Electrical Engineer's Reference Book

Fourteenth Edition

Electrical Engineer's Reference Book

Fourteenth Edition

Edited by
M A Laughton
BASc, PhD, DSc(Eng), FIEE
and
M G Say
PhD, MSc, CEng, FRSE, FIEE, FIERE, ACGI, DIC

with specialist contributors

Butterworths
London · Boston · Durban · Singapore
Sydney · Toronto · Wellington

All rights reserved. No part of this publication may be
reproduced or transmitted in any form or by any means,
including photocopying and recording without the written
permission of the copyright holder, application for which
should be addressed to the Publishers. Such written
permission must also be obtained before any part of this
publication is stored in a retrieval system of any nature.

This book is sold subject to the Standard Conditions of Sale
of Net Books and may not be resold in the UK below the net
price given by the Publishers in their current price list.

First published in 1945 by George Newnes Ltd
Fourteenth edition 1985
  Reprinted 1986

© Butterworth & Co (Publishers) Ltd 1985

British Library Cataloguing in Publication Data

Electrical Engineer's Reference Book—14th ed.
  1 Electric Engineering
  I. Laughton, M. A.   II. Say, M. G.
  621.3   TK 145

ISBN 0-408-00432-0

Typset by Mid-County Press,
2a Merivale Road, London SW15 2NW
Printed and bound in England by
Robert Hartnoll (1985) Ltd, Bodmin, Cornwall

Contents

Preface

List of Contributors

1 Units, Mathematics and Physical Quantities
International unit system . Mathematics . Physical quantities . Physical properties . Electricity

2 Electrotechnology
Nomenclature . Thermal effects . Electrochemical effects . Magnetic field effects . Electrical field effects . Electromagnetic field effects

3 Network Analysis
Introduction . Basic network analysis . Power-system network analysis

4 Control System Analysis
Introduction . Laplace transforms and the transfer function . Block diagrams . Feedback . Generally desirable and acceptable behaviour . Stability . Classification of system and static accuracy . Transient behaviour . Root-locus method . Frequency-response methods . Conclusion

5 Materials
Conductors and superconductors: Conducting materials . Superconductors . Semiconductors, thick and thin films . Silicon . *Insulators:* Insulating materials . Properties and testing . Gaseous dielectrics . Liquid dielectrics . Semi-fluid and fusible materials . Varnishes, enamels, paints and lacquers . Solid dielectrics . Irradiation effects . Ferroelectrics . *Magnetic materials:* Ferromagnetics . Soft irons and relay steels . Ferrites . Nickel-iron alloys . Permanent magnetic materials . *Special alloys and applications:* Fuses . Contacts . Special alloys . Solders . Rare and precious metals . Temperature-sensitive bimetals . Nuclear reactor materials

6 Electrical Metrology and Instrumentation
Introduction . Terminology, traceability and physical standards . Traceability and quality assurance data . Physical reference standards . DC and industrial-frequency analogue instrumentation . Resistance and conductance instrumentation . Integrating (energy) metering . Electronic instrumentation . Digital wattmeters . Electronic oscillography . Potentiometers and bridges . Measuring and protection transformers . Magnetic measurements . Transducers

7 Steam Generating Plant
Introduction . Combustion . Sources of chemical energy . Thermodynamics and hydrodynamics of steam generating plant . Types of steam generator . Fuel handling and storage . Pollution control

8 Turbines and Diesel Plant
Steam turbine plant: Cycles and types . Turbine technology . Turbine construction . Turbine support plant . Turbine operation and control . *Gas turbine plant:* Open-cycle plant . Closed-cycle plant . Combined-cycle plant . Cogeneration plant . *Hydro-electric plant:* Catchment area . Reservoir and dam . Pipelines . Power station . Turbines . Electrical plant . Discharge and tail race . Operational features . Pumped storage . *Diesel engine plant:* Theory and general principles . Engine features . Engine primary systems . Engine ancillaries . AC generators . Switchgear and controls . Operational aspects . Plant layout . Economic factors . Total energy

9 Nuclear Reactor Plant
Introduction . The nucleus as an energy source . Nuclear fission . Reactor components . Types of nuclear reactor . Nuclear reactor plant for ship propulsion . The nuclear fuel cycle

10 Alternative Energy Sources
Wind energy . Geothermal energy . Wave energy . Ocean thermal energy . Solar energy . Tidal energy . Direct conversion . Fuel cells . Nuclear fusion . Combined heat and power . Total energy . Economics . Heat pumps

11 Alternating-Current Generators
Introduction . Airgap flux distribution and no-load e.m.f. . Stator windings . Insulation . Temperature rise . Output equation . Armature reaction . Reactances and time-constants . Steady-state operation . Synchronising . Operating charts . On-load excitation . Sudden 3-phase short circuit . Excitation systems

12 Overhead Lines
General . Conductors . Conductor fittings . Electrical characteristics . Insulators . Supports . Lightning . Loadings

13 Cables
Introduction . Cable components . General wiring cables and flexible cords . Supply distribution cables . Transmission cables . Current-carrying capacity . Jointing and accessories . Cable fault location

14 Power Transformers
General . Magnetic circuit . Windings and insulation . Connections . Three-winding transformers . Quadrature booster transformers . On-load tap changing . Cooling . Fittings . Parallel operation . Auto-transformers . Special types . Maintenance . Testing . Surge protection . Purchasing specifications

15 Switchgear and Protection
Arc mechanism . Fusegear . Switchgear . Circuit-breakers . Disconnectors . Reactors and capacitors . Switchgear testing . Overcurrent and earth leakage protection . Application of protective systems . Testing and commissioning . Overvoltage protection

16 Supply and Control of Reactive Power
Introduction . General considerations of reactive volt-amperes . Management of reactive power . Reactive compensation and transient stability . Variable var compensators . Special aspects of compensation application . Future prospects

17 Power System Operation and Control
Introduction . Objectives and requirements . System description . Data acquisition and telemetering . Decentralised control: excitation systems and control characteristics of synchronous machines . Decentralised control: electronic turbine controllers . Decentralised control: substation automation . Decentralised control: pulse controllers for voltage control with tap-changing transformers . Centralised control

18 Power System Planning and Economics
Load forecast . Reliability of supply . Legislation . Choice of generating plant . Plant scheduling . Revenue . British electricity supply systems

19 Power Electronics
Rectification . Controlled rectification

20 Installation
Layout . Regulations and specifications . High-voltage supplies . Fault currents . Substations . Wiring systems . Lighting and small power . Floor trunking . Standby and emergency supplies . Special buildings . Switchgear . Transformers . Power-factor correction . Earthing . Inspection and earthing

21 Electroheat
Introduction . Direct resistance heating . Indirect resistance heating . Infrared heaters . Ultraviolet processes . Electric ovens and furnaces . Ultraviolet processes . Electric ovens and furnaces . Induction heating . Metal melting . Dielectric heating . Plasma torches . Glow discharge processes . Lasers

22 Welding
Arc welding . Resistance welding

23 Electrochemical Technology
Introduction . Cells and batteries . Primary cells . Secondary cells and batteries . Anodising . Electrodeposition . Hydrogen and oxygen electrolysis

24 Microcontrollers and Their Applications
Introduction . General requirements . Dedicated controller . Operator interface . Communications with other controllers . User-designer interface . Maintenance facilities . Applications . Serial communication

25 Analogue, Hybrid and Digital Computation
Digital and programmable electronic systems: Digital systems . Processing binary information . Sequential logic . Programmable systems . *Digital computers and computer programming:* Digital computer structure . Interrupts . Input/Output . Instruction execution . Computer instruction types . Computer instruction format . Computer program execution . Computer programming . Software design and specification . Machine language . Program development . High-level language types . Real time operation . *Analogue computation:* Operational amplifier . Operational amplifier integrated circuits . Analogue functional elements . Complex functions . Voltage follower . Function generators . Comparator . Diode function generator . Multifunction converter . Analogue computer . Applications . *Hybrid computation:* Signal switches . Sample and hold . Digital-to-analogue converter (DAC) . Analogue-to-digital converter (ADC) . Analogue multiplexers . Hybrid computer . Hybrid computer applications . *Digital simulation:* Introduction . System models . Integration schemes . Organisation of problem input . Illustrative examples . Conclusions

26 Electromagnetic machines
Energy conversion . Electromagnetic devices . Industrial rotary and linear motors

27 Illumination
Light and vision . Quantities and units . Photometric concepts . Lighting design technology . Lamps . Choice of lamps . Lighting design . Applications

28 Environmental Control
Introduction . Environmental comfort . Energy requirements . Heating and warm air systems . Control . Energy conservation . Interfaces

29 Transportation—Roads
Electrical equipment of road transport vehicles . Light rail transit . Trolleybuses . Battery vehicles

30 Transportation—Railways
Railway electrication . Diesel-electric traction . Railway signalling and control

31 Transportation—Ships
Introduction . Regulations . Conditions of service . DC installations . AC installations . Earthing . Machines and transformers . Switchgear . Cables . Emergency power . Steering gear . Refrigerated cargo spaces . Lighting . Heating . Watertight doors . Ventilating fans . Radio interference and electromagnetic compatibility . Deck auxiliaries . Remote and automatic control systems . Tankers . Steam plant . Generators . Diesel engines . Electric propulsion

32 Mining
General . Power supplies . Winders . Underground transport . Coalface layout . Power loaders . Heading machines . Flameproof and intrinsically safe equipment . Gate-end boxes . Flameproof motors . Cables, couplers, plugs and sockets . Drilling machines . Underground lighting . Monitoring and control

33 Agriculture and Horticulture
Introduction . Supply . Installation . Electrical equipment . Farming processes . Horticultural processes . Information

34 HVDC Transmission
Introduction . Application of h.v.d.c. . Principles of h.v.d.c. . Transmission arrangements . Converter station design . Insulation co-ordination . Thyristor valves . Harmonic filters . Reactive power consideration . Control of h.v.d.c. . AC damping controls . Interaction between a.c. and d.c. systems . Multi-terminal systems . Future trends

35 Education and Training
Introduction . Summary of the three levels . Chartered engineers . Technician engineers . Engineering technicians . Bridges and ladders . Addresses for further information

36 Standards
Introduction . United Kingdom . International standards . USA standards . World standards . Further reading

INDEX

Preface

The *Electrical Engineer's Reference Book* was first published in 1945, when some fields of electrical power engineering, now of great importance, were unknown or only speculative. The original concept of the *Reference Book* and its numerous subsequent editions testify to the aim of reflecting the modern state of the art.

In this fourteenth edition, the *Reference Book* has again been completely revised and rewritten with the help of a team of specialist contributors. Further, the format has been considerably improved: larger pages with double columns make for easier reading, paragraphs are assigned numerics and the facilities of up-to-date printing techniques have been applied. The text has been carefully devised with the convenience of the reader given prime consideration.

The basic plan of the *Reference Book* has been arranged to group its subjects in a logical sequence, as follows.

FUNDAMENTALS (Chapters 1–6)   Units, mathematical formulae, electrophysics, electro-technology, network solution applied to basic circuits and to power-supply networks, control-system analysis and instrumentation. The text here covers the common background of electrical power engineering technology and gives a number of reference tables.

ENERGY SUPPLY (Chapters 7–20)   Conversion of energy into the electrical form, prime movers, generators, transformers, lines, economics. The essential features of planning, operation and control of electrical power-supply systems are described.

POWER PLANT (Chapters 19, 21–23, 26)   Application of electrical energy in electro-mechanical drives, electro-thermal plant, electro-welding and electro-chemistry. In all these areas the advent of solid-state power electronics has had an important effect.

APPLICATION (Chapters 24, 25, 27–34)   Automatic control methods, which have changed radically with the advent of micro digital control computers (Chapters 24 and 25). Specific fields of application, such as illumination, transportation, mining, agriculture and installation technology (Chapters 27–34).

STANDARDS (Chapter 36)   Guide to Standards relevant to the subjects of the preceding Chapters.

The *Electrical Engineer's Reference Book* brings within the scope of a single volume a very wide range of material for the information of the electrical engineer. As a practical complement to academic studies, or as a guide to areas outside the personal experience of the reader, it constitutes both a database and a knowledge base of interest and importance to engineers at all stages of their careers.

The text is consistently cast in terms of the international unit system (SI). For approaching a half-century, electrical engineering units have been based on the metre-kilogram-second (m.k.s.) system, which is now incorporated in SI, a coherent system that covers all branches of applied science. With power in watts, energy in joules and force in newtons, the technologies of allied branches of engineering are now interconsistent. This is of great advantage to the electrical engineer, who is often required to cope with electro-mechanical, electro-thermal and electro-chemical interconversions. It has, nevertheless, been felt to be helpful to give, on occasion, the equivalent non-SI values of quantities with which an engineer might, for the time being, be more familiar.

Professional engineers have written from time to time with suggestions for improvements or additions. Such communications are welcomed, and all are given careful consideration.

1985

MAL   MGS
London   Edinburgh

List of Contributors

D I Alabaster BTech
GEC Electrical Projects Ltd

D R Andrews PhD, CEng, FIM, MIProdE
University of Aston in Birmingham

D S Armstrong MIEE, MIMechE
British Railways Research Division

A C Bailey BSc, FIEE

J A Bailey
GEC Traffic Automation Ltd

L Banbury
BICC General Cables Ltd

M Barak MSc, DPhil, FRSC, CChem, CEng, FIEE, FRSA
Consulting chemist and engineer

H Barber BSc
Loughborough University of Technology

D J Barrow
Lucas Electrical Ltd

P Beckley BSc, PhD, CEng, FIM, FIEE
British Steel Corporation, Newport

J W Beeston BSc(Eng), PhD
GEC Electrical Projects Ltd

R I Bell
Thorn Lighting Ltd

G L Bibby BSc, CEng, MIEE
University of Leeds

G Brundrett
Electricity Council Research Centre

S W Bullett BSc
GEC Electrical Projects Ltd

J Burley MIEE
Brown Boveri Cie, Baden

A G Clegg MSc, PhD
Sunderland Polytechnic

J S Cliff MBE, CEng, FIEE
Engineering Consultant

A Cox
Lucas Electrical Ltd

I G Crow BEng, PhD, CEng, MIMechE, FIMarE, MemASME
NEI Projects Ltd

N G Dovaston BSc, PhD, MPRI
Sunderland Polytechnic

T A C Dulley MA, MIEE
Merz and McLellan

M D Dwek BSc, FIEE
Merz and McLellan

H E Evans BSc, PhD, CEng, MIM
Berkley Laboratories, Gloucestershire

B D Field BSc(Eng)(Met), CEng, MIM
Johnson Matthey Ltd, Harlow

J O Flower BSc(Eng), PhD, DSc(Eng), CEng, FIEE, FIMarE
University of Exeter

R W Flux
NEI-Peebles Ltd

P F Gale BTech, PhD, CEng, MIEE
Biccotest Ltd

A Gavrilović
GEC, Stafford

H Glavitsch Dr-Ing
ETH, Zurich

T G F Gray BSc, PhD, CEng, MIMechE, MWeldI
University of Strathclyde

D Grieve BSc
Welwyn Microelectronics, Bedlington

P H Hammond FIEE, MInstMC
Computer Aided Design Centre, Cambridge

J E Harry BSc(Eng), PhD
Loughborough University of Technology

R Hartill BSc(Hons), CEng, FIEE, FIMEMME, FRSA
Trolex Products Ltd

D J Holding
University of Aston

D R Holmes MA, DPhil, FInstP, FICST
National Corrosion Service, NPL, Teddington

A Hunt BSc(Eng), CEng, FIEE
Engineering Consultant

D Inman
Imperial College, London

P J King
Control Systems Engineering, Tosco Corporation, USA

J R Kirkman BSc, PhD, MIERE, CEng
Sunderland Polytechnic

P A Kurn DFH, CEng, MIEE
ERA Technology Ltd, Leatherhead

M A Laughton BASc, PhD, DSc(Eng), FIEE
Queen Mary College, London

D C Leslie MA(Oxon), DPhil, CEng, FINucE
Queen Mary College, London

D Lush BSc(Eng), CEng, MIEE, MCIBS
Ove Arup Partnership

B J Maddock MA, FInstP, CEng, MIEE
Central Electricity Research Laboratories, Leatherhead

L L J Mahon CEng, FIEE, FBIM
Engineering Consultant

R V Major MSc, PhD, FInstP
Telcon Ltd, Crawley

D McAllister
Formerly BICC Power Cables Ltd

J D McColl
GEC Rectifiers Ltd

G McHamish
Boving & Co Ltd

J McTaggart
GEC Industrial Controls Ltd

J P Milne BSc(Eng), CEng, FIEE, FCIBSE
Engineering Consultant

S Muckett BSc, CEng, MIM
Agnet Ltd, Reading

W T Norris ScD, CEng, FIEE, SrMemIEEE
Central Electricity Generating Board, London

T E Norris BSc, FIMechE, FIEE
Merz and McLellan

G Orawski BSc(Eng)(Hons), CEng, FIEE, SMIEEE
Balfour Beatty Power Construction Ltd

A J Pearmain BSc(Eng), PhD, CEng, MIEE
Queen Mary College, London

F Peneder Dipl-Ing
Brown Boveri Cie, Baden

A M Plaskett BSc(Eng), MIEE
GEC Electrical Projects Ltd

B F Pope MSc, AMBIM
GEC Industrial Controls Ltd

G S Powell BA, MBCS
GEC Electrical Projects Ltd

B J Prigmore MA, MSc, DIC, CEng, FIEE
Consultant Electrical Engineer

K Reichert Dr-Ing
ETH, Zurich

W Rizk CBE, MA, PhD, FEng, FIMechE
GEC Power Engineering Ltd

M G Say PhD, MSc, CEng, FRSE, FIEE, FIERE, ACGI, DIC
formerly Heriot-Watt University, Scotland

A Sensicle BSc, CEng, MIEE, MInstMC
Institute of Measurement and Control

P R Smith BSc, PhD, CEng, FINucE, FInstP
Queen Mary College, London

E C Snelling BSc(Eng), CEng, FIEE
Philips Research Laboratories, Redhill

R H Taylor
Central Electricity Generating Board

H L Thanawala PhD, BE(Elect), BE(Mech)
GEC, Stafford

H Watson BSc, CEng, FIMechE, FIMarE, MemASME, FFB
Engineering Consultant

W H Whitehouse
British Railway Board

W P Williams BSc, CEng, FIEE, SMIEEE
GEC, Stafford

R K Wood
University of Alberta

A Wright PhD, DSc, CEng, FIEE
University of Nottingham

Units, Mathematics and Physical Quantities

M G Say, PhD, MSc, CEng, ACGI, DIC, FIEE, FRSE
Heriot-Watt University

Contents

1.1 International unit system 1/3
    1.1.1 Base units 1/3
    1.1.2 Supplementary units 1/3
    1.1.3 Notes 1/3
    1.1.4 Derived units 1/3
    1.1.5 Auxiliary units 1/4
    1.1.6 Conversion factors 1/4
    1.1.7 CGS electrostatic and electromagnetic units 1/4

1.2 Mathematics 1/4
    1.2.1 Trigonometric relations 1/4
    1.2.2 Exponential and hyperbolic relations 1/6
    1.2.3 Bessel functions 1/8
    1.2.4 Series 1/9
    1.2.5 Fourier series 1/9
    1.2.6 Derivatives and integrals 1/10
    1.2.7 Laplace transforms 1/10
    1.2.8 Binary numeration 1/12
    1.2.9 Power ratio 1/12

1.3 Physical quantities 1/12
    1.3.1 Energy 1/13
    1.3.2 Structure of matter 1/16

1.4 Physical properties 1/24

1.5 Electricity 1/24
    1.5.1 Charges at rest 1/24
    1.5.2 Charges in motion 1/24
    1.5.3 Charges in acceleration 1/26

This reference section provides (a) a statement of the International System (SI) of Units, with conversion factors; (b) basic mathematical functions, series and tables; and (c) some physical properties of materials.

1.1 International unit system

The International System of Units (SI) is a metric system giving a fully coherent set of units for science, technology and engineering, involving no conversion factors. The starting point is the selection and definition of a minimum set of independent 'base' units. From these, 'derived' units are obtained by forming products or quotients in various combinations, again without numerical factors. For convenience, certain combinations are given shortened names. A single SI unit of energy (joule = kilogram metre-squared per second-squared) is, for example, applied to energy of any kind, whether it be kinetic, potential, electrical, thermal, chemical..., thus unifying usage throughout science and technology.

The SI system has seven *base* units, and two *supplementary* units of angle. Combinations of these are *derived* for all other units. Each physical quantity has a quantity symbol (e.g. m for mass, P for power) that represents it in physical equations, and a unit symbol (e.g. kg for kilogram, W for watt) to indicate its SI unit of measure.

1.1.1 Base units

Definitions of the seven base units have been laid down in the following terms. The quantity symbol is given in italic, the unit symbol (with its standard abbreviation) in roman type. As measurements become more precise, changes are occasionally made in the definitions.

Length: l, metre (m) The metre was defined in 1983 as the length of the path travelled by light in a vacuum during a time interval of 1/299 792 458 of a second.

Mass: m, kilogram (kg) The mass of the international prototype (a block of platinum preserved at the International Bureau of Weights and Measures, Sèvres).

Time: t, second (s) The duration of 9 192 631 770 periods of the radiation corresponding to the transition between the two hyperfine levels of the ground state of the caesium-133 atom.

Electric current: i, ampere (A) The current which, maintained in two straight parallel conductors of infinite length, of negligible circular cross-section and 1 m apart in vacuum, produces a force equal to 2×10^{-7} newton per metre of length.

Thermodynamic temperature: T, kelvin (K) The fraction 1/273.16 of the thermodynamic (absolute) temperature of the triple point of water.

Luminous intensity: I, candela (cd) The luminous intensity in the perpendicular direction of a surface of $1/600\,000$ m^2 of a black body at the temperature of freezing platinum under a pressure of 101 325 newton per square metre.

Amount of substance: Q, mole (mol) The amount of substance of a system which contains as many elementary entities as there are atoms in 0.012 kg of carbon-12. The elementary entity must be specified and may be an atom, a molecule, an ion, an electron..., or a specified group of such entities.

1.1.2 Supplementary units

Plane angle: α, β ..., radian (rad) The plane angle between two radii of a circle which cut off on the circumference of the circle an arc of length equal to the radius.

Solid angle: Ω, steradian (sr) The solid angle which, having its vertex at the centre of a sphere, cuts off an area of the surface of the sphere equal to a square having sides equal to the radius.

1.1.3 Notes

Temperature At zero K, bodies possess no thermal energy. Specified points (273.16 and 373.16 K) define the Celsius (centigrade) scale (0 and 100°C). In terms of *intervals*, 1°C = 1 K. In terms of *levels*, a scale Celsius temperature θ corresponds to $(\theta + 273.16)$ K.

Force The SI unit is the newton (N). A force of 1 N endows a mass of 1 kg with an acceleration of 1 m/s^2.

Weight The weight of a mass depends on gravitational effect. The standard weight of a mass of 1 kg at the surface of the earth is 9.807 N.

1.1.4 Derived units

All physical quantities have units derived from the base and supplementary SI units, and some of them have been given names for convenience in use. A tabulation of those of interest in electrical technology is appended to the list in *Table 1.1*.

Table 1.1 SI base, supplementary and derived units

Quantity	Unit name	Derivation	Unit symbol
Length	metre		m
Mass	kilogram		kg
Time	second		s
Electric current	ampere		A
Thermodynamic temperature	kelvin		K
Luminous intensity	candela		cd
Amount of substance	mole		mol
Plane angle	radian		rad
Solid angle	steradian		sr
Force	newton	kg m/s^2	N
Pressure, stress	pascal	N/m^2	Pa
Energy	joule	N m, W s	J
Power	watt	J/s	W
Electric charge, flux	coulomb	A s	C
Magnetic flux	weber	V s	Wb
Electric potential	volt	J/C	V
Magnetic flux density	tesla	Wb/m^2	T
Resistance	ohm	V/A	Ω
Inductance	henry	Wb/A, V s/A	H
Capacitance	farad	C/V, A s/V	F
Conductance	siemens	A/V	S
Frequency	hertz	s^{-1}	Hz
Luminous flux	lumen	cd sr	lm
Illuminance	lux	lm/m^2	lx
Radiation activity	becquerel	s^{-1}	Bq
Absorbed dose	gray	J/kg	Gy
Mass density	kilogram per cubic metre		kg/m^3
Dynamic viscosity	pascal-second		Pa s
Concentration	mole per cubic metre		mol/m^3
Linear velocity	metre per second		m/s
Linear acceleration	metre per second-squared		m/s^2
Angular velocity	radian per second		rad/s
Angular acceleration	radian per second-squared		rad/s^2

cont'd

Table 1.1 (continued)

Quantity	Unit name	Derivation	Unit symbol
Torque	newton metre		N m
Electric field strength	volt per metre		V/m
Magnetic field strength	ampere per metre		A/m
Current density	ampere per square metre		A/m^2
Resistivity	ohm metre		Ω m
Conductivity	siemens per metre		S/m
Permeability	henry per metre		H/m
Permittivity	farad per metre		F/m
Thermal capacity	joule per kelvin		J/K
Specific heat capacity	joule per kilogram kelvin		J/(kg K)
Thermal conductivity	watt per metre kelvin		W/m K)
Luminance	candela per square metre		cd/m^2

Decimal multiples and submultiples of SI units are indicated by prefix letters as listed in *Table 1.2*. Thus, kA is the unit symbol for kiloampere, and μF that for microfarad. There is a preference in technology for steps of 10^3. Prefixes for the kilogram are expressed in terms of the gram: thus, 1000 kg = 1 Mg, not 1 kkg.

1.1.5 Auxiliary units

Some quantities are still used in special fields (such as vacuum physics, irradiation, etc.) having non-SI units. Some of these are given in *Table 1.3* with their SI equivalents.

1.1.6 Conversion factors

Imperial and other non-SI units still in use are listed in *Table 1.4*, expressed in the most convenient multiples or submutiples of the basic SI unit [ ] under classified headings.

1.1.7 CGS electrostatic and electromagnetic units

Although obsolescent, electrostatic and electromagnetic units (e.s.u., e.m.u.) appear in older works of reference. Neither system is 'rationalised', nor are the two mutually compatible. In e.s.u. the electric space constant is $\varepsilon_0 = 1$, in e.m.u. the magnetic space constant is $\mu_0 = 1$; but the SI units take account of the fact that $1/\sqrt{(\varepsilon_0\mu_0)}$ is the velocity of electromagnetic wave propagation in free space. *Table 1.5* lists SI units with the equivalent number n of e.s.u. and e.m.u. Where these lack names, they are expressed as SI unit names with the prefix 'st' ('electrostatic') for e.s.u. and 'ab' ('absolute') for e.m.u. Thus, 1 V corresponds to $10^{-2}/3$ stV and to 10^8 abV, so that 1 stV = 300 V and 1 abV = 10^{-8} V.

1.2 Mathematics

Mathematical symbolism is set out in *Table 1.6*. This subsection gives trigonometric and hyperbolic relations, series (including Fourier series for a number of common wave forms), binary enumeration and a list of common derivatives and integrals.

1.2.1 Trigonometric relations

The trigonometric functions (sine, cosine, tangent, cosecant, secant, cotangent) of an angle θ are based on the circle, given by $x^2 + y^2 = h^2$. Let two radii of the circle enclose an angle θ and form the sector area $S_c = (\pi h^2)(\theta/2\pi)$ shown shaded in *Figure 1.1* (left): then θ can be defined as $2S_c/h^2$. The *right-angled* triangle with sides h (hypotenuse), a (adjacent side) and p (opposite side)

Table 1.2 Decimal prefixes

10^{18} exa E	10^9 giga G	10^2 hecto h	10^{-3} milli m	10^{-12} pico p
10^{15} peta P	10^6 mega M	10^1 deca da	10^{-6} micro μ	10^{-15} femto f
10^{12} tera T	10^3 kilo k	10^{-1} deci d	10^{-9} nano n	10^{-18} atto a
		10^{-2} centi c		

Table 1.3 Auxiliary units

Quantity	Symbol		SI		Quantity	Symbol		SI	
Angle					Mass				
degree	(°)		$\pi/180$	rad	tonne	t		1000	kg
minute	(')		—	—	Nucleonics, Radiation				
second	(")		—	—	becquerel	Bq		1.0	s^{-1}
Area					gray	Gy		1.0	J/kg
are	a		100	m^2	curie	Ci		3.7×10^{10}	Bq
hectare	ha		0.01	km^2	rad	rd		0.01	Gy
barn	barn		10^{-28}	m^2	roentgen	R		2.6×10^{-4}	C/kg
Energy					Pressure				
erg	erg		0.1	μJ	bar	b		100	kPa
calorie	cal		4.186	J	torr	Torr		133.3	Pa
electron-volt	eV		0.160	aJ	Time				
gauss-oersted	Ga Oe		7.96	μJ/m^3	minute	min		60	s
Force					hour	h		3600	s
dyne	dyn		10	μN	day	d		86 400	s
Length					Volume				
Ångstrom	Å		0.1	μm	litre	l or L		1.0	dm^3

Table 1.4 Conversion factors

Length [m]	
1 mil	25.40 μm
1 in	25.40 mm
1 ft	0.3048 m
1 yd	0.9144 m
1 fathom	1.829 m
1 mile	1.6093 km
1 nautical mile	1.852 km

Area [m^2]	
1 circular mil	506.7 μm^2
1 in^2	645.2 mm^2
1 ft^2	0.0929 m^2
1 yd^2	0.8361 m^2
1 acre	4047 m^2
1 mile^2	2.590 km^2

Volume [m^3]	
1 in^3	16.39 cm^3
1 ft^3	0.0283 m^3
1 yd^3	0.7646 m^3
1 UKgal	4.546 dm^3

Velocity [m/s, rad/s] Acceleration [m/s^2, rad/s^2]	
1 ft/min	5.080 mm/s
1 in/s	25.40 mm/s
1 ft/s	0.3048 m/s
1 mile/h	0.4470 m/s
1 knot	0.5144 m/s
1 deg/s	17.45 mrad/s
1 rev/min	0.1047 rad/s
1 rev/s	6.283 rad/s
1 ft/s^2	0.3048 m/s^2
1 mile/h per s	0.4470 m/s^2

Mass [kg]	
1 oz	28.35 g
1 lb	0.454 kg
1 slug	14.59 kg
1 cwt	50.80 kg
1 UKton	1016 kg

Energy [J], Power [W]	
1 ft lbf	1.356 J
1 m kgf	9.807 J
1 Btu	1055 J
1 therm	105.5 kJ
1 hp h	2.685 MJ
1 kW h	3.60 MJ
1 Btu/h	0.293 W
1 ft lbf/s	1.356 W
1 m kgf/s	9.807 W
1 hp	745.9 W

Thermal quantities [W, J, kg, K]	
1 W/in^2	1.550 kW/m^2
1 Btu/(ft^2 h)	3.155 W/m^2
1 Btu/(ft^3 h)	10.35 W/m^3
1 Btu/(ft h °F)	1.731 W/(m K)
1 ft lbf/lb	2.989 J/kg
1 Btu/lb	2326 J/kg
1 Btu/ft^3	37.26 KJ/m^3
1 ft lbf/(lb °F)	5.380 J/(kg K)
1 Btu/(lb °F)	4.187 kJ/(kg K)
1 Btu/(ft^3 °F)	67.07 kJ/m^3 K

Density [kg/m, kg/m^3]	
1 lb/in	17.86 kg/m
1 lb/ft	1.488 kg/m
1 lb/yd	0.496 kg/m
1 lb/in^3	27.68 Mg/m^3
1 lb/ft^3	16.02 kg/m^3
1 ton/yd^3	1329 kg/m^3

Flow rate [kg/s, m^3/s]	
1 lb/h	0.1260 g/s
1 ton/h	0.2822 kg/s
1 lb/s	0.4536 kg/s
1 ft^3/h	7.866 cm^3/s
1 ft^3/s	0.0283 m^3/s
1 gal/h	1.263 cm^3/s
1 gal/min	75.77 cm^3/s
1 gal/s	4.546 dm^3/s

Force [N], Pressure [Pa]	
1 dyn	10.0 μN
1 kgf	9.807 N
1 ozf	0.278 N
1 lbf	4.445 N
1 tonf	9.964 kN
1 dyn/cm^2	0.10 Pa
1 lbf/ft^2	47.88 Pa
1 lbf/in^2	6.895 kPa
1 tonf/ft^2	107.2 kPa
1 tonf/in^2	15.44 MPa
1 kgf/m^2	9.807 Pa
1 kgf/cm^2	98.07 kPa
1 mmHg	133.3 Pa
1 inHg	3.386 kPa
1 inH$_2$O	149.1 Pa
1 ftH$_2$O	2.989 kPa

Torque [N m]	
1 ozf in	7.062 nN m
1 lbf in	0.113 N m
1 lbf ft	1.356 N m
1 tonf ft	3.307 kN m
1 kgf m	9.806 N m

Inertia [kg m^2], Momentum [kg m/s, kg m^2/s]	
1 oz in^2	0.018 g m^2
1 lb in^2	0.293 g m^2
1 lb ft^2	0.0421 kg m^2
1 slug ft^2	1.355 kg m^2
1 ton ft^2	94.30 kg m^2
1 lb ft/s	0.138 kg m/s
1 lb ft^2/s	0.042 kg m^2/s

Viscosity [Pa s, m^2/s]	
1 poise	9.807 Pa s
1 kgf s/m^2	9.807 Pa s
1 lbf s/ft^2	47.88 Pa s
1 lbf h/ft^2	172.4 kPa s
1 stokes	1.0 cm^2/s
1 in^2/s	6.452 cm^2/s
1 ft^2/s	929.0 cm^2/s

Illumination [cd, lm]	
1 lm/ft^2	10.76 lm/m^2
1 cd/ft^2	10.76 cd/m^2
1 cd/in^2	1550 cd/m^2

Table 1.5 Relation between SI, e.s. and e.m. units

Quantity	SI unit		Equivalent number n of			
			e.s.u.		e.m.u.	
Length	m		10^2	cm	10^2	cm
Mass	kg		10^3	g	10^3	g
Time	s		1	s	1	s
Force	N		10^5	dyn	10^5	dyn
Torque	N m		10^7	dyn cm	10^7	dyn cm
Energy	J		10^7	erg	10^7	erg
Power	W		10^7	erg/s	10^7	erg/s
Charge, electric flux	C		3×10^9	stC	10^{-1}	abC
density	C/m^2		3×10^5	stC/cm^2	10^{-5}	abC/cm^2
Potential, e.m.f.	V		$10^{-2}/3$	stV	10^8	abV
Electric field strength	V/m		$10^{-4}/3$	stV/cm	10^6	abV/cm
Current	A		3×10^9	stA	10^{-1}	abA
density	A/m^2		3×10^5	stA/cm^2	10^{-5}	abA/cm^2
Magnetic flux	Wb		$10^{-2}/3$	stWb	10^8	Mx
density	T		$10^{-6}/3$	stWb/cm^2	10^4	Gs
Mag. fd. strength	A/m		$12\pi \times 10^7$	stA/cm	$4\pi \times 10^{-3}$	Oe
M.M.F.	A		$12\pi \times 10^9$	stA	$4\pi \times 10^{-1}$	Gb
Resistivity	Ω m		$10^{-9}/9$	stΩ cm	10^{11}	abΩ cm
Conductivity	S/m		9×10^9	stS/cm	10^{-11}	abS/cm
Permeability (abs)	H/m		$10^{-13}/36\pi$	—	$10^7/4\pi$	—
Permittivity (abs)	F/m		$36\pi \times 10^9$	—	$4\pi \times 10^{-11}$	—
Resistance	Ω		$10^{-11}/9$	stΩ	10^9	abΩ
Conductance	S		9×10^{11}	stS	10^{-9}	abS
Inductance	H		$10^{-12}/9$	stH	10^9	cm
Capacitance	F		9×10^{11}	cm	9×10^{11}	abF
Reluctance	A/Wb		$36\pi \times 10^{11}$	—	$4\pi \times 10^{-8}$	Gb/Mx
Permeance	Wb/A		$10^{11}/36\pi$	—	$10^9/4\pi$	Mx/Gb

Gb = gilbert; Gs = gauss; Mx = maxwell; Oe = oersted.

give ratios defining the trigonometric functions

$\sin\theta = p/h$ $\qquad \operatorname{cosec}\theta = 1/\sin\theta = h/p$
$\cos\theta = a/h$ $\qquad \sec\theta = 1/\cos\theta = h/a$
$\tan\theta = p/a$ $\qquad \cot\theta = 1/\tan\theta = a/p$

In any triangle (*Figure 1.1*, right) with angles A, B and C at the corners opposite, respectively, to sides a, b and c, then $A + B + C = \pi$ rad (180°) and the following relations hold:

$a = b\cos C + c\cos B$
$b = c\cos A + a\cos C$
$c = a\cos B + b\cos A$
$a/\sin A = b/\sin B = c/\sin C$
$a^2 = b^2 + c^2 + 2bc\cos A$
$(a+b)/(a-b) = (\sin A + \sin B)/(\sin A - \sin B)$

Other useful relationships are:

$\sin(x \pm y) = \sin x \cdot \cos y \pm \cos x \cdot \sin y$
$\cos(x \pm y) = \cos x \cdot \cos y \mp \sin x \cdot \sin y$
$\tan(x \pm y) = (\tan x \pm \tan y)/(1 \mp \tan x \cdot \tan y)$
$\sin^2 x = \tfrac{1}{2}(1 - \cos 2x) \quad \cos^2 x = \tfrac{1}{2}(1 + \cos 2x)$
$\sin^2 x + \cos^2 x = 1 \quad \sin^3 x = -\tfrac{1}{4}(3\sin x - \sin 3x)$
$\cos^3 x = \tfrac{1}{4}(3\cos x + \cos 3x)$

$\sin x \pm \sin y = 2 \begin{bmatrix} \cos \\ \sin \end{bmatrix} \tfrac{1}{2}(x-y) \cdot \begin{bmatrix} \sin \\ \cos \end{bmatrix} \tfrac{1}{2}(x+y)$

$\cos x \pm \cos y = \mp 2 \begin{bmatrix} \cos \\ \sin \end{bmatrix} \tfrac{1}{2}(x-y) \cdot \begin{bmatrix} \sin \\ \cos \end{bmatrix} \tfrac{1}{2}(x+y)$

$\tan x \pm \tan y = \sin(x \pm y)/\cos x \cdot \cos y$
$\sin^2 x - \sin^2 y = \sin(x+y) \cdot \sin(x-y)$
$\cos^2 x - \cos^2 y = -\sin(x+y) \cdot \sin(x-y)$
$\cos^2 x - \sin^2 y = \cos(x+y) \cdot \cos(x-y)$

$d(\sin x)/dx = \cos x \qquad \int \sin x \cdot dx = -\cos x + k$
$d(\cos x)/dx = -\sin x \qquad \int \cos x \cdot dx = \sin x + k$
$d(\tan x)/dx = \sec^2 x \qquad \int \tan x \cdot dx = -\ln|\cos x| + k$

Values of $\sin\theta$, $\cos\theta$ and $\tan\theta$ for $0° < \theta < 90°$ (or $0 < \theta < 1.571$ rad) are given in *Table 1.7* as a check list, as they can generally be obtained directly from calculators.

1.2.2 Exponential and hyperbolic relations

Exponential functions For a positive datum ('real') number u, the exponential functions $\exp(u)$ and $\exp(-u)$ are given by the summation to infinity of the series

$\exp(\pm u) = 1 \pm u + u^2/2! \pm u^3/3! + u^4/4! \pm \ldots$

with $\exp(+u)$ increasing and $\exp(-u)$ decreasing at a rate proportional to u.
If $u = 1$, then

$\exp(+1) = 1 + 1 + 1/2 + 1/6 + 1/24 + \ldots = e = 2.718\ldots$
$\exp(-1) = 1 - 1 + 1/2 - 1/6 + 1/24 - \ldots = 1/e = 0.368\ldots$

In the electrical technology of transients, u is most commonly a negative function of time t given by $u = -(t/T)$. It then has the

Table 1.6 Mathematical symbolism

Term	Symbol
Base of natural logarithms	e $(=2.71828\ldots)$
Complex number	$\mathbf{C}=A+jB=C\exp(j\theta)$
	$=C\angle\theta$
argument; modulus	$\arg\mathbf{C}=\theta$; $\mod\mathbf{C}=C$
conjugate	$\mathbf{C}^*=A-jB=C\exp(-j\theta)$
	$=C\angle-\theta$
real part; imaginary part	$\operatorname{Re}\mathbf{C}=A$; $\operatorname{Im}\mathbf{C}=B$
Co-ordinates	
cartesian	x, y, z
cylindrical; spherical	$r, \phi, z; r, \theta, \phi$
Function of x	
general	$f(x), g(x), F(x)$
Bessel	$J_n(x)$
circular	$\sin x, \cos x, \tan x \ldots$
inverse	$\arcsin x, \arccos x,$
	$\arctan x \ldots$
differential	dx
partial	∂x
exponential	$\exp(x)$
hyperbolic	$\sinh x, \cosh x, \tanh x \ldots$
inverse	$\operatorname{arsinh} x, \operatorname{arcosh} x,$
	$\operatorname{artanh} x \ldots$
increment	$\Delta x, \delta x$
limit	$\lim x$
logarithm	
base b	$\log_b x$
common; natural	$\lg x; \ln x$
	(or $\log x; \log_e x$)
Matrix	A, B
complex conjugate	A^*, B^*
product	AB
square, determinant	$\det A$
inverse	A^{-1}
transpose	A^t
unit	I
Operator	
Heaviside	$p \; (\equiv d/dt)$
impulse function	$\delta(t)$
Laplace $L[f(t)]=F(s)$	$s \; (=\sigma+j\omega)$
nabla, del	∇
rotation $\pi/2$ rad;	j
$2\pi/3$ rad	h
step function	$H(t), u(t)$
Vector	$\mathbf{A}, \mathbf{a}, \mathbf{B}, \mathbf{b}$
curl of A	$\operatorname{curl} \mathbf{A}, \nabla \times \mathbf{A}$
divergence of A	$\operatorname{div} \mathbf{A}, \nabla \cdot \mathbf{A}$
gradient of ϕ	$\operatorname{grad} \phi, \nabla\phi$
product: scalar; vector	$\mathbf{A}\cdot\mathbf{B}; \mathbf{A}\times\mathbf{B}$
units in cartesian axes	$\mathbf{i}, \mathbf{j}, \mathbf{k}$

Table 1.7 Trigonometric functions of θ

θ deg	rad	$\sin\theta$	$\cos\theta$	$\tan\theta$
0	0.0	0.0	1.0	0.0
5	0.087	0.087	0.996	0.087
10	0.175	0.174	0.985	0.176
15	0.262	0.259	0.966	0.268
20	0.349	0.342	0.940	0.364
25	0.436	0.423	0.906	0.466
30	0.524	0.500	0.866	0.577
35	0.611	0.574	0.819	0.700
40	0.698	0.643	0.766	0.839
45	0.766	0.707	0.707	1.0
50	0.873	0.766	0.643	1.192
55	0.960	0.819	0.574	1.428
60	1.047	0.866	0.500	1.732
65	1.134	0.906	0.423	2.145
70	1.222	0.940	0.342	2.747
75	1.309	0.966	0.259	3.732
80	1.396	0.985	0.174	5.671
85	1.484	0.996	0.097	11.43
90	1.571	1.0	0.0	∞

graphical form shown in *Figure 1.2* (left) as a time dependent variable. With an initial value k, i.e. $y=k\exp(-t/T)$, the rate of reduction with time is $dy/dt=-(k/T)\exp(-t/T)$. The initial rate at $t=0$ is $-k/T$. If this rate were maintained, y would reach zero at $t=T$, defining the *time constant T*. Actually, after time T the value of y is $k\exp(-t/T)=k\exp(-1)=0.368k$. Each successive interval T decreases y by the factor 0.368. At a time $t=4.6T$ the value of y is $0.01k$, and at $t=6.9T$ it is $0.001k$.

If u is a quadrature ('imaginary') number $\pm jv$, then

$$\exp(\pm jv)=1\pm jv-v^2/2! \mp jv^3/3! + v^4/4! \pm \ldots$$

because $j^2=-1, j^3=-j1, j^4=+1$, etc. *Figure 1.2* (right) shows the summation of the first five terms for $\exp(j1)$, i.e.

$$\exp(j1)=1+j1-1/2-j1/6+1/24$$

a complexor expression converging to a point P. The length OP is unity and the angle of OP to the datum axis is, in fact, 1 rad. In general, $\exp(jv)$ is equivalent to a shift by $\angle v$ rad. It follows that $\exp(\pm jv)=\cos v \pm j\sin v$, and that

$$\exp(jv)+\exp(-jv)=2\cos v \quad \exp(jv)-\exp(-jv)=j2\sin v$$

For a complex number $(u+jv)$, then

$$\exp(u+jv)=\exp(u)\cdot\exp(jv)=\exp(u)\cdot\angle v$$

Hyperbolic functions A point P on a rectangular hyperbola $(x/a)^2-(y/a)^2=1$ defines the hyperbolic 'sector' area $S_h=\frac{1}{2}a^2\ln[(x/a)-(y/a)]$ shown shaded in *Figure 1.3* (left). By analogy

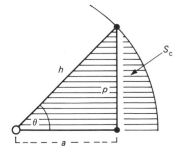
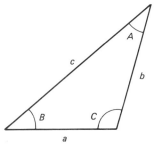

Figure 1.1 Trigonometric relations

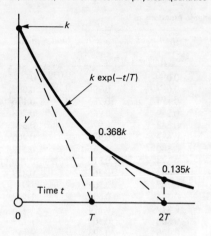

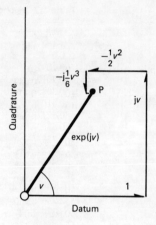

Figure 1.2 Exponential relations

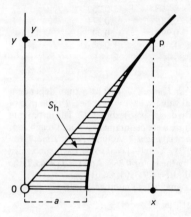

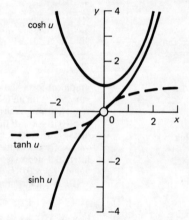

Figure 1.3 Hyperbolic relations

with $\theta = 2S_c/h^2$ for the trigonometrical angle θ, the hyperbolic entity (not an angle in the ordinary sense) is $u = 2S_h/a^2$, where a is the major semi-axis. Then the hyperbolic functions of u for point P are:

$\sinh u = y/a$ $\operatorname{cosech} u = a/y$
$\cosh u = x/a$ $\operatorname{sech} u = a/x$
$\tanh u = y/x$ $\coth u = x/y$

The principal relations yield the curves shown in the diagram (right) for values of u between 0 and 3. For higher values $\sinh u$ approaches $\pm \cosh u$, and $\tanh u$ becomes asymptotic to ± 1. Inspection shows that $\cosh(-u) = \cosh u$, $\sinh(-u) = -\sinh u$ and $\cosh^2 u - \sinh^2 u = 1$.

The hyperbolic functions can also be expressed in the exponential form through the series

$\cosh u = 1 + u^2/2! + u^4/4! + u^6/6! + \ldots$
$\sinh u = u + u^3/3! + u^5/5! + u^7/7! + \ldots$

so that

$\cosh u = \tfrac{1}{2}[\exp(u) + \exp(-u)]$ $\sinh u = \tfrac{1}{2}[\exp(u) - \exp(-u)]$
$\cosh u + \sinh u = \exp(u)$ $\cosh u - \sinh u = \exp(-u)$

Other relations are:

$\sinh u + \sinh v = 2 \sinh \tfrac{1}{2}(u+v) \cdot \cosh \tfrac{1}{2}(u-v)$
$\cosh u + \cosh v = 2 \cosh \tfrac{1}{2}(u+v) \cdot \cosh \tfrac{1}{2}(u-v)$
$\cosh u - \cosh v = 2 \sinh \tfrac{1}{2}(u+v) \cdot \sinh \tfrac{1}{2}(u-v)$
$\sinh(u \pm v) = \sinh u \cdot \cosh v \pm \cosh u \cdot \sinh v$

$\cosh(u \pm v) = \cosh u \cdot \cosh v \pm \sinh u \cdot \sinh v$
$\tanh(u \pm v) = (\tanh u \pm \tanh v)/(1 \pm \tanh u \cdot \tanh v)$
$\sinh(u \pm jv) = (\sinh u \cdot \cos v) \pm j(\cosh u \cdot \sin v)$
$\cosh(u \pm jv) = (\cosh u \cdot \cos v) \pm j(\sinh u \cdot \sin v)$
$d(\sinh u)/du = \cosh u$ $\int \sinh u \cdot du = \cosh u$
$d(\cosh u)/du = \sinh u$ $\int \cosh u \cdot du = \sinh u$

Exponential and hyperbolic functions of u between zero and 6.908 are listed in *Table 1.8*. Many calculators can give such values directly.

1.2.3 Bessel functions

Problems in a wide range of technology (e.g. in eddy currents, frequency modulation, etc.) can be set in the form of the Bessel equation

$$\frac{d^2 y}{dx^2} + \frac{1}{x} \cdot \frac{dy}{dx} + \left[1 - \frac{n^2}{x^2}\right] y = 0$$

and its solutions are called Bessel functions of order n. For $n=0$ the solution is

$$J_0(x) = 1 - (x^2/2^2) + (x^4/2^2 \cdot 4^2) - (x^6/2^2 \cdot 4^2 \cdot 6^2) + \ldots$$

and for $n = 1, 2, 3 \ldots$

$$J_n(x) = \frac{x^n}{2^n n!} \left[1 - \frac{x^2}{2(2n+2)} + \frac{x^4}{2 \cdot 4(2n+2)(2n+4)} - \ldots \right]$$

Table 1.9 gives values of $J_n(x)$ for various values of n and x.

Table 1.8 Exponential and hyperbolic functions

u	exp(u)	exp(−u)	sinh u	cosh u	tanh u
0.0	1.0	1.0	0.0	1.0	0.0
0.1	1.1052	0.9048	0.1092	1.0050	0.0997
0.2	1.2214	0.8187	0.2013	1.0201	0.1974
0.3	1.3499	0.7408	0.3045	1.0453	0.2913
0.4	1.4918	0.6703	0.4108	1.0811	0.3799
0.5	1.6487	0.6065	0.5211	1.1276	0.4621
0.6	1.8221	0.5488	0.6367	1.1855	0.5370
0.7	2.0138	0.4966	0.7586	1.2552	0.6044
0.8	2.2255	0.4493	0.8881	1.3374	0.6640
0.9	2.4596	0.4066	1.0265	1.4331	0.7163
1.0	2.7183	0.3679	1.1752	1.5431	0.7616
1.2	3.320	0.3012	1.5095	1.8107	0.8337
1.4	4.055	0.2466	1.9043	2.1509	0.8854
1.6	4.953	0.2019	2.376	2.577	0.9217
1.8	6.050	0.1653	2.942	3.107	0.9468
2.0	7.389	0.1353	3.627	3.762	0.9640
2.303	10.00	0.100	4.950	5.049	0.9802
2.5	12.18	0.0821	6.050	6.132	0.9866
2.75	15.64	0.0639	7.789	7.853	0.9919
3.0	20.09	0.0498	10.02	10.07	0.9951
3.5	33.12	0.0302	16.54	16.57	0.9982
4.0	54.60	0.0183	27.29	27.31	0.9993
4.5	90.02	0.0111	45.00	45.01	0.9998
4.605	100.0	0.0100	49.77	49.80	0.9999
5.0	148.4	0.0067	74.20	74.21	0.9999
5.5	244.7	0.0041	122.3	$\cosh u$	$\tanh u$
6.0	403.4	0.0025	201.7	$= \sinh u$	$= 1.0$
6.908	1000	0.0010	500	$= \tfrac{1}{2}\exp(u)$	

1.2.4 Series

Factorials In several of the following the factorial ($n!$) of integral numbers appears. For n between 2 and 10 these are

2! =	2	1/2! = 0.5	
3! =	6	1/3! = 0.1667	
4! =	24	$1/4! = 0.417 \times 10^{-1}$	
5! =	120	$1/5! = 0.833 \times 10^{-2}$	
6! =	720	$1/6! = 0.139 \times 10^{-2}$	
7! =	5 040	$1/7! = 0.198 \times 10^{-3}$	
8! =	40 320	$1/8! = 0.248 \times 10^{-4}$	
9! =	362 880	$1/9! = 0.276 \times 10^{-5}$	
10! =	3 628 800	$1/10! = 0.276 \times 10^{-6}$	

Progression
Arithmetic $a + (a+d) + (a+2d) + \ldots + [a+(n-1)d]$
$= \tfrac{1}{2}n$(sum of 1st and nth terms)

Geometric $a + ar + ar^2 + \ldots + ar^{n-1} = a(1-r^n)/(1-r)$

Trigonometric See Section 1.2.1.
Exponential and hyperbolic See Section 1.2.2.
Binomial

$$(1 \pm x)^n = 1 \pm nx + \frac{n(n-1)}{2!}x^2 \pm \frac{n(n-1)(n-2)}{3!}x^3 + \ldots$$

$$+ (-1)^r \frac{n!}{r!(n-r)!}x^r + \ldots$$

$(a \pm x)^n = a^n[1 \pm (x/a)]^n$

Binomial coefficients $n!/[r!(n-r)!]$ are tabulated:

Term r =	0	1	2	3	4	5	6	7	8	9	10
n = 1	1	1									
2	1	2	1								
3	1	3	3	1							
4	1	4	6	4	1						
5	1	5	10	10	5	1					
6	1	6	15	20	15	6	1				
7	1	7	21	35	35	21	7	1			
8	1	8	28	56	70	56	28	8	1		
9	1	9	36	84	126	126	84	36	9	1	
10	1	10	45	120	210	252	210	120	45	10	1

Power If there is a power series for a function $f(h)$, it is given by

$$f(h) = f(0) + hf^{(i)}(0) + (h^2/2!)f^{(ii)}(0) + (h^3/3!)f^{(iii)}(0) + \ldots$$
$$+ (h^r/r!)f^{(r)}(0) + \ldots \quad \text{(Maclaurin)}$$

$$f(x+h) = f(x) + hf^{(i)}(x) + (h^2/2!)f^{(ii)}(x) + \ldots$$
$$+ (h^r/r!)f^{(r)}(x) + \ldots \quad \text{(Taylor)}$$

Permutation, combination

$^nP_r = n(n-1)(n-2)(n-3)\ldots(n-r+1) = n!/(n-r)!$

$^nC_r = (1/r!)[n(n-1)(n-2)(n-3)\ldots(n-r+1)] = n!/r!(n-r)!$

Bessel See Section 1.2.3.
Fourier See Section 1.2.5.

1.2.5 Fourier series

A univalued periodic wave form $f(\theta)$ of period 2π is represented by a summation in general of sine and cosine waves of fundamen-

Table 1.9 Bessel functions $J_n(x)$

n	$J_n(1)$	$J_n(2)$	$J_n(3)$	$J_n(4)$	$J_n(5)$	$J_n(6)$	$J_n(7)$	$J_n(8)$
0	0.7652	0.2239	−0.2601	−0.3971	−0.1776	0.1506	0.3001	0.1717
1	0.4401	0.5767	0.3391	−0.0660	−0.3276	−0.2767	−0.0047	0.2346
2	0.1149	0.3528	0.4861	0.3641	0.0466	−0.2429	−0.3014	−0.1130
3	0.0196	0.1289	0.3091	0.4302	0.3648	0.1148	−0.1676	−0.2911
4	—	0.0340	0.1320	0.2811	0.3912	0.3567	0.1578	−0.1054
5	—	—	0.0430	0.1321	0.2611	0.3621	0.3479	0.1858
6	—	—	0.0114	0.0491	0.1310	0.2458	0.3392	0.3376
7	—	—	—	0.0152	0.0534	0.1296	0.2336	0.3206
8	—	—	—	—	0.0184	0.0565	0.1280	0.2235
9	—	—	—	—	—	0.0212	0.0589	0.1263
10	—	—	—	—	—	—	0.0235	0.0608
11	—	—	—	—	—	—	—	0.0256
12	—	—	—	—	—	—	—	—
13	—	—	—	—	—	—	—	—
14	—	—	—	—	—	—	—	—
15	—	—	—	—	—	—	—	—

n	$J_n(9)$	$J_n(10)$	$J_n(11)$	$J_n(12)$	$J_n(13)$	$J_n(14)$	$J_n(15)$
0	−0.0903	−0.2459	−0.1712	0.0477	0.2069	0.1711	−0.0142
1	0.2453	0.0435	−0.1768	−0.2234	−0.0703	0.1334	0.2051
2	0.1448	0.2546	0.1390	−0.0849	−0.2177	−0.1520	0.0416
3	−0.1809	0.0584	0.2273	0.1951	0.0033	−0.1768	−0.1940
4	−0.2655	−0.2196	−0.0150	0.1825	0.2193	0.0762	−0.1192
5	−0.0550	−0.2341	−0.2383	−0.0735	0.1316	0.2204	0.1305
6	0.2043	−0.0145	−0.2016	−0.2437	−0.1180	0.0812	0.2061
7	0.3275	0.2167	0.0184	−0.1703	−0.2406	−0.1508	0.0345
8	0.3051	0.3179	0.2250	0.0451	−0.1410	−0.2320	−0.1740
9	0.2149	0.2919	0.3089	0.2304	0.0670	−0.1143	−0.2200
10	0.1247	0.2075	0.2804	0.3005	0.2338	0.0850	−0.0901
11	0.0622	0.1231	0.2010	0.2704	0.2927	0.2357	0.0999
12	0.0274	0.0634	0.1216	0.1953	0.2615	0.2855	0.2367
13	0.0108	0.0290	0.0643	0.1201	0.1901	0.2536	0.2787
14	—	0.0119	0.0304	0.0650	0.1188	0.1855	0.2464
15	—	—	0.0130	0.0316	0.0656	0.1174	0.1813

Values below 0.01 not tabulated.

tal period 2π and of integral harmonic orders n $(=2, 3, 4, \ldots)$ as

$$f(\theta) = c_0 + a_1 \cos\theta + a_2 \cos 2\theta + \ldots + a_n \cos n\theta + \ldots$$
$$+ b_1 \sin\theta + b_2 \sin 2\theta + \ldots + b_n \sin n\theta + \ldots$$

The mean value of $f(\theta)$ over a full period 2π is

$$c_0 = \frac{1}{2\pi} \int_0^{2\pi} f(\theta) \cdot d\theta$$

and the harmonic-component amplitudes a and b are

$$a_n = \frac{1}{\pi} \int_0^{2\pi} f(\theta) \cdot \cos n\theta \cdot d\theta, \quad b_n = \frac{1}{\pi} \int_0^{2\pi} f(\theta) \cdot \sin n\theta \cdot d\theta$$

Table 1.10 gives for a number of typical wave forms the harmonic series in square brackets, preceded by the mean value c_0 where it is not zero.

1.2.6 Derivatives and integrals

Some basic forms are listed in *Table 1.11*. Entries in a given column are the integrals of those in the column to its left and the derivatives of those to its right. Constants of integration are omitted.

1.2.7 Laplace transforms

Laplace transformation is a method of deriving the response of a system to any stimulus. The system has a basic equation of behaviour, and the stimulus is a pulse, step, sine wave or other variable with time. Such a response involves integration: the Laplace transform method removes integration difficulties, as tables are available for the direct solution of a great variety of problems. The process is analogous to evaluation (for example) of $y = 2.1^{3.6}$ by *transformation* into a logarithmic form $\log y = 3.6 \times \log(2.1)$, and a subsequent *inverse transformation* back into arithmetic by use of a table of antilogarithms.

The Laplace transform (L.t.) of a time-varying function $f(t)$ is

$$L[f(t)] = F(s) = \int_0^\infty \exp(-st) \cdot f(t) \cdot dt$$

and the inverse transformation of $F(s)$ to give $f(t)$ is

$$L^{-1}[F(s)] = f(t) = \lim \frac{1}{2\pi} \int_{\sigma-j\omega}^{\sigma+j\omega} \exp(st) \cdot F(s) \cdot ds$$

The process, illustrated by the response of a current $i(t)$ in an electrical network of impedance z to a voltage $v(t)$ applied at $t=0$,

Table 1.10 Fourier series

Wave form	Series

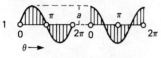

1. Sine: $a \sin \theta$ Cosine: $a \sin \theta$

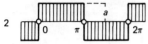

2. Square:
$$a\frac{4}{\pi}\left[\frac{\sin\theta}{1}+\frac{\sin 3\theta}{3}+\frac{\sin 5\theta}{5}+\frac{\sin 7\theta}{7}+\ldots\right]$$

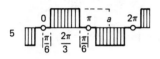

5. Rectangular block:
$$a\frac{2\sqrt{3}}{\pi}\left[\frac{\sin\theta}{1}-\frac{\sin 5\theta}{5}-\frac{\sin 7\theta}{7}+\frac{\sin 11\theta}{11}+\frac{\sin 13\theta}{13}-\frac{\sin 17\theta}{17}-\ldots\right]$$

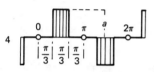

4. Rectangular block:
$$a\frac{4}{\pi}\left[\frac{\sin\theta}{2\cdot 1}-\frac{\sin 3\theta}{3}+\frac{\sin 5\theta}{2\cdot 5}+\frac{\sin 7\theta}{2\cdot 7}-\frac{\sin 9\theta}{9}+\frac{\sin 11\theta}{2\cdot 11}\right.$$
$$\left.+\frac{\sin 13\theta}{2\cdot 13}-\frac{\sin 15\theta}{15}+\frac{\sin 17\theta}{2\cdot 17}+\ldots\right]$$

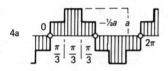

4a. Stepped rectangle:
$$a\frac{3}{\pi}\left[\frac{\sin\theta}{1}+\frac{\sin 5\theta}{5}+\frac{\sin 7\theta}{7}+\frac{\sin 11\theta}{11}+\frac{\sin 13\theta}{13}+\frac{\sin 17\theta}{17}+\ldots\right]$$

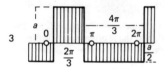

3. Asymmetric rectangle:
$$a\frac{3\sqrt{3}}{2\pi}\left[\frac{\sin\theta}{1}-\frac{\sin 5\theta}{5}-\frac{\sin 7\theta}{7}+\frac{\sin 11\theta}{11}+\frac{\sin 13\theta}{13}-\ldots\right.$$
$$\left.-\frac{\cos 2\theta}{2}-\frac{\cos 4\theta}{4}+\frac{\cos 8\theta}{8}+\frac{\cos 10\theta}{10}-\ldots\right]$$

Triangle:
$$a\frac{8}{\pi^2}\left[\frac{\sin\theta}{1}-\frac{\sin 3\theta}{9}+\frac{\sin 5\theta}{25}-\frac{\sin 7\theta}{49}+\frac{\sin 9\theta}{81}-\frac{\sin 11\theta}{121}+\ldots\right]$$

Sawtooth:
$$a\frac{3}{\pi}\left[\frac{\sin\theta}{1}-\frac{\sin 2\theta}{2}+\frac{\sin 3\theta}{3}-\frac{\sin 4\theta}{4}+\frac{\sin 5\theta}{5}-\ldots\right]$$

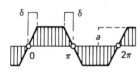

Trapeze:
$$a\frac{4}{\pi\delta}\left[\frac{\sin\delta\cdot\sin\theta}{1}+\frac{\sin 3\delta\cdot\sin 3\theta}{9}+\frac{\sin 5\delta\cdot\sin 5\theta}{25}+\ldots\right]$$
$$a\frac{6\sqrt{3}}{\pi^2}\left[\frac{\sin\theta}{1}-\frac{\sin 5\theta}{25}+\frac{\sin 7\theta}{49}-\frac{\sin 11\theta}{121}+\ldots\right] \text{ for } \delta=\pi/3$$

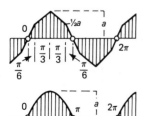

Trapeze-triangle:
$$a\frac{9}{\pi^2}\left[\frac{\sin\theta}{1}+\frac{\sin 5\theta}{25}-\frac{\sin 7\theta}{49}-\frac{\sin 11\theta}{121}+\frac{\sin 13\theta}{169}+\ldots\right]$$

Rectified sine (half-wave):
$$a\frac{1}{\pi}+a\frac{2}{\pi}\left[\frac{\pi\sin\theta}{4}-\frac{\cos 2\theta}{1\cdot 3}-\frac{\cos 4\theta}{3\cdot 5}-\frac{\cos 6\theta}{5\cdot 7}-\ldots\right]$$

cont'd

Table 1.10 (continued)

Wave form	Series
	Rectified sine (full-wave): $$a\frac{2}{\pi} - a\frac{4}{\pi}\left[\frac{\cos 2\theta}{1\cdot 3} + \frac{\cos 4\theta}{3\cdot 5} + \frac{\cos 6\theta}{5\cdot 7} + \frac{\cos 8\theta}{7\cdot 9} + \dots\right]$$
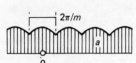	Rectified sine (m-phase): $$a\frac{m}{\pi}\sin\frac{\pi}{m} + a\frac{2m}{\pi}\sin\frac{\pi}{m}\left[\frac{\cos m\theta}{m^2-1} - \frac{\cos 2m\theta}{4m^2-1} + \frac{\cos 3m\theta}{9m^2-1} - \dots\right]$$
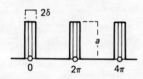	Rectangular pulse train: $$a\frac{\delta}{\pi} + a\frac{2}{\pi}\left[\frac{\sin\delta\cdot\cos\theta}{1} + \frac{\sin 2\delta\cdot\cos 2\theta}{2} + \frac{\sin 3\delta\cdot\cos 3\theta}{3} + \dots\right]$$ $$a\frac{\delta}{\pi} + a\frac{2\delta}{\pi}\left[\frac{\cos\theta}{1} + \frac{\cos 2\theta}{2} + \frac{\cos 3\theta}{3} + \dots\right] \quad \text{for } \delta \ll \pi$$
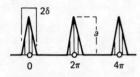	Triangular pulse train: $$a\frac{\delta}{2\pi} + a\frac{4}{\pi\delta}\left[\frac{\sin^2(\frac{1}{2}\delta)}{1}\cos\theta + \frac{\sin^2 2(\frac{1}{2}\delta)}{4}\cos 2\theta + \frac{\sin^2 3(\frac{1}{2}\delta)}{9}\cos 3\theta + \dots\right]$$ $$a\frac{\delta}{2\pi} + a\frac{\delta}{\pi}[\cos\theta + \cos 2\theta + \cos 3\theta + \dots] \quad \text{for } \delta \ll \pi$$

is to write down the transform equation

$$I(s) = V(s)/Z(s)$$

where $I(s)$ is the L.t. of the current $i(t)$, $V(s)$ is the L.t. of the voltage $v(t)$, and $Z(s)$ is the *operational* impedance. $Z(s)$ is obtained from the network resistance R, inductance L and capacitance C by leaving R unchanged but replacing L by Ls and C by $1/Cs$. The process is equivalent to writing the network impedance for a steady state frequency ω and then replacing $j\omega$ by s. $V(s)$ and $Z(s)$ are polynomials in s: the quotient $V(s)/Z(s)$ is reduced algebraically to a form recognisable in the transform table. The resulting current/time relation $i(t)$ is read out: it contains the complete solution. However, if at $t=0$ the network has initial energy (i.e. if currents flow in inductors or charges are stored in capacitors), the equation becomes

$$I(s) = [V(s) + U(s)]/Z(s)$$

where $U(s)$ contains such terms as LI_0 and $(1/s)V_0$ for the inductors or capacitors at $t=0$.

A number of useful transform pairs is listed in *Table 1.12*.

1.2.8 Binary numeration

A number N in decimal notation can be represented by an ordered set of binary digits $a_n, a_{n-1}, a_{n-2}, \dots, a_2, a_1, a_0$ such that

$$N = 2^n a_n + 2^{n-1} a_{n-1} + \dots + 2a_1 + a_0$$

where the as have the values either 1 or 0. Thus, if $N = 19$,

$$19 = 16 + 2 + 1 = (2^4)1 + (2^3)0 + (2^2)0 + (2^1)1 + (2^0)1 = 10011$$

in binary notation. The rules of addition and multiplication are

$0+0=0, 0+1=1, 1+1=10; 0\times 0=0, 0\times 1=0, 1\times 1=1$

1.2.9 Power ratio

In communication networks the powers P_1 and P_2 at two specified points may differ widely as the result of amplification or attenuation. The power ratio P_1/P_2 is more convenient in logarithmic terms.

Neper [Np] This is the natural logarithm of a voltage or current ratio, given by

$$a = \ln(V_1/V_2) \quad \text{or} \quad a = \ln(I_1/I_2) \text{ Np}$$

If the voltages are applied to, or the currents flow in, identical impedances, then the power ratio is

$$a = \ln(V_1/V_2)^2 = 2\ln(V_1/V_2)$$

and similarly for current.

Decibel [dB] The power gain is given by the common logarithm $\lg(P_1/P_2)$ in bel [B], or most commonly by $A = 10\log(P_1/P_2)$ decibel [dB]. With again the proviso that the powers are developed in identical impedances, the power gain is

$$A = 10\log(P_1/P_2) = 10\log(V_1/V_2)^2 = 20\log(V_1/V_2) \text{ dB}$$

Table 1.13 gives the power ratio corresponding to a gain A (in dB) and the related identical-impedance voltage (or current) ratios. Approximately, 3 dB corresponds to a power ratio of 2, and 6 dB to a power ratio of 4. The decibel equivalent of 1 Np is 8.69 dB.

1.3 Physical quantities

Engineering processes involve energy associated with physical materials to convert, transport or radiate energy. As energy has several natural forms, and as materials differ profoundly in their

Decimal	1	2	3	4	5	6	7	8	9	10	100
Binary	1	10	11	100	101	110	111	1000	1001	1010	1100100

actual value have been matched automatically. The correcting variable of the run-up controller with proportional–integral (PI) response passes to the electro-hydraulic transducer over a smallest-value selector and the valve-position controller. It then influences, via the control oil system, the position of the turbine control valves. The critical-speed ranges are passed through with the maximum admissible acceleration.

Once nominal speed is reached, the frequency controller, set to nominal frequency, takes over from the run-up controller, which then ceases to intervene since the speed set-value n_s continues to increase up to 106%.

Load controller with frequency response. Load-frequency set-value The signal for the desired load is formed by the target-load set-value P_z and the adjustable loading rate of change (dP/dt) or rate of change of load shedding $(-dP/dt)$. The load set-value P_s is varied accordingly, this taking place via a transfer unit actuated by push-button.

The load component dependent on frequency is formed by a comparison of the frequency set-value f_s and the actual value f; the control deviation is gained by adjusting the frequency droop. If desired, the influence of the frequency controller can be suppressed within an adjustable range f_T.

The frequency-dependent load component and the load set-value are coupled via a summing junction. The output signal of this junction is thus representative of the load to be delivered by the turbine. All units connected beyond this output are either limiters or subsidiary controllers capable of temporary intervention. These are described individually below.

Maximum-load limiter P_{max} An analogue-value generator P_{max} transmits a value that, for the turbine or the steam generator, represents a load limit not to be exceeded under any circumstances.

Load limiter, admissible rate of change of load Neither the steam turbo-set nor the steam generator can contend with sudden large load changes. These occur particularly when control is according to a frequency–load characteristic and when large frequency fluctuations (e.g. line faults) occur. The step-change and rate-of-change limiter limits these load changes to a variable step change ($\pm$%) and the remainder to a likewise variable rate of change ($\pm dP/dt$). This limiter can be influenced by a controller in relation to the thermal stresses of the turbine, thus overriding the load set-value.

Minimum-load setting P_{min} This prevents the turbine from being erroneously tripped by the reverse-power relay. This could take place after parallel connection of the generator and the line by the synchroniser. The setting is authorised as soon as the generator breaker closes, and causes the generator to deliver a fixed minimum output. As soon as the load set-value exceeds the minimum-load setting, the latter automatically ceases to intervene. It remains, however, on stand-by until the preselected target-load set-value has been attained.

'Unit load demand' signals (ULD 1, 2, 3) These signals can be employed for operation in conjunction with the steam generation control system. A signal relevant to the given application is used, e.g. as a reference variable or as a signal for corrective action.

Load control with live-steam pressure influence For the live-steam pressure controller, a load-controlled turbine represents a controlled object without inherent feedback. The pressure control can be additionally stabilised by applying the control deviation Δp, within certain limits, to the load controller of the turbine. Thus for minor pressure control fluctuations a controlled object with inherent feedback is achieved.

Valve-position controller, turbine control valves and turbine master station The output signal of the lowest-value selector for all controllers represents the reference variable for the valve-position controller and is limited to fixed minimum and maximum values.

To ensure co-ordination of the various PI action controllers (run-up, main and live-steam pressure controllers) during normal operation, the integral part of each controller continually tracks the controller having the smallest registered proportional component.

The turbine master station TLS allotted to the valve-position controller is not used during normal operation but only in the event of faults in, for example, the main controller. In this event the correcting variable received from the main controller is suppressed and the turbine master set to 'manual'. The correcting variables received from the acceleration limiter and the safety-system coordinator always remain on stand-by. When the turbine master is set to 'manual', any desired valve position can be commanded. It should, however, be noted that with this type of operation the operating staff are responsible for ensuring that the admissible turbine limits are observed.

Live-steam pressure controller To prevent an excessive drop in the live-steam pressure when faults occur in the steam generator, or when there are rapid positive changes in load, the live-steam pressure controller intervenes on the adjustable limit p_T being attained. This reduces the turbine load in accordance with the available live-steam flow. The limit can, if desired, be set to zero, after which the Turbotrol acts as an initial pressure controller.

Acceleration limiter When load-shedding to house load, this limiter closes the turbine control valves immediately. Simultaneously, the load set-value P_s is brought to zero within less than 1 s. As soon as the acceleration of the turbine fades out, the limiter ceases to intervene. The frequency set-value alone now determines, via the main controller, the speed of the turbine by controlled opening of the valves (operation feeding only the house load).

Safety system When the turbine trips, the pressure switch MS responds. The signal is led via the evaluator logic to the turbine controller and causes all control values to close immediately.

17.6.6.3 Control room operation and display

Basically, the number and type of units required in a control room for the Turbotrol system vary according to the installation. The functions are actuated by means of non-arresting push-buttons. These may be adapted for use as illuminating buttons. Either solid-state output units or coupling relays should be provided for the alarm signals, depending on the type of warning system.

17.6.6.4 Operating behaviour and maintenance

The philosophy applied to safety, availability and reliability is expounded here without details of the theories of system safety and reliability.

Safety A solid-state Turbotrol control system as described is a single-channel arrangement. In assessing the safety of the entire turbo-set, however, it should be borne in mind that, parallel to the electronic control system, a completely independent hydraulic safety system can be provided with redundant monitors (speed, pressure, vacuum etc.) in a 1-out-of-2 arrangement and independent actuators (main and reheat stop valves).

The safety system is then coupled to the control valves both hydraulically via a relay, and electro-hydraulically via function

semi-automatically with the aid of the run-up controller (424) and starts with the turbine in the no-load condition. Initiated manually, the run-up controller accelerates the turbo-set at a defined rate until the first overspeed monitor responds at 110% of nominal speed. A further command, likewise manually authorised, causes the run-up controller to accelerate the turbo-set up to 112% of nominal speed, at which the second overspeed monitor responds. The turbine is then tripped. A fault in the system causing the set-point value to increase to 114% of nominal speed trips the turbine automatically, i.e. independently of the overspeed monitors.

17.6.6.2 Principle of operation

Run-up controller Providing that the change-over unit is set to 'auto', a speed programme unit SFH (*Figure 17.38*) adjusts the speed set-value by a certain rate of change (dn/dt). Change-over to 'auto' can only be authorised when the speed set-value and the

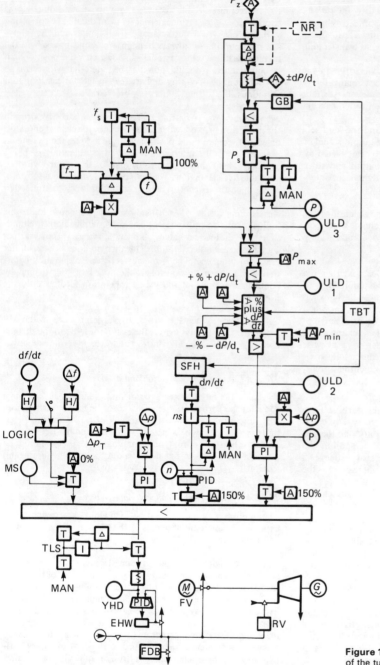

Figure 17.38 Block diagram of the overall arrangement of the turbine controller

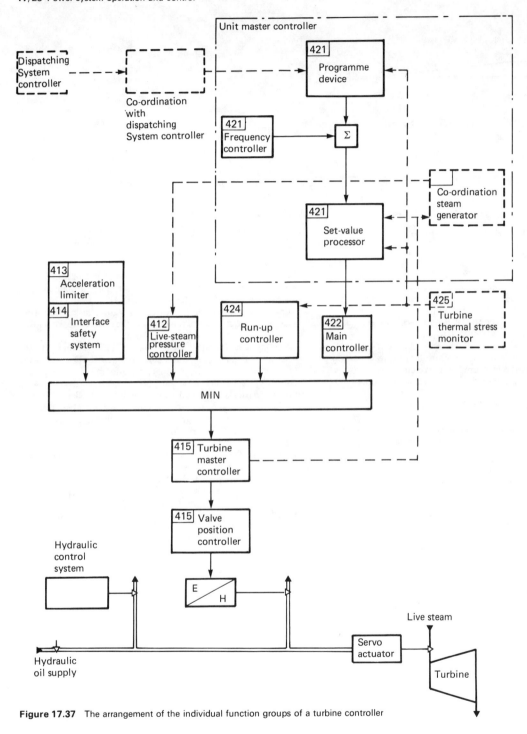

Figure 17.37 The arrangement of the individual function groups of a turbine controller

operating staff to ensure that the admissible turbine limits are observed.

Tripping of the turbine Direct link-up of the turbine hydraulic safety system with the control system, and an electro-hydraulic link between the Turbotrol and the safety system, ensures closure of the control valves each time the turbine is tripped. In addition, the Turbotrol is made ready for start-up, i.e. the turbine master station changes to 'manual' and the 'zero' correcting variable is dispatched to the control valves.

Overspeed test When standard tests have been completed and the necessary preparations made in the hydraulic safety circuit, the overspeed test should be carried out. This takes place

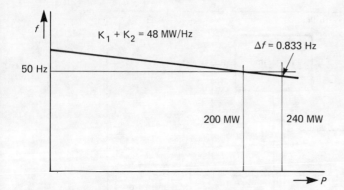

Figure 17.36 Two-machine system, combined characteristic

17.6.6 Steam-turbine control system (Turbotrol)

The Turbotrol electro-hydraulic control system has to fulfil the following objectives:

(1) Automatic variation of the speed set-value for speed-controlled run-up of a turbo-set with respect to critical speed and temperature conditions.
(2) Speed control during no-load operation.
(3) Frequency control when feeding house load.
(4) Accurate rapid load control in accordance with an adjustable linear frequency–load characteristic.
(5) Automatic load set-value variation for loading or load shedding, taking into consideration all measures necessary to prevent the turbo-set from being overstressed.
(6) Maintenance of the speed within admissible limits on load shedding.
(7) Co-ordinated action of the steam generator controller and turbine controller during load operation.
(8) Processing of the signals of an overriding control system, e.g. from the load dispatching system controller.

The Turbotrol system also must ensure (i) high operational safety and availability, (ii) high-response sensitivity when changing specific control variables and (iii) ease of servicing and minimum maintenance.

17.6.6.1 Structure of control system and types of operation

The arrangement of the individual function groups of a turbine controller is shown by the block diagram in *Figure 17.37*. For clarity the numbers assigned to these groups have been inserted in the text. The control system must be designed for the following types of operation:

Normal conditions: start-up, no load, synchronising, load operaton.
Special condition: control of the live-steam pressure by the turbo-set.
Fault conditions: manual control with the turbine master station, turbine trips.
Testing: overspeed, simulation.

Start-up During this phase the run-up controller (424) is in action, the turbine master station is set to 'auto' and the valve-position controller (415) is switched on. The controller begins the run-up from any given starting speed to the nominal speed, full use being made of the admissible operating range of the turbo-set. On reaching nominal speed, the frequency controller, which is set to nominal frequency, takes over from the run-up controller. The operation arrived at is 'no load'.

No load During no-load operation the speed of the turbine is determined by the frequency set-point via the control unit (421), the main controller (422), the turbine master station and the valve-position controller (415). No-load operation at nominal speed forms the basis for both the on-line synchronisation of the generator and the overspeed test.

Synchronisation This type of operation is authorised when a signal is received from the synchroniser. The turbine speed is determined by the frequency set-value, this being varied by pulses from the synchroniser until the frequency and phase of the turbo-set coincide with those of the line, and the circuit-breaker closes.

Load operation After synchronising and closing of the generator breakers, loading can commence. The following function groups are then active: programme unit, frequency controller, set-value processor (421), main controller (422), turbine master station and valve-position controller (415).

The programme unit forms a reference variable from the predetermined target load set-value and the adjustable rate of change of load. This variable is added to the output signal of the frequency controller and then reprocessed in the set-value processor. The frequency controller compares the actual frequency value with the predetermined frequency set-value and delivers the amplified difference as an output signal. The set-value processor contains limit circuits which prevent the turbo-set from being overstressed by rapid changes of set-values. The output signal of the set-value processor is compared with the actual value of the active power in the main controller.

Thus, during load operation, continuous load–frequency control is accomplished under the direction of the load programme device, full use being made of the permissible operating range of the turbo-set.

Turbo-set control of live-steam pressure The Turbotrol system and the steam-generator control system here operate in conjunction. The live-steam pressure controller (412) intervenes when a fault occurs in the steam generation system, thereby taking over from the main controller (422). The system (412) overrides the load set-value and controls the live-steam pressure. The turbine output is then determined by the available live-steam flow.

Manual control The turbine master station is an auxiliary unit which is usually set to position 'auto', but which automatically changes over to 'manual' as soon as an appropriate monitor in the electronic control system has responded. The reference variable for the valve position existing prior to the fault is retained by the turbine master station, while faulty controllers that are not absolutely necessary for the safety of the turbo-set cease to intervene. Function groups 413 and 414 remain permanently on stand-by. Emergency operation may be carried out by manual control for as long as the fault is present.

During manual operation it is the responsibility of the

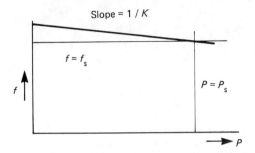

Figure 17.33 Static characteristic: the operating point is confined to the sloping straight line

where D is the *droop* of the system (in Hz/MW). Together with the intersection $(f_s + P_s/K)$, it determines the position of the straight-line characteristic of the speed-governing system. Discussion of the role of the parameters gives an insight into the operation of the system.

17.6.4.1 No-load operation and f_s

On no load the output is zero and the speed is given by the intersection of the straight line with the ordinate. Thus f_s is used to set the no-load speed. P_s is set to zero.

17.6.4.2 Synchronous operation and P_s

When the no-load frequency has reached the system frequency and the synchronising conditions are met, the generator is first connected to the power system, then loaded by raising P_s to the desired value. The characteristic is thereby raised in parallel. The frequency remains unchanged. Should the generator breaker be opened inadvertently, the frequency will assume the value given by the intersection of the straight line with the ordinate. This involves a rise in frequency that is governed by K.

Consider a generator which has the following settings and parameters:

$f_s = 50$ Hz $P_s = 200$ MW $K = 80$ MW/Hz

The generator is operating at $f = 50$ Hz, $P = 200$ MW. After disconnection of the generator, the no-load frequency will become $f = 52.5$ Hz. The droop is $D = 1/K = 0.0125$ Hz/MW. It can, however, also be expressed as a percentage:

$D = 100(\Delta f/f_0)(P_0/P)$

where f_0 and P_0 are nominal values. In this example, the droop is 5%. Thus the droop is a determining factor for the no-load frequency after load shedding.

17.6.4.3 Isolated operation

Although the set-point f_s is maintained at 50 Hz, the frequency f will deviate from f_s because of load changes. The intersection of the characteristic with the load characteristic (e.g. vertical line) fixes the frequency.

Consider the situation given by *Figure 17.34* (not drawn to scale). The set-points are $f_s = 50$ Hz, $P_s = 160$ MW and $K = 100$ MW/Hz. The load is 180 MW. Hence $f = 49.75$ Hz. In order to adjust the frequency to its nominal value, P_s has to be raised by 20 MW. Load shedding would result in a frequency rise according to $f_s + DP_s$.

Frequency control in isolated operation means either letting the operating point vary along the characteristic or readjusting the set-point whenever a frequency deviation arises.

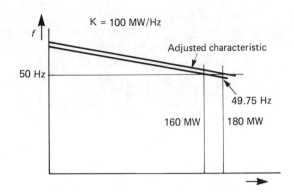

Figure 17.34 Isolated operation of one generator: load increase by 20 MW, and readjustment of characteristic in order to raise frequency to 50 Hz

17.6.5 Parallel operation of generators

Consider a two-machine system to serve a common variable load. Each set has a speed governor. The system schematic is given in *Figure 17.35*; it is assumed that the generators supply 200 MW shared in the ratio 80/120. The frequency is at the nominal value of 50 Hz.

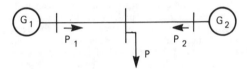

Figure 17.35 Two-machine system: G_1, G_2, generators operating at the same frequency serving P; $P_1 = 80$ MW, $P_2 = 120$ MW

How will an increase of load by 40 MW be shared? The problem is readily solved by calculation or graph. The required data for the two generators G_1 and G_2 is

G_1: $f_{s1} = 50.0$ Hz $P_1 = 80$ MW $P_{s1} = 80$ MW $K_1 = 16$ MW/Hz
G_2: $f_{s2} = 50.0$ Hz $P_2 = 120$ MW $P_{s2} = 120$ MW $K_2 = 32$ MW/Hz

The new frequency f and the power increments ΔP_1 and ΔP_2 are to be determined, with $\Delta P_1 + \Delta P_2 = 40$ MW. Then

$P_{s1} - P_1 - \Delta P_1 + K_1(f_{s1} - f) = 0$

$P_{s2} - P_2 - \Delta P_2 + K_2(f_{s2} - f) = 0$

But $f_{s1} = f_{s2} = f_s$, and f is the common frequency; hence these equations in combination give

$(P_{s1} + P_{s2}) - (P_1 + P_2) - (\Delta P_1 + \Delta P_2) - (K_1 + K_2)(f_s - f) = 0$

a resultant characteristic with a set-point of $P_{s1} + P_{s2}$ and a gain $K_1 + K_2$, or a droop $1/(K_1 + K_2)$ (*Figure 17.36*). The solution is $\Delta P_1 = 13.3$ MW, $\Delta P_2 = 26.7$ MW.

This principle of combining generators with equivalent generators, with a composite droop, is quite general and can be extended to multi-machine systems. Any generator can be taken as generator G_1 and the rest as generator G_2. The combined system has a composite characteristic called a power–frequency characteristic. Thus in an interconnected system, control of power and frequency is realised by adjusting the set-points of the various generators either manually or by automatic means.

The logical extension of frequency control at this level (primary control) to a regional level is load–frequency control or automatic generation control.

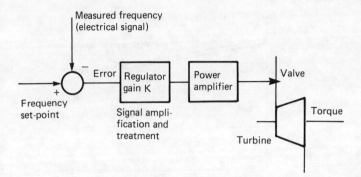

Figure 17.31 Basic arrangement of speed governor (frequency controller: the basic functions of error detection, signal amplification and power amplification are shown)

with a reference signal (also called the 'set-point'). The difference between the measured signal and the set-point is converted to the valve position, which involves a gain. This gain has a two-fold meaning. First, it is a pure signal gain relating a deviation, based on nominal quantities, and a deviation of the valve position which is also referred to a nominal position. Second, the gain means also a power amplification, i.e. the conversion of a weak electrical signal to a strong mechanical torque. A schematic of the basic arrangement is given in *Figure 17.31* where the functional relations are shown.

Assuming that the stability of such a system is guaranteed, transient and steady-state conditions can easily be calculated. The basic functions can best be understood by considering steady-state or quasi-steady-state operating points. If the need arises to deliver more power to the generator, the valve of the turbine has to be opened. This is realised by forming a difference between the set-point and the measured speed signal either by raising the set-point or by a drop in frequency. Changes in the governing system will take place until a balance is reached. Steady-state conditions are found from the relations

$$\Delta f K = M_e \qquad \Delta f = f_s - f$$

where f is the frequency, Δf if the frequency deviation, f_s is the set-point, M_e is the electrical or load torque and K is the gain of the speed governor in units of torque per unit of frequency.

The gain determines the amount of frequency deviation. The higher K is, the smaller Δf will be. However, the gain cannot be chosen to be arbitrarily high. There are problems of stability and certain restrictions due to the allocation of changes of power. The relation $\Delta f K = M_e$ also shows that the speed governor has a double function. It carries out both load control, i.e. the matching of prime-mover torque to the load torque, and frequency control. The relation between these two functions is given by the gain K. Its dimensions in practice are megawatts per hertz; i.e. instead of torque, it is the power which is measured at the output.

Hence, in isolated operation, frequency changes follow a load change so long as the set-point is fixed. A change in set-point will cause a rise in frequency when the load remains constant.

In interconnected operation, the frequency remains practically unchanged. Therefore, a change in loading cannot be affected by the power system. The output of the generator changes without any variation of the frequency, only when the set-point is changed. In this case the speed governor is a pure load controller.

In practice the speed governor is always ready to take on either function. It is the boundary conditions that determine its momentary role, i.e. the governor controls the output power when the frequency is imposed and it controls the frequency when the load is imposed. In a more complex situation, when load and frequency are dependent, the governor varies its output until an equilibrium is reached.

17.6.4 Static characteristic

Modern speed governors have amplifiers that include an integrating property. This choice is made for both reasons of principle and technical reasons, the main one of which is the control of servomotors by valves or signal converters. The proportional behaviour is still maintained by adding the output power as a measured quantity. Such a governor acting on a turbine–generator is commonly represented by a block diagram as shown in *Figure 17.32*.

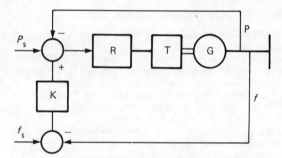

Figure 17.32 Schematic of a power–frequency-controlled generator: G, generator; T, turbine; R, regulator (integrating); P, measured power; P_s, set-point power; f, measured frequency; f_s, frequency set-point; K, gain

In this schematic, two signals measured at the output are compared with set-points, i.e. frequency and power. All deviations are combined in one summing junction, the frequency deviation being weighted by K. The summing junction produces an error which drives the regulator as an integrator. The regulator, which is at the same time a power amplifier, drives the valve.

The steady-state behaviour of the system is described by a characteristic graph with a few parameters. The starting point is the zero value of the error e when the system (*Figure 17.32*) has reached a stationary operating point. Then

$$e = P_s - P + K(f_s - f) = 0$$

where P is the measured power output and P_s is the power set-point; the other parameters have already been defined. P and f are variables, represented by the axes in *Figure 17.33*. The frequency f can be expressed by $f = f_s + (P_s - P)/K$, a linear relation between f and P. The slope is

$$df/dP = -1/K = -D$$

swing depends on the site of the short-circuit, because torque M_{B2} incorporates the network reactances between the machine and short-circuit location. In some instances, therefore, a short circuit across the terminals may be less of a stability problem than a distant short circuit.

Direct account is taken of all the braking torques M_{B1}–M_{B3} only in the case of models 1 and 2. M_{B1} is present with all the models. With model 3 there is also torque M_{B2}.

The following conclusions can be drawn:

(1) The transformation terms in the Park equations (models 1 and 2) and in the network equations must be taken into account (i) when power is low, (ii) when the short circuit is in the vicinity of the generator, and (iii) if the braking torque M_{B3} brought about by the transformation terms cannot be introduced into the mechanical equations.
(2) The backswing effect occurs only when the site of the short circuit is close to the generator. From this it follows that under certain circumstances a short circuit remote from the generator can be more dangerous, as regards stability, than a short circuit nearby.
(3) It is known that the individual synchronous machine models behave very differently.

In many instances, especially in the case of transient stability problems, the simpler models (4 and 5) exhibit the same behaviour as model 2 or 1 with transformation effect. Nevertheless, with the simple models it is particularly important to represent damping correctly. Models 4 and 5 are therefore particularly suitable for simulating synchronous machines with constant excitation voltage that are located far away from the short circuit. Investigation shows also that the widely recommended model 6 (constant voltage beyond transient reactance) yields results that are pessimistic.

If it is assumed that the constant H decreases, and reactances X_d, X_d' and X_t increase, with rising generator capacity, then the critical short-circuit duration will be smaller as generator capacity rises. Studies show, however, that the behaviour of even large synchronous machines can be described with the known models of synchronous machines, and that there is usually no need to make allowance for the transformation terms.

References

1. BONANOMI, P., GÜTH, G., BLASER, F. and GLAVITSCH, H., 'Concept of a Practical Adaptive Regulator for Excitation Control', Paper No. 79 453-2, IEEE Summer Power Meeting (1979)
2. PENEDER, F. 'Modern Excitation Equipment for Power Station Generators', *Brown Boveri Review*, **67**, 173–179, March (1980)
3. PENEDER, F. and BERTSCHI, R., 'Slip Stabilization Equipment, Transfer Functions and the Relevant Experience', *Brown Boveri Review*, **65**, 724–730, November (1978)
4. PENEDER, F. and BERTSCHI, R., 'Slip Stabilization', *Brown Boveri Review*, **61**, 448–454, September/October (1974)

17.6 Decentralised control: electronic turbine controllers

17.6.1 Introduction

The turbine equipped with a controller represents an important item of decentralised control as its controller governs the active power input to the power system. For many reasons, both historical and technical, this controller is kept separate from excitation control, although a computer would be able to fulfil the combined function.

Modern turbine controllers are electronic and have to meet a series of functions and requirements. Frequency control is just one of these; safety functions, monitoring, limit checking etc. are equally important. The most important requirements are the following:

Automatic start-up and shut-down of the turbine
Application of a load–frequency control system
Application of an external reference input
Adjustable droop of the speed controller
Good adaptability to the given operating condition
High-response sensitivity
Co-ordinated operation of the turbine–generator set and all other systems in the power-plant
Short reaction time between controller and control valves
Short closing time of the control valves

Nowadays turbine controllers are composed of electronic modules or functional groups. Most of these can be used in the control system of steam, gas or water turbines. However, the control schemes of such turbines vary widely. We therefore describe both a modern steam-turbine control system and a system for a Francis water turbine. First, however, we explain in simple terms the basic functions of frequency and load control. We derive the elementary block diagram of a frequency controller, explain set-point control, and set out the basic concept of droop. Our treatment concludes with a consideration of frequency control in a multi-machine system.

17.6.2 The environment

The turbine driven by water or steam is connected to a synchronous generator. The synchronous generator has the important property of being able to align itself with other generators when the armatures are interconnected and the rotors are excited. When this alignment is achieved, i.e. when peaks and zero crossings of the induced voltages occur at the same instant, 'synchronous' operation is realised. In greater detail, there are small differences between the mechanical positions of the rotors due to the individual loadings, which may reach 45°. We do not consider different numbers of poles, but assume that all generators have the same number of poles.

In synchronous operation all generators have the same frequency or speed. When the frequency changes, all generators change their speed. Hence, frequency or speed control in synchronous or interconnected operation means control of one common frequency. Small deviations and transients with respect to this frequency are corrected by the inherent ability of the synchronous machines to maintain alignment.

A single generator can also be operated in a system. However, in this case there is no other generator to which it can be aligned. The operation is similar to that of a d.c. generator. Speed control and angular position are purely a matter of the balance between prime-mover torque and load torque acting on the inertia of the shaft. Following a load change, there will be a change in frequency. The speed governor, which is the frequency controller, has an important bearing on the frequency behaviour.

In contrast, a change in loading of a single generator in interconnected operation does not necessarily affect the frequency because it is maintained by the other generators. Hence it is important to know the environment in which a speed governor has to function. Modern governing systems are being designed to cope with a wide range of system conditions.

17.6.3 Role of the speed governor

The speed governor can best be understood by considering a steam engine with a generator serving a local load in isolation, the turbine being controlled by a fly-ball governor. The basic function is realised by a proportional control, whereby the valve position is the actuating signal. Speed or frequency is the input from the system, i.e. the controlled quantity which is compared

$I_N = S_N / \sqrt{3} V_N$ for currents i_d and i_q (17.26)

$M_N = S_N / \omega_B$ for torque M (17.27)

I_{fd0} for excitation current i_{fd}

V_{fd0} for the excitation voltage

In Equations (17.1), R_e and L_e are respectively the resistance or the inductance of the positive-sequence system between generator and infinite network, i.e. $\omega L_e = X_{tr} + X_N$.

If the site of the short circuit is between the generator and the infinite network, there are ten first-order differential equations to integrate.

Model 2 This model has only one damper winding in the quadrature axis, i.e. $i_{tq} = 0$. The relationships $\psi_{tq} = f(i_q, i_{tq}, i_{kq})$ in Equations (17.2) and $d\psi_{tq}/dt = R_{tq} i_{tq}$ in Equations (17.3) do not apply. Instead of Equations (17.16) and (17.19) we have

$$X_{kq} = \frac{(X_q - X_p)(X_q'' - X_p)}{X_q - X_q''}$$ (17.16a)

$$R_{kq} = \frac{(X_q - X_p)^2 X_q''}{(X_q - X_q'') T_q'' X_q \omega_B}$$ (17.19a)

Equations (17.20), (17.23) and (17.24) do not apply.

Model 2 also takes account of the transformation terms, but has only two damper windings. In this case only nine differential equations have to be integrated.

Model 3 This, like model 2, has one damper winding in each of the direct and quadrature axes. The transformation terms in the stator equations do not apply, i.e. Equations (17.3) become

$V_d^G = \psi_q \omega - R i_d$

$V_q^G = -\psi_d \omega - R i_q$

$d\psi_{fd}/dt = V_{fd} - R_{fd} i_{fd}$ (17.3a)

$d\psi_{kd}/dt = -R_{kd} i_{kd}$

$d\psi_{kq}/dt = -R_{kq} i_{kq}$

Model 3 also has two damper windings. The transformation terms are disregarded. There are only five differential equations to integrate. It is found from experience that with this configuration the time step can be much larger than with models 1 and 2.

The equations of motion (17.7)–(17.9) remain unchanged. In addition to M_B in Equation (17.4), however, the braking torque M_{Bs}, caused by the transformation terms when switching operations occur, must be introduced into Equation (17.7):

$$2H(ds/dt) = M - M_B - M_{Bs}$$

if the movement process is to be represented correctly.

Model 4 This model has no damper windings. The basic equations therefore become

$\psi_d = (X_p + X_{ad}) i_d - X_{ad} i_{fd}$
$\psi_{fd} = -X_{ad} i_d + (X_{ad} + X_{fd}) i_{fd}$ (17.2b)
$\psi_q = (X_p + X_{aq}) i_q$

$V_d^G = \psi_q \omega - R i_d$

$V_q^G = -\psi_d \omega - R i_q$ (17.3b)

$d\psi_{fd}/dt = V_{fd} - R_{fd} i_{fd}$

In the absence of damper windings, the asynchronous damping torque (M_{B2}) also has to be included in equation of motion (17.7):

$$2H(ds/dt) = M - M_B - M_{Bs} - M_{B2}$$

Instead of Equations (17.10)–(17.24) we have the expressions

$X_{ad} = X_d - X_p$ (17.10)

$X_{aq} = X_q - X_p$ (17.11)

$$X_{fd} = \frac{(X_d - X_p)(X_d' - X_p)}{X_d - X_d'}$$ (17.13c)

$$R_{fd} = \frac{X_d - X_p}{X_d - X_d'} \frac{X_d'}{T_d' X_d \omega_B}$$ (17.17c)

The transformation terms are disregarded. Thus only three differential equations need to be integrated.

Model 5 This model has no field winding and no damper winding. The base equations for this model are:

$\psi_d = X_d' i_d - E_f$

$\psi_q = X_q i_q$ (17.2d)

$V_d^G = \psi_q \omega - R i_d$

$V_q^G = -\psi_d \omega - R i_q$ (17.3d)

The mechanical behaviour is described by Equations (17.4) and (17.7)–(17.9). E_t is calculated in terms of the initial values of i_d, i_q and V_q^G:

$$E_t = V_q^G + R i_q + X_d' i_d$$

On the stator, conditions are represented by one winding each in the direct and quadrature axes, their reactances being different. The effect of excitation appears as a current source in the equivalent circuit of the direct-axis winding. Only the two differential equations for the mechanical system have to be integrated.

Model 6 This model, like model 5, has neither field winding nor damper windings. Equations (17.2d) contain X_d' instead of X_q. Thus Equation (17.4) becomes

$$M_B = E_t i_q$$

Model 6 resembles model 5 except that it has identical reactances in the direct and quadrature axes.

17.5.10.2 Comparison of the models

The essential difference between models 1 and 2 on the one hand, and models 3–6 on the other, is the allowance made for the transformation effect. Models 1–3 have damper windings, while models 4–6 do not. The stability of a synchronous machine after disconnection of the short circuit is determined chiefly by the accelerating torque M_A, or braking torque M_B, that was acting on the rotor during the short circuit. Accelerating torque M_A is composed of driving torque M and braking torque M_B: i.e. $M_A = M - M_B$. The braking torque M_B arising during the short circuit and oscillations must therefore be correctly represented in each model.

The braking torque M_B acting after the fault has occurred consists of the following component parts:

(1) a braking torque M_{B1} corresponding to losses in the stator and network resistance R;
(2) an asynchronous braking torque M_{B2} corresponding to losses in the rotor field and damper windings;
(3) a decaying braking torque M_{B3} introduced by the transformation terms in the machine and network equations. This torque becomes effective at every change of state.

M_{B3} is mainly responsible for the well-known 'backswing' effect. The synchronous machine is then braked initially in the first 20 ms after the short circuit. However, the extent of the back-

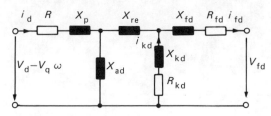

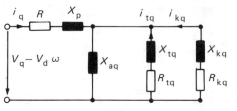

Figure 17.30 Equivalent circuit for model 1 of a synchronous machine: for symbols, see text

Saturation is disregarded and the same equations of motion are used for all models.

Model 1 This model has six windings (three damper windings). Its equivalent circuit diagram is shown in *Figure 17.30*. The corresponding equations are as follows.

The network equations are

$$V_d^G = V_d^N + R_e i_d + L_e(di_d/dt) - \omega L_e i_q$$
$$V_q^G = V_q^N + R_e i_q + L_e(di_d/dt) + \omega L_e i_d \quad (17.1)$$

in which $\omega L_e = X_{tr} + X_N$.

The following equations are used for the synchronous machine:

Flux–current relationships

$$\psi_d = (X_p + X_{ad})i_d + X_{ad}i_{kd} - X_{ad}i_{fd}$$
$$\psi_{kd} = X_{ad}i_d + (X_{ad} + X_{re} + X_{kd})i_{kd} - (X_{ad} + X_{re})i_{fd}$$
$$\psi_q = (X_p + X_{aq})i_q + X_{aq}i_{tq} + X_{aq}i_{kq} \quad (17.2)$$
$$\psi_q = (X_p + X_{aq})i_q + X_{aq}i_{tq} + X_{aq}i_{kq}$$
$$\psi_{tq} = X_{aq}i_q + (X_{aq} + X_{tq})i_{tq} + X_{aq}i_{kq}$$
$$\psi_{kq} = X_{aq}i_q + X_{aq}i_{tq} + (X_{aq} + X_{kq})i_{kq}$$

Voltage equations

$$d\psi_d/dt = -V_d^G + \psi_q\omega - Ri_d$$
$$d\psi_q/dt = -V_q^G - \psi_d\omega - Ri_q$$
$$d\psi_{fd}/dt = V_{fd} - R_{fd}i_{fd} \quad (17.3)$$
$$d\psi_{tq}/dt = -R_{tq}i_{tq}$$
$$d\psi_{kd}/dt = -R_{kd}i_{kd}$$
$$d\psi_{kq}/dt = -R_{kq}i_{kq}$$

Mechanical equations

Torque:

$$M_B = \psi_q i_d - \psi_d i_q \quad (17.4)$$

Power:

$$P = V_d^G i_d + V_q^G i_q \quad (17.5)$$
$$Q = V_q^G i_d - V_d^G i_q \quad (17.6)$$

Equation of motion:

$$2H(ds/dt) = M - M_B \quad (17.7)$$
$$d\theta/dt = s \quad (17.8)$$
$$\omega = 1 + s \quad (17.9)$$

The quantities L, X, R, ψ, i, V, M and ω are per-unit quantities.
From the input data

$R, X_p, X_c, X_d, X_q, X'_d, X''_d, X''_q, T'_d, T'_q, T''_d, T''_q, S_N, V_N, f, I_{fd0}, V_{fd0}, p, H$

of which T, S_N, V_N, f, I_{fd0}, V_{fd0} and H are not in p.u., the constants X_{ad} and X_{aq} of the synchronous machine can be calculated in the following manner:

$$X_{ad} = X_d - X_p \quad (17.10)$$
$$X_{aq} = X_q - X_p \quad (17.11)$$
$$X_{re} = \frac{X_{ad}(X_{re} - X_p)}{X_d - X_{re}} \quad (17.12)$$
$$X_{td} = \left(\frac{X_d - X_p}{X_d - X_{re}}\right)^2 \frac{(X_d - X_e)(X'_d - X_e)}{X_d - X'_d} \quad (17.13)$$
$$X_{kd} = \left(\frac{X_d - X_p}{X_d - X_{re}}\right)^2 \frac{(X'_d - X_e)(X''_d - X_e)}{X'_d - X''_d} \quad (17.14)$$
$$X_{tq} = \frac{(X_q - X_p)(X'_q - X_p)}{X_q - X'_q} \quad (17.15)$$
$$X_{kq} = \frac{(X'_q - X_p)(X''_q - X_p)}{X'_q - X''_q} \quad (17.16)$$
$$R_{fd} = \left(\frac{X_d - X_p}{X_d - X_{re}}\right)^2 \frac{(X_d - X_e)^2}{X_d - X'_d} \frac{1}{T_{fd}\omega_B} \quad (17.17)$$
$$R_{kd} = \left(\frac{X_d - X_p}{X_d - X_{re}}\right)^2 \frac{(X'_d - X_e)^2}{X'_d - X''_d} \frac{X''_d}{T''_d X'_d \omega_B} \quad (17.18)$$
$$R_{kq} = \frac{(X'_q - X_p)^2}{X'_q - X''_q} \frac{X''_q}{T''_q X'_q \omega_B} \quad (17.19)$$
$$R_{tq} = \frac{(X_q - X_p)^2}{X_q - X'_q} \frac{1}{T_{tq}\omega_B} \quad (17.20)$$
$$T_{td} = T'_d \frac{X_d}{X'_d} - \left(T_{kd} - T''_d \frac{X'_d}{X''_d}\right) \quad (17.21)$$

where

$$T_{kd} = T''_d \frac{X'_d}{X''_d}$$
$$\times \frac{(X_d - X_{re})(X'_d - X''_d) + (X'_d - X_{re})(X''_d - X_{re})}{(X'_d - X_{re})^2} \quad (17.22)$$

$$T_{tq} = T'_q \frac{X_q}{X'_q} - \left(T_{kq} - T''_q \frac{X'_q}{X''_q}\right) \quad (17.23)$$

$$T_{kq} = T''_q \frac{X'_q}{X''_q}$$
$$\times \frac{(X_q - X_p)(X'_q - X''_q) + (X'_q - X_p)(X''_q - X_p)}{(X'_q - X_p)^2} \quad (17.24)$$

$$\omega_B = 2\pi f \quad (17.25)$$

The reference values are:

V_N for voltages V_d^G and V_q^G

S_N for powers P and Q

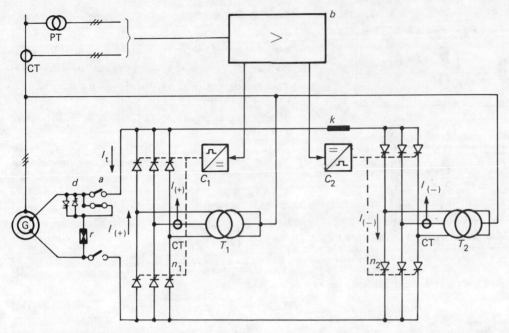

Figure 17.29 Circuit diagram for excitation device: CT, current transformer; G, synchronous generator; I_f, field current of synchronous generator; PT, voltage transformer; T_1, T_2 transformers; a, field suppression switch; b, control electronics; C_1, C_2, firing-angle devices; d, crowbar; k, reactor; n_1, static converter for positive excitation current; n_2, static converter for negative excitation current; r, non-linear de-energising resistor

If the operating conditions now change specifically with regard to the frequency, the limitations on the use of negative excitation currents must be carefully investigated, since the design of the means for compensating the transmission system will be affected. For instance, with a synchronous machine operating on a long line, if a sudden drop occurs in the active load at the consumer end of the line then, in addition to the charging power for the network, a frequency rise also occurs. This leads both to a rise in the quadrature-axis reactance of the machine and to a fall in the capacitive network impedance. This implies, however, that maintaining the voltage is rendered more difficult by the square of the frequency change. The maximum negative excitation that can be used must in this case be reduced in inverse proportion to the frequency.

In order to decouple the two d.c. circuits (to absorb the different voltage–time areas) and to avoid short-circuit balancing currents, a reactor is incorporated in the d.c. intermediate circuit. The balancing current between the two bridges is regulated to a minimum, but it does provide a guarantee that the two converter sets always carry current; therefore change-over from negative to positive excitation current can be effected virtually without loss of time.

17.5.10 Machine models for investigating stability

Generally, when investigating stability, one examines the consequences of the following disturbances:

(1) Short circuit (1- to 3-phase) in the network with subsequent complete (3-phase) or partial interruption of the load flow between generator and network (rapid reclosure).
(2) Partial or complete interruption of load flow between generator and network (load rejection).
(3) Small fluctuations in the load flow (static stability).

It is always assumed that the generators are connected to an infinite power system via a transformer and a network reactance X_n (single-machine problem).

The major parameters for studying stability are:

(a) the generator parameters $H, X_d, X'_d, X''_d, X'_q, (X'_q), X''_q, T'_d, T''_d, (T'_q), T''_q, X_p, X_{re}, E = f(I_F)$;
(b) the transfer functions of the voltage regulator and turbine controller;
(c) the unit-transformer reactance X_{tr} and the network;
(d) the nature of the fault (1-phase, 3-phase, short circuit etc.), the fault location and fault time t_F.

The generator parameters are determined by the construction of the machine. Evaluation of the data for a variety of machines reveals the following trend: for a given frequency f, the inertia constant H diminishes with increase in rating S and a decreasing number p of pole-pairs:

$$H = \frac{GD^2}{4} \frac{\omega_s^2}{2S}$$

where $GD^2/4$ is the moment of inertia referred to diameter, and $\omega_s = 2\pi f/p = 2\pi n_s$ is the synchronous angular velocity. Only a lower limit of H can be specified, because its magnitude depends considerably on the turbine inertia.

The reactances X_d and X'_d, the stator resistance R and the short-circuit time-constant T'_d show a tendency to become larger with increasing power rating. However, these relationships can be very strongly influenced by structural features and design layout.

Experience shows that the method of calculation and the machine model influence the results more strongly with high-speed transients than with slow ones.

17.5.10.1 Synchronous machine models

Six different models of synchronous machines are generally used.

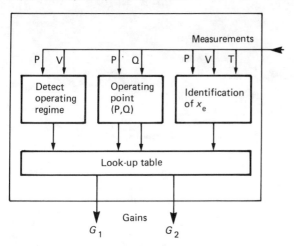

Figure 17.27 The adaptation scheme

adaptation. The identification of X_e is based on a simple curve-fitting procedure. The adaptation scheme (*Figure 17.27*) is implemented on a digital microprocessor. The gains G_1 and G_2 are then applied to the analogue regulator of *Figure 17.26*.

The need for adaptation at critical operating points is demonstrated in *Figure 17.28*. Two sets of gains are used, each designed for a specific operating point. The graphs show that the

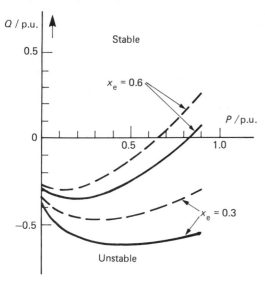

Figure 17.28 Steady-state stability limits of the generator: ---, unadapted regulator; ——, adapted regulator

gains designed for one case may not be used in the other. When the wrong gains are used, instability occurs after a disturbance or in the form of self-induced oscillations. *Figure 17.28* shows the region of stable operating points for two typical values of X_e. The stability boundary is evaluated using the linearised model.

17.5.9 Static excitation systems for positive and negative excitation current

Hydroelectric power-plants are usually remote from the load centre, and the power produced is transmitted over long 3-phase transmission lines to the consumer. Depending upon the active power to be transmitted (light- or heavy-load operation), the transmission line will supply or consume reactive power, which must be absorbed or supplied by generators at two ends of the line or by reactive current compensators (var sources). Owing to the great distance of transmission along the line, the excitation equipment of such synchronous machines must be designed for high ceiling voltages and negative excitation currents in conjunction with stability of operation. Static excitation equipments with controlled static converters connected in antiparallel and subject to circulating current are capable of satisfying these requirements for hydroelectric generators and synchronous compensators; they improve the capacity of these machines for absorbing reactive load, while retaining optimum control and operating characteristics not merely for steady-state operation but also for dynamic and transient conditions in the network. The circuit diagram for an excitation device is shown in *Figure 17.29*.

The maximum continuously permissible positive excitation current is defined by the maximum continuous load of the synchronous machine for a given power factor. The short-term loadings due to ceiling current and load-independent short-circuit current at the rotor in the event of a fault must also be superimposed upon this value. The maximum continuous current of an excitation device for hydroelectric generators is therefore about 1.5 times the rated excitation current, if redundancy design is ignored. On the other hand, the maximum negative excitation current that must be applied to maintain the voltage of the synchronous machine under extreme capacitive load is determined solely by the design, and therefore by the machine parameters, of the synchronous machine.

A salient-pole generator of terminal voltage V operating at a load angle δ with an internal field electromagnetic force (e.m.f.) E_p supplies a reactive power Q given by

$$Q = \frac{E_p V}{X_d} \cos \delta - \frac{V^2}{X_d}\left(1 + \frac{X_d + X_q}{X_q} \sin \delta\right)$$

where X_d and X_q are respectively the direct- and quadrature-axis reactances. If the load angle δ and the excitation E_p aer zero, the synchronous machine takes continuously from the network a reactive power, the value of which is determined by the direct-axis reactance X_d: i.e. $Q = -V^2/X_d$. The supply of reactive power by the network is then just equal to the absorption capability of the synchronous machine. If no negative excitation current can be supplied and the capacitive network load increases further, self-excitation will occur, and the machine must be disconnected. Since, however, the synchronising torque $M_e = dP/d\delta$ begins to decrease only with negative excitation beyond the value corresponding to $E = -V[(X_d - X_q)/X_q]$, negative excitation up to this value can be introduced with quick-acting regulators, the reactive power absorption rising to $Q = -V^2/X_q$.

The ratio X_d/X_q for salient-pole hydro-generators lies in the range 1.3–1.4, and the no-load excitation current I_{f0} corresponds to the generator terminal voltage. The maximum negative exciting current is therefore $I_{f0}(X_d/X_q - 1)$, which approximates to $0.4 I_{f0}$. Thus the equipment for imposing negative excitation needs to be designed for only about 40% of the excitation on no load; it does, however, ensure an increase of about 40% in the reactive power absorption at rated load and frequency. If the reactive load supplied by the connected network rises above $Q_c = V^2/X_q$, the synchronous machine will no longer be capable of holding the voltage, and the definitive self-excitation condition, which cannot be controlled by any excitation current or regulator, will be initiated.

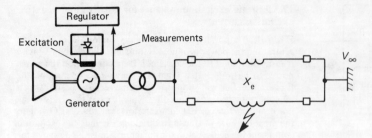

Figure 17.25 Power system model

transmission lines with circuit-breakers. V_∞ is an ideal 3-phase voltage source, and the lines are represented by pure reactances. X_e denotes the equivalent reactance between the generator and V_∞. This configuration corresponds to the realistic case of a remote power-station supplying a distant load centre.

The control quality of the synchronous machine, is assessed in terms of the steady-state behaviour and the dynamic performance in the presence of disturbances. Two kinds of *disturbances* must be considered:

(a) short circuits in the network, line switchings;
(b) changes of operating point (power, voltage, network parameters).

Therefore in the design of the voltage regulator the following performance criteria must be considered:

(1) Regulation of the generator terminal voltage with regard to: (i) smoothness (no ripple during steady-state operation); (ii) speed of response (to avoid overvoltages after load or topology changes); (iii) accuracy.
(2) Ability to keep synchronism after a fault. This requirement means simply that the excitation voltage should be at the maximum during faults and during dangerous rotor accelerations. The excitation may return to normal after the first peak of the rotor swing.
(3) Damping of rotor oscillations.
(4) Criteria (1)–(3) for a wide region of operating points of the generator (power loading, voltage).

The voltage regulator used here corresponds to the present-day conventional regulator as described earlier. However, the stabilising signals derived from power and frequency measurements are introduced via variable gain factors (*Figure 17.26*). P- and ω-feedback are regarded as additional signals, the purpose of which is to produce damping (stabilisation) of rotor oscillations.

The gains for the power and frequency feedback are determined by using the D-decomposition technique. D-decomposition (or domain separation) is a method based on the characteristic equation of the linearised model. It produces curves of constant damping in the plane of the gains G_1 and G_2, at a certain operating point (P=const., Q=const.) and constant line impedance X_e. The method is combined with an optimisation procedure to generate optimal gains (optimal damping of the dominant poles). However, as already stated, the gains that are adequate for the best damping of rotor oscillations depend on the operating point of the generator. Owing to the non-linearities in the system, the operating point itself may vary in time, depending on the loading, voltage and network topology (X_e).

Three quantities are sufficient to define the operating point: active power P, reactive power Q and the reactance X_e of the network. P and Q are measurable directly, while X_e can be identified from local measurements by system response. When the three quantities are known, the linearised model of the generator is fully defined (assuming $V_\infty = 1$ p.u.) and the adequate gains can be adjusted automatically.

However, two situations must be considered: steady-state operation and transient operation (short circuit, rotor oscillations). In the steady state, the gains G_1 and G_2 should be reduced for better voltage control, but they may be larger during transient operation.

The resulting adaptation scheme is shown in *Figure 17.27*. The entries of the look-up table are computed off-line. G_1 and G_2 are adjusted whenever major events occur or the parameters have drifted significantly. Updating of the gains is only permitted once every few seconds in order to secure the stability of the

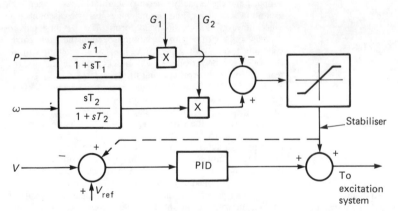

Figure 17.26 The analogue regulator (present-day Brown Boveri (BBC) regulator): ---, alternative connection required by some customers

The relationship between change in torque and change in rotor angle in a closed voltage-regulator loop is therefore

$$\frac{\Delta M_{dE2}}{\Delta \delta} = \frac{k_1 k_3 k_5 + k_2 k_4 (1 + p T_E)}{1/k_2 + k_6 k_E + p(T_E/k_2 + T'_{d0}) + p^2 T_E T'_{d0}}$$

The factors, $k_1, \ldots, k_6$ have a considerable influence on the damping of the synchronous machine and its behaviour, as follows:

k_1 is generally positive and can assume negative values only at correspondingly large network reactances, but it is then no longer possible to transmit stable power. In the normal range, however, this factor has a stabilising effect, i.e. damping.

k_2 is always positive and has a stabilising effect.

k_3 is independent of the rotor angle and responds only to changes in excitation.

k_4 is negative and reduces damping in the system, an effect that can be markedly reduced by high gain in the voltage regulator. The negative electric torque generated by k_4 is synchronous with the torque achieved with k_1 and is compensated by this.

k_5 comprises two components. With large system reactances x_e, i.e. where the rotor angle is large anyway, the expression can become negative. A torque that reduces damping is generated and is further enlarged by the gain at the voltage regulators. Active-load hunting commences with large amplitudes and leads to the machine losing synchronism and being shut down. Only with slip-stabilisation equipment can the effects of this component be eliminated and stable power transmission restored.

k_6 is independent of rotor angle and therefore of no significance as far as hunting is concerned.

It can be seen from the block circuit diagram that the damping-reducing influence is introduced through the difference between set-point and actual values at the voltage regulator. Provided that the synchronous machine is not equipped with static excitation, it is recommended to feed the stabilising signal to the mixing point of the control amplifier. Where static excitation equipment is provided, mixing is also possible direct at the entrance of the gate control system of the power stage of the excitation equipment. To damp the rotor oscillations, it is essential that more active power is delivered at the stator terminals when the rotor is accelerated, and less when it is braked.

There are various means of measuring speed and acceleration, and their corresponding effect on the rotor motion. One simple method is to measure the active power. Assuming that the prime-mover power is constant, changes in the active power cause corresponding acceleration or deceleration of the rotor. Measuring the active power fluctuations gives the first derivative of speed. The actual speed can be determined by integration. The accuracy of the method is adequate for this purpose. However, the reaction to load changes at the drive end is entirely false. Any danger to operation can safely be eliminated by limiting the effect of the stabilising equipment on the voltage control and by introducing an additional signal that is also proportional to frequency.

Thus the use of Δf, i.e. the angular frequency ($\Delta \omega$), instead of integration of ΔP represents a considerable advance in this field. The base signal combination ($\Delta \omega$ plus ΔP) arrangement therefore gives the best arrangement for damping of oscillations (*Figure 17.24*). Test results have fulfilled all expectations in respect of the stabilising signal. Peaks in the signal, which occur when breaker operations take place in the network, are of such short duration that they have no effect upon the excitation. The load/frequency system combines the advantages of load and frequency measurement without any detrimental side-effects.

17.5.8 Adapted regulator for the excitation of large generators

The excitation of a generator is in principle controlled by automatic voltage regulators, and improvements have been achieved in several respects. The original purpose of the feedback arrangement was *voltage regulation*. This basic requirement is important for maintaining stable synchronous operation of the generators in the power system and for controlling the voltage supplied to customers. Further advantages have been obtained since *static* (power semiconductor) *excitation* systems have been used. Through these fast-acting devices, the regulator may contribute to the *transient stability* after faults (keeping the generator from falling out of step) and to the *damping* of electromechanical rotor oscillations. Furthermore, stabilising signals have been introduced with pre-set amplification gains as a compromise to various operational points. These features are now provided by many commercial regulators. The adapted regulator improves the performance one step further. The main result is the good damping provided for a wide region of operating points, and the smooth voltage regulation.

For better understanding of the benefits of an adaptive control on a network, the one-machine system is presented in *Figure 17.25*. The generator is connected to the 'infinite bus' V_∞ over two

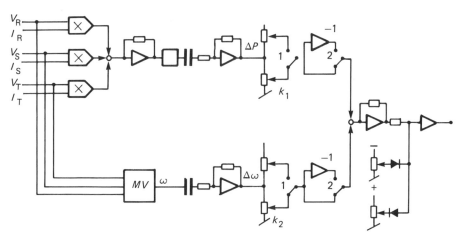

Figure 17.24 $\Delta \omega + \Delta P$ base signal combination arrangement for damping of oscillations: MU,

set-point results in a drooping characteristic. In a steady-state condition the effect is the same as that of a reactance. A subtractive current bias causes a rising characteristic, i.e. a compounding. The effect of a true reactance can be reduced by compounding. In simple circuits the reactive current droop is accompanied by a slight compounding in the active current, but in most cases this is useful and under no circumstances is it a disturbing influence.

It should also be noted that the current bias curve is not entirely linear but slightly progressive. This is of virtually no significance. The current bias can be varied between zero and maximum by altering the load resistance. Nearly all voltage regulators are issued with this equipment.

As mentioned above, the current phasor is chosen in most cases so that an almost purely reactive load effect is created. However, an active-load effect or a mixture of the two effects can be achieved by selecting a different phase angle. Consequently, in medium-voltage networks, additional active current compounding has shown itself to be a suitable means of improving the operating behaviour.

17.5.7 Slip stabilisation

Power transmission between synchronous machines and the load centre must remain stable even over long distances and under complex conditions. Experience has shown that certain network conditions can cause excessive hunting. Analysis indicates that hunting can be effectively reduced only by the introduction into the voltage control circuit of transient stabilising signals derived from the machine speed/frequency or produced by a change in the electrical output of the generator. Knowledge of the origin of the instability and the means of intervening in the controlled synchronous machine are essential for applying additional signals.

If the synchronous machine is connected to a rigid network through a reactance, it can be easily seen from the phasor diagram that the terminal voltage of the machine and the electrical torque alter with the rotor angle and with the main flux. The changes in terminal voltage, torque and main flux are given by differential quotients. In order to be able to make a quantitative statement for a given duty point, small changes are observed and the transmission functions and their representative factors are linearised about the duty point (index 0). This gives simple expressions, and a block circuit diagram (*Figure 17.23*) can be drawn which characterises the response of the synchronous machine at the rigid network. Let

$$\Delta M_{dE} = k_1 \Delta \delta + k_2 \Delta E'_q$$

$$\Delta E'_q = k_3 \Delta E_p + k_4 \Delta \delta$$

$$\Delta V_t = k_5 \Delta \delta + k_6 \Delta E'_q$$

$$k_1 = \frac{\Delta M_{dE}}{\Delta \delta} = \frac{V_0 \cos \delta_0}{x_q + x_e} E_{q0}$$
$$+ \frac{x_q - x'_d}{x'_d + x_e} V_0^2 \sin \delta_0 \frac{1}{x_q + x_e}$$

$$k_2 = \frac{\Delta P_d}{\Delta E'_q} = \frac{V_0 \sin \delta_0}{x'_d + x_e}$$

$$k_3 = \frac{\Delta E'_q}{\Delta E_p} = \frac{x'_d + x_e}{x_d + x_e}$$

$$k_4 = -\frac{x_d - x'_d}{x'_d + x_e} V_0 \sin \delta_0$$

$$k_5 = \frac{\Delta V_t}{\Delta \delta} = \frac{V_d}{V_t} V_0 \cos \delta_0 \frac{x_q}{x_q + x_e}$$
$$- \frac{V_q}{V_t} V_0 \sin \delta_0 \frac{x'_d}{x'_d + x_e}$$

$$k_6 = \frac{\Delta V_t}{\Delta E'_q} = \frac{V_q}{V_t} \frac{x_e}{x'_d + x_e}$$

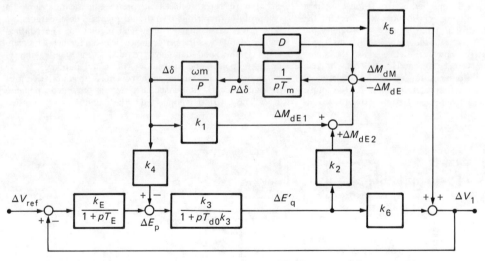

Figure 17.23 Block circuit diagram of a synchronous machine connnected to a rigid network: *D*, damping constant of synchronous machine; *k*, *k*$_E$, proportionality factors; *M*$_{dE}$, electric torque; *M*$_{dM}$; mechanical torque; *T*$'_{d0}$, *T*$_E$, time-constants; *T*$_m$, moment of inertia; ω_m, angular velocity of synchronous machine; *p*, operator d/d*t*; other symbols, as for *Figure 17.17*. For linearisation of transfer functions assuming small changes,

$$k_1 = \frac{\Delta M_{dE1}}{\Delta \delta} \quad E'_q = \text{constant}$$

$$k_2 = \frac{\Delta M_{dE2}}{\Delta E'_q} \quad \delta = \text{constant}$$

$$k_3 = \frac{\Delta E'_q}{\Delta E_p} \quad \delta = \text{constant}$$

$$k_4 = \frac{\Delta E'_q}{\Delta \delta} \frac{1}{k_3}$$

$$k_5 = \frac{\Delta V_t}{\Delta \delta} \quad E'_q = \text{constant}$$

$$k_6 = \frac{\Delta V_t}{\Delta E'_q} \quad \delta = \text{constant}$$

17.5.6.2 Principle of current bias

The principle of phasor addition of a current-proportional voltage to the generator voltage has been known in various forms for some time (*Figure 17.22*). All 3-phase voltages should be measured in each case to keep ripple and filter time-constants small. As shown in *Figure 17.22*, a 1-phase current bias can be applied. In the case of controllers for large generators, the extra cost for symmetrical 3-phase bias is justified.

The summation is made such that the over-excited reactive component of the current increases the actual voltage. This method of falsifying the actual value in relation to an unchanged

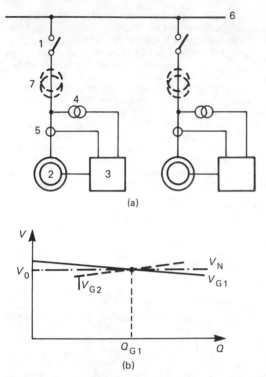

Figure 17.21 Parallel operation in a power-station: (a) basic circuit diagram; (b) characteristics; 1, generator breaker; 2, generator; 3, voltage regulator; 4, voltage transformer; 5, current transformer; 6, busbar; 7, unit-connected transformer; V_N, system voltage; V_{G1}, V_{G2}, generator voltage characteristics; V_0, no-load voltage; Q_{G1}, reactive power output of generator 1

17.5.6.1 Parallel operation in a power-station

Parallel operation in a power-station is illustrated in *Figure 17.21*. As virtually all large generators are operated in unit connection, the transformer reactance between the terminals and the h.v. busbar causes a natural reactive current droop of 8–12%. This accurately defines the reactive power output. Conditions are very different in the case of low-voltage (l.v.) generators, which must operate in parallel direct at the machine terminals. Here there is no natural droop, and even the smallest difference in the voltage values will lead to undesirable mutual exchange of reactive power between the machines, without resulting in a defined distribution of demand. It is not until an artificial droop is introduced by applying reactive current in the measuring loop of the voltage regulator that the reactive load distribution becomes stable; initially it is immaterial whether the generators themselves define the voltage at a purely consumer network, or coincide with the voltage given by a live network.

Electronic controllers have no dead-band and permit closed-loop gains of between 100 and 200 in accordance with a proportional range of 1–0.5%. Consequently, in isolated duty a satisfactory distribution over the parallel generators is achieved even with a 3–4% artificial droop. For reasons discussed later, there are various factors that make higher droop values more suitable for interconnected systems. The polygon connection used previously in some power-stations for distributing the reactive load has therefore lost its significance to a large extent because total elimination of changes in the steady-state voltage had to be bought at the cost of complicated circuitry.

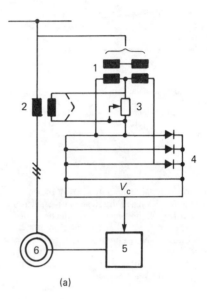

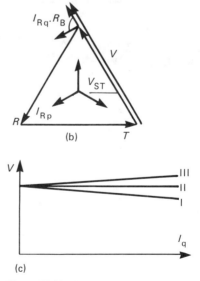

Figure 17.22 Current bias: (a) basic circuit diagram; (b) phasor diagram; (c) characteristics; 1, 3-phase voltage transformer; 2, current transformer; 3, load resistance R_B; 4, rectifier; 5, voltage regulator; 6, generator; V_C, V, r.m.s. voltage; I_{Rp}, active current in phase R; I_{Rq}, reactive current in phase R; I_q, reactive current, super-excited; I, reactive current droop; II, no effect; III, reactive current compounding

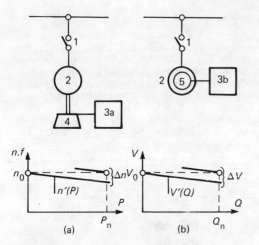

Figure 17.19 Characteristics for parallel operation: (a) speed/active-power control; (b) voltage/reactive-power control; 1, generator breaker; 2, generator; 3a, speed regulator; 3b, voltage regulator; 4, turbine; 5, rotor winding; 6, network; P, active power; Q, reactive power; n, speed; f, frequency; V, voltage

isolated duty, the appropriate value is controlled by the corresponding control system, i.e. the speed (frequency) or generator voltage. This is no longer the case, however, in parallel operation with a live network. Frequency and voltage are already present and can be changed by the set only to a limited extent. Secondary controlled variables are involved here, and it is imperative that they are controlled in parallel operation. These secondary variables are the active and the reactive power.

On what is this twofold nature of the control system based? In isolated duty, only one control system is acting on any control loop. In parallel operation, however, many loops are coupled through the common controlled variable. The primary control task is distributed over a large number of control points. Selective stable distribution is of decisive importance. The overall task here is to maintain frequency and to produce active load, or to maintain voltage and to produce reactive load. The familiar solution is to provide the regulator with a drooping characteristic, i.e. the set-point of the primary controlled variable drops as the secondary controlled variable rises. The point of intersection of this characteristic of the network rating defines the corresponding operating conditions. Any set-point can also be set for the secondary controlled variable by parallel displacement of the characteristic. The gradient of the characteristic is usually expressed as a droop of the form $S_n = \Delta n/n$ or $S_v = \Delta V/V$, i.e. as the percentage deviation in the primary controlled variable that is necessary to bring the secondary controlled variable from zero to the rated value.

While the frequency of the network is the same throughout, in the case of voltage this applies only to the imaginary voltage of an infinite busbar. The true network represents roughly a variable 'mountain landscape' of voltages defined by its line impedances and feed-in or feed-out at a large number of junction points. This true network must first be accessible by reducing it to a simple equivalent circuit (*Figure 17.20*).

Each individual generator feeds through transformer and line impedances into the 'finite busbar' formed by the sum of all the other generators. As can be seen from the phasor diagram in *Figure 17.20*, which may be considered as qualitative, the products of the reactances and reactive components of the current are decisive for the magnitude of the voltage drop, $|\Delta V|$. The drop in active voltage and the phase displacement of the voltage phasor can here be ignored. The reactances between the generator terminals and the infinite network thus produce a natural drop in the reactive current, whereas the drop of the speed-control system is always synthetic. The generator reactance X_D is within the control loop and therefore need not be taken into account for quasi-steady-state phenomena. The effect of the natural drop in reactive current, and its increase or decrease by artificial means, is the main theme to be discussed from various viewpoints in the following.

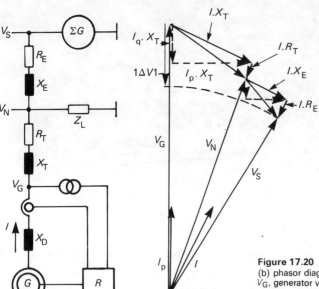

Figure 17.20 Impedances and voltage drops: (a) equivalent circuit diagram; (b) phasor diagram; V_S, voltage of infinite system; V_N, system voltage; V_G, generator voltage; I, generator current; I_p, active component; I_q, reactive component; X, external (short-circuit) reactance; X_T, transformer reactance; X_D, direct-axis reactance of generator; E_E, R_T, active resistances; Z_L, load impedance; G, generator; ΣG, sum of all other generators; R, voltage regulator

However, when because of, for example, breaker failure the fault is not cleared, or when generated reactive power no longer suffices to maintain the voltage level, over-excitation will continue to be present. The limit controller now aims to reduce the excitation current before any of the protection gear is tripped. A further important aspect is that this also causes the short-circuit power of the system to be reduced, which, in turn, will diminish the extent of the damage resulting from failure of a breaker or other protective gear.

The condition in which the maximum continuous excitation permissible is present might persist for a long time, and experience has shown that further short circuits appear relatively frequently during this time. When these secondary disturbances occur, it becomes necessary to deliver the maximum reactive current to the system once more, especially to retain system stability. This means that limitation has to be cancelled again for a short time. The characteristic which trips the reset is the steep drop in voltage, dV/dt.

A further requirement results from the fact that today most of the large excitation equipment is fed from the generator terminals. The necessary ceiling excitation current is normally already available at 90% of the rated voltage. This means that at 120% of the rated voltage, a ceiling current which is 33% higher than the necessary value will flow when an instantaneously acting excitation limiter is not provided.

The limitation signal can be introduced as a reference-value variation of the voltage regulator. However, two facts suggest applying the signal to the voltage regulation output for use as dominant parallel limitation. These are: (i) the limitation is absolute and thus fully effective even when the actual value for the voltage regulator is lost (voltage-transformer fuses blown); (ii) the limit controller can be adpted to the dynamic conditions of the field-current controller and become independent of voltage-regulator response.

As has been mentioned, limitation to the permissible continuous excitation current should be delayed in order to first give the voltage the maximum support possible. The permissible continuous current in the excitation circuit is determined by the thermal stressing of the field winding, the supply transformer, or the thyristors or diodes. The overload capacity of these elements can usually be represented by integration of the current. A slight overrun of the set limit for the continuous current is then tolerated correspondingly longer. It can be argued, however, that the ceiling excitation will be required only as long as fault clearance is still taking place, in which case a fixed timing element is provided. In both cases, it is usually desired that ceiling excitation should be available again in the event of secondary disturbance.

Based on previous experience, improved rotor-current limit controllers were developed to cater for the many varied applications met with in practice; these controllers can be easily adapted to the particular protection method used.

A maximum-current limit controller, instantaneous-acting and always available, ensures that for terminal-fed excitation the desired ceiling current is not exceeded over the entire operating voltage range.

In the case of the simple integrator mode, the speed at which reduction to the limit value takes place is relative to the magnitude of the overcurrent. The integrator is reset only when the limit for the continuous rotor current is underrun.

With the switching mode, reduction to the limit begins after a set delay T_v of several seconds. Every time there is a sharp drop in voltage, maximum-current limitation begins anew and the timing element is reset.

A method frequently used is the combined mode: here, integration of the overcurrent is combined with resetting of the integrator by each succeeding voltage drop.

The actual current can usually be measured with a current transformer in the alternating-current (a.c.) power supply prior to rectification by the thyristors or diodes. A d.c./d.c. converter makes connection to a shunt also possible.

When partial failure of components in the excitation circuit causes the permissible continuous current to be reduced, change-over to a pre-set second current limit set-point is possible.

17.5.5.2 Load-angle limitation

For a number of years it has been standard practice to provide protection gear that detects excitation failure or loss of generator synchronism and trips the generator breaker. Inadequate excitation can, for example, also be caused by a change in the system configuration due to a fault, by reference-value failure or by other faults in the voltage regulator. The load-angle limiter obviates unnecessary disconnection by the under-excitation protection device in all these cases.

The limit angle is set intentionally to a value considerably smaller than the pull-out limit. For turbo-generators, for example, the angle lies between 70° and 85°. For salient-pole machines the actual stability must be taken into account and a suitable response limit determined. Thus, an adequate angle difference exists for the acceleration of the generator during clearance of a nearby short circuit. This reserve is directly related to the time allowed for clearing the fault.

The principle of analogue simulation of the phasor diagram using stator voltage and current is applied. This method is notably simpler than the direct measuring processes. In the event of a transmitter on the shaft providing the rotor position, the same device may be employed for further evaluation. The angle between the two simulated phasors, rotor voltage E_p and the infinite bus voltage V_s, is converted into a d.c. signal in a solid-state angle discriminator. When transients appear, the angle generated by the simulation leads the true load angle slightly. This is advantageous as this limiter should intervene at an early stage. Such a procedure has been shown to be reliable in practice.

17.5.5.3 Stator-current limitation

The stator-current limit, which is governed by thermal considerations, normally lies beyond the operating range of the synchronous generator. Thus, stator-current limitation is employed only in special cases. To illustrate this point, actual applications of a stator-current limit controller, usually in conjunction with limit control of the load angle and excitation current, are given below:

(1) With gas turbo-sets, for example, the useful output can be raised at times to cover peak demands, so that the limit is governed partly by the stator current.
(2) In coastal areas it is necessary to reduce the system voltage at certain times of the year because of salt vapour; there is thus an increase in stator current for the same power.
(3) In the case of reactive power compensators the under-excitation limit can be attained only with a stator-current limiter.
(4) With synchronous motors, a practical utilisation limit can appear that will be detected approximately by a stator-current limit controller.

17.5.6 Control characteristics for synchronous machines

As far as the system of synchronous generator and network is concerned, there is a strong similarity between the speed and active-power control system on the one hand, and the voltage and reactive-power control system on the other. To understand control by characteristic it is useful to compare the two systems (*Figure 17.19*). As long as the set is operating at no load, or on

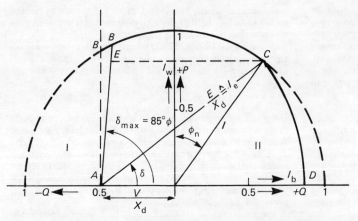

Figure 17.17 Current and power diagram of a turbo-generator: I=1 p.u., $V=V_k$=1 p.u., $X_d=X_q$=2 p.u., cos ϕ=0.8; I, under-excitation; II, over-excitation; AB, practical stability limit; AB', steady-state stability limit; BC, limit of stator temperature rise; CD, limit of rotor temperature rise; CE, active power limit; E, rotor e.m.f.; I_b, reactive current; I_e, excitation current; I_w, active current; P, active power; Q, reactive power; V_k, terminal voltage; X_d, direct-axis reactance; X_q, quadrature-axis reactance; δ, load angle; ϕ_n, rated phase angle

introduced is the requirement that these do not interfere with the normal operation of the voltage regulator, and that neither intervention nor return to voltage regulation leads to disturbances.

Electronic load-angle and current-limit controllers were introduced some time ago, and these are now standard components of all good voltage regulation equipments. Practical experience has made possible the addition of improvements and refinements.

Consider the static excitation equipment commonly used for large generators (*Figure 17.18*). Normally, the voltage regulator holds the generator voltage (and thus, indirectly, the reactive power output) at a constant level. The static converter adjusts the excitation current so that the reference voltage remains a function of the reactive current. At the same time, the rotor-current limit controller measures the excitation current in the static converter supply and compares it with the relevant reference limit. The load-angle limit controller continuously forms from the generator current and voltage a signal that is proportional to the angle, and compares it with the limit.

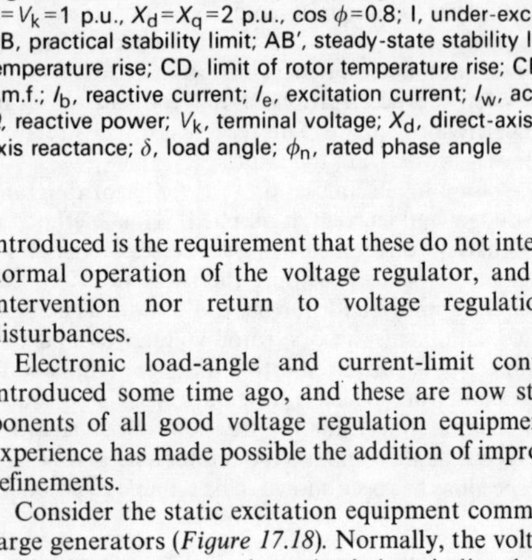

The two limiters operate as parallel controllers, i.e. their signals completely replace the voltage as controlled variable when their output variables become smaller or greater than the voltage control signal. Thus, the change in specific dynamic behaviour of the control circuit for each limit function can be fully taken into account. It will be shown that unwanted side-effects occurring on signal take-over can be avoided entirely.

17.5.5.1 Rotor-current limitation

Every exciter is capable of delivering a maximum current appreciably greater than the continuous value based on thermal considerations. This capacity for over-excitation is necessary to provide the additional reactive power demanded during selective clearance of faults and to maintain a synchronous torque even when the voltage level has fallen.

Although it safeguards the winding insulation against damage from prolonged over-excitation, rotor protection gear (such as overcurrent and overtemperature relays) disconnects the generator even when disconnection is not desired. Use of a controller to limit the excitation current to an acceptable value during operation would substantially improve the availability of the generator. For safety reasons, however, a separate protective device must be retained to safeguard the winding against overloading.

The demands to be met by the rotor-current limiter are best explained with reference to the voltage and excitation current when a short-line fault occurs. The voltage regulator reacts to the drop in voltage with surge excitation. The controller is not intended to impede this, but merely to determine the ceiling current I_{fm}. Although the clearance normally takes place in a few hundred milliseconds, the longest duration (back-up protection) of 2–3 s will be assumed. When, after the fault has been cleared, nominal voltage is attained with a permissible continuous excitation current, the thermal controller will not intervene.

Figure 17.18 Block diagram of the automatic voltage regulation system for a synchronous generator: 1, voltage regulator; 2, gate control unit; 3, static converter; 4, rotor-current limiter; 5, load-angle limiter; 6, voltage transformer; 7, generator-current transformer; 8, excitation-current transformer; G, synchronous generator; V_g, generator voltage; I_g, generator current; I_f, rotor current; δ, rotor angle

The capacity of the excitation system is dependent on the generator specification and on the requirements of the power system. The rating of the excitation transformer or exciter in particular is determined by the maximum continuous current rating of the rotor winding and the required ceiling voltage. In designing the converter circuits, which have much shorter thermal time-constants, the ceiling current and short-circuit current capacity must also be considered.

The voltage of the transformer secondary (or of the auxiliary generator) is determined by the maximum ceiling voltage, V_c of the excitation system, and its current by the maximum continuous current I_{fm} of the generator field winding. A simple approximation to the rating of a 3-phase excitation transformer is $1.35 V_c I_{fm}$.

The static converter comprises one or more fully controlled thyristor bridges, which are almost always cooled by forced air. The number of parallel bridges and the number of series elements in each bridge branch are determined by the technical data (current rating and inverse voltage) of the thyristor bridges, the maximum exciting current, the transient overload (due, for example, to a system fault) and the ceiling voltage. An extra redundant bridge is normally included (in parallel) so that, even if one bridge fails, the full operational requirements can still be met. To ensure selectivity, it is necessary to have at least three bridges in parallel. This is because a faulty thyristor together with a healthy branch can form a two-pole short circuit that persists until the fuse in series with the defective element blows. If only two parallel bridges are provided then the short-circuit current in the healthy branch is shared by only two thyristors; the fusing integral of both the corresponding fuses may then be exceeded, resulting in an excitation trip.

The rectifier can have either a block structure or be arranged phase by phase. In the block arrangement, one or more bridges are grouped together and gated from a common pulse amplifier. Blocks are normally provided with individual fans, but common cooling is also possible.

The arrangement of the converter in phase groups reduces the risk of an interphase short circuit. All parallel thyristors of one phase are brought together in a single rack. A separate cooling system for each rack is inappropriate in this arrangement and common cooling is employed, consisting of two fans each capable of supplying the full cooling-air requirement.

With the phase-by-phase arrangement, the excitation transformer is usually composed of three 1-phase units; this will always provide a higher degree of safety against interphase short circuits.

The field suppression system comprises essentially a two-pole field-breaker and a non-linear field discharge resistor. The component ratings are so chosen that neither the permissible arcing voltage at the breaker nor the maximum permissible field-circuit voltage is exceeded, even with the maximum possible field current (following a terminal short circuit on the generator). A 'crowbar' is included in the discharge-equipment cubicle as additional overvoltage protection. This comprises two anti-parallel thyristor groups which are triggered by suitable semiconductor elements such as breakover diodes (BODs). If, for instance, a voltage is induced in the rotor field circuit owing to generator slip, this voltage can rise only to the threshold of the BOD element. If this value is reached, the element triggers the thyristors, and the rotor circuit is connected to the discharge resistance, allowing a current to flow which immediately causes the induced voltage to break down.

17.5.4 Automatic voltage regulator and firing circuits for excitation systems

The voltage regulator for brushless exciters as well as for static excitation systems constitutes the central control element in a modern plant. This unit uses a voltage transformer to measure the actual value, rectifies it and compares it with the reference value. The difference-voltage is fed to an amplifier with a lead–lag filter (the response characteristics of which must be carefully adjusted to suit the particular generator and grid parameters) and the amplified value is brought to the gate control unit. Using additional reactive current compensation it is possible to adjust the reactive current behaviour of the generator with regard to the network.

The voltage regulator has normally to be supplemented with parallel-connected limiters which function before corresponding generator protection relays are activated. Depending on requirements, rotor-current, load-angle and stator-current limiters may be employed. Their features are described later.

If the limitation control causes a reduction in the excitation current (on over-excited operation), time-delay elements are incorporated to allow high transient currents. For under-excited operation there is no time delay.

Stabilising equipment (2) can be used to damp power oscillations that may arise under exceptional network conditions.

The voltage regulator may be supplemented with an over-riding system for controlling the power factor or the reactive power flow, as required. All voltage regulators are provided with manual control (change-over from closed-loop to open-loop control). A separate redundant gate control unit is often provided (double-channel type). In case of a fault in the 'automatic' channel, a follow-up control system ensures a smooth transition to manual operation.

The amplified difference signal is transformed into pulses with the appropriate firing angle by the gate control unit. The firing pulses are amplified in an intermediate stage and led to the individual output stages which are assigned to the various converter units. The output stages shape the pulses to the steep slopes necessary to ensure simultaneous firing of all parallel thyristors. The pulses from the output stages are fed via impulse transformers to the thyristor gates.

The transfer function of the automatic voltage regulator (AVR) including the gate control system is

$$\frac{\Delta V_f}{\Delta V_g} = A \frac{(1+pT_1)(1+pT_2)}{(1+pT_3)(1+pT_4)} \frac{1}{1+pT_M} \frac{1}{1+pT_E}$$

where T_M is the measuring time-constant, T_E is the exciter time-constant, $T_1, \ldots, T_4$ are the equivalent lead–lag time-constants, A is the amplification, and p is the operator d/dt.

17.5.5 Limiting the excitation of synchronous machines

Ever-increasing demands on power system reliability led to the development of ancillary equipment (limiters), to inhibit the tripping of protection gear when this was not wanted and to make, within the permissible limits, better use of the synchronous machine. These limits are clearly depicted in the power chart of a turbo-generator shown in *Figure 17.17*.

The limit of active power output (line CE) is determined by the prime mover, and is disregarded. A factor connected directly with excitation current, however, is the permissible temperature rise in the rotor winding, represented by the arc CD with centre A corresponding to the non-excited condition. This thermal limit is defined by the ageing of the insulation. It may therefore be exceeded for a short time—a requirement essential for the stability of the power system.

In the under-excited mode, operation of the synchronous machine is limited by the airgap torque necessary for the active power transfer, the steady-state condition being represented by a definite, permissible rotor displacement angle (line AE). This limit has mechanical and dynamic characteristics, and calls for instantaneous intervention as soon as it is exceeded, to prevent the generator from falling out of step.

Another important condition governing the limits to be

also field suppression, when required, by change-over to inverter mode.

17.5.2 Brushless excitation systems

Brushless excitation systems are mainly employed either where the control requirements are not too stringent or where the atmosphere is chemically aggressive. The main generator excitation is provided by an auxiliary exciter generator mounted on the main generator shaft.

This exciter is a salient-pole synchronous generator with a 3-phase winding on the rotor and a d.c. excitation winding on the stator. The alternating currents produced by the rotor winding are rectified by rotating diode bridges mounted on the shaft, the resulting d.c. being fed to the field winding of the main generator. The excitation for the exciter stator winding is supplied by a thyristor regulator which comprises voltage regulators and limit-value controllers. Field suppression of the exciter machine is also carried out by the regulator (*Figure 17.15*).

The controlled rectifiers and the electronic circuitry of the regulator are supplied either from a p.m. generator on the main shaft or through a transformer fed from the main generator terminals. In the latter case, compounding equipment is available for the required selective switching (short-circuit duration in excess of 0.5 s) which is then combined with the standard regulators.

In its basic form the regulation system contains the continuously active elements for automatically controlled operation, which are:

A voltage regulator
A reference-value potentiometer for remote adjustment
A supply transformer
A field-breaker and discharge resistors
Diode monitoring consisting of:
    The manual control device, i.e.
    a variac transformer
    a diode bridge
    a change-over switch (auto to manual)

17.5.3 Static excitation systems

On generators with difficult regulation requirements a rotating exciter should not be used. Here the excitation is supplied either directly from the generator terminals via a transformer and a controlled rectifier bridge (*Figure 17.16*) or by an auxiliary generator.

The main components of a static excitation system are:

An excitation transformer (or auxiliary generator)
A static converter
A field suppression system
Control and firing circuits
A protection and monitoring system

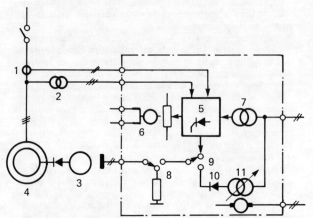

Figure 17.15 Voltage regulation system with manual control facility: 1, current transformer; 2, voltage transformer; 3, exciter; 4, generator; 5, voltage regulation; 6, reference value; 7, supply transformer for the voltage regulator; 8, field switch/discharge resistor; 9, change-over switch; 10, rectifier; 11, variac with motor drive

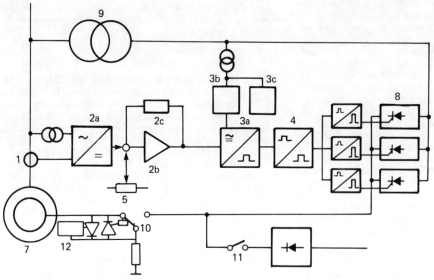

Figure 17.16 Block diagram of a voltage regulator with static excitation (without manual control); 1, instrument transformer; 2a, transducer; 2b, voltage regulator; 2c, PID filters; 3a, firing-angle control unit; 3b, filter; 3c, voltage relay for control of excitation field flashing; 4, pulse amplifier; 5, reference-value setting unit; 7, generator; 8, main excitation rectifier; 9, excitation transformer; 10, field suppression device; 11, field flashing device; 12, overvoltage protection

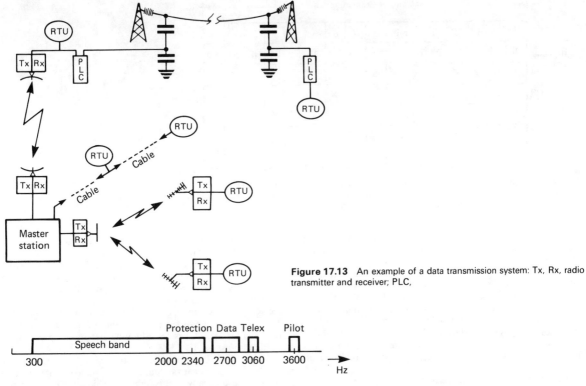

Figure 17.13 An example of a data transmission system: Tx, Rx, radio transmitter and receiver; PLC,

Figure 17.14 Allocation of communication functions to 4 kHz channel

each having measurands would have a measurand and indication interrogation scan time (with no changes of state to transmit) of approximately 6.5 s, using a 200-baud channel. To decrease this time, either the number of RTUs in the part system would have to be reduced or the speed of transmission would have to be increased.

17.4.5 Communication channels

The transmission of the frequency-shift keying signals requires a 4-wire circuit communication channel, which can be telephone-type cable, power line carrier equipment, radio transmission or any combination of these media in series or parallel. *Figure 17.13* gives an example of a possible combination of different types of communication channels.

Often the telecontrol signal must share the v.f. band with other transmissions such as speech, telex, protection or even another telecontrol part-system transmission. An example of the frequency multiplexing of the v.f. band is given in *Figure 17.14*. The multiplexing equipment is normally available with the power line carrier. The frequency-shift keying takes place within the frequency bandwidth indicated in *Figure 17.14*: e.g. for a 200-baud modem the lower keying frequency is −90 Hz and the upper +90 Hz of the channel centre frequency. The total bandwidth required is 360 Hz.

References

1 FUNK, G. and SODER, G., 'Indactic 13 and 33 Telecontrol Systems based on ED 1000 Modules', *Brown Boveri Review*, **61**, 393–398, August (1974)

17.5 Decentralised control: excitation systems and control characteristics of synchronous machines

17.5.1 Introduction

A synchronous generator operating on an interconnected grid requires a fast-response excitation system to ensure its stable operation. This system must be able to adjust the level of magnetic flux in the generator according to the grid network requirements. Since the parameters of the generator will fall into a fairly narrow range according to the type of generator (salient-pole or turbo), the excitation system must be designed to bring the generator flux (i.e. excitation current) to the required level in the shortest possible time, notwithstanding the long time-constants of the generator transfer functions.

Brushless excitation equipment has been developed to meet these requirements on various types of synchronous generator— particularly on smaller turbo-generators—and has given excellent results in service. The control characteristics of this system, however, are dictated largely by the exciter and the rotating diodes rather than by the voltage regulator (d.c. exciters are not used in modern excitation systems).

For large generators a static excitation system represents the best solution since, in principle, it imposes no limit on the ceiling voltage. In the per-unit system the ceiling voltage is the ratio of the maximum excitation voltage to the excitation voltage for nominal open-circuit voltage at the generator terminals. Values of 10 or 15 p.u. are quite often used. A great advantage of static excitation is that the total field-voltage range is available with practically no time delay. This speeds up not only field forcing but

Data acquisition and telemetering

Bit no.	15	14	13	12	11	10	9	8	7	6	5	4	3	2	1	0
Bit significance	M2	M1	L	D7	D6	D5	D4	$\overline{H4}$	D3	D2	D1	H3	D0	H2	H1	H0
$\overline{H4}$	X	X	X	X	X	X	X	●								
H3	X	X	X	X					X	X	X	○				
H2	X	X			X		X		X	X			X	○		
H1	X		X		X		X		X		X		X		○	
H0	X		X		X		X		X		X		X		X	○

Figure 17.11 An example of message telegram organisation: ●, odd parity check; ○, even parity check; ×, bits supervised by Hamming bits; D, data bit; L, message sequence complete; M, message type definition; H, Hamming bit

converter module reassembles the message into a parallel word including the check bits. The message can then be checked for errors induced during transmission.

There are various methods of checking for errors in a message telegram. The error-detecting capability of a checking system is often described as a Hamming distance. The Hamming distance between two binary words is the number of bit positions by which they differ that it is undetectable by the error-checking system. For most purposes in telecontrol, the minimum Hamming distance is 4, i.e. all 3-bit errors are detected. An example of a modified Hamming code error-checking system is shown in *Figure 17.11*.

The advantage of this system is that in addition to detecting the same number of error patterns as equivalent error-checking systems of the same size, the introduction of odd parity in one check-bit position and even parity in the others forces at least two changes in bit settings in the telegram. This considerably reduces the risk of loss of synchronism when the word is being read. The Hamming distance of this code is 4: i.e. all patterns of 3-bit errors are detected, as well as all odd numbers of bit error patterns and over 92% of all even numbers of bit error patterns.

17.4.4 Transmission system

The data transmission system must be designed to allow all the RTUs to transmit their data to the control centre in a reasonable time without imposing impractical demands on the communication system. The data transmission can be divided into several parts, each of which can comprise one or more RTUs. Each part system is completely independent of the others; thus they effectively operate in parallel.

Three basic systems are possible: point-to-point, radial or multi-point (see *Figure 17.6*). Combinations of radial and multi-point are also possible.

With point-to-point systems, the transmission can be simplex (where the RTU sends data continuously to the master station) or half- or full-duplex (where data can be transmitted in both directions).

With radial or multi-point systems, the dialogue between the master station and the RTU is entirely controlled by the master station to avoid two or more RTUs sending data simultaneously. The master station transmits instructions to the RTU, which, in complying with the instruction, acknowledges it. The RTU takes no initiative in transmitting data and sends data only after receiving a specific instruction to do so from the master station. Thus all data in a part system is transmitted from each RTU in turn and response times are a limiting feature of a part system configuration.

There are two basic forms of data to transmit: event-controlled data which appears at random intervals, and data that must be transmitted cyclically as it is liable to be continually changing. The cyclically transmitted data, e.g. measurands which require frequent updating at the control centre, takes up most of the transmission time; however, as the RTU is quiescent, periodically it must be asked whether it has any spontaneous data to transmit. To save transmission time, this is made in a 'broadcast' interrogation to all RTUs in the part system simultaneously. If no RTU answers positively, the normal cyclic sequence is resumed; however, if one or more RTU answers that it has spontaneous data to send, the normal cycle is interrupted and each RTU is interrogated in turn until all the spontaneous data have been transmitted and received at the master station. In order that the detection of spontaneous data (e.g. a breaker trip) is transmitted in a reasonable time, the spontaneous data interrogation is interjected between the transmission of groups of measurands (say 16). An example of a transmission sequence is given in *Figure 17.12*. The spontaneous data can be sent in chronological order of events taking place and also with the time at which each event happened.

The response times are dependent on the transmission speeds, and a choice of speed is available. Standard speeds of 50, 100, 200, 600 and 1200 bits/s are available within the v.f. band. The choice of speed is then dependent on the response times required and the bandwidth available on the communication channel. As an example of response times, a part system comprising four RTUs

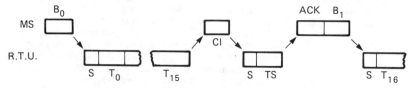

Figure 17.12 A typical transmission sequence: B_n, call for transmission of a group of measurands; S, start character; T_n, values; CI, call for transmission of a change of state; T_s, change-of-state message; ACK, acknowledgement of receipt of message; MS, master station

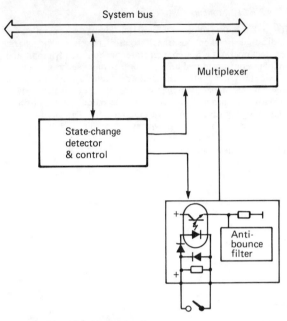

Figure 17.9 Digital input circuit

17.4.2.2 Analogue inputs

The electrical measurements are converted from current-transformer and voltage-transformer levels to analogous direct-current (d.c.) milliampere signals by suitable transducers. These convert voltage, current, active and reactive power, frequency, temperature etc. into proportional d.c. signals which are fed into an analogue input module. This module (*Figure 17.10*) has a multiplexer for scanning multiple inputs (usually 16) and an analogue-to-digital converter (ADC) which converts the d.c. signal into a digital code (usually binary or BCD). The local scanning rate of the measurands is less than 0.5 s, with an accuracy of conversion for normal requirements of 1% for bipolar (7 binary bits + sign) values and of 0.5% for unipolar values. For higher accuracy requirements an accuracy of 0.1% is used (11 bits).

17.4.2.3 Pulse inputs

Energy or fuel consumed can be measured by integrating meters giving a proportional pulse output. The pulses are integrated in a pulse input module which is periodically scanned, e.g. at 15 min intervals. The consumption over that period is then the difference between the last two readings. Special measures are taken to ensure that the reading does not interfere with the pulse integration or vice versa.

17.4.2.4 Digital output

Digital output modules convert a coded message, received from the control centre, into a contact output, e.g. to trip or close a circuit-breaker or to raise or lower a tap-changer. Obviously, the transmission of these signals must be very secure, and checking that the correct output is given must be rigorous. Echo checks are made to ensure that the correct code has been received by the module, and additional circuits can be added to detect stuck contacts and to prevent more than one command being output at a time.

17.4.2.5 Analogue output

Set-points are output via analogue output modules that receive a coded message which they convert into a d.c. analogue signal; this is output e.g. as a turbine-governor setting.

17.4.2.6 Clock

An optional feature of the RTU is a clock module, which enables the events in the substations to be arranged and transmitted in chronological order and the time at which an event happened to be added. The clock is capable of giving a 10 ms time resolution in event sequence recording and the time in hours, minutes, seconds and hundredths of a second.

17.4.2.7 Printer

A further option can be included to print the alarm and event sequences locally at the substation by adding memory, serial–parallel interface a printer and additional program. If the memory capacity is limited, the output is either coded or abbreviated; however, with adequate memory the print-out can be in clear text.

17.4.3 Data transmission

The substation data is locally scanned and memorised in the RTU ready to be transmitted to the control centre. When an instruction is received from the master station, the processor prepares to send the data to the control centre. To every message containing data, it adds check bits that enable the master station to ensure that the message has not been corrupted during transmission. Similarly, to any message output from the master station, check bits are added to enable the RTU to verify that the received message has not been corrupted. The data are then output to the communication channel via the parallel-to-serial converter and the modulator–demodulator (modem) which converts d.c. pulses into keying frequencies in the voice-frequency (v.f.) range. Normally two v.f.-range frequencies are used to represent '1's and '0's of the message code and to modulate the carrier of the transmission channel. At the receiving end, the modem converts the keying frequencies into d.c. pulses and the serial–parallel

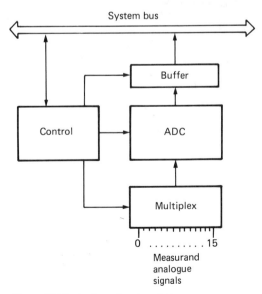

Figure 17.10 Analogue input module

read-only memory (PROM) and random-access memory (RAM). The data-acquisition program and the transmission program are loaded into the PROM, which is non-volatile (not corrupted during a power supply failure). All the system modules are connected to the system bus (see *Figure 17.7*) which is structured to facilitate data transfer from the information source to its destination, under the control of the program. The microprocessor function is monitored by the watch-dog module. The RTU is configured to suit the substation requirements. A typical configuration is shown in *Figure 17.8* with the input/output for one high-voltage (h.v.) circuit only. The input modules fall into three main categories: digital input, analogue input and pulse input.

17.4.2.1 Digital inputs

Digital input modules are used for inputting contact states indicating breaker positions, isolator positions and alarm contact operation. The contact states are continually monitored but are not transmitted unless a contact state has been changed. Under program control, the module is cyclically interrogated to see whether it has a change of state to report: if the reply is negative, the program moves to another module; if it is positive, the change of state is written to the memory. The input circuits are decoupled from the power system equipment by electromagnetic relays or opto-couplers and are filtered to eliminate the effect of contact bounce. A digital input circuit is illustrated in *Figure 17.9*.

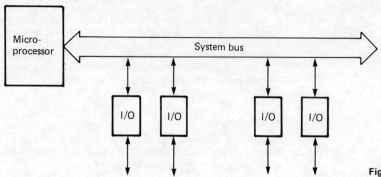

Figure 17.7 RTU bus structure: I/O, input/output

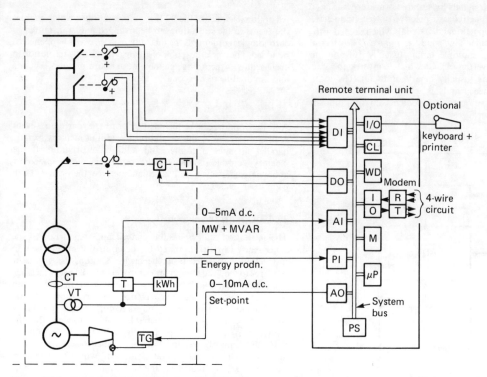

Figure 17.8 Example of the configuration of an RTU: DI, digital input; DO, digital output; AI, analogue input; PI, pulse input; AO, analogue output; μP, microprocessor; M, memory; CL, clock (optional); I/O, serial input/output int3rface; WD, watch-dog; PS, power supply; TG, turbine governor; CT, current transformer; VT, voltage transformer; T, transducer; CT, close, trip; R/T, receiver/transmitter

impedances. Then the following description, which is machine-readable, is possible:

Connection	Impedance	Connection	Impedance
1–2	0.05 + j0.68	3–6	0.06 + j0.75
2–3	0.06 + j0.75	3–5	0.03 + j0.35
1–4	0.03 + j0.35	4–5	0.04 + j0.50
1–5	0.04 + j0.50	5–6	0.05 + j0.65
2–5	0.05 + j0.70		

The first column specifies the nodal connection, the second the complex impedance. Each line of the list corresponds to a connection or a line of the network.

The generation of the nodal admittance matrix can be explained by a multiplication of the primitive admittance matrix Y_p (diagonal matrix) by the corresponding incidence matrix and its transpose from left and right respectively:

$$Y = A^T Y_P A$$

The nodal admittance matrix contains all the information necessary (topology, impedance) to describe the network.

Loads and generators are connected to the nodes of the network. The description of the nodal constraints depends on the way the network is employed in an analysis or decision-making process.

For the purposes of load-flow calculations the specification is done in terms of PQ- or PV-nodes, where PQ means that the active power P and reactive power Q at the node are constant (i.e. independent of voltage), and PV means constant active power and constant voltage.

For dynamic analysis more complex models for the nodal description have to be added. On the load side, the frequency dependence or, if necessary, a set of different equations is needed. For the generating unit, differential equations are comprehensive, but not always transparent. Hence, a block diagram is used to specify forward and feedback paths, wherever appropriate. As an example, the signal flow and the generation of torque within a turbo-generator are given by the block diagram of *Figure 17.5*. The dynamics of a synchronous machine would also be amenable to such a description, but in practice a description by differential equations (two-reaction theory) is usually preferred. Details of controllers and regulations are described by block diagrams, as is common in control engineering.

Reference

1 KOLLER, H. and FRÜHAUF, K., 'PRIMO Data Base Management System', *Brown Boveri Review*, **66**, 204, March (1979)

17.4 Data acquisition and telemetering

17.4.1 Introduction

To be able to supervise and control a power system, the network control engineer requires reliable and current information concerning the state of the network. This he obtains from the power flows, bus voltages, frequency, and load levels, plus the position of circuit-breakers and isolators; the source of these data being in the power-stations and substations of the network. As these are spread over a wide geographical area, the information must be transmitted over long distances. Thus the electrical parameters and switch states must be transduced into a suitable form for transmission to the control centre. A telecontrol system is also required to transmit these data using communication channels which are often limited in capacity and must be shared with other facilities such as speech, protection and, perhaps, also telex communication. Speed of data transmission is therefore often limited, and the telecontrol network must be configured to optimise the use of the available bandwidth on the carrier channel.

The telecontrol system comprises a master station communicating over communication channels with remote terminal units (RTUs) located in the power-stations and switching stations. Normally the RTUs are quiescent, i.e. they only send data after a direct interrogation from the master station; thus more than one RTU can be connected to a transmission channel as the channel can be time-shared. A telecontrol network can be configured as a point-to-point, star (radial) or multi-point (party-line) system as illustrated in *Figure 17.6*. The transmission can be either duplex, half-duplex, or simplex (in the case where data traffic is unidirectional).

17.4.2 Substation equipment

The data-acquisition equipment in a substation is usually a microprocessor-based RTU, equipped with both programmable

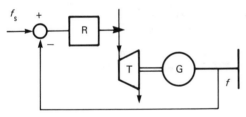

Figure 17.5 Schematic of a generator and a speed governor: G, generator; T, turbine; R, regulator/speed governor; f, measured frequency; f_s, set-point of frequency

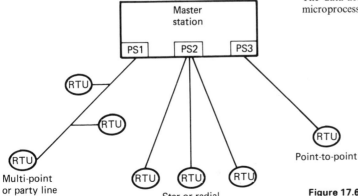

Figure 17.6 Transmission network configuration: PS, part system

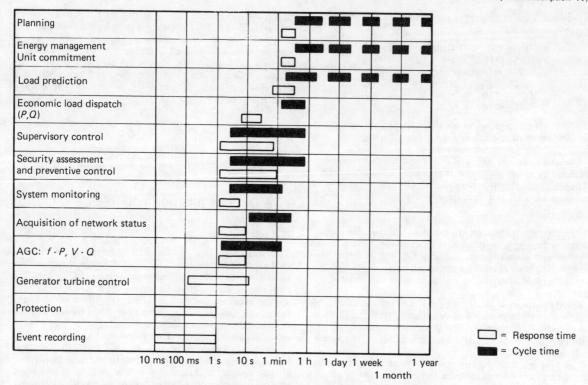

Figure 17.2 Response times and cycle times for plant management, monitoring, protection and control functions: AGC, automatic generation control; P, active power; Q, reactive power; f, frequency; V, voltage

representation, whereas the incidence matrix is a mathematical formulation which can be interpreted by a computer. An example will illustrate the correspondence between these two descriptions. *Figure 17.3* shows a graph derived from an electrical network. The connections between nodes have been marked by arrows; this fixes the way of counting and thereby orientates the graph. *Figure 17.4* is the incidence matrix corresponding to the graph. The way of counting is indicated by the sign of the entries. Both the graph and the matrix describe the same network. The graph, however, is much easier to grasp.

The incidence matrix is the base for two processes, both of which are important for systematic description. The first is the reading of network data, wherein the way in which lines are connected is already implied. The second is the establishment of admittance and impedance matrices.

Consider the graph in *Figure 17.3* where the nodes have a particular numbering. This numbering permits the specification not only of a connection between two nodes but also of the type of connection. Assume that the connections consist of simple series

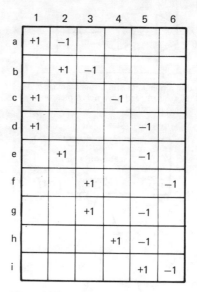

	1	2	3	4	5	6
a	+1	−1				
b		+1	−1			
c	+1			−1		
d	+1				−1	
e		+1			−1	
f			+1			−1
g			+1		−1	
h				+1	−1	
i					+1	−1

Figure 17.4 Nodal incidence matrix **A** corresponding to the graph in *Figure 17.3*. The matrix establishes relations between line and nodal quantities (currents, voltages). The arrow leaving a node determines the positive sign of the entry (+1)

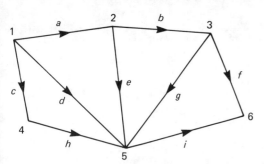

Figure 17.3 Orientated graph describing the structure of a 6-node network: 1, 2, ..., 6, nodes; a, b, ..., i, lines

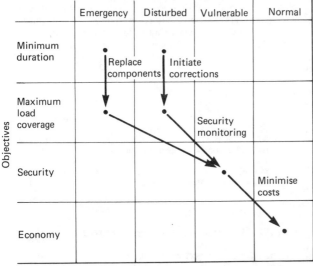

Figure 17.1 Security states and objectives: state diagram (above) and controlled transitions (below) following given objectives

(4) *Emergence (E) state:* The system cannot supply all of its loads, overloads are present, there is a low voltage profile and part of the system is disconnected.

There are inadvertent transitions between the states caused by faults, human error and dynamic effects. However, control actions will return the system to the normal state. A thorough understanding of system operation will show that appropriate objectives and corresponding control functions will effect this return in a logical manner. These mechanisms are illustrated in the schematic of *Figure 17.1*, which shows (upper part) a state diagram with various transitions. The lower part of the figure gives transitions associated with a number of objectives. These pertain to control actions only. It is clear that the objective of full load coverage has to be met first before the vulnerable state can be reached. Further, all vulnerable conditions have to be eliminated before economic dispatching can be initiated.

At the decentralised level the objectives are much simpler and easier to understand. A state concept is not necessary since the objectives are expressed in terms of errors or time sequences in a straightforward manner. As an illustration, let us consider protection and excitation control.

In protection the objective is to keep fault duration or outage to a minimum. The fault itself cannot be avoided, but the adverse effects, possibly leading to a disturbed or emergency state, can be. Thus protection supports the aim of security on the decentralised level.

Excitation control maintains the voltage at a given location of the system and supports the stability. It is the stable voltage, with all its implications, that is at stake. A stable voltage is a very important prerequisite for system security. For its realisation many detailed considerations are required since it is system dynamics that determines the performance of excitation control. Beyond that, an excitation system has many monitoring functions, i.e. limit-checking and even protective functions. The objectives of decentralised control, which become recognisable in terms of set-points, time periods and the like, must be co-ordinated. This co-ordination is performed either in the planning or operational planning phase (e.g. for protective relays), or in real time (e.g. for economic dispatching).

In power system operation and control it must be recognised that the control functions have a certain time range in which they are effective. Thus the corresponding objective has its validity within this time range only. As *Figure 17.2* shows, there is a complete hierarchy of functions ordered in terms of their effective time range. This hierarchy in time is also responsible for the functional hierarchy of the control system.

Reference

1 Special issue 'Load Dispatching Systems', *Brown Boveri Review*, **66**, March (1979)

17.3 System description

In describing an electric power system we must distinguish the network (responsibile for transmission and distribution) and the generating stations, as well as the loads.

The network is amenable to description by topological means such as graphs and incidence matrices as long as the connection of two nodes, by a line or cable, is of importance. Thereby it is implied that the connection is made by a 3-phase line which is itself described by differential equations. This topological description is always inherent and necessary. In routine work, however, it may not always be obvious what the true background is. Hence it is worthwhile to consider these topological means as a starting point.

Graphs and incidence matrices are equivalent and describe the way in which nodes are connected. A graph is a pictorial

17.1 Introduction

Power system operation and control are guided by the endeavour of the utility to supply electric energy to the customer in the most economic and secure way. This objective is underlined by the fact that the electric energy system is a coherent conductor-based system in which the load effects the generation without delay. Energy storage can only be realised in non-electrical form in dedicated power-stations, so that each unit of power consumed must be generated at the same time. It is therefore necessary to maintain all the variables and characteristics designating the quality of service, such as frequency, voltage level, wave-form etc.

The power system is operated continuously and extends geographically over wide areas, even over continents. The number of generators supplying the system and the number of components contributing to the objective are extremely high. However, the task of system operation is alleviated by the fact that there are not too many different types of component (if the intricacies of power-stations are excluded). In the transmission system the components that are quite easily managed are lines, cables, breakers, disconnectors, busbars, transformers and compensators. In the power-stations the generator commonly used is the synchronous machine. The prime mover is a turbine driven by steam or by water. Overlooking the details, there are similarities in the operation and control of turbines and generators that permit a certain uniformity of approach; however, many details have to be considered when it comes to design, failure modes, actual performance, quality of service etc.

The mechanism of power flow in the system, which is a key issue in all considerations of operation under normal and disturbed conditions, is governed by the Ohm and Kirchhoff laws. Higher voltage at an appropriate phase angle will cause more power flow when a suitable path is available. To draw power from a system is extremely simple: the consumer's load is just connected to a 3-phase busbar, where the voltage is regulated so that the load can be supplied. The system includes control actions by which the power is allocated to various generators, but there is little interaction in the transmission system.

For economic and secure operation, the actions are quite sophisticated. Control can no longer be performed in a single location by observing a single variable. The control system becomes a multivariable, multi-level system with real-time and prophylactic interventions, either manual or automatic. Thus, the control system is a hierarchical system where the levels and corresponding functions can be characterised as follows:

Decentralised control
(1) In the power-station: control of voltage, frequency, active and reactive power.
(2) In the substation: control of voltage (tap-changer on transformers), switching of lines and cables, protection.

Centralised control
(1) Regional control centre: switching, start-up and shut-down of generating units, unit commitment, reactive power control, load–frequency control.
(2) Utility control centre: economic load dispatching, security assessment, load–frequency control (automatic generator control), power exchange with other areas.

A hierarchical control system needs extensive communication. Hence there is an extensive telemetering and telecontrol system which provides data to the various control levels and executes control actions at a particular location that have been initiated in a control centre.

The following material is organised in such a way that the objectives and requirements of system operation and control are given and discussed first (Section 17.2): they underline the motivation for the development of modern complex systems. A way of presenting components and subsystems is given in Section 17.3. Data acquisition and telemetering are then treated (Section 17.4); these are prerequisites for power system control.

Decentralised control is divided into *excitation* control (Section 17.5), *turbine* control (Section 17.6) and *substation* automation (Section 17.7), which are the most important control functions at a local level. Pulse controllers for tap-changers are also considered (Section 17.8).

Centralised control is dealt with in Section 17.9, where the hardware and software aspects of computer-based system operation are considered. With the various systems and functions to hand, present-day system operation is characterised in Section 17.10. Finally, the reliability of system control is considered in Section 17.11.

References

1 DYLIACCO, T. E. 'The Adaptive Reliability Control System', *IEEE Trans. PAS*, **86**(5) (1967)
2 GLAVITSCH, H., 'Power System Control and Protection—The Interaction of New and Existing Concepts in Providing Economic and Reliable Supply', keynote address given to the IFAC Symposium on Automatic Control and Protection of Electric Power Systems, Melbourne, Australia, February 21–25 (1977)

17.2 Objectives and requirements

The often-cited objectives of economy and security have to be considered in various time-scales and in different system conditions in order to achieve a systematic approach to power systems operation, particularly when computer-based systems are involved. Before we go into details, it should be noted that system operation as considered here is a problem within the framework of a given power system. Planning problems and problems of procuring the primary energy are omitted.

Any objective or requirement is derived from the basic task of supplying electric energy to the consumer with the least expenditure of economic effort, measured over a long period. Hence a variety of cost items and even some intangibles such as environmental effects are involved. These items range from generating costs, losses, the cost of outages and the cost of damages, to risks and the amount of emissions, etc. Stated thus, these items are still too general to be converted immediately into an objective upon which a control function could be built. Different objectives apply for decentralised and centralised control; moreover, within a centralised control system it is necessary to distinguish various conditions to which particular objectives apply.

The most widely used concept for the realisation of a systematic control approach in centralised control is the concept of *states*. A state of the power system is characterised by reserves, by the ability of the system to override a disturbance, by the presence or absence of overloads, etc. The starting point is a series of considerations concerning the security of the system. However, considerations of economy can be easily added.

The four states with some rough characterisations are as follows:

(1) *Normal (N) state:* The system has no overloads and a good voltage profile; it can withstand a line or generator outage and is stable.
(2) *Vulnerable (V) state:* The system has no overloads and a good voltage profile; line or generator outage causes overloads and/or voltage droop; there is a low stability margin.
(3) *Disturbed (D) state:* The system still supplies its loads, but overloads are present and there is a low voltage profile.

17 Power System Operation and Control

H Glavitsch, Dr-Ing
ETH, Zurich

K Reichert, Dr-Ing
ETH, Zurich

F Peneder, Dipl-Ing
Brown Boveri Cie., Baden

J Burley, MIEE
Brown Boveri Cie., Baden

Contents

17.1 Introduction 17/3

17.2 Objectives and requirements 17/3

17.3 System description 17/4

17.4 Data acquisition and telemetering 17/6
    17.4.1 Introduction 17/6
    17.4.2 Substation equipment 17/6
    17.4.3 Data transmission 17/8
    17.4.4 Transmission system 17/9
    17.4.5 Communication channels 17/10

17.5 Decentralised control: excitation systems and control characteristics of synchronous machines 17/10
    17.5.1 Introduction 17/10
    17.5.2 Brushless excitation systems 17/11
    17.5.3 Static excitation systems 17/11
    17.5.4 Automatic voltage regulator and firing circuits for excitation systems 17/12
    17.5.5 Limiting the excitation of synchronous machines 17/12
    17.5.6 Control characteristics for synchronous machines 17/14
    17.5.7 Slip stabilisation 17/17
    17.5.8 Adapted regulator for the excitation of large generators 17/18
    17.5.9 Static excitation systems for positive and negative excitation current 17/20
    17.5.10 Machine models for investigating stability 17/21

17.6 Decentralised control: electronic turbine controllers 17/24
    17.6.1 Introduction 17/24
    17.6.2 The environment 17/24
    17.6.3 Role of the speed governor 17/24
    17.6.4 Static characteristic 17/25
    17.6.5 Parallel operation of generators 17/26
    17.6.6 Steam-turbine control system (Turbotrol) 17/27
    17.6.7 Water-turbine control system (Hydrotrol) 17/31

17.7 Decentralised control: substation automation 17/35
    17.7.1 Introduction 17/35
    17.7.2 Hardware configuration 17/35
    17.7.3 Software configuration 17/36
    17.7.4 Applications 17/36

17.8 Decentralised control: pulse controllers for voltage control with tap-changing transformers 17/37

17.9 Centralised control 17/38
    17.9.1 Hardware and software systems 17/38
    17.9.2 Hardware configuration 17/38
    17.9.3 Man–machine interface 17/39
    17.9.4 Wall diagram 17/39
    17.9.5 Hard copy 17/39
    17.9.6 Software configuration 17/39
    17.9.7 Memory management 17/41
    17.9.8 Input/output control 17/41
    17.9.9 Scheduling 17/41
    17.9.10 Error recovery 17/41
    17.9.11 Program development 17/41
    17.9.12 Inter-processor communication 17/41
    17.9.13 Data base 17/41
    17.9.14 System software structure 17/42

commutated devices. Predictable timing of thyristor conduction can be achieved only if the a.c. voltage waveform is reasonably sinusoidal when it is essential for TCR to operate in a controlled manner.

With forced or artificial commutation using commutating capacitors, (1) the control of an SVC is almost independent of the system voltage and therefore its operation would be less influenced by high system impedance, and (2) it is possible to swing smoothly from absorption to generation of vars. In addition, the capacitor-commutated SVC could use smaller total capacitor rating. Although the present experimental circuits suffer from the fact that the SVC needs a relatively long time to restart after an a.c. system short-circuit, future improvements using better controls may be developed.

16.7.3.4 Gate turn-off thyristors

Transistors have been generally used in electronic circuits to control very small amounts of power smoothly and continuously. Recent developments have resulted in larger power transistors being available. Even so, they are able to control only up to 200–300 kW per device, far too low for SVC application.

A gate turn-off thyristor (GTO) is a thyristor capable of being switched off by application of relatively large current pulse of short duration at its gate; it does not therefore require an external forced-commutation circuit. An SVC based on self commutated converter circuits (SCC) using GTOs will be less dependent on the a.c. system voltage condition and provide better assistance to a weak system compared with the TCRs and TSCs at present available.

A GTO has about one-third of the mean current capability of a thyristor of similar size (e.g. 75 mm silicon slice) and costs twice as much. Better and cheaper devices may be available in the future and GTOs may advantageously replace even the largest 100 mm thyristors in static var compensators.

16.8 References

1 *Supplies to arc furnaces*, Engineering Recommendation P7/2 (1970), Electricity Council, London, England
2 *Static shunt devices for reactive power control*, CIGRE Working Group 31-01, CIGRE Paper 31-08 (1974)
3 *Limits for harmonics in the UK electricity supply system*, Engineering Recommendation G5/3, September 1976, London, England
4 *Modelling of static shunt var systems (SVS) for system analysis*, CIGRE Working Group 31-01, *Electra* No. 51 (March 1977)
5 *Proceedings of EPRI Seminar on Transmission Static Var systems*, EPRI, USA, 1978
6 International Symposium on Controlled reactive compensation, IREQ/EPRI, 1979, Canada
7 *Proposed terms and definitions for subsynchronous oscillations*, IEEE Committee Report, IEEE Transactions on Power Apparatus and Systems, Vol. PAS-99, No. 2, March/April 1980, USA
8 *Thyristor and variable static equipment for ac and dc transmission*, IEE Conference Publication No. 205, 1981, England
9 MILLER, T. J. E. (ed.), *Reactive power control in electric systems*, John Wiley, New York (1982)
10 *UIE flickermeter, functional and design specifications*, Disturbances study committee, International Union for Electroheat, 1983, Paris
11 MATHUR, R. M. (ed.), *Static compensators for reactive power control*, Canadian Electrical Association, Canada (1984)
12 *Var management, problem recognition and control*, IEEE Committee Report, IEEE Transactions on Power Apparatus and Systems, Vol. PAS-103, No. 8, Aug. 1984, USA
13 *AC and DC Power Transmission*, IEE Conference Publication No. 255, 1985, England

differential equations of the complete electrical–mechanical system. For studying such phenomena more economically, however, digital computer programs have been developed, based on either eigen-value analysis or frequency-response analysis of the complete system of linearised machine and network equations, including any shaft torsional equations of the turbine prime movers.

Various methods have been proposed to damp sub-synchronous oscillations: these include the use of SVCs, damping filters and generator-excitation-control signals.

The awareness of the possible danger of SSR, the improved methods of study and the available methods of damping such oscillations will enable the full use of series compensation.

The sub-harmonic oscillation phenomena are associated with the use of series capacitors in conjunction with non-linear reactive devices using saturable iron in the magnetic path, such as transformers, gapped-core 'linear' reactors or saturable reactors. These phenomena are amenable to non-linear or linearised analysis, but only with a limited accuracy except for the harmonic-compensated saturated reactors, whose performance can now be analysed fairly accurately, and, where necessary, appropriate sub-harmonic damping filters can be designed accurately. Model studies using transient network analysers or simulators are employed in assessing individual applications involving transformers or 'linear' reactors. Theoretical analysis of ferroresonance phenomena has not resulted in generally applicable conclusions.

16.7 Future prospects

16.7.1 Long-distance transmission

Several trends in reactive power requirements of the system should be noted. Increase of the line voltage to the maximum practical value reduces the load current and consequently the var consumption. The transiently needed peak vars can be met by rapidly switching in capacitor banks (possibly by thyristors) and so maintaining stability. However, this relies on obtaining timely signals during transient conditions and the total time delay may sometimes be too long to save the system. Owing to the possibility of sudden total load rejection, adequate linear or saturated shunt reactors must be permanently connected.

Series capacitor compensation is an economic way of increasing the power transfer capacity of a line, but some of the potential gain in additional capacity is lost by the need to have permanently connected linear shunt reactors, and sub-synchronous resonance conditions must be evaluated at design stage.

Shunt compensation using SVCs would appear to be simple, as it provides good voltage control along the line as well as increased transmission capacity. Unfortunately, it is not always the most economic method of achieving the required performance if the total 'dynamic' or 'swing' range of the continuously variable vars is high. The use of variable reactors instead of linear reactors has already been implemented in service on a limited scale. The advantage is obvious: at light or no load the variable reactor consumes maximum vars and at full load it consumes none, but it is available to prevent severe overvoltages in case of load rejection. The need for very fast action is obvious and, hence, the harmonic-compensated saturated reactor (SR) may be more suitable than a TCR; in addition, the former does not rely on controls to change its var consumption. The SR has an overload capability that will allow time for additional linear reactors to be switched in, so reducing the cost of the more expensive SR and its slope-correction capacitor.

There is likely to be a need in series compensated lines to provide variable reactive shunt compensation at the receiving end of the line to assist regulation and voltage stability of the load network. A possible further option, not yet fully exploited in practice, although discussed in published papers, is the use of a combination of series capacitors and SVCs, not only at the termination of the line, but also at intermediate points.

16.7.2 General application

For general system application, equipment is available to fulfil duties similar to those for long lines. It has been mentioned that SVCs can be used for system swing damping. Another type of variable static equipment, h.v.d.c., has been used successfully to damp a.c. system phase angle oscillations by modulation of the power fed into the a.c. system. It has also been proposed to use h.v.d.c. to provide a.c. system voltage control by controlling the converters to consume reactive power from the network in response to a system signal.

These recently developed methods of controlling reactive power by variable static equipment (SVC and h.v.d.c.) indicates considerable advance considering that the only methods available up to the late 1960s were the generator itself and the synchronous compensator.

16.7.3 Future developments in var compensation equipment

16.7.3.1 Thyristor tapchangers and combination of SR with TCR

The technique of using thyristors instead of mechanical switches is available for controlling transformer tap ratios, but the considerably higher cost has so far not justified their use. However, it is a method available for fast control of voltage and vars. The magnitude as well as the phase angle of the voltage can be made rapidly controllable.

The use of slope-correcting series capacitors adds to the cost of harmonic-compensated saturated reactors. In addition, the SR is intrinsically a constant-voltage device. If a relatively small thyristor controlled valve, or thyristor controlled tap-changer, is used in conjunction with the saturated reactor, the superior overload capacity of the SR combined with controllability of a TCR may find application in transmission schemes.

16.7.3.2 Present limitations

What are the future needs? First, critically consider the SVC characteristics available at present. Static var systems rely on capacitors for var generation. This may, on occasion, be inconvenient from two points of view.

The vars generated by capacitors vary with the square of the applied voltage, and therefore reduce when there is a particular need for vars to boost the voltage.

SR and TCR depend for their correct operation on a good a.c. voltage waveform, as in a 'strong' system. The term 'weak' refers to an a.c. system in which the trip of an individual load or loss of a line can transiently affect the voltage waveform of the remaining system which has to continue to operate. The impedance of the system in such cases is likely to be relatively high with respect to the rating of the SVC. Moreover, in weak systems there is normally a need for large shunt capacitance to boost the voltage. Unfortunately, shunt capacitors increase the effective system impedance and tend to cause resonance on low-order harmonics. These resonances are controlled by specially designed filters, which add to cost and loss.

16.7.3.3 Forced commutation

In h.v.d.c. converters and TCRs the current is commutated from one thyristor to another by the system voltage—i.e. they are line-

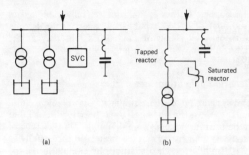

Figure 16.21 Arc furnace compensation: (a) busbar compensator; (b) tapped reactor/saturated reactor compensator

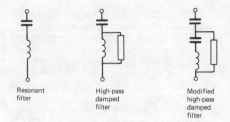

Figure 16.22 Harmonic filters

The harmonic content has undesirable disturbing effects on the system and load equipments and on adjoining telecommunication circuits, which require its control and reduction by means of system design and, where necessary, by installation of shunt harmonic filter equipment at appropriate system busbars. Inductive or capacitive coupling between telephone and power lines will induce harmonic interference over the audiofrequency range, and appropriate weighting factors are applied to the individual frequencies when assessing total distortion effect using measures such as Telephone Interference Factor (TIF). Rotating generators and induction motors experience increased power loss due to the flow of harmonic currents. Waveform distortion can sometimes cause maloperation of electromagnetic as well as static protective relaying equipment. Shunt capacitor banks primarily for power factor correction are particularly affected by harmonics, as their impedance falls inversely with frequency so that they tend to act as 'sinks' of harmonic currents and may overheat by increased loss. Shunt capacitors form a resonant circuit with respect to the rest of the system at some frequency above the fundamental (i.e. at a supersynchronous frequency), and if the resonant frequency is close to that of a local source of harmonic current, large and potentially damaging harmonic overvoltages and current magnifications can occur.

In order to minimise harmonic problems, supply authorities apply harmonic current and voltage distortion limits. The limits are, in general, based on observed disturbances and on a desire to protect other consumers and telecommunication services. Compliance with the limitations at the point of common coupling with other loads may require reduction of system impedance by addition of lines or transformers, or by connection of the distorting load to a different point in the network. In case of converter loads the pulse number may have to be increased from 6 to 12, or even 24.

Connection of a properly designed harmonic filter consisting of a combination of inductive, capacitive and resistive elements at a suitable busbar will provide a low-impedance shunt path to the flow of harmonic currents generated by the distorting loads. The filter and the system will share the harmonic current injected by the load in the inverse ratio of their respective impedances. The desired impedance pattern of the filter circuit can be achieved by an appropriate combination of single-frequency tuned arms and broad-band damped arms (*Figure 16.22*).

The system impedance varies with harmonic frequency in a complex manner, as the system is not, in general, purely inductive. The system will have intrinsic resonant frequencies, and if these occur close to a particular harmonic, there will be dramatic changes in impedance from inductive to capacitive for only $\pm 10\%$ variation in frequency. The supply system will, in general, have many different operating configurations, and to take account of the worst conditions the system impedance vector may be assumed to lie within a circular segment limited by maximum impedance angles, except in simpler industrial supply circuits, where the harmonic impedance is dominated by a fixed supply transformer. The latter assumption can also be made in the absence of other data if the system up to the point of connection of the filters is not primarily composed of long lines. The system may contain several harmonic sources other than the particular known loads, or compensators, or h.v.d.c. converters; in the assessment of the voltage distortion levels and filter component current ratings the effects of such 'pre-existing' harmonics should not be ignored.

The high-pass damped filter shown in *Figure 16.22* is sometimes advantageous compared with the tuned arms if the filter is to be satisfactory over a wide range of system operating frequency (e.g. from 49 to 51 Hz); but such damped arms can have greater losses. The modified damped filter is of particular use if the frequencies of the harmonics to be passed are low, as the loss in the resistor can then be excessive at the fundamental frequency unless it is shunted by the series L–C circuit tuned to the fundamental frequency. Such a filter is specifically useful for suppressing unintended resonances at 'non-characteristic' frequencies (e.g. 2nd harmonic) which are inherently low in current magnitude and thus would not give large loss in the resistor at these harmonic frequencies.

The optimum design of a combination of different filter arms (tuned, damped, etc.) to achieve a particular harmonic impedance pattern requires an economic assessment of the effects of choosing different relative proportions (in terms of fundamental Mvar) of the filter arms. The choice is governed not only by the harmonic performance requirements, but also by the fundamental plus harmonic ratings of the filter components—in particular, the capacitor banks. The effects of variations in the filter impedance with system frequency, temperature and component tolerances should be taken into account. The choice of the filter combination should also ensure that the non-characteristic harmonics (e.g. 2nd, 3rd or 4th order), which may be relatively small, are not subject to excessive magnifications due to unintended resonances of the selected filter itself with the system. The need to minimise fundamental plus power-frequency losses in the filters also plays an important role in the final choice of the design.

16.6.3 Subsynchronous/sub-harmonic resonance damping

Series capacitors in large power systems require some caution at the system design stage, as they can resonate with the generator and line inductances at frequencies below the power frequency, and such subsynchronous oscillations can lead to self-excitation of generators, to rotor hunting or to shaft oscillations. The possible danger of this phenomenon was illustrated in early 1970s when the shaft of a large turbo-generator was twice damaged in service, owing to subsynchronous resonance (SSR) excited by series capacitors. The SSR phenomena is now well understood as being due to near coincidence of the electrical and shaft torsional resonance frequencies, and can be fully analysed using non-linear

The spectral density of voltage fluctuations produced by an arc furnace is approximately in inverse proportion to the square root of the frequency, while the reaction of people subject to light flicker is subjective; generally, human sensitivity peaks just below 10 Hz. As seen in *Figure 16.19*, a weighted combination of these characteristics shows that the frequencies most liable to cause visual annoyance lie in a band from 2 to 25 Hz. If the voltage

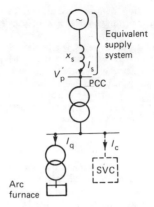

Figure 16.20 Simplified arc furnace supply circuit: PCC = point of common coupling (with other consumers)

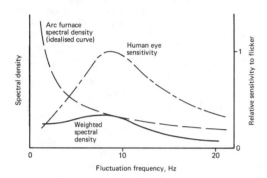

Figure 16.19 Eye sensitivity to arc furnace induced flicker

fluctuations at 10 Hz are greater than 0.2%, then they are likely to give a noticeable flicker on a 230 V filament lamp. For an 110 V lamp, which has a heavier filament with a greater thermal capacity, the limiting voltage fluctuation is somewhat lower.

A circuit supplying an arc furnace can be simplified to that shown in *Figure 16.20*, where the point of common coupling is the point in the network at which other consumers are connected. If V_p is the voltage dip at this point due to the variation of arc furnace var demand, then by neglecting the power component of the furnace current (which is an acceptable approximation if the X/R ratio of the supply system is not particularly low) and if there is no SVC installed) we get:

$$V_p = I_s \cdot X_s = I_q X_s$$

It is thus relatively easy to estimate the voltage dip. It is, however, more difficult to estimate the effect of varying voltage dips on the human eye. A flicker meter that measures numerically the flicker perception level for different voltage fluctuations has been developed. It has been accepted by I.E.C. but work is continuing to determine its application to system design. Until then a simple estimating procedure, that has been obtained empirically and has proved to be valid on many installations, uses a 'short-circuit voltage depression' (s.c.v.d.) criterion. This is the steady state voltage drop at the point of common coupling due to change in furnace var demand from no load to that due to a 3-phase short-circuit on the electrodes. If it is greater than 2%, consumers are likely to complain about light flicker.

Whereas this criterion can be used to determine the maximum furnace rating that should be connected to a given system, it cannot be used to determine the rating of a compensator for flicker reduction unless the compensator is capable of reducing all flicker frequencies in the visual annoyance range reasonably equally. Where a compensator has an acceptably linear fluctuation frequency versus speed of response characteristic up to about 25 Hz, then, if connected as shown in *Figure 16.20*, a steady state calculation of s.c.v.d. can determine its rating—i.e. the compensator current jI_c makes up the difference between the allowable $-jI_s$ and the value of $-jI_q$.

A high speed of response is essential for flicker reduction. It has been shown that if the compensator has a control-time delay of 10 ms, no matter what its rating, it can give very little reduction of flicker; and with a time delay of 20 ms or greater, certain frequencies within the visual annoyance range will be accentuated.

The harmonic-compensated saturated reactor without a slope-correction circuit can give flicker reduction of up to 3:1. It has been employed successfully in many installations as a busbar compensator (*Figure 16.21(a)*), having been designed on the basis of the s.c.v.d. criterion. The slope-corrected saturated reactor has also been used successfully in many cases; although its response at the higher fluctuation frequencies is less rapid, making it less effective, the reduction of its steady state slope down to zero, if required, makes it more effective at the lower fluctuation frequencies.

The thyristor-controlled reactor used as a busbar compensator can be made suitable for arc-furnace compensation for flicker reduction of about 2 or 3:1 if it is given a sufficiently fast control system and also if it has a low per-unit reactance; the latter feature may, however, lead to more expensive thyristor valves as well as higher harmonic currents of the SVC. The thyristor switched capacitor compensator cannot, however, give the necessary speed of response to reduce the arc-furnace flicker in the frequency range above 5 Hz, where the human eye is most sensitive.

To obtain a flicker reduction of greater than 4:1 the tapped-reactor/saturated-reactor scheme (*Figure 16.21(b)*) can be used, for a single arc-furnace. In this scheme the saturated reactors are 1-phase devices and the slope correction is achieved by the tapped reactor winding ratios; this compensator inherently compensates for the unbalance loadings of the arc furnace and gives an instantaneous response. It produces considerable harmonic distortion and requires adequate filtering.

16.6.2 Harmonic filter design

Many types of plant draw a distorted, non-sinusoidal current from the supply system, as, for instance, large rectifiers (aluminium smelters), controlled converters (traction drives, mine winders, rolling mills), discharge lamps, light dimmers and television receivers. Power transformers (due to magnetising currents), h.v.d.c. converter stations, static var compensators (using thyristor controlled reactors and saturated reactors) are major sources of waveform distortion in the network itself. These current and consequent voltage waveform distortions can be described in terms of fundamental and harmonic frequency components using Fourier analysis.

thyristors. The actual relative positions of these constant-current and continuous-conduction parts of the characteristic may be changed to suit the design requirements.

A speed of response to small disturbances of 1–3 periods is achievable with this type of control system, particularly if the TCR design employs a linear reactor of substantially lower impedance than the typical values given above. A faster response can be obtained by using an open-loop current control; this can be usable for certain industrial applications where the highest speed is required with some sacrifice of control accuracy.

16.5.4.4 Thyristor switched capacitor compensator (TSC)

The TSC is the only static compensator that can give directly the effect of a variable capacitance, although for reasons given below the variation is in steps. A.C. thyristor valves consisting of back-to-back connected thyristors, similar to those used for the TCR, are used to give rapid switching of blocks of capacitors (*Figure 16.18*).

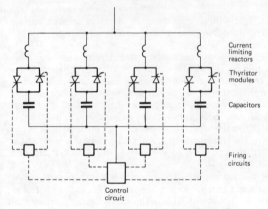

Figure 16.18 One phase of a thyristor switched capacitor compensator

After the capacitor current through the thyristor ceases at current zero, unless re-gating occurs the capacitor will remain charged at peak voltage while the supply voltage peaks in the opposite polarity after a half-period. This imposes a doubled voltage stress on the non-conducting thyristor, so that an increase is necessary in the number of thyristors in series in a TSC compared with a TCR valve of equivalent voltage, although normally this increase is less than double. Further, because of the very high inrush currents that can occur on energising capacitors, it is necessary to choose the point-on-wave for gating the thyristor with accuracy. Ideally gating should take place when the voltage across the thyristor is zero and the supply voltage is at its peak (i.e. $dv/dt=0$), when the energisation will be transient-free; such conditions cannot always be achieved, owing to mismatch between the capacitor voltage, left slowly leaking its charge from a previous energisation, and the system voltage, which may have changed differently over the same interval. The function of the control system is therefore to choose the gating instant for minimum transient.

There will be occasions when severe transients occur (for instance during initial energisation of an uncharged capacitor); hence, series reactors must be installed to reduce the magnitude of the inrush current under the worst case to a value within the thyristor capability. With the exceptions of these transients, however, the TSC does not produce harmonics.

Once gated, the thyristor will conduct for a half-period and unless again gated the current will then cease. The TSC, therefore, can only give a variation of capacitance in discrete steps for a discrete period of an integral number of half-periods. The number of steps and, hence the fineness of control are dictated by economics in the cost of separate valves for each step and the complexity of controls. The effective number of steps can be increased by using different sizes; for instance, by having one block of capacitors half the size of the rest, the effective number can be doubled. But this increases the control complexity.

Thyristors can be used to switch-on a large bank of capacitors faster than is possible with a mechanical switch; therefore a potential application of a TSC is for fast boosting of a transmission-line voltage to maintain system stability. A TSC having several steps could also be used on its own to provide swing damping in a transmission system.

TSCs are used in conjunction with TCRs to give much reduced losses at zero var output (float condition) compared with the losses of a scheme using a fixed capacitor and a larger TCR, and to provide an increased operating range in the var generation region where speed of capacitor energisation is of importance or a high frequency of operation is required. In these applications the TSC provides the coarse steps, seldom exceeding four, whereas the TCR gives a fine continuous control in between. As the cost of the combination of TSC and TCR is substantially higher than the combination of mechanically switched capacitors and a TCR or an SR, the overall system requirements must be carefully evaluated.

TSCs have been used for power-factor control of loads with frequently varying var demand. Here the capacitor banks and, hence, the a.c. thyristor valves are at a low voltage and a large number of them are used in parallel, giving a close rapid control. The cost effectiveness of such applications must be carefully examined, as their effectiveness in reducing light flicker and voltage fluctuation is very limited compared with that of other types of SVCs.

16.6 Special aspects of compensation application

16.6.1 Light flicker compensation

The power demand of industrial loads, such as arc furnaces or thyristor drives for rolling mills and mine winders, varies rapidly in both magnitude and power factor. In an arc furnace the impedance of the arc changes half-cycle to half-cycle and, particularly during the early part of a melt, the arc may be short-circuited by scrap metal, when the furnace current is limited only by the circuit reactance. This gives a current demand which very rapidly fluctuates from zero to short-circuit in a random manner.

In thyristor drives very large changes in energy are required during acceleration and deceleration: for instance, in a rolling mill, as an ingot enters the rolls, the drive motor is running at low speed and therefore at a low voltage; the current then increases very rapidly to maximum. Under such conditions the thyristors must be phased back until the motor has accelerated, and this will produce a high current demand at about 0.2–0.3 power factor lagging. Unlike the situation in the arc furnace, the changes in power and var requirements are predictable, as they occur in a regular cycle and are relatively slow.

Both these loads can give voltage fluctuations caused by fast var changes in the system reactance which may be unacceptable unless the network is strengthened or reactive compensation applied. The limitations arise from either the production of light flicker from filament lamps or interference with electronic control circuits of other equipment in service. Since the former when caused by arc furnaces is the more difficult to deal with, it will be emphasised in this section; but the general principles are applicable to the compensation of other fluctuating loads.

Figure 16.16 Thyristors and assemblies

applications such thyristors can be rated at up to 5 kV (peak), but derating factors of 2 or more have to be applied in a series string of thyristors to determine the actual operating voltage of each thyristor.

Differences in the thyristor leakage currents, stray capacitances and stored charge all tend to result in an uneven voltage distribution between series-connected thyristors (*Figure 16.15(c)*). A resistor–capacitor a.c. grading network is used in parallel with a d.c. grading resistor across each thyristor level. These grading networks reduce the voltage unbalance, and the R–C network also serves to damp the transient overshoots in the reverse voltage which occur when thyristors come out of conduction (twice per cycle per valve). Individual thyristor levels within a module are protected against overvoltage by breakover diodes or similar devices, which provide a gate pulse for forward firing of the thyristors above a preset overvoltage level. Special care must be taken in the design of the gating electronics and the signal transmission circuits to ensure coherent (simultaneous) firing of all the series-connected thyristors.

The thyristor valves are air insulated, and air or water cooled (deionised pure water being used in the latter case), and are suitable for indoor installation. Only pure demineralised water having high resistivity is allowed to flow through the thyristor heat sinks, which would be at high voltage with respect to ground, and through insulating water pipes. The water cooling plant is mounted separately at ground potential and the pipework from ground to the thyristor stacks is made of insulating material. Air cooling is much simpler in this respect, but it generally permits less efficient utilisation of the current capability.

Figure 16.16 illustrates thyristors (silicon slices and individual units) and their modular assemblies; the left-hand side shows a water cooled assembly of four 75 mm thyristors and heat sinks and that on the right an air cooled assembly of two 56 mm thyristors, the former assembly being capable of more than four times the MVA handling capability of the latter.

The control system of a TCR can produce any desired voltage–current characteristic derived from various input signals. A basic characteristic for a transmission compensator having a closed-loop voltage control, with current compounding, is shown in *Figure 16.17*. When the valve reaches maximum conduction, the characteristic follows the slope set by the impedance value of the linear reactors plus coupling transformer (typically 0.7–1.0 p.u. on the basis of full-load rating) up to the point of the thyristor current thermal limit, when by phasing back the firing angle the current is held constant up to the maximum voltage limit of the

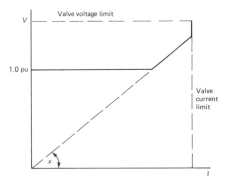

Figure 16.17 Voltage–current characteristic for thyristor controlled reactor

Variable var compensators **16**/13

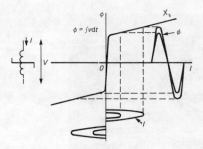

Figure 16.14 Basic self-saturated reactor operation

Treble Tripler reactor (i.e. within the range of 8–15% on rating) would have an inherent speed of response, discounting control delays, equal to the Treble Tripler without slope correction.

The simplest design of TCR (*Figure 16.15(a)*) uses three 1-phase valves connected in delta giving a 6-pulse unit: hence, it produces substantial 5th and 7th harmonic currents and in most cases will require the use of harmonic filters. Although for some ratings it may be possible to connect the reactor direct to the power system busbar, for most applications an interconnecting transformer will be required to match the optimised valve design voltage to the system voltage.

For large installations it may be economic to use a step-up transformer with two phase-displaced secondary windings connected star and delta, each with a delta-connected thyristor valve and reactors (*Figure 16.15(b)*). This 12-pulse design gives better harmonic compensation, the principal harmonics being now 11th and 13th, and filters are not normally required.

The a.c. thyristor valve is based on a technology similar to that developed for the valve of higher voltage rating used in h.v.d.c. transmission. The first transmission SVCs used modified d.c. designs. With the back-to-back connection of the thyristors for the a.c. valve (*Figure 16.15(c)*) the valves are of modular construction using similar elements in series to obtain a wide range of ratings with voltages up to 40 kV and currents up to 4 kA. The cost of a thyristor valve is influenced not only by the number of thyristors, but also by the number of voltage levels in series. Large-diameter silicon wafers used in the latest designs of thyristor can give valve current ratings of up to 4 kA; taking into account the maximum desirable rating of the reactors and connections, this means that paralleling of thyristors, as in the earlier generation of valves, is no longer necessary. For TCR

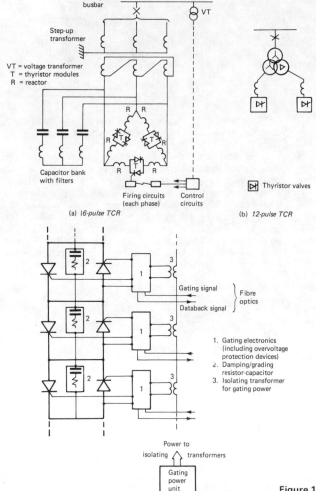

Figure 16.15 Features of thyristor controlled reactor compensators

Figure 16.11 Treble Tripler saturated reactors

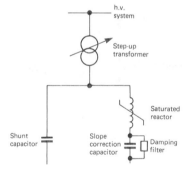

Figure 16.12 Saturated reactor compensator

A significant feature of the saturated-reactor types of SVC is that, being of transformer-like construction, they have very low maintenance/operational requirements and also a high short-time-rated overload capability of up to 4–5 times rated load for 1 s, or longer if necessary. This can be of use in networks with long transmission lines, where the Ferranti effect following a loss of load can produce excessively high temporary overvoltages. The reactor, by swinging rapidly into a high var absorption mode, can balance the var production of the line and so reduce the inherent overvoltage.

16.5.4.3 Thyristor-controlled reactor compensator (TCR)

The thyristor-controlled reactor comprises a linear reactor, connected in series with what is termed an 'a.c. thyristor valve' made up of back-to-back connected pairs of high-power high-voltage thyristors, which are themselves connected in series and/or parallel to obtain the necessary rating. Variation of the current is obtained by control of the thyristor conduction duration in each half-cycle, from a 90° firing angle delay as measured from the applied voltage zero for full conduction to 180° delay for no conduction.

The operation of the self-saturated reactor (SR) may appear to be quite different from that of a TCR. However, there are similarities. One phase of a TCR is shown in *Figure 16.13*. The

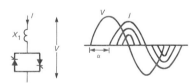

Figure 16.13 Basic TCR operation

current conduction (and therefore its value) is varied by varying the delay angle. A single phase of a saturated reactor is shown in *Figure 16.14*; while the switching function in a TCR is performed by thyristors, in this case it is performed by core saturation. If the series reactance X_1 of the TCR is made equal to the saturated (slope) reactance X_s and if the gating of the TCR thyristors is controlled by a signal proportional to the integral of the applied voltage, the operation of the two devices would be identical. *Figure 16.10* shows the practical circuit of a Treble Tripler saturated reactor which can be considered as analogous to an 18-pulse converter circuit as regards harmonic content. Theoretically an 18-pulse TCR with a reactance equal to that of a

saturated reactors, the first a d.c.-controlled transductor and the second a self-saturated harmonic-compensated device. During the next decade many SVCs were commissioned on industrial systems. The first thyristor-controlled reactor was put into service on a transmission system in 1978 and this coincided with a rapid growth in such applications of both thyristor and saturated reactor (SR) types of compensators (*Figure 16.8*). This growth resulted mainly from two factors: first, the rapid increase in the cost of power made utilities very conscious of the need to minimise transmission losses; and second, the growing opposition in industrialised countries to the construction of new transmission lines encouraged the maximisation of power transfer down existing rights of way.

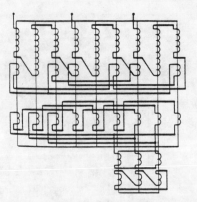

Figure 16.10 Circuit of Treble Tripler reactor

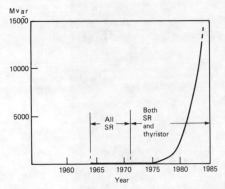

Figure 16.8 SVC installed capacity (data approximate only)

16.5.4.2 Harmonic-compensated self-saturated reactor compensator (SR)

This is the only type of SVC that gives an inherent variation of reactive current with voltage without an imposed control system. The variable element comprises an iron-cored inductor operating in the saturated region to give an r.m.s. voltage–current characteristic as shown in *Figure 16.9*, the operating current adjusting itself to give a predetermined voltage as defined by the system load line.

An iron-cored inductor with a simple winding is preferably not used in this manner on a power network, as it would generate large harmonic currents. However, by interconnecting sections of all three phase windings it is possible to cancel out most of these harmonics. The Twin Tripler and Treble Tripler harmonic compensated saturated reactors, invented and developed by Dr E. Friedlander of the General Electric Company, England, use this principle and have, respectively, six and nine active iron-cored limbs with inter-star connected windings (*Figure 16.10*). These harmonic-compensated reactors have virtually superseded the d.c.-controlled transductor-type saturable reactors for most applications. *Figure 16.11* illustrates the core and windings of an 114 MVA 18 kV Treble Tripler saturated reactor. In addition to reducing the harmonic content of the reactor current to negligible proportions the designs give a voltage–current characteristic with a sharp knee point and a slope that remains linear to within about 1% for currents down to 7 or 8% of full load. The slope reactance is normally within the range of 8–15% on the basis of nominal full-load rating.

The saturated reactor is a fixed-voltage device. If it is to be connected to a system whose voltage level must be adjustable, then a regulating transformer is inserted between the reactor and system busbars (*Figure 16.12*). Where a step-up transformer is necessary to match the reactor voltage (normally 6.6–69 kV) to the system voltage, then the regulation can be obtained by using an on-load tap changer on the step-up transformer.

The *V*–*I* slope typically required at the terminals of an SVC for transmission applications is around 3%, whereas that obtained from a saturated reactor plus step-up transformer may be 15–25%. To compensate for this, a slope-correcting capacitor is connected in series with the saturating reactor as shown in *Figure 16.12*. To avoid the possibility of sub-harmonic oscillations after switch-on, a damping filter is fitted across this capacitor.

The speed of response of a saturated reactor of the Twin Tripler or Treble Tripler design is very fast. Owing to the polyphase construction and the interconnection of its windings, these reactors operate in a manner different from that in a linear shunt reactor. The latter absorbs the capacitive energy of the system as electromagnetic energy and returns it all every half-cycle, phase by phase. In contrast, the Treble Tripler reactor, by virtue of its nine active cores and interconnected windings, transfers about 90% of its energy from phase to phase as the magnetic flux is transferred from one core to the next every 1.1 ms at 50 Hz ($2 \times 9 = 18$ transitions per period). From a different viewpoint, in the saturated reactor a change in current is caused by a supply voltage change acting on its slope reactance which is in the range of 8–15% of the reactance of a linear reactor of the same rating: the response is therefore almost instantaneous. The inclusion of the slope-correcting capacitor does, however, slow its response slightly and a slope-corrected Treble Tripler reactor SVC responds effectively to a step change within one to two cycles, dependent on the system impedance and on the rating of its sub-harmonic damping circuit. An alternative method of achieving the slope-correction effect without time delays is sometimes employed for light flicker compensation as illustrated in *Figure 16.21(b)*.

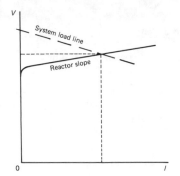

Figure 16.9 Saturated reactor characteristic

points of distribution systems with long overhead feeders, such as those supplying rural areas; the capacitors, often pole-mounted, part-way down the feeders would be permanently connected, with those at the substations being switchable. In the UK, however, owing partly to tariff incentives, load power factors tend to be high and no general application of this type has been carried out.

Capacitors are not generally required for cable distribution networks; but where these include substantial lengths at higher voltages, shunt reactors are sometimes installed at bulk supply points to compensate for the excess of vars under light-load conditions. To reduce the cost of losses inherent in all compensation devices, these are normally switched off as the load builds up.

An important application for shunt capacitor banks is in power factor correction for industrial consumers. Here the economics of reducing var demand is straightforward. The installation of such a bank may, however, cause resonance between its capacitance and the system reactance at a prevalent harmonic, with possible overvoltages or excessive currents. The capacitor bank itself (which will have a low impedance at harmonic frequencies) may act as a sink for harmonic currents produced by pre-existing voltage distortion in the network, and so become overloaded. Before installing power factor correction banks, measurements or estimates of pre-existing harmonic voltages in the network should be made and the effects of the capacitors on the remainder of the system should be studied.

It is worth noting that since shunt capacitor banks are made up of a number of small units of up to 300 kvar each, it is possible to increase a bank size in steps without the necessity for replacing previously installed units. This usually allows a good economic choice to be made in their rating.

Any switchgear installed to control a shunt capacitor bank or a reactor, be it circuit-breaker or load switch, must be carefully chosen, as the duty, even with the low load currents normally encountered, is not easy. The number of operations is often high. Capacitor switching produces high transient inrush currents which may severely stress both the switch and the capacitor, and with reactor switching it is necessary to ensure that excessive overvoltage cannot be generated.

16.5.3 Synchronous compensators

All synchronous machines can give continuously variable var compensation, and in industrial systems where large synchronous motors are installed they are normally used in this way for load power factor correction in addition to their main driving duty. Self-driven synchronous machines not connected to any mechanical load (synchronous compensators) can be used anywhere in a system either to generate or to absorb vars, on a balanced 3-phase basis. When overexcited, the machine generates vars and is operating in a stable condition; when underexcited, it absorbs vars with a reducing stability down to zero excitation. Most machines operate with an automatic excitation control which requires a rapid response to assist the machine in maintaining stability through system disturbances or to follow rapid reactive load changes.

The synchronous compensator has the most positive action of the static var compensator. It has, however, now been superseded for most applications by static designs owing to its inherent limitations: notably a slower speed of response, typically 0.2 s; higher losses; inertia that can cause underdamped oscillations or loss of stability; higher maintenance requirements; and higher capital costs.

The cynchronous compensator has the most positive action of all the types of compensators in supporting a weak system, as it is the only one with a generated voltage and an inertia acting transiently as a prime mover. As opposed to a compensator using capacitors (where, when it has reached the limit of its variable range, the output current decreases with decreasing voltage), the synchronous compensator can maintain full rated current independent of its terminal voltage and may for a short time increase its var output to transiently assist the system. This is of value in certain applications, for example, where power is supplied by h.v.d.c. to a system having minimal generation, the receiving system has to be strengthened by a synchronous type of compensator.

16.5.4 Static var compensators (SVCs)

16.5.4.1 Basic features

The main elements of the SVC are a capacitor to generate vars and an inductor to absorb vars. To operate in both the generation and absorption modes, both elements must be used, and at least one of them must be variable (*Figure 16.7(a)*). A variable capacitive element cannot give a stepless variation: therefore, if a smooth continuous variation is needed over the whole range A–A' in *Figure 16.7(b)*, then a capacitor bank having a rating to

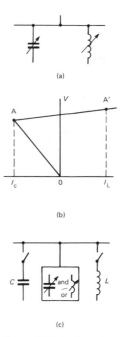

Figure 16.7 Basic static var compensator (SVC)

give at least the current I_c must be connected and the variable element must be inductive with a rating to give a current at least equal to $I_c + I_L$.

In some system applications the SVC must be capable of operating under particular system conditions at the extremes of its range (i.e. at A and A'), but it need not be capable of swinging in a continuous manner from one extreme to the other. Its 'dynamic' or 'swing' range, which is the range over which it can give a near-instantaneous response, may be only a proportion of its total range. In such a design the size of the variable element, which is by far the most expensive per unit of rating, can be reduced to deal only with the swing; and mechanically switchable L and C elements (*Figure 16.7(c)*) can be controlled automatically to give the total range. The overall cost of the SVC is then reduced.

The first static var compensator for a transmission system was installed in 1967, followed by another in 1969: both used

Table 16.1 Comparison of the characteristics of different var compensators

Item	A	B	C	D
Type of compensator / Features	Synchronous	Harmonically compensated self-saturated reactor/fixed capacitors	Thyristor-controlled reactor/fixed capacitor	Thyristor-switched capacitors/ thyristor-controlled reactors
Steady-state characteristic	AVR easily adjustable	Restricted adjustment possible on site	Controller easily adjustable	Controller easily adjustable
Harmonic content of current	Negligible	Internally compensated in balanced 3-phase system	6 pulse design usually requires filters, particularly if reactance < 100% on rating	TSC, negligible TCR, smaller than in C
Fault infeed	Approximately 4–6 × rating	Nil	Nil	Nil
Overload capability	Gives both generation and absorption up to approximately 2 × normal during swings	Inherent very high absorption (e.g. 3 × normal) Generation overload limited to capacitor bank rating unless special star-delta switching used	Absorption limited by full conduction of thyristors	
Losses	High at full output	High at zero var output (float condition)		Low at zero var output
Maintenance requirements	As for any large rotating plant	Small; as for conventional outdoor equipment	Moderate; as for special electronic (indoor) equipment if properly designed	
Approximate response time in typical complete system	0.2 s	0–0.05 s	0.02–0.06 s	0.02–0.06 s
Response to rapidly fluctuating loads	Relatively slow	Inherently fast	Normally less rapid than B, but some control improvement possible	Slower than C
Voltage control under load rejection (i.e. potential over-voltage condition)	Poor to fair, liable to self-excitation and loss of synchronism below minimum excitation limit	Very good with inherent fast response	Fair, controlled absorption limited by full conduction of thyristors	Poor, limited by the low rating of TCR reactor
Voltage control under line outage (i.e. potential undervoltage condition)	Good, but relatively less rapid	May require switched capacitors to support voltage		
Behaviour following system fault	Machine inertia could cause swinging and loss of synchronism	Inherent response as a constant voltage device	Auxiliary controls could be used to damp load swings	

voltage transmission systems down to the correction of power factor on a works distribution system.

The factors that can influence the decision to install switched reactive compensation units are both technical and economic. Of these factors, the latter type is often difficult to assess. In many cases from operational considerations the units could be permanently connected, but this will cause unnecessary loss. The installation of a frequently operated switch, on the other hand, brings additional capital and maintenance costs. If the compensation is permanently connected to the load or the line, the latter's protective circuit-breaker will also serve for clearing faults in the compensation plant. In assessing the var balance it is necessary to know for what proportion of the time a unit could be switched out, a figure often difficult to obtain from the existing loading information and even more so to predict for the future.

Typical applications are considered below.

In transmission systems with long lines, shunt reactors are installed at the ends of the lines to control their voltage under light-load conditions. To increase transmission capability and minimise system losses, it would be desirable to switch these reactors out during heavy loading when the line var gain due to the shunt capacitance is offset by its series var loss. However, this requires transmission-voltage circuit-breakers, which are expensive. Moreover, in the event of a large loss of load it would leave the system with a large excess of vars, leading possibly to excessively high voltage levels which could be unacceptable even for the short time necessary to reconnect the shunt reactors. As described later, a combination of fixed and continuously variable reactors may be a solution in these circumstances.

It is the practice to install shunt capacitor banks at critical

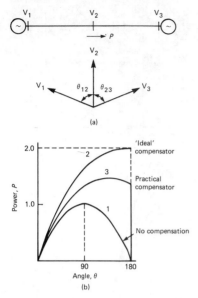

Figure 16.5 Power–angle features of a transmission line using a shunt compensator

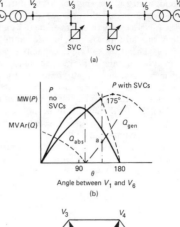

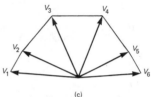

Figure 16.6 Long transmission line with multiple static var compensators

Without the support at the intermediate point the power will be as in (b), curve 1:

$P = (V_1 V_3 / X_{13}) \sin \theta_{13}$

$P_{max} = (V_1 V_3) / X_{13}$

With the voltage support at the point 2 the condition will be as in curves 2 and 3. If the reactive power were to be supplied instantaneously, the power transfer would be doubled (curve 2). In practice this cannot be achieved, owing to the response delay of a practical SVC (curve 3). However, it is possible to provide SVC support at several places along the line and so make up to an extent for the lack of the 'ideal compensator'.

Figure 16.6 illustrates an example of conditions to be expected on a very long high-voltage transmission line with voltage stabilisation at two intermediate substations.

For each section there are two limiting conditions which can be responsible for the maximum power to be reached. Assume that the stabilisers can hold the voltage rigidly constant irrespective of load. In this case the stability limit will be reached if the phase angle between two adjacent SVCs reaches 90° theoretically; in practice the resistance of the line and the SVC response delays limit the total angle at maximum P with SVCs to well below $3 \times 90°$, e.g. 175° in Figure 16.6(b). Figure 16.6(c) illustrates the phase angles of the line at such steady-state power limit.

The second limiting condition might be that any one of the SVCs reaches the limit of its var generation capacity; the required vars from the SVC are illustrated in Figure 16.6(b) by the curve Q (generation or absorption). When the power P and angle θ reach the condition of this limit, as for point 'a' in Figure 16.6(b), the SVC ceases to be effective in controlling the line voltage at its point of connection and the stability limit is reached even if the total angle has not reached the first limit (e.g. 175°).

16.5 Variable var compensators

16.5.1 General

The load carried by a power system is continuously changing. Major variations in total load occur slowly over a period of tens of minutes or longer in a daily cycle, which itself changes with a weekly and yearly variation. Normally any rapid changes are a small proportion of the total and do not affect the overall pattern.

For a system operating under normal conditions the overall var compensation requirements also change relatively slowly; and since in a well proportioned system the var flows are not sensitive to small load changes, it is sometimes possible to use fixed shunt reactors or capacitor banks. Often, however, all or some portion of these shunt compensation elements have to be switchable to give better control and reduce unnecessary loss. These discretely variable devices are the simplest var compensators and by far the most common.

Continuously variable var compensators can be used for particular categories of application, as follows:

(1) to compensate for rapidly varying loads that could otherwise cause unacceptable voltage disturbances on the power network;
(2) to give a continuous control at a weak point of a network where voltage variations due to normal load changes could become excessive and pose a threat of voltage instability;
(3) to damp out load swings on a tie-line between two power systems which otherwise could lead to dynamic instability;
(4) to give a continuous fast control where transient stability is important, particularly on long-distance transmission;
(5) to give as near as possible instantaneous overvoltage control in the event of a major load rejection on a long-distance transmission system.

These different applications call for different characteristics of the variable var compensator. The types of compensator are reviewed in this section, with particular reference to their operating characteristics and applications, a comparative summary of which is given in Table 16.1.

16.5.2 Mechanically switched capacitors and reactors

Mechanically switched capacitors and reactors are the most common. They are used at all voltage levels from the highest

stability limit corresponding to a phase angle of 90° is bound to become unstable following a major disturbance due to the increase of phase angles. It is necessary to design a line to be able to transmit more than pre-fault power up to the point of maximum angular swing. Therefore sufficient capacitive vars must be available to compensate for the increased var consumption by line current at increased phase angles. In consequence, for transient stability to be secured, some excess capacitive power must be available. This can be achieved either by operating the line sufficiently below its surge impedance power before the fault or by adding capacitive power transiently for the period in which the line phase angle will be abnormally increased. However, without adequately fast voltage control along the line, either solution can lead to dangerous overvoltages, particularly under the condition of a severe 'back-swing' (i.e. a transient phase condition when the power transmitted is much less than the pre-fault level) or during a load-rejection situation consequent to loss of stability.

For a line of series inductance L and shunt capacitance C, the surge impedance is $Z_s = \sqrt{(L/C)}$, and the surge-impedance load at rated voltage V_1 is $P_s = V_1^2/Z_s$. Then the total reactive power Q to be absorbed from the line when operating at a voltage V and power P approximates to

$$Q = Q_0[V^2 - (P/V)^2]$$

where Q_0 is the total reactive power in C at rated voltage V_1. Here P, Q and Q_0 are expressed in per-unit of P_s, and V in per-unit of V_1.

For $V = 1$ p.u., Q is equal to Q_0 at $P = 0$, and falls to zero for $P = 1$—i.e. a power equal to P_s. Power can be increased above P_s if Q can be made negative—i.e. if shunt capacitors are added.

16.4.2 Series capacitor compensation

Series capacitors inserted in one or two locations along a transmission line have the following important characteristics:

By reducing the effective series inductive reactance of a transmission line they allow a higher power transfer capability. This may be a higher power over a given distance or a longer distance for a given power; alternatively, it may be a smaller steady-state phase angle for a given power and distance.

By increasing the capacitive vars generated as the load current increases, the series capacitors improve the var balance in a line and, hence, reduce its voltage regulation, thereby reducing shunt capacitive compensation requirements at the line terminals.

Figure 16.4 shows an illustrative theoretical example of the power angle characteristic of a line including its terminal impedances; the line is represented as three π-sections of about 200 km length each, with shunt reactors (R) at each line section terminal and series capacitors (C) at two intermediate locations. The curves of *Figure 16.4* are calculated neglecting power losses, and the values are given in per-unit terms on the base of the surge impedance load (P_s).

Even if long h.v. and e.h.v. lines are series compensated, shunt reactors would be required to absorb part of the shunt vars of the line. Typically between 40% and 100% of line shunt capacitive power may be thus compensated at light load to prevent overvoltages on the line. To restrict insulation stresses caused by overvoltages following sudden load rejection, a substantial part of the shunt reactive compensation is usually left permanently connected, reducing the maximum power limit of a long line. As the degree of shunt compensation is increased, the line requires a greater amount of series compensation to maintain the power-angle characteristics of the line. A comparison of curves 1 and 6 and the table of *Figure 16.4* shows that with 100% shunt reactor compensation the line requires almost 50% series compensation to reach the same maximum stable power limit as for the case

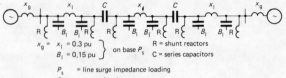

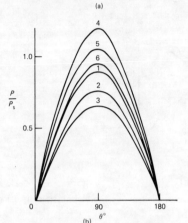

Curve	Compensation of line shunt capacitance (%)	Compensation of line series reactance (%)	$\dfrac{P_{max}}{P_s}$
1	0	0	0.91
2	50	0	0.77
3	100	0	0.66
4	0	50	1.20
5	50	50	1.06
6	100	50	0.95

(c)

Figure 16.4 Transmission line with series and shunt compensation

without shunt and series compensation. The maximum allowable limit of series capacitor compensation and choice of its location are governed not only by economic consideration, but also from subsynchronous resonance aspects, discussed later, and by the series capacitor short-circuit overcurrent protection needs. The latter requires spark-gaps or non-linear resistors. The line distance protection relaying must also be co-ordinated with these features.

16.4.3 Variable static shunt compensation

Series capacitors increase the amount of power that can be transmitted over a line by effectively reducing its series reactance and so reducing the phase angle. Another way of reducing the critical phase angle for stability consideration is to subdivide the line.

In *Figure 16.5(a)* an assumption is made that the point 2 is connected to a virtual infinite bus which will always provide the right amount of vars to keep the voltage magnitude constant (but not the phase angle). In this context, the 'infinite bus' is not required to supply active power, but must supply the required vars as an 'ideal' compensator. A static var compensator, SVC, can be designed to supply any required vars very quickly (in 20–50 ms), but not instantaneously.

The angles between V_1 and V_2 and between V_2 and V_3 now each become the critical angle for stability, and therefore can be allowed to increase separately so that the total angle between V_1 and V_3 can be greater than the conventional critical value of 90°.

i.e. they have a positive var balance. However, under emergency conditions following a loss of one or more lines, the positive balance may turn into a deficit which under extreme conditions can give problems of voltage instability as well as possible loss of synchronism due to the network's low power-transfer capability.

A reactive power planning study involves the balancing of the vars in the complete network. Then, after allocating an amount of var production to generating plant, it is necessary to decide on the best method to deal with the resultant surplus or deficit. For a large system this can require a protracted and major design exercise, and over the last two decades much effort has gone into the development of mathematical and computational techniques to aid system designers in this task.

The second area of var management is in system operation, where the facilities provided must be used in the most efficient manner to maximise system operational security and minimise system losses. Again much work has been done on the development of both off-line and on-line computer programs. Various studies have shown that, in general, minimising network losses also gives maximum security and minimum unnecessary circulation of vars in the network.

The third area is the commercial one of tariff policy, in which financial incentives are adopted to persuade the consumer to maintain a power factor close to unity.

16.3.3 Var management by the consumer

The induction motor is by far the most common driving machine and in consequence the natural power factor of most industrial loads is in the 0.7–0.8 lagging region in the absence of reactive compensation. Such a load requires a var supply of between 75 and 100% of its power, increasing the loading on the plant (e.g. transformers and cables) and causing significant losses and voltage drops. By generating the required vars close to the loads, these disadvantages can be largely overcome and for this reason supply authorities impose tariffs on industrial consumers to encourage them to install power-factor correction plant within their systems. This can often be achieved in a relatively inexpensive way by shunt capacitors, either direct on the motor terminals or as larger banks on supply voltage busbars. Often a combination is the optimum solution.

Fixed capacitors or switched banks may not always be adequate where the var requirements fluctuate widely and rapidly, producing greater voltage variations at a point of common coupling with other consumers than the supply authority allows. Such loads are found, for example, in the steel industry, where, with thyristor driven rolling mills, the sudden impact of the billet meeting the rolls produces a rapid increase in low-power-factor current; and with electric arc furnaces, where frequent and random short-circuits occur between electrodes, producing short bursts of maximum current, again at a very low power factor.

Most supply authorities have regulations covering allowable limits for voltage fluctuations, particularly in respect of lamp flicker; as the human eye is most sensitive to fluctuation frequencies in the region of 8–10 Hz, these limits vary with frequency. Typically they might allow only 0.25% voltage dips at 10 Hz but 1–2% at 1 Hz. To compensate for such conditions requires a fast and continuously variable reactive source such as some types of static var compensator.

Thyristor power conversion equipment (which is increasingly being used in large ratings) and electric arc furnaces generate harmonic currents that flow into the supply network. These can affect other plant, causing losses, possible overheating or interference with electronic and telecommunication circuits. Supply authorities must therefore limit the harmonic distortion on their network and many stipulate maximum permissible values. To avoid exceeding these, it may be necessary to prevent a proportion of the harmonic currents generated by the consumer from penetrating the supply system. It is common to bypass these currents into harmonic filters. It is often possible to convert capacitors banks for power-factor correction into harmonic filters, giving significant cost savings. In any case the harmonic filters will generate vars at the power supply frequency and these must be accounted for in the system var balance.

16.4 Reactive compensation and transient stability

16.4.1 Transmission characteristics

The use of reactive compensation to improve transient stability and so increase the amount of power which can be transmitted over given transmission lines is discussed here.

Transient instability can occur following a sudden disturbance such as a short-circuit in the transmission system. The power output of a generator located near a fault may be greatly reduced, thereby increasing its speed, while the output of a remote generaror may be hardly affected by the event. As the acceleration of the two generators will differ, the phase angle between their internal e.m.f.s will change. In a simple system the two generators will drop out of synchronism if the fault persists long enough to allow the phase angle between their e.m.f.s to swing too far above 90°. The loss of synchronism takes place usually within 1 s of the disturbance. It is clear that the larger the series inductive reactance of the transmission lines and system transformers in the network interconnecting the generators, the greater attention must be paid to the possibility of phase-angle instability. In a well-interconnected system not having long distances between generators, such as the system of Central Electricity Generating Board in the UK, transient and dynamic instability is not a difficult problem with the fast clearance times obtained using modern relays and circuit-breakers.

On the other hand, slow swings or oscillations, possibly initiated by minor disturbances, between two parts of the system or possibly affecting only one remote generator, may take up to 20 or 30 s to build up to unacceptable levels. Modern generator automatic excitation controllers or static var compensators, both with auxiliary stabilising signals, can be used to damp such oscillations. Dynamic stability and swing damping will not be discussed further.

There has been a tendency to increase the length of transmission lines typically for the following requirements: supplying large amounts of power to conurbations and industrial centres from economic but distant energy sources—e.g. hydroelectric or mine-mouth thermal stations; supplying moderate amounts of power to areas remote from the power grid of the central region—i.e. by radial subtransmission lines, having loads tapped at one or several locations along each line.

A number of performance features are associated with long distance transmission:

At light loads which are only a small fraction of SIL (surge impedance load), or at energisation of the line from one end only, the distributed inductance and capacitance of the line can cause a large overvoltage due to the Ferranti effect, which must be countered by either (a) cancelling the line inductance by series capacitor compensation or (b) cancelling the line capacitance by shunt reactor compensation.

The maximum power which can be transmitted down a long uncompensated line is limited by its series inductance and by the line surge impedance. Increase in transmitted power can be achieved by series or shunt capacitor compensation. (The shunt capacitors are required to boost the line voltage at high loads.)

Owing to the inertia of rotating machines connected at ends of transmission lines, any system operating near its steady-state

banks located in load and distribution networks should also be taken into account.

Loads that can produce network disturbances or distortions require special attention. These are in the main found in a.c. traction systems, and in the metal and mining industries, where arc furnaces and large thyristor drives can cause problems due to large and rapid fluctuations in load, particularly var demands, as well as problems due to the generation of harmonics.

16.2.2 Network voltage considerations

It is important for the stable operation of a system that voltages at load points should not be allowed to drop below certain limits, for if this occurs, the condition can become progressively worse, leading to a complete collapse of voltage. Such voltage instability is more likely to occur in systems with long transmission lines and insufficient variable reactive compensation, and can be exacerbated by loads that tend to consume constant power and vars irrespective of the magnitude of their supply voltage. Such loads can be those supplied by transformers with on-load tap-changers having an automatic control to maintain a constant secondary voltage.

Figure 16.3(b) shows a set of voltage–load curves for the system (a). The voltage (V_R) applied to the loads is plotted against the total power delivered to these loads for a number of different power factors. For a particular power factor curve and for any power less than the maximum there are two operating points, points A, A_1 ... being stable (i.e. dV_R/dP is negative) and points D, D_1 ... unstable. The upper stable values represent possible system operating conditions; however, if the load power factor is decreased and the operating point moves round the bend of the curve, then there will be a progressive reduction in voltage, leading to a complete collapse of the system.

This could arise in the network in *Figure 16.3(c)*, where the loads are fed from both the remote generator A and the local generator B, the latter supplying most of the load vars. If B is suddenly disconnected, there will be an immediate increase in demand for both power and vars in the transmission line, and though generator A may have sufficient power rating, the increase in vars could well depress the voltage V_R onto the lower part of the curve in (b), causing a system collapse. If the deficit in vars could be corrected rapidly following the loss of generator B, say by the switching in of a capacitor bank C to return the voltage to the upper slope of the curve, then either the deficit in power could be made up by increasing output from generator A or time could be gained to reduce the deficit by load shedding. In either case the system would be saved from complete collapse.

The phenomenon of voltage instability has been the cause of, or a contributory factor in, a number of major system failures; the adequate provision and the correct control of the network var resources can play a vital role in its prevention.

16.3 Management of reactive power

16.3.1 Objectives

Adequate provision of vars at all times can be crucial for the security of a system. In the foregoing example it was assumed that all var requirements of the load were supplied from the generators during normal system conditions. In an actual system this may not be within the operating capabilities of the generators, it may not be the most economical method in terms of either capital costs or operating costs, and, as has been demonstrated, it certainly is not the best for system security. It is necessary to balance various factors to arrive at an optimum solution for management of reactive power.

The objectives in such management must be both economic and technical:

(i) to optimise the capital investment in plant for the generation and absorption of vars;
(ii) to minimise system losses;
(iii) to optimise plant utilisation, so postponing the need for system reinforcement;
(iv) to obtain adequate system security;
(v) to maintain adequate quality of supply;
(vi) to control system overvoltages.

To attain such objectives the system designers, operators and consumers can use some or all of the following types of plant or procedures: generators; shunt reactors and capacitor banks, switched or unswitched; on-load tap-changing transformers; synchronous compensators; static var compensators; series capacitors; and switching of transmission circuits.

16.3.2 Var management by the supply authority

The benefits of good reactive power management by the supply authority can be significant.

Three areas of management can be identified.

The first is in the system planning and design stage, where a choice must be made on the rating and type of any reactive compensation equipment to be used, where it should be connected in the network, and when is the optimum time for installation. The characteristics of any power system, particularly the proportions of generation connected at the different transmission voltage levels and the ratio between the system maximum and minimum loading, greatly influence the decisions to be made.

When a transmission line is operating at its surge impedance load (SIL), the var absorption in its series reactance equals the var generation in its shunt capacitance, and the voltage along the line is the rated voltage (ignoring line resistance). Most transmission networks, even at peak demand, operate below their surge impedance load and, hence, generate more vars than they absorb:

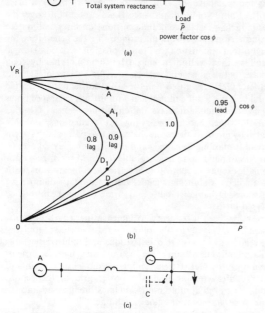

Figure 16.3 Transmission line voltage/load characteristics

(SCR, which is approximately the reciprocal of the synchronous reactance). Where transient stability or the use of var control in damping system oscillations is to be examined, then the inertia of the turbine–generator combination is required also.

The frame size of a generator depends upon its MVA rating; hence, the cheapest design for a given MW output would be one rated to operate at unity power factor. However, for stability such an operating point may not be acceptable. Operation at a lagging power factor requires a higher level of field excitation, which reduces the generator load angle for a given power output. To prevent a generator losing synchronism during angular swings following a system disturbance, it is desirable to operate it in steady state with an adequate margin of load angle from its stability limit. For a typical cylindrical-rotor turbogenerator, which will have an SCR in the range of 0.45–0.6 (see Chapter 11); this would require operation around 0.9–0.95 lagging p.f., as shown in the capability chart of *Figure 16.2*, if it is feeding into a reasonably strong system. Under these conditions the generator will be supplying vars to the system amounting to 0.48–0.33 Mvar/MW of output.

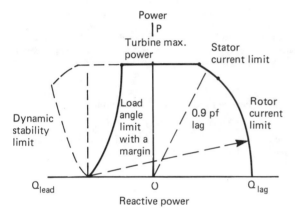

Figure 16.2 Generator capability chart

When it is necessary for the generator to absorb vars (i.e. operate at a leading power factor), as occurs during light load conditions for generating stations supplying remote load centres through long high-voltage lines, the field excitation must be lower than that at unity p.f. The generator must be designed to maintain stability, which requires that positive excitation is needed under all conditions. This gives a design with an SCR of, say, 1.0–1.5. Such high values are common in salient-pole low-speed generators in hydroelectric stations, which account for most such applications. So high a value is unusual in round-rotor turbo-generators and would require an unusually large machine for the MW output. It may be cheaper to provide var absorption equipment within the network in these circumstances.

To obtain economies of standardisation, many turbogenerator makers will offer a machine from a standard range of frame sizes; different combinations of MW, rated power factor and short-circuit ratio are possible with each design of the range. This must be allowed for in considering the overall var balance for a projected system.

The proportion of the system var requirements supplied by generators varies from system to system. In most systems the high-voltage transmission network has a surplus of vars—i.e. its shunt capacitance generates more vars than are absorbed by the I^2X losses in its series reactance, even under maximum-demand conditions. This allows the generators to operate at high power factors, even if the system loads themselves have low power factors. The operating power factor in practice may be dictated more by stability than load requirements, and it is quite usual for generators to be operated at a lagging power factor higher than their rated value.

16.2.1.2 Transmission lines and cables

Transmission lines and cables have an inherent capability to generate vars by their shunt capacitance but they also absorb vars in their series inductance. The higher the voltage the higher the var generation per unit length. High voltages are used for transmission in order to achieve higher power transmission capability per right of way and to minimise losses, a balance being struck between increased equipment cost and reduced operating costs.

Most countries have two or three transmission voltage levels within their system, the highest normally being in the 400–500 kV range, with a few countries in North and South America and the USSR operating or installing networks in the 750 kV range. Although the use of u.h.v. of 1000–1500 kV is technically feasible, its commercial application is likely to be limited in the forseeable future.

From the analytical point of view, cables are indistinguishable from overhead lines except that the ratio between shunt capacitance and series reactance is considerably higher. The use of cables in high-voltage networks is increasing as higher voltages are being brought into urban areas, and to a more limited extent for a.c. underwater links; but it is still very small compared with overhead lines. Where cables are used in any quantity, their high shunt capacitance reduces the need for additional var generation at peak demand periods but increases the need for var absorption during low loading.

16.2.1.3 Transformers

In closely integrated networks, apart from the generator reactances, the transformer leakage reactances have the greatest significance. So, although power loss and magnetising current can be neglected when considering var flows, the I^2X absorption in its leakage reactance is important. During temporary network overvoltages transformer cores may become saturated, drawing abnormally high magnetising current. Although this increased var consumption will assist in reducing the overvoltages, the harmonics so generated may produce undesired resonances which themselves can cause overvoltages.

On-load tap-changing of transformers is a useful function in the co-ordination of var control between various voltage levels within a system and for balancing var flows within a single voltage level. Two distinct practices occur with transformers used to interconnect different transmission voltage levels in a single system; a number of authorities, with a view to using transformers having the highest reliability, install fixed ratio transformers, whereas others use on-load tap-changers having a total range of 20–30%. In interconnecting transmission networks to sub-transmission or distribution networks, the use of tap-changing transformers is almost universal.

16.2.1.4 Loads

In general, loads are inductive and their power factor can vary widely, dependent on *type*, from industrial through commercial to domestic loads, and on *location* (commercial and domestic loads in hot climates have an increasing proportion of motors for air conditioning). Although for most studies load representation as constant impedances (or constant P and Q) is adequate, when studying dynamic conditions involving wide variations in voltage a more detailed representation which allows for changing power factors will be required. Power-factor correction capacitor

16.1 Introduction

When designing transmission or distribution systems, the engineer must take into account not only the power requirement of loads, but also the fact that they consume reactive power; and, equally important, that the networks include inductive and capacitive elements which themselves absorb or generate reactive power. Whereas power is generated and consumed in a controlled manner only at specific points in the networks (ignoring losses), reactive power is generated and absorbed throughout the network in significant quantities which vary with the system loading and configuration. For brevity the term 'vars' (volt-amperes reactive) will be used for reactive power. The convention for generation and absorption of reactive power by inductors and capacitors is described in Chapter 3.

Electricity supply networks have predominantly reactive series impedances; therefore, to a first approximation, power is transmitted through a network by a difference in phase angle between voltage vectors at two points. For the same reason the flow of vars between the two points is dependent primarily on a difference in voltage magnitudes. Put simply, phase angle controls power flow and voltage magnitude controls var flow. This is a simplified description of the first principles in a.c. power transmission and reference should be made to Chapter 3 for the theoretical background.

If a power system is to operate efficiently and securely, the importance of the correct and co-ordinated provision and control of vars cannot be overemphasised. It is necessary to examine these needs for both steady-state and dynamic operation, and it is normal to consider them separately.

The objectives for steady-state operation are: optimisation of the network voltage profile under varying load conditions to maximise power transfer and minimise losses; reduction of the unnecessary circulation of vars within the network.

For dynamic operation emphasis is on: control of over- and under-voltages in a network during and after transient disturbances; provision of voltage support at an intermediate point of a transmission line, so increasing the transmitted power limit by allowing a larger phase angle across the complete line without loss of stability; provision of series capacitors in long transmission lines to reduce the effective series reactance and improve the var balance; in special cases, compensation of loads having large and rapid fluctuations to prevent 'pollution' of other parts of the system.

It is possible to separate the needs of system var control and load var control, since, although the objectives may be the same (i.e. the efficient operation of the system) and similar control principles may be used, their application methods will be different.

16.2 General considerations of reactive volt-amperes

Figure 16.1 shows a system where different components and reactive power devices have been combined to illustrate the main features of reactive power supply and demand in a network. The characteristics of individual components and the methods of network analysis are covered in other chapters. Reference will be made here only to those characteristics or parameters relevant to the management and supply of vars, network voltage control and reactive compensation.

16.2.1 Characteristics of system components

16.2.1.1 Generators

The main generator parameters relevant to the var balance in a system are the transient reactance and the short-circuit ratio

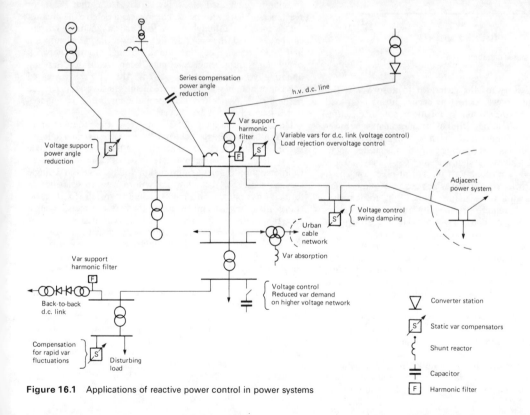

Figure 16.1 Applications of reactive power control in power systems

16 Supply and Control of Reactive Power

A Gavrilović
GEC, Stafford, UK

W P Williams, BSc, CEng, FIEE, SMIEEE
GEC, Stafford, UK

H L Thanawala, PhD, BE(Elect), BE(Mech)
GEC, Stafford, UK

Contents

16.1 Introduction 16/3

16.2 General considerations of reactive volt-amperes 16/3
    16.2.1 Characteristics of system components 16/3
    16.2.2 Network voltage considerations 16/5

16.3 Management of reactive power 16/5
    16.3.1 Objectives 16/5
    16.3.2 Var management by the supply authority 16/5
    16.3.3 Var management by the consumer 16/6

16.4 Reactive compensation and transient stability 16/6
    16.4.1 Transmission characteristics 16/6
    16.4.2 Series capacitor compensation 16/7
    16.4.3 Variable static shunt compensation 16/7

16.5 Variable var compensators 16/8
    16.5.1 General 16/8
    16.5.2 Mechanically switched capacitors and reactors 16/8
    16.5.3 Synchronous compensators 16/10
    16.5.4 Static var compensators (SVCs) 16/10

16.6 Special aspects of compensation application 16/15
    16.6.1 Light flicker compensation 16/15
    16.6.2 Harmonic filter design 16/17
    16.6.3 Subsynchronous/sub-harmonic resonance damping 16/18

16.7 Future prospects 16/18
    16.7.1 Long-distance transmission 16/18
    16.7.2 General application 16/18
    16.7.3 Future developments in var compensation equipment 16/18

Figure 15.57(b) shows a lock-open sequence, the open-circuit times being adjustable between 0.25 and 1 s. If the fault clears and operation is stopped during the cycle, the mechanism returns to the starting position. The 'instantaneous' trips are rapid (2–7 cycles) to avoid fuse deterioration: they give two chances for fault clearance. The time delay switching occurs twice to blow the fuses on a permanently faulted section. If the sequence continues, the operation terminates with the recloser locked open. In *Figure 15.57(c)* is shown a sequence ending in a hold-closed condition. A permanent fault is cleared by other means, after which the mechanism automatically resets. Transient faults are cleared as with type (*b*).

Where adequate fuse co-ordination is not obtainable, the alternative is the use of sectionalisers, as at (ii) in *Figure 15.57(a)*, associated with automatic reclosers. A sectionaliser comprises a normally closed oil switch latched in against spring loading. When a fault occurs, the current passes through a series coil which actuates the mechanism and counts the fault current impulses during the recloser sequence, opening the sectionaliser automatically during a pre-set open-circuit period. The mechanism resets if the fault clears prior to the end of the sequence, but if the sectionaliser reaches the open condition, it must be reset by hand.

Auto-reclosing is also sometimes used on transmission systems to assist in maintaining stability by reclosing before synchronous machines have had time to lose synchronism. In some cases, particularly on single-circuit lines, the added complication of selecting, tripping and reclosing only the faulty phase is justified.

When the loss of supply may have serious consequences, duplicate supplies or a ring main may be installed.

Further reading

JOHN, M. N., 'Electricity supply—problems and possibilities', *IEE Electronics and Power*, **29**, No 10, 702–704.
FORREST, J. S., 'Electricity supply—present and future', *IEE Electronics and Power*, **29**, Nos 11/12, 796–800
FLURSCHEIM, C. H. (ed), *Power Circuit Breaker Theory and Design*, IEE Power Engineering Series No. 1, Peter Peregrinus, 602
WRIGHT, A. and NEWBERRY, P. G., *Electric Fuses*, IEE Power Engineering Series No 2, Peter Peregrinus, 229
ROSEN, P. et al., 'Recent advances in HRC fuse technology', *IEE Electronics and Power*, **29**, No 6, 495–8
Electricity Distribution (CIRED 1981), IEE Conference Publication No 197, Peter Peregrinus, 371 (1981)
TEDFORD, D. J. et al., 'Modern research and development in SF_6 switchgear', *IEE Electronics and Power*, **29**, No 10, 719–723
KEINERT, L. et al., *Service Experience With, and Development of SF_6 Gas Circuit-Breakers Employing the Self-Extinguishing Principle*, IEE Conference Publication No 197, 36–40 (1981)
DONON, J. and VOISIN, G., *Factors Influencing the Ageing of Insulating Structures in SF_6*, CIGRE Paper 15-04, 11 (1980)
BOGGS, S. A. et al., *Coupling Devices for the Detection of Partial Discharges in Gas-Insulated Switchgear*, IEEE PES Paper 81 WM 139-5, 5
MALLER, V. N. et al., *Advances in H.V. Insulation and Arc Interruption in SF_6 and Vacuum*, Oxford: Pergamon, 282 (1981)
BLOWER, R. W. et al., *Experience with Medium Voltage Vacuum Circuit-breaker Equipment*, IEE Conference Publication No 197, 51–55 (1981)
LATHAM, R. V., *High-voltage Vacuum Insulation—the Physical Basis*, Academic Press, 245 (1981)
SLEMECKA, E., *Interruption of Small Inductive Currents*, Electra No 72, 73–103 (Oct. 1980)
MILLER, T. J. E., *Reactive Power Control in Electric Systems*. Wiley Interscience, 381 (1982)
KRIECHBAUM, K., *Progress in H.V. Circuit-breaker Engineering*, SRBE Bulletin, Vol. 96, Nos 3–4, pp. 147–158 (1980)
Developments in Design and Performance of EHV Switching Equipment, IEE Conference Publication No 182, Peter Peregrinus, 134
GARDNER, G. E. et al., *Development Testing Techniques for Air-blast and SF_6 Switchgear*, IEE Proceedings, Vol. 127, Part C, No 5, 285–293 (Sept. 1980)
BASAK, P. K. et al., *Survey of TRV conditions on the CEGB 400 kV System*, IEE Proceedings, Vol. 128, No 6, 342–350
Developments in power system protection, IEE Conference Publication No 185, Peter Peregrinus, 303
Power system protection, Edited by the Electricity Council
Power system protection I, Principles and Components, Peter Peregrinus, 544
Power system protection II, Systems and Methods, Peter Peregrinus, 352
Power system protection III, Application, Peter Peregrinus, 450
CHAMIN, M., *Transient Behaviour of Instrument Transformers and Associated High-speed Distance and Directional Comparison Protection*, Electra No 72, 115–139 (Oct. 1980)
MADDOCK, B. J. et al., *Optical Fibres in Overhead Power Transmission Systems for Communication and Control*, Proceedings 29th International Wire and Cable Symposium, USA, 402–409 (1980)
DAVY, R. A. et al., *State of the Art in Protection and Control for E.H.V. Transmission*, Electrical Review. Vol. 207, No 19, 27–29 (21 Nov. 1980)
Protective Relays—Application Guide, GEC Measurements, 447
CHISHOLM, W. A. et al., *Lighting Performance of Overhead Lines*, Ontario Hydro Review No 3, 19–24 (Jan. 1981)
LAT, M. V. et al., *Distribution Arrester Research*, IEEE PES paper H WM 199-9, 9 (1981)
The Use of ZnO Surge Diverters in the Austrian E.H.V. Network, Oesterrich Z Electr. Vol. 33 No 12 (Dec. 1980), 433–437, In German
RAGALLER, K. (ed.), *Surges in H.V. Networks*, Proceedings of Brown Boveri Symposium 423 (1979)

as is practicable to the terminals of the plant. An ideal arrester (a) takes no current at normal system voltage, (b) establishes an instantaneous path to earth for any voltage abnormally high, (c) is capable of carrying the full discharge current and (d) inhibits subsequent power-frequency current. The modern device comprises an assembly of small gaps and non-linear resistors in series, the whole being contained in a cylindrical porcelain housing. The use of multiple rather than single gaps gives the most rapid breakdown. The resistor elements offer a low resistance to surge currents (limiting the voltage across the arrester) and a much higher resistance to the power-frequency follow current. Careful gap design ensures that the follow current does not restrike.

Rod gap A plain air-gap is cheap, and it satisfies requirements (a), (b) and (c) above. However, it does not fulfil condition (d), and although gap breakdown protects the plant from overvoltage, there may be a power-frequency follow current which must be cleared by a circuit-breaker operation, involving an outage. A plain rod gap, connected between line and earth, has the following typical gap lengths giving breakdown at about 80% of the plant impulse level:

system voltage (kV):	36	72.5	145	300	420
gap length (m):	0.23	0.35	0.66	1.25	1.70

15.11.2.4 Prevention of outages

About 80% of the earth faults on overhead lines are of a transient nature resulting from lightning, birds or other causes, and no damage to the equipment is caused; however, the ionised path caused by the transient flashover permits a power-frequency follow current to flow which must be cleared by a circuit-breaker operation. Such outages can, however, be avoided in certain cases by the use of arc suppression coils (Petersen coils) or auto-reclose circuit-breakers.

Arc suppression coil For reasons of insulation economy and/or safety it is customary to earth the neutral point of a system either directly or through a resistor. Any earth fault results in fault current and necessitates a circuit outage to clear it. Were the neutral point isolated, it might be expected that an earth fault would result in no fault current and that the system could continue to operate, provided that the raising of the two sound phases to line voltage above earth potential were acceptable. However, owing to the system capacitance a leading current will flow through the fault as shown in *Figure 15.56(a)*. If the fault involves an arc, any such current, if it exceeds a few amperes, is likely to cause damaging voltage surges and to be very persistent. Operation with a completely isolated neutral point is therefore impracticable except on very small isolated systems.

If, however, the neutral is earthed through a high inductive reactance X (several hundred ohms), the resulting lagging current can, neglecting losses, precisely neutralise the capacitance current, as shown in *Figure 15.56(b)*. There will thus be no fault current and the system can be operated with the fault until such time as it can conveniently be repaired. To secure precise balance the value of the arc suppression coil reactance must be $X = 1/3\omega C$.

Arc suppression coils are effective on 12 kV, 36 kV and occasionally up to 245 kV systems, but at the higher voltages and with long lines the resistive and other losses prevent precise phase opposition of I_{fl} and I_{fc} so that there is a resultant current in the fault which, if it involves an arc, is inextinguishable and damaging.

Auto-reclose circuit-breaker A transient flashover rarely causes damage if the fault is cleared by the normal protective equipment.

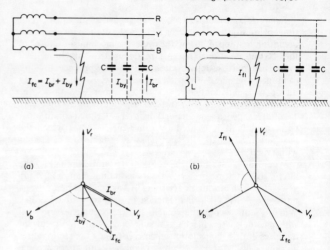

Figure 15.56 Action of arc-suppression coil. (a) isolated neutral, (b) neutral earthed through arc-suppression coil

After an interval of 0.4–0.8 s, sufficient for the natural de-ionisation of the arc path, the circuit-breaker can be reclosed with safety.

Circuit-breakers with automatic reclosure are widely used on 12 kV and 36 kV radial distribution networks. They are required to provide time delay switching in the event of a permanent fault; permit of normal fuse discrimination to limit the area of interruption; operate rapidly on the initial passage of a fault current. *Figure 15.57(a)* shows three applications. In (i) all

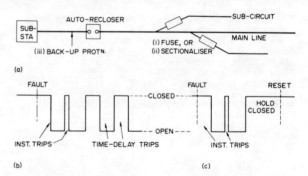

Figure 15.57 Auto-reclosing. (a) system, (b) lock-open sequence, (c) hold-closed sequence

transient faults are cleared by the recloser, and in the case of permanent faults the recloser sequence provides a time delay trip to blow the fuse only on the faulty subcircuit. In (ii) sectionalisers provide the disconnecting means for faulty subcircuits. Method (iii) operates as (i), except that on permanent fault the recloser holds closed and back-up protection operates.

A typical current operated recloser comprises a normally closed oil circuit-breaker for pole mounting. The breaker is held closed by springs. When the current exceeds, for example, twice full-load value, movement of the plunger of a series solenoid causes the recloser to open. The plunger is then spring-reset and the reclosing is automatic. Relay features control the reclosing and opening times, and a mechanism is provided to enable the recloser to lock open or hold closed at the end of its operating sequence.

Table 15.1 Insulation co-ordination, 1–36 kV

V_0 (kV r.m.s.)	v_{lp}		V_1 (kV r.m.s.)
	(kV peak)	(kV peak)	
Series I	List 1	List 2	
3.6	20	40	10
7.2	40	60	20
12	60	75	28
17.5	75	95	38
24	95	125	50
36	145	170	70
	500 kVA and below	Above 500 kVA	
Series II			
4.4	60	75	19
13.2–14.5	95	110	34
26.4	150	150	50
36.5	200	200	70

Table 15.2 Insulation co-ordination, 52–245 kV

V_0 (kV r.m.s.)	v_{lp} (kV peak)	V_1 (kV r.m.s.)
52	250	95
72.5	325	140
123	450, 550	185, 230
145	550, 650	230, 275
170	650, 750	275, 325
245	750, 850, 950, 1050	325, 360, 395, 460

Table 15.3 Insulation co-ordination, 300–765 kV

V_0 (kV r.m.s.)	v_{sp} (kV peak)	v_{lp} (kV peak)
300	750	850, 950
300	850	950, 1050
362	850	950, 1050
362	950	1050, 1175
420	950	1050, 1175
420	1050	1175, 1300, 1425
525	1050	1175, 1300, 1425
525	1175	1300, 1425, 1550
765	1300	1425, 1550, 1800
765	1425	1550, 1800, 2100
765	1550	1800, 1950, 2400

Canada, and applicable to other countries within the sphere of influence of (I) or (II). There are two lists in (I), the choice between them being made on consideration of the degree of exposure to lightning and switching impulse voltages, the type of system neutral earthing and (where applicable) the type of overvoltage protective devices. In (II) the data are based on the apparent-power rating of the plant protected.

Middle range In *Table 15.2* more than one level of v_{lp} is given, the highest being for plant in systems where the earth fault factor exceeds 1.4.

Higher range *Table 15.3* specifies no power-frequency values, the switching overvoltage having priority. The rated values of v_{lp} associated with a standard v_{sp} have been chosen in accordance with: (a) for plant protected by surge arresters the two lowest values apply, and (b) for plant not—or not effectively—protected by arresters only the highest value applies.

Several insulation levels may exist in the one power network, appropriate to different situations or to the characteristics of different protective equipments.

15.11.2 Protective equipment

15.11.2.1 Lightning

Underground cables are virtually immune from direct lightning strokes, so that preventive measures apply in general only to overhead lines.

Earth wire Shielding the line conductors by an earth wire is reasonably effective in preventing a direct stroke to the conductors, provided that the conductors lie within a segment subtending an angle of about 45° (or preferably 25° for towers above 50 m high) from the earth wire to the ground. Owing to the complex (and not fully understood) nature of the leader stroke as it approaches the earth, such protection sometimes fails; in cases of particular importance, e.g. near a major substation or where lightning is particularly prevalent, two earth wires may be installed.

Tower footing resistance A lightning stroke to the earth wire can produce a back-flashover to the line conductors resulting in a line surge voltage unless the tower footing resistance is very low, e.g. not greater than 1 Ω per 100 kV of impulse level. In difficult situations a buried *counterpoise* earthing system, comprising wires radiating from the foot of the tower to a distance of 30–60 m, or continuous wires from tower to tower, help to lower the footing resistance.

15.11.2.2 Switching surges

While switching surges cannot be entirely avoided, it is desirable on lines of 300 kV upward to limit them by shunting across circuit-breaker contacts, during closure, resistors of value approximating to the surge impedance of the line (300–400 Ω).

15.11.2.3 Damage to plant

Overhead-line damage by overvoltage can occasionally occur, but is generally repairable fairly quickly, but damage to substation plant (particularly to transformers) is costly and results in a long outage. Overvoltage protection is thus aimed primarily at the defence of transformers.

Surge attenuation A travelling wave of voltage attenuates (mainly by corona loss) as it is propagated along a line, and its magnitude and wavefront steepness are mitigated. Inductive coupling with the earth and with earth or counterpoise wires assists in this process. A reduction by one-half may thus occur in a line length of 5–8 km.

A length of underground cable between the terminal of an overhead line and the substation plant also reduces the magnitude of an incoming voltage surge, owing to the lower surge impedance of the cable. Additional reflections from the junctions may, however, partially offset this advantage and the cable is itself vulnerable. Although such a section of cable is often desirable for amenity or other reasons, it is rarely installed solely for protective purposes.

Surge arrester The aim is to direct a surge to earth before it reaches a vulnerable plant. The arrester must be located as near

be used for threading through the transformers with current supplied from a heavy current testing transformer. The cable should be as central as possible, as small errors are introduced if the cable is asymmetrically located. For balancing purposes it is noted that if the relative position of each cable in each transformer is the same, the same error will be introduced and the effect neutralised.

15.10.3 Secondary injection tests

The preceding methods utilise primary current, but use can be made of secondary injection. The effect of primary current in a c.t. is to develop a secondary voltage; in secondary injection, voltage is applied to the secondary terminals, usually from an injection transformer, so applying the 'output' voltage. The method has application for checking relays, calibration, commissioning and maintenance. It is not in itself sufficient for balanced schemes, and must in such cases be supplemented by tests on load; however, it is possible in most cases to simulate through-faults to earth, and phase–phase conditions, by rearrangement of the secondary c.t. connections. One limitation is that the 'load' current is dependent on the load on the unit, usually determined by network conditions, and on balanced systems the stability check may be at a low primary current.

The various types of injection transformers generally have a tapped primary winding, with a secondary resistance controlled output covering a range of voltage. The method gives fine control for calibrating and enables a low power input to be used.

When using injection transformers for timing tests on induction-type relays with inverse time characteristics, the wave form characteristics should be carefully ascertained. Such relays are very sensitive to wave form (having an inherent, saturable iron circuit), and a more satisfactory and reliable method for checking timing is to supply the test current from a low voltage with resistance control. In this way the saturating feature of the relay is absorbed by the controlling resistance, the source of supply being treated as a transformer of relatively high capacity. It must be noted that the saturation of the injection transformer itself is important.

15.10.4 Fault location

Rapid location of a fault, together with some idea of its nature, is clearly an essential preliminary to its quick repair. With overhead lines visual inspection either from the ground or from a helicopter may be possible, but if this is not practicable, and almost always in the case of underground cables, inspection of the flag indicators on protective relays and simple Megger or continuity tests will usually provide useful evidence and enable a suitable method of more detailed investigation to be selected.

15.10.4.1 Loop tests

Details of the basic Murray loop test and its variants are given in Section 6. If the fault is of low resistance, a low-voltage (car) battery may suffice for the test supply. For higher-resistance faults about 500 V from, say, a Megger may be required, while for very high resistance, where it is necessary to break down the fault, a high-voltage rectifier set may be used.

15.10.4.2 Other tests

Fall-of-potential, capacitance and pulse reflection tests are more sophisticated methods. Such tests, with loop tests, may be accurate to a location within 20 m or so of the fault point. In the case of underground cables a more precise location is desirable before excavation. Typical are induction and discharge methods.

15.11 Overvoltage protection

Lightning, switching and other less common phenomena produce overvoltages on transmission and distribution systems. Precautions against consequential damage and system outage must be taken, by (a) preventing overvoltages from being impressed on the system and (b) protecting vulnerable equipment from voltage surges.

Insulation levels are based on combinations of short-duration power-frequency overvoltages, and impulse voltages arising from lightning or switching. Protective devices are (1) non-linear resistor surge arresters, (2) expulsion surge arresters (up to voltages not exceeding 36 kV) and (3) spark-gaps. These do not provide the same degree of protection, and the choice between them depends on such factors as the relative importance of the plant to be protected, the consequence of an outage, the system layout, the probable lightning activity and the system voltage.

The breakdown of plant insulation varies with the time for which an overvoltage is maintained. *Figure 15.55* indicates typical voltage–time relations for plant insulation and overvoltage protection devices. A surge arrester has a characteristic

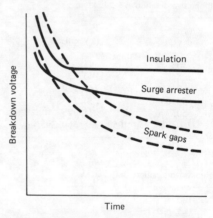

Figure 15.55 Typical voltage/time relations for plant insulation and overvoltage protection devices

similar to that of the plant, and is usually arranged to spark-over at a voltage slightly lower. A spark-gap has a more nearly hyperbolic characteristic, so that the plant insulation may break down first on overvoltages of short duration; but if, to avoid this, the spark-gap breakdown voltage is reduced, many unnecessary outages may result.

15.11.1 Insulation co-ordination

International and national standards for phase-to-earth insulation are referred to three system voltage ranges, viz. 1–36, 52–245 and 300–765 kV r.m.s. (Standards for other than phase-to-earth insulation are under consideration.) *Tables 15.1–15.3* give the standard voltage levels in terms of:

V_0 highest system voltage (in kV r.m.s.) at which the plant and the protective equipment have to operate normally,
v_{lp} rated lightning-impulse withstand voltage (in kV peak),
v_{sp} rated switching-impulse withstand voltage (in kV peak),
V_1 rated power-frequency short-duration withstand voltage (in kV r.m.s.).

Lower range Table 15.1 gives voltage values in two series based on current practice (I) in Europe and (II) in the USA and

sequence components of the line currents by means of a sequence filter, arranged to allow for the greater relative heating effect of negative sequence current components.

15.9.3.2 Rectifiers

The majority of rectifier duties are now performed by semiconductors: these have special demands for protection. Faults can be classified as: (a) cell overload and failure; (b) operational faults such as backfires; (c) d.c. faults on the busbars and cables; (d) a.c. faults on the supply transformer and cables.

Because of their low thermal mass and consequent low fault-current withstand, the series- or parallel-connected cells need to be individually protected, usually by fuses shunted by a small indicating fuse which, on blowing, operates a microswitch to trip the supply or give an alarm.

Semiconductor cells are susceptible to overvoltage failures caused by switching or lightning surges, arc voltages on fuse clearance, chopping of load current, etc. Protection usually takes the form of surge diverters or resistor/capacitor networks connected to earth (or between phases) and across the d.c. output. Rectifier transformers are often provided with earthed screens between the windings to limit the surge transfer.

15.10 Testing and commissioning

The testing of protective schemes has always been a problem, because protective gear is concerned only with fault conditions and cannot readily be tested under normal system operating conditions. The problem has been aggravated in recent years by the complexity of protective schemes and relays.

Protective gear testing may be divided into three stages: (1) factory tests, (2) commissioning tests, (3) periodic maintenance tests.

The first two stages prove the performance of the protective equipment during its development and manufacture, and in its operational environment. The last stage, properly planned, ensures that this performance is maintained throughout its life.

The relay manufacturer must provide adequate testing of protective gear before it is accepted and commissioned. The tests performed are (a) tests in which the operating parameters of the relays, etc., are simulated; (b) conditions such as temperature range, vibration, mechanical shock, electrical impulse withstand, etc., which might affect the operation in service. In some cases tests of both groups are conducted simultaneously to check performance.

With the advent of static relays the use of semiconductors and other electronic components has posed new problems to the manufacturer, because the production of such devices is not within his control. Quality control testing procedures have been established with the object of identifying the device that may fail early in its life, the usual method being to determine which critical parameters will show a substantial change when subjected to accelerated life testing and to examine these.

15.10.1 Commissioning tests

The object of the commissioning tests is to ensure that the connections are correct, that the performance of current transformers and relays agrees with the expected results, and that no components have been damaged by transport or installation. This performance test includes correct current transformer ratio, correct calibration of relays and tests proving that the tripping, intertripping and indication of the scheme are in order.

Although the details of every protective scheme vary, the main points to be checked on every scheme are: (a) stability of the system under all conditions of 'through' current; (b) that the sensitivity of all relays with reference to the primary current is correct.

Prior to making these final checks on site a very careful preliminary check on the protective gear scheme should be made. Such tests would include:

(1) Examination of all small wiring connections, and insulation resistance tests on all circuits.
(2) Identification of all pilot cables—insulation resistance tests and loop resistance tests on balanced pilot-wire schemes.
(3) Polarity and ratio check (if possible) on all current transformers.
(4) Tests on d.c. tripping, intertripping and operation of all breakers connected with the unit, including alarm circuits and relay signals.
(5) Checking of calibration of all relays by secondary injection.
(6) Miscellaneous tests depending on special details of a particular scheme.

15.10.1.1 Saturation curve

Considerable use on site may be made of a c.t. saturation curve using a low-voltage local supply; it compares closely with results obtained by primary current. A low-voltage a.c. supply is fed to the secondary winding through a control rheostat, and a curve of voltage–current for the secondary is plotted. The equivalent primary current is inferred from the turns ratio. The test is useful for comparing the performance of c.t.s required to have matched characteristics.

Such preliminary tests ensure that components are correct. The commissioning process thereafter must depend upon site facilities. The following notes give a general outline.

15.10.2 Primary current tests

A *generator* is isolated with the unit under test and by means of primary short-circuits and earth faults stability figures, on balanced systems, under full load conditions and operation tests with internal faults may be carried out. It is important that the makers of the generator be consulted before it is used for steady unbalanced conditions, e.g. testing with one phase earthed, as the distortion of flux combined with armature reaction, if prolonged, may result in excessive heating. Earthing resistors and transformers should be short-circuited or by-passed during current testing, as they are normally short time rated. An *inverted transformer* is a means of stepping-up testing current where a transformer is available between the unit under test and the source of test supply. For example, if the transformer has a normal step-up ratio of 6.6 kV to 66 kV, then by connecting the testing supply (isolated generator or low-voltage source) to the 66 kV windings and supplying the test current from the output of the 6.6 kV windings a 10/1 step-up of current is obtained. This 'inversion' can readily be carried out by flexible cabling when the connections to the transformer are made through open-type bushings, and is applicable to any power transformer.

A heavy current *testing transformer* may be designed with an input suitable for a general range of low-voltage supply. The types available include 3-phase and single-phase units with resistance or induction regulator control. For the purpose of producing primary current the output terminals are connected across the primary of the circuit under test, to test windings or test bar primaries embodied in the current transformers. On metal clad switchgear, connection to the primary can be made through the circuit spouts with testing plugs; potential transformer spouts may also be used, but the current-carrying capacity should be investigated.

When using externally mounted current transformers around cables, e.g. in core balance types of protection, flexible cables may

improvement in continuity of supply and system stability. The choices of dead time, of reclose time, and of whether to use a single- or a multi-shot scheme, are of fundamental importance. With a 3-phase scheme the tripping of all three phases on the occurrence of a fault means that no interchange of synchronising power can take place during the dead time. If only the faulty phase is tripped during earth fault conditions (which account for the majority of faults on overhead lines), synchronising power can still be interchanged through the healthy phases, but this entails the provision of circuit-breakers provided with tripping and closing mechanisms on each phase, and the reclosing circuitry is more complicated. In the event of a multi-phase fault, all three phases are tripped and locked out. It is important to ensure simultaneous tripping of the circuit-breakers at all ends of the faulty section, and distance protection imposes some difficulties in this respect, so that a signalling channel is often used between ends. On highly interconnected transmission systems where the loss of a single line is unlikely to cause two sections of the systems to drift apart and lose synchronism, 3-phase auto-reclosing can be delayed for 5–60 s with an increase in successful reclosures.

Electromechanical, semi-static and fully static schemes using printed circuit boards for the mounting of the majority of the components give efficient utilisation of spaces. Further developments in the static field are likely to come from the application of computers or processors for auto-reclose applications.

15.9.2.1 Busbars

Busbars have often been left without specific protection for one or more of the following reasons:

(1) The busbars and switchgear have a high degree of reliability and are often regarded as intrinsically safe.
(2) It was feared that accidental operation of busbar protection might cause widespread dislocation which, if not quickly cleared, would cause more loss of supply than the very infrequent busbar faults.
(3) It was expected that system protection, or back-up protection, would provide sufficient busbar protection if needed.

The reasons are applicable only to small indoor metal enclosed stations: with outdoor switchgear the case is less clear, for although the likelihood of a fault is higher, the risk of damage is much less. In general, busbar protection is required when the system protection does not cover the busbars, or when, to maintain system stability, high-speed fault clearance is necessary.

To maintain the high order of stability needed for busbar protection, it is now an almost invariable practice to make tripping dependent on two independent measurements of fault quantities. Methods include: (a) two similar differential systems; (b) a single differential system (*Figure 15.54*), checked by a frame earth system (*Figure 15.53*); (c) earth fault relays energised by c.t.s in the neutral earth connection; (d) overcurrent relays; (e) a frame earth system checked by earth fault relays. Separate current transformers are normally used with separate tripping relays series-connected in pairs to provide a single tripping output, so that independence is maintained throughout. The essential principle is that no single occurrence of a secondary nature shall be capable of causing an unnecessary trip of a busbar circuit-breaker.

15.9.3 Motors and rectifiers

15.9.3.1 Motors

Protection is required against (a) imposed external conditions, and (b) internal faults.

Category (a) includes unbalanced supply voltages, under-

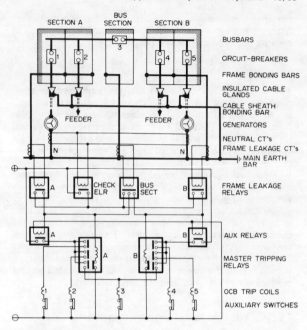

Figure 15.53 A single differential system for busbar protection

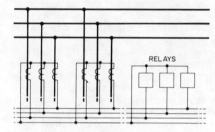

Figure 15.54 A frame earth system

voltage, single-phasing, and reverse-phase-sequence starting; (b) includes bearing failures, internal earth faults and overloads.

Motors up to about 400 kW usually have ball or roller bearings which may fail very quickly: the only remedy is to disconnect the motor as rapidly as possible to avoid the overcurrent damaging the windings. Larger machines usually have sleeve bearings, incipient failure of which may be detected by a temperature device embedded in the bearing. As the current increase may be only 10–20%, it cannot be detected by the overload relays.

The majority of winding failures are caused by overload, leading to deterioration of the insulation followed by electrical fault in the winding.

The wide diversity of motor designs makes it impossible to cover all types and ratings with a simple characteristic curve. Motors with fluctuating loads where shutdown would affect the whole process might be left running by giving the overload relay a higher current setting. Motors on a steady load can be tripped more quickly, as overload will probably be due to a mechanical fault.

A temperature compensated static relay can follow changes of the working temperature of the motor more accurately, so that it will not be shut down on overload unless it is overheated. By including a number of alternative operating time–current characteristics it is possible to cover a wide range of motor designs and applications. Protection of motors operating on unbalanced voltages is provided by separating the positive and negative

necessary. Shunt reactors for compensating line capacitance are usually oil immersed and protected by time-delayed overcurrent relays, with instantaneous restricted earth fault relays where the system neutrals are earthed. Buchholz relays are often used on oil immersed reactors.

15.9.2 Feeders

Many schemes are available: therefore, only general guidance can be given. Scope still exists for adapting schemes to individual requirements. Fuses and/or overcurrent protection with graded time lags are commonly used for 11 kV and 33 kV distribution systems with balanced opposed-voltage schemes for the more important 33 kV lines.

For higher-voltage and longer-distance lines, the protection can be by either electromagnetic or static relays, both of which may be switched by fault detectors, and for earthed systems can incorporate earth fault relays.

The length of feeder frequently results in additional circuits being necessary between the relays at each end. Pilot wires are often used for this purpose, but for distances above about 20 km the pilot wires may behave as a transmission circuit, and may need shunt reactors to compensate for the circuit capacitance. Pilot wires are also liable to electrical interference, manual disturbance and limits as to overvoltage and current, so that many schemes are provided with continuous supervision to warn that the overall protection is unsound and relay adjustment may be necessary. By using the line itself as a link and injecting high-frequency signals (70–700 kHz), the necessary end-to-end protective coupling can be achieved. Information is impressed on the signal by a modulation process. Attenuation of the signal carrier makes pilots unsuitable for transmitting the amplitude of a measured quantity; however, the phase can be transmitted satisfactorily. Frequency modulation can convey all the characteristics of the modulating quantity but involves frequency–amplitude conversion. The best use of carrier involves simple on/off switching: this method of modulation has been extensively used in conjunction with polyphase directional relays by transmitting a locking signal over the line for through-faults.

With extensive power systems and interconnection to ensure continuity of supply and good voltage regulation, the problems of combining fast fault clearance with protective gear co-ordination have become increasingly important. To meet these requirements high-speed protective systems suitable for use with automatic reclosure of the circuit-breakers are under continuous development and are already very widely used. The distance scheme of protection is comparatively simple to apply and offers high speed, primary and back-up facilities, and needs no pilot wires; but by combining it with a signal channel (such as carrier) it is particularly suitable for use with high-speed auto-reclosing for the protection of important transmission lines.

Since the impedance of a transmission line is proportional to its length, it is appropriate to use a relay capable of measuring the impedance of a line up to a given point. The basic principle of measurement (Section 15.8.7) involves the comparison of the fault current seen by the relay with the voltage of the relaying point, whence it is possible to measure the impedance of the line up to the fault. The plain impedance relay takes no account of the phase angle between the current and voltage applied to it, and is therefore non-directional. It has three important disadvantages:

(1) It is inherently non-directional and therefore needs a directional element to give it discrimination.
(2) It is affected by arc resistance.
(3) It is highly sensitive to power swings, because of the large area covered by the impedance circle.

The *admittance relay*, by addition of a polarising signal, combines the characteristics of the impedance and directional relays. It is satisfactory on long lines carried on steel towers with overhead earth wires, where the effect of arc resistance can be neglected, but for lines on wooden poles without earth wires the earth fault resistance reduces the effective zone to such an extent that most faults take longer to detect. This problem can usually be overcome by using either reactance or fully cross-polarised admittance relays for the detection of earth faults. In theory, any increase in the resistive component of the fault impedance has no effect upon a reactance relay: however, in practice, when the fault resistance approaches that of the load, then the relay characteristics are modified by the load and its power factor and it may cover either more or less of the line. This can be overcome by the use of a fully cross-polarised admittance relay, obtained by the use of a phase comparator circuit which needs two input signals for a $\pm 90°$ comparison angle.

A distance protection scheme comprises starting, distance measuring, auxiliary, zone timer and tripping relays. To cater for the economic and technical requirements of a particular power system, schemes are available using either a plain distance measurement with several steps of protection, or a combination of distance measurement and a high-speed signalling channel or power-line carrier to form a unit system of protection over the whole of the protected line and to provide its own back-up protection to the adjacent lines. Full-distance, or switched-distance, schemes are applied according to the system voltage and the importance of the lines to be protected. The main difference between the two arrangements is that the full-distance scheme uses six measuring units (three for phase and three for earth faults), whereas the switched-distance scheme uses only one measuring unit for all types of fault, this being switched to the correct fault loop impedance by means of a suitable set of starting units of the overcurrent or underimpedance type.

One of the main disadvantages of conventional time stepped distance protection is that the instantaneous zone of protection at each end of the protected line cannot be set to cover the whole of the feeder length. It usually covers about 80%, leaving two end zones in which faults are cleared instantaneously at one end of a feeder but in much longer time at the other end. In some applications this cannot be tolerated: either the delay in fault clearance may cause the system to become unstable, or, where high-speed auto-reclosing is used, the non-simultaneous opening of the circuit-breakers interferes with the auto-reclosing cycle. These objections can be overcome by interconnecting the distance protections at each end by a signalling channel to transmit information about the system conditions at one end to the other end of the protected line. It can also initiate, or prevent, tripping of the remote circuit-breaker. The former arrangement is a 'transformer trip scheme', while the other is a 'blocking scheme'.

If two circuits are supported on the same tower or are otherwise in close proximity over the whole of their length, there is mutual coupling between them. The positive and negative sequence coupling is small and usually neglected. The zero sequence coupling cannot be ignored. Types of protection that use current only (e.g. power-line carrier phase comparison and pilot-wire differential systems) are not affected, whereas types using current and voltage, especially distance protection, are influenced by the phenomenon and its effect requires special consideration.

The protective schemes described previously for the protection of two-ended feeders can also be used for multi-ended feeders with load or generation at any terminal. However, the application of these schemes to multi-ended feeders is much more complex and requires special attention.

As transient faults, such as an insulator flashover, make up 80–90% of all faults on h.v. and e.h.v. systems, and are cleared by the immediate tripping of one or more circuit-breakers, they do not recur when the line is re-energised. Such lines are frequently provided with auto-reclosing schemes, and result in a substantial

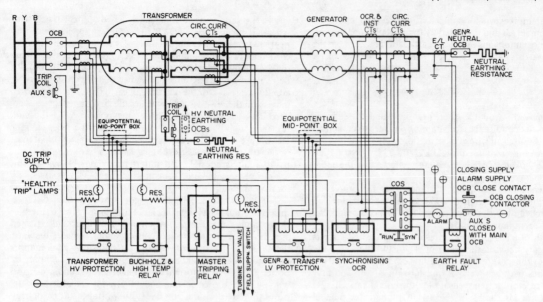

Figure 15.52 Generator–transformer protection

Where generator–transformers are used, the differential protection will always need to be biased, using either harmonic bias or special attention to settings. Earth fault protection will cover the generator and the transformer primary winding; in addition, the transformer h.v. winding will usually be provided with restricted earth fault protection and Buchholz protection (see *Figure 15.52*).

The field of a faulty generator should be suppressed as quickly as possible. For machines above about 5 MVA this is usually done by connecting a field discharge resistor of about five times the rotor field-winding resistance in parallel with the generator field when the field circuit-breaker opens. This reduces the field time constant, but it may still be more than 1 s.

15.9.1.2 Transformers

The power transformer is one of the most important links in a transmission system, and as it has the greatest range of characteristics, complete protection is difficult. These conditions must be reviewed before detailed application of protection is settled. The ratings of units used in transmission and distribution schemes range from a few kVA to several hundred MVA. The simplest protection, such as fuses, can be justified only for the small transformers; those of the highest ratings have the best protection that can be afforded.

A fault in a transformer winding is controlled in magnitude not only by the source and neutral earthing impedance, but also by the leakage reactance of the transformer, and the fact that the fault voltage may differ from the system voltage according to the position of the fault in the winding. Star-connected windings may be earthed either solidly or through an impedance, presenting few protection problems; but delta-connected windings require close consideration, as the individual phase currents may be relatively low. Faults between phases within a transformer are comparatively rare, and interturn faults in low-voltage transformers are unlikely unless caused by mechanical force on the windings due to external short-circuits. Where a high-voltage transformer is connected to an overhead transmission line, a steep-fronted voltage surge due to lightning may be concentrated on the end turns of the winding, resulting in 70–80% of such transformer failures unless some form of voltage grading is employed. The bolts that clamp the core together are insulated to prevent eddy currents. Should this insulation fail, then severe local heating may damage the winding. As the oil is usually broken down, the gas produced can be used to operate a Buchholz relay. Tank faults are rare, but oil sludging can block cooling ducts and pipes, leading to overheating.

One of the major problems in the protection of large transformers is the phenomenon of magnetising current inrush when switching on. This is not a fault condition and the protection relays must not trip, although the inrush current appears superficially as an internal fault and may have a long time constant. To avoid long time delays and, hence, damage to important transformers, it is essential to clear all faults rapidly. Use is made of the fact that magnetising inrush currents contain a unidirectional component plus second, third and higher harmonics. The second harmonic is the most useful as a stabilising bias against inrush effects, and by combining this (extracted through a filter) with the differential current through a static device, a setting of 15% can be obtained, with an operating time of 45 ms for all fault currents of two or more times rated current. The relay will restrain when the second harmonic component exceeds 20% of the current.

Most of the usual transformer arrangements such as star/delta, auto-, earthing, etc., can be protected by differential relays, sometimes with restricted earth fault relays.

Sometimes transformers are connected directly to transmission lines without circuit breakers. In such cases the transformer and feeder are both protected by differential protection plus intertripping of the remote circuit-breaker. An alternative to intertripping is to detect earth faults on a feeder connected to a non-earthed transformer winding by measuring the residual voltage on the feeder, using either voltage transformers or capacitors to detect the neutral displacement.

15.9.1.3 Reactors

The most common are the series and tie-bar reactors, for limiting overcurrents. On earthed systems with air insulated reactors the protection usually consists of differential relays or time-delayed overcurrent with time-delayed earth fault relays. For insulated-neutral systems the earth fault relays are un-

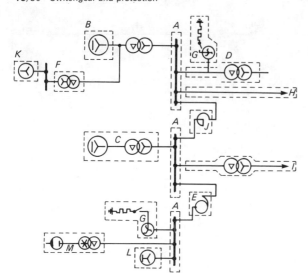

Figure 15.51 Zones of protection. A Busbars; B Generator; C Generator-transformer; D Power transformer; E Auto-transformer; F Unit transformer; G Earthing transformer; H Transmission line of feeder; I Transformer feeder; J Reactor; K Induction motor; L Synchronous motor; M Rectifier

for example, a triple-pole overcurrent relay can be used for both overcurrent and earth fault protection, and the following combinations are commonly found in practice: (1) inverse time overcurrent and earth fault; (2) inverse time with instantaneous high-setting overcurrent, with or without inverse time earth fault; (3) thermal overcurrent, with instantaneous overcurrent and earth fault.

The power system neutral is considered to be earthed when the neutral point is connected to earth directly or through a resistor. Earth fault protection is applicable in all such cases. The neutral point is considered to be insulated when the neutral point is not connected to earth, or is earthed through a continuously rated arc suppression coil or through a voltage transformer.

Arc suppression coils are a form of protection applicable where the majority of the earth faults are expected to be of a transient nature only. They can be continuously or short-time rated (normally 30 s), with one phase of the system faulted to earth. The continuously rated coils can be provided with an alarm to indicate the presence of a persistent fault, and in some cases directional earth fault relays are used to indicate the location of an earth fault. The short-rated coil is short-circuited, either directly or through a resistor after a short time, so that persistent earth faults can be cleared by normal discriminative earth fault protection.

15.9.1 Plant

15.9.1.1 Generators

The core of an electrical power system is the generator, requiring a prime mover to develop mechanical power from steam, gas, water or diesel engines. The range of size extends from a few hundred kVA up to turbine driven sets exceeding 600 MVA in rating. Small sets may be directly connected to the distribution system, while the larger units are associated with an individual step-up transformer through which the set is coupled to the transmission system.

A modern generating unit is a complex system consisting of the generator stator, the rotor with its field winding and exciters, the turbine and associated condenser, the boiler with auxiliary fans and pumps, and possibly an associated or unit transformer. Faults of many kinds can occur in such a system, for which diverse protective means are needed. The amount of protection applied will be governed by economic considerations. In general, the following faults need to be considered:

stator insulation earth faults	failure of prime mover
overload	low vacuum in condenser
overvoltage	lubrication oil failure
unbalanced loading	overspeeding
rotor fault	rotor distortion
loss of excitation	differential expansion
loss of synchronism	excessive vibration

The neutral point of a generator is normally earthed, with some impedance inserted in the earthing lead to limit the magnitude of earth fault current to values from a few amperes to about rated full load current. Phase–phase faults clear of earth are unusual, as are interturn faults; and they usually involve earth in a short time. Circulating current biased differential protection is the most satisfactory way of protecting a generator stator or a generator–transformer unit.

Generators on a large system under continuous supervision are not usually subject to prolonged accidental overloading and are often protected only by built-in temperature measuring devices on both stator and rotor. For the smaller machines i.d.m.t. overcurrent relays are used (often forming a back-up feature) and may be voltage restrained to give better discrimination.

Power frequency overvoltages are usually dealt with by instantaneous relays with fairly high settings, and transient overvoltages by surge diverters and capacitors connected to the generator terminals if the incoming surges are not sufficiently reduced by transformer interconnections.

Any unbalanced load produces negative sequence components, the resulting reaction field producing double-frequency currents in the field system and rotor body, which can result in serious overheating. For turbo sets the negative sequence continuous rating is only 10–15% of the continuous mean rating value for positive sequence. The unbalance is detected by relays in a negative sequence network, providing first an alarm, and finally tripping the generator load.

Rotor faults due to earthing of the winding, or partial short-circuited turns, can be detected by methods involving a potentiometer across the winding, or a.c. or d.c. injection. Failure of the field system results in a generator losing synchronism, running above synchronous speed and operating as an induction generator with the main flux produced by reactive stator current drawn from the system. In general, a 60 MW machine with conventional cooling will not be overheated by asynchronous operation at full load for 5 min; but a hydrogen cooled 500 MW set should not be run like this for more than 20 s. A generator may also lose synchronism because of a severe system fault, or operation at a high load with leading power factor and, hence, a relatively weak field. This subjects the generator and prime mover to violent oscillations of torque, but synchronism can be regained if the load is reduced in a few seconds: otherwise, it is necessary to isolate the machine. Alternatively, the excitation may be removed so that the generator runs asynchronously, thereby removing the violent power oscillations. Reclosing the excitation at a low value will then cause the machine to resynchronise smoothly. The range of possible positions of saving curves makes detection by simple relays relatively difficult, but by the use of two impedance relays, pole slipping can be detected and corrected by varying the excitation.

Various faults may occur on the mechanical side of the set, such as failure of the prime mover, overspeed, boiler failure, loss of vacuum, lubricating oil failure, rotor distortion, etc. These can all be protected, the inclusion being a matter of economics.

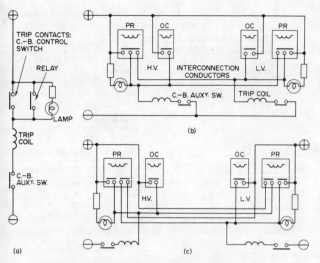

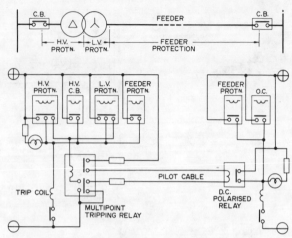

Figure 15.49 Tripping circuits. (a) Single circuit breaker; (b) common d.c. supply and (c) separate d.c. supplies for intertripping of transformer h.v. and l.v. circuit-breakers

Figure 15.50 Intertripping of feeder transformer

respective breakers only. Buchholz protectors or high-temperature relays, where used, should be connected so as to intertrip. When transformers are feeding distribution networks, three overcurrent relays on the h.v. side only with intertrip to the l.v. breaker are sometimes used, i.e. with no overcurrent relays on the l.v. side. The disadvantage of this arrangement is apparent, for the tripping of the l.v. breaker depends solely on the intertripping wire or cable. Also, the overcurrent relays are inoperative with the transformer charged from the l.v. side only, and with balanced earth leakage on each winding there would be no phase fault protection, obviously a dangerous condition. It is considered better practice to install overcurrent relays on each breaker so that, in the event of damage or bad connection on the interconnecting pilots, the 'back-up' value of overcurrent relays will trip each breaker independently, as well as giving complete protection with either side energised.

Distance intertripping is applicable to feeder transformer protection with breakers situated in different stations, or to special schemes in some forms of busbar zone protection. Basically, feeder transformer protection is treated as a combination of (1) feeder protection, generally a balanced protection employing either circulating current or balanced voltage principles; (2) transformer protection in the form of restricted earth leakage or circulating current, both sections being independent on the a.c. side, but the arrangements necessitate the additional tripping of the breaker at the remote end of the feeder in the event of a fault on either side of the transformer.

Figure 15.50 shows an arrangement embodying a d.c. polarised relay which is energised over two pilot cables by the operation of the transformer protection relays. It is a 'dead pilot' system, as the pilots are connected to the d.c. supply through limiting resistors only when the protection operates—an obvious advantage. The scheme counters the induction of e.m.f.s in the pilots by earth fault currents in neighbouring main cables. Special relays, sensitive to d.c. but strongly biased against a.c., are also available for use in this arrangement.

Another method, applicable to certain types of balanced voltage protection, operates by interrupting one pilot cable and applying d.c. or a.c. injection across the break.

Alarm systems Modern protective relays are fitted with relay signals, but it is also necessary in attended stations to have audible and visual indication that a breaker has tripped. The methods are:

(1) Electrically operated alarm systems in which an alarm bell and lamp circuit are energised by a relay, manually reset or with self-holding contacts; in some makes a shunt or series coil closing suitable contacts is embodied in the protective relay. Separate contacts on a multipoint tripping relay, which are bridged when the relay trips, may be utilised.
(2) Mechanically operated free-handle type, in which an auxiliary switch is operated during the closing of the breaker, the alarm lamp and audible circuit being completed by an auxiliary switch on the circuit-breaker when it opens; this scheme gives the alarm not only when the breaker is tripped from protective relays, but also if the breaker slips or opens through mechanical vibration.

In all cases (with the exception of hand-reset alarm relays) cancellation of the alarm system is effected by turning the circuit-breaker controller to the open position.

D.C. supply In large stations it is the practice to employ a trickle-charged 'floating' battery with charging from the local a.c. supply through a rectifier. Where no charging supplies are available, a replacement routine must be employed.

To reduce fire hazard, modern switchgear is sectionalised. The d.c. control circuits should be similarly sectionalised and fed from the main d.c. panel by separate cables.

15.8.9 Efficacy of protection scheme

The true measure of the efficacy of a protective unit may be expressed in terms of the number N of operations, of which n are incorrect, as $(N-n)/N$. The number N includes both through-faults and internal faults; n includes failure to trip on faults in the protected zone and false operation under through-fault conditions. Many factors influence this efficacy, such as imperfect design, application or commissioning, or failure of the equipment from damage and other causes. Protective equipment must therefore be carefully selected and meticulously maintained.

15.9 Application of protective systems

There are usually several ways of protecting any given equipment, and the more usual zones of protection are shown in *Figure 15.51*. One relay can often be used for several functions;

Typical arrangements for the three steps of a stepped-time scheme are also shown in *Figure 15.47*. At (*g*) the directional feature is given by the 'mho' relays used for the first two stages, while at (*h*) the mho relay for the third stage is used as a starting element and also prevents the reactance relays operating under load conditions.

Inaccurate operation can result if the relay voltages are low, owing to a high source/line impedance ratio, e.g. 30/1 or more. A polarising winding on the relay fed (1) through a memory circuit so that it retains pre-fault voltage or (2) from the sound phases through a phase shifting circuit, may be used. Method (2) is not effective for 3-phase faults.

15.8.7.7 Accelerated distance protection

It can be seen from *Figure 15.44*(*b*) that faults near the remote end of the protected section AB will only be cleared after an appreciable time lag. Such a fault will, however, be cleared 'instantaneously' by the right-to-left equipment at B: and if this equipment is also made to transmit a signal to A over a pilot circuit (usually carrier over the line), it can initiate the immediate tripping of the circuit breaker at A, thus giving almost instantaneous protection over the whole line.

15.8.8 Miscellaneous equipment

15.8.8.1 Negative sequence

Any fault other than a symmetrical 3-phase fault develops negative sequence currents, so that a network responsive to n.p.s. but not to p.p.s. components will indicate the presence of a fault and can be made to operate a relay. Two such circuits are shown in *Figure 15.48*. The cross-connection of the current transformers

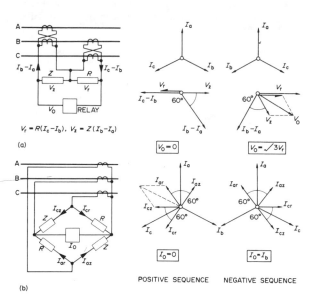

Figure 15.48 Negative-phase-sequence protection. (a) High impedance relay; (b) low-impedance relay

in (*a*) eliminates the z.p.s. components: the latter are earth leakage currents and are detected in other ways. The n.p.s. schemes illustrated employ impedance elements of resistance R and inductive impedance $Z = R \angle 60°$, but capacitive impedances $Z = R \angle -60°$ give the same results if the positions of R and Z are interchanged.

15.8.8.2 Neutral displacement

Displacement of the neutral point potential from its normal earth potential is indicative of a fault condition, and may be used to initiate tripping.

15.8.8.3 Buchholz relay

The Buchholz relay is used for the protection of large oil immersed transformers or shunt reactors with oil conservators, and is fitted in the pipe from the tank to the conservator. Any arcing fault causes the oil to decompose, generating gas which passes up the pipe to the conservator and is trapped in the relay. In the case of a large fault a bulk displacement of the oil takes place. In a two-float relay the upper float responds to the slow accumulation of gas due to the mild or incipient faults, while the lower float is deflected by the oil surge caused by a major fault. The floats control contacts—in the first case, to give an alarm; in the second case to isolate the unit.

Such relays also incorporate a petcock at the top for the removal of the gas; its subsequent analysis can show the origin and severity of the internal fault.

15.8.8.4 Tripping of circuit-breakers

Direct-operated trip coils The circuit-breaker trip coil may be energised directly from the current in the main circuit for lower voltage circuit-breakers, or from a current transformer with the trip coil shunted by a fuse which blows on overcurrent to cause tripping. With c.t.s it is possible to use the residual current to operate the tripping at a much lower earth fault current than for overcurrents. The advent of solid state relays enables much better fault discrimination to be achieved, and also obviates the need for external auxiliary supplies (usually d.c.) for tripping.

15.8.8.5 D.C. Tripping and operating circuits

The d.c. circuits brought into operation by protective relays are of importance, as they are responsible for circuit-breaker actuation.

Single tripping The tripping of one local circuit-breaker ('unit tripping') by the operation of relays is applicable to overcurrent or balanced protection of feeders having relays at each end. The essentials are shown in *Figure 15.49*(*a*): it includes a 'healthy trip circuit' indicating lamp with external resistance of such value that, if the lamp is short-circuited, the current is less than that required for tripping. The lamp also serves as a 'circuit-breaker closed' indicator, and proves the continuity of the trip circuit when the breaker is closed.

Intertripping By this is meant the necessary tripping of circuit-breakers identified with the unit protected; for example, the tripping between h.v. and l.v. breakers on a transformer with (1) overall circulating current protection or (2) balanced earth leakage and overcurrent protection on each side, it being necessary to trip *both* breakers for a fault on either side. Intertripping may be local and relatively straightforward, or it may require to be performed on circuit-breakers at separated points in the network.

Local intertripping When two circuit-breakers are involved, use can be made of a standard 'three-point' scheme, either with a three-point relay, or by the addition of an interposing relay which gives a positive tripping supply to two independent circuits. *Figure 15.49*(*b*) shows an arrangement with three-point relays and a common d.c. supply, and *Figure 15.49*(*c*) the modification for separate supplies. In both cases the overcurrent relays, whether for 'back-up' or phase fault protection, trip their

neous' operation for faults occurring within the nearest 80% of feeder AB, i.e. over distance X_1. Relay 2 operates, after a time lag, for faults occurring up to a point just beyond circuit-breaker B; it thus acts as a back-up for relay A over the distance X_1. Relay 3 operates with a still longer time lag for faults beyond the cut-off point of relay 2; it thus acts as a further back-up for relay A, and also for relay 2 over the last 20% of feeder AB.

15.8.7.5 Fault impedance

The impedance 'seen' by a relay depends upon the type of fault. Two sets of relays, differently connected to the system, are required (although a single set with a switching device is also practicable).

Earth fault For a simple earth fault fed with current I_e and phase-to-neutral voltage V_a of the faulted phase, the earth loop impedance presented to the relay is

$$V_a/I_e = Z_1 + Z_e = Z_1(1+k)$$

where Z_1 is the normal positive-sequence line impedance from the relay to the fault, Z_e is the corresponding impedance of the earth path, and $k = Z_e/Z_1$.

Phase–phase fault The relay is fed with the phase–phase voltage and with the difference $(I_a - I_b)$ between the faulted phase currents, so that the impedance is

$$V_{ab}/(I_a - I_b) = V_a/I_a = Z_1$$

Three-phase fault Relays connected as for the phase–phase fault see the same impedance Z_1, because

$$V_{ab}/\sqrt{3}I_a = V_a/I_a = Z_1$$

and consequently operate correctly with the same setting.

Earth current compensation If, as is common, the system is earthed at more than one point, a proportion of the earth fault current will return by a path other than that at which the relays are located. The impedance seen is thus in error: it will be too low. This can be corrected in two ways.

Sound-phase compensation A fraction $p = Z_e/(Z_1 + Z_e)$ of the sound-phase currents is added to the faulted-phase current, as shown in *Figure 15.45(a)*. Typical values for a 132 kV line are $Z_1 = 0.44$ Ω/km and $Z_e = 0.21$ Ω/km, so that $p = 0.34$.

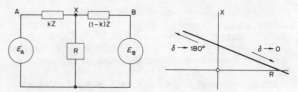

Figure 15.46 Impedance during power swing

may result. Typical values of fault resistance lie between 0.5 and 3 Ω, the higher values for lower currents.

Power swing If a power swing occurs, i.e. phase swinging of the terminal voltages of the section of the system concerned, the impedance seen by a relay may fall to a low value even if there is no fault in or near its protected zone. In *Figure 15.46* the impedance seen by the relay at X and looking towards B is given by

$$Z_p = Z[A/(A-1) - k]$$

where $A = (E_A/E_B) \angle \delta$, with δ representing the phase angle between E_A and E_B, and Z the total impedance of the line. The expression gives approximately the linear impedance locus shown, part of which around $\delta = 90°$ may well fall within the operating zone of the relay and cause false tripping unless special precautions are taken.

15.8.7.6 Relay characteristics

Characteristics for various relay constructions are given in *Figure 15.47*, the relays operating if the impedances seen by them fall within the shaded area. The impedance for which a relay is required to operate is given by OZ_1, OZ_2 or OZ_3 in *Figure 15.47*, so that this value should lie within the shaded area and any other impedance seen by the relay (due to load, arc resistance, power swings, etc.) should lie outside it. The quadrilateral characteristic (*f*) is desirable, but it can be obtained only by electronic means.

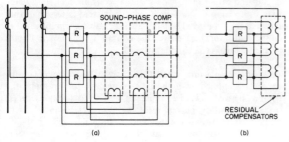

Figure 15.45 Sound-phase and residual compensation

Residual compensation A fraction q of the residual current I_r is added to the relay operating current, as in *Figure 15.45(b)*. If $q = Z_e/Z_1$, the measured impedance is Z_1, and the relays must be set for Z_1 instead of for Z_e to give correct operation. With the typical values above, $q = 0.52$.

Arc resistance If the fault has an arc resistance, the relay will 'see' the fault as located too far away, and errors in discrimination

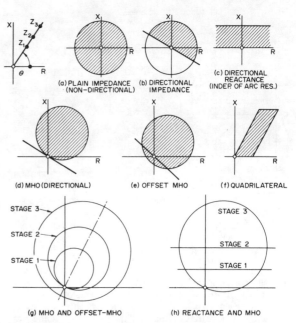

Figure 15.47 Distance relay characteristics

about 0.1 µF. The c.t.s operate, even on internal fault, with a high impedance burden; on a through-fault they are virtually on open circuit, and unless precautions are taken, the voltage between pilot wires can reach 1 kV, making magnetic balance of the transformers difficult and also giving rise in the pilot wires to capacitance currents which are likely to cause maloperation. Of the many schemes devised to overcome these difficulties, that mentioned below is typical.

This scheme employs induction relays so arranged that only two pilot wires are required, their capacitance currents being used to give a restraining effect. The pilot wire voltage is limited to about 130 V. In *Figure 15.42* consider a fault F between the R

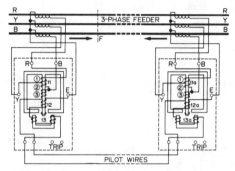

Figure 15.42 Biased balanced voltage protection

and Y phases: the currents in sections (1) of the relay primary windings 11 and 11a induce e.m.f.s in 12 and 12a which, being now additive, circulate a current in the operating coils 13 and 13a and in the two pilot wires. Both upper and lower magnets are energised, and if the fault current exceeds the scale setting, the relays operate to trip their associated circuit-breakers. A fault between phases Y and B energises section (2) of windings 11 and 11a, while a fault between phases R and B energises both sections (1) and (2), the fault setting being one-half of that in the former cases.

In the event of an earth fault on line R, the resultant secondary current from the phase R c.t. flows through sections (1), (2) and (3) of windings 11 and 11a. A Y phase earth fault energises sections (2) and (3); and a B phase fault, only section (3).

Pilot wires It is often desired, especially for feeder protection, to reduce the number of pilot wires to two. One method is the *discriminating delta* connection of *Figure 15.43(a)*, in which the

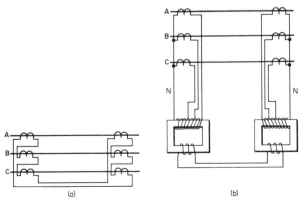

Figure 15.43 Balanced protection principles. (a) Discriminating delta; (b) summation transformers

c.t.s have different ratios (e.g. N, 3 N/4 and N/2). A more common method employs a *summation transformer* (b). If the primary turns of the summation transformer are in the ratios 1 between A and B, 1 between B and C, and n between C and N, then the relative sensitivities for various types of fault are in the following relative proportions as secondary outputs:

Fault	A–N	B–N	C–N	A–B	B–C	C–A	A–B–C
Proportion	$2+n$	$1+n$	n	1	1	2	$\sqrt{3}$

Stability ratio High *sensitivity* is desirable to ensure operation of the relays on minor or incipient faults, while high *stability* is required to prevent operation on heavy through-faults. These requirements are in conflict and it is usual to restrain ('bias') the relays by an additional winding carrying a current proportional to the fault current, thus desensitising the response to heavy fault conditions. The ratio (maximum through-fault stable current)/ (minimum fault operating current) can thus be made as great as 100.

Distance (impedance) protection This type of protection, used chiefly for overhead lines, involves the measurement of the impedance of the circuit as 'seen' from the relay location; if this impedance falls below a specified value, a fault is present. The impedance observed is approximately proportional to the distance of the fault from the relay and discrimination is obtained by introducing time delays for the more distant faults. Two schemes are available, *impedance time* and *stepped time*, the latter being more widely used on lines above 33 kV, on account of the shorter operating times obtainable.

15.8.7.3 Impedance time scheme

Relays (usually induction) having a directional characteristic give an operating time proportional to the impedance presented, i.e. proportional to the distance of the fault from the relay. By locating such relays at each circuit-breaker on a succession of feeders the characteristics of *Figure 15.44(a)* are obtained. It can

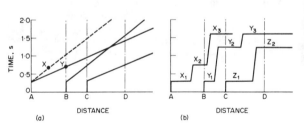

Figure 15.44 Distance (impedance) protection principles

be seen that should a relay, e.g. that at B, fail to operate, it will be backed up by that at A after a time interval. With feeders of different lengths the slopes of the characteristics must be adjusted to give correct discrimination; thus, the relay at A must have characteristic X (dotted) rather than Y, as the latter would give incorrect back-up discrimination for a fault near the end of section CD. The relays cater for fault currents flowing from left to right; a similar set would be required to cater for fault currents from right to left.

15.8.7.4 Stepped time scheme

Three separate relays are required for each direction of fault current at every circuit-breaker location. The relays are of the induction or (more rarely) of the beam type. In *Figure 15.44(b)* relay 1 of the group at circuit-breaker A is set to give 'instanta-

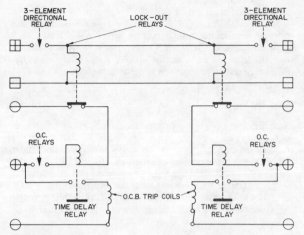

Figure 15.39 Locking signal protection using pilot wires

initiates a signal on the pilot circuit which energises the lock-out relays and prevents tripping at both ends. A failure of the pilot circuit will cause the sending end to trip on a through fault but will not interfere with operation on an internal fault.

Either private pilot wires, telecommunication pilots or carrier current over the line itself may be used, the equipment for the latter being similar to that for carrier current phase comparison.

15.8.6 Earth leakage protection

A current-to-earth on an h.v. system invariably indicates the presence of a fault: if large enough, it will operate the o.c. relays, but it is necessary to detect earth currents even if they are limited by resistance earthing of the neutral point. Separate earth leakage (e.l.) protective equipment is thus usually desirable. The arrangements of *Figure 15.40* are available, usually with induction relays, or linear or rotary attracted-armature types for high sensitivity. (An overhead conductor falling to earth of dry soil may produce no more than 5 A of earth fault current, but the associated potential field may be dangerous to human and animal life.)

Schemes (a)–(e) in *Figure 15.40* have no discrimination and operate for any fault beyond the equipment. If, however, the e.l. relays are given a time lag, any serious fault can be cleared discriminatively by o.c. or other protection. In (c) the transformer tank, switchgear frame or metallic busbar covering must be lightly insulated from earth: an insulation resistance of a few ohms is adequate, but it must never be short-circuited, e.g. by a length of conduit or a metal tool leant against it. Scheme (d) employs a 'core balance' c.t. in which the core flux is produced only by earth leakage currents. Scheme (a) is commonly extended to include two o.c. relays, as in (e).

Discrimination can be provided by balancing the e.l. currents at two points in the network. A common application is for the protection of the star-connected winding of a transformer, (f), in which the e.l. relay will operate only for internal faults in the transformer.

A directional feature can be given to an e.l. relay, the second input being usually provided from open-delta voltage transformers.

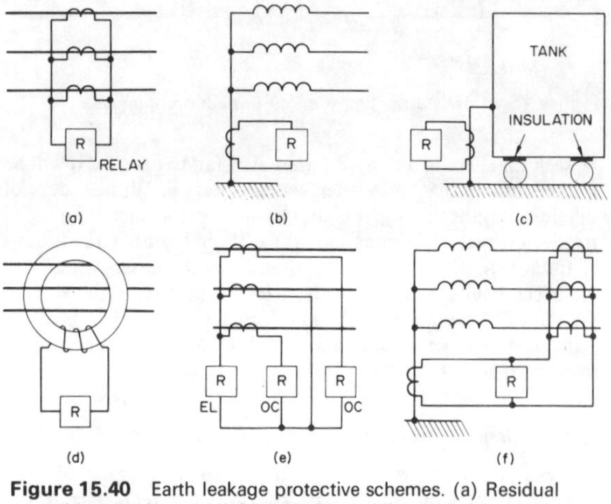

15.8.7 Balanced (differential) protection

Balanced protection is based on the principle that the current entering a sound circuit is equal to the current leaving it, whereas, if faulted, the detectable current difference can be used to trip the appropriate circuit-breakers.

15.8.7.1 Circulating current schemes

The c.t.s in circulating current schemes operate with low-impedance secondary burdens, provided that the pilot wires are short, so that identical c.t. characteristics at the two ends are not difficult to achieve even under conditions of heavy fault; but a large d.c. fault current component may cause dissimilarity and spurious unbalance.

The relays must be located at the electrical mid-point of the pilot wires. This introduces no difficulty where the ends of the protected zone are not far apart, but is less suitable for feeders (although some are in fact so protected). Short pilot circuits obviate the need for summation transformers to reduce their number, and the arrangement of *Figure 15.41* is typical. Circulating current protection is widely used for reactors, transformers, generators and busbars. In appropriate circumstances it has the advantages of simplicity, sensitivity and rapid response (e.g. 10–20 ms).

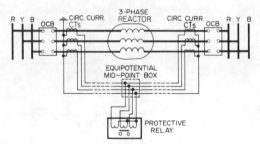

Figure 15.41 Circulating current protection on a reactor

15.8.7.2 Opposed voltage schemes

As relays can readily be installed at each end of the protected zone, an opposed voltage scheme is suitable for feeder protection. Summation transformers or other means reduce pilot wires to one pair, but even so the pilot problems limit the practicable feeder length.

Per kilometre loop, the resistance of the typical pilot circuit with 7/0.075 mm conductors is 12.5 Ω and the capacitance is

Figure 15.40 Earth leakage protective schemes. (a) Residual c.t.s; (b) neutral c.t.s; (c) frame leakage; (d) core balance; (e) overcurrent and earth leakage; (f) balanced (restricted earth leakage)

The relay has two adjustable settings. The *plug setting* (current setting) comprises a tapped winding for the current coil, the tappings being brought out to a plug bridge on which, by insertion of the plug, current settings of 50, 75, 100, 125, 150, 175 and 200% of the full-load output of the current transformer (5 A) can be obtained; removal of the plug automatically short-circuits the current transformer terminals. A *time setting* is achieved by varying the amount of travel of the contacts, giving time multipliers between 1.0 and 0.1. The times on *Figure 15.36* are for a time setting of 1.0 and the currents are multiples of the plug setting current.

15.8.5.1 Discrimination

Graded time lags On simple radial distribution systems, discrimination by graded time lags is effective, the lags of the relays being set successively longer for relays nearer to the supply source. The minimum time difference at all parts of the relay characteristic may be estimated as: error of relay 1 (near to fault), 0.1 s; circuit-breaker operating time, 0.1 s; overshoot of relay 2 (nearer to supply), 0.04 s; error of relay 2, 0.1 s; total, 0.34 s. The greater low-current times of the extremely inverse time characteristic may be suitable for grading with fuses, and also are less likely to cause tripping on switching-in after a supply interruption, when there may be heavy overloads resulting from plant left connected.

Graded time lags and directional relays If the overcurrent relays embody a directional feature (DOC), discrimination on more complicated networks can be obtained. *Figure 15.37(a)* shows a *ring main*: the arrows indicate the current direction for which the relays will operate, and the figures show typical time settings. The directional feature ensures that time grading can be obtained in each direction around the ring.

For the two *parallel transformers* in *Figure 15.37(b)* directional overcurrent relays are employed at the transformer ends on the l.v. side to detect phase faults on the line. Thus, a fault at F would

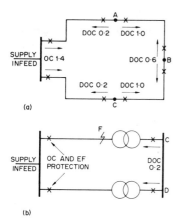

Figure 15.37 Directional overcurrent protection

be cleared by relays at C, while those at D would restrain. For a fault on the l.v. system both relays at C and D would restrain.

A more complicated network is shown in *Figure 15.38*. Suppose that the interconnector and feeder 6 are open and that there is a fault F on feeder 5. Fault current will then flow from the supply infeed direction into feeder 5 through 2. Now if feeder 2 is provided at each end with current operated relays having definite time delay, and if similar relays for feeder 5 are set lower than

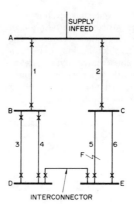

Figure 15.38 Network time-grading

those for 2, then the former will operate first and clear the fault, after which the relays on 2 reset and leave the feeder still connected to the system. It will be apparent that a fault on feeder 2 will be cleared in a longer time than a fault on feeder 5.

Suppose now that feeder 6 is closed and that its relays are set the same as those of 5; then for fault F on 5 the relays on 5 and 6 both operate to trip both feeders. This unwanted result is avoided if the relays at the E end of 5 and 6 are set to a shorter time than those at the C end and provided with a directional feature; then for fault F the first operation will be the feeder 5 relays at E, and the second will be those at C, which are set in time to discriminate with the relays for feeder 2. Such a scheme is the simplest form of directional overcurrent protection, and at each point the relays provided may be (1) one overcurrent (o.c.) relay in each phase together with an earth fault (e.f.) relay, or (2) two o.c. relays and one e.f. relay, or (3) three o.c. relays only, where the system is solidly earthed. In the case of earthing through a resistor, the o.c. and e.f. relays would require different settings.

Consider the effect of closing the interconnector with a fault F on feeder 5. If the two radial systems are identical in length, feeder size and relay time settings, then, although fault current will divide between feeders 1 and 2, the first relays to operate will be those at E on feeder 5. However, when this occurs the fault is fed through 2, 6 and the interconnector; therefore the feeder 6 relays at E will operate first if their operating time plus that of the feeder 5 relays at E is less than the operating time of the feeder 5 relays at C. Further, the interconnector cannot be proved with DOC relays satisfactorily for its own protection unless it is regarded as a 'loose link', in which case they would have to be set for rapid tripping to cut off the second infeed to a fault elsewhere on the system. If it is required that the interconnector should not trip, it must be provided with balanced protection and the o.c. relays set high for back-up.

A disadvantage of time graded systems is that the relays on feeders nearest to the supply point must have the longest operating times. As fault currents at such points are greatest, this is a drawback in the case of phase faults, though less so for single phase-earth faults, where fault current may be limited by earthing resistors.

Locking signals To avoid the long time lags associated with overcurrent equipment, a scheme involving transmission of *locking signals* from the remote end of a line may be employed. Consider *Figure 15.39*, in which only the locking signal pilot and relay contact circuits are shown: if the o.c. relays at, say, the left-hand end detect a fault current, they will operate to close the trip-coil circuit of the circuit-breaker at that end; if, however, the directional relay at the right-hand end detects a fault current going *out* of the feeder (indicating a fault on another section), it

produce maximum torque. Although the relay may be inherently wattmetric, its characteristic can be varied by the addition of phase shifting components to give maximum torque at the required phase angle. Several different connections have been used, and examination of the suitability of each involves determining the limiting conditions of the voltage and current applied to each phase element of the relay for all fault conditions, taking into account the possible range of source and line impedances.

Beam A balanced beam with two electromagnets can be used as a simple double-input device and has been widely applied to distance protection (*Figure 15.34b*). It can also serve as a directional relay.

Definite time Devices employing a clockwork escapement released by the operation of an attracted-armature relay, or an escapement driven by a disc or drum, are available. More usually, however, induction relays are used because of the long periods of inactivity in service.

15.8.4 Solid state equipment

Thermionic valve equipment has been in use for a half-century in carrier current protective signalling. The advent of solid state devices (transistors and thyristors) has made possible alternatives to electromagnetic relays.

Solid state equipment differs markedly from conventional electromagnetic equipment in several ways, particularly in the reduction in overall size.

The new flexibility and characteristics enable schemes to be devised that are difficult to achieve conventionally, such as the distance relay with a quadrilateral characteristic, *Figure 15.47(f)*. For most applications the transistor has the advantage of high speed, high sensitivity and simple amplifying and switching properties. Where heavier currents are necessary, e.g. in a tripping circuit, the thyristor is required. Solid state devices can, if desired, readily be associated with electromagnetic relays if the time delays (10–20 ms) associated with the latter can be accepted; however, the lighter and faster (1–2 ms) reed relay is applicable in many cases.

Components are subject to damage by moisture, but encapsulation can obviate the trouble.

Static components have some resilience and very low mass; they can withstand mechanical shock and vibration, significantly reducing risk of damage during transport and erection.

All components, but particularly transistors and thyristors, are sensitive to transient voltages, even of only a few microseconds' duration. Such transients arise from switching, and can be injected into the protective equipment directly through current and voltage transformers, or by electric or magnetic field induction. Equipment must be impulse tested, e.g. by a 5 kV 1/50 μs impulse wave.

The burden imposed by static relay equipment on c.t.s is much less than that of comparable electromagnetic equipment. As the cost of instrument transformers is 20–40% of that of the complete protective scheme, smaller and improved types are being developed which may offset the higher cost of the static relay itself.

Even with conventional equipment, the use of 1 A secondary currents (instead of 5 A) is sometimes desirable, especially where long interconnecting leads are necessary. With static equipment even smaller currents can be used. The ultimate in this respect is a very small line mounted high-voltage current transformer feeding a laser system; the electromagnetic power (about 2 W) is converted to light power and transmitted to earth potential equipment by light-guides, the resulting power (a few μW) then being amplified and fed to static relay equipment.

The range of solid state relays now available covers many applications hitherto dealt with by electromechanical devices.

The static equivalent cannot at present compete on economic terms with such equipment as a multi-contact hinged armature unit, but the static approach offers greater flexibility in the design to meet given protection functions. Typical characteristics are shown in *Figure 15.35* for static protection, which needs no auxiliary supply and provides a high reliability and accuracy.

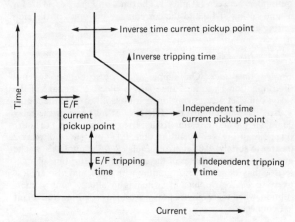

Figure 15.35 Static protection unit characteristics

The future of the static relay lies probably in new characteristics and concepts for protection schemes. The exploitation of integrated circuits, particularly in custom-built equipment, will increase as the price of these components falls. In the future it may be possible to accommodate the logic and amplifying functions for a protective relay on a single chip, greatly reducing the size. This may lead to the use of redundancy techniques, such as the two-out-of-three arrangement, which will help to provide more reliable protection.

15.8.5 Overcurrent protection

The most common overcurrent relay is the induction type, typical characteristics of which are given in *Figure 15.36*. The standard characteristic follows no simple law, but the *very-inverse-time* characteristic is $(I-1)t=k$, and the *extremely-inverse-time* law is $(I^2-1)t=k$.

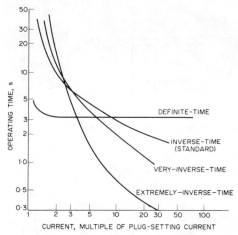

Figure 15.36 Overcurrent relay characteristics

forms are available as follows:

(1) attracted armature
(2) moving coil
(3) induction
(4) thermal
(5) motor-operated
(6) mechanical
(7) magnetic amplifier
(8) thermionic
(9) semiconductor
(10) photoelectric

Of these main group types, (7)–(10) are commonly known as 'static' relays.

15.8.3.1 Single-input relays

In the *repeater* type a small signal can be multiplied by 10^3 or more to operate one or more secondary devices, e.g. the trip coil of a circuit-breaker. Operation is rapid (within 20 ms), although a time delay can be incorporated. The *magnitude indicator* type operates instantaneously (or after a fixed time delay) when the magnitude of the signal exceeds or falls below a specified value. In the *time-dependent* relay the operating time depends on the magnitude of the signal; the most usual characteristic gives an operating time inversely proportional to the magnitude.

15.8.3.2 Double-input relays

In the *amplitude comparator* one input, I_1, tends to operate the relay, while the other, I_2, tends to restrain it. Operation takes place when I_1/I_2 is less than the specified value. When drawn on a complex plane R–X, the characteristic is a circle (*Figure 15.33a*).

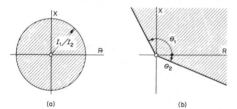

Figure 15.33 Comparator relay characteristics. (a) Amplitude; (b) phase

Operation is independent of the phase angle between the currents and occurs within the shaded area.

In the *phase comparator* operation takes place when the phase angle between the currents lies within specified limits (*Figure 15.33b*).

15.8.3.3 Multiple-input relays

By using more than two inputs, various forms of operating zone can be obtained, as required for distance feeder protection.

15.8.3.4 Types of relay

Attracted armature This single-input relay comprises a small coil carrying the signal current: it attracts a magnetic armature with one or more pairs of contacts which control the secondary circuits, and is suitable for both d.c. and a.c. operation. The setting may be based on minimum pickup or maximum drop-out current, with ranges usually selected by tappings to secure a constant value of operating m.m.f. Tapping changes the impedance and, therefore, the burden on associated c.t.s. The speed of operation is a function of the m.m.f., the length of armature travel and the inertia of the moving parts. The latter may be partly offset by counterweighting. When the relay coil is intermittently energised, it may be uprated to increase the operating speed: a set of contacts may be provided which, when energised by the operation of the initiating relay contacts, gives a holding effect. The current in the coils of tripping relays is then usually broken by an auxiliary switch on the circuit-breaker. Magnetic sticking of the armature in attracted-armature relays is prevented by stops of non-magnetic inserts on the working face.

Rotary This is a rotary version of the foregoing. The armature is usually mounted on a vertical spindle, the rotating parts are light and well-balanced, and the relay is sensitive at a relatively low burden. Adjustment and accurate setting is by torsion head and helical spring control. In some types the armature carries a light silver contact, the coil operated contacts being energised when the silver contacts close. The relay can be made to operate with a few milliamperes, and both linear and rotary forms of the attracted-armature relay can be arranged for double-input working.

Induction The time-dependent induction protective relay is probably the most widely used of all. It employs essentially the same construction as an integrating a.c. energy meter. For a single-input relay (*Figure 15.34a*) torque is produced by a phase difference between the flux of the main coil on the central limb of

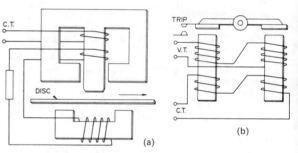

Figure 15.34 (a) Induction and (b) beam relays

the upper magnet and that in the lower (secondary) magnet, an effect secured by impedance compensation in the secondary. The torque is proportional to the square of the exciting current and is unidirectional. The speed of the disc may be governed by a control spring and a permanent-magnet brake. Slots are cut in the disc to modify the operating time–current characteristic, and the shape of this curve at higher currents can be made almost flat if the magnetic circuit is arranged to saturate at two or three times the setting.

The driving disc usually has a vertical spindle, but one make of relay has a horizontal drum driven about its axis, the time–current characteristic being similar to that of the disc type.

The *single-input* relay is widely used for overcurrent protection. By feeding a separate signal to the secondary coil the relay becomes a *double-input* device with a torque proportional to $I_1 I_2 \sin \alpha$, where α is the phase angle between the currents. In this form it can be given directional properties, and is also applicable to certain forms of distance protection.

Directional Directional relays operate in a manner similar to that of an electrodynamic wattmeter or, more commonly, an induction energy meter. In the latter the torque on the disc is proportional to $VI \cos \beta$, where β is the phase angle between the voltage and current inputs.

The maximum-torque angle is the angle by which the current applied to the relay must be displaced from the applied voltage to

Rated secondary values are normally 1 or 5 A (c.t.), and 110 V (v.t.).

Relays These, operating from c.t. and v.t., detect the presence of faults and energise the trip circuits of the appropriate circuit-breakers. Electronic (solid state) devices are now available as alternatives to the electromagnetic types, and give shorter operating times and improved performance.

Pilot circuits These carry signals between points in the system for comparison, for feeder protection and other purposes. The circuits may be provided in a variety of ways:

(1) Conductors specially installed for the purpose.
(2) Rented telecommunication lines; the current and voltage limitations are 60 mA and 130 V peak, and the pilots must be insulated from all power system equipment for 15 kV.
(3) Power transmission lines, using carrier signals at frequencies between 70 and 700 kHz.
(4) Radio links using microwave transmission between line-of-sight terminals.
(5) Optical links, usually embedded in the earth wire.

The choice requires very careful consideration, particularly for long distances. Even apparently simple conductors, for which the resistance and capacitance between ends can be measured, often act as a transmission line operating in conjunction with relay equipment. The performance of the pilot circuit may be complex, and in operation is made even more so by inductive interference effects.

It is possible to obtain high-speed feeder protective schemes using voice frequencies of 600–3000 Hz, which are suitable for pilot-wire channels, or pilot-line carrier or radio channels.

The overall fault clearance time is made up of the signalling time, and the operating times of the protection relay, trip relay and circuit-breaker opening. This total time must be less than the maximum for which the fault can remain on the system for minimum plant damage, loss of stability, etc. In recent years practical minimum times have been achieved in reducing protection times from 60 to 20 ms, trip relay times from 10 to 3 ms, circuit-breaker times from 60 to 40 ms, leaving protection signalling times to be reduced from 70–180 ms down to 15–40 ms in the UK, and to 5 ms in certain parts of the world where system stability is critical.

General details and classifications of current and voltage transformers are given in Section 6. Protection current transformers may have to maintain accuracy up to 30 per unit overcurrent, although often the error is less important than that c.t. characteristics should match.

Current transformers The errors are due to the exciting current, and they vary with the phase angle of the secondary burden. An increased burden demands a corresponding increase in core flux, and as the exciting current is a non-linear function of the flux, it is subject to a more than proportional rise accompanied by a greater harmonic content, so that the composite error is increased.

Protective equipment is intended to respond to fault conditions, and is for this reason required to function at current values above the normal current rating. Current transformers for this application must retain a reasonable accuracy up to the largest relevant current, known as the 'accuracy limit current', expressed in primary or equivalent secondary terms. The ratio between the accuracy limit current and the normal rated current is known as the 'accuracy limit factor'. Standard values are 5, 10, 15, 20 and 30. Protective c.t. ratings are expressed in terms of rated burden, class and accuracy limit factor (e.g. 10 VA class 10P 10). Classes 5P and 10P are useful only for overcurrent protection; for earth fault protection, in particular, it is better to refer directly to the maximum useful e.m.f. which can be obtained from the c.t. In this context the 'knee point' of the excitation curve is defined as that point at which a further increase of 10% secondary e.m.f. would require an increment of exciting current of 50%. Such current transformers are designated class X.

Current transformers may have primaries wound or of bar ('single-turn') form for mounting on bushings. For high-voltage systems the c.t.s may be separately mounted with oil or SF_6 insulation.

Voltage transformers The main requirement in protection is that the secondary voltage should be a true reflection of the primary voltage under all conditions of fault. It is usual with electromagnetic v.t.s to apply additional delta-connected windings (*Figure 15.32*) to give a measure of the residual voltage at the point of connection. The three voltages of a balanced

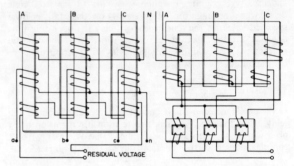

Figure 15.32 Voltage transformers with residual-voltage windings

system summate to zero, but this is not so when the system is subject to a single-phase earth fault. The residual voltage in this case is of great value for protective gear practice as a means of detecting or discriminating between earth fault conditions. The residual voltage of a system is measured by connecting the primary windings of a 3-phase v.t. between the three phases and earth, and connecting the secondary windings in series or 'open delta'. The residual voltage is three times the zero sequence voltage. To measure this component it is necessary for a zero sequence flux to be set up in the v.t., and for this to be possible there must be a return path for the resultant summated flux: the v.t. core must have unwound limbs linking the yokes (*Figure 15.32*). If three single-phase units are used (as is common for e.h.v. systems), each phase unit has a core with a closed magnetic circuit, so that the above consideration does not arise.

An alternative, avoiding the cost of an h.v. voltage transformer, is to use the secondary voltages from a v.t. on the l.v. side of a power transformer, the voltage drops in which are compensated by the addition or subtraction of voltages developed by c.t.s in the delta of the power transformer. The v.t.s and c.t.s must be provided with tappings if the power transformer is equipped with tap-changing gear, and arranged for automatic selection with the main tappings.

For high-voltage systems the *capacitor voltage divider* gives a cheaper (but less accurate) device than its electromagnetic counterpart. A typical divider for 400 kV has a total capacitance of 1500–2000 pF, with about 34 000 pF between the tapping point and earth, to give about 13.5 kV across the primary of the intermediate transformer, the secondary of which gives 63.5 V (phase to neutral).

15.8.3 Relays

Relays may be classified according to the number of current inputs and their operating function. Types and constructional

reactor is an arc suppression coil, point E is raised to A, the sound phase voltages to earth become $\sqrt{3}$ times normal, and the residual voltage at the fault will be 3 times normal phase-to-neutral value.

Earthed through resistor The conditions (*Figure 15.29b*) are essentially similar to those for reactor earthing except that there is unbalance in the voltages to earth of the sound phases. The volt drop across the resistor is E″N.

15.8.1.4 Double phase-to-earth fault

With *solid earthing* at one point only, the voltages to earth at a double phase-to-earth fault will, as shown in *Figure 15.30*, be zero, zero and C′E, and at some distance away will be A′E′, B′E′ and C′E′. The voltage to earth of the sound phase may be

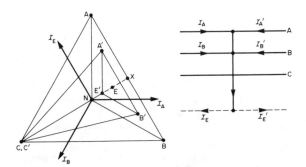

Figure 15.30 Double phase-to-earth fault

increased at the point of fault. The angle α between the phase currents I_A and I_B (or I_A and I_B') may be less than 60°, depending on the ratio of balanced to zero-sequence impedance; the angle varies considerably, according to the earthing arrangements, and in the limit with the neutral point isolated it becomes 180°. The residual current (i.e. the earth current) is the vector sum of the currents in the faulted phases and that (if any) in the sound phase. The residual voltage at the point of fault is the normal phase-to-neutral voltage, and it decreases with distance from the fault position.

With the system neutral earthed through a low-valued reactor, the drop across the latter is NE′ and the voltage to earth of the sound phase is CE, increasing to a possible maximum of 1.5 times normal value if the reactor is an arc suppression coil.

With system earthing through a *resistor*, the point E in *Figure 15.30* is no longer on the line CX.

The system Z_0/Z_1 ratio is defined as the ratio of zero sequence and positive sequence impedances viewed from the fault. It is a variable ratio dependent on the method of earthing, fault position and system operating arrangement. When assessing the distribution of residual quantities through a system, it is convenient to use the fault point as the reference, as it is the point of injection of unbalanced quantities into the system. The residual voltage is measured in relation to the normal phase-neutral system voltage, and the residual current is compared with the 3-phase fault current at the fault point. It can be shown that the character of these quantities can be expressed in terms of the ratio Z_0/Z_1 of the system; see *Figure 15.31*. The residual current in any part of the system can be obtained by multiplying the current from the curve by the appropriate zero sequence distribution factor. Similarly, the residual voltage is calculated by subtracting from the voltage curve three times the zero sequence drop between the measuring point in the system and the fault.

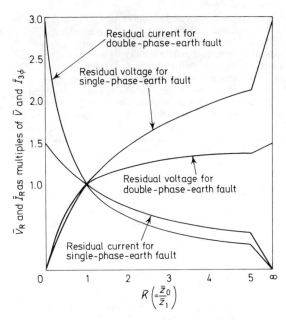

Figure 15.31 Variation of residual quantities at fault point

15.8.2 Protective equipment

The satisfactory operation of protective equipment (which extends from simple domestic subcircuit fuses to sophisticated electronic system devices) depends largely on *co-ordination*. The influencing factors include the network layout and characteristics, fault MVA levels, earthing, the availability of pilot cables and the physical extent of the system. Generally, the higher the voltage and fault levels the more necessary is quick-acting discriminative protection covering lines and plant.

Protective equipment may be broadly classified into (a) *restricted zone*, giving full discrimination, (b) *semi-restricted zone*, (c) *unrestricted zone*, with no discrimination. Type (a) can be considered as applying to a network of such magnitude that the disconnection of each item of plant is necessary if an internal fault develops. Types (b) and (c) apply to less important installations and also as 'back-up' in restricted-zone schemes.

15.8.2.1 Modes of operation

An alternative classification based on the operating mode is as follows:

Overcurrent equipment operates if the current exceeds a pre-set value. Restrictions regarding the direction of the overcurrent are often included to improve discrimination.

Balanced (differential) equipment operates if an out-of-balance occurs between currents or voltages which under normal conditions are balanced.

Distance (impedance) equipment operates if the impedance (proportional to distance), as viewed at a circuit-breaker supplying a feeder, falls below a specified value.

Miscellaneous equipment is designed for special purposes.

15.8.2.2 Equipment elements

The chief items of plant making up a complete protective installation are:

Current and voltage transformers These provide convenient levels of current or voltage proportional to the system values.

may be simple or complicated, and involve conductor breakage or short-circuited turns on transformers or generators.

The incidence of faults on cables and lines depends on the installation and the climatic conditions. For a typical system in a temperate zone the distribution of faults as percentages of the total is approximately: overhead lines, 60; cables, 15; transformers, 12; switchgear, 13. The causes of overhead-line faults, and the number of faults per 100 km of line per year, are, typically: lightning, 1.0; gales, 0.15; fog and frost, 0.1; snow and ice, 0.06; salt spray, 0.06. The total is 1.37 faults per 100 km per year: in tropical countries lightning faults may be markedly more numerous. In general, the higher the voltage the lower the number of lightning faults.

15.8.1 Fault conditions

Below are given circuit and vector diagrams relevant to the behaviour of protective equipment under the simpler fault conditions.

15.8.1.1 Three-phase fault

A 3-phase fault usually develops first as a phase-earth fault, and it may be unbalanced. Even when a circuit-breaker closes on to a 3-phase fault, one phase may momentarily be faulted before the other two, a matter of importance in high speed protection. *Figure 15.26* shows the relevant vector diagram. At the fault the

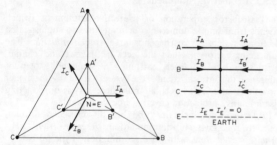

Figure 15.26 Three-phase fault

voltages to neutral are zero; and if AN, BN, CN are the phase-to-neutral pre-fault voltages, the voltages at some distance from the fault (e.g. at a relay controlling a circuit-breaker) are A′N, B′N, C′N. With a symmetrical short-circuit the conditions are independent of the system earthing arrangements.

15.8.1.2 Phase-to-phase fault

Two phases are faulted clear of earth, an unusual kind of fault even less likely on cables than on overhead lines. *Figure 15.27* shows AN, BN and CN as the normal voltages at the site of the fault. On occurrence of the fault, the voltages there are XN, XN and CN, and the voltages at some specified distance are A′N, B′N and C′N, with the locus of A′ and B′ on the line AB. The voltage to earth of the sound phase is not affected. The conditions are independent of the system earthing, but the fault current will be reflected into the further sides of associated transformers in accordance with their connections.

15.8.1.3 Single phase-to-earth fault

This type of fault is considered for a system with the generator or transformer having its neutral earthed (a) solidly, (b) through an reactor, (c) through a resistor.

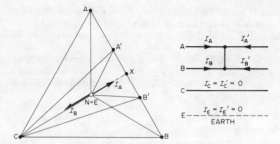

Figure 15.27 Phase-to-phase fault

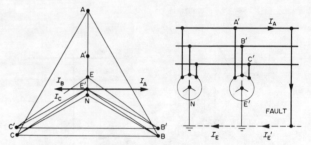

Figure 15.28 Single phase-to-earth fault (solid earthing)

Neutral solidly earthed The voltages to earth at the point of fault with phase A earthed are zero, B′E and C′E, as in *Figure 15.28*, and at some distance away the voltages are A′E′, B′E′ and C′E′. The locus of A′ of the line AN, and the voltages to earth of the sound phases are slightly changed to B′E and C′E by reason of the currents in these phases between the two infeeds. With the system earthed at one point only (apart from the fault), the total earth fault current divides between the two infeeds. If there are two (or more) system earth points and the infeeds are I_A and I'_A, there may be currents in the sound phases of direction decided by the preponderance of zero- over positive- and negative-sequence impedance of one infeed compared with the other. In this case the residual or earth current on each side will be $I_A + I_B + I_C = I_E$; the residual voltage will be $0 + B'E + C'E$ at the fault, and $A'E' + B'E' + C'E'$ some distance away.

Earthed through reactor With a low-valued reactor (small compared with an arc suppression coil), the earth point potential on the occurrence of a fault will be E (*Figure 15.29a*), and the drop in the reactor will be E′N. The fault position voltages are zero; voltages to earth of the sound phases are increased. If the

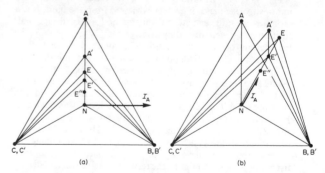

Figure 15.29 Single phase-to-earth fault. (a) Through reactor; (b) through resistor

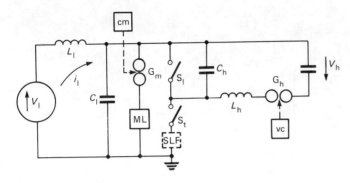

V_l = lower voltage, supplying power-frequency current.
L_l = inductance to give correct short-circuit current i_l from V_l.
C_l = capacitance of power-frequency current circuit and with L_l controlling injected voltage characteristic.
S_t = test circuit-breaker.
S_l = auxiliary circuit-breaker.
G_h = gap for closing high-voltage supply to test circuit-breaker.
G_m = gap for closing multi-loop re-ignition circuit.
vc = voltage dependent control circuit for closing gap G_h after current zero.
cm = current dependent control circuit for closing gap G_m.
V_h = higher voltage, supplying main part of transient recovery voltage.
L_h = inductance of voltage circuit.
C_h = capacitance of voltage circuit and with L_h controlling main part of the transient recovery voltage characteristic.
ML = multi-loop re-ignition circuit (step-by-step method).
SLF = short-line fault circuit, when testing under short-line fault conditions.

Figure 15.25 Voltage injection scheme with voltage circuit in parallel to the auxiliary circuit-breaker, simplified diagram

relating to the circuit-breaker itself, but also other data relating to short-circuit conditions, such as instability, interference with communication systems, earth currents and the behaviour of protective equipment. Detailed preparation for field tests may thus require many months of work.

15.7.5 Miscellaneous tests

Interrupting capability tests occupy the major time and expense of a testing programme, but other important tests are also necessary.

Making The circuit-breaker is closed on to a fault with the maximum asymmetry to check that the contacts, mechanism and connections will properly withstand the electromagnetic forces. As possible arcing before the contacts close must also be checked, the test must be carried out at normal voltage.

Switching Tests, both making and breaking, for the transient voltages set up when switching transformers or unloaded lines are desirable, although the currents involved are quite small.

Impulse voltage The circuit-breaker as a whole must be subjected to the lightning and switching impulse voltage tests appropriate to the system into which it is to be connected.

Temperature rise The breaker and associated switching resistors, current transformers, etc., must be subjected to temperature rise tests at rated full-load current; also at maximum short-circuit current, which may last for 3 s and may involve the equivalent of two or three successive operations.

Switching resistor Special tests may be desirable for the switching resistors, as these may be subject to abnormal transient voltages arising from differences (up to 1 ms) in the closing and opening instants of the series breaks in one phase.

Environment Circuit-breakers may have to operate in any part of the world in conditions very different from those normally existing at the manufacturer's works. Temperatures between $-50°C$ in arctic regions and $70°C$ or more in tropical sunlight may be encountered; the former may involve testing under severe icing and with special low-freeze oils. Very fine dust may occur in some tropical areas and earthquake conditions may also have to be simulated. The effect of height and atmospheric pollution in insulator flashover has also to be considered. A large and well-equipped environmental test chamber is thus required.

15.8 Overcurrent and earth leakage protection

The main effects of fault current in a power system are:

(1) Disturbance of the connected load.
(2) Overheating at the fault point and in associated plant.
(3) Electromagnetic forces of abnormal magnitude, with consequent mechanical damage.
(4) Loss of synchronous stability.

The function of the protective equipment is to isolate the faulty plant from the running system by initiating tripping signals for appropriate circuit-breakers. It should therefore *discriminate* between faulty plant and sound plant carrying through-fault current. The whole process must be effected with the minimum of delay and disturbance.

Exceptionally, some faults may be allowed to persist: an example is an earth fault on a system earthed through an arc suppression (Petersen) coil, or one having an isolated neutral, where the fault may be located and alternative feeds provided before the fault is isolated.

Faults are due to insulation breakdown by deterioration, overvoltage, mechanical damage or short-circuit effects; they

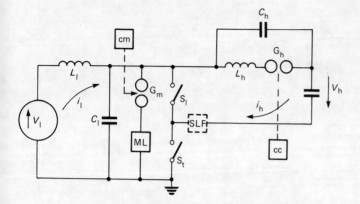

- V_l = lower voltage, supplying power-frequency current.
- L_l = inductance to give correct short-circuit current i_l from V_l.
- C_l = capacitance of power-frequency current circuit.
- S_t = test circuit-breaker.
- S_l = auxiliary circuit-breaker.
- G_h = gap for closing high-voltage supply to test circuit-breaker.
- G_m = gap for closing multi-loop re-ignition circuit.
- cc = current dependent control circuit for closing gap G_h before current zero.
- cm = current dependent control circuit for closing gap G_m.
- V_h = higher voltage, supplying injected current i_h and main part of transient recovery voltage.
- L_h = inductance of voltage circuit.
- C_h = capacitance of voltage circuit and with L_h controlling main part of the transient recovery voltage characteristic.
- ML = multi-loop re-ignition circuit (step-by-step method).
- SLF = short-line fault circuit, when testing under short-line fault conditions.

Figure 15.23 Current injection scheme with voltage circuit in parallel to the auxiliary circuit-breaker, simplified diagram

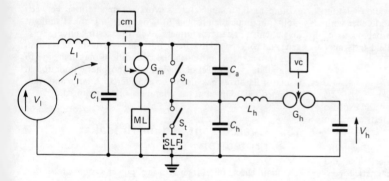

- V_l = lower voltage, supplying power-frequency current.
- L_l = inductance to give correct short-circuit current i_l from V_l.
- C_l = capacitance of power-frequency current circuit and with L_l controlling injected voltage characteristic.
- S_t = test circuit-breaker.
- S_l = auxiliary circuit-breaker.
- C_a = capacitance in parallel to auxiliary circuit-breaker.
- G_h = gap for closing high-voltage supply to test circuit-breaker.
- G_m = gap for closing multi-loop re-ignition circuit.
- vc = voltage dependent control circuit for closing gap G_h after current zero.
- cm = current dependent control circuit for closing gap G_m.
- V_h = higher voltage, supplying main part of transient recovery voltage.
- ML = multi-loop re-ignition circuit (step-by-step method).
- L_h = inductance of voltage circuit.
- C_h = capacitance of voltage circuit and together with C_a, C_l and L_h controlling main part of the transient recovery voltage characteristic.
- SLF = short-line fault circuit, when testing under short-line fault conditions.

Figure 15.24 Voltage injection scheme with voltage circuit in parallel to the test circuit-breaker, simplified diagram

made with the appropriate fraction of the recovery voltage, but the rate of rise of the transient recovery voltage is, of course, that for the breaker as a whole. Provided that the voltage distribution over the several modules is uniform (as it is in modern designs), unit tests can be carried out with confidence.

15.7.3 Synthetic testing

For the testing of modern circuit-breakers having a short-circuit rating many times that available in the largest direct test plants, it was essential to introduce some form of synthetic testing in which the short-circuit current is obtained from one source at low voltage, and the transient recovery voltage (which occurs separately a few microseconds later) from another source. IEC 427-1973 covers the requirements for breaking tests, and deals with two methods of applying the high voltage:

(1) Current injection, in which the high-voltage source is switched into the test circuit *before* current zero, thus providing the current through the test circuit-breaker during the current zero period.
(2) Voltage injection, in which the high-voltage source is switched into the test circuit *after* current zero, the power-frequency current circuit providing the current through the test circuit-breaker during the current zero period.

In both cases the high-voltage circuit may be switched in parallel with the test or the auxiliary breaker: see *Figures 15.22–15.25*.

Where the breaker being tested may arc for more than one loop of current, the arcing may be prolonged by various methods, including the provision of a multi-loop re-ignition circuit and the step-by-step method of testing.

A variety of transient recovery voltage waveforms, including those for short-line faults, and the four parameter waveforms for high-voltage circuit-breakers are possible using a variety of *RLC* values.

Comparative tests made on both current and voltage injection synthetic circuits have verified their equivalence to direct test circuits within the limits of experimental error for air-blast, small-oil-volume, and SF_6 single-pressure breakers provided that the circuits are carefully chosen and controlled to give the correct stresses.

Since 1973 there have been developments both in breakers and in synthetic testing circuits, so that suitable tests can now be made for many of the newer designs, covering: make–break duty cycles; air-blast breakers with low ohmic breaking resistors; a.c. recovery voltages; tests on complete metal clad breakers; breaking tests at very low frequency recovery voltages associated with u.h.v. breakers; 3-phase breaking tests; asymmetrical breaking current tests; and capacitive breaking tests.

For the higher ratings synthetic testing is the only economical method and it has been widely adopted, with facilities for breaking currents up to 140 kA and injected voltages up to 1500 kV peak.

15.7.4 Field testing

While field testing on an actual system may be the only way in which sufficient short-circuit MVA for a full-scale test can be achieved, it may also have the advantage that certain tests can be carried out that are not easily reproducible in a test plant, e.g. unloaded transmission lines, long transformer feeders, etc. Such testing is, however, naturally not favoured by operating engineers.

Very careful preliminary calculations must be carried out to ensure that the staged faults will not interfere with the security of the system; rearrangement of system connections and alterations to protective equipment settings will almost certainly be required. In order to secure the maximum of information from such tests, elaborate instrumentation must be provided. The opportunity is usually taken to obtain not only information

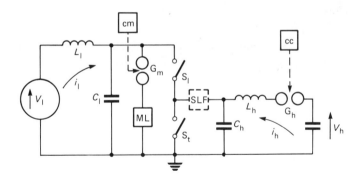

V_1 = lower voltage supplying power-frequency current.
L_1 = inductance to give correct short-circuit current i_1 from V_1.
C_1 = capacitance of power-frequency current circuit.
S_t = test circuit-breaker.
S_1 = auxiliary circuit-breaker.
G_h = gap for closing high-voltage supply to test circuit-breaker.
G_m = gap for closing multi-loop re-ignition circuit.
cc = current dependent control circuit for closing gap G_h before current zero.
cm = current dependent control circuit for closing gap G_m.
V_h = higher voltage, supplying injected current i_h and transient recovery voltage.
L_h = inductance of voltage circuit.
C_h = capacitance of voltage circuit and with L_h controlling the transient recovery voltage characteristic.
ML = multi-loop re-ignition circuit (step-by-step method).
SLF = short-line fault circuit, when testing under short-line fault conditions

Figure 15.22 Current injection scheme with voltage circuit in parallel to the test circuit-breaker, simplified diagram

parallel star-delta winding connections enable various voltages up to 15 or 22 kV to be generated.

The generator is run up to speed by a d.c. or induction motor of 400–700 kW, the motor being disconnected immediately prior to the test to avoid supply system disturbance. The energy required for the short-circuit is provided from the stored kinetic energy without significant speed drop.

A special excitation system is required, the separately driven exciter being capable continuously of 600–1000 A at 100–1200 V. At the instant of short-circuit the excitation current is boosted to several times these values by short-circuiting a series resistor, in order to avoid excessive decrement of the test current during the period of the test. The energy for the boost is provided by a flywheel (6–10 t) mounted on the exciter shaft.

The transformers are 1-phase units of low leakage reactance (0.03–0.05 p.u.) and, by series-parallel and star-delta winding arrangements, give a voltage range up to about 275 kV 3-phase or 400 kV single-phase.

For testing equipment rated at medium voltage a.c. step-down transformers are used; again these usually consist of 1-phase units of low leakage reactance which may be connected in 3- or 1-phase with series-parallel, star-delta connections. At these lower voltages very high currents can be obtained which are useful for short-time current tests on cables, current transformers, bushings, etc.

For d.c. tests it is usual to obtain this from rectifiers connected to the step-down transformers, voltages of 750–3000 V being obtainable at currents of 140–35 kA. Tests at voltages below 750 V are usually obtained by reducing the excitation of the generators.

15.7.1.2 Network parameters

Resistors and inductors are necessary to adjust the magnitude and power factor of the short-circuit current, the inductors (usually air insulated) comprising a number of small units that can be appropriately connected. Capacitors are necessary to control the transient recovery voltage and also to test the circuit-breaker capability in clearing capacitance currents. An artificial line, or the equivalent, is installed in order to test under short-line fault conditions.

15.7.1.3 Control switches

The main closing switch for applying the short-circuit is of heavy air insulated construction. It can be closed at any point in the generated voltage wave. A 'master' circuit-breaker is included to protect the plant in case of failure of the test circuit-breaker.

15.7.1.4 Sequence control

Control of the various operations necessary for initiating and controlling the test may be provided by (1) a pendulum with numerous adjustable contacts along its sweep, or (2) a drum driven at constant speed and provided with adjustable contact-operating strikers.

For time delays less than about 10 ms, the above methods are supplemented by electronic timing for control of the point-on-wave of voltage and synthetic testing, etc.

15.7.1.5 Instrumentation

The chief items required for the test plant are:

(1) A multi-element electromagnetic oscillograph for recording the normal-frequency phase voltages and currents, and the arc voltage.
(2) Cathode ray oscillographs for recording transient recovery voltages in each phase.
(3) Shunts and voltage dividers for supplying the oscillographs (instrument transformers being unsuitable for transients).
(4) Electromechanical transducers for sensing contact travel, tank pressure, etc.
(5) Miscellaneous indicating instruments for pre-setting the fault conditions.

Records from a typical test are shown in *Figures 15.20* and *15.21*, from which the currents and voltages and transient recovery voltage handled by the test breaker can be determined. For accuracy, the measuring equipment is usually calibrated prior to a test, and again at its conclusion.

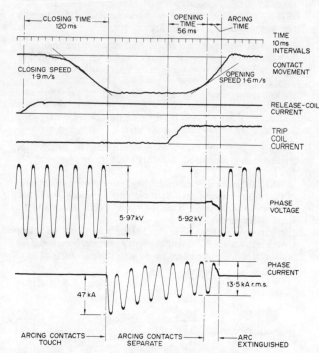

Figure 15.20 Electromagnetic oscillogram of typical make-break test. Voltage and current shown for one-phase only

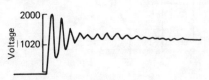

Figure 15.21 Cathode ray oscillogram of typical transient recovery voltage

15.7.2 Single-phase and unit testing

If the phases of a circuit-breaker are separate (i.e. not in a common tank) and independently operated, *single-phase testing* can be adopted. The full line voltage of the 3-phase test plant can be applied, or special generator and transformer connections used, so that a major part of the available capacity of the test plant is concentrated on one phase of the breaker under test. True *unit testing*, however, refers to the separate testing of one or more units of a circuit-breaker of modular construction. The tests are

that negligible flux pass into the tank walls; this is usually ensured by surrounding the coreless coils by a laminated iron shield. An alternative is to surround the coils by a screen of copper rings; flux tending to pass outside the screen is suppressed by currents induced in the rings. If the screen diameter is about twice the coil diameter, these currents reduce the effective inductance by about 10%.

15.6.2 Shunt reactors

Shunt reactors are invariably oil immersed. A major air-gap is necessary in the magnetic circuit to limit saturation so that the inductance is substantially constant up to 30% above rated voltage. The *magnetic shield* construction (as for series reactors) is common, the flux density in the shield being kept to about 1 T or less to minimise noise. An alternative is the *gapped core* construction, in which the core comprises a number of laminated iron sections separated by non-magnetic spacers; fringing at the edges of the gaps may cause noise and losses, and the stamping and assembly of the small packets of laminations may cause difficulties. Although the gapped core construction may be slightly cheaper at lower ratings (20 MVAr at 132 kV), the coreless type is generally preferred. The active power loss, about 0.002–0.003 p.u. of the reactive power rating, can usually be dissipated by natural cooling.

15.6.3 Capacitors

Most power system capacitors comprise five or more layers of impregnated paper between aluminium foil electrodes, the whole being wound on a mandrel as a continuous length to form an element of circular or elliptical section. Impregnation in a synthetic oil (trichlordiphenyl) gives a relative permittivity of 4–6 and a working stress of about 8 kV/mm. Several elements are assembled in a welded steel tank to give units of 50–100 kVAr at 1.5 kV of capacitance 70–140 µF. Units are rack-mounted and can be connected in series parallel to give any required voltage and rating. Outdoor mounting is common. The space occupied is 0.6–0.9 m²/MVAr and the mass is 600–800 kg/MVAr. The dielectric loss is from 0.02 to 0.002 per unit.

Discharge resistors must be provided to ensure safety after de-energisation, the voltage remaining on the capacitor not being permitted to exceed 50 V after 1 min for units up to 660 V and 5 min for units of higher voltage. The power loss in such resistors is negligible.

Protection of each individual unit of a bank may be by a fuse located either internally to the tank or externally; the latter, although more expensive, has the merit that a faulty unit can quickly be detected. Protection of the complete bank is normally carried out by balancing the currents in similar sections of the bank, an unbalance usually operating an alarm.

Series capacitors require, in addition, a protective spark-gap in parallel to prevent excessive voltage when carrying currents in excess of twice full-load current. The gap should be self-extinguishing in order that the compensation is restored after clearance of the fault.

Harmonic voltages can seriously overload capacitors. Series rejector networks can be included to suppress them.

15.6.4 Variable reactive power

Variable reactive powers can be obtained from reactors and capacitors if the units are separately switched. Alternatively, the bank can be fed from a tap-changing transformer. A reactor can itself be fitted with taps, or designed as a transductor with d.c. bias to vary the magnetic saturation. Usually, however, where variable control is required, it involves both generation and absorption of lagging reactive power, so that a combination of

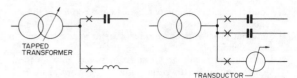

Figure 15.18 Static reactive power supply

reactors and capacitors is necessary; two possible schemes are shown in *Figure 15.18*, the transductor scheme having the merit of a quicker response.

15.6.5 Resonant link

To provide a connection between two power systems without significantly increasing the short-circuit level on either, a network comprising a series-connected reactor and capacitor in resonance has recently been developed. The network offers negligible impedance at normal frequency, but under fault conditions, when the voltage across each element rises, the overvoltage protection on the capacitor short-circuits it, leaving the full impedance of the reactor in the system. After fault clearance the conditions of resonance are restored. The overvoltage equipment could be a protective gap, or preferably a transductor–resistor arrangement in which the transductor saturates just below full-load current.

15.7 Switchgear testing

The high ratings of modern switchgear raise difficult testing problems. Full-scale testing requires expensive plant and is impracticable for ratings above about 8 GVA at 275 kV. Single-phase testing, the testing of individual units, a simulation by 'synthetic' testing or (in a few special cases) field testing must then be adopted.

15.7.1 Direct testing

A large switchgear testing station comprises generators and transformers to give the required voltage and apparent-power rating; adjustable resistors, inductors and capacitors to reproduce the specified circuit conditions; control circuit-breakers; and extensive instrumentation and sequence control gear. A schematic diagram of a typical testing station is given in *Figure 15.19*.

15.7.1.1 Supply equipment

The generating plant comprises one or more motor-driven 3-phase generators running at 1500 or 3000 rev/min. The generators are designed with a low transient reactance (0.01–0.02 p.u.), a short-time thermal rating and heavily braced windings to withstand repeated short-circuit forces. The dimensions of a machine for 1–2 GVA short-circuit rating are comparable with those of a 60–80 MVA turbo-generator. Two or three machines can be paralleled for higher short-circuit ratings, and series-

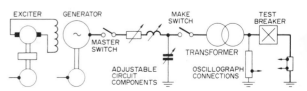

Figure 15.19 Basic circuit for short-circuit tests

(2) Good thermal conductivity to assist rapid cooling of the arc roots.
(3) Sufficient metal vapour from low-current arcing to control current chopping.
(4) Limitation of metal vapour and thermionic emission from high-current arcing to permit voltage recovery at current zero.
(5) Low weld and cold adhesion strengths at the contacting surfaces to give easy and consistent separation.
(6) Separation of the contacting surfaces with only small roughness to preserve the electric strength of the gap.
(7) Low and uniform erosion of contacts to give long operating life.

Contacts having a normal current rating of up to 3 kA are available, and indicators can be fitted to show contact wear. The short stroke and low inertia of moving parts, operated by mechanisms designed to have short stroke, high speed and freedom from bounce on closing, have resulted in an operating energy much less than that of the equivalent oil circuit-breaker, so that a modern 10 Ah battery will provide over 1000 close/open operations. Alternatively, spring closing mechanisms can be provided.

As the interrupter needs no maintenance, it need not be withdrawable, so that with switch disconnectors and selectors, isolation, earthing and testing can be achieved.

With static relays giving high-speed protection, and auto-reclose features, such circuit-breakers are now used for many distribution circuits, 25 kV single-phase railway supply protection, and underground mining supplies at voltages up to 11 kV.

15.5 Disconnectors

To carry out maintenance on items of plant, it is necessary to provide disconnectors on each of the circuits to which the plant is connected. These normally operate only under no-load conditions ensured by suitable interlocks. Earthing switches are also needed, commonly in preference to temporary earth connections, and these may have to close and carry the fault current of the system.

With indoor cubicle mounted or metal clad switchgear the disconnectors may be incorporated in the structure and interlocked with the circuit breaker; alternatively, with metal clad gear, they may take the form of plugs through which the circuit-breaker is connected to the busbars and the feeder, the circuit-breaker being drawn out horizontally or lowered vertically to effect isolation.

Outdoor disconnectors are built as separate units and have the merit of providing clear visual evidence that a particular item of plant is isolated; interlocks are also, of course, essential. At the lower voltages the isolators are usually simple knife-switches operated manually through a linkage mechanism from ground level. At the higher voltages either a single-break vertical movement or a double-break horizontal movement switch, with the switch blade rotating about a centre support, can be used. Vertical break disconnectors are common in the USA, but British practice favours the horizontal movement at 132–400 kV. In all cases it is essential for safety of personnel that the impulse strength of the open gap exceed by 10–20% the breakdown strength to earth of the support insulators under all foreseeable conditions. It must also be remembered that the disconnectors may have to operate under conditions of icing and pollution.

15.5.1 Switches

Switches have to be capable of closing and opening circuits up to their normal current rating. At medium voltages they are often either self-contained or combined with fuses or starters, and thus also act as disconnectors for maintenance purposes. At the higher voltages they have often to break the magnetising current of transformers or the charging current of lines or cables. Again they are sometimes combined with disconnectors and may incorporate simple aids for clearing by the use of gas-producing enclosures, or simple puffer-type SF_6 interrupters.

15.6 Reactors and capacitors

Reactors are used on a power system as *series reactors* for limiting short-circuit currents and as *shunt reactors* to compensate for system capacitance that might otherwise cause excessive voltage rises, especially at light loads.

Capacitors may be used as *series capacitors* in order to neutralise wholly or partially (not usually above about 60%) the line inductance, thereby reducing the voltage drop and, in the case of large systems, increasing the power limits and improving stability. *Shunt capacitors* are used for compensating the lagging reactive power of a system; on both large high-voltage and small low-voltage distribution systems the purpose is to improve voltage conditions and reduce losses by improving the power factor. Industrial consumers commonly find it economic to introduce such equipment in order to reduce their electricity costs.

On a typical large system the rating of reactive power compensation required, excluding industrial equipment, is commonly between 10 and 20% of the system maximum load.

15.6.1 Series reactors

Most commonly, series reactors are located in the busbars of a substation or generating station (*Figure 15.17*) rather than in a

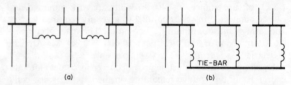

Figure 15.17 Location of reactors. (a) Ring system; (b) tie-bar system

feeder circuit, where they would have to carry the full feeder current, with consequent loss and volt drop. Busbar reactors of rating up to 0.1 or 0.15 per unit (based on the station rating) can give effective fault current limitation.

Although the terminal voltage of the reactor is low, it must be insulated for full system voltage to earth. An essential feature is that the magnetic circuit is all (or largely) of air, so that it does not saturate under heavy current. Attention must also be paid to the high forces due to short-circuit currents which tend to burst the coil radially outwards and crush axially.

15.6.1.1 Air insulated

In air insulated reactors the conductors are moulded into concrete supports, the whole structure being mounted on insulators. Such reactors are unsuitable for systems above 33 kV and must be installed under cover. The magnetic leakage field may heat adjacent metalwork.

15.6.1.2 Oil immersed

Most reactors are of the oil immersed type, with insulation and cooling arrangements like those of a transformer. It is necessary

alumina filter through which exhaust gas is returned to the compressor, which starts automatically when a break operation occurs.

Advantages of the SF_6 compared with the air-blast breaker include quiet operation, freedom from chopping and suitability for use in city areas with metal clad gear. Ratings of 132 kV 3.5 GVA and 275 kV 15 GVA have been used in the UK for both metal clad and outdoor arrangements.

A disadvantage of the two-pressure SF_6 design was the cost of producing the blast of gas for arc extinction by using a compressor and high-pressure reservoir which, owing to the characteristics of the gas, require heating when temperatures are below about 10°C. As the flow of gas was only required during arc extinction, later designs were produced whereby the insulation was obtained from gas at 0.65 MN/m^2, the blast being produced by movement of the opening contacts. Such designs have now superseded both the air-blast and two-pressure SF_6 circuit-breakers for metal clad and outdoor equipments.

There are basically two arrangements for obtaining a gas flow during contact opening (*Figure 15.15a, b*). In (*a*), a duo-blast

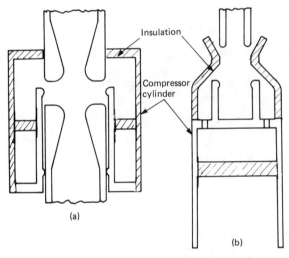

Figure 15.15 Schematic of types of single-pressure breaker

arrangement, the arc contact gap (and general design) is fixed and is bridged by the moving contact attached to the insulation compression cylinder. In (*b*), a mono-blast arrangement, the arc contact gap increases in length as the circuit-breaker opens, the gas flow being directed round and through the fixed arc contact by an insulating guide.

Using arrangement (*a*) with bridging contacts suitable for a normal current rating of up to 4000 A, and graphite arc contacts giving rated breaking currents of 31.5–63 kA, the numbers of breaks used for a voltage range are: 100–170 kV, one; 170–420 kV, two; 360–525 kV, three. For the multi-break units the division of the transient recovery voltage between breaks is achieved by capacitors in parallel with each break. These breakers are available in either metal-clad or outdoor arrangements.

To achieve the high opening speed necessary for a two-cycle total break time a hydraulic mechanism is used at 31 MN/m^2 (4500 lbf/m^2) backed up by bottled nitrogen.

For the metal clad arrangement the busbars are enclosed and insulated by SF_6 gas. They can be supplied either with each phase segregated or with the three phases within a single enclosure. Also within the circuit-breaker enclosure are SF_6 insulated discon-

nectors and fault-making earthing switches so that any of the usual single or double busbar, one-and-a-half switch and four-switch mesh, circuit arrangements can be achieved. The outgoing circuits can also be either by overhead line through bushings or by cables. Test bushings can be provided for either testing the cables or testing the SF_6 insulation in the metal clad substation.

So far the highest commercial voltage is 765 kV and only outdoor switchgear has been used. Recent developments of SF_6 single-pressure circuit-breakers have now made them available for this voltage. The design of the interrupter is based on *Figure 15.15(a)* (duo-blast arrangement), except that to speed up the SF_6 flow during opening, the piston, instead of being fixed, is also moved mechanically in the direction opposite to that of the opening contacts. Improved electrohydraulic operating mechanisms give a total break time of two cycles (33 ms at 60 Hz), which assists system stability. At present the maximum short-circuit current is about 25 kA, but future development may raise this to 40–50 kA, so that the 63 kA test current that the circuit-breaker has withstood should suffice for many years. This is achieved using four breaks per phase. At voltages above 500 kV the switching-in of unloaded lines is a particularly critical operation, owing to switching overvoltages, even more so when the lines carry residual charges following an auto-reclose to clear a single-phase to earth fault. To control these overvoltages, the circuit-breakers are fitted with mechanically operated closing resistors of value approximately equal to the surge impedance of the line. The resistors close about 15 ms before the main contacts touch, have sufficient thermal capacity for repeated closing and keep the switching overvoltage well below permitted levels.

15.4.4 Vacuum circuit-breakers

The vacuum interruption process has been developed for voltages up to 36 kV per break. It would seem to be the ideal process for high-performance circuit-breakers. The interrupting chamber (*Figure 15.16*) is a sealed porcelain or vitrified glass vessel, and maintenance of the contacts with a gap of about 1 cm

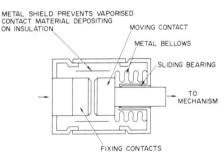

Figure 15.16 Vacuum circuit-breaker

is possible. The life, governed by contact erosion, is expected to be about 20 years, provided that the vacuum of about 10^{-5}–10^{-6} Torr is maintained. Commercial development has been made possible by the design of suitable seals, and by materials that contain no gas that might be occluded and released into the chamber during arcing.

Current chopping has been contained by developments in contact material to a level of about 5 A. Contact welding has also been reduced, so that currents up to 40 kA can be interrupted, using magnetic rotation of the arc by cutting slots in the contacts to reduce their local heating.

The contact material used is a compromise between the following requirements:

(1) Good mechanical strength and electrical conductivity.

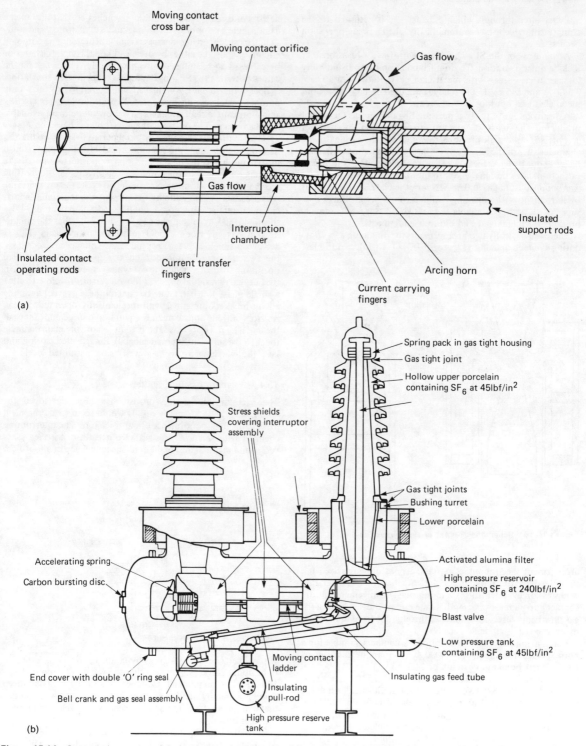

Figure 15.14 Contact arrangement of dual pressure sulphur hexafluoride circuit-breaker. (b) General arrangement of dual pressure sulphur hexafluoride circuit-breaker

The original construction of the interrupter itself was similar to that of an air-blast unit, but the gas was not exhausted to atmosphere on account of its cost. A closed-circuit arrangement was therefore used, as shown in *Figure 15.14*. The equipment comprised a compressor; a storage vessel holding gas at about 1.65 MN/m^2 (240 lbf/in^2); a blast valve which admits gas to the interrupting chamber (the latter being normally held at about 310 kN/m^2 (45 lbf/in^2) for insulation purposes); and an activated

remove the ionised gas. No further arc control device is used, although care and skill are required in designing the contacts and air passages to ensure adequate air flow. Air-blast circuit-breakers are built for voltages up to the highest so far required (765 kV).

In earlier (and still extant) high-voltage designs the contacts are held closed by means of a spring as shown in *Figure 15.12*.

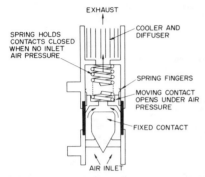

Figure 15.12 Air-blast interrupting head (spring closed)

Admission of air to the interrupting head forces the contacts open and then, as soon as the arc is extinguished, opens the series disconnecting switch. Making the circuit is effected by closing the disconnecting switch by compressed air at high speed to avoid a damaging arc; the contacts must be substantial, as the switch may have to close on a short circuit.

A modular construction with 40–100 kV per break and up to 12 breaks in series is necessary for higher voltages. Earlier designs for up to 220 kV had interrupting heads stacked vertically, but although simple and requiring minimum insulation path to earth, this construction was structurally weak, and had differing air-passage lengths for the several breaks.

Continuously pressurised type Recent designs dispense with series isolators by arranging for the moving contacts to be held open after a trip, and by sealing the exhaust valve. This ensures that when the contacts are open, the breaker is pressurised up to the exhaust valve to maintain full electric strength. Air consumption is reduced, and external moving parts eliminated. Such breakers are suitable for ratings up to 40 GVA at 765 kV. A 132 kV 6-break version is shown in *Figure 15.13*; the overall

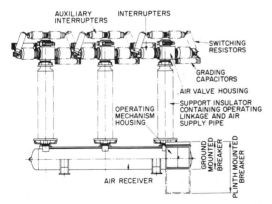

Figure 15.13 One phase of continuously pressurised air-blast circuit-breaker

height from the ground is about 5.5 m and the horizontal distance between supports about 2.5 m. The horizontal air receiver for each phase is used as a base for three vertical insulators carrying air pipes and resin-bonded glass operating rods. At the top of each column are two interrupters.

The breaker is operated by a pneumatic closing mechanism with spring trip, similar in principle to that used in oil circuit-breakers. When a trip is initiated, this mechanism applies tension to the bonded glass rods, which operate servo-valves in the blast-heads, causing the exhaust valves to open. This produces a pressure differential on pistons which retract the moving contacts, and these are arranged to reach a high speed before the arc contacts part. High-pressure air exhausts through the nozzle to atmosphere extinguishing the arc. The contact piston also operates the servomechanism, which causes the exhaust valves to close and air pressure to build up again. An over-centre spring biasing device holds the moving contacts in the open or closed positions if air pressure is removed. The breaker is closed by a reversal of the pressure differential on the moving contact pistons, and the valves for this are also operated by tension rods.

An advantage of this arrangement is that the full operating pressure is available at the contacts, and the speed of operation is very high. The breaker incorporates resistors for control of the rate of rise of transient recovery voltage with separate interrupters. When the breaker is tripped, the resistor interrupter closes first. When the exhaust valve resets after the arc is extinguished, the resistor interrupter reopens automatically, breaking the resistor current. Capacitance control of voltage distribution is also applied. For voltages of 500 kV upwards, switching-in overvoltages are controlled by fitting closing resistors to each phase.

The action of the air-blast circuit-breaker, unlike that of the arc controlled oil circuit-breaker, is independent of the magnitude of the current being interrupted: the powerful blast necessary for interrupting high currents may, at low currents, cause chopping, with the consequent high transient voltages. A resistor of the order of 10–20 kΩ is, as noted above, inserted into the circuit before final interruption. This resistor is additional to the voltage-equalising capacitor units.

The air released during an opening operation is typically 7 m^3 in a time of 0.2 s for a 132 kV 3.5 GVA circuit-breaker, and 50 m^3 in 0.1 s for a 400 kV 35 GVA unit. The exhaust to atmosphere gives rise to noise levels of 120 dB or more at a distance of 20 m, and silencers must be incorporated to reduce the level to 80 dB or less.

A substation with air-blast breakers must have a reliable, comprehensive compressed air supply. The standard arrangement in British 400 kV substations is typical: duplicate 100 kW four-stage compressors are installed, each giving 4.5 m^3/min at a pressure of 20 MN/m^2 (3000 lbf/in^2). The air is filtered by activated carbon, dried and passed to two storage vessels. Reducing valves lower the pressure and feed the air into a double ring main from which supplies to circuit-breakers are tapped. The air consumption per make–break operation is 45 m^3 and the storage can give ten operations.

15.4.3 Sulphur hexafluoride circuit-breakers

An alternative to air as an interrupting medium is sulphur hexafluoride (SF$_6$), a colourless non-toxic gas with a density five times that of air, a high thermal conductivity and a high electronegative attraction for electrons. Its insulating properties at 206 kN/m^2 (30 lbf/in^2) are equivalent to those of switchgear oil and its arc interrupting characteristics at 1379 kN/m^2 (200 lbf/in^2) are equivalent to those of air at 4826 kN/m^2 (700 lb/in^2). A single break can recover its dielectric strength at about 10 kV/µs.

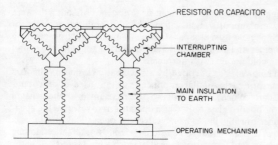

Figure 15.9 Modular 'live tank' circuit-breaker

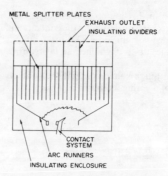

Figure 15.10 Metal plate arc chute

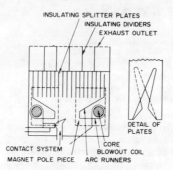

Figure 15.11 Insulated plate arc chute

avoiding the virtual impossibility of testing complete units of high rating in test plant.

An essential feature of modular construction is to ensure that each break of the series carries its proper share of the duty. While care must be taken to ensure that all breaks occur simultaneously, it is of prime importance that the recovery voltage components be divided equally across the breaks during the process of interruption. The natural division depends upon stray capacitance (500–1000 pF), to earth and between contacts, and results in a very uneven voltage distribution. This is corrected by connecting resistors or capacitors in parallel with the interrupting heads (*Figure 15.9*). These swamp the effect of stray capacitance.

Although the modular construction is essential for high ratings, it has, compared with the bulk-oil arrangement, more complicated operating mechanisms and a greater number of paths to earth over insulator surfaces. Again current transformers must be mounted as separate units.

15.4.2 Air circuit-breakers

Circuit-breakers using air as the interrupting medium avoid the fire hazard associated with oil. *Air-break* units use atmospheric air, while *air-blast* units employ compressed air.

15.4.2.1 Air-break circuit-breakers

Air-break circuit-breakers use high-resistance interruption, the arc being lengthened and presented to a large cooling surface within an arc chute.

For the control of medium-voltage circuits, air-break circuit-breakers are now universal, from the miniature and earth leakage circuit-breakers for domestic use, to the large moulded-case or metal clad circuit-breakers for the higher currents and breaking capacities. In some cases, by using the expulsion due to the short-circuit current on the contacts, some cutoff of the peak current can be achieved at the higher values with very short opening times. The circuit-breaker is tripped open at lower overload currents by magnetic or thermal means, which take much longer.

Air-break circuit-breakers are used at voltages up to 15 kV. Movement of the arc into the chute results from natural convection of the hot gas; from magnetic action (by the arc current itself or by a magnetic blowout coil carrying the arc current); or from a mechanically produced puff of air to assist the extinction of small overcurrents.

Arc chutes In the *metal plate* type (*Figure 15.10*) the arc is driven on to a series of metal plates, which split it into a number of short arcs each with its characteristic anode and cathode volt drops, so reducing the effective arc column voltage and raising the effective resistance.

The arc in the *insulated plate* type (*Figure 15.11*) is forced into a labyrinth of insulator plates of temperature-resistant ceramic, thus increasing its length and resistance. In both cases it is usual to mount exhaust baffles at the top of the arc chute to prevent external flashover.

Hard-gas unit Here the arc space is surrounded by a material that is volatilised by the arc heat, the resulting gas (mainly hydrogen) being forced across the arc path by the pressure set up. Such circuit-breakers are not suitable for breaking inductive circuits, because the rate of rise of electric strength is low; but the principle is sometimes applied to load-breaking switches. Disadvantages are the possibility of flashover due to atmospheric pollution and the presence of ionised gas following an arc.

D.C. circuit-breakers Breakers for d.c. interruption are invariably of the air-break type. They are made in many forms fitted with automatic tripping for overcurrent, reverse current and undervoltage. Most are of very simple construction for small currents at up to 600 V; but more complex units with strong magnetic blowout action and high-speed tripping mechanisms are used, primarily for traction service, with continuous ratings of several kiloamperes at 1.5–3 kV. For higher voltages circuit-breakers become impracticable, and none is commercially available for use on high-voltage d.c. transmission schemes.

All air-break circuit-breakers give easy access to contacts, facilitating maintenance. Transients are damped by the high resistance of the arc and its gradual extinction, eliminating current chopping. The effects of failure are likely to be less serious than with an oil circuit-breaker.

15.4.2.2 Air-blast circuit-breakers

The low-resistance arc principle is applied, with the arc length minimised to a few centimetres and interruption at a current zero. A blast of air at 2.5–7.5 MN/m² (350–1000 lbf/in²) is directed across the arc path either axially or by a cross-blast, to cool and

15.3.1.3 Metal clad

All live parts are completely enclosed by a substantial metal casing. Although normally used indoors, the structure can readily be made weatherproof and used outdoors. The circuit-breaker itself is usually arranged for withdrawal, horizontally or vertically, with plug-and-socket connections between the busbars and outgoing circuits. Sometimes the circuit-breaker is fixed, with disconnectors between it and the busbars.

15.3.1.4 Outdoor

Each of the main items of equipment is mounted separately in an open air substation, relays and instruments being housed in a building near the site. This construction can be used for any unit size from that for a small rural installation to the largest required.

15.4 Circuit-breakers

The forms of circuit-breaker in common use can be grouped according to construction and to the mode of arc interruption:

(1) Oil
    (a) plain-break
    (b) arc controlled: (i) bulk-oil; (ii) low-oil-content
(2) Air
    (a) plain-break
    (b) air-blast: (i) with isolator; (ii) continuously pressurised
(3) Sulphur hexafluoride (SF_6)
(4) Vacuum

15.4.1 Oil circuit-breakers

In about 1890 it was found that immersion of a.c. circuit-breaker contacts in oil was effective in arc extinction. The arc energy decomposes the hydrocarbon oil around the arc into constituent gases (e.g. 80% hydrogen, 22% acetylene, 5% methane, 3% ethylene), to an amount of about 60 cm^3/kJ. The arc thus takes place in a bubble of gas surrounded by insulating oil.

As an arc extinguishing medium, oil has the following advantages: (1) absorption of arc energy by decomposition; (2) arc cooling, chiefly by hydrogen gas, which has a high diffusion rate and good heat absorption in changing from the diatomic to the monatomic state; (3) cooling by oil around the gaseous arc path; (4) the ability of cold oil of high electric strength to flow into the arc space after a current zero; (5) the natural insulation value of the oil, enabling clearances to be minimised.

The chief disadvantage of oil is its flammability, the possible formation of an explosive air–gas mixture, and the maintenance (changing and purifying) necessary to keep it in good condition.

15.4.1.1 Plain break circuit-breakers

The contacts, two in series, are contained in an earthed metal weatherproof tank containing oil.

The rather low post-zero resistance of the arc space tends to damp the transient recovery voltage and reduce its rate of rise, so that the performance of this type of circuit-breaker has often been more satisfactory than might have been expected. Although simple and inexpensive, it has arc interruption times of several periods and is somewhat inconsistent in operation. The plain-break construction for medium voltages has been superseded by air-break circuit-breakers, and for high voltages by the fitting of arc control devices.

15.4.1.2 Arc control devices

Great improvement in interruption capacity and in operating consistency is achieved by enclosing each contact in an arc

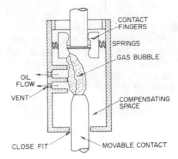

Figure 15.8 Arc-control 'pot'

control 'pot' (*Figure 15.8*). A high pressure is set up by the gas formed by the arc inside the pot and this forces oil out across the arc path, thus causing rapid extinction. Careful design of the pot is required, since the pressure generated depends on the magnitude of the current being interrupted; a compromise must be effected between ensuring sufficient pressure to extinguish the arc at low currents and too great a pressure (which could burst the pot) at high currents.

All oil circuit-breakers employ some form of arc control device, and various forms have been developed. Many designs use only a single break up to 132 kV, while others use several breaks in series, depending upon the short-circuit MVA rating and voltage.

15.4.1.3 Bulk-oil circuit-breakers

The 'earthed' or 'dead tank' construction is widely used at the higher ratings. Above 66 kV, separate tanks are normally used for each phase, and shaped around the interrupter structure to minimise the bulk of oil required. Even so, up to 70 m^3 of oil may be needed for the three tanks of a 275 kV circuit-breaker. A significant advantage of the bulk-oil construction is that current transformers can readily be accommodated in the terminal bushings; also, the exposed insulation comprises only two bushings per phase.

Although the moving parts within the tank are simple with the conventional two-break arrangement, economy can be effected by use of a single-break form. Multiple breaks (e.g. six in series per phase for 275 kV) have been employed, but structural complication imposed by short-circuit forces is considerable at higher ratings. The practical economic limit for bulk-oil circuit-breakers at 275 kV is therefore about 15 GVA.

15.4.1.4 Low-oil-content circuit-breakers

In the bulk-oil construction the oil is both an arc extinguisher and a main insulation. To reduce the oil content to that for the arc alone (about 10% of the bulk-oil quantity), the low-oil-content construction has been developed. The arc control device is enclosed in a 'live' insulated housing supported by a porcelain insulator. Single-break units of this type are in use at 132 kV. For higher voltages, multiple breaks in series can be obtained using the 'modular' construction, giving ratings up to 40 GVA at 400 kV or 500 kV, with up to ten breaks.

Current transformers must be mounted as separate units.

15.4.1.5 Modular construction

Connection of two or more interrupting heads in series, in the manner shown for a four-break unit in *Figure 15.9*, yields two main advantages. First, a standard unit can be employed, enough units (up to twelve) being series connected as appropriate to the system voltage; second, each unit can be separately tested,

15.2.3 Discrimination

Two or more fuses are often in series between the power source and a fault; such fuses are referred to as *major* and *minor* fuses, the latter being more remote from the source. Discrimination is correctly achieved when the major fuse is unaffected by interruption of the minor fuse under maximum fault conditions. The total I^2t passed by the minor fuse must thus be less than the pre-arcing I^2t of the major fuse. A ratio of 2/1 in the ratings of successive fuses usually meets this requirement.

15.2.4 Fuse applications

15.2.4.1 Distribution

Fuses are in common use on the high-voltage side of 11 kV and 6.6 kV distribution transformers. In rural areas, however, they are liable to operate on transient overhead line faults, so that auto-reclose circuit-breakers and sectionalisers may be preferred.

On the low-voltage (415 V) side, HRC fuses are widely applied in distribution pillars to protect the transformers. The fuses discriminate correctly with consumers' fuses so that no other protection is required.

Switch-fuse units for distribution are common in circuits at voltages of 6.6 kV or 11 kV. The fuses are in series with oil immersed switches and so arranged that the blowing of a single fuse operates a striker pin to initiate automatic tripping of the switch.

The fuses may be mounted in either air or oil (the air mounted fuse having a normal current rating slightly lower than when in oil), with ratings in oil from 5 to 140 A, and breaking capacities at 6.6 kV and 11 kV of 250 MVA.

15.2.4.2 Motors

A common fuse application is in the short-circuit protection of motors. The normal control gear (usually contactors and overcurrent relays) provides overcurrent protection, but does not have short-circuit interruption capability. Care is needed in fixing the change-over current between the two devices to prevent fuse operation during starting but at the same time prevent short-circuit current damage to contactors and, particularly, thermal overload devices.

For medium-voltage motor circuits the wiring regulations permit fuses to have a normal current rating twice that of the cable, on the basis that the motor starter overload protection will protect the cable. Hence, for this application the fusing factor is not a relevant parameter. For the larger motors having direct on-line starting, special motor-circuit fuse links are available.

To avoid single-phasing of induction motors, which might cause serious overheating, special overload relays are available which open the associated contactor starter.

For high-voltage motors (3.6–7.2 kV) the fuses must have the ability to withstand without deterioration the repeated current surges associated with motor starting. The selection of fuse rating depends on the run-up times, the frequency of starts per hour and the starting current. At 3.6 kV the current ratings are from 100 A to 630 A, with a breaking capacity of 250 MVA; and at 7.2 kV the current ratings are from 50 A to 630 A, with a breaking capacity of 500 MVA. All ratings can be supplied with striker pins for tripping associated starters to prevent single-phasing.

15.2.4.3 Voltage transformers

The fuse in the high-voltage side of a voltage transformer is to protect the system in the event of a voltage transformer fault. The current rating is low (1–3 A) but the interrupting capacity must be high (e.g. up to 1000 MVA at 11 kV), and the fuse must not blow on magnetising current inrush. Modern practice is to use a silver-wire fuse, which gives a high pre-arc I^2t and thus reduces the tendency to operate on switching transients. Such fuses are suitable for mounting in air or oil.

15.2.4.4 Solid state devices

The high power/size of diodes and thyristors severely limits their thermal and voltage withstand levels. The fuse is the only device that can give adequate protection, and the rating of the solid state equipments may indeed be determined by the protection provided. Specially designed fuses are required, with I^2t values about one-tenth of those for standard industrial fuse-gear, and with particular attention to the limitation of arc voltage.

To obtain these characteristics, several fuses are often connected in parallel with special connections to ensure satisfactory division of the total current. A wide range of a.c./d.c. voltages are available from 20 to 200 V a.c./200 to 1500 V d.c. and currents up to 700 A. Where indication of a blown fuse is required, this can be obtained from a trip indicator fuse mounted in parallel with the main fuses and operating a small microswitch.

15.2.4.5 D.C. circuits

The limited demand has made it uneconomic to design fuses specially for d.c. circuits. The interrupting duty is more onerous than for a.c. for overload rather than for short-circuit conditions; but satisfactory performance can generally be achieved by an a.c. fuse with, say, 40 kA short-circuit rating rather than the 80 kA short-circuit rating which would be used in a comparable a.c. circuit.

15.3 Switchgear

The term 'switchgear' includes not only circuit-breakers, but also auxiliary equipment such as current and voltage transformers, switches, disconnectors, motor starters, operating mechanisms and comparable devices.

15.3.1 Mounting

Four general methods of mounting switchgear are in common use. Three of them are widely used for industrial switchgear.

15.3.1.1 Cubicle

The circuit-breaker and its associated equipment are mounted in a cubicle, usually of sheet steel, a number of cubicles being arranged side by side with busbars running along their upper chambers. Access is through front- or back-opening doors suitably interlocked to prevent opening when the circuits are alive. Operating handles, indicating lamps, relays, measuring instruments and auxiliary circuit fuses are mounted on the front of the cubicle. Such units are generally suitable for medium-voltage switchgear.

15.3.1.2 Truck

The truck is a modified form of cubicle mounting in which all the equipment is mounted on a 'truck' that can be withdrawn from the cubicle for maintenance. Connections to the busbars and feeder circuits are made by plugs and sockets, protective shields automatically falling over the sockets as the truck is moved. Interlocks prevent movement of the truck unless the circuit-breaker is open.

15.2.1.2 I^2t value

The I^2t value (in A^2 s or J/Ω) is obtained from the r.m.s. current I during the total interruption time t, and is therefore related to the shaded area in *Figure 15.4(a)*. Values of I^2t for a range of typical fuses are given in *Figure 15.5(b)*. These represent the maximum value of I^2t that the fuse will permit to pass into the circuit being protected.

15.2.1.3 Current/time characteristic

A typical characteristic is shown in *Figure 15.6*. The time in a particular case depends upon circuit conditions, as the arcing time is a function of the energy stored in the circuit at the moment of interruption, all of which has to be dissipated in the arc. For the 500 A fuse with the characteristics illustrated, melting will not occur below 750 A, giving a *fusing factor* (minimum fusing current in terms of rated current) of 1.5. This ensures that the fuse temperature at a rated current does not cause gradual deterioration.

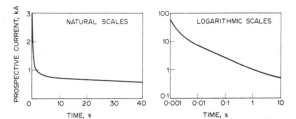

Figure 15.6 Current/time characteristics for a 500 A fuse

15.2.2 Types of fuse

15.2.2.1 Semi-enclosed rewirable fuse

Semi-enclosed fuses are suitable for use on d.c. or a.c. circuits having a maximum voltage to earth not exceeding 250 V. Normal current ratings up to 100 A were available but, in general, these fuses are now used only up to 30 A, with rated breaking currents of 1000 A for the 5 A size and up to 4000 A for the 30 A size. The fusing factors range up to 2.0 and are therefore higher than for the corresponding cartridge fuses.

Although inexpensive and easily and cheaply rewirable—too often with oversize wire, thus impairing protection—considerations of safety should preclude itw use except for minor subcircuits in domestic and commercial installations, fed from small transformers and with lengths of cable between the transformer and the fuse.

15.2.2.2 Low-voltage high-rupturing-capacity (HRC) fuse

The HRC fuse is the common type for general power system practice and has ratings up to 1600 A and up to 660 V. The fuse comprises a closed cylindrical ceramic body, the fusible link and appropriate terminations. The fusible link is of silver wire or strip giving a low resistance. To secure modification of the conventional current/time characteristic, an insert of longer heating time constant can be introduced. Alternatively, a fusible metal may be added which, on melting, penetrates the crystal structure of the underlying silver (M effect), increasing the resistance. As this phenomenon is time-dependent, it enables the fuse to withstand short heavy-current pulses.

The body is filled with silica sand or granulated quartz having a high heat-absorbing capacity and through which the fusible link passes. An indicator to show that the fuse has operated may be provided; this may comprise a fine wire in parallel with the fusible element and passing through a mild explosive charge in a visible pocket at the side of the body. The terminal connections must provide a low contact resistance and carry away the heat developed on rated current. The fuse carrier must be designed and located with a view to cool running under all normal conditions.

The characteristics in *Figures 15.5* and *15.6* refer to fuses of the HRC type.

15.2.2.3 High-voltage current-limiting fuse

The high-voltage current-limiting fuse is similar in operation to the low-voltage HRC fuse, and its construction is also similar, except that a much greater length of fusible element is required in order to build up an arc voltage sufficiently large to dissipate the energy of the circuit. The fusible element is wound helically on a porcelain former within the powder-filled ceramic container. A typical fuse, of 85 A rating and 250 MVA breaking capacity at 11 kV, may have a container about 0.35 m long and of 60 mm diameter: it can be air mounted or oil immersed.

15.2.2.4 High-voltage non-current-limiting fuse

For higher voltages, up to 33 kV, fuses with an action similar to that of a circuit-breaker are available, the arc interruption occurring at or near a current zero.

The *expulsion fuse* is contained in a cylindrical vessel of insulating material open at the lower end; arc quenching is assisted by a jet of vapour originating from the arc and possibly also from a volatile solid, e.g. boric acid, which vaporises because of the heat. Such fuses are effective up to 33 kV but are only suitable for outdoor use, owing to the fairly violent discharge that takes place.

The *liquid quenched fuse* is enclosed in a glass vessel (*Figure 15.7*); the vessel contains the arc quenching liquid, e.g. carbon tetrachloride, and the length of break is increased by the action of the spring. Such fuses are often re-wirable.

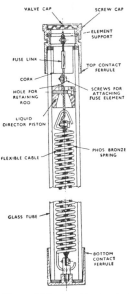

Figure 15.7 Liquid-quenched fuse

phase angle of the short-circuit current approaches 90° lagging, so that the normal-frequency recovery voltage is at a peak at the instant of a current zero. The sudden appearance of this voltage across the breaker contacts (and therefore across C) gives rise to a high-frequency transient of frequency $1/[2\pi\sqrt{(LC)}]$ in the LC circuit: this transient recovery voltage is superposed on the normal-frequency component as shown in *Figure 15.3(b)*. The consequent rate of voltage rise across the circuit-breaker contacts may be of the order 1–6 MV/μs, and so comparable with the rate of growth of the electric strength of the gap. Low stray capacitance gives a high rate, and vice versa.

If the circuit-breaker interrupts the current before a current zero (a condition of current chopping), additional overvoltage components may be induced because of the circuit inductance.

In making an inductive circuit, as in closing on to a fault, the peak current may be two or more times the symmetrical peak short-circuit current, depending upon the power factor of the circuit and the point on the voltage wave at which the circuit is closed, thus causing severe mechanical forces of electromagnetic origin on switchgear, busbars and generator or transformer windings.

Short-line fault   If inductance and capacitance are present on both the line and the supply sides of the circuit-breaker in *Figure 15.3(a)*, two LC combinations are concerned, imposing a double-frequency transient recovery voltage on the breaker. A particularly severe condition arises when a fault at a short distance (up to about 6 km) from the circuit-breaker is interrupted, as successive voltage reflections may lead to very high initial rates of rise.

15.1.2.3 Capacitive condition

With a 90° leading phase angle and interruption at a current zero the network capacitance remains charged to the peak system voltage after the breaker has opened, leaving instantaneously no voltage across the contacts and, apparently, making interruption very easy. But as the source voltage swings in the following half-period to reversed peak, the voltage between contacts reaches double peak value. This may initiate a restrike and result in an overshoot, giving a transient voltage to earth of up to three times system peak. Further extinction and restrike may occasionally cause even higher voltages.

The making of a capacitive circuit can cause a high, but very short, transient current as the system capacitance takes an impulsive charge.

15.1.2.4 Asynchronous condition

A circuit-breaker may be called upon to open a circuit connecting two parts of the system which have fallen out of step. The normal-frequency recovery voltage then reaches twice rated value, and interruption of a heavy fault current may be very difficult.

15.2 Fusegear

The fusible link was the earliest form of cutout. Modern developments have resulted in fuses for power circuits, the main feature being the very short operating time (0.3–3 ms) under heavy fault currents. These are suitable for d.c. voltages up to 1500 V and for a.c. voltages up to 1000 V. Fuse links are also used for voltages up to 36 kV a.c. For the lower voltages the fuse-links are usually mounted in insulating holders and may be assembled in fuse-boards or fuse-switches. The rated breaking capacities are 80 kA at a rated voltage of, or exceeding, 415 V a.c., and/or 40 kA at a rated voltage of up to 500 V d.c.

15.2.1 Fuse action

With currents of little more than the fuse rating the fusible element melts slowly, starting usually midway between its terminals. The arc which forms at the break elongates until the system voltage can no longer maintain it.

With high currents heating is rapid and uniform along the length of the element. Melting occurs at a number of points, forming globules with arcs between them; although these quickly merge into a single arc, they introduce a significant resistance into the circuit and reduce the current. The arc voltage across the fuse during the interruption process may, however, exceed the normal system voltage, owing to the rapidly falling current in the circuit inductance. Maximum values of 1000–3500 V, depending on the normal current rating of the fuse-link, are specified.

15.2.1.1 Cutoff value

Typical oscillograms of fuse action are given in *Figure 15.4*. *Figure 15.4(a)* shows that the current is cut off before it reaches

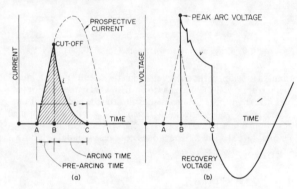

Figure 15.4   Fuse current and voltage transients

the full prospective value determined by the parameters of the faulted circuit. The transient arc voltage across the fuse at cutoff and during the arcing period is shown in *Figure 15.4(b)*.

The relation between the cutoff and prospective values of current for a typical range of fuses of high rupturing capacity is shown in *Figure 15.5(a)*. These cutoff values limit the electro-magnetic stresses in the equipment protected as well as easing the duty of the fuse itself. Typical values for a 500 A fuse interrupting a 415 V circuit with a prospective current of 80 kA r.m.s. are: peak asymmetrical current, 180 kA; cutoff current, 52 kA; maximum arc voltage, 2500 V.

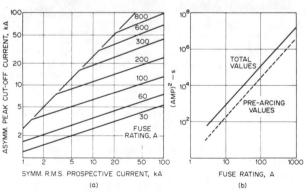

Figure 15.5   Fuse characteristics

of bringing it more nearly into phase with the voltage, so that when the current falls to zero at the end of the half-cycle, the voltage is also near zero and there is little tendency for the arc to restrike at the beginning of the next half-cycle.

15.1.1.2 Low-resistance (zero point) interruption

The arc resistance is minimised by keeping its length short, the aim then being to prevent its restriking after going out at a current zero. A major factor in the behaviour is therefore the state of the arc space at, and immediately following, the current zero; at the moment when the current ceases, the arc space contains ions and electrons that have not yet recombined to form neutral molecules, although this process is occurring naturally and very rapidly.

There are two main theories of zero point interruption—the *recovery rate theory* and the *energy balance theory*.

In the recovery rate theory the rate at which the ions and electrons in the arc space recombine to form, or are replaced by, neutral molecules (i.e. the rate at which the gap recovers its dielectric strength) is compared with the rate at which the voltage across the contacts rises. If the voltage rises more rapidly than the dielectric strength, the space breaks down and the arc persists until the next current zero.

The energy balance theory involves the resistance of the contact space following the current zero (the post-zero resistance) and relates the power put into this resistance from the circuit with the power dissipated from the contact space by convection or other means. At current zero the power delivered from the circuit to the contact space is clearly zero. As the voltage across the contact space rises, a current flows in it, causing power generation; however, the action of the circuit-breaker is to cause the post-zero resistance to increase towards infinity, so that the current and power again become zero. Between these two points the power generated in the contact space becomes a maximum, and if it is greater than the rate at which energy can be dissipated from the contact space, the temperature rises and an arc will form again.

Much research work is being done on conditions in the contact space at, and about, the current zero. It would seem that the energy balance theory is more appropriate for explaining the behaviour of circuit-breakers with low post-zero resistance, while the recovery rate theory is preferable for those in which the post-zero resistance is high.

The rapid increase of dielectric strength, or post-zero resistance, that is necessary can be obtained in the following ways:

Lengthening of the gap The dielectric strength, or the post-zero resistance, is proportional to the length of the gap, so that lengthening, by rapid opening of the contacts, is an obvious procedure. However, the permissible gap length is limited by other considerations, e.g. arc energy and the possibility of transient voltages due to current chopping.

Cooling Natural recombination of ions and electrons takes place more rapidly if they are cool. Cooling by conduction to adjacent parts, e.g. baffles, or by the use of a gas such as hydrogen which has a high diffusion and heat absorption rate is therefore effective.

Blast effect Cool non-ionised gas or liquid can be forced into the contact space, thereby cooling the ions and electrons and also sweeping them away.

15.1.1.3 Vacuum arc interruption

The use of a high vacuum as an interrupting medium has the advantage of a high dielectric strength (about 700 kV/cm) and very fast radial diffusion of ions and electrons from the arc. The degree of vacuum necessary to set up a 'vacuum arc' must be such that the mean free path of any particles is large compared with the discharge path; pressures must thus be lower than about 0.01 N/m^2 (or 10^{-4} mm Hg), giving a mean free path of about 50 cm. The arc ions and electrons result from electron emission and vaporisation of the contacts as they part. Once away from the cathode spots, collisions are rare and the arc voltage is only about 0.05 V/cm; however, the voltage drop at the cathode may be 15–20 V, about half of which is used in accelerating electrons which subsequently dissipate this energy at the anode.

The vacuum circuit-breaker can interrupt only small direct currents; its ability to deal with much greater alternating currents depends upon the prevention of cathode spots occurring on the negative contact immediately after a current zero. The cathode spots disappear at (or immediately before) current zero, leaving ionised vapour in the contact space which rapidly diffuses, giving a rate of increase of electric strength of up to about $10 \text{ kV/}\mu\text{s}$.

15.1.2 Circuit conditions

The voltage across the open contacts immediately following a current zero during the interruption process has a decisive effect on the behaviour of the circuit-breaker. This voltage, which is characteristic of the network to which the breaker is connected, has normally two components: (1) the *normal-frequency recovery voltage* of the network, and (2) the *transient recovery voltage* (or 'restriking voltage'), a high-frequency transient superimposed on (1). The transient voltage has a high rate of rise across the contacts, tending to restrike the arc.

For a 1-phase network the normal-frequency recovery voltage is the e.m.f. of the effective source, which may be somewhat less than the prefault value, owing to the demagnetising effect on the generators concerned. On earthed 3-phase systems the recovery voltage is the phase voltage V_p for earth faults, $1\frac{1}{2} V_p$ for the first phase to clear of a 3-phase fault, and $\sqrt{3} V_p$ for a phase-to-phase fault.

The conditions of interruption are closely dependent upon the effective network parameters, which can be considered as approximating to resistive or inductive or capacitive conditions.

15.1.2.1 Resistive condition

The resistive condition includes load circuits having normally high power factors. The recovery voltage zero coincides with the current zero and there is no transient. The voltage rise follows the normal-frequency sine wave and there is little tendency to restrike. On making such a circuit the rate of rise of current may be high, but there is no transient or overshoot.

15.1.2.2 Inductive condition

The circuit is highly inductive (*Figure 15.3a*), but there is inevitable stray capacitance C due to cables, bushings, etc. The

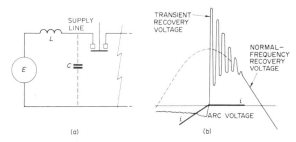

Figure 15.3 Interruption of inductive circuit. (a) Basic circuit; (b) recovery voltage components

The control of a power system involves the use of switchgear for making and breaking circuits, and protective equipment for detecting the presence of fault currents and overvoltages and for isolating the faulty plant with the least disturbance to the system. The methods and equipment used for this are being constantly improved in performance, and the British Standards Institution (BSI) and the International Electrotechnical Commission (IEC) publications, with which they must comply, are also the subject of constant revision, to reflect the latest practice.

British Standards, hitherto mainly domestic, are now based on those of the IEC with the object of harmonising standards throughout the world and facilitating international trade. Many characteristics of power station equipment and performances are specified in BSI and IEC publications, and are an essential part of understanding the design and behaviour of many system components. The more important publications are listed under 'References'. Complete sets of the latest BS are kept for reference purposes at most of the major libraries throughout the UK.

In addition to the relevant standards, there are two bi-annual conferences dealing with switchgear and protection: the Conférence Internationale de Grands Réseaux Electriques (CIGRE), held in Paris every even year, issues a monthly publication *Electra*; and Congrès International des Réseaux Electriques de Distribution (CIRED), which meets every odd year. CIRED covers switchgear up to 36 kV, while CIGRE covers generation and high-voltage transmission problems. (See also Chapter 36.)

15.1 Arc mechanism

When a current-carrying circuit is broken, an arc is inevitably formed between the opening contacts, prolonging the interruption process for a time in the range 10–100 ms or more. The arc participates in dissipating the stored circuit energy, and if it did not occur naturally, some equivalent phenomenon would have to be invoked.

The arc consists of a column of ionised gas, i.e. gas in which the molecules have lost one or more of their electrons, leaving positive ions. The electrons are attracted towards the positive contact and move towards it very rapidly; the ions are attracted towards the negative contact but, as they comprise almost the whole mass of the molecule, move relatively slowly. The electron movement thus constitutes the current flow.

Once the arc has started and a voltage gradient is maintained along it, the rapidly moving electrons collide with molecules of the gas and dislodge further electrons by collision. More and more current can thus flow and the resistance of the arc drops with increasing current so that the voltage across it drops only slightly, as shown in *Figure 15.1(a)*. If external conditions in the circuit cause the current to drop to a low value (5–10 A) such that there are not sufficient electrons to maintain the ionisation process, the arc changes to a glow discharge and dies out. Ionisation is accompanied by the emission of light and heat, the power p dissipated at any instant being given by the product of the instantaneous voltage v and current i, i.e. as shown in *Figure 15.1(b)* for an arc carrying a sinusoidal current; the energy dissipated is represented by the shaded area.

Three regions of the arc are important:

Cathode Here there is a voltage drop associated with electron emission from the cathode surface. The drop may be 20–50 V.

Arc column The column of ionised gas has, for a given current and ambient physical conditions, a well-defined cross-section. The temperature may reach 6000 K, and the voltage gradient is approximately constant. Increase of temperature reduces the gradient by increasing the kinetic energy of the electrons, so facilitating ionisation of the gas in the column. However, the gradient increases with gas pressure, by reason of the reduction in the mean-free path of the electrons. If the arc is constrained in cross-section, the gradient is again increased.

Anode At the anode surface there is a voltage drop of 10–20 V.

15.1.1 Arc extinction

Three general methods of arc extinction are available for use in circuit-breakers. In (1) *high-resistance interruption* the arc is controlled in such a way that its resistance is caused to increase rapidly, thus reducing the current until it falls to a value that is insufficient to maintain the ionisation process. In (2) *low-resistance interruption* the arc resistance is kept low, in order to limit the arc energy, and use is made of a natural or artificial current zero when the arc extinguishes and is then prevented from restriking. In (3) *vacuum interruption* the arc takes place in a vacuous chamber, and the current carriers are ions and electrons emitted from the contacts. The first method is normally employed for d.c. and air-break a.c. cases; the second, for a.c. oil, air-blast and sulphur hexafluoride (SF_6) circuit-breakers. Commercial development of the vacuum method is recent, but circuit-breakers of this type are available for a.c. at working voltages up to 132 kV using several vacuum bottles in series per phase.

15.1.1.1 High-resistance interruption

The arc resistance can be raised by (a) lengthening, (b) cooling and (c) splitting. The effect of increase of arc resistance is shown by oscillograms in *Figure 15.2*. In the d.c. case (*Figure 15.2a*) the

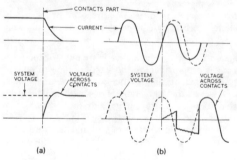

Figure 15.2 High resistance interruption. (a) D.c. circuit; (b) a.c. circuit

increasing resistance reduces the current until it falls to such a low value, e.g. about 10 A, that the arc cannot maintain itself; the voltage across the contacts simultaneously rises to normal circuit voltage, the high resistance preventing any inductive rise to values appreciably higher than this. In the a.c. case (*Figure 15.2b*) the current to be interrupted will, if it is a short-circuit current, be lagging the voltage by nearly 90°; the arc resistance then has the effect not only of reducing the magnitude of the current, but also

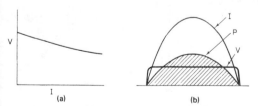

Figure 15.1 Arc characteristics. (a) Voltage-current characteristics; (b) arc energy in a.c. circuit

15 Switchgear and Protection

J S Cliff, MBE, CEng, FIEE
Engineering Consultant

Contents

15.1 Arc mechanism 15/3
    15.1.1 Arc extinction 15/3
    15.1.2 Circuit conditions 15/4

15.2 Fusegear 15/5
    15.2.1 Fuse action 15/5
    15.2.2 Types of fuse 15/6
    15.2.3 Discrimination 15/7
    15.2.4 Fuse applications 15/7

15.3 Switchgear 15/7
    15.3.1 Mounting 15/7

15.4 Circuit-breakers 15/8
    15.4.1 Oil circuit-breakers 15/8
    15.4.2 Air circuit-breakers 15/9
    15.4.3 Sulphur hexafluoride circuit-breakers 15/10
    15.4.4 Vacuum circuit-breakers 15/12

15.5 Disconnectors 15/13
    15.5.1 Switches 15/13

15.6 Reactors and capacitors 15/13
    15.6.1 Series reactors 15/13
    15.6.2 Shunt reactors 15/14
    15.6.3 Capacitors 15/14
    15.6.4 Variable reactive power 15/14
    15.6.5 Resonant link 15/14

15.7 Switchgear testing 15/14
    15.7.1 Direct testing 15/14
    15.7.2 Single-phase and unit testing 15/15
    15.7.3 Synthetic testing 15/16
    15.7.4 Field testing 15/16
    15.7.5 Miscellaneous tests 15/18

15.8 Overcurrent and earth leakage protection 15/18
    15.8.1 Fault conditions 15/19
    15.8.2 Protective equipment 15/20
    15.8.3 Relays 15/21
    15.8.4 Solid state equipment 15/23
    15.8.5 Overcurrent protection 15/23
    15.8.6 Earth leakage protection 15/25
    15.8.7 Balanced (differential) protection 15/25
    15.8.8 Miscellaneous equipment 15/28
    15.8.9 Efficacy of protection scheme 15/29

15.9 Application of protective systems 15/29
    15.9.1 Plant 15/30
    15.9.2 Feeders 15/32
    15.9.3 Motors and rectifiers 15/33

15.10 Testing and commissioning 15/34
    15.10.1 Commissioning tests 15/34
    15.10.2 Primary current tests 15/34
    15.10.3 Secondary injection tests 15/35
    15.10.4 Fault location 15/35

15.11 Overvoltage protection 15/35
    15.11.1 Insulation co-ordination 15/35
    15.11.2 Protective equipment 15/36

The minimum spacing of the gap electrodes, or the discharge characteristic of the surge arrester, is chosen to ensure that a flashover will not occur under normal steady state or transient power-frequency operating conditions. The voltage at which the gap will flashover following a lightning surge is known, and the insulation level of the transformer is chosen to withstand this gap voltage together with a margin of safety. (The nominal margin is 20%, but in practice it is somewhat higher, owing to the fact that the actual strength of the transformer insulation is necessarily greater than its specified level.)

Whether surge arresters are used instead of the simple discharge gap depends on such factors as the severity of lightning in the area and the importance of ensuring that the individual supply circuit remains intact.

A lightning surge that causes a gap discharge with either form of protection is inevitably followed by a power arc sustained by normal operating voltage. With surge arresters the non-linear characteristic causes the resistance to rise immediately the lightning discharge current (which may amount to several kiloamperes) ceases to flow. The higher resistance reduces the power-frequency follow current to a value such that the arcs across the multiple gap are self-extinguishing and steady state voltage conditions are restored. With simple arcing horns, however, a flashover of the gap due to lightning is followed by a power arc which is not self-extinguishing and which forms a direct earth fault on the phase in question. This leads to disconnection by the normal earth fault protection, although if the associated switchgear is arranged for automatic reclosing, there is only a momentary interruption.

In addition to ensuring continuity of supply, surge arresters have a further advantage over open gaps for transformer protection in areas where lightning is particuarly intense. This is because the voltage/time characteristic of a multiplicity of small gaps is flatter than that for a single large gap. In the event of a steep-fronted surge (resulting from a lightning strike to the line within a short distance of the transformer), the surge arrester will flash over at a lower voltage than will the open-type gap and thus provide better protection for the transformer.

In Britain it has been found that, in general, surge arresters are unnecessary largely because of the relative freedom from severe lightning storms and the small statistical probability of a lightning strike close to a transformer. Open-type gaps (usually referred to as 'co-ordinating gaps') give perfectly adequate protection against lightning surges with relatively slow 'fronts' (of the order of 5 µs or greater), although necessarily with the disadvantage described above in respect of continuity of supply following a gap flashover. In other parts of the world where lightning storms are intense, it is normal practice to provide surge arresters because of the better protection that they provide.

Lightning arresters are used in Britain on certain lower-voltage lines (as an alternative to the use of automatic reclosing circuit-breakers) and on some 132 kV wood pole lines without earth wires. Such lines have an inherently high insulation level to ground (compared with a steel tower line, in which a lightning surge is usually immediately discharged by a flashover of a line insulator on the nearest tower) which results in high-amplitude surges travelling along the line to the transformer.

Surge arresters are also used in certain special cases (e.g. in association with shunt inductors) where circuit characteristics are such that it is desirable to limit the magnitude of switching surges by installing an arrester to provide a discharge path. Conversely, some circuit configurations (e.g. the complex busbar system of a main switching station) are such as to reduce to an insignificant level any likelihood of a co-ordinating gap flashover being caused by lightning. In such cases surge arresters may be found unnecessary, even on such important installations as main generator step-up transformers.

14.16 Purchasing specifications

The essential basic information to be given is the following:

Standard (national or international)	Nominal rating
Rated winding voltages	Number of phases
Frequency	
Cooling medium (external)	Phase connection
Cooling medium (internal)	Phase relation
Ambient temperature of cooling media	Terminal arrangement
Transport limitations (weight, dimensions)	Altitude (if > 1000 m)

Impulse levels for lightning and switching surges (if applicable)
Impedance (with acceptable minimum or maximum values for limitation of fault level or voltage regulation)
Characteristics of transformers with which parallel operation is required
Tapping range (number and size of steps, whether on-load or off-load, restriction of impedance variation over tap range)
Performance requirements for cyclic or emergency overloading
Special requirements (fittings, paint finish, etc.)

In addition to the technical information above, the manufacturer should be advised on the basis of purchase (minimum first cost, maximum efficiency or capitalised loss). If a loss-capitalising formula is to be used, details should be provided to minimise the work of preparing and evaluating bids.

Specifications should set down a clear definition of technical requirements and operating conditions, and exclude detailed clauses on constructional points better left to the maker. The highly competitive transformer market, together with the normal practice of buying on the lowest tender that complies with the specification, means that the manufacturer must put forward the minimum-cost design with no extra capability not specified. It is thus important to specify every abnormal requirement.

resistance), and the net power taken from a supply is equal to the total power loss in the two transformers under test. If tappings are not available, or are unsuitable to circulate approximately rated full-load current, it is possible to inject the requisite circulating current through a small booster transformer connected in the leads running between the two main transformers.

The back-to-back connection, or direct loading, must be employed when making temperature tests on dry-type transformers, because the heat transfers from the core and windings to the cooling medium (air) are largely independent. The test conditions must therefore represent actual conditions in service when the core is heated by the magnetisation loss and the windings are heated by the load current.

In oil filled transformers the heat generated in both core and windings is transferred to the oil, and the total heat then has to be dissipated from the oil to the cooling medium. Because the loss in the core is usually appreciably less than the loss in the windings, it is possible to employ the so-called short-circuit method of test. The connections are identical with those shown in *Figure 14.29*.

Under short-circuit the loss in the core is almost negligible and the current in the windings is adjusted to a value slightly above the rated figure, so that the total loss in the windings is equal to the sum total of the separately measured load and core losses. The procedure during a short-circuit temperature test is to load the transformer with this increased current until such time as the observed oil temperature rise becomes sensibly constant, or is not rising at more than 1°C/h. This part of the test proves that the cooling equipment of the transformer is adequate to dissipate the total losses under normal full-load conditions and the oil temperature rise is recorded accordingly.

The increased current obviously results in the temperature gradient between windings and oil being higher than when the current is at the rated value. The thermal inertia of the windings is relatively low, however, and any change in current is quickly followed by a corresponding change in the temperature gradient between windings and oil.

After the oil temperature rise has been recorded, current is reduced from the increased to the normal rated value and maintained at this level for 30 min. The supply is then disconnected and the d.c. resistance of the windings measured in a manner similar to that employed prior to the loss measurement tests when the transformer was cold. By taking three temperature measurements in succession and plotting a graph against time from shut-down and extrapolating back to the time of shut-down, the resistance of the windings at the instant of shut-down can be determined. The winding temperature rise is then calculated by comparing the cold and hot resistances, with an allowance being made for any fall in oil temperature during the last half-hour at rated current. Full detailed of the methods are given in BS 171.

14.14.2.3 Sound level measurement

Because of the importance of noise as an environmental factor, the specification of a sound level limit for power transformers is increasingly common. The British Electricity Board's Specifications for various sizes of transformers stipulate maximum acceptable sound levels for each size of transformer purchased. The sound level tests, normally carried out at the manufacturer's works, can only be satisfactorily made at times when factory noise (including that of running test plant) can be kept to a level well below that of the noise emitted by a transformer on test. It is also important that the transformer on test is well clear of walls or other large areas which would reflect sound and cause a build-up of noise which would give a false reading. These important practical difficulties have led to considerable delay in including noise measurement as part of the national standard specification. BS 171: Part 1: 1978 includes the measurement of acoustic sound level as a special test which is carried out according to IEC Publication 551.

14.14.2.4 Switching surge test

At high system voltages (300 kV and above) the voltage surges resulting from circuit-breaker operation tend to become of greater significance than those due to lightning. A modified form of impulse test in which the shape of the applied wave is more representative of a switching surge than is the normal 1.2/50 µs wave is therefore frequently specified.

The standard wave shape for a switching surge test on air insulated equipments is of the order of 250/2500 µs, i.e. a relatively slow rise of front followed by a tail of long duration. The practical difficulty in producing such a wave from an impulse generator connected to a transformer winding (related to the generator capacity and the transformer-core saturation) has led to a relaxation of the requirements for the switching surge wave specified for application to transformer windings. Limiting features are a wavefront rise time of at least 90 µs, a time above 90% of the specified amplitude of at least 200 µs, and a total duration from virtual origin to first zero passage of at least 500 µs.

14.14.3 Commissioning tests at site

Commissioning tests vary considerably with the size and importance of the installation.

For a small or medium-sized distribution transformer the minimum requirements would be a visual examination for transport damage and an insulation test with a portable instrument. Preferably there should be a check of the ratio (by applying a medium voltage to the high-voltage terminals and measuring the induced voltage at the low-voltage terminals), and on the oil level and condition to confirm that ingress of moisture has not occurred. Measurements of ratio and of polarity are essential if a transformer is to be connected into a circuit where it will operate in parallel with other transformers.

On large units which are normally despatched either without oil or only partially filled, checks must be made on the filling procedure and of the condition of the oil prior to filling, in addition to ensuring that the insulation has not become wet during transport.

Auxiliary equipment such as on-load tap changing gear and any protective relays and current transformers associated with the main transformer must also be checked for correct operation.

In general, a repetition of high-voltage tests carried out at the works is not considered to be necessary. Where the transformer is subjected to retesting on site at high voltage, the test voltage level is normally restricted to 75% of that applied during tests at the works.

14.15 Surge protection

Any transformer connected to an overhead transmission line must be protected against surges resulting from lightning striking the line conductors. National and international standards exist for the insulation level of lines, switchgear and transformers at all usual transmission voltages, in the context of *insulation co-ordination*.

Transformers are protected against lightning surges by discharge gaps which may be in the form of simple arcing horns attached to the transformer bushings, or more sophisticated surge arresters. The later consist of a number of small gaps in series connected together through non-linear resistor material in which the resistance varies inversely as some power of the current.

14.14.2.1 Impulse test

The impulse test simulates the conditions that exist in service when a transformer is subjected to an incoming high-voltage surge due to lightning or other disturbances on the associated transmission line.

Impulse tests were introduced originally solely as type tests and, because of fears that they might cause undetected damage to the insulation of a transformer, the tests were largely confined to specially built prototype assemblies and were not applied to production units prior to going into service. Gradually it was realised that with increasing sensitivity of the equipment provided to detect insulation breakdown during the test, there was little risk of a service transformer suffering undetected damage. Still later it was appreciated that the sensitivity of the failure detection equipment was such that it would disclose hitherto unsuspected damage which might have occurred during a preceding power-frequency over-potential test. In Britain, therefore, common practice is for the impulse test to follow the power-frequency overvoltage tests, and the impulse test is now regarded as a routine test on all large and important transformers.

Details of the impulse wave shapes are specified in BS 171 for transformers for various system voltages. The precise form of a complete impulse test varies with customers' preferences and on whether the test is being applied as a basic type test on the first unit of a new design, or as a routine check on insulation strength following power-frequency test as described above.

The normal sequence for impulse tests in Britain is:

(1) One reduced-level full-wave (voltage between 50% and 75% of the full-wave voltage test level).
(2) One full-wave at specified test level.
(3) Two chopped waves with a crest value not less than the specified full-wave test level.
(4) One full-wave at the test level.

Evidence of insulation failure during an impulse test primarily depends on oscillograph records of the impulse voltage wave and either of the current passing through the winding under test or of the voltage induced by inductive transfer in another winding of the phase under test.

The normal sweep time for the oscillograph recording the voltage wave is 100 μs for a full-wave and a shorter time for a chopped-wave test, e.g. 10 μs. Current oscillograms are taken with the same sweep time, or preferably simultaneous records are taken of current with three different sweep times, e.g. 10, 100 and 500 μs.

The current and voltage wave shapes as shown on the oscillographic records are carefully compared with each other and particularly with the traces taken during the preliminary reduced voltage shots. Any discrepancy, however slight, in the wave shape of any one of the records compared with the others indicates a change in the conditions either in the test equipment or in the transformer under test. Such discrepancies must be investigated and satisfactorily explained, and any failure to do so involves the risk of a transformer with damaged insulation being put into service. BS 171 states that in cases of any doubt as to the interpretation of discrepancies three subsequent 100% full-wave shots shall be applied, and that if the discrepancies are not enlarged by these tests, the impulse test is deemed to have been withstood.

Before the development of the sensitive failure detection techniques now available the principal criteria for determining whether an impulse test had been successfully withstood were that there should be no noise from within the transformer during the test and no sign of smoke or bubbles in the oil. These remain as accepted criteria, but are now of secondary significance compared with the oscillographic procedures.

The chopped wave test is not usually included when impulse tests form part of the routine tests on a transformer, because of the time involved in setting up the chopping gap and some doubt as to whether the chopped wave test is useful for the detection of defects of workmanship or material. The main purpose of the chopped wave test as part of the impulse-type test series is to increase the stress in the insulation within and between the coils adjacent to the transformer line terminal. This is normally regarded as a feature of the design rather than of the individual unit. The chopped wave test is, nevertheless, representative of the conditions that arise when an incoming surge is suddenly chopped because of a flashover of an insulator in or near the substation, and it is prudent to ensure that the transformer is capable of safely withstanding events of this nature.

14.14.2.2 Temperature test

Each new design of transformer should be subjected to a test to determine that the temperature rise at rated load will not exceed the guaranteed values. It is uneconomic, if not totally impossible, to test a large transformer at the maker's works with both full voltage applied and full-load current in the windings, as the total output would have to be supplied and dissipated in some way.

On small transformers where the rating of the available test plant is equal to or greater than twice the rating of the transformer under test, it is possible to arrange two units connected back-to-back in the manner shown in *Figure 14.31*. If the ratio of both units is identical, no current will flow in the windings, but if the ratio is deliberately unbalanced (by connecting the two units on different tappings), a circulating current will flow through the two units, of magnitude governed by the out-of-balance voltage and the total impedance of the two units in series. The current is largely reactive (since the reactance of a transformer is normally considerably greater than the

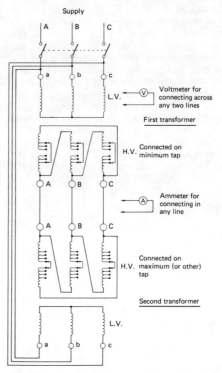

Figure 14.31 Method of connection for a back-to-back heat run

14.14.1.5 Core loss and magnetising current

The fundamental principle of this test is that normal rated voltage is applied to one winding while the other is left open circuit. The current flowing in the winding to which the supply is connected is the magnetising current and this is recorded as part of the test records. This magnetising current is normally a small percentage of the full-load current and the I^2R loss is negligible compared with the core loss.

The two-wattmeter method of measurement is again employed and the diagram is identical with that in *Figure 14.29*, except that the secondary winding is not short-circuited. To avoid unnecessarily high voltages in the test circuit during the core loss test, it is normal practice to connect the supply to the lower-voltage winding of the transformer.

14.14.1.6 Applied high-potential tests

Applied high-potential tests are normally made, in turn, between each winding, and the core and all other windings connected to earth.

Figure 14.30 shows the arrangement with the high-voltage winding under test and the low-voltage winding and core connected together and to earth.

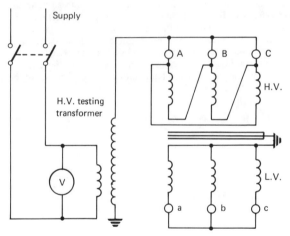

Figure 14.30 Connections for applied-high-potential test

The magnitude of the applied potential test depends on the rated voltage of the winding in question and on whether the major insulation between it and earth is uniform or graded. For a uniformly insulated winding the applied voltage test provides the principal dielectric test of the main insulation. It is usually of the order of $(2E+1)$ kV, where E is the 'highest system voltage' for the winding in question. Full details are given in BS 171. On a 3-phase transformer an applied voltage test of $2E$ raises the line terminals to 3.46 × their normal operating voltage to earth.

For graded insulation windings the applied voltage test is at a value appropriate to the insulation level at the neutral point and therefore does not adequately prove the strength of the line-end insulation.

14.14.1.7 Induced overvoltage test

The induced overvoltage test involves exciting the transformer on open circuit at a voltage higher than normal for a short period. On transformers with uniform insulation on which the applied high-potential test provides the principal check on the strength of the major insulation, the purpose of the induced test is to prove the strength of the insulation between turns and between other parts of the transformer operating at different potentials. The magnitude of the test is usually twice rated voltage, and to prevent overexcitation of the core the frequency of supply also has to be increased to at least twice normal.

On transformers with graded insulation the induced overvoltage test constitutes the main test of the major insulation. The magnitude of the test is fixed so that the potential to earth of each of the high-voltage terminals in turn is raised to the appropriate test voltage for the system on which the transformer is to operate. The magnitude of the test may be as high as 3.46 times normal rated voltage, and interturn and other insulation is obviously tested to this degree at the same time.

The duration of the test is 60 s for any test frequency up to and including twice rated frequency. When the test frequency exceeds twice rated frequency, the duration of the test is for 6000 periods (i.e. 1 min at 100 Hz) or a minimum of 15 s, whichever is the greater. The magnitude of the test voltages for different system operating voltages and conditions are given in BS 171.

The traditional fixed relation between specified impulse and power-frequency test voltage levels is operationally illogical. The former is determined by insulation co-ordination, surge arrester and circuit-breaker characteristics, earthing, etc. A trend for e.h.v. units (300 kV upwards) is for impulse and switching surge test levels to be specified with regard to expected operational surges, and for the induced voltage test to be directly related to the system voltage (e.g. 1.5 times) and to be applied for a much longer duration (30 min or 1 h) than in the 6000 period test at higher voltage, to give a reliable check on the capability of the insulation in normal service. It is likely that a single international standard on these lines will be agreed.

14.14.2 Type tests at works

Type tests consist of:

Impulse test.
Temperature test.
Noise level measurement.
Switching surge test (when specified by customer).

Until recently the distinction between type and routine tests was clearly defined, but operational experience, particularly on very large e.h.v. transformers, seems to be leading towards a compromise arrangement being adopted on transformers other than on the first of a new design. In particular, a simplified impulse test is frequently specified for application to all transformers purchased by certain customers. The logic is that many weaknesses disclosed by impulse test have been found to be due to poor materials or workmanship and not to fundamental errors of design. In the light of this experience it is obviously prudent to adopt the highly sensitive impulse test, even in modified form, as a routine check on materials and workmanship.

Another recent requirement is for the manufacturer to demonstrate during the induced overpotential test that the transformer is free from harmful internal discharge within the insulation structure. This measurement is frequently referred to as a 'corona' test. While there is little doubt that the corona test provides a valuable check on the integrity of the high-voltage insulation, there is a considerable difference of opinion among Supply Authorities and manufacturers on the method to be used to measure the discharge and on the limits to be regarded as tolerable and above which the transformer is deemed not to be suitable for service.

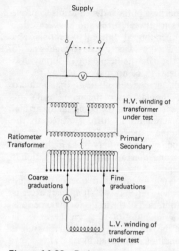

Figure 14.28 Ratio test

Because of the inductive effect of the core, care must be taken to ensure that a steady d.c. value is reached before voltage and current readings are recorded.

14.14.1.3 Insulation resistance

The insulation resistance between windings and from each winding to earth is measured by a special instrument such as the Megger or Metrohm.

The insulation resistance is commonly used as one of the criteria for determining that the transformer has been properly dried out. It varies widely and inversely with temperature, and care is necessary to ensure that the readings are correctly interpreted.

14.14.1.4 Load loss and impedance

Load loss and impedance are measured by short-circuiting the terminals of one winding of the transformer and applying a low voltage to the other winding sufficient to cause rated full-load current to flow. Because the applied voltage, and, hence, the magnetisation of the core, is extremely low, the core loss can reasonably be neglected and the measured input power represents the total load loss at rated load on the complete transformer.

Figure 14.29 shows a typical connection diagram for the I^2R loss and impedance test. It will be noted that the short-circuit is applied to the l.v. winding and the supply is connected to the h.v. winding, at which side all readings are taken. In principle, the same result would be obtained if the h.v. winding were short-circuited and the supply connected to the l.v. winding, but this involves measuring the heavier l.v. rated currents, which may be too large for convenience.

On 3-phase transformers the two-wattmeter method of measuring the power input is adopted, and the diagram shows the switching arrangement used to permit the requisite measurements to be made with one set of instruments. A double bank, double-pole switch is used to connect the measuring instruments into two of the three phases in turn. When the wattmeter switch is in the *off* position, the supply is connected directly to the h.v. terminals of the transformer under test, and the instruments do not indicate. With the wattmeter switch in position 1, the current coil of the wattmeter and the ammeter carry the current in phase A, and the voltage coil of the wattmeter and the voltmeter are connected between phase A and phase B.

With the switch in position 2, the current coil of the wattmeter and the ammeter carry the current in phase C, and the voltage coil and the voltmeter are connected between phase C and phase B. The switch is first closed in one direction and after the voltage has been adjusted to give full-load rated current on the ammeter, the readings of the voltmeter and wattmeter are taken. This is repeated with the switch in position 2. The total load loss is given by the sum of the two wattmeter readings and the impedance voltage is given by either voltmeter reading.

It is important that the test is performed at normal frequency to ensure the correct proportioning of the I^2R and stray losses in the windings and structure of the transformer, which are dependent on frequency.

In transformers with exceptionally high reactance the voltage that has to be applied to circulate rated full-load current through the windings, even under short-circuit, may be sufficient to magnetise the core to a level at which the loss therein may be significant. In such cases the core loss during the short-circuit test may be determined by removing the short-circuit and measuring magnetising loss when the transformer is excited on open circuit at the previously measured voltage required to circulate full-load current. The true load loss is then the difference between the two successive measurements.

It is important that the short-circuiting connections be substantial and applied in such a manner that the loss therein does not represent a significant fraction of the loss within the transformer. It is also important that the temperature be measured at the time that the load loss measurements are made, so that the necessary correction can be made to deduce the copper loss at the temperature (75°C) at which guarantees normally apply. BS 171 details the manner in which the load loss measurements at any given temperature can be corrected to the equivalent figure at 75°C.

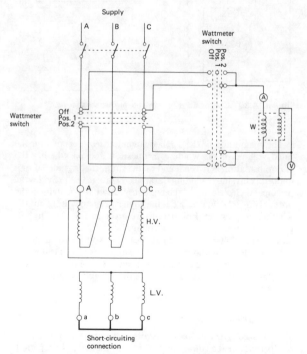

Figure 14.29 Connections for short-circuit test on a transformer for measurement of load loss, impedance and temperature rise

which in turn have a deleterious effect on the solid insulation. Poor ventilation in a transformer chamber (b) results in condensation inside the transformer, which similarly is liable to promote acidity and sludging. The electrical strength of the oil is considerably reduced by included moisture or fibres and particularly by a combination of both (c).

Deterioration can be greatly reduced or even arrested by attention to operating conditions and by routine precautions. Samples of the oil should, therefore, be taken from the transformer at regular intervals and the characteristics checked. BS 5730 *Code of Practice for Maintenance of Insulating Oil* should be consulted.

With large transformers, generally speaking, little trouble is experienced with acidity, mainly owing to the lower operating temperatures. Standard CEGB practice is to specify conservator-type transformers, and apart from one experimental installation associated with a group of generator transformers, it has not been necessary to consider the use of 'inhibited' oil. Similarly, any form of 'sealing' has in the past been considered unnecessary, for the same reasons. More sophisticated oil preservation equipment is now being specified for 400 kV transformers.

In small transformers there is a greater tendency for the development of acidity, but with modern oil (and provided that reasonable precautions are taken) no serious inconvenience should be experienced in this respect.

Discharge under oil may result in the flash point of the oil being reduced, although after a relatively short period of time it may recover. Similar reduction in flash point can occur as the result of abnormal local heating such as may be experienced during the development of an incipient fault, e.g. a core fault: involving circulating currents within the core itself due to a breakdown of interlamination resistance or failure of core bolt insulation.

14.13.2 Insulation

The standard method for checking the state of insulation is by measurement of the insulation resistance. It should be noted, however, that a transformer with relatively 'wet' insulation may have a high insulation resistance when this quantity is measured with the transformer cold, but the value may drop rapidly as maximum operating temperatures are approached. A hot insulation resistance reading below 1 MΩ per 1000 V rating of the tested windings is generally indicative that drying out is necessary.

When suitable equipment is available, measurement of dielectric power factor gives a reliable check on the state of the internal solid insulation. As the value of the p.f. will depend on the transformer design, a reference value taken on the actual transformer during works test is necessary for comparison before any useful assessment can be made.

A useful method mainly used in the factory for checking the state of dryness in solid insulation is the 'dispersion' test, based on the fact that in dry insulation the distribution of elements is such that individual shunt paths of time constant greater than 3 ms are effectively absent and that the presence of moisture introduces time constant within the range 3–300 ms. The method of test involves the application of a 300 ms pulse followed by a 3 ms short-circuiting pulse. Any shunt paths having time constants greater than 3 ms retain their charge, which is measured as a voltage by a suitable measuring device, the measurement indicating the moisture content.

The windings of a transformer should be inspected at long-term intervals. Any slackness due to insulation shrinkage or to the falling out of packing can then be remedied either by the adjustable coil clamping screws or by packing out the winding. This operation is particularly advisable in the case of transformers subject to heavy load surges, such as furnace transformers. Large transformers are usually built with preshrunk windings, so that slackening in service is almost entirely eliminated.

14.13.3 On-load tap changing equipment

From an operational aspect, an important factor is the period of time that can be allowed to elapse before attention to the switch contacts becomes necessary. If practicable, this period should be such that it can be co-ordinated with other outages of plant. This is of particular importance with generator transformers, where outage of the unit means the non-availability of generating plant. Until comparatively recently it was normal practice to carry out maintenance after every 10 000 operations, reasonably corresponding to a normal 12 month period between generator overhauls. The modern high-speed resistor tap changer, however, requires diverter switch maintenance only after 100 000 or more operations, and tap changer maintenance is no longer a limiting factor from the operational point of view.

14.14 Testing

The normal practice for testing power and distribution transformers is to carry out a comprehensive set of tests at the maker's works — the number and nature of which depend on whether the transformer is the first of a new design or otherwise — and a few relatively simple tests after installation at site to prove that the transformer is ready for service. The two classes of works tests are referred to as 'type' and 'routine', respectively. The first transformer of a particular design is subjected to both type and routine tests, while routine tests only are applied to later units.

14.14.1 Routine tests at works

Routine tests are:

Ratio, polarity and interphase connections checked.
Winding resistance measured.
Insulation resistance measured.
Load loss and impedance measured.
Core loss and magnetising current measured.
Applied high-potential tests of windings and of core insulation.
Induced overvoltage test.

14.14.1.1 Ratio, polarity and interphase connections

Figure 14.28 shows the type of circuit used for measuring ratio. This involves the use of a 'ratiometer', which basically consists of a multiratio transformer from which tappings are taken to coarse and fine adjusting switches. The ratiometer and the transformer under test are connected in opposition. When the ratiometer is adjusted to give a ratio exactly equal to that of the transformer under test, no current will flow in the secondary circuit. The ratio can then be read directly from the dial readings on the ratiometer.

Polarity and interphase connections are checked by measuring voltages between various terminals when the transformer is energised at a low voltage.

14.14.1.2 Winding resistance

The d.c. resistance of each phase of each winding is measured separately by the voltammeter method and is recorded together with the temperature of the winding at the time. This information is required for use in connection with later measurements of the load loss and the temperature rise of the transformer under rated load.

being based upon a 100 A line-to-neutral load. Each balancer winding carries one-third of the neutral or out-of-balance current, and has one-third of the line voltage impressed across it. The rating of the balancer as a 3-phase transformer is therefore $2(\sqrt{3}V/3)(I/3)(3/2) = 0.58VI$, where V is the line-to-neutral voltage, I is the neutral current and VI is the 1-phase load being balanced.

(1) To supply a 1-phase load. A balancer will be found to be cheaper than a 1-phase transformer connected across two lines, and much cheaper than a 3/1-phase transformer. Overload protection is provided by a fuse in the load circuit as shown in *Figure 14.27(a)*, and the balancer neutral current corresponds to the load current.
(2) To improve the voltage regulation of 4-wire networks. Balancers have in the past been used chiefly for this application. The improved regulation has been useful on rural distribution systems with isolated 1-phase loads. The balancer rating depends upon the out-of-balance current of the system, which usually is taken as 20% of the 3-phase line current.
(3) To transform a 3-wire system into a 4-wire one. Fuses, or even a circuit breaker, must not be employed in the balancer line. If one fuse were to blow, the neutral current would have to flow through the high-impedance paths offered by the sound balancer limbs and there would be a rise in the line-to-neutral voltage on the sound phases.

14.12.2 Welding transformers

Owing to their simplicity, economy and efficiency, transformer welding sets predominate over their rotary machine d.c. counterpart. Stick electrodes have been specially developed for a wide range of applications for use on a.c. sources, but even where a d.c. source is imperative, preference is shown for transformer/rectifier equipment over rotary machines.

The basic requirement for a.c. welding is a low-voltage power source (70–100 V), with an adjustable series inductor to ensure stability of welding current and provide phase shift between the source voltage and the welding current, enabling the arc to be restruck in each half-period after the current has passed through zero.

Power sources are supplied for use either by individual operators when part of the series inductive reactance may be incorporated in the transformer, or by groups of operators when a single multiphase transformer of relatively low impedance provides low-voltage (90 V) supply through a number of separate adjustable inductors. In the latter case advantage can be taken of the diversity factor in minimising the power rating of the transformer. For general-purpose applications standard a.c. single-operator welding sets are also available with inbuilt rectifiers and a smoothing inductor enabling the operator to use a wider range of electrodes.

The inherently high reactance of the source produces a very low p.f. load on the supply which, owing to its variability and intermittency, cannot be continuously corrected by capacitors. However, some correction is both possible and desirable, and most single-operator sets are designed to house a capacitor of a size recommended by the supplier.

The performance of welding power sources, specified in BS 638, covers the basic requirements for the majority of applications, but for other uses such as consumable electrode shielded gas systems, the characteristics of the welding set play a fundamental part in determining the quality of the weld. Typically, in transformer/rectifier power sources the internal impedance of the sets is very low and the open-circuit voltage is little more than the arc voltage; this produces a high rate of change of welding current when the arc length varies, and automatically adjusts the burn-off rate.

In the past most welding sets supplied for the British market have been oil cooled, and these are most suitable for the onerous conditions found, for instance, in shipyards; but the development of new insulating materials has enabled air cooled dry-type sets increasingly to take their place in less onerous conditions and where light weight and portability are valuable.

14.12.3 Mining transformers

The operating conditions in coal mines impose special requirements on transformers for use underground. There must be no possibility of a defect in the equipment itself causing an explosion of the gaseous atmosphere in the mine, and headroom is normally extremely limited.

Until the early 1950s, special low-height oil filled non-flameproof mining transformers were used underground up to within 300 m of the coal face. The switchgear directly mounted on these transformers was of certified flameproof construction. As the size of the load increased, the problems of voltage drop in the l.v. cables between the transformer and coal face machinery became greater, leading to the need for a completely flameproof transformer (and associated switchgear) which could be taken close to the coal face. The modern flameproof underground transformer is an air insulated unit constructed with class C insulation and contained in a flameproof case.

14.12.4 Small transformers

Small transformers are made in large numbers for electronic apparatus and like needs. The open construction has been superseded by a hermetically sealed arrangement in air filled or oil filled metal containers. The connections are brought out through metal/glass or metal/ceramic seals soldered to the container.

Transformers for mobile equipment, which may be operated at 400–1600 Hz for the sake of the saving in weight and size, are made of relatively costly materials to obtain larger magnetic and electric loadings. Thus, very thin cold-rolled silicon steel or thin nickel-iron cores may be used with coils insulated by high-temperature dielectrics such as glass fibre, silicones, etc. In this way a 30 VA, 1.6 kHz, fully sealed transformer weighs about 0.1 kg compared with its open-type equivalent of 1940 weighing 1 kg or more.

14.13 Maintenance

Maintenance can be described as the measures adopted to ensure that equipment is kept in a fully serviceable and reliable condition. Of necessity it is therefore mainly a routine involving attention at regular intervals to particular features based on service experience and manufacturers' recommendations. In the former, the measures involved tend to be general, whereas the latter tend to cover, in addition to general points, particular measures depending on the constructional characteristics of the individual manufacturer's designs. Consideration here is limited to two particular issues, namely insulating oil and solid insulation.

14.13.1 Insulating oil

Oil forms part of the main insulation of most transformers, but it tends to deteriorate in service owing to (a) operating temperature, (b) atmospheric conditions (applicable to unsealed, non-conservator-type transformers) and (c) presence of moisture or fibres.

In (a) deterioration is accelerated by prolonged high operating temperatures leading to the development of acidity and sludging,

leads are suitably transposed. Parallel operation is not possible between a 0° or 180° group and a ±30° group.

Some typical examples of connections are:

Connection	BS Reference No.	Displacement
Delta/star	Dy. 11	+30°
Star/delta	Yd. 1	−30°
Star/star	Yy. 0	0
Star/inter-star	Yz. 11	+30°
Delta/inter-star	Dz. 0	0

From this it will be seen that parallel operation is not possible between star/star and delta/star units. If the turns ratios of the transformers are not identical, a circulating current traverses the transformer windings, increasing the no-load losses. The magnitude of this circulating current for the case of two transformers A and B is

$$I_s = (V_A - V_B)/(Z_A + Z_B)$$

where V is the no-load secondary voltage for a common primary voltage and Z is the leakage impedance, all quantities being complexors.

The relative values of the percentage (or per-unit) impedance determine the proportion of the total load shared by each transformer. Thus, when all the percentage impedances are identical, each transformer will take its fair share of the load. Although it is advisable to have this condition, dissimilar impedance units may be connected in parallel in an emergency, provided that the current carried by any transformer does not exceed its normal rating or an acceptable overload value.

The percentage resistance drops of the transformers need not be the same. A difference in resistance drop, when the percentage impedances are numerically equal, results in an angular displacement of the individual transformer currents and reduces slightly the maximum permissible output. This is not normally of serious consequence.

14.11 Auto-transformers

The great advantage of the auto-connection, as distinct from the usual double-winding arrangement, is that the transformer physical size and losses are much smaller, provided that the primary to secondary turns ratio is not large. The amount of apparent reduction is termed the auto fraction

$$n = (V_1 - V_2)/V_1$$

where V_1 is the higher and V_2 is the lower voltage. Thus, the equivalent frame rating of an auto is equal to n times the load or throughput rating. For a 2/1 ratio this means that the auto is half the size of a double-wound transformer for the same duty. A requirement for tappings can have a marked effect on the apparent economy of using an auto-transformer. Unfortunately this economy is not obtained without certain liabilities, so that care is required in specifying auto-transformers unless the conditions are known and appreciated.

The calculated reactance of an auto-transformer on a frame kVA base has to be multiplied by n to obtain the reactance on a throughput base, e.g. a 2/1 ratio auto-transformer with a frame reactance of 4% would present an impedance of 4/2 = 2% to through faults. Given a high fault MVA infeed on the system, this could lead to short-circuit currents of more than the maximum permitted value of 25 × normal. System operating conditions must be clearly specified and, if necessary, additional impedance introduced to limit fault currents. It is the joint responsibility of purchaser and manufacturer to ensure that the transformer will not be subjected to excessive stresses.

The common electrical connection between the primary and secondary sides is a potential source of danger. The position of the earth connection with a 3-phase star-connected auto-transformer is important, and it is normally preferable to connect the supply neutral (assumed earthed) to the auto-neutral, and not to have the auto-neutral floating.

The use of the auto arrangement on transformers interconnecting different voltage levels (e.g. 400/275 kV) of a transmission system enables significant cost savings to be achieved. Small units are most useful as voltage regulating devices: they lend themselves readily to the provision of tappings, and as the loads generally have constant impedance characteristics, a small unit can control a large load. Consider a 10 kW, 400 V heating load, taking 25 A in an equivalent resistance of 16 Ω. It is required to control the heat in five steps by adjustment of the secondary load voltage E_s, using an auto-transformer. The secondary current for each tap is $I_s = E_s/16$, and the corresponding primary current is $I_p = I_s(E_s/400)$. The winding currents are I_p in the part corresponding to $(400 - E_s)$, and $(I_s - I_p)$ in the remainder. The winding could be graded to carry the maximum current in each portion, and it should be noticed that the 100 V tapping currents are less than those already determined for any portion of the winding. It is an axiom that, for constant impedance autotransformers, any tappings below half the supply voltage do not influence the transformer size. The equivalent kVA is the sum of the winding kVA divided by 2, and for the example this is 1.65 kVA.

14.11.1 Auto-starters

The principles above are applied to auto-starter transformers for 3-phase induction motors. The value of the equivalent motor impedance (assumed constant) is

$$Z = 1.25V/\sqrt{3}kI = 0.72V/kI$$

where V is the supply voltage, I is the full-load current of the motor and k is its ratio between short-circuit current and full-load current. The winding currents are determined according to the number of tappings. These currents are of short duration.

14.12 Special types

14.12.1 Static balancer

The static balancer is a simple apparatus which in its 3-phase form comprises an ordinary 3-phase transformer core carrying two windings per limb in zig-zag connection. Normally, when connected to a 3-phase line the balancer draws only a small magnetising current. When a load is connected between one line and neutral, however, so that the current balance of the feeders is upset, the load current flows through the balancer windings. This condition is illustrated in *Figure 14.27*, the current distribution

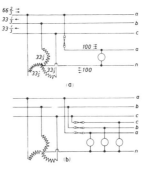

Figure 14.27 A.c. static-balancer applications

before any refinements and extras required for satisfactory operation are specified.

Terminals and bushings Transformer terminals must of necessity cover a wide range of voltages, currents and operating conditions. Outdoor type bushings are standardised in Britain and are detailed in ESI Standards 35-1 and 35-2 and BEB Specifications T1, T2 and T3. Some outdoor bushings are fitted with arcing horns which provide a safety gap to discharge an incoming surge which might otherwise damage the transformer windings. It is important that there is the correct correlation between the insulation strength of the transformer winding and the flashover characteristics of the gap. The setting of the gap must be small enough to protect the windings without causing too frequent interruption of the supply due to power arcs following gap flashover due to surges on the system.

Cable boxes These are essential when paper or plastics insulated cables require connection to the transformer. Cable box arrangements depend on the number of cores in the cable and the number of cables connected in parallel. Several factors must be considered in designing a cable box: adequate electrical and mechanical clearances are necessary; the compound must not ooze out of the box nor any transformer oil leak in; and voids must not easily occur in the compound, either during pouring operations or during normal thermal expansion and contraction. In order to make transformers interchangeable the box flanges are standardised (ESI 35-1 and 35-2).

Oil conservator Transformer oil has a coefficient of thermal expansion of 0.000 725 per degC at 0°C, equivalent to a 7.25% volume change over an oil temperature range of 0–100°C. The expansion may be accommodated in a free-breathing or sealed space at the top of the tank, or in a conservator tank mounted on the tank cover. It is desirable to avoid oxidation of the oil, which causes sludging and acidity to develop. With either the sealed tank or the conservator a nitrogen cushion can be maintained above the surface of the oil, thus preventing oxidation. A free-breathing conservator is preferable to an unsealed tank, as the temperature of the oil in contact with the air is lower and this in itself reduces the rate of oxidation. Some form of oil conservation arrangement is essential for large transformers and for those working at 33 kV and above; it is desirable also for small transformers, especially those subjected to heavy periodic peak load.

Breather This device allows ingress and egress of air to compensate for changes in oil volume. Except on small transformers, the breather should incorporate either a chemical or a refrigerating system for removing moisture from the air entering the transformer.

Oil gauge This can be a direct-reading glass window, or a dial instrument operated by magnetic coupling from a float on the oil surface. The dial gauge can be fitted with contacts to give a low-oil-level alarm.

Oil temperature indicator Normally of the dial type, oil thermometers can be arranged for remote electrical indication of oil temperature.

Winding temperature indicator The so-called winding temperature indicator is basically an oil temperature thermometer in which the bulb is associated with a heater coil which carries a current proportional to the load on the transformer. The heater coil introduces an increment of temperature above that of the oil, to correspond to the gradient between winding and oil temperatures. The instrument thus indicates a figure which, although a reasonable analogue of the temperature of the windings, is not a direct measurement thereof.

Buchholtz relay Any electrical fault occurring inside a transformer is accompanied by an evolution of gas. Appreciable quantities of gas may be produced before the fault develops to such an extent that it can be detected by the normal electromagnetic protection equipment. The Buchholtz relay, connected between the transformer tank and the conservator, contains two elements: (1) a float which operates a gas alarm device to give warning of gas discharge from within the transformer, and (2) a surge element connected to trip the transformer out of circuit in the event of a massive surge of oil and gas resulting from a major internal fault. In the event of a gas alarm being given, it is sometimes possible to deduce the nature of the defect within the transformer by observing whether the gas emission is at a constant rate (voltage dependent) or varies with load (current dependent).

The Buchholtz relay provides sensitive protection against certain types of fault (e.g. flashover between tapping leads) to which the normal electromagnetic protection is relatively insensitive, and may not operate until extensive damage has been caused.

Relief or explosion vents Many users specify relief or explosion vents, which are intended to act as safety valves to reduce internal pressure in the event of a major fault within the transformer and thus to protect the tank from damage. The vents can be spring loaded or fitted with thin non-metallic material to fracture under pressure. Provided that the build-up of pressure is relatively slow, the relief vent can operate satisfactorily and prevent the tank from bursting, but in the event of a violent fault leading to a shock wave of pressure, the tank may be burst before the relief vent has time to operate.

Tapping switches (off-load) Almost every distribution and medium-sized power transformer is fitted with voltage adjusting tappings, usually for five positions corresponding to $\pm 2\frac{1}{2}\%$ and $\pm 5\%$ of the supply voltage. To obviate opening up the transformer to change tapping links, a tapping switch is necessary. Such a switch is often fitted on top of the transformer core and is gang operated on all phases by means of a hand wheel on the tank end.

14.10 Parallel operation

The following information is required when the parallel operation of transformers is considered:

(1) Output and temperature rise of the transformers.
(2) Polarity for 1-phase units; angular displacement for 3-phase units; BS group reference or phasor diagram.
(3) Turns ratio on all tappings.
(4) Percentage impedance at 75°C.
(5) Percentage resistance drop at 75°C, or the load loss.

The polarity is basically determined by the direction of the primary and secondary windings and the position of the transformer line leads with respect to the start and finish of the windings. British transformers usually have a subtractive polarity. In the case of 3-phase units it is the angular phase displacement, i.e. the angle between the primary and secondary phase to neutral voltages, which has to be considered. This angle may be 0, 180° or $\pm 30°$, depending on the direction of the windings and the interphase connections.

All groups of transformers having the same angular displacement may be connected in parallel. Those having $+30°$ displacement may be paralleled with $-30°$ units, provided that the line

14.8.1.2 Forced air cooling (AF)

The temperature conditions are the same as for AN, but the improved heat transfer properties resulting from the forced air stream enables the current density in the windings and the flux density in the core to be increased and greater output to be obtained from a given size of unit.

14.8.2 Oil immersed, air cooled

BS 171 recognises two maximum oil temperatures: 60°C when the transformer is sealed or equipped with a conservator, and 55°C when the transformer is not so equipped.

Winding temperature rise by resistance for oil immersed transformers with ambient air temperatures of 40°C and a daily average of not more than 30°C is limited to 65°C, irrespective of the type of cooling or the cooling medium.

The various types of cooling and the new symbols are as follows.

14.8.2.1 Natural oil circulation, natural air flow (ONAN)

The great majority of power transformers up to ratings of 5 MVA are of ONAN type.

A plain sheet-steel tank radiates about 13 W/m^2 per degC rise. Above about 25 kVA, 3-phase increased cooling surface becomes necessary. This extra surface may be obtained by fins or corrugations, but the most common method is to employ a tubed tank. The tubes are usually 40–50 mm in diameter, of welded steel construction, having a wall thickness of about 1.5 mm. For medium sizes (2–5 MVA) tubes of elliptical section are preferred, as a greater number can be accommodated on a given tank. For transformers larger than 5 MVA it becomes necessary to employ radiator banks of elliptical tubes, or banks of corrugated radiators.

Transformer tanks have been constructed with finned or pilled tubes in order to augment the surface. They are difficult to paint, and are liable to collect water if employed out of doors.

The power dissipated by a tubular tank is a function of the ratio between tank envelope and total surface, for radiation is a function of the envelope, while convection depends upon the whole surface.

14.8.2.2 Natural oil circulation, air blast (ONAF)

By directing an air blast on to an ordinary tubular tank or on to separate radiators the rate of heat dissipation is increased; thus, while the transformer itself is not reduced in size, less external cooling surface is required.

14.8.2.3 Forced oil circulation, natural air flow (OFAN)

OFAN is an uncommon system, but is useful where for reasons of space the coolers have to be well removed from the transformer. The oil is pumped round the cooling system, from which heat dissipation is by natural air. The forced oil circulation permits high current densities to be employed in the windings, so that there is a reduction in transformer size.

14.8.2.4 Forced oil circulation, air blast (OFAF)

The OFAF system is employed for most large transformers. The forced oil enables the windings and core to be economically rated, while the forced air blast reduces the size of the radiating surfaces, an important point for transformers of 30 MVA upward. Depending upon the type and disposition of the radiators and on the purchaser's requirements, *mixed* cooling is employed in which the transformer operates as an ONAN unit up to 50% of its forced cooled rating (66% in the USA). As the load increases further, temperature sensitive elements start the pumps and fans of the forced cooling equipment.

14.8.3 Oil immersed, water cooled

Cooling-water inlet temperature limits as defined in BS 171 must not exceed 25°C. Oil temperature and winding temperature limits are as for air cooled units.

14.8.3.1 Natural oil, water (internal cooler) (ONWF)

A copper cooling coil is mounted above the transformer core in the upper portion of the transformer tank.

14.8.3.2 Forced oil, water (external cooler) (OFWF)

The OFWF system uses oil/water heat exchangers external to the transformers. The arrangement has a number of advantages over ONWF cooling:

(1) The transformer is smaller in size, as the windings can be more highly rated owing to forced oil flow and the tank is not required to accommodate a large cooling coil.
(2) Condensation troubles are non-existent.
(3) Water leakage into oil is improbable with the oil pressure maintained higher than that of the water. In cases where the cooling water has a high head at the transformer plinth level, it is necessary to employ two heat exchangers in series, i.e. oil/water and water/water; the water in the intermediate circuit between the two coolers is separate from the main water supply and at a low head.
(4) The cooling tubes may be easily cleaned.

Water cooling of transformers is common at generating stations (particularly hydro), where ample cooling water supplies are available.

14.8.4 Overload capability

The temperature limits for oil, windings and insulation laid down in BS 171 are chosen to ensure that a transformer operating within these limits will have a satisfactory life of 20 or more years. The relationship between operating temperature and life is complex. Experimental work indicates that each 8°C increase in operating temperature halves the life of the insulation, but there is little information available to enable an operator to determine precisely the actual life expectation under any given operating conditions. This obviously depends on other factors, including the incidence and severity of short-circuit forces to which the ageing insulation may be subjected. To some extent also the loss of insulation life due to overload operation during an emergency can be offset, provided that for some further and more protracted period operating temperatures are kept well below the specified limits.

Because of the considerable thermal inertia of the mass of oil and metal in a transformer, appreciable overloads can be carried for short periods without endangering insulation life. BS Publication CP 1010, *Guide to Life of Oil-immersed Transformers to BS 171*, gives details of permitted overloads. These are, however, normally regarded as conservative and have been exceeded in service without adverse results.

14.9 Fittings

The number of fittings and their siting on the transformer tank constitute major problems of transformer standardisation. It is recommended, therefore, that the fittings enumerated in BS 171 be used, careful consideration being given to the essential points

transformers in parallel, although the length of time to complete an initial tap change may be considered to be excessive where more than two transformers are involved. The voltage relay initiates the movement of the 'master' tap changer: when this has completed one tap step, the auxiliary switches operate to cause the second transformer of the chain to come into line with the first. This is followed in sequence by the movement of the remaining units. The time for completing one tap step on all units in the bank is, therefore, the sum of the individual operating times. As in (1), this scheme is suitable only for substantially identical tap changers.

(3) Master-follower scheme with multiple voltage-regulating relays, a more complicated version of (2) in that each transformer is equipped with its own voltage-regulating relay and any one of the group may be selected to act as master, the remaining units following in any desired sequence.

(4) Circulating-current control schemes. The foregoing schemes all require multipoint switches in each tap changer, interconnected by multicore cable. The circulating-current schemes depend on the fact that if two transformers operating in parallel are out of step, a circulating current will flow between them in a direction depending on the relative ratio of transformation. Each transformer is controlled by its own voltage-regulating relay, and as individual characteristics are not absolutely identical, it follows that, when a change of voltage occurs, one relay of the group will initiate a tap change on its associated transformer earlier than the others. As soon as the first transformer of a group has completed a tap change, there will be an imbalance of ratio and a circulating current will flow in the main circuit connections between the transformers. This circulating current is used to control auxiliary relays in each tap changer of the group, so that no further movement can occur to increase the imbalance, i.e. the second and later transformers in the group can change step to come into line with the first unit which has already moved; alternatively, the first unit can move in the reverse direction to bring itself back into its original position and therefore directly in step with the others. This type of scheme allows transformers to operate indefinitely in a 'one step out' condition. The wiring is relatively simple, as the only interconnection between units are the secondary leads of the circulating-current c.t.'s. It permits automatic parallel control of transformers of different rating, impedance, tapping range and number and size of tapping steps, as the tap changers take up positions to minimise the circulating current between units. Provided that the c.t's are selected to correspond to the rating of each main transformer, optimum loading of the group is achieved.

(5) Parallel control by reverse compounding. All of the schemes mentioned above necessitate secondary connections between the transformers that are to operate in parallel, and all except (4) suit only transformers with near-identical characteristics. Where parallel operation is required between transformers with differing characteristics, or where the transformers are situated some distance apart, it is possible to achieve stable operation with each being controlled by its own voltage-regulating relay by introducing negative compounding of the reactance element of a *line-drop compensator*. This tends to give a negative compounding characteristic (i.e. output voltage drops as load increases), but compensation can be provided by increasing the positive compounding of the in-phase element. Unless the negative reactance characteristic is introduced, any two tap changers controlled by independent voltage-regulating relays which are not positively locked together will inevitably move quickly to opposite extremes of their range.

14.7.2 Line-drop compensation

This device permits the output voltage of the transformer to be compounded so that it rises with load to compensate for voltage drop in the cables connected to the secondary side, and to maintain approximately constant voltage at a remote point. The line-drop compensator comprises an 'artificial line' circuit consisting of adjustable resistances and reactances connected into the voltage relay operating-coil circuit (*Figure 14.26*). A current transformer in the main secondary connection injects current into the adjustable resistance and reactance coils and thus biases the voltage applied to the voltage sensitive element of the regulating relay.

14.8 Cooling

Small transformers are air-cooled and insulated. For units of larger rating and higher voltage, oil cooling becomes economical, because oil provides greater insulation strength than air for a given clearance, and augments the rate of removal of heat from the windings. With the exception of certain special installations, such as in coal mines or within occupied buildings where mineral oil is undesirable because of the fire risk involved, almost all power transformers are oil immersed. The combination of oil and paper insulation has been used in transformers for many years and there appears little likelihood of its being superseded by any modern synthetic material. A principal reason for this is that both materials can be operated safely at approximately the same maximum temperature, approximately 105°C. Any alternative materials would have to show significant advantages in either or both insulating and heat transfer properties compared with the combination of oil and paper.

The most common types of cooling arrangements, detailed below, are identified in BS 171 by a system of symbols which indicate the cooling medium in contact with the windings; the cooling medium in contact with the external cooling system; and the kind of circulation for each. The symbols for oil, water and air are O, W and A, respectively; those for the kind of circulation are N and F for 'natural' and 'forced', respectively. BS and IEC specifications stipulate temperature limits for windings (measured by resistance) and insulation and define normal standard values for the temperature of the cooling medium.

14.8.1 Air insulated, air cooled

14.8.1.1 Natural air cooling (AN)

The temperature rise by resistance is limited by the class of insulation used. Typical figures are 60°C for class A, 90°C for class B and 150°C for class C materials; all above a maximum ambient temperature of 40°C and a daily average of not more than 30°C. Type AN cooling is generally limited to relatively small units, although the development of high-temperature insulation, such as glass and silicon resins, has resulted in its use on transformers up to 1500 kVA and for special application as in mines.

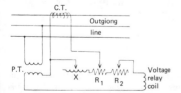

Figure 14.26 Line-drop compensation (R_1, X = adjustable to suit line characteristics; R_2 = adjustment for output-voltage level)

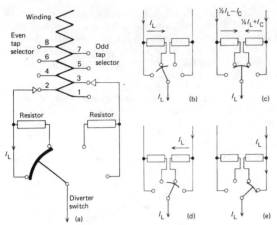

Figure 14.25 Resistor transition. (a) Outgoing tap operating; (b) Load current in resistor; (c) Resistors in parallel to load current and in series to circulating current; (d) Load current in resistor; (e) Incoming tap operating

superseded by high-speed resistor transition. The switching sequence is shown in *Figure 14.25* and is similar to that with inductors, except that two resistors are used. Back-up main contacts are provided which short-circuit the resistors for normal running conditions.

Advantages of inductor transition were: the inductor could be continuously rated and a failure of auxiliary supply during a tap change did not necessitate the main transformer being taken out of service; also, that the intermediate or bridging position could be used as a running position, giving a voltage equivalent of one-half tap step. The main disadvantage is that the circulating current between taps during the bridging condition is at low power factor, adversely affecting diverter-switch contact life. The inductor itself was costly and occupied a significant amount of space in the transformer tank.

Resistor transition is now used almost exclusively by British and European tap changer manufacturers, although inductor transition is still used in the USA, possibly because it is common practice there to specify a large number of small tap steps, a requirement met by using the bridging position as a running tap.

Resistor transition requires one winding tap for each operating position. The basic disadvantage is that the resistors cannot be continuously rated, if their physical size is to be kept small. It is essential to minimise the period during which they are in circuit. Some form of energy storage has, therefore, to be incorporated in the driving mechanism to ensure that a tap change, once initiated, is completed irrespective of failure of auxiliary supply. Early resistor tap changers operated at low speed and the stored-energy mechanism was a flywheel or a falling mass. All modern tap changers use springs for energy storage, and the total time that a resistor is in circuit during a tap change is limited to a few periods. The advantages of the high-speed resistor tap changer are its compactness and lack of wear of diverter-switch contacts because of the high speed of break and because the circulating current is at unity power factor. Contact life of 250 000 operations is common, compared with the 10 000–20 000 for reactor tap changers.

Irrespective of the form of transition, all on-load tap changers fall into one of three categories in respect of the switching arrangement. These are as follows:

(1) The oldest (and now least common) arrangement is for a separate contactor to be connected to each winding tap. Contactors are operated by a camshaft to ensure the correct sequencing. A later development was to use mercury switches instead of open-type contactors, giving the advantage of freedom from carbonisation of the coil.

(2) The winding tappings are connected to a series of fixed contacts of a selector switch of either linear or circular form, and an associated pair of moving contacts operates to provide the required switching sequence. Current making and breaking occurs at the selector switch contacts and some degree of oil carbonisation and contact wear is inevitable. This type of tap changer, usually called 'single-compartment', is now common for transformers up to about 20 MVA rating.

(3) For the largest and most important transformers the tap selector switches do not move when carrying current. Current making and breaking is carried out by two separate diverter switches, usually in a separate compartment of the tap changer to minimise the amount of oil contaminated by carbon. The diverter switches operate to make and break the current and are mechanically interlocked with the selector switches, which move only when not carrying current to provide the correct sequence of connection to the winding taps.

At one stage of development of medium-size tap changers, large mercury switches were used as diverter switches; these could be mounted in a selector switch compartment as there was no risk of oil contamination. A recent development along the same lines is the application of vacuum-switch diverters, capable of many thousand operations without attention and with freedom from contamination. Trials are being made with thyristors to provide 'contactless' switching: while this might yield marginal advantage in minimising maintenance and perhaps improving reliability, the cost and complexity of the control arrangements are likely to inhibiting factors.

14.7.1 Tap changer control

Control gear can vary from simple local push-button control to a complex scheme for the automatic control of as many as four transformers operating in parallel. The objective of automatic tap change control is to maintain output voltage either constant or with a compound characteristic rising with load.

The main component is the automatic voltage-regulating device, which consists of a voltage governor, a time-delay relay and compounding elements. The time-delay element prevents tap changes occurring due to minor short-time fluctuations of voltage. It can be set for delay periods of up to a minute.

Tap change control circuits necessarily involve auxiliary switches mounted within the driving mechanism of the tap changers themselves, and this has led to a proliferation of types of control schemes each designed to operate with a particular type of tap changer. These variations have made it extremely difficult for the British Electricity Supply Authorities to develop a national standard control scheme. Some of the differences arise in the method adopted to arrange simultaneous or near-simultaneous tap changes of transformers operating in parallel. A further complication arises because of differences between one transformer and another in the tapping range and the number and size of tapping steps. Some of the various types of parallel control schemes are as follows:

(1) Simultaneous operation of two or more tap changers initiated from one voltage regulating relay. Closing of the 'raise' or 'lower' contact in the voltage-regulating relay closes the appropriate motor contactor in each tap changer, and auxiliary switches lock out further movement until all the tap changers have completed a single change of position. This scheme is, in general, only applicable to tap changers with identical main and auxiliary characteristics.

(2) Master-follower operation initiated from one voltage-regulating relay. This is also suitable for two to four

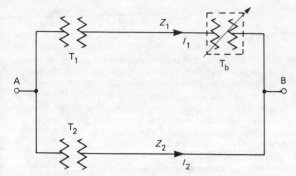

Figure 14.22 Power flow along two parallel lines of different impedance

continuous rating. The tertiary must withstand the effects of a short-circuit fault across its external terminals, as well as those due to earth faults on the main windings. The conductor sections and the bushings are designed for the most onerous operating condition.

14.6 Quadrature booster transformers

Consider two lines in parallel transmitting power between A and B (*Figure 14.22*): T_1 and T_2 are conventional transformers, either or both having on-load tap changing equipment. The division of load currents I_1 and I_2 between the lines is in inverse proportion to the impedances Z_1 and Z_2. For better load sharing it may be desirable to adjust the division of current. Tap changing the transformers to give a voltage difference V between T_1 and T_2 will cause a circulating current $I_c = V/(Z_1 + Z_2)$ to flow around the loop; and I_c will lag V by the natural phase angle of $Z_1 + Z_2$, increasing the current and $I^2 R$ loss in the branches without achieving the desired change in load sharing.

If a booster transformer T_b is introduced into the loop to inject a small voltage in quadrature with the supply voltage, the resulting circulating current will be approximately cophasal with the active-power component of the load current. It will therefore add arithmetically to the active current in one branch and subtract from that in the other, thus controlling the relation between I_1 and I_2. The adjustment permits an increase in the total load that can be transmitted over the system.

The quadrature booster is an important element for power control in parallel circuits. A simple method of deriving the necessary injected voltage is shown for one phase in *Figure 14.23*. A single transformer unit has a delta connected main winding and a tapped series winding to provide an adjustable voltage in quadrature with the phase voltage.

For large high-voltage quadrature boosting, two transformers are commonly used in order to provide the necessary dielectric and short-circuit strengths. The first 'excitation' transformer supplies the second 'injection' transformer. By suitable choice of connections either quadrature or in-phase regulation can be obtained. By use of two tap changers and more complex tapping windings, separate control of both in-phase and quadrature components can be achieved.

14.7 On-load tap changing

The essential feature of all methods of tap changing under load is that circuit continuity must be maintained throughout the tap stepping operation. The general principle of operation used in all forms of on-load tap changer is that, momentarily at least, a connection is made simultaneously to two adjacent taps on the transformer during the transition period from one tap to the next. Impedance in the form of either resistance of inductive reactance is introduced to limit the circulating current between the two tappings. The circulating current would represent a short-circuit between taps if not so limited.

Figure 14.24 shows in diagrammatic form the use of a centre-tap inductor or autotransformer as the transition impedance. In *Figure 14.24(a)* the load current is shown passing from the maximum tap through the halves of the inductor in opposition, and, hence, non-inductively. In *Figure 14.24(b)* one of the two tap-selector switch contacts has opened and the load current is carried inductively through one half of the inductor. In *Figure 14.24(c)* the inductor is shown bridging the two adjacent tappings. The load current is shared equally between the two tappings and passes non-inductively in opposition through the halves of the inductor. The tap step voltage is applied to the whole of the winding of the inductor and the circulating current is limited by the total impedance. In this position, in which the tap changer can remain indefinitely, the effective voltage is equivalent to the mean of the two individual tap voltages. *Figure 14.24(d)* shows the momentary condition where one half is connected inductively to the incoming tap position, and *Figure 14.24(e)* shows the final stage of the transition where both selector switch contacts are connected to the incoming tap and the inductor is non-inductively connected.

Circulating-current limitation by centre-tapped inductor was common in the late 1940s, but has since been almost entirely

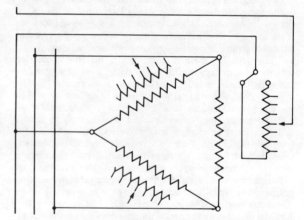

Figure 14.23 Connections for quadrature regulation

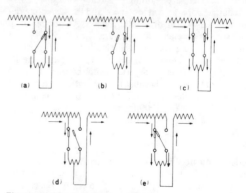

Figure 14.24 Inductor transition

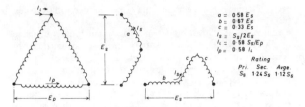

Figure 14.20 Leblanc 3/2-ph transformation ($S_s = kVA$ rating of two-phase load)

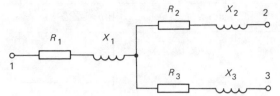

Figure 14.21 Equivalent circuit of three-winding transformer

Three-phase to single-phase transformation can be achieved by an open delta connection which involves a standard, 3-limb, 3-phase core, outer limbs wound. Alternatively, an arrangement similar to the Scott connection can be used. With the open delta connection the output voltages from each limb are 120° apart, so that the voltage applied to the load is $\sqrt{3}$ times the voltage across one winding, and in the case of the T connection the output voltages are 90° apart, so that the voltage applied to the load is $\sqrt{2}$ times the voltage across one winding. With either form of connection, if a 3-wire 1-phase supply is required and the loads on each side of the mid-point are liable to unbalance, the windings should be subdivided and interconnected to distribute more evenly the out-of-balance current and to avoid excessive voltage regulation.

It is important to realise that, although a 3/1-phase connection can be used to transform a line–neutral load on the secondary side to a line–line current on the primary side, it does not result in a balanced primary load. The same degree of unbalance must appear on the primary; or in terms of symmetrical components, all zero- and negative-sequence currents on the secondary will also flow on the primary side.

14.5 Three-winding transformers

Typical applications of 3-winding transformers are:

(1) Feeding two secondary networks, of different voltage or phase relationship, from a common primary supply.
(2) Connecting two generators to the same high-voltage system while maintaining a relatively high impedance between them to limit cross-feed of fault energy.
(3) Feeding two parts of a sectionalised low-voltage network so as to limit the fault level of each part without too high an impedance between the high-voltage and the low-voltage sides.
(4) Providing a rectifier with a multiphase (e.g. 12- or 24-phase) supply.

Provided that there is at least one delta connected winding to permit the flow of third-harmonic currents, there is freedom to adopt star or delta connection to meet phasing or earthing requirements.

14.5.1 Impedance characteristics

Two-winding technology does not apply. The essence of the procedure for a 3-winding unit is that the leakage impedance can be represented by assuming each of the three windings to have an individual resistance and leakage reactance, and mutual impedance effects (other than those that result from these individual values) to be absent. The equivalent circuit can be represented by the star network in *Figure 14.21*. The leakage impedance values (in per-unit form to a common kVA base) are given in terms of the conventional 2-winding impedances: for resistances

$$R_1 = \tfrac{1}{2}(R_{12} + R_{31} - R_{23}) \quad R_2 = \tfrac{1}{2}(R_{23} + R_{12} - R_{31})$$
$$R_3 = \tfrac{1}{2}(R_{31} + R_{23} - R_{12})$$

X being substituted for R to give the leakage reactances. These for the individual arms are then combined to give the effective values between any pair of terminals: e.g. $R_{12} = R_1 + R_2$, $X_{23} = X_2 + X_3$, and so on. (As the equations for the individual arms include a negative term, some particular evaluations may be found to be negative.)

14.5.2 Tertiary windings for harmonic suppression

The most common 3-winding transformer is star/star connected with a delta tertiary to provide a path for z.p.s. currents. If third-harmonic distortion of the main flux (and, hence, of the secondary voltage wave form) is to be avoided, a tertiary must be used on any configuration of core that provides a low-reluctance path for third-harmonic components of the flux. This applies to shell types, and to 3-phase core types having more than three limbs; a 3-phase 3-limb core-type transformer has no ferromagnetic return path for third-harmonic flux components, so suppressing z.p.s. distortion of the flux, and it will operate satisfactorily provided that the load is not significantly unbalanced. For an unbalance exceeding 10% it would be prudent to include a tertiary. However, a 2-winding transformer called upon to supply zero-sequence loads could alternatively be provided with an external source of zero-sequence power (a) by a direct connection between the transformer in question and the neutral of a standby unit having a delta winding or (b) by the installation of a zigzag connected 'earthing transformer'. If several star/star units are acquired and undue unbalanced loading is not expected, it would be preferable on grounds of cost to specify them without tertiaries, subsequent remedial action being taken if and when it becomes desirable.

Whether or not third-harmonic problems arise depends on the complete installation and how it is operated. The transformer characteristics do not alone determine the issue, for other factors are (a) the level of the transformer core flux density (and, hence, the magnitude of the third-harmonic component of magnetising current); (b) the expected degree of load unbalance; (c) the impedance of external sources of zero-sequence current; (d) the proximity of telecommunication circuits to the external zero-sequence current path.

Tertiary windings fitted only for harmonic suppression and not connected to external terminals must be rated to withstand the effects of primary and secondary earth fault currents: the tertiary current then depends on the p.p.s., n.p.s. (negative phase sequence) and z.p.s. impedances of both the supply and the transformer. As such fault currents persist for only a few seconds, the temperature rise of the tertiary winding is determined by its thermal capacity.

14.5.3 Tertiary windings for external loads

Tertiaries providing auxiliary power circuits, or reactive power compensating capacitors or inductors, must have an appropriate

Three-winding transformers 14/13

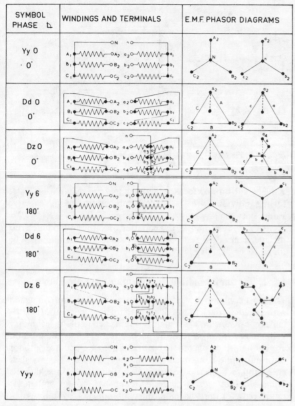

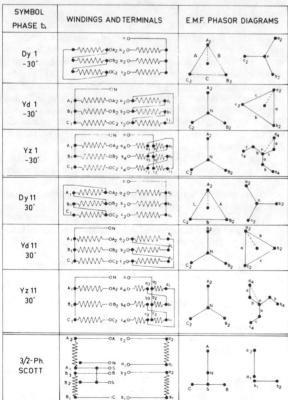

Figure 14.18 Standard connections for three-phase transformers

even 24-phase connections may be needed in rectifier transformers.

Single-phase transformers can be independent; arranged to provide a 2-wire or 3-wire supply, centre-point earthed; or combined to form a 3-phase bank. In the case of 3-phase windings, three forms of connection are possible; star, delta and interconnected-star (zigzag). When combined on the same core, a delta winding and an interconnected-star winding can be arranged to provide zero phase displacement, and when either of these is combined with a star winding, a 30° phase displacement results, either leading or lagging. By reversal of one winding with respect to the other when a combination of the same connection is involved, or where the combination is of connections giving the same phase displacement, a 180° phase displacement is produced. A summary of the combinations detailed in BS 171 and the corresponding e.m.f. diagrams are given in *Figure 14.18*.

In a star/star connection unbalanced load may result in neutral displacement and third-harmonic currents may circulate between lines and earth. These difficulties may be overcome by providing a delta connected stabilising (tertiary) winding with a rating sufficient to take short-circuit fault currents. The need for this winding depends, however, on the core construction. Where, as in banks of three 1-phase units or in 5-limb 3-phase core-type transformers, there is an independent iron path for the zero-sequence flux in each phase, the z.p.s. impedance is consequently high, making a stabilising winding essential. With 3-phase 3-limbed cores the z.p.s. fluxes are forced out of the core and the z.p.s. impedance is lower; consequently, a tertiary winding may not be necessary.

14.4.1 Phase conversion

The Scott connection is the most common 2/3-phase conversion. Two 1-phase transformers are generally used, with one pair of windings arranged to form a T connection for the 3-phase supply. The 'main' unit is wound for the phase-to-phase voltage with a mid-point tapping, and forms the head of the T. The other unit, the 'teaser', is wound for 0.866 times the line voltage and both are designed to carry full line current. A neutral point can be provided by a tapping on the teaser winding at a point 0.577 of the turns from the line end; see *Figure 14.19*. The two units, for operational convenience, can be made interchangeable. A winding arrangement which results in a minimum leakage reactance between the winding halves of the main unit (for example, an interleaved winding) is essential to ensure correct current distribution and to avoid excessive voltage regulation. In the place of two single-phase units, a 3-limb core, outer legs wound, may be used if the unwound centre limb is proportioned to suit the flux conditions, i.e. given 41.5% greater section.

As an alternative to the Scott connection, which is basically a 1-phase arrangement, a Leblanc connection can be used. It employs a standard, 3-limb, 3-phase core and a standard 3-phase delta connected winding (*Figure 14.20*). Compared with the Scott arrangement, the Leblanc connection requires a smaller core but involves a greater winding section.

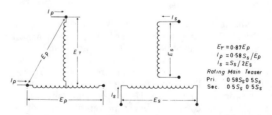

Figure 14.19 Scott 3/2-ph transformation

cooling, it is common for the transformer to operate without the pump up to some proportion of its forced cooled rating. It is, therefore, important to ensure that the natural thermal circulation of the oil is such that there is a flow over the main cooling surface of the coils. This is a simple matter where the main cooling surfaces are vertical, but when (as in disc windings and particularly on large transformers) the major cooling surfaces are horizontal, special means must be taken to ensure that the oil will flow across the horizontal surface under both natural and forced cooling conditions. Oil-flow barriers are introduced for this purpose.

14.3.6 Short-circuit conditions

Under conditions of system fault, mechanical and thermal stresses of considerable magnitude can be imposed on a transformer winding.

Mechanical forces of magnetic origin may be resolved into two components: (1) *radial*, due to the coil currents lying in the axial component of the leakage flux, tending to burst the outer and crush the inner winding; (2) *axial*, due to the radial component of the leakage flux arising from ampere-turn unbalance, tending to displace the windings (or parts thereof) with respect to each other. Axial forces may increase the unbalance that produces them, and repeated short-circuits may have a cumulative effect. Windings must be designed and built to withstand the mechanical forces, an important consideration being ampere-turn unbalance between windings, especially that caused by tappings.

Thermal stresses arise from the temperature attained by the windings when carrying sustained fault current. The limits of temperature rise permitted (BS 171) for copper windings are: 250°C for class A insulation in oil and 350°C for class B in air. Aluminium windings are limited to 200°C for both classes.

Mechanical forces have to be considered relative to the asymmetrical peak current, whereas thermal stresses are governed by the symmetrical r.m.s. value and duration. The current itself is controlled by the leakage impedance of the transformer plus the impedance of the Supply System. BS 171: Part 5: 1978 states that for transformers rated 3150 kVA or less, the system impedance shall be neglected in the total impedance if it is equal to or less than 5% of the leakage impedance of the transformer.

The duration for which a transformer can withstand short-circuit current depends on thermal considerations. BS 171: Part 5: 1978 states that the duration of the short-circuit current to be used for calculation of the thermal ability to withstand short-circuit is two seconds unless otherwise specified by the purchaser. The temperature attained by the winding can be calculated by the method described in BS 171 on the assumption that, during the period of short-circuit, all heat developed by the loss in the windings is stored in the conductor material.

14.3.7 Impedance voltage

The impedance voltage of a transformer can be defined as that voltage required to circulate full-load current in one winding with the other winding(s) short-circuited. It comprises a component to supply the IR drop and another to overcome the e.m.f. induced in leakage inductance. The leakage flux is a function of the winding ampere-turns and of the area and length of the paths of the leakage flux. By adjustment of these parameters the transformer can be designed for a range of reactances. The most economical arrangement of core and windings results in a 'natural' value of reactance. This value can be varied to some limited extent without any great influence on the cost and performance of the transformer. Interleaving the windings and so reducing the effective area of the leakage paths will reduce reactance. High reactance requirements usually result in greater stray load loss, because of the necessarily greater leakage flux.

It is usual to express the leakage impedance Z of a transformer as a percentage (or per-unit) value. The per-unit value is IZ/V, and the percentage value is $100\,(IZ/V)$, where I and V refer to the full-load current and rated voltage of one of the windings, and IZ is the voltage measured at rated current during a short-circuit test on the transformer. In the case of a transformer with tappings, the impedance is conventionally expressed in terms of the rated voltage for the tapping concerned.

14.3.7.1 Efficiency

Efficiency is the ratio between power output and power input. Its actual value is less important than that of the magnitude of the losses, which determine heating, cooling, rating and the cost of supplying the losses under given loading conditions in a system.

For a given system voltage and transformer tapping, the flux density has constant peak value, and the core loss p_i is considered to be constant. The load loss p_c varies with the square of the loading, and because of the change of conductor resistivity with temperature is commonly stated at 75°C. If the power input is P_1 and the power output is P_2, the efficiency is

$$P_2/P_1 = P_2/(P_2 + p_i + p_c)\ \text{per unit}$$

14.3.7.2 Regulation

The regulation is the difference between the no-load and full-load secondary voltages expressed in terms of the former, with constant primary voltage. The difference is the result of voltage drops due to resistance and leakage reactance; it is proportional to load but is strongly influenced by the load power factor. *Figure 14.17* gives a diagram to show the voltage drop IZ in the leakage impedance Z of the transformer (as obtained from a short-circuit test). The regulation is a maximum when the load phase angle ϕ is

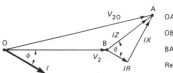

Figure 14.17 Regulation

equal to the angle θ in the impedance triangle of R, X and Z. If the load is leading reactive, the regulation is reduced; it may even be reversed so that the secondary voltage rises on load. Formulae (BS 171) for the regulation ε of a two-winding transformer on full rated load, in terms of the % IR drop voltage ε_r and the % reactance voltage ε_x are:

(1) $\varepsilon = \varepsilon_r \cos\phi + \varepsilon_x \sin\phi + (\varepsilon_x \cos\phi - \varepsilon_r \sin\phi)^2/200\%$

(2) $\varepsilon = \varepsilon_r \cos\phi + \varepsilon_x \sin\phi\%$

where (1) applies to transformers having impedance voltages up to 20%, and (2) is a simplified expression for cases in which the impedance voltage does not exceed 4%. For unity p.f. load, (1) reduces to $\varepsilon = \varepsilon_r + \varepsilon_x^2/200$, and (2) to $\varepsilon = \varepsilon_r$.

14.4 Connections

Transformer windings can be constructed for connection to a 1-phase, 2-phase or 3-phase power supply. Combinations are also possible, namely 3/2-phase or 3/1-phase conversion. Six-, 12- and

remote end (with x expressed as a fraction of the winding length) is:

isolated-neutral winding $e_x = E (\cosh \alpha x/\cosh \alpha)$
earthed-neutral winding $e_x = E (\sinh \alpha x/\sinh \alpha)$

for a surge voltage peak of E.

If the surge voltage is maintained, an approximately uniform voltage distribution will be reached; but between these two states (if they differ) a complex array of damped oscillations will occur, creating abnormal stresses in the insulation (*Figure 14.14c*).

A chop occurring between 3 and 10 μs after application of the impulse voltage will result in augmented interturn and intercoil stresses, although the voltages to earth are normally reduced. The stressing again depends on the initial distribution and on the actual chopping instant. The chop may be regarded as a unit-function voltage, of polarity opposite to that of the incoming surge, and superimposed on the conditions existing within the winding at the instant of chopping.

With the front-of-wave test a higher voltage is applied, but its duration is shorter and the increased stresses are generally confined to the entrance insulation (particularly the bushing). Internal voltage stresses are usually less than those arising from a chopped wave test.

The ratio α is high for a small transformer. For transformers of higher rating and especially for higher line voltage, C_g decreases because it is determined largely by clearances, while C_s increases because of the greater radial depth of the winding. Thus, $\alpha = \sqrt{C_g/C_s}$ is lower and the initial voltage distribution approximates more closely to the uniform. To reduce stress concentrations at the ends, disc-coil windings are provided with stress rings that act as radial shields, although they do not materially improve the axial distribution. Axial improvement can be gained by the addition of rib shields to the line-end turns (*Figure 14.15*) or by several other means of controlling the electric field distribution.

Compared with a disc winding, a multilayer coil has a higher series capacitance and therefore a superior transient distribution, although electric field shielding at the ends of the layers may be necessary. The choice between disc-coil and multilayer windings depends, therefore, on the specified impulse level and on the rating.

14.3.4 Losses

The load (or 'copper') loss comprises two components; a direct I^2R loss due to ohmic resistance of the windings, and a stray loss arising from eddy currents in the conductors due to their own flux, influenced by the tank and by steel clamping structures. The eddy loss is negligible when the section of the conductors is small. When the current is too great for a single conductor without excessive eddy loss, a number of strands must be used in parallel. Because the parallel components are joined at the ends of the coil, steps must be taken to circumvent the induction of different e.m.f.s in the strands, which would involve circulating currents and further loss. Forms of conductor transposition have been devised for this purpose.

Ideally, each conductor element should occupy every possible position in such a way that all elements have the same resistance and the same induced e.m.f. Transposition, however, involves some sacrifice of winding space. If the winding depth is small, one transposition halfway through the winding is sufficient; or in the case of a two-layer winding, at the junction of the layers. Windings of greater depth need more transpositions.

Typical forms of transposition are shown in *Figure 14.16*. The methods apply mainly to helical coils. In disc windings where

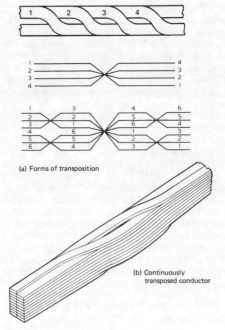

Figure 14.16 Conductor transposition

there are two or more conductors in parallel, the connections between the discs can be arranged to give the necessary effect.

Stray loss may also be produced by radial components of leakage flux, but can be minimised by careful ampere-turn balance of the windings.

14.3.5 Cooling

The majority of transformer windings are cooled by thermal circulation of oil. Even where pumps are fitted to provide forced

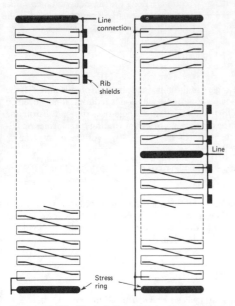

Figure 14.15 Disc windings fitted with stress rings and rib shields

Table 14.4 Insulation level test voltages

Highest system voltage kV r.m.s.	Power-freq. test kV r.m.s.*	Impulse test voltage, kV		Power-frequency test voltage, kV	
		Standard 1	Standard 2	Standard 1	Standard 2
less than 1.1	2.5				
1.1	3.0				
3.6	8.0	45		16	
7.2	15.0	60		22	
12.0	25.0	75		28	
17.5	36.0	95		38	
24.0		125		50	
36.0		170		70	
52.0		250		95	
72.5		325		140	
100.0		450	380	185	150
123.0		550	450	230	185
145.0		650	550	275	230
170.0		750	650	325	275
245.0		1 050	900	460	395
300.0		—	1 050	—	460
362.0		—	1 175	—	510
420.0		—	1 425	—	630

* No impulse test.

transformers with graded insulation. The value depends on the method of earthing. For solid earthing the separate-source power frequency test voltage is 38 kV, although 45 kV is required for transformers connected to the British 132, 275 and 400 kV systems.

14.3.3.1 Surge voltage distribution

The voltage distribution in power frequency voltage tests is substantially uniform between turns and coils, and corresponds to the normal service condition in respect of voltage to earth. Under impulse test, however, the distribution can be far from uniform. An impulse voltage wave has a steep front and a long tail. The standard form, defined in BS 923:1980, has a front rising from zero to peak value in 1.2 μs, and falling thereafter on the tail to 50% of peak value in 50 μs. This is termed a 1.2/50 μs wave (*Figure 14.13a*). Impulse voltage tests on transformer windings include the application of a *full wave* and a *chopped wave* impulse, the latter being a full wave shortened by sparkover of a rod gap or its equivalent. American practice adds a front-of-wave test, in which a voltage rising at approximately 1 MV/μs is chopped on the wavefront. The three forms of impulse test voltage are shown in *Figure 14.13(b)*.

A full wave application tests the ability of the insulation structure to withstand voltage surges; a chopped wave test simulates stresses that occur on collapse of a surge tail by operation of a rod gap or a flashover to earth. Transformers for systems operating at 300 kV or above may be also required to demonstrate ability to withstand switching surges, typically by a 250/2500 μs wave: see under 'Testing'.

Ignoring resistance, a transformer winding and its surroundings may be represented by an inductor–capacitor network (*Figure 14.14a*). When a steep-fronted surge is applied to the line terminal, the initial voltage distribution is governed solely by the capacitance network, in particular by the ratio $\alpha = \sqrt{(C_g/C_s)}$ of the capacitance to earth and in series. The greater the value of α the greater is the divergence from uniform voltage distribution from line to ground, as shown in *Figure 14.14(b)*. For a uniform winding of identical sections having the same capacitance values C_s and C_g, the initial voltage to earth at any point x from the

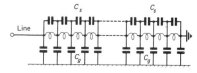

(a) Equivalent circuit of transformer winding

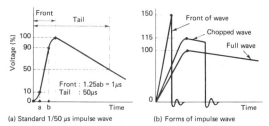

(a) Standard 1/50 μs impulse wave   (b) Forms of impulse wave

Figure 14.13 Impulse testing

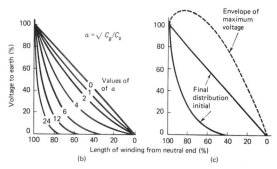

Figure 14.14 Voltage distribution

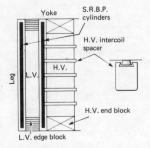

Figure 14.11 Typical insulation for small and medium-sized transformers

of pressboard or paper wound directly over the inner winding, the h.v. winding being assembled directly over the outer layer, with cooling ducts transferred from the inside turns to some point in the winding. In this case the pressboard or paper wraps are extended beyond the axial winding length and flanged at the ends to assist with the insulation to earth (*Figure 14.12b*). A combination of s.r.b.p. cylinders and pressboard wrappings can also be employed, the wrappings located over the outer cylinder again flanged at the ends. Further, the s.r.b.p. tube can be manufactured such that the bonding is limited to the central portion, leaving the ends plain so that these in turn can be flanged.

With multilayer helical high-voltage windings, the interlayer insulation is provided in the form of a combination of paper or pressboard wrappings, and bars for cooling ducts, with the paper between the individual layers flanged at the ends (*Figure 14.12c*). When internal interlayer connections are involved (*Figure 14.8c*), the interlayer paper wrappings may be formed in tapered halves so that the maximum thickness of insulation is provided at appropriate places between the actual connection and the layers.

Paper is used almost exclusively for the interturn insulation of high-voltage windings.

14.3.3 Winding design

The main factors to be taken into account in winding design are:

(1) To provide adequate insulation strength to withstand (a) power frequency applied or induced overvoltage tests to prove the insulation to earth and between windings; (b) an induced voltage test for internal winding insulation (interturn, intercoil and between phases); (c) an impulse voltage test to prove the ability of the insulation structure to withstand transient overvoltages such as may result from atmospheric surges; and (d), in some cases, a switching-surge test to prove the strength to resist surges arising from switching operations. At system voltages of 500 kV and above switching surges rather than lightning control the design.
(2) To ensure that the load loss (i.e. the sum of the I^2R and stray losses) does not exceed the guaranteed performance figure.
(3) To provide adequate cooling to meet guaranteed temperature rise limits.
(4) To ensure adequate short-circuit strength.
(5) To achieve the required impedance characteristic.

Transformer windings may be fully insulated, where the insulation to earth at all points will withstand the separate source voltage test specified for the line terminals; or have graded insulation, where the insulation to earth is reduced from that required for the line terminals to a smaller amount at the neutral end, and which, therefore, will withstand only a separate-source voltage test corresponding to the insulation level at the neutral. In the latter case induced voltage is employed to test not only the internal insulation and that between phases, but also the insulation to earth at the line end.

The required level of insulation at the line end is normally designated as the *basic insulation level* (b.i.l.). This is specified by the purchaser of the transformer and is determined by taking into account the maximum expected atmospheric or switching surges which may be imposed on the transformer, plus a small margin of safety.

The b.i.l. for any particular installation at a given system voltage is governed by the type and effectiveness of the protection against lightning and the combination of line and circuit-breaker characteristics that control the switching surge limit. National and international standards of insulation levels are strictly observed in some countries such as Britain, but in others (such as the USA) are frequently varied to achieve the most economic overall transmission system cost.

For systems where the neutral is isolated, or earthed through an impedance (e.g. an arc-suppression coil) such that, during a line-to-earth fault, the voltage to earth of the unfaulted lines can exceed 80% of the normal line-to-line voltage, fully insulated systems are essential. They are called *non-effectively earthed* systems. Graded insulation is permitted in *effectively earthed* systems, where the method of earthing at each transformer is such that the 80% value is not exceeded for any operating condition. The requirement is met when the zero phase sequence/positive phase sequence (z.p.s./p.p.s.) reactance ratio of the system is less than 3, and the z.p.s./p.p.s. resistance ratio is less than 1.

Uniform insulation must be provided for all delta-connected windings, and for star-connected windings where the neutral is not earthed. (Grading may be applied in the latter case if the neutral is earthed.) Where earthing conditions permit grading, they also, in general, allow forms of overvoltage protection for which test voltage levels are lower. Two series of test voltages are given for oil immersed transformers in BS 171, namely Standard 1 for non-effectively earthed and Standard 2 for effectively earthed systems. Typical values, together with those for dry-type transformers and oil immersed transformers not designed to withstand impulse testing, are given in *Table 14.4*.

With uniform windings, the power-frequency test voltage values in the Table apply to the separate-source overvoltage test, the voltage for the induced test being such as to produce, between any two parts of the winding, twice the voltage normally between them. For graded windings the induced-voltage test is required to raise the line terminals above earth to the appropriate value listed.

Further voltage tests are specified for the neutral ends of

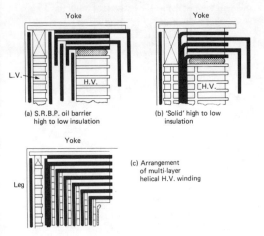

Figure 14.12 Insulation design

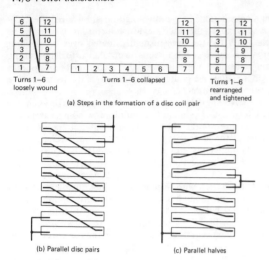

Figure 14.10 Disc coil windings

the coil from the lathe. The continuous coil avoids assembly and joining and saves joint space, an advantage with an inside winding. For large currents the coils can be wound with multiple conductors, or constructed as separate parallel connected pairs (*Figure 14.10b*). Alternatively, two half-windings, one half reversed with respect to the other, can be stacked as in *Figure 14.10(c)* with the line connection taken from the centre of the stack; this method is common for large high-voltage transformers having a directly earthed neutral. Horizontal cooling ducts are provided by spacers between individual discs and between disc pairs.

The disc coil is a general-purpose winding element, applied to the higher-voltage windings of transformers above 1 MVA up to the highest ratings and voltages, and also for the lower-voltage windings (normally 33 kV upwards) of medium and large power transformers. *Table 14.3* gives a survey of typical applications of the various forms of winding.

14.3.2 Insulation

Materials of classes A, E, B, F, H and C (in temperature sequence) are recognised in BS 171 as being suitable for transformers. A, B, H and C are most common, but F is used to an increasing extent. All these are employed in dry-type transformers, the silicon-treated materials being advantageous (but at extra cost) by reason of their water-repellent properties. For windings immersed in hydrocarbon oil (BS 148: 1972) or in synthetic liquids, the coil insulation is usually a class A material. Cotton is confined to small units, with paper and paper derivatives used for most purposes up to the largest ratings; but there is a growing application of synthetic enamel as interturn insulation in small transformers and in the low-voltage windings (up to about 17.4 kV) of large transformers.

Most power transformers are immersed in hydrocarbon oil. To reduce fire risk, chlorinated biphenyls (e.g. Pyroclor) have been used, but as these liquids are toxic (and difficult to dispose of safely), they have virtually been banned on environmental grounds. Synthetic silicone-based liquids are now available without these disadvantages, but they are relatively expensive; and the varnishes and binders usable with hydrocarbon oil may not be suitable. Where fire risk is unacceptable, air-cooled 'dry'-type designs with glass and epoxy resin insulation are usually preferred. Gas filled transformers using electronegative sulphur hexafluoride gas (SF6) at a pressure of about 1 atm eliminate fire risk, but the cost is high, the force-cooled heat exchangers are complicated, and the transformers have short thermal time constants, giving reduced overload capability.

14.3.2.1 Insulation design

For transformers of small and medium size, the inner (l.v.) winding is insulated from the core by pressboard or a synthetic resin-bonded paper (s.r.b.p.) cylinder, with axial bars of pressboard or equivalent material arranged round it to form cooling ducts for the inside surface of the winding. With disc or disc-helix coils, the bars have a wedge section on which intercoil or interturn dovetail-slotted spacers can be threaded. Similar bars are placed over the outer surface of the inner winding and, if necessary, between layers of a helix winding. The main h.v./l.v. insulation is provided by another pressboard or s.r.b.p. cylinder. The arrangement of bars and spacers is repeated for the outside (h.v.) winding. Insulation to earth at the ends of the windings takes the form of blocks keyed to the axial bars in line with the spacers, to form a series of columns round the windings by which the windings can be effectively clamped. The cross-section of a typical arrangement is shown in *Figure 14.11*.

Insulation arrangements for dry-type transformers are similar to those described, except that the materials chosen are appropriate to the insulation class. For example, the cylinders and bars can be made from suitably impregnated glass fibre, ceramic materials may be used for the coil spacers and glass tape for interturn insulation.

When operating voltages are such that thick major insulation is required between windings, it is customary to use a number of concentric thin-walled cylinders, spaced by axial bars, the insulation being carried round the ends of the outer (h.v.) winding by means of flanged collars interfitting with the ends of the tubes (*Figure 14.12a*). Alternatively, with h.v. windings at above 110 kV, so-called *solid* insulation can be used, comprising layers

Table 14.3 Transformer windings

Service	MVA	H.V. winding		L.V. winding	
		kV	Type	kV	Type
Distribution	up to 1	11, 33	Foil, crossover or multilayer	0.43	Helix
System	1–30	33, 66	Disc	11	Disc, helix or disc-helix
Transmission	30 upward	132–500	Disc or multilayer	11, 33, 66	Disc or disc-helix
Generator	30 upward	132–500	Disc or multilayer	11–22	Disc-helix

14.3.1 Types of coil

The simplest *helix* coil consists of a single layer, formed by turns lying directly side by side, extending over the axial winding length. Each turn may comprise a single conductor or a number of conductors in parallel, the helix at each end of the coil being supported by a suitably shaped edge block to give adequate mechanical strength in an axial direction (*Figure 14.8a*). This type of coil, single- or double-layer, is used for the low-voltage windings of small and medium-sized transformers.

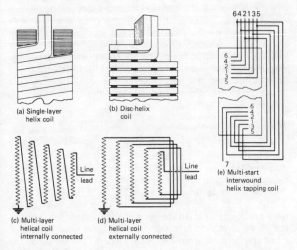

Figure 14.8 Forms of coil

When the current rating necessitates a large number of conductors in parallel for each single turn, the individual strips can be laid one above the other in a radial direction, as either a single column or two columns in parallel, each turn or column separated by spacers to provide sufficient cooling surface (*Figure 14.8b*). This type of coil, sometimes termed a *disc helix*, is used for the low-voltage windings of large transformers.

Multilayer helix coils can be employed for the high-voltage windings of large transformers. Because of the high capacitance between the individual layers, this type of winding has good inherent strength against incoming surge voltages. It is fairly simple to calculate the surge response of the multilayer winding, but difficult to ensure that the long thin layers of conductors have adequate strength to resist axial forces. As ratings increase for a given primary service voltage, a *disc* coil (see below) becomes relatively less difficult to design from the point of view of predicting surge strength and more satisfactory because of its high inherent strength against axial forces. The use of the multilayer helical coil is thus generally confined to the smaller MVA ratings of any given voltage class above about 200 kV.

This type of coil shows particular advantage when the transformer neutral is directly earthed. The inside layer of the winding is then near earth potential, thus reducing the major insulation between high- and low-voltage windings. The length of the individual layers decreases progressively towards the outer layer to provide increased insulation to earth at the line end. Interlayer connections, usually top-to-bottom, can either be arranged internal to the coil between the layers, or formed by external joints (*Figure 14.8c, d*). Multilayer coils are often used for the high-voltage windings of small distribution transformers, in this case wound as continuous layers, with top-to-top and bottom-to-bottom interlayer connections.

Another form of layer coil is a *multistart interwound helix* employed for the separate tapping windings of large transformers when a considerable range of voltage variation is required. The coil is so arranged that each single conductor forms one tapping section, with the requisite turns distributed over the winding length to provide axial ampere-turn balance for the various tapping positions. The individual conductors are physically located so that the voltage between them is reduced to a minimum: see *Figure 14.8(e)*.

A *crossover* or *bobbin* coil is wound on a former between side cheeks. It is, in effect, similar to a multilayer winding, except that the layers are short. The interlayer insulation needs to be extended at the ends to guard against failure by creep at the edges; alternatively, it can be folded round the last turn of each layer, or the overhang crimped to form a trough of depth equal to the thickness of the conductor. The assembly of separate coils is connected in series, with horizontal cooling ducts provided by spacers between the individual coils. Intercoil connections are usually made back-to-back and front-to-front, with adjacent coils reversed. Back-to-front connection requires insulation of the intercoil connections corresponding to the voltage developed across the coil: see *Figure 14.9*. Crossover coils, generally of round wire, are employed for the high-voltage windings of distribution transformers up to about 1000 kVA.

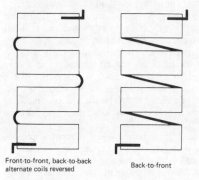

Figure 14.9 Crossover coils: intercoil connections

Because of increase in the cost of copper, more attention is being paid to the use of aluminium for transformer windings. Aluminium is a less effective conductor than copper, but despite this intrinsic disadvantage (which results in a larger transformer for a given efficiency) there is a net saving in overall cost. Aluminium strip can be used to wind any of the types of coil described above, but it is also uniquely used in *foil* form. This is because aluninium can be rolled to a thinner and more flexible foil than copper. The foil is used in bobbin-type coils of one turn per layer with intercoil connections as shown in *Figure 14.9*.

The *disc* coil differs from other windings in that adjacent turns, consisting of strip conductor, are wound one above the other in a radial direction from the centre outwards; thus, the coil might be compared with a crossover coil comprising one turn per layer. To achieve the required disposition of turns, the coils are formed in pairs, the requisite turns for one disc being loosely wound so that the conductor finishes in a position to provide the start of the inside turn of the second disc, which can then be wound from the inside outwards, the turns of the first disc being rearranged by folding one inside another in such a manner that the start of the coil is located as an outside turn (*Figure 14.10a*). The coil is then tightened to remove slack from the re-formed disc.

Disc coils can be wound either as individual pairs, which are then assembled and connected in series by external joints, or as a continuous winding where, following formation of one pair, the procedure is repeated without cutting the conductor or removing

unit value. For distribution transformers in the range 500–1000 kVA and built of c.r.o.s. the no-load current is of the order of 1.5% (0.015 p.u.). Values for large high-voltage transformers may be less than 0.75%.

The saturation characteristic of the steel is such that the no-load current increases rapidly when the transformer is over-excited. If with c.r.o.s. the normal peak density is 1.6 T, the no-load current will be at least doubled by a 10% overvoltage. Further, the harmonics in the current wave form will be substantially increased. Provided that the circuit connections permit, the most pronounced harmonic is the third, although fifth and seventh harmonics can become significant at high densities. If the circuit connections do not permit third-harmonic currents to flow (in a 3-phase transformer there must either be a neutral connection to a source of zero sequence current, or a delta connection within the transformer in which third harmonics can circulate), the flux wave form will distort because of the lack of the component. The distortion appears as a third-harmonic ripple in the secondary line-to-earth voltage.

The magnetic path length associated with the central phase of a 3-phase 3-limbed core-type transformer (*Figure 14.3c*) is significantly shorter than that of either of the outer phases. The configuration shows that the outer path lengths include two half-yokes in addition to the limb. As a consequence, the magnetising current and core loss values are asymmetric, to an extent depending on the path length ratio. If the central path length is one-half that of either outer, then its magnetising current is likely to be about 30% less, and this is independent of the peak flux density level. Measurement of the total core loss by the two-wattmeter method is made difficult by the effect of the low power factor and the small difference between two large readings: the measurement also depends upon which phase currents pass through the wattmeter current circuits, and even on the phase sequence if the wattmeter volt circuits have inductance errors. The most satisfactory core loss assessment is made with one wattmeter for each phase.

14.2.3.1 Magnetising inrush current

When a transformer is 'switched-in' on no load, it may take an initial *inrush* current greatly exceeding normal no-load value, and sometimes greater than full-load current. The inrush transient decays to normal no-load level within a few periods. The first peak depends on the voltage at the instant of switching, and on the magnetic state of the core as left after the previous switching-out.

If the instant of switch-on corresponds to a voltage zero, the flux must, in the first half-period, produce a complete change of $2\Phi_m$ from zero, as shown in *Figure 14.7*. The peak flux therefore rises to twice normal peak Φ_m. The maximum flux reached will be increased to $(2\Phi_m + \Phi_r)$ if there is a residual flux Φ_r already present in the core in the same direction as that to be taken by the first half-cycle of flux growth. These high-flux conditions demand very high peaks of exciting current with large harmonic content. As the core steel will saturate, much of the flux will follow an 'air'

path, and the peak inrush current is consequently influenced by the area enclosed by the winding excited.

Under adverse conditions the magnitude and asymmetry of inrush currents may cause maloperation of overcurrent or balanced forms of protection, but in practice the worst conditions are statistically unlikely, and terminal voltage drop will reduce the peaks. In a 3-phase transformer the inrush conditions must differ for each phase.

14.2.3.2 Magnetostriction

Magnetostriction is a property of magnetic material whereby a small change in linear dimensions (usually an elongation) follows the flux cycle in a complex pattern. In transformer steel the linear change is a few parts in a million. It causes vibration of the core at twice supply frequency and at multiples thereof, producing sound waves. Magnetostriction in the material of the core is the main source of transformer noise.

14.2.3.3 Noise

The unremitting hum of a transformer installed in a residential area can lead to complaints and in extreme cases to legal action. Careful design and manufacture is necessary to ensure that the noise emitted is within the level normally accepted as reasonable for a given size of transformer. The Electricity Supply Industry Standards (ESI 35-1 for Distribution Transformers up to 1000 kVA and ESI 35-2 for Emergency Rated System Transformers) and British Electricity Board Specification (BEBS) T2 for the large units contain curves relating MVA rating and noise level which form part of the manufacturer's contractual obligations. BEAMA publication No. 227, *Guide to Transformer Noise Measurement*, is a useful reference work on the subject.

If it is expected that the noise emitted from a particular installation may create a nuisance, various mitigating measures are possible, including the following:

(1) Specifying a noise level less than the normal standard. Usually not economic (or even feasible) if reductions of more than about 10 dB below standard are likely to be necessary.
(2) Concealing the transformer behind a screen of trees or a wall. Sometimes the psychological effect is greater than the actual measured reduction in noise level at the point of complaint.
(3) Completely enclosing the transformer in a 'sound-proof' housing. This is expensive, as obviously special measures have to be taken to emit heat without emitting noise.
(4) A combination of (2) and (3) in which specially designed coolers form a screen wall completely surrounding the transformer. This is a useful development from the point of view of amenity in respect of both appearance and noise reduction.

14.3 Windings and insulation

Because of their direct association with power systems, the windings and associated insulation are the most vulnerable parts of a transformer. They must be designed and constructed to withstand the voltage stresses and thermal conditions of normal service, the mechanical and thermal stresses resulting from system faults and short-circuits, and transient overvoltages such as those generated by lightning and switching. The core and windings together must meet specified impedance and loss requirements.

In view of the almost total predominance in Britain of core-type transformers, the following section relates only to windings for this type.

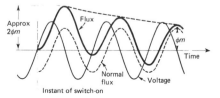

Figure 14.7 Flux transient for switching at voltage zero (no residual)

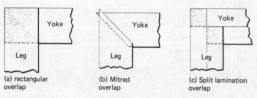

Figure 14.4 Types of joint in cores (shaded portions indicate overlap areas)

directional properties of c.r.o.s., the area overlap has been minimised by means of mitred joints (*Figure 14.4b*). The affected area can also be reduced by using split laminations (*Figure 14.4c*), with the split butted in small cores, or separated to form a cooling duct in the case of large cores.

By eliminating core joints, or by arranging that joints do not disturb the optimum flux path, maximum advantage can be taken of c.r.o.s. Such conditions are achieved with the wound core (*Figure 14.5*). This construction is widely used for transformers

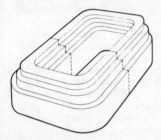

Figure 14.5 Strip-wound core (cuts in legs for butt-jointed core are shown dotted)

up to 25 kVA, the limitation in rating being imposed by special coil-winding requirements which involve a split former so that the coils can be wound on the completed core. A variation called a C core, of slightly reduced efficiency, is a wound core cut across the centre of the leg section, thus involving a butt joint in each limb.

14.2.2.4 Core building factor

The total core loss divided by the total mass of the completed magnetic circuit gives the specific loss of the built core. If this figure is divided by the specific loss of the material used (from tests on samples or from the steelmaker's guarantee), the result is the *core building factor*, a measure of the effectiveness of the magnetic circuit design. Factors vary with the size of core, smaller cores generally giving factors near to unity. Factors for large cores can be kept to less than 1.2 if full attention is paid in the design to make the best use of the directional properties of the steel.

14.2.2.5 Cutting and punching

Cold-rolled grain-oriented steel is produced in the form of strip about 750–800 mm wide and is supplied in coils of approximately 2 t weight. Individual laminations are produced by slitting to the widths required, and subsequent cropping to length by guillotine. Where bolt-holes are employed, it is preferable to use a single hydraulic press to crop the strip material to length and punch bolt-holes simultaneously. Following cutting and punching, the individual laminations may be dressed to remove any edge burrs, but as deburring may harm the magnetic properties of the

material, it is preferable that the cutting tools be maintained in good condition so that deburring is not necessary.

Cutting and punching adversely affect the magnetic properties of the material, and it is desirable that the finished laminations (or for small units the complete cores) should be stress-relief annealed to remove cutting strains. Various types of annealing furnace have been used, including the *batch* furnace and the *continuous belt* furnace, in which small stacks of laminations (up to 10 or 12 plates deep) pass slowly through a heating zone at about 800°C in an inert atmosphere of nitrogen. In the *single-sheet roller hearth* furnace, single laminations pass relatively rapidly through the heating zone; with this furnace the laminations are in the heated zone for a comparatively short time and it is unnecessary to provide an inert atmosphere to prevent oxidation. Annealing ensures uniformity of material and may reduce the core loss by 5–15% and the magnetising vars by even greater amounts.

Another form of annealing (which can rarely be economically utilised by the individual transformer manufacturer because of the relatively small quantities of material of any given width) is the *catenary* annealing method: the steel is annealed in continuous strip form after it has been cut to the required width. The steel is allowed to hang as an unsupported loop (hence the name) as it passes through the heated zone of the furnace. It is thus maintained under constant tension throughout the heating and subsequent cooling process. The residual tension left in the steel after it has cooled results in improved magnetostriction properties. Some steel manufacturers sell annealed strip to transformer manufacturers in ready-for-use widths.

14.2.3 Magnetic circuit characteristics

The magnetic circuit characteristics are the no-load core loss and the magnetising current. The former is commonly divided into hysteresis and eddy current components. The hysteresis loss depends on the cycle of flux density and the frequency, while the eddy current loss depends on the r.m.s. density and on the degree of subdivision (i.e. lamination) of the core. The specific loss components are given by

$$p_h = k_h f B_m^n \text{ and } p_e = k_e f^2 B^2 t^2$$

where B_m is the peak and B the r.m.s. flux density, f is the frequency, t is the lamination thickness, and k_h and k_e are constants for a given material. The exponent n is empirical, with a value generally not very different from 2.

The measured no-load loss includes I^2R loss due to the no-load current, and dielectric loss (especially for large high-voltage transformers); these components can often be neglected.

The no-load current I_0 can be taken as comprising an active component I_{0a} in phase with the applied voltage V and accounting for the no-load power input P_0, together with a reactive or magnetising component I_{0r} (*Figure 14.6*). The two components are not physically separable, as both are concerned together in the magnetisation of the core. The no-load current is normally expressed in terms of the full-load current as a percentage or per-

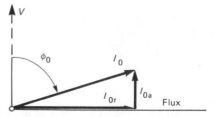

Figure 14.6 Phasor diagram of no-load current

alternate high-voltage/low-voltage groups is dependent on the required reactance characteristics of the transformer.

In Great Britain the core-type transformer is used for all power system applications. The shell form is used only for special applications, usually where very heavy current low-voltage outputs are required for such purposes as electric arc furnaces or short-circuit testing stations. The shell form is used in the USA and elsewhere for power transformers of the largest sizes and highest voltages. The fact that both core- and shell-type transformers have existed side by side for many years indicates that neither has any significant economic advantage. There is, however, a discernible trend away from the shell towards the core type of construction, due entirely to economic considerations.

Core-type transformers (circular coils on circular limbs) are built in various forms, as shown in *Figure 14.3*. The choice is

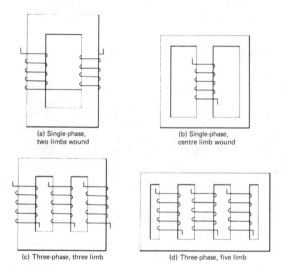

Figure 14.3 Core arrangements

dictated by the type of winding or the need to meet transport loading restriction. Thus, a 5-limb core will have a smaller overall height than a 3-limb core for a 3-phase transformer. A very large high-voltage 1-phase transformer might also have to be designed with a multi-limb core.

The basic parameters of the core are (a) the core circle diameter, (b) the yoke height, (c) the limb centres and (d) the limb length. The normal practice in a design office is that the first three are related to a range of fixed standards, while a limited variation on core size is secured by change of limb length. The relation between the core circle diameter D and the rating S of a transformer is given approximately by

$$D = k(S)^{1/4}$$

where k is a constant depending on the type of transformer. Thus, if a 500 kVA transformer has core circle diameter of 190 mm, then a 1000 kVA transformer of the same voltage ratio and service conditions would have $D = 220$ mm approximately. The actual diameter adopted would depend on the nearest standard diameter available.

The e.m.f. per turn E_t is a function of the frequency f, the peak flux density B_m and the net core area A_i:

$$E_t = 4.44 f B_m A_i.$$

14.2.2.1 Limb space factor

To obtain the most economical use of core steel and copper it is essential to utilise as much as possible of the available core circle. Consequently, the limb sections are built in steps; the greater the diameter, the larger the number of steps. Optimum utilisation factors for cores with 1–7 steps are given in *Table 14.2*.

Table 14.2 Core utilisation factors

Number of steps	1 (square)	2	3	4	5	6	7
Fraction of core circle areas	0.64	0.78	0.85	0.89	0.91	0.92	0.94

In practice these areas are reduced by the lamination stacking factor, normally between 0.9 and 0.95, depending on the thickness of the interlaminar insulation and the tightness of the core clamping. With large transformers allowance must be made for loss of area due to cooling ducts in the core, and the area will also be affected by the relative size of the individual core steps, which can depend on the width necessary to accommodate core clamping plates.

Cores having more than 7 steps (up to as many as 15 or more) are common, but the improvement in utilisation decreases with an increasing number of steps and is offset from the production point of view by the increasing complexity involved in producing a greater number of widths.

14.2.2.2 Yoke dimensions

As the yoke height is not restricted by a winding, the area selected may be larger than that of the corresponding limbs. Dissimilar steps between the limb and yoke packets, however, can give rise to cross-flux, and the resultant increase in core loss may offset benefits accruing from the greater yoke area; this is a factor of particular significance with c.r.o.s. Apart from very small cores, modern practice is to step the yoke and limb sections in a similar, or preferably identical, manner.

When the magnetic circuit is completed by a single-path yoke, e.g. a 1-phase 2-limb core (*Figure 14.3a*), or a 3-phase 3-limb core (*Figure 14.3c*), yoke areas at least equal to the limb section are necessary. Where the yoke path is split, the yoke area can be appropriately reduced, e.g. with a 1-phase centre-limb-wound core (*Figure 14.3b*), the yoke area can be halved, and with a 3-phase 5-limb core (*Figure 14.3d*), the yoke area can be reduced to about 0.55–0.6 of that of the corresponding limb.

14.2.2.3 Core clamping and core joints

Bolt-holes cause local flux deviation and crowding, which results in increased noise and loss. These effects have led to the development of boltless cores, which are held together by circumferential bands. These bands are sometimes of steel with suitable insulated sections, but synthetic-resin-impregnated glass fibre tape is preferable, as it eliminates any risk of the band becoming involved in an electrical failure or becoming overheated owing to eddy currents. Banded cores are now used even in the largest transformers. Other methods of clamping include bolts passing through oil ducts in the core to avoid holes in the sheet, and reduction of the bolt area by the use of high-tensile steel.

The introduction of c.r.o.s. necessitated a modification of the traditional rectangular overlap (*Figure 14.4a*) between yoke and leg laminations commonly used with hot-rolled strip cores. To achieve minimum loss by taking maximum advantage of the

14.1 General

A transformer consists essentially of two or more electric circuits in the form of windings magnetically interlinked by a common magnetic circuit. An alternating voltage applied to one of the windings produces, by electromagnetic induction, a corresponding e.m.f. in the other windings, and energy can be transferred from the primary circuit to the other circuits by means of the common magnetic flux.

The British Standard Specification for power transformers is BS 171: 1970 and 1978 to which reference should be made for all information relating to the standard characteristics, guaranteed performance and tolerances, testing and operation, introduced to conform with the International Electrotechnical Commission (IEC) Publication No. 76 (1967, 1976 et al.), *Power Transformers*. Other useful references are Electricity Supply Industry (ESI) Standard 35-1 for Distribution Transformers up to 1000 kVA and 35-2 for Emergency Rated System Transformers; and British Electricity Board Specification (BEBS) T2 for Transformers and Reactors.

14.2 Magnetic circuit

The magnetic circuit, or core, provides a closed path for the flux. To prevent excessive eddy current loss within the metal of the core itself, it must be laminated in a plane parallel to the flux path. Individual laminations are insulated with a thin layer of phosphate applied during the manufacturing of the steel itself. For cores having a lamination width greater than about 250 mm, it is usual to apply an additional thin coat of varnish to each lamination.

14.2.1 Core steel

For many years power transformer cores were made up with laminations cut from sheets of special 4% silicon alloy steel produced by a hot-rolling process. During the late 1940s an improved material was developed known as cold-rolled grain-oriented strip (c.r.o.s.). This material has a 3% silicon content and is produced in strip form, usually 0.30 mm thick. Because of the effect of the cold-rolling process on the grain formation, it has magnetic properties in the direction of rolling that are far superior to those in other directions.

The basic requirements for transformer core steel are laid down in BS 601. *Figure 14.1* shows the directional properties of

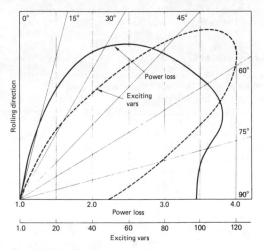

Figure 14.1 Directional properties of c.r.o.s.

c.r.o.s.: the actual specific loss (W/kg) and excitation (r.m.s. var/kg) for a particular grade of material, when magnetised parallel to the direction of rolling (i.e. 0°), at a flux-density of 1.5 T, are referred to unity, and the change in these quantities with magnetisation at other angles is shown as a multiple. For example, the loss with magnetisation at 45° to the rolling direction is increased by a factor of 3 and the excitation vars by a factor of 85.

The 'knee' of the magnetisation curve of c.r.o.s. occurs at about 1.7 T, and provided that the directional properties are properly exploited, it can be operated at appreciably higher flux densities than the older hot-rolled steel cores. Whereas the latter were typically operated at 1.35 T, modern c.r.o.s. cores usually operate at nominal densities of 1.6–1.8 T. The development of c.r.o.s. is a major factor in the remarkable recent improvement in transformer output per unit of active material.

Table 14.1 shows the various grades commonly used. The original grades are now designated M4, M5 and M6. The M2H grade is a more recent development having an improved degree of orientation.

Table 14.1 Typical core loss in W/kg at 50 Hz for flux densities of 1.5 Tesla and 1.7 Tesla

Grade	Thickness (mm)	Watts per kilogram	
		1.5 Tesla	1.7 Tesla
35M6	0.35	1.02	1.44
30M5	0.30	0.90	1.32
28M4	0.28	0.85	1.26
30M2H	0.30	0.83	1.12

Although c.r.o.s. is now used for virtually all 'power' transformer cores from 1 kVA upwards, there are other special low-loss steels (e.g. Mumetal) used for the cores of instrument transformers. These materials have markedly superior magnetic properties (i.e. magnetising vars and power loss) at low densities, but they saturate at much lower levels than c.r.o.s. They are therefore not economic for power transformers, where the advantage of a high operating flux density overrides all other factors.

14.2.2 Magnetic circuit design

The two fundamental types of construction used, shown diagrammatically in *Figure 14.2*, are known as *core* and *shell*, respectively. The normal arrangement of a core-type transformer is for circular primary and secondary windings to be arranged concentrically around the core leg of substantially circular cross-section. In the shell type the magnetic circuit is of rectangular cross-section formed by a stack of laminations of constant width. The coils are straight-sided and the primary and secondary windings are interleaved in a sandwich fashion. The number of

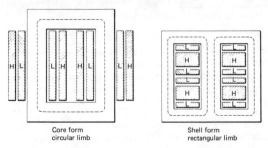

Figure 14.2 Basic cell and magnetic circuit arrangements

14 Power Transformers

R W Flux
NEI — Peebles Ltd

Contents

14.1 General 14/3

14.2 Magnetic circuit 14/3
    14.2.1 Core steel 14/3
    14.2.2 Magnetic circuit design 14/3
    14.2.3 Magnetic circuit characteristics 14/5

14.3 Windings and insulation 14/7
    14.3.1 Types of coil 14/7
    14.3.2 Insulation 14/8
    14.3.3 Winding design 14/9
    14.3.4 Losses 14/11
    14.3.5 Cooling 14/11
    14.3.6 Short-circuit conditions 14/12
    14.3.7 Impedance voltage 14/12

14.4 Connections 14/12
    14.4.1 Phase conversion 14/13

14.5 Three-winding transformers 14/14
    14.5.1 Impedance characteristics 14/14
    14.5.2 Tertiary windings for harmonic suppression 14/14
    14.5.3 Tertiary windings for external loads 14/14

14.6 Quadrature booster transformers 14/15

14.7 On-load tap changing 14/15
    14.7.1 Tap changer control 14/16
    14.7.2 Line-drop compensation 14/17

14.8 Cooling 14/17
    14.8.1 Air insulated, air cooled 14/17
    14.8.2 Oil immersed, air cooled 14/18
    14.8.3 Oil immersed, water cooled 14/18
    14.8.4 Overload capability 14/18

14.9 Fittings 14/18

14.10 Parallel operation 14/19

14.11 Auto-transformers 14/20
    14.11.1 Auto-starters 14/20

14.12 Special types 14/20
    14.12.1 Static balancer 14/20
    14.12.2 Welding transformers 14/21
    14.12.3 Mining transformers 14/21
    14.12.4 Small transformers 14/21

14.13 Maintenance 14/21
    14.13.1 Insulating oil 14/21
    14.13.2 Insulation 14/22
    14.13.3 On-load tap changing equipment 14/22

14.14 Testing 14/22
    14.14.1 Routine tests at works 14/22
    14.14.2 Type tests at works 14/24
    14.14.3 Commissioning tests at site 14/26

14.15 Surge protection 14/26

14.16 Purchasing specifications 14/27

MIRANDA, F. J. and GAZZANA PRIOROGGIA, P., 'Recent advances in self-contained oil-filled cable systems', *IEE Electron. Power*, 136–140, February (1977)

ARKELL, C. A. *et al.*, 'Design and construction of the 400 kV cable system for the Severn Tunnel', *Proc. IEE*, **124**(3), 303–316, March (1977)

ARNAUD, U. G. *et al.*, 'Development and trials of the integral pipe cooled e.h.v. cable system', *Proc. IEE*, **124**(3), 286–293, March (1977)

ARKELL, C. A., HUTSON, R. B. and NICHOLSON, J. A., 'Development of internally oil-cooled cable systems', *Proc. IEE*, **124**(3), 317–325, March (1977)

ARKELL, C. A., GREGORY, B. and SMEE, G. J., 'Self-contained oil-filled cables for high power circuits', Paper F 77 547-3, *IEEE PES Summer Meeting*, Mexico City (1977)

IWATA, Z., ICHIYANAGI, N. and KAWAI, E., 'Cryogenic cable insulated with oil-impregnated paper', *IEEE Winter Meeting* (1977)

SHIMSHOCK, J. F., *Installed Cost Comparison for Self-contained and Pipe-type Cable*, EPRI Report EL-935, November (1978)

ARKELL, C. A. *et al.*, 'The design and installation of cable systems with separate pipe water cooling', CIGRE Paper 21-01, Paris (1978)

MAINKA, A. G., BRAKELMANN, H. and RASQUIN, W., 'High power transmission with conductor cooled cables', CIGRE Paper 21-10, Paris (1978)

OCCHINI, E. *et al.*, 'Self-contained oil-filled cable systems for 750 kV and 1100 kV. Design and tests', CIGRE Paper 21-08, Paris (1978)

BEALE, H. K., 'Underground cables for HV power transmission', *CEGB Res.*, 24–32, June (1979)

DONAZZI, F., OCCHINI, E. and SEPPI, A., 'Soil thermal and hydrological characteristics in designing underground cables', *Proc. IEE*, **126**(6), 505–516, June (1979)

HEAD, J. G., GALE, P. S. and LAWSON, W. G., 'Effects of high temperatures and electric stresses on the degradation of OF cable insulation', *IEE 3rd Int. Conf. on Dielectric Materials*, 10 September (1979)

ROSEVEAR, R. D. and VECELLIO, B., 'Cables for 750/1100 kV transmission', *2nd IEE Int. Conf. Progress Cables for 200 kV and Above*, September (1979)

ALEXANDER, S. M. *et al.*, 'Rating aspects of the 400 kV West Ham–St John's Wood cable circuits', *2nd IEE Int. Conf. Progress Cables for 220 kV and Above*, September (1979)

ITOH, H., NAKAGAWA, M. and ICHIMO, T., 'EHV self-contained oil-filled cable insulated with composite paper, DCLP', *2nd IEE Int. Conf. Progress Cables 220 kV and Above*, September (1979)

HEUMANN, H. *et al.*, '380 kV oil-filled cable for municipal power supply in Vienna', *2nd IEE Int. Conf. Progress Cables for 220 kV and Above*, September (1979)

BALL, E. H. *et al.*, 'Connecting Dinorwic pumped storage power station to the grid system by 400 kV underground cables', *Proc. IEE*, **126**(3), March (1979)

FAVRIE, E. and AUCLAIR, H., '225 kV low density polyethylene insulated cables', *2nd IEE Int. Conf. Progress Cables 220 kV and Above*, September (1979)

MITZUKAMI, T. *et al.*, 'Prototype tests of EHV cryoresistive cable', *IEEE Trans.*, **PAS-99**, March/April (1980)

PEORMAN, A. J. *et al.*, 'Preliminary ageing tests on a superconducting cable dielectric', *IEEE Symp. on Electrical Insulation*, 132–135 (1980)

ARKELL, C. A. *et al.*, 'Development of polypropylene paper laminate OF cable UHV systems', CIGRE Paper 21-04, Paris (1980)

ALLAM, E. M. and McKEAN, A. L., 'Design of an optimised ±600 kV d.c. cable system', *IEEE Trans.*, **PAS-99**, September (1980)

SPENCER, E. M. *et al.*, 'Research and development of a flexible 362 kV compressed gas insulated transmission cable', CIGRE Paper 21-02, Paris (1980)

KUSANO, T. *et al.*, 'Practical use of "Siolap" insulated OF cables', IEE Paper WM 114-8 (1981)

ALLAM, E. M. and McKEAN, A. L., 'Laboratory experiments of a ±600 kV d.c. pipe type cable', *IEEE Trans.*, **PAS-100**, March (1981)

KOJIMA, K. *et a..*, 'Development and commercial use of 275 kV XLPE insulated power cables', *IEEE Trans.*, **PAS-100**, January (1981)

Ratings

BUCKINGHAM, G. S., 'Short-circuit ratings for mains cables', *Proc. IEE*, **108**(A), June (1961)

GOSLAND, L. and PARR, R. G., 'A basis for short-circuit ratings for paper insulated cables up to 11 kV', *Proc. IEE*, **108**(A), June (1961)

BUCKINGHAM, G. S., BOWIE, G. A. and PARR, R. G., 'Current ratings of PVC and polythene insulated cables', *IEE/ERA Symp. on Plastic Insulated Cables* (1962)

PARR, R. G., *Bursting Currents of 11 kV 3-core Screened Cables (Paper-insulated Lead-sheathed)*, ERA Report F/T 202 (1962)

MOCHLINSKI, K., 'Assessment of the influence of soil thermal resistivity on the ratings of distribution cables', *Proc. IEE*, **123**(1), January (1976)

GOSDEN, J. H. and KENDALL, P. G., 'Current ratings of 11 kV cables', *IEE Conf. on Distribution Cables and Jointing Techniques*, May (1975)

Jointing and accessories

CROSSLAND, J., 'Joints on 3-core 11 kV paper insulated cables', *IEE Conf. on Distribution*, May (1976)

McALLISTER, D. and RADCLIFFE, W. S., 'Joints incorporating mechanical connectors and cast resin filling for 600/1000 V cables', *IEE Conf. on Distribution*, May (1976)

RADCLIFFE, W. S. and ROBERTS, B. E., 'Resin-filled joints for 11 kV paper insulated cables', *IEE Conf. on Distribution*, May (1976)

ROSS, A., 'Jointing trials and tests on 11 kV aluminium sheathed cables', *IEE Conf. on Distribution*, May (1976)

WHYTE, D. H., 'A resin jointing system for all voltages', *Electr. Equip.*, November (1979)

JORGENSEN, J. and NIELSEN, O. J., *Straight Joints for Solid Dielectric Insulated Cables of 12–170 kV*, CIRED, Liege (1979)

TODD, D., 'Electric cable for signalling and track to train communications', *I. Mech. E.R.E.J.*, September (1975)
ANON., 'Wiring cables and flexible cords—new harmonised standards', *BSI News*, February (1976)
LATHAM, W. B., 'Mineral insulated cables in hazardous areas', *Electr. Times*, 2 April (1976)
BUNGAY, E. W. G. and HOLLINGSWORTH, P. M., 'Progress with harmonisation of cable standards with CENELEC', *Electr. Times*, 2 April (1976)
ANON., 'Selecting cables for the hostile conditions on construction sites', *Electr. Times*, 2 December (1977)
ANON., 'Compatibility of PVC cables', *Electr. Times*, 4 August (1978)
WILSON, I. O., 'Magnesium oxide as a high temperature cable insulant', *Proc. IEE*, Pt A, **128**(3), 159–164, April (1981)

Power cables (general)
REYNOLDS, E. H. and ROGERS, E. C., 'Discharge damage and failure in 11 kV belted cables', *Trans. S. Afr. Inst. Electr. Engrs*, **52**, Pt 10, October (1961)
GOSDEN, J. H., *Reliability of Overhead Line and Cable Systems in Great Britain*, CIRED, Liege (1975)
SWARBRICK, P., 'Developments in 11 kV underground cable systems. Paper-insulated aluminium sheathed cables and resin filled joints', *Electr. Rev.*, 14 December (1973)
ROSS, A., 'Cable practice in (UK) Electricity Board distribution networks: 132 kV and below', *Proc. IEE*, **121**, No. 11R, 1307–1344, November (1974)
BLECHSCHMIDT, H. H. and GOEDECKE, H. P., *Cables with Synthetic Insulation in the Federal Republic of Germany*, CIRED, Liege (1975)
BUNGAY, E. W. G., *et al.*, *The Development of 11 kV Cable Systems*, CIRED, Liege (1975)
SCHMELTZ, J., 'EDF distribution cables', 8th Cables Seminar, Hanover, 28 September (1976) (Electricity Council translation OA 2116)
GOSDEN, J. H. and WALKER, A. J., 'The reliability of cable circuits for 11 kV and below', *IEE Conf. on Distribution*, May (1976)
BALDOCK, A. T. and HAMBROOK, L. G., 'Regulations relevant to the design and utilisation of distribution cables for the Electrical Supply industry and the consumer', *IEE Conf. on Distribution*, May (1976)
BUNGAY, E. W. G. and PHILBRICK, S. E., 'Paper-insulated 11 kV aluminium-sheathed cables', *IEE Conf. on Distribution*, May (1976)

CNE cables
HENDERSON, J. T. and SWARBRICK, P., 'The Consac cable system', *IEE/ERA Conf. on Distribution*, Edinburgh (1970)
HUGHES, O. I. and BROMLEY, G. E. A., 'Development and production of a PME elastomeric insulated MV cable', *IEE/ERA Conf. on Distribution*, Edinburgh (1970)
ROCKCLIFFE, R. H., HILL, E. and BOOTH, D. H., 'Protective earthing practices in the UK and their associated underground cable systems', Paper 71C 42-PWR, *IEEE Conf. on Power Distribution* (1971)
McALLISTER, D. and COX, E. H., 'Behaviour of MV power distribution cables when subjected to external damage', *Proc. IEE*, **119**(4), 479–486, April (1972)
RADCLIFFE, W. S. and McALLISTER, D., 'Cables and joints for PME distribution systems', *Electr. Rev.*, 24 March (1972)
GEER, P. K. and SLOMAN, L. M., 'Cables for PME distribution systems', *IEE Conf. on Distribution*, May (1976)
BURTON, J. M., 'Consac cable system development in the Midlands Electricity Board', *IEE Conf. on Distribution*, May (1976)

Polyethylene and XLPE insulated cables
TABATA, T., NAGAI, H., FUKUDU, T. and IWATA, Z., 'Sulphide attack and treeing of polyethylene insulated cables—cause and prevention', *IEEE Summer Meeting*, Paper 71-TP 551-PWR (1971)
VAHLSTROM, W., 'Investigation of insulation deterioration, in 15 kW and 22 kV polyethylene cables removed from service', *IEEE Trans.*, **PAS-91**, 1023–1035, May/June (1972)
LAWSON, J. H. and VAHLSTROM, W., 'Investigation of insulation deterioration in 15 kV and 22 kV cables removed from service—Part 2', *IEEE Trans.*, **PAS-92**, 824–835, March/April (1973)
TANAKA, T., FUKUDU, S., SUZUKI, Y. and NITTA, H., 'Water trees in cross-linked polyethylene power cables', *IEEE Trans.*, **PAS-93**, 693–702, March/April (1974)
BAHDER, G., KATZ, C., LAWSON, J. H. and VAHLSTROM, W., 'Electrical and electrochemical treeing effect in polyethylene and cross-linked polyethylene cables', *IEEE Trans.*, **PAS-93**, 977–990, May/June (1974)
LACOSTE, A., ROYERE, A., LEPERS, J. and BENART, P., 'Experimental construction prospects for the use of 225 kV, 600 MVA links, using polyethylene insulated cables with forced external water cooling', CIGRE Paper 21-12, Paris (1974)
McKEAN *et al.*, *Investigation of Mechanism of Breakdown in XLPE Cables*, EPRI Report 7809-1 (1976)
SWARBRICK, P., 'Developments in XLPE Cables', *Electr. Rev.*, 7 January (1977)
SWARBRICK, P., 'Developments in the manufacture of XLPE cables', *Electr. Rev.*, 28 January (1977)
JACOBSON, C. T., ATTERMO, R. and DELLBY, B., 'Experience of dry-cured XLPE-insulated high voltage cables', CIGRE Paper 21-06, Paris (1978)
BAHDER, E., KATZ, C., GARCIA, F. C., WALLDORF, F., CHAROY, M., FAVRIE, E. and JOCTEUR, R., 'Development of extruded cables for EHV applications in the range 138–400 kV', CIGRE Paper 21-11, Paris (1978)
BERNSTEIN, B. S., *Research to Determine the Acceptable Emergency Operating Temperatures for Extruded Dielectric Cables*, EPRI Report EL-938, November (1978)
LANCTOA, T. P., LAWSON, J. H. and McVEY, W. L., 'Investigation of insulation deterioration in 15 kV and 22 kV polyethylene cables removed from service—Part 3', *IEEE Trans.*, **PAS-98**(3), 912–925, May/June (1979)
FERREN, J. and PINET, A., 'Development of a new 20 kV cable with synthetic insulation and of its fittings', CIRED Paper 31, Liege (1979)
JORGENSON, J. and NEILSON, O. K., 'Straight-through joints for extruded solid dielectric insulated cables, 12–170 kV', CIRED Paper 33, Liege (1979)
HYDE, H. B. *et al.*, 'Earth fault spiking tests at system voltage on 11 kV polymeric cables', *IEE Conf. on Distribution*, May (1976)
NAYBOUR, R. D., 'The growth of water trees in XLPE at operating stresses and their influence on cable life', *IEE 3rd Int. Conf. on Dielectric Materials*, 10 September (1979)
EICHHORN, R. M., 'Treeing in solid extruded electrical insulation', *IEEE Trans.*, **EI-12**(1), February (1976)
MATSUBA, H. and KAWAI, E., 'Water tree mechanism in electrical insulation', IEEE Paper F 75 (1975)
DENSLEY, R. J., 'An investigation into the growth of electrical trees in XLPE insulation', *IEEE Trans.*, **EI-14**(3), 148–158, June (1979)
LAWSON, J. H. and THUE, W. A., 'Summary of service failure of high voltage extruded dielectric insulated cables in the USA', *IEEE Symp. on Electrical Insulation*, June (1980)
NUNES, S. L. and SHAW, M. T., 'Water treeing in polyethylene—a review of mechanisms', *IEEE Trans.*, **EI-5**(6), 437–450, December (1980)

Transmission cables
ROGERS, E. C., SLAUGHTER, R. J. and SWIFT, D. A., 'Design for a superconducting a.c. power cable', *Proc. IEE*, **118**, No. 10, October (1971)
ENDACOTT, J. D., 'Underground power cables', *Phil. Trans. R. Soc. London*, **A275**, 193–203 (1973)
RAY, J. J., ARKELL, C. A. and FLACK, H. W., '525 kV self-contained oil-filled cable systems for Grand Coulee third powerplant—Design and development', IEEE Paper T 73, 492–6 (1973)
BROOKS, E. J., GOSLING, C. H. and HOLDUP, W., 'Moisture control of cable environment with particular reference to surface troughs', *Proc. IEE*, **120**, (1), 51–60, January (1973)
ARKELL, C. A., ARNAUD, U. G. and SKIPPER, D. J., 'The thermomechanical design of high power, self-contained cable systems', CIGRE, Paper 21-05, Paris (1974)
ENDACOTT, J. D., *et al.*, 'Progress in the use of aluminium in duct and direct buried installations of power transmission cable', *IEEE Underground Transmission and Distribution Conf.*, 466–474 (1974)
MIRANDA, F. J. and GAZZANA PRIOROGGIA, P., 'Self-contained oil-filled cables. A review of progress', *Proc. IEE*, **123**(3), 229–238, March (1976)
ARKELL, C. A., 'Self-contained oil-filled cable: installation and design techniques', *IEEE Underground and Transmission Conf.*, 497–502 (1976)

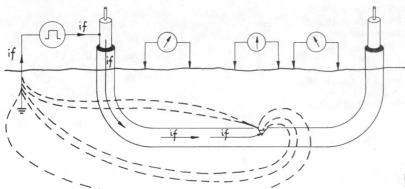

Figure 13.35 Pool of potential method for serving fault location

technique is in cable route tracing when an artificial core-to-core fault is applied at the far end of the cable.

Both the acoustic and the induction pinpointing techniques require a complete metallic path in which the signal currents can flow. An alternative approach, used in pinpointing serving faults on insulated sheath transmission cables, is to apply a signal between the metallic sheath and the general mass of ground so that a voltage gradient is established in the earth in the vicinity of the fault. The voltage gradient or pool of potential is detected using a sensitive voltmeter connected to a pair of probes and the fault position is pinpointed accurately (*Figure 13.35*). Probing over the complete length of a long tranmission cable is time consuming, and many serving faults are sectionalised, using a magnetometer to trace the current flowing in the sheath up to the fault point.

When cables are accessible, being installed above ground or still on the drum prior to installation, it is often possible to pinpoint faults by exploiting their 'microphonic' characteristics. A high-gain amplifier, a.c. coupled to a faulty cable, will pick up the small 'noise' voltages generated when the fault is subjected to mechanical vibration. Partially broken conductors can be detected and pinpointed by circulating a constant direct current through the cable; insulation faults can be detected using a high-voltage source to pass a small polarising current through the fault resistance.

Acknowledgements

The author would like to express his appreciation and grateful thanks to Dr P. F. Gale of BICCOTEST Instruments for the section on 'Fault Location' and to Mr L. G. Banbury and colleagues of BICC General Cables Ltd for the section on 'General Wiring Cables'.

Further reading

General books
HEINHOLD, L., *Power Cables and their Applications*, Siemens Aktiengesellschaft (1970)
GRANEAN, P., *The Science, Technology and Economics of High Voltage Cables*, Wiley (1980)
WEEDEY, B. M., *Underground Transmission of Electric Power*, Wiley (1980)
McALLISTER, D., *Electric Cables Handbook*, Granada Publishing (1982)

Conductors
GRIESSER, E. E., 'Sodium as an electrical conductor', Wire, 2006–2014, December (1966)

HUMPHREY, L. E. *et al.*, 'Insulated sodium conductors', *IEEE Trans.*, **PAS-86** (1967)
BALL, E. H. and MASCHIO, G., 'The a.c. resistance of segmented conductors used in power cables', *IEEE Trans.*, **PAS-87** (1968)
ANON., 'New wiring cable has copper-clad aluminium conductors', *Elect. Times*, 43–45, 2 July (1970)
McALLISTER, D., 'Aluminium cables accepted by industry', *Elec. Times*, 10 September (1971)
McALLISTER, D., 'Terminations for aluminium conductor power cable', *BSI News*, August (1976)
CHATTERGEE, S., 'Failures of terminations on aluminium conductor cables in domestic wiring installations', Paper No. 6, Electrical Contractors Association of Eastern India, August (1977)
KALSI, S. S. and MINNICH, S. H., 'Calculation of circulating current losses in cable conductors', *IEEE Trans.*, **PAS-99** (1980)

Polymeric Materials
BLOW, C. M., *Rubber Technology and Manufacture*, Newnes-Butterworth, London (1975)
BRYDSON, J. A., *Plastics Materials*, 4th edn, Butterworth Scientific, London (1982)
TOWN, W. L., 'Using extrudable materials for cable insulation', *Electr. Times*, 23 February (1979)

Cables in fires
LINDSTROM, R. S. *et al.*, 'Effects of flame and smoke additives in polymer systems', *JFF/Fire Retardant Chem.*, **1**, 152, August (1954)
DAY, A. G., 'Oxygen index tests: temperature effects and comparison with other flammability tests', *Plast. Polym.*, **43** (106), 64 (1975)
WHITE, T. M., 'Cable fires in power stations', *IEE Electron. Power*, February (1977)
BENNETT, H. R., 'Cables for limited fire performance', *Electr. Times*, 11 March (1977)
NATIONAL MATERIALS ADVISORY BOARD, *Flammability, Smoke, Toxicity and Corrosive Gases of Cable Materials*, Publication NMAB-342, National Academy of Sciences, Washington (1978)
SULLIVAN, T. and WILLIS, A. J., 'Reducing the hazards of cables in fires', *Electr. Times*, 17 November (1978)
NESS, D. E. M. and BLACK, R. M., 'New materials for cable sheathing' *IEE Electron. Power*, 698 (1982)

General wiring type cables
HOLLINGSWORTH, P. M., 'The effect of application on the choice of cable materials', *Electr. Rev.*, 24 April (1964)
TOWN, W. L., 'Wiring cables', *Electr. Times*, 22 June, 6 July (1972)
RODGERS, J. A., 'Good wiring guide', *Electr. Wholesaler*, October (1972)
TAYLOR, F. G., 'Cables for electronics', *Electrotechnology*, March (1973)
HOLLINGSWORTH, P. M. and TOWN, W. L., 'Trends in wiring cable design and installation', *Electr. Rev.*, 11 May (1973)
TOWN, W. L., 'A guide to the selection of electrical and electronics wires and cables', *OEM Design*, April (1974)
SECCOMBE, G. H., 'EEC and cable standards: where are we going?', *Electr. Times*, 2 January (1975)

the cable terminals are therefore functions of the length of cable between that terminal and the fault. Both current and voltage transients are created, but the safest and most economical method is to detect the current transient using a 'linear coupler' (*Figure 13.31*).

Special digital electronic instruments are now available which allow these extremely short-duration transients to be recorded and subsequently displayed on an oscilloscope screen where they can be easily and accurately measured (*Figure 13.32*). In addition

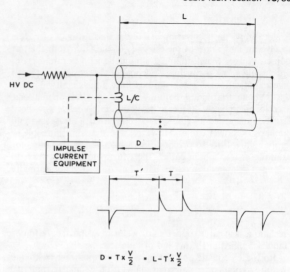

Figure 13.33 Impulse current method for EHV faults

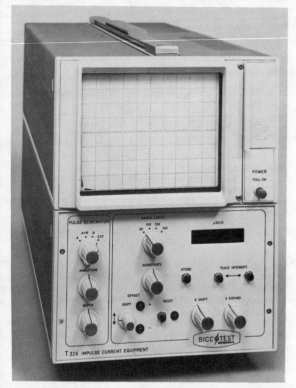

Figure 13.32 Impulse current equipment

to being used to locate high-resistance and flashing faults the 'impulse current' method is also applicable to low-resistance and open-circuit faults, making it a versatile power cable fault location technique. The 'impulse current' equipment and 'linear coupler' can also be used to prelocate flashing and intermittent faults on extra-high-voltage transmission cables by using a high-voltage d.c. test set to bring about flashover at the fault (*Figure 13.33*).

13.8.4 Pinpointing

Pinpointing is essential on direct buried cables if the location and repair of a fault is to be accomplished with a single excavation. The majority of faults on high-voltage power cables are pinpointed by detecting the acoustic signal generated when the fault spark-gap breaks down either from the application of a voltage impulse from a surge generator or a high direct voltage from a pressure test set. In some cases the acoustic signal can be detected without any special equipment but, in general, it is an advantage to use a 'ground microphone' and amplifier to pick up the mechanical shock wave. The importance of the pinpointing stage cannot be overemphasised, and anything which might jeopardise

the generation of the acoustic signal, such as prolonged pre-conditioning, should be avoided.

Once a fault develops into a very-low-resistance 'welded' condition, it will 'short out' the spark-gap, making it impossible to generate an acoustic signal. Multiple excavations may then be unavoidable unless the fault is between two conductors or both conductors are welded to the cable sheath. If a low-resistance path exists between one conductor and another, it is possible to pinpoint the fault using the 'A.F. induction' or 'Bimec' method (*Figure 13.34*). Between the signal generator and the fault the signal induced in the search coil will exhibit a characteristic rise and fall owing to the lay of the cable cores, while beyond the fault the signal will either disappear completely or, more likely, will become constant. The lay effect is the only positive means of identifying that the received signal is emanating from the faulty cable and is not caused by re-radiation from other adjacent buried metallic services. The induction method of pinpointing can be applied, subject to the necessary fault conditions, without any prior knowledge of the cable route and without prelocating the fault. As the appropriate fault conditions for the 'induction' method occur relatively infrequently, the main use of the

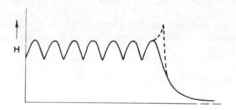

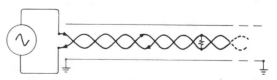

Figure 13.34 Induction method for core-to-core faults (and cable route tracing)

13.8.3 Prelocation

Prelocation is the application of a test at the terminals of a cable to give an indication of the distance of the fault from the test point. While the measurement should be as accurate as conditions will allow, the primary purpose of the terminal tests is to give an indication, as quickly as possible, of the vicinity in which to commence the final pinpointing tests. Bridge or loop methods, used for many years to locate insulation faults, have now been largely superseded by the 'pulse echo' and 'impulse current' methods on power cables. However, where it is not possible to apply high voltage to precondition a high-resistance fault—for example, on a telephone cable—the 'Murray loop' (*Figure 13.29*) or one of its many derivatives, is still used. As it is not normally

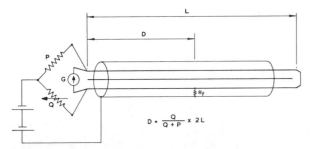

Figure 13.29 Murray loop test

possible to pinpoint telephone cable faults, the 'bridge' test must be carried out with care, and detailed cable records are essential, particularly if the route consists of a mixture of cable sizes.

Cable routes of mixed cross-sections are more common on power systems than on telephone circuits. The use of travelling-wave methods has eliminated the complications of equivalent-length calculations necessary with 'bridge' methods, as the velocity of wave propagation is a function only of the cable dielectric, and therefore does not vary with changes in either conductor material or cross-section.

The 'pulse echo' method (*Figure 13.30*) uses an oscilloscope to measure the time interval T taken for a low-voltage pulse of 10–100 V to travel from the cable terminal to the fault and return by reflection. The distance D (in m) for a time T (in μs) is given by

$$D = \tfrac{1}{2} u T$$

where $u = 300/\varepsilon_r$ is the speed of propagation (in m/μs) and ε_r is the relative permittivity of the cable dielectric. Typical values of u are:

Dielectric	u
Air	294
Dry paper (telephone)	210–240
Oil impregnated paper	168
Polyethylene	200
Cross-linked polyethylene	156–174

The 'pulse echo' method depends for its success on recognising the reflection, or 'echo', from the fault among the other reflections caused by non-fault mismatches (e.g. joints) along the cable length. The amplitude of the fault reflection is determined by the relative value of R_f and the cable surge impedance Z_0 and for insulation faults the practical limit is generally taken as $R_f = 10 Z_0$, which represents a 5% reflection. However, with Z_0 on power cables lying between 10 and 100 ohms, the '$10 Z_0$' limit means that most faults cannot be prelocated without first being preconditioned. By comparison, most conductor faults (which are completely open-circuit, at least at low levels of applied voltage) produce 100% reflection, making them readily identifiable. Although some improvement in detection sensitivity can be obtained by using 'phase comparison' or 'differential' techniques, the 'pulse echo' method is directly applicable only to low-resistance insulation faults and open-circuit conductor faults.

The need to precondition high-resistance and flashing faults is eliminated with the 'impulse current' method, in which the spark-gap of the fault equivalent circuit is caused to flashover by injecting a high-voltage impulse from a surge generator. The flashover launches travelling waves into the lengths of cable on either side of the fault which propagate to the cable terminals, where, depending on the impedance of any connected equipment, they are reflected back towards the fault. Once flashover has occurred at the fault, the ionised gas forms a low-impedance path between the two spark-gap 'electrodes' and, hence, any returning waves are re-reflected from the fault. The transients appearing at

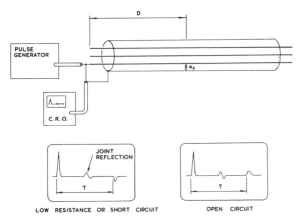

Figure 13.30 Pulse echo method

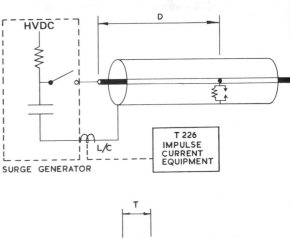

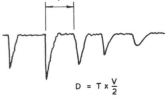

Figure 13.31 Impulse current method

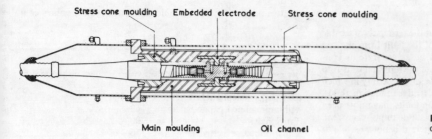

Figure 13.26 Single-core 132 kV OF cable stop joint

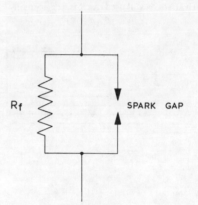

Figure 13.28 Equivalent circuit of a cable fault

Table 13.12 Fault classification

Type of fault	R_f	Spark-gap breakdown voltage
Conductor		
Open-circuit	∞	Usually low; may be high for pulled joints
Insulation		
Low-resistance	$<10Z_0$	Can readily be broken down
High-resistance	$>10Z_0$	using impulse voltage
Flashing	∞	Breakdown under impulse or direct voltage; cannot be burnt down
Intermittent	∞	Occasional breakdown during direct voltage test

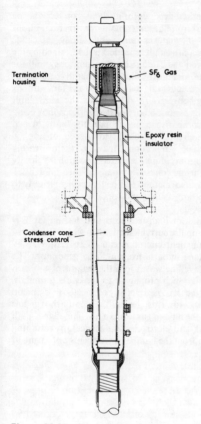

Figure 13.27 SF_6 sealing end for 132 kV OF cable

systematic approach if time and cost are to be minimised, and the modern approach is to follow the 4-stage programme of diagnosis, preconditioning, prelocation and pinpointing.

13.8.1 Diagnosis

Diagnosis is used to confirm the existence of a fault and to characterise it according to *Table 13.12*. The dividing line between low- and high-resistance faults is based on the assumption that prelocation will be by a modern technique based on the propagation and reflection of travelling waves in cables, rather than by any of the classical bridge methods. It is essential that the fault resistance be measured using an ohmmeter and not a 'Megger'. Cable continuity can be checked using a 'pulse echo' set rather than an ohmmeter, with the advantage that breaks in the cable sheath as well as in the cable cores will be detected.

13.8.2 Preconditioning

Preconditioning (often referred to as 'fault burning') is used to change high-resistance into low-resistance faults which can be prelocated using the pulse echo method. 'Fault burners' are usually designed to give various combinations of voltage and current output at ratings of 5 kVA or more. While reasonably successful on paper insulated cables (except on flashing and intermittent faults), fault burning has not proved to be effective on cables employing polymeric insulation, particularly XLPE. Preconditioning may also be attempted by re-energising a faulty cable in the hope that the current may change the fault into one more amenable to prelocation. One application of this technique is on low-voltage distribution cables where the presence of consumers' loads precludes the application of any voltage other than normal system voltage during the fault location exercise.

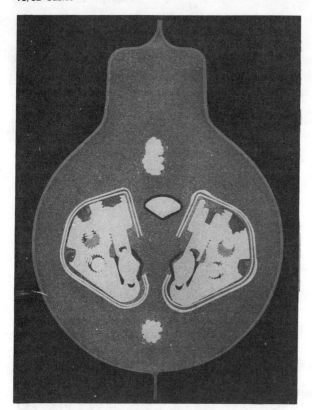

Figure 13.24 Section of cast resin service joint on Waveconal cable, showing the insulation piercing phase conductor connector

Straight joints The main requirements for joints' connecting adjacent lengths of cables are: (a) to provide electrical continuity for the cable conductors; (b) to maintain the electrical insulation; (c) to provide an oil-tight connection between the sheaths; (d) in the case of cross-bonded single-core installations, to provide an insulating barrier between adjacent cable sheaths with facilities for cross connection; and (e) to maintain the insulation of the sheaths in the case of cross-bonded cable installations.

The electrical connection between conductors is usually provided by a compression ferrule for copper conductors and MIG welding for aluminium. As the factory-applied insulation has had to be removed to permit access to the conductors, it has to be replaced by hand-applied impregnated paper tapes and rolls. These are not electrically as strong as the factory-applied insulation and must be built up to a greater diameter. The different diameters for the cable and hand-applied insulation introduce longitudinal electric stresses in the joint. The stresses (and, hence, the shape of the boundaries) must be carefully controlled, as the electric strength of paper insulation along the laminations is only about 1/15 of that normal to the surface of the paper. *Figure 13.25* illustrates a typical straight joint for 33 kV 3-core cable.

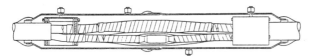

Figure 13.25 Straight joint for 33 kV, 3-core, OF cable

Trifurcating joint This type of joint resembles a straight joint but is used to connect a 3-core cable to three single-core cables, usually for terminating purposes.

Stop joint A stop joint provides all the functions of a straight joint but, in addition, separates the two adjacent cables hydraulically. It is used to limit the hydraulic pressure in a cable system installed when there are significant changes in elevation. This requirement has a major effect on the design, as it is necessary to introduce a different material in the electrical insulation to be capable of withstanding the hydraulic pressure difference. In modern designs of stop joint, this function is carried out by a moulded barrier of epoxy resin with a mineral filler. A stop joint for 132 kV single-core cable is shown in *Figure 13.26*.

Outdoor termination The termination is enclosed in a porcelain housing, which retains the oil pressure and provides protection against climatic conditions. The porcelain has a long creepage path to allow for contamination by dirt, rain, fog and snow. As for straight joints, care must be taken to control the longitudinal stresses in the paper insulation. In very-high-voltage cable systems this is usually provided by means of a 'capacitor cone' similar to that used in high-voltage bushing.

Oil immersed termination This is similar to an outdoor termination but is used to make connection to oil immersed equipment. As there is no surface contamination from the atmosphere, the length of the termination is appreciably shorter than the equivalent outdoor form.

SF_6 immersed termination Increasing use is now being made of SF_6 gas instead of oil for insulation. *Figure 13.27* shows a termination for SF_6 insulated metal clad switchgear.

13.8 Cable fault location

Properly designed and manufactured cables have a high reliability, but once installed and operational they are exposed to hazards that may result in failure. The commonest cause of failure on buried cables is accidental disturbance during excavation on, or adjacent to, the cable route. Frequently the damage is only slight and some time may elapse before the insulation deteriorates sufficiently to cause the circuit protection to operate. While most cable faults arise from failure of the insulation due either to ingress of moisture or overvoltage surges caused by lightning or switching transients, faults in joints and terminations, caused by, for example, ground subsidence or overheating at bad contacts, often result in loss of continuity in either the conductor(s) or cable sheath. Frequently conductor faults become apparent only after the insulation is affected and a path created in which fault current can flow.

Figure 13.28 shows a simplified equivalent circuit of a cable fault in which the fault resistance R_f is shunted by a spark-gap to represent the non-linear voltage characteristic exhibited by most faults. The fault resistance can vary from zero to infinity, while the flashover voltage of the spark-gap can vary from a few hundred to several thousand volts. A fault may be located either by detecting the discontinuity created by the fault resistance or by analysis of the transient(s) generated in the cable when the spark-gap breaks down. The method adopted in a particular case is governed more by the operating and/or permissible test voltage of the faulty cable than by the characteristics of the fault. Techniques which can be applied to an isolated high-voltage cable are not normally applicable to low-voltage distribution, pilot or telephone cables, where service must be maintained on the non-faulty phases or pairs. Irrespective of the type of cable, fault location demands a

more bulky. The publication of test requirements such as BS 4579, Part 3, has been an important step forward.

Reference has been made to the problems with aluminium conductors in the tunnel type terminations of house-wiring fittings and these emphasise the care necessary in designing any mechanical fittings for aluminium. Suitable approved fittings are available for specific requirements such as the concentric neutral conductor in Waveform distribution cable, and for aluminium conductors in house-service cut-outs. Alternatively, for terminations, fittings are available for connecting a short piece of copper to an aluminium conductor.

Where maximum strength and minimum volume of joint are essential, welding techniques have been developed, but they have only achieved widespread use for oil filled cables. There are doubts about the creep strength of soldered joints and jointing is made more difficult by the oil flow which must be maintained. The MIG (metal inert gas) welding procedure, which is essentially a casting process, proved to be practical and reliable.

13.7.2 Joints for distribution cables

During the last decade the traditional practices requiring highly skilled jointers for soldering and plumbing have been revolutionised. The heavy, bulky and cumbersome practices involving filling metal sleeves and/or cast-iron boxes with hot bitumen have largely been dispensed with, even with paper insulated cables. By the use of mechanical joints and cast resin filling into simple shell type plastics moulds, all the components can be packed as a convenient complete kit and only simple tools are needed on site. The resin is poured while cold, and when mechanical fittings are used, no heating of any kind is required. At the time of mixing on site, the resin is fluid and penetrates well to fill all cavities, but it subsequently sets quickly to a hard solid mass. Most joints can be completed in $1-1\frac{1}{2}$ h and can be energised immediately. The basic principles of cast resin jointing apply to all types of cable for all voltages up to and including 10 kV. *Figure 13.23* shows two typical examples. Where 'live' jointing is practicable, e.g. for 240/415 V services, it can equally well be carried out with these designs. The main advantages, however, accrue from the lower skill demanded and the short time required for jointer training.

Figure 13.24 shows a cross-section of a service joint on CNE cable taken through one of the phase connectors. With this type of connector it is not necessary to remove the insulation from the phase conductor, even if it consists of XLPE, because knife edges on the connector pierce the insulation and establish firm contact with the conductor.

13.7.3 Joints for transmission cables

Special fittings are required to joint and terminate transmission cables, the detailed design of which depends on the type of cable system employed. The following description of those used for low-pressure OF cables gives an indication of the important categories.

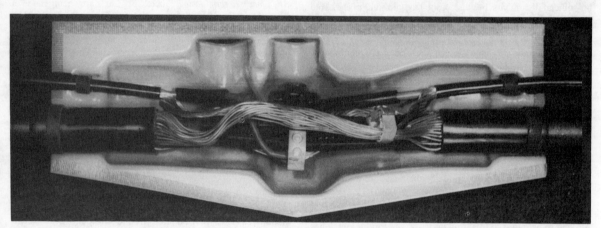

(a)

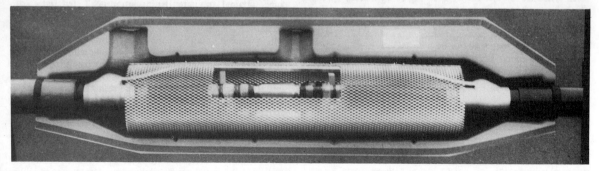

(b)

Figure 13.23 Cast resin type joints before resin filling: (a) service joint on Waveconal cable; (b) straight joint on paper insulated cable

that in time a layer of silt may build up, and investigations in canals have shown that such layers can have high thermal resistivity.

13.6.4 Short-time and cyclic ratings

A cable on load will show an exponential temperature rise/time relationship and, if starting from a low temperature, may take many hours to reach stable condition at maximum temperature. It can, therefore, carry more than maximum continuous rating for a limited time, the factor for overload depending on the extent of initial loading.

For cyclic loadings some increase of rating, compared with continuous, may be applied to an extent which will vary with the shape of the load curves. Calculations to take advantages of this possibility are rather tedious, but guidance may be obtained from ERA Report F/T 186.

13.6.5 Short-circuit ratings

Often the conductor size necessary is related to short-circuit current rather than continuous current requirements. The short-circuit current, which may be 20 or more times normal, produces thermal and electromagnetic effects proportional to the square of the current. So far as the cable insulation itself is concerned, much higher conductor temperatures can be allowed because the heating and cooling are very rapid and the full temperature will not be sustained for significant time by the insulation. Figures are included in *Table 13.9* and *13.10*.

Short-circuit ratings are not published for individual cables in any official documents, because of the large number of the cable types and sizes involved and the fact that they have to be related to the duration of short circuit which applies to the particular circuit. *Figure 13.22* illustrates a typical example of the graphs available from manufacturers to provide information on a basis of a maximum conductor temperature with a range of durations.

Other factors may dictate a lower rating for a particular design of cable or installation condition. A short-circuit in a cable produces electromagnetic forces which could burst the cable if the cores are not adequately bound together. The accessories must also be designed to withstand both electromagnetic and thermomechanical forces; and accessories must be compatible with the cable in this respect. Soldered joints impose limitation of the short-circuit temperatures to 160°C.

The method of installation may also limit permissible short-circuit current. Local pressure due to clamping may lead to high forces, with deformation of cable components. Longitudinal expansion can also be considerable and has to be absorbed uniformly. When cables are buried, the cable is restrained and these forces must be accommodated by joints and terminations.

13.6.6 Voltage drop

Voltage drop may be of great significance for 0.6/1 kV cables but is not usually important at higher voltages. A typical requirement in the IEE Regulations is that the voltage drop in a cable run should be such that the total drop in the circuit, of which the cable forms a part, does not exceed 2.5% of the nominal voltage.

As the actual power factor of the load is seldom known, a practical approach is to assume the worst condition, i.e. where the phase angle of the load is equal to that of the cable. Cable manufacturers issue tabulated figures for volt drop, as in *Table 13.11*, based on this assumption. If the actual current differs greatly from the tabulated current the figure may be approximate only. From a table such as this, a suitable cable size may be selected but it must also be able to carry the current.

13.6.7 Protection against overload current

For some types of cable, particularly wiring cables, the required current rating of the cable must be determined by the overload protective device rather than the circuit current. The rating of the device must not be less than the circuit current and, of course, such ratings are in discrete steps. In the 15th edition of the IEE Wiring Regulations the protective device must satisfy the requirements of:

$I_B \leqslant I_n \leqslant I_z$

$I_2 \leqslant 1.45 I_z$

where (a) the nominal or current setting I_n shall not be less than the design current I_B of the circuit; (b) I_n does not exceed the lowest of the current-carrying capacities I_z of any of the conductors in the circuit; (c) the current causing effective operation of the protective device I_2 does not exceed 1.45 times the lowest of the current-carrying capacities I_z of any of the conductors of the circuit.

13.7 Jointing and accessories

13.7.1 Aluminium conductor jointing

The technique involving a flux for soldering with aluminium is essentially the same as for copper, but more care is required and strict observance of temperature limits is important. A special flux is necessary to remove the oxide skin and its composition should be without dermatitic risk. Solid conductors have an advantage in that a tinned layer can more readily be produced by the abrasion tinning procedure.

Problems with aluminium conductors, however, have largely been overcome by development of improved and simple compression methods suitable for both solid and stranded conductors. Probably the only remaining difficulty is that, when making straight joints on multicore cables, the larger separation necessary between cores for inserting the tool head makes the joints

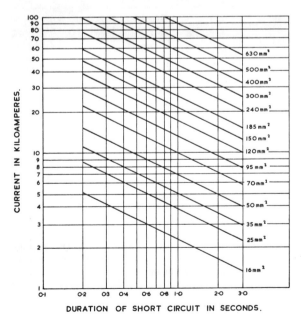

Figure 13.22 Short-circuit ratings for copper conductor XLPE insulated cables (temperature rise 90 to 250°C)

Table 13.9 Paper cables: conductor temperature limits

Rated voltage, U_0/U (kV)	Design	Temperature* (°C)
0.6/1, 1.8/3 and 3.6/6	Belted	80
6/10	Belted	65
6/10	Screened	70
8.7/15	Screened	70
12/20 and 18/30 MIND	Screened	65

* For continuous operation. Temperature for short-circuit conditions is 160°C, except for 0.6/1 kV cable, for which the limit is 250°C, subject to the accessories being suitable.

Table 13.10 Polymeric cables: conductor temperature limits

Insulating compound	*Temperature (°C)	
	Continuous	Short-circuit
Polyvinyl chloride	70	160†
Polyethylene	70	130
Butyl rubber	85	220
Ethylene propylene rubber	90	250
Cross-linked polyethylene	90	250

* Temperature limits are based on intrinsic properties and do not take account of variations in cable and accessory design. Short-circuit ratings are affected by (a) reduction of thickness of PVC and PE by thermomechanical forces; (b) conductor and core screens; (c) design of accessories (e.g. soldered conductor joints are unsuitable).
† 140°C for conductors above 300 mm².

withstand much higher temperatures without undue deformation. Limitation by ageing effects is a factor with natural rubber compounds.

13.6.2.2 Conductor losses

With the exception of some higher voltage transmission cables, the I^2R conductor losses represent the major source of heat produced. These also have to be dissipated through the longest radial path in the cable. When the conductors are large, the effective resistance may also be increased because of skin effect. The increase is negligible for sizes up to about 185 mm² and can be reduced by the use of the Milliken construction described earlier.

Proximity effects may be caused by the interaction of magnetic fields associated with adjacent current-carrying conductors and these, too, can cause further redistribution, as with skin effect. The proximity effect occurs with small spacing and so is most significant for low-voltage cables of large conductor size.

13.6.2.3 Dielectric losses

Dielectric losses are reasonably negligible for paper and XLPE cables up to about 60 kV and for PVC cables up to 6 kV. They are mainly of importance for high-voltage transmission cables. One reason that PVC has not found much application in the 10–20 kV field is that the losses are high in comparison with paper and XLPE. Even so, they represent only around 6–8% of the conductor losses in 11 kV cables. With XLPE the figure is around 0.1%.

13.6.2.4 Sheath and armour losses

Losses in metallic sheaths are of great importance for large conductor single-core cables bonded and earthed at both ends. As explained in Section 13.5.5, they can be avoided by cross-bonding. Although they make some contribution to total losses, the effect is not very significant for multicore cables. Similar remarks apply to armour, but losses due to magnetic effects are dominant for single-core cables and it is usually necessary to use non-magnetic armour material.

13.6.2.5 Internal thermal resistance

Thermal resistance within the cable is related (a) to cable design and construction, e.g. the number of separate layers and the volume; and (b) to the thermal resistivities and thicknesses of the individual materials. Values in °C m/W included in IEC 287 are:

Impregnated paper, solid cables	6.0
O.F. cables	5.0
Polyethylene	4.0
Polyvinyl chloride	7.0
Bituminous textiles	6.0

13.6.2.6 External thermal resistance

For cables in free air the heat dissipation is related to the degree of exposure and to the surface emissivity, which depends on surface condition. Published ratings assume shading from the sun, and if this is not provided, derating may be necessary.

13.6.3 Sustained ratings

Table 13.11 indicates a typical example of published ratings for one type of cable installed in air, in ducts and buried directly in the ground. When cables are installed in ducts, two other thermal

Table 13.11 Sustained current ratings and volt drop for 3-core copper, 0.6/1 kV, XLPE insulated, armoured cables

Conductor area (mm²)	Rating, dg (A)	Rating, sd (A)	Rating, air (A)	Volt drop (mV m/A)
16	115	94	105	2.6
25	150	125	140	1.6
35	180	150	170	1.2
50	215	175	205	0.87
70	265	215	260	0.61
95	315	260	320	0.45
120	360	300	370	0.36
150	405	335	430	0.30
185	460	380	490	0.25
240	530	440	580	0.21
300	590	495	660	0.19

Depth of laying, 0.5 m.
Soil thermal resistivity, 1.2°C m/W.
Ground temperature, 15°C.
Ambient air temperature, 25°C.
Maximum conductor temperature, 90°C.
Rating: dg, direct in ground; sd, in single-way ducts; air, in free air.

resistances are introduced—namely, air space between cable and duct and the duct itself. As will be seen from *Table 13.11*, these cause a heavy rating penalty. *Table 13.11* also illustrates that while ratings for cables in air and buried direct in the ground are broadly similar, in air they are lower for small conductor sizes and higher for larger sizes. These differences are related to heat dissipation as a function of surface area.

Cables installed under water have the lowest external thermal resistance and highest ratings. However, there is always a danger

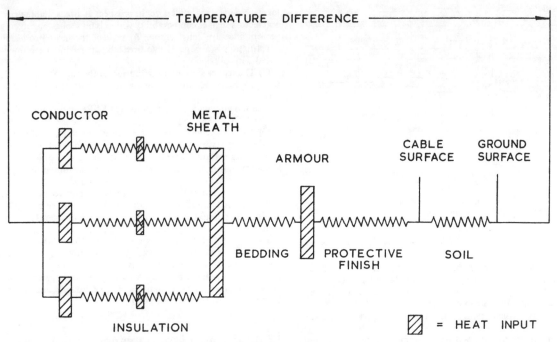

Figure 13.21 Equivalent circuit diagram for heat flow in 3-phase belted cable

(2) Report ERA 69-30 (4 parts), published by ERA Technology Ltd, provides ratings for paper cables up to 33 kV, PVC cables up to 3.3 kV and thermosetting insulated cables up to 3.3 kV.

(3) Manufacturers' catalogues.

(4) Lower ratings may be selected for specialised installations because of particular environmental conditions. For example, the IEE Regulations for the Electrical and Electronic Equipment of Ships stipulate an ambient temperature of 45°C and somewhat lower maximum temperatures for continuous operation (80°C for thermosetting insulation).

(5) The above cover installations based on British cable practice. When USA types of cable and system are applicable, reference may be made to IPCEA publications.

(6) IEC Publications 448 and 287 (*Table 13.1*): 448 provides ratings for unarmoured cables and 287 gives the basic methods for calculating ratings using the standard data included. Values prepared by all other bodies are almost always derived in accordance with this specification.

In general, these documents provide tabulated figures for copper and aluminium conductor cables installed in air, in ducts and buried directly in the ground. The data quoted are for standard conditions, and multiplying factors are given for variations in the conditions.

A feature of USA practice is that data provision is made for limited periods of operation with emergency overload for a specified number of hours per year to a higher temperature. While it is recognised that such operation could affect the life of the cable, the conditions are chosen to ensure that only limited ageing is likely to occur. British practice has not yet included this feature in published recommendations.

Another important aspect relates to the fact that the published ratings are quoted for 'continuous' or 'sustained' operation. Few cables are loaded for the whole of their life to full rating, and allowance is made for this. Nevertheless, the derivation of ratings is a most complex subject and many large users, such as the UK Electricity Boards, have developed ratings which allow for their own circumstances, such as cyclic operation and the emergency conditions which can arise with 11 kV cables normally installed as open rings (Electricity Council Engineering Recommendation P 17—Current Rating Guide for Distribution Cables).

A standard rating for the particular cable and specified installation conditions having been determined, factors have then to be applied to obtain the actual rating for the individual conditions. The references quoted provide these factors for variations such as ambient or ground temperature, depth of laying, thermal resistivity of soil and mutual heating due to cables being installed close together.

13.6.2 Factors in cable ratings

13.6.2.1 Temperature

As previously mentioned, ratings are governed primarily by the permissible temperature rise from a declared base temperature to a maximum for the particular cable design. The base temperature is normally 15°C for buried cables and either 25 or 30°C for cables in air. At the maximum continuous temperature the heat generated in the cable equates with the heat dissipation from it, which is dependent on the thermal resistance of the cable components and the surroundings.

The internationally recognised limits for conductor temperatures with the common types of insulation and cable design are shown in *Tables 13.9* and *13.10*.

In the case of the insulation materials, it is not usually chemical degradation which is the main aspect. With paper insulation the permissible temperature is reduced with increasing voltage and this ensures that there is not undue expulsion of impregnating compound for the duty required. Similarly, screened paper cables are more independent of compound effects in the filler spaces and can be operated to a higher temperature. Thermoplastic insulation softens significantly with increasing temperature and the limit is governed by deformation. Thermosetting materials can

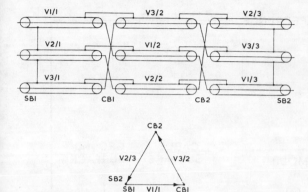

Figure 13.20 Cross-bonding of cable sheaths to provide transposition for reducing sheath losses

earth at one end only, but modern practice is to employ a transposition method (*Figure 13.20*) in which the metallic sheaths are interrupted every few hundred metres with cross-connection at jointing positions. Voltages are then balanced at every third joint and usually kept below 65 V under full-load conditions. High transient voltages can occur; and to check that the sheath insulation is satisfactory, a 10 kV d.c. test is carried out after laying.

When two circuits are laid on a common route, the current rating will be reduced by mutual heating unless thermal independence can be obtained by a separation of about 2 m at 132 kV, with progressive increases for higher voltages. To obtain the most economic solution, it is necessary to examine the cost of larger conductor cable, extra trenching and external cooling. In the case of a single-core circuit the mutual heating effects between the phase cables must also be taken into account. With single-point bonding and cross-bonding to eliminate sheath circulating currents, a flat formation with spacing between cables of 150–300 mm is usually beneficial to avoid unduly high sheath voltages. If single-core cables are bonded and earthed at both ends, it is necessary to install the cables in trefoil formation, because with wide spacing the increase in sheath losses would more than offset the reduction in mutual heating.

13.5.6 Future development

It was mentioned earlier that the low-pressure OF cable was nearing its limit of performance at the present maximum service voltage of 525 kV. However, overhead lines will soon be in operation at around 1000 kV, and at some future date there will be a need for cables to match them.

Problems arise because the power loss in a cable dielectric is proportional to the square of the phase voltage multiplied by the capacitance and the power factor. A level of voltage is soon reached, therefore, at which long circuits of conventional design can carry little more than their own charging currents unless cooling is used. Capacitance is dependent on the conductor diameter together with the thickness and permittivity of the dielectric material. Of these, only the permittivity offers limited scope for improvement, and overall the most important factor for reducing dielectric losses is the adoption of material with lower dielectric power factor. Some of the possibilities are:

(1) Further refinement of the OF cable by improvement of dielectric materials and increase of oil pressure to 15 bar together with forced cooling of the cable.
(2) Insulation comprising either paper/plastics laminates or layers of a synthetic film alone under oil or gas (sulphur hexachloride) pressure, with the object of reducing the dielectric losses. No ideal material is yet in prospect at a reasonable price, and factors such as compatability with oils plus long-term ageing performance are considerable. At present the most promising construction is a paper/polypropylene/paper laminate impregnated with oil.
(3) The use of sulphur hexafluoride as the primary dielectric with spaces to maintain the conductors in the centre of a flexible tube.
(4) Solid polymeric insulation such as XLPE, at present limited to the 130–225 kV range.
(5) A completely different approach is to make use of superconductivity: very large currents can be carried by small conductors without generation of much heat. To attain such a low temperature, it is necessary to use liquid helium in the cable and provide thermal insulation from the ground. An alternative is to use a 'cryo-resistive' principle with more conventional conductor metals and a temperature such as 77 K, which can be attained with liquid nitrogen. The engineering problems are not inconsiderable, but experimental cables have been made and trials carried out to demonstrate the practicability of such schemes. It has, nevertheless, been established that the economics are such that this form of transmission can only be justified for ratings of the order of 5–10 GVA. In the UK such requirements are several times what can immediately be envisaged, and further development work has been postponed.

13.6 Current-carrying capacity

The continuous current rating of a cable is dependent on the way heat generated in the conductor, insulation and metallic components is transmitted through the cable and then dissipated from its external surface. For convenience the conductor temperature is taken as the reference datum for the cable. A notional maximum cable rating can then be calculated from the permissible temperature rise from a standard base ambient or ground temperature to the maximum temperature that the particular type of insulation will withstand with a reasonable margin of safety. Adjustments to this notional rating have to be applied to cover many factors, which include a different base temperature and variations in heat dissipation from the cable surface: e.g. dissipation from a cable in a duct is lower than from a cable in free air.

The difference between conductor temperature and external or ambient temperature is directly related to the total heat losses and the law of heat flow, using a conduction current analogue. This analogy may be extended into the type of circuit diagram in *Figure 13.21*, which shows how the heat input at several positions has to flow through various layers of different thermal resiatance. To make calculations, values of thermal resistivity have to be measured for all the materials involved. Thermal resistivity is defined as the difference in degrees Kelvin between opposite faces of a metre cube caused by the transference of one joule per second of heat; the SI unit is K m/W.

13.6.1 Availability of continuous ratings

Cable users frequently need to ascertain the rating for a particular type of cable at a given voltage and with a range of copper or aluminium conductor sizes. The most common sources of reference are:

(1) The IEE Regulations for the Electrical Equipment of Buildings. This source covers cables of all standard types (up to 1 kV). A difference from the others listed below is that the ratings quoted are lower, as they are calculated from a base ambient temperature of 30 °C (compared with 25 °C) and, hence, a lower permissible temperature rise. Only 'in air' ratings are included.

determined by the geometry and the permittivity of the dielectric. It is usual to assume a uniform permittivity, as this property is affected only to a minor degree by changes in cable temperature and voltage. In d.c. cables, however, the steady state stress distribution is dependent on the geometry and resistivity of the dielectric. If the latter remains uniform, the stress distribution is the same as that for a.c. However, the resistivity of the dielectric is highly dependent on the dielectric temperature and to a lesser degree on the applied stress. When the cable is carrying load, there is a temperature gradient across the dielectric, the effect of which is to reduce the stress adjacent to the conductor and to increase it at the outside. It is possible to arrive at the conditions where the stress at the outside exceeds that at the conductor and the insulation must be designed to cater for these changing stress conditions. *Figure 13.18* illustrates the principles involved.

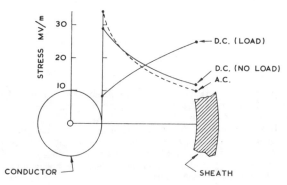

Figure 13.18 Stress distribution in d.c. and a.c. cables

Nevertheless, as pressurising of the insulation is not so necessary and does not give much advantage in direct voltage and impulse breakdown strength, it is possible to use mass impregnated solid type cables for much higher equivalent direct voltages. This assumes that the dielectric is not weakened by migration of impregnating compound, i.e. the insulation is 'non-draining'. For the highest voltage levels, OF cables are used and, as with a.c. but for a different reason, clean paper obtained by the use of deionised water is desirable. Excessive conductivity due to ionic impurities can lead to thermal instability and breakdown.

A power of 500 MW can be transmitted by *three* single-core a.c. cables with 1000 mm^2 conductors at 275 kV. The same power can be conveyed by *two* single-core 800 mm^2 d.c. cables at ± 250 kV. To transmit 1500 MW would require a double circuit comprising *six* naturally cooled 2000 mm^2 a.c. cables, but still only *two* d.c. cables, reducing by two-thirds the number of substantially identical cables required.

13.5.5 Cable ratings and forced cooling

The considerations in the general section on current carrying capacity are applicable also to transmission cables, but because of the much greater power carried, the effects of heat dissipation in the ground are of particular importance. First, it is necessary to inspect the soil to determine its thermal resistivity: 1.2 °C m/W is taken as a representative value, but it may be much higher in sand, shingle or made-up ground, or if the soil is likely to be permanently dry. The moisture content is a significant factor in ground thermal resistivity; this became apparent when cables were loaded continuously so that moisture could not seep back during reduced load periods. If the ground surrounding the cable reaches a temperature of around 50 °C, there is a considerable danger, with certain types of soil, of reaching a 'runaway' condition: complete drying out, high thermal resistivity, excessive temperature rise in the cable and breakdown.

When there is doubt about the thermal properties of the backfill, it is safer to surround the cable with imported material having known thermal resistivity in the dry condition. This means creating a dense mass with little air space by a controlled mixture of sand and gravel, particle sizes being blended to obtain good packing. Laboratory control of composition and compaction is important. An alternative is to use a mixture of selected sand and cement in the proportion of 14:1.

During the early 1960s, following the first failure due to ground drying out, several installations were completed in the UK with cooling pipes laid adjacent to each cable, water being circulated through a closed heat exchanger. The latter was air cooled or in some locations water cooled by supplies from bore-holes. Initially, aluminium was used for the pipe but this was later changed to high-density polyethylene. In these later installations four pipes were used with an internal bore of approximately 66 mm. Specially selected sand was used around the cables and pipes. Some 160 circuit km of 275 kV with a winter rating of 760 MVA was installed in this way. It was later found that this rating, together with 1100 MVA at 400 kV, could be achieved by naturally cooled cables with a stabilised backfill and more realistic assumption of ambient ground parameters.

Separate pipe cooling came back into prominence in 1977 when overhead line ratings were further increased by raising the operating temperature to 65 °C. This required a winter rating of 2038 A, equivalent to 970 MVA at 275 kV and 1410 MVA at 400 kV, which could not be achieved with the 2500 mm^2 maximum conductor size and stabilised backfill. A similar need for additional cooling also arose with the 400 kV cables in the Dinorwic pumped storage scheme in North Wales. Improved water pipe systems were adopted, the emphasis being on the use of larger pipe diameters and special arrangements for water cooling of the joints (*Figure 13.19*).

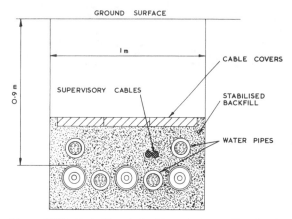

Figure 13.19 Typical layout of cables and water cooling pipes in a trench

A rating factor which is particularly important with high-voltage single-core cables is to prevent the very high losses in metallic sheaths and reinforcement if these were solidly bonded at both ends of a feeder. The losses accrue from currents induced in the low-impedance sheath circuit and are related to the conductor current and separation between phases. Without elimination of such losses the use of aluminium sheaths would not generally be economic. In some cases it may be possible to bond

Table 13.8 D.C. cable schemes

Link	Rating (MW)	Voltage (kV)	Commissioning date	Type of cable	Conductor (mm^2)	Route length (km)
Götland (Sweden)	20	100	1954	Solid	90	100
Cross-Channel (UK–France)	160	±100	1961	Solid	340/595	52
Cook Strait (New Zealand)	600	±250	1965	Gas filled	520/800	40
Konti Skan (Sweden–Denmark)	250	250	1965	Solid/flat OF	620/800	90
Sardinia–Corsica (Italy–France)	200	±100	1966	Solid	420	121
Vancouver Island (Canada)	360	±300	1969	Solid	400	30
London area (UK)	640	±266	1971	OF	800	82
Vancouver Island (Canada)	400	±300	1977	OF	400	31
Skagerrak (Norway–Denmark)	500	±250	1977/78	Solid	800	130
Tsugoru (Japan)	600	±260	1978	OF	600/900	45

1969. EDF also claimed that the overall economics were favourable. This is despite the use of a lead sheath and limitation with straight polyethylene of the operating temperature to 70°C compared with 85°C with paper insulation.

Following consolidation of satisfactory experience at 33 kV, it is at around 132 kV that large-scale experience will first be obtained and it is here that the need or otherwise for a metallic sheath will be determined. Many small installations at 132 kV are in service with a design stress of 7–9 MV/m. The problems have been enumerated in Section 13.4.6 and are primarily concerned with the production of clean insulation and screen interfaces plus possible incidence of water treeing in service. The French experience with straight polyethylene indicates that the former can be overcome, and only time will prove whether a metal sheath is necessary to prevent degradation by water tree mechanisms.

Much development work is proceeding on the optimum method of curing XLPE. Although dry-cured material has a lower content of microvoids than steam-cured material, there is little difference between the two levels in water tree growth if water is in contact with the insulation. If kept dry, treeing ceases to be a problem and the improved short-term breakdown strength of the dry-cured material enables it to be operated at a higher stress.

13.5.3 Submarine power cables

The engineering of submarine cable links is complex. Apart from the choice between a.c. and d.c. transmission, cable design has to take account of the maximum depth on the projected route, the potential hazards caused by shipping, corrosion and possibly marine borers. There are many submarine power cable installations giving satisfactory service at voltages up to ±300 kV d.c. and 420 kV a.c.

The choice of cable for a submarine crossing is influenced by the system voltage, the maximum depth and the length of the crossing. Solid type paper insulated cables are suitable for voltages up to 300 kV d.c. and 33 kV a.c. For deep-water installations special design features are necessary to enable this type of cable to resist the external water pressure. Solid type cable has been used successfully at 550 m depth on a d.c. link between Norway and Denmark. For higher voltages self-contained pressure assisted cables (oil filled and gas filled) are used, the internal pressure being maintained above the external water pressure at the deepest part of the route. Although not so far used for major submarine transmission schemes, polymeric insulated cables may be attractive for future a.c. links.

It is preferable that the cable be manufactured in continuous lengths without joints. As this is not always possible, proven techniques have been developed for the construction of flexible joints in all types of power cable, to facilitate laying in a continuous operation. Manufacture has to be arranged so that the cable can be coiled down on land and then reloaded directly into the cable laying vessel.

Experience on the ±100 kV d.c. cable circuit between England and France (1961) and on the Sweden–Denmark ('Konti-Skan') ±250 kV d.c. link (1965), indicates that cables laid directly on the sea-bed across busy shipping lanes or fishing zones are liable to suffer frequent impact damage caused by dragging anchors and trawls. A significant increase in circuit security can then be obtained by embedding the cables. Techniques have been developed for cutting trenches in the sea-bed and for accurate positioning of the cables within the trenches. On many cable routes adequate security can be obtained by burying the cable at the shore approaches only.

13.5.4 D.C. transmission

Table 13.8 indicates the main d.c. schemes in operation, all but one being submarine. There are advantages in cable cost, but expensive terminal conversion stations may make such schemes uneconomic unless there are other overriding considerations. These arise when large national systems need to be interconnected and occasionally when large blocks of power have to be transmitted within a network, such as in the CEGB d.c. link in South London. Most of the existing schemes are for submarine links where the charging currents for a.c. cable systems would be excessive.

D.C. cables can be operated at much higher design stresses than a.c. cables. For example, the 266 kV OF cable supplied for the CEGB link in South London had a maximum design stress of 33 MV/m, whereas a comparable 275 kV a.c. cable would have a design stress of 15 MV/m. Although the partial discharge in voids mechanism does not apply in d.c. operation, there are other factors, such as stress distribution and transient voltages arising from rectifier malfunction, which have to be taken into account.

In a.c. cables the stress distribution in the insulation is

water used in papermaking and washing for very-high-voltage cables may be deionised. For electrical stress reasons, and to obtain good bending performance without disturbance of the dielectric by wrinkling, etc., the papers tapes are graded from thin adjacent to the conductor (where the electrical stress is highest) towards much thicker and wider on the outside (to withstand the higher mechanical stress). With the large insulation thickness (up to 30 mm) required for the highest-voltage cables it is necessary to control the design and paper lapping parameters to allow individual paper layers to slide over one another on bending. This is done by shrinking the papers by predrying before lapping, and carrying out the lapping in a low-humidity atmosphere with careful control of tension.

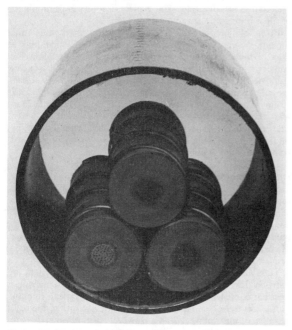

Figure 13.17 Pipe type 230 kV high-pressure OF cable

13.5.2.3 High-pressure oil filled (HPOF) cable

The high-pressure oil filled cable is a type of cable (*Figure 13.17*) developed in the USA and used extensively only in a few countries. It evolved from the predominant American practice of installing cables in buried ducts, the steel pipe being essentially a duct which can be installed a short length at a time without need for long trenches to be kept open. Either the cable cores have a temporary lead sheath which is stripped off as the cable is pulled into the pipe, or the unsheathed cable is delivered to site on a specially sealed and protected drum. So that the cores are not damaged during the pulling operation, D-shaped skid wires are applied helically over the insulation. After jointing, the pipe is evacuated and filled with oil to a pressure of 14 bar and the pressure is maintained by automatic pumping stations. The relatively large volume of oil and the high pressure enable a viscous impregnant to be used.

Pipe type HPOF cables tend to be more expensive than self-contained OF cables laid directly in the ground, but in built-up inner city areas, or where robustness is desirable, they can be advantageous. Except for terminations, they always consist of three single-core cables pulled into a single pipe. The proximity of the cores and the high electrical losses in the pipe impose lower ratings than for self-contained cables.

13.5.2.4 Mollerhoj type oil filled cable

In the Mollerhoj type design the cores are laid side by side in a flat formation under a lead sheath. By providing longitudinally applied corrugated tapes along the flat faces of the sheath and binding them with helical wires and brass tapes, an elastic diaphragm is created. After jointing, the internal oil pressure is raised and the cable then sealed off from the pressure source. The Mollerhoj cable has been used extensively in Denmark, particularly for submarine transmission to off-shore islands, but has found little application elsewhere.

13.5.2.5 Gas pressure cables

During the 1930s–1940s, many designs became established to utilise the principle of gas pressure to suppress partial discharge in voids.

In the 'internal gas pressure' cable and in one form of pipe type cable the gas was admitted directly into the cable insulation and held by the metal sheath or steel pipe. The 'gas compression' cable worked on a different principle. Insulated oval conductors were sheathed with either lead or polyethylene and the individual cores or the three laid-up cores were then covered with a pressure retaining metallic sheath or pulled into a steel pipe. The space between the inner sheath and the outer pressure retaining member was filled with high-pressure gas, usually nitrogen. Expansion and contraction of the relatively viscous impregnating compound was compensated for by the inner sheath acting as a diaphragm.

With one exception, gas pressure designs are obsolete mainly because they cannot match the technical performance of oil filled cables throughout the voltage range. At 275 kV and above, low power factor and high breakdown strength (a.c. and impulse) become increasingly important, and OF cables can be operated to higher design stresses.

The exception is the 'pre-impregnated gas filled cable', useful for applications where problems exist in creating practicable oil sections, e.g. on hilly and undersea routes. Other advantages are: (a) there is no need for specialised oil equipment and (b) long continuous lengths suitable for submarine use can be manufactured. The total length for installation is determined by what can be coiled down in a ship, as joints between lengths can be made either in the factory or on the ship. In this gas filled cable system the paper is impregnated with a special greasy compound before being lapped on to the conductor. With modern designs the impregnated cores are covered by a smooth aluminium sheath and gas is admitted directly into the insulation after installation. There is a minimum of impregnating compound, and although voids do exist from the outset, the high nitrogen pressure provides good electrical strength. As with other forms of gas cable, however, operating voltages are usually limited to 132 kV.

13.5.2.6 Cables with polymeric insulation

Mention has already been made of the increasing use of polymeric insulation, largely XLPE, for distribution cables up to 33 kV. The low power factor of around 0.0005 which is attainable with such cables is also clearly attractive in comparison with the minimum of about 0.002 which is possible with the best OF cables. It seems likely that a new phase of cable transmission is emerging. Incentives are simpler jointing techniques, and reduced maintenance as pressurising equipment and oil leaks are avoided. An economic consideration relates to the voltage limit above which it is desirable to have a metal sheath over the insulation to prevent contact with water.

It was reported in 1978 that 188 km of lead sheathed 225 kV cable with thermoplastic polyethylene insulation was in satisfactory service in France, the first lengths having been installed in

has to be compared with the service performance requirement of between 10.2/1 and 6.2/1 according to voltage, i.e. cables must be designed on an impulse breakdown stress basis and then they will have a large safety margin for a.c. performance. The reverse would soon lead to breakdown. *Table 13.7* also illustrates that, because the impulse/a.c. ratio reduces with increasing voltage, higher design stresses can be adopted as voltage increases: typical values are 7.5 MV/m at 33 kV; 12 MV/m at 132 kV; and 15 MV/m at 275–400 kV.

13.5.2.2 Low-pressure oil filled (LPOF)

Right from the beginning the LPOF cable has been well to the fore and has been the only type of cable widely used in the UK at 275 and 400 kV. Single- and 3-core designs are available from 33 to 132 kV, but because of diameter limitations, only single-core cables can be produced for the higher voltages. *Figure 13.15* illustrates how oil channels are provided within the cable. In

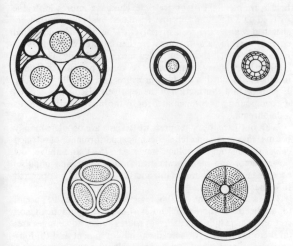

Figure 13.15 Cross-sections of typical oil filled cables

single-core cables the oil flow is normally through a duct in the centre of the conductor, but for short lengths used to terminate 3-core cables the design may incorporate an annulus formed by the provision of longitudinal ribs on the inside of the lead sheath. In 3-core lead sheathed cables and aluminium sheathed cables with circular conductors, a duct is placed in the fillers between the cores. Alternatively, with a corrugated aluminium sheath (CAS) it is possible to omit the ducts and fillers. At 33 kV the conductors may be of oval shape and the construction is known as ductless shaped oil filled (DSO), whereas for higher voltages with circular conductors it becomes DCO (ductless circular oil).

As the cable heats, the oil expands and is forced out of the cable through pipes at joints or terminations into a tank reservoir having internal pressurised capsules so designed that, on cooling, there is a feedback of oil into the cable. *Figure 13.16* illustrates the system. Tanks are of sizes to suit the route length and volume of oil in the cable. They are pressurised to take into account variations in height along the cable route. By the inclusion of stop joints between lengths of cable the circuit may be split into several oil sections. The designed static pressure within the cable is 5.25 bar, but transient pressures up to 8 bar can occur during periods of rapid heating due to increasing load. Optimum planning of the oil feed and sectionalising arrangements is a very important part of the economic design of a cable system.

From the time the cable is filled with oil during manufacture, the oil pressure must be continually maintained. A small tank is fitted on the cable drum; it remains connected during cable laying and even during jointing a flow of oil is maintained. When cables are installed in vertical shafts, e.g. for pumped storage stations, special arrangements are necessary. The Cruachan 275 kV pumped storage scheme in Scotland has a vertical head of 325 m with a consequent hydrostatic pressure of approximately 30 bar. The cable had to be partially drained under vacuum to limit the oil flow while making the lower stop joint, and reimpregnated before making the upper sealing end.

Lead sheaths will not withstand significant internal pressures, and are reinforced to withstand a continuous pressure of 5.25 bar for normal installations by the helical application of bronze tapes. In spite of the reinforcement, the lead sheath is subject to some expansion under creep stress. British practice favours the use of $\frac{1}{2}$C alloy (0.2% tin, 0.075% cadmium). Aluminium sheaths have technical and economic advantages and nowadays they are of corrugated design. This enables the thickness to be reduced and also provides greater flexibility for handling.

To maximise the efficiency of oil flow and the length of individual oil sections, the viscosity of the oil needs to be as low as possible, consistent with low power factor and electrical strength. Until recently, mineral oils were used with a viscosity of about 12 centistokes maximum at 20°C. However, current practice is to use synthetic alkylates of dodecylbenzene type which have better gas absorbing properties under electric stress. Such impregnants are intermediates in detergent manufacture.

Apart from the impregnant, the insulation for OF cables also differs significantly from that for lower-voltage paper cables. To keep the power factor as low as possible, the paper needs to be more thoroughly treated to remove impurities. For example, the

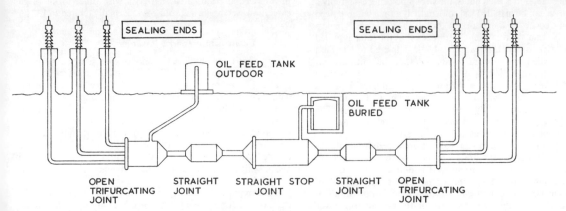

Figure 13.16 Diagrammatic layout of typical 3-core OF cable system

around 200 mm². With the growth in the usage of electricity during the next 30 years, cable voltages and ratings had to keep pace with and match the overhead-line circuits. Developments quickly proceeded to find solutions for the higher voltages with conductors up to 2500 mm².

Bulk transmission in the UK began in the 1930s with a circuit requirement of 110 MVA at 132 kV; 275 kV followed in the late 1950s with a winter circuit rating of 760 MVA; and by the late 1960s the circuit demand had increased to 2600 MVA at 400 kV. To obtain this from a single cable circuit meant that design had to be pushed towards the limit for paper insulation in relation to electrical features, diameter and coiling on drums. To match the increases in overhead-line ratings, it has become necessary for the heat generated in cables to be removed by more sophisticated engineering means involving cooling pipes.

13.5.1.4 Alternatives to impregnated paper insulation

The OF cable has been most successful in meeting all requirements up to 525 kV. At some future date (possibly not until the next century in the UK, but earlier in some other countries) there will be a need for undergrounding parts of transmission lines operating at 800–1000 kV. To produce cables within a diameter (say 160 mm) which is practicable for manufacture and handling, and to keep losses within an economic limit, it seems essential to use higher stresses (possibly 25 MV/m) and dielectrics with lower power factor and permittivity than can be achieved at present with impregnated paper. Possibilities are discussed later.

There has been much recent interest in materials such as polyethylene and XLPE, which have good potentialities for very high voltages and other advantages for the lower voltage range. To date, the enthusiasm has stemmed not primarily from low losses, but from the possibility of a much simpler cable construction, the promise of less complicated jointing requirements and, above all, the elimination of pipework and pressurising equipment. In some overseas countries, where installation and maintenance skills are not readily available, this is an important factor which could well justify a higher intrinsic cable cost. Location and repair of oil leaks can be troublesome. However, some of the problems with polymeric insulation for distribution cables have already been discussed, and for 130 kV upwards the usage so far of PE and XLPE cables has been relatively small.

13.5.2 Types of cable

13.5.2.1 Basic requirements

Apart from absolute consistency and freedom from defects, the essential requirements of high-voltage dielectrics are:

(1) High impulse strength, because this is the ultimate design stress requirement and determines dimensions.
(2) Low dielectric power factor in order to keep the heat generation to a minimum. When conductor size is at the maximum possible, much expense may have to be devoted to means of cooling the cable to obtain an economic circuit.
(3) Low permittivity to reduce both the electrical losses and charging current.
(4) Ease of bending during installation without sustaining damage which could affect service life.

Impregnated paper under oil pressure is the only dielectric which so far has met all these requirements up to about 525 kV. In relation to electrical losses, however, it is reaching its limit at this voltage without forced cooling. Impregnated paper with gas pressure is technically satisfactory up to 275 kV but is not generally economically competitive with OF cables.

Low-pressure OF cable (*Figure 13.14*) is used almost universally in the UK throughout the voltage range and it is

Figure 13.14 400 kV single-core oil filled cable

predominant throughout the world. High-pressure OF cable is favoured in the USA and the flat Mollerhoj OF type of cable is used in Denmark.

The influence of the impulse strength requirement on design stress can be seen from *Table 13.7*.

Conventional OF cable has a safe impulse stress of around 100 MV/m and a.c. stress of 30 MV/m, a ratio of about 3/1. This

Table 13.7 Working and impulse voltages

System voltage (kV)	Working voltage (kV)	Impulse-test voltage (kV)	Impulse/working ratio
33	19	194	10.2
66	38	342	9.1
132	76	640	8.4
275	160	1050	6.6
400	230	1425	6.2

13.4.7.2 Tests of completed cables at works

Tests of completed cables at works comprise the following:

(1) Measurement of the thickness of insulation and other prescribed components.
(2) Conductor resistance test.
(3) An a.c. test for 5 min at a voltage which is usually $2.5U_0 + 2$ kV for cables rated up to 3.6/6 kV, and $2.5U_0$ for cables of 6/10 kV and above. For multicore non-screened cables the test is required between conductors and also between any conductor and sheath. For cables with individually screened cores it is from conductor to sheath only.
(4) For paper insulated cables with rated voltage U_0 of 8.7 kV and above, a dielectric power factor/voltage test is required to determine compliance with prescribed limits for maximum power factor at 0.5 times U_0 and maximum difference in power factor from 0.5 to 1.25 times U_0 and from 1.5 to 2.0 times U_0.
(5) A partial discharge test is required for cables insulated with butyl rubber, PE and XLPE of rated voltages above 1.8/3 kV and on cables insulated with EPR and PVC of rated voltages above 3.6/6 kV. The magnitude of discharge at $1.25U_0$ must not exceed 20 pC for butyl, EPR, PE and XLPE and 40 pC for PVC.

13.4.7.3 Tests after installation

(a) Paper cables: A 15 min d.c. test at a voltage of 70% of the values given in (3) above.
(b) Polymeric cables: A 15 min d.c. test at a voltage of $4V_0$.

13.4.7.4 Special and type tests

(1) A bending test at a radius much more severe than stipulated for installation, followed by a voltage test. For paper cables the diameter of the test cylinder varies, according to the cable rated voltage and type, from 12 to 25 times the diameter of the cable plus the diameter of the conductor $(D+d)$. Three cycles of bending are required and maximum limits are stipulated for tearing of individual paper tapes. For polymeric cables two cycles of bending are required over a test cylinder of 20 $(D+d)$ for single-core cables and 15 $(D+d)$ for multicore cables.
(2) A drainage test for non-draining paper cables at the maximum continuous operating temperature for the cable. The maximum permissible drainage is 2.5–3% of the internal volume of the metal sheath.
(3) A dielectric security test for paper cables comprising sequential bending, impulse and a.c. tests. The impulse withstand requirement is 95 kV for $U_0 = 8.7$ kV, 125 kV for $U_0 = 12$ kV and 170 kV for $U_0 = 18$ kV. The a.c. application is $4U_0$ for oil/resin impregnation and $3U_0$ for non-draining impregnants.
(4) A power factor/temperature test for paper cables of $U_0 = 8.7$ kV and above to a temperature 10°C above rated temperature. Limits are 20–60°C, 0.0060; 70°C, 0.0130; 75°C, 0.160; 80°C, 0.0190; 85°C, 0.0230.
(5) An electrical test for butyl, PE and XLPE cables above 1.8/3 kV and PVC or EPR cables above 3.6/6 kV. This requires sequential application and/or measurement of insulation resistance, partial discharge, bending, power factor/voltage, power factor/temperature, load cycles, partial discharge, impulse withstand and a.c. high voltage. The impulse requirement is $U_0 = 3.6$ kV, 60 kV; $U_0 = 6$ kV, 75 kV; $U_0 = 8.7$ kV, 195 kV; $U_0 = 12$ kV, 125 kV; and $U_0 = 18$ kV, 170 kV. The a.c. test comprises 4 h at $3U_0$.
(6) Tests on the component materials before and after ageing and, in the case of polymeric cables, on the complete cables after ageing.

13.5 Transmission cables

13.5.1 Historical development sequences for a.c. transmission

13.5.1.1 Problems due to partial discharges within paper insulation

Reference has already been made to the work by Hochstädter which led to the 'H' type or 'screened' radial field design. Such constructions were quite satisfactory at 33 kV and to a limited extent at 66 kV. Failures at 66 kV and higher voltage were found to be due to discharges in minute vacuous voids formed by expansion of the impregnating compound with insufficient subsequent contraction to fill all the space available. Emanueli in the later 1920s pioneered the first solution, which was the oil filled cable. The basic requirement was either to eliminate completely the possibility of voids being created, as in the pressurised oil filled cable, or to ensure that they were always under a high gas pressure. Gas may be admitted directly into the insulation or exerted externally on a flexible sheath over the insulation, in which case the void suppression principle is more akin to that of the OF cable.

As the void formation mechanism was also clearly related to the temperature excursions of the insulation, the operating temperature limit of the solid (i.e. non-pressure) insulation could be raised from 65 to 85°C, with consequently much improved cable ratings. Even more important was the fact that the a.c. operating electrical stress of impregnated paper insulation could be increased from 4 MV/m to about 16 MV/m and reductions in dielectric power factor were achieved.

13.5.1.2 Types of paper-insulated pressurised cables

Many different types of pressuring are possible and may be classified into the two basic constructions indicated in *Table 13.6*, which lists those currently in service.

Table 13.6 Pressure cables and voltage in commercial service

Design	Voltage range (kV)
Fully oil impregnated	
Lead or aluminium sheath:	
Low-pressure OF	30–525
Mollerhoj flat	30–132
Steel pipe:	
High-pressure OF	30–500
External gas pressure with	
diaphragm sheath	30–275
Internal gas pressure	
Lead or aluminium sheath	30–275
Steel pipe	30–132

A fact which emerges from *Table 13.6* is that oil pressure can be used up to the highest voltages at present required (525 kV). Gas pressure has a limitation primarily related to a lower electrical breakdown strength. Although gas pressure cables have some advantages in terms of the associated accessories and equipment, the oil pressure cables are usually more economic.

13.5.1.3 Transmission system requirements

When Emanueli first developed the pressure cable technique, he was fulfilling a need for the requirement of the early 1930s to transmit in the range of 30–132 kV with conductor sizes of

extruded bedding, galvanised steel wire armour and PVC oversheath. A single-core cable would have a PVC bedding over the concentric copper earth wires, then aluminium wire armour and PVC oversheath.

Bearing in mind the faults that can be experienced due to contact between groundwater and insulation, and acknowledging the fact that PVC oversheaths may be damaged during installation or subsequently, new designs with components or special layers to restrict movement of water within cables are being considered.

13.4.6.6 Dielectric deterioration by treeing phenomena

No discussion on polymeric insulation at high voltage would be complete without some reference to deterioration caused by treeing mechanisms. These are related to a prebreakdown characteristic which gradually spreads through the dielectric under electrical stress through paths which, when visible or made visible, resemble the branch structure of trees. Trees are of two basic types:

(1) *Electrical trees.* These are trees in a dielectric consisting of permanent channels having dendritic or branching patterns due to partial discharges during application of a.c., d.c. or impulse electrical stresses. The channels originate at sites of high stress due to non-uniform electrical fields from imperfections such as protrusions at an insulation interface, a void or a contaminant.
(2) *Electrochemical trees.* This is a class of tree generated in a dielectric during application of electrical stress in the presence of liquid water or water vapour—hence, often known as 'water trees'. They consist of fine water channels which can be seen under a microscope after staining. They disappear if the sample is dried, but reappear after boiling in water. Electrochemical trees are formed at stresses which are much lower than those required to produce electrical trees, and the rate of growth may be very slow. The tree patterns appear generally at opaque areas in the translucent polyethylene. If the dielectric or screen is in contact with soil water containing such minerals as sulphides, the water may have a characteristically coloured stain. The initiation of electrochemical trees is at the same types of site as indicated above for electrical trees. Characteristic names are often given to them according to origin, e.g. 'bow-tie' trees from contaminants (*Figure 13.13a*) and 'bush' or 'broccoli' from surface imperfections. *Figure 13.13(b)* shows an electrical tree which is developing in an area where electrochemical treeing has become extensive.

It is this treeing phenomenon which is the important reason for the insulation to be free from all irregularities and for the surfaces to be smooth and in good contact with the screens. Cables may operate for many years before a tree size is generated which will contribute to ultimate breakdown. The presence of water is a requisite for treeing to be initiated, but a very small amount suffices, and for the highest voltages it is desirable to ensure that all moisture is excluded, e.g. by provision of a metallic sheath.

13.4.7 Cable tests

Full details of the tests and procedures required are given in the cable Standards listed at the beginning of the section. IEC 55 and IEC 502 are the most relevant documents and values quoted below are taken from these Standards. A complete summary would be lengthy and it is only possible to give a brief outline to illustrate the general basis for the more important tests.

13.4.7.1 Manufacturing tests

Tests during manufacture are restricted to those which are not possible on finished cables and comprise a.c. spark tests on polymeric insulation and sheaths.

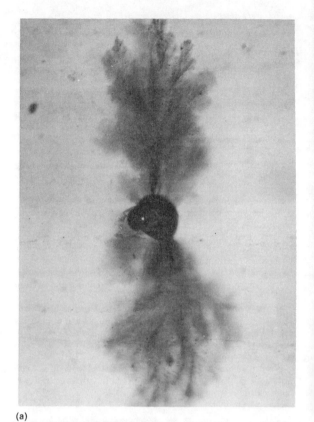

Figure 13.13 (a) Bow-tie tree at an inclusion; (b) electrical tree in an area of extensive water tree development

without adequate testing to ensure freedom from discharge. It was also not until the mid-1970s that ideal forms of screening were developed which could be readily removed for jointing and adequately deal with thermal expansion and contraction. Then, in the succeeding years, the final problem was to identify and find solutions to problems caused by effects of water in contact with the insulation. Water has minute solubility in PE and XLPE, but 'tree-like' structures were found in the insulation and it was eventually established that these could lead to electrical breakdown.

Although UK manufacturers supplied some of the first cables used in the early 1960s, the uses at home have largely been for specialised applications, notably for cables in CEGB generating stations. Three main reasons have accounted for this. First, for the types of cable used, there has not yet been a clear economic incentive in comparison with paper insulation. Second, in the countries where polymerics have been adopted, one of the prime reasons has been to enable jointing to be carried out with less skilled resources. In the UK this has not been necessary. Third, the higher operating temperature of cross-linked insulation has particular benefit in reducing the derating penalty in countries having high ambient temperatures. Nevertheless, cable manufacturers export a large proportion of their production and so have been obliged to produce competitively.

In the problems that have arisen, polyethylene has no advantages over XLPE and, because it is a thermoplastic material, has great disadvantages in current ratings. Even in the USA, where the use of PE has been substantial, it is now rapidly giving way to XLPE. The remainder of this section refers to XLPE only.

No British Standards have yet been prepared for high-voltage polymeric cables, but IEC Specification 502 forms the basis for most national requirements.

13.4.6.1 Conductors

Circular conductors in either solid or stranded form have been standard to date. Uniform application of insulation and screens is somewhat more difficult with shaped conductors, but there is a limited commercial production. Because of the importance of the screen between conductor and insulation, a smooth conductor surface is desirable and stranded conductors need to be well compacted.

13.4.6.2 Conductor screens

Many of the early cable failures were due to imperfections resulting from the use of semiconducting fabric tapes as conductor screens. A thin layer of extruded semiconducting polymeric material is now mandatory, and to ensure a clean interface it is normally extruded in tandem with the main insulation and cured with it. In the case of stranded conductors, a semiconducting tape may be applied between the conductor and extruded screen to prevent penetration between the wires and facilitate removal for jointing.

13.4.6.3 Insulation

Extrusion and curing can be carried out by a variety of processes, but a cardinal feature of all of them is that good material handling to avoid dirt and contamination is vital. The most common is the continuous catenary vulcanising (CCV) method, in which the curing is carried out by steam pressure and the cooling is in water. For cables of the voltage range under consideration the small water content induced by steam curing does not seem to be of importance, but it can be avoided, with the same equipment, by using radiant heating and nitrogen under pressure. The Monosil process, to which reference has been made in Section 13.2.2, is also suitable for 10–20 kV cables, and, although the curing is in water, the insulation performance is equivalent to that of 'dry cured' cable.

13.4.6.4 Insulation screen

One of the main factors concerning the dielectric screen is that it should be easily removed for jointing. A layer of semiconducting polymeric material, compatible with the insulation, can readily be extruded and cured in the same operation, but it then tends to be integrally bonded and can only be removed with special tools. Great care is necessary to produce a smooth surface with no damage to the insulation.

The most common type of screen to date has consisted of a primary screening layer of semiconducting varnish followed by an easily removed semiconducting tape and a copper tape. However, extruded semiconducting screens are now available which are well bonded to the insulation but are readily 'strippable', and these are likely to become of standard use.

The amount of metal in the insulation screen has to be related to what is required for earth fault current-carrying capacity. For 3-core cables copper tapes applied over the semiconducting layer may be adequate and can be supplemented as necessary by copper wires in the filler spaces. For single-core cables a concentric layer of copper wires may be needed.

13.4.6.5 Finish

A typical 3-core cable is shown in *Figure 13.12*, in which the copper taped cores are laid-up, then provided with a PVC

Figure 13.12 Three-core 8.7/15 kV XLPE insulated steel wire armoured cable

As PVC insulated cables are little affected by moisture, no metal sheath is required, and this contributes greatly to ease of handling as well as simplifying jointing and terminating procedures. No precautions have to be taken to prevent entry of moisture.

BS 6346 caters for conductors of stranded copper or solid aluminium, but not stranded aluminium. The solid form was chosen because it provides the most economic cable construction and is particularly suitable for manufacture with PVC insulation. Solid conductors are also very much better for either soldering or mechanical jointing techniques. Stranded aluminium conductors are often preferred by overseas users and are also used for power supply cables in coal mines, as they facilitate coiling for taking the cables down the mine shafts.

Except for the smallest sizes, the conductors are shaped, and uniform thickness of insulation is obtained by extruding the PVC as a slightly oversize tube which is made a snug fit on the conductor by a combination of conductor feed speed control and internal vacuum. For multicore cables, the cores fit tightly together, leaving few gaps, but when these are of larger size, non-hygroscopic fillers are included so that the laid-up cores are reasonably circular.

For the armour bedding, there is a choice between PVC tapes and a layer of extruded PVC. The latter is more expensive but provides a robust cable which is preferred for cables with circular conductors, for cables laid underground, and when it is desirable for terminating glands to provide a seal on to the bedding.

While any form of armour can be supplied (e.g. steel tape or strip, aluminium strip or galvanised wire (GSW)), BS 6346 covers only GSW or aluminium strip. GSW is normally preferred, as it gives optimum mechanical protection and adequate earth conductivity. Aluminium strip armour is now usual only when extra earth conductance is required and it is then important that suitably designed aluminium terminating glands be used. Aluminium armour is also necessary for single-core cables, because steel armour, being magnetic, increases losses, with an adverse effect on ratings.

A PVC sheath is usually applied overall and no bitumen is normally included over the armour. When this was done in the early years, following conventional practice with textile servings, it was found that the bitumen extracted plasticisers from the PVC, creating a mobile black liquid which would bleed from the cables at terminations below vertical runs.

The early choice of PVC by the National Coal Board for mining cables operating at 3.3, 6.6 and, to some extent, at 11 kV was associated with the fact that the resilience of the insulation was found to provide better resistance than paper insulation to damage by rock falls. Although PVC is also satisfactory for higher voltages and has been used extensively in Germany at 20 kV, the electrical losses tend to be high. Better materials such as XLPE are now available when polymeric insulation is preferred.

Although the relative hardness of PVC at ambient temperature can be modified considerably by the choice and proportion of the platicisers used, these cannot exert a significant effect on deformation at maximum operating temperatures. Heat resisting grades are defined in BS 6746, and such grades can even be formulated to allow PVC to operate for limited periods up to around 100°C without serious degradation due to chemical factors. However, they do little to improve deformation resistance and not much use has been made of them for power cables.

13.4.5 XLPE insulated cables up to 3.3 kV

Polyethylene has never found much application outside the USA for power cables, largely because PVC became established and polyethylene suffered from the same disadvantage of thermal deformation. XLPE completely overcomes this problem, and in the voltage range up to 3.3 kV it provides an advantageous alternative with cable constructions which are essentially identical. The main difference is that, as it is a much tougher material, the insulation thickness can be reduced, in the case of 1 kV cables to the minimum which can be extruded satisfactorily.

XLPE has not yet made much impact on the established use of PVC for industrial cables in the UK, largely for price reasons, but this situation could well change in the future. XLPE has positive advantages because it is a better insulating material with much lower dielectric loss factor; more particularly, it can be operated satisfactorily to 90°C, with corresponding improvement in cable ratings. These factors have clearly provided an incentive for XLPE to be considered as a competitive material throughout the whole range of power cables up to the highest voltages, and there is little doubt that, because of its universal sphere of application, it will be well to the fore in the future. There are competitors, such as ethylene propylene rubber, which may have advantages for specific cable types, but are unlikely to be economic over the whole range.

Up to 3.3 kV, therefore, XLPE is competing both with PVC and with paper insulation. In comparison with PVC, the continuous current rating advantage is usually more apparent than real, because cable size is dictated by voltage drop rather than current rating. The short-circuit rating based on 250°C instead of 160°C is likewise a bonus only infrequently required. Where XLPE does gain is in that, when ambient temperature is high, such as in tropical countries, the benefit from a smaller derating factor can be substantial. XLPE is not flame retardant, as is PVC, but as flame retardancy is governed more by the oversheath than the insulation, this is not normally significant.

In comparison with paper insulation, XLPE also has a small continuous rating benefit, but the main advantage is the absence of a metallic sheath and the availability of cable which is much cleaner and easier to handle in laying and jointing, together with lower permissible bending radii. The simpler jointing techniques, without any need for plumbing, provide strong attraction for developing countries where such skills are not readily available; and, to date, this is probably the area where XLPE has made the greatest impact.

Another field for XLPE is for self-supporting 240/415 V cables for overhead distribution, as a replacement for bare conductors. Four insulated circular stranded aluminium conductors are twisted together with a long lay and used with special fittings. Following widespread use in Europe, this application has now proved to be an economic alternative in the UK.

13.4.6 PE and XLPE cables for 11 kV to 45 kV

The excellent dielectric properties of PE and XLPE brought these materials into prominence in the early 1960s for higher-voltage applications and an increasing scale of effort has been devoted to them ever since. In some countries, particularly the USA, they came into regular use at 10–20 kV at an early stage, and, in spite of a very poor initial service performance in comparison with paper insulation, they have since virtually replaced it for many years at voltages up to 45 kV.

The most important single factor which has caused problems is that, as with paper insulation, internal partial discharges occur at voltages of 5 kV upwards at any irregularities within or at the surface of the insulation. However, whereas paper insulation has fairly good resistance to such discharges and the effects in butt gap spaces can be minimised by oil or gas pressure, polyolefines such as PE and XLPE are particularly weak. Both PVC and EPR are rather better, but have other limitations.

While it was recognised that the insulation must be extremely clean and free from voids, and that screening at both surfaces of the insulation was necessary, many cables were put into service

techniques involving the use of mechanical fittings and cast resin filling.

One of the important features in the design of any CNE cable is that, in the event of cable failure, there should be no loss of the important protective neutral conductor. It has also to be recognised that, with the growing use of mechanical excavating equipment, the main source of cable failures is now third-party damage. If the PVC oversheath on Consac is damaged, local corrosion of the aluminium sheath will follow and water entering the insulation will produce detectable cable failure before there is any severe reduction in conductance of the neutral.

The Waveform cable type (*Figure 13.10*) is also known as Waveconal and Alpex. Although introduced some years after

Figure 13.10 Waveconal cable

Consac, and a little more expensive, it has become more extensively used than Consac, because of its simple jointing techniques. The XLPE insulation represented the first departure by the UK Electricity Boards from paper insulation for mains cables. In Waveform cable the neutral conductor comprises aluminium wires applied with a sinusoidal lay, and, as with all modern cable designs, there is an outer PVC oversheath. For making service joints, no cutting of the neutral is involved and the wires can readily be formed into two bunches for mechanical jointing. A key point in the design is that the wires are spaced and encapsulated between two layers of unvulcanised rubber so that each wire is separately embedded in the rubber. In the event of local damage to the oversheath, the entry of groundwater is thus limited: again an important factor in preventing loss of the neutral conductor. In Germany this form of sinusoidal lay neutral construction is known as 'Ceander', but has only been employed with copper wires and without the rubber bedding.

In Scotland some use has been made of 'Districable', a type which is also found in France. This is a 4-core construction, again with XLPE insulation on the phase cores, but the neutral/earth circular or shaped conductor has a lead sheath to protect it from corrosion. Two thin galvanised steel tapes are applied as a binder and to provide a metal screen in contact with the neutral/earth conductor. A PVC sheath is applied overall.

Ultimate simplicity in CNE cable design and ease of jointing can be achieved by the use of four shaped aluminium conductors, insulated with XLPE and then provided with a PVC oversheath with no outer metallic protection. This type has already replaced all other public supply mains cable types in Germany. In the UK it has been rejected, however, even though a metallic envelope is not mandatory below 650 V and spiking tests have shown that danger from flash or shock is little greater than with the Waveconal, Consac, Districable or lead sheathed paper cables. The reason is that, in the event of mechanical damage, there could be exposure of the aluminium in the neutral to groundwater, with the possibility of undetected loss of this conductor. Damage to phase conductor insulation could also give rise to currents in the ground, flowing through other buried metalwork.

13.4.3 Service cable

Prior to the introduction of CNE mains cables, the service cables were of the 3-conductor split-concentric type. This design is still used for consumers where it is not practicable to provide a PME earth terminal connected to the supply neutral/earth. The single-core (or multicore for 3-phase supply) phase conductors are insulated with PVC. In a helically applied concentric layer around the phase core or cores, some of the copper wires are bare to form the earth conductor and some have a thin layer of PVC coating to comprise the neutral conductor, the two portions being separated by PVC strings.

For a PME system, the construction is further simplified to a 2-conductor design, the concentric layer consisting of bare copper wires.

As an alternative design of service cable, some users of Waveform mains cable prefer to adopt the same construction, i.e. sinusoidal lay aluminium wires embedded in rubber, as the neutral/earth conductor.

In addition to use for house service, all these cables find applications for such requirements as street lighting, traffic signs and complete individual routes for motorway lighting.

13.4.4 PVC insulated power cables

Although used for public supply cables in some overseas countries, PVC insulation has never been adopted in the UK for this purpose, other than for the service cables described above. The reason is associated with its thermoplastic nature and resultant softening at elevated temperatures. Thus, at 1 kV, ratings are restricted by a maximum temperature of 70°C, whereas paper can be operated to 80°C and XLPE to 90°C. More important, however, is that, in the event of a short overload, severe thinning may occur due to deformation by conductor thrust at bends, whereas paper or XLPE insulation would be relatively unaffected.

Close fusing to give cable protection is usually impracticable in public supply systems, but presents no great problem in industrial applications. Since the late 1950s, therefore, PVC insulated cables have been almost universally applied in this sector for voltages up to 3.3 kV. Close fusing was defined by the 14th edition of IEE Wiring Regulations as an excess-current operating device which operates within 4 h at 30% excess rated value for cables direct in the ground, or 50% excess for cables in ducts or in air. In the 15th edition the whole concept has been changed (see Section 13.6.7).

Figure 13.11 shows typical 1 kV cable and 6.6 kV cable, the latter as used in coal mines.

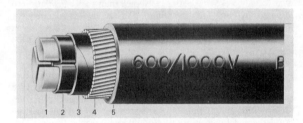

(a)

(b)

Figure 13.11 PVC insulated cables: (a) 3-core 1 kV, SWA for industrial use; (b) National Coal Board 3-core 6.6 kV, DWA

However, if full load is to be carried regularly, it is desirable with smooth aluminium sheathed cable to employ joints filled with cast resin to overcome possible problems at the plumbs and buckling of cores within the joint sleeve. Most of the 11 kV cable installed by the public supply authority in the UK operates at less than maximum rating because of factors of which the most pertinent is that the cable network is in open rings and is only required to carry full load when the ring is closed because of a fault near to the transformer.

Belted and screened types are available (*Figure 13.7*) the belted construction being more widely used. When the sheath is corrugated, there is a large space between the outside of the insulation and the inside of the sheath. It is important that this should be partially (but not completely) filled with impregnating compound.

13.4.2 CNE cables for PME systems

In this heading CNE denotes a 'combined neutral and earth' conductor in cable construction and PME signifies 'protective multiple earth' applied to a network.

13.4.2.1 PME systems

Traditional practice in UK buried systems involved earthing of the neutral conductor at one point only, at the substation. This meant that the supply cables along the streets required five conductors, three phases, one neutral and one earth (the lead sheath). Consumers normally obtained a satisfactory earth by connection to buried lead water pipes, but when conditions were difficult, as with overhead line sections in rural areas, a practice of multiple earthing of the neutral (MEN) was introduced in the 1940s by burying suitable metal adjacent to poles. This gradually extended to the PME concept, which basically implies that consumers are provided with an earth terminal connected to the supply neutral conductor.

In the 1950s further problems arose when lead water pipes began to be replaced by plastic pipes. One solution was to earth consumers' plant to the lead sheath of the supply cable, but this was only satisfactory if across all straight and branch joints the lead sheaths were plumbed to the jointing sleeves. In many cases, however, the joints were of the mechanical grip type in cast-iron boxes, and these had such a high resistance that the earth path to the substation was inadequate. By additional earthing of the neutral conductor, nowadays usually only at the remote end of the run, and by using this conductor also as the protective earthing conductor, the consumer earthing problems were overcome. Moreover, the supply cable required one fewer conductor and by developing a new range of cables very considerable savings were obtained. *Figure 13.8* shows the much lower material utilisation achieved with one form of CNE cable.

Figure 13.7 Smooth (a) and corrugated (b) aluminium sheathed, 11 kV, paper insulated cables

Figure 13.8 Comparative dimensions of 4-core PILS/STA and Consac CNE cables of equal rating

Initially, PME systems with CNE cables were kept separate from the existing networks, but by appropriate and simple bonding between neutral conductors and lead sheaths, all existing systems can be converted to PME. The UK network has largely been modified in this way. CNE cables can, therefore, be installed indiscriminately for replacements and additions. A point requiring attention is that, if consumers are given a PME earthing facility, all exposed earthed metal within the installation which may be touched must be suitably bonded to provide an equipotential background.

13.4.2.2 CNE cable types

The Consac cable, introduced in the mid-1960s and shown in *Figure 13.9*, maintains the use of paper insulation. Aluminium sheathing provides the neutral. The use of solid aluminium conductors with 1 kV paper insulation had already become established with 4-core PILS cables before the development of

Figure 13.9 Consac cable

Consac. Because of the total use of aluminium as conductor metal and the small amount of other material, Consac has a very economic construction, but in the early days some of the cost advantage was lost by the extra difficulty of plumbing the aluminium sheath. This was later overcome by the simple

spaces. An even more important effect, to be seen in *Figure 13.6*, is that, in addition to the radial stresses through the layers of paper, there is also a tangential stress component along the paper surface. In the tangential direction the electric strength of impregnated paper is only one-tenth of that radially.

When supply voltages were increased to 22 kV and 33 kV in the 1920s, many cable failures occurred due to lack of appreciation of this fact. Hochstädter identified the need for an earthed metallic layer over the insulation to create a purely radial field, a construction subsequently known as 'H'-type or screened. Very little metal is required for the purpose, and while thin copper tapes have been used, the most common form nowadays for multicore cables is a layer of thin aluminium tape or of metallised paper, consisting of aluminium foil on a paper backing. The latter is usually pinpricked to facilitate passage of oil during impregnation. The cores and fillers are held together by a binder of 'copper woven fabric tape' (CWF) containing a few thin wires woven into the weft. This gives protection against scuffing and provides electrical contact between the screen and the metallic sheath. Another construction, used mainly in continental Europe, is the 'HSL' or 'HSA' type, which denotes three lead or aluminium sheathed single cores laid up together and then armoured overall.

The screened construction is optional at 11 kV, but mandatory for higher voltages. Because the dielectric has much better electric field distribution the operating temperature can be increased and higher current ratings obtained. Some 11 kV users find that these factors justify the somewhat greater expense and the extra skill required in jointing.

13.4.1.2 Insulation

The insulation comprises layers of paper tapes, of thickness in the range of 0.7–1.9 mm, carefully applied to maintain controlled butt gap spacings and optimum registration between layers. The stress is highest at the conductor surface and may be increased locally, owing to the conductor profile or lack of smoothness. To improve this situation at voltages of 6.35/11 kV and above, a layer of semiconducting carbon paper is applied over the conductor to exclude from the field the small spaces between the wires of the outer layer which otherwise could be sites for discharge.

The thickness of insulation has to be determined by both mechanical and electrical requirements, the former being dominant at the lower voltages, e.g. to withstand bending and to resist damage due to impact. Similarly, at 11 kV, while impregnated paper itself has an a.c. breakdown strength of the order of 10 MV/m, the actual cable design stress is only 2 MV/m, the effects in butt gap spaces being one of the most important factors.

The impregnation of the paper is carried out before application of the metallic sheath by the 'mass-impregnation' process. The cores, on drums or rewound into trays, are inserted into large tanks. These are first evacuated to remove all the moisture in the paper. Hot impregnating compound is then admitted, and the tank is maintained under pressure for a period which depends on voltage, and then cooled slowly to ensure that contraction voids are not present within the insulation.

To obtain good impregnation, the compound viscosity at 120 °C should be low, but in the operating temperature range of the cable it needs to be as high as practicable, so that no drainage occurs into the inevitable space under the metallic sheath and into joints. Traditionally, the compound has consisted of mineral oil thickened with gum rosin. A problem with such compounds is that the viscosity at maximum operating temperature is not high enough to prevent drainage when cables were installed vertically or on hilly routes, thus leaving the already relatively weak butt gap spaces devoid of impregnant. In the 1950s, BICC developed the 'mass-impregnated non-draining compounds' (MIND) which subsequently became standardised in the UK. In these compounds the viscosity control is obtained by the addition of such materials as microcrystalline waxes and polyethylene to mineral oil.

When single-core or 3-core cables are operated at 11 kV or higher, some discharge may occur in the space between the insulation and the inside of the metallic sheath. Although this is not unduly detrimental, it is eliminated by the inclusion of a carbon paper over the insulation.

13.4.1.3 Lead sheath

Unalloyed lead is suitable for the majority of armoured cables but is prone to fatigue cracking if subjected to vibration or to high expansion and contraction, as when cables are suspended on hangers or are in manholes. In the UK, when moderate improvement of fatigue strength is required, it is usual to adopt Alloy 'E' to BS 801 (0.4% tin, 0.2% antimony). Alloy 'B' (0.85% antimony) has higher fatigue strength and is desirable for conditions involving severe flexing, such as aerial cable installations and cables on bridges. Other alloys are available and are preferred in some countries.

The use of very high-purity lead is detrimental because it can give rise to large grain size and low fatigue strength. Hence, it is always preferable to use lead with impurities up to the limit of 0.1% as permitted by BS 801. Tin and antimony are frequently added for this reason.

13.4.1.4 Armour

The use of armour fulfils a variety of functions, primarily to supply mechanical protection during cable handling and installation, and subsequently in service. Steel taping is the most common, but in the UK a layer of galvanised steel wire is applied for 11 kV and higher to increase the longitudinal strength of the cable. Galvanised steel tape is popular in tropical countries to provide greater resistance to corrosion; narrow steel strips are often preferred in continental Europe. In general, the resistance to damage is proportional to the armour thickness, and steel strips or tapes are less effective than steel wire.

13.4.1.5 11 kV aluminium sheathed cables

The replacement of a lead sheath by an aluminium sheath with good corrosion protection, such as an extruded plastics oversheath, provides a very economic cable construction eliminating armour. For cables of a type offering other advantages, e.g. to provide a concentric conductor as in the Consac CNE type cable described later, and HV cables operating under internal pressure, aluminium sheaths have been widely used since the mid-1960s.

For other types of paper insulated cables, however, aluminium sheaths have not found favour, one factor being the somewhat greater skill required for sheath plumbing when joints other than the cast resin type are used. An exception is that in the UK the public supply Boards have almost universally standardised on 11 kV aluminium sheathed paper cables since the mid-1970s. This resulted in a very considerable cost saving and has made XLPE insulated cables unable to compete in price. Initially, some Boards favoured a smooth aluminium sheath, while others preferred the corrugated form. With a weight reduction of 50%, both types of cable are as easy to handle as lead sheathed cables, but in the larger sizes of 240 and 300 mm^2 cables with smooth sheaths are more difficult to pull into ducts and to bend at terminations when space is limited. The preference is now for the corrugated sheath.

A significant factor is the effect of thermomechanical forces at straight joints. With the flexible corrugated sheath the position does not greatly differ from that with lead sheathed cables.

insulation by heavy additions of aluminium trihydrate in EPR and EVA. For beddings and oversheaths similar addition may be made to ethylene acrylic elastomer, but such compounds do not have the toughness and oil resistance of CSP and PCP compounds.

For aircraft engine components, where cable weight is important, the cable construction is based on silicone rubber plus quartz with PTFE coverings.

For ships' cables, silicone rubber is also used, and where a glass braid is also included, the silica ash enables the IEC 331 test at 750°C to be met. By use of special EPR compounds, the withstand temperature may be increased to 1000°C. The use of mica/glass tapes on conductors provides good high-temperature insulation which is cost effective in comparison with silicone/glass and mineral insulated designs.

13.4 Supply distribution cables

For underground public supply systems and mains distribution in factories, the cables have traditionally been of the paper insulated type. While this still largely applies in the UK for public supply, changes have been occurring progressively during the last decade. For industrial use up to 6 kV, PVC insulation ousted paper in the early 1960s. XLPE has come in at 1 kV for some types of public-supply cables and has now exerted a strong challenge for all the higher-voltage cables.

In Europe and elsewhere, consumers are supplied at around 240 V single-phase and 240/415 V 3-phase, as required. From the outset, the system for urban areas has been underground with direct burial of multicore cables, and 3-phase transformers feeding large groups of consumers through cables along the whole length of every road. The step-down from the transmission grid has moved towards voltages of 19/33 kV and 6.35/11 kV, but international standardisation for cable specifications caters for a full range of 0.6/1, 1.8/3, 3.6/6, 6/10, 8.7/15, 12/20 and 18/30 kV r.m.s. The rounding off of voltages to whole numbers allows for the fact that the designs cater for 20% variation of voltage.

In the USA, and other countries following American practice, cable designs and voltage standards are the same, but the types favoured and the practical utilisation tend to be very different. The supply to the consumer caters for both 110 and 220 V or thereabouts, and except for the innermost areas of cities the distribution is largely by overhead lines. Instead of 3-phase transformers, the local supply is from single-phase units at 10–15 kV or higher, using appropriate transformers to obtain the dual consumer voltages. Conventionally, such transformers are pole mounted, and small, as they feed only a few consumers.

Undergrounding on the American system tends to be a replica of the overhead practice by continuation of the use of similar small transformers and merely adding insulation to the overhead line conductors. Extruded polyethylene or XLPE is convenient for this purpose for both low-voltage and high-voltage requirements, and this is why the interest first developed on the American continent. Very simple single-core cable constructions meet the requirement and much high-voltage cable has been installed in which the neutral conductor comprises copper wires applied over the insulation with no outer sheath. In recent years a concerted effort has been made to get away from the unsightly poles and overhead distribution lines with emphasis on 'Underground Residential Distribution' (URD). This embraces the concept of small single-phase transformers outlined above but more emphasis is now being placed on direct burial of cables instead of installation in ducts. With the dual voltage requirement for consumers, the URD concept, together with the use of single-core cables, provides a way of undergrounding overhead networks at minimum cost. It seems unlikely that it will be adopted in countries where systems have long been geared to other practices.

13.4.1 Paper insulated cables

From its introduction at the end of the last century, impregnated paper has given excellent service to the cable industry. Under normal conditions, users have been able to install the cables and then forget about them. Ultimate lives of 50–60 years are common, and the majority of cables have been replaced only because they became too small for the load. The UK supply industry depreciates paper cables over a 40 year life—surely towards the maximum for any industrial plant.

While the basic dielectrics have changed little throughout this century, there have been considerable improvements in quality of materials and manufacturing techniques, and these have led to successive reductions in thickness over the years.

13.4.1.1 Belted and screened constructions

In multicore cables a greater insulation thickness is required between conductors than from conductor to metal sheath. The most economic construction, therefore, is to apply part over the individual conductors and then a small thickness as a 'belt' over the laid-up cores (*Figure 13.5*). The spaces between the cable

Figure 13.5 Four-core 1 kV, paper insulated, lead sheathed cable

cores under the belt are filled with jute or paper, but whereas the main insulation consists of paper tapes applied in a controlled manner, the filler insulation has to be softer and less dense to be compressed into the space available. It is therefore weaker electrically, and it will be seen from the pattern of flux distribution in *Figure 13.6* that significant stresses arise in the filler

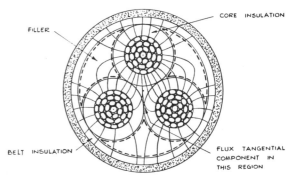

Figure 13.6 Flux distribution in paper insulated belted cable with top conductor at peak potential

they need to withstand all the rigours of service in a rugged and rough environment. The range of cables, given in BS 6116, incorporates ethylene propylene rubber (EPR) as insulation, a rubber undersheath, a layer of stranded galvanised steel wires applied as an armour, and a tough weather resistant outer sheath of polychloroprene (PCP). Cables are available for 600/1000 V, 3.3 kV and 6.6 kV systems. Some higher-voltage installations have been made using cables with individually screened cores and a thick overall PCP sheath, but their use is subject to permission from the relevant authorities.

Cables for use for equipment connection in underground coal mines are manufactured to specifications issued by the National Coal Board and use an EPR compound insulation specially formulated to give good impact strength and crush resistance. Individual core screening is the norm as part of the safety measures necessary for operation at the coal face. The sheath is of PCP, which has excellent mechanical properties and is flame retardant. Low-voltage pliable armoured cables similar to those in quarries are also used for portable supply cables to conveyor loaders, etc., and similar cables suitable for 3.3 kV and 6.6 kV systems are used for connection to transformers.

Besides these elastomeric insulated types, thermoplastic insulated cables are also widely used for power, lighting and signalling purposes.

An increasing demand for improved communication services in mines and quarries is reflected in the expanding use of two-way mobile radio, radio paging and radio control systems. However, mines and quarries often present situations where free-space propagation is not possible: for instance, propagation in tunnels can be restricted to only a few hundred metres. One solution is the radiating cable or leaky feeder in which signals radiate from the cable rather than from a conventional aerial. Radiating cables have special screens which provide and control the electromagnetic field around the cable so that signals can be picked up by nearby mobile receivers, with communication in the reverse direction also possible. These cables use low-permittivity dielectrics, which, with the special screens, ensure good radiation and transmission characteristics unaffected by the arduous external environmental conditions encountered in mines.

13.3.10 Mineral insulated metal sheathed cables

Mineral insulated metal sheathed cables consist of copper or aluminium conductors insulated with highly compressed magnesium oxide (MgO) and enclosed in a copper or stainless steel tube. The manufacturing process, to ensure compression of the dielectric and its complete enclosure to exclude moisture, consists of building up large-diameter short sections containing the requisite proportion of conductor, dielectric and sheath, and then drawing them down through a succession of dies until the required final dimensions are obtained. Annealing is carried out at regular intervals during the drawing process in order to restore ductility to the cable, lost through work-hardening. Normally, MI cables need no further protection over the metal sheath, will withstand high service temperatures and are impervious to oil and water. Being composed of inert inorganic materials, they are incombustible and non-ageing. They are made in 1-, 2-, 3-, 4-, 7-, 12- and 19-core constructions, and in light-duty (600 V) and heavy-duty (1000 V) grades.

The current rating of MI cables is determined chiefly by the permitted sheath temperature. For normal applications and where the cable is PVC covered, this temperature is 70°C. The PVC oversheath is recommended for MI cables buried underground or installed in aggressive industrial environments. However, bare MI cable with copper sheath can operate continuously at sheath temperatures up to 250°C, and when so rated, the current carrying capacity is greatly increased. Above 250°C significant oxidation of the copper sheath occurs, although the cable can function for limited periods with the sheath temperature in the region of 1000°C.

As for other types of cables installed in buildings, the published current ratings are based on an ambient temperature of 30°C and adjustment may need to be applied as detailed in the section on current carrying capacity.

13.3.11 Cables in fire hazard

An area of electric cable technology where much research and development work has been concentrated in recent years is that of the behaviour of cables in fires. Although they may overheat when subject to current overloads or mechanical damage, electric cables in themselves do not present a primary fire hazard. However, cables are frequently involved in outbreaks of fire from other causes which can eventually ignite the cables. The result can be the propagation of flames and production of noxious fumes and smoke. This result, added to the fact that cables can be carrying power control circuits which it is essential to protect during a fire to ensure an orderly shutdown of plant and equipment, has led to a large amount of development work by cablemakers. This work has included investigations on a wide range of materials and cable designs, together with the establishment of new test and assessment techniques.

Although PVC is essentially flame retardant, it has been found that, where groups of cables occupy long vertical shafts and there is a substantial airflow, fire can be propagated along the cables. Besides delaying the spread of fire by sealing ducts at spaced intervals, an additional safeguard is the use of cables with reduced flame propagating properties. Attention has also been focused on potential hazards in underground railways, where smoke and toxic fumes could distress passengers and hinder their rescue. Compounds with reduced acidic products of combustion have been incorporated into cables which have barrier layers to significantly reduce the smoke generated. Other cablemaking materials have been developed which contain no halogens and which also produce low levels of smoke and toxic fumes as well as having reduced flame propagating properties.

A different requirement in many installations, such as in ships, aircraft and the petrochemical industry (both on and off-shore), is that critical circuits should continue to function during and after a fire. Cables with good fire-withstand performance include mineral insulated metal sheathed cables for emergency lighting systems and industrial installations where 'fire survival' is required. As fire survival requirements on oil rigs and petrochemical plants become more severe, new control cable designs have been developed to meet fire tests at 1000°C for 3 h with impact and water spray also applied, and also to have low smoke and low toxic properties.

Another novel approach to fire protection in power stations and warehouses is the use of fire detector cables (*Figure 13.4*).

Figure 13.4 Heat sensor cable

These are used in a system which both detects and initiates the extinction of a fire in the relatively early stages of its growth. These cables have also been installed in shops, offices and public buildings, where the cables can be used to operate warning lights or alarms.

The present position in relation to materials is that problems due to smoke and objectionable fumes are dealt with in the case of

Figure 13.2 Typical ship wiring cable

The operating temperature of 85°C provided by this combination has been proved satisfactory over a number of years and has been used extensively for applications on North Sea oil platforms, although in this case braid armoured cable with an outer CSP sheath has been the main standard.

EPR has excellent electrical properties, good corona resistance and good low-temperature flexibility. The CSP sheath provides a tough outer surface which is resistant to weather, oil resistant and flame retardant.

The increasing use of higher power generation on board ship has meant that systems with voltages of 3.3 and 6.6 kV are now in operation. Again the use of suitable designs of EPR insulated CSP sheathed types have proved fully adequate for this service. For oil rig platforms 13.8 kV cables are also in regular operation.

The conductors of ship wiring cables are of a more flexible construction than comparable land-based cables because of the problem of installing them through complex structures characteristic of shipboard installations. The reactance of a cable operating in an a.c. system depends on many factors, including, in particular, the axial spacing between conductors, and the proximity and magnetic properties of adjacent steelwork. This latter point is of crucial importance in ships and oil rigs. It is desirable to minimise the effect of magnetic induction by means of adequate spacing between cables and steelwork, minimum spacing between conductors, and the avoidance of magnetic materials between single cores in the same circuit.

There are many Classification Authorities that will approve cables to international and to some national standards. When a ship installation is contemplated, the local surveyor of the chosen Classification Authority should be consulted before a decision is taken on the particular design to be used.

13.3.7 Aircraft cables

Cables used for wiring aircraft are continuously under development and, as technology improves, new materials are used in this most exacting of applications. The ambient temperature variation within an aircraft is wide and provision has to be made for cables to operate down to −75°C and up to 260°C. In addition, various special fire resisting cables are needed for use in aircraft-engine fire zones.

For the lower temperature a combination of special PVC, glass braid and nylon has been used where a conductor temperature of 105°C is deemed appropriate. These designs are detailed in BS G177, which covers the 'Nyvin' range of cables. In the mid-1960s development was undertaken and a miniature range of cables was introduced and subsequently detailed in BS G195. This range has the generic name 'Minyvin' and is available in multicore and screened versions. Greater attention has recently been paid to the effect of cable weight on the performance of the aircraft and, as a result, cables with polyimide insulation are being widely used in the aircraft industry for general airframe wiring. This material has been chosen because of its excellent mechanical and electrical properties and its resistance to the various chemical contaminants present in aircraft. Depending upon the type of conductor and coating used for colouring purposes, polyimide insulated cables are approved to operate at conductor temperatures of 150 and 210°C.

The higher temperature ranges of cable consist largely of combinations of PTFE, glass and polyimide, and when using a nickel-plated conductor, are approved for operation at conductor temperatures up to 260°C.

Special fire resisting cables for engine-bay wiring are available and comprise combinations of silicone rubber, glass-fibre, quartz fibre, polyimide and PTFE to ensure that circuit integrity can be maintained for a short period during a fire.

13.3.8 Cables for railways

The main cable application for railways is for track signalling. Multicore signalling cables are laid along the trackside. The insulation is a combination of natural and synthetic rubber. A natural rubber layer is applied next to the conductor to provide electrical integrity, and the outer layer is of polychloroprene (PCP) to give each insulated conductor some oil resistant properties. The cores are collected together and covered overall with a thick sheath of heavy-duty PCP compound specially chosen for toughness, and weather and abrasion resistance.

Cables for track power feeds on electric systems at medium and low voltage are insulated with EPR and sheathed with chlorosulphonated polyethylene (CSP). Cables for traction and rolling stock are conventionally of EPR/CSP composite insulation, but recent developments have resulted in new materials having low smoke properties and greater resistance to the oils and fluids used in traction and rolling stock.

In the underground system operated by London Transport Executive particular emphasis has been placed on minimising hazards resulting from fire. Designs are now approved which, in a fire, give off fewer toxic products and far less smoke than did previous designs (*Figure 13.3*).

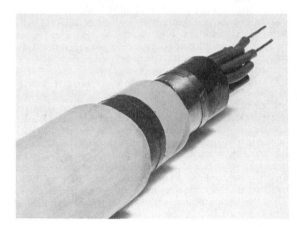

Figure 13.3 Multi-pair signalling cable with materials and construction for optimum flame retardance and freedom from smoke and fumes in a fire

Two-way communication between a control centre and moving trains is now feasible even when the trains are in tunnels. It is accomplished by installing suitably designed electric cables near to the track, either at ground level or overhead, to act as elongated aerials. The most common cable for this purpose is the so-called radiating coaxial or leaky feeder.

13.3.9 Cables for mines and quarries

Cables for metalliferous mines and quarries have been standardised for many years. They are essentially flexible and tough, as

Several national standards exist for the colour identification of insulation and sheath. Unfortunately, there is not yet a recognised international standard.

To prevent electrical interference in both control and thermocouple circuits within and between the cables, metallic screens (usually in the form of tapes) are applied over the individual pairs and/or the laid-up cores.

The finish of any type of cable to be buried in a petrochemical environment has to be given special consideration because of the presence of hydrocarbons in the soil. The steel wire armour which is normally applied as a mechanical protection on underground cables is not a barrier to the ingress of hydrocarbons into the heart of the cable: this is best achieved by applying a lead sheath. Thus, such control and instrumentation cables contain a PVC bedding, lead sheath, another PVC bedding, single wire armour and PVC oversheath. BS 5308 gives details of control cables with this type of protection. North American practice is to use an aluminium sheath.

Safety in hazardous areas has to be carefully considered. Intrinsic safety is a protective technique which ensures that any electrical sparking which may occur is incapable of causing an ignition of gas or vapour. Although a cable itself will rarely cause an explosion, it is possible for gas or vapour to percolate along the interstices of cable from a hazardous zone to a non-hazardous one. This problem can be cured by use of a stopper box or sealing gland. Reference should be made to BS 5345 for specific details. The British Approvals Service for Electrical Equipment in Flammable Atmospheres (BASEEFA) test and certify intrinsically safe and flameproof equipment.

13.3.4 Cables for electronic applications

Cables for electronic applications are single-core and multicore equipment wires, and radiofrequency cables. Equipment wires are usually regarded as insulated single conductors with or without a screen and sheath. They can also be supplied in flat formation as multicores or multipairs, often with a transparent backing. The space between conductors is precisely controlled to ensure consistent electrical characteristics and to assist in termination. Multicore equipment wires generally have PVC or PTFE insulation and sheath in a range of conductor sizes and number of cores, unscreened and screened.

Radiofrequency cables transmit high-frequency signals at minimum loss. Calculation of the optimum cable design involves operating frequency, capacitance, velocity ratio, impedance and attenuation. With the exception of a few twin and special designs, RF cables are normally *coaxial*; they have an inner conductor, insulation, outer concentric conductor forming the screen, and sheath. Also used for RF applications are PE and PTFE insulated multipair cables for use in computer interfacing.

For conductors, plain annealed copper is most common but even this is available in several grades, each conferring special properties. It is possible to draw copper to extremely fine sizes because of its good ductility. By leaving copper in the hard-drawn condition, additional tensile strength may be obtained at the expense of elongation. Where d.c. conductivity is of secondary importance, composite conductors incorporating a high-tensile-steel core are employed in miniature cables.

Insulants used in electronic cables vary between PVC, PE, PPE, PVF_2, PVF, ETFE, FEP and PTFE (see *Table 13.4*). The choice of insulant is an optimisation of performance and economics. In RF cables, where electrical performance is paramount, the insulation is usually either PE or PTFE. However, chemical, mechanical and thermal performance must also be considered.

Screens are applied to prevent electrical interference between circuits or to control the amount of pickup by, or leakage from, a cable. Braided and lapped wires, tapes, tubes, foils and films are among the screening materials used, depending on the application.

Sheaths are applied to act as protection. A thermoplastic or fluorocarbon material is most common. In more demanding environments it may be necessary to have additional protection over the sheath in the form of steel wire braids or armour.

Although there are many specifications covering electronic cables, manufacturers have their own standards.

In the optical fibre cable, the signals are transmitted optically rather than electrically. The conductor is made of a high-quality glass-fibre which transmits light. The main advantages of optical fibre cables compared with conventional metallic conductor cables include low weight, small volume, increased system capacity, freedom from electromagnetic interference and improved security.

13.3.5 Arc welding cables

British Standard 638, Part 4, covers three basic designs: (1) a single layer of 85°C heat, oil and flame retardant (HOFR) sheath; (2) as (1) but with 85°C rubber insulation as an inner layer; (3) a single layer of 85°C rubber sheath of Type GP3 which is usually of a specially compounded EPR.

The conductor is made up of a large number of small copper or aluminium wires, usually with a separator between conductor and insulation to make the cable supple. Single-conductor cables meet the majority of requirements for connection to electrode holders, arc-welding guns or leads for both manual and automatically controlled metal arc-welding equipment, or to form extension or return leads. Multicore cables are sometimes required for connections to the distribution boxes of multi-operator equipment.

Because of the variable periods of operation, current ratings have to be derived specifically for arc-welding cables and are contained in BS 638, Part 4. The period during which current flows varies from periodic to continuous, according to the application. The longer the period of use, the greater the conductor heating effect, so that current ratings are reduced as the operating cycle lengthens. The operating (or duty) cycle is defined as the time a cable operates in each 5 min period expressed as a percentage: e.g. up to $1\frac{1}{2}$ min operation in the 5 min period is 30%. Duty cycles are classified as follows:

Duty cycle	%
Automatic	up to 100
Semi-automatic	30–85
Manual	30–65
Intermittent or occasional	up to 30

Excessive voltage drop can occur when long cable lengths are required between the set and the electrode; conductors of larger current rating must then be selected. For flexibility, the final length of cable to the electrode can revert to the area appropriate to the current rating.

13.3.6 Ship's cables

A wide variety of installation conditions and extremes of temperarure are experienced in tankers, refrigerated vessels, ferries, trawlers, tenders, passage vessels, dry cargo ships, etc. In the UK the cables used are largely standardised as ethylene propylene rubber (EPR) insulated and chlorosulphonated polyethylene (CSP) sheathed and produced to BS 6883 (*Figure 13.2*).

Although rubber was for many years the insulant for wiring cables, it has been superseded almost entirely by PVC, or for some applications by mineral insulated cables. Although several alternatives have been tried, plain annealed copper remains the sole conductor material for wiring cables up to 16 mm^2.

PVC was originally developed in Germany in the 1930s. It took many years for PVC to be universally accepted for use in wiring cables. As PVC compounds improved, the thickness of insulation was progressively reduced. It now approaches half of that originally used and is acceptable for twice the operating voltage. The insulation is designed to have a higher tensile strength, resistance to deformation and a higher insulation resistance than the sheath. Sheathing compounds are usually formulated to provide good abrasion resistance and yet have easy-tear properties to facilitate stripping at terminations.

When vulcanised rubber insulation is used with copper conductors, it is necessary to tin the copper to prevent chemical reaction with the rubber. With PVC insulation, plain conductors have become universally accepted. The introduction of cables to metric standards in 1969 achieved a greater degree of international alignment.

PVC wiring cables in common use in the UK are of two basic designs. One is the single-core unsheathed cable, used in conduit or trunking, the numbers of cores varying from 2 up to as many as 38. The other is the insulated and sheathed cable, available in single-core, 2-core and 3-core versions, the 2-core and 3-core cables (made in a flat formation) having the option of a bare earth continuity conductor.

Apart from the flat-twin festoon lighting cables, the only existing wiring system cable with elastomeric insulation is an ethylene propylene rubber (EPR) insulated, textile braided and compounded single-core conduit wire. This cable has an upper temperature rating of 85°C and is in demand for applications where the upper temperature limit of PVC is restrictive.

13.3.2 Flexible cords

Most flexible cords are designed for and used in domestic premises as the supply lead from the socket outlet to portable, and some fixed, appliances. The range of domestic appliances covers such items as can openers, food freezers, microwave ovens, towel rails and washing machines. There is a flexible cord to suit each one.

The major cable insulants used in flexible cords are natural or synthetic rubber and PVC. Where flexibility is a prime requirement (for instance, on an electric iron), a rubber insulated type is most suitable. However, PVC is less expensive, has an attractive surface finish and is available in a larger range of colours. It is therefore the first choice for the supply lead to most domestic appliances.

Nowadays, many appliances carry the approval mark of the British Electrotechnical Approvals Board (BEAB) for Household Equipment, which means that in most cases the supply lead should meet the requirements of BS 6500: Insulated flexible cords. Independent auditing of flexible cords to BS 6500 by the British Approvals Service for Electric Cables (BASEC) provides an assurance of the integrity and reliability of the cablemakers' products. This standard specifies cords having elastomeric, thermoplastic and glass-fibre insulants. These types are designed for a wide range of applications with upper temperature limits varying from 60 to 185°C. The natural rubber insulated types are rated at 60°C and cover the twin-twisted and textile braided cord in use for many years for pendant flexibles. The recommendations for this application are now 85 and 150°C rubber or glass-fibre insulated types. The unkinkable domestic flexible cord is used on electric irons and consists of rubber insulation and a thin rubber sheath, over which is applied a semi-embedded textile braid.

The PVC insulated and sheathed cords in BS 6500 are rated at 70°C and are available in light- and ordinary-duty versions for such applications as table lamps, TV sets, washing machines and refrigerators. For night-storage and immersion heaters, and for situations involving contact with oil and grease, there is an elastomeric insulated and sheathed flexible cord, rated at 85°C and having an oil resisting sheath. The glass-fibre insulated types in BS 6500 are primarily intended for use with light fittings (luminaires) and other situations where the cord is not subject to mechanical damage or continuous flexing. They can be used at temperatures up to 185°C.

Some flexibles are available in cut and trimmed form as an aid to the handyman, i.e. the cores are cut to length and the sheath and insulation are stripped. A few are also available in coiled extensible form, the most common type being for electric shavers.

There are also flexible cables designed for more harsh industrial environments. Three common types have a copper wire or galvanised steel wire braid, or a spiral steel strip, over the inner sheath, and an outer oversheath. The copper wire braided type is mainly used for portable hand lamps and where a flexible cable is required in certain flameproof installations. The steel wire braided and steel strip types are utilised where both flexibility and mechanical protection are required.

Some, but not all, designs of flexible cord used in the UK have been harmonised in accordance with the CENELEC procedure and for these the same design is standardised throughout the Common Market countries.

13.3.3 Control and instrumentation cables

Whereas power cables are the 'arteries' of industry, control and instrumentation cables are its 'nerves' and are used for the control of equipment and data collection. They range from single-core cables used in the wiring of control panels and switchgear, to the complex control and instrumentation cables used in power stations and petrochemical sites.

At one end of the scale are the single-core cables used within machine tools and switchgear. Where normal ambient temperatures are involved, PVC insulation is employed. At the other end of the scale is, for example, a North Sea oil terminal utilising 500 km of cable and connecting as many as 2000 instruments measuring flow rates or liquid levels in storage tanks. Control cables have copper conductors and are laid up in multicore or multipair formation, each core being separately identified.

Thermocouple cables (*Figure 13.1*) are used for connecting the thermocouple to its measuring instrument. The term 'ther-

Figure 13.1 Multi-pair thermocouple cable with PVC insulation, pair screening, lead sheath and wire armour

mocouple cable' is often used to describe both extension and compensating cables. Extension cables utilise conductors of the same alloys or metals as the thermocouple itself, while compensating cables utilise conductors of cheaper material although having similar thermoelectric characteristics. The normal conductor materials or alloys used are constantan, copper, iron, copper-nickel, nickel-chromium and nickel-aluminium.

Table 13.5 Electrical properties of polymeric materials

Material	Type	Volume resistivity (min) at 20°C (ohm m)	Permittivity at 50 Hz	tan δ at 50 Hz
Thermoplastic *				
Polyvinyl chloride	TI 1	2×10^{11}	6–7	0.1
Polyvinyl chloride	2	1×10^{12}	4–6	0.08–0.1
Polyvinyl chloride	TI 2	2×10^{11}	6–7	0.09–0.1
Polyvinyl chloride	4	1×10^{9}	5–6	0.07–0.13
Polyvinyl chloride	5	5×10^{11}	6	0.9
Polyethylene LD	PE 03	1×10^{16}	2.35	0.0003
Polyethylene LD	PE 2	1×10^{16}	2.35	0.0003
Polyethylene HD		1×10^{16}	2.35	0.0006
Polypropylene		1×10^{16}	2.25	0.0005
Elastomeric †				
General-purpose GP rubber	EI 1	2×10^{12}	3–4.5	0.01–0.03
Heat resisting GP rubber	GP 1	7×10^{12}	3–4	0.01–0.02
Heat resisting GP rubber	GP 2	1×10^{13}	3–4	0.01–0.02
Heat resisting MEPR rubber	GP 4	7×10^{12}	3–4	0.01–0.02
Flame retardant rubber	FR 1	5×10^{12}	4.5–5	0.02–0.04
Flame retardant rubber	FR 2	1×10^{13}	4–5	0.015–0.035
OFR rubber	OR 1	1×10^{10}	8–11	0.05–0.10
Silicone rubber	EI 2	2×10^{12}	2.9–3.5	0.002–0.02
Ethylene vinyl acetate	EI 3	2×10^{12}	2.5–3.5	0.002–0.02
Hard ethylene propylene rubber		2×10^{13}	3.2	0.01
Cross-linked polyethylene		1×10^{14}	2.3–5.2	0.0004–0.005
Fluorocarbons				
Polytetrafluoroethylene		1×10^{16}	2	0.0003

* BS 6746 for PVC types and BS 6234 for polyethylene types.
† BS 6899 for GP rubber types, BS 5469 for EPR types and BS 5468 for cross-linked polyethylene.

and polyimide/FEP tapes are often used. These are light in weight, and are resistant to abrasion and cut-through, even when applied with small radial thicknesses.

13.2.4.2 Distribution cables and transmission cables

Even when armoured cables are installed indoors above ground, it is unusual for the armour to be left bare, because (a) there are few environments where corrosion will not occur, (b) without an outer covering the armour layer may become disturbed during installation, and (c) few cables are above ground for their whole length. Nowadays, to make cables easy to handle and provide a clean finish, a PVC sheath is usually applied overall.

Where cables are buried, the soils can be aggressive and cable life may depend on the degree of protection provided. Traditionally, lead sheathed cables have depended on bituminised textiles as a bedding *under* the armour, and a serving of two layers of hessian or a layer of helically applied jute strings *over* the armour. Bitumen is provided over each layer, as well as to flood the armour, and use of the optimum grade of bitumen at each stage is important. Today extruded PVC is being used to an increasing extent to replace bituminous finishes, even on lead sheathed cables, and all new designs of cable introduced during the last two decades have had PVC oversheaths. The successful introduction of aluminium for sheathing depends entirely on adequate protection with PVC.

For exposure to sunlight, PVC oversheaths should be black, but in other situations colours are sometimes used as a means of identification. To this end the sheaths are also embossed with the words 'Electric Cable' and the voltage. Even though very tough, PVC oversheaths can be damaged during installation, and if they protect an aluminium sheath, it is important to carry out an inspection before backfilling. In the case of expensive transmission cables it is usual to test for damage by applying a graphite coating over the PVC and then to carry out a 10 kV d.c. test between this electrode and the metal sheath. As an alternative to PVC the tougher high-density polyethylene is frequently used for such cables, especially at high ambient temperatures.

One of the disadvantages of PVC concerns cables in buildings or in tunnels. Although PVC is basically flame retardant, if a serious fire develops, it can transmit flame and will decompose, with evolution of noxious acidic fumes and dense smoke. Alternative PVC compositions with special fillers and materials based on other polymers are now available to overcome this hazard, but are more expensive (see later).

13.3 General wiring cables and flexible cords

13.3.1 Wiring system cables

A wiring system cable is usually regarded as the final link in the transmission network which begins in the power station and ends at the socket outlet in the home, office or factory work-bench.

Table 13.4 Physical properties of polymeric materials

Material	Type	Tensile strength (min) (N/mm^2)	Elongation at break (min) (%)	Limiting temperature* Rating (°C)	Limiting temperature* Installation (°C)
Thermoplastic†					
Polyvinyl chloride	TI 1	12.5	125	70	0
Polyvinyl chloride	2	18.5	125	70	0
Polyvinyl chloride	TI 2	10	150	70	−10
Polyvinyl chloride	4	7.5	125–150	85	0
Polyvinyl chloride	5	12.5	125	85	0
Polyethylene LD	PE 03	7	300	70	−60
Polyethylene LD	PE 2	7	300	70	−60
Polyethylene HD		37	500	80	−40
Polypropylene		37	400	80	−10
Elastomeric§					
General-purpose GP rubber	EI 1	5.0	250	60	−45
Heat resisting GP rubber	GP 1	4.2	200	85	−45
Heat resisting GP rubber	GP 2	4.2	200	85	−45
Heat resisting MEPR rubber	GP 4	6.5	200	90	−45
Flame-retardant rubber	FR 1	5.5	200	85	−30
Flame-retardant rubber	FR 2	5.5	200	85	−30
OFR rubber	OR 1	7.0	200	85	−30
Silicone rubber	EI 2	5.0	150	150	−55
Ethylene vinyl acetate	EI 3	6.5	200	105	−25
Hard ethylene propylene rubber		8.5	200	90	−40
Cross-linked polyethylene		12.5	200	90	−40
Fluorocarbons					
Polytetrafluoroethylene		24	300	260	−75

* Maximum temperature for sustained operation and minimum temperature for installation.
† BS 6746 for PVC types and BS 6234 for polyethylene types.
§ BS 6899 for GP rubber types, BS 5469 for EPR types and BS 5468 for cross-linked polyethylene.

cables this is provided by SWA. Cables with aluminium sheaths seldom require armour. Cables with lead sheaths may be armoured with steel tape, which is cheaper, but SWA is preferred in the UK for the heavier higher-voltage cables for 10 kV and upwards because it increases corrosion resistance and the longitudinal strength of the cable for installation purposes. Steel tape is normally protected against corrosion by bitumen, but if better corrosion resistance is required, the tape may be galvanised.

When additional mechanical or tensile strength is needed, as for river crossings, coal mines or long vertical runs, a double layer of steel wires may be employed. Single-core a.c. cables are rarely armoured, but if armour is necessary, it can be provided by non-magnetic tape or wire, normally of aluminium.

13.2.4 Oversheaths and protective finishes

13.2.4.1 General wiring cables

The choice of sheathing material depends on its environmental performance. Matters for consideration are: ambient temperature; flexibility; resistance to abrasion, water, oil and other chemicals; and compatibility with other materials with which a cable is in contact during its operational life. The sheathing material must also be chemically compatible with the other materials used in the cable both during and after processing. While insulants are chosen primarily for their electrical characteristics, sheaths are selected on their physical properties. Thus, not all insulants are suitable sheaths. However, in general, insulation and sheath materials are similar: e.g. a thermoplastic sheath protecting thermoplastic insulation.

Elastomeric sheathing materials include natural rubber, used for ordinary-duty domestic flexible cords; and synthetic rubbers such as NBR/PVC, PCP(OFR), CSP(HOFR). PCP is classed as an OFR material (oil resistant and flame retardant), and CSP as an HOFR material (heat resistant, oil resistant and flame retardant). These materials can be specially formulated to meet special requirements, e.g. improved water resistance, extra flame retardance, improved mechanical properties. Most elastomeric materials have a wider operational temperature range than thermoplastics and their superior performance under adverse environments (as found in mines and quarries) makes them first choice for such applications. Not only are they abrasion resistant, but also they are flexible over a wide range of temperatures.

In thermoplastic insulated cables, e.g. general wiring cables and radiofrequency cables, the predominant sheathing material is PVC in various formulations. Polyethylene is chosen where cables are in water or operate at subzero temperatures. For more specialised applications nylon and polyurethane are also used. Nylon has application where the cable is likely to be attacked by hydraulic fluids and is also a termite barrier. Polyurethane is being introduced into designs which call for a cable having good flexibility, abrasion and impact properties under arduous low-temperature conditions.

Cables with high-performance insulation materials can be sheathed with conventional thermoplastic or elastomeric materials but more commonly they have sheaths similar to the insulation composition. PFTE, FEP and a combination of PTFE

vulcanisation, generally by chemical methods, to provide them with the familiar characteristics of rubber compounds.

Examples of elastomeric materials are natural rubber (NR), ethylene propylene rubber (EPR), polychloroprene (PCP), chlorosulphonated polyethylene (CSP) and silicone rubber (SR). Of these, EPR is the most common because it combines the flexibility and electrical properties of natural rubber with a higher operating temperature limit and easier strippability.

Examples of thermoplastic materials are polyvinyl chloride (PVC), polyethylene (PE) and polypropylene (PP). PVC is the most usual insulant because of its uncomplex processability, good general-purpose performance and economic advantage. By adjustment of formulation, PVC compounds can meet a variety of requirements. Their robustness, relative chemical inertness, good ageing and attractive appearance in a range of colours have led to wide use not only as an insulatnt, but also as bedding for armour wires and for sheathing. At temperatures below 0°C PVC hardens but will recover its flexibility on returning to normal ambient temperatures. General-purpose PVC compounds are limited to a maximum conductor operating temperature of 70°C.

Where electrical properties are paramount, e.g. radiofrequency cables, polyethylene is the preferred insulant. It is also a more effective water barrier where a water-resistant property is important.

Elastomeric compounds are of advantage for long-term operation at temperatures higher than those that PVC can tolerate. EPR can operate at 85°C continuously and silicone rubber at 150°C continuously. Elastomers are also the first choice where flexibility combined with mechanical ruggedness is required. Applications for this type of material range between flexible cords for domestic flat-irons and flexible trailing cables for mines and quarries.

Where light weight and high operating temperatures are of paramount importance, fluorocarbon tapes or extrusions are adopted, particularly in the aircraft industry. Among the materials used in such high-performance cables are polytetrafluoroethylene (PTFE), fluorinated ethylene propylene (FEP), ethylene tetrafluoroethylene (ETFE) and polyimide/FEP tapes. They are characterised by low coefficient of friction, excellent electrical properties, resistance to chemical attack and stability at elevated temperatures. Besides use in aircraft applications, these materials have also been used in specialised radiofrequency cables and equipment wires.

Glass fibre in the form of lappings and braids is the insulation in a range of cables and cords for use, for example, in luminaires.

Tables 13.4 and 13.5 provide data on physical and electrical properties for a range of thermoplastic and elastomeric insulating materials.

13.2.2.2 Thermoplastic and elastomeric materials for power distribution cables

In the distribution field PVC is also the most widely used thermoplastic material, but as flexibility is not important, cross-linked polyethylene (XLPE) is more favoured than EPR as a thermosetting material. (It is more usual to refer to cross-linking rather than curing or vulcanising; and to thermosetting rather than to elastomeric materials.)

Although EPR can be produced in a hard grade (known as HEPR) with properties similar to those of XLPE, it is more expensive than XLPE and, as is common with rubber compounds, it contains a large number of ingredients. XLPE comprises merely polyethylene, an antioxidant and a cross-linking agent. The cross-linking can be accomplished by a variety of methods. Until recently the most common has been to mix an organic peroxide (dicumyl peroxide) with the PE and to extrude the insulated conductor into a large catenary tube containing steam under high pressure. For high-voltage cables, where minimum moisture content of the insulation may be important, radiant electrical heating may be substituted for steam and the CCV tube filled with nitrogen.

In the CCV extrusion process there is a wastage when starting and stopping; increasing use is being made of a process by which the extrusion can be in a conventional line, as for thermoplastic materials, and the cross-linking can be accomplished by a different chemical process. A silane and an accelerator are blended with the polyethylene and cross-linking is achieved by immersion of the insulated conductor in hot water. One method requires a separate process for preparing the graft polymer (which has a limited storage life), but in the Monosil process all the ingredients are blended in the hopper of the extrusion machine.

13.2.2.3 Impregnated paper

Layers of paper tapes are lapped around the conductor and the cable is dried and impregnated before application of the metal sheath which is required to keep the insulation dry and undamaged in service. The paper consists of a felted matt of long cellulose fibres derived from wood pulp. Washing of the fibres, both at the pulp stage and after formation of the sheet, is an important factor in the control of the properties of paper for cables. Large quantities of water are used, and for paper intended for the highest-voltage cables this water has to be deionised to ensure minimum power factor.

Impregnants have traditionally been based on mineral oils thickened with gum rosin to limit drainage from the insulation at service operating temperatures and to provide resistance to oxidation. It is now more usual to substitute materials such as microcrystalline waxes to obtain improved non-draining performance. For high-voltage internal pressure cables of the oil filled type it is necessary to use an impregnant with very low viscosity; hence, the highly refined mineral oils formerly used have been replaced by synthetic alkylates of dodecylbenzene type.

The electrical properties of the dielectric are not critical for voltages up to about 10 kV, but, at this level, ionisation in any air spaces becomes important and the impregnation process must ensure that butt gap spaces between paper tapes are well filled with impregnant. Impregnated paper itself, in sheet form, has a high electric strength, around 10 MV/m in short-time a.c. tests.

13.2.3 Armour

13.2.3.1 General wiring cables

When cables are not installed in conduit or trunking, they may require armour, most commonly provided by galvanised steel wire (GSW) helically applied in a single layer and known as 'SWA' (single-wire armour).

Pliable wire armour finds application for portable cables in quarries and mines. It consists of stranded 7-wire bunches of GSW applied helically in a similar manner to SWA, but with a shorter lay length, thus providing good mechanical protection with improved flexibility, enabling the cable to be moved without affecting performance.

Braided GSW armour is mainly used in cables for ships and off-shore applications. It has the advantage of easier installation in complex cable runs. For single-core cables in a.c. circuits, where magnetic effects could cause high losses with GSW, tinned phosphor-bronze wires are normally used.

13.2.3.2 Supply distribution cables

Most types of power cable require mechanical protection and/or an earth conductor to carry fault currents. For most distribution

but more because conductivity decreases significantly with degree of working. Impurities also affect conductivity and are kept to a maximum of 0.01%.

Tinned copper wires have been used for wiring and flexible cables, partly to improve solderability but mainly to prevent interaction between copper and the sulphur present to produce vulcanised rubber insulation. With the substitution of PVC for natural rubber the use of tinned conductors has diminished.

Aluminium Although having only 61% of the conductivity of copper and, hence, for equal conductance requiring a conductor area 1.6 times that of copper, the low density of aluminium results in the actual weight of a comparable conductor being only half that required with copper; i.e. the current-carrying capacity of an aluminium conductor is 78% of that of a copper conductor of equal area, and one ton of aluminium does the work of two tons of copper, at a lower price per ton. However, as the size is larger, the amounts of all the other materials in the cable are increased. The economic advantage, therefore, varies with the relative metal prices and with the type of cable. When the conductor metal is a small fraction of overall volume, aluminium is uneconomic.

Another difference from copper is that whereas solid copper conductors become difficult to handle above about 16 or 25 mm^2, solid aluminium conductors can be handled easily up to 240 or 300 mm^2, which further keeps dimensions to a minimum. Aluminium in a soft temper is quite suitable for these solid conductors but lacks strength in wire for stranded conductors. However, as aluminium does not suffer the same penalty as copper in loss of conductivity with work-hardening, it is satisfactory to use a broad ¾-hard temper for aluminium wire.

Certain impurities lower the conductivity of aluminium but the effect is not as great as with copper. Subject to control, the basic grade of 99.5% purity produced by electrolytic refining is satisfactory and appropriately defined in Cable Standards.

Apart from the economic factor, aluminium has no real advantage over copper for cables and it also suffers from a positive disadvantage. This relates to the protective oxide skin which is always present and which requires somewhat greater care to be taken when making soldered joints. To a large extent, compression joints have replaced soldering and have been designed to deal satisfactorily with this aspect.

Copper clad aluminium At times when the price ratio has been favourable to aluminium, attempts have been made to introduce it for house wiring. The major problem has been the effect of the oxide skin in the terminals of wiring accessories in which the conductor is held by a grub screw. With many designs the aluminium may also yield slowly under the screw, with consequent overheating. Improved fittings have been developed, but another solution is to provide the aluminium with a substantial coating of copper, commonly 10% by volume, i.e. 27% by weight. While this overcomes installation and jointing problems, extra cost is incurred.

13.2.1.2 Constructions

Metric conductor sizes, standard in the UK since 1969, are used in all countries other than the USA. *Table 13.3* shows comparisons.

Stranded conductors are available in circular form up to 2000 mm^2 and, for the lower voltage ranges, in sector-shaped contour up to 630 mm^2. The minimum number of wires is defined in IEC 228. Shaped conductors are normally prespiralled so that the cores fit easily together without applied twist in the laying-up operation. To provide a smooth surface and reduce the dimensions, it is now the practice to compact stranded conductors by a rolling process.

In the UK, aluminium conductors for cables up to 1.8/3 kV

Table 13.3 Conductor data

Standard metric size (mm^2)	Equivalent Imperial size (in^2)	Maximum d.c. resistance at 20°C	
		Aluminium (Ω/km)	Copper (Ω/km)
1.5	0.0023	—	12.1
2.5	0.0038	—	7.41
4	0.0061	7.41	4.61
6	0.0092	4.61	3.08
10	0.016	3.08	1.83
16	0.025	1.91	1.15
25	0.038	1.20	0.727
35	0.053	0.868	0.524
50	0.072	0.641	0.378
70	0.104	0.443	0.268
95	0.144	0.320	0.193
120	0.182	0.253	0.153
150	0.224	0.206	0.124
185	0.281	0.164	0.0991
240	0.369	0.125	0.0754
300	0.463	0.100	0.0601
400	0.592	0.0778	0.0470
500	0.746	0.0605	0.0366
630	0.963	0.0469	0.0283
800	1.23	0.0367	0.0221
1000	1.55	0.0291	0.0176

have largely been solid and of sector shape for multicore cables. Sector corner radii are fairly sharp to produce a compact construction. In Europe larger radii have been used and this form is likely to be adopted for high-voltage applications. Four 90° sector conductors laid-up together are used for 380–960 mm^2 solid-sectoral circular cables to obtain increased flexibility.

The Milliken construction is frequently used for circular conductors in sizes of 900 mm^2 and above to reduce skin effect and improve flexibility. Four or six individual sectors are used with a layer of insulation tape over alternate sectors.

The comments above refer largely to wiring and distribution cables. Different conditions apply to conductors for transmission cables, particularly of the oil filled type. Single-conductor sizes extend to 2500 mm^2 and an oil duct is required in the centre. The duct may be formed by laying the conductor strands round an open metal helix or, more usually, by creating a self-supporting centre by the use of curved segmental sections. Succeeding layers may consist of wires or flat strips, the latter improving both compactness and flexibility. For 18/30 kV cables, shaped conductors are oval rather than sectoral; circular conductors only are used at all higher voltages.

13.2.2 Insulation

13.2.2.1 Thermoplastic and elastomeric materials for wiring cables

In the general wiring-type cable field and for many power cables the major insulants in general use are either thermoplastic or elastomeric materials. In making the choice, several factors have to be considered. No insulant is ideal: a compromise is sought between processability, performance and economics.

An elastomeric material is one which returns rapidly to approximately its initial dimensions and shape after deformation at room temperature by a weak stress. Under such conditions a thermoplastic material shows permanent deformation. Conventional elastomeric compounds need to be cross-linked by

Table 13.2 Physical properties of metals used in cables (20°C)

Property	Copper	Aluminium	Lead
Density (kg/m^3)	8890	2703	11 370
Resistivity ($\mu\Omega$ m)	0.01724	0.02826	0.214
Res.-temperature coefficient (per °C)	0.0039	0.0040	0.0040
Thermal expansion coefficient (per °C)	17×10^{-6}	23×10^{-6}	29×10^{-6}
Melting point (°C)	1083	659	327
Thermal conductivity (W/m K)	380	240	34
Ultimate tensile strength			
soft temper (MN/m^2)	225	70–90	—
$\frac{3}{4}$H to H (MN/m^2)	—	125–205	—
Elastic modulus (GN/m^2)	260	140	—
Hardness			
soft (DPHN)	50	20–25	5
$\frac{3}{4}$H to H (DPHN)	—	30–40	—
Stress fatigue endurance limit (MN/m^2)	±65	±40	±2.8

Standards into conformity with them. There must be no extra requirements or deviations, except in special circumstances and subject to general agreement, and then only on a temporary basis.

To date, the two most important harmonisation documents for cables which have come into effect are:

HD 21 Polyvinyl Chloride (PVC) Insulated Cables and Flexible Cords of Rated Voltage up to and Including 450/750 V

HD 22 Rubber Insulated Cables and Flexible Cords of Rated Voltage up to and Including 450/750 V

but the intention is to extend harmonisation to all types of mains cables.

Among the features arising from this structure are the following:

(1) The policy is to use IEC standards, if suitable and available, as a basis for harmonisation.
(2) When CENELEC begins work on a particular type of cable (or a subject having an important bearing on cable standards), a 'standstill' or 'status quo' arrangement comes into effect. Changes in relevant national standards cannot then be made until harmonisation has been agreed or permission obtained from CENELEC, to avoid prejudicing the harmonisation.
(3) While the number of cable designs and types will be kept to a minimum, it will still be possible to have a national Standard for a type of cable not of interest to other member countries of CENELEC.
(4) Individual customers can still obtain cables made to their own specification, but it is hoped that this will be kept to a minimum. Because they have to be produced as 'specials' they may suffer from a delivery and/or a price penalty.
(5) Although implementation for flexible cables and some wiring cables has been effected, several years may elapse before all mains cables are harmonised. PVC cables to BS 6346 will probably be the first mains cables involved, but the changes are likely to be relatively minor.

Another important facet of CENELEC activity relates to certification and associated marks. These are issued by national approval organisations (NAOs) to indicate independent assurance of manufacture to specification. A common marking scheme, utilising the 'HAR' mark, has been devised. There is a reciprocal agreement accepted by most of the CENELEC countries' national approval organisations, under which for cables to harmonised standards, each NAO will recognise the common marking as superseding its own national mark. The procedures for granting a licence to a manufacturer to use the mark are identical in all the participating countries, based on initial approval, from inspection of manufacturing and testing facilities and testing of samples, and subsequent surveillance in the form of periodic testing of samples. In the UK the approval organisation is British Approvals Service for Electric Cables (BASEC). Under the Low Voltage Directive, of which the requirements have to be incorporated in the national laws of the Common Market countries, it is required that electrical equipment for voltages up to 1000 V should be accepted in each country as meeting the safety requirements of the directive if it conforms with a harmonised standard. Moreover, it should be presumed to conform with the relevant harmonised standard if the manufacturer qualifies for the use of the common mark.

13.2 Cable components

13.2.1 Conductors

13.2.1.1 Conductor materials

Materials and the form in which they are used comprise normally: (a) copper in solid form up to 2.5 mm^2 for wiring cables, 25 mm^2 for power cables and 150 mm^2 for mineral insulated cables, and in stranded form up to 2000 mm^2; (b) tinned copper similarly but in a narrower range for wiring cables; (c) solid aluminium up to 300 mm^2 and stranded aluminium up to 2000 mm^2; (d) copper clad aluminium in the smaller sizes of single-wire and stranded conductors for wiring; (e) an aluminium sheath as a concentric neutral conductor; and (f) lead and aluminium sheaths and steel wire or strip as an earth conductor.

In the USA some use was made of sodium by filling it into an insulating polyethylene tube. Although technically satisfactory, handling difficulties were found to outweigh the economic advantage. Another, more novel, application, which has been proved technically but not yet brought to commercial fruition, is the use of niobium alloys with superconducting properties at very low temperatures. Further reference to this subject is made in Section 13.5.

Some typical physical and electrical properties of the metals used in cables are given in *Table 13.2*.

Copper Because of its excellent conductivity, reasonable price and ease of working into rod and wire, copper has always been the basic material of the cable industry. Until the 1950s it was virtually without any challenger. Except when tensile strength is important, notably for self-supporting overhead line cables, it is always used in the annealed condition, partly to obtain flexibility,

Table 13.1 (*continued*)

Cables and Flexible Cords (*continued*)

IEC 502	Extruded solid dielectric insulated power cables for rated voltages from 1 kV to 30 kV
IEC 541	Comparative information on IEC and North American flexible cord types

Conductors

BS 2627	Wrought aluminium for electrical purposes—Wire
BS 3988	Wrought aluminium for electrical purposes—Solid conductors
BS 4109	Copper for electrical purposes: wire for general electrical purposes and insulated cables and flexible cords
BS 4990	Copper-clad conductors in insulated cables
BS 5714	Method of measurement of resistivity of metallic materials
BS 6360	Conductors in insulated cables
IEC 228	Conductors of insulated cables
	228A (Supplement). Guide to the dimensional limits of circular conductors

Insulation and Sheathing (*non-metallic*)

BS 5468	Crosslinked polyethylene insulation of electric cables
BS 5469	Hard ethylene propylene rubber insulation of electric cables
BS 6234	Polyethylene insulation and sheath of electric cables
BS 6746	PVC insulation and sheath of electric cables
BS 6899	Rubber insulation and sheath of electric cables
IEC 173	Colours of the cores of flexible cables and cords
IEC 304	Standard colours for PVC insulation for low-frequency cables and wires
IEC 391	Marking of insulated conductors
IEC 446	Identification of insulated and bare conductors by colour

Tests on Cables and Materials

BS 903	Methods of testing vulcanised rubber (in 53 parts)	
BS 5099	Spark testing of electric cables	
IEC 55	See above	
IEC 60	High voltage test techniques:	
	60-1, Part 1.	General definitions and test requirements
	60-1, Part 2.	Test procedures
	60-3, Part 3.	Measuring devices
IEC 141	Tests on oil-filled and gas pressure cables and their accessories.	
	141-1, Part 1.	Oil-filled, paper-insulated, metal-sheathed cables for alternating voltages up to and including 400 kV
	141-2, Part 2.	Internal gas-pressure cables and accessories for alternating voltages up to and including 275 kV
	141-3, Part 3.	External gas-pressure (gas compression) cables and accessories for alternating voltages up to 275 kV
	141-4, Part 4.	Oil-impregnated, paper-insulated high pressure, oil-filled, pipe-type cables and accessories for a.c. voltages up to 400 kV

Tests on Cables and Materials (*continued*)

IEC 229	Tests on anti-corrosion protective coverings for metallic cable sheaths
IEC 230	Impulse tests on cables and their accessories
IEC 270	Partial discharge measurements
IEC 330	Method of test for PVC insulation and sheath of electric cables
IEC 538	Electric cables, wires and cords: methods of test for polyethylene insulation and sheath
IEC 540	Test methods for insulations and sheaths of electric cables and cords (elastomeric and thermoplastic compounds)

Jointing and Accessories

BS 4579	Performance of mechanical and compression joints in electric cable and wire connectors:	
	Part 1.	Compression joints in copper conductors
	Part 2.	Compression joints in nickel, iron and plated copper conductors
	Part 3.	Mechanical and compression joints in aluminium conductors
BS 5372	Cable terminations for electrical equipment	
BS 6081	Fittings for mineral insulated cables:	
	Part 1. Copper	
	Part 2. Aluminium	
BS 6121	Mechanical cable glands for elastomer and plastics insulated cables	

Miscellaneous

BS 801	Lead and lead alloy sheaths of electric cables
BS 1441	Galvanised steel wire for armouring submarine cables
BS 1442	Galvanised mild steel wire for armouring cables
BS 4066	Tests on electric cables under fire conditions
	Part 1. Method of test on a single vertical wire or cable
IEC 38	IEC Standard voltages
IEC 183	Guide to the selection of high-voltage cables
IEC 287	Calculation of the continuous current rating of cables (100% load factor)
IEC 331	Fire resisting characteristics of electric cables
IEC 332	Tests on electric cables under fire conditions:
	332-1, Part 1. Test on a single vertical wire or cable
IEC 448	Current-carrying capacities of conductors for electrical installations in buildings
IEC 724	Guide to the short-circuit temperature limits of electric cables with a rated voltage not exceeding 0.6/1 kV

consequently a time lag in taking account of new developments, but British Standards have been more comprehensive in requirements and in updating revisions.

However, a fundamental change affecting the British Standards arises from the activities of the European Committee for Electrotechnical Standardisation (CENELEC), of which the membership consists of the Standards organisations of the European Common Market and the EFTA countries. The aim of CENELEC is to remove technical barriers to trade among the member countries, and to this end it is engaged in the harmonisation of their national standards.

The mechanism is the preparation of Harmonisation Documents; and after these are issued, the member countries are required to bring the technical requirements of their national

13.1 Introduction

The essential components of a cable are a metallic conductor of low resistivity to carry the current and insulation to provide a dielectric medium for isolating conductors from one another and from their surroundings. The conductor may consist of solid metal or of wires or segments stranded together. A single-core wiring cable for installation in conduit represents this basic construction. For other applications, two or more single-core units may be assembled together with overall protective coverings to prevent moisture ingress, and provide resistance to mechanical damage and to other external influences such as corrosion and fire.

In general, the voltage range extends from automobile cables at 6–12 V to the highest transmission voltages, which now are reaching towards 760 kV. In order to specify suitable insulation and construction for the required service performance, the design voltages are quoted in the form of U_0/U, i.e. (voltage to earth)/(voltage between phases). Cables are not manufactured, however, for every individual voltage requirement—e.g. although the most common supply voltage in the UK is 240/415 V, the cables actually used are designated as 0.6/1 kV. This is largely related to the fact that the minimum thickness of insulation which can be economically applied meets the higher voltage.

It has been traditional practice to describe cables in categories of low-voltage (LV), medium-voltage (MV), high-voltage (HV), supertension (ST), extra-high-voltage (EHV) and ultra-high-voltage (UHV). However, the exact demarcations have never been very precise, because they vary between different countries, among different groups of engineers and with the passing of time. Not so long ago, the conventional supply cables operating at 240/415 V were known as MV, but now they are LV; 11 kV cables have also changed from HV to MV. Thus, there can be confusion especially at international meetings, and it is better to refer only to the actual voltage designation.

There is also some overlapping when cables are classified into the three major groups of usage: (1) wiring and general, (2) power distribution and (3) transmission. When heavy power cables were essentially of the paper-insulated type, there was little problem, because they tended to be made in different factories from wiring cables and the type of insulation governed the usage group. When paper insulation operated with internal or external fluid or gas pressure, as required for voltages above 33 kV, the cable came into the transmission grouping. Nowadays, the basic insulation materials and constructions used for wiring cables can, with apppropriate insulation thickness, be utilised across the three major groupings and it is becoming more complex to define specific cable categories. For public supply there is no problem, but for ships' cables, off-shore supplies and large factory distribution there can be considerable overlap between the wiring and the distribution categories. At the top end of the voltage range we now have 132 kV cables for distribution rather than transmission, and polyethylene or cross-linked polyethylene is finding acceptance across the whole spectrum from 1 to 275 kV.

In this edition, the section on power distribution cables is aimed at public supply networks together with the larger power cables used in factories, etc., from 1 to 33 kV. Where the broad wiring cable category encroaches on this field, reference is made in the text. Cables for voltages above 33 kV are covered in the section on transmission cables.

13.1.1 Standards

Table 13.1 lists most of the British (BS) and International (IEC) Standards applicable to cables and cable systems, including tests. While most national standards are in accordance with IEC requirements, the IEC Standards represent a consensus of national opinion and take many years to prepare. There is

Table 13.1 British and IEC Standards

Cables and Flexible Cords

BS 638	Arc welding plant (includes cables)
BS 4553	PVC-insulated split concentric cables with copper conductors for electricity supply
BS 5055	PVC-insulated and elastomer-insulated cables for electric signs and HV luminous discharge tube installations

BS 5308	Instrumentation cables intended for intrinsically safe systems
	    Part 1. Polyethylene insulated cables
	    Part 2. PVC-insulated cables

BS 5467	Armoured cables with thermosetting insulation for electricity supply
BS 5593	Impregnated paper insulated cables with aluminium sheath/neutral conductor and three shaped solid aluminium phase conductors (Consac) for electricity supply
BS 6004	PVC-insulated cables (non-armoured) for electric power and lighting
BS 6007	Rubber-insulated cables for electric power and lighting
BS 6116	Elastomer-insulated flexible trailing cables for quarries and miscellaneous mines
BS 6195	Insulated flexible cables and cords for coil leads

BS 6207	Mineral insulated cables:
	    Part 1. Copper sheathed cables with copper conductors
	    Part 2. Aluminium sheathed cables with copper conductors and aluminium sheathed cables with aluminium conductors

BS 6231	PVC-insulated cables for switchgear and control wiring
BS 6346	PVC-insulated cables for electricity supply
BS 6480	Impregnated paper-insulated cables for electricity supply: Part 1. Lead and lead alloy sheathed cables for working voltages up to 33 kV
BS 6500	Insulated flexible cords
BS 6708	Trailing cables for mining purposes
BS 6862	Cables for vehicles: Part 1. Cables with copper conductors
BS 6883	Elastomeric insulated cables for fixed wiring in ships
BS 6977	Braided travelling cables for electric and hydraulic lifts

BS Series	Cables for aircraft use:
	    G177 (Nyvin), G189 and G227 (Tersil),
	    G192 and G222 (Efglas),
	    G195 and G221 (Minyvin), G206 (Fepsil),
	    G210 (PTFE), G212 (General requirements)

IEC 55	Paper insulated metal sheathed cables for rated voltages up to 18/30 kV (with copper or aluminium conductors and excluding gas pressure and oil-filled cables)
55.1	Part 1. Tests
55.2	Part 2. Construction
IEC 92	Electrical installations in ships: 92-352 Choice and installation of cables for low-voltage power systems
IEC 227	Polyvinyl chloride insulated flexible cables and cords with circular conductors and a rated voltage not exceeding 750 V
IEC 245	Rubber-insulated flexible cables and cords with circular conductors and a rated voltage not exceeding 750 V

13 Cables

Laurence Banbury
BICC General Cables Ltd

P F Gale, BTech, PhD, CEng, MIEE
Biccotest Ltd

D McAllister
Formerly BICC Power Cables Ltd

Contents

13.1 Introduction 13/3
    13.1.1 Standards 13/3

13.2 Cable components 13/5
    13.2.1 Conductors 13/5
    13.2.2 Insulation 13/6
    13.2.3 Armour 13/7
    13.2.4 Oversheaths and protective finishes 13/8

13.3 General wiring cables and flexible cords 13/9
    13.3.1 Wiring system cables 13/9
    13.3.2 Flexible cords 13/10
    13.3.3 Control and instrumentation cables 13/10
    13.3.4 Cables for electronic applications 13/11
    13.3.5 Arc welding cables 13/11
    13.3.6 Ships cables 13/11
    13.3.7 Aircraft cables 13/12
    13.3.8 Cables for railways 13/12
    13.3.9 Cables for mines and quarries 13/12
    13.3.10 Mineral insulated metal sheathed cables 13/13
    13.3.11 Cables in fire hazard 13/13

13.4 Supply distribution cables 13/14
    13.4.1 Paper insulated cables 13/14
    13.4.2 CNE cables for PME systems 13/16
    13.4.3 Service cable 13/17
    13.4.4 PVC insulated power cables 13/17
    13.4.5 XLPE insulated cables up to 3.3 kV 13/18
    13.4.6 PE and XLPE cables for 11 kV to 45 kV 13/18
    13.4.7 Cable tests 13/20

13.5 Transmission cables 13/21
    13.5.1 Historical development sequences for a.c. transmission 13/21
    13.5.2 Types of cable 13/22
    13.5.3 Submarine power cables 13/25
    13.5.4 D.C. transmission 13/25
    13.5.5 Cable ratings and forced cooling 13/26
    13.5.6 Future development 13/27

13.6 Current-carrying capacity 13/27
    13.6.1 Availability of continuous ratings 13/27
    13.6.2 Factors in cable ratings 13/28
    13.6.3 Sustained ratings 13/29
    13.6.4 Short-time and cyclic ratings 13/30
    13.6.5 Short-circuit ratings 13/30
    13.6.6 Voltage drop 13/30
    13.6.7 Protection against overload current 13/30

13.7 Jointing and accessories 13/30
    13.7.1 Aluminium conductor jointing 13/30
    13.7.2 Joints for distribution cables 13/31
    13.7.3 Joints for transmission cables 13/31

13.8 Cable fault location 13/32
    13.8.1 Diagnosis 13/33
    13.8.2 Preconditioning 13/33
    13.8.3 Prelocation 13/34
    13.8.4 Pinpointing 13/35

22 'CIGRE, survey of the lightning performance of EHV transmission lines', prepared by E. R. Whitehead, *ELECTRA*, No. 33, March (1974)
23 MARTIN, D., 'Design of uplift foundations', *ELECTRA*, No. 38, January 1979—prepared within scope of WGO7 of CIGRE SC 22
24 VANNER, M. J., 'Foundations and the effect of the change in ground conditions over the seasons', *2nd Int. Conf. Progress in Cables and Overhead Lines for 220 kV and Above*
25 MANUZIO, C. and PARIS, L., 'Statistical, determination of wind loading effects on overhead line conductors', CIGRE Paper 231, Paris (1964)
26 ARMITT, J., COJAN, M., MANUZIO, C. and NICOLINI, P., 'Calculation of wind loadings in components of overhead lines', *Proc. IEE*, **122** (11), November (1975)
27 EPRI, *Transmission Line Reference Book: 345 kV and Above*, 2nd edition (1982)
28 ADAM, J. F., BRADBURY, J., CHARMAN, W. R., ORAWSKI, G. and VANNER, M. J., 'Overhead lines – some aspects of design and construction'. *Proc. IEEE*, **131** (5), September (1984)

12.8 Loadings

The mechanical loadings on an overhead line are due to conductor tension, wind, icing and temperature variation. In line design an early decision must be made on the loadings to be accepted. Statistical techniques[25] have been suggested; more recently methods[26] for calculating wind loads, including wind characteristics, topography, height and shape of components, etc., have been put forward. The concept of a 'working load' multiplied by a 'factor of safety' is unsatisfactory, and international work is proceeding in an effort to rationalise the problem.

Accepting that failures *do* occur, a 'rational' transmission-line design could be based on the minimum of installation cost + capitalised annual cost of operation + annual cost of failures. The last-named term is the product of the estimated cost of each failure and the average number of failures expected per year.

Meteorological stations collect data on wind, ice, temperature, storms, etc. The distribution of wind pressure can be represented by a 'law of extreme values', as shown by $\phi(w)$ in *Figure 12.12*. L_w is the load with the highest occurring frequency and $L_{w'}$ is a higher wind load with a smaller frequency.

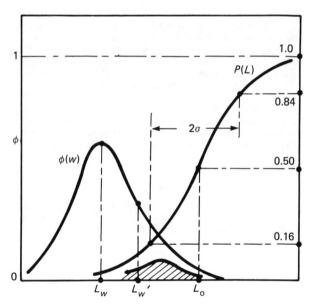

Figure 12.12

Each constructional element of an overhead line has inherently a statistical distribution of strength around an average, assumed to be Gaussian with a standard deviation σ that can be obtained from tests. It is represented in *Figure 12.12* by the function $P(L)$, with L_0 the average strength value. Its shape depends on the standard deviation σ. Then the shaded area is a measure of the risk of failure. The technique depends upon the provision of all cost data at the design stage; apart from capital cost, however, not all the required data may be to hand.

A useful concept in deciding on wind pressure is that of 'return period' of a given wind speed. The return period is the mean time between the occurrences of a phenomenon equal to (or greater than) a specified value. As return period T is related to probability p by the expression $Tp = 1$, it follows that a gale for which $T = 50$ years has a 2% probability of being exceeded in any one year.

The foregoing design approach does not require 'factors of safety' in the traditional sense. Loads are derived on the basis of statistical safety and the associated statistical risk of failure. The loads so derived can be considered as ultimate loads, with overload factors (1.2–1.5) to allow for the dispersion in the strength of each line element. Work on the relationship between loadings and strength is still in progress.

References

1. Statutory instruments, 1970 No. 1355—Electricity. The Electricity (Overhead Lines) Regulations 1970, HMSO
2. ASH, D. O., DEY, P., GAYLARD, B. and GIBBON, R. R., 'Conductor systems for overhead lines: some considerations in their selection, *Proc. IEE*, **126** (4), April (1979)
3. BUTTERWORTH, S., *Electrical Characteristics of Overhead Lines*, ERA Report O/T4 (1954)
4. PRICE, S. J., ALLNUTT, J. G. and TUNSTALL, M. J., 'Subspan oscillations of bundled conductors', *IEE Conf. Progress in Cables and Overhead Lines for 220 kV and Above*, September (1979)
5. Study Committee 22 Report, 'Aeolian vibration on overhead lines', CIGRE Paper 22–11, Paris (1970)
6. WATES, R. H., JACKSON, G. B., DAVIS, D. A., ERSKINE, A., BROWN, R. C. and ORAWSKI, G., 'Major high voltage long span river crossings in Great Britain', CIGRE Paper 226, Paris (1964)
7. BRADBURY, J., KUSKA, G. F. and TARR, D. J., 'Sag and tension calculations in mountainous terrain', *IEE Conf. Progress in Cables and Overhead Lines for 220 kV and Above*, September (1979)
8. BRADBURY, J., DEY, P., ORAWSKI, G. and PICKUP, K. H., 'Long term creep assessment for overhead line conductors', *Proc. IEE*, **122** (10), October (1975)
9. *Specification for CISPR Radio Interference Apparatus for the Frequency Range 0.15 MHz to 30 MHz*, CISPR Publication No. 1 from IEC
10. CIGRE, *Interferences Produced by Corona Effect of Electric Systems*—Description of Phenomena, Practical Guide for Calculation (document established by Working Group 36–01 (Interferences) (1974)
11. Anon., Survey of extra high voltage transmission line radio noise', *ELECTRA*, January (1972)
12. *Insulation Co-ordination*, IEC Standard—Publications 71–1 and 71–2
13. *BS 5622: Guide for Insulation Co-ordination*. Part I, Terms Definitions, principles and rules. Part II, Affiliation guide
14. *Tests on Insulators of Ceramic Material or Glass for Overhead Lines with a Nominal Voltage Greater than 1000 v*, IEC Standard—Publication 383
15. *BS 137: Specification for Insulators of Ceramic Material or Glass for Overhead Line with a Nominal Voltage Greater than 1000 V*. Part 1, Tests. Part 2, Requirements
16. PARIS, L. and CORTINA, R., 'Switching and lightning impulse discharge characteristics of large air gaps and long insulator strings', *IEEE Trans.*, **PAS-87** (4), April (1968)
17. MORRIS, THOMAS A. and OAKESHOTT, D. F., *Choice of Insulation and Surge Protection of Overhead Transmission Lines of 33 kV and Above*, ERA Report O/T14
18. *Artificial Pollution Tests on High Voltage Insulators to be Used on a.c. Systems*, IEC Report, Publication 507
19. WHITEHEAD, E. R., 'Lightning protection of transmission lines', Chapter 22, *Lightning* (ed. R. H. Golde), Academic Press
20. CLAYTON, J. H. and YOUNG, F. S., 'Estimating lightning performance of transmission lines', *IEEE Trans.*, **PAS-83** (Paper No. 64–138)
21. GILMAN, D. W. and WHITEHEAD, E. R., 'The mechanism of lightning flashover on HV and EHV transmission lines', *ELECTRA*, No. 27

is of the form $M = kbd^3$ for a width b and a depth d. The coefficient k depends on the shape of the bloc and the properties of the soil.

Uplift foundations In the UK, where 'pad and chimney' foundations are favoured, it is customary to assume that the resistance to uplift is due to (a) the dead weight of the foundation and (b) the weight of the earth contained in a volume of a frustum defined by the base of the foundation and an angle to the horizontal reflecting the soil properties. In addition, the 'chimney' is designed in bending. A general expression for the uplift force[24] is

$$F_t = A[Ck_1 + d\gamma(k_2 + k_3) + qk_4] + p_c + p_s$$

where A is the lateral area rising from the base of the foundation to the surface of the soil, d is the depth, p_c is the weight of concrete and p_s the weight of soil above the foundation, and q is the overburden pressure for foundations set below the critical depth. The constants k depend on the type of soil. In a pad and chimney foundation the pad can be shaped by formers or cast on site into an excavated hole. When access is difficult and concrete expensive (because of lack of available water or aggregates), grillage pads may be considered, as individual bars can be transported to site. In bad soils pile or raft foundations may be used; in good rock the foundation may employ rock anchors.

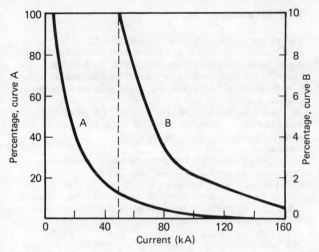

Figure 12.11

12.7 Lightning

Lightning results from a phenomenon in which the clouds in the ambient atmosphere acquire substantial electric charges, with corresponding potentials which may be several megavolts. When the limit breakdown value of the air is reached, a lightning discharge takes place between clouds, or between clouds and earth. Because of their height, overhead transmission lines 'attract' lightning discharge. Data are available for the percentage of days when thunder is heard[17] and the number of flashes to earth per annum per unit area. The 'keraunic level' is the number of days thunder is heard in a year.

12.7.1 Mechanisms of insulation flashover

12.7.1.1 Induced voltage

A lightning stroke to ground near a line discharges the energy in the cloud very rapidly, initiating by travelling wave action an induced voltage rise across the insulators. The problem arises only at line voltages of 66 kV downward, for which lines it would be uneconomic to provide a complete guard against this hazard.

12.7.1.2 Shielding failure

Shielding failure occurs when the leader stroke of the lightning discharge bypasses the earth wire (which is usually at ground potential) and strikes a conductor. Some protection against this hazard is provided by positioning the earth wire(s) above the conductors. Analytical models[19] have been evolved for the calculation of the appropriate position(s). A shielding angle of 30° is generally adequate: the angle is that between the vertical through the earth wire and the line joining the earth wire and the protected conductor. However, in areas of high lightning intensity, zero (or even negative) angles have been used.

12.7.1.3 Back-flashover

Figure 12.11 shows that about 10% of lightning strokes involve a current i exceeding 50 kA of very steep wave front. When lightning strikes (a) a tower or (b) an earth wire, their potential may be raised to a value well in excess of the insulation level, possibly high enough to cause flashover to the line conductors.

(a) *Tower* If the tower resistance (mainly that of the footing) is R and the inductance as a vertical circuit element is L, then the potential of the tower top will be raised from zero to $v = Ri + L(di/dt)$. This simple approach illustrates the effect of the basic parameters: a more critical study would be required to take account of capacitive and inductive couplings between conductors, the effects of multiple reflections, etc.

(b) *Earth wire* Given that the surge impedance of the earth wire is Z_0, a lightning stroke raises the earth wire potential from zero to $v = \frac{1}{2}Z_0 i$. Two travelling waves are propagated in opposite directions from the point of strike, until each reaches a tower top and is partially discharged through the tower to ground, as in (a).

12.7.2 Lightning performance

Various methods (ERA[17], IEEE[20], EPRI[27]) have been evolved, and refined in the light of experience, for the estimation of the lightning performance of overhead lines. The basic step is the calculation of the potential rise of the tower top or earth wire as a function of the lightning current discharged to ground. It is then necessary to develop an analytical model to include the statistical distribution of current wave shapes, grounding parameters, leader approach angles, ground flash density, back-flashover and shielding failure[21,22]. The ERA method employs a step-by-step procedure clearly demonstrating the factors affecting line performance; the IEEE method, easier to apply, is based on generalised graphs. The EPRI method is very explicit.

In overhead-line design, some parameters will depend on lightning as they affect the number and positioning of earth wires, the length of insulator strings and tower-footing resistance. Further considerations relate to tower height, phase configurations and span lengths. It is now accepted that an earth wire with 10% smaller sag than that of the associated conductor minimises the likelihood of mid-span flashover between them. The main purpose of an earth wire is to shield phase conductors from direct strokes, which would almost always result in insulator flashover.

Increased tower height attracts more lightning strokes. Thus, the lightning performance of a line is improved by shorter towers (and consequently shorter spans). Short towers with wide bodies have lower surge impedance. But in many cases it is economic considerations that settle the final choice.

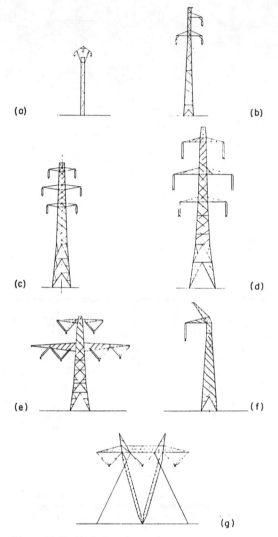

Figure 12.10 Typical steel towers

12.10(*b*) or rigid type, in which each leg of the support has a separate foundation, or of the *narrow base* or flexible type, with only a single foundation (*a*). The latter is, of course, cheaper, but is less resistant to the twisting moment caused by a broken conductor; on the other hand, it gives some flexibility in the direction of the line which tends to relieve the forces to some extent. The narrow base may also simplify wayleave problems in cultivated areas on account of the smaller area occupied.

Double-circuit lines (*c*) are more economical than two single-circuit lines, although for a given voltage the tower height is greater. There is the possibility of both circuits being affected simultaneously by lightning, but wayleave problems are eased.

The cat's head horizontal configuration (*g*) minimises height and reduces the possibility of conductor clashing, but line erection is complicated by having to thread one phase through the steelwork.

The tower (*f*) for a single-circuit d.c. line is simpler, smaller and 20–30% cheaper than a 3-phase single-circuit tower to carry the same electrical load.

The 400 kV double-circuit tower (*d*), widely used in Britain, is 50 m high and may interfere with amenity. To avoid this the lower type (*e*) may be used but is 10–15% more expensive.

The use of insulated cross-arms has been suggested for use even at 400 kV. It would reduce tower height by the length of one insulator string.

Tubular steel poles These are being increasingly used in the UK, because of the reduction in timber supplies. They are usually formed from three tapered, hollow sections, which rest inside one another during transportation and storage, and which are assembled on the site by means of a special winch, or are driven together by a sledgehammer or wooden maul.

12.6.3 Tower geometry

Considerations of system design will decide whether a line should carry a single or a double circuit. The choice is influenced by the problem of wayleave. In the UK most grid lines carry a double circuit; elsewhere a multiple-circuit construction to accommodate circuits of different voltages is sometimes used.

The geometry of the tower top is controlled essentially by electrical considerations:

(1) Height to bottom cross-arm: the sum of the clearance to ground, the sag and the insulator-set length.
(2) Vertical spacing: the sum of the electrical clearance plus the insulator length.
(3) Horizontal spacing: (a) in still air, a clearance at least equal to the impulse strength of the insulation; (b) in wind conditions (on the assumption that a fault is unlikely to occur during maximum wind speed), a clearance corresponding at least to the power frequency withstand voltage of the air-gap; the deflected position of the insulator string determines the cross-arm length.

It is generally found that horizontally placed conductors have sufficient clearance when they are separated by earthed steelwork. For exceptionally long spans, or when two conductors at the same level are on the same side of the structure, their minimum separation should be

$$D = K_1 \sqrt{(s+l)} + K_2 V$$

where s is the maximum sag of the conductor in the span; l is the length of a suspension-insulator set; V is the line voltage (kV); $K_1 = 0.6 \times K_3$ (usually $0.7 < K_1 < 0.9$); $K_3 = \dfrac{\sqrt{(W^2 + p^2)}}{W}$, where W is unit weight of conductor (kgf/m) and p is 50% of design pressure (kgf/m) per unit length; and $K_2 = 0.0025 \times \sqrt{3}$. The expression is empirically based on the performance of long spans.

The phase geometry of single-circuit lines is either triangular or horizontal. Double-circuit lines are more likely to have a vertical formation. From the system viewpoint, a triangular arrangement results in minor imbalance of the phase impedances (see also Section 12.7).

12.6.4 Foundations

Foundations must be adapted to the loadings that they have to withstand and the properties of the soil, i.e. cohesion C, angle of internal friction ϕ, and density γ.

For single poles and narrow-based towers, the foundations are designed on overturning; for broad-based towers, on uplift/compression. Compression effects are well documented, but uplift has been the subject of lengthy investigation[23].

Overturning (monobloc) foundations Resistance to overturning

Table 12.4 Voltage distribution over insulator string

No. of units	Unit number from live end	Voltage across unit (%)
4	1, 2, 3, 4	32, 24, 21, 23
6	1, 2, 3, 4, 5, 6	25, 18, 14, 13, 14, 16
8	1, 2, 3, 4, 5, 6, 7, 8	24, 15, 13, 11, 9, 8, 9, 11
12	1, 2, 3, 4, 5, 6, 7, 8, 9, 10, 11, 12	23, 13, 10, 8, 7, 6, 5.5, 5, 5, 5.5, 6, 7
16	1, 2, 3, 4, 5 …	20, 13, 9, 7.5, 5.5, …

12.6 Supports

The supports of an overhead line have the greatest impact on visual amenity. They cannot be hidden, but their appearance may be made less objectionable.

12.6.1 Materials

Common structural materials are wood, concrete, tubular or rolled-section steel and aluminium alloys. Choice is governed by economics, availability, resistance to deterioration (e.g. by termites in wood, corrosion in steel), and sometimes by problems of transportation and erection. When helicopters are used in regions of difficult access, aluminium alloy structures may show a lower cost of installation that counterbalances their greater prime cost.

Wood poles are usually preferred for l.v. distribution lines; they have been used for 345 kV lines in the USA and 132 kV in the UK, but normally wood poles are unlikely for voltages over 66 kV.

Concrete (reinforced, prestressed or spun) has for poles the drawbacks of weight and fragility during transport, but have been employed for lines up to about 220 kV. The material is unlikely to be competitive unless local production facilities are available or there are constraints on imported material.

Steel, the material most preferred, can be fabricated in convenient lengths for transport after rolling into angles, beams and rounds. Parts are readily bolted together, facilitating unit-type construction. Steel for overhead lines is normally galvanised by the hot-dip process (BS 729) as a protection against corrosion. In clean air, galvanising can ensure 30–50 years of trouble-free life. Where the atmosphere is potentially corrosive, it may be necessary to paint steel structures after 10 years or less. Two grades of steel are commonly applied to line towers, viz. grade 43A and grade 50B (high tensile) to BS 4360, the latter grade showing advantage in weight reduction.

Aluminium alloy is advantageous in special cases (sites difficult of access or with corrosive atmospheres). A structure of this material is 0.6–0.75 of the weight of an equivalent steel tower, but its cost may be three times as much.

12.6.2 Configurations

There is a considerable variety in configuration: the more usual forms are shown diagrammatically in *Figure 12.9* for wood poles and *Figure 12.10* for steel towers.

Wood poles These are more economical than lattice-steel supports for lines having spans less than about 200 m and are widely used for distribution in rural areas at voltages up to and including 33 kV. Where ample supplies of wood are available, e.g. in Sweden and parts of America, they have been used for lines up to 220 kV; at these high voltages the portal type of construction is essential to keep the length of pole to a minimum. For lower-

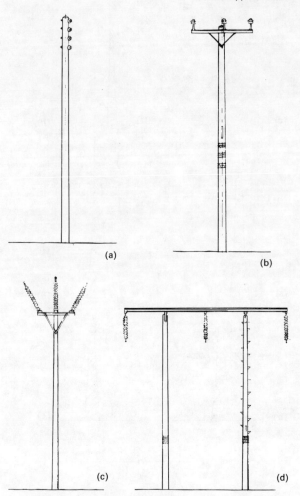

Figure 12.9 Typical wood pole constructions—415 V to 132 kV. (a) Low voltage single member pole; (b) 11 kV single member pole; (c) 132 kV support with composite (polymeric) insulators; (d) 132 kV portal structure (H) with standard insulator sets

voltage lines the single pole is generally used, although its transverse strength can be increased three or four times by using two poles arranged in A or H formation as shown in *Figure 12.9*. The cross-arms in both the single and the portal types may be of wood or steel, the latter being more usual. A zinc or aluminium cap covers the top of the poles to protect the end grain. The life of a red fir pole, if properly creosoted, is 25–30 years, although British fir or larch has a shorter life.

Wood is an insulator, so that the pole adds to the insulation strength between the conductors and earth and renders flashover due to lightning less likely. On the other hand, if a pole is struck, it may be shattered, causing complete failure of the line. The probability of shattering can be reduced by earthing all metal supports for the insulators either individually at each pole or by connection to a continuous earth wire.

Steel towers These are employed where long spans and high supports are needed. Normal spans of 200–500 m are usual, with special cases (such as river crossings) requiring spans up to 2 km. Lines of 66 kV and over usually have lattice steel supports, especially if they are double-circuit lines.

The lattice-steel support may be of the *broad base* (*Figure*

(g) Insulator set.
(h) Vee insulator set.

The elements (d, e, f) can be assembled into one, two (g) or more strings in parallel. Vee types (h) are often favoured for e.h.v. lines, as they show economic advantage.

12.5.2 Selection

As insulators must meet specified criteria, the choice of type, configuration, etc., is complex. Units are characterised by lightning impulse voltage; power frequency voltage when wet; electromechanical failing load, mechanical load and puncture voltage (when applicable); and dimensions, of which creepage path is the most important parameter for selection.

In addition, strings or sets are characterised by: lightning withstand voltage (dry); switching impulse withstand voltage (wet); and power frequency voltage (wet). The wave shapes for switching surge and lightning impulse tests have been standardised in terms of time to crest and time to half-value on the tail. These, in microseconds, are

Switching surge: $(250 \pm 100)/(2500 \pm 100)$
Lightning impulse: 1,2/50

The performance of an air-gap depends on the electrode shapes and on the applied voltage wave shape. *Figure 12.8* shows typical values for a rod/plane gap. Comparable information (or correction factors) is needed for other configurations[16]. They apply to lines as follows:

Rod/plane: live ends of insulators and fittings to structure and ground.

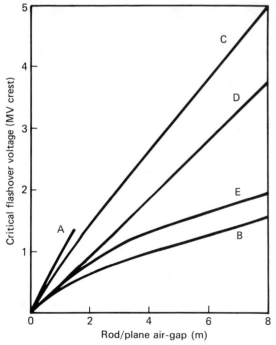

Figure 12.8 Insulation strength of rod/plane air-gaps.
A 200/3000 μs, negative ⎫ switching surge
B 200/3000 μs, positive ⎭
C 1.2/50 μs, negative ⎫ impulse
D 1.2/50 μs, positive ⎭
E 50 Hz, power frequency

Conductor/structure: inside or outside a tower window, outer phases to all types of tower.
Conductor/plane: clearance at mid span to ground, or conductor jumper to cross-arm.
Conductor/conductor: flashover between phases or between conductor and earth wire.

For insulation co-ordination, the electric strength of the air-gaps must be about 10% greater than that of the insulators[17].

Up to about 400 kV, line insulation is governed by the requirements of creepage distance. For higher voltages, switching-surge considerations predominate.

12.5.3 Pollution

A first approximation to the number of units in a string is by an assessment of the intensity of pollution in the area, adopting a creepage distance of 18 mm/kV of line voltage for clean air and up to 30 mm/kV or more for severe pollution.

Insulator surfaces may be contaminated by dust and dirt, whose composition varies according to the area where the line is built. If frequent rains occur, these layers may be washed, but occasionally insoluble layers are formed in the presence of mist or dew; and in the absence of heavy downpours, they reduce the insulation resistance of the insulators. Under certain conditions, flashover can occur which ionises a path in air, providing a channel for follow-currents at power frequency.

To assess the insulation performance of insulators under pollution, two series of tests have been devised[18]:

(1) Salt-fog method: generally representative of coastal areas or of zones where rain can wash the insulators. In this test, insulators are submitted to electrical stresses in an artificial fog of variable salinity.

(2) Solid-layer method: generally representative of industrial zones or desert areas where a solid crust of insoluble deposits can be formed. In this test, the insulators are covered with a special mixture attempting to reproduce natural pollution.

If measurements are made of leakage currents under natural pollution (e.g. at an open-air station), these can be reproduced in the laboratory by varying either the degree of salinity or the composition of the solid layer. Thus, a correlation can be found and a rational assessment can be made as to the ultimate behaviour of the insulators.

Work has been done on palliative measures such as coatings of hydrocarbon or silicon grease, or live-line working. These measures are normally restricted to specific areas.

12.5.4 Voltage distribution over insulator strings

When assessing radio noise from an insulator string, it is desirable to know how the voltage along the string is distributed. This can be found by test: (1) a small spark-gap of known flashover voltage is connected across the cap and pin of one insulator element; (2) the voltage across the string is raised until flashover at the spark-gap, giving the voltage across the element as a fraction of the test voltage. *Table 12.4* gives the results of such a test for a string of porcelain elements of diameter 254 mm, the string having no modifying devices. It will be noted that for strings with eight or more units, the live-end unit carries 0.20–0.24 of the string voltage, and the second 0.13–0.15.

The application of grading rings and horn fittings modifies the voltage distribution by introducing capacitance in shunt with the line-end units to reduce the voltage across them.

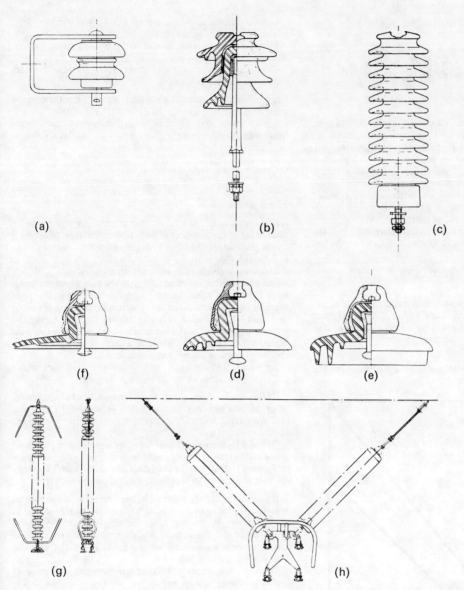

Figure 12.7 Typical insulator units and insulator sets. (a) Shackle insulator; (b) pin insulator; (c) post insulator; (d) normal cap and pin unit; (e) antifog cap and pin unit; (f) aerodynamic cap and pin unit; (g) insulator set (single string); (h) 500 kV Vee insulator set

power frequency overvoltage caused by faults or abnormal conditions, and (c) impulse overvoltage (switching); (2) externally applied stresses due to lightning. Insulation requirements strongly affect the support tower geometry.

The insulation of an overhead line cannot be determined in isolation. It has to be co-ordinated with the system of which it is a part[12,13] and meet specific requirements[14,15].

12.5.1 Types

The most common materials are porcelain (ceramic) and glass. Porcelain insulators are surface-glazed, usually in brown but occasionally in other colours (e.g. blue) for reasons of amenity. With glass, the toughening process develops two zones, an outer layer in compression, an inner layer in tension. Damage to the outer skin usually results in an 'explosion' of the unit. This feature is sometimes regarded as a maintenance advantage, as a damaged unit is readily observed by foot or helicopter patrols.

Composite insulators, comprising a high-strength core of parallel glass fibres with sheds of synthetic rubber or cycloaliphatic resin, are gaining acceptance: they show economic advantages for voltages above 500 kV.

Besides withstanding electric stress, insulators must deal with mechanical stresses. *Figure 12.7* shows some common types of insulator unit. These are

(a) Shackle: for low-voltage distribution lines.
(b) Pin: used for voltages up to 33 kV (occasionally 66 kV).
(c) Post: used up to 66 kV, mainly for the support of air insulated busbars.
(d) Cap and pin: as elements in insulator strings.
(e) Anti-fog cap and pin: for regions of high pollution level, especially if humid.
(f) Aerodynamic cap and pin: for desert regions.

negligible, but skin effect may produce a small increase in the effective resistance.

Inductance For a 3-phase line with asymmetrical phase spacing and transposition, each phase comprising (a) a single conductor of radius r and geometric mean radius r' ($\simeq 0.78r$) or (b) a bundle of geometric mean radius $r' = r_i$, the line-to-neutral inductance is

$$L = [0.20 \ln (s'/r') + K] \text{ mH/km}$$

where $s' = (s_{ab} \cdot s_{bc} \cdot s_{ca})^{1/3}$ is the geometric mean spacing between phases ab, bc and ca. K is a correction factor for the steel core and is equal to zero when there is no steel[3].

Capacitance An isolated 3-phase asymmetric transposed line with single conductors per phase of overall radius r has a line-to-neutral capacitance

$$C = 1/[18 \ln (s'/r)] \; \mu\text{F/km}$$

As the earth is a conductor, it influences the line capacitance. An adjusted value of C is obtained by modifying the logarithmic term to $[\ln (s'/r) - \ln (s'/s_i)]$, where $s_i = (s_{aa'} \cdot s_{bb'} \cdot s_{cc'})^{1/3}$ is the geometric mean of the respective distances between conductors abc and their images a'b'c' across the earth plane. A similar (but much more complex) expression is required for a bundle-conductor assembly[9].

12.4.3 Voltage gradient effects

For a single conductor of radius r (in cm) operating at a phase voltage V_{ph} (in kV), the surface electric field strength is

$$E = 1.8 V_{ph} C/r \text{ kV/cm}$$

where C is the capacitance of the conductor (in pF/cm). In a bundle conductor the voltage gradient is not the same at all points on a subconductor surface[9]. For n subconductors each of radius r (in cm) spaced a distance a apart, the maximum voltage gradient (in r.m.s. kV/cm) is given by

$$E = \frac{1.8 V_{ph} C}{nr}\left[1 + \frac{2(n-1) \sin (\pi/n)}{a/r}\right]$$

Corona The corona onset voltage gradient[3] can be estimated from the Peek formula

$$E_c = 32\delta[1 + 0.308/\sqrt{(\delta r)}] \text{ kV/cm}$$

where δ is the relative air density $3.92b/(\theta + 273)$, b is the barometric pressure (in cmHg), θ is the temperature (in °C) and r is the conductor radius (in cm). The numeric 32 is an average value of the peak breakdown electric strength of air, corresponding to about 22 r.m.s. kV/cm. Thus, a surface voltage gradient of 18–19 kV/cm r.m.s. represents an acceptable upper limit.

Radio noise Electromagnetic fields, including interference fields, are generally expressed either in microvolts per metre (μV/m), in millivolts per metre (mV/m) or in decibels, as follows:

field in decibels = 20 log (field in μV/m)

Thus, an interference level expressed as 46 dB corresponds to an electric field of 200 μV/m (i.e. $46 = 20 \log \dfrac{200}{1}$). Consequently, the radio noise from overhead lines is expressed in decibels above one microvolt per metre (dB/1 μV/m). The main source of radio interference (RI) is conductor corona which depends on voltage gradient. Other factors come into play. There are many methods for calculating RI, both analytical and empirical, but the following empirical formula is often accepted.

$$RI = (RI)_0 + 3.8(E_a - E_0) + 40 \log \frac{r}{r_0} + 10 \log \frac{n}{n_0}$$

$$+ 30 \log \frac{D_0}{D} + 20 \log \frac{1 + f_0^2}{1 + f^2}$$

$$+ \frac{h - h_0}{300}$$

where RI = radio noise to be calculated, in dB/1 μV/m; E_a = average surface gradient; r = radius of subconductor; n = number of subconductors in bundle; h = altitude of line; D = distance of the line at which the noise is measured (typically 15–30 m); f = frequency at which noise is calculated (typically 0.5 or 1.0 MHz in the medium-wave band). The suffix '0' refers to the same parameters as obtained from a known line with fairly close characteristics[11]. For example, a 500 kV line is designed so that $E_a = 16$ kV/cm; $r = 1.6$ cm; $n = 4$; $D = 30$ m; $f = 1.0$ MHz; $h = 600$ m.

Analysis of data for a comparable construction yields $(RI)_0 = 57$; *$E_0 = 17$ kV/cm; $r_0 = 2.14$; $n_0 = 2$; $D_0 = 15$; $f_0 = 0.5$ MHz; $h_0 = 300$ m.

Thus,

$$RI = 57 + 3.8(16 - 17) + 40 \log \frac{1.6}{2.14} + 10 \log \frac{4}{2}$$

$$+ 30 \log \frac{15}{30} + 20 \log \frac{1 + 0.5^2}{1 + 1^2} + \frac{600 - 300}{300}$$

$$= 57 - 3.8 - 5.05 + 3 - 9.03 - 4.1 + 1$$

$$= 39.02 \text{ dB above 1 } \mu\text{V/m}.$$

Before the design of the line is accepted, it is necessary to obtain the strength of the radio signal received at the position of the line. The decision will be governed by the desired quality of reception, on the basis of the following table.

Class of reception	Quality of reception	Approximate signal/noise ratio
A	Background undetectable	32
B	Background detectable	27
C	Background evident	22
D	Background evident but speech still understood	16

If, for example, a class of reception B is desired, in the case of the line mentioned above, the signal strength received at 30 m from the axis of the line should be $39.02 + 27 = 66.02$ dB/1 μV/m under fair weather conditions. If this is not the case, another look at the design or at the route of the line would be justified.

Under foul weather conditions, the radio noise from the line would be some 10 dB/1 μV/m greater.

12.5 Insulators

Air is ubiquitous and is an excellent insulator. The performance of air-gaps is well known as regards breakdown as a function of electrode geometry, so that clearances between phase conductors and from conductors to ground can be determined. However, at line supports it is necessary to provide man-made insulators to maintain adequate air-gaps and to provide mechanical support.

The electric stresses applied to an overhead line are: (1) internal stresses resulting from (a) normal power frequency voltage, (b)

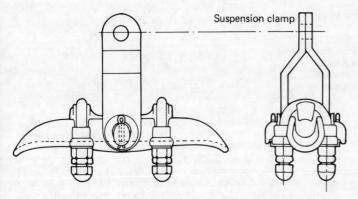

Figure 12.4 Suspension clamp

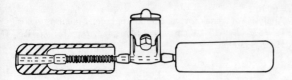

Figure 12.5 Stockbridge vibration damper

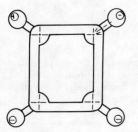

Figure 12.6 Schematised spacer-damper

12.4 Electrical characteristics

Electrical characteristics are concerned with voltage regulation and current carrying capability. The electric field strength (or voltage gradient) at the conductor surfaces affects corona and radio interference phenomena.

Voltage regulation This must be maintained within specified limits (normally 5–12%). The voltage drop depends on the electrical line parameters of resistance, inductance and capacitance, the two latter being influenced by the geometry of the support structures and the frequency of the system. For power frequency lines of length up to about 80 km the capacitance is usually ignored in electrical calculations.

Current rating Aluminium is subject to annealing at temperatures of 75 °C and above, and 75 °C is therefore taken as the upper limit of conductors in still air. The current rating is assessed from a heat balance equation, i.e. I^2R loss + heat absorption from solar radiation = heat loss by radiation and convection. As the current rating (which has no unique value) depends on local meteorological conditions, a knowledge of these is paramount. In some cases techniques are being introduced by system operators to adapt the thermal rating of the conductors to the immediate ambient conditions.

Radio interference Considerations of amenity require the radio noise from overhead lines to be kept within acceptable limits of signal/noise ratio in the vicinity of the line.

12.4.1 Bundle conductors

At operating frequency the inductive reactance of an overhead line with an equivalent spacing s' and a conductor system of geometric mean radius r' has the form $X = K \ln(s'/r')$. An increase of r' reduces the inductance (and therefore increases the capacitance) of the line; it also lowers the voltage gradient for a given working voltage. High-voltage lines can utilise bundles of two or more spaced subconductors to give an effective increase in r'. For a bundle of n subconductors each of radius r, arranged symmetrically around the circumference of a circle of radius R, the geometric mean radius of the assembly is

$$r_i = [nrR^{n-1}]^{1/n}$$

Bundle conductors have the following advantages:

(1) Reduced inductance and increased capacitance, improving 'surge impedance loading', i.e. raising the power transmission capability of a long line.
(2) Greater thermal rating because of the greater cooling surface area compared with that of a single conductor with the same total cross-sectional area.
(3) Lower surface voltage gradient, so reducing corona loss and radio interference.

The main disadvantage is the sometimes unsatisfactory aerodynamic performance: interbundle oscillations can occur between subconductor spacers. The effect can be controlled by judicious arrangement of spacers and spacer–dampers[4].

12.4.2 Electrical parameters

Resistance Because of the lay in strands, the current flow in a conductor is helical, developing an axial magnetic field. If the conductor has a steel core, the magnetic loss therein may increase the effective resistance by 4–5%. The proximity effect is usually

Table 12.3 Creep coefficients for ACSR conductors

No. of strands		Ratio Area Al/Steel	Process*	Coefficients				
Al	Steel			κ	ϕ	β	γ	δ
54	7	7.71	HR	1.1	0.0175	2.155	0.342	0.2127
			EP	1.6	0.0171	1.418	0.377	0.1876
48	7	11.4	HR	3.0	0.0100	1.887	0.165	0.1116
30	7	4.28	EP	2.2	0.0107	1.375	0.183	0.0365
26	7	6.16	HR	1.9	0.0235	1.830	0.229	0.08021
24	7	7.74	HR	1.6	0.0235	1.882	0.186	0.00771
18	1	18	EP	1.2	0.0230	1.503	0.332	0.1331
12	7	1.71	HR	0.66	0.0115	1.884	0.273	0.1474

*Industrial processing of Al rod: HR, hot-rolled; EP, extruded or Properzi.

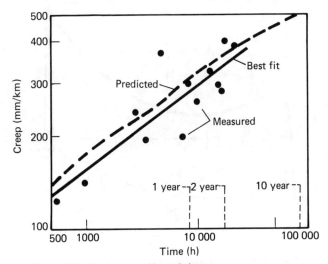

Figure 12.3 Typical creep/time relations

Here E is the modulus of elasticity, θ is the temperature (°C) and α is the coefficient of linear thermal expansion per °C.

The maximum permissible stress f_2 (or tension $f_2 a$) that will occur under the most onerous conditions and at a low temperature θ_2 is usually known. The stress f_1 at which the line must be strung can thus be calculated. A 'stringing chart' for a specified line relating span, tension and sag at a given temperature is provided for line erection. With the extended use of computers, the stringing tables are now automatically printed and supplied. They supersede the previous charts. For spans up to 300 m the parabolic assumption is adequate for most cases. Longer spans and abnormal conditions demand the use of the catenary equations, which can readily be solved by computer.

12.2.4.2 Equivalent span

As an actual overhead line comprises a series of spans of not necessarily equal length, supported by suspension-insulator sets and with tension insulators at the ends of a section, it is generally assumed that the behaviour of a section is that given by a series of equal equivalent spans, each given by

$$L_{eq} = \sqrt{[\Sigma L_i^3 / \Sigma L_i]}$$

where L_i is an individual span length. It is possible to improve on this estimate by use of a complex computer program.

12.2.4.3 Creep

Active research[8] into the problem of creep has resulted in the empirical expression

$$\varepsilon = K \cdot f^\beta \cdot \exp(\phi \theta) \cdot t^{\gamma/f\delta}$$

for the creep extension ε (in mm/km) in terms of conductor stress f (in kg/mm^2) and temperature θ (in °C), time t (in h), a constant K and creep indices β, ϕ, γ and δ. Typical values of K and the creep indices are given in *Table 12.3* for ACSR conductors. Evaluation of the expression for the range of operating conditions requires a sophisticated computer program. *Figure 12.3* shows a predicted creep–time curve, a series of measured values and a 'best fit' for them, in the case of a Zebra 54/7/3.18 conductor.

12.3 Conductor fittings

Suspension clamp *Figure 12.4* illustrates a typical UK design, light and suitably profiled to limit the effects of combined static tension, compression and bending, and stresses due to dynamic bending. Positioning the axis of rotation on the axis of the conductor is of help in these aspects.

Tension and mid-span joints A bolted clamp or a cone-type grip is usually adequate for monometallic conductors of small size. For composite conductors (especially when greased) compression clamps must be used for joints and conductor ends.

Vibration dampers These absorb vibration energy. The Stockbridge damper (*Figure 12.5*) is well known. Other types, such as 'bretelles', festoons, Elgra, etc., have also proved successful.

Spacers and spacer–dampers These maintain bundled conductors in proper configuration. The term is now applied to fittings that damp all types of induced vibration: *Figure 12.6*.

Anti-galloping devices When fitted to bundled conductors, significant reduction in gallop amplitude is achieved, reducing outages due to interphase flashover.

Table 12.2 Characteristics of conductor materials at 20 °C

Property		Annealed copper	Hard-drawn copper	Cadmium-copper	Hard-drawn aluminium	Aluminium alloy (BS 3242)	Galvanised steel
Relative conductivity	(%)	100	97 (avg)	79.2 (min.)	61 (min.)	53.5	—
Volumetric resistivity	(Ω mm^2/m)	0.017 24 (std)	0.017 71 (avg)	0.021 77 (max.)	0.028 26 (max.)	0.032 2 (std)	—
Mass resistivity	(Ω kg/km)	0.153 28	0.157 41	0.194 72	0.076 40	0.086 94	—
Resistance at 20 °C	(Ω mm^2/km)	17.241	17.71	21.77	28.26	32.2	7780
Density	(kg/m^3)	8890	8890	8945	2703	2700	7.78
Mass	(kg/mm^2 km)	8.89	8.89	8.945	2.703	2.70	
Resistance temperature coefficient at 20 °C	(per °C)	0.003 93	0.003 81	0.003 10	0.004 03	0.0036	
Coefficient of linear expansion	(per °C)	17 × 10^{-6}	17 × 10^{-6}	17 × 10^{-6}	23 × 10^{-6}	23 × 10^{-6}	11.5 × 10^{-6}
Ultimate tensile stress (approx.)	(MN/m^2)	255	420	635	165	300	1350
Modulus of elasticity	(MN/m^2)	100 000	125 000	125 000	70 000	70 000	200 000

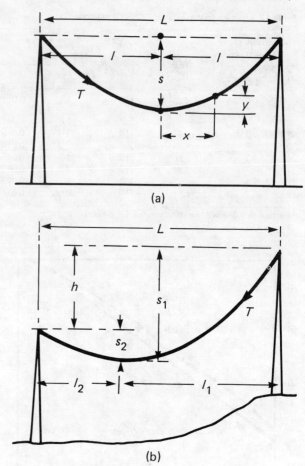

Figure 12.2 Sag and tension

and the length of conductor in a half-span is

$$l_c = l + w^2 l^3 / 6 T_h^2 = l + 2s^2/3l$$

For supports at different heights (*Figure 12.2b*) that differ by h, the distances of the supports from the point of maximum sag are

$$l_1, l_2 = \tfrac{1}{2}[L \pm 2hT/wL]$$

with the parabolic assumption. For large values of h the value of l_2 may be negative, indicating that the lowest point of the conductor is outside the span, on the left of the lowest support. Such a case would involve an upward component of pull on the lower support, not admissible with suspension insulators.

12.2.4.1 Change of state

If after erection the conductor temperature rises (because of I^2R loss or of a rise in ambient temperature), the conductor expands, increasing the sag; but at the same time reduction of the tension allows the conductor to contract elastically. Further, if the loading increases (owing, e.g., to wind pressure and/or ice), the tension rises and the conductor stretches. Analysis of these opposing tendencies leads to a cubic equation relating tension, temperature, loading and elasticity. For two sets of conditions (subscripts 1 and 2)

$$\frac{w_2^2 L^3}{24 f_2^2 a^2} = \frac{w_1^2 L^3}{24 f_1^2 a^2} + \frac{f_2 - f_1}{E} - (\theta_1 - \theta_2)\alpha$$

Table 12.1 Typical properties of ACSR conductors

Code name	Stranding	Alu area (mm^2)	Steel area (mm^2)	Diameter (mm)	Mass (kg/km)	Breaking load (kN)	Resistance, 25 °C (Ω/km)
Horse	12/7/2.79	73.4	42.8	13.95	538	61.2	0.3936
Lynx	30/7/2.79	183.4	42.8	19.53	842	79.8	0.1441
Zebra	54/7/3.18	428.9	55.6	28.62	1621	131.9	0.0674
Dove	26/3.72+7/2.89	282	45.9	23.55	1137	99.88	0.1024

loading and ambient temperature), and on internal conditions (such as stranding, modulus of elasticity, thermal expansion and creep). For lines of 33 kV and over, hard-drawn AAC is not likely to be adopted; economics and creep behaviour lead to the choice of ACSR. As is seen from *Figure 12.1*, a wide range of breaking strength/weight ratio can be achieved by modifying the aluminium and steel content, and by using aluminium alloy strands where conditions demand. *Table 12.1* gives typical properties of ACSR conductors.

Modulus of elasticity For a complete ACSR the modulus of elasticity can be estimated from

$$E = [(70r + 200)/(r + 1)] \times 10^3 \text{ MN/m}^2$$

where r is the ratio aluminium area/steel area.

Linear expansion The coefficient of linear expansion of an ACSR conductor per °C is given by

$$\alpha = 10^{-6}[(23r + 32.6)/(r + 2.83)]$$

Some useful reference data are given in *Table 12.2*.

Oscillation phenomena Exposed to wind and ice loading, overhead line conductors are subject to oscillations that affect the design and application of the conductor fittings.

Aeolian vibrations are typically in the range 8–40 Hz, occasionally higher, and are generated by wind speeds of 2–40 km/h. As a laminar air flow is involved, certain regions of terrain are prone to develop the aeolian phenomenon. The consequent vibration causes strand breakdown by fatigue. The subject is well documented[5].

Subspan oscillations affect only bundled conductors (Section 12.4.1). They occur at a frequency of 0.5–5 Hz[4] and wind speeds of 15–65 km/h. Their intensity is considered to be reduced if the ratio between subconductor spacing and conductor diameter exceeds 16.

Galloping involves complete spans in a fundamental mode. Amplitudes reaching the value of the sag have been recorded. Bundled conductors may be more prone to galloping than single conductors, but the latter have been known to gallop[6]. It is now generally accepted that near-freezing temperatures are required for galloping to occur with conductor diameters up to 35 mm. Large conductors can gallop under specific conditions due to aerodynamic instability, even in the absence of ice.

Creep Aluminium based conductors are subject to a permanent non-elastic elongation ('creep'), which must be predicted with adequate precision in order that ground clearance regulations may be satisfied. Creep changes the conductor length, tension and sag, and this in turn affects the creep, as does also the conductor temperature.

12.2.4 Sag and tension

Almost all the elements of an overhead line are based on the mechanical loading of the conductors. The loading includes the effect of stringing tension, wind pressure, temperature and ice formation.

The configuration of a conductor between supports approximates a catenary curve, but for most purposes it may be taken as a parabola. The catenary has a constant mass per unit length along the *conductor*; the parabolic curve assumes a constant mass along the *straight line* between the points of support. The following nomenclature is employed.

Quantity	Symbol	Unit SI	Unit Imperial
Span length	L	m	ft
Half span length	l	m	ft
Conductor diameter	d	m	ft
Conductor section	a	m^2	ft^2
Thickness of ice	t	m	ft
Tension			
total	T	N	lbf
horizontal	T_h	N	lbf
Stress	f	N/m^2	lbf/ft^2
Maximum sag (at midspan)	s	m	ft
Weight			
conductor	w_c	N/m	lbf/ft
ice	w_i	N/m	lbf/ft
Wind pressure	p	N/m^2	lbf/ft^2
Wind loading	w_w	N/m	lbf/ft
Total loading	w	N/m	lbf/ft

The stress is $f = T/a$, the wind loading is $w_w = p(d + 2t)$ and the total loading is $w = \sqrt{[(w_c + w_i)^2 + w_w^2]}$ as the conductor and ice loadings are directed vertically downward, whereas the wind loading is directed horizontally and acts on the projected area of the conductor and its ice loading (if applicable).

The catenary equation for a conductor between supports at the same height is expressed in terms of the distance x from the centre of the span (point of maximum sag) and of y the conductor height above this point (*Figure 12.2a*). Then

$$y = (T_h/w)[\cosh(wx/T_h) - 1]$$

is the catenary equation. Putting $x = L$ gives y at either support and therefore the sag s. The hyperbolic cosine term can be expanded into a series. For typical low-sag cases all terms but the first are negligible. Further, with the assumption of a uniform conductor tension throughout, the equation reduces to $y = wx^2/2T_h$, representing a parabola. The sag is therefore

$$s = wL^2/8T_h \text{ for a span } L.$$

12.1 General

The overhead line is the cheapest form of transmission and distribution of electrical energy. Line design and construction involve several engineering disciplines (electrical, civil, mechanical, structural, etc.), and must conform to national and international specifications, regulations and standards. These refer to conductor size and tension, minimum clearance to ground, stresses in supports and foundations, insulation levels, etc. Lines must operate in conditions of large temperature change and in still air and gales, and (in non-tropical climates) may have ice formation on the conductors. In the UK, the CEGB has accepted loading conditions more onerous than those stipulated in the 'Statutory Instruments'[1].

The structural design of a d.c. line does not differ essentially from that of an a.c. line, but electrical features related to frequency (such as inductive, capacitive and skin effects) do not apply in the d.c. case under normal operating conditions.

12.2 Conductors

12.2.1 Materials

Many years of operating experience, and the costs as affected by metal market trends, have combined to favour aluminium-based conductors[2]. Copper and cadmium-copper are seldom used, even for distribution lines.

The intractability of large solid conductors has led to the almost exclusive use of stranded conductors (in spite of their greater cost), which have a larger diameter than equivalent solid circular conductors. With strands of diameter d, the outside diameter of uniformly stranded conductors is

No. of strands	3	7	19	37	61
Overall diameter	$2.15d$	$3d$	$5d$	$7d$	$9d$

Aluminium-based conductors in normal use are categorised as

AAC	All-Aluminium Conductors (BS 215, IEC 207)
ACSR	Aluminium Conductors Steel Reinforced (BS 215, IEC 209)
AAAC	All-Aluminium Alloy Conductors (BS 3242, IEC 208)
AACSR	All-Aluminium Alloy Conductors Steel Reinforced (IEC 210)
ACAR	Aluminium Conductor Alloy Reinforced

Figure 12.1 illustrates typical strandings of ACSR. The conductor with an outer layer of segmented strands has a smooth surface and a slightly reduced diameter for the same electrical area.

12.2.2 Nomenclature

There is as yet no international agreement on nomenclature. In the UK conductors were referred to the approximate area in (inches)^2 of a copper conductor having the same conductance. Nowadays, aluminium based conductors are referred to their nominal aluminium area. Thus, ACSR with 54 Al strands surrounding 7 steel strands, all strands of diameter $d = 3.18$ mm (designated: 54/7/3.18; alu. area = 428.9 mm^2, steel area = 55.6 mm^2) is described as 400 mm^2 nominal aluminium area. In France the total area (485 mm^2) is quoted; in Germany the aluminium and steel areas are quoted (429/56); whereas in Canada and the USA the area is stated in thousands of circular mils (1000 circular mils = 0.507 mm^2).

Code names using animal, bird or flower, etc... names are also used. Thus, the 400 mm^2 nominal aluminium area ACSR is known as 'Zebra' . . . .

12.2.3 Mechanical characteristics

The choice of a conductor from the mechanical viewpoint depends on external loading conditions (such as wind speed, ice

Conductor sections

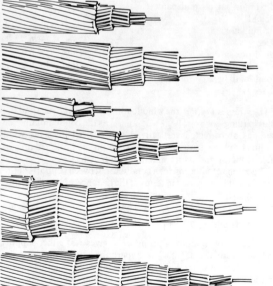

Figure 12.1 ACSR conductor stranding

12 Overhead Lines

G Orawski, BSc(Eng)Hons, CEng, FIEE, SMIEEE
Transmission Consultant
Balfour Beatty Power Construction Ltd

Contents

12.1 General 12/3

12.2 Conductors 12/3
    12.2.1 Materials 12/3
    12.2.2 Nomenclature 12/2
    12.2.3 Mechanical characteristics 12/3
    12.2.4 Sag and tension 12/4

12.3 Conductor fittings 12/6

12.4 Electrical characteristics 12/7
    12.4.1 Bundle conductors 12/7
    12.4.2 Electrical parameters 12/7
    12.4.3 Voltage-gradient effects 12/8

12.5 Insulators 12/8
    12.5.1 Types 12/9
    12.5.2 Selection 12/10
    12.5.3 Pollution 12/10
    12.5.4 Voltage distribution over insulator strings 12/10

12.6 Supports 12/11
    12.6.1 Materials 12/11
    12.6.2 Configurations 12/11
    12.6.3 Tower geometry 12/12
    12.6.4 Foundations 12/12

12.7 Lightning 12/13
    12.7.1 Mechanisms of insulation flashover 12/13
    12.7.2 Lightning performance 12/13

12.8 Loadings 12/14

5 *Transmission and Distribution Handbook*, Westinghouse Electrical Corporation, Pittsburgh, USA
6 BROWN, J. G. (Ed.), *Hydroelectric Engineering Practice*, Vol. 2, Chs 7, 8, 9, Blackie and Son
7 CENTRAL ELECTRICITY GENERATING BOARD, *Modern Power Station Practice*, Vol. 4, (1971)
8 SAY, M. G., *Introduction to the Unified Theory of Electromagnetic Machines*, Pitman Books Ltd (1971)
9 ADKINS, B. and HARLEY, R. G., *The General Theory of A.C. Machines*, Chapman and Hall (1975)
10 SAY, M. G., *Alternating Current Machines*, 5th ed., Pitman Books Ltd (1983)
11 GUILE, A. E. and PATERSONS, W., *Electrical Power Systems*, Vol. 1, Ch. 7, 'Synchronous machines', Pergamon Press (1977)
12 SARMA, M. S., *Synchronous Machines: Their Theory, Stability and Excitation Systems*, Gordon and Breach (1979)
13 WALKER, J. H., *Large Synchronous Machines*, Clarendon Press (1981)
14 FITZGERALD, A. E., KINGSLEY, C. Jr. and UMANS, S. D., *Electric Machinery*, McGraw Hill (1983)

There is a very large literature about a.c. generators, their theory, design and operation, and associated equipment, in the journals of learned societies, e.g.

IEE: *Proceedings*
IEEE: *Transactions on Power Apparatus and Systems* (P.A.S.)
CIGRE: *Electra*

in the house journals of major manufacturers, and in other technical journals. It is impracticable to give here a comprehensive reference list. The reader is referred to:

1 *Electrical and Electronic Abstracts*, published monthly by IEE as part of the Inspec Information Service (IEE, Savoy Place, London WC2R 0BL)
2 *List of CEGB Bibliographies*, Ref. no. 196 (Central Electricity Generating Board Technical Information Unit, Sudbury House, 15 Newgate St., London EC1A 7AU)
3 *Bibliography of Rotating Electrical Machinery*, IEEE paper T 73 011-4 (IEEE Service Dept. 445 Hoes Lane, Piscataway, New Jersey 08854, USA)

A few review papers or reports on particular subjects are:

1 VICKERS, V. J., 'Recent trends in turbogenerators', *IEE Proceedings*, **121**, No. 11R, Nov. (1974)
2 SEONI, R. M. *et al.*, 'Review of trends of large hydroelectric generating equipment', *IEE Proceedings*, **123**, No. 10R (1976)
3 'IEEE working group report of problems with hydrogenerator thermoset stator windings', *P.A.S.*, **100**, No. 7.
4 KURTZ, M. and LYLES, J. F., 'Generator insulation diagnostic testing', *P.A.S.*, **90**, No. 5
5 'IEEE rotating machinery committee report on torsional oscillations and loss of life in turbogenerators', *P.A.S.*, **101**, Sept. (1982)
6 IEEE Standard No. 421-A: 1978, *Guide for Identification, Testing and Evaluation of the Dynamic Performance of Excitation Control Systems*
7 *Excitation System Models for Power System Stability Studies*, IEEE Committee Report: F80 258-4
8 IEEE Tutorial, *Power System Stabilisation via Excitation Control*, 81 EHO 175-0-PWR

control, after a long period of idleness, and can deliver full load within 2 or 3 min of start-up.

Generators may be mounted on sole-plates or on common baseplates with the prime movers. Most generators have a bearing at the non-drive end and have the drive end of the rotor bolted solidly to the shaft of the prime mover or gearbox.

The machines operate in a variety of environments which dictate the type of enclosure and method of cooling. Open ventilation with the appropriate class of protection is simplest where an adequate supply of clean air is available, or an open machine in an enclosure supplied with filtered air may be suitable. In dirtier surroundings a totally enclosed machine with its air circulated through an air-to-air or air-to-water cooler would be needed. Modern machines are almost invariably built with Class F insulation, but temperature rises may be limited to Class B or less, depending upon the application.

For generators driven by diesel engines, in addition to design checks on critical speeds it is necessary to check torsional natural frequencies of the shaft systems in order to avoid resonance with the frequencies in the engine torque. Other technical problems can arise from transient loads, which produce sudden voltage or speed changes; in this connection, motor starting duties can be particularly onerous. Rectifier loads need careful consideration as they draw harmonic currents which produce additional heating in the generator, especially in solid pole-shoes.

The most common form of excitation is the brushless a.c. exciter with a rotating armature and rectifier. An alternative is a static electronic regulator feeding the generator field through slip-rings.

11.19 Synchronous compensators

Synchronous compensators are synchronous motors running without mechanical load; they are used to generate or absorb reactive power, in order to control the voltage of a power system. Hence they are usually installed near a load, or part way down a long transmission line to support the voltage at the intermediate point.

At times of heavy load, compensators run overexcited to supply the magnetising power demanded by the load (transformers and induction motors) or the inductive $I^2 X$ losses of the line. At light load they must run underexcited to take reactive power from the line to offset the capacitive line-charging current and so avoid excessive voltage rise. In many h.v. systems (200 kV and above), the line capacitive power exceeds the load magnetising power, even at times of heavy load.

Static inductors and capacitors, switched to suit the system conditions, can be used for the same purpose, but the synchronous compensator has the advantage of providing continuously variable control, and with thyristor excitation it can have a response fast enough for many contingencies.

Thyristor-controlled static compensators give rapid and continuously variable control, but require filters to limit harmonic generation in the power system. Lower maintenance and running costs give them an advantage over rotating machines in most new installations.

Synchronous compensators up to 300 Mvar are in service. Air cooling has been used up to ratings of 40 Mvar, but hydrogen cooling is now normal to reduce the size and the light-load losses, the latter by reducing windage. As the shaft-end need not emerge from the hydrogen-tight casing, no shaft seals are needed. Losses at full load (overexcited) are in the range 0.01–0.016 MW per Mvar of rating.

The underexcited capability is usually about half the overexcited rating; for this a short-circuit ratio of about 0.75 is desirable, to ensure that at the underexcited capability the rotor has sufficient positive excitation to maintain stability. A short-circuit ratio of 1.3–1.5 will provide an underexcited capability level equal to the overexcited level, but the machine is larger and has higher losses than the design with the lower short-circuit ratio. Water cooling has been used for stator and/or rotor windings at ratings of 200 Mvar or more.

The compensator may be run up to speed as an induction motor through a step-down transformer, by means of a direct-coupled pony motor, or by using a variable-frequency inverter.

11.20 Induction generators

If an induction motor is driven above synchronous speed it will deliver power to the system, with a slip of about −0.05 at full load. It has the advantage of simple construction, and needs no excitation, speed governing, or synchronising. This makes it cheaper than a synchronous machine and operationally more convenient, e.g. for unattended hydrostations or wind-driven generators. The disadvantage is that it must draw from the power system magnetising power of 0.5–0.75 of its rated active power output, and this has limited the size to about 5 MW.

Research in the USA has shown that, by using static var compensators to supply the reactive power, it should be practicable to run induction generators of a few hundred megawatts output either in parallel with synchronous generators or even as a separate supply system. Speed control would then be essential to fix the frequency of the separate system.

Connection to the power network can be made merely by closing the breaker when the machine is up to synchronous speed. To reduce the current surge in large machines, the machine can be allowed to build up to normal voltage by first connecting a capacitor and then synchronising in the usual way. By suitable design of the machine and the static compensator, efficiency and stability can be comparable to those of a synchronous machine.

The most suitable locations appear to be where transmission by h.v. cable is required, for the cable capacitance will contribute to the reactive-power requirement, and at points in the system where substantial var support is installed anyway, the generator being run when active power is also required.

To maintain stability following system faults requires considerable reactive-power capacity beyond that needed for steady full-load operation. Where the system is strong enough, however, this is not too costly, and the total cost of the induction generator installation can be less than that of a synchronous unit.

11.21 Acknowledgements

The author is indebted to the General Electric Co. p.l.c. (UK) for permission to reproduce *Figures 11.16*, *11.22* and *11.23* and for information on which *Figures 11.18–11.21* are based. The author wishes to record his gratitude to colleagues for their assistance, especially to Messrs N. Foster, G. K. Ridley and F. J. Parker for their contribution to the section on hydrogenerators, and to A. B. J. Reece and Dr R. D. M. Whitelaw for valuable discussion and suggestions.

11.22 References

1 LANGSDORF, A. F., *Theory of A.C. Machinery*, 2nd edn, McGraw-Hill (1955)
2 KIMBARK, E. W., *Power System Stability, Vol. 3, Synchronous Machines*, Chapman and Hall (1956)
3 SAY, M.G., *Performance and Design of A.C. Machines*, Pitman Books Ltd (1958)
4 JAIN, G. C., *Design, Operation and Testing of Synchronous Machines*, Asia Publishing House, London (1966)

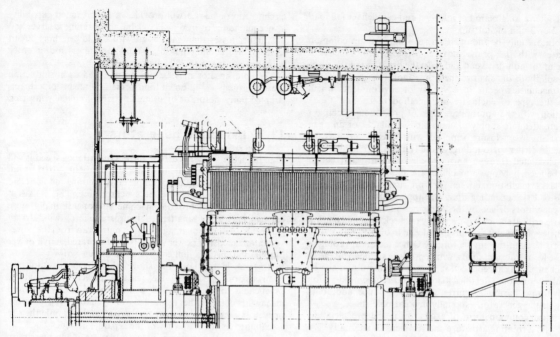

Figure 11.23 Section of a pumped-storage motor/generator

larger size may be needed to attain the desired inertia constant, or the lower capitalised cost of its lower losses may be enough to offset the lower first cost of the smaller frame. In some circumstances, water-cooled stator and rotor windings may be economically justified at ratings as low as 150 MVA; conversely, water-cooled stator windings with some form of improved air cooling for the rotor may be preferred on grounds of adequacy and simplicity for ratings as high as 700 MW.

11.17.3 Pumped-storage units

Pumped-storage units were originally installed as peak-levelling units, running as generators at times of high system load, and as motors pumping water up to the top reservoir during light-load periods. A unidirectional set has separate pump and turbine, whereas a reversing set uses the same hydraulic machine either as a pump or as a turbine, depending on the direction of rotation.

The original purpose has been extended to provide spinning reserve and to deliver power into the system at a 'few seconds' notice to assist in maintaining stability if other generation, or a system interconnection, is suddenly lost. This introduces particular problems of thermal cycling and mechanical fatigue, especially with reversing sets, which may be required to go from full-speed pumping to reversed full-speed generating within seconds, and to do this perhaps several times a day. For reversing units, separately driven fans are usually provided because rotor-mounted fans designed for both directions of rotation have low efficiency. Water cooling may be applied, as for generators, and may be particularly valuable for the damper cage if this is used for 'induction-motor' starting.

Figure 11.23 shows a cross-section of one half of the motor/generator of a reversible pumped-storage unit rated at 330 MVA, 18 kV, 0.95 p.f., 500 rev/min. The design at this output and speed approaches the limit achievable with present materials and air cooling. The outside diameter of the stator core is 6.2 m, the rotor diameter 4.5 m and the active core length 3.6 m. The mechanical design of the rotor is dominated by centrifugal stresses and the fatigue effects of reversals. The rotating parts weigh about 440 tonne, and the hydraulic thrust raises the load on the thrust bearing to almost 600 tonne.

11.17.3.1 Starting

Methods available for starting and run-up of a machine in the pumping mode are: (i) by a direct-coupled auxiliary starting turbine or pony motor; (ii) by back-to-back connection with another machine driven by its own turbine and acting as a generator; (iii) from the power network through a step-down transformer, using the rotor solid pole-shoes or the damper winding as a cage for an 'induction' start; (iv) from the power network through a variable-frequency thyristor converter controlled to give an output over the range of a few hertz up to normal system frequency. In (iii) the pole-shoes or damping windings must be designed to carry the induced currents without excessive rise in temperature. In (iv) the machine moves from rest by induction torque, but at a low frequency it synchronises with the converter and thereafter remains in synchronism up to normal frequency, to be then synchronised with the power network. The starting equipment is expensive, but run-up is more readily supervised, and the damping cage (or pole-shoe) design is not constrained. The method is preferred for large units.

11.18 Engine-driven generators

Medium-sized generators with outputs up to 30 MW are used for industrial and marine applications, including transportable units, drilling-rig supplies, small power-houses and emergency stand-by supplies. Sets driven by steam turbine or gas turbine usually have generators that are gear driven at speeds between 1000 and 1800 rev/min. Diesel-engine-driven sets cover a much wider speed range from 1800 to 125 rev/min. Gas-turbine-driven sets, and espeically diesel-driven sets, are valuable for emergency power supplies because they can start quickly, often by remote

For forced-air cooling, the flow is usually from one end of the run to the other along one phase, half the flow returning along each of the other two phases. The air circuit is totally enclosed, with an air-to-water heat-exchanger extracting the loss. If the air circulation fails, the naturally cooled rating is about 60% of the forced-flow rating.

The use of water-cooled stator windings led to the development of water-cooled connections in solid tube or cable form. The water-cooling circuit is similar to, but usually separate from, that for the generator stator. The smaller dimensions are an advantage where space is limited, and the greater loss is not usually economically unacceptable. The system has not been widely adopted, however, because there is usually room for air-cooled connections, while the extra water-cooling auxiliaries introduce additional maintenance and extra complications of duplication and control to guard against shut-down of the generator if an auxiliary item fails.

11.17 Hydrogenerators

The design of generators for hydroelectric plant is largely determined by mechanical considerations. Their low speeds (50–1000 rev/min), fixed by the water head available and the size and type of turbine, result in large physical dimensions, often making it necessary to transport them in sections. Either turbine-governing and speed-regulation requirements or the transient stability of the associated power system decide the amount of flywheel effect of the set. The contribution of the turbine to the flywheel effect is small, so that the generator inertia required is frequently greater than that inherent in the machine on a basis of the electrical specification alone. This may well be a deciding factor in the design, demanding a large diameter to achieve the flywheel effect. In extreme cases a separate flywheel is incorporated.

Water turbines are subject to high overspeeds of the order of twice normal speed in the event of failure of the governor gear or water-regulating gates, and it is usual to design generators to withstand the consequent stresses. Critical speeds are an additional consideration and care is taken at the design stage to ensure that the first critical speed is well above the maximum overspeed.

11.17.1 Construction

Both horizontal- and vertical-shaft arrangements are employed, the former usually for impulse turbines or smaller reaction units, the latter for large reaction and Kaplan turbines. The bearings of horizontal generators follow normal practice, except that a thrust bearing is provided to resist axial thrust due to unbalanced hydraulic forces.

Three different bearing arrangements are used for vertical machines. The slowest machines, being of relatively large diameter and short core length, have a combined thrust and guide bearing below the rotor, often referred to as 'umbrella' construction. At higher speeds, requiring a smaller diameter and longer core length, an additional guide bearing is fitted above the rotor—a semi-umbrella arrangement. For the highest speeds, the 'top-bracket' construction is usual; for this, a combined thrust and guide bearing is mounted above the rotor and a second guide bearing is provided underneath.

The hydraulic thrust from the turbine may be several hundred tonnes, exceeding the dead weight of the rotating masses; both of these must be supported by the thrust bearing. Bearing oil is cooled usually by oil-to-water heat-exchangers mounted either within the oil chamber or externally. In the latter case a motor-driven pump is generally necessary to ensure adequate circulation; otherwise the natural pumping action of the bearing is relied upon. Sometimes cooling water is circulated directly through the bearing pads.

Vertical-shaft machines are provided with pneumatically operated brakes to bring the set to rest without protracted running at the low speeds for which lubrication of the thrust bearing may be inadequate. To reduce wear on the brake linings, the brakes are not usually applied at full speed, except in emergency; dynamic electric braking is frequently preferred to friction brakes. The friction brakes, operated hydraulically, are used as jacks when the machine is stationary, either for flooding of the surfaces of the thrust bearing with oil after prolonged standing or for maintenance such as the removal or inspection of the thrust-bearing pads.

The construction of the active magnetic and electrical parts of the generator follows established practice for synchronous machines. The rotor is of salient-pole construction, the poles being either solid forgings or built up from punched steel laminations clamped between solid steel end-plates. In the latter case, copper damper windings are inserted in the pole faces and may be connected between poles to form a complete cage winding. With solid poles, sufficient damping is obtained from eddy currents induced in the pole faces. The poles are keyed either to a laminated rim or to a solid rotor body which may be forged or built up from steel plates.

11.17.2 Excitation

Excitation may be by main and pilot exciters directly coupled to the generator shaft. Separate motor-driven exciters are also used, in conjunction with a flywheel to maintain excitation during momentary interruptions or reduced voltage of the driving-motor supply under system fault conditions. More frequently, static excitation equipment is used. To supply the turbine-governor driving motor, a small auxiliary generator (usually with a permanent-magnet field) is mounted on the generator shaft.

Generation is usually within the voltage range 6.6–18 kV. Hydro power stations may have to be situated at considerable distance from their load centres; the long transmission lines then necessitate machines of a high short-circuit ratio to supply the line-charging currents.

Except for small machines, closed-circuit air cooling is usually employed, the heat being removed by water coolers mounted on the back of the stator frame. About 4 m^3/s of air per kilowatt of loss is required. In cold climates, a proportion of the hot exhaust air from generators may be bled off for station heating. The generators may be mounted in pits below station-floor level or enclosed by sheet-steel covers.

Hydrogen cooling has not been applied to hydrogenerators, mainly because of the cost and practical difficulties of making an explosion-proof casing. However, when inertia is not a controlling factor, a useful reduction in physical size can be achieved by water-cooling the stator or rotor windings, or both. Water-cooling of only the stator winding introduces a multiplicity of joints which must not leak, but greatly increases the current capacity, or conversely allows the slot size to be reduced for a given rating; this slightly reduces the outside diameter of the core and significantly reduces the leakage reactance.

Water-cooling of the rotor winding introduces more constructional difficulties and affects the design more profoundly. The excitation capability is increased without risk of exceeding the specified temperature limits, so a higher short-circuit ratio (longer airgap) is possible, improving the underexcited (line-charging) capability. Alternatively a smaller machine with higher electric loading is possible with the stator design suitably adjusted. However, it may be necessary to adopt a size larger than the smallest determined from purely thermal considerations. This

the hydrogen- and air-side flows are kept separated.

A wide range of indicators fitted with audible and visible alarms is necessary to indicate any departure from normal operation of the various parts of the gas and oil systems.

11.15.7.2 Direct cooling

Hydrogen The winding conductors are much more effectively cooled by passing the coolant through them in direct, or almost direct, contact with the copper. Hence much higher current density is possible with an acceptable temperature rise, and the output per unit volume of active material can be greatly increased. Furthermore, significant improvement in performance with hydrogen cooling is obtainable up to operating pressures of 5 or even 6 bar (abs.) (500–600 kN/m^2).

In the rotor, several flow arrangements are used:

(a) *Axial flow*. The gas enters tubular, or axially grooved, conductors in the end-winding region and leaves radially through a group of holes through the conductors and wedges at the mid-length of the rotor.
(b) *Axial flow*. This is as for (a), but for longer rotors. The middle quarter (approximately) is fed from subslots, while the two end-portions are fed as in (a).
(c) *Radial*. The gas enters each end of a subslot cut beneath the winding slot and flows radially outwards through holes punched through the flat copper strips and insulation, distributed throughout the rotor length.
(d) *Axial–radial*. This is a combination of (a) and (c) using axially grooved conductors in which the radial exit holes are displaced axially from those that feed gas from the subslots.
(e) *Gap pick-up*. Specially shaped holes in the wedges 'scoop up' gas from the gap, and others eject it after it has passed through cooling ducts formed in the copper by punched holes or transverse grooves.

Types (a) and (b) require high-pressure axial-flow blowers, which may have three to seven stages of blades; (c) and (d) rely almost wholly on the self-ventilating action of the rotor; (e) may be applied by dividing the rotor into axially adjacent inlet and outlet zones, which may be co-ordinated with the stator ventilation zones. All these schemes are used for hydrogen-cooled machines up to the largest ratings. Type (c) is now common for air-cooled machines with totally enclosed air circuits.

Directly hydrogen-cooled stator coils are used by some makers up to ratings of 600 MW or more. The gas flows down thin-walled bronze or stainless steel tubes that are bonded among the conductor strips and lightly insulated from them. Entry and exit has to be at the ends of the coils, so a moderately high pressure differential is needed. The system co-ordinates well with rotor ventilation of type (a) or (b).

Water The high heat-removal capacity of water and its low viscosity allow it to be used in tubular subconductors that are still small enough to keep the eddy-current losses low.

At each end of the conductor the subconductors (tubes and strips, or all tubes) are brazed into a water box. The coil-to-coil current-carrying connection may be tubular to carry the water also, or separate connections may be made. The water boxes are connected to the inlet and outlet manifolds by insulating pipes of polytetrafluoroethylene (PTFE) or some other synthetic material. Water taken from the boiler make-up system is circulated by pumps around a closed pipework system containing the winding, coolers, filters, control valves and monitoring instruments. Water conductivity is easily maintained by using a demineraliser unit, usually of the resin-bead type; less than 10 µS-cm is easily attained, and the leakage currents down the water columns are insignificant.

The temperature difference between copper and water is only about 2 °C, and the water temperature rise between inlet and outlet is usually about 25–30 °C. The inlet water pressure is kept below the hydrogen pressure; if leakage does occur, it is leakage of hydrogen into the water (easily detected), and water does not enter the machine.

Water-cooling of the rotor winding increases the output available from a given frame size; the slots can be smaller, leaving more room for magnetic flux, so the machine can be shorter provided the stator design is adjusted to suit. Constructional problems occur because of the need to convey water to and from the rotor, to accommodate water manifolds on the shaft, to support the water-pipes against centrifugal force, and to design them and the winding to withstand internal pressures (produced by rotation) up to around 15 MN/m^2. Nevertheless, such rotors are in successful service, though they have not yet been widely adopted.

11.16 Generator–transformer connections

For small ratings up to 3 MW, single-core cables are used. For larger ratings, solid copper or aluminium busbars have been used, supported on insulating cleats or ceramic post insulators, within a concrete duct. They were spaced and supported to suit the operating voltage and the electromagnetic forces produced by short-circuit currents, and the phases were often segregated by insulating barriers. Each phase conductor usually consisted of two angle- or channel-section bars mounted face to face to form an open diamond or box-shaped section. These gave lower skin-effect losses than flat rectangular bars of the same weight, and natural air cooling was adequate, up to at least 200 MW (around 10 kA).

To avoid the possibility of phase-to-phase faults, however, phase-isolated busbars were adopted for ratings above about 200 MW, and now above 60 MW. Each line connection consists of two angle- or channel-section bars supported by post insulators inside an aluminium tube which is physically and electrically continuous along its length. The tubes of the three phases are connected together electrically at each end of the run and are joined mechanically to the generator and transformer frames. Thus only phase-to-earth faults are possible. Eddy currents induced in the aluminium tubes cause additional losses, but they confine the magnetic field largely within the tube, so that heating of the foundation steelwork or other structures is avoided.

For the higher ratings, say above 12 kA, each conductor bar may be of semi-hexagon or semi-octagon shape, so that the complete conductor approximates to the circular cross-section that gives minimum skin effect and minimum loss.

Natural air cooling is practicable up to about 20 kA, but a significant reduction in cross-section or an increase in current rating is made possible by forced-air cooling. A maximum conductor temperature rise of 55 °C above the ambient air is usual. With this, and a continuous rating of 19.5 kA (660 MW at 23 kV, 0.85 p.f.), typical dimensions are as given in *Table 11.3*.

Table 11.3

Cooling	Natural air	Forced air	Pumped water
Conductor shape	[ ]	[ ]	○
Diameter of circumscribing circle (m)	0.9	0.5	0.13
Diameter of isolating trunking (m)	1.5	1.1	0.75
Total loss in conductor and trunking, for 3 phases (kW/m)	2.0	4.5	7.5

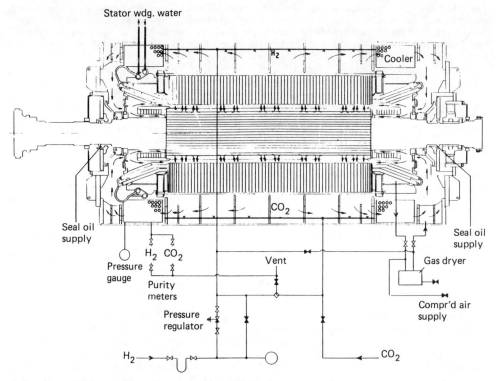

Figure 11.22 Section of a turbogenerator with simplified hydrogen-gas cooling

surface of the windings. Thus heat generated in the winding passes through the main insulation to the rotor and stator teeth respectively, and is picked up by the cooling gas mainly from the iron surfaces.

With air cooling, closed-circuit ventilation is universal except in the very smallest sizes, and coolers are separately mounted — usually in a basement beneath the generator, but occasionally above or at the side of the machine. With hydrogen cooling, however, there is no alternative but to build coolers within the gas-tight explosion-proof structure of the machine itself. *Figure 11.22* shows a simplified diagram of a hydrogen-cooled machine with its gas system.

The cooling gas is usually circulated by a fan at each end of the rotor, though many air-cooled generators (30 to 60 MW or so) had motor-driven fans mounted in the basement with the air coolers. The rotor fans may be of the centrifugal or the aerofoil (axial-flow) type. Their main purpose is to establish the gas flow through the stator frame, core and coolers; the flow through the rotor results mainly from its own rotation. Most rotors have axial ventilation slots in the teeth that are closed by wedges except near the middle of the rotor body where the flows from each end emerge into the gap and then pass through the radial ducts in the stator core to the back of the frame.

Pure hydrogen has a density approximately 1/14 that of air, while its specific heat is 14 times that of air; it has a higher heat transfer coefficient and much better thermal conductivity. In service there may be about 1% impurity consisting of air and carbon dioxide; this increases the density by about 13%, but has no significant effect on the cooling properties listed in *Table 11.2*. Windage losses are proportional to density, but even at an operating pressure of 5 bar (abs.) they are still only 40% of what they would be in air at atmospheric pressure.

Early hydrogen-cooled machines were designed to operate at just above atmospheric pressure, but raising the pressure to 2 bar (abs.) and then to 3 bar raised the output available from a given frame size by approximately 15% and then by a further 10% respectively. No worthwhile improvement occurs above 4 bar (abs.) because the temperature gradient across the winding insulation is a large part of the permissible temperature rise.

The auxiliary equipment for hydrogen-cooled generators divides into two main groups; gas control and seal-oil treatment. The gas control system maintains the pressure and purity of hydrogen in the casing within predetermined limits and provides means for filling and emptying the casing without risk of forming an explosive hydrogen–air mixture.

The shaft seals, which prevent leakage of hydrogen along the shaft to the bearings, are supplied with oil maintained at a pressure above the gas pressure. Ring seals which encircle the shaft are simple and allow free axial expansion of the shaft; however, they also allow a significant rate of oil flow towards the hydrogen side of the seal, and rather more to the air side. The gas-side flow absorbs hydrogen, and would release air or moisture into the machine if it contained these in solution. To avoid the consequent pollution, and loss, of the hydrogen, the oil is vacuum treated before being fed to the seals, and the hydrogen-side oil passes through detraining tanks to allow entrained hydrogen to return to the frame before the oil is vacuum treated and recirculated.

The face, or thrust, type of seal is a ring which operates against the radial face of a collar on the shaft. The hydrogen-side oil flow is insignificant, so vacuum treating is unnecessary. The extremely thin hydrogen-side oil film (say 60 μm) makes the seal rather vulnerable to dirt particles, and if the ring does not slide freely to follow shaft expansion it will leave the collar and allow leakage or will suffer excessive face pressure and damage to the white metal. Both types of seal are in satisfactory operation. A doubly fed ring-type seal offers the advantages of both types, provided that the pressures of the two systems are kept accurately balanced so that

from the rotor; they must be thick enough to provide adequate creepage distance to withstand the specified h.v. tests.

The end-windings are packed, partly or wholly, with blocks of insulating material to avoid distortion and the consequent risk of short circuits between turns.

11.15.4 Stator core

The stator core is built up of segments of electrical sheet steel, usually 2–3% silicon, 0.35 mm thick, cold rolled and non-oriented. To minimise weight, the core is worked at the highest flux density consistent with resonable losses. In a 2-pole machine the magnetic force across the airgap subjects the core to an elliptical distortion that rotates with the rotor, so producing a double-frequency ($2f$) vibration. The core depth must be chosen so that its natural frequency of vibration in this elliptical mode is well away from $2f$; usually $3f$ or more is practicable without excessive depth of core. Grain-oriented steel has better permeability and lower losses than non-oriented steel, but as it has a lower modulus of elasticity its other advantages cannot be realised without accepting higher vibration. This, and its higher cost, severely limit its use in turbogenerators.

11.15.5 Stator casing

Air-cooled machines may have bearing pedestals on a bedplate; the stator frame then merely supports the core and forms the ventilation enclosure. Alternatively, it may be a more rigid box frame with the rotor bearings carried in end-brackets.

A hydrogen-cooled machine must have a totally enclosed and gas-tight construction; the end-bracket bearing arrangement is adopted to minimise the bearing span and to raise the critical speeds. Hence the frame must be rigid enough to provide proper support for the bearings and to contain the gas pressure that might occur in the unlikely event of a hydrogen–air explosion inside the frame. This could produce pressures up to about 1400 kN/m^2; therefore this pressure, rather than the continuous working pressure of hydrogen, becomes the design criterion.

In large 2-pole machines (which are invariably hydrogen cooled) some form of flexible mounting is needed between the core and the casing, and the casing should not have any natural frequency near to $2f$. This is to avoid excessive magnetic noise and the risk of unacceptable vibration on the casing, coolers or pipework.

Where transport facilities are inadequate for handling the complete stator, the core and windings must be made separately from the outer casing, separately transported, and assembled on site before the rotor is inserted. By contrast, smaller machines can be transported complete to some sites, with the rotor clamped in temporary supports; this arrangement facilitates erection.

11.15.6 Stator winding

For small machines the voltage is usually fixed at a standard network voltage (e.g. 6.6 or 11 kV), but for large machines, where a generator–transformer is used, the designer has a free choice. A high voltage avoids difficulties due to high currents, but valuable space in the slots has to be sacrificed to insulation; a compromise is thus about 15 kV for 100 MW and 200 MW machines, and up to 22 or 25 kV for the larger sets. Even so, generators of more than about 50 MW rating will have two circuits in parallel per phase; for more than 1000 MW it may be necessary to use special winding arrangements to have four parallel paths in a 2-pole machine. In 4-pole generators, four circuits occur naturally and can be in parallel or in series–parallel.

In the slots, the conductors are subjected to pulsating forces by the interaction between the current and the slot leakage flux, so the slot wedges and packings must be designed to prevent looseness developing.

The winding is usually of the two-layer basket type, the coils spanning about 5/6 of the pole-pitch. Each conductor is made up of small subconductors, each lightly insulated with, for example, glass braid and resin, and transposed usually by the Roebel method to limit eddy currents to an acceptable level. For 2-pole machines of 400 MW or more each conductor will usually have four, not two, stacks of subconductors.

The main insulation is of mica paper, or a mixture of this and mica flakes, bonded with epoxy resin. It is cured by heat and pressure to form a solid thermosetting covering that has a thickness appropriate to the voltage and is as far as possible free from voids, with good mechanical and dielectric strength. It is most important that the end-windings are well supported against two effects: continuous though fairly low vibratory forces when on load, and the very much larger transient forces that occur if a short circuit occurs on the system, the transformer or the generator connections. The end-windings are secured to a strong structure of insulating materials (resin-bonded wood laminate, fibre-glass rings and conformable resin-impregnated packings are commonly used), all bolted securely to the end of the stator core structure. Some designs, however, permit the end-winding structure to move axially to accommodate differential expansion between the coils and the core.

11.15.7 Cooling

Efficiencies between 98% and 99% are achieved; however, the losses will be 5–10 MW, appearing as heat that must be removed by circulation of an appropriate cooling medium: oil for removing bearing-friction losses, and air, hydrogen or water for other losses. Details of the cooling media used for stator and rotor windings are given in *Table 11.2*. The heat transfer coefficients are typical, but depend considerably on velocity and duct size.

11.15.7.1 Indirect cooling

The cooling medium may be air for machines up to 30 or even 60 MW, and hydrogen for larger machines, the coolant being blown along the airgap, through ducts in the core, and over the

Table 11.2 Properties of cooling media

Medium	Absolute pressure (bar)	Specific heat (kJ/(kg K))	Density (kg/m^3)	Relative volume flow	Relative thermal capacity	Relative heat transfer coefficient
Air	1.0	1.0	1.1	1.0	1.0	1.0
Hydrogen	1.0	14.3	0.076	1.0	0.99	1.45
	2.0	14.3	0.152	1.0	1.98	2.52
	4.0	14.3	0.304	1.0	3.96	4.38
Water	1.0	4.2	1000	0.01	38	60

to offset its higher capital cost compared with a high-speed unit. Design constraints are less exacting in the low-speed turbine and generator.

11.15.1 Main dimensions

The output coefficient ranges typically from about 0.6 MVA-s/m^3 for a rating of 20 MVA to about 2.0 MVA-s/m^3 for a 1000 MVA unit. For 3000 rev/min machines these figures correspond to D^2L of 0.84 and 11.5 m^3 respectively. Economic diameters for these outputs range from approximately 0.75 m to 1.3 m, the latter being a limit set by centrifugal stresses in the rotor teeth and end-rings. Hence outputs range from approximately 17 to about 170 MVA per metre of core length. The higher values of output coefficient are made possible by enhanced cooling techniques. Typical dimensions for a 660 MW 2-pole hydrogen-cooled machine may be: rotor diameter, 1.15 m; core length, 6.8 m; overall shaft length, 13.5 m; core outside diameter, 2.7 m; outer casing, 4.8 m in diameter and 10.3 m long; total weight, 480 tonne.

11.15.2 Rotor body

The output available from a turbogenerator is largely determined by the excitation m.m.f. that can be carried on the rotor with acceptable winding temperatures. The high centrifugal stresses make cylindrical (i.e. non-salient-pole) construction essential. Within the chosen diameter the number, shape, size and spacing of the winding slots have to be optimised to obtain the maximum m.m.f. capability with acceptable stresses in the teeth and slot wedges, with adequate insulation, with acceptable magnetic flux densities and with ducts for ventilation that enable temperature guarantees to be met. For air-cooled machines of medium output the manufacturing simplicity afforded by parallel-sided slots and solid copper conductors of rectangular cross-section may outweigh the loss of optimum performance and provide the cheapest design. For larger ratings tapered slots are used to accommodate more copper, while giving approximately constant mechanical stress and magnetic flux density along the radial length of the teeth.

A rotor is forged from a single steel ingot, the largest of which approach 500 tonne in weight; this would produce a rotor weighing 250 tonnes; enough for a 4-pole machine of about 1250 MW at 1500 rev/min. The forgings contain the alloying elements nickel, chromium, molybdenum and vanadium; according to size and speed, ultimate tensile strengths of the forgings range from 650 to 800 MN/m^2, while their 0.2% proof stresses range from 550 to 700 MN/m^2. The forgings are inspected with ultrasonics and magnetic-particle ink before use. Many generator makers now rely on these examinations and do not bore the forging axially along its centre line, except for large forgings or if ultrasonics reveals defects that can be removed by boring.

The end-windings (i.e. the parts projecting beyond the ends of the slots) must be supported against centrifugal forces by end-rings (retaining rings). These are steel forgings, usually shrunk on to the ends of the rotor body and insulated from the rotor end-windings. Usually they are non-magnetic austenitic steel (18% manganese, 4–5% chromium, 0.3% carbon) warm worked to give high strength, up to 1100 MN/m^2 proof stress and 1220 MN/m^2 ultimate for the largest rotors. A recently developed alloy contains 18% manganese and 18% chromium, and has much better resistance to stress corrosion.

Rotor vibration at running speed must be low—typically about 50 μm peak-to-peak measured on the shaft near the bearings, though up to twice this is commercially acceptable. Hence balance weights must be carefully positioned, axially as well as circumferentially, and the design of the rotor, its bearings and its supports must ensure that its critical speeds are sufficiently far from rated speed. Small 3000 rev/min rotors will have one critical speed below 3000 rev/min, say about 1700–2000 rev/min, but as ratings (and therefore the bearing span) increase, two or even three criticals will occur below 3000 rev/min (typically around 650, 1750 and 2500 rev/min). When the rotor is coupled to the turbine, the critical speeds are usually raised slightly, so that behaviour in both the coupled and the uncoupled condition must be acceptable, for site running and works testing (without the turbine) respectively.

Electrical faults on the power system or on the machine itself produce abnormally high oscillatory torques on the rotor, at system frequency and often at twice this frequency also. Complex torsional oscillations develop in the shaft system, with components determined by the inertias and stiffnesses of the several shafts of the turbine, generator and exciter. The shaft dimensions must be chosen to avoid a serious loss of fatigue life during such incidents as well as to satisfy the critical speed criteria mentioned above.

Bearings are of the white-metalled cylindrical type, with forced oil lubrication and, except on small sets, high-pressure oil jacking also. Jacking allows the set to run slowly (typically 3–20 rev/min) on the turning gear to cool off the turbine and generator rotors before the unit is finally stopped. Without this, the rotors would bend because temperature gradients would occur across the diameter of each rotor, and vibration would occur on the next run-up.

All turbogenerator rotors bend under their own weight, and with 2-pole rotors the amount of bend may be more when the pole axis is horizontal than when it is vertical. Hence a vibration occurs at a frequency corresponding to twice the running speed; it is caused by the changing stiffness of the rotor in the vertical plane, and therefore cannot be removed by mechanical balancing. The stiffness about the polar (direct) axis and the slot (quadrature) axis must be made as nearly equal as possible under running conditions (when centrifugal force on the windings increases the stiffness in the plane of the quadrature axis). This is done either by cutting axial slots along the pole areas (and filling them with magnetic steel if necessary to avoid magnetic saturation) or by cutting narrow arcuate grooves circumferentially across the poles, sufficient grooves being spaced down the length of the rotor to reduce the stiffness to match that in the quadrature plane.

11.15.3 Rotor winding

Excitation currents range from say 400 A at 200 V for a 30 MW generator to 5.7 kA at 640 V for a 1000 MW 2-pole machine. For rotors using indirect cooling each coil is wound with a continuous length of copper strap, bent on edge at the four corners. The copper contains about 0.1% of silver and is hard drawn to increase its strength and so avoid the coil-shortening effect that occurred with plain soft copper as a result of heating while part of the copper was prevented (by centrifugal force) from expanding axially.

Directly cooled coils are usually made of larger-section conductors of silver-bearing copper containing grooves and holes to provide gas passages. Half-turns are brazed together in the end regions after they have been positioned in the rotor slots. The slots may be parallel-sided, or tapered to contain more copper without increased tooth stress.

Insulation between turns is usually provided by interleaves of resin-bonded glass fabric or some other synthetic material. The coils are insulated from the rotor body by U- or L-shaped troughs of resin-bonded fibre-glass, nomex or melamine, or combinations of such materials. Similar insulating strips insulate the top conductor from the slot wedge; in direct-cooled rotors the strips must have through-holes to allow the cooling gas to escape

speeds below about 95% of normal. This 'constant volts-per-cycle' control is needed also for generators that have to operate over a speed range, e.g. for ship propulsion.

11.14.8.4 System stability controls

To achieve high accuracy of voltage control and rapid response, the excitation system should have a high gain (i.e. change of generator excitation voltage divided by change of error voltage). Too high a gain inherently causes oscillation within the AVR circuits; however, this can be avoided by suitable stabilising feedback or by incorporating phase-advance circuits in the AVR forward path. More importantly, a high gain can encourage dynamic instability of the machine in relation to the power system, causing unacceptable fluctuations of voltage, power and load angle, and eventually loss of synchronism. In some system situations it is necessary to make the excitation responsive to signals such as power output, rate of change of power, load angle or angular acceleration, in order to maintain stability. Such conditions will arise if a generator (or a group) has to operate underexcited, especially if it feeds through a relatively high system reactance.

11.14.9 Overall voltage response

The overall voltage response is defined in terms of steady-state and transient behaviour with the generator on open circuit and on no load, and under the control of the excitation system. For a generator operating *alone*, the following conditions may be relevant, their importance depending on the duty required of the generator:

1. *Steady state:* accuracy of voltage control over the range of load and power factor.
2. *Transient:* 1. Response of the open-circuit generator voltage to a step change in reference voltage; 2. voltage response when a sudden increase or decrease of load occurs.

Conditions of importance in 2.2 are the voltage rise and recovery time of the generator voltage when load is suddenly removed, and the voltage dip and recovery time when a large motor is switched on to the generator terminals. For conditions 1 and 2.2, various 'grades' of voltage regulation are specified in BS 4999 (Pt. 40, Characteristics of Synchronous Machines). Accuracy of voltage control may conform to $\pm 1\%$ or $\pm 2.5\%$ or $\pm 5\%$. Voltage dip must not exceed 0.15 p.u. when a current of 0.35 or 0.6 or 1.0 p.u. of rated generator current is suddenly demanded. The voltage is required to recover to 0.94 or 0.97 p.u. within 1.5 or 1.0 or 0.5 s. Values of the temporary voltage rise that occurs when rated load at p.f. 0.8 is thrown off are specified with a range 0.35–0.15 p.u. Appendix 40 of BS 4999 shows that, the more severe the conditions, the lower must be the generator subtransient reactance to limit the voltage dip or rise. Values of X_d'' of 0.25–0.12 p.u. are indicated. There is a consequential increase in the short-circuit current and in the generator frame size; both these increases raise the cost of the generator and, perhaps, of its switchgear.

The response defined in 2.1 above, though not directly applicable to service conditions, is a convenient way of expressing the overall performance and of testing it during commissioning. The terms used are

V_1	initial voltage
V_2	final voltage
$V_1 - V_2$	voltage step
v	voltage overshoot beyond V_2
t_1	rise time, that in which V_2 is first reached (and passed)
t_2	settling time, the shortest time after which v does not differ from V_2 by more than say 0.5% of rated voltage

Normally, either V_1 or V_2 is the rated voltage V_r, depending on whether a 'step-up' or a 'step-down' change is being tested. The voltage–time curve is a well-damped transient, settling at V_2 after a few oscillations. Typical values of the quantities defined above are

$V_2 - V_1$	0.1 p.u. of V_r	v	not more than $(V_2 - V_1)/2$
t_1	0.2–0.6 s	t_2	1–5 s

For a given step change $(V_2 - V_1)$, t_1 is reduced by increasing the ceiling voltage V_C, by increasing the excitation system gain, and/or by reducing the system time-constants. The parameters of the generator and exciter cannot be changed once the machines are made, but the AVR parameters are designed to be adjustable. Changes that reduce t_1 will increase the overshoot v and the settling time t_2; hence settings of AVR gain, time-constants and feedback signals (if any, e.g. exciter voltage or exciter field current) are calculated to achieve the desired compromise, and performance is checked over a range of values of step change during commissioning (say steps of 1%, 5%, 10% and 20% of V_r).

When generators operate in parallel, as most do, the excitation systems must be designed and adjusted to achieve the best compromise between highly accurate voltage control, steady-state stability, and transient stability following a system disturbance. Thus some step-change tests, or tests by injecting low-frequency sinusoidal voltage into the reference circuit, are desirable, to confirm the calculated performance.

11.14.9.1 AVR fault protection

Failure of AVR components, or of other components in the excitation system, may cause excessive or insufficient excitation for safe operation. Either condition trips the AVR to manual control and alerts the operator.

Voltage-transformer fuse failure is detected by comparing the voltages from two voltage transformers. If the AVR voltage transformer fails, the system trips to manual control; if the comparison circuit fails, an alarm shows that fuse failure protection is no longer working.

11.14.9.2 Double-channel AVR

To enhance reliability of operation, large or vital generators frequently use regulators in which the automatic and manual control circuits are duplicated; often the thyristor output stage supplying the exciter field is duplicated too. Occasionally the much larger thyristor bridge that feeds the generator field directly may be duplicated. Each channel is able to perform the full excitation duty; the two channels may operate in parallel or in main and stand-by mode. If one channel fails, the other maintains the excitation unchanged. If the second channel fails subsequently, it trips to manual control. Alarms indicate the abnormal conditions.

11.15 Turbogenerators

The characteristic features of turbogenerators are their high speed to meet steam-turbine requirements and their large outputs to provide economy of capital and operating costs for the power station. Most are 2-pole units running at 3000 or 3600 rev/min, but 4-pole generators at 1500 or 1800 rev/min have become common for large ouptuts (800 MW or more) from nuclear reactors of the boiling-water or pressurised-water type. These reactors deliver large volumes of steam at temperatures and pressures that are lower than those provided by fossil-fired boilers or some gas-cooled reactors. The low-speed turbine may handle these conditions with a greater efficiency that is sufficient

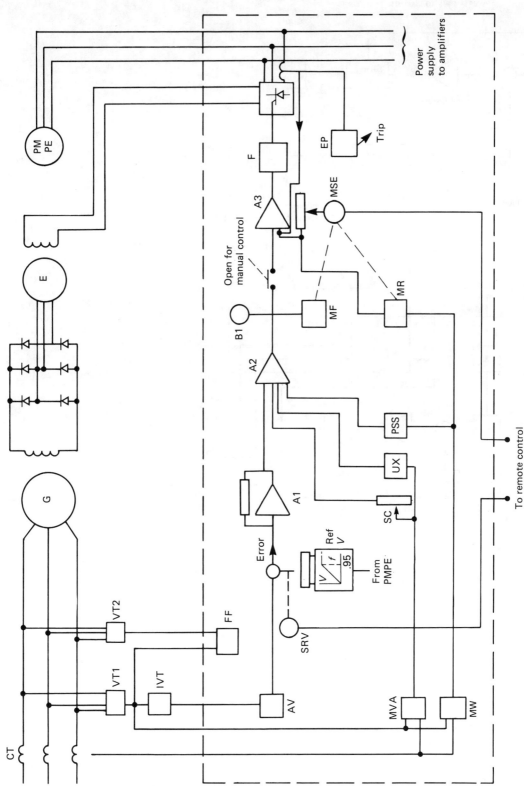

Figure 11.21 Automatic voltage regulator with additional control features: A1, A2, A3, amplifiers; AV, average-voltage circuit; BI, balance indicator; CT, current transformers; E, exciter; EP, excitation protection circuit (operates on high or low excitation current); F, thyristor firing gear; FF, voltage-transformer fuse failure detector and alarm; G, generator; IVT, isolating voltage transformer; MF, manual follow-up circuit; MR, manual restrictive circuit; MSE, manual set-excitation control; MVA, MW, circuits providing signals proportional to MVA and MW; PMPE, permanent-magnet pilot exciter; PSS, power-system stabiliser control; Ref V, reference voltage; SC, set-current compounding control; SRV, set-reference voltage control; UX, underexcitation (Var limit)

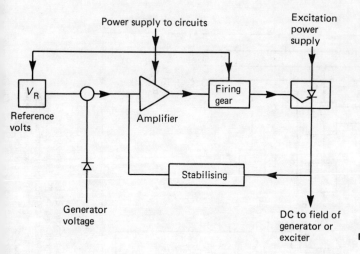

Figure 11.19 Basic circuit of an AVR

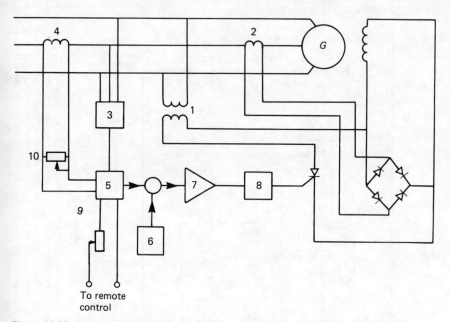

Figure 11.20 Self-excitation with a 1-phase thyristor: 1, excitation transformer; 2, excitation current transformer; 3, voltage transformer; 4, compounding current transformer; 5, voltage-measuring circuit; 6, reference voltage; 7, amplifier; 8, firing gear; 9, voltage-setting rheostat; 10, compounding adjustment

11.14.8 Additional control features

11.14.8.1 Parallel operation

To ensure satisfactory sharing of reactive load between generators paralleled at their terminals, the AVR can arrange for the terminal voltage to fall with increasing reactive load, usually by 2.5–4.0% at full load. For generators paralleled on the h.v. side of step-up transformers, the AVR can either add to or partly compensate for the transformer impedance drop, as desired.

11.14.8.2 Excitation limits

Fault conditions on the power system will cause the excitation to rise to ceiling value to try to maintain normal voltage. An adjustable timer is used to return to normal excitation after several seconds in order to avoid overheating the machines if the fault persists.

A reactive-power-limiting circuit can be used to prevent the excitation falling so low that the generator will not remain in step with the system. The reactive power (underexcited) at which this circuit operates is automatically varied in response to machine voltage and power output to maintain an adequate stability margin.

11.14.8.3 Overfluxing protection

It is operationally desirable to be able to leave the AVR in control when the generator is shut down or run up. To avoid overfluxing the machine and its associated transformer (if any) the reference voltage is arranged to decrease in proportion to frequency at

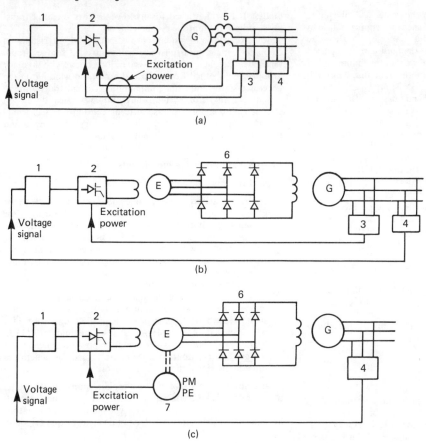

Figure 11.18 Excitation systems: (a) direct self-excitation; (b) self-excitation through an exciter; (c) separate excitation. 1, control circuits; 2, power-output stage (AVR); 3, excitation power transformer; 4, voltage transformer; 5, current transformers for excitation power; 6, diode rectifier (static or rotating); 7, permanent-magnet pilot exciter

to the desired level and the error voltage returns to near zero. The set level is obtained by adjusting the proportion of the machine voltage that is compared with the reference voltage or by adjusting the reference voltage itself. The basic circuit (*Figure 11.19*) is incorporated in *Figures 11.20* and *11.21*.

11.14.7.1 Control range

Generators are usually designed to deliver any load from zero to rated output over a voltage range of $\pm 5\%$, at any power factor between rated (usually 0.8–0.9 lag) and say 0.95–0.9 lead. The AVR setting controls must provide the corresponding range of excitation, and also provide say 85% of rated voltage on no load. Accuracy of control is usually within ± 2.5 or $\pm 1.0\%$ of the set value over the load range.

11.14.7.2 Manual control

Manual control is usually provided for use if the automatic control fails or for convenience when the set is being commissioned. In small units the manually controlled system may be entirely independent of the automatic, but (especially for economy on large units) it uses the thyristor output stage and the associated firing circuits of the regulator.

Some regulators, when in auto control, drive the manual control so that, if it were in use, it would give the same excitation as the auto circuit. However, this follow-up must be prevented from driving the manual control down to excitation levels that are stable with continuously acting control but unstable with fixed excitation.

Some systems use the manual circuits continuously to control the steady-state excitation. The auto circuits continuously trim this to suit minor fluctuations of load, voltage etc.; larger disturbances will cause rapid automatic changes of excitation voltage, up to full boost or full buck if necessary. If changed conditions persist for more than a few seconds, the follow-up circuit adjusts the manual control to the new steady state and the auto-circuit output falls to its usual low level.

11.14.7.3 Manual-to-auto change-over

Whichever system is in control, it is necessary to adjust the other, automatically or manually, so that a change-over can be made without causing a significant change in excitation. A balance meter is provided so that the outputs of the two systems can be matched before making the change. The manual rheostat and the voltage-setting rheostat are often motorised for control from a remote control room, and the balance-meter reading must be repeated there too, unless automatic matching is provided.

arm, failure of one diode or fuse leaves the exciter with one phase unloaded; exciters are usually designed to supply full-load excitation in this condition without damage, so that the generator can remain in service until the fault can be repaired conveniently. However, experience shows that the failure rate of diodes is extremely low and that more often fuse links fail mechanically. Hence some makers supply salient-pole generators, up to say 25 MW, with no fuses at all, but use generously rated diodes to provide a large margin. These generators would use up to three diodes in parallel per bridge arm. If one diode fails, the short-circuit current in the exciter armature induces a.c. into its field winding, and this is used to operate a protective relay to trip the unit. For turbogenerators up to about 70 MW, some designs use two diodes in series—each of full duty, with one, two or more series pairs in parallel per bridge arm—and no fuses. On large units, redundant parallel paths, individually fused, are provided as in static equipments.

For units that use fuses, the striker-pin indicator type can still be used, the pin being observed by causing it to interrupt a light beam falling on to a photoelectric cell, or it can be observed visually with a stroboscope. Alternatively a neon lamp is connected across the fuse and glows when the fuse blows.

More elaborate indication, perhaps coupled with measurements of current and voltage and indication of earth fault, can be arranged by telemetry, but the telemetry may be less reliable than the diodes. Frequently instrument slip-rings are used, with solenoid-operated brushes that make contact only when readings are required.

11.14.4 Thyristor excitation

Direct control of the field current of the synchronous machine by thyristors gives quicker response than can be obtained by controlling the exciter field current, because the time delay in the exciter is eliminated and the machine field current can be forced down by using the thyristors to reverse the machine field voltage. (By contrast, with a diode bridge, the machine field voltage can only be reduced to zero by reversing the exciter field voltage.) This is valuable for generators and synchronous compensators in certain power-system situations: for example, to minimise the voltage dip caused by large and possibly frequent load changes; to maintain transient stability of a generator under short-circuit conditions on the power system; to enable a synchronous compensator to maintain close control of the system voltage by rapid change in its reactive load; to minimise the voltage rise following sudden load rejection; to reduce more quickly the current resulting from a fault between the generator and its nearest protective circuit-breaker when field suppression is the only means available.

The synchronous machine requires slip-rings and brushes, and this is a disadvantage, especially for large machines for which brushgear maintenance may become a significant inconvenience.

The excitation power may be supplied by direct-coupled main and pilot exciters, the main exciter being designed to work continuously at ceiling voltage. Alternatively, the excitation power may be supplied via a step-down transformer from the machine terminals (or, unusually, from another connection to the power system). To maintain adequate excitation when the system and machine voltages are depressed by a fault on the h.v. side, the transformer must provide a high ceiling voltage under normal conditions, perhaps up to twice that needed for full-load output. A lower ceiling is adequate if power-rated current transformers are added in order to derive some excitation from the machine output current, so boosting excitation during the fault. The set is shortened by the absence of the exciter, and this may save costs on foundations and building. For very-low-speed generators the scheme may well be cheaper than a direct-coupled exciter and diodes. It is widely used for hydrogenerators on long transmission lines to improve system stability.

Some excitation systems use diodes and thyristors in combination, e.g. in a full-wave half-controlled bridge circuit. One patented scheme uses a full-wave diode bridge with thyristor 'trimmer' control fed from a special excitation winding on the generator stator and from compounding current transformers.

Rotating thyristor systems have not yet been developed commercially, mainly because of technical difficulties in transferring control signals from the stationary equipment and problems concerning the reliability of rotating control circuitry.

11.14.5 Excitation system circuits

Typical systems are shown in *Figure 11.18*.

(a) *Self-excitation* provides a simple and inexpensive scheme for generators up to about 3 MVA, using a 1-phase thyristor output stage. With a 3-phase thyristor bridge the scheme is applied for the highest ratings. The bridge rectifier may be half-controlled, with thyristors and diodes in combination. Another variant has a diode bridge that provides more exciting current than is demanded and thyristors to divert part of this current from the field winding.

(b) *Self-excitation through an exciter* is convenient for brushless sets where the diodes are mounted on the generator–exciter shaft, and where the cost or mechanical complication of a pilot exciter is undesirable. A typical rating limit is 10 MVA.

(c) *Separate excitation* provides excitation power independent of the generator output. It is commonly used for generators rated at 10 MVA up to the maximum.

Scheme (a) is capable of the most rapid response. In (b) and (c) some delay is introduced by the exciter time-constant; consequently a high exciter ceiling voltage and a large output from the pilot exciter are needed to obtain a more rapid response.

11.14.6 Excitation control

In normal operating conditions, control of the excitation maintains the terminal voltage required and controls the generation of reactive power. Under conditions of fault, the control must as far as possible maintain the voltage and keep the generator stable on the power network. Manual control of a field rheostat in the main or exciter field circuits is inadequate: automatic control is essential.

Electromechanical voltage regulators, in use only in old installations, may be of the carbon-pile, vibratory-contact (Tirrill) or rolling-sector (Brown Boveri) type. These have been superseded, initially by magnetic amplifiers, with or without amplidynes, and now by solid-state control systems using transistor amplifiers at low power levels, with thyristors for field-circuit power. The new systems are continuously acting (i.e. have no dead band) and can be arranged to respond to many control signals besides that from the terminal voltage, so the term 'automatic excitation controller' is more logical than 'automatic voltage regulator' (AVR). Power supply for the control circuits is derived from the machine terminals or the pilot exciter.

11.14.7 Basic principles of voltage control

A direct voltage proportional to the generator average terminal voltage is derived via voltage transformers and a diode rectifier circuit. This voltage is compared with a stable reference voltage generated within the regulator. Any difference (the 'error voltage') is amplified and used to control the firing of a thyristor circuit which supplies the excitation, either to the field of the synchronous machine or to its main exciter field winding. Thus the excitation is raised or lowered to restore the machine voltage

11.14 Excitation systems

A synchronous machine requires an excitation system to provide the field current for magnetising the machine to the desired voltage and, when it is running in parallel with others, determining the lagging reactive power generated or received. It is customary for each generator to have its own self-contained excitation system, which provides the power required to supply the I^2R loss in the field circuit. This varies between about 10 kW/MVA for small machines and 4 kW/MVA for very large units.

Excitation voltage and currents are chosen (i) to give field winding conductors that are mechanically robust in small machines and not too massive in large ones, (ii) to suit the ratings of available diodes or thyristors, and (iii) to give convenient designs of exciter, and also of slip-rings where these are used. Values range from a few score amperes and volts on very small machines up to say 6 kA at 600 V on the largest turbogenerators. At no load or with leading power factor, control of the exciting current is needed down to about one-third of the value for rated load.

The excitation system must respond to applied control signals quickly enough to provide the desired control of generator voltage. For example, the control should hold the generator voltage to within 1% or less of the set voltage over long periods, despite load changes, should minimise short-term voltage fluctuations caused by sudden changes of load or by power system disturbances, and should restore the set voltage quickly after such sudden changes. In order to change the field current quickly enough to do this, the excitation system must have a ceiling voltage sufficiently higher than the full-load field voltage.

Exciter performance is usually judged by the response rate or response ratio. Starting with the open-circuit exciter voltage equal to V_f, the generator field voltage at rated load with the field winding at a mean temperature of 75 °C, the excitation control is suddenly changed to cause the exciter to reach ceiling voltage. The mean rate of change of exciter voltage over the first 0.5 s is the response rate in volts per second. This divided by V_f is the response ratio.

A response ratio of 0.5 is adequate for many situations, but 2 or more may be needed if, for example, a generator has to start a large motor with a small voltage dip and quick recovery. Hence the ceiling voltage is usually in the range 1.5–3 p.u. (i.e. 1.5–3 times the full-load field voltage). The exciter response ratio is determined by (a) the time-constant of its field circuit, (b) the change of voltage applied to it by the pilot exciter, and (c) the output voltage at which the exciter saturates. It may be possible, and cheaper, to attain a desired exciter response ratio by using a high-output pilot exciter rather than a very large exciter with a higher saturation voltage.

The same response definition is often used for a thyristor system but, as ceiling voltage can be reached in a few milliseconds, a ceiling voltage of 1.5 p.u. gives a response ratio of about 2.

Any excitation system which reaches 95% of ceiling voltage in 0.1 s or less, starting from V_f on open circuit, is called a high-initial-response system, according to the *IEEE Standard Criteria and Definitions for Excitation Systems for Synchronous Machines, Standard no. 421.*

11.14.1 D.C. exciters

Until diode rectifiers were adopted in the United Kingdom (in about 1960), the exciter was a d.c. generator coupled to the shaft of the synchronous machine, feeding its output to the main field through slip-rings. For high-speed generators of more than about 50 MW, the exciter had to be driven at a lower speed (typically 1000 or 750 rev/min) through gears, or separately by a motor, in order to avoid difficulties of construction and of commutation. On very-low-speed hydrogenerators a directly coupled exciter would be excessively large, so a higher-speed exciter, driven by a motor or perhaps a small water turbine, was used.

For small ratings, the exciter was shunt excited; however, most were separately excited from a directly coupled shunt-excited pilot exciter. Control of the generator excitation was provided by controlling the field current of the main exciter.

11.14.2 A.C. exciters with static rectifiers

The advent of solid-state rectifiers made it possible to avoid commutators by using an a.c. exciter, directly coupled to the generator and feeding its output via floor-mounted rectifiers to the generator field winding through slip-rings. The exciter can operate at any economically convenient frequency, usually between 50 and 250 Hz, and the system is suitable for generators up to the largest ratings.

The diode cubicles may be cooled by natural convection or forced air flow. Alternatively, especially for large ratings, the diodes may be mounted on water-cooled busbars; this greatly reduces the size of the cubicles so that they can, if desired, be mounted on the sides of the main exciter frame, thus avoiding the need for long runs of a.c. and d.c. busbars or cables.

The main exciter field is supplied by an a.c. pilot exciter, often a permanent-magnet generator. The excitation of the main generator is controlled by controlling the main exciter field current via the automatic voltage regulator.

The main exciter is usually 3-phase, and the diodes are connected in the 6-arm bridge circuit, usually with a fuse in series with each diode to interrupt the fault current should a diode break down (which almost always causes it to conduct in both directions, i.e. to act as a short circuit). Diodes are available with current ratings up to 400 A (mean d.c.) and 5 kV peak inverse voltage. For other than small ratings, each bridge arm has several diodes in parallel; all the diodes are fused and have sufficient current margin to enable the bridge to carry full-load excitation continuously with one or more of the diodes failed and isolated by their fuses. Hence the generator can remain in service until maintenance can be carried out conveniently. Diodes and fuses are reliable; however, some installations continue the practice that was originally used on large generators of fitting a.c. and d.c. isolators to allow parts of the bridge to be worked on without the generator being taken out of service.

The diodes must be able to withstand induced transient currents and voltages resulting from system short circuits, asynchronous running, pole-slip and faulty synchronising, as well as from faults in the excitation system itself. Their continuous duty rating must leave some margin for imperfect sharing between parallel paths and for the possible loss of one or more paths, as noted above.

The fuse characteristic is co-ordinated with that of the diode, so that fuses should not blow unless a diode fails or there is a short circuit on the d.c. output. The fuses must clear the fault current under the most onerous condition, which is usually that of a failure with the exciter at ceiling voltage. Fuse blowing is easily indicated by a microswitch operated by a striker pin that is ejected from an indicator fuse in parallel with the main fuse.

11.14.3 Brushless excitation

The a.c. exciter has a rotating armature and stationary field system; the diodes and fuses are mounted on the rotor, and the rectified output is led directly to the generator field winding without need of brushes and slip-rings. The diodes are mounted on well-ventilated heat sinks, and special designs of fuse are used to withstand the centrifugal force on the fusible link.

On units small enough to require only one diode and fuse per

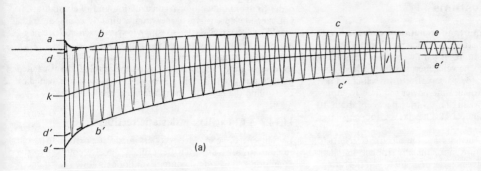
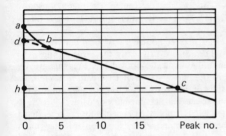
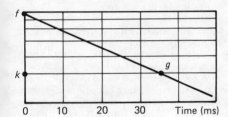

Figure 11.17 Analysis of a short-circuit current oscillogram: (a) current envelope; (b) logarithmic plot

the envelopes at each current peak and subtract ee' from each. Plot the results as ordinates on a logarithmic scale, to a linear base of time. This gives the curve abc in *Figure 11.17(b)*.

(6) Project cb by a straight line to d at $t=0$. Then

[Od (cm)/$(2\sqrt{2})$] × current scale of oscillogram + $I_d = I'_d$

whence

$$X'_d = E_o/I'_d$$

(7) The subtransient r.m.s. current at $t=0$ is $I''_d = I'_d +$ the rapidly decaying component represented by da. Thus

$$I''_d = I'_d + \frac{\text{intercept da}}{2\sqrt{2}} \text{ (cm)} \times \text{current scale}$$

and

$$X''_d = E_o/I''_d$$

The intercepts between ab and db are drawn to extended current and time scales in the lower part of *Figure 11.17(b)*. Point f on the ordinate scale corresponds to da.

(8) Time-constants are obtained from the slopes of the current–time plots. With reference to the line dc, the current I_t at time t is related to I'_d at $t=0$ by

$$I_t - I_d = (I'_d - I_d)\exp(-t/T'_d)$$

Consequently

$$T'_d = \frac{t}{\ln(I'_d - I_d) - \ln(I_t - I_d)} = \frac{hc}{\ln d - \ln h}$$

where hc is in seconds and d and h are in centimetres. T''_d is obtained similarly from the extended-scale plot in *Figure 11.17(b)*. The $T''_d = kg/(\ln f - \ln k)$.

(9) The procedure 5–8 is repeated for the other two phases and the mean values are obtained.

(10) The reactance values decrease with increasing short-circuit current (and therefore with increasing open-circuit voltage E_o). Rated-current values are obtained when at $t=0$ the transient current I'_d is equal to rated current. A range of short-circuit tests spanning the expected X'_d will give a plot of reactance to a base of transient current I'_d from which the appropriate value can be found. A rated-voltage value, if required, is obtained by testing at 1 p.u. voltage for a small machine without a transformer, but the electromechanical forces on the stator end-windings would be excessive in a large generator. Tests up to 0.7 p.u. voltage simulate a fault on the h.v. side of the transformer in a generator–transformer unit and are more relevant to the service conditions of a large machine.

(11) The armature short-circuit time-constant T_a is determined from the decay of the d.c. components of stator current or from the decay of the a.c. component of induced field current. The latter method is simple: it requires only a log–linear plot of the a.c. component to a base of time, similar to the plot in *Figure 11.17(b)*. The stator d.c. component is represented by the median line kl of the current envelopes in *Figure 11.17(a)*. However, if there are significant even-order harmonics then the median is displaced and a wave-form analysis is necessary to find the harmonic effect.

11.13 Sudden 3-phase short circuit

If a 3-phase generator is initially excited to a phase e.m.f. E_o on open circuit, and then the three phases are suddenly short-circuited together, the stator winding carries balanced 3-phase currents of up to several times full-load value, depending on the magnitude of E_o. These currents produce an m.m.f. that rotates synchronously with the rotor, with its axis along the main pole axis, tending to reduce the mutual flux from its initial value. Change of flux linkage in a closed circuit induces therein a current opposing the change. Hence large direct currents are induced in the rotor damper circuits and field winding.

The combined effect of the large and opposing stator and rotor currents is to produce large leakage fluxes round the stator winding, the damper circuits and the field winding, while the mutual flux along the main flux axis decreases correspondingly, so that the *total* flux linkage with each winding remains momentarily unchanged. I^2R losses rapidly dissipate stored magnetic energy and the damper currents rapidly decay. Typically, the induced field current reaches its peak a period or two after the short-circuit instant and then decays relatively slowly (*Figure 11.16*).

In the stator, each phase current is asymmetric to an extent depending on how near the phase voltage was to zero at the instant of short circuit; zero instantaneous voltage produces full asymmetry. Thus each phase carries a d.c. component which, at the instant of short circuit, is equal and opposite to the instantaneous a.c. component. These d.c. components produce a stationary m.m.f. sufficient to hold the stator flux linkage momentarily unchanged, i.e. fixed relative to the stator in the position the stator flux linkage occupied at the short-circuit instant. The d.c. rapidly decays, and with it the stationary field; while it persists, however, the rotation of the rotor within it induces rotational-frequency currents in the rotor damper and field circuits. The a.c. in the field is clearly seen in *Figure 11.16*; it is the greater, the less effective the damper circuits are. With no damper at all, the induced d.c. and the zero-to-peak amplitude of the a.c. would initially be equal.

The path taken in the rotor by the stationary airgap flux has a greater permeance when the d-axis coincides with the stationary flux axis than when the q-axis coincides. Hence the flux fluctuates at twice fundamental frequency, and double-frequency current is induced in the stator winding. This current and the a.c. in the rotor decay as the stator d.c. component decays. The magnitude of the 2nd-harmonic stator current depends on the difference between X''_d and X''_q; it is small in turbogenerators and in salient-pole machines with good interconnected damper windings.

In summary, the a.c. components in the stator give rise to the d.c. components in the rotor circuits; after the subtransient period, stator a.c. and field d.c. decay together with the transient short-circuit time-constant, T'_d. The stator d.c. components produce the rotor a.c., and these decay together with the armature short-circuit time-constant T_a.

Direct-axis reactances and time-constants are derived from the oscillograms of a 3-phase short circuit as follows (see *Figure 11.17* and IEEE Publication 115). The oscillograms should record for not less than 0.5 s; I_d can be measured by instruments or by taking a second oscillogram after the steady state has been reached. Speed and field current should be constant throughout. E_o is the open-circuit phase e.m.f. corresponding to the rotor excitation. The analysis is then as follows:

(1) Draw the envelopes abc and a'b'c' of one phase-current oscillogram. Then aa' is the double-amplitude of the prospective current at the instant $t = 0$ of short circuit. (The first current peak is slightly less than $\frac{1}{2}$aa' because of the rapid subtransient decrement.) Taking aa' *as scaled in per-unit terms*, the r.m.s. current is $I''_d = \text{aa}'/(2\sqrt{2})$ and

$$X''_d = E_o / I''_d$$

(2) Project the envelopes in the transient region, cb and c'b', back respectively to d and d', ignoring the initial rapid subtransient decrement (this cannot be done with great accuracy). Then $I'_d = \text{dd}'/(2\sqrt{2})$ and $X'_d = E_o / I'_d$.

(3) Repeat steps 1 and 2 for the other two phases and derive the mean values of X''_d and X'_d.

(4) For a closer estimate, the equation relating the r.m.s. value of the a.c. short-circuit current I_t to time may be used:

$$I_t = I_d + (I'_d - I_d)\exp(-t/T'_d) + (I''_d - I'_d)\exp(-t/T''_d)$$

where $I_d = \text{ee}'/(2\sqrt{2})$ is the sustained steady-state short-circuit current, and I'_d and I''_d are the transient and subtransient r.m.s. currents respectively, corresponding to dd' and aa' respectively, at $t = 0$.

(5) Measure (e.g. in centimetres) the double-amplitude between

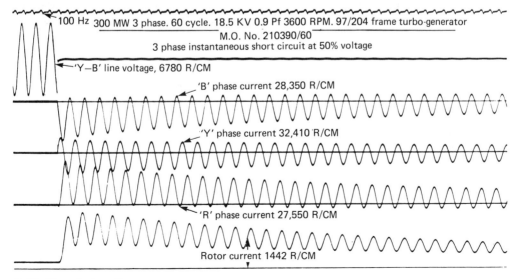

11.16 Short-circuit oscillogram

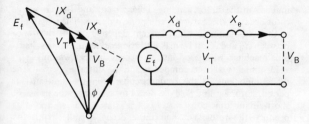

Figure 11.12 Effect of system reactance

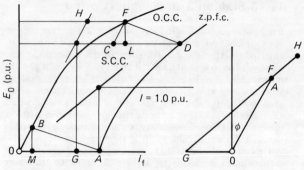

Figure 11.14 Potier-reactance excitation diagram

11.12.2 The ANSI Potier-reactance method (*Figure 11.14*)

ANSI Potier-reactance method (IEEE Publication 115, Sect. 4) requires the tested o.c. and z.p.f. characteristics, and is limited to machines that can be loaded for a z.p.f. test. In *Figure 11.14(a)*, A is the rated-current short-circuit point and D is the rated-current rated-voltage point: to reach D the field current exceeds the rated-load excitation level. Draw DC=AO, and draw CF parallel to the airgap line OBH. Drop FL perpendicular to DC, and draw triangle OBAM similar to CFDL. Then FL is the Potier voltage drop IX_P, and DL is the armature-reaction m.m.f.

The argument is that for a given stator current the armature-reaction and leakage-reactance volt drops are constant, but the latter requires more excitation at the higher levels of saturation.

OG, OA and FH in *Figure 11.14(b)* are excitation currents as in (*a*), with OG for rated voltage on the *airgap* line. GH is the total excitation required.

The z.p.f. test may have to be performed at less than rated stator current; the z.p.f.c. is then closer to the o.c.c., and DL and FL are smaller. Nevertheless, the Potier reactance is still considered to be $X_P = FL/I$ up to rated value.

Given only the o.c.c. and s.c.c., and no facility for adequately loading the machine, one procedure is to measure the d-axis subtransient reactance from a sudden-short-circuit test (or by the IEEE method below) and to use it as X_P.

The test method for X_d'' in IEEE 115, clause 7.30.25, is to apply a voltage E of normal frequency to each pair of stator terminals in turn and to observe the current I with the rotor stationary. Let the three quotients of E/I be A, B and C. Then with E and I in per-unit values of rated phase voltage and current, $X_d'' = (A+B+C)/6$ to an approximation. To avoid rotor overheating, the duration of the test should not exceed the maker's recommendations (e.g. 0.2 p.u. for a time sufficient to read the meters).

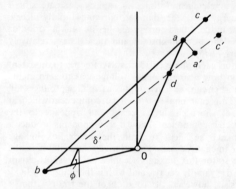

Figure 11.15 M.M.F. excitation diagram from design calculations

11.12.3 Use of design calculation (*Figure 11.15*)

Methods can be refined by allowing for the rotor pole-to-pole leakage. From a knowledge of the airgap flux, the m.m.f.s required for the gap, stator teeth and stator core are calculated. The m.m.f. for the rotor pole and body are calculated from the gap flux and pole leakage. The component phasors are added as in the diagram, where the total rotor m.m.f. per pole bc is obtained from Oa (armature-reaction), Ob (gap, teeth and core) and ac (pole and body). Saliency can be allowed for by dividing Oa at d such that $Od/Oa = X_q/X_d$. Then the total rotor m.m.f. is bc'. This differs very little from bc, but the load angle δ' is more accurate.

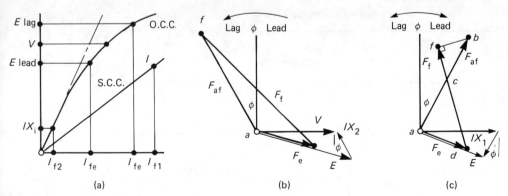

Figure 11.13 M.M.F. excitation diagrams: (a) open- and short-circuit characteristics; (b) cylindrical-rotor, power factor lagging; (c) salient-pole, power factor leading

component of stator current; considered as delivered to the power system, Oa leads the terminal voltage V.

Stable operation in practice is not possible up to the theoretical steady-state limit line Om. It is usual to construct a practical stability line such as Ofk on which, from each point such as k_1, operation for a given excitation $Os = Ok_1$ is not permissible in the region rk_1. The load increment rs may be a fixed fraction of rated load or may change progressively between no load and full load. Often, a minimum acceptable field current is defined to avoid pole-slip at low load when a system voltage drop occurs (for the synchronising torque depends on VI_f). The minimum excitation is typically 20% of that needed on no load, giving the limiting arc ef.

With a constant active-power output, the stator leakage flux in end-winding spaces increases as the power factor becomes more leading. This increases the loss, and the temperature in the end core packets and clamping plates will rise. Hence an end-heating limit line ln may be specified. Either the practical stability or the end-of-core temperature may set the limit on the reactive power that can be absorbed.

The operating chart can form the dial of a $P-Q$ meter having a pointer moving parallel to each axis. Where the pointers cross indicates the load point, and the margins between the output and the several limits are readily observed.

If the $P-Q$ diagram is scaled in MW and Mvar, changing the operating voltage (but not the *rated* power, power factor and voltage) has the following effects:

(1) Points a, b and d remain unchanged.
(2) Point O moves to a new value of Mvar, in proportion to V^2.
(3) The current and voltage scales for triangle Oab change in inverse proportion to V.
(4) Movement of the stability-limit line through O emphasises the change in the stability margin that exists for any given operating point defined by the P and Q values.
(5) The change in the length of Ob (to the MVA scale) shows how the maximum power P_m at the stability limit rises with increasing voltage, and conversely.

11.11.2 Salient-pole generator

From the phasor diagram it can be shown that the $P-Q$ chart for a salient-pole generator has the form of *Figure 11.11*. Here Oa, ab, ad and ag have the same significance as in *Figure 11.10*, and $as = 1/X_q$; $ac/ab = X_q/X_d$. The e.m.f. $E_f = Oe = lb$ (these two lines

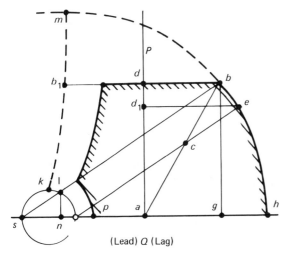

Figure 11.11 Operating chart for a salient-pole generator

being parallel) contributes the power ad_1, and d_1d is the reluctance power. The circle of diameter $Os = V(1/X_q - 1/X_d)$ is the zero-excitation circle, within which ordinates such as $ln = d_1d$ also represent the reluctance power.

If between the generator and the fixed-voltage busbars there is a reactance X_e introduced by, for example, a transformer or power line, the phasor diagram of *Figure 11.7* is modified to that in *Figure 11.12*. The reactive power at the stability limit of a cylindrical-rotor machine becomes $V/(X_d + X_e)$. For a salient-pole machine, the intercepts of the zero-excitation circle lie at $V/(X_d + X_e)$ and $V/(X_q + X_e)$ on the reactive-power axis. In brief, the stability-limit line moves toward the $P-Q$ origin.

In many installations the generator transformer has an on-load tap changer, and reactive-power control is achieved by tap-changing, so reducing the voltage range required of the generator.

11.12 On-load excitation

Use of the constant unsaturated value of X_d leads to values of E_f higher than those that occur for a practical machine. Thus a cylindrical-rotor generator with $X_{du} = 2$ p.u. and carrying rated load at p.f. 0.85 lagging would have $E_f = 2.67$ p.u., while a practical machine would saturate at around 1.5 p.u. If I_f were read from the airgap line for $E_f = 2.67$ p.u. it would be about 15% low. At leading power factor, for which saturation levels are low, the error would be small. However, a more accurate estimate of the field current I_f is needed in the design of the excitation system and its cooling, and to determine the open-circuit e.m.f. that would be reached if the automatic voltage regulator failed to limit the excitation on sudden load rejection.

Computer programs facilitate detailed calculation of mutual and leakage fluxes, allowing for field-current and armature-reaction m.m.f.s, and adjusting magnetic permeability to suit local flux conditions. Such techniques are valuable for detailed design, but are elaborate and expensive for routine calculation. Hence methods based on phasor diagrams and adjusted reactances, and making separate allowance for saturation, are still of use. They are necessary, too, for calculating exciting currents from test results on built machines.

Such methods add the rotor m.m.f. phasor needed to generate the no-load voltage to that needed to balance armature reaction. They differ in the choice of the voltage and in the way allowance is made for the effects of saturation.

The open-circuit short-circuit and zero-power-factor characteristics (o.c.c., s.c.c. and z.p.f.c.) are required, either by test or from design calculations. All the methods should give the full-load excitation to within ±5%, or closer at leading power factor. We consider three methods here.

11.12.1 M.M.F. phasor diagram (*Figure 11.13*)

The phasor diagram (b) is that of *Figure 11.7* turned to a more convenient position. The e.m.f. E behind the leakage reactance X_l requires a field m.m.f. F_e, read from the calculated or tested o.c.c. F_{af} is the armature-reaction m.m.f. in rotor terms, obtained by using the equations in Section 11.7 or (if X_l is known) by calculation from the tested s.c.c. as follows. To circulate stator current I on short circuit, excitation I_{f1} is needed; to generate an e.m.f. to balance IX_l drop, I_{f2} is needed; hence the armature-reaction m.m.f. is $F_{af} = I_{f1} - I_{f2}$. F_f is the excitation m.m.f. required for the load current I at terminal voltage V and p.f. angle ϕ. *Figure 11.13(c)* shows the diagram for a salient-pole machine with a leading p.f. load.

(preferably about 0.2% high); (2) its r.m.s. voltage should equal the busbar voltage within ±5%; (3) the machine and busbar voltages must be momentarily in phase, or within ±5° of phase coincidence.

In manual synchronising, condition (3) involves the use of a synchroscope to indicate to the operator the relative phase positions. Allowance must be made by the operator for the 0.2–0.4 s time-lag between initiation of switch closure and the actual closure of the switch contacts.

Alternatively, the process may be carried out by automatic means which monitor the conditions and initiate switch closure at the proper instant.

11.10.2 Synchronising power and torque

If a machine running in parallel with others is disturbed from its steady-state condition, i.e. if its load angle changes, a synchronising torque and corresponding power are developed, tending to restore the machine to normal synchronous running. For small displacements $\Delta\delta$ from the steady-state power-angle δ, the synchronising power coefficient is

$$P_s = \frac{dP}{d\delta} = \frac{VE_f}{X_d}\cos\delta + \frac{V^2(X_d - X_q)}{X_d X_q}\cos 2\delta$$

and the power restoring synchronism is $P_s \Delta\delta$, per phase. The term in 2δ is usually negligibly small for cylindrical-rotor machines. The corresponding synchronising torque coefficient is

$$T_s = 3P_s/2\pi n$$

where n is the synchronous speed (r/s). The machine remains in synchronism unless the disturbance is a major load change or a system fault causing a severe fall in voltage.

11.10.3 Oscillation frequency

The restoring torque acting on the torsional inertia J of the generator/prime-mover combination gives the set a natural angular frequency of oscillation ω_n. If $T_m = pT_s$ is the synchronising torque per mechanical radian for a $2p$-pole machine, then

$$\omega_n = \sqrt{(T_m/J)} \text{ mech. rad/s} \quad \text{or} \quad f_n = \sqrt{(T_m/J)}/2\pi \text{ Hz}$$

For most machines, regardless of speed, the oscillation frequency f_n lies between 1 and 3 Hz. If f_n is within 10% (or less in a well-damped system) of some cyclic disturbing torque (e.g. with reciprocating engine drive), parallel operation may be difficult, or impossible, by reason of excessive cyclic variation in load angle, power, current and voltage.

It is usual to fit damping windings in the pole-shoes of salient-pole machines to reduce oscillation, the eddy currents induced dissipating oscillation energy. Dampers also increase the synchronising torque, and raise f_n typically by about 10%. The increased f_n should be taken into account when determining whether there is a margin between f_n and the cyclic disturbance. It may be necessary to increase the inertia J to keep f_n below the disturbing frequency. Additional inertia may also be necessary to reduce speed fluctuations even when electromechanical resonance is not a problem. For example, natural frequencies of shaft oscillation will produce voltage fluctuation if the generator operates alone or as a substantial part of a small power system. As little as 0.5% voltage fluctuation at 10 Hz (typical of low-speed engines) may cause lighting flicker. However, this can be reduced by excitation control.

11.11 Operating charts

Operating charts based on *Figure 11.5* or *11.8* define the operating limits imposed by the prime mover, excitation, load and stability. Saturation is ignored and the reactances are deemed constant.

11.11.1 Cylindrical-rotor generator

In *Figure 11.10*, Oab is the synchronous-reactance triangle. The terminal phase voltage V is Oa to a scale of v (V/unit length); to this scale ab represents IX_d. The voltage E_f behind the

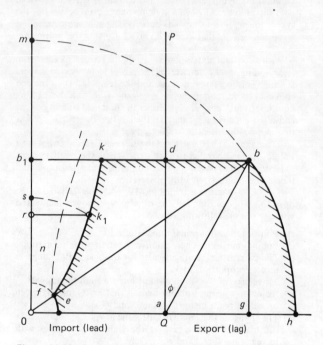

Figure 11.10 Operating chart for a cylindrical-rotor generator

synchronous reactance is Ob, read from the airgap line of *Figure 11.4* for a field current I_f. Angle dab is the power-factor (p.f.) angle ϕ. To a scale v/X_d (A/unit length), aO represents $V/X_d = I_{sc}$, the stator current drawn from the supply for zero excitation; ad is the active and bd the reactive component of the stator current ab. Hence to a scale of $3Vv/X_d$ (W or var/unit length), ad is the active, ag the reactive and ab the apparent power. To a scale deduced from the airgap line of the open-circuit characteristic (*Figure 11.4*), Ob is the field current required for the load conditions represented by point b.

The vertical P axis and the horizontal Q axis of the diagram are commonly marked in MW or Mvar, or in per-unit terms where, with $V = 1$ p.u., point b represents the rated load. The chart gives

ab = rated MVA = 1 p.u.   ad = rated active power = $\cos\phi$ p.u.
aO = rated MVA/X_d = $1/X_d$ p.u.

Operation must be so controlled that the operating point is within the boundary set by (i) an arc of centre a and radius ab representing rated stator current, (ii) an arc bh of centre O representing rated field current, and (iii) the line bdb_1 representing the rated active power output of the prime mover.

The line Om, corresponding to a load angle $\delta = 90°$, shows the theoretical maximum power (since the output falls for $\delta > 90°$), while Ob_1 is the lowest field current for which rated power can be delivered, the corresponding stator current then being ab_1.

Instability at all loads occurs when the generator (being underexcited) absorbs the reactive power Oa of value $1/X_d$ p.u. The line Oa to the current scale is the zero-power-factor

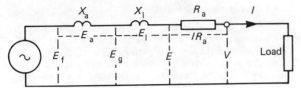

Figure 11.6 Equivalent circuit for a cylindrical-rotor generator

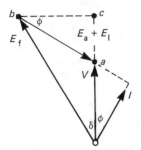

Figure 11.7 Simplified phasor diagram pertaining to output power for a cylindrical-rotor generator on load

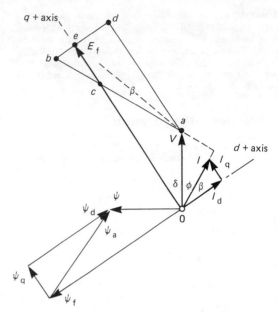

Figure 11.8 Phasor diagram for a salient-pole generator on load

If, as is usually permissible, R_a is neglected, then V and E coincide, simplifying the diagram to that in *Figure 11.7*. Here ab on a voltage scale represents IX_d. If the scale is divided by X_d then ab represents I. The angle abc is ϕ, the power-factor angle; bc represents the active-power component of I; but bc is also $E_f \sin \delta$, and hence the power per phase is $(VE_f/X_d)\sin\delta$.

At no load, V and E_f coincide. Hence δ is the *power* (or *load*) *angle*, i.e. the angle by which the rotor must be driven forward relative to the resultant flux (i.e. forward from its no-load position) to deliver active power. If V and E_f are fixed, then the active-power output P is proportional to $\sin\delta$, reaching a maximum for $\delta = 90°$.

For a *salient-pole* machine, account must be taken of the differing d- and q-axis reluctances. The q-axis component current I_q has a lower flux-producing effect than in a cylindrical-rotor machine, i.e. X_q is smaller than X_d. *Figure 11.8* shows the phasor diagram for an output at lagging power factor, with Ψ_l regarded as part of Ψ_a and E_l as part of IX_d. The position of the q-axis and the length $Oe = E_f$ are found by dividing ab at c such that $ab/ac = X_d/X_q$. The voltage triangle abd is similar to the current triangle I, I_d, I_q, each side being the appropriate current multiplied by X_d. As $ac/ab = de/db = X_q/X_d$, it follows that $de = I_q X_q$, whence

$$Oe = V + I_d X_d + I_q X_q = E_f$$

Again δ is the load angle. It is less than that for a cylindrical-rotor machine of the same X_d delivering the same active power at the same voltage and excitation. From the geometry of the phasor diagram it can be shown that the active power output is given by

$$P = VE_f \frac{1}{X_d}\sin\delta + V^2 \frac{X_d - X_q}{2X_d X_q}\sin 2\delta$$

The second term is the power that is available with zero field excitation ($E_f = 0$) and which is developed as a reluctance torque and power that depend on the different axis reluctances X_d and X_q. *Figure 11.9* shows power–angle curves for a salient-pole machine with typical values of X_d and X_q and for different excitation levels. The cylindrical-rotor machine may have a small difference of axis reactances, but the saliency effect is usually only about 0.05 p.u.

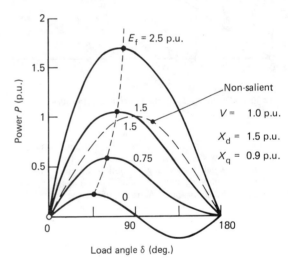

Figure 11.9 Power–angle relations for a salient-pole generator

11.10 Synchronising

Almost all large a.c. generators operate in parallel with others. This raises the problem of switching a machine safely into service ('synchronising') and ensuring that it subsequently remains in synchronism. It is here assumed that a generator is to be connected to a system large enough to fix its voltage and frequency regardless of changes in load and excitation on an individual generator.

11.10.1 Synchronising procedure

The following conditions must be satisfied by the incoming machine with respect to the network busbars: (1) the speed must be such that its frequency is close to that of the busbars

Measurements of the behaviour of machines and systems when they are subjected to test disturbances have shown that for some stability calculations the simpler circuits, even with the subtransient effects neglected, are adequate. However, for calculating short-circuit torques, the subtransients must be included; and for detailed analysis of the effects of excitation control, more elaborate models are needed.

11.9 Steady-state operation

Prediction of the operation of a generator in the steady state is based on the open- and short-circuit characteristics in *Figure 11.4*.

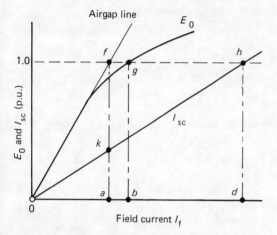

Figure 11.4 Open-circuit and short-circuit characteristics

With the stator winding on *open circuit*, the field current I_f produces a mutual flux linking the stator and rotor windings, plus a relatively small rotor leakage flux linking the rotor winding only. The corresponding stator winding flux linkage per phase Ψ_o generates the stator e.m.f. At rated speed, the value of I_f represented by Ob generates rated e.m.f. E_o, represented by bg; Oa is the current needed to overcome the reluctance of the airgap, and ab is required for the iron parts of the magnetic circuit.

With the stator winding *short-circuited*, field current Od circulates rated stator current I_{sc}, represented by hd. Armature-reaction m.m.f. produced by I_{sc} is wholly demagnetising, and the difference between it and the field m.m.f. (Od) develops a mutual flux Φ_{sc} sufficient to induce an e.m.f. E_{sc} equal to the stator leakage reactance drop $I_{sc}X_l$. This neglects the stator winding resistance R_a and any harmonic fluxes developed by the stator and rotor m.m.f.s.

The flux Φ_{sc} is too small to cause magnetic saturation; hence I_{sc} is proportional to I_f. As both E_{sc} and X_l are proportional to frequency, the short-circuit characteristic is almost independent of speed; nevertheless, it is usual to obtain it at rated speed.

Still neglecting saturation and armature resistance, a field current I_f = Oa gives E_o = af on open circuit and I_{sc} = ak on short circuit. Thus on short circuit the stator appears to present a reactance $X_{du} = E_o/I_{sc}$ = af/ak, a constant representing the *unsaturated* d-axis synchronous reactance $X_{ad} + X_l$. X_{du} is usually defined in terms of I_f as the ratio Od/Oa in *Figure 11.4*. The short-circuit ratio is defined as Ob/Od, which from the geometry is (Ob/Oa)(1/X_{du}). Here Ob/Oa is a saturation factor, usually in the range 1.1–1.2.

11.9.1 Steady load conditions

The resultant m.m.f. of the stator current I and the field current I_f develops an airgap flux Φ which induces the stator phase e.m.f. E_{ph}. The stator leakage flux Φ_l induces the e.m.f. E_l.

Detailed analysis allowing for local flux distribution and the variation of magnetic permeability in ferromagnetic parts of the magnetic circuit is necessary in design. However, the performance of a generator on load and under fault conditions can be examined conveniently (and, for many purposes, adequately) by combining the e.m.f.s considered to be produced by I_f alone and I alone, taken separately. Neglecting saturation, the machine can be represented by an equivalent circuit of constant reactances, and with e.m.f.s proportional to their respective currents. Usually the effect of stator-winding resistance can be neglected. Further, by assuming the airgap flux distribution to be sinusoidal, phasor diagrams can be employed. Finally, for a cylindrical-rotor machine the uniform gap length makes it permissible to assume that the d- and q-axes have equal reluctances.

Adopting the conventions of IEC Publication 34-10, the basis is generator action, with power positive when it flows from generator to load. An induced e.m.f. is $e = -d\Psi/dt$, where Ψ is the linkage between flux and stator winding. This means that the e.m.f. phasor lags the flux (or linkage) phasor by 90° electrical. Then, since the flux linking a phase winding produced by stator current is in time phase with the current, the phasor diagram in *Figure 11.5* applies for a generator on inductive load, and *Figure 11.6* is the corresponding equivalent circuit. Here Ψ_f is the stator linkage produced by the field current I_f alone; it would induce on open circuit the e.m.f. E_f. The load current I produces the linkage Ψ_a (the armature-reaction effect) reducing E_f to the airgap e.m.f. E_g. Stator leakage flux produces the linkage Ψ_l and the e.m.f. E_l, and the total induced e.m.f. is E. Subtraction of the volt drop IR_a gives the terminal voltage V. With the convention that the inductive volt drops leads the current by 90°, the two equivalent equations in phasor terms are

$$V = E_f + E_a + E - IR_a \quad \text{or} \quad V = E_f - I(X_a + X_l) - IR_a$$

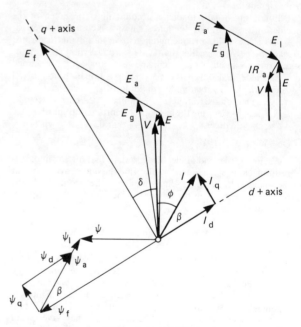

Figure 11.5 Phasor diagram for a cylindrical-rotor generator on load

Table 11.1 Synchronous generators: typical reactances (p.u.) and time-constants (s)

Parameter	Symbol	Turbogenerator	Salient-pole generator		Compensator
			with dampers	without dampers	
Synchronous reactance					
d-axis	X_d	1.0–2.5	1.0–2.0		0.8–2.0
q-axis	X_q	1.0–2.5	0.6–1.2		0.5–1.5
Armature leakage reactance	X_l	0.1–0.2	0.1–0.2		0.1–0.2
Transient reactance					
d-axis	X'_d	0.2–0.35	0.2–0.45	0.2–0.45	0.2–0.35
q-axis	X'_q	0.5–1.0	0.25–0.8	0.5–1.0	0.25–0.8
Subtransient reactance					
d-axis	X''_d	0.1–0.25	0.15–0.25	0.15–0.3	0.1–0.25
q-axis	X''_q	0.1–0.25	0.2–0.8	0.5–1.0	0.2–0.8
Negative-sequence reactance	X_2	0.1–0.25	0.15–0.6	0.25–0.65	0.15–0.5
Zero-sequence reactance	X_0	0.01–0.15	0.4–0.2		0.03–0.2
Time-constants					
d.c.	T_a	0.1–0.2	0.1–0.2		0.1–0.2
transient	T'_d	1.0–1.5	1.5–2.0		1.5–2.5
subtransient	T''_d	0.03–0.1	0.03–0.1		0.03–0.1
open-circuit transient	T'_{do}	4.5–13	3–8		5–8

currents: if we neglect the effect of R_f, the time constant is

$$T''_{do} = \left(X_{kd} + \frac{X_{ad}X_f}{X_{ad}+X_f}\right)\frac{1}{\omega R_{kd}}$$

11.8.8.2 Short-circuit

With the stator short-circuited, current and flux changes are influenced by X_l, but R_a is usually negligible. The transient time-constant is

$$T'_d = \left(X_f + \frac{X_l X_{ad}}{X_l + X_{ad}}\right)\frac{1}{\omega R_f}$$

The subtransient short-circuit time-constant depends primarily on damper-circuit parameters: it is

$$T''_d = \left(X_{kd} + \frac{X_{ad}X_f X_l}{X_{ad}X_f + X_f X_l + X_l X_{ad}}\right)\frac{1}{\omega R_{kd}}$$

The armature short-circuit time-constant relates to the rate of decay of d.c. components of stator current that occur from the beginning of a sudden short circuit: it is

$$T_a = (X''_d + X''_q)/\omega R_a$$

11.8.8.3 Q-axis

Only subtransient time-constants are concerned. In the simple model with one q-axis damper circuit

$$T''_{qo} = (X_{aq} + X_{kq})/\omega R_{kq} \quad \text{(open circuit)}$$

$$T''_q = \left(X_{kq} + \frac{X_{aq}X_l}{X_{aq}+X_l}\right)\frac{1}{\omega R_{kq}} \quad \text{(short circuit)}$$

11.8.8.4 Other relations

From the foregoing, it is found that

$$T'_d/T'_{do} = X'_d/X_d \quad T''_q/T''_{qo} = X''_q/X_q \quad T''_d/T''_{do} = X''_d/X'_d$$

11.8.9 Potier reactance

The Potier reactance X_P is an estimate of armature leakage reactance deduced from the open-circuit and zero-power-factor (z.p.f.) curves; it is used in one method of calculating the on-load field current allowing for saturation. X_P is slightly higher than the true X_l, especially for salient-pole machines, where pole saturation is greater than on normal load because of the greater pole-to-pole leakage flux for the z.p.f. conditions. Hence using X_P somewhat overestimates the load excitation.

11.8.10 Frequency-response tests

The d- and q-axis parameters can be measured by injecting 1-phase current into the stator winding over a frequency range (typically 1 mHz to 1 kHz), the rotor being stationary with its d- and q-axes in turn aligned with the stator field. By fitting an expression in the Laplace form

$$X_d(s) = \frac{(1+sT'_d)(1+sT''_d)}{(1+sT'_{do})(1+sT''_{do})}$$

to the curve of d-axis reactance against frequency, X_d and the time-constants can be found, and the corresponding X'_d and X''_d values derived, appropriate to a machine model with one damper circuit on the d-axis.

To fit the q-axis reactance–frequency curve reasonably closely, it is necessary to assume two damper circuits, which gives an expression similar to the d-axis expression quoted, but with q-axis quantities. Thus X'_q and T'_q values are deduced, as well as X''_q and T''_q, despite the absence of a field winding on the q-axis.

More accurate fits to the measured frequency-response curves may be obtained by using an equivalent with more damper circuits and corresponding time-constants. Techniques have been developed for taking frequency-response curves on the machine in service in order to obtain values more appropriate to the load condition. There is an extensive literature in the *IEEE Journal (Power Apparatus and Systems)*. A recent review of the subject, and some specific papers, are contained in IEEE Publication 83TH0101-6-PWR, Symposium on Synchronous Machine Modelling for Power System Studies, February 1983.

maintain the required stator flux linkage. Thus the induced rotor currents, and the stator current, decrease in unison, at first rapidly as the damper currents decay, then for a time more slowly, until eventually the induced current in the field winding has disappeared, the flux is fully established along the main flux paths and the stator current is settled at its magnetising value $I = V/X_d$. The main decay time is called the *transient* period, and the brief initial decay period is the *subtransient*.

The machine as seen from the supply system can be represented by the equivalent circuit shown in *Figure 11.3(a)*, where I_{kd} and I_f are the components of I_d needed to balance the induced damper and field currents respectively. The effective impedance increases

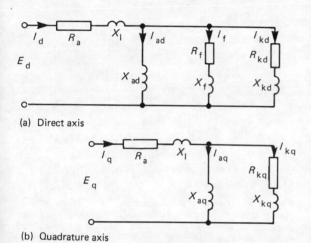

(a) Direct axis

(b) Quadrature axis

Figure 11.3 Equivalent circuits for d- and q-axis reactances

progressively from its initial value of X_l in series with the other three circuits in parallel, through X_l in series with X_{ad} and the field circuit in parallel when I_{kd} has reached zero, to $X_l + X_{ad}$ in the steady state. Thus the transient reactance X'_d and subtransient reactance X''_d are

$$X'_d = X_l + \frac{1}{1/X_{ad} + 1/X_f} \qquad X''_d = X_l + \frac{1}{1/X_{ad} + 1/X_f + 1/X_{kd}}$$

If the moment of switching is such that flux is established along the quadrature axis, similar arguments apply except that q-axis flux has no net linkage with the field winding. However, damper circuits are present, and conditions pass rapidly through the subtransient to the steady state. The equivalent circuit is that in *Figure 11.3(b)*, and the q-axis subtransient reactance is

$$X''_q = X_l + \frac{1}{1/X_{aq} + 1/X_{kq}}$$

There is no transient reactance analogous to X'_d.

The equivalent circuits in *Figure 11.3*, with fixed parameters and simple exponential decay, cannot exactly represent the conditions in the machine, but they are adequate in some cases, e.g. for salient-pole machines with laminated poles and specific damper windings. In principle a more accurate model could be devised by use of two damper circuits on each axis, although more sophisticated design analysis and testing would be necessary to obtain suitable values for the greater number of parameters.

Traditionally, X''_d and X'_d have been deduced from the oscillograms of a sudden symmetrical 3-phase short-circuit test; however, the values obtained are low because of magnetic saturation and they are not appropriate for less severe system short circuits or for sudden load changes. No comparable test is available for the q-axis parameters, but IEEE Publication 115 gives test methods.

11.8.5 Negative-sequence reactance

Unbalanced load or fault conditions are usually analysed by the method of symmetrical positive, negative and zero phase-sequence components (p.p.s., n.p.s. and z.p.s. components). Currents of n.p.s. in the stator produce an m.m.f. rotating at synchronous speed in a direction opposite to that of the rotor. This m.m.f. acts upon the d- and q-axes in turn, inducing double-frequency currents in any available rotor circuit. The stator presents a low reactance to n.p.s. current, taken to be the mean of the subtransient d- and q-axis values, i.e.

$$X_2 = \tfrac{1}{2}(X''_d + X''_q)$$

11.8.6 Zero-sequence reactance

Zero-sequence currents in the three phases are equal and in time phase, and their combined effect is to produce a stationary field alternating at supply frequency, therefore inducing a stator e.m.f. of that frequency. The m.m.f. and therefore the flux are small compared with the p.p.s. and n.p.s. components, and they depend heavily on the coil-span, falling from a maximum value to zero as the span decreases from full pitch to 2/3 pitch. Accordingly X_0 is usually quite small.

11.8.7 Reactance values

Ranges of typical values of reactances and time-constants are given in *Table 11.1*. Since the ranges all depend on the details of the machine design, there will be exceptions to the following generalisations, which do however indicate normal trends.

For a given output and speed, the physically smaller machine will have a higher current loading, so all reactances will be higher than those of a larger, and therefore 'slacker', design. At a given speed, reactances tend to rise with increasing rated output, since higher electrical loading and more intensive cooling are needed to attain more output per unit volume of active material. For a given output, low-speed machines are physically larger, and tend to have higher reactances, than high-speed designs.

Having chosen the electrical loading and gap length to attain the output and the desired X_d on an economic frame size, some limited adjustment of the transient and subtransient reactances is possible by changing the stator slot design. These two reactances differ only because of the damper-circuit reactance. It is rarely justifiable, or even practicable, to adopt an abnormal damper design in order to obtain an unusual relationship between them.

11.8.8 Reactances and time-constants

The reactances and time-constants are based on equivalent circuits, such as those in *Figure 11.3*. Time-constants are given by inductance/resistance ratios, i.e. the ratio of $X/2\pi f$ to R. In the formulae below, $2\pi f$ is written as ω, the angular frequency. All time-constants are in seconds if all X and R values are expressed consistently in per-unit or ohmic values.

11.8.8.1 Open-circuit

With the stator winding open-circuited, the leakage impedance (X_l and R_a) has no influence and the transient behaviour is determined by the inductance and resistance of the field winding. The open-circuit transient time-constant is

$$T'_{do} = (X_{ad} + X_f)/\omega R_f$$

The subtransient duration depends primarily on damper-circuit

delivered (with the rotor underexcited) within the limit of steady-state stability.

11.8 Reactances and time-constants

In order to evaluate the steady-state behaviour of a synchronous generator or its response to changes of load, excitation and system disturbances, a mathematical model of the machine is required. Reactances have been defined which, with winding resistances where significant, can form an appropriate equivalent circuit for which behaviour equations can be written; these equations can then be solved to determine the performance of the machine.

The reactances commonly employed are described below. They are associated with a two-axis model, represented on the d-axis by an equivalent stator winding, the field winding and a damper winding, and on the q-axis by a second equivalent stator winding and one damper winding. Circuit impedances can normally be taken as reactances because the resistances are comparatively small; however, the resistance values directly influence the time-constants.

Voltages, currents and reactances are usually expressed in per-unit (p.u.) terms of rated voltage and current; other values are put in p.u. by dividing actual voltage and current by the rated values. The unit reactance is the ratio rated-voltage/rated-current. Thus a reactance having a voltage drop of 0.2 p.u. when carrying 0.5 p.u. current has a value of 0.4 p.u. Stator voltages, currents and reactances are all per-phase values.

Each reactance is associated with a particular component of flux produced by either the d- or the q-axis component of current in the stator winding. A d-axis current produces a d-axis flux: as shown in *Figure 11.1*, the conductors in which it flows are near the q-axis but they form coils magnetising on the d-axis. Similarly a q-axis current produces a q-axis flux. The numerical value of each reactance is then the fundamental-frequency e.m.f. per phase generated by the associated flux, divided by the corresponding component of current. Usually reactances are defined with rated current in the d- or q-axis and are termed 'rated' or 'unsaturated' values. With the heavy currents occurring under short-circuit conditions, saturation reduces the flux per ampere, and the saturated reactance values are lower.

The reactance values can be derived from the machine geometry by calculating first the permeance of the associated flux path and then the inductance L, i.e. the flux linkage with the stator winding per ampere of stator current in the required axis. The reactance (in ohms) is $2\pi f L$. Difficulties arise in defining exactly the flux paths and permeances and in allowing for saturation at high flux densities, even though a large part of the leakage flux paths is in air.

11.8.1 Armature leakage reactance

The armature leakage reactance X_l results from stator leakage flux that crosses the stator slots, flux that passes circumferentially from tooth to tooth round the airgap without entering the rotor and flux linking the stator end-winding. X_l is a component of all the positive- and negative-sequence reactances. Since the flux paths are independent of the rotor, the reactance has the same value for both axes; and since the flux paths are the same for positive- and negative-sequence currents, X_l has the same value for both.

11.8.2 Magnetisation (armature-reaction) reactances

The magnetisation (armature-reaction) reactances are associated with the synchronously rotating flux set up by positive-sequence current in the stator winding, i.e. by balanced phase currents. If the field winding is unexcited, enough stator current flows to set up a flux sufficient to induce into the stator winding an e.m.f. equal to the rated value (ignoring resistance and leakage). The d-axis reactance X_{ad} applies to a flux coincident with the pole axis (where the radial airgap is a minimum in a salient-pole machine), while the q-axis value X_{aq} is smaller as the gap reluctance in the interpolar zone is considerably higher. Thus in a salient-pole machine the ratio X_{aq}/X_{ad} is typically 0.6. In a cylindrical-rotor machine the ratio is rarely less than 0.9; the airgap is uniform, but the presence of rotor slots slightly increases the effective airgap.

11.8.3 Synchronous reactances

The synchronous reactances are the total reactances presented to the applied stator voltage when the rotor is running synchronously but unexcited:

$$X_d = X_{ad} + X_l \quad \text{and} \quad X_q = X_{aq} + X_l$$

The magnetising and synchronous reactances are steady-state values applicable with balanced phase currents of constant r.m.s. value. They are defined by considering the machine to be excited by stator current only. The two axis reactances can be represented by the equivalent circuits in *Figure 11.2*.

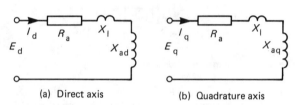

(a) Direct axis   (b) Quadrature axis

Figure 11.2 Equivalent circuits for d- and q-axis armature-reaction reactances

11.8.4 Transient and subtransient reactances

The transient and subtransient reactances relate to conditions that arise when the m.m.f. on the magnetic circuit of the machine is suddenly changed. Consider the conditions following the sudden application of a 3-phase supply of voltage V to the stator winding, with the rotor running synchronously and its field winding closed but unexcited. The 3-phase stator currents develop an m.m.f. that rotates synchronously with the rotor. (There are direct-current (d.c.) components as well, but they are not relevant here.) Suppose that the instant of switching is such that the stator m.m.f. is impressed on the pole axis. The flux linkages of the stator winding must induce an e.m.f. that balances V (neglecting resistance). If there were no current paths on the rotor, the stator would immediately carry currents necessary to magnetise the machine to the required flux level, i.e. $I = V/X_d$. However, if there are closed rotor circuits available (i.e. the field winding, damper windings and solid iron pole-shoes), currents are induced in them, inhibiting the rise of flux through the rotor poles and so forcing the flux into rotor leakage paths of high reluctance. Hence the initial stator current must be larger than V/X_d. The leakage paths are largely circumferential in the pole-faces, from pole to pole in the gap between adjacent salient poles, and across the wedges and tooth-tips in a turbogenerator rotor. This path adds only a small permeance to that of the stator leakage paths alone, so the effective reactance is not much more than X_l.

I^2R losses in the damper circuits cause these currents to decay rapidly, enabling the flux to penetrate past the dampers into the pole and field winding region. The permeance of the available flux path therefore increases, decreasing the stator current needed to

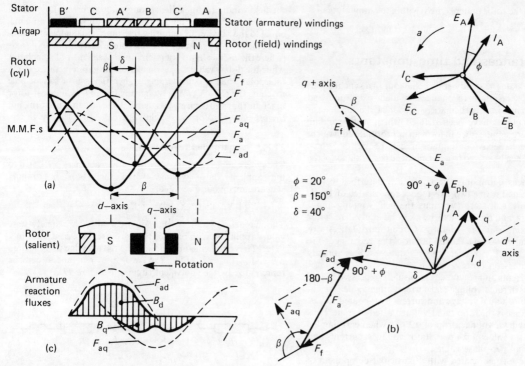

Figure 11.1 Airgap m.m.f. distribution

the e.m.f.s E_f, E_a and E_{ph} in *Figure 11.1(b)* are respectively induced by F_f, F_a and F. Then E_{ph} is the terminal e.m.f. for a representative stator phase.

11.7.2 Salient-pole rotor machine

The gap reluctance is not uniform, and a given stator m.m.f. acting on the q-axis produces a smaller flux than it would if acting on the lower reluctance of the d-axis. This is indicated in *Figure 11.1(c)*. It is therefore necessary to resolve F_a into the components F_{ad} and F_{aq} and to determine separately the fluxes they produce. The d-axis flux is distributed approximately sinusoidally, but the q-axis flux has significant space-harmonic content, mainly the 3rd harmonic. For analysis it is usual to evaluate the fundamentals of B_d and B_q. There are several techniques for solving this problem: Wieseman, for example, gives coefficients in terms of the pole profile and the gap reluctance at the pole centre.

11.7.3 Equivalent field m.m.f.

To evaluate the rotor excitation required on load it is necessary first to express the armature reaction m.m.f. F_a as an equivalent rotor m.m.f. F_{af} ('equivalent' here means that each m.m.f. would produce the same fundamental flux if acting alone along the d-axis). For a 3-phase winding of 60° phase spread the peak of the fundamental m.m.f. is

$$F_a = 1.35 I_c K_w N/p \text{ A-t/pole}$$

where I_c is the r.m.s. current in each stator conductor and N is the total number of turns per phase.

For most slotting arrangements in a *cylindrical rotor* the peak fundamental m.m.f. is approximately $F_f = 1.05 I_f N_f$ when a rotor current of I_f A flows in N_f turns/pole. Equating F_a and F_f, we obtain the rotor equivalent, F_{af}, of F_a:

$$F_{af} = I_f N_f = F_a/1.05 = 1.29 I_c K_w N/p \text{ A-t/pole}$$

In a *salient-pole machine*, with distributed stator winding, concentrated field winding and non-uniform airgap, it is necessary to calculate and equate the fundamental fluxes, rather than equating m.m.f.s. For most geometries

$$F_{af} = I_f N_f = (0.8-0.85)F_a = (1.08-1.15)I_c K_w N/p \text{ A-t/pole}$$

Figure 11.13 shows how F_{af} is combined with F_e to calculate F_f, the on-load rotor ampere-turns per pole.

11.7.4 Short-circuit ratio

The short-circuit ratio (s.c.r.) is a measure of the strength of the armature reaction, and is defined as

$$r_{sc} = F_{oc}/F_{sc}$$

i.e. the ratio of the excitation for rated e.m.f. on open circuit at rated speed to the excitation for rated stator current on short circuit. The s.c.r. largely determines (i) the reactive power (at leading power factor) at the limit of steady-state stability and (ii) by how much the required excitation changes with variation of output and power factor. The chosen s.c.r. thus influences the design of the field winding and the cooling it needs. In designing up to the thermal limit for a given output, a high s.c.r. necessitates a larger machine (or more sophisticated cooling) to keep the field-winding temperature within its thermal limit. The desired s.c.r. is achieved by choice of the airgap length because, with reasonable saturation, the airgap demands about 0.85 of the no-load excitation.

Evidently a low s.c.r. permits of a larger output from a given physical size of machine within the thermal limit set by the rotor. Typical s.c.r. values are: turbogenerator, 0.45; medium-speed machines up to about 25 MW, 0.7–1.0; hydrogenerators and synchronous compensators, 1.0–1.5. The last-named may have to absorb large per-unit reactive powers from the power network, and the high s.c.r. increases the leading reactive power that can be

conductor. The 3-phase winding normally has six groups of coils in each double pole-pitch. With an integral number of slots per pole and phase, all coil groups are identical. In multipolar machines it is often necessary to use fractional-slot windings; the groups then have unequal numbers of coils. The same *overall* pattern must, however, be obtained for all phases to provide symmetry.

11.4 Insulation

Winding insulation must retain its dielectric and mechanical strength over its many years of service life. High temperature is usually the most significant cause of deterioration, and insulation systems have been classified according to the maximum steady operating temperature for which a satisfactory lifetime can be expected. Those used in a.c. generator windings are Classes B (130 °C) and F (155 °C). Class H (180 °C) is used for special high-rated (and usually small) machines.

Classes B and F employ inorganic materials (mica, glass fibre, synthetic film) bonded with a thermosetting synthetic resin. Class H also includes silicone elastomers. These temperature classifications depend on the thermal capabilities of the resins used to bond the materials.

For high-voltage (h.v.) stator windings the permissible dielectric stress is a design criterion, and the deleterious effects of vibration and (in long machines) differential expansion must be withstood. Field windings operate at low voltage (e.g. 100–600 V) and present no electric-strength problems, but thermal expansion and centrifugal force are significant factors.

11.5 Temperature rise

Limits of temperature rise or, for direct-cooled machines, the total temperature are specified in national and international standards such as BS 4999 and 5000, IEC Publication 34-1, American National Standards Institute (ANSI) C.50 and the West German VDE rules 0530. Operation within the prescribed limits should give a service life of at least 20 years. Sustained operation at 10°C higher may halve the life.

11.6 Output equation

The apparent power rating of a generator is a function of the stator bore diameter D, the active axial length L, the speed n, the stator winding factor K_w, the number of pole-pairs p, and the specific magnetic and electric loadings B and A. The specific loadings give the mean gap-flux density and the mean circumferential stator current density:

$$B = 2p\Phi/\pi DL \qquad A = I_c N_c/\pi D$$

where Φ is the flux per pole, I_c is the current per conductor and N_c is the total number of stator conductors. The apparent power rating is then

$$S = 11 K_w B A D^2 L n \text{ (VA)}$$

The actual values of voltage and current depend on the design of the stator winding.

The magnetic loading B is governed by the permissible saturation of the magnetic circuit and does not vary widely with size. Typical values lie in the range 0.5–0.8 T, corresponding to maximum tooth flux densities of 1.5–2.0 T.

The current loading A depends chiefly on the cooling system: it ranges from 15 kA/m in small natural-air-cooled machines to about 70 kA/m for salient-pole machines of 20 MVA, and up to over 200 kA/m for large turbogenerators with hydrogen and water cooling.

The expression above for S leads to the output coefficient given by

$$G = S/D^2 L n = 11 K_w B A \text{ [VA/(m}^3 \text{ s)]}$$

a figure representing the output per unit volume per rev/s. Given the output and speed for a projected design, and having selected appropriate values of B and A, one can derive an approximate product $D^2 L$. An economic design usually results with L/D large; however, excessive length may introduce problems of critical speed or of the ventilation of the middle bulk of the machine. The ratio of length L to pole-pitch $\pi D/2p$ may be 0.8 for a hydrogenerator if a high inertia is required, 1.5–3 for the general run of industrial generators, up to 4 for salient-pole machines where stresses limit the peripheral speed to about 130 m/s, and 5 for large 2-pole turbogenerators where peripheral speeds up to 220 m/s are permissible. These latter machines will run through two or three critical speeds on run-up from rest to the rated speed of 3000 or 3600 rev/min.

11.7 Armature reaction

Balanced 3-phase sinusoidal currents in the stator winding develop a magnetomotive force (m.m.f.) that rotates synchronously with the rotor. The m.m.f. distribution around the airgap is approximately sinusoidal, with a wavelength spanning a double pole-pitch. Actually the m.m.f. wave-shape changes between peaked and flat-topped distributions every 1/12 period, but both shapes have the same *fundamental* component and can be considered as that fundamental with space-harmonics of order $6m \pm 1$, where m is a positive integer. Harmonics of order $6m+1$ (i.e. the 7th, 13th, ...) rotate forward in the same direction as the fundamental, while those of order $6m-1$ (i.e. the 5th, 11th, ...) rotate backward, in both cases at peripheral speeds inversely proportional to their orders. The harmonics induce e.m.f.s in the stator winding, but these of small magnitudes that do not affect the r.m.s. value of the phase e.m.f. However, the harmonic fluxes induce eddy currents in solid pole-faces, damper windings and end-shields; the 5th and 7th can give rise to significant loss unless the stator coils are short-chorded to 5/6 pitch.

11.7.1 Cylindrical-rotor machine

The gap flux wave developed by the fundamental stator ('armature') m.m.f. acting alone is nearly sinusoidally distributed because of the uniform airgap. The gap flux developed by the rotor ('field') m.m.f. acting alone has a trapezoidal distribution. On load the gap flux results from the stator and rotor m.m.f.s in combination. The fundamental components of the distributed m.m.f.s can be represented by phasors of peak values F_a and F_f ampere-turns per pole respectively, each directed along the corresponding axis of maximum m.m.f. The fundamental component of gap flux on load is proportional to the phasor sum F of F_a and F_f (neglecting magnetic saturation).

F_f is centred on the pole axis (the direct or d-axis) to which the interpolar axis (the quadrature or q-axis) is in electrical space quadrature. In general, the axis of F_a is displaced from the d-axis by an angle β depending on the power factor. F_a can be resolved into components F_{ad} and F_{aq} respectively on the d- and q-axes.

Figure 11.1(a) shows, for a balanced 3-phase cylindrical-rotor machine, the stator and rotor current-sheet patterns for an instant of zero current in stator phase C. (The black areas represent outward, and the cross-hatched areas inward, current direction.) The m.m.f. phasors F_a and F_f are displaced by angle β. The resultant m.m.f. acting on the airgap is the phasor sum $F_f + F_a = F$. Assuming each m.m.f. to develop an individual flux,

11.1 Introduction

The alternating-current (a.c.) generator is almost invariably a 3-phase machine operating at a standard frequency of 50, 60 or 25 Hz. It has two functions: (i) converting the prime-mover mechanical power into electrical active power; (ii) generating or receiving electrical reactive power.

An isolated generator must supply the active and reactive power demand of the load to which it is connected: changing the excitation changes the load terminal voltage and the load current. Most generators, however, are synchronous machines connected to an extensive supply network, and changing the excitation of one machine does not affect the power output or the speed: it changes only the reactive power, and a compensating change in reactive power is shared by the generators operating on the same or neighbouring busbars. System engineers, for brevity, refer to the flow of 'power' and 'vars', meaning respectively active power and reactive power. By convention, inductors and induction machines, and underexcited synchronous machines, demand positive vars from the system. Capacitors and over-excited synchronous machines supply positive vars to the system.

The constructional forms of a.c. generators are basically dependent on the speed of the prime mover. The number of pole-pairs p of a machine driven at a speed n (rev/s) and operating at a frequency f (Hz) is

$p = f/n$

which defines the *synchronous speed*. Generators may be classified as follows.

Synchronous generators. Turbogenerators These are cylindrical-rotor machines driven by steam turbine at 60, 50, 30 or 25 rev/s (3600, 3000, 1800 or 1500 rev/min), with ratings up to 1200 MW. In general, hydrogen cooling is used for ratings above 100 MW, while above 200 MW the stator winding is usually water cooled.

Hydrogenerators These are salient-pole machines driven by water turbines at speeds in the range 50–1000 rev/min, depending on the type of turbine which in turn depends on the head and flow-rate of the water available. Ratings up to 800 MW have been achieved.

Engine-driven generators These are salient-pole machines driven directly by gas or diesel engines or through gearboxes by gas (or occasionally steam) turbines with ratings up to about 25 MW.

Synchronous compensators These are self-driven machines for dealing only with reactive power, they are normally salient-pole machines with six or eight poles and ratings up to 350 Mvar. Hydrogen cooling is normal for ratings of 60 Mvar upward. The demand for synchronous compensators has been reduced by the development of static equipments (Chapter 16).

Induction generators These are asynchronous generators that are occasionally employed, in ratings up to 5 MW and speeds up to 1000 rev/min. They require reactive power from the associated network for magnetisation.

The power output of any generator is controlled by the governor setting of its prime mover. The governor responds to small speed changes to adjust the mechanical input. With synchronous machines, the voltage and speed are held almost constant over the load range by the combination of the governor and the automatic voltage regulator. In the induction generator, the synchronous speed is determined by the supply network to which it is connected, and the rotor speed varies over a small range above synchronous speed (i.e. with negative slip) to control the power output.

Field windings for synchronous machines are always carried on the rotor, in slots for a turbogenerator or on salient poles for the other types. The induction generator has no separate field windings as such, and the rotor carries a cage winding.

In the following, Sections 11.2–11.14 deal with synchronous generators operating on 'infinite busbars', Sections 11.15–11.18 with turbo-, hydro-, and engine-driven synchronous machines, Section 11.19 with synchronous compensators and Section 11.20 with the induction generator.

11.2 Airgap flux distribution and no-load e.m.f.

The distributed field winding and the uniform airgap length of a turbogenerator result in a gap-flux density of trapezoidal shape around the surface of the stator. The salient-pole machine has a more rectangular distribution with short arcs of low flux density between the poles. Pole-shoes are usually chamfered, i.e. shaped so that the gap length increases progressively from the pole-centre towards the pole-tips, rounding the field form. In both types the no-load field form can be regarded as sinusoidal (with one pole-pitch equal to 180° electrical) together with a number of odd-order harmonics, particularly the third and fifth. Normally the amplitude of the fundamental flux-density component is a few percent greater than that of the field form, but the root-mean-square (r.m.s.) values of the actual wave and its fundamental component are almost identical. Further, the distribution of the stator winding around the stator bore surface and the use of coil-spans less than a pole-pitch (usually a 5/6 pitch) make the harmonic content of the stator electromotive force (e.m.f.) much smaller than that in the field form. Thus in calculating the e.m.f., the assumption of a sinusoidal flux distribution gives negligible error. The r.m.s. value of the phase e.m.f. is then

$E_{ph} = \sqrt{2} \pi f K_w N_s \Phi = 4.44 f K_w N_s \Phi$ volts

where f is the frequency, N_s the number of turns in series per phase, K_w the winding factor and Φ the fundamental component of the flux per pole in webers.

The winding factor K_w is the product of k_d, the *distribution factor*, and k_c, the *chording factor*. Normally each phase band has a spread of 60° electrical, and k_d is close to 0.96; $k_c = \cos(\frac{1}{2}\alpha)$ for the fundamental and $\cos(\frac{1}{2}n\alpha)$ for the nth harmonic, where α is the electrical angle by which the coils are short-pitched. For the common span of 5/6 of a pole-pitch, $\alpha = 30°$ and k_c has the following values:

Harmonic order:	Funda-mental	3	5	7	9	11, 13
Coil-span factor:	0.966	0.707	0.259	0.259	0.707	−0.966

Thus the 5th and 7th harmonics in the e.m.f. wave are reduced. The 3rd and 9th do not appear between lines in a 3-phase winding.

The practicable and economic line voltage increases with the rated output, typically 400 V to 11 kV (line-to-line) for industrial salient-pole generators up to 22–28 kV for large turbo or hydro units.

11.3 Stator windings

To provide a balanced 3-phase voltage a double-layer winding is used, with formed coils (or half-coils) wedged in open slots. The numbers of slots per pole, turns per coil and parallel circuits per phase are chosen to provide the required phase voltage and current with appropriate values of voltage and current per

11 Alternating-current Generators

A Hunt BSc(Eng), CEng, FIEE
Engineering Consultant

Contents

11.1 Introduction 11/3
11.2 Airgap flux distribution and no-load e.m.f. 11/3
11.3 Stator windings 11/3
11.4 Insulation 11/4
11.5 Temperature rise 11/4
11.6 Output equation 11/4
11.7 Armature reaction 11/4
    11.7.1 Cylindrical-rotor machine 11/4
    11.7.2 Salient-pole rotor machine 11/5
    11.7.3 Equivalent field m.m.f. 11/5
    11.7.4 Short-circuit ratio 11/5
11.8 Reactances and time-constants 11/6
    11.8.1 Armature leakage reactance 11/6
    11.8.2 Magnetisation (armature-reaction) reactance 11/6
    11.8.3 Synchronous reactances 11/6
    11.8.4 Transient and subtransient reactances 11/6
    11.8.5 Negative-sequence reactance 11/7
    11.8.6 Zero-sequence reactance 11/7
    11.8.7 Reactance values 11/7
    11.8.8 Reactances and time-constants 11/7
    11.8.9 Potier reactance 11/8
    11.8.10 Frequency-response tests 11/8
11.9 Steady-state operation 11/9
    11.9.1 Steady load conditions 11/9
11.10 Synchronising 11/10
    11.10.1 Synchronising procedure 11/10
    11.10.2 Synchronising power and torque 11/11
    11.10.3 Oscillation frequency 11/11
11.11 Operating charts 11/11
    11.11.1 Cylindrical-rotor generator 11/11
    11.11.2 Salient-pole generator 11/12
11.12 On-load excitation 11/12
    11.12.1 M.M.F. phasor diagram 11/12
    11.12.2 The ANSI Potier-reactance method 11/13
    11.12.3 Use of design calculation 11/13
11.13 Sudden 3-phase short circuit 11/14
11.14 Excitation systems 11/16
    11.14.1 D.C. exciters 11/16
    11.14.2 A.C. exciters with static rectifiers 11/16
    11.14.3 Brushless excitation 11/16
    11.14.4 Thyristor excitation 11/17
    11.14.5 Excitation system circuits 11/17
    11.14.6 Excitation control 11/17
    11.14.7 Basic principles of voltage control 11/17
    11.14.8 Additional control features 11/19
    11.14.9 Overall voltage response 11/21
11.15 Turbogenerators 11/21
    11.15.1 Main dimensions 11/22
    11.15.2 Rotor body 11/22
    11.15.3 Rotor winding 11/22
    11.15.4 Stator core 11/23
    11.15.5 Stator casing 11/23
    11.15.6 Stator winding 11/23
    11.15.7 Cooling 11/23
11.16 Generator–transformer connections 11/25
11.17 Hydrogenerators 11/26
    11.17.1 Construction 11/26
    11.17.2 Excitation 11/26
    11.17.3 Pumped-storage units 11/27
11.18 Engine-driven generators 11/27
11.19 Synchronous compensators 11/28
11.20 Induction generators 11/28

11 MASTERS, J. and PEARSON, J., 'Automotive gas engines power air conditioning systems, heat pump installations and heat and power units', *Int. Gas Engineers*, January (1981)
12 JENSEN, W., 'Criteria for the use of compression heat pumps in industry', *ASVE*, **6**, 22–28 (1981)
13 PAUL, J., 'Schrauben und Kolbenverdichter in Vergleich', *Kalte Klimatechnik*, **12**, 1–6 (1981)
14 IOFFE, A. F., *Semiconductor Thermoelements and Thermoelectric Cooling*, Infosearch Ltd (1957)
15 SPANKE, D., 'Air conditioning using heat pumps and Peltier cells', *Elektrowarme Int.*, **26** (6), 220–227, June (1968). Electricity Council OA Translation 1152
16 YANKOV, V. S. and FILKOV, V. M., *Soviet Research on Large Heat Pump Stations for Centralised Heat Supply*, World Energy Conference Working Party Report (1979)
17 Swedish State Power Board, *Heat Pumps and Solar Energy* (1982)
18 NEI Projects Ltd, *The Templifier Heat Pump*, Publication NP2 (1983)
19 LAWTON, J., MACLAREN, J. E. T. and FRESHWATER, D. C. 'Heat pumps in industrial processes', in *The Rational Use of Energy*, 47–56, Watt Committee London (1977)
20 BRAHAM, D., 'The energy factor', *51st Annual Conference Baths & Recreational Management*, 1–27, September (1981)
21 FESSEL, E., 'Public indoor swimming pool with heat recovery', *Elektrowarme Int.*, **33** (5), 230–234 (1975)
22 HODGETT, D. L., 'Dehumidifying evaporators for high temperature heat pumps', *UF-Int. Inst. Refrigeration Conference*, Belgrade (1977)
23 PERRY, E. J., *Drying by Cascaded Heat Pumps*, Inst. of Refrigeration (1981)
24 GEERAERT, B. 'Air drying by heat pumps with special reference to timber drying', in Camatini, E. and Kester, T. (Eds.), *Heat Pumps and their Contribution to Energy Conservation*, Noordhoff (1982)
25 SETA, P., BERHONDO, P. and ROBIN, P., 'Heat pump tests in new individual homes and block apartments: first results and observations', *Comité Français d'Electrothermie: Versailles Symposium* (1979). Electricity Council Trans. OA 2378
26 JACKSON, A. and STERLINI, P. A., 'The performance of air to water heat pumps in domestic premises', *Domestic Heating*, **14**, 12–14 (1981)
27 KALISCHER, P., 'Operating experience with dual fuel heat pump systems', *Elektrowarme Int.*, **37**, 266–271 (1979)
28 KALISCHER, P., 'The heat pump for hot water supply in the residential sector', *Unipede Workshop on Domestic Electric Hot Water Supply*, EBES, Antwerp, April (1982)
29 BRUNDRETT, G. W. and BLUNDELL, C. F., 'An advanced dehumidifier for Britain', *Heating Vent. Engr*, 6–9, November (1980)

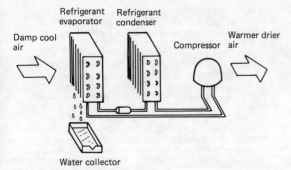

Figure 10.24 A domestic dehumidifier removes water vapour and provides heat

10.13.5 Conclusions

Heat pump applications are very diverse. The expertise required varies widely with the different types of application. The very large plants (~1 MW) are tailormade for specific tasks which need much design analysis for successful integration. As the plant size decreases, the heat pump technology is built into packaged units and the application skills needed become those of a building service engineer. Illustrations of the performance and size of different applications are summarised in *Figure 10.25*.

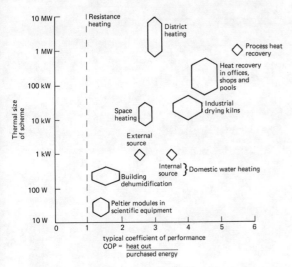

Figure 10.25 Illustrative Coefficients of Performance for heat pumps in different applications

In general, the heat pump is an advanced piece of engineering which saves energy by extracting it from a low-temperature source and making it available at a higher and more useful temperature. The three key factors for its successful use are: (1) where both heating and cooling are required, preferably simultaneously; (2) where moisture has to be removed, and preferably where some heating is needed simultaneously; (3) where the hours of use are long each year, so that the revenue savings can justify the increased initial cost which the heat pump incurs.

Professional guides

ARI (Air Conditioning and Refrigeration Institute, USA):
   ARI 240-77 Air Source Unitary Heat Pump Equipment
   ARI 260-75 Application, installation and servicing of unitary systems
   ARI 320-76 Water source heat pumps
   ARI 340-80 Commercial and industrial heat pump equipment
   ARI 270-80 Sound rating of outdoor unitary equipment
   ANSI/ARI 310-76 Packaged terminal heat pumps

UL (Underwriters Laboratories Inc., USA):
   ANSI/UL—559-1976 Heat Pumps (A standard for safety)

NBS (National Bureau of Standards, USA):
   NBSIR 76-1029 Unitary heat pump specification for military family housing. C. W. Phillips, B.A. Peavy and W. J. Milroy (1976). Prepared for Family Housing Division; US Air Force
   NBSIR 80-2002 Method of testing, rating and estimating the seasonal performance of heat pumps. W. H. Parken, G. E. Kelly and D. A. Didion (1980)

ASHRAE (American Society for Heating Refrigerating and Air Conditioning Engineers, USA):
   ANSI/ASHRAE 37-38 Method of testing for rating unitary air conditioning and heat pump equipment

German Standards (Deutsche Normen):
   DIN 8900 Heat Pumps; heat pump units with electric driven compressors
      Pt. 1. Concepts (April 1980)
      Pt. 2. Rating conditions, extent of testing, marking (October 1980)
      Pt. 3. Testing of water/water and brine/water heat pumps (August 1979)
      Pt. 4. Testing of air/water heat pumps (June 1982)

British Air Conditioning Approvals Authority, 30 Millbank, London, SW1P 4RD:
   Interim standard for performance and rating of air to air and air to water heat pumps up to 15 kW capacity. B. J. Hough (1982)

Electricity Council, 30 Millbank, London, SW1P 4RD:
   Heat pumps and air conditioning: A guide to packaged systems, 1982.

References

1. AMBROSE, E. R., *Heat Pumps and Electric Heating*, Wiley (1966)
2. HEAP, R. D., *Heat Pumps*, Spon (1979)
3. BERNIER, J., *La pompe de chaleur*, Pyc Edition (1979)
4. REAY, D. A. and MACMICHAEL, D. B. A., *Heat Pumps—Design and Application*, Pergamon Press (1979)
5. VON CUBE, H. L. and STEIMLE, F. (English edn., E. G. A. Goodall), *Heat Pump Technology*, Butterworths (1981)
6. GROFF, G. C. (Ed.), *Heat Pump and Space Conditioning Systems for the 1990's*, International Symposium by Carrier Corporation, USA (1979)
7. LOPEZ-CACICEDO, C. L., 'Electrically driven heat pumps: current research and future prospects', *Energy World* Heat Pump Supplement, 19, October (1981)
8. COOPER, K. W. and SUMNER, L. E., 'An open cycle study using moist air thermodynamics', *ASHRAE J.*, 68–71, January (1978)
9. VAUTH, R., 'A new heat pump operating on the cold air principle', *Heat Pump Conf.*, Essen 1977, Electricity Council OA Trans., 1775
10. JESINGHAUS, J., 'Development trends in absorption heat pumps', *Sonnenergie Warmepumpe*, **6** (5), 31–34, September/October (1981). British Gas, Watson House Translation WH 738

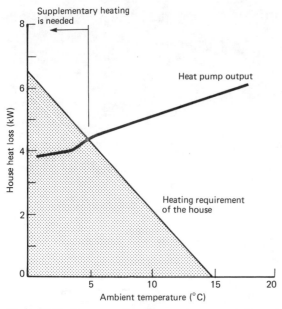

Figure 10.21 The output of an air source heat pump declines with colder conditions. The small step decline around 5°C is due to de-icing the evaporator

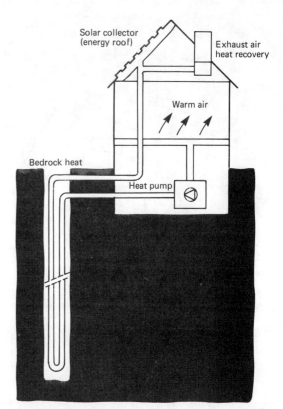

energy, provided that a sufficient area of brine filled pipes are buried in it (*Figure 10.22*).

10.13.4.4 1–10 kW (thermal) packaged units

Two types of equipment are available, one domestic and one industrial. The domestic unit is a heat pump water heater and it is often combined with a hot water cylinder[28] (*Figure 10.23*). The heat source is air from inside or outside the building. Such applications are very attractive if interior cooling is needed, e.g. cooling a beer cellar and using the reject heat to provide hot water for washing the glasses. Conventional equipment has a maximum temperature of 55 °C and therefore the water storage volume has to be a little higher than with other forms of water heating.

The industrial application is dehumidification. Portable units are available to dry out damp or freshly built buildings. Fixed units are now being used to maintain low humidities in warehouses, particularly as warehouses are becoming automated and no longer require heating for the occupants.

10.13.4.5 100 W–1 kW (thermal) units

Damp, cold conditions characterise Britain's winter climate. Small heat pump dehumidifiers extract moisture and translate the latent heat into sensible heat. Cool, damp air enters the evaporator and is chilled, depositing much of its moisture. The same air is then reheated over the condenser and returned to the room. Present equipment has a coefficient of performance which varies from 1.1 to 2.0, the higher value being associated with warmer and damper conditions[29] (*Figure 10.24*).

10.13.4.6 10–100 W (thermal) modules

Peltier modules are effective ways of providing very small heat flows which can heat or cool. Their main application is to provide stable reference temperatures in scientific equipment.

Figure 10.22 Advanced house design uses the heat pump in conjunction with heat recovery. The 'chimney' is now a ventilator, the roof a solar collector and the floor a low-temperature heat emitter. The ground provides the extra energy[17]

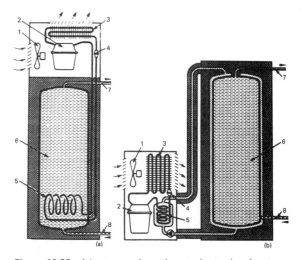

Figure 10.23 A heat pump domestic water heater. Local water regulations may prohibit the direct immersion of a refrigerant heat exchanger into the water cylinder[28]. 1 = fan, 2 = compressor, 3 = evaporator, 4 = expansion valve, 5 = condenser, 6 = storage, 7 = hot water, 8 = cold water

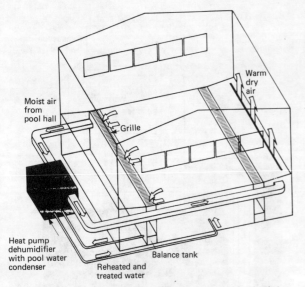

Figure 10.17 Heat pump dehumidifiers can recover much of the sensible and latent heat from the warm moist air and return it to the pool water. Not only does this recover energy, but also it enables the conventional ventilation rate to be reduced[20]

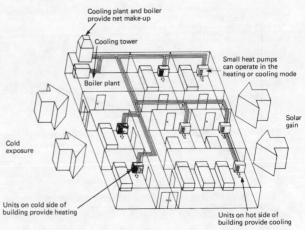

Figure 10.19 Linked room units can provide heating in some areas and cooling in others

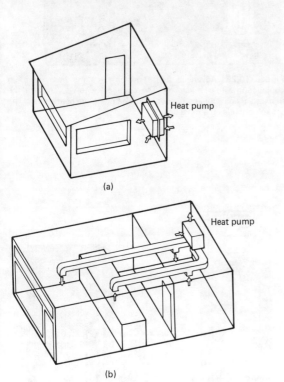

Figure 10.18 Reversible air to air room units can either heat or cool. (a) Reversible air to air room unit; (b) Self-contained ducted roof mounted heat pump

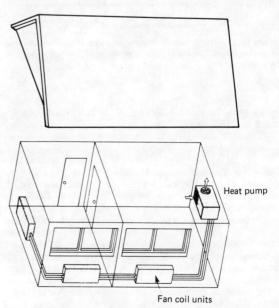

Figure 10.20 An air-to-water domestic heat pump. The heat pump is shown inside the roof space but it can be sited in the garden

The second and less common type is the air/water space heating heat pump[25-27]. They usually use outside air as the heat source and supply the heat to the house through conventional water radiators (*Figure 10.20*). They are often used in conjunction with supplementary heating because both the effectiveness and the output of such machines fall when the outdoor temperature falls and heating need is greatest (*Figure 10.21*). They also have a maximum water temperature of 55 °C, which is lower than the figure for the conventional boiler of 80 °C. Care has to be taken, therefore, to ensure that the area of radiator is sufficient, when operating at 55 °C. There are two other cautionary points—noise and starting current surges. Air source heat pumps can be noisy and must therefore be selected and sited so that noise levels immediately outside the bedroom window are below 45 dB(A). Compressors with single-phase electrical drives greater than 1 kW rating have to be checked with the local Electricity Board to see whether the electrical network would be unduly disturbed by the connection of such a pump. Soft start units are now available on single-phase domestic units. Such devices are particularly helpful if the heat pump is switching on and off frequently.

While ambient air is the usual energy source, groundwater can be used. Even the earth around the building can supply the

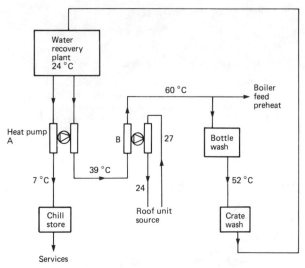

Figure 10.14 Two McQuay centrifugal compressors operate up to 70°C to provide 1 MW of heat recovery in the British Milk Marketing Board's Dairy at Bamber Bridge. The Coefficient of Performance under these conditions is 5.5

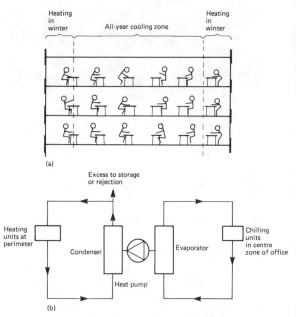

Figure 10.15 Heat pump recovery in deep plan offices. (a) Section of a deep plan office; (b) The cooling circuit provides the perimeter heating in winter

Process heat treatment can be illustrated by the 1 MW Milk Marketing Board Dairy Plant at Bamber Bridge, England. This dairy is a bottling and cartoning depot serving retail outlets. Two McQuay Templifier centrifugal compressors are employed in series (*Figure 10.14*). Recycled effluent from the bottle washers is stored and then pumped through both the evaporator and the condenser sections of the first heat pump. The outlet water from the evaporator at 7 °C is directed into the dairy supply tank as chilled water for dairy services. The water leaving the condenser passes into the condenser of the second heat pump, where it is heated to 60 °C and provides a boiler feed preheat and a crate washing unit. The overall COP is 5.5.

10.13.4.2 100 kW–1MW (thermal) schemes

The three main applications are commercial buildings, small industrial batch drying plant and swimming pools[20,21]. The compressors are multicylinder piston or rotating vane types.

Deep open plan offices require winter heating at the perimeter to combat the outdoor climate but require permanent cooling of the central core. The heat pump enables the energy to be redistributed within the building (*Figure 10.15*). Such installations halved the energy cost of air conditioned buildings in Britain, and formed the new concept of integrated environmental design. Coefficients of performance are 3–4.

The second type of machine is usually a factory packaged system for batch industrial drying[22–24]. Compact dehumidifiers which operate up to 80 °C are now available for timber drying. Such machines are particularly suitable for the controlled drying required for hardwood to avoid timber splitting (*Figure 10.16*).

Swimming pools are particularly energy intensive. Internal design conditions must not exceed 70% relative humidity of the pool hall air if condensation and mould growth are to be avoided. The conventional technique is to ventilate at a high rate and lower the moisture content of the air by dilution with air from outside. Heat pump dehumidification enables the moisture in the pool air to be controlled without losing great amounts of heat with high ventilation rates. The latent heat recovered from the moisture is used to heat up the pool water (*Figure 10.17*). The warm, moist air conditions mean that the heat pump can operate at a COP of 5–6.

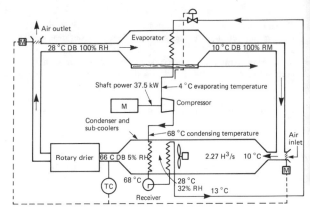

Figure 10.16 Single-stage heat pump dryer with sub-cooler has a Coefficient of Performance of 4.4[23]

10.13.4.3 10–100 kW (thermal) packaged units

Two types of machine of this size are factory packaged[22]. The most common is the reversible air-to-air space conditioner. It is reversible because by an arrangement of valves the unit can interchange evaporator and condenser by cooling in summer or heating in winter. The equipment is usually installed 'through the wall' in offices and shops, with each unit controlling a small zone within the building (*Figure 10.18*). It is also commonly sited on the flat roof of shops. A modification of this principle is applied to large buildings, particularly older office blocks where the glazing area is large. In such buildings the individual heat pumps are attached to a ring main of recirculating tepid water. For those parts of the building needing cooling the local heat pumps reject the heat to the ring main (*Figure 10.19*). Those parts of the building needing heating use the heat from the ring main as their heat source. Any net heating or cooling is provided from the central boiler or the central chiller.

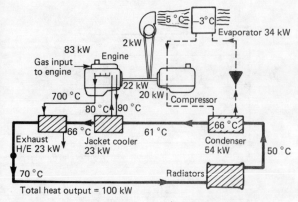

Figure 10.10 A gas engine driven air-to-water heat pump with heat recovery from the engine. (Coefficient of Performance = 2.7)

sor. Piston compressors, where the lubricating oil is in intimate contact with the working fluid, are constrained by oil degradation. Dry compressors, where the bearings are sealed from the working fluid, can operate at higher temperatures. However, the chemical and physical stability of the working fluid itself then provides the working temperature limits[12,13].

A range of operating temperatures for different working fluids is illustrated in *Figure 10.11*.

10.13.3.3 Thermoelectric heat pump: Peltier device

When a direct electric current passes round a circuit incorporating two different metals, one junction of the two metals is heated and the other cooled[14,15]. To be effective these Peltier couples must have a high thermoelectric coefficient α, a low thermal conductivity κ and low electrical resistivity ρ. The high thermal conductivity of metals normally makes the units very inefficient (COP 1.01). However, recent progress in semiconductors has improved α, enabling much more effective units to be made.

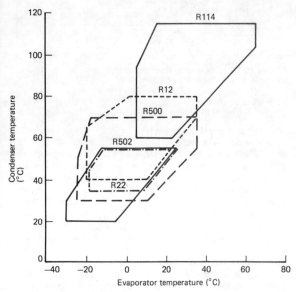

Figure 10.11 Operating temperature range for different working fluids

The Peltier effectiveness (z) is given by

$$z = \frac{\alpha^2}{\kappa\rho}$$

Present-day materials have $z = 0.003/\kappa$.

The overall performance of such devices is still short of that achieved by vapour compression cycles but the small size, reliability and ease of making low-capacity modules gives them a special market. A typical module layout is illustrated in *Figure 10.12*.

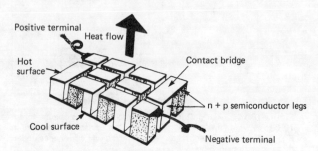

Figure 10.12 Peltier thermoelectric module (d.c. electric)

10.13.4 Scale

The size and complexity of heat pump applications is very wide, with appropriate specialist techniques for each application. For convenience we shall examine applications by size.

10.13.4.1 1–10 MW (thermal) schemes

Large-scale heat pump investments are attractive when the running time is long each year. Two types of application meet this requirement. These are base load space and water heating for district heating schemes[16,17] and heat recovery techniques in large continuous industrial processes[18,19]. The compressors are usually high-speed centrifugal types or screw compressors.

Groundwater, sea-water, lakes or sewage treatment can provide the heat source for district heating schemes. Preliminary results from the Swedish Sala Municipal district heating network show that a screw compressor can provide 3.2 MW thermal energy at an annual COP of 2.7 (*Figure 10.13*). Availability in its first year was 80%. This heat pump operates throughout the year, providing the base heating load in winter and the hot water heating in summer. Supplementary heating which is needed in the depth of winter is provided by a conventional oil burning boiler.

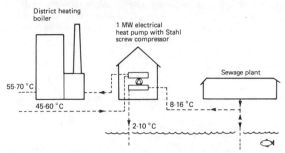

Figure 10.3 Town sewage provides the heat for Sala, central Sweden, with an annual Coefficient of Performance of 2.7

replacement with electricity produced on-site in generating plant integrated with heating (or cooling) plant.

10.13 Heat pumps

10.13.1 Introduction

Heat pumping is the use of a thermodynamic cycle to extract heat from a lower-temperature source and supply it to a higher-temperature sink where it is useful. In doing so the purchased energy needed to drive the cycle is less than that usefully supplied. The ratio between useful heat delivered and the energy purchased is the Coefficient of Performance (COP). It should always be greater than unity.

Thermodynamic cycles have been well developed over the last century for refrigeration but it is only in recent years that the heating application has developed. They were first applied in the USA, where the coastal regions required air conditioned cooling in summer and space heating in winter. Over two million such heating/cooling units have been sold.

In the last 10 years there has been a rapid widening of heat pump applications for winter space heating, industrial heat recovery and dehumidification[1-7].

10.13.2 Thermodynamics

The ideal thermodynamic cycle for heat pumps, developed by Carnot, assumes a perfect working fluid operating in perfect conditions:

$$\text{ideal COP} = \frac{T_{\text{hot}}}{T_{\text{hot}} - T_{\text{cold}}}$$

where the heat is extracted from a cold source and supplied to a hot sink. The temperatures are absolute temperature (i.e. °C + 273) (*Figure 10.8*).

This ideal cycle shows that the usefulness of a heat pump cycle depends mainly on the temperature difference between heat source and heat sink. The heat pump performance improves as this temperature difference narrows. The heat pump also improves slightly with increasing temperature.

$$\text{Coefficient of Performance} = \frac{\text{heat out}}{\text{compressor + fan energy in}}$$

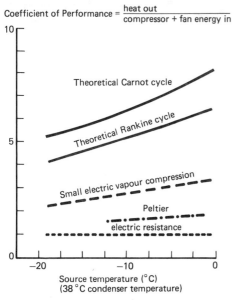

Figure 10.8 Coefficients of performance in theory and practice

In practice, with real fluids and real equipment the best performance is obtained from the vapour compression cycle, which achieves about a third of the ideal Carnot efficiency.

13.13.3 Practical cycles

There are three basic cycles:

10.13.3.1 Air cycle

When air is compressed, it becomes warmer. Heat can be extracted and the cooled pressurised air expanded down to its original pressure. The expander can be a turbine which drives the compressor. This open cycle can be used for heating buildings. Unfortunately, the equipment is very bulky and its efficiency is very senstiive to the inefficiencies of both the compressor and the expander[8,9]. It is not commercially attractive except for special applications where compressed air is readily available, such as aircraft air conditioning.

10.13.3.2 Vapour compression cycle

The vapour compression cycle relies on the condensation temperature increasing with increase in pressure. A vapour from the evaporator when compressed will condense at a higher temperature, corresponding to the new higher pressure. Successful working fluids must have a high latent heat of condensation so that the bulk of the heat can be extracted at the highest possible temperature. This principle applies to all conventional heat pumps (*Figure 10.9*).

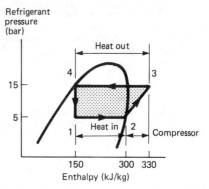

Figure 10.9 The vapour compression cycle (R22). From 1–2 the refrigerant vapour absorbs heat; from 2–3 the compressor compresses the gas; from 3–4 the gas is condensed and its latent heat released; from 4–1 the liquid expands to a vapour at the lower pressure

The change in pressure between the evaporator and the condenser can be created by any mechanically driven compressor or by a physical absorption cycle[10]. At input evaporating temperatures between −10 and +10 °C, typical of space heating heat pumps, the absorption cycle achieves 24–40% of the efficiency of a mechanical drive cycle. Almost all the equipment in use is mechanically driven and electric motors are the driving units. They are favoured because of their cost, simplicity, silence, efficiency, long life and reliability. However, there is an increasing use of fossil fuel driven engines because the waste heat from such an engine can often be incorporated into the heating scheme[11] (*Figure 10.10*).

The selection of heat pump working fluids is complex but a critical factor is the range of permissible condensing temperatures. The upper temperature limits depend upon the compres-

Table 10.2 Combined heat and power plant

Prime mover	Ratio of useful heat to power	Heat source	Available at temperature (°C)
Steam turbine:			
back-pressure	6:1 to 10:1	steam	100–300
pass-out condensing	up to 5:1		
Diesel or dual-fired engine:			
unfired waste-heat boiler	0.5:1 to 1.5:1	jacket-water	80
fired waste-heat boiler	up to 3:1	steam from waste-heat boiler	200*
Gas turbine:			
unfired waste-heat boiler	3:1	steam from waste-heat boiler	250*
fired waste-heat boiler	up to 15:1		

* Higher temperatures are attainable but are not normally required for process heating.

met or to produce surplus electricity when the maximum demand for process heat is being satisfied. In the former case the surplus heat must be dumped, which reduces the overall thermal efficiency of the scheme; in the latter, the surplus electricity can be exported to the public electricity supply system.

10.11 Total energy

There is no official definition of the term 'total energy'. It was coined in the 1960s, apparently in the USA, and has been taken to imply a scheme in which a single energy conversion centre provides electricity and heat to a system which is independent of any other sources of electricity and heat. It is, therefore, a particular form of CHP. Total-energy schemes which fall under this definition are typically small, with combined electricity and heat loads less than 30 MW.

There are numerous examples of total-energy schemes, some commercial and some industrial, though none appears to have been built in recent years. All, or nearly all, are based on a clean and easy to use fuel—natural gas. Indeed such schemes were first promoted in the USA to provide a market for the sale of natural gas. Both gas turbines and dual-fuel compression ignition engines have been used as prime movers, with exhaust heat recovery in boiler plant, sometimes with supplementary firing.

Perhaps the best-known total-energy scheme in the UK is that at the John Player tobacco factory in Nottingham. The factory requires electricity, steam and refrigeration (the latter for air-conditioning). The installation comprises eight Ruston 1.1 MW gas turbines operating on natural gas (of which six are intended to be available at any given time), four boilers operated on the exhaust from the gas turbines and fitted with auxiliary gas firing, and a steam turbine driven compressor together with two absorption machines for the refrigeration. Cooling towers are provided also, to reject to the atmosphere low-grade heat from the air-conditioning plant. The overall thermal efficiency of the scheme has been stated to be about 63%.

A total-energy scheme, since it is to be independent of public electricity supplies, must include spare plant capacity to cover planned maintenance and breakdowns. The amount of spare capacity will be influenced both by the degree of reliability desired and by the nature of the loads.

Transport systems—for example, ships and aeroplanes—might be regarded as special forms of total energy systems.

10.12 Economics

As in the case of all forms of investment, the decision whether to proceed with CHP or to deal separately with the requirements for heat and power will be made on the basis of an economic analysis. CHP schemes are likely to require a greater initial capital expenditure and to offer the prospect of a reduced annual expenditure on fuel or on electricity. For example, in a scheme using a back-pressure steam turbine, some or all of the electricity requirements are produced in-house and there is, therefore, a saving in the cost of electricity which otherwise would be purchased from the public electricity supply. In comparison with a heat-only installation, however, a more expensive boiler is required (to produce steam at a higher temperature and pressure) and the back-pressure steam turbogenerator must be installed. Thus, a comparison must be made between (a) the additional capital expenditure and the higher annual fuel costs and (b) the annual savings in the cost of purchased electricity.

The willingness of the public electricity supply authority to pay for surplus electricity exported to its system and the amount that it is willing to pay can have an important bearing on the economic attractiveness of CHP schemes.

The choice of CHP scheme will itself be influenced by the availability and cost of fuel. For example, coal may be plentiful and cheap. Coal is suitable only for use in boiler plant; thus, for CHP cheap coal implies the use of steam turbine plant.

Gas turbines have to be operated on clean fuels such as gas or distillate. Since these fuels are nowadays expensive, the scope for the use of gas turbines in CHP schemes is diminishing.

Many schemes for district heating have been investigated and it is generally accepted that, owing to the high cost of heat distribution, a combination of a densely populated area and a long heating season provides economic circumstances most favourable to district heating. These circumstances occur relatively frequently in continental Europe. The winter in the UK is not consistently cold enough for there to be a continuous high heating load. In continental Europe this is not the case, but even there district heating on any scale is limited to those countries where the supply of electricity is arranged on a municipal basis.

Most total-energy schemes in Europe and the USA were conceived at a time when natural gas was plentiful and cheap. The situation now is different: there is no longer surplus gas, nor is it relatively cheap. The economics of total-energy schemes depend on the elimination of purchased electricity and its

controlled by regulating the amount of steam which passes through the remainder of the turbine to the condenser. With the pass-out turbine, therefore, it is possible to control both the electrical output and the pass-out pressure. An electrical output of up to 5 MW and a steam flow rate at turbine inlet of 30 000 kg/h at 45 bar abs. and 400 °C with pass-out of 13 500 kg/h at 3 bar abs. might be considered typical. Because some of the heat produced in the boiler is rejected to waste in the condenser, the overall thermal efficiency of a pass-out condensing cycle will be lower than that of a back-pressure cycle, ranging from 20% or less at minimum pass-out to 70% at maximum pass-out.

10.10.2 Diesel engines

The diesel engine is a relatively efficient prime mover, but 55% or more of the energy in the fuel is rejected in the form of heat: about 25% to the jacket and charge air cooling water (which is raised in temperature thereby to 80 °C) and about 30% to the atmosphere via the exhaust gases, which have a temperature of 400 °C or more. If the exhaust gas is passed through a waste-heat boiler, some of this energy can be used to raise steam, albeit at modest pressure, which can be used for process or space heating. Where use can be made of this steam and of the heat in the jacket and charge-air cooling water, overall efficiencies of 70% and more can be achieved.

In order to avoid corrosion in the boiler from acids in the exhaust gases from the diesel engine, the gas temperature at the outlet of the waste heat boiler is not allowed to fall below the corresponding dew-point. Depending on the fuel, an outlet gas temperature of about 175 °C is normally considered to be a reasonable minimum. The maximum steam temperature will be determined by the minimum practical temperature difference between the exhaust gas and the water or steam. This minimum value occurs at the pinch-point, which is usually the point where the gas leaves the evaporative section of the boiler. Typically, saturated steam is produced at a pressure of 8 bar abs.

The exhaust gases from a diesel engine contain a significant proportion of oxygen—up to twice as much air is passed through a diesel engine as is theoretically required to support combustion—and it is possible, therefore, to burn additional fuel in the exhaust gas. The recoverable heat in the exhaust can in this way be increased up to sixfold at the expense, of course, of an increase in fuel consumption. As an alternative, or in addition, separate oil fired boiler plant may be installed to augment the output from the waste-heat boilers.

In general, the capital cost of diesel generator plant equipped with waste-heat boiler plant is lower than that of a conventional boiler and steam turbogenerator. In addition, some diesel engines are capable of burning residual oils of the type used in boiler plant. The diesel generator, therefore, provides an attractive alternative for some schemes, although maintenance is usually more expensive and reliability is lower than for steam plant. The output of a diesel generator used in a CHP scheme would typically lie within the range 1–5 MW. Depending on the scheme, a single engine or a number of engines might be employed. The diesel generator plant is usually operated in parallel with the public electricity supply.

A CHP scheme which has recently been put into service at Hereford is built around two 7.5 MW diesel generator sets which operate on residual fuel. The electricity is fed directly into the public electricity supply system. The diesel generator sets are provided with economisers and waste-heat boilers in which the exhaust gas is employed to raise steam at 20.5 bar with 11 °C superheat for use in two nearby factories. The boiler feedwater is passed through heat exchangers to recover heat from the engine lubricating oil and jacket water. Separate oil fired boiler plant supplies steam during peak heat load periods and when the diesel plant is shut down. The thermal efficiency of the scheme is said to be 76%. The Hereford scheme is operated by the Midlands Electricity Board.

10.10.3 Gas turbine plant

The gas turbine is a relatively inefficient prime mover, having an efficiency of between 18 and 30%, according to size and complexity, and most of the energy in the fuel appears in the exhaust gas, the temperature of which can be 500 °C or more.

Since the temperature of the exhaust gas is higher from a gas turbine than from a diesel engine steam can be raised, in a waste heat boiler, to a higher temperature.

The air/fuel ratio of a gas turbine is perhaps 70:1 or twice as high as that of a diesel engine. There is therefore a greater proportion of oxygen in the exhaust gas and more scope, with supplementary combustion, for increasing the ratio of useful heat to power.

Gas turbines have generally to be operated on clean fuels such as gas or distillate.

10.10.4 District heating

District heating is a particular application of CHP. As generally understood, district heating combines the production of heat for domestic and commercial space heating and for commercial process heating with the generation of electricity in large power stations. The power plant for the larger schemes is usually of the pass-out condensing steam turbine type, a typical machine having an electrical output of 150 MW with a heat output of 200 MW. The heating steam passed out of the turbine is employed in a heat exchanger at the power station to heat water, usually from 50 to about 95 °C. The hot water is pumped through insulated mains, sometimes several kilometres long, to residential and commercial areas, where the heat is transferred, usually via heat exchangers, to the consumers' heating circuits.

10.10.5 Desalination

The combination of electricity generation with the production of fresh water in distillation plants is another example of a particular application of CHP. In principle, the scheme is similar to district heating. The extracted steam, however, is employed, usually in multiple-stage flash distillation plants, to produce fresh water from salt or brackish water.

Some large combined power/desalination plants have been installed in recent years, notably in the Middle East. In such an installation a pass-out condensing machine with an electrical output of 150 MW might be employed to provide steam at 2 bar abs, to a desalination plant with an output of 27 500 m^3/day of distilled water.

10.10.6 Ratio between heat and power

It has already been noted that the ratio between the requirement for heat and the requirement for electrical power, and the temperature (or temperatures) at which the heat is required, are important elements when designs for CHP schemes are under consideration. The requirements will exert an influence on the choice of prime mover. An indication of the ratio between useful heat and electric power output (measured in the same units) of the different types of prime mover and the temperature at which the heat is available is given in *Table 10.2*.

It has been noted earlier that there will be occasions when the output of the CHP plant cannot be matched exactly to requirements for heat and power. The mismatch will be in the form of either a surplus or a deficit of one of the requirements when the other is being met. The plant can be designed either to produce a surplus of heat when the maximum demand for electricity is being

turn, will have a bearing on the choice of prime mover in a CHP scheme. The application of different forms of prime movers to CHP schemes is discussed briefly below.

10.10.1 Steam plant

In steam power stations the greater part of the energy in the fuel which is not converted into electricity is dissipated as low-grade heat in cooling water, large quantities of which are discharged at a temperature in the range 25–45 °C. Unfortunately, not much use can be made of copious volumes of tepid water, except perhaps for fish farming. The cooling water is drawn initially from natural sources such as rivers, lakes and the sea, and either is returned to these sources or, where this is not practical, recirculated through cooling towers in which the heat absorbed by the water in the power station is given up to the atmosphere mostly by evaporation.

Back-pressure steam turbines. If useful heat is to be obtained from steam plant, it must be derived from the steam at temperature higher than that prevailing at the exhaust in power station condensing plant. The expansion of steam in the turbine must be halted and the steam brought out at some suitable temperature and pressure. In the simplest arrangement a back-pressure turbine (without a condenser) will be used in which all the steam passing into the turbine will be exhausted at the chosen back-pressure (*Figure 10.6*).

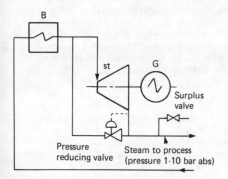

Figure 10.6 Back-pressure steam turbine providing electricity and process heat

If the latent heat in the steam exhausted from the back-pressure turbine is employed for process heating and if the hot condensate from the process heat exchangers is returned to the boiler house, such a scheme can produce heat and electricity at an overall thermal efficiency of up to 80%—i.e. the scheme may be nearly as efficient as a heat-only industrial plant.

The amount of electricity that can be generated in a back-pressure scheme depends on: (a) the steam temperature and pressure at the turbine inlet, (b) the steam pressure at the turbine exhaust, (c) the efficiency of the steam turbine and (d) the rate of flow of the steam. The steam pressure at the turbine exhaust is usually determined by the temperature required in the industrial process. It is normal practice to allow about 10 °C of superheat at the turbine exhaust to avoid wetness in the last stages of the turbine and condensation in the process steam supply pipework. Excess of superheat is generally avoided because, in comparison with dry saturated steam, superheated steam has a poor coefficient of heat transfer and because superheated steam cannot give up all its heat at a constant temperature. The steam flow rate is determined by the process requirement for heat. The steam conditions at the turbine inlet are selected so that, after expansion through the turbine, the steam at the exhaust is at the desired condition. A number of combinations of inlet temperature and pressure are possible for any particular requirement. The metallurgical properties of boiler tubing present an upper limit but this is rarely approached in industrial CHP. The inlet conditions may be selected so as to obtain a particular electrical output from the steam, but it is more likely that the conditions will be chosen after consideration of practical matters such as turbine design, reliability and ease of operation.

Thus, the scope for adjusting the heat/power ratio, for given process heating requirements, is limited. In many factories, however, the required heat/power ratio fluctuates widely, both daily and seasonally, and some means must be available for reconciling the required heat/power ratio with that available. If the requirements for both are to be met in full, there must be means for controlling the plant over the complete range of requirements. In practice this is rarely possible and arrangements have to be made outside the controllable range for the disposal of surplus electricity or heat. A number of methods are available, the most convenient of which is to operate the back-pressure turbogenerator in parallel with the public electricity supply: the demand for process steam determines the output of electricity, and electricity is imported from or exported to the public electricity supply system according to factory requirement and the output of the back-pressure turbogenerator. If for any reason it is not possible to operate the back-pressure plant in parallel with the public electricity supply system, the heat and power demands can still be balanced by means of a pressure reducing valve and a surplus valve, as shown in *Figure 10.6*.

A typical back-pressure set would produce 2 MW, lower and upper limits for industrial purposes being perhaps 500 kW and 15 MW, respectively. Steam conditions at the turbine inlet might range from 28 bar abs. and 400 °C for the smaller machines up to 63 bar abs. and 480 °C for the larger ones.

A typical back-pressure turbogenerator set might produce 2 MW of electrical output when expanding 17 500 kg/h of steam from inlet conditions of 42.5 bar abs. and 370 °C to exhaust conditions of 2.75 bar abs. and 140 °C.

Pass-out condensing steam turbines. Pass-out condensing steam turbines (in which some steam is taken from the turbine before the expansion is complete and some is expanded fully and taken to a condenser) are sometimes used when the demand for heating steam is small in relation to the demand for electricity or when the demand for heating steam is liable to wide fluctuations (*Figure 10.7*). The steam pressure at the pass-out is usually automatically

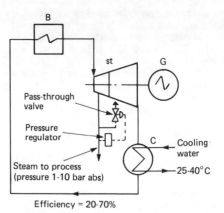

Figure 10.7 Pass-out condensing steam turbine providing electricity and process heat

The research programmes of the major nations in the fusion field (USA, USSR, Japan and Euratom) are closely co-ordinated. Major new fusion research facilities planned are:

TFTR	tokamak	1982	USA
JT 60	tokamak	1980	Japan
T	tokamak	1985	USSR
T M	tokamak	–	USSR
Jet	tokamak	1982	Europe-Culham
Shiva	Nd – glass laser	1980	USA

It is likely that a number of these will achieve ignition of a D–T fuel.

There is still a long way to go before commercial fusion reactors could be built. It is unlikely that they would be operational much before about 2020. A reactor would probably use the kinetic energy of fast neutrons to heat the tritium producing coolant of liquid lithium in a 'blanket' around the reaction chamber, and this liquid would pass through a series of heat exchangers to drive a turbine. Major engineering problems need to be overcome in the design of wall materials to withstand radiation densities of 4 MW/m^2 or more, the tritium breeding blanket of lithium, superconducting magnets, refuelling mechanisms, the reactor core to withstand vacuum and magnetic forces and the 100 MW, 300 keV neutron beams required for plasma heating, and their associated energy storage.

Tokamaks look, at present, to be the most favoured route to achieve fusion, although other devices, such as the stellerator, may also be successful. It is hoped that the Shiva laser in the USA will answer many of the outstanding questions on laser fusion, but so far laser systems are unsuited to power station application, since they have efficiencies below 1% and low repetition rates (up to 10 per hour). Efficiencies need to be improved by an order of magnitude and short-wavelength lasers need to be developed which can deliver pulses with peak power densities in excess of 10^{19} W/m^2 at a frequency of 10 Hz.

In the USSR particular interest has focused on hybrid fission–fusion devices in which fast neutrons from a fusion core are used to breed plutonium in a U-238 blanket for use in conventional thermal reactors. The fusion does not have to achieve break-even conditions in this configuration. The Russians hope to use the T-20 device as a hybrid system around 1990.

10.10 Combined heat and power

Because of the laws of nature rather than the limitations of engineering science, the conversion of heat into mechanical energy is inherently an inefficient process. The overall thermal efficiency of the most modern conventional thermal power station is not much more than 40% and 60% of the energy in the fuel appears as useless heat (*Figure 10.4*). In contrast, the conversion of energy in the fuel into useful heat is a relatively efficient process. Many industries use boiler plant to raise steam for product processing, and if the condensate is returned to the boiler house, an overall efficiency of conversion of primary energy into usable energy of over 80% can be achieved (*Figure 10.5*).

Since, in general, we require both electricity and heat, meeting the requirements for both from the same plant is (in concept) attractive. The major prospective gain of such a combined heat and power (CHP) scheme is the more efficient use of primary energy resources. The major difficulty is that of matching the supply of electricity and heat simultaneously to the demands. In practice the choice between CHP and the separation of heat from power is determined by their relative costs.

While the voltage at which electricity is generated can readily be transformed to the voltage (higher or lower) required by the load, the temperature at which heat is available from a particular plant cannot be increased unless more energy is supplied to the system. Heat content can be quantified only with reference to some base temperature. Thus, in the selection of suitable CHP plant for a particular application the temperature at which heat is required is an important factor. Equally important is the heat/power ratio.

The usual applications for CHP plant are in industry and for district heating. Only in a few countries (e.g. Russia, Sweden and Denmark) is district heating commonplace: there is, however, a current upsurge of interest in CHP/district heating in the UK. There are industries in which CHP is the established practice. In some, use is made of a waste product from the process; for example, in the processing of sugar cane the crushed cane (bagasse) remaining after extraction is burned in boilers to raise steam for use in turbogenerators, steam turbine mechanical drives or process heating. There are also industries which, although they use bought-in fossil fuels, have a history of the use of CHP; examples are the brewing, leather working, paper making and dyeing industries.

In general, process heat is required at constant temperature and thus steam, which has high latent heat, is the most suitable heat transfer medium. For district heating applications hot water is usually considered to be a more suitable medium for transferring heat from the CHP plant to the load.

Perhaps the most important technical consideration in the planning of a CHP scheme is the matching of the heat and electricity loads. These loads are liable to vary, more or less independently, from hour to hour, day to day and during the year, and clearly the plant must be capable of meeting the maximum and the minimum demands for heat and electricity and all intermediate demands. For many processes, the quality of the produce is unaffected by the source of the heat energy input. For example, cooking might be carried out by steam, electric hotplate or gas ring; and a compressor might be driven by an electric motor or a steam turbine. The form in which energy is employed, where there is a choice, will affect the heat/power ratio and this, in

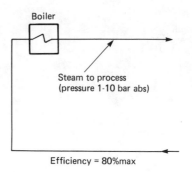

Figure 10.5 Industrial steam raising/process heating

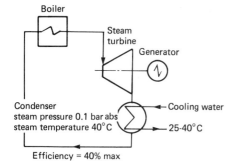

Figure 10.4 Steam power station

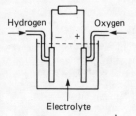

Figure 10.3 Hydrogen/oxygen fuel cell

Hydrogen/oxygen cell. The action of a fuel cell is illustrated by the H/O cell (*Figure 10.3*). Two porous nickel electrodes, incorporating suitable catalysts, are immersed in a potassium hydroxide electrolyte (30–40% concentration) which has good conductivity and relatively small corrosion. Hydrogen fed to one electrode reacts with hydroxyl ions of the electrolyte to form water and free electrons ($H_2 + 2OH^- \rightarrow 2H_2O + 2e$). Oxygen fed to the other electrode is reduced to form the hydroxyl ions, which migrate to the hydrogen electrode ($\frac{1}{2}O_2 + H_2O + 2e \rightarrow 2OH^-$). The cell operates at about 5 bar and 200 °C.

If the cell is on open circuit, the electrons accumulate on the hydrogen (negative) electrode with a corresponding depletion on the oxygen (positive) electrode to give a potential of just over 1 V. When the circuit is closed, the electrons travel around the circuit to take part in the reactions. The electrodes and the electrolyte are not consumed and the waste product is water. The potential is then reduced to 70–80% of its open-circuit value by the resistance of the electrolyte and 'polarisation' losses at the electrodes at typical current densities of 100–200 mA/cm² of electrode area.

A disadvantage of the alkaline cell is that, for use in bulk generation, CO_2 must be removed from the feed gases (e.g. air and H_2 from gasified coal or steam reformed natural gas or oil) in order to prevent the formation of relatively insoluble carbonate which clogs the porous electrodes. The use of an acid electrolyte overcomes this problem but the electrode reactions become appreciably slower and, hence, significant quantities of expensive catalyst are required to give acceptable current densities. Nevertheless such a system is being developed using 60% H_3PO_4 with platinum-catalysed carbon electrodes and operating at 180 °C with a current density of 150 mA/cm². A 4.5 MW demonstration unit, with integral reformers to convert truckable naphtha fuel to H_2, is at present under construction by UTC in a residential district in Manhattan, USA, and should operate with an overall conversion efficiency approaching 39%.

An alternative approach, still at the laboratory stage, is to use a molten carbonate electrolyte supported on ceramic tiles (e.g. Li_2CO_3/K_2CO_3, supported on $LiAlO_2$) and operating at a much higher temperature (650 °C) with nickel and nickel oxide electrodes. The reject heat from this device would be available at a sufficiently high temperature to allow a conventional steam cycle to be used as a 'bottoming' process to improve the overall efficiency to more than 50%. An additional advantage is that CO as well as H_2 can be converted directly at the electrodes and a combined fuel cell/steam cycle scheme would thus be very suitable for integration with a coal gasification system.

Finally, some attention has been directed towards a fuel cell using a solid electrolyte in which, at a sufficiently high temperature, oxide ions (O^{2-}) have a conductivity nearly as great as that of liquid electrolytes. The bulk of effort has centred around thin-film devices employing MgO or Y_2O_3 doped zirconia (ZrO_2) electrolytes and operating at 1000 °C with porous Ni and indium oxide (In_2O_3) as anode and cathode materials, respectively. Fabrication of devices which can withstand thermal cycling and maintain leak-tight interconnections is, however, very difficult, and the materials mentioned suffer degradation after prolonged operation at 1000 °C. Research is now moving towards the identification and synthesis of novel solid oxide electrolyte materials which would enable operation at the reduced temperature of 700–800 °C.

The low-temperature hydrogen/oxygen cell has been used for small vehicles, fork-lift trucks and navigational beacons as well as in spacecraft, where the waste product, water, is valuable. Large units are under active development, with possible future uses to regenerate electricity from hydrogen produced by nuclear power stations at off-peak times.

10.9 Nuclear fusion

As with fission, the fusion of deuterium (D) and tritium (T) nuclei results in a net mass loss and a release of energy of about 22 MeV. The D–T reaction

$$D_1^2 + T_1^3 \rightarrow He_2^4 + \text{neutron} + 17.6 \text{ MeV}$$

is, of a number of exothermic fusion reactions, the easiest to 'ignite'. D is available naturally (e.g. in sea-water), but T has to be made by absorbing the neutrons produced from the D–T reaction in a blanket of liquid lithium around the plasma:

$$Li_3^6 + \text{neutron} \rightarrow T_1^3 + He_2^4 + 4.6 \text{ MeV}$$

Net fusion-power production may be achieved when the Lawson criteria are met. For the D–T reaction above these are (a) temperature $\rightarrow 10^8$ K (10 keV), (b) $n\tau \rightarrow 10^{20}$, where n is the ion density per m³ and τ the containment time. Experiments in hand, or planned, are likely to approach these criteria. Power reactors would have to reach temperatures 50% higher and densities an order of magnitude higher.

Two main techniques are employed to attempt controlled nuclear fusion.

In the first—magnetic confinement—the D–T plasma is heated by electrical discharge or neutron beams and confined within a magnetic 'bottle'. Unfortunately, plasmas are not adequately confined by simple toroidal or magnetic mirror configurations. Owing to plasma oscillations and magnetohydrodynamic instabilities, more complex arrangements have to be used. The currently favoured toroidal system is the tokamak, in which the plasma is heated by pulsed currents (many MA are necessary for a reactor-size device), the confining magnetic field being produced by superconducting magnets. The combination of the toroidal magnetic field and the plasma current magnetic field (poloidal) leads to a resultant field which spirals around the plasma torus. The relative strengths of the toroidal and poloidal magnetic fields (safety factor, q) and the ratio between plasma kinetic pressure and the magnetic pressure exerted by the toroidal field on the plasma (β) are two important design parameters.

The second confinement technique—inertial confinement—relies upon heating a small 'pellet' of D and T to the temperatures required, while the material is confined by its own inertia. The method relies upon high-powered pulsed lasers, although plans to use ion or electron beams are under consideration. The detailed physics of how laser energy is transferred to the pellet is the subject of much research. It is generally agreed that short laser wavelengths (0.3–0.6 μm) are desirable and that peak powers of about 100 TW are required to achieve adequate compression. Most experiments are concentrated on the so-called 'explosive pusher' heating mode, in which the D–T gas at a pressure of several atmospheres is held in a small glass ampoule. The laser light is largely absorbed in the glass and compression of the D–T is achieved by the resulting shock wave. So far, insufficient compression has been achieved by this technique and work is now proceeding to produce spatially and temporally shaped laser pulses which may lead to significant improvements.

the energy of its electrons to a level enabling them to escape from the surface and flow to the anode: at the anode their energy appears partially as heat (removed by cooling) and partially as electrical energy delivered to the circuit. Although the distance between anode and cathode is only about a millimetre, the negative space charge with such an arrangement hinders the passage of the electrons and must be reduced—e.g. by introducing positive ions into the inter-electrode space, caesium vapour being a valuable source of such ions. Anode materials should have a low work-function (e.g. barium oxide and strontium oxide), while that of the cathode should be considerably higher, tungsten impregnated with a barium compound being a suitable material. With these materials temperatures up to 2000 °C will be needed to secure, for the generator itself, efficiencies of 30–35%, although higher overall efficiency can be obtained by utilising the heat from the coolant. Electrical outputs of about 6 W/cm^2 of anode surface have been suggested.

Developments of thermionic generators using radioactive isotopes as the heat source have taken place for space applications. Thermionic devices, in general, do not appear to offer significant potential as power sources.

10.7.3 Magnetohydrodynamic (m.h.d.) generators

In the magnetohydrodynamic generator a partially conducting gas is heated by a fuel fired or nuclear reactor, allowed to expand through a nozzle to convert the heat energy to kinetic energy, and then passed between the poles of an electromagnet, the field of which converts some of the kinetic energy to electrical energy which can be collected from electrodes situated in the gas channel (*Figure 10.2*). The generator is thus not quite a direct heat-to-

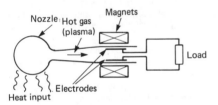

Figure 10.2 Magneto-plasma-dynamic generator

electricity device, as are the thermoelectric and thermionic devices, for there is an intermediate kinetic energy stage; also, largely owing to the power required for the electromagnets and other losses, it is unlikely to give a useful output unless built in sizes of 50 MW or more.

Provided that the gas is conducting and moving at right angles to the magnetic field, an e.m.f. will be set up at right angles to the direction of motion and of the magnetic field, being proportional to the velocity of the gas and to the magnetic flux density. This e.m.f. can be collected from suitable electrodes located in the gas stream and can supply power to an external circuit. The power output is proportional to the square of the velocity, to the square of the flux density and to the conductivity of the gas between the electrodes.

The field density should therefore be as high as possible, making superconducting magnets to give fields of 4–5 T a great advantage over conventional magnets. Gas velocities up to 1000 m/s are practicable. The electrical conductivity of gases even at temperatures of 2000–3000 °C is too low to give practicable powers. Ionisation must therefore be artificially increased by *seeding* the gas with an easily ionisable element such as caesium or potassium. Either must be recycled for economic operation, but caesium is so expensive and corrosive that a closed-cycle system is essential.

The possibility of using a liquid metal, sodium or potassium, is being investigated; such a fluid would have a much higher electrical conductivity but a lower velocity, the major problem being that of producing a high velocity with sufficient liquid density to give an adequate conductivity.

With gaseous conductors a complication is introduced by the Hall effect—i.e. by the fact that the current flow between the electrodes is not in the same direction as the field: the Hall angle may reach 80°, increasing with low pressures, high magnetic fields and high electron mobilities. The resulting axial component of current flow leads to inefficiency. To counter the Hall effect the number and configuration of the electrodes is more complex. In the Faraday generator a single pair of electrodes, or several pairs connected to separate load circuits, are used, but these arrangements are not appropriate for Hall angles of more than 45°. In the Hall generator use is made of the axial component by a more complex electrode arrangement in which current is collected from axially spaced electrodes; Hall angles up to 80° are appropriate with this type.

Extensive programmes of work are in progress in the USA and USSR to develop the open-cycle fossil fired system. In this the fuel combustion products at temperatures over 2000 °C, achieved by preheating the combustion air, are seeded with potassium carbonate and passed through an m.h.d. duct. The waste gases are then used to heat a conventional steam cycle. The m.h.d. process is therefore a topping unit increasing the overall efficiency of power generation to about 45% and potentially over 50%. The potassium carbonate also combines with the sulphur in the coal to form potassium sulphate which is removed from the boiler with the ash, the sulphur being removed and the potassium recycled. The system therefore has an added advantage where sulphur emissions must be controlled, as is required in the USA. The main problems are the development and cost of a suitable gas duct and electrode system to withstand the high temperature and corrosive effect of the gas for long periods, and the effective recovery and recycling of the seed material. In addition, large air heaters and superconducting magnets would be needed. The USSR has a 20 MW prototype station which has operated continuously for up to 250 h at 10 MW on natural gas. A larger unit is proposed in which the duct life problem is avoided by having two ducts which are used alternately and regularly refurbished. The US programme is also being directed towards the evaluation of large-scale components for open-cycle m.h.d. but using coal as the fuel.

10.8 Fuel cells

The conversion of the chemical energy of a fuel to electrical energy without passing through the heat state removes the Carnot efficiency limitation, so that theoretical efficiencies approaching 100% are attainable. The simple primary battery is an example of this process, but its application is limited to very small outputs on account of its high cost and short life.

Research on the fuel cell, to which a continuous supply of fuel is given, has been going on for over 100 years but only within the last decade have commercially practicable cells been developed; even so, costs are still high and outputs limited.

Of the available fuels, hydrogen has so far given the most promising results, although cells consuming coal, oil or natural gas would be economically more useful for large-scale applications. Some of the possible reactions are:

Hydrogen/oxygen	1.23 V	$2H_2 + O_2 \rightarrow 2H_2O$
Hydrazine	1.56 V	$N_2H_4 + O_2 \rightarrow 2H_2O + N_2$
Carbon (coal)	1.02 V	$C + O_2 \rightarrow CO_2$
Methane	1.05 V	$CH_4 + 2O_2 \rightarrow CO_2 + 2H_2O$

useful as a source of low-temperature building heat. Significant research on these lines is being carried out in Israel.

10.5.6 Biological conversion

Photosynthesis in plants, algae and some bacteria uses solar energy to convert carbon dioxide and water into starch, cellulose or sugars with the evolution of oxygen. This solar energy is stored in the biomass. Several routes are available to utilise this process for energy conversion, including the traditional wood burning, and chemical and biological methods of producing oils, alcohols and gases.

Since most plant material contains a high moisture content, the use of wet processes is preferable to direct burning. Several techniques are available with an efficiency of conversion often exceeding 50%. They include: anaerobic digestion; chemical reduction (hydrogenation); pyrolysis (of 'dry' biomass); and hydrogasification and gasification. Of these, chemical reduction and particularly hydrogasification and gasification are still at an early stage of development.

10.6 Tidal energy

Sea-level varies approximately sinusoidally on a 12.4 h cycle at most places. The peak-to-peak amplitude of the tidal variation is the 'tidal range'. In mid-ocean this range is only about 1 m but it is often amplified in coastal areas by a complex interplay of geographical features. The greatest amplitude occurs in estuaries where the tide is in a resonance condition. The range also varies sinusoidally over as much as 3 to 1 from spring to neap tide on a 14 day cycle. Smaller seasonal variations also occur.

There are essentially two methods for extracting energy from a tidal scheme. The first is to build a single dam across an estuary, with turbines in the dam, to generate electricity while the tide is ebbing as result of the head developed from water stored behind the dam during the flood tide. In some schemes generation may also take place during the flood tide or in both directions. This conceptually simple way of generating power does so, however, only according to Moon time, since little power can be developed close to high tide and low tide. The second approach is to build two basins allowing (at much greater cost) the whole operation to be retuned and continuous power to be extracted. One basin—the high-level basin—is filled roughly between mid-tide and high tide; the other—the low-level basin—is emptied roughly between mid-tide and low tide. Generation is achieved by turbines situated between the high-level and the low-level basins. Many variants on this basic principle are possible, including the ability to incorporate pumped storage.

Tidal barrage schemes may have important environmental effects, and need to be studied for each system. Effects on the ecology of plants and animals, silting patterns, sewage and effluent disposal, shipping, land drainage and similar matters may be important, as is the effect of the barrage itself on the tidal range.

Assessments of technical feasibility and economics are very site-specific. France has a pilot scheme in the Rance Estuary near St. Malo giving a peak output of 240 MW. The highest tidal range in the world is thought to occur in the Bay of Fundy in Canada, the second highest in the Severn Estuary in England. Proposals have been put forward for tidal schemes in both these locations and they have recently been the subject of detailed assessment. Studies of the Severn Estuary, for example, have suggested that a scheme costing £5600 M (1980 prices), taking 12 years to complete, could generate 13 TWh per year from an installed capacity of 7.2 GW. This would produce electricity for about 3.1 p/k Wh. Design and acceptability studies would have to be carried out before construction could begin.

10.7 Direct conversion

Some methods of extracting electrical energy from renewable sources do not rely upon a heat stage, and the limitations of Carnot efficiency are avoided. Some other methods, still largely small-scale and experimental, also eliminate machinery, relying instead on direct conversion processes. Like solar cells, the direct conversion processes described below produce electricity at low direct voltage; many units need to be interconnected and coupled to inverter systems to give a.c. output. Direct conversion processes generally need to be operated at high temperatures, and difficulties are encountered in finding suitable materials to withstand such temperatures over a long lifetime.

10.7.1 Thermoelectric generators

If two dissimilar materials are joined in a loop with the two junctions maintained at different temperatures, an e.m.f. $E = \alpha \theta$ is set up around the loop, where θ is the temperature difference and α is the Seebeck coefficient (itself depending to some extent on temperature). The phenomenon, long used in thermocouples, enables generators with semiconductor junctions to supply up to 5 kW for radionavigation beacons and satellites. A useful figure of merit is $Z = \alpha^2 \sigma / K$, where σ is the electrical conductivity (which should be high, to reduce $I^2 R$ loss) and K is the thermal conductivity (which should be low, to limit heat transfer between junctions).

If the thermoelectric generator works between absolute temperatures T_1 and T_2, the efficiency as a fraction of the Carnot efficiency is

$$\eta = [(1 + ZT)^{1/2} - 1] / [(1 + ZT)^{1/2} + T_2 / T_1]$$

where $T = \frac{1}{2}(T_1 + T_2)$. Thus, the efficiency depends on the product ZT, which is a convenient assessment for possible thermoelectric materials. In practice no single combination maintains a high ZT over a wide temperature range: most practical designs use a series of stages with n- and p-type semiconductor junctions that have a high ZT over their relevant temperature differences.

Taking into account mechanical characteristics, stability under operating conditions and ease of fabrication, bismuth telluride appears one of the most suitable; it can be alloyed with such materials as bismuth selenide, antimony telluride, lead selenide and tin telluride to give improved properties, is suitable for temperatures up to about 180°C, and can give efficiencies up to about 5%. A silicon–germanium alloy with phosphorus and boron impurities can be used up to 1000 °C and might give efficiencies up to 10%.

10.7.1.1 Practical developments

A typical thermoelectric couple could be designed to give about 0.1 V and 2 A (i.e. about 0.2 W), so that a 10 W device suitable for a navigational beacon or unattended weather station would require about 50 couples in series. Various methods have been utilised to provide a source of heat for the hot junction. These include small oil or gas burners, isotopic heating and solar radiation. Although some progress has been made in developing suitable materials, theoretical studies seem to show that the scope for improvement in ZT values is rather limited. For this reason, interest in thermoelectric devices has declined over the last decade.

10.7.2 Thermionic generators

In its simplest form the thermionic converter comprises a heated cathode (electron emitter) and an anode (electron collector) separated in a vacuum, the electrical output circuit being connected between the two. Heat supplied to the cathode raises

Europe through a European consortium (EUROCEAN), and as part of the French solar programme.

The major technical problems to be overcome before this technology can contribute to energy supply in the tropics are the long cold-water pipes, the advanced heat exchangers, large pumps and overcoming biofouling.

10.5 Solar energy

Direct use of solar radiation is exploited in a number of energy conversion processes. Solar radiation is indirectly the driving force behind wind and wave energy. The resource (1 kW/m² at some points on the Earth's surface) is immense; it is estimated that the Earth receives from the Sun each year about 10 000 times as much energy as is consumed.

10.5.1 Photovoltaic conversion (solar cells)

When light strikes certain semiconducting materials, such as silicon, gallium arsenide or cadmium sulphide, fabricated in the form of a p–n junction, an electric current can flow through an externally connected circuit. Efficiencies are limited to around 25% because the band gaps in the materials are matched only to one wavelength in the incident radiation. Thus, photons with energy less than the band gap (typically 1.1 eV for silicon, corresponding to a wavelength of about 1100 nm) have insufficient energy to create an electron–hole pair in the material, while those with higher energy than the critical value (shorter wavelength) may nonetheless have insufficient energy to create two electron–hole pairs, and much of the incident energy is wasted in both cases.

Solar cells operate satisfactorily in diffuse sunlight, although their power output falls with decreasing intensity. Since a single silicon cell exposed to an irradiance of 1 kW/m² will produce an open-circuit voltage of about 0.5 V, a short-circuit current of 400 A/m² and a maximum power output of about 150 W/m², such cells need to be interconnected in series-parallel to provide a current/voltage relation suitable for the load.

Cells on land must be protected from corrosion, erosion, icing and thermal cycling. They have been successful in a number of small-scale applications, however; in the USA a large programme is currently aimed at improving the economics for large-scale power production. At present, costs are an order of magnitude greater than target costs, but series production of silicon cells may have a dramatic effect in reducing cost. In some parts of the world (such as the UK) the limited land areas available, low solar radiation levels and the seasonal mismatch between output and demand are likely to mean that terrestrial solar cells will be unattractive for other than small-scale remote application.

Solar cells on Earth can operate only during daytime and the atmosphere filters much useful energy. An ambitious scheme to overcome these problems was proposed by Glaser in 1969. He suggested that 10 GW(e) orbiting solar power stations could be built in synchronous orbit about the Earth. The solar irradiance of 1350 W/m² would be collected by a 60 km² solar array and the power beamed back to a receiving station on Earth with a collector area of 90 km², using microwaves. Such schemes are still being considered in the USA, but are unlikely to be pursued in the near future.

10.5.2 Photoelectrochemical conversion

In the photoelectrochemical conversion process (sometimes referred to as 'photogalvanic conversion') irradiation of an electrode/electrolyte system results in a current flow in an external circuit. The current may be generated by a photochemical reaction in the electrolyte or by a photosensitive electrode. Devices based upon the effect may be used either to produce electric power directly or to produce (photoelectrolysis) a chemical product which stores the energy and regenerates the reactants on subsequent conversion to electricity. It is this capacity for energy storage that makes devices based upon the electrochemical effect particularly attractive for solar energy conversion. So far, however, only low efficiencies have been obtained and the process is at present almost entirely experimental.

10.5.3 Solar thermal generation

A solar thermal power generation system consists of up to five elements: a solar concentrator to collect the sunlight, an absorption system to convert the concentrated radiation into a heat fluid, a transfer system for the fluid and a power generation system. There is also an opportunity to incorporate heat storage in the system. Two basic plant types can be envisaged:

(1) A distributed system in which many receivers each convert the sunlight to thermal energy and transfer the heat to a central generating facility. Sunlight is usually concentrated by a factor of up to 100.
(2) A central system in which Sun tracking flat-plate reflectors (heliostats) transfer energy on to a central collector system on a tower. This has the advantage over (1) of disposing of the need to conduct a working fluid through a pipe system, but requires accurate tracking and reflection of the sunlight on to the power plant. High concentration ratios (200–1000) and thus very high temperatures can be achieved.

Solar thermal generation is of little interest to countries such as the UK, because much of the incident solar radiation is diffuse and cannot be focused. Several major prototype installations have been built in the USA, France, Italy, Spain and Japan. These include a 10 MW tower installation at Barstow in California employing 2200 heliostats; an EEC installation of 1 MW (Eurelios) in Sicily; two 500 kW plants (one of each design) in Almeria, Spain, sponsored by the International Energy Agency; and two 1 MW systems in Japan. Present costs are high, but it is hoped to reduce them by series production.

10.5.4 Flat-plate collectors for space and water heating

A metal plate backed with heat insulation, painted black on the front and covered with one or two sheets of glass, acts as a heat absorber. If made hollow or attached to pipes at the back through which water is circulated, the heat may be removed and utilised. Such collectors are mounted in a fixed position facing the general direction of the Sun. The water temperature so obtained can hardly exceed 100 °C, so that application is limited to space heating and the provision of hot water. The collecting efficiency is about 50%.

10.5.5 Solar ponds

A shallow pond about 1 m deep, with a blackened bottom, can be used as a solar collector, provided that convection can be prevented by ensuring that the density of the water increases towards the bottom. As the lower layers of water are the warmer, this is contrary to the natural density gradient and a salt concentration must be introduced into the lower layers to increase their density. Heat can be removed by circulating the lower layer of water and passing it through a heat exchanger; experience shows that this can be done without disturbing the upper layers. Water temperatures up to 100 °C can be achieved with a collecting efficiency of 15–20%. Such ponds may prove

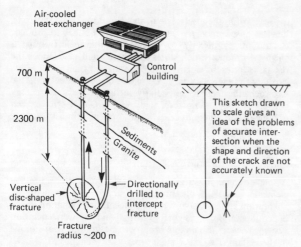

Figure 10.1 Schematic diagram of a hot rock geothermal system (courtesy Los Alamos Scientific Laboratory)

extracting heat from hot dry rocks. The granites studied have a thermal gradient with depth of 50–60 °C/km. Experiments have achieved an output of 5 MW (thermal) over a period of more than 200 days but more recent experiments aim to achieve 20–50 MW. Two boreholes have recently been drilled to a depth of more than 5 km where temperatures in excess of 300 °C have been achieved. The system is being designed for a 10 year lifetime.

The Los Alamos studies have involved hydrofracturing of hot rocks. At Camborne School of Mines in Cornwall, UK, a major project to achieve the required heat exchange surface between rock and water is being carried out using controlled explosive charges to open up incipient fractures in the rock. These are then developed by hydrofracturing.

Although the Los Alamos thermal gradients are very high by world standards, figures of 30–40 °C/km are not uncommon (in the UK, for example in Cornish granite) and the experience built up at Los Alamos and Camborne may be important in the long term for exploring the geothermal potential of countries without natural hydrogeothermal resources. The economics of such plant are far from certain, but some estimates in the range 3–5 p/kWh make hot rock geothermal energy promising. The major costs are in drilling deep boreholes in hard rock.

10.3 Wave energy

Wave power is a form of wind power in that waves are generated by the wind. The main areas of interest for the exploitation of wave energy are those where a prevailing wind develops a deep ocean swell with a very long attenuation distance. The north-west coast of Britain is such a location, and average power densities of 50 kW/m of wave front have been measured. Because of its favourable location, much of the work on wave power has been carried out in the UK.

A wide range of devices has been suggested for extracting energy from the waves. They include:

Salter duck. Mr S. Salter of Edinburgh University proposed a design of rocking or nodding 'ducks' or cams with a leading edge shaped so that it moves with the free motion of the water, but with a trailing edge which does not disturb the water behind. The motion is converted to electrical energy through a hydraulic system housed in a long spine on which the individual ducks are mounted. Recent designs employ gyroscopes as torque converters.

Cockerell raft. This device, proposed by Sir Christopher Cockerell, consists of a number of coupled rafts or pontoons which follow the wave contour. Power is taken off at the hinges between the rafts.

Oscillating water column. A number of designs, including the Japanese *Kaimei* wave-power ship moored in the Sea of Japan, and the UK National Engineering Laboratory device, attempt to make use of a vertical column of water oscillating in phase with the incoming waves. Air trapped above the water column can then drive an air turbine, either by a system of flaps and valves to give unidirectional flow, or by a self-rectifying system such as the Wells turbine.

Lanchester clam. In early designs the clam device consisted of a series of hinged plates mounted on a long hull or spine, held perpendicular to the incoming waves. The hinged plates respond to the wave motion, and compress and expand flexible bags mounted in the hull. Recent designs omit the hinged plate and rely upon the compression and expansion of the bags alone to drive air round a circuit and through an air turbine.

Lancaster flexible bag. Air-filled bags are attached along the top of a submerged hull lying end-on to the sea. The hull contains high- and low-pressure air ducts connected to the bags by non-return valves. As a wave train runs along the device, the bags act as bellows pumping air from the low- to the high-pressure ducts, driving an air turbine.

Bristol submerged cylinder. This device exploits the fact that waves incident on a cylinder moored just beneath the surface with its axis perpendicular to the incoming waves cause the axis of the cylinder to move orbitally about an offset axis. In principle, this motion can be used to activate a hydraulics system on the sea-bed.

Norwegian device. The devices so far described are either terminators, moored with their axis perpendicular to the incoming waves, or attenuators (e.g. flexible bag), moored with their hulls parallel to the waves. In Norway much work has centred on oscillating water columns acting as point absorbers. Ideas for controlling the motion of the devices to be in phase with the waves are being studied.

These devices together with others such as the 'rectifier', 'flounder', and 'triplate' systems have now been studied as part of the UK programme. Some have been tested at 1/10 scale. Major problems include survivability under storm conditions, mooring, repair and maintenance, and detailed problems of power take-off. Typical efficiencies for converting wave motion to mechanical motion of the device in a real sea are in the range 30–50% and typical power-chain efficiencies are ~60%. Cost estimates for full-scale devices lie in the range 5–20 p/kWh at 1979 prices.

10.4 Ocean thermal energy

Ocean thermal energy conversion (OTEC) uses the temperature difference between the sea at the surface and at depths of 1000 m or more to extract energy. In tropical areas this can be as great as 25 °C. In some designs cold water is drawn up a vertical pipe to condense a working fluid—such as ammonia—in a closed cycle. Warm water at the surface is used to evaporate the fluid. By pumping the working fluid around a closed circuit, the latent heat can be extracted through heat exchangers and used to generate electricity. The maximum efficiency is determined by the temperature difference and is unlikely to exceed a few per cent.

Prototype OTEC devices were first constructed in Cuba in 1929 and on the Ivory Coast in 1956. Both projects ran into severe problems. In 1979 a 50 kW plant (mini-OTEC) was sited off Hawaii as a research tool to design a 1 MW unit (OTEC-1). Larger units up to 40 MW are planned in the USA for operation in the mid-1980s. Other projects are proceeding in Japan (1 MW),

Table 10.1 Wind power plant in 1981 (38 m diameter and above)

Location	Operational date	Turbine	Diameter (m)	Power (MW)	Generator
USA					
Ohio	1975	MOD O	38	0.1	S
New Mexico	1977	MOD OA	38	0.2	S
Puerto Rico	1978	MOD OA	38	0.2	S
Rhode Island	1979	MOD OA	38	0.2	S
Hawaii	1980	MOD OA	38	0.2	S
North Carolina	1979	MOD 1	61	2.0	S
Washington	1980	MOD 2	91	2.5	S
Washington (2)	1981	MOD 2	91	2.5	S
Wyoming	–	WTS 4	80	4.0	S
Denmark					
Nibe	1979	Nibe A	40	0.63	A
Nibe	1980	Nibe B	40	0.63	A
Tvind	1978	Tvind	54	2.0	S
Sweden					
Malmö	–	KKV/HS	78	3.0	S
Gotland	–	KMW/ERNO	75	2.2	A
West Germany					
Hamburg	–	Growian 1	100	3.0	A
Stuttgart	–	Voith	52		S
UK					
Orkney	–	WEG	60	3.0	S

S, synchronous; A, asynchronous.

10.2 Geothermal energy

The production of electricity from the heat of the Earth, in principle, has much in its favour. The Earth is an almost infinite source of heat which, unlike wind or wave, can deliver a constant source of power. There are two possible techniques for exploiting geothermal energy: the first is to utilise existing hydrothermal sources; the second, to attempt to extract heat from hot dry impermeable rocks at depth below the surface. The first is now well established. The second is still at the research stage.

10.2.1 Hydrothermal sources

Although steam and hot water come naturally to the surface of the Earth in some locations, for large-scale use boreholes are normally sunk with depths up to 3000 m, releasing steam and water at temperatures up to 200 or 300 °C at pressures up to 3000 kN/m² (450 lbf/in²). Many installations are in use in various parts of the world: three of the largest are near Lardarello, Italy (500 MW); Wairakei, New Zealand (250 MW); and the Geysers, California (1000 MW). Mexico, Japan and the Phillipines also have sizeable (and expanding) geothermal programmes.

Flowing well-head steam pressures vary between 200 and 1500 kN/m² (between 30 and 200 lbf/in²). From the well-head the steam is transmitted by pipe lines up to 1 m in diameter over distances up to about 3 km to the power station. Water separators are usually required, as superheating the steam to minimise wetness is uneconomic; care must also be taken in the choice of materials for the plant, as the steam may contain solid and gaseous impurities. Steam may come from several boreholes at different pressures, so that the steam layout of the generating station may involve two or three different pressures in cascade. Direct use of the steam in turbines is possible in most cases, jet condensers being used with the condensate discharged to waste along with the cooling water. The only major electrical plant required in the generating station is the cooling-water pump. Owing to the low pressure, station efficiency is only 10–15%, so that large quantities of exhaust heat are available if a local use can be found; otherwise this must be suitably discharged.

At Lardarello the geothermal field provides superheated dry steam at 200–500 kN/m² (30–75 lbf/in²) and 140–200°C, containing significant impurities. Some of the fields are of limited capacity, so that attention has been given to demountable temporary stations. This is also being planned in Mexico.

At Wairakei the field produces hot water and steam in a ratio of about 5 to 1, so that extensive water separation at the well-head is necessary. Shallow wells giving steam at 600 kN/m² (90 lbf/in²) as well as deeper wells giving 1500 kN/m² (210 lbf/in²) are available. A three-pressure system with turbine pressure of 1200 kN/m² (180 lbf/in²), 350 kN/m² (55 lbf/in²) and atmospheric pressure has proved satisfactory.

The Geysers field in California now employs some of the largest geothermal units in the world (in excess of 100 MW). It is a dry-steam field. Environmental legislation may inhibit its further rapid expansion.

Heat from aquifers at lower temperatures, unsuitable for electricity generation, may be used for district heating or agricultural purposes. Several such aquifers with temperature in the range 60–80°C are thought to exist in the UK and these are currently being explored.

10.2.2 Hot dry rocks

A possible technique for extracting heat from hot dry rock several kilometres down is illustrated in *Figure 10.1*. Hydrostatic pressure with the controlled use of explosive charges is used to fracture rock at the bottom of twin boreholes and water is circulated in a closed cycle to extract the heat from the rock. In practice, there are many difficulties in achieving a low impedance path with a high heat-transfer area between the rock and water, and in ensuring satisfactory intersection of the borehole systems. The technique is still experimental.

The University of California at the Los Alamos Laboratories in New Mexico have been carrying out field trials since 1975 on

Renewable energy sources (wind, wave, solar, tidal) do not employ finite and dwindling reserves of fossil fuel; their technology is costly; and as their availability is subject to the vagaries of weather or the lunar cycle, they cannot fully replace conventional plant. Geothermal energy is a vast resource, often considered as 'renewable'. Direct-conversion processes not requiring rotating machinery are small-scale and experimental. Nuclear fusion offers an alternative to the established nuclear fission process. Although not strictly 'alternative', schemes for the better use of available heat can be obtained by combined heat and power and by total energy plants.

10.1 Wind energy

Wind power has been used for millenia for agricultural uses such as grinding corn and pumping water. Recently, it has been the subject of considerable research and development for large-scale electricity generation.

The power available to a windmill is proportional to the area swept out by the blades and to the cube of the wind velocity. In a wind of 10 m/s (a typical yearly mean for a very good site in the UK) a modern wind turbine generator might produce more than 10 GW-h per annum operating at a peak efficiency of more than 40%. Power is generated when the wind speed reaches a cut-in speed typically of a few metres per second. The power output then increases roughly according to a cube law relationship up to a value known as the rated wind speed. Above this, the output is kept nearly constant, either by blade pitch control or by progressive blade stalling. At very high wind speeds machines are shut down for safety reasons. The maximum power is largely a design choice. Wind turbine generators developing full power at a high ratio between rated and mean wind speed achieve high power outputs but for a relatively small fraction of the year. Those rated to a peak output at lower wind speeds do so for longer periods and thus have a higher load factor.

A wide range of windmill designs has been suggested and built. They generally fall into two classes: horizontal-axis propellor-type and vertical-axis designs. In the horizontal-axis type, two or three blades are designed to rotate at typical speeds of 15–100 rev/min, with blade-tip: wind speed ratios 6–10. Some machines operate at constant speed, using blade pitch control; others operate at variable speeds depending upon the wind strength. The blades are mounted at the top of a tower on a nacelle which contains the main shaft, gear-box and generator. The nacelle assembly and blades are free to rotate about a vertical axis to keep the blades at right angles to the wind direction. Generators are generally either synchronous or induction-type, depending upon whether the blades rotate at constant speed and on the strength of the electrical system into which the machine is connected. Whereas most designs concentrate on ensuring that resonant frequencies for the structure lie well above the rotational frequency, much recent research in the USA has concentrated on 'soft' or 'compliant' designs using smaller amounts of structural material (thus achieving cost savings), with resonant frequencies below that of rotation. Concepts such as teetering hubs, 'soft' drives and multispeed generators are also being explored.

Vertical-axis designs, in principle, have the advantage that they do not need to be turned into the wind and that their gear-box and generator can be located at the base of the tower. They suffer from the disadvantage that torque and electrical output vary cyclically over a revolution. Two designs—the Darrieus or 'egg-whisk' machine and the Musgrove variable geometry hinged design—show considerable promise.

Until recently, the largest machine was the 60 m diameter horizontal-axis Smith–Putnam machine built in the USA in the early 1940s. This windmill, which generated a peak output of 1.25 MW, was damaged by strong winds shortly after entering commercial service. The fact that it was not repaired was largely due to difficulties during wartime. Currently there are many giant windmills being designed and tested, and wind energy is one of the most attractive of the renewable energy technologies. In the USA a series of wind turbine generators has been built ranging from 40 m diameter machines generating a few hundred kilowatts to 100 m diameter designs generating a few megawatts. The largest prototypes in operation at present are the MOD 2, 90 m diameter, 2.5 MW machines at Goldendale in Washington State. An array of three such machines is producing power for a local utility. Even larger machines generating 6–7 MW are being designed. It is predicted that when series-produced and built on sites with mean wind speeds of just over 6 m/s, advanced designs could generate electricity in the USA for 2–3 cents/kWh at 1979 costs.

Britain, Sweden, Denmark and Germany are building megawatt-size modern wind turbine generators as part of government funded programmes. The machines completed or under construction in mid-1981 are listed with details of their performance in *Table 10.1*.

In the UK there are projects under way to examine the possibility of utilising wind power on island, lowland and offshore sites. Novel designs are also being studied.

The Orkney project—involving the construction of 250 kW and 3 MW prototype wind turbines in a collaboration between government, the Wind Energy Group and the NSHEB—will seek to test the feasibility of using wind energy at such locations. The 250 kW machine will be completed in early 1983 and will be one of the most advanced designs of intermediate-size machine then built. It will be used to study the integration of wind energy into the Orkneys' electricity supply and will be heavily instrumented. The larger machine is due to be completed in late 1984 and is currently being redesigned to incorporate a number of advanced features such as a soft tower and a teetering hub. Both of these machines are horizontal-axis designs. Another UK consortium hopes to construct, with government backing, a novel 25 m diameter version of a vertical-axis machine based on an original design by Musgrove (University of Reading). This type of machine offers considerable promise.

If wind turbines sited on land are to make a major contribution to UK electricity supply in the future, machines which are reliable and economic on lower wind speed sites must be developed. In 1981 the Central Electricity Generating Board announced a step-by-step ordering programme to examine the technical feasibility of, economics of and public reaction to wind turbines on such sites.

A large potential resource is available from wind energy if machines are sited in shallow offshore waters. Detailed assessments in a number of countries have suggested that offshore siting is feasible and that high wind speeds may partly offset the increased civil engineering and maintenance costs of building at sea. Costs are thought to be about 5p/kWh at 1979 prices, but there is potential for reducing these if large, advanced machines can be constructed offshore.

Small machines typically for use on farms have been developed vigorously in the USA and in some European countries. The economics of small machines depends on capital costs, machine reliability (affecting lifetime and operation and maintenance costs), government subsidies, tax concessions, rates for buying back excess electricity, the costs of interconnection and the cost of the energy source with which they must compete. This varies strongly from country to country. There is thought to be a large potential for such machines in developing countries, particularly for pumping.

10 Alternative Energy Sources

R H Taylor
Central Electricity Generating Board

T A C Dulley, MA, MIEE
Merz & McLellan

G Brundrett
Electricity Council Research Centre

Contents

10.1 Wind energy 10/3

10.2 Geothermal energy 10/4
    10.2.1 Hydrothermal sources 10/4
    10.2.2 Hot dry rocks 10/4

10.3 Wave energy 10/5

10.4 Ocean thermal energy 10/5

10.5 Solar energy 10/6
    10.5.1 Photovoltaic conversion (solar cells) 10/6
    10.5.2 Photoelectrochemical conversion 10/6
    10.5.3 Solar thermal generation 10/6
    10.5.4 Flat-plate collectors for space and water heating 10/6
    10.5.5 Solar ponds 10/6
    10.5.6 Biological conversion 10/7

10.6 Tidal energy 10/7

10.7 Direct conversion 10/7
    10.7.1 Thermoelectric generators 10/7
    10.7.2 Thermionic generators 10/7
    10.7.3 Magnetohydrodynamic (m.h.d.) generators 10/8

10.8 Fuel cells 10/8

10.9 Nuclear fusion 10/9

10.10 Combined heat and power 10/10
    10.10.1 Steam plant 10/11
    10.10.2 Diesel engines 10/12
    10.10.3 Gas turbine plant 10/12
    10.10.4 District heating 10/12
    10.10.5 Desalination 10/12
    10.10.6 Ratio between heat and power 10/12

10.11 Total energy 10/13

10.12 Economics 10/13

10.13 Heat pumps 10/14
    10.13.1 Introduction 10/14
    10.13.2 Thermodynamics 10/14
    10.13.3 Practical cycles 10/14
    10.13.4 Scale 10/15
    10.13.5 Conclusions 10/19

The highly active fission products constitute a small proportion of the total volume of material recovered from the spent fuel, about 5 m³ of highly active liquid waste being produced per ton of fuel processed. At present it is general practice to store the concentrate in high-integrity tanks, housed in thick-walled concrete cells with cooling arrangements to remove fission product decay heat. Eventual disposal is discussed below. Medium-activity wastes are concentrated by evaporation and stored for a number of years to allow the short-lived activity to decay. It may then be safe to discharge the residues to the environment; alternatively, a bituminised waste can be formed for long-term storage. Low-activity wastes can be safely dispersed to the environment local to the reprocessing plant in accordance with national and international regulations. They may contain a variety of radioactive elements; an important example is tritium, which may be safely discharged to the sea after dilution — at inland sites, however, wastes containing tritium may have to be evaporated and discharged to the environment as water vapour.

At present the high-level wastes in the UK are stored in acid solution in double-skinned stainless steel tanks which are cooled by water circulation. In the longer term it is intended to convert these wastes to solids in the form of glass which will be stored in steel containers; artificial cooling will still be required for some years before disposal can be made to a final repository in which natural cooling will be adequate to maintain acceptable temperatures. The volume of high-activity waste for eventual disposal in this form is relatively small; each GW(e)-year of nuclear power production will give rise to about 4 m of solidified wastes containing 15% by weight of fission product oxides. If this is encapsulated in the proposed containers, which are 0.5 m diameter and 3 m long, each GW(e)-year leads to a requirement for ten additional containers. A UK nuclear programme building up to an installed capacity of 40 GW(e) by the year 2000 would produce in total a quantity of high-level wastes sufficient to fill 3500 standard containers, and a repository to house this number would occupy a volume of rock measuring 150 m × 400 m × 150 m.

The sequence of events in the management of high-activity wastes can be summarised as follows: on discharge from a reactor, the fuel rods are stored in ponds for a period of months or years before reprocessing; after reprocessing, the high-activity liquor is stored in cooled tanks, probably for some years; the active liquor is then solidified in glass blocks, housed in steel containers, and placed in a retrievable store which is artificially cooled, again for some years; the containers are then placed in a final repository with no human surveillance.

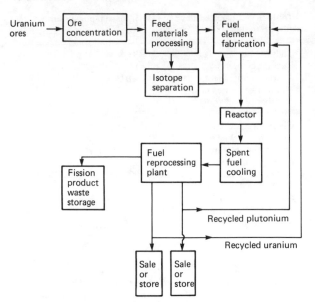

Figure 9.12 The reactor fuel cycle

UF$_4$ by reaction with HF gas. The uranium tetrafluoride is used as feed material for the production of uranium metal, uranium hexafluoride and fuel grade uranium dioxide.

For metal production the UF$_4$ is blended with magnesium metal turnings; the mixture is then pelletised and reduced to uranium metal in an electric furnace. For uranium hexafluoride production the UF$_4$ is mixed with CaF$_2$ and caused to react with fluorine; for uranium dioxide production the liquid UF$_6$ is vaporised and caused to react with steam in a rotary kiln.

9.7.1 Enrichment plant

Apart from MAGNOX and CANDU reactors, which use natural uranium, all thermal reactor systems require enriched uranium fuel. Natural uranium contains one part in 140 of U-235, the remainder, for all practical purposes, being U-238. The process of enrichment requires the proportion of U-235 to be increased, and since the isotopes are chemically identical, a non-chemical process has to be used. The two methods currently in use employ, respectively, gaseous diffusion and gaseous centrifugation.

In the gaseous diffusion process uranium hexafluoride is passed through a porous barrier; the separation factor achieved depends upon the square root of the molecular mass ratio, which is 1.0043, so that many stages are required to change appreciably the proportion of U-235. Typically, a cascade of some 1000 stages is necessary to produce an enrichment of 2–4%, and the investment required for such plant is large. Substantial energy is required to heat the uranium hexafluoride to the gaseous state and to pump it through the porous barriers.

The high-speed centrifuge method also uses the mass difference between U-235 and U-238 to effect a separation, but the separation factor is much greater, typically between 1.25 and 2, and fewer stages are required to produce fuel of a given enrichment. The energy input to the centrifugation process is appreciably less than that required for enrichment by gaseous diffusion.

Other techniques with promise for the future are laser enrichment and the jet nozzle process. Lasers are used to produce intense beams of monochromatic light which selectively ionise atoms of one of the uranium isotopes or molecules bearing that isotope, so that they can subsequently be separated easily by chemical or physical means. The possible advantages of this method are a larger separation per stage and lower energy requirements, giving lower costs of the enriched uranium product. The jet nozzle process, which has been under development in Germany for many years, depends upon the centrifugal action already described in combination with a gas scattering effect. The latter depends upon preferential scattering of the lighter component from a well-collimated beam of two gases, and involves pumping uranium hexafluoride gas down cylindrical elements with axial slit nozzles and flow dividing edges. South Africa has developed a related process, described as a stationary wall centrifuge.

9.7.2 Fuel element fabrication

Uranium, either natural or enriched, is the most common fuel, as regards both availability and use. It can be employed in the form of pure metal, as a constituent of an alloy and as an oxide, carbide or other suitable compound. The principal fuels in current use are uranium metal, uranium oxide, plutonium oxide and uranium carbide.

Uranium metal can be fabricated by conventional means, including casting, rolling, extrusion, forging, swaging, drawing and machining. Cold pressing followed by sintering or hot pressing is used to produce high density with random orientation of grain structure to give greater radiation stability. Uranium dioxide is used extensively as a fuel material, having the advantages of high temperature stability and improved resistance to radiation. It is also chemically inert to attack by hot water, which makes it particularly suitable for use in water cooled and moderated systems. Uranium dioxide fuel pins are produced from a powder which may be prepared from uranyl nitrate solution. Conventional ceramic procedures (cold pressing, extrusion, slip casting, etc.) have been used to form uranium dioxide pellets which are sintered at temperatures ranging from 1300 to 2000°C. More unconventional methods, such as hot and cold swaging and vibratory compaction, can also be used.

9.7.3 Reprocessing plant and waste disposal

The fuel elements which are discharged from a nuclear power reactor are highly radioactive, owing to the presence of accumulated fission products, but contain valuable quantities of uranium and plutonium. Reprocessing plant is required to separate the uranium and plutonium for recycling as fresh reactor fuel and to concentrate the radioactive fission products in a form convenient for economical storage.

The processes and plant employed have developed in size and complexity since reprocessing on an industrial scale commenced in the USA some 30 years ago; the early carrier precipitation methods applied to low-burn-up metallic natural uranium fuel have given way to highly developed countercurrent solvent extraction systems dealing with enriched high-burn-up oxide fuels. Before chemical separation of the fuel constituents can proceed, it is necessary to separate physically the cladding or canning material. This can be effected mechanically for MAGNOX fuel by pressing the can through a die. A system of mechanical chopping is commonly used for oxide fuels, the fuel pins being broken into lengths of about 50 mm, and the uranium, the plutonium and the fission products are then leached out in a strong nitric acid solution. This solution is then subjected to a process of solvent extraction with tributyl phosphate (the Purex process), which allows the separation of the uranium and plutonium from the fission products. The fission products are removed in high-, medium- and low-activity waste streams, for each of which there are appropriate waste management techniques.

able in the UK is often quoted as having an energy equivalent of more than 100 000 million barrels of oil if used in fast reactors. The annual arisings of reactor depleted uranium from UK reprocessing plant is equivalent, if fissioned in fast reactors, to 6250 million barrels of oil, which is almost twice the present annual oil output of Saudi Arabia.

9.6 Nuclear reactor plant for ship propulsion

The applications of reactor plant so far described have been related to electrical power generation; this account would not be complete without reference to the other major application of nuclear reactor plant, which is in the field of ship and submarine propulsion.

The use of a nuclear reactor as a power source for propulsion has obvious advantages for naval vessels, since it removes the requirement for refuelling at relatively short intervals which is a characteristic of oil burning ships, and gives an enormously increased range. For submarines the removal of the need for oxygen for combustion permits submerged operation for almost indefinite periods. It is no surprise, then, to find that the main maritime application of nuclear power reactors has been military.

Jane's Fighting Ships lists over 300 nuclear powered ships, the majority being submarines but with a substantial number of surface ships, notably in the US Navy, which has 3 aircraft carriers and 8 cruisers powered by nuclear reactors. The principal nuclear navies are those of the USA and USSR, both of which have over 150 nuclear powered vessels; the UK has 19 nuclear submarines in commission or under construction and France has 4 submarines in use, 5 building and at least 4 more proposed, and also intends to build a nuclear powered aircraft carrier.

Details of the nuclear propulsion units are not available, but, apart from one sodium cooled system (Seawolf) installed in 1957 and replaced by a PWR 2 years later, they are all PWRs; the output required is typically of the order of 12 000 shaft kW per reactor, which indicates that they are considerably smaller than the reactors used for electrical power generation.

The advantages of an extended range and higher speeds might be thought to make nuclear propulsion equally attractive for civil maritime purposes, and indeed there have been many design studies which have shown an economic advantage for large high-speed nuclear powered ships on long-haul operation. There has, however, been little progress in this direction.

The first civil nuclear ship to be built was the NS *Savannah*, a US cargo–passenger vessel of 20 000 tons which had a 74 MW(th) PWR giving a speed of 21 knots and an endurance of 350 000 miles. The maiden voyage was in 1962; the ship operated a demonstration commercial service from 1965 to 1969 and was then laid up in 1971, having sailed 470 000 miles and visited 77 ports in 26 countries. The objective, to demonstrate the technical viability of nuclear marine propulsion, had been achieved.

The *Otto Hahn* was the next to be launched, in 1964; this was a German ore carrier with an improved integral reactor design (see *Table 9.11*), using a 38 MW(th) PWR to provide 8000 shaft kW and a cruising speed of 17 knots. The ship sailed 600 000 miles without incident, but in 1979 it was decided not to proceed with a new core and the reactor is now being decommissioned.

A Japanese design of nuclear reactor was installed in the *Mutsu*, a freighter/survey vessel of 8350 tons powered by a 36 MW(th) PWR. This project had an unfortunate history in that, construction having been completed in 1972, environmentalist opposition focused through local fishermen made it impossible to provide docking and fuel handling facilities for the ship, so that no further progress has been possible.

Table 9.11 Characteristics of a reactor for ship propulsion (*Otto Hahn*)

Thermal output	38 MW(th)
Fuel type	UO_2
Average enrichment	4.03%
Number of fuel elements	16
Number of fuel rods	3144
Fuel rod diameter	11 mm
Cladding material	Cr–Ni–Nb
Cladding thickness	0.35 mm
Operating pressure	6.23 MN/m^2 (62.3 bar)
Inlet temperature	267°C
Outlet temperature	278°C
Core diameter	1150 mm
Active core height	1120 mm
Number of control rods	12
Absorber material	B_4C

Finally, the USSR has three large nuclear powered ice-breakers in operation and a fourth is building. The first, *Lenin*, was commissioned in 1959 and has seen long and successful service; *Arktika* was brought into operation in 1975 and *Sibir* in 1977. The latter two have displacements of 19 300 tons and use twin PWRs to generate 55 000 shaft kW to give a speed of 21 knots. Canada has recently indicated an intention to build a large nuclear ice-breaker to open up large areas of the Arctic Ocean.

The slow progress towards a more widespread adoption of nuclear power for civil ship propulsion must be attributed to the associated problems of port access and liability in the event of a nuclear accident, neither of which are at present satisfactory, and perhaps to some extent to the paucity of facilities for the special operations involved in the periodic servicing and refuelling of ship-borne nuclear plant.

9.7 The nuclear fuel cycle

'Fuel cycle' is the phrase used to describe the path followed by nuclear reactor fuel in the many stages of a complex operation, which includes the production of the fuel material, the fabrication of fuel elements, the reprocessing of spent fuel for recovery of fissile and fertile material, and the disposal of radioactive wastes. *Figure 9.12* shows the interrelationship between these processes.

The process, from the extraction of uranium from its ores to the production of a refined metal, oxide or fluoride, consists of two main stages. In the first stage the uranium is extracted from the ore and converted to a concentrate which contains 60–65% uranium. This is carried out at the uranium mill, which is located close to the mine. The concentrate, commonly known as yellow cake, is then converted to the desired final product in a refinery which is closely associated with the factories at which the reactor fuel is produced.

The conventional extraction process for recovering uranium from ore comprises a number of operations, including leaching, solid–liquid separation, precipitation of uranium concentrates and production of refined products.

The processing of uranium yellow cake in the UK is carried out at the British Nuclear Fuels plant at Springfields. Here uranium concentrates are first dissolved in nitric acid and any insoluble material is filtered off. The uranyl nitrate solution is purified by solvent extraction, dibutyl phosphate in kerosene being used as the extractant. The purified solution is concentrated and decomposed to UO_3 in a fluidised bed reactor; the uranium oxide product is then reduced with hydrogen to UO_2, again in a fluidised bed reactor, and the uranium dioxide is converted to

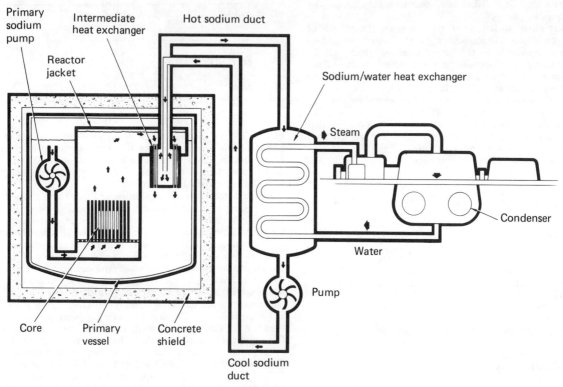

Figure 9.11 Schematic layout for a Liquid Metal Fast Breeder Reactor

mercial system of 1300 MW(e) is now planned as the next step in the programme. France also has a prototype LMFBR, known as PHENIX; this is a 250 MW(e) plant which achieved power in 1973 and is being followed by a 1200 MW(e) commercial system, SUPERPHENIX, which is now being built in partnership with West Germany and Italy. The USA have a fast flux test facility (FFTF) and an experimental breeder reactor (EBR), and are constructing a 380 MW(e) liquid metal fast breeder demonstration plant at Clinch River. The USSR has a substantial fast reactor programme; the first unit to produce power, the BOR 60, was commissioned in 1969, some 10 years after the first Soviet experimental fast reactor, the BR 5. It was followed by the BN 350, which produces 120 MW(e) and 50 000 m³/day of desalinated water. A full-size reactor — the BN 600, of 600 MW(e) — was commissioned in April 1980.

Figure 9.11 shows a schematic layout for an LMFBR system. The liquid sodium coolant is confined within a large pool and exchanges its heat with a second, intermediate circuit of liquid sodium. This intermediate circuit, in turn, transfers its heat to steam generators which produce steam to drive the turbine, thus avoiding the possibility of the release of radioactive sodium as a result of steam generator problems and also the possibility of penetration of water into the reactor core. The reactor core contains mixed uranium–plutonium oxides and is surrounded by a blanket of uranium oxide. The uranium in both core and blanket is U-238, which is a waste product from both enrichment and reprocessing plant. Fission occurs primarily in the core region and conversion in both core and blanket; conversion in the blanket is essential, since the core is a net consumer of plutonium. Regular reprocessing is required to recover the bred fissile material.

The core and blanket are made up of a compact arrangement of vertical rods of 5.8 mm diameter, clad in stainless steel. The rods forming the radial blanket are of the same composition throughout; the rods constituting the core and axial blanket have fissile and fertile material in the core portion but only fertile material in the blanket region. The rods are arranged in assemblies, with 271 rods per assembly and 318 fuelled assemblies in a particular large commercial design.

The pool type of design, in which the core, the primary coolant pumps and the intermediate heat exchangers are all contained within the reactor vessel, offers a number of safety features: the primary coolant is retained within the reactor vessel, which can be located below ground, and there are no penetrations of the reactor vessel below the sodium level; in the event of primary pump failure, there is a degree of natural circulation available to effect decay heat removal. This contrasts with designs in which the primary coolant is circulated in pipes between primary vessel and heat exchanger.

One of the attractions of the fast reactor concept is that in the longer term a programme of fast reactor development can be self-sufficient in fissile fuel. The initial inventory, both of Pu-239 and of depleted uranium (U-238), will come in the early stages of the programme from the existing stocks produced in thermal reactors and recovered in the reprocessing plant; subsequently the inventory of fissile material, both for refuelling existing fast reactors and for the initial charge for new fast reactors, will be produced by breeding. The conversion of U-238 to Pu-239 in fast reactors conserves the relatively limited stocks of uranium, effectively allowing an increase by a factor of 60 in the energy which can be extracted from natural uranium, compared with thermal reactors. The existing stock of depleted uranium avail-

and light water coolant in the SGHWR design permits a judicious choice of relative volumes, so that a change in voidage will have little or no effect upon neutron multiplication and reactor power.

Other attractive features of the design are the absence of moving mechanical components within the reactor core; a shutdown mechanism which is independent of the primary circuit; and an emergency core cooling system which injects water along the whole length of the fuel channel, a feature not possible if the fuel is segmented, as in CANDU.

In 1977 it was proposed that the SGHWR would form the basis of the next stage of the UK nuclear power programme, but detailed design studies for a commercial size SGHWR revealed complications leading to increased costs which had not been evident in the Winfrith prototype design.

Characteristics of a proposed commercial design are listed in *Table 9.8*.

Table 9.8 Characteristics of a proposed commercial SGHWR design

Core thermal power	2026 MW(th)
Plant efficiency	30.02%
Plant electrical output	660 MW(e)
Cladding material	Zircaloy
Cladding thickness	0.69 mm
Number of fuel channels	584
Fuel material	UO_2
Fuel enrichment	1.8% U-235
Number of fuel pins per bundle	36 (or 60 in advanced design)
Total fuel inventory (UO_2)	113.2 t
Lattice pitch (square)	26 cm
Fuel power density	17.9 MW/t
Coolant type	light water
Steam drum operating pressure	6.48 MN/m^2 (64.8 bar)
Final feed water temperature	199°C
Pressure tube mean thickness	5 mm
Pressure tube internal diameter	13 cm
Design fuel burn-up	21 500 MW-day/t

9.5.6 Fast breeder reactors

All the reactors so far described have been *thermal* reactors, in which a moderator is used to slow down the fast neutrons emitted in the fission process in order to take advantage of the greatly increased probability of fission at lower neutron energies.

In all thermal reactors which use natural or low-enrichment uranium as fuel there is a conversion of U-238, which cannot be fissioned by thermal neutrons, to Pu-239, which can be fissioned by neutrons of any energy. Some of this Pu-239 will, in turn, be fissioned, but the production rate exceeds the *in situ* fission rate and there will be an accumulation. The amounts which accumulate in various types of thermal reactor are shown in *Table 9.9*; the figures in each case refer to a 1000 MW(e) nuclear power station operating for 1 year.

Thermal reactors thus provide a source of plutonium which, after extraction from the spent fuel, can be used to fabricate fresh fuel elements. Pu-239 has the property that the number of neutrons liberated per neutron absorbed in fission by a fast neutron is substantially greater than the corresponding number for a thermal neutron induced fission. This leads to the concept of a *fast* reactor with Pu-239 fuel, having no moderator, so that the mean energy of the neutrons is higher, with a correspondingly high fission neutron yield. In a modern commercial fast reactor the mean neutron energy is typically 100 keV, compared with less than 0.1 eV in a thermal reactor. Many of the neutrons which escape from the core are then absorbed in a surrounding *blanket* which contains *fertile* elements such as U-238 or Th-232, and neutron absorption in these elements produces additional fissile material, respectively Pu-239 or U-233.

Table 9.9 Production of plutonium expressed as weight of Pu-239 for each GW-year of electricity

Reactor type	Net production of Pu-239
PWR	270 kg
MAGNOX	617 kg
CANDU	493 kg
AGR	173 kg

Although there is a net production of Pu-239 in a thermal reactor, there is no net production of fissile material when the burn-up of the original fuel is taken into account. In a fast reactor, however, it is possible to design the core and blanket assembly so that more atoms of fissile material are formed than are consumed. There is then a net gain in fissile material which leads to the concept of a fast *breeder* reactor, which produces more fuel than it consumes. The breeding rate is relatively slow; the *doubling time*, which is the time taken to breed enough fissile material to start another similar fast reactor, is typically in the range 10–30 years.

The probability of fission is lower for high-energy neutrons than it is for slow neutrons, so that a higher concentration of fissile material is necessary in the core of a fast reactor, typically 15% or more, compared with 2–3% for many thermal reactors. The absence of a moderator, however, makes for a compact core with higher power density, requiring an effective coolant with good heat transfer properties. The liquid metal fast breeder reactor (LMFBR) uses sodium, a metal that is liquid over a large range of temperature and one that can be used at essentially atmospheric pressure while still allowing operation at temperatures which will give a high thermal efficiency. Characteristics are listed in *Table 9.10*.

In the UK development of LMFBR technology began with an experimental system known as the Dounreay fast reactor, a 14 MW(e) system which began operation in 1959. This was followed somewhat belatedly by a 250 MW(e) prototype, also at Dounreay, which was commissioned in 1974, and a full-scale com-

Table 9.10 Characteristics of a liquid metal fast breeder reactor (commercial design)

Thermal power	3800 MW(th)
Overall plant efficiency	39.5%
Net electrical power	1500 MW(e)
Core diameter	310 cm
Core height	120 cm
Cladding material	SS 316
Fuel material	PuO_2/UO_2
Fuel rod diameter	6.9 mm
Number of fuel rods per assembly	271
Number of core assemblies	318
Number of blanket assemblies	234
Vessel diameter	7.16 m
Vessel length	18.14 m
Reactor inlet temperature	385°C
Reactor outlet temperature	528°C
Peak fuel burn-up	150 000 MW-day/t
Doubling time	12–15 years

is required compared with an LWR; typically, bulk light water is injected into the primary circuit from a storage tank and subsequently recycled from the well of the reactor building as necessary.

The pressure tube design with the bulk moderator at low pressure means that there is no requirement for a massive pressure vessel, and has the additional safety advantage that the failure of a single pressure tube is much less serious than the failure of a pressure vessel. (It is of course necessary to prove that pressure tube failure cannot propagate.) The CANDU system is economical in respect of fuel costs, since natural uranium can be used, but this is to some extent offset by the high cost of the heavy water inventory. The fuel burn-up achieved in CANDU reactors is appreciably higher than in MAGNOX systems, but this is also offset by the higher costs of oxide fuel compared with metallic uranium.

9.5.5.2 Steam generating heavy water reactor

The steam generating heavy water reactor differs from the CANDU design in two important respects: first, although the moderator is still heavy water, the coolant is now light water; second, a direct cycle is used, steam being produced in the core and fed, after separation from the liquid phase in a steam drum, to the turbines. The only SGHWR in commission is a 100 MW(e) prototype at Winfrith, Dorset, UK, which achieved full design power in January 1968.

The reactor consists of banks of zirconium alloy pressure tubes which are located within, but separated from, the tubes of a calandria which contains the heavy water moderator. About 30% of the moderation in the reactor takes place in the light water coolant, which reduces the heavy water inventory. The heavy water temperature does not exceed 80°C and, with the low pressures in the moderator circuit, leakage losses of the expensive heavy water can be maintained at a very low level.

The fuel is slightly enriched (~2%) uranium dioxide pellets of diameter about 12.5 mm, clad in zirconium alloy cans of thickness 0.69 mm and arranged in 36 rod clusters of length 3.66 m. Fuel clusters are positioned within the vertical pressure tubes and the fission heat causes partial boiling of the upward-flowing coolant. The steam produced is separated from the water, in steam drums, and passes directly to the turbine. The water passes to the lower part of the drum and mixes with the condensate returning from the turbine condenser before recirculation to the core. A flow diagram is shown in *Figure 9.10*. An emergency cooling system is provided to cool the fuel elements in the event of a rupture of the primary circuit and a loss of normal cooling. The system injects water into each fuel element through a central sparge pipe which provides a spray of water to the fuel rods.

Normal regulation of reactor power is by adjustment of the level of the heavy water in the calandria, which affects the neutron moderation and the neutron chain reaction. Long-term reactivity changes due to fuel burn-up are compensated for by variation of the boric acid concentration in the heavy water moderator. The reactor is normally shut down by dumping the heavy water from the calandria, but for rapid shut-down a fast-acting liquid shutdown system is also provided, borated liquid being injected under gas pressure into tubes within the calandria.

Like CANDU, the pressure tube design of the SGHWR permits factory fabrication of pressure components, leading to a high integrity of the primary coolant circuit and a short site construction time, together with the demonstrated ability to replace major pressure retaining components if necessary. Single-channel access is available for refuelling; rapid off-load refuelling is utilised for the prototype and proposed for the commercial design.

In a natural uranium system which is both cooled and moderated by heavy water, there is a positive void coefficient of reactivity; that is, an increase in voidage within a coolant channel gives rise to an increase in neutron multiplication and, consequently, in fission rate and power. In a light water cooled and moderated system the opposite is true and the void coefficient is strongly negative. The combination of heavy water moderator

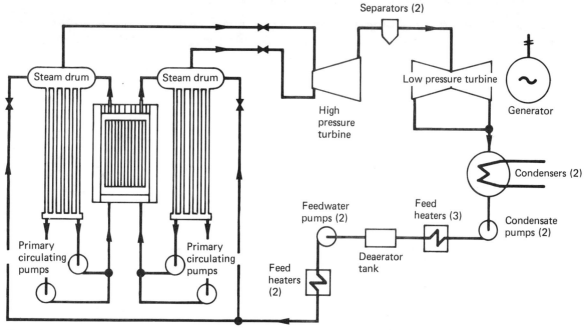

Figure 9.10 Steam generating heavy water reactor (SGHWR)

water as both moderator and coolant. There are now 10 CANDU reactors in operation and another 15 under construction or planned; the range of CANDU reactor systems includes a boiling light water design at Gentilly, Quebec, and an organic cooled design at Whiteshall, Manitoba. It seems that no more examples of these types will be built. The use of a light water coolant with a heavy water moderator distinguishes a UK design known as the steam generating heavy water reactor (SGHWR). A prototype SGHWR has been operating successfully at Winfrith since 1967, but no progress has been made towards a full-scale commercial reactor of this type apart from design studies.

9.5.5.1 CANDU

The fuel of a CANDU reactor (characteristics listed in *Table 9.7*) consists of pellets of uranium dioxide sealed into Zircaloy clad fuel pins which are then arranged in fuel *bundles*, each of 37 pins. The pellets are 1.21 cm in diameter, the cladding is 0.04 cm thick and the cylindrical fuel bundles are typically 50 cm long and 10 cm in diameter. The calandria which contains the heavy water moderator is a cylinder 7.6 m in diameter and 7.6 m in length, with stainless steel walls 3 cm thick and ends 5 cm thick. The horizontal calandria tubes in which the fuel is loaded contain individual pressure tubes fabricated from zirconium alloy, with a gas space between the pressure tube and the surrounding calandria tube. The tubes are arranged in a square lattice of 28 cm pitch.

The pressure tubes have a header at either end and may be opened individually during reactor operation for refuelling. Coolant from a primary pump passes through a distribution header to the individual pressure tubes, then horizontally through the core and on to the header at a steam generator and thence through the U tube steam generator back to the primary pump. The coolant flow in alternate channels is in opposite directions, so maintaining a more uniform temperature distribution in the core. Twelve fuel bundles are loaded end-to-end in each channel, and on-load refuelling is effected by inserting fresh bundles and pushing through the spent fuel; refuelling is also in opposite directions for alternate channels, and this helps to achieve a more uniform burn-up distribution in the core. The adoption of horizontal fuel channels in the CANDU design is thus seen to result in both improved temperature and improved burn-up distributions. *Figure 9.9* shows a schematic layout for a CANDU reactor system.

Reactivity control is provided by solid absorber rods containing cadmium which are used for shut-down, by light water control absorbers for zonal adjustments and by the addition of boron and gadolinium compounds to the moderator for long-term compensation. The moderator temperature is maintained constant by circulation through a separate heat exchanger, with a moderator clean-up system which controls impurities and includes the capability for removing the boron and gadolinium. The moderator provides a heat sink during a loss of coolant accident, so that a less elaborate emergency core coolant system

Table 9.7 Characteristics of a CANDU reactor system

Core thermal power	2140 MW(th)
Plant efficiency	28%
Plant electrical output	600 MW(e)
Core diameter	6.3 m
Core active length	5.9 m
Cladding material	Zircaloy
Cladding thickness	0.04 cm
Fuel material	natural uranium oxide
Number of fuel pins per bundle	37
Number of fuel bundles per channel	12
Number of fuel channels	380
Total fuel inventory (UO_2)	95 t
Fuel power density	22.5 MW/t
Coolant type	heavy water
Coolant pressure (mean)	10.7 MN/m^2 (107 bar)
Coolant inlet temperature	267°C
Coolant outlet temperature	312°C
Moderator type	heavy water
Moderator pressure	approximately atmospheric
Moderator inlet temperature	43°C
Moderator outlet temperature	71°C
Design fuel burn-up	7500 MW-day/t
Refuelling sequence	continuous, on-line

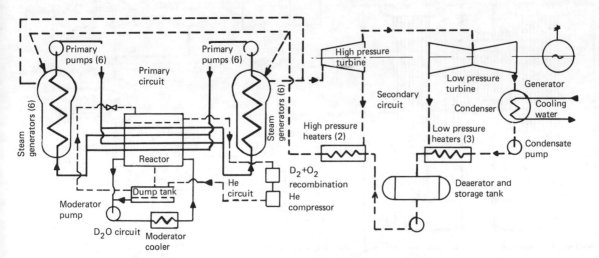

Figure 9.9 CANDU system schematic (by courtesy the International Atom Energy Agency)

The boiler and circulator units, of which there are eight per reactor in the later AGRs, are located within the pressure vessel in five of the AGR Stations; at Hartlepool and Heysham A, they are contained in vertical cavities or *pods* in the cylindrical wall of the pressure vessel.

It is intended that fuel replacement shall be carried out while the reactor is at power by means of a fuelling machine which can access each fuel channel through individual standpipes located in the top cap of the pressure vessel. The fuelling machine is gantry mounted and traverses and serves both the reactors in an AGR station.

Vertical control rod channels are located at selected interstitial positions in the graphite lattice; the control rods are used to compensate for reactivity changes and during start-up and shut-down of the reactor. The control rods are of stainless steel tubular construction with boron steel inserts and are partially filled with graphite. Secondary and tertiary shut-down systems are also available to give rapid and complete shut-down capability independent of the control rod system. The secondary system injects nitrogen gas into interstitial holes in the core; the tertiary system uses a solid absorber in the form of borated glass beads.

A comparison with the MAGNOX data indicates the relative compactness of the AGR design, with higher fuel power density, higher outlet gas temperature and an improved thermal efficiency.

9.5.4 High-temperature reactors (HTRs)

The high-temperature reactor concept represents a major departure from the technology of the MAGNOX and AGR systems in that, although neutron moderation is still achieved by the use of graphite, the coolant is now helium and the fuel is in the form of particles of oxides or carbides of uranium or thorium, coated with layers of carbon and silicon carbide. The absence of metallic cladding avoids wasteful neutron absorption, and the use of helium removes the restriction on coolant temperature imposed by the reaction between CO_2 and graphite. To realise this advantage in practice, coolant impurities must be kept to levels usually associated with laboratory apparatus. In particular, lubricant ingress could not be tolerated, and it has been necessary to develop gas bearing circulators.

The chief disadvantage of gas cooling is that heat transfer coefficients are much lower than with liquid coolants. With MAGNOX and AGRs, this implies that the fuel rating (or fuel power density) must be kept low in order to keep the can surface temperature and the fuel centre temperature within the permissible limits. In HTRs this disadvantage is overcome by the very high surface temperatures which are permissible for graphite in contact with helium, and the small temperature drop across the coated particle fuel. The volumetric ratings of HTRs can, therefore, be comparable with those of water reactors.

The very high coolant outlet temperature possible with HTR (950°C has been demonstrated and 1100°C may be achievable) is another potential advantage. With outlet temperatures as high as this, one can contemplate both direct cycle operation using helium gas turbines and using the reactor as a source of process heat. In processes such as steel making and the production of methane by the water–gas reaction, fossil carbon is used both as a chemical and as a source of heat. If the necessary heat can be taken from an HTR instead, the reaction yield can be much improved. The many paper studies have emphasised the difficulty of both these technologies. The medium-term future of the HTR must be as an electricity generator, and here its advantages are the high power density referred to above and the excellent uranium utilisation made possible by eliminating metal from the core.

HTR fuel may be either cylindrical (prismatic) or in the form of balls (pebbles), typically 60 mm in diameter: in either case, it consists of a dispersion of coated particles in graphite. The prismatic fuel is made up into a normal core, which differs from MAGNOX and the AGR in that the moderator is completely removable. Pebble fuel is stacked into a pebble bed and again the moderator is completely removable. In the AGR, reactor life is likely to be limited by erosion of the moderator by the CO_2 coolant, so that this is another advantage of the HTR.

The reactor may be operated on either the U–Pu or the U–Th fuel cycle. On the U–Pu fuel cycle the enrichment is typically 10%, while on the U–Th cycle the uranium is highly enriched (93%). The U–Th cycle gives much better uranium utilisation, but it requires a higher initial uranium investment and special fuel cycle facilities.

The first HTR was the Dragon experimental prototype at Winfrith, Dorset, UK, which was built and operated by the HTR or Dragon project, a collaborative effort by a number of West European countries. This reactor has operated on both fuel cycles with prismatic fuel: it was shut down in 1976.

The only country in Western Europe with its own HTR programme is the Federal Republic of Germany (which was also a partner in Dragon). The Republic has a special problem of heat rejection. Her coastline is short and far from the main industrial centres, and her rivers are overloaded. The HTR is attractive, because its high efficiency reduces the amount of heat which must be rejected for each unit of electricity generated. In the longer term, the Republic is also interested in process heat applications. She has pursued the pebble bed type of reactor, using the U–Th fuel cycle. The AVR experimental reactor was commissioned at Julich in 1967 and will be kept in operation until the mid-1980s. The THTR 300 MW(e) industrial prototype is now under construction at Uentrop and should be commissioned shortly.

The USA has developed the U–Th cycle with prismatic fuel. The Peach Bottom experimental reactor shut down in 1974, having given excellent service, and the construction of a 330 MW(e) industrial prototype at Fort St. Vrain in Colorado started in 1968. This reactor has suffered manifold difficulties and delays and has not yet reached full power. The combination of this experience with environmentalist attacks on nuclear power makes it unlikely that there will be a US HTR programme in this century: the four commercial HTR power stations ordered by utilities have all been cancelled.

9.5.5 Heavy water reactors

Heavy water is an alternative reactor moderator material, having good slowing-down properties and very low thermal neutron absorption, which means that, as with graphite, a critical chain reaction can be sustained with natural uranium fuel; the reactivity of a heavy water/natural uranium lattice is in fact higher than for a comparable graphite/natural uranium lattice. The ratio between moderator volume and fuel volume in a heavy water system is appreciably greater than in a light water system, and this leads to the possibility of having individually cooled fuel channels with a separate heavy water moderator surrounding the channels. This, in turn, leads to a *pressure tube* design in which the coolant (which may be light or heavy water) is pressurised while the heavy water moderator, separately contained in a *calandria*, is at near-atmospheric pressure. (The term 'calandria' is used to describe a cylindrical vessel whose plane surfaces are joined by cylindrical penetrations, each containing a pressure tube; these may be horizontal or vertical.) This reduces the volume and consequently the fabrication costs of the high pressure primary circuit; it also reduces the loss by leakage of the expensive heavy water under normal operating conditions and the seriousness of moderator loss in the event of a breach of the calandria.

The main advances in heavy water reactor technology have been in Canada, where the CANDU system (CANadian Deuterium Uranium) has been developed, initially using heavy

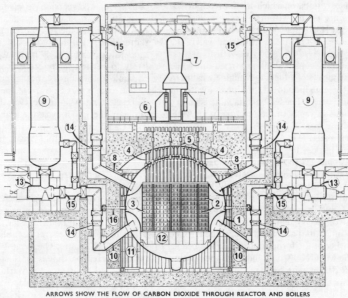

Figure 9.7 Layout of a MAGNOX Reactor system (by courtesy the Central Electricity Generating Board). 1 Reactor pressure vessel; 2 Fuel elements; 3 Graphite-moderator; 4 Charge tubes; 5 Control rod standpipes; 6 Charge floor; 7 Charge/discharge machine; 8 Hot gas outlets from reactor; 9 Boilers (six per reactor); 10 Cool gas inlets to reactor; 11 Thermal shield; 12 Diagrid; 13 Main circulators; 14 Gas isolating valves; 15 Hinged expansion bellows; 16 Can-failure detection standpipes

Table 9.6 Characteristics of a typical advanced gas cooled reactor system

Core thermal power per reactor	1510 MW(th)
Plant efficiency	41.2%
Plant electrical output per reactor	622 MW(e)
Core diameter	9.3 m
Core active length	8.23 m
Cladding material	stainless steel
Cladding thickness	0.38 mm
Fuel material	UO_2
Fuel enrichment	1.4 and 1.6%
Number of fuel pins per element	36
Number of fuel elements per channel	8
Number of fuel channels	324
Control rod type	boron steel tubes
Number of control rods	81
Total fuel inventory (UO_2)	120 t (UO_2)
Fuel power density	12.5 MW(th)/t (UO_2)
Coolant type	CO_2
Coolant pressure	4.14 MN/m^2 (41.4 bar)
Coolant inlet temperature	318°C
Coolant outlet temperature	651°C
Design fuel burn-up	18 000 MW-day/t
Refuelling sequence	continuous, on-load

being slightly higher to flatten the radial power distribution; in some designs axial variation of enrichment is also used, in this case to flatten the axial power distribution and improve burn-up. The cylindrical core is about 8.25 m high, with a diameter of 9 m.

The integral reactor system is contained within a prestressed concrete pressure vessel in the form of a vertical cylinder, typically of height 29.25 m and outside diameter 25.9 m. This is lined internally with mild steel plate about 25 mm thick and insulated with stainless steel foil and mesh layers protected by 6.35 mm thick steel cover plates. The pressure vessel is prestressed circumferentially by means of concentrated bands of wire wound under tension around the outer periphery of the vessel, while vertical prestressing is achieved by means of vertical tendons. *Figure 9.8* shows the layout of Hunterston B AGR.

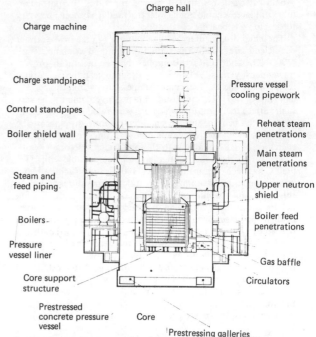

Figure 9.8 Advanced Gas-Cooled Reactor layout (by courtesy South of Scotland Electricity)

Table 9.4 Details of the UK AGR programme

Location	Electrical output (MW(e))	No. of reactors	Mean power rating (kW(th)/kgU)
Dungeness B	1200	2	9.5
Hinkley Point B	1250	2	13.1
Hunterston B	1250	2	13.1
Hartlepool	1250	2	12.5
Heysham	1250	2	12.5

as high as 2500°C. It is also outstandingly resistant to radiation damage. The original choice of cladding was beryllium, which has a high melting point and very low thermal neutron absorption; it is, however, very brittle and has poor resistance to radiation damage, and stainless steel was selected in its place. While the required temperatures can be accommodated satisfactorily, the neutron absorption of stainless steel is greater than that of beryllium and results in a higher fuel enrichment.

Two additional AGR stations have now been ordered and will be sited at Heysham and Torness.

The third generation of graphite moderated reactors, known as high temperature reactors (HTRs), represents a further advance in technology, but apart from a 32 MW(th) prototype known as DRAGON, commissioned at Winfrith in 1964, no progress has been made in the UK beyond design studies. In the USA General Atomic has built a 330 MW(e) HTR but no progress has been made with larger commercial versions. In the Federal Republic of Germany a pebble-bed HTR is being built and is expected to be commissioned in 1983.

9.5.3.1 MAGNOX reactors

The core of a MAGNOX reactor (characteristics listed in *Table 9.5*) is built up from accurately machined graphite blocks, typically of 20 cm × 20 cm cross-section, each having a cylindrical channel of about 100 mm diameter to accommodate the fuel. The number of fuelled channels varies from 1696 for Calder Hall to 3000–4000 for the later designs and 6150 for the larger Wylfa reactors. The fuel, which is in the form of natural uranium metal rods about 90 cm in length and 2.5 cm in diameter, is enclosed in magnesium alloy cans to form individual fuel elements; eight to ten of these elements are loaded end-to-end in each channel. To improve the heat transfer from the uranium bar through the cladding to the coolant, the heat transfer surface is increased by the use of spiral fins. The fuel elements are cooled by circulating carbon dioxide gas upwards through the channels; the gas then gives up its heat in a number of steam raising units or boilers before returning to the core to repeat the cycle. The steam from the boilers is used to generate electricity as in conventional plant. Control of the neutron chain reaction is by means of boron control rods which are raised or lowered in vertical channels at selected interstices of the graphite blocks. A layout schematic for one of the Hunterston A reactors is shown in *Figure 9.7*. Provision is made for emergency shut-down by a system which injects balls containing boron into the core.

A typical MAGNOX reactor core is approximately cylindrical, with a radius of about 6 m and a height of about 9 m, although there is some variation in size between different designs. In the early systems the core is enclosed in a cylindrical or spherical steel pressure vessel, but in the later Oldbury and Wylfa designs this is replaced by prestressed concrete. In the case of Oldbury this is an upright cylinder of internal diameter 23.5 m and height 18.3 m, with a wall thickness of 4.57 m. Prestressing is by layers of cable laid to a helical pattern, with alternate layers twisting in opposite directions. Each stressing cable is made from 12 high-tensile 7-wire strands 1.5 cm in diameter, terminated in one of four annular stressing galleries. The design of a MAGNOX reactor allows continuous refuelling, in contrast to the batch schemes which are obligatory for the PWR and BWR. Individual fuel channels are refuelled while the reactor is at power, by use of a charge–discharge machine mounted on a travelling carriage which makes a sealed connection with a fuel standpipe and then effects fuel changes under remote control. In most designs charge and discharge is from above the core, although at Hunterston A the charge machine works from below the reactor where the working temperatures are lower.

Gas temperatures and plant thermal efficiency are limited in the MAGNOX design by a phase change which occurs in natural uranium metal at 665°C. Fuel which is cycled through this temperature undergoes a volumetric change and is consequently subject to considerable distortion which would result in can failure and release of fission products to the primary circuit. The burnup is also restricted for this fuel by reactivity considerations to about 5000 MW-day/t; the fuel has been improved to the point where metallurgical factors no longer limit burn-up. In the AGR design these limitations are avoided by the use of uranium oxide fuel.

9.5.3.2 Advanced gas cooled reactors

The graphite core of an AGR (characteristics listed in *Table 9.6*) contains about 300 identical fuel channels (Dungeness B has, exceptionally, over 400). Each channel contains a stringer of eight fuel elements, each of which comprises a cluster of 36 stainless steel clad fuel pins containing low-enrichment uranium dioxide pellets, the cluster being surrounded by a graphite sleeve, and supported by a central tie-bar or 'strongback' made of stainless steel. The cylindrical UO_2 pellets have a diameter of 14.48 mm; the stainless steel cladding is 0.4 mm thick; and the fuel element is 1041 mm long and 191 mm in diameter. The fuel channels are arranged in a square lattice of 457.2 mm pitch, each channel having a diameter of 269.9 mm. The coolant is carbon dioxide gas.

Fuel enrichment, which is required to compensate for the higher neutron absorption in the stainless steel cladding, varies in the range 2–3% U-235. A two-region differential enrichment scheme is usually employed, the enrichment in the outer region

Table 9.5 Characteristics of a typical MAGNOX reactor system

Core thermal power per reactor	892 MW(th)
Plant efficiency	33.6%
Plant electrical output per reactor	300 MW(e)
Core diameter	14.2 m
Core active length	9.75 m
Cladding material	MAGNOX A12
Cladding thickness	2.54 mm
Fuel material	natural uranium metal
Number of fuel rods per channel	8
Number of fuel channels	3308
Control rod type	boron steel
Number of control rods	101
Total fuel inventory (U)	293 t
Fuel power density	3.07 kW(th)/KgU
Coolant type	CO_2
Coolant pressure	2.5 MN/m² (25 bar)
Coolant inlet temperature	245°C
Coolant outlet temperature	412°C
Design fuel burn-up	4000–5000 MW-day/t
Refuelling sequence	continuous, on-load

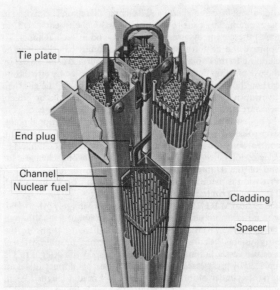

Figure 9.5 Boiling Water Reactor fuel assembly (General Electric)

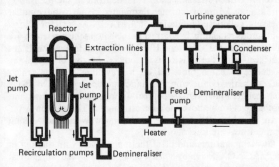

Figure 9.6 Layout schematic for a boiling water reactor system (General Electric)

in diameter, 22 m high, with a wall thickness of 15–18 cm. The head of the vessel and the steam separators and dryers are removable to allow access for refuelling the core. Like the PWR, the BWR is refuelled off-load, typically every 18 months in a three-batch cycle.

Emergency core cooling systems are provided. A high-pressure core spray system (early BWRs) or a high-pressure coolant injection system (latest BWRs) ensures adequate cooling of the core in the event of a small leak; these systems are initiated if the feedwater pumps fail to maintain the water level at a preset height above the core, or if high pressure is detected in the drywell containment which surrounds the pressure vessel. In the event of a more serious leak an automatic depressurisation system would operate to discharge steam through pressure relief valves into a suppression pool system, so lowering the pressure in the reactor vessel in order that operation of the low-pressure emergency cooling system could be initiated. This comprises a low-pressure core spray backed up by a low-pressure coolant injection system.

Current BWR designs usually provide both primary and secondary containment. The former is a steel pressure vessel surrounded by reinforced concrete and is designed to withstand peak transient pressures for the most serious loss of coolant accident. The secondary containment system is the building which houses the reactor and its primary containment system; it is designed for low leakage with sealed joints and interlocked double door entries.

9.5.3 Graphite moderated reactors

The technology of graphite moderated power reactors has been developed mainly in the UK. The first-ever critical assembly, built in Chicago in 1942, used graphite as a moderator with natural uranium fuel, and this important characteristic of graphite systems, that a critical chain reaction can be sustained without the need for expensive enriched uranium fuel, was exploited in the development of what has become known as the MAGNOX programme. This began with Calder Hall, commissioned in 1956 with four reactors producing a total of 200 MW(e). After a similar power station at Chapelcross, a series of MAGNOX reactors was built with progressively improved technology and increased power output, ending with a two-reactor station at Wylfa in Anglesey, commissioned in 1970 with a nominal electrical output of 1180 MW(e). Details are shown in *Table 9.3*.

The MAGNOX reactors derive their name from the cladding which encloses the natural uranium metal fuel and in which metallic magnesium is the chief constituent; the coolant is CO_2.

The second generation of UK graphite moderated reactors are known as advanced gas cooled reactors (AGRs). Following the commissioning of a 33 MW(e) prototype at Windscale in 1962, a programme of five stations was initiated, and these are currently at or near completion. AGRs use enriched uranium oxide fuel with stainless steel cladding and CO_2 coolant.

The AGR system (see *Table 9.4*) was evolved to take better advantage of the high-temperature possibilities of the gas–graphite combination. UO_2 does not undergo a phase change as does metallic uranium, and can be used with a centre temperature

Table 9.3 Details of the UK MAGNOX reactor programme

Location (all UK)	Commissioned	Electrical output (MW(e))	No. of reactors	Mean power rating (kW(th)/kgU)
Calder Hall	1956	200	4	2.40
Chapelcross	1959	200	4	2.40
Berkeley	1962	276	2	2.42
Bradwell	1962	300	2	2.20
Hunterston	1964	320	2	2.27
Trawsfynydd	1965	500	2	3.11
Hinkley Point A	1965	500	2	2.78
Dungeness A	1965	550	2	2.78
Sizewell A	1966	580	2	2.96
Oldbury	1967	600	2	3.04
Wylfa	1970	1180	2	3.16

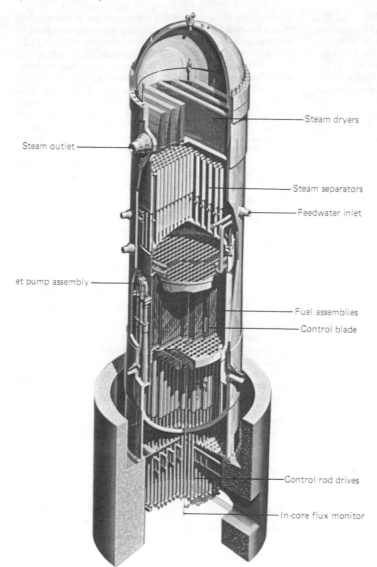

Figure 9.4 General arrangement of a boiling water reactor pressure vessel (General Electric)

this pressure water boils and forms steam at 285°C. The general arrangement within the pressure vessel is shown in *Figure 9.4*.

The fuel used in the BWR is uranium oxide in the form of cylindrical pellets of diameter about 10 mm, which are enclosed in tubes of Zircaloy-4 to form fuel pins. The fuel pins are then arranged in assemblies, typically of 8×8 pins, with an enclosing sheath which constrains the flow of water within the assembly, unlike the PWR assembly, which is open. A large BWR may contain over 700 assemblies with a total number of pins approaching 50 000. The active core height is typically 380 cm and the diameter 460 cm. The fuel is enriched to a few per cent U-235; the core is usually arranged in two roughly cylindrical zones with a slightly higher enriched fuel in the outer zone to flatten the radial power distribution. A typical BWR fuel assembly is shown in *Figure 9.5*.

Control is effected by cruciform control elements containing neutron absorbers usually in the form of boron carbide; these traverse vertically in the interstitial gaps between four adjacent fuel assemblies. The neutron density in a BWR is reduced towards the top of the core, owing to the relatively poor moderation of the steam–water mixture, and since control absorbers are more effective in a high neutron density, it is necessary for the control elements to enter the core from below. Consequently, the control rod drive mechanisms for a BWR are located at the bottom of the reactor vessel and space has to be provided below the core for the retraction of the control elements. The space above the core is occupied by steam separators and steam dryers, so that the overall height of a BWR vessel is appreciably greater than that for a PWR of comparable power output. Soluble poison cannot be used for reactivity control in a BWR, since the concentrations required to control excess reactivity at the start of life would make the void coefficient of reactivity very positive. Burnable poisons are used, often in the form of plates or curtains located in the gaps between fuel assemblies which are not occupied by control elements.

About a third of the coolant water is continuously by-passed out of the vessel through two external recirculation loops and returned to the vessel through internal jet pumps (*Figure 9.6*) to provide the motive force for core circulation, the rate of which determines the amount of heat that can be extracted from a reactor core of a given size.

Typically, a 1000 MW(e) BWR has a reactor vessel about 6.4 m

coolant loops. When the plant electric load is decreased, the temperature of the primary coolant rises and a positive pressure surge in the primary system results in the automatic operation of the spray system in the top of the pressuriser; this condenses some of the steam and decreases the pressure. During a negative pressure surge caused by an increased plant electrical load the electric heaters are turned on and generate sufficient steam within the pressuriser to maintain primary system pressure. A schematic layout of a typical PWR plant is shown in *Figure 9.3*. (Note that a large PWR system may have up to four coolant loops, each with its own steam generator and pump: the loops normally work into a single turbine.)

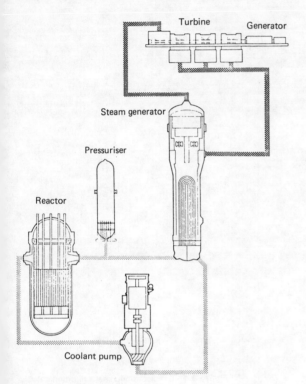

Figure 9.3 Layout schematic for a pressurised water reactor system (Westinghouse)

Each nuclear plant has many features that are provided to compensate for operator errors, equipment defects, abnormal occurrences and a variety of accidents. In particular, provision is made in a PWR system for the removal of decay heat in the event of a break that permits water to leak from the primary system. The emergency core cooling systems consist of several independent subsystems. The first is a passive accumulator injection system which consists of two or more large tanks connected through check valves to the primary coolant pipes; these tanks contain cool borated water stored under nitrogen gas at a pressure of 1.3–4.5 MN/m^2 (13–45 bar). If a large break were to occur in the reactor primary system, with consequent rapid decrease of pressure, the check valves would open to discharge a large volume of water into the reactor vessel. Two active systems also provide emergency core cooling; one is a low-pressure system which comes into action in the event of a large break as soon as the accumulator tanks have emptied; the other is a high-pressure pumped system which is activated if the break is small and pressure remains high within the primary loop.

The entire primary system of a PWR is enclosed in a containment building which, in the event of a break in the primary system, is capable of preventing the release of radioactive material into the environment. Most modern PWR containments are constructed of reinforced concrete with a steel liner. In some PWRs the containment space is kept slightly below atmospheric pressure so that any leakage through the containment walls would, under normal conditions, be inward. Other systems have double barriers. In some plants cold water sprays are provided within the containment to condense the steam resulting from a major escape of primary system coolant; in other plants stored ice is used for this purpose. Another safety measure is the provision of blowers to recirculate the containment atmosphere through filters and absorption beds to remove airborne radioactive materials.

9.5.2 Boiling water reactors

Although the spread of boiling water reactor (BWR) technology (see *Table 9.2*) is less extensive than that of the PWR, there are now about 50 BWR systems in operation in some 11 countries.

Table 9.2 Characteristics of a typical boiling water reactor system

Core thermal power	3579 MW(th)
Plant efficiency	34%
Plant electrical output	1220 MW(e)
Core diameter	460 cm
Core active length	388.6 cm
Cladding material	Zircaloy-2
Cladding thickness	0.813 mm
Fuel material	UO$_2$
Pellet diameter	10.4 mm
Assembly array of fuel pins	8 × 8
Number of fuel assemblies	748
Control rod type	cruciform blades (boron carbide)
Number of control rods	177
Total fuel inventory (UO$_2$)	155 t
Fuel power density	23.09 MW/t
Coolant flow rate	47.2 × 10 t/h
Coolant pressure	7.17 MN/m^2 (71.7 bar)
Coolant inlet temperature	215°C
Coolant outlet temperature	288°C
Design fuel burn-up	28 400 MW-day/t
Refuelling sequence	1/3 of fuel every 18 months
Refuelling time	1 week

About a third of the LWRs operating or under construction in the USA are of this type, the supplier being General Electric. Boiling water reactor technology has been developed in other countries, notably Sweden and the Federal Republic of Germany, where the suppliers are ASAE–Atom and Kraftwerk Union, respectively. US designs have been exported to a number of countries, including Japan. In the direct or single-cycle BWR system water is boiled within the pressure vessel, producing saturated steam that passes through internal steam separators and dryers before being fed directly into the steam turbine. The operating pressure is considerably below that for a PWR and the design of the nuclear steam supply system less complex. Water from the steam separators and water returned from the turbine condenser mix and flow downwards through the annulus between the core shroud and the reactor vessel to the bottom of the core before passing upwards through the core to complete the cycle. The pressure in a typical BWR is 6.9 MN/m^2 (690 bar). At

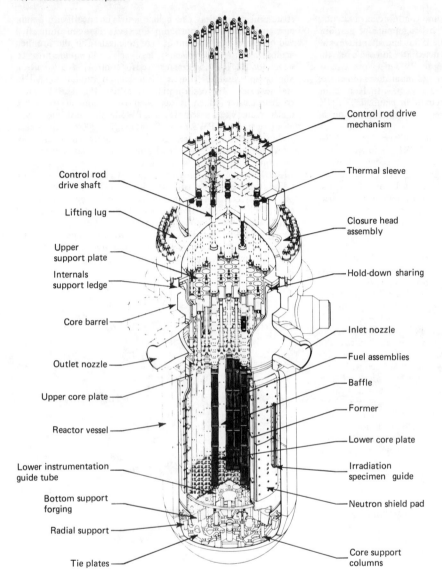

Figure 9.2 General arrangement of a Pressurised Water Reactor pressure vessel and internals (Westinghouse)

core, and the path of coolant flow within the reactor vessel is downward around the core barrel and then upward through the fuel assemblies. The downflow provides a reflector region which flattens the radial neutron flux distribution in the core. The pressure of a typical large power reactor system is sustained at about 15.5 MN/m^2 (155 bar), at which pressure water can be heated to about 343°C without boiling. PWR reactor vessels are cylindrical in shape, some being over 12 m high, with an inside diameter of over 4.25 m and wall thickness in excess of 20 cm.

The high-pressure water, heated to an average temperature around 315°C, is piped from the reactor vessel to two or more steam generators, in which the high-pressure reactor coolant water circulates through tubes whose outer surfaces are in contact with a second stream of water returning from the turbine condenser and from steam separated within the vessel. Heat transferred from the hot water inside the tubes causes the water of the secondary stream to boil and produce steam for the turbine. A typical vertical U tube heat exchanger contains 3260 U tubes, has an overall height of 19.2 m, with an upper shell diameter of 4.42 m, a lower shell diameter of 3.43 m and a wall thickness of 215–250 mm. Since the steam generators are so large and operate under high pressure, structural strength requires fabrication from thick steel plate and makes them some of the most massive objects in the plant.

The primary coolant pumps circulate the water through the primary coolant system, and high reliability and performance are demanded of them. They must operate at high temperature and pressure and circulate water which is radioactive; the stringent design and manufacturing criteria used for other primary circuit components apply equally to them. Each of the four primary coolant pumps in a typical 1000 MW(e) PWR is about 9 m high, is driven by a 4.5 MW electric motor and has a pumping capacity of about 520 m^3/h. The pump casings are designed for operation at 343°C at a pressure of 17.25 MN/m^2 (172.5 bar).

PWR steam supply systems are equipped with pressurisers; a typical pressuriser contains immersion-type electric heaters of about 2000 kW capacity, multiple safety and release valves, and a spray nozzle. The lower portion of the pressuriser contains water and the upper portion contains steam. The pressuriser is connected by a surge line to the hot side of one of the reactor

either complete or near completion, and two more have recently been ordered. France also started a programme of graphite moderated reactor development in the 1950s, and after three low-power demonstration plants, constructed four full-scale reactor systems of this type before switching to PWR development.

Canada has been responsible for the main development of heavy water reactors in the CANDU system, with 10 reactors now operating and 15 under construction or planned. The UK has a single prototype heavy water reactor in operation at Winfrith.

Developments in fast reactor technology began in the USA with an experimental breeder reactor, EBRI, commissioned in 1951 and in the UK with the Dounreay fast breeder prototype, completed in 1959, and are now progressing on a broad front in France, the USSR, the USA and the UK. The USSR was the first country to commission a fast reactor of more than prototype size, in April 1980.

9.5.1 Pressurised water reactors

There are more pressurised water reactors (PWRs) in use throughout the world than any other type of power reactor; the major suppliers are Westinghouse, Babcock and Wilcox, Combustion Engineering, Framatome and Kraftwerk Union. Westinghouse alone list about 130 PWR systems operating or under construction, of which 30 are under licence. PWR plant is installed or under construction in some 30 countries; in addition, all the propulsion units for nuclear ships and submarines are of this type and this currently represents more than 300 power reactors (although these are typically small compared with land-based plant), principally in the navies of the USA and the USSR but with significant numbers in those of the UK and France. Characteristics are listed in *Table 9.1*.

The PWR uses uranium dioxide fuel pellets, typically enriched to a few per cent U-235, sealed in a helium atmosphere in zirconium alloy cladding which is chosen for its low neutron absorption, resistance to corrosion by light water and good structural properties. The pellets are cylindrical, with height equal to diameter of about 9 mm. The clad pellets constitute a fuel rod or pin, typically about 365 cm in length. The pins are then arranged in *fuel assemblies*, typically a 17×17 square array in which some 24 locations are reserved for control pins of a neutron absorbing material such as silver–indium–cadmium, clad in stainless steel. The control pins are ganged together to form a control cluster which is activated by a control rod drive mechanism. A fuel assembly for a PWR is shown in *Figure 9.1*.

A large modern PWR may have as many as 200 fuel assemblies within the reactor core, arranged in two roughly cylindrical zones with higher-enrichment fuel in the outer zone to flatten the power distribution. The core is enclosed in a massive steel pressure vessel. Refuelling can only be effected by removing the pressure vessel head and the control rod drive mechanisms which penetrate it; consequently, refuelling is a lengthy operation which can only be carried out with the reactor shut down, so that all PWRs are operated in batch mode, a third or a quarter of the fuel being replaced at intervals of one or two years. The layout of a typical PWR pressure vessel is shown in *Figure 9.2*. The coolant inlet and outlet ducts are situated above the level of the top of the

Table 9.1 Characteristics of a typical pressurised water reactor system

Core thermal power	3820 MW(th)
Plant efficiency	34%
Plant electrical output	1300 MW(e)
Core diameter	335 cm
Core active length	365 cm
Cladding material	Zircaloy-4
Cladding thickness	0.62 mm
Fuel material	UO_2
Pellet diameter	8.19 mm
Assembly array of fuel pins	17×17
Number of fuel assemblies	193
Control rod type	silver–indium–cadmium pin clusters
Number of control rod assemblies	57 full-length, 8 part-length
Number of control rods per assembly	24
Total fuel inventory (UO_2)	89 t
Fuel power density	38.33 MW/t
Coolant flow rate	65.45×10^3 t/h
Coolant pressure	15.5 MN/m^2 (155 bar)
Coolant inlet temperature	292°C
Coolant outlet temperature	325°C
Design fuel burn-up	32 000 MW-day/t
Refuelling sequence	1/3 of fuel per year
Refuelling time	17 day (minimum)

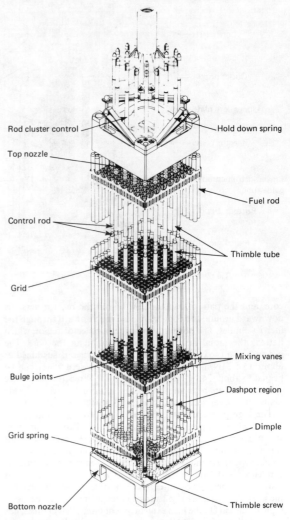

Figure 9.1 Fuel assembly for a pressurised water reactor (Westinghouse) showing rod cluster control assembly

from the fuel reprocessing plant; this contains an increased proportion of U-238 and could be used to surround a fast reactor core as a *blanket* in which escaping fast neutrons are absorbed to produce Pu-239. By suitable design of core and blanket it can be arranged that more fissile atoms of Pu-239 are produced in the blanket than are destroyed in the core, which leads to the concept of a *breeder* reactor which produces more fissile material than it consumes.

U-233 is also formed as a result of neutron absorption, in this case by Th-232 and the subsequent radioactive decay of Th-233. Thus, Th-232 can also be used as a blanket material for a fast reactor.

Most of the energy released in the fission process appears initially as kinetic energy of the fission fragments, and this is rapidly converted, by collision, into thermal energy: some of this energy is deposited in the moderator rather than in the fuel. Some 8%, however, is associated with the radioactive decay of the fission fragments and does not appear immediately, but with a delay determined by the half-lives of the associated radioactive processes, which range from a fraction of a second to thousands of years. This means that when a nuclear power reactor is 'shut down' (i.e. the fission chain reaction is stopped), the reactor will continue to generate significant amounts of thermal energy from the radioactive decay of the fission fragments. This residual energy, which for a 1000 MW(e) reactor would be of the order of 240 MW(Thermal) immediately after shut-down, is known as *decay heat*. If the shut-down followed a prolonged period at full power, the decay heat would still be 40 MW after 1 day. In the design of reactor plant provision has to be made for the safe removal of this decay heat under both normal and accident conditions.

9.4 Reactor components

Nuclear fuel in the form of rods, plates or pins is *clad*, or canned, to prevent the escape of fission products and arranged in a regular lattice to form the reactor *core*. The fuel material may be natural or enriched uranium or plutonium; natural uranium may be used in a metallic form but this imposes serious temperature limitations, and enriched uranium is normally used in the less restrictive oxide form; carbides and cermets have also been considered, the latter comprising a ceramic in a metallic matrix. Plutonium is also used in oxide form.

In a thermal reactor part of the space between the fuel locations is occupied by a *moderator* which serves to slow down the fast neutrons emitted in the fission process. Provision is also made for the flow of a *coolant* within the lattice to carry away the heat generated in the fuel. In a *direct cycle* system the fluid is usually light water and steam is raised in the core, separated and fed to a turbine; a gas cooled direct cycle system, in which the primary fluid is fed to a gas turbine, is also possible. In an *indirect cycle* system for a thermal reactor the primary fluid, which may be liquid or gas, gives up its energy in a *heat exchanger* or steam generator, to raise steam in a secondary circuit to feed to a turbine; for a fast reactor there is an additional intermediate coolant circuit and heat exchanger.

In some systems the reactor core is enclosed within a massive *pressure vessel*; in others individual fuel elements or clusters of fuel elements are positioned within *pressure tubes*. In an *integral* design the main primary components — i.e. reactor core, heat exchanger and coolant pumps — are all located within the pressure vessel or, in the case of fast reactors, the primary vessel.

Pressurised-water-reactor steam supply systems are equipped with a *pressuriser* to maintain required primary coolant pressure during steady state operation and to limit pressure changes caused by coolant thermal expansion and contraction as plant loads change.

A control system allows the plant operator to maintain design operating conditions. Reactivity control can be effected by a system of *control rods* or control pin clusters which are inserted into or withdrawn from the core; in one design tubes within the core are filled with a neutron absorbing liquid to reduce reactivity and shut down the reactor. In some systems longer-term reactivity compensation is achieved by 'fixed' absorber rods inserted within the lattice; in some water cooled systems *soluble poisons*, usually in the form of boric acid, are dispersed within the coolant for this purpose. *Burnable poisons*, often in the form of boron or gadolinium, can be added homogeneously to the fuel or heterogeneously to the lattice; the neutron absorption due to this form of absorber decreases as the poison is 'burned up'.

Shielding is provided to give protection to the reactor operating staff against the high levels of radioactivity which arise within the reactor core. The shield immediately surrounding the core is termed the *primary* shield. The materials used will vary from system to system and many combinations are effective in reducing γ and neutron radiation to acceptable levels. Concrete or lead is commonly used to attenuate γ radiation; water provides an effective shield against fast neutrons. In some reactor systems the primary fluid becomes radioactive under neutron irradiation and the whole of the primary circuit must then be surrounded by a *secondary* shield. This is particularly true of pressurised water reactors in which the coolant water becomes activated while passing through the reactor core.

9.5 Types of nuclear reactor

There are many possible combinations of materials which can sustain a neutron chain reaction and, correspondingly, many possible different reactor types. Nuclear power reactors, however, are commonly classified into five groups, four of which have names derived from one or more of the materials used in their construction, and one whose name derives from the neutronic processes involved. The groups are: (1) pressurised water reactors; (2) boiling water reactors; (3) graphite moderated reactors; (4) heavy water reactors; and (5) fast breeder reactors. Within each of these categories are to be found considerable variations in size, both in terms of physical dimensions and power output, and in the types and arrangements of material; these variations will be considered under the appropriate heading. Nuclear reactors are used for a variety of purposes, including materials testing, neutronic research and the production of plutonium, but discussion will be restricted here to the types of reactor whose function is the generation of thermal power for the production of electricity or for ship propulsion.

Light water reactor (LWR) technology developed initially in the USA, starting with the US Naval Reactors programme leading to the commissioning of the first nuclear powered submarine, the USS *Nautilus*, in 1955, and the land-based prototype Shippingport pressurised water reactor (PWR) in 1957. Boiling water reactor (BWR) technology advanced in parallel and Dresden I BWR was completed in 1960. By 1973 there were 30 LWRs in operation in the USA with more than 150 reactor years of operating experience. Competence in LWR design and development has now crossed the Atlantic to the UK, the Federal Republic of Germany, France and Sweden, all of which countries have substantial LWR programmes, civil or military.

The first graphite moderated reactor was also built in the USA; this was CP1, the first reactor to achieve criticality, built in Chicago in 1942, using natural uranium fuel. Subsequent developments in graphite technology were principally in the UK, where Calder Hall was completed in 1956 and Chapelcross in 1959, followed by a programme of nine graphite moderated reactor power stations ending with Wylfa in 1970. Five more advanced power stations (AGRs) using graphite moderators are

9.1 Introduction

Nuclear power reactors provide a means of power generation as an alternative to systems which depend upon oil, coal or hydro sources. Oil reserves are depleting rapidly and oil should be completely replaced as a fuel for electrical power generation soon after the turn of the century; while there are considerable resources of coal, its mining in substantially increased quantities presents problems both human and environmental; hydroelectric power can make only a limited contribution and has little capacity for expansion, except in some developing countries. Alternative energy sources such as fusion, solar, wind, wave and tidal power may in due course offer some promise for the future, but none of these can at present provide an effective source of bulk electrical power. Nuclear reactors have been used to generate electrical power for more than 20 years, and current reactor technology is based upon this extensive experience of design, construction and operation. In the UK between 15 and 20% of electrical power comes from this source, and the proportion is expected to increase as new plant is ordered and commissioned. The total world nuclear generating capacity is about 120 GW(e) in 22 countries, a total which will rise to over 400 GW(e) in 37 countries by about 1990 as plant under construction or planned comes into operation.

9.2 The nucleus as an energy source

In a nuclear power reactor the energy liberated within the fuel is carried away by a coolant and used to generate steam to drive either a conventional turbo-generator to generate electricity or a propeller for ship propulsion. The energy derives from the conversion of mass to energy which has been described by Einstein's well-known relationship $E = mc^2$ (where E is energy, m is mass and c is the velocity of light). There are two distinct nuclear processes which depend upon this relationship, known, respectively, as fusion and fission. In the fusion reaction two light nuclides (isotopes of hydrogen) are combined to form a single heavier nuclide, with a consequent decrease of mass and release of energy. This procedure, while promising an abundant energy source for the future, has not yet been developed beyond the research stage, and a fully engineered system is not expected for at least 20 years. The fission reaction is the basis of all current nuclear power reactor technology, and depends upon the splitting, or fission, of one of a small group of heavy nuclides (of which U-235 is the best-known example) into two fission fragments of roughly equal mass; in this process there is also a decrease of mass which appears as a release of energy. The fission fragments are, in general, unstable and radioactive, usually releasing β and/or γ radiation in order to regain stability. In the design of nuclear plant, account must be taken of this source of radioactivity, to protect the operating staff who work in the vicinity of the reactor, to protect the general public against the possibility of an accidental release of radioactive material to the environment, and to ensure that releases during normal operation are acceptably small.

Nuclear fuel provides a highly compact source of energy; the complete fission of 1 g of U-235 would release approximately 1 MW-day of thermal energy, so that a 1000 MW(e) nuclear reactor operating at one-third efficiency would consume only 3 kg of U-235 a day. It is not, in fact, possible to fission *all* the atoms of U-235 before the fuel is removed from a reactor; but it is perhaps of interest to note that for every tonne of uranium required to fuel a nuclear reactor (of the types currently in general use) the equivalent amount of fuel for a coal fired power station would be 20 000 t of coal. The development of fast reactors, discussed below in more detail, will increase this ratio to 10^6. It should also be noted that the compactness of the power source implies a correspondingly small volume of spent fuel and radioactive waste, since the latter is formed within the fuel. It has been calculated, for example, that when all the plant at present under construction is commissioned, the total volume of high-activity wastes for final disposal arising from the UK nuclear power programme will be no more than 30 m^3/year.

9.3 Nuclear fission

In theory energy can be released by the fission of any one of a group of heavy nuclides, but in practice there are only three which need to be considered as fuels for nuclear power reactors. These three are distinguished by the fact that fission will occur upon absorption of a slow neutron, while all the rest require additional neutron kinetic energy to be available. The three nuclides are U-235, Pu-239 and U-233. Reactors which depend upon fission induced by slow neutrons are termed *thermal* reactors, in contrast with *fast* reactors, in which the average neutron energy is considerably higher. In a thermal reactor a *moderator* is used to slow down, by collision, the fast neutrons which are liberated in the fission process until they are in thermal equilibrium with the moderator atoms; the probability of neutron absorption leading to fission is then exceptionally high. A fast reactor has no moderator. An average fission releases about 200 MeV of energy and two or three fast neutrons. Some of these neutrons may escape from the reactor and some will be parasitically absorbed in reactor materials, but if at least one on average can be made to produce a further fission, a self-sustaining *neutron chain reaction* can be achieved. Since the fraction of neutrons lost by leakage decreases as the size of the reactor is increased, there will be a *critical size* at which this self-sustaining chain reaction occurs for a particular combination of materials. As a reactor operates, variations in neutron absorption may be induced by changes in isotopic concentrations within the fuel due to a wide variety of neutron induced reactions or by the deliberate insertion or removal of neutron absorbing materials to exercise control over the fission rate and, consequently, over the power released by producing a convergent or divergent chain reaction. Departure from criticality is indicated by the *reactivity* of a reactor, positive or negative reactivities corresponding to divergent and convergent chain reactions, respectively. Of the three fissile nuclides, U-235 is the only one which occurs in nature, being present with an abundance of 0.7% in uranium deposits. Thus, *natural uranium* fuel has a proportion of one part in 140 of U-235, the remainder being U-238. *Enriched uranium* fuel contains a higher proportion of U-235, typically a few per cent; uranium is enriched by gaseous diffusion or centrifugation, which adds substantially to the fuel costs. Almost all current power reactors use U-235 as the primary fissile fuel material.

The fuelled region in a reactor is referred to as the reactor *core*. Pu-239 is formed by the radioactive decay of U-239, which results from the absorption of a neutron in U-238. This process is significant in all reactors fuelled with natural or low-enrichment uranium, and fission of the Pu-239 so produced makes a contribution to the power of such reactors. Spent fuel contains valuable amounts of this nuclide which, being chemically dissimilar from uranium, can be separated by chemical processes. This recovery of plutonium from spent fuel is an important function of fuel reprocessing plant. Spent fuel contains other plutonium isotopes, generated by successive neutron absorptions; Pu-240 is a parasitic absorber of neutrons in thermal reactors, but Pu-241, which is produced in relatively small amounts, is also fissile by slow neutrons. The plutonium recovered from spent fuel from thermal reactors could be used to provide fresh fuel for similar systems; its preferred use, however, is for initial fuel charges for fast reactors.

Uranium depleted in the fissile isotope U-235 is also recovered

Nuclear Reactor Plant

D C Leslie, MA (Oxon), DPhil, CEng, FINucE
Queen Mary College

P R Smith, BSc, PhD, CEng, FINucE, FInstP
Queen Mary College

Contents

9.1 Introduction 9/3

9.2 The nucleus as an energy source 9/3

9.3 Nuclear fission 9/3

9.4 Reactor components 9/4

9.5 Types of nuclear reactor 9/4
    9.5.1 Pressurised water reactors 9/5
    9.5.2 Boiling water reactors 9/7
    9.5.3 Graphite moderated reactors 9/9
    9.5.4 High-temperature reactors 9/12
    9.5.5 Heavy water reactors 9/12
    9.5.6 Fast breeder reactors 9/15

9.6 Nuclear reactor plant for ship propulsion 9/17

9.7 The nuclear fuel cycle 9/17
    9.7.1 Enrichment plant 9/18
    9.7.2 Fuel element fabrication 9/18
    9.7.3 Reprocessing plant and waste disposal 9/18

(2) *Peak lopping*
   —independent of utility
   —in parallel with utility
(3) *Standby to utility sumply.*

Running costs are related to the pattern of operation, the number, rating and type of engines, the fuel type used, the overhaul intervals and the hours of maintenance per year. The following cost indicators will enable reasonable estimates to be made at the first stage of any appraisal, when alternative schemes are being investigated:

8.29.1 Fuel costs

The cost of fuel to generate electricity only can be estimated as one-eighth of the cost of purchased electricity expressed in unit currency per kilocalorie (e.g. p/kcal, where 1 kWh = 0.863 kcal). Assume a consumer purchased 150 000 kWh of electricity over a given period at a total price of 1.5 p/kWh (= 1.74 p/kcal). This would equate to a cost of purchased power of £2610 over the period. The cost of fuel to privately generate 150 000 kWh would then be £326.

8.29.2 Lubricant costs

Lubricating oil consumption may be taken to be 1.5% of the fuel-oil consumption at full-load. To this must be added the quantities representative of any oil changes at routine service intervals. This may vary from between 250 h to every 5000 h of running, depending upon the size and speed of the engine. The sump capacity of a 1.2 MW engine is of the order of 800 l. For first estimates of lubricating oil costs, one may work on the basis of 5% of fuel costs for the same period.

8.29.3 Maintenance costs

Estimate on the basis of 5% of the combined costs of fuel and lubricating oil over the given period. This covers spares and labour.

8.29.4 Depreciation, interest, insurance and rates

Engineers do not always appreciate how substantial these costs can be. A recent economic appraisal for a private generating scheme, employing two 1.2 MW sets (with a third in reserve) operating at a 92% load factor and written down over a 7 year period, revealed that the annual sum of these costs was 70% of the combined total of the other running costs, i.e. those attributable to fuel, lubricant, maintenance, spares and labour.

8.29.5 Capital costs

Table 8.4 offers a very rough guide for first estimates on capital costs (including installation), using the cost of the engine-generator and its auxiliaries as base 100.

Table 8.4 Estimates on capital costs

Cost item	Size of unit		
	1 MW	2 MW	4 MW
Engine-alternator and auxiliaries	100	100	100
Civil works	20	19.5	16.5
Cranes and services in station	15.5	13.5	10
Installation	18.5	16.5	12.5

8.30 Total energy

A total energy system implies on-site power generation in which the energy input from either liquid fuel (diesel engines) or a combination of gaseous and liquid fuels (dual-fuel engines) is maximised by recovering the waste heat from the generating process. By so doing the overall thermal efficiency of generation may be raised from 37% to about 80%.

Compared with gas turbines the quality of waste heat is of a relatively low grade. Nevertheless, sizeable combined heat and power installations have been commissioned both in the UK and in Europe in recent years, most favouring multi-engined systems employing 1.5–2 MW unit sizes with dual-fuel operation.

Table 8.5 shows typical heat balances for a 2 MW 750 rev/min diesel-generator giving 1.8 MW in the dual-fuel mode.

Table 8.5 Heat balances for a 2 MW 750 rev/min diesel-generator

Heat balance at full load	Diesel (%)	Dual-fuel (%)
To electricity	38.5	39
To exhaust	36	34
To jacket water	11	10
To lubricating oil	4.5	4
To charge air	4	2.5
To radiation, etc.	6	6
To unmixed gases	—	4.5
	100	100

Engines of this type and size give about 1 MW of recoverable heat from jacket water and lubricating oil in the form of l.p. water at 80 °C and another 1 MW recovered from the exhaust gases to give about 1400 kg/h of steam at 850 kPa.

Recoverable heat from a diesel engine is of the order of 250 000 kcal/h/MW from the exhaust gases and 350 000 kcal/h/MW from the jacket-water.

Using exhaust heat recovery alone, 0.5 kg of steam at 850 kPa/kWh is possible. With full jacket, oil and exhaust heat recovery this figure doubles to 1 kg of steam per kWh and can be raised still further to 2.5 kg//kWh if pre-heating of the jacket-water into an automatic boiler, using the same fuel as the engine, is employed.

Below a unit size of 500 kW it is considered to be uneconomic to fit heat-recovery systems as the heat balance is low and the cost of recovery disproportionately expensive.

Table 8.3 The major classes of failure at diesel and gas engined plants

Class of failure	Percentage of total stoppages
Fuel injection equipment and fuel supply	26
Water leakages and cooling	16
Valve systems and seatings	13
Bearings	7
Governor gear	6
Turbocharger/lubrication/piston assemblies/gearing and drives	4 each

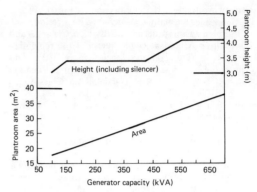

Figure 8.37 Floor area and height for standby generators

8.28 Plant layout

The size of a plant room is largely determined by the number and the rating of the generators installed and by their requirements for ancillary equipment. In addition to the engine-generator assemblies there will be switchgear, distribution and control gear, engine starting equipment, fuel service tanks, provisions for fuel and lubricating oil storage, engine cooling systems, exhaust silencing equipments, and (on the larger engines) lubrication and fuel systems external to the engine.

The internal layout should be such that the basic requirement is to construct a station building around the machinery. It is prudent to provide for future expansion. Growth may be in the form of a larger generator unit to replace the original or additional units, in multi-generator plant, to cater for increased load demand. A removable end wall offers one way of providing for future plant room expansion.

A minimum space of 2 m should be allowed around each set to facilitate maintenance. On multi-engined stations there should be sufficient headroom for overhead, installation and servicing cranes—one for large lifts of between 10–30 tonnes capacity and a smaller unit of about 2–3 tonnes. Height to the underside of the common crane rail girder should be such that the distance between the bottom of the hook on the larger crane, when fully raised, and the floor level, is about 6 m for the biggest engines installed.

Most generator-set manufacturers provide an advisory service on foundation requirements. Where sets are not supplied with anti-vibration mountings a concrete foundation block, preferably 'isolated' from the main building structure to minimise vibration (and therefore noise) transmission, is necessary. A good empirical estimate for foundation mass is that it should be at least 1.5 times the dynamic mass of the associated diesel-alternator. On new civil works a basement can be provided at little extra cost to house engine auxiliaries. Similarly, a gallery can be constructed to accommodate fuel service and water make-up tanks. This reduces the need for pipework trenches within the generator hall itself. Access to basement auxiliaries for replacement or maintenance could be by removable open-mesh gratings.

Where radiators are fitted to generating sets they should have pusher fans and be installed near to and facing an outside wall, with air ductwork and control louvres to regulate the plant room temperatures. Space between rear wall and the generator end of each set should be sufficient to allow for end-removal of major components, as required.

For ventilation purposes it can be assumed that approximately 8% of a generator nameplate kW rating is radiated as heat from the engine and generator carcases. The combustion air required by an engine may be taken to be approximately 9.5 m^3/h per nameplate kW. Plant-room air exchange required is then the sum of the ventilation air and engine combustion air requirements. Exhaust gases should be piped to atmosphere through insulant-caulked apertures in an outside wall and silencers should be mounted external to the plant room, if at all possible.

Figure 8.37 gives floor area and height requirements, recommended by the Building Services Research and Information Association (BSRIA) in their *Technical Note TN4/79*, for individual standby generators in the range 50–625 kVA.

Where noise is likely to be a community problem, it is necessary to identify all the noise contributors in the plant and obtain octave-band frequency analyses for each so as to calculate the total generated noise level, in worst-case conditions. This must then be related to any noise level limits imposed by communal interests or local legislation in order to determine the noise reduction required. The appropriate noise control treatments should then be selected to restrict the noise transmitted and radiated from the plant room to a value below the promulgated level. Acoustic barriers, partial enclosures, vibration damping materials, vibration isolation, inertial blocks, lined duct work and splitter silencers in ventilation inlets and hot air discharge outlets are but some of the noise control techniques that could be considered. Any treatment applied should not prejudice the operation, maintenance and safety of the plant.

8.29 Economic factors

When considering the installation of private generating plant it is important that proposals are not only technically suitable-for-purpose but are also economically defensible. The actual or predicted costs of purchased power should be compared with the projected costs of privately generated power for clearly defined electrical and heat load cycles, before any management decision is sought.

Factors to be considered are:

Capital costs, embracing land; site preparation and access; foundations; buildings; workshops and tools; fuel storage; the power plant and its ancillary equipments; cranes; stores and non-consumable spares and heat recovery equipment (if applicable).
Installation costs.
Operating costs; fuel; lubricating oil; service spares; wages of operating and maintenance staff; insurance; depreciation, interest on capital and rates.
Costs of any consequent supply outages.
The size of the installation and its mode of operation:
(1) *Base load*
 —independent of utility supply
 —supplemented by utility
 —with utility as standby
 —total energy

with a viscosity greater than 150 s Redwood No. 1 (expressed as 150 SR1) at 37.8 °C is normally considered to be a residual oil.

BS 2869 lists nine classes of fuel oil, two of which, Classes A and B, are specifically produced as engine fuels. Of the remaining seven, Classes E, F, G and H are classified as industrial and marine fuels: E and F are light residuals or boiler fuels, G and H are heavy residuals or bunker fuels. The distillates, Class A (gas oil) and Class B (diesel oil) are the most widely used fuels on medium- and high-speed engines. Their American ASTM diesel fuel classification equivalents are *D 975-66T* Nos. 1-D and 2-D. Grade 4-D of *ASTM.D 975-66T*, although considered a distillate oil, is intolerable to some high-speed engines and is best suited to low- and medium-speed units.

Almost all medium-speed engines will operate on the Class E and F light residual fuels, whilst crosshead 2-stroke low-speed engines and some of the slower running medium-speed 4-stroke engines can successfully operate on Class G and, to a lesser extent, Class H heavy residual fuels. Residual fuels must be pre-treated, i.e. settled, heated, separated by centrifuge and filtered or waterwashed before being transferred to an engine fuel system. Heavy fuels should not be mixed with gas, distillate or light residual fuels.

The effects that fuel constituents have on engine performance and their influence on maintenance may be briefly summarised as follows:

Viscosity The lubricating property of the fuel falls as the viscosity reduces. Fuel injection equipment is most affected by any reduction in fuel lubricity or excessive water content in the fuel. Special precautions are necessary when 'dry' fuels such as Avtar, Avtag and Avcat are used with high-speed engines.

Sulphur content Sulphur in a fuel forms a corrosive acid when it combines with exhaust gas condensates. This is wear-inducing and may be minimised by ensuring that high jacket-water temperatures are maintained.

Conradson carbon residue value This is a measure of the tendency of a fuel to form 'coke' when heated. A fuel with high Conradson value reduces combustion efficiency whilst increasing the rate of carbon build-up in the engine. This, if coincident with a high sulphur content, greatly increases wear rate.

Ash content The effect on the engine is similar to that described for the Conradson value.

Vanadium or sodium content Their presence in any appreciable quantity affects engine exhaust valves and turbochargers. Contents below 50 parts per million are recommended if frequent incidences of valve seat burning and turbine blade failure are to be avoided.

8.27.3 Lubricating oil

Selection of lubricant depends on whether the engine has separately lubricated bearings and cylinders (low-speed and some medium-speed engines) or whether it has combined lubrication of bearings and cylinders (all high-speed and most medium-speed engines).

Where separate lubrication is used, specially formulated and refined mineral oils containing anti-oxidant additives are specified for bearings. Some engine operators favour detergent additives despite their tendency to thicken relatively rapidly. Cylinder lubrication calls for a heavier oil than that suitable for bearings, but on many engines the same viscosity grade is satisfactorily applied to both.

For combined lubrication, oils need: a viscosity applicable to both cylinders and bearings; high oxidation stability; sufficient detergency and corrosion inhibition properties. A big proportion of medium- and high-speed engines use a straight mineral oil of the type applied to separately lubricated bearings; but others, especially the highly rated high-speed engines, require heavy duty oils with detergent and contaminant dispersive properties, e.g.

MIL-L-46152 specification for naturally aspirated engines and MIL-L-2104C for turbocharged units. If there is any possibility of an engine being subjected to prolonged light-load running, such as on many telecommunication transmitter applications, it is advisable to employ a heavy duty oil even though the engine might not normally require it.

Engine makers will always specify the oil to be used consistent with the duty and rating of an application and the fuel which it is intended to use. Mono-grade oils are normally recommended but makers will countenance the use of multi-grade types, provided that the operator obtains the oil supplier's assurance that the type proposed meets the certified performance level of the equivalent and acceptable mono-grade oil.

8.27.4 Maintenance

The diesel engine and the gas turbine are both i.c. heat engines but since the former employs reciprocating motion it is mechanically the more complex machine and demands a greater skill in its maintenance. Also its wear rate is higher. However, repair services for diesel engines are more readily available worldwide than they are for gas turbines.

Plant operating personnel should be fully familiar with all its component parts and be capable of maintaining it in its optimum condition. Operation must be continuously monitored and accurate records kept so that incipient faults are timeously detected and repair effected before major, unscheduled and costly stoppages occur.

The starting point of any preventive maintenance programme must be the manufacturer's Operator and Service Manuals. They indicate what needs to be checked and how frequently. Planned schedules should at first rigidly adhere to the maker's recommended frequencies for inspection checks and maintenance tasks. Only after sufficient operational experience has been accumulated should one contemplate any modification to fit the particular installation and its operating conditions, and certainly not before the first general overhaul (stripping to crankshaft bearing level) has been undertaken. This gives the opportunity to assess achievement against the maker's wear and renewal limit schedules.

Whilst the service intervals recommended by manufacturers are conservative and based on average experience and temperate conditions, periodicity of inspection and service may have to be increased where, for instance, the quality of fuel is in question or where corrosive and very dusty environments pertain, or where less than 50% loading conditions are initially expected. Over the last decade, engine and component manufacturers have achieved impressive improvements in times between overhauls. Typically, major overhauls on well-maintained high-speed engines many now occur at 15 000 h intervals, whilst those on medium-speed engines could be at between 20 000 and 30 000 h—the longer interval applying to the slower running engines.

Careful consideration must be given to plant spares stocks. A more comprehensive holding of both non-consumable and consumable items would be necessary at remote sites or in certain developing countries, where there are no local accredited spares stockists. At locations where there is ready access to good stockists, a much-reduced inventory on consumables is justified.

Even with the most vigilant monitoring, problems can arise at any time. Running plants are usually most vulnerable immediately after commissioning and after general overhauls. *Table 8.3* based on four years of analyses of the excellent *Annual Working Cost and Operational Reports* published by the Institution of Diesel and Gas Turbine Engineers, shows the major classes of failure (expressed as rounded-off percentages of total unscheduled stoppages) reported by over 100 diesel and gas engined plants, worldwide.

Recording instruments provide useful information for reconstruction of events, for registering operational patterns and indicating trends. In their multi-channelled form they give a time-sequenced record of events during disturbances and faults, particularly when they simultaneously monitor closing and tripping of key circuit breakers in the system.

8.26.4 Protection

The parameters to be monitored on the prime-mover have been dealt with in the preceding text. Protection related to the generator for which provision should be made, might in addition to system-fitted protective devices safeguarding against external short-circuits and overcurrent, include the following:

Restricted-earth-fault protection, combined if necessary with over-current protection in the one relay.
Differential (circulating-current) protection, where access is possible to the stator windings before the star point. Where a generator and step-up transformer are directly connected, an overall-biased differential protection scheme could be considered.
Stator earth-fault protection. It is possible to combine this with differential protection in the one relay.
Rotor earth-fault protection, by field or stator monitoring.
Generated overvoltage and overfrequency protection.
Unbalanced loading protection, with negative-phase sequence protective gear.
Load-shedding protection.

Additionally for generators running in parallel:

Reverse-power protection. Relays should be set to operate about 10% reverse power and be time delayed.
Check synchronising relays.

Voltage surge protection should be considered where vacuum breakers or contactors are used, particularly with larger and higher voltage rated generators, if surges induced by lightning discharges on lines external to the diesel generator plant or initiated by switching, are likely.

8.26.5 Control gear

Control gear should be designed to give a comprehensive indication of the state of a generator plant at all times and provide the means for modifying that state. Equipment for this purpose would include: measuring instruments, condition indicators, alarm annunciators, prime-mover and generator regulating controls and command signalling devices to engines and switchgear.

On the simplest single-generator installations, instrumentation and logic controls are usually incorporated within the switchgear cubicle(s) to give convenient operation of the plant from one location. On the more complex multi-generator plants, since adjustments on one machine affect all others in parallel with it, simultaneous observation of all the effects of any one action is essential. Where only two or three generators are involved, individual switchboards arranged as for the single generator may suffice, provided that they are placed as close together as possible and preferably in a quiet location away from the noise of running machinery.

But where adequate observation of a greater number of switchboards is difficult from one vantage point, it is advisable to use a control desk placed some distance in front of the composite switchboard. This may be used to house the generator regulating and control equipments whilst instrumentation and switchgear indicators are retained on the individual switchboards. Where size and cost of an installation warrants, a mimic diagram, representative of the complete generator plant and its feeder networks, may be incorporated into this control desk. Alternatively, it may be fitted into a separate diagram panel surrounding the composite switchboards. The diagram should incorporate pilot lights and semaphores to represent circuit breakers, bus couplers, isolators and selector switches with miniature indicating instruments to cover feeders and generator output conditions.

8.27 Operational aspects

8.27.1 Engine ratings

Since the i.c. engine is an air-aspirating machine its output is affected by changes in the temperature and the pressure of the air it breathes. Power ratings should be quoted to the Standard Reference Conditions (S.R.C.) contained in the applicable National Standards of the engine maker concerned. For British built engines this is *BS 5514 Part 1*, which is equivalent to *ISO 3046/1*. Although corresponding German (*DIN 6270*), American (*SAE J-243*) and Japanese National Standards cover the same technical subject matter, their treatment of it differs particularly with regard to the standard reference conditions adopted and definitions of kinds of power.

BS 5514, which covers reciprocating i.c. engines using both liquid and gaseous fuels, applies the following standard reference conditions:

Barometric pressure	100 kPa (750 mm Hg)
Corresponding altitude at 0.88 mechanical efficiency	110 m
Temperature	300 K (27 °C)
Relative humidity	60%
Charge air coolant	300 or 350 K (27 or 77 °C)

The ISO standard power rating of the engine has to be adjusted for ambient conditions, falling outside the standard reference condition, to arrive at a predicted on-site or service power rating. Section 10 of *BS 5514* lists the formulae to be applied for this purpose and Annexes B to O offer tables and nomograms to help simplify such calculations. As it is rare, in any part of the world, for high humidity to be combined with very high temperature, a de-rating exceeding 6% for humidity is seldom, if ever, warranted.

The power categories of importance to the diesel generator user are 'continuous power' and 'overload power', as defined in *BS 5514*: a footnote states that it is customary to permit an overload of 110% power 'for periods and speed corresponding to the engine application'. Engine makers almost universally permit 10% overload for 1 h in any 12 of continuous operation on generator application. In so doing they rightly perpetuate a requirement of *BS 649*, which was withdrawn on publication of *BS 5514* in 1977.

Prospective engine users should be wary of rating classifications other than those above, for generator applications, e.g., standby rating, continuous duty with time-limitation rating, reserve rating, intermittent rating and maximum rating, to name but a few of those cited. American, European and Japanese manufacturers will usually declare ISO standard power ratings, when asked.

8.27.2 Fuel oils

Liquid fuels used in compression-ignition engines are categorised as either distillate fuels or residual (blended) fuels. These are further subdivided into classes based upon viscosity, measured in seconds, using either the Redwood or the Saybolt systems. A fue

Voltages of 2.4, 4.2, 6.9, 13.2, 13.8 and 15 kV are encountered in American and European practice.

BS 4999 requires the output voltage to be held to within $\pm 5\%$, but most a.v.r's will control to $\pm 2\frac{1}{2}\%$ over the load range. Regulation down to $\pm 1\%$ is possible with solid-state devices in closed-loop control systems.

Standard classification for insulating materials is given in Chapter 5. Most machines up to 3 MVA employ class E or F insulation, or combinations thereof, for stator and rotor windings.

8.25.4 Parallel operation

Active power load-sharing is a function of the engine and its governing system. The sharing of reactive power is determined by excitation and synchronous impedance: proper sharing is obtained by applying quadrature current compensation (q.c.c.) to give automatic controlled droop of the output-voltage/reactive-current relation. The amount of droop should be the same for each generator paralleled into a power system: it is typically about 5%.

8.25.5 Short-circuit performance

Fast-response brushless and compounded machines can have subtransient short-circuit current levels of 6 to 10 times full-load current. But where excitation is derived from the output voltage, a short-circuit removes the excitation supply and the generator voltage collapses.

In distribution networks it is essential that an adequate 'permanent' short-circuit current is maintained by the energy source in order to allow discriminative operation between protective devices such as circuit breakers, fuses and over-current releases. This is necessary if faults on final circuits are to be cleared as quickly as possible. Maintenance of short-circuit current at a level 2 to 3 times full-load current may be achieved by excitation power derived from current transformers in series with the generator output leads. This power is fed to the exciter field through a relay which closes when the output voltage collapses.

8.25.6 Generator selection

In selecting the right size of generator for a specific application the following factors, all influencing the rating, should be considered.

Application The mode of operation, e.g., standby or continuous; single running or paralleled with a utility supply or with similar generators; load power factors.
Location Altitude, ambient temperature, humidity and other environmental conditions.
Dynamic loading Any limitations imposed on voltage transient performance in the starting of large induction motors or on waveform characteristics where static converter equipments or thyristor drives form part of the load.

8.26 Switchgear and controls

8.26.1 Planning

Factors effecting the choice and design of switchgear will be:

The size and nature of initial and future loads. For example, spare panel positions should be considered in an initial switchgear layout, to cater for future extensions.
In-service conditions. These relate to such aspects as temperature, humidity, air conditioning and ventilation, dust, corrosion or pollution conditions and affect not only choice of components but also enclosure design.

Foreknowledge of standard equipments available in the market. Custom-built equipments are not justified when standard and competitive units may be readily adapted.
Requirements for reliability and security of supply. For example, the degree of security required influences the choice of busbar systems both in terms of numbers installed and their sectionalising.
Requirements for maintenance and safety, consistent with the skills of operating personnel.

8.26.2 Fault considerations

Overcurrent protective devices must operate to isolate short-circuit faults safely, minimise damage to circuit elements and avoid, if possible, shutdown of plant. An accurate knowledge of prospective fault currents throughout the system is essential for the correct application of protective devices and the design of busbars and terminal arrangements to withstand consequential mechanical and thermal stresses.

Generator short-circuit fault current decreases from a high initial value determined by the subtransient reactance X_d'' of the machine, through a lower value determined by the transient reactance X_d', settling after 0.6–2.0 s to a steady-state level determined by the synchronous reactance X_d. Circuit-breakers and fuses should operate before the steady-short-circuit condition is reached.

System faults are also fed by synchronous and induction motors, which may generate by release of kinetic energy by their rotating masses. The fault contribution from an induction motor ceases after a few periods. Typical values of subtransient and transient reactance, in per-unit on a machine-rating base, are

	X_d''	X_d'
Salient-pole generators:		
up to 12-pole	0.16	0.33
14 poles and upwards	0.21	0.33
Synchronous motors:		
4 and 6 poles	0.15	0.25
Induction motors (low-voltage)	0.20	–

In the calculation of system fault levels it is sufficient to employ reactances only, except that in low-voltage (l.v.) systems the resistance of cables cannot be ignored. Arcing impedances should be included in l.v. system calculations. Typical values of per-unit arcing-fault current on 0.415 kV 3-ph systems are: 0.70 for a 3-ph arc, 0.57 for line-line and 0.14 for line-neutral.

8.26.3 Instrumentation and metering

Single running sets require an ammeter in each line (or an ammeter and a selector switch), a frequency indicator and a wattmeter. Optionally, a watthour meter and a power recorder may be fitted.

The instrumentation on each of any parallel running sets should include: an ammeter in each line (or a single ammeter with selector switch) a wattmeter and a reactive volt-ampere (VAr) meter. Power-factor meters may sometimes be substituted for VAr meters but as they develop low torque at low load they are prone to reading errors below 25% of rated current. Power factor calculated from kW and kvar values provides a more reliable and accurate indicator. Additionally, a set of synchronising instruments, perhaps mounted on a swivelling frame, is required, comprising a synchroscope, a double movement voltmeter and a double movement frequency indicator. Any incoming generator before being paralleled to the busbar should be connected by plug or switch to the 'incomer' movements of this voltmeter and frequency indicator, the second movements of both instruments being permanently connected to the busbars.

detect the gear teeth on the flywheel rim. The signal so derived is compared with a potentiometer-set speed reference. Any detected difference is amplified within a control unit to adjust the signal to an hydraulic or electrical actuator fitted to the engine fuel rack to correct the fuel and return the engine to its preset speed. Response to speed changes is much faster than with the mechanical-hydraulic types and fully isochronous load sharing is possible on paralleled generators with these types of governor. Moreover they may be used with a wide selection of control modules to provide fully automated and integrated multi-generator installations.

Where a governor is independently mounted from fuel injector pump(s) it is critical that the engine be fitted with some form of overspeed shutdown device, either acting directly on the fuel pump rack(s) or completely cutting off the intake air flow to the engine. This is necessary to prevent over-fuelling of the engine should any casual jamming of racks and fuel control levers take place or should any of the elements within the governing system fail.

8.24.2 Engine monitoring

It is essential that the strategic temperature, pressure, speed and flow parameters of an engine system are regularly monitored to interpret its behaviour and performance and relay this information to planned maintenance operations.

Maker's instruction books give a good indication of the degree of instrumentation required. Much also depends upon the build specification for the particular engine but parameters such as those listed below may be monitored by instruments, wherever applicable.

Jacket and raw water temperatures, pressures and flows.
Lubricating oil temperature and pressure.
Differential pressures across fuel and lubricating oil filters.
Jacket water temperature at outlets from individual cylinders.
Exhaust temperatures at individual cylinders and before or after turbochargers.
Charge air temperatures and pressures.
Starting air pressure.
Engine, pumps, compressors, and other relevant speeds.
Fuel temperature and pressure.
Fuel and lubricating oil tank levels.
Mechanically driven or transducer operated cylinder pressure indicators.

Abnormal operating conditions may be detected by sensors whose output signals are fed into alarm/shutdown logic controls. Various combinations of indicative and protective action are possible.

Two-stage alarm and shutdown.
Simultaneous alarm and shutdown.
Alarm only (visible and audible).

The conditions to be covered may include:

High jacket and raw water temperatures.
Low lubricating oil pressure.
High exhaust temperature before turbocharger.
High differential pressure across both fuel and lubricating oil filters.
High charge air temperature.
Excessive vibration.
Engine overspeed.

This list is by no means exhaustive and the engine maker's advice should be sought on the extent of protective insurance to be taken.

8.25 A.C. generators

Generator technology is dealt with in Chapter 11. Engine-driven generators for outputs of 20 kVA and above have direct-compound or brushless excitation.

Compounded generators use the load current to provide part of the excitation. They have good overload capacity and rapid voltage recovery, and may be preferred for marine application where large induction motors have to be started.

Brushless excitation is common for diesel-generator plant. The exciter is a 3-ph machine connected to a shaft-mounted rectifier diode assembly the output from which is fed to the main generator field. The brushless machine may be self-excited from its output terminals through a solid-state automatic voltage regulator (a.v.r.). The d.c. output from the a.v.r. feeds the stator field of the a.c. exciter to control the output voltage of the main generator. Alternatively, the exciter may obtain its field supply from a shaft-mounted permanent-magnet pilot exciter in combination with an a.v.r.

For medium- and low-speed generators the rotor is usually of laminated form with fully interconnected damper windings. In some high-speed sets the poles have solid bolted-on shoes, which provide eddy-current damping.

8.25.1 Construction

Types of construction and mounting arrangements are designated in *BS 4999 Part 22 (IEC347)*. The equivalent German Standard is *DIN42950*.

Almost all diesel-generator sets employ direct coupling of the prime-mover and generator. The latter may be treated either as a separate machine flexibly or solidly coupled to the engine, or flange-mounted to the flywheel housing and close-coupled.

Most 2-, 4- and 6-pole machines incorporate end-shield-mounted, grease-lubricated ball and roller bearings. Above about 3 MW, plain bearings are sometimes employed, self- or oil-scavenge lubricated. The large slow running machines tend to use single or dual pedestal outer bearings, mounted on a common baseframe with the generator stator casing.

8.25.2 Protection

Where generators are housed in buildings or canopies and enclosures, protection to IP23 of *BS 4999 Part 20* is sufficient. (Technically equivalent standards are *IEC 345* and *DIN 40040*.) Ventilation is provided by an internal shaft-mounted fan drawing air from the non-driving end and discharging it at the driving end. Dust filters may be fitted to this type of enclosure when the inlet air contains fine dust, sand, moisture or oil vapour. An output power reduction factor of about 0.95 should be applied to compensate for the resulting restricted cooling air flow. It is advisable to use thermistor temperature-sensing probes, either embedded in the stator windings or placed in the cooling air flow, to guard against damage by overheating, if the filters are not cleaned at regular intervals. For operation in extremely dirty conditions and on outdoor sites such as quarries, a totally enclosed construction, using closed-air circuit ventilation, should be specified. The closed-circuit air is directly cooled by air-to-air (TECACA) or air-to-water (TECACW) heat exchangers, which may be attached to the generator casing. These enclosures are defined as IP45.

8.25.3 Voltage

In the UK the preferred output voltages for 3-ph 50 Hz supply, and the appropriate ratings of engine-driven generators, are:

Voltage (kV)	0.415	3.3	6.6	11
Rating (MVA)	up to 1.5	0.5–6	0.8–10	1–20

maintenance is not always possible: heavy-duty paper element filters with highly efficient (90–95%) centrifugal pre-cleaner stages and preferably with self emptying or dust unloading arrangements. Filters of this type usually offer safety elements as an optional feature. Their purpose is to protect the engine in the event of main element perforation or act as temporary substitutes when the main elements are being serviced.

It is good practice to fit air restriction indicators to dry type filters to warn when the elements have reached a pre-set limit of fouling.

8.24 Engine ancillaries

8.24.1 Starting equipment

The two main energy sources for engine starting are batteries for electric-start systems and air receivers for air-start systems. On small high-speed engines hydraulic energy or spring type inertia systems are occasionally employed. The simplest method of starting is of course manual cranking, but it is practicable only with the smallest engines. On the very largest low-speed engines small pony i.c. engines are sometimes employed to give crankshaft rotation at starting through a clutched pinion engaging with the engine fly-wheel gear ring.

Most electric-starting systems use motors fitted with Bendix-type pinions on their armature shafts to mesh with a toothed rim on the engine flywheel. Lower output engines in the high-speed range may require only one such motor whereas larger engines in that range, and medium-speed engines, need two motors with a common synchronising control.

Starter motors are either of the axial or co-axial type. In the former the complete armature assembly and pinion move forward axially to engage with the flywheel teeth. On the latter type only the pinion dome moves forward to engage under reduced power. This minimises engagement shock and reduces wear on the gear teeth. Starter windings may be of the 'hold-on' or 'non-hold-on' type. The non-hold variety is usually applied to motors for remotely or automatically started engines. The most frequently used voltages are 12 and 24 V, but 6 V for small engines and 32, 48 or 64 V for the top end of the electrically started range of engines are not unusual.

A cost effective alternative to starter motors may be applied on generating sets below 15 kW rating, making use of a special d.c. starting winding within the generator to motor the engine. The same winding is used for charging the starter battery when the engine is running.

Starter batteries are either of the heavy-duty lead–acid or alkaline type. The latter, in its nickel–cadmium form, is favoured for standby generators because it retains its charge better over long periods without use. Disadvantages are that it is bulkier than the lead–acid battery of similar capacity, it is more expensive and it tends to have a larger terminal voltage drop with heavy starting current drain.

As a typical illustration the capacity of a lead–acid battery required to give six consecutive 20 s cranking periods with 5 s rest periods between each at ambient temperatures down to $-7\,°C$ on an 8-cylinder vee-form engine rated at 600 kW is 236 A-h at the 5 h rate for a 25% discharged condition. At this temperature the steady cranking current demand at 80 rev/min is of the order of 1 kA corresponding to an engine cranking torque of 1 kN-m. Breakaway current and torque figures are as much as 150% above these values.

Starting difficulties, particularly in low ambient temperatures, are more usual on the smaller high-speed engines because their large surface area-to-cylinder volume ratios tend to dissipate the heat of compression. Moreover, restrictions on the size and weight of starting equipment limit the amount of starting torque available. Various proprietary starting aids are available: devices such as electric glow plugs or those using ether–air mixtures pumped into the air inlet manifolds to give access to the engine combustion chambers to promote combustion. Decompression mechanisms may also be used to hold-off either the inlet or the exhaust valve on each cylinder, during the initial starting period. Once the engine is up to steady cranking speed, full compression is restored. The engine should then fire and run up to self-sustaining speed.

Compressed air for engine starting may be expanded either in an air motor engaging with the flywheel or directly within the engine cylinders, to move the pistons downwards on their working strokes until firing occurs. As in the electric starter the drive of the air motor is through a sliding pinion. Air motors may be applied to the range of engines that otherwise uses electric starter motors.

Direct air starting applies to the larger medium-speed engines and to low-speed units. Air is directed to each of the engine cylinders, in their proper firing sequence, through non-return valves either from a camshaft driven distributor or through mechanically operated valves.

Compressed air for either form of starting is stored in one or more receivers at pressures between 100–300 kPa (1–3 kg/cm^2). Air charge is maintained by a small single-stage auxiliary compressor driven by an electric motor, an i.c. engine or by the main engine itself. On large or critical installations it is usual to provide back-up auxiliaries; the primary compressor being perhaps electric-motor-driven with an i.c. engine-driven standby unit.

8.24.2 Governors

Diesel generators use variable speed governors set to operate at the predetermined synchronous speed. Choice of governor is dictated by:

(1) The engine type and its application. For example: independent operation feeding an isolated load or parallel operation with similar generators or with a utility supply.
(2) The standard of governing required, i.e. defining the limits for speed regulation (speed droop), steady-state stability and dynamic behaviour. (Classes of governing accuracy and their parameters are defined in *BSS 5514 Part 4* and its corresponding International Standard ISO 3046 Part 4.)
(3) The available inertia or flywheel effect of the combined engine and alternator. It may be possible to employ a relatively simple mechanical governor in conjunction with a higher inertia flywheel to meet a tight governing specification in an economical manner without recourse to a more sophisticated and expensive governing system. Engine makers will calculate the minimum generating set inertia required for each eligible type of governor to fulfil the requirements of any governing system.

Governors vary from the simple all-speed mechanical type through various forms of mechanical-hydraulic and electro-hydraulic types to all-electric or electronic types.

The mechanical types use rotating flyweights to measure engine speed. The weights move radially and assume a position related to the speed of the engine. In the straight mechanical governor this position is directly translated to the fuel pump rack. Speed setting is usually fixed (or adjustable through only a very narrow speed range) and speed droop is non-adjustable. Pre-selection of speeder springs provides a choice of droop settings— between 4 and 12%. Most high-speed diesel generators up to 1 MW capacity use block type fuel pumps with an all-speed mechanical governor fitted to one end of the injector pump housing. Mechanical–hydraulic governors amplify flyweight movement through a hydraulic servo system to the fuel rack(s). Droop is readily adjustable from 0 to about 8%.

In the electronic governor, engine speed is measured through a magnetic pick-up usually mounted in the flywheel housing to

the circulating pump, which may be either of the gear or the multi-lobed rotor type; a pressure relief valve; oil filter(s) and an oil-to-engine coolant heat exchanger fitted between the feed pump and the filters. Delivery pressure is normally in the range of 50–200 kPa (0.5–2 kg/cm^2) but it may even be as much as 400 kPa in high-speed engines.

In the so-called dry-sump system two pumps are employed. Oil that has circulated through the engine oilways returns by gravity to the crankcase pan. The task of the first pump is to transfer this oil from the pan to a reservoir tank external to the engine. The second pump draws oil from this tank and delivers it via the heat exchanger and filters to the bearings, etc.

The arrangement, mainly used on high-speed engines, in which the crankcase oil pan is in itself the reservoir is known as a wet-sump system.

Probably the most severe cylinder wear conditions occur just after an engine is started, when piston lubrication is at its poorest. For this reason pre-priming systems are incorporated into both the manual start procedures and automatic start controls of low-speed and higher-rated medium-speed engines. It is also advisable to use periodic priming of lubricant on the larger high-speed engines, when these are operating in the automatic standby mode. Oil fed from a separate electric-motor-driven pump is circulated at periodic intervals through the oilways to flush the liners and generally wet the engine moving parts in readiness for an automatic start. Periodic priming also has the advantage of reducing the severe wear that can occur on cylinders due to condensed combustion products when an engine is at standstill.

8.23.3 Cooling

Engines may be either air-cooled, using a mixture of air and oil as cooling medium, or water-cooled with water and oil as the cooling media. Air cooling is simpler and is satisfactorily applied to high-speed engines up to approximately 400 kW output. Air is drawn into an impellor (usually secured to the engine flywheel or vee-belt—driven off the crankshaft) and discharged through shrouding across the finned external surfaces of the cylinder and cylinder heads. Whilst very low output engines may not require separate oil-to-air lubricating oil heat exchangers, they are standard features on larger engines.

Good installation of air-cooled engines is critical especially in high ambient conditions and in confined spaces. Care must be taken with the design and application of air intake and hot air outlet trunking to avoid, in particular, the possibility of hot discharged air being recirculated within the generator housing.

Water-cooling by circulation of water through cylinder jackets is the cooling method most frequently applied to engines in generating plants. Detailed arrangements vary considerably but most installations employ some form of closed-circuit jacket cooling system to transfer the engine heat to a heat exchanger. This may be a fan-cooled radiator or a shell-and-tube type heat exchanger. Thermostatic elements are incorporated within the systems to by-pass the heat exchanger when starting from cold so as to allow the engine to attain its operating temperature more quickly.

Radiators may either be mounted on the same baseframe as their engine-generator assemblies, or remote mounted. Set-mounted radiators usually have engine-driven cooling fans, whilst remote units incorporate single or multiple fans driven by electric motors. At ratings below 1 MW, single or double sectioned radiators are employed. On larger engines multi-sectioned units are used to provide separate circuits for jacket water, lubricating oil and charge air cooling.

Since most modern medium- and high-speed engines are designed for high-temperature coolant conditions they employ pressurised closed circuit cooling systems. System pressures vary between 30–70 kPa (0.3–0.7 kg/cm^2) in practice. They equate to system boiling points at sea level of 107 °C and 115 °C, respectively, so that engine makers may design for corresponding maximum operating temperatures at engine outlet of 80 °C and 90 °C.

Where a plentiful and cheap raw water supply of good quality is available at less than 30 °C, it may be used as a secondary fluid to circulate through an engine's tubular heat exchangers before it is run to waste. Pre-knowledge of the character and quality of the water is necessary to ensure that correct selection of materials for the tubes and shells of the heat exchangers is made.

Where a raw water supply is of dubious quality, clean (or treated) water may be used in a secondary closed circuit, which is in its turn cooled by the raw water in, for example, a cooling tower. *Figure 8.36* illustrates diagrammatically an arrangement for a system of this kind. Whatever the system, it is advisable to use a high rate of jacket circulation with a small temperature difference between engine inlet and outlet, rather than a slow circulation and a large temperature rise.

8.23.4 Induction

Every engine should have air-intake filtration/silencing equipment. The induction system must be designed to supply clean dry air to the engine at as near ambient temperature as is possible and with the minimum of restriction. Engine makers stipulate the maximum permissible restriction at induction manifold or turbocharge inlet. The quality and quantity of the air supply has a direct bearing upon the engine output, fuel consumption and life.

Choice of filter depends upon the plant environment and the service life required. The following suggest the duty category in which the various filter types may be applied. The engine maker's recommendations should be sought and complied with.

(1) For plant installed in sheltered and low dust concentration conditions: oil bath or paper element (dry) type filters, both types without pre-cleaner stages.
(2) For installations in temperate, relatively dry and moderately dusty conditions: oil bath or dry element filters with centrifugal pre-cleaner stages and with greater dust holding capacity than those in (1);
(3) For severe dust concentration applications or where regular

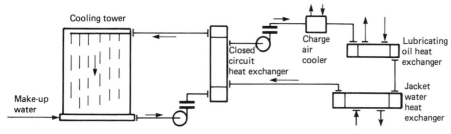

Figure 8.36 A typical closed-circuit secondary cooling circuit

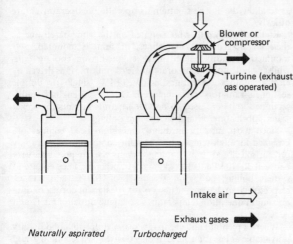

Figure 8.35 Four-stroke engine with exhaust gas-operated turbocharger

Air-to-air charge cooling is usually carried out in a section of the engine's radiator. An alternative water-cooled arrangement uses either a separate radiator circuit for charge-air cooling water or, in marine installations, sea-water-cooled heat exchangers with finned tubes carrying the sea-water and over which the charge air passes.

8.22 Engine features

8.22.1 Basic classification

In generator drives, the compression-ignition engine is usually a multi-cylinder unit classified by the synchronous speed required, the type of fuel to be used, and the mechanical arrangement (i.e. the geometric arrangement of the cylinders).

8.22.2 Synchronous speed

Consistent with the output power required, the operating speed N (rev/min) is determined by the frequency f (Hz) and the number p of generator pole-pairs in accordance with the relation $N = 60f/p$. Thus a 2-pole generator must be driven at 3000 rev/min for 50 Hz or 3600 rev/min for 60 Hz; and for an 8-pole generator 750 or 900 rev/min.

8.22.3 Fuels and operating modes

(1) In its compression-ignition form the internal combustion (i.c.) engine may be run on liquid fuels using either distillate or light or heavy residual oils.
(2) As a high compression unit in its compression-ignition form, the i.c. engine may operate in a dual-fuel mode using a mixture of gas and air ignited in the cylinders by the injection of a small pilot charge of liquid distillate fuel. The pilot fuel consumption is between 5 and 10% of the normal full load quantity required for straight diesel operation and it remains fairly constant throughout the load range unless there is a gas shortage, in which case any input energy shortfall is made up by an increase in liquid fuel injected. Should the gas supply at any time fail or become inadequate for the load demanded, the engine automatically reverts to the straight diesel mode. Because of the need to modulate both oil and gas flows, the control and protection system needed is more complex than for straight diesel or gas engines.

The engine may be switched at any load from dual-fuel to diesel operation and vice versa. The same output rating must therefore be selected for both diesel and gas operation.
(3) As a spark ignition unit the i.c. engine may operate on gaseous fuels such as natural gas, propane or sewage gas.
(4) Finally, in an alternative-fuel form, an i.c. engine may be designed to incorporate both fuel-injection and spark-ignition systems to give operation on either liquid or gaseous fuels. More often than not change to either type of fuel requires engine stoppage, but link mechanisms can be fitted to disconnect the fuel pump drive and close the air intake whilst simultaneously energising the electrical ignition and turning on the gas input. Changeover may then be effected with the engine running, much as in the dual-fuel mode described above. It must be appreciated that output rating will vary with the type of fuel used; it will be lower for gas operation owing to the change in compression ratio.

8.22.4 Mechanical arrangements

Perhaps the most widely applied engine format is the vertical design in which the cylinder axes are perpendicular and in-line. In the medium-speed range vee-form engines offering high power in a small bulk are attractive, particularly for trailer-mounted and transportable generator plants. The included angle of the vee may range from about 35–90°. Other arrangements occasionally used are horizontal designs, opposed piston engines of various forms and vertical twin crankshaft or double-bank engines.

8.23 Engine primary systems

8.23.1 Fuel injection

Most modern compression ignition engines have a mechanical or airless fuel injection system embodying jerk-pumps. Medium-speed engines tend to use individual camshaft-actuated pumps for each cylinder. Higher-speed engines employ block pumps within which all the jerk-pump elements are incorporated and driven from a self-contained camshaft which, in turn, is coupled to an auxiliary drive from the engine. Certain two-stroke engines use the common rail system, wherein fuel is maintained at a constant pressure by a pump and a hydraulic accumulator. A fuel valves at each cylinder, driven and timed from the main camshaft, delivers the fuel to the engine.

The injector, the device that introduces the fuel by spray into the combustion space, is in essence a spring-loaded needle valve, whose tip covers the injector nozzle hole(s). The number of holes, their angles and the angles of spray are largely dependent upon the shape of the combustion chamber. Since fuel systems must have small passages and nozzle holes, it is of paramount importance that the fuel within these systems is well filtered.

8.23.2 Lubrication

The lubricant in an engine performs many tasks. In addition to reducing friction (and the potentially considerable power loss due to it) and minimising wear its purpose is to provide:

cooling (either under-crown and/or in piston ring areas);
cleaning and flushing of impurities;
absorption of shocks and impacts between bearings and other engine parts.

It also affords a seal between piston rings and cylinder walls to reduce the seepage of gas that passes between the piston and cylinder walls from the combustion chamber into the crankcase.

Most engines use a pressurised or force-fed system to circulate the lubricant from an external drain tank or from a sump in the base of the crankcase. The main components of any system are:

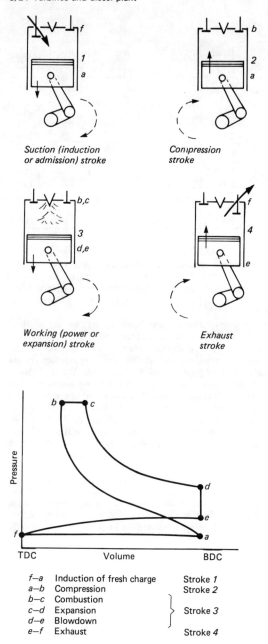

Suction (induction or admission) stroke

Compression stroke

Working (power or expansion) stroke

Exhaust stroke

f–a	Induction of fresh charge	Stroke 1
a–b	Compression	Stroke 2
b–c	Combustion	
c–d	Expansion	Stroke 3
d–e	Blowdown	
e–f	Exhaust	Stroke 4

Figure 8.33 Four-stroke cycle

the crankshaft) are available, but since they deprive the engine of a portion of its shaft output, they are not as economical as the turbocharger which utilises the otherwise wasted energy of the engine's exhaust gases. The turbocharger very simply consists of a gas turbine, driven by the exhaust gas flow, mounted on a common spindle with a blower or compressor placed in the air intake path. *Figure 8.35* illustrates in schematic form the application of the turbocharger to a four-stroke engine. Further engine upratings are now being achieved in some medium-speed, four-stroke engines by employing two-stage turbocharging to give higher air intake densities.

The full potential of this increase in air inlet density by pressure charging is, however, marginally offset by an increase of air temperature due to adiabatic compression in the turboblower. This loss is recoverable by the use of charge air coolers (intercoolers) placed downstream of the turboblower, which have the effect of increasing the fuel/air ratio, allowing more fuel to be injected into the cylinder and so raising the engine's power output. The lower air intake temperature has the further effect of reducing not only the maximum cylinder pressure but also the exhaust temperature, and with it the engine's thermal loading. Increase in engine power over a straight turbocharged model is usually of the order of 20–25% and thermal efficiencies of over 40% are obtainable.

increased without altering crankshaft speed or cylinder volume. This is effectively what pressure charging does. It can increase output by as much as 50% over an equivalent naturally aspirated engine of similar speed and dimensions. Furthermore, appreciable reductions are achieved at all loads in specific fuel consumption rates and the less arduous working conditions at the cylinders give increased engine reliability and reduced maintenance. On the debit side, however, it is impractical to expect a pressure charged engine to accept more than about 85% of its full load capability in one step in less than 10 s from crank initiation.

Several types of compressor (driven by chain or gearing from

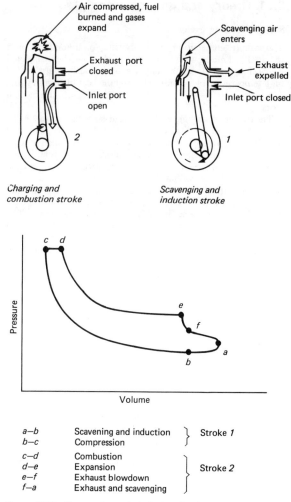

Charging and combustion stroke

Scavenging and induction stroke

a–b	Scavenging and induction	Stroke 1
b–c	Compression	
c–d	Combustion	
d–e	Expansion	Stroke 2
e–f	Exhaust blowdown	
f–a	Exhaust and scavenging	

Figure 8.34 Two-stroke cycle (valveless form)

Table 8.2 Pumped-storage installations
S = separate pump and turbine R = reversible pump/turbine

Scheme and completion date	System	Head (m)	No.	Sets (MW)
Herdecke, Germany (1930)	S	165	4	35
Niagara, Canada (1955)	R	25	6	35
Hiwassee, USA (1957)	R	62	1	56
Festiniog, Britain (1961)	S	305	4	75
Vianden, Luxembourg (1965)	S	290	9	100
Cruachan, Britain (1965)	R	375	4	100
Shiroyama, Japan (1965)	R	180	4	65
Azumi, Japan (1969)	R	135	4	110
Foyers, Britain (1974)	R	180	2	150
Shin Toyone, Japan (1972)	R	230	4	225
Coo Trois Ponts, Belgium (1970)	R	269	3	150
Dinorwic, Britain (1981)	R	500	6	300

operational economy. Security of supply is essentially a function of the availability of engines and the number of units and spare capacity installed in relation to the average load demand.

8.21 Theory and general principles

8.21.1 Working cycles

The compression ignition engine operating on liquid fuels, or in the dual-fuel mode, works on the principle of fuel being injected into a charge of compressed air and spontaneously ignited by the high temperature of the induced air by the heat of compression. The process converts the heat energy of the fuel into mechanical work.

The two basic working cycles are: four-stroke and two-stroke. These are diagrammatically represented in *Figures 8.33* and *8.34* together with the appropriate indicator diagrams, which portray the events within the engine cylinder during each cycle.

In the two-stroke engine the working stroke occurs in each revolution of the crankshaft whereas in the four-stroke engine it occurs once in every two revolutions. It does not follow that, because the two-stroke engine has twice as many power strokes as the four-cycle engine, it will produce twice the power. The down stroke of the two-cycle engine (*Figure 8.34*) combines both power and exhaust strokes. As the intake and exhaust ports are cleared by the piston some mixing of fresh air charge and burned gases takes place (scavenging). Not all the burned gases are exhausted, which prevents a larger fresh charge of air being induced into the cylinder. The resulting power stroke has, therefore, less thrust.

In the four-stroke engine, however, nearly all the burned gases are forced out of the combustion chamber by the up-stroking piston. This allows almost a full air/fuel mixture to enter the cylinder since a complete piston stroke is devoted to induction of the mixture. The power stroke therefore produces relatively more power than its two-cycle counterpart.

8.21.2 Combustion

The major advantage of the reciprocating internal combustion engine is that its design is not limited by the properties of the materials of its construction, since none of its parts is required to work continuously at maximum-cycle temperature. This allows high maximum-cycle temperatures to be used, which results in a high thermal efficiency—this latter is a measure of the efficiency with which the fuel is burned during the combustion process to produce engine power.

Whilst compression-ignition engines are generally about 5% more efficient than their prime mover competitors, there is appreciable variation amongst them in thermal efficiency. Much depends upon the size of engine and the type of combustion chamber.

Combustion chambers are basically of two types: those designed for indirect injection and those for direct injection. The former employ pre-combustion chambers in the cylinder head into which a relatively coarse fuel spray is injected at low pressure. They are popular with European and American engine manufacturers and have the advantage of being able successfully to handle a wide range of fuels. When required to operate over a wide band of environmental conditions, however, they compare unfavourably with direct-injection chambers on fuel consumption performance. Also, since heat loss from the pre-combustion chamber is high, cold starting can be difficult without prolonged cranking or recourse to external heating (such as glow plugs).

In direct-injection systems the underside of the cylinder head is usually flat and clearance volume on compression is mainly contained within the piston crown. Crown depressions are so shaped as to effectively induce swirled air turbulence, as the piston rises on its compression stroke. Fuel is then injected in the same direction as this flow of swirling air. The direct-injection principle is almost universally employed in modern medium-speed and on many high-speed engines.

Small-bore engines tend to have lower thermal efficiencies because their high surface area-to-cylinder volume ratios give larger heat losses. Again, the greater heat losses from the larger exposed surfaces of indirect-injection systems means that they give lower thermal efficiencies than direct injection ones. For these reasons, small indirect injection engines may have thermal efficiencies as low as 28% whilst larger engines, particularly those using direct injection techniques, may have efficiencies as high as 40%.

8.21.3 Pressure charging and inter-cooling

In the naturally aspirated four-stroke engine the working cylinder is almost (but not fully) charged with fresh air at atmospheric temperature and pressure at the end of the suction stroke. The density of this aspirated air regulates the weight of fuel which can be burned during the working stroke and this in turn determines the maximum power that can be developed. If a compressor were to be employed to supply the engine with intake air at a pressure higher than atmospheric, the mean effective pressure (and therefore the power output) of the engine would be

Table 8.1 Significant hydro-electric schemes

Scheme and initial commissioning date	Head (m)	Capacity MW Present	Capacity MW Ultimate
Grand Coulee, USA (1941)	—	2025	9480
Sayansk, USSR (—)	236	—	6300
Guri, Venezuela (1967)	100	527	6000
Churchill Falls, Canada (—)	313	—	5225
Iron Gate, Rumania (—)	—	—	2160
Tarbela, Pakistan (—)	115	—	2100
High Aswan, Egypt (1967)	57	1750	2100
Kariba, Rhodesia (1959)	86	600	1500
El Chacon, Argentina (1973)	58	1200	1200
Sharavathy, India (1968)	440	540	900
Murray I, Australia (1965)	521	800	800

8.20.1 Layout

Water is pumped from a lower to an upper reservoir at times of light load and allowed to run down, generating power at subsequent peak-load periods. The generators can also be used, when appropriate, as synchronous compensators. The upper reservoir may also be arranged to collect additional water from its surrounding catchment area as in a conventional hydro-electric station. In some cases, notably in Japan, the sea is used as the lower reservoir but British investigations on these lines have indicated possible difficulties due to sea-water permeation of the land surrounding the upper reservoir.

The civil-engineering works are similar to those of conventional stations. The only special features are associated with the pump and turbine units.

8.20.2 Pump/turbine plant

In the earlier schemes separate pump and turbine units mounted on the same shaft and driving a single generator/motor unit were used; this gives a simple starting and changeover procedure and permits the pump and the turbine each to be designed for maximum efficiency. A clutch enables the pump to be disconnected when generating.

In the early 1950s a single reaction machine was developed which can operate satisfactorily as either a pump or turbine, thereby reducing the capital cost of the set as well as the building to contain it. Most installations now use such machines, although the following complications are introduced:

(1) The set runs in opposite directions for pumping and for turbine operation; rapid changeover is thus more difficult as the set must be braked to standstill and its direction reversed.
(2) Starting for pumping is more complicated.
(3) The most economical speed for turbine operation is about 20% above that for pumping, so that either a compromise design must be used (the usual solution) or a two-speed synchronous machine must be employed. Pump efficiency is not appreciably affected by the compromise design, but turbine efficiency is reduced by 1–1.5%.

8.20.3 Starting and changeover

Starting a set for *generation* simply involves, after starting the essential auxiliaries, admitting water to the turbine. Where separate pump and turbine units are installed, starting for *pumping* involves running the set up to speed by the turbine and synchronising; the turbine guide vanes are then closed so that the set is motoring; the turbine casing is then dewatered by compressed air and the pump discharge valve is opened. With combined pump/turbine units, the situation is more difficult as, owing to its wrong direction of rotation, the unit cannot be used for starting. In order to reduce the starting torque required the unit is dewatered and then started by using the synchronous machine as a motor (usually at reduced voltage) or by running up with a separate pony motor or small water turbine. The starting power required is 7–8% of the full-load rating, so that starting by the synchronous machine may impose an excessive load on the system, especially with large units, so that the separate starting machine is more common. Changeover times in either direction can, with automatic control, generally be effected within 5–10 min, the change from pumping to generation being usually the more critical and quicker.

8.20.4 Economics

The overall energy efficiency of a pumped-storage scheme is usually about 70%; the losses arise chiefly from the double energy conversion and from loss of water due to evaporation from the upper reservoir.

The running cost of the station comprises the cost of the pumping energy (including the above losses), wages and maintenance, and the cost of any transmission losses arising from the remoteness of the station. Pumping energy can be purchased during off-peak times at 0.15–0.2 p/kWh, so that it can be returned to the system at 0.2–0.3 p/kWh plus the cost of wages and maintenance (e.g. 0.01 p/kWh).

Capital costs depend on the topography of the situation, but, provided they can be kept below those of an alternative form of peak-load station (e.g. a gas turbine station), the scheme can be economic.

Some typical stations are listed in *Table 8.2*.

In East and West Europe many pumped-storage plants have been built and are being planned and will provide 6–8% of the maximum load. Japan has nearly 20 plants totalling over 2500 MW in use, and extensive future plans. USA has a relatively small number of plants, although a 2000 MW plant is under-construction on the Hudson River.

DIESEL ENGINE PLANT

Generators powered by diesel engines are employed in three main roles:

(1) On primary or base-load duty in locations where there is no utility supply or as an independent power source to ensure security of supply where a public supply system is available.
(2) For peak-lopping (or peak-shaving) duty to supplement and/or reduce the cost of supply from a utility source.
(3) As standby to a power supply from a utility.

The speed of crankshaft rotation basically determines the weight, size and cost of an engine in relation to its output power. Engines are generally accepted as being divided into three classes:

High speed	Over 1000 rev/min
Medium speed	400–1000 rev/min
Low speed	Up to 400 rev/min.

The maximum size of diesel plant for primary power generation is for all practical purposes between 150 and 200 MW per station. Output ratings available are in unit sizes from 1 kW to 30 MW. The most significant range for generating plant in all three utilisation categories (continuous, peaking and standby) lies between 250 kW and 3.5 MW unit sizes, in the medium and high-speed classes.

The choice between low, medium or high speed engines must be related to evaluation of power supply security against

Figure 8.31 Runner for Kaplan turbine

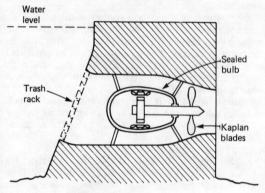

Figure 8.32 Bulb turbine

tages are that building and civil engineering costs are low. units of this type are installed at the Rance tidal scheme (France) where 38 sets each of 9 MW are installed working at a head between 5 and 10 m.

8.16 Electrical plant

Since speeds are not greater than 1000 or 1200 rev/min, salient-pole synchronous generators are almost universally employed, although small induction generators up to 4 MW are used in isolated cases. Significant features in the design of synchronous hydro-generators are a vertical shaft, a high runaway speed, the common need to incorporate high mechanical inertia to improve electrical stability, and the often difficult transport problems.

8.17 Discharge and tail race

The water leaving the power station will discharge into a tail race and the design of the civil works must avoid damage to masonry or the gradual washing away of the banks of the river by the action of h.p. water jets. If it is desired to shut the station down, or where a deflector governing is employed, the water jet must, in general, be diffused before being allowed to emerge into the tail race. This requires the employment of diffusing discharge regulators, or valves so arranged as to disperse the water in a hollow cone, which quickly loses its energy.

8.18 Economics

Capital costs of hydro-electric plants are, on account of the extensive civil-engineering works, usually high. The life of much of the plant, however, may well be 50–80 years, partly because of its durability and partly because the problem of obsolescence hardly arises. Running costs are very low, comprising only staff and small maintenance costs.

The high capital and low running cost (as with nuclear plant) might seem to justify base-load operation. In fact, design for peak-load operation is usually more economic since the amount of energy available per year is determined by the rainfall on the catchment area and a fixed sum must be expended on the catchment works and reservoir. Having expended this money the power station may be designed to use the water at a steady rate (100% load factor) or, by installing a greater megawatt rating of plant, it may use it at a much higher rate for a shorter period; the incremental cost of the plant for the latter design is, of course, less than for the scheme as a whole and usually much less than the cost of any alternative peak-load plant.

Hydro-electric stations are usually operated in conjunction with fuel-fired or nuclear stations; having built such a hydro-electric plant it is essential (to minimise fuel costs for the system as a whole) so to operate it that use is made of all the available water. Detailed calculations are thus necessary to ensure a correct balance between water used and water stored, say during the winter period, to be certain that during the ensuing summer period there will be enough water available to meet foreseeable requirements both for emergency and for best economy.

Where several stations are located on the same river and are successively using the same water, it is highly desirable to design and operate them in combination to ensure that overall usage and storage give optimum economy.

In costing a hydro-electric scheme, allowance must be made for the transmission lines to the load area; for remote stations these may well cost more than the station itself.

8.19 Operational features

Hydro-electric stations are commonly located away from population centres, so that, although personnel requirements are small, difficulties due to isolation can arise. Remote or fully automatic control can, however, readily be achieved for stations up to 50 or 100 MW.

Maintenance is small, outage time due to planned maintenance or failures being about 0.5%, as compared to 10% or more for fuel-fired stations. Starting from cold is rapid, full load being obtainable after a few minutes even with the largest sets, a factor which adds to the usefulness of hydro-electric plant for peak-load operation.

Although in many countries such as Sweden, Switzerland, Japan and Britain, most of the economically viable hydro-electric resources have been developed, there are important schemes recently built or under construction, particularly in developing countries. Some of the more significant of these are listed in *Table 8.1*.

8.20 Pumped storage

As in many countries the available water-power resources are becoming fully utilised, modern hydro-electric practice is tending towards pumped-storage schemes for peak loads.

the energy then imparted to the runner wheel is rapidly destroyed. By this means, very small forces are enabled to govern exactly the largest water wheels of this type; and regulation, in cases of a complete dropping of load from 100% to zero, can be as close as 3%. Braking of wheels of this nature is achieved by a counter jet which throws a stream of water on to the back of the buckets by the use of an additional nozzle for this purpose. The majority of impulse turbines are designed with the shaft horizontal. In some cases, two wheels are mounted at the ends of a common shaft, with the generator between.

Although reaction turbines are being built for higher heads, several large impulse turbines are in use or under construction, for example: Sharavathi (India), 440 m head, 105 MW at 300 rev/min; New Colgate (USA), 412 m head, 170 MW at 180 rev/min; Evanger (Norway), 770 m head, 110 MW at 500 rev/min.

The runners for the New Colgate plant have a diameter of 5.5 m with 22 twin buckets, weighing 44.5 tonnes and being cast in stainless steel as a single unit; each of the six nozzles gives a jet of water 360 mm in diameter which, in emergency, can be deflected from the buckets in 3 s. The head at this station is, however, such that either a reaction or an impulse turbine could appropriately have been chosen. Better part-load efficiency can be achieved with an impulse turbined by varying the number of nozzles in use.

8.15.3 Reaction (Francis) turbines

This type is used for medium heads up to about 600 m. The water flows through a spiral casing inwards through the wheel and acts on a series of blades running between the hub and an outer wheel ring, *Figure 8.29*. It is thus enabled to make use of its kinetic and its potential energy. The complete installation provides a casing for a ring of movable guide vanes running around the circumference of the runner. Mechanical linkages, operated by servo-mechanisms, open or close these gates to regulate the flow of water to the turbine. Once it has passed the wheel, the water is taken away through a draught tube and the design of this tube forms an integral part of the design of the turbine itself. The setting of the machine centre line below tailwater level must be chosen so as to avoid the formation of vapour bubbles on the trailing edges of the water-wheel blades. The implosion of these bubbles can result in serious pitting (cavitation) under these conditions. Stainless steel is welded to the water-wheel edges in many cases, to prolong their life.

The reaction turbine is usually built (in the larger sizes) with the shaft vertical and remarkable economy of power station building space is attained in this fashion. By the installation of suitable cranes, maintenance operations involving the lifting out of the runner are greatly facilitated by the vertical arrangement.

Reaction turbines can be built in sizes up to 600 MW, e.g. the 508 MW, 93.8 rev/min, 93 m head units for Krasnoyarsk (USSR) with welded steel runners 7.5 m diameter carrying 14 blades and weighing 14 tonnes. The specific speed is 230.

A modification of the Francis turbine is the Deriaz machine, first used on a large scale in 1957 for the six 30 MW, 90 m head reversible units at the Adam Beck station, Niagara. The blades are movable about an inclined axis, *Figure 8.30*, such movement giving a better part-load efficiency. The number of blades, 10–12, is less than for a normal Francis turbine, the maximum head is about 200 m, the specific speed usually lies between 160 and 250 and the runaway speed is about twice the normal running speed.

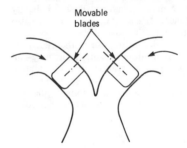

Figure 8.30 Deriaz turbine

8.15.4 Propeller turbines

The propeller turbines, used for lower heads, is a development of the reaction wheel. The runner takes the form of a ship's propeller and comprises a number of blades attached to a heavy hub which revolves below the spiral casing. This latter component and the draught tube are of the same general design as that employed in the reaction type of installation. A particular type of this propeller turbine is known as the Kaplan, and here the blades have an adjustable pitch enabling them to accommodate themselves to a wide range of operating conditions. Such a runner is shown in *Figure 8.31*.

Kaplan turbines are used in large sizes for heads as low as 6 m. The governor servo-mechanism may be situated at the top of the shaft, or it may be housed in the hub itself. The low speed of propeller and Kaplan units results in large physical size for a given output, which may make the development of very low heads uneconomic. For such cases a recent development is the *bulb* turbine, *Figure 8.32*, in which both turbine and generator are mounted horizontally in the water passage, the generator being contained in a watertight bulb. Such units may operate with forward or reverse flow in generator or motor modes. Advan-

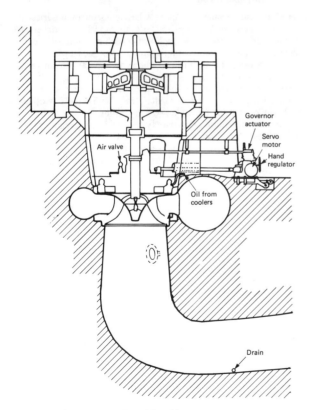

Figure 8.29 Reaction (Francis) turbine

silt behind the dam over a period of 20 years or more may also constitute a problem.

8.13 Pipelines

From the dam the water is led to the power station by one or more concrete or steel pressure pipes laid on or near the ground surface, or as a concrete-lined tunnel. In some case, depending on the topography, an open flume leads from the dam to a forebay at the head of the pressure pipe, while in others the pressure pipe may be within the dam itself.

Intake valves may be installed at the head of the pipeline; to permit maintenance, these, and other outdoor valves, may need to be electrically heated to prevent freezing.

Excessive pressure (water-hammer) caused by the sudden shut down of a turbine may be avoided by relief valves by-passing the turbine and automatically opening within a few seconds, or a surge tank or surge shaft built to a height corresponding to the head so that water can rise up in it, thus dissipating the excess pressure.

8.14 Power station

The building housing the turbines and electrical plant may be above or below ground, the latter frequently being more economical as well as advantageous from scenic and war-time safety considerations. Most turbines can be effectively designed to operate with their axis vertical, the generators being mounted above the turbine. This arrangement leads to a compact station (0.3–0.6 m³/kW) with the generating hall usually at, or slightly above, ground level, and the turbine plant in the basement; it also raises the electrical equipment well above likely flood levels.

Since there is no ash and negligible dirt or smell, there is a trend towards architectural beautification of the generating-hall interior by terrazo flooring, mural decoration, attractive lighting and, in at least one case (Tysse II, Norway), a carpeted floor. Such treatment not only improves operating conditions for the staff but also, as power plants are often tourist attractions, improves the public image of the industry.

8.15 Turbines

Three types of turbines are in general use, the impulse, the reaction (Francis) and the propellor (Kaplan) types, the choice depending largely on the head, h.

8.15.1 Specific speed

Performance of turbines of various types and sizes operating under different conditions can be compared by referring to their *specific speed*, defined as the speed at which an exactly similar turbine would run if it were designed to give unit output (1 kW) at unit head (1 m). The specific speed n_s is related to the actual speed n by the relation

$$n_s = n(P^{1/2}/h^{5/4})$$

In selecting an appropriate specific speed for the turbines of a particular installation, use may be made of the empirical expression $n_s = 1800/\sqrt{h}$; this leads to the choice of impulse turbines for high heads, reaction turbines for medium heads and propellor turbines for low heads. General details of the three basic types are shown in *Figure 8.27*. A further factor influencing the choice is the relation between the water velocity ($\sqrt{2gh}$) and the peripheral velocity of the turbine runner; where water velocity

Head (m)	Low 4–50	Medium 20–600	High 200–1000
Storage	Pondage or run-of-river	Storage or pondage	Ample storage
Turbine type	Propeller (Kaplan)	Reaction (Francis or Deriaz)	Impulse (Pelton)
Speed (rev/min)	50–250	150–600	200–1000
Runner	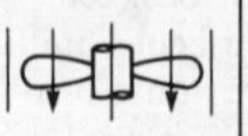	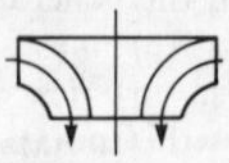	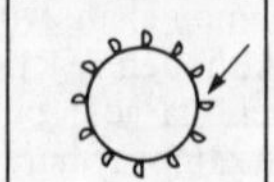
Specific speed (kW–m units)	40–80	60–400	300–900
Peripheral speed of runner / Water speed	2–3	0.75–0.9	0.5
Runaway speed / Normal speed	2–2.2	1.8–2.1	1.8

Figure 8.27 Basic types of water turbine

is high, as in high-head stations, low ratio is desirable, indicating the use of an impulse turbine.

Design developments tend to increase the specific speed of turbines to minimise weight, size and cost, but too high a value leads to low part-load efficiencies and to cavitation (erosion of turbine runners).

8.15.2 Impulse turbines

The impulse turbine or Pelton wheel consists of a series of buckets placed round the perimeter of the wheel, two buckets of hemispherical shape being employed at each position with space between them, thus preventing obstruction to the jet of water by the succeeding pair of buckets. The general arrangement for a machine with a single jet is shown in *Figure 8.28*. Up to six jets may be employed on a single runner, or two runners may be coupled together on the same shaft, each with one or two jets.

Many modern turbines employ a governing device which does not involve deflection or by-pass systems and which regulates the power delivered to the wheel by interpolation into the jet orifice of a diffuser, consisting of vanes on the conical needle which closes the mouth of the jet. When the mechanism forces these vanes to protrude, they break up the jet into a hollow cone and

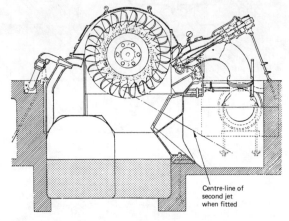

Figure 8.28 Impulse (Pelton) turbine

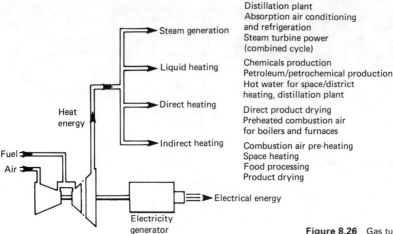

Figure 8.26 Gas turbine cogeneration heat utilisation

pressures, say between an absolute pressure of 274–620 kPa. Since it is economic to generate steam in an HRSG at much higher pressures, say 1825–6310 kPa, the opportunity arises to expand the steam from its generation pressure to the required process pressures using a back pressure steam turbine, with steam extraction for intermediate pressure requirements. Thus, valuable extra electrical megawatts can be generated, thereby improving the economics of an installation, while serving the process steam requirement. Such plant would be properly referred to as combined cycle cogeneration equipment, although frequently only the cogeneration term is applied.

Figure 8.25(c) shows the energy utilisation of a typical combined cycle cogeneration plant. The figure shows that overall energy utilisation efficiencies well in excess of 80% can be achieved with this plant configuration.

HYDRO-ELECTRIC PLANT

Water-power stations vary widely in capacity and type of load service. Some are associated with water resources permitting a steady full load, others have seasonal fluctuations, and many schemes can give full output only for a few peak-load hours in each day. Hydro schemes may incorporate pumped-storage, in which water is returned to reservoirs at times when the flow is in excess of that required for power generation. For this purpose, pumps are connected to the generators, which run during the pumping operation as synchronous motors; and when the main water flow is inadequate, this water can be run back through the turbines, often for peak-load provision.

The varied terrains amidst which water power exists call for such widely diverse schemes that it is scarcely practicable to speak of a 'typical station'; yet if such a station were postulated, it could be said to consist of the following elements:

catchment area, reservoir, and dam,
pipe lines, tunnels and surge tank,
power station,
turbines,
electrical plant,
discharge arrangements and the tail race.

8.11 Catchment area

In planning the extent of country to be regarded as the catchment area, calculations from rainfall statistics are apt to be fallacious. This is due to the varying amounts of water held back by the surface of the land over which the rain falls. For instance, the ideal catchment area surface, such as sheer rock, would obviously lose practically no water by infiltration into the land itself; whereas at the other extreme, vegetation and fissures leading to underground reservoirs draining in a different direction would absorb up to 50% of the water dispersed from the clouds on the area in question. The only reliable records are actual measurements of stream and river flows taken over a period which some authorities suggest should be not less than the sunspot cycle of eleven years. Calculations made on theoretical and statistical grounds are often widely at variance with the figures obtained by these practical tests; but in the case of unexplored territory it is obvious that calculations will have to be on a much shorter period of accurately measured statistics.

8.12 Reservoir and dam

Since rainfall is variable, a reservoir or natural lake is normally necessary to store water collected from the catchment area. The storage may be sufficient to cater for day-to-day, month-to-month or even year-to-year variations; the energy stored is 9.81 kJ $(2.72 \text{ Wh})/\text{m}^3/\text{m}$ head. Storage is thus more effective, and usually more readily available, in high-head stations giving month-to-month storage, e.g. from the summer months (melting snows) to the winter months. Storage with lower-head stations is more likely to be on a day-to-day, or even hour-to-hour basis ('pondage'). In the limit, on large even-flowing rivers, there may be negligible storage ('run-of-river' plant). Year-to-year storage, to cater for possible dry years, is usually only economic where use can be made of a large natural lake.

The choice of dam construction lies between the reinforced-concrete arch, mass-concrete gravity, buttress concrete, pre-stressed concrete and earth- or rock-fill types. In some cases it is necessary to provide locks for navigation, log shutes for the passage of timber or fish passes (usually a fish ladder). Association of the power project with irrigation may also be desirable and in many cases scenic amenity must be considered.

The world's highest dam (310 m) is at Nurek (USSR) and is of the rock-fill type; but over twenty, mostly of concrete construction, exceed 200 m. Artificial reservoirs holding up to $160 \times 10^9 \text{ m}^3$ are in use (e.g. Kariba in Zimbabwe and High Aswan in Egypt). Only the upper 10–15% of the water in most reservoirs is normally available for withdrawal; accumulation of

PERFORMANCE DATA

OPEN CYCLE SYSTEMS

Gas turbine type	EM85	EAS-1	ERB-1	ELM-125	ELM-150	EM610
Base load output, kW	7110	14 900	24 500	22 000	33 000	65 000
Overall thermal efficiency, %	22.5	28.1	35.3	37.0	37.2	28.9

COMBINED CYCLE SYSTEMS

Gas turbine type	EM85	EAS-1	ERB-1	ELM-125	ELM-150	EM610
Gas turbine base load output, kW	6800	14 500	24 030	21 640	32 520	61 296
Steam turbine base load output, kW	4600	7000	8600	7625	9220	40 974
Total base load power output, kW	11 400	21 500	32 630	29 265	41 740	102 270
Overall thermal efficiency, %	36.4	40.6	47.0	49.2	46.9	45.9

COGENERATION SYSTEMS

Gas turbine type	EM85	EAS-1	ERB-1	ELM-125	ELM-150	EM610
Gas turbine base load output, kW	6800	14 500	24 030	21 640	32 520	61 296
Total heat in exhaust above 120°C, kW	17 470	28 540	34 831	29 490	41 190	134 300
Overall thermal efficiency, %	77.4	81.3	84.8	86.0	82.8	87.8
Steam production from and at 100°C, kg/s	7.59	12.39	15.12	12.80	17.88	58.29
Tonnes/h	27.29	44.59	54.44	46.08	64.36	209.84

COAL-BURNING COGENERATION SYSTEM

Gas turbine base load power output, kW	6295
Heat to process and steam turbine, kW	25 000 – 60 000
Overall thermal efficiency, %	70 – 83

Basis for nominal performance listings:
  15°C, sea level, no intake or exhaust losses for open cycle, 254 mm H_2O back-pressure loss in exhaust heat recovery systems. Performance at power turbine coupling. Steam production figures assume 98% boiler efficiency.

Cogeneration systems are tailored for specific projects and the above examples are typical.

Figure 8.25 (*continued*)

most (if not all) of the exhaust heat can be credited to the overall thermal efficiency of the cogeneration plant. Accordingly, cogeneration installations can achieve very high overall efficiencies, up to 90%, depending on the nature of the heat utilisation.

From the point of view of economically effective plant installations, the two most interesting forms of heat use shown in *Figure 8.26* are the direct use of the gas turbine exhaust gases and the use of exhaust heat for steam generation. The first is of interest because no expensive intermediate heat exchange equipment is required to recover the exhaust heat; this possibility arises because the gas turbine exhaust gases are relatively clean, particularly when the gas turbine is operating on high-quality natural gas, and they contain a large percentage of oxygen which can be used in further combustion processes downsteam of the gas turbine. Accordingly, the exhaust heat can be used directly for duties such as product drying or preheating of combustion air for product heating in fired heaters (petrochemical industry), kilns (brick and ceramics industry), coke ovens (steel industry), etc.

The steam generation function is of interest because it is often found that the steam is required at relatively low process

The main disadvantage of the closed cycle is the air heater. It is very difficult to attain high rates of heat transfer to a gaseous fluid like air, and excessively high tube temperatures are easily developed with consequent failure. Increasing the air velocity through the tubes improves the heat transfer but increases the pressure drop of the air in passing through the tubes. A similar problem occurs in the design of heat exchangers where improvements in heat transfer by higher velocities and greater turbulence result in excessive pressure drop. The solution is always a compromise between these conflicting factors.

8.8 Indirect cycle

There is a notable growing interest in externally heated gas turbines exploiting the high transfer properties of fluidised bed combustion systems. This is widely predicted to be the answer to the long-standing question of how to run the gas turbine on cheap dirty fuels which has eluded solution for so many years.

8.9 Combined-cycle plant

The combined cycle, in its conventional power generation form, recovers much of the gas turbine's unutilised exhaust heat in a heat recovery steam generator (HRSG). The h.p. steam generated by the HRSG is used in a conventional Rankine cycle where the steam is expanded in a condensing steam turbine/generator set to provide an additional conversion of fuel energy to electrical energy. Unfortunately, because not all the exhaust energy can be recovered by the HRSG and the efficiency of the Rankine cycle is unlikely to exceed 38%, considerable energy is still eventually lost to atmosphere as low temperature heat, principally from the condenser. (The Rankine cycle thus acts as a bottoming cycle for the gas turbine's higher temperature cycle). Two typical arrangements are shown in *Figure 8.24*.

Figure 8.25(b) illustrates the use of a high efficiency simple gas turbine in combined cycle mode. The situation has considerably improved and overall power generation efficiencies in the range 44–50.5% are currently available. Application is, however, still restricted due to the need for clean but expensive fuels such as distillates and natural gas. However, the above-mentioned application of fluidised bed combustion systems capable of burning all coal types, peat, refinery bottoms and cokes, is anticipated to change this. The increasing influence of emission controls also favour fluidised bed combustors which can absorb sulphur oxides (SO_x) and produce inherently low-nitrogen oxides (NO_x) without performance penalties. The addition of flue gas scrubbers to conventional steam plant to absorb SO_x and NO_x causes a reduction in plant efficiency of 2–3 percentage points and 25–30% increases in capital cost. This will give a considerable incentive to utilities to give serious consideration to the new generation of fluidised bed combined cycles that is now available.

8.10 Cogeneration plant

If instead of using recovered exhaust heat solely for additional power generation in a bottoming cycle, it is used directly as heat for a process requirement, further substantial improvements in energy utilisation can be achieved. This is known as cogeneration (of both heat and power), or sometimes as combined heat and power (CPH).

Figure 8.26 shows diagrammatically some of the heat utilisation forms that are possible with gas turbine cogeneration plant. In each case exhaust heat is used to eliminate, or significantly reduce the requirements for additional combustion of fuel in separate heat generation plant (boilers, kilns, combustors, fired heaters, etc.). In addition, since a number of the applications involve eventual heat rejection at temperatures close to ambient,

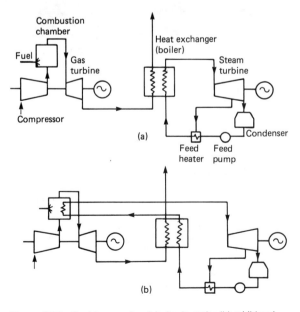

Figure 8.24 Gas/steam cycles: (a) simple cycle; (b) additional evaporator in combustion chamber

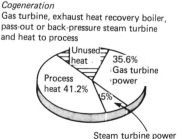

Figure 8.25 Energy utilisation for various gas turbine systems

8.6.2 More complex plant

The efficiency of the plant described above is somewhat low (18–25%). It has advantages of simplicity, low cost and independence of water. Higher efficiencies (25–37%) may be attained at the expense of greater complication and cost.

The addition of an efficient heat exchanger has a most pronounced beneficial effect on efficiency. *Figure 8.22* shows a gas turbine plant in which a heat exchanger is included. Heat in the turbine exhaust is transferred to the air delivered by the

8.6.3 Aircraft-type gas turbines

Since the 1960s there has been widespread application of aerojet gas turbines in which the propulsion jet is replaced by a heavy duty free power turbine. The development of this type of turbine makes use of the extensive research carried out by the aero-engine industry and employs one or more aircraft jet engines to discharge into a gas turbine. Such units are reasonably inexpensive, compact and have exceptionally quick starting properties, e.g. full load from a cold start in 2 min.

The jet engine on aircraft has frequent skilled maintenance and normally operates at a high altitude with low atmospheric pressure. By de-rating the engine and incorporating certain small modifications it can be operated successfully at sea level at a power between its normal flight and take-off ratings and can give 20–25 000 h operation between overhauls. Its compactness also makes it suitable for mobile power plants.

8.6.4 Free-piston gas generator

Instead of the compressor and combustion chamber of the conventional gas turbine, one or more free-piston gas generators may be used to produce the hot gas for discharge into the turbine. This comprises a cylinder containing two free pistons; fuel is admitted between the pistons and, by compression ignition and appropriate operation of valves, hot gas is emitted from the exhaust ports. The chief advantage is a higher overall efficiency (30–35%) than the conventional gas turbine, but cost and maintenance requirements are generally greater so that applications are limited. A few successful plants of between 1 and 10 MW are, however, in operation.

8.7 Closed-cycle plant

In the open-cycle plant a continuous supply of air from the atmosphere is drawn into the compressor. The intake pressure and density of the air are, therefore, fixed at atmospheric conditions. By using the closed system shown in *Figure 8.23*, in which the same air circulates continuously, the pressure and density of the air may be increased. More power is obtained from

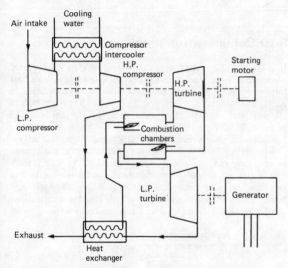

Figure 8.22 Compound cycle gas turbine

compressor so that less heat is required from the combustion of fuel. The output of the plant is not increased but the reduction in the amount of fuel burned improves the efficiency. To justify its space, weight and cost the heat exchanger must have a high thermal ratio. It usually consists of a shell containing nests of small-bore tubes over which the turbine exhaust passes giving up heat to the compressor delivery air passing through the tubes on its way to the combustion chamber; however heat exchangers have not been adopted widely for reasons of cost, bulk and reliability except in automotive gas turbines.

For electric power generation the gas turbine is essentially a constant-speed machine. The simple open-cycle plant is not flexible and operates with highest efficiency at full load and constant speed. More flexible operation and better efficiencies at part load can be obtained, but only by increasing the complexity of the plant. There are many possible combinations of compressors, intercoolers, turbines, combustion chambers and heat exchangers. *Figure 8.22* shows a plant in which the compressor and turbine are each in two stages. Interstage cooling of the compressed air increases the efficiency of compression. Another combustion chamber may be interposed between the h.p. turbine and l.p. turbine to increase the output of the plant. The separation of the turbine producing useful output from that driving the compressors makes the plant more adaptable to load variations with less reduction of efficiency at part load. The cost of more complex plant is, of course, justified only when the load factor and fuel costs are high.

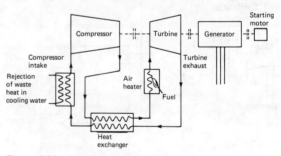

Figure 8.23 Closed-cycle gas turbine

a given size of plant and output may be varied by changing the pressure and mass flow of air, the pressure ratio and speed being retained at the optimum values.

Combustion must be external to the air stream and the heat transferred to the air in an air heater. Waste heat is extracted from the system in a cooler placed before the compressor intake.

A further advantage of the closed-cycle plant is that the air is not contaminated by products of combustion and dust drawn in from the atmosphere. Other gases than air might be used if their physical properties were superior for the purpose, e.g. helium has been used for nuclear reactor cycles.

8/14 Turbines and diesel plant

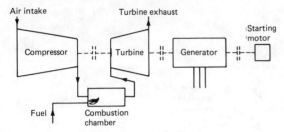

Figure 8.19 Simple open-cycle, single shaft gas turbine. Note that the turbine may also be split into high-pressure and low-pressure elements to form a two-shaft gas turbine

8.6 Open-cycle plant

The simplest form of gas turbine consists of a compressor, a combustion chamber and a turbine as shown in *Figure 8.19*. This is the 'single shaft' gas turbine which is generally used for electricity generation applications. Driven by the turbine, the compressor delivers compressed air to the combustion chamber where there is continuous combustion of the injected fuel at constant pressure. In order that the temperature after combustion may not be too high, the ratio of fuel to air, by weight, is of the order of 0.017. This ratio is too low for combustion, so the air from the compressor is divided at entry to the combustion chamber. Part of it is used in the combustion zone and part acts as a coolant. The two streams are thoroughly mixed before entering the nozzles of the turbine at the permissible temperature. *Figure 8.20* shows the outline of a gas turbine combustion chamber. The combustion of the fuel must be as nearly 100%

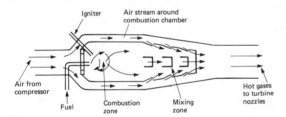

Figure 8.20 Schematic layout of combustion chamber

efficient as possible to ensure a high plant efficiency and to reduce carryover of carbon matter to the turbine blades. The quantity of fuel injected must control the output of the plant and it is essential that flame stability shall be maintained over a wide range of fuel injection rates.

The heated air from the combustion chamber, including the products of combustion, is expanded in the turbine and is then exhausted to the atmosphere. The power developed in the turbine is partly used up in driving the compressor and the remainder is available for the generation of electricity or for other useful purposes.

8.6.1 Simple power relations

Figure 8.21 shows the ideal pressure volume diagram representing the ideal cycle work derived from the turbine and the ideal work necessary to drive the compressor. In practice, the work done in the turbine is less than this ideal and the isentropic efficiency of the turbine, η_t, is less than unity. Similarly, the work required in the compressor is greater than this ideal and the isentropic efficiency η_c of the compressor is less than unity. The

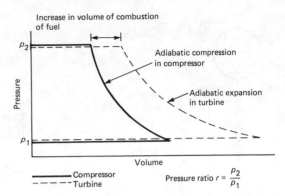

Figure 8.21 Pressure/volume diagram for ideal gas turbine cycle

useful work per unit mass of air is the difference between the work done by the turbine and the work required by the compressor, namely

$$W = c_p[T_2(1-1/k)\eta_t - T_1(k-1)/\eta_c]$$

where $k = r^{(\gamma-1)/\gamma}$, r is the pressure ratio of compression and expansion, γ is the ratio of specific heats, c_p is the specific heat at constant pressure, T_1 is the absolute temperature at compressor inlet, and T_2 is the absolute temperature of the air after combustion and before entry to the turbine.

It is evident that if η_t, η_c and T_2 were too low this expression could have a negative value. The turbine output would not be sufficient to drive the compressor and there would be no useful work. Hence the success of the gas turbine depends on the attainment of high isentropic efficiencies in the compressor and turbine and the ability to operate at a high turbine inlet temperature.

The gas turbine must operate at very high temperatures to develop efficiencies comparable with those attained in steam turbines and diesel engines. The turbine blades must withstand both high stresses due to rotation and the high temperature of the combustion gases passing over them. The success of the gas turbine has been made possible by the development of steels which will permit stressing at high temperatures without excessive creep. To allow the use of higher gas temperatures the turbine blades are often cooled by fluid circulating through passages in them. The designed maximum temperature of operation is determined by the required 'life' of the turbine. A lower temperature of operation will reduce the creep rate of the stressed parts and increase the number of hours of running life of the turbine. Temperature gradients in rotors and blades also induce stresses and alternating stresses due to blade vibration could cause failure by fatigue. With the materials and design experience now available, efficient and reliable gas turbine plants are being built and many are in successful operation.

The gas turbine is not a self-starting machine; it must be brought up to speed by a starting motor until the compressor is running fast enough to attain an adequate pressure ratio and component efficiencies for a self-sustaining or viable cycle. After ignition of the fuel the machine will increase in speed until the turbine drives the compressor and provides the useful output. If the gas turbine drives a generator with separate exciter, the exciter may be used as a starting motor.

As in the operation of a steam turbine, starting up and shutting down must take place with due precaution to avoid excessive temperature gradients, particularly in the rotors.

Contemporary gas turbines have compressor pressure ratios (delivery pressure/inlet pressure) as much as 30:1 resulting in combustion chamber pressures up to 3000 kN/m² (435 lbf/in² gauge).

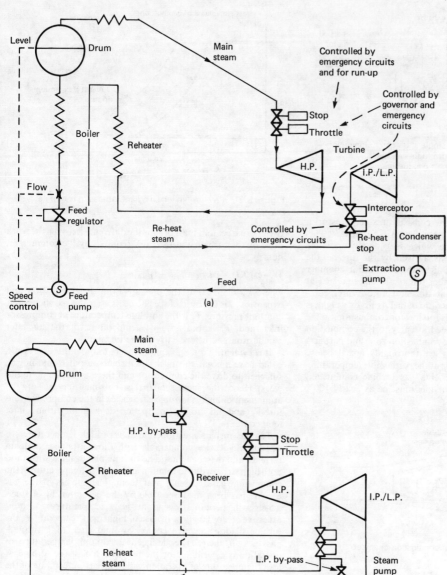

Figure 8.18 (a) Basic steam supply and control system; (b) system with steam by-pass

GAS TURBINE PLANT

During recent years many electric supply authorities have installed gas turbines, in unit sizes up to 120 MW, primarily for peak-load or stand-by, but occasionally (where fuel is cheap) for base-load operation. Many industrial plants have also used gas turbines in connection with total energy schemes.

The major advantages of a gas turbine over steam plant are low capital cost, quick starting, small erection time and the facility for using a wide range of fuels from heavy oil to natural gas; also remote control or fully automatic operation is readily achieved.

The disadvantages are the low efficiency (18–25%) unless considerable complications are included, and the limitation on size imposed by the non-availability of materials to withstand the severe temperature and other conditions.

In the *open-cycle* plant, air is taken from the atmosphere, burnt with the fuel in a combustion chamber, passed through the turbine and finally exhausted to the atmosphere. In the *closed-cycle* plant the air (or gas) is circulated around a closed circuit, the heat being supplied to the air through a heat exchanger from an external combustion chamber. The former is the more common; its lower capital cost and adaptability to the relative importance of capital cost and efficiency are cogent factors.

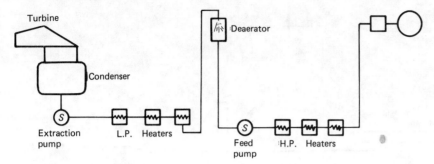

Figure 8.16 Typical feed system

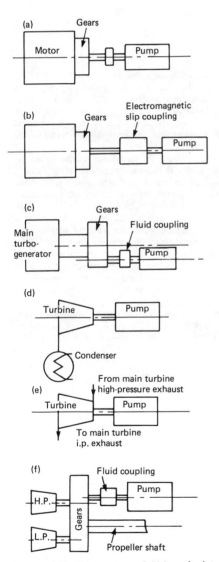

Figure 8.17 Different ways of driving a feed pump to gain economy (a) constant-speed drive; (b) electric drive with some speed variation but with coupling losses; (d) separate turbine drive, condensing—fully variable speed; (e) turbine drive integrated into the steam circuit of the main turbine—fully variable speed; (f) a marine arrangement

thermal energy. If they are closed too soon, the machine may be damaged. Automatic steam traps can be used but they need very careful selection and arrangement to ensure that they function reliably.

8.5 Turbine operation and control

There are two basic ways of arranging and controlling the steam supply. Where boilers have substantial storage volume in their drums, the steam is arranged to pass direct from the boiler to the turbine, *Figure 8.18(a)*. This is common practice in Britain. A system more frequently met overseas is *Figure 8.18(b)*, which uses by-passes. This is essential for supercritical boilers, for no drum is used and the boiler has but little steam storage capacity. But the by-pass system is also advantageous where only small drums are used. It allows the boiler to operate on part-load and pass steam via the by-pass system to the condenser without its going through the turbine. Thus, should the turbine lose full load, rather than blow the safety valves the by-pass will open and take steam from the boiler until the furnace heat has been adjusted to the load condition.

The by-pass system has other advantages. For instance, the i.p. stage can be warmed through in parallel with the h.p. rather than in series with it, so applying hot steam sooner to the large i.p. cylinder and quickening the start of the unit as a whole.

When on load in either system, the control of the turbine is by the governing system operating the throttle valves. Emergency signals of overspeed, oil loss, vibration, vacuum loss, etc. can close or reduce both stop and throttle valves by over-riding or acting through the governor. Instruments record the more important parameters of steam pressure, temperature, eccentricity, sliding, differential expansion, etc. and alarms operate from selected parameters.

Starting the machine is mostly automatic, or can be made so. There is a set sequence for taking each action, starting the oil system, barring the shaft, raising vacuum, etc. The most difficult operation and the one that requires most judgement is the admission of steam. The walls of the turbine valve chests and cylinders are thick and steam must not be admitted so quickly that severe thermal stresses develop. Many machines are equipped with instruments to measure the temperature difference across critical sections of chest or casing and so guide the operator on whether the warming through and starting is proceeding safely.

The operator must watch other factors too during a start. Most machines require special care to ensure that axial clearances are not taken up and these and the ease of sliding of the critical guidance and expansion surface are shown by appropriate instrumentation.

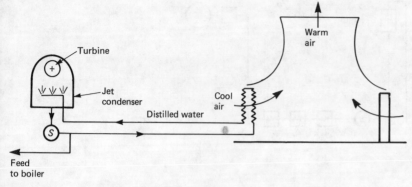

Figure 8.14 Dry cooling tower with jet condenser

In a sea-water system the turbine exhaust temperature is set by the temperature of the sea plus the temperature difference required by the condenser to transfer heat. In cooling-tower plants a further temperature rise occurs because the air is normally warmer than the sea; thus the station efficiency is lower and the cooling plant more costly.

Some water is lost by a wet cooling tower. About 1.4% of the total throughput is carried away by the air—about 27 000 tonne/day for a 660 MW plant. At some sites a reasonable source of water is too remote for economic use and a dry cooling system must be adopted.

Figure 8.14 shows the principle of the dry cooling tower. There is no direct contact between the water coolant and air, so no water is lost. Cooling takes place in large heat exchangers at the base of the tower across which air is drawn either by natural draught or by fans. The cooled water, which must be pure enough to use in the boiler, is sprayed into the steam entering a jet condenser. Such a system was installed at Rugeley in England in 1961 as an experiment at 120 MW and others have been built overseas since. An alternative arrangement is to use a surface condenser instead of a jet condenser. Raw water is then used as a coolant instead of feed water.

The dry-tower system is less efficient than the wet because the terminal temperature differences are higher so poorer vacua are obtained. Also the tower itself is much larger and so the system costs more. It can, however, enable large amounts of power to be produced in a dry terrain.

Where lower-power units are required, an air-cooled condenser, *Figure 8.15*, may be more economical than a dry tower. In the air-cooled condenser, the turbine exhausts to a heat exchanger mounted some distance above it. Steam goes through tubes and is condensed by air drawn or blown across them. The choice between air condensers and dry towers depends on many factors, but currently the former is used below about 100 MW and dry towers above about 200 MW.

8.4.2 Feed plant

The feed water produced in the condenser is taken back to the boiler through a train of devices. First the extraction pump raises it to well above atmospheric pressure and drives the water through a series of l.p. heaters, which warm it with steam drawn from the turbine. The heaters are usually tubular-type, the direct contact type is out of favour after accidents in which water spilled back into the turbine.

After leaving the l.p. heaters, the feed water is lifted to the de-aerator where it is mixed with more steam bled from the turbine, and releasing dissolved air that would otherwise a.pear in the boiler. Sometimes the de-aerator is incorporated in the condenser itself.

Water is then passed through a network of further tubular heaters and the feed pump, before passing to the boiler via flow regulating valves. A typical arrangement is shown in *Figure 8.16*.

The feed-heating system is complicated, expensive and often takes up almost as much station space as the turbine itself. In spite of the fuel saved by it (and even for a small station and medium steam conditions this can amount to over 10%) there is a trend to simplify by reducing the number of feed heating stages, more particularly on stations for developing countries where simple robustness is desirable and where fuel is not expensive.

The feed pumps of a large power station absorb considerable power—up to about 15 MW for a 600 MW generator. There is much to be gained by driving them in an economical way. *Figure 8.17* shows several contending methods. Most large units use some form of variable-speed steam drive. Practice in Britain is usually as shown in *Figure 8.17(e)*. Overseas, the condensing system, *Figure 8.17(d)*, is more often used where power units are large.

8.4.3 Other auxiliaries

The lubrication, control, feed water, cooling water and main and reheat steam systems have been mentioned with their main auxiliaries. The other important systems to accomodate in the station and to supply with power and controls, are the gland steam, the drain and, sometimes, the flange-warming systems. Usually the most troublesome during the project stage and in operation is the drain system, perhaps because it is rarely given the depth of engineering which it deserves. Taking drains from a turbine can involve a complicated mechanism of two-phase flow, with water near to boiling point, flashing to steam as pressure is dropped, e.g. through a drain valve. It then assumes a volume of many times that of the original water. Great care is required in sizing and routing pipes, valves and drain traps. Also, control needs careful thought: if drains are left open too long they waste

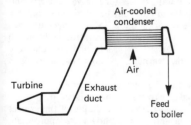

Figure 8.15 Air-cooled condenser

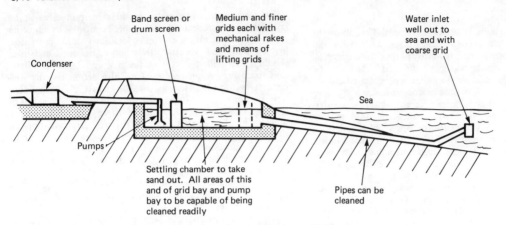

Figure 8.12 Sea-water cooling system where weed and sand are a problem

Thin tubes of titanium are now becoming more usual. The evidence so far is that they offer much greater resistance to corrosion and erosion and often promise economic advantage.

With the high-steam conditions now used, it is vital to keep salts from the boiler and tube fixtures usually involve expansion or welding. Packings are rare. Sometimes anti-leak measures extend to the use of double tubeplates, with pressurised distilled water filling the space between the two plates. This ensures that, if there is a leak into the steam space, it will be of clean water and can be monitored. Large condensers usually have arrangements for on-load cleaning.

Ejectors are used to remove air from the condenser. Currently, steam-jet ejectors appear to be in favour but there are various alternative electrically driven ejectors.

The cooling water system can be quite elaborate and usually involves a considerable amount of both mechanical and civil engineering. In a typical British coastal 660 MW unit, for every 1 kg of steam that enters the condenser, about 30 kg of cooling water must be pumped through it. This means a flow of some 1000 m^3/min. A 1300 MW nuclear unit on the Californian coast uses as much as 3000 m^3/min of cooling water.

There is no standard cooling system using sea-water: the arrangement and equipment are made to suit the circumstances. Most have one or all of the elements shown *Figure 8.12*. The system shown is elaborate because it is designed for the extreme conditions met, for instance, in the Mediterranean, where pollution causes growth of sea-grass. Periodic disturbance by storm can bring large quantities of this into the cooling system of a power station. The choking which results can take the turbines off load and is always a worry as regards chemical attack of the tubes. The weed has to be trapped and extracted before the condenser is reached if trouble is to be avoided.

Sand can also be a hazard, especially if it is hard and sharp. Settling tanks are often used to take sand out of the system.

In such systems the water inlet is taken well out to sea. This usually avoids the areas of higher weed density. The inlet is kept well above the seabed to avoid intake of sand, yet well beneath the surface to avoid oil slick. Water inlet positions must be planned most carefully with all these factors in mind and with others, such as avoiding warm water from the discharge re-entering the system. The work involves a careful on-the-spot survey with seasonal observations. It usually entails a hydraulic model.

Power stations are often sited on rivers. In some estuaries the water may be seriously short of oxygen or have aggressive pollutants, which promote tube damage. In others, the water is warm through use by other power stations or may be in seasonal short supply. Cooling towers then become necessary. Their essential principle is shown in *Figure 8.13*. Warm cooling water is sprayed into the atmosphere within the tower and passes its heat to the air, which in turn becomes warm and rises. The tower is thus filled with air which is warmer than that outside and the thermal syphon effect creates a through-draught. Wind across the top of the tower can stimulate the process.

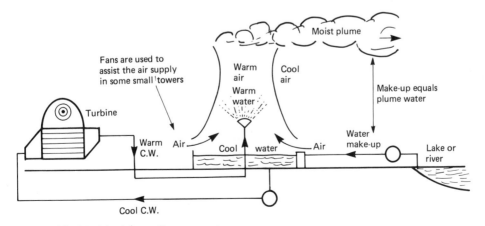

Figure 8.13 Principle of wet cooling tower system

8.3.3 Foundations

The large turbine generator is supported some 10–12 m above the basement, a space that accommodates pipes, condensers and drainage. Supports were formerly mostly of reinforced concrete, but steel supports are increasingly preferred. They can be accommodated more readily into a building programme and are more predictable in their dynamic characteristics.

The block usually takes the form of an entablature supported on columns. It must be designed so that the shaft runs smoothly even though it may be slightly out-of-balance. The foundations must sometimes accommodate earth tremors. Both these requirements are better served if the structure is in steel than in concrete, because with steel the stresses, loadings and masses commercially used give a lower and more predictable natural frequency. Vertical frequencies of about 10 Hz and sway frequencies somewhat lower are typical of such designs. These frequencies are substantially less than the shaft frequency of 50 Hz or half-shaft frequency of 25 Hz, both of which must be avoided if resonance is not to occur.

Concrete foundations tend to give calculated frequencies nearer 50 Hz, either above or below it. If the softness of the earth beneath the foundations is allowed for, these fundamental frequencies fall usually to well below 50 Hz but not so low as with steel. While concrete foundations have given good service, their response is more difficult to predict and they are less likely to accommodate shaft and earth dynamics so acceptably as do steel foundations. Also, steel is usually more amenable to a quick station-building programme. However, in many countries concrete may be more economic and the choice has to be made on the circumstances.

8.3.4 Lubrication

The security of the turbine depends on a continuity of oil supply to its bearings and on the safe functioning of its hydraulic control gear. Further, the fire record in power stations shows that the oil system may play a part in starting or aggravating a fire. It pays to invest thought and money to make the oil system completely reliable and safe. Oil systems involve quite large flows with large turbines, partly because the bearings (especially the thrust block) have turbulent conditions in them unlike the less power-absorbent laminar conditions in former 30 MW sets. Special measures aim to reduce these power losses but they can still reach 3.5 MW for a 660 MW set.

Most machines have shaft-driven main oil pumps, usually centrifugal, primed by jet pumps or pumps driven by a hydraulic motor. The main pumps are backed by a.c. full-duty pumps and by a battery-operated d.c. emergency pump to provide oil until the shaft stops should there be an a.c. failure. Sometimes a small a.c. pump is fitted for use on barring. Shafts are generally jacked by high-pressure oil to help starting.

In many arrangements the main oil pump also supplies the control gear with fluid; in others, separate pumps are used and feed control gear at much higher pressures than are used in the lubrication system. A fire-resistant fluid is sometimes used for control but is rarely considered for lubrication. The lubricant fluid is still oil, which can burn: it is therefore essential to design the oil system and fire-fighting plant to reduce fire risk. Attention is needed especially to pipe flange and coupling design. Any item that may fracture or leak oil should be eliminated or designed to reduce risk.

8.3.5 Governing

The governing and protection system operates control valves both to keep the speed steady within narrow limits and to prevent overspeed on sudden load loss. The large turbine has stop and throttle valves at its inlet, and stop and interceptor valves at the i.p. inlet from the reheater. Usually, speed control is by adjusting the h.p. throttles. Overspeed on sudden load loss is prevented, first by closing the throttle valves and interceptors and then, should speed still rise, closing the h.p. and i.p. stop valves. Overspeed prevention requires fast valve action, usually well under 0.5 s closure and in some cases down to 0.15 s. The modern large machine has a high power/mass ratio and a sudden loss of full load can cause accelerations of 10–15% of full speed per second. This allows only a short time to cut off the steam supply, a situation aggravated by the stored energy of steam already in the pipes and spaces downstream of the valves. It is an important aspect of machine arrangement to minimise such spaces, e.g. using short steam loop pipes.

Other protective circuits besides overspeed operate the valves. The list usually includes low bearing-oil pressure, high vibration and poor vacuum.

Mechanical governors of large machines operate the steam control valves via hydraulic relays and power amplifiers and, eventually, by a servo-motor whose position is determined by speed level and held by a pilot valve and feedback. There are variations on this theme, some having an acceleration element.

Electric governing is becoming more general. Speed accelerations are sensed electrically and translated to an error signal in some adjustable way within the electric governor. The final movement and positioning of the steam valve is still hydraulic.

Some small machines use mechanical relays between speed-sensor and valve, but most use hydraulics or electrical signalling and hydraulic power for valve movement. Lubricating oil is normally used, with pressures taken from the main oil pump at 0.35–2 MPa (59–300 lbf/in^2) but occasionally separate pumps are used for control and with pressures as high as 17 MPa. Fire-resistant fluids are used in the control systems of some large machines. In USA practice, double piping is often called for if lubricating oil is used. The reason is that oil leaks are retained within the pipes rather than spill to atmosphere and cause a fire hazard.

8.4 Turbine support plant

8.4.1 Condensing plant

Condenser arrangements have already been described. The condenser itself is large and costly consisting essentially of tubes through which cooling water is pumped and about which steam and air move. The steam is condensed and the air extracted. The condensate is drawn out by extraction pumps and in an efficient condenser it is nearly at the same temperature as the steam. It is returned to the boiler to use again. For every 1 °C by which the exhaust temperature is lowered in a typical 100 MW set, up to 0.5% extra fuel is burnt. Thus, a larger surface area of tubing allows the exhaust to be brought down more nearly to the coolant temperature; but, of course, this means a higher plant cost.

Having decided the surface area, however, it is important to use it properly. A particular offender is air, which can increase the exhaust temperature if it enters the condenser in large quantities, blanketing the tubes from steam. Excluding air is important: the condenser must extract such air as does enter without its forming pockets and blankets.

In the past, tubes have usually been made of a brass. Where there is a severe corrosion risk, cupro-nickel is sometimes used. Dosing is usually employed, e.g. by ferrous sulphate injections, especially where there is a risk of corrosion or of erosion. Sometimes a cleaning gear is used in which shoals of floating balls pass through the tubes, are recollected and put back into the tubes.

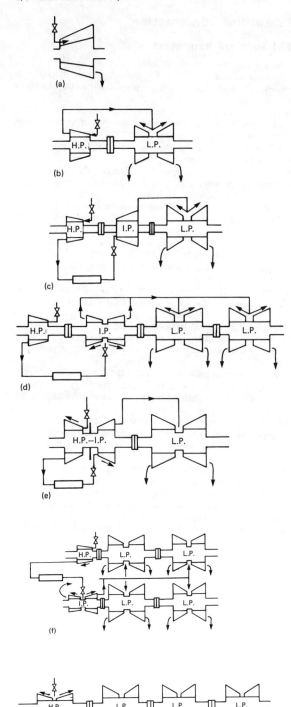

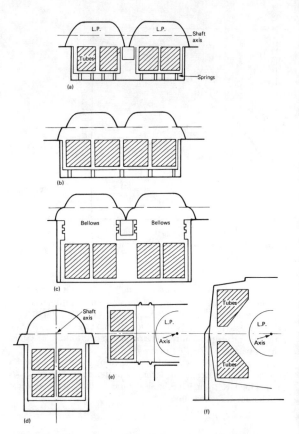

Figure 8.11 Condenser and exhaust arrangements: (a) two condensers, tubes transverse underslung; (b) unit condenser, Bridge underslung, transverse tubes; (c) floor-mounted condenser; (d) axial underslung; (e) pannier half-view, axial arrangement, end-on views; (f) radial or integral half-view

Figure 8.10 Forms of turbine: (a) simple cylinder, non-reheat; (b) two cylinder, non-reheat; (c) three cylinder, reheat; (d) four cylinder, reheat; (e) reheat with combined high power and interim power; (f) cross compound; (g) nuclear tandem for PWR

practice. It gives least load on the sliding feet of the l.p. In USA practice it is not uncommon to mount the condenser direct on the floor and connect it to the l.p. by bellows, *Figure 8.11(c)*. The arrangement separates the turbine and condenser contracts more clearly, though it means that the l.p. structure has to slide on its support under very considerable loads from the atmosphere (about 10 tonne/m² of plan area under vacuum).

A number of the British 500 MW and comparable machines had underslung condensers with tubes arranged axially, as in *Figure 8.11(d)*. Others use pannier condensers as in *Figure 8.11(e)*, while in some the tubes are brought close in to the turbine and spread around it as in the radial design, *Figure 8.11(f)*. (A very similar arrangement is sometimes termed an 'integral' design.) Designs in *Figure 8.11(f)* have compactness and should give a good tube array, but they make turbine maintenance a little more difficult and axial steam distribution requires more thought than with, say, the design of *Figure 8.11(b)*. The radial (or integral) and the pannier require cooling water to be pumped to a greater height than in the underslung designs: this can be uneconomic in coastal or river sites where no cooling towers are used. A factor in the choice of such designs is the extra efficiency they provide to the exhaust, offset against the extra pumping cost. Overall present trends appear to prefer underslung arrangements, usually with transverse tubes, and either spring-mounted as in *Figure 8.11(a)* or floor mounted as *Figure 8.11(c)*.

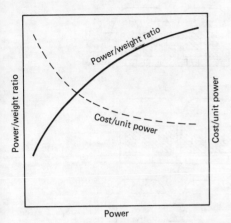

Figure 8.7 Power/weight and cost/unit power trends

8.3 Turbine construction

8.3.1 Form and arrangement

It would be ideal to have all turbines as simple as the single-cylinder machine of *Figure 8.4*, shown diagrammatically in *Figure 8.10(a)*. In fact, these are built for outputs up to 80 or 90 MW. Above these, the longer blades make two or more exhausts necessary and these need a separate shaft. The next step is to *Figure 8.10(b)*, in which in the non-reheat form provides outputs of 250 MW or more.

When reheat is adopted there are two hot inlet flows, one at high pressure and a second at an intermediate pressure. Some makers mount each on a separate shaft, giving the high-pressure, intermediate-pressure (i.p.) and low pressure arrangement of *Figure 8.10(c)*. Others take both inlets to a common h.p.–i.p rotor, *Figure 8.10(e)*. The latter is common American practice.

The arrangements so far described are for rated speeds of 3000 or 3600 rev/min. An important design factor concerns the last blade size: increase in blade height permits the number of exhausts to be reduced. Current UK designs for 3000 rev/min turbines include blading heights up to 0.95 m on a mean diameter of 2.65 m. For 300 MW these use two exhausts as in *Figure 8.10(c)* and four for 600 MW, as in (*d*), or alternatively six where the higher efficiency obtained is economically justified.

Developments may lead to even longer blades, capable of 600 MW with two exhausts as in *Figure 8.10(c)* provided that low coolant temperatures are feasible. In some areas of the world these cannot be obtained, and only low area exhausts are justified.

Where the power rating is such that the exhaust area is beyond the capacity of the largest blades, the necessary area can be provided by a two-shaft or cross-compound arrangement, e.g. *Figure 8.10(f)*, or by halving the speed as in the 1800 rev/min 1300 MW machines such as that in *Figure 8.10(g)*. These arrangements accept, as most do, that eight exhausts on a single high-speed shaft line would not be a preferred solution.

There is a compromise between the two methods shown by using a cross-compound with a high-speed, h.p.–i.p., line and a low-speed, low-pressure line, either 3000/1500 or 3600/1800 rev/min depending on grid frequency. Inherent in these schemes is the fact that the lower speeds make it possible to use much longer blades and so enlarge the exhaust area and machine output. However, they do so at the expense of more metal for the low-speed rotors and casings, which are massive in comparison with those for a high-speed set.

Engineering solutions to the problem of increased output are always evolving and much larger turbine sizes than the current 1300 MW units could, no doubt, be devised. The limit tends to be imposed not by technical aspects of turbines, but by economics and by factors other than the turbine itself. An influence that the turbine contributes to this limiting of growth, however, is that the return tends to level off as size increases and can easily be nullified or even reversed if availabilities of the large machines fall even slightly beneath those of the smaller ones.

8.3.2 Exhaust and condenser arrangements

In large machines, the l.p. stage, its exhaust and the condenser it feeds, comprise a substantial part of the structure of the machine. A number of arrangements of condenser l.p. stages are in vogue for these multiple-exhaust machines. *Figure 8.11(a)* shows an arrangement in which each l.p. turbine has itw own condenser, both underslung and with much of their weight carried by springs from the floor. In a more compact arrangement, *Figure 8.11(b)*, the condenser and l.p. are combined into a single unit slung as a bridge between the h.p.–i.p. support block and the generator block.

Underslinging and support by springs is usual in British

efficient on part load: energy is wasted by throttling down the pressure in valves. The steam jets are slow and do not strike the blades at the right angle, and the exhaust is too hot. If a turbine has to be used frequenctly at part load, it can pay to adopt nozzle rather than throttle control. The arrangement, *Figure 8.8* could show advantage over that in *Figure 8.9*.

In nozzle control, part load is achieved by using fewer nozzles and un-throttled steam. The technique is more common overseas than in the UK, though mechanical difficulty tends to make it less suitable for large machines.

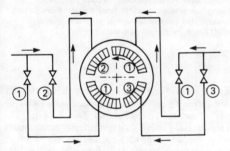

Figure 8.8 Steam inlet for nozzle control

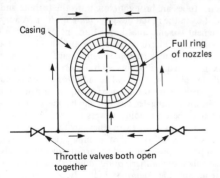

Figure 8.9 Steam inlet arrangement for throttle control

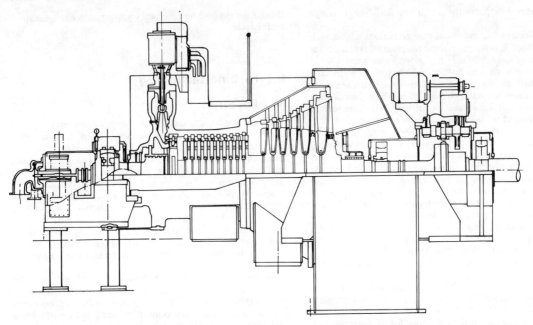

Figure 8.4 Section through a single-cylinder pressure-compounded turbine. (Courtesy GEC Ltd)

Terry turbine is used only for very low powers, e.g. for driving small auxiliaries.

Both these types are now rare. Most modern machines have axial flow with either reaction blading or a combination using h.p. stages of mainly impulse and l.p. stages of mainly reaction design. Such an arrangement is shown in *Figure 8.4*.

8.2.2 Size effects

The steam turbine is notable in that higher outputs can be achieved without a pro-rata increase in mass and cost. A power increase of 20% is achieved by using blades 20% longer and this increases the mass of the machine by much less than 20%. So the mass/output ratio is a curve such as in *Figure 8.5*. Also, parasitic losses such as gland loss, bearing loss, etc. do not increase at the same rate as power does when greater outputs are used. Thus, increased size produces some gain in efficiency, *Figure 8.6*.

Taking these two factors together there is a financial gain to be got from increased size, *Figure 8.7*. Two important points to be noted, however, are:

(1) If size is achieved at the expense of reliability, the financial gain may well become a loss.
(2) The gain due to higher outputs is best achieved if blades are lengthened and the change confined to this and to directly related issues. It will be only partially realised if, say, increased exhaust area is achieved by using more exhausts and hence jore cylinders, rather than if the change is confined to longer blades.

8.2.3 Part load

The steam turbine performs best at full load, where the whole system of nozzles and blades operate with proper steam velocities, with minimal throttle losses in valves. It will be less

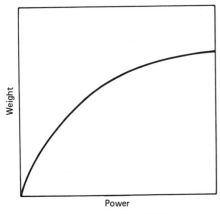

Figure 8.5 Weight of turbines of higher output

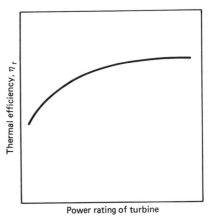

Figure 8.6 Trend of efficiency as outputs grow

normally higher than that of the gas turbine, for which T_1 is high but so is T_2.

The combined cycle aims to optimise conditions with a cycle that uses the high T_1 of the gas turbine and the cool exhaust of the steam turbine, *Figure 8.2(f)*. The cycle works as follows.

Fuel is burnt in the combustion chamber of a gas turbine, which generates electricity and exhausts to a heat exchanger. Here the hot gases boil water to raise steam which then drives a steam turbine which also generates electricity. The steam turbine exhaust is changed to boiler feed water in a condenser.

A number of such plants are in service and more are being built because of their high efficiency. They also have the quick-starting ability that the straight steam turbine lacks. Also they can offer power early in the building programme by installing the gas turbine first and the steam turbine later.

They have as yet a reliability somewhat lower than that of simple steam plant, and the gas turbine has a shorter life. The availability is more acceptable if natural gas is used rather than oil. This position is, however, likely to improve as more experience is gained: the cycle is certainly promising. A common arrangement is to take about two-thirds of the power output from the gas turbines and one-third from the steam turbine; say, two gas turbines and one steam turbine per unit.

While this cycle is as yet limited to gas and carefully prepared oil, there are prospects of its use with coal. The pollution problem is creating so strong an emphasis on clean burning that new combustion methods are being developed for coal. Pressurised, fluidized bed combustion is one; gasification is another. Such arrangements promise to produce a gas clean enough for gas turbines to use. So there is a prospect of using the combined cycle to give thermal efficiencies of around 40% using coal, which is the profuse fossil fuel.

Combined cycles are not necessarily arranged as in *Figure 8.2(f)*. For instance, fuel can also be burnt in the gas-turbine exhaust to raise more steam.

8.1.7 Alternative fluids

From time to time research is done into fluids alternative to steam. The physical properties of Freon and Ammonia have certain advantages, but so far the disadvantages have overruled them. Energy costs, however, may yet impose a change from the steam cycle.

8.2 Turbine technology

8.2.1 Blade types

There are two basic types of blade—impulse and reaction. In the impulse type, *Figure 8.3(a)*, all the expansion of steam is done in fixed nozzles and the high velocity jets so created drive the rotor blade. In most machines the expansion is in successive stages, of which there are usually as many as 20 from inlet to exhaust, each comprising a row of nozzles and a row of moving blades.

Some turbines use a 'Curtiss' stage, *Figure 8.3(b)*, to drop the temperature quickly at inlet or to shorten the machine. This stage takes a more than usual pressure drop. An exceptionally high jet velocity results and this is converted to work by driving two or more rows of blades from the one jet. The sketch shows how this is done using guide blades to re-direct the steam leaving the first row.

In reaction blading, *Figure 8.3(c)*, the moving blades are driven not only by the steam jets from preceding nozzles: they act as nozzles themselves and the jet they create helps to propel them. Therefore, pressure is lost both in the fixed and moving blades and power is produced both by impulse and jet-reaction.

In practice, few machines are purely impulse and none are purely reaction. In the modern so-called impulse machine all blades have a degree of reaction and generally this increases towards the exhaust. In some designs the amount of reaction is more marked, depending on the background experience of the manufacturer. Properly designed, there is little difference in efficiency between the impulse and reaction types.

In all but a few small machines, the steam flows along the shaft, parallel to the axis. There are two unusual designs in which this is not so. One is the Ljungström turbine, in which steam flows radially outwards through concentric rings of counter-rotating blades. The other is the Terry turbine where the steam jets travel at an angle inwards to drive against hollows in a rotating disc. The Ljungström turbine is confined to medium outputs, while the

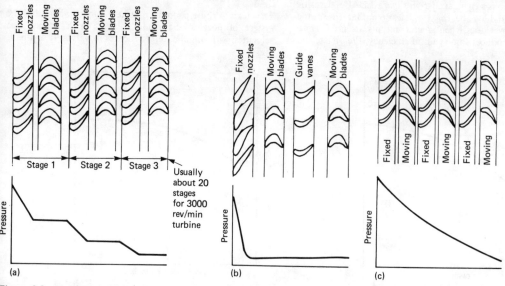

Figure 8.3 Blade types: (a) impulse, pressure compounded; (b) Curtiss stage; (c) reaction

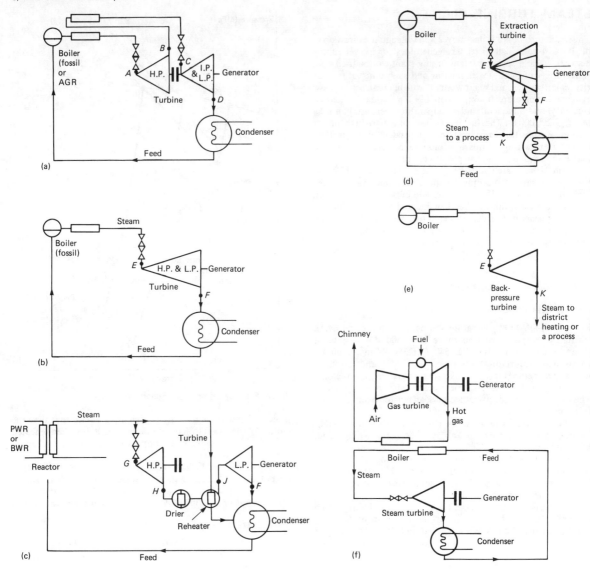

Figure 8.2 Various turbine cycles: (a) reheat cycle, commonly used in fossil and AGR power stations; (b) non-reheat cycle, used for lower power; (c) cycle for PWR and BWR nuclear; (d) extraction cycle; (e) back-pressure or 'total energy' cycle; (f) combined cycle, one of various steam and gas turbine arrangements

Some factories need steam for process work, e.g. at upwards of 150 °C and at a pressure useful enough to drive the steam to and through the process plant—say 350–1000 kPa. District heating schemes usually require water at 65 °C and obtain it from steam at around these conditions.

The turbine can be arranged to exhaust at these higher pressures and to pass all its exhaust steam to the process or to the district-heating scheme. Although the higher back pressure of, say, 700 kPa makes the machine much less efficient than with a more normal condensing turbine, the cycle as a whole is efficient because the turbine exhaust heat is not lost. Thus, a back-pressure plant generating electricity and exhausting to a factory process can have a thermal efficiency of 70% compared with 35% for a normal power station.

The basic problem with back-pressure units is that the heat rate required near to the turbine-generator rarely keeps pace with the electrical power required locally. Basically, it is easier to distribute electricity to more distant parts than to send heat: back-pressure units are therefore usually small. Confining their power and heat output to nearby use, they are rarely of more than 20 MW rating and often much less. There may well be a future for such cycles in the conservation of increasingly costly fuel, and the potential of such 'total energy' schemes must inevitably attract more attention.

8.1.6 Combined cycle

The ideal engine has a high inlet thermodynamic (absolute) temperature T_1 and a low exhaust temperature T_2. The highest attainable Carnot efficiency is $(T_1 - T_2)/T_1$. The steam turbine has only a moderate T_1 but a very low T_2, so that its efficiency is

STEAM TURBINE PLANT

The steam turbine is a prime-mover well suited to the direct drive of 2- or 4-pole high-speed a.c. generators for power-supply networks. The associated steam-raising plant can be fired by a wide variety of fuels—fossil, nuclear and such unusual fuels as city refuse, sawdust and sugar waste. The turbine can accept a low exhaust temperature (making possible a reasonable thermal efficiency) and a moderate inlet temperature (permitting a long life, e.g. 200 000 h). The turbine is best suited to high power ratings, as parasitic losses tend to be independent of size; further, output ratings can be raised without a pro-rata increase in materials, so giving a higher power cost ratio.

Steam turbines are constructed in the range 5–1300 MW. At the highest ratings the only comparable prime-mover is the hydraulic turbine, while at the lower end of the range the steam turbine competes with the diesel engine and the gas turbine.

Besides generator drive, the steam turbine is applied to ship propulsion and to rotary compressors. This chapter deals exclusively with electric power generator drive.

8.1 Cycles and types

8.1.1 Reheat cycle

Most fossil-fuel-fired generating stations use the reheat cycle. Currently, outputs span the range 200–1300 MW. Common steam conditions are 16 MPa, 540 °C (2350 lbf/in², 1000 °F). Occasionally supercritical conditions, e.g. 22 MPa and above, are met, and sometimes there are two stages of reheat; but the one-stage reheat subcritical cycle here described is more common. The British advanced gas-cooled nuclear-reactor plant also uses this cycle, with outputs currently of 660 MW.

In *Figure 8.1*, ABCD shows the conditions of steam throughout the cycle, and *Figure 8.2(a)* gives the plant arrangement. Steam is expanded in the high-pressure (h.p.) stage down to near-saturation and is then returned to the boiler for reheat to the original temperature but at lower pressure. It then expands in the low-pressure (l.p.) stage until it becomes as wet (usually not more than 13% wet) as the final blades can tolerate.

The cycle is efficient and eases the design of last blading, which is usually the critical problem in turbine development. By using reheat, less steam per MWh is needed and this means shorter blades at exhaust. The cycle usually gives an exhaust somewhat drier than others, reducing the blade erosion hazard. However, all this is at the expense of some complication and extra boiler size and the smaller units—say under 100 or 200 MW—use the simpler non-reheat cycle.

8.1.2 Non-reheat cycle

This is shown by the line EF, in *Figure 8.1*, and the plant layout in *Figure 8.2(b)*. Steam usually at 5–8.5 MPa (900–1250 lbf/in²) is expanded through the turbine to exhaust from the condenser without return to the boiler. This simple and compact machine is usually constructed in a single cylinder. The plant has a lower first cost that that for reheat, but is 8% or more higher in fuel consumption.

8.1.3 Pressurised-water and boiling-water reactor cycle

The steam conditions are given by line GHJF and the layout in *Figure 8.2(c)*. Water-cooled reactors require a special cycle because steam is supplied at an unusually low temperature, about 280 °C (540 °F). The steam is saturated and when expanded, quickly becomes wet. Water in the high-pressure stages can cause damage, so the steam is taken out of the turbine after the h.p.

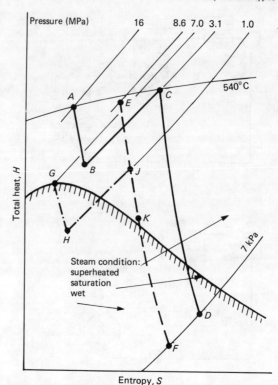

Figure 8.1 Cycles on Mollier chart

stage and put through a drier—usually a device that uses the motion of the steam to separate water from it. It is then reheated by steam bled from the main supply to a surface heat exchanger alongside the turbine.

The cycle is not efficient and turbine design must take care of the potentially damaging effects of water in the steam. However, it suits the p.w.r. and b.w.r. reactors and their economics. Machines of up to 1300 MW are in service using such cycles.

8.1.4 Extraction cycle

This cycle, line EKF in *Figure 8.1* with layout in *Figure 8.2(d)*, is particularly economic where steam is used to supply a process as well as to generate: a typical example is the desalination plant fed with steam from a turbo-generator, an arrangement in demand in countries where industry and population develop in desert areas. Here, power units of 20–120 MW are in operation, and pass-out steam feeds multi-flash sea-water desalination plants.

Steam is expanded in the h.p. stage of a turbine and is then withdrawn. A control system acts to take off as much steam as the process requires. The steam that the process does not need is then returned to the turbine to expand and do work in the l.p. stages and ultimately to exhaust to the condenser under vacuum.

8.1.5 Back-pressure ('total energy') cycle

Figure 8.1 shows the cycle by line EK, and *Figure 8.2(e)* shows the plant diagram. While the turbine is efficient in itself, with stage efficiencies of around 80%, the basic cycle has a much lower efficiency (usually 30–38% overall for power stations), mainly because over half the energy in the steam is lost in the exhaust. With the exhaust pressure usually as low as 5 kPa (0.75 lbf/in²) the steam is at only 33 °C (92 °F). It is difficult to find ways to use energy at such a relatively low temperature and so the very considerable heat in the exhaust is discarded to a condenser.

        8.20.2 Combustion 8/23
        8.20.3 Pressure charging and inter-cooling 8/23

8.21 Engine features 8/25
        8.21.1 Basic classification 8/25
        8.21.2 Synchronous speed 8/25
        8.21.3 Fuels and operating modes 8/25
        8.21.4 Mechanical arrangements 8/25

8.22 Engine primary systems 8/25
        8.22.1 Fuel injection 8/25
        8.22.2 Lubrication 8/25
        8.22.3 Cooling 8/26
        8.22.4 Induction 8/26

8.23 Engine ancillaries 8/27
        8.23.1 Starting equipment 8/27
        8.23.2 Governors 8/27
        8.23.3 Engine monitoring 8/28

8.24 A.C. generators 8/28
        8.24.1 Construction 8/28
        8.24.2 Protection 8/28
        8.24.3 Voltage 8/28
        8.24.4 Parallel operation 8/29
        8.24.5 Short-circuit performance 8/29
        8.24.6 Generator selection 8/29

8.25 Switchgear and controls 8/29
        8.25.1 Planning 8/29
        8.25.2 Fault considerations 8/29
        8.25.3 Instrumentation and metering 8/29
        8.25.4 Protection 8/30
        8.25.5 Control gear 8/30

8.26 Operational aspects 8/30
        8.26.1 Engine ratings 8/30
        8.26.2 Fuel oils 8/30
        8.26.3 Lubricating oil 8/31
        8.26.4 Maintenance 8/31

8.27 Plant layout 8/32

8.28 Economic factors 8/32
        8.28.1 Fuel costs 8/33
        8.28.2 Lubricant costs 8/33
        8.28.3 Maintenance costs 8/33
        8.28.4 Depreciation, interest, insurance and rates 8/33
        8.28.5 Capital costs 8/33

8.29 Total energy 8/33

Turbines and Diesel Plant

H Watson, BSc, CEng, FIMechE, FIMarE, MemASME, FFB
Engineering Consultant

W Rizk, CBE, MA, PhD, FEng, FIMechE
GEC Gas Turbines Ltd

G McHamish
Boving & Co Ltd

LLJ Mahon, CEng, FIEE, FBIM, CDipAF
Engineering Consultant

Contents

STEAM TURBINE PLANT

8.1 Cycles and types 8/3
    8.1.1 Reheat cycle 8/3
    8.1.2 Non-reheat cycle 8/3
    8.1.3 Pressurised-water and boiling-water reactor cycle 8/3
    8.1.4 Extraction cycle 8/3
    8.1.5 Back-pressure (total energy) cycle 8/3
    8.1.6 Combined cycle 8/4
    8.1.7 Alternative fluids 8/5

8.2 Turbine technology 8/5
    8.2.1 Blade types 8/5
    8.2.2 Size effects 8/6
    8.2.3 Part load 8/6

8.3 Turbine construction 8/7
    8.3.1 Form and arrangement 8/7
    8.3.2 Exhaust and condenser arrangements 8/7
    8.3.3 Foundations 8/9
    8.3.4 Lubrication 8/9
    8.3.5 Governing 8/9

8.4 Turbine support plant 8/9
    8.4.1 Condensing plant 8/9
    8.4.2 Feed plant 8/11
    8.4.3 Other auxiliaries 8/11

8.5 Turbine operation and control 8/12

GAS TURBINE PLANT

8.6 Open-cycle plant 8/14
    8.6.1 Simple power relations 8/14
    8.6.2 More complex plant 8/15
    8.6.3 Aircraft-type gas turbine units 8/15
    8.6.4 Free-piston generator 8/15

8.7 Closed-cycle plant 8/15

8.8 Combined-cycle plant 8/16

8.9 Cogeneration plant 8/16

HYDRO-ELECTRIC PLANT

8.10 Catchment area 8/18

8.11 Reservoir and dam 8/18

8.12 Pipelines 8/19

8.13 Power station 8/19

8.14 Turbines 8/19
    8.14.1 Specific speed 8/19
    8.14.2 Impulse turbines 8/19
    8.14.3 Reaction (Francis) turbines 8/20
    8.14.4 Propeller turbines 8/20

8.15 Electrical plant 8/21

8.16 Discharge and tail race 8/21

8.17 Economics 8/21

8.18 Operational features 8/21

8.19 Pumped storage 8/21
    8.19.1 Layout 8/22
    8.19.2 Pump/turbine plant 8/22

DIESEL ENGINE PLANT

8.20 Theory and general principles 8/23
    8.20.1 Working cycles 8/23

discharge to atmosphere. A simple limestone wet-scrubbing device is shown in *Figure 7.14*. Here, again, these devices, comparable in size with the boiler, increase costs and reduce plant efficiencies, but are necessary in order to comply with pollution regulations which are becoming increasingly severe. They also produce large amounts of sludge, which also requires ecological disposal. An alternative arrangement is the system of precombustion cleaning, in which the sulphur content of the coal is reduced prior to milling.

With regard to NO_x control, nitrogen oxides are produced as a by-product of fuel combustion in air. Control of excess air amounts and of the temperature of the combustion process (as in the fluid bed boiler) both contribute to the reduction of NO_x. Hence, the control of this pollutant must be primarily achieved by the design of the boiler furnace and of the fuel burners.

Care must therefore be taken in the selection of any boiler plant to ensure that it is capable of complying with the current, or proposed, air pollution legislation.

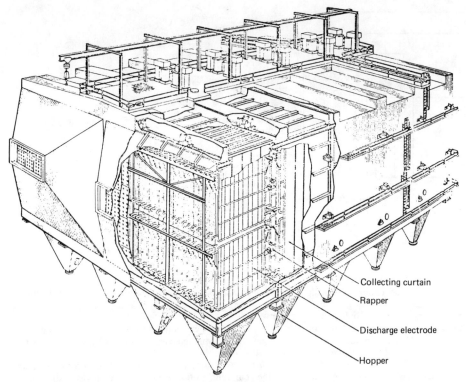

Figure 7.13 Electrostatic precipitator – parallel plate type (from *Steam, its Generation and Use*, 39th edition, Library of Congress Catalog Card No. 77-90791)

Coal poses the more difficult problem for pollution control. However, the simpler forms of combustion systems, such as the stoker and the cyclone furnace, present the easiest situation, as the coal is not so finely crushed (or pulverised) as it is in the larger PF boilers. Hence, grit arrestors or mechanical collectors can operate with efficiencies above 90%, as particle sizes are above 90 μm. However, their collection efficiency falls rapidly as the dust size decreases. Increasingly, therefore, particularly in the USA, stoker fired boilers are having to be equipped with devices similar to those installed on PF boilers.

Pulverised fuel boilers produce essentially two types of ash: that dropped out of the furnace and collected from below, and flyash, most of which is capable of passing right through the boiler. Flyash constitutes up to 85% of the ash make of a PF boiler and it must be collected. Two devices have found general application for this duty; the electrostatic precipitator and the bag-house. The former device imposes an electric charge on the particles passing through it, and they are then collected on plates. Periodically the collecting plates are rapped and the dust collected on them is dropped into hoppers, as shown in *Figure 7.13*. Collection efficiencies, in favourable conditions, can be in excess of 99%. Coals producing very fine ash particles, but particularly those producing a high-resistivity ash (such as low-sulphur coals), are found difficult to collect, and so the size and cost of the precipitator will increase if acceptable collection efficiencies are to be achieved. The alternative, the bag-house system employs fabric filters in the form of bags to trap the dust, and periodically these are cleaned, possibly by reversing the gas flow. These devices are less successful in collecting very fine ashes but have the advantage of being independent of ash conductivity. In general, they produce a higher pressure drop than do precipitators and are more temperature-sensitive, but nonetheless are gaining acceptance.

As to the disposal of the particulates collected, it is common on large installations for ash lagoons to be provided. Such disposal must, however, guard against the pollution of any groundwater courses, as the ash does contain high levels of chemicals which are water-soluble. The linings of the lagoons must therefore be carefully prepared to prevent seepage.

More recently it has become necessary to remove, or prevent the formation of, the oxides of sulphur or nitrogen discharged from the boiler stack. To-day, particularly in highly industrialised countries (such as the USA, Japan and parts of Europe), scrubbers are fitted to remove SO_2 from the flue gases prior to

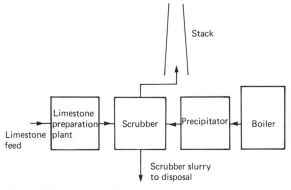

Figure 7.14 A wet scrubbing system for flug-gas desulphurisation

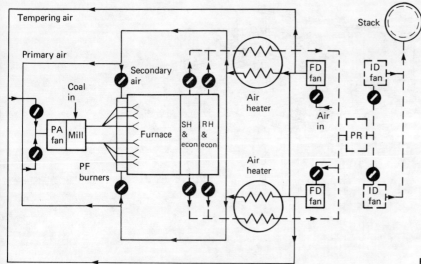

Figure 7.12 Air-gas flow diagram

wagon unloading facilities, which may be either to British practice using bottom discharging wagons and 'merry-go-round' rail system or to the American system of railcar tippling. The latter system is particularly applicable when there is a chance of receiving frozen or partially frozen coal. However, in the UK this problem never arises and so the bottom-opening wagons are preferred.

For the smaller industrial installations, which would expect to receive their fuel supplies by road, the systems of pneumatic transport are appropriate. However, difficulties have been reported with some of these systems in respect to their tendency to segregate the coal, so causing dusting and other problems.

For coal delivered over long distances coal–water slurry pipelines are gaining popularity. In this case, however, dewatering facilities must be provided and the used water also disposed of in an acceptable fashion. The 'dried' coal can be expected still to contain 25% free moisture and allowances for this must therefore be made in designing any preparation equipment associated with the boilers. These latter factors may pose severe disadvantages both economically and ecologically, and may prohibit the use of slurry transport from areas where water is in short supply.

Marine coal unloading (and loading) is becoming increasingly important as the world coal market develops. Consequently, it is of major importance to develop reliable systems. Self-unloading vessels may prove to be most economical, but for other vessels some type of wharfside unloaders must be provided. Again, a major design consideration for marine schemes should be to minimise double handling. In the field of power generation the only type of equipment generally accepted is the clamshell-type crane unloader. However, for the larger ships, continuous unloaders using bucket chains or bucket wheels are available and are already used in the materials handling market.

The third system which has been used on materials other than coal is the screw-type unloader. Its flexibility and lightness make it particularly attractive. The presence of tramp and dirt in the coal is the major factor which continuous unloaders must demonstrably overcome before they will gain ready acceptance.

As a consequence of the combustion of fuel in any boiler, certain amounts of ash remain and must be disposed of in a proper manner. Coal fired boilers produce most ash, which may be a significant proportion of the fuel burned, as seen in *Table 7.4*. The following types of ash are obtained from a boiler burning pulverised coal (PF firing):

Furnace bottom ash. The material dropped out of the furnace combustion zone. Collected by hydraulic or mechanical scraper conveyors.

Flyash. Fine particles which pass right through the boiler in the flue gas stream. Must be collected by specialised equipment (see next section).

Mill rejects. Coarse materials, stone, slate, iron pyrites rejected by the coal mills. Collected and removed periodically.

For the smaller industrial plants, ash collection into silos and subsequent removal from site is the usual means of disposal. For large installations such as power stations it is common for ash lagoons to be constructed on site or close to it. Ash from stoker fired boilers will usually find ready use in the construction industry, while for PF fired boilers, the flyash may find use in the cement industry or as a land-fill medium.

7.7 Pollution control

The control of pollution arising from the activities associated with the combustion of fossil fuels has become increasingly important, owing to the size and number of the plants both at present in use and projected. Predictions suggest that before the turn of the century the rate of use of coal alone will triple; hence, there is an increasing concern with regard to environmental protection.

The combustion of fossil fuels creates by-products, both gaseous and solid, which can find their way into the atmosphere and into groundwater courses. For a typical 300 MWe coal fired public utility boiler generating some 930 000 kg/h of steam, the amount of flue gas produced will be in excess of 1300 t/h, the amount of ash produced will be of the order of 20 t/h and the sulphur dioxide may well be in excess of 5 t/h. In order to prevent such large quantities of potential pollutants from causing nuisance, substantial pollution control precautions need to be taken.

For oil fired installations, the amount of ash produced is small and the problem is primarily contained within the boiler setting, so that the flue gases are usually sufficiently clear of particulates to be within any statutory limits. As for the gaseous pollutants, it has for many years been the practice to disperse these by the use of high stacks, and chimneys as high as 370 m are not uncommon, particularly for public utility sites.

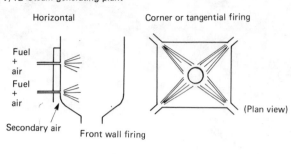

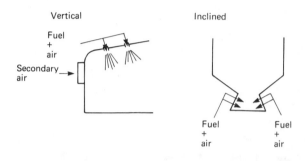

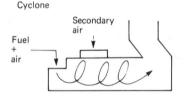

Figure 7.11 Some common burner and furnace arrangements

Table 7.7 Typical flue gas velocities and convection zone tube spacings

Fuel	Flue gas velocity (m/s)	Typical tube spacing (mm)	
		Superheater	Economiser
Gas and distillates oil	30	50	25
Heavy oil	30	150	25
Coal	15–20	400–150	50–25

industrial user to 8 weeks' for power stations. In extreme cases some power station reserves extend to a full year's supply. Hence, coal handling and storage systems are a significant feature of all coal fired boiler plant installations.

For large consumers, such as power stations, coal stacking and reclaiming equipment has been in use for many years. However, following the construction of large units with capacities of 500 MW and above, and the advent of mechanisation in the mines, much larger and more reliable means of coal handling and storage had to be sought, because now coaling rates of up to 3000 t/h with coal containing a high proportion of fines and increasing quantities of dirt are being burned.

In order to arrive at a stockyard system which will fulfil the essential requirement that an uninterrupted flow of coal can be maintained to the boiler bunkers, it is necessary to optimise a number of interrelated factors, viz.: boiler capacity; the area of land required for the coal store; the rates and means of delivery of coal to the site; the mode of working and degree of automation in the stockyard; the rate of consumption of coal by the boilers (with seasonal variations); and climatic and environmental effects. The derived capacity of the coal store therefore reflects all these aspects and must also be compatible with the geographical nature of the stockyard site, which will, in turn, influence its form and layout if double handling is to be minimised. For a stockyard receiving a steady daily supply of coal by any one or a combination of road transport, rail or barge, a stockyard system would be expected to be capable of receiving and unloading the coal deliveries and then to be able to supply directly the boiler bunkers from this newly arriving coal. Sufficient supplies must be maintained during the working periods to enable a live stock of coal to be built up for use at weekends or holiday periods. This live stock would be expected to be used within about 10 days of delivery; otherwise it must be moved to the permanent storage area, and compacted and protected to avoid problems of spontaneous combustion and degradation. Any coal surplus to these two requirements (perhaps that arriving during periods of low consumption) must be used to build up the permanent store, and the stockyard system must be arranged for this.

In the above situation a major proportion of fuel by-passes the yard. At the other extreme, for cases where fuel arrives by sea-going colliers, it is necessary to unload the ships rapidly in order to avoid excessive demurrage charges. Ship sizes of up to 100 000 dwt can be accommodated at the most modern power stations in Europe, the Middle East and the Far East, and so the capacity and layout of the stockyard must reflect the fact that there may be long periods between deliveries and that high ship unloading rates are required. In this case it can be expected that for long periods the flow of coal would be from the stockyard to the boilers.

As to the stockyard itself, experience shows that good drainage is essential and it must be designed so as to avoid possible damage by the action of stockyard equipment. Further, in many installations rainwater run-off must be collected and treated prior to discharge from the site in order to comply with local pollution regulations. The foundations of any rail-mounted equipment are particularly important and must be of sufficient quality to ensure no subsequent subsidence, as all such equipment is level-sensitive. With regard to climatic conditions such as heat, high winds and rain, appropriate action may include such precautions as screen walls or the covering of certain portions of the stockyard. For permanent stores good compaction and sealing are most important. In certain areas which are particularly sensitive to noise and dust, stockyards have been partially covered, while in other instances where typhoon conditions are to be expected, protection of the permanent stores has been by means of an asphalt covering.

For coal fired installations requiring coal storage capacities of less than 10 000 t, alternative storeyard installations may be appropriate, and the use of storage bins or silos may be economically considered. In such instances coal would be transferred from reception into silos with discharge facilities which may include rotary plough feeders or a pneumatic conveying system. The significant factors here in deciding whether silo storage is economic are the civil engineering costs, which rise in proportion with increases in storage capacity. For storage capacities up to a nominal 20 000 t a polar or conical stockpile incorporating underground discharge feeders can be economical. These lower capacities are appropriate to industrial or small public utility installations.

The transport of materials around sites of any size must be properly designed if major problems are to be avoided and double handling minimised. Coal reception by rail necessitates

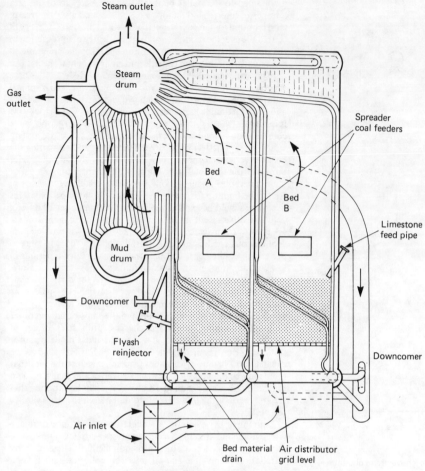

Figure 7.10 An atmospheric fluidised-bed boiler (from Foster Wheeler Power Products Ltd, Brochure No. SP80.6)

pass into an air heater, where incoming combustion air is heated. For the case of a coal fired utility boiler, an air–gas flow diagram is shown in *Figure 7.12*. Air heaters fall into two categories, recuperative and regenerative. In a *recuperative* design heat is transferred directly from hot gases on one side of a membrane to the air on the other. In a *regenerative* design a heat storage medium is used to transfer heat indirectly to the air from the hot gas. Recuperative heaters are generally tubular, while the regenerative principle invokes a rotation whereby plates are alternatively heated by flue gases and cooled by the incoming air. Designs are available in which the plates rotate within a stationary split casing or in which the plates are stationary and the gas streams are progressively redirected by rotating hoods.

In some instances separately fired air heaters or steam coils may be used. For those utilising flue gas heat, however, it is important that the gas leaving temperature should be sufficiently high to prevent condensation of flue gas contaminants which may otherwise cause 'back-end' corrosion. Typical gas temperature limits for oil fired boilers would be 150°C and for coal fired units 126°C. Clearly it is important to keep these leaving temperatures as low as possible if high boiler efficiencies are to be maintained.

7.6 Fuel handling and storage

Fuel oil, because of its high heating value per unit volume and the fact that it can become (if necessary by heating) a liquid capable of being pumped, has become a most attractive fuel. It can be transported readily by sea, rail, truck and pipeline, and can be stored in tanks at the consuming site. One of the main hazards of storage is from tank failure or leakage, which is usually overcome by the use of bund walls (or cofferdams) placed around the groups of tanks. For the heavier oils heating must be provided in the tanks and the pipelines must be trace-heated so that the oil may be pumped. Tanks, piping and heaters used on these heavy oils require regular cleaning to keep them free of sludge.

For gas installations, particularly natural gas, transportation by pipeline is attractive, while for overseas transport the gas is usually liquefied under pressure (LNG) to facilitate transportation. Storage systems are normally water-seal tanks or pressurised insulated steel tanks.

By contrast, coal is undoubtedly the most difficult of the fossil fuels to transport and store. For these reasons it is common for coal fired boiler installations to maintain a reserve of fuel on the site which may amount to some 2 weeks' reserve at full load for an

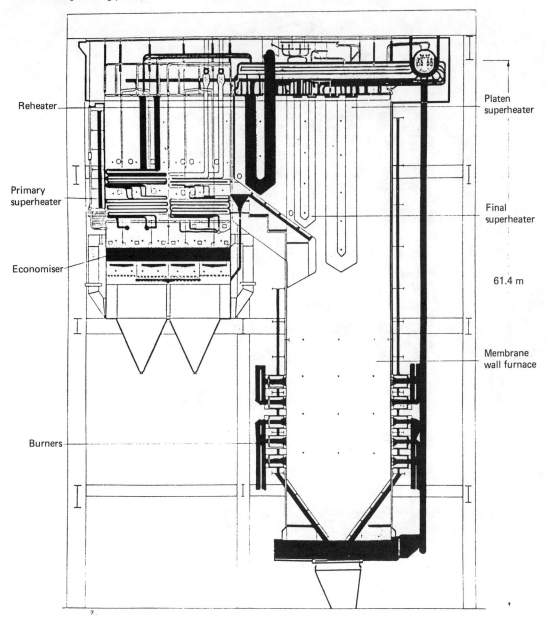

Figure 7.9 Large coal-fired sub-critical drum boiler which provides steam for a 660 MWe turbine (from Babcock Power Ltd Publicity No. 3003/2k/0979)

completely burned. By contrast, oil firing produces only minor quantities of ash, and its preparation for firing is limited to the preheating of the heavier residual oils in order to facilitate pumping.

Combustion gases leaving the furnace then enter the convection zone, and, for the case of coal firing, it would be expected that the gas temperature would be below the ash deformation temperature in order that the particles carried with the gas stream do not form deposits on the tube banks (which now extend across the gas flow path in order to absorb the available heat). High gas velocities, while providing good mixing, should be avoided if tube erosion by ash particles is to be controlled. Current trends are to reduce previously accepted gas velocities and to increase tube pitches in order to provide high tolerance to the widest possible range of coals. *Table 7.7* gives some typical gas velocities for various fuels. As a means of controlling the build up of gas side deposits on tube banks, soot blowing equipment is usually installed. This consists of a series of lances which direct streams of air or steam onto the tube banks and thereby dislodge the deposits and improve heat transfer rates. For difficult fuels these devices may be put into an operational cycle every 4 h.

On larger boilers, as the flue gases cool they pass into the economiser banks, where water incoming from the feed pump is heated prior to admission to the furnace. It is not uncommon for such tubes to be of the extended surface type, having fins to improve their heat transfer properties. Finally, the gases may

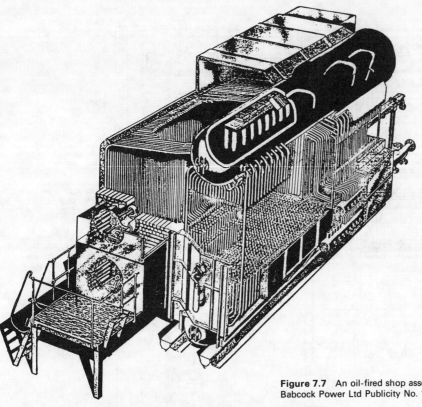

Figure 7.7 An oil-fired shop assembled package boiler (from Babcock Power Ltd Publicity No. 19488/2k/1274)

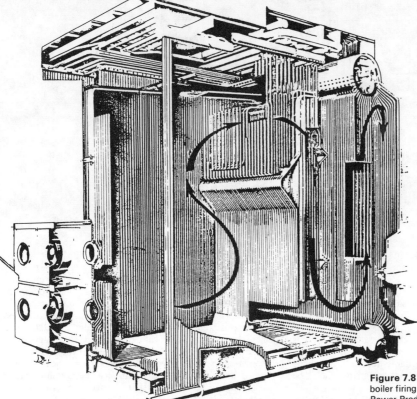

Figure 7.8 A medium-sized bi-drum water-tube boiler firing either oil or gas (from Foster Wheeler Power Products Ltd, Brochure No. 150.12.76)

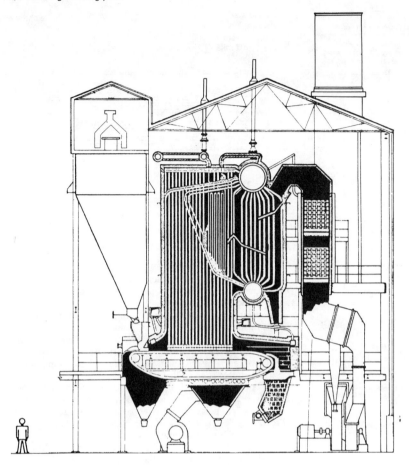

Figure 7.6 A stoker-fired water-tube boiler of 36 300 kg/h steam capacity at 28 bar, 385°C (from NCB publication *Boiler House Design for Solid Fuel*, (1975))

systems are based on the atmospheric fluidised bed system, which involves the combustion of the fuel (generally coal) in a turbulent bed of inert particles (sand, limestone or coal ash). The bed is kept turbulent by the injection of air, uniformly through the base. By this means very high heat transfer rates from bed to water can be achieved in any water tubes passing through the bed: more importantly, the bed temperature is usually below 900°C. This means that few if any of the gaseous pollutants formed in conventional furnaces are formed, or escape into the flue gases. For example, high-sulphur coals can be burned and almost all the sulphur can be retained in the bed if limestone is used as the bed material. Further, low-grade and variable coal mixtures can also be burned successfully and cleanly, this latter point being a major consideration with regard to pollution control.

7.5.1 Design options and objectives

Modern industrial and public utility boilers are expected to be capable of operating continuously for extended periods without distress and with little, if any, loss efficiency. Such reliability can only be achieved by matching the fuel and its combustion characteristics with the expected operating demands for the steam supply, in turn dictated by the generating or process plant to which the boiler is connected.

The furnace configuration of a water tube boiler must be such as to induce conditions under which the complete combustion of the fuel can take place. Fuel and combustion air must, therefore, be introduced into the furnace and ignited, and an example of the simplest option for solid fuels, stoker firing, may be seen in *Figure 7.6*. In this system coal is introduced onto a slowly moving grate where it burns, combustion air also being introduced into the furnace at suitable points. For larger boilers, or those burning liquid fuels, burners penetrating the tube walls are used. In this situation the burner is designed to mix the fuel and some combustion (or primary) air and to introduce this mixture into the furnace, where it meets more air (secondary air). Some common furnace configurations using burner systems are shown in *Figure 7.11*. For combustion, liquid fuels are vaporised or atomised in the combustion air. Atomising burners may either spray fuel from a nozzle at pressures up to 20 bar or atomise it with air or steam at pressures up to some 14 bar.

For coal fired boilers fuel preparation by pulverisation and drying in mills is necessary, before it is ducted along with the primary air (about 10–20% of the total combustion air) to the burner. Ball, roller and attrition mills have all been used successfully for coal preparation; however, they do introduce a further complication, as does the presence of significant quantities of ash in the coal. Of major concern in the furnace of coal fired boilers is the formation of slag deposits (formed from coal ash) on the water walls. Designs must arrange to minimise these undesirable effects while at the same time providing sufficient residence time within the furnace for all fuel particles to be

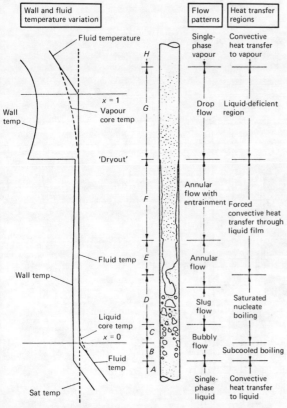

Figure 7.4 Regions of heat transfer in convective boiling for a straight vertical tube (from J. G. Collier, *Convective Boiling and Condensation,* McGraw-Hill (1972))

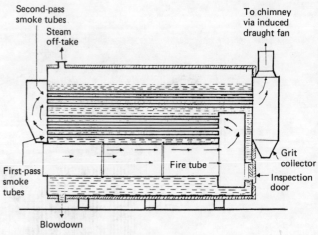

Figure 7.5 Triple-pass wet-back shell boiler

Table 7.6 Some flow velocities common in steam generators

Type of service		Flow velocity (m/s)
Boiler furnace tubes		0.35–3.5
Steam lines	High-pressure	40–60
	Low-pressure	60–75
Superheated tubes		10–25
Forced draft ducts (air)		7.5–18
Induced draft flues (flue gas)		10–18

a unit is shown in section in *Figure 7.5.* Shell boilers are capable of firing solid, liquid and gaseous fuels, with their capacity mainly limited by the fabrication problems associated with the shell. For oil fired units, capacities of up to 27 250 kg/h are available, but for coal fired installations this would be reduced by about half. Pressures normally extend up to 17 bar, and to 38 bar for special units, while by the introduction of superheater tubes (sometimes placed in the final smoke pass) some superheat is available. In almost every case shell boilers are 'package units', i.e. they are constructed and tested in the factory and delivered to the site as working units. Typical shell boiler installations would be seen in breweries and textile mills.

Water tube boilers overcome the limitations of the shell boiler. They are available in sizes up to the highest capacity and can be designed to fire any of the fuels listed in the previous section. Consequently, there is a very wide range of designs available and some of the more significant are noted below.

The simplest, and oldest, type of water tube boiler is the coal fired spreader stoker boiler, which is found to be economic up to capacities of about 13 500 kg/h of steam at pressures up to some 60 bar. Numerous types of stoker arrangements have been developed, such as dumping, oscillating and travelling, and a typical example is given in *Figure 7.6.*

For liquid or gaseous fuels, particularly, it is common for shop assembled or packaged boilers to be used in capacities up to 130 000 kg/h at pressures up to 79 bar at temperatures to 480°C (*Figure 7.7*). The principal limitation to their size is the capacity of transport facilities, particularly by road.

For the larger industrial applications the bi-drum boiler of the type shown in *Figure 7.8* is common, while for the largest capacities a single-drum radiant boiler of the type shown in *Figure 7.9* is usually selected. As alternatives to this option, supercritical boilers can be selected, which, by operating at pressures above the critical point of steam, can eliminate the need for drums (or steam/water separators) and are effectively once-through boilers. In subcritical drum boilers the water entering the furnace tubes is only partially converted to steam and so the resulting mixture requires separation, with water returning to the furnace and the steam generated passing on to the superheaters. This separation is achieved in the steam/water drum, where water is also added to compensate for the quantity of steam removed. In general, it will be found that about 50% by volume of steam is present in the average steam–water mixture. The mass of steam generated in the furnace is dependent upon pressure, and the ratio between the steam generated in a single pass through the furnace and the amount of water entering the furnace (known as the 'circulation ratio') will fall with increasing pressure.

It is possible to arrange for a boiler to operate in the once-through mode at pressures below the critical. Such a boiler is usually termed a Benson boiler. In this design no steam/water drums are required when the unit is operating in its normal mode, and in order to maintain good flow rates on the water side of the furnace, the tubes are spirally wound instead of being vertical, as they are in a drum boiler. Once-through boilers are popular with public utilities in Western Europe; however, the single-drum circulating boiler is favoured by nearly all American and UK utilities.

A recent development concerning a concept new to the boiler industry is that of the fluidised bed boiler, and an example of such a unit is shown in *Figure 7.10.* Essentially, the currently available

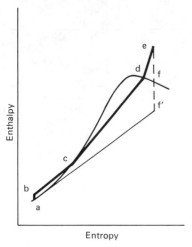

Figure 7.2 The Mollier or enthalpy-entropy diagram

transformed into dry-saturated steam. If the boiler is equipped with superheaters, further heat can be added, to achieve a final condition at e. The processes between c and d occur at an essentially constant temperature (the saturation temperature equivalent to the pressure in the boiler), while in the superheat region the added heat causes the steam temperature to rise.

At point d or e the steam can be used to provide heat for process plant or be used to drive a work producing engine or turbine. If steam leaves the boiler at d, which is a dry-saturated condition, it is evident that the removal of energy from it by heat transfer or by work in an engine will cause it to start to condense, and x will fall. However, if the steam is initially superheated to e, it can provide an amount of work approximately equivalent to the drop in enthalpy $(h_e - h_f)$ before any moisture is formed: this is the great advantage of superheat. Nonetheless, many applications, particularly in the process industries, require only heat and the added complications of providing large amounts of superheat are avoided, providing that process temperature demands are compatible with available boiler pressures.

By contrast, in cases of power generation, high pressures and temperatures are thermodynamically attractive, as is the expansion of the steam to vacuum in a condenser (point f' on *Figures 7.1 and 7.2*), so making the term $(h_e - h_f)$ as large as possible. The boiler may then be said to be working as part of a regenerative cycle with steam under vacuum at f' being condensed along f'a to a pump at a, where it is pressurised and returned to the boiler at b.

Mixtures of these two systems are common, particularly in industry, where some electrical power as well as process steam is required. However, in all cases the steam generator must provide the steam at the desired conditions, and to achieve this the boiler designer must be fully aware of the processes involved within the working fluid as it moves from b to e.

For the simplest type of industrial boiler, the shell or fire tube boiler, where only modest steam pressures are available and little or no superheat is produced, the combustion occurs in furnace and fire tubes and the products of combustion pass down further tubes (called smoke tubes). The tubes are contained within a vessel (or shell) which contains essentially non-flowing water, heated by the tubes passing through it. The rate at which heat can be transferred to the water is controlled essentially by the properties of the water. If the rate of heat transfer is poor or a very high flux is provided by the furnace, overheating of the tubes may occur and the material may fail. Information on these heat transfer rates, either by experience or by experiment, are therefore required by the designer.

For higher pressures and capacities and for applications calling for appreciable amounts of superheat, the roles are reversed, in that water occupies the tubes, which now form the enclosure of the furnace. Such a boiler is known as a water tube boiler. In such designs, water flows in the tubes and it is induced to do so either by means of a pump or by the 'natural' effects induced by heating water in inclined or vertical tubes so that the steam so formed is free to rise.

The heat flux in the boiler must therefore be controlled to match the ability of the working fluid to absorb heat. The effects of varying heat fluxes on the working fluid and on the tube wall temperatures are illustrated in *Figure 7.3*. The region between

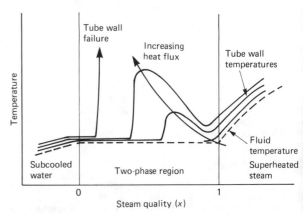

Figure 7.3 Working fluid and tube wall temperatures for tubes in the furnace of a water-tube steam generator

$x = 0$ and $x = 1$ (c and d in *Figure 7.1*) is known as the convective boiling zone and has been the subject of much research, as the optimisation of this region can be fundamental to boiler reliability. The regions of heat transfer in the convective boiling region are shown in *Figure 7.4*, from which it will be seen that the boiling process is complex, with many regimes being identified (A–H in the diagram). It should be remembered that the two phases are flowing in the tube, so that the regimes will move backwards and forwards along the tube, owing to fluctuations in heat flux, pressure and flow rate.

Figures 7.3 and *7.4* both give indications of the variation of the tube wall metal temperatures for both water- and steam-swept surfaces. They illustrate the fact that the resistance to heat flow for water-swept surfaces comes almost totally from the gas (or flame side), but where surfaces are steam-swept the steam side resistance constitutes a significant amount in the overall resistance, which is shown in the increase in tube wall temperatures. Hence, it is important for the designer to ensure adequate flow rates and good distribution of the working fluid to all parts of the boiler if problems of tube overheating and failure are to be avoided. This is done by use of the established hydrodynamic relationships for fluid flow in ducts and pipes. *Table 7.6* illustrates some typical velocities found appropriate in steam generators, both on the water side and on the gas side.

7.5 Types of steam generator

The simplest type of boiler commonly encountered in industrial situations that require little or no superheat and modest quantities of steam at low pressures is the shell or fire tube boiler. Such

Table 7.4 Examples of some hard coals available

Country	Typical analysis				Higher heating value (kcal/kg)
	Moisture (%)	Volatile matter (%)	Fixed carbon (%)	Ash (%)	
Australia	0.5–2.6	30–37	50–65	11–15	6 500–7 300
Canada	1.4–4.0	25–34	51–62	10–12.5	7 000
China	2.5–5.4	25–30	40–45	20–31	5 200–6 000
Great Britain	0.7–7.5	11–36	57–85	2–4	7 400–8 000
Poland	4–17	21–32	40–59	6–22	4 500–7 500
South Africa (export)	2.0–5.5	16–27	50–70	9–16	5 900–6 800
USA (export)	1.5–3.5	20–40	—	5–9	7 700

Table 7.5

Fuel	Heating value (kcal/kg)	Ash (%)	Moisture (%)	Volatile matter (%)
Peat (air-dry)	2600–3000	1–2	37–45	38–40
Bagasse	2200–4860	1.315	40	45
Wood (air-dry)	4200–5400	0.5–3	10–20	75
Wood waste with bark	3300	2	15	70

Anthracitic coals have a high heating value, are non-agglomerating, have low volatility and very high fixed carbon content (in excess of 85%). Bituminous coals have a heating value not less than 5700 kcal/kg, and are agglomerating, while subbituminous coals can have heating values as low as 4600 kcal/kg. Lignitic coals are characterised by high moisture contents (up to 60%) and heating values lower than 4600 kcal/kg.

To provide an indication of the variations of coals world-wide, some of the more significant producers are listed in *Table 7.4*, together with corresponding typical hard coal properties.

7.3.4 Other solid fuels

Other fuels, mainly vegetable or vegetable waste, have been found to be useful. Peat is, in fact, burned by certain public utilities. Wood and wood wastes as well as bagasse waste are also well-known fuels, while, more recently, trials have been conducted into the use of industrial and domestic refuse as a boiler fuel. Some indication of the properties of these fuels is given in *Table 7.5*.

7.4 Thermodynamics and hydrodynamics of steam generating plant

In a conventional steam generator the working fluid (water) receives heat from the chemical reactions occurring between its fuel and the air, and in an ideal situation the working fluid is at constant pressure. In a real boiler there is a pressure drop between inlet and outlet due to the effects of friction; however, the process can reasonably be represented by the line bcde in *Figures 7.1* and *7.2*. Initially, the incoming water or feedwater to the boiler has to be pumped into the boiler, and this is represented by the line ab. At inlet to the boiler (b) the water enthalpy will be subcooled and will first need to be heated to saturation at c. At c its mass quality (x) is zero. Mass quality may be defined as,

$$x = W_g/(W_g + W_f) \tag{7.1}$$

where W_g and W_f are the flow rates of steam and water, respectively. A mass quality definition can also be stated as

$$x = (h - h_f)/h_{fg} \tag{7.2}$$

where h is the enthalpy of the working fluid, h_f the enthalpy of saturated water at the same pressure and temperature, and h_{fg} the latent heat at the same conditions.

In Equation (7.1) x can vary between zero and unity, while in Equation (7.2) x can assume negative values if the fluid is subcooled and values greater than unity when the fluid is superheated. In the range $0 \leqslant x \leqslant 1$, the values of x derived from Equations (7.1) and (7.2) are identical if thermodynamic equilibrium exists.

Between c and d the mass quality increases from zero to unity as steam is produced, until at d the working fluid is completely

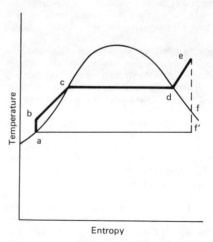

Figure 7.1 The temperature-entropy diagram

Table 7.2 Examples of some fuel oils available

Grade of fuel	Composition % (by weight)					Gross heating value (kcal/kg)	Description
	C	H_2	O_2+N_2	S	Ash		
Distillate (amber)	86.5	12.7	0.2	0.7	Trace	10 800	For general-purpose domestic heating
Very light residual (black)	86	12	0.5	1.5	0.02	10 500	For installation not equipped with preheating
Residual (black)	85.75	10.5	0.9	3	0.08	10 150	For installations equipped with preheating
Crude oil (Mexican)	83.3	10.9	2.2	3.6	—	9 665	
Coal-tar	89.5	6.5	3.5	0.6	—	9 020	

7.3.1 Liquid fossil fuels

Fuel oil, owing to its relative ease of handling and storage, the absence of large quantities of residual ash and its high calorific values, had become by 1970 a most attractive fossil fuel. Since then, however, its relative cost with regard to other fuels and the uncertainty of its long-term availability have both tended to reduce its competitiveness. Nonetheless, particularly in the case of smaller boiler installations, oil remains a viable fuel.

The type of oil burned for steam generation can vary from light distillates having low viscosity and low specific gravity to the heaviest residual black fuel oils with very high viscosities. Occasionally crude (or unrefined) oil may be considered for use directly in the boiler furnace; however, the presence of highly volatile fractions (normally removed in a refining process) may prove troublesome. Typical characteristics of fuel oils are shown in *Table 7.2*. Despite the low percentages of ash indicated in *Table 7.2*, trace elements such as vanadium, sodium and sulphur can be responsible for a number of operating problems.

A further source of liquid fuel is oil derived from oil-shale, which is a fine-grained, compact, sedimentary rock containing an organic material known as kerogen. Shale oil is obtained by heating the rock to 470°C. (The yield is normally about 115 litres per tonne of rock.) Substantial reserves are known to exist in North America, in some parts of the Middle East, in China and in Australia. The use of shale oil in steam generators may be restricted by cost, but consideration is currently being given to the direct combustion of oil-shale in boilers.

7.3.2 Gaseous fuels

In many ways gas can be considered the most easily used of all the chemical fuels. It is capable of easy transportation from the producing or gathering plant to the consumer, and the use of pipelines can eliminate storage problems, particularly for very small installations. Care is required, however, in large boilers to avoid the occurrence of explosions during unit ignition or at other times when ignition may be lost. As may be expected, for clean gases complete combustion with a low excess air requirement is possible and the combustion does not produce smoke; additionally, it is substantially free of ash. Typical analyses of gases burned in steam generators are shown in *Table 7.3*. Blast-furnace gas is included in *Table 7.3* for completeness, although it is usually available only as a by-product of a steelworks. It is normally heavily contaminated with dust, and great care must be taken to avoid plugging of fuel pipes or furnace fouling. Clean-up is usually achieved by a washing process.

7.3.3 Solid fossil fuels

Coal is a major fossil fuel throughout the world and has been in use for many decades by the industrialised nations. It is found in many forms, ranging from anthracitic through bituminous and subbituminous to lignitic. With so many sources and with such variations of type and quality, it has been found impossible to produce any generalised analysis for this fuel which will predict its behaviour in the boiler. As opposed to oil and gas, coal contains relatively high ash levels, which in turn will contain trace compounds (such as SiO_2, AlO_2, TiO_2, Fe_3O_4, CaO, MgO, Na_2O, K_2O, Mn_3O_4, P_2O_5) whose presence and effects need to be recognised if efficient and reliable coal combustion is to be achieved.

With these reservations in mind, however, it is possible to divide coals into certain classifications. In Europe the International System of Classification is generally used, while in the USA the ASTM System is used. Both systems have attempted to group coals together by their heating values, moisture contents and volatile matter contents.

Table 7.3 Examples of some gaseous fuels available

Gaseous fuel	Composition								Higher heating value (kcal/kg)
	CO	H_2	CH_4	CnHm	CO_2	N_2	O_2		
Coal gas	4.7	2.5	38.5	35.9	10.3	7.3	0.8		8 640
Water gas	71	6.4	0.3	—	14	8.3	—		3 560
Producer gas	31.3	0.9	0.2	—	8.5	59.1	—		985
Natural gas	—	—	63–88	6–29	0.5–8	0.5–7.3	—		9 000–11 050
Blast furnace gas	30	0.2	1.3	—	14	55	—		5 900

7.1 Introduction

The production of steam at conditions suitable for supplying an engine or a turbine, or of providing steam for heating or process plant, is achieved by the transfer of heat from a primary energy source to water contained in a boiler, or steam generator, which in its most rudimentary form may be little more than a vessel heated from below. Currently, there are two major energy sources capable of supplying heat at a sufficiently high temperature and in a controlled and economically acceptable fashion. These are nuclear fission and the combustion of fossil fuels, such as coal, oil, natural gas and their derivatives. It is the latter, or 'conventional' energy source and its associated plant which will be considered here.

7.2 Combustion

The combustion process for a fossil fuel is a rapid exothermic chemical reaction between the oxygen in the air and the combustible elements in the fuel. There are two such principal elements, carbon and hydrogen, which react with oxygen, thus:

$C + O_2 = CO_2 + $ heat

$2H_2 + O_2 = 2H_2O + $ heat

The heat release for carbon is approximately 7830 kcal/kg of carbon burned, and the equivalent figure for hydrogen is 33 940 kcal/kg. The requirements for the above reactions to proceed to completion are first, a sufficiently high temperature to ignite the constituents; second, adequate mixing of the constituents; and third, sufficient time for the reaction to be completed. With regard to the ignition temperature, *Table 7.1* gives an indication of these values for a number of fuels.

Table 7.1 Ignition temperatures for some common fuels in atmospheric air

Fuel	Ignition temperature (°C)
Charcoal	343
Bituminous coal	407
Anthracite	450–600
Ethane (C_2H_6)	470–630
Ethylene (C_2H_4)	480–550
Hydrogen (H_2)	575–590
Methane (CH_4)	630–750
Carbon monoxide (CO)	610–657

The mixing of the constituents and the reaction times are both major elements in the design of the furnace of a steam generator, which must be large enough and provide sufficient turbulence to ensure that these two requirements are met.

The absorption of heat in the boiler is achieved in two ways. In the furnace, where all the combustion takes place, heat is transferred to the water by *conduction* through the walls of its containing vessel or tubes, which receive heat by radiation from the burning fuel. At the outlet of the furnace the products of combustion are still very hot (probably in excess of 1000°C), so that they remain capable of transferring heat; this time by *convection*. This heat can be used in non-superheating boilers to further heat the water, while in superheating boilers, as the gas passes through the convection zone, it may progressively be used to superheat the steam first produced in the furnace, and eventually the cold incoming feedwater. Finally, the gaseous products of combustion are ducted to a chimney (sometimes via flue gas clean-up devices) and so any residual heat left in these gases is lost. Nonetheless, the efficiency of steam generating plant, as defined by the ratio between the heat energy of the steam leaving the boiler and the chemical energy of the fuel entering the furnace, can be high, with typical figures in excess of 85% for coal fired plant, rising to as high as 98% for compact oil fired boilers.

With regard to efficiency definitions, care should always be taken to establish the basis of the efficiency calculation, as the heat content of the fuel is generally assessed differently in Europe and the United States. In the USA the higher heat of combustion (Q_H) is used which incorporates the heat of vaporisation of water (that is, it is assumed that all water vapour formed by combustion is condensed). The lower heat of combustion (or lower calorific value, Q_L) is derived by assuming that the products of combustion remain in the gaseous state. It may be shown that

$Q_L = Q_H - 578.24W$ kcal/kg

where W is the weight of water formed per kilogram of fuel. European practice tends to favour the use of Q_L.

Calorific values, are obtained experimentally by means of a bomb calorimeter in which combustion occurs at constant volume and the derived value is Q_H. Q_L is obtained from the above relationship and, in fact, refers to combustion occurring at constant pressure.

The major heat loss from a boiler has already been identified as the heat loss up the stack, or chimney. Other losses may arise due to incomplete combustion of the fuel, leaving carbon in the ash or by producing CO rather than CO_2. In practical installations it is always necessary to use more than the theoretical air requirements to be sure of complete combustion. However, in order to keep the stack loss to a minimum this excess must be carefully controlled. For fuel oil fired installations, excess air is typically in the 5–10% by weight range, referred to the theoretical air requirements. Corresponding figures for a pulverised coal fired unit would be 15–20%, while for a stoker fired boiler the figures would be in the range 30–60%. The remaining inherent loss for a boiler arises due to the moisture content of the fuel and the production of water vapour in the combustion process, while the remaining avoidable loss arises from radiation from the boiler setting. This latter loss can, of course, be controlled by providing the unit with good insulation.

7.3 Sources of chemical energy

The fossil fuels available as suitable heat sources occur naturally in the earth and are the remains of organic materials once found living on the Earth's surface. It is not surprising, therefore, that the properties of individual fuels vary markedly from place to place and that the fuels contain many trace elements which can in some cases profoundly influence their combustion characteristics. It is necessary, therefore, that the boiler plant designer knows the type of fuel that will be burned and also fully understand the effects the properties of the fuel will have upon all his design parameters. In consequence, much effort has been directed towards understanding the combustion of conventional fuels which are commonly used, and considerable data are available on the fossil fuels (oils, natural gases and coals).

As well as fossil fuels, less important but common vegetable fuels such as peat, wood, wood bark, bagasse (sugar cane residue), grain hulls and residues from coffee grounds and tobacco stems are useful sources of heat. Sometimes gases, tars and chars produced as by-products in steel making or oil refining processes are used. In many instances when waste products are to be used they will be burned in conjunction with the major fuels (oil, coal and gas).

7 Steam Generating Plant

I G Crow, BEng, PhD, CEng, MIMechE, FIMarE, MemASME
NEI Projects Ltd

Contents

7.1 Introduction 7/3

7.2 Combustion 7/3

7.3 Sources of chemical energy 7/3
    7.3.1 Liquid fossil fuels 7/4
    7.3.2 Gaseous fuels 7/4
    7.3.3 Solid fossil fuels 7/4
    7.3.4 Other solid fuels 7/5

7.4 Thermodynamics and hydrodynamics of steam generating plant 7/5

7.5 Types of steam generator 7/6
    7.5.1 Design options and objectives 7/8

7.6 Fuel handling and storage 7/11

7.7 Pollution control 7/13

Most strip-chart recorders are panel mounted, and some accept plug-in amplifiers and attenuators to extend the range of application. Thus, a linear output can be obtained from a thermocouple (which has a non-linear e.m.f./temperature characteristic for a wide temperature range) by use of a plug-in amplifier to provide both impedance matching and non-linearity compensation. As they condition the signal by impedance matching and gain control, these amplifiers are sometimes referred to as 'conditioning amplifiers'.

Many portable solid state battery-driven electronic recorders are available for desk-top or vertical mounting. They are normally restricted to one- or two-channel inputs and have smaller widths of chart. Miniature recorders (mains-, battery- or clockwork-driven) have been considerably developed for use in large schematic process-control display panels.

References

1 Glossary of Terms Used in Metrology, *BS 5233.1975*. Terms Common to Power, Telecommunications and Electronics, *BS4727*
2 Direct Acting Electrical Indicating Instruments, *BS89: Part 1: 1970*
3 HEALY, J. T., *Automatic Testing and Evaluation of Digital Integrated Circuits*, Reston, Prentice-Hall (1981)
4 LECLERC, G., 'Etalons représentatifs de l'ohm et du volt', *Metrologia*. **14**, 171–174 (1978)
5 GALAKHOVA, O. P. et al., *IEEE Trans. on Instrumentation and Measurement*, **IM-29**, 171–174 (1978)
6 (a) JOSEPHSON, B. D., *Phys. Letters*, **1**, 251 (1962)
   (b) LANGENBERG, D. N. et al., 'The Josephson effects', *Scientific American*, May, 30–39 (1966)
7 THOMPSON, A. M. and LAMPARD, D. G., *Nature*, **177**, 888 (1956)
8 LAMPARD, D. G., *Proc. IEE, Monograph 216M* (1957)
9 IEE Wiring Regulations, 15th Edition (1981)
10 JONES, L. T. et al., *Hewlett-Packard Journal*, **32**, No. 4, 23 (1981)
11 STOCKTON, J. R., *IEE Electronics Letters*, **13**, No. 14, pp. 406–407 (1977)
12 KNIGHT, R. B. D. and STOCKTON, J. R., *NPL Reports DES60* (1981), also STOCKTON, J. R. and CLARKE, F. J. J., *DES71* (August 1981)
13 IKEDA, Y., *NPL DES. Memorandum No. 21* (1976)
14 TURGEL, R. S., *NBS Tech. Note 870* (June 1975)
15 TURGEL, R. S., *IEE Trans. on Instrumentation and Measurement*, **IM-23**, 337–341 (1974)
16 EMMENS, T., *IEE Electronics and Power*, p. 166 (February 1981)
17 Yokogawa Electric Limited, Model 2885
18 Yokogawa Electric Limited, Model 2552

transient, requiring a linear or logarithmic time-base X amplifier, with the Y channel drawn by low-frequency signals derived from a digital transient storage recorder of the high-frequency one-shot transient phenomenon.

Two-pen $X-Y$ recorders are available to relate three parameters by means of a pair of separate Y input channels which print in sequence against a common X input signal.

The writing process may be a fibre pen with ink cartridges on paper, or an inkless pressure-sensitive method using specially treated paper. For small portable roll-paper instruments the usual paper width is 10 cm, while desk-top instruments have a 30 cm width. Large-display fixed-sheet instruments are available with paper sizes up to 75 cm square.

The response is relatively slow. A full-scale traverse of the Y axis takes about 0.2 s, and corresponding X-axis chart speeds lie between 10 and 0.01 mm/s. The inherent sensitivity varies between 0.1 and 5 mV/mm and the best uncertainty is about $\pm 0.2\%$ of full-scale deflection.

6.14.3 Analogue and digital strip-chart recorders

The principles of measurement and display used in most analogue strip-chart recorders are based either on an unbalanced bridge and galvanometer or on a null-seeking potentiometric servo system. An alternative technique employs analogue/digital conversion of the measurement with the appropriate selection of one of many fixed printing styli (thermal, electrostatic or pressure-sensitive) which are equispaced across the Y axis of the chart. The time base in all cases is developed by driving the paper mechanically past the recording heads at some selected constant speed.

6.14.3.1 Galvanometer recorders

A high-torque galvanometer is driven directly by the signal source or indirectly by the out-of-balance voltage from an active or passive bridge, one arm of which is the signal transducer. The pen can be attached to a long radial arm, normally fixed to the axis of the galvanometer. Such displays, whether mechanical or optical, exhibit some non-linearity towards maximum deflection; the error is not usually important, and could in principle be compensated by an appropriate scale on the graph paper.

6.14.3.2 Potentiometric recorders

Potentiometric recorders are always associated with servo-motor pen drive, running along a linear guide track in the Y direction. The transducer voltage to be recorded is compared with a voltage proportional to the pen position, and the error voltage applied to a high-gain amplifier and servo-motor combination. The pen is driven to a position of zero error. The response is fast enough to produce a continuous trace of a single slowly changing phenomenon.

With several signals fed into separate matched amplifiers and presented in sequence to a single pen, up to 16 separate traces can be drawn. The individual zeros can be offset as required. It is important that the variable gain and linearity of each amplifier should be known from recent recalibration. As recording must be inoperative until the null balance is reached, direct-contact writing methods require the pen to be raised during the nulling process. All such multiple traces therefore appear as discrete (but closely spaced) dots.

6.14.3.3 Digital recorders

Digital recorders accept conventional analogue input signals, which are then converted by normal analogue/digital logic networks into decimal readout signals. The stationary recorder arm, which spans the paper in the Y direction, comprises many equispaced printing styli. One stylus is selected at a deflection proportional to the particular value of the readout signal, and actuated to print a dot—e.g. by a thermal or electrostatic process. Except for the time base generated by the moving paper, the measurement and writing system has no moving parts.

Thermal-writing methods employ about 500 equispaced thermal styli directed towards heat-sensitive white paper to produce a single dark spot by a heated stylus. Electrostatic methods include some 1000 wire-element styli and when a voltage pulse is applied to a stylus it transfers a pinpoint charge to a special (low-cost) paper; the 'spot' is preserved by electrostatic attraction of positively charged ink particles and, after cleaning, the spot becomes bonded to the paper immediately after exposure to air. Thermal and electrostatic 'spots' are created sequentially often at 200 and 1000 per second, respectively; successive spots are jointed by straight lines, particularly when the input signal changes between spots. The maximum frequency is about 15 kHz but with a very segmented representation of wave forms, square-wave response appearing to be superior to sine waves. The digital nature of these devices provides all the corresponding convenience of application, processing and interfacing with digital systems; in general they appear to be very useful for low-frequency signals. The galvanometer-type ultraviolet recorder could provide a more faithful reproduction of complex wave forms up to a higher frequency for an appropriately selected galvanometer.

6.14.3.4 Instrument selection

The main distinction between graphical display instruments is between those for continuous monitoring and those for occasional use for particular experiments. The recorders described above can be satisfactory for either purpose, but for special applications their properties should be compared with $X-Y$ time-domain instruments and ultraviolet galvanometer recorders.

The null principle is superior to the unbalanced-bridge method, because at balance its high input impedance virtually leaves the signal source unloaded. However, for many routine monitoring tasks the bridge-type instrument rather than the more expensive null-seeking servo instrument is adequate.

Characteristic features of strip-chart recorders are listed below:

Type		*Potentiometer*	*Galvanometer*
Response time for full-scale deflection	ms	500	2
Linearity based on full-scale deflection	%	0.1	1
Range of time base speeds:			
upper	cm/s	40	10
lower	cm/h	3	3
Maximum number of channels		16	8
Maximum chart width	cm	30	30
Maximum operating frequency	Hz	5	150

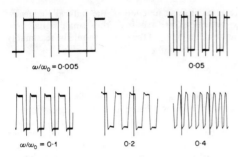

Figure 6.71 Galvanometer response to square wave

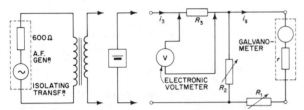

Figure 6.72 Networks for galvanometer parameters

audiofrequency generator through an isolating transformer with a low-impedance secondary. The results from a typical low-frequency magnetically damped instrument with a coil resistance $r = 40\,\Omega$, tested at 40 Hz and giving a 10 cm peak-to-peak deflection, were: $i_3 = 1.6$ mA, $R_1 = 350\,\Omega$, $R_2 = 10\,\Omega$, $R_3 = 600\,\Omega$; whence $i_g = 40\,\mu$A and the a.c. sensitivity is $8\sqrt{2}\,\mu$A/cm.

Tests over a range of frequencies identify the limits of flat frequency response, using input signals that produce a given maximum display on the recording paper.

6.14.1.8 Coil resistance

The coil resistance r can be measured by Wheatstone bridge provided that the bridge current is less than the peak of the a.c. continuous rating. The bridge method *may* be satisfactory for a.c. galvanometers but might destroy a d.c. one.

An elementary indirect measurement of r is to replace the a.c. source transformer in *Figure 6.72* by a 2 V battery. Suppose that R_1 is a variable 10 kΩ resistor, $R_2 = 1\,\Omega$ and $R_3 = 10$ kΩ; then i_3 is effectively constant at about 200 μA, and for a steady deflection θ

$$i_g = 200 R_2 / [(1+r) + R_1] = \theta(K/G)$$

A plot of $1/\theta$ to a base of R_1 yields a linear graph having an extrapolated value of $-(1+r)$ at $1/\theta = 0$. Alternatively $1+r$ can be computed from widely different values of R_1. It is not necessary to know r to an uncertainty better than $\pm 1\%$.

6.14.1.9 Damping factor

The d.c. form of *Figure 6.72* is used. Let $R_1 = 0$ and R_2 be adjusted to give maximum deflection (taking care of the coil current limits). The source is quickly disconnected and the deflection of the first overshoot is noted. If R_2 is readjusted until the overshoot is 25%, the damping factor is $\delta = 0.4$, for a known total damping resistance $R_d = R_2 + r$. For any *one* galvanometer the product δR_d is constant, so that R_d can now be calculated for any required δ such as 0.64 as quoted by the manufacturer. Any significant changes in δ will call for readjustment of the damping networks.

6.14.1.10 Instrument selection

In choosing an instrument it is necessary to assess (1) the largest uncertainty (distortion) that can be tolerated in the output display with due regard to the use intended; (2) the type, quality, linearity of the transducer–bridge–amplifier system to be used to drive the galvanometer; (3) the dynamic impedance matching of the galvanometer and signal source. Selection will often involve a compromise between the conflicting requirements of high sensitivity and wide flat-frequency range.

Ultraviolet recorders are restricted to audiofrequency signals, and the galvanometer response need be no better than the quality of the transducer–amplifier driving it. The high-frequency attenuation can be an advantage in filtering high-frequency noise; thus, the top frequency requirement could be reduced to obtain higher sensitivity, an advantage in very-low-frequency applications such as those involving thermocouple or strain gauge bridges or accelerometers.

Phase distortion is important for non-sinusoidal and square wave forms. For applications involving d.c. transients the selection is based on the acceptable degree of overshoot, the rise time and the settling time to a steady finite or zero reading. The best compromise response may be obtained from a damping ratio of about 0.6, which would also suit mixed a.c. and d.c. transient work.

6.14.1.11 Applications

Ultraviolet recorders are widely applicable to multiple d.c. and audiofrequency phenomena requiring the time correlation of numerous signals coupled with an immediate capability for investigating the fine detail of each. Recorder facilities and special galvanometers available include the following:

Integrating galvanometers, often used with velocity transducers to give a spot deflection proportional to displacement: they are very heavily damped (e.g. $\delta = 20$).

Filter galvanometers, heavily damped to attenuate signals at above twice natural undamped frequency.

Special galvanometers for computing the sum, difference or product of signals—e.g. instantaneous audiofrequency power, or average active and apparent power; or for event marking and sequential momentary blanking of the trace, the optical recording of grid and base lines and time-scale oscillations.

6.14.2 X–Y recorders

The continuous graphs to cartesian co-ordinates drawn by an X–Y recorder can represent interrelated d.c. functions $y = f(x)$ both of which change slowly with time, or time-dependent functions $y = f(t)$ when a time-base signal is applied to the X input. The recorder uses two separate servo-control systems, each based on a null potentiometric measurement:

Y-deflection is produced by the movement of a pen along a guided linear track.

X-deflection is generated either (1) by transporting the complete Y assembly across fixed chart paper, or (2) by impressing the same X displacement on a movable-roll chart which is driven fore and aft across the path of the fixed Y carriage.

Input signals to the servo systems can be obtained from a range of amplifiers with linear or non-linear gain characteristics, capable of transforming their inputs into some output function more suitable as a final display. At the same time the output signal levels and impedances are matched to the servo-control inputs. As an example, separate X and Y linear and logarithmic frequency/voltage conversion amplifiers provide for the display of gain/log-frequency system response characteristics. Another application is the low-frequency display of a high-frequency

The predominant damping can be electromagnetic and/or viscous-frictional. Both convert kinetic energy into heat.

Electrodynamic The motional e.m.f. in the coil drives current through a matched resistance network. The external resistance normally quoted by makers is that value required for a damping factor of 0.64; it is made up by the internal resistance of the current source and a suitable series-parallel combination of resistors for the remainder.

Fluid Oil-damped movements are used for higher frequencies. A small electromagnetic damping is also present. The manufacturer's quoted external resistance is made up as in the foregoing.

Special applications may require galvanometer characteristics significantly different from the 'optimum' discussed, and may combine electromagnetic with up to 35% viscous damping. Provided that the external resistance R for the $\delta = 0.64$ damping condition is known, then the external resistance R_1 for some other overall damping factor δ_1 can be found from

$$(\delta_1 - \delta_v)/(0.64 - \delta_v) = (r + R)/(r + R_1)$$

where r is the resistance of the galvanometer coil and δ_v is the viscous damping.

Flat frequency response This term refers to the bandwidth throughout which all signals are faithfully transmitted within stated limits. A typical instrument with $\delta = 0.64$ may have a ± 0.02 per-unit flat frequency response up to $0.67 f_0$; see *Figure 6.70*. With a ± 0.08 per-unit flat response the upper limit would be $0.83 f_0$. Hence, for $f_0 = 20$ kHz a given galvanometer may be variously quoted as having a flat frequency response up to a frequency of 13 or 17 kHz, depending on the specified limits.

6.14.1.3 Phase distortion

Phase distortion β_n can have two adverse effects: (1) the time correlation between simultaneous phenomena displayed by two different galvanometers may be lost; (2) complex waveforms will not be faithfully reproduced. The curves in *Figure 6.70(b)* show the relation between phase change and frequency, for various values of the damping factor δ. The relative phase change between the fundamental and the harmonics is eliminated if the phase change is a linear function of frequency: such a condition is most nearly achieved with $\delta = 0.75$, but there is a fair linearity still with $\delta = 0.64$, the value usually adopted for optimum amplitude response.

6.14.1.4 Response to step function

The response is governed by the damping factor δ. With underdamping ($\delta < 1$) the response is fast but contains an overshoot and a decaying oscillation. With overdamping ($\delta > 1$) the time to reach a final deflection is excessive. The compromise damping factor (0.64) used for a.c. signals provides a reasonable value for d.c. transients, although a slightly lower value may be adopted in particular cases.

Figure 6.71 shows typical responses to a square waveform, which combines both periodic and transient characteristics.

6.14.1.5 Sensitivity

Two sensitivity ratios are normally quoted: (1) *current* sensitivity, the ratio between coil current and linear deflection of the spot; (2) *voltage* sensitivity, the ratio between the deflection and the total voltage required to produce this deflection on an optimally damped system. As the reciprocal of these may also be quoted by the makers, and as each sensitivity is in any case a function of the optical magnification length, it is important to assess comparative sensitivities with care.

Sensitivity is an inverse function of the natural frequency. When large deflections are required from small signals, the undamped natural frequency of the system should be reasonably low.

6.14.1.6 Errors

The amplitude, phase change and transient limitations of the galvanometer are due to the inherent electromechanical/optical properties of any ideal instrument. Superimposed upon these limitations will be a group of errors that may be different for instruments having the same nominal specification and, collectively, will contribute to the makers' declared uncertainty in the product.

These imperfections include mechanical asymmetry and non-linear higher modes of oscillation, temperature–viscosity changes, non-uniform magnetic flux density surrounding the rotating coil and changes in this density with time; all these variations cause non-uniform damping effects, of which the flux density change probably constitutes the main component. Additional non-uniformity may be caused by optical imperfections and incorrect location of the light beam perpendicularly to the paper, together with slight oscillations around the average take-off speed of rotation. Hysteresis (drift of the zero after a deflection) can be present, and electrical noise may enter the system despite high insulation levels.

Modern galvanometer recorders are high-quality instruments and their cumulative errors are small. It is nevertheless prudent to recalibrate instruments occasionally to ensure that matching and damping factors are unchanged and that the correct sensitivity factors are being used to interpret the results.

Some details of the ranges and tolerances of ultraviolet recorders are given below:

Type of damping		Fluid	Magnetic
Flat frequency response (± 0.08 per unit)	Hz	1000–20 000	6–500
Undamped d.c. sensitivity*	µA/cm	200–20 000	1–150
Voltage sensitivity*	mV/cm	15–2000	0.03–15
Damping resistance	Ω	20–60	25–400
Terminal resistance	Ω	2–130	40–180
Maximum safe current (5 s overload)	mA	20–200	5–50

Percentage tolerances:	
Speed stability (time base linearity) after 0.3 s	3
Mean magnetic flux density	10
Flux density variation along magnetic block	10
Voltage sensitivity for a given flux density	5
Coil resistance for damping factor 0.64	10
Timing lines with crystal time control	0.02
Hysteresis in terms of full-scale deflection	1
Natural undamped frequency	10
Total non-linearity based on full-scale deflection	1

* For an optical arm 0.3 m long.

6.14.1.7 Calibration

The sensitivity factors and response characteristics can readily be checked with conventional laboratory equipment.

R_1 in *Figure 6.72* is selected to provide 0.64 damping factor for the galvanometer of coil resistance r. Most of the total current i_3 is shunted by R_2, leaving a small current i_g to give a maximum peak-to-peak readout, say 10 cm. The supply is taken from an

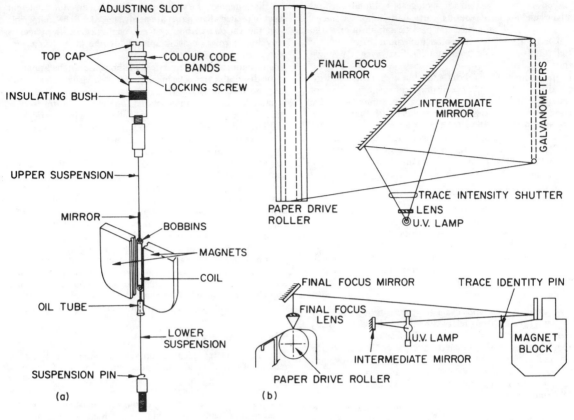

Figure 6.69 Ultraviolet recorder (Southern Measuring Instruments Ltd)

condition is met, then very nearly the ratio between the deflection for the nth harmonic and that for the fundamental (for the same current amplitude in each case, i.e. per unit) is

$$\frac{\theta_n}{\theta_1} = \frac{1}{\sqrt{[\{1-(n\alpha)^2\}^2 + (2\delta n\alpha)^2]}}$$

where $\alpha = \omega_1/\omega_0$. The deflection θ_n lags on the current i_n by an angle β_n on its own angular scale, or β_n/n on the scale of the fundamental, given by

$$\tan \beta_n = 2\delta/[(n\alpha) - (1/n\alpha)]$$

6.14.1.2 Amplitude distortion

The ratio θ_n/θ_1 would be constant and independent of n if the second (damping) term in the denominator compensated at all harmonic frequencies for the reduction in the first term—i.e. for a damping coefficient such that

$$\delta = \tfrac{1}{2}\sqrt{[2-(n\alpha)^2]}$$

For example, suppose that the natural frequency of the galvanometer moving system is $f_0 = \omega_0/2\pi = 1000$ Hz and that the fundamental current has the frequency $f_1 = \omega_1/2\pi = 0.5$ Hz; then $\alpha = 0.5/1000$ and the damping factors necessary to obtain uniform response are:

Harmonic frequency (Hz)	0.5	50	200	400	600
order n	1	100	400	800	1200
Damping factor δ	0.71	0.71	0.70	0.68	0.64

For most instruments it is the practice to adopt $\delta = 0.64$ as a compromise giving the best 'flat' sensitivity/frequency response. The curves in *Figure 6.70(a)* illustrate the importance of proper damping conditions.

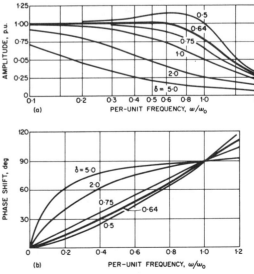

Figure 6.70 Galvanometer response

Photodiodes can be selected for a range of optimum spectral sensitivites and are fast-operating, particularly so for the silicon planar p–i–n type. Phototransistors provide amplification at the cost of a relatively slow response (minimum 20 μs). Photomultiplier tubes provide high amplification, even for single photons.

Integrated-circuit chips about 4 mm² in area, produced with progressively more complex networks, can include scores of photodiode arrays, with amplifiers, logic counting networks, scanning circuits, etc., to give such outputs as digital pulse rates proportional to illumination, logic for character recognition and punched-card reading, and as replacements for photomultipliers. These devices find a wide range of application in transducer instrumentation for control and communication.

6.13.12 Nuclear radiation sensors

Radiation sensors are required in particle physics research, for monitoring reactions in nuclear power stations, as gauges for industrial processes, and for measurements employing radioactive isotopes. The emanations used for industrial processes include α, β and γ rays, with X-rays for special applications such as the examination of welds; the sensor characteristics largely determine the radiation to be processed. The devices include the following.

Low-pressure gas ionisation chambers to measure α-particles by accumulating the electrons released by their collisions with gas molecules, and without subsequent multiplication.

Proportional counters, similar to the foregoing but with a higher accelerating voltage (500–800 V) to give cumulative gas amplification (Townsend avalanche) and an electron gain of 10^5–10^6. The use is for counting α particles in the presence of β particles and γ rays.

Geiger counters operating at 800–1500 V, or at low voltage, for β and γ ray detection. The complete ionisation must be quenched between counts, and the dead time restricts the counting rate to about 1000/s.

Scintillation sensors (or 'counters') for α, β or γ detection. There are also semiconductor sensors for the same application and for X-rays.

Alpha, beta and gamma ray sensors require to be followed by electronic pulse amplifiers and counters. A counter may accumulate a total/fractional scaled count, or it may give a rate-meter display. These outputs, together with pulse height analysers, are used to assess the energy spectra of the received signals. Coincidence counters have multiple input sensors so arranged that only the required type of emanation is measured during any selected period.

The detectors can form the bases for industrial gauges and monitors when used with radioisotope sources. The selective absorption of materials enables β radiations through paper to indicate the uniformity of the product; liquid-level gauges give correcting signals when the rise of level absorbs the radiation; radioisotope tracers with short half-lives can be introduced into liquid and gas channels to indicate flow rate, uniformity of mixing and the detection of leakages.

6.14 Data recording

Chart recorders responsive to signals at frequencies up to a maximum of 25 kHz include (1) a.c. ultraviolet, (2) X–Y and (3) analogue and digital strip-chart forms. All provide records of related phenomena that need to be preserved in graphical form.

The visual discrimination of the best-quality trace, when stated as a fraction of the chart width, should not be significantly different from the accuracy and linearity of the display; this condition dictates the width of the chart paper. Many different types of treated chart paper are needed in order to be compatible with the various writing systems mentioned below. Paper is provided in rolls or overlapping folds, the latter being very convenient for quick retrieval of data.

Conventional pens for ink-feed methods use felt, ball-point or nylon disposable tips. The ink cartridges are easily replaced, and the supply is drawn to the tip by capillary action. The nylon pen is suitable for higher writing speeds; to retain a uniform trace, a variable pressure is automatically applied to the ink cartridge. In multiple-display applications, the possibility of easy trace separation by colour is a unique convenience of ink writing.

Many alternative writing methods have been developed in an attempt to overcome the inertia (response time) limitation of conventional ink writing, as well as to improve the quality of the trace. These include: (1) a heated stylus and sensitive plastic-coated paper; (2) mechanical pressure on chemically treated paper; (3) ultraviolet light beams directed on to photographic paper, the trace appearing and becoming fixed within a few seconds of exposure to natural light; (4) the passage of current through treated metallised paper causing the chemical reduction process to develop a black imprint; (5) electrostatic copying, in which a flat capacitor is formed from conductive paper with a dielectric coating—a very small area can be charged, and particles are attracted to the charged area to provide the visible trace, which may require treatment to make it permanent.

6.14.1 Ultraviolet recorder

The ultraviolet recorder can provide permanent time domain records of periodic and transient currents in up to 50 galvanometers. A typical galvanometer is shown in *Figure 6.69(a)* and the principle of a complete instrument in *Figure 6.69(b)*. Each galvanometer deflects (ideally) through an angle corresponding to the instantaneous current in its coil. A high-pressure mercury vapour 100 W light source is used to illuminate the galvanometer mirrors, the reflected beams being focused on to paper sensitive to ultraviolet light. A time base is introduced by rotating the paper at uniform speed in the range 1 mm/s to 10 m/s.

6.14.1.1 Response to periodic waveform

The response of a galvanometer depends on its physical dimensions and inherent electrical properties. The electromechanical behaviour must be compatible with both the characteristics of the current to be recorded and the signal source impedance.

The moving-coil system has, for an instantaneous angular deflection θ, a restoring torque $K\theta$ due to the suspension; a damping torque $F(d\theta/dt)$ due to electromagnetic effects in the coil or to its immersion in a viscous oil, and taken as approximately proportional to angular speed; and an inertial torque $J(d^2\theta/dt^2)$, where J is the polar moment of inertia of the moving system. The system is in consequence inherently oscillatory, with a natural undamped angular frequency $\omega_0 = \sqrt{(K/J)}$ and a damping factor $\delta = F/2\sqrt{(KJ)}$.

An instantaneous coil current i produces a proportional deflecting torque Gi, so that the equation of motion is given by

$$Gi = J(d^2\theta/dt^2) + F(d\theta/dt) + K\theta$$

A time-varying coil current i of complex waveform and of fundamental angular frequency ω_1 can be represented by a Fourier series of harmonics each of the form $i_n \sin(n\omega_1 t + \phi_n)$. The faithful reproduction of the waveform depends upon the galvanometer response being the same for any order n of harmonic in the current. In general, the nearer the harmonic frequency $n\omega_1$ is to the natural frequency ω_0 the greater is the deflection per unit current. The galvanometer should therefore have a natural frequency well above the highest frequency to be recorded. If this

is a submultiple $1/n$, then the object appears stationary in n different positions. It is desirable for the repetition frequency to exceed 30 Hz in order to avoid visual flicker.

From the flash frequency and observation, the character of velocity perturbations can be viewed and counted and dynamic distortion effects seen. Electrical stroboscopes may involve special neon lamps connected to the secondary side of an induction coil, the primary of which is interrupted by a driven tuning fork. In modern instruments the lamp is a xenon gas discharge tube emitting white light. The flashing frequency is adjusted by means of a transistorised multivibrator-shaper circuit producing pulses of constant energy and fast rise time. Each trigger pulse initiates the discharge of a capacitor to create the flash. Capacitor recharging limits the upper frequency to about 1.5 kHz with a 200 lx illumination level. High frequencies up to 10 kHz can be obtained from a Z-modulated c.r. tube as a low-level light source.

6.13.11 Photo sensors

In this context, 'light' usually means radiation at wavelengths covering the visible spectrum from the near infrared to the near ultraviolet. Light transducers can be identified by three basic forms of physical behaviour: photovoltaic, photoresistive/photojunction and photoemissive.

6.13.11.1 Photovoltaic

Photovoltaic sensors are p–n junction devices with the normal barrier layer potential present between the two materials. When illuminated, the higher-energy protons raise electrons from the valence into the conduction band to create electron–hole pairs. The consequent continual charge separation is enough to drive current through a resistive load without any external source.

Selenium cell. This has a linear current/illumination characteristic for load resistances up to $100\,\Omega$, but the relation is progressively more non-linear for higher resistance, and reaches an open-circuit e.m.f. of 0.6 V. The peak spectral response is in the centre of the visible band ($0.57\,\mu m$), but the energy efficiency is only 0.5%.

Silicon p–n junction cell. The energy conversion is 10–15%, which is adequate for solar cell use in space vehicles. As the time response (a few μs) is very fast, silicon photocell arrays are used for reading punched-card and tape, and for optical tracking. The output of a single sensor is typically 70 μA for 5000 lx, the current being constant up to 200 mV. When they are reverse-biased, silicon photosensors behave as photoresistive devices.

6.13.11.2 Photoresistive

A photoresistive sensor consists of a single homogeneous semiconductor material such as n-type cadmium sulphide or cadmium selenide. The material is doped to permit of a large electron charge amplification by the selective absorption of holes, and a spectral response that can simulate that of the eye. This type of sensor is well known as a photographic exposure meter. The CdS sensor is not quite linear over wide illumination ranges, and its response time (100 ms), is slow compared with that (10 ms) of the CdSe type.

The sensors can be used in series with a source and a relay, the latter operating when illumination reduces the sensor resistance to about 1 kΩ. The power dissipation by the sensor can be minimised if the slow resistance changes are used to alter the state of a conventional Schmitt trigger circuit (*Figure 6.68*). The required illumination level is set by R_1; when the illumination is reduced, the cell resistance rises to switch on T_1, which causes T_2 to switch off and T_3 then saturates. The relay operates when the

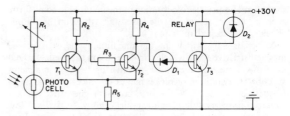

Figure 6.68 Photo-resistive cell in Schmitt trigger circuit

Zener diode conducts at about 12 V. The sensor is useful for alarm circuits, street-lighting control and low-rate counting.

6.13.11.3 Photojunction

Normal diodes and transistors are light-shielded. In the photojunction devices controlled light is admitted to enhance the light effect.

Photodiodes These are normally of silicon unless the greater infrared response of germanium is needed. Time constants of 10–100 ns are normal. Silicon planar p-type/intrinsic/n-type photodiodes can respond to laser pulses of 1 ns, and are usable at very low light level equivalent to a current of about 100 pA. Special low-noise transistor amplifiers are required.

Phototransistors The normal form is the n–p–n silicon planar transistor, with conventional bias except for the reverse-biased photodiode between base and collector. It behaves as a common-collector transistor amplifier with a photocurrent generator between base and collector. The maximum cut-off frequency is 50 kHz. A derivative of the field effect transistor, the 'photofet', is more sensitive and the gain–bandwidth product is higher.

6.13.11.4 Photoemissive

Photoemissive sensors are vacuum or gas-filled phototubes. The emission of electrons from the cathode occurs when light falls on it, to be collected by a positive anode. Vacuum phototubes have been superseded for light measurement by photodiodes. In the *photomultiplier* tube the vacuum emission is enhanced by secondary emission at a succession of anodes to give an electron multiplication of 10^8 or 10^9, providing significant output for flashes of very short duration. The photomultiplier is used as a *scintillation* sensor by viewing the weak light emissions that occur in selected phosphors exposed to α, β or γ radiation. These scintillation 'counters' have wide applications in spectrophotometry, flying-spot scanning, photon counting and whenever a very weak light signal is to be amplified. An alternative device is the Bendix magnetic multiplier. This consists of a single layer of resistive film with the emissive cathode at one end; by suitable accelerating p.d.s applied along the film, together with shaped magnetic fields, the electrons are multiplied by secondary emission and bounced along the strip to accumulate on an anode to give an overall electron gain of 10^9. With different emissive materials, ultraviolet rays and X-rays can be amplified for subsequent analysis.

Gas filled phototubes are low-frequency devices but have a high signal/noise ratio. They are used for film sound-track sensing, and have been applied to monitor natural gas flames (which emit considerable ultraviolet radiation) to avoid danger from unburnt gas.

6.13.11.5 Selection

Photovoltaic cells do not require an external supply, but they are generally less sensitive than photoresistive types, which do.

the natural frequencies of vibration of the material in some form, such as a bar, resonance increases the vibration amplitude considerably to perhaps 1 part in 10^3.

Materials vary widely in the magnitude and sign of the magnetostrictive effect, which also depends on the magnetic field intensity. Pure iron may have positive or negative magnetostriction: the addition of nickel makes the effect positive at all frequencies. With 30% Ni the longitudinal magnetostriction falls to zero, a property utilised in invar–nickel alloys to give very low coefficients of thermal expansion, because the latter is neutralised by the magnetostrictive contraction. In nickel alloys there is a peak dimensional change when the Ni content is about 45%, and at about 63% the most sensitive magnetostrictive condition is reached.

6.13.8.1 Oscillator

The magnetostriction oscillator is a frequency-selective device in which a magnetised rod is set in vibration by compression stresses transmitted from end to end. A coil is wound on a ferrous alloy bar. The length and volume of the bar follow the changes in current, and induce a voltage in a pick-up coil as input to an untuned amplifier that maintains the bar excitation. The oscillations at magnetostrictive resonance frequency provide a stable source in a frequency range not duplicated by a quartz crystal. Frequencies up to 100 kHz are possible, but the size of the magnetostrictive element is inversely proportional to frequency and limits the upper range. Eddy losses increase as the square of the frequency, so that the element should be constructed of annealed and insulated laminations bonded to form a rigid assembly. Thin tubes have been used, but their power capability is low. Digitally divided high-frequency crystal output is more stable and would be preferred today.

6.13.8.2 Sensors

For the purpose of detecting surface and submarine ships, magnetostriction transducers have been employed using underwater supersonic vibrations. The compression stresses generated by the transmitter are reflected by the submerged object and are picked up by a detector utilising the inverse effect—i.e. change of magnetic properties under mechanical stress.

The presence of flaws, air pockets or impurities can similarly be detected in some opaque substances (e.g. rubber) which should normally be homogeneous. The phase displacement between the incident and reflected waves can be interpreted to indicate the position of the obstruction.

Liquid level can be controlled by a sensor probe just above the required level and connected to an oscillator network. As the level of the liquid reaches the probe, the oscillation ceases because of the greatly increased damping. For the measurement of mechanical stress (*Figure 6.66*) the bridge is balanced when the sensor and compensating elements are equally stressed, the compensating element eliminating thermal errors. If the stress in the sensor element is increased, its permeability is reduced, and the out-of-balance current is indicated on the detector D, which can be calibrated in stress units. If the stress is rapidly varying, D can be replaced by a c.r.o.

6.13.9 Reactance sensors

Many transducers use the properties of inductive and capacitive reactance. The inductance of an iron-cored inductor varies very rapidly with the length of an included air-gap. The basic network in *Figure 6.67* employs a bridge with inductors 1 and 2, one with a fixed gap and the other with a gap variable in accordance with

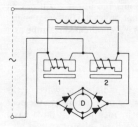

Figure 6.67 Reactance bridge electric gauge

some physical quantity such as pressure, strain, thickness, acceleration, etc. The detector D is calibrated appropriately for direct indication. Reliable readings of displacement down to about 3×10^{-5} mm are readily secured with a properly designed instrument. If the displacement fluctuates, it will modulate the supply waveform, and the modulation waveform can be filtered out from the carrier.

Capacitive reactance can also be used, but as the magnitude of its change with displacement is very small, the capacitor sensor is made to be part of the capacitance in an oscillatory circuit, the frequency of which can be related to the displacement. One type of capacitor pressure sensor consists of two flexible parallel plates with a 'vacuum' dielectric. A cylindrical capacitor suitable for sensing high pressures has as electrodes an inner deforming cylinder and an outer ring section: dimensional changes in the radial gap occur when the axial force on the inner cylinder causes a change in its diameter. The change of capacitance is well within the range of commercial strain-gauge bridges.

6.13.9.1 Angular velocity sensors

Measurements can be read from the output of a permanent-magnet d.c. tachometer for speeds up to 3000 rev/min. The instrument is direct-reading, but a proportional direct output voltage can be used as an electrical transducer. For low speeds the commutator ripple can be excessive and an a.c. tachometer with full-wave rectification will provide a more uniform output, particularly if the instrument is of the multi-tooth variable-reluctance type. These instruments act as loads on the source and they are unable to assess the fine detail of the angular velocity such as superimposed small amplitude oscillations, or non-uniform accelerations and retardations due to changes in load. Simple optical solutions to the problem involve the use of a stroboscope or photosensor coupled with digital counters, and each method avoids any loading of the machine.

6.13.10 Stroboscope

The stroboscope generates high-intensity impulsive flashes of light at controllable repetition frequency. If the repetition frequency corresponds to the time of one revolution of a rotating object or to one complete excursion of a reciprocating object (or any multiple thereof), the object appears to be at rest. If the period

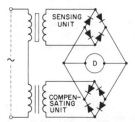

Figure 6.66 Magnetostriction stress sensor circuit

6.13.7.1 Resistance-wire, p–n junction and carbon gauges

A grid of resistance wire is cemented between two insulating films. The grid is usually smaller than 30 mm × 15 mm, and has a resistance in the range 60–2000 Ω. When the gauge is cemented to the surface under test, the change ΔR in total resistance R due to a displacement ΔL in a length L is converted into strain ε by means of the gauge factor

$$k_g = (\Delta R/R)(L/\Delta L) = (\Delta R/R)/\varepsilon$$

If F represents the force and E the elastic modulus (i.e. the stress/strain ratio), then

$$F = k_g E a R/\Delta R$$

where a is the area normal to the direction of the applied force. Resistance-wire strain gauges have been in use for many years. Recent developments include metal-foil strain gauges based upon printed circuit and photoetching techniques: these can be made smaller. Thin-film techniques have been developed to apply the gauges directly to very small surface areas.

Semiconductor *silicon junction* strain gauges are perhaps 30 times more sensitive than the resistive gauge; however, their stress and temperature ranges are more restricted, and they are more difficult to 'cement' in position. The inherent non-linearity of the response by sensors can be offset by special bridge techniques using compensation networks.

Gauges consisting of a layer of *carbon*, which responds to strain in a manner similar to that of a carbon microphone, have a gauge factor typically of 20, considerably higher than for metals but much less than for p–n junctions. The inherent disadvantage is sensitivity to temperature and humidity.

6.13.7.2 Strain-gauge measurement techniques

Strain An 'active' gauge is cemented to the workpiece, and a similar 'compensating' gauge is left free. The gauges form two arms of a bridge (*Figure 6.62(a)*), which is inherently self-compensating for temperature. Balance is achieved initially by

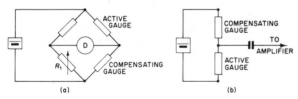

Figure 6.62 Strain-gauge networks

adjusting R_1, the load is applied and the bridge rebalanced. For multiple-gauge arrangements, 100-channel instruments with zero-balancing facilities are available, giving a printed readout. Alternatively, after initial balance, the residual bridge voltage is measured, or applied to a c.r.o. for recording. Dynamic strain may be measured by the circuit of *Figure 6.62(b)*, usable with a flat-response amplifier over the range 0.5–1000 Hz. As in any structural stress analysis it is important to know the time relation as well as the amplitude of stress, the amplifier must have either a negligible phase shift or one directly proportional to frequency over the required working range.

Torque For a circular-section shaft in torsion, the principal stresses lie at 45° to the axis. If four gauges are applied as in *Figure 6.63*, and connected to a bridge through suitable slip-ring gear (usually silver ring surfaces and silver graphite brushes), the bridge unbalance is four times that of a single gauge. The

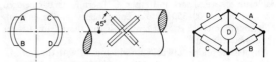

Figure 6.63 Strain-gauge torque measurement

arrangement is inherently temperature-compensated. For engine testing it is normal to make up a special length of shaft with gauges and slip-rings, to be fitted between the engine and the brake or driven unit. With automatic recorders it is possible to record simultaneous data such as torque, pressure variation, temperature, etc., to give all phenomena on a time base display.

Thrust Various other measurements of mechanical power transmission can be measured. The arrangement in *Figure 6.64* is for thrust. Four axial (A) and four circumferential (C) gauges are

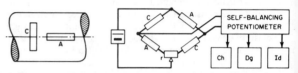

Figure 6.64 Strain-gauge thrust measurement

disposed symmetrically and connected as indicated in a bridge network, with r as balancing adjustment at zero load. The monitoring system (which can be used with other strain-gauge applications) consists of a self-balancing potentiometer with chart (Ch), digital (Dg) and indicator (Id) readouts. The speed of a self-balancing potentiometer is restricted to about one reading per second.

Principal stresses The direction and magnitude of principal stresses can be determined from a *rosette* of three or more strain gauges arranged with suitable orientations (*Figure 6.65*). Com-

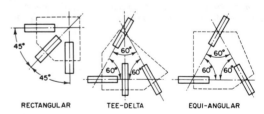

Figure 6.65 Typical strain-gauge rosettes

mercial rosettes are available in several geometrical formations. As it is unlikely that the principal strain will be on the axis of any one gauge, note must be taken of the sensitivity of the gauges in a direction at right angles to the axis if accurate results are required. The correction is not likely to exceed 3%.

6.13.8 Magnetostrictive transducers

Magnetostriction is the dimensional change that occurs in a ferromagnetic material when subjected to magnetisation, and an inverse effect of change in magnetisation when the physical dimensions are altered by the application of external force. The dimensional changes are very small (e.g. 30 parts in 10^6), but if the magnetic field alternates at a frequency corresponding to one of

coefficient of several parts in 10^2 per °C. Thermistors can be used over the range 200–550 K. One consequence of the current/voltage characteristic is self-heating if the thermistor carries appreciable current.

Wheatstone bridge networks are widely used, with the thermistor in one arm and the out-of-balance detector current as an indication of temperature. Although the characteristics are strongly non-linear, it is possible to obtain thermistors in matched pairs, which can be applied to differential temperature measurement. Another application, not directly a temperature measurement, is to the indication of air flow in pipes, or as anemometers: one thermistor is embedded in a metal bloc, acting as a thermal reservoir, the other senses the air speed as a cooling effect.

6.13.3 p–n junctions

Certain semiconductor diodes are suitable for temperature measurement over a wide range; Ge, for example, is useful below 35 K. One differential thermometer has two matched p–n diodes which, when linearly amplified, give a discrimination better than 0.0001 °C. Such instruments have several indirect applications, such as sensing thermal gradients in 'constant–temperature' enclosures or vats, monitoring load changes, and displaying the input–output liquid temperature differences in fluid pumps.

6.13.4 Pyrometers

Radiation thermometers respond to the total radiation (heat and light) of a hot body, while optical thermometers make use only of the visible radiation. Both forms of *pyrometer* are specially suited to the measurement of very high temperatures, because they do not involve contact with the source of heat.

Radiation thermometers Both the Fery variable-focus and the Foster fixed-focus types comprise a tube containing at the closed end a concave mirror which focuses the radiation on to a sensitive thermocouple. In use the tube is 'sighted' on to the hot body. The sighting distance is not critical, provided that the image formed is large enough to cover the thermocouple surface. Calibration is by direct sighting on to a body of known surface temperature.

Optical thermometers The commonest form is the disappearing-filament type, which consists of a telescope containing a lamp, the filament brightness of which can be so adjusted by circuit resistance that, when viewed against the background of the hot body, the filament vanishes. The lamp current passes through an ammeter scaled in temperature. The telescope eyepiece contains a monochromatic glass filter to utilise the phenomenon that the light of any one wavelength emitted by the hot body depends on its temperature. Although calibration is based on the assumption that the hot body is a uniform radiator, departure from this condition involves less error than in the radiation pyrometer.

6.13.5 Pressure

The piezoelectric effect is the separation of electronic charge within a material when applied pressure deforms the crystal structure. Conversely, the application of charge to the crystal will cause changes in the dimension of the crystal. Quartz is the only *natural* crystal in general use as a sensor. It is important that the crystal be prepared by slicing along the plane most sensitive for the particular application. Quartz is used in oscillators as the stable frequency-reference element. The self-resonant frequency is temperature-dependent, so constant temperature (e.g. 35 °C) ovens are required for very stable oscillators: this application of quartz is not as a sensor, but the same temperature dependence of resonant frequency is applied for very-low-temperature measurement (below 35 K) where the temperature change can be interpreted by changes in frequency. Numerous ceramic crystals have been developed based upon barium titanate with controlled added impurities; the piezoelectric properties have to be applied by a special polarising treatment during the cooling period following the sintering process in a kiln.

Active four-arm strain gauge bridges, diffused into a single crystal silicon diaphragm, form the basic sensor for a wide range of sensitive pressure elements which are encapsulated in solid state transducers/transmitters (e.g. Druck Limited, Groby, Leicestershire, England). Minimum to maximum pressures can be measured, from venous or arterial values in physiology, to aerospace applications in engines and satellites, and to high-pressure industrial processes. The sensors have 10 V d.c. or a.c. input, with $15\,mV \rightarrow 10\,V$ outputs, 0.1% non-linearity working into $4\frac{1}{2}$ digit multichannel indicator/BCD-recorder which, when coupled with $5\frac{1}{2}$ digit, 0.04% calibrator, provides traceability through a low-cost primary standard.

6.13.6 Acceleration

Accelerometers can be used for velocity measurement by integration of the output signal. The acceleration force of a mass is made to increase or decrease the spring pressure on a ceramic or quartz crystal to produce proportional piezoelectric effect. The device is insulated and hermetically sealed, with care taken to avoid loss of signal through parallel insulation paths of resistance comparable with that of the crystal itself (e.g. 1000 MΩ). Conditioning of the signal involves matching to the much lower input impedances of conventional amplifiers by use of an emitter–follower network. 'Integrated circuit' forms of the network can be located actually within the transducer unit.

The signal amplifier can be a voltage amplifier or a charge amplifier (i.e. an operational amplifier with capacitive feedback). The latter is preferred because the equivalent feed-back capacitive effect is large and dominates the input capacitance associated with varying lengths of coaxial cable.

Other accelerometers are based either on changes in the reluctance of differential transformers or on variations in the resistance of a strain gauge. Slow changes in acceleration can be sensed by connecting the seismic mass to a wiper on a resistance voltage divider.

The upper frequency of the output is limited by the natural frequency of the transducer, while the minimum is almost zero. If the lowest useful frequency of an a.c. electronic amplifier is 20 Hz, the analysis of very-low-frequency signals may be achieved by first recording them on a precision frequency-modulated tape-recorder, which is then replayed at high speed for amplification and analysis.

Constant bandwidth frequency analysers are convenient for stable periodic complex waveforms. Constant per cent bandwidth types suit cases such as machines in which there is some small fluctuation in the nominal periodic behaviour. Truly random vibrations of a stochastic nature may more profitably be analysed in terms of the power spectral density function. Analysers are available for measuring many properties of random signals—e.g. 'probability density function' devices based on instantaneous signal amplitudes (see Electronic Analysers', page 6/27).

6.13.7 Strain gauges

Electrical strain gauges are devices employed primarily for detecting and measuring small dimensional variations in the surfaces to which they are attached, particularly where direct measurement is difficult. The gauge essentially converts mechanical displacement into a change in some electrical quantity (usually resistance). The essential feature is that strain shall be communicated to the gauge without fatigue or inertia effects.

6.13.1 Resistive transducers for temperature measurement

A resistive sensor based on a substantially linear resistance/temperature coefficient should have small thermal capacity for rapid response and avoidance of local temperature gradient. Hence, materials of high resistivity (giving small volume) and temperature coefficient are desirable. Thermistors are made to fulfil these requirements.

6.13.1.1 Pure-metal sensors

Pure conductors such as Pt, Ni and W have low resistance/temperature coefficients but these are stable and fairly linear. Pt is generally adopted for precision measurements and, in particular, for the definitive experiment within the range 100–903.5 K of the International Practical Temperature Scale. Platinum and other metals can be deposited on ceramics to produce metal-film resistors suitable (if individually calibrated) for temperature measurement.

6.13.1.2 Platinum resistance thermometer

To a reasonable approximation the resistance R of a platinum wire at temperature θ (in °C) in terms of its resistance R_0 at 0 °C is

$$R = R_0(1 + k\theta) \quad \text{where} \quad k = (R_{100} - R_0)/100 R_0$$

The temperature θ_p for a given value R is

$$\theta_p = 100(R - R_0)/(R_{100} - R_0)$$

and is known as the *platinum temperature*. Conversion to true temperature is found from the difference formula

$$\theta - \theta_p = \delta[(\theta/100)^2 - (\theta/100)]$$

The value of δ depends upon the purity of the metal. It is obtained by measurement of the resistance of the sensor at 0, 100 and 444.67 °C, the last being the boiling point of sulphur. The value of δ is typically 1.5.

A practical sensor usually consists of a coil of pure platinum wire wound on a mica or steatite frame, the coil being protected by a tube of steel or refractory material. The resistance measurement is carried out by connecting the coil in a Wheatstone bridge network or to a potentiometer. In the former case the bridge usually has equal ratio-arms and a pair of compensating leads is connected in the fourth arm. These leads run in parallel to the actual leads from the sensor coil and compensate for their resistance changes. If the initial resistance and coefficient k of the sensor are large, the compensating leads may not be required. *Figure 6.61* shows the method of connecting three such coils in turn in a Wheatstone bridge network, so as to measure the temperatures at three different locations. In this case the bridge is not balanced; the out-of-balance current through the galvanometer gives the temperature directly. Initial setting is done by adjustment of the battery current until some definite deflection is obtained, when a standard resistance replaces the sensors in the bridge circuit.

6.13.1.3 Thermocouple sensors

Thermocouple sensors are active transducers exploiting the Seebeck effect developed between two dissimilar metals when two junctions are at different temperatures. The International Practical Temperature Scale is defined in terms of a (0.9 Pt + 0.1 Rh)/Pt thermocouple over the range 903.5–1336 K. Readings between 100 and 3000 K are possible with Au/CoCu at the lower and W/Re at the higher end. Conventional materials for intermediate ranges include

copper/constantan (670 K)   iron/constantan (1030 K)
chromel/constantan (1270 K)   chromel/alumel (1640 K)

the figures giving the upper limits. The thermo-e.m.f. varies between 10 and 80 µV/K. To make an instrument direct-reading for the temperature at one junction, the second ('cold') junction must be kept at a reference temperature. The cold junction can be maintained at 0 °C by immersion within chipped, melting ice in a thermos flask, or by using a commercial ice-point apparatus working upon the Peltier effect. When high temperatures are being measured, the terminals of the detector can sometimes be used as the 'cold' junction, but compensating leads should be connected between the thermocouple and the detector. The most accurate method for measuring the e.m.f. is with a d.c. potentiometer.

The e.m.f. e for a hot-junction temperature θ (in °C) with the cold junction at 0 °C is of the form $\log(e) = A \log(\theta) + B$, where the constants A and B have, for example, the values 1.14 and 1.36 for copper/constantan, with e in microvolts.

The small thermal capacity of thermocouples makes them suitable for the measurement of rapidly changing temperatures and for the temperatures at particular points in a piece of apparatus. One useful application is the measurement of surface temperatures, in which case the thermocouple consists of the two metals in the form of a flat flexible strip with a welded junction at the centre, this strip being applied to the surface under test.

A microprocessor-controlled digital readout thermometer, sensed by thermocouples (type 6000 series, Comark, Littlehampton, England), is a mains-driven instrument with a 1000 h battery for off-mains use. Storage in memory of the characterisation parameters includes those of six thermocouple specifications (to ± 0.1 °C) defined in BS 4937, HBS125, NFC42321 and DIN 43710 (-200 °C to $+1767$ °C). Measurements are effected, without change of range, to an overall uncertainty (including characterisation) of less than $\pm 0.2\%$. The cycle of measurements includes short-circuit, cold junction and internal 'calibration', the display being refreshed every 0.25 s. Scales displayed are µV, °C, °F or K, and facilities include programmable multiplying factors; alarms; dwell times for up to 99 min per channel, with pause between scans up to 4 days; 10 input channels; paralleled BCD or high-resolution analogue outputs.

Many portable and bench digital thermometers are available (e.g. Avo types AT1, AT2) and usable with a range of thermocouple probes giving 0.1 °C resolution, auto-ranging and analogue output; cold junction compensation is based upon a precision thermistor and the overall uncertainty is about $\pm 0.3\%$ of reading.

6.13.2 Thermistors

Thermistors are semiconductor sensors which are made from the sintered compounds of metallic oxides of Cu, Mn, Ni and Co formed into beads, rings and discs. A high resistivity is achieved with a large resistance/temperature coefficient. The units can have resistances at 20 °C from 1 Ω to several megohms. The thermal-inertia time-constant is not more than 1 s for small beads, but rather more for discs and coated specimens. The outstanding property is the negative resistance/temperature

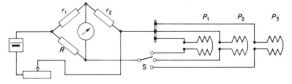

Figure 6.61 Temperature sensor network

With ferrite and powder cores, as with Q-value tests of a coil with and without a slug core, r.f. resonance methods are convenient.

Production and quality control test equipment for core-loss and coil-turns instruments include a 'coil-turns tester' (Tinsley 5812D) with a four-digit display of the number of turns in non-uniform and uniform windings, by using an inductive comparative measurement against a standard winding: the same range of 'magnetic type' instrumentation includes a 'shorted-turns tester' which will detect single and multiple shorted circuited turns in coils of any shape or number of windings; while a 'core tester' can check transformer core characteristics of EI stacks, C cores and toroids including hysteresis loop output for a c.r.o. display.

6.12.3 Bridge methods

Two examples of bridges for audiofrequency tests are given. The non-linear magnetic characteristics of materials introduce harmonics at higher flux densities, imposing limitations because the bridge must be balanced at fundamental frequency.

6.12.3.1 Campbell method

The specimen, shown as a toroid, is uniformly wound with N_2 secondary turns, overlaid with N_1 primary turns (*Figure 6.60(a)*). The magnetising current I_1 passes through the primary of the

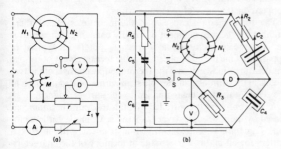

Figure 6.60 Bridges for magnetic measurements. (a) Campbell; (b) Owen

mutual inductor M and the resistor r. At balance indicated on detector D, the power factor and power loss are given by

$$\cos\phi = r/\sqrt{(r^2 + \omega^2 M^2)} \quad \text{and} \quad P = I_1^2 r(N_1/N_2)$$

The corresponding magnetising force and flux density are found from

$$H = (I_1 N_1/l)\sin\phi \quad \text{and} \quad B_m = V/4faN_2$$

where l is the mean length of the magnetic path in the toroid, a is its cross-sectional area, and V is the *mean* voltage as measured on the high-impedance *average* voltmeter V.

6.12.3.2 Modified Owen method

The toroidal specimen (*Figure 6.60(b)*), has N_1 a.c. magnetising turns, and an additional winding N_2 through which a steady polarising d.c. excitation can be superposed. The d.c. supply is taken from a battery through a high-valued inductor to minimise induced alternating current from N_1. The Wagner earth arrangement $R_5 C_5 C_6$ may be added to eliminate earth capacitance errors from the main bridge arms by balancing in both positions of the switch S. The high-impedance voltmeter V measures the r.m.s. voltage V across R_3. The a.c. magnetising force H_a due to the current V/R_3 in N_1, the d.c. magnetising force H_d due to I_d in N_2,

the power loss in the specimen (including the $I^2 R$ loss in the winding N_1 which can be allowed for separately) and the peak flux density B_m are given by

$$H_a = VN_1/R_3 l; \quad H_d = I_d N_2/l; \quad P = V^2(C_4/C_3 R_3)$$
$$B_m = \sqrt{2}V(C_4/\omega C_2 a N_1)\sqrt{(1 + \omega^2 C_2^2 R_2^2)}$$

where a is the cross-sectional area of the specimen, and it is assumed that all time-varying quantities have sinusoidal waveform and angular frequency ω.

6.13 Transducers

In instrumentation systems a transducer is a device that can sense changes of one physical kind and transpose them systematically into a different physical kind compatible with a signal processing system. The compatible signals considered here are generally electrical or magnetic. Transducers can sense most non-electrical quantities (e.g. humidity, pressure, temperature, force, etc.) and, so far as the electrical response is concerned, can be classified as 'active' or 'passive', as follows:

Passive. The output response produces proportional changes in a passive network parameter such as resistance, inductance and capacitance.

Active. Transducers act as generators, the class including piezoelectric, magnetoelectric and electrochemical devices.

The 'Code for Temperature Measurement' is included in BS 1041. The sensors and instruments for the purpose are classified according to temperature range:

(1) Specialised thermometers for measurements near absolute zero. One such, based on the magnetic susceptance of certain paramagnetic salts, is suitable for temperatures below 1.5 K; another, employing an acoustic resonant-cavity technique, can be used up to 50 K. The associated instruments are a mutual inductance bridge for the former, and electronic measurement of length for the latter.
(2) Ge, Si and GaAs p–n junction diode and carbon resistor thermometers are used with d.c. potentiometers and Wheatstone bridges for temperatures up to 100 K.
(3) Vapour pressure thermometers are used for standard readings below 5 K and for general measurements up to 370 K.
(4) Thermistor (resistance) and quartz (resonance) methods, with Wheatstone bridge and electronic counter, respectively, cover the range 5–550 K.
(5) Electrical resistance sensors (usually of Pt) are used with d.c. bridges throughout the range 14—1337 K.
(6) Thermocouple e.m.f.s are measured by d.c. potentiometer. The method, widely employed in industry, has ranges between 100 and 3000 K with various combinations of metals.
(7) The expansion of mercury in glass or steel capillary tubes is applied to measurements from 230 to 750 K. The range is 80–300 K if the mercury is replaced by toluene.
(8) Radiation and optical thermometers employ thermopiles, photodiodes or photomultipliers for measurements up to 5000 K, using d.c. potentiometric or optical balance methods. These provide the only practical methods for very high temperatures.

Mercury-in-glass and optical thermometers are indicating instruments only, but the remainder can be made to furnish graphical records. The electrical resistance and thermoelectrical thermometers are especially suitable for multi-point recording as for heated solid surfaces, points in a mass, or inaccessible places in electrical machines.

6.12.1 Instruments

The most simple fundamental standard of magnetic flux density is based on the accurately known dimensions of a long, uniformly wound solenoid carrying a known current. Alternatively, a uniformly wound toroid can be used, although in this case the flux density will not be quite uniform across the core section. The measurement of flux and flux density can be carried out by means of the ballistic galvanometer or fluxmeter for the former, and the Hall effect instrument for the latter.

6.12.1.1 Ballistic galvanometer

The moving coil of a galvanometer with a long periodic time gives a first swing proportional to the time integral of the current through it (i.e. the charge) provided that the duration is very short. The charge results from a time integral of voltage (i.e. magnetic linkage) impressed on a known resistance. The moving coil is connected in series with a search coil and a resistor, and is immersed in the magnetic field due to a known current. When this current is *rapidly* reversed, the total change of linkage is presented to the galvanometer as an impulsive charge, and the first deflection of the instrument is used for calibration against a known (or calculable) linkage change, or for measuring an unknown linkage. The main limitations to accuracy are the observation of the scale and the uncertainty in the current measurement. A typical high sensitivity ballistic glavanometer has a period of 20 s, a moving-coil resistance of 850 Ω, and a sensitivity of 8.5 m/μC on a scale distant 1 m from the coil using a lamp and mirror technique.

6.12.1.2 Fluxmeter

The fluxmeter is a permanent-magnet instrument with a moving coil of low inertia and negligible control torque. Its damping is made high by use of a relatively thick aluminium former, so that the period is long. It is immaterial whether the time taken to change the linkage in a search coil connected to the moving coil is long or short, and in consequence the instrument is useful in iron testing, where the time taken for a flux to collapse or reverse may be several seconds. The deflection is read from the initial position of a pointer on a quadrant scale when the pointer reaches its maximum deflection; after this the pointer drifts slowly back to a zero position. A typical full-scale deflection would be given by a change of 10 μWb-t.

Strong air-gap flux densities can be measured by an alternative method in which a small coil is rotated at a high and known speed, the induced e.m.f. being proportional to the local flux density.

6.12.1.3 Hall effect instrument

The Hall effect applies to conductors generally, but its application is normally associated with semiconductor materials. These should be of low resistance so that, even for small signals, the thermal (Johnson) noise effects will be low. Bulk materials, indium arsenide and indium antimonide, have low resistances and good output voltage coefficients, with the InSb voltage larger than that of InAs; however, InAs is usually selected, as its temperature coefficient is one-tenth that of InSb. Thin-film InSb is used for switching applications where the higher output e.m.f. is the prime consideration. The Hall effect response is very fast, being usable up to the MHz range. Bulk material InAs elements are more effective than thin-film elements, owing to the much lower resistances obtainable.

In commercial instruments the current I_c in the sensing element can be a direct current chopped at audio frequency; the resultant a.f. Hall voltage E_H, after linear amplification and demodulation, can be displayed on a taut-band d.c. moving coil indicator against a calibrated scale of flux density. Typical ranges have full-scale deflections between 0.1 mT and 5 T for steady fields. Pulsating fields can be measured with d.c. instruments at frequencies up to 500 Hz with a c.r.o. detector. Instruments for alternating fields only are available for frequencies of about 30 kHz, with ranges similar to those of d.c. instruments.

An instrument and probe can be checked by use of stable reference magnets, which may have uncertainties of 0.5–1.0%. The overall uncertainty of a scale reading is typically $\pm 3\%$ of the full-scale deflection for the range.

The sensing elements are protected by epoxy glass-fibre or similar enclosures, and are available in a variety of forms for probing transverse, coaxial or tangential fields. The flat form is most common, typically with the dimensions 0.5 mm × 4 mm × 25 mm. Incremental measurements, using two matched probes and backing-off networks, enable perturbations as small as 0.1 μT to be observed either in the presence of, say, a given 0.1 T field or as a difference between two separate 0.1 T field systems. Hall effect instruments have many obvious applications where the magnitude and direction of the field distribution is required such as air gaps of machines and instruments, magnetron magnets, mass spectrometer fields, residual interference fields, etc.: in addition, the Hall effect principle can be used in a variety of transducer applications by making I_c and B proportional to separate physical functions, and measuring the instantaneous or average results of this product. One example is a wideband, 50 kHz wattmeter in which I_c and B are made, respectively, proportional to the scalar values of the load current and voltage.

6.12.2 Magnetic parameters

The d.c. magnetisation curve, hysteresis loop and relative permeability can be obtained by use of a fluxmeter or calibrated ballistic galvanometer, using a method substantially the same as for calibration. The magnetic test material constitutes the core of the toroid, or is built in strips to form a hollow square. The a.c. cyclic magnetisation loop differs from the 'static' loop, and is still further modified when a d.c. bias m.m.f. is superposed. The a.c. loop can be viewed on a c.r.o. displaying the B and H functions on the Y and X axes, respectively. The H function can be obtained from the volt drop across a small resistor carrying the magnetising current, and the B function from the time integral of the search-coil voltage obtained with the aid of an operational amplifier.

6.12.2.1 Core loss

In a low-frequency test the loop area of the B/H relation is directly proportional to the core loss per cycle of magnetisation, which provides a simple comparative test for specimen material. A more convenient method employs low-p.f. wattmeters, but care is needed to exclude instrument and connection losses. Selected samples of sheet steel intended for the cores of inductors and power transformers are prepared for testing in the form of strips (e.g. 30 cm × 3 cm) assembled in bundles and butted or interleaved at the corners to form a hollow square. The sides are embraced by magnetising and search coils of known dimensions. Input power and current, and mean and r.m.s. voltages, are read at various frequencies and voltages to assess the overall power loss within the material and (if required) an approximate indication of the eddy and hysteresis loss components. The Lloyd–Fisher and Epstein squares are commonly used forms, the latter being referred to in BS 601, *Sheet Steel*.

The magnetic properties of bars, forgings and castings are derived from galvanometer or fluxmeter measurements (BS 2454). A.C. potentiometer and bridge methods are also used, especially for low-loss material at frequencies in the audio range.

6.11.2.1 Burden and output

The preferred rated output burdens lie between 10 and 200 VA/ph, and the rated burden is normally the limit for which stated accuracy limits apply.

6.11.2.2 Errors

Five classes are listed in descending order of accuracy, namely AL, A, B, C and D. The voltage ratio error limit varies between 0.25 and 5% and applies for *small* voltage changes ($\pm 10\%$) around the rated voltage; and the phase error limits are ± 10 min to ± 60 min from class AL to class C. The phase error of D is not defined.

For dual purpose v.t. (i.e. for both measuring and protective application) additional classes E and F are defined which quote the permitted limits of error for classes A, B and C when used at voltages between 0.05 and 1.9 times rated voltage. The v.t. is then denoted by both relevant class letters. The extended error limits are ± 3–5% and ± 2–$5°$.

6.11.2.3 Measurement

By convention, a phase error is positive when the secondary voltage phasor V_s leads the primary voltage phasor V_p. The ratio error is $(k_r V_s - V_p)/V_p$, where k_r is the nominal rated transformation ratio. The rated output is stated in volt-amperes at unity power factor for the specified class accuracy.

6.11.2.4 Dannatt method

This deduces the error in terms of network parameters (*Figure 6.58*). The v.t. primary is fed from a supply to which is also

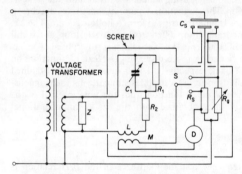

Figure 6.58 Dannatt method for measurement of voltage transformer errors

connected the standard air capacitor C_s in series with a low-valued resistor R_s. The v.t. secondary is loaded with the rated burden Z in parallel with a series-parallel network comprising high-valued resistors R_1 and R_2, a capacitor C_1 and the primary of a mutual inductor M. The secondary voltage of the mutual inductor is balanced against the volt drop across R_s. To ensure that the guard and guarded electrodes of C_s are held at the same potential, a variable resistance R_g is connected so that balance is obtained with either position of the switch S. The balance condition is such that the ratio and phase errors are given very closely by

$$M/C_s R_s (R_1 + R_2) \quad \text{and} \quad \omega(L - C_1 R_1^2)/(R_1 + R_2)$$

where L is the self inductance of the mutual inductor primary.

6.11.2.5 Kusters method

The a.c. comparator bridge can be applied. The capacitance ratio of two gas-filled h.v. capacitors supplied by a common secondary voltage is measured with the currents passing through the separate balancing windings. If one h.v. capacitor and its series-connected balancing winding are now fed from the primary supply of the v.t., then the voltage ratio can be found by rebalancing the comparator when the two capacitors are connected, respectively, to the primary and secondary voltages.

6.11.2.6 Capacitor-divider voltage transformer

For h.v. transformers for 100 kV and above, a more economical and satisfactory voltage division for measurement and protective purposes is given by use of the capacitor-divider system (*Figure*

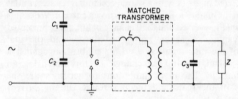

Figure 6.59 Capacitor-divider voltage transformer

6.59). The reduced intermediate voltage appears across the protecting gap G as the input to a tuned transformer. Changes in the secondary burden Z, which would adversely affect the errors, are minimised by adjusting the transformer leakage reactance and/or that of a series inductor L to resonate with the input capacitance $(C_1 + C_2)$ effectively across the primary. There is a consequent sensitivity to frequency.

The classification is the same as for magnetic transformers, except that AL does not apply and there is an additional limitation on frequency. The transient performance is naturally significantly different from that of the normal v.t.

6.12 Magnetic measurements

Because of the non-homogeneity of bulk ferromagnetic materials, magnetic metrology is relatively inaccurate and imprecise. The fields of interest lie in the basic physics of magnetism, the distribution of the magnetic field, the assessment of magnetic parameters, and the measurement of core loss.

Physical basis. The origin of magnetic phenomena lies in the statistical quantum-mechanical behaviour of electrons and particles. Additional information can be obtained by photography of domain formation.

Field distribution. Two-dimensional field distributions in air-gaps are readily traced by current-field analogues such as conducting paper or the electrolytic tank. The tank can also be employed for restricted three-dimensional fields, and tests around models or actual equipments can be based on various types of small sensor working into calibrated instruments (fluxmeter or ballistic galvanometer) or a Hall effect device.

Parameters. The 'static' magnetisation characteristic, the relative permeability and the hysteresis loop are the principal parameters of interest.

Core loss. The loss in a core is a matter of considerable technological importance. It is, however, very difficult to measure directly the losses in such a way that the results can be directly applied to machine constructions other than transformers.

core and secondary encloses the bar primary, which is equivalent to a single turn.

For lower primary currents it is necessary to provide a conventional primary winding. This gives a better primary/secondary coupling and a greater accuracy.

Current transformers must be insulated to withstand the service voltage, which may be up to 400 kV. Oil immersion may be necessary, in which case the core and coils are clamped to the top cover to facilitate removal as a complete unit.

6.11.1.1 Burden and output

Primary currents have preferred values between 1 A and 75 kA, with rated secondary currents of 5 A and 1 A (also 2 A). The *rated burden* is the ohmic impedance Z_T of the secondary circuit when carrying the rated current I_s at a stated power factor. The *rated output* is the volt-ampere product $(I_s Z_T)I_s$. The rated output, the limiting value for which accuracy statements apply, appears in selected preferred values between 1.5 and 30 VA. The normal operating mode of a c.t. is as a low-impedance device; hence, a short-circuiting switch is provided for the secondary winding for use when a low-impedance load (e.g. an ammeter) is not connected to the secondary terminals.

6.11.1.2 Errors

In an ideal c.t. the primary/secondary current ratio is precisely the same as the secondary/primary turns ratio, and the currents produce equal m.m.f.s in exact antiphase. In a practical c.t. the current ratio diverges from the turns ratio, and the phase angle differs by a small defect from opposition. These *ratio* and *phase angle errors* arise from that component of the primary current required to magnetise the core and provide for core loss, and from the e.m.f. necessary to circulate the secondary current through its burden. In transformers intended for accurate measurement and for metering, these errors must be small.

6.11.1.3 Classification

The limits of error define the 'class' of transformer. BS 3938 defines six classes for measuring and three for protective c.t.

Measuring c.t. The classes in descending order of accuracy are designated AL, AM, BM, CM, C and D. The limits of ratio error vary from $\pm 0.1\%$ for AL to $\pm 5\%$ for D; and the phase error from ± 5 min in AL to ± 120 min ($2°$) in types CM and C, with no limits quoted for type D.

Protective c.t. The classes are termed S, T and X. The limits of ratio error for S and T are, respectively, ± 3 and 5%, and the corresponding phase limits are $3°$ and $6°$ approximately. Limits for type X are not stated explicitly. In testing, S, T and X classes it is necessary to distinguish between low- and high-impedance types.

6.11.1.4 Measurement

Errors are normally measured by comparison with a c.t. of higher class. By convention, a phase error is positive when the secondary current phasor leads that of the primary current.

6.11.1.5 Arnold method

In *Figure 6.56* C is the c.t. under test and B is a standard of the same nominal ratio and having known, very small errors. The working burden of the test c.t. is Z. The network is such that the difference between the secondary currents I_b and I_c flows in the non-reactive resistor R and is measured by means of the mutual inductor M and slide wire r fed by the auxiliary c.t. F (con-

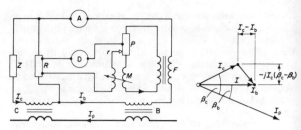

Figure 6.56 Arnold method for measurement of current transformer errors

veniently of ratio 5/5 A). Phase balance is achieved by adjusting M (positive or negative), and ratio balance by selection of r around the centre-tap of P. From the phasor diagram, the components of the difference current I are approximately $I_b - I_c$ and $-jI_c(\beta_c - \beta_b)$ because the phase angles are actually very small. At balance the condition is given by $I_b(r - j\omega M) = RI$. Equating the 'real' parts gives $I_c/I_b = 1 - (r/R)$; whence in terms of the current ratios $k_c = I_p/I_c$ and $k_b = I_p/I_b$, where the primary current is common to both c.t.s,

$$k_c = k_b(I_c/I_b) = k_b/[1 - (r/R)] \simeq k_b[1 + (r/R)]$$

provided that $r \ll R$. Equating the 'imaginary' parts gives

$$\beta_c - \beta_b = (\omega M/R)(I_b/I_c) \simeq \omega M/R$$

The bridge readings give the errors directly by comparison with those of the reference transformer.

6.11.1.6 Kusters method

This is a comparator method (*Figure 6.57*) in which C is the c.t. under test with its rated burden Z, and B is an a.c. comparator

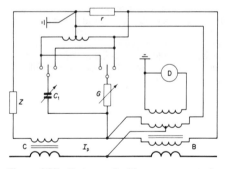

Figure 6.57 Kusters method for measurement of current transformer errors

with the same nominal ratio as C but possessing negligible errors. The errors of C can be deduced directly from the balance settings of the arms of the parallel $G - C_1$ network.

6.11.2 Voltage transformers

Magnetically coupled voltage transformers (v.t.) resemble power transformers in basic principle, and have recommended primary voltages up to $396/\sqrt{3}$ kV. Oil-immersion is necessary for these levels. The preferred secondary voltages lie between $110/\sqrt{3}$ and 220 V/ph.

transformer calibration, as it enables an external burden B to be supported under balance conditions. Any change in the capacitive error due to B can be neutralised by W_c connected on one side of the separate centre-tapped transformer W_b. By careful design the capacitance distribution is uniform. The internal p.d. of the major winding can be neutralised by a parallel compensation winding W_c wound within the magnetic shield, so reducing capacitive currents, as it can be held at earth or reference ground potential.

The commercial range of comparators includes d.c. potentiometers, direct voltage and current ratios, d.c. and a.c. 4-terminal resistance comparison, voltage- and current-transformer errors, and high-voltage capacitor comparison. All the instruments derived from d.c. and a.c. comparators possess very linear properties, excellent discrimination and good stability. When certified they can be used as comparative reference standards.

6.10.4.14 H.V. capacitance comparator bridge

Standard capacitors used with h.v. bridges (such as the Schering and comparator types) use compressed gas construction up to a maximum of 1000 pF, although 100 pF is more common. Measurement of such capacitors is based on the principle that a stable alternating voltage to two capacitors produces a current ratio equal to the capacitance ratio. The variable turns ratio of the comparator bridge can be used to balance an unknown capacitor up to 1000 times as large as the standard. By cascading, the range can be extended still further.

6.10.4.15 Substitution

In any active network the equivalent parameters of a component may be measured if variable reference standards can be either substituted for the component or connected in parallel with it—provided always that the standards can be adjusted so that no final change occurs in the voltage, current, phase or harmonic content of the waveforms in the network. The substitution principle can be applied to any type of bridge, or other class of network; resonant networks often provide the best discrimination between the two conditions. It is important that the stray capacitance, inductance and resistance paths, to earth and between components, should be unchanged by the act of substitution; hence, for the substitution test there should be (ideally) no change in the geometrical layout of leads and components. This can be easily achieved with low-loss change-over switches or coaxial connectors.

For the measurement of radiofrequency components, where shunt capacitance effects are prominent, the substitution principle is particularly valuable. Typical applications are discussed below.

6.10.4.16 Parallel-T network

Figure 6.54(a) shows a general parallel-T network, with an a.c. source as input and a detector across the output terminals. The balance condition (of zero output) in terms of the impedance operators is

$$Z_1+Z_2+(Z_1Z_2/Z_3)+Z_4+Z_5+(Z_4Z_5/Z_6)=0$$

Suppose that Z_1 and Z_2 are pure reactances of like sign, e.g., $-jX_1$ and $-jX_2$, and that Z_3 is purely resistive: then

$$Z_1Z_2/Z_3 = -(X_1X_2/R_3)$$

which is equivalent to a negative resistance and capable of balancing out a positive resistance in Z_4 or Z_5. An example is shown in *Figure 6.54(b)*. The source is a modulated signal generator and the detector is, in effect, a radio receiver tuned to the 'carrier' frequency. Balance is achieved by variation of C_3 and C_6 with the test terminals open-circuited. If an unknown admittance $G_x + jB_x$ is now connected across the test terminals and the disturbed balance restored by altering C_3 and C_6 to C'_3 and C'_6, respectively, then

$$B_x = \omega(C_3 - C'_3) \quad \text{and} \quad G_x = \omega^2 C^2 (C'_6 - C_6) P/C_4$$

Bridge screening is simplified because the common source and detector terminal can be earthed. The use of a variable resistor as a balance arm can often be avoided, a considerable simplification for radiofrequency measurement.

6.10.4.17 Bridged-T network

This can be used at frequencies up to about 50 MHz. *Figure 6.55(a)* shows the schematic arrangement, and *Figure 6.55(b)* illustrates a common application. If the unknown impedance

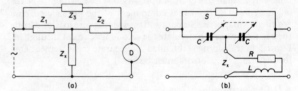

Figure 6.55 Bridged-T networks

$Z_x = R + j\omega L$ is such that R^2 is negligible compared with $\omega^2 L^2$, then for balance

$$1/\omega^2 LC = 2 \quad \text{and} \quad R = S/4$$

whence R and L can be obtained if the angular frequency ω is known.

6.11 Measuring and protection transformers

Measuring (instrument) transformers are used primarily for changing currents and voltages in power networks to values more suited to the range of conventional indicators. Protection transformers are employed in systems of fault protection. Both types are dealt with in BS 3938 (*Current Transformers*) and BS 3941 (*Voltage Transformers*).

6.11.1 Current transformers

Air-cooled current transformers (c.t.) are used in circuits of voltage up to 660 V and currents up to 75 kA. The *bar primary* form employs the actual cable or busbar of the main circuit as the primary winding. The core is built of high-grade steel laminations in rectangular or circular shape. The secondary has the appropriate number of complete turns uniformly around the annular core or on all four sides of the rectangular core. The assembly of

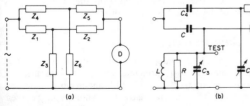

Figure 6.54 Parallel-T networks

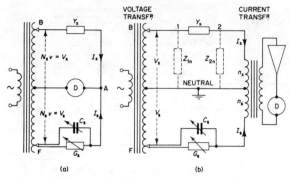

Figure 6.52 Inductive ratio-arm bridge. (a) Basic principle; (b) With isolated detector

6.10.4.11 Inductive ratio-arm bridge

The principle is explained by reference to the simplified diagram in *Figure 6.52(a)*, the essential features being the coupled windings of, respectively, N_x and N_s turns. If the voltage per turn is v, then the two windings have the voltages $V_x = N_x v$ and $V_s = N_s v$, respectively. Let $V_x = V_s$; then the current I_x through the unknown capacitive admittance Y_x can be made equal to the current I_s by variation of the parallel standards (conductance G_s and capacitance C_s), until the detector shows a null reading. For this condition the standard admittances represent the *parallel equivalent network values* of Y_x. If Y_x is inductive, the C_s connection at F is transferred to B to avoid the use of standard inductors, which are very inferior to standard capacitors.

The voltage linearity of the input transformer is of fundamental importance to this class of bridge. With modern techniques and magnetic core materials, it is possible to wind toroidal cores to produce an extremely uniform reactive distribution (leakage inductive reactance and turn-to-turn capacitance) with a resultant non-linearity of about 1 in 10^7 in the magnitude of the mutually induced e.m.f. per turn.

In one well-known bridge (Wayne Kerr) the primary winding of a current transformer is inserted at A and the detector removed to the secondary winding as in *Figure 6.52(b)*. The two parts of the primary winding (n_s and n_x turns) are wound in opposite senses and the detector can indicate the zero m.m.f. condition in the windings.

In general, N_s differs from N_x and n_s from n_x, so that $I_x n_x = I_s n_s$ for balance; further, the primary of the current transformer is at the ground (or earth) potential of the neutral if the trivial resistance drop of the windings is ignored, so that $I_x = (N_x v) Y_x$ and $I_s = (N_s v) Y_s$. Combining these results gives

$$Y_x = Y_s (N_s/N_x)(n_s/n_x)$$

The double turns ratio product can be used to permit *one* accurate standard resistor or capacitor to be switched to any N_s turns of value 1–10 to simulate *ten* accurate standard components by the precise selection of turns. Moreover, wide-range multiplication of the simulated standard is achieved by decade tappings on N_x and n_x.

An important advantage with this class of bridge is the possibility of connecting the neutral to earth; then small leakage currents from Y_x to earth are almost completely eliminated, since the current transformer is at earth potential and any *small* leakage currents from the high-voltage side of Y_x merely causes a slight *uniform* reduction in v but does not alter the turns ratio. Hence, determinations such as the following are possible:

(1) One-port low-capacitance measurement in the presence of a large shunt capacitance due to connecting leads.
(2) One-port *in situ* impedance measurement, the effect of the remainder of the network (shown dotted in *Figure 6.52*) is neutralised.
(3) Three-terminal impedance transfer functions for correctly terminated networks.

The results of measurements appear in *parallel admittance* form, and may require elementary computation to render them into *series impedance* form, with due regard to frequency. With medium to high values of impedance, resistance and capacitance are unaffected, but for inductance measurement (involving a 'resonance' balance) a reciprocal ω^2 correction is needed. For low-impedance networks, resistance and inductance are unchanged but capacitance readings require correction.

6.10.14.12 Characteristics

Universal bridges of the inductive ratio-arm type measure over a wide range of values, at an angular frequency of, e.g., 10 krad/s from a built-in oscillator, or at other frequencies up to 10 MHz using external signal generators and tuned electronic detectors. The measurement uncertainty can be within ± 0.1–1.0%, provided that the standard component values are trimmed against superior standards and that the actual frequency is measured when frequency sensitive parameters are measured.

The Wayne Kerr precision bridge operates at 1591.55 Hz ($\omega = 10$ krad/s) and the 6-figure display has a minimum uncertainty of 0.01%. The instrument uses an electronic null-seeking process to assist the *automatic* balancing procedure. With external source the bridge has a frequency range of 0.2–20 kHz with manual balancing.

Admittance bridges are commercially available to operate up to 100 MHz. At this frequency lumped parameter measurements are being displaced by distributed-parameter network characteristics.

6.10.4.13 Current comparator bridge

The a.c. current comparator instrument uses current transformers only in the load network. It is based on the principle of m.m.f. (i.e. ampere-turn) balance on an ideal magnetic circuit, in which the current ratio is the inverse turns ratio.

The two basic current coils, P and S in *Figure 6.53*, are wound in opposing senses on a common magnetic core with a high initial permeability. The null m.m.f. condition is sensed by winding M

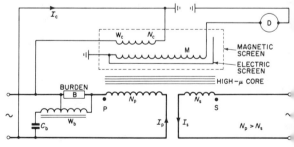

Figure 6.53 Current comparator with capacitive and burden compensation

and manual selection of the turns ratio N_p/N_s. The ideal balance condition can be closely approached in practice by shielding the detector from magnetic leakage by a laminated magnetic screen and by a compensation winding W_c. The shield must not form a short-circuited turn; but it does aid magnetic energy transfer between windings P and S. This is advantageous in current-

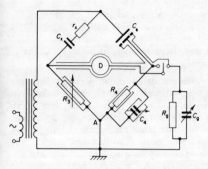

Figure 6.47 Schering bridge

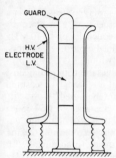

Figure 6.48 Air capacitor

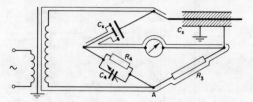

Figure 6.49 Schering bridge and earthed bushing

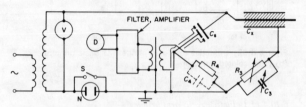

Figure 6.50 Discharge bridge

the high-voltage standard capacitor C_s) is calculable, and with an air or gas dielectric can be assumed to have zero loss. A typical standard capacitor for 150 kV is shown in *Figure 6.48*; for higher voltages the clearances necessary for avoiding corona and breakdown are large, but the dimensions can be reduced by a construction in which the dielectric gas is under pressure.

The test capacitor in *Figure 6.47* is represented by C_x and a series loss resistance r_x. R_3 is a variable resistor, and the fourth arm comprises a variable low-voltage capacitor C_4 in parallel with a resistor R_4 (which may be fixed). Then, for balance

$$C_x = C_s(R_4/R_3) \quad \text{and} \quad \tan \delta_x = \omega C_4 R_4$$

The loss tangent is a valuable and sensitive indication of the quality of the test insulation. If periodical tests reveal a gradual increase in the loss tangent, then deterioration is occurring, the loss power is increasing and breakdown in service is probable.

To avoid electric field coupling between bridge components—in particular, between the h.v. electrodes and the connecting leads—that would affect accuracy, components R_3, R_4 and C_4 are enclosed in metal screens connected to the earthed point A. Again, to avoid errors due to intercapacitive currents between the centre l.v. electrode and the guard electrode of C_s, these electrodes should be brought to the same potential by aid of the auxiliary branches C_g and R_g. Balance is achieved with the detector switched to this combination, in addition to the main balance. If the test capacitor C_x is a cable with an earthed sheath, or a capacitor bushing on site and not readily insulated from earth, then the bridge node A must be isolated: careful screening is now even more necessary. *Figure 6.49* shows the arrangement for a portable Schering bridge equipment for testing an earthed bushing. It will be seen that an earthed screen isolates the bridge from the transformer primary, and a second screen connected to point A keeps capacitive currents from the high-voltage connections out of the bridge arms.

Discharge bridge This has been developed to detect the onset of void breakdown in cables and bushings. The network (*Figure 6.50*) can be used to test an unearthed component—for example,

the bushing C_x; the arrangement resembles that of the Schering bridge. The stray capacitances of the guards and screening of C_s act as the capacitance C_4, which, however, at the discharge frequencies concerned, is greater than that required for balance. Hence, it is necessary to provide the variable capacitor C_3. The bridge output is fed to a rectifier milliammeter D through a filter and amplifier designed to pass, e.g., the band 10–20 kHz, so that the indication is a measure of the discharge current. On the supply side of the bridge is connected some artificial source of discharge voltage, such as the neon tube N, which serves initially to balance the detector D.

6.10.4.10 Wagner earth

Capacitance between bridge arms and earth in the Schering bridge causes disturbing currents to flow so that false balance conditions may result. If, as in *Figure 6.51*, the bridge supply is not earthed but the nodes c and d are at earth potential, then the

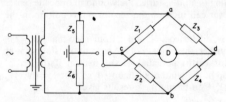

Figure 6.51 Wagner earth arrangement

detector is also at earth potential and no stray capacitance currents can flow in it. Further, all branch capacitances to earth are stabilised. The Wagner earth method secures this condition by the use of the additional impedances Z_5 and Z_6, the junction of which is solidly earthed. Z_5 and Z_6 must be of the same type as either Z_1 and Z_2, or Z_3 and Z_4. If balance is achieved for both positions of the switch, then nodes c and d must have the same potential as the earthed junction, which is zero. Stray capacitance at a and b is innocuous as it merely 'loads' the supply.

The Wagner earth is usually applied to commercial Schering bridges, but the technique is applicable to any bridge and can be particularly helpful at higher frequencies.

at the proper potentials with respect to the equipment and the environment.

Continuously variable calibrated air-dielectric capacitors with ranges between 20 and 1200 pF are used for small adjustment, and ranges of fixed or variable mutual inductors are available. For high-voltage bridges the standard capacitors are of fixed value, and designed with air or compressed gas as dielectric for 300 kV and above.

6.10.4.5 Low-frequency and audiofrequency bridges

A selection is given of the more important bridges, many of which form the basis of commercial instruments; for example, the inductance bridges of Maxwell and Hay, the modified de Sauty bridges for capacitance, the Schering bridge for capacitance and loss tangent at low and high voltages, and the Wien bridge for frequency and a.c. resistance. These bridges are commonly made to give results with a $\pm 1\%$ uncertainty, but can be designed for $\pm 0.1\%$.

All bridges should be checked occasionally to show that the reference standard components have not changed and that the range and ratio division has not drifted; also, if the bridge is frequency-sensitive, the frequency of the oscillator and the drift rate should be investigated. Naturally, superior standard components must be available which have recent 'traceability' certification, in addition to accurately known decade-ratio measurement techniques. Where bridge measurements are implicit in contract specifications, it may be prudent or necessary to justify the claims by referring the instrument to a calibration service standards laboratory.

6.10.4.6 Mutual inductance

Figure 6.44(a) shows the simple Felici–Campbell bridge for the measurement of an unknown mutual inductance M_x in terms of a variable standard M. For balance on the null detector D, then

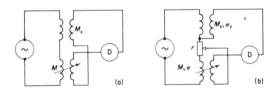

Figure 6.44 Mutual inductance bridges. (a) Felici-Campbell; (b) Hartshorn

$M_x = M$, on the assumption that the inductors are perfect, i.e. that the secondary e.m.f. is in phase quadrature with the primary current. Each, however, has a small in-phase component that can be represented by an equivalent resistance σ. The modification in Figure 6.44(b) shows Hartshorn's arrangement, which enables the impurities to be measured: at balance by adjustment of M and r, the conditions are $M_x = M$ and $\sigma_x = r + \sigma$.

6.10.4.7 Inductance

Inductance may be measured by Wien's modification of the Maxwell bridge (Figure 6.45(a)). If the unknown inductor has an inductance L_x and an equivalent series resistance R_x, balance gives

$L_x = R_2 R_3 C_1; \quad R_x = R_2 R_3/R_1; \quad Q_x = \omega L_x/R_x = \omega C_1 R_1$

The advantage is that inductance is measured in terms of a high-quality and almost loss-free capacitor. The bridge can measure a

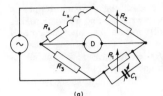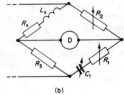

Figure 6.45 Inductance bridges. (a) Maxwell-Wien; (b) Hay

wide range of inductance values with Q factors less than 10. High Q factors require excessively large values of R_1, and by creating the same effective phase characteristic by means of a low-valued resistor R_1 in series with C_1, the Hay bridge of Figure 6.45(b) is obtained. The balance conditions for it are

$Q_x = 1/\omega C_1 R_1; \quad R_x = R_2 R_3/R_1 (1 + Q_x^2);$

$L_x = R_2 R_3 C_1/(1 + 1/Q_x^2)$

A disadvantage is the need to measure the frequency.

A commercial instrument using the Maxwell and Hay bridges has built-in supplies at frequencies of 50 Hz, 1 kHz and 10 kHz, and an inductance range from $0.3\,\mu\text{H}$ to 21 kH with an accuracy within $\pm 1\%$. The loss resistance is known to about $\pm 5\%$, and it is possible to introduce a d.c. bias.

6.10.4.8 Capacitance at low voltage

Relatively pure capacitors can be measured by the de Sauty bridge (Figure 6.46(a)), the balance condition for which is $C_x = C_s(R_4/R_3)$. Imperfect capacitors can be compared by the arrangement shown in Figure 6.46(b), in which the imperfection is

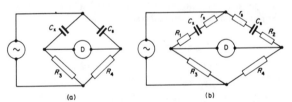

Figure 6.46 Capacitance bridges. (a) De Sauty; (b) Modified de Sauty

represented by a series resistance r. To obtain balance, adjustment of all four resistors R is necessary:

$C_x = C_s(R_4/R_3) = C_s[(R_2 + r_s)/(R_1 + r_x)]$

For the loss tangent,

$\tan \delta_x = \tan \delta_s + \omega C_s [R_2 - (R_1 R_4/R_3)]$

Portable commercial capacitance testers, with battery-driven 1 kHz oscillators and rectified d.c. or headphone detectors, use a modified de Sauty bridge network so that loss tangent can be estimated and the capacitance measured to an uncertainty of about $\pm 0.25\%$.

6.10.4.9 Capacitance at high voltage

Dielectric tests at high voltages, particularly of the loss tangent, can be made with the Schering bridge (Figure 6.47). The bridge is in wide use for precision measurements on solid and liquid dielectrics, and for insulation testing of cables, high-voltage machine windings and capacitor bushings. The capacitance of

Good-quality commercial bridges claim an uncertainty of about ±0.05% for unknowns between 1 Ω and a few microhms. Near the latter level a small constant resistance becomes a limitation. As $R_1, \ldots, R_4$ constitute a Wheatstone bridge network, a Kelvin bridge is often extended to measure 2-terminal resistances up to 1 MΩ with an uncertainty of ±0.02%.

6.10.3.3 Digital d.c. low resistance instruments

Digital milliohmeter/micro-ohmeter instruments have built-in d.c. supplies; they are auto-ranging and make 4-terminal measurements by internal comparator-ratio techniques to give, e.g., $4\frac{1}{2}$-digit displays in the case of the Tinsley 5878 microohmmeter with 0.1 μΩ best resolution and uncertainty ±(0.1% → 0.3%) of reading. Also, IEEE 488 bus attachment is available for use in automated test systems. Thermal e.m.f. balance is incorporated as well as lead compensation.

6.10.3.4 Comparator resistance bridge

The comparator resistance bridge is a variant of the comparator potentiometer (*Figure 6.37*). In this case the network connections and adjustments are such that, at balance,

$$I_m R_m = I_b R_b \quad \text{and} \quad I_m N_m = I_b N_b, \quad \text{whence} \quad R_m = N_m(R_b/N_b)$$

where R_b is the known standard 4 terminal resistor and N_b has been adjusted to make R_b/N_b an exact decade value. The unknown 4-terminal resistance R_m is found in terms of a number of turns N_m and a decade factor.

The two resistors can be compared at different current levels, so that, for example, the rated current can be used with the test resistor, while the standard resistor can be used at a current well below the level at which self-heating effects would change its resistance value. The actual comparison uncertainty with this class of bridge is less than 1 in 10^6, to which must be added the basic uncertainty in R_b, probably 2 in 10^6 if it is a class 'S' (or 'Wilkins') standard 4-terminal resistor *known* to be stable.

6.10.4 A.C. bridge networks

A.C. network parameters (impedance, admittance, phase angle, loss angle, etc.) can be measured in the following ways:

(1) Single-purpose or multipurpose a.c. bridge networks, with four or six arms or with inductive ratio arms.
(2) Analogue and digital networks developed for special applications and consisting of frequency-selective or resonant or filter networks; they are associated usually with measurements at radiofrequencies and high audiofrequencies.
(3) Digital multimeter instruments for measuring the modulus of the unknown parameter (but not other qualities).

The treatment in this subsection is confined largely to (1).

6.10.4.1 Balance conditions

In the four-arm bridge of *Figure 6.43*, the arms comprise impedance operators Z_1, Z_2, Z_s and Z_x. The network is supplied at

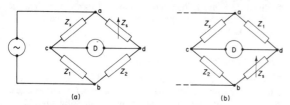

Figure 6.43 Basic four-arm a.c. bridge. (a) Ratio bridge; (b) Product bridge

nodes a and b from a signal generator, and a sensitive null detector is connected between nodes c and d. If Z_1 and Z_2 (fixed standards) and Z_s (variable standard) are so adjusted that the p.d. between c and d is zero, then the unknown impedance Z_x depends on whether Z_1 and Z_2 are adjacent as in *Figure 6.43(a)*, or in opposite arms as in *Figure 6.43(b)*; in either case the balance condition is that the product of pairs of opposite arms is equal, giving

(a) ratio bridge: $Z_x = Z_s(Z_1/Z_2)$;
(b) product bridge: $Z_x = Y_s(Z_1 Z_2)$

where $Y_s = 1/Z_s$. These are *complex* expressions, so that the equality involves both magnitude and phase angle; alternatively, both phase ('real') and quadrature ('imaginary') components. Thus, Equation (a) above can be written in terms of *magnitudes* in either of the following ways:

$$Z_x \angle \phi_x = Z_s(Z_1/Z_2) \angle (\phi_s + \phi_1 - \phi_2)$$
$$R_x + jX_x = (R_s + jX_s)(R_1 + jX_1)/(R_2 + jX_2)$$

The bridge arms will include intentional or stray inductance and capacitance, and the corresponding reactances are functions of frequency. Most practical bridges are so chosen that only two or three final adjustments are necessary to obtain the optimum null condition of the detector.

6.10.4.2 Signal source

The bridge supply can be from a screened power-frequency instrument transformer, but more probably from a fixed- or variable-frequency signal generator (see 'Electronic oscillators', p. **6**/26). A voltage up to 30 V may be required, but a few volts is normally adequate, especially for radiofrequency bridges. The maximum power of a signal generator is of the order of 1000 mW, so that to avoid overload and distortion it is important that the appropriate voltage be set and a reasonable match be obtained between the generator and the effective input impedance of the bridge network.

6.10.4.3 Detector

A moving-coil vibration galvanometer may be used at discrete frequencies over the range 10 Hz–1.2 kHz. The galvanometer has a taut suspension mechanically tuned to the operating frequency. The method is sensitive, and effectively filters harmonics, but it is often more convenient to employ the electronic detectors included in most commercial bridges. These include tuned-resonance amplifier networks or broad band high-gain electronic amplifiers; both give a rectified out-of-balance display on a d.c. instrument, with 'magic eye' or c.r.o. as possible alternatives. In some portable instruments it was the practice to use head-telephone sets, which are highly sensitive in the audio range of 0.8–1.0 kHz.

6.10.4.4 Bridge-arm components

Precision a.c. resistance boxes can be designed with very low time constants. They are available with switched or plugged groups of ten resistors in decade values between 0.1 Ω and 1 MΩ. Careful construction and screening enable some of these devices to be used at frequencies up to 1.2 MHz before the residual reactance becomes significant. An overall uncertainty of ±0.05% is possible, but 0.1–1.0% is more conventional. (See also 'Wilkins' resistor, paragraph 6.10.1.8).

Multi-dial switched mica capacitors are available with good stability and low loss tangent. Single- and double-screen capacitors have, respectively, 3 and 4 terminals, and care is needed to avoid unwanted stray capacitance, the screens being maintained

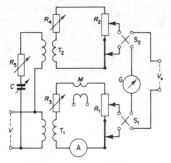

Figure 6.40 Gall co-ordinate a.c. potentiometer

6.10.2.2 Gall potentiometer

The Gall potentiometer (*Figure 6.40*) provides two quadrature voltage components, summed and balanced against the unknown V_x. The supply voltage V is applied to the primaries of isolating transformers T_1 and T_2. The secondary of T_1 supplies a current, approximately in phase with V, to R_1 which consists of a tapped resistor and slide wire in series. The current is adjusted by resistor R_3 and read on an ammeter A, and passes through the primary of a fixed mutual inductor M. Any desired value of in-phase voltage (within the available range) is obtained from tappings on R_1. The primary of transformer T_2 has a series resistor R_5 and a variable capacitor C, and may be so adjusted that the current in R_2 is in quadrature with that in R_1. The exact phase angle, and the calibration of the quadrature circuit, are achieved by balancing the secondary e.m.f. of the mutual inductor M against a fraction of the p.d. across R_2. Thus, variable voltages, phase and quadrature, can be obtained from R_1 and R_2, respectively. These, in series, are balanced against the unknown V_x through the vibration galvanometer G. Reversing switches S_1 and S_2 facilitate balance and provide for a phase angle over a full 360°.

6.10.3 D.C. bridge networks

A bridge network has as its distinctive feature a 'bridge' connection between two nodes, with a null detector to sense the balance condition of zero p.d. between the nodes.

6.10.3.1 Wheatstone bridge

The Wheatstone bridge is an arrangement in very wide use for the determination of one unknown resistance in terms of three known resistances. The network is shown in *Figure 6.41*, where R_1, R_2, R_3 and R_4 are resistors connected at the nodes a and b through a reversing switch S to a d.c. supply. The galvanometer G, with a shunting resistor to control its sensitivity, and the key K are connected to the nodes c and d as shown. R_1 and R_3 may be set at known fixed values. R_2 is a variable resistor and R_4 is the unknown resistance to be measured. The bridge is balanced by adjusting the value of R_2 until the deflection of the galvanometer, set for maximum sensitivity, is brought to zero. The condition for balance is easily seen to be $R_4/R_3 = R_2/R_1$ and is independent of the voltage applied to the bridge: whence the unknown is $R_4 = (R_3/R_1)R_2$. The bridge may be built up as a single unit with, for example, R_1 and R_3 each able to be set at one of the resistance values 1, 10, 100 or 1000 Ω. The bridge ratio R_3/R_1 may therefore have a range of values from 1000/1 to 1/1000. In this case R_2 might conveniently be adjustable, by means of dial switches, in steps of 1 Ω from 1 to 11 110 Ω.

The bridge resistors are wound from manganin wire, since manganin is an alloy with a low temperature coefficient of resistance and a low thermo-electric e.m.f. to copper. Any thermo-electric e.m.f.s in the bridge connections which did not on average cancel themselves out could give a false result. Errors from this cause are eliminated by taking a balance for each position of the changeover switch S and using the mean of two observed values of R_2.

High-precision Wheatstone bridges are capable of measuring resistances between 1 Ω and 100 MΩ with uncertainties which vary between 5 and 100 in 10^6 of the reading over this range of values. Modern resistors are very stable devices and they should not change in value by more than a few parts in 10^6 per year; however, it is prudent to check the linearity of resistive bridges regularly, particularly if they are being used to the limit of the original specification. Some bridges have trimming facilities available with the principal resistors, although corrections can always be applied which although tedious, can avoid imposing new resistance drifts on the bridge as a consequence of changing *stable* resistors.

Various forms of high-quality bridges are used for measuring the changes which occur in resistive transducers. The Smith and Mueller bridges, used for measurements with precision platinum resistance thermometers, can yield 0.001 °C discrimination in readings. For strain gauge transducers there are portable bridges that can detect changes of 0.05% or less.

6.10.3.2 Kelvin double bridge

The Kelvin double bridge is an adaptation of the Wheatstone bridge which may be used for the accurate measurement of very low resistances, such as 4-terminal shunts or short lengths of cable. In the network (*Figure 6.42*) R_x is the unknown and R_s is a standard of approximately the same ohmic value. The two are

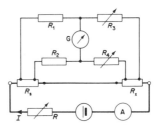

Figure 6.42 Kelvin double bridge

connected in series in a heavy-current circuit (e.g. 50 A for $R_s = 10$ mΩ, giving 0.5 V drop and adequate sensitivity). Suitable values for R_1, R_2, R_3 and R_4 are in each case a few hundred ohms. To use the bridge, R_1 and R_2 are made equal; then R_3 and R_4 are adjusted (keeping $R_3 = R_4$) until balance is achieved. Then the balance condition is $R_x/R_s = R_4/R_2 = R_3/R_1$. Provided that it is short and of adequate cross-section, the connection between R_x and R_s does not affect the measurement.

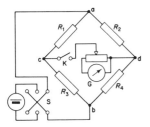

Figure 6.41 Wheatstone bridge

opposition to the p.d. across R_m, and balance achieved by use of the adjustable direct-current controller C_m, which is then left to maintain this current level.

The equivalent cell e.m.f. $I_m R_m$ is now connected (through S2) in opposition to $I_b R_b$, and N_m turns are selected to be numerically equal to E_r. The m.m.f. is balanced automatically by the feedback control system and by a small adjustment to N_b so that $I_m R_m = I_b R_b = E_r$, represented numerically by N_m. With switch S set to S_0 the unknown p.d. V_x is presented through S_x and galvanometer G_1 in opposition to $I_b R_b$ and balance achieved by manual adjustment of N_m. Then N_m is numerically the unknown voltage.

Instruments of this kind are probably the most linear d.c. potentiometers available. They incorporate additional windings that enable the linearity to be checked by the operator. While the actual values of R_m and R_b are not of first importance, it is vital that they should be constant. Resistors are incorporated for subdivision of I_m for lower decades of N_m in order to simulate the 10^7 turns implied by a discrimination of 1 in 10^7, but they do not contribute unduly to the overall non-linearity, which is about $\pm 0.3 \times 10^{-6}$ of full scale. The system uncertainty must include that of the standard cell, at least ± 1–$2\,\mu$V.

6.10.1.7 Pulse width modulation potentiometer[17]

This method subdivides a known stable d.c. voltage in terms of a time-period ratio. If a standard voltage source E_r is repeatedly connected to a low-pass filter through a very-high-speed semi-conductor switch for a time t_1, and then disconnected for an additional time t_2, the high-frequency rectangular, chopped waveforms will, after smoothing, give an output $E_0 = E_r \cdot t_1 / (t_1 + t_2)$ with $t_1 + t_2 = T$, a constant period. The time *ratio* can be precisely set using a variable time-interval counter for t_1 and a fixed-period counter for T, with each counter being driven from a common highly stable crystal-controlled oscillator. An unknown voltage V_x, in opposition to E_0 through a galvanometer, can be measured by variation of E_0 (through the voltage-scaled setting of the t_1 counter) until galvanometer zero balance is achieved. The e.m.f. E_r of the solid state source of steady voltage must be known accurately and the source must be capable of delivering small current surges into the low-pass filter without any significant change in voltage. Conversely, V_x and E_r could be interchanged when $V_x > E_r$.

One d.c. variable-voltage source uses the accurately determined voltage as input to constant gain d.c. amplifiers (designed for decade steps) to give a wide-range d.c. voltage standard source with ± 50 p.p.m. uncertainty (3 months' stability) up to 1.2 kV with a $25 \rightarrow 10$ mA (at 1 kV) current capability and $0.5 \rightarrow 3.0$ s settling times.

6.10.1.8 Applications

The d.c. potentiometer may be used to determine temperature by measuring the thermo-e.m.f.s in calibrated thermocouples. In conjunction with standard 4-terminal resistors and shunts it is applied to the calibration of voltmeters, ammeters and wattmeters, and for the comparison of resistors. In *Figure 6.38* the values V_x, I_x, R_x and P_x are unknowns, and E_p is measured by the potentiometer. Resistors $R_1, \ldots, R_7$ have known values. At (a) a voltage divider is used to measure V_x, and at (b) a shunt is employed to measure I_x. At (c) R_x is compared with R_4 by means of their respective p.d.s. The network (d) enables the power $P_x = V_x I_x$ to be determined from V_x and I_x. The resistance of 4-terminal resistors is known between the potential (inner) terminals, and *not* the current (outer) terminals, which minimises errors due to non-uniform current densities and contact resistances.

The NPL designed 'Wilkins' resistor is a high-stability, 4-terminal, a.c./d.c. resistor available in various decade and other

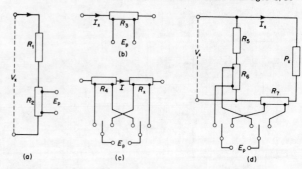

Figure 6.38 D.c. potentiometer measurement of voltage current, resistance and power

values (Tinsley Co., London) which has an exceptionally low time constant for accurate a.c./d.c. comparator networks, etc.

6.10.2 A.C. potentiometers

The a.c. potentiometer may be used to measure the magnitude of an alternating voltage and its phase relative to a datum waveform. The measurements are normally restricted to sinusoidal waveforms, as the presence of harmonics makes a null balance impossible; and operation is usually confined to a single frequency. One recent instrument (Yorke) is capable of measurements up to audio-frequency, with an uncertainty approaching $\pm 0.05\%$. The general principles of the Larsen and Gall potentiometers are set out below.

6.10.2.1 Larsen potentiometer

The primary winding of a variable mutual inductor is connected in series with a tapped resistor to a source of voltage V. The unknown voltage V_x is normally derived from the same source, because identity of frequency is essential. The supply current I is controlled by resistor R_1 and indicated on an ammeter. The inductor secondary e.m.f. in series with a tapped fraction of the volt drop across R is balanced against the unknown V_x (*Figure 6.39*) by means of the vibration galvanometer G, the sensitivity of which can be varied by resistor R_2. If M and R are the mutual inductance and resistance values at balance, then

$$V_x = I\sqrt{|R^2 + \omega^2 M^2|} \quad \text{and} \quad \phi = \arctan(\omega M / R)$$

where ϕ is the phase angle between V_x and I. In practice it may be necessary to reverse either or both of the component voltages to secure balance, and reversing switches are incorporated for this purpose. The mutual inductor should be astatic, to avoid pick-up errors.

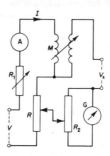

Figure 6.39 Larsen a.c. potentiometer

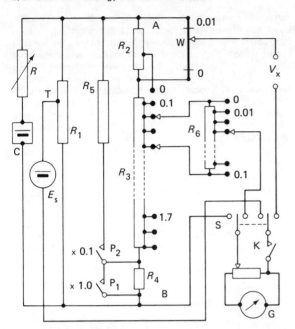

Figure 6.36 Practical potentiometer

One terminal of the standard cell is connected to a tapping on the parallel resistor R_1, and the other terminal may be connected (when the switch S is in the left-hand position and the key K is depressed) to the point B through galvanometer G. A balance may be obtained with the standard cell by adjusting the rheostat R, and the resistance values are such that the p.d. between successive studs on R_3 is then precisely 0.1 V, and the voltage between A and B a little over 1.8 V. In parallel with the shunting resistor R_2 is the slide wire W. The tapping on R_2 enables a true zero, and small negative readings, to be obtained on W. By suitable choice of the resistance of R_2 and W the voltage range covered by the whole slide wire is a little over 0.01 V.

R_6 is a tapped resistor with 10 equal sections and 11 studs. The total resistance of R_6 is made equal to the resistance of two sections of R_3, and R_6 always spans two sections by means of sliding contacts on the R_3 switch as shown. The total resistance of R_6 and two sections of R_3 in combination is therefore always equal to a single section of R_3 and the potential difference between successive tappings on R_6 is then 0.01 V.

The unknown voltage V_x is applied to terminals connected through the galvanometer switch S to the sliding contact on W and to the tapping on the R_6 switch. If the slide-wire dial has 100 divisions, each division represents 0.0001 V. Thus, for the switches in the positions shown, V_x might be 0.2376 V. The maximum V_x readable, assuming that there are some graduations provided beyond the 0.01 position on the slider dial, is $1.7 + 0.1 + 0.0100 = 1.8100$ V.

The resistors R_4 and R_5 are introduced into the circuit by opening switch P_1 and closing P_2; this reduces the p.d.s in the potentiometer circuits to one-tenth of those marked. One division on the slider dial would then correspond with 0.00001 V or $10\,\mu$V.

Medium- to high-precision potentiometers include a 'Standard Cell Setter' so that the actual e.m.f. of the standard cell can be used for standardisation and, subsequently, for rechecking the standardisation during use *without* altering the measuring dials: this is more convenient, although *less* accurate than standardising against the dials. The contact T could be the slider of a 10 turn voltage divider with an indicated range, e.g., of 1.018250 to 1.018750 V in 1 μV steps.

D.C. potentiometers of very high precision have a ratio non-linearity of ± 5 parts in 10^6. A stable current of 50 mA must be provided by an electronic controller, and a high-gain low-noise detector is necessary: this may be a photo-cell galvanometer amplifier with a display sensitivity of 5000 mm/μV in a 10 Ω circuit.

6.10.1.5 Comparator potentiometer (using m.m.f. balance)

The comparator potentiometer is one of a group of precision bridges employing mutual-coupled inductive ratio arms. The bridges exploit the high m.m.f. linearity and discrimination of a constant current in a variable number of turns.

A simplified network is shown in *Figure 6.37*. The measuring loop M is magnetically coupled to the balancing loop B by the turns N_m and N_b, respectively, on a magnetic core, which also carries a feedback winding with N_f turns. The feedback network includes an a.c. modulated sensor which actuates an a.c./d.c. electronic control network C_b to change the direct current I_b until a zero m.m.f. condition is achieved in the core.

6.10.1.6 Operation and standardisation

If I_m and N_b are constant, then for an automatically maintained m.m.f. balance $N_m I_m = N_b I_b$ it follows that linear changes in N_m result in corresponding linear changes in I_b. As N_m has in effect a discrimination and linearity of about 1 in 10^7, these properties are imposed also on I_b. To convert this linear current scale to voltage across R_b, the instrument has to be standardised against a standard cell. The cell, of e.m.f. E_r, is connected (through S1) in

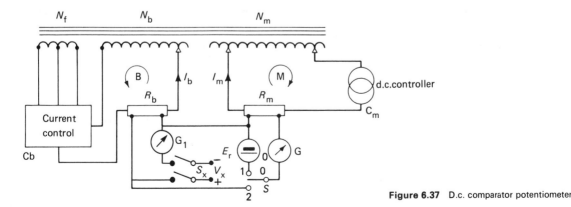

Figure 6.37 D.c. comparator potentiometer

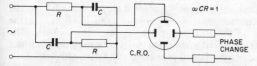

Figure 6.34 Connections for circular time-base

circular time base is made 10 times that of the next higher count rate section. With the latter used to modulate, 10 equal bright and dark peripheral sectors should appear on the trace.

6.10 Potentiometers and bridges

6.10.1 D.C. potentiometers

D.C. potentiometers may be used to measure a direct voltage by comparison with the known ratio of the e.m.f. of a Weston standard cell or electronic solid state reference device. The methods of achieving the linear ratio between the unknown p.d. and the standard-cell e.m.f. are by successive approximation to the balance condition (a) using groups of matched resistors, (b) using opposed m.m.f.s between matched windings and current-ratio resistors (comparator potentiometer) and (c) pulse width modulation (ratio of time interval). The d.c. potentiometer is a ratio device of high precision with the uncertainty of the standard voltage source limiting the ultimate accuracy.

6.10.1.1 Weston standard cell

The highest quality saturated mercury/cadmium–mercury cell has an e.m.f. of about 1.018 620 V at 20 °C. The net e.m.f./temperature coefficient is about $-40\,\mu V/°C$, arising from the two different coefficients of the limbs, each of the order of $350\,\mu V/°C$. It is, therefore, important to avoid a temperature differential across the cell by mounting it in oil or air within a constant temperature enclosure (e.g. for 20 °C ± 10 m °C, the uncertainty can be only $\pm 0.4\,\mu V$).

Standard cells are stable, the e.m.f. falling by, perhaps, $1\,\mu V$ per year. For accurate work, *regular* confirmation of the e.m.f. should be made at a Standards Laboratory.

The internal resistance is 600–800 Ω. The current drawn from a cell should ideally be zero: in practice it should be limited to a few nanoamperes for a few seconds.

Should a cell be compared with an electronic solid state voltage reference (below), the cell must be presented and removed while the electronic source is switched on.

6.10.1.2 Electronic solid state e.m.f. reference

These devices use the constant voltage property of Zener diodes to deliver highly stable d.c. voltages, usually at 1 V and 10 V levels for d.v.m. calibrations, and 1.018 61 V to simulate a Weston cell for d.c. potentiometric standardisation. Up to four diodes (each stabilising at about 6.3 V) are used, having been carefully selected for stability, low noise and low temperature/voltage coefficient properties. After further ageing, these temperature-stabilised monolithic devices may be used either separately with a carefully designed operational amplifier, or connected in cascade in specialised networks to achieve the required output p.d. by potential division using highly stable resistors.

Cropico Ltd (London) and Tinsley Co. (London) claim reproducible e.m.f.s to within 1 or 2 p.p.m. well within an hour from switch-on. A recent experiment with a Tinsley Type 5646, calibrated in London and measured in California, showed agreement within $1\,\mu V$ (after allowing for the different sizes of the British and US volts).

It would appear that these devices are excellent transportable standards of voltage (which should be reliable to within $2\,\mu V/year$) and they are much more convenient than Weston cells in this respect. Except for the most extreme requirements of accuracy, they could be used as laboratory standards, provided that at least three separate units are available for inter-comparison. With more experience and/or minor refinement to these devices, they could shortly replace the Weston cell completely, since at present the Weston cell can be used as a transportable standard to less than $0.5\,\mu V$, but only if very great care is exercised.

Typical electronic devices can deliver, say, 5 mA, 10 V from 2 mΩ and 1.0186 V from 2 kΩ.

6.10.1.3 Potentiometer principle (*using groups of matched resistors*)

In *Figure 6.35* AB is a manganin wire of length a little greater than 1 m, through which the current I from a 2 V secondary cell may be varied by means of R. The current is adjusted so that the p.d.

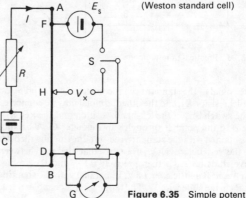

Figure 6.35 Simple potentiometer

between points F and D, 1.0186 m apart, balances the e.m.f. of the Weston standard cell, the balance condition being obtained by use of the switch S and galvanometer G. The volt drop along the wire is now 1 mV per mm length. The switch S may now be set to apply the unknown voltage V_x, and a new balance point H found. Then the length DH in mm represents the unknown voltage in mV.

This very simple arrangement is inconvenient in practice. It suffers also from the disadvantage that the precision of the voltage measurement depends upon the precision with which the point of contact of the slider is known. For example, if the length DH is known only to the nearest millimetre, the precision of the voltage measurement would, in general, be less than 1 in 1000. Much greater precision than this is often required, and to obtain it (and to secure a more compact form of apparatus) some elaboration and refinement in the circuit elements is required. Many practical forms of d.c. potentiometer have been devised.

6.10.1.4 Practical potentiometer

The potentiometer in *Figure 6.36* is supplied across AB from a 2 V secondary cell through a rheostat R. Resistors R_2, R_3 and R_4 are connected in series between A and B, but the conditions first considered are with R_4 short-circuited by the switch at P_1. The resistor R_3 is divided into equal sections with 19 tappings brought out to studs, and there is an additional stud connected to a tapping on R_2.

used to match the source to the standard 50 Ω input impedance of a high-frequency c.r.o. (100 MHz and above), an important matter in that a quarter wavelength at 100 MHz is about 75 cm, and severe standing-wave effects may otherwise occur to invalidate the measurements.

6.9.4.2 Earths and grounds

Ground planes (e.g. extended screens) and earth planes (e.g. trunking) constitute low-impedance paths through which interfering signals can circulate, of a wide variety of frequencies and waveforms. The arbitrary connection of screens to earth and ground terminals can have the unwelcome result of increasing noise, due to earth-loop currents. The latter must not enter the input signal lines of either the system or the c.r.o.; and the interference c⁻⁻ be minimised by using a single ground/earth connection to se; 'e the whole equipment, if this is possible. The common connection should be made at the zero-potential input signal lead.

6.9.5 Calibration

Operational specifications will include performance statements, typically: voltage accuracy, $\pm 3\%$; time-base frequency $\pm 3\%$; Y-amplifier linearity, $\pm 3\%$, phase shift, 1°; pulse response, $\pm 5\%$ after 2 ns; sweep delay accuracy, $\pm 2\%$. The calibration facility provided usually consists of a 1 or 2 kHz square waveform with a 3 μs rise time, with both the frequency and voltage amplitude within $\pm 1\%$.

The practical application of the c.r.o. as a test instrument includes (a) the general diagnostic display of waveforms, and (b) the testing of a product during development or manufacture in relation to its accuracy and quality control. For (a) the built-in calibration facility is adequate; but for (b) it is prudent (or essential) to make regular checks of the actual parameters of the self-calibration facility and of the basic characteristics of the instrument. Such tests would include assessments of the accuracy on alternating voltages, and the ratio and phase angle of the amplifiers and attenuators over the frequency range. The most important and critical test for a c.r.o., which will inevitably be used with modern digital networks, is its response to h.f. square waveforms. The rise time is quite easily measured, but the initial resonant frequency of the 'flat' part of the waveform and its rate of attenuation are important in assessing the actual duration of this part.

The quality and usefulness of these tests is ultimately related to the readability of the display, which would be no better than $\pm 0.3\%$ in a 10 cm trace. The quality of the standard reference devices would, for example, be $\pm 0.1\%$ for a variable-frequency alternating voltage source; hence, there would be a total measurement uncertainty of about 0.4% in the $\pm 1\%$ waveform. It follows that the actual c.r.o. voltage source must be correct to within 0.6% if it is to be used with confidence as a $\pm 1\%$ source. It is a fundamental requirement that the quality of all the reference standards be known from recent linearity and traceability tests.

6.9.6 Applications

The principal application is for diagnostic testing. In addition, the c.r.o. is used to monitor the transfer characteristics of devices and systems, as well as for numerical measurements. Among special uses are to be found the television camera, electron diffraction camera and electron microscope.

Single-deflection measurements. The tube functions with a single pair of deflecting plates as a voltmeter or ammeter, with the advantages of being free from damping, unaffected by change of frequency or temperature or over-deflection, and imposing only a minute load on the test circuit. Examples are the monitoring of signals in broadcast and recording studios, modulation checks, indication of motor peak starting currents, thickness measurement, null detection and the measurement of current and voltage in networks of very low power.

Differential measurements. Similar or related phenomena are compared by use of both pairs of deflecting plates, the resulting Lissajous figure being observed. In this way phase difference can be measured, frequencies accurately checked against a standard, armature windings tested, modulation indicated by a stationary waveform, and the distortion measured in receivers, amplifiers and electro-acoustic equipment. *Figure 6.32* is a Lissajous figure comparing two frequencies by counting the horizontal and vertical loops, giving the ratio 4/5. The diagram in *Figure 6.33* shows the connections for testing the distortion of an amplifier:

Figure 6.32 Lissajous figure

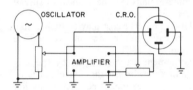

Figure 6.33 Distortion test on amplifier

the oscillator is set to a known frequency, the two pairs of deflectors connected, respectively, to the input and output (with suitable attenuators). A straight-line display indicates absence of distortion, while curvature or looping indicates distortion.

Repetitive time base. Using a linear time base enables sustained waveforms to be displayed for machines, transformers and rectifiers, rapid operations to be timed, and surge phenomena to be shown with the aid of a recurrent-surge generator.

Single sweep. Non-repetitive transients, such as those arising by lightning or switching, are most readily traced by use of the continuously evacuated tube, but sensitive sealed-off tubes can also be employed. In either case a non-repetitive time base is used consisting of the voltage of a capacitor discharged through a resistor and triggered by a gap. The transient discharge is normally obtained by use of a surge generator for power system plant testing, but other single transients, such as the small voltages of cardiac origin, may also require to be recorded.

Independent bases. The c.r.o. is well suited to the display of quantities related by some variable other than time. The current due to an impressed voltage can be displayed on a frequency base as a resonance curve, radio receivers aligned, and B/H loops taken for steel samples. Transistor curve tracers are useful applications capable of giving a complete family of characteristics in a single display.

Modified time bases. There are several modified forms of time base. If two voltages of the same amplitude and frequency are displaced in time phase by a quarter-period and applied, respectively, to the X and Y plates as in *Figure 6.34*, they will produce a circular trace. One application, in decade digital logic circuits, involves a Z modulation of the c.r.o. grid. To check that a frequency divider section is working correctly, the period of the

frequency range; the pulse response must be fast (e.g. 2 ns); and the drift and noise must be small (e.g. 3 µV).

Amplifiers have d.c. and a.c. inputs. The d.c. coupling is conventional, to minimise the use of capacitors (which tend to produce low-frequency phase distortion). The d.c. inputs also permit the use of a reference line. The a.c. input switch introduces a decoupling capacitor.

Two separate Y amplifiers are often provided; in conjunction either with one switched or with two separate X amplifiers, they permit the examination of two voltages simultaneously. It is important to distinguish the dual-trace oscilloscope (in which a single beam is shared by separate signal amplifiers) and the true double-beam device (with a gun producing two distinct beams).

6.9.3 Instrument selection

6.9.3.1 (a) Analogue c.r.o.

A general purpose c.r.o. must compromise between a wide frequency range and a high amplification sensitivity, because high gain is associated with restricted bandwidth. An economically priced c.r.o. with solid state circuitry may have a deflection sensitivity of 5 mV/cm, bandwidth 10 MHz and a 3 kV accelerating voltage for a 10 cm useful display. For greater flexibility, many modern c.r.o.s comprise a basic frame unit with plug-in amplifiers; the frame carries the tube, all necessary power outlets and basic controls; the plug-in units may have linear or special characteristics. The tube must be a high-performance unit compatible with any plug-in unit. The cost may be higher than that of a single function c.r.o., but it may nevertheless be economically advantageous.

Many classes of measurement are catered for by specially designed oscilloscopes. In the following list, the main characteristic is given first, followed by the maximum frequency, deflection sensitivity and accelerating voltage, with general application notes thereafter.

Low-frequency (30 MHz, or 2 MHz at high gain; 0.1 mV/cm–10 V/cm; 3–10 kV). General purpose oscillography, for low-frequency system applications. Usually, high X-sweep speeds are needed for transient displays and complex transducer signals. Two traces available with two-beam or chopped single-beam tube. Single-shot facility. Y-amplifier signal delay useful for 'leading-edge' display. Rise times 10–30 ns.

Medium-frequency (100 MHz; 0.1 mV/cm; 8–15 kV). General-purpose, including wide-band precision types with a rise time of 3–10 ns.

High-frequency (non-sampling) (275 MHz; 10 mV/cm; 20 kV). May include helical transmission-line deflecting plates. Rise time 1.5 ns for testing digital system and computer displays.

Variable persistence storage c.r. tube for h.f. transients.

Very-high-frequency (sampling) (12 000 MHz; 5 mV/cm; 20 kV). The v.h.f. waveforms are 'instantaneously' measured and optically stored, n times at n successively retarded instants during the period of each of n successive waveforms (with n typically 1000). The n spots accumulate in the correct time sequence as a single waveform derived from the n waveforms sampled.

Large-screen (20 MHz; 60 mV/cm; 28 kV). Screen area up to 0.25 m². Useful for industrial monitoring, classroom display and computer graphics. (Magnetic deflection gives a shorter overall length of tube, probably with improved focusing towards the edges but with a lower maximum frequency.)

6.9.3.2 (b) Digital c.r.o.

The display will usually be formed from a sequence of closely connected dots, although, in one instrument, analogue-type networks are used to connect the dots as continuous lines. The digital c.r.o., having digitised the analogue input data, is then able to perform many of the analytical or statistical functions of a specialist analyser while still providing the routine applications of the conventional analogue c.r.o. Since any waveform, e.g. a fast transient, can be stored in digital form, the instrument has the inherent property of a storage-type c.r.o. (although of much lower maximum frequency) while using a normal c.r. tube phosphor: close examination of part of the trace can be achieved by withdrawing data from store at any selected slow rate (i.e. time base expansion property)—also permitting photography with normal film speeds. A useful design feature is that the stored wave can be redisplayed for inter-comparison with the extant waveform resulting from modifications to the network under test. The high-speed stored waveform can be permanently recorded on a slow-speed pen or galvanometer-type $X-Y$ recorder by withdrawing the data from store at a slow rate. Digital c.r.o.s have maximum bandwidths of about 7 MHz; resolution, linearity and accuracy better than 0.5% f.s.d.; maximum speed time per point 50 ns (20 MHz digitising rate): data points per waveform 100–4000 and storage capacity 4096 words by 12 bits.

6.9.3.3 Other features

With all c.r.o.s, the $X-Y$ cartesian co-ordinate display can be augmented by Z-modulation. This superimposes a beam-intensity control on the $X-Y$ signals to produce the illusion of three dimensions. The beam control may brighten one spot only, or gradually 'shape' the picture.

Oscilloscopes are supplemented by such accessories as special probes for matching and/or attenuating the signal to the signal/cable interface. Anti-microphony cables are available which minimise cable-generated noise and capacitance change. Many types of high-speed camera can be used, including single-shot cameras that can employ ultraviolet techniques to 'brighten' a record from a trace of low visual intensity.

6.9.4 Operational use

To avoid degradation of the signal source, a c.r.o. has a high input impedance (1 MΩ, 20 pF) and a good common-mode rejection (50 dB). The quality can be considerably impaired by impedance mismatch and incorrect signal/earth (or ground) connections.

6.9.4.1 Impedance matching

Voltage signals are often presented to a c.r.o. through wires or coaxial cables terminated on crocodile clips. Although this may be adequate at low frequencies, it can behave like a badly matched transmission line, causing attenuation distortion and phase distortion. The earth screen of a coaxial cable is not fully effective, and in consequence the cable can pick up local radiated interference. In a normal cable the central conductor is not bonded to the insulation, and flexure of the cable may distort the signal waveform because of changes in the capacitance distribution, or generated noise. The cable capacitance of about 50 pF/m is additive to the c.r.o. input capacitance, and will distort the signal, particularly if the frequency and source impedance are high.

These adverse effects can be mitigated by use of fully screened probes and cables of low and constant capacitance associated with low loss. The probes may be *active* (including an impedance matching unity-gain amplifier) or *passive* (including resistance with adjustable capacitance compensation); attenuators are incorporated in each to reduce both the signal level and the capacitive loading of the source, the compensation being most effective for high-impedance sources. The active probe can be

Table 6.3 Classification of phosphors

Phosphor	Persistence	Colour of trace	Relative brightness	Application
P2	medium/short	yellow/green	6.5	low repetition rates (general oscillography)
P4	medium	white	5.0	high-contrast displays (monochrome TV)
P7	long	white or yellow/green	4.5	long-persistence low-speed (radar)
P11	medium/short	blue	2.5	high writing speeds (photography)
P31	medium/short	green	10	high-brightness, general use

the normal 'writing' properties of the tube. Storage, at reduced brightness, is possible for up to 8 h.

Phosphor materials are insulators. The current circuit for all the incident beam electrons is completed by an equal number of low-energy electrons being attracted along various paths through the vacuum to the positive conducting-graphite coating. The result of secondary emission of electrons is to leave the screen with a positive charge only a little lower in potential than that of the graphite coating; thus, the surplus secondary emission electrons return to the phosphor-coated screen surface. This mechanism works perfectly at the voltages usually employed in oscillography and television (3–20 kV), also in small projection-tube sealed-off high-speed oscillographs (10–25 kV). Above 25 kV the mechanism is unsatisfactory and the dispersion of charge has to be supplemented by other means.

The power concentration in the luminous spot of a c.r. tube is very high. In direct-vision television tubes the momentary luminance may exceed 10 cd/mm². A 3 mA 80 kV beam of a large-screen projection tube may be concentrated in an area of 0.1 mm², representing a power density of 2 kW/mm². (By comparison, melting tungsten radiates about 10 W/mm².) The width of the trace in a conventional 10 kV c.r. tube with a beam current of 10 μA is only 0.35 mm.

6.9.2 Deflection amplifiers (analogue c.r.o.)

Amplifiers are built into c.r. oscilloscope equipments to provide for time base generation and the amplification of small signals. The signals applied to the vertical (Y) deflecting plates are often repetitive time functions. To display them on a time base, the vertical displacements must be moved horizontally across the screen at uniform speed repeatedly at intervals related to the signal frequency, and adjusted to coincidence by means of a synchronising signal fed to the time base from the Y amplifier. An additional constraint may be the requirement to start the time base from the left side of the screen at the instant the Y-plate signal is zero: a synchronising trigger signal is fed to the time base circuit to initiate each sweep. Finally, the spot must return (flyback) to the origin at the end of each sweep time. Ideally, flyback should be instantaneous: in practice it takes a few score nanoseconds. During flyback the beam is normally cut off to avoid return-trace effects, and this is the practice for television tubes.

Linear time bases are described in terms either of the sweep time per cm measured on the horizontal centre line of the graticule, or of the frequency. The repetition frequency range is from about 0.2 Hz to 2.5 MHz.

6.9.2.1 Ramp voltage (saw-tooth) generator

If a constant current I is fed to a capacitor C, its terminal voltage increases linearly with time. An amplified version of the voltage is applied to the X plates. Various sweep times are obtained by selecting different capacitors or other constant current values.

The method shown in *Figure 6.31* employs a d.c. amplifier of high gain $-G$ with a very large negative feedback coupled through an ideal capacitor C_f. The amplifier behaves in a manner dictated by the feedback signal to node B rather than by the small direct input voltage E_s. The resultant signal v_i is small enough for node B to be considered as virtually at earth potential. Analysing the network based on the 'virtual earth' principle, then $v_i = 0$ and $i_i = 0$, the input resistance R_i being typically 1 MΩ. Then $i_s + i_f = 0$, whence

$$(E_s/R_s) + C_f(dv_o/dt) = 0 \text{ giving } v_o = -(1/C_f R_s) \int E_s \cdot dt$$

The output voltage v_o increases linearly with time to provide the ramp function input to the X amplifier.

6.9.2.2 Variable delay

When two similar X amplifiers, A and B, are provided, B can be triggered by A; if B is set for, say, five times the frequency of A and this had a variable delay facility, then any one-fifth of the complete A waveform display can be extended to fill the screen. Both complete and expanded waveforms can be viewed simultaneously by automatic switching of A and B to the X plates during alternate sweeps.

6.9.2.3 Signal amplifiers

The X- and Y-plate voltages for maximum deflection in a modern 10 cm c.r. tube may lie between 30 and 200 V. As many voltage phenomena of interest have amplitudes ranging between 500 V and a few millivolts, both impedance matched attenuators and voltage amplifiers are called for. The amplifiers must be linear and should not introduce phase distortion in the working

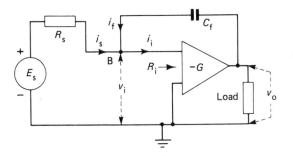
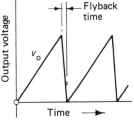

Figure 6.31 Time-base operational amplifier

a constant. If i_b is the beam current, then the apparent diameter of the spot is given approximately by $\sqrt{(i_b/J_s)}$.

The voltage V_d applied to the deflection plates produces an angular deflection δ given by

$$V_d = \tfrac{1}{2}\delta V d / L_d \qquad (6.2)$$

Deflection is possible, however, only up to a maximum angle α, such that

$$\alpha L_d + \theta L = \tfrac{1}{2} d \qquad (6.3)$$

From these three expressions, certain quantities fundamental to the rating of a c.r. tube with electric field deflection can be derived.

6.9.1.3 Definition

This is the maximum sweep of the spot on the screen, $D = 2\alpha L$, divided by the spot diameter, i.e. $N = D/s$. In television N is prescribed by the number of lines in the picture transmission. In oscillography N is between 200 and 300. Better definition can be obtained only at the expense of deflection sensitivity.

6.9.1.4 Specific deflection sensitivity

This is the inverse of the voltage that deflects the spot by a distance equal to its own diameter, i.e. which produces the smallest perceptible detail. The corresponding deflection angle is $\delta_0 = s/L$. Use of Equations (6.1)–(6.3) gives for the specific deflection sensitivity

$$S = k_1 L / [(L/L_d + \alpha/\theta)\sqrt{(Vi_b)}] \qquad (6.4)$$

where k_1 is a constant. Thus, the larger the tube dimensions the better the sensitivity at the smallest beam power consistent with adequate brightness of trace.

6.9.1.5 Maximum recording speed

The trace remains visible so long as its brightness exceeds a certain minimum. The brightness is proportional to the beam power density VJ, to the luminous efficiency ς of the screen, and to the time s/U during which the beam sweeps at speed U over one element of the screen. Hence,

$$U = k_2 s J V \varsigma \qquad (6.5)$$

Recording speeds up to 40 000 km/s have been achieved with sealed-off tubes, permitting single-stroke photography of the trace.

6.9.1.6 Maximum specified recording speed

Unless compared with the spot size s, the value of U does not itself give adequate information. The specific speed $U_s = U/s$ is more useful: it is the inverse measure of the shortest time-interval in which detail can be recorded. Combining Equations (6.5) and (6.1)

$$U_s = k_2 J V \varsigma = k_3 \varsigma (V\theta)^2 \qquad (6.6)$$

The product $\varsigma(V\theta)^2$ is the essential figure in the speed rating of a c.r. tube.

6.9.1.7 Limiting frequency

A high recording speed is useful only if the test phenomena are recorded faithfully. The c.r. tube fails in this respect at frequencies such that the time of passage of an electron through the deflection field is more than about one-quarter of the period of an alternating quantity. This gives the condition of maximum frequency f_m:

$$f_m = 1.5 \times 10^5 \sqrt{V}/L_d \qquad (6.7)$$

which means that the smallest tube at the highest acceleration voltage has the highest frequency limit. This conflicts with the condition for deflection sensitivity. Combining the two expressions gives the product of the two:

$$(Sf_m) = k_4/[1 + (L_d/L)(\alpha/\theta)]\sqrt{i_b} \qquad (6.8)$$

Compromise must be made according to the specific requirement. It may be observed that the optimum current i_b is not zero as might appear, because for such a condition the convergence angle θ is also zero. In general, the most advantageous current will not differ from that for which the electron beam just fills the aperture.

Limitation of c.r. tube performance by space charge is an effect important only at low accelerating voltages and relatively large beam currents. For these conditions the expressions above do not give practical ratings.

6.9.1.8 Fluorescent screen

The kinetic energy of each incident electron is converted into light owing to the exchange of energy which occurs within the phosphors coated on the inside of the screen. The result, *luminescence*, is usually described as either fluorescence or phosphorescence. *Fluorescence* is the almost instantaneous energy conversion into light, whereas *phosphorescence* is restricted to the phosphors which 'store' the converted energy and release it, as light, after a short delay. The *afterglow* can last many seconds. The respective terms are applied to the phosphors in accordance with their dominant light-emitting behaviour; however, each property is always present in a phosphor. In most c.r. tubes fluorescent phosphors are used with a very short afterglow (40 µs); a longer afterglow would blur a quickly changing display.

The most common fluorescent substances are silicates and sulphides, with calcium tungstate for photographic work. An excellent zinc silicate occurs naturally in the mineral willemite, but all other substances are prepared artificially. After purification, the substances are activated by the addition of very small quantities of suitable metals, and then ground to a powder of grain size 5–10 µm (for sulphides 10–50 µm).

In applying the powder to the glass several methods are used. Very finely ground silicates can be applied without a binder, by setting the powder from water, or an electrolyte. Sulphides are usually compounded with the glass wall by means of a binder such as sodium silicate; good results have been obtained also by baking them directly into the glass. In choosing the method two conflicting considerations must be taken into account. Baking or compounding gives good thermal contact, and prevents overheating of the powder particles by the electron beam. But at the same time good optical contact is established with the glass, with the result that light from the powder enters the glass also at angles exceeding the total reflection angle. This part of the light cannot escape at the viewing side, but is reflected and forms a luminous 'halo' which blurs the picture.

While silicates fluoresce usually green or blue-green, white fluorescence can be obtained by mixing cadmium sulphide and zinc cadmium sulphide in suitable proportions. The luminous efficiency of modern fluorescent powders is very high. Zinc silicate emits up to 2.5 cd/W of beam power, and some sulphides are even more effective. The luminous efficacy increases with voltage up to about 10 kV, beyond which it decreases, partly because fast electrons are not stopped by the crystal grains.

The range of phosphors is classified by 'P' numbers, as in *Table 6.3*.

The light persistance of the most widely used phosphor (P31) is 40 µs; however, there are tubes available, using this phosphor, which have both variable persistence and 'mesh storage' facilities. Adjustable persistence enables very-low-frequency signals, lasting for a minute, to be viewed as a complete trace while retaining

6.9.1 Cathode-ray tube

The c.r. tube takes its name from Pluecker's discovery in 1859 of 'cathode rays' (a better name would be 'electron beam', for J. J. Thomson in 1897 showed that the 'rays' consist of high-speed electrons). The developments by Braun, Dufour and several other investigators enabled Zworykin in 1929 to construct the essential features of the modern c.r. tube, namely a small thermionic cathode, an electron lens and a means for modulating the beam intensity. The block diagram (*Figure 6.28*) shows the principal components of a modern cathode-ray oscilloscope. The tube is

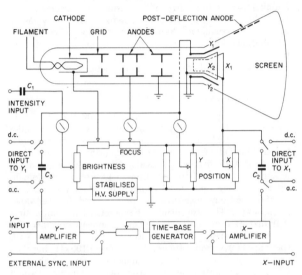

Figure 6.28 Analogue cathode-ray oscilloscope circuits

designed to focus and deflect the beam by means of structured electric field patterns. Electrons are accelerated to high speed as they travel from the cathode through a potential rise of perhaps 15 kV. After being focused, the beam may be deflected by two mutually perpendicular X and Y electric fields, locating the position of the fluorescing 'spot' on the screen.

6.9.1.1 Electron gun and beam deflection

The section of the tube from which the beam emerges includes the heater, cathode, grid and the first accelerating anode; these collectively form the *electron gun*. A simplified diagram of the connections is given in *Figure 6.29*. The electric field of the relatively positive anode partly penetrates through the aperture of the grid electrode and determines on the cathode a disc-shaped area (bounded by the $-15\,\text{kV}$ equipotential) within which electrons are drawn away from the cathode: outside this area

they are driven back. With a sufficiently negative grid potential the area shrinks to a point and the beam current is zero. With rising grid potential, both the emitting area and the current density within it grow rapidly.

The diagram also shows the electron paths that originate at the emitting area of the cathode. It can be seen that those which leave the cathode at right angles cross the axis at a point not far from the cathode. This is a consequence of the powerful lens action of the strongly curved equipotential surfaces in the grid aperture. But as electrons are emitted with random thermal velocities in all directions to the normal, the 'cross-over' is not a point but a small *patch*, the image of which appears on the screen when the spot is adjusted to maximum sharpness.

The conically shaped beam of divergent electrons from the electron gun have been accelerated as they rise up the steep potential gradient. The electrical potentials of the remaining nickel anodes, together with the post-deflection accelerating anode formed by the graphite coating, are selected to provide additional acceleration and, by field shaping, to refocus the beam to a spot on the screen. To achieve sharply defined traces it is essential that the focused spot should be as small as possible; the area of this image is, in part, dictated by the location and size of the 'point source' formed in front of the cathode. The rate at which electrons emerge from the cathode region is most directly controlled by adjustment of the negative potential of the grid; this is effected by the external *brightness control*. A secondary result of changes in the brightness control may be to modify, slightly, the shape of the point source. The slight de-focusing effect is corrected by the *focus control*, which adjusts an anode potential in the electron lens assembly.

6.9.1.2 Performance (electrostatic deflection)

If an electron with zero initial velocity rises through a potential difference V, acquiring a speed u, the change in its kinetic energy is $Ve = \frac{1}{2}mu^2$, so that the speed is

$$u = \sqrt{(2Ve/m)} \simeq 6 \times 10^5 \sqrt{V} \,\text{m/s}$$

using accepted values of electron charge, e, and rest-mass, m. The performance is dependent to a considerable extent on the beam velocity, u.

The deflection sensitivity of the tube is increased by shaping the deflection plates so that, at maximum deflection, the beam grazes their surface ($Y_1 Y_2$, *Figure 6.28*). The sensitivy and performance limits may be assessed from the *idealised* diagrams (and notation) in *Figure 6.30*. Experience shows that a good c.r. tube produces a beam-current density J_s at the centre of the luminous spot in rough agreement with that obtained on theoretical grounds by Langmuir, i.e.

$$J_s = K J_c \theta^2 V / T \tag{6.1}$$

where J_c is the current density at the cathode, V the total accelerating voltage, T the absolute temperature of the cathode emitting surface, θ one-half of the beam convergence angle, and k

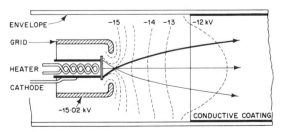

Figure 6.29 Electron beam in electron gun

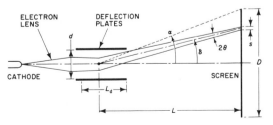

Figure 6.30 Deflection parameters

generally cover a more restricted range. As voltage sources, the output power is low (100 mW–2 W). The total harmonic distortion (i.e. r.m.s. ratio between harmonic voltage and fundamental voltage) must be low, particularly for testing quality amplifiers, e.g. 90 dB rejection is desirable and 60 dB (10^3:1) is normal for conventional oscillators. Frequency selection by dials may introduce $\pm 2\%$ uncertainty, reduced to 0.2% with digital dial selection which also provides 0.01% reproducibility. Frequency-synthesiser function generators should be used when a precise, repeatable frequency source is required. Amplitude stability and constant value over the whole frequency range is often only 1–2% for normal oscillators.

6.8.8 Electronic analysers

Analysers include an important group of instruments which operate in the frequency domain to provide specialised analyses, in terms of energy, voltage, ratio, etc. (a) to characterise networks, (b) to decompose complex signals which include analogue or digital intelligence, and (c) to identify interference effects and non-linear cross-modulations throughout complex systems.

The frequency components of a complex waveform (Fourier analysis) are selected by the sequential, or parallel, use of analogue to digital filter networks. A frequency-domain display consists of the simultaneous presentation of each separate frequency component, plotted in the x-direction to a base of frequency, with the magnitude of the y-component representing the parameter of interest (often energy or voltage scaled in dB).

Instruments described as network, signal, spectrum or Fourier analysers as well as digital oscilloscopes have each become so versatile, with various built-in memory and computational facilities, that the distinction between their separate objectives is now imprecise. Some of the essential generic properties are discussed below.

6.8.8.1 Network analysers

Used in the design and synthesis of complex cascaded networks and systems. The instrument normally contains a swept-frequency sinusoidal source which energises the test system and, using suitable connectors or transducers to the input and output ports of the system, determines the frequency-domain characteristic in 'lumped' parameters and the transfer function equations in both magnitude and phase: high-frequency analysers often characterise networks in 'distributed' s-parameters for measurements of insertion loss, complex reflection coefficients, group delay, etc. Three or four separate instruments may be required for tests up to 40 GHz. Each instrument would have a wide dynamic range (e.g. 100 dB) for frequency and amplitude, coupled with high resolution (e.g. 0.1 Hz and 0.01 dB) and good accuracy, often derived from comparative readings or ratios made from a built-in reference measurement channel.

Such instruments have wide applications in R & D laboratories, quality control and production testing—and they are particularly versatile as an automatic measurement testing facility, particularly when provided with built-in programmable information and a statistical storage capability for assessment of the product. Data display is by X–Y recorder and/or c.r.o. and the time-base is locked to the swept frequency.

6.8.8.2 Spectrum analysers

Spectrum analysers normally do not act as sources; they employ a swept-frequency technique which could include a very selective, narrow-bandwidth filter with intermediate frequency (i.f.) signals compared, in sequence, with those selected from the swept frequency; hence, close control of frequency stability is needed to permit a few per cent resolution within, say, a 10 Hz bandwidth filter. Instruments can be narrow range, 0.1 Hz–30 kHz, up to wide range, 1 MHz–40 GHz, according to the precision required in the results; typically, an 80 dB dynamic range can have a few Hz bandwidth and 0.5 dB amplitude accuracy. Phase information is not directly available from this technique. The instrument is useful for tests involving the output signals from networks used for frequency mixing, amplitude and frequency modulation or pulsed power generation. A steady c.r.o. display is obtained from the sequential measurements owing to digital storage of the separate results when coupled with a conventional variable-persistence c.r. tube (a special c.r. tube is not required for a digital oscilloscope): the digitised results can, in some instruments, be processed by internal programs for (a) time averaging of the input parameters, (b) probability density and cumulative distribution, and (c) a comprehensive range of statistical functions such as average power or r.m.s. spectrum, coherence, autocorrelation and cross-correlation functions, etc.

6.8.8.3 Fourier analysers

Fourier analysers use digital signalling techniques to provide facilities similar to the spectrum analysers but with more flexibility. These 'Fourier' techniques are based upon the calculation of the Discrete Fourier Transform using the Fast Fourier Transform Algorithm (which calculates the magnitude and phase of each frequency component using a group of time-domain signals from the input signal variation of the sampling rate); it enables the long measurement time needed for very-low-frequency ($\ll 1$ Hz) assessments to be completed in a shorter time than that for a conventional swept measurement—together with good resolution (by digital translation) at high frequencies. Such specialised instruments are usable only up to 100 kHz but they are particularly suitable for examining low-frequency phenomena such as vibration, noise transmission through media and the precise measurement of random signals obscured by noise, etc.

6.8.9 Data loggers

Data loggers consist of an assembly of conditioning amplifiers which accept signals from a wide range of transducers or other sources in analogue or digital form; possibly a small computer program will provide linearisation and other corrections to signals prior to computational assessments (comparable to some features of a spectrum or other class of analyser) before presenting the result in the most economical manner (such as a graphical display). The data logger, being modular in construction, is very flexible in application; it is not limted to particular control applications and by the nature of the instrument can appear in various guises.

6.9 Electronic oscillography

The cathode-ray oscilloscope (c.r.o.) is one of the most versatile instruments in engineering. It is used for diagnostic testing and monitoring of electrical and electronic systems, and (with suitable transducers) for the display of time-varying phenomena in many branches of physics and engineering. Two-dimensional functions of any pair of repetitive, transient or pulse signals can be developed on the fluorescent screen of a c.r.o.

These instruments may be classified by the manner in which the *analogue* test signals are conditioned for display as (a) the 'analogue c.r.o.', using analogue amplifiers, and (b) the 'digital c.r.o.', using A/D converters at the input with 'all-digit' processing.

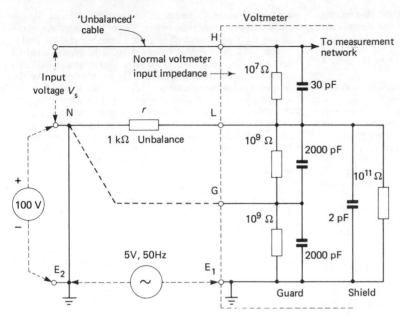

Figure 6.27 Typical guard and shield network for digital voltmeter

(2) High input impedance, and the effect on likely sources of the *dynamic* impedance.
(3) Electrical noise rejection, assessed and compared with (a) the common-mode rejection ratio based on the input and guard terminal networks and (b) the actual inherent noise rejection property of the measuring principle employed.
(4) Requirements for binary-coded decimal, IEEE-488, storage and computational facilities.
(5) Versatility and suitability for use with a.c. or impedance converter units.
(6) Use with transducers (in which case 3 is the most important single factor).

6.8.4.14 Calibration

It will be seen from *Table 6.2* that digital voltmeters should be recalibrated at intervals between 3 and 12 months. Built-in self-checking facilities are normally provided to confirm satisfactory operational behaviour, but the 'accuracy' check cannot be better than that of the included standard cell or Zener diode reference voltage and will apply only to *one* range. If the user has available some low-noise stable voltage supplies (preferably batteries) and some high-resistance helical voltage dividers, it is easy to check logic sequences, resolution, and the approximate ratio between ranges. The accuracy of the 1 V range can be tested with an external Weston standard cell, provided that the cell voltage is known to within $\pm 3\,\mu V$ by recent NPL calibration and that the digital logic voltage noise which appears at the input terminals will not harm the cell.

6.8.5 Digital wattmeters[13,14,15]

Digital wattmeters employ various techniques for sampling the voltage and current waveforms of the power source to produce a sequence of instantaneous power representations from which the average power is obtained. One new NPL method uses high-precision successive-approximation A/D integrated circuits with the multiplication and averaging of the digitised outputs being completed by computer[11,12]. A different 'feedback time-division multiplier' principle is employed by Yokogawa[17,18].

The new NPL prototype digital wattmeter uses 'sample and hold' circuits to capture the two instantaneous voltage (v_x) and current (i_x) values, then it uses a novel double dual-slope multiplication technique to measure the average power from the mean of numerous instantaneous ($v_x i_x$) power measurements captured at precise intervals during repetitive waveforms. Referring to the single dual-ramp (*Figure 6.25a*) the a.c. voltage v_x (captured at T_0) is measured as $v_x T_1 = V_r(T_2 - T_1)$. If a second voltage (also captured at T_0) is proportional to i_x and integrated for $T_2 - T_1$, then, if it is reduced to zero (by V_r) during the time $T_3 - T_2$ it follows that $i_x(T_2 - T_1) = V_r(T_3 - T_2)$.

The instantaneous power (at T_0) is $v_x i_x = K(T_3 - T_2)$ and, by scaling, the mean summation of all counts such as $T_3 - T_2$ equals the average power. The prototype instrument measures power with an uncertainty of about $\pm 0.03\%$ f.s.d. (and $\pm 0.01\%$, between 50 and 400 Hz, should be possible after further development).

The 'feed-back time division multiplier' technique develops rectangular waveforms with the pulse height and width being proportional, respectively, to the instantaneous voltage (v_x) and current (i_x); the average 'area' of all such instantaneous powers ($v_x i_x$) is given (after l.p. filtering) as a d.c. voltage measured by a DVM scaled in watts; such precision digital wattmeters, operating from 50 to 500 Hz, have $\pm 0.05\%$ to $\pm 0.08\%$ uncertainty for measurements up to about 6 kW.

6.8.6 Energy meters

A *solid state* single-phase meter, Sangamo model S720, claims to fulfil the specifications of its traditional electromechanical counterpart while providing a reduced uncertainty of measurement coupled with improved reliability at a similar cost. The instrument should provide greater flexibility for reading, tariff and data communication of all kinds, e.g. instantaneous power information, totalised energy, credit limiting and multiple tariff control. The instrument is less prone to fraudulent abuse.

6.8.7 Electronic oscillators

Oscillators usually provide sinusoidal voltage output (occasionally ramp and square wave also) between 5 Hz and a few MHz, although the more accurate and stable instruments

6.8.4.10 Noise limitation

The information signal exists as the p.d. between the two input leads; but each lead can have unique voltage and impedance conditions superimposed on it with respect to the basic *reference* or *ground* potential of the system, as well as another and different set of values with respect to a local *earth* reference plane.

An elementary electronic instrumentation system will have at least one ground potential and several earth connections—possibly through the (earthed) neutral of the main supply, the signal source, a read-out recorder or a cathode ray oscilloscope. Most true earth connections are at different electrical potentials with respect to each other, owing to circulation of currents (d.c. to u.h.f.) from other apparatus, through a finite earth resistance path. When multiple earth connections are made to various parts of a high-gain amplifier system, it is possible that a significant frequency spectrum of these signals will be introduced as electrical noise. It is this interference which has to be rejected by the input networks of the instrumentation, quite apart from the concomitant removal of any electrostatic/electromagnetic noise introduced by direct coupling into the signal paths. The total contamination voltage can be many times larger (say 100) than the useful information voltage level.

Electrostatic interference in input cables can be greatly reduced by 'screened' cables (which may be 80% effective as screens), and electromagnetic effects minimised by transposition of the input wires and reduction in the 'aerial loop' area of the various conductors. Any residual effects, together with the introduction of 'ground and earth-loop' currents into the system, are collectively referred to as *series* and/or *common-mode* signals.

6.8.4.11 Series and common-mode signals

Series-mode (normal) interference signals, V_{sm}, occur in series with the required information signal. Common-mode interference signals, V_{cm}, are present in both input leads with respect to the reference potential plane: the required information signal is the difference voltage between these leads. The results are expressed as *rejection ratio* (in dB) with respect to the *effective* input error signal, V_e, that the interference signals produce, i.e.

$$K_{sm} = 20 \log (V_{sm}/V_e) \text{ and } K_{cm} = 20 \log (V_{cm}/V_e)$$

where K is the rejection ratio. The series-mode rejection networks are *within* the amplifier, so V_e is measured with *zero* normal input signal as V_e = output interference voltage ÷ gain of the amplifier appropriate to the bandwidth of V_e.

Consider the elementary case in *Figure 6.26*, where the input-lead resistances are unequal, as would occur with a transducer input. Let r be the difference resistance, C the cable capacitance, with common-mode signal and error voltages V_{cm} and V_e, respectively. Then the common-mode numerical ratio is

$$V_{cm}/V_e = V_{cm}/ri = 1/2\pi f C r$$

assuming the cable insulation to be ideal, and $X_C \gg r$. Clearly, for a common-mode statement to be complete, it must have a stated frequency range and include the resistive unbalance of the source. (It is often assumed in c.m.r. statements that $r = 1 \,\text{k}\Omega$.)

The c.m.r. for a digital voltmeter could be typically 140 dB (corresponding to a ratio of $10^7/1$) at 50 Hz with a $1 \,\text{k}\Omega$ line unbalance, and leading consequently to $C = 0.3 \,\text{pF}$. As the normal input cable capacitance is of the order of 100 pF/m, the situation is to inhibit the return path of the current i by the introduction of a guard network. Typical guard and shield parameters are shown in *Figure 6.27* for a six-figure digital display on a voltmeter with $\pm 0.005\%$ uncertainty. Consider the magnitude of the common-mode error signal due to a 5 V, 50 Hz common-mode voltage between the shield earth E_1 and the signal earth E_2:

N–G not connected. The a.c. *common*-mode voltage drives current through the guard network and causes a change of 1.5 mV to appear across r as a *series*-mode signal; for $V_s = 1 \,\text{V}$ this represents an error V_e of 0.15% for an instrument whose quality is $\pm 0.005\%$.

N–G connected. The *common*-mode current is now limited by the shield impedance, and the resultant *series*-mode signal is 3.1 µV, an acceptably low value that will be further reduced by the noise rejection property of the measuring circuits.

6.8.4.12 Floating-voltage measurement

If the d.c. voltage difference to be measured has a p.d. to E_2 of 100 V, as shown, then with N–G open the change in p.d. across r will be 50 µV, as a *series*-mode error of 0.005% for a 1 V measurement. With N–G connected the change will be 1 µV, which is negligible.

The interconnection of electronic apparatus must be carefully made to avoid systematic measurement errors (and short circuits) arising from incorrect screen, ground or earth potentials.

In general, it is preferable, wherever possible, to use a *single* common reference node, which should be at zero signal reference potential to avoid leakage current through r. Indiscriminate interconnection of the shields and screens of adjacent components can increase *noise* currents by short-circuiting the high-impedance stray path between the screens.

6.8.4.13 Instrument selection

A precise seven-digit voltmeter, when used for a 10 V measurement, has a discrimination of ± 1 part in 10^6 (i.e. $\pm 10 \,\mu\text{V}$), but has an uncertainty ('accuracy') of about ± 10 parts in 10^6 (i.e. $\pm 100 \,\mu\text{V}$). The distinction is important with digital read-out devices, lest a higher quality be accorded to the number indicated than is in fact justified. The quality of any reading must be based upon the time stability of the total instrument since it was last calibrated against *external* standards, and the cumulative evidence of previous calibrations of a like kind.

Selection of a digital voltmeter from the list of types given in *Table 6.2* is based on the following considerations:

(1) No more digits than necessary, as the cost per extra digit is high.

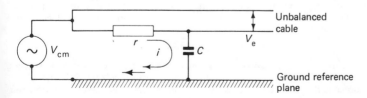

Figure 6.26 Common-mode effect in unbalanced network

Table 6.2 Typical characteristics of digital multimeters

Type (a) ± Uncertainty of maximum reading (b) Resolution (c)	D.C. ranges			A.C. ranges		C.M.R. at 50 Hz (dB) (minimum)	S.M.R. at 50 Hz (dB) (minimum)	Parallel input R and C	A.C. operating time (ms)	Recalibration period (months)*
	Voltage	Current	Resistance	Voltage and frequency min ≃ 40 Hz	Current and frequency min ≃ 40 Hz					
(a) 3½ digits general-purpose (mean sensing)	0.1 V–1 kV	0.1 A–1 A	1.1 kΩ–11 MΩ	0.1 V–1.0 kV	0.1 A–1.0 A	100	10 W	10 MΩ	350	3
(b)	0.3%	0.8%	0.5%	1.5%→8% (2 kHz→10 kHz)	2%→3.5% (2 kHz→5 kHz)			30 pF		
(c)	100 μV	100 μA	1 Ω	100 μV	100 μA					
(a) 4½ digits general purpose (mean sensing)	20 mV–1 kV	200 μA–2 A	200 Ω–20 MΩ	200 mV–100 V	200 μA–100 mA	120	60	1 MΩ	100	12
(b)	0.03%	0.1%→0.6%	0.02%→0.1%	0.15%→0.5% (10 kHz→20 kHz)	0.3%→0.7% (10 kHz→20 kHz)			100 pF		
(c)	1 μV	10 nA	10 mΩ	10 μV	10 nA					
(a) 5½ digits (r.m.s.)	0.1 V–1 kV	—	0.1 kΩ→15 MΩ	r.m.s. 1 V–1 kV	—	160	60	d.c. 10^{10} Ω →10^7 Ω	350	3
(b)	0.01%	—	50→1000 p.p.m.	0.1%→9% (20 kHz→1 MHz)	—			2 MΩ		
(a) 5½ digits (mean)				mean 1 V–1 kV 0.2%→0.8% (100 Hz→250 kHz)				100 pF		
(b)										
(c)	1 μV		1 mΩ	10 μV						
(a) 6½ digits precision (r.m.s. up to 200 V)	10 mV–1 kV	—	14 Ω–14 MΩ	0.1 V–1 kV	—	150	50	1 MΩ 150 pF	1 s	6
(b)	20 p.p.m.		50 p.p.m.	0.15%–0.9% (10 kHz→100 kHz)	—					
(c)	1 μV		1 mΩ	1 μV	—					
(a) 8½ digits High precision solarton 7081	0.1 V–1 kV	—	0.1 kΩ–1.4 GΩ	0.1 V–1 kV r.m.s.	—	140	40	a.c. 1 MΩ 150 pF d.c. 10 GΩ to 10 MΩ 10 MΩ (0.1→1 kV)	51 s to 1.7 ms	3
(b)	15 p.p.m.		17–46 p.p.m. (0.1% 1 GΩ range)	0.02%→2% 40 Hz→1 MHz	—					
(c)	10 nV		10 μΩ	1 μV	—					

* Specifications often include 12, 6, 3-month periods each with progressively smaller uncertainties of measurement down to 24-hour statements for high precision instruments.

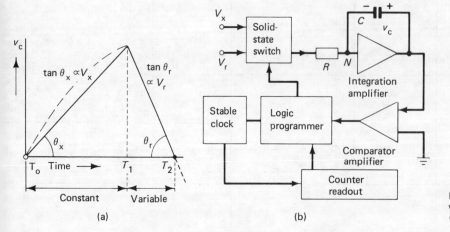

Figure 6.25 Dual-slope digital voltmeter. (a) Ramp voltages; (b) Block diagram

network based on the ramp technique. During the integration (*Figure 6.25*) the unknown voltage V_x is switched at time zero to the integration amplifier, and the initially uncharged capacitor C is charged to V_x in a known time T_1, which may be chosen so as to reduce noise interference. The ramp part of the process consists in replacing V_x by a *reversed* biased direct reference voltage V_r, which produces a constant-current discharge of C; hence, a known linear voltage/time change occurs across C. The total voltage change to zero can be measured by the method used in the linear-ramp instrument, except that the slope is negative and that counting begins at maximum voltage from the diagram in *Figure 6.25(a)* it follows that $\tan\theta_x/\tan\theta_r = (T_2 - T_1)/T_1 = V_x/V_r$, so that V_x is directly proportional to $T_2 - T_1$. The dual-slope method is seen to depend ultimately on time-base linearity and on the measurement of time difference, and is subject to the same limitations as the linear-ramp method, but with the important and fundamental quality of inherent noise rejection capability.

Noise interference (of series-mode type) is principally due to e.m.f.s at power supply frequency being induced into the d.c. source of V_x. When T_1 equals the periodic time of the interference (20 ms for 50 Hz), the charging p.d. v_c across C would follow the dotted path (*Figure 6.25(a)*) without changing the final required value of v_c, thus eliminating this interference.

6.8.4.6 (f) Multislope

At time T_1 (above) the maximum dielectric absorption in C will coincide with v_c maximum; this effect degrades (a) the linearity of the run-down slope during $T_2 - T_1$ and (b) the identification of zero p.d. One improved 'multislope' technique reduces dielectric absorption in C by inserting various reference voltages during the run-up period and, after subtraction, leaves a lower p.d. v'_c, prior to run-down while storing the most significant digit of V_x during run-up period. During run-down, C is discharged rapidly to measure the next smaller digit; the residue of v'_c, including overshoot, is assessed to give the remaining three digits in sequence using three v_r/t functions of slope $+1/10$, $-1/100$, $+1/1000$ compared with the initial rapid discharge. Measurement time is reduced, owing to the successive digits being accumulated during the measurement process[10].

6.8.4.7 (g) Mixed techniques

Several techniques can be combined in the one instrument in order to exploit to advantage the best features of each. One accurate, precise, digital voltmeter is based upon precision inductive potentiometers, successive approximation, and the dual-slope technique for the least significant figures. An uncertainty of 10 parts in 10^6 for a 3-months period is claimed, with short-term stability and a *precision* of about 2 parts in 10^6.

6.8.4.8 Digital multimeters

Any digital voltmeter can be scaled to read d.c. or a.c. voltage, current, immittance or any other physical property, provided that an appropriate transducer is inserted. Instruments scaled for alternating voltage and current normally incorporate one of the a.c./d.c. converter units listed in a previous paragraph, and the quality of the result is limited by the characteristics inherent in such converters. The digital part of the measurement is more accurate and precise than the analogue counterpart, but may be more expensive.

For systems application, programmed signals can be inserted into, and binary-coded or analogue measurements received from, the instrument through multiway socket connections, enabling the instrument to form an active element in a control system (e.g. IEC bus, IEEE-488 bus).

Resistance, capacitance and inductance measurements depend to some extent on the adaptability of the basic voltage-measuring process. The dual-slope technique can be easily adapted for two-, three- or four-terminal ratio measurements of resistance by using the positive and negative ramps in sequence; with other techniques separate impedance units are necessary. See *Table 6.2*.

6.8.4.9 Input and dynamic impedance

The high precision and small uncertainty of digital voltmeters make it essential that they have a high input impedance if these qualities are to be exploited. Low test voltages are often associated with source impedances of several hundred kilohms: for example, to measure a voltage with source resistance 50 kΩ to an uncertainty of $\pm 0.005\%$ demands an instrument of input resistance 1 GΩ, and for a practical instrument this must be 10 GΩ if the loading error is limited to one-tenth of the total uncertainty.

The dynamic impedance will vary considerably during the measuring period, and it will always be lower than the quoted null, passive, input impedance. These changes in dynamic impedance are coincident with voltage 'spikes' which appear at the terminals owing to normal logic functions; this noise can adversely affect components connected to the terminals, e.g. Weston standard cells.

Input resistances of the order of 1–10 GΩ represent the conventional range of good-quality insulators. To these must be

6.8.4.2 (b) Charge balancing

This principle[16] employs a pair of differential-input transistors used to charge a capacitor by a current proportional to the unknown d.c. voltage, and the capacitor is then discharged by a large number of small $+\delta q$ and $-\delta q$ quantities, the elemental discharges being sensed and directed by fast comparator/flip-flop circuits and the total numbers being stored. Zero is in the *middle* of the measurement range and the displayed result is proportional to the difference between the number of $+\delta q$, $-\delta q$ events.

The technique which, unlike dual-ramp methods, is inherently bipolar, claims some other advantages, such as improved linearity, more rapid recovery from overloads, higher sensitivity and reduced noise due to the averaging of thousands of zero crossings during the measurement, as well as a true *digital* auto-zero subtraction from the next measurement, as compared with the more usual capacitive-stored analogue offset p.d. being subtracted from the measured p.d.

Applications of this digital voltmeter system (which is available on a monolithic integrated circuit, Ferranti ZN 450) apart from d.c. and a.c. multimeter uses, include interfacing it *directly* with a wide range of conventional transducers such as thermocouples, strain gauges, resistance thermometers, etc.

6.8.4.3 (c) Successive approximation

As it is based on the potentiometer principle, this class produces very high accuracy. The arrows in the block diagram of *Figure 6.23* show the signal-flow path for one version; the resistors are selected in sequence so that, with a constant current supply, the test voltage is created within the voltmeter.

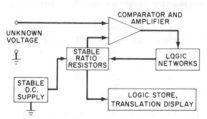

Figure 6.23 Successive-approximation digital voltmeter

Each decade of the unknown voltage is assessed in terms of a sequence of accurate stable voltages, graded in descending magnitudes in accordance with a binary (or similar) counting scale. After each voltage approximation of the final result has been made and stored, the residual voltage is then automatically re-assessed against smaller standard voltages, and so on to the smallest voltage discrimination required in the result. Probably four logic decisions are needed to select the major decade value of the unknown voltage, and this process will be repeated for each lower decade in decimal sequence until, after a few milliseconds, the required voltage is stored in a coded form. This voltage is then translated for decimal display. A binary-coded sequence could be as follows, where the numerals in **bold** type represent a logical *rejection* of that number and a progress to the next lower value:

Unknown analogue voltage	3	9	.	2	0	6
Logic decisions in vertical binary sequences and in descending 'order'.	8 4 2 1	8 4 2 1		8 4 2 **1**	8 **4** 2 1	8 4 **2** 1
Decoded decimal display	3	9	.	2	0	6

Voltages obtained from residual (difference) currents across high-stability resistors

The actual sequence of logical decisions is more complicated than is suggested by the example. It is necessary to sense the initial polarity of the unknown signal, and to select the range and decimal marker for the read-out; the time for the logic networks to settle must be longer for the earlier (higher-voltage) choices than for the later ones, because they must be of the highest possible accuracy; offset voltages may be added to the earlier logic choices, to be withdrawn later in the sequence; and so forth.

The total measurement and display takes about 5 ms. When noise is present in the input, the necessary insertion of filters may extend the time to about 1 s. As noise is more troublesome for the smaller residuals in the process, it is sometimes convenient to use some different techniques for the latter part. One such is the voltage frequency principle (see below), which has a high noise rejection ratio. The reduced accuracy of this technique can be tolerated as it applies only to the least significant figures.

6.8.4.4 (d) Voltage–frequency

The converter (*Figure 6.24*) provides output pulses at a frequency proportional to the instantaneous unknown input voltage, and the non-uniform pulse spacing represents the variable frequency output. The decade counter accumulates the pulses for a predetermined time T and in effect measures the average frequency during this period. When T is selected to concide with the longest period time of interfering noise (e.g. mains supply frequency), such noise averages out to zero.

Instruments must operate at high conversion frequencies if adequate discrimination is required in the final result. If a 6-digit display were required within 5 ms for a range from zero to 1.000 00 V to $\pm 0.01\%$, then 10^5 counts during 5 ms are called for, i.e. a 20 MHz change from zero frequency with a 0.01% voltage-frequency linearity. To reduce the frequency range, the measuring time is increased to 200 ms or higher. Even at the more practical frequency of 0.5 MHz that results, the inaccuracy of the instrument is still determined largely by the non-linearity of the voltage–frequency conversion process.

In many instruments the input network consists of an integrating operational amplifier in which the average input voltage is 'accumulated' in terms of charge on a capacitor in a given time.

6.8.4.5 (e) Dual-slope

This instrument uses a composite technique consisting of an integration (as mentioned above) and an accurate measuring

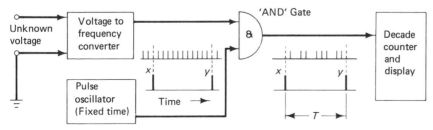

Figure 6.24 Voltage-to-frequency digital voltmeter

6.8.2.2 Distortion

The degree of distortion to be expected in the voltage waveform may dictate the choice of a.c./d.c. converter, particularly when the distortion is related to the quality of the result (or the uncertainty) required. For near-sine inputs the peak-responsive instrument is likely to give a good-quality result over a wide frequency range when used with a probe. With greater distortion of waveform the average-response instrument may be better, but there may be some sacrifice in bandwidth. The true r.m.s. instrument should provide the most accurate result for any waveform, and will be particularly valuable when power (proportional to voltage squared) is to be calculated. In general, the r.m.s. instrument is more expensive than the other types.

6.8.2.3 Range

Conventional solid state voltmeters are obtainable for measurements between $3\mu V$ and $1.5\,kV$ f.s.d. over a frequency range from zero to 1 GHz, with inaccuracies between ± 1 and $\pm 5\%$. No single instrument has all these properties. Measurements of alternating current are made using 'clip-on' probes rated at about 100 mA for 10 MHz, or 25 A maximum for audio frequencies.

D.C. electronic voltmeters provide ranges and uncertainties similar to those of their a.c. counterparts, and are often used as multipurpose instruments. As such they provide current readings from 10 pA to 10 A, with 2–3% uncertainty, and resistance from a few ohms to 100 MΩ, with 2–5% uncertainty.

6.8.3 Differential instruments

Precision d.c. differential electronic voltmeters are based upon the fundamental potentiometric principle of measurement, in which the unknown voltage is balanced against a known reference voltage derived from stable resistance ratios and a Zener diode/Weston cell reference potential. These instruments require an electronic impedance converter of impedance at least 10 GΩ to isolate the low-impedance range and measurement networks of the potentiometer from all values of the unknown voltage. These voltmeters can be very accurate, reading up to 1 kV with uncertainties of 20 parts in 10^6 of maximum reading. Although the stability is good, the quality degrades within a few months of manufacture.

6.8.3.1 D.C. precision sources

These include a high-quality differential voltmeter as reference and a control element that can provide up to 1 kV and 50 mA. These d.c. sources can be used for the testing of digital voltmeters as well as for driving the d.c. measurement side on an a.c./d.c. thermocouple transfer standard.

The adjective 'differential' is used synonymously with 'difference' in amplifiers and voltmeters having the ability to reject many common-mode interference signals, and may also be applied to operational amplifiers capable of performing the mathematical operation of differentiation. Care is therefore needed in the interpretation of the word.

6.8.4 Digital voltmeters

These provide a digital display of d.c. and/or a.c. inputs, together with coded signals of the visible quantity, enabling the instrument to be coupled to recording or control systems. Depending on the measurement principle adopted, the signals are sampled at intervals over the range 2–500 ms. The basic principles are (a) linear ramp; (b) charge balancing; (c) successive-approximation/potentiometric; (d) voltage to frequency integration; (e) dual-slope; (f) multislope; (g) some combination of the foregoing.

Modern digital voltmeters often include a multiway socket connection (e.g. B.C.D.; I.E.C. bus; IEEE-488 bus, etc.) for data/interactive control applications.

6.8.4.1 (a) Linear ramp

This is a voltage/time conversion in which a linear time-base is used to determine the time taken for the internally generated voltage v_S to change by the value of the unknown voltage V. The block diagram (*Figure 6.22(b)*) shows the use of comparison networks to compare V with the rising (or falling) v_S; these networks open and close the path between the continuously running oscillator, which provides the counting pulses at a fixed clock rate, and the counter. Counting is performed by one of the binary-coded sequences; the translation networks give the visual decimal output. In addition, a binary-coded decimal output may be provided for monitoring or control purposes.

Limitations are imposed by small non-linearities in the ramp, the instability of the ramp and oscillator, imprecision of the coincidence networks at instants y and z, and the inherent lack of noise rejection. The overall uncertainty is about $\pm 0.05\%$ and the measurement cycle would be repeated every 200 ms for a typical 4-digit display.

Linear 'staircase' ramp instruments are available in which V is measured by counting the number of equal voltage 'steps' required to reach it. The staircase is generated by a solid state diode pump network, and linearities and accuracies achievable are similar to those with the linear ramp.

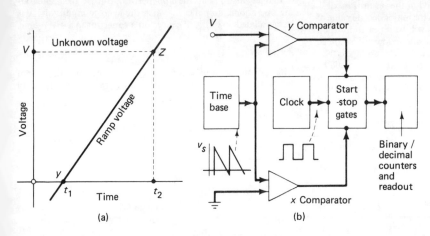

Figure 6.22 Linear ramp digital voltmeter. (a) Ramp voltage; (b) Block diagram

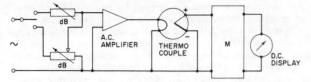

Figure 6.19 Thermo-couple electronic voltmeter. M is a cascaded group of chopper + modulator + demodulator + machine amplifier

squared signal; i.e. the device must have a square law response coupled with an inertial property. One of the best such devices is the thermocouple with a non-reactive heater carrying the current to be measured (*Figure 6.19*). The thermal inertia of the heater and thermocouple effectively averages the cyclic changes of I^2R to provide a steady thermo-e.m.f. This is amplified, e.g. by a conventional modulator, chopper amplifier, demodulator, and an output matching emitter-follower amplifier chain feeding the d.c. moving-coil display indicator. Modern multiple-thermocouple network design has overcome many of the difficulties arising from slow response (excessive thermal inertia), non-uniform squaring with large signals, and fragility.

For relatively small peak-to-peak signal changes, square law characteristics can be obtained from suitably biased transistor amplifiers and some solid state devices. Shaping networks are possible in which an approximate square law can be achieved by a few short linear segments of successively increasing slope; none of these is a true square law device. In the case of these instruments relying on bias, it is only the *change* in signal level that will be proportional to the input; and as the instrument must read zero with no input, the correct bias adjustment *may* need to be trimmed externally as a 'set zero' correction.

6.8.1.3 Average

As the mean value of a symmetrical waveform is zero, the term 'average' is applied either to the mean of the positive half-period or to the mean of the positive and inverted-negative half-periods. The half-period mean of a non-sinusoidal waveform is given by

$$V_{av} = (2\sqrt{2}/\pi)[V_1 + V_3/3 + \ldots + V_n/n + \ldots]$$

where $V_1 \ldots V_n$ are the r.m.s. values of the fundamental ... nth harmonic voltage components, and the true mean (i.e. the direct voltage component) is zero or has been blocked by an input capacitor.

Instruments responding to average have an inherent torque (and deflection) corresponding to V_{av}. The scale graduations are, however, in r.m.s. values, on the assumption that the harmonics are zero. True r.m.s. values cannot be inferred from the scaled values unless the form factor is known. The rectification process may be a conventional full-wave bridge diode network or a wide-band operational amplifier of the type shown in *Figure 6.20*.

6.8.1.4 Peak

Converters for peak indication employ diode rectification to charge capacitive-resistive networks of very low time-constant to the *peak* value of the input waveform (*Figure 6.21*). Subsequent d.c. amplification is then introduced for display on a moving-coil instrument. For the r.m.s. scaling the initial calibration is made

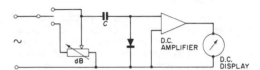

Figure 6.21 Peak-sensing electronic voltmeter

using a sine input; hence, the actual r.m.s. indication is always $1/\sqrt{2}$ times the peak value. Care is necessary in interpreting the results with *non*-sinusoidal signals. The current/voltage characteristic of a rectifier is non-linear for small signals, for which incremental changes may be r.m.s. responsive; the indication becomes a peak response for larger signals.

The input capacitor C of the network in *Figure 6.21* is charged to the peak of the input signal, and as modern solid state diodes are small and have low charge storage, they can operate successfully up to frequencies of several hundred megahertz. For such frequencies the diode or transistor can be mounted in a 'probe', which can be taken close to the network being measured and which develops a d.c. signal transmitted to the amplifier by screened leads. This technique avoids r.f. standing waves, mismatch and incorrect voltages, which become troublesome at frequencies above about 5 MHz.

6.8.2 Instrument selection

6.8.2.1 Sensitivity

All electronic voltmeters respond to a wide range of voltage magnitude and frequency. 'Noise' is a fundamental limitation on the sensitivity (minimum voltage) capability. A broad-band amplifier has more self-generated noise than one for a restricted bandwidth, and is more responsive to external noise. Thus, a very sensitive $3\,\mu V$ voltmeter would require to be associated with a relatively narrow band (e.g. 5 Hz–500 kHz) to obtain a tolerable signal/noise ratio.

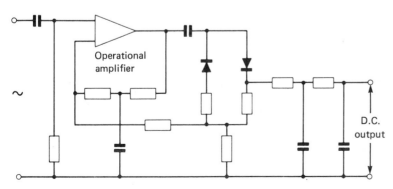

Figure 6.20 Wide-band operational amplifier: rectification type

6.7.4.4 Data acquisition system

A more versatile electronic summator system is provided by the Ferranti Series-70 Data Acquisition System, developed to deal with complicated multiple tariffs and for faster and more systematic data handling. It can accommodate 64 channels and provide a memory; also available for each channel are the total integrations and maximum-demand outputs presented in decimal printed figures, as coded punched tape, or in serial form for transmission to a remote print-out.

The CEGB National Control Centre has been designed to have sufficient flexibility and capacity to ensure economic and reliable supply. Its functions are:

(1) Load forecasting, based on continuous weather data, seasonal variations, past experience and the occurrence of special national events; the predictions for economic generation during the following 24 h are off-line computerised studies.
(2) Cross-country power transfers (including those to or from the French and southern Scottish networks). Real-time computation.
(3) Alternative route-finding for rapid and economical power supply during fault conditions.

The control room evidence for extant conditions is provided by electronic summators.

6.8 Electronic instrumentation

The rapid development during the last decade of large-scale integrated (l.s.i.) circuits, as applied to analogue and especially digital circuitry, has allowed both dramatic changes in the format of measurement instrumentation by designers and in the measurement test systems available for production or control applications.

A modern instrument is often a multipurpose assembly, including a microprocessor for controlling both the sequential measurement functions and the subsequent programmed assessment of the data retained in its store; hence, these instruments can be economical in running and labour costs by avoiding skilled attention in the recording and statistical computation of the data. Instruments are often designed to be compatible with standardised interfaces, such as the IEC bus or IEEE-488 bus, to form automatic measurement systems for interactive design use or feed-back digital-control applications; typically a system would include a computer-controlled multiple data logger with 30 inputs from different parameter sensors, statistical and system analysers, displays and storage. The input analogue parameters would use high-speed analogue-to-digital (A/D) conversion, so minimising the front-end analogue conditioning networks while maximising the digital functions (e.g. filtering and waveform analyses by digital rather than analogue techniques). The read-out data can be assembled digitally and printed out in full, or presented through D/A converters in graphical form on X–Y plotters using the most appropriate functional co-ordinates to give an economical presentation of the data.

The production testing of the component subsystems in the above instruments requires complex instrumentation and logical measurements, since there are many permutations of these minute logic elements resulting in millions of high-speed digital logic operations–which precludes a complete testing cycle, owing to time and cost. Inspection testing of l.s.i. analogue or digital devices and subassemblies demands systematically programmed measurement testing sequences with varied predetermined signals; similar automatic test equipments (ATE) are becoming more general in any industrial production line which employs these modern instruments in process or quality-control operations. Some companies are specialising in computer-orientated measurement techniques to develop this area of ATE; economic projections indicate that the present high growth rate in ATE-type measurement systems will continue or increase during the next few years.

The complementary nature of automated design, production and final testing types of ATE leads logically to the interlinking of these separate functions for optimising and improving designs within the spectrum of production, materials handling, quality control and costing to provide an overall economic and technical surveillance of the product.

Some of the more important instruments have been included in the survey of electronic instrumentation in the following pages, but, owing to the extensive variety of instruments at present available and the rapid rate of development in this field, the selection must be limited to some of the more important devices.

Analogue instruments normally provide as output a continuous function of a smoothly changing input quantity. A pointer indicator is a typical analogue instrument.

Digital instruments usually provide an output in visual decimal or coded digit form as a discontinuous function of a smoothly changing input quantity. In practice the precision of a digital instrument can be extended by adding digits to make it better than an analogue instrument, although the final (least significant) digit must always be imprecise to ± 0.5.

6.8.1 Electronic analogue voltmeters

A fundamental tenet of metrology is that the process of measurement must not affect the phenomenon being assessed. Despite the very high input impedance of electronic voltmeters, most instruments take some power, even though it may be of the order of $1\,\mu\text{W}$. The input impedance is commonly represented by a parallel network of a capacitance (2–50 pF) and a resistance (several MΩ). The high-impedance property, together with high-frequency amplification up to more than 1 GHz, is the predominant virtue of this type of voltmeter, although many instruments are restricted to about 5–10 MHz, since capacitance can be the limiting factor in high-frequency applications.

To use these voltmeters over a wide range of input voltage, the signal amplitude must be 'conditioned' to the range limits of the instrument by high-impedance attenuators for high-voltage and high-gain high-impedance amplifiers for low-voltage signals. In each case the output impedance of the final amplifier must be low, so that it behaves as a current source for the a.c./d.c. converter network required to drive the d.c. instrument used to display the output.

The output scale of the indicator may be divided to read r.m.s. logarithmic, average or peak values of the input voltage; but the accuracy of the scale reading can be interpreted only on the basis of the inherent characteristics of the a.c./d.c. converter, together with a knowledge of the signal waveform. Some single-range logarithmic voltmeters display a wide *dynamic range* of voltage on a single scale in dB, e.g. $100\,\text{V} \rightarrow 316\,\mu\text{V}$, quoted as $20\log_{10}(100\,\text{V} \div 316\,\mu\text{V}) = 110\,\text{dB}$ dynamic range.

6.8.1.1 A.C./D.C. voltage converter characteristics

Converters can produce d.c. outputs representing true r.m.s. average or peak of the input waveform. All three types of instrument usually have scales calibrated in r.m.s. volts: this is satisfactory for 'true r.m.s.' readings, but for 'average' and 'peak' readings the indication is correct only if the input waveform is sinusoidal.

6.8.1.2 True r.m.s.

The requirement is that a true r.m.s. device must be able (a) to square a complex signal voltage and (b) deliver the average of the

6/18 Electrical metrology and instrumentation

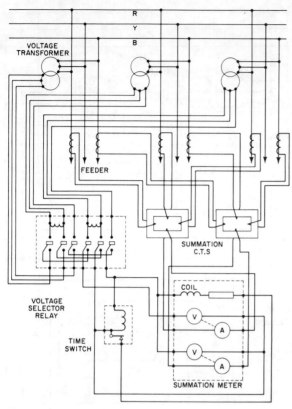

Figure 6.17 Summation with summation transformers

of the discs, they are the seat of induced e.m.f.s, which are rectified and fed to a multivibrator, triggering it to produce an output pulse.

The electronic pulse-forming networks can produce, typically, 18 000 impulses during any half-hour maximum-demand integration period, and so can provide more precise information than is available from a mechanical contactor. Each impulse, representing a known value of the quantity being recorded, is fed to the summator.

6.7.4.3 Summator

Mechanical A device at the end of each signal path translates the electrical impulses into mechanical movements (e.g. by d.c. stepper motor); the rotary motions are added (e.g. by differential gearing); and the resultant output operates a summation clock, a maximum-demand indicator or a print-out. Mechanical summators have been used for small numbers of circuits and parameters, but the more recent electronic summators operate faster and are more adapted to computer control systems.

Electronic This can deal with the relatively high pulse rates of electronic transmitters. Input signals derived from up to 16 channels determine the states of separate bistable networks. The latter are reset sequentially by a scanner (oscillator) network; and when reset, the gate permits a pulse to be passed into the channel register as well as to a total register path. Thus, pulses can arrive simultaneously (in parallel) but are accumulated in series. Conventional digital networks can readily be incorporated for summation, subtraction of exported power and the logical interrogation of particular parameters for any special requirement. As the actual register in a conventional maximum-demand meter may be driven by a stepper motor, the high impulse rate of the electronic signals needs to be successively divided (down to 10 impulses/s) to enable the stepper motor to index correctly. Low-rate pulses can also be used to drive a printometer (at 4 impulses/s) or to telemeter a more complex data recorder.

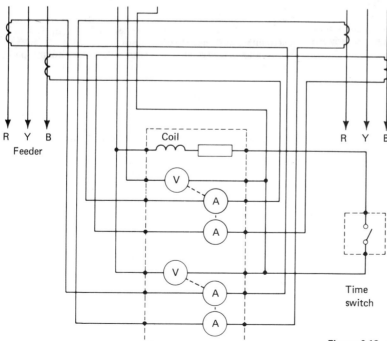

Figure 6.18 Summation with split-winding meter coils. The upper part of the diagram is identical with that in *Figure 6.16*

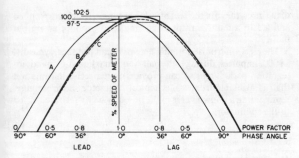

Figure 6.15 Kilovolt ampere-hour measurement over limited p.f. range constant VA input: A at u.p.f. normal speed and volt-coil flux; B = volt-coil flux lagging 18° by magnetic shunt and series resistance; C = increased speed 2½% by adjustment of voltage transformer

−3% for all p.f. up to 50% overload, even after allowing for reasonable variations in the frequency and supply voltage.

6.7.3.2 Maximum-demand kilovar-hour indicator

This measurement can be made with a standard kWh meter by either (a) supplying the meter volt coils from a separate 'quadrature' transformer or (b) cross-connecting the volt coils internally. In each case the effect is to retard the volt coil fluxes by 90°.

6.7.4 Summation metering

Factories, and other large undertakings covering a wide area, are often supplied over a number of feeders. The overall tariff will depend on the diversity of the total import of active, reactive and apparent power, so that these parameters must be monitored for each feeder and collectively for all. Commercial summation consoles display all individual and the summated quantities, together with a printed record of the demand integrations over the appropriate successive periods (e.g. of 30 min).

Collation of information from individual feeders may be direct or indirect, in the latter case by proportional impulse signals. Direct methods employ current transformers or special meters.

6.7.4.1 Direct summation

Current transformers (c.t.s) The direct method using current transformers in parallel is shown in *Figure 6.16*. Two or more networks can be summated in this manner, provided that the feeder voltages are both equal and in phase and that the loads are comparable. The c.t. ratios must all be the same. The voltage selector relay is fitted to ensure that, in the event of a supply failure on a feeder, there is a maintained supply to the volt coils of the meter.

Summation transformers can be used if there are several feeders. These, shown in *Figure 6.17*, include as many primary windings as there are circuits to be summed, and there is one common secondary. The method is more flexible than that with paralleled c.t.s as it can cope with feeders of substantially different ratings and current ratios. This means that for feeders remote from the summation meter it is possible to use 1 A secondary currents; but the transformers must be of high quality.

Special meters This method is restricted to one pair of networks. For the highest accuracy the two voltages should be in phase and the two sets of measuring transformers identical. *Figure 6.18* shows that the two R and B load currents are each transformed into four electrically separate coils, so wound that the m.m.f.s for the R load are additive in one current coil of the meter and the B

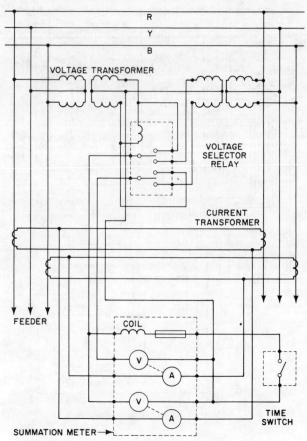

Figure 6.16 Summation with parallel-connected current transformers

loads in the second current coil. These precision integrating meters conform to BS 37 and are designed for statistical and accounting work, and the errors must not exceed ±0.5% over a wide range of load conditions. The instrument can be of the split-series double-element type with two discs on a common spindle.

6.7.4.2 Indirect summation

Methods involve impulsing techniques, so adapted that transmitted pulses from rotating meters can be collated by a common summator to display the average and total values of all the relevant parameters. Impulsing adaptors are fitted to summation instruments and to all the instruments already discussed, which may be monitoring a single network parameter. The impulsing devices can be any of the three types described below.

Mechanical contactor This can be cam-operated from an auxiliary spindle driven from the main spindle. The gear ratio provides typically for 2000 impulses as a maximum during any half-hour integration period.

Phototransistors Two phototransistors, illuminated by a small lamp fitted between two slotted discs driven from the main spindle, give a maximum impulsing rate dependent on the number of slots.

Inductive transmitter The method now in general use also uses two slotted discs, but the lamp between the discs is replaced by a coil fed from an oscillator, and the phototransistors are replaced by pickup coils. When the pickup coils are unshielded by the teeth

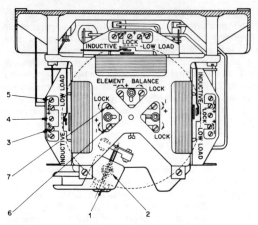

Figure 6.12 Polyphase energy meter (Ferranti Ltd). Adjustments: 1. Full load; 2. Lock; 3. Inductive load; 4. Lock; 5. Light load; 6. Equalising; 7. Lock

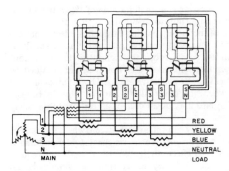

Figure 6.13 Polyphase energy meter connections (Ferranti Ltd). Rating, over 650 V and 50 A; M=mains; L=load current; S=voltage

error limits are claimed by manufacturers (with the approval of government departments) as applying up to three or four times nominal maximum current. Polyphase two-rate meters are available with two registers, selected by an internal electromagnet actuated from an external timing mechanism.

6.7.3 Polyphase maximum-demand indicator

Energy tariffs are commonly based on energy used, together with charges for the average maximum active or apparent or reactive power demand in (typically) successive half-hour periods. Instruments are required to integrate these quantities over the specified periods and to record permanently the maximum demand. The indicators are essentially 3-phase energy meters modified to record the integrated values on six small dials, or cyclometer rollers, fitted within a circular dial scaled in the maximum-demand quantity (*Figure 6.14*). The driving pointer A pushes a slave pointer B ahead of it. B remains at maximum reading after A returns automatically to zero at the end of the integration period. The reset mechanism can be impulsed from an external time-switch or an internal synchronous-motor drive. External monitoring of the maximum-demand quantity involves either a two-way mechanical contactor which is suitable for sending pulses to drive a remote maximum-demand indicator and a printing recorder, or an electronic transmitter feeding an electronic summator.

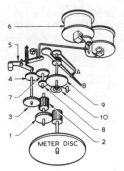

Figure 6.14 Mechanism of a maximum demand indicator

The resetting mechanism (*Figure 6.14*) shows the drive pointer A driven by the meter disc through the pinion and wheels 1, 2, 3, 4, 7, 8 and 9. At the end of the integration period, electromagnet 6 is energised through a time-switch and the rotary armature is moved across the poles. The movement is transmitted by lever 5 to the shaft carrying wheels 3 and 4, causing 4 to disengage from 7. As soon as this occurs, spiral spring 10, which has been wound up, returns to its original position carrying with it pointer A. The position of pointer B is left unchanged, and it will remain in position unless advanced further in a subsequent integration period.

6.7.3.1 Maximum-demand kilovolt-ampere-hour indicator

Tariffs based on active power demand are most usual, but charging on an apparent power basis encourages load power factor improvement. The term 'kVA maximum demand' is not precise, and instruments operating on different principles may give widely different indications on the same load.

Integrating instruments are available using the arithmetic sum (a.s.) or vector sum principles: the latter is usually preferred because the a.s. instruments give higher indications if the power factor varies during the integration period, or if the load is unbalanced.

Vector sum instruments may be of the full-range pattern compensated for all power factors, or of the restricted-range pattern compensated for a limited p.f. range such as 0.8–1.0 or 0.5–0.9. There is little difference, if any, between the indications of a restricted- and a full-range instrument, provided that the load p.f. at maximum demand is within the range for which the former is compensated. The power factor at lower loads is immaterial, because it cannot affect the reading. The p.f. of most industrial loads at the time of maximum demand is rarely less than 0.5, nor is it greater than 0.9 unless p.f. improvement plant is installed. In these circumstances the higher cost of the full-range instrument may be unjustified.

Typical characteristics (*Figure 6.15*) show how a limited p.f. range (1.0–0.8 lagging) can be achieved with a $\pm 2\tfrac{1}{2}\%$ uncertainty in registration by adjustments of the speed as well as the phase of the volt coil flux. A similar technique can give limited lagging p.f. ranges of 0.98–0.67 and 0.9–0.5.

One development in maximum-demand meter design employs electronic techniques for measuring the arithmetic sum of polyphase kVAh or kVA quantities. D.C. outputs are obtained by rectification of the secondary currents of the three current transformers which carry the load currents of the network. The total direct current, controlled by a transistorised circuit, supplies the series coils of a single-element, single-disc, induction meter. The voltage coil of the meter is energised directly from the line voltage, rather than the phase voltage, to reduce errors due to phase voltage imbalance. The cumulative error is typically +2 to

high-power-factor loads and slower on lagging reactive loads: an increase of frequency has the opposite effect (*Figure 6.10(c)*). The errors are cumulative if both voltage and frequency variations occur together.

6.7.1.11 Testing of single-phase meters

Comparison is made with a reference instrument, which must have been calibrated *recently* against a superior standard having a known traceability to the national standards. The three recognised test methods are described below.

6.7.1.12 (A) Dial testing

A batch of meters is run on the same load as a substandard meter and the dial readings compared. Allowing for the certified error of the substandard, the 'true' errors of the batch can be determined.

6.7.1.13 (B) Rotating substandard

The rotating substandard (r.s.s.) is connected to the same load as the meters under test, but can be stopped and started at will by means of a switch in its volt circuit. It is also provided with a zero reset of its revolution-counter pointers. The procedure is to zero the r.s.s. and observe the disc of the test meter, the edge of which has a marked spot. As the spot passes a fixed point, the r.s.s. is started. After a predetermined number of disc revolutions, the r.s.s. is stopped as the test disc passes the fixed point. If the test meter had the same error as the r.s.s., then the ratio between the revolution counts should be the ratio between the revolutions per energy unit (r.p.u.); hence,

r.s.s. count = (r.s.s. r.p.u./meter r.p.u.) × meter count

The difference between the actual r.s.s. revolutions and the calculated value enables the error difference between the r.s.s. and the meter to be found. If the certified error of the r.s.s. is added, then the 'true' error of the meter is obtained.

6.7.1.14 (C) Wattmeter and stop-watch

The load P (in kilowatts) monitored by the meters under test is held constant by electronic and/or manual trimming, and measured by a wattmeter: the connections are shown in *Figure 6.11*. The tester observes the meter disc, and starts his stop-watch as the spot passes the fixed point. He then times N revolutions of the disc, which takes, say, T seconds. Every meter has a revolutions-per-kWh constant A (a value stamped on the nameplate) appropriate to its counter gearing. In a *perfect* meter the disc speed is related directly to energy with a constant ratio k = energy (kJ) per rev = $3600/A$, and the time duration for a power P and N disc revolutions is $T_0 = kN/P$. The difference between T_0 and the actual time T enables the percentage error to be calculated. Any wattmeter error must be added: in practice it is often allowed for by adjusting the timing T by a suitable percentage. Wherever possible, P and N are selected so as to simplify the arithmetic.

6.7.1.15 Run-off method

This is used when at least six meters are to be tested simultaneously. It is similar to method (A) above, except that the disc of the reference meter is rotated through a fixed number of revolutions (e.g. 25). All meters are started with the disc spot properly located, and the error of each meter is assessed from the total number of complete and fractional revolutions made by its disc.

6.7.1.16 Choice of method

The cost of meter testing is a small fraction of 1% of the annual revenue, and it is important to ensure that the proper revenue is earned. If most of the meters passing through the test room are new, it may be convenient to make dial tests and reject those outside the limits, if test room throughput is the only consideration. The most reliable results, however, are obtained by individual rather than by collective testing. This involves method (B) or (C), or, where the 'cover effect' is not variable, by run-off. The run-off standard may be an r.s.s. or an ordinary (but newly calibrated) meter.

The batch method by run-off is quick, except where the batches comprise meters with inconsistent 'cover effect'. Older patterns of meter may have cover effect differences of 1–3% and may require method (B) or (C). Ordinary meters used as standards are calibrated by method (B) or (C), while rotating substandards are tested by method (C) using instruments with *valid* traceability certificates.

6.7.2 Polyphase meter

A polyphase meter comprises a number of 1-phase meters with their rotor torques combined mechanically to drive a common register. *Figures 6.12* and *6.13* show features of the layout and the essential connections of a three-element meter driving a single disc. The adjustments for full-load, light-load and inductive-load accuracy are provided for each element. There is an additional feature to balance the torque between the three elements: it consists of three mild-steel springs, which run on rack-and-pinion devices to give micrometer screw adjustment and which are arranged to act as magnetic shunts to the three volt coil fields. This meter, with its single disc, has reduced wall area, but some increase in forward projection. Three-element meters having two discs, one disc being driven by two elements, are also made.

Polyphase meters for load currents above 150 A can be combined with their current transformers, with a saving in space, ease of installation and connection, and permanent association of each meter element with the transformer with which it was originally calibrated.

6.7.2.1 Accuracy

Three-phase whole-current energy meters have ±2% limits of error for 1.0 p.f. loads throughout most of the current range as defined in BS 37. For extended- or long-range instruments, these

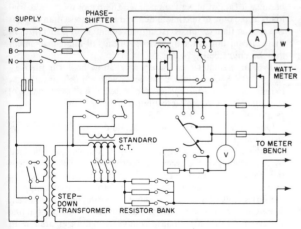

Figure 6.11 Single-phase meter-testing circuits

kilowatt-hour are readily obtainable, and usually are simple in construction, reliable in operation and quickly replaceable.

Some manufacturers have adopted a price change mechanism consisting of a circular disc on the meter casing, concentric with the operating handle, which is rotated by the consumer after the insertion of a coin in a slot. Around the edge of the disc are engraved a number of figures indicating the number of kilowatt-hours obtainable for the insertion of one coin. By varying the position of the disc on its axis, the amount of energy obtainable by the consumer for each coin can be adjusted. This variation is readily made by the meter reader, if necessary without having access to the interior of the meter, and no skill is necessary in making the adjustment. As the device is sealed, it cannot be altered by unauthorised persons.

The scaling of the disc is in terms of 'units per coin' and the advances are in uniform steps. In practice, energy is sold in terms of 'price per unit', and when the conversion from one system to the other is made, it will be found that a large number of the steps provided are of no value, because they represent prices which are not met with in practice. On the other hand, some prices which might be required are not available and thus the apparent usefulness of the device may be limited.

A mechanism that enables various price-per-unit selections to be made is shown in *Figure 6.9*, excluding the two-part tariff device on the right of the diagram. A cone of 17 wheels, one for

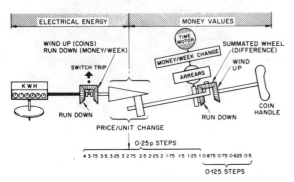

Figure 6.9 Principle of two-part tariff mechanism (Ferranti Ltd)

each price per unit, is engaged by a sliding pinion which has a locked carriage to ensure positive meshing. A slotted selector plate ensures that the correct wheel is engaged. Each price is marked on a price change dial plate. Change of price is effected by opening a separately sealed door on the meter cover and moving the selector knob to the price required. When the door is opened, a spring-loaded plunger automatically locks the credit and prevents loss of credit during the changeover. This plunger can be unlocked independently to free the coin storage when the cover is removed for testing: it must be restored before the cover is replaced. The price-per-unit mechanism is on the coin side of the main differential, thus giving constant gearing between the meter and the switch tripping device and ensuring constant loading on the meter element regardless of the price setting.

6.7.1.7 Two-part tariff meter

On most two-part tariff prepayment meters the fixed-charge mechanism includes a small mains-operated synchronous motor as a timing element (*Figure 6.9*), geared to the prepayment mechanism so as to supplement the run-down of the debit side by the meter. The latter, of course, increases the debit proportionally to the energy taken; the timing element, however, increases the debit at a constant rate, as determined by its gear train. This debit (charge per week) can be adjusted in steps of 1 p up to 130 p, or in 2 p steps up to 260 p. Arrears of payment are accumulated on a dial up to a maximum of £20.

6.7.1.8 Performance of single-phase meters

The limits of permissible error are defined in BS 37. In general, the error should not exceed $\pm 2\%$ over most of the working range for credit meters. For prepayment meters, $+2$ and -3% for loads above 1/30 full load at unity power factor, at any price setting to be used, are the specified limits. Tests for compliance with this requirement at and below 1/10 marked current must be made with the coin condition that not less than one nor more than three coins of the highest denomination acceptable by the meter are prepaid.

It is usual for manufacturers to adjust their meters to less than one-half of the permissible error over much of the working range. The mean error of an individual meter is, of course, less than the maximum observable error, and the mean collective error of a large number of meters is likely to be much less than $\pm 1\%$ at rated voltage and frequency. If no attempt is made during production testing to bias the error in one direction, the weighted mean error of many meters taken collectively is probably within 0.5%, and is likely to be positive (see *Figure 6.10(a)*).

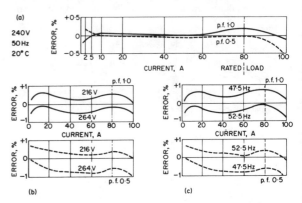

Figure 6.10 Typical performance of 240 V, 50 Hz, 20–80, A house service meter (Ferranti Ltd)

6.7.1.9 Temperature

The temperature error of a.c. meters with no temperature compensation is negligible under the conditions normally existing in the UK. In North America it is common practice to fix meters on the outside of buildings, where they are subjected to wide temperature ranges (e.g. 80 °C), which makes compensation desirable. However, even in the UK it is common to provide compensation (when required) in the form of a strip of nickel-rich alloy encircling the stator magnet. Its use is advantageous for short-duration testing, as there is an observable difference in error of a meter when cold and after a 30 min run.

6.7.1.10 Voltage and frequency

Within the usual limits of $\pm 4\%$ in voltage and $\pm 0.2\%$ in frequency, the consequent errors in a meter are negligible. Load-shedding, however, may mean substantial reductions of voltage and frequency. A significant reduction in voltage usually causes the meter to read high on all loads, as shown in *Figure 6.10(b)*. Reduction of the supply frequency makes the meter run faster on

The accuracy of the meter is affected on low loads by pivot friction. A *low-load adjustment*, consisting of some device that introduces a slight asymmetry in the volt magnet working flux, is fitted to mitigate frictional error. There are several methods of producing the required asymmetry: one is the insertion of a small magnetic vane into the path of the working flux. The asymmetry results in the development at zero load of a small forward torque, just sufficient to balance friction torque without causing the disc to rotate.

6.7.1.1 Rotor bearings

The wear of sapphire jewels depends on the direction of cut. With unlubricated jewels, minimum wear is obtained when the optic axis of the sapphire is at 90° to the normal to the tangent plane of the base of the jewel cup.

Proper lubrication of the pivot and jewel with suitable oil is as important as orientation. Excellent results have been obtained with lubricated bearings irrespective of orientation, the wear being so reduced that the low-load accuracy is not seriously impaired even after many millions of revolutions of the rotor. The working surfaces of pivots and jewels are liable to damage by mutual impact due to careless handling or transport. The severity of impact can be reduced by lubrication. Many makers provide a spring seating to the jewel to reduce impact damage, and a recommendation to this effect is included in BS 37.

An alternative rotor support is the magnetic suspension assembly illustrated in *Figure 6.7*. The meter employs a pair of

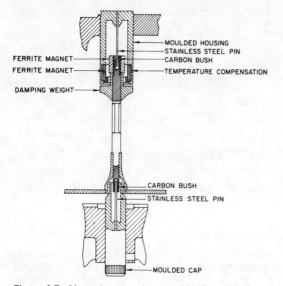

Figure 6.7 Magnetic suspension assembly (Ferranti Ltd)

concentric magnets of partially anisotropic ferrite material located at the upper end of the rotor spindle and magnetised each with three sets of poles arranged to be in repulsion. The rotor is constrained against lateral movements by fixed stainless steel pins in carbon bushes mounted into the rotor spindle. The design produces a high vertical *stiffness factor*, a criterion of the effort exerted by the rotor system to restore itself to its correct operating position when disturbed. The magnetic suspension is remote from the induction magnets, and is temperature-compensated.

6.7.1.2 Magnet materials

Brake magnets are of cobalt steel, Alni or Alnico. Use of these alloys has contributed to improved overcurrent performance of a.c. meters by increased braking torque and reduced disc speed.

The cores of the volt and current magnets are assembled from laminated precision cold-reduced steel.

6.7.1.3 Register

The register can comprise six separate decade clock dials, or a roller 'cyclometer' mechanism with a five-digit read-out and a units dial duplicated in analogue form to display the final digit more precisely.

The internal view of an assembled meter is shown in *Figure 6.8*.

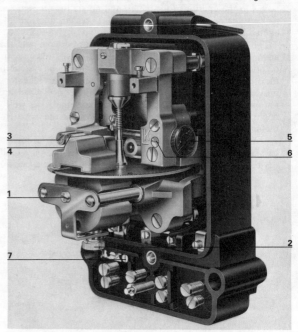

Figure 6.8 Single-phase energy meter (Ferranti Ltd) 1. Full load adjustment; 2. Full load adjustment lock; 3. Inductive load adjustment; 4. Inductive load adjustment lock; 5. Low load adjustment; 6. Low load adjustment lock; 7. Testing link

6.7.1.4 Single-phase two-rate meter

A special tariff provides for off-peak energy at a cheaper cost. The energy can be registered by a conventional meter with external time-switch suitably programmed to connect the supply only during off-peak periods; alternatively, an uninterrupted supply can be provided through a two-rate meter with a separate programmed time-switch. The two-rate meter has two separate registers. An electromagnet transfers the drive to one or other of the registers: it is of semi-rotary construction, giving positive action for a low power demand. The changeover operates through a moulded differential gear. If the time-switch is out of circuit, only the normal-rate register continues to record.

6.7.1.5 Single-phase prepayment meter

The meter incorporates a device to vary the energy available per coin according to the ruling tariff.

6.7.1.6 Single-rate meter

The simplest device consists of a change wheel or a small gear train arranged as a unit in such a manner that it can easily be detached from the prepayment mechanism and another unit substituted in its place. Units giving any desired price per

Supply source and function	Test voltage (kV)	Resistance (MΩ)
Mains: heavy-duty for cables, switchgear, etc.	5	10^5
	10	2×10^5
Mains or heavy-duty battery: electronic (h.v.)	5	5×10^4
Mains: electronic (high-resistance)	0.5	2×10^8
Hand-driven generator: general purpose	0.25	50
	5	10^4
Light-duty battery: sealed, intrinsically safe for mines, etc.	1	100

Instruments with test voltages up to 1 kV often incorporate bridges for resistance measurements between 10 mΩ and 1 MΩ, suitable for earth fault location on cables.

After testing at high voltage it is important that stored capacitive energy be discharged through a resistor. Many instruments include this facility.

6.6.2 Earth resistance tests

Simple earth rod systems rarely have a resistance below 0.5 Ω, but complex earthing systems may have a resistance of 0.1 Ω or less, and require special measurement techniques.

The 'Megger' earth tester is used to determine soil resistivity during site surveys as well as the actual electrode resistance for existing installations. The instrument includes a double-coil ratiometer and a special a.c. generator. To avoid the effects of electrolytic e.m.f. in the soil, and of p.d. arising from earth currents, the instrument is designed to pass a.c. through the soil but d.c. through the indicator. Indications on various ranges lie between 50 mΩ and 20 kΩ. Null-balance earth testers are usable down to 10 mΩ.

6.6.3 Conductivity tests

Measurements of resistance down to a few microhms (i.e. conductance up to several megasiemens) at 100 A are required for the assessment of conductivity. The 'Ducter' is an example of a low-resistance test set for switch contacts, machine windings, soldered joints and the like. The four-terminal measurement is made with the aid of two pairs of 'spike' probes; the outer probes are arranged to inject direct current into the test piece, and the inner probes select the p.d. across the resistance of interest. The injected current and that from the selected region pass through the control and deflecting coils, respectively, mounted in the usual ratiometer dual-coil assembly to give a deflection scaled in ohms. Internal and external heavy-duty batteries, or rectified mains supplies, are used to provide test currents between 0.2 and 100 A, to give maximum and minimum readings of 10 Ω and a few microhms, respectively.

6.7 Integrating (energy) metering

Integrating meters record the time-integral of active, reactive and apparent power as a continuous summation. The integration may be limited by a specified total energy (e.g. prepayment meters) or by time (e.g. maximum demand). Meters for a.c. supplies are all of the induction type, with measuring elements in accordance with the connection of the load (1-phase or 3- or 4-wire 3-phase). Manufacturing and testing specifications are given in BS 37. Instrument transformers for use with meters are listed in BS 3938 and BS 3941.

6.7.1 Single-phase meter

The rotor is a light aluminium disc on a vertical spindle, supported in low-friction bearings. The lower bearing is a sapphire cup, carrying the highly polished hemispherical hardened steel end of the spindle. The rotor is actuated by an induction driving element and its speed is controlled by an eddy current brake. The case is usually a high-quality black phenolic moulding with integral terminal block. The frame is a rigid high-stability iron casting which serves as the mounting, as part of a magnetic flux path, and as a shield against ambient magnetic fields.

The driving element has the basic form shown in *Figure 6.6*. It has two electromagnets, one voltage-excited and the other current-excited. The volt magnet, roughly of E-shape, has a

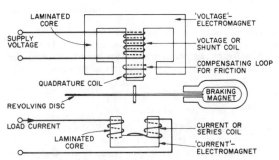

Figure 6.6 Essential parts of induction meter

nearly complete magnetic circuit with a volt coil on the centre limb. Most of the flux divides between the outer limbs, but the *working* flux from the central limb penetrates the disc and enters the core of the current magnet. The latter, of approximate U-shape, is energised by a coil carrying the load current. With a condition of *zero* load current, the working flux from the volt magnet divides equally between the two limbs of the current magnet and returns to the volt magnet core through the frame, or through an iron path provided for this purpose. If the volt magnet flux is symmetrically disposed, the eddy current induced in the disc does not exert any net driving torque and the disc remains stationary. The volt magnet flux is approximately in phase quadrature with the applied voltage.

When a load current flows in the coil of the current magnet, it develops a cophasal flux, interacting with the volt magnet flux to produce a resultant that 'shifts' from one current magnet pole to the other. Force is developed by interaction of the eddy currents in the disc and the flux in which it lies, and the net torque is proportional to the voltage and to that component of the load current in phase with the voltage—i.e. to the active power. The disc therefore rotates in the direction of the 'shift'.

The disc rotates through the field of a suitably located permanent isotropic brake magnet, and induced currents provide a reaction proportional to speed. Full-load adjustment is effected by means of a micrometer screw which can set the radial position of the brake magnet.

The volt magnet working flux lags the voltage by a phase angle rather less than 90° (e.g. by 85°). The phase angle is made 90° by providing a closed *quadrature coil* on the central limb, with its position (or resistance) capable of adjustment.

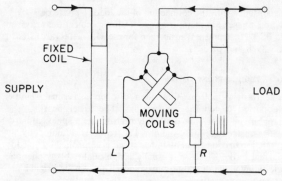

Figure 6.4 Single-phase electro-dynamic power factor indicator

between any two voltages (of the same frequency) over a wide range of frequencies. Either the conventional solid state digital instruments generate a pulse when the two voltages pass zero, and measure and display the time between these pulses as phase difference, or the pulses trigger a multivibrator for the same period, to generate an output current proportional to phase (time) difference. These instruments provide good discrimination and, with modern high-speed logic switching, the instant for switch-on (i.e. pulse generation) is less uncertain: conversely, the presence of small harmonic voltages around zero could lead to some ambiguity.

6.5.8 Phase sequence and synchronism

A small portable form of *phase sequence* indicator is essentially a primitive three-phase induction motor with a disc rotor and stator phases connected by clip-on leads to the three-phase supply. The disc rotates in the marked forward direction if the sequence is correct. These instruments have an intermittent rating and must not be left in circuit.

The *synchroscope* is a power factor instrument with rotor slip-rings to allow continuous rotation. The moving-iron type is robust and cheap. The direction of rotation of the rotor indicates whether the incoming 3-phase system is 'fast' or 'slow', and the speed of rotation measures the difference in frequency.

6.5.9 Frequency

Power frequency monitoring frequency indicators are based on the response of reactive networks. Both electrodynamic and induction instruments are available, the former having the better accuracy, the latter having a long scale of 300° or more.

One type of indicator consists of two parallel fixed coils tuned to slightly different frequencies, and their combined currents return to the supply through a moving coil which lies within the resultant field of the fixed coils. The position of the moving coil will be unique for each frequency, as the currents (i.e. fields) of the fixed coils are different unique functions of frequency. The indications are within a restricted range of a few hertz around nominal frequency. A ratiometer instrument, which can be used up to a few kilohertz, has a permanent-magnet field in which lie two moving coils set with their planes at right angles. Each coil is driven by rectified a.c., through resistive and capacitive impedances, respectively, and the deflection is proportional to frequency.

Conventional solid state counters are versatile time and frequency instruments. The counter is based on a stable crystal reference oscillator (e.g. at 1 MHz) with an error between 1 and 10 parts in 10^8 per day. The displays have up to eight digits, with a top frequency of 100 MHz (or 600 MHz with heterodyning). The

resolution can be 10 ns, permitting pulse widths of 1000 ns to be assessed to 1 per cent: but a 50 Hz reading would be near to the bottom end of the display, giving poor discrimination. All counters provide for measurements giving the period of a waveform to a good discrimination.

6.6 Resistance and conductance instrumentation

A wide range of portable instruments has been devised to fulfil the need for safety measurements, and specialised conductive tests, as required by IEE Wiring Regulations[9], various Electricity Board and Factory Inspectorate specifications, the Mines and Quarries Act, etc.

6.6.1 Insulation tests

The test instrument requires a source of direct test voltage between 250 V and 10 kV. The prime source is often a hand-driven a.c. generator. For prolonged testing the generator may be motor-driven for voltages up to 2.5 kV, or a mains feed through a transformer and static rectifier up to 10 kV. Battery-fed instruments use a transistor oscillator, Zener diode and stabilising circuits, step-up transformer and rectifier; these sets are limited to a few kilovolts.

In most instruments the measurement is based upon the ratiometer principle, using two moving coils connected mutually at right angles within a permanent-magnet field (*Figure 6.5*). The

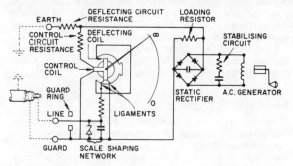

Figure 6.5 Bridge Megger tester (Evershed & Vignoles Ltd)

control coil of the movement is in series with a constant resistance, the deflecting-coil in series with the insulation resistance to be measured. No control spring is fitted; hence, the position of the movement is determined by the ratio between the currents in the two coils. This deflection is a function of resistance, any change in the applied voltage affecting the two coils in the same proportion. The inherent (reciprocal) resistance scale is opened out by means of suitably biased non-linear diodes. As an alternative, some electronic instruments use a conventional microammeter, either to measure the current through the insulation or to detect the out-of-balance signal in a bridge network.

The list below gives a general classification of supply source, function, maximum test voltage and maximum measurable insulation resistance:

switching surge. Thermocouple voltmeters are low-impedance devices (taking, e.g. 500 mW at 100 V) and may be unsuitable for electronic circuit measurements.

6.5.4 Medium and high direct and alternating voltage

6.5.4.1 Electrostatic voltmeter

This is in effect a variable capacitor with fixed and moving vanes. The power taken (theoretically zero) is in fact sufficient to provide the small dielectric loss. The basically square-law characteristic can be modified by shaping the vanes to give a reasonably linear upper scale. The minimum useful range is 50–150 V in a small instrument, up to some hundreds of kilovolts for large fixed instruments employing capacitor multipliers. Medium-voltage direct-connected voltmeters for ranges up to 15 kV have CI ratings from 1.0 to 5.0. The electrostatic indicator is a true primary a.v./d.v. transfer device, but has now been superseded by the thermocouple instrument. The effective isolation property of the electrostatic voltmeter is attractive on grounds of safety for high-voltage measurements.

6.5.5 Power

6.5.5.1 Electrodynamic indicator

The instrument is essentially similar to the electrodynamic voltmeter and ammeter. It is usually 'air-cored', but 'iron-cored' wattmeters with high-permeability material to give a better torque/mass ratio are made with little sacrifice in accuracy. The current (fixed) coils are connected in series with the load, and the voltage (moving) coil with its series range resistor is connected either (a) across the load side of the current coil, or (b) across the supply side. In (a) the instrument reading must be *reduced* by the power loss in the volt coil circuit (typically 5 W), and in (b) by the current coil loss (typically 1 W).

The volt coil circuit *power* corrections are avoided by use of a *compensated wattmeter*, in which an additional winding in series with the volt circuit is wound, turn-for-turn, with the current coils, and the connection is (a) above. The m.m.f. due to the volt-coil circuit current in the compensating coil will cancel the m.m.f. due to the volt circuit current in the current coils.

The volt circuit terminal marked $\pm$ is immediately adjacent to the voltage coil; it should be connected to the current terminal similarly marked, to ensure that the wattmeter reads positive power and that negligible p.d. exists between fixed and moving coils, so safeguarding insulation and eliminating error due to electric torque.

When a wattmeter is used on d.c., the power should be taken from the mean of reversed readings; wattmeters read also the active (average) a.c. power. For the measurement of power at very low power factor, special wattmeters are made with weak restoring torque to give f.s.d. at, e.g., a power factor of 0.4. Range extension for all wattmeters on a.c. can be obtained by internal or external current transformers, internal resistive volt-range selectors or external voltage transformers. Typical self-contained ranges are between 0.5 and 20 A, 75 and 300 V. The usable range of frequency is about 30–150 Hz, with a best CI of 0.05.

Three-phase power can be measured by single dual-element instruments.

6.5.5.2 Thermocouple indicator

The modern versions of this instrument are more correctly termed electronic wattmeters. The outputs from current- and voltage-sensing thermocouples, multiplied together and amplified, can be displayed as average power. Typical d.c. and a.c. ranges are 300 mV–300 V and 10 mA–10 A, from pulsed and other non-sinusoidal sources at frequencies up to a few hundred kilohertz. The interaction with the network is low; e.g. the voltage network has typically an impedance of 10 kΩ/V.

6.5.5.3 Induction indicator

The power-reading instrument includes both current and voltage coils. The energy meter is referred to later. The instruments are frequency-sensitive and are used only for fixed-frequency switchboard applications. The CI number is 1.0–5.0, usually the latter.

6.5.5.4 Electrodynamic reactive-power indicator

The active-power instrument can read reactive power if the phase of the volt circuit supply is changed by 90°, to give $Q = VI \sin \phi$ instead of $P = VI \cos \phi$ (for sinusoidal conditions).

6.5.6 Maximum alternating current

6.5.6.1 Maximum-demand instrument

The use of an auxiliary pointer carried forward by the main pointer and remaining in position makes possible the indication of maximum current values, but the method is not satisfactory, because it demands large torques, a condition that reduces effective damping and gives rise to overswing. A truer maximum demand indication, and one that is insensitive to momentary peaks, is obtainable with the aid of a thermal bimetal. Passing the current to be indicated through the bimetal gives a thermal lag of a few minutes. For long-period indication (1 h or more) a separate heater is required.

Such instruments have been adapted for 3-phase operation using separate heaters. A recent development permits a similar device to be used for two phases of a 3-phase supply. The currents are summed *linearly*, and the maximum of the combined unbalanced currents maintained for 45 min is recorded on a kVA scale marked in terms of nominal voltage.

6.5.7 Power factor

Power factor indicators have both voltage and current circuits, and can be interconnected to form basic electrodynamic or moving-iron direct-acting indicators. The current coils are fixed; the movement is free, the combined voltage- and current-excited field producing both deflecting and controlling torque. The electrodynamic form has the greater accuracy, but is restricted to a 90° deflection, as compared with 360° (90° lag and lead, motoring and generating) of the moving-iron type.

One electrodynamic form comprises two fixed coils carrying the line current, and a pair of moving volt coils set with their axes mutually at almost 90° (*Figure 6.4*). For one-phase operation the volt-coil currents are nearly in quadrature, being, respectively, connected in series with an inductor L and a resistor R. For three phase working, L is replaced by a resistor r and the ends of the volt-coil circuits taken, respectively, to the second and third phases. A three-phase moving-iron instrument for balanced loads has one current coil and three volt coils (or three current and one voltage); if the load is unbalanced, three current and three volt coils are required. All these power factor indicators are industrial frequency devices.

The one-phase electrodynamic wattmeter can be used as a phase-sensing instrument. When constant r.m.s. voltages and currents from two, identical-frequency, *sinusoidal* power supplies are applied to the wattmeter, then the phase change in one system produces power scale changes which are proportional to $\cos \phi$.

Many electronic phase-meters of the analogue and digital type are available which, although not direct-acting, have high impedance inputs, and they measure the phase displacement

readings to be obtained from supplies with distorted current waveforms, although excessive distortion will cause errors. Nevertheless, instruments can be designed for use in the lower audio range, provided that the power loss (a few watts) can be accepted. Ranges are from a few milliamperes and several volts upward. The power loss is usually significant and may have to be allowed for.

6.5.3.3 Split indicators

Figure 6.3 illustrates a typical indicator of the 'clip-on' form, with a split and hinged magnetic circuit. It obviates the need for a

Figure 6.3 Split electromagnet instrument

current transformer, and by eliminating windings minimises electrical failure. Such indicators are light, convenient and operable by one hand; they can be used for both a.c. and d.c.; and the range can be altered by substituting the detachable movement units. In use it is advisable to press and release the trigger immediately before taking a reading, in order to reduce hysteresis error.

6.5.3.4 Electrodynamic indicator

In this the permanent magnet of the moving-coil instrument is replaced by an electromagent (field coil). For ammeters the moving coil is connected in series with two fixed field coils; for voltmeters the moving and the fixed coils each have series-connected resistors to give the same time-constant, and the combination is connected in parallel across the voltage to be measured. With ammeters the current is limited by the suspension to a fraction of 1 A, and for higher currents it is necessary to employ non-reactive shunts or current transformers.

6.5.3.5 Precautions and corrections

The flux density is of the same order as that in moving-iron instruments, and similar precautions are necessary. The torque has a square-law characteristic and the scale is cramped at the lower end. By using the mean of reversed readings these instruments can be d.c. calibrated and used as adequate d.c./a.c. transfer devices for calibrating other a.c. instruments such as moving-iron indicators. The power loss (a few watts) may have to be allowed for in calculations derived from the readings.

6.5.3.6 Induction indicator

This single-frequency instrument is robust, but limited to CI number from 1.0 to 5.0. The current or voltage produces a proportional alternating magnetic field normal to an aluminium disc arranged to rotate. A second, phase-displaced field is necessary to develop torque. Originally it was obtained by pole 'shading', but this method is obsolete; modern methods include separate magnets one of which is shunted (Ockenden), a cylindrical aluminium movement (Lipman) or two electromagnets coupled by loops (Banner). The induction principle is most generally employed for the measurement of energy.

6.5.3.7 Moving-coil rectifier indicator

The full-wave rectification process is discussed under 'Analogue/ Digital Instrumentation'. The place of the former copper oxide and selenium rectifiers has been taken by silicon diodes. Used with taut-band suspension, full-wave instrument rectifier units provide improved versions of the useful rectifier indicator.

6.5.3.8 Precautions and corrections

Owing to its inertia, the polarised moving-coil instrument gives a mean deflection proportional to the mean rectified current. The scale is marked in r.m.s. values on the assumption that the waveform of the current to be measured is sinusoidal with a form factor of 1.11. On non-sinusoidal waveforms the true r.m.s. value cannot be inferred from the reading: only the true *average* can be known (from the r.m.s. scale reading divided by 1.11). When true r.m.s. values are required, it is necessary to use a thermal, electrodynamic or square-law electronic instrument.

6.5.3.9 Multirange indicator

The rectifier diodes have capacitance, limiting the upper frequency of multirange instruments to about 20 kHz. The lower limit is 20–30 Hz, depending on the inertia of the movement. Indicators are available with CI numbers from 1.0 to 5.0, the multirange versions catering for a wide range of direct and alternating voltages and currents: a.c., 2.5 V–2.5 kV and 200 mA–10 A; d.c. 100 µV–2.5 kV and 50 µV–10 A. Non-linear resistance scales are normally included; values are derived from *internal* battery-driven out-of-balance bridge networks.

6.5.3.10 Moving-coil thermocouple indicator

The current to be measured (or a known proportion of it) is used as a heater current for the thermocouple, and the equivalent r.m.s. output voltage is steady because of thermal inertia—except for very low frequency (5 Hz and below). With heater current of about 5 mA in a 100 Ω resistor as typical, the possible voltage and current ranges are dictated by the usual series and shunt resistance, under the restriction that only the upper two-thirds of the square-law scale is within the effective range. The instrument has a frequency range from zero to 100 MHz. Normal radio-frequency low-range self-contained ammeters can be used up to 5 MHz with CI from 1.0 to 5.0. Instruments of this kind could be used in the secondary circuit of an r.f. current transformer for measurement of aerial current. The minimum measurable voltage is about 1 V.

The thermocouple, giving a true r.m.s. indication, is a primary a.c./d.c. transfer device. Recent evidence[5] indicates that the transfer uncertainty is only a few parts in 10^6. Such devices provide an essential link in the traceability chain between microwave power, r.f. voltage and the primary standards of direct voltage and resistance.

6.5.3.11 Precautions and corrections

The response is slow, e.g. 5–10 s from zero to f.s.d., and the overload capacity is negligible; the heater may be destroyed by a

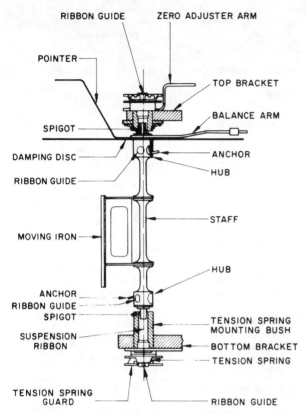

Figure 6.2 Taut-ribbon suspension assembly (Crompton Parkinson Ltd)

current. The rectifier instrument, although an a.c. instrument, has a scale which is usually linear, as it depends on the moving-coil characteristic and the rectifier effect is usually negligible. (In low-range voltmeters, the rectifier has some effect and the scale shape is between linear and square-law.) Thermal ammeters and electrodynamic voltmeters and ammeters usually have a true square law scale, as any scale shaping requires extra torque and it is already low in these types. Moving-iron instruments are easily designed with high torque, and scale shaping is almost always carried out in order to approach a linear scale. In some cases the scale is actually contracted at the top in order to give an indication of overloads that would otherwise be off-scale. The best moving-iron scale shape is contracted only for about the initial 10% and is then nearly linear. Logarithmic scales have the advantage of equal percentage accuracy all over the scale, but they are difficult to read, owing to sudden changes in values of adjacent scale divisions. Logarithmic scales are unusual in switchboard instruments, but they are sometimes found in portable instruments, such as self-contained ohmmeters.

BS 3693[2] deals with scales. Following the revision of BS 89[2] there has been the requirement for new instruments to identify the effective range. Further, for some years nearly 'circular-scale' instruments have been made, and the advent of taut-band suspensions has made a scale covering 250° generally available.

6.5.2 Direct voltage and current

6.5.2.1 Moving-coil indicator

This instrument comprises a coil, usually wound on a conducting former to provide eddy-current damping, with taut-band or pivot/control-spring suspension. In each case the coil rotates in the short air-gap of a permanent magnet. The direction of the deflection depends on the polarity, so that unmodified indicators are usable only on d.c., and may have a centre zero if required. The error may be as low as $\pm 0.1\%$ of f.s.d.; the range, from a few microamperes or millivolts up to 600 A and 750 V, which makes possible multirange d.c. test sets. The scale is generally linear (equispaced) and easily read. Non-linear scales can be obtained by shaping the magnet poles or the core to give a non-uniform air-gap.

6.5.2.2 Corrections

The *total* power taken by a normal-range voltmeter can be 50 µW/V or more. For an ammeter the *total* series loss is 1–50 mW/A. Such powers may be a significant fraction of the total delivered to some electronic networks: in such cases electronic and digital voltmeters with trivial power loss should be used instead.

6.5.2.3 Induced moving-magnet indicator

The instrument is polarised by a fixed permanent magnet and so is only suitable for d.c. A small pivoted iron vane is arranged to lie across the magnet poles, the field of which applies the restoring torque. One or more turns of wire carry the current to be measured around the vane, producing a magnetic field at right angles to that of the magnet, and this drive torque deflects the movement to the equilibrium position. It is a cheap, low-accuracy instrument which can have a nearly linear centre-zero scale, well suited to monitoring battery charge conditions.

6.5.3 Alternating voltage and current

6.5.3.1 Moving-iron indicator

This is the most common instrument for a.c. at industrial frequency. Some (but not all) instruments may also be used on d.c. The current to be measured passes through a magnetising coil enclosing the movement. The latter is formed either from a nickel–iron alloy vane that is *attracted* into the field, or from two magnetic alloy members which, becoming similarly polarised, mutually *repel*. The latter is more usual but its deflection is limited to about 90°; however, when combined with attraction vanes, a 240° deflection is obtainable but can be used only on a.c. on account of the use of material of high hysteresis.

The inherent torque deflection is square law and so, in principle, the instrument measures true r.m.s. values. By careful shaping of vanes when made from modern magnetic alloys, a good linear scale has been achieved except for the bottom 10%, which readily distinguishes the instrument from the moving-coil linear scale. Damping torque is produced by eddy-current reaction in an aluminium disc moving through a permanent-magnet field.

6.5.3.2 Precautions and corrections

These instruments work at low flux densities (e.g. 7 mT), and care is necessary when they are used near strong ambient magnetic fields despite the magnetic shield often provided. As a result of the shape of the B/H curve, a.c./d.c. indicators used on a.c. read high at the lower end and low at the higher end of the scale. The effect of hysteresis with d.c. measurements is to make the 'descending' reading higher than the 'ascending'.

Moving-iron instruments are available with class index values of 0.3–5.0, and many have similar accuracies for a.c. or d.c. applications. They tend to be single-frequency power instruments with a higher frequency behaviour that enables good r.m.s.

Table 6.1 Comparison of instruments

Measurement parameter	Type	Advantages	Disadvantages
Direct voltage and current	Moving-coil	Accurate: wide range	Power loss 50–100 mW
	Induced moving-magnet	Cheap	Inaccurate; high power loss
Alternating voltage and current	Moving-iron	Cheap; reasonable r.m.s. accuracy	No low range; high power loss
	Electrodynamic	Adequate a.c./d.c. comparator; more accurate r.m.s.	Expensive; high power loss; square law scale
	Induction	Long scale	Single frequency; high power loss; inaccurate
	Moving-coil rectifier	Low power loss (a few mW); wide range: usable up to audio frequency	Non-sinusoidal waveform error
	Moving-coil thermocouple	True r.m.s.; good a.c./d.c. comparator; usable up to radio frequency	No overload safety factor; slow response; power loss (e.g. 1 W)
Direct and alternating voltage	Electrostatic	Negligible power; good a.c./d.c. comparator; accurate r.m.s.; wide frequency range	High capacitance; only for medium and high voltage; poor damping
A.C. power (active)	Electrodynamic	Accurate; linear scale	Expensive; high power loss
	Thermocouple electronic	High frequency; pulse power; true r.m.s.	Expensive
	Induction	Cheap; long scale	Single frequency; high power loss
A.C. power (reactive)	Electrodynamic	Accurate; linear scale	Power loss; needs phase-rotation network
A.C. maximum current	Thermal maximum demand	Long adjustable time-constant; thermal integration; cheap	Relatively inaccurate

(1) *Electromagnetic torque*: moving-coil, moving-iron, induction and electrodynamic (dynamometer).
(2) *Electric torque*: electrostatic voltmeter.
(3) *Electrothermal torque*: maximum demand.

A comparison is made in *Table 6.1*, which lists types and applications.

6.5.1.1 Torque effects

Instrument dynamics is discussed later. The relevant instantaneous torques are: *driving* torque, produced by means of energy drawn from the network being monitored; *acceleration* torque; *damping and friction* torque, by air dashpot or eddy current reaction; *restoring* torque, due usually to spring action, but occasionally to gravity or opposing magnetic field.

The higher the driving torque the better, in general, are the design and sensitivity; but high driving torque is usually associated with movements having large mass and inertia. The torque/mass ratio is one indication of relative performance, if associated with low power demand. For small instruments the torque is 0.001–0.005 mN-m for full-scale deflection; for 10–20 cm scales it is 0.05–0.1 mN-m, the higher figures for induction and the lower for electrostatic instruments.

Friction torque, always a source of error, is due to imperfections in pivots and jewel bearings. Increasing use is now made of taut-band suspensions (*Figure 6.2*); this eliminates pivot friction and also replaces control springs. High-sensitivity moving-coil movements require only 0.005 mN-m for a 15 cm scale length: for moving-iron movements the torques are similar to those for the conventional pivoted instruments.

6.5.1.2 Scale shapes

The moving-coil instrument has a linear scale owing to the constant energy of the permanent magnet providing one of the two 'force' elements. All other classes of indicators are inherently of the double-energy type, giving a square-law scale of the linear property (voltage or current) being measured; for a wattmeter, the scale is linear for the average scalar product of voltage and

6/6 Electrical metrology and instrumentation

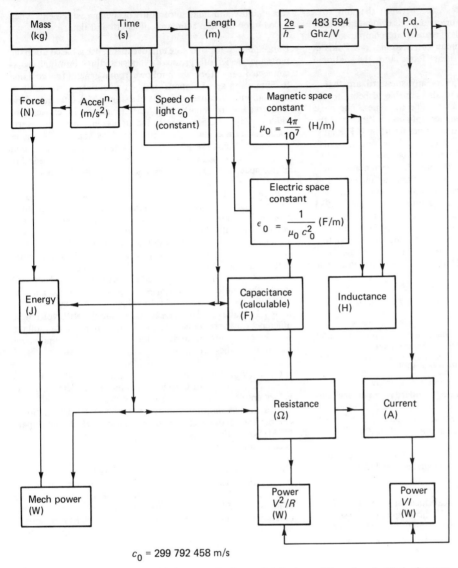

Figure 6.1 Dimensional relationship between absolute and derived quantities using electrical references based upon (i) voltage (Josephson Junction)[6] and (ii) calculable capacitor (Lampard)[7,8]

unit. The input arrows to a given block indicate, collectively, all the more primitive dimensional components of that block.

6.5 D.c. and industrial-frequency analogue instrumentation

The principal specification for the characteristics and testing of low-frequency instruments is BS 89. Instruments are classified into nine groups conforming closely to the 'Class Index' (CI) notation of the IEC (Publication No. 51). The class index numbers are 0.05, 0.1, 0.2, 0.3, 0.5, 1.0, 1.5, 2.5 and 5.0, and they correspond to the percentage error at *full-scale* deflection for most (but not all) instruments when used under closely specified operational conditions. The 'effective range' is that over which the class index limit applies: the lower limit of this range is dictated by the inherent torque characteristic of the instrument and the length of the scale, and can vary from 10% of the fiducial value (often full-scale deflection) to 33% for square-law instruments. The scales of most instruments manufactured prior to 1971 include markings below the lower limit of effective range, but after that date the effective range *should* be unambiguous, either by omission of most of the out-of-range marks or by arrows on a completely marked scale.

Class indices 0.05 and 0.1 apply to instruments for use as reference indicators; and these instruments, when justified by *recent* traceability evidence, could be used to check bench and semi-portable indicators with higher CI values (0.2, 0.3 and 0.5). The remainder (CI 1.0–5.0) are some bench, and all portable, panel and switchboard instruments.

6.5.1 Direct-acting indicators

These can be described in terms of the dominating torque-production effects:

Very small agreed differences exist between the unit sizes maintained by national laboratories, and this evidence has been available since 1927 as the result of regular intercomparisons[4] of the important units by the Bureau International des Poids et Mésures (BIPM) at Sèvres, near Paris and the national laboratories[5].

The most desirable properties of any standard are that it should be completely stable, and also capable of being reproduced exactly in different places at different times. The most stable, precise and (by definition) accurate unit is 'time' and the reciprocal value of 'frequency' (events per unit time). The most accurate set of standard unit sizes could be one in which frequency plays a dominant, if not a fundamental, role in the determination of each unit size. Progress has been made in this direction by employing various aspects of nuclear and electronic resonance phenomena. Current (gyromagnetic ratio of particles) is susceptible to this approach. The metre was defined in 1983 using interferometric/laser techniques in terms of the velocity of light in a vacuum. The d.c. volt has been maintained since 1972 by means of the Josephson effect experiment, which is sensitive to a particular fixed-frequency microwave radiation: the calculable cross-capacitor (Thompson and Lampard) depends upon a *single* (interferometric) measurement of length.

6.4.2.1 Fundamental standards

Abbreviated definitions of the S.I. base units are given in Chapter 1. The notes below concern the metrological aspects.

Kilogram (mass) This is a comparison unit.

Metre (length) The basic experiment employs an interferometric technique. Many working standards used in mechanical metrology are precision gauge blocks of stainless steel which are reproducible to 0.25 μm. Some commercial laser interferometers provide resolutions of about 1 part in 2×10^9.

Second (time) The first continuously operating standard 'atomic clock' based on energy transitions in caesium-133 was made at the NPL in 1955. The uncertainty in modern caesium clocks is about seven parts in 10^{12}. The duration of the second was deduced from complex astronomical calculations of the Earth's orbital motion around the Sun, and it has supplanted the second of Universal Time (UT) based on the mean time of rotation of the Earth about its own axis. UT corrected for perturbations due to polar motion and seasonal variations is termed UT 2, and this is longer than the standard second, because of the decreasing rate of rotation of the Earth. Atomic Time (AT) is the international reference standard. The progressive phase difference between AT and UT 2 is corrected by the occasional (perhaps annual) addition of a 'leap second' to the AT scale. AT and UT 2 seconds are radiated in terms of time and frequency signals by radio; the most stable signals are of low frequency, e.g. 16 kHz (GBR, Rugby, UK) and 20 kHz (WWVL, Fort Collins, USA), because the groundwave is well defined.

Monthly bulletins are issued by NPL (for use by relevant R & D sections in industry) which state the daily traceable deviations in the radiated frequency and/or time-interval information which is available from various transmissions: e.g. BBC Radio 4 (Droitwich) 200 kHz carrier wave (until 1986); GBR 16 kHz; MSF 60 kHz; Loran emissions; BBC TV channels 26, 33, 46, 40. Channels 27, 31 have line and frame synchronisation signals which are generally controlled by an atomic frequency standard.

Speed of light c_0 in a vacuum Now defined, by international agreement, to be the physical constant $c_0 = 299\,792\,458$ m/s. This definition for the speed of electromagnetic radiation (or light) in free space was possible owing to the high precision, excellent repeatability and small uncertainty of such measurements when based upon modern standards of length and time.

Capacitance The absolute unit of capacitance is based upon the particular design of a group of cylindrical, three-terminal, cross-connected capacitors which follows from a general theorem due to Thompson and Lampard[7,8]. The uncertainty in the value of this computable capacitor is due to a single axial, interferometric measurement of length. Practical designs yield capacitance values of about 0.5 pF which, inevitably, includes the electric space constant ε_0 (i.e. the 'permittivity of free space'), where $\varepsilon_0 = c_0^{-2} \times \mu_0^{-1} = 8.854\,188$ pF/m, since, by definition in rationalised MKS units, the magnetic space constant μ_0 (i.e. the 'permeability of free space') $= 4\pi \times 10^{-7}$ H/m and $c_0 = 299\,792\,458$ m/s.

Voltage The definitive experiment is based upon the physics of superconductivity. When a current-carrying Josephson junction[6] is irradiated at a particular microwave frequency, a d.c. voltage (typically 5 mV) appears across the junction having a sufficiently high degree of stability and reproducibility to constitute the internationally accepted fundamental experimental determination of voltage. The BIPM recommended value of $2e/h = 483\,594$ GHz/V is employed in this derivation at NPL and most other national laboratories—except in the USA and the USSR, whose $2e/h$ values yield d.c. voltage standards which differ by $-1.2\,\mu V$ and $+3.3\,\mu V$, respectively, from the NPL volt: these differences are observable with modern transportable d.c. voltage sources and must be allowed for when assessing equipment calibrated in these countries.

A.C./D.C. transfer of voltage The basic standard method to transfer d.c. voltage or current values into equivalent r.m.s. a.c. values employs one or more thermocouples to sense the two equivalent temperature changes. A recent (1980) international comparison[5] of these a.c./d.c. transfer devices (UK, USSR, Japan, USA) showed that a discrepancy between a.c. and d.c. quantities is less than 10 parts per million up to 50 kHz.

Kelvin (temperature) The fundamental fixed point is the triple point of water, to which is assigned the thermodynamic temperature of 273.16 K exactly: on this scale the temperature of the ice point is 273.15 K (or 0 °C). The platinum resistance and thermocouple thermometers are the primary standard instruments up to 1336 K (gold point), above which the standard optical pyrometer is used to about 4000 K.

Candela (luminance) This has nearly the same magnitude as the former International Candle but can be more readily reproduced. It is not an absolute unit.

6.4.2.2 Derived electric and magnetic standards

Unit sizes are based upon the farad and volt. The national standard of d.c. resistance (the ohm) is derived from a 0.5 pF calculable capacitor using various precision a.c. bridges and frequency-sensitive techniques which determine in stages, 1000 pF, $10^5 \,\Omega$ a.c. to the 1 Ω d.c. reference value. The ampere follows from the volt and ohm, the coulomb from the ampere-second (or volt-farad) and the henry from the farad, and the weber and tesla are derived by using a standard solenoid.

6.4.2.3 Dimensional relationships

The *dimensional* relationships between the absolute and the derived units (not necessarily the *metrological* sequences above) are shown diagrammatically in *Figure 6.1*. The arrows indicate the direction from the *more* fundamental towards the derived

(2) The conventional analogue or digital display from voltmeters, counters, bridges: or a microprocessor-based automatic test equipment when associated with a display or digital record.

During the last decade the advent of the microprocessor has led to the design of many automatic test systems. Automatic test equipment is used, for example, to evaluate complex digital networks at each economically appropriate stage of production[3]. Likewise, general electronic instrumentation has improved dramatically in the facilities now available to the user—in addition to the pre-programmed assessment of large amounts of high-speed data prior to automatic onward transmission. The ephemeral nature of this fast-developing technology makes it impracticable to deal with systems in this section. The most important assessment to be made in the planning of a complex automatic test equipment is the total development cost of all the software facilities (including the specialist operators), since the hardware specification is not now the dominant economic element.

6.3.1 Traceability and uncertainty in the product

Imperfections in the performance of any system—whether through a communication channel, a closed-loop control assembly, a bridge or a single-instrument display—will inevitably depend upon the cumulative malfunctions and errors of the constituent parts. Any assessments of this kind must be limited by the known uncertainties in both the measurement technique and the test instruments.

The quality control of any product, or the total uncertainty which can be assigned to any numerical result in research, requires the evidence of recent, systematic *traceability* evidence to domestic standards and, hence, also to national standards. When the term 'traceability' is used, it implies that systematic measurement procedures are used and that the appropriate domestic standards of reference are available, whereby the numerical uncertainty of the reported result can be known in terms of the national standards. 'Traceability' can be applied to measurements of the most modest accuracy, since it is only the logical inter-relationship of given components to other components of higher quality which is essential in order to justify the traceability claim.

Traceability of product assessment figures has become a more general contractual requirement, particularly as a feature of the procurement policies of various governments. In general, the traceability process is a prudent activity for maintaining the quality of a product, and in promoting confidence in the results published in research bulletins and specifications.

6.3.1.1 United Kingdom Electrical Standards Laboratories

In the UK the National Physical Laboratory (NPL) is responsible for maintaining the fundamental standard unit values, which have been agreed through the International Electrotechnical Commission (IEC) and the Bureau International des Poids et Mésures (BIPM) in Paris. Traceability to these agreed values may be achieved to a known uncertainty of measurement, by regular reference to British Calibration Service (BCS) laboratories; this service has been in operation since 1968. The headquarters of BCS is at the NPL, Teddington, England. At present (1983) selected existing non-BCS industrial quality control laboratories and 'test houses' are being organised (at NPL) into a 'National Testing Laboratory Accreditation Scheme (NATLAS) in order to establish widespread national recognition of the competence of accredited testing laboratories based upon traceable measurements to be derived, generally, from BCS laboratories.

The UK Department of Trade and Industry is the controlling agency of the national laboratories.

6.4 Physical reference standards

The first International Electrotechnical Congress (IEC), Paris (1881), accepted an 'absolute' system of CGS electromagnetic units as a basis on which others could be derived. Several memorial names for unit sizes (volt, ohm, coulomb, farad(ay), etc.) were agreed. The absolute units of the ampere and the henry enabled all other electrical unit sizes to be realised through various precision bridge networks but, as the absolute unit sizes could not be determined with adequate consistency, the IEC adopted (London, 1908) an 'International Unit' system based on refined physical comparisons such as a mercury column for resistance and electrochemical deposition for electric quantity. Rapid advances in metrology led to a return to absolute units in 1940; the implementation was delayed by war until 1948. Two years later the IEC adopted rationalised MKS absolute units.

6.4.1 Previous electric and magnetic standards

The values were based upon d.c. current and mutual inductance.

6.4.1.1 Ampere (electric current)

The definition enabled 'current balances' to be constructed. When one pair of mutual-coupled coils carrying a current can move, the change in force between the coils, when the current direction in one is reversed, can be 'weighed'. The 'weight' can also be calculated in terms of geometry and the square of the current. Thus, current was measured in terms of force, and this ephemeral current 'stored' as the ratio of the e.m.f. of a standard cell to the resistance of a stable resistor.

A determination of the ampere, in 1974 at NPL, used a redesigned current balance to achieve an uncertainty of ± 1 part per million. A smaller uncertainty of about ± 0.2 part per million has now been obtained through the volt and farad, as discussed below.

6.4.1.2 Mutual inductance

The calculation of mutual inductance is part of the practical realisation of the ampere by current balance, as it is explicitly a component of the magnetising force computation from the Biot–Savart law. The ohm used to be measured in terms of frequency and mutual inductance by the differential method first devised by Lorentz. The unit of potential difference followed from the ohm × ampere product. The coulomb is the ampere-second, and the farad was obtained from the potential difference per coulomb. The weber is derived from the henry and the ampere, using standard solenoids.

6.4.2 Modern standards

Basic physical units of mass, length, time, thermodynamic temperature, luminance and quantity of substance have been defined as well as voltage and capacitance, which have in recent years replaced current in the experimental determination of electrical units. These and many related quantities are agreed as the result of systematic and carefully controlled experiments conducted between national laboratories. The absolute electrical units are re-created in national laboratories when required, as the result of very exacting experiments lasting some months, to yield results with a minimum uncertainty. The absolute values are ephemeral and must be preserved by standard cells and resistors to retain the national volt and ohm between determinations.

6.1 Introduction

Modern analogue and digital commercial instruments can provide measurement facilities and accuracies that approach the quality of a standards laboratory. If this new range of high-cost high-quality instrumentation is to be used profitably and integrated into systems which are economic, engineers need to be more aware of those principles of systematic measurement science which were, until recently, rather the domain of their colleagues in standards laboratories.

The subjects of electrical and electronic instrumentation, electrical metrology, automatic test equipment and general data acquisition are so extensive that it is possible to give only a brief selection of the more important devices likely to be used by the electrical engineer. The operational behaviour of instruments has been coupled with the need to develop a critical assessment of the device, in terms of the principles of electrical metrology, and the total uncertainty (i.e. total error) of the instrumentation system.

6.2 Terminology

Some of the technical terms used to describe various aspects of metrology give rise to ambiguous interpretations. The following brief list of definitions, with comments, has been compiled in order to provide a consistent usage throughout this section. British Standard definitions are given in Reference 1.

Metrology. The theory and practice of the science of measurement.

Quantity. By general usage the term now refers to the property of a physical phenomenon (e.g. mass), rather than the number of the basic unit amounts (e.g. 5 kg).

Dimension. Describes the essential quality of a phenomenon by reference to the fundamental dimensions of mass $|M|$, length $|L|$, time $|T|$ and quantity of electric charge $|Q|$ as an arbitrary fourth dimension for electromagnetic phenomena. Other selected dimensions for non-electromagnetic phenomena could include luminance and temperature.

Absolute measurement. The measurement of the unit size of a quantity, in terms of the unit sizes allocated to the fundamental dimensions of the quantity.

Absolute electrical unit. The unit size defined by reference to its own physical laws: the realisation of this unit value is often referred to as an 'absolute electrical measurement'.

Absolute electrical measurement. The measurement uses non-electrical units, each of which is a fundamental dimension (e.g. mass, length or time), together with a defined fourth electrical 'dimension' (e.g. quantity of charge Q). The electrical 'dimension' is required to link, or balance, the collected dimensions of the non-electrical absolute units with those of the actual physical electromagnetic laws of the particular electrical unit.

Traceability. The systematic process by which the quality of any measurement may be assessed by reference to superior standards. Each standard, in sequence, is compared with more accurate standards up to the ultimate International Reference Standards.

Traceable standard. A stable device, maintained under known controllable environmental conditions, which has been subjected to regular systematic traceability processes. The uncertainty in the numerical value of a traceable standard must be justifiable, ultimately, with respect to the national standard values.

Calibration. The experimental traceable process for determining and recording the *difference* between the indicated and 'true' values of a quantity. It is not necessarily the act of adjusting an instrument or device to a 'correct' value.

Self-calibration system. An instrumentation system, which includes a microprocessor, so programmed that, in effect, corrections are made to various sensitivie *internal* measurement parameters by repeated comparisons with *external* traceable standard values. The system (e.g. an automatic test facility) may include numerous digital test-pattern generators to ensure correct internal sequential operations of the digital logic networks, as well as analogue drift-voltage-to-zero adjustments etc.

Self-correcting system. Similar in apparent performance to a self-calibrating system—except that the overall uncertainty in the output value is referred to internal standard comparative networks. Self-adjusting (i.e. self-correcting) systems of this kind are very often referred to as 'self-calibrating' systems even though the output value is referred to *internal* standard comparative nettruly traceable process of measurement.

Systematic error. That part of the total error which is assumed to be due to an accumulation of constant causes. The algebraic value may be computed or assessed; however, the result cannot be included in statistical calculations.

Random error. That part of the total error which is not of a systematic kind. The random error may be treated in a statistical manner, since the sources of this error may, for repeated tests, yield results which are scattered around a mean value.

Uncertainty. The numerical values, assigned to the result of a measurement, which express the upper and lower limits of the *corrected* mean result within which the 'true value' should lie. To justify these limits, the uncertainty should include all systematic errors, together with a defined and stated statistical assessment of the random error.

Total error. From the measured value subtract the reference (i.e. 'true') value.

Correction. The negative of the total algebraic error. The number which must be added, algebraically, to the measured value to achieve the 'true' value.

Precision and repeatability. 'Precision' is a comparative term. High precision implies the ability to reproduce the measurement of a stable phenomenon with only a minor scatter in the results. High precision is *not* synonymous with a small uncertainty (i.e. high accuracy), although a small uncertainty requires high precision. The numerical assessment of scatter is the repeatability.

Indicated value. The measured number read from an instrument after application of nominal range factors.

Fiducial value. A value to which reference is made in order to specify the accuracy of an instrument. The value is usually the full-scale deflection (f.s.d.) value of a direct-acting instrument as defined in BS 89.

Effective range (of direct-acting instruments). That proportion of the total scale length of an instrument dial for which the stated accuracy limits of the instrument apply. The effective range varies from 10–100 to 33–100 per cent of f.s.d., depending on the type and class of instrument. This term applies to older types of analogue instruments.

Class index[2]. The number that indicates the accuracy of an indicator; it is numerically equal to the limit of error expressed as a percentage.

6.3 Traceability and quality assurance of data

The quality assurance of data falls, broadly, into two categories:

(1) Those areas in which a significant proportion of modern electronic measurements are effected between instruments within complex systems; such designs are generically described as 'signal processing systems'; often an analogue or digital read-out display is not required.

6.14 Data recording 6/51
    6.14.1 Ultraviolet recorder 6/51
    6.14.2 $X-Y$ recorders 6/54
    6.14.3 Analogue and digital strip-chart recorders 6/55

6 Electrical Metrology and Instrumentation

G L Bibby, BSc, CEng, MIEE
University of Leeds

Contents

6.1 Introduction 6/3

6.2 Terminology 6/3

6.3 Traceability and quality assurance of data 6/3
    6.3.1 Traceability and uncertainty in the product 6/4

6.4 Physical reference standards 6/4
    6.4.1 Previous electric and magnetic standards 6/4
    6.4.2 Modern standards 6/4

6.5 D.c. and industrial-frequency analogue instrumentation 6/6
    6.5.1 Direct-acting indicators 6/6
    6.5.2 Direct voltage and current 6/8
    6.5.3 Alternating voltage and current 6/8
    6.5.4 Medium and high direct and alternating voltage 6/10
    6.5.5 Power 6/10
    6.5.6 Maximum alternating current 6/10
    6.5.7 Power factor 6/10
    6.5.8 Phase sequence and synchronism 6/11
    6.5.9 Frequency 6/11

6.6 Resistance and conductance instrumentation 6/11
    6.6.1 Insulation tests 6/11
    6.6.2 Earth resistance tests 6/12
    6.6.3 Conductivity tests 6/12

6.7 Integrating (energy) metering 6/12
    6.7.1 Single-phase meter 6/12
    6.7.2 Polyphase meter 6/15
    6.7.3 Polyphase maximum-demand indicator 6/16
    6.7.4 Summation metering 6/17

6.8 Electronic instrumentation 6/19
    6.8.1 Electronic analogue voltmeters 6/19
    6.8.2 Instrument selection 6/20
    6.8.3 Differential instruments 6/21
    6.8.4 Digital voltmeters 6/21
    6.8.5 Digital wattmeters 6/26
    6.8.6 Energy meters 6/26
    6.8.7 Electronic oscillators 6/26
    6.8.8 Electronic analysers 6/27
    6.8.9 Data loggers 6/27

6.9 Electronic oscillography 6/27
    6.9.1 Cathode-ray tube 6/28
    6.9.2 Deflection amplifiers 6/30
    6.9.3 Instrument selection 6/31
    6.9.4 Operational use 6/31
    6.9.5 Calibration 6/32
    6.9.6 Applications 6/32

6.10 Potentiometers and bridges 6/33
    6.10.1 D.C. potentiometers 6/33
    6.10.2 A.C. potentiometers 6/35
    6.10.3 D.C. bridge networks 6/36
    6.10.4 A.C. bridge networks 6/37

6.11 Measuring and protection transformers 6/41
    6.11.1 Current transformers 6/41
    6.11.2 Voltage transformers 6/42

6.12 Magnetic measurements 6/43
    6.12.1 Instruments 6/44
    6.12.2 Magnetic parameters 6/44
    6.12.3 Bridge methods 6/45

6.13 Transducers 6/45
    6.13.1 Resistive transducers for temperature measurement 6/46
    6.13.2 Thermistors 6/46
    6.13.3 P–n junctions 6/47
    6.13.4 Pyrometers 6/47
    6.13.5 Pressure 6/47
    6.13.6 Acceleration 6/47
    6.13.7 Strain gauges 6/47
    6.13.8 Magnetostrictive transducers 6/48
    6.13.9 Reactance sensors 6/49
    6.13.10 Stroboscope 6/49
    6.13.11 Photo sensors 6/50
    6.13.12 Nuclear radiation sensors 6/51

In gas-cooled reactors, graphite is used as a moderator. In practice, each pile consists of a collection of fuel elements penetrating a large block of graphite which has been fabricated from graphite bricks. The moderator, in this case, thus has a structural role also. It is because of this that graphite oxidation rates in the CO_2 coolant are limited by additions of CO, CH_4 and H_2O.

A similar dual-role philosophy is usually used in water-cooled reactors also where the moderator (either light water or heavy water) can also serve as the coolant.

5.26.5 Pressure vessel

All thermal reactors use a pressurised coolant so that the integrity of the primary system is a principal factor in reactor operation and safety. Most designs rely on a single, large pressure vessel but 'pressure tube' reactors enclose each fuel channel in a minature vessel or pressure tube.

The most widespread material for the construction of large pressure vessels is mild steel, manufactured to a wall thickness which varies with reactor type, e.g. 120 mm for Magnox and 215 mm, for PWR. A problem with this material is the need always to demonstrate that brittle fracture of the vessel will not occur. A further limitation is that only relatively simple vessel shapes can be fabricated. These disadvantages can be overcome with the use of pre-stressed concrete vessels as used in the CAGR and later Magnox designs.

Pressure-tube reactors are less popular and are confined to water-cooled systems. The tube material may be either one of the zircaloy series or a Zr-$2\frac{1}{2}\%Nb$ alloy.

5.26.6 Shield

The reactor shield must attenuate neutrons and gamma radiation. In most cases, the cheapest material is heavy concrete such as barytes concrete in which crude barium sulphate replaces the normal aggregate.

Table 5.38 Thermostatic bimetals

Bimetal type*	Deflection constant $K(°C^{-1})$	Range of maximum sensitivity (°C)	Modulus of elasticity (kg/mm²)	Electrical resistivity ($\mu\Omega \cdot m$)
200	19.3×10^{-6}	−25–200	13 500	1.11
140	14×10^{-6}	0–175	16 000	0.76
400	11.8×10^{-6}	0–310	16 000	0.70
188†	8.8×10^{-6}	0–130	17 500	0.87
200 R17‡	18.9×10^{-6}	−25–200	13 500	0.16
R 5 M‡	13.4×10^{-6}	−20–200	16 000	0.06

* Telcon Metals Limited
† Corrosion resistant type
‡ Trimetals

5.26.1 Fuels

The only naturally occurring fuel is the isotope ^{235}U, present in natural uranium to about 0.7%. Most modern nuclear plants use fuel containing an enrichment of ^{235}U up to about 4%. Additionally, ^{239}Pu can be produced from ^{235}U, and ^{233}U from ^{232}Th.

The extraction of U is by ball-mill crushing and dissolution in nitric acid to yield an aqueous solution of uranyl nitrate. UO_3 is then obtained usually by thermal decomposition of the nitrate.

This higher oxide may then be reduced partially to UO_2 by high-temperature exposure to H_2-bearing gases or may be reduced completely to U metal by fluoridation to UF_4 and subsequent reduction by Mg.

Enrichment of U with the isotope ^{235}U is achieved by further fluoridation of UF_4 to UF_6 and then making use of the different gaseous diffusion rates of $^{235}UF_6$ and $^{238}UF_6$ to achieve partial separation of the isotopes. An alternative technique of separation relies on the differing masses of the isotopes and their response to centrifugal action.

Early commercial reactors in the UK relied on U metal as fuel but this technology has now been superseded by the worldwide use of compacted UO_2 pellets. The overriding advantage of the latter fuel is its improved efficiency of burn-up of the ^{235}U isotope. Both types of fuel, however, are enclosed in metal-walled cans in the reactor in order to contain the products of the nuclear reaction and to provide mechanical support and stability.

^{235}U fissions on interaction with a neutron having energy in the thermal range, say <1 MeV—hence, its use as a fuel in so-called 'thermal reactors'. In the future, breeder or 'fast' reactors may become of commercial importance. In such a system, highly-energetic neutrons are used to react with ^{238}U to produce ^{239}Pu as further fuel. An alternative fuel cycle involving the transmutation of ^{232}Th to ^{233}U in fast reactors and the subsequent use of this U isotope as fuel in thermal reactors is also a possibility.

5.26.2 Fuel cladding

The fuel cladding must provide structural support to the fuel stack and contain the fission products. Cladding materials must have a relatively low neutron absorption (for thermal reactors), have suitable low- and high-temperature strength and ductility, have good chemical compatibility with both the fuel and reactor coolant and have a resistance to property degradation as a result of neutron irradiation. Even though these requirements are particularly onerous, a number of cladding alloys have found service in commercial reactors.

The first generation of gas-cooled reactors in the UK used a Mg–0.8% Al cladding alloy, known as Magnox. This name has been adopted as the generic description of this type of reactor. Because of the low melting temperature of the alloy ($\sim$923 K), its use imposed limitations on thermal efficiency in the steam cycle.

Accordingly, the second generation of UK reactors, with their associated high $T2$ temperatures of up to 950 K, employ a highly alloyed stainless steel. This material contains 20% Cr (for oxidation resistance), 25% Ni and is stabilised with Nb. The alloy is used in the fully-annealed state and is entirely austenitic. This reactor system, termed the Advanced Gas-cooled Reactor (CAGR), is currently employed in base-load operation but may be required to load follow around the turn of the century. This change of operation will require the use of fuel cladding of higher creep strength; to this end, a modified version of the 20 Cr/25 Ni stainless steel, containing a dispersion of titanium nitride particles is available.

Fuel cladding in water reactors experiences appreciably lower temperatures of $\sim$600 K and simple alloys based on Zr have been used generally. That currently in favour, is termed Zircaloy-4 and consists of Zr, containing $\sim$1.5% Sn, 0.15% Cr and 0.15% Fe. Under normal operating conditions, this material has high creep strength and should allow load-following operation.

The fast-reactor programme has not yet reached a stage of commercial deployment but the prototype stations in France and the UK use fuel cladding based on Type 316 stainless steel.

5.26.3 Coolant

Commercial gas-cooled reactors use a CO_2-based coolant pressurised to $\sim$4 MNm^{-2} to aid in heat transfer. Additions of CO, CH_4 and H_2O are made to this in a closely controlled manner to optimise gaseous reaction within the primary circuit.

Water-cooled reactors may employ either light water (the LWRs) or heavy water (HWRs) as coolant. In each case, pressurisation is required to maintain the liquid phase at reactor operating temperature. Again, close control of minor constituents such as O_2 and also of pH level is essential for satisfactory performance.

The generation of large quantities of heat in the relatively small core of a fast reactor necessitated the use of a liquid sodium coolant to achieve adequate heat transfer in the early conceptual designs of the fast reactor. This choice has been maintained in present prototypes. An advantage of such a coolant is that pressurisation is unnecessary, but there are obvious disadvantages in the necessity to avoid water ingress and in the need for separate heating circuits to avoid freezing at room temperature. With recent advances in gas-circulator technology, it is now feasible to design a gas-cooled fast reactor using a pressurised CO_2 coolant.

5.26.4 Moderator

Thermal reactors require moderators to slow down fast neutrons to a sufficiently low energy to permit a fission reaction to occur with the ^{235}U isotope.

Table 5.37 Rare and precious metals used for contacts (Johnson Matthey data)

Metal or alloy	Melting point (°C)	Vickers hardness (annealed)	Density (kg/m³)	Resistivity Ω-m $\times 10^8$ at 20°C
Light duty contacts				
Gold	1064	20	19 300	2.2
Platinum	1769	65	21 450	10.6
10% Iridium–platinum	1780	120	21 600	24.5
20% Iridium–platinum	1815	200	21 700	30.0
25% Iridium–platinum	1845	240	21 700	32.0
30% Iridium–platinum	1885	285	21 800	32.3
25% Iridium–ruthenium–platinum	1890	310	20 800	39.0
7% Platinum–silver–gold	1100	60	17 100	16.8
30% Silver–gold	1025	32	16 600	10.4
30% Silver–copper–gold	1014	95	14 400	14.0
10% Silver–copper–gold	861	160	13 700	12.5
Rhodium	1963	40	12 400	4.9
Iridium	2447	220	22 600	5.1
Palladium	1554	40	12 000	10.8
40% Silver–palladium	1290	95	11 900	35.8
40% Copper–palladium	1200	145	10 400	35.0
Medium duty contacts				
10% Gold–silver	965	30	11 400	3.6
20% Palladium–silver	1070	55	10 700	10.1
10% Palladium–silver	1000	40	10 600	5.8
5% Palladium–silver	965	33	10 500	3.8
Fine silver	962	26	10 500	1.6
0.2% Magnesium–0.2% nickel–silver	961	140	10 400	2.8
1% Graphite–silver	961	40	9 900	1.8
2% Graphite–silver	961	40	9 700	2.0
Standard silver	778	56	10 300	1.9
10% Copper–silver	778	60	10 300	2.0
10% Cadmium oxide–silver	850	50	9 800	2.1
10% nickel–silver	961	40	10 300	2.0
15% Cadmium oxide–silver	850	60	10 000	2.3
20% Copper–silver	778	85	10 200	2.1
20% Nickel–silver	961	48	10 100	2.1
Cadmium–copper–silver	800	65	10 100	4.2
50% Copper–silver	778	95	9 700	2.1
Heavy duty contacts				
10% Cadmium oxide–silver	850	55	10 000	2.1
15% Cadmium oxide–silver	850	65	9 800	2.3
40% Tungsten carbide–silver	960	90	11 900	2.5
45% Tungsten carbide–silver	960	95	12 200	2.8
50% Tungsten–silver	960	125	13 600	2.8
50% Tungsten carbide–silver	960	160	12 500	3.0
55% Tungsten–silver	960	140	13 400	3.0
60% Tungsten carbide–silver	960	200	13 200	4.8
65% Tungsten–silver	960	185	14 800	3.3
73% Tungsten–silver	960	220	15 600	4.0
78% Tungsten–copper	1080	240	15 200	6.1
68% Tungsten–copper	1080	160	13 600	5.3
60% Tungsten–copper	1080	140	12 800	4.3

as overload circuit breakers, where the bimetal is heated by the direct passage of current. These include a number of trimetals in which the centre component is a low resistivity metal such as copper or nickel.

A straight bimetal strip of length L, thickness t, width w, fixed at one end and free to move at the other, will produce a free end deflection of $d = 1.1 K(\Delta T)L^2/t$ for a temperature change ΔT.

The force developed, if the free end is restrained from moving, is $1.1 KE(\Delta T)wt^2/4L$, where E is its modulus of elasticity. Similarly, for a bimetal spiral or helical coil of radius r, the angular deflection is $130KL/(\Delta t)/t$ and the restrained force is $0.19KE(\Delta T)wt^2/r$.

The properties of a representative range of bimetals are given in *Table 5.38*.

5.26 Nuclear reactor materials

The chief components of a nuclear reactor are: (1) the fuel; (2) the fuel cladding; (3) the coolant, which transports heat from the fuel to the heat exchangers; (4) the moderator; (5) the pressure vessel, which contains the high coolant pressure and (6) the shield and blanket materials.

tailored to give the same expansion coefficient as various types of glass for use as metal-to-glass or ceramic seals for TV tubes, integrated circuits and fluorescent lights. The expansion coefficient of these alloys will be in the range 4–10 parts per million/°C.

5.22.4 Heat-resisting alloys

A range of nickel–chromium based alloys has been specifically developed to meet strict limitations on the permissible creep of vital components in gas turbines in severe conditions of time, mechanical stress and working temperature. A 43/37/18/2% iron/nickel/chromium/silicon alloy is heat-resisting in oxidising conditions up to 950 °C or higher if the atmosphere is reducing. Developed originally for wire-woven conveyor belts for electric furnaces, it is now used also for a wide range of high-temperature applications.

5.23 Solders

Soldering is a process whereby metal components are joined together using a low-temperature filler metal, which is usually a tin-containing alloy. To assist in the wetting of the basis metal by molten solder, a flux, which is a weak acid, must be present to dissolve the thin oxide films already present on the surface of the components and to prevent further oxidation during heating of the joint.

5.23.1 Fluxes

Soldering fluxes are liquid or solid materials which, when heated, are capable of promoting or accelerating the wetting of metals by molten solder. Fluxes are usually divided into three groups by a classification based on the nature of their residues, namely corrosive, intermediate and non-corrosive fluxes. The National Standard for soft-soldering fluxes, BS 5625 (1980), incorporates a larger number of categories which gives an indication of the chemical nature of each flux type and their application.

For electrical components and other applications where corrosive residues could be difficult to remove, a non-corrosive rosin flux is used. Solder wire with a continuous core, or cores, of rosin flux can be used for manual soldering operations. The National Standard outlining the requirements of such material is BS 441 (1980).

5.23.2 Solders

A selection of solder alloys are available which melt at temperatures ranging between 60 and 310 °C. British Standard grades of solder, their maximum levels of impurities that are permissible and typical applications are listed in BS 219 (1977). For the soldering of electrical connections and high-quality sheet metal work, an alloy containing 60% tin 40% lead (Grade K) is often used. For the machine soldering of electronic assemblies a solder of equivalent alloy composition but with a lower level of impurities is recommended (Grade KP). Tin–solder alloys with lower tin contents are used for general engineering and the joining of copper conductors and lead sheathing, etc.

The shear strengths of soldered joints are generally within the range of 20–60 Nmm^{-2} at room temperature. As the temperature is increased the strength of joints made with tin–lead solders can decrease significantly. For this reason several solder alloy compositions, such as 95% tin 5% antimony (Grade 95A) and 96.5% tin 3.5% silver (Grade 96S) are recommended for use at service temperatures in excess of 100 °C. There are various methods of mechanically attaching two components prior to soldering in order to give added joint strength.

Depending on the soldering method employed solder can be used in the form of a bath of molten metal, sticks, solid wire, flux-cored wire, powder, solder creams, solder paint, or as preforms stamped out of thin foil. Solder paint and solder creams, which are a mixture of flux and solder powder, can be pre-placed into the area to be soldered by brush, syringe or other convenient means prior to heating.

5.24 Rare and precious metals

One of the most important and widespread electrical uses of the rare and precious metals is for contacts in applications, ranging from everyday electrical appliances to heavy duty switchgear and contact breakers, as well as in scientific and precision instruments, and communication equipment. Contacts can be broadly divided into light, medium and heavy duty and in *Table 5.37* the materials within these groups are approximately arranged in order of descending cost.

Light duty contacts require that the surfaces do not corrode appreciably, so the more noble metals and alloys are often used, while currents are low, so that resistivity is less important.

Medium duty contacts handle heavier currents, so that low resistivity is important and, since contact forces are normally high, slight corrosion/tarnishing is less important, but higher hardness becomes desirable.

For heavy duty applications severe arcing and heavy mechanical wear must be expected, so that higher resistivity can be accepted, in the interests of high hardness and arc resistance.

The choice of contact material is very much a compromise between the intrinsic initial cost, the ease and cost of replacement and the electrical and mechanical properties of the alloy.

Platinum and rhodium–platinum alloys are extensively used for high temperature thermocouples, which are accurate and particularly stable, as well as for the elements for high-temperature furnaces. Iridium–platinum, and rhodium–platinum are also used as electrodes in cathode ray tubes. Caesium salts are used in the manufacture of photo electric cells.

Pure silver is commonly used for electrical fuses and also in certain types of batteries and in capacitors, while a wide range of precious metals and alloys are used for thermal fuses acting as overtemperature cut-outs in electric furnaces.

A number of precious metal alloys are used for precision variable resistances, where the contact resistance at the wiper brush must be minimised.

5.25 Temperature-sensitive bimetals

Temperature-sensitive bimetals, commonly known as thermostatic bimetals, are produced by bonding together two metals having different coefficients of expansion and cold rolling the composite into strip. When subjected to a temperature change, the strip alters curvature in a precise and calculable manner. The bimetals can be used in forms such as the deflection of a straight strip, the rotation of spirals or helices and the snap action of dished discs. Applications include temperature indicators, thermostatic controls, energy regulators, temperature compensation and automotive fuel control devices.

The alloys used for the low-expansion components are normally Invars, 36% or 42% nickel–iron. The high-expansion components are mainly alloys based on manganese, iron or nickel. Alloys have been developed for special applications, such as shower temperature control units and steam traps, where the corrosion resistance is specially important. A range of bimetals with closely controlled resistivities are available for devices, such

transfer from one contact to another in a unidirectional manner and advantage can be taken of this by the use of dissimilar contact materials or even dissimilar contact sizes. With alternating current, the reversal of the arcing current will average the erosion so that the contacts erode equally. The unidirectional transfer by d.c. generates a 'pip and crater' condition which may reach the state where the 'pip' wedges into the crater effectively locking the contacts together. This effect is noticeable on small contacts which have too small a separation force to break the pipe clear of the crater. Very low force contacts as in reed switches are particularly susceptible and gold contacts may even cold weld when left closed for a long period. Precautions are necessary to reduce these effects and for d.c. low current applications RC suppression and resistance to limit current pulses through the contacts is generally used.

5.22 Special alloys

Many alloys have been developed for special applications either at elevated temperatures for heating elements or as heat resisting materials or at room temperature where a minimum change of resistance or dimensions is required.

5.22.1 Heating alloys

There is a considerable range of alloys which are used for heating elements for a wide range of applications including electric fires, storage heaters and industrial and laboratory furnaces. These alloys usually contain nickel together with chromium, copper and iron in varying proportions and often with small amounts of other elements. Similar alloys are also used for the construction of fixed and variable resistors. For heating elements, a considerable resistivity is required to limit the bulk of wire required. In addition the temperature coefficient of resistivity should be small so that the current remains reasonably constant at constant applied voltage. *Table 5.36* gives the properties and trade names of a range of resistance heating alloys.

The operating temperature of these alloys is dependent on the cross section of the wire or strip and on the atmosphere in which the material is to be used. The manufacturer's literature should be consulted before any application is finalised.

For higher temperatures, ceramic rods are used. Silicon carbide may be used in applications ranging from below 600 °C up to 1600 °C in either air or controlled atmospheres, although the type of atmosphere will determine the recommended element temperature. For even higher temperatures up to 1800 °C, cermets consisting of molybdenum disilicide ($MoSi_2$) with additions of a ceramic glass phase may be used, but again the maximum temperature depends on the type of atmosphere in which it is to be used.

5.22.2 Resistance alloys

Alloys for standard and fixed resistors are required to have a low-temperature coefficient of resistivity in the region of room temperature.

Manganin (84% Cu, 4% Ni, 12% Mn) This has been the traditional material for high-grade standard resistors. Its resistivity is about 0.40 $\mu\Omega$-m and its temperature coefficient is about 1×10^{-5}/°C.

Karma and Evanohm Trade names for quarternary alloys (73% Ni, 21% Cr, 2% Al, 2% Fe or Cu) which are being used increasingly for standard resistors, especially those of high value. The resistivity is about 1.30 $\mu\Omega$-m and the temperature coefficient is $\pm 0.5 \times 10^{-5}$/°C. Each of the above alloys have a low thermo-e.m.f. against copper. Normally joining the above alloys to copper should be by argon arc welding or if this is not possible hard soldering may be used.

Constantan, Eureka Advance and Ferry Proprietary names for copper–nickel alloys (55% Cu, 45% Ni) which are used for heavy duty and fixed resistors, potentiometers and strain gauges. They have a resistivity of about 0.50 $\mu\Omega$-m and the temperature coefficient varies between $\pm 4 \times 10^{-5}$/°C. The high thermo-e.m.f. against copper (-40 μV/°C) is a disadvantage for d.c. resistors but the effect is usually negligible in a.c. resistors. These alloys may be soft soldered satisfactorily.

5.22.3 Controlled expansion alloys

These give a range of thermal expansion required in precision parts, control devices and glass-to-metal seals. The lowest expansion alloy is a 36% nickel iron alloy and is variously called Invar, Nilex and Nilo. The expansion coefficient of these alloys can be less than 1 part in a million/°C although this can only be attained over a limited temperature range. Other alloys in the nickel–iron series with additions of cobalt or chromium can be

Table 5.36 Resistance heating alloys

Trade name	Nichrome Brightray C/S	Nichrome 3 Brightray B	Nichrome 1 Brightray F	Alferon 20 Brightray ICA	Alferon Y Fecralloy	Kanthal
Nominal composition, %						
Ni	Balance	59	38	—	—	—
Cr	20	16	18	20	16	22
Fe	1	Balance	Balance	Balance	Balance	Balance
Y	—	—	—	—	0.4	—
Al	—	—	—	5	6	4.5 to 5.8
Si	1.5	0.35	2.2	0.3	—	—
Maximum cycling temperature, °C	1150	1100	1000	Up to 1300	Up to 1350	Up to 1375
Resistivity						
$\mu\Omega$-m at 20 °C	1.08	1.12	1.06	1.37	1.37	1.30
at 1000 °C	1.15	1.26	1.30	1.46	1.45	1.50

Some or all of these alloys can be supplied by British Driver Harris, Stockport; Wiggin Alloys, Hereford; Kanthal, Stoke on Trent and B.S.C. Stocksbridge

The contact shape has importance if the contacts are not expected to erode during their lifetime and thus change their original shape.

5.21.3 Contact design

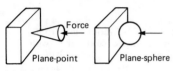
(a) Point contacts

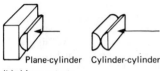
(b) Line contacts

(c) Plane contacts

Figure 5.21 Basic forms of contact

Table 5.35 Constants for copper contacts
$r = k/f^n$ with r in ohms, f in newtons

Form	Surface condition	n	k
Point	Normal	0.5	0.0007
Line	Normal	0.7	0.0015
Plane	Normal	1.0	0.004
(160 mm^2)	Lubricated	1.0	0.003
	Tinned	1.0	0.012
	Fine-ground, new	2.0	5

The passage of current at the contacting face will cause heating and this may cause a local softening of the material with a resulting increase in contacting area and reduced contact resistance. This would appear to be an advantage but represents a dangerous condition since welding may occur. In the minimal case the contacts may be separated mechanically and the weld broken but in the worst case the contacts become permanently joined! Welding may also occur due to contacts arcing and particularly so when contacts 'bounce' while carrying current.

To discourage welding, contact 'alloys' are available which contain low resistance silver and a hard material such as nickel or tungsten or an oxide of cadmium, tin or zinc. Graphite may also be included to reduce welding but at the expense of increased contact resistance.

5.21.3.1 Medium voltage (up to 660 V)

Make and break contacts for this industrial range are required to provide a useful life of many thousands of operations. The duty may be a motor load where the starting current is typically six times the running current. For this duty (AC3), a long life can be expected since little erosion occurs at contact close. However, bounce can cause welding so contacts needs to be rated according to the duty. Contacts which make and break equal currents (AC4) would erode more rapidly. At high currents shaped contacts may be used so that a defined area of the contact breaks last and carries the eroding arc. The remaining area, by staying clean, provides a low resistance for the continuous load current. The best combination of contact force, size, shape and material has to be made for a contactor to control a wide range of loads; capacitive loads (fluorescent lights) are particularly prone to weld problems. Silver–cadmium oxide is a common choice of contact material because of its good erosion and weld resistance properties. Silver–tin oxide is a recent alternative. The granular structure of these sintered materials has a great influence on their performance. The ratio of Silver to Cadmium/Tin/Nickel is high and thus acceptably low contact resistance is obtained. However, the thermal rating of the contacts normally decides their dimensions.

5.21.3.2 High voltage–high power (3 kV+)

This application is generally required for a low duty since the operating rate is slower. The higher voltages may require a multiple contact break and external influences are used to extend the arc into a shute where it is cooled and extinguished. Sulphur Hexafluoride Gas and oils may be used to assist in deionizing the arc. (The special behaviour of an arc in a vacuum is used in the Vacuum Contactor which is capable of interrupting current at a non-current zero (current chopping).) Contact materials for this high-power duty have to withstand higher temperatures and tungsten alloys are common. In oil, copper is acceptable for the load carrying part of the contact since oxidation does not occur and the arc tips only are fitted with tungsten alloy.

The effect of a current passing through closed contacts is represented by $I^2 rt$. The contact resistance, r, is considered low and stable, but t is large so the resulting heat needs to be dispersed. This is performed by conduction into the mass of the supporting contact backing and, in many constructions, into the cable connections. During the contact make and break, r is larger but t is small. The heat is dispersed only into the contact area so the thermal capacity and thermal conduction of the immediate contact assembly is most significant. Good ventilation also assists in reducing contact temperature.

When separating contacts which are passing a current, the final contact is small so the resultant resistance produces considerable heat and high temperatures are reached. $\theta = kV^2/\gamma\rho$ where V is contact voltage, γ is thermal conductivity and ρ is electrical resistivity of the contact materials. V has a major influence and temperatures in excess of 10 000 K are readily attained. Such temperatures cause vapourisation, thermionic emission and together with electromagnetic forces these cause destruction of the contact surfaces evidenced as erosion.

When the contacts are subjected to an arc during circuit interruption, it is advantageous to minimise the duration of the arc. An a.c. circuit carries current which passes through zero twice per cycle and at this instant there is no energy to support the arc which then extinguishes. The rise in voltage across the contacts may restrike the arc if any ionisation remains. The contacts must therefore be separated to a sufficient distance and the gap cooled and ventilated. For a d.c. circuit, no such current zero occurs and the contact separation is required to break the arc. This may be assisted by multiple break contacts, extending the arc by magnetic fields, forcing the arc against cooling plates and generally by a very fast separation of the contacts. The use of blow-out coils to create a magnetic field is commonly used in large units but these are not effective at low currents and a permanent magnet may be added to assist the low current performance.

The life of contacts is a function of the current and the number of operations there being both mechanical wear and electrical surface disruption. With d.c. there is a tendency for material to

Table 5.34 (continued)

Material	Properties	Area of application
Ag/Ni 70/30...60/40	Contact properties similar to 10–20% Ni, but higher contact resistance and lower wear (increasing with increasing Ni-content).	Circuit breakers for d.c. and a.c. Automotive horn switches. Controllers
Ag/CdO 90/10...85/15	Contact resistance somewhat higher than for Ag/Ni 90/10. No welding up to peak currents of 3000 A. Low arc erosion in the range 100 to 3000 A. Very good arc extinction properties. Unfavourable arc movement properties. Limited workability	Low voltage contactors, motor- and motor-protection-switches with ratings from 10 A. Low voltage circuitbreakers with ratings up to about 100 A. Miniature circuit breakers and earth leakage circuit breakers with peak currents up to 3000 A. Lighting switches
Ag/C 97/3...95/5	Low contact resistance. Very high reliability against welding (increasing with increasing C-content). Good friction properties. High wear. Bad arc-mobility properties. Bad workability	Miniature circuit breakers and earth leakage circuit breakers. Low voltage circuit-breakers (unequal pairs with Ag/Ni). Capacitor protective relays. Sliding contacts with self-lubrication
Ag/ZnO 92/8	Similar properties to Ag/CdO, but arc erosion in the current range 100–3000 A somewhat larger and in the range 3000–5000 A smaller	Low voltage circuitbreakers with ratings up to 200 A. Earth leakage circuit breakers
Base: palladium		
Pd	Highly resistant to corrosion, but prone to catalytic reaction with organic materials (brown powder). Highly resistant to arc erosion. Low electrical conductivity	Switching contacts at voltages $U = 20...60$ V
Ag/Pd 70/30...50/50	Generally resistant to corrosion, but worse than Au-alloys. For Ag compared to Ag/Pd 70/30 about 7 times faster and compared with Ag/Pd 50/50 about 100 times faster formation of surface films. Highly wear-resistant	Switching contacts at voltages $U = 20...60$ V, e.g. for telephone relays and selectors. Usual material in telecommunications. Sliding contact in precision-potentiometers
Pd/Cu 85/15 and 60/40	Corrosion behaviour similar to Pd, but at 40% Cu thin oxide layers form high resistance to arc erosion. Low tendency to transfer	Switching contacts at voltages $U = 6...60$ V. High switching currents
Base: platinum Pt/W 95/5 Pt/Ni 91.5/8.5	Resistance to corrosion better than for Pd-alloys, but also formation of 'brown powder'. Low, even transfer. Very highly wear resistant	Switching contacts at high load currents and very long life
Base: gold Au, Au/Pt 90/10	Highest resistance to corrosion, contact resistance constant over long periods. Prone to cold welding. Material transfer	Opening contacts for very small currents and voltages (dry circuits), e.g. in measuring devices, reed switches
Au with hardening additives, electrolytically produced (hard gold)	Similar to Au, but slightly higher contact resistance and less prone to cold welding	Plugs, slide rails, rotary- and sliding-switches. PCB edge connections
Au/Ag 92/8...70/30 Au/Ag/Pt 69/25/6	Good resistance to corrosion. Higher hardness and resistance to wear and less prone to transfer than Au	Switching contacts with voltage <24 V and for small currents, e.g. circuits. Plugs for frequent operation
Au/Co 95/5	Resistance to corrosion similar to, hardness and wear resistance higher than Au/Ag alloys. Slight tendency to transfer. Less malleable	Switching contacts for long life, e.g. for flashers, measuring devices, clocks. Plugs with long life
Au/Ag/Cu 70/25/5, 70/20/10 Au/Ag/Ni 71/26/3	Good resistance to corrosion but slightly less than Au/Ag, decreasing with higher base metal content. Transfer worse than for Au/Ni and Au/Co	Switching contacts, e.g. telegraph relays at voltages <24 V. Plugs for normal life at contact forces of about 0.5 N

Table 5.33 Physical properties of contact materials (*continued*)

Material		Density $kg/m^3 \times 10^{-3}$	Melting point[1] °C	Boiling point[2] °C	Hardness soft HV	Hardness hard HV	Tensile strength $\frac{N}{(mm)^2}$ soft	Tensile strength $\frac{N}{(mm)^2}$ hard	Elongation (%) soft	Elongation (%) hard	Thermal conductivity $\frac{W}{K-m}$ at 20°C	Electric conductivity $\frac{m}{\Omega mm^2}$
Gold–Silver–Copper	20% Ag 10% Cu	15.1	865	2200	125	230	480	820	20	1	68	7.3
	25% Ag 5% Cu	15.2	940	2200	90	185	400	700	25	2	67	8.1
Gold–Silver–Nickel	26% Ag	15.4	990	2200	80	120	350	570	20	1		8.3
Gold–Nickel	5% Ni	18.2	995	2370	105	160	380	640	25	1	52	7.3
Gold–Platinum	10% Pt	19.5	1100	2970	45	160	260	410	20	1	54.5	8
	25% Pt	19.9	1220	2970	80	155						3.6
Gold–Silver	8% Ag	18.1	1035	2200	30	100	150	320	25	1	147	15.8
	20% Ag	16.5	1035	2200	35	90	190	390	25	1	75	10.5
	30% Ag	15.4	1025	2200	40	95	220	380	25	1		9.8
Gold–Silver–Platinum	25% Ag 6% Pt	16.0	1050	2200	30	150	300	500	25	1		6.1
Platinum–Tungsten	5% W	21.4	1830	4400	160	250						2.3
Platinum–Nickel	8.5% Ni	19.2	1670	2370	180	260						3.7

[1] For alloys the solidus point is given, for sintered materials the melting point of the lowest melting component
[2] The boiling point of the lowest boiling component is given
[3] The composition is given in weight %
Data from Johnson Matthey (UK) and Degussa (Germany)

Table 5.34

Material	Properties	Area of application
Base: tungsten or molybdenum		
W/Ag 80/20…20/80	Very low wear, decreasing with increasing W-content. High contact resistance, increasing with increasing W-content. Resistance increases during life. High contact forces necessary. Bad arc mobility properties. Not workable	Low voltage and high voltage circuit-breakers, miniature circuitbreakers (in particular American systems). Railway switches
WC/Ag 80/20…40/60	Slightly better than W/Ag for erosion. suppression of forming of tungstate	As W/Ag, in particular for simple pairs of contacts (no special arcing contacts)
Mo/Ag 80/20…50/50	Similar to W/Ag	Similar to W/Ag
W/Cu 85/15…50/50	Similar properties to W/Ag, but more prone to forming of oxide	High voltage-load breaking switches and circuitbreakers (contacts in air, oil, SF_6); transformer tap-changers (contacts under oil). Electrodes for spark-erosion, electrolytic removal and welding
Base: silver		
Ag	Highest electrical and thermal conductivity. Oxidation-resistant but formation of sulphide. Material transfers. Easily worked	Control switches, microswitches, regulators and selector switches: Voltages $U > 60$ V. Currents $I < 10$ A
Ag/Ni 99.85/0.15 (Fine grain silver) Ag/Cu 97/3…90/10 (Hard silver)	Similar to Ag, but lower erosion. Contact resistance increases with increasing base metal content. Welding tendency low for peak currents below 100 A	Control-, micro-, selector-, relax switches, regulators, miniature circuit-breakers with ratings up to 5 A
Ag/Ni 90/10…80/20	Contact resistance similar to hard silver, but less increase in resistance during life. Lower erosion. No welding for current peaks up to 100 A. Low and flat material transfer when switching d.c. Erosion debris on insulating materials non-conducting. Good arc extinguishing properties	Control switches, regulators, selector switches for d.c. and a.c. up to 100 A. Switches for domestic appliances. Miniature circuitbreakers up to 25 A rating. Motor control switches, contactors up to 25 A rating. Automotive switches. M.c.b.s for d.c. and a.c. (unequal pairs with Ag/C). Controllers

(*continued*)

Table 5.33 Physical properties of contact materials

Material		Density kg/m³ ×10⁻³	Melting point[1] °C	Boiling point[2] °C	Hardness soft HV	Hardness hard HV	Tensile strength $\frac{N}{(mm)^2}$ soft	Tensile strength hard	Elongation (%) soft	Elongation (%) hard	Thermal conductivity $\frac{W}{K-m}$ at 20°C	Electrical conductivity $\frac{m}{\Omega\,mm^2}$
Pure metals												
Silver	Ag	10.5	962	2170	30	80	200	360	30	2	419	62
Gold	Au	19.3	1064	2850	25	60	140	240	30	1	297	44
Platinum	Pt	21.5	1769	3820	40	95	140	400	50		72	9.5
Palladium	Pd	12.0	1554	3000	40	100	200	480	44	2	72	9
Rhodium	Rh	12.4	1963	3700	130	280	420		9		88	22
Iridium	Ir	22.5	2447	4530	220	350					59	19
Copper	Cu	8.9	1085	2590	50	100	200	450	33	2	394	58
Tungsten	W	19.3	3422	5700	250	450	1000	5000			167	18
Molybdenum	Mo	10.2	2623	4630			600	2500			142	19
Iron	Fe	7.9	1540	2760	90	150					75	10
Nickel	Ni	8.9	1455	2900	80	200	450	900	50	2	92	14
Power engineering materials[3]												
Fine grain silver	0.15% Ni	10.5	960	2200	55	100	220	360	25	1	415	58
Silver–Copper	3% Cu	10.4	900	2200	65	120	250	470	25	1	372	52
(Hard silver)	5% Cu	10.4	850	2200	70	125	270	550	20	1	335	51
	10% Cu	10.3	780	2200	75	130	280	550	15	1	335	50
	20% Cu	10.2	780	2200	85	150	320	650	15	1	335	49
Silver–Nickel	10% Ni	10.3	961	2200	50	90	220	400	20	1		54
	15% Ni	10.2	961	2200	55	92	240	420	17	1		
	20% Ni	10.1	961	2200	60	95	280	450	15	1		47
	30% Ni	10.0	961	2200	65	105	330	530	8	1		42
	40% Ni	9.8	961	2200	70	115	370	580	6	1		37
Silver–Cadmium oxide	10% CdO	10.2	961	2200	70	100	300	460				48
(internally oxidised)	15% CdO	10.1	961	2200	80	125						42
Silver–Cadmium oxide	10% CdO	10.2	961	2200	50	80	230	450				48
(sintered)	12% CdO	10.2	961	2200	60	95						47
	15% CdO	10.1	961	2200	65	115	280	420				45.5
Silver–Zinc oxide	8% ZnO	10.2	961	2200	60							49
Silver–Graphite	3% C	9.1	961	2200	40							47
	5% C	8.6	961	2200	40							43.5
Tungsten–Silver	20% Ag	15.4	961	2200	180	240					245	26–28
	35% Ag	14.8	961	2200	100	130					280	34–36
	50% Ag	13.5	961	2200	900	100						40
	65% Ag		961	2200	80	90						
	80% Ag		961	2200	70	80						42
Tungsten Carbide–	20% Ag	13.3	961	2200	400	470						25–35
Silver	60% Ag	11.2	961	2200	100	130						
Tungsten–Copper	15% Cu	16.0	1083	2300	190	260						
	20% Cu	14.7	1083	2300	180	240						24–26
	25% Cu	14.3	1083	2300	170	220						
	30% Cu	13.1	1083	2300	160	200						28–30
	40% Cu	12.7	1083	2300	140	170						
	50% Cu	11.8	1083	2300	130	150						
Molybdenum–Silver	25% Ag		961	2200	190	210						
	35% Ag	10.3	961	2200	160	180						22
	50% Ag	10.3	961	2200	130	150						24
Light current materials[3]												
Fine grain silver	0.15% Ni	10.5	960	2200	55	100	220	360	25	1	415	58
Silver–Palladium	60% Pd	11.4	1330	2200	100	170	380	720			29.3	2.4
	50% Pd	11.2	1290	2200	90	160	360	700	15	1	34	3.3
	40% Pd	11.1	1225	2200	70	140	350	630	20	1	46	4.9
	30% Pd	10.9	1155	2200	65	120	300	600	20	1	59	6.7
Palladium–Copper	15% Cu	11.4	1370	2300	100	260	400	800	42	1		2.5
	40% Cu	10.5	1200	2300	120	280	450	850	39	1	37.7	2.7

(continued overleaf)

speed fuses this principle is employed to a high degree. The extent to which the mass of the heat sink can be increased while reducing the length of the relatively thin element is determined by the requirement that the fuse should withstand the system voltage after the current has been interrupted (i.e. must not restrike). Considerable ingenuity has reconciled these two mutually incompatible requirements. More than one restriction may be used in series along the length of an element to cater for increased voltage, but this aggravates the problem of dissipating heat from the elements. The solution lies in an increase in the transfer of radial heat through the surrounding media. Thus the fuse element must not be looked at in isolation, but as a composite whole with the rest of the assembly.

The fashioning of fuse elements to produce elaborate shapes is economically limited by the means available to achieve the shape required. A variety of means are employed and these often influence the choice of material as regards its physical constants, e.g. the purity of silver as a factor in hardness, etc.

5.20.4 'M' effect

'M' effect, deriving from an exposition by Metcalf, refers to exploiting the thermal reactions of dissimilar metals in the control of time/current characteristics. The thermally most stable fuse element is a simple homogeneous metal. Such an element provides the highest degree of non-deterioration and reliability with adequate breaking capacity at higher overcurrents, but it may be insensitive at lower overcurrents. A lower melting temperature metal with higher resistivity and, therefore, greater thermal mass, can be made to respond more sensitively to lower overcurrents, but may be unreliable at higher currents.

'M' effect is a means by which these extremes can be combined to produce a desired characteristic, but it needs to be used with care in design to avoid compromising non-deterioration properties. An element incorporating 'M' effect is shown in *Figure 15.20(a)*.

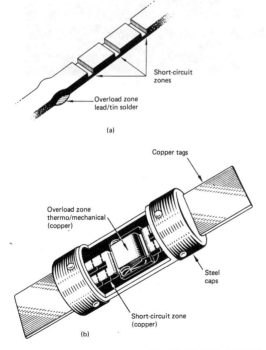

Figure 5.20 (a) Fuse element (English Electric); (b) Dual-element fuse

5.20.5 Composite or dual-element fuses

Satisfactory operation throughout the overcurrent and short-circuit ranges is sometimes obtained in the same package by combining what are, in effect, two fuses connected in series in the same cartridge., *Figure 5.20(b)*. Typical of these is the so-called dual-element design common in the USA. The short-circuit zone is similar to the homogeneous element used in single purpose h.r.c. fuses. The overload zone may take the form of a massive slug of low melting-point alloy, or some electromechanical device, e.g. two copper plates soldered together and stressed by a spring so that when the solder melts the plates spring apart to interrupt the current. The variables in such designs are considerable and many ingenious ideas have been exploited with some success.

5.21 Contacts

Contacts may be classified according to the load they control, and are here discussed under four basic headings

(1) Low voltage, light current.
(2) Low voltage, high current.
(3) Medium voltage (<660 V) and power levels.
(4) High voltage, high power.

An indication of the physical properties of contact materials and the performance and application of contact alloys is given in *Tables 5.33* and *5.34*.

5.21.1 Low voltage, low current contacts

These contacts are required to make and break a very low electrical duty so contact erosion is not a problem. Ideally, the contacts should have a low contact resistance which does not introduce electrical noise by electrothermal or electromechanical means. Importance is therefore placed on surface contamination and deterioration in use and in storage. Silver–nickel alloys are commonly used in low-current control circuits, but for low noise a plated gold surface may be applied which survives a signal level service but rapidly exposes the silver–nickel for higher current applications. Rhodium or palladium may also be used in combination with gold for higher mechanical duty applications.

5.21.2 Low voltage, high current contacts

Separation of contacts carrying a current will create an arc. Since an arc requires a minimum voltage to maintain itself, the arc rapidly extinguishes for low voltage. However, inductance in the load will cause an arc of longer duration and results in contact burning. The high current requires a low contact resistance and large contacts with high contact force are used. Contact materials may be silver–nickel though some automotive applications will use copper-alloys for economy.

Contact resistance is a function of the contact materials, the force applied and, to some extent, the shape of the contacts. The resistance of a pair of contacts may be expressed as $r = k/f^n$ where f is the force applied, and k and n are constants dependent on the contact materials and shape.

Typical values for f and n for copper are given in *Table 5.35*.

At the instant of contact closure a single contact point only may be considered; increasing force distorts the surface and more contact points are established in parallel with the first. With a 'soft' material and with a high force, contact is eventually established over an area which may represent a large proportion of the available contact area, resulting in a low contact resistance. This system would soon cause mechanical erosion of the contact surfaces and so such techniques are only applied to contacts which are only occasionally separated such as plugs and sockets or isolators. Sliding of the contacts may also be permitted to ensure clean surfaces are presented at each contact 'make'.

SPECIAL ALLOYS AND APPLICATIONS

5.20 Fuses

Fuses are the commonest form of protection used in electric circuits and they have been used since 1864.

In theory any conducting metal can be used as a fusible element. In practice a variety of metals are used, ranging from the cheaper base metal alloys to the expensive rarer metals, the choice depending upon the precise function of the fuse and the performance characteristics required.

A fuse by BS definition is the complete assembly. This in its simplest form consists of a piece of base metal wire between two terminals on a suitable support; and at its most complex, of elaborate cartridge assemblies fitted into carrying handles and fuse bases. A typical cartridge fuselink is shown in *Figure 5.18*.

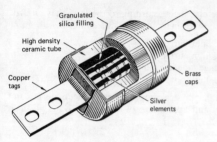

Figure 5.18 Materials in HRC fuse link

Many present-day applications require cartridge fuses, which are made up of fusible elements contained in rigid tubes filled with suitable exothermal and arc-quenching powders, usually sealed by metal end-caps which also carry the conducting tags or end connections. The metal parts, other than the fusible elements, while complying with exacting thermal, mechanical and electrical requirements, are invariably copper, brass, steel or composites. The choice of fusible metal is both wider and more critical; it has to be designed for predictable performance under a wide range of conditions, from normal thermal cycling to the violent changes of state that occur when the element is subjected to transient arcing during fault interruption.

5.20.1 Fuse technology

Fuses perform two basic functions: the *passive* function of carrying current during normal conditions in the circuit and the *active* function of interrupting overcurrents during fault clearance.

The passive function requires that a fuse should be able to carry normal load currents and even transient overloads (and the thermal cycling which accompanies them) for a service life of 20 years or so, without any change of state that might affect its electrical performance. This property of 'non-deterioration' implies that the fusible element is both thermally and chemically compatible with the ambient media.

The active function requires that a fuse should respond thermally to overcurrents by melting and subsequently interrupting the circuit. The melting of a fusible element is followed by arcing, a manifestation of circuit energy which in power circuits can be very high, and the magnitude and duration of which is a function of the circuit. Successful fault interruption implies that the arcing is properly and wholly contained within the fuse cartridge: this capability is the *breaking* or *rupturing capacity*. Inadequate rupturing capacity can result in disastrous damage and explosion in high energy circuits. The operating time of a fuse varies inversely with the level of the overcurrent and discrimination can be obtained in networks by choosing fuses with the necessary time/current characteristics and current ratings.

An important property of the modern high rupturing capacity (h.r.c.) fuse is its ability to limit fault energy, i.e. to melt and quench the arc long before the fault current can rise to the 'prospective' values which the circuit is capable of producing under fault conditions. This property requires considerable sophistication in fusible element design and involves complicated shaping of the elements quite apart from the choice of the basic material. A high degree of energy limitation is achievable with most well-designed h.r.c. fuses, but the degree may vary, as for instance, between industrial power fuses and high-speed fuses for the protection of solid-state devices.

5.20.2 Materials

Materials for fusible elements are silver, copper, tin, lead, zinc or composite alloys of these. Rarer metals are used for special applications. The most common material for high-quality, high-performance fuses is silver. For less precise performance, copper is widely used. The other base materials are introduced to provide varying characteristics in the low-power field.

Silver is a practical choice because it gives a balance of physical properties within practical economic limits, is reasonably free from corrosion in normal atmospheric conditions, and is chemically compatible with silica sand and other media by which it is normally surrounded in fuse construction. Even when prone to oxidation at elevated temperatures its conductance remains relatively unimpaired because the conductivity of the silver oxide is not dissimilar from that of the parent metal. It is ductile and easily fashioned into a variety of shapes by stamping, cutting, or swaging, *Figure 5.19*. It is easily joined within the circuit and

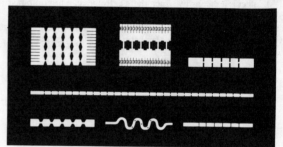

Figure 5.19 Shaped silver fuse elements

hard enough to be mechanically self-supporting. It can be combined with other dissimilar metals (e.g. 'M' effect) to produce eutectic zinc, without impairing its stability during thermal cycling. The physical break-up which follows melting when 'operating' (or 'blowing') is regular and predictable for elements of prescribed purity. The vaporised metal can be made to disperse within the cooling and arc quenching media to combine and condense so that the resulting 'fulgurite' becomes completely insulated.

A silver element may be heated almost to melting and then allowed to cool without its state significantly changing its performance as a fuse. This condition often arises in service when dangerous overcurrents are interrupted elsewhere in the circuit or allowed to subside before the fuse has time to melt. It is important under these circumstances that the fuse should not weaken or otherwise change its designed time/current characteristic.

5.20.3 High-speed fuses

All modern h.r.c. fuses consist of fusible elements, usually with restrictions, of small sectional area connected between relatively massive end connections which act as heat sinks. With high-

Table 5.32 Names and applications of permanent magnet materials

Material	Names	Applications
Al-Ni-Co-Fe Anisotropic	Alcomax II, II, IV Ticonal G Magloy Alnico V	Moving coil instruments, watt-hour meters, loudspeakers (TV), telephone receivers, weighing machines, relays
Columnar	Columax Ticonal GX Alnico V Col	Nuclear magnetic resonance, large motors, loudspeakers (TV)
High coercivity	Hycomax III Ticonal X Alnico XI	Centre core instruments, magnetic bearings, thermostats and relays
Ferrites Isotropic and bonded	Feroba I	Holding catches, refrigerator doors, toys, miniature d.c. motors
Anisotropic	Feroba II, III	High-fidelity loudspeakers, mineral separators, d.c. motors and starter motors, NMR body scanners, holding devices and chucks
Rare earth cobalt Anisotropic		Travelling wave tubes, klystrons, gyroscopes, centrifuges, stepper motors, clutches and frictionless bearings
Neodymium iron boron		Vehicle starter motors and robotic motors

from cracking. Their magnetic properties are not so good as the unbonded materials, at the best the BH_{max} is about 50%.

5.19.6 Other materials

There are a number of other permanent materials with minority applications. Lodex is theoretically interesting because it is made by compacting fine elongated particles of iron and cobalt, the particles having been produced by electrolysis with a mercury cathode. Although its properties are not outstanding it has been found convenient for making certain small sizes with reasonable accuracy.

Other permanent magnet alloys mainly find application because they can be mechanically formed to shape, which is not possible for Alnico, Ferrites or rare earth alloys. It must be emphasised that these other alloys are only available in a limited range of sizes and shapes.

Steels which were the original permanent magnets are now almost obsolete because their properties are inferior to those of the materials mentioned above. They do, however, find applications in hysteresis motors where intermediate coercivities are required.

Vicalloy (V, Co, Fe) and Cunife (CuNFe) are alloys which can be rolled and drawn into wire and the anisotropy is produced by this rolling or drawing. More recent developments are in a CrFeCo alloy which can be rolled and can have properties similar to those of anisotropic Alnico, the anisotropy being produced by the rolling process. Another development is in MnAlC alloys which have the attraction of being free of cobalt, the properties are obtained by extrusion.

Table 5.31 gives the range of properties for each class of material. This range includes deliberate variations in properties obtained by small changes in composition and heat treatment and usually listed by manufacturers with different numbers or letters such as Alcomax II, III and IV or Ticonal G and X.

Table 5.32 lists a few of the most common applications of each broad class of material, also the trade names most likely to be met.

5.19.7 Further reading

WOHLFARTH, E. P. (ed.), *Ferromagnetic Materials*, 3 vols. published, North Holland (1980 to 1983)
BRAILSFORD, F. *Magnetic Materials*, 3rd edn., Methuen (1960)
BERKOWITZ, A. and KNELLER, E. (eds.), *Magnetism and Metallurgy*, 2 vols., Academic Press (1969)
BOLL, R. (ed.), *Soft Magnetic Materials*, Vacuumschmelze Handbook, Heyden (1977)
SMIT, J. and WIJN, H. P. J. *Ferrites*, pp. 6–9, Philips Technical Library, Eindhoven (1959)
STREET, B. G. *Ferrite component manufacture*, Powder Metall. (GB), (1982), **25**, No. 3, pp. 173–176
BROESE VAN GROENOU, A., BONGERS, P. F. and STUIJTS, A. L., *Magnetism, microstructure and crystal chemistry of spinel ferrites*, Mater. Sci. Eng. (1968/9), **3**, pp. 325–331
SNELLING, E. C. *Soft Ferrites, Properties and Applications*, Butterworths, London (1969)
SNELLING, E. C. and GILES, A. D. *Ferrites for inductors and transformers*, Research Studies Press, London (1983)
WATONABE, H., IIDA, S. and SUGIMOTO, M. (eds.), *Ferrites (Proceedings of ICF3)*, Centre for Academic Publications, Japan (1981)
MCCAIG, M., *Permanent Magnets in Theory and Practice*, Pentech Press, London and Plymouth (1977)
This will be found to be useful for reference as it refers to all the more important books and papers and gives details of manufacturers throughout the world.
GOULD, J. E., *Permanent Magnets*, Proc. IEE, v125, No. 11R (Nov. 1978) p. 1137–1151
WRIGHT, W. and MCCAIG, M., *Permanent Magnets*, (Design Council, B.S.I. and IEE), Oxford (1977)
MOSKOWITZ, L. R., *Permanent Magnet Design and Application Handbook*, Cahners Books, Boston, Mass. (1976)
For more recent information it will be found useful to refer to:
IEEE Trans on Magnetics which contains the proceedings of Intermag Conferences.
Proceedings of International Workshops on Rare Earth Cobalt Permanent Magnets, from the University of Dayton, Dayton, Ohio, U.S.A.
IEC Standards on Magnetic Materials, No. 404 parts 1 to 5 and 7 have been published on the classification and methods of measurement of magnetic properties. It is expected that these will all be published as part of British Standard 6404.

low price per unit of available magnetic energy, the high coercivities, the high resistivities and the low density. The isotropic grades are the least expensive to manufacture and may be magnetised into complex pole configurations; they are used for a wide range of relatively small-scale applications. The largest application in terms of market volume is for magnets for loudspeakers; here the high remanence anisotropic grades are used. The other high-volume application is for field magnets in small d.c. motors, particularly those used in the automotive industry, e.g. for windscreen wipers, blowers, etc. In this application the high coercivity anisotropic ferrite is formed into an arc-shaped segment, magnetised radially; a pair of segments embrace the armature and provide the stator field. Other applications of anisotropic hard ferrites include magnetic chucks and magnetic filters and separators.

5.19.3 Rare earth cobalt

This is a relatively recent development and the main commercial material is $SmCo_5$. The alloy powder is prepared either by reducing the rare oxide powder together with cobalt powder or by preparing the alloy and powdering it down. The powder is compacted and pressed almost invariably in a magnetic field to produce anisotropic magnets. The properties are due to a very strong magnetocrystalline anisotropy related to its hexagonal crystal structure. The range of sizes which can be produced is from about 1 mm cubes up to blocks of $50 \times 50 \times 25$ mm, the short dimension being the preferred direction.

A current development in this series is the Sm_2Co_{17} alloy which has a higher remanence and $(BH)_{max}$ than the $SmCo_5$ but it seems to be much more difficult to produce with consistent properties.

5.19.4 Neodymium iron boron

This is a recent development the basic alloy being $Nd_{15}B_8Fe_{77}$. It has risen as a result of interest in amorphous alloys obtained by very rapid cooling from the melt. When alloys of the above composition are recrystallised, an alloy with very good permanent magnet properties is obtained. An alternative route is to prepare the alloy by melting neodymium with ferroboron in the correct proportions. The crystal structure of this alloy is tetragonal the long axis being the preferred direction. The alloy is powdered down and pressed in a magnetic field to produce an anisotropic compact. This compact is then sintered. Throughout the processing a protective atmosphere must be used to protect the neodymium from oxidation.

The basic alloy suffers from a rapid change of magnetic properties with temperature in particular coercivity changes. It has been reported that this can be reduced by an addition of a small amount of dysprosium.

5.19.5 Bonded materials

Each of the above materials can in the powdered form be mixed with a bond. These bonds can be rubbers, polymers or plastics, they may be flexible or rigid. The flexible rubber bonded ferrites have found wide application in holding and display devices and the best quality anisotropic material is used in small motors. Alnico and rare earth cobalt are also made in bonded forms, they have the advantages of a uniform level of properties and freedom

Table 5.31 Characteristics of permanent magnet materials
B_r remanence, T
$(BH)_{max}$ energy product, J/m^3
H_{cB} coercivity, kA/m
μ_r relative recoil permeability

Material	B_r	$(BH)_{max}$	H_{cB}	μ_r
Alnico				
normal anisotropic	1.1–1.3	36–43	46–60	2.6–4.4
high coercivity	0.8–0.9	32–46	95–150	2.0–2.8
columnar	1.35	60	60	1.8
Ferrites (Ceramics)				
Barium isotropic (a)	0.22	8	130–155	1.2
anisotropic (a)	0.39	28.5	150	1.05
Strontium anisotropic (a)	0.36–0.42	24–33	240–290	1.05
Bonded ferrite				
isotropic (a)	0.14	3.2	90	1.1
anisotropic (a)	0.25	11.2	176	1.05
Rare earth				
$SmCo_5$ sintered (b)	0.9	160	640–700	1.05
$SmCo_5$ bonded (b)	0.5–0.6	56–64	400–460	1.1
Sm_2Co_{17} (provisional) (c)	1.1	190–220	720	1.05
NdFeB (provisional) (d)	1.3	290	800	1.1
Others				
Lodex	0.4–0.8	10–27	56–100	1.5–3.8
Cunife	0.54	10	40	—
Vicalloy	0.9–1.1	8–10.5	18–24	—
MnCAl	0.5–0.6	40–56	176–216	—
CrFeCo	1.3	40–46	46	3–4

Intrinsic coercivity H_{cJ}, kA/m: (a) 160–340 kA/m; (b) 800–1500 kA/m; (c) 1700 kA/m
The Curie temperature of Alnico is 800–850 °C, of ferrite 450 °C, of SmCo over 700 °C and for NdFeB it is 300 °C. The Alnicos have a resistivity of about 50×10^{-8} Ω-m, for the ferrites it is about 10^4 Ω-m and for the rare earths $90–140 \times 10^{-8}$ Ω-m

working point will fall to say, the point P. If now a magnetising field is applied to the magnet, the B value will recoil up the line PQ depending on the amount of the field. The slope of this line is $\mu = \mu_0 \mu_r$ where μ_r is the relative recoil permeability and μ_0 is the magnetic constant ($=4\pi \times 10^{-7}$). This is an important consideration in dynamic applications such as motors, generators and lifting devices where the working point of the magnet changes as the magnetic circuit configuration changes. It is also important when considering premagnetising before assembly into a magnetic circuit. If μ_r is nearly equal to unity as it is for ferrites and for rare earth cobalt alloys then the magnets can be premagnetised before assembly into a magnetic circuit without much loss of available flux. However, if μ_r is considerably greater than unity, care must be taken to magnetise the magnet in the assembled magnetic circuit.

The dotted curve in *Figure 5.17* is the J–H curve where $B = \mu_0 H + J$. For materials with B_r much greater than $\mu_0 H_{cB}$, e.g. Alnico, the two curves are almost identical, but for ferrites and rare earth cobalt alloys where B_r and $\mu_0 H_{cB}$ are close in value the curves become quite different. The J–H curve has only a very gradual slope from B_r, this indicates that the magnetisation in the material is nearly uniform and a new parameter H_{cJ} is introduced. This intrinsic coercivity H_{cJ} is a measure of the difficulty or ease of demagnetisation. As the value of H_{cJ} is increased the more the material will resist demagnetisation due to stray fields, etc. However, as a field of at least three times H_{cJ} must be applied to the magnet for magnetisation, difficulties may be encountered in attaining the full magnetisation if H_{cJ} is too high.

Magnets in the three main groups Alnico, ferrites and rare earth cobalt alloys are generally made anisotropic with the properties in one direction considerably better than in the other directions. This best direction is called the preferred direction of magnetisation. The curves and parameters supplied for such a material refer to the properties in this preferred direction.

To choose between these three materials for a particular application, Alnico must be used where magnetic stability is a requirement and this would apply in any type of instrument application. Ferrite will be used where cost is the main consideration and rare earth magnets will be used if the highest possible strength is required or if miniaturisation is needed.

5.19.1 Alnico alloys

A wide range of alloys with magnetically useful properties is based on the Al-Ni-Co-Fe system. They are characterised by high remanence, available energy and moderately high coercivity. They are very stable against vibration and have the widest useful temperature range (up to over 500 °C) of any permanent magnet material. But they are mechanically hard, impossible to forge and difficult to machine except by grinding and special methods such as spark erosion.

The preferred form of magnet is a relatively simple shape made by casting or sintering, precision ground on essential surfaces only. Soft iron pole pieces may be clamped on or held by a suitable adhesive such as Araldite. Soldering, diecasting and even brazing (with precautions against overheating) may be used.

The older isotropic forms of these alloys should be used only for shapes that cannot be made in the anisotropic material. Although the isotropic alloys are slightly cheaper weight for weight, the anisotropic alloys have a much greater energy, so that a smaller volume is required. It is always cheaper to use the anisotropic alloys when this is possible.

The commonest alloys contain 23–25% Co, 12–14% Ni, about 8% Al, a few per cent Cu and sometimes small additions of Nb and Si, with the balance Fe. They are cooled at a controlled rate in a magnetic field applied in the direction in which the magnets are to be magnetised. The properties are much improved in this direction at the expense of those in other directions. There is finally a fairly prolonged and sometimes rather complicated treatment at temperatures in the range 650–550 °C. With the aid of the electron microscope it has now been shown conclusively that after this treatment the magnets have a two-phase structure consisting of fine elongated magnetic particles separated by a non-magnetic phase, this structure being oriented by the magnetic field applied during cooling. It was already known that an assembly of fine elongated magnetic particles should exhibit permanent magnet properties because of their shape anisotropy, i.e. their tendency to remain magnetised in their long direction.

The field cooling is much more effective in producing anisotropic properties if the magnets are cast with a columnar structure. Basically a columnar structure is produced by casting the alloy into a hot mould with a cold chill. The mould may be made of special materials which can be preheated to at least 1250 °C in a furnace, or alternatively, contain chemicals which heat it by an exothermic reaction. Either method is inconvenient and expensive; there are also limitations on the lengths and shapes of magnets that can be made columnar. Columax is therefore regarded as a rather expensive, quality material.

The coercivity of Alnico can be improved by a factor of 2 to 3 times by increasing the cobalt content to 30–40% and adding 5–8% of Ti with possibly some Nb. For the best properties it is necessary to hold these alloys for several minutes in a magnetic field at a constant temperature, accurately controlled to ± 10 °C, instead of the usual field cooling process. This meticulous heat treatment as well as the high cobalt content makes these alloys expensive and they are only used where the higher coercivity is required. Columnar versions of these alloys can also be made but these are only used for very specialised applications.

The Alnico alloys can be produced as castings of up to 100 kg and down to a few grams in weight but it is found more economical to produce the smaller sizes of 50 grams and less by the sintering process. In all cases it is advisable to contact the manufacturers before any design is considered.

5.19.2 Ferrite

The permanent magnet ferrites (also called ceramics) are mixed oxides of iron (ferric) oxide with a divalent heavy metal oxide usually either barium or strontium. These ferrites have a hexagonal crystal structure, the very high anisotropy of which gives rise to high values of coercivity, e.g. 150–250 kA/m (compared with about 110 kA/m for the best Alnico alloys). The general formula is $M0.5.9 Fe_2O_3$ where M is either Ba or Sr, the crystal structure is called magnetoplumbite as it was originally found in the equivalent lead oxide compound. These ferrites are made by mixing together barium or strontium carbonate with iron oxide in the correct proportions. The mixture is fired in a mildly oxidising atmosphere and the resulting mixture is milled to a particle size of about 1 micron. This powder is then pressed in a die to the required shape (with a shrinkage allowance), anisotropic magnets are produced by applying a magnetic field in the direction of pressing. After pressing the compact is fired. This material is a ceramic and can only be cut by high-speed slitting wheels. Ferrite magnets are produced in large quantities in a variety of sizes for different applications. Flat rings are made for loudspeakers ranging up to 300 mm in diameter with a thickness of up to 25 mm. Segments are made for motors with ruling diameters from 40–160 mm and rectangular blocks are made for separators with dimensions of up to $150 \times 100 \times 25$ mm. These blocks can be built up into assemblies or cut down into smaller pieces for a variety of applications. In each of these cases the preferred direction of magnetisation is through the shortest dimension. Generally the magnetic length does not exceed 25 mm and if a longer length is required, this is built up with magnets in series.

The great success of permanent magnet ferrites is due to the

Table 5.30 Iron–cobalt alloys

Alloy	Saturation flux density (T)	Initial permeability	Maximum permeability	Coercive force (A/m)	Remanence (T)	Resistivity ($\mu\Omega$-m)	Curie temperature (°C)
24% Co (24 Permendur*) (Hiperco 27[+])	2.35	250	2000	130	1.5	0.2	925
49/49/2 Co/Fe/V (49 Permendur*) (Hiperco 50[+])	2.35	800	7000	950	1.6	0.4	975

* Telcon Metals Ltd., UK
[+] Carpenter Steel Co; USA
Forms of supply—Rod, Bar, Forgings, Cold Rolled Strip 1.6–0.025 mm
Heat treatment to develop magnetic properties — 2 h at 760°C in pure dry hydrogen

5.17.5 Forms of supply

The alloys are produced in the form of bar, rod, wire and cold rolled strip down to 0.003 mm thick. Laminations or toroidal cores are produced from the strip. High temperature heat treatment in an atmosphere of pure dry hydrogen at 1000°C to 1200°C, is necessary to develop the full magnetic performance and all forms of mechanical strain should be avoided following this operation. The thin strip toroidal cores are normally protected by means of a case, before winding, to avoid the introduction of such strains.

5.18 Iron–cobalt alloys

The addition of cobalt to iron results in an increase in saturation, up to a maximum flux density of 2.45 Tesla at 35% cobalt. The high cost of cobalt, relative to that of iron, restricts the widespread use of these alloys, although they are widely employed in the aircraft and associated industries. Their high Curie temperature and magnetostriction are also of special importance. Typical properties of a representative range of alloys are given in *Table 5.30*.

5.18.1 24/27% Cobalt iron

The ductility drops very considerably when the addition of cobalt exceeds 27% and compositions in this area are chosen for applications such as magnet pole tips where a combination of good ductility, ease of machining and high magnetic saturation flux density are required. The permeability, coercive force and loss characteristics are, however, inferior to the 50% cobalt alloy.

5.18.2 50% Cobalt iron

The optimum combination of magnetic properties is obtained in the 50/50 cobalt–iron composition. The ductility of the binary alloy is so low that it is not possible to fabricate the alloy to any extent. The addition of 2% vanadium greatly increases the ductility and fortunately, from the eddy current loss viewpoint, also substantially increases the resistivity. The 49/49/2 Fe/Co/V alloy has been extensively used for stators in lightweight electrical generators for many years. Recent developments, which have enhanced its mechanical properties, have extended its use also to the rotor laminations. Other applications include lightweight/small-volume transformers, special relays, diaphragms, loudspeakers and magnet pole tips.

5.19 Permanent magnet materials

The properties of a permanent magnet material are given by the demagnetisation curve, *Figure 5.17*, the second quadrant of the B–H hysteresis loop. This extends from the remanence B_r to the

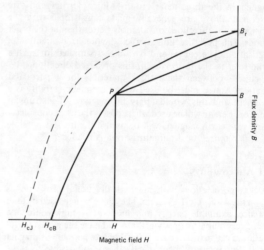

Figure 5.17 B–H relation

coercivity H_{cB}. It can be shown that when a piece of permanent magnet material is put into a magnetic circuit, the magnetic field generated in a gap in the circuit is proportional to BHV, where B and H are the corresponding points at a point on the demagnetisation curve and V is the volume of permanent magnet. So that to obtain a given field with the minimum volume of magnet material we require the product BH to be a maximum. The magnet is then designed so that its B, H value is as close as possible to the $(BH)_{max}$ value. It is also useful to use the $(BH)_{max}$ value to compare the characteristics of materials. Generally the material with the highest $(BH)_{max}$ will be chosen but this has to be weighed against such considerations as cost, shape, manufacturing problems and stability.

When the magnet is fully magnetised in a completed magnetic circuit, the working point is the remanence B_r. When a gap is made in the circuit the magnet will be partly demagnetised and its

e.m. wave; devices based on such materials will exhibit discrimination as to the directional properties of such a wave.

This phenomenon gives rise to a family of waveguide and stripline devices used in microwave engineering, e.g. unidirectional attenuators (isolators) and energy routing devices such as circulators.

5.17 Nickel–iron alloys

The nickel–iron alloys, ranging in composition from 30–80% nickel are the most versatile of soft magnetic materials. The design engineer can select from a wide range of available properties such as saturation flux density, permeability, coercive force and loss. The alloys can be subdivided into four main groups. Typical properties of a representative range of alloys are given in *Table 5.29*.

5.17.1 30% Nickel

The 30% nickel–iron alloy has a Curie temperature of the order of 60 °C and at high fields it possesses a linear permeability/temperature characteristic over the range −30 °C to +40 °C. This property is frequently utilised in instruments employing permanent magnets or electromagnets, such as domestic watt-hour meters and speedometers. The 30% nickel–iron alloy is used as a magnetic shunt to compensate for changes with temperature of the magnet or the conductivity of associated copper or aluminium parts. The 32% nickel alloy, which has a Curie temperature of 150 °C, is used for devices operating over a wider temperature range.

5.17.2 36% Nickel

The alloy has a lower saturation flux density, 1.2 T, and permeability than the 50% alloy but its lower cost and high resistivity make it an attractive alternative. The main applications are relays, high frequency transformers and inductors. The alloy is also employed in devices, such as inductive displacement transducers, where its very low coefficient of expansion, $1 \times 10^{-6}/°C$, is a great advantage.

5.17.3 50% Nickel

This alloy has the maximum saturation in the nickel–iron range, 1.6 T, and is used where a higher permeability and/or a better corrosion resistance than that of silicon–iron is required. Another advantage of the alloy is its high incremental permeability over a wide range of polarising d.c. fields. Applications include chokes, relays, small motors and synchros.

Special processing techniques enable a wide range of properties to be developed in this alloy. A cold reduction, in excess of 99%, produces a cube texture in the annealed strip and a square hysteresis loop. Applications utilising this property are magnetic amplifiers, inverters and pulse transformers. Annealing in a magnetic field, below the Curie temperature, can also alter the magnetic characteristics. If the field is applied in the conventional manner, both the initial and maximum permeabilities are substantially increased, as in the case of Satmumetal. When the field is applied in a transverse direction a very flat hysteresis loop with a low remanence is produced, as demonstrated by Permax F. These properties are ideal for thyristor firing circuits where a unipolar pulse is required and the transformer only operates between remanence and saturation.

5.17.4 80% Nickel

The high permeability and low coercivity of this group of alloys is due to the fact that both the magnetostriction and crystalline anisotropy are essentially simultaneously zero at this composition. Until the very recent discovery of the cobalt–boron metal glass alloys with a similar combination of properties, the 80% nickel–iron alloys were unique for their ultra-high permeability performance.

The commercial alloys have additions of small percentages of molybdenum, copper or chromium to give improved magnetic performance over the binary alloy. Applications include sensitive relays, pulse and wideband transformers, current transformers, current balance transformers for sensitive earth leakage circuit breakers and magnetic recording heads. Magnetic shielding is also a major area of application either in the form of fabricated cans or screens for transformers and cathode ray tubes. Annealed tape is also used in the manufacture of special cables.

Table 5.29 Nickel–iron based alloys

Alloy	30% Ni R 2799*	32% Ni R 2800*	36% Ni[1] Radiometal 36*	45% Ni[2] Radiometal 4550*	50% Ni[3] Super Radiometal*	50% Ni[4] HCR*	54% Ni Satmumetal*	65% Ni Permax+ F Z	77% Ni[5] Mumetal*	77% Ni Super Mumetal 180
Initial permeability $\times 10^{-3}$	—	—	3	6	11	0.5	50	— —	60	200
Maximum permeability $\times 10^{-3}$	—	—	20	40	100	100	120	— —	240	400
Saturation flux density (T)	0.2	0.7	1.2	1.6	1.6	1.54	1.5	1.3 1.3	0.8	0.8
Coercive force (A/m)	—	—	10	10	3	10	2.5	10 2	1.0	0.5
Remanence (T)	—	—	0.5	1.0	1.1	1.5	0.7	0.2 1.25	0.45	0.5
Resistivity (Ωμ-m)	0.85	0.83	0.8	0.45	0.4	0.4	0.45	0.6	0.6	0.6
Density kg/m³	8000	8000	8100	8300	8300	8300	8300	8500	8800	8800
Curie temperature (°C)	60	150	280	530	530	530	550	520	350	350
Expansion coefficient/°C $\times 10^6$	10	5	1	8	10	10	11	12	13	13

* Telecon Metals Limited
+ Vacuumschmelze GmbH
1, 2, 3, 4, 5—BS 2875, 1976. Classes C, B1, B2, D and A respectively.

In the case of MnZn ferrites the finite conductivity gives rise to some eddy current core loss at higher frequencies. This loss depends on the core size and shape but it is usually a small proportion of the total loss. The low-amplitude hysteresis loss is expressed in terms of a coefficient, η_B, such that the hysteresis loss factor $(\tan \delta_h)/\mu_i = \eta_B \hat{B}$.

At the high flux densities that characterise power applications, the hysteresis core loss is expressed as power per unit volume, i.e. the power loss (volume) density, P_h. It is found that $P_h = k\hat{B}^n f^m$ where n is the Steinmetz exponent. For MnZn ferrites intended for power applications, n usually lies between 2 and 3 (typically 2.5) and m is approximately 1.3 (at $\theta \approx 100\,°C$). Typical curves for P_h are given in *Figure 5.16*. The ferrite composition is usually chosen so that the minimum power loss density occurs at the typical operating temperature of the core, e.g. 85 °C.

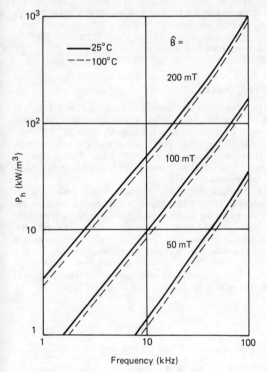

Figure 5.16 Hysteresis power loss density of a typical MnZn ferrite intended for power applications (courtesy Mullard Ltd)

Given the permitted core loss and operating frequency, the maximum permitted flux density in a core of given volume is readily obtained from the power loss density curves and, from the induction formula, the required number of turns may be determined. In practice, it may be necessary to allow for a small additional core loss due to eddy currents.

Also, in power applications, it is important that the available flux density should be as high as possible at the operating temperature. In the design of a power transformer, a flux density excursion into the saturation region is usually to be avoided since this causes a steep rise in the magnetisation current.

5.16.1.2 Applications

Magnetically soft ferrites are used in large quantities as cores for inductors and transformers in telecommunications and in industrial and consumer electronics. Referring to *Table 5.28*, the first three columns indicate typical properties of MnZn ferrites. The first is a low-loss ferrite intended for high-quality inductors used in filters. It is usually made in the form of circular or square pot cores, the cores being gapped to provide specified effective permeabilities. There is usually provision for adjusting the inductance over a range of about $\pm 10\%$. Such components combine potentially high Q-factors (300–1000) and good stability with temperature and time. The second ferrite, (ii), has high initial permeability and is intended for low-power wideband or pulse transformers. It is usually made in the form of toroids and E cores as well as various types of pot core shape. Low hysteresis is also an important characteristic. The third column refers to ferrite specifically developed for high-power applications, such as cores for power transformers operating in the 10–100 kHz range. Such transformers have become very important in electronic switched-mode power supplies and they are manufactured in large quantities. The form of these cores is usually the U core or the E core. The latter type of core often has a cylindrical centre limb and is designed to have proportions that are optimum for the application. Typical power handling capabilities range from 10–750 W. Similar ferrites are used for deflection yokes in TV receivers, although for this application the magnetic properties are not critical. The production of deflection yokes exceeds by weight that of any other soft ferrite core type. Finally, column (iv) refers to a typical NiZn ferrite intended for cores of high-quality inductors operating in the frequency range 2–12 MHz. It is usually made in the form of pot cores, toroids, rods and beads.

In addition to the main applications outlined above, soft ferrites are employed in a variety of other applications. They are used, for example, as rods for radio antennas, as recording heads in audio and video tape recorders, as cores in radio interference suppressors, for magnetic proximity detectors and, occasionally, for the massive coupling cores in large particle accelerators.

5.16.2 Other ferrite types

5.16.2.1 Rectangular loop ferrites

A variety of mixed ferrites, notably manganese–magnesium, manganese–copper and lithium–nickel ferrites, exhibit substantially rectangular hysteresis loops. Small toroids of such materials may be used as passive devices having two stable states (the positive and negative remanences) and may thus be used to store and process binary information. Many thousands of millions of minute toroids, so-called memory cores, have been used to provide the main frame memories in large computers. This application has now been largely taken over by integrated circuit memories but for certain applications where environmental conditions make such memories unreliable, the core memory with its planes of four or sixteen thousand cores, each core threaded with address and sense conductors, are still used in significant quantities.

5.16.2.2 Microwave ferrites

The application of ferrites at microwave frequencies depends on the fact that the spinning electron (the magnetic elements of the lattice) will, if disturbed, precess in a direction dependent only on the direction of the static field aligning the spins and at a frequency dependent on the static field strength. An e.m. wave, having positive circular polarisation relative to the static field direction and a frequency equal to the precession frequency will couple to the spin to produce a resonance. There is no corresponding coupling or resonance for a negatively polarised wave. Thus, a material such as a ferrite saturated by a static magnetic field and thus constituting an array of aligned spins will have properties that depend on the sense of polarisation of an incident

facturing process. For example, increasing the proportion of zinc lowers the Curie Point and influences the saturation flux density, and the presence in MnZn ferrites of a small proportion of the iron in divalent form critically affects the temperature coefficient of permeability and the magnetic losses. By varying these factors a range of ferrite grades can be made, each specifically matched to a particular application.

Generally the MnZn ferrites have the higher permeabilities, lower losses at frequencies up to about 1 MHz and resistivities in the range 0.05–5 Ωm. The NiZn ferrites have lower permeabilities (dependent on the Ni/Zn ratio) and higher losses below about 1 MHz but they maintain these properties up to much higher frequencies; their resistivities are about three orders of magnitude higher than those for MnZn ferrites. Thus, in general, MnZn ferrites find applications at operating frequencies below about 2 MHz; above this frequency the NiZn ferrites can give better performance where low loss resonance applications are concerned.

The initial permeability, μ_i, is the permeability at very low field strengths. High values of μ_i are generally desirable but the higher the permeability the lower is the frequency at which the magnetic properties deteriorate due to the approach of ferromagnetic (spin) resonance. This causes a large decrease in permeability with frequency. Therefore ferrites intended for higher frequency operation invariably have lower permeabilities.

The temperature coefficient of permeability is of great importance for inductor applications. As a material property it is usually expressed as a temperature factor $\Delta\mu_i/(\mu_i^2 \Delta\theta)$. If a core is gapped so that it has an effective permeability, μ_e, then the temperature coefficient of effective permeability (and therefore of the inductance of a winding on the gapped core) is μ_e times the temperature factor. The temperature coefficient of inductance may thus be determined by the choice of the air gap length.

At the low flux densities appropriate to signal applications, the core loss in ferrites is expressed in terms of a loss factor which is the tangent of the loss angle divided by the initial permeability, i.e. $(\tan \delta)/\mu_i$. The loss tangent of a gapped core equals the loss factor multiplied by the effective permeability. In many applications the flux density is sufficiently low that the hysteresis loss is negligible. The main loss is then represented by the residual loss factor, $(\tan \delta_r)/\mu_i$; this remains low over the lower frequency region and then rises rapidly towards a frequency that is inversely proportional to the initial permeability. This frequency marks the onset of ferromagnetic resonance. *Figure 5.15* shows the residual loss factor as a function of frequency for three typical ferrites. Other properties of the same ferrites are given in *Table 5.28*.

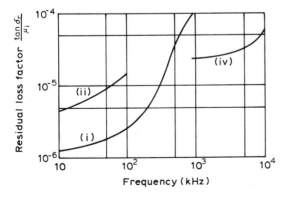

Figure 5.15 Residual loss factor as a function of frequency (see Table 5.28 for identification of materials)

Table 5.28 Typical properties of some magnetically soft ferrites*

Parameter	Conditions			Ferrite grade				Units
	f (kHz)	$\hat{B}$ (mT)	Miscellaneous	(i)	(ii)	(iii)	(iv)	
Approximate composition:								
$MnO/ZnO/Fe_2O_3$				27/20/53	25/22/53	34/14/52	—	mol %
$NiO/ZnO/Fe_2O_3$				—	—	—	32/18/50	mol %
Initial permeability, μ_i	<10	<0.1		1200–2500	3800–10000	1000–3000	70–150	
Temperature factor, $\Delta\mu_i/\mu_i^2 \Delta\theta$	<10	<0.1	5 to 55 °C	0.5–1.5	5–10	—	0–8	$°C^{-1} \times 10^{-6}$
	30			0.8–2.0	10–50	—	—	
	100			1.5–3.0	—	—	—	
Residual loss factor, $(\tan \delta_r)/\mu_i$	10^3	<0.1		—	—	—	20–40	10^{-6}
	10×10^3			—	—	—	60–100	
Hysteresis coefficient, η_B	10	1 to 3		0.5–1.0	0.1–1.0	—	2–10	$mT^{-1} \times 10^{-6}$
Saturation flux density, B_{sat}			$H = 1$ kA/m	0.35–0.45	0.35–0.45	0.4–0.5	0.3–0.4	T
Curie Point, θ_c				130–250	90–220	180–280	250–400	°C
Resistivity	d.c.	<0.1		1–5	0.05–0.5	1–5	$>10^3$	Ωm

* at 25°C unless otherwise stated.

the supply of relay steels and soft iron, but this can lead to wide variations of magnetic properties within a batch which meets the chemical specification.

The British Steel Corporation is now offering steel to a magnetic specification within three coercive force grades:

(1) Less than 0.7 Oe (56 A/m)
(2) 0.7–1.0 Oe (56–80 A/m)
(3) Greater than 1.0 Oe (80 A/m)

(There is still much use made of the CGS unit, the oersted for soft iron grading)

Material supplied to a magnetic guarantee is much more satisfactory to the relay designer and quality control engineer.

It is expected that an international product standard will emerge in accord with the outline specification given in IEC Standard 404-1 (1979).

5.15.7 Heat treatment

It is important that iron should be in a fully annealed stress-free state in its final form in a magnetic circuit. Most material is punched, sheared, upset or machined in one way or another during fabrication, all of which induces high stress. An appropriate heat treatment of a piece-part will allow the material to give the best performance in service.

Optimum heat treatments are best worked out in conjunction with the steel supplier—a service which British Steel offers from its laboratories at Orb Works in Newport. Broadly a 950 °C anneal in a neutral atmosphere followed by a slow cool over several hours is satisfactory.

Where material is fairly thin (up to 1 or 2 mm) the final heat treatment can, with advantage be decarburising so that lowest coercive force is obtained as residual carbon is removed. Decarburising anneals employ furnace atmospheres of wet hydrogen and again require specialist advice.

5.15.8 Ageing

A final anneal takes residual carbon into solution and some of this may precipitate over a period of years leading to increases in coercive force as domain wall motion becomes impeded by carbon precipitates. Ageing may be minimised by keeping carbon contents low and by use of special stabilising additions to the steel. When necessary modern irons can be made ageing-free.

5.15.9 Test methods

British Standard 5884 deals with test methods appropriate to relay steels. It covers the more lengthy procedures in which a welded ring is used as sample (or stamped rings from sheet), or the use of a d.c. permeameter with flux closure yokes. BS 5884 also gives details of the Vibrating Coil Magnetometer with which it is possible to obtain a coercive force reading in a few seconds. This is a very simple procedure compared with the more lengthy permeametric methods which require ballistic galvanometers, etc.

All varieties of such test apparatus are available from the British Steel Corporation who have pioneered the production of the Vibrating Coil Magnetometer for sale to industry in conjunction with the National Physical Laboratory.

IEC Standard 404-4 deals with test methods for solid steels and IEC Standard 404-7 specially covers coercive force measurements in an open magnetic circuit.

Study of the UK and IEC Test Standards is of the greatest assistance in understanding how to grade soft irons for quality.

5.15.10 Other applications

Where relays must operate at the highest speeds it is necessary to raise the electrical resistivity of the magnetic circuit. This is necessary so that flux can penetrate the core iron rapidly. A low electrical resistivity gives rise to very vigorous induced eddy currents which oppose flux penetration to the centre of the material. The use of laminations is not often suitable so resistivity can best be raised by the addition of silicon, up to some 3% as required. This produces a four-fold increase in resistivity.

Thin iron is often used for screening. This may be applied to cables to reject outside interference or to load them inductively. Whole areas may be screened with iron enclosures where steady field or field-free conditions are essential. For this purpose screens consisting of layers of iron and air are the most effective.

5.16 Ferrites

In the present context, ferrites may be defined as compounds of metal oxides formulated to have specific magnetic properties. They generally have the form of polycrystalline ceramic materials and are used extensively as magnetic core materials in a wide variety of applications in communication and electronic engineering. In the crystal lattice of these materials the metal ions are separated by oxygen ions and this results in high electrical resistivities which suppress the effects of eddy currents. The oxygen is also associated with antiparallel alignment of the magnetic moments of the metal ions on the different sub-lattices so that the net available magnetisation is the difference between the magnetisations of the sub-lattices. This, together with the dilution due to the non-magnetic oxygen ions, inherently limits the saturation flux density to about 0.4–0.6 T (compared with 2 T for some magnetic alloys).

The usual manufacturing process consists of the following stages: mixing of raw materials (oxides, carbonates, etc.) in the required proportions, calcining at about 1000 °C, crushing, milling, powder granulation, forming to the required shape by pressing the powder in a die or by extrusion, and, finally, sintering the piece parts at about 1250 °C for up to 12 h in a controlled atmosphere. During sintering the crystallites are formed by solid state reaction and this is accompanied by a shrinkage of linear dimensions of between 10 and 25%. The product is a black brittle ceramic component having a density of about 4800 kg/m^3. Any subsequent shaping operations, e.g. pole-face finishing, have to be done by grinding.

There are two types of ferrite—soft with high permeability and hard with permanent magnet properties. The hard ferrites are described later.

5.16.1 Magnetically soft ferrites

The magnetically soft ferrites have the cubic crystal structure of the mineral spinel. This structure is characterised by a very low anisotropy which leads to the ferrite having low coercivity, high permeability and low magnetic losses at frequencies extending up to 1–100 MHz depending on composition. The chemical formulation is represented by $MeFe_2O_4$ where Me is most commonly a combination of manganese and zinc (MnZn ferrite) or nickel and zinc (NiZn ferrite).

These ferrites are almost invariably manufactured as specific core shapes, e.g. rings, E cores, U cores, pot cores, etc. Apart from the rings, the complete cores are usually assembled from pairs of half cores, the mating surfaces being ground flat to ensure a low reluctance joint. Often, one of the poles faces of the centre limb (the limb embraced by the winding) is ground back to form an air gap in order to modify the properties of the wound assembly.

5.16.1.1 Properties

The properties are strongly influenced by the composition, the presence of minor constituents and the details of the manu-

Carbon levels of 0.005% are advantageous, but expensive to produce, and material of intermediate quality can be obtained with carbon ranging from this low level up towards 0.05% and higher. At 0.1% Carbon the material becomes a common 'mild steel' and while it can still be processed to have useful magnetic properties it is no longer a specialist material. Sulphur levels are kept as low as possible and <0.01% is a good aim.

5.15.2 Applications

Soft iron (or 'relay steel' as it is often called) is used largely for the magnetic circuits of electromechanical relays and solenoids, for the pole pieces of d.c. magnets and parts of the magnetic circuits of small generators. Quite a considerable amount of soft iron is used in the construction of particle accelerators for nuclear research where huge magnets are required to guide accelerated particle beams.

5.15.3 Physical forms

Soft irons come in four main physical forms:

(1) Flat rolled, mainly in the thickness range 0.3–2.0 mm.
(2) Round rod in the diameter range 2.0–20 mm.
(3) Thick plate from 2.0–25 mm thick.
(4) Solid block up to 100 mm to 150 mm thick.

5.15.4 Metallurgical state

To achieve the best magnetic properties the metal should be free from non-magnetic inclusions, have large grains and be free from internal stress.

5.15.5 Magnetic characteristics

The primary grading property of soft iron is coercive force—that is the magnitude of reverse field required to demagnetise the steel fully after a previous magnetisation to a specified value of magnetic flux density. This is shown as the distance PQ on the H axis of the B–H loop in *Figure 5.12*.

Coercive force is important as it indicates the amount of hysteresis which a material exhibits.

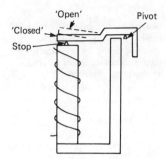

Figure 5.13 A simple electromechanical relay

In an electromechanical relay, *Figure 5.13*, the magnetic circuit has quite a large air gap in it when the relay armature is in the relaxed 'open' state. When 'closed' the air gap is less but still of finite size (often set by a non-magnetic stop pip). It is well known that the magnetic reluctance of the air gap in a magnetic circuit produces 'shearing' of the B–H loop of the circuit material (see *Figure 5.14*). The coercive point X stays the same under loop shear, but the remnant point Y moves to Y'. It is apparent that material having a 'thin' loop with a small H_c will show a greater fall in remanence for a similar amount of 'shear' than will a material of high H_c and having a 'fat' loop. For this reason coercive force is used as a quality guide for material which ideally should have as low a hysteresis as possible.

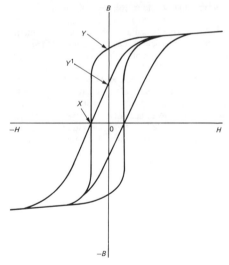

Figure 5.14 Shearing of the B–H loop

Besides coercive force, permeability is important. Depending on the application the low-, mid- or high-field permeability may be important and strict specifications may be placed on the magnetic induction (tesla) achieved for a given applied magnetic field strength.

5.15.6 Grades and specifications

At the present there is no UK standard for grades of soft iron and much of such material is supplied against the house specifications of users. Traditionally a chemical specification has been used for

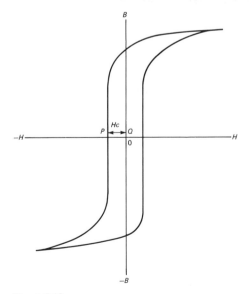

Figure 5.12

5.14.2.4 Stacking factor

This is the ratio of the height of a stack of laminations calculated from the volume of metal present to the measured height of the stack of laminations under a specified clamping pressure, for the same cross-sectional area of material. Stacking factor is expressed as a percentage and typical figures range from 96–98%.

5.14.2.5 Ductility

This is expressed as the number of 180° bends the material can sustain over a specified radius without cracking (typically > 10 bends over a radius of 5 mm). Full details of specifications are given in British Standard 601 (1973) Parts 1–4, in Euronorm 106 (1971) for non-oriented steel and in Euronorm 107 (1975) for grain-oriented steel.

These European specifications are in the process of revision and IEC Grade Standards are in preparation.

5.14.3 Chemistry and production

Electrical steels are produced by a complex process of casting, hot rolling, cold rolling and heat treatment. The end product must be very low in carbon and sulphur and free from non-metallic inclusions—that is metallurgically very 'clean'.

The magnetic properties are much improved if the grain size of the metal can be enlarged, up to a maximum of some 1 cm diameter for some grades and considerable metallurgical skill is employed to procure an end product that is chemically appropriate, metallurgically clean and of optimum grain size.

5.14.4 Physical form

Electrical steels are normally sold in coil form in coil weights up to several tonnes and in widths up to 1 m wide, or in sheets cut to lengths of up to 3 m. Thickness normally ranges from 0.27 mm up to 0.65 mm (1.6 mm if high tensile grades are included). Strip can be supplied fully finished in a state of final anneal and carrying an insulating coating or semi finished so that users can cut or punch the material and apply a final heat treatment and perhaps coating before building into a core. Electrical steels are susceptible to stress and mechanical damage so that they must be handled with care and, if appropriate, re-annealed to recover properties damaged by stress. Appropriate heat treatments are normally an anneal at 800°C in a neutral atmosphere (non-oxidising) or a decarburising anneal at some 800°C in wet hydrogen.

When supplied as slit, narrow strip steel can be used to wind 'clockspring' cores directly without the need for punching.

5.14.5 Coatings

The insulative coatings applied to electrical steels fall into two main classes, organic and inorganic. Organic coatings are inexpensive but cannot be used if further annealing is to be applied to the steel.

Inorganic coatings are resistant to annealing and some may be formed during the processing of the steel for the development of its magnetic properties.

5.14.6 Grain orientation

The grains in electrical steel of the so-called 'unoriented' type are, for the most part, randomly arranged so that the magnetic properties of sheet are broadly isotropic. This is convenient for many applications where flux must flow in varying directions in the steel.

However, if the grains are specially oriented so that an easy direction of magnetisation of the crystal lies along one direction of the sheet (normally the production rolling direction) then the magnetic properties in this direction are much enhanced at the expense of those at other angles to the rolling direction.

This, so called, 'grain orientation' requires a costly and complex regime of rolling and heat treatments and the high temperature anneal stage of production produces a glass film on the surface of the steel which acts not only as an insulating coating but as a further means of enhancing magnetic properties. When steel bearing a glass coating cools from a high temperature the steel contracts more than the glass so that at room temperature the glass coating is holding the steel in a state of tension. This tension improves the magnetic properties of the steel, so that the material at its best functions as a composite system of a precisely grain oriented steel substrate held in a state of beneficial tension by a very thin (some 1–2 microns) layer of glass which also acts as an inter-laminar insulant. A final phosphate coating is usually applied to complete the insulation and tensioning process.

5.14.7 Test methods

The test methods for electrical steels are fully described in BS 6404 part 2 (1985), in Euronorm 118 (1975) and in IEC Standards 404-2 and 404-3. These tests seek to simulate the service condition of electrical steels.

One of the best-known test methods is the Epstein test in which a set of strips of steel each 30.5 cm long and 3 cm wide are arranged in four limbs with double lapped corners to form a hollow square. Four double sets of windings enclose the strips and these are energised to produce a situation analogous to that in a small transformer at operating levels of magnetic flux density and under 'no load' conditions. Careful measurements of magnetic properties are carried out using a precision wattmeter and ancillary instrumentation.

There is a growing trend towards the use of single plates some 50 cm square for tests, and the increasing use of such plate tests will reduce the labour involved in the lengthy Epstein test.

The British Steel Corporation is a leading manufacturer of electrical steels and also produces specialist test equipment for the measurement and quality control of electrical steels.

Equally important is the customer advisory service which is offered as well as the services of their Standards Laboratory which holds British Calibration Service Approval for power loss testing at Newport in Gwent. This is the only BCS Laboratory certified for power loss.

5.14.8 Various

Electrical steels cover a wide range of applications, some are made to have especially high strength for use where heavy mechanical loads arise. Some are of specially precise grain orientation where very low power loss and the highest permeability are vital. Some are simple but inexpensive for use in applications where cost and ease of manufacture of the final article may be paramount.

Electrical steels used inside nuclear reactors must behave well under conditions of high temperature and neutron flux.

Wherever alternating magnetic fields must be managed efficiently and economically, electrical steels of one variety or another can assist. The British Steel Corporation produces in the UK a complete range of electrical steels to cover the needs of the electrical industry.

5.15 Soft irons and relay steels

5.15.1 Composition

The primary aim of soft irons is to be a close approximation to pure iron. Efforts are made to keep contaminant elements to a low level—particularly carbon and sulphur.

Table 5.27 Selection of electrical steel grades produced by the British Steel Corporation—Typical properties

Grade identity	Thickness mm	Typical specific total loss at $\hat{B}=1.5T$, 50 Hz (or 1.7T, 50 Hz where stated) (W/kg)	Typical specific apparent power at $\hat{B}=1.5T$, 50 Hz (or 1.7T, 50 Hz where stated) (VA/kg)	Nominal silicon content %	Resistivity $\Omega m \times 10^8$	Stacking factor	$\hat{B}$ at $\hat{H}=5000\ A/m$ (or 1000 A/m where stated) (T)	Typical applications
Grain oriented: magnetic properties measured in direction of rolling only								
Unisil-H								
30M2H	0.30	1.12 (1.7, 50)	1.55 (1.7, 50)	2.9	45	96.5	1.93 (1000)	High efficiency power transformers
Unisil								
27M4	0.27	0.84	1.16	3.1	48	96	1.85 (1000)	
30M5	0.30	0.89	1.32	3.1	48	96.5	1.85 (1000)	
35M6	0.35	1.00	1.39	3.1	48	97	1.84 (1000)	
Non-oriented: magnetic properties measured on a sample comprising equal number of strips taken at 0° and at 90° to the direction of rolling								
Transil 300/35	0.35	2.95	30	2.9	48	98	1.65	Large rotating machines
Losil 400/50	0.50	3.60	19	2.4	44	98	1.69	Small transformers/chokes
Losil 800/65	0.65	6.50	14	1.6	34	98	1.75	Motors and FHP motors
Non-oriented: grades supplied in the 'semi-processed' condition and which require a decarburising anneal after cutting/punching to attain full magnetic properties								
Newcor 800/65	0.65	7.00	10	Nil	12	97	1.77	Motors and FHP motors
Newcor 1000/65	0.65	8.80	11.5	Nil	12	97	1.74	
Tensile grades								
Tensiloy 250	1.60	—	—	Nil	—	—	1.66	Pole pieces large rotating machines

particularly injurious and the presence of these has to be reduced to the lowest possible limits. The presence of small proportions of silicon, up to approximately 5%, is beneficial. Small proportions of manganese are not injurious, but when the manganese content reaches 12%, a particularly non-magnetic steel is obtained. To neutralise the effect of impurities and to confer special properties, iron is alloyed with other elements, of which the following are the most important: nickel, cobalt, aluminium, silicon, copper, chromium, tungsten, molybdenum and vanadium.

The magnetic properties of ferromagnetic materials depend not only on their chemical composition, but also on the mechanical working and heat treatment that they have undergone. For practical purposes magnetic materials fall into two main groups, high permeability or soft magnetic materials and permanent magnet or hard magnetic materials.

There are a number of groups of soft magnetic materials including soft iron, mild steel, silicon steels, nickel iron alloys and soft ferrites. In static d.c. applications mild steel is the main bulk material, but where better properties are required with higher magnetic permeability, soft iron is used. Silicon steel is the major bulk material for low frequency alternating field applications and the nickel irons are used for more specialised applications with the soft ferrites being applied at the higher frequencies because of their high resistivity. There are four main classes of permanent magnet materials, steels, Alnicos, hard ferrites (also called ceramics) and rare earth cobalt alloys. Of these the steels are almost obsolete because of their inferior properties and during the last twenty years the ferrites have gradually taken over from the Alnicos as the main bulk material principally because of their cost advantage. The rare earth alloys, which are a more recent development are finding increasing numbers of applications because of their much superior properties enabling considerable weight and space savings to be made where cost is not of paramount importance.

5.14 Electrical steels including silicon steels

5.14.1 General

Electrical steels form a class of sheet material used for the flux-carrying cores of transformers, motors, generators, solenoids and other electromechanical devices. Almost always alternating magnetisation is employed and the material is designed to keep power losses due to eddy currents and magnetic hysteresis to a minimum.

5.14.1.1 Eddy current losses

Minimisation of eddy currents is achieved by using thin laminations. The e.m.f. generated in a lamination is proportional to the cross section of the lamination for a given peak magnetic flux density and frequency; while the path length (and resistance) varies only slightly for a given lamination width, so that lamination of the core reduces the V/R ratio for the eddy currents (see *Figure 5.11*).

The current is proportional to (area of cross section)/(path length).

For (i) this is $\dfrac{ab}{2b+2a} \sim \dfrac{ab}{2b} = \dfrac{a}{2}$

For (ii) this is $\dfrac{2ab}{2b+4a} \sim \dfrac{2ab}{2b} = a$

Power rises as $I^2 R$ so that loss per unit volume rises rapidly as the sheet thickness increases. By increasing the resistivity of the steel, eddy currents may be further restrained and this is usually done by adding silicon as an alloying element. Silicon additions range from zero up to some $3\tfrac{1}{4}\%$.

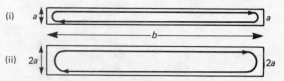

Figure 5.11 V/R ratio for eddy currents

Over this range the resistivity may vary from about 12 to about 48–$50\ \Omega\text{-m} \times 10^{-8}$.

The addition of silicon has the effect of reducing the saturation flux density of the steel so that a larger core cross sectional area is needed to carry a given magnetic flux.

5.14.1.2 Hysteresis losses

Magnetic hysteresis is another source of power loss which can be reduced by the addition of silicon as an alloying element. While the area of the very low frequency B–H loop is often taken as a measure of hysteresis loss, effects which impede magnetic domain wall motion alter with frequency so that at 50 or 60 Hz the overall loss mechanism is a complex mixture of macro eddy currents relating to lamination thickness, micro eddy currents associated with the movement of the internal domain walls and energy dissipative hysteretical effects due to impediments to free domain wall motion. Such impediments may be non-metallic inclusions, grain boundaries or regions of stress within the crystal lattice of the material.

Table 5.27 gives an outline of the principal types of electrical steel produced in the UK, and some of their properties and applications.

Key properties As the percentage of added silicon increases, the steel becomes more difficult and expensive to process. Thinner material is more costly because it requires more rolling and processing. The key points of a material specification are given below.

Primary grading This is by power loss in W/kg at a peak magnetic flux density of 1.3, 1.5 or 1.7 tesla 50 Hz (60 Hz in USA).

5.14.2 Ancillary properties

5.14.2.1 Specific apparent power (VA/kg)

Specific apparent power gives an indication of the RMS ampere-turns which must be available to maintain a given state of alternating magnetisation, e.g. a peak value of 1.5 tesla at a specified frequency. Most of the current drawn is 90° out of phase with the supply voltage and so is non-dissipative, but copper or aluminium windings of sufficient cross section must be used to keep copper losses low in the face of the current demanded.

5.14.2.2 Permeability

There are many varieties of permeability, but for electrical steels those of most interest are $\hat{B}_{1000}$, $\hat{B}_{2500}$, $\hat{B}_{5000}$ and $\hat{B}_{10000}$, that is the peak value of magnetic flux density attained under the influence of an applied field of 1000 A/m, 2500 A/m, 5000 A/m and 10 000 A/m respectively.

5.14.2.3 Surface insulation resistance

This is a measure of the goodness of the surface insulating coating (expressed in $\Omega\text{-m}^2$) used to restrain the flow of current between laminations.

upon the resistance of the basic materials to the effects of radiation. Cellulose derivatives, such as cellulose acetate, have poor resistance, mechanical deterioration being rapid; hence lacquers, adhesives and moulded or extruded parts made from them also deteriorate rapidly under irradiation. On the other hand, diphenyl silicone and products made from this resin, have relatively good resistance. The combination of mineral materials, whether as powdered or fibrous fillers or as sheet reinforcement (e.g. woven glass cloth), with the organic base materials usually results in products having superior radiation resistance.

Synthetic resins: thermoplastic Of the thermoplastic synthetic resins *polystyrene* is one of the most stable. Styrene co-polymers are generally poorer in resistance to effects of radiation, and high-impact-strength polystyrene loses some of its impact strength when irradiated. A good deal of work has been done on the effects of irradiating *polyethylene*, the principal change being the elimination of its melting point due to cross-linkage: this can be beneficial. Some increase in tensile strength occurs at first but, with continued irradiation at high dosages, the tensile strength decreases and the material ultimately becomes brittle and cheesy. The high-density varieties are slightly better. *Nylon* behaves somewhat similarly when irradiated, cross-linkage occurring with consequent increase in tensile strength in the case of sheet, but rapid reduction of strength occurs in nylon fibres, with decrease of elongation and impact strength.

Of the *vinyl polymers* and *co-polymers*, polyvinyl carbazole has good resistance to radiation; *polyvinyl chloride* (*p.v.c*) has radiation resistance equivalent to that of polyethylene but liberates hydrogen chloride when irradiated; *vinyl chloride acetate* has a lower radiation resistance than p.v.c., softening and turning black at low dosages.

Irradiation of *polyethylene terephthalate* in fibrous form causes rapid loss in strength, and the films become brittle and darken. *Cellulose plastics* such as cellulose acetate deteriorate mechanically to a serious extent by low dosages, but electrical properties of the latter are not appreciably affected. *Acrylic resins* also have comparatively low radiation resistance.

The thermoplastic resins most seriously affected by radiation are the fluoroethylene polymers such as *polytetrafluoroethylene* and *monochlorotrifluoroethylene*, the latter being superior. Decrease of tensile strength and elongation of these is rapid and they become very brittle with only moderate dosages.

Synthetic resins: thermosetting Of the amino resins, the *melamine-formaldehyde* type is slightly superior to the *urea-formaldehyde*, but with cellulose fillers both deteriorate rapidly and become brittle when irradiated. The radiation resistance of *epoxy* resins is above the average for plastics but it depends largely upon the hardener used. *Epoxyphenolic* resins are better than ordinary epoxies and phenolics for radiation resistance. *Phenolic* and *polyester* resins without fillers have low resistance but this is increased considerably by the addition of fillers, especially minerals such as asbestos.

Silicone resins are more resistant to radiation than silicone fluids and elastomers and do not rapidly deteriorate physically, the major electrical properties of most resins being maintained even after subjection to high dosage.

Solid materials: inorganic In general, inorganic materials are much more resistant than organic materials to irradiation. The principal effects are to cause colour changes and to induce conductivity in glasses, ceramics, fused silica and mica, followed by disintegration of natural mica irradiated by high dosages at elevated temperatures (e.g. 150 °C). Synthetic mica seems to be more resistant than the natural form to mechanical and electrical degradation. Built-up mica products made from flake mica or reconstituted mica paper are affected according to the bonding medium—which is usually organic. Wires insulated with swaged magnesium oxide have shown high resistance to radiation at a temperature of 815 °C.

5.12 Ferroelectrics

Ferroelectrification, by analogy with ferromagnetism, is a state of spontaneous dielectric polarisation due to displacement of the positive and negative electrical centres in atoms, displacement of positive and negative ions in molecules, and the alignment of permanent electric dipoles in materials. Rather less than twenty compounds are known to exhibit electric flux without the application of an external electric field, the most important being barium titanate ($BaTiO_3$). A 'domain theory' resembling that for ferromagnetic materials has been advanced, and the analogy includes hysteresis effects and high permittivities. Unlike ferromagnetics, however, the ferroelectrics exhibit their optimum properties near the Curie point which, for barium titanate, is about 120 °C: in this region a relative permittivity of 10 000 may be approached.

5.12.1 Manufacture

Barium titanite is formed from the reaction of a mixture of $BaCO_3$ and TiO_2 heated to 1250 °C. The product is powdered and then worked by means of common ceramic techniques. Admixtures of other oxides are employed to modify the dependence of permittivity on temperature.

5.12.2 Applications

Permittivity Ceramic capacitors of large capacitance/bulk ratio at high and low voltages are made for the range 500–10 000 pF, normally of the two-plate type, for electronic equipment.

Hysteresis effect The dependence of permittivity on polarising voltage (analogous to incremental permeability) permits timing control or carrier-frequency modulation by means of a voltage control. The effect is also applicable to computers using binary counting.

Transducer effect Barium titanate has a piezo-electric property, and although mechanically inferior to quartz it has a compensating advantage in that it can be fabricated in a variety of complex shapes.

MAGNETIC MATERIALS

5.13 Ferromagnetics

The choice of materials having ferromagnetic properties is very wide. Besides iron, a number of elements such as nickel and cobalt as well as some manganese alloys are ferromagnetic.

It has so far been found impossible to obtain absolutely pure iron, in the true sense, even under laboratory conditions; but iron in which the impurities have been reduced to a mere trace has been found to have a maximum relative permeability of 50 000 and a very low hysteresis loss. But even if it were possible to produce an iron of this degree of purity commercially, its low resistivity would be a disadvantage with alternating fluxes because of eddy currents.

The common impurities in iron are carbon, manganese, silicon, copper, sulphur, phosphoros and oxygen and the general effect of these impurities is to decrease the permeability and to increase the hysteresis loss. Sulphur, phosphorus and oxygen are

(3) Van de Graaf electron accelerator.
(4) Microwave linear electron accelerator.

The most convenient is the linear accelerator, in which a small diameter beam can be made to scan the parts to be irradiated with uniformity of dosage. The physical conditions—temperature, ambient atmosphere (air, carbon dioxide, nitrogen, etc.), humidity—must be carefully selected and controlled.

5.11.2 Irradiation effects

The effects of radiation are usually assessed in the first place by visual examination, as many materials discolour, crack, disintegrate or melt, while flexible or soft materials may become hard and brittle.

More advanced work relies on the determination of changes in measured properties after certain periods of irradiation. Results of tests on unirradiated samples are compared with those on similar samples after subjection to known energies for different times. Non-destructive tests (loss tangent, permittivity, resistivity, changes in weight and changes in dimensions) are particularly useful because they can be repeated on the same specimen as a function of the total dose. Other tests used to determine the effects of irradiation are electric strength; tensile, shear and impact strength; elongation; hardness; flexibility; and water absorption.

In general, organic insulating materials deteriorate mechanically and electrically as the result of irradiation, the mechanical properties usually being impaired at a greater rate than electrical ones. These results are mainly due to chemical changes which occur in the materials consequent upon certain rearrangements of their molecular structure, usually with the evolution of one or more gases, such as methane, carbon monoxide, carbon dioxide or hydrogen. The probable life-dose for typical insulating materials is given in *Table 5.26*. The dose figure is that at which marked deterioration is observable, there is a 50% reduction in mechanical and electric strength, or some similar factor. The life-dose is approximate: it depends on the actual formulation of the material and irradiation conditions such as dose-rate and ambients.

Gases The spark-gap breakdown voltage of air is reduced by about 20% under intense nuclear radiation. A gas having good radiation stability is sulphur hexafluoride (SF_6), whereas halogenated hydrocarbons slowly polymerise with evolution of corrosive products, and other gases, such as perfluoropropylene, polymerise rapidly to form liquids when irradiated.

Liquids The viscosity of hydrocarbon oils increases and most liquid dielectrics polymerise when irradiated, the electrical properties being lowered considerably. Gases are usually evolved by liquids during irradiation and can create difficulties due to increase of pressure in containers of capacitors, transformers, etc. Silicone oils polymerise to form elastomers, those of high molecular weight yielding elastomers of low tensile strength, but the solids produced from the low molecular weight (e.g. 300–20 000) oils crumble on handling.

Semi-fluid and fusible materials Silicone compounds made from fluids with mineral fillers react to irradiation in the same manner as the fluids; they rapidly harden and then gell. Petroleum greases are affected similarly. Dimethyl silicone fluids of low viscosity are to be preferred for use where doses do not exceed about 100 Mrad.

Organic solids Most of the organic solid insulating materials in general use are synthetic resins or are based upon such resins and similar plastics. The performance of the materials in the form in which they are used under irradiation conditions depends largely

Table 5.26 Radiation resistance of organic insulating materials

Probable useful-life dose in Mrad

Material	Mrad	Material	Mrad
Gases		*Liquids*	
Sulphur hexafluoride	5000	Polyphenyls	5000
Difluorodichloromethane	1000	Radn. resistant petroleum oil	2000
Trifluoromonochloroethylene	500	Transformer oil (naphthenic)	1000
Perfluoropropylene	100	Transformer oil (paraffinic)	500
		Silicone oil	200
Moulded of laminated plastics (filled)		Pyrochlor	100
Diphenyl silicone/glass	10 000		
Mineral-filled epoxyphenolics	10 000	*Resins, bitumens, etc. (unfilled)*	
Mineral-filled phenolics	4000	Diphenylsilicone	5000
Epoxy/glass cloth	4000	Polystyrene	5000
Cellulose-filled phenolics	1000	Polyvinyl carbazole	4000
Cellulose-filled urea-formaldehyde	1000	Bituminous compounds	2000
		Nylon	2000
Elastomers		Polyethylene	2000
Polyvinyl chloride (plasticised)	500	High-impact polystyrene	2000
Polyurethane rubber	400	Polyurethane	1000
Butadiene-styrene + antirad.	300	Alkyd resins	500
Phenylmethyl silicone	200	Phenol-formaldehyde resins	500
Polychloroprene	150	Polyethylene terephthalate	500
Natural rubber	150	Cellulose nitrate	100
Acrylonitrile	100	Cellulose butyrate	50
Polysulphide	80	Cellulose acetate	50
Dimethylsilicone	30	Methyl methacrylate	50
Polyisobutylene	20	Polytetrafluoroethylene	5

Table 5.25 Properties of typical organic and inorganic moulded and formed materials

Test methods of BS 488 and 771 (PF Phenol formaldehyde; UF Urea formaldehyde)

Organic		Binder: Filler:	PF Wood filler	PF Wood filler	PF Asbestos	PF Mineral powder	UF Wood	Rubber Mineral	Cellulose- acetate
Density	kg/m^3		1400	1360	1900	1880	1500	1700	1300
Plastic yield temperature	°C		>100	>140	>180	>140	>100	80	60–80
Coefficient of expansion × 10^6	per °C		40	40	30	30	45	80	160
Water absorption†	%		0.3–0.4	0.3	0.05	0.02–0.05	0.5	<0.01	1.5–3
Elastic modulus	GN/m^2		5	5	7	3	7	3	2
Tensile strength	MN/m^2		48	48	35	35	63	24	31
Cross-break strength	MN/m^2		63	70	49	52	105	67	49
Crushing strength	MN/m^2		240	240	160	120	230	140	140
Impact strength	Nm		0.3	0.3	0.2	0.3	0.3	0.17	0.14
Resistivity	Ω-m		10^9	5 × 10^{10}	10^9	5 × 10^8	10^6	10^{14}	5 × 10^6
Surface resistivity	MΩ/sq.		5 × 10^4	10^6	10^4	2 × 10^7	5 × 10^5	3 × 10^5	4 × 10^6
Relative permittivity	—		7–10	4–8	8–18	5	9	4.1	4–6.5
Loss tangent, 50 Hz	—		0.25	0.04	0.2–0.6	—	0.08	0.016	0.016
1 kHz			0.2	0.04	0.1–0.4	0.02	0.06	0.012	0.03
1 MHz			0.15	0.035	0.1	—	0.04	0.01	0.06
Electric strength*	kV/mm		1.2–4	9	3–4	15	3	15	12

Inorganic			Porcelain	Steatite	Al-oxide	Glass	Glass- mica	Asbestos cement	Glass- ceramic
Density	kg/m^3		2400	2750	3650	2250	2680	1600	2600
Plastic yield temperature	°C		>1200	1400	—	600	450	>700	1250
Coefficient of expansion × 10^6	per °C		4	6–8	6.2	3.2	9.8	—	5.7
Water absorption (20°C, 24 h)	%		0	0.01	—	0	0	10–15	0
Specific heat	J/kg		900	840	750	840	840	—	750
Tensile strength	MN/m^2		35	56	75	—	42	7	150
Flexural strength	MN/m^2		70	100	330	—	93	28	140
Crushing strength	MN/m^2		420	840	1670	970	270	55	—
Resistivity (20°C)	Ω-m		10^{16}–10^{13}	10^{13}	10^{14}	>10^{12}	10^{11}	—	10^{14}
Relative permittivity (20°C), 1 MHz	—		5–7	4.1–6.5	10	4.5–4.9	6–7.5	—	5.6
Loss tangent (20°C), 1 kHz	—		—	0.005	0.008	0.005	0.007	—	0.0025
1 MHz			0.006	0.0045	0.0006	0.003	0.002	—	0.0015
Electric strength**	kV/mm		6–16	8–15	48	14	20	0.8–4	—

* R.M.S., 3 mm at 90 °C
** R.M.S. 20 °C
† Cellulose acetate, 24 h; remainder 7 days

radiation, often at high energy levels. Such radiation can change the characteristics of many materials and may cause severe deterioration; on the other hand, radiation can have beneficial effects, especially by causing synthetic polymers to cross-link. Particular cases are the irradiation of polyethylene and other thermoplastic materials giving new products with improved properties, especially resistance to heat and mechanical failure; the vulcanisation of rubber; the polymersiation of plastics; the curing of resin and varnish films and the manufacture of heat-shrinkable films and sleevings.

5.11.1 Type of radiation

The more common forms of radiation encountered are: neutron and gamma from nuclear reactors, neutron and gamma from isotopes and electrons and X-rays from particle accelerators. As a unit of absorbed radiation the rad (= 100 erg/g) or the megarad (= 10^8 erg/g of material). The megarad is equivalent to 10 kJ/kg.

Although the effects of radiation on materials are cumulative and therefore dependent on the total dose, the dose rate (generally expressed as M rad/h) may have some further effect. This point must be considered when experimental work is being carried out. Fortunately, it has been found that changes in most materials due to irradiation are practically independent of the type of radiation encountered. Hence, various sources of irradiation may be used for experimental work and those normally employed are:

(1) 'Hot' fuel elements and other parts of nuclear reactors.
(2) Radioactive isotopes (e.g. Cobalt 60).

resin products such as polyvinyl chloride, nylon, polystyrene; cellulose derivatives such as cellulose acetate.

(2) *Organic—thermo-setting* They are, chiefly: cured shellac products including micanite (for simple shapes, e.g. commutator cones); rubber-sulphur and similar vulcanisable compositions; compounds made from synthetic resins of the phenol-formaldehyde type, urea-formaldehyde, silicones, polyesters, alkyds, epoxies and polyurethanes.

(3) *Inorganic* The principal materials are:
  (*a*) *Asbestos–cement* compositions, mainly for high heat resistance, particularly for parts exposed to arcs;
  (*b*) *Concrete*, cast or moulded, for inductors and switchgear where strength and fire-resistance are needed;
  (*c*) *Porcelain*, mainly for use out-of-doors and other cases where dust and moisture collect readily, also special grades for high temperatures;
  (*d*) *Steatite*, for uses similar to those of porcelain;
  (*e*) *Special ceramics* for radio capacitors, sparking plugs, etc.;
  (*f*) *Fire-clay* for holding electric heating elements;
  (*g*) *Glass* for out-door insulators, lamp bulbs, valves and high frequency insulation;
  (*h*) *Mica-glass* composition for high-temperature applications requiring good electrical properties, especially at high frequency.

5.10.8 Methods of moulding and forming organic materials

The principal methods of manufacturing parts from the organic materials are:

(1) *Forming* of laminated and other sheet materials in open moulds or other forming tools with moderate pressure, the material being made plastic by a suitable liquid or, more usually, by heat. Such forming is generally restricted to relatively simple shapes such as channels, cones, tubes, collars, spools and wrappings. Heat is usually applied during the forming operation, and the material sets either by cooling, heat-treatment in the mould, or subsequent drying, baking, etc. (e.g. vulcanisation).
(2) *Moulding under pressure* in closed metal moulds ('compression moulding'), using a powder, dough, treated paper, treated fabric or other form of moulding material and heat treating during or subsequent to moulding.
(3) *Injecting* material, made plastic by heat or other means, into moulds under pressure and setting by cooling, or heat processing.
(4) *Extruding* plastic material as in (3) but not into moulds, the shape being determined by the orifice or die through which the material is extruded and the setting being generally due to cooling, but may be followed by a further process such as vulcanisation.
(5) *Casting* a liquid or molten material into moulds (without pressure) and conversion to a solid condition by cooling or heating (thermo-setting materials), or by the action of chemicals with or without heat.
(6) *Coating* with polymers in powder form, spread on heated metal components by several methods including electrostatic fields or fluidised bed techniques. Thick coatings with good electrical properties can be applied overall. Conversion to the solid is by cooling with thermoplastic materials, or by further heat treatment for thermosetting materials. Some polymers can be applied to surfaces (which need not be metal) in the form of dispersions in a suitable liquid carrier, followed by curing by heating.

Properties of typical organic moulding compounds are listed in *Table 5.25*.

5.10.9 Methods of manufacture of inorganic materials (ceramics, etc.)

The methods used for producing articles from the inorganic materials are, briefly, as follows:

Asbestos-cement compounds Made from asbestos and other minerals, e.g. powdered silica, mixed wet with lime or Portland cement and moulded *cold* by compression moulding. After removal from the mould the parts are cured in live steam.

Concrete The large mouldings for inductors are made of high-grade concrete, made from specially selected Portland cement, sand and aggregate, the wet mixture being poured and 'puddled' into suitable moulds and, after preliminary setting, the parts being cured in live steam.

Porcelain Made from china clay (*kaolin*), ball clay, quartz and felspar, finely powdered and mixed with water. Small parts (e.g. tumbler switch bases) are made by the *dry process* (or *die-pressing*), in which a slightly damp mixture is compressed in steel moulds. Many high-voltage parts, particularly large pieces such as transformer bushings, are made by the *wet process* from a wet plastic mixture, shaped on a potter's wheel and turned on a lathe. Others, such as overhead line insulators, are made by pouring a *creamy* mixture into plaster moulds in which partial drying occurs. Parts formed by the foregoing processes are then dried and usually coated with glaze, after which they are fired in a kiln at temperatures such as 1200–1400°C.

Steatite Consists of powder soapstone (talc) die-pressed dry or moulded wet, similar to porcelain, and finally fired at about 1400 °C.

Special ceramics A number of ceramics are made, similar to porcelain and steatite, from such materials as rutile (a form of titanium dioxide)—having low losses and high permittivity, and suitable for high-frequency capacitors—and aluminium oxide, mainly for sparking plugs. A ceramic composed of barium-titanate has exceptionally high permittivity.

Fire-clay refractory ceramics Made from special grades of clay, usually die-pressed and fired, somewhat as in the case of porcelain.

Glass Made from powdered silica mixed with metallic bases (soda or potash) and a flux (e.g. borax). The mixture is fused at temperatures of the order of 1200–1400 °C, and the molten mass 'blown' into moulds, or forced into moulds under pressure, the parts being removed when cool.

Recent developments have been made in which the glass instead of remaining in its usual super-cooled liquid state is devitrified, with the result that a fine-grain structure develops and the mechanical properties are much improved. These new materials are known as glass-ceramics and can have specially controlled properties. In particular the coefficient of thermal expansion can be selected from high positive to negative values.

Mica–glass composition Produced from powdered mica and glass, heated to a semi-molten condition and moulded in steel moulds by compression or injection at high pressure.

Properties of typical ceramics and other inorganic formed or moulded compositions are summarised in *Table 5.25*.

5.11 Irradiation effects

Electrical equipment is being used increasingly in situations where it will be exposed to the effects of nuclear and other types of

impregnated with a silicone elastomer are available straight or bias cut in thicknesses from 0.07–0.5 mm. Other fabric tapes, coated on one side with partially cured silicone elastomer, are used for taping bars and coils and the curing subsequently completed by baking. The cured taping forms a homogeneous coating having good resistance to moisture, discharges and heat, e.g. for continuous use at about 180 °C and up to 250 °C for short periods, and possessing high electric strength and low dielectric losses. Most grades of silicone rubbers remain flexible at temperatures down to $-60\,°C$ and some can be used down to $-90\,°C$.

5.10.4.7 Composite sheet insulations

These combine two or more materials, especially varnished cloth, press papers, vulcanised fibre, plastic films bonded with special adhesives. Particularly widely used have been cellulose fibres combined with polyester or cellulose acetate film, polyester fibre paper combined with polyester film and various combinations of mica, varnished glass fabrics and polyester film. Among the newer developments are combinations of aromatic polyamide paper with polyimide film, giving a material suitable for use at temperatures above 220 °C. Another combination is aromatic polyamide paper with polyester film, having good class B performance.

5.10.5 Sleevings, flexible tubings and cords

Tubular sleevings made of cotton, polyester, asbestos, glass, ceramic, quartz, silica, polyamide and other fibres supplied untreated in flat tubular form are suitable for low-voltage insulation of connections of coils, etc. Glass sleeving given a high-temperature treatment before use has the fibres set in place, thus preventing unravelling during assembly. Varnishing of all these sleevings during treatment of the coils is the usual practice.

Similar sleevings may be varnished or treated with resins and polymers before use; these types are known as coated sleevings. Most colours can be produced, thus helping in identification of circuits. The coating materials in general use for cotton and rayon sleevings are natural or synthetic varnishes. For glass fibres, high-temperature polyvinylchlorides, polyurethanes and silicone elastomers are widely used as well as high-temperature varnishes such as those based on acrylics, polyesters, silicones or polyimide resins. Coatings of fluorinated resins on glass fibres are available.

Flexible tubings can be extruded from most of the materials mentioned in Section 5.10.4.6, but as many of these materials are thermoplastic and also liable to oxidation problems, care is required at temperatures above 90 °C.

Other methods of making tubing are by helical winding with adhesives as a bond, when very thin walls (e.g. 0.025 mm) are required or only small quantities are needed. Many of the extruded tubings and some of the helix types can be caused to shrink by applying heat after the tubes have been assembled. Most shrinkage takes place in diameter but there is sometimes a small amount in length. These sleeves are useful for insulating connections and joints, for covering capacitors, resistors, diodes, etc., for colour-coding, sealing and moisture proofing. The two main shrinkage processes are: (i) Thermoplastics stretched during manufacture, and allowed to cool so that the mechanical strains are frozen in, shrink next time the material is heated. (ii) When certain materials are exposed to high-energy radiation, molecular cross-linking occurs, and they shrink when the temperature is subsequently raised.

Cords for lashing bundles of switchboard wiring, holding leads in position and for tying down coils in machines and transformers, may be made from yarns laid together and twisted or braided, or from narrow woven or slit tape. The trend is away from cellulose materials (linen, hemp) to synthetics, especially glass and polyester fibres. Because of the poor abrasion resistance of glass fibres, glass cords are often pre-treated with epoxy or silicone resins or with various polymers. Normally the type of treatment depends on the application.

For high-voltage machines, a useful material is a round cord 3–5 mm diameter comprising a central core of straight polyester fibres surrounded by a braided polyester sheath. This material shrinks slightly on heating, and tightens. For smaller windings, polyester mat can be slit into tapes and several tapes twisted together. Polyester fibre cords of all kinds have the advantage over glass cords that knots can be made without appreciable loss in strength. Glass and polyester bands can be treated after application with varnishes and both have good resistance to abrasion and to mould growth. For rotors straight unidirectional glass fibre bands pre-treated with epoxy, polyester, or acrylic resins are used. These bands must be applied under controlled tension to ensure that the windings are held against centrifugal force. Such bands can be placed safely near voltage carrying parts and (unlike wire bands) are not affected by magnetic fields.

For switchboard wiring, cable harnesses, etc., neat lashings can be made with small diameter extruded polyvinylchloride threads. These materials do not burn readily and are obviously well suited for use with p.v.c. insulated switchboard wiring.

5.10.6 Wire coverings

Wires for coils and armature conductors are insulated with lappings of cotton, asbestos, glass fibre and polyamide fibre, all of which are hygroscopic and require treatment with oils, compounds or varnishes, usually after winding. Relevant specifications are: BS 1497, 1933, 2479, 2480 and 2776. A recent addition to the range of wire and strip coverings is lapped polyimide film, bonded to itself and to the conductor with f.e.p. polymer: the films are thin and flexible, electrically good and workable at 220 °C.

Orthodox oleo-resin (BS 156) and polyvinyl-formal and acetal enamels (BS 1844) are tough, resistant to abrasion and to softening by varnishes. Enamelled wires have a better space factor than those with fibre covering.

Some wires are covered in separate operations with synthetic enamels of widely differing characteristics, giving a range of finely balanced properties. The inner coat can be considered as the insulant; the outer coat can give improved resistance to solvents and abrasion, better cut-through resistance, heat-bonding of turns into a solid coil, etc. Wires coated with polyurethane polymers are useful because they can be soldered without first removing the enamel (BS 3188).

For high-temperature working, the performance of p.t.f.e. coverings is limited by cold flow of the insulant under mechanical pressure. Silicone resin gives a high-temperature covering but the solvent resistance and mechanical properties are not very good. Better mechanical characteristics are given by polyimide, polyamide-imide, polyester-imide and polyhydantoin, but these materials are still uncommon.

5.10.7 Moulded and formed compositions, plastics, ceramics, etc.

Articles of various shapes, often quite complicated, which cannot readily or economically be matched or built up from sheet, rod and tube materials may be obtained by forming, moulding, coating or casting one of the 'plastics' (which are all essentially organic), or a ceramic such as porcelain, or glass or other inorganic materials. The materials can be grouped approximately as follows:

(1) *Organic—thermo-plastic* Examples are: compounds made of natural gums, bitumen, etc., not specially processed; synthetic

temperatures. Both materials suffer from flow at high pressures even at moderate temperatures. Operating temperatures are otherwise similar to those given for f.e.p. above.

P.V.F. and P.V.F.2 Polyvinyl fluoride and polyvinylidene fluoride are produced by substituting fluorine for some of the chlorine in p.v.c. Both have the excellent temperature and chemical stability shown by fluorine-substituted materials. Electrical properties are not as good as those of p.t.f.e. and f.e.p.

Polyethylene This is not widely used as electrical insulation, mainly because of the low softening temperature. Controlled radiation with high-energy electrons produces a film which has similar properties to the original film but because cross-linking has occurred, there is no sharp melting point. This enables the material to be used at higher temperatures without damage; also, with this treatment, the material can be made heat-shrinkable.

Polyethylene terephthalate These films are widely used for slot insulation in motors, as the dielectric in capacitors, for coils, etc. For slot insulation the film is frequently combined with polyester fibre paper but may be used alone in smaller equipments. It is suitable for Class E temperatures although some manufacturers claim it can be used at Class B temperatures. When protected from oxygen by films or varnishes with higher temperature resistance, some users claim temperatures up to Class F are suitable for this material.

Polyamide (nylon) Films show moisture absorption in humid atmospheres, but are strong, solvent-resistant and have a high melting point; they cannot be used at temperatures exceeding 80 °C in air because of oxidation.

Polycarbonate The film is strong, flexible and highly stable against temperatures up to 130 °C. Electrical properties are excellent and the film is finding use as a capacitor dielectric.

Polypropylene This dielectric film has a higher softening point than polyethylene film and the dielectric properties are similar. The material is resistant to hot mineral oils and chlorinated polyphenols and is finding widespread use in low-voltage and power capacitors.

Polyimide An outstanding film of recent development. The material has no melting point and can withstand exposure at temperatures above 500 °C for several minutes. Life at 250 °C is over 10 years, and more than 10 h at 400 °C. It remains flexible down to the temperature of liquid helium. The film is replacing glass and mica insulation in motors especially for traction. As thin layers can be obtained, considerable space saving (often as high as 50%) can be achieved. The film cannot be bonded easily, but can be supplied with a very thin layer of thermoplastic f.e.p. on one or both sides, and layers can be bonded at temperatures approaching 300 °C.

Polyvinyl chloride These films have low water absorption and resist most chemicals and solvents. The high elongation makes it possible to apply tapes neatly and tightly over irregular shapes. This class of material is often used as a pressure-sensitive tape.

A summary of the properties of materials available in the form of thin films is given in *Table 5.24*.

Silicone elastomers Silicone rubbers or elastomers are a range of heat stable elastic silicone materials used for electrical insulation as sheet, tape, wire and cable coverings, extruded sleevings and mouldings, unsupported, but more extensively as coated glass fibre cloths, tapes and braided glass sleevings. Such fabric tapes

Table 5.24 Properties of materials as thin films

CA	Cellulose acetate
PA	Polyamide
PVF	Polyvinyl fluoride
PVF2	Polyvinylidene fluoride
PTFE	Polytetrafluorethylene
PTFCE	Polytrifluorochloroethylene
FEP	Fluorinated ethylene propylene
PET	Polyethylene terephthalate
CT	Cellulose triacetate
PI	Polyimide
PVC	Polyvinyl chloride (plasticised)

	Material			CA	CT	FEP
1	Density		kg/m^3	1290	1290	2150
2	Tensile strength		MN/m^2	50–80	60–110	21
3	Elongation at break		%	15–60	10–40	300
4	Resistivity (20 °C)		Ω-m	10^{13}	10^{13}	$>10^{16}$
5	Relative permittivity (20 °C),	1 kHz	—	3.6–5	3.2–4.5	2.0
6		1 MHz	—	3.2–5	3.3–3.8	2.0
7	Loss tangent (20 °C),	1 kHz	—	0.015–0.03	0.015–0.025	0.0002
8		1 MHz	—	0.025–0.05	0.03–0.04	0.0002
9	Electric strength (20 °C), r.m.s.		kV/mm	60	60	200

	PTFE	PTFCE	PA	PVC	PET	PI	PVF	PVF2
1	2200	2100	1140	1260	1400	1420	1380–1570	1750
2	20	40	50–80	13–30	170	170	100–140	40–45
3	200–300	200–350	250–500	150–350	100	70	100–200	150–500
4	$>10^{10}$	10^{16}	$>3 \times 10^{11}$	10^9–10^{12}	10^{16}	10^{16}	$>10^{11}$	2×10^{12}
5	2.1	2.8	3.8	3–7	3.2	3.5	8.5–9	8.0
6	2.1	2.5	3.4	—	—	—	—	6.6
7	0.0005	0.016	0.01	0.01–0.02	0.005	0.003	0.015	0.018
8	0.0002	—	0.025	—	—	—	—	0.17
9	60	120–200	60–70	10–40	280	280	160	50

Table 5.22 Properties of typical flexible sheets

VF	BS 2768 Vulcanised Fibre
PP	BS 231 and 3255 Presspaper Grade II
CP	BS 698 Cellulose Paper (dry, or impregnated in mineral oil)
AP	BS 3057 Asbestos Paper
PF	Polyamide Fibre Paper
MP	Muscovite Mica Paper

Material		VF	PP	CP dry	CP oil	AP	PF	MP
Density	kg/m^3	1260	1050	1050	—	700	950	1600
Thickness	mm	0.25	0.25	0.25	0.25	0.15	0.25	0.15
Maximum operating temperature	°C	90	90	90	120	300	220	600
Tensile strength m	kN/m	31	19	12	12	2	30	1.5
c	kN/m	14	8	4	4	1	18	1.2
Tearing strength m	kg	0.42	0.38	0.26	0.26	—	0.55	—
c	kg	0.46	0.41	0.29	0.29	—	0.9	—
Resistivity (20 °C) (dry)	Ω-m	—	—	10^{16}	10^{12}	—	10^{14}	10^{13}
Relative permittivity (20 °C, 50 Hz)		—	2.8	3.1	1–2.5	3–4	—	2.6	—
Loss tangent (20 °C, 50 Hz)		—	0.05	0.012	0.0025	0.0025	—	0.01	—
Electric strength* in oil	kV/mm	16	60	—	50–80	—	—	—
in air (90 °C)	kV/mm	15	11	9	—	7	30	20

m = in machine direction; C = cross direction
* r.m.s.

Table 5.23 Properties of typical varnished textiles

Test methods of BS 419
P.E.T.P. = Polyethylene terephthalate; Y = yellow; B = black

Property		Cotton Y/B	Nylon Y/B	P.E.T.P. Y*/B	Glass Y*/B
Thickness	mm	0.25	0.15	0.15	0.15
Tensile strength, wp	kN/m	8/9	9	9	21
wf	kN/m	6/7	5	5	16
Tearing strength, wp	kg	0.25/0.22	0.12	0.1	0.15
wf	kg	0.26/0.23	0.3	0.21	0.34
Electric strength (1 min):†					
large electrodes, 20 °C	kV/mm	—	36/38	34/40	44/42
90 °C	kV/mm	24/32	32/32	30/34	32/40
150 °C	kV/mm	—	—	28/30	18/19
6.35 mm diameter electrodes (20 °C), tapes stretched by					
0%	kV/mm	—	52/58	52/62	42
5%	kV/mm	—	50/56	50/60	40
10%	kV/mm	—	50/54	50/54	28
20%	kV/mm	—	26/48	22/20	—

* Special heat-resistant varnish
† Root–mean–square

thicknesses of about 0.008 mm up to about 0.5 mm are usual. Most are strong, have good electrical properties and good resistance to moisture, but some have certain limitations in operating temperature. The thermoplastics often soften or melt at comparatively low temperatures and some films which melt at very high temperatures are damaged by oxidation processes at much lower temperatures.

Cellulose acetate, triacetate and similar acetates These films have been available for a long time and still find use in machines and in coils at temperatures up to Class Y conditions.

F.E.P. This copolymer of tetrafluoroethylene and hexafluoropropylene (which is hardly affected by any known chemicals and solvents) has a service temperature range from about −250 °C up to more than 200 °C. It can be bonded to itself and to other materials by heat and pressure and has excellent high-frequency characteristics.

P.T.F.E. and P.T.F.C.E. Polytetrafluoroethylene and polytrifluorochloroethylene have excellent chemical stability and electrical properties. The former material does not soften, but degrades above 300 °C; the latter is thermoplastic at high

5.10.2 Rods

Many of the board materials can be obtained in the form of rods produced by: (i) *Machining from sheet*: e.g. vulcanised fibre, s.r.b. paper, cotton fabric, glass, phenolic laminated wood, etc.; there may be a plane of weakness due to the laminar structure. (ii) *Compression moulding*: e.g. tubes of s.r.b. paper and cotton or glass fabric wound on very small mandrels and subsequently compressed in a split mould: 'flash' at the split lines must be removed, sometimes giving local weakness. (iii) *Extrusion*: e.g. ebonite and many plastics; a type of extrusion can be used by drawing fibres and resins simultaneously through a die with heat applied to cure the resin. (iv) *Casting*: e.g. phenolic resins and polymethylmethacrylate; by laying up glass or other fibres and applying epoxy or polyester resins (often in vacuum) reinforced rods of great strength can be produced.

5.10.3 Tubes and cylinders

Tubes and cylinders can be produced by several methods. Materials capable of being moulded can be treated by compression or casting techniques. Many plastics can be extruded. Vulcanised fibre tubes are wound from paper treated with zinc chloride solutions. Laminated materials are generally wound on special machines with heated rollers while tension and pressure are applied to consolidate the layers. In this way s.r.b. paper and fabric tubes and cylinders can be produced by rolling the treated material convolutely on heated mandrels. Some of the smaller tubes are moulded in a split mould while still on the mandrel. Cylinders (considered as tubes with internal diameters above 76 mm) are made by similar techniques. These tubes which are generally for use in large transformers may have diameters as high as 2 m and lengths of 4 m. S.R.B. cotton fabric tubes and cylinders are produced by similar methods; s.r.b. glass-fibre tubes and cylinders can be rolled from most of the rigid-board materials.

Pressboard tubes and cylinders (mainly for oil-immersed transformers) are produced by winding presspaper on mandrels under tension, applying adhesives resistant to transformer oil (gum arabic, casein, phenolic resins, etc.). Tubes previously rolled from shellac bonded micanite for high voltage use are now widely superseded by tubes rolled from mica paper bonded with epoxy or silicone resins.

In addition to the convolutely wound tubes, a wide range of products can be produced by helical winding of strips (papers, fabrics, film, etc.), adhesives being applied meanwhile. The edges of the strip are generally butted together and this technique makes it possible to produce tubes the length of which is limited only by the requirements of transport.

Yet another method of winding tubes, chiefly with glass fibres, is known as filament winding. Strands of resin treated glass fibre are applied to mandrels in special winding machines. Restrictions on shape are less stringent, and very good mechanical properties are obtained.

5.10.4 Flexible sheets, strips and tapes

In very many applications a certain amount of flexibility is necessary, mainly to enable the materials to be readily applied to conductors, coils and various shapes, often irregular. The following are the principal flexible insulating materials, used for such purposes as wrappings on conductors and connections of machines and transformers, busbars and other parts; interlayer and connection insulation of coils; and slot linings of armatures.

5.10.4.1 Micanite

Tapes and sheets of micanite (micafolium) are widely used for high-voltage and high-temperature machine windings. It consists of mica splittings bonded with gum, bitumen or synthetic adhesive, often backed with thin paper or fabric (especially glass-fibre) to assist taping and to give mechanical support for micanite slot liners and coil insulation. Synthetic bonds, especially epoxies and silicones, are used for the higher temperature applications, i.e. for Classes B, F and H. Similar sheet and tape materials are made from mica-paper produced by a paper-making process using minute particles of mica, the sheet being treated subsequently with shellac or synthetic resins; these mica-paper products are like micanite but are more adaptable and uniform.

5.10.4.2 Vulcanised fibre and presspaper

These are useful flexible materials for coils, slot liners and many uses in machines, transformers and other apparatus.

5.10.4.3 Papers

Chiefly made from wood-pulp, cotton or manila fibres, these are employed in capacitor and cable manufacture, as insulations in coils and in the manufacture of s.r.b.p. boards, tubes and bushings, also as backing material for flexible micanite products. Asbestos paper has uses for high temperature conditions.

Papers are produced from other bases, such as ceramic, silica and glass fibre. A polyamide paper will not melt nor support combustion, and has good electrical and mechanical properties; as with most cellulose papers it can be supplied in creped form to facilitate the taping of irregular forms, and it can be obtained in combination with mica platelets to give greater resistance to h.v. discharges. Papers made from polyester fibres are used normally in combination with materials such as polyester film.

Properties of some flexible sheets are given in *Table 5.22*.

5.10.4.4 Fabrics

Cotton, nylon and polyethylene terephthalate cloths are used mostly as bases for varnished fabrics for small coils where flexible dielectrics 0.1–0.25 mm thick are wanted. Woven asbestos and glass fibre cloths are applied where temperatures are too high for organic textiles. The introduction of p.e.t.p. fabrics with heat-resisting (e.g. alkyd and polyurethane) varnishes has provided a range for class E or even class B insulation. Varnished glass fibre cloths and tapes—for which special coating materials have been formulated, including silicone resins and elastomers—are able to meet most of the high temperature requirements of Classes B, F and H apparatus. Fabrics of polyamide fibre can be coated with high-temperature resins and elastomers to give a material suitable for use at temperatures above 200 °C.

5.10.4.5 Tapes

Pressure-sensitive adhesive tapes are used in the construction of all types of equipment and are invaluable for holding and positioning conductors, preventing relative movement, identification of parts, exclusion of moisture, etc. Many types of such material are available based on most of the papers, fabrics, films and metal foils together with adhesives which can be thermoplastic or can be made to cure on the application of heat. Extra care has to be taken to reduce corrosiveness since these tapes are often used in fine-wire coils where electrolysis can cause erosion of a conductor.

Table 5.23 gives some typical properties.

5.10.4.6 Films

Many plastic materials are available in film form and many are used for the production of composite materials. Films with

Table 5.19 Properties of synthetic resin bonded glass fabric laminates
EP, Epoxy resins; MF, Melamine-formaldehyde resins; PF, Phenol-formaldehyde resins; PR, Polyester resins; SIL, silicone resins.
Test methods of BS 2782 and 3953.

Property		EP1	EP2	MF1	PR1	PR2	SIL1	SIL2	SIL3
Water absorption‡		10	20	118	20	75	9	11	47
Tensile strength	MN/m²	173	207	103	138	173	84	117	110
Crossbreak strength	MN/m²	240	310	103	207	138	90	117	90
Impact strength†	Nm	2.8	4.8	4.1	4.1	5.5	2.8	4.1	5.5
Ins. resistance (wet)	MΩ	100	100	1	10	—	1000	100	10
Relative permittivity at 1 MHz	—	5.5	5.5	7.5	4.5	4.9	4.0	4.3	4.8
Loss tangent at 1 MHz	—	0.035	0.035	0.025	0.04	0.05	0.003	0.004	0.01
Electric flatwise	kV/mm	6.3	6.3	2.8	8.3	7.1	—	—	—
Strength* edgewise	kV/mm	30	30	15	35	30	30	25	20

* R.M.S. for 25 mm length at 90 °C
† Per 12.7 mm width 3 mm thick

Table 5.20 Properties of typical high-temperature laminates
A/AP, acrylic + asbestos paper; S/AP, silicone + asbestos paper; P/GF, polyimide + glass fabric; PI/GF, polyamide-imide + glass fabric; PEP/GF, PEP/AF, polyaralkye-ether/phenol + glass or asbestos fabric

Property		A/AP	S/AP	P/GF	PI/GF	PEP/GF	PEP/AF
Density	kg/m³	1700	1720	1600	1800	1770	1650
Maximum operating temperature	°C	180	>220	300	280	250	250
Tensile modulus (r)	GN/m²	—	—	19	24	37	14
Tensile strength (r)	MN/m²	150	120	360	380	435	130
(250°C)	MN/m²	—	—	275	—	300	—
Crossbreak strength (r)	MN/m²	280	190	440	450	690	190
(288°C)	MN/m²	—	—	296	340	—	—
Impact strength† (r)	Nm	0.6	—	11	—	14	2.5
Relative permittivity (r, 1 MHz)	—	—	—	3.6	—	4.8	—
Loss tangent (r, 1 MHz)	—	—	—	0.012	—	0.011	—
Electric strength (r)*	kV/mm	9–20	10.5	—	—	27–34	—

(r) = room temperature
* r.m.s.
† Per 12.7 mm width

Table 5.21 Properties of typical rigid sheets and boards

VF	BS 216, Grey Vulcanised Fibre
PB	BS 231, Absorbent Pressboard Grade II
SP	BS 1137, Synthetic Resin Bonded Paper Type I
SW	BS 2572, Synthetic Resin Bonded Wood
SF	BS 2966, Synthetic Resin Bonded Fabric Type 3B
BA	BS 3503, Self-extinguishing Bitumen-treated Asbestos-cement

Property		VF	PB	SP	SW	SF	BA
Density	kg/m³	1300	1000	1330	1320	1330	1580
Water absorption	%	20–50	150	0.3	0.8	0.7	0.03
Elastic modulus	GN/m²	5	low	10	17	6	5.5
Tensile strength	MN/m²	80	40	60	100*	75	10
Shear strength	MN/m²	75	—	36	40*	90	28
Cross-breaking strength	MN/m²	—	—	100	150*	140	28
Crushing strength	MN/m²	—	—	240	200	240	90
Compression at 70 MN/m²	%	10	30	1.3	—	3.2	20
Relative permittivity	—	—	2.5–5	3.2†	4.5	4.5	10	30
Loss tangent at 50 Hz	—	—	—	0.015‡	0.02	0.02	0.3	0.3
Electric strength (1-min) r.m.s.							
through laminae	kV/mm	1.4–4	8.8	8	4	1	1.5
along laminae (25 mm length)	kV/mm	1.2	2	2	1.4	0.6	0.8

* Varies considerably with relation of grain to lamination
† Dry: 4 when dried and oil-impregnated
‡ Dry: 0.04 when dried and oil-impregnated

Table 5.18 Properties of rigid mica-paper sheets and tubes

Type of mica and resin		P Shellac	M Epoxy	M Epoxy	M Silicone
Sheets					
Density	kg/m³	2270	2300	2200	2000
Bond content	%	5	13	20	9
Maximum operating temperature	°C	130	180	180	600–700
Water absorption	%	14	0.7	0.05	0.7
Tensile strength (15 °C)	MN/m²	125	275	310	158
Flexural strength (15 °C)	MN/m²	—	440	380	100
Coefficient of expansion × 10⁶:					
normal to sheet	per °C	60	60	60	60
plane of sheet	per °C	10–12	10–12	10–12	10–12
Thermal conductivity	W/(mK)	0.2–0.3	0.2–0.3	0.2–0.3	0.2–0.3
Electric strength (15 °C)					
normal to laminae r.m.s.	kV/mm	30	32	32	32

Type of mica and resin		P Silicone	M or P Epoxy
Tubes			
Density	kg/m³	1800	1600
Maximum operating temperature	°C	350	200
Water absorption	mg/cm²	1.5	0.8
Cohesion between layers, wall 1.6 mm	N	44	267
3.2 mm	N	133	890
Electric strength, r.m.s.:			
normal to laminae in oil (90 °C), wall 1.6 mm	kV/mm	12	20
3.2 mm	kV/mm	10	16
in air (15 °C), wall 1.6 mm	kV/mm	10	16
3.2 mm	kV/mm	10	16
along laminae in oil (90 °C), 25 mm length	kV	26	40

M = Muscovite mica; P = Phlogopite mica. Test methods of BS 2782 and 1314

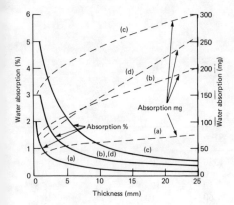

Figure 5.10 Water absorption of typical synthetic resin bonded paper and fabric boards. (a) S.R.B.P. Type I; (b) S.R.B.P. Type II; (c) S.R.B.P. Type III; (d) S.R.B.P. Type IIIA

mechanical properties in certain planes. Such boards bonded with epoxy resins can have flexural strengths as high as 1300 MN/m² and flexural moduli approaching 5 GN/m².

In addition to the resins mentioned, new high-temperature materials are now available and thoroughly suitable for use with all forms of glass, asbestos and other reinforcement. Some of these materials give thermal lives as good as the silicones but with mechanical properties more nearly equal to the epoxies. Typical materials in this class are resins based on acrylics, polyimides, polyamide–imides and combinations of polyaralkyl ether with phenols, etc. Also, resins are now available for all types of laminate especially phenolic, epoxy and polyester where, in order to decrease the fire risk, the resin has flame-retardant properties: see *Table 5.20*.

S.R.B. laminates of various types are made with a thin layer of copper (or other metals such as nickel, cupronickel, etc.) bonded to one or both surfaces. These materials are known as metal (or specifically copper-) clad laminates and are used for printed circuit board applications.

A comparatively new material in rigid form is produced by bonding high-density polyamide paper piles by heat and pressure to give a tough, strong board with a long life at temperatures up to 220 °C.

Other materials from the wide range of plastics are available as rigid boards or sheets although most of these materials are thermoplastic. Often such materials are moulded to produce finished products but it is possible to cut shapes and panels from sheets. Typical materials are based on polyvinyl chloride, acrylonitrile–butadiene–styrene, polyolefin, polymethylmetahacrylate, etc. Many are flame-retardant or self-extinguishing; and reinforcing materials such as glass and asbestos fibres, glass spheres and mineral fillers can be added to improve mechanical properties and resistance to high temperatures. The well-known material ebonite, a mineral-filled rubber-based material, was a forerunner of this class.

The properties of typical sheets and boards are given in *Table 5.21*.

Table 5.17 Properties of typical varnishes

Air-drying
A1: BS 634, Type 1; spirit shellac/methylated spirit
A2: BS 634, Type 2; oil and resin/petroleum spirit

Baking
B1: BS 2778, Grade F; oil/petroleum spirit
B2: BS 2778, Grade G; black bitumen/petroleum spirit
B3: BS 2778, Grade H; synthetic resin and oil/toluol

Air-drying and baking
AB; —; pigmented oil/white spirit

Silicone
SA: cured for 16 h at 250 °C
SB: cured for 16 h at 150 °C

Property		A1	A2	B1	B2	B3	AB
Density	kg/m^3	935	920	910	890	990	1400
Body content	%	38	62	54	44	63	73
Viscosity	c.g.s.	0.66	5.1	2.5	1.9	4.9	—
Drying time at 15 °C	h	0.5	6–8	—	—	—	4–6
105 °C	h	—	—	2	2.5	4	2
Electric strength at 90 °C, r.m.s.	kV/mm	16	60	44	73	64	27
Electric strength damp at 20 °C, r.m.s.	kV/mm	13	24	21	31	24	11

Property		SA		SB	
		dry	*wet**	*dry*	*wet**
Relative permittivity (25 °C)	100 Hz	3.0	3.0	3.0	3.0
	1 MHz	2.9	2.9	2.9	2.9
Loss tangent (25 °C)	100 Hz	0.0077	0.0069	0.0084	0.0085
	1 MHz	0.0039	0.0053	0.0043	0.0047
Electric strength r.m.s.	kV/mm	60	40	60	40

* After 24 h immersion in distilled water

must be removed and baking must then be carried out at temperatures in the range 150–260 °C.

Insulating varnishes are the subjects of BS 634 and 2778. Typical properties are listed in *Table 5.17*.

5.10 Solid dielectrics

The many solids that can be used for insulating purposes are considered in groups according to their form.

5.10.1 Rigid boards and sheets

Panels and simple machined parts Asbestos-cement boards (3–100 mm) are incombustible and arc-resistant, and are used as barriers and arc-chutes. *Glass-bonded mica*, in sheets 0.5 × 0.4 m of thickness 3–30 mm, are specially good for high-frequency, high-voltage and high-temperature (400 °C) application. *Micanite*, mica splittings bonded with shellac or synthetic resins, in thicknesses up to 25 mm or more; also *mica-paper* materials, comprising layers of mica-paper bonded under heat and pressure with a wide variety of resins, including high-temperature types: see *Table 5.18*.

Pressboard This is a paper product, widely used in oil-immersed transformers in thicknesses up to 10 mm. Thicker boards are built up by bonding piles together with adhesives such as phenolic resins or casein. In the USA, pressobards are available with special treatments which are claimed to permit operation at higher temperatures than untreated materials. In one case the material is treated with an amine; in another the cellulose molecule is modified chemically by cyanoethylation.

Laminates These are sheets of paper, fabric, etc., bonded with gums, shellac or synthetic resins under heat and pressure usually in hydraulic presses. Resins can be phenol–formaldehyde, melamine–formaldehyde, polyester, epoxy, silicone, polyimide, polyamide–imide, etc. These are some of the most useful insulating materials for panels, terminal boards, coil flanges, packings, cleats, slot wedges and many other uses. The principal varieties are: *Synthetic resin bonded paper* in thicknesses from 0.2–50 mm in several grades. *Synthetic resin bonded cotton* fabric (0.2–100 mm), the most common being bonded with phenolic resin, but an epoxy bonded type has been developed; both are tough and have good machinability.

Curves for water absorption for some types of s.r.b. paper and cotton fabric are given in *Figure 5.10*. *Synthetic resin bonded asbestos* paper, fabric and felt, used for low-voltage work at somewhat higher temperatures, e.g. 130–150 °C. *Synthetic resin bonded glass-fibre*, bonded under heat and pressure with synthetic resins of the melamine, epoxy, phenolic, polyester and silicone types in the thickness range up to 12.7 mm. Most of these materials have good electrical and mechanical properties and low water absorption: see *Table 5.19*. Similar materials are available where the reinforcement is a web of random laid glass fibres — known as mat. These materials are generally cheaper than those based on fabrics, although some of the properties are not as good. By arranging for a preponderance of the fibres to be in one direction, it is possible to produce boards with very good

and freedom from corona discharges. Good adhesion at the terminals is ensured by the nature of the resin system, but this also means that it sticks to the mould unless a release agent is used. Various preparations are used, including mixtures of high polymers, silicones, greases and waxes. The agent must be confined to the mould and not allowed to contaminate the components inside the casting.

The requirements of low shrinkage and low exotherm have been stated. The effect of the former is to damage components by compressional forces and is particularly severe on some nickel–iron alloys used in making inductors; thin-film resistors and capacitors are also vulnerable to this form of damage. Isolating the components from the resin by means of low modulus materials can appreciably reduce this defect. The heat of reaction (exotherm) is also damaging to organic materials and to semi-conductors; this often results in a loss of volatile matter causing shrinkage and the effects already noted above.

All adhesives contain polar bonds in their molecular structure. These give rise to changes in dielectric properties as the frequency and temperature varies, causing the electrical behaviour of the components inside the casting to change as the frequency and temperature changes.

Another disadvantage of potted circuits is that they are irreparable: they must be designed as 'throw away' sub-units and made at an economical price. For this reason they are rarely used in domestic applications such as radio or television, but are of particular use in military electronics, for machine control and similar purposes where the utmost reliability is essential and first costs are relatively unimportant.

5.9 Varnishes, enamels, paints and lacquers

Numerous liquid materials, which form solid films, are used extensively in the manufacture of insulating materials and for protecting windings, etc.

5.9.1 Air-drying varnishes, paints, etc.

One class of air-drying varnishes and lacquers consists of plain solutions of shellac, gums, cellulose derivatives or resins which dry (e.g. in 5–30 min) and deposit films by evaporation of the solvent. Other air-drying varnishes and paints form films which harden by evaporation of solvent accompanied by oxidation, polymersiation, or other chemical changes which harden and toughen the film, these processes taking several hours.

5.9.2 Baking varnishes and enamels

Where the toughest and most resistant coatings are required, baking varnishes and enamels are used, the evaporation of solvents and hardening of the material being effected by the appplication of heat. Typical varnishes of this class require baking at, say, 90–110 °C in a ventilated oven for 1–8 h. During the baking (or 'stoving') the hardening is usually caused by oxidation, but in some cases polymerisation takes place. The latter process does not require oxygen, and provided that the solvents are first removed, drying can take place within the interior of coils. In consequence these varnishes are preferred to the oxidising types which skin over and leave liquid varnish underneath. Such thermo-setting impregnating varnishes are used extensively for treating coils, and the windings of small machines and transformers.

5.9.3 Solventless varnishes

Thermo-setting synthetic resins are used for impregnating windings and, owing to the manner of hardening during which little or no volatile matter is evolved, spaces in the interior of windings can be filled completely with non-porous resin. One type consists mainly of oil-modified phenolic resins, similar to the oleo-synthetic resinous varnishes but without solvents; they are consequently termed *solventless varnishes*. When used for impregnating they are in a hot liquid condition, the material solidifying within the winding by baking after impregnation.

Specially formulated low-viscosity resins of the polyester or epoxy type are also used for impregnation. It is possible to use solventless resins in conveyorised plants where the parts are dipped in the resin, allowed to drain and then passed through a heated tunnel to cure the resin. In other cases, particularly for tightly wound apparatus, the parts are treated in the resin using a vacuum-pressure process to ensure that a high level of impregnation is attained.

The most recent development in treating industrial machine windings is the 'trickle' process. The wound part is heated and mounted at a slight angle so that it can be rotated slowly about its axis. A metered quantity of the resin is allowed to trickle on to the winding and under the action of gravity and the rotational forces, the resin penetrates to all parts of the winding and completes the impregnation. It is possible to completely fill the interstices of the winding without loss of resin by draining. Radiant heat may be applied to complete the cure while the parts are rotating, or heating currents may be circulated through the winding. Trickle impregnation can be performed automatically and the process can be completed in a few minutes without removing the wound parts from the production line.

5.9.4 Silicone varnishes

Varnishes based upon silicone resins are in general use. Some include other resins, etc., such as alkyd resins. Silicone varnishes are used for impregnating and coating cloths, tapes, cords, sleevings, papers, etc., made from glass fibres or asbestos, and polyethylene-terephthalate fibre; for bonding mica, glass-cloths, asbestos cloths and papers, e.g. for slot insulation; for coating and bonding glass or asbestos coverings on conductors; for 'enamelling' wires for windings; and for all manner of impregnating, bonding, coating and finishing purposes such as the treatment of windings for Classes B, F and H.

5.9.5 Properties of varnishes, etc.

The properties of the solid films formed after drying and hardening varnishes, paints, etc., naturally depend mainly on the principal basic materials, e.g. gums, resins and oxidising oils. Other materials added are: 'driers' to accelerate drying; 'plasticisers' to improve the flexibility; and pigments to provide the required colour and improve the hardness and filling capabilities.

Treatments of windings and insulation parts, by varnishes, enamels, lacquers or paints, take the form of (*a*) application of external coatings—chiefly for providing protection against moisture and oils, or (*b*) impregnation of windings and absorbent materials, for rendering them less susceptible to moisture and improving their electrical and heat-resisting properties. Both air-drying and baking materials are used for (*a*), but only baking varnishes and enamels are suitable for (*b*). In practically all cases, thorough drying prior to application of the varnish is essential as for compound treatment, but with varnishes, extra care is required to remove excess varnish by proper draining and to extract solvents thoroughly before hardening the material by baking. This is usually done in well-ventilated ovens at temperatures of from 80–150 °C and sometimes by the application of vacuum for a period to assist the removal of solvent. These processes are not required when solventless varnishes are used. In the case of silicone varnishes, the solvents

electrical properties maintained up to 200 °C and higher. Some silicones can work continuously at 200 °C and intermittently to 300 °C: they can be applied to insulation in classes F, H and C. Silicone resins are used for bonding mica, asbestos and glass-fibre textiles and for producing compounds, varnishes, micanite, wire coverings, etc.

A number of silicone compounds can be used for filling and sealing where heat and moisture resistance are required. One such, of the consistency of petroleum jelly, is of use as a waterproof seal in high-voltage ignition systems: it protects cable insulation from moisture, oxidation and electrical discharges.

Polyesters Alkyd resins are more rubbery than phenolic resins, have good adhesion and do not readily track; they are therefore of use for finishes and for varnishing glassfibre and similar material to produce heat-resisting varnished cloths. *Unsaturated polyesters* are useful for casting and potting, as solventless varnishes, and in the manufacture of laminates with glass fabric or mineral fillers.

Epoxies The epoxy resins have become important as casting, potting, laminating, adhesive and solventless varnish agents. They have good electrical properties and resistance to heat, moisture and tracking, and adhere well to metal parts. They have been applied for h.v. insulation in switchgear and for casting, in which case they are often mixed with mineral fillers. Earlier epoxy resins showed damage when subjected to severe weather and to high electric stress on creepage surfaces. New types have been based on cycloaliphatic resins which, because of the different molecular structure, produce less carbon during the passage of surface discharges and leakage currents under polluted conditions. Further improvements have been made in this application by using specially selected and treated mineral fillers which also reduce the effects of weathering and surface tracking. It is claimed that these newer products are suitable for use on high-voltage equipment out of doors.

Polyurethanes, isocyanates These are used mainly for coating fabrics (such as glass- and polyethylene-terephthalate fibre) to produce heat-resisting flexible sheet insulation and for coating wires.

Polyimides These have been specially developed for use at high temperature as mouldings, films, wire enamels and laminate bondings. The materials can be used continuously at temperatures in the 200–240 °C region; and for very short periods temperatures up to 500 °C can be withstood without apparent damage. Mechanical and electrical properties are good and resistance to most chemicals, solvents and nuclear radiation is excellent. *Polyamide-imide* resins are similar to the polyimides and are available in the same forms. The performance at high temperatures is marginally lower but the resins are simpler to use in the manufacture of laminates, and have longer shelf life.

Polyaralkyl ether/phenols These have a high-temperature performance not quite as good as that of the polyimides, but are cheaper. The resins can be used as bonds for glass and asbestos laminates and mineral filled moulding powders are available. A high proportion of room temperature strength is retained at temperatures up to 250–300 °C. Long-term operation at temperatures of 220–240 °C is possible. Resistance to most chemicals and solvents is excellent.

5.8.10 Encapsulation

When electronic or electrical components and circuits must resist the effects of climate, industrial atmospheres, shock or vibration they may be encapsulated. They are then generally known as 'potted circuits'.

Certain thermosetting synthetic resins are of the greatest use as they can be easily poured from low-viscosity liquids and made to set without the use of pressure and in some instances, with very little heat. A suitable material must (i) be a good insulator over a wide range of temperatures (volume resistivity say 10^6 Ω-m); (ii) polymerise or set without splitting off water or other products; (iii) have low viscosity at pouring temperature, low vapour pressure, and freedom from deleterious side effects on personnel who are using it; (iv) shrinkage must be small, especially when changing from the liquid to the solid state; (v) must adhere to all materials commonly found in electrical equipments, e.g. to brass, solder, steel, insulating boards, etc. Only the epoxy resins possess all the essential requirements for successful encapsulation and even these need an inorganic filler to obtain the best heat resistance, low shrinkage, good electrical properties and high thermal conductivity.

The epoxide resins of use in potted circuits are derived from a condensation reaction between epichlorhydrin and bisphenol A. They are cross-linked with aliphatic or aromatic amines, acid anhydrides and a few other chemical compounds to give thermally, electrically and mechanically stable resins. The addition of inorganic fillers improves them and reduces their shrinkage, tendency to crack at low temperatures and cost. The mixture of resin and cross-linking agent (known as a hardener) is called a system. An accelerator may be added, as well as various diluents, both reactive and non-reactive. Accelerators and promotors alter the speed of reaction and the pot life.

Typical potting formulations, in parts by weight, are:

A: resin 100, hardener 10, mica flour filler, 15.
B: resin 100, hardener 82, quartz flour filler 375, accelerator 1.

Formulation A is satisfactory for small units of about 0.1 kg of mixture. It sets at room temperature, but is post-cured at an elevated temperature dependent on the heat resistance of the included components; 18 h at 65 °C is usually satisfactory. Such a unit is suitable for small electronic packages containing small components, including semiconductors. Larger masses, depending on their geometry, may exhibit a strong exotherm (i.e. generate heat) which may damage the included components.

Formulation B is used hot (usually at about 65 °C) and is very fluid at this temperature besides possessing a long pot life. This makes it suitable for the potting of transformers. Vacuum impregnation is essential to eliminate voids, which would result in ionisation and corona discharges. The large amount of filler greatly improves the thermal and mechanical properties and allows a larger casting to be produced without a high exotherm. Again, a post-curing cycle is essential to bring out the best properties; in this case about 18 h at 120 °C is satisfactory and will produce a material with high volume resistivity and heat resistance. The pot life of this mixture is about two hours at 65 °C, which may be compared with formulation A whose pot life is less than half an hour at room temperature.

In use the resin and dried filler of A are mixed together and stored under vacuum or in a dessicator until used. The hardener is added and thoroughly mixed just before pouring into the mould or other article to be potted. It stands at room temperature to set or 'gel' and is subsequently post-cured; this latter process may be carried out after removal from the mould if required.

For formulation B the resin, hardener and dried filler are all mixed and stored as before, and the accelerator is added to the mixture just before use and heated to 65 °C. It is poured into the mould or other article. For transformers this is usually done under a vacuum of about 1 mm–1 cm Hg. Although a large amount of inorganic filler is used, this is filtered to some extent, by the windings of the transformer so that the insulation between turns is mostly of unfilled resin, giving high breakdown strength

Table 5.16 Properties of unfilled phenol-formaldehyde resins

Property		Varnish type	Casting type
Density	kg/m^3	1260	1300
Softening point	°C	60–80*	—
Plastic yield temperature	°C	120	85
Hardening time at 105 °C	min	45*	—
150 °C	min	6*	—
Linear expansion × 10^6	per °C	20	28
Water absorption	%	0.1	0.07
Elastic modulus	GN/m^2	5–7	3
Tensile strength	MN/m^2	35	28–56
Compressive strength	MN/m^2	170	100
Resistivity	Ω-m	10^{10}	10^9
Relative permittivity	—	4–7	7–11
Loss tangent:			
50 Hz	—	0.05	0.10
1 kHz	—	0.03	0.20
1 MHz	—	0.02	0.25
Electric strength			
(90 °C, 3 mm thick)	kV r.m.s./mm	8–20	12

* In the stage (1); other properties are for stage (3)

extruded. Highly resistant to moisture and chemicals, withstands temperatures up to 250 °C. Low-loss high-frequency application.

Polystyrene Softens at 70 °C. Can be compression or injection moulded and may be used with a mineral filler to improve heat resistance.

Polyvinylacetate and co-polymers Polyvinyl *acetates* are obtained from acetylene and acetic acid: they are used as adhesives and enamels.

Polyvinyl chlorides These are obtained from the combination of acetylene and hydrochloric acid as a white powder used with stabilisers, plasticisers, etc., to produce various rubber-like materials that can be extruded as tubes for wire protection. Sheet and moulded p.v.c. has a loss tangent too high for h.f. use. Co-polymers of the acetate and chloride forms are tough, rigid and water-resistant and can be injection moulded.

Acrylates The most important product is poly-methyl-methacrylate, a rigid glass-clear material with good optical, electrical and mechanical qualities. It can be obtained in sheet, rod and tube form, and as a moulding powder. Its low softening point (60 °C) limits its application to moderate temperatures.

Polyesters Some non-hardening *alkyds* have limited use as adhesives.

Polyethylene terephthalate This has a sharp melting point at about 260 °C and is formed into filaments for textile manufacture. Also extruded to form films. It has a high resistance to temperature and ageing and to water absorption. The textiles are suitable for class E insulation and, when suitably varnished, may withstand temperatures greater than class E.

Cellulose acetate and *triacetate* These are also esters. Produced as lacquers, textiles, sheet, rod, film and moulding powder, they are suitable for machine windings. The acetate softens at 60–80 °C, the triacetate at 300 °C. The materials are available as fibrous cotton or paper tapes.

Polyamides Super-polyamides are known as 'nylon': they produce monofilaments and yarns, with very good mechanical properties. The electrical properties are not outstanding, but nylon gives tough and flexible synthetic 'enamel' covering for wires. Films and mouldings can also be produced.

Polyacetal The material has good dimensional stability and is tough and rigid. Can be injection moulded and extruded, and is replacing metal parts in relays.

Polypropylene This material has a low density, dielectric loss and permittivity. Special stabilisers may be necessary when the material is extruded on to copper conductors.

4-Methylpentene-1 Similar to polyethylene and polypropylene and with similar resistance to chemicals and solvents. It is the lightest known thermoplastic. The high melting point (above 240 °C) cannot be fully exploited because of softening and oxidation. Permittivity and loss are low and remain fairly constant over a wide frequency and temperature range.

Polycarbonate Approaches thermo-setting materials in retention of stability up to 130 °C. Self-extinguishing and useful for structural parts, housings and containers for hand tools and domestic appliances.

Polyphenylene oxide Stiff and resistant to comparatively high temperatures. Permittivity and loss tangent fairly constant at frequencies up to 1 MHz.

Acrylonitrile butadiene styrene Has good dimensional stability and mechanical strength from −40 to 100 °C.

5.8.9 Thermo-setting synthetic resins

Phenol-formaldehyde These versatile resins are available in varnishes, adhesives, finishes, filling and impregnating compounds, laminated materials (boards, tubes, wrappings, sheets), moulding powders, and cast-resin products. The principal resins (Bakelite) are made by reacting phenolic material with formaldehyde. The final polymerising ('curing') time, which vitally affects the use, varies at 150 °C from a few seconds to an hour or more. The resins are normally solids of softening point between 60 and 100 °C. They are readily soluble in methylated spirit for coating papers, fabrics, etc., in the manufacture of laminates. Varieties are suitable for pouring molten into moulds followed by polymerisation. The most extensive use is as moulding powders with fillers (wood flour, powdered mica, fibres and colourings) to give mechanical strength and suitable electrical properties.

Phenol-furfural Produced by the reaction of phenol with furfural, an aldehyde obtained by acid treatment of bran and fibrous farm-waste. They are suitable for injection moulding.

Urea-formaldehyde The main use is as the binder of cellulose, wood flour or mineral powder in mouldings, made by compression at 115–160 °C. Mouldings can be delicately coloured.

Melamine-formaldehyde Properties superior to those of the urea-formaldehydes. Suitable for mouldings for ignition equipment. Good resistance to tracking.

Silicones These are organic compounds of silicon. By variation of the basic silicon–oxygen structure and of the attached organic groups, many different products can be made, including fluids, resins, elastomers and greases. Main properties are water-repellency, stability to heat, cold and oxidation, and good

Table 5.15 Properties of thermoplastic and casting resins

Thermoplastic resins		Poly-ethylene	Poly-styrene	Poly-methyl-methacrylate	Poly-amide (Nylon 6.6)	Polyacetal	Polypropylene	Poly-carbonate	Poly-phenylene-oxide	4-methyl-pentene-1	Acrylo-nitrile-butadiene-styrene
Density	kg/m³	920	1050	1190	1140	1420	920	1200	1200	830	900–1000
Softening temperature	°C	95	70–95	80–85	180	158	145	135	100–150	178	85
Melting point	°C	110	—	—	260	163	164	230	—	240	—
Linear expansion × 10⁶	per °C	220	75	80	100	95	110	70	35	120	60–120
Water absorption	%	0	0.05	0.4	<8	0.2	0.03	0.35	0.07	0.01	0.1–1
Elastic modulus	GN/m²	1–2	3	3.3	3	2.6	1–1.4	2.4	2.5–8.4	1.4	0.7–2.8
Tensile strength	MN/m²	12	41	59	45	69	31–38	93	69–120	27	17–62
Flexural strength	MN/m²	low	77	100	87	96	—	—	90–138	—	27–84
Impact strength	kgf-m	5–30	0.05	0.05	0.15–0.3	0.32	0.08–0.8	—	0.2–0.3	0.05–0.1	0.4–1.7
Resistivity (20°C)	Ω-m	3 × 10¹⁵	10¹⁵–10¹⁶	>10¹³	10¹³	5 × 10¹²	>10¹⁴	2 × 10¹⁴	10¹⁵	>10¹⁴	10¹¹–10¹⁴
Relative permittivity (20°C)	—	2.3	2.5–2.7	2.8	3.5–6	3.7	2.23	3.1	2.65	2.12	2.7–4
Loss tangent (20°C):	50 Hz	0.0001	0.0002	0.06	0.015	0.004	0.0002	0.0009	0.0005	0.0001	0.004–0.07
	1 kHz	0.0001	0.0002	0.03	0.020	—	0.0002	—	—	0.00005	—
	1 MHz	0.0001	0.0002	0.02	0.02–0.06	0.004	0.0005	0.01	0.001	0.0002	0.007–0.02
Electric strength*	kV/mm	15	20	10	15–19	20	30–32	16	20	28	12–15

Casting resins		Polyester		Epoxy (Bisphenol type)		Epoxy (Cycloaliphatic type)			
		Unfilled	Mineral filled	Unfilled	Mineral filled	Unfilled	Mineral filled:		
							Anti-track	Mech. strength	
Density	kg/m³	1100–1400	1600–1800	1100–1200	1600–2000	1100–1200	1700–1800	1700–1800	
Linear expansion × 10⁶	per °C	75–120	60–70	45–65	20–40	90–95	38–43	38–43	
Water absorption	%	0.15–0.6	0.1–0.5	0.08–0.15	0.04–0.1	0.04–0.05	0.02–0.04	0.02–0.04	
Elastic modulus	GN/m²	2–4	2.5–3	2–2.5	2.5–3	3.4–3.6	17–18	18–20	
Tensile strength	MN/m²	42–70	20–35	60–80	50–75	40–50	30–40	50–60	
Compressive strength	MN/m²	90–250	120–200	95–140	100–270	120	130–150	180–200	
Flexural strength	MN/m²	56–120	50–100	90–140	56–100	80–100	60–70	80–100	
Resistivity (20°C)	Ω-m	10¹¹	10¹⁰–10¹¹	10¹⁰–10¹⁵	10¹¹–10¹⁴	5 × 10¹⁴	6 × 10¹²	5 × 10¹³	
Relative permittivity (20°C):	1 kHz	3.2–4.3	3.8–4.5	3.5–4.5	3.2–4	3.5–3.6	4.6–4.7	4.3–4.5	
	1 MHz	2.8–4.2	3.6–4.1	3.34	3–3.8	—	—	—	
Loss tangent (20°C)	1 kHz	0.006–0.04	0.008–0.05	0.002–0.02	0.008–0.03	0.01	0.02	0.035–0.039	
	1 MHz	0.015–0.03	0.015–0.03	0.03–0.05	0.02–0.04	—	—	—	
Electric strength*	kV/mm	20	15–20	16–22	16–22	20–21	15–17	19–21	

* R.M.S. for 3 mm thickness.

used for sealing over the tops of primary batteries and accumulators. Compounds containing beeswax are useful impregnants for small coils not exposed to heat, e.g. on telephone apparatus; and sulphur, 'sealing wax' and 'Chatterton's compound' are examples of materials finding uses for miscellaneous applications where, say, the heads of screws in insulating panels and mouldings require to be sealed over.

5.8.6 Treatments using fusible materials

Treatments with bitumen, waxes, etc., usually consist of thorough vacuum drying of the coils, capacitors or other parts to be treated, followed by complete immersion in the compound while in a molten condition and at a temperature such that the viscosity is low enough to facilitate penetration; the molten compound generally being admitted to the impregnating vessel under vacuum. Pressure (up to 10 atm) is often applied during the immersion period to assist penetration. Such treatments enable spaces in windings to be filled thoroughly, and absorbent materials such as papers and fabrics are often well saturated with the impregnants, especially in the case of waxes. These treatments provide good resistance to moisture absorption and improve transference of heat from the interior of coils, also eliminating discharges in high-voltage windings and capacitors by the filling of air spaces.

5.8.7 Synthetic resins

An increasing number of the well-known synthetic resins, which form the basis of the principal 'plastics', are of great use to electrical engineers on account of their fusibility or softening characteristics at elevated temperatures, which enables them to be converted to desired shapes. The synthetic resins can be divided into two groups as follows:

Thermo-plastic	*Thermo-setting*
Polyethylene	Phenol-formaldehyde
Polystyrene	Phenol-furfural
Polyvinyl acetate	Urea-formaldehyde
Polyvinyl chloride	Melamine-formaldehyde
Acrylates	Silicones
Polyesters, alkyds, etc. (non-hardening)	Polyesters, alkyds, etc. (thermo-hardening)
Polyamides	Epoxy (epoxide)
Polyacetal	Polyurethanes
Polypropylene	Polyimide
Polycarbonate	Polyimide-amide
Polyphenylene oxide	Polyaralkyl ether/phenols
4-Methylpentene-1	
Acrylonitrile-butadiene-styrene	

Thermo-plastic synthetic resins In the case of most of the thermoplastic materials of this type, heating to temperatures within a certain range causes considerable softening and sometimes melting of the material to a viscous liquid. This enables them to be cast, formed, moulded or extruded into various required shapes by virtue of re-solidification on cooling again to normal temperatures. Some of the resins (e.g. acrylates and alkyd resins) have good adhesive properties and can therefore be used for bonding purposes either in the form of a solution or, more usually, by the application of heat. Layers of sheet materials, such as paper and fabric, can thus be bonded together into boards, simple mouldings, tubes, etc. The resins are often used alone, but more usually mixed with materials such as fillers and plasticisers, and in both varieties these synthetic materials are usually capable of being formed to all manner of shapes by the usual moulding processes (see *Table 5.15*).

Thermo-setting synthetic resins These enable useful compositions to be made, and withstand temperatures in excess of 100 °C. The most widely used are the phenol-formaldehyde type. The materials pass through three stages of physical condition:

(1) in which the resins are fusible at temperatures such as 80 °C, and are soluble in suitable solvents,
(2) results from heating the stage (1) resin until it becomes relatively infusible and insoluble,
(3) the infusible and insoluble state reached by continued heating after stage (2); no further change occurs and the materials are 'fully cured' or 'completely polymerised'.

These physical stages make thermo-setting resins suitable for three main uses:

(a) In spirit solutions; as ingredients in varnishes for impregnating purposes and the production of surface finishes; as enveloping, potting or encapsulating materials, and as ingredients in filling compounds.
(b) As adhesives for bonding layers of wood, paper, fabrics, etc., together to form laminated sheets, tubes, wrappings, and other simple shapes.
(c) As the basic material in moulding compositions for use in making articles by compression or injection moulding, extrusion or casting.

Properties of typical thermo-setting resins of the phenol-formaldehyde type, unfilled, are given in *Table 5.16*.

Many of the thermo-setting resins, e.g. phenol-formaldehyde and melamine-formaldehyde, require heavy pressure during the heating and hardening processes (b) and (c) above. Several new resins requiring little or no pressure (polyesters, epoxies and polyurethanes) have been developed as 'low-pressure', 'contact' or 'casting' resins, or as 'solventless varnishes'. These resins are initially in a low-viscosity liquid state, to which a 'hardener' or catalyst (e.g. a peroxide) is added. In some cases polymerisation sets in at normal room temperatures, or at temperatures of only 80–100 °C, the hardening process taking place more rapidly as the temperature is increased. Thus the resins can be readily cast to required shapes in 'moulds', and can also be used for impregnating and coating windings as they readily fill interstices and do not leave voids on hardening owing to the fact that no volatile constituents evaporate—hence their use as 'solventless varnishes'. Mixed with suitable fillers (e.g. glass-fibres, asbestos or other minerals) or applied to fabrics, papers and other sheet materials (usually of glass fibres), they are used extensively for producing castings, mouldings and laminates of varying degrees of mechanical and electrical strength, sometimes in very large pieces which could not readily be made by normal moulding methods; they are usually referred to as 'reinforced plastics'.

A brief description of the principal synthetic resins in electrical use is given below. Some are suitable for moulding with or without fillers, some for the preparation of laminated materials. Rod, sheet and tube forms are available in certain cases.

5.8.8 Thermo-plastic synthetic resins

Polyethylene Waxy, translucent, tough and flexible, with a sharp melting point at about 110 °C. For high-voltage and high-frequency application. Readily injection-moulded, extruded as wire coverings, and in sheet, rod and film form.

Polytetrafluorethylene A white powder that can be moulded or

5.8.1 Bitumens

Highly refined bitumens, which are usually steam-distilled, and of numerous grades, varying from semi-liquids to hard bitumens of melting-point over 120 °C, are used extensively for filling cable boxes, transformers and switchgear. These have high electric strength and are very inert and stable. As the coefficient of expansion is high, care has to be taken in filling large spaces to prevent voids and cracks on cooling. Some of the bitumens, especially those of high melting point, are rather brittle; all are soluble in oil, but they have excellent resistance to moisture. BS 1858 deals with bitumen-base filling compounds for electrical purposes. Properties of typical bituminous compounds are given in *Table 5.13*. Some bitumens are used as ingredients in varnishes and paints, rendering these very resistant to moisture and chemical attack. A few impregnating compounds contain bitumens, especially those used for treating high-voltage machine bars and coils.

5.8.2 Mineral waxes and blends

Various mineral waxes such as paraffin, ceresine, montan and ozokerite—including micro-crystalline waxes—also blends and gels of these, having melting points ranging from 35–130 °C, are used for impregnating capacitors, radio coils and transformers, also for other purposes such as cable manufacture. Properties of mineral waxes are given in *Table 5.14*.

5.8.3 Synthetic waxes

A few synthetic waxes—principally chlorinated naphthalene—with melting points up to 130 °C have certain advantages over natural waxes, particularly higher permittivity which enables smaller paper-insulated capacitors to be made. Properties of a typical synthetic wax of this type are given in *Table 5.14*.

5.8.4 Natural resins or gums

These materials, which may be classified broadly as shellac, rosin (colophony), copals and gum-arabic, are used principally as ingredients in varnishes or liquid adhesives. In some cases they are used direct, e.g. as powders, for a bonding medium between layers of mica which are hot pressed, but they are usually dissolved in spirit solvents, e.g. methylated spirits (or water in the case of gum-arabic) and applied as a solution to mica, paper, etc. for subsequent laminating and hot rolling, pressing or moulding (see *Table 5.14*).

5.8.5 Miscellaneous fusible compounds

Numerous compounds of bituminous and other types are used for all manner of cavity-filling purposes, some (mainly rosin) are oil resisting and are employed where bituminous compounds cannot be used owing to the presence of oil, e.g. for bushings of oil-circuit breakers and oil-immersed transformers. Others are

Table 5.13 Properties of fusible bituminous compounds

Class (BS 1858)		I	II	III	IV	V
Density	kg/m^3	960	1030	1040	1050	1027
Softening point (R & B)	°C	—	55–60	85	118	143
Pouring temperature	°C	—	175	171	193	204
Flash point	°C	>200	260	308	260	312
Viscosity at 100 °C	Rdwd.-s	750	—	—	—	—
Solubility in CS$_2$	%	99.5	99.8	99.5	>99	99.8
Acidity	g KOH/kg	1–2	2	1.5–4	4	4
Cubic expansion	per °C	0.000 65 for all classes				
Electric strength at 60 °C	kV r.m.s.	15–25	30–40	25	25–30	25–40

Table 5.14 Properties of fusible waxes, resins and gums

Property		Natural shellac	Natural copal gum (kauri)	Non-bituminous filling compound	Paraffin wax	Hydro-carbon wax	Synthetic chloro-naphthalene wax
Density	kg/m^3	1000–1100	1040	1100	900	800–1000	1550
Softening point	°C	50–70	60–90	70	45–50	—	90
Melting point	°C	80–120	120–180	80	50–60	40–130	123
Flash point	°C	—	—	230	200	275	—
Mineral ash	%	0.5–1	3	5	0	0	—
Acid value	g KOH/kg	60–65	70–85	—	0	<0.1	—
Saponification value	—	200–225	80	—	0	<0.5	—
Iodine value	%	9	90	—	0	—	—
Relative permittivity (20 °C)	—	2.3–3.8	—	—	2.2	2–2.5	5
Resistivity (20 °C)	Ω-m	10^{14}	—	—	10^{13}–10^{17}	10^{14}–10^{16}	$>5 \times 10^{11}$
Electric strength* (20 °C)	kV/mm	16–23	14–18	>30†	12	>50‡	6
Resistance to mineral oils		fair	good	good	poor**	fair	—

* R.M.S. for 3 mm thickness, except for † 1.2 mm gap and ‡ 4 mm gap between 13 mm electrodes
** Dissolves

Table 5.12 Properties of typical oils and fluids

Property	Unit	M	S
Density at 15 °C	kg/m^3	880	970
Viscosity, 21 °C	cSt	35	21
60 °C	cSt	6	11
Boiling range	°C	170–200	>200
Evaporation loss at 110 °C	%	0.7	0
Flash point (closed)	°C	149	271
Pour point (max.)	°C	−31	−50
Sludge value	%	0.8	0
Acidity	gKOH/kg	0.01	—
Breakdown voltage, r.m.s.*	kV	45–70	40–60
Relative permittivity at 20 °C, 50 Hz	—	2.1	2.7
Loss tangent at 20 °C, 50 Hz	—	0.0002	0.0002
Coefficient of cubic expansion per °C	—	0.0008	0.001

M: mineral oil to BS 148
S: liquid methyl silicone
* In standard test cell

properties are similar to mineral oils, but the visocisity–temperature characteristics show much higher low-temperature viscosity than comparable polybutenes. However, the gas-absorbing characteristics are good. Electrical properties are generally similar to mineral oils but the permittivity is somewhat higher for dodecyl benzene.

Polychlorinated biphenyls (also called askarels) have been used as high permittivity (3–6) fire-resistant insulating liquids since the 1930s but their effect as an ecological poison has limited their use to sealed equipment in recent years and all use of these liquids is being discouraged.

Silicone fluids (poly-dimethyl siloxanes) have been used as alternative fire-resistant insulating liquids, but their fire resistance is inferior to the askarels. They are generally gas evolving and their arc products can cause problems. However, they are very stable and have good electrical properties and are used in transformers and capacitors.

Several liquids have been developed as alternative high permittivity insulating liquids to replace the askarels for capacitor dielectrics. One possible group of materials is the organic esters. They have good viscosity temperature characteristics and are less flammable than mineral oil and can be either gas producing or gas absorbing, depending on their composition. When carefully purified, their electric strengths are about 20 kV/mm and dissipation factors average 0.001 at 20 °C. Diesters have relatively high permittivities (4.3 for Di- 2 ethylhexylphthalate). Other liquids that may be suitable for electrical insulation are phosphate esters, halogenated hydrocarbons, fuoroesters and silicate esters. Castor oil is a good insulation material for d.c. stress with a permittivity of 4.7, but it has a high dissipation factor of 0.002 that makes it unsuitable for most a.c. applications.

5.8 Semi-fluid and fusible materials

A few semi-fluid or semi-plastic compounds, and various fusible materials which are solids at normal temperature and melt to liquids of low viscosity or soften considerably with heat, are used principally in the following ways:

(1) For filling small cavities and large spaces, e.g. in metal-clad switchgear, transformers, cable-boxes and capacitors.
(2) For impregnating absorbent materials and windings.
(3) As the bond in laminated materials.
(4) As the basic material in moulding compounds.
(5) For external coverings of parts and apparatus (i.e. envelopment and encapsulation).

The materials most commonly used for these purposes are bitumen, natural waxes, shellac, synthetic waxes and synthetic resins; with the exception of many of the latter, and shellac, these are all *thermo-plastic* materials, i.e. they soften and melt on heating and solidify again on cooling without any substantial chemical change, and they can be re-softened or re-melted. In the case of some of the synthetic resins, especially those of the phenolic type, gradual hardening takes place as they are heated, and the melting point rises, so that, after being melted, solidification takes place on further heating, the material then becoming infusible: i.e. the process of melting and solidification on cooling is not repeatable; they are therefore known as *thermo-setting* materials. Shellac also has thermo-setting properties, but it requires longer heating to effect marked rise of melting point than in the case of many synthetic resins.

The properties of chief importance in such materials are: mechanical strength; electric strength; freedom from impurities; softening and melting temperatures; viscosity at pouring or impregnating temperatures; coefficient of expansion, and chemical effects on other materials.

Several materials which are *semi-fluid* at normal temperatures are used for filling and sealing purposes. For example: good grades of petroleum jelly of the Vaseline type are preferred to oil for filling apparatus and components where a liquid is undesirable, or where molten compounds cannot readily be poured or may affect other materials which are present (e.g. rubber and thermo-plastic materials).

Various '*semi-plastic*' compounds or cements, of a putty-like consistency, are also used for plugging and filling purposes, where semi-fluid and molten compounds cannot readily be applied. Some of these are almost permanently plastic and are therefore preferred where, for example, a certain amount of flexibility is required (e.g. where leads of coils may be moved slightly in assembly or service). Others may harden gradually in course of time (as in the case of ordinary putty), or they may harden quickly by chemical action (e.g. litharge and glycerine cement) or by heating—the latter usually being necessary with synthetic-resin compounds.

particles of such solid impurities—particularly organic fibres—is present, the breakdown voltage being reduced considerably by even small quantities.

Typical values of electric strength in different conditions are given in *Table 5.11*. Water and other impurities can be removed from oil by means of filter presses, centrifuges or (where high voltages are concerned) the application of vacuum. In addition to removing moisture, the vacuum will remove dissolved gases, but it is necessary to heat the oil and to spread it out over a very large surface area to facilitate the process. Once oil has been treated in this way, it must be stored out of contact with air and for preference at a temperature higher than the ambient.

Table 5.11 Effect of contaminants on electric strength of mineral insulating oil
Breakdown voltage, kV r.m.s. (between 13 mm diameter spheres, 4 mm gap)

Contaminant	(g/m³)	Water present, parts/10 000			
		0.2	1.0	2.5	5
Clean oil	0	86	80	80	80
Cotton	0.02	68	36	30	28
	0.28	33	11	10	10
Pressboard	0.08	82	64	56	54
fibres	0.37	56	30	29	28
	1.4	26	12	11	11
Carbon	1.9	83	80	80	79
	35.0	73	70	70	69

Viscosity This property, particularly at low temperature, is of great importance in oils used primarily for cooling in transformers and rheostats, it being necessary for the viscosity to be sufficiently low to ensure the necessary convection at the operating temperatures. This property is usually determined by methods such as those described in BS 188, and the viscosity is expressed in centistokes. Oil to BS 148 has a maximum viscosity (kinematic) of 37 centistokes at 21.1 °C (70 °F). This is approximately equivalent to 151 s at 21.1 °C and 200 s at 15.5 °C (60 °F) obtained with a Redwood No. 1 viscometer. *Figure 5.9* gives typical viscosity/temperature characteristics.

Flash point For standard oil this is not less than 146 °C, and may be as high as 240 °C for rheostats: these values refer to a *closed* flash-point tester.

Thermal properties The specific heat is about 1900 J/(kg K) at 15 °C and 2200 at 80 °C. The thermal conductivity is of the order of 0.15 W/(m K).

Chemical stability Insulating oils should be stable and not liable to deteriorate other materials or cause corrosion. The *acidity* is therefore closely controlled and oils are tested to ensure that they do not cause discoloration of copper. The worst feature of oils in this connection is the formation of sludge. This is mainly due to the oxidation of unsaturated hydrocarbons, particularly at high temperatures, and is accelerated by exposure of the oil to air and light, and (due to catalytic action) to copper. BS 148 includes tests for acidity, discoloration of copper, tendency to sludge formation and development of acidity.

Useful guidance on means of maintaining insulating oils in service is given in the British Standard Code of Practice, CP 1009 (1959), for the Maintenance of Insulating Oil with Special Reference to Transformers and Switchgear. This refers to oil supplied to BS 148 and describes the nature of deterioration or contamination likely to occur in storage, or in the course of handling or in service. It also gives recommendations for routine methods of sampling and testing to enable the suitability of oil for further service to be determined.

The properties of a typical mineral oil, complying with BS 148, are shown in *Table 5.12*.

5.7.2 Inhibited transformer oil

Oils operating at comparatively high temperatures, in the presence of oxygen and various catalytic materials, develop sludges and high acidity. These effects can be alleviated by adding various inhibiting substances to the oil, the most widely known being di-tertiary-butyl-paracresol used in quantities generally less than about 0.5% of the oil by weight. Materials of this type delay the point at which sludge and acid formation begin; but once the inhibitor has been used up, deterioration will proceed at the same rate as if no inhibitor had been used. For large power transformers, it has not been found necessary to use these inhibiting substances because of improvements in the construction which have reduced access of oxygen by conservators, hermetic sealing or the use of a nitrogen blanket above the oil surface. Another improvement has been the covering of copper surfaces so reducing the catalytic effect. For transformers operating under more adverse conditions of temperature such as distribution or pole-mounted units, a better case can be made for using inhibited oils.

5.7.3 Synthetic insulating liquids

Synthetic hydrocarbons are fairly widely used for power-cable insulation and as capacitor dielectrics. These are more expensive than petroleum oils but they generally have better electrical properties because of their lower contamination levels and they can have better gas-absorbing properties. A commonly used synthetic oil is poly-iso-butylene, commonly known as polybutene. Different polymer chain lengths can be produced giving a wide range of viscosities from low viscosity liquids to sticky semi-solids. The high molecular weight tacky and rubbery materials can be used mixed with oil, resin, bitumens, polyethylene and inorganic fillers to produce non-draining and potting compounds. Another synthetic oil that is used for cable insulation is dodecyl benzene, an aromatic. The physical

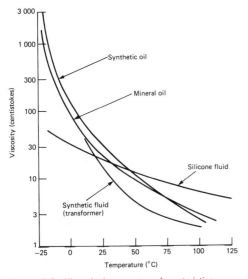

Figure 5.9 Viscosity/temperature characteristics

Table 5.10 Breakdown voltage of nitrogen and sulphur hexafluoride (kV peak)
Direct voltage and uniform field

Gas	Pressure (atm)	Positive polarity gap (mm)					Negative polarity gap (mm)				
		5	10	15	20	25	5	10	15	20	25
Nitrogen	1	—	30	—	56	—	—	30	—	56	—
	2	30	55	—	100	—	30	54	—	100	—
	3	41	76	114	147	180	42	77	113	147	178
80% nitrogen	1	38	74	111	145	178	38	74	111	146	178
20% sulphur	2	72	143	212	—	—	72	142	210	—	—
hexafluoride	3	111	220	—	—	—	111	221	—	—	—
Sulphur	1	44	88	133	175	213	44	88	134	176	208
hexafluoride	2	85	171	252	—	—	84	171	251	—	—
	3	132	260	—	—	—	131	258	—	—	—

5.6.4 Hydrogen

This gas is used as a cooling medium in some large turbo-generators and synchronous motors; the main advantages are the efficient removal of heat and reductions in windage loss. Although there is a fire and explosion risk, troubles of this kind have been few during the 30 or more years that the gas has been used commercially for electrical machines. The electric strength of hydrogen at atmospheric pressure is about 65% that of air but most machines operate at pressures of 2–5 atm, and over this range the electric strength is higher than for air at atmospheric pressure. High-voltage discharges are thus not likely to be any more severe, and as discharges in hydrogen do not produce ozone or oxides of nitrogen, injurious effects are considered to be negligibly small.

5.6.5 Vacuum

Considerable investigation has been made into the utilisation of high vacua both for the insulation of equipment and as the interrupting medium in vacuum circuit breakers and contactors. The major advantage is due to the fact that very high electric field strengths can be achieved with a maximum operating pressure of 1 atm (negative), whereas with gases, very high operating pressures are generally essential and this complicates the mechanical design of the tank or other containing structure. Vast improvements in high-vacuum technology together with the need to solve problems in the construction of very large high-voltage equipment such as cyclotrons, Van de Graaf generators, etc., will obviously result in advances in this field.

5.7 Liquid dielectrics

The liquids which are most commonly used for electrical insulation are petroleum oils. For some applications these are being replaced by synthetic hydrocarbon oils, particularly as impregnant for oil-impregnated paper insulated power cables. Polychlorinated biphenyls (askarels) were widely used where non-flammable insulation was required for transformers, and for capacitor dielectrics. However, these have now been withdrawn for most applications because of environmental pollution effects and silicone oils are now used for non-flammable transformer insulation. Capacitors often use silicone liquids or synthetic hydrocarbons as dielectrics, but various esters are now being introduced that offer a higher permittivity and hence a higher capacitance value for the same dimensions. An insulating liquid that is sometimes used in the less developed countries is castor oil.

The principal uses of liquid dielectrics are:

(1) As a filling and cooling medium for transformers and some electronic equipment, and as a filling medium for capacitors, bushings, etc.
(2) As an insulating and arc-quenching medium in switchgear.
(3) As an impregnant of absorbent insulation, e.g. paper, porous polymers and pressboard. These are used in transformers, switchgear, capacitors and cables.

The important properties of the liquid used vary with the application, but they include electric strength, permittivity, chemical and thermal stability, gassing characteristics, fire resistance and viscosity.

5.7.1 Insulating oils

The insulating oils used extensively are highly refined hydrocarbon mineral oils obtained from selected crude petroleum, and have densities ranging from 860–890 kg/m^3 at 15°C. Oil for transformers and switchgear is dealt with in BS 148. A number of special mineral oils are employed for impregnated paper capacitors and cables and others—usually of higher viscosity and flash point—for rheostats and for filling busbar chambers in switchgear. Typical properties are given in *Tables 5.5* and *5.12*.

Electric strength This is a property involving similar phenomena to sparkover in gases. On raising the voltage between two electrodes in oil, electrical discharges may first appear in the space surrounding the electrodes—particularly at sharp corners—and at a higher voltage, sparks pass across the intervening space between the conductors: these are often intermittent ('pilot') sparks, and, on raising the voltage further, a continuous stream of sparks usually occurs and may develop into an arc, with complete breakdown of the oil.

The electric strength is generally tested with electrodes consisting of two metal spheres of about 13 mm diameter separated by a gap of 4 mm. For clean, dry oil the breakdown voltage should be in the region of 100 kV r.m.s. or more, but careful treatment, storage and handling are needed to maintain this level. For the oil to comply with BS 148, it should withstand for 1 min without breakdown 40 kV r.m.s. applied between the spherical electrodes under conditions laid down in the Specification.

The electric strength of insulating oil is strongly affected by impurities, especially water and particles of fibrous material. The latter are attracted to the testing gap by the electric field and readily align themselves across the shortest space. The presence of moisture in oils is shown by electric strength tests when

this is largely due to the absence of corona prior to flash-over if the spacing does not exceed the radius of the spheres. Sphere-gaps are suitable also for the measurement of impulse voltages. Voltages of about 2 kV and upwards can be measured reliably. BS 358 gives detailed information on the effects of humidity, air density (or barometric pressure), etc. The effect of density is pronounced in the case of equipment used at high levels above 1000 m, in aircraft where altitudes up to 15 km may be met, or in spacecraft where outer space is an almost perfect vacuum.

Needle gaps Humidity has here a strong influence on breakdown voltage where the electrode shape leads to field concentration. For this reason, as well as that of a degree of frequency-dependence, needle gaps are unreliable for high-voltage measurements. Rod gaps (e.g. 12 or 16 mm square-section rods with sharp corners) are used for chopping impulse voltages, but with these too a humidity correction is necessary.

Corona This term is used to describe the glow or 'brush' discharge around conductors when the air is stressed beyond the ionisation point without flashover developing, it is of more or less serious consequence according to the application concerned. It causes a certain amount of energy loss with alternating current, which may become appreciable on high-voltage transmission lines; it produces radio interference; it may initiate surface deterioration and breakdown on solid insulation surfaces; and it produces secondary chemical effects.

In thin films, particularly in spaces between layers of sheet insulation, air can readily become ionised due to the electric stress across such spaces exceeding the critical value. This is often due to the fact that, with dielectrics in series, the stress in each section is inversely proportional to its permittivity. When the critical stress in the air or gas is exceeded, discharges occur (often called corona, ionisation, glow or brush discharges) and this causes splitting up of the gas molecules. In air this leads to the formation of ozone and nitrogen oxides which in the presence of moisture produce nitric acid. The ozone has, of course, a strong oxidising effect, but the more serious chemical effects of ionisation are those due to the nitrogen products, as the nitric acid attacks most of the organic insulating materials and causes corrosion of metal parts. The action of either or both the ozone and the nitrogen oxides on many materials is to cause decomposition and often the formation of acids; for example, oxalic acid by the oxidation of cellulose materials, and acetic acid from the decomposition of cellulose acetate.

In addition to the chemical effects, discharges in spaces, films or cavities within dielectrics can have serious consequences mainly due to the high energy in some of the individual discharges. Mechanical, electrical and thermal damage can occur and breakdown in service may result after long periods. There has been a considerable growth in methods of detecting the presence of such discharges in various types of equipment especially where oil-impregnated paper dielectrics are used. Discharges within air or gas films in such material can cause severe damage often followed by complete breakdown.

Compressed air This is used as the arc-extinsuishing medium and dielectric insulation in air-blast circuit breakers.

5.6.2 Nitrogen

Instead of air, which is a mixture of approximately 21% oxygen and 79% nitrogen, nitrogen alone is sometimes used when there is a risk of oxidation of another material such as insulating oil. Nitrogen is often used in gas-filled high-voltage cables, as an inert medium to replace air in the space above the oil in some transformers, in low-loss capacitors for high-voltage testing, etc. There is no appreciable difference between the electric strength of nitrogen and that of air. Some results relating to the electric strength of nitrogen for uniform fields at pressures above atmospheric up to 20 atm are given in *Table 5.9*. Included are some similar results for carbon dioxide.

Table 5.9 Breakdown voltages of nitrogen and carbon dioxide (kV peak)

Gas	Gap (mm)	Pressure (atm (abs))				
		3	5	10	15	20
Nitrogen	1	10	15	27	35	45
	8	90	123	180	220	—
Carbon dioxide	1	13	17	27	38	52
	8	85	115	200	260	—

5.6.3 Sulphur hexafluoride

Sulphur hexafluoride is an electronegative gas which has come into wide use during recent years both as a dielectric (in X-ray equipment, in waveguides, coaxial cables, transformers, etc.) and as an arc-quenching medium in circuit-breakers. Its electric strength is of the order of 2.3 times that of air or nitrogen, and at a pressure of 3–4 atm it has an electric strength similar to that of transformer oil at atmospheric pressure. The gas sublimes at about $-64\,°C$ and it may be used at temperatures up to about $150\,°C$. Although the gas is considered to be non-toxic, non-flammable and chemically inert, under the influence of arcs or high-voltage discharges, there may be some decomposition with consequent attack on certain insulating materials and metals. In circuit-breakers this problem is overcome by careful selection of materials (e.g. polytetrafluoroethylene for interrupter nozzles) and by the use of filters and absorbents to remove the products of decomposition after circuit interruption. Some figures relating to the electric strength of sulphur hexafluoride and mixtures of this gas with nitrogen are given in *Table 5.10*.

Numerous other electronegative gases such as perfluoro-propane C_3F_8, octafluorocyclobutane C_4F_8 and perfluoro-butane C_4F_{10} have been developed, but few have found such widespread use as sulphur hexafluoride, The main interest for these gases is as dielectrics in transformers, waveguides, capacitors, etc., but one difficulty is that the temperature at which condensation occurs may not be sufficiently low for safety in outdoor equipment likely to remain un-energised for long periods. This problem can be overcome partly by fitting heaters or by using admixtures with a more volatile gas (such as nitrogen). Addition of nitrogen often improves some of the characteristics, while at the same time reducing the overall cost. Some of these gases can be used at temperatures well above $200\,°C$.

Most of the fluorinated gases have an electric strength between two and five times that of air or nitrogen under the same conditions but as with sulphur hexafluoride, care must be taken to prevent high-voltage discharges or arcs in the gas because of the dangers of producing decomposition products.

Some electric machines and special devices have to operate in a gas other than air—for example most refrigerator compressor motors operate in gaseous refrigerants mostly based on chloro-fluoro-hydrocarbons (such as 'Arcton', 'Freon', etc.). These materials can act as solvents for some of the components used in insulating materials with consequent failure of the equipment due to blocked tubes and valves in the refrigerator circuit. Careful selection of materials for resistance to these fluids is essential.

voltage gradient in a particular region exceeds the critical value for air, this happening readily at points of electric flux concentration, e.g. sharp edges of metal parts. If this local breakdown becomes unstable—as it will when the voltage between conductors is increased sufficiently—spark-over will occur. This may be an isolated spark from one conductor to the other, the intervening air then re-healing itself; if the voltage is maintained (or increased), the spark may be followed by a continuous stream of sparks.

Typical values of breakdown voltages for gaps of different forms and sizes of electrodes (at normal temperature and air pressure) are given in *Table 5.6*.

Table 5.6 Typical breakdown voltages in air at N.T.P. (kV peak at 50 Hz)

Two-electrode system	Spacing or gap (mm)				
	10	50	100	200	300
Spheres, diameter 1.0 m	—	137	266	503	709
0.25 m	31	137	243	363	—
Needle points	13	50	78	127	178
Parallel wires, diameter 8.25 mm	—	38	57	83	117
Concentric cylinders: outer/inner radius, mm	38/1.3	38/11	67/17	67/2	
Breakdown voltage, kV	26	55	88	103	

Partial or complete breakdown of air in gaps can be influenced by suspending sheets of material at particular places in the electric field. In some cases, the sheets may be of metal in others of insulating materials; even woven fabrics can have an effect. The effect is generally greater the more divergent the field. This solution can be useful where clearances are limited inside equipment, or particularly in high-voltage test areas.

For plane gaps, all gases exhibit a minimum breakdown voltage known as the Paschen minimum; this occurs at a given value of the product Pd of absolute gas pressure and gap length. For air this minimum occurs at $Pd \simeq 6$ (in torr-mm). At this point the voltage is approximately 330 V direct, or 250 V r.m.s. for alternating voltages of sine-wave form. Thus as the value of Pd is reduced from a higher region the breakdown voltage falls to the minimum value quoted, and further decrease in either P or d results in an increase in the voltage required to break down the gap. This explains why quite small gaps under conditions of high vacuum can sustain very high voltages. *Table 5.7* gives the values of Pd and the minimum voltages for several gases.

Sphere gaps As the electric strength of air is dependable, standard sphere gaps can be used as reliable and accurate means for measuring high voltages (*Table 5.8*), particularly where peak voltages are to be measured, as it is, of course, the *peak* value which determines the breakdown. Standard sizes of spheres are generally used as electrodes, as, provided the size is appropriate and proper precautions are taken (e.g. to avoid effects such as those due to the proximity of other objects and uncontrolled irradiation of the gap by other discharges), clean, smooth, metal spheres are most reliable as a means of determining high voltages;

Table 5.7 Minimum breakdown voltage for gases. 1 torr -1 mm Hg -1/760 atm

Gas	CO_2	Air	O_2	N_2	H_2	Ar	He	Ne
$P \times d$ (torr-mm)	5	6	7	7.5	12.5	15	25	30
Direct voltage (V)	420	330	450	275	295	265	150	244

Table 5.8 Sphere-gap breakdown voltages (kV peak)

Gap (mm)	Sphere diameter (m)							
	0.02	0.0625	0.125	0.25	0.5	0.75	1.0	1.5
0.5	2.8	—	—	—	—	—	—	—
1	4.7	—	—	—	—	—	—	—
1.5	6.4	—	—	—	—	—	—	—
2	8.0	—	—	—	—	—	—	—
4	14.4	14.2	—	—	—	—	—	—
5	17.4	17.2	16.8	—	—	—	—	—
6	20.4	20.2	19.9	—	—	—	—	—
8	25.8	26.2	26.0	—	—	—	—	—
10	30.7	31.9	31.7	31.7	—	—	—	—
15	(40)	45.5	45.5	45.5	—	—	—	—
20	—	58.5	59.0	59.0	—	—	—	—
30	—	79.5	85.0	86.0	86	86	86	—
40	—	(95)	108	112	112	112	112	—
50	—	(107)	129	137	138	138	138	138
100	—	—	(195)	244	263	265	266	266
150	—	—	—	(314)	373	387	390	390
200	—	—	—	(366)	460	492	510	510
300	—	—	—	—	(585)	665	710	745
400	—	—	—	—	(670)	(800)	875	955
500	—	—	—	—	—	(895)	1010	1130

In air at 20 °C, 1013 mb. One sphere earthed. For alternating voltages of either polarity; and for standard negative impulse voltages (50% breakdown value). Figures in brackets not reliable.

about 20 kV r.m.s. for 25 mm distance between two 38 mm diameter electrodes with fairly sharp edges.

Tracking Leakage along the surface of a solid insulating material, often a result of surface contamination and moisture or of discharges on or close to the surface, may result in carbonisation of organic materials and conduction along the carbonised path. This is known as 'tracking'. It is usually progressive, eventually linking one electrode to another and causing complete breakdown along the carbonised track. The method of test recommended in IEC Publication 112 is very similar to that given in BS 3781, although the methods of expressing the results are slightly different.

Permittivity This property is specific to a material under given conditions of temperature, frequency, moisture content, etc. When two or more dielectrics are in series and an electric stress is applied across them, the voltage gradient across each individual dielectric is inversely proportional to its permittivity. This is particularly important when air spaces exist in solid and liquid dielectrics, as the permittivities of these are always higher than that of air, hence the air is liable to have the higher stress and may fail and cause spark-over through the air space in consequence.

Values of permittivity for some insulating materials are given in *Table 5.5*.

Dielectric loss A capacitor with a perfect dielectric material between its electrodes and with a sinusoidal alternating voltage applied takes a pure capacitive current $I = \omega CV$ with a leading phase-angle of 90°. In a practical case, conduction and hysteresis effects are present, the phase-angle is less than 90° by a (normally) small angle δ. The power factor, no longer zero, is given by $\cos(90° - \delta) = \sin \delta \simeq \tan \delta$: the latter is called the *loss tangent*. The power loss is, to a close approximation, $P = V^2 \omega C \tan \delta$ where $\omega = 2\pi f$: it is proportional to the square of the voltage and to the product $\varepsilon \tan \delta$, because the absolute permittivity ε determines the capacitance of a system of given dimensions and configuration.

The loss tangent varies, sometimes considerably, with frequency, also with temperature, values of $\tan \delta$ usually increasing with rise of temperature, particularly when moisture is present, in which case the permittivity also rises with the temperature, so that total dielectric losses are often liable to a considerable increase as the temperature rises. This is very often the basic cause of electric breakdown in insulation under a.c. stress, especially if it is thick, as the losses cause internal temperature rise with consequent increase in the dielectric loss, this becoming cumulative and resulting in thermal instability and, finally, breakdown.

Permittivity and loss tangent are usually determined by means of a Schering bridge (BS 903). For power devices such as cables and bushings, the test is made at 50 Hz; but for high-frequency equipments it is necessary to determine loss tangent and permittivity at much higher frequencies. BS 2067 covers such measurements by the Hartshorn and Ward method at frequencies between 10 kHz and 100 MHz. Other methods are available for other frequencies (see IEC Publication 250). Typical values of loss tangent and permittivity for some of the principal insulating materials used for high voltages and for high frequencies are given in *Table 5.5*.

5.5.4 Chemical properties

The chemical and related properties of insulating materials of importance may be grouped as follows:

(1) Resistance to external chemical effects.
(2) Effects on other materials.
(3) Chemical changes of the insulating material itself.

Under (1) there are such properties as resistance to:

(a) The effect of oil on materials liable to be used in oil (in transformers and switchgear), or to be splashed with lubricating oil.
(b) Effects of solvents used with varnishes employed for impregnating, bonding and finishing.
(c) Attack by acids and alkalis, e.g. nitric acid resulting from electrical discharge, acid and alkali vapours and sprays in chemical works, and deposits of salts from sea spray.
(d) Oxidation, hydrolysis and other influences of atmospheric conditions, especially under damp conditions and in direct sunlight.
(e) Effects of irradiation by high-energy nuclear radiation sources, e.g. neutrons, beta particles and gamma rays.

In group (2), typical effects of the insulating materials on other substances with which they may be used are:

(i) Direct solvent action, e.g. of oils and of spirits contained in varnishes, on bitumen and rubber; corrosion of metals in contact with the insulation; and attack on other materials by acids and alkalis contained in the insulating materials in a free state.
(ii) Effects of impurities contained in the insulation.
(iii) Effects resulting from changes in the material, for example acids and other products of decomposition and oxidation affecting adjacent materials.

These effects are generally referred to under the heading 'Compatability'. If meaningful test results are to be obtained, all components of an insulation system must be present and they must have been treated in the same way as will be used in manufacture.

Group (3) includes such features as:

(1) Oxidation resulting from driers included in varnishes.
(2) Deterioration due to acidity (e.g. in oils, papers and cotton products).
(3) Chemical instability of synthetic resins.
(4) Self-polymerisation of synthetic compounds.
(5) Vulcanisation of rubber-sulphur mixtures.

Most of these chemical properties are determined by well-known methods of chemical analysis and test. The principal tests are for acidity and alkalinity, pH value, chloride content in vulcanised fibre, and conductivity of aqueous extract (for presence of electrolytes). Some of these are dealt with in BS 2689, 2782 and 3266.

Increasing attention is being paid to chemical features of the raw materials and processes used in the manufacture of insulating materials—particularly varnishes, synthetic resins and all manner of plastics—and much research work is being carried out on these features and on the correlation of chemical structure of dielectrics with their physical, electrical and mechanical properties.

5.6 Gaseous dielectrics

5.6.1 Air

Air is the most important gas used for insulating purposes, having the unique feature of being universally and immediately available at no cost. The resistivity of air can be considered as infinite under normal conditions when there is no ionisation. There is, therefore, no measurable dielectric loss, negligible $\tan \delta$, and a relative permittivity (for all practical purposes) of unity. The electric strength at s.t.p. is 3 MVp/m for a uniform field. In a practical airgap the voltage gradient is a maximum at the electrode surfaces. The sparkover (breakdown) voltage of an airgap is therefore a non-linear function of its length.

Partial breakdown of air, locally, often occurs when the

Table 5.5 Representative properties of typical insulating materials

Volume resistivity $\rho = 10^n$ Ω-m; the value of n is tabulated. Relative permittivity ε_r. Loss tangent tan δ.

Insulant	n	ε_r	Tan δ 50 Hz	1 KHz	1 MHz
Vacuum	∞	1.0	0	0	0
Gases					
Air	∞	1.0006	0	0	0
Sulphur hexafluoride	∞	1.002	0	0	0
Liquids					
Mineral insulating oil	11–13	2–2.5	0.0002	0.0001	—
Dodecyl benzene	12–13	2.1–2.5	0.0002		
Organic esters	10–12	2.9–4.3	0.001		
Polybutenes	12–14	2.1–2.2	0.0005		
Silicone fluids	12	2.7	0.0001		
Solids					
Paraffin wax	14	2.2		0.0003	0.0001
Bitumen	12	2.6	0.008		
Pressboard	8	3.1	0.013		
Bitumen-asbestos	10–11				0.08
Paper: dry	10	1.9—2.9	0.005	0.007	
oil-impregnated	14–16	3.2–4.7	0.002		
Cloth: varnished cotton	13	5	0.2	0.15	
Ethyl cellulose	11	2.5–3.7	0.02	0.03	0.02
Cellulose acetate film	13	4–5.5	0.023	0.04	
Cellulose acetate moulding	10	4–6.5	0.016	0.03	0.06
Synthetic-resin (phenol) bonded paper	11–12	4–6	0.02	0.03	0.04
Mica	10	5.5–7	0.0005	0.0005	0.0005
Nylon	11	3.8			0.03
Phenol—formaldehyde	10	4–7	0.05	0.03	0.02
cast	9	7–11	0.1	0.2	0.25
Polystyrene	15	2.6	0.0002	0.0002	0.0002
Polyethylene	15	2.3	0.0001	0.0001	0.0001
Polypropylene	15	2.3	0.003		0.0003
Polytetrafluoroethylene	15	2.1	0.0002		0.0002
Methyl methacrylate	13	2.8	0.06	0.03	0.02
Synthetic—resin compounds					
Phenol formaldehyde mineral filled	10–12	5	0.015	0.015	0.01
Urea formaldehyde mineral filled	10	5–8	0.1	0.1	0.038
Polyvinyl chloride	11	5–7	0.1	0.1	

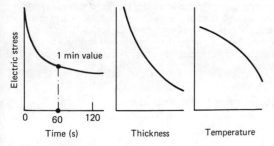

Figure 5.8 Effect of time, thickness and temperature

factors, and the data from comparative tests. *Figure 5.8* gives typical results of the variation of electric strength with thickness of specimen and with temperature.

Surface breakdown and flashover When a high-voltage stress is applied to conductors separated only by air where they are closest together, and the stress is increased, breakdown of the intermediate air will take place when a certain stress is attained, being accompanied by the passage of a spark from one conductor to the other, i.e. the electric strength of the air has been exceeded. This may also be followed by a continuous arc. The voltage at which this occurs is the sparkover or flashover value. Similar conditions obtain with oil as the insulant when a spark passes through the oil between the conductors.

In most electrical assemblies the live parts are separated by both solid insulation and the ambient air, and failure may take place either by breakdown of the solid material or by flashover through the air. Often the process involves surface leakage, deterioration and surface flashover. This phenomenon is generally due to the nature and design of the metal parts, as sharp edges of nuts and washers (for example) give local concentrations of stress. In addition, the onset of surface discharges at metal edges (which can initiate breakdown) is influenced by the permittivity of the dielectric material. The higher the permittivity, the lower the voltage at which flashover is likely to occur. Insulating materials are sometimes tested for surface breakdown or flashover between two electrodes on a typical surface but, unless the material itself or its surface is poor electrically, flashover in air takes place in preference, usually at values of

5.5.1 Physical properties

Density This is of importance for varnishes and oils. The density of solid insulants varies widely; in a few cases it is the measure of relative quality (as in pressboard).

Moisture absorption This usually causes serious depreciation of electrical properties, particularly in oils and fibrous materials. Swelling, warping, corrosion and other effects often result. Under severe conditions of humidity, such as occur in mines and in tropical climates, moisture sometimes causes serious deterioriation; products made from linseed oil varnishes, for example, are prone to complete destruction of the varnish film in a damp atmosphere. Fungus growth and electrolysis are other examples of effects due to moisture.

It is usual to determine the absorbency of solid materials by ascertaining the weight of water absorbed by a standard specimen when immersed for a specified period: however, the quantity of water absorbed is not a reliable criterion of the *electrical* performance of a material if taken in isolation. Some British Standard methods require that electrical tests especially for insulation resistance and loss tangent be carried out immediately after the samples have been removed from water following a period of immersion of 24 h.

Thermal effects These often seriously influence the choice and application of insulating materials, important features being freezing point (of gases and liquids); melting point (e.g. of waxes); softening or plastic yield temperatures; flash point of liquids; ignitability, flammability, ability to self-extinguish if ignited; resistance to electric areas; liability to carbonise or track; specific heat; thermal resistivity or conductivity; coefficient of expansion, etc.

Ageing This is concerned with the maximum temperature that a material or combination of materials will withstand for long times without serious degradation of properties. It is important that during tests to determine this statistic all components of an insulation system be present because of the possibility of *compatability* problems. Testing of this type is generally carried out on models made to reproduce as far as possible the conditions under which the materials will operate in service. Such model investigations, often called 'functional testing', are generally accelerated by using temperatures considerably above those envisaged for service; but, provided that agreed procedures are used, it is often possible by extrapolation to obtain long-term results from comparatively short-term tests.

Miscellaneous characteristics These include viscosity (of liquids such as molten bitumen), moisture content (of wood, pressboard, etc.), uniformity of thickness and porosity (of papers, porcelain, etc.).

5.5.2 Mechanical properties

The usual mechanical properties of solid materials are of varying significance in the case of those required for insulating purposes, *tensile* strength, *cross-breaking* strength, *shearing* strength and *compressive* strength often being specified. Owing, however, to the relative degree of inelasticity of most solid insulations and the fact that many are quite brittle, it is frequently necessary to pay attention to *compressibility*, *deformation under bending* stresses, *impact* strength and *extensibility*; *tearing* strength and ability to fold without damage are important properties of thin sheet insulations such as papers, pressboards and varnished cloths.

Methods of test for the above properties are given in British Standards.

Many other mechanical features of insulating materials have to be considered, for example: machinability (especially as regards drilling and punching) and resistance to splitting, the latter being of particular importance in the case of laminated materials, wood and pressboards.

5.5.3 Electrical properties

The essential property of a dielectric, is, of course, that it shall insulate. But there are properties other than resistivity determining the insulation value: these are the electric strength, permittivity and loss tangent.

Resistivity This concerns volume resistivity (a bulk property) and surface resistivity (concerning leakage current across the insulator surface between electrodes having a potential difference). The former is specified in ohm-metres (or megohm-metres) and the latter in ohms per square: the surface resistance between opposite sides of a square surface is independent of the size of the square. The properties are affected by surface or bulk moisture, so that measurements of insulation resistance of pieces of material or of insulated systems are often used to assess the state of dryness. Values of volume resistivity are given in *Table 5.5*.

Electric strength Electric strength (or dielectric strength) is the property of an insulating material which enables it to withstand electric stress without failure. It is usually expressed in terms of the minimum electric stress (i.e. potential difference per unit distance) that will cause failure or 'breakdown' of the dielectric under specified conditions, e.g. shape of electrodes, temperature and method of application of voltage, as these and several other features all influence the liability of the material to fail under electric stress. It is, therefore, important to state most of these conditions when quoting values of electric strength, and they have been standardised accordingly by BSI and others. The standard method for testing oils for electric strength is given in BS 148, and that for proof tests on bitumen-base filling compounds in BS 1858. Details of the standard method for proof tests on solid insulations, such as moulded compounds and sheet materials, are given in BS 1539, BS 1137 and others, based upon technique developed by the ERA.

The electric strength of most materials falls with rise in temperature and it is usual to carry out tests for this property at suitably elevated temperatures.

Other features which vitally affect the apparent electric strength are: the sharpness or radius of edges of electrodes; the wave form of the voltage (as breakdown is dependent upon the *peak* value); the rate of increase in voltage and the time any voltage stress is maintained; the moisture content of the material; the thickness of specimen tested and the medium (usually air or oil) in which the test is made. Comparisons of electric strength are made generally by determining the electric stress that will cause failure one minute after its application. Specifications frequently call for a *Proof Test*, the material being required to withstand for, say, one minute, a specified electric stress under controlled conditions.

In view of all the features which affect the apparent electric strength of dielectrics it is preferable to obtain comparative values, say, at a range of temperatures, thicknesses and test durations. Tests may be made with alternating or direct voltages; and it is now becoming more usual to test with impulse or switching-surge voltages if the material is liable to sustain transient voltages in operation such as occur with overhead-line insulators, switchgear, power transformers and some machine windings. The object is to determine the highest stress that a material or assembly will withstand indefinitely. An indication can be obtained from a voltage/time curve, *Figure 5.8*, plotted from the stresses that cause breakdown in measured periods. The safe operating stress is then settled by experience, the use of safety

protected inside hermetic packages. Thick or thin film can be used, depending on resistor values and tolerances. The semiconductor dice are attached by conductive epoxy or eutectic bonding, then they are wire-bonded to the conductor pads. This tends to be a relatively slow manual job, but automatic bonders are now available with pattern recognition cameras which can locate the bonding pads and make the bonds.

5.3.10.4 Multilayers

The interconnect density of a hybrid can be multiplied several times by using multilayered conductors. It is possible to fit into a 30-pin package 40×25 mm (1.6×1 in) about 15–20 integrated circuit chips, which would normally need a printed circuit board at least 100 mm (4 in) square. Gold conductors are used because of their excellent conductivity, with gold wire bonds to the chips. Capacitors are attached with conductive epoxy and often discrete thick or thin film resistor chips are added by wire bonding.

Copper multilayers with solder attach are gaining in importance. The metal/dielectric systems have been available for several years, but compatible resistor systems are still being developed, because of the difficulties in firing resistors in a nitrogen atmosphere.

5.3.10.5 Microwave circuits

These circuits are made chiefly in thin film, because of the better control of line sizes and edge definition with this process, and the substrate material can be varied to exploit different dielectric constants.

5.3.11 Further reading

MAZDA, F. (ed.), *Electronic Engineers Reference Book*, 5th edn., Butterworths (1983), chapters 28 to 31
MCKELVEY, J. P., *Solid State and Semiconductor Physics*, Harper Row (1966)
KITTEL, C., *Introduction to Solid State Physics*, 5th edn., Wiley (1976)
COLCLASER, R. A., *Microelectronics Processing and Device Design*, Wiley (1980)
MORGAN, D. V. and BOARD, K., *Introduction to Semiconductor Microtechnology*, Wiley (1983)

INSULATORS

5.4 Insulating materials

Electrical insulating materials can be solid, liquid or gaseous, often in combination such as oil-impregnated paper. The materials may be organic or inorganic and natural or synthetic. Both the electrical and the mechanical properties of the materials are important and the changes in the properties with temperature are of the first importance.

5.4.1 Classification

Insulating materials, especially those used in generators, motors, transformers and switchgear, are often classified on the basis of their thermal stability according to the scheme described in BS 2757:1956 and IEC 85. This uses seven temperature classes, allocating materials to a class with 'temperature limits that will give acceptable life under usual industrial conditions of service'.

Temperature limits

Class	Y	A	E	B	F	H	C
Temperature (°C)	90	105	120	130	155	180	>180

Examples of materials in each class:

Class Y Unimpregnated paper, cotton or silk, vulcanised natural rubber, various thermoplastics that have softening points that would only permit their use up to 90 °C. Aniline and urea formaldehydes.

Class A Paper, cotton or silk impregnated with oil or varnish, or laminated with natural drying oils and resins or phenol formaldehyde. Polyamides. A variety of organic varnishes and enamels used for wire coating and bonding.

Class E Polyvinyl formal, polyurethane, epoxy resins and varnishes, cellulose triacetate, polyethylene terephthalate, phenol formaldehyde and melamine formaldehyde mouldings and laminates with cellulosic materials.

Class B Mica, glass and asbestos fibres and fabrics bonded and impregnated with suitable organic resins such as shellac bitumen, alkyd, epoxy, phenol- or melamine-formaldehyde.

Class F As Class B but with resins that are approved for Class F operation such as alkyd, epoxy and silicone-alkyd.

Class H As Class B but with silicone resins or other resins suitable for Class H operation. Silicone rubber.

Class C Mica, asbestos, ceramics and glass alone or with inorganic binders or certain silicone resins. Polytetrafluoroethylene.

The allocation of materials to classes such that life will be adequate under usual industrial conditions means that the materials may not give adequate life if the service is unusually severe, e.g. equipment normally operated very near to full load. Conversely, it may be economic to use materials of a lower temperature class for equipment operated infrequently or normally operated at very low load.

5.4.2 Temperature index

Various suggestions have been made for alternative temperature classification systems, especially as techniques such as cross-linking enable the thermal stability of materials to be significantly improved and new high-temperature materials are constantly being developed. The IEEE have suggested that a temperature index related to a particular property of the material should be assigned to a material on the basis of experience or comparison with materials that have established indices. The index would be based on the life of the material in particular environmental conditions and would preferably be a number chosen from the series 90, 105, 130, 155, 180, 200.

5.5 Properties and testing

The properties of insulating materials fall into the following categories:

(1) Physical
(2) Mechanical
(3) Electrical
(4) Chemical

Insulating materials may have to operate in the vicinity of apparatus producing high intensity radiation such as nuclear reactors, isotopes, microwave and electron generators, and considerable work has been done on the properties of materials under these conditions.

5.3.8.1 Thick film processes

There are many stages between the customer's original drawing and the final circuit.

The circuit elements are laid out using computer aided design (CAD), from which photopositives are made with a photoplotter for each different ink to be printed on the substrate. The CAD will also prepare all the necessary drawings and instructions for production personnel. Typical conductor line width is 0.25–0.75 mm (0.010–0.030 in) and resistors are usually not less than 1.25 mm (0.050 in) square.

The photopositive is exposed and developed on a screen coated with u.v.-sensitive emulsion. The screen mesh is stainless steel or synthetic fabric, with between 2.4 and 13 meshes/mm (60 and 325/in). A thick film paste placed on the screen is printed on to the substrate by a moving rubber squeegee blade.

Printing machines can be loaded by hand or fitted with mechanical feed systems, which take substrates from magazines, and transfer printed substrates to belt driers. The dried pastes are fired in multizone belt furnaces through a controlled temperature–time profile. A typical profile for conductors/resistors is 60 min through-time with 10 min at a peak temperature of 850 °C.

Resistors are trimmed by laser energy, which has the advantages of speed, precision and cleanliness. Circuit density could not have reached its present level without this compatible trimming method. Substrates 50 mm (2 in) square with multiple circuits are scribed using a more powerful laser.

Components attached by solder joints are placed on printed solder cream pads. The cream is reflowed on non-stick belts moving over heated platens, or by condensing fluorocarbon vapour in a vapour phase reflow machine.

Semiconductor chips are usually glued in position with epoxy. This can be made conductive with silver or gold powder. This material is also used for capacitors on gold conductors as solder dissolves gold very quickly. Bare chips are bonded with aluminium or gold wires of about 0.025 mm (0.001 in) diameter. Aluminium wire is not used with pure gold conductors because the bond strength deteriorates rapidly.

Terminals are supplied on reels with perhaps 50 000 on a continuous strip. Insertion machines crop unwanted terminals and insert substrates into the strip, cutting it into convenient lengths for soldering, after which the tie bar is cut off, leaving the finished circuit.

Metal packages are closed by soldering or welding a lid or cover to the header on which the substrate is mounted. A dry nitrogen atmosphere is maintained inside the package, to preserve the circuit from corrosion. The packages themselves are protected by plated films of tin, solder, nickel or gold.

5.3.9 Thin film materials

The preferred substrate materials are glass or 99.6% alumina because the surface finish has to be very smooth to allow the deposition of a uniform metal film. The film itself has a resistivity between 50 and 500 Ω/sq with TCR ± 10 p.p.m., and many materials are in use, such as nichrome or tantalum nitride. The resistive film is overlaid by a conductive metal film. Conductors for wire bonding use gold, whereas if solderable conductors are required, nickel covered with a gold flash is used. To optimise conductor properties, there is a complex layered structure, with titanium and palladium layers under the gold to minimise diffusion effects, especially of chromium to the surface of the conductor.

Substrates can be metallised in-house or bought-in with a film of known resistivity from an outside supplier. The add-on components are the same as for thick film.

5.3.9.1 Thin film processes

Typically, the smallest resistor line width is 10 microns and conductor lines can be 50 microns. Designs can avoid crossovers by wirebonding over several lines, using the fine line capability to crowd the conductor tracks. Designs cannot be photoplotted at this line width, so must be produced at a magnified scale and reduced photographically. The photopositives are produced on glass because of the dimensional stability required.

A thin film of liquid photoresist is spun on to the substrate and the pattern is exposed and developed. Selective etchants remove the metal layers as required to produce the circuits.

Thin film circuits produced in a matrix on glass substrates are separated by dicing with a diamond wheel. Thereafter, individual circuits are assembled using the same methods as for thick film, although generally they are used in hermetic packages, which protect the glass substrates.

5.3.10 Applications of hybrids

With the wide range of materials and components available, the hybrid process can be used to produce any kind of circuit, from simple resistor networks to complex multilayers which can cost one thousand times the price of the simple unit. The following examples illustrate the capability of hybrids to produce circuits of increasing complexity.

5.3.10.1 Resistor networks

These networks are usually fabricated on ceramic substrates using thick film techniques, when the wide range of paste resistivities can be used to achieve any mix of resistor values. The package can be single-in-line (SIL) with the substrate vertical, or dual-in-line (DIL) with two rows of terminals and the substrate horizontal. With a substrate thickness of 0.063 mm (0.025 in), SIL circuits can be easily fitted to printed circuit boards on 2.5 mm (0.1 in) pitches, and the height above the board can be between 3.75 mm (0.150 in) and the maximum the customer will allow.

The pin-out configuration is very flexible, using terminals at 2.54 or 1.27 mm (0.1 in or 0.050 in) pitches. Both sides of the substrate may be printed with resistors, provided the total dissipation does not exceed the substrate capability. If the network has to withstand power pulses from lightning strikes, a thicker substrate is used, which is stronger and withstands the sudden heating effect.

Divider networks made with the same thick film pastes have a good tracking performance, but to obtain the ultimate performance in this application, thin film is best.

5.3.10.2 Networks with surface mounted devices

It is a small step to convert a passive resistor network into an active circuit by incorporating semiconductor devices and capacitors. Transistors, diodes, capacitors, resistors, inductors, integrated circuits and transformers have been produced in miniaturised forms for surface mounting. The components are usually soldered to mounting pads on the circuit, or occasionally conductive epoxy may be used. For large-volume applications, pick and place machines are now available which take components from loaded reels and place them in the correct position on the substrate. The circuit may be protected by printed or dipped organic resin, or the circuit may be left in its overglazed state, since the add-on components have their own protection.

5.3.10.3 Networks with wire-bonded chips

By using bare chips, a substantial improvement in packing density can be realised, which is valuable for military and aerospace applications, but the wire bonds themselves must be

5.3.7 Thick and thin film microcircuits

Thick and thin films are used in a variety of microcircuits and systems replacing many individual components such as resistors and capacitors and thus saving both space and weight. Where a mixture of thick and thin films is used together with discrete active devices, the total system is called a hybrid microcircuit. The first application of microcircuit techniques was for the wartime production of proximity fuses, using printed carbon composition resistors on ceramic substrates. The basic methods are recognisable forty years later, even though modern materials have greatly improved performance.

Screen printing using precision stainless steel or synthetic mesh screens is the method by which thick film patterns of conducting pastes are deposited on ceramic substrates. High temperature firing fuses the pastes to the substrates, forming circuit elements less than one thousandth of an inch in thickness. Even so, these are still 'thick' films, compared with the film thickness of 'thin' film circuits. The process here is completely different, using photolithography and selective etching on purely metallic films to generate the circuits.

Microcircuits will be discussed in terms of materials and processes, and how they are applied to a range of circuit applications.

5.3.8 Thick film materials

The materials used for hybrids are being continually up-graded and extended, to improve the final product. The basic properties of the materials are as follows:

The substrate material must be insulating, flat, non-reactive and thermally stable to the firing temperature (about 850 °C). High purity (96%) alumina is the most widely used material, used in plates from 50 mm (2 in) square to 150×100 mm (6×4 in), and usually 0.63 mm (0.025 in) in thickness, although any thickness between 0.25 and 2.5 mm (0.01 in and 0.100 in) can be obtained.

Beryllia substrates are occasionally used because of better thermal conductivity, and recently, metal substrates coated with glaze (with compatible conductor and resistor systems) have become available.

For any particular application, a conductor is chosen which will give adequate performance in terms of adhesion, solderability, etc., at an economic cost. The air-firing conductors are based on the precious metals silver, gold, palladium and platinum. Copper and nickel systems have been developed in recent times because of the lower cost of the metal, but these conductors must be fired in a non-oxidising atmosphere (nitrogen, with oxygen content less than 20 p.p.m.). Common alloys and their properties are given in Table 5.4.

Modern resistor pastes are based upon ruthenium, either as the dioxide or as bismuth ruthenate. Resistivities between 1 Ω/sq and 10 MΩ/sq can be obtained, and intermediate values can be made by blending. The temperature coefficient of resistance (TCR) is ± 100 p.p.m. or better, and load stability is good. Fired resistors are often overglazed with low-temperature glaze (550 °C) which contributes to the mechanical and environmental protection of the resistors.

Multilayer hybrids are built up from layers of conductor tracks, separated by layers of dielectric glaze and interconnecting through windows (called vias) in the glaze. The glaze must tolerate multiple re-fires as each layer is added and have a low dielectric constant, to minimise stray capacitance between crossing tracks. Resistors can be fired on top of glazes loaded with ceramic, giving results similar to resistors on alumina substrates.

Screen-printable protections are printed over trimmed resistors and cured at 200 °C or less. Complete circuits can be dip-coated in liquid resin suspensions, powder-coated, moulded or potted. The usual resin types which are used are silicones, epoxies or phenolics.

To attach components with solder, solder cream is printed over conductor pads on the substrate, into which the component feet are placed prior to reflow. The solder cream consists of small solder balls, flux and solvent and it is printed through a coarse mesh screen, often with a metal foil stencil to boost the thickness of deposit.

The components which can be attached to hybrids are of all shapes and sizes. The name 'hybrid circuit' is derived from the ability of this technology to mix all kinds of component to achieve the desired circuit characteristics.

Semiconductor chips are used in many packaging styles. Bare chips without any protective coating are wire-bonded into circuits and chips with moulded plastic protection are soldered.

Complex chips are packaged in chip carriers, which are small hermetic packages made from ceramic or glass fibre laminate. These chip carriers permit the electrical parameters to be fully tested before soldering into a circuit.

Of passive devices, multilayer ceramic capacitors are used in the greatest numbers. They consist of alternate layers of dielectric and conductor, sintered into a solid block, and with metallised terminations. The capacitors range in size from 1.25×1 mm to 6×5.5 mm (0.050 in $\times$ 0.040 in to 0.220 in $\times$ 0.240 in) and in value from 1 pF to 1 μF. High-value capacitors are miniaturised tantalum capacitors with metal end caps, or moulded with metal tabs. Chip resistors can be made in thick or thin film and are useful for multilayer circuits, when resistor printing may be difficult to control. Thin film chips are used when precision matching of resistors is needed, for example, in digital-to-analogue convertors. Small wire-wound components such as transformers and inductors have also become available for hybrid use.

Table 5.4 Common alloys and their properties

Material	Resistivity (milliohm/sq)	
Ag	2–5	Inexpensive. Poor leach resistance in molten solder. Little used
Pd/Ag	20–35	Most widely used. Good adhesion, leach resistance. Wire bondable
Pd/Au	50–90	More expensive. Better adhesion, leach resistance. Wire bonding fair
Pt/Au	70–100	Very expensive. Excellent adhesion, leach resistance. Wire bonding poor
Au	3–4	Most widely used multilayer conductor. Excellent conductivity, good wire bonding, not solderable
Au/2% Pd	5–7	Modified gold for aluminium wire bonding
Cu	2–4	Good solderability but needs non-oxidising furnace atmosphere

unable to complete the covalent bonding and a vacancy or hole is produced. This vacancy can be filled by the thermal motion of neighbouring electrons so that a mobile hole is produced which can drift in an applied field to produce current flow. The impurity atom will be left with a negative charge which is immobile.

The introduction of the Group 3 atoms give rise to mobile holes, fixed negative charges and a reduction in the number of mobile electrons in the p-type material. The equality $np = n_i^2$ still holds. Typical p-type impurities are boron, aluminium, gallium and indium.

The energy band diagrams show impurity levels within the forbidden energy gap either just below the conduction band (n-type) or just above the valence band (p-type). These levels are within 0.04 eV of the respective bands with the consequence that at room temperature the impurities are fully ionised.

As the number of levels of energy at the bottom of the conduction band is large ($\sim 10^{25}$ m^{-3}) it is usual to regard the number of electron states N_c as all existing at the same energy level E_c. Similarly, the number of electron states N_v at the top of the valence band are regarded as existing at energy level E_v.

The actual number of electrons in the conduction band is determined from the number of possible electron states and the probability that these states are filled. The probability, pr, that an electron state is filled is given by the Fermi–Dirac distribution

$$pr = (1 + \exp(E - E_F)/kT)^{-1}$$

This distribution function and its variation with temperature are shown in *Figure 5.7(a)*. *Figures 5.7(b)* and *(c)* indicate the application of the function to the energy band representation to give the actual number of electrons and holes in the conduction and valence bands of n-type and p-type material. Note that, in the n-type material, the Fermi level has been shifted upward to give a greater propability of finding an electron in the conduction band so that the actual number of electrons here is increased. This shift gives a corresponding decrease in the number of holes in the valence band as indicated by equation $np = n_i^2$.

A similar situation exists in the p-type material but the shift of the Fermi level is towards the valence band.

5.3.6 Single crystal growth and device manufacture

Although silicon solar cells are now made from polycrystalline material, the production of most other semiconductor devices usually requires the growth of large single crystals of silicon. One method of growing single crystals is the Czochralski technique of 'pulling' from the melt. Silicon, in a quartz crucible set in a graphite cup, is held in an inert atmosphere and brought to its melting point by radio frequency heating the graphite holder. A small single crystal seed, cut at the correct crystallographic orientation is lowered to the melt and, after wetting, is slowly withdrawn at a controlled rate. Silicon from the melt grows on the cooler seed and takes up the orientation of the seed. Usually both seed and melt are rotated slowly to achieve symmetrical growth. Single crystals of approximately 40 mm diameter and around 100–150 mm length can be obtained.

A shortcoming of the method, apart from the limit to the crystal size, arises from the fact that impurities have a greater solubility in the molten material than in the solid. This means that impurities tend to be concentrated in the melt as the crystal is grown giving rise to an increase in conductivity towards the bottom end of the crystal.

This greater solubility in the molten material is utilised in an alternative method of single crystal production, i.e. zone-refining. A molten zone is produced in the polycrystalline silicon by radio frequency heating and this zone is moved along the length of the material, carrying impurities with it. Repeated sweeps of the solid material by the molten zone are used to reduce the impurity level in the solid material.

To produce the individual semiconductor devices, the single crystal is cut into thin slices perpendicular to its length. These slices are accurately lapped to a thickness of the order of 0.25 mm and etched to give a mirror finish. An oxide layer is produced on the silicon surface and, using photolithographic techniques, part of this layer is removed to expose the silicon surface into which can be diffused either n- or p-type impurities. This oxide masking procedure is used repeatedly to diffuse different patterns of impurity into the wafer and build up the multilayer structure of the semiconductor device. The patterns are repeated across the whole of the silicon slice to give devices with almost identical characteristics.

Other techniques include epitaxial growth, i.e. the growth from the vapour stage of layers of silicon, containing the required impurities, on to silicon substrates. The thickness of these layers can be controlled accurately and they take up the orientation of the silicon substrate thus preserving the single crystal properties.

Diffusion of impurities into silicon at high temperatures (1250 °C) is a widely used technique in device production but an alternative method of localised doping is ion bombardment. Here the impurity to be introduced into the silicon lattice is formed into an ion beam which is used to bombard selected areas of the silicon wafer.

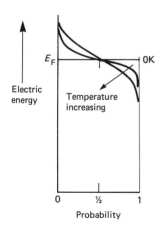
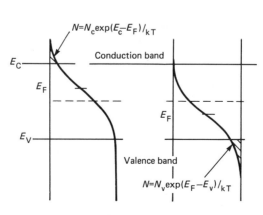

Figure 5.7 Electron distribution function

5.3.2 Electrical conduction

In the perfect crystal, all the valence electrons are taken up by the covalent bonding and are not available for the conduction of electric current. However, under external influences such as heat or illumination, energy is imparted to the electrons causing some of the covalent bonds to be disrupted. These electrons are then free to migrate through the crystal as negative charge carriers. The holes which are left in the bonds behave as mobile positive charges and they are able to move through the lattice by the

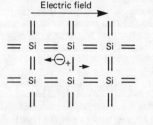

Figure 5.4 Intrinsic conduction

reciprocal motion of other bonding electrons required to fill the holes. Application of an electric field as in *Figure 5.4*, gives rise to drift currents of electrons and holes, both of which effectively carry negative charge in the same direction so that the conductivity of the material is given by:

$$\sigma_i = n_i e \mu_n + p_i e \mu_p$$

where n_i is the number of mobile electrons per cubic metre and p_i is the number of mobile holes per cubic metre. In this pure (intrinsic) material $n_i = p_i$.

The distinction between mobile and valence electrons is illustrated in the energy level scheme shown in *Figure 5.5*.

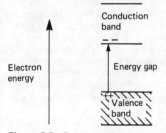

Figure 5.5 Energy band scheme for intrinsic semiconductor

Valence (bonding) electrons occupy a set of energy levels which, in the ideal crystal at absolute zero, are all filled. These levels lie within a band called the valence band and, because the levels are all filled, the electrons in this band make no contribution to electrical conduction. Above the valence band and separated from it by an energy gap in which few energy levels exist, is the conduction band. It is the presence of electrons in this band that gives rise to electrical conduction; the conductivity is proportional to the number of electrons present. Within this conduction band the separation of the energy levels is so small (approximately 10^{-19} eV) that the band may be regarded as a continuum of unfilled electron energy levels. Thus, when an electric field is applied to an electron in this band it is able to move up into a higher energy level consistent with the increase in energy it has acquired as kinetic energy, i.e. electrical conduction requires there to be vacant energy levels immediately above those occupied by the electron which is to carry the conduction current. The energy gap for silicon is 1.106 eV at 300 K (27 °C) and at this temperature the average thermal energy of the electron is 0.025 eV. The probability therefore, of an electron acquiring sufficient energy to move up from the valence to the conduction band by thermal means is quite small. Nevertheless, at this temperature, the number making this transition (the intrinsic electron and hole concentration, n_i and p_i) is $1.5 \times 10^{16}/m^3$, giving rise to a conductivity of 4.4×10^{-4} $(\Omega \, m)^{-1}$.

As the temperature is increased, electrons acquire greater thermal energies and hence their number in the conduction band (and correspondingly the number of holes in the valence band) will increase giving rise to an increase in conductivity.

The major practical application of silicon in the semiconductor field depends not on its intrinsic conductivity but rather on its extrinsic (or impurity) conductivity. The addition of suitable impurities into the silicon lattice enable a conductivity to be achieved anywhere between that of the intrinsic material and that of some metals (Sb, As, Bi).

5.3.3 Extrinsic conduction

The conductivity of the intrinsic material can be enhanced by the substitution into the silicon lattice of relatively small numbers of atoms of elements in Group 3 or Group 5 of the periodic table. The extrinsic material so produced is known as p-type or n-type respectively where the prefix indicates the sign of the excess charge carrier.

5.3.4 n-type conduction

The substitution of a Group 5 element into the lattice is indicated in *Figure 5.6*. Four of the five valence electrons of the Group 5 impurity are needed to complete the covalent bonding with the neighbouring Si atoms, whilst the fifth valence electron is superfluous and can easily be detached by thermal vibrations.

Figure 5.6 Extrinsic (impurity) conduction

Once detached from the impurity atom, the excess electron is free to conduct electricity in an applied field. The loss of the excess electron by the impurity atom leaves this atom with a positive charge which is fixed in the lattice.

The addition of the Group 5 impurity into the lattice gives rise to

(1) excess electrons which make the major contribution to the electrical conductivity,
(2) fixed positive charges which make no contribution to conductivity but which can give rise to space charges when the mobile electrons are drawn away.

A further, but important, consequence is the suppression of the thermally generated holes by the excess electrons. This arises because of the equality

$$np = n_i^2$$

where n is the density of excess electrons, p is the density of mobile holes and n_i is the density of electrons in the intrinsic material $(1.5 \times 10^{16}/m^3)$. n-type impurities are phosphorus, arsenic and antimony.

5.3.5 p-type conduction

A similar process occurs with the substitution of a Group 3 element into the silicon lattice. The trivalent atom is, however,

high-speed computers which must be compact to reduce signal transit times. The need for liquid helium cooling is not seen as a serious disadvantage.

As a logic gate, currents in control lines overlying the SQUID induce transitions between the zero voltage and resistive states by altering the flux linking the device. For memory cells, superconducting loops may be used with the 1 and 0 binary values being represented by the presence or absence of a persistent circulating current. One Josephson junction in series with the loop and one inductively linked to it are required to 'write' and 'read' respectively. Arrays of junctions may be deposited on a substrate using many of the processes developed for fabricating silicon microcircuits.

5.2.6 Further reading

BRANDES, A. E. (ed.), *Smithells Metals Reference Book*, 6th edn., Butterworths (1983)
Engineering Alloys, 5th edn., Van Nostrand Reinhold, New York (1973)
DAVIDSON, H. W. *Manufactured Carbon*, Pergamon (1968)
ROSE-INNES, A. C. and RHODERICK, E. H., *Introduction to Superconductivity*, Pergamon, 2nd edition (1978)
WILSON, M. N., *Superconducting Magnets*, Oxford University Press (1983)
FONER, S. and SCHWARTZ, B. B. (eds.), *Superconducting Machines and Devices. Large Systems Applications*, NATO Advanced Study Institutes Series Vol. 1, Plenum Press (1974)
FONER, S. and SCHWARTZ, B. B. (eds.), *Superconducting Applications: SQUIDs and Machines*, NATO Advanced Study Institutes Series Vol. 21, Plenum Press (1977)
MATISOO, J., *The Superconducting Computer*, Scientific American **242**, 38–53 (May 1980)
Proceedings of the biennial Applied Superconductivity Conferences, e.g. *IEEE Transactions on Magnetics MAG-15*, no. 1, Jan. 1979; *MAG-17*, no. 1, Jan. 1981; *MAG-19*, no. 3, (2 parts) (May 1983)
MATULA, R. A., Electrical resistivity of copper, gold, palladium and silver. *J. Phys. Chem. Ref. Data*, Vol. 8, no. 4, 1147–1298 (1979)
Copper Development Association publications from C.D.A., Potters Bar, Herts., England

SEMICONDUCTORS, THICK AND THIN FILMS

5.3 Silicon

The early development of semiconductor devices was based largely on germanium as the semiconductor material. The transition to silicon took place as it became obvious that the supply of germanium was restricted and that this material had a relatively low intrinsic resistivity which in turn lead to low breakdown voltages and low power in germanium devices. For a short while germanium devices persisted in the high frequency field, where the high mobility of carriers in germanium is approximately three times that in silicon, but the rapid improvements in manufacturing techniques soon matched the higher frequency response of the germanium devices. Under ideal conditions of chemical purity, physical perfection and at very low temperatures, silicon is a near-insulator; at higher temperatures it conducts electricity by electron flow. This conductivity can be enhanced by the introduction into the crystal lattice of specific impurities: other electrical properties, such as the minority carrier lifetime, are strongly influenced by other impurities as well as imperfections in the crystal lattice.

5.3.1 Crystal structure

Silicon (Si) of atomic number 14, is in Group 4 of the periodic table of the elements, i.e. it is tetravalent and its atomic structure consists of four valence electrons surrounding an inert core of net charge +4 electronic units. The valence electrons are responsible for chemical binding and the tetravalent property gives rise to the hard and brittle properties of this crystalline solid.

Knowledge of the crystal structure is a fundamental requirement in the understanding of electrical conduction in this

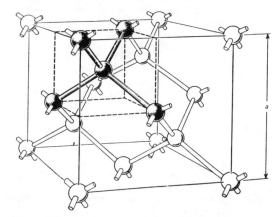

Figure 5.3 Crystal structure of silicon. a is the lattice constant

material. *Figure 5.3* illustrates how the atoms of silicon are arranged in the silicon lattice. Every atom is bound to its four nearest neighbours which are symmetrically placed at the corners of a regular tetrahedron; each bond is formed by a pair of valence electrons—one electron being donated by each atom. This is termed electron-pair or covalent bonding and the structure is of the diamond form. Linear dimensions of the unit cell are specified by the lattice constant, a, which equals the length of the side of the unit cube shown in *Figure 5.3*. The value of this constant and other parameters are given in *Table 5.3*.

Table 5.3 Properties of silicon

Property		Si
General Properties		
Atomic number/atomic weight		14/28.08
Density	kg/m^3	2328
Melting point	°C	1420
Boiling point	°C	2600
Specific heat (0–100 °C)	J/(kg K)	755
Coefficient of thermal expansion $\times 10^6$	°C^{-1}	4.2
Thermal conductivity (25 °C)	W/(m K)	83
Intrinsic Material at 300 K (27 °C)		
Lattice constant, a	m	5.42×10^{-10}
Atomic density	atoms/m^3	5.0×10^{25}
Energy gap	eV	1.106
Electron and hole concentrations, n_i	per m^3	1.5×10^{16}
Electron drift mobility, μ_n	m^2/Vs	0.135
Hole drift mobility, μ_p	m^2/Vs	0.048
Intrinsic conductivity, σ_i	S/m	4.4×10^{-4}
Relative permittivity	—	12

is a factor of 100 or even 1000 less. This advantage can extend to lower fields, especially when the field is required over a large volume. All the other applications of high-field superconductors listed below have so far been explored less extensively and often in only a few places in the world. The economics of the complete systems have not yet been sufficiently attractive to provoke substantial development.

5.2.3.1 Laboratory magnets

Fields in excess of 10 T can be generated readily and standards of stability and uniformity hitherto unobtainable achieved. The use of nuclear magnetic resonance both for high resolution analysis and for body scanning (tomography) depends on these features.

5.2.3.2 High-energy physics

Many superconducting magnets are used for particle accelerators, for handling the extracted beams and for analysing particle reactions in bubble chambers. One large bubble chamber coil has a diameter of 4.6 m and a central field of 3.5 T. Radio-frequency cavities employing surface superconductors are also used in some accelerators. The good performance of all these devices and the associated refrigeration plant has shown that equipment using superconductors can run economically and reliably. If magnetically confined thermonuclear fusion is to be a future power source, the windings needed will have to be superconducting. Such magnets have already been built for experimental investigations.

5.2.3.3 Separation processes

Superconducting magnets can be used for extracting weakly magnetic materials. This is being considered, for example, for extracting haemoglobin from blood, for removing impurities from china clay, coal and water and for separating minerals.

5.2.3.4 Trains

A superconducting coil mounted in the bottom of a wagon moving over a conducting surface will provide lift if the wagon is moving fast enough. The principle offers a possible efficient alternative to wheels on very high-speed trains.

5.2.3.5 D.C. machines

Homopolar motors using superconducting exciter coils have been built and work satisfactorily. A generator and motor operated back-to-back can form a quiet efficient 'gearbox' for ship propulsion.

5.2.3.6 A.C. generators

A superconducting exciting winding on the rotor offers a more efficient a.c. generator than one with a conventional copper winding: about 0.5–1% more output is achievable for a given mechanical input.

5.2.3.7 Supertransductor

The conventional transductor intended for fault current limitation in electric power systems has a d.c. bias winding to saturate the core of an inductor, and so reduce the a.c. impedance of a coil on the same core. The core can come out of saturation on one-half cycle if the a.c. current attempts to rise excessively: the impedance of the transductor then increases. Superconductors offer an attractive method of providing the d.c. bias.

5.2.3.8 Other power applications

If magnetohydrodynamic generators are to be used to produce electricity efficiently, their windings will have to be superconducting. For energy storage, gigantic inductors holding perhaps 5000 MWh are being discussed. Such a store could retain electricity generated cheaply at night for use during the day. A small superconducting energy store for controlling stability in a power system is being used on an experimental basis. Because superconductors that carry a.c. do not function satisfactorily in strong fields, they convey no benefit to power transformer design.

5.2.4 Power transmission

High power superconducting cables, either a.c. or d.c., appear more economic than conventional ones only for power transfers in excess of a few GW; such power transfers in single circuits are not at present required. A.C. cables would use superconductors in a surface mode to avoid excessive hysteresis losses.

5.2.5 Electronic devices

When two pieces of superconductor are connected together by a 'weak' link, many intriguing phenomena occur which form the basis for a variety of electronic devices. Niobium and lead are commonly used and the link, normally rather poorly conducting or even insulating, may consist of a point contact, a fine constriction or a thin (2–5 nm) layer of oxide. The resulting junction can carry supercurrents (typically up to a few hundred μA) by the tunnelling of electron pairs through the barrier without any voltage drop. This is the d.c. Josephson effect. If the current exceeds some critical value, which depends on the magnetic flux in the barrier, the junction switches to a resistive state with a voltage drop of typically a few mV. When there is a voltage across the junction, the supercurrent component oscillates at a very high frequency determined by the voltage (483.6 GHz/mV). This is the a.c. Josephson effect; it can be used as a precise voltage standard.

5.2.5.1 SQUIDs

An extremely sensitive measuring device may be made from a pair of Josephson junctions connected in parallel, by superconductor, and known as a SQUID (superconducting quantum interference device). The total critical current then depends periodically on the flux (the period being the flux quantum) within the loop formed by the two junctions. Field changes as low as 10^{-14} T can be detected. Careful screening, together with differential techniques, are needed to eliminate unwanted responses. Small voltages in low-resistance circuits (e.g. 10^{-10} V in 1 Ω) and small currents may be measured by using the SQUID to determine the associated magnetic field. Sometimes just one junction is used in a superconducting loop with the flux being sensed by its effects on an inductively coupled r.f. bias circuit. With sufficiently high bias frequencies, measurements into the microwave region are possible.

Instruments incorporating SQUIDs are finding application particularly in geophysics—studies of rock magnetism, exploration based on magnetic anomalies, determination of deep earth conductivity—and in biomagnetism—investigation of electrical activity in the heart, brain and muscles for instance.

5.2.5.2 Computer elements

Many groups are exploring the possibility of using SQUIDs and related devices in computers. Their bistable characteristics, small size, ultrashort switching times (as low as 10 ps) and very low-power consumption make them attractive components for large

Table 5.2 Some superconducting materials and their properties

Material	Critical temperature (K)	Upper critical magnetic field at 4.2 K (Boiling point of liquid helium) (T)	Critical current density (in superconductor) at 4.2 K and 5 T (GA/m^2)
Lead	7.2	0.06	—
Niobium	9.2	0.4	—
Niobium-60 at% titanium	9.3	11	1–2
Niobium tin (Nb$_3$Sn)	18.1	23	5–10
Niobium germanium (Nb$_3$Ge)	23	37	5–15

All the superconductors listed are type II except lead (type I). Critical current densities are very dependent on the way the material is prepared

superconducting current only in a surface layer and only in low magnetic fields. The magnetic field at the surface is equal to the current per unit width of surface. In general, the resistance of superconductors to alternating currents is not zero, though small: that of surface superconductors can be virtually zero. The small power losses that do then occur arise mainly from factors such as surface irregularities.

Bulk superconductors can carry currents throughout their section. Losses under changing current are relatively high and bulk superconductors are not used at power or higher frequencies. The loss is effectively a hysteresis loss. Some bulk superconductors, often referred to as 'hard' superconductors, can carry large current densities in strong magnetic fields (see *Table 5.2*). Materials with good superconducting properties tend to have high normal resistivities.

Physicists divide superconductors into types I and II. Type I materials are perfectly diamagnetic below the critical magnetic field: they carry surface currents only and exclude all magnetic flux from their bulk. The depth of current and field penetration is determined by quantum mechanical considerations and is typically 10^{-7} m, say 1000 atoms. Above the critical field they lose all diamagnetic properties. Type II materials also show diamagnetism but lose the property gradually as the field is raised above a first critical field. The magnetic flux penetrates the bulk in quantised flux bundles or fluxons (2×10^{-15} Wb per bundle) which can be detected with sensitive techniques. Electrical resistance returns above the upper critical field.

The quantum theory of superconductivity envisages a low-temperature condensed state of electrons in which they interact, through the atomic lattice, to form temporary pairs. Each member of the pair distorts the lattice in such a way as to attract the other. Because of their lower energy state, these pairs are not readily scattered and flow unimpeded. Without detailed prior knowledge of the electronic and atomic structure of a material, the theory is not sufficiently quantitative to predict whether or not that material will exhibit superconducting properties, let alone what the properties will be. Analogy with known superconductors has proved the best guide to new ones.

Typical materials used for engineering purposes are shown in the table. Niobium is the surface superconductor most commonly envisaged for use in alternating current equipment such as power cables. Lead is cheaper but not usually preferred unless for special reasons a low-field capacity is satisfactory. Bulk superconductors can be used in low fields as surface superconductors, e.g. niobium tin (Nb$_3$Sn). Of the bulk superconductors, niobium–titanium is a ductile alloy which needs to be coldworked to produce good current-carrying capacity. Niobium tin is a compound and very brittle: it is usually made by diffusion of tin into niobium, sometimes from a bronze, sometimes in ribbon form to reduce mechanical stresses during handling. The diffusion is often carried out by heat treatment of the magnet coil after it has been wound. Niobium germanium has the highest critical temperature of all, but properties vary from sample to sample and it cannot yet be considered for engineering applications. It too is brittle. The critical temperatures of Nb$_3$Sn as well as Nb$_3$Ge are sufficiently high to make cooling by liquid hydrogen appear physically feasible but it is not yet an economic engineering proposition.

5.2.2 Stabilisation of magnet conductors

Bulk superconductors, especially when used near the critical conditions, suffer from electrothermal instability. A small increase in temperature lowers the critical current density which can lead to a further increase in temperature ultimately driving the material into its normal resistive state; heat generation then becomes very rapid and the whole coil goes 'normal'. Stabilisation can be achieved in several ways. All methods use a composite conductor in which ordinary conductors, frequently of copper, are in intimate contact with the superconductor. The copper has a much lower electrical resistance than the superconductor in its normal state. The most robust stabilisation system is the so-called cryostatic method. The design provides plenty of copper in parallel with the superconductor and good cooling of the composite. It envisages some agent causing the temperature to rise and current to transfer into the low resistance copper over a short length of conductor. The thermal conditions are designed to be such that the composite conductor will cool so that once more the superconductor element regains its superconducting capacity. This method requires coolant access to most of the winding.

A more compact scheme for coils aims to limit localised temperature rises. Magnetically induced electrical losses in the superconductor during current changes are reduced by using a composite of fine strands of superconductor coprocessed in a matrix, often of copper. The strands can be twisted and the matrix incorporate components with high resistance, such as cupro-nickel, to control circulating currents and losses. The need for additional metal means that current density in terms of the total conductor cross section is up to ten or more times lower than that in the superconductor itself. This is especially so for Nb$_3$Sn and Nb$_3$Ge. Perhaps most importantly, the winding can be encapsulated in resin to prevent frictional heating as the winding takes up mechanical loads; preventing cracking of the resin is an important part of the technology.

5.2.3 High-field electromagnets

To produce magnetic fields greater than about 2 T using conventional copper windings can require powers measured in megawatts. The outstanding success of superconductors has been in providing strong fields for laboratory experiments and for high energy nuclear physics equipment without a massive power burden. The power to drive the helium refrigerators or liquefiers

5.1.3.3 Carbon contacts

Electrical contacts include components for switches, circuit breakers, contactors, relays and sliding contacts. The properties of carbon and graphite of importance in such applications include: non-welding characteristics, a melting temperature in excess of 3500 °C and sublimation at atmospheric pressure. It also has arc resisting characteristics because a higher voltage is required to maintain a carbon arc than for many metal arcs. The material does not tarnish giving a constant contact resistance in the absence of any surface films.

The material is selected in accordance with the required contact resistance, the metal-impregnated and metal–graphite classes giving the lowest and the carbon class the highest values.

5.1.3.4 Resistance brazing and welding

Using the relatively high resistance characteristic of carbon it is possible to obtain a heating effect that can be used for the joining of metals. The major application is in resistance brazing where one component in the system is melted, or a low melting alloy solder is introduced to complete the bond.

The high melting and boiling points of carbon prevent it from sticking or welding to the workpieces and prevent distortion of the tips of the tool even at white heat under pressure. Design of the carbon tips, shape and area of contact, is used to control the heat in the joint. The lower strength of carbon compared with water-cooled metal electrodes limits use in resistance welding and spot welding applications. Plain carbon or electrographite material can be used, the latter generally giving the better life.

5.1.3.5 Arc welding

Electrodes of carbon are very suitable for arc welding. The weld is achieved either by a fusion process (i.e. fusion of butting edges) or by feeding a weld rod into the arc formed between the workpiece and the electrode. Carbon arcs find limited use in metal cutting typical applications being in cutting of risers from castings or grooving of metal plates. Carbon arcs are widely used for brazing of thin mild steel sheet.

5.1.3.6 Granules

Carbon granules for microphone applications are made in sizes from 60–700 μm, the largest being suitable for maximum response but developing more background noise. The smallest granules give good quality but a lower response. For normal telephone work, granules of the smaller range of sizes are employed to give good response and acceptable noise. Carbon is superior to metal powders in such applications due to high specific and contact resistance, the absence of any tendency to oxidise or tarnish leading to a stable and reproducible performance necessary in audio applications.

Telephone granular carbon is a crushed product made from petroleum coke; particle size and shape are important in determining performance characteristics. Special non-ageing grades are available.

5.1.3.7 Fibres

Fibres are the latest form in which carbon is manufactured; they have a near-perfect oriented structure. Fibres have a high strength and elastic modulus which may be controlled by the maximum temperature attained during heat treatment. Low density leads to high specific strength and stiffness particularly important in aerospace applications. Selective use of carbon fibres in composites is employed to maximise properties in desired directions. Carbon fibres are graphitic structures produced by special degradation procedures involving polyacrylonitrile or cellulose fibres. More recently petroleum or coaltar pitch have been used.

5.2 Superconductors

Superconducting materials lose all electrical resistance below a sufficiently low temperature called the 'critical temperature' which is different for different materials. The phenomenon was first observed in mercury in 1911. Niobium germanium (Nb_3Ge) is the material with the highest critical temperature (23 K) known up to 1985.

Engineering interest in superconductors became really significant in the early 1960s when materials capable of carrying high current densities ($\sim 10^9$ A/m^2) in high magnetic fields ($\sim$ several tesla) were discovered. These opened the way for high field electromagnets. The interest was reinforced by advances (partly spurred by superconductor development) in the technology of large-scale helium refrigeration (hundreds of watts cooling at 4 K) which could produce cold gaseous or liquid helium for cooling purposes. The discovery of the Josephson effect (1962) has led to small, low-field, superconducting electronic devices (see Section 5.2.5).

Once cooled below their critical temperature, T_c, superconductors can carry resistanceless current up to a maximum determined by the ambient magnetic field and the temperature. If the limiting combination of current, magnetic field and temperature is exceeded, the material reverts to an electrically resistive or 'normal' condition. In a plot of temperature, magnetic field and current density, the surface which divides the superconducting from the normal condition is a key characteristic of the material. *Figure 5.2* illustrates a typical surface for a superconductor with high bulk current-carrying capacity. The magnetic field at which the superconducting current-carrying capacity vanishes is known as the critical field: it depends upon the temperature.

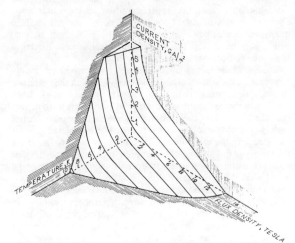

Figure 5.2 Limiting values of temperature, magnetic field and current density (curtailed at top) for a superconducting niobium-titanium alloy

5.2.1 Varieties of superconductor

For engineering purposes there are two varieties of superconductor—surface superconductors and bulk superconductors—determined as much by the way they are used as by their intrinsic properties. Surface superconductors carry

Table 5.1 Carbon brush grades

Grade	Morganite Grade coding	
C/S	For commutator/slip-ring application	v Contact volt drop per brush
C*	For traction commutator	μ Coefficient of friction
J	Normal current density, mA/mm²	u Normal maximum operating speed
p	Normal pressure, mN/mm²	

The characteristics v, μ and u are indicated by numbers:

Number	Contact voltage drop, V/brush	Coefficient of friction	Surface speed, m/s
1	<0.4	<0.1	20
2	0.4–0.7	0.1–0.15	30
3	0.7–1.2	0.15–0.2	50
4	1.2–1.8	>0.2	60
5	>1.8	—	>60

Grade	C/S	J	p	v	μ	u	Properties
Electrographite class							
EG0	C	100	18	3	2	1	Low-voltage, high-current
	S	115	18	3	2	3	Steel and bronze slip-rings
EG3	C	85	14	3	2	1	Small and medium industrial motors
	C*	85	20–50	3	2	2	Traction a.c. motors
EG260	C	100	21	3	2	2	Wide range of d.c. machines, and for corrosive atmospheres
	S	115	21	3	2	2	
EG8101	C	85	21	4	1	2	Difficult a.c. commutating conditions: fractional motors
EG12	C	100	18	3	3	3	For prolonged overloading
EG14D	C*	100	20–50	3	1	3	For d.c. and 16⅔ Hz a.c. traction
EG17	C	100	21	3	2	4	Long life: stable current collection
EG116	C*	110	20–50	4	1	3	A.C. and d.c. traction motors
EG133	C	100	18	3	3	3	For high-rated welding generators
EG224	C	95	21	3	2	3	For rolling-mill motors
EG236	C	110	18	4	2	4	For rolling-mill motors and mine-hoist generators
EG236S	C	110	21	2	3		For industrial d.c. machines and traction generators
EG251	C	95	21	4	1	3	For peaky overloads, light-load running
EG6345	C	95	18	3	2	2	For d.c. machines in ships
EG6749	C*	95	20–50	3	2	3	For d.c. traction motors
Natural graphite class							
HM2	C	100	14	3	2	3	Low friction
HM6R	C	100	14	3	3	4	Low inertia, good collection
	S	100	14	3	3	5	Suitable for turbogenerator rings
HM100	C	100	14	3	3	4	Good load-sharing
	S	100	14	3	3	5	For helical-grooved steel rings
Resin-bonded class							
IM6	C	25	21	5	1	2	High contact drop: for d.c. and universal fractional motors
IM23	C	95	21	5	1	2	For Schrage motors
Carbon and carbon–graphite class							
A2Y	C	70	14	3	3	1	For fractional machines with flush mica
B	C	85	14	3	4	1	High conductivity, for easy commutation conditions
C4	C	65	14	3	4	1	Dense, for machines with hard mica
H100	C	85	84	3	1	2	Contains MoS₂; quiet, low friction: for car generators
PM9	C	65	50–20	4	4	1	For fractional machines with flush mica
PM70	C	40	21	4	4	1	For difficult fractional machines with recessed mica
Metal-graphite class							
CM+O	C	170	130–70	1	3	1	Low loss; for car starter motors only
CM	S	230	14	1	3	1	Very low loss; for heavily loaded slip-rings
CM3H	C	125	21	1	2	2	High graphite content: for car starters and d.c. machines up to 12
	S	155	21	1	2	2	
CM5B	C	115	21	2	1	2	For d.c. machines up to 50 V
	S	140	21	2	1	2	For totally enclosed slip-rings, car accessory motors
CM5H	C	115	21	2	1	2	For d.c. machines up to 30 V, induction-motor slip rings
Metal-impregnated graphite class							
DM4A	C	115	—	2	2	1	Short-period current densities up to 2300 mA/mm²; pressure according to application; for automobile and battery-vehicle motor and contacts
DM5A	C	115	—	3	2	1	

and new applications are constantly being developed. The former include mechanically alloyed materials, high-strength materials prepared by powder metallurgical methods, materials strengthened by reinforcement with steel, silicon carbide, carbon, boron or alumina fibres or refractory oxide dispersions, directionally and rapidly solidified materials and amorphous or glassy alloys with unusual properties. New applications are in memory alloys for temperature sensing and control, in solar energy collector systems, in underground cables both as conductors and sheathing, in battery and welding cables and in back-up conductors for superconductors in addition to its evergrowing role in transport. Magnetic alloys of aluminium are dealt with in Section 5.19.

More details of the properties of aluminium and its alloys can be obtained in the publications of the Aluminium Federation, Birmingham, England, from *Metals Reference Book* (6th Edn, Butterworths, London, 1983) and from *Engineering Alloys* (5th Edn, Van Nostrand Reinhold, New York, 1973).

5.1.3 Carbon

Carbon occurs in nature in two crystalline forms, graphite and diamond, the former being thermodynamically more stable at ambient temperatures and pressures. Synthetic carbons prepared from coke or other precursors may be predominantly graphitic or amorphous depending on the conditions prevailing during manufacture. Diamonds may be prepared from graphite using high-pressure techniques but the resulting products are usually suitable only for industrial applications; new techniques have been reported for the production of gem quality stones by vapour deposition on to small specimens.

In engineering applications synthetic carbons have the advantage of closely controlled impurity levels which can lead to highly predictable properties.

The engineering applications of carbon exploit the following properties: it is thermally stable and in the absence of oxidising atmospheres retains most of its mechanical strength to a temperature of 3500 °C at which, under atmospheric pressure, it sublimes; its oxides are gases and leave no surface film; it is dimensionally stable, does not swell in water and can be machined to close tolerance; it has low expansion coefficient and low density (each about one-quarter that of steel); is a good conductor of heat, with a high specific heat and a great resistance to thermal shock; it is not wetted by molten metals; it is chemically inert; it is self lubricating under normal atmospheric conditions; it is electrically conductive and has high contact resistance with metals.

5.1.3.1 Carbon brushes

Current collection in moving contact forms one of the most important electrical applications of carbon. Brushes must carry heavy current without excessive overheating or wearing of the parts contacted. Low friction, high contact resistance and infusibility are amongst the most desirable characteristics. *Table 5.1* gives data on some widely used brush grades.

Commutators Contact resistance is of major importance in commutation. High contact resistance reduces losses due to high circulating currents minimising problems due to overheating and sparking. Commutator machines use (non-metallic) 100% carbon brushes except in low voltage applications where metal-graphite grades may be used to reduce voltage drop and hence losses. Current density and surface speed must be considered in selection of suitable grades. Low friction materials and good design serve to reduce chatter maintaining good contact between brush and commutator bars. Friction characteristics may be affected by a number of factors including chemical and mechanical properties of commutator metal as well as humidity, temperature, contaminants and abrasives. Ash content of brush material is important in determining friction and abrasion characteristics and in some types of commutator may serve to wear down insulation between commutator segments. The performance of a brush on a commutator machine is influenced by its position on the commutator, i.e. the circumferential and axial stagger.

Slip-rings When commutation phenomena are absent, brush grades with low contact drops, particularly in the metal–graphite class, can be employed. Again, current density and surface speed must be taken into account: at the highest surface speeds it is necessary to select a grade from the natural graphite class. For applications in instrumentation, such as pick-up from thermocouples and strain gauges, silver-graphite brushes on pure silver rings are needed to give minimal and constant contact drop.

Wear The rate of wear of brushes is not directly related to hardness of the brush material but more to the grade, the current density and quality of mechanical features (surface roughness, eccentricity, stability of brush-holders and brush arms and angle of brush relative to the pick-up surface).

Spring-pressure The pressure specified for a particular brush grade is determined from laboratory and field tests to give the optimum performance and life, and should be carefully adhered to. Pressures are normally of the order of 18–21 mN/mm^2, but in conditions of considerable vibration (as in traction and in aircraft) pressures may be as high as 50–70 mN/mm^2.

Brush materials Brushes utilising lampblack base are widely used for medium- and high-voltage machines, d.c. motors and generators and universal motors. These materials possess high contact resistance necessary for good commutation. Graphitised petroleum coke is used for non-metallic slip-ring brushes where high contact resistance is not required. Metal powders, copper or silver, blended with graphite may be used for slipping or low-voltage commutating machinery. The bond in most non-metal brush grades is carbon from pyrolysis of the coal-tar pitch binder. Some graphite grades are bonded with synthetic resins. Final properties are controlled by use of impregnating agents combined with heat-treatments which are designed to control contact resistance, filming action and friction characteristics.

Brush design For standards of brush design, reference should be made to BS 4392 Section A1, and to IEC Publications 136-1 and 136-2, on which the British Standard is based.

5.1.3.2 Linear current collection

This is the reverse of the machine condition in that the conductor is stationary and the collector (equivalent to the brush) is moving. Current collection of this type is found on rail and trolley traction systems, cranes and line process plants. In many applications the collector and conductor are open to aggresive environments including adverse weather conditions. Arcing which results from such situations roughens the conductor and accelerates wear in the collector. Carbon as a contact material provides the real benefits of a very low rate of conductor wear, long collector life and good contact stability. The carbon element, which needs no applied lubrication is very hard and strong. Carbon shoes are in competition with brass or bronze wheels and copper shoes. Radio interference is minimised by use of carbon shoes; the grades most generally used are Link CY3TA and Link MY7D, the latter being metal impregnated.

assembly the contacting joints should be coated with a recommended high-temperature grease to exclude air and moisture from the joint.

Problems may arise with differential thermal expansion effects causing creep of the relatively soft aluminium and relaxation of the joint, but these can be mitigated by the use of Belleville spring washers or a systematic retightening schedule. Aluminium-to-copper joints should be avoided if possible but when mechanical methods are used to join these two materials, the same procedures as for joints between aluminium components are adopted, but exclusion of moisture from the dissimilar metal contact is vital to prevent corrosion and subsequent overheating. Similar greases are used, sometimes with added inhibitors and for additional protection binding with inhibited waterproof tape or covering the entire joint with a moisture-impermeable paint, mastic, or plastic coating is recommended. Alternatively, one may interpose a sandwich material of aluminium-copper foil with its exposed edge extending well beyond the electrical contact area of the joint, or a profiled transition washer which has recently been developed. Prior plating or tinning of mechanically jointed conductors is not currently considered desirable or necessary. In long busbar runs allowance must be made for the high coefficient of expansion of aluminium ($23 \times 10^{-6}/°C$) relative to those of other constructional materials by the incorporation of standard expansion devices.

When aluminium and especially aluminium alloys are used in severe industrial or marine environments they require more protection than that afforded by their usual formation of stable, continuous and generally self-repairing films. This can be provided by anodising in a suitable electrolyte followed by sealing or painting. Recommended procedures and anodising thicknesses are given in BS 1615 (1972, 1982). For some applications aluminium alloys clad with pure aluminium can be obtained, as the latter material has the better corrosion resistance.

5.1.2.3 Other applications of aluminium and its alloys

Switchgear, generators, motors and transformers Aluminium casting alloys containing 0–12% silicon with small additions of iron, copper, magnesium and nickel are widely used for busbar castings and for switchgear and generator casings where lightness is essential. Tensile strengths range up to 345 MN/m^2 and conductivities to 50% of that of IACS copper. With this excellent conductivity and strength, the alloys are suitable for induction motor cage rotors which are fabricated either by integral casting techniques or by brazing pre-cast slot bars to the end rings. Aluminium strip and wire is also used for transformer windings and casting alloys have recently been used in place of steel for a large transformer casing with advantages in reducing the associated magnetic losses.

Heat sinks and heat exchangers Because of its high specific heat (917 J/kg K) and excellent thermal conductivity (238 W/m K) aluminium and its alloys with copper, silicon, magnesium and other metals are commonly used as heat sink and heat exchanger materials. Applications are in heat sinks and cooling fins for semiconductor power rectifiers, in fuel cladding for water-cooled nuclear reactors and in aluminium bronze (up to 14% Al) condenser tubing in power stations.

Transport The widest use of aluminium and its alloys is in transport vehicles because of its low density and the exceptionally high strength-to-weight ratio of some of its alloys. (For example, Alloy 7075 containing 5.6% Zn, 2.5% Mg, 1.6% Cu can be heat treated to give tensile strengths up to 600 MN/m^2, 2–3 times that of standard structural steels.) This alloy and related less highly alloyed materials are finding increasing application for structural and chassis members in electric traction vehicles, in addition to their already common use for cladding. Some of these alloys, e.g. 7075 and 7045, exhibit the practically important property of superplasticity allowing them to be deep drawn by up to 300–400% to intricate shapes. A similar increased use of aluminium and its alloys is also occurring in the engines, bodies, structural members and transmissions of fossil fuel powered vehicles because of the energy savings resulting from their use. In space vehicles, weight saving becomes even more vital to reduce launching costs, so use of 5–6 Mtonnes of aluminium alloys in the first half of the 21st century has been projected for extra-terrestrial solar power generation.

Primary and secondary batteries Although developments are not yet commercially viable there is strong interest in the use of aluminium and its compounds in both primary (including mechanically rechargeable) and secondary cells, because of the ready availability and cheapness of the pure metal and its high free energy of oxidation, with associated high electromotive force. Current developments include its use in place of zinc in primary and secondary metal/air or oxygen batteries employing aqueous alkaline or phosphoric acid electrolytes and either alone or alloyed with lithium in secondary cells with Al or LiAl/FeS or TiS_2 electrodes and molten chloride electrolytes consisting of $AlCl_3$, LiCl, NaCl, KCl eutectics. Another important development close to commerical exploitation which employs aluminium compounds is the sodium/sulphur secondary battery. This has liquid sulphur and sodium electrodes and a solid β', β''-alumina electrolyte. β',β''-alumina is a layer compound (β' is $Na_2O, 11Al_2O_3$, β'' is $Na_2O, 5.33Al_2O_3$) formed by doping pure alumina with sodium oxide, which has a remarkable ionic conductivity for sodium ions. Such batteries provide energy densities up to 100 Wh/kg compared with 10–20 Wh/kg for conventional and 30–40 Wh/kg for advanced lead/acid batteries, coupled with storage efficiencies of 80% and good charge/discharge characteristics.

Capacitors High purity aluminium is widely used as a capacitor plate material. High-capacity, low-volume capacitors can be fabricated using anodically formed oxide films on the metal as the dielectric or alternatively, in a recent development high-capacity compact capacitors for use up to 600 V are made from coils of aluminised polypropylene film. An interesting application for special configuration capacitors formed from anodic oxide films on the metal is their use as secondary instruments for detecting and determining gaseous water concentrations, as the dielectric constant of the oxide is dependent on the water content of the gas in its pores.

Aluminium for protective purposes Aluminium is now used as commonly as zinc for protecting less corrosion resistant materials such as steels and aluminium alloys as it is the most electronegative of all the common cheap materials except magnesium. Thus aluminised coatings are used for protecting steels from atmospheric exposure and gas turbine blades fabricated from advanced nickel and cobalt alloys from corrosion by hot fuel impurities. In marine and industrial applications sacrificial aluminium or aluminium alloy anodes are used in galvanic cathodic protection systems.

5.1.2.4 Miscellaneous applications and future developments

Aluminium's excellent combination of properties—good thermal and electrical conductivity, good corrosion resistance, lightness and exceptional strength/weight ratio when alloyed—has given its use a faster growth rate than other metals. It has already replaced copper in many electrical applications and is now second only to steel in overall annual usage. New types of alloys

5.1.2.1 Resistivity

Pure aluminium has a resistivity of $2.64\ \Omega\text{-m} \times 10^{-8}$ at $20\,°\text{C}$ with a mean temperature coefficient of resistivity over the range $0\text{–}100\,°\text{C}$ of $4.2 \times 10^{-3}/°\text{C}$. Thus it has about 64% of the conductivity of pure copper or 66% of that of the International Annealed Copper Standard (IACS) at $20\,°\text{C}$, but on an equal weight basis the conductivity at this temperature is 2.11 times that of copper and exceeds that of all known materials except the alkali metals.

5.1.2.2 Mechanical properties and electrical conductor applications of aluminium and its alloys

Overhead-line conductors All-aluminium and steel-cored aluminium conductors (ACSR) for overhead lines are currently made to BS 215 (1970)* with a maximum permitted resistivity for the aluminium of $2.83\ \Omega\text{-m} \times 10^{-8}$ at $20\,°\text{C}$ and a constant mass temperature coefficient of resistance of $4.03 \times 10^{-3}\,°\text{C}$ at this temperature. Requirements for the steel core wires, including the new high-strength reinforcing wires are given in BS 4565 (1982) and for the aluminium in BS 2627 (1980). The corrosion, fatigue and creep resistance of ACSR conductors is good and, except in marine or industrially polluted locations, service lives up to 50 years are expected. Complete greasing of both central steel core and of all the inner aluminium wires is now recognised to be necessary to obtain maximum life in polluted conditions and is accepted practice in the UK. Recent studies have shown that the properties of the aluminium wires, particularly their corrosion resistance, are significantly affected by their fabrication route and consequent metallographic structure, with superior material being produced by the extrusion route. Minimum breaking stresses for the aluminium wires used in stranding the overhead lines are specified in BS 215 and 2627 and lie in the range 160–180 MN/m^2 for 2.06–4.65 mm diameter wire. Breaking loads for all-aluminium stranded overhead lines are taken as 95% of the total of those for the individual wires if there are not more than 37 of them or 90% if this number is exceeded.

With ACSR overhead lines practical experience has demonstrated the importance of adequate galvanising of the reinforcing steel wires to avoid galvanic corrosion of the aluminium wires and consequent premature failure. Suitable specifications are given in BS 4565 (1982), 443 (1982) and 215, Pt 2 (1970) which also cover the strength requirements of the steel. Zinc coating weights should lie in the range 0.185–0.305 kg/m^2, dependent on wire diameter. In calculating the resistance of ACSR conductors the BS recommend that the conductivity of the steel wires should be neglected. The breaking load is taken as the sum of the strengths of the aluminium wires (taken from the tables of specified minimum tensile strengths) plus the sum of the strengths of the steel wires calculated from the specified minimum stress at 1% elongation. As the aluminium and steel in ACSR conductors have widely different mechanical properties, the fittings necessary for jointing and anchoring them have to be of special design. These generally take the form of compression fittings applied by a hydraulic or mechanical crimping tool, although bolted clamps are still used for special purposes. In the former a steel sleeve is first compressed over the stranded steel core and then an aluminium sleeve is compressed over the overlying aluminium strands. The joints must meet severe electrical and mechanical requirements specified in BS 3288 (1973, 1979), including heating cycle tests. All steel or cast iron fittings used with any type of aluminium-containing conductor, ACSR, all-aluminium or all-aluminium alloy conductor (AAAC) that are subject to atmospheric exposure should be galvanised, at least to BS 729 (1971) specification. Otherwise severe galvanic corrosion of the aluminium may occur.

All-aluminium alloy conductors, made of alloys containing 0.3–1.0% silicon, 0.4–0.7% magnesium with small amounts of iron and manganese, are being used increasingly for high voltage (400 kV or greater) overhead lines. These alloys can be precipitation hardened by heat treatment to give doubled tensile strengths (relative to aluminium) in the range 310–415 MN/m^2 while retaining conductivities 52–60% of that of IACS copper. Requirements for their manufacture and properties are given in BS 3242 (1970) which specifies a maximum resistivity of $3.28\ \Omega\text{-m} \times 10^{-8}$ (53% IACS conductivity), a constant mass temperature coefficient of resistivity of $3.6 \times 10^{-3}/°\text{C}$ at $20\,°\text{C}$ and breaking stresses in the range 295–335 MN/m^2 dependent on wire diameter. The rules for calculating the breaking stresses of stranded conductors are the same as those for aluminium conductors. Use of AAACs in place of ACSRs can provide useful economic advantages. Their higher strength-to-weight ratios mean that transmission lines can be strung with less sag in spite of their greater coefficients of thermal expansion and, thus, lower lighter towers can be employed. Their higher electrical conductances reduce power losses and joints and end connections are simpler, involving only one compression operation.

Aluminium alloy reinforced aluminium conductors (ACAR), employing a reinforcing core of aluminium alloy wires, are also being used since they offer a good combination of conductivity, strength and weight.

Busbars Aluminium and aluminium alloys of similar compositions to those used in overhead lines are widely used for busbars in the form of round and rectangular section bar and extruded channel sections. Requirements are covered by BS 2898 (1980) and are generally similar to those for overhead line conductor wires except that specified strengths for both types of material are lower (60–85 and 200 MN/m^2 respectively for the commonly used materials designated* 1350 and 6101A). A more stringent lower resistivity of less than $3.133 \times 10^{-8}\ \Omega\text{-m}$ is specified for alloy busbar material. The temperature coefficient of resistivity of the alloy material is less ($3.6 \times 10^{-3}/°\text{C}$) than that of aluminium so at $85\,°\text{C}$ its current carrying capacity is 97% of that of aluminium busbar.

Aluminium busbar has a greater cross-sectional area (by 64%) than copper busbar of equal resistance per unit length (as well as being approximately half the weight) and thus has a greater heat dissipation and current loading. For the same temperature rise and current loading on a circular cross section the diameter of an aluminium bar needs to be only 18% greater than that for a copper bar and the weight will be 58% less. Skin effect currents can set limits to permissible diameters for a.c. busbars; the a.c. rating ($I_{\text{a.c.}}$) is less than the d.c. rating ($I_{\text{d.c.}}$) by the ratio:

$$I_{\text{a.c.}}/I_{\text{d.c.}} = \sqrt{R_{\text{d.c.}}/R_{\text{a.c.}}}$$

where $R_{\text{d.c.}}/R_{\text{a.c.}}$ is the skin effect ratio. Electrical connection (jointing) of aluminium or aluminium alloy busbar can be effected by standard aluminium brazing or soldering techniques but more preferably by inert gas arc welding or by mechanical means (bolting or clamping). Use of the latter method requires special care and even in entirely satisfactory joints resistances can vary by a factor of two. The overlap should be 1.25–2.5 times the width of the bar, with an overall uniform clamping pressure of 10–14 MN/m^2, if necessary maintained by the use of steel back-up plates or substantial washers. Steel or aluminium alloy bolts, washers and plates are best employed. If steel components are used they should be galvanised or aluminised to guard against electrolytic corrosion effects. Filing, scratch brushing or machining under an inert grease immediately prior to assembly is desirable for the production of satisfactory joints. Before

---

* The date given in brackets after the BS No. is that of the latest amendment traced

* The internationally agreed numerical designation scheme for aluminium alloys is described in BS 2627 (1980)

Copper alloys containing 0.7–1.0% cadmium have greater strengths under both static (up to 750 MN/m^2) and alternating stresses and greater resistance to wear, making them useful for contacts and telephone wires, although there is little improvement in strength retention at elevated temperatures. Conductivity is between 80 and 97% of that of IACS material, dependent on the degree of coldwork.

Alloys containing 0.77% chromium can be heat treated to retain their enhanced hardness and tensile strength (up to 480 MN/m^2) even after exposure for more than a 1000 h at 340 °C when the tensile strength of OFHC or tough pitch copper would be reduced to 170–200 MN/m^2. Both tensile strength and conductivity depend on the heat treatment, high strength (solution) treated materials having typical values of 230 MN/m^2 and 45% IACS while for annealed (precipitation hardened) material the values are 450 MN/m^2 and 80%. These materials are used in electrical engineering for welding electrodes and for light current-carrying springs. Copper beryllium and copper zirconium alloys have similar properties and applications with superior notch fracture resistance.

For applications where the highest conductivity is essential and enhanced creep strength at high temperatures is required (e.g. for the rotor conductors in large turbogenerators and for components which have to be tinned, soldered or baked during fabrication) copper alloys containing up to 0.15% silver are used. These have the same conductivity as IACS copper and retain their mechanical properties to 300 °C.

Alloys containing tellurium (0.3–0.7%), sometimes with small amounts of nickel and phosphorus, have machining properties approaching those of free-cutting brass and retain their tensile strength (275 MN/m^2) to 315 °C with improved oxidation resistance. Tellurium additions alone produce only small reductions in conductivity as the solubility of tellurium in copper is only about 0.003% at 600 °C. Copper sulphur (0.4%) and copper lead (0.8%) alloys are also finding applications because of their easy machining and electroplating properties.

For castings electrolytically refined copper is sometimes used as the raw material but more commonly deoxidised tough pitch copper is employed. If high electrical conductivity is to be retained the common deoxidant phosphorus cannot be used because of its high solubility in copper (about 0.6%), so more expensive deoxidants such as silicon, lithium, magnesium, beryllium, calcium or boron are required. Cadmium and chromium copper alloys are also used for castings.

5.1.1.3 Other applications of copper and its alloys

Springs The material selected depends on whether the spring itself is required to carry current. For low conductivity spring materials, phosphor bronze (3.75–6.75% tin, 0.1% phosphorus) and nickel silver (10–30% nickel, 10–35% zinc) are widely used and have conductivities in the ranges 12–27% and 6–8% IACS respectively. Beryllium copper is used in more critical applications: a 2% beryllium alloy can be coldworked and heat-treated to give a tensile strength of 1000–1500 MN/m^2 while retaining a conductivity of 25–35% IACS. With other alloy compositions and heat treatments even higher conductivities can be obtained.

Resistance and magnetic materials Copper alloys containing manganese (9–12%) and aluminium (0–5%) and manganese (10–72%) and nickel (0–10%) are widely used as resistance materials because they have high resistivities (38–48 and 44–175 Ω·m $\times 10^{-8}$ respectively), low or zero temperature coefficients of resistance (-0.03 to $+1.4 \times 10^{-3}$/°C over the range 0–100 °C) and low thermal EMFs relative to those of copper. Copper is also a constituent of many magnetic materials both soft and hard (see Sections 5.13–5.19).

Contacts Palladium copper and silver copper (7.5–50% copper) are suitable for light contacts. For heavy-duty contacts, sintered copper tungsten materials are tough and durable and similar materials are used for electric discharge machining electrodes.

Electroplating alloys Copper alloys containing nickel (7–23%) and zinc (10–35%), known as German or nickel silvers are widely used for electroplating as they form durable corrosion resistant coatings with reflectivities equal to that of standard silver.

Heat transfer materials Copper alloys are extensively used in nuclear and fossil fuel steam plant for electric power generation for heat exchangers such as feedheaters and condensers, although there are increasing tendencies towards the substitution of mild steel for the former and of titanium for the latter. The principal condenser alloys are Admiralty brass (70% Cu, 30% Zn with small arsenic and tin additions to prevent dezincification) and aluminium brass (76% Cu, 2% Al, 0.4% As, balance Zn) where improved corrosion resistance is required. In more erosive conditions caused by the presence of suspended solids in the cooling water, cupronickel alloys (70–90% Cu, 30–10% Ni, with 1–2% Fe, 1–2% Mn) are used at higher initial capital cost and with some penalty in heat transfer. For feedheater applications the cupronickel alloys were most widely used in older plant.

Memory effect alloys Some copper zinc aluminium alloys (8–14% Al, 0–14% Zn, 0–3% Ni) have the useful property of existing in two distinct shapes above and below a critical transformation temperature (within the range -70 °C to $+130$ °C) due to a structural change. They are therefore finding applications as temperature-sensitive actuating devices, backing-up or replacing bimetallic strip or thermistor controlled relay devices.

Superconducting alloys Dendritic copper niobium alloys (20–30% Nb) may be plated or diffused with tin and reacted *in situ* to form fine superconducting Nb$_3$Sn filaments intimately incorporated in a copper matrix. Such materials have high critical current densities in the 8–14 tesla range and improved tolerance to strain (see section 5.2.1).

Future developments The versatility and range of application of copper and its alloys seems almost unlimited. Interesting new developments include amorphous or glassy copper titanium alloys, alloys with enhanced mechanical properties formed by rapid solidification techniques or by powder metallurgical compounding of copper with refractory oxides giving useful strength retention in the range 400–600 °C and superplastic alloys containing aluminium or nickel and zinc which can be drawn 300–400%.

More comprehensive information on the properties and electrical engineering applications of copper and its alloys can be obtained from the publications of the Copper Development Association, from *Metals Reference Book* (6th Edn, Butterworths, London 1983) and from *Engineering Alloys* (5th Edn, Van Nostrand Reinhold, New York, 1973).

5.1.2 Aluminium

Aluminium and its alloys are widely used in the electrical industry because of their good thermal and electrical conductivity, generally excellent mechanical properties and corrosion resistance, ease of fabrication, low density and non-magnetic properties. Aluminium extraction and refining take 3–5% of the electrical power generated in advanced countries, so it has been suggested that bulk elemental aluminium can usefully be regarded as a form of stored energy.

CONDUCTORS AND SUPERCONDUCTORS

5.1 Conducting materials

5.1.1 Copper

Copper, the most commonly used conducting metal, has high electrical and thermal conductivity, good mechanical properties and resistance to corrosion, easy jointing, ready availability and high scrap value. Copper is also the basis of an important family of alloys.

5.1.1.1 Conductivity

Pure copper has a resistivity at 20 °C of 1.679 Ω-m $\times 10^{-8}$, lower than that of any known material except silver. The relative conductivity at 20 °C of other metals compared to that of copper ($=100$) is (Smithells *Metals Reference Book*, 6th edition) silver 104, aluminium 60, nickel 25, iron 17, platinum 16, tin 13, lead 8. The resistivity of a standard purity of copper (IACS, International Annealed Copper Standard) was established in 1913 by the International Electrotechnical Commission as 0.174 21 Ω for the resistance of a metre length of 1 mm² cross-section at 20 °C. The conductivity of all other grades and purities of copper and its alloys is commonly referred to this value taken as 100.

Impurities in high conductivity copper must be closely controlled because of their deleterious effect (*Figure 5.1*). Phosphorus in particular drastically lowers the conductivity, as little as 0.05%* being sufficient to reduce the value to about 70% IACS. Apart from silver, cadmium and zinc, all other impurities that copper may contain cause appreciable reductions in conductivity.

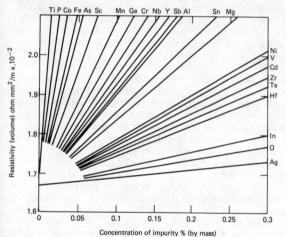

Figure 5.1 Effect of added elements upon the electrical resistivity of copper (courtesy of Copper Development Association)

The resistivity of copper, like that of all pure metals, varies with temperature. A recent comprehensive analysis of the data for pure copper showed that over the temperature range 200–400 K, the resistivity, corrected for changes in linear dimensions, increased linearly with temperature T by 6.78×10^{-11} Ωm/K. Over this temperature range, the temperature coefficient of resistivity, α_T, corrected for thermal expansion is given by $1/(T(\text{K})-45.723)$ or $1/(T(\text{C})+227.457)$ i.e. $4.396 \times 10^{-3}/°\text{C}$ at 0 °C and $4.04 \times 10^{-3}/°\text{C}$ at 20 °C. A slightly different value is recommended by the Copper Development Association for 100% IACS copper of $3.95 \times 10^{-3}/°\text{C}$ at 20 °C.

* All concentrations are in weight per cent

Note that for accurate work, the true temperature coefficient of resistivity will differ from the temperature coefficient at constant mass due to the change in dimensions of the conductor with temperature, but for copper the difference is small, less than $0.02 \times 10^{-3}/°\text{C}$.

The resistivity of copper at liquid helium temperatures (4 K) is also of practical importance because of its increasing use as a thermal and electrical back-up material for superconductors. The resistivity at this temperature is more radically affected by impurities than it is at room temperature and further refining of the copper to purities greater than those of tough pitch (see below) or oxygen free high conductivity copper may be necessary to obtain the required factors of improvement of about 200 or more. (The factor of improvement is the ratio of the resistivities at 0 °C and 4 K.)

5.1.1.2 Mechanical properties and electrical conductor applications of copper and its alloys

Copper as cast has a tensile strength of 150–170 MN/m²; subsequent rolling, drawing or other hot and cold working can raise the tensile strength to 230 for annealed material and to a maximum of 450 MN/m² for hard-drawn wire. Over this range of strengths, tensile moduli increase from 110 to 130 GN/m², Vickers hardness numbers from 50 to 110–130 and ductilities decrease from 45–60% to 5–20%. Cold drawn materials start to recrystallise and lose their strength at temperatures in the range 110–200 °C. The increase in strength due to cold working is associated with a loss of conductivity. For hard-drawn material of tensile strength T MN/m² in the range 300–450 MN/m², there is a loss of conductivity of $0.007T\%$ relative to IACS copper.

For applications in which conductors are not required to withstand severe mechanical stresses or elevated temperatures, it is usual to employ a commercial purity grade (known as tough pitch (TP) copper) containing controlled amounts of oxygen (0.02–0.10%) which produces some hardening and strengthening by precipitation of a fine cuprous oxide/copper eutectic mixture. Because oxygen has negligible solubility in the copper matrix (less than 0.002%) the conductivity is only slightly reduced by the presence of oxygen (1% IACS for 0.08%). Some electrical applications require the casting of intricate shaped conductors and tough pitch copper is not satisfactory for these or for situations in which it is liable to hydrogen embrittlement (e.g. for torch brazing or for hydrogen cooled turbogenerators). For these applications the slightly more expensive oxygen-free high conductivity (OFHC) copper containing less than 0.01% oxygen is specified as it is immune to hydrogen embrittlement. These two materials are most widely used for wire and strip conductors in the electrical industry, for the windings of a.c. and d.c. motors and generators and of transformers. Heavier gauge bar, strip and channel section material of these grades are used for busbars. Large quantities of high conductivity wire and strip are also used in both insulated telephone and power cables and in bare solid and stranded overhead conductors. Another form of cable, the mineral insulated copper-clad cable, employs copper for the sheathing material as well as for the conductors and is invaluable for resisting fire and mechanical damage in critical control or supply circuits where high integrity is essential.

Alloying additions of cadmium, chromium, silver, beryllium and zirconium are most commonly used to enhance the mechanical properties of copper-based materials and to ensure their retention to higher temperatures, but all reduce the electrical conductivity to a greater or lesser extent (*Figure 5.1*). Copper materials hardened by mechanical dispersions of refractory oxides and composite materials strengthened by various types of fibres are currently being developed.

        5.6.4   Hydrogen 5/23
        5.6.5   Vacuum 5/23

5.7   Liquid dielectrics 5/23
        5.7.1   Insulating oils 5/23
        5.7.2   Inhibited transformer oil 5/24
        5.7.3   Synthetic insulating liquids 5/24

5.8   Semi-fluid and fusible materials 5/25
        5.8.1   Bitumens 5/26
        5.8.2   Mineral waxes and blends 5/26
        5.8.3   Synthetic waxes 5/26
        5.8.4   Natural resins or gums 5/26
        5.8.5   Miscellaneous fusible compounds 5/26
        5.8.6   Treatments using fusible materials 5/27
        5.8.7   Synthetic resins 5/27
        5.8.8   Thermo-plastic synthetic resins 5/27
        5.8.9   Thermo-setting synthetic resins 5/29
        5.8.10   Encapsulation 5/30

5.9   Varnishes, enamels, paints and lacquers 5/31
        5.9.1   Air-drying varnishes, paints, etc. 5/31
        5.9.2   Baking varnishes and enamels 5/31
        5.9.3   Solventless varnishes 5/31
        5.9.4   Silicone varnishes 5/31
        5.9.5   Properties of varnishes, etc. 5/31

5.10   Solid dielectrics 5/32
        5.10.1   Rigid boards and sheets 5/32
        5.10.2   Rods 5/35
        5.10.3   Tubes and cylinders 5/35
        5.10.4   Flexible sheets, strips and tapes 5/35
        5.10.5   Sleevings, flexible tubings and cords 5/38
        5.10.6   Wire coverings 5/38
        5.10.7   Moulded and formed compositions, plastics, ceramics, etc. 5/38
        5.10.8   Methods of moulding and forming organic materials 5/39
        5.10.9   Methods of manufacture of inorganic materials (ceramics, etc.) 5/39

5.11   Irradiation effects 5/39
        5.11.1   Type of radiation 5/40
        5.11.2   Irradiation effects 5/41

5.12   Ferroelectrics 5/42
        5.12.1   Manufacture 5/42
        5.12.2   Applications 5/42

MAGNETIC MATERIALS

5.13   Ferromagnetics 5/42

5.14   Electrical steels including silicon steels 5/43
        5.14.1   General 5/43
        5.14.2   Ancilliary properties 5/43
        5.14.3   Chemistry and production 5/45
        5.14.4   Physical form 5/45
        5.14.5   Coatings 5/45
        5.14.6   Grain orientation 5/45
        5.14.7   Test methods 5/45
        5.14.8   Various 5/45

5.15   Soft irons and relay steels 5/45
        5.15.1   Composition 5/45
        5.15.2   Applications 5/46
        5.15.3   Physical forms 5/46

        5.15.4   Metallurgical state 5/46
        5.15.5   Magnetic characteristics 5/46
        5.15.6   Grades and specifications 5/46
        5.15.7   Heat treatment 5/47
        5.15.8   Ageing 5/47
        5.15.9   Test methods 5/47
        5.15.10 Other applications 5/47

5.16   Ferrites 5/47
        5.16.1   Magnetically soft ferrites 5/47
        5.16.2   Other ferrite types 5/49

5.17   Nickel–iron alloys 5/50
        5.17.1   30% Nickel 5/50
        5.17.2   36% Nickel 5/50
        5.17.3   50% Nickel 5/50
        5.17.4   80% Nickel 5/50
        5.17.5   Forms of supply 5/51

5.18   Iron–cobalt alloys 5/51
        5.18.1   24/27% Cobalt iron 5/51
        5.18.2   50% Cobalt iron 5/51

5.19   Permanent magnet materials 5/51
        5.19.1   Alnico alloys 5/52
        5.19.2   Ferrite 5/52
        5.19.3   Rare earth cobalt 5/53
        5.19.4   Neodymium iron boron 5/53
        5.19.5   Bonded materials 5/53
        5.19.6   Other materials 5/54
        5.19.7   Further reading 5/54

SPECIAL ALLOYS AND APPLICATIONS

5.20   Fuses 5/54
        5.20.1   Fuse technology 5/55
        5.20.2   Materials 5/55
        5.20.3   High-speed fuses 5/55
        5.20.4   'M' effect 5/56
        5.20.5   Composite or dual-element fuses 5/56

5.21   Contacts 5/56
        5.21.1   Low voltage, low current contacts 5/56
        5.21.2   Low voltage, high current contacts 5/56
        5.21.3   Forms of contact 5/60

5.22   Special alloys 5/61
        5.22.1   Heating alloys 5/61
        5.22.2   Resistance alloys 5/61
        5.22.3   Controlled expansion alloys 5/61
        5.22.4   Heat-resisting alloys 5/62

5.23   Solders 5/62
        5.23.1   Fluxes 5/62
        5.23.2   Solders 5/62

5.24   Rare and precious metals 5/62

5.25   Temperature-sensitive bimetals 5/62

5.26   Nuclear reactor materials 5/63
        5.26.1   Fuels 5/64
        5.26.2   Fuel cladding 5/64
        5.26.3   Coolant 5/65
        5.26.4   Moderator 5/65
        5.26.5   Pressure vessel 5/65
        5.26.6   Shield 5/65

5 Materials

D R Holmes MA, DPhil, FInstP, FICST, National Corrosion Service, NPL, Teddington (Sections 5.1.1–5.1.2)

N G Dovaston BSc, PhD, MPRI, Sunderland Polytechnic (Section 5.1.3)

B J Maddock MA, FInstP, CEng, MIEE, Central Electricity Research Laboratories, Leatherhead (Section 5.2)

W T Norris ScD, CEng, FIEE, Sr Mem IEEE, CEGB, Walden House, London (Section 5.2)

J R Kirkman BSc, PhD, MIERE, CEng, Sunderland Polytechnic (Sections 5.3.1–5.3.6)

D Grieve BSc, Welwyn Microelectronics, Bedlington, Northumberland (Sections 5.3.7–5.3.10)

A J Pearmain BSc (Eng), PhD, CEng, MIEE, Queen Mary College, London (Sections 5.4–5.12)

A G Clegg MSc, PhD, Magnet Centre, Sunderland Polytechnic (Sections 5.13, 5.19, 5.22, and general editor)

P Beckley BSc, PhD, CEng, FIM, FIEE, British Steel Corporation, Newport (Sections 5.14–5.15)

E C Snelling BSc (Eng), CEng, FIEE, Philips Research Laboratories, Redhill, Surrey (Section 5.16)

R V Major MSc, PhD, FInstP, Telcon Limited, Crawley (Sections 5.17–5.18, 5.25)

A Wright PhD, DSc, CEng, FIEE, University of Nottingham (Section 5.20)

P A Kurn DFH, CEng, MIEE, ERA Technology Limited, Leatherhead (Section 5.21)

S Muckett BSc, CEng, MIM, Agnet Limited, Reading (Section 5.23)

B D Field BSc (Eng) (Met), CEng, MIM, Johnson Matthey Ltd, Harlow (Section 5.24)

H E Evans BSc, PhD, CEng, MIM, Berkley Laboratories, Gloucestershire (Section 5.26)

Contents

CONDUCTORS AND SUPERCONDUCTORS

5.1 Conducting materials 5/3
    5.1.1 Copper 5/3
    5.1.2 Aluminium 5/4
    5.1.3 Carbon 5/7

5.2 Superconductors 5/9
    5.2.1 Varieties of superconductor 5/9
    5.2.2 Stabilisation of magnet conductors 5/10
    5.2.3 High-field electromagnets 5/10
    5.2.4 Power transmission 5/11
    5.2.5 Electronic devices 5/11
    5.2.6 Further reading 5/12

SEMICONDUCTORS, THICK AND THIN FILMS

5.3 Silicon 5/12
    5.3.1 Crystal structure 5/12
    5.3.2 Electrical conduction 5/12
    5.3.3 Extrinsic conduction 5/13
    5.3.4 n-type conduction 5/13
    5.3.5 p-type conduction 5/13
    5.3.6 Single crystal growth and device manufacture 5/14
    5.3.7 Thick and thin film microcircuits 5/15
    5.3.8 Thick film materials 5/15
    5.3.9 Thin film materials 5/16
    5.3.10 Applications of hybrids 5/16

INSULATORS

5.4 Insulating materials 5/17
    5.4.1 Classification 5/17
    5.4.2 Temperature index 5/17

5.5 Properties and testing 5/17
    5.5.1 Physical properties 5/18
    5.5.2 Mechanical properties 5/18
    5.5.3 Electrical properties 5/18
    5.5.4 Chemical properties 5/20

5.6 Gaseous dielectrics 5/20
    5.6.1 Air 5/20
    5.6.2 Nitrogen 5/22
    5.6.3 Sulphur hexafluoride 5/22

improving the performance of systems, i.e. compensating the system, can be more easily appreciated.

Practice in using root locus, Bode diagrams and Nichols charts is very necessary if confidence in design work is to be achieved. To obtain this practice the reader should satisfy himself that he could obtain the diagrams shown in *Table 4.2*.

Further reading

The author has found the following books useful for basic control engineering studies. This list is by no means exhaustive.

Anand, D. K., *Introduction to Control Systems*, Pergamon Press (1974)
Chen, C-F. and Haas, I. J., *Elements of Control Systems Analysis*, Prentice-Hall (1968)
DiStefano, J. J., Stubberud, A. R. and Williams, I. J., *Theory and Problems of Feedback and Control Systems*, Schaum's Outline Series, McGraw-Hill (1967)
Dorf, R. C., *Modern Control Systems*, Addison-Wesley (1980)
Douce, J. L., *The Mathematics of Servomechanisms*, English Universities Press (1963)
Elgard, O. I., *Control Systems Theory*, McGraw-Hill (1967)
Healey, M., *Principles of Automatic Control*, Hodder and Stoughton (1975)
Jacobs, O. L. R., *Introduction to Control Theory*, Oxford University Press (1974)
Langill, A. W., *Automatic Control Systems Engineering*, Vol. I and Vol. II, Prentice-Hall (1965)
Marshall, S. A., *Introduction to Control Theory*, Macmillan (1978)
Power, H. M. and Simpson, R. J., *Introduction to Dynamics and Control*, McGraw-Hill (1978)
Raven, F. H., *Automatic Control Engineering*, McGraw-Hill (1961)
Saucedo, R. and Schiring, E. E., *Introduction to Continuous and Digital Control Systems*, Macmillan (1968)

damping (say $\zeta < 0.5$) an asymptotic plot can be in considerable error around the break frequency and more careful evaluation may be required around this frequency. The phase goes from minus a few degrees at low frequencies towards $-180°$ at high frequencies, being $-90°$ at $\omega = 1/\tau$.

(g) *Quadratic lead term* $(1 + 2\tau\zeta s + \tau^2 s^2)$ This is argued in a similar way to the lag term with the gain curves inscribed and the phase going from plus a few degrees to $180°$ in this case.

Example Plot the Bode diagram of the open-loop frequency-response function

$$G(j\omega) = \frac{10(1 + j\omega)}{j\omega(j\omega + 2)(j\omega + 3)}$$

and determine the gain and phase margins (see *Figure 4.22*). Note: *Table 4.2* shows a large number of examples and also illustrates the gain and phase margins.

4.10.3 The Nichols chart

This is a graph with the open-loop gain in dB as co-ordinate and the phase as abscissa. The open-loop frequency response is for a particular system and is plotted with frequency ω as parameter.

Now the closed-loop frequency response is given by

$$W(j\omega) = \frac{G(j\omega)}{1 + G(j\omega)}$$

and corresponding lines of constant magnitude and constant phase of $W(j\omega)$ are plotted on the Nichols chart as shown in *Figure 4.23*.

When the open-loop frequency response of a system has been plotted on such a chart, the closed-loop frequency response may be immediately deduced from the contours of $W(j\omega)$.

4.11 Conclusion

In this summary of classical control system theory we have outlined the salient features of system characterisation and shown how to view such systems in transfer function form. Considerable attention was paid to the second-order underdamped system, since its characteristic behaviour may often be used in assessing the behaviour of higher-order systems, at least to a first approximation.

We have presented a brief introduction to the three main tools of this type of control system design work, i.e. root locus, Bode diagrams and Nichols charts. From such plots, ways of

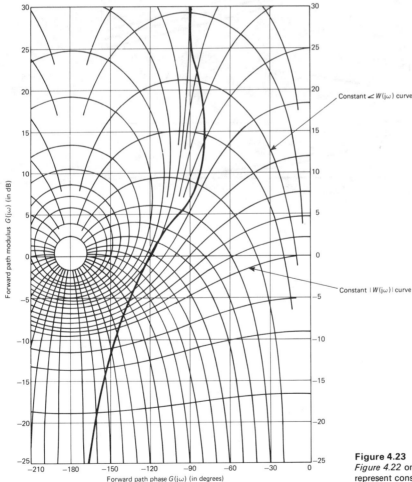

Figure 4.23 Nichols chart and plot of system shown in *Figure 4.22* on this chart. Orthogonal families of curves represent constant $|W(j\omega)|$ and constant $\angle W(j\omega)$

0 dB, and when $\omega^2\tau^2$ is large compared with unity, the gain will be $-20 \log \omega\tau$. With logarithmic plotting this specifies a straight line having a slope of -20 dB/decade of frequency (6 dB/octave) intersecting the 0 dB line at $\omega = 1/\tau$. The actual gain at $\omega = 1/\tau$ is -3 dB and so the plot has the form shown in plot 1 of *Table 4.2*. The frequency at which $\omega = 1/\tau$ is called the *corner* or *break frequency*. The two straight lines, i.e. those with 0 dB and -20 dB/decade, are called the asymptotic approximation to the Bode plot. These approximations are often good enough for not too demanding design purposes.

The phase plot will lag a few degrees at low frequencies and fall to $-90°$ at high frequency, passing through $-45°$ at the break frequency.

(e) *First-order lead term* $(1 + \omega\tau)$ The lead term properties may be argued in a similar way to the above but the gain, instead of falling, rises at high frequencies at 20 dB/decade and the phase, instead of lagging, leads by nearly $90°$ at high frequencies.

(f) *Quadratic-lag term* $[1/(1 + 2\tau\zeta s + \tau^2 s^2)]$ The gain for the quadratic lag is given by

$$-10 \log\left[\left(1 - \left(\frac{\omega}{\omega_n}\right)^2\right)^2 + \left(2\zeta\frac{\omega}{\omega_n}\right)^2\right]$$

and the phase angle

$$\angle G(j\omega) = \arctan\left[-\frac{2\zeta(\omega/\omega_n)}{1 - (\omega/\omega_n)^2}\right]$$

where $\tau = 1/\omega_n$. At low frequencies the gain is approximately 0 dB and at high frequencies falls at -40 dB/decade. At the break frequency $\omega = 1/\tau$ the actual gain is $20 \log(1/2\zeta)$. For low

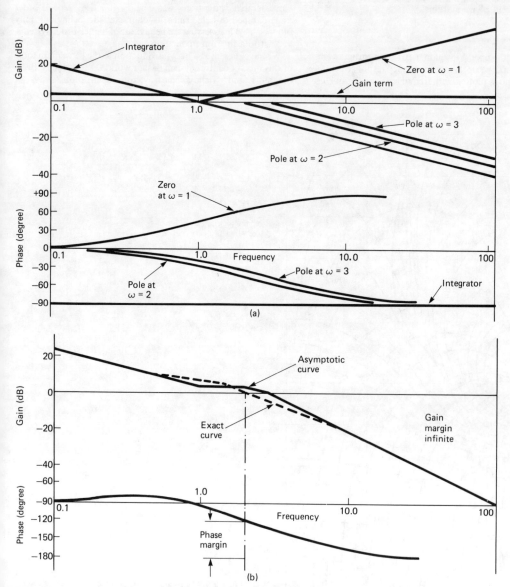

Figure 4.22 (a) Gain and phase curves for individual factors (see *Figure 4.18*); (b) composite gain and phase curves. N.B. phase margin about 60°; gain margin infinite, since phase tends asymptotically to $-180°$

Table 4.2 continued

	$G(s)$	Polar plot	Bode diagram	Nichols diagram	Root locus	Comments
5.	$\dfrac{K}{s(sr_1+1)}$					Elementary instrument servo; inherently stable; gain margin $=\infty$
6.	$\dfrac{K}{s(sr_1+1)(sr_2+1)}$					Instrument servo with field-control motor or power servo with elementary Ward-Leonard drive; stable as shown, but may become unstable with increased gain
7.	$\dfrac{K(sr_a+1)}{s(sr_1+1)(sr_2+1)}$ $r_a < \dfrac{r_1 r_2}{r_1+r_2}$					Elementary instrument servo with phase-lead (derivative) compensator; stable
8.	$\dfrac{K}{s^2}$					Inherently unstable; must be compensated

Table 4.2 Transfer function plots for typical transfer functions

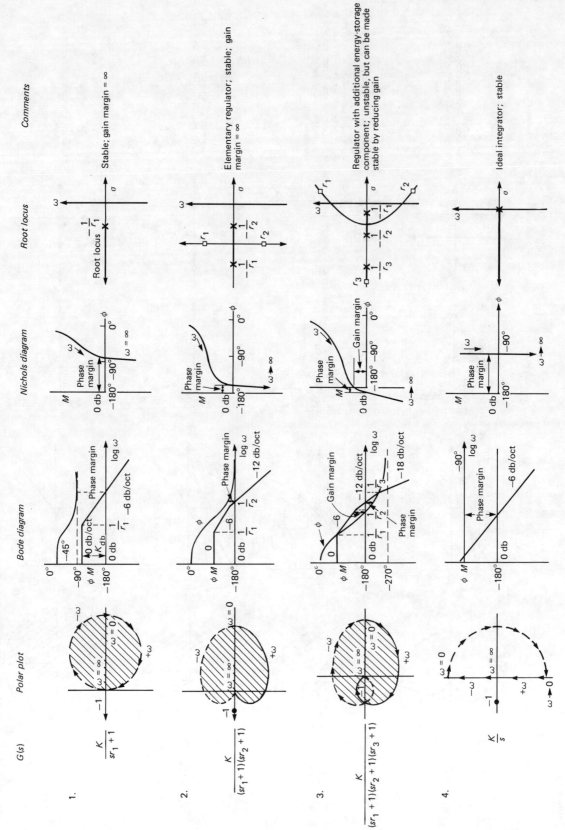

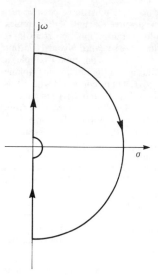

Figure 4.20 Modification of the mapping contour to account for poles appearing at the origin

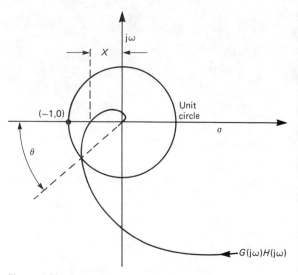

Figure 4.21 Gain and phase margin illustration. Gain margin = $1/X$; phase margin = θ

axis, e.g. integrator $1/s$, then the contour is indented as shown in *Figure 4.20* and the rule above still applies to this modification.

4.10.1.1 Relative stability criteria

Obviously the closer the $H(j\omega)G(j\omega)$ locus approaches the critical point the more critical is the consideration of stability, i.e. we have an indication of relative stability, given a measure by the gain and phase margins of the system.

If the modulus of $H(j\omega)G(j\omega) = X$ with a phase shift of 180°, then the *gain margin* is defined as

gain margin = $1/X$

The gain margin is usually specified in decibels, where we have

gain margin (dB) = $20 \log(1/X) = -20 \log X$

The *phase margin* is the angle which the line joining the origin to the point on the open-loop response locus corresponding to unit modulus of gain makes with the negative-real axis. These margins are probably best appreciated diagrammatically (*Figure 4.21*). They are useful, since a rough working rule for reasonable system damping and stability is to shape the locus so that a gain margin of at least 6 dB is achieved and a phase margin of about 40°.

Examples of the Nyquist plot are shown in *Table 4.2*. Although from such plots the modifications necessary to achieve more satisfactory performance can be easily appreciated, precise compensation arrangements are not easily determined, since complex multiplication is involved and an appeal to the Bode diagram can be more valuable.

4.10.2 Bode diagram

As mentioned above, the Bode diagram is a logarithmic presentation of the frequency response and has the advantage over the Nyquist diagram that individual factor terms may be added rather than multiplied, the diagram can usually be quickly sketched using asymptotic approximations and several decades of frequency may be easily considered.

Now suppose that

$H(s)G(s) = H(s)G_1(s)G_2(s)G_3(s)\ldots$

i.e. the composite transfer function may be thought of as being composed of a number of simpler transfer functions multiplied together, so

$|H(j\omega)G(j\omega)| = |H(j\omega)||G_1(j\omega)||G_2(j\omega)||G_3(j\omega)|\cdots$

$20 \log|H(j\omega)G(j\omega)| = 20 \log|H(j\omega)| + 20 \log|G_1(j\omega)|$
$\qquad + 20 \log|G_2(j\omega)| + 20 \log|G_3(j\omega)| + \cdots$

This is merely each individual factor (in dB) being *added* algebraically to a grand total. Further:

$\angle H(j\omega)G(j\omega) = \angle H(j\omega) + \angle G_1(j\omega) + \angle G_2(j\omega) + \angle G_3(j\omega) + \cdots$

i.e. the individual phase shift at a particular frequency may be *added* algebraically to give the total phase shift.

It is possible to construct Bode diagrams from elemental terms including gain (K), differentiators and integrators (s and $1/s$), lead and lag terms ($(as+1)$ and $(1+as)^{-1}$), quadratic lead and lag terms ((bs^2+cs+1) and $(bs^2+cs+1)^{-1}$), and we consider the individual effects of their presence in a transfer function on the shape of the Bode diagram.

(a) *Gain term K* The gain in dB is simply $20 \log K$ and is *frequency independent*; it merely raises (or lowers) the combined curve $20 \log K$ dB.

(b) *Integrating term, $1/s$* Now $|G(j\omega)| = 1/\omega$ and $\angle G(j\omega) = -90°$ (a constant) and so the gain in dB is given by $20 \log(1/\omega) = -20 \log \omega$. On the Bode diagram this corresponds to a straight line with slope -20 dB/decade (or -6 dB/octave) of frequency and passes through 0 dB at $\omega = 1$; see plot 4 in *Table 4.2*.

(c) *Differentiating term, s* Now $|G(j\omega)| = \omega$ and $\angle G(j\omega) = 90°$ (a constant) and so the gain in dB is given by $20 \log \omega$. On the Bode diagram this corresponds to a straight line with slope 20 dB/decade of frequency and passes through 0 dB at $\omega = 1$.

(d) *First-order lag term $((1+s\tau)^{-1})$* The gain in dB is given by

$20 \log\left(\dfrac{1}{1+\omega^2\tau^2}\right)^{1/2} = -10 \log(1+\omega^2\tau^2)$

and the phase angle is given by $\angle G(j\omega) = -\tan^{-1}\omega\tau$. When $\omega^2\tau^2$ is small compared with unity, the gain will be approximately

$$\frac{Y(s)}{F(s)} = G(s) = \frac{1}{ms^2 + bs + k}$$

If $f(t) = F_0 \exp(j\omega t)$, then

$$y_{ss}(t) = \frac{F_0 \exp(j\omega t)}{(k - \omega^2 m) + j\omega b}$$

whence

$$y_{ss}(t) = \frac{F_0 \exp[j(\omega t - \phi)]}{\sqrt{(k - \omega^2 m)^2 + (b\omega)^2}}$$

where $\phi = \arctan b\omega/(k - m\omega^2)$.

Within the area of frequency-response characterisation of systems three graphical techniques have been found to be particularly useful for examining systems and are easily seen to be related to each other. These techniques are based upon:

(1) The *Nyquist plot*, which is the locus of the frequency-response function plotted in the complex plane using ω as a parameter. It enables stability, in the closed-loop condition, to be assessed and also gives indication of how the locus might be altered to improve the behaviour of the system.
(2) The *Bode diagram*, which comprises two plots, one showing the amplitude of the output frequency response (plotted in decibels dB) against the frequency ω (plotted logarithmically) and the other of phase angle θ of the output frequency response plotted against the same abscissa.
(3) The *Nichols chart*, a direct plot of amplitude of the frequency response (again in decibels) against the phase angle, with frequency ω as a parameter, but further enables the closed-loop frequency response to be read directly from the chart.

In each of these cases it is the *open-loop* steady state frequency response, i.e. $G(j\omega)$, which is plotted on the diagrams.

4.10.1 Nyquist plot

The closed-loop transfer function is given by

$$\frac{C(s)}{R(s)} = \frac{G(s)}{1 + H(s)G(s)}$$

and the stability is determined by the location of the roots of $1 + H(s)G(s) = 0$, i.e. for stability no roots must have positive-real parts and so must not lie on the positive-real half of the complex plane. Assume that the open-loop transfer function $H(s)G(s)$ is stable and consider the contour C, the so-called Nyquist contour shown in *Figure 4.19*, which consists of the imaginary axis plus a semicircle of large enough radius in the right half of the s-plane such that any zeros of $1 + H(s)G(s)$ will be contained within this contour. This contour C_n is mapped via $1 + H(s)G(s)$ into another curve γ into the complex plane s'. It follows immediately from complex variable theory that the closed loop will be stable if the curve γ does not encircle the origin in the s'-plane and unstable if it encircles the origin or passes through the origin. This result is the basis of the celebrated Nyquist stability criterion. It is rather more usual to map not $1 + H(s)G(s)$ but $H(s)G(s)$; in effect this is merely a change of origin from $(0, 0)$ to $(-1, 0)$, i.e. we consider curve γ'_n.

The statement of the stability criterion is that the closed-loop system will be stable if the mapping of the contour C_n by the open-loop frequency-response function $H(j\omega)G(j\omega)$ does not enclose the so-called critical point $(-1, 0)$. Actually further simplification is normally possible, for:

(1) $|H(s)G(s)| \to 0$ as $|s| \to \infty$, so that the very large semicircular boundary maps to the origin in the s'-plane.
(2) $H(-j\omega)G(-j\omega)$ is the complex conjugate of $H(j\omega)G(j\omega)$ and so the mapping of $H(-j\omega)G(-j\omega)$ is merely the mirror image of $H(j\omega)G(j\omega)$ in the real axis.
(3) Note: $H(j\omega)G(j\omega)$ is merely the frequency-response function of the open loop and may even be directly measurable from experiments. Normally we are mostly interested in how this behaves in the vicinity of the $(-1, 0)$ point and therefore only a limited frequency range is required for stability assessment.

The mathematical mapping ideas stated above are perhaps better appreciated practically by the so-called *left-hand rule* for an open-loop stable system, which reads as follows: if the open-loop sinusoidal response is traced out going from low frequencies towards higher frequencies, the closed loop will be stable if the critical point $(-1, 0)$ lies on the left of all points on $H(j\omega)G(j\omega)$. If this plot passes through the critical point, or if the critical point lies on the right-hand side of $H(j\omega)G(j\omega)$, the closed loop will be unstable.

If the open loop has poles that actually lie on the imaginary

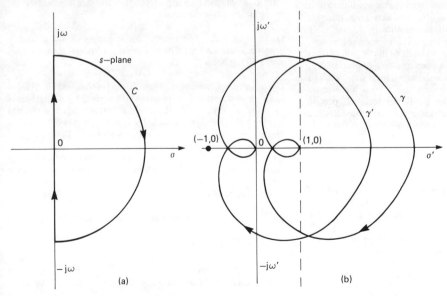

Figure 4.19 Illustration of Nyquist mapping: (a) mapping contour on s-plane; (b) resulting mapping of $1 + H(s)G(s) = 0$ and shift of origin

and become equal when K has the maximum value that will enable them both to be real and, of course, coincident. Thus, an evaluation of K around the breakaway point will rapidly reveal the breakaway point itself.

Example Draw the root locus for

$$KG(s) = \frac{K(s+1)}{s(s+2)(s+3)}$$

Procedure (*Figure 4.18*):

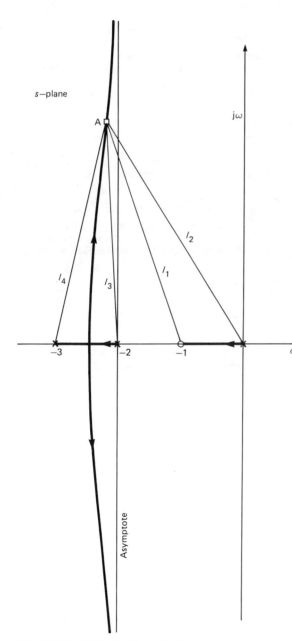

Figure 4.18 Root-locus construction for

$$KG(s) = \frac{K(s+1)}{s(s+2)(s+3)}$$

(1) Plot poles of open-loop system (i.e. at $s=0$, $s=-2$, $s=-3$).
(2) Plot zeros of system (i.e. at $z=-1$).
(3) Determine sections on the real axis at which closed-loop poles can exist. Obviously these are between 0 and -1 (this root travels along real axis between these values as K goes from $0 \to \infty$), and between -2 and -3 (two roots moving towards each other as K increases and of course will break away).
(4) Angle of asymptotes

$$\phi_1 = \tfrac{1}{2} \times 180° = 90°$$
$$\phi_2 = \tfrac{3}{2} \times 180° = 270°$$

(5) Centroid σ_A located at

$$\sigma_A = \frac{-2-3+1}{2} = -2$$

(6) Breakaway point, σ_B

σ_B	-2.45	-2.465	-2.48
K	0.418	0.4185	0.418

(7) Modulus. For a typical root situated at, for example, point A, the gain is given by $K = l_2 l_3 l_4 / l_1$.

After a little practice the root locus can be drawn very rapidly and compensators can be designed by pole-zero placement in strategic positions. A careful study of the examples given in the table will reveal the trends obtainable for various pole-zero placements.

4.10 Frequency-response methods

Frequency-response characterisation of systems has led to some of the most fruitful analysis and design methods in the whole of control system studies. Consider the situation of a linear, autonomous, stable system, having a transfer function $G(s)$, and being subjected to a unit-magnitude sinusoidal input signal of the form $\exp(j\omega t)$, starting at $t=0$. The Laplace transformation of the resulting output of the system is

$$C(s) = G(s)/(s-j\omega)$$

and the time domain solution will be

$$c(t) = G(j\omega) \exp(j\omega t) + \begin{cases} \text{terms whose exponential terms} \\ \text{correspond to the} \\ \text{roots of the denominator of } G(s) \end{cases}$$

Since a stable system has been assumed, then the effects of terms in braces will decay away with time and so after sufficient lapse of time the steady state solution will be given by

$$c_{ss}(t) = G(j\omega) \exp(j\omega t)$$

The term $G(j\omega)$, obtained by merely substituting $j\omega$ for s in the transfer function form, is termed the *frequency-response function*, and may be written

$$G(j\omega) = |G(j\omega)| \angle G(j\omega)$$

where $|G(j\omega)| = \text{mod } G(j\omega)$ and $\angle G(j\omega) = \text{phase } G(j\omega)$. This implies that the output of the system is also sinusoidal of magnitude $|G(j\omega)|$ and with a phase-shift of $\angle G(j\omega)$ with reference to the input signal.

Example Consider the equation of motion

$$m\ddot{y} + b\dot{z} + ky = f(t)$$

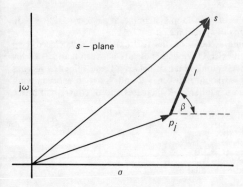

Figure 4.16 Representation of $(s-p_j)$ on the s-plane
$[l=|s-p_j|, \beta = \angle(s-p_j)]$

plane, is illustrated in *Figure 4.16*, where the $\text{mod}(s-p_j)$ and $\text{phase}(s-p_j)$ are also illustrated. The determination of the magnitudes and phase angles for all the factors in the transfer function, for any s, can therefore be done graphically.

The complete set of all values of s_r constituting the root locus may be constructed using the angle condition alone; once found, the gain K giving particular values of s_r may be easily determined from the magnitude condition.

Example Suppose that $G(s) = K/[(s+a)(s+b)]$, then it is fairly quickly established that the only sets of points satisfying the angle condition

$$-\text{phase}(s_r+a) - \text{phase}(s_r+b) = 180 + n360°$$

are on the line joining $-a$ to $-b$ and the perpendicular bisector of this line (*Figure 4.17*).

Rules for construction of the root locus

(1) The angle condition must be obeyed.
(2) The magnitude condition enables calibration of the locus to be carried out.
(3) The root locus on the real axis must be in sections to the left of an odd number of poles and zeros. This follows immediately from the angle condition.
(4) The root locus must be symmetrical with respect to the horizontal real axis. This follows because complex roots must appear as complex conjugate pairs.

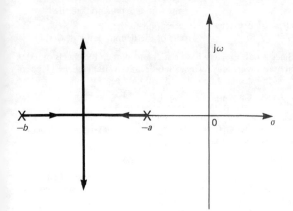

Figure 4.17 Root-locus diagram for
$$KG(s) = \frac{K}{(s+a)(s+b)}$$

(5) Root loci always emanate from the poles of the open-loop transfer function where $K=0$. Consider $a(s) + Kb(s) = 0$; then $a(s) = 0$ when $K = 0$ and the roots of this polynomial are the poles of the open-loop transfer function. Note: this implies that there will be n branches of the root locus.

(6) m of the branches will terminate at the zeros for $K \to \infty$. Consider $a(s) + Kb(s) = 0$, or $(1/K)a(s) + b(s) = 0$, whence as $K \to \infty$, $b(s) \to 0$ and, since this polynomial has m roots, these are where m of the branches terminate. The remaining $n-m$ branches terminate at infinity (in general, complex infinity).

(7) These $n-m$ branches go to infinity along asymptotes inclined at angles ϕ_i to the real axis, where

$$\phi_i = \frac{(2i+1)}{n-m} 180°, \quad i = 0, 1, \ldots, (n-m-1)$$

For consider a root s_r approaching infinity: then $(s_r - a) \to s_r$ for all finite values of a. Thus, if ϕ_i is the phase s_r, then each pole and each zero term of the transfer function term will contribute approximately ϕ_i and $-\phi_i$, respectively. Thus,

$$\phi_i(n-m) = 180° + i360°$$

$$\phi_i = \frac{(2i+1)}{n-m} 180°, \quad i = 0, 1, \ldots, (n-m-1)$$

(8) The centre of these asymptotes is called the asymptote centre and is (with good accuracy) given by

$$\sigma_A = \left(\sum_{i=1}^{n} p_i - \sum_{j=1}^{m} z_i \right) \bigg/ (n-m)$$

This can be shown by the following argument. For very large values of s we can consider that all the poles and zeros are situated at the point σ_A on the real axis. Then the characteristic equation (for large values of s) may be written

$$1 + \frac{K}{(s+\sigma_A)^{n-m}} = 0$$

or approximately by using the binomial theorem

$$1 + \frac{K}{s^{n-m} + (n-m)s^{n-m-1}\sigma_A} = 0$$

Also, the characteristic equation may be written

$$1 + \frac{K \prod_{i=1}^{m}(s+z_i)}{\prod_{i=1}^{m}(s+p_i)} = 0$$

Expanding this for the first two terms results in

$$1 + \frac{K}{s^{n-m} + (a_{n-1} - b_{m-1})s^{n-m-1}} = 0$$

where

$$b_{m-1} = \sum_{i=1}^{m} z_i \quad \text{and} \quad a_{n-1} = \sum_{i=1}^{m} p_j$$

whence

$$(a_{n-1} - b_{m-1}) = (n-m)\sigma_A$$

$$\sigma_A = \frac{a_{n-1} - b_{m-1}}{n-m}$$

as required.

(9) When a locus breaks away from the real axis, it does so at the point where K is a local maximum. Consider the characteristic equation $1 + K[b(s)/a(s)] = 0$; then we can write $K = p(s)$, where $p(s) = -[a(s)/b(s)]$. Now, where two poles approach each other along the real axis they will both be real

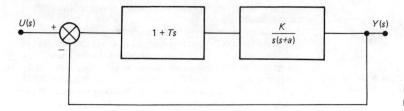

Figure 4.14 Schematic diagram of proportional plus derivative control system

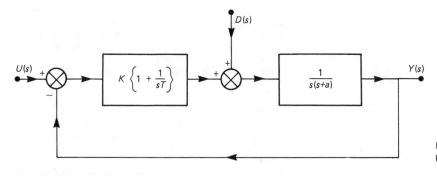

Figure 4.15 Schematic diagram of proportional plus integral control

Assuming that $d(t)$ is a unit step, $D(s)=1/s$, and using the final-value theorem, $\lim_{t\to\infty} y(t)$ is obtained from

$$\lim_{t\to\infty} y(t) = \lim_{s\to 0}\left[\frac{s}{[s(s+a)+K]s}\right] = \frac{1}{K}$$

and so the effect of this disturbance will always be present. By incorporating an integral control as shown in *Figure 4.15* the output will, in the steady state, be unaffected by the disturbance, viz.

$$Y(s) = \frac{Ts}{Ts^2(s+a)+K(1+Ts)} D(s)$$

and so

$$y_{ss} \to 0$$

This controller is called a proportional-plus-integral controller.

An unfortunate side-effect of incorporating integral control is that it tends to destabilise the system, but this can be minimised by careful choice of T. In a particular case it might be that *proportional-plus-integral-plus-derivative control* (PID) may be called for, the amount of each particular type being carefully proportioned.

In the foregoing discussions we have seen, albeit by using specific simple examples, how the behaviour of a plant might be modified by use of certain techniques. It is hoped that this will leave the reader with some sort of feeling for what might be done before embarking on more general tools which tend to appear rather rarefied and isolated unless a basic physical feeling for system behaviour is present.

4.9 Root-locus method

The root locus is merely a graphical display of the *variation of the poles of the closed-loop system* when some parameter, often the gain, is varied. The method is useful, since the loci may be obtained, at least approximately, by straightforward application of simple rules and possible modification to reshape the locus can be assessed.

Considering once again the unity-feedback system with open-loop transfer function $KG(s)=Kb(s)/a(s)$, where $b(s)$ and $a(s)$ represent mth- and nth-order polynomials, respectively, and $n>m$, then the closed-loop transfer function may be written

$$W(s) = \frac{KG(s)}{1+KG(s)} = \frac{Kb(s)}{a(s)+Kb(s)}$$

Note: the system is nth order and the zeros of the closed loop and the open loop are identical for unity feedback. The characteristic behaviour is determined by the roots of $1+KG(s)=0$ or $a(s)+Kb(s)=0$. Thus, $G(s)=-(1/K)$ or $b(s)/a(s)=-(1/K)$.

Let s_r be a root of this equation; then

$$\mod\left[\frac{b(s_r)}{a(s_r)}\right] = \frac{1}{K}$$

and

$$\text{phase}\left[\frac{b(s_r)}{a(s_r)}\right] = 180° + n360°$$

where n may take any integer value, including $n=0$. Let $z_1,\ldots,z_m$ be the roots of the polynomial $b(s)=0$, and $p_1,\ldots,p_n$ be the roots of the polynomial $a(s)=0$. Then

$$b(s) = \prod_{i=1}^{m}(s-z_i)$$

and

$$a(s) = \prod_{i=1}^{n}(s-p_i)$$

Therefore

$$\frac{\prod_{i=1}^{m}|s_r-z_i|}{\prod_{i=1}^{n}|s_r-p_i|} = \frac{1}{K}, \quad \text{the magnitude condition}$$

and

$$\sum_{i=1}^{m}\text{phase}(s_r-z_i) - \sum_{i=1}^{n}\text{phase}(s_r-p_i) = 180° + n360°,$$

the angle or phase condition

Now, given a complex number p_j the determination of the complex number $(s-p_j)$, where s is some point in the complex

The *percentage overshoot* is defined as:

$$\text{percentage overshoot} = \frac{100(\text{max. value of } y(t) - \text{steady state value})}{\text{steady state value}}$$

$$= 100\exp(-\zeta\pi/\gamma) = 100\exp[-a\pi/\sqrt{(4K-a^2)}]$$

i.e. the percentage overshoot increases as the gain K increases.

The *frequency of oscillation* ω_r is immediately seen to be

$$\omega_r = \omega_n\gamma = \sqrt{[K-(a/2)^2]}$$

i.e. the frequency of oscillation increases as the gain K increases.

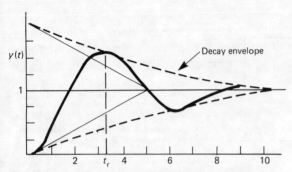

Figure 4.12 Step response of system shown in *Figure 4.11*.
Note: (1) rise time t_r, i.e. time taken to reach maximum overshoot; (2) overshoot displacement of maximum overshoot from newly demanded position; (3) predominant time constant indicated by tangents to the envelope curve

The *predominant time constant* is the time constant associated with the envelope of the response (*Figure 4.12*) which is given by $\exp(-\zeta\omega_n t)$ and thus the predominant time constant is $1/\zeta\omega_n$,

$$\frac{1}{\zeta\omega_n} = \frac{1}{(a/2\sqrt{K})\sqrt{K}} = \frac{2}{a}$$

Note that this time constant is unaffected by the gain K and is associated with the 'plant parameter a', which will normally be unalterable, and so other means must be found to alter the predominant time constant should this prove necessary.

The *settling time* t_s is variously defined as the time taken for the system to reach 2–5% (depending on specification) of its final steady state and is approximately equal to four times the predominant time constant.

It should be obvious from the above that characteristics desired in plant dynamical behaviour may be conflicting (e.g. fast rise time with small overshoot) and it is up to the skill of the designer to achieve the best compromise. Overspecification can be expensive.

A number of the above items can be directly affected by the gain K and it may be that a suitable gain setting can be found to satisfy the design with no further attention. Unfortunately, the design is unlikely to be as simple as this, in view of the fact that the predominant time constant cannot be influenced by K. A particularly important method for influencing this term is the incorporation of so-called *velocity feedback*.

4.8.3 Velocity feedback

Given the prototype system shown in *Figure 4.11*, suppose that this is augmented by measuring the output $y(t)$, differentiating to form $\dot{y}(t)$, and feeding back in parallel with the normal feedback a signal proportional to $\dot{y}(t)$: say $T\dot{y}(t)$. The schematic of this arrangement is shown in *Figure 4.13*. Then by simple manipulation the modified transfer function becomes

$$W'(s) = \frac{K}{s^2 + (a+KT)s + K}$$

whence the modified predominant time constant is given by $2/(a+TK)$. The designer effectively has another string to his bow in that manipulation of K and T is normally very much in his command.

A similar effect may be obtained by the incorporation of a *derivative* term to act on the error signal (*Figure 4.14*) and in this case the transfer function becomes

$$W'(s) = \frac{K(1+Ts)}{s^2 + (a+KT)s + K}$$

It may be demonstrated that this derivative term when correctly adjusted can both stabilise the system and increase the speed of response. The control shown in *Figure 4.14* is referred to as *proportional-plus-derivative* control and is very important.

4.8.4 Incorporation of integral control

Mention has previously been made of the effect of using integrators within the loop to reduce steady state errors; a particular study with reference to input/output effects was given. In this section consideration is given to the effects of disturbances injected into the loop, and we consider again the simple second-order system shown in *Figure 4.11* but with a disturbance occurring between the amplifier and the plant dynamics. Appealing to superposition we can, without loss of generality, put $U(s) = 0$ and the transfer function between the output and the disturbance is given by

$$\frac{Y(s)}{D(s)} = \frac{1}{s(s+a)+K}$$

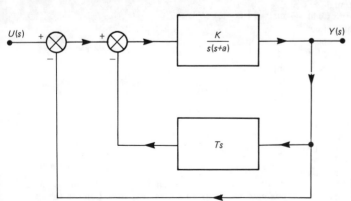

Figure 4.13 Schematic diagram showing incorporation of velocity feedback

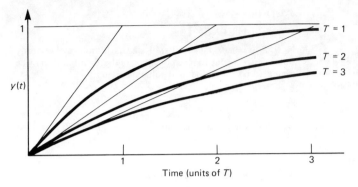

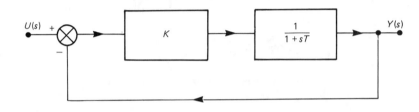

Figure 4.8 First-order lag response to a unit step (time constant = 1, 2, 3 units)

Figure 4.9 First-order lag incorporated in feedback loop

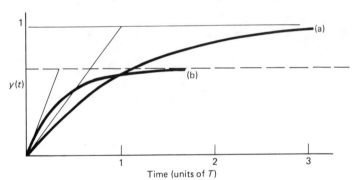

Figure 4.10 Response of first-order lag: (a) open-loop condition ($T=1$); (b) closed-loop condition ($T=1$, $K=2$)

provides the ability to control the effective time constant by altering the gain of an amplifier, the original physical system being left unchanged.

4.8.2 Second-order system

The behaviour characteristics of second-order systems are probably the most important of all, since many systems, of seemingly greater complexity, may often be approximated by a second-order system because certain poles of their transfer function dominate the observed behaviour. This has led to system specifications often being expressed in terms of second-order system behavioural characteristics.

In Section 4.2 the importance of the generator behaviour of the second-order was mentioned, and this subject is now taken further by considering the system shown in *Figure 4.11*.

The closed-loop transfer function for this system is given by

$$W(s) = \frac{KG(s)}{1+KG(s)} = \frac{K}{s^2+as+K}$$

and this may be rewritten in general second-order terms in the form

$$W(s) = \frac{\omega_n^2}{s^2+2\zeta\omega_n s+\omega_n^2}$$

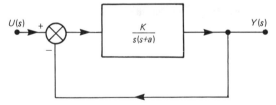

Figure 4.11 Second-order system

where $K=\omega_n^2$ and $\zeta=a/(2\sqrt{K})$. The unit-step response is given by

$$y(t) = 1 - \exp(-\zeta\omega_n t)[\cos(\gamma\omega_n t) - (\zeta/\gamma)\sin(\gamma\omega_n t)]$$

where $\gamma = \sqrt{(1-\zeta^2)}$. This assumes, of course, that $\zeta<1$, so giving an oscillating response decaying with time.

The *rise time* t_r will be defined as the time to reach the first overshoot (note: other definitions are used and it is important to establish which particular definition is being used in a particular specification).

$$t_r = \pi/(\gamma\omega_n) = \pi/\sqrt{[K-(a/2)^2]}$$

i.e. the rise time decreases as the gain K is increased.

Thus,
$$e_{ss} = \frac{1}{1+\infty} \to 0$$

i.e. there is no steady state error in this case and we see that this is due to the presence of the integrator term $1/s$. This is an important practical result, since it implies that steady state errors can be eliminated by use of integral terms.

(b) *Velocity-error coefficient K_v*
Let us suppose that the input demand is a unit ramp, i.e. $u(t) = t$, so $U(s) = 1/s^2$. Then

$$e_{ss} = \lim_{s \to 0}[sE(s)] = \lim_{s \to 0}\left[\frac{1}{s+sKG(s)}\right] = \frac{1}{\lim_{s \to 0}[sKG(s)]} = \frac{1}{K_v}$$

where $K_v = \lim_{s \to 0}[sKG(s)]$ is called the *velocity-error coefficient*.

Examples For a type 0 system $K_v = 0$, whence $e_{ss} \to \infty$.
 For a type 1 system $K_v = K(b_0/a_0)$ and so this system can follow but with a finite error.
 For a type 2 system

$$K_v = \lim_{s \to 0}\left[\frac{K}{s}\frac{b_0}{a_0}\right] \to \infty$$

whence $e_{ss} \to 0$ and so the system can follow in the steady state without error.

(c) *Acceleration-error coefficient K_a*
In this case we assume that $u(t) = t^2/2$, so $U(s) = 1/s^3$ and so

$$e_{ss} = \lim_{s \to 0}[sE(s)] = \lim_{s \to 0}\left[\frac{1}{s^2+s^2KG(s)}\right]$$

$$= \frac{1}{\lim_{s \to 0}[s^2KG(s)]} = \frac{1}{K_a}$$

where $K_a = \lim_{s \to 0}[s^2KG(s)]$ is called the *acceleration-error coefficient* and similar analyses to the above may be performed.

These error-coefficient terms are often used in design specifications of equipment and indicate the minimum order of the system that one must aim to design.

4.7.3 Steady state errors due to disturbances

In *Figure 4.7* the prototype unity-feedback closed-loop system is shown modified by the intrusion of a disturbance, $D(s)$, being allowed to affect the loop. For example, the loop might represent a speed-control system and $D(s)$ might represent the effect of changing the load. Now, since linear systems are under discussion, in order to evaluate the effects of this disturbance on $Y(s)$ (denoted $Y_D(s)$) we may tacitly assume $U(s) = 0$ (i.e. invoke the superposition principle)

$$Y_D(s) = D(s) - KG(s)Y_D(s)$$
$$Y_D(s) = D(s)/[1+KG(s)]$$

Now $E_D(s) = -Y_D(s) = -D(s)/[1+KG(s)]$, and so the steady state error, e_{ssD}, due to the application of the disturbance, may be evaluated by use of the final-value theorem as

$$e_{ssD} = -\lim_{s \to 0}\left[\frac{sD(s)}{1+KG(s)}\right]$$

Obviously the disturbance may enter the loop at other places but its effect may be established by similar analysis.

4.8 Transient behaviour

Having developed means of assessing stability and steady state behaviour, we turn our attention to the transient behaviour of the system.

4.8.1 First-order system

It is instructive to examine first the behaviour of a first-order system (a first-order lag with a time constant T) to a unit-step input, *Figure 4.8*.
 Now

$$\frac{Y(s)}{U(s)} = G(s) = \frac{1}{1+sT}$$

where $U(s) = 1/s$

$$Y(s) = \frac{1}{s(1+sT)} = \frac{1}{Ts[s+(1/T)]} = \frac{1}{s} - \frac{1}{[s+(1/T)]}$$

or $y(t) = 1 - \exp(-t/T)$; note also that $dy/dt = (1/T)\exp(-t/T)$.
 Figure 4.8 shows this time response for different values of T where it will be noted that the corresponding trajectories have slopes of $1/T$ at time $t = 0$ and reach approximately 63% of their final values after T.

Suppose now that such a system is included in a unity-feedback arrangement together with an amplifier of gain K (*Figure 4.9*); therefore

$$\frac{Y(s)}{U(s)} = \frac{K/(1+sT)}{1+K/(1+sT)} = \frac{K}{(1+K)\left(1+s\dfrac{T}{1+K}\right)}$$

For a unit-step input the time response will be

$$y(t) = \frac{K}{1+K}[1-\exp\{-(1+K)(t/T)\}]$$

This expression has the same form as that obtained for the open loop but the effective time constant is modified by the gain and so is the steady state condition (*Figure 4.10*). Such an arrangement

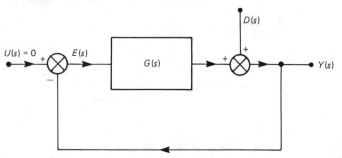

Figure 4.7 Schematic diagram of a disturbance entering the loop

There are cases that arise which need a more delicate treatment.

(1) Zeros occur in the first column, while other elements in the row containing a zero in the first column are non-zero.

In this case the zero is replaced by a small positive number, ε, which is allowed to approach zero once the array is complete.

For example, consider the polynomial equation $s^5 + 2s^4 + 2s^3 + 4s^2 + 11s + 8 = 0$:

s^5:   1   2   11
s^4:   2   4   8
s^3: ε 5   0
s^2: α_1 8
s^1: α_2 0
s^0:   8

where $\alpha_1 = \dfrac{4\varepsilon - 10}{\varepsilon} \simeq -\dfrac{10}{\varepsilon}$ $\alpha_2 = \dfrac{5\alpha_1 - 8\varepsilon}{\alpha_1} \simeq 5$

Thus, α_1 is a large negative number and we see that there are effectively two changes of sign and, hence, the equation has two roots which lie in the right-hand half of this plane.

(2) Zeros occur in the first column and other elements of the row containing the zero are also zero.

This situation occurs when the polynomial has roots that are symmetrically located about the origin of the s-plane, i.e. it contains terms such as $(s+j\omega)(s-j\omega)$ or $(s+v)(s-v)$.

This difficulty is overcome by making use of the auxiliary equation which occurs in the row immediately before the zero entry in the array. Instead of the all-zero row the equation formed from the preceding row is differentiated and the resulting coefficients are used in place of the all-zero row.

For example, consider the polynomial $s^3 + 3s^2 + 2s + 6 = 0$.

s^3:   1   2
s^2:   3   6   (auxiliary equation $3s^2 + 6 = 0$)
s^1:   0   0

Differentiate the auxiliary equation giving $6s = 0$ and compile a new array using the coefficients from this last equation, viz.

s^3:   1   2
s^2:   3   6
s^1:   6   0
s^0:   1

Since there are no changes of sign, the system will not have roots in the right-hand half of the s-plane.

Although the Routh method allows a straightforward algorithmic approach to determination of stability, it gives very little clue as to what might be done if stability conditions are unsatisfactory. This consideration will be taken up later.

4.7 Classification of system and static accuracy

4.7.1 Classification

The discussion in this section is restricted to unity-feedback systems (i.e. $H(s) = 1$) without seriously affecting generalities. We know that the open-loop system has a transfer function $KG(s)$, where K is a constant and we may write

$$KG(s) = \frac{K(s-z_1)(s-z_2)\cdots(s-z_m)}{s^l(s+p_1)(s+p_2)\cdots(s-p_3)} = \frac{K\sum_{k=0}^{m} b_k s^k}{s^l \sum_{k=0}^{n-1} a_k s^k}$$

and for physical systems $n \geq m+1$.

The *order* of the system is defined as the degree of the polynomial in s appearing in the denominator, i.e. n.

The *rank* of the system is defined as the difference in the degree of the denominator polynomial and the degree of the numerator polynomial, i.e. $n - m \geq 1$.

The *class* (or *type*) is the degree of the s-term appearing in the denominator (i.e. l), and is equal to the number of integrators in the system.

Example

(1) $G(s) = \dfrac{s+1}{s^4 + 6s^3 + 9s^2 + 3s}$

implies order 4, rank 3 and type 1.

(2) $G(s) = \dfrac{s^2 + 4s + 1}{(s+1)(s^2 + 2s + 4)}$

implies order 3, rank 1 and type 0.

4.7.2 Static accuracy

When a demand has been made on the system, then it is generally desirable that after the transient conditions have decayed the output should be equal to the input. Whether or not this is so will depend both on the characteristics of the system and on the input demand. Any difference between the input and output will be indicated by the error term $e(t)$ and we know that for the system under consideration

$$E(s) = \frac{U(s)}{1 + KG(s)}$$

Let $e_{ss} = \lim_{t \to \infty} e(t)$ (if it exists), and so e_{ss} will be the steady state error. Now from the final-value theorem we have

$$e_{ss} = \lim_{t \to \infty} e(t) = \lim_{s \to 0} [sE(s)]$$

Thus,

$$e_{ss} = \lim_{s \to 0} \left[\frac{sU(s)}{1 + KG(s)} \right]$$

(a) *Position-error coefficient K_p*

Suppose that the input is a unit step, i.e. $R(s) = 1/s$; then

$$e_{ss} = \lim_{s \to 0} \left[\frac{1}{1 + KG(s)} \right] = \frac{1}{1 + \lim_{s \to 0} [KG(s)]} = \frac{1}{1 + K_p}$$

where $K_p = \lim_{s \to 0} [KG(s)]$ and this is called the *position-error coefficient*.

Example For a type 0 system

$$KG(s) = \left[K \sum_{k=0}^{m} b_k s^k \right] \bigg/ \left[\sum_{k=0}^{n} a_k s^k \right]$$

Therefore $K_p = K(b_0/a_0)$ and $e_{ss} = 1/(1 + K_p)$.

It will be noted that after the application of a step there will always be a finite steady state error between the input and the output but this will decrease as the gain K of the system is increased.

Example For a type 1 system

$$KG(s) = \left[K \sum_{k=0}^{m} b_k s^k \right] \bigg/ \left[s \sum_{k=0}^{n-1} a_k s^k \right]$$

and

$$K_p = \lim_{s \to 0} \left[K \sum_{k=0}^{m} b_k s^k \right] \bigg/ \left[s \sum_{k=0}^{n-1} a_k s^k \right] \to \infty$$

4.5 Generally desirable and acceptable behaviour

Although specific requirements will normally be drawn up for a particular control system, there are important general requirements applicable to the majority of systems. Usually an engineering system will be assembled from readily available components to perform some function, and the choice of these components will be restricted. An example of this would be a Diesel–alternator set for delivering electrical power, in which normally the most convenient Diesel engine–alternator combination will be chosen from those already manufactured.

Even if such a system were assembled from customer-designed components, it would be fortuitous if it performed in a satisfactory self-regulatory way without further consideration of its control dynamics. Hence, it is the control engineer's task to take such a system and devise economical ways of making the overall system behave in a satisfactory manner under the expected operational conditions.

For example, a system may oscillate, i.e. it is unstable; or, although stable, it might tend to settle after a change in input demand to a value unacceptably far from this new demand, i.e. it lacks static accuracy. Again, it might settle to a satisfactory new steady state but only after an unsatisfactory transient response. Alternatively, normal operational load disturbances on the system may cause unacceptably wide variation of the output variable, e.g. voltage and frequency of the engine–alternator system.

All these factors will be normally quantified by an actual design specification, and fortunately a range of techniques is available for improving the behaviour. But the application of a particular technique to improve the performance of one aspect of behaviour often has a deleterious effect on another, e.g. improved stability with improved static accuracy tends to be incompatible. Thus, a compromise is sought which gives the 'best' acceptable all-round performance. We now discuss some of these concepts and introduce certain techniques useful in examining and designing systems.

4.6 Stability

This is a fairly easy concept to appreciate for the types of system under consideration here. Equation (4.2) with the right-hand side made equal to zero governs the free (or unforced, or characteristic) behaviour of the system, and because of the nature of the governing LCCDE it is well known that the solution will be a linear combination of exponential terms, viz.

$$y(t) = \sum_{i=1}^{n} A_i \exp(\alpha_i t)$$

where the α_is are the roots of the so-called characteristic equation.

It will be noted that should any α_i have a positive real part (in general, the roots will be complex), then any disturbance will grow in time. Thus, for stability no roots must lie in the right-hand half of the complex plane or s-plane. In a transfer function context this obviously translates to 'the roots of the denominator must not lie in the right-hand half of the complex plane'. For example, if $W(s) = G(s)/[1 + H(s)G(s)]$, then the roots referred to are those of the equation

$$1 + H(s)G(s) = 0$$

In general, the determination of these roots is a non-trivial task, and as, at this stage, we are interested only in whether the system is stable or not, we can use certain results from the theory of polynomials to achieve this without the necessity of evaluating the roots.

A preliminary examination of the location of the roots may be made using the *Descartes rule of signs*, which states: if $f(x)$ is a polynomial, the number of positive roots of the equation $f(x) = 0$ cannot exceed the number of changes of sign of the numerical coefficients of $f(x)$, and the number of negative roots cannot exceed the number of changes of sign of the numerical coefficients of $f(-x)$. 'A change of sign' occurs when a term with a positive coefficient is immediately followed by one with a negative coefficient, and vice versa.

Example Suppose $f(x) = x^3 + 3x - 2 = 0$; then there can be at most one positive root. Since $f(-x) = -x^3 - 3x - 2$, the equation has no negative roots. Further, the equation is a cubic and must have at least one real root (complex roots occur in conjugate pairs); therefore the equation has one positive-real root.

Although Descartes' result is easily applied, it is often indefinite in establishing stability or not, and a more discriminating test is that due to Routh, which we give without proof.

Suppose that we have the polynomial

$$a_0 s^n + a_1 s^{n-1} \cdots a_{n-1} s + a_n = 0$$

where all coefficients are positive, which is a necessary (but not sufficient) condition for the system to be stable, and we construct the following so-called Routh array:

s^n:	a_0	a_2	a_4	a_6	$\cdots$
s^{n-1}:	a_1	a_3	a_5	a_7	$\cdots$
s^{n-2}:	b_1	b_2	b_3	$\cdots$	
s^{n-3}:	c_1	c_2	c_3	$\cdots$	
s^{n-4}:	d_1	d_2	$\cdots$		

where

$$b_1 = \frac{a_1 a_2 - a_0 a_3}{a_1}, \quad b_2 = \frac{a_1 a_4 - a_0 a_5}{a_1}, \quad b_3 = \frac{a_1 a_6 - a_0 a_7}{a_1}, \ldots$$

$$c_1 = \frac{b_1 a_3 - a_1 b_2}{b_1}, \quad c_2 = \frac{b_1 a_5 - a_1 b_3}{b_1}, \ldots$$

$$d_1 = \frac{c_1 b_2 - b_1 c_2}{c_1}, \ldots$$

This array will have $n+1$ rows.

If the array is complete and *none* of the elements in the first column vanishes, then a sufficient condition for the system to be stable (i.e. the characteristic equation has all its roots with negative-real parts) is for all these elements to be positive. Further, if these elements are not all positive, then the number of changes of sign in this first column indicates the number of roots with positive-real parts.

Example Determine whether the polynomial $s^4 + 2s^3 + 6s^2 + 7s + 4 = 0$ has any roots with positive-real parts. Construct the Routh array:

s^4:	1	6	4
s^3:	2	7	
s^2:	$\frac{(2)(6)-(1)(7)}{2} = 2.5$	$\frac{(2)(4)-(1)(0)}{2} = 4$	
s:	$\frac{(2.5)(7)-(2)(4)}{2.5} = 3.8$		
s^0:	4		

There are five rows with the first-column elements all positive, and so a system with this polynomial as its characteristic would be stable.

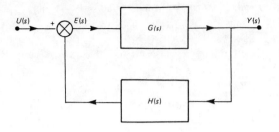

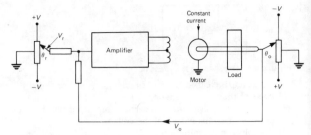

Figure 4.3 Block diagram of prototype feedback system

Figure 4.4 Reduction of diagram shown in *Figure 4.3* to single block

$$\frac{Y(s)}{U(s)} = \frac{G(s)}{1+H(s)G(s)} = W(s)$$

In block diagram form we have *Figure 4.4*. If we eliminate $Y(s)$ from the above equations, we obtain

$$\frac{E(s)}{U(s)} = \frac{1}{1+H(s)G(s)}$$

4.4 Feedback

The last example is the basic feedback conceptual arrangement, and it is pertinent to investigate it further, as much effort in dealing with control systems is devoted to designing such feedback loops. The term 'feedback' is used to describe situations in which a portion of the output (and/or processed parts of it) are fed back to the input of the system. The appropriate application may be used, for example, to improve bandwidth, improve stability, improve accuracy, reduce effects of unwanted disturbances, compensate for uncertainty and reduce the sensitivity of the system to component value variation.

As a concrete example consider the system shown in *Figure 4.5*, which displays the arrangement for an angular position control system in which a desired position θ_r is indicated by tapping a voltage on a potentiometer. The actual position of the load being driven by the motor (usually via a gear-box) is monitored by θ_o,

Figure 4.5 Schematic diagram of simple position control system

indicated, again electrically, by a potentiometer tapping. If we assume identical potentiometers energised from the same voltage supply, then the misalignment between the desired output and the actual output is indicated by the difference between the respective potentiometer voltages. This difference (proportional to error) is fed to an amplifier whose output, in turn, drives the motor. Thus, the arrangement seeks to drive the system until the output θ_o and input θ_r are coincident (i.e. the error is zero).

In the more general block diagram form the above schematic will be transformed to that shown in *Figure 4.6*, where $\theta_r(s)$, $\theta_o(s)$ are the Laplace transforms of the input, output position; $K_1(s)$ and $K_2(s)$ are the potentiometer transfer functions (normally taken as straight gains); $V_r(s)$ is the Laplace transform of reference voltage; $V_o(s)$ is the Laplace transform of output voltage; $G_m(s)$ is the motor transfer function; $G_l(s)$ is the load transfer function; and $A(s)$ is the amplifier transfer function.

Let us refer now to *Figure 4.3* in which $U(s)$ is identified as the transformed input (reference or demand) signal, $Y(s)$ is the output signal and $E(s)$ is the error (or actuating) signal. $G(s)$ represents the *forward transfer function* and is the product of all the transfer functions in the forward loop, i.e. $G(s) = A(s)G_m(s)G_l(s)$ in the above example.

$H(s)$ represents the *feedback transfer function* and is the product of all transfer functions in the feedback part of the loop.

We saw in Section 4.3 that we may write

$$\frac{Y(s)}{U(s)} = \frac{G(s)}{1+H(s)G(s)} \quad \text{and} \quad \frac{E(s)}{U(s)} = \frac{1}{1+H(s)G(s)}$$

i.e. we have related output to input and the error to the input.

The product $H(s)G(s)$ is called the *open-loop transfer function* and $G(s)/[1+H(s)G(s)]$ the *closed-loop transfer function*. The open-loop transfer function is most useful in studying the behaviour of the system, since it relates the error to the demand. Obviously it would seem desirable for this error to be zero at all times, but since we are normally considering systems containing energy storage components, total elimination of error at all times is impossible.

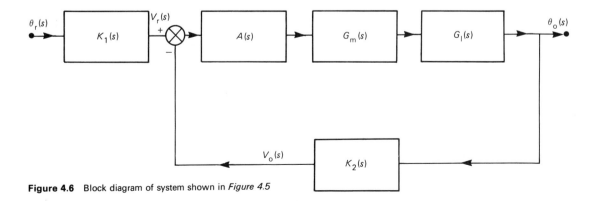

Figure 4.6 Block diagram of system shown in *Figure 4.5*

Figure 4.1 Input–output representation

$z_1, \ldots, z_m$ are called the *zeros* and $p_1, \ldots, p_n$ are called the *poles* of the transfer function.
(3) It is not an explicit function of input or output but depends entirely upon the nature of the system.
(4) The block diagram representation shown in *Figure 4.1* may be extended so that the interaction of composite systems can be studied (provided that they do not load each other); see below.
(5) If $u(t)$ is a delta function $\delta(t)$, then $U(s) = 1$, whence $Y(s) = G(s)$ and $y(t) = g(t)$, where $g(t)$ is the *impulse response* (or weighting function) of the system.
(6) Although a particular system produces a particular transfer function, a particular transfer function does not imply a particular system, i.e. the transfer function specifies merely the input–output relationship between two variables and, in general, this relationship may be realised in an infinite number of ways.
(7) Although we might expect that all transfer functions will be ratios of finite polynomials, an important and common element which is an exception to this is the pure-delay element. An example of this is a loss-free transmission line in which any disturbance to the input of the line will appear at the output of the line without distortion, a finite time (say τ) later. Thus, if $u(t)$ is the input, then the output $y(t) = u(t - \tau)$ and the transfer function $Y(s)/U(s) = \exp(-s\tau)$. Hence, the occurrence of this term within a transfer function expression implies the presence of a pure delay; such terms are common in chemical plant and other fluid-flow processes.

Having performed any manipulations in the Laplace transformation domain, it is necessary for us to transform back to the time domain if the time behaviour is required. Since we are dealing normally with the ratio of polynomials, then by partial fraction techniques we can arrange $Y(s)$ to be written in the following sequences:

$$Y(s) = \frac{K(s - z_1)(s - z_2) \cdots (s - z_m)}{(s - p_1)(s - p_2) \cdots (s - p_n)}$$

$$Y(s) = K\left[\frac{A_1}{s - p_1} + \frac{A_2}{s - p_2} + \cdots + \frac{A_n}{s - p_n}\right]$$

and by so arranging $Y(s)$ in this form the conversion to $y(t)$ can be made by looking up these elemental forms from *Table 4.1*.

Example Suppose

$$Y(s) = \frac{5(s^2 + 4s + 3)}{s^3 + 6s^2 + 8s} = \frac{5(s^2 + 4s + 3)}{s(s + 2)(s + 4)}$$

$$= 5\left[\frac{3}{8s} + \frac{1}{4(s + 2)} + \frac{3}{8(s + 4)}\right]$$

i.e.

$$y(t) = \frac{5}{4}\left[\frac{3}{2}\{1 + \exp(-4t)\} + \exp(-2t)\right]$$

4.2.3 Certain theorems

A number of useful transform theorems are quoted without proof.

(1) *Differentiation*
If $F(s)$ is the Laplace transformation of $f(t)$, then

$$L[d^n f(t)/dt^n] = s^n F(s) - s^{n-1} f(0) - s^{n-2} f'(0) - \cdots - f^{n-1}(0)$$

For example, if $f(t) = \exp(-bt)$, then

$$L\left[\frac{d^3}{dt^3} \exp(-bt)\right] = \frac{s^3}{s + b} - s^2 + bs - b^2$$

(2) *Integration*
If $L[f(t)] = F(s)$, then

$$L\left[\int_0^t f(t)\, dt\right] = \frac{F(s)}{s} + f(0)$$

Repeated integration follows in similar fashion.

(3) *Final-value theorem*
If $f(t)$ and $f'(t)$ are Laplace transformable and if $L[f(t)] = F(s)$, then if the limit of $f(t)$ exists as t goes towards infinity, then

$$\lim_{s \to 0} sF(s) = \lim_{t \to \infty} f(t)$$

For example,

$$F(s) = \frac{b - a}{s(s + a)(s + b)}$$

then

$$\lim_{s \to 0} \frac{s(b - a)}{s(s + a)(s + b)} = \frac{b - a}{ab} = \lim_{t \to \infty} f(t)$$

(4) *Initial-value theorem*
If $f(t)$ and $f'(t)$ are Laplace transformable and if $L[f(t)] = F(s)$, then

$$\lim_{s \to \infty} sF(s) = \lim_{t \to 0} f(t)$$

(5) *Convolution*
If $L[f_1(t)] = F_1(s)$ and $L[f_2(t)] = F_2(s)$, then

$$F_1(s) \cdot F_2(s) = L\left[\int_0^\infty f_1(t - \tau) \cdot f_2(\tau)\, d\tau\right]$$

4.3 Block diagrams

It is conventional to represent individual transfer functions by boxes with an input and output (see note (4), Section 4.2.2). Provided that the components represented by the transfer function do not load those represented by the transfer function in a connecting box, then simple manipulation of the transfer functions can be carried out. For example, suppose that there are two transfer functions in cascade (see *Figure 4.2*); then we may write $X(s)/U(s) = G_1(s)$ and $Y(s)/X(s) = G_2(s)$. Eliminating $X(s)$ by multiplication, we have

$$Y(s)/U(s) = G_1(s)G_2(s)$$

which may be represented by a single block. This can obviously be generalised to any number of blocks in cascade.

Another important example of block representation is the prototype feedback arrangement shown in *Figure 4.3*. We see that $Y(s) = G(s)E(s)$ and $E(s) = U(s) - H(s)Y(s)$. Eliminating $E(s)$ from these two equations results in

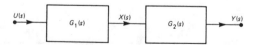

Figure 4.2 Systems in cascade

Table 4.1 Laplace transforms and z-transforms. $\delta(t)$ is the unit impulse function; $h(t)$ is the unit step function $u_T(t)$ is the unit step function followed by a unit negative step at $t=T$. T is the sampling period

$f(t)$	$F(s)$	$F(z)$
0	0	0
$f(t-nT)$	$\exp(-nsT)F(s)$	$z^{-n}F(z)$
$\delta(t)$	1	1
$\delta(t-nT)$	$\exp(-nsT)$	z^{-n}
$\sum_{n=0}^{\infty} \delta(t-nT)$	$(1-\exp(-sT))^{-1}$	$z(z-1)^{-1}$
$h(t)$	s^{-1}	$z(z-1)^{-1}$
$u_T(t)$	$(1-\exp(-sT))s^{-1}$	—
A	As^{-1}	$Az(z-1)^{-1}$
t	s^{-2}	$Tz(z-1)^{-2}$
$f(t)t$	$-dF(s)/ds$	—
$(t-nT)h(t-nT)$	$\exp(-nsT)s^{-2}$	$Tz^{-(n-1)}(z-1)^{-2}$
t^2	$2s^{-3}$	$T^2 z(z+1)(z-1)^{-3}$
t^n	$n!\, s^{-(n+1)}$	—
$\exp(\alpha t)$	$(s-\alpha)^{-1}$	$z(z-\exp(\alpha T))^{-1}$
$f(t)\exp(\alpha t)$	$F(s-\alpha)$	$F(z\exp(-\alpha T))$
$\delta(t)+\alpha\exp(\alpha t)$	$s(s-\alpha)^{-1}$	—
$t\exp(\alpha t)$	$(s-\alpha)^{-2}$	$Tz\exp(\alpha T)(z-\exp(\alpha T))^{-2}$
$t^n \exp(\alpha t)$	$n!\,(s-\alpha)^{-(n+1)}$	—
$\sin \omega t$	$\dfrac{\omega}{s^2+\omega^2}$	$\dfrac{z\sin\omega T}{z^2-2z\cos\omega T+1}$
$\cos \omega t$	$\dfrac{s}{s^2+\omega^2}$	$\dfrac{z(z-\cos\omega T)}{z^2-2z\cos\omega T+1}$
$\dfrac{t}{2\omega}\sin\omega t$	$\dfrac{s}{(s^2+\omega^2)^2}$	—
$\dfrac{1}{2\omega}(\sin\omega t - \omega t\cos\omega t)$	$\dfrac{\omega^2}{(s^2+\omega^2)^2}$	—
$\dfrac{1}{\cos\delta}\sin(\omega t+\delta)$	$\dfrac{A}{s^2+\omega^2}\left(s+\dfrac{\omega}{A}\right)$ where $\tan\delta=A$	—
$\dfrac{1}{\cos\delta}\cos(\omega t+\delta)$	$\dfrac{1}{s^2+\omega^2}(s-A\omega)$	—
$\exp(\alpha t)\sin\omega t$	$\dfrac{\omega}{(s-\alpha)^2+\omega^2}=\dfrac{\omega}{(s-\alpha+j\omega)(s-\alpha-j\omega)}$	$\dfrac{z\exp(\alpha T)\sin\omega T}{z^2-2z\exp(\alpha T)\cos\omega T+\exp(2\alpha T)}$
$\exp(\alpha t)\cos\omega t$	$\dfrac{s-\alpha}{(s-\alpha)^2+\omega^2}$	$\dfrac{z(z-\exp(\alpha T)\cos\omega T)}{z^2-2z\exp(\alpha T)\cos\omega T+\exp(2\alpha T)}$
$\dfrac{t}{2\omega}\exp(\alpha t)\sin\omega t$	$\dfrac{s-\alpha}{[(s-\alpha)^2+\omega^2]^2}$	—
$\dfrac{1}{2\omega}\exp(\alpha t)(\sin\omega t - \cos\omega t)$	$\dfrac{\omega^2}{[(s-\alpha)^2+\omega^2]^2}$	—
$\dfrac{1}{\cos\delta}\exp(\alpha t)\sin(\omega t+\delta)$	$\dfrac{A}{(s-\alpha)^2+\omega^2}\left(s-\alpha+\dfrac{\omega}{A}\right)$ where $\tan\delta=A$	—
$\dfrac{1}{\cos\delta}\exp(\alpha t)\cos(\omega t+\delta)$	$\dfrac{1}{(s-\alpha)^2+\omega^2}(s-\alpha-A\omega)$	—
$\sinh\omega t$	$\omega(s^2-\omega^2)^{-1}$	—
$\cosh\omega t$	$s(s^2-\omega^2)^{-1}$	—
$f'(t)$	$sF(s)-f(0-)$	—
$f''(t)$	$s^2 F(s)-sf(0-)-f'(0-)$	—
$f^n(t)$	$s^n F(s)-s^{n-1}f(0-)-s^{n-2}f'(0-)\cdots - f^{n-1}(0-)$	—
$f^{-1}(t)$	$\dfrac{F(s)}{s}+\dfrac{f^{-1}(0-)}{s}$	—
$f(t)\big|_{t\to 0}$	$sF(s)\big|_{s\to\infty}$	$F(z)\big|_{z\to\infty}$
$f(t)\big|_{t\to\infty}$	$sF(s)\big|_{s\to 0}$	$(z-1)z^{-1}F(z)\big|_{z\to 1}$

4.1 Introduction

Examples of the conscious application of control ideas have appeared in technology since very early times; certainly the float-regulator schemes of ancient Greece were notable landmarks in this area. Much later came the automatic direction-setting of windmills and the Watt governor and its derivatives. The first third of the present century saw applications in the areas of automatic ship-steering, process control in the chemical industry, and so forth. For some of these latter applications serious analytical attention was given to the dynamical behaviour of such systems and attempts were made to account for their sometimes seemingly capricious behaviour.

However, it was not until during and immediately after World War II that the fundamentals of the above somewhat disjointed control studies were subsumed into a coherent body of knowledge which was recognised as a new engineering discipline in its own right. The main antecedents of the great thrust forward in control at this time may be traced to the theoretical developments in the burgeoning electronics industry in the decade or so before World War II. In this area considerable efforts were expended with great success in formulating and understanding the pervasive concept of feedback which is so important to the control engineer.

Since that time tremendous efforts have been made to extend the boundaries of control engineering knowledge. For example, ideas from classical mechanics and the calculus of variations have been taken up, extended and reformulated from a control-theoretic viewpoint. However, it must be admitted that the practical usefulness of many of these developments has yet to be demonstrated. Most control system design still relies heavily on the work and techniques developed during World War II, aided and extended, it must be stressed, by the computational abilities of computers (from main-frame to micro versions); and, indeed, computers themselves are now fairly common elements in actual control systems.

Techniques from this 'classical' period of control development have proved themselves to be easily understood, wide-ranging in application and (probably most important) capable of coping with deficiencies in detailed knowledge about the systems to be controlled. Further work in recent years has led to new research interest in these 'old' methods, for they appear to be capable of systematic extension into the difficult area of multi-variable system control design.

With the above comments in mind, we present the classical conventional approach to control theory, secure in the knowledge that a basic understanding of these ideas and methods has stood practitioners in good stead for several decades.

4.2 Laplace transforms and the transfer function

In most engineering analysis it is usual to produce mathematical models (of varying precision) to predict the behaviour of physical systems. Often such models are manifested by a differential equation description. This appears to fit in with the causal behaviour of idealised components, e.g. Newton's law relating the second derivative of displacement to the applied force. It is possible to model such behaviour in other ways (for example, using integral equations), although these are much less familiar to most engineers. All real systems are non-linear; however, it is fortuitous that most systems behave approximately like linear ones, with the implication that superposition holds true to some extent. We shall further restrict the coverage in that we shall be concerned particularly with systems whose component values are not functions of time—at least over the time-scale of interest to us.

In mathematical terms this latter point implies that the resulting differential equations are not only linear, but also have constant coefficients, e.g. many systems behave approximately according to the equation

$$\frac{d^2x}{dt^2} + 2\zeta\omega_n \frac{dx}{dt} + \omega_n^2 x = \omega_n^2 f(t) \tag{4.1}$$

where x is the dependent variable—displacement, voltage, etc.; $f(t)$ is a forcing function—force, voltage source, etc.; and ω_n^2 and ζ are constants whose values depend upon the size and interconnection of the individual physical components making up the system—spring-stiffness constants, inductance values, etc.

Equations having the form of Equation (4.1) are called linear constant coefficient ordinary differential equations (LCCDE) and may, of course, be of any order. There are several techniques available for solving such equations but the one of particular interest here is the method based upon the Laplace transformation. This is treated in detail elsewhere but it is useful to outline the specific properties of particular interest here.

4.2.1 Laplace transformation

Given a function $f(t)$, then its Laplace transformation, $F(s)$, is defined as

$$L[f(t)] = F(s) = \int_0^\infty f(t)\exp(-st)\,dt$$

where, in general, s is a complex variable and of such a magnitude that the above integral converges to a definite functional value.

A list of Laplace transformation pairs is given in *Table 4.1*.

The essential usefulness of the Laplace transformation technique in control engineering studies is that it transforms LCCDE and integral equations into algebraic ones and, hence, makes for easier and standard manipulation.

4.2.2 The transfer function

This is a central notion in control work and is, by definition, the Laplace transformation of the output of a system divided by the Laplace transformation of the input, with the tacit assumption that all initial conditions are at zero.

Thus, in *Figure 4.1*, where $y(t)$ is the output of the system and $u(t)$ is the input, then the transfer function $G(s)$ is

$$L[y(t)]/L[u(t)] = Y(s)/U(s) = G(s)$$

Supposing that $y(t)$ and $u(t)$ are related by the general LCCDE

$$a_n \frac{d^n y}{dt^n} + a_{n-1} \frac{d^{n-1} y}{dt^{n-1}} + \cdots + a_0 y$$

$$= b_m \frac{d^m u}{dt^m} + b_{m-1} \frac{d^{m-1} u}{dt^{m-1}} + \cdots + b_0 u \tag{4.2}$$

then, on Laplace transforming and ignoring initial conditions, we have (see later for properties of Laplace transformation)

$(a_n s^n + a_{n-1} s^{n-1} + \cdots + a_0)Y(s)$
$= (b_m s^m + b_{m-1} s^{m-1} + \cdots + b_0)U(s)$

whence

$$\frac{Y(s)}{U(s)} = G(s) = \sum_{i=0}^{m} b_i s^i \bigg/ \sum_{i=0}^{n} a_i s^i$$

There are a number of features to note about $G(s)$:

(1) Invariably $n > m$ for physical systems.
(2) It is a ratio of two polynomials which may be written

$$G(s) = \frac{b_m(s-z_1)\cdots(s-z_m)}{a_n(s-p_1)\cdots(s-p_n)}$$

Control System Analysis

J O Flower, BSc(Eng), PhD, DSc(Eng),
CEng, FIEE, FIMarE
University of Exeter

Contents

4.1 Introduction 4/3

4.2 Laplace transforms and the transfer function 4/3
    4.2.1 Laplace transformation 4/3
    4.2.2 The transfer function 4/3
    4.2.3 Certain theorems 4/5

4.3 Block diagrams 4/5

4.4 Feedback 4/6

4.5 Generally desirable and acceptable behaviour 4/7

4.6 Stability 4/7

4.7 Classification of system and static accuracy 4/8
    4.7.1 Classification 4/8
    4.7.2 Static accuracy 4/8
    4.7.3 Steady state errors due to disturbances 4/9

4.8 Transient behaviour 4/9
    4.8.1 First-order system 4/9
    4.8.2 Second-order system 4/10
    4.8.3 Velocity feedback 4/11
    4.8.4 Incorporation of integral control 4/11

4.9 Root-locus method 4/12

4.10 Frequency-response methods 4/14
    4.10.1 Nyquist plot 4/15
    4.10.2 Bode diagram 4/16
    4.10.3 The Nichols chart 4/20

4.11 Conclusion 4/20

kinetic energy acquired by the rotor during this period is represented by area A. At δ_s the power–angle relation becomes P_1 corresponding to a single healthy link. This reverses ΔP and the rotor decelerates. At the angle δ_2 such that area B (representing kinetic energy returned from the rotor) is equal to A, the rotor speed is again synchronous. However, ΔP is now reversed and the rotor will begin to swing back. The range of rotor-angle excursions is stable, but there is a critical value of the fault-clearance angle δ_s that, if exceeded, will result in instability. If all the power–angle relations are true sinusoids, the critical angle can be found analytically.

Conditions other than that shown in *Figure 3.49* can be dealt with if the relevant power–angle relations can be drawn. It is to be noted that a *swing* curve may be required to relate rotor angle to *time*, as it is the time of fault clearance (or other event)—a quantity based on the delay of switch opening—that is normally specified.

the broken curve, the angle cannot immediately change because of the inertia of the generator and prime-mover rotors. The electrical power drops, and a power difference ΔP appears, accelerating the rotating members towards the new balancing angle δ_1. Overshoot takes the rotor angle to δ_2. If the disturbance is not severe, the rotor assumes the angle δ_1 after some rapidly decaying oscillations of frequency 1 or 2 Hz. The angle–time relation is that of curve A in *Figure 3.48*. However, if ΔP is large, the overshoot may cause loss of synchronism—the unstable curve B. A comprehensive investigation of stability thus involves the calculation of *swing* curves for the machines concerned.

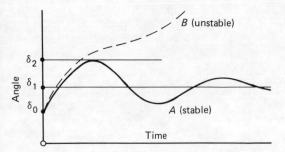

Figure 3.48 Swing curves

Equation of motion The equation of motion for a *single* machine is

$$M(d^2\delta/dt^2) + K_1(d\delta/dt) + K_2 = \Delta P$$

where M is the angular momentum. If the damping coefficients K_1 and K_2 are ignored, the equation of motion reduces to $d^2\delta/dt^2 = \Delta P/M$.

A mass of inertia J rotating at angular speed ω stores a kinetic energy $W = \tfrac{1}{2}J\omega^2$. The momentum $M = J\omega$ can be usefully related to the machine rating S by the *inertia constant*

$$H = W/S = \tfrac{1}{2}M\omega_1/S = \tfrac{1}{2}J\omega_1^2/S \simeq 20 J n_1^2/S$$

in which ω_1 is the synchronous angular speed (rad/s) and $n_1 = \omega_1/2\pi = f/p$ is the corresponding rotational speed (r/s) for a machine with $2p$ poles operating at a frequency f. The magnitude of H (in joules/volt-ampere, or MJ/MVA, or seconds) has the typical values given in *Table 3.6*.

A direct solution of the equation of motion is not normally possible, and a *step-by-step* process must be adopted. For this, a succession of time intervals (e.g. 50 ms) is selected and the rotor acceleration ($d^2\delta/dt^2$) is calculated at the beginning of each. Assuming the acceleration to be constant throughout a time interval, the angular velocity and the movement δ during the interval can be found. At the end of the first interval, the new ΔP is obtained from the power–angle curve and used to calculate the acceleration during the second interval, and so on. The complete *swing curve* can thus be obtained. The method can be extended (if the relevant data are available) to include damping, changes in excitation, saliency, prime-mover governor action and other factors that affect the swing phenomenon.

Multi-machine system A *two-machine* system with a transmission link can be represented by a single machine feeding an infinite busbar and having an equivalent momentum $M = M_1 M_2/(M_1 + M_2)$.

A *group* of machines 1, 2,... paralleled on the same busbar can be treated as a single machine of rating $S = S_1 + S_2 + \ldots$ and of equivalent momentum $M = (S_1/S)M_1 + (S_2/S)M_2 + \ldots$.

For a *multi-machine* network, a separate equation of motion must be set up for each generator and a step-by-step solution undertaken. Determination of ΔP for each machine at the end of a time interval involves a comprehensive loadflow calculation by computer.

Equal-area criterion Neglecting damping, governor action and changes in excitation, the stability of a simple generator/link/infinite-busbar system can be checked graphically using power–angle relations. Consider the system in *Figure 3.49*, with the generator operating at a load angle δ_0 on the power–angle curve P_2 with a prime-mover input P_m and both transmission links intact. A fault occurs on one link, changing the power–angle relation to P_f and giving a power difference ΔP between P_m and the electrical output. As a result the rotor accelerates until the angle δ_s is reached, when the faulted link is switched out. The

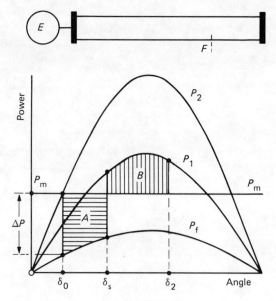

Figure 3.49 Equal-area stability criterion

Table 3.6 Inertia constants of 50 Hz synchronous machines

Machine	n (r/s)	H (s)	Machine	n (r/s)	H (s)
Turbogenerators	50	3–7	Hydrogenerators	8.3	2–4
	25	5–10		5	2–3.5
Compensators	—	1–1.25		2.5	2–3
Motors	—	2–2.25		1.7	15–2.5

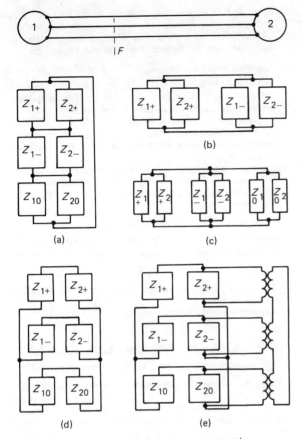

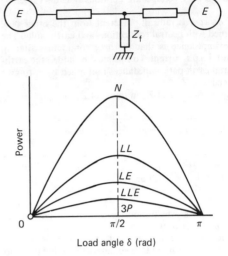

Figure 3.46 Power–angle relations

Figure 3.45 Interconnected phase-sequence networks

as the voltage falls, and beyond the maximum-power conditions the machines will stall, and will draw heavy 'pick-up' currents after a restoration of the voltage.

(2) *Synchronous stability* The receiving-end active and reactive powers in terms of V_s and V_r, and the parameters $ABCD$, are given for a transmission link in Section 3.2.13.1. For a short line, $A = 1 \angle 0°$ and $B = Z \angle \beta$, where $Z = R + jX$ and $\beta = \arctan(X/R)$, conditions shown in *Figure 3.27*.

If the resistance R can be neglected (as is often the case, especially where the link includes terminal transformers), the receiving-end active power P_r and its maximum P_{rm} become

$$P_r = (V_s V_r / X) \sin \theta \quad P_{rm} = V_s V_r / X$$

To attain maximum active power, the receiving end must also accept a leading reactive power $Q_r = V_r^2 / X$.

Interpreting the angle θ between V_s and V_r as that between the generator rotor (indicated by the e.m.f. E_t) and V_r, and including the appropriate generator reactance in X, the angle is now the *load angle* δ, and maximum active power transfer will occur for a load angle $\delta = \pi/2$ rad (90° elec.). The relation for normal conditions is marked N in *Figure 3.46*.

Although a system does not operate under continuous steady-state conditions with a system fault, the power–angle relation is important in the assessment of transient stability. The network for which the curve is calculated is obtained by connecting a 'fault shunt' Z_f at the point of fault. The value of Z_f is in terms of Z_- and Z_0, respectively the total impedance to n.p.s. and z.p.s. currents up to the point of fault. These values are given below for line–line

(LL), single-earth (LE), double-earth (LLE) and 3-phase (3P) faults, while the corresponding power–angle relations are shown in *Figure 3.46*.

Fault	LL	LE	LLE	3P
Z_f	Z_-	$Z_- + Z_0$	$Z_- Z_0 / (Z_- + Z_0)$	zero

3.3.5.2 Transient conditions

If a system in a steady state is subjected to a sudden disturbance (e.g. short circuit, load change, switching out of a loaded circuit) the power demand will not immediately be balanced by change in the prime-mover inputs. To restore balance the rotors of the synchronous machines must move to new relative angular positions; this movement sets up angular oscillations, with consequent oscillations of current and power that may be severe enough to cause loss of synchronism. The phenomenon is termed *transient instability*.

Rotor angle For a single machine connected over a transmission link to an infinite busbar, the simple system of *Figure 3.47* applies. The mechanical input P_m is, in the steady state, balanced by the electrical output for the angle δ_0 on the full-line power–angle relation. If an electrical disturbance occurs such that the power–angle relation is suddenly changed to that indicated by

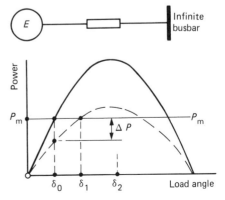

Figure 3.47 Single-machine system

source e.m.f.s: the n.p.s. voltages are 'fictitious' ones developed by the fault.

The *z.p.s. network* is radically different from the other two, being concerned with neutral connections and earth faults. The effective line impedance is that of three conductors sharing equally the total n.p.s. current. To this must be added the earth-connection and earth-path impedances multiplied by 3, to give the z.p.s. impedance Z_0.

Typical values of Z_- and Z_0 in terms of Z_+ are

Ratio	Generator	Transformer	Transmission link	
			Overhead	Cable
Z_-/Z_+	0.6–0.7	1.0	1.0	1.0
Z_0/Z_+	0.1–0.8	1.0 or ∞	3–5	1–3

The value of Z_0 for a synchronous generator depends on the arrangement of the stator winding.

To evaluate the system when faulted, it is necessary to determine the fault currents and the voltage of the sound line(s) to earth. If the voltages and currents at the fault are V_a, V_b, V_c and I_a, I_b, I_c respectively, the following expressions always apply:

$$V_a = E_a - I_+ Z_+ - I_- Z_- - I_0 Z_0 \qquad I_a = I_+ - I_- + I_0$$
$$V_b = h^2 E_a - h^2 I_+ Z_+ - h I_- Z_- - I_0 Z_0 \qquad I_b = h^2 I_+ + h I_- + I_0$$
$$V_c = h E_a - h I_+ Z_+ - h^2 I_- Z_- - I_0 Z_0 \qquad I_c = h I_+ + h^2 I_- + I_0$$

where h is the 120° rotation operator (see Section 3.2.12). From the boundary conditions at the fault concerned it is possible to write three equations and to solve them for the symmetrical components I_+, I_- and I_0.

3.3.4.2 Boundary conditions

Three simple cases are shown in *Figure 3.44*. It is assumed that only fault currents are concerned, and that in-feed to the fault is from one direction.

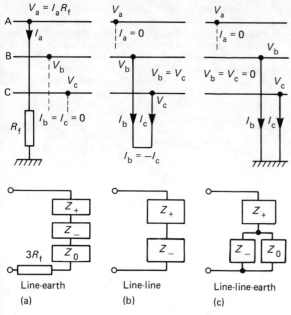

Figure 3.44 Boundary conditions at line faults

(a) *Earth fault of resistance R_f on line A.* At the fault, $V_a = I_a R_f$ and $I_b = I_c = 0$. This leads to $I_{a0} = I_{a+} = I_{a-}$, so that the three sequence currents in phase A are identical. It follows that I_b and I_c are zero, as required. From the basic equations

$$I_{a+} = I_{a-} = I_{a0} = \tfrac{1}{3} I_a = E_a/(Z_+ + Z_- + Z_0 + 3R_f) = E_a/Z$$

and the fault current is $I_a = 3E_a/Z$. The three sequence networks are, in effect, connected in series. The component currents, and the voltages V_b and V_c, are obtained from those in phase A by application of the basic relations in Section 3.3.4.1. Each sequence current divides in the branches of its network in accordance with the configuration and impedance values.

(b) *Short circuit between lines B and C.* The boundary conditions are $I_a = 0$, $I_b = -I_c$ and $V_b = V_c$. As there is no connection to earth at the fault, the z.p.s. network is omitted. The p.p.s. and fault currents are

$$I_{a+} = E_a/(Z_+ + Z_-) = E_a/Z \qquad I_b = -I_c = -j\sqrt{3} E_a/Z$$

where $Z = Z_+ + Z_-$ is the impedance of the p.p.s. and n.p.s. networks in series. The voltages to neutral at the fault are

$$V_a = E_a - I_+ Z_+ - I_- Z_- = 2E_a(Z_-/Z) \qquad V_b = V_c = E_a(Z_-/Z)$$

(c) *Double line–earth fault on lines B and C.* Here the boundary conditions are $I_a = 0$, and $V_b = V_c = 0$. The sequence components of the fault current are

$$I_{a+} = E_a/[Z_+ + Z_- Z_0/(Z_- + Z_0)]$$
$$I_{a-} = -I_{a+}[Z_0/(Z_- + Z_0)] \qquad I_{a0} = -I_{a+}[Z_-/(Z_- + Z_0)]$$

The sequence networks are connected in series–parallel.

Figure 3.45 shows the interlinked phase-sequence networks where both ends feed the fault F. Conditions in (*a*), (*b*) and (*c*) correspond to those in *Figure 3.44*. Networks for a broken-conductor condition are shown in (*d*) and (*e*): the former is for a case in which both ends at the break remain insulated, while the latter applies where the conductor on side 2 falls to earth, the additional constraint involving ideal 1/1 transformers in side 2 of the combined sequence network. In more complicated cases, ideal transformers with phase-shift or with ratios other than 1/1 may be required. Evaluation of these, and of conditions involving simultaneous faults at different points and/or phases, requires matrix analysis and computer programs.

3.3.5 Network power limits and stability

Steady-state a.c. power transfer over transmission links has limitations imposed by terminal voltages and link impedance. Transient conditions for stable operation are dynamic, and more complicated.

3.3.5.1 Steady-state conditions

Two typical cases concern a link transferring power from a sending-end generator at busbar voltage V_s to a receiving end of voltage V_r where there is either (1) a static load only or (2) a generator. Case (2) is the more important, as loss of synchronism is possible.

(*1*) *Load stability* The power taken by a static load of constant power factor is proportional to the square of the voltage. As the load power is increased the voltage falls, at first slightly but subsequently more rapidly until maximum power is attained. Thereafter both load voltage and power decrease, but the system is still stable, though overloaded. The condition could occur following the clearance of a system fault.

The load is rarely purely static: it usually contains motors. With induction motors the reactive-power requirements increase

Synchronous machines such as generators and motors are represented by a voltage source in series with an appropriate admittance. For example, at busbar k where the node voltage is V_k, the current could be represented by $I_k = y_k''(E_k'' - V_k)$, using the subtransient e.m.f. and admittance. The term $y_k''V_k$ can be transferred to the other side of the nodal-admittance equation in such a way that y_k'' joins any load-admittance term in the diagonal element Y_{kk}.

The network equations have now been modified to the form

$$Y''V = \begin{pmatrix} y''E'' \\ 0 \end{pmatrix}$$

where Y'' is the admittance matrix Y with diagonal elements supplemented by equivalent load admittances and machine subtransient admittances, and the right-hand-side elements are either of the type $y_k''E_k''$ or zero.

3.3.3.2 Method of solution

In 3-phase short-circuit conditions the voltages V will differ from the steady-state load values, but the right-hand-side elements will, with the exception of the element corresponding to the faulted busbar, remain constant. Let busbar m be short-circuited; then $V_m = 0$. Solving the remaining equations for the voltages V_i (with $i = 1, 2, \ldots, n, i \neq m$) by the Gauss–Seidel procedure and then substituting the voltage values obtained in the mth equation yields a new right-hand-side value of one or other of the forms

$$y_m''E_m'' + I_{msc} \quad \text{or} \quad 0 + I_{msc}$$

Here I_{msc} is the 3-phase per-unit short-circuit current injected into busbar m to make $V_m = 0$. The fault level (in MVA) is then

$$V_{m(\text{prefault})} \times I_{msc}^* \times \text{MVA(base)}$$

A preferred alternative uses the superposition theorem (Section 3.2.2.1). The injection of I_{msc}, when acting alone, superposes a change $\Delta V_m (= -V_{m(\text{prefault})})$ at busbar m. The equations to be solved become

$$Y'' \begin{pmatrix} \Delta V_1 \\ \vdots \\ \Delta V_m \\ \vdots \\ \Delta V_n \end{pmatrix} = \begin{pmatrix} 0 \\ \vdots \\ I_{msc} \\ \vdots \\ 0 \end{pmatrix} \quad \text{or} \quad Y''\Delta V = I_{sc}$$

Inversion of Y'' gives $V = Z''I_{sc}$, in which the mth equation is known to be $-V_m = Z_{mm}''I_{msc}$. If we assume nominal prefault voltage, i.e. $V_m = 1 + j0$ p.u., then the value of the 3-phase short-circuit current at busbar m is $I_{msc} = -1/Z_{mm}''$. The voltage at any other busbar k can then be found from

$$V_k = V_{k(\text{prefault})} + \Delta V_k = V_{k(\text{prefault})} + Z_{km}''I_{msc}$$

By shifting the short circuit from busbar to busbar, the fault level for each can be found from the inverses of the appropriate diagonal elements of matrix Z''.

3.3.4 System-fault analysis

The analysis of *unbalanced* faults in 3-phase power networks is an important application of the *symmetrical-component* method (Section 3.2.12). The procedure for given fault conditions is:

(1) Obtain the sequence impedance values for all items of the plant, equipment and transmission links concerned.
(2) Reduce all ohmic impedances to a common line-to-neutral base and a common voltage.
(3) Draw a single-line connection diagram for each of the sequence components, simplifying where possible (e.g. by star–delta conversion—see Section 3.2.4.5).
(4) Calculate the z.p.s., p.p.s. and n.p.s. currents, tracing them through the network to obtain their distribution with reference to the particular values sought.

Impedance in the neutral connection to earth, and in the earth path itself, must be multiplied by 3 for z.p.s. currents when the z.p.s. connection diagram is being set up in (3) because the three z.p.s. component currents are co-phasal and flow together in the z.p.s. path.

In general, a network offers differing impedances Z_+, Z_- and Z_0 to the sequence components. In static plant (e.g. transformers and transmission lines) Z_- may be the same as Z_+, but Z_0 is always significantly different from either of the other impedances. The presence of z.p.s. currents implies that a neutral connection is involved.

3.3.4.1 Sequence networks

As an example, *Figure 3.43* shows transmission lines 4 and 5–6 linking a generating station with generators 1 (isolated neutral) and 2 (solid-earthed neutral) to a second station with generator 3 (neutral earthed through resistor R_n). The numerals are used to

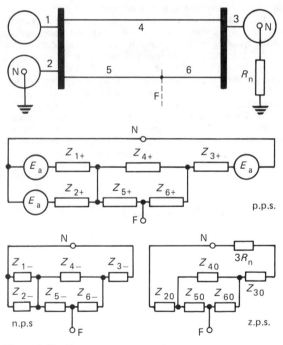

Figure 3.43 Phase-sequence networks

indicate position: e.g. Z_{1+} is the p.p.s. impedance per phase of generator 1, and Z_{60} is the z.p.s. impedance of line 6 between generator 3 and a fault at F.

The *p.p.s. network* is identical with the physical set-up of the original network (which operates with p.p.s. conditions when normally balanced and unfaulted). Each generator is a source of p.p.s. voltages only. It is here assumed that all the generators develop the same e.m.f. E_a.

The *n.p.s. network* is similar in configuration (but not usually in impedance values) to the p.p.s system. There are, however, no

the following:
$$V_2^{(p+1)} = -(Y_{21}/Y_{22})V_1 + 0 - (Y_{23}/Y_{22})V_3^{(p)}$$
$$\quad -(Y_{24}/Y_{22})V_4^{(p)} + S_2^*/Y_{22}V_2^{*(p)}$$
$$V_3^{(p+1)} = -(Y_{31}/Y_{33})V_1 - (Y_{32}/Y_{33})V_2^{(p+1)} + 0$$
$$\quad -(Y_{34}/Y_{33})V_4^{(p)} + S_3^*/Y_{33}V_3^{*(p)}$$
$$V_4^{(p+1)} = -(Y_{41}/Y_{44})V_1 - (Y_{42}/Y_{44})V_2^{(p+1)}$$
$$\quad -(Y_{43}/Y_{44})V_3^{(p+1)} + 0 + S_4^*/Y_{44}V_4^{*(p)}$$

Note that as each new estimate becomes available, it is used in the succeeding equations. Being iterative, the process of convergence can usually be assisted by the use of 'acceleration'. If $\Delta V_k^{(p)} = V_k^{(p+1)} - V_k^{(p)}$, then a new estimate can be obtained from $V_k^{(p+1)} = V_k^{(p)} + \omega \Delta V_k^{(p)}$; here ω is an accelerating factor, optimally a complex number but usually taken as real, with typical values in the range 1.0–1.6.

To terminate the successive-estimation process, various convergence tests are applied. The simplest is to examine the difference between successive voltage estimates and to stop when the maximum of $|V_k^{(p+1)} - V_k^{(p)}|$ for $k = 1, 2, \ldots, n$ is less than ε, a suitable small number such as 0.00001 p.u. However, the preferred test is

$$V_k^{(p)} \Sigma_m Y_{km}^* V_m^{*(p)} - S_k < \gamma$$

where γ is a measure of the maximum allowable apparent-power mismatch at any busbar, with a value typically 0.01 p.u.

During iteration, other calculations (e.g. voltage magnitude corrections at generator busbars and changes in transformer tap settings) can be included. If busbar 2 in the example above is a generator busbar, then the reactive power Q_2 can be assigned an initial value, say $Q_2^{(0)} = 0$, and $V_2^{(1)}$ obtained therefrom. The voltage estimate can then be scaled to agree with the specified magnitude $|V_2|$, and $Q_2^{(1)}$ immediately calculated, prior to proceeding to the next equation.

The Gauss–Seidel procedure is well suited to the microcomputer, in which core space is limited. However, matrix-inversion techniques (as required in the Newton–Raphson procedure following) for large networks demand too much core space.

3.3.2.4 Newton–Raphson procedure

The Newton–Raphson procedure is at present the most generally adopted method. It has strong convergence characteristics and suits a wide range of problems. The method employs the preliminary terms in the Taylor series expansion (Section 1.2.4) of a set of functions of variables V. The kth function is defined as

$$\mathbf{f}(V) = \mathbf{f}[V^{(p)} + \mathbf{J}(V^{(p)})(V - V^{(p)})]$$

The true set of values V is taken as given by $V = V^{(p)} + \gamma^{(p)}$, i.e. by the sum of an approximate set $V^{(p)}$ and a set of error terms $\gamma^{(p)}$. Then, taking the first two terms of the Taylor expansion,

$$\mathbf{f}(V) = \mathbf{f}[V^{(p)} + \mathbf{J}(V^{(p)})(V - V^{(p)})]$$

whence

$$\mathbf{J}(V^{(p)})\gamma^{(p)} = -\mathbf{f}(V^{(p)})$$

Matrix $\mathbf{J}(V^{(p)})$ is the Jacobian matrix of first derivatives of the functions $\mathbf{f}(V)$. Voltage estimates $V^{(p)}$ are used to evaluate specific matrix elements of $\mathbf{J}$; and $\gamma^{(p)}$ is the column matrix of voltage differences $\Delta V^{(p)}$ to be evaluated, these being the difference between the true and approximate values of the voltages V. Likewise, the term $-\mathbf{f}(V^{(p)})$ is the set of per-unit apparent-power differences $\Delta S^{(p)}$ between specified and calculated values, where

$$\Delta S_k^{(p)} = S_k - V_k^{(p)} \Sigma_m Y_{km}^* V_m^{*(p)}$$

The loadflow equation to be solved becomes

$$\mathbf{J}(V^{(p)})\Delta V^{(p)} = \Delta S^{(p)}$$

When $\Delta V^{(p)}$ is determined, the voltages are updated to $V^{(p+1)} = V^{(p)} + \Delta V^{(p)}$.

The polar form of the equations is most usually employed, so with $V_k = |V_k| \angle \delta_k$ and $Y_{km} = |Y_{km}| \angle \psi_{km}$ the function $\mathbf{f}_i = 0$ becomes becomes

$$(V_i Y_{i1} V_1 \cos \beta_{i1} + V_i Y_{i2} V_2 \cos \beta_{i2} + \ldots - P_i)$$
$$+ \mathrm{j}(V_i Y_{i1} V_1 \sin \beta_{i1} + V_i Y_{i2} V_2 \sin \beta_{i2} + \ldots - Q_i) = 0 + \mathrm{j}0$$

where $\beta_{i1} = \delta_i - \delta_1 - \psi_{i1}$ and similarly for $\beta_{i2}, \ldots$. Partial differentiation to form the terms of the Jacobian matrix, and then separation of the real and imaginary parts, gives the matrix equations to be solved. For generator busbars the voltage magnitudes are fixed, so that only equations in reals are needed to evaluate $\angle \delta$.

Generator busbars are often termed 'P, V' and load busbars referred to as 'P, Q' busbars, reflecting the values specified.

The Jacobian equations for the Newton–Raphson method are thus of the form

$$\begin{pmatrix} \partial \mathbf{f}_1/\partial \delta_1 & \partial \mathbf{f}_1/\partial \delta_2 & \ldots & \partial \mathbf{f}_1/\partial V_1 & \partial \mathbf{f}_1/\partial V_2 & \ldots \\ \partial \mathbf{f}_2/\partial \delta_1 & \partial \mathbf{f}_2/\partial \delta_2 & \ldots & \partial \mathbf{f}_2/\partial V_1 & \partial \mathbf{f}_2/\partial V_2 & \ldots \\ \vdots & \vdots & & \vdots & \vdots & \\ \partial \mathbf{f}_1/\partial \delta_1 & \partial \mathbf{f}_1/\partial \delta_2 & \ldots & \partial \mathbf{f}_1/\partial V_1 & \partial \mathbf{f}_1/\partial V_2 & \ldots \\ \partial \mathbf{f}_2/\partial \delta_1 & \partial \mathbf{f}_2/\partial \delta_2 & \ldots & \partial \mathbf{f}_2/\partial V_1 & \partial \mathbf{f}_2/\partial V_2 & \ldots \\ \vdots & \vdots & & \vdots & \vdots & \end{pmatrix} \begin{pmatrix} \Delta \delta_1 \\ \Delta \delta_2 \\ \vdots \\ \Delta V_1 \\ \Delta V_2 \\ \vdots \end{pmatrix} = \begin{pmatrix} \Delta P_1 \\ \Delta P_2 \\ \vdots \\ \Delta Q_1 \\ \Delta Q_2 \\ \vdots \end{pmatrix}$$

Written in abbreviated form, this is

$$\begin{pmatrix} \mathbf{J}_{11} & \mathbf{J}_{12} \\ \mathbf{J}_{21} & \mathbf{J}_{22} \end{pmatrix} \begin{pmatrix} \Delta \delta \\ \Delta V \end{pmatrix} = \begin{pmatrix} \Delta P \\ \Delta Q \end{pmatrix}$$

To save computer memory space, it is usual to omit $\mathbf{J}_{12}$ and $\mathbf{J}_{21}$, an approximation that leaves two *decoupled* sets of equations. This approach, called the 'fast decoupled Newton–Raphson loadflow', is in wide use.

For any set of estimates of the voltage, the elements of the Jacobian matrix are evaluated, and the set of equations solved (using space-saving sparse-matrix programming techniques) for $\Delta \delta$ and ΔV; the values of V, ΔP and ΔQ are updated, and so on. Convergence is achieved for most networks in a few iterations.

A further development is to extend the Taylor series to the second-derivative term, when the series will terminate if Cartesian co-ordinates are employed. Iteration is more lengthy, but the convergence characteristics are more powerful. This 'second-order Newton–Raphson procedure' is gaining popularity.

3.3.3 Fault-level analysis

The calculation of 3-phase fault levels in large power networks again involves solution of the nodal-admittance equations $YV = I$ subject to constraints.

3.3.3.1 System representation

Representation of generators and loads by fixed $P, |V|$ and P, Q requirements is not valid because of the large and sudden departure of the busbar voltages from their nominal values.

Passive loads are usually represented by a constant admittance, implying that the load power is proportional to the square of the busbar voltage. Relations $P \propto |V|^{1.2}$ and $Q \propto |V|^{1.6}$ would be more likely, but the $|V|^2$ proportionality affords a measure of demand variability and is more easily represented in the admittance matrix Y.

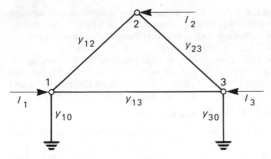

Figure 3.41 Sample network

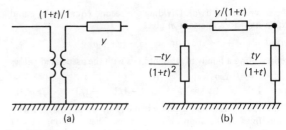

Figure 3.42 Transformer equivalent circuit

Cast into matrix form, these equations become

$$\begin{pmatrix} y_{10}+y_{12}+y_{13} & -y_{12} & -y_{13} \\ -y_{21} & y_{21}+y_{23} & -y_{23} \\ -y_{31} & -y_{32} & y_{30}+y_{31}+y_{32} \end{pmatrix} \begin{pmatrix} V_1 \\ V_2 \\ V_3 \end{pmatrix} = \begin{pmatrix} I_1 \\ I_2 \\ I_3 \end{pmatrix}$$

This can be abbreviated to the nodal-admittance matrix equation $YV = I$, matrix Y being known as the nodal-admittance matrix. The relation between the branch elements and the corresponding matrix elements can be seen by inspection.

For some purposes the alternative impedance matrix equations are used: viz. $ZI = V$, where $Z = Y^{-1}$ is the nodal-impedance matrix. However, evaluating the inverse of Y is more complicated than finding Y, and the admittance form may be considered as the primary (or given) form.

3.3.2 Loadflow analysis

Loadflow analysis is the solution of the nodal equations, subject to various constraints, to establish the node voltages. At the same time generator power outputs, transformer tap settings, branch power flows and powers taken by voltage-sensitive loads (including r.p. compensators) are determined.

3.3.2.1 Problem description

As indicated in *Figure 3.41*, the elements of a power system can be represented either as equivalent branches with appropriate admittance incorporated into the matrix Y, or as equivalent current sources added to the matrix I.

Transmission lines and cables are represented by their series admittance and shunt (charging) susceptance. These parameters are actually distributed quantities, but are taken into account by Π (or, occasionally, T) equivalent networks (Section 3.2.5).

Transformers are modelled by equivalent circuits with an ideal transformer in series with a leakage admittance (*Figure 3.42(a)*). With two terminals connected to a common reference point (earth) the circuit reduces to that in *Figure 3.42(b)*: the ideal transformer with a turns-ratio $(1+t)/1$ is replaced by an equivalent Π. The tap setting t represents the per-unit of nominal turns-ratio: e.g. $t = \pm 0.05$ for $\pm 5\%$ taps.

Loads can be represented either as equivalent admittances Y_{k0} connected between busbars and earth, or as current sources. If the load demand at busbar k is $S_k = P_k + jQ_k$, then the equivalent admittance at voltage V_k is found from

$$S_k = V_k I_k^* = |V_k|^2 Y_{k0}^* \quad \text{or} \quad Y_{k0} = S_k^*/|V_k|^2$$

where $|V_k|$ is assumed to be 1.0 p.u. (i.e. rated voltage). The equivalent admittance is incorporated into the corresponding diagonal element of the nodal-admittance matrix. In the current-source representation, the current I_k is substituted into the current column-matrix I by a fixed power requirement S_k and the (unknown) voltage V_k, where $I_k = S_k^*/V_k^*$.

Generator units or stations can likewise be represented by current sources, but usually the busbar to which a generator is connected has a controlled voltage. At a busbar m the requirement would be for specified values of active power P_m and voltage $|V_m|$, with the reactive power Q_m to be determined.

Slack busbar In a loadflow study, the total active power supplied cannot be specified in advance because the loss in the supply network will not be known. Further, in an n-busbar network, there are n complexor equations involving $2n$ real-number equations, however, there are $2n+2$ unknowns. To reduce this number to $2n$ it is the practice to specify the voltage of one busbar in both magnitude and phase angle. This is termed the *slack busbar*, to which the chosen *slack generator* is connected. The slack-busbar equation can now be removed from the solution process and, when all other voltages have been determined, the slack-busbar generation can be found. For a slack busbar k, for example, the generation S_k is found from

$$S_k = V_k \Sigma_m (Y_{km}^* V_m^*)$$

3.3.2.2 Network solution process

Solution of the matrix equation is now sought. As it embodies several simultaneous equations, the solution has to be iterative. The two main procedures are the Gauss–Seidel and the Newton–Raphson methods. An important consideration in the computation process is the rate of convergence.

3.3.2.3 Gauss–Seidel procedure

This early (and still effective) technique resembles the over-relaxation method used in linear algebra. Consider a four-busbar network described by the equations

$$Y_{11}V_1 + Y_{12}V_2 + Y_{13}V_3 + Y_{14}V_4 = S_1^*/V_1^*$$
$$Y_{21}V_1 + Y_{22}V_2 + Y_{23}V_3 + Y_{24}V_4 = S_2^*/V_2^*$$
$$Y_{31}V_1 + Y_{32}V_2 + Y_{33}V_3 + Y_{34}V_4 = S_3^*/V_3^*$$
$$Y_{41}V_1 + Y_{42}V_2 + Y_{43}V_3 + Y_{44}V_4 = S_4^*/V_4^*$$

where $Y_{12} = -y_{12}$ etc. Let busbar 1 be chosen as the slack busbar, and let V_1 be $1 + j0$. Then the remaining three equations are to be solved for V_2, V_3 and V_4. The method adopted is one of successive estimation.

First, the equations are rearranged by extracting the diagonal terms $Y_{11}V_1$, $Y_{22}V_2$, ... and transferring all other terms to the right-hand side. Each equation is then divided by the diagonal admittance element (Y_{11}, Y_{22}, ...).

If $V_k^{(p)}$ denotes the pth estimate of V_k and the equations are solved in the sequence 2–3–4–2–3–..., then an iterative process is

if z is written for dy/dt. Dividing the second equation by z and equating it to a constant m gives

$m = dz/dy = -(z+y)/z$ or $z/y = -1/(1+m)$

representing a family of straight lines with the associated values

m	-4	-2	-1	0	1	2	4	∞
z/y	$1/3$	1	∞	-1	$-1/2$	$-1/3$	$-1/5$	0

Draw the z/y axes on the phase plane (*Figure 3.39*) marked with short lines of the appropriate slope m. Then, starting at any arbitrary point, a trajectory is drawn to cross each axis at the indicated slope. With no drive, all trajectories approach, and finally reach, the origin after oscillations in a counter-clockwise direction; for a steady drive V, the only difference is to shift the vortex to V on the y-axis. The approach to O or V represents the decaying oscillation of the system and its final steady state. Because $dt = dy/z$, the finite difference $\Delta t = \Delta y/z$ gives the time interval between successive points on a trajectory.

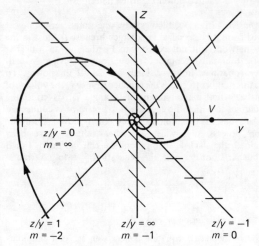

Figure 3.39 Phase-plane trajectories

3.3 Power-system network analysis

3.3.1 Conventions

Modern power-system analyses are based mainly on nodal equations scaled to a per-unit basis, with a particular convention for the sign of reactive power.

3.3.1.1 Per-unit basis

The total apparent power in a 3-phase circuit ABC with phase voltages V_a, V_b, V_c, currents I_a, I_b, I_c and phase angles between the associated voltage and current phasors of $\theta_a, \theta_b, \theta_c$ is

$S = (P_a + P_b + P_c) + j(Q_a + Q_b + Q_c) = P + jQ$

where for phase A the active power is $P_a = V_a I_a \cos \theta_a$ and the reactive power is $Q_a = V_a I_a \sin \theta_a$. Corresponding expressions apply for phases B and C.

If the phases are balanced, all three have the same scalar voltage V and current I, and all phase angles are θ. Then $S = 3(VI \cos \theta + jVI \sin \theta)$. This can be written in the form

$$\frac{S}{k} = \left(\frac{3}{a}\frac{V}{b}\frac{I}{c}\frac{\cos\theta}{d}\right) + j\left(\frac{3}{a}\frac{V}{b}\frac{I}{c}\frac{\sin\theta}{d}\right)$$

where k, a, b, c and d are scaling factors. It is customary to choose $a = 3$ and $d = 1$, leaving c and b, one of which is assigned an independent value while the other takes a value depending on the overall scaling relation $k = abcd$.

In normal operating conditions the scalar voltage V approximates to the rated phase voltage V_R; hence b is taken as V_R so that V/b is the voltage in per-unit of V_R. Defining the scaled variables as S_{pu}, V_{pu} and I_{pu}, the *total* apparent power is

$S_{pu} = (V_{pu} I_{pu} \cos \theta) + j(V_{pu} I_{pu} \sin \theta)$

which is an equation in 1-phase form, justifying the use of single-line schematic diagrams to represent 3-phase power circuits.

The scaling factors are termed *base* values; i.e. k is the base apparent power, b is the base phase voltage and c is the base current. These definitions imply further base values for impedance and admittance, namely $Z_{base} = b/c$ and $Y_{base} = c/b$.

If the line-to-line voltage V_l is used, then in the foregoing scaling equation $3/a$ and V/b becomes $\sqrt{3}/a'$ and V_l/b'. Choosing $a' = \sqrt{3}$ and b' as rated line voltage leaves S/k unchanged. Note, however, (i) that θ remains as the angle between *phase* voltage and current, and (ii) that in the per-unit equation $S = VI^*$ the voltage V is the per-unit *phase* voltage, not the line-to-line voltage (although numerically both have the same per-unit value).

3.3.1.2 Reactive power convention

Reactive power may be lagging or leading. The common convention is to consider lagging reactive power flow to be positive, as calculated from the product of the voltage and current-conjugate phasors: thus $S = VI^* = P + jQ$, with Q a positive number for a lagging power factor condition. For a leading power factor, Q has the same flow direction but is numerically negative. As a consequence, an inductor *absorbs*, but a capacitor *generates*, lagging reactive power, as shown in *Figure 3.40*.

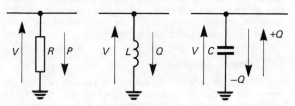

Figure 3.40 Power taken by resistive, inductive and capacitive loads

Note System engineers refer, for brevity, to the flow of 'power' and 'vars', meaning respectively active power and reactive power.

3.3.1.3 Nodal-admittance equations

The nodal-admittance equations derive from the node-voltage method of analysis in Section 3.2.3. For the network in *Figure 3.41*, currents I_1, I_2 and I_3 are injected respectively into nodes 1, 2 and 3. The loads are linked by branches of admittance y_{12}, y_{23} and y_{31}, and to the earth or external reference node r by branches of admittance y_{10} and y_{30}. Assuming that the node voltages V_1, V_2 and V_3 are expressed with reference to r and that $V_r = 0$, then writing the Kirchhoff current equations and simplifying gives

$(y_{10} + y_{12} + y_{13})V_1 - y_{12}V_2 - y_{13}V_3 = I_1$

$-y_{21}V_1 + (y_{21} + y_{23})V_2 - y_{23}V_3 = I_2$

$-y_{31}V_1 - y_{32}V_2 + (y_{30} + y_{31} + y_{32})V_3 = I_3$

made from oxides of the iron group of metals with the addition of small amounts of ions of different valency, and are applied to temperature measurement and control. Thermistors with a positive resistance–temperature coefficient made from monocrystalline barium titanate have a resistance that, for example, increases 100-fold over the range 50–100 °C; they are used in the protection of machine windings against excessive temperature rise.

Voltage-sensitive resistors, made in disc form from silicon carbide, have a voltage–current relation approximating to $v = ki^\beta$, where β ranges from 0.15 to 0.25. For $\beta = 0.2$ the power dissipated is proportional to v^6, the current doubling for a 12% rise in voltage.

Inductors The current in a load fed from a constant sinusoidal voltage supply can be varied over a wide range economically by use of a series inductor carrying an additional d.c.-excited winding to vary the saturation level and hence the effective inductance. The core material should have a flux–m.m.f. relation like that in *Figure 3.36(f)*. Grain-oriented nickel and silicon irons are suitable for the inductor core. A related phenomenon accounts for the inrush current in transformers.

Describing function In a non-linear system a sinusoidal drive does not produce a sinusoidal response. The describing-function technique is devised to obtain the *fundamental-frequency* effect of non-linearity under steady-state (but not transient) conditions.

Consider a stimulus $x = h \cos \omega t$ to give a response $y = f(t)$. As non-linearity inevitably introduces harmonic distortion, y can be expanded as a Fourier series (Section 1.2.5) to give

$$y = c_0 + a_1 \cos \omega t + a_2 \cos 2\omega t + \ldots$$
$$+ b_1 \sin \omega t + b_2 \sin 2\omega t + \ldots$$

The components $a_1 \cos \omega t$ and $b_1 \sin \omega t$ are regarded as the 'true' response, with a gain factor $(a_1 + jb_1)/h$, the other terms being the distortion. The gain factor is the describing function. Let $y = ax + bx^2$ with $x = h \cos \omega t$; applying the expansion gives the fundamental-frequency term $y = (a + \tfrac{3}{4}bh^2)h \cos \omega t$. The describing function is therefore $a + \tfrac{3}{4}bh^2$, which is clearly dependent on the magnitude h of the input. Thus the technique consists in evaluating the Fourier series for the output wave-form for a sinusoidal input, and finding therefrom the magnitude and phase angle of the fundamental-frequency response.

Ferroresonance The individual r.m.s. current–voltage characteristics of a pure capacitor C and a ferromagnetic-cored (but loss-free) inductor L for a constant-frequency sinusoidal r.m.s. voltage V are shown in *Figure 3.37*. With L and C in series and carrying a common r.m.s. current I, the applied voltage is $V = V_L - V_C$. At low voltage V_L predominates, and the $I-V$ relation is the line OP, with I lagging V by 90°. At P, with $V = V_0$ and $I = I_0$, the system is at a limit of stability, for an increase in V results in a reduction in $V_L - V_C$. At a current level Q the difference is zero. The current therefore 'jumps' from I_0 to a higher level I_1 (point R), still with $V = V_0$. During the rapid rise there is an interchange of stored energy, and for $V > V_0$ the circuit is capacitive. When V is reduced from above to below V_0, a sudden current jump from I_1 to I_0 occurs. A comparable jump phenomenon takes place for a parallel connection of C and L.

Phase-plane technique 'Phase' here means 'state' (as in the solid, liquid and vapour 'phases' of water). The phase-plane technique can elucidate non-linear system behaviour graphically. *Figure 3.38* shows a circuit of series R, L and C with a drive having the

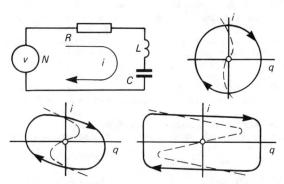

Figure 3.38 Phase-plane trajectories for an oscillatory circuit

voltage–current relation $v = -ri + ai^3$. Then, with constant circuit parameters,

$$L(di/dt) + (R - r + ai^2)i + q/C = 0$$

where q is the time-integral of i. The presence of L and C indicates the possibility of oscillation. The middle ('damping') term can be negative for small currents (increasing the oscillation amplitude) but positive for larger currents (reducing the amplitude). Hence the system seeks a constant amplitude irrespective of the starting condition. The $q-i$ phase-plane loci show the stable condition as related to the degree of drive non-linearity (indicated by the broken curves). With suitable scales the locus for minor non-linearity is circular, indicating near-sinusoidal oscillation; for major drive non-linearity, however, the locus shows abrupt changes and an approach to a 'relaxation' type of wave-form.

Isoclines A non-linear second-order system described by

$$d^2y/dt^2 + f(y, dy/dt) + g(y, dy/dt)y = 0$$

can be represented at any point in the phase plane having the co-ordinates y and dy/dt, representing, for example, position and velocity, or charge and current. By writing $dy/dt = z$ and eliminating time by division, we obtain a first-order equation relating z and y:

$$dz/dy = -[f(y,z)z + g(y,z)y]/z$$

Integration gives phase-plane trajectories that everywhere satisfy this equation, starting from any initial condition. If dz/dy cannot be directly integrated, it is possible to drawn the trajectories with the aid of *isoclines*, i.e. lines along which the *slope* of the trajectory is constant. Make $dz/dy = m$, a constant; then $-mz = f(y,z)z + g(y,z)y$. Since for $z = 0$ the slope m is infinite (i.e. at right angles to the y-axis) the trajectories intersect the horizontal y-axis normally, except at singular points.

Consider a linear system with an undamped natural frequency $\omega_n = 1$ and a damping coefficient $c = 0.5$. For zero drive

$$d^2y/dt^2 + dy/dt + y = 0 \quad \text{or} \quad dz/dt = -(z+y)$$

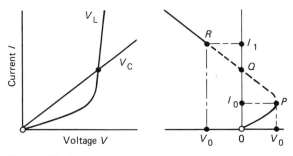

Figure 3.37 Ferroresonance

position. The effect of the constant K is to lift the whole plot upward by $20 \lg (K)$. The summed phase angles approach $-90°$ at zero frequency and $-180°$ at infinite frequency.

Nichols diagram The Nichols diagram resembles the Nyquist diagram in construction, but instead of phasor values the magnitudes are the log moduli. The point $(-1+j0)$ of the Nyquist diagram becomes the point (0 dB, $\angle -180°$). The Nichols diagram is used for determining the closed-loop response of systems.

3.2.16 Non-linearity

A truly linear system, in which *effect* is in all circumstances precisely proportional to *cause*, is a rarity in nature. Yet engineering analyses are most usually based on a linear assumption because it is mathematically much simplified, permits of superposition and can sometimes yield results near enough to reality to be useful. If, however, the non-linearity is a significant property (such as magnetic saturation) or is introduced deliberately for a required effect (as in rectification), a non-linear analysis is essential. Such analyses are mathematically cumbersome. No general method exists, so that *ad hoc* techniques have been applied to deal with specific forms of non-linearity. The treatment depends on whether a steady-state or a transient condition is to be evaluated.

3.2.16.1 Techniques

Some of the techniques used are: (i) step-by-step solution, graphical or by computation; (ii) linearising over finite intervals; (iii) fitting an explicit mathematical function to the non-linear characteristic; (iv) expressing the non-linear characteristic as a power series.

Step-by-step solution Consider, as an example, the growth of the flux in a ferromagnetic-cored inductor in which the inductance L is a function of the current i in its N turns. Given the flux magnetomotive-force (m.m.f.) characteristic, and the (constant) resistance r, the conditions for the sudden application of a constant voltage V are given by

$$V = ri + d(Li)/dt \simeq ri + N(\Delta\Phi/\Delta t)$$

which is solved in suitable steps of Δt, successive currents i being evaluated for use with the magnetic characteristic to start the next time-interval.

Linearising A non-linear characteristic may be approximated by a succession of straight lines, so that a piecemeal set of linear equations can be applied, 'matching' the conditions at each discontinuity.

For 'small-signal' perturbations about a fixed quiescent condition, the mean slope of the non-linear characteristic around the point is taken and the corresponding parameters derived therefrom. Oscillation about the quiescent point can then be handled as for a linear system.

Explicit function For the resistance material in a surge diverter the voltage–current relation $v = ki^x$ has been employed, with x taking a value typically between 0.2 and 0.3. The resistance–temperature relation of a thermistor, in terms of the resistance values R_1 and R_2 at corresponding absolute temperatures T_1 and T_2 takes the form

$$R_2 = R_1 \exp(k/T_2 - k/T_1)$$

Several functions, such as $y = a \sinh(bx)$, have been used as approximations to magnetic saturation excluding hysteresis. A static-friction effect, of interest at zero speed in a control system, has been expressed as $y = k(\operatorname{sgn} x)$, i.e. a constant that acts against the driving torque.

Series A typical form is $y = a_0 + a_1 x + a_2 x^2 + \ldots + a_n x^n$, where the coefficients a are independent of x. Such a series may have a restricted range, and the powers limited to even orders if the required characteristic has the same shape for both negative and positive y. A second-degree series $y = a_0 + a_1 x + a_2 x^2$ can be fitted through any three points on a given function of y, and a third-degree expression through any four points. However, the prototype characteristic must not have any discontinuities.

Rational-fraction expressions have also been developed. The open-circuit voltage of a small synchronous machine in terms of the field current might take the form $v = (27 + 0.006i)/(1 + 0.03i)$. Similarly, the magnetisation curve of an electrical sheet steel might have the B–H relation

$$H = B(426 - 760B + 440B^2)/(1 - 0.80B + 0.17B^2)$$

with hysteresis neglected. An exponential series

$$B = a[1 - \exp(-bH)] + c[1 - \exp(-dH)] + \ldots + \mu_0 H$$

has been suggested to represent the magnetisation characteristic of a machine, the final term being related to the airgap line.

Non-linear characteristics Figure 3.36 shows some of the typical relations $y = f(x)$ that may occur in non-linear systems. Not all are analytic, and some may require step-by-step methods.

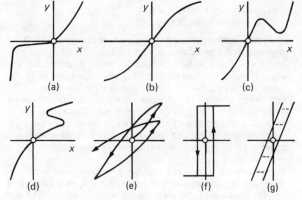

Figure 3.36 Typical non-linear characteristics

The *simple* relations shown are: (*a*) response depending on direction, as in rectification; (*b*) skew symmetry, showing the effect of saturation; (*c*) negative-slope region, but with y univalued.

The *complex* relations are: (*d*) negative-slope region, with y multivalued; (*e*) build-up of system with hysteresis, unsaturated; (*f*) toggle characteristic, typical of idealised saturated hysteresis; (*g*) backlash, with y taking any value between the characteristic limit-lines.

3.2.16.2 Examples

A few examples of non-linear parameters and techniques are given here to illustrate their very wide range of interest.

Resistors Thermally sensitive resistors (thermistors) may have positive or negative resistance–temperature coefficients. The latter have a relation between resistance R and absolute temperature T given by $R_2 = R_1 \exp[b(1/T_2 - 1/T_1)]$. They are

angle $\alpha=0$. For $\omega=\infty$, the terminal capacitor effectively short-circuits the output terminals so that $V_o/V_i=0$. At intermediate frequencies the gain $|M|$ rises to a peak at $\omega=3.3$ rad/s and thereafter falls toward zero. The phase angle α is small and positive below $\omega=1$, being always negative thereafter, to become $-180°$ at infinite frequency.

Nyquist diagram The Nyquist diagram is a polar plot of $|M|\angle\alpha$ over the frequency range (*Figure 3.34(b)*), for an input $V_i=1+\text{j}0$. The plot is particularly useful for feedback systems. If the open-loop transfer function is plotted, and in the direction of increasing ω it encloses the point $(-1+\text{j}0)$, then when the loop is closed the system will be unstable as the output is more than enough to supply a feedback input even when $V_i=0$. The Nyquist criterion for stability is therefore that the point $(-1+\text{j}0)$ shall not be enclosed by the plot.

Bode diagram The Bode diagram for the system of *Figure 3.33* is *Figure 3.34(a)* redrawn with logarithmic ordinates of $|M|$ and a logarithmic scale of ω. Normally the ordinates are expressed as a *gain* $20\lg|M|$ in decibels. For the example being considered, $M=0.5$ for very low frequencies, so that $20\lg|M|=-6$ dB; for $\omega=3.3$ the amplitude of M is 1.4 and the corresponding gain is $+2.9$ dB; and at the two frequencies when the output and input magnitudes are the same, $M=1$ and $20\lg(1)=0$ dB. All these are shown in the Bode diagram (*Figure 3.34(c)*). On the logarithmic frequency scale, equal ratios of ω are separated by equal distances along the horizontal axis. If successive values $0.5, 1, 2, 4, \ldots$ are marked in equidistantly, their successive ratios $1/0.5, 2/1, \ldots$ are all equal to 2, so that each interval is a *frequency octave*. Correspondingly the equispaced frequencies $0.1, 1, 10, \ldots$ express a *frequency decade*.

The phase-angle plot is drawn in degrees to the same logarithmic scale of frequency.

An advantage of the Bode plot is the ease with which system functions can be built up term by term. The product of complex operators is reduced to the addition of the logarithms of their moduli and phase angles; similarly the quotient is reduced to subtraction. If the system function can adequately be expressed in *simple* terms, the Bode diagram can be rapidly assembled. Such terms are:

jω: represented by a line through $\omega=1$ and rising with frequency at 6 dB per octave or 20 dB per decade, and with a constant phase angle $\alpha=90°$.

$1/\text{j}\omega$: as for jω, but falling with frequency, and with $\alpha=-90°$.

$1+\text{j}\omega T$: a straight line of zero gain for frequencies up to that for which $\omega T=1$, and thereafter a second straight line rising at 6 dB per octave; the change of direction occurs at the *break point* (*Figure 3.35(a)*).

$1/(1+\text{j}\omega T)$: as for $1+\text{j}\omega T$, except that after the break point the gain drops with frequency at 6 dB per octave.

In *Figure 3.35(a)* the true gain shown by the broken curve is

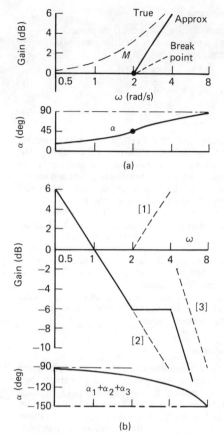

Figure 3.35 Bode diagrams

approximated by the two straight lines meeting at the break point. The approximate and true gains, and the phase angles, are given in *Table 3.5* for the term $1+\text{j}\omega T$. The error in the gain is 3 dB at the break point, and 1 dB at one-half and twice the break-point frequency, making correction very simple.

The uncorrected Bode plot for the system function

$$KG(\text{j}\omega)=K\frac{(1+\text{j}\omega 0.5)}{\text{j}\omega(1+\text{j}\omega 0.25)^2}=K\frac{[1]}{[2][3]}$$

is shown in *Figure 3.35(b)*. Term [1] is the same as in *Figure 3.35(a)*. Term [2] is a straight line running downward through $\omega=1$ with a slope of 6 dB per octave. Term [3] has a break point at $\omega=4$, but as it is a squared term its slope for $\omega>4$ is 12 dB per octave. The full-line plot of gain is obtained by direct super-

Table 3.5 Gain and phase angle for $1+\text{j}\omega T$

ωT	Gain (dB)		Angle (deg)	ωT	Gain (dB)		Angle (deg)
	approx.	true			approx.	true	
0	0	0.0	0	2	6	7.0	63.5
0.01	0	0.00	0.5	4	12	12.3	76
0.1	0	0.04	5.7	8	18	18.1	83
0.25	0	0.26	14	10	20	20.0	84
0.5	0	1.0	26.5	16	24	24.0	86.5
1.0	0	3.0	45	100	40	40.0	89.5

3.2.15.2 System performance

In general, a system function takes the form numerator/denominator, each a polynomial in s, relating response to input stimulus. Two forms are

$$KG(s) = \frac{b_m s^m + b_{m-1} s^{m-1} + \ldots + b_0}{a_n s^n + a_{n-1} s^{n-1} + \ldots + a_0} \quad \text{(i)}$$

$$= \frac{b_m}{a_n} \frac{(s-z_1)(s-z_2)\ldots(s-z_m)}{(s-p_1)(s-p_2)\ldots(s-p_n)} \quad \text{(ii)}$$

The response depends both on the system and on the stimulus. Performance can be studied if simple formalised stimuli (e.g. step, ramp or sinusoidal) are assumed; an exponential stimulus is even more direct because (in a linear system) the transient and steady-state responses are then both exponential. With the system function expressed in terms of the complex frequency $s = \sigma + j\omega$ it is necessary to express the stimulus in similar terms and to evaluate the response as a function of time by inverse Laplace transformation. The response in the *frequency domain* (i.e. the output/input relation for sustained sinusoidal stimuli over a frequency range) is obtained by taking $s = j\omega$ and solving the complexor $KG(j\omega)$. Another alternative is to derive the poles (p) and zeros (z) in Equation (ii) above.

Thus there are several techniques for evaluating system functions. Some are graphical and give a concise representation of the response to specified stimuli.

3.2.15.3 Poles and zeros

In Equation (ii) above, the numbers z are the values of s for which $KG(s) = 0$; for, if s is set equal to z_1 or $z_2 \ldots$, the numerator has a zero term as a factor. Similarly, if s is set equal to p_1 or $p_2 \ldots$, there is a zero factor in the denominator and $KG(s)$ is infinite. Then the 'z's are the *zeros* and the 'p's are the *poles* of the system function. Except for the term b_m/a_n the system function is completely specified by its poles and zeros.

Consider the network of *Figure 3.33*, the system function required being the output voltage v_o in terms of the input voltage v_i. This is the ratio of the paralleled branches $R_2 L_2 C$ to the whole impedance across the input terminals. Algebra gives

$$KG(s) = 8(s+1)/(s^3 + 3s^2 + 14s + 16)$$

$$= 8 \frac{s+1}{(s+1.36)[s+(0.82+j3.33)][s+(0.82-j3.33)]}$$

by factorising numerator and denominator. Thus there is one zero for $s = -1$. There are three poles, with $s = -1.36$, and $-0.82 \pm j3.33$. These are plotted on the complex s-plane in *Figure 3.33*. Poles on the real axis σ correspond to simple exponential variations with time, decaying for negative and increasing indefinitely for positive values. Poles in conjugate pairs on the $j\omega$ axis correspond to sustained sinusoidal oscillations. If the poles occur displaced from the origin and not on either axis, they refer to sinusoids with a decay or a growth factor, depending on whether the term σ is negative or positive.

3.2.15.4 Harmonic response

This is the steady-state response to a sinusoidal input at angular frequency ω. When a sine signal input is applied to a linear system, the steady-state response is also sinusoidal and is related to the input by a relative magnitude M and a phase angle α. The system function is $KG(j\omega)$.

Consider again the network of *Figure 3.33*. Writing $s = j\omega$ and simplifying gives the phasor expression for V_o/V_i as

$$KG(j\omega) = \frac{8(j\omega + 1)}{(16 - 3\omega^2) + j\omega(14 - \omega^2)} = |M| \angle \alpha$$

Plots of $|M|$ and $\angle \alpha$ are shown in *Figure 3.34(a)*. For $\omega = 0$ the network is a simple voltage divider with $V_o/V_i = 0.5$ and a phase

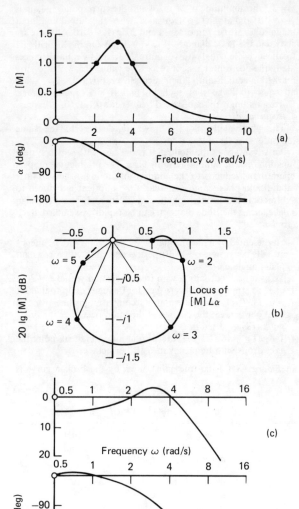

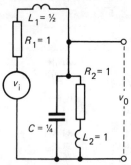 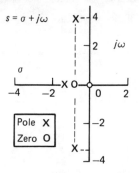

Figure 3.33 Poles and zeros

Figure 3.34 Harmonic response

Table 3.4 System transfer functions [the relation $f_2(t)/f_1(t)$ of output to input quantity in terms of the Laplace transform $F_2(s)/F_1(s)$]

v	voltage	L	inductance	θ	angular displacement
i	current	k_e	e.m.f. coefficient	ω	angular velocity
Z	impedance	c	damping coefficient	a	acceleration
R, r	resistance	T	time-constant	k_a	acceleration coefficient

M	mass	J	inertia
F	viscous friction coefficient	K	stiffness

System		Transfer function	
1 Electrical network		$\dfrac{V_2(s)}{V_1(s)} = \dfrac{Z_2(s)}{Z_1(s)+Z_2(s)}$	
2 Electrical network		$\dfrac{V_2(s)}{I_1(s)} = \dfrac{Z_1(s)Z_2(s)}{Z_1(s)+Z_2(s)}$	
3 Feedback amplifier		$\dfrac{V_2(s)}{V_1(s)} = -\dfrac{Z_2(s)}{Z_1(s)}$	
4 Second-order system		$\dfrac{\theta_2(s)}{\theta_1(s)} = \dfrac{1}{1+2csT+s^2T^2}$	$T=\sqrt{(J/K)}$ $c=F/2\sqrt{(JK)}$
5 Accelerometer		$\dfrac{V_2(s)}{A_1(s)} = \dfrac{k_a}{1+2csT+s^2T^2}$	$T=\sqrt{(M/K)}$ $c=F/2\sqrt{(MK)}$
6 Permanent-magnet generator		$\dfrac{V_2(s)}{\omega_1(s)} = k_e$	
7 Separately excited generator		$\dfrac{V_2(s)}{V_1(s)} = \dfrac{k_e\omega}{R(1+sT)}$	$T=L/R$
8 Motor: armature control		$\dfrac{\theta_2(s)}{V_1(s)} = \dfrac{K_c}{s(1+sT)}$	$K_c=k_e/(FR+k_e^2)$ $T=JR/(FR+k_e^2)$
9 Motor: field control		$\dfrac{\theta_2(s)}{V_1(s)} = \dfrac{k_e}{s(1+sT_1)(1+sT_2)}$	$T_1=J/F,\ T_2=L/R$

3.2.15.1 Closed-loop systems

In *Figure 3.32*, (*a*), (*b*) and (*c*) are *open-loop* systems. However, the output can be made to modify the input by feedback through a network $K_f G_f(s)$ as in (*d*). The signal

$$\theta_f(s) = [K_f G_f(s)]\theta_o(s)$$

is combined with $\theta_i(s)$ to give the modified input.

For *positive feedback*, the resultant input is $\sigma(s) = \theta_i(s) + \theta_f(s)$, and the effect is usually to produce instability and oscillation.

For *negative feedback*, the resultant input is the difference $\varepsilon(s) = \theta_i(s) - \theta_f(s)$, an 'error' signal. With the main system $KG(s)$ now relating ε and θ_o, the output/input relation is

$$\frac{\theta_o(s)}{\theta_i(s)} = \frac{KG(s)}{1+[KG(s)][K_f G_f(s)]}$$

Suppose that there is unity feedback, $K_f G_f(s) = 1$; then, if $KG(s)$ is large, there results

$$\theta_o(s)/\theta_i(s) = KG(s)/[1+KG(s)] \simeq 1$$

and the output closely follows the input in magnitude and wave-shape, a condition sought in servo-mechanisms and feedback controls.

where $a = R/L = 1/T$, the inverse Laplace transform gives

$$i(t) = (V/aL)[1 - \exp(-at)] = (V/R)[1 - \exp(-t/T)]$$

which is the complete solution. More complex problems require the development of partial fractions to derive recognisable transforms which are then individually inverse-transformed to give the terms in the solution of $i(t)$.

3.2.15 System functions

It is characteristic of linear constant-coefficient systems that their operational solution involves three parts: (i) the excitation or stimulus, (ii) the output or response and (iii) the system function. Thus in the relation $I(s) = V(s)/Z(s)$ for the current in Z resulting from the application of V, $1/Z(s)$ is the *system function* relating voltage to current. For the simple electrical system shown in *Figure 3.32(a)* the system function $Y(s)$ relating $V(s)$ to $I(s)$ in $I(s) = V(s)Y(s)$ is $Y(s) = 1/(R + Ls + 1/Cs)$. Different functions could relate the capacitor charge or the magnetic linkage in the inductor to the transform $V(s)$ of the stimulus $v(t)$.

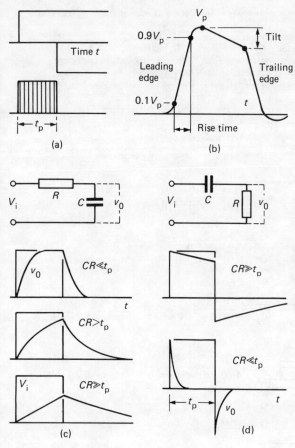

Figure 3.31 Pulse drive

For $CR \gg t_p$ the response shows a tilt; for $CR \ll t_p$ the capacitor charges rapidly and the output v_o comprises positive- and negative-going spikes that give a measure of the time-differential of V_i.

3.2.14.3 Laplace transform method

Application of the Laplace transforms is the most usual method of solving transient problems. The basic features of the Laplace transform are set out in Section 1.2.7 and *Table 1.4*, which gives transform pairs. The advantages of the method are that (1) any stimulus, including discontinuous and pulse forms, can be handled, (2) the solution is complete with both steady-state and transient components, (3) the initial conditions are introduced at the start and (4) formal mathematical processes are avoided.

Consider the system in *Figure 3.28(a)*. The applied direct voltage V has the Laplace transform $V(s) = V/s$; the operational impedance of the circuit is $Z(s) = R + Ls$. Then the Laplace transform of the current is

$$I(s) = \frac{V(s)}{Z(s)} = \frac{V}{s} \cdot \frac{1}{R+Ls} = \frac{V}{L} \cdot \frac{1}{s(s+R/L)}$$

The term V/L is a constant unaffected by transformation. The term in s is almost of the form $a/s(s+a)$. So, if we write

$$I(s) = \frac{V}{aL} \cdot \frac{a}{s(s+a)}$$

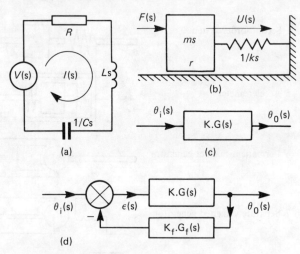

Figure 3.32 System functions

The mechanical analogue (*Figure 3.32(b)*) of this electrical system, as indicated in Section 1.3.1, has a system transfer function to relate force $f(t)$ to velocity $u(t)$ of the mass m and one end of the spring of compliance k in the presence of viscous friction of coefficient r. Then $F(s)$ and $U(s)$ are the transforms of $f(t)$ and $u(t)$, and the operational 'mechanical impedance' has the terms ms, $1/ks$ and r. In general an input $\theta_i(s)$ and an output $\theta_o(s)$ are related by a system transfer function $KG(s)$ (*Figure 3.32(c)*), where K is a numeric or a dimensional quantity to include amplification or the value of some physical quantity (such as admittance). The transform of the integro-differential equation of variation with time is expressed by the term $G(s)$. The system is then represented by the block diagram in *Figure 3.32(c)*; i.e. $\theta_o(s)/\theta_i(s) = KG(s)$.

A number of typical system transfer functions for relatively simple systems are given in *Table 3.4*.

The output of one system may be used as the input to another. Provided that the two do not interact (i.e. the individual transfer functions are not modified by the connection) the overall system function is the product $[K_1 G_1(s)] \times [K_2 G_2(s)]$ of the individual functions. If the systems are paralleled and their outputs are additatively combined, the overall function is their sum.

subsequently the decay of i_t allows the current to approach the steady-state condition.

If $\omega L \gg R$, then approximately $\phi = \tfrac{1}{2}\pi$. Let the switch be closed at $v=0$ for which $\alpha=0$. Then the current is

$$i = (v_m/\omega L)[\sin(\omega t - \tfrac{1}{2}\pi) + 1]$$

which raises i to twice the normal steady-state peak when t reaches a half-period: this is the *doubling effect*. However, if the switch is closed at a source-voltage maximum, the current assumes its steady-state value immediately, with no transient component.

Summary for an RL circuit. The transient current has a decaying exponential form, with a value of k such that, when it is added to the final steady-state current, the initial current flowing in the circuit at $t=0$ is obtained. (In both of the cases in *Figure 3.28* the initial current is zero.) Thus if the initial circuit current is 10 A and the final current is 25 A, the value of k is -15 A.

For the *CR* circuit of *Figure 3.29*, the form of the transient is found from $Ri + q/C = 0$; differentiating, we obtain

$$R(di/dt) + (1/C)i = 0 \quad \text{or} \quad R\lambda + 1/C = 0$$

from which $\lambda = -1/CR = -1/T$, where $T = CR$ is the *time-constant*. Thus $i_t = k\exp(-t/T)$. With the capacitor initially uncharged and a source direct voltage V switched on at $t=0$,

$$i = i_s + i_t = 0 + k\exp(-t/T)$$

As this must be V/R at $t=0+$, then $k=V/R$, as shown in *Figure 3.29(a)*. In *Figure 3.29(b)* the initiation of a *CR* circuit with a sine voltage is shown.

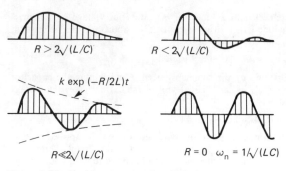

Figure 3.30 Double-energy transient forms

The resulting transient depends on the sign of the quantity in parentheses, i.e. on whether $R/2L$ is greater or less than $1/\sqrt{(LC)}$. Four wave-forms are shown in *Figure 3.30*.

(1) *Roots real*: $R > 2\sqrt{(L/C)}$. The transient current is unidirectional and results from two simple exponential curves with different rates of decay.
(2) *Roots equal*: $R = 2\sqrt{(L/C)}$. This has more mathematical than physical interest, but it marks the boundary between unidirectional and oscillatory transient current.
(3) *Roots complex*: $R < 2\sqrt{(L/C)}$. The roots take the form $-\alpha \pm j\omega_n$, and the transient current oscillates with the interchange of magnetic and electric energies respectively in L and C; but the oscillation amplitude decays by reason of dissipation in R. With $R=0$ the oscillation persists without decay at the undamped natural frequency $\omega_n = 1/\sqrt{(LC)}$.

Pulse drive The response of networks to single pulses (or to trains of such pulses) is an important aspect of data transmission. An ideal pulse has a rectangular wave-form of duration ('width') t_p. It can be considered as the resultant of two opposing step functions displaced in time by t_p as in *Figure 3.31(a)*.

In practice a pulse cannot rise and fall instantaneously, and often the amplitude is not constant (*Figure 3.31(b)*). Ambiguity in the precise position of the peak value V_p makes it necessary to define the *rise time* as the interval between the levels $0.1V_p$ and $0.9V_p$. The *tilt* is the difference between V_p and the value at the start of the trailing edge, expressed as a fraction of V_p.

The response of the output network to a voltage pulse depends on the network characteristics (in particular its time-constant T) and the pulse width t_p. Consider an ideal input voltage V_i of rectangular wave-form applied to an ideal low-pass series network (*Figure 3.31(c)*), the output being the voltage v_o across the capacitor C. Writing p for d/dt, then

$$\frac{v_o}{V_i} = \frac{1/pC}{R + 1/pC} = \frac{1}{1+pCR} = \frac{1}{1+pT}$$

where $T = CR$ is the network time-constant. This represents an exponential growth $v_o = V_i[1 - \exp(-t/T)]$ over the interval t_p. The trailing edge is an exponential decay, with t reckoned from the start of the trailing edge. Three typical responses are shown. For $CR \ll t_p$ the output voltage reaches V_i; for $CR > t_p$ the rise is slow and does not reach V_i; for $CR \gg t_p$ the rise is almost linear, the final value is small and the response is a measure of the time-integral of V_i.

With C and R interchanged as in *Figure 3.31(d)* to give a high-pass network, the whole of V_i appears across R at the leading edge, falling as C charges. Following the input pulse there is a reversed v_o during the discharge of the capacitor. The output/input voltage relation is given by

$$\frac{v_o}{V_i} = \frac{R}{R + 1/pC} = \frac{pCR}{1+pCR} = \frac{pT}{1+pT}$$

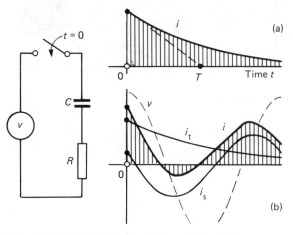

Figure 3.29 Transients in a capacitive–resistive circuit

Summary for an RC circuit. The transient current is a decaying exponential $k\exp(-t/T)$. The initial current is determined by the voltage difference between the voltage applied by the source and that of the capacitor. (In *Figure 3.29* the capacitor is in each case uncharged.) If this p.d. is V_0, then the initial current is V_0/R.

Double energy A typical case is that of a series *RLC* circuit. The transient form is obtained from $L(di/dt) + Ri + q/C = 0$, differentiated to

$$d^2i/dt^2 + (R/L)(di/dt) + (1/LC)i = 0$$

Thus $\lambda^2 + (R/L)\lambda + 1/LC = 0$ is the required equation, with the roots

$$\lambda_1, \lambda_2 = -\frac{R}{2L} \pm \left(\frac{R^2}{4L^2} - \frac{1}{LC}\right)^{1/2}$$

connected to a source of instantaneous voltage v, the corresponding rates of energy input, dissipation (in R) and storage (in L and C) are related by

$$p = vi = Rii + [L(di/dt)i + (q/C)i]$$

Dividing by the common current i and writing the capacitor charge q as the time-integral of the current gives the voltage equation

$$v = Ri + L(di/dt) + (1/C)\int i\,dt$$

and any changes in the parameters or in the applied voltage demand changes in the distribution of the circuit energy. The integro-differential equation can be solved to yield both steady-state and transient conditions.

In practical circuits the system may be too complex for such a direct solution; the following methods may then be attempted:

(i) formal mathematics for simple cases with linear parameters;
(ii) simplification, e.g. by linearising parameters or by neglecting second-order terms;
(iii) writing a possible solution based on the known physical behaviour of the system, with a check by differentiation;
(iv) setting up a model system on an analogue computer;
(v) programming a digital computer to give a solution by iteration.

3.2.14.1 Classification

Where the system has only one energy-storage component, *single-energy* transients occur. Where two (or more) different storages are concerned, the transient has a *double-* (or *multiple-*) *energy* form. Transients may occur in the following circumstances:

(1) *Initiation*: a system, initially dead, is energised.
(2) *Subsidence*: an initially energised system is reduced to a zero-energy condition.
(3) *Transition*: a change from one state to another, where both states are energetic.
(4) *Complex*: the superposition of more than one disturbance.
(5) *Relaxation*: transition between states that, when reached, are themselves unstable.

Further distinctions can be made, e.g. between linear and non-linear parameters, neglect or otherwise of propagation time within the system, etc. Attention here is mainly confined to simple electric networks with constant parameters and, by analogy (Section 1.3.1), to corresponding mechanical systems.

3.2.14.2 Transient forms

During transience, the current i for an impressed voltage stimulus $v(t)$ is considered to be the superposition of a transient component i_t and a final steady-state current i_s, so that at any instant $i = i_s + i_t$. Alternatively, the voltage v for an impressed current stimulus $i(t)$ is the summation $v = v_s + v_t$. The quantities i_s or v_s are readily derived by applying the appropriate steady-state technique. The *form* of i_t or v_t is characteristic of the system itself, is independent of the stimulus and comprises exponential terms $k \exp(\lambda t)$ where k depends on the boundary conditions. This is the case because of the fixed proportionality between the stored energy $\frac{1}{2}Li^2$ and the rate of energy dissipation Ri^2 in an RL circuit; and similarly for $\frac{1}{2}Cv^2$ and v^2/R in an RC circuit. Hence the transient form can be obtained from a case in which the final steady state is of zero energy, i.e. a subsidence transient.

The subsidence transient in a *single-energy* (first-order) system having the general equation $dy/dt + ay = 0$ can be found by substituting λ for d/dt to give $\lambda y + ay = 0$, whence $\lambda = -a$. Then the solution is

$$y = k \exp(\lambda t) = k \exp(-at)$$

a simple exponential decay as in *Figure 1.2* of Section 1.2.2. For a *double-energy* (second-order) system the basic equation is

$$d^2y/dt^2 + a(dy/dt) + by = 0 \quad \text{or} \quad \lambda^2 + a\lambda + b = 0$$

The quadratic in λ has two roots, λ_1 and λ_2, and the solution has a pair of exponential terms that depend on the relation between a and b. For a *multiple-energy* (nth-order) system there will be n roots. From Section 1.2.2 it will be seen that exponential terms can represent oscillatory as well as decay forms of response.

Single-energy Consider the RL circuit in *Figure 3.28*, subsequent to closure of the switch at $t = 0$. The transient current

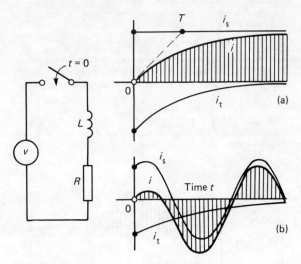

Figure 3.28 Transients in an inductive–resistive circuit

form is obtained from $L(di/dt) + Ri = 0$, or $L\lambda i + Ri = 0$, giving $\lambda = -R/L$. Then

$$i_t = k \exp[-t(R/L)] = k \exp(-t/T)$$

where $T = L/R$ is the *time-constant*. The final steady-state current depends on the source voltage v. In *Figure 3.28(a)*, with $v = V$, a constant direct voltage, $i_s = V/R$. Immediately after switching, with $t = 0+$, the current i is still zero because the inductance prevents any instantaneous rise. Hence

$$i = i_s + i_t = V/R + k \exp(-0) = V/R + k$$

so that $k = -(V/R)$. From $t = 0$ the current is therefore

$$i = i_s + i_t = (V/R)[1 - \exp(-t/T)]$$

The two terms and their summation are shown in *Figure 3.28(a)*.

If, as in *Figure 3.28(b)*, the source voltage is sinusoidal expressed by $v = v_m \sin(\omega t - \alpha)$ and again switching occurs at $t = 0$, the form of the transient current is unchanged, but the final steady-state current is

$$i_s = (v_m/Z)\sin(\omega t - \alpha - \phi)$$

where $Z = \sqrt{(R^2 + \omega^2 L^2)}$ and $\phi = \arctan(\omega L/R)$. At $t = 0+$

$$i = i_s + i_t = (v_m/Z)\sin(-\alpha - \phi) + k = 0$$

which gives $k = -(v_m/A)\sin(-\alpha - \phi)$. The final steady-state and transient current components are shown in *Figure 3.28(b)* with their resultant. Initially the current is asymmetric, but

angle α: thus $V_r > V_s$, the *Ferranti effect*. For the loaded condition, $I_r B$ is added to $V_r A$ to give V_s. Similarly $V_r C$ is added to $I_r D$ to obtain I_s.

The product $V_r I_r = I_r (V_s - V_r A)$ is the receiving-end complex apparent power S_r. Let V_s lead V_r by angle θ; then the receiving-end load has the active and reactive powers P_r and Q_r given by

$$P_r = (V_s V_r / B) \cos(\theta - \beta) - (V_r^2 A/B) \cos(\beta - \alpha)$$
$$Q_r (V_s V_r / B) \sin(\theta - \beta) + (V_r^2 A/B) \sin(\beta - \alpha)$$

where α and β are the angles in the complexors A and B. The importance of B (roughly the overall series impedance) is clear.

Line chart Operating charts for a transmission circuit can be drawn to relate graphically V_s, V_r, P_r and Q_r, using the appropriate overall $ABCD$ parameters (e.g. with terminal transformers included).

Receiving-end chart A receiving-end chart gives active and reactive power at the receiving end for V_r constant (*Figure 3.26(a)*).

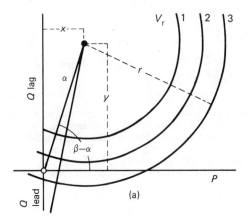

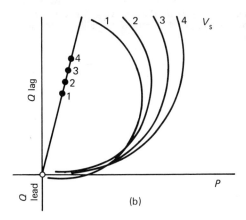

Figure 3.26 Line charts

The co-ordinates (x, y) and the radius (r) of the constant-voltage circles are

$$x = -V_r^2 (A/B) \cos(\beta - \alpha) \quad y = -V_r^2 (A/B) \sin(\beta - \alpha)$$
$$r = V_s V_r / B$$

where A and B are scalar magnitudes, and α and β the angles in A and B. For a given V_r the chart comprises a family of concentric circles, each corresponding to a particular V_s. If a given receiving-end load is located by its active and reactive power components, V_s is obtained from the corresponding V_s circle.

Sending-end chart For a given V_s the sending-end chart comprises a family of circles as shown in *Figure 3.26(b)*, each circle corresponding to a particular V_r. Load points outside the envelope of these circles cannot be supplied at the V_s for which the chart is drawn.

3.2.13.2 Short line

For an overhead interconnector line the capacitive shunt admittance is neglected, reducing the general parameters to $A = D = 1$, $B = Z = R + jX$ and $C = 0$. The operating conditions are those in *Figure 3.27*, with a receiving-end voltage V_r (taken as reference phasor), a sending-end voltage V_s and a load current I at

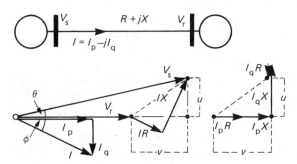

Figure 3.27 Operating conditions for a short transmission line

a lagging phase angle ϕ with respect to V_r and having active and reactive components respectively I_p and I_q. Then

$$V_s = V_r + (I_p - jI_q)(R + jX)$$
$$= V_r + (I_p R + I_q X) + j(I_p X - I_q R) = V_r + v + ju$$

To a close approximation, v is the difference of the voltages V_s and V_r, while u determines their phase difference (or transmission angle). The regulation and angle are therefore v/V_s p.u. and $\theta = \arctan(u/V_s)$ rad.

Suppose that $V_r = V_s$; then $v = 0$ giving $I_q = -I_p(R/X)$, and $u = I_p X [1 + (R/X)^2]$ giving $\theta = \arctan(I_p/V_s)[1 + (R/X)^2]$. The consequences are that (i) for a receiving-end active power P the load must be able to absorb a leading reactive power $Q = P(R/X)$; and (ii) the transmission angle is determined largely by X. If $R/X = 0.5$, typical of an overhead line, then $Q = 0.5P$ and $\theta = \arctan[1.25 X(I_p/V_s)]$. With interconnector cables the R/X ratio is usually greater than unity and shunt capacitance current is no longer negligible.

Independent adjustment of V_s and V_r is not feasible, and effective load control requires adjustment of v (e.g. by transformer taps) and of u (e.g. by quadrature boosting).

3.2.14 Network transients

Energy cannot be instantaneously converted from one form to another, although the time needed for conversion can be very short and the conversion rate (i.e. the power) high. Change between states occurs in a period of *transience* during which the system energies are redistributed in accordance with the energy-conservation principle (Section 1.3.1). For example, in a simple series circuit of resistance R, inductance L and capacitance C

The components are evaluated from the arbitrary identities

$$E_a = Z + P + N \quad E_b = Z + h^2 P + hN \quad E_c = Z + hP + h^2 N$$

where

$$Z \doteq E_a + E_b + E_c)/3 \quad P = (E_a + hE_b + h^2 E_c)/3$$
$$N = (E_a + h^2 E_b + hE_c)/3$$

Figure 3.24 is drawn for an asymmetric system with voltages $E_a = 200$, $E_b = 100$ and $E_c = 400$ V, and phase-displacement angles $\alpha = 90°$, $\beta = 120°$ and $\gamma = 150°$. In phasor terms,

$$E_a = 200 \angle 0° = 200 + j0 \text{ V}$$
$$E_b = 100 \angle (-90°) = 0 - j100 \text{ V}$$
$$E_c = 400 \angle 150° = -346 + j200 \text{ V}$$

Then

$$Z = (200 - j100 - 346 + j200)/3 = -49 + j33 = E_{a0}$$
$$P = (200 + 87 + j50 + 347 + j200)/3 = 211 + j83 = E_{a+}$$
$$N = (200 - 87 + j50 - j400)/3 = 38 - j117 = E_{a-}$$

The summation $E_{a0} + E_{a+} + E_{a-} = 200 + j0 = E_a$. The p.p.s. and n.p.s. components of E_b and E_c are readily obtained.

3.2.12.1 Power

In linear networks there is no interference between currents of different sequences. Thus p.p.s. voltages produce only p.p.s. currents, etc. The total power is therefore

$$P = P_a + P_b + P_c$$
$$= 3(V_0 I_0 \cos\phi_0 + V_+ I_+ \cos\phi_+ + V_- I_- \cos\phi_-)$$

This is equivalent to the more obvious summation of phase powers

$$P = V_a I_a \cos\phi_a + V_b I_b \cos\phi_b + V_c I_c \cos\phi_c$$

Symmetrical-component techniques are useful in the analysis of power-system networks with faults or unbalanced loads: an example is given in Section 3.3.4. Machine performance is also affected when the machine is supplied from an asymmetric voltage system: thus in a 3-phase induction motor the n.p.s. components set up a torque in opposition to that of the (normal) p.p.s voltages.

3.2.13 Line transmission

Networks of small physical dimensions and operated at low frequency are usually considered to have a zero propagation time; a current started in a closed circuit appears at every point in the circuit simultaneously. In extended circuits, such as long transmission lines, the propagation time is significant and cannot properly be ignored.

The basics of energy propagation on an ideal loss-free line are discussed in Section 2.6.3. Propagation takes place as a voltage wave v accompanied by a current wave i such that $v/i = z_0$ (the surge impedance) travelling at speed u. Both z_0 and u are functions of the line configuration, the electric and magnetic space constants ε_0 and μ_0, and the relative permittivity and permeability of the medium surrounding the line conductors. At the receiving end of a line of finite length, an abrupt change of the electromagnetic-field pattern (and therefore of the ratio v/i) is imposed by the discontinuity unless the receiving-end load is z_0, a termination called the *natural load* in a power line and a *matching impedance* in a telecommunication line. For a non-matching termination, wave reflection takes place with an electromagnetic wave running back towards the sending end. After many successive reflections of rapidly diminishing amplitude, the system settles down to a steady state determined by the sending-end voltage, the receiving-end load impedance and the line parameters.

3.2.13.1 A.c. power transmission

The steady-state condition considered is the transfer of a constant balanced apparent power per phase from a generator at the sending end (s) to a load at the receiving end (r) by a sinusoidal voltage and current at a frequency $f = \omega/2\pi$. The line has uniformly distributed parameters: a conductor resistance r and a loop inductance L effectively in series, and an insulation conductance g and capacitance C in shunt, all per phase and per unit length. The series impedance, shunt admittance and propagation coefficient per unit length are $z = r + j\omega L$, $y = g + j\omega C$ and $\gamma = \sqrt{(yz)}$ respectively. For a line by a length l the overall parameters are $zl = Z$, $yl = Y$ and $l\sqrt{(yz)} = \sqrt{(YZ)} = \gamma l$. The solution for the receiving-end terminal conditions is in terms of $\sqrt{(YZ)}$ and its hyperbolic functions as a two-port:

$$V_s = V_r A + I_r B = V_r \cosh(\sqrt{YZ}) + I_r z_0 \sinh(\sqrt{YZ})$$
$$I_s = V_r C + I_r D = V_r (1/z_0) \sinh(\sqrt{YZ}) + I_r \cosh(\sqrt{YZ})$$

Using the hyperbolic series (Section 1.2.2) and writing $z_0 = \sqrt{(Z/Y)}$, we obtain for a symmetrical line

$$A = 1 + YZ/2 + (YZ)^2/24 + \ldots = D$$
$$B = Z[1 + YZ/6 + (YZ)^2/120 + \ldots]$$
$$C = Y[1 + YZ/6 + (YZ)^2/120 + \ldots]$$

The significance of the higher powers of YZ depends (i) on the line configuration, (ii) on the properties of the ambient medium and (iii) on the physical length of the line in terms of the wavelength $\lambda = u/f$. For air-insulated overhead lines and inductance is large, the capacitance small: the propagation velocity approximates to $u = 3 \times 10^5$ km/s (corresponding to a wavelength $\lambda = 6000$ km at 50 Hz), with a natural load z_0 of the order of 400–500 Ω. Cable lines have a low inductance and a large capacitance: the permittivity of the dielectric material and the presence of armouring and sheathing result in a propagation velocity around 200 km/s, a surge impedance below 100 Ω, and the possibility (in high-voltage cables) that the charging current may be comparable with the load current if the cable length exceeds 25–30 km.

For balanced 3-phase power transmission, the general equations are applied for the line-to-neutral voltage, line current and phase power factor. Phasor diagrams for the load and no-load ($I_r = 0$) receiving-end conditions for an overhead-line transmission are shown in *Figure 3.25*, with V_r as datum. On no load, $V_s = V_r A$, and as A has a magnitude less than unity and a small positive angle α, the phasor $V_r A$ is smaller than V_r and leads it by

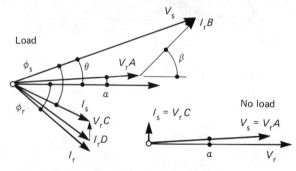

Figure 3.25 Transmission-line phasor diagram

summates the power automatically; with separate instruments, one will tend to read reversed under certain conditions, given below.

(4) *Three-phase, 3-wire, load balanced* With sinusoidal voltage and current the conditions in *Figure 3.23(c)* obtain. Wattmeters W_1 and W_2 indicate powers P_1 and P_2 where

$$P_1 = V_{ab}I_a \cos(30° + \phi) = V_l I_l \cos(30° + \phi)$$
$$P_2 = V_{cb}I_c \cos(30° - \phi) = V_l I_l \cos(30° - \phi)$$

The total active power $P = P_1 + P_2$ is therefore

$$P = V_l I_l [\cos(30° + \phi) + \cos(30° - \phi)] = \sqrt{3}\, V_l I_l \cos\phi$$

where $\cos\phi$ is the *phase* power factor. The algebraic difference is $P_1 - P_2 = V_l I_l \sin\phi$, whence the reactive power is given by

$$Q = \sqrt{3}\, V_l I_l \sin\phi = \sqrt{3}\,(P_1 - P_2)$$

and the phase angle can be obtained from $\phi = \arctan(Q/P)$. For $\phi = 0$ (unity power factor) both wattmeters read alike; for $\phi = 60°$ (power factor 0.5 lag) W_1 reads zero; and for lower lagging power factors W_1 tends to read backwards.

The active power of a single phase has a double-frequency pulsation (*Figure 3.18*). For the asymmetric 2-phase system under balanced conditions and a phase displacement of 90°, and for all symmetric systems with $m = 3$ or more, the total power is constant.

3.2.11.6 Harmonics

Considering a symmetrical balanced system of 3-phase non-sinusoidal voltages, and omitting phase displacements (which are in the context not significant), let the voltage of phase A be

$$v_a = v_1 \sin\omega t + v_2 \sin 2\omega t + v_3 \sin 3\omega t + \dots$$

Writing $\omega t - \tfrac{2}{3}\pi$ and $\omega t - \tfrac{4}{3}\pi$ respectively for phases B and C, and simplifying, we obtain

$$v_a = v_1 \sin\omega t \quad + v_2 \sin 2\omega t \quad + v_3 \sin 3\omega t + \dots$$
$$v_b = v_1 \sin(\omega t - \tfrac{2}{3}\pi) + v_2 \sin 2(\omega t - \tfrac{2}{3}\pi) + v_3 \sin 3\omega t + \dots$$
$$v_c = v_1 \sin(\omega t - \tfrac{4}{3}\pi) + v_2 \sin 2(\omega t - \tfrac{2}{3}\pi) + v_3 \sin 3\omega t + \dots$$

The fundamentals have a normal $2\pi/3$ rad (120°) phase relation in the sequence ABC, as also do the 4th, 7th, 10th,... harmonics. The 2nd (and 5th, 8th, 11th,...) harmonics have the $2\pi/3$ rad phase relation but of reversed sequence ACB. The *triplen* harmonics (those of order a multiple of 3) are, however, *cophasal* and form a zero-sequence set.

The relation $V_l = \sqrt{3}\, V_{ph}$ in a 3-phase star-connected system is applicable only for sine wave-forms. If harmonics are present the line- and phase-voltage wave-forms differ because of the effective phase angle and sequence of the harmonic components. The nth harmonic voltages to neutral in two successive phases AB are $v_n \sin n\omega t$ and $v_n \sin n(\omega t - \tfrac{2}{3}\pi)$, and between the corresponding line terminals the nth harmonic voltage is $2v_n \sin n(\tfrac{1}{3}\pi)$. For triplen harmonics this is zero; hence no triplens are present in balanced line voltages because, in the associated phases, their components are equal and in opposition. In a balanced delta connection, again no triplens are present between lines: the delta forms a closed circuit to triplen circulating currents, the impedance drop of which absorbs the harmonic e.m.f.s.

3.2.12 Symmetrical components

Figure 3.21(a) shows the sine waves and phasors of a balanced symmetric 3-phase system of e.m.f.s of sequence ABC. The magnitudes are equal and the phase displacements are $2\pi/3$ rad. In *Figure 3.21(b)*, the asymmetric sine wave-forms have also the sequence ABC, but they are of different magnitudes and have the phase displacements α, β and γ. Problems of asymmetry occur in the unbalanced loading of a.c. machines and in fault conditions on power networks. While a solution is possible by the Kirchhoff laws, the method of *symmetrical components* greatly simplifies analysis.

Any set of asymmetric 3-phase e.m.f.s or currents can be resolved into a summation of three sets of symmetrical components, respectively of positive phase-sequence (p.p.s.) ABC, negative phase-sequence (n.p.s.) ACB, and zero phase-sequence (z.p.s.). Use is made of the operator h, resembling the 90° operator j (Section 3.2.9.1) but implying a counter-clockwise rotation of $2\pi/3$ rad (120°). Thus

$$h = 1 \angle 120° = \tfrac{1}{2}(-1 + j\sqrt{3}) \quad h^2 = 1 \angle 240° = \tfrac{1}{2}(-1 - j\sqrt{3})$$
$$h^3 = 1 \angle 360° = 1 + j0 \qquad 1 + h + h^2 = 0$$

A symmetric 3-phase system has only p.p.s. components

$$E_a = E_{a+} \quad E_b = h^2 E_{a+} \quad E_c = h E_{a+}$$

whereas an asymmetric system (*Figure 3.24*) comprises the three sets

z.p.s. $\quad E_{a0} \quad E_{b0} = E_{a0} \quad E_{c0} = E_{a0}$

p.p.s. $\quad E_{a+} \quad E_{b+} = h^2 E_{a+} \quad E_{c+} = h E_{a+}$

n.p.s. $\quad E_{a-} \quad E_{b-} = h E_{a-} \quad E_{c-} = h^2 E_{a-}$

where the subscripts 0, + and − designate the z.p.s., p.p.s. and n.p.s. components respectively. The p.p.s. and the n.p.s. components sum individually to zero. Therefore, if the originating phasors E_a, E_b, E_c also sum to zero there are no z.p.s. components; if they do not, their residual is the sum of the three z.p.s. components.

The asymmetrical phasors have now been reduced to the sum of three sets of symmetrical components:

$$E_a = E_{a0} + E_{a+} + E_{a-} \qquad E_b = E_{b0} + E_{b+} + E_{b-}$$
$$E_c = E_{c0} + E_{c+} + E_{c-}$$

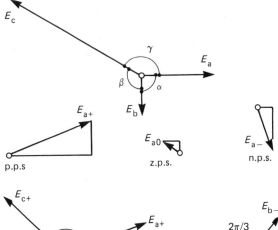

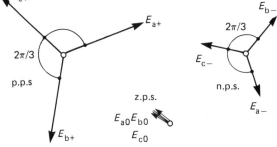

Figure 3.24 Symmetrical components

and its e.m.f. E_{xy} is represented by an arrow with its point at X. The e.m.f. E_{yx} between terminals Y and X is therefore $-E_{xy}$. Further, when two windings XN and YN have a common terminal N, the e.m.f.s are

X to Y: $E_{xy} = E_{xn} - E_{yn}$ Y to X: $E_{yx} = E_{yn} - E_{xn} = -E_{xy}$

Common phase interconnections are shown in *Figure 3.22*.

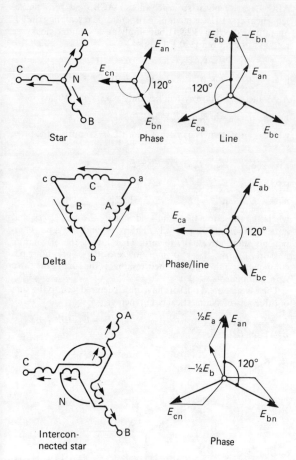

Figure 3.22 Three-phase interconnections

3.2.11.2 Star

Let the phase e.m.f.s be E_{an}, E_{bn} and E_{cn} with an arbitrary positive direction outward from the star-point N. Then the line e.m.f.s are

$E_{ab} = E_{an} - E_{bn}$ $E_{bc} = E_{bn} - E_{cn}$ $E_{ca} = E_{cn} - E_{an}$

These are of magnitude $\sqrt{3}$ times that of a phase e.m.f., and provide a symmetric 3-phase system of line e.m.f.s, with E_{ab} leading E_{an} by 30°. Thus $E_l = \sqrt{3}\, E_{ph}$ and $I_l = I_{ph}$, the subscripts l and ph referring to line and phase quantities respectively.

3.2.11.3 Delta

The line-to-line e.m.f. is that of the phase across which the lines are connected. The line current is the difference of the currents in the phases forming the line junction, so that the relations for symmetric loading are $E_l = E_{ph}$ and $I_l = \sqrt{3}\, I_{ph}$.

3.2.11.4 Interconnected star

A line-to-neutral e.m.f. comprises contributions from successive half-phases and sums to $\tfrac{1}{2}\sqrt{3}$ of a complete phase e.m.f. The line-to-line e.m.f. is $1\tfrac{1}{2}$ times the magnitude of a complete phase e.m.f. and the line current is numerically equal to the phase current.

3.2.11.5 Power

The total power delivered to or absorbed by a polyphase system, be it symmetric and balanced or not, is the algebraic sum of the individual phase powers. Consider an m-phase system with instantaneous line currents $i_1, i_2, \ldots, i_m$, the algebraic sum of which is zero by the Kirchhoff node law. Let the voltages of the input (or output) terminals, with reference to a common point X, be $v_1 - v_x, v_2 - v_x, \ldots, v_m - v_x$; then the instantaneous powers will be $(v_1 - v_x)i_1, (v_2 - v_x)i_2, \ldots, (v_m - v_x)i_m$, which together sum to the total instantaneous power p. There is no restriction on the choice of X; it can be any of the terminals, say M. In this case $v_m - v_x = v_m - v_m = 0$, and the power summation has only $m - 1$ terms. The average power over a full period T is therefore

$$P \doteq (1/T)\int_0^T \left[(v_1 - v_m)i_1 + \ldots + (v_{m-1} - v_m)i_{m-1}\right] dt$$

The first term of the sum in brackets represents the indication of a wattmeter with i_1 in its current-circuit and $v_1 - v_m$ across its volt-circuit, i.e. connected between terminals 1 and M. It follows that three wattmeters can measure the power in a 3-phase 4-wire system, and two in a 3-phase 3-wire system. Some of the common cases are listed below.

(1) *Three-phase, 4-wire, load unbalanced* The connections are shown in *Figure 3.23(a)*. Wattmeters W_1, W_2 and W_3 measure the phase powers separately. The total power is the sum of the indications:

$P = P_1 + P_2 + P_3$

(2) *Three-phase, 4-wire, load balanced* With the connections shown in *Figure 3.23(a)*, all the meters read the same. Two of the wattmeters can be omitted and the reading of the remaining instrument multiplied by 3.

(3) *Three-phase, 3-wire, load unbalanced* Two wattmeters are connected with their current-circuits in any pair of lines, as in *Figure 3.23(b)*. The total power is the algebraic sum of the readings, regardless of wave-form. A two-element wattmeter

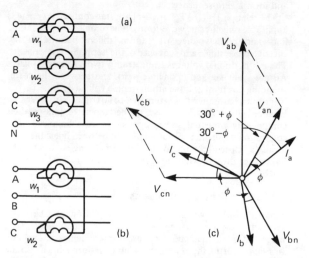

Figure 3.23 Three-phase power measurement

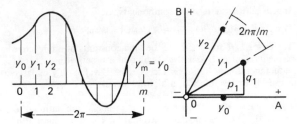

Figure 3.20 Graphical harmonic analysis

A semi-graphical method is shown in *Figure 3.20*. The base of a complete period (2π) is divided into m parts, the corresponding ordinates being $y_0, y_1, y_2, \ldots, y_m$. Construct axes OA and OB; set out the radii y_0 to $y_m (=y_0)$ at angles $0, 2n\pi/m, 4n\pi/m, \ldots$ from the axis OA. Then project the extremities horizontally (p) and vertically (q), and take the sum of the two sets of projections with due regard to their sign. Then for the nth harmonic

$$a_n = \frac{2}{\pi} \sum_0^{m-1} p \qquad b_n = \frac{2}{\pi} \sum_0^{m-1} q$$

The labour is reduced if $2\pi/m$ is a simple fraction of 2π, for then some groups of radii are coincident.

3.2.10.3 Power

The r.m.s. value of a current

$$i = I_0 + i_1 \sin(\omega t + \alpha_1) + i_2 \sin(2\omega t + \alpha_2) + \ldots$$

is obtained from the square root of the average squared value, resulting in

$$I = \sqrt{(I_0^2 + \tfrac{1}{2}i_1^2 + \tfrac{1}{2}i_2^2 + \ldots)} = \sqrt{(I_0^2 + I_1^2 + I_2^2 + \ldots)}$$

where $I_1 = i_1/\sqrt{2}$, $I_2 = i_2/\sqrt{2}$ etc. are the r.m.s. values of the individual harmonic components. The r.m.s. voltage is obtained in a similar way.

Power The instantaneous power p in a circuit with an applied voltage

$$v = v_1 \sin(\omega t + \alpha_1) + v_2 \sin(2\omega t + \alpha_2) + \ldots$$

producing a current

$$i = i_1 \sin(\omega t + \alpha_1 - \phi_1) + i_2 \sin(2\omega t + \alpha_2 - \phi_2) + \ldots$$

is the product vi: this includes (i) a series of the form

$$v_n i_n \sin(n\omega t + \alpha_n) \sin(n\omega t + \alpha_n - \phi_n)$$

all terms of which have a fundamental-period average $\tfrac{1}{2} v_n i_n \cos \phi_n$; and (ii) a series of the form

$$v_p i_q \sin(p\omega t + \alpha_p) \sin(q\omega t + \alpha_q - \phi_q)$$

which, over a fundamental period, averages zero. Power is circulated by a voltage and a current of different frequencies, but the circulation averages zero. The mean (active) power is therefore

$$P = \tfrac{1}{2}v_1 i_1 \cos\phi_1 + \tfrac{1}{2}v_2 i_2 \cos\phi_2 + \ldots + \tfrac{1}{2}v_n i_n \cos\phi_n + \ldots$$
$$= V_1 I_1 \cos\phi_1 + V_2 I_2 \cos\phi_2 + \ldots + V_n I_n \cos\phi_n + \ldots$$

where the capital letters denote component r.m.s. values. Thus the harmonics contribute power separately.

Power factor The ratio of the active power P to the apparent power S is

$$P/S = (V_1 I_1 \cos\phi_1 + \ldots + V_n I_n \cos\phi_n + \ldots)/VI$$

This may be less than unity even with all phase angles zero if the ratio V_n/I_n is not the same for each component. Where the applied voltage is a *pure sinusoid* of fundamental frequency there can be no harmonic powers; the active power is $P = V_1 I_1 \cos\phi_1$. Then

$$P/S = (V_1 I_1 \cos\phi_1)/V_1 I = (I_1/I) \cos\phi_1$$

where $I_1/I = I_1/\sqrt{(I_1^2 + I_2^2 + \ldots + I_n^2)} = \delta$, the *distortion factor*. The overall power factor is consequently $\delta \cos\phi_1$. This is typical of circuits containing non-linear elements.

3.2.11 Three-phase systems

A symmetrical m-phase system has m source e.m.f.s, all of the same wave-form and frequency, and displaced $2\pi/m$ rad or $1/m$ period in time; m is most commonly 3, but is occasionally 6, 12 or 24.

Symmetric 3-phase system In *Figure 3.21(a)* the symmetric sinusoidal 3-phase system has source e.m.f.s in phases A, B and C given by

$$e_a = e_m \sin\omega t \quad e_b = e_m \sin(\omega t - 2\pi/3) \quad e_c = e_m \sin(\omega t - 4\pi/3)$$

The instantaneous sum of the phase e.m.f.s (and also the phasor sum of the corresponding r.m.s. phasors E_a, E_b and E_c) is zero.

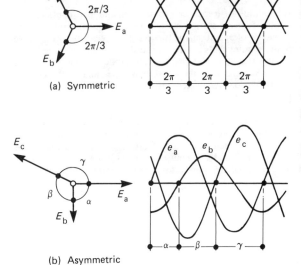

Figure 3.21 Three-phase systems

Asymmetric 3-phase system The asymmetric system in *Figure 3.21(b)* has, in general, unequal phase voltages and phase displacements. Such asymmetry may occur in machines with unbalanced phase windings and in power supply systems when faults occur; the usual method of dealing with asymmetry is described in Section 3.2.12. Attention here is confined to the basic symmetric cases.

3.2.11.1 Phase interlinkage

While individual phase sources can be used separately, they are generated in the same machine and are normally interlinked. Using r.m.s. phasors, let the e.m.f. in a generator winding XY be such as to drive positive current out at X: then X is positive to Y

and load at a *peak* rate

$Q = \tfrac{1}{2} v_m i_m \sin \phi = VI \sin \phi$

The power conditions thus summarise to the following:

Active power P: the *mean* of the instantaneous power over an integral number of periods giving the mean rate of energy transfer from source to load in watts (W).

Reactive power Q: the *maximum* rate of energy interchange between source and load in reactive volt-amperes (var).

Apparent power S: the product of the r.m.s. voltage and current in volt-amperes (VA).

Both P and Q represent real power. The apparent power S is not a power at all, but is an arbitrary product VI. Nevertheless, because of the way in which P and Q are defined, we can write

$P^2 + Q^2 = (VI)^2(\cos^2\phi + \sin^2\phi) = (VI)^2$

whence $S = \sqrt{(P^2 + Q^2)}$, a convenient combination of mean active power with peak power circulation.

Complex power: the active and reactive powers can be determined for voltage and current phasors by

$S = P \pm jQ = VI^*$ or $S = V^*I$

using the conjugate of either I or V.

Power factor: the ratio of active to apparent power, $P/S = \cos \phi$ for sinusoidal conditions.

3.2.9.4 Resonance

A condition of resonance occurs when the load contains two forms of energy-storing element (L and C) such that, at the frequency of operation, the two energies are equal. The reactive power requirements are then satisfied internally, as the inductor releases energy at the rate that the capacitor requires it. The source supplies only the active power demand of the energy-dissipating load components, the load externally appearing to be purely resistive.

Acceptor resonance The series RLC circuit in *Figure 3.19(a)* has, at angular frequency ω, the impedance $Z = R + jX$, where X is $\omega L - 1/\omega C$, which for $\omega = \omega_0 = 1/\sqrt{(LC)}$ is zero. The impedance is then $Z = R$ and the input current has a maximum $I_0 = V/R$, conditions of *acceptor* resonance. Internally, large voltages appear across the reactive components, viz.

$V_L = I_0 \omega L = V \omega_0 (L/R)$ $V_C = I_0 (1/\omega_0 C) = V/\omega_0(CR)$

The terms L/R and $1/CR$ are the time-constants of the reactive elements, and $\omega_0 L/R$ is the Q-value of a practical inductor of inductance L and loss-resistance R. The Q-value may be large (e.g. 100) for resonance at a high frequency.

Rejector resonance This occurs in a parallel combination of L and C, the energies circulating around the closed LC loop. If in *Figure 3.19(b)* the resistance R is zero, the terminal input admittance vanishes at angular frequency $\omega_0 = 1/\sqrt{(LC)}$, with $\omega_0 C = 1/\omega_0 L$ and an input susceptance $B = 0$. Where the circuit contains resistance R the resonance conditions are less definite. Three possible criteria are (i) $\omega_0 = 1/\sqrt{(LC)}$, (ii) the input admittance is a minimum and (iii) the input admittance is purely conductive. All three criteria are satisfied simultaneously in the simple acceptor circuit, but differ in rejector conditions; however, where resonance is an intended property of the circuit, the differences are small.

Some expressions for resonance are given for six circuit arrangements in *Table 3.3*.

3.2.10 Non-sinusoidal alternating quantities

Periodic but non-sinusoidal currents occur (i) with non-sinusoidal e.m.f. sources, (ii) with sinusoidal sources applied to non-linear loads and (iii) with any combination of (i) and (ii).

3.2.10.1 Fourier series

Any univalued periodic wave-form can be represented as a summation of sine waves comprising a *fundamental*, where frequency is that of the periodic occurrence, and a series of *harmonic* waves of frequency $2, 3, \ldots, n$ times that of the fundamental. The Fourier series for a periodic function $y = f(x)$ takes either of the following equivalent forms:

(1) $y = c_0 + c_1 \sin(x + \alpha_1) + c_2 \sin(2x + \alpha_2) + \ldots$

(2) $y = c_0 + a_1 \cos x + a_2 \cos 2x + \ldots + a_n \cos nx$
        $+ b_1 \sin x + b_2 \sin 2x + \ldots + b_n \sin nx$

where c_0 is a constant, $c_n = \sqrt{(a_n^2 + b_n^2)}$ and $\alpha_n = \arctan(a_n/b_n)$. The coefficients of the terms are given by

$c_0 = (1/2\pi) \int_0^{2\pi} f(x) \, dx =$ mean of the wave over one period

$a_n = (1/\pi) \int_0^{2\pi} f(x) \cos nx \, dx$ $b_n = (1/\pi) \int_0^{2\pi} f(x) \sin nx \, dx$

These can be evaluated mathematically for simple cases. The work may sometimes be reduced by inspection: thus $c_0 = 0$ for a wave symmetrical about the baseline; or only odd-order harmonics may be present.

3.2.10.2 Analysis

The series for a range of mathematically tractable wave-forms are given in *Table 1.10*. For experimentally derived wave-forms there are several methods, but none yields the amplitude of higher-order harmonics without considerable labour, unless a computer program is available.

A particular harmonic, say the nth, may be found by super-posing n copies of the wave, displaced relatively by $2\pi/n, 4\pi/n, \ldots$, and adding corresponding ordinates. The result is a wave of frequency n times that of the harmonic sought (with the addition, however, of harmonics of orders kn where k is an integer). The method gives also the phase angle α_n.

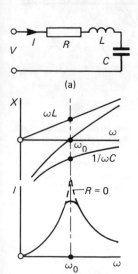
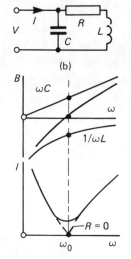

Figure 3.19 Resonance

Table 3.3 Impedance of network elements at angular frequency ω (rad/s)

Impedance: $Z = R + jX = |Z| \angle \theta$ $\qquad$ $|Z| = \sqrt{(R^2 + X^2)}$ $\qquad$ $\theta = \arctan(X/R)$
Admittance: $Y = 1/Z = |Y| \angle (-\theta)$ $\qquad$ $|Y| = \sqrt{[(R/Z^2)^2 + (X/Z^2)^2]}$ $\qquad$ $\theta = -\arctan(X/R)$

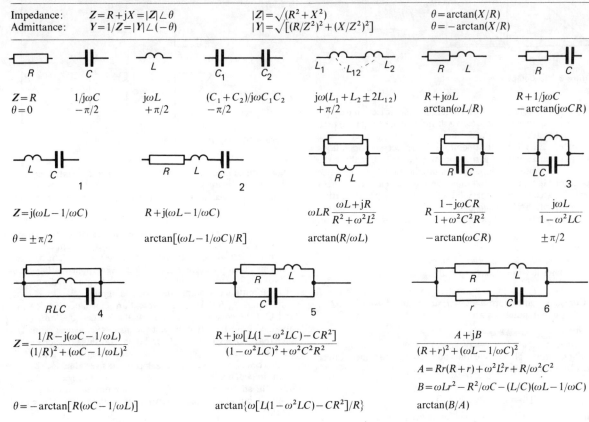

$Z = R$ $\qquad$ $1/j\omega C$ $\qquad$ $j\omega L$ $\qquad$ $(C_1 + C_2)/j\omega C_1 C_2$ $\qquad$ $j\omega(L_1 + L_2 \pm 2L_{12})$ $\qquad$ $R + j\omega L$ $\qquad$ $R + 1/j\omega C$
$\theta = 0$ $\qquad$ $-\pi/2$ $\qquad$ $+\pi/2$ $\qquad$ $-\pi/2$ $\qquad$ $+\pi/2$ $\qquad$ $\arctan(\omega L/R)$ $\qquad$ $-\arctan(j\omega CR)$

1 $\qquad$ 2 $\qquad$ 3

$Z = j(\omega L - 1/\omega C)$ $\qquad$ $R + j(\omega L - 1/\omega C)$ $\qquad$ $\omega LR \dfrac{\omega L + jR}{R^2 + \omega^2 L^2}$ $\qquad$ $R \dfrac{1 - j\omega CR}{1 + \omega^2 C^2 R^2}$ $\qquad$ $\dfrac{j\omega L}{1 - \omega^2 LC}$

$\theta = \pm \pi/2$ $\qquad$ $\arctan[(\omega L - 1/\omega C)/R]$ $\qquad$ $\arctan(R/\omega L)$ $\qquad$ $-\arctan(\omega CR)$ $\qquad$ $\pm \pi/2$

RLC 4 $\qquad$ 5 $\qquad$ 6

$$Z = \frac{1/R - j(\omega C - 1/\omega L)}{(1/R)^2 + (\omega C - 1/\omega L)^2}$$

$$\frac{R + j\omega[L(1 - \omega^2 LC) - CR^2]}{(1 - \omega^2 LC)^2 + \omega^2 C^2 R^2}$$

$$\frac{A + jB}{(R + r)^2 + (\omega L - 1/\omega C)^2}$$

$A = Rr(R + r) + \omega^2 L^2 r + R/\omega^2 C^2$

$B = \omega L r^2 - R^2/\omega C - (L/C)(\omega L - 1/\omega C)$

$\theta = -\arctan[R(\omega C - 1/\omega L)]$ $\qquad$ $\arctan\{\omega[L(1 - \omega^2 LC) - CR^2]/R\}$ $\qquad$ $\arctan(B/A)$

Resonance conditions for LC networks numbered 1–6 above, for $\omega = \omega_0 = 1/\sqrt{(LC)}$:
(1) $|Z| = 0$, $\theta = 0$ (2) $|Z| = R$, $\theta = 0$ (3) $|Z| = \infty$, $\theta = 0$ (4) $|Z| = R$, $\theta = 0$
(5) $|Z| = L/CR$, $\theta = -\arctan(\omega CR)$ for $R \ll \omega L$ (6) $|Z| = R$ (const.) for $R = r = \sqrt{(L/C)}$

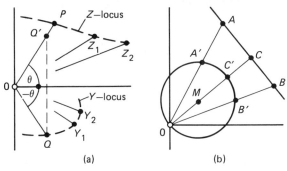

Figure 3.17 Inversion

Now resolve i into the active and reactive components

$i_p = (i_m \cos \phi) \sin \omega t$ $\qquad$ $i_q = (i_m \sin \phi) \sin(\omega t - \tfrac{1}{2}\pi)$

as in *Figure 3.18(b)*; then the instantaneous power can be written

$p = (v_m (i_m \cos \phi) \sin^2 \omega t - v_m (i_m \sin \phi) \sin \omega t \cos \omega t$

Over a whole number of periods the average of the first term is

$P = \tfrac{1}{2} v_m i_m \cos \phi = VI \cos \phi$

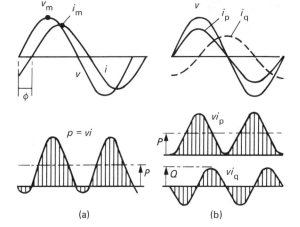

Figure 3.18 Active and reactive power

giving the *average* rate of energy transfer from source to load. The second term is a double-frequency sinusoid of average value zero, the energy flow changing direction rhythmically between source

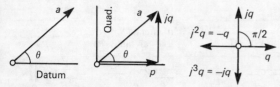

Figure 3.15 Complexors

$a = p + jq$, the *rectangular* form, in terms of its projection p on the datum and q on a quadrature axis at right angles thereto: q (as a scalar magnitude along the datum) is rotated counter-clockwise by angle $\tfrac{1}{2}\pi$ rad (90°) by the operator j. Two successive operations by j (written as j²) give a rotation of π rad (180°), making the original $+q$ into $-q$, in effect a multiplication by -1. Three operations (j³) give $-jq$ and four give $+q$. Thus any complexor can be located in the complex datum–quadrature plane. Further obvious forms are the *trigonometric*, $a = a(\cos\theta + j\sin\theta)$ and the *exponential*, $a = a\exp(j\theta)$. Summarising, the four descriptions are:

Polar $\qquad a = a\angle\theta$
Rectangular $\qquad a = p + jq$
Exponential $\qquad a = a\exp(j\theta)$
Trigonometric $\qquad a = a(\cos\theta + j\sin\theta)$

where $a = \sqrt{(p^2 + q^2)}$ and $\theta = \arctan(q/p)$.

Consider complexors $a = p + jq = a\angle\alpha$ and $b = r + js = b\angle\beta$. The basic manipulations are:

Addition $\qquad a + b = (p + r) + j(q + s)$
Subtraction $\qquad a - b = (p - r) + j(q - s)$
Multiplication   The exponential and polar forms are more direct than the rectangular or trigonometric:

$$ab = (pr - qs) + j(qr + ps)$$
$$= ab\exp[j(\alpha + \beta)] = ab\angle(\alpha + \beta)$$

Division   Here also the angular forms are preferred:

$$a/b = [(pr + qs) + j(qr - ps)]/(r^2 + s^2)$$
$$= (a/b)\exp[j(\alpha - \beta)] = (a/b)\angle(\alpha - \beta)$$

Conjugate   The conjugate of a complexor $a = p + jq = a\angle\alpha$ is $a^* = p - jq = a\angle(-\alpha)$, the quadrature component (and therefore the angle) being reversed. Then

$$ab^* = ab\angle(\alpha - \beta)$$
$$a^*b = ab\angle(\beta - \alpha)$$
$$a^*a = aa^* = a^2 = p^2 + q^2$$

The last expression is used to 'rationalise' the denominator in the complexor division process.

3.2.9.2 Impedance and admittance operators

Sinusoidal voltages and currents can be represented by phasors in the expressions $V = IZ = I/Y$ and $I = VY = V/Z$. Current and voltage phasors are related by multiplication or division with the complex operators Z and Y. Series resistance R and reactance jX can be arranged as a right-angled triangle of hypotenuse $Z = \sqrt{(R^2 + X^2)}$ and the angle between Z and R is $\theta = \arctan(X/R)$. The relation between Z and Y for the same series network elements with $Z = R + jX$ is

$$Y = \frac{1}{Z} = \frac{1}{R + jX} = \frac{R - jX}{(R + jX)(R - jX)} = \frac{R - jX}{R^2 + X^2}$$
$$= R/Z^2 - j(X/Z^2) = G - jB$$

where G and B are defined in terms of R, X and Z. The *series* components R and X become *parallel* branches in Y, one a pure conductance, the other a pure susceptance. Further, a positive-

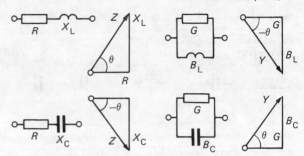

Figure 3.16 Impedance and admittance triangles

angled impedance has, as inverse equivalent, a negative-angled admittance (*Figure 3.16*).

The impedance and phase angle of a number of circuit combinations are given in *Table 3.3*.

Impedance and admittance loci   If the characteristics of a device or a circuit can be expressed in terms of an equivalent circuit in which the impedances and/or admittances vary according to some law, then the current taken for a given applied voltage (or the voltage for a given current) can be obtained graphically by use of an admittance or impedance locus diagram.

In *Figure 3.17(a)*, let OP represent an impedance $Z = R + jX$ and OQ the corresponding admittance $Y = G - jB$. Point Q is obtained from P by finding first the *geometric* inverse point Q' such that $OQ' = 1/OP$ to scale, and then reflecting OQ' across the datum line to give OQ and thus a reversed angle $-\theta$, a process termed *complexor* inversion. If Z has successive values $Z_1, Z_2, \ldots$ on the impedance locus, the corresponding admittances $Y_1, Y_2, \ldots$ lie on the admittance locus. The inversion process may be point-by-point, but in many cases certain propositions can reduce the labour:

Inverse of a straight line   The geometric inverse of a straight line AB about a point O not on the line is a circle passing through O with its centre M on the perpendicular OC from O to AB (*Figure 3.17b*). Then A' is the geometric inverse of A, B' of B, etc.; also, A is the inverse of A', B of B', etc.
Inverse of a circle   From the foregoing, the geometric inverse of a circle about a point O on its circumference is a straight line. If, however, O is not on the circumference, the inverse is a second circle between the same tangents; but the distances OM and OM' from the origin O to the centres M and M' of the circles are not inverses of each other.

The choice of scales arises in the inversion process: for example, the inverse of an impedance $Z = 50 \angle 70°\ \Omega$ is $Y = 0.02 \angle (-70°)$ S. It is usually possible to decide on a scale by taking a salient feature (such as a circle diameter) as a basis.

3.2.9.3 Power

The instantaneous power delivered to a load is the product of the instantaneous voltage v and current i. Let $v = v_m \sin\omega t$ and $i = i_m \sin(\omega t - \phi)$ as in *Figure 3.18(a)*; then the instantaneous power is

$$p = \tfrac{1}{2}v_m i_m[\cos\phi - \cos(2\omega t - \phi)]$$

This is a quantity fluctuating at angular frequency 2ω with, in general, excursions into negative power (i.e. that returned by the load to the source). Over an integral number of periods the mean power is

$$P = \tfrac{1}{2}v_m i_m \cos\phi = VI\cos\phi$$

where V and I are r.m.s. values.

for $R = 0$ and $V = 0$ when the source power is dissipated entirely in r. The maximum-power condition is utilised only with sources whose power capability is very small.

3.2.8 Steady-state a.c. networks

An alternating current flows alternately in the specified *positive* direction and then in the *negative* direction in a circuit, repeating this cycle continuously. A graph of current or voltage to a time base shows the *wave-form* as a succession of *instantaneous values*. In general there will be a maximum or *peak* value in both positive and negative half-periods where the current or voltage is greatest. The time for one complete *cycle* is the *period T*. The number of periods per second is the *frequency* $f = 1/T$.

An alternating current is produced by an alternating voltage. Two such quantities may have a difference of *phase*, to which a precise meaning can be given only when the quantities are both sinusoidal functions of time.

3.2.8.1 Root-mean-square (r.m.s.) value

The numerical value assigned to an alternating current or voltage is normally defined in terms of mean power in a pure resistor. An *alternating current* of 1 A is that which produces heat energy at the same mean rate as a direct current of 1 A in the same non-reactive resistor. If i is the instantaneous value of an a.c. in a pure resistance R, the heat developed in a time element dt is $dw = i^2 R \, dt$. The mean rate (i.e. the mean power) over a complete period T is

$$P = (1/T) \int_0^T dw = (1/T) \int_0^T i^2 R \, dt = I^2 R$$

and I is the r.m.s. value of the current. An *alternating voltage* is defined in a similar way; the instantaneous power is v^2/R, and the mean is V^2/R where V is the square root of the mean v^2.

In some cases the *peak* or the *mean* value of the current or voltage wave-form is more significant, particularly with asymmetric, pulse or rectified wave-forms. The value to be understood by the term 'mean' is then obvious. In the case of a symmetrical wave, the *half-period mean* value is intended, as the mean over a complete period is zero. *Table 3.2* gives the mean and r.m.s. values for a number of typical wave-forms, together with the values of

form factor, K_f = r.m.s./mean
peak factor, K_p = peak/r.m.s.

The techniques developed for the solution of steady-state a.c. networks depend on the wave-form. One technique applies to purely sinusoidal quantities, another to periodic but non-sinusoidal waveforms. In each case the network is assumed to be linear so that the principle of superposition is valid.

3.2.9 Sinusoidal alternating quantities

For pure sinusoidal wave-forms, a current can be expressed as a function of time, $i = i_m \sin(2\pi f t) = i_m \sin(\omega t)$, completing f cycles

Table 3.2 Values of alternating quantities (peak = a)

Wave-form	r.m.s.	mean	K_f	K_p
Sinusoidal	$a/\sqrt{2}$	$a(2/\pi)$	1.11	1.41
Half-wave rectified sine	$a/2$	a/π	1.57	2.0
Full-wave rectified sine	$a/\sqrt{2}$	$a(2/\pi)$	1.11	1.41
Rectangular	a	a	1.0	1.0
Triangular	$a/\sqrt{3}$	$a/2$	1.16	1.73

in 1 s with a period $T = 1/f$. The quantity $2\pi f$ is contracted to ω, the *angular frequency*. The sine-wave shape has the advantages that (i) it is mathematically simple and its integral and differential are both cosinusoidal, (ii) it is a wave-form desirable for power generation, transmission and utilisation, and (iii) it lends itself to phasor and complexor representation.

The graph of a sinusoidal current or voltage of frequency f can be plotted to a time-angle base ωt by use of trigonometric tables. Alternatively it can be represented by the projection of a line of length equal to the peak value and rotating counter-clockwise at angular speed ω about one end O. A *stationary* line can represent the sine wave, particularly in relation to other sine waves of the same frequency but 'out of step'. Two such waves, say v and i with peak values v_m and i_m respectively, can be written

$$v = v_m \sin \omega t \quad \text{and} \quad i = i_m \sin(\omega t - \phi)$$

and drawn as in *Figure 3.14*, the phase difference or phase angle between them being ϕ rad. Then two lines, OA and OB, having an angular displacement ϕ, can represent the two waves in peak magnitude and relative time phase.

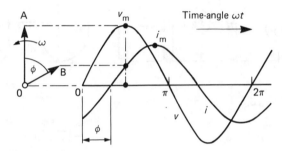

Figure 3.14 Phasors

Although developed from rotating lines of peak-value length, it is more convenient to change the scale and treat the lengths as r.m.s. values. The processes of addition and subtraction of r.m.s. values are performed as if the lines were coplanar vector forces in mechanics. Physically, however, the lines are not vectors: they substitute for scalar quantities, alternating sinusoidally with time. They are termed *phasors*. Certain associated quantities, such as impedance, admittance and apparent power, can also be represented by directed lines, but as they are not sinusoids they are termed *complexors* or *complex operators*. Both phasors and complexors can be dealt with by application of the theory of complex numbers. The definitions concerned are:

Complexor A generic term for a non-vector quantity expressed as a complex number.
Phasor A complexor (e.g. voltage or current) derived from a time-varying sinusoidal quantity and expressed as a complex number.
Complex operator A complexor derived for the ratio of two phasors (e.g. impedance and admittance); or a complexor which, operating on a phasor, gives another phasor (e.g. $V = IZ$, in which V and I are phasors but Z is a complex operator).

3.2.9.1 Complexor algebra

The four arithmetic processes for complexors are applications of the theory of complex numbers. Complexor a in *Figure 3.15* can be expressed by its magnitude a and its angle θ with respect to an arbitrary 'datum' direction (here taken as horizontal) as the simple *polar* form $a = a \angle \theta$. Alternatively it can be written as

Table 3.1 (cont.)

Network	Matrix
ABCD 1 / ABCD 2 (parallel)	Parallel networks $$\begin{pmatrix} (A_1B_2+A_2B_1)/(B_1+B_2) & B_1B_2/(B_1+B_2) \\ C_1+C_2+\dfrac{(A_2-A_1)(D_1-D_2)}{B_1+B_2} & \dfrac{B_1D_2+B_2D_1}{B_1+B_2} \end{pmatrix}$$
Symmetrical lattice (Z_1, Z_2)	Symmetrical lattice network $$\begin{pmatrix} (Z_1+Z_2)(Z_1-Z_2) & 2Z_1Z_2/(Z_1-Z_2) \\ 2/(Z_1-Z_2) & (Z_1+Z_2)/(Z_1-Z_2) \end{pmatrix}$$

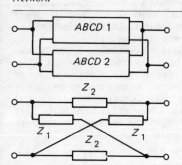
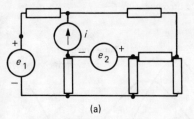
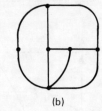

(a) (b) (c) (d)

Figure 3.10 Network topology

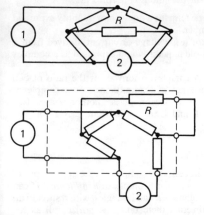

Figure 3.11 Conversion to a passive network

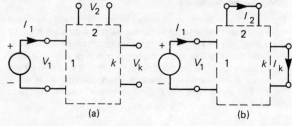

Figure 3.12 A multiport network

(b) All ports but one are short-circuited: a current I_1 (requiring a voltage V_1) is fed in at port 1. Then I_1/V_1 is the *short-circuit (s.c.) driving-point admittance* at port 1, I_k/V_1 is the s.c. *transfer admittance* from port 1 to port k, and I_k/I_1 is the s.c. *current ratio* of port k to port 1.

3.2.7 Steady-state d.c. networks

The steady state implies that energy storage in electric and magnetic fields does not change, and only the resistance is significant. In *Figure 3.13* a source of constant e.m.f. E and internal resistance r provides a current I at terminal voltage V to a network represented by an equivalent resistance R. On open circuit ($R=\infty$), $I=0$ and $V=E$. As R is reduced the source provides a current $I=E/(R+r)=(E-V)/r$. The greatest output power $P=VI$ occurs for the condition $R=r$; further reduction of R reduces the network power, down to a short-circuit condition

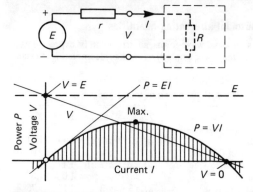

Figure 3.13 D.C. system

Table 3.1 General *ABCD* two-port parameters

Network	Matrix	Network	Matrix
Direct connection	$\begin{pmatrix} 1 & 0 \\ 0 & 1 \end{pmatrix}$	Loaded network	$\begin{pmatrix} A+BY_0 & B \\ C+DY_0 & D \end{pmatrix}$
Cross-connection	$\begin{pmatrix} -1 & 0 \\ 0 & -1 \end{pmatrix}$	Shunted network	$\begin{pmatrix} A & B \\ C+AY_1 & D+BY_1 \end{pmatrix}$
Series impedance	$\begin{pmatrix} 1 & Z \\ 0 & 1 \end{pmatrix}$	Mutual inductance	$\begin{pmatrix} 0 & -j\omega L_{12} \\ 1/j\omega L_{12} & 0 \end{pmatrix}$
Shunt admittance	$\begin{pmatrix} 1 & 0 \\ Y & 1 \end{pmatrix}$	Mutual inductance	$\begin{pmatrix} 0 & j\omega L_{12} \\ -1/j\omega L_{12} & 0 \end{pmatrix}$
L-network	$\begin{pmatrix} 1+YZ & Z \\ Y & 1 \end{pmatrix}$	Ideal transformer	$\begin{pmatrix} N_1/N_2 & 0 \\ 0 & N_2/N_1 \end{pmatrix}$
L-network	$\begin{pmatrix} 1 & Z \\ Y & 1+YZ \end{pmatrix}$		
T-network	$\begin{pmatrix} 1+YZ_1 & Z_1+Z_2+YZ_1Z_2 \\ Y & 1+YZ_2 \end{pmatrix}$		
Symmetrical T-network	$\begin{pmatrix} 1+YZ/2 & Z(1+YZ/4) \\ Y & 1+YZ/2 \end{pmatrix}$		
Π-network	$\begin{pmatrix} 1+Y_2Z & Z \\ Y_1+Y_2+Y_1Y_2Z & 1+Y_1Z \end{pmatrix}$		
Symmetrical Π-network	$\begin{pmatrix} 1+YZ/2 & Z \\ Y(1+YZ/4) & 1+YZ/2 \end{pmatrix}$		
Cascaded networks	$\begin{pmatrix} A_1A_2+B_1C_2 & A_1B_2+B_1D_2 \\ A_2C_1+C_2D_1 & B_2C_1+D_1D_2 \end{pmatrix}$		

Multiplication shows that $AD - BC = 1$.

For the symmetric T with $Z_1 = Z_2 = \frac{1}{2}Z$,

$A = 1 + \frac{1}{2}YZ = D \quad B = Z + \frac{1}{4}YZ^2 \quad C = Y$

3.2.5.3 Π-network

In a similar way, the general parameters for the asymmetric case are

$A = 1 + Y_2 Z \quad B = Z \quad C = Y_1 + Y_2 + Y_1 Y_2 Z \quad D = 1 + Y_1 Z$

which reduce with symmetry to

$A = 1 + \frac{1}{2}YZ = D \quad B = Z \quad C = Y + \frac{1}{4}Y^2 Z$

The values of the *ABCD* parameters, in matrix form,

$\begin{pmatrix} A & B \\ C & D \end{pmatrix}$

are set out in *Table 3.1* for a number of common cases.

3.2.5.4 Characteristic impedance

If the output terminals of a two-port are closed through an impedance $V_2/I_2 = Z_0$, and if the input impedance V_1/I_1 is then also Z_0, the quantity Z_0 is the *characteristic impedance*. Consider a symmetrical two-port ($A = D$) so terminated: if V_1/I_1 is to be Z_0 we have

$$\frac{V_1}{I_1} = \frac{V_2 A + I_2 B}{V_2 C + I_2 A} = \frac{V_2(A + B/Z_0)}{I_2(A + CZ_0)} = Z_0 \frac{A + B/Z_0}{A + CZ_0}$$

which is Z_0 for $B/Z_0 = CZ_0$. Thus the characteristic impedance is $Z_0 = \sqrt{(B/C)}$. The same result is obtainable from the input impedances with the output terminals first open-circuited ($I_2 = 0$) giving Z_{oc}, then short-circuited ($V_2 = 0$) giving Z_{sc}: thus

$Z_{oc} = A/C \quad Z_{sc} = B/A \quad Z_0 = \sqrt{(Z_{oc} Z_{sc})} = \sqrt{(B/C)}$

3.2.5.5 Propagation coefficient

The parameters *ABCD* are functions of frequency, and Z_0 is a complex operator. For the Z_0 termination of a symmetrical two-port (for which $A^2 - BC = 1$) the input/output voltage or current ratio is

$V_1/V_2 = I_1/I_2 = A + \sqrt{(BC)} = A + \sqrt{(A^2 - 1)}$
$= \exp(\gamma) = \exp(\alpha + j\beta)$

The magnitude of V_1 exceeds that of V_2 by the factor $\exp(\alpha)$ and leads it by the angle β, where α is the *attenuation* coefficient, β is the *phase* coefficient and the combination $\gamma = \alpha + j\beta$ is the *propagation* coefficient.

3.2.5.6 Alternative two-port parameters

There are other ways of expressing two-port relations. For generality, both terminal voltages are taken as *applied* and both currents are *input* currents. With this convention it is necessary to write $-I_2$ for I_2 in the general parameters so far discussed. The mesh-current and node-voltage methods (Section 3.2.4) give $V_1 = I_1 z_{11} + I_2 z_{12}$ etc. and $I_1 = V_1 y_{11} + V_2 y_{12}$ etc., respectively. A further method relates V_1 and I_2 to I_1 and V_2 by hybrid (impedance and admittance) parameters. The four relations then appear as follows:

General

$\begin{pmatrix} V_1 \\ I_1 \end{pmatrix} = \begin{pmatrix} A & B \\ C & D \end{pmatrix} \begin{pmatrix} V_2 \\ -I_2 \end{pmatrix}$

Impedance

$\begin{pmatrix} V_1 \\ V_2 \end{pmatrix} = \begin{pmatrix} z_{11} & z_{12} \\ z_{21} & z_{22} \end{pmatrix} \begin{pmatrix} I_1 \\ I_2 \end{pmatrix}$

Admittance

$\begin{pmatrix} I_1 \\ I_2 \end{pmatrix} = \begin{pmatrix} y_{11} & y_{12} \\ y_{21} & y_{22} \end{pmatrix} \begin{pmatrix} V_1 \\ V_2 \end{pmatrix}$

Hybrid

$\begin{pmatrix} V_1 \\ I_2 \end{pmatrix} = \begin{pmatrix} h_{11} & h_{12} \\ h_{21} & h_{22} \end{pmatrix} \begin{pmatrix} I_1 \\ V_2 \end{pmatrix}$

If the networks are passive, then $z_{12} = z_{21}$, $y_{12} = y_{21}$ and $h_{12} = -h_{21}$. If in addition the networks are symmetrical, then $A = D$, $z_{11} = z_{22}$ and $y_{11} = y_{22}$. If the networks are active (i.e. they contain sources), then reciprocity does not apply and there is no necessary relation between the terms of the 2×2 matrix.

3.2.6 Network topology

In multibranch networks the solution process is aided by representing the network as a graph of nodes and interconnections. The topology is the scheme of interconnections. A network is planar if it can be drawn on a closed spherical (or plane) surface without cross-overs. A non-planar network cannot be so drawn: a single cross-over can be eliminated if the network is drawn on a more complicated surface resembling a doughnut, and more cross-overs require closed surfaces with more holes.

The nomenclature employed in topology is as follows.

Graph A diagram of the network showing all the nodes, with each branch represented by a plain line.
Tree Any arrangement of branches that connects all nodes together without forming loops. A *tree branch* is one branch of such a tree.
Link A branch that, added to a tree, completes a closed loop.
Tie set A loop of branches with one a link and the others tree branches.
Cut set A set of branches comprising one tree branch, the other branches being tree links. A cut set dissociates two main portions of a network in such a way that replacing any one element destroys the dissociation.

Before setting up the equations for network solution, some guide is necessary in forming the proper number of independent equations. Given the network (*a*) in *Figure 3.10*, the first step is to draw the graph (*b*). Two of its possible trees are shown in (*c*). The trees are then used to set up the equations.

3.2.6.1 Network equations

Voltage The network diagram for the upper tree in *Figure 3.10(c)* is drawn in (*d*). Specifying the tree-branch voltages specifies also the voltages across the links. It is convenient to choose r as a reference node, leaving $n = 6$ independent nodes requiring $n = 6$ voltage equations.

Current A tree has no closed paths. As the links are added with specified currents, each creates one loop. Then the sum of the links m is the number of currents to be evaluated. For a network of b branches and n independent nodes, the number of independent meshes is $m = b - n$.

3.2.6.2 Ports

It is often helpful to place the sources outside the network and to regard their connections to the (now passive) remainder as ports. Again, branches of interest can be taken outside and used to terminate ports, as in *Figure 3.11*.

A multiport network (*Figure 3.12*) has the following characteristic definitions:

(a) All ports but one are open-circuited: a voltage V_1 is applied to port 1 and a current I_1 flows into it. Then V_1/I_1 is the *open-circuit* (*o.c.*) *driving-point impedance* at port 1, V_k/I_1 is the o.c. *transfer impedance* from port 1 to port k, and V_k/V_1 is the o.c. *voltage ratio* of port k to port 1.

connected in parallel to a common load of impedance Z is $V = I_{sc}Z_p$, where I_{sc} is the sum of the short-circuit currents of the individual source branches and Z_p is the effective impedance of all the branches in parallel, including the load Z. If E_1 and E_2 are the e.m.f.s of two sources with internal impedances Z_1 and Z_2 connected in parallel to supply a load Z, and if I_1 and I_2 are the currents contributed by these sources to the load Z, then their common terminal voltage V must be

$$V = (I_1 + I_2)Z = [(E_1 - V)/Z_1 + (E_2 - V)/Z_2]Z$$

whence

$$V(1/Z + 1/Z_1 + 1/Z_2) = E_1/Z_1 + E_2/Z_2$$

The term in parentheses on the left-hand side of the equation is the effective admittance of all the branches in parallel. The right-hand side of the equation is the sum of the individual source short-circuit currents, totalling I_{sc}. Thus $V = I_{sc}Z_p$. The theorem holds for any number of sources.

3.2.4.7 Helmholtz–Thevenin (*Figure 3.7*)

The current in any branch Z of a network is the same as if that branch were connected to a voltage source of e.m.f. E_0 and internal impedance Z_0, where E_0 is the p.d. appearing across the branch terminals when they are open-circuited and Z_0 is the impedance of the network looking into the branch terminals with all sources represented by their internal impedance.

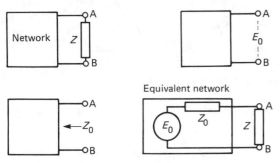

Figure 3.7 The Helmholtz–Thevenin theorem

In *Figure 3.7*, the network has a branch AB in which it is required to find the current. The branch impedance Z is removed, and a p.d. E_0 appears across AB. With all sources replaced by their internal impedance, the network presents the impedance Z_0 -to AB. The current in Z when it is replaced into the original network is

$$I = E_0/(Z_0 + Z)$$

The whole network apart from the branch AB has been replaced by an equivalent *voltage* source, resulting in the simplified condition of *Figure 3.1(a)*.

3.2.4.8 Helmholtz–Norton

The Helmholtz–Norton theorem is the dual of the Helmholtz–Thevenin theorem. The voltage across any branch Y of a network is the same as if that branch were connected to a current source I_0 with internal shunt admittance Y_0, where I_0 is the current between the branch terminals when short-circuited and Y_0 is the admittance of the network looking into the branch terminals with all sources represented by their internal admittance. Then across the terminals AB in *Figure 3.7* the voltage is

$$V = I_0/(Y_0 + Y)$$

Thus the whole network apart from the branch AB has been replaced by an equivalent *current* source, i.e. the system in *Figure 3.1(b)*.

3.2.5 Two-ports

In power and signal transmission, input voltage and current at one port (i.e. one terminal-pair) yield voltage and current at another port of the interconnecting network. Thus in *Figure 3.8* a voltage source at the input port 1 delivers to the passive network a voltage V_1 and a current I_1. The corresponding values at the output port 2 are V_2 and I_2.

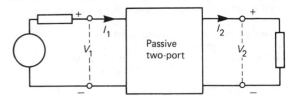

Figure 3.8 Two-port network

3.2.5.1 Lacour

According to the theorem originated by Lacour, any passive linear network between two ports can be replaced by a 2-mesh or T network, and in general no simpler form can be found. Such a result is obtained by iterative star–delta conversion to give the T-equivalent; by one more star–delta conversion the Π-equivalent is obtained (*Figure 3.9*). In general the equivalent networks are

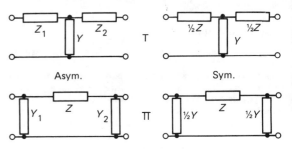

Figure 3.9 T and Π two-ports

asymmetric; in some cases, however, they are symmetric. It can be shown that a passive two-port has the input and output voltages and currents related by

$$V_1 = AV_2 + BI_2 \quad \text{and} \quad I_1 = CV_2 + DI_2$$

where $ABCD$ are the general two-port parameters, constants for a given frequency and with $AD - BC = 1$. The conventions for voltage polarity and current direction are those in *Figure 3.8*.

3.2.5.2 T-network

Consider the asymmetric T in *Figure 3.9*. Application of the Kirchhoff laws gives

$$I_1 = I_2 + (V_2 + I_2Z_2)Y = V_2Y + I_2(1 + YZ_2)$$
$$V_1 = V_2(1 + YZ_1) + I_2(Z_1 + Z_2 + Z_1Z_2Y)$$

Hence in terms of the series and parallel branch components

$$A = 1 + YZ_1 \quad B = Z_1 + Z_2 + Z_1Z_2Y \quad C = Y \quad D = 1 + YZ_2$$

3.2.3.2 Node-voltage equations

Of the network nodes, one is chosen as a reference node to which all other node voltages are related. The sources are represented by current generators feeding specified currents into their respective nodes and the branches are in terms of admittance Y. Then for the n independent nodes

$$I_a = V_a Y_{aa} + V_b Y_{ab} + \ldots + V_n Y_{an}$$
$$I_b = V_a Y_{ba} + V_b Y_{bb} + \ldots + V_n Y_{bn}$$
$$\vdots$$
$$I_n = V_a Y_{na} + V_b Y_{nb} + \ldots + V_n Y_{nn}$$

Here $Y_{aa}, Y_{bb}, \ldots, Y_{nn}$ are the *self-admittances* of nodes $a, b, \ldots, n$, i.e. the sum of the admittances terminating on nodes $a, b, \ldots, n$; and $Y_{ab}, Y_{pq}, \ldots$ are the *mutual admittances*, those that link nodes a and b, p and q, ... respectively and which are usually negative.

The mesh-current and node-voltage methods are general and basic; they are applicable to all network conditions. Simplified and auxiliary techniques are applied in special cases.

3.2.3.3 Techniques

Steady-state conditions Transient phenomena are absent. For d.c. networks the constant current implies absence of inductive effects, and capacitors (having a constant charge) are equivalent to an open circuit. Only resistance is taken into account, using the Ohm law.

For a.c. networks with sinusoidal current and voltage, complexor algebra, phasor diagrams, locus diagrams and symmetrical components are used, while for a.c. networks with periodic but non-sinusoidal waveforms harmonic analysis with superposition of harmonic components is employed.

Transient conditions Operational forms of stimuli and parameters are used and the solutions are found using Laplace transforms.

3.2.4 Network theorems

Network theorems can simplify complicated networks, facilitate the solution of specific network branches and deal with particular network configurations (such as two-ports). They are applicable to linear networks for which *superposition* is valid, and to any form (scalar, complexor or operational) of voltage, current, impedance and admittance. In the following, the Ohm and Kirchhoff laws, and the reciprocity and compensation theorems, are basic; star–delta transformation and the Millman theorem are applied to network simplification; and the Helmholtz–Thevenin and Helmholtz–Norton theorems deal with specified branches of a network. Two-ports are dealt with in Section 3.2.5.

3.2.4.1 Ohm (Figure 3.2(a))

For a branch of resistance R or conductance G,

$$I = V/R = VG \quad V = I/R = I/G \quad R = V/I = 1/G$$

Summation of resistances $R_1, R_2, \ldots$ in series or parallel gives

Series: $\quad R = R_1 + R_2 + \ldots \quad$ or $\quad G = 1/(1/G_1 + 1/G_2 + \ldots)$
Parallel: $\quad R = 1/(1/R_1 + 1/R_2 + \ldots) \quad$ or $\quad G = G_1 + G_2 + \ldots$

The Ohm law is generalised for a.c. and transient cases by $I = V/Z$ or $I(p) = V(p)/Z(p)$, where p is the operator d/dt.

3.2.4.2 Kirchhoff (Figure 3.4)

The node and mesh laws are

Node: $\quad i_1 + i_2 + \ldots = \sum i = 0$
Mesh: $\quad e_1 + e_2 + \ldots = i_1 Z_1 + i_2 Z_2 + \ldots \quad$ or $\quad \sum e = \sum iZ$

3.2.4.3 Reciprocity

If an e.m.f. in branch P of a network produces a current in branch Q, then the same e.m.f. in Q produces the same current in P. The ratio of the e.m.f. to the current is then the *transfer* impedance or admittance.

3.2.4.4 Compensation

For given circuit conditions, any impedance Z in a network that carries a current I can be replaced by a generator of zero internal impedance and of e.m.f. $E = -IZ$. Further, if Z is changed by ΔZ, then the effect on all other branches is that which would be produced by an e.m.f. $-I\Delta Z$ in series with the changed branch. By use of this theorem, if the network currents have been solved for given conditions, the effect of a changed branch impedance can be found without re-solving the entire network.

3.2.4.5 Star–delta (Figure 3.5)

At a given frequency (including zero) a 3-branch star impedance network can be replaced by a 3-branch delta network, and conversely. For a star Z_a, Z_b, Z_c to be equivalent between

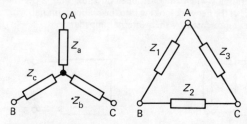

Figure 3.5 Star–delta conversion

terminals AB, BC, CA to a delta Z_1, Z_2, Z_3, it is necessary that

$Z_a = Z_3 Z_1/Z \qquad Z_1 = Z_a + Z_b + Z_a Z_b/Z_c$
$Z_b = Z_1 Z_2/Z \qquad Z_2 = Z_b + Z_c + Z_b Z_c/Z_a$
$Z_c = Z_2 Z_3/Z \qquad Z_3 = Z_c + Z_a + Z_c Z_a/Z_b$

where $Z = Z_1 + Z_2 + Z_3$. The *general* star–mesh conversion concerns the replacement of an n-branch star by a mesh of $\frac{1}{2}n(n-1)$ branches, but *not* conversely; and as the number of mesh branches is greater than the number of star branches when $n > 3$, the conversion is only rarely of use.

3.2.4.6 Millman (Figure 3.6)

The Millman theorem is also known as the *parallel-generator* theorem. The common terminal voltage of a number of sources

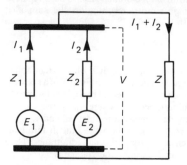

Figure 3.6 The Millman theorem

A more concise representation of the behaviour of pure parameters uses the differential operator p for d/dt and the inverse $1/p$ for the integral operator: then

(a) Resistance $v = Ri = v/G$ $i = Gv = (1/R)v$
(b) Self-inductance $v = Lpi$ $i = (1/Lp)v$
    Mutual inductance $e_1 = L_{12}pi_2$ $e_2 = L_{21}pi_1$
(c) Capacitance $v = (1/Cp)i$ $i = Cpv$

For the steady-state direct-current (d.c.) case, $p = 0$. For steady-state sinusoidal alternating current (a.c.), $p = j\omega$, giving for L and C the forms $j\omega L$ and $1/j\omega C$ where ω is the angular frequency. In general, Lp and $1/Cp$ are the *operational* impedance parameters.

3.2.1.3 Configuration

The assembly of sources and loads forms a network of branches that interconnect nodes (junctions) and form meshes. The 7-branch network in *Figure 3.3* has 5 nodes (a, b, c, d, e) and 4 meshes (1, 2, 3, 4). Branch ab contains a voltage source; the other

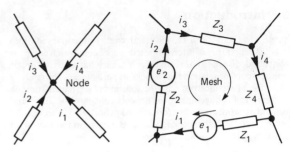

Figure 3.4 The Kirchhoff laws

polarity, is in fact the normal condition, but where possible the minor non-linearities are ignored in order to permit the use of greatly simplified analysis and the principle of superposition.

3.2.2.1 Superposition

In a strictly linear network, the current in any branch is the sum of the currents due to each source acting *separately*, all other sources being replaced meantime by their internal impedances. The principle applies to voltages and currents, but not to powers, which are current–voltage products.

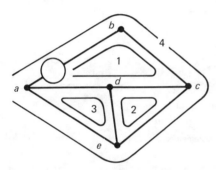

Figure 3.3 Network configuration

branches have (unspecified) impedance parameters. Inspection shows that not all the meshes are independent: mesh 4, for example, contains branches already accounted for by meshes 1, 2 and 3. Further, if one node (say, e) is taken as a *reference* node, the voltages of nodes a, b, c and d can be taken as their p.d.s with respect to node e. The network is then taken as having $b = 7$ branches, $m = 3$ independent meshes and $n = 4$ independent nodes. In general, $m = b - n$.

3.2.2 Network laws

The behaviour of networks (i.e. the branch currents and node voltages for given source conditions) is based on the two Kirchhoff laws (*Figure 3.4*).

(1) *Node law*   The total current flowing into a node is zero, $\sum i = 0$. The sum of the branch currents flowing into a node must equal the sum of the currents flowing from it; this is a result of the 'particle' nature of conduction current.
(2) *Mesh law*   The sum of the voltages around a closed mesh is zero, $\sum v = 0$. A rise of potential in sources is absorbed by a fall in potential in the successive branches forming the mesh. This is the result of the nature of a network as an energy system.

The Kirchhoff laws apply to all networks. Whether the evaluation of node voltages and mesh currents is tractable or not depends not only on the complexity of the network configuration but also on the branch parameters. These may be active or passive (i.e. containing or not containing sources), linear or non-linear. Non-linearity, in which the parameters are not constant but depend on the voltage and/or current magnitude and

3.2.3 Network solution

A general solution presents the voltages and currents everywhere in the network; it is initiated by the solution simultaneously of the network equations in terms of voltages, currents and parameters.

The Kirchhoff laws can be applied systematically by use of the *Maxwell circulating-current* process. To each mesh is assigned a circulating current, and the laws are applied with due regard to the fact that certain branches, being common to two adjacent meshes, have net currents given by the superposition of the individual mesh currents postulated. Generalising, the network can be considered as either (i) a set of independent nodes with appropriate node-voltage equations or (ii) a set of independent meshes with corresponding mesh-current equations.

3.2.3.1 Mesh-current equations

This is a formulation of the Maxwell circulating-current process. If source e.m.f.s are written as E, currents as I and impedances as Z, then for the m independent meshes

$$E_1 = I_1 Z_{11} + I_2 Z_{12} + \ldots + I_m Z_{1m}$$
$$E_2 = I_1 Z_{21} + I_2 Z_{22} + \ldots + I_m Z_{2m}$$
$$\vdots$$
$$E_m = I_1 Z_{m1} + I_2 Z_{m2} + \ldots + I_m Z_{mm}$$

Here $Z_{11}, Z_{22}, \ldots, Z_{mm}$ are the *self-impedances* of meshes $1, 2, \ldots, m$, i.e. the total series impedance around each of the chosen meshes; and $Z_{12}, Z_{pq}, \ldots$ are the *mutual impedances* of meshes 1 and 2, p and $q, \ldots$, i.e. the impedances common to the designated meshes.

The mutual impedance is defined as follows. Z_{pq} is the p.d. per ampere of I_q in the direction of I_p, and Z_{qp} is the p.d. per ampere of I_p in the direction of I_q. The sign of a mutual impedance depends on the current directions chosen for the meshes concerned. If the network is coplanar (i.e. it can be drawn on a diagram with no cross-over) it is usual to select a single consistent direction—say clockwise—for each mesh current. In such a case the mutual impedances are *negative* because the currents are oppositely directed in the common branches.

3.1 Introduction

In an electrical network, electrical energy is conveyed from *sources* to an array of interconnected *branches* in which energy is converted, dissipated or stored. Each branch has a characteristic voltage-current relation that defines its *parameters*. The analysis of networks is concerned with the solution of source and branch currents and voltages in a given *network configuration*. Basic and general network concepts are discussed in Section 3.2. Section 3.3 is concerned with the special techniques applied in the analysis of power-system networks.

3.2 Basic network analysis

3.2.1 Network elements

Given the sources (generators, batteries, thermocouples etc.), the network configuration and its branch parameters, then the network solution proceeds through network equations set up in accordance with the Kirchhoff laws.

3.2.1.1 Sources

In most cases a source can be represented as in *Figure 3.1(a)* by an electromotive force (e.m.f.) E_0 acting through an internal series impedance Z_0 and supplying an external 'load' Z with a current I at a terminal voltage V. This is the Helmholtz-Thevenin *equivalent voltage generator*. As regards the load voltage V and current I, the source could equally well be represented by the Helmholtz-Norton *equivalent current generator* in *Figure 3.1(b)*,

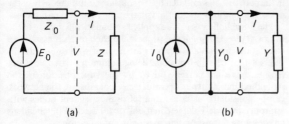

Figure 3.1 Voltage and current sources

comprising a source current I_0 shunted by an internal admittance Y_0 which is effectively in parallel with the load of admittance Y. Comparing the two forms for the same load current I and terminal voltage V in a load of impedance Z or admittance $Y = 1/Z$, we have:

Voltage generator
$V = E_0 - IZ_0$
$I = (E_0 - V)/Z_0$
$\quad = E_0/Z_0 - V/Z_0$
$\quad = I_0 - VY_0$

Current generator
$I = I_0 - VY_0$
$V = (I_0 - I)/Y_0$
$\quad = I_0/Y_0 - I/Y_0$
$\quad = E_0 - IZ_0$

These are identical provided that $I_0 = E_0/Z_0$ and $Y_0 = 1/Z_0$. The identity applies only to the *load* terminals, for internally the sources have quite different operating conditions. The two forms are *duals*. Sources with $Z_0 = 0$ and $Y_0 = 0$ (so that $V = E_0$ and $I = I_0$) are termed *ideal* generators.

3.2.1.2 Parameters

When a real *physical* network is set up by interconnecting sources and loads by conducting wires and cables, all parts (including the connections) have associated electric and magnetic fields. A resistor, for example, has resistance as the prime property, but the passage of a current implies a magnetic field, while the potential difference (p.d.) across the resistor implies an electric field, both fields being present in and around the resistor. In the *equivalent* circuit drawn to represent the physical one it is usual to lump together the significant resistances into a limited number of *lumped* resistances. Similarly, electric-field effects are represented by lumped capacitance and magnetic-field effects by lumped inductance. The equivalent circuit then behaves like the physical prototype if it is so constructed as to include all significant effects.

The lumped parameters can now be considered to be free from 'residuals' and *pure* in the sense that simple laws of behaviour apply. These are indicated in *Figure 3.2*.

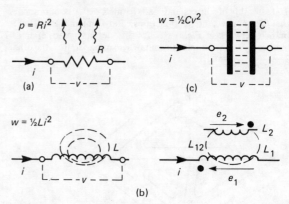

Figure 3.2 Pure parameters

(a) *Resistance* For a pure resistance R carrying an instantaneous current i, the p.d. is $v = Ri$ and the rate of heat production is $p = vi = Ri^2$. Alternatively, if the conductance $G = 1/R$ is used, then $i = Gv$ and $p = vi = Gv^2$. There is a constant relation

$$v = Ri = v/G \quad i = Gv = v/R \quad p = Ri^2 = Gv^2$$

(b) *Inductance* With a *self*-inductance L, the magnetic linkage is Li, and the source voltage is required only when the linkage changes, i.e. $v = d(Li)/dt = L(di/dt)$. An inductor stores in its magnetic field the energy $w = \frac{1}{2}Li^2$. The behaviour equations are

$$v = L(di/dt) \quad i = (1/L)\int v\, dt \quad w = \tfrac{1}{2}Li^2$$

Two inductances L_1 and L_2 with a common magnetic field have a *mutual* inductance $L_{12} = L_{21}$ such that an e.m.f. is induced in one when current changes in the other:

$$e_1 = L_{12}(di_2/dt) \quad e_2 = L_{21}(di_1/dt)$$

The direction of the e.m.f.s depends on the change (increase or decrease) of current and on the 'sense' in which the inductors are wound. The 'dot convention' for establishing the sense is to place a dot at one end of the symbol for L_1, and a dot at that end of L_2 which has the same polarity as the dotted end of L_1 for a given change in the common flux.

(c) *Capacitance* The stored charge q is proportional to the p.d. such that $q = Cv$. When v is changed, a charge must enter or leave at the rate $i = dq/dt = C(dv/dt)$. The electric-field energy in a charged capacitor is $w = \frac{1}{2}Cv^2$. Thus

$$i = C(dv/dt) \quad v = (1/C)\int i\, dt \quad w = \tfrac{1}{2}Cv^2$$

It can be seen that there is a duality between the inductor and the capacitor. Some typical cases of the behaviour of pure parameters are given in *Figures 2.3, 2.21* and *2.28*.

Network Analysis

M G Say PhD, MSc, CEng, FRSE, FIEE, FIERE, ACGI, DIC
Heriot-Watt University

M A Laughton BSc, PhD, DSc(Eng), FIEE
Queen Mary College, London

Contents

3.1 Introduction 3/3

3.2 Basic network analysis 3/3
    3.2.1 Network elements 3/3
    3.2.2 Network laws 3/4
    3.2.3 Network solution 3/4
    3.2.4 Network theorems 3/5
    3.2.5 Two-ports 3/6
    3.2.6 Network topology 3/7
    3.2.7 Steady-state d.c. networks 3/9
    3.2.8 Steady-state a.c. networks 3/10
    3.2.9 Sinusoidal alternating quantities 3/10
    3.2.10 Non-sinusoidal alternating quantities 3/13
    3.2.11 Three-phase systems 3/14
    3.2.12 Symmetrical components 3/16
    3.2.13 Line transmission 3/17
    3.2.14 Network transients 3/18
    3.2.15 System functions 3/21
    3.2.16 Non-linearity 3/25

3.3 Power-system network analysis 3/27
    3.3.1 Conventions 3/27
    3.3.2 Loadflow analysis 3/28
    3.3.3 Fault-level analysis 3/29
    3.3.4 System-fault analysis 3/30
    3.3.5 Network power limits and stability 3/31

facilitated by the use of Bewley's 'lattice diagram': a horizontal scale (of distance along a system in the direction of the initial surge) is combined with a downward vertical scale of time. A lattice of distance–time lines occupies the plane, the lines being sloped to correspond with the velocities of propagation in the system. The lines are marked with their surge voltage values in terms of v_1 and the various transmission and reflection coefficients.

The process is illustrated by the example in *Figure 2.33*. A 500 kV steep-fronted surge reaches the junction P of an overhead transmission line with a cable PQ, 1 km long. At Q the cable is connected to a second overhead line having a short circuit at S, distant 2 km from Q. The overhead lines have a propagation velocity of $u = 0.30$ km/μs and surge impedances z_0 respectively of 500 and 600 Ω; corresponding values for the cable are 0.20 km/μs and 70 Ω. Assuming the surge impedances to be purely resistive and neglecting attenuation, the Bewley lattice diagram is to be drawn for the system and the surge–voltage distribution found for an instant 19 μs after the surge wavefront reaches P.

Five sets of transmission and reflection factors are required, one at S and two each at P and Q for the two directions of propagation. The reflection factor is $\alpha = (z - z_0)/(z + z_0)$ and the transmission (or absorption) factor is $\beta = (1 + \alpha)$. Then

Direction left to right:
At P: $\alpha = (70 - 500)/(70 + 500) = -0.75$ $\beta = +0.25$
At Q: $\alpha = (600 - 70)/(600 + 70) = +0.79$ $\beta = +1.79$
At S: $\alpha = (0 - 600)/(0 + 600) = -1.0$ $\beta = 0$

Direction right to left:
At P: $\alpha = (500 - 70)/(500 + 70) = +0.75$ $\beta = +1.75$
At Q: $\alpha = (70 - 600)/(70 + 600) = -0.79$ $\beta = +0.21$

The lattice diagram has distance plotted horizontally. The distance scale for the cable section PQ is enlarged by the factor 3/2 to take account of its lower propagation velocity. Time is plotted vertically with time zero at the instant that the surge voltage reaches P. Starting from the upper left-hand side, a sloping straight line is drawn to show the position at any instant of the surge wavefront. Reflection occurs at each junction, so that corresponding lines of the same slope (but running downward to the left) are drawn. The junctions are now marked with the appropriate voltages, using the calculated reflection and transmission factors α and β. Thus the incident surge at P is reflected with $\alpha = -0.75$ to give -375 kV; the transmitted voltage, with $\beta = +0.25$, is 125 kV. The former is turned back at P and is superimposed on the incident surge, while the latter proceeds in the direction PQ, to be split at Q into reflected and transmitted components. At any junction the marked voltages must sum on each side to the same total. The surge–voltage distribution for 19 μs is found by summing the voltages marked on the sloping lines up to this instant.

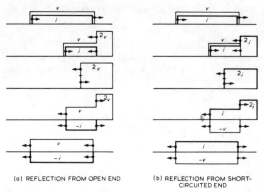

(a) REFLECTION FROM OPEN END (b) REFLECTION FROM SHORT-CIRCUITED END

Figure 2.32 Surge reflection

doubling of the current and reflection with reversed voltage (*Figure 2.32b*).

If the sending-end generator has zero impedance and remains closed on to the line, reflections take place repeatedly with an increase of i in the current each time, building up eventually to an infinite value—as would be expected in a lossless line with its end short-circuited. Under realisable conditions the current rises with steps of reducing size to a value limited by the series line impedance.

Terminated line If a line is terminated on an impedance z, only partial reflection will take place, some of the incident surge energy being dissipated as heat in the resistive part of z. Let the incident surge be $v_1 i_1$ such that $v_1/i_1 = z_0$; and let the reflected wave be $v_2 i_2$, with $v_2/i_2 = -z_0$. The negative sign is required algebraically to take account of the reversal of either current or voltage by reflection. The voltage v and current i at the termination z during the reflection are such that $v = v_1 + v_2$ and $i = i_1 + i_2$. But $v = iz$, so that, for the reflected current and voltage,

$$v_2 = v_1 \frac{z-z_0}{z+z_0} = \alpha v_1 \quad \text{and} \quad i_2 = -i_1 \frac{z-z_0}{z+z_0}$$

where α is the *reflection factor*. At the termination itself

$$v = v_1 \frac{2z}{z+z_0} = \beta v_1 \quad \text{and} \quad i = i_1 \frac{2z_0}{z+z_0}$$

where $\beta = 1 + \alpha$ is the *absorption factor*. The characteristic impedance is resistive, and reflection takes place unless $z = z_0$ and is also resistive. In the latter case v_2 and i_2 vanish, there is no reflection, and the energy in the incident surge $v_1 i_1$ is absorbed in z at the same rate as that at which it arrives.

Line junction or discontinuity A line having an abrupt change of surge impedance from z_0 to z_0' (as, for example, at the junction of a branch line, or at line-cable connection) can be treated as above, substituting z_0' for z. The voltage $v = \beta v_1$ is then characteristic of the *transmitted surge*, and β becomes the transmission factor.

Writing $v = 2v_1 \cdot z/(z+z_0)$, it is seen that v can be considered as the voltage across the load z of a voltage generator of e.m.f. $2v_1$ and internal impedance z_0. Any line termination of reasonable simplicity can now be dealt with. The discontinuity z may be a shunt load, or any combination of loads and extension lines reducible to an equivalent z. If z is resistive, the calculation of the terminal v and i is straightforward; but if z contains a time function (i.e. if it contains inductive and/or capacitive elements), the expressions $v = \beta v_1$ and $v_2 = \alpha v_1$ become integro-differential equations, to be solved by the methods given in Section 3. For z resistive, reflection and transmission (or absorption) of surges takes place without change of shape: for z containing terms in L and C, the shape of the incident surge is modified.

Single reflections Applying the equivalent circuit, a line on open-circuit has $z = \infty$, giving $v = 2v_1$ and $v_2 = v_1$. For a short-circuited line, $z = 0$, $v = 0$ and $v_2 = -v_1$. These cases are shown in *Figure 2.32*.

Shunt inductor termination Because L offers infinite opposition to infinite rate of current rise, it acts as an open circuit at the instant of surge arrival, degenerating with time constant L/z_0 to a condition of short-circuit. Thus the voltage at the termination is

$$v = 2v_1 \exp(-t \cdot z_0/L)$$

Shunt capacitor termination C is initially the equivalent of a short-circuit, but charges as the surge continues to arrive, and eventually approaches the charged state, for which it acts as an open circuit. The voltage across it is consequently

$$v = 2v_1[1 - \exp(-t/Cz_0)]$$

Series inductor termination If an inductor is inserted into a line of surge impedance z_0, the equivalent circuit is made up of the generator of voltage $2v_1$, loaded with z_0, L and z_0 in series. The voltage transmitted into the line extension is then

$$v' = v_1[1 - \exp(-t \cdot 2z_0/L)]$$

showing the reduction of wave front steepness produced. This roughly represents the action of a series surge modifier.

Multiple reflections Surges on power networks will be subject to repeated reflections at terminations, junctions, towers and similar discontinuities. Such cases are handled by developing the appropriate reflection and transmission (or absorption) factors α and β for each discontinuity and each direction. Procedure is

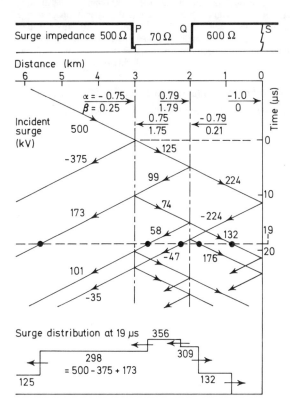

have $f = quB = ma = mu^2/R$, the particle being constrained by the force to move in a circular path of radius $R = (u/B)(m/q)$. For an electron $R = 5.7 \times 10^{-22}(u/B)$.

Combined electric and magnetic fields The two effects described above are superimposed. Thus, if the E and B fields are coaxial, the motion of the particle is helical.

The influence of static (or quasi-static) fields on charged particles is applied in cathode ray oscilloscopes and accelerator machines.

2.6.2 Free space propagation

In Section 1.5.3 the Maxwell equations are applied to propagation of a plane electromagnetic wave in free space. It is shown that basic relations hold between the velocity u of propagation, the electric and magnetic field components E and H, and the electric and magnetic space constants ε_0 and μ_0. The relations are:

$$u = 1/\sqrt{(\mu_0 \varepsilon_0)} \simeq 3 \times 10^8 \text{ m/s}$$

is the free space propagation velocity. The electric and magnetic properties of space impose a relation between E (in V/m) and H (in A/m) given by

$$E/H = \sqrt{(\mu_0/\varepsilon_0)} = 377 \, \Omega.$$

called the *intrinsic impedance* of space. Further, the energy densities of the electric and magnetic components are the same, i.e.

$$\tfrac{1}{2}\varepsilon_0 E^2 = \tfrac{1}{2}\mu_0 H^2$$

Propagation in power engineering is not (at present) by space waves but by guided waves, a conducting system being used to direct the electromagnetic energy more effectively in a specified path. The field pattern is modified (although it is still substantially transverse), but the essential physical propagation remains unchanged. Such a guide is called a transmission line, and the fields are normally specified in terms of the inductance and capacitance properties of the line configuration, with an effective impedance $z_0 = \sqrt{(L/C)}$ differing from $377\,\Omega$.

2.6.3 Transmission line propagation

If the two wires of a long transmission line, originally dead, are suddenly connected to a supply of p.d. v, an energy wave advances along the line towards the further end at velocity u (*Figure 2.31*). The wave is characterised by the fact that the advance of the voltage charges the line capacitance, for which an advancing current is needed: and the advance of the current establishes a magnetic field against a counter-e.m.f., requiring the voltage for maintaining the advance. Thus, current and voltage are propagated simultaneously. Let losses be neglected, and L and C be the inductance and capacitance per unit length of line. In a brief time interval dt, the waves advance by a distance $u \cdot dt$.

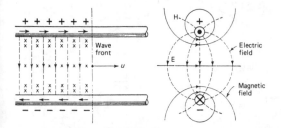

Figure 2.31 Transmission-line field

The voltage is established across a capacitance $Cu \cdot dt$ and the rate of charge, or current, is

$$i = vCu \cdot dt/dt = vCu$$

The current is established in an inductance $Lu \cdot dt$ producing the magnetic linkages $iLu \cdot dt$ in time dt, and a corresponding counter-e.m.f. overcome by

$$v = iLu \cdot dt/dt = iLu$$

These two expressions, by simple manipulation, yield

Propagation velocity $u = 1/\sqrt{(LC)}$
Surge impedance $z_0 = v/i = \sqrt{(L/C)}$
Energy components $\tfrac{1}{2}Li^2 = \tfrac{1}{2}Cv^2$

For a line in air consisting of parallel conductors of radius r spaced a between centres, the inductance (neglecting internal linkage) and capacitance per unit length are

$$L = (\mu_0/\pi)\ln(a/r); \quad C = \pi\varepsilon_0/\ln(a/r)$$

whence $u = 1/\sqrt{(\mu_0 \varepsilon_0)} \simeq 3 \times 10^8$ m/s, exactly as for free space propagation. For lines in which the relative permittivity (and/or, rarely, the relative permeability) of the medium conveying the electromagnetic wave is greater than unity, the speed is reduced by the factor $1/\sqrt{(\mu_r \varepsilon_r)}$. Line loss and internal linkage slightly reduce the speed of propagation. With the L and C values quoted, the surge impedance is

$$z_0 = 120 \ln(a/r)$$

which is usually in the range 300–600 Ω. For cables the different geometry and the relative permittivity give a much lower value.

2.6.3.1 Reflection of surges

The relation of p.d. and current direction in a pair of wires forming a long transmission line is determined by the direction in which the energy is travelling (*Figure 2.31*). The current flows in the direction of propagation in the positive conductor, and returns in the negative. Two waves travelling in opposite directions on a line must have *either* the currents *or* the p.d.s in opposite senses. If two such waves meet, either the currents are subtractive and the voltage additive, or the reverse. In each case the natural ratio $v/i = z_0$ for each wave is not apparent: in fact, if the resultant voltage/current ratio is not z_0, the actual distribution of current and voltage must be due to two component waves having opposite directions of propagation. Such conditions arise when a surge is *reflected* at the end of a line.

Consider a steep-fronted surge which reaches the open-circuited end of its guiding line. At the point the current must be zero, which requires an equal reflected surge current of opposite sense. The voltages have the same sense and combine to give a doubling effect (*Figure 2.32a*). Reversal of the energy flow imposed by the line discontinuity, i.e. reflection, is thus accompanied by voltage doubling and an elimination of current. In unit length of a surge, the total energy is $\tfrac{1}{2}Li^2 + \tfrac{1}{2}Cv^2$; when two surges (one incident, one reflected) are superimposed, the total energy is electrostatic and of value $\tfrac{1}{2}C(2v)^2 = 2Cv^2$, which, of course, is equal to the total energy per unit length of the two waves: $2(\tfrac{1}{2}Cv^2 + \tfrac{1}{2}Li^2) = Cv^2 + Li^2 = 2Cv^2$.

If the sending-end generator has zero impedance and remains closed on to the line, the returning surge is again reflected at voltage v and the to-and-fro 'oscillation' is maintained indefinitely. In practice, the presence of the losses reduces the reflections and the voltage settles down finally to the steady value v, with a current determined by line admittance.

An incident surge which reaches the *short-circuited* end of a line is presented with a condition of termination across which no p.d. can be maintained. The voltage collapses to zero, initiating a

2.5.3.4 Conduction and absorption

Solid dielectrics in particular, and to some degree liquids also, show conduction and absorption effects. Conduction appears to be mainly ionic in nature. Absorption is an apparent storing of charge *within* the dielectric. When a capacitor is charged, an initial quantity is displaced on its plates due to the *geometric* capacitance. If the p.d. is maintained, the charge gradually grows, owing to *absorptive* capacitance, probably a result of the slow orientation of permanent dipolar molecules. The current finally settles down to a small constant value, owing to conduction.

Absorptive charge leaks out gradually when a capacitor is discharged, a phenomenon observable particularly in cables after a d.c. charge followed by momentary discharge.

2.5.3.5 Grading

The electric fields set up when high voltages are applied to electrical insulators are accompanied by voltage gradients in various parts thereof. In many cases the gradients are anything but uniform: there is frequently some region where the field is intense, the voltage gradient severe and the dielectric stress high. Such regions may impose a controlling and limiting influence on the insulation design and on the working voltage. The process of securing improved dielectric operating conditions is called *grading*. The chief methods available are:

(1) The avoidance of sharp corners in conductors, near which the gradient is always high.

(2) The application of high-permittivity materials to those parts of the dielectric structure where the stress tends to be high, on the principle that the stress is inversely proportional to the permittivity: it is, of course, necessary to correlate the method with the dielectric strength of the material to be employed.

(3) The use of intersheath conductors maintained at a suitable intermediate potential so as to throw less stress on those parts which would otherwise be subjected to the more intense voltage gradients.

Examples of (1) are commonly observed in high-voltage apparatus working in air, where large rounded conductors are employed and all edges are given a large radius. The application of (2) is restricted by the fact that the choice of materials in any given case is closely circumscribed by the mechanical, chemical and thermal properties necessary. Method (3) is employed in capacitor bushings, in which the intersheaths have potentials adjusted by correlation of their dimensions.

2.5.4 Electromechanical effects

Figure 2.29 summarises the mechanical force effects observable in the electric field. In (*a*), (*b*) and (*c*) are sketched the field patterns for cases already mentioned in connection with the laws of electrostatics. The surface charges developed on high-ε materials are instrumental in producing the forces indicated in (*d*). Finally, (*e*) shows the forces on pieces of dielectric material immersed in a gaseous or liquid insulator and subjected to a non-uniform electric field. The force direction depends upon whether the piece has a higher or lower permittivity than the dielectric medium in which it lies. Thus, pieces of high permittivity are urged towards regions of higher electric field strength.

2.6 Electromagnetic field effects

Electromagnetic field effects occur when electric charges undergo acceleration. The effects may be negligible if the rate of change of velocity is small (e.g. if the operating frequency is low), but other

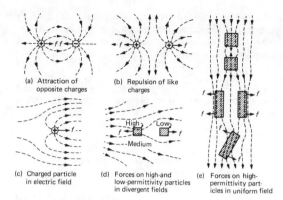

Figure 2.29 Electromechanical forces

conditions are also significant, and in certain cases effects can be significant even at power frequencies.

2.6.1 Movement of charged particles

Particles of small mass, such as electrons and protons, can be accelerated in vacuum to very high speeds.

Static electric field The force developed on a particle of mass m carrying a positive charge q and lying in an electric field of intensity (or gradient) E is $f = qE$ in the direction of E, i.e. from a high-potential to a low-potential region (*Figure 2.30a*). (If the charge is negative, the direction of the force is reversed.) The

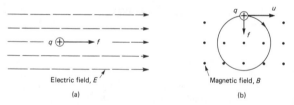

Figure 2.30 Motion of charged particles

acceleration of the particle is $a = f/m = E(q/m)$; and if it starts from rest its velocity after time t, is $u = at = E(q/m)t$. The kinetic energy $\frac{1}{2}mu^2$ imparted is equal to the change of potential energy Vq, where V is the p.d. between the starting and finishing points in the electric field. Hence, the velocity attained from rest is

$$u = \sqrt{[2V(q/m)]}$$

For an electron ($q = -1.6 \times 10^{-19}$ C, $m_0 = 0.91 \times 10^{-30}$ kg) falling through a p.d. of 1 V the velocity is 600 km/s and the kinetic energy is $w = Vq = 1.60 \times 10^{-19}$ J, often called an electron-volt, 1 eV.

If $V = 2.5$ kV, then $u = 30\,000$ km/s; but the speed cannot be indefinitely raised by increasing V, for as u approaches $c = 300\,000$ km/s, the free-space electromagnetic wave velocity, the effective mass of the particle begins to acquire a rapid relativistic increase to

$$m = m_0 / [1 - (u/c)^2]$$

compared with its 'rest mass' m_0.

Static magnetic field A charge q moving at velocity u is a current $i = qu$, and is therefore subject to a force if it moves across a magnetic field. The force is at right angles to u and to B, the magnetic flux density, and in the simple case of *Figure 2.30(b)* we

The electric field strength (voltage gradient) E is inversely proportional to the radius, over which it is distributed hyperbolically. The maximum gradient occurs at the surface of the inner conductor and amounts to

$E_m = V/r \ln(R/r)$

At any other radius x, $E_x = E_m(r/x)$. For a given p.d. V and gradient E_m there is one value of r to give minimum overall radius R: this is

$r = V/E_m$ and $R = 2.72r$

For the cylindrical capacitor (d) with two dielectrics, of permittivity ε_1 between radii r and ρ and ε_2 between ρ and R, the maximum gradients are related by $E_{m1}\varepsilon_1 r = E_{m2}\varepsilon_2 \rho$.

Parallel cylinders (Figure 2.27e) The calculation leads to the value

$C = \pi\varepsilon/\ln(a/r)$ farad/metre

for the capacitance between the conductors, provided that $a \gg r$. It can be considered as composed of two series-connected capacitors each of

$C_0 = 2\pi\varepsilon/\ln(a/r)$ farad/metre

C_0 being the *line-to-neutral* capacitance. A 3-phase line has a line-to-neutral capacitance identical with C_0, the interpretation of the spacing a for transposed asymmetrical lines being the same as for their inductance.

The voltage gradient of a 2-wire line is shown in *Figure 2.27(e)*. If $a \gg r$, the gradient in the immediate vicinity of a wire may be taken as due to the charge thereon, the further wire having little effect: consequently,

$E_m = V/r \ln(a/r)$

is the voltage gradient at a conductor surface.

2.5.2.3 Connection of capacitors

If a bank of capacitors of capacitance $C_1, C_2, C_3, \ldots$ be connected in *parallel* and raised in combination each to the p.d. V, the total charge is the sum of the individual charges $VC_1, VC_2, VC_3 \ldots$, whence the total combined capacitance is

$C = C_1 + C_2 + C_3 + \ldots$

With a *series* connection, the same displacement current occurs in each capacitor and the p.d. V across the series assembly is the sum of the individual p.d.s:

$V = V_1 + V_2 + V_3 + \ldots = q[(1/C_1) + (1/C_2) + (1/C_3) + \ldots] = q/C$

so that the combined capacitance is obtained from

$C = 1/[(1/C_1) + (1/C_2) + (1/C_3) + \ldots]$

2.5.2.4 Voltage applied to a capacitor

The basis for determining the conditions in a circuit containing a capacitor to which a voltage is applied is that the p.d. v across the capacitor is related definitely by its capacitance C to the charge q displaced on its plates: $q = Cv$.

Let a direct voltage V be suddenly applied to a circuit devoid of all characteristic parameters except that of capacitance C. At the instant of its application, the capacitor must accept a charge $q = CV$, resulting in an infinitely large current flowing for a vanishingly short time. The energy stored is $W = \frac{1}{2}Vq = \frac{1}{2}CV^2$ joules. If the voltage is raised or lowered uniformly, the charge must correspondingly change, by a constant charging or discharging current flowing during the change (*Figure 2.28a*).

If the applied voltage is sinusoidal, as in (b), such that

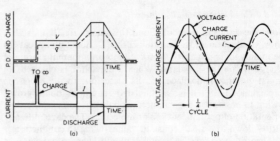

Figure 2.28 Voltage applied to a pure capacitor

$v = v_m \cos 2\pi ft = v_m \cos \omega t$, the same argument leads to the requirement that the charge is $q = q_m \cos \omega t$, where $q_m = Cv_m$. Then the current is $i = dq/dt$, i.e.

$i = -\omega C v_m \sin \omega t$

with a peak $i_m = \omega C v_m$ and an r.m.s. value $I = \omega C V = V/X_c$, where $X_c = 1/\omega C$ is the *capacitive reactance*.

2.5.3 Dielectric breakdown

A dielectric material must possess (a) a high insulation resistivity to avoid leakage conduction, which dissipates the capacitor energy in heat; (b) a permittivity suitable for the purpose—high for capacitors and low for insulation generally; (c) a high electric strength to withstand large voltage gradients, so that only thin material is required. It is rarely possible to secure optimum properties in one and the same material.

A practical dielectric will break down (i.e. fail to insulate) when the voltage gradient exceeds the value that the material can withstand. The breakdown mechanism is complex.

2.5.3.1 Gases

With gaseous dielectrics (e.g. air, hydrogen), ions are always present, on account of light, heat, sparking, etc. These are set in motion, making additional ionisation, which may be cumulative, causing glow discharge, sparking or arcing unless the field strength is below a critical value. A field strength of the order of 3 MV/m is a limiting value for gases at normal temperature and pressure. The dielectric strength increases with the gas pressure.

The polarisation in gases is small, on account of the comparatively large distances between molecules. Consequently, the relative permittivity is not very different from unity.

2.5.3.2 Liquids

When very pure, liquids may behave like gases. Usually, however, impurities are present. A small proportion of the molecules forms positive or negative ions, and foreign particles in suspension (fibres, dust, water droplets) are prone to align themselves into semiconducting filaments: heating produces vapour, and gaseous breakdown may be initiated. Water, because of its exceptionally high permittivity, is especially deleterious in liquids such as oil.

2.5.3.3 Solids

Solid dielectrics are rarely homogeneous, and are often hygroscopic. Local space charges may appear, producing absorption effects; filament conducting paths may be present; and local heating (with consequent deterioration) may occur. Breakdown depends on many factors, especially thermal ones, and is a function of the time of application of the p.d.

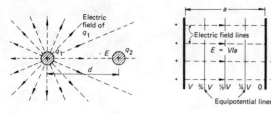

Figure 2.25 Electric fields

2.5.2 Capacitance

On the configuration and geometry of the conductor system in *Figure 2.24* depends the relation between the charge q on each conductor and the p.d., V, between them. Then $q = CV$, where C is the capacitance of the *capacitor* formed by the system. A capacitor has unit capacitance (1 F) if (a) the energy stored in the associated electric field is $\tfrac{1}{2}$ J for a p.d. of 1 V, or (b) the p.d. is 1 V for a charge of 1 C. Definition (a) follows from the energy storage property: if the p.d. across a capacitor is raised uniformly from zero to V_1, a charge $q_1 = CV_1$ is established at an average p.d. $\tfrac{1}{2}V_1$, so that the energy input is

$$W = \tfrac{1}{2}V_1 \cdot q_1 = \tfrac{1}{2}V_1 \cdot CV_1 = \tfrac{1}{2}CV_1^2$$

2.5.2.1 Dielectrics

Up to this point it has been assumed that the electric field between the plates of a capacitor has been established through vacuous space. If a material insulator be used—gas, liquid or solid—the electric field will exist therein. It will act on the molecules of the *dielectric* material in accordance with electrostatic principles to 'stretch' or 'rotate' them, and so to orientate the positive and negative molecular charges in opposite directions. This *polarisation* of the dielectric may be imagined to take place as in *Figure 2.26*. Before the p.d. is applied, the

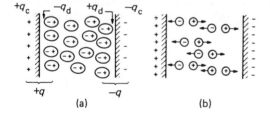

Figure 2.26 Dielectric polarisation and breakdown

molecules of the dielectric material are neutral and unstrained. As the p.d. is raised from zero as in (a), the electric field acts to separate the positive and negative elements, the small charge displacement forming a *polarisation* current.

The effect of the application of a p.d. to a capacitor with a material dielectric, then, is to displace a surface charge q_d of polarisation, having a polarity opposite to that of the adjacent capacitor plate q_c. The electric field in the dielectric is due to the resultant or net charge $q = q_c - q_d$. The field strength (and therefore the p.d.) is less than would be expected for the charge q_c on the plates: the relative reduction is found to be approximately constant for a given dielectric. It is called the *relative permittivity*, symbol ε_r. The same field strength as for a vacuum will exist in the dielectric for ε_r times as much charge on the capacitor plates, so that the capacitor has ε_r times the capacitance of a similar capacitor having free space between its plates.

Permittivity effects can thus be taken into account either by assigning to a dielectric a relative permittivity or by considering its polarisation. The latter is of use where internal dielectric forces are concerned, and in dielectric breakdown (*Figure 2.26b*).

2.5.2.2 Calculation of capacitance

The capacitance of capacitors of simple geometry can be found by assigning, respectively, charges of $+1$ C and -1 C to the plates or other electrodes, between which the total electric flux is 1 C. From the field pattern the electric flux density D at any point is found. Then the electric field strength at the point is $E = D/\varepsilon$, where $\varepsilon = \varepsilon_r \varepsilon_0$ is the absolute permittivity of the insulating medium in which the electric flux is established. Integration of E over any path from one electrode to the other gives the p.d. V, whence the capacitance is $C = 1/V$.

Parallel plates (*Figure 2.27a*) The electric flux density is uniform except near the edges. By use of a *guard ring* maintained at

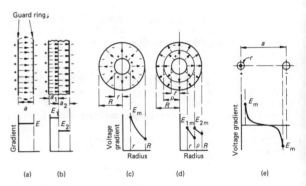

Figure 2.27 Capacitance and voltage gradient

the potential of the plate that it surrounds, the capacitance of the inner part is calculable on the reasonable assumption of uniform field conditions. With a charge of 1 C on each plate, and plates of area S spaced a apart, the electric flux density is $D = 1/S$, the electric field intensity is $E = D/\varepsilon = 1/S\varepsilon$, the potential difference is $V = Ea = a/S\varepsilon$, and the capacitance is therefore

$$C = q/V = \varepsilon(S/a)$$

A case of interest is that of a parallel plate arrangement (*Figure 2.27b*), with two dielectric materials, of thickness a_1 and a_2 and absolute permittivity ε_1 and ε_2, respectively. The voltage gradient is inversely proportional to the permittivity, so that $E_1\varepsilon_1 = E_2\varepsilon_2$. The field pattern makes it evident that the difference in polarisation produces an interface charge, but in terms of the charge q_c on the plates themselves the electric flux density is constant throughout. The total voltage between the plates is $V = V_1 + V_2 = E_1a_1 + E_2a_2$, from which the total capacitance can be obtained.

Concentric cylinders (*Figure 2.27c*) With a charge of 1 C per metre length, the electric flux density at radius x is $1/2\pi x$, whence $E_x = 1/2\pi x\varepsilon$. Integrating for the p.d. gives

$$V = (1/2\pi\varepsilon)\ln(R/r)$$

The capacitance is consequently

$$C = 2\pi\varepsilon/\ln(R/r) \text{ farad/metre}$$

(a) the energy stored in the common magnetic field is 1 J when the current in each circuit is 1 A, or (b) the e.m.f. induced in one is 1 V when the current in the other changes at the rate 1 A/s, or (c) the secondary linkage is 1 Wb-turn when the primary current is 1 A.

Figure 2.23(a) shows two coils on a common magnetic circuit: it is assumed that *all* the flux due to a primary current also links the secondary (a condition approached in a transformer). Let 1 A

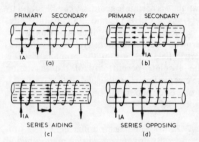

Figure 2.23 Mutual inductance

in the two-turn primary produce 2 Wb. The primary self-inductance is consequently $L_1 = 2 \times 2 = 4$ Wb-t/A $= 4$ H. The secondary linkage is $4 \times 2 = 8$ Wb-t, so that the mutual inductance is $L_{21} = 8$ H. Let now a current of 1 A circulate instead in the secondary (*b*): it develops double the m.m.f. of that developed by the primary in (*a*) and twice the flux, i.e. 4 Wb. The self-inductance is $L_2 = 4 \times 4 = 16$ H. The primary linkage is $2 \times 4 = 8$, i.e. $L_{12} = 8$ H. Thus, $L_{12} = L_{21}$.

In *Figure 2.23(c)* the coils are connected in *series aiding*. With a current of 1 A, the total m.m.f. is $2+4 = 6$, the common flux is 6 Wb and the total inductance is $6 \times 6 = 36$ H, which can be shown to be

$$L = L_1 + L_2 + L_{12} + L_{21} = 4 + 16 + 2(8) = 36 \text{ H}$$

For the *series opposing* connection (*d*) the m.m.f.s oppose with a net value $4 - 2 = 2$, and the resulting flux is 2 Wb. The linkages oppose, amounting to $(4 \times 2) - (2 \times 2) = 4$. The total inductance is 4 H, obtained by

$$L = L_1 + L_2 - L_{12} - L_{21} = 4 + 16 - 2(8) = 4 \text{ H}$$

The example shows that $L_{12} = L_{21} = \sqrt{(L_1 L_2)}$. Normally the linkages are less complete, and the ratio $L_{12}/\sqrt{(L_1 L_2)} = k$, the *coefficient of coupling*.

2.4.4.4 Connection of inductors

In the absence of mutual inductance, the total inductance of a circuit consisting of inductors $L_1, L_2 \ldots$ is $L = L_1 + L_2 + \ldots$ if they are in series, and $L = 1/[(1/L_1) + (1/L_2) + \ldots]$ if they are in parallel. When there is mutual inductance, it is necessary to set up a circuit equation including the mutual inductance coefficients L_{12}, L_{13}, $L_{23} \ldots$. For two inductors the inductance, as already discussed, is

$$L = L_1 + L_2 \pm 2L_{12}$$

With coils associated on a common ferromagnetic circuit, L_{12} may differ from L_{21} because of saturation effects.

2.5 Electric field effects

When two conductors are separated by a dielectric medium and are maintained at a potential difference, an electric field exists between them. Consider two such conductors A and B (*Figure 2.24*): the application of a p.d. causes a transfer of conduction

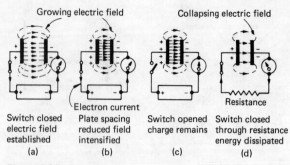

Figure 2.24 Capacitor charge and discharge

electrons from A to B, leaving A positive and making B negative, and setting up the electric field (*Figure 2.24a*). The positive charge on A prevents more than a given number of electrons leaving this conductor, depending on the p.d., the size and configuration, and the spacing. Similarly, the surplus of electrons on B repels others arriving, so that here, too, an equilibrium is established. If, as in (*b*), the spacing is reduced, a further electron transfer takes place until equilibrium is again reached; the charge and the electric field have been intensified by the increase in *capacitance*. If now the switch is opened, the charge, p.d. and field remain as a store of electric field energy (*c*). Let the supply be removed (*d*) and the switch closed through a resistor; the charges on the conductors are dissipated by an electron current from B to A, and the field energy is converted into heat in the resistor.

2.5.1 Electrostatics

The lines used to depict the pattern of an electric field begin on a positive charge and terminate on an equal negative one. Two similar point charges of q_1 and q_2 coulombs, spaced d metres apart in free space (or air) develop a force

$$f = q_1 q_2 / 4\pi\varepsilon_0 d^2 \text{ newton}$$

of repulsion if the charges have the same polarity, of attraction if they have opposite polarity, ε_0 being the electric space constant. The force on q_2 can be considered as due to its immersion in the electric field E_1 of q_1, i.e. $f = E_1 q_2$; whence

$$E_1 = q_2 / 4\pi\varepsilon_0 d^2 \text{ volt/metre}$$

defining the electric field strength at distance d from a concentrated charge q_1. Thus, a unit charge (1 C) is that which repels a similar charge at unit distance (1 m) with a force of $1/4\pi\varepsilon_0$ newton. Similarly, a field of unit strength (1 V/m) produces a mechanical force of 1 N on a unit charge placed in it.

To charge a system like that in *Figure 2.24*, a quantity of electricity has been moved under an applied electric force—i.e. work has been done being measured by the charge transferred and the p.d. Unit p.d. (1 V) exists between two points in an electric field when unit work (1 J) is done in moving unit charge (1 C) between them. The two conductors are equipotential surfaces, and potential levels or equipotential lines can be drawn at right angles to the field lines (*Figure 2.25*). Equipotential lines resemble contour lines on the map of a hill: the closer they are, the greater is the voltage gradient. The change of potential in a given direction is

$$V = -\int E \cdot dx$$

where E is the electric field strength, or potential gradient, in the direction x; whence $E = -dV/dx$.

which remains constant. The current therefore continues to circulate indefinitely at value I_1.

Suppose that V is applied for a time t_1, then reversed for an equal time interval, and so on, repeatedly. The resulting current is shown in *Figure 2.21(b)*. During the first period t_1 the current rises uniformly to $I_1 = (V/L)t_1$, and the stored energy is then $\frac{1}{2}LI_1^2$. On reversing the applied voltage the current performs the same rate of change, but negatively so as to reduce the current magnitude. After t_1 it is zero and so is the stored energy, which has all been returned to the supply from which it came.

If the applied voltage is sinusoidal and alternates at frequency f, such that $v = v_m \cos 2\pi f t = v_m \cos \omega t$, and is switched on at instant $t=0$ when $v = v_m$, the current begins to rise at rate v_m/L (*Figure 2.21c*); but the immediate reduction and subsequent reversal of the applied voltage require corresponding changes in the rate of rise or fall of the current. As $v = L(di/dt)$ at every instant, the current is therefore

$$i = \int \frac{v}{L} dt = \frac{v_m}{\omega L} \sin \omega t$$

The peak current reached is $i_m = v_m/\omega L$ and the r.m.s. current is $I = V/\omega L = V/X_L$, where $X_L = \omega L = 2\pi f L$ is the *inductive reactance*.

Should the applied voltage be switched on at a voltage zero (*Figure 2.21d*), the application of the same argument results in a sine-shaped current, unidirectional but pulsating, reaching the peak value $2i_m = 2v_m/\omega L$, or twice that in the symmetrical case above. This is termed the *doubling effect*. Compare *Figure 2.21(b)*.

2.4.4.2 Calculation of inductance

To calculate inductance in a given case (a problem capable of reasonably exact solution only in cases of considerable geometrical simplicity), the approach is from the standpoint of definition (c). The calculation involves estimating the magnetic field produced by a current of 1 A, summing the linkage ΦN produced by this field with the circuit, and writing the inductance as $L = \Sigma \Phi N$. The cases illustrated in *Figures 2.10* and *2.22* give the following results.

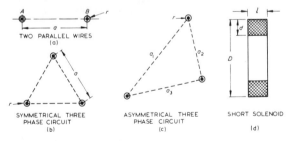

Figure 2.22 Parallel conductors

Long straight isolated conductor (*Figure 2.10a*) The magnetising force in a circular path concentric with the conductor and of radius x is $H = F/2\pi x = 1/2\pi x$; this gives rise to a circuital flux density $B_0 = \mu_0 H = \mu_0/2\pi x$. Summing the linkage from the radius r of the conductor to a distance s gives

$$\psi = (\mu_0/2\pi)\ln(s/r)$$

weber-turn per metre of conductor length. If s is infinite, so is the linkage and therefore the inductance: but in practice it is not possible so to isolate the conductor.

There is a magnetic flux following closed circular paths within the conductor, the density being $B_i = \mu x/2\pi r^2$ at radius x. The effective linkage is the product of the flux by that proportion of the conductor actually enclosed, giving $\mu/8\pi$ per metre length. It follows that the internal linkage produces a contribution $L_i = \mu/8\pi$ henry/metre, regardless of the conductor diameter on the assumption that the current is uniformly distributed. The absolute permeability μ of the conductor material has a considerable effect on the internal inductance.

Concentric cylindrical conductors (*Figure 2.10b*) The inductance of a metre length of concentric cable carrying equal currents oppositely directed in the two parts is due to the flux in the space between the central and the tubular conductor set up by the inner current alone, since the current in the outer conductor cannot set up internal flux. Summing the linkages and adding the internal linkage of the inner conductor:

$$L = (\mu/8\pi) + (\mu_0/2\pi)\ln(R/r) \text{ henry/metre}$$

Parallel conductors (*Figure 2.22*) Between two conductors (a) carrying the same current in opposite directions, the linkage is found by summing the flux produced by conductor A in the space a assuming conductor B to be absent, and doubling the result. Provided that $a \gg r$, this gives for the loop

$$L = (\mu/4\pi) + (\mu_0/\pi)\ln(a/r) \text{ henry/metre}$$

It is permissible to regard one-half of the linkage as associated with each conductor, to give for each the *line-to-neutral* inductance

$$L_0 = (\mu/8\pi) + (\mu_0/2\pi)\ln(a/r)$$

A 3-phase line (b) has the same line-to-neutral inductance if the conductors are symmetrically spaced. If, however, the spacing is asymmetric but the conductors are cyclically transposed (c), the expression applies with $a = \sqrt[3]{(a_1 a_2 a_3)}$, the geometric mean spacing.

Toroid (*Figure 2.10c*) For a core of permeability μ the inductance is

$$L = \mu N^2 A/2\pi R = \mu N^2 A/l \text{ henry}$$

where $l = 2\pi R$ is the mean circumference and A is the effective cross-sectional area of the core.

Solenoid A solenoid having a ratio length/diameter of at least 20 has an inductance approximating to that of the toroid above. A short solenoidal coil of overall diameter D, length l, radial thickness d and N turns has an inductance given approximately by

$$L = \frac{6.4\mu_0 N^2 D^2}{3.5D + 8l} \cdot \frac{D - 2.25d}{D}$$

For the best ratio of inductance to resistance, $l = d$, giving a square winding cross-section, and $D = 4.7d$. Then $L = (0.8 \times 10^{-6})N^2 D$.

Inductor with ferromagnetic circuit (*Figure 2.10d*) Saturation makes it necessary to obtain the m.m.f. F for a series of magnetic circuit fluxes Φ. Then with an N-turn exciting coil the inductance is $L = N\Phi/I$, where $I = F/N$. Thus, L is a function of I, decreasing with increase of current. The variation can be mitigated by the inclusion in the magnetic circuit of an air-gap to 'stiffen' the flux.

2.4.4.3 Mutual inductance

If two coils (primary and secondary) are so oriented that the flux developed by a current in one links the other, the two have mutual inductance. The pair have unit mutual inductance (1 H) if

Types of machine For unidirectional torque, the axes of the pole centres and armature m.m.f. must remain fixed relative to one another. Maximum torque is obtained if these axes are at right angles. The machine is technically better if the field flux and armature m.m.f. do not fluctuate with time (i.e. are d.c. values): if they do alternate, it is preferable that they be co-phasal.

Workable machines can be built with (1) concentrated ('field') or (2) phase windings on one member, with (A) commutator or (B) phase windings on the other. It is basically immaterial which function is assigned to stator and which to rotor, but for practical convenience a commutator winding normally rotates. The list of chief types below gives the type of winding (1, 2, A, B) and current supply (d or a), with the stator first:

D.C. machine: 1d/Ad The arrangement is that of *Figure 2.19(d)*. A commutator and brushes is necessary for the rotor.

Single-phase commutator machine: 1a/Aa The physical arrangement is the same as that of the d.c. machine. The field flux alternates, so that the rotor m.m.f. must also alternate at the same frequency and preferably in time phase. Series connection of stator and rotor gives this condition.

Synchronous machine: Ba/1d The rotor carries a concentrated d.c. winding, so the rotor m.m.f. must rotate with it at corresponding (synchronous) speed, requiring a.c. (normally 3-phase) supply. The machine may be inverted (1d/Ba).

Induction machine: 2a/Ba (Figure 2.19e) The polyphase stator winding produces a rotating field of angular velocity ω_1. The rotor runs with a slip s, i.e. at a speed $\omega_1(1-s)$. The torque is maintained unidirectional by currents induced in the rotor winding at frequency $s\omega_1$. With d.c. supplied to the rotor (2a/Bd) the rotor m.m.f. is fixed relatively to the windings and unidirectional torque is maintained only at synchronous speed ($s=0$).

All electromagnetic machines are variants of the above.

2.4.3.7 Magnetohydrodynamic generator

Magnetohydrodynamics (m.h.d.) concerns the interaction between a conducting fluid in motion and a magnetic field. If a fast-moving gas at high temperature (and therefore ionised) passes across a magnetic field, an electric field is developed across the gaseous stream exactly as if it were a metallic conductor, in accordance with Faraday's law. The electric field gives rise to a p.d. between electrodes flanking the stream, and a current may be made to flow in an external circuit connected to the electrodes. The m.h.d. generator offers a direct conversion between heat and electrical energy.

2.4.3.8 Hall effect

If a flat conductor carrying a current I is placed in a magnetic field of density B in a direction normal to it (*Figure 2.20*), then an electric field is set up across the width of the conductor. This is the Hall effect, the generation of an e.m.f. by the movement of conduction electrons through the magnetic field. The Hall e.m.f. (normally a few microvolts) is picked off by tappings applied to the conductor edges, for the measurement of I or for indication of high-frequency powers.

2.4.4 Inductance

The e.m.f. induced in an electric circuit by change of flux linkage may be the result of changing the circuit's own current. A magnetic field always links a current carrying circuit, and the linkage is (under certain restrictions) proportional to the current. When the current changes, the linkage also changes and an e.m.f. called the *e.m.f. of self-induction* is induced. If the linkage due to a current i in the circuit is $\psi = \Phi N = Li$, the e.m.f. induced by a change of current is

$$e = -d\psi/dt = -N(d\Phi/dt) = -L(di/dt)$$

L is a coefficient giving the linkage per ampere: it is called the *coefficient of self-induction*, or, more usually, the *inductance*. The unit is the henry, and in consequence of its relation to linkage, induced e.m.f., and stored magnetic energy, it can be defined as follows:

A circuit has unit inductance (1 H) if (a) the energy stored in the associated magnetic field is $\tfrac{1}{2}$J when the current is 1 A, or (b) the induced e.m.f. is 1 V when the current is changed at the rate 1 A/s, or (c) the flux linkage is 1 Wb-turn when the current is 1 A.

2.4.4.1 Voltage applied to an inductor

Let an inductor (i.e. an inductive coil or circuit) devoid of resistance and capacitance be connected to a supply of constant potential difference V, and let the inductance be L. By definition (b) above, a current will be initiated, growing at such a rate that the e.m.f. induced will counterbalance the applied voltage V. The current must rise uniformly at V/L amperes per second, as shown in *Figure 2.21(a)*, so long as the applied p.d. is maintained.

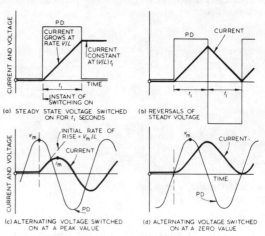

Figure 2.21 Voltage applied to pure inductor

Simultaneously the circuit develops a growing linked flux and stores a growing amount of magnetic energy. After a time t_1 the current reaches $I_1 = (V/L)t_1$, and has absorbed a store of energy at voltage V and average current $I_1/2$, i.e.

$$W_1 = V \cdot \tfrac{1}{2}I_1 \cdot t_1 = \tfrac{1}{2}VI_1 t_1 = \tfrac{1}{2}LI_1^2 \text{ joules}$$

since $V = I_1 L/t_1$. If now the supply is removed but the circuit remains closed, there is no way of converting the stored energy,

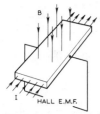

Figure 2.20 Hall effect

2.4.3.5 Ideal transformer

An ideal transformer comprises two resistanceless coils embracing a common magnetic circuit of infinite permeability and zero core loss (*Figure 2.18*). The coils produce no leakage flux: i.e. the

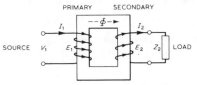

Figure 2.18 Ideal transformer

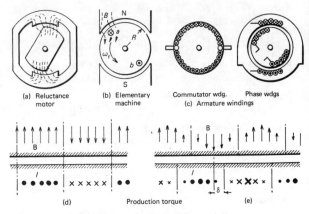

Figure 2.19 Electromagnetic machines

whole flux of the magnetic circuit completely links both coils. When the primary coil is energised by an alternating voltage V_1, a corresponding flux of peak value Φ_m is developed, inducing in the N_1-turn primary coil an e.m.f. $E_1 = V_1$. At the same time an e.m.f. E_2 is induced in the N_2-turn secondary coil. If the terminals of the secondary are connected to a load taking a current I_2, the primary must accept a balancing current I_1 such that $I_1 N_1 = I_2 N_2$, as the core requires zero excitation. The operating conditions are therefore

$$N_1/N_2 = E_1/E_2 = I_2/I_1; \quad \text{and} \quad E_1 I_1 = E_2 I_2$$

The secondary load impedance $Z_2 = E_2/I_2$ is reflected into the primary to give the impedance $Z_1 = E_1/I_1$ such that

$$Z_1 = (N_1/N_2)^2 Z_2$$

A practical power transformer differs from the ideal in that its core is not infinitely permeable and demands an excitation $N_1 I_0 = N_1 I_1 - N_2 I_2$; the primary and secondary coils have both resistance and magnetic leakage; and core losses occur. By treating these effects separately, a practical transformer may be considered as an ideal transformer connected into an external network to account for the defects.

2.4.3.6 Electromagnetic machines

An electromagnetic machine links an electrical energy system to a mechanical one, by providing a reversible means of energy flow between them in the common or 'mutual' magnetic flux linking stator and rotor. Energy is stored in the field and released as work. A current carrying conductor in the field is subjected to a mechanical force and, in moving, does work and generates a counter e.m.f. Thus the force–motion product is converted to or from the voltage–current product representing electrical power.

The energy-rate balance equations relating the mechanical power p_m, electrical power p_e, and the energy stored in the magnetic field w_f, are:

Motor: $p_e = p_m + dw_f/dt$

Generator: $p_m = p_e + dw_f/dt$

The mechanical power term must account for changes in stored kinetic energy, which occur whenever the speed of the machine and its coupled mechanical loads alters.

Reluctance motors The force between magnetised surfaces (*Figure 2.15b*) can be applied to rotary machines (*Figure 2.19a*). The armature tends to align itself with the field axis, developing a *reluctance torque*. The principle is applied to miniature rotating-contact d.c. motors and synchronous clock motors.

Machines with armature windings Consider a machine rotating with constant angular velocity ω_r and developing a torque M. The mechanical power is $p_m = M\omega_r$: the electrical power is $p_e = ei$, where e is the counter e.m.f. due to the reaction of the mutual magnetic field. Then $ei = M\omega_r + dw_f/dt$ at every instant. If the armature conductor a in *Figure 2.19(b)* is running in a non-time-varying flux of local density B, the e.m.f. is entirely rotational and equal to $e_r = Blu = Bl\omega_r R$. The tangential force on the conductor is $f = Bli$ and the torque is $M = BliR$. Thus, $e_r i = M\omega_r$ because $dw_f/dt = 0$. This case applies to constant flux (d.c., three-phase synchronous and induction) machines.

If the armature in *Figure 2.19(b)* is given two conductors a and b they can be connected to form a turn. Provided the turn is of full pitch, the torques will always be additive. More turns in series form a winding. The total flux in the machine results from the m.m.f.s of all current carrying conductors, whether on stator or rotor, but the torque arises from that component of the total flux at right angles to the m.m.f. axis of the armature winding.

Armature windings (*Figure 2.19c*) may be of the commutator or phase (tapped) types. The former is closed on itself, and current is led into and out of the winding by fixed brushes which include between them a constant number of conductors in each armature current path. The armature m.m.f. coincides always with the brush axis. Phase windings have separate external connections. If the winding is on the rotor, its current and m.m.f. rotate with it and the external connections must be made through slip-rings. Two (or three) such windings with 2-phase (or 3-phase) currents can produce a resultant m.m.f. that rotates with respect to the windings.

Torque *Figure 2.19(d)* shows a commutator winding arranged for maximum torque: i.e. the m.m.f. axis of the winding is displaced electrically $\pi/2$ from the field pole centres. If the armature has a radius R and a core length l, the flux has a constant uniform density B, and there are Z conductors in the $2p$ pole pitches each carrying the current I, the torque is $BRlIZ/2p$. This applies to a d.c. machine. It also gives the mean torque of a single-phase commutator machine if B and I are r.m.s. values and the factor $\cos\phi$ is introduced for any time phase angle between them.

The torque of a phase winding can be derived from *Figure 2.19(e)*. The flux density is assumed to be distributed sinusoidally, and reckoned from the pole centre to be $B_m \cos\alpha$. The current in the phase winding produces the m.m.f. F_a, having an axis displaced by angle δ from the pole centre. The total torque is then

$$M = \pi B_m F_a lR \sin\delta = \tfrac{1}{2}\pi \Phi F_a \sin\delta$$

per pole pair. This case applies directly to the 3-phase synchronous and induction machines.

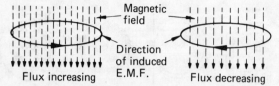

Figure 2.16 Faraday-Lenz law

linkage ψ with time. The flux linkage is the summation of products of magnetic flux with the number of turns of the circuit linked by it. Then

$$e = -d\psi/dt = -\Sigma N(d\Phi/dt)$$

the negative sign being indicative of the direction of the e.m.f. as specified in Lenz's law.

Consider a circuit of N turns linked completely with a flux Φ. The linkage $\psi = \Phi N$ may change in a variety of ways:

(1) Supposing the flux constant in value, the circuit may move through the flux (relative motion of flux and circuit: the *motional* or *generator effect*).

(2) Supposing the coil stationary with reference to the magnetic path of the flux, the latter may vary in magnitude (flux pulsation: the *pulsational* or *transformer effect*).

(3) Both changes may occur simultaneously (movement of coil through varying flux: combination of the effects in (1) and (2)).

The generator effect is associated with conversion of energy between the electrical and mechanical form, using an intermediate magnetic form; the transformer effect concerns the conversion of electrical energy into or from magnetic energy.

2.4.3.1 Generator effect

In simplified terms applicable to heteropolar rotating electrical machines (*Figure 2.17*) the instantaneous e.m.f. due to rate of change of linkage resulting from the motion at speed u of an N-turn fullpitch coil of effective length l is $e = 2NBlu$, where B is the flux density in which the coil sides move at the instant considered.

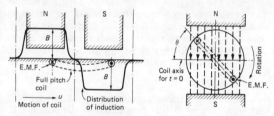

Figure 2.17 Motional e.m.f.

On this expression as a basis, well-known formulae for the motional e.m.f.s of machines can be derived. For example, consider the arrangement (right) in *Figure 2.17*, where the flux density B is considered to be uniform: let the coil rotate at angular velocity ω_r rad/s corresponding to a speed $n = \omega_r/2\pi$ rev/s. Then the peripheral speed of the coil is $u = \omega_r R$ if its radius is R. Let the coil occupy a position perpendicular to the flux axis when time $t = 0$. At $t = \theta/\omega_r$ it will be in a position making the angle θ. Its rate of moving across the flux is $\omega_r R \sin\theta$, and the instantaneous coil e.m.f. is

$$e = \omega_r RNBl \sin\theta = \omega_r N\Phi_m \sin\theta$$

where $\Phi_m = 2BlR$ is the maximum flux embraced, i.e. at $\theta = 0$. The e.m.f. is thus a sine function of frequency $\omega_r/2\pi$ and the values

peak: $e_m = \omega_r N\Phi_m$ r.m.s.: $E = (1/\sqrt{2})\omega_r N\Phi_m$

The same result is obtained by a direct application of the Faraday law. At $t = 0$ the linked flux is Φ_m; at $t = \theta/\omega_r$ it is $\Phi_m \cos\theta = \Phi_m \cos\omega_r t$. The instantaneous e.m.f. is

$$e = -d\psi/dt = -N\Phi_m d(\cos\omega_r t)/dt = \omega_r N\Phi_m \sin\theta$$

as before.

2.4.3.2 Transformer effect

The practical case concerns a coil of N turns embracing a varying flux Φ. If the flux changes sinusoidally with time it can be expressed as

$$\Phi = \Phi_m \cos\omega t = \Phi_m \cos 2\pi f t$$

where Φ_m is its time maximum value, f is its frequency, and $\omega = 2\pi f$ is its angular frequency. The instantaneous e.m.f. in the coil is

$$e = -N(d\Phi/dt) = \omega N\Phi_m \sin\omega t$$

This relation forms the basis of the e.m.f. induced in transformers and induction motors. The e.m.f. and flux relationship is that of *Figures 2.16* and *2.21(c)*.

2.4.3.3 Calculation of induced e.m.f.

The two methods of calculating electromagnetically induced e.m.f.s are (1) the change-of-flux law and (2) the flux-cutting law.

Flux change This law has the basic form

$$e = -N(d\Phi/dt)$$

and is applicable where a circuit of constant shape links a changing magnetic flux.

Flux cutting Where a conductor of length l moves at speed u at right angles to a uniform magnetic field of density B, the e.m.f. induced in the conductor is

$$e = Blu$$

This can be applied to the motion of conductors in constant magnetic fields and when sliding contacts are involved.

Linkage change Where coils move in changing fluxes, and both flux-pulsation and flux-cutting processes occur, the general expression

$$e = -d\psi/dt$$

must be used, with variation of the linkage ψ expressed as the result of both processes.

2.4.3.4 Constant linkage principle

The linkage of a *closed* circuit cannot be changed instantaneously, because this would imply an instantaneous change of associated magnetic energy, i.e. the momentary appearance of infinite power. It can be shown that the linkage of a closed circuit of *zero resistance* and no internal source cannot be changed at all. The latter concept is embodied in the following theorem.

Constant linkage theorem The linkage of a closed passive circuit of zero resistance is a constant. External attempts to change the linkage are opposed by induced currents that effectively prevent any net change of linkage.

The theorem is very helpful in dealing with transients in highly inductive circuits such as those of transformers, synchronous generators, etc.

In designing a magnet it is necessary to allow for leakage by use of an m.m.f. allowance F_a (normally not more than 1.25) and a flux allowance Φ_a, which may be anything from 2 to 20, being greater for a high ratio between gap length and gap section.

If H_g = field strength in gap, l_g = gap length, A_g = gap section, B_m = working density in magnet, H_m = working demagnetising field strength in magnet, l_m = magnet length and A_m = magnet section, then it may be shown that

$$H_g = \sqrt{[(B_m H_m / F_a \Phi_a)(A_m l_m / A_g l_g)]}$$

i.e. it is greatest when $B_m H_m$ is a maximum. This occurs for a working point at $(BH)_{max}$. The magnet length and section must be proportioned to suit the alloy and the gap dimensions to secure the required condition. The section is $A_m = (H_g A_g \Phi_a / B_m)$ and the length $l_m = (H_g l_g F_a / H_m)$. To calculate these the B_m and H_m values at the $(BH)_{max}$ point must be known. Alternatively, if the three points corresponding to the remanent flux density B_r, the $(BH)_{max}$ and the coercive force H_c are given, the working values can be calculated from

$$B_m = \sqrt{[(BH)_{max}(B_r/H_c)]} \quad \text{and} \quad H_m = \sqrt{[(BH)_{max}(H_c/B_r)]}$$

2.4.2 Magnetomechanical effects

Mechanical forces are developed in magnetic field systems in such a way that the resulting movement increases the flux linkage with the electric circuit, or lowers the m.m.f. required for a given flux. In the former case an increase of linkage requires more energy from the circuit, making mechanical energy available; in the latter, stored magnetic energy is released in mechanical form.

Systems in which magnetomechanical forces are developed are shown diagrammatically in *Figure 2.15*.

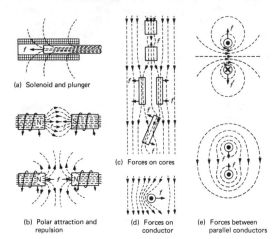

(a) Solenoid and plunger

(b) Polar attraction and repulsion

(c) Forces on cores

(d) Forces on conductor

(e) Forces between parallel conductors

Figure 2.15 Magnetomechanical forces

(a) *Solenoid and magnetic core* The core is drawn into the solenoid, increasing the magnetix flux and, in consequence, the circuit's flux linkages.

(b) *Attraction or repulsion of magnetised surfaces* The attraction in (b) increases the flux by reducing the reluctance of the intergap. Repulsion gives more space for the opposing fluxes and again reduces the reluctance.

(c) *Forces on magnetic cores in a magnetic field* Cores in line with the field attract each other, cores side by side repel each other, and a core out of line with the general field direction experiences a force tending to align it.

(d) *Electromagnetic force on current carrying conductor* A current carrying conductor lying in an externally produced (or 'main') field tends to move so as to increase the flux on that side where its own field has the same direction as the main field.

(e) *Electromagnetic force between current carrying conductors* Two parallel conductors carrying currents in opposite directions repel, for in moving apart they provide a greater area for the flux between them. If carrying currents in the same direction, they attract, tending to provide a shorter path for the common flux.

It is worth noting as a guide to the behaviour of magnetic field problems (although not a physical explanation of that behaviour) that the forces observed are in directions such as would cause the flux lines to shorten their length and to expand laterally, as if they were stretched elastic threads.

The calculation of the force developed in the cases is based on the movement of the force-system by an amount dl and the amount of mechanical energy dW thereby absorbed or released. Then the force is

$$f = dW/dl$$

With energy in joules and displacement in metres, the force is given in newtons. The calculation is only directly possible in a few simple cases, as below. The references are to the diagrams in *Figure 2.15*.

Case (c) The energy stored per cubic metre of a medium in which a magnetising force H produces a density B is $\frac{1}{2}BH = \frac{1}{2}B^2/\mu_r\mu_0$ joules. At an iron surface in air, a movement of the surface into a space originally occupied by air results, for a constant density B, in a reduction of the energy per cubic metre from $B^2/2\mu_0$ to $B^2/2\mu_r\mu_0$, since for air $\mu_r = 1$. The force must therefore be

$$f = \frac{B^2}{2\mu_0}\left(1 - \frac{1}{\mu_r}\right) \simeq \frac{B^2}{2\mu_0} \quad (\text{N/m}^2)$$

the latter expression being sufficiently close when $\mu_r \gg 1$.

Case (d) This case is of particular importance, as it is the basic principle of normal motors and generators, and of moving-coil permanent magnet instruments. Consideration of the mechanical energy gives, for a current I of length l lying perpendicular to the main field, a force

$$f = BIl$$

where B is the flux density. The force, as indicated in *Figure 2.15(d)*, is at right angles to B and to I. Suppose the conductor to be moved in the direction of the force (either with it or against it): the work done in a displacement of x is

$$fx = W = BIlx = \Phi I \text{ joules}$$

where $\Phi = Blx$ is the total flux cut across by the conductor.

2.4.3 Electromagnetic induction

A magnetic field is a store of energy. When it is increased or decreased, the amount of stored energy increases or decreases. Where the energy is obtained from, or restored to, an associated electric circuit, the energy delivered or received is in the form of a current flowing by reason of an *induced e.m.f.* for a time (specified by the conditions), these three being the essential associated quantities determining electrical energy. The relative directions of e.m.f. and current depend on the direction of energy flow. This is described by *Lenz's law (Figure 2.16)*, which states that the direction of the e.m.f. induced by a change of linked magnetic field is such as would oppose the change if allowed to produce a current in the associated circuit.

Faraday's law states that the e.m.f. induced in a circuit by the linked magnetic field is proportional to the rate of change of flux

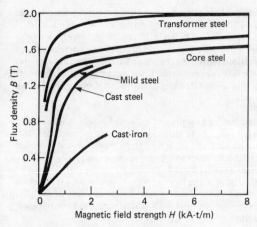

Figure 2.11 Magnetisation curves

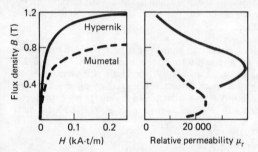

Figure 2.12 Magnetisation and permeability curves

2.4.1.1 Permeability

Certain *diamagnetic* materials have a relative permeability slightly less than that of vacuum. Thus, bismuth has $\mu r = 0.9999$. Other materials have μ_r slightly greater than unity: these are called *paramagnetic*. Iron, nickel, cobalt, steels, Heusler alloy (61% Cu, 27% Mn, 13% Al) and a number of others of great metallurgical interest have *ferromagnetic* properties, in which the flux density is not directly proportional to the magnetising force but which under suitable conditions are strongly magnetic. The more usual constructional materials employed in the magnetic circuits of electrical machinery may have peak permeabilities in the neighbourhood of 5000–10 000. A group of nickel–iron alloys, including *mumetal* (73% Ni, 22% Fe, 5% Cu), *permalloy 'C'* (77.4% Ni, 13.3% Fe, 3.7% Mo, 5% Cu) and *hypernik* (50% Ni, 50% Fe), show much higher permeabilities at low densities (*Figure 2.12*). Permeabilities depend on exact chemical composition, heat treatment, mechanical stress and temperature conditions, as well as on the flux density. Values of μ_r exceeding 5×10^5 can be achieved.

2.4.1.2 Core losses

A ferromagnetic core subjected to cycles of magnetisation, whether alternating (reversing), rotating or pulsating, exhibits *hysteresis*. *Figure 2.13* shows the cycle *B–H* relation typical of this phenomenon. The significant quantities *remanent flux density* and *coercive force* are also shown. The area of the *hysteresis loop* figure is a measure of the energy loss in the cycle per unit volume of material. An empirical expression for the *hysteresis loss* in a core taken through f cycles of magnetisation per second is

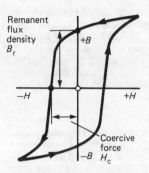
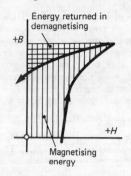

Figure 2.13 Hysteresis

$p_h = k_h f B_m^x$ W per unit mass or volume

Here B_m is the maximum induction reached and k_h is the hysteretic constant depending on the molecular quality and structure of the core metal. The exponent x may lie between 1.5 and 2.3. It is often taken as 2.

A further cause of loss in the same circumstances is the *eddy current loss*, due to the $I^2 R$ losses of induced currents. It can be shown to be

$p_e = k_e t^2 f^2 B^2$ W per unit mass or volume

the constant k_e depending on the resistivity of the metal and t being its thickness, the material being laminated to decrease the induced e.m.f. per lamina and to increase the resistance of the path in which the eddy currents flow. In practice, curves of loss per kilogram or per cubic metre for various flux densities are employed, the curves being constructed from the results of careful tests. It should be noted that hysteresis loss is dependent on the maximum flux density B_m, while the eddy current loss is a function of r.m.s. induced current and e.m.f., and therefore of the r.m.s. flux density B, and not the maximum density B_m.

2.4.1.3 Permanent magnets

Permanent magnets are made from heat treated alloys, or from ferrites and rare earths, to give the material a large hysteresis loop. *Figure 2.14* shows the demagnetisation B/H quadrant of the loop of a typical material. In use, a magnet produces magnetic

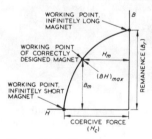

Figure 2.14 Ideal permanent magnet conditions

energy in the remainder of the magnetic circuit derived from a measure of self-demagnetisation: consequently, the working point of the magnet is on the loop between the coercive force/zero flux point and the zero force/remanent flux point. Different parts of the magnet will work at different points on the loop, owing to leakage, and the conditions become much more complex if the reluctance of the external magnetic circuit fluctuates.

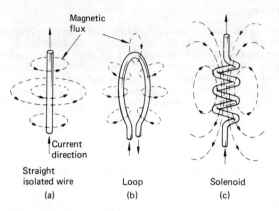

Figure 2.9 Magnetic fields

2.4.1 Magnetic circuit

By analogy with the electric circuit, the magnetic flux produced by a given current in a magnetic circuit is found from the magnetomotive force and the circuit reluctance. The m.m.f. produced by a coil of N turns carrying a current I is $F = NI$ ampere-turns. This is expended over any closed path linking the current I. At a given point in a magnetic field in free space the m.m.f. per unit length or magnetising force H gives rise to a magnetic flux density $B_0 = \mu_0 H$, where $\mu_0 = 4\pi/10^7$. If the medium in which the field exists has a relative permeability μ_r, the flux density established is

$$B = \mu_r B_0 = \mu_r \mu_0 H = \mu H$$

The summation of $H \cdot dl$ round any path linking an N-turn circuit carrying current I is the total m.m.f. F. If the distribution of H is known, the magnetic flux density B or B_0 can be found for all points in the field, and a knowledge of the area a of the magnetic path gives $\Phi = Ba$, the total magnetic flux.

Only in a few cases of great geometrical simplicity can the flux due to a given system of currents be found precisely. Among these are the following.

Long straight isolated wire (Figure 2.10a) This is not strictly a realisable case, but the results are useful. Assume a current of 1 A. The m.m.f. around any closed linking path is therefore 1 A-t. Experiment shows that the magnetic field is symmetrical about, and concentric with, the axis of the wire. Around a closed path of radius x metres there will be a uniform distribution of m.m.f. so that

$$H = F/2\pi x = 1/2\pi x \text{ (A-t/m)}$$

Consequently, in free space the flux density (T) at radius x is

$$B_0 = \mu_0 H = \mu_0/2\pi x$$

In a medium of constant permeability $\mu = \mu_r \mu_0$ the flux density is $B = \mu_r B_0$. There will be magnetic flux following closed circular paths within the cross-section of the wire itself: at any radius x the m.m.f. is $F = (x/r)^2$ because the circular path links only that part of the (uniformly distributed) current within the path. The magnetising force is $H = F/2\pi x = x/2\pi r^2$ and the corresponding flux density in a non-magnetic conductor is

$$B_0 = \mu_0 H = \mu_0 x/2\pi r^2$$

and μ_r times as much if the conductor material has a relative permeability μ_r. The expressions above are for a conductor current of 1 A.

Concentric conductors (Figure 2.10b) Here only the inner conductor contributes the magnetic flux in the space between the conductors and in itself, because all such flux can link only the inner current. The flux distribution is found exactly as in the previous case, but can now be summed in defined limits. If the outer conductor is sufficiently thin radially, the flux in the inter-conductor space, per metre axial length of the system, is

$$\Phi = \int_r^R \frac{\mu_0}{2\pi x} dx = \frac{\mu_0}{2\pi} \ln \frac{R}{r}$$

Toroid (Figure 2.10c) This represents the closest approach to a perfectly symmetrical magnetic circuit, in which the m.m.f. is distributed evenly round the magnetic path and the m.m.f. per metre H corresponds at all points exactly to the flux density existing at those points. The magnetic flux is therefore wholly confined to the path. Let the mean radius of the toroid be R and its cross-sectional area be A. Then, with N uniformly distributed turns carrying a current I and a toroid core of permeability μ,

$$F = NI; \quad H = F/2\pi R; \quad B = \mu H; \quad \Phi = \mu FA/2\pi R$$

This applies approximately to a long solenoid of length l, replacing R by $l/2\pi$. The permeability will usually be μ_0.

Composite magnetic circuit containing iron (Figure 2.10d) For simplicity practical composite magnetic circuits are arbitrarily divided into parts along which the flux density is deemed constant. For *each* part

$$F = Hl = Bl/\mu = BlA/\mu A = \Phi S$$

where $S = l/\mu A$ is the *reluctance*. Its reciprocal $\Lambda = 1/S = \mu A/l$ is the *permeance*. The expression $F = \Phi S$ resembles $E = IR$ for a simple d.c. circuit and is therefore sometimes called the *magnetic Ohm's law*.

The total excitation for the magnetic circuit is

$$F = H_1 l_1 + H_2 l_2 + H_3 l_3 + \ldots$$

for a series of parts of length $l_1, l_2 \ldots$, along which magnetic field intensities of $H_1, H_2 \ldots$, (A-t/m) are necessary. For free space, air and non-magnetic materials, $\mu_r = 1$ and $B_0 = \mu_0 H$, so that $H = B_0/\mu_0 \simeq 800\,000 B_0$. This means that an excitation $F = 800\,000$ A-t is required to establish unit magnetic flux density [1 T] over a length $l = 1$ m. For ferromagnetic materials it is usual to employ B-H graphs (*magnetisation curves*) for the determination of the excitation required, because such materials exhibit a *saturation* phenomenon. Typical B-H curves are given in *Figures 2.11* and *2.12*.

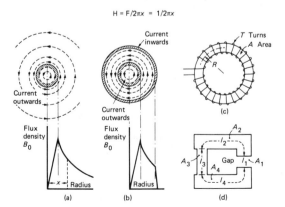

Figure 2.10 Magnetic circuits

2.3.2 Cells

2.3.2.1 Primary cells

An elementary cell comprising electrodes of copper (positive) and zinc (negative) in sulphuric acid develops a p.d. between copper and zinc. If a circuit is completed between the electrodes, a current will flow, which acts in the electrolyte to decompose the acid, and causes a production of hydrogen gas round the copper, setting up an e.m.f. of polarisation in opposition to the original cell e.m.f. The latter therefore falls considerably. In practical primary cells the effect is avoided by use of a *depolariser*. The most widely used primary cell is the Leclanché. It comprises a zinc and a carbon electrode in a solution of ammonium chloride, NH_4Cl. When current flows, zinc chloride, ZnCl, is formed, releasing electrical energy. The NH_4 positive ions travel to the carbon electrode (positive), which is packed in a mixture of manganese dioxide and carbon as depolariser. The NH_4 ions are split up into NH_3 (ammonia gas) and H, which is oxidised by the MnO_2 to become water. The removal of the hydrogen prevents polarisation, provided that the current taken from the cell is small and intermittent.

The *wet* form of Leclanché cell is not portable. The *dry* cell has a paste electrolyte and is suitable for continuous moderate discharge rates. It is exhausted by use or by ageing and drying up of the paste. The 'shelf life' is limited. The *inert* cell is very similar in construction to the dry cell, but is assembled in the dry state, and is activated when required by moistening the active materials. In each case the cell e.m.f. is about $1\frac{1}{2}$ V.

2.3.2.2 Standard cell

The Weston Normal Cell has a positive element of mercury, a negative element of cadmium, and an electrolyte of cadmium sulphate with mercurous sulphate as depolariser. The open-circuit e.m.f. at 20°C is about 1.018 30 V, and the e.m.f./temperature coefficient is of the order of -0.04 mV/°C.

2.3.2.3 Secondary cells

In the *lead–acid* storage cell or accumulator, lead peroxide reacts with sulphuric acid to produce a positive charge at the anode. At the cathode metallic lead reacts with the acid to produce a negative charge. The lead at both electrodes combines with the sulphate ions to produce the poorly soluble lead sulphate. The action is described as

```
Charged      PbO₂   +  2H₂SO₄  +  Pb      Discharged
              Brown    Strong acid   Grey
      =   PbSO₄  +   2H₂O    +  PbSO₄
          Sulphurate  Weak acid    Sulphate
```

Both electrode reactions are reversible, so that the initial conditions may be restored by means of a 'charging current'.

In the *alkaline* cell, nickel hydrate replaces lead peroxide at the anode, and either iron or cadmium replaces lead at the cathode. The electrolyte is potassium hydroxide. The reactions are complex, but the following gives a general indication:

```
Charged                                      Discharged
2Ni(OH)₃ + KOH + Fe  = 2Ni(OH)₂ + KOH + Fe(OH)₂
or 2Ni(OH)₃ + KOH + Cd = 2Ni(OH)₃ + KOH + Cd(OH)₂
```

2.3.2.4 Fuel cell

The only practicable fuel cell for direct conversion of fuel into electrical energy is the hydrogen–oxygen cell (*Figure 2.8*). Microporous electrodes serve to bring the gases into intimate contact with the electrolyte (potassium hydroxide) and to provide the cell terminals. The hydrogen and oxygen reactants are fed continuously into the cell from externally, and electrical energy is available on demand.

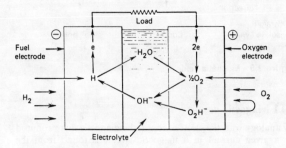

Figure 2.8 Fuel cell

At the fuel (H_2) electrode, H_2 molecules split into H atoms in the presence of a catalyst, and these combine with OH^- ions from the electrolyte, forming H_2O and releasing electrons e. At the oxygen electrode, the oxygen molecules (O_2) combine (also in the presence of a catalyst) with water molecules from the electrolyte and with pairs of electrons arriving at the electrode through the external load from the fuel electrode. Perhydroxyl ions (O_2H^-) and hydroxyl ions (OH^-) are produced: the latter enter the electrolyte, while the more resistant O_2H^- ions, with special catalysts, can be reduced to OH^- ions and oxygen. The overall process can be summarised as:

Fuel electrode	$H_2 + 2OH^- = 2OH_2O + 2e$
Oxygen electrode	$\frac{1}{2}O_2 + H_2O + 2e = 2OH^-$
Net reaction	$H_2 + \frac{1}{2}O_2 \rightarrow 2e$ flow $\rightarrow H_2O$

In a complete reaction 2 kg hydrogen and 16 kg oxygen combine chemically (not explosively) to form 18 kg of water with the release of 400 MJ of electrical energy. For each kAh the cell produces 0.33 litre of water, which must not be allowed unduly to weaken the electrolyte. The open-circuit e.m.f. is 1.1 V, while the terminal voltage is about 0.9 V, with a delivery of 1 kA/m^2 of plate area.

2.4 Magnetic field effects

The space surrounding permanent magnets and electric circuits carrying currents attains a peculiar state in which a number of phenomena occur. The state is described by saying that the space is threaded by a *magnetic field of flux*. The field is mapped by an arrangement of *lines of induction* giving the strength and direction of the flux. *Figure 2.9* gives a rough indication of the flux pattern for three simple cases of magnetic field due to a current. The diagrams show the conventions of polarity, direction of flux and direction of current adopted. Magnetic lines of induction form closed loops in a *magnetic circuit* linked by the circuit current wholly or in part.

2.2.2.7 Thermoelectric devices

If a current flows through a thermocouple (*Figure 2.6b*), with one junction in thermal contact with a heat-sink, the other removes heat from a source. The couple must comprise conductors with positive and negative Seebeck coefficients, respectively. The arrangement is a *refrigerator* with the practical advantages of simplicity and silence.

A heat source applied to a junction develops an e.m.f. that will circulate a current in an external load (*Figure 2.6d*). If semiconductors of low thermal conductivity are used in place of metals for the couple elements, a better efficiency is obtainable because heat loss by conduction is reduced. The couples in *Figure 2.7(a)* of a thermoelectric power generator are constructed with p- and n-type materials. The efficiency, limited by Carnot cycle considerations, does not at present exceed 10%.

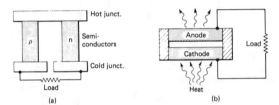

Figure 2.7 Thermoelectric devices

A thermocouple generator in which one element is an electron stream or plasma is the *thermionic generator*, in effect a diode with flat cathode and anode very close together. By virtue of their kinetic energy, electrons emitted from the cathode reach the anode against a small, negative anode potential, providing current for an external circuit (*Figure 2.7b*). The work function of the anode material must be less than for the cathode. The device is a heat engine operating over the cathode–anode temperature fall, with electrons providing the 'working fluid'.

Outputs of 2 kW/m² at an efficiency of 25% may be reached when the device has been fully developed and the space charge effects overcome. Cathode heating by solar energy is a possibility.

2.3 Electrochemical effects

2.3.1 Electrolysis

If a liquid conductor undergoes chemical changes when a current is passed through it, the effect is ascribed to the movement of constituent parts of the molecules of the liquid *electrolyte*, called *ions*, which have a positive or negative electric charge. Positive ions move towards the negative electrode (*cathode*) and negative ions to the positive electrode (*anode*). The ionic movement is the reason for the current conduction. Ions reaching the electrodes have their charges neutralised and may be subject to chemical change. Hydrogen and metal ions are electropositive: non-metals of the chlorine family (Cl, Br, I and F) and acid radicals (such as SO_4, NO_3) form negative ions in solution. As examples, hydrochloric acid, HCl, forms H^+ and Cl^- ions; sulphuric acid forms $2H^+$ and SO_4^- ions; and sodium hydroxide, NaOH, yields Na^+ and OH^- ions. The products of electrolysis depend on the nature of the electrolute. Basic solutions of sodium or similar hydroxides produce H_2 and O_2 gases at cathode and anode, respectively. Acid solutions give products depending on the nature of the electrodes. Solutions of metal salts with appropriate electrodes result in electrodeposition.

The mass of the ion of an element of radical deposited on, dissolved from or set free at either electrode is proportional to the quantity of electricity passed through the electrolytic cell and to the ionic weight of the material, and inversely proportional to the valency of the ion; whence the mass m_e in kilogram-equivalents is the product (z in kilogram-equivalents per coulomb) × (Q in coulombs). The value of z is a natural constant 0.001 036. Representative figures (for convenience in milligrams per coulomb) are given in Table 2.4.

Table 2.4 Electrochemical equivalents z (mg/C)

Element	Valency	z	Element	Valency	z
H	1	0.010 45	Zn	2	0.338 76
Li	1	0.071 92	As	3	0.258 76
Be	2	0.046 74	Se	4	0.204 56
O	2	0.082 90	Br	1	0.828 15
F	1	0.196 89	Sr	2	0.454 04
Na	1	0.238 31	Pd	4	0.276 42
Mg	2	0.126 01	Ag	1	1.117 93
Al	3	0.093 16	Cd	2	0.582 44
Si	4	0.072 69	Sn	2	0.615 03
S	2	0.166 11	Sn	4	0.307 51
S	4	0.083 06	Sb	3	0.420 59
S	6	0.055 37	Te	4	0.330 60
Cl	1	0.367 43	I	1	1.315 23
K	1	0.405 14	Cs	1	1.377 31
Ca	2	0.207 67	Ba	2	0.711 71
Ti	4	0.124 09	Ce	3	0.484 04
V	5	0.105 60	Ta	5	0.374 88
Cr	3	0.179 65	W	6	0.317 65
Cr	6	0.089 83	Pt	4	0.505 78
Mn	2	0.284 61	Au	1	2.043 52
Fe	1	0.578 65	Au	3	0.681 17
Fe	2	0.289 33	Hg	1	2.078 86
Fe	3	0.192 88	Hg	2	1.039 42
Co	2	0.305 39	Tl	1	2.118 03
Ni	2	0.304 09	Pb	2	1.073 63
Cu	1	0.658 76	Bi	3	0.721 93
Cu	2	0.329 38	Th	4	0.601 35

To pass a current through an electrolyte, a p.d. must be applied to the electrodes to overcome the drop in resistance of the electrolyte, and to overcome the e.m.f. of *polarisation*. The latter is due to a drop across a thin film of gas, or through a strong ionic concentration, at an electrode.

Every chemical reaction may be represented as two electrode reactions. The algebraic p.d. between the two is a measure of the reactivity. A highly negative p.d. represents a spontaneous reaction that might be utilised to generate a current. A high positive p.d. represents a reaction requiring an external applied p.d. to maintain it.

2.3.1.1 Uses of electrolysis

Ores of copper, zinc and cadmium may be electrolytically treated with sulphuric acid to deposit the metal. Copper may be deposited by use of a low voltage at the cathode, while oxygen is emitted at the anode. Electrorefining by deposition may be employed with copper, nickel, tin, silver, etc., produced by smelting or electrowinning, by using the impure metal as anode, which is dissolved away and redeposited on the cathode, leaving at the bottom of the cell the impurities in the form of sludge. Electroplating is similar to electrorefining except that pure metal or alloy is used as the anode.

2.2.2.3 Measurement of temperature rise

The temperature rise of a device developing heat can be measured (a) by a thermometer placed in contact with the surface whose temperature is required, (b) by resistance-temperature detectors or thermocouples on the surface of, or embedded in, the device, or (c) by the measurement of resistance (in the case of conducting circuits), using the known resistance-temperature coefficient. These methods measure different temperatures, and do not give merely alternative estimates of the same thing.

2.2.2.4 Heating and cooling cycles

In some cases a device (such as a machine or one of its parts) developing internal heat may be considered as sufficiently homogeneous to apply the exponential law. Suppose the device to have a temperature rise θ after the lapse of a time t. In an element of time dt a small temperature-rise $d\theta$ takes place. The heat developed is $p \cdot dt$, the heat stored is $Gh \cdot d\theta$, and the heat dissipated is $A\theta\lambda \cdot dt$. Since the heat stored and dissipated together equal the total heat produced,

$$Gh \cdot d\theta + A\theta\lambda \cdot dt = p \cdot dt$$

the solution of which is

$$\theta = \theta_m [1 - \exp(-t/\tau)]$$

where θ_m is the final steady temperature rise, calculated from $\theta_m = p/A\lambda$, and $\tau = Gh/A\lambda$ is called the *heating time constant*. For the lapse of time t equal to the time constant

$$\theta = \theta_m [1 - \exp(-1)] = 0.632\theta_m$$

When a heated body cools owing to a reduction or cessation of internal heat production, the temperature–time relation is the exponential function

$$\theta = \theta_m \exp(-t/\tau_1)$$

where τ_1 is the *cooling time constant*, not necessarily the same as that for heating conditions.

Both heating and cooling as described are examples of *thermal transients*, and the laws governing them are closely analogous to those concerned with transient electric currents, in which exponential time relations also occur.

2.2.2.5 Fusing currents

For a given diameter d, the heat developed by a wire carrying a current I is inversely proportional to d^3, because an increase of diameter reduces the current density in proportion to the increase of area, and the emitting surface is increased in proportion to the diameter. The temperature rise is consequently proportional to I^2/d^3. If the temperature is raised to the fusing or melting point, $\theta = aI^2/d^3$ and the fusing current is

$$I = \sqrt{(\theta d^3/a)} = k d^{3/2}$$

This is *Preece's law*, from which an estimate may be made of the fusing current of a wire of given diameter, provided that k is known. The exponent $3/2$ and the value of k are both much affected by enclosure, conduction of heat by terminals, and similar physical conditions.

It is obvious that any rule regarding suitable current densities giving a value regardless of the diameter is likely to be uneconomically low for small wires and excessive for large ones. Further, the effects of length and enclosure make a direct application of Preece's law unreliable. For small wires the exponent x in the term d^x may be 1.25–1.5, and for larger wires it may exceed 1.5.

2.2.2.6 Thermo-e.m.f.s

An effect known as the *thermo-electric effect* or *Seebeck effect* is that by which an e.m.f. is developed due to a difference of temperature between two junctions of dissimilar conductors in the same circuit. The *Thomson effect* or *Kelvin effect* is (a) that an e.m.f. is developed due to a difference of temperature between two parts of the same conductor, and (b) that an absorption or liberation of heat takes place when a current flows from a hotter to a colder part of the same material. The *Peltier effect* describes the liberation or absorption of heat at a joint where current passes from one material to another, whereby the joint becomes heated or cooled.

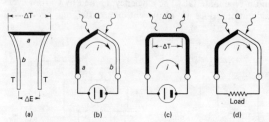

Figure 2.6 Thermo-e.m.f.

In *Figure 2.6(a)–(c)* the symbols are absolute temperature T, thermo-e.m.f. E and rate of heat production or absorption Q. The Seebeck coefficient (a) is the e.m.f. per degree difference between hot and cold junctions:

$$\alpha_S = \Delta E / \Delta T$$

Typical e.m.f.s for a number of common junctions are given in *Table 2.3*. In the Peltier effect (b) a rate of heat generation (reversible, and distinct from the irreversible I^2R heat) results

Table 2.3 Thermocouple e.m.f.s (mV): cold junction at 0°C

Hot-junction temperature (°C)	Platinum/ ^{87}Pt–^{13}Rh	Chromel/ Alumel	Iron/ Eureka	Copper/ Eureka
100	0.65	4.1	5	4
200	1.46	8.1	11	9
400	3.40	16.4	22	21

from the passage of a current i through the different conductors A and B. The *Peltier coefficient* is

$$\alpha_P = Q/i$$

The Thomson effect (c) concerns the rate of reversible heat when a current i flows through a length of homogeneous conductor across which there is a temperature difference. The *Thomson coefficient* is

$$\alpha_T = \Delta Q / i \cdot \Delta T$$

The relation between the Seebeck, and Peltier coefficients is important: it is

$$\alpha_S = \alpha_P / T$$

The Seebeck coefficient is the more easily measured, but the Peltier coefficient determines the cooling effect of a thermo-electric refrigerator.

useful cross-section of the conductor is less than the actual area, and the effective resistance is consequently higher. This is called the *skin effect*. An analogous phenomenon, the *proximity effect*, is due to mutual inductance between conductors arranged closely parallel to one another. The effects depend on conductor size, frequency f of the current, resistivity ρ and permeability μ of the material. For a circular conductor of diameter d the increase of effective resistance is proportional to $d^2 f \mu / \rho$. At power frequencies and for small conductors the effect is negligible. It may, however, be necessary to investigate the skin and proximity effects in the case of large conductors such as busbars.

2.2.1.6 Non-linear resistors

Prominent among non-linear resistors are electric arcs; also silicon carbide and similar materials.

Arcs An electric arc constitutes a conductor of somewhat vague dimensions utilising electronic and ionic conduction in a gas. It is strongly affected by physical conditions of temperature, gas pressure and cooling. In air at normal pressure a d.c. arc between copper electrodes has the voltage–current relation given approximately by $V = 30 + 10/I + l[1 + 3/I]10^3$ for a current I in an arc-length l metres. The expression is roughly equivalent to 10 V/cm for large currents and high voltages. The current density varies between 1 and 1000 A/mm^2, being greater for large currents because of the *pinch* effect.

Silicon carbide Conducting pieces of this material have a current–voltage relation expressed approximately by $I = KV^x$, where x is usually between 3 and 5. For rising voltage the current increases very rapidly, making silicon carbide devices suitable for circuit protection and the discharge of excess transmission-line surge energy.

2.2.2 Heating and cooling

The *heating* of any body such as a resistor or a conducting circuit having inherent resistance is a function of the losses within it that are developed as heat. (This includes core and dielectric as well as ohmic I^2R losses, but the effective value of R may be extended to cover such additional losses.) The *cooling* is a function of the facilities for heat dissipation to outside media such as air, oil or solids, by radiation, conduction and convection.

2.2.2.1 Rapid heating

If the time of heating is short, the cooling may be ignored, the temperature reached being dependent only on the rate of development of heat and the thermal capacity. If p be the heat developed per second in joules (i.e. the power in watts), G the mass of the heated body in kilograms and c its specific heat in joules per kilogram per kelvin, then

$$Gc \cdot d\theta = p \cdot dt, \quad \text{giving} \quad \theta = (1/Gc) \int p \cdot dt$$

For steady heating, the temperature rise is p/Gc in K/s.

Standard annealed copper is frequently used for the windings and connections of electrical equipments. Its density is $G = 8900$ kg/m^3 and its resistivity at 20°C is 0.017 $\mu\Omega$-m; at 75°C it is 0.021 $\mu\Omega$-m. A conductor worked at a current density J (in amperes per square metre) has a specific loss (watts per kilogram) of $\rho J^2/8900$. If $J = 2.75$ MA/m^2 (or 2.75 A/mm^2), the specific loss at 75°C is 17.8 W/kg, and its rate of self-heating is $17.8/375 = 0.048$ °C/s.

2.2.2.2 Continuous heating

Under prolonged steady heating a body will reach a temperature rise above the ambient medium of $\theta_m = p/A\lambda$, where A is the cooling surface area and λ the specific heat dissipation (joules per second per square metre of surface per degree Celsius temperature rise above ambient). The expression is based on the assumption, roughly true for moderate temperature rises, that the rate of heat emission is proportional to the temperature rise. The specific heat dissipation λ is compounded of the effects of *radiation*, *conduction* and *convection*.

Radiation The heat radiated by a surface depends on the absolute temperature T (given by $T = \theta + 273$, where θ is the Celsius temperature), and on its character (surface smoothness or roughness, colour, etc.). The Stefan law of heat radiation is

$$p_r = 5.7 e T^4 \times 10^{-8} \text{ watts per square metre}$$

where e is the coefficient of radiant emission, always less than unity, except for the perfect 'black body' surface, for which $e = 1$. The radiation from a body is independent of the temperature of the medium in which it is situated. The process of radiation of a body to an exterior surface is accompanied by a re-absorption of part of the energy when re-radiated by that surface. For a small spherical radiating body inside a large and/or black spherical cavity, the radiated power is given by the Stefan–Boltzmann law:

$$p_r = 5.7 e_1 [T_1^4 - T_2^4] 10^{-8} \text{ watts per square metre}$$

where T_1 and e_1 refer to the body and T_2 to the cavity.

The emission of radiant heat from a perfect black body surface is independent of the roughness or corrugation of the surface. If $e < 1$, however, there is some increase of radiation if the surface is rough.

Conduction The conduction of heat is a function of the thermal or temperature gradient and the thermal resistivity, the latter being defined as the temperature difference in degrees Celsius across a path of unit length and unit section required for the continuous transmission of 1 W. Thus, the heat conducted per unit area along a path of length x in a material of thermal resistivity ρ for a temperature difference of θ is

$$p_d = \theta/\rho x \text{ watts}$$

Resistivities for metals are very low. For insulating materials such as paper, $\rho = 5$–10; for *still* air, $\rho = 20$ W per °C per m and per m^2, approximately.

Convection Convection currents in liquids and gases (e.g. oil and air) are always produced near a heated surface unless baffled. Convection adds greatly to heat dissipation, especially if artificially stimulated (as in force cooling by fans). Experiment shows that a rough surface dissipates heat by convection more readily than a smooth one, and that high fluid speeds are essential to obtain *turbulence* as opposed to *stream-line* flow, the former being much more efficacious.

Convection is physically a very complex phenomenon, as it depends on small changes in buoyancy resulting from temperature rise due to heating. Formulae for dissipation of heat by convection have a strongly empirical basis, the form and orientation of the convection surfaces having considerable influence.

Cooling coefficient For electrical purposes the empirically derived coefficient of emission λ, or its reciprocal $1/\lambda$, are employed for calculations on cooling and temperature rise of wires, resistors, machines and similar plant.

tics are the same or different for the two current-flow directions. Rectifiers are an important class of non-linear, asymmetrical resistors.

A hypothetical device having the current–voltage characteristic shown in *Figure 2.4(c)* has, at an operating condition represented by the point P, a current I_d and a p.d. V_d. The ratio $R_d = V_d/I_d$ is its *d.c. resistance* for the given condition. If a small alternating voltage Δv_a be applied under the same condition (i.e. superimposed on the p.d. V_d), the current will fluctuate by Δi_a and the ratio $r_a = \Delta v_a/\Delta i_a$ is the *a.c.* or *incremental resistance* at P. The d.c. resistance is also obtainable from $R_d = \cot \theta$, and the a.c. resistance from $r_a = \cot \alpha$. In the region of which Q is a representative point, the a.c. resistance is *negative*, indicating that the device is capable of giving a small output of a.c. power, derived from its greater d.c. input. It remains in sum an energy dissipator, but some of the energy is returnable under suitable conditions of operation.

2.2.1.3 D.C. or ohmic resistance: linear resistors

The d.c. or ohmic resistance of linear resistors (a category confined principally to metallic conductors) is a function of the dimensions of the conducting path and of the *resistivity* of the material from which the conductor is made. A wire of length l, cross-section a and resistivity ρ has, at constant given temperature, a resistance

$R = \rho l/a$ ohms

where ρ, l and a are in a consistent system of dimensions (e.g. l in metres, a in square metres, ρ in ohms per 1 m length and 1 m^2 cross-section—generally contracted to ohm-metres). The expression above, though widely applicable, is true only on the assumption that the current is uniformly distributed over the cross-section of the conductor and flows in paths parallel to the boundary walls. If this assumption is inadmissible, it is necessary to resort to integration or the use of current-flow lines. *Figure 2.5* summarises the expressions for the resistance of certain arrangements and shapes of conductors.

Resistivity The resistivity of conductors depends on their composition, physical condition (e.g. dampness in the case of non-metals), alloying, manufacturing and heat treatment, chemical purity, mechanical working and ageing. The *resistance-temperature coefficient* describes the rate of change of resistivity with temperature. It is practically 0.004Ω per °C at 20°C for copper. Most pure metals have a resistivity that rises with

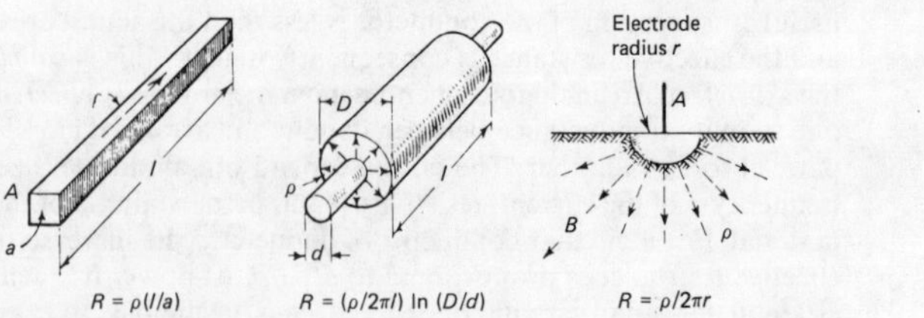

Figure 2.5 Resistance in particular cases

$R = \rho(l/a)$ $R = (\rho/2\pi l) \ln (D/d)$ $R = \rho/2\pi r$

temperature. Some alloys have a very small coefficient. Carbon is notable in that its resistivity decreases markedly with temperature rise, while uranium dioxide has a resistivity which falls in the ratio 50:1 over a range of a few hundred degrees. *Table 2.2* lists the resistivity ρ and the resistance-temperature coefficient α for a number of representative materials. The effect of temperature is assessed in accordance with the expressions

$R_1 = R_0(1 + \alpha\theta_1)$; $R_2/R_1 = (1 + \alpha\theta_2)/(1 + \alpha\theta_1)$;

or $R_2 = R_1[1 + \alpha(\theta_2 - \theta_1)]$

where R_0, R_1, R_2 are the resistances at temperatures 0, θ_1 and θ_2, and α is the resistance-temperature coefficient at 0°C.

2.2.1.4 Liquid conductors

The variations of resistance of a given aqueous solution of an electrolyte with temperature follow the approximate rule:

$R_\theta = R_0/(1 + 0.03\theta)$

where θ is the temperature in °C. The conductivity (or reciprocal of resistivity) varies widely with the percentage strength of the solution. For low concentrations the variation is that given in *Table 2.2*.

2.2.1.5 Frequency effects

The resistance of a given conductor is affected by the frequency of the current carried by it. The simplest example is that of an isolated wire of circular cross-section. The inductance of the central parts of the conductor is greater than that of the outside skin because of the additional flux linkages due to the internal magnetic flux lines. The impedance of the central parts is consequently greater, and the current flows mainly at and near the surface of the conductor, where the impedance is least. The

Table 2.2 Conductivity of aqueous solutions (mS/cm)

(a) NaOH = caustic soda
(b) NH$_4$Cl = sal ammoniac
(c) NaCl = common salt
(d) NaNO$_3$ = Chilean saltpetre
(e) CaCl$_2$ = calcium chloride
(f) ZnCl$_2$ = zinc chloride
(g) NaHCO$_3$ = baking soda
(h) Na$_2$CO$_3$ = soda ash
(j) Na$_2$SO$_4$ = Glauber's salt
(k) Al$_2$(SO$_4$)$_3$K$_2$SO$_4$ = alum
(l) CuSO$_4$ = blue vitriol
(m) ZnSO$_4$ = white vitriol

Concentration (%)	a	b	c	d	e	f	g	h, j	k	l, m
1	40	18	12	10	10	8	6	4	3	3
2	72	35	23	20	20	16	12	8	6	6
3	102	51	34	30	30	24	18	12	9	8
4	130	65	44	39	39	32	23	16	11	10
5		79	55	48	47	39	28	20	13	11
7.5		110	79	69	67	54	39	29	18	16
10			99	90	85	69		31	22	20

form of a complex number.

Polarisation: The change of the electrical state of an insulating material under the influence of an electric field, such that each small element becomes an electric dipole or doublet.

Potential: The electrical state at a point with respect to potential zero (normally taken as that of the earth). It is measured by the work done in transferring unit charge from potential zero to the point.

Potential difference: A difference between the electrical states existing at two points tending to cause a movement of positive charges from one point to the other. It is measured by the work done in transferring unit charge from one point to the other.

Potential gradient: The potential difference per unit length in the direction in which it is a maximum.

Power: The rate of transfer, storage, conversion or dissipation of energy. In sinusoidal alternating current circuits the *active power* is the mean rate of energy conversion; the *reactive power* is the peak rate of circulation of stored energy; the *apparent power* is the product of r.m.s. values of voltage and current.

Power factor: The ratio between active power and apparent power. In sinusoidal alternating current circuits the power factor is $\cos\phi$, where ϕ is the phase angle between voltage and current wave forms.

Quantity: The product of the current and the time during which it flows.

Reactance: In sinusoidal alternating current circuits, the quantity ωL or $1/\omega C$, where L is the inductance, C is the capacitance and ω is the angular frequency.

Reactor: A device having reactance as a chief property; it may be an inductor or a capacitor. A *nuclear reactor* is a device in which energy is generated by a process of nuclear fission.

Reluctance: The ratio between the magnetomotive force acting around a magnetic circuit and the resulting magnetic flux. The reciprocal of permeance.

Remanence: The remanent flux density obtained when the initial magnetisation reaches the saturation value for the material.

Remanent flux density: The magnetic flux density remaining in a material when, after initial magnetisation, the magnetising force is reduced to zero.

Residual magnetism: The magnetism remaining in a material after the magnetising force has been removed.

Resistance: That property of a material by virtue of which it resists the flow of charge through it, causing a dissipation of energy as heat. It is equal to the constant potential difference divided by the current produced thereby when the material has no e.m.f. acting within it.

Resistivity: The resistance between opposite faces of a unit cube of a given material.

Resistor: A device having resistance as a chief property.

Susceptance: The reciprocal of reactance.

Time constant: The characteristic time describing the duration of a transient phenomenon.

Voltage: The same as potential difference.

Voltage gradient: The same as potential gradient.

Wave form: The graph of successive instantaneous values of a time-varying physical quantity.

2.2 Thermal effects

2.2.1 Resistance

That property of an electric circuit which determines for a given current the rate at which electrical energy is converted into heat is termed *resistance*. A device whose chief property is resistance is a *resistor*, or, if variable, a *rheostat*. A current I flowing in a resistance R develops heat at the rate

$P = I^2 R$ joules/s or watts

a relation expressing *Joule's law*.

2.2.1.1 Voltage applied to a resistor

In the absence of any energy storage effects (a physically unrealisable condition), the current in a resistor of value R is I when the voltage across it is V, in accordance with the relation $I = V/R$. If a *steady* p.d. V be suddenly applied to a resistor R, the current instantaneously assumes the value given, and energy is expended at the rate $P = I^2 R$ watts, continuously. No transient occurs. If a constant frequency, constant amplitude *sine wave* voltage v is applied, the current i is at every instant given by $i = v/R$, and in consequence the current has also a sine wave form, provided that the resistance is linear. The instantaneous rate of energy dissipation depends on the instantaneous current: it is $p = vi = i^2 R$. Should the applied voltage be non-sinusoidal, the current has (under the restriction mentioned) an exactly similar wave form. The three cases are illustrated in *Figure 2.3*.

In the case of alternating wave form, the average rate of energy dissipation is given by $P = I^2 R$, where I is the *root-mean-square* current value.

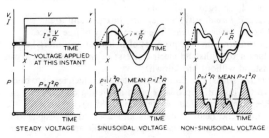

Figure 2.3 Voltage applied to a pure resistor

2.2.1.2 Voltage–current characteristics

For a given resistor R carrying a constant current I, the p.d. is $V = IR$. The ratio $R = V/I$ may or may not be invariable. In some cases it is sufficient to assume a degree of constancy, and calculation is generally made on this assumption. Where the variations of resistance are too great to make the assumption reasonably valid, it is necessary to resort to less simple analysis or to graphical methods.

A constant resistance is manifested by a constant ratio between the voltage across it and the current through it, and by a straight-line graphical relation between I and V (*Figure 2.4a*), where $R = V/I = \cot\theta$. This case is typical of metallic resistance wires at constant temperature.

Certain circuits exhibit *non-linear* current–voltage relations (*Figure 2.4b*). The non-linearity may be *symmetrical* or *asymmetrical*, in accordance with whether the conduction characteris-

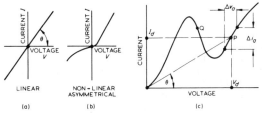

Figure 2.4 Current–voltage characteristics

Conductance: For steady direct currents, the reciprocal of the resistance. For sinusoidal alternating currents, the resistance divided by the square of the impedance.

Conductivity: The reciprocal of resistivity.

Core loss (iron loss): The loss in a magnetic body subject to changing magnetisation, resulting from hysteresis and eddy current effects.

Current: The flow or transport of electric charges along a path or around a circuit.

Current density: The current per unit area of a conductor, or per unit width of an extended conductor.

Diamagnetic: Having a permeability less than that of free space.

Dielectric loss: The loss in an insulating body, resulting from hysteresis, conduction and absorption.

Displacement current: The current equivalent of the rate of change of electric flux with time.

Eddy current: The current electromagnetically induced in a conductor lying in a changing magnetic field.

Electric field: The energetic state of the space between two oppositely charged conductors.

Electric field strength: The mechanical force per unit charge on a very small charge placed in an electric field. The negative voltage gradient.

Electric flux: The electric field, equal to the charge, between oppositely charged conductors.

Electric flux density: The electric flux per unit area.

Electric space constant: The permittivity of free space.

Electric strength: The property of an insulating material which enables it to withstand electric stress; or the stress that it can withstand without breakdown.

Electric stress: The electric field intensity, which tends to break down the insulating property of an insulating material.

Electromagnetic field: A travelling field having electric field and magnetic field components and a speed of propagation depending on the electrical properties of the ambient medium.

Electromagnetic induction: The production of an electromotive force in a circuit by a change of magnetic linkage through the circuit. The e.m.f. so produced is an *induced e.m.f.*, and any current that may result therefrom is an *induced current*.

Electromotive force (e.m.f.): That quality which tends to cause a movement of charges around a circuit. The direction is that of the movement of positive charges. E.m.f. is measured by the amount of energy developed by transfer of unit positive charge. The term is applied to sources that convert electrical energy to or from some other kind (chemical, mechanical, thermal, etc.).

Ferromagnetic: Having a permeability much greater than that of free space, and varying with the magnetic flux density.

Force: The cause of the mechanical displacement, motion, acceleration and deformation of massive bodies.

Frequency: The number of repetitions of a cyclically time-varying quantity in unit time.

Hysteresis: The phenomenon by which an effect in a body depends not only on the present cause, but also on the previous state of the body. In *magnetisation* a flux density produced by a given magnetic field intensity depends on the previous magnetisation history. A comparable effect occurs in the *electrification* of insulating materials. In cyclic changes hysteresis is the cause of energy loss.

Immittance: A circuit property that can be either impedance or admittance.

Impedance: The ratio between voltage and current in r.m.s. terms for sinusoidally varying quantities.

Impedance operator: The ratio between voltage and current in operational terms.

Inductance: The property of a circuit by virtue of which the passage of a current sets up magnetic linkage and stores magnetic energy. If the linkage of a circuit arises from the current in another circuit, the property is called mutual inductance.

Inductor: A device having inductance as a chief property.

Insulation resistance: The resistance under prescribed conditions between two conductors or conducting systems normally separated by an insulating medium.

I^2R *loss (copper loss):* The loss (converted into heat) due to the passage of a current through the resistance of a conductor.

Line of flux (line of force): A line drawn in a field to represent the direction of the flux at any point.

Linkage: The summation of the products of elements of magnetic flux and the number of turns of the circuit they embrace in a given direction.

Loss angle: The phase angle by which the current in a reactor fails to lead (or lag) the voltage by $\frac{1}{2}\pi$ rad under sinusoidal conditions. The tangent of this angle is called the *loss tangent*.

Magnetic circuit: The closed path followed by a magnetic flux.

Magnetic field: The energetic state of the space surrounding an electric current.

Magnetic field strength: The cause at any point of a magnetic circuit of the magnetic flux density there.

Magnetic flux: The magnetic field, equal to the summation of flux density and area, around a current. A phenomenon in the neighbourhood of currents or magnets. The magnetic flux through any area is the surface integral of the magnetic flux density through the surface. Unit magnetic flux is that flux, the removal of which from a circuit of unit resistance causes unit charge to flow in the circuit; or in an open turn produces a voltage–time integral of unity.

Magnetic flux density: The magnetic flux per unit area at a point in a magnetic field, the area being oriented to give a maximum value to the flux. The normal to the area is the direction of the flux at the point. The direction of the current produced in the electric circuit on removal of the flux, and the positive direction of the flux, have the relation of a right-handed screw.

Magnetic leakage: That part of a magnetic flux which follows such a path as to make it ineffective for the purpose desired.

Magnetic potential difference: A difference between the magnetic states existing at two points which produces a magnetic field between them. It is equal to the line integral of the magnetic field intensity between the points, except in the presence of electric currents.

Magnetic space constant: The permeability of free space.

Magnetising force: The same as magnetic field strength.

Magnetomotive force: Along any path, the line integral of the magnetic field strength along that path. If the path is closed, the line integral is equal to the total magnetising current in ampere-turns.

Paramagnetic: Having a permeability greater than that of free space.

Period: The time taken by one complete cycle of a wave form.

Permeability: The ratio of the magnetic flux density in a medium or material at a point to the magnetic field strength at the point. The *absolute permeability* is the product of the *relative permeability* and the *permeability of free space* (magnetic space constant).

Permeance: The ratio between the magnetic flux in a magnetic circuit and the magnetomotive force. The reciprocal of reluctance.

Permittivity: The ratio between the electric flux density in a medium or material at a point and the electric field strength at the point. The *absolute permittivity* is the product of the *relative permittivity* and the *permittivity of free space* (electric space constant).

Phase angle: The angle between the phasors that represent two alternating quantities of sinusoidal wave form and the same frequency.

Phasor: A sinusoidally varying quantity represented in the

Table 2.1 Electrotechnical symbols and units

Quantity	Quantity symbol	Unit name	Unit symbol
Admittance	Y	siemens	S
Ampere-turn	—	ampere-turn	At
Angular frequency	$\omega = 2\pi f$	radian/second	rad/s
Capacitance	C	farad	F
Charge	Q	coulomb	C
Conductance	G	siemens	S
Conductivity	γ, σ	siemens/metre	S/m
Current	I	ampere	A
Current density, linear	A	ampere/metre	A/m
Current density, surface	J	ampere/metre-square	A/m^2
Electric field strength	E	volt/metre	V/m
Electric flux	Q	coulomb	C
Electric flux density	D	coulomb/metre-square	C/m^2
Electric space constant	ε_0	farad/metre	F/m
Electromotive force	E	volt	V
Force	F, f	newton	N
Frequency	f	hertz	Hz
Impedance	Z	ohm	Ω
Inductance, mutual	L_{jk}, M	henry	H
Inductance, self-	L	henry	H
Linkage	ψ	weber-turn	Wbt
Loss angle	δ	radian	rad
Magnetic field strength	H	ampere/metre	A/m
Magnetic flux	Φ	weber	Wb
Magnetic flux density	B	tesla	T
Magnetic space constant	$\mu_0 = 4\pi/10^7$	henry/metre	H/m
Magnetomotive force	F	ampere (-turn)	A, At
Period	T	second	s
Permeability, absolute	$\mu = \mu_r \mu_0$	henry/metre	H/m
Permeability, free space	μ_0	henry/metre	H/m
Permeability, relative	μ_r	—	—
Permeance	Λ	weber/ampere (-turn)	Wb/A, Wb/At
Permittivity, absolute	$\varepsilon = \varepsilon_r \varepsilon_0$	farad/metre	F/m
Permittivity, free space	ε_0	farad/metre	F/m
Permittivity, relative	ε_r	—	—
Phase angle	ϕ	radian	rad
Potential	V	volt	V
Potential difference	V, U	volt	V
Potential gradient	E	volt/metre	V/m
Power, active	P	watt	W
Power, apparent	S	volt-ampere	VA
Power, reactive	Q	var	var
Quantity	Q	coulomb	C
Reactance	X	ohm	Ω
Reluctance	S	ampere (-turn)/weber	A/Wb, At/Wb
Resistance	R	ohm	Ω
Resistivity	ρ	ohm-metre	Ωm
Susceptance	B	siemens	S
Time constant	τ	second	s
Voltage	V	volt	V
Voltage gradient	E	volt/metre	V/m

Admittance operator: The ratio between current and voltage in operational terms.

Ampere-turns: The product of the number of turns of a circuit and the current flowing in them.

Angular frequency: The number of periods per second of a periodically varying quantity multiplied by 2π.

Capacitance: The property of a conducting body by virtue of which an electric charge has to be imparted to it to produce a difference of electrical potential between it and the surrounding bodies. The ratio between the charge on a conductor and its potential when all neighbouring conductors are at zero (earth) potential. The ratio between the charge on each electrode of a capacitor and the potential difference between them.

Capacitor: A device having capacitance as a chief property.

Charge: The excess of positive or negative electricity on a body or in space.

Coercive force: The demagnetising force required to reduce to zero the remanent flux density in a magnetic body.

Coercivity: The value of the coercive force when the initial magnetisation has the saturation value.

Complexor: A non-vectorial quantity expressible in terms of a complex number.

Electrotechnology concerns the electrophysical and allied principles applied to practical electrical engineering. A completely general approach is not feasible, and many separate *ad hoc* technologies have been developed using simplified and delimited areas adequate for particular applications.

In establishing a technology it is necessary to consider whether the relevant applications can be dealt with (a) in *macroscopic* terms of physical qualities of materials in bulk (as with metallic conduction or static magnetic fields); or (b) in *microscopic* terms involving the microstructure of materials as an essential feature (as in domain theory); or (c) in molecular, atomic or *subatomic* terms (as in nuclear reaction and semiconduction). There is no rigid line of demarcation, and certain technologies must cope with two or more such subdivisions at once. Electrotechnology thus tends to become an assembly of more or less discrete (and sometimes apparently unrelated) areas in which methods of treatment differ widely.

To a considerable extent (but not completely), the items of plant with which technical electrical engineering deals—generators, motors, feeders, capacitors, etc.—can be represented by *equivalent circuits* or *networks* energised by an electrical *source*.

For the great majority of cases within the purview of 'heavy electrical engineering' (that is, generation, transmission and utilisation for power purposes, as distinct from telecommunications), a *source* of electrical energy is considered to produce a *current* in a conducting *circuit* by reason of an *electromotive force* acting against a property of the circuit called *impedance*. The behaviour of the circuit is described in terms of the energy fed into the circuit by the source, and the nature of the conversion, dissipation or storage of this energy in the several circuit components.

Electrical phenomena, however, are only in part associated with conducting circuits. The generalised basis is one of magnetic and electrical fields in free space or in material media. The fundamental starting point is the conception contained in Maxwell's electromagnetic equations (Section 1.5.3), and in this respect the voltage and currents in a circuit are only representative of the fundamental field phenomena within a restricted range. Fortunately, this range embraces very nearly the whole of 'heavy' electrical engineering practice. The necessity for a more comprehensive viewpoint makes itself apparent in connection with problems of long-line transmission; and when the technique of ultra-high-frequency work is reached, it is necessary to give up the familiar circuit ideas in favour of a whole-hearted application of field principles.

2.1 Nomenclature

2.1.1 Circuit phenomena

Figure 2.1 shows in a simplified form a hypothetical circuit with a variety of electrical energy sources and a representative selection of devices in which the energy received from the source is converted into other forms, or stored, or both. The forms of variation of the current or voltage are shown in *Figure 2.2*. In an actual circuit the current may change in a quite arbitrary fashion as indicated at (*a*): it may rise or fall, or reverse its direction, depending on chance or control. Such random variation is inconveniently difficult to deal with, and engineers prefer to simplify the conditions as much as possible. For example (*Figure 2.2b*), the current may be assumed to be rigidly constant, in which case it is termed a *direct current*. If the current be deemed to reverse cyclically according to a sine function, it becomes a *sinusoidal alternating current* (*c*). Less simple wave forms, such as (*d*), may be dealt with by application of Fourier's theorem, thus making it possible to calculate a great range of practical cases—

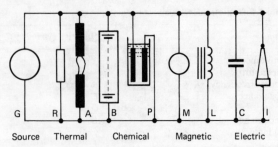

Figure 2.1 Typical circuit devices. G = Source generator; R = resistor; A = arc; B = battery; P = plating bath; M = motor; L = inductor; C = capacitor; I = insulator

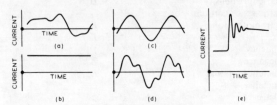

Figure 2.2 Modes of current (or voltage) variation

such as those involving rectifiers—in which the sinusoidal wave form assumption is inapplicable. The cases shown in (*b*), (*c*) and (*d*) are known as *steady states*, the current (or voltage) being assumed established for a considerable time before the circuit is investigated. But since the electric circuit is capable of storing energy, a change in the circuit may alter the conditions so as to cause a redistribution of circuit energy. This occurs with a circulation of *transient* current. An example of a simple oscillatory transient is shown in *Figure 2.2(e)*.

The calculation of circuits in which direct currents flow is comparatively straightforward. For sine wave alternating current circuits an algebra has been developed by means of which problems can be reduced to a technique very similar to that of d.c. circuits. Where non-sinusoidal wave forms are concerned, the treatment is based on the analysis of the current and voltage waves into fundamental and harmonic sine waves, the standard sine wave method being applied to the fundamental and to each of the harmonics. In the case of transients, a more searching investigation may be necessary, but there are a number of common modes in which transients usually occur, and (so long as the circuit is relatively simple) it may be possible to select the appropriate mode by inspection.

Circuit *parameters*—resistance, inductance and capacitance—may or may not be constant. If they are not, approximation, linearising or step-by-step computation is necessary.

2.1.1.1 E.M.F. sources

Any device that develops an e.m.f. capable of sustaining a current in an electric circuit must be associated with some mode of energy conversion into the electrical from some different form. The modes are (1) mechanical/electromagnetic, (2) mechanical/electrostatic, (3) chemical, (4) thermal, (5) photoelectric.

2.1.2 Electrotechnical terms

The following list includes the chief terms in common use. The symbols and units employed are given in *Table 2.1*.

Admittance: The ratio between current and voltage in r.m.s. terms for sinusoidally varying quantities.

2 Electrotechnology

M G Say, PhD, MSc, CEng, ACGI, DIC, FIEE, FRSE
Heriot-Watt University

Contents

2.1 Nomenclature 2/3
    2.1.1 Circuit phenomena 2/3
    2.1.2 Electrotechnical terms 2/3

2.2 Thermal effects 2/6
    2.2.1 Resistance 2/6
    2.2.2 Heating and cooling 2/8

2.3 Electrochemical effects 2/10
    2.3.1 Electrolysis 2/10
    2.3.2 Cells 2/11

2.4 Magnetic field effects 2/11
    2.4.1 Magnetic circuit 2/12
    2.4.2 Magnetomechanical effects 2/14
    2.4.3 Electromagnetic induction 2/14
    2.4.4 Inductance 2/17

2.5 Electric field effects 2/19
    2.5.1 Electrostatics 2/19
    2.5.2 Capacitance 2/20
    2.5.3 Dielectric breakdown 2/21
    2.5.4 Electromechanical effects 2/22

2.6 Electromagnetic field effects 2/22
    2.6.1 Movement of charged particles 2/22
    2.6.2 Free space propagation 2/23
    2.6.3 Transmission line propagation 2/23

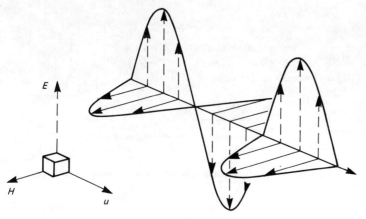

Figure 1.11 Electromagnetic wave

$u = 1/\sqrt{(\varepsilon_0 \mu_0)} = c \simeq 3 \times 10^8$ m/s

$E/H = \sqrt{(\mu_0/\varepsilon_0)} = Z_0 \simeq 377\ \Omega$

The velocity of propagation in free space is thus fixed; the ratio E/H is also fixed, and is called the intrinsic impedance. Further, $\tfrac{1}{2}\varepsilon_0 E^2 = \tfrac{1}{2}\mu_0 H^2$ [J/m³], showing that the electric and magnetic energy densities are equal.

Propagation is normally maintained by charge acceleration which results from a high-frequency alternating current (e.g. in an aerial), so that waves of E and H of sinusoidal distribution are propagated with a wavelength dependent on the frequency (*Figure 1.11*). There is a fixed relation between the directions of E, H and the energy flow. The rate at which energy passes a fixed point is EH [W/m²], and the direction of E is taken as that of the wave polarisation.

Plane wave transmission in a perfect homogeneous loss-free insulator takes place as in free space, except that ε_0 is replaced by $\varepsilon = \varepsilon_r \varepsilon_0$, where ε_r is the relative permittivity of the medium: the result is that both the propagation velocity and the intrinsic impedance are reduced.

When a plane wave from free space enters a material with conducting properties, it is subject to attenuation by reason of the I^2R loss. In the limit, a perfect conductor presents to the incident wave a complete barrier, reflecting the wave as a perfect mirror. A wave incident upon a general medium is partly reflected, and partly transmitted with attenuation and phase-change.

Table 1.24 gives the wavelength and frequency of free space electromagnetic waves with an indication of their technological range and of the physical origin concerned.

Table 1.24 Electromagnetic wave spectrum

Free space properties:

Electric constant	$\varepsilon_0 = 8.854 \times 10^{-12}$	F/m	Intrinsic impedance	$Z_0 = 376.8$	Ω
Magnetic constant	$\mu_0 = 4\pi \times 10^{-7}$	H/m	Velocity	$c = 2.9979 \times 10^8$	m/s

The product of wavelength λ [m] and frequency f [Hz] is $f\lambda = c \simeq 3 \times 10^8$ [m/s].

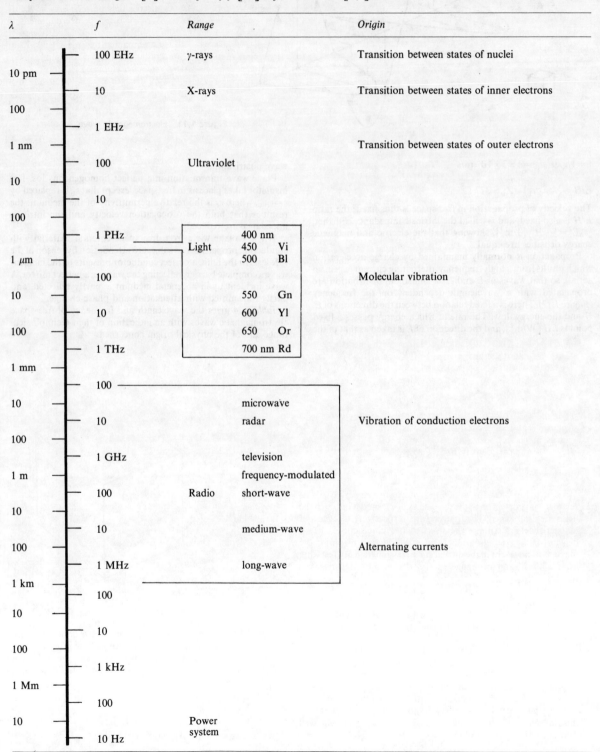

strength and flux density, also surface and path-length elements, are vectorial.

Field	Electric		Magnetic		Conduction	
Potential	V	[V]	F	[A]	V	[V]
Field strength	E	[V/m]	H	[A/m]	E	[V/m]
Flux	Q	[C]	ϕ	[Wb]	I	[A]
Flux density	D	[C/m^2]	B	[T]	J	[A/m^2]
Material property	ε	[F/m]	μ	[H/m]	σ	[S/m]

The total electric flux emerging from a charge $+Q$ or entering a charge $-Q$ is equal to Q. The integral of the electric flux density D over a closed surface s enveloping the charge is

$$\int_s D \cdot ds = Q \tag{1.1}$$

If the surface has no enclosed charge, the integral is zero. This is the Gauss law.

The magnetomotive force F, or the line integral of the magnetic field strength H around a closed path l, is equal to the current enclosed, i.e.

$$\int_o H \cdot dl = F = i_c + i_d \tag{1.2}$$

This is the Ampère law with the addition of displacement current.

The Faraday law states that, around any closed path l encircling a magnetic flux ϕ that changes with time, there is an electric field, and the line integral of the electric field strength E around the path is

$$\int_o E \cdot dl = e = -(d\phi/dt) \tag{1.3}$$

Magnetic flux is a solenoidal quantity, i.e. it comprises a structure of closed loops; over any closed surface s in a magnetic field as much flux leaves the surface as enters it. The surface integral of the flux density B is therefore always zero, i.e.

$$\int_s B \cdot ds = 0 \tag{1.4}$$

To these four laws are added the *constitutive equations*, which relate the flux densities to the properties of the media in which the fields are established. The first two are, respectively, electric and magnetic field relations; the third relates conduction current density to the voltage gradient in a conducting medium; the fourth is a statement of the displacement current density resulting from a time rate of change of the electric flux density. The relations are

$$D = \varepsilon E; \quad B = \mu H; \quad J_c = \sigma E; \quad J_d = \partial D/\partial t$$

In electrotechnology concerned with direct or low-frequency currents, the Maxwell equations are rarely used in the form given above. Equation (1.2), for example, appears as the number of amperes (or ampere-turns) required to produce in an area a the specified magnetic flux $\phi = Ba = \mu Ha$. Equation (1.3) in the form $e = -(d\phi/dt)$ gives the e.m.f. in a transformer primary or secondary turn. The concept of the 'magnetic circuit' embodies Equation (1.4). But when dealing with such field phenomena as the eddy currents in massive conductors, radio propagation or the transfer of energy along a transmission line, the Maxwell equations are the basis of analysis.

1.5.3.2 Electromagnetic wave

The local 'induction' field of a charge at rest surrounds it in a predictable pattern. Let the position of the charge be suddenly displaced. The field pattern also moves, but because of the finite rate of propagation there will be a region in which the original field has not yet been supplanted by the new. At the instantaneous boundary the electric field pattern may be pictured as 'kinked', giving a transverse electric field component that travels away from the charge. Energy is propagated, because the transverse electric field is accompanied by an associated transverse magnetic field in accordance with the Ampère law.

Consider a unit cube of *free space (Figure 1.10)* approached by a transverse electric field of strength E at a velocity u in the

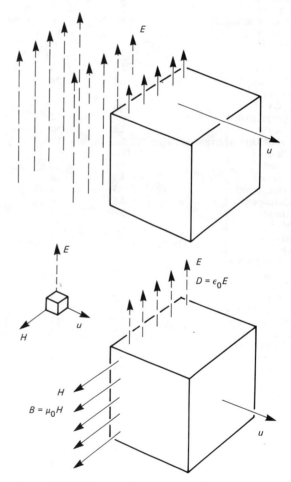

Figure 1.10 Electromagnetic wave propagation

specified direction. As E enters the cube, it produces therein an electric flux, of density $D = \varepsilon_0 E$ increasing at the rate Du. This is a displacement current which produces a magnetic field of strength H and flux density $B = \mu_0 H$ increasing at the rate Bu. Then the E and H waves are mutually dependent:

$$Du = \varepsilon_0 Eu = H \tag{1.5}$$

$$Bu = \mu_0 Hu = E \tag{1.6}$$

Multiplication and division of (1.5) and (1.6) give

Table 1.23 Electrical properties of insulating materials

Typical approximate values (see also Section 1.4):

ε_r	relative permittivity	
E	electric strength	[MV/m]
tan δ	loss tangent	
θ	maximum working temperature	[°C]
k	thermal conductivity	[mW/(m K)]
G	density	[kg/m³]

Material	ε_r	E	tan δ	θ	k	G
Air at n.t.p.	1.0	3	—	—	25	1.3
Alcohol	26	—	—	—	180	790
Asbestos	2	2	—	400	80	3000
paper	2	2	—	250	250	1200
Bakelite moulding	4	6	0.03	130	—	1600
paper	5	15	0.03	100	270	1300
Bitumen						
pure	2.7	1.6	—	50	150	1200
vulcanised	4.5	5	—	100	200	1250
Cellulose film	5.8	28	—	—	—	800
Cotton fabric						
dry	—	0.5	—	95	80	—
impregnated	—	2	—	95	250	—
Ebonite	2.8	50	0.005	80	150	1400
Fabric tape, impregnated	5	17	0.1	95	240	—
Glass						
flint	6.6	6	—	—	1100	4500
crown	4.8	6	0.02	—	600	2200
toughened	5.3	9	0.003	—	—	—
Gutta-percha	4.5	—	0.02	—	200	980
Marble	7	2	0.03	—	2600	2700
Mica	6	40	0.02	750	600	2800
Micanite	—	15	—	125	150	2200
Oil						
transformer	2.3	—	—	85	160	870
castor	4.7	8	—	—	—	970
Paper						
dry	2.2	5	0.007	90	130	820
impregnated	3.2	15	0.06	90	140	1100
Porcelain	5.7	15	0.008	1000	1000	2400
Pressboard	6.2	7	—	95	170	1100
Quartz						
fused	3.5	13	0.002	1000	1200	2200
crystalline	4.4	—	—	—	—	2700
Rubber						
pure	2.6	18	0.005	50	100	930
vulcanised	4	10	0.01	70	250	1500
moulding	4	10	—	70	—	—
Resin	3	—	—	—	—	1100
Shellac	3	11	—	75	250	1000
paper	5.5	11	0.05	80	—	1350
Silica, fused	3.6	14	—	—	—	—
Silk	—	—	—	95	60	1200
Slate	—	0.5	—	—	2000	2800
Steatite	—	0.6	—	1500	2000	2600
Sulphur	4	—	0.0003	100	220	2000
Water	70	—	—	—	570	1000
Wax (paraffin)	2.2	12	0.0003	35	270	860

along a copper conductor carrying an alternating current, but the conduction current is vastly greater even at very high frequencies. In poor conductors and in insulating materials the displacement current is comparable to (or greater than) the conduction current if the frequency is high enough. In free space and in a perfect insulator only displacement current is concerned.

Equations The following symbols are used, the SI unit of each appended. The permeability and permittivity are absolute values ($\mu = \mu_r \mu_0, \varepsilon = \varepsilon_r \varepsilon_0$). Potentials and fluxes are scalar quantities; field

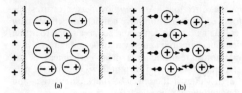

Figure 1.9 Polarisation and breakdown in insulator

electric field exerts opposite mechanical forces on the negative and positive charges and the atoms become more and more highly strained (*Figure 1.9a*). On the left face the atoms will all present their negative charges at the surface: on the right face, their positive charges. These surface polarisations are such as to account for the effect known as *permittivity*. The small displacement of the atomic electric charges constitutes a *polarisation current*. *Figure 1.9(b)* shows that, for excessive electric field strength, conduction can take place, resulting in insulation breakdown.

The electrical properties of metallic conductors and of insulating materials are listed in *Tables 1.22* and *1.23*.

1.5.2.6 Convection current

Charges can be moved mechanically, on belts, water-drops, dust and mist particles, and by beams of high-speed electrons (as in a cathode ray oscilloscope). Such movement, independent of an electric field, is termed a *convection current*.

1.5.3 Charges in acceleration

Reference has been made to the emission of energy (photons) when an electron falls from an energy level to a lower one. Radiation has both a particle and a wave nature, the latter associated with energy propagation through empty space and through transparent media.

1.5.3.1 Maxwell equations

Faraday postulated the concept of the field to account for 'action at a distance' between charges and between magnets. Maxwell (1873) systematised this concept in the form of electromagnetic field equations. These refer to media *in bulk*. They naturally have no direct relation to the electronic nature of conduction, but deal with the fluxes of electric, magnetic and conduction fields, their flux densities, and the bulk material properties (permittivity ε, permeability μ and conductivity σ) of the media in which the fields exist. To the work of Faraday. Ampère and Gauss, Maxwell added the concept of displacement current.

Displacement current Around an electric field that changes with time there is evidence of a magnetic field. By analogy with the magnetic field around a conduction current, the rate of change of an electric field may be represented by the presence of a *displacement current*. The concept is applicable to an electric circuit containing a capacitor: there is a conduction current i_c in the external circuit but not between the electrodes of the capacitor. The capacitor, however, must be acquiring or losing charge and its electric field must be changing. If the rate of change is represented by a displacement current $i_d = i_c$, not only is the magnetic field accounted for, but also there now exists a 'continuity' of current around the circuit.

Displacement current is present in any material medium, conducting or insulating, whenever there is present an electric field that changes with time. There is a displacement current

Table 1.22 Electrical properties of conductors

Typical approximate values at 293 K (20°C):
 g conductivity relative to I.S.A.C. [%]
 ρ resistivity [nΩ m]
 α resistance–temperature coefficient [mΩ/(Ω K)]

Material		g	ρ	α
International standard annealed copper (ISAC)		100	17.2	3.93
Copper				
annealed		99	17.3	3.90
hard-drawn		97	17.7	3.85
Brass (60/40)				
cast		23	75	1.6
rolled		19	90	1.6
Bronze		48	36	1.65
Phosphor-bronze		29–14	6–12	1.0
Cadmium-copper, hard-drawn		82–93	21–18	4.0
Copper-clad steel, hard-drawn		30–40	57–43	3.75
Aluminium				
cast		66	26	3.90
hard-drawn		62	28	3.90
duralumin		36	47	—
Iron				
wrought		16	107	5.5
cast				
grey		2.5	700	—
white		1.7	1000	2.0
malleable		5.9	300	—
nomag		1.1	1600	4.5
Steel				
0.1% C		8.6	200	4.2
0.4% C		11	160	4.2
core				
1% Si		10	170	—
2% Si		4.9	350	—
4% Si		3.1	550	—
wire				
galvanised		12	140	4.4
45 ton		10	170	3.4
80 ton		8	215	3.4
Resistance alloys*				
80 Ni, 20 Cr	(1)	1.65	1090	0.1
59 Ni, 16 Cr, 25 Fe	(2)	1.62	1100	0.2
37 Ni, 18 Cr, 2 Si, 43 Fe	(3)	1.89	1080	0.26
45 Ni, 54 Cu	(4)	3.6	490	0.04
20 Ni, 80 Cu	(5)	6.6	260	0.29
15 Ni, 62 Cu, 22 Zn	(6)	5.0	340	0.25
4 Ni, 84 Cu, 12 Mn	(7)	3.6	480	0.0
Gold		73	23.6	3.0
Lead		7.8	220	4.0
Mercury		1.8	955	0.7
Molybdenum		30	57	4.0
Nickel		12.6	136	5.0
Platinum		14.7	117	3.9
Silver				
annealed		109	15.8	4.0
hard-drawn		98.5	17.5	4.0
Tantalum		11.1	155	3.1
Tungsten		31	56	4.5
Zinc		28	62	4.0

* Resistance alloys: (1) furnaces, radiant elements; (2) electric irons, tubular heaters; (3) furnace elements; (4) control resistors; (5) cupro; (6) German silver, platinoid; (7) Manganin.

very slow drift (measurable in mm/s) of an immense number of electrons.

A current may be the result of a two-way movement of positive and negative particles. Conventionally the direction of current flow is taken as the same as that of the positive charges and against that of the negative ones.

1.5.2.2 Metals

Reference has been made above to the 'electron atmosphere' of electrons in random motion within a lattice of comparatively rigid molecular structure in the case of copper, which is typical of the class of good metallic conductors. The random electronic motion, which intensifies with rise in temperature, merges into an average shift of charge of almost (but not quite) zero continuously (*Figure 1.7*). When an electric field is applied along the length of a

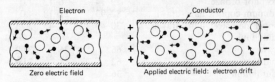

Figure 1.7 Electronic conduction in metals

conductor (as by maintaining a potential difference across its ends), the electrons have a *drift* towards the positive end superimposed upon their random digressions. The drift is slow, but such great numbers of electrons may be involved that very large currents, entirely due to electron drift, can be produced by this means. In their passage the electrons are impeded by the molecular lattice, the collisions producing heat and the opposition called *resistance*. The conventional direction of current flow is actually opposite to that of the drift of charge, which is exclusively electronic.

1.5.2.3 Liquids

Liquids are classified according to whether they are *non-electrolytes* (non-conducting) or *electrolytes* (conducting). In the former the substances in solution break up into electrically balanced groups, whereas in the latter the substances form ions, each a part of a single molecule with either a positive or a negative charge. Thus, common salt, NaCl, in a weak aqueous solution breaks up into sodium and chlorine ions. The sodium ion Na^+ is a sodium atom less one electron; the chlorine ion Cl^- is a chlorine atom with one electron more than normal. The ions attach themselves to groups of water molecules. When an electric field is applied, the sets of ions move in opposite directions, and since they are much more massive than electrons, the conductivity produced is markedly inferior to that in metals. Chemical actions take place in the liquid and at the electrodes when current passes. Faraday's Electrolysis Law states that the mass of an ion deposited at an electrode by electrolyte action is proportional to the quantity of electricity which passes and to the *chemical equivalent* of the ion.

1.5.2.4 Gases

Gaseous conduction is strongly affected by the pressure of the gas. At pressures corresponding to a few centimetres of mercury gauge, conduction takes place by the movement of positive and negative ions. Some degree of ionisation is always present due to stray radiations (light, etc.). The electrons produced attach themselves to gas atoms and the sets of positive and negative ions drift in opposite directions. At very low gas pressures the electrons produced by ionisation have a much longer free path before they collide with a molecule, and so have scope to attain high velocities. Their motional energy may be enough to *shock-ionise* neutral atoms, resulting in a great enrichment of the electron stream and an increased current flow. The current may build up to high values if the effect becomes cumulative, and eventually conduction may be effected through a *spark* or *arc*.

In a *vacuum* conduction can be considered as purely electronic, in that any electrons present (there can be no *molecular* matter present in a perfect vacuum) are moved in accordance with the force exerted on them by an applied electric field. The number of electrons is small, and although high speeds may be reached, the conduction is generally measurable only in milli- or microamperes.

Some of the effects are illustrated in *Figure 1.8*, representing part of a vessel containing a gas or vapour at low pressure. At the

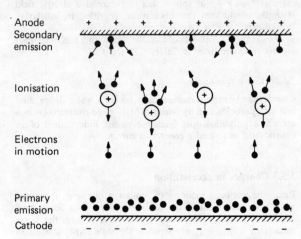

Figure 1.8 Conduction in low-pressure gas

bottom is an electrode, the *cathode*, from the surface of which electrons are emitted, generally by heating the cathode material. At the top is a second electrode, the *anode*, and an electric field is established between the electrodes. The field causes electrons emitted from the cathode to move upward. In their passage to the anode these electrons will encounter gas molecules. If conditions are suitable, the gas atoms are ionised, becoming in effect positive charges associated with the nuclear mass. Thereafter the current is increased by the detached electrons moving upwards and by the positive ions moving more slowly downwards. In certain devices (such as the mercury arc rectifier) the impact of ions on the cathode surface maintains its emission. The impact of electrons on the anode may be energetic enough to cause the *secondary emission* of electrons from the anode surface. If the gas molecules are excluded and a vacuum is established, the conduction becomes purely electronic.

1.5.2.5 Insulators

If an electric field is applied to a perfect insulator, whether solid, liquid or gaseous, the electric field affects the atoms by producing a kind of 'stretching' or 'rotation' which displaces the electrical centres of negative and positive in opposite directions. This polarisation of the dielectric insulating material may be considered as taking place in the manner indicated in *Figure 1.9*. Before the electric field is applied, the atoms of the insulator are neutral and unstrained; as the potential difference is raised the

to produce variants of crystal structure and consequent magnetic characteristics.

1.4 Physical properties

The nature, characteristics and properties of materials arise from their atomic and molecular structure. Tables of approximate values for the physical properties of metals, non-metals, liquids and gases are appended, together with some characteristic temperatures and the numerical values of general physical constants.

1.5 Electricity

In the following paragraphs electrical phenomena are described in terms of the effects of electric charge, at a level adequate for the purpose of simple explanation.

In general, charges may be at rest, or in motion, or in acceleration. *At rest*, charges have around them an electric (or *electrostatic*) field of force. *In motion* they constitute a current, which is associated with a magnetic (or *electrodynamic*) field of force additional to the electric field. In *acceleration*, a third field component is developed which results in energy propagation by *electromagnetic waves*.

1.5.1 Charges at rest

Figure 1.5 shows two bodies in air, charged by applying between them a potential difference, or (having been in close contact) by forcibly separating them. Work must have been done in a

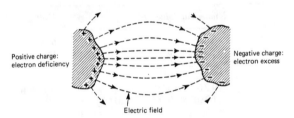

Figure 1.5 Charged conductors and their electric field

physical sense to produce on one an excess and on the other a deficiency of electrons, so that the system is a repository of potential energy. (The work done in separating charges is measured by the product of the charges separated and the difference of electrical potential that results.) Observation of the system shows certain effects of interest: (1) there is a difference of electric potential between the bodies depending on the amount of charge and the geometry of the system; (2) there is a mechanical force of attraction between the bodies. These effects are deemed to be manifestations of the *electric field* between the bodies, described as a special state of space and depicted by *lines of force* which express in a pictorial way the strength and direction of the force effects. The lines stretch between positive and negative elements of charge through the medium (in this case, air) which separates the two charged bodies. The electric field is only a concept—for the lines have no real existence—used to calculate various effects produced when charges are separated by any method which results in excess and deficiency states of atoms by electron transfer. Electrons and protons, or electrons and positively ionised atoms, attract each other, and the stability of the atom may be considered due to the balance of these attractions and dynamic forces such as electron spin. Electrons are repelled by electrons and protons by protons, these forces being summarised in the rules, formulated experimentally long before our present knowledge of atomic structure, that 'like charges repel and unlike charges attract one another'.

1.5.2 Charges in motion

In substances called *conductors*, the outer shell electrons can be more or less freely interchanged between atoms. In copper, for example, the molecules are held together comparatively rigidly in the form of a 'lattice'—which gives the piece of copper its permanent shape—through the interstices of which outer electrons from the atoms can be interchanged within the confines of the surface of the piece, producing a random movement of free electrons called an 'electron atmosphere'. Such electrons are responsible for the phenomenon of electrical conductivity.

In other substances called *insulators*, all the electrons are more or less firmly bound to their parent atoms, so that little or no relative interchange of electron charges is possible. There is no marked line of demarcation between conductors and insulators, but the copper group metals, in the order silver, copper, gold, are outstanding in the series of conductors.

1.5.2.1 Conduction

Conduction is the name given to the movement of electrons, or ions, or both, giving rise to the phenomena described by the term *electric current*. The effects of a current include a redistribution of charges, heating of conductors, chemical changes in liquid solutions, magnetic effects, and many subsidiary phenomena.

If at a specified point on a conductor (*Figure 1.6*) n_1 carriers of electric charge (they can be water-drops, ions, dust particles, etc.)

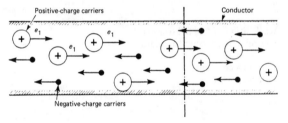

Figure 1.6 Conduction by charge carriers

each with a positive charge e_1 arrive per second, and n_2 carriers (such as electrons) each with a negative charge e_2 arrive in the opposite direction per second, the total rate of passing of charge is $n_1 e_1 + n_2 e_2$, which is the charge per second or *current*. A study of conduction concerns the kind of carriers and their behaviour under given conditions. Since an electric field exerts mechanical forces on charges, the application of an electric field (i.e. a potential difference) between two points on a conductor will cause the movement of charges to occur, i.e. a current to flow, so long as the electric field is maintained.

The discontinuous particle nature of current flow is an observable factor. The current carried by a number of electricity carriers will vary slightly from instant to instant with the number of carriers passing a given point in a conductor. Since the electron charge is 1.6×10^{-19} C, and the passage of one coulomb per second (a rate of flow of *one ampere*) corresponds to $10^{19}/1.6 = 6.3 \times 10^{18}$ electron charges per second, it follows that the discontinuity will be observed only when the flow comprises the very rapid movement of a few electrons. This may happen in gaseous conductors, but in metallic conductors the flow is the

paramagnetic. The spin effect may, in certain cases, be very large, and high magnetisations are produced by an external field: such materials are *ferromagnetic*.

An iron atom has, in the $n=4$ shell (N), electrons that give it conductive properties. The K, L and N shells have equal numbers of electrons possessing opposite spin directions, so cancelling. But shell M contains 9 electrons spinning in one direction and 5 in the other, leaving 4 net magnetons. Cobalt has 3, and nickel 2. In a solid metal further cancellation occurs and the average number of unbalanced magnetons is: Fe, 2.2; Co, 1.7; Ni, 0.6.

In an iron crystal the magnetic axes of the atoms are aligned, unless upset by excessive thermal agitation. (At 770 °C for Fe, the Curie point, the directions become random and ferromagnetism is lost.) A single Fe crystal magnetises most easily along a cube edge of the structure. It does not exhibit spontaneous magnetisation like a permanent magnet, however, because a crystal is divided into a large number of *domains* in which the various magnetic directions of the atoms form closed paths. But if a crystal is exposed to an external applied magnetic field, (a) the electron spin axes remain initially unchanged, but those domains having axes in the favourable direction grow at the expense of the others (domain wall displacement); and (b) for higher field intensities the spin axes orientate into the direction of the applied field.

If wall movement makes a domain acquire more internal energy, then the movement will relax again when the external field is removed. But if wall movement results in loss of energy, the movement is non-reversible—i.e. it needs external force to reverse it. This accounts for hysteresis and remanence phenomena.

The closed-circuit self-magnetisation of a domain gives it a mechanical strain. When the magnetisation directions of individual domains are changed by an external field, the strain directions alter too, so that an assembly of domains will tend to lengthen or shorten. Thus, readjustments in the crystal lattice occur, with deformations (e.g. 20 parts in 10^6) in one direction. This is the phenomenon of *magnetostriction*.

The practical art of magnetics consists in control of magnetic properties by alloying, heat treatment and mechanical working

Table 1.20 Characteristic temperatures
Temperature T [kelvin] corresponds to $\theta_c = T - 273.15$ [degree Celsius] and to $\theta_f = \theta_c(9/5) - 32$ [degree Fahrenheit].

Condition	T	θ_c	θ_f
Absolute zero	0	−273.15	−459.7
Boiling point of oxygen	90.18	−182.97	−297.3
Zero of Fahrenheit scale	255.4	−17.78	0
Melting point of ice	273.15	0	32.0
Triple point of water	273.16	0.01	32.02
Maximum density of water	277.13	3.98	39.16
'Normal' ambient	293.15	20	68
Boiling point of water	373.15	100	212
Boiling point of sulphur	717.8	444.6	832
Freezing point of silver	1234	962	1762
Freezing point of gold	1336	1064	1945

Table 1.21 General physical constants (approximate values, to five significant figures)

Quantity	Symbol	Numerical value	Unit
Acceleration of free fall (standard)	g_n	9.8066	m/s^2
Atmospheric pressure (standard)	p_0	1.0132×10^5	Pa
Atomic mass unit	u	1.6606×10^{-27}	kg
Avogadro constant	N_A	6.0220×10^{23}	mol^{-1}
Bohr magneton	μ_B	9.2741×10^{-24}	J/T, A m^2
Boltzmann constant	k	1.3807×10^{-23}	J/K
Electron			
charge	$-e$	1.6022×10^{-19}	C
mass	m_e	9.1095×10^{-31}	kg
charge/mass ratio	e/m_e	1.7588×10^{11}	C/kg
Faraday constant	F	9.6485×10^4	C/mol
Free space			
electric constant	ε_0	8.8542×10^{-12}	F/m
intrinsic impedance	Z_0	376.7	Ω
magnetic constant	μ_0	$4\pi \times 10^{-7}$	H/m
speed of electromagnetic waves	c	2.9979×10^8	m/s
Gravitational constant	G	6.6732×10^{-11}	N m^2/kg^2
Ideal molar gas constant	R	8.3144	J/(mol K)
Molar volume at s.t.p.	V_m	2.2414×10^{-2}	m^3/mol
Neutron rest mass	m_n	1.6748×10^{-27}	kg
Planck constant	h	6.6262×10^{-34}	J s
normalised	$h/2\pi$	1.0546×10^{-34}	J s
Proton			
charge	$+e$	1.6022×10^{-19}	C
rest mass	m_p	1.6726×10^{-27}	kg
charge/mass ratio	e/m_p	0.9579×10^8	C/kg
Radiation constants	c_1	3.7418×10^{-16}	W m^2
	c_2	1.4388×10^{-2}	m K
Rydberg constant	R_H	1.0968×10^7	m^{-1}
Stefan–Boltzmann constant	σ	5.6703×10^{-8}	J/(m^2 K^4)
Wien constant	k_W	2.8978×10^{-3}	m K

Table 1.19 Physical properties of gases

Values at 0°C (273 K) and atmospheric pressure:
- δ density [kg/m³]
- v viscosity [μPa s]
- c_p specific heat capacity [kJ/(kg K)]
- c_p/c_v ratio between specific heat capacity at constant pressure and at constant volume
- k thermal conductivity [mW/(m K)]
- T_m melting point [K]
- T_b boiling point [K]

Gas		δ	v	c_p	c_p/c_v	k	T_m	T_b
Air	—	1.293	17.0	1.00	1.40	24	—	—
Ammonia	NH_3	0.771	9.3	2.06	1.32	22	195	240
Carbon dioxide	CO_2	1.977	13.9	0.82	1.31	14	216*	194
Carbon monoxide	CO	1.250	16.4	1.05	1.40	23	68	81
Chlorine	Cl_2	3.214	12.3	0.49	1.36	7.6	171	239
Deuterium	D	0.180	—	—	1.73	—	18	23
Ethane	C_2H_6	1.356	8.6	1.72	1.22	18	89	184
Fluorine	F_2	1.695	—	—	0.75	—	50	85
Helium	He	0.178	18.6	5.1	1.66	144	1.0	4.3
Hydrogen	H_2	0.090	8.5	14.3	1.41	174	14	20
Hydrogen chloride	HCl	1.639	13.8	0.81	1.41	—	161	189
Krypton	Kr	3.740	23.3	—	1.68	8.7	116	121
Methane	CH_4	0.717	10.2	2.21	1.31	30	90	112
Neon	Ne	0.900	29.8	1.03	1.64	46	24	27
Nitrogen	N_2	1.251	16.7	1.04	1.40	24	63	77
Oxygen	O_2	1.429	19.4	0.92	1.40	25	55	90
Ozone	O_3	2.220	—	—	1.29	—	80	161
Propane	C_3H_8	2.020	7.5	1.53	1.13	15	83	231
Sulphur dioxide	SO_2	2.926	11.7	0.64	1.27	8.4	200	263
Xenon	Xe	5.890	22.6	—	1.66	5.2	161	165

* At pressure of 5 atm.

can escape from the surface it must be endowed with an energy not less than $\phi = W - \varepsilon^*$, called the *work function*.

Emission occurs by *surface irradiation* (e.g. with light) of frequency v if the energy quantum hv of the radiation is at least equal to ϕ. The threshold of photoelectric emission is therefore with radiation at a frequency not less than $v = \phi/h$.

Emission takes place at *high temperatures* if, put simply, the kinetic energy of electrons normal to the surface is great enough to jump the potential step W. This leads to an expression for the emission current i in terms of temperature T, a constant A and the thermionic work function ϕ:

$$i = AT^2 \exp(-\phi/kT)$$

Electron emission is also the result of the application of a *high electric field intensity* (of the order 1–10 GV/m) to a metal surface; also when the surface is bombarded with electrons or ions of sufficient kinetic energy, giving the effect of *secondary emission*.

Crystals When atoms are brought together to form a crystal, their individual sharp and well-defined energy levels merge into energy *bands*. These bands may overlap, or there may be gaps in the energy levels available, depending on the lattice spacing and interatomic bonding. Conduction can take place only by electron migration into an empty or partly filled band; filled bands are not available. If an electron acquires a small amount of energy from the externally applied electric field, and can move into an available empty level, it can then contribute to the conduction process.

1.3.2.5 Insulators

In this case the 'distance' (or energy increase Δw in electron-volts) is too large for moderate electric applied fields to endow electrons with sufficient energy, so the material remains an insulator. High temperatures, however, may result in sufficient thermal agitation to permit electrons to 'jump the gap'.

1.3.2.6 Semiconductors

Intrinsic semiconductors (i.e. materials between the good conductors and the good insulators) have a small spacing of about 1 eV between their permitted bands, which affords a low conductivity, strongly dependent on temperature and of the order of one-millionth that of a conductor.

Impurity semiconductors have their low conductivity raised by the presence of minute quantities of foreign atoms (e.g. 1 in 10^8) or by deformations in the crystal structure. The impurities 'donate' electrons of energy level that can be raised into a conduction band (n-type); or they can attract an electron from a filled band to leave a 'hole', or electron deficiency, the movement of which corresponds to the movement of a positive charge (p-type).

1.3.2.7 Magnetism

Modern magnetic theory is very complex, with ramifications in several branches of physics. Magnetic phenomena are associated with moving charges. Electrons, considered as particles, are assumed to possess an axial spin, which gives them the effect of a minute current turn or of a small permanent magnet, called a Bohr *magneton*. The gyroscopic effect of electron spin develops a precession when a magnetic field is applied. If the precession effect exceeds the spin effect, the external applied magnetic field produces less magnetisation than it would in free space, and the material of which the electron is a constituent part is *diamagnetic*. If the spin effect exceeds that due to precession, the material is

Table 1.17 Physical properties of non-metals

Approximate general properties:
δ density [kg/m³]
e linear expansivity [μm/(m K)]
c specific heat capacity [kJ/(kg K)]
k thermal conductivity [W/(m K)]
T_m melting point [K]
ρ resistivity [MΩ m]
ε_r relative permittivity [−]

Material	δ	e	c	k	T_m	ρ	ε_r
Asbestos (packed)	580	—	0.84	0.19	—	—	3
Bakelite	1300	30	0.92	0.20	—	—	7
Concrete (dry)	2000	10	0.92	1.70	—	0.1	—
Diamond	3510	1.3	0.49	165	4000	10^7	—
Glass	2500	8	0.84	0.93	—	10^6	8
Graphite	2250	2	0.69	160	3800	10^{-11}	—
Marble	2700	12	0.88	3	—	10^3	8.5
Mica	2800	3	0.88	0.5	—	10^8	7
Nylon	1140	100	1.7	0.3	—	—	—
Paper	900	—	—	0.18	—	10^4	2
Paraffin wax	890	110	2.9	0.26	—	10^9	2
Perspex	1200	80	1.5	1.9	—	10^{14}	3
Polythene	930	180	2.2	0.3	—	—	2.3
Porcelain	2400	3.5	0.8	1.0	1900	10^6	6
Quartz (fused)	2200	0.4	0.75	0.22	2000	10^{14}	3.8
Rubber	1250	—	1.5	0.15	—	10^7	3
Silicon	2300	7	0.75	—	1690	0.1	2.7

Table 1.18 Physical properties of liquids

Average values at 20 °C (293 K):
δ density [kg/m³]
v viscosity [mPa s]
e cubic expansivity [10^{-3}/K]
c specific heat capacity [kJ/(kg K)]
k thermal conductivity [W/(m K)]
T_m melting point [K]
T_b boiling point [K]
ε_r relative permittivity [−]

Liquid		δ	v	e	c	k	T_m	T_b	ε_r
Acetone	$(CH_3)_2CO$	792	0.3	1.43	2.2	0.18	178	329	22
Benzine	C_6H_6	881	0.7	1.15	1.7	0.14	279	353	2.3
Carbon disulphide	CS_2	1260	0.4	1.22	1.0	0.14	161	319	2.6
Carbon tetrachloride	CCl_4	1600	1.0	1.22	0.8	0.10	250	350	2.2
Ether	$(C_2H_5)_2O$	716	0.2	1.62	2.3	0.14	157	308	4.3
Glycerol	$C_3H_5(OH)_3$	1270	1500	0.50	2.4	0.28	291	563	56
Methanol	CH_3OH	793	0.6	1.20	1.2	0.21	175	338	32
Oil		850	85	0.75	1.6	0.17	—	—	3.0
Sulphuric acid	H_2SO_4	1850	28	0.56	1.4	—	284	599	—
Turpentine	$C_{10}H_{16}$	840	1.5	0.10	1.8	0.15	263	453	2.3
Water	H_2O	1000	1.0	0.18	4.2	0.60	273	373	81

them to 'rebound' with a velocity of random direction but small compared with their average velocities as particles of an electron gas. Just as a difference of electric potential causes a drift in the general motion, so a difference of temperature between two parts of a metal carries energy from the hot region to the cold, accounting for thermal conduction and for its association with electrical conductivity. The free electron theory, however, is inadequate to explain the dependence of conductivity on crystal axes in the metal.

At absolute zero of temperature (zero K = −273 °C) the atoms cease to vibrate, and free electrons can pass through the lattice with little hindrance. At temperatures over the range 0.3–10 K (and usually round about 5 K) the resistance of certain metals, e.g. Zn, Al, Sn, Hg and Cu, becomes substantially zero. This phenomenon, known as *superconductivity*, has not been satisfactorily explained.

Superconductivity is destroyed by moderate magnetic fields. It can also be destroyed if the current is large enough to produce at the surface the same critical value of magnetic field. It follows that during the superconductivity phase the current must be almost purely superficial, with a depth of penetration of the order of 10 μm.

Electron emission A metal may be regarded as a potential 'well' of depth $-V$ relative to its surface, so that an electron in the lowest energy state has (at absolute zero temperature) the energy $W = Ve$ (of the order 10 eV): other electrons occupy levels up to a height ε^* (5–8 eV) from the bottom of the 'well'. Before an electron

Table 1.16 Physical properties of metals

Approximate general properties at normal temperatures:

- δ density [kg/m^3]
- E elastic modulus [GPa]
- e linear expansivity [μm/(m K)]
- c specific heat capacity [kJ/(kg K)]
- k thermal conductivity [W/(m K)]
- T_m melting point [K]
- ρ resistivity [nΩ m]
- α resistance–temperature coefficient [mΩ/(Ω K)]

Metal	δ	E	e	c	k	T_m	ρ	α
Pure metals								
4 Beryllium	1840	300	120	1700	170	1560	33	9.0
11 Sodium	970	—	71	710	130	370	47	5.5
12 Magnesium	1740	44	26	1020	170	920	46	3.8
13 Aluminium	2700	70	24	900	220	930	27	4.2
19 Potassium	860	—	83	750	130	340	67	5.4
20 Calcium	1550	—	22	650	96	1120	43	4.2
24 Chromium	7100	25	8.5	450	43	2170	130	3.0
26 Iron	7860	220	12	450	75	1810	105	6.5
27 Cobalt	8800	210	13	420	70	1770	65	6.2
28 Nickel	8900	200	13	450	70	1730	78	6.5
29 Copper	8930	120	16	390	390	1360	17	4.3
30 Zinc	7100	93	26	390	110	690	62	4.1
42 Molybdenum	10 200	—	5	260	140	2890	56	4.3
47 Silver	10 500	79	19	230	420	1230	16	3.9
48 Cadmium	8640	60	32	230	92	590	75	4.0
50 Tin	7300	55	27	230	65	500	115	4.3
73 Tantalum	16 600	190	6.5	140	54	3270	155	3.1
74 Tungsten	19 300	360	4	130	170	3650	55	4.9
78 Platinum	21 500	165	9	130	70	2050	106	3.9
79 Gold	19 300	80	14	130	300	1340	23	3.6
80 Mercury	13 550	—	180	140	10	230	960	0.9
82 Lead	11 300	15	29	130	35	600	210	4.1
83 Bismuth	9800	32	13	120	9	540	1190	4.3
92 Uranium	18 700	13	—	120	—	1410	220	2.1
Alloys								
Brass (60 Cu, 40 Zn)	8500	100	21	380	120	1170	60	2.0
Bronze (90 Cu, 10 Sn)	8900	100	19	380	46	1280	—	—
Constantan	8900	110	15	410	22	1540	450	0.05
Invar (64 Fe, 36 Ni)	8100	145	2	500	16	1720	100	2.0
Iron, soft (0.2 C)	7600	220	12	460	60	1800	140	—
Iron cast (3.5 C, 2.5 Si)	7300	100	12	460	60	1450	—	—
Manganin	8500	130	16	410	22	1270	430	0.02
Steel (0.85 C)	7800	200	12	480	50	1630	180	—

and continues to Ni, becoming again regular with Cu, and beginning a new irregularity with Rb.

The electron of a hydrogen atom, normally at level 1, can be raised to level 2 by endowing it with a particular quantity of energy most readily expressed as 10.2 eV. (1 eV = 1 electron-volt = 1.6×10^{-19} J is the energy acquired by a free electron falling through a potential difference of 1 V, which accelerates it and gives it kinetic energy.) The 10.2 V is the *first excitation potential* for the hydrogen atom. If the electron is given an energy of 13.6 eV, it is freed from the atom, and 13.6 V is the *ionisation potential*. Other atoms have different potentials in accordance with their atomic arrangement.

1.3.2.4 Electrons in metals

An approximation to the behaviour of metals assumes that the atoms lose their valency electrons, which are free to wander in the ionic lattice of the material to form what is called an electron gas. The sharp energy levels of the free atom are broadened into wide bands by the proximity of others. The potential within the metal is assumed to be smoothed out, and there is a sharp rise of potential at the surface which prevents the electrons from escaping: there is a potential-energy step at the surface which the electrons cannot normally overcome: it is of the order of 10 eV. If this is called W, then the energy of an electron wandering within the metal is $-W + \tfrac{1}{2}mu^2$.

The electrons are regarded as undergoing continual collisions on account of the thermal vibrations of the lattice, and on Fermi–Dirac statistical theory it is justifiable to treat the energy states (which are in accordance with Pauli's principle) as forming an energy continuum. At very low temperatures the ordinary classical theory would suggest that electron energies spread over an almost zero range, but the exclusion principle makes this impossible and even at absolute zero of temperature the energies form a continuum, and physical properties will depend on how the electrons are distributed over the upper levels of this energy range.

Conductivity The interaction of free electrons with the thermal vibrations of the ionic lattice (called 'collisions' for brevity) causes

Table 1.15 Elements: periodic table

Periods	1a	2a	3b	4b	5b	6b	7b	8b	8b	8b	1b	2b	3a	4a	5a	6a	7a	0
I	1 H				Metals										Non-metals			2 He
II	3 Li	4 Be											5 B	6 C	7 N	8 O	9 F	10 Ne
III	11 Na	12 Mg			Transitions								13 Al	14 Si	15 P	16 S	17 Cl	18 Ar
IV	19 K	20 Ca	21 Sc	22 Ti	23 V	24 Cr	25 Mn	26 Fe	27 Co	28 Ni	29 Cu	30 Zn	31 Ga	32 Ge	33 As	34 Se	35 Br	36 Kr
V	37 Rb	38 Sr	39 Y	40 Zr	41 Nb	42 Mo	43 Tc	44 Ru	45 Rh	46 Pd	47 Ag	48 Cd	49 In	50 Sn	51 Sb	52 Te	53 I	54 Xe
VI	55 Cs	56 Ba	57 La	72 Hf	73 Ta	74 W	75 Re	76 Os	77 Ir	78 Pt	79 Au	80 Hg	81 Tl	82 Pb	83 Bi	84 Po	85 At	86 Rn
VII	87 Fr	88 Ra	89 Ac															
		Rare earths			58 Ce	59 Pr	60 Nd	61 Pm	62 Sm	63 Eu	64 Gd	65 Tb	66 Dy	67 Ho	68 Er	69 Tm	70 Yb	71 Lu
		Actinides			90 Th	91 Pa	92 U	93 Np	94 Pu	95 Am	96 Cm	97 Bk	98 Cf	99 Es	100 Fm	101 Md	102 No	103 Lr

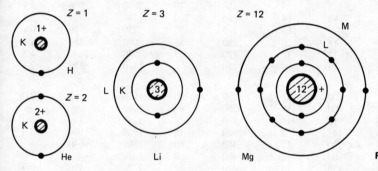

Figure 1.4 Atomic structure

1.3.2.3 Electrons in atoms

Consider the hydrogen atom. Its single electron is not located at a fixed point, but can be anywhere in a region near the nucleus with some probability. The particular region is a kind of shell or cloud, of radius depending on the electron's energy state.

With a nucleus of atomic number Z, the Z electrons can have several possible configurations. There is a certain radial pattern of electron probability cloud distribution (or shell pattern). Each electron state gives rise to a cloud pattern, characterised by a definite energy level, and described by the series of quantum numbers n, l, m_l and m_s. The number $n(=1, 2, 3 \ldots)$ is a measure of the energy level; $l(=0, 1, 2, \ldots)$ is concerned with angular momentum; m_l is a measure of the component of angular momentum in the direction of an applied magnetic field; and m_s arises from the electron spin. It is customary to condense the nomenclature so that electron states corresponding to $l=0, 1, 2$ and 3 are described by the letters s, p, d and f and a numerical prefix gives the value of n. Thus boron has 2 electrons at level 1 with $l=0$, two at level 2 with $l=0$, and one at level 3 with $l=1$: this information is conveyed by the description $(1s)^2(2s)^2(2p)^1$.

The energy of an atom as a whole can vary according to the electron arrangement. The most stable state is that of minimum energy, and states of higher energy content are *excited*. By Pauli's *exclusion principle* the maximum possible number of electrons in states $1, 2, 3, 4, \ldots, n$ are $2, 8, 18, 32, \ldots, 2n^2$, respectively. Thus, only 2 electrons can occupy the $1s$ state (or K shell) and the remainder must, even for the normal minimum-energy condition, occupy other states. Hydrogen and helium, the first two elements, have, respectively, 1 and 2 electrons in the 1-quantum (K) shell; the next, lithium, has its third electron in the 2-quantum (L) shell. The passage from lithium to neon results in the filling up of this shell to its full complement of 8 electrons. During the process, the electrons first enter the $2s$ subgroup, then fill the $2p$ subgroup until it has 6 electrons, the maximum allowable by the exclusion principle (see *Table 1.14*).

Very briefly, the effect of the electron-shell filling is as follows. Elements in the same chemical family have the same number of electrons in the subshell that is incompletely filled. The rare gases (He, Ne, Ar, Kr, Xe) have no uncompleted shells. Alkali metals (e.g. Na) have shells containing a single electron. The alkaline earths have two electrons in uncompleted shells. The good conductors (Ag, Cu, Au) have a single electron in the uppermost quantum state. An irregularity in the ordered sequence of filling (which holds consistently from H to Ar) begins at potassium (K)

Table 1.14 (continued)

Z	Name and symbol		A	Shells			
				KLM	N	O	P
57	Lanthanum	La	138.9	28	18	9	2
58	Cerium	Ce	140.1	28	19	9	2
59	Praseodymium	Pr	140.9	28	21	8	2
60	Neodymium	Nd	144.3	28	22	8	2
61	Promethium	Pm	147.0	28	23	8	2
62	Samarium	Sm	150.4	28	24	8	2
63	Europium	Eu	152.0	28	25	8	2
64	Gadolinium	Gd	157.3	28	25	9	2
65	Terbium	Tb	159.2	28	27	8	2
66	Dysprosium	Dy	162.5	28	28	8	2
67	Holmium	Ho	163.5	28	29	8	2
68	Erbium	Er	167.6	28	30	8	2
69	Thulium	Tm	169.4	28	31	8	2
70	Ytterbium	Yb	173.0	28	32	8	2
71	Lutecium	Lu	175.0	28	32	9	2
72	Hafnium	Hf	178.6	28	32	10	2
73	Tantalum	Ta	181.4	28	32	11	2
74	Tungsten	W	184.0	28	32	12	2
75	Rhenium	Re	186.3	28	32	13	2
76	Osmium	Os	191.5	28	32	14	2
77	Iridium	Ir	193.1	28	32	15	2
78	Platinum	Pt	195.2	28	32	17	1
79	Gold	Au	197.2	28	32	18	1
80	Mercury	Hg	200.6	28	32	18	2
81	Thallium	Tl	204.4	28	32	18	3
82	Lead	Pb	207.2	28	32	18	4
83	Bismuth	Bi	209.0	28	32	18	5
84	Polonium	Po	210.0	28	32	18	6
85	Astatine	At	211.0	28	32	18	7
86	Radon	Rn	222.0	28	32	18	8

Z	Name and symbol		A	Shells			
				KLMN	O	P	Q
87	Francium	Fr	223.0	60	18	8	1
88	Radium	Ra	226.0	60	18	8	2
89	Actinium	Ac	227.0	60	18	9	2
90	Thorium	Th	232.0	60	18	10	2
91	Protoactinium	Pa	231.0	60	20	9	2
92	Uranium	U	238.0	60	21	9	2
93	Neptunium	Np	237.0	60	22	9	2
94	Plutonium	Pu	239.0	60	24	8	2
95	Americium	Am	243.0	60	25	8	2
96	Curium	Cm	247.0	60	25	9	2
97	Berkelium	Bk	247.0	60	26	9	2
98	Californium	Cf	251.0	60	28	8	2
99	Einsteinium	Es	254.0	60	29	8	2
100	Fermium	Fm	257.0	60	30	8	2
101	Mendelevium	Md	257.0	60	31	8	2
102	Nobelium	No	254.0	60	32	8	2
103	Lawrencium	Lr	256.0	60	32	9	2
104	Kurchatovium	Ku	—				
105	Hahnium	Ha	—				

two protons, the two electrons occupying the K shell which, by the Pauli exclusion principle, cannot have more than two. The next element in order is lithium ($Z = 3$), the third electron in an outer L shell. With elements of increasing atomic number, the electrons are added to the L shell until it holds a maximum of 8, the surplus then occupying the M shell to a maximum of 18. The number of 'valence' electrons (those in the outermost shell) determines the physical and chemical properties of the element. Those with completed outer shells are 'stable'.

Isotopes An element is often found to be a mixture of atoms with the same chemical property but different atomic masses: these are isotopes. The isotopes of an element must have the same number of electrons and protons, but differ in the number of neutrons, accounting for the non-integral average mass numbers. For example, neon comprises 90.4% of mass number 20, with 0.6% of 21 and 9.0% of mass number 22, giving a resultant mass number of 20.18.

Energy states Atoms may be in various energy states. Thus, the filament of an incandescent lamp may emit light when excited by an electric current but not when the current is switched off. Heat energy is the kinetic energy of the atoms of a heated body. The more vigorous impact of atoms may not always shift the atom as a whole, but may shift an electron from one orbit to another of higher energy level within the atom. This position is not normally stable, and the electron gives up its momentarily acquired potential energy by falling back to its original level, releasing the energy as a light quantum or photon.

Ionisation Among the electrons of an atom, those of the outermost shell are unique in that, on account of all the electron charges on the shells between them and the nucleus, they are the most loosely bound and most easily *removable*. In a variety of ways it is possible so to excite an atom that one of the outer electrons is torn away, leaving the atom *ionised* or converted for the time into an *ion* with an effective positive charge due to the unbalanced electrical state it has acquired. Ionisation may occur due to impact by other fast-moving particles, by irradiation with rays of suitable wavelength and by the application of intense electric fields.

1.3.2.2 Wave mechanics

The fundamental laws of optics can be explained without regard to the nature of light as an electromagnetic wave phenomenon, and photoelectricity emphasises its nature as a stream or ray of corpuscles. The phenomena of diffraction or interference can only be explained on the wave concept. *Wave mechanics* correlates the two apparently conflicting ideas into a wider concept of 'waves of matter'. Electrons, atoms and even molecules participate in this duality, in that their effects appear sometimes as corpuscular, sometimes as of a wave nature. Streams of electrons behave in a corpuscular fashion in photoemission, but in certain circumstances show the diffraction effects familiar in wave action. Considerations of particle mechanics led de Broglie to write several theoretic papers (1922–1926) on the parallelism between the dynamics of a particle and geometrical optics, and suggested that it was necessary to admit that classical dynamics could not interpret phenomena involving energy quanta. Wave mechanics was established by Schrödinger in 1926 on de Broglie's conceptions.

When electrons interact with matter, they exhibit wave properties: in the free state they act like particles. Light has a similar duality, as already noted. The hypothesis of de Broglie is that a particle of mass m and velocity u has wave properties with a wavelength $\lambda = h/mu$, where h is the Planck constant, $h = 6.626 \times 10^{-34}$ J s. The mass m is relativistically affected by the velocity.

When electron waves are associated with an atom, only certain fixed-energy states are possible. The electron can be raised from one state to another if it is provided, by some external stimulus such as a photon, with the necessary energy difference Δw in the form of an electromagnetic wave of wavelength $\lambda = hc/\Delta w$, where c is the velocity of free space radiation (3×10^8 m/s). Similarly, if an electron falls from a state of higher to one of lower energy, it emits energy Δw as radiation. When electrons are raised in energy level, the atom is *excited*, but not ionised.

continuous motion. In a *solid* the molecules are relatively closely 'packed' and the molecules, although rapidly moving, maintain a fixed mean position. Attractive forces between molecules account for the tendency of the solid to retain its shape. In a *liquid* the molecules are less closely packed and there is a weaker cohesion between them, so that they can wander about with some freedom within the liquid, which consequently takes up the shape of the vessel in which it is contained. The molecules in a *gas* are still more mobile, and are relatively far apart. The cohesive force is very small, and the gas is enabled freely to contract and expand. The usual effect of heat is to increase the intensity and speed of molecular activity so that 'collisions' between molecules occur more often; the average spaces between the molecules increase, so that the substance attempts to expand, producing internal pressure if the expansion is resisted.

Molecules are capable of further subdivision, but the resulting particles, called *atoms*, no longer have the same properties as the molecules from which they came. An atom is the smallest portion of matter than can enter into chemical combination or be chemically separated, but it cannot generally maintain a separate existence except in the few special cases where a single atom forms a molecule. A molecule may consist of one, two or more (sometimes many more) atoms of various kinds. A substance whose molecules are composed entirely of atoms of the same kind is called an *element*. Where atoms of two or more kinds are present, the molecule is that of a chemical *compound*. At present over 100 elements are recognised (*Table 1.14*: the atomic mass number A is relative to 1/12 of the mass of an element of carbon-12).

If the element symbols are arranged in a table in ascending order of atomic number, and in columns ('groups') and rows ('periods') with due regard to associated similarities, *Table 1.15* is obtained. Metallic elements are found on the left, non-metals on the right. Some of the correspondences that emerge are:

Group 1a: Alkali metals
      (Li 3, Na 11, K 19, Rb 37, Cs 55, Fr 87)
  2a: Alkaline earths
      (Be 4, Mg 12, Ca 20, Sr 38, Ba 56, Ra 88)
  1b: Copper group (Cu 29, Ag 47, Au 79)
  6b: Chromium group (Cr 24, Mo 42, W 74)
  7a: Halogens (F 9, Cl 17, Br 35, I 53, At 85)
  0 : Rare gases
      (He 2, Ne 10, Ar 18, Kr 36, Xe 54, Rn 86)
3a–6a: Semiconductors
      (B 5, Si 16, Ge 32, As 33, Sb 51, Te 52)

In some cases a horizontal relation obtains as in the transition series (Sc 21 ... Ni 28) and the heavy-atom rare earth and actinide series. The explanation lies in the structure of the atom.

1.3.2.1 Atomic structure

The original Bohr model of the hydrogen atom was a central nucleus containing almost the whole mass of the atom, and a single *electron* orbiting around it. Electrons, as small particles of negative electric charge, were discovered at the end of the nineteenth century, bringing to light the complex structure of atoms. The hydrogen nucleus is a *proton*, a mass having a charge equal to that of an electron, but positive. Extended to all elements, each has a nucleus comprising mass particles, some (*protons*) with a positive charge, others (*neutrons*) with no charge. The atomic *mass number A* is the total number of protons and neutrons in the nucleus; the *atomic number Z* is the number of positive charges, and the normal number of orbital electrons. The nuclear structure is not known, and the forces that bind the protons against their mutual attraction are conjectural.

The hydrogen atom (*Figure 1.4*) has one proton ($Z=1$) and one electron in an orbit formerly called the K shell. Helium ($Z=2$) has

Table 1.14 Elements (Z, atomic number; A, atomic mass; KLMNOPQ, electron shells)

Z	Name and symbol		A	Shells			
				K	L		
1	Hydrogen	H	1.008	1	—		
2	Helium	He	4.002	2	—		
3	Lithium	Li	6.94	2	1		
4	Beryllium	Be	9.02	2	2		
5	Boron	B	10.82	2	3		
6	Carbon	C	12	2	4		
7	Nitrogen	N	14.01	2	5		
8	Oxygen	O	16.00	2	6		
9	Fluorine	F	19.00	2	7		
10	Neon	Ne	20.18	2	8		
				KL	M	N	
11	Sodium	Na	22.99	10	1	—	
12	Magnesium	Mg	24.32	10	2	—	
13	Aluminium	Al	26.97	10	3	—	
14	Silicon	Si	28.06	10	4	—	
15	Phosphorus	P	31.02	10	5	—	
16	Sulphur	S	32.06	10	6	—	
17	Chlorine	Cl	35.46	10	7	—	
18	Argon	Ar	39.94	10	8	—	
19	Potassium	K	39.09	10	8	1	
20	Calcium	Ca	40.08	10	8	2	
21	Scandium	Sc	45.10	10	9	2	
22	Titanium	Ti	47.90	10	10	2	
23	Vanadium	V	50.95	10	11	2	
24	Chromium	Cr	52.01	10	13	1	
25	Manganese	Mn	54.93	10	13	2	
26	Iron	Fe	55.84	10	14	2	
27	Cobalt	Co	58.94	10	15	2	
28	Nickel	Ni	58.69	10	16	2	
29	Copper	Cu	63.57	10	18	1	
30	Zinc	Zn	65.38	10	18	2	
31	Gallium	Ga	69.72	10	18	3	
32	Germanium	Ge	72.60	10	18	4	
33	Arsenic	As	74.91	10	18	5	
34	Selenium	Se	78.96	10	18	6	
35	Bromine	Br	79.91	10	18	7	
36	Krypton	Kr	83.70	10	18	8	
				KLM	N	O	
37	Rubidium	Rb	85.44	28	8	1	
38	Strontium	Sr	87.63	28	8	2	
39	Yttrium	Y	88.92	28	9	2	
40	Zirconium	Zr	91.22	28	10	2	
41	Niobium	Nb	92.91	28	12	1	
42	Molybdenum	Mo	96.0	28	13	1	
43	Technetium	Tc	99.0	28	14	1	
44	Ruthenium	Ru	101.7	28	15	1	
45	Rhodium	Rh	102.9	28	16	1	
46	Palladium	Pd	106.7	28	18	—	
47	Silver	Ag	107.9	28	18	1	
48	Cadmium	Cd	112.4	28	18	2	
49	Indium	In	114.8	28	18	3	
50	Tin	Sn	118.7	28	18	4	
51	Antimony	Sb	121.8	28	18	5	
52	Tellurium	Te	127.6	28	18	6	
53	Iodine	I	126.9	28	18	7	
54	Xenon	Xe	131.3	28	18	8	
				KLM	N	O	P
55	Caesium	Cs	132.9	28	18	8	1
56	Barium	Ba	137.4	28	18	8	2

cont'd

Table 1.13 Decibel gain: power and voltage ratios

A	P_1/P_2	V_1/V_2	A	P_1/P_2	V_1/V_2
0	1.000	1.000	9	7.943	2.818
0.1	1.023	1.012	10	10.00	3.162
0.2	1.047	1.023	12	15.85	3.981
0.3	1.072	1.032	14	25.12	5.012
0.4	1.096	1.047	16	39.81	6.310
0.6	1.148	1.072	18	63.10	7.943
0.8	1.202	1.096	20	100.0	10.00
1.0	1.259	1.122	25	316.2	17.78
1.2	1.318	1.148	30	1000	31.62
1.5	1.413	1.189	35	3162	56.23
2.0	1.585	1.259	40	1.0×10^4	100.0
2.5	1.778	1.333	45	3.2×10^4	177.8
3.0	1.995	1.413	50	1.0×10^5	316.2
3.5	2.239	1.496	55	3.2×10^5	562.3
4.0	2.512	1.585	60	1.0×10^6	1 000
4.5	2.818	1.679	65	3.2×10^6	1 778
5.0	3.162	1.778	70	1.0×10^7	3 160
6.0	3.981	1.995	80	1.0×10^8	10 000
7.0	5.012	2.239	90	1.0×10^9	31 620
8.0	6.310	2.512	100	1.0×10^{10}	100 000

'lumped'. An ideal transformer, in which the primary m.m.f. in ampere-turns $i_1 N_1$ is equal to the secondary m.m.f. $i_2 N_2$ has as analogue the simple lever, in which a force f_1 at a point distant l_1 from the fulcrum corresponds to f_2 at l_2 such that $f_1 l_1 = f_2 l_2$.

A simple series circuit is described by the equation $v = L(di/dt) + Ri + q/C$ or, with i written as dq/dt,

$$v = L(d^2q/dt^2) + R(dq/dt) + (1/C)q$$

A corresponding mechanical system of mass, compliance and viscous friction (proportional to velocity) in which for a displacement l the inertial force is $m(du/dt)$, the compliance force is l/k and the friction force is ru, has a total force

$$f = m(d^2l/dt^2) + r(dl/dt) + (1/k)l$$

Thus the two systems are expressed in identical mathematical form.

1.3.1.2 Fields

Several physical problems are concerned with 'fields' having stream-line properties. The eddyless flow of a liquid, the current in a conducting medium, the flow of heat from a high- to a low-temperature region, are fields in which representative lines can be drawn to indicate at any point the direction of the flow there. Other lines, orthogonal to the flow lines, connect points in the field having equal potential. Along these equipotential lines there is no tendency for flow to take place.

Static electric fields between charged conductors (having equipotential surfaces) are of interest in problems of insulation stressing. Magnetic fields, which in air-gaps may be assumed to cross between high-permeability ferromagnetic surfaces that are substantially equipotentials, may be studied in the course of investigations into flux distribution in machines. All the fields mentioned above satisfy Laplacian equations of the form

$$(\partial^2 V/\partial x^2) + (\partial^2 V/\partial y^2) + (\partial^2 V/\partial z^2) = 0$$

The solution for a physical field of given geometry will apply to other Laplacian fields of similar geometry, e.g.

The ratio I/V for the first system would give the effective conductance G; correspondingly for the other systems, q/θ gives the thermal conductance, Q/V gives the capacitance and Φ/F gives the permeance, so that if measurements are made in one system the results are applicable to all the others.

It is usual to treat problems as two-dimensional where possible. Several field-mapping techniques have been devised, generally electrical because of the greater convenience and precision of electrical measurements. For two-dimensional problems, conductive methods include high-resistivity paper sheets, square-mesh 'nets' of resistors and electrolytic tanks. The tank is especially adaptable to three-dimensional cases of axial symmetry.

In the electrolytic *tank* a weak electrolyte, such as ordinary tap-water, provides the conducting medium. A scale model of the electrode system is set into the liquid. A low-voltage supply at some frequency between 50 Hz and 1 kHz is connected to the electrodes so that current flows through the electrolyte between them. A probe, adjustable in the horizontal plane and with its tip dipping vertically into the electrolyte, enables the potential field to be plotted. Electrode models are constructed from some suitable insulator (wood, paraffin wax, Bakelite, etc.), the electrode outlines being defined by a highly conductive material such as brass or copper. The metal is silver-plated to improve conductivity and reduce polarisation. Three-dimensional cases with axial symmetry are simulated by tilting the tank and using the surface of the electrolyte as a radial plane of the system.

The conducting-*sheet* analogue substitutes a sheet of resistive material (usually 'teledeltos' paper with silver-painted electrodes) for the electrolyte. The method is not readily adaptable to three-dimensional plots, but is quick and inexpensive in time and material.

The *mesh* or resistor-net analogue replaces a conductive continuum by a square mesh of equal resistors, the potential measurements being made at the nodes. Where the boundaries are simple, and where the 'grain size' is sufficiently small, good results are obtained. As there are no polarisation troubles, direct voltage supply can be used. If the resistors are made adjustable, the net can be adapted to cases of inhomogeneity, as when plotting a magnetic field in which permeability is dependent on flux density. Three-dimensional plots are made by arranging plane meshes in layers; the nodes are now the junctions of six instead of four resistors.

A stretched elastic membrane, depressed or elevated in appropriate regions, will accommodate itself smoothly to the differences in level: the height of the membrane everywhere can be shown to be in conformity with a two-dimensional Laplace equation. Using a rubber sheet as a membrane, the path of electrons in an electric field between electrodes in a vacuum can be investigated by the analogous paths of rolling bearing-balls. Many other useful analogues have been devised, some for the rapid solution of mathematical processes.

Recently considerable development has been made in point-by-point computer solutions for the more complicated field patterns in three-dimensional space.

1.3.2 Structure of matter

Material substances, whether solid, liquid or gaseous, are conceived as composed of very large numbers of *molecules*. A molecule is the smallest portion of any substance which cannot be further subdivided without losing its characteristic material properties. In all states of matter molecules are in a state of rapid

System	Potential	Flux	Medium
current flow	voltage V	current I	conductivity σ
heat flow	temperature θ	heat q	thermal conductivity λ
electric field	voltage V	electric flux Q	permittivity ε
magnetic field	m.m.f. F	magnetic flux ϕ	permeability μ

Table 1.12 (continued)

Definition	$f(t)$ from $t=0+$	$F(s) = L[f(t)] = \int_{0-}^{\infty} f(t) \cdot \exp(-st) \cdot dt$	
16. Exponentially decaying sine	$\exp(-\alpha t)\sin \omega t$	$\dfrac{\omega}{(s+\alpha)^2 + \omega^2}$	
17. Cosinusoidal	$\cos \omega t$	$\dfrac{s}{s^2 + \omega^2}$	
18. Phase-advanced cosine	$\cos(\omega t + \phi)$	$\dfrac{s \cos \phi - \omega \sin \phi}{s^2 + \omega^2}$	
19. Offset cosine	$1 - \cos \omega t$	$\dfrac{\omega^2}{s(s^2 + \omega^2)}$	
20. Cosine × t	$t \cos \omega t$	$\dfrac{s^2 - \omega^2}{(s^2 + \omega^2)^2}$	
21. Exponentially decaying cosine	$\exp(-\alpha t)\cos \omega t$	$\dfrac{(s+\alpha)}{(s+\alpha)^2 + \omega^2}$	
22. Trigonometrical function $G(t)$	$\sin \omega t - \omega t \cos \omega t$	$\dfrac{2\omega^3}{(s^2 + \omega^2)^2}$	
23. Exponentially decaying trigonometrical function	$\exp(-\alpha t) \cdot G(t)$	$\dfrac{2\omega^3}{[(s+\alpha)^2 + \omega^2]^2}$	
24. Hyperbolic sine	$\sinh \omega t$	$\dfrac{\omega}{s^2 - \omega^2}$	
25. Hyperbolic cosine	$\cosh \omega t$	$\dfrac{s}{s^2 - \omega^2}$	
26. Rectangular wave (period T)	$f(t)$	$\dfrac{1 + \tanh(sT/4)}{2s}$	
27. Half-wave rectified sine ($T = 2\pi/\omega$)	$f(t)$	$\dfrac{\omega \exp(sT/2)\operatorname{cosech}(sT/2)}{2(s^2 + \omega^2)}$	
28. Full-wave rectified sine ($T = 2\pi/\omega$)	$f(t)$	$\dfrac{\omega \coth(sT/2)}{s^2 + \omega^2}$	

f	force [N]	M	torque [N m]	v	voltage [V]
m	mass [kg]	J	inertia [kg m²]	L	inductance [H]
r	friction [N s/m]	r	friction [N m s/rad]	R	resistance [Ω]
k	compliance [m/N]	k	compliance [rad/N m]	C	capacitance [F]
l	displacement [m]	θ	displacement [rad]	q	charge [C]
u	velocity [m/s]	ω	angular velocity [rad/s]	i	current [A]

The force necessary to maintain a uniform linear velocity u against a viscous frictional resistance r is $f = ur$; the power is $p = fu = u^2 r$ and the energy expended over a distance l is $W = fut = u^2 rt$, since $l = ut$. These are, respectively, the analogues of $v = iR$, $p = vi = i^2 R$ and $W = vit = i^2 Rt$ for the corresponding electrical system. For a constant angular velocity in a rotary mechanical system, $M = \omega r$, $p = M\omega = \omega^2 r$ and $W = \omega^2 rt$, since $\theta = \omega t$.

If a mass is given an acceleration du/dt, the force required is $f = m(du/dt)$ and the stored kinetic energy at velocity u_1 is $W = \tfrac{1}{2}mu_1^2$. For rotary acceleration, $M = J(d\omega/dt)$ and $W = \tfrac{1}{2}J\omega_1^2$. Analogously the application of a voltage v to a pure inductor L produces an increase of current at the rate di/dt such that $v = L(di/dt)$ and the magnetic energy stored at current i_1 is $W = \tfrac{1}{2}Li^2$.

A mechanical element (such as a spring) of compliance k (which describes the displacement per unit force and is the inverse of the stiffness) has a displacement $l = kf$ when a force f is applied. At a final force f_1 the potential energy stored is $W = \tfrac{1}{2}kf_1^2$. For the rotary case, $\theta = kM$ and $W = \tfrac{1}{2}kM_1^2$. In the electric circuit with a pure capacitance C, to which a p.d. v is applied, the charge is $q = Cv$ and the electric energy stored at v_1 is $W = \tfrac{1}{2}Cv_1^2$.

Use is made of these correspondences in mechanical problems (e.g. of vibration) when the parameters can be considered to be

1/14 Units, mathematics and physical quantities

Table 1.12 Laplace transforms

Definition	$f(t)$ from $t=0+$	$F(s) = L[f(t)] = \int_{0-}^{\infty} f(t) \cdot \exp(-st) \cdot dt$	
Sum	$af_1(t) + bf_2(t)$	$aF_1(s) + bF_2(s)$	
First derivative	$(d/dt)f(t)$	$sF(s) - f(0-)$	
nth derivative	$(d^n/dt^n)f(t)$	$s^n F(s) - s^{n-1} f(0-) - s^{n-2} f^{(1)}(0-) - \ldots - f^{(n-1)}(0-)$	
Definite integral	$\int_{0-}^{T} f(t) \cdot dt$	$\dfrac{1}{s} F(s)$	
Shift by T	$f(t-T)$	$\exp(-sT) \cdot F(s)$	
Periodic function (period T)	$f(t)$	$\dfrac{1}{1-\exp(-sT)} \int_{0-}^{T} \exp(-sT) \cdot f(t) \cdot dt$	
Initial value	$f(t), t \to 0+$	$sF(s), s \to \infty$	
Final value	$f(t), t \to \infty$	$sF(s), s \to 0$	

Description	$f(t)$	$F(s)$	$f(t)$ to base t
1. Unit impulse	$\delta(t)$	1	Area = 1
2. Unit step	$H(t)$	$\dfrac{1}{s}$	
3. Delayed step	$H(t-T)$	$\dfrac{\exp(-sT)}{s}$	
4. Rectangular pulse (duration T)	$H(t) - H(t-T)$	$\dfrac{1-\exp(-sT)}{s}$	
5. Unit ramp	t	$\dfrac{1}{s^2}$	
6. Delayed ramp	$(t-T)H(t-T)$	$\dfrac{\exp(-sT)}{s^2}$	
7. nth-order ramp	t^n	$\dfrac{n!}{s^{n+1}}$	
8. Exponential decay	$\exp(-\alpha t)$	$\dfrac{1}{s+\alpha}$	
9. Exponential rise	$1 - \exp(-\alpha t)$	$\dfrac{\alpha}{s(s+\alpha)}$	
10. Exponential × t	$t \exp(-\alpha t)$	$\dfrac{1}{(s+\alpha)^2}$	
11. Exponential × t^n	$t^n \exp(-\alpha t)$	$\dfrac{n!}{(s+\alpha)^{n+1}}$	
12. Difference of exponentials	$\exp(-\alpha t) - \exp(-\beta t)$	$\dfrac{\beta - \alpha}{(s+\alpha)(s+\beta)}$	
13. Sinusoidal	$\sin \omega t$	$\dfrac{\omega}{s^2 + \omega^2}$	
14. Phase-advanced sine	$\sin(\omega t + \phi)$	$\dfrac{\omega \cos \phi + s \sin \phi}{s^2 + \omega^2}$	
15. Sine × t	$t \sin \omega t$	$\dfrac{2\omega s}{(s^2 + \omega^2)^2}$	

Table 1.11 Derivatives and integrals

$d[f(x)]/dx$	$f(x)$	$\int f(x) \cdot dx$
1	x	$\tfrac{1}{2}x^2$
nx^{n-1}	x^n $(n \neq -1)$	$x^{n+1}/(n+1)$
$-1/x^2$	$1/x$	$\ln x$
$1/x$	$\ln x$	$x \ln x - x$
$\exp x$	$\exp x$	$\exp x$
$\cos x$	$\sin x$	$-\cos x$
$-\sin x$	$\cos x$	$\sin x$
$\sec^2 x$	$\tan x$	$\ln(\sec x)$
$-\cosec x \cdot \cot x$	$\cosec x$	$\ln(\tan \tfrac{1}{2}x)$
$\sec x \cdot \tan x$	$\sec x$	$\ln(\sec x + \tan x)$
$-\cosec^2 x$	$\cot x$	$\ln(\sin x)$
$1/\sqrt{(a^2-x^2)}$	$\arcsin(x/a)$	$x \arcsin(x/a) + \sqrt{(a^2-x^2)}$
$-1/\sqrt{(a^2-x^2)}$	$\arccos(x/a)$	$x \arccos(x/a) - \sqrt{(a^2-x^2)}$
$a/(a^2+x^2)$	$\arctan(x/a)$	$x \arctan(x/a) - \tfrac{1}{2}a \ln(a^2+x^2)$
$-a/x\sqrt{(x^2-a^2)}$	$\arccosec(x/a)$	$x \arccosec(x/a) + a \ln|x + \sqrt{(x^2-a^2)}|$
$a/x\sqrt{(x^2-a^2)}$	$\arcsec(x/a)$	$x \arcsec(x/a) - a \ln|x + \sqrt{(x^2-a^2)}|$
$-a/(a^2+x^2)$	$\arccot(x/a)$	$x \arccot(x/a) + \tfrac{1}{2}a \ln(a^2+x^2)$
$\cosh x$	$\sinh x$	$\cosh x$
$\sinh x$	$\cosh x$	$\sinh x$
$\sech^2 x$	$\tanh x$	$\ln(\cosh x)$
$-\cosech x \cdot \coth x$	$\cosech x$	$-\ln(\tanh \tfrac{1}{2}x)$
$-\sech x \cdot \tanh x$	$\sech x$	$2 \arctan(\exp x)$
$-\cosech^2 x$	$\coth x$	$\ln(\sinh x)$
$1/\sqrt{(x^2+1)}$	$\arsinh x$	$x \arsinh x - \sqrt{(1+x^2)}$
$1/\sqrt{(x^2-1)}$	$\arcosh x$	$x \arcosh x - \sqrt{(x^2-1)}$
$1/(1-x^2)$	$\artanh x$	$x \artanh x + \tfrac{1}{2}\ln(1-x^2)$
$-1/x\sqrt{(x^2+1)}$	$\arcosech x$	$x \arcosech x + \arsinh x$
$-1/x\sqrt{(1-x^2)}$	$\arsech x$	$x \arsech x + \arcsin x$
$1/(1-x^2)$	$\arcoth x$	$x \arcoth x + \tfrac{1}{2}\ln(x^2-1)$
$u\dfrac{dv}{dx} + v\dfrac{du}{dx}$	$u(x) \cdot v(x)$	$uv - \int v \dfrac{du}{dv} dv$
$\dfrac{1}{v}\dfrac{du}{dx} - \dfrac{u}{v^2}\dfrac{dv}{dx}$	$\dfrac{u(x)}{v(x)}$	—
$r \exp(xa) \times \sin(\omega x + \phi + \theta)$	$\exp(ax) \times \sin(\omega x + \phi)$	$(1/r)\exp(ax)\sin(\omega x + \phi - \theta)$ $r = \sqrt{(\omega^2 + a^2)}$ $\theta = \arctan(\omega/a)$

physical characteristics, separate technologies have been devised around specific processes; and materials may have to be considered macroscopically in bulk, or in microstructure (molecular, atomic and subatomic) in accordance with the applications or processes concerned.

1.3.1 Energy

Like 'force' and 'time', energy is a unifying concept invented to systematise physical phenomena. It almost defies precise definition, but may be described, as an aid to an intuitive appreciation.

Energy is the capacity for 'action' or *work*.
Work is the measure of the change in energy *state*.
State is the measure of the energy condition of a *system*.
System is the ordered arrangement of related physical entities or processes, represented by a *model*.
Mode. is a description or mathematical formulation of the system to determine its *behaviour*.
Behaviour describes (verbally or mathematically) the energy processes involved in changes of state. Energy *storage* occurs if the work done on a system is recoverable in its original form.

Energy *conversion* takes place when related changes of state concern energy in a different form, the process sometimes being reversible. Energy *dissipation* is an irreversible conversion into heat. Energy *transmission* and *radiation* are forms of energy transport in which there is a finite propagation time.

In a physical system there is an identifiable energy input W_i and output W_o. The system itself may store energy W_s and dissipate energy W. The energy conservation principle states that

$$W_i = W_s + W + W_o$$

Comparable statements can be made for energy changes Δw and for energy rates (i.e. powers), giving

$$\Delta w_i = \Delta w_s + \Delta w + \Delta w_o \quad \text{and} \quad p_i = p_s + p + p_o$$

1.3.1.1 Analogues

In some cases the *mathematical* formulation of a system model resembles that of a model in a completely different physical system: the two systems are then analogues. Consider linear and rotary displacements in a simple mechanical system with the conditions in an electric circuit, with the following nomenclature:

group 414. Loss of pressure in the hydraulic fail-safe circuit causes the control valves to receive closing commands over two independent channels.

The Turbotrol system itself is designed to ensure the highest possible degree of safety whilst, with regard to the signal paths, still upholding the principles of simplicity and transparency. The most important peripheral units and transducers are actively redundant.

Two separate d.c. power sources reduce the danger of a turbine trip as a result of failure of the supply voltage. Failure of internal power-supply units simply causes clearly defined subcircuits to be disconnected, fault indication taking place simultaneously. Generally, it is still possible to carry out restricted operation with the Turbotrol. When a fault occurs, some of the internal monitoring elements set the turbine master station immediately to 'manual', so that the turbo-set can at least still be operated manually. Even during such operation, the subcircuits important for the safety of the installation (413, 414) remain active. Response by either of these, or a fault in one of them, causes an immediate trip.

The subcircuits and functions can, to a large extent, be tested during operation.

Availability Regular checks, and the modular design of the equipment, permit faults that occur to be recognised early and eliminated by replacing the appropriate module before it causes operational failure. If a breakdown should nevertheless occur, the fault may be quickly located by means of the many indication, test and simulation devices, and promptly eliminated. The availability of the installation is thus ensured.

Reliability The individual parts of the control system, its components and, finally, the complete system unit should be subjected to rigorous testing. Combined with a simple logically designed system that has the minimum number of component parts, these measures contribute to the high degree of reliability of the entire control and safety system.

Final designs of various kinds, depending on the respective importance attached to the terms safety, availability and reliability, can be achieved.

Operation with the hydraulic control system This control system serves as back-up when the Turbotrol is in operation. If a fault occurs in either the turbine master station or the electro-hydraulic transducer (415), the respective hydraulic lines must be blocked manually. Renewed turbine start-up then takes place manually via the hydraulic control system. The variable readings required for this are displayed in the control room, provided that the d.c. power supply for the Turbotrol is intact.

17.6.6.5 Fast valving

Power export is reduced by the loss of voltage in the event of short circuits in the transmission systems. Acceleration of the turbine rotors occurs in the turbine–generator units involved by a change in the balance of the mechanical-drive and the electrical-load torques. In the turbine–generator unit closest to the short-circuit point, the rotor angle reaches a critical value that depends upon the generated load and the duration of the short circuit; this causes the unit to lose synchronisation.

By briefly closing the turbine control valves ('fast valving'), the drive torque of the turbine–generator unit is rapidly reduced to control a sustained short circuit without loss of synchronism. Depending upon the load in the transmission system, it may be necessary before reconnecting the load to set a load level lower than that before the short circuit.

The subassembly 'Fast valving' (functional group 470—see *Figures 17.39* and *17.40*), takes action in the turbine control system in the event of failures in the power transmission system by brief closure of the control valves in conjunction with a reduction in the load setting. This takes place if the load on the turbine–generator unit is greater than the adjustable maximum value (approximately 60% load). Failure detection in the power transmission system and the initiation of fast valving are performed by the grid supervision system.

Both the main control and the intercept valves take part in fast valving. While the intercept valves are briefly closed (T_{FV}) via solenoid valves, the main control valves are actuated by the Turbotrol electronic control system.

The electrical position signal YHP for the main control valves is set to 0% by a minimum selection gate. The turbine master of the Turbotrol is switched to 'manual'. After the time T_{FV}, YHP is switched to a preselectable value $X\%$. Simultaneously, a 'lower' instruction for tracking the turbine master is issued. The integrators of the controllers also follow the tracking of the turbine master. The actual load value then appears according to $X\%$. The load set-point is tracked in the unit master.

The load set-point drops until the controlled variable YR is smaller than YFV, i.e. until the main controller is ready to take over the control task. The turbine master then automatically switches back to 'auto'.

The load target set-point has remained stationary during the entire fast-valving process so that it is immediately possible to reconnect the load by switching on the automatic loading system.

It is only possible to test the subassembly with a load of less than an adjustable minimum value (about 30%).

17.6.7 Water-turbine control system (Hydrotrol)

The Hydrotrol (*Figure 17.41*) performs, in conjunction with the electro-hydraulic transducers and the hydraulic servo-motors, the following tasks:

(1) Automatic provision of an opening for run-up of the turbine until the speed controller takes over.
(2) Speed control during no-load operation.
(3) Frequency control during isolated grid operation and when station auxiliaries are being supplied.
(4) Opening control according to an adjustable linear frequency–opening characteristic.
(5) Arresting of overspeed after load shedding.
(6) Connection and processing of signals of a higher-order control system with or without load feedback (e.g. from the water-level controller or dispatching system controller).
(7) Position control of the guide-vane apparatus in Francis turbines.
(8) Position control of the guide vane and impeller blades (with a pre-set relationship between guide-vane and impeller-blade angles) in Kaplan turbines.
(9) Position control of the needle and deflector (with co-ordination between jet diameter and deflector position) in Pelton turbines.
(10) Redundant measurement of speed. The speed-measurement device is equipped with outputs for indicator devices, control and monitoring.

The Hydrotrol 4 ensures (i) high operating reliability and availability, (ii) high-response sensitivity in respect of changes in specific control variables and (iii) easier servicing with low maintenance requirements.

17.6.7.1 Design and operation of the controller

Figure 17.41 shows the basic design and the way in which the function groups of the Hydrotrol are interconnected. The numbers of the functional units are referred to in the text. The controller is designed for the following methods of operation:

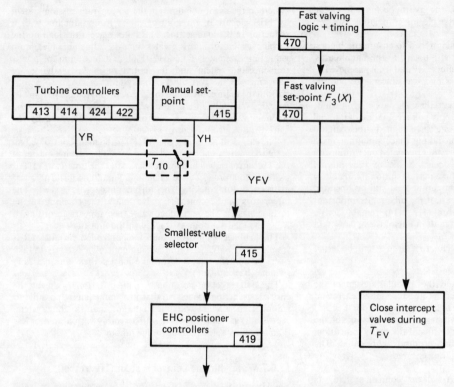

Figure 17.39 Block diagram showing action of the fast-valving functional group (470)

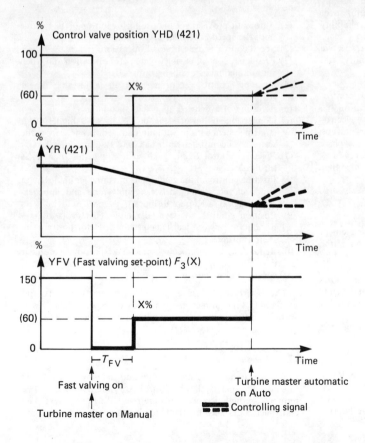

Figure 17.40 The response of the signals YHP (421), YR (421) and YFV during fast valving

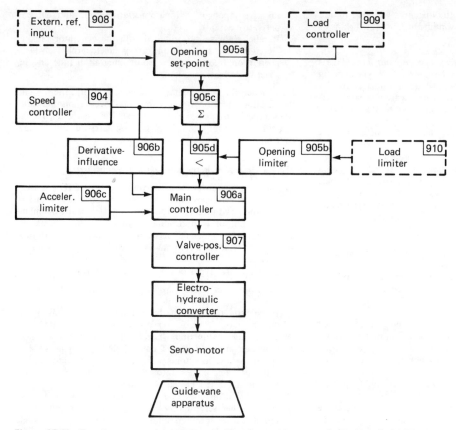

Figure 17.41 Function groups of the Hydrotrol with valve-position controller for Francis turbines

Normal:
   start-up (with opening limiter);
   no-load operation;
   synchronising;
   load operation with opening feedback;
   load operation with opening limiter in action;
   load operation with dispatching system controller connected;
   load operation with external reference input;
   shut-down.
Disturbance:
   load shedding;
   turbine tripping.
Testing:
   overspeed;
   simulated operation.

Start-up As soon as the starting command is given, the opening limiter (905b) indicates the starting opening (*Figure 17.41*). In the smallest-value selector (905d) the control deviation of the speed controller (904) is limited by the start opening and is supplied to the main controller (906a). The servo-motor opens the guide-vane equipment via the valve-position controller (907) to the value specified by the opening limiter (905b). The machine accelerates until the speed controller (904) comes into action via the derivative action (906b) below the nominal frequency. Acceleration consequently ceases and the speed controller (904) controls the turbine at nominal frequency.

No-load operation The turbine speed can be set by the frequency set-point value. The control deviation of the speed controller is formed by comparing the set-point with the actual frequency, and passes via the smallest-value selector (905d) to the main controller (906a). In this method of operation the main controller is switched to proportional–integral–differential (PID) response for reasons of stability. No-load operation at nominal speed is the starting point for synchronising the generator with the network.

Synchronising The synchronising device acts on the frequency set-point. Its pulses adjust the set-point until the frequency and phase position of the unit coincide with the network, and the generator breaker then closes.

Load operation with opening feedback After synchronising, the plant changes to load operation with the generator breaker closed. The following function units of the control system are now in operation: speed controller (904); opening set-point (905a); summing junction (905c); smallest-value selector (905d); main controller (906a); valve-position controller (907).

The turbine is loaded by run-up of the opening set-point. This is compared with the opening of the guide-vane equipment; the resultant deviation, after evaluation by the pre-set droop variable, is added to the control deviation of the speed controller. The acceleration influence (derivative action) on the main controller is blocked. In this mode of operation the main controller is switched to PI response.

Load operation with opening limiter in action As soon as the opening of the guide-vane device exceeds the value set on the opening limiter (905b), the latter begins to act via the smallest-value selector (905d) and limits the opening. The load limiter (910) automatically limits the maximum output of the turbine or generator via the opening limiter.

Load operation with applied dispatching system controller This is initiated by actuating a button on the control desk. The opening set-point is controlled by a superposed frequency–load controller (909). The load set-point can be pre-selected by a separate setpoint generator or an external dispatching system controller.

Load operation with external reference input This method is preselected by a button on the control desk, which causes the opening set-point (905a) to approach the pre-selected reference input (908) from a water-level controller or dispatching system controller. As soon as the two signals agree, the external reference input is applied directly as an opening set-point. The opening setpoint tracks the external reference intput so that, when this mode of operation is switched off, the controller continues to retain the last-occurring reference input.

Shut-down After the load has been reduced by decreasing the opening set-point, the generator breaker is opened; the controller again operates in the no-load mode. The turbine can then be fully shut down by automatically switching the opening limiter to a closure command of about 10%.

Load shedding To arrest the speed rise of the turbine on load shedding as rapidly as possible, an acceleration limiter (906c) is incorporated. As soon as the acceleration exceeds a limiting value, this device, acting via the servo-motor, causes immediate closure of the guide vanes. When the overspeed vanishes, the speed controller again takes over control at nominal frequency.

Turbine tripping All the emergency-closure and rapid-closure criteria simultaneously have the effect of causing a switch-over to a closure command of -10%, as with the normal shut-down operation.

Overspeed test For this, the speed controller (904) is rendered ineffective. By increasing the speed with the opening limiter (905b), the function of the overspeed protection device can be tested.

Simulated operation This is initiated by inserting a plug in the Hydrotrol 4; it facilitates functional testing of the electronic controller with the plant shut down, by means of incorporated simulation equipment.

17.6.7.2 Operation of the main controller

In order that the control deviation does not remain permanently in the stabilised state, the main controller is constructed with a PI circuit. With the circuit selected, the main controller governs for fairly small control deviations, and the servo-motor operates in its linear rapid-operating range. For large control deviations, the servo-motor can no longer follow the correcting variable (output of the main controller), since the maximum speed of the servomotor is limited by hydraulic orifice plates. The main controller will disengage, resulting in pronounced overshoot of the servomotor position (in the extreme case, event instability). However, electronic simulation of these hydraulic orifice plates limits the integral component of the main controller in such a way that, as soon as the servo-motor again reaches the linear range, the main controller immediately corrects itself and can take up optimised control.

Two types of stabilisation are possible in the main controller:

(1) No-load stabilisation, initiated with the generator breaker open (no-load stabilisation corresponds to a reduced proportional component, a long integration time and effective derivative action).
(2) Parallel stabilisation, initiated with the generator breaker closed (parallel stabilisation corresponds to a relatively large proportional component, a short integration time and ineffective derivative action).

If the generator is supplying an isolated network, the no-load stabilisation must be switched on. This occurs automatically when a specific frequency band is exceeded, and is signalled in the control room. It must also be possible to carry out this changeover manually.

17.6.7.3 Operation of the valve-position controller

The output signal from the main controller (with PID response) acts upon the valve-position controller. The latter has the task of making the servo-motors of the control devices track the maincontroller output (proportional response). It represents the actual connection device between controller and hydraulic system, and serves as a command device of the electro-hydraulic transducer.

There are three types of valve-position controller:

(1) For Francis turbines. In this variant, the position of the guidevane servo-motor is regulated.
(2) For Pelton turbines. In this variant, the positions of the needle and deflector servo-motors are regulated. By simulating a function, 'needle movement to water-jet diameter' in the 'deflector' control circuit, it is possible to ensure that the deflector is positioned above the water jet. The deflector then comes into action only if rapid control movements in the 'closed' direction occur.
(3) For Kaplan turbines. Here the positions of the guide-vane and impeller servo-motors are regulated. Function transmitters ensure optimum tuning between the guide-vane angle and impeller-blade angle as a function of the head. The corresponding device is situated in the 'impeller' control circuit.

17.6.7.4 Structure of the speed-measuring equipment

The Hydrotrol is equipped with a redundant speed-measuring device, with outputs for indicator devices, control and monitoring. The basic design of the device is shown in *Figure 17.42*. References to the units and outputs are given in the text in parentheses.

The speed is measured with three channels. The speed is picked up from a wheel (e) by ferrostat transmitters ($a1, a2, a3$). The generated frequency (proportional to speed) is transmitted via the pre-amplifier ($b1, b2, b3$) and the speed-measuring device ($c1, c2, c3$). The speed measurement generates an analogue voltage ($n1, n2, n3$) which is accurately proportional to the speed. The speed transmitters are monitored: a fault at speed transmitter $a1$ (such as a voltage failure, short circuit or line fault), is signalled via monitoring channel $u1$ and is annunciated by the alarm 'speed transmitter faulty' (output s). Simultaneously, the speed signal for the speed controller (i), the speed limiting-values (j) and the indicator (k) are switched over from speed transmitter $a1$ to transmitter $a2$.

The overspeed protection (p) consists of a double-channel '2-out-of-3 logic' (l). This triggers an emergency stop (output q) and a quick shut-down (output r) as soon as two or three speed limiting-values ($d1, d2, d3$) respond or faults occur at two or three speed transmitters. The direction indicator (t) serves for determining the direction of rotation in pump turbines.

The speed transmitters must be mounted at angular intervals $\beta = 360(n + \frac{2}{3})/z$ (degrees), where z is the number of gearwheel teeth and n is an integer. The outputs v and w indicate the

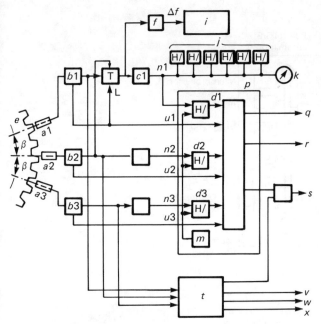

Figure 17.42 Block diagram of a speed-measuring device with speed limiting-values, overspeed protection and direction indicator: D/A, digital/analogue

direction of rotation, and output x indicates when the shaft is stationary.

17.6.7.5 Testing

Component testing Every component approved for use in electronic systems must be subjected to stringent tests, carried out in compliance with IEC publications.

Testing of the controller The controllers must be individually tested and adjusted in the laboratory on the basis of the particular turbine characteristics. Before delivery, every Hydrotrol must operate for a week in the closed-loop of an analogue turbine simulator, with all control variables and the most important control parameters continuously recorded. In this way, changes in control behaviour and premature failure, for example of semiconductor components, can be detected.

17.6.7.6 Operational reliability and availability

Reliability An electronic water-turbine Hydrotrol is, in general, of single-channel construction. The important peripheral devices and transmitters (e.g. for speed) are designed with redundancy. A failure in the supply voltage or internal supply equipment causes rapid closure and simultaneous signalling of the fault. Faults in internal monitoring devices such as those for the speed transmitters or valve-position controllers lead to an emergency closure.

Availability Regular testing and the modular construction of the equipment should permit any faults that occur to be detected at an early stage and to be rectified by replacement equipment before they cause operating faults.

References

1 COHN, N., *Control of Generation and Power Flow on Interconnected Systems*, Wiley (1966)
2 GLAVITSCH, H. and STOFFEL, J., 'Automatic Generation Control', *Journal on Electrical Power and Energy Systems*, **2**(1), January (1980)
3 ANDRES, W., SCHAIBLE, W. and SCHATZMANN, G., 'Turbotrol 4 Electronic Control System for Steam Turbines based on PC 200 Equipment', *Brown Boveri Review*, **62**, 377, September (1975)
4 MÜHLEMANN, M., 'The Electronic Water Turbine Controller Hydrotrol 4', *Brown Boveri Review*, **66**, October (1979)

17.7 Decentralised control: substation automation

17.7.1 Introduction

With the increasing complexity of power systems, the difficulties of manning substations and the requirement for shorter restoration times after power failures, more tasks are being automated in the substation. The conventional solution to automating these tasks has been by dedicated hard-wired logic. Though effective, this is inflexible and difficult to test off-line. Hard-wired logic is being superseded by systems based on mini- or micro-processors which have advantages over their hard-wired counterparts with regard to flexibility, space requirements, test and commissioning facilities. A further and important advantage is their ability to self-monitor and detect an internal fault almost immediately, and not only when operating. However, they have the disadvantage of intolerance to voltages in the connecting cables induced by switching of h.v. apparatus or thunderstorms. Special precautions must therefore be taken to decouple the plant connections from the electronic circuits.

Several tasks can be assembled in one set of hardware, and thus a centralised system can be used. The speed of transmission and the availability of transmission channels are limiting factors in achieving reasonable response times for some substation tasks from a remote control centre; many of these tasks are therefore performed by substation equipment.

17.7.2 Hardware configuration

A simplified substation automation hardware configuration is illustrated in *Figure 17.43*, its connections to one feeder circuit of an h.v. substation.

The central logic is microprocessor with the automation routines stored in a non-volatile programmable read-only memory (EPROM) which can be erased and modified using special equipment. Thus, by triggering a power-failure restart routine held in the EPROM, the equipment can recover from power failures without reloading. The variable data are held in a random-access memory (RAM) which can be corrupted on power failure. It must be capable of replacement after a power failure either by local input or by down-loading from a higher-level control system. Similarly, the variable data, such as limits, tripping priorities, etc., must be capable of modification, e.g. from a keyboard.

The data acquisition and automation programs are executed by the microprocessor, which communicates with the input/output equipment via the system bus. The input and output modules are similar to those already described for the telecontrol remote terminal unit (TRU). The connection to the higher-level control system can be via a direct link or a telecontrol TRU.

An important decision to be made in the layout of an automation system in substations is the location of the transducers and the method of transmitting the h.v.-circuit data to the central processor unit. Economically, the transducers are better

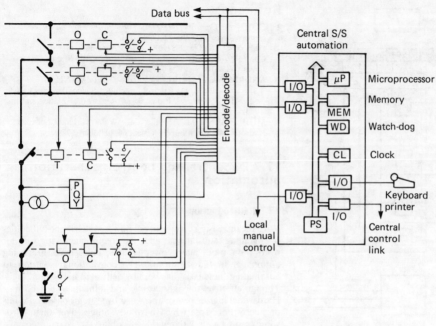

Figure 17.43 An example of a decentralised automation system; O, Open; C, Close; T, Trip; P, active power; Q, reactive power; V, voltage; PS, power supply; S/S, substation

located near to the measurement point to avoid long instrument-transformer connections. However, this means that light-current connections are long and vulnerable to interference and high induced voltages. A solution is to install encoding equipment at the data source and serial transmission equipment to send the messages to the central logic over a serial data bus which can be common to a number of h.v. circuits, as in the telecontrol system transmission already described. As the technique develops, this transmission can be over a fibre-optic line.

17.7.3 Software configuration

The software is modular to allow the system to be easily extended and developed. The individual application programs run independently of each other, but share data and input/output routines. For small systems the programs can run cyclically; however, for larger systems a system of priorities must be established and the resources allocated by an executive program. The program structure is illustrated in *Figure 17.44*.

17.7.4 Applications

17.7.4.1 Sequence control (switching programs)

In closing circuit-breakers on to live circuits, several criteria have to be met to ensure a successful operation; for example, in energising a feeder, the breaker must first be connected to the required busbar, the line isolator closed, the line–earth switch open, and the synchronising criteria satisfied. It must be ensured that no sanctions are valid that would inhibit the operation. Similarly, when energising a transformer circuit, manually or automatically, it is normal to make similar checks and circuit preselections before closing the l.v. breaker and then the h.v. breaker. In changing a circuit from one busbar to another, a switching sequence involving the bus coupler and perhaps the bus section switches must be performed.

These sequences can be programmed and performed by local automation equipment, provided that all switches are motorised. The sequence can be initiated by one command (e.g. 'close line 123 on to bus 3'). The responsibility from there on is taken from the operator and the instruction is performed by the local automation equipment which, in carrying it out, will respect all the interlocking requirements and inhibits.

17.7.4.2 Switching interlock supervision

Manual switch operation in a substation is required either as the normal mode of operation or as a back-up to the automatic mode. For these operations, interlocking is required to prevent incorrect switching that might endanger equipment or the stability of the power network. The conventional solution for interlocking is a device with auxiliary contacts and inter-device wiring forming a hard-wired supervision logic. Interlocking can also be achieved by the local automation equipment using programmed logic accessing the shared data base. This latter solution considerably reduces the inter-device cabling as well as the number of device auxiliary switches or repeater relays used and their discrepancy supervision.

17.7.4.3 Load shedding and restoration

With a falling frequency, at certain levels of frequency it may be necessary to reduce load without waiting for remote intervention or relying on transmission channels. The load shedding can be performed automatically by local automation equipment in progressive steps, starting with the tripping of storage heating or cooling loads via ripple control signals, and progressing to more drastic action by tripping feeder circuits at the substation.

These tasks can be performed by substation equipment that scans the frequency measurand and, at various pre-set frequency thresholds, instigates load-shedding action in pre-arranged steps and programs. The programs can be arranged to cycle the load shedding so that each consumer is tripped in turn and no particular load is singled out to be always first on the tripping list.

Once the frequency has returned to predetermined levels,

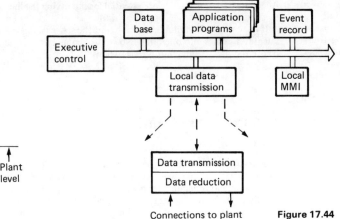

Figure 17.44 Software configuration: MMI,

supplies to loads can be progressively restored in a predetermined order until the situation is normal.

Similar actions can be initiated on a feeder overload when the current exceeds the thermal limit of the cable.

17.7.4.4 Automatic reclosing

The majority of overhead line faults are transient, caused by falling trees, birds, lightning, etc., and they clear after the breaker interrupts the fault current. To reduce the duration of the outage, the breaker can be automatically reclosed, after a time delay to allow for deionisation of the fault path. If after the first interruption the fault has not cleared, either the breaker is locked open or further reclosing attempts are made before locking the breaker open. Tripping and reclosing can be either 1-phase or 3-phase, depending on the selectivity of the protection and the construction of the breaker.

Conventionally, these reclosing sequences are performed by timing relays and hard-wired logic requiring dedicated breaker auxiliary switches. Settings of the number of reclosures, the dead time between reclosures, and the reclaim time are adjustable. Equally, these functions can be performed by common logic using a shared data base to reduce wiring and auxiliary contacts on the breaker. An additional advantage appears with microprocessor-based logic in that parameters such as dead times and 1-phase or 3-phase reclosing can be selected and down-loaded from the control centre, depending on the network conditions.

17.7.4.5 Event recording

Event recording in a substation can be divided into two basic categories: one that records alarms, contact closures or device operations, and another that records wave-forms of currents and voltages prior to, during and immediately after a fault on the power system.

The first category detects changes of state of contacts with a time resolution of 10 ms and records the operation on a printer. Thus, in post-event analysis, the maintenance engineer has a record of the times at which the events occurred and their chronological sequence.

The second category is an 'oscillo-perturbo-graph', which continuously records, temporarily memorises and then erases the wave-forms of the currents and voltages in a circuit. After the detection of a fault, the memorised pre-fault, fault and post-fault data (which can include contact operation) is permanently recorded for subsequent analysis. The conventional equipment for faulting recording is an electro-mechanical device that has a galvanometer logger writing the wave-forms on to a rotating inked cylinder. At the end of the cylinder rotation, the marks made by the recording pen are erased and replaced with a later value. When a fault is detected, paper is brought into contact with the rotating cylinder to print the recorded parameters. This device requires periodic maintenance, but it has been effective for many years. It can be superseded by a microprocessor-based digital system but, as the scanning speed required to record the fault parameters is of the order of 1000 Hz, it is unlikely that this function can be incorporated in a central substation logic. One advantage of digital recording is that the data can be transmitted to a remote control centre for subsequent post-fault analysis.

17.7.4.6 Protection

Protection is a vast subject, the principles of which have been covered elsewhere. It suffices here to mention it as a part of substation automation. Because of the high scanning speeds required, it is unlikely that primary protection can be integrated into a central substation automation system. Thus, even though micro-processor-based protection will be available, it will be dedicated to a particular task or circuit as with the present solutions.

References

1 LEUZINGER, J. and BAUMANN, R., 'BECOS 10—A Software Package for Out-Stations and Small Dispatching Centres', *Brown Boveri Review*, **66**, 175–180, March (1979)
2 JERABEK, A. and RISCHEL, H., 'BECOS 20—A Software System for Regional Dispatching Centres', *Brown Boveri Review*, **66**, 181–187, March (1979)

17.8 Decentralised control: pulse controllers for voltage control with tap-changing transformers

As a result of ever-increasing automation and rationalisation, electronic pulse controllers are becoming widely used in the field of tap-changer controls. In interconnected operation, maintenance of frequency characterises the equilibrium between generation and consumption of active power, while voltage control determines the control of reactive power in the system. The essential difference between the frequency/active-power and the voltage/reactive-power characteristics is that the frequency has the same value throughout the system while the voltage forms a system of ever-varying peaks and valleys (assuming a constant rated voltage) which in turn decides the direction and magnitude of reactive power flow.

Tap-changing transformers are the variable step functions in this range of peaks and valleys, and this introduces additional freedom at certain points in the interconnected system. Depending on their place of application, tap-changing transformers can be generator or interconnecting or consumer-load units.

The task of the pulse control unit is automatically to maintain the voltage in the system or to direct the reactive power flow. *Figure 17.45* illustrates an example of application–the control of a consumer network.

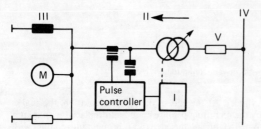

Figure 17.45 Consumer network regulated by tapped transformer: I, tap-changer; II, reactive power flow; III, consumer network; IV, supply network; V, line reactance

The voltage to be regulated is compared in the regulator with an adjustable reference value. Depending on the polarity and magnitude of the difference between these values, 'higher' or 'lower' impulses are given, resulting in the necessary adjustment of the tap-changer. The impulse sequence is inversely proportional to the difference signal. The integration time t_1 and the pulse duration t_2 are adjustable. As long as the difference signal is smaller than the set sensitivity, the impulses are blocked. A further adaptation of the impulse sequence is possible owing to a time factor so that a stable and quasi-steady regulation may always be obtained. *Figure 17.46* shows a typical characteristic of a final control element with a stepwise mode of operation.

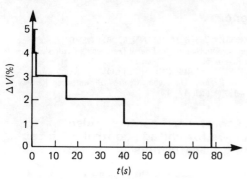

Figure 17.46 Typical characteristic of a final control element with steps of 1%: $\Delta V = 5\%$, $\alpha = 0.25$, $\varepsilon = 0.8\%$, $t_1 = 50$ s, $t_2 = 0.2$ s, $t_3 = 0.5$ s

It is important that the controller does not cause the tap-changer to carry out unnecessary switching operations. If, for example, a short-time fault were to occur on the system, no purpose would be served by the tap-changer responding. The fault would be rectified before the switching operation could be completed and the controller would then immediately issue commands for the tap-changer to return to the original position.

Adequate damping is therefore essential for preserving the life of the tap-changer.

If the control deviation tends to exceed the set sensitivity, an integrator comes into operation. No control command is issued before the pre-set integration time expires.

17.9 Centralised control

17.9.1 Hardware and software systems

In a control centre, the basic requirement of the operator is information about the power system, presented to him in a clear and unambiguous manner. He needs an overview of the network, usually in the form of a large wall diagram giving the network configuration and perhaps coarse line load levels. He also needs detailed diagrams of individual parts of the network showing the load of each h.v. circuit plus its switching configuration by isolator, circuit-breaker and earth-switch positions. These circuit diagrams are selectable on to colour visual display units (VDUs). *Figure 17.48* gives an overview of a typical control room.

In addition to the basic requirements, the operator requires other information to be calculated and automatic controls to be performed, as discussed later. These requirements can be realistically met only by an on-line real-time computer system interfacing with the telecontrol system to provide up-to-date network data and to accept commands.

The availability of such a system must be of a very high order (99% plus); a single-processor system would not be adequate. Multi-processor systems must therefore be considered, the classic configuration of which is the dual main computer system with front-end processors for the telecontrol system, and an independent wall-diagram control.

17.9.2 Hardware configuration

The main computer hardware is configured as in *Figure 17.49* to allow either computer to operate in an on-line, hot stand-by or

Figure 17.48 Overview of a typical control room

off-line mode. In the on-line mode, all (or most of) the peripheral devices are connected to the on-line machine. The on-line computer communicates with the remote terminals via the front-end systems and with the man–machine interface equipment in the control room.

In the 'hot stand-by' mode, the second computer is available and updated with all the power-system data, ready to take over control if the on-line computer fails. To enable the stand-by computer to assume control in the shortest possible time, it should be updated, preferably simultaneously with the main computer. Thus, on fail-over, time is not lost on a general interrogation of all the RTUs. Any state changes that could not be accepted during the switching operation will be held in the front-end system until they can be transferred.

The simultaneous updating of both on-line and stand-by computers is possible with a data-bus connection between all computers of the control system. The data from the front-end computers is output for the bus and is accepted by all partners on the bus. Similarly, commands and configuration instructions are accepted by the front-ends and noted by the stand-by computer; thus it can keep itself fully informed of the power network and telecontrol network configuration. The input from the console keyboard can be transferred either via the data bus or via an additional communication channel.

17.9.3 Man–machine interface

The operation peripherals forming the man–machine interface equipment can be switched either individually or as a group to either computer. Normally, they are switched to the on-line machine. The main man–machine interfaces are the console keyboards and the graphic or semi-graphic VDUs. Network diagrams, alphanumeric information, trend curves and histograms can be selected for display on the VDU screens (see *Figure 17.50*). The displays consist of both fixed and variable data, the latter being updated as necessary. For example, when a station diagram is displayed, the switch-position indications are automatically updated after the information of the change of state has been received from the substation. Measurand values are also updated on the screen at the same rate as the telecontrol cycle. For trend curves, the screen is used rather like a multi-pen chart recorder, using colours to identify the curves.

The keyboards contain both functional keys for operations that are repeated frequently, and alphanumeric keys for inputting numerical data and text. The keyboards are interactive with the displays on the VDUs, which allows parameter changes and device control by identification of the object to be addressed by device or position reference input, or via the function and alphanumeric keys, or by the positioning of a cursor. The cursor movement is controlled by tracker-ball, joystick or direction keys. Thus a dialogue is possible between the operator and the computer system to select displays, to give commands, and to input data for limits, set-points, calculation parameters etc. Text, for tagging or recording or sanctions, can be temporarily added to the displays, entered via the alphanumeric keyboard.

The VDU screens can be either full-graphic (where the cathode-ray tube spot can be moved in any direction under program control) or the more economic point-graphic system which operates on a scanning system where characters and symbols can be created from a pattern of dots on the screen. From the three primary colours (red, green and blue) plus combinations of these phosphors, seven colours can be used. The background can be changed, or the characters flashed, for high-lighting.

17.9.4 Wall diagram

The wall diagram serves to give the operator an overview of the power system network; however, few details need to be given as these are available on the VDU screens. Nevertheless, the wall diagram can be used as a back-up to the VDU system and, in addition to showing the network topology, it can also display the loading of the network with line load indicators. In their simplest form, the line load indicators give direction of power flow and load level in quartiles, by lamps or light-emitting diodes. The wall diagram can also be used to display some alarms if, for example, a substation has an alarm-state current or if it has been blocked from the telecontrol scan.

The drive unit for the wall diagram can be the main computer or a separate unit deriving the data direct from the front-end computers, for example over the data bus. In the latter case, the wall-diagram system can serve as a back-up control system in the event of both the main computer being unavailable. If a separate console is added, commands can be sent and a limited number of measurands can be displayed under emergency conditions.

17.9.5 Hard copy

Events and alarms are recorded for subsequent analysis on event and alarm printers. If a printer fails, the message is output to the alternative unit. The printers operate at a speed of say 100 characters/s. They are therefore suitable for small amounts of data; however, output from a study program producing large quantities of data would be tedious on a small machine. For this purpose, high-speed printers are used, operating at speeds of say 180 characters/s; alternatively, line printers, printing at say 300 lines/min, are used.

Other hard-copy devices are available that can copy either alphanumeric or graphic data displayed on the VDU screen, to make a permanent record of a display existing at a particular moment.

17.9.6 Software configuration

The software system of a computer-based control system consists of many individual tasks which fall into three basic categories of real-time, extended real-time and batch. The programs of the real-time group, such as data acquisition, man–machine interface, automatic generator control etc., have to respond to external events within a given time. The response time of extended real-time programs, such as state estimation or economic dispatch control, is not so critical, though these programs work with real-time data. Extended real-time programs, and batch programs, have no stringent response-time definition.

In a centralised network control system, there may be several hundred programs and subroutines all competing for the limited hardware resources; they must therefore share the computer time and store. Consequently, an overall co-ordination system is required to allocate the processor time and the memory space and access. A real-time multi-programming operating system and a data-base management system ensure that all programs can share the computer system resources and all data, without mutual exclusions and inter-program or data corruption.

The heart of an operating system is the real-time executive which allocates the computer resources, in order of priority, to the programs requiring them at a particular moment. When real-time or extended real-time programs are required to run, they need high priority on the system resources including main memory space and processor time. The executive allocates the resource in order of priority by assigning partitions of main memory to a program and the appropriate input/output handler routines. When the higher-priority programs do not require the resources, the batch (off-line) programs can run in turn. Programs can be interrupted at check-points in their sequence and be delayed while a program of higher priority is executed.

The priority control is triggered by hardware and software interrupts which can be externally or internally generated. The interrupts are assigned levels of priority. Several interrupts can be assigned to the same level, in which case they are queued to run when the resources are available to that priority level.

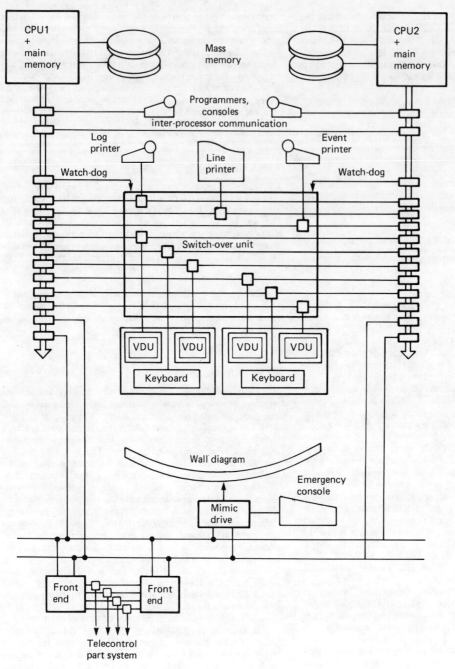

Figure 17.49 Central control system configuration: CPU 1, CPU 2, central processor units

17.9.7 Memory management

The limited main memory must be shared between many programs which cannot all reside simultaneously in the main memory. Thus they must be loaded into a partition of the main memory, from the mass memory (usually disc) when they are required to run. Once a program has been executed, the partition can be liberated for another program. The memory management system governs the allocation of the main memory, which may also require rearrangement of the space already allocated to obtain a contiguous area for a partition large enough for a program that has been called to run. A very large program may be too large to run in the available space, in which case the program itself must be divided into partitions which must be called into a sub-partition of the allocated space.

Programs that run frequently and need fast response times are normally in the main memory, as this avoids loading time from disc, which may be between 40 and 120 ms for a program of say

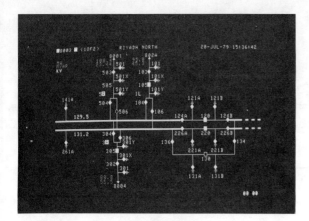

Figure 17.50 A typical VDU display

5000 words. It is obviously desirable to keep the disc access to a minimum, but it becomes an unavoidable overhead for programs that overflow their partition allocation.

17.9.8 Input/output control

The operating system incorporates peripheral-device drive routines which are available for all application programs and can be called to read data from, or write data to, the peripheral devices. The driver conducts the communication with the device and reports malfunction or non-availability of the device. The operating system temporarily allocates the driver to an application program requesting it.

17.9.9 Scheduling

Some programs (e.g. Automatic Generation Control, AGC) must run periodically, others must run at certain times of the day (e.g. daily log), and yet others after a certain delay (e.g. time-outs). All timers for this program scheduling are available in the operating system, which also keeps track of the date. The computer time can be periodically corrected by reference to a 'standard' time unit external to the computer system.

17.9.10 Error recovery

An important part of the operating system lies in the detection of errors in hardware and software, and in taking subsequent action to ensure the security of the system. The errors may be device errors which, if the device is redundant, do not jeopardise the satisfactory running of the system. In the case of serious errors from which the system cannot recover (e.g. disc-drive or a recurrent parity failure), the errors must initiate a switch-over to the stand-by machine.

At the detection of a power failure, the executive arranges that the contents of the volatile registers are stored in the non-volatile memory before the system halts. When the power supply is restored, the system can restart automatically under control of the executive, which restores the system to its former state. However, where program operation (e.g. data acquisition or AGC) is affected by an extended power failure, re-initialising is necessary.

17.9.11 Program development

Throughout the life of the control system, power-network extensions will have to be added to the data base, and program development will be required without interference with the on-line systems. The operating system must permit and assist this work; hence it must include language compilers (for languages such as FORTRAN, Coral, Pascal and Assembler) as well as editing and debugging facilities.

17.9.12 Inter-processor communication

If the control centre is a part of a hierarchical network control system and must be capable of exchanging data with its neighbour, a method of processor-to-processor communication is necessary. The communication system should have minimal effect on the computer system; thus it should have direct memory access to perform cyclic redundancy encoding and checking (CRC) with separate hardware logic and buffers. The security of the code should have a minimum Hamming distance of 4 and automatic repetition of messages if errors are detected.

17.9.13 Data base

The many programs and subroutines that make up the control-centre software require access to data that, in many cases, are common to several programs. For example, the display update routines and the state estimation program both require the same measurands supplied by the data acquisition program. To avoid the necessity of passing data, created or acquired by one program, to many different programs that may require it, it is more convenient, secure and economical to have the data stored once only. If the central data storage is organised and manipulated by a data management system, the data storage becomes independent of any application program. Data reorganisation then necessitates no modifications to the programs using the data, an essential requirement if developments and extensions are to take place economically and with the minimum interference to the working system.

The management system organises and keeps track of all system data and makes them available to any authorised program, through standard access routines. Thus, in a multi-programming environment, a certain discipline is imposed on the data users, which prevents accidental corruption of the system data. The management system also guarantees consistency of the data: for example, when an interdependent set of data are being modified, access by other programs must be prevented until modification is complete. To access the data, the user program holds the data access references and, when calling for the data, presents these references to the access routines. The access routines refer to a 'schema' that contains a unique definition of each data item in the data base. Thus, as long as the references do not change, even though the data have changed their physical position, the user program is unaffected by data-base extensions or reorganisation.

17.9.14 System software structure

The power system network is regularly expanding as new lines and substations are commissioned. The software structure must allow the control system to expand in step with the network to include not only additional data but also the new network control facilities demanded as the complexity of the network increases. The software structure must therefore be flexible so that modifications or additions have minimal effect on the rest of the system.

The ideal method of achieving this goal lies in the modularity of all application programs, whose execution is organised by the operating system and by co-operating with other programs via the data base (see *Figure 17.51*). However, with well-established co-operating programs, in the interest of response time this is not always necessary or even desirable. Thus, in certain cases, it is advantageous to exchange data directly between programs.

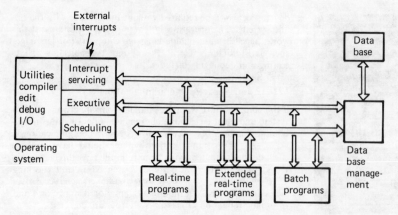

Figure 17.51 Software configuration

References

1 REICHERT, K., 'Application Software for Power System Operation', *Brown Boveri Review*, **66**, 197–203, March (1979)
2 FROST, R., HUYNEN, M. and STAHL, U., 'BECOS 30 Software System for Large Dispatching Centres', *Brown Boveri Review*, **66**, 188–196, March (1979)
3 REICHERT, K., 'Systems Engineering for Power System Operation', *Brown Boveri Review*, **66**, 225–233, March (1979)
4 SCHAFFER, G., 'Power Application Software for the Operation of Power Supply Systems', *Brown Boveri Review*, **70**, 28–35, January/February (1983)

17.10 System operation

The various controls and the incidence of a vast number of disturbances make the operation of a system a very complex task. It is manageable only by the allocation of appropriate sub-tasks to different hierarchical levels; this breaks down the problems, which can then be handled effectively by man or machine.

So long as changes in load and topology, and fault incidence, appear as foreseen in the planning stage, the power system is able to maintain stable conditions through the action of decentralised control. This is true for protective devices and voltage, load and frequency control. However, whenever topological changes take place or become necessary, limits are reached or potential risks appear, control actions that cannot be handled at the decentralised level are imminent. This is where centralised control has to take on its important role.

To what degree these control actions can be executed by automatic means is still an open question. The only centralised automatic control system is that for load–frequency control (automatic generator control) together with economic dispatching. Switching operations, and start-up and disconnection of generators, are still subject to manual intervention. However, extensive support for the human operation is provided by the computer-based control system. Primarily, it provides the real-time information about incidents in the power system; next, it forecasts load development and the effects of possible faults, facilitating efficient decision-making. Beyond that there is a host of control functions, which can be conceived as a black box, having a set of inputs and a set of outputs. They can be called upon by manual intervention or by another function. The output is available after the lapse of a certain response time. The mode of using these functions is either repetitive or event-driven.

As seen, most of these system control functions are implemented in terms of digital programs, i.e. by software. However, some functions are realised by hardware.

In order to illustrate typical software necessary for the control of a power system, the structure of the framework of functions is given in *Figure 17.52*. It comprises centralised control only, and is typical for the control centre of a utility. It shows distinct categories of functions that are characterised by data acquisition (level A); dynamic control (e.g. frequency) (B), security assessment and optimisation (C), and adaptation (D). A common link is established by the data bases (real-time, operation planning). Commands generated by the functions or originating from the human operator are executed via the telecontrol system).

Reference

1 MILLER, R., *Power System Operation*, McGraw Hill (1970)

17.11 Reliability considerations for system control

17.11.1 Introduction

The overall objective of power system operation, namely the provision of power to the customer at all times (i.e. with a high availability), places certain obvious requirements on the reliability and performance of power system components and their controls. Accidental events in the outside world (e.g. the atmosphere) that influence the system, failures of system components and imperfections of the control system including the operator can all affect the performance of the overall system and the final objective, and must therefore be taken account of. The positive or negative contributions of the generation, transmission and distribution systems should also be distinguished. For these subsystems, statistics on the failure rates, unavailabilities etc. are given, and the question arises as to what degree system control in the widest sense can improve the performance of the system, i.e. the availability of power and energy.

In a more detailed analysis of an overall system with the objective as outlined above, a closer look has to be taken at the following:

(1) the meaning of availability in a power system, particularly at the consumer's end;
(2) the availability of components and subsystems;
(3) the specification of reliability characteristics for the control functions;
(4) the method of reliability analysis and of availability evaluation;
(5) availability optimisation considering certain cost constraints.

It should be implicitly understood that improvements in system performance can be achieved via both the heavy equipment and the control system. Since certain similarities exist in the

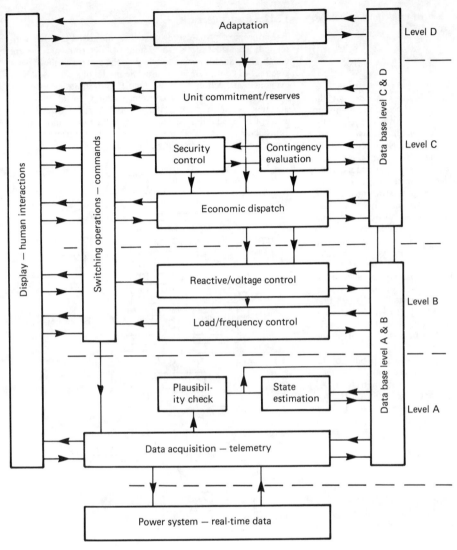

Figure 17.52 Framework of functions: software structure valid for a system control centre

procedures for both domains; the emphasis here is placed on the control side, where hardware and software have to be considered. In addition to hardware considerations, which are very often put in the foreground, we consider the treatment of data and manual interventions.

The main line of the given approach is the establishment of a functional relationship between the supply of power at the consumer's end and the characteristics of the control functions. Bearing this in mind, one can evaluate controls at all levels from protection to the system control centre.

17.11.2 Availability and reliability in the power system

Before working out various details, we discuss the meaning of the terms 'availability' and 'reliability' as applied to a power system. How these concepts are understood depends largely on the object, the location etc. to which they are applied.

Here three points of view are considered; namely those of (i) the consumer, (ii) the utility and (iii) the system specialist. Quite different considerations and requirements apply to the three domains.

17.11.2.1 The consumer

At the consumer's end, i.e. at a single load point or at a supply point to an l.v. system, an availability in terms of up times and down times can be defined:

$$A = \frac{\text{MUT}}{\text{MUT} + \text{MDT}} = 1 - \frac{\text{MDT}}{\text{MUT} + \text{MDT}} = 1 - \bar{A}$$

where MUT is the mean up time, MDT is the mean down time, A is the availability and $\bar{A}$ is the unavailability. A represents an average fraction (per unit) of the time during which power and energy could have been delivered to the load point. In developed systems this figure reaches values of up to 0.9995–0.9997, so that the average down time per year is as low as 2–4 h. This applies to single supply points and customers. The figure may vary from point to point and cannot be transferred to a complete voltage level or to the transmission system.

17.11.2.2 The utility

Single outages at supply points and unavailabilities at the consumer's end are undesirable and should be kept to a minimum that is determined by economic considerations. Technically, however, the disconnection of single consumers has no detrimental effects on the overall system. There is simply a certain reduction of the load, which is balanced by the various control mechanisms. The utility must be interested to reduce these outages in accordance with its legal obligations, but its real concern lies in the continuous operation of the transmission and generating system. There, any disturbance that might endanger the overall system is to be avoided. There are concepts that measure the amount of non-served energy accumulated over a given period, e.g. one year, in terms of the maximum load multiplied by 'system minutes', where the system minutes constitute a measure of the unavailability of the system. However, this figure in minutes is of secondary importance to the utility as long as no complete breakdowns of the transmission system occur. This consideration is further supported by the fact that many utilities employ load shedding in order to avoid emergency situations. Load shedding also causes outages, so that consumers are not served, but it is done for the benefit of the integrity of the overall system.

So, what finally counts for the utility is the availability of the transmission system measured in terms of up and down times. Many utilities have remarkable records, i.e. availabilities of 100% over tens of years. However, there have also been catastrophic black-outs lasting for hours in utilities all over the world; these have received considerable attention and have motivated significant research efforts.

17.11.2.3 The control specialist

Here, we consider the reliability aspects of control levels in a power system from a technical point of view. These aspects include protection and decentralised and centralised control. Clearly, improving the availability of control functions will improve the availability of the supply of power and of the system itself. However, the control specialist differentiates between flat improvement of a characteristic of control functions and augmentation of a parameter which might have significant effects on the system. Detailed investigation reveals that the power system is quite tolerant, as it can maintain its function in the absence of certain control functions, at least over a certain period. Hence, the control specialist is interested in identifying those elements and controls that contribute most significantly to the goal of system availability. There are, however, other elements which tolerate a lower availability or a reduced performance, and the control specialist must certainly take a broad view, i.e. he must also consider events and failures with low probability. In the end, it is the concerted effort of all types of control over a long period which constitutes success, i.e. the high performance of the power system under acceptable economic conditions.

To illustrate the interaction between disturbances and control actions on various levels, a cause-and-consequence chart (CCC) for a power system with its primary controls and its system control centre is shown in *Figure 17.53*, and explained in *Table 17.1*. The CCC is the basis for any reliability analysis of a controlled power system. It reveals the causal relations between events and faults on the one hand and effects on the system on the other.

17.11.3 System security

The concept of system security is often discussed together with reliability considerations. It is primarily a deterministic concept which gives an answer as to whether the system can survive a given set of contingencies. It uses the model of system states as explained in Section 17.2 and load-flow techniques to check contingencies. In the basic concept, nothing is said about the duration of the various states (probabilities). However, the concept is amenable to extension to include this, and various approaches to the reliability analysis of the transmission and generation system have followed this direction.

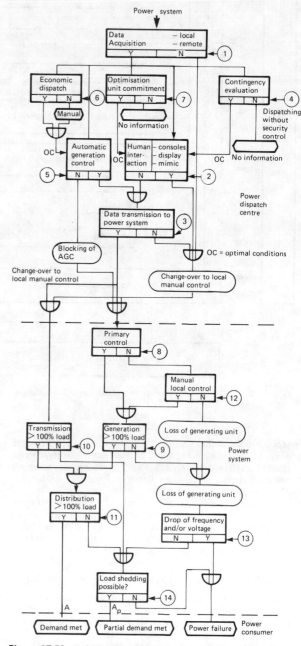

Figure 17.53 A CCC for a power system, its primary control and its system control centre

Table 17.1 Details of the CCC in *Figure 17.53*

Circle	Cause	Consequences
1	Loss of transducers, telemetry system, front-end equipment	Loss of information on system state
2	Failure of computer or peripherals	Loss of information on status of substations, line loading, alarms
3	As for 2	No updating of set-points in power-stations
4	As for 2, loss of application program	No contingency evaluation, no security check
5	As for 4	No automatic generation control
6	As for 4	No economic dispatch control
7	As for 4	No operating planning unit commitment
8	Loss of auxiliary equipment in speed governor	No telemetered change in set-point
9	Loss of generation	Not sufficient reserve
10	Loss of transmission	Not sufficient transmission capacity
11	Loss of distribution	Not sufficient distribution capacity
12	Loss of generation	Reduction of reserve
13	Load demand not met	Load shedding
14	Load shedding not possible	Power failure

Basically, a secure system is understood to be one that can withstand a number of outages, mostly single outages. This leads to the idea of $n-1$ security, in which, out of n components, the disconnection of any component alone does not endanger the operation of the system.

17.11.4 Functions

To assess the various control functions and their contributions to the availability of the power system, it is necessary to know their structure and framework as given in *Figure 17.52*, though with much additional detail and an allocation to the various hardware components. On the basis of such a structure and its functional relations, the possible contribution of the control system could be assessed by a simulation. In principle the following relations must hold.

The contribution of a *perfect* control system to the availability A is given by

$$A = A' + \Delta A_0$$

where A' is the availability of the uncontrolled power system and ΔA_0 the possible contribution of the control system. A *real* control system, however, has itself got a finite availability A_c; hence

$$A = A' + A_c \Delta A_0$$

The availability A_c of the control system can be derived from the various functions f_j and their individual functional availabilities A_{cj} under the assumption that certain stationary probabilities g_j are known which give a rate at which the function f_j are called upon. The probabilities g_j must add up to unity:

$$\sum_j g_j = 1$$

then

$$A_c = \sum_j g_j A_{cj}$$

A_{cj} is a functional availability that applies directly to a so-called RT function (real-time function). Its contribution comes from its on-line operation. It is called upon whenever an event bearing some risk for the power system arises. In contrast, so-called PR functions (preventive functions) condition the system for the event in advance. Their operation does not coincide with the appearance of the event, and the control function is not needed in the event. Thus, there are less stringent requirements for the actual availability A_{aj} of such a function. For a more detailed discussion of this subject, see references 1 and 2.

17.11.5 Impact of system control

At this point, the question arises as to the possible contribution of system control to the availability of the power system. The answer would be of great value for the design of control systems, telecontrol equipment, regulators and control centres. However, a complete assessment for a real system seems impossible. With the help of a mathematical model, though, a limited answer can be given, at least in terms of the relative merits of the various functions. The treatment of such a model requires a Monte-Carlo simulation of the system behaviour over sufficiently long time periods.

Such simulations prove that the performance of the power system can actually be improved, although the degree of improvement will depend upon the inherent performance, the loading, redundancy etc. (for details see reference 3). It can also be shown that the distinction between RT functions and PR functions is well justified. It turns out that repair plays an important role for the PR functions, which is further supported by the fact that the requirements for control change with the daily load cycle. Stress situations appear two or three times per day. In

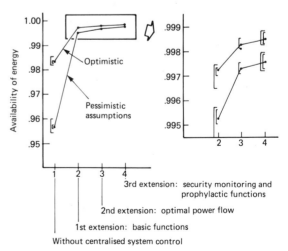

Figure 17.54 Effectiveness of different functions:
+ mean value (five simulation runs)
[range of results assuming small uncertainty within the failure data
[range of results assuming large uncertainty within the failure data

between, the system can be conditioned for a possible event. A PR function need not be available at a specified time but can be delayed. Thus, repair is possible.

In order to give an idea of such a result, an example from reference 3 is presented in *Figure 17.54*. The figure and its enlarged section show the increase in the availability of energy as a function of the categories of control functions. Clearly, the most significant improvement can be achieved by the basic functions of system control like SCADA. Further improvements are harder to realise, but some gain is still possible. The remaining unavailability is due to the particular structure of the test system. It is an isolated system with a peak load of about 4000 MW. The system was assumed to be heavily loaded. Thus, control had a limited effect because of the lack of reserves in the system.

17.11.6 Conclusions

As far as the performance of a power system as expressed by its availability is concerned, three domains have to be considered: (i) the generation system and its reserves; (ii) the transmission system; (iii) the control system. Assuming that the first two systems are fixed, it has been shown that system control will improve the performance. Quite detailed studies are necessary in order to evaluate the contributions of the various control functions. In such a treatment the final question concerns the configuration of the computer system.

Over the years one basic set-up has evolved which has not changed very much and seems to prove its validity as time goes on. It is the multi-computer concept on three levels having several front-end computers, a double system on the main level and a background computer. The configuration is shown in *Figure 17.55*.

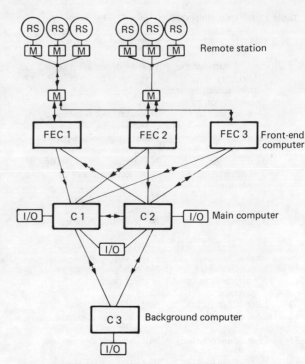

Figure 17.55 Hierachical and redundant multi-computer system with three levels: I/O, input/output; FEC, front-end computer; C, main processor; M, modem; RS, remote station

References

1 FREY, H., GLAVITSCH, H. and WAHL, H., 'Availability of Power as Affected by the Characteristics of the System Control Center, Part I: Specification and Evaluation', *Proc. IFAC Symposium on Automatic Control and Protection of Electric Power Systems, Melbourne, February 21–25 (1977)*

2 FREY, H., GLAVITSCH, H. and WAHL, H., 'Availability of Power as Affected by the Characteristics of the System Control Centre, Part II: Realization and Conclusions', *Proc. IFAC Symposium on Automatic Control and Protection of Electric Power Systems, Melbourne, February 21–25 (1977)*

3 GLAVITSCH, H. and KAISER, W., 'Assessment of Reliability Parameters of the Power System and their Dependence on Control Functions', *Paper No. 81 SC05*, CIGRE SC32 Study Committee Meeting, Rio de Janeiro, September 21–24 (1981)

18 Power System Planning and Economics

T E Norris, BSc, FIMechE, FIEE
Merz and McLellan Consulting Engineers

T A C Dulley, MA, MIEE
Merz and McLellan Consulting Engineers

M D Dwek, BSc, FIEE
Merz and McLellan Consulting Engineers

Contents

18.1 Load forecasts 18/3
    18.1.1 Forecasting for development 18/3
    18.1.2 Forecasting for operation and maintenance 18/4

18.2 Reliability of supply 18/4
    18.2.1 Availability of plant and equipment 18/4
    18.2.2 Reserve capacity 18/5
    18.2.3 System design 18/5

18.3 Legislation 18/6

18.4 Choice of generating plant 18/6
    18.4.1 Load curves 18/6
    18.4.2 Base-load and peak-load plant 18/7
    18.4.3 Types of plant 18/7
    18.4.4 Standby plant 18/9
    18.4.5 Load shedding 18/9

18.5 Plant scheduling 18/9
    18.5.1 Thermal plant 18/9
    18.5.2 Hydro-electric plant 18/9
    18.5.3 Mixed thermal and hydroelectric plant 18/10
    18.5.4 Pumped storage plant 18/10

18.6 Revenue 18/10
    18.6.1 Tariff structure 18/10
    18.6.2 Forms of tariff 18/11
    18.6.3 Connection charges 18/11
    18.6.4 Typical tariffs 18/12
    18.6.5 Metering 18/13

18.7 British electricity supply systems 18/13
    18.7.1 History 18/13
    18.7.2 275 kV network 18/14
    18.7.3 400 kV network 18/14
    18.7.4 Substations 18/15
    18.7.5 Direct current transmission 18/15

For an electricity supply undertaking planning is essential if a cheap and reliable supply of electricity is to be made available to those who desire it. This planning ranges from the ability to meet the expected load a few minutes ahead to the optimum development programme to meet the expected needs of the consumers during the next decade or more. Failure to plan adequately sooner or later leads to a supply which is unreliable or expensive, or both. Reliable information must be available to enable planning engineers and operators to make assessments of the future requirements. Care must be taken to ensure that the information is correctly interpreted if these assessments are to be realistic. Clearly the future requirements are unlikely to match exactly the forecasts, and the effects and costs of these deviations must be assessed, not least for the determination of the preferred direction of deviation.

In all areas where decisions have to be taken both the technical and economic effects of alternative plans have to be considered. Where decisions have to be made on the minute-by-minute operation of the system, the rules must already be known and codified.

18.1 Load forecasts

18.1.1 Forecasting for development

18.1.1.1 Past statistics

In any expanding system a first approach to forecasting the immediate future is to extrapolate from the immediate past load growth. The growth in the recent past may, however, have been the result of exceptional circumstances, such as a succession of very cold winters or the implementation of a large industrial development on a comparatively small system. The usual statistical method of reducing the effect of such exceptional circumstances is to consider a longer period of time in the past, though not so long that the data obtained are unrepresentative of the current trend. A period of 5 or 6 years is generally accepted as the best compromise, although a slightly longer period can be analysed to compare the more recent with earlier trends. In general, forecasting by extrapolation of past load growth should only be done after a thorough analysis and understanding of the past.

Where sufficient statistical information is available, analysis can be made of, for example, (a) the sales of energy to different consumer or tariff groups — domestic, commercial, industrial, agricultural, etc.; (b) the growth in the numbers of consumers in the different groups and, hence, the specific energy consumption in each; (c) the growth of power demand and its daily, monthly, seasonal and annual variations; (d) the distribution of load between geographical zones, generating sources and substations; (e) the effect of weather on demand and the consequent desirability of correcting data to constant weather conditions; (f) the losses in the system.

In analysing past statistics it is important to isolate from the general body of consumers large individual consumers (say those taking 1% or more of the total energy sales) whose requirements may distort the analysis.

18.1.1.2 Extrapolation

From these past statistics the initial (or current) load and rate of growth for individual consumer groups or for the whole can be determined and almost any method of mathematical extrapolation yields a forecast of requirements for the immediate future, say 1 year, with a high probability of accuracy, provided that there are no exceptional circumstances.

Beyond this immediate future, however, the accuracy obtained by direct extrapolation inevitably decreases and it is important, therefore, to establish a preferred direction of error for any inaccuracies that may arise in the forecast. In the shorter term, load forecasts form the basis for the procurement of plant and equipment, for which a period of 3, 4 or more years may be required. If the timing of such a project were based on a forecast which proved to be low, the plant might not be available in time to meet the load, and load shedding might result. In the short term therefore, a load forecast should err, if at all, on the high side. Longer-term forecasts, on the other hand, form the basis not for the immediate ordering of plant but for the economic planning of system development, and the economics are generally regarded as being on a sounder basis if the load forecast is not too high. Long-term forecasts should in any case be reviewed regularly, and ordering of plant and equipment advanced sufficiently to cater for loads higher than originally forecast.

The most generally accepted form of growth is based on the exponential (or compound growth) law under which the initial (or current) electricity load E_0 (expressed in terms of power or energy) increases at an annual growth rate g to $E_0(1+g)^n$ in n years' time.

Until recently it was common to find that the rate of annual load growth on long-established systems remained at about 6 to 8% over many years. On the other hand, for developing systems and particularly those with an element of suppressed load initially, growth rates as high as 20% are equally common in the first few years after the imbalance between supply and demand of electricity has been rectified. It is uncommon, however, for high rates of growth to be maintained in the longer term, one reason being that continued substantial capital investment would have to be made available for electricity supply in competition with other services, and it is unlikely that the national economy would provide for this. It is generally accepted that, while a growth rate as high as indicated by optimistic assessment of the electricity requirements of such developing systems can be used for the short-term forecast — thus possibly erring on the high side — some regression (or gradual reduction in growth rate) must be allowed for in the longer-term forecast. Indeed, such regression has occurred recently on many long-established systems, as a result of world economic recession.

A common form of regression adopted for longer-term forecasts is represented by the so-called straight line law under which the load increases by a constant annual amount, which in turn represents a gradually smaller rate of growth.

It is believed in some quarters that growth in electricity requirements is subject to cyclical variations, as generally is economic activity to which electricity requirements are related. Under this theory high growth rates are followed by lower rates for only a few years, after which growth rates will increase again.

18.1.1.3 Direct enquiries

As a check on load forecasts based on past load statistics and as an aid to reducing their inaccuracy, it is important to make enquiries from and about existing and potential consumers, or possibly to carry out a full market survey, and to adjust the forecasts accordingly. Such enquiries should include existing and potential large consumers, whose estimated requirements should be added to the forecasts for the general body of consumers.

18.1.1.4 Economic indicators

It can also be worthwhile to analyse statistics of population, household income and expenditure, and production and consumption of different products, all of which bear some relationship to the increasing demand for electricity. Attempts are also made to correlate electricity usage with gross domestic product (GDP) and, hence, to base the forecasts of electricity on forecasts of GDP. This correlation can change significantly with time, and forecasts of GDP are notoriously inaccurate.

18.1.1.5 Comparisons with other places

Where there is less statistical information on past electrical loads, such as in new areas of supply, it becomes necessary to base load estimates more on direct enquiry, demographic and economic statistics and national or world-wide comparisons with other places at a comparable stage of development.

18.1.1.6 Maximum and minimum forecasts

As an alternative to preparing a single best estimate of future electricity requirements, the limits of accuracy of the load forecast can themselves be estimated by preparing maximum and minimum forecasts, the latter being used mainly for long-term system planning, subject to periodic review, and the former mainly for short-term procurement of plant and equipment. The use of alternative load forecasts will also facilitate comparison and demonstrate the effect of different rates of growth on the merits of alternative developments.

18.1.2 Forecasting for operation and maintenance

Prediction up to a year ahead is necessary for planning the withdrawal of plant for maintenance. The more important predictions, however, are of day-to-day and hour-to-hour loading: these are to ensure that plant is available to meet any expected demand and that sufficient spinning reserve is provided, and to enable the optimum economic choice of the available plant to be made.

A basic load for a given time and a particular day in the week is estimated from records, and this is weighted by factors depending on temperature, cloud cover, wind, visibility and rainfall, all of which must be predicted from weather reports. This procedure, largely manual, depends on the skill of the operator, but experience in the UK indicates that a temperature change of 1 °C will produce a change in load of about 1%; an overcast sky as compared with a clear sky will increase load by 3–4%; thick fog may increase it by 10–12%; and wind will increase loads by about 1% for every 4 km/h of its velocity. There is, however, usually a time-lag in any change: for instance, on two successive days at an equally low temperature the load will be significantly higher on the second day.

18.2 Reliability of supply

Absolute reliability of supply cannot be guaranteed, but a very high level of reliability can be, and in developed countries is, achieved. The reliability of supply achieved depends upon (a) the availability of the generating plant and of the transmission and distribution equipment on the system; (b) the margin, known as the reserve capacity, between the installed supply capacity and the expected maximum load; and (c) the design of the system.

18.2.1 Availability of plant and equipment

The overall availability of generating plant and transmission and distribution equipment itself depends upon whether any of the plant or equipment is out of service (a) deliberately, to allow planned maintenance to be carried out (a 'scheduled outage') or because the magnitude of the load does not require it to be in service, or (b) accidentally, as a result of breakdown (a 'forced outage').

The probability of any of the plant or equipment being out of service as a result of either kind of outage can be calculated from the statistical data of performance of existing plant and equipment; but, since maintenance is a matter of planning rather than of statistics, it is often preferable to draw up maintenance programmes when considering scheduled outage in relation to availability of plant and equipment.

18.2.1.1 Scheduled outage

The most effective way of minimising forced outages is to carry out strictly a system of regular planned maintenance. Unless maintenance is carried out regularly, the plant or equipment will ultimately fail, perhaps at the most inopportune time — say when and because it is fully loaded; its repair following such failure may well also take a much longer time to complete than the time required for its routine planned maintenance. Inadequate maintenance, as well as impairing reliability of supply, can also seriously affect efficiency and increase operating costs.

On systems where there are large seasonal variations in demand the maintenance of plant and equipment of the highest capacity should generally be scheduled for the season of low demand. The maintenance of hydro-electric generating plant should generally be scheduled for the dry season (if there is one) when there may not be sufficient water for all the plant anyway.

As implied above, generating plant should normally be overhauled once every year. The extent of this annual maintenance on some plant (for example, hydro-electric generating sets and steam turbo-generators) may be small and take 2 or 3 weeks to complete in most years, or be of major proportion, taking 1–3 months perhaps only every 4 or 5 years. For other items of generating plant there may be statutory requirements: for example, that boilers as pressure vessels should be inspected at intervals of not more than 29 months in the UK.

On systems where there is little or no seasonal variation in demand, the statistical approach in estimating the probability of scheduled outage is suitable, but the effect of simultaneous outages of plant requires consideration.

Maintenance of static equipment such as overhead lines and transformers consists largely of visual inspection. Regular foot patrolling is usual, but in difficult terrain tracked vehicles or helicopters are necessary; radio communication is, of course, essential, and where very adverse weather conditions may be expected refuge huts must be provided, together with a rescue organisation in case of injury. In carrying out the inspection, various sophisticated techniques are used such as infra-red detectors for indicating hot joints and discharge detectors for indicating excessive corona. Outages to replace damaged insulators, joints or spacers can be minimised by using 'live-line' working. Underground cables cannot be inspected visually; periodic tests, particularly of dielectric power factor, should be made and any adverse change fully investigated.

Maintenance of switchgear and protective gear consists of periodically checking its proper operation, and the condition of contacts in circuit-breakers and of the oil in oil circuit-breakers.

18.2.1.2 Forced outage

The statistical probability of any component of a supply system being forced out of service by its own failure (its 'forced outage probability' or 'forced outage rate') is defined as the proportion of a given period, termed the exposure time, during which it is forced out of service, i.e.

$$\text{forced outage rate } p = \frac{\text{period on forced outage}}{\text{exposure time}}$$

(where exposure time = period on forced outage + period

available for service, i.e. excluding any period on scheduled outage). The corresponding probability of the same component being available for service is sometimes referred to as its service probability or

service rate $q = \dfrac{\text{period available for service}}{\text{exposure time}}$

and clearly $p + q = 1$.

Considerable work has been done, particularly in the USA over the past 40 years, to collect and analyse availability and outage statistics of generating plant. The results of this analysis are now published annually by the North American Electric Reliability Council and include average forced outage rates for different types of plant (steam turbine, nuclear, gas turbine, diesel, hydro, etc.), different components of the plants (boilers, turbines, condensers, etc.) and different plant capacities. These statistics indicate that the average forced outage rates are approximately as follows:

Steam turbine plant:	2–3% for units up to 100 MW, increasing to about 10% at 600 MW and 15% or more for the largest units
Nuclear plant:	10%
Gas turbine and diesel plant:	30%
Hydro-electric plant:	1%

Applying the rules of the theory of probability, if there are n units on a system, all of which have the same forced outage (and therefore service) rate(s) p (and q), the probability of all n units being forced out of service is p^n.

The probability of $n-1$ units being forced out (and therefore of 1 being in service) is $np^{n-1}q$.

The probability of $n-2$ units being forced out (and therefore of 2 being in service) is ${}_nC_2 p^{n-2} q^2$.

In general, therefore, the probability of r units being forced out and $n-r$ units being in service is ${}_nC_r p^{n-r} q^r$. This is the general term of the binomial expansion $(p+q)^n$, by the successive terms of which the probabilities of having $n, n-1, n-2$, etc., units forced out are given.

If, on the other hand, as is normally the case, of the total of n units there are b units with forced outage and service rates p_1 and q_1, respectively, c units with p_2 and q_2, etc., the probabilities of having $n, n-1, n-2$, etc., units forced out are given by the successive terms of the product of the separate binomial expansions relating to each group: $(p_1 + q_1)^b \times (p_2 \times q_2)^c \times \ldots$.

The application of the binomial theorem makes it possible, therefore, to calculate the statistical probability (expressed as a fraction of exposure time) that the capacity of generating plant available for service will be less than the system maximum demand. Such calculations are much facilitated by the use of a computer. If the shape of the system load curve is known, the period during which the load cannot be met in full can be calculated.

Although generating plant has been considered in the above discussion, the same principles apply to transmission and distribution equipment. The forced outage rates of overhead lines are calculated from statistics of the frequencies of fault occurrences, which in the UK range between 1 or 2% per year per 100 km for high-voltage transmission lines, through 5–10% per year for subtransmission and primary distribution lines to 20% per year or more for low-voltage distribution lines. The frequency of faults on underground cables is about one-third to one-half of that on overhead lines.

18.2.2 Reserve capacity

The desired degree of security varies considerably between electricity supply authorities. Since higher security of supply generally results in an increase in cost, security of supply should ideally be fixed at the point where the benefits obtained are equal to the extra cost; but for an economic criterion to be used in assessing the benefits of security of supply, a value has to be assigned to such intangible factors as lost production, lost leisure time and inconvenience caused by loss of supply. It is very difficult to assign values to these factors, since opinion and judgement play a predominant role in their evaluation.

Furthermore, any attempt at basing supply reliability on the value of energy not delivered involves a very detailed investigation of the whole generation, transmission and distribution systems over many years. As a result security of supply and, hence, reserve capacity, have come to be fixed by many undertakings in accordance with rules based on experience, although statistical methods may be used as an aid in determining these rules. Some of these rules are:

(1) *Reserve capacity not less than the capacity of the largest supply unit.* This is the smallest amount of reserve that should be provided if frequent load shedding is to be avoided. This criterion is frequently applied to systems with a small number of supply units. In practical terms, it allows for the largest supply unit, e.g. the largest generator, to be out of service on either scheduled or forced outage, but quickly results in a decrease in the security of supply as the system grows.

(2) *Reserve capacity not less than the combined capacities of the largest and second-largest supply units.* This is a logical step after (1), above. It becomes applicable as the number of supply units on the system increases and its effectiveness depends on the actual sizes of the largest and second-largest units. In practical terms, it allows for the forced outage of the second-largest supply unit simultaneously with scheduled outage of the largest, or vice versa. As the system continues to increase in size, however, the continued used of this rule also leads to a steady decrease in the security of supply.

(3) *Reserve capacity not less than a fixed percentage of the system annual maximum demand forecast.* This is used by many undertakings. The percentage to be used can be assessed by comparison with other systems with comparable characteristics. The fixed percentage method is, however, most applicable to systems in which the capacity of the largest supply unit represents only a small proportion of the maximum demand.

(4) *Reserve capacity such that the statistically determined probability of load shedding is no greater than a chosen value.* In using statistical methods as described above, it is desirable to have statistics of plant availability and forced outage rates and of daily and seasonal load variations, and estimates of the precision of the basic load forecast. It is particularly important to note that the results can only be as good as the statistical data available.

The rule used in any particular case will depend upon the type and size of system. Whichever is in use, care must be taken that, as the system grows and the nature of the load and the capacities of supply units changes, the method used to assess the minimum desired reserve capacity is still the most suitable for the changed system.

UK generating plant margins The planned margin of reserve capacity in the UK is currently about 28% of the estimated future maximum demand on the system.

18.2.3 System design

There are many ways in which the design of the electricity system may affect the reliability of supply. For example, an arrangement in which a number of boilers in a power station supply a number of steam turbo-generators through a common steam header or main is inherently more reliable than one in which each boiler and turbo-generator form a separate unit; in the former case the loss of any single boiler need not result in the loss of the whole output of a generator, provided that the respective boiler and generator capacities are correctly co-ordinated. On the other hand, an arrangement of two generators connected to the system through a single transformer is inherently less reliable than one of generators connected through their own individual transformers. Two single-circuit overhead transmission lines on separate poles or towers are more reliable than a double-circuit line on shared supports. A ring circuit is more reliable than a radial feeder, a closed ring more reliable than an open ring, and a meshed or grid system more reliable still, but very extensively meshed systems can be less reliable because it is difficult to provide discriminative protection on such systems.

Minimum levels of security in the UK The minimum level of security adopted by the electricity supply industry in the UK depends on the demand of the load and is classified as shown in *Table 18.1*.

The level of demand met after circuit outages is determined largely by the design of the supply system. For example, under Class B an alternative supply may be made available within 3 h of an outage by manual and perhaps by extensive switching operations on the system. Under Class C two normally closed circuits, or one circuit with automatic switching of an alternative circuit following an outage, may be provided. Class E may involve three separate supply circuits. Class F, not shown in *Table 18.1*, applies to loads in excess of 1500 MW connected to the 275 kV or 400 kV supergrid and subject to special conditions of supply.

18.3 Legislation

The power system engineer has to operate within the provisions of many Acts of Parliament and other regulations designed to ensure the safety of operators, consumers and the general public; to establish the rights and responsibilities of those concerned; to minimise disturbance to scenic and other amenities; and to ensure a proper standard of workmanship. The more important Acts and Regulations applicable in Britain are listed below.

Electricity Supply Regulations, 1937, for securing the safety of the public and for ensuring a proper and sufficient supply of electrical energy.

Hydro-Electric Development (Scotland) Act, 1943, to establish the North of Scotland Hydro-Electric Board for the development of supplies of electricity in the north of Scotland; and to authorise the Board to generate and supply electricity.

Electricity (Factories Act) Special Regulations, 1944, which are primarily to safeguard operators in factories, including substations. The first regulation is all-embracing and provides that 'all apparatus and conductors shall be sufficient in size for the work they are called upon to do and so constructed, installed, protected, worked and maintained as to prevent danger so far as is reasonably practicable'. The remaining regulations give a more detailed code applicable to particular circumstances.

Electricity Act, 1947, to establish the British Electricity Authority with responsibility for generation and transmission, and of 14 Area Electricity Boards with responsibility for distribution, throughout England, Wales and the south of Scotland.

Public Utilities and Street Works Act, 1950, which requires electricity supply authorities to obtain approval from highway authorities before opening a street for cable laying.

Electricity Reorganisation (Scotland) Act, 1954, to establish the South of Scotland Electricity Board and to transfer to it the responsibilities in the south of Scotland for generation and transmission from the British Electricity Authority, renamed Central Electricity Authority, and for distribution from the two Scottish Area Electricity Boards.

Electricity Act, 1957, to dissolve the Central Electricity Authority, to establish the Central Electricity Generating Board and the Electricity Council, and to transfer to them, respectively, the executive (i.e. generation and transmission) and supervisory functions of the Central Electricity Authority.

Health and Safety at Work Act, 1974, to secure the health, safety and welfare of persons at work, to protect others against risks to health or safety in connection with the activities of persons at work, to control the keeping and use and prevent the unlawful acquisition, possession and use of dangerous substances, and to control certain emissions into the atmosphere.

Town and Country Planning Acts, under which before a generating station, substation or transmission line can be erected the approval of the local planning authority must be sought; if objections are raised, there may be a public enquiry, after which the Minister of State will agree to the proposal, impose any conditions he may think proper or reject it in its entirety.

IEE Regulations, which deal with the wiring of buildings and with permanent and temporary wiring in general, with emphasis on the safety of the user; although not legally enforceable, they are frequently quoted as the example of recommended practice.

British Standards, issued by the British Standards Institution, which lay down dimensions, performance and quality requirements of a great variety of electrical (and other) equipment ranging from turbo-alternators to domestic appliances and wiring details.

18.4 Choice of generating plant

18.4.1 Load curves

A typical daily load curve for a system is shown in *Figure 18.1*, with the minimum load, or trough, occurring in the middle of the night and the maximum, or peak, load in the early evening. The

Table 18.1 Minimum security levels, UK

Class of supply	Group demand (MW)	Minimum demand to be met after first circuit outage	second circuit outage
A	up to 1	RT: GD	nil
B	1+ to 12	(a) Wn 3 h: (GD−1) (b) RT: GD	nil
C	12+ to 60	(a) Wn 15 min: smaller of (GD−12) and $\frac{2}{3}$GD (b) Wn 3 h: GD	nil
D	60+ to 300	(a) Im: (GD−20) (automatically disconnected) (b) Wn 3 h: GD	(c) Wn 3 h: for GD>100, smaller of (GD−100) and $\frac{1}{3}$GD (d) Wn time to restore arranged outage: GD
E	300+ to 1500	(a) Im: GD	(b) Im: $\frac{2}{3}$GD (c) Wn time to restore arranged outage: GD

GD = group demand (MW); Im = immediate; RT = in repair time; Wn = within.

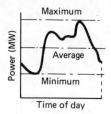

Figure 18.1 Typical system daily load curve

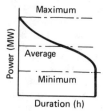

Figure 18.2 System load-duration curve

corresponding system load–duration curve is shown in *Figure 18.2*; the abscissa at any point on this curve represents the total time during that day when the load given by the ordinate or a higher load has to be supplied. Similar system load–duration curves can be produced for any period, the most common being a week, a month or a year.

The ratio between the average and the maximum load on the system is the *system load factor*, while the area under the load-duration curve represents the *system energy requirements*, for the period considered. Typical values of system annual load factors can be as low as 10% for developing systems with only a domestic lighting load, and up to 50–75% for highly developed systems with mixed domestic, industrial and other loads, the higher end of this range being more common on systems with little seasonal variation in load.

The way in which the system load can in theory be supplied by the generating plant is illustrated by the plant load–duration curve in *Figure 18.3*, in which the system load–duration curve of *Figure 18.2* is shown dotted. The peak level on the plant load-

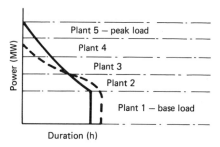

Figure 18.3 Plant load-duration curve

duration curve is the sum of the capacities of all the generating plant available during the period considered. The difference between the peak levels on the plant and system load–duration curves represents the reserve capacity available to offset unscheduled loss (forced outage) of generating plant. Typical values of reserve capacities lie between 10 and 40% of system maximum demands, depending upon the desired security of supply.

In *Figure 18.3* it is assumed that all the available plant, including that providing the reserve capacity, operates at least some of the time during the period considered. It is also assumed that no plant operates at its full capacity continuously for the period considered, i.e. at a capacity factor of 100%; this is in practice true if the period considered is, say, a year, but may not be true if the period is a day.

For the lowest total operating costs it is axiomatic that the plant with the lowest individual operating cost should run for the longest and that with the highest cost for the shortest possible time.

The plant with the lowest operating cost (Plant 1 in *Figure 18.3*) therefore operates whenever it is available, to supply the base load at the maximum plant capacity factor; the plant (5) with the highest operating cost operates only infrequently and for short periods to supply the peak load at the minimum plant capacity factor; and the remaining plants (2, 3 and 4) supply the intermediate load at capacity factors which are gradually higher the lower their operating costs.

18.4.2 Base-load and peak-load plant

In large systems with many generating plants of various types the choice of plant to supply the different tranches of load as shown in *Figure 18.3* is of considerable economic significance.

For base-load operation the chief requirements are low specific operating cost (i.e. cost/kWh supplied) and high availability. Since the capital cost of the plant is spread over a large amount of energy, high capital cost is generally acceptable, as also is high incremental capital cost (for instance, of a refinement or improvement) if it results in lower specific operating cost.

It is more difficult to secure an economic solution to the problem of supplying peak load. The desirable requirements are: (a) ability to start and take full load within, say, 30 min; (b) low capital cost in view of the small annual output, with operating cost only a secondary consideration; (c) a siting near the load centre, or accessible transmission facilities to minimise transmission costs.

It is advantageous if the peak-load plant can, in emergency, be used as a back-up to supply the base load.

18.4.3 Types of plant

18.4.3.1 Fossil-fuelled plant

Large coal-fired steam turbine power stations (say with 4×500 MW units) have a capital cost at mid-1983 price levels of about £500/kW. With oil firing the cost is somewhat lower. The annual efficiency of a large steam turbine power station is about 35% and in the UK the operating cost, i.e. the cost excluding capital charges, is about 2.2 p/kWh supplied. The overall specific cost lies in the range 3.0 p/kWh supplied (at 80% capacity factor) to 3.5 p/kWh (at 50%). Large, modern power stations are operated as far as possible to supply the base load, not only because this leads to lowest system costs but also because operating difficulties may arise (especially with boiler plant) if the plant is operated with varying loads. Thus, the annual capacity factor is not usually less than 50%. On the other hand, the maximum annual capacity factor is not much greater than 80% in practice.

As new plants are built, they supply the increasing base load and replace older base-load plants which can then be used as two-shift, one-shift and eventually peak-load plants. However, the large plants now coming into this category are less suitable for operation at very low capacity factors, and other means must be sought to supply the peak load.

One possibility is steam plant designed specifically for peak loads, with a simple thermal cycle, forced-circulation boilers and rapid-start turbines. The economics of building new peak-load rather than new base-load steam turbine plant (with consequent downgrading of the capacity factors of existing plants) must be carefully considered.

Gas turbines, particularly of the open-cycle type, can start and be loaded within a few minutes, are amenable to remote or automatic control, and because of the small space needed can be sited near a load centre or in the precincts of a major steam turbine station. They are therefore ideally suited to peak-load operation. Their rating is limited at present to about 100 MW. Gas turbines, particularly those used in a peak-load role, suffer from the requirement to operate on clean fuels, such as expensive distillates. In general, the use of gas turbines can be justified for periods of operation of up to a few hundred hours per year.

As stated above, therefore, medium-load stations which supply the intermediate load between base and peak loads (plants 2, 3 and 4 in *Figure 18.3*) are fossil-fuelled stations displaced from base-load operation on an expanding system by the introduction of newer power stations with lower specific operating costs. In future years, however, this progression may not be realised: first, because a plateau of development has been reached and there is no immediate prospect of a reduction in specific operating costs; secondly, because in many countries the growth in demand for electricity is sluggish; and thirdly, because large modern steam turbine plant is not well suited to intermittent duty, as is required to meet the medium ranges of load.

In these circumstances, consideration can be given to the construction of power stations specifically designed for two-shift operation with a relatively low specific operating cost. Among the types of plant are simple steam turbine plant without reheat, and combined-cycle plant incorporating gas and steam turbines. The latter plant suffers the disadvantage that a clean fuel, such as expensive distillates, must be used in the gas turbine. Proposals have been made to integrate coal gasification with gas and steam turbine generation, but it seems that it will be some years before the reliability of such an integrated scheme can be proved, and in any case gasification plant is not suited to operation at varying loads.

Where there is no coal and where steam turbine units of economic capacity are precluded by the magnitude of the load, as is the case on many developing systems, diesel generating plant is used to supply the base load.

18.4.3.2 Nuclear plant

Nuclear power stations cost over twice as much per kW to build as fossil-fuelled stations, but their specific operating costs are up to 50% lower; for minimum costs, therefore, it is even more desirable that nuclear stations, where they are available, be operated at a high annual capacity factor, to supply the base load. Recent experience suggests, however, that it is unrealistic to expect nuclear power stations to operate at annual capacity factors much in excess of 80% on average.

18.4.3.3 Hydro-electric plant

The firm power available from run-of-river hydro-electric stations is determined by the minimum reliable flow of water. Additional power may or may not be available at any time according to the flow. The additional power cannot be relied on and back-up plant, usually fossil-fuelled, must be provided to make this power firm. The amount of firm power can be increased by providing storage and, in the limit, the firm power is then determined by the long-term average rate of flow of water past the site of the storage dam. The cost of the civil engineering works associated with a dam to provide complete regulation of flow is usually high; some lesser amount of regulation is more likely to be provided. If the capacity of the hydro-electric plant is such that the annual plant capacity factor is 100%, the capital cost per kW installed will be relatively high. Hydro-electric plant has the advantage of high availability and may run for years with only a few brief shut-downs for inspection.

The capital cost of a hydro-electric project is not easy to estimate, but once it is built there is little subsequent cost; in contrast, the capital cost of a fossil-fuelled plant can be forecast fairly accurately but future fuel costs are uncertain. Thus hydro-electric plant can be used to supply base load.

Alternatively, its quick-starting facility and ability to handle rapidly varying loads make hydro-electric plant ideal for peak-load operation. However, such plants are often remote from the load centre, and the high capital cost and low operating cost appear, at first sight, to be the reverse of what is required. On the other hand, if the civil engineering works required to handle and store the water of a given catchment area are sufficient to give a continuous output of, say, 10 MW and it is decided to design the plant for operation at a capacity factor of, say, 10%, the only additional expense involved in providing the additional 90 MW of peak-load capacity will be that of the additional generating units and pipeline, together with a transmission system capable of a greater power transfer. Thus, the incremental cost of the peak-load plant in such a case is quite small. Of course, such a plant could never operate continuously at its full capacity, as insufficient water would be available. It is desirable that the tailrace of such a peak-load plant should be a lake, otherwise the discharge, in 1 or 2 h, of a day's flow may cause difficulties lower downstream.

If a hydro-electric plant is designed and built at the outset for high capacity factor operation, it cannot subsequently be moved up the load–duration curve as may be done with a steam turbine plant. To do so would necessitate increasing the generating capacity, since the full use of the available water is axiomatic: generating sets could be added fairly cheaply, but the cost of additional pipelines and other civil engineering works might be prohibitive, unless suitable provision had been made in the initial installation.

18.4.3.4 Pumped storage plant

The operational merits of hydro-electric generating plant for peak-load operation can be combined with pumped storage in a high-level reservoir during off-peak periods.

If the pumping energy is supplied by conventional hydro-electric plant, the specific operating cost of pumping is low and the efficiency of pumping is of little importance; but if it is supplied by thermal (fossil fuelled or nuclear) plant, the specific operating cost is higher and pumping efficiency is important.

18.4.3.5 Compressed-air storage plant

In compressed-air storage schemes a motor-driven compressor is used to charge an underground cavern, making use of excess electricity at off-peak periods. The compressed air, after additional input of energy in the form of fuel, is expanded in a gas turbine to generate electricity at peak-load periods.

18.4.3.6 Wind, tidal and solar power plants

Wind, tidal and solar power plants generate electricity at unpredictable or uncontrollable times. Such plants must, therefore, if operating alone or if supplying the major part of the load on a network, operate in conjunction with some form of storage. Battery storage is appropriate for small units (e.g. a 4 kW wind generator) but pumped-water storage must be used with larger installations such as a tidal scheme.

If such a plant is connected to a large network which is able to absorb its maximum output whenever it may occur, storage facilities need not be provided. Without storage the output cannot be relied on and there is no saving in plant capacity elsewhere on the system; there is, however, a saving in fuel (or water) consumption elsewhere. If storage is provided, some or all of the output can be made reliable, but the cost may exceed that of providing the same reliable capacity in a conventional power station.

18.4.4 Standby plant

In the UK, where there is a closely interconnected system with numerous generating stations, standby plant is not necessary as part of the main supply system. Individual large consumers, e.g. important manufacturers, hospitals, pumping installations, etc., may, however, consider it desirable to install standby plant to guard against a supply failure. The required outputs do not usually exceed a few hundred kW, and diesel plant, which can be started almost instantaneously, is usually provided for this duty.

There are circumstances where a load centre is supplied by an overhead transmission line from a distant hydro-electric station, and standby plant may then form part of the main system. Diesel plant, or steam turbine plant with quick-starting forced circulation boilers, has been used in such cases.

18.4.5 Load shedding

A method of dealing with (though not supplying) peak load is by load shedding, i.e. by reducing the system voltage or frequency, or by switching out certain loads. A 1% drop in voltage results in a maximum drop in load of 2%, depending on the nature of the load. A 5% voltage drop leads to complaints by consumers, while a 10% drop may cause small motors to overheat or burn out. Further reduction would affect power station auxiliaries and probably lead to a complete loss of supplies.

A frequency reduction of 1% (the voltage being normal) may reduce the load by 1–2% as a result of the lower speed of industrial motors, but increases inductive (magnetising) load by upwards of 5%. The wide use of frequency- or speed-sensitive equipment precludes frequency reduction as a general means of load shedding.

Disconnection of loads is a last resort. Considerable planning is necessary to minimise interruption to essential services such as hospitals, transport systems, etc.

18.5 Plant scheduling

To minimise the cost of supply it is necessary to make a proper selection, hour by hour, of plant to be added to the system, or shut down, as the demand changes.

The selection is made primarily on the basis of the operating costs associated with each item of generating plant — fuel, operation and maintenance. Capital cost, an important factor in the original choice of the generating plant to be installed on a system, is not relevant to the scheduling of plant to meet the load once the plant has been installed.

18.5.1 Thermal plant

18.5.1.1 Order-of-merit commitment

In a simple and widely used procedure for a system in which thermal plant predominates, the specific operating cost (i.e. the cost/kWh of energy supplied from each generating set) is calculated and the sets arranged on a commitment list in order of increasing specific operating cost. As the system load rises, sets are added in this order, due allowance being made for spinning reserve (plant which is running but not fully loaded) and other special factors.

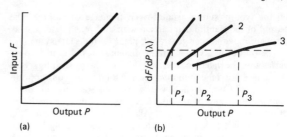

Figure 18.4 Generating plant characteristics. (a) Input-output curve; (b) Incremental input-output curves

18.5.1.2 Equal incremental cost loading

With equal incremental cost loading, the load allocated to each set is decided from a consideration of the incremental costs of the sets supplying the load. *Figure 18.4(a)* shows for a typical set how the input F varies with the output P. The input F might be coal or steam or heat consumption; the input cost is derived from one of these. Output P will typically be the net power output of the set (in MW). Such a curve is termed an input–output curve. Alternatively, the way that P varies with F can be shown on an efficiency curve. From the input–output curve can be derived a curve showing the rate of incremental input per incremental output, dF/dP, against P. Curves of this nature, derived from the input–output curves of three generating sets, are shown in *Figure 18.4(b)*. These curves show the slope of the input–output curve for any value of P. In its simplest form the theory of equal incremental cost loading may be explained as follows.

If two generating sets supplying P_1 and P_2 have costs F_1 and F_2, respectively, then the value of P_1 for lowest overall cost $F = F_1 + F_2$ can be found from

$$(dF/dP_1) = (dF_1/dP_1) + (dF_2/dP_1) = (dF_1/dP_1) - (dF_2/dP_2)$$

because $dP_2 = -dP_1$. For a minimum $(dF/dP_1) = 0$, and hence $(dF_1/dP_1) = (dF_2/dP_2)$, which enables P_1 and P_2 to be determined.

In a similar way it can be shown that the input to a system containing any number of generating sets operating in parallel to meet a given load is minimised when the given load is shared between the sets such that the incremental input rate for each set is the same, i.e. when the loads are P_1, P_2 and P_3 (on sets 1, 2 and 3) in *Figure 18.4(b)*. The common incremental input rate for the given system load, then, is λ, where

$$\lambda = \frac{dF_1}{dP_1} = \frac{dF_2}{dP_2} = \frac{dF_3}{dP_3} = \cdots = \frac{dF_i}{dP_i} = \cdots = \frac{dF_n}{dP_n}$$

Since ΣP_n must be equal to the total generation required to meet the load and since λ varies with P_i, the optimum division of the load between generating sets is best obtained by use of computers which have made it possible to solve problems of this type rapidly. Many modifications of this basic procedure have been developed for use in load despatching centres.

In practice, on many systems a number of power stations supply the load via transmission tie-lines and grid systems. When line losses are appreciable, the load allocation procedure must be modified to take them into account.

18.5.2 Hydro-electric plant

The basic requirement for economic operation of hydroelectric plant is that the maximum utilisation be made of the water available. Ideally, therefore, no water should be wasted over spillways. In an all-hydro system, however, some spillage is almost inevitable. The operating procedure will vary, depending on the type of station (run-of-river, pondage or storage), the output of the plant in comparison with the system load, and its association with other generating plant on the same system.

The output of a water turbine depends on the head, and so varies with the water storage level. With several generating sets in the same station the minimum specific total water consumption (say in m^3/kWh) is achieved when the incremental water consumption rates of all sets are equal, in a manner similar to optimised operation of thermal plant, as discussed above. It is rarely possible to apply this principle to load-sharing between stations, as factors involving storage will take precedence. With hydraulically independent stations the operator has little control over the total system storage, and must confine his efforts to conserving impounded water. Where plants are in cascade on the same river, there is more scope for economy.

Hydro-electric generating stations are often remote from load centres, and transmission losses may assume importance.

18.5.3 Mixed thermal and hydro-electric plant

Almost invariably, hydro-electric plant operates in parallel with thermal plant, and the plant must be operated in such a way as to minimise fuel costs. With a run-of-river station the water must be used whenever it is available, and there is little the operator can do to cater for loads later in the day, week or year.

Where storage is provided, a further requirement is to ensure that sufficient water is stored at all times to meet future loads, e.g. later in the day or week where small-scale storage, or pondage, is provided, or later in the year where large-scale dams and reservoirs are built to provide seasonal or over-year regulation. Thus, the factors involved are the foreseeable load variations, the seasonal water availability, the storage capacity and the ratio between the capacities of hydro and thermal plants. The first two of these factors necessitate the collection and processing of data from previous years.

The assessment of probable water availability on a seasonal basis can be carried out by computer simulations of the operation of the hydraulic system comprising the reservoir, waterways and turbines, using hydrological data from past records spanning as many years as possible. From such simulations the firm water flow (or alternatively water flows with specified probabilities of failure) and the corresponding values of potential hydro-electric generation for the amount of storage available can be determined.

By use of these values of potential hydro-electric generation, the capacity (or capacities) of the hydro-electric plant can be fitted on the plant load–duration curve to supply the appropriate tranche (or tranches) of load, as illustrated in *Figure 18.3*. The remaining tranches of load can then be allocated to the thermal plant in order of merit. The whole process of producing a predicted plant load–duration curve for the period considered and of fitting the capacities of hydro and thermal plants on the curve can be done by computer.

18.5.4 Pumped storage plant

The economic principle of pumped storage is the replacement of energy generated in high-incremental-cost thermal plant at peak load times by energy generated in low-incremental-cost plant during off-peak periods. In a typical case the efficiencies of the component parts of the system might be: pump/turbine 82%, motor/generator 96%, generator transformer 98%, transmission 97% and pipeline 98%. Allowance must also be made for power station auxiliaries and for evaporation from the reservoirs, and the overall conversion efficiency will be of the order of 70%, so that the operating cost of energy delivered from the installation is about 1.4 times that of the energy used at off-peak periods for pumping. Pumped storage plant generation is worth while if the incremental operating cost of other plant at peak periods is more than about 1.4 times the incremental operating cost during the pumping period.

Pumped storage plant has other advantages. It is typically capable of rapid response to load change and this property enables the plant to be used for system frequency control, which eases the duty of thermal generating plant. The plant can also be suitable for spinning unloaded and ready to take up load in the event of failure of other plant on the system.

18.6 Revenue

The costs of owning and operating an electricity supply system must normally be recovered by charges made to electricity consumers. Ancillary services such as the sale of electrical appliances or electrical installation contracting work rarely account for more than a few per cent of the total revenue. In theory, each individual consumer should be charged in relation to the actual cost of supplying him. In practice this is impossible to implement and it is normally sufficient to charge a group of generally similar consumers at a common rate such that the total group revenue is related to the cost of supplying that group.

Consumers are normally classified into such groups, each charged at a common rate, by the type of use to which they put the electricity (e.g. industrial, commercial, domestic), their size and sometimes their location (e.g. urban or rural).

18.6.1 Tariff structure

The cost of supplying electricity has three main components:

(1) A component related to the supply (generation, transmission and distribution) capacity provided to meet the power demand, i.e. a component to cover mainly the capital charges (depreciation and interest) on generating plant and transmission and distribution equipment. This component also covers other fixed costs, including the fixed components of operating and maintenance costs, administration costs, taxes and other overheads related to the system capacity.
(2) A component related to the connection of each consumer's premises to the system, to cover the costs of service cabling and metering.
(3) A component related to the quantity of energy supplied, to cover the costs of fuel, the energy-related costs of operation and maintenance and losses in the system.

Components (1) and (2) are termed the *fixed costs*, and (3) is termed the *variable, running or operating cost*.

Other factors that influence the tariff structure are:

(a) The time of day and/or of year, since the supply capacity is related to the demand during the peak-load periods and any incremental demand during those periods involves the supply authority in providing additional supply capacity, whereas incremental demand during off-peak periods can be met by the available capacity which is at that time working below rating.
(b) The power factor, since a low power factor also necessitates additional supply capacity and incurs higher system losses.
(c) The voltage of supply, since the lower this is the more transformation steps are required and the higher are the system losses.
(d) The continuity of supply, since a guarantee of such continuity involves the supply authority in providing additional standby supply capacity. Some consumers may be prepared to accept, by prior arrangement, reductions or even total loss of supply during periods of potential peak loads, thus obviating the necessity for additional standby supply capacity.

The structure of most tariffs, therefore, incorporates components which primarily reflect the fixed and variable costs of supply, with refinements or adjustments to cater for the other factors listed above. The fixed component may be truly a simple standing charge or a charge related directly to the power demand or supply capacity (in kW say, or in kVA to take account of power factor, although this involves more costly metering equipment), while the variable component is always related to the energy consumed (in kWh); in such cases the tariff is termed a *two-part* tariff. Alternatively, a single-part *block* energy tariff may be used, with successive blocks of energy consumption charged at progressively lower rates, the lowest rate being termed the follow-on rate: in such a tariff the blocks at the higher rates are designed to recover the fixed and variable costs of supply, and that at the follow-on rate the variable costs only.

Refinements of the above two simple structures of tariffs include:

(1) Combinations of the two, e.g. a *two-part block* energy tariff incorporating a maximum demand charge and successive blocks of energy charges.
(2) Demand or capacity related block energy charges, e.g. incorporating successive blocks of energy per kW or kVA of demand or capacity at progressively lower rates.
(3) Three-part tariffs incorporating standing, power and energy components.

The simplest forms of tariffs are those in which no metering is involved either because metering (for example, of public lighting) would be impractical or (as in some developing countries) because the electricity requirements of the lowest-income domestic consumers may be so small that the costs of metering would alone far exceed any charges which might justifiably be made to them. When supplies are not metered, charges are usually made at a fixed rate determined by estimating the average power demand and energy consumption.

While it is undesirable for tariffs to be altered frequently, circumstances may increase the costs of supply and therefore the need to increase revenue. In some cases it may be sufficient merely to impose a temporary surcharge on all rates. When the costs of supply are subject to frequent adjustment, usually upwards, as has happened over the past few years as a result of fuel price increases, then appropriate adjustments to electricity charges may be made by imposing successive and progressively larger surcharges, or in the case of fuel by incorporating a fuel cost adjustment whereby the energy rate (or rates) in the basic tariffs is adjusted in relation to the fuel price. With two-part tariffs different surcharges can be imposed on, or adjustments made to, the fixed and variable components to reflect the appropriate increases in costs. Such surcharges and adjustments should, however, periodically be incorporated in the basic tariffs by reviewing both their structure and rates. In general, changes in structure are made to shape the pattern of usage (or reshape it when, for example, a particular tariff is being abused), while changes in rates are made to produce financial viability.

18.6.2 Forms of tariff

18.6.2.1 Bulk supply

Generation and transmission authorities (such as the CEGB in England and Wales) frequently sell electricity in bulk to distribution authorities (such as the Area Boards) or directly to large individual consumers such as steelworks or electric railways. The form of tariff is usually at least two-part, with power demand and energy charges, and sometimes three-part to include a standing or service charge additionally made to cover specifically the costs of providing and maintaining the bulk supply points.

Adjustments to the demand and energy charges for time of day, time of year, power factor and interruptible supplies are also sometimes incorporated, as is fuel cost adjustment of energy charges.

18.6.2.2 Industrial and commercial

Two-part tariffs are usual for industrial and commercial consumers with loads in excess of about 40 kW. Again either or both of the demand and energy rates may be dependent on the time of day and/or year, and adjustments for power factor, interruptible supplies and fuel price may also be made.

The cost of metering maximum power demand makes it uneconomic to apply it to loads of less than 20–40 kW, so that the smaller industrial and commercial consumers are usually charged on a block energy tariff, the fixed costs of supply being recovered more or less by the higher-rate blocks.

For very small industrial and commercial consumers domestic tariffs may be used.

18.6.2.3 Domestic

For loads of little more than lighting in some developing countries, a simple single energy rate per kWh may be used. More simply still where, as mentioned above, the costs of metering cannot be justified a restricted supply may be provided through a fuse or load limiter, and a fixed charge made depending on the fuse or limiter rating. For larger loads two-part tariffs with a fixed charge related to floor area, rateable value, installed total load, etc., which used to be quite common, have largely been superseded by block energy tariffs.

18.6.2.4 Farming and irrigation

For general farming consumers without large loads block energy tariffs are usual. With large loads (for example, irrigation pumping) two-part tariffs or demand- or capacity-related block energy tariffs are usual, due allowance being made for coincidence or non-coincidence of peak farming and peak system loads.

18.6.2.5 Off-peak

Lower rates are commonly offered for electricity taken during off-peak periods, e.g. during the night, or during the summer in Britain. The lower rates can be charged either (a) on the consumer's whole electricity requirements, by the provision of separate time-switched on-peak and off-peak meters (e.g. the British 'Economy 7' tariffs which incorporate a 7 h off-peak charging period), or (b) on the requirements of only certain of the consumer's appliances, e.g. storage heaters or water heaters, the supply to which is restricted by time switches to specific off-peak periods.

18.6.2.6 Public lighting

Metering of public lighting is generally impractical, and so, as the load is normally fixed and controlled by time-switches or comparable devices, the chargeable power demand and energy consumption are both calculated.

18.6.3 Connection charges

In rural areas, if the cost of running a distribution line to an isolated farm is substantial, it is usual for the consumer to have to contribute to this cost or to guarantee a minimum consumption for a number of years, or both. A similar problem arises with new housing estates, and it is then usual for the developer to meet some of the cost of both distribution and service cabling.

18.6.4 Typical tariffs

Table 18.2 gives examples of typical tariffs.

Table 18.2 Typical tariffs

INDUSTRIAL/COMMERCIAL/FARMING
Annual or differential monthly maximum demand (m.d.) tariff for supplies from HV substations.
(1) *Monthly charge*: a monthly charge of £45.
(2) *Service capacity charge*: 35p/month or £4.20/year for each kVA of service capacity.
(3) *Maximum demand charge*:
    (a) Where metered at high voltage:
        (i) Annual tariff: £/kW of m.d.—17.40 up to 3000 kW, 16.80 for each additional kW.
        (ii) Differential monthly tariff: £/month per kW of monthly m.d.—Apr. to Oct., 0.05; Nov. and Mar., 1.20; Dec. to Feb., 5.00 up to 3000 kW, 4.80 for each additional kW.
    (b) Where metered at low voltage:
        (i) Annual tariff: £18.60 per kW of m.d.
        (ii) Differential monthly tariff: £/month per kW of monthly m.d.—Apr. to Oct., 0.05; Nov. and Mar., 1.40; Dec. to Feb., 5.30.
(4) Adjustment for power factor: m.d. charge (3) increased by p/kW for each complete 0.01 by which the average lagging power factor is less than 0.9;
    (i) Annual tariff, 10.00.
    (ii) Differential monthly tariff, 0.83.
(5) Energy, p/kWh:
    (a) Where metered at high voltage:
        (i) 'Any time' unit charge, 3.09.
        (ii) 'Day/night' unit charges, 0030–0730 hours, 1.49; other times, 3.43.
    (b) Where metered at low voltage:
        (i) 'Any time' unit charge, 3.19.
        (ii) 'Day/night' unit charges, 0030–0730 hours, 1.53; other times, 3.54.
(6) Fuel cost adjustment: energy charge (5) increased (decreased) by, p/kWh
    (a) Where metered at high voltage, 0.00044
    (b) Where metered at low voltage, 0.00042
for each 1p by which the cost of fuel is more (less) than £45/tonne.

INDUSTRIAL/COMMERCIAL/FARMING
General purpose and maximum demand tariffs for supplies from the LV system.

Tariff name	Ordinary general purpose	Economy 7* general purpose	Evening and weekend	Quarterly maximum demand	Economy 7 maximum demand
Upper limits	50 kW or 60 000 kWh/year				
Quarterly charge (£)	7.54	9.39	9.39	14.95	14.95
m.d. charge (£)	—	—	—	8.20	8.20
Energy charges (p/kWh)					
Weekday daytime	5.80	6.20	7.50	3.55	3.80
Evening and weekend	5.80	6.20	3.55	3.55	3.80
Night-time	5.80	1.90	3.55	3.55	1.90

* Upper limits apply to daytime only.

INDUSTRIAL/COMMERCIAL/FARMING

General purpose time of day tariff
(1) Fixed monthly charge:
    (a) On supplies from high voltage substations, £45.
    (b) On supplies from low voltage distribution system, £15.
(2) Service capacity charge, p/month for each kVA of service capacity:
    (a) On supplies from high voltage substations, 35.
    (b) On supplies from low voltage distribution system, 50.
(3) Adjustment for power factor:
        (i) HV supplies: service capacity charge (2) increased by 0.7 p/month per kVA for each complete 0.01 by which the average lagging power factor is less than 0.9.
        (ii) LV supplies: nil (power factors less than 0.9 not permitted).
(4) Energy, p/kWh:

Table 18.2 continued

	HV supplies metered at HV	HV supplies metered at LV	LV supplies
Dec.–Feb. (1630–1830 h weekdays)	29.00	30.00	32.00
Nov.–Mar. (0900–2000 h weekdays)	5.20	5.35	6.12
Year round (0030–0730 h all week)	1.49	1.53	1.76
Any other time	3.04	3.14	3.47

(5) Fuel cost adjustment: energy charge (4) increased (decreased) by, p/kWh
    (a) HV supplies metered at HV, 0.00042
    (b) HV supplies metered at LV, 0.00044
    (c) LV supplies, nil
for every 1p by which the cost of fuel is more (less) than £45/tonne.

DOMESTIC
(a) General purpose
    (1) Fixed quarterly charge: £6.76.
    (2) Energy: 5.07 p/kWh.
(b) Economy 7
    (1) Fixed quarterly charge: £8.61.
    (2) Energy, p/kWh: 0700–2400 hours, 5.37; 2400–0700 hours, 1.90.

18.6.5 Metering

Metering equipment must be installed for each consumer and must comply with legal limits of accuracy. Meters may be of the credit type, for later charging and payment, or of the prepayment type ('slot meter'), through which a supply can be obtained only after insertion of the appropriate coins. Credit meters should be recertified after 20 years, prepayment meters after 15 years. Typical metering costs (relative to a single-phase credit type) are listed in Table 18.3.

Bulk-supply consumers will, in addition to energy meters, require kVAr metering to assess m.d. averaged over $\frac{1}{2}$ h periods. Summation metering may be needed to obtain energy and m.d. with two or more supply circuits. For *industrial* and large *commercial* consumers a 3-phase supply is usual, with current transformers having 5 A secondaries to supply the meters where the load exceeds 50 kVA. Maximum demand metering is normally required for loads above 20 MVA. *Domestic* and small *commercial* consumers must have credit or prepayment energy meters. Ratings of 40 A and 80 A (overcurrent capabilities 6.25 A and 120 A) are common.

18.6.5.1 Meter reading

The cost of reading a meter, and of preparing and posting the account, may represent a significant proportion of the total cost

Table 18.3 Relative metering costs

Single-phase metering:		Three-phase metering:	
credit type	1	energy	8–12
prepayment	2.5–3.0	energy and m.d.	15–20
certification	0.02–0.03	1-feeder bulk	
changing	0.2–0.3	supply	400–500
Test and repair		2-feeder summation	
1-phase		electromechanical	70–80
credit	0.05–0.25	electronic	120–150
1-phase			
prepayment	0.08–1.25		
3-phase	1.25–1.75		

of supplying a domestic consumer. A meter reader can visit 120–150 consumers per day in an urban area, but only one-half as many in a semi-rural district. Reply-paid cards (on which the consumer records his own reading), and 'estimated readings' by the supply authority, can reduce the number of visits (including abortive ones when the consumer is absent), but can cause annoyance. Meter panels mounted outside the premises have been used with moderate success. Centralised metering has been devised, but is expensive.

18.6.5.2 Prepayment

Despite some social advantages, prepayment meters cannot be justified on technical or financial grounds. Their cost is high, and only 30–40 consumers per day can be visited for collection: the cost of this service is about three times that for reading credit meters. Theft and abuse are common, and the cost of special visits to clear a stuck coin or a full box is heavy.

18.7 British electricity supply systems

18.7.1 History

In the early days of electricity supply in the UK, each town possessed a local generating station publicly or privately owned. By 1918 there were 600 such undertakings, and the heavy demands of war industries had shown the weakness of such parochialism. As a result, Parliament passed the Electricity Supply Act, 1919, to set up the Electricity Commissioners 'to promote, regulate and supervise the development of the industry'. The hope that small undertakings would be combined into larger and more economic units was not realised, and the work of the Weir Committee of 1925 resulted in the 1926 Act which authorised the setting up of the Central Electricity Board to control all generation and main transmission in the UK except in the north of Scotland, and to standardise the frequency of all supplies at 50 Hz.

Of the 500 generating stations then in use, 126 of the largest and most efficient were selected and interconnected by a 'grid' of high-voltage (132 kV) transmission lines. The remaining stations were closed down, the supply undertakings concerned then

Table 18.4 British supply systems (1982/83)

System description		Authority		
		CEGB	SSEB	NSHEB
Number of generating stations		100	16	66
Generating station installed capacity				
fossil fuelled steam turbine plant	GW	49.4	6.5	1.3
gas turbine plant	GW	2.7	0.2	—
diesel plant	GW	—	—	0.1
nuclear plant	GW	5.2	1.5	—
hydro-electric plant	GW	0.1	0.1	1.1
pumped storage plant	GW	0.7	—	0.7
total	GW	58.1	8.3	3.2
Generating station sent out capacity	GW	54.8	6.4	3.1
System annual maximum demand	GW	42.1	4.0	1.5
Annual energy supplied to system*	TWh	206.7	19.3	6.6
Length of overhead transmission lines				
400 kV	circuit km	9414	336	—
275 kV	circuit km	3775	1490	1690
132 kV	circuit km	459	1530	3350
Length of underground cables				
400 kV	circuit km	126	4	—
275 kV	circuit km	526	70	5
132 kV	circuit km	202	152	50
Number of 400 kV and 275 kV substations		207	46	11
Capacity of main transformers				
400 kV	GVA	89	8	—
275 kV	GVA	63	11	7
132 kV	GVA	4	7	4

* Including interchanges between systems.

obtaining a bulk supply from the grid through a suitable substation. The total cost of the transmission lines and associated works was estimated at £30 million, and it was calculated that the savings resulting from the reduction in spare plant required and the increased economy of generation would more than offset this cost: these estimates proved correct. All the lines in the original scheme were in operation by 1932, although additions were continually being made to meet increased demands. Several large new generating stations were built and some stations originally scheduled for closing down were retained.

In 1943 the North of Scotland Hydro-Electric Board was set up to cover the Highlands of Scotland north-west of a line from Glasgow to Inverness. As most of the district covered is very sparsely populated, many of the distribution schemes are uneconomic and are subsidised by the transmission of large blocks of electricity southwards for sale in the industrial areas of central Scotland.

On 1 April 1948, as a result of the enactment of a Bill to nationalise the electricity supply industry, the ownership and control of all supply authorities south of a line approximately from Glasgow to Dundee passed into the hands of the British Electricity Authority (BEA) and 14 Area Boards. Apart from some change in its area, the functions of the North of Scotland Hydro-Electric Board were not appreciably affected. In 1955 the South of Scotland Electricity Board took over responsibility for generation and supply in the rest of Scotland, so that the NSHEB and the SSEB between them then held the authority for all Scotland. As a result of this severance, the industry in England and Wales comprised the Central Electricity Authority and 12 Area Boards. The CEA had authority to generate electricity for bulk supply to the Area Boards by the main grid of transmission lines, to co-ordinate the work of the Boards and to determine policy. The Area Boards purchased energy from the CEA (and in certain cases from other sources) and were responsible for distribution systems formerly owned by local authorities or private companies.

In 1957, as a result of the recommendations of the Herbert Committee, further changes were made, setting up the present organisation. The supervisory functions of the CEA were transferred to the Electricity Council and its executive functions to the Central Electricity Generating Board. The two Scottish Boards continued as before, as did the Area Boards of England and Wales. The many sub-areas and districts based on the former patchwork of separate authorities has been replaced by a 'two-tier' structure comprising, for each Area Board, an area organisation and 10–15 districts.

18.7.2 275 kV network

Increasing load led to the need for a 275 kV transmission 'supergrid' superimposed on the former 132 kV 'grid'. The first line was commissioned in 1953. The system comprised double-circuit overhead lines with twin-wire bundle conductors, and some lines were built for ultimate operation at 400 kV.

18.7.3 400 kV network

Rapid load growth in the 1950s, especially in areas away from the older industrial centres, and the construction of nuclear stations in remote sites, led the CEGB in 1960 to introduce a 400 kV network, partly with new lines and partly by upgrading 275 kV sections. The 275 kV lines with twin 256 mm^2 (0.4 in^2) conductors can be operated at 400 kV by increasing the number of suspension–insulator units to 18 or 19, the line rating becoming 700–1100 MVA. The new 400 kV lines have quadruple 256 mm² conductors and carry 1400–2200 MVA per circuit. Oil-filled cables have been built for 400 kV, with electric stress of about 15 kV/mm and a charging current of about 35 A/km.

The 400 kV network is now complete, and extends northwards to Carlisle and Newcastle in England. There is also a small 400 kV network in the Clyde and Glasgow area of Scotland. Although it is doubtful whether the closely knit British system will need a voltage higher than 400 kV, experimental work is being carried out at 750 kV.

Table 18.4 gives some details of the British system.

18.7.4 Substations

18.7.4.1 Transformers

All transformers are designed for outdoor operation and most are of the 3-phase core type with mixed cooling for larger ratings. On-load tap changing gear giving $\pm 10\%$ regulation is provided. The 132 kV windings are star-connected with solidly earthed star point. Transformers stepping down from 132 kV have outputs up to 300 MVA and are normally double-wound with a solidly earthed star high-voltage winding and a delta-connected low-voltage winding with earthing through an earthing transformer or arc-suppression coil. Transformers stepping down from 275 kV or 400 kV are auto-connected in star with solidly earthed star point and a delta tertiary. Their ratings are up to and include 1000 MVA.

18.7.4.2 Circuit-breakers

For 132 kV circuit-breakers a 3-phase short-circuit breaking current rating of 20 kA and a 1-phase rating of 25 kA are required. The ratings are 31.5 kA and 40 kA at 275 kV, and 50 kA and 57.5 kA at 400 kV, respectively.

At 132 kV all types of circuit-breaker are installed, namely bulk-oil, small-oil-volume, air-blast and SF_6, with the latter two types predominating in recent years. At 275 kV the bulk-oil designs have also been overtaken by air-blast and SF_6 types. Only air-blast and SF_6 circuit-breakers are installed at 400 kV. SF_6 insulated metal clad units also exist at 275 kV and 400 kV.

18.7.4.3 Protective gear

On 400 kV and 275 kV feeders, two full discriminative, high-speed systems of protection using either distance protection or phase comparison (with either power line carrier or voice frequency over Telecom pilots) are provided. For shorter lines differential protection utilising either private or Telecom solid core pilots is used. Operating times are typically 40 ms. On 132 kV feeders one high-speed system selected from the above types of protection is provided, supplemented by back-up IDMTL protection.

Transformer protection comprises differential protection with back-up provided by Buchholz, overcurrent, winding temperature and standby earth fault protection.

Busbar and circuit-breaker fail protection is installed at all important switching stations.

18.7.5 Direct current transmission

A link between the British and French systems was proposed in the early 1950s (a) to conserve coal by using in Britain excess French hydroelectric energy, (b) to conserve stored water in France by using in France excess British off-peak energy, and (c) to exchange peak-load energy, as the two system peaks occur at different times. As a result of developments in high-voltage d.c. transmission it was decided to interconnect the two systems with a ± 100 kV d.c. submarine cable system of 160 MW capacity. The cables were brought into service in 1961, and in spite of some outages caused by cable or transformer faults the scheme has justified itself.

Governmental approval in principle was given in 1978 to the installation of a second submarine d.c. link between the British and French systems, with a capacity of 2000 MW.

Largely to gain experience in the use of direct current as a part of the main a.c. network, an 84 km, 640 MW, ± 266 kV d.c. underground cable was installed between Kingsnorth generating station in North Kent and two substations, Willesden and Beddington, in the outer London area, and was brought into operation in 1973. Each substation, connected between one of the outer cables and a neutral cable, receives 320 MW at 255 kV.

19 Power Electronics

J D McColl
GEC Rectifiers Ltd

Contents

19.1 Rectification 19/3
    19.1.1 Ideal rectifier element 19/3
    19.1.2 Practical rectifier elements 19/3
    19.1.3 Basic rectifier circuits 19/3
    19.1.4 Multiphase rectification 19/4
    19.1.5 Bridge connection 19/4
    19.1.6 Interphase inductor 19/5
    19.1.7 Power rectifier connections 19/5
    19.1.8 Short-circuit conditions Z9/7
    19.1.9 Overlap 19/7

19.2 Controlled rectification 19/8
    19.2.1 Voltage control 19/8
    19.2.2 Inversion 19/8
    19.2.3 Reversing motor drives 19/8
    19.2.4 Frequency conversion 19/9
    19.2.5 Thyration 19/9
    19.2.6 Ignitron 19/9
    19.2.7 Polycrystalline rectifiers 19/10
    19.2.8 Monocrystalline rectifiers 19/10

19.1 Rectification

Rectification is the process by which unidirectional voltages and currents are developed from an alternating voltage supply, by means of a switching device having the property of asymmetrical conductivity. Although a.c. supply is almost universal, d.c. is preferred for some applications (e.g. motor speed control over wide ranges) and essential for others (e.g. electrochemistry). Methods of rectification of a.c. to d.c. are therefore of interest, and equally the reverse process, inversion from d.c. to a.c., is required.

19.1.1 Ideal rectifier element

A rectifier element (in general terms, a *diode*) is represented by the graphical symbol in *Figure 19.1 (a)*. An ideal rectifier element has the current–voltage relation (*b*), in which the resistance offered to

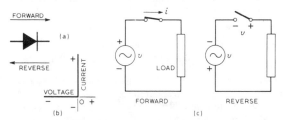

Figure 19.1 Rectification process

the passage of conduction current in one (the *forward*) direction is zero, and in the other (the *reverse*) direction is infinite. The element thus acts as if it were an ideal switch, closed in the forward and open in the reverse direction (*c*). The circuit conditions for the two modes of operation are then:

Forward: the current is determined by the supply voltage and the load impedance; the voltage across the rectifier element is zero.

Reverse: the current is zero; the whole supply voltage is impressed across the rectifier element.

These two conditions bring out the essentials of successful rectification. In the forward direction there must be a load impedance to limit the current, while in the reverse mode the element must be able to withstand the peak reverse voltage.

19.1.2 Practical rectifier elements

No homogeneous conductor can exhibit the rectification property. Some form of interface between conducting electrodes is necessary. There are several available, some of which have been extensively developed for power rectification over a wide range of current and voltage ratings. Typical current–voltage relations are shown in *Figure 19.2*. The limitations applicable to the ideal element must be observed even more stringently here: (a) there is a finite forward resistance (non-linear) so that I^2R loss occurs,

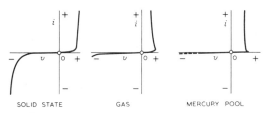

Figure 19.2 Asymmetric current/voltage characteristics

heat is developed, and, since its performance is very temperature sensitive, the element must be adequately cooled; and (b) there is a 'breakdown' limit on the magnitude of the peak reverse voltage that can be applied, beyond which the element becomes reverse conducting. To satisfy given current and voltage requirements it may be necessary to connect elements in parallel and/or series, a process that may demand means for ensuring that the elements properly share both forward current and reverse voltage.

19.1.3 Basic rectifier circuits

Consider a 1-phase supply connected to an inductive-resistive load through an ideal rectifier element (*Figure 19.3a*). The load

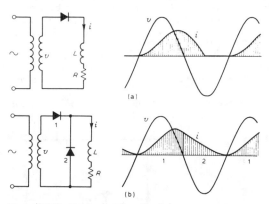

Figure 19.3 Single-phase rectification

voltage and current waveforms show that the load inductance has a marked influence on the behaviour of the circuit, to the extent of delaying the rise of current at the start of the forward half-period, and of maintaining forward current into the half-period of supply voltage reversal, which lowers the average output voltage. The current is discontinuous: one method of reducing its fluctuation is to provide a second or *freewheel* rectifier element to keep the d.c. load circuit closed during the negative half-periods of the supply (*Figure 19.3b*). The periodic transfer of current from one rectifier element to the other is called the *commutation* process: it occurs at the instant when the respective element voltages are the same.

A capacitor can also serve as a means of smoothing the load current, but the action is very different. The capacitor takes an impulsive charging current for a fraction of the positive half-period, and sustains the whole load current for the remainder of the full period. The method is commonly employed for low-power rectification.

As a means of isolating the d.c. load from the a.c. supply, and also of providing a required level of direct output voltage, a transformer is necessary. The a.c. component of the output current is balanced by a corresponding primary current, but the d.c. component cannot be so balanced, and causes magnetic saturation of the transformer core. For this reason, as well as for the undue output current fluctuation, the simple arrangement of *Figure 19.3 (a)* is not employed for power rectification. Half-wave rectification is rejected in favour of the full-wave action (*Figure 19.4*: small letters *v*, *i* refer to instantaneous values, capitals to mean or r.m.s. values).

By using two rectifier elements and a single-phase transformer with a centre-tapped secondary, each half-secondary conducts during alternate half-periods so that the effective secondary m.m.f. has a symmetrical a.c. wave form, balanced by a corresponding primary a.c. and avoiding a d.c. saturating com-

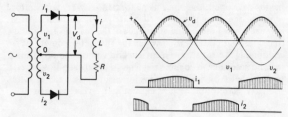

Figure 19.4 Full-wave rectification

ponent. The output current has less fluctuation than in the half-wave case. The more common bridge connection (*Figure 19.5*) gives the same overall effect, a simpler transformer, but requires two more rectifier elements. The current wave form is dependent entirely upon the relative amount of L and R in each case.

A basic arrangement for 3-phase supply is shown in *Figure 19.6*. The secondary starpoint is the equivalent of the centre tap in *Figure 19.5*. There is now a condition such that two elements have a forward voltage applied to them simultaneously. Conduction will take place through that element subjected to the greater positive (forward) voltage, and during the time that the two elements have the same voltage, they conduct in parallel. Conduction is commutated naturally from one element to the next, as the voltage of the former diminishes and that of the latter rises. The output voltage v_d therefore consists of the 'tops' of the successive secondary voltage sine waves, and the resultant d.c. voltage ripple has, compared with 1-phase operation, clearly been mitigated.

19.1.4 Multiphase rectification

Still less ripple in the output voltage v_d is obtained with six diodes supplied from a 3/6-phase transformer. Each diode conducts for only one-sixth of a period per cycle. With p diodes fed from a star-connected p-phase transformer secondary, and operation in the manner shown for $p=3$ in *Figure 19.6*, the following (ideal) conditions hold:

Voltage. Let the r.m.s. voltage of each secondary transformer phase be V_a; then the mean direct output voltage is

$$V_d = \sqrt{2} V_a (p/\pi) \sin(\pi/p)$$

The ripple voltage superimposed on the mean direct voltage has components of harmonic orders p, $2p$, $3p$, ..., of the supply frequency.

Current. Assuming that the d.c. load possesses an inductance sufficient to make the direct output constant at I_d (so that the diode currents in *Figure 19.6* become 'rectangular blocks'), then the transformer secondary phase currents I_a become

$$I_a \text{ (r.m.s.)} = I_d/\sqrt{p} \qquad I_a \text{ (mean)} = I_d/p$$

Transformer rating. The output is $P_d = V_d I_d$. The loading of each of the p transformer secondary phases is $V_a I_a$. The full secondary rating is therefore

$$S_2 = p V_a I_a = V_d I_d / \sqrt{(2p)} \cdot (1/\pi) \sin(\pi/p)$$

As the case is ideal, the transformer secondary rating might be expected to equal the output: but the discontinuous character of the diode currents renders the transformer secondary windings idle for considerable parts of each period, and the transformer requires more conductor material than one working on a comparable a.c. load.

The relations above are evaluated in *Table 19.1* for various typical secondary phase numbers. The figures are per unit values

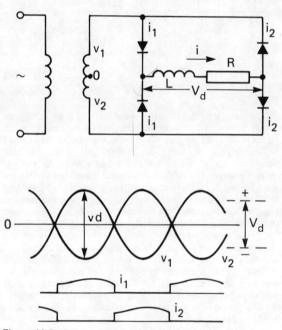

Figure 19.5 Full-wave bridge rectification

Table 19.1 Voltage and current relations in simple rectifiers

Number of secondary phases	p	2	3	6
Transformer secondary r.m.s. voltage	V_a	1.11	0.84	0.74
Diode r.m.s. current	I_a	0.71	0.58	0.41
Transformer secondary rating	S_2	1.57	1.46	1.82

in terms of $V_d = 1.0$ p.u. and $I_d = 1.0$ p.u. As p is increased, the direct output voltage approaches a constant value with a reducing ripple; but the conduction time of each diode shortens, and the transformer secondary rating rises. The worsening of the transformer utilisation is such as to make the simple star connection less economic when $p > 3$.

19.1.5 Bridge connection

Bridge connection is shown for a 3-phase rectifier in *Figure 19.7*. The bridge connection consists of two commutating groups in series, each active in alternate half-periods so that each transformer secondary phase winding conducts two-way; that is, once in

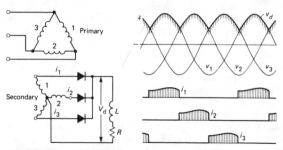

Figure 19.6 Three-phase three-pulse rectifier

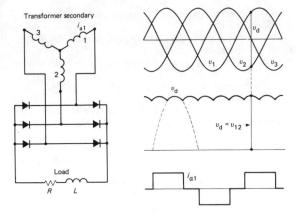

Figure 19.7 Three-phase six-pulse bridge rectifier

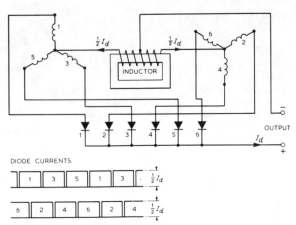

Figure 19.8 Six-pulse rectifier with interphase inductor

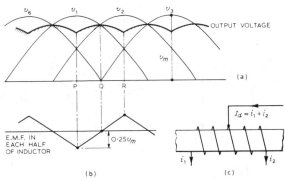

Figure 19.9 Interphase inductor

each half-period. The phase current is therefore not unidirectional but alternating (although not sinusoidal), so doubling the utilization time.

One set of diodes is connected to the positive output terminal, while a second group is connected to the negative. The load has an impressed voltage equal to the sum of the instantaneous part-voltages. As the two groups operate at opposite polarity, their ripple voltages are displaced, giving the direct output a 6-pulse ripple. The a.c. peak voltage of each group is $2V_a$, and of the combination is $\sqrt{2}\sqrt{3}V_a = \sqrt{6}V_a$.

Each transformer secondary phase has a $2\pi/3$ pulse duration in each half-period, so that the primary current is also devoid of d.c. component and the primary phases may be connected in star or delta. The latter is preferred because it eliminates harmonic currents of order 3, 6, 9, ..., from the supply. The mean output voltage is

$$V_d = \sqrt{6}V_a(6/\pi)\sin(\pi/6) = 2.34 V_a$$

The r.m.s. diode current is $I_d/\sqrt{3}$, and the transformer secondary r.m.s. current is $I_d/\sqrt{(3/2)}$. As a result the secondary load is

$$S_2 = 3V_a I_a = 1.05 V_d I_d$$

which applies also to the primary. Thus, the ripple content has been reduced (by making the pulse number = 6) and the transformer utilisation markedly raised: there are, however, twice as many diodes as for the simple 3-pulse arrangement of *Figure 19.6*.

19.1.6 Interphase inductor

Another method of securing a smooth voltage output at the same time as effective transformer utilisation is to split the halves of a 6-phase transformer into two sets of three, and to connect the two star points through an inductor. The interphase inductor (often called an interphase transformer) performs the function of combining two separate 3-phase systems (as *Figure 19.3*). At any instant two diodes, one from each group, are conducting in parallel; consequently, the output voltage is the mean of the respective phase voltages. Since now each diode carries only one-half of the output current, and does so for one-third of each period, the transformer phases are more effectively utilised. *Figure 19.8* shows the essential connections.

The operating principle is that, to cause two diodes to conduct simultaneously in parallel, their voltages must be equalised. The interphase inductor secures this condition and at the same time divides the output current equally between the conducting diodes. Consider instant P in *Figure 19.9 (a)*. Diode 1 is conducting and diode 2 is taking over a share in the load current from diode 6. The transformer phase voltages v_1 and v_2 are quite different at instant P; v_1 is at a peak, while v_2 is at one-half peak value. To equalise the voltages it is therefore necessary to subtract 0.25 of peak voltage from phase voltage v_1 and add a similar amount to v_2. This transfer is accomplished by the interphase inductor shown in *Figure 19.9 (c)*: it may be regarded as a centre-tapped auto-transformer operating at 3 times the fundamental frequency. Neglecting magnetising current, the m.m.f. balance between the two halves of the winding is maintained only if $i_1 = i_2 = \frac{1}{2}I_d$, i.e. the two currents are constrained to be equal.

At instant Q the voltages v_1 and v_2 become instantaneously equal and no voltage is impressed across the interphase inductor. Subsequently the polarity reverses and increases to a maximum at R. The e.m.f. in each half of the inductor varies with time in accordance with the waveform (b), from which it is seen that the peak e.m.f. is 0.25 of the peak transformer phase voltages in this (6-phase) case.

Table 19.2 summarises the voltage and current relations for 3-phase, 6-pulse centre-tap, 6-pulse bridge, and 6-pulse interphase inductor connections. In each case $V_d = 1.0$ p.u. and $I_d = 1.0$ p.u.

19.1.7 Power rectifier connections

The arrangements normally employed for large power rectifiers are summarised below and shown diagrammatically in *Figure 19.10*. The various connections are classified (a) in accordance with the number of pulses (or commutations per period); (b) as single- or double-way (i.e. transformer secondary phase current

Table 19.2 Voltage, current and rating

		6-phase star	3-phase bridge	3+3 phase inductor
Number of pulses	p	6	6	6
Transformer secondary r.m.s. voltage	V_a	0.74	0.43	0.84
Diode r.m.s. current	I_a	0.41	0.48	0.29
Transformer secondary rating	S_2	1.82	1.05	1.46
Transformer primary rating	S_1	1.29	1.05	1.03

respectively uni- or bidirectional); (c) whether or not a bridge connection is used, or an interphase inductor. The idealised waveforms at various parts of the network are shown in each case. The nine arrangements are:

(a) *Two pulse*; single-way; 1-phase primary, centre-tapped secondary.
(b) *Two pulse*; double-way; 1-phase primary and secondary bridge.
(c) *Three-pulse*; single-way; delta or star* primary, zigzag secondary.
(d) *Six-pulse*; single-way; delta or star* primary, double-star secondary; interphase inductor, $3f$.
(e) *Six-pulse*; single-way, star primary with isolated star point, diametral secondary.
(f) *Six-pulse*; double-way; delta or star primary and secondary; bridge.
(g) *Twelve pulse*; single-way; delta or star primary, quadruple zigzag star secondary; two $3f$ and one $6f$ inductors. (The $6f$ interphase inductor may be omitted.)
(h) *Twelve-pulse*; double-way; delta or star primary, delta and star secondaries; two 3-phase bridges in parallel; interphase inductor ($6f$).
(j) *Twelve-pulse*; double-way; delta or star primary, delta and star secondaries; two 3-phase bridges in series.
*Line current same shape as phase current.

19.1.7.1 Applications

The multi-anode mercury arc rectifier with a common cathode pool is now virtually obsolete: it was usually connected in a single-way arrangement (c, d, e, g). All these connections can be used with silicon diodes, which are now preferred.

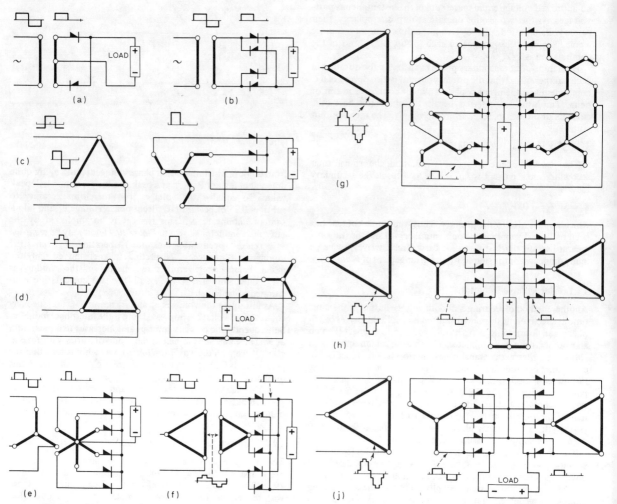

Figure 19.10 Power rectifier connection

For 1-phase 2-pulse equipments, the average transformer rating for (a) is 1.2 times that for (b), but the number of diodes is only one-half, which not only reduces the diode conduction loss, but also may more than offset the extra cost of the transformer. However, the diodes for (a) must have twice the voltage withstand capability of (b): thus (b) is generally preferred for higher voltage systems.

In 6-pulse cases (d, c) the ratio of transformer ratings is 1.2:1.4:1.0, and for (d) an additional requirement is an interphase inductor of relative rating 0.06. Connection (e) has a greater regulation.

The relative transformer ratings for the 12-pulse arrangements (g, h, j) are 1.3:1.0:1.0; also, (g) requires interphase inductors of rating 0.06 (3f) and 0.01 (6f), and (h) an interphase inductor rating of 0.01 (6f). Connection (g) was common with large-current low-voltage mercury-arc equipments. For relatively high output voltages (1.5 kV and over) connection (j) may be used, with diodes in series in each path of each bridge arm if necessary (200 kV rectifiers are built thus).

The non-sinusoidal transformer primary currents produce voltage waveform distortion on supply systems. For installations of 100 kW upwards it is desirable to limit distortion by increasing the number of pulses. The minimum number is related to the supply network impedance expressed in terms of the short-circuit fault level S at the point where other consumers likely to be affected by supply distortion are connected. In the UK a rough relation between rectifier rating P and fault level S is

Minimum ratio S/P:	200	50	30
Number of pulses:	6	12	24

Harmonic loading is extensively covered in Engineering Recommendation G5/3 (1976) of the Electricity Council.

A 24-pulse system may be provided by two separate 12-pulse equipments with outputs in parallel, with the necessary 15 degree shift in phase angles provided by a $7\frac{1}{2}$ degree shift on each primary winding. Alternatively, four separate 6-pulse equipments suitably phase-shifted may be employed.

19.1.7.2 Current and voltage relations

Basic ideal relations for connections (a–j) of *Figure 19.10* are given in *Table 19.3*. For the open-circuit condition the mean direct output voltage is $V_d = 1.0$: for the load condition $V_a = 1.0$ and the steady direct output current is $I_d = 1.0$. The assumptions made in *Table 19.3* are that the supply voltage is sinusoidal, the primary/secondary turns ratio of the main transformer is unity, and the output current is perfectly smoothed. The effects of transformer leakage impedance and magnetising current are neglected and there is no 'overlap'.

Connections (d), (g) and (h) are seen in *Table 19.3* to require a greater phase voltage on load than on open-circuit to obtain a given output voltage $V_d = 1$. These connections all employ interphase inductors, which are not effective when the output current is zero. As soon, however, as an output current is taken by the load, the inductor can provide the averaging effect already described. Thus, for a given phase voltage V_a the output voltage drops from open-circuit value V_{do} to the load value V_d, as shown for an ideal system in *Figure 19.11* (a). Practical inductors require an equivalent r.m.s. current I_t for full magnetisation, as in (b). The regulation there shown results from transformer impedance, loss, the diode forward volt drop and overlap.

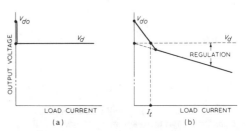

Figure 19.11 Regulation effect of interphase transformer

19.1.8 Short-circuit conditions

Silicon diodes have a smaller short-time overload margin than do mercury arc rectifiers, and the estimation of the current peaks that can occur in a given equipment and type of connection must be carefully checked in order that suitable diodes can be selected.

Apart from the impedance of the a.c. supply network only the transformer impedance—which depends intimately on the connections and the leakages—exerts a limitation on the d.c. short-circuit current level.

19.1.9 Overlap

Up to this point it has been assumed that the diode currents have wave forms resembling rectangular blocks, implying that one diode drops its current instantaneously and the next diode instantaneously picks it up. This is not, in fact, possible, because of the leakage inductance of the transformer and of the secondary connections. The inductance delays extinction and growth of current. *Figure 19.12* illustrates the effects for a simple 3-pulse rectifier. At point A diodes 1 and 2 have the same supply voltage; immediately thereafter, the voltage of diode 2 rises above that of diode 1. At and after point A, therefore, the current carried by diode 1 tends to collapse, current 2 to rise. A time duration corresponding to angle θ_u is required for the current transfer to be completed. The diode current waveforms lose their rectangular shape; and during the overlaps voltages 1 and 2 must be the same. A forward e.m.f. is induced in transformer phase 1 inductance by the collapse of the current, and an equal counter-e.m.f. in the next phase is produced by the current rise in diode 2; these force the effective voltages to equality. Briefly, during overlap the output voltage of the two phases in parallel is the mean of the respective transformer voltages. Comparison of the output wave form of the voltage with that in *Figure 19.6* shows that it has been reduced in mean value by an amount proportional to the overlap. As θ_u is a function of the load current to be transferred, the output voltage falls with increase of load. The relation is linear, i.e. the

Table 19.3 Current and voltage relations

Load	Open-circuit			
Connection	V_a	V_h	v_p	V_a
a	1.11	0.48	3.14	1.11
b	1.11	0.48	1.57	1.11
c	0.84	0.18	2.09	0.84
d	0.74	0.04	2.09	0.85
e	0.74	0.04	2.09	0.85
f	0.74	0.04	1.05	0.74
g	0.74	0.01	2.09	0.85
h	0.71	0.01	1.05	0.74
j	0.37	0.01	0.52	0.37

V_h = r.m.s. harmonic voltage
v_p = peak inverse voltage

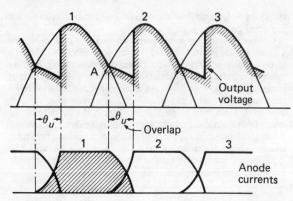

Figure 19.12 Overlap

voltage–current graph is a straight line, as indicated in *Figure 19.11(b)*.

19.2 Controlled rectification

Means are available in certain rectifier elements (such as thyristors, transistors, thyratrons and mercury pool rectifiers) to inhibit natural commutation and to delay the initiation of conduction. The process can be used (a) to control output voltage of a rectifier over wide limits and (b) to effect inversion, i.e. d.c. to a.c. conversion.

19.2.1 Voltage control

Figure 19.13(a) shows the output voltage v_d of mean value V_d formed from the peaks of successive transformer phase voltages in a simple *p*-pulse/rectifier. Ignoring overlap, point A represents the instant at which anode 2 takes over conduction from anode 1, the instant of *natural commutation*. In (*b*) the initiation of element/anode conduction has been delayed by angle α: the element which would, under natural commutation conditions, take over conduction is not 'fired' into conduction, allowing the preceding element to continue conduction on a falling voltage. The output wave form has enlarged harmonic content and a lower mean value. Thus, firing delay produces voltage reduction

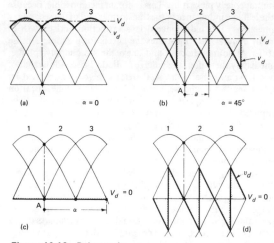

Figure 19.13 Delay angle

(accompanied by a commensurately poorer power factor due to the delay of anode current with respect to anode voltage).

If the d.c. load is devoid of energy storage (i.e. no back-e.m.f., such as inductors, capacitors and running machines), the output voltage is reduced to zero by a delay $\alpha = 2\pi/p$, as in *Figure 19.13 (c)*. If, however, the d.c. load is highly inductive in a real or equivalent sense, then a delay of approximately π/p is enough to make the mean voltage zero, as in (*d*).

If V_d is the mean voltage output with natural commutation, then with a delay angle α the mean output voltage for load conditions with no back-e.m.f. is

$$V_d = v_m(p/\pi) \sin(\pi/p) \cdot \cos \alpha = V_d \cos \alpha$$

The amplitude of output voltage harmonics of order $n = kp$, where $k = 1, 2, 3, \ldots$, is given by

$$v_n = [2V_d/(n^2 - 1)] \sqrt{(1 + n^2 \tan^2 \alpha)}$$

and these clearly become large for the higher values of delay angle. There is the further effect that sudden impulsive rises of inverse voltage are applied to the rectifier elements.

In *Figure 19.13* the effects of overlap have been omitted. In practice, leakage inductance causes delay in commutation by a further angle θ_u during which successive anodes conduct in parallel. The effect on the output voltage wave form in *Figure 19.13 (b)* is shown in *Figure 19.14*. The result is a further minor drop in output voltage.

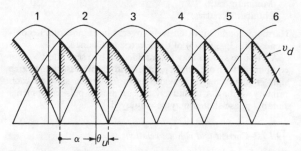

Figure 19.14 Delay and overlap angles

19.2.2 Inversion

If the angle of delay α is increased beyond $2\pi/3 = 120°$ up to a maximum of about $5\pi/6$ (150°), the conditions in *Figure 19.15* are obtained. The mean 'rectified' voltage V_d is negative. As the direction of current flow through the converter can only be forward, two conditions must be satisfied to permit operation: (1) there must be a live d.c. supply of appropriate polarity, and (2) its voltage E must exceed V_d by an amount adequate to supply the IR volt drop of the circuit. When these conditions obtain, current can flow in the normal forward direction and the rectifier operates as an inverter, with power flow from the d.c. to the a.c. side. Decreasing the delay angle α (equivalent to increasing the advance angle β) will decrease V_d and permit of a greater current I_d.

For inversion it is clearly necessary for the a.c. side of the inverter to be connected to a live a.c. network to provide an appropriate operating voltage and frequency.

19.2.3 Reversing motor drives

For duties such as mine winders and reversing rolling-mills it is now normal to use a controlled converter to feed a d.c. motor instead of a Ward Leonard set. This can result in appreciable saving in running costs, particularly if there is much running at

19.2.5 Thyratron

The thyratron is a gas-filled triode. The presence of the gas gives the grid a control characteristic substantially identical with that of the grid controlled mercury arc rectifier. The grid, if biased sufficiently negative with respect to the cathode, can prevent the passage of anode current; but once conduction starts, the grid can exert no further control.

The gas filled triode has been developed in a wide variety of types and sizes, with nominal anode current ratings upwards from about 100 mA, but is now superseded by the power transistor and thyristor.

19.2.5.1 Applications

Thyratron control was widely applied in the past to several types of control system, notably the speed control of d.c. motors. Many of these (small) equipments are still in service. As with the grid controlled mercury arc rectifier, it can be applied to the inversion (d.c. to a.c.) process. An example is shown in *Figure 19.17*, where an alternating control voltage is fed into the grid circuits through

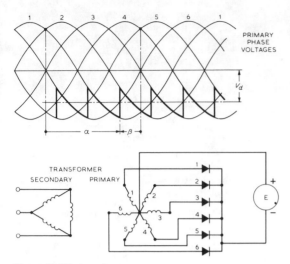

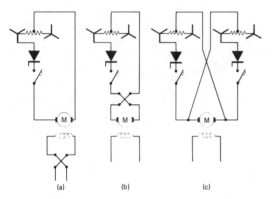

Figure 19.15 Inversion

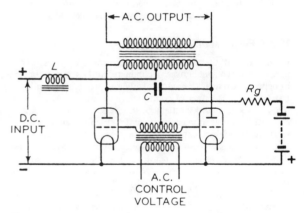

Figure 19.16 Reversing mill drives

Figure 19.17 Separately excited parallel-type inverter

a transformer having a centre-tapped secondary to which a suitable negative bias is applied with respect to the tube cathodes. The voltages at the respective ends of the secondary to which the grids are connected are in phase opposition, so that when one tube is made to conduct by a positive grid-voltage pulse, the other is biased off. The capacitor C is important. When either tube conducts, C is charged to a voltage equal to that across one-half of the output transformer primary, and the capacitor terminal connected to the anode of the conducting tube is negative with respect to the other terminal. When the second tube strikes, the capacitor polarity reverses, applying a reverse voltage to quench the first tube. Today such a system would use power transistors.

19.2.6 Ignitron

The ignitron is a 3-electrode valve with an anode, a mercury pool cathode and an ignitor. It differs from the mercury arc rectifier in that the arc is initiated each cycle at the required firing instant.

If a rod (ignitor) of high resistivity material (e.g. carborundum, silicon carbide) is partly immersed in a mercury pool *in vacuo*, and a current is passed through the combination, an arc is formed at the junction of the rod (anode) and pool (cathode) after a few microseconds. The mercury does not 'wet' the ignitor rod, so that actual contact is confined to a number of minute areas. As a result of this and of the high resistivity of the rod, large potential

light load, and it also avoids the necessity for heavy foundations. Since the converter can only pass current in one direction, it is necessary to make provision for feeding the motor so that it can be reversed. Three possible methods (*Figure 19.16*) are (a) motor field reversal; (b) motor armature reversal; (c) cross-connection. Both (a) and (b) have the disadvantage that for a short time during reversal the converter exercises no control over the motor, either because the motor field is near zero or because the reversing switch is in the process of changing over. The cross-connection (c) overcomes this difficulty by providing continuous control from forward to reverse, at the expense of having to provide two converters instead of one. If (a) is chosen, the field is normally fed by cross-connected auxiliary converters as in (c).

19.2.4 Frequency conversion

It is possible to transfer power between two a.c. systems, either of differing frequencies or whose frequencies cannot be maintained in synchronism, by employing two converters with an intermediate d.c. link. The cross-Channel d.c. transmission system constitutes such a link between the British and French national grid systems. The additional advantage is that fault conditions can always be modified by firing angle control, obviating the need for the larger a.c. circuit-breakers that would be necessary with an a.c. link.

differences occur at the contact region, causing electron emission from the mercury and very rapid local vaporisation. The action resembles arc formation when a pair of electrodes are separated. The initiation current required is quite small and its duration is very limited (e.g. 10 A, 100 µs): the ignition pulse is consequently of low energy.

The electrodes are housed in an evacuated steel envelope which is provided with a water jacket for cooling. The ignitor is conical in shape and is permanently immersed tip downwards in the mercury pool cathode. If the ignitor is pulsed and the anode has a potential positive with respect to that of the cathode, anode current will flow so long as the anode voltage is maintained. Output voltage control is obtained by firing the ignitor and, hence, the anode at selected points in the a.c. positive half-period. Currents up to several kiloamperes can be handled. The cathode spot is not continuously maintained as in the mercury arc rectifier, but is formed in each conducting half-period.

Ignitrons are commonly fitted with thermostats, used either to switch on the cooling water when the upper limit of operating temperature is reached or to switch off the output load if the ignitron overheats through overload or shortage of cooling water. Many small ignitrons are in service in antiparallel pairs used as timing controllers for electric welding plant. They are being rapidly superseded by thyristor antiparallel pairs.

19.2.7 Polycrystalline rectifiers

In the range of metal rectifiers, the rectification process takes place at a dry contact between a conductor and a semiconductor, or between two semiconductors of differing properties, forming a barrier layer. The area of contact may be minute, as in high-frequency silicon and germanium rectifiers, to as much as 200 cm^2 in a single selenium heavy-current power-frequency rectifier.

There is no inherent limit to the forward current, but a practical limit is set by heating because the barrier layer may be damaged by excessive temperature. The natural thermal capacity, however, permits of large short-period overloads and transient peaks.

The safe reverse voltage is limited in two ways. Excessive reverse voltage may puncture the barrier layer, or damage may result from overheating if the voltage is sustained.

The first metal rectifier to find really wide application in the field of power rectification was the copper oxide type, but this was largely superseded by the selenium rectifier for power applications. The copper oxide rectifier, however, is still in use in measurement applications such as instrument rectifiers in a.c.–d.c. test instruments, and where a high and stable reverse resistance is necessary at low operating power levels. The selenium rectifier is by far the most widely used metal rectifier for power applications (from 1 W to several kW) on account of its ability to operate at a higher element voltage and temperature.

19.2.7.1 Characteristics

The static current–voltage characteristics of a copper oxide element are shown in *Figure 19.18*, with the current scales expressed in terms of current density. Appreciable forward conduction begins at about 0.25 V. The intercept on the baseline formed by an extrapolation of the substantially straight part of the characteristic ('bottom-bend voltage') is about 0.3 V. The maximum reverse voltage is limited to 6–8 V, but transient peaks up to 15 V can be sustained. The safe maximum element temperature is 55 °C.

Considerable increase in the peak-reverse voltage capability of selenium rectifiers has been obtained, as indicated for 'double-voltage' and 'quadruple-voltage' designs in *Figure 19.19*, without undue sacrifice of forward conductivity. For the 'double-voltage' type, appreciable forward conduction does not occur below

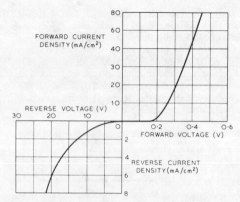

Figure 19.18 Current/voltage characteristic of copper-oxide rectifier elements

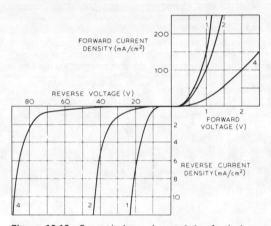

Figure 19.19 Current/voltage characteristic of selenium rectifier element.
(1) Single voltage; (2) Double voltage; (4) Quadruple voltage

about 0.5 V, but this is not a serious limitation for power applications in view of the high reverse working voltage. The safe working temperature is about 70 °C. The elements are subject to an ageing process that is accelerated if the operating temperature is high.

19.2.7.2 Capacitance

The self-capacitance is of the order of 3–5 nF/cm^2 of active area. When forward current flows, the self-capacitance effectively disappears. At power frequencies the effect of the capacitance is negligible, but it limits use at high frequencies. High-voltage selenium rectifiers may be used at frequencies up to 50 kHz; copper oxide instrument rectifiers begin to lose accuracy above about 100 kHz.

19.2.8 Monocrystalline rectifiers

An ideal rectifier offers zero impedance to current flow in one direction and infinite impedance in the other. A germanium or silicon rectifier approximates to this definition: the forward resistance is very low and the reverse resistance extremely high. A single germanium cell may drop only 0.5 V for a forward current of 150 A, yet pass only a few milliamperes when blocking a reverse voltage of 300 V. Similarly, a silicon cell might have a forward drop of 1.2 V at 600 A and a reverse current of 1 mA at 1500 V. In consequence, the power loss is smaller and the rectifying efficiency higher than for other kinds of rectifier.

The base element consists of a thin slice of material cut from a single crystal. One surface is alloyed with a donor and produces n-type material, while the other is alloyed with an acceptor impurity, such as indium, and produces p-type material. The junction between the n-type and p-type materials lies inside the single crystal, and is called a p–n junction. Forward biased, it conducts current; reverse biased, the p–n junction is an insulator and withstands the inverse voltage without permitting reverse current to flow, except for a small leakage due to charge carriers introduced by thermal agitation.

The permissible current density in the forward direction is very high for semiconductor rectifiers. Both silicon and germanium elements can operate continuously with a mean current density of 100 A/cm^2. With the usual transformer connection, the corresponding r.m.s. current density is 150 A/cm^2. This implies that the active junction area of the rectifier is often no greater than the cross-section of the connecting copper bars.

High efficiency and small size are advantages that have encouraged the application of monocrystalline rectifiers to almost all power rectification. Semiconductor diodes have virtually replaced copper oxide and selenium rectifiers in the low- and medium-power ranges, and mercury arc rectifiers in the high-power range. In the field of control, the silicon controlled rectifier or thyristor has considerable advantages for fast voltage control and inversion.

19.2.8.1 Diode

Characteristics The forward characteristics of both germanium and silicon diodes (*Figure 19.20*) show low volt drops for very high current densities. For a peak density of 250 A/cm^2 the forward volt drop is aout 0.5 V for germanium and 1.1 V for silicon.

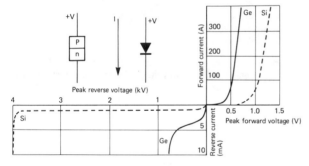

Figure 19.20 Current/voltage characteristics of germanium and silicon diodes at 20 °C

The reverse characteristics show very small currents over a wide range of voltage. These 'saturation currents' are independent of voltage but dependent on temperature. For germanium the value could be 1–2 mA at room temperature, while for silicon it is even less. At a voltage which varies from device to device with the method of manufacture, the reverse characteristic bends over as the phenomenon of avalanche breakdown commences. The bend-over is usually sharper for silicon and occurs at a much higher voltage than for germanium.

The characteristics are modified by temperature rise. As the temperature at the rectifier junction increases, so the forward power loss decreases almost linearly, while the reverse power loss increases approximately exponentially. Reverse voltage-withstand capability thus reduces.

Rating The voltage rating of a p–n junction is determined by its reverse characteristic. The maximum permissible operating voltage is fixed by the 'breakdown' voltage, at which the inverse power exceeds a predetermined safe value, taken together with a derating safety factor to ensure that accidental circuit overvoltages do not damage the device element.

The current rating is determined by the size of the junction area, the maximum permissible junction temperature, and the normal and abnormal (fault) overloads.

The junction size is determined largely by the economics of large-scale manufacture and by the technical difficulties associated with the manufacture of single crystals of very large diameter.

The maximum permissible temperature of the p–n junction to ensure long life is fixed by the results of tests and operating experience as well as by the details of device design. The continuous operating temperature of the junction is designed to be less than the maximum permissible value, to allow for overloads and fault currents. Typical junction temperatures during test are 75 °C for germanium and 120–200 °C for silicon. It is difficult to measure directly the temperature at the junction, and very often temperatures are measured at the heat-sink. From a knowledge of the thermal resistance of the device and cooling fin, and the loss generated, it is possible to calculate the corresponding junction temperature.

Connections Large-power converters always consist of a number of single devices connected in series and/or parallel to each transformer phase. In order to ensure that no device exceeds its maximum permissible junction temperature, and also to obtain the maximum possible output from the converter, it is essential to secure good current sharing.

Semiconductor diodes can be made with comparatively small variation in their forward characteristics, but a minor voltage difference may represent a current difference ten times as great. Further, the forward resistance has a negative temperature coefficient, so that an overheated element tends to take even more current. Manufactured diodes may be graded into limited spreads of forward volt drop for parallel connection, and circuits so designed as to equalise resistance and inductance. Thyristors are less easy to grade, but current sharing is often facilitated by a series inductor.

When devices are connected in series, they divide the circuit reverse voltage between them in accordance with their respective reverse resistances. At low or normal circuit voltages the division is determined by the individual element characteristics in the saturation current region; at high or excess circuit voltages the division depends on the characteristics in the avalanche region. Saturation current varies sufficiently from element to element to cause unequal division of voltage, which varies with both the total voltage and the temperature. Although in theory it appears self-protecting, nevertheless it is sometimes preferred to force equal voltage division by the use of resistors connected across each device.

Protection Although the normal working reverse voltage permitted across each device is generally restricted to about 50% of its breakdown value by circuit design, it is necessary to protect against overvoltages. The devices can themselves produce dangerous voltages at every commutation, due to the sudden cessation of the reverse current which flows for a short time after forward conduction. The interruption of this 'hole storage' or 'carrier storage' current snap-off in the inductive transformer circuit causes high-frequency voltage spikes and oscillations; but these are of relatively low energy and are easily damped by small capacitors, often connected across the a.c. terminals. Overvoltage surges can also be produced by circuit-breaker switching, by lightning discharges and sometimes by the load itself. These surges can generally be attenuated by *RC* circuits connected appropriately. Spark gaps, non-linear resistors and Zener diodes are other devices sometimes used to give co-ordinated voltage protection. Generally, when devices in series are employed, the 'hole storage capacitors', if connected across each device, can serve the dual purpose of 'hole storage protection' and transient voltage sharing.

The scheme of protection against overcurrents varies with the method of rating. When the converter is rated generously on a continuous load basis, it can withstand maximum fault currents for 0.1 s or more. Conventional circuit-breakers are then used, capable of being tripped by an overload relay fed from current transformers reflecting the converter current. Where no one relay characteristic completely matches the rectifier overload–capability curve, it is possible by means of both instantaneous and inverse-time elements to protect against most faults. Device temperature thermostats may sometimes be used to give protection over a restricted range of overloads.

When the converter is assigned a higher continuous current, it can then withstand maximum fault currents only for shorter periods. Conventional breakers alone are not fast enough, and high-speed circuit-breakers are employed in the d.c. side. The transformer fault current is arranged to trip the a.c. breaker.

Additional protection against overheating is often given by means of thermostats and flow switches in the air or cooling fluid circuit.

Regulation The inherent regulation of a semiconductor converter equipment depends principally on the transformer connection used, the effective leakage reactance of the transformer and the resistance of the circuit. The reactance is usually designed as a compromise, to limit the fault current when a d.c. short-circuit occurs and yet maintain an acceptable regulation.

Voltage drop across the semiconductor devices alone is very small indeed, and is not linearly related to the current flowing. (At 50 A a germanium element drops about 0.3 V at 10 A and 0.5 V at 150 A. A 200 A silicon element drops about 0.7 V at 10 A, and 1.2 V at 600 A.)

19.2.8.2 Thyristor

The thyristor, or s.c.r. (silicon controlled rectifier), resembles the semiconductor diode except that a four-layer structure is employed. (There are no commercial germanium thyristors.) *Figure 19.21* shows diagrammatically (a) a diode and (b) a thyristor. The core of the thyristor is a wafer of n-type silicon with two p-type layers diffused into its surfaces. The cathode connection forms a p–n junction, with the trigger or gate electrode attached to the diffused p-layer. The anode connection is made to the base, which provides a rigid platform for the whole device. The 'hockey puck'

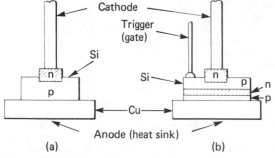

Figure 19.21 Silicon devices. (a) Normal diode with positive heat sink (reverse polarities are available); (b) Thyristor

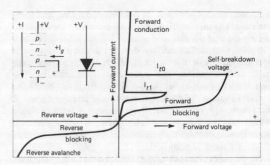

Figure 19.22 Current/voltage characteristic of silicon thyristor

design of housing permits heat to be taken from anode and cathode face', and inherently allows reversal of polarity by upside-down mounting. Great care is required in making the joint, and manufacturers provide full information.

Characteristics *Figure 19.22* shows the static characteristic. In the reverse direction the characteristic is the same as for an uncontrolled diode. In the forward direction the characteristic is the same as for the reverse direction up to the turn-on voltage, at which point avalanche multiplication begins, and the total current increases rapidly until the device becomes highly conductive. The thyristor is then in a switched-on state and will remain in this condition until the forward current is reduced below a small value known as the holding current. Below this value the action again reverts to the forward blocking condition.

Increasing the gate current I_g reduces both the forward breakdown voltage and the holding current. In the fully-on state the thyristor behaves like an ordinary semiconductor diode, but is generally limited to a junction temperature at 125 °C.

The thyristor can be made to fire at a given forward voltage by controlling the gating instant (angle α). In this manner the direct output voltage from a number of suitably connected devices can be varied and even reversed in polarity for inversion, as previously described (Sections 19.2.1, 19.2.2).

Rating The parameters of importance are the forward current, reverse voltage and forward voltage in the non-conducting state; and also the requirements of the trigger or gate. Voltages of transient nature, also surge and reverse currents, must fall within the maker's 'absolute maximum' limits, and as these depend on the circuit, the design must consider the system as a whole. There are low as well as high limits on the junction temperature: the lower limit of −55 °C can be achieved by some devices.

The heat producing losses are: forward, reverse, switching (turn-on and turn-off) and gate losses, together with forward losses in the non-conducting state. Heat dissipation is improved by provision of a heat-sink.

The gate ratings (forward and reverse voltage and current, peak and mean power) are used principally in the design of trigger circuits.

Switching (1) *Turn-on* When a gate pulse is applied to a thyristor having a positive anode, there is a brief delay after which the thyristor switches rapidly to the conducting state. Typical values are 0.2 μs and 0.1 μs for the delay and rise times; but these are affected by the characteristics of the trigger pulse, the duration of which has to be adapted to the type of load. The thyristor remains conducting after the pulse has been removed (provided that its load current exceeds the latching current, typically twice the holding current).

(2) *Turn-off* Once the thyristor is conducting, it can be switched off only by reducing the load current to a value below

the minimum holding current, typically 10 mA, for a discrete turn-off time. The turn-off time is usually between 15 and 100 μs. It is important that the rate of change dv/dt of the anode voltage reapplied after turn-off should not exceed the value given by the makers: otherwise the thyristor will regain its conducting state.

Control The methods used are substantially those mentioned earlier in this section. The four most common configurations are the half-controlled and fully controlled bridges, and the half- and fully controlled a.c. circuits. The half-controlled bridge for 1-phase and 3-phase systems is employed for motor speed control, magnetic clutches and brakes; the fully controlled version, for regenerative applications of a similar type. The a.c. controllers are employed for furnaces and similar applications, with the fully controlled type used where a lower ripple content is desired. All these employ phase control of the triggering angle (*Figure 19.13*).

The *integral cycle* control technique is one in which complete successive periods of conduction occur for a determined time, and then the supply is blocked off for a further duration. By varying the conduction and blocking times, the average power to the load can be varied. Integral cycle triggering is used where the load has a long time constant, as in furnaces and thermal storage systems. The thyristors work at minimum average loss for a given mean output current, and the rate of rise of current is low because triggering occurs only at low instantaneous voltage. In phase-control systems the rate of rise of current may produce difficulties when triggering takes place at an anode voltage peak.

Commutation The load current of a thyristor can be turned off by forcing it to a value below the minimum holding current (*forced commutation*). The methods are by impulse, harmonic or capacitor commutation. Most thyristor switching applications can work satisfactorily with capacitor commutation.

The essentials are shown for a simple case in *Figure 19.23*. Initially the thyristor is conducting, and has a forward volt drop v_f. The capacitor C is previously charged to a voltage V_c of the

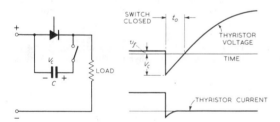

Figure 19.23 Essentials of capacitor commutation

polarity shown, and isolated. When the switch is closed, the thyristor is reverse-biased by V_c, its current ceases, and the capacitor discharges through the load. The capacitor voltage falls to zero, then rises with polarity reversed towards the supply voltage level. Provided that t_0 exceeds the discrete turn-off time for the particular thyristor, conduction stops.

In multiphase cases, and those in which the load is inductive, both the action and the circuit arrangements are much more complex.

Protection Because of their low heat capacity, semiconductor devices are sensitive to overcurrents. In a short-term overload a rapid surge of current may damage the device within one half-period (10 ms in a 50 Hz system). Protection can be provided by special fuses with an ultra-fast fusing characteristic. A voltage surge may occur when the fuse blows, and it is necessary to ensure that the permissible non-repetitive peak reverse voltage of the thyristor (or diode) is not exceeded. This protection is often afforded by the hole storage capacitors.

In general, a thyristor control system can be considered as a thyristor stack, the output current of which can be controlled by gating. The stack is connected to a mains a.c. supply through a circuit-breaker. There are several sources of voltage transient, viz. the supply itself, the circuit-breaker operation, the gating, hole storage, lightning, etc. Each may produce a rate of rise of voltage dv/dt exceeding the value that the thyristors can withstand without turning on.

System transients comprising voltage 'spikes' require input RC filter suppressors. If the circuit has (due, for example, to transformer leakage) a substantial inductance, it may be necessary to consider the possibility of high-frequency oscillations. Thyristor turn-on and turn-off with an inductive output load can produce voltage transients with a high rate of change of voltage; these may be induced into control circuitry and may wrongly trigger other thyristors in the stack, so that overall control is lost.

The thyristor, while being a highly sensitive device demanding protection against phenomena that may in more traditional equipment have negligible importance, is nowadays applied with full confidence. Important factors to be remembered are:

(1) The provision of adequate cooling for limitation of temperature on load, to avoid thermal runaway.
(2) The limitation of current by the load, or otherwise, should spurious thyristor triggering occur.
(3) The consequential effect of overcurrent on the load.
(4) The magnitude of transient voltages and currents, and their effect on the functioning of diodes and thyristors.

Basic circuits The most generally used low-power 3-phase thyristor control arrangements are shown in *Figures 19.24* and *19.25*.

Three-phase a.c./a.c. controller These controllers are common in the control of the transformer primary side of rectifier systems designed to deliver high-voltage/low-current or low-voltage/high-current outputs, because the primary quantities have more convenient levels.

A half-controlled arrangement (*Figure 19.24*) has a thyristor T_a in parallel with a reversed diode D in each of the phase connections. When T_a is triggered in phase A, current flows to the load through it, returning to the supply through diodes D_a or D_c or both together, depending on the instant in the cycle. Thus, the load current in any phase line is the combination of the forward and reverse current components.

Conditions for phase A are shown with various gate-firing angles and a pure resistance load: the wave forms for $\alpha = 0$ are for triggering at the start of the positive half-period of phase A, giving a whole-wave result. For $\alpha = 90°$ the conduction through T_a is delayed by the quarter-period. For a trigger angle $\alpha = 210°$ all thyristors are non-conducting and the output current is zero.

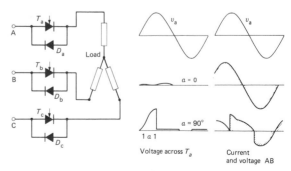

Figure 19.24 Three-phase half-controlled a.c./a.c. controller

A fully controlled arrangement is obtained if each diode D in *Figure 19.24* is replaced by a thyristor. For current to flow, it is necessary to gate at least two thyristors at a time. This form of control is used when the harmonic content of the output is to be considered.

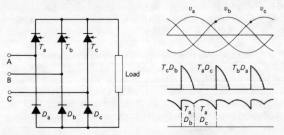

Figure 19.25 Three-phase half-controlled bridge a.c./d.c. converter

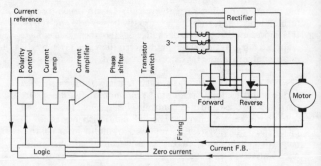

Figure 19.26 Control for reversing mill drive

Three-phase a.c./d.c. converter *Figure 19.25* shows a half-controlled bridge connection, and wave forms for the line-to-neutral voltage and current for a resistive load. Consider thyristor T_a. For high α (short conduction) gate angles, T_a conducts until it is commutated by the voltage of phase C rising above that of phase A. For low gate angles T_a conducts into phase B, which is most negative, switching subsequently to phase C.

The converter is fully controlled if the diodes are replaced by thyristors. Cross-coupling of the gate pulses is necessary.

In all these arrangements, *natural* commutation occurs by reason of the variation of the relative phase voltages from instant to instant.

D.C. short circuits can be interrupted by removing the gate pulses, and inrush currents reduced by large α angle firing delay in phase control. A.C./D.C. converters (or controlled rectifiers) often take the form of the half-controlled bridge. The thyristors can be connected in either arm of the bridge. Where the load is highly inductive (e.g. magnetic clutches and motor field systems), it is essential to commutate the thyristor current at the end of each half-period. This is done by using separate flywheel diodes, or by similar action within the bridge. If regeneration is necessary, a fully controlled bridge connection is essential.

Applications Provided that its special operating conditions are fully met, the thyristor is able to meet the requirements of all duties formerly serviced by mercury arc equipments and many that the mercury arc could not handle.

Simple motor *speed control* systems have been developed for shunt commutator machines, with the armature fed from a device stack arranged as a half-controlled 1-phase bridge, a protective filter separating the stack from the mains. The field winding is energised through diodes (or thyristors, if control is required). Feedback information for stabilising the speed is derived from the motor armature e.m.f. controlling the gate pulse timing. Alternatively, a tachometer generator can be used to give a speed proportional control voltage.

A basic electric furnace *temperature control* system uses a trigger module receiving one signal from a temperature sensor, and a second from a current-limit module, monitoring the heater current. These two signals determine the gate pulse angle of the thyristor stack.

Large d.c. power blocks, sometimes exceeding 100 MW in one unit, are required in large *electrochemical plants* such as those for the production of aluminium (1 kV, 100 kA). The modern unit is invariably a silicon rectifier bank of 12- or 24-pulse equipments operating in parallel. The commonest control is a transformer with on-load tap changing and a saturable inductor for inter-tap adjustment, but thyristor control is already of major importance in electrochemical machining, and in tinning and galvanising lines.

Thyristor supply has been applied to all types of *rolling mills* using drives up to 15 MW. Power supply to the drive motors must be continuously controllable throughout the rolling cycle of threading, acceleration, full-speed running, retardation, in either direction from standstill. Variable voltage d.c. motors have the appropriate torque–speed characteristics for such a duty; and as they are required to regenerate during retardation, they must operate also as generators and the thyristor equipment as an inverter. The three functions required are: (1) to provide continuous rolling power, (2) to provide a short-time peak power for acceleration and (3) to accept a short-time peak power for retardation and reverse jogging.

Functions (1) and (2) are performed by a rectifying thyristor converter, and (3) by an inverting thyristor converter, one form of which is an antiparallel connection where the gate pulses are switched to select the converter required. The dead-time involved in a changeover has to be substantially less than 50 ms on strip mills, for which there would be a noticeable variation in strip tension. The control method is shown in block schematic form in *Figure 19.26*.

During normal operation only one of the two bridges is required to conduct. A current feedback signal from current transformers is rectified to produce a d.c. proportional to the instantaneous motor armature current, its direction being determined by the logic circuits. When reverse current is demanded, gate pulses are moved to maximum delay, current zero is detected, and after a short time delay, firing pulses are released to the reverse bridge, the polarity of the current reference is reversed and the gate pulses advanced to pick up current.

20 Installation

J P Milne BSc(Eng), CEng, FIEE, FCIBSE
Engineering Consultant

Contents

20.1 Layout 20/3
    20.1.1 System supply 20/3
20.2 Regulations and specifications 20/3
20.3 High-voltage supplies 20/3
20.4 Fault currents 20/4
20.5 Substations 20/4
    20.5.1 Low-voltage equipment 20/6
    20.5.2 Packaged substations 20/6
    20.5.3 Low-voltage distribution 20/6
20.6 Wiring systems 20/7
    20.6.1 Steel conduit 20/7
    20.6.2 Plastics conduit 20/8
    20.6.3 Trunking 20/8
    20.6.4 Cabling 20/9
20.7 Lighting and small power 20/9
    20.7.1 Lighting circuits 20/9
    20.7.2 Small power circuits 20/9
20.8 Floor trunking 20/10
    20.8.1 Underfloor trunking 20/10
    20.8.2 Open-top trunking 20/10
    20.8.3 Cavity floor system 20/11
    20.8.4 Under-carpet system 20/11
20.9 Standby and emergency supplies 20/11
20.10 Special buildings 20/11
    20.10.1 High-rise buildings 20/11
    20.10.2 Public buildings 20/11
    20.10.3 Domestic premises 20/11
20.11 Switchgear 20/12
    20.11.1 Moulded-case and miniature circuit breakers 20/12
    20.11.2 Prospective fault current 20/12
    20.11.3 Discrimination 20/12
    20.11.4 Motor control gear 20/14
20.12 Transformers 20/14
    20.12.1 Installation 20/15
    20.12.2 Transformer protective devices 20/15
20.13 Power-factor correction 20/16
    20.13.1 Capacitor rating 20/16
20.14 Earthing 20/16
    20.14.1 Earth electrode 20/16
    20.14.2 Soil resistivity 20/17
    20.14.3 Electrode installation 20/17
    20.14.4 Resistivity and earth resistance measurement 20/18
    20.14.5 Protective conductors 20/18
    20.14.6 System earthing 20/18
20.15 Inspection and earthing 20/19
    20.15.1 Visual inspection 20/19
    20.15.2 Testing 20/19
    20.15.3 Test gear 20/22

20.1 Layout

This chapter deals with the installation of a power supply to a consumer. One example might be a large industrial concern with a factory complex spreading over several hectares; another the single tower-block with one supply entry point in the basement but with additional substations at intermediate floors. At the other extreme is the small dwelling-house.

The general rules of safety of personnel and equipment, continuity of supply and ease of operation and maintenance apply to all consumers, but the layout naturally varies with the individual establishment.

20.1.1 System supply

When an electricity supply is to be provided to a factory or building complex, there are several features that must be taken into account. Safety is one that cannot be measured in terms of capital cost, but it should not be difficult to achieve an acceptable standard, provided that good-quality equipment is purchased, properly installed and well maintained, and operated by experienced staff.

Loss of supply may result in loss of output from a factory or danger to life in a hospital. The system designer must take these matters into account. To eliminate all risk of outage is likely to be very costly: the alternative of accepting some measure of risk is often preferable, bearing in mind the present-day reliability of equipment and of the public electricity supply.

Losses occur in cables, overhead lines and transformers, and the system should be designed to minimise these, by locating substations as near as possible to the centres of load. The system power factor is also relevant, for it can affect equipment sizing, I^2R losses and the cost of electrical energy.

In few installations is maintenance afforded the importance it deserves. The design and installation engineer cannot control this, but he can ensure that the equipment supplied allows for proper isolation and regular testing with the minimum interruption of supply.

Few installations remain unchanged throughout their life. Whether it be the provision of a spare way on a distribution board or space for an additional high-voltage ring-main circuit, thought should be given to the problem at the time of design and installation.

In determining the cost of installation it must be borne in mind that, in addition to the capital cost of the equipment, attention must be given to the running cost, and also to the cost of providing accommodation for the equipment. For example, it may be better to provide outdoor equipment rather than less costly indoor plant when provision of a building to house it is taken into consideration.

20.2 Regulations and specifications

In the United Kingdom the supply of electricity to premises is governed by the Electricity Supply Regulations, which require that all electrical equipment in the premises shall be maintained in an efficient state and that, as regards high-voltage equipment, authorised persons are available to cut off supply in emergency. The actual erection and installation should comply with the Institution of Electrical Engineers Wiring Regulations for Electrical Installations: compliance with these Regulations will produce a high standard of work and allow a good factor of safety against possible breakdown.

In 1981 the new edition (15th) of these Regulations (referred to as W.R.15) marked a radical change in both style and approach. It is based on internationally agreed installation rules, the two bodies concerned being the International Electro-technical Commission (IEC) and the European Committee of Electrotechnical Standardisation (CENELEC).

Every IEC Standard issued expresses, as nearly as possible, an international consensus on its subject, and the intention is that every member country should adopt the text of the Standard, as far as the individual country's conditions will permit. Additionally, the members of the EEC are committed to removing trade barriers, and CENELEC aims to help in this by attempting to 'harmonise' the corresponding national standards. In 1968 IEC started work to formulate standard rules. It was soon realised that combining the existing rules of member countries was not feasible, and so IEC decided to go back to fundamentals. The work is far from complete, but some sections have been published, and the IEE decided to make use of the IEC plan and the technical content already agreed by IEC in revising W.R.15. This has resulted in a set of IEE Wiring Regulations very different from any previously published.

These new IEE Regulations W.R.15 are aimed directly at the designer of electrical installations rather than the site installer. They demand an analytical, mathematical approach to the design of the installation. Additional design time will be required, but the designer is given a greater degree of freedom to produce an economic design.

W.R.15 has already spawned two books of explanatory notes. But what is not in doubt is that an installation fully in accordance with W.R.15 will be well-designed, safe in operation, and with minimum risk from fire, shock or burns. Additionally, it is to be hoped that the customer will get better value for money.

The intention here is to provide guidance on installation matters, but it must be pointed out that casual and occasional reference to W.R.15 is not enough to inform readers of the requirements. This can only be achieved by a study of them in depth.

20.3 High-voltage supplies

The general method of supplying bulk power to factories or other complexes is by means of a high-voltage supply, usually at 11 kV, occasionally at 6.6 kV. The installation comprises a main substation at the point of entry, with h.v. cables to supply subsidiary substations located near load centres. Where desirable, it may be possible for the consumer to obtain two supplies from the power authority, and in special cases where loss of supply could give rise to a particular hazard (e.g. the danger to life in a hospital) the two supplies may be provided from separate sources.

The distribution system generally comprises either radial feeders or a ring main using underground cables to supply the subsidiary substations. The emphasis today is towards ring-main supplies, and this is considerably helped by the ready availability of competitively priced ring-main switchgear. Typical is the example shown in *Figure 20.1*. This is the basic unit, consisting of two oil switches, capable of making on to a fault and of breaking load, and a fuse-switch. Other units are available, assembled into various configurations. They can include circuit-breakers.

Four distribution arrangements are illustrated in *Figure 20.2(a–d)*, although variants are also available.

Diagram (*a*) is a *radial-feeder* arrangement with each feeder controlled by its own automatic circuit-breaker. A cable or transformer fault will result in loss of supply only to the one substation, but supply cannot be restored till the repair is effected.

Diagram (*b*) is a simple *ring-main* arrangement which allows for any individual transformer substation to be isolated for inspection or test without interrupting the supply to the other substations Any fault occurring on the system will result in a total shutdown, but the fault can be isolated and the supply restored to

Figure 20.1 Ring-main switchgear

all substations but one, unless the fault is in the cable between substations C and D, in which case all substations can be re-energised.

In diagram (c) the distribution substations have their outgoing cables controlled by switches and the transformer by a fuse-switch. Hence, every transformer can be independently switched, and should a fault occur and one of the fuses blow, the fuse-switch is arranged to open all three phases automatically, thus disconnecting the transformer from the system. The system would normally be operated with the ring open, but any cable fault will result in part of the system being shut down until the fault is located, when power can be restored to all the substations.

By using automatic circuit-breakers throughout, as shown in diagram (d), a greater degree of security of supply is obtained. The system can be operated as a closed ring, and even greater security is obtained by duplicating the transformers as shown in the diagram.

Modern transformers and switchgear are very reliable, and provided that cables are installed carefully, the number of faults that will occur will be extremely small. Since cost will always be of importance, it is common to use the simplest ring-main switchgear that is acceptable for the particular application, and switches with a fuse-switch tee-off represent very good value for money and are widely used both by public electricity authorities and by industry. The situation is helped by the availability of fault monitoring and locating equipment. Hence, if a fault does occur, the faulted section can be rapidly located and isolated for repair. This equipment comprises a current transformer on the outgoing feeder cable to recognise an unbalanced condition caused by any fault current. With the advent of microprocessors the system can be extended so that equipment can be installed to identify exactly where the fault lies on a network. In the future, it can be expected that auto-reclose of the healthy circuits will be possible.

20.4 Fault currents

A relatively simple calculation determines the required interrupting capacity of switchgear. This is particularly useful to the installation engineer when it is necessary to provide extensions to an existing system and he must ensure that the changes will not result in the rupturing capacity being exceeded. There is less risk of this today, most equipment being manufactured with a minimum rupturing capacity of 250 MVA at 11 kV, but where old gear is already in use, problems could arise.

The technique involves using the reactances of the equipment. In the case of underground cables, the resistance is high compared with the reactance and so for calculation purposes the impedance of the cable is used. For generators the sub-transient reactance is used, as it is effective for the first few cycles following a fault.

The ratio between the phase voltage drop and the normal line-to-neutral voltage for a stated current (or kVA) is the per-unit (or per cent) impedance at that current. For example, if a 1000 kVA transformer has a 0.05 p.u. (or 5%) reactance, then its voltage drop on rated load is 0.05 of the line-to-neutral voltage. It follows that if the transformer output terminals are short-circuited, the volt-drop is 1.0 p.u. (100%) and the short-circuit kVA loading is $1000 \times (1/0.05) = 20\,000$ kVA, and this is the maximum that the transformer can pass. Again, consider a length of 11 kV cable of impedance $1\,\Omega$/ph and carrying 1000 kVA. The current is $I = 1000/(11\sqrt{3}) = 52.5$ A, the phase voltage is $E_{ph} = 11/\sqrt{3} = 6.35$ kV, the volt-drop is $v = 52.5 \times 1.0 = 52.5$ V, and the impedance of the cable is $(52.5/6350) = 0.0083$ p.u. (0.83%).

Calculation of the interrupting capacity S required at a point in a system is by assuming an arbitrary kVA S_0 in the system, and evaluating the voltage required to pass this load as a fraction x of the line-to-neutral voltage. Then $S = S_0/x$. The effective reactance of the system configuration involves summing branches in series and parallel and, in some cases, the application of the star/delta transformation technique. Details are given in Section 20.1. *Table 20.1* gives typical values of reactance for generators and transformers.

20.5 Substations

The space requirements of a substation depend on the equipment to be housed, and on whether a new building can be erected for it or it has to be fitted into an existing building. In the latter case it may be difficult to achieve an ideal solution, but where no severe limitations are imposed the layout in *Figure 34.4* would prove satisfactory. This is suitable for a main 11 kV substation, also supplying local l.v. distribution, and it will be seen that it meets most of the following requirements:

There is adequate clearance around the equipment, and space to withdraw circuit-breakers for maintenance.

Equipment operating at different voltages is segregated (advisable but not mandatory).

There is sufficient space for drawing in and connecting cables, and for the delivery and erection of additional switchgear: doorways are high and wide enough.

The walls can act as fire barriers. If an h.v. bus-section switch is included, a decision must be made as to the need for fire barrier walls between the sections. Even with oil circuit breakers, this risk is small and many engineers disregard it.

Transformers are usually housed in open or semi-open compounds; but if an indoor location is essential, particular

Substations 20/5

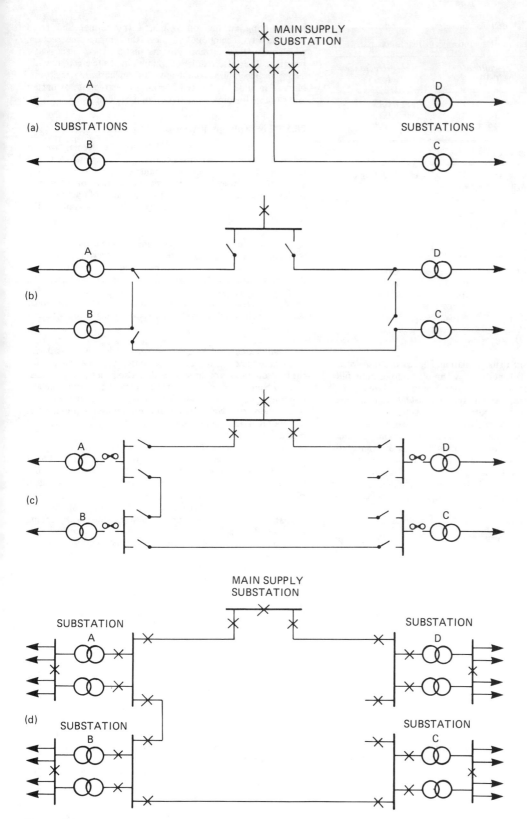

Figure 20.2 Alternative supply system designs

Table 20.1 Typical per-unit reactances

(a) Synchronous generators: subtransient reactance, p.u.

Rating (MVA)	Voltage (kV)	Reactance (p.u.)
Up to 20	11	0.12
20–30	11	0.135
Over 30	22, 33	0.17, 0.20

(b) Transformers: leakage reactance, p.u.

Rating (MVA)	6.6/0.4 or 11/0.4 kV p.u.	33/11 kV p.u.	132/11 kV p.u.
0.05	0.034	—	—
0.1	0.036	—	—
0.2	0.040	—	—
0.5	0.043	—	—
1	0.045	—	—
2	0.056	0.06	0.075
5	0.07	0.075	0.09
10	—	0.09	0.10
20	—	0.10	0.10
45	—	0.10	0.11

attention must be paid to its ventilation. The risk of a transformer fire is extremely small: nevertheless, an oil sump (usually filled with pebbles) should be provided to trap burning oil which could escape following a fire. Alternatively air insulated or cast resin transformers can be used, but are more expensive.

To simplify the stock of spares and to ensure ready interchangeability between gear in different substations, as much standard equipment as possible should be used, even at the expense of varying from the ideal installation. In general, substations should be limited to a capacity of about 2000 or 3000 kVA, with individual transformers no larger than 1500 kVA, to allow for the use of commercial l.v. switchgear of about 31 MVA rupturing capacity.

20.5.1 Low-voltage equipment

At substations the low-voltage distribution gear consists usually of a circuit-breaker for each transformer with circuit-breakers, or switches and fuses for the outgoing feeders. Low-voltage feeders are usually supplied with ammeters and often with meters to measure the energy consumption of the feeder. In many cases maximum-demand meters are included so that this feature can be periodically monitored—a useful facility in large industrial complexes.

Feeders radiate to the various sectional distribution centres, where they terminate at switchboards to which smaller sub-main cables are connected, supplying power to the various departments or shops by means of other smaller distribution boards. The most important feeders will again be duplicated or interconnected to some extent, to safeguard the supply, so that an l.v. breakdown will result only in a temporary shutdown of one section. With duplicate cables, one may be entirely spare or both cables may share the load, provided that they are both of sufficient capacity to carry the total current independently in emergency.

20.5.2 Packaged substations

In recent years the 'packaged' substation has become increasingly popular. *Figure 20.4* shows a typical design, incorporating an h.v. oil switch, a transformer and fused outgoing ways. Their popularity lies in the compact construction, which makes them attractive for installation where space is at a premium. They also require the minimum of site erection work. They can be supplied complete with a prefabricated enclosure, often of moulded reinforced fibreglass construction, or unclad and suitable for direct installation in a building.

20.5.3 Low-voltage distribution

From the substations described earlier, l.v. feeders run to subsidiary substations or load centres. These can include multi-motor starter boards, air-conditioning control boards, lighting and heating power distribution boards, vertical or horizontal busbars, street lighting supplies, etc. The range of alternatives is wide and the following paragraphs outline only some of the solutions available.

Multi-motor starter boards Large motors are usually supplied with independent separate feeders. This is advisable because of the heavy fluctuating load, which might otherwise cause disturbance to other plant on the same supply. However, where a number of small or medium-sized motors are in reasonable proximity, it is convenient to group the starters in a multi-motor starter panel, which often includes one or more distribution boards for lighting.

Air-conditioning and ventilation control boards These are variants of the multi-motor starter board. As well as starters for the fans and pumps of the system, they also include the specialist control equipment needed for the automatic control of the air-conditioning and ventilation. Similar developments are found in many industries where plant process control equipment is incorporated into combined motor starter and small-power boards.

Busbar systems Generally, wherever there are continuous rows of machine tools to be fed, an overhead busbar system provides the required degree of flexibility. Used vertically, a busbar system offers simple and flexible provision of power in a high-rise building. Several manufacturers market copper/aluminium busbars enclosed in steel trunking and provided with tapping points at intervals. Standard tees, bends and other accessories are available. At the tapping points it is possible to insert a tapping box, which can be safely applied or removed with the internal busbars live. When positions of machines have to be changed or new machines installed, it is easy to insert a tapping box at the appropriate point in the run of the busbar trunking. The maintenance cost is low, depreciation is minimal and there is a high recovery value if the trunking has to be demounted and rerun. Earth continuity can be provided by an external link bolted between lengths of trunking, but for large ratings a separate earth conductor is necessary. No cables other than the busbars themselves may be included within the trunking. Fire barriers are required for vertical runs, and for long runs it is advisable to provide busbar expansion joints.

From the individual fuse (or circuit-breaker) units on the overhead busbar, cables in conduit or flexible metallic tubing are run to the machine tools. In the case of tools with more than one motor, distribution boards can be mounted at a convenient point on the equipment.

Vertical busbars are usually run in a vertical duct within the building. A lock-proof cupboard is formed at each floor level and the distribution boards and switches are accommodated therein, with the outgoing supplies fed to meet the lighting and power needs of the floor occupants. Where possible, it is preferable to use only one phase on each floor of the building, as this eliminates

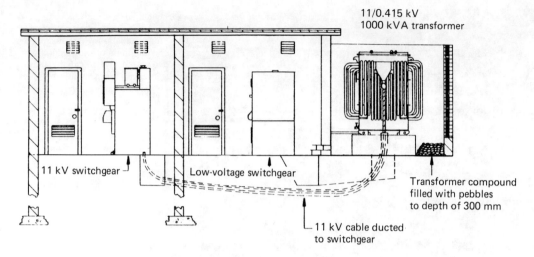

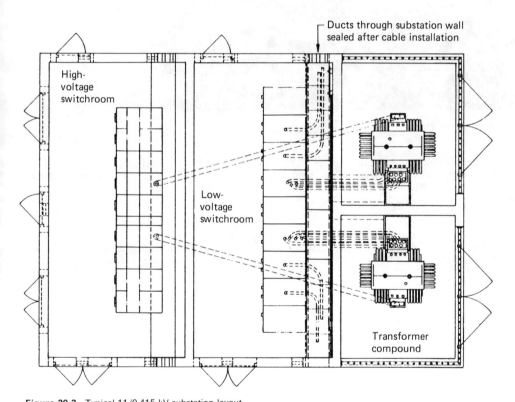

Figure 20.3 Typical 11/0.415 kV substation layout

risk of voltages in excess of 240 V being encountered by personnel.

20.6 Wiring systems

The choice of wiring system for industrial, commercial and domestic installations depends on the structure, application, supply voltage, load, appearance and cost. An essential feature is a proper earthing system giving an earth continuity conducting path for protection and safety.

20.6.1 Steel conduit

Screwed steel conduit provides a potentially good earth continuity, and makes it possible to enclose cables throughout their length with strong mechanical protection. The conduit is normally of a heavy-gauge welded construction.

Skill is required to achieve a proper system of good appearance. The conduit should be erected completely before cables are drawn in; tube ends should be reamered to prevent damage to cables; draw-in boxes should be amply proportioned and accessible; the conduit should be adequately supported by

Figure 20.4 'Packaged' substation equipment

saddles or clips. There should be no more than two right-angle bends (or their equivalent) between successive draw-in boxes. The whole system should be mechanically continuous throughout, including connections to switches, fittings, distribution boards, motor control gear and the metal cases of other equipment; and it must be properly earthed.

Most conduit systems in factories are run on the surface, but in offices, or where conduit would be affected by corrosive fumes, it may be sunk in concrete in the floor or in walls behind plaster or cement. The layout must be carefully preplanned, and joints must be tight, to prevent the ingress of wet concrete, cement or plaster. A draw wire should be pulled through as soon as possible after pouring or plastering, to remove any intrusive material before it hardens. The actual wiring is left until all is dry, and all wiring in one conduit must be drawn in at the same time.

20.6.2 Plastic conduit

In recent years there has been an upsurge in the use of plastics conduit. PVC components are generally unaffected by water, acids, oxidising agents, oils, aggressive soils, fungi and bacteria, and they can be buried in concrete or plaster. They provide slight resistance to heat and flame, and in this respect are inferior to steel conduit. Their mechanical strength is also much less.

They are not electrically conductive and therefore separate earth continuity conductors have to be run, which on occasion can require a larger diameter than for a steel conduit. Expansion couplings may need to be included in long straight runs, as linear expansion of PVC gives it an extension of about 1.5 mm/m for a temperature rise of 20 °C. On the other hand, the conduit can be readily cut by hacksaw and bent when gently heated. PVC conduit can be screwed using suitable dies, but jointing of lengths is commonly by a solvent welding compound. This is easier, quicker and provides an entirely watertight joint.

PVC conduit and fittings are marginally cheaper than the equivalent steel conduit components, but PVC is a by-product of the oil industry and its price comparison with steel conduit is likely to vary.

20.6.3 Trunking

Trunking is used to bunch numbers of cables or wires which follow the same route, with branches to motors, switchgear and lighting circuits by spurs in trunking or conduit. Trunking is made in various sections from 50 mm × 50 mm upward and in standard lengths. Tees, reducers, crossings, elbows and other fittings are made as shown on *Figure 20.5*.

Both steel and plastic trunking are used. With the former, low-resistance joints between lengths of trunking and fittings are essential. If the trunking is itself used as a 'protective conductor' under the requirements of the Wiring Regulations, its bonding and jointing are specified. Where conduits connect to the trunking, a clearance hole is drilled in the trunking wall and the conduit connected with a socket and male bush. Connections between trunking and switchboards and similar panels are generally effected by flanges securely bonded to the trunking and bolted to the switchgear.

On a switchboard or distribution panel in which lengths of trunking enclose the various interconnecting cables, the units should be bonded to the trunking by a copper earth tape. In the case of vertical runs of trunking, internal fire barriers must be fitted at each floor level. These barriers must also be fitted in horizontal runs where the trunking passes from one zone of fire protection to another.

Plastics trunking is made in sizes similar to those for steel trunking, small sizes being extruded and large ones formed from PVC laminated sheet. Common fittings are available, but as PVC is easily 'formed', special fittings can be 'welded' with the help of a

In industrial installations where cost is a prime consideration, it may sometimes be acceptable to run PVC sheathed cabling fixed by cleats spaced about 1 m apart. Cables should preferably not be concealed and must be protected where they run through floors, and also on walls if fixed at a level below about 1.6 m from the floor.

20.7 Lighting and small power

Whatever type of electrical installation is involved—be it for a factory, a public or commercial building, or domestic premises—supplies for electric lighting and small power will be needed.

20.7.1 Lighting circuits

There are two main ways of running lighting circuits, 'loop-in' and 'junction box'. Both can be used in the same installation. In the loop-in system (a) in *Figure 20.6* the terminals for joining the cable ends form part of the ceiling rose. The junction-box system (b) is used when the lighting fittings have no loop-in terminals, or to save cable when the lamp and the switch are far apart. The connections are similar to those in the loop-in system.

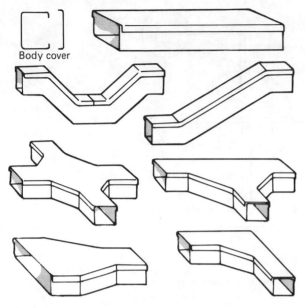

Figure 20.5 Cable trunking

20.7.1.1 Two-way switching

The two-way switch is a single-pole changeover switch. When interconnected in pairs, two-way switches provide control from two positions and are therefore installed on landings and staircases, in long halls and in any room with two doors. The arrangement is as shown diagrammatically at (a) in *Figure 20.7*.

20.7.1.2 Intermediate switches

Where there are long halls, corridors or passageways with several doors, it may be convenient to introduce additional switching positions. The 'intermediate' switch enables any number of additional switching points to be introduced to the two-way switch circuit, as in (b) of *Figure 20.7*.

20.7.1.3 Master control switch

A master control switch is sometimes provided to give overall control to a number of lamps which are, additionally, independently switched, as in *Figure 20.7(c)*. A typical installation would be an hotel bedroom or suite, where a double-pole switch at the entrance to the room will switch off all lights on the occupant's leaving the room.

hot-air welding gun. The characteristics of PVC trunking make it particularly suitable for fume-laden atmospheres, humid tropical conditions and salt-laden coastal air. The material has the physical characteristics mentioned in Section 20.6.2 and for the same reason requires a bare copper conductor to be included to provide an effective earth. Screwed or snap covers may be used.

20.6.4 Cabling

Details of insulated cables or conductors are given in Chapter 13. The following cables have applications special to the wiring of buildings.

30.6.4.1 Mineral insulated metal sheathed cables (MIMS)

A copper MIMS cable consists of high-conductivity single-strand copper conductors inside a seamless copper sheath packed with magnesium oxide powder as the insulant. As an alternative, aluminium conductors and sheath are also available. The sheath acts as a robust and malleable conduit, capable of withstanding considerable mechanical abuse. It is strongly fire resistant, and when covered overall with a PVC sheathing can withstand chemical corrosion.

Because magnesium oxide is hygroscopic, the exposed cable ends must be sealed. Normally the sealants are suitable for operating temperatures up to 70 °C, but special oil-resistant and high-temperature seals are available for special applications.

Although dearer than the equivalent steel conduit installation, there are applications where MIMS cabling is superior and preferred. For example, some fire authorities insist on MIMS cables for all fire protection, detection and alarm circuits.

20.6.4.2 PVC sheathed cables

For domestic installations, PVC insulated and sheathed cables are often used, run on the surface or buried in plaster. In down-runs to switches, it is preferable to enclose the cables in PVC conduit. To comply with regulations, the cable must embody an earth wire, which must be connected at lighting points, switches, socket boxes and ceiling roses to the earth terminal.

20.7.2 Small power circuits

The socket outlet is a safe and convenient means by which free-standing or portable apparatus can be connected to an electric supply. It is false economy to limit the number of socket outlets installed. Sufficient should be provided to match the predicted needs of the consumer, and should be located adjacent to the most convenient point of use of the apparatus.

Both radial and ring final subcircuits may be used to connect socket outlets, but it is suggested that for general use the 30 A ring-main system using BS 1363 fused plugs approaches the ideal. With this system, the current-carrying and earth-continuity conductors are in the form of loops, both ends of which are connected to a single way in the distribution board. The conductors pass unbroken through socket outlets or junction boxes (or must be joined in an approved manner). The ring may feed an unlimited number of socket outlets or fixed appliances but the floor area covered shall not exceed 100 m². In domestic

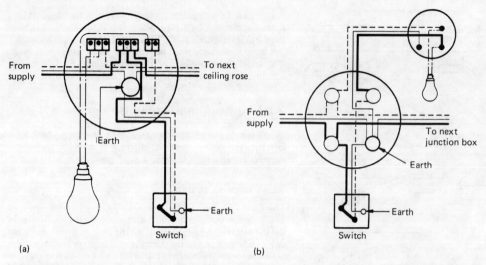

Figure 20.6 Ceiling-rose loop-in and junction-box wiring: (a) Loop-in system; (b) junction-box system. Key: ———, red, always live; ---, black, neutral except where returning from switch; ———, earth; —·—, flex to lamp

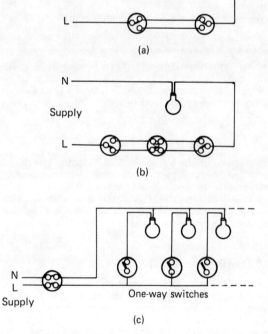

Figure 20.7 Lamp switching: (a) Two-way switching; (b) two-way and intermediate switching; (c) master control switching

premises particular consideration should be given to the loading in the kitchen, which may require an additional circuit.

In the UK an unlimited number of fused spurs may be installed in a ring circuit, but the number of *non*-fused spurs must not exceed the number of socket outlets and items of fixed equipment in the circuit. Also a non-fused spur must feed only one single or twin-socket outlet, or one permanently connected item of equipment. A typical ring circuit is shown in *Figure 20.8*.

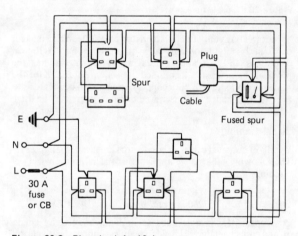

Figure 20.8 Ring circuit for 13 A outputs

20.8 Floor trunking

Floor trunking is used mainly as a flexible means of providing small-power and telephone facilities throughout an open floor area. The basic methods are underfloor, open-top and cavity-floor systems. All must have continuity between metal sections and be adequately earthed.

20.8.1 Underfloor trunking

The several proprietary brands available consist basically of a closed trunking with one, two or three compartments, laid in a grid pattern on the floor slab. Specially designed intersection boxes take care of the crossovers in the lines of trunking, and fixed outlet boxes from the trunking are installed to meet all possible future needs. When the trunking is screeded over, the intersection and outlet box lids are flush with the final floor level.

20.8.2 Open-top trunking

Open-top trunking is a more recent concept of floor trunking. As with the underfloor trunking, it is laid on the floor slab before

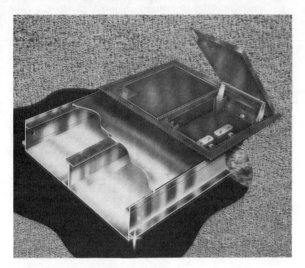

Figure 20.9 Open-top trunking

screeding, but in this case the heavy-duty cover plates of the trunking are flush with the final floor finish (*Figure 34.11*), which makes it easy for outlet boxes to be installed anywhere along the trunking are flush with the final floor finish (*Figure 20.9*), which changing requirements and additional boxes can be added, so giving considerable flexibility.

20.8.3 Cavity floor system

This flexible system comprises wood or metal square plates resting on support jacks. The services are run in conduit or trunking on the surface of the floor slab under the floor plates. Flush surface boxes can be installed into holes cut almost anywhere in the syspended floor, and wiring to power and telephone outlets is routed via conduits or trunking to central junction boxes.

20.8.4 Under-carpet system

This recently introduced system is laid on the office floor, dispensing with trunking. It comprises a flat cable capable of loading up to 30 A, with three (1-phase) or five (3-phase) colour-coded conductors sealed between two polyester layers, then sandwiched between a plastics abrasion shield (for protection against floor irregularities) and a copper earthing shield which is bonded to the earth conductor at 0.7 m spacing. Installation is simple: the cable is unreeled in position, then covered with a protective steel tape. The thickness of the assembly is about 0.9 mm, thin enough to be folded where necessary to turn corners. Supply is tapped through small pedestals where and when necessary. Transition boxes enable the flat cable to be joined to conventional cable at wall sockets or within distribution boards.

20.9 Standby and emergency supplies

Provision for supply failure has sometimes to be considered, most commonly for emergency lighting in the form of (i) escape lighting (to ensure the provision of adequate illumination of emergency escape routes) and (ii) standby lighting (to enable normal activities to continue during the period of mains failure). The two most common systems of emergency lighting are: *maintained*, where normal lighting is fully maintained by automatic switching to an alternative source should mains failure occur; *non-maintained*, where the emergency lamps are energised only when the normal lighting has failed. Both systems are used extensively. The maintained system is essential in certain public buildings, and is the system preferred for 'standby' lighting.

Standards of emergency lighting have proved hard to define, but in the UK the CIBS Lighting Code and BS 5266 do now provide some useful guidance.

In the past, most emergency lighting systems were supplied from a central battery and charger unit, but self-contained emergency lighting luminaires have become readily available for both maintained and non-maintained systems. Each unit comprises a lamp, battery, charger and control equipment, and is wired conventionally into the normal lighting system. Usually a neon or diode indicator is incorporated to provide a visual indication that the battery is under charge. In general, where fewer than 20 fittings are involved, the self-contained luminaires are likely to give the more cost-effective solution.

20.10 Special buildings

20.10.1 High-rise buildings

With large high-rise buildings, the heavy electrical and mechanical services plant is usually located in the basement. However, there are occasions when air-conditioning chillers, standby generating plant and even boilers are located on the roof, and this can influence the positioning of the main electrical switchboards. Nevertheless, since the main power supply enters at basement level, it is usual to locate the main transformer and the primary substation in the basement. Basement-mounted equipment will normally provide supply for 10–12 floors. For taller buildings, additional substations are required every 10–15 floors, and these are often of the packaged type.

When equipment is installed in buildings to be used as offices or apartments, particular attention must be given to preventing unnecessary vibration. Rotating plant should be provided with anti-vibration mountings. In a new building the architect and the structural engineer should co-operate in providing a 'floating' floor to support equipment liable to vibration.

20.10.2 Public buildings

In installations for large buildings and special establishments open to the public (such as theatres, cinemas, etc.) special precautions have to be taken in the layout of the electrical equipment. Regulations framed by local public authorities must be strictly observed in planning the scheme. These deal more especially with the requirements necessary to avoid danger to the public from fire, explosion, electric shock, etc., and are also concerned with the question of dual supply or emergency lighting in the event of a breakdown of the power mains.

20.10.3 Domestic premises

Domestic premises are commonly connected to the street supply by a 1-phase low-voltage PVC cable, often with aluminium conductors. The incoming supply is rated at between 60 and 100 A and is capable of supplying the lighting, heating and cooking demands of the house or flat.

The incoming cable terminates in a service cut-out and connects via the house meter to a consumer's unit. There is an increasing desire with new housing to locate the meter in a position where the meter reader can read it without access to the house. The consumer unit is equipped with an incoming supply switch and outgoing ways using HRC fuses or miniature circuit-breakers. One or more 30 A ring mains for socket outlets will be provided. Immersion heaters fitted to hot-water cylinders in excess of 15 litres capacity should be supplied from an independent circuit. Electric cookers are fed by a 30 A final subcircuit through a cooker control unit which may also include a 13 A socket outlet.

When the ratings of the protective devices or the quality of the earthing is such that an earth fault would not be cleared within the prescribed time, a residual-current device (r.c.d.) should be fitted. It could cover a whole domestic installation, but there can then be a risk of spurious tripping caused by the normal leakage currents of healthy equipment.

An alternative would be to use one r.c.d. on a socket outlet circuit. Certainly, if a socket outlet is used to connect garden tools, this should be protected by an r.c.d. The r.c.d. can be incorporated in the consumer unit, or a unit comprising one socket outlet and an r.c.d. unit can be contained in a twin-socket box.

20.11 Switchgear

Heavy switchgear is dealt with in Chapter 15. The term 'industrial switchgear' is generally applied to that used on voltages not exceeding 600 V and controlling power from the l.v. side of a transformer to subdistribution boards which may include automatic circuit-breakers, or to fuse-switches and contactors for motor-starting equipments. Consideration must be given to the prospective fault current at the point of installation, particularly as it may directly affect a greater number of personnel, often unskilled and unaware of the danger should the apparatus fail to perform its function.

Oil circuit-breakers for operation at 415 V are out of favour. Air circuit-breakers are still preferred by some engineers to moulded circuit-breakers, and for the higher current and fault levels they are essential.

20.11.1 Moulded-case and miniature circuit-breakers

Of particular interest to the installation engineer are the miniature and moulded-case circuit-breakers widely used in domestic, commercial and industrial applications. They are designed to operate manually for normal switching functions and automatically under overload and short-circuit conditions. The m.c.b. enables circuits to be restored quickly after a transient fault, and gives instant indication of tripping.

The primary function of an m.c.b. is to protect an installation or appliance against dangerous sustained overloading and short-circuit faults, but it will also give protection against earth faults, provided that the earth fault-loop impedance is low. As a secondary function, an m.c.b. may also be used as an isolating switch or for local control switching.

A circuit-breaker used to protect a subcircuit should have a current rating matched to its load; the circuit-breaker current rating must not exceed that of the cable. If the breaker is protecting an appliance with a fixed load, say an electric heater, the current rating of the breaker should be at least equal to that of the appliance. Also, the current causing effective operation of the breaker must not exceed 1.45 times the lowest current-carrying capacity of any of the conductors in the circuit. (This requirement of W.R.15 also applies to other protective devices, such as fuses.)

If the appliance imposes a fluctuating load on the supply and, in particular, if heavy starting surges are involved (as with motors or discharge lighting circuits), it may be necessary to select a current rating greater than the nominal current of the appliance to be protected.

Miniature circuit-breakers usually embody both overload and short-circuit tripping devices, the overload usually by a thermal device, the short-circuit by a magnetic one. A trip-free mechanism is incorporated so that the contacts cannot be held closed against a fault, and the thermal element prevents continuous rapid reclosing of a circuit when a fault or overload persists. Most types must be derated when operating at high ambient temperatures, which may also affect the tripping times.

Moulded-case circuit-breakers are primarily intended for applications such as protecting main feeder cables, or acting as main circuit-breakers controlling banks of other circuit-breakers. As such, their major function is to provide back-up protection of the subcircuit protective devices. Overcurrent protection is a relatively secondary function. The moulded-case breaker is a more robust device and can often be provided with auxiliary items such as extended operating handles for assembly in multi-switch cubicles, motor-operated mechanisms, mechanical interlocking with an adjacent m.c.c.b., and shunt trip coils. Miniature circuit-breakers have current ratings from 0.4 A to about 100 A at 440 V and with short-circuit capability up to 9 kA. Moulded case circuit-breakers have ratings from 15 A to about 1500 A at 660 V and short-circuit capability of the order of 40 kA.

20.11.2 Prospective fault current

The prospective fault current at any point in a system is the current that would flow if there were a solid short-circuit at that point, and it represents the maximum current the protective device would have to interrupt. In practice, the actual current is usually much less than the prospective fault current because seldom does a solid 3-phase short-circuit occur at the terminals of the protection device, and even short lengths of intervening cable reduce the fault current considerably. Nevertheless, knowledge of prospective fault current is necessary to ensure that the circuit-breakers, fuses, etc., in the system can deal with the fault current. If the fault level at the point is in excess of the rupturing capacity of the device, damage can be done to the installation and to the protective device itself, because of the amount of energy that passes before the current is completely interrupted. In such cases the protective device must be backed up by another breaker, or by fuses capable of interrupting the fault and before the 'let-through' energy has built up. The total 'let-through' energy is identified by $I^2 t$, where I is the fault current and t is the time for which the current flows before complete interruption.

When a fault current in a.c. circuit starts to rise, the fuse element heats and begins to melt. This takes a finite *pre-arcing* time. As the element ruptures, arcing occurs: very shortly after, the break is complete, the arcing ceases and the current drops to zero. This is the *arcing time*. The sum of these two times is termed the *total operating time*.

The high-rupturing-capacity (HRC) fuse is a valuable device for use in distribution circuits; it is capable of clearing very high fault currents rapidly. *Figure 20.10(a)* shows how the 'let-through' energy is restricted. This has led to HRC fuses being widely used as protective devices in distribution circuits in their own right, and also as back-up devices where the primary device is incapable of handling the prospective fault current.

If one fuse is backing up another, the total $I^2 t$ for the minor fuse must not exceed the pre-arcing $I^2 t$ of the back-up fuse. From *Figure 20.10(b)* it will be seen that a 50 A fuse cannot be used as back-up to a 40 A fuse. A 60 A fuse would be adequate, but tolerances suggest that an 80 A fuse would be preferred.

20.11.3 Discrimination

A major consideration in designing a distribution system is the problem of ensuring effective discrimination. Ideally, the protective devices should be so graded that, when a fault occurs, only the device nearest the fault operates. The other devices should remain intact and continue supplying the healthy circuits.

theoretically, is the ideal condition, but in practice—to allow for the discrepancies mentioned above—it is usually advisable to aim for back-up protection taking over at a fault-current level not exceeding about 70% of the circuit-breaker breaking capacity. The need to comply with this requirement will automatically set the upper limit to the current rating of the back-up device. The lower limit of current rating is usually set by the need to avoid loss of discrimination.

Where m.c.b.s in subcircuit distribution boards are concerned, the m.c.b. should be the device to operate for all fault levels up to 1000 A, i.e. the great majority of subcircuit fault currents. If the zone in which the back-up fuse tends to take over from the circuit-breaker is below 1000 A, discrimination troubles may be experienced, but provided that it is not below about 700 to 800 A, operating experience indicates that little trouble will be met on this score. If the zone is above 1300 A, the installation should be relatively free, in practice, from any form of discrimination trouble.

If fuse time/current characteristics are available, the probable position of the take-over zone can be assessed quite readily. The total time taken by an m.c.b. to clear a short-circuit fault is usually about 10 ms. If, therefore, the prearcing or melting time of the back-up fuse—at a given level of fault current—is less than 10 ms, it is reasonable to assume that at that level of fault current an m.c.b. will not consistently discriminate against the back-up fuse. Thus, the current at which the fuse prearcing time/current characteristic crosses the 10 ms line will give a working guide to the position of the take-over zone.

The family of time/current characteristics shown in *Figure 20.11* is for a typical range of quick-acting HRC fuses. From this diagram it will be seen that the 63 A characteristic crosses the 10 ms line at about 1000 A. From this it is reasonable to assume that the take-over zone, i.e. the range of current over which there is a possibility of loss of discrimination, will probably extend from about 800 to 1000 A. For fault currents less than 800 A there

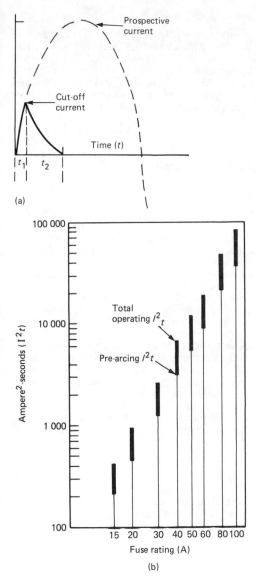

Figure 20.10 (a) Oscillogram of current: t_1, pre-arcing time; t_2, arcing time. (b) I^2t characteristic for HRC Type T fuse links

It is usually possible to assess, with fair accuracy, how effective the discrimination will be between combinations of circuit-breakers and fuses, but there are several practical considerations to be taken into account: (i) manufacturing tolerances in components, and (ii) operating conditions such that the devices do not conform to the published data. Over a period of time a fuse may be subjected to high currents causing it to 'age', and this can have an adverse effect on its behaviour.

On a normal installation there will inevitably be other devices between any circuit-breaker and the source of supply, owing to the methods of subdividing the supply for distribution. At the other end of the scale, as in the normal domestic ring circuit, there may be fuses further down the line backed by m.c.b.s. The primary objective is to ensure that the circuit-breaker will deal with faults up to its breaking capacity, the back-up device taking over above this level. (These other protective devices may be circuit-breakers, or fuses of rewireable or cartridge type.) This,

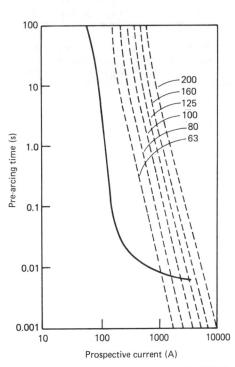

Figure 20.11 Time/current characteristics for HRC fuses: HRC fuse; ——, 30 A, MCb

should be no trouble with the back-up fuse. With an 80 A fuse the zone will probably extend from about 1100 to 1500 A, and with the 100 A rating from about 1700 to 2500 A. Hence, if a quick-acting h.r.c. fuse is used for back-up, ideally it should not be rated at less than 80 A, and preferably at not less than 100 A.

20.11.4 Motor control gear

A motor starter may be an individual item of equipment, or one of several in a multi-motor starter panel. Methods and conditions of starting, speed control, reversal, etc., are given in another Chapter. In general, the supply authority lays down the limits of starting current permissible. Small 1-phase motors are normally switched direct-on-line, but larger machines may require starting methods that limit the starting current and power factor.

20.11.4.1 Isolating switches

The Factories Act and the IEE Regulations W.R.15 require that every motor be provided with a means of disconnecting the motor and all its auxiliary equipment, including any automatic circuit-breaker. The isolating switch should be mounted nearby and it is often incorporated in the same case as the starter itself. The switch must be capable of making and breaking the stalled current of the motor and under normal conditions it remains closed. The switch often incorporates auxiliary contacts which (among other functions) isolate the control circuitry in the starter panel. It is usual for the isolating switch to be mechanically interlocked with the access door to the starter, to ensure that the starter and its control circuit are isolated before the door is opened and maintenance work begun.

20.11.4.2 Contactor starter

For the smallest motors, a manual mechanical switch is acceptable; however, the Regulations require that means must be provided to prevent automatic restarting after stoppage (due to drop in voltage or failure of supply) where unexpected restarting might cause danger. Where such a device is provided, it is logical that it should also be used to operate the switching contacts, and here is the reason for the popularity of electromagnetically operated contactor starters. Many other advantages, such as easy control by relays, limit switches and other light current devices, follow the adoption of contactor starters.

There are two basic types of contactor—the electrically held-in and the latched-in type. With both, the contactor should close when the voltage is as low as 85% of nominal. With the electrically held-in type, should there be a transient failure of the supply, the contactor will drop out. With certain types of drive this can be embarrassing, and a latched-in contactor may be preferred. With this type, the 'close' coil energises the contactor, which then latches in, and the 'close' coil is de-energised. The contactor is opened by energising an 'open' coil or by the operation of a protective device.

20.11.4.3 Motor protection

Motor starters are usually fitted with a trip device which deals with overcurrents from just above normal running current of the motor to the stall current. The aim should be for the device to match the characteristics of the motor so that full advantage may be taken of any overload capacity. Equally, the trip device must open the starter contactor before there is any danger of permanent damage to the motor.

Contactors are not normally designed to cope with the clearance of short-circuit conditions, and it is therefore usual for the contactor to be backed up by h.r.c. fuses or by circuit-breaker.

20.11.4.4 Remote control of motors

When the starter is mounted remote from its motor, care must be taken in positioning the control push-buttons. Accidents can occur when the operator cannot view the motor from the control position; it may then be necessary to mount pilot lights adjacent to the motor to indicate to local personnel that it is about to start. In some cases, particularly with large complex machines, it may be necessary to give audible warning that the equipment is about to be started.

When an emergency stop button is located adjacent to the motor, it should be of the lock-off variety, so that when operated, not only does the motor come to a standstill, but also it cannot restart until the lock-off push-button has been released.

20.11.4.5 Braking and stopping machines

When it is necessary to stop a machine very quickly, some form of braking must be employed. An ordinary friction brake, the simplest device, should be held off against a spring or counterweight, so that it is applied automatically if power should fail. Electrodynamic braking is ineffective at slow speeds: it is therefore a useful addition rather than an alternative to a friction brake. Torque reversal (plugging) is very effective, but in many cases a reverse rotation cut-out will be required.

Care must be taken with the connection to the brake, Some motors tend to generate during run-down, and the brake circuit must therefore be interrupted positively to ensure that the friction brake is applied. A particular danger arises when rectified current is used for the brake solenoid on large machines. It may be found that the rectifier acts as a low-resistance shunt across the brake coil and prevents the magnetic flux collapsing quickly, making the brake action sluggish. In this case it is necessary for the main contactor to have separate contacts in the brake-magnet circuit.

When heavy machines are controlled by plugging, a problem will arise if the power supply should fail, for this would leave the machine uncontrolled, unless there is back-up emergency braking.

20.11.4.6 Limit switches

Limit switches are generally used to initiate a control sequence at the correct point in the mechanical duty cycle of a machine. For example, they may signify the end of a crane travel and initiate the stop sequence. Limit switches must be of robust construction—well protected against the ingress of fluids, dust or dirt. They are frequently used on machines expected to have a long life, and it is essential that the limit switches be equally generously designed. The current-carrying capacity of the electrical contacts should also be generously rated, since they may well have to handle inrush current.

20.11.4.7 Inching control

When frequent inching is required, it is preferable that a separate contactor be employed. However, the motor circuit should still be taken through the overload protection device, to ensure that the motor is not allowed to overheat. At the same time the use of the separate contactor does restrict wear on the contacts of the main contactor.

20.12 Transformers

Most transformers used in industrial and commercial applications are oil-filled, natural-cooled and suited for mounting out of doors. The transformer is contained in a tank, plain or with cooling tubes or fins. Tanks may be equipped with

Table 20.2 Preferred ratings (kVA)

5	6.3	8	10	12.5	16	20	25	31.5	40
50	63	80	100	125		160	200	250	315	400
500	630	800	1000, etc.						

small wheels, but more usually have a flat bottom for mounting on a concrete plinth. *Table 20.2* gives the preferred ratings of 3-phase transformers.

The modern transformer is reliable and tolerant of its location. Although fires are infrequent, there is still some reservation with regard to siting oil-filled units indoors. For such locations use may be made of air-insulated encapsulated units, or replacing the normal hydrocarbon oil by bio-degradable, non-toxic dielectric fluid, such as MIDEL. The use of PCBs (Askarels) is not to be recommended because of their toxicity. A new development is the use of SF_6 as an insulant.

The heat loss from a transformer makes ventilation important. *Figure 20.3* shows a suitable arrangement for a substation transformer.

It is usual for a transformer to be delivered ready for service, and it is recommended that the installation engineer does not remove the lid unless there is evidence of some abnormality. If, however, there is doubt, the check testing below is carried out.

Voltage ratio. Apply a low-voltage 3-phase supply to the h.v. winding and measure the l.v. output voltage.

Phase grouping. Connect one pole of the h.v. and l.v. windings together (*Figure 20.12*); then by applying a 3-phase l.v. supply to the h.v. windings, voltages can be measured and the grouping determined from the results.

It will be readily seen that in both the above tests it is imperative that the l.v. supply be connected to the h.v. winding.

It is seldom that a small or medium-sized transformer is shipped without oil; but if this should happen, or if the transformer has been in store, it may be necessary to dry out the windings before commissioning the unit. If carried out at the manufacturer's works, this would be done in a drying chamber, but on site it should be done by circulating currents in the transformer windings. It is then preferable that the oil be circulated through a filter plant at the same time. During the drying-out test, regular readings of the insulation resistance of the winding should be taken and checks made on the electric strength of the oil. The drying-out process is likely to take several days.

Where large transformers are concerned (say above 10 MVA), it would be foolish to proceed with any precommissioning or commissioning tests without enlisting the aid of the manufacturers.

20.12.1 Installation

Before connecting the transformer to the supply, several precautions are necessary.

First, the tank should be efficiently earthed. In the case of a transformer with off-load taps, the correct tappings should be chosen.

With self-cooled units fitted with radiators, the valves at the top and bottom of the headers should be open. In the case of artificially cooled units, all the valves must be open and the correct quantities of oil and water should be circulating. Where the transformer is of the type employing forced oil circulation with a water-cooled cooler, the pressure of oil in the cooler must be greater than that of the water, so that, if there is any leakage, water will not be forced into the oil. In the case of transformers with fan coolers, these should be checked and tested, and the setting for the cutting-in of the fan checked.

Breathers, where used, should be fully charged with the drying agent. The jointing of the cable boxes should be checked, and where the transformer is fitted with bushings, these should be examined for cracks or chips, and if fitted with arcing horns, the gaps should be checked and set.

If the transformer is required to operate in parallel with another unit, it should be 'phased in', i.e. the phase sequence on the secondary, with the primary excited, should be identical with that of the other unit before the secondary is connected to the common outgoing supply. In a number of 3-phase transformer connections there is a phase displacement of the secondary line voltage, as, e.g., in a delta/star connected transformer. If the primaries of this be paralleled with the primaries of a star/star transformer, then it is not possible to run the secondaries in parallel, for the line voltages in the two cases have a phase displacement of 30°. A delta/star group may be paralleled with a star/delta or a star/interconnected star group, but not with a star/star or delta/delta group. A grouping commonly found in industrial and commercial work is the delta/star, either a Dn 1 or a Dn 11. In normal circumstances it is not possible to parallel Dn 1 and Dn 11 transformers, since, when the primaries are connected to the h.v. supply, there would be a 60° phase displacement between the l.v. connections. However, this can be corrected by crossing two of the h.v. connections on one of the transformers and the corresponding pair of l.v. connections.

After switching in and applying full voltage successfully, it is desirable to have the transformer operating in this no-load condition for as long as is practicable. The heat due to core loss warms the coils and the oil gradually; this minimises absorption of moisture and also allows trapped air to be removed by the circulation of the oil.

Inspection of naturally cooled transformers should be made annually; artificially cooled units should be inspected every 6 months. It is always advantageous to take a sample of oil and test it for electric strength. This gives an indication of the presence of moisture or other impurities in the oil.

At all times oil levels should be maintained correctly and breathers should be recharged regularly.

20.12.2 Transformer protective devices

The practicable indication of the temperature rise of a transformer in service is the top oil temperature. There is a temperature differential between the winding and oil temperatures, and the winding temperature responds much more quickly to changes of load, but the winding temperature cannot readily be monitored.

It is usual to fit an instrument on the tank to read the sum of the

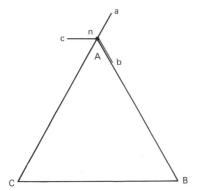

Figure 20.12 Checking the phaser group of a transformer

oil temperature and an analogue of the winding gradient. The instrument comprises a normal dial thermometer, the bulb of which is surrounded by a heating coil, with thermal characteristics similar to those of the transformer windings and fed from the secondary of a current transformer. In this manner the dial reading gives an indication approximating to the temperature of the windings.

Winding temperature indicators are fitted with alarm contacts, so that warning can be given should the windings reach dangerous temperature.

The Buchholz relay is a mechanical device fitted in the pipe between the transformer tank and the conservator. It usually consists of two floats with contacts: one to an alarm circuit, the other to a trip circuit. Any breakdown in transformer insulation is accompanied by the generation of gas in the oil. A serious fault results in the rapid generation of gas, and as this rushes through the pipe, it operates the float and closes the trip contacts. Alternatively, if the fault develops slowly, gas is generated slowly but is sufficient to operate the alarm float and contacts.

20.13 Power-factor correction

Most industrial loads have a lagging power factor. The common method of correction is by means of static capacitors, generally oil-impregnated and oil-cooled and with a paper dielectric. The loss in a capacitor is less than 2 W/kVAr, and the temperature rise does not exceed about 15 °C above ambient. Maintenance is negligible, and it is unnecessary to filter or replace the oil during the life of the capacitor. No special foundations are needed. Capacitors are available for direct connection to systems up to 33 kV at power frequency, and can be installed indoors or outdoors.

In many cases capacitors can be connected directly to the terminals of induction motors. Individual correction of large numbers of small motors (e.g. for looms) can be economic.

An alternative p.f. correction method is the use of synchronous machines with overexcitation. Operated at a leading p.f., the machine can correct the lagging p.f. of the rest of the system. However, the method is applicable only when the synchronous machine is required for a specific duty and when system conditions are such that the motor is not shut down while the rest of the system is still in operation.

20.13.1 Capacitor rating

In order to evaluate the capacitor requirements, the system p.f. must be known: it can be obtained from a p.f. meter, or calculated from active, reactive and apparent power indicators.

Let a load of active power P, reactive power Q_1 (lag), apparent power S_1 and p.f. $\cos\phi_1$ require correction to a lagging p.f. $\cos\phi_2$, corresponding to P, Q_2 and S_2. The correction is shown in *Figure 20.13* with specific reference to a 3-phase 415 V 50 Hz load of $P = 100$ kW and a p.f. $\cos\phi_1 = 0.5$, to be raised to $\cos\phi_2 = 0.9$, both lagging. For the two conditions

p.f. 0.5: $P = 100$ kW $S_1 = 200$ kVA $Q_1 = 173$ kvar
p.f. 0.9: $P = 100$ kW $S_2 = 111$ kVA $Q_2 = 48$ kvar

The required capacitor rating is therefore $Q_a = (Q_1 - Q_2) = 125$ kvar. With a delta-connected 3-phase bank, each branch is rated at $125/3 = 42$ kvar. *Table 20.3* gives the rating (in kvar per kW of load) to raise a given p.f. to a selection of higher values. To obtain p.f.s approaching unity, the capacitor rating tends to be large, and it is usually uneconomic to correct to p.f.s greater than about 0.9.

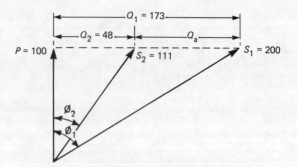

Figure 20.13 Power-factor correction

Table 20.3 Capacitor rating, kvar per kW of load

$\cos\phi_1$	Correction to $\cos\phi_2$				
	0.85	0.90	0.95	0.98	1.0
0.40	1.67	1.80	1.96	2.08	2.29
0.45	1.36	1.50	1.66	1.78	1.99
0.50	1.11	1.25	1.40	1.53	1.73
0.55	0.90	1.03	1.19	1.32	1.52
0.60	0.71	0.85	1.00	1.13	1.33
0.65	0.55	0.68	0.84	0.97	1.17
0.70	0.40	0.54	0.69	0.81	1.02
0.75	0.26	0.40	0.55	0.67	0.88
0.80	0.13	0.27	0.42	0.54	0.75
0.85	—	0.14	0.29	0.42	0.62
0.90	—	—	0.15	0.28	0.48

For a rating Q_a (kvar) the capacitance C (μF) required is

$$C = (Q_a / 2\pi f V^2) \, 10^9$$

where f is the frequency (Hz) and V is the line voltage (V).

20.14 Earthing

Earthing an electrical system is the process of connecting all metalwork (other than the electrical power conductors themselves) to the main body of earth. The aim is to convey to earth any leakage of electrical energy to the metalwork without hazard to personnel or equipment. An earthing system has two distinct but related parts: (i) a low-resistance conductor bonding the metalwork, connected to (ii) an electrode or array of electrodes buried in the ground. A 'good' earth may be difficult to achieve. The main factors in (ii) are the form and configuration of the electrode assembly, and the electrical resistivity of the soil.

20.14.1 Earch electrode

The ideal electrode is a conducting hemisphere. Current entering the ground from the hemispherical surface flows in radial lines, with concentric hemispherical equipotential surfaces decreasing in potential from the electrode. Such an electrode, of radius r in soil of resistivity ρ, has a resistance to the main body of earth given by

$$R = \rho / 2\pi r$$

Most of this resistance is located near the electrode surface, where the current density is greatest. A relatively small leakage current, say 5 A, for $R = 50\,\Omega$ would raise the potential of the electrode to 250 V above the general body of earth and produce a steep voltage gradient in its vicinity. A man standing close to the electrode would probably experience a shock; a cow would almost certainly be electrocuted.

Practical electrodes are usually rods, plates or grids, and have a non-uniform current distribution in close proximity, but the more remote from the electrode system the closer does the current distribution resemble that of the hemisphere. Formulae for estimating the value of R for typical configurations are given below. In each case ρ is the soil resistivity.

Plate of radius r buried with its centre at a depth h:

$$R = \rho \frac{1}{8r}\left[1 + \frac{r}{2.5h + r}\right]$$

For a rectangular plate of area a, take $r = \sqrt{(a/\pi)}$.

Rectangular grid with a buried conductor of length l and a grid area a:

$$R = \rho[(1/4r) + (1/l)]$$

where $r = \sqrt{(a/\pi)}$.

Single rod of diameter d and buried length l:

$$R = \rho(1/2\pi l)[\ln(8l/d) - 1]$$

Multiple rods. Arrays of rods have practical relevance in that they are commonly used for substation earthing. The following formula is due to G. F. Tagg, *Earth Resistance* (Newnes, London, 1964). For a group of n rods, each of resistance R_1, in a hollow-square configuration with a spacing s between rods:

$$R = (R_1/n)(1 + k\alpha)$$

where $\alpha = r_h/s$, where r_h is the radius of the equivalent hemisphere for one rod, and k is a factor depending on n, as indicated in *Figure 20.14*.

20.14.2 Soil resistivity

The quality of the soil (sand, clay, gravel, rock,...) and its homogeneity determine its resistivity. Typical values (Ω m) are

Rich arable soil, wet or moist	50
Poor arable soil, gravel	500
Rocky ground, dry sand	3000

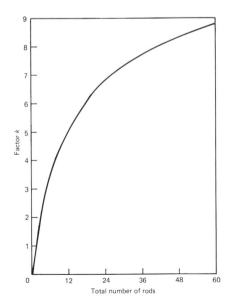

Figure 20.14 Factor k for hollow-square multiple-rod earthing array

The resistivity is greatly affected by (i) moisture content, (ii) temperature and (iii) the presence of chemical contaminants. Thus, the resistivity rises substantially when the moisture content is below 30% and also when the temperature falls below about 10 °C; but the presence of salts in the soil lowers the resistivity. It is evident that when soil is subject to considerable seasonal changes of climate, earthing electrodes should be deeply buried. In practical terms this suggests the adoption of rod electrodes, which can be driven to a depth where the moisture content and the temperature are more stable.

20.14.3 Electrode installation

For toughness and resistance to bending, driven electrodes are of steel rod, either galvanised or with copper molecularly bonded to the surface (*Figure 20.15*). Rods of diameter 15–20 mm can

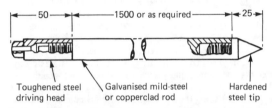

Figure 20.15 Typical earth-rod electrode

normally be satisfactorily driven. For adequate depth, rods supplied in standard lengths can be connected by screwed couplers or special connectors. In an installation requiring few rods, manual driving is easy and economic, but for hard compacted soil, and where many rods are concerned, it is preferable to use a power hammer.

Each rod should be driven until a minimum resistance reading has been reached, as described later. If the overall resistance of a given rod assembly is not low enough, additional rods may be driven, spaced typically at double the depth.

An earth electrode system must resist corrosion. Both galvanised and copper-clad steel rods have good corrosion resistance. The system must also retain its ability to carry large currents repeatedly throughout its life (30 years or more). Individual rods can deal with currents up to the fusing values, indicated in *Table 34.4*, for three typical durations.

Plate Electrode. This has lost favour. It is expensive to install, more liable to corrosion, and is buried in top layers of soil where stable soil-resistivity conditions are not readily achieved.

Grid Electrode. Where the ground is rocky and the soil resistivity high, the only practicable method of obtaining an

Table 20.4 Earthing-rod fusing currents

Rod sectional area (mm²)		Rod fusing current (kA)		
Copper-clad	Equivalent steel	0.25 s	0.5 s	1.0 s
62	72	15	11	8
98	115	24	17	12
113	132	28	20	14
126	147	31	22	16
158	183	39	27	19
235	271	57	41	29
422	522	110	78	55

adequately low earth resistance may be by installing a grid of buried conductors.

Addition of Chemicals. Generally this is of limited application, as chemicals tend to leach out. But where earthing is difficult, gypsum (a form of calcium sulphate) may be employed, mixed into the top soil or spread on the surface. Its value lies in its relatively low solubility in water. With moderate rainfall, gypsum percolates through the soil and continues to keep down the earth resistivity for the life of the electrodes. It has minimal corrosive effect, but sulphates do attack concrete.

20.14.4 Resistivity and earth resistance measurement

The soil resistivity must be known for the design of an earthing system to be determined and the earthing resistance of the system must be checked after installation. Both measurements involve passing current through the soil and measuring the volt drop. Composite instruments are available. *Figure 20.16(a)* shows the measurement of *resistivity*. The test prods are pushed into the ground to a depth not exceeding $s/20$, where s is the spacing between prods. The greater the spacing s, the greater the volume of soil concerned in the measurement. By repeating the test with various values of s, the uniformity of the soil can be assessed: reasonably constant values indicate good soil homogeneity.

Testing *electrodes*, individually or in an array, is shown in *Figure 20.16(b)*, using the same test set with C1 connected to P1. Prods C2 and P2 should be far enough from the electrode system for accurate readings of resistance to be obtained. Provided that there is an adequate distance between C1 and C2, there should be a zone within which the potential prod P2 gives a constant reading.

20.14.5 Protective conductors

The earthing conductor connects the consumer's main earth terminal to the earth-electrode system. It should be capable of disconnection for testing the electrode system, but such disconnection must require the use of tools.

Under the IEE Regulations W.R.15, the *circuit protective conductors* (c.p.c.; formerly the 'earth-continuity conductors') are those that join the consumer's main earth terminal to all the exposed conductive parts of the system. The c.p.c. provides a low-impedance path for fault current, but must also ensure that no dangerous voltage can occur in metalwork in the vicinity of the fault. This is achieved by the connection of all extraneous metalwork to the main earth terminal: the provision covers such items as water and gas pipes, accessible structural steelwork, and central heating pipework, a requirement termed *main equipotential bonding*. The connections should be made as close as possible to the point of entry to the building concerned. Additionally, *supplementary bonding* is the connection of extraneous conductive parts that are accessible simultaneously with other conductive parts but are not electrically connected to the equipotential bonding: e.g. water taps, sink and bath waste outlets, central heating radiators.

The cross-sectional area (mm²) of the c.p.c. may be calculated from

$$S = (1/K)\sqrt{(I^2 t)}$$

where I is the current (A) for a fault of negligible impedance, t is the operating time (s) of the disconnecting device, and K is a factor dependent on the material of the c.p.c., the insulation, and the initial and final temperatures. The operating time is usually $t = 5$ s for fixed and $t = 0.4$ s for portable equipment. A quick rule-of-thumb alternative to the formula is: S is the same as the cross-sectional area of the phase conductor where the latter is 16 mm² or less. $S = 16$ mm² for phase conductor sections between 16 and

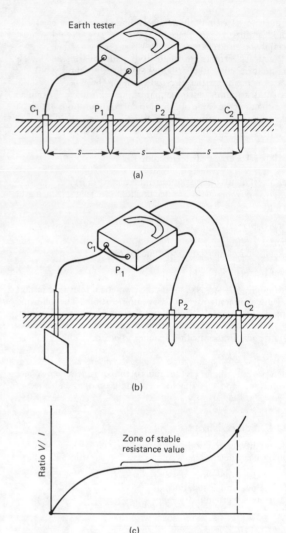

Figure 20.16 Measurement of (a) soil resistivity and (b) earthing resistance

35 mm². S is one-half of the phase conductor section for larger areas.

For main equipotential bonding conductors, the area shall be not less than one-half of the main earthing conductor of the installation, with a minimum of 6 mm². Generally, supplementary bonding using 4 mm² bare or 2.5 mm² mechanically protected conductors will be accepted.

20.14.6 System earthing

W.R. 15 has introduced a new designation for electricity supply systems to identify the type of earthing in use. A three- or four-letter code is employed, as follows.

First letter: 'supply' earthing arrangement. T indicates that one or more points of the supply are directly earthed. I indicates that the supply is nowhere earthed, or is earthed through a fault-limiting impedance.

Second letter: 'installation' earthing arrangement. T indicates that all exposed conductive metalwork is connected directly to earth. N indicates that all exposed conductive metalwork is connected directly to the earthed supply conductor (usually the neutral).

Third and fourth letters: earthed supply conductor arrangement. S indicates separate neutral and earthed conductors. C indicates combined neutral and earth in a single conductor.

In a TT system the exposed conductive parts of the installation are connected to an earthing electrode which is independent of the supply earth.

In a TN–S system the consumer's earth terminal is connected to the supply authority's protective conductor, usually the cable sheath and armouring, and so provides a continuous metallic path back to their earth electrode. The majority of installations with an underground supply are of this type.

In a TN–C system the neutral and protective functions are combined in a single conductor throughout the system. Typical is a system using earthed concentric wiring. If fed from a public supply authority, this system can only be used under special conditions and with the authority's permission.

In a TN–C–S system the supply neutral and protective functions are combined and earthed at several points, often with an earth electrode at or near the consumer's incoming terminals. The exposed conductive parts of the installation are connected to the neutral and to the supply neutral/earth at the consumer's main earthing terminal. This system is also referred to as *protective multiple earthing.*

In an IT system the exposed conductive parts of the installation are connected to the consumer's earth electrode, while the supply is either isolated from earth or connected to earth through an impedance. This form of supply cannot be used for public supply in the UK.

20.15 Inspection and earthing

On completion and before energising, every installation must be inspected and tested to ensure that it complies with the appropriate Regulations.

20.15.1 Visual inspection

A visual inspection is required to ensure that all electrical equipment is in compliance with the appropriate national standard; has been correctly selected and installed; and is in a safe working condition. The inspector/tester should understand how the installation functions, and to this end he should ensure that his inspection covers such items as the identification and checking of the size of conductors, and the labelling of circuits, fuses, distribution boards and other switchgear. Finally, he should insist on the availability of diagrams, charts or tables describing the system under inspection.

20.15.2 Testing

Testing is required to ensure that the installation has been designed and installed (i) to give protection against electric shock during normal operation and during fault conditions, (ii) to ensure that it provides protection against overloads and short circuits, and (iii) to provide protection against thermal effects—in particular, against fire and burns.

The IEE Regulations W.R.15 are more specific than formerly on the methods of testing to be employed. In particular, they cite the following items which, where relevant, shall be tested, and in the sequence indicated:

Continuity of ring final circuit conductors.
Continuity of protective conductors, including main and any supplementary equipment bonding.
Earth-electrode resistance.
Insulation resistance.
Insulation of site-built assemblies.
Protection by electrical separation.
Protection by barriers or enclosures provided during erection.
Insulation of non-conducting floors and walls.
Polarity.
Earth-fault loop impedance.
Operation of residual-current and fault-voltage operated protective devices.

Tests must be carried out in the correct sequence. For example, it is important that the continuity (and therefore the effectiveness) of the protective conductors be checked before insulation tests are carried out, because an open-circuited protective conductor coupled with a low insulation resistance reading could make the whole system live at the test voltage during the insulation test.

In general, the application of any voltage to a system prior to connection to the mains must be done with care, to ensure that no danger to persons, property or equipment can occur, even if the tested circuit is defective. There is an increasing use of electronic components in present-day installations, and these can be damaged by relatively low voltages. Therefore the insulation resistance tests should be undertaken before the components are fitted, or they should be removed before testing commences.

20.15.2.1 Continuity

To check continuity it is not sufficient that a measurement be made across the end terminals. In the case of final ring circuits two measurements need to be made, and in practice W.R.15 proposes two acceptable methods. Both assume that an outlet is installed at or near the mid-point of the ring.

In Method 1 the resistance of each pair of like conductors is measured with the two ends separated. After reconnecting the two conductors, a measurement is taken between the distribution board and the mid-point. After measuring and deducting the resistances of the test lead, it should be found that the first reading is four times the second reading; see *Figure 20.17(a)*.

In Method 2 no long lead is needed to make contact with the mid-point of the ring. The continuity of each conductor of the ring circuit is measured between the ends of the conductor when separated, as for Method 1. The conductors are then reconnected at the origin, all the connections at the mid-point outlet are short-circuited and the continuity is measured between the phase and neutral terminals at the distribution board. This value should be one-half of the previous reading. Finally, the continuity is measured between the phase and the earth; see *Figure 20.17(b)*.

Continuity tests in protective conductors, particularly steel conduit, require the application of a current approximately 1.5 times the normal design current, using a voltage not exceeding 50 V a.c. or d.c. (If d.c. is used, verify that no inductor is included in the circuit.)

It will be appreciated that the resistances being measured in the continuity tests will be small (e.g. between 0.1 and 0.5 Ω), and it is suggested that a testmeter with a range between 0.005 and 2.0 Ω would be useful.

20.15.2.2 Earth electrode

Methods of testing earth electrodes are described in Section 20.14.4 and illustrated in *Figure 20.18*.

20.15.2.3 Insulation resistance

Insulation resistance tests are carried out to verify that the insulation of the conductors and electrical accessories is satisfactory. A low reading would indicate a deterioration in the insulation, and a very low reading a total failure.

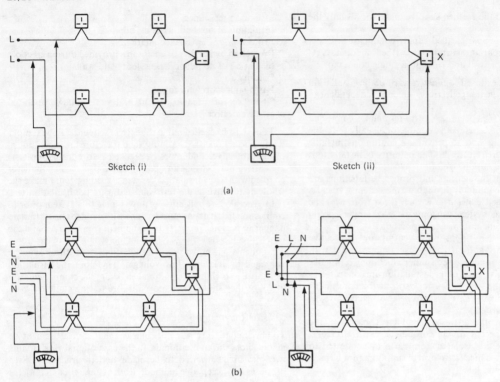

Figure 20.17 Measurement of continuity of ring circuit: (a) method 1; (b) method 2

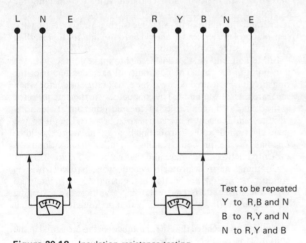

Figure 20.18 Insulation resistance testing

The test meter should provide a d.c. voltage of not less than twice the nominal voltage of the circuit under test (r.m.s. value for an a.c. circuit), but the test voltage need not exceed 500 V d.c. for installations rated up to 500 V, or 1000 V d.c. for installations between 500–1000 V.

Insulation resistance testing involves tests of insulation resistance between conductors of different polarity and between all conductors and earth, and the value must not be less than 1 MΩ. A large installation may be tested by dividing it into groups, each containing not less than 50 outlets; the minimum insulation resistance demanded of each group is 1 MΩ. The term 'outlet' includes every point and every switch, except where the switch is incorporated in a lighting fitting, socket outlet or power-consuming apparatus.

During the insulation test to earth of a whole installation, all fuses should be in place and all switches closed, including the consumer's main switch (if practicable). For the test (*Figure 20.18*) all conductors are bonded together at the load terminal of the main switch by means of a copper wire. The bonding wire is then connected to the 'line' terminal of the tester, the other terminal being solidly connected to the consumer's earthing terminal. (When it is required to test to earth a live installation, the consumer's main switch must be opened and the supply from the mains disconnected.) Portable apparatus should not be connected to the system, as it can be tested separately. Any item of fixed apparatus may be disconnected if desired and tested separately. With the test set connected as described above, the insulation resistance is measured. Where two-way switches are included in the circuits, two test readings should be taken, one switch in each pair being changed over for the second test to include the strapping wires in the test.

If the insulation resistance value exceeds the specified minimum, the system can be passed; but if not, further sectional tests will have to be made to locate the faulty section.

For insulation testing between conductors, if practicable all switches and circuit-breakers should be closed, and all lamps removed, together with all fixed appliances.

Any equipment disconnected before the tests are done, and having exposed conductive parts, should itself have an insulation value of not less than 0.5 MΩ.

20.15.2.4 Polarity

It must be checked that all fuses and single-pole circuit-breakers and switches are in the 'line' side, that the wiring of plugs and socket outlets is correct, and that the outer contacts of concentric bayonet and screw lampholders are in the neutral (or earth) side.

20.15.2.5 Earth-fault loop impedance

If a fault of negligible impedance should occur from a phase conductor to earth, the supply voltage will provide a fault current through the earth-fault loop. The significance of the earth-fault loop impedance is that it determines the current, and this, in turn, determines whether or not the protective device will operate quickly enough to meet the requirements of the relevant regulations. Take, for example, the arrangement shown in *Figure 20.19* and assume a fault to occur in an item of consumer's equipment. The fault current I_F will flow from the transformer, through the fuse and then via the casing of the equipment to the consumer's earth, and thence back through the earth path to the transformer, such that $I_F = V_s/Z$, where Z is the loop impedance. From Z calculations can be made for the overall performance of the system under fault conditions. For example, W.R.15 calls for all equipment connected to socket outlets to be cleared in 0.4 s or less when faulted. Other circuits supplying only fixed equipment are acceptable if the fault is cleared within 5 s. The Regulations give tables of Z to meet these conditions.

In practical terms, a phase-to-earth loop can be tested by connecting a known resistor R between the phase and protective conductors when the system is live, with R typically $10\,\Omega$. Measuring the volt drop v across it gives the current $I = v/R$. If the supply voltage is V_s, then $V_s = I(Z+R)$, or $Z = (V_s/I) - R$, giving the required loop impedance.

The loop impedance must be tested using a *phase-earth loop tester*, which gives a direct reading in ohms. (The neutral-earth loop tester previously specified is no longer acceptable.) A dual-scale instrument covering 0–2 and 0–100 Ω is suitable.

20.15.2.6 Residual-current devices

Where the loop impedance is such that the required tripping times cannot be achieved, and in special cases (e.g. where the socket outlet in a house is used to supply gardening tools, or where the supply authority does not provide an earth terminal), a *residual-current* device must be fitted. It usually comprises coils on a magnetic circuit to carry the phase and neutral currents in opposing directions. In balanced conditions no magnetic flux is set up; but if a fault occurs on the system, even one of high resistance, the phase and neutral current imbalance induces an e.m.f. in a third coil (*Figure 20.20*), tripping the circuit. To test the device, an a.c. voltage not exceeding 50 V r.m.s., obtained from a mains-fed double-winding transformer, is applied across the neutral and earth terminals. The device should immediately trip.

Fault-voltage operated circuit-breakers may alternatively be used, but are subject to tripping when fault occur outside the protected zone. If used, however, the earth-fault loop impedance including earth-electrode resistance must not exceed 500 Ω.

Whether or not this simple go/no-go test is sufficient is debatable. Special test sets are already available that will measure both the current and the time to trip.

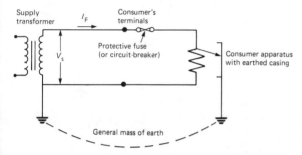

Figure 20.19 Earth-fault loop impedance

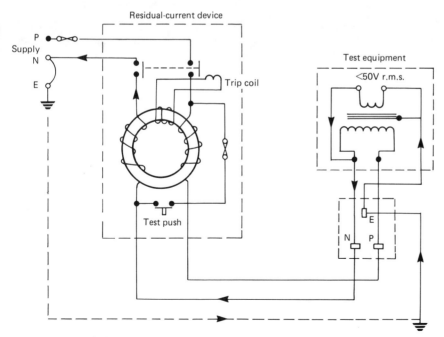

Figure 20.20 Residual-current device, showing test connections. The arrows show the path of the test current

20.15.3 Test gear

Because the testing requirements of W.R.15 are stringent, the UK Electrical Contractors' Association has suggested that the tester should equip himself with the following instruments, incorporating the features itemised:

Insulation-resistance/continuity tester: with 500/1000 V selector, analogue scale down to 0.01 Ω, battery power, push-button operation.

Digital milliohmmeter: with battery power, push-button operation, voltage protection indicator.

Polarity/earth-fault loop-impedance tester: with mains power, polarity indicator lamps, analogue low-resistance scale.

Residual-current-device/fault-voltage-device tester.

Earth-electrode/soil-resistivity tester.

21

Electroheat

J E Harry, BSc(Eng), PhD
Loughborough University of Technology

H Barber, BSc
Loughborough University of Technology

Contents

21.1 Introduction 21/3

21.2 Direct resistance heating 21/3
    21.2.1 Metals 21/3
    21.2.2 Glass 21/4
    21.2.3 Water 21/4
    21.2.4 Salt baths 21/4

21.3 Indirect resistance heating 21/4
    21.3.1 Resistance heaters 21/4

21.4 Infrared heaters 21/6

21.5 Ultraviolet processes 21/6

21.6 Electric ovens and furnaces 21/6
    21.6.1 Heating element construction for ovens and furnaces 21/6
    21.6.2 Ovens 21/7
    21.6.3 Furnaces 21/8

21.7 Induction heating 21/9
    21.7.1 Power sources for induction heating 21/10
    21.7.2 Load coupling 21/10
    21.7.3 Coil design 21/13
    21.7.4 Through heating of billets and slabs 21/13
    21.7.5 Surface and localised heating 21/13
    21.7.6 Semiconductor manufacture 21/13

21.8 Metal melting 21/13
    21.8.1 Resistance melting 21/13
    21.8.2 Channel induction furnaces 21/14
    21.8.3 Coreless induction furnaces 21/14
    21.8.4 Electric arc furnaces 21/15

21.9 Dielectric heating 21/17
    21.9.1 Radiofrequency power sources and applicators 21/17
    21.9.2 Microwave power sources and applicators 21/18

21.10 Plasma torches 21/19
    21.10.1 Direct-coupled plasma torch 21/19
    21.10.2 Induction-coupled plasma torch 21/19
    21.10.3 Plasma furnaces 21/19

21.11 Glow discharge processes 21/20

21.12 Lasers 21/20

21.1 Introduction

Electroheat is defined here as the use of electrical energy for industrial heating processes, including such different techniques as the resistance heating of glass, metal melting and dielectric drying, but excluding welding and space heating. The subject has grown in importance since the first introduction of electric melting in the nineteenth century and today represents a substantial part of the heating methods in industry and a major part of the total electric load.

The processes vary widely and utilise the entire frequency range from zero frequency (d.c.) to wavelengths in the visible region of the spectrum (*Table 21.1*) at power levels ranging from a few watts to several hundred megawatts.

Table 21.1 Electric heating processes

Technique	Frequency range	Power range
Direct resistance	0–50 Hz	0.01–30 MW
Indirect resistance	50 Hz	0.5–5 kW
Oven, furnace	50 Hz	0.01–1 MW
Arc melting	50 Hz	1–100 MW
Induction heating	50 Hz–450 kHz	0.02–10 MW
Dielectric heating	1–100 MHz	1–500 kW
Microwave heating	0.5–25 GHz	1–100 kW
Plasma torch	4 MHz	0.001–1 MW
Laser	30 THz	0.1–60 kW
Electron beam	2–3 THz	1–500 kW
Infrared	1–1.5 Thz	1–20 kW
Ultraviolet	0.5–0.75 Thz	1 kW

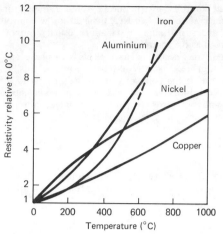

Figure 21.1 Variation of resistivity with temperature (relative to 0°C) of aluminium, copper, nickel and iron

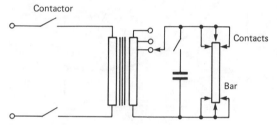

Figure 21.2 Electric circuit for direct resistance billet heater

21.2 Direct resistance heating

Direct resistance heating is used in the iron and steel industry for heating rods and billets prior to rolling and forging; for ferrous and non-ferrous annealing; either alone or in combination with other fuels for melting glass; in electrode boilers, for water heating and steam raising; and in salt baths, for the surface heat treatment of metallic components.

21.2.1 Metals

Figure 21.1 shows the variation of resistivity with temperature of some of the metals of interest. The relatively high resistivity of steel allows billets of large cross-section to be heated efficiently, provided that the length is several times greater than the diameter. Heating is very rapid, so that heat losses (e.g. radiation from the surface and thermal conduction through the contacts) are small. The efficiency of the process can be of the order of 90% or better. The workpiece resistance is normally low, and for these efficiencies to be achieved, the supply resistance must be lower. The low resistivity of copper and similar materials implies that the length/diameter ratio should be considerably higher than 6 if the process is to be successful, and with these materials the more common application is the annealing of wire and strip. In all cases the cross-sectional area of the current flow path must be uniform, otherwise excessive heating, with the possibility of melting, will occur at the narrower sections. With the normal 50 Hz supplies non-uniform current distribution caused by skin effect leads to higher heating rates at the surface, but this is counteracted by surface heat loss.

A typical supply circuit is shown in *Figure 21.2*. The high-current transformer will usually have off-load tappings to take account of the major changes in workpiece resistivity that occur over the heating cycle. The reactance of this transformer and its associated connections should be less than 20% of that of the workpiece and the whole arrangement must be mechanically robust to withstand the magnetic forces at the high currents involved, which can, in the larger units, be in excess of 100 kA. Although this is a resistance heating application, the inductance in the workpiece and connections may result in a power factor as low as 0.4 lagging at the start of the process, and power factor correction capacitors are often incorporated. As the temperature increases, the resistance rises; with ferrous materials there is a considerable decrease in inductance once the Curie temperature (at which the material becomes non-magnetic) is exceeded, resulting in a sudden increase in current.

A high contact resistance between the supply and the workpiece leads to excessive voltage drop and local overheating. Large-volume contacts behave as heat sinks with the possibility of cold ends, and excessive pressure can cause welding of the contact to the bar. Allowance needs to be made for the longitudinal thermal expansion during heating. One commercial arrangement uses a number of hemispherical contacts of copper–tungsten alloy at each end of the bar and applied by hydraulic pressure. Liquid contacts, using mercury or water, have been employed for the continuous heating of wire and strip but a sliding or rolling metal/metal contact is preferred.

Process control can be achieved on the basis of a simple timed cycle with the transformer tappings adjusted during the cycle to maintain the required current. A refinement involves weighing the billet before the process is started and using this information to adjust the cycle time to give the required temperature. Other installations are controlled by an optical pyrometer to monitor the surface temperature and feed back a signal to the control system. Wire and strip heating installations usually employ throughput speed as the control variable.

The direct resistance heating of bar or billets is a 1-phase load that is switched at frequent intervals. This results in possible voltage unbalance at the point of common coupling and in transient voltage disturbances. The load can be phase-balanced by inductive and capacitive components. In large units the switch-on disturbances may be compensated by a soft-start arrangement.

21.2.2 Glass

Glass at temperatures above 1100°C, is molten and has a resistivity low enough (*Figure 21.3*) for direct resistance heating.

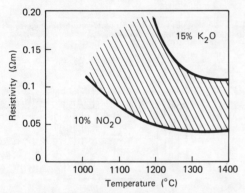

Figure 21.3 Resistivity with temperature of glasses

In the UK electricity is used in mixed melting units where, typically, electrodes are added to a fuel furnace to increase the output for a relatively low capital outlay. Current is passed between electrodes immersed in the molten glass. These electrodes must withstand the high temperatures involved and the movement of molten glass across their surface. Contamination of the glass by pick-up of electrode material must also be avoided and either molybdenum or tin oxide electrodes are used with current densities of the order of 1500 A/m^2. A 1-, 2- or 3-phase electrode system may be employed; the 2-phase connection uses Scott-connected transformers. The 3-phase arrangement is preferred for phase balance and low cost, and also produces electromagnetic forces in the glass which, together with the thermal forces, lead to significant movement in the melt. Provided that it is not excessive, the movement improves quality and melting rate.

Existing electric furnaces have usually followed a traditional design pattern with a rectangular tank, but circular designs have now been developed which (it is claimed) give improved stirring in the melting zone.

The power input to the furnace and, hence, the melting rate is controlled by varying the input voltage by use of tapped transformers or saturable inductors. An alternative method is to change the effective surface area of the electrodes by raising them from the melt.

21.2.3 Water

The generation of steam and hot water by passing current between electrodes in water is now common practice. As with molten glass, the conductivity is dependent on the ions that the water contains, but in many districts it is sufficient to use normal tap-water. Electrode boilers range in size from a few kilowatts up to 20 MW. The larger-capacity units operate at high voltages up to 6 kV, and the water is sprayed on to the cast-iron electrodes. The heating rate is regulated by moving a porcelain shield over the jets to divert the water. Output control in the smaller units is achieved by varying the surface area of the electrode in contact with the water, either by vertical movement of the electrodes or, more simply, by changing the water level in the boiler. It is important to ensure that the conductivity of the water is maintained at the appropriate value: too low a conductivity reduces the power, while if too high, the result is a deposition of insoluble salts on the electrodes. In large units the conductivity and pH values are continuously monitored and their values automatically controlled.

21.2.4 Salt baths

Salt baths can be used for heat treatment of metal components. The heated salt reacts chemically with the surface layer of the workpiece to give the required surface properties. At temperatures above about 800°C direct resistance heating, as opposed to the use of sheathed elements or external heat sources, is the only usable method. Although the salt is a good conductor when molten, the bath must be started up from cold by using an auxiliary starting electrode to draw a localised arc. The electrodes have to withstand the corrosive effects of the salt and are manufactured from graphite or a corrosion-resistant steel alloy. The currents involved can be up to 3 kA at voltages of 30 V. Both 1- and 3-phase units are available.

21.3 Indirect resistance heating

In indirect resistance heating electrical energy produces heat by use of a resistance heating element. Applications include surface heaters, immersion heaters and open heating elements used in ovens and furnaces. The factors affecting the selection of element materials for this purpose are quite different from those where the temperature rise may be only a few tens of degrees and power densities are low. Large and non-linear changes in resistance may occur over the operating range and must be taken into account in the design of power supplies. Resistance to creep, thermal shock and oxidation, and the presence of any other gases or vapour, are important. Good ductility combined with mechanical strength at high temperatures and a resistivity such that an adequate cross-section for mechanical strength can be obtained at normal temperatures are also desirable.

The principal available materials are listed in *Table 21.2* and include nickel and iron alloys; the so-called refractory metals such as platinum, tungsten and tantalum, which have melting points above 1500°C; and some non-metallic conductors, including silicon carbide, molybdenum disilicide and lanthanum chromite. The resistivity–temperature characteristics of these materials are shown in *Figure 21.4* while *Figure 21.5* gives the vapour pressure, an important characteristic if the elements are to be used in vacuum furnaces.

21.3.1 Resistance heaters

Resistance heaters (*Figure 21.6*) are used for many applications ranging from localised heat sources in plastics sealing machines to melting soft metals. The maximum temperature required is not normally high and nickel-chrome or ferrous resistance wire is used. However, the choice of insulating support material has a major effect on the design. Insulating materials and power densities used for resistance heaters, with their upper operating temperatures, are listed in *Table 21.3*.

Table 21.2 Materials for resistance-heating elements

Material	θ	ρ	α	Principal applications
Copper	350	1.73	—	Low-power surface heaters
Nickel-based alloys*:				
80Ni/20Cr	1200	108	6.0	Furnaces; resistance heaters
80Ni/20Cr+Al	1250	124	−2.0	Lower-cost furnaces
60Ni/15Cr/25Fe	1100	112	13	Lower-cost furnaces
50Ni/18Cr/32Fe	1075	111	17	Lower-cost furnaces
37Ni/18Cr/43Fe/2Si	1050	105	24	Lower-cost furnaces
44Ni/56Cu	400	49	—	Resistance heaters; domestic appliances
Iron-based alloys*:				
72Fe/22Cr/4Al	1050	139	4.7	Furnaces
72Fe/22Cr/4Al+Co	1375	145	3.2	Furnaces
78Fe/18Cr/4Al+Yt, C	1300	134	12	Furnaces
Refractory materials:				
platinum	1300	11	392	Small muffle furnaces
90Pt/10Rh	1550	19.2	200	Small muffle furnaces
60Pt/40Rh	1800	17.4	200	Small muffle furnaces
molybdenum	1750†	5.7	550	Vacuum furnaces: small muffle furnaces in hydrogen atmosphere
tantalum	2500	12.5	320	Vacuum furnaces
tungsten	1800†	5.6	594	Infrared lamps; vacuum furnaces
Non-metals:				
graphite	3000†	1000	−26.6	Vacuum furnaces and reducing atmospheres
molybdenum disilicide	1800	40	1200	Small glass-melting furnaces; forehearths
silicon carbide	1600	110 000	−0.26	Furnaces; oxidising and reducing atmospheres
lanthanum chromite	1800	2100	−2.77	Furnaces; oxidising and reducing atmospheres

* Composition approximate.
† Not in air.
θ maximum operating temperature in dry air (°C); ρ resistivity at 20°C ($\mu\Omega$ m × 10^{-2}); α mean resistance/temperature coefficient over operating range (× 10^{-5}).

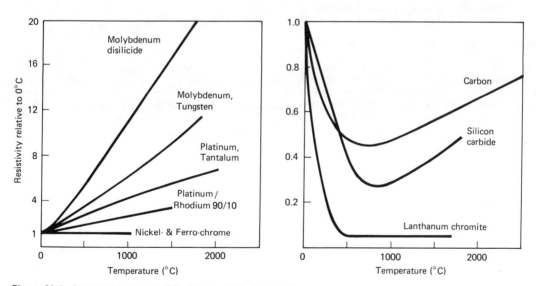

Figure 21.4 Resistivity (relative to 0°C) of some resistance materials

The metal sheathed heating element comprises a wire helix enclosed in a metal sheath and electrically insulated from it with magnesium oxide powder. The latter has a resistivity of 10^7 Ω-m at 1000°C. Seamless tube and flat heaters constructed in this way can be shaped after manufacture. Ring and cartridge heaters can be constructed for specific applications. The sheath material depends on the application: special materials are used for corrosive environments such as salt baths, low-melting-point metals and acids. Electrical connections include connection to both ends; single-ended, 1- or 3-phase connection, and parallel connection.

The maximum surface temperature of the heater is limited to about 750°C by the thermal conductivity of the electrical insulation. The power density is governed by the thermal conductivity of the insulation and by the heat transfer at the heater surface. Very high localised power densities, up to 10

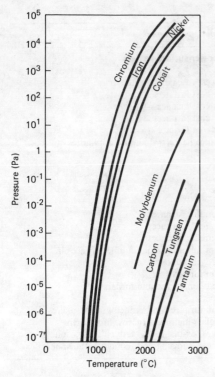

Figure 21.5 Vapour pressure and temperature of resistance element materials

kW/m², can be achieved with the cartridge heater, which is designed as an interference fit for a high heat transfer coefficient at the heated surface.

A more flexible construction is obtained by incorporating the heater wire in a woven or mesh insulating support. Applications include the heating of pipes and vessels in the chemical and food industries and annealing large metal structures. Another form employs a resistance film or wire helix between a pair of parallel conductors; an advantage is the ability to cut to any length. A similar construction, using graphite powder between the conductors, is claimed to have a self-limiting temperature due to the high positive resistance/temperature coefficient eliminating the need for temperature control.

21.4 Infrared heaters

The intensity distribution of an infrared source varies with temperature and proportionately more power is radiated at the shorter wavelengths as the temperature increases (*Figure 21.7*). Power densities obtainable from different radiant sources are given in *Table 21.4*. As a result infrared sources with relatively low temperatures are normally used for the heating and drying of non-metallic materials, which generally absorb effectively at long wavelengths; but high-temperature sources with most of their output at short wavelengths (but which reflect at long wavelengths) are more suitable for heating metals which reflect at long wavelengths.

21.5 Ultraviolet processes

Ink and plastics coatings can be cured at high rates with ultraviolet sources. The coatings contain monomers and photo-activators which enable very rapid polymerisation to take place when exposed to ultraviolet radiation. Although this is not strictly a heating process, it has much in common, and is often in direct competition with, infrared heating for drying or cross-linking; however, the energy levels are normally much lower, since the process requires only the specific amount of energy to stimulate the reaction.

Ultraviolet lamps with fused quartz envelopes have a higher transmission than glass at the ultraviolet end of the spectrum, the spectral emission being concentrated mainly between 250 and 450 nm. The radiated energy from an electrical discharge is not governed by the black body laws: the intensity can be very high and limited to a few bands of wavelength. A typical output spectrum of a high-pressure mercury vapour discharge lamp is shown in *Figure 21.8*.

Power ratings of these lamps are in the range 1–4 kW over active lengths of 400–1400 mm. Low-pressure actinic lamps, with output radiations at short wavelengths and suitable phosphors to obtain emission in the region 300–500 nm, are used for curing polyesters. Typical power ratings are 120 W with an active length of about 1.5 m. The tube is made of glass, which has good transmission properties down to about 300 nm.

21.6 Electric ovens and furnaces

Electric ovens and furnaces are used for a great variety of different processes, ranging from sintering ceramic materials at temperatures up to 1800°C to drying processes close to ambient temperature at power ratings varying from a few kilowatts to more than 1 MW. A substantial part of the electric heating load is taken by electric ovens and furnaces of conventional design used for heat treatment (*Table 21.5*) and similar processes at temperatures up to about 1100°C.

21.6.1 Heating element construction for ovens and furnaces

The materials used for resistance heating elements have been listed in *Table 21.2*. Forms of construction of furnace heating elements are shown in *Figure 21.9*. Metal resistance heating elements for furnaces are normally in the form of wire, strip or tube. Heavy-section low-voltage high-current elements are made with an alloy casting or with corrugated or welded tube. Helically wound wire heating elements are manufactured with a wire-to-mandrel diameter ratio of between 3:1 and 8:1, limited by collapse of the helix. The helix is then expanded so that its length is about three times the close-wound length. Coiled-wire or strip heating elements may be inserted in ledges or grooves supported at intervals by nickel alloy or ceramic pegs. The end connections of the heating element, normally made of a material different from that of the element to reduce attack from oxidisation and chemical reaction with the refractories, have a lower resistance to reduce heat dissipation where the leads pass through the furnace wall; this lower resistance is achieved by making the diameter of the ends greater than that of the heating zone or by using a material of higher electrical conductivity.

Silicon carbide rod or tubular heating elements are mounted vertically at the side or arranged to span the roof of the furnace. The cold ends of these rod elements are impregnated with silicon or made by joining end sections of higher conductivity. The hot zone of a tubular element is obtained by cutting a helix so that the current path length is increased and the cross-section is reduced, which increases the resistance. Single-ended double-spiral and 3-phase rod heaters can be suspended vertically from the roof of the furnace.

Molybdenum disilicide is normally available only in the hairpin or W form of construction for suspending vertically or

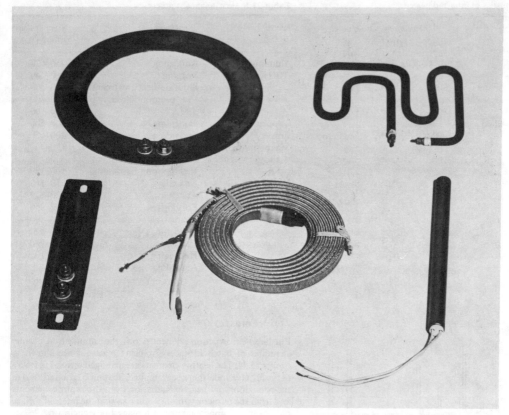

Figure 21.6 Types of resistance heaters. Top left: metal sheathed ring heating element. Top right: tubular mineral insulated heating element. Bottom left: flat mineral insulated heating element. Centre: flexible surface heater. Bottom right: cartridge heater

Table 21.3 Maximum operating temperatures and power densities of insulation and support materials for flexible heating elements

Insulation/support material	Max. workpiece temperature (°C)	Max. power density (kW/m^2)
Polytetrafluoroethylene	170	2
Silicone rubber	200	3
Asbestos	400	—
Glass fibre	425	8
Mineral insulation (MgO):		
copper–nickel sheath	350	6
stainless steel sheath	600	13
Stainless steel mesh	600	13
Quartz fibre	800	12
Ceramic beads	1000	40

supporting in a horizontal plane. The heating zone is of reduced cross-section to increase the resistance. Graphite in the form of machined rods or slabs, or as tubes, is used in muffle furnaces.

21.6.2 Ovens

An oven is usually defined as having an upper temperature limit of about 450°C. Ovens, using natural or forced convection, are widely employed for drying and preheating plastics prior to

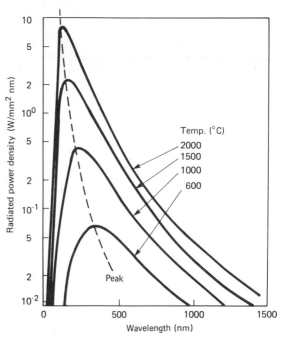

Figure 21.7 Radiated power intensity of a black body with temperature and wavelength (peak intensity lies on broken line)

Table 21.4 Characteristics of infra-red heaters

Type	Maximum surface temperature (°C)	Maximum power density (kW/m²)	Heater wavelength
Embedded ceramic heater	800	68*	Long
Mineral insulated element	800	40*	Long
Tubular quartz-sheathed element	900	50*	Medium
Circular heat lamp (375 W)	2100	30*	Short
Linear heat lamp (1 kW):			
parabolic reflector	2100	50*	Short
elliptical reflector	2100	90†	
Quartz-halogen linear heat lamp (12 kW):			
parabolic reflector	2700	200*	Short
elliptical reflector	2700	3000†	Short

* Average power density.
† Power density at focus.

Table 21.5 Heat-treatment processes

Metal	Treatment process	Temperature range (°C)
Aluminium and alloys	Annealing	250–520
	Forging	350–540
	Solution heat treatment	400–500
	Precipitation hardening	100–200
	Stress relief	100–200
Copper	Annealing	200–500
Brass	Annealing	400–650
Magnesium and alloys	Annealing	180–400
	Solution heat treatment	400
	Precipitation hardening	100–180
Nickel and alloys	Annealing	650–1100
Carbon steel (0.6–1.5% C)	Annealing	720–770
	Forging	800–950
	Tempering	200–300
	Hardening	760–820
Stainless steels	Tempering	175–750
	Hardening	950–1050
Cast-iron	Annealing	500

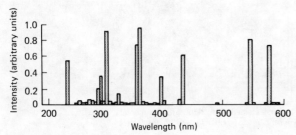

Figure 21.8 Spectrum of output wavelengths and intensity from an ultraviolet lamp

forming, curing, annealing glass and aluminium, baking and a host of other applications. Coiled nickel-chrome wire or mineral-insulated metal sheathed heating elements are distributed around the oven so as to obtain as uniform a temperature distribution as possible. Heat transfer rates may be increased by using a fan to circulate air over the heating elements onto the workpiece, the air being recirculated through ducts. An important advantage of convection ovens is that the operating temperature is normally the element temperature and the maximum temperature is never exceeded, as the process is self-limiting, which prevents overheating if the material is left in the oven too long. This is particularly important for temperature sensitive materials such as plastics.

High heating rates can be achieved by direct radiation from heating elements in infrared ovens. The oven walls are made of sheet metal which reflects the radiation, no thermal insulation is necessary and vapour is easily removed. Two forms of continuous infrared oven are illustrated in *Figure 21.10*, the low thermal mass and weight allowing a light-weight structure and a very high power density. The high-intensity lamps employed have a low thermal inertia and can be switched to a low level of stand-by power, full power being used only when the workpiece is inside the oven. Infrared heating processes are advantageous when only the surface layer is required to be heated, as when curing coatings, so that the overall efficiency may be very high compared with other methods which heat also the substrate.

21.6.3 Furnaces

Furnace construction depends on the application. Some examples of batch furnaces for heat treatment are shown in *Figure 21.11*. The heating elements are arranged around the sides of the furnace and, where very uniform heating is required, also in the roof, door and hearth. Radiation from the heating elements and from the refractory lining occurs, so that the internal surface of the furnace approximates to a black body enclosure.

Forms of work handling used with batch furnaces include bogie or car-bottom hearths and elevated hearths raised into the furnace on hydraulic rams to facilitate loading and unloading. The box furnace is normally used at temperatures where radiation is dominant. Where it is necessary to avoid exposure to air during the quenching process, the sealed quench furnace is used. Forced convection furnaces allow high heating rates and, with careful design, good temperature uniformity can be achieved; but usually these are limited to maximum temperatures of 700–900°C. In the pit furnace, capable of operating at higher temperatures, radiation is the dominant mode of heat transfer and, by using a retort, the process can be carried out in a controlled atmosphere.

The bell furnace may be used as a hot-retort vacuum furnace. By reducing the pressure inside the bell very low partial pressures of oxygen are obtained. Since heat losses by convection are greatly reduced at low pressures, the bell can be raised when the required temperature is reached, one bell then being used to heat several retorts.

The rectangular bell furnace is a form of box furnace that allows unimpeded access to the furnace hearth. It is useful for heat treating large components such as mill rolls or fragile items such as ceramics. The cold-retort vacuum furnace allows very high temperatures to be achieved. It is used for heat treatment and brazing in controlled low-pressure atmospheres using vacuum interlocks to achieve high throughputs. The hydrogen muffle furnace is used for sintering alumina at 1700°C, zone refining where a precise temperature profile is required, and the continuous annealing of wire and strip in controlled atmospheres by use of an open-ended muffle.

Continuous furnaces use a conveyor mechanism enabling in-line processes to be carried out combining heating and cooling at a controlled rate (*Figure 21.12*). Some examples of different

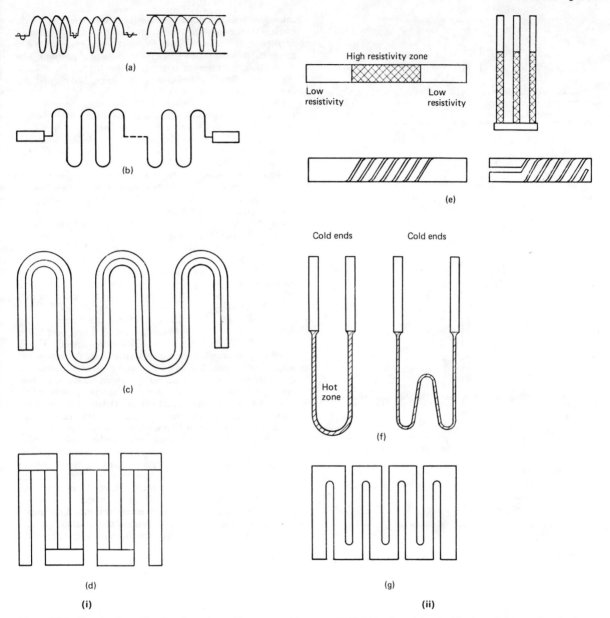

Figure 21.9 Construction of heating elements used in ovens and furnaces. (i) Metal heating elements: (a) wire coil supported on hooks or in grooved-tile; (b) strip; (c) cast element; (d) tubular heating elements. (ii) Non-metallic heating elements; (e) rod and tubular silicon carbide elements; (f) molybdenum disilicide; (g) graphite

conveyor mechanisms are illustrated in *Figure 21.13*. The choice depends on the nature of the process and the size of the workpiece. Applications range from bright annealing of fasteners to normalising steel billets.

21.7 Induction heating

Induction heating makes use of the transformer effect. The workpiece is placed in the alternating magnetic field of a coil. The field produces eddy currents in the workpiece, which is heated as a result of I^2R losses. The workpiece must have a low resistivity and therefore the technique is confined mainly to the metals industry, both ferrous and non-ferrous. Hysteresis loss heating also occurs in ferrous metals; it is normally small compared with the eddy current effect but it is applicable in heating metal powders at high frequencies. The current density in the workpiece is a maximum nearest to the surface, and thus the heating effect is non-uniform and is a function of the material and supply parameters (resistivity, permeability and frequency). A suitable frequency allows the heating pattern to be controlled. In particular, if the frequency is high, most of the heat is developed in a thin surface layer, while lower frequencies give a more uniform distribution. Induction techniques are used for both through-heating and surface heating of metallic materials at frequencies in the range 50 Hz–1 MHz. They are used for melting and also, at

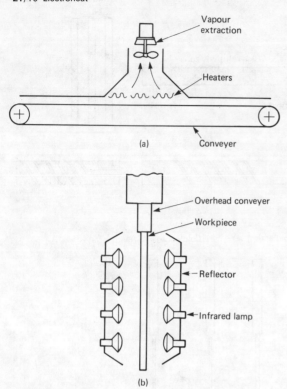

Figure 21.10 Infrared conveyor ovens: (a) horizontal conveyor; (b) vertical

very high frequencies, in the manufacture of semiconductor materials.

The eddy current power P per unit length of a cylindrical workpiece of radius R, resistivity ρ and absolute permeability μ, set in a coil of length substantially greater than that of the workpiece and energised at a frequency f, is

$$P = 2H^2 \rho R K Q$$

where H is the magnetic field strength at the surface of the workpiece and $K = (2/\delta)^{1/2}$ is a function of the skin depth

$$\delta = (2\rho/2\pi f \mu)^{1/2}$$

Q, a factor depending on δ and the shape of the workpiece, is plotted in *Figure 21.14* for a cylinder in terms of a normalised (dimension/skin depth) parameter. Values of δ, assuming where appropriate a relative permeability of 800 (i.e. $\mu = 800\,\mu_0$), are given for a number of materials in *Table 21.6*. The curves show that, for maximum eddy current power in a given workpiece, the frequency f is of major significance, although it may be necessary to compromise if a particular heat distribution is required. High frequencies (i.e. small skin depths) are needed if workpieces of small dimensions are to be heated.

21.7.1 Power sources for induction heating

There is an obvious economic advantage in using 50 Hz from the mains, in spite of the relatively poor power transfer obtained from conventional coil designs at this frequency. Depending on the size of the load, the power input is controlled by either an off-load tap changing transformer or an auto-transformer. The major problems are the low power factor, the effects of the large 1-phase loads on the power supply network and the voltage transients caused by switching large units. Power factor correction is with conventional power capacitors, often water cooled. Soft-start arrangements can be employed to reduce voltage transients and phase balancing networks lessen the imbalance problem. The applications include metal melting, the through heating of slabs, sheet and billets, surface heating of large cylinders and the heating of metal tanks.

Frequency triplers provide a 150 Hz supply for through heating slabs prior to rolling, derived from three 1-phase saturated transformers with secondaries connected in open delta (*Figure 21.15*). Harmonics generated in the tripler are filtered by an arrangement of inductors and capacitors. A 20 MW installation of this type is operated in Canada.

Higher frequencies give faster heating rates and, because of skin effect phenomena, enable the effective induction heating of a smaller cross-section of workpiece: the melting of small-diameter charge material is also possible. Power at these frequencies, before the advent of the thyristor, was obtained from rotating machines designed to produce power at frequencies up to 10 kHz. They are normally vertically mounted and, in the larger sizes, water cooled. Control of power output uses conventional techniques and a balanced load is presented to the supply. These machines do, however, have a constant frequency output, the efficiency is comparatively low (particularly at reduced power) and they are noisy.

The availability of high-power thyristors capable of operating at frequencies up to 10 kHz has opened up a new field to the designers of medium-frequency heating equipment. Sources based on these devices are commercially available with power outputs up to about 2 MW. The major applications are to through heating and melting, and this source has the advantage over the machine of a flexible operating frequency. The efficiency is generally higher, even on partial load. The operating principle is familiar: the rectified 50 Hz supply is chopped by thyristors and fed into a resonant load circuit formed by the resistance and inductance of the loaded work coil together with a tuning capacitor. The firing pulses may be applied at a preset frequency; alternatively, the control signal may be derived by feedback from the load circuit.

Oscillatory sources in the frequency range 10–200 kHz have been commercially developed in which the resonant frequency is determined by the load. The tank circuit is supplied through a transformer which enables the power supply to be remote from the load, avoiding large circulating currents in the connecting leads.

At frequencies above 50 kHz the power source uses a vacuum triode feeding into a tank circuit of which the load forms a part (*Figure 21.16*) either directly or through a coupling transformer. Water cooled inductors are used in the tank circuit and, in all but the smallest ratings, the valve is also water cooled. Conventional industrial valves have been and still are used for this application, in ratings up to 500 kW operating in the class C mode with conversion efficiencies greater than 50%. Even higher efficiencies are obtained with the magnetically beamed triode, which has an inherent low grid dissipation. This valve is robust and offers more flexible control than the conventional unit, owing to the lower grid power requirements which enable semiconductor control circuits to be used. Closed-loop control of the process variables, i.e. power and temperature, is possible.

21.7.2 Load coupling

The matching of the loaded work coil to the source is extremely important in the successful application of induction heating processes. Variable switched tuning capacitors are used at the lower frequency range, and at higher values a coupling transformer may be necessary. The impedance of the load will

Induction heating 21/11

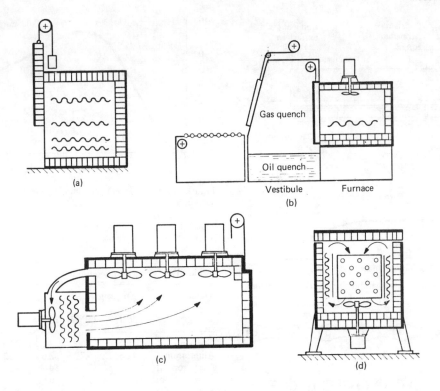
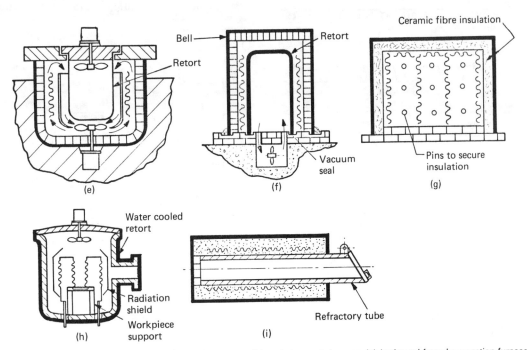

Figure 21.11 Batch furnaces for heat treatment: (a) box furnace; (b) sealed quench furnace; (c) horizontal forced convection furnace; (d) vertical forced convection furnace; (e) pit furnace; (f) bell (hot retort) vacuum furnace; (g) rectangular bell (low thermal mass) furnace; (h) cold retort vacuum furnace; (i) muffle furnace

21/12 Electroheat

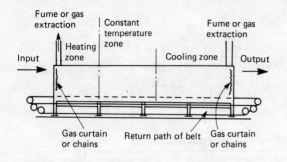

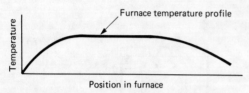

Figure 21.12 Conveyor furnace and typical temperature profile

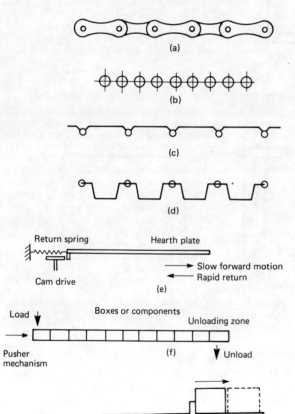

Figure 21.13 Some examples of conveyor furnace mechanisms: (a) cast link conveyor; (b) roller hearth; (c) slat belt conveyor; (d) slat pan conveyor; (e) shaker hearth; (f) pusher furnace mechanism; (g) walking beam furnace

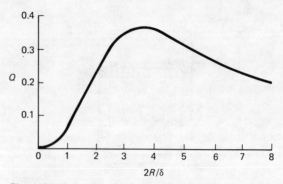

Figure 21.14 Variation of Q with the normalised diameter for a solid cylinder

Table 21.6 Typical skin depth (mm) and frequency relation

Metal		Frequency				
		50 Hz	150 Hz	1 kHz	10 kHz	450 kHz
Copper		9.2	5.3	2.1	0.65	0.096
Aluminium		11.6	6.7	2.6	0.82	0.12
Grey iron	(a)	62	37	14	4.5	0.89
	(b)	2.2	1.3	0.51	0.16	0.021
Steel	(a)	22	13	5.1	1.60	0.23
	(b)	0.78	0.45	0.18	0.057	0.008
Nickel	(a)	18	11	4.3	1.3	0.19
	(b)	0.85	0.38	0.15	0.05	0.007

(a) Below Curie temperature ($\mu_r = 800$)
(b) Above Curie temperature ($\mu_r = 1$)

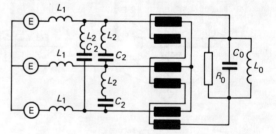

Figure 21.15 Three-phase transformer with open delta secondary for generation of 3rd harmonic

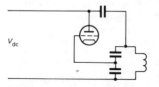

Figure 21.16 Basic oscillator circuit for r.f. induction heaters

depend on its geometry and physical parameters, while the number of turns on the coil is governed by the size of the area required to be heated: furthermore, this load impedance may change considerably during the heating cycle, making it necessary to retune the circuit.

21.7.3 Coil design

Coil designs depend on the operating frequency and the application. The basic objectives are to produce a high magnetic field strength over that section of the workpiece to be heated and to insulate the coil both to prevent electrical breakdown and to reduce heat transfer from the workpiece to the coil. The heat generated by the I^2R loss in the coil implies some form of forced cooling, and it is customary to use hollow water cooled copper tubular conductors. There is usually a considerable proximity effect resulting in the concentration of current at the inside surface of the coil nearer to the workpiece, and at high frequencies skin effect also influences the current distribution. Single-layer designs have usually been preferred.

21.7.4 Through heating of billets and slabs

Induction heating is used extensively for the through heating of both ferrous and non-ferrous metal billets prior to rolling or forging. The billet, of circular or rectangular cross-section, is passed through a series of current-carrying water cooled coils. The frequency of the current, the power input and the length of time in the coil are chosen to give the required throughput rate with an acceptable temperature distribution over the cross-section of the workpiece. A compromise is sought between high heating rates produced by high frequencies, and an acceptable skin depth. A typical example is a 350 kW, 1 kHz unit for heating steel transmission forgings to a temperature of 1150 °C; the energy consumption is of the order of 400 kWh/t and the heating time can be less than 5 min. Heaters with power outputs of up to 4 MW are readily available and, although the frequency for most applications is in the range 0.5–3 kHz, higher values up to 10 kHz are occasionally employed. The efficiency of the process is a function of the coil design and the coupling between it and the workpiece. In some instances a tapered coil has been used to allow for the changing parameters of the workpiece material as the temperature increases during the process. The workpiece resistivity at the final working temperature may be four or five times that at 20 °C and, in the case of ferromagnetic materials, the relative permeability will fall to unity when the Curie temperature is reached.

Metal slabs are also heated by induction processes. One of the largest single installations is in the USA, where a 200 MW, 1 kHz power supply heats large steel slabs prior to rolling. The heating of a thin sheet using this technique will present difficulties due to interaction of the eddy currents in the narrow cross-section unless very high frequencies are used. An alternative technique is the transverse flux method. The sheet is situated between two sets of magnetic poles energised at 50 Hz, the flux passes transversely through the sheet and currents are induced in the plane of the sheet. The winding and pole arrangement must be designed to give uniform current and heat distribution over the sheet. Development is also being carried out on the travelling-wave induction heater, which utilises a 3-phase winding operating at 50 Hz, similar in construction to that of a linear motor with the secondary (workpiece) held stationary. Applications of interest are the heating of metal cylinders and tanks; this technique could also be used for the heating of sheets and slabs.

21.7.5 Surface and localised heating

Induction heating at high frequencies, up to 500 kHz, is used extensively in the engineering manufacturing industries for the

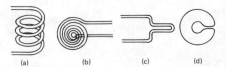

Figure 21.17 Coils for surface hardening by induction heating: (a) cylindrical coil for shaft hardening; (b) pancake coil for heating flat surfaces; (c) hair-pin coil for localised heating; (d) current concentrator

hardening of bearing surfaces, welding and brazing and similar surface heat treatment processes. The workpiece is placed in a coil (*Figure 21.17*) designed to cover the area to be heated. Heating takes place within a few seconds and, owing to the high frequencies, is confined in the earlier stages to a thin surface layer. The heated area is then cooled rapidly, so that the required surface hardening effect is obtained while still leaving the centre ductile. Contoured shapes can be heated, as can the internal surfaces of tubes, etc., with suitably designed coils. Flat spiral coils are used for the treatment of localised areas in sheet materials. An extension of the application is to soldering and brazing, in which case the field distribution is modified by the use of a field intensifier, a suitably shaped piece of magnetic material such as a ferrite core. The technique is also used for seam welding of steel tubes from flat strip, bent into shape and the two edges welded. The field, and therefore the heat, is concentrated at the two surfaces which are to be welded. In this latter application high throughput rates are achieved by using power sources with ratings as high as 1 MW at 450 kHz.

21.7.6 Semiconductor manufacture

Since the energy from the heating coil can be generated in the workpiece without any heat transfer medium, the whole process can be carried out in a vacuum with the coil either inside or outside the vacuum chamber. This is particularly useful in the manufacture of semiconductor materials. In one technique the material is placed in a conducting crucible in the vacuum space and is heated by induction. Semiconductor materials can also be heated directly by induction without the need for the crucible: in this case very high frequencies up to 4 MHz are used.

21.8 Metal melting

Electricity is widely used in the UK for metal melting. Applications range from the tonnage production of steel to the melting of small quantities of precious metals. Induction and arc furnaces are both extensively employed, and resistance melting finds a limited application for small quantities of low-melting-point metals.

21.8.1 Resistance melting

Immersion heaters with metal sheathed heating elements are employed for melting soft metals (lead, tin and type-metal) with melting points below about 650 °C. Some aluminium alloys may also be melted in this way, but many are corrosive and rapidly erode the sheath material. Zinc and zinc alloys can be melted with silicon carbide heating elements in cast-iron sheaths.

Small furnaces employing a crucible heated by external elements are used for melting or holding soft metals, including aluminium and most copper alloys. The heating elements are usually chromium-iron or silicon carbide, with heat transfer by radiation to the outside of the crucible. An earthing screen is fitted between the crucible and the elements to prevent electrical breakdown and the whole assembly is contained in a refractory-

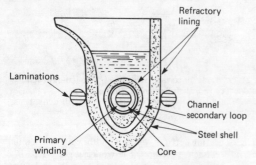

Figure 21.18 Wyatt channel furnace, single-coil, single-loop

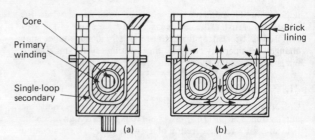

Figure 21.19 Channel furnace inductor box designs: (a) single-coil, single-loop; (b) double-coil, double-loop

lined vessel. Alternatively, heating elements may be mounted above the melt, or surface heaters can be clamped to the sides of the crucible.

21.8.2 Channel induction furnaces

In the channel induction furnace (*Figure 21.18*) the molten metal loop in the channel forms a single-turn secondary winding, the circuit of which is completed by the metal in the furnace above the channel. Typically the voltage induced in the single-turn secondary of molten metal is of the order of 10 V at 1 kA. The channel insulation and furnace structure reduce the coupling coefficient but the power factor is still relatively high. It is necessary to leave a heel of molten metal in the furnace to complete the secondary circuit, limiting flexibility when a range of alloys and continuous operation of the furnace is necessary. It also reduces the amount of metal that can be taken from the furnace.

The diameter of the channel is, in part, governed by the skin depth δ at 50 Hz, the value usually adopted being about twice the skin depth at the metal melting temperature. If the channel is made larger than 2δ, the power density in the channel and the total power density will decrease rapidly. The reactance of the channel is dependent on its geometry and the thickness of the refractory lining which decreases with refractory erosion; thus, the increase in power factor gives a direct indication of channel wear. The channel furnace in this form is limited to relatively low powers and small tonnages and is used almost exclusively for copper, aluminium and their alloys using magnesia or alumina linings.

The interaction of the currents in the V-shaped channel results in a reduction in magnetohydrodynamic pressure owing to the magnetic pinch effect at the base of the V. This causes a pressure gradient which opposes that of the static head of molten metal and increases the effects of convection, thus assuring circulation of hot metal through the channel. Outside the channel the magnetic flux interacts with the current loop, causing a motor effect which improves the stirring action. A W-shaped channel results in the highest magnetic flux density being produced in the central channel, which in practice is narrower than the outer channels, and creates a pressure gradient forcing the molten metal around the outer channels of the inductor box and the upper part of the furnace itself, so improving the flow and enabling higher power outputs to be employed.

Various inductor designs include single and twin loops of molten metal and single and double cores. Today most inductors are of the single-core/single-coil or single-core/twin-coil types (*Figure 21.19*). Separate inductor boxes enable them to be used in pairs either side of the metal bath, or in line. Pairs of inductor boxes so connected may be supplied from a 3-phase system using a Scott-connected transformer.

The depth of the furnace bath is governed partly by the degree of mixing by convection required and partly by the power density. A typical bath depth is of the order of 1 m, although smaller baleout furnaces have been built for aluminium diecasting. The difficulty of tilting large furnaces, which in this form have a high mass-centre, has led to the development of the drum furnace, which can be tilted about its axis. Three inductor boxes can be used in line to provide a balanced 3-phase load.

21.8.3 Coreless induction furnaces

Coreless furnaces, in which the metal in the crucible forms a single-turn secondary with the coil acting as the primary (*Figure 21.20*), are used for melting ferrous and non-ferrous scrap, nickel and aluminium refining, duplexing and vacuum melting. Outputs

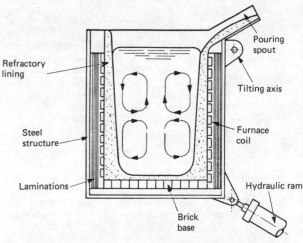

Figure 21.20 Coreless melting furnace showing stirring patterns

range from less than 1 kg to more than 1 t at power inputs up to 30 MW at 50 Hz and 5 MW at 500 Hz. The furnace coil is constructed from D-section tube insulated with impregnated tape or spacers and compressed to form a rigid structure.

Ideally, for maximum power dissipation the ratio D/δ should be 3.5; however, in practice such a furnace would be inconveniently small and values of $D/\delta > 3.5$ are normal. The ratio L/D is governed by the static head, which, if too large, prevents uniform electromagnetic stirring and hinders good mixing. If L/D is too small, the end effects of the coil result in excessive reactance. Increasing the furnace diameter reduces the coupling loss due to leakage flux, since the insulation thickness remains the same; it also reduces the power density. In practice, a ratio L/D between 1:1 and 3:1 is normally adopted. The power factor of a coreless furnace may be as low as 0.1 and coil voltages are correspondingly high. Power factor correction and a 3-phase balancing network are normally necessary for large loads.

As the frequency is increased, the optimum furnace diameter decreases. Commonly used frequencies are 50 Hz from the mains supply, and harmonics at 150 Hz, 450 Hz and 550 Hz. Frequencies from 50 Hz to 50 kHz are now obtained from static inverters. Frequencies of 50–450 kHz from radio frequency oscillators are also used in the melting of small quantities (a few kilograms) of precious metals. At 50 Hz a molten heel of one-quarter to one-third of the furnace volume is often left, to ensure good coupling when the furnace is recharged: this is particularly useful when fine scrap, e.g. turnings and borings, is used. At frequencies above 50 Hz the furnace is easier to start from cold and it is not necessary to use a starting plug or to leave a molten heel.

The flow pattern in the melt is shown in *Figure 21.20*. Stirring is due to the interaction of the current in the metal with the magnetic field. The effect is greatest at 50 Hz, decreasing as the frequency is raised. Where medium frequencies are used for melting, a low-frequency supply may be connected to the melting coil when the charge is molten, or separate stirring coils used.

Coil design ratings are limited to about 80 A-T/mm by the power dissipation in the coil itself and in the furnace. The surface power density limitation, 0.3–0.5 W/mm^2, of the circumferential area of the melt is imposed by pinch effect and by the possibility of refractory damage caused by excessive turbulence. This is overcome in the horizontal coreless furnace, where the higher static head enables the power densities (in terms of circumferential area) to be increased to 1.25 W/mm^2, giving high continuous outputs.

Metals can be melted under vacuum with the coreless furnace. Pouring can also take place under vacuum, so enabling a high degree of degassing to be achieved. In these circumstances lower operating voltages and careful insulation are necessary to prevent low-pressure corona and glow discharges which might lead to electrical breakdown.

Small versions of the coreless furnace with capacities of up to 200 kg have been designed in which the crucible can be removed and the coil used to service several crucibles. The crucible is manufactured from ceramic, graphite or silicon carbide bonded materials. In the case of a crucible made from conducting material, such as silicon carbide or graphite, some current flow and therefore power dissipation occurs within the crucible. This may be advantageous in melting materials of high conductivity, such as copper, for which the skin depth is large.

21.8.4 Electric arc furnaces

The electric arc furnace, dating back to the end of the last century, is an early example of electric heating. Heating occurs primarily by radiation from the arc and from the ends of the electrodes. Of the various designs of furnace developed, the 3-phase direct arc furnace is most widely used.

21.8.4.1 Direct arc furnace

The direct arc furnace, used for producing low-carbon steels from scrap, has almost totally replaced the open-hearth furnace. It is of very robust construction, with the hearth dish-shaped and shallow (*Figure 21.21*), to enable high heat transfer rates and effective slag reactions to be obtained. Bath diameters up to 7 m and charge capacities of more than 400 t are used. Electromagnetic stirring may be incorporated by using a non-magnetic steel shell and incorporating a low-frequency stirring coil below the furnace bath. The entire structure, including electrodes, masts, etc., is normally mounted on a hydraulically operated rack and pinion enabling it to be tilted in one direction for pouring and

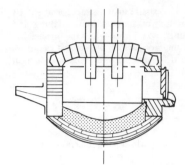

Figure 21.21 Direct arc furnace

in the reverse direction for slagging. The roof structure is pivoted so that it can be swung aside (with the electrodes raised) for charging. The electrodes can be slipped in the clamp to allow for electrode wear. The electrode arms can also be raised and lowered individually by hydraulic or winch systems, and the electrode height above the melt is controlled by feedback signals derived from the arc voltage and current. The electrodes are made of coated graphite and can be up to 0.6 m diameter, with connections to the busbars made using water-cooled flexible conductors so that the roof can be moved.

The substation for a large arc furnace is normally adjacent to the furnace itself and contains the furnace transformer, which normally has a star connected primary and an input voltage of 33 kV. The transformer must withstand very large electromechanical forces produced by the high short-circuit currents; it is oil cooled and has terminations brought out to which the flexible cables are connected. The furnace power is varied by on-load tap changing. Electrical contact to the electrodes is made by a large copper pad contained in the electrode clamp connected in water cooled busbars which rise and fall with the furnace electrodes. Various configurations have been adopted to ensure that the geometry remains as nearly symmetrical as possible, independent of the busbars, to minimise out-of-balance currents. The furnace electrodes are normally connected in delta, and where very high currents are used, the delta is closed at the electrode clamp in order to minimise the effects of reactance in the transformer secondary circuit.

Economies of scale have resulted in progressively larger arc furnaces with ratings in excess of 100 MVA and increasing power inputs per tonne to reduce cycle times; the capacities of these units are more than 150 t. Further increase in size is limited by the need to tilt the furnace, the electrode diameter required, the inductive reactance of the circuit and the difficulties in raising and lowering the large electrodes independently. Operation with d.c. is claimed to give increased utilisation of the furnace, reduced noise and flicker and less wear of refractories and electrodes: it is not affected by the inductance of the furnace connections. This technique is currently being studied. Continuous feeding of pelletised prereduced iron or fragmented scrap is also being investigated as a method of overcoming some of the limitations involved in scaling up conventional arc furnaces. This type of feed eliminates the need to remove the roof for charging and to tilt the furnace for pouring.

21.8.4.2 Indirect arc furnace

The indirect arc furnace relies on radiation for heat transfer to the molten metal and the refractory lining from an arc drawn between two electrodes on the axis of a cylindrical horizontal shell. The furnace rocks about its axis so that the refractory lining is washed with molten metal; this assists heat transfer to the melt and cools the lining. The largest size manufactured is about 1 t capacity, with a power rating of about 1 MW. Melting and refining can be carried out in the same unit, and the furnace has been used mainly for melting non-ferrous metals and cast-iron, but it has been superseded by channel and coreless induction furnaces.

21.8.4.3 Submerged arc processes

The submerged arc process is not essentially an arc process, as heating occurs also by direct resistance with, perhaps, some limited heating from arcs and sparks during interruption of the current path. The principal applications are for reducing highly endothermic ferroalloys of high melting point, such as ferro-manganese, nickel, chrome, silicon, tungsten and molybdenum, which are subsequently remelted in arc furnaces to produce special alloys.

The design of a submerged arc furnace depends on its application. In principle it is a dished vessel (*Figure 21.22*), brick lined, as with the arc furnace. But there the similarity ends: the

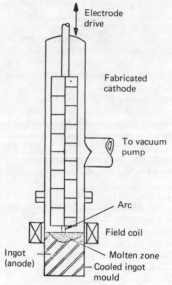

Figure 21.23 Vacuum arc furnace

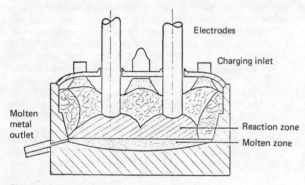

Figure 21.22 Submerged arc smelting furnace

dish and the roof are axially fixed, although the roof, together with the electrodes, may rotate. The furnace is charged through ports in the roof, and molten metal and slag flow from the furnace continuously. The electrodes are of the Soderberg type, formed *in situ* by pouring a mixture of pitch and tar plus anthracite into a steel tubular shell. The process is carried out several metres above the furnace, and as the electrode is lowered, it bakes, so driving off the volatile binding. By the time it enters the furnace it is a solid mass. Electrodes capable of carrying very high currents, up to 120 kA, can be produced in this way.

21.8.4.4 Vacuum arc furnace

The vacuum arc furnace (*Figure 21.23*) is used primarily for remelting metals of very high quality, including titanium, tantalum, niobium, hafnium, molybdenum, tungsten, zirconium and some steel and nickel alloys. Ingots of up to 100 t can be produced. The furnace operates at low pressure, down to about 0.01 Pa, and very effective degassing of the droplets of molten metal (which have a high surface area) occurs. The ingot forms a molten 'skull' which freezes in contact with the copper mould, thus eliminating contamination from refractory linings and minimising thermal stresses and piping at the ends. Impurities are carried through the ingot and collect in the molten pool on the surface. The electrode is either prefabricated or melted first in a vacuum induction furnace; the arc is d.c. and operates with a current of 10–25 kA and a voltage of 20–30 V. A low voltage is used to prevent the arc attaching to the walls of the vessel, and an additional field coil, which interacts with any radial component of arc current, tends to help the stabilising effect and produces a strong stirring action.

21.8.4.5 Electroslag refining

Electroslag refining (*Figure 21.24*) is directly competitive with vacuum arc processes for materials not unduly reactive in air. A high degree of refining, not possible with the vacuum arc process, can be obtained, since the droplets of molten metal penetrate through the molten slag, enabling desulphurising to be carried out and oxide inclusion to be reduced. The process, like the vacuum arc furnace, forms a molten skull and has similar advantages.

The electroslag refining process is essentially one of resistance heating, since it relies on electrical conduction in the molten slag. Single- or 3-phase operation (using three electrodes over one ingot) is possible. The operating voltage is kept to the range 40–60 V to prevent formation of arcs, and currents of up to 3 kA are used. Operation is controlled by limiting the voltage and keeping the electrode immersed in the electrically conducting slag in order to avoid drawing an arc. The slag composition is carefully controlled to maintain its electrical conductivity and to allow it to refine the molten metal droplets. Cylindrical and slab

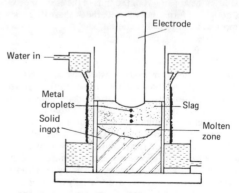

Figure 21.24 Electroslag refining furnace

ingots can be produced: typical ingot diameters are 350 mm (single-phase), 900 mm (3-phase), with a melting rate of 180–360 kg/h at 360 kVA (single-phase) or up to 6 MVA at 180 V and 18 kA (3-phase); thus ingots of up to 15 t can be produced.

21.8.4.6 Electron beam furnace

Electron beams are used for welding, melting and the production of evaporated coatings. The beam is obtained from a heated filament or plate and is accelerated in an electron gun by a high electric field produced by one or more annular anodes. Electrons on the axis of the gun pass through the final anode at very high velocities (e.g. 85×10^6 m/s at 20 kV). The electron gun and chamber are kept at a low pressure of around 0.001 Pa and, as little energy is lost from scatter or production of secondary electrons, practically all the kinetic energy of the beam is converted to heat at the workpiece; thus, the conversion efficiency of electrical energy input to thermal energy in the workpiece is very high. The electron beam furnace (*Figure 21.25*) utilises a cooled ingot mould in the same way as the vacuum and electroslag furnaces.

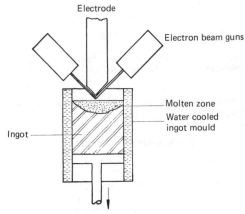

Figure 21.25 Electron-beam melting furnace

Ingots, slabs, tubes, castings, pellets and powder can be produced. One system, shown in *Figure 21.25*, comprises one, two or three guns arranged around a consumable electrode. Individual power ratings up to 400 kW are possible, which enables total power inputs of up to 1.2 MW and melting rates of 500 kg/h to be obtained.

21.9 Dielectric heating

Process heating of non-metal materials can present a difficult production problem, especially when the material is a poor thermal conductor. In these circumstances high heating rates using conventional methods of thermal radiation, conduction or convection imply high surface temperatures: as these may damage the material, heating must proceed slowly. The electrical conductivity of the materials is inevitably low, making heat generation by I^2R effects impracticable. It many instances dielectric loss mechanisms can be used for heating. As the heat is not conducted through the surface, high throughputs can be achieved without damage to the product.

The rate P of energy dissipation by dielectric loss in a non-conducting material of absolute permittivity $\varepsilon = \varepsilon_r \varepsilon_0$ and a loss tangent $\tan \delta$ is

$$P = 2\pi f E^2 \varepsilon \tan \delta \quad \text{W/m}^3$$

at a frequency f. For a high heating rate the frequency ω, the electric field strength E and the loss factor $\varepsilon \tan \delta$ must all be large.

Although all forms of dielectric polarisation are effective, the most significant is usually that which occurs with dipolar materials. Dielectric heating is therefore important in the processing of materials which have a high loss factor. It can also be used in the selective heating of one compound in a mixture of two materials which have substantially different values of the parameter: a typical example is in drying processes. *Table 21.7* indicates the loss factor for some typical materials, and illustrates the wide differences which can occur between substances which appear similar. *Table 21.7* also shows the influence of frequency

Table 21.7 Typical loss factor and frequency relation

Material	Temperature (°C)	Frequency (MHz)			
		1.0	10	100	3000
Ice	—	0.50	0.067	—	0.003
Water	1.5	1.6	0.17	0.61	25
	15	2.5	—	—	16
	65	5.6	—	—	4.9
	95	7.9	0.72	0.17	2.4
Porcelain	25	0.015	0.013	0.016	0.028
Glass (borosilicate)	25	0.002	0.003	0.004	0.004
(soda-silica)	25	0.07	—	0.051	0.066
Nylon (610)	25	0.07	0.06	0.06	0.033
	84	0.76	0.43	0.23	0.10
PVC (QYNA)	20	0.046	0.033	0.023	0.016
	96	0.24	0.14	0.086	—
(VG5904)	25	0.60	0.41	0.22	0.10
(VU1900)	25	0.29	0.17	0.087	0.015
Araldite (E134)	25	0.34	0.41	0.48	0.15
(adhesive)	25	0.11	0.12	0.11	0.07
Rubber (natural)	25	0.004	0.008	0.012	0.006
Neoprene (GN)	25	0.54	0.94	0.54	0.14
Wood (fir)	25	0.05	0.06	0.06	0.05
Paper (royal-grey)	25	0.11	0.16	0.18	0.15
	82	0.08	0.14	0.19	0.23
Leather (dry)	25	0.09	0.09	0.12	—
(15% water)	25	0.78	0.49	0.45	—

and temperature on the loss factor. The former is due to the well-known dependence of both real and complex parts of the permittivity on frequency, and the latter to the fact that an increase in temperature increases the material's internal energy, affecting the polarisation mechanism.

The frequency should theoretically be chosen to maximise the product of frequency and loss factor. However, practical limitations dictate that unless the equipment is adequately screened (often difficult in an industrial situation), the frequencies which can be used are limited to those in the ISM Band (*Table 21.8*).

21.9.1 Radiofrequency power sources and applicators

The power source for heating in the range 10–500 MHz is a class C amplifier/oscillator. At these frequencies cavity construction is normally used for the tank circuit (*Figure 21.26*). Industrial radiofrequency triodes are used, currently available in power ratings up to 500 kW. The load is loosely coupled to the source, typical values of loaded Q being of the order of 100, and the need to observe the allowable bandwidths means that a high degree of

Table 21.8 Frequency allocations for ISM purposes in the UK

Frequency (MHz)	Wavelength (m)	Tolerance (+/−)
13.56*	22	0.05%
27.12*	11	0.6%
40.68*	7.5	0.05%
42, 49, 56, 61, 68	7.1, 6.1, 5.4, 4.9, 4.4	0.2%
84, 168	3.6, 1.8	0.005%
896	0.33	10 MHz
2450*	0.12	50 MHz
5800*	0.052	75 MHz
24 125*	0.012	125 MHz
40 680	0.0073	—

* International ISM frequency bands

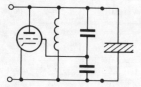

Figure 21.26 Basic oscillator circuit for r.f. dielectric heaters

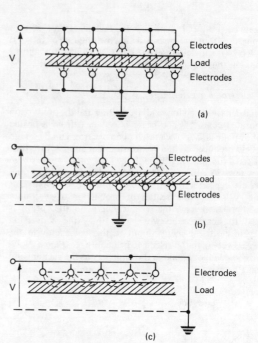

Figure 21.27 Applicators for r.f. dielectric heating: (a) through field; (b) staggered through field; (c) stray field

frequency stability must be achieved even under the varying load conditions to which most heaters are subjected in operation. An important consideration in applicator design is the voltage distribution over the electrode system. The length of the applicator should be considerably less than a quarter-wavelength at the fundamental frequency in order to obtain a reasonable degree of uniformity of field strength in the workpiece. Thus, the physical size of the workpiece affects the operating frequency. The simplest applicator is a pair of parallel plates with the load completely filling the space between them. There is no air-gap and the field strength in the load is relatively high (although it must be well within the breakdown strength of the material). Pressure can be applied to the plates if desired. The technique is used for the heating and welding of plastics materials. For loads of varying geometry and for continuous-process systems an air-gap is required: the field strength in the load is reduced and heating proceeds at a slower rate. The load is carried between the electrodes on a belt made from low-loss material such as PTFE. The physical size and weight of the electrodes becomes a problem with large applicators and, if the process involves the evaporation of water or solvent from a substrate, there is little ventilation to remove the vapour. Perforated-plate systems can be employed, or a rod applicator, which is available in a variety of configurations (*Figure 21.27*). The throughfield system is a simple equivalent of the parallel plates. The staggered throughfield arrangement gives a significant field component in the plane of the load; this is particularly helpful in the drying of materials such as paper, where, owing to its fibrous nature, the loss factor is greater along the plane of the web than transverse to it. In some applications, such as the gluing of joints in wooden structures (e.g. doors), it is difficult to arrange for a throughfield and instead the strayfield arrangement has to be used.

Most commercial heaters have a simple anode-voltage or grid-current control which is fixed at the start of the process. Thereafter heating takes place over a timed cycle.

Dielectric heating at these frequencies is used in many industries, including the welding of PVC materials for articles ranging from small containers to large sheets and car seats; in the textile industries for drying after washing and dyeing in the manufacture of paper; and in the food manufacturing and woodworking industries.

21.9.2 Microwave power sources and applicators

For industrial heating applications using microwaves, the usual power source is the magnetron, chosen for its reliability and long life. Occasionally, when high power from a single source is required, the klystron is used. The largest magnetron commercially available at 896 MHz is 25 kW, and this reduces to 5 kW at 2450 MHz. When higher powers are needed, several magnetrons are fed into one applicator. One form of applicator is a waveguide in which the load is placed. Unabsorbed energy is reflected from the closed end back into the load, but is prevented from being reflected back into the source by means of a directional coupler. A more sophisticated arrangement is the slow-wave, or serpentine, applicator (*Figure 21.28*), in which the energy, after being transmitted once through the load, is fed into a second, parallel, section of waveguide, where it again interacts with the load, which in this case is a laminar material fed through

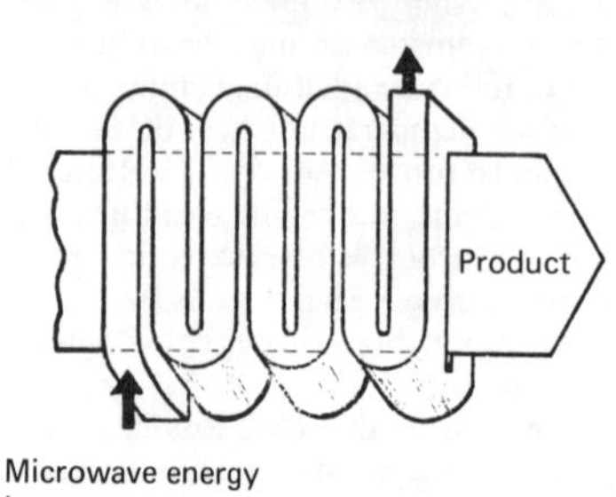

Figure 21.28 Slow-wave or serpentine microwave applicator

a slot in the waveguide wall, and so on through several passes where the application demands it. The waveguide arrangement can be adapted for both batch and continuous processing. Cuboid arrangements, i.e. microwave ovens, designed to give as uniform an energy distribution as possible within the cavity, have been in use for some time. A recent development is a cylindrical cavity in which the energy is concentrated on the central axis; it can be used to heat materials such as liquids which are passed through a loss-free tube at the centre of the cavity. It can also be employed to heat groups of textile threads.

There is an obvious thermal danger from radiated microwave fields, and the permitted level of radiation is strictly controlled. In continuous systems this may mean fitting quarter-wavelength chokes at the inlet and outlet ports, and providing doors of cavity applicators with an adequate seal, a quarter-wavelength choke in the form of a high-loss strip.

Although microwave heating is used mainly in the rubber industry for curing and preheating prior to moulding, it is also employed in the leather industry for curing, and in the food and pharmaceutical industries.

21.10 Plasma torches

21.10.1 Direct-coupled plasma torch

Plasma torches use a stabilised and normally constricted arc at temperatures up to 20 000 °C. This constriction can be achieved by the proximity of cooled surfaces, such as a water cooled nozzle, axial and vortex gas or liquid flows, or movement at high velocity in a transverse magnetic field. Breakdown occurs between the cathode and the anode in the non-transferred mode (*Figure 21.29*) and the plasma torch is used in this way for spraying cermets and for carrying out certain chemical reactions.

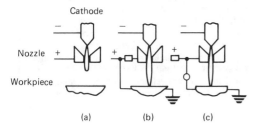

Figure 21.29 Direct-coupled plasma torch: (a) non-transferred mode; (b) transferred mode; (c) superimposed mode

In the transferred mode the arc is transferred from the nozzle to a second anode outside the torch. Such torches are principally used for cutting and welding. Some plasma reactors have been developed at power levels up to 1 MW. The plasma torch can be used to cut all grades of steel alloys and non-ferrous metals, the principal application being the cutting of stainless steel above about 6 mm thick. Other applications include welding (including low-current precision welding) and the machining of hard metals by softening the metal in front of the cutting tool. Most materials which are available in powder form and have a liquid phase, including metals, ceramics and plastics, can be sprayed using a plasma torch. By using a nozzle with a rectangular slit across which a transverse magnetic field is applied, the arc can be made to oscillate in the nozzle. A large area can be scanned in this way during one traverse of the torch without local overheating, and a very uniform temperature distribution can be achieved in the workpiece. This technique is potentially suitable for surface heating processes.

21.10.2 Induction-coupled plasma torch

The induction-coupled plasma torch utilises an electrical discharge to break down a small volume of ionised gas. Electrical energy from a radiofrequency power source is then coupled into this, by use of an induction heating coil. An annulus of ionised gas is formed and stabilised by a vortex gas flow inside a fused quartz tube, the depth being limited by the skin depth at high frequency. Very high temperatures are obtained (>25 000 °C), but the temperature distribution is highly non-uniform and the conversion efficiency from input power to power in the heated gas is limited to about 60% by the conversion efficiency of the radiofrequency power source. A power supply for an induction-coupled plasma torch is shown schematically in *Figure 21.30*.

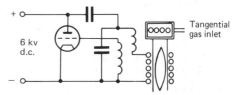

Figure 21.30 Radio frequency induction-coupled discharge and power supply

High-frequency operation is necessary because of the poor coupling between the induction heater and the initiating ionisation. The frequency is normally above 1 MHz, although operation down to 10 kHz has been achieved at power inputs of about 1 MW. Examples of applications include the fusion of high-purity optically transparent quartz, the manufacture of ultra-fine silica particles for thixotropic fluids, metallic boron and zirconia.

21.10.3 Plasma furnaces

Selective non-equilibrium reactions can be carried out by utilising either the high excitation levels available or selective dissociation and recombination (including quenching metastable states) rather than by using thermal energy to increase the energy of all the molecules. Large reaction zones and small temperature gradients in the heated gas are important in chemical reactors. The quenching stage may also be critical in order to enable the required phases to be trapped; these must include metastable states if high process yields are to be obtained.

Several plasma reactors have been developed (*Figure 21.31*) at power levels up to about 1 MW in an attempt to meet the above requirements. An early form of torch using a long arc stabilised by a gas vortex in a cylindrical reactor was developed for the manufacture of acetylene. Subsequent torch designs based on this concept have been used for re-entry simulation, for heating oxygen to 1500 °C in the commercial manufacture of titanium dioxide, and for melting iron and steel alloys. One or more torches in transferred or superimposed modes of operation have been used to produce a relatively large volume of ionised gas in the reaction zone.

A large reaction zone can be obtained by rotating the arc at several thousand revolutions per second between coaxial electrodes, an axial magnetic field being used, so that the gas flowing between the electrodes experiences several rotations of the arc. High heat transfer rates and relatively uniform treatment of the material flowing between the electrodes can be achieved. Systems have now been produced capable of output powers up to several megawatts. Some examples of promising chemical processes are given in *Table 21.9*.

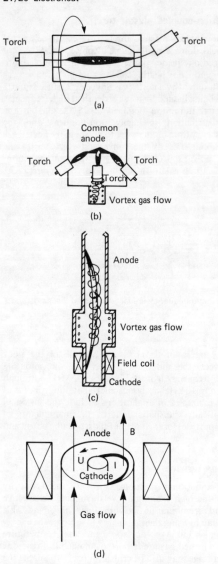

Figure 21.31 Plasma reactors: (a) centrifugal furnace; (b) superimposed arcs; (c) vortex stabilised torch; (d) magnetically rotated arc

21.11 Glow discharge processes

Glow discharges can be used for surface heat treatment and to carry out chemical reactions (such as polymerisation processes involving non-equilibrium selective transitions) with high efficiency. The glow discharge is characterised by a high voltage gradient in the column and the discharge roots. It is not in thermal equilibrium (unlike the arc discharge) and can be developed at any frequency between zero and microwave. One application is for heat treatment, in which wire is passed continuously through a beam of electrons by use of annular electrodes. Another application which is in commercial development is ion nitriding. In this process a low-pressure glow discharge using an electron beam gun produces a stream of electrons and ions which bombard the workpiece, which serves as an anode. The ions penetrate several atomic layers and react with the substrate material to produce a hard layer. The process is carried out in a vacuum.

Table 21.9 Examples of potential plasma processes in chemical synthesis and extractive metallurgy

Starting product	Reaction products
$N_2 + O_2$	2NO
$N_2 + 2H_2$	N_2H_4
$TiCl_4 + 4Na$	$Ti + 4NaCl$
$2FeCl_3 + 3H_2$	$2Fe + 6HCl$
$TiCl_4 + O_2$	$TiO_2 + Cl_4$
$ZrCl_3 + 3Na$	$Zr + 3NaCl$
MoS_2	Mo Z
$Fe_2O_3 + CH_4$	$2Fe + 2H_2O + CO$
$FeOTiO_3$	ferrotitanium + TiO_2
chromite	ferrochrome + SCr
$2CH_4$ + oil feedstock	$C_2H_2 + 3H_2$
$V_2O_5/V_2O_3 + C + Fe$	FeVC + CO
SiO	surface active SiO_2
coal	coal gas
steel scrap	alloy steel
tin slag	tin oxide
$UO_2 + C$	UC + CO

21.12 Lasers

Since its discovery in 1960 the laser has found extensive application in industry in cutting, welding, material removal and heat treatment processes requiring power densities above 100 W/mm². The principal features of lasers in terms of heat treatment sources are their monochromatic output, low divergence and high intensity, which enable the parallel monochromatic beam to be focused to a smaller diameter than is possible with other light sources. The minimum focused diameter for a Gaussian intensity distribution approaches the diffraction limited diameter, $d = 1.22 \lambda/F$, where λ is the emission wavelength and F is the aperture of the optics. Very high power densities can be obtained in this way. Ruby, argonion, neodymium and CO_2 lasers are used in industrial processes. The neodymium laser is excited with xenon or krypton discharge lamps; the high pulse energy combined with the smaller focused spot size obtainable at the shorter wavelength of 1060 nm has resulted in its being used primarily for drilling, microwelding and scribing and trimming electronic circuits. Continuous-wave neodymium lasers are used for the spiral cutting of resistors. The CO_2 laser utilises an electric glow discharge to excite the laser transitions in a mixture of carbon dioxide, helium and nitrogen at efficiencies of up to 28% at a wavelength of 10 600 nm. Below 50 W a sealed tube construction can be used, but at higher powers a continuous-flow system is required. Outputs up to about 500 W are obtained with a low-velocity axial gas flow, but at powers in the range 1–60 kW high-speed gas flow is used, either along the axis or transverse to it. Higher cutting speeds of thick materials are made possible by using a gas jet after the focusing lens and parallel with the laser beam to blow away molten or evaporated material. In the case of metals, if the gas is chosen to react with the metal, as in the iron–oxygen reaction, considerable increase in cutting effectiveness is obtained.

Industrial applications for CO_2 lasers include dividing and perforation of substrates for integrated circuits and electronic components; perforation of elastomeric materials; manufacture of dies for cutting cartons; ready-to-wear suits and thin glass tube for fluorescent lamps, for quartz tubes and borosilicate glass and for the manufacture of flexigravure plates. Deep-penetration welds, where the ratio of the weld depth to the heat affected zone may be as high as 10/1 compared with those attainable with electron beam welding, can also be obtained at power levels above about 1 kW. Selective heat treatment is also possible.

22 Welding

T G F Gray BSc, PhD, CEng, MIMechE, MWeldI
Department of Mechanics of Materials, University of Strathclyde

D R Andrews PhD, CEng, FIM, MIProdE
Department of Mechanical and Production Engineering, University of Aston in Birmingham

Contents

22.1 Arc welding 22/3
    22.1.1 Manual welding with flux-coated electrodes 22/3
    22.1.2 Power supply for MMA welding 22/5
    22.1.3 Submerged-arc welding 22/7
    22.1.4 Electroslag welding 22/8
    22.1.5 Tungsten inert-gas welding 22/8
    22.1.6 Plasma-arc welding 22/11
    22.1.7 Gas-shielded metal-arc welding 22/11
    22.1.8 Metal cutting and gouging 22/12

22.2 Resistance welding 22/12
    22.2.1 Scope 22/13
    22.2.2 Welding equipment 22/13
    22.2.3 Spot welding 22/13
    22.2.4 Projection welding 22/17
    22.2.5 Seam welding 22/17
    22.2.6 Butt and flash welding 22/19
    22.2.7 Resistance-welding electrodes 22/20

22.1 Arc welding

The welding arc consists of a high-current-density plasma maintained by passage of an electric current which keeps the ionised gas at a high temperature (of the order of 15 000 K). The body of the arc is electrically neutral, i.e. there are equal numbers of positively and negatively charged particles.

Voltage distribution across the arc from cathode to anode will depend to a large extent on the cathode material. In a tungsten arc the voltage distribution is as shown in *Figure 22.1*. Three

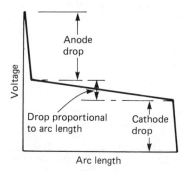

Figure 22.1 Voltage distribution across a tungsten arc

distinct voltage drops are involved: the cathode drop, the column drop and the anode drop. Both the cathode and anode drops are to a first approximation constant, but the column drop is a function of arc length.

The arc characteristics are shown in *Figure 22.2*. There are three regions: at low currents the voltage–current relation has a negative slope (as the current increases the voltage decreases), the flat portion is essentially a constant voltage irrespective of current, and at high currents the slope of the voltage–current relation is positive.

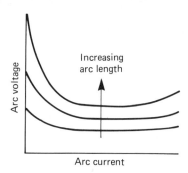

Figure 22.2 Typical voltage–current arc characteristics

22.1.1 Manual welding with flux-coated electrodes

A high proportion of all welding carried out in the United Kingdom (UK) is still done manually, using covered electrodes.

22.1.1.1 Electrodes

Electrode wires are coated with suitable fluxes to stabilise the arc: to provide a gas shield through which molten metal is transferred; to provide a slag which, floating on the top of the molten-metal weld pool, protects it from atmospheric contamination and improves its profile; to supply deoxidants (e.g. Al, Si, Ti) and alloying elements to give the weld the required mechanical properties. It is more convenient to introduce deoxidants and alloying elements via the flux coating, as different properties can then be obtained from a single wire formulation.

22.1.1.2 Classification and coding of electrodes

The coding for manual metal-arc (MMA) electrodes to BS 639:1976 describes the following characteristics: (i) the tensile and yield strengths of the deposited weld; (ii) the elongation and impact properties of the deposited weld; (iii) the type of electrode coating; (iv) the efficiency of deposition, i.e. the ratio (mass deposited)/(mass of core wire); (v) the welding position in which the electrode can be used; (vi) electrical characteristics; (vii) the limit on diffusible hydrogen. Manufacturers are not required to give the information (iv)–(vii), so sometimes the code is shortened.

As an example of the coding, consider an MMA electrode with basic covering, high efficiency, low hydrogen deposit, and minimum properties of yield strength 380 N/mm^2, tensile strength 560 N/mm^2, elongation 22%, impact strength 47 J at $-20\,°C$, efficiency 158%. In terms of (i)–(vii) above, the coding is

Characteristic (i) (ii) (iii) (iv) (v) (vi) (vii)
Code   E51 33   B   160 2   0   (H)

Details of the code letters and numbers are given below.

Mechanical properties Two grades of ultimate tensile strength and minimum yield strength are specified:

Grade	E43	E51
Tensile strength (N/mm^2)	430–550	510–650
Minimum yield strength (N/mm^2)	330	360

The next two digits specify elongation and impact properties: for the first digit,

First digit	0	1	2	3	4	5
Minimum elongation (%)						
E43	n.s.	20	22	24	24	24
E51	n.s.	18	18	20	20	20
Temperature for 28 J impact (°C)	n.s.	+20	0	−20	−30	−40

and for the second digit,

Second digit	0	1	2	3	4	6
Minimum elongation (%)						
E43	n.s.	22	22	22	–	–
E51	n.s.	22	22	22	18	18
Impact value (J)						
E43	n.s.	47	47	47	–	–
E51	n.s.	47	47	47	47	47
Temperature (°C)	n.s.	+20	0	−20	−30	−50

In both tables 'n.s.' indicates that a value is not specified.

Covering The types of electrode coating are designated by a letter symbol:

A acid (iron oxide)   AR acid (rutile)
B basic   R   rutile (medium-coated)
C cellulosic   RR rutile (heavy-coated)
O oxidising   S   other

Rutile-coated electrodes are commonly used for general-purpose welding of mild steel; they are used in the vertical and overhead positions. Cellulosic coatings give highly penetrating all-position electrodes and are often used for pipeline and storage-tank welding. Oxidising electrodes are used mostly in flat positions and give a smooth appearance. Electrodes of classes A and AR need material of good weldability but can be operated at high rates with easy slag removal. Basic electrodes are also referred to as 'low-hydrogen' electrodes and are used where high resistance to cracking is required on hardenable steels or highly restrained structures. Special care needs to be taken to keep these electrodes dry.

Efficiency and welding position The efficiency of weld deposition as defined above is quoted directly in the coding as a percentage. Note that efficiencies greater than 100% are common, as the flux covering contributes material to the deposit. The positional capability is identified next in the coding by a single digit:

1 all positions
2 all positions except vertical down
3 flat, and horizontal–vertical for fillet welds
4 flat
5 flat, vertical down, and horizontal–vertical for fillet welds
6 any other position or combination

Electrical characteristics The recommended direct-current (d.c.) and alternating-current (a.c.) welding conditions are defined by the next single digit:

Code	0	1	2	3	4	5	6	7	8	9
D.C.: recommended polarity	r	+	− or −	+	+ or −	−	+	+ or −	−	+
A.C.: minimum open-circuit voltage (V)	d	50	50	50	70	70	70	90	90	90

r, as recommended, d, not suitable for a.c.

Hydrogen control The designation (H) terminates the coding if the electrodes are capable of producing a deposit containing not more than 15 ml of diffusible hydrogen per 100 g of weld metal.

22.1.1.3 Electrode sizes

Table 22.1 gives metric sizes with their alternative inch dimensions and approximate wire-gauge (s.w.g.) numbers, and typical ranges of welding currents.

Table 22.1 Electrode sizes

Electrode diameter		S.W.G. No.	Current A
mm	in		
2	0.080 or 5/64	14	45–80
2.5	0.104 or 3/32	12	60–95
3.15*	0.128 or 1/8	10	80–125
4	0.160 or 5/32	8	115–180
5	0.192 or 3/16	6	140–260
6	0.232	4	190–340
6.3	0.25 or 1/4	—	200–370
8	0.324 or 5/16	—	250–500
10	0.375 or 3/8	—	320–600

* 3.15 mm is the ISO recommended size; the 3.2 mm diameter is, however, widely used and is to be retained temporarily.

22.1.1.4 Electrode compositions for various jobs

Low-alloy steels Alloying elements are present in steels to increase certain properties such as tensile strength, toughness, corrosion resistance and creep resistance.

In general, MMA welding electrodes are available with weld-metal properties to match all low-alloy steels in general use. It is not always necessary that the electrodes have the same chemical analysis as the material to be welded; however, where this necessity does exist, for instance with Cr–Mo creep-resistant steels, these electrodes are available.

Some low-alloy steels are readily welded with the minimum of precautions, but others may require carefully controlled preheating before welding and a post-weld heat treatment such as stress relieving immediately after welding if cracking is to be avoided.

Corrosion- and heat-resisting steels A fairly wide range of electrodes is available for the welding of materials in this category, which includes stainless steels. BS 2926: 'Chromium–Nickel, Austenitic and Chromium Steel Electrodes for Manual Metal Arc Welding' gives the range of electrode types available. In this Standard, electrodes are classified according to their chemical composition and type of flux coating. The first figure indicates the chromium content, the second figure indicates the nickel content and the third figure indicates the molybdenum content. Next comes either an L indicating a low carbon content, an H indicating a high carbon content, the symbol Nb indicating the presence of niobium as a stabilising element or W indicating the presence of tungsten. Following these there is a suffix, B indicating a basic lime–fluorspar covering, R indicating a rutile-type covering, or MP indicating that the core wire is not of the chemical composition indicated and that the bulk of the alloying elements are added via metal powder in the coating. Electrodes bearing the suffix B are generally used on d.c. electrode-positive only; those with suffix R can be used on either d.c. electrode-positive or a.c. according to the manufacturer's recommendations.

Welding of chromium–nickel corrosion- and heat-resisting steels is generally carried out with the minimum of difficulty provided that the weld metal contains a nominal amount of ferrite. Fully austenitic weld metals are susceptible to micro-cracking and in some cases even gross cracking of the weld metal. The use of a weld metal containing a small amount of ferrite will generally avoid the trouble; however, this is not always acceptable, and in such cases small runs and interpass cooling may be necessary.

The plain chromium steels present more of a problem and in some cases preheat and the use of a chromium–nickel electrode may be necessary.

Copper and copper alloys The high conductivity of copper is disadvantageous with regard to its weldability, and joining of thicker sections by the MMA process is generally carried out using a bronze electrode. Most electrode manufacturers recommend tin bronzes containing between 4% and 10% tin and up to 0.5% phosphorus. Silicon bronzes and aluminium bronzes will in most instances also give satisfactory results. Where it is necessary to make joints in copper with an electrical conductivity matching that of the parent material, the tungsten inert-gas process is generally used.

Good welds with excellent corrosion resistance and mechanical properties can be made by the MMA process in both brasses and bronzes; however, owing to the volatilisation of zinc which occurs when brass electrodes are used, both types of alloy are generally welded with bronze electrodes.

Nickel and nickel alloys The metal-arc welding of nickel and nickel alloys is widely practised, and in principle the techniques used differ only slightly from those used for mild steel. For instance, it is generally important to ensure that the joint is stress-free; this is accomplished by using annealed material and avoiding joint geometries which induce restraint. It is advisable to use small electrodes and to prevent overheating of the workpiece by allowing it to cool down between weld runs. Direct current is essential and the electrode should be connected to the positive pole.

A fairly comprehensive range of electrodes is available for the welding of nickel and its alloys. Some of these are also ideally suited to the making of dissimilar-metal joints, e.g. the welding of stainless steel to mild steel.

22.1.2 Power supply for MMA welding

In the metal-arc welding process the electrode is consumed in the arc and deposited in the weld pool. As this system is normally a manual operation there can be variations in the length of the arc, and consequently also the arc voltage and current. A change in current is undesirable as it leads to a higher burn-off rate. The consequence of this is that the current supply for MMA welding is required to be reasonably constant; the power source is required to have a drooping voltage–current characteristic.

22.1.2.1 D.C. equipment

Mobile equipments are mainly engine-driven generators, frequently fitted with voltage and current control. The adjustment of these two controls enables the welder to obtain a voltage–current characteristic that gives a measure of control over the welding conditions. The modern tendency with respect to a.c. generators is to use 'brushless' types. It is of course necessary to have a rectifier system to convert the output of the generator to d.c. If generators are used in fixed positions or where mains supply is available, they are powered by electric motors.

Rectifiers Rectifiers are most commonly of the 3-phase bridge configuration (*Figure 22.3*); they produce a lower ripple-voltage content than 1-phase rectifiers.

Silicon diodes have largely superseded selenium metal rectifiers; they have the advantages of greater efficiency and smaller size. They are, however, more sensitive to surges and overvoltage, and must be adequately rated and have suitable protection. Special care must be taken to protect silicon diodes where high-frequency (h.f.) arc starting systems are employed.

Forced-air cooling is generally employed, which reduces size of the heat sink required for the rectifier bank.

Current control The basic principle of *transductor* control, employed in both 1-phase and 3-phase welding equipments, is that an iron-cored inductor is placed in each rectifier supply line, and its inductance controlled by a d.c. winding which varies the conditions of core saturation. Alternatively, a *moving-core inductor* may be used, in which the inductance is a function of the airgap. Both methods introduce an adjustable inductive impedance into the a.c. side.

22.1.2.2 A.C. equipment

Where a.c. mains are available the equipment comprises essentially a transformer for single- or multiple-arc operation. The transformer may be of either the air-cooled or the oil-cooled types. The former have some advantage for light work and are usually more closely rated than the oil-immersed types; they are therefore of lighter weight and more suitable for intermittent work where a small inexpensive unit is required. For production welding or work out of doors oil-immersed sets are usually preferred. In such cases the windings are totally enclosed in a tank which protects them from damage due to entry of moisture, dirt etc. and therefore renders them very reliable even under conditions of rough usage.

The availability of high-temperature insulants is gradually extending the use of air-cooled transformer sets into the heavier classes of welding duty.

There is a British Standard covering electric arc-welding equipment: BS 638. It lays down the rating for both air- and oil-cooled single-operator transformers, and may be summarised as follows: the continuously rated load shall be 70% of the immersed transformers, in each case for single-arc welding equipment.

22.1.2.3 Single-operator transformer equipment

The most common form of single-operator equipment comprises a transformer wound on the primary side for normal low-voltage (l.v.) mains and on the secondary side for 60–100 V, together with a tapped inductor (*Figure 22.4*) or moving-core inductor. It is preferred to connect the primary across two lines of 3-phase supply rather than between line and neutral.

Sets are produced with T or open-delta connection to 3-phase supplies, but as the 1-phase load cannot actually be balanced over three phases no great advantage can be obtained.

22.1.2.4 Two-operator equipments

Transformer welding sets are also available for two welders, who can work simultaneously. These may consist of two 1-phase units included in a single tank (*Figure 22.5*) or comprise a single transformer to which two variable inductors are connected in parallel on the secondary side. Such sets, of either type, have the advantage that the two secondary circuits can be connected together to a single arc to give a high current for large electrodes or heavy work.

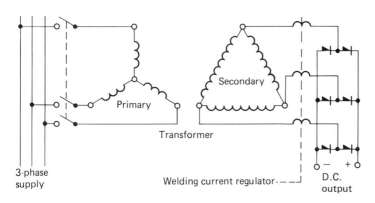

Figure 22.3 Rectified a.c. supply for metal-arc welding

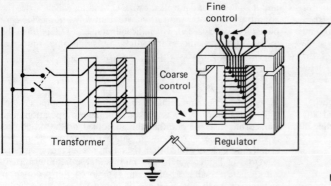

Figure 22.4 Single-phase transformer and tapped inductor

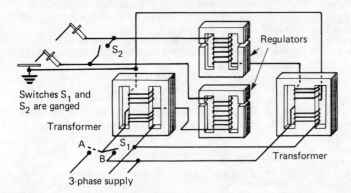

Figure 22.5 Double-operator welding equipment

In order to avoid the possibility of shocks when the electrode holder on a.c. transformer welding equipment is handled, a device can be fitted which will reduce the voltage automatically after the arc is broken.

The unit is a timed relay device which operates in such a manner that within a fraction of a second after the termination of a welding run the voltage at the electrode holder is automatically reduced from the normal 80 or 100 V to approximately 30 V.

The unit is introduced into the secondary circuit of the transformer, i.e. between the output side of the welding equipment and the electrode holder.

22.1.2.5 Power-factor correction

In view of the fact that the load taken by a welding transformer is highly inductive, the power factor will necessarily be low. For single-operator equipment it is of the order of 0.3–0.5 lagging, depending on the design of the set and the type of electrodes used, arc length, etc. It is well known that the greater the value of the inductance of the welding circuit, the better the conditions. A welding set which is not provided with a capacitor or other means of power-factor correction cannot, therefore, operate at a high power factor.

The simple 1-phase transformer welding set lends itself readily to power-factor correction by connection of a capacitor on the primary side. Also there are available sets in which the capacitor is incorporated in or attached to the welding transformer.

Where many single-operator sets are connected, care must be taken not to overcorrect the power factor. It must be borne in mind that a welding load is intermittent and probably for 50% of the time the set is connected no welding will be in progress. If, therefore, each welding equipment is provided with a separate capacitor, there is the danger of a leading load being taken if the diversity factor of the welding load is small. This difficulty may be overcome by:

(1) Automatic switching on of the capacitors when load is applied.
(2) A central bank of capacitors to correct the load of whole works based on an average observed load.
(3) Careful instructions to operatives to switch off capacitors as soon as they stop welding.

22.1.2.6 Multi-operator transformer equipment

Multi-operator transformer equipment consists of a single transformer, to the secondary windings of which a number of arcs can be connected. Each welding circuit must be provided with a current regulator which can be a variable resistor or inductor. The latter gives higher efficiency and improved arc stability.

The most common form of installation of a 3-phase transformer having a delta-connected primary winding and an interconnected-star secondary (*Figure 22.6*). It will be noted that the inductive current regulators are connected to the line terminals and the material to be welded to the star or neutral, which should be connected to earth.

The number of reactors and welders' circuits on the 3-phase system must be a multiple of three, so that they can be distributed evenly on the three secondary windings.

This type of plant will be found to be economical for installations where work is concentrated in one shop, and also for outside constructional work where it is desirable to safeguard welders from possible shock from the primary voltage of the supply. This precaution is particularly necessary in the case of shipbuilding, bridge and storage-tank construction.

BS 638 indicates a method of connecting the welding equip-

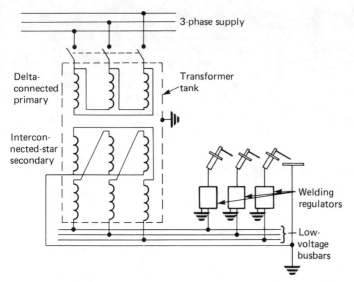

Figure 22.6 Three-phase multi-operator welding equipment

ment and specifies special plug and socket connections and distribution boxes. The leading dimensions of these are laid down, so components of different manufacture are interchangeable.

The sizes of transformers are limited to four, viz. 57, 95, 128 and 160 kVA continuous rating, which are capable of providing one, two, three or four welding operators per phase each at a maximum continuous hand-welding current of 350 A, or a lesser number at a hogher current. The three standard sizes of current regulators are designed to give a maximum welding current of 350, 450 or 600 A, at 90 V.

22.1.3 Submerged-arc welding

In this process a bare wire is used and the flux is added in the form of a powder which covers the weld pool and the end of the electrode; thus the arc formed is completely submerged (*Figure 22.7*). The current range for submerged-arc welding is from 200 to 2000 A and the corresponding arc-voltage range is 25 to 45 V. Welding currents far in excess of 2000 A have been used with this process, but the mechanical properties of the weld deposit are adversely affected as the volume of weld deposited in one run is increased.

The flux in the immediate vicinity of the welding arc is melted during welding and approximately equal proportions by weight of wire and flux are used in the making of a weld run. On completion of the weld run, the unfused flux is removed for subsequent reuse, leaving the fused flux, which is often self-lifting, to be removed from the weld deposit. Providing that the welding parameters have been correctly set, the resulting weld bead is of neat and regular appearance.

Fluxes and wires for submerged-arc welding are usually developed together in order to give welds with the required mechanical properties or chemical compositions, and they are not necessarily interchangeable. Flux–wire combinations are available for a wide range of mild and low-alloy steels, corrosion- and heat-resisting steels, and nickel alloys.

The submerged-arc welding head can be stationary (in which case the workpiece will move along or revolve beneath it), or it can be moved over the work on a self-propelled carriage or gantry. For applications such as the welding of stiffeners to a deck plate, machines are available with twin heads which will weld both sides of a T fillet simultaneously.

A development of the submerged-arc process is multiple-electrode welding, where two or more electrodes are used, in parallel where wider welds are required, or in tandem where higher welding speeds are desirable. In either case a considerable increase in the rate of deposition of weld metal can be realised. If the arcs are supplied from separate power supplies, at least one is a.c. so that the magnetic fields produced around the electrodes do not interfere with the smooth running of the arcs.

Where it is required to use the submerged-arc system for overlaying or surfacing (for instance, a mild or low-alloy steel shaft may be overlayed with a corrosion-resistant layer of weld metal which could well reduce the overall cost of the component considerably), techniques can be employed which reduce the penetration of the process, thus minimising the amount of mixing of base metal and weld metal. These techniques include: a two-wire tandem electrode system where each electrode is connected to a separate pole of a single d.c. power source; the use of a strip electrode which tends to spread the arc over a wider area, thus reducing the penetration considerably. This latter technique is often referred to as 'strip cladding'.

A wide variety of a.c. or d.c. power sources can be used for submerged-arc welding, but the most common choice is a d.c.

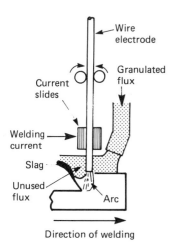

Figure 22.7 Submerged-arc process

source with a drooping characteristic, as in MMA practice. The arc length is controlled by sensing the arc voltage and using the signal to vary the wire-feed motor speed to hold the arc length constant. This methd is common for welding systems that use an automatic feed of a heavy-section wire.

22.1.4 Electroslag welding

Electroslag welding, originally developed for the welding of thick mild and low-alloy steels, is restricted to welding in the vertical or near-vertical position. *Figure 22.8* shows the basic arrangement.

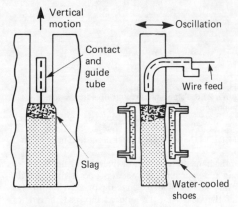

Figure 22.8 Electroslag process

Once established, the electroslag process is basically arcless, the heat required to melt the wire and fuse the parent material being supplied by resistive heating of the molten slag bath. To ensure uniform fusion of the joint faces, it is necessary with thick material either to have more than one electrode or to traverse the electrode(s) to and fro across the joint. As the weld progresses, the level of the weld pool rises and the welding head and water-cooled copper shoes are moved slowly up the joint. A typical welding speed for electroslag welding of 76 mm thick mild steel would be 1 m/h at 550 A and 44 V. Since the process is only stable once a slag bath of appropriate depth has been established, it is essential that a run-on plate is provided so that the defective start portion of the weld can be removed on completion of the joint.

As the electroslag process is virtually a continuous casting process, the resulting welds have a large grain size and a tendency towards columnar growth. Because of this the weld metal and heat-affected zone generally have extremely poor impact properties. While this can be mitigated by a post-weld normalising heat treatment, such treatment may cancel out the economic advantages that the process could otherwise offer.

Although electroslag welding is mainly used for the joining of mild and low-alloy steels, satisfactory welds have been made in high-alloy steels, stainless steels, titanium and aluminium.

Consumable-guide electroslag welding is a much simplified version of electroslag welding which is rapidly gaining popularity. This simplification stems mainly from the use of a wire guide which is progressively melted into the weld. The consumable guide (or guides) eliminates the need for a moving welding head and also results in faster welding speeds. The two moving water-cooled copper shoes used for electroslag welding are generally replaced by two pairs of shoes which are leap-frogged up each side of the joint as welding progresses.

Electrogas welding is very similar in principle to electroslag welding, the main difference being that heat for welding is generated by an arc which is formed between a flux-cored electrode and the molten weld pool. The flux from the electrode forms a thick protective layer over the weld pool, but additional protection in the form of a gaseous shield (usually CO_2) is used. The electrogas process is generally faster than electroslag welding when used on relatively thin sections. The resulting weld metallurgy is similar to a high-current submerged-arc weld.

22.1.5 Tungsten intert-gas welding

Tungsten inert-gas (TIG) welding employs a heat source in the form of an arc which is struck between a tungsten electrode and the material to be welded. The tungsten electrode, arc and weld area are all protected from atmospheric contamination by a shield of inert gas.

The polarity of the electrode is important, since the arc heat is not evenly divided between the anode and cathode. *Figure 22.9* indicates the way in which the arc heat is distributed with the d.c. system and also shows the relative penetration characteristics.

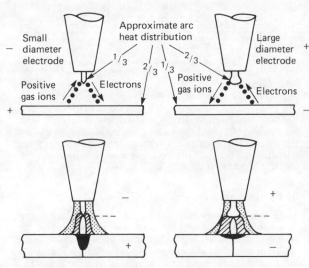

Figure 22.9 Effect of d.c. electrode polarity in TIG welding

The majority of materials can be readily welded using the d.c. electrode-negative system. This, however, has a disadvantage in the welding of aluminium or magnesium, or alloys in which either material is predominant, in that the refractory oxide always present with these materials floats on the surface of the molten weld pool to such an extent that no pool is in fact visible to the welder. With the electrode-positive system, the oxide is broken down or dissipated leaving a clean and bright weld pool. Owing to the distribution of heat, discussed earlier, the welding of thicker materials with d.c. electrode-positive is impracticable, since very-large-diameter electrodes would be required to carry the currents involved. For this reason, the a.c. system is used for the welding of these materials. The a.c. arc combines the cleaning action of d.c. electrode-positive and the cooler running of the electrode of d.c. electrode-negative systems.

22.1.5.1 Electrodes

While it is possible to use a pure tungsten electrode, it is general practice to use electrodes with 1% or 2% thoria and 1% or 2% zirconia for d.c. and a.c. welding respectively. These small additions of rare earth oxides assist with arc striking and tend to reduce the risk of tungsten inclusions in the weld.

22.1.5.2 Gases

Most of the TIG welding in the UK is carried out using argon as the shielding gas. For certain applications, however, other gases or gas mixtures can offer advantages. One instance would be the use of an argon–hydrogen mixture for welding of corrosion- and heat-resisting steels and nickel alloys. The hydrogen addition (generally 2–5%) to the argon has a twofold effect: firstly, the hydrogen tends to combine with the oxides which form with these materials, resulting in a much cleaner weld; secondly, the hydrogen (a diatomic gas) is broken down to its monatomic state under the influence of the arc and subsequently recombines with a resultant increase in the effective heat of the arc, giving faster welding speeds. Another instance is the use of helium or argon–helium mixtures for the welding of copper. The arc voltage with the gases is increased, with a resulting rise in arc energy for a given current level. This again increases the welding speed and also minimises the need for preheating.

22.1.5.3 Applications

The practical working range of the TIG process with the d.c. electrode-negative system can be from as low as 1 A up to 700–800 A; it is therefore a very versatile welding process. Material as thin as 0.1 mm (0.004 in) can be welded at the low end of the range; the high currents are employed particularly for the welding of thick copper. The working range for a.c. is from about 25 to 350 A. At currents lower than 25 A the arc tends to be unstable, and at currents above 350 A the risk of tungsten particles being transferred across the arc increases.

Although high current levels have been quoted for the TIG process, the rate at which metal can be deposited is slow in comparison to other processes; it is therefore primarily a process for the welding of sheet materials and small components. The process may be used in any welding position. When welding vertically, however, the vertical-up technique is generally recommended. In addition, the excellent control of heat input makes the process especially useful for making controlled penetration root runs in butt joints in thicker materials, especially pipework, where access is only available from one side of the joint.

The process is readily mechanised and gives high-quality precision joints. It is therefore quite widely used in the aircraft, atomic energy and instrument industries.

22.1.5.4 Electrical requirements

Generally the arc is initiated by a series of h.f. high-voltage (h.v.) sparks, which ionise the gas between the electrode and workpiece. The voltage and frequency are approximately 10–20 kV at 2–3 MHz; hence it can be seen that care must be taken to prevent insulation breakdown of the welding equipment. Any measuring equipment used must be in the workpiece line, and must be suitably protected by inductors and capacitors. H.F. feedback into the welding power source is eliminated by an air-cored inductor between the h.f. generator and transformer/rectifier unit; it may be built into the h.f. transformer as shown in *Figure 22.10*. Meters in the power source are protected by series inductors and shunt capacitors. Care must be taken to ensure that the equipment is properly earthed, and that all welding leads are kept as short as possible, as an appreciable amount of radio-frequency interference is generated.

Shutting off the arc at full welding current tends to leave a crater at the end of the weld; this is undesirable in terms of stress, and may lead to cracking. It is therefore usual to incorporate a control that automatically tapers the current to a low level before breaking the supply to the arc.

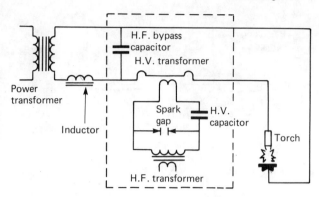

Figure 22.10 H.F. arc-starting unit for TIG welding

22.1.5.5 D.C. process

The arc is formed between a tungsten (or tungsten alloy) electrode and the workpiece. Tungsten and its alloys are thermionic emitters, and are therefore used as cathodes.

As in the metal-arc system, the arc voltage is determined by the arc length. It is also required that the current is eventually constant, so the power supply must have a drooping voltage–current characteristic, as shown in *Figure 22.11*.

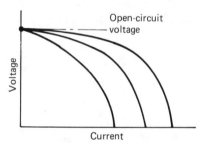

Figure 22.11 Static voltage–current characteristic

22.1.5.6 A.C. process

The cyclic nature of the current introduces certain difficulties. When the tungsten electrode changes from positive to negative polarity a smooth transition takes place because the tungsten electrode (being a thermionic emitter) has an electron cloud available fro reignition as an arc cathode. When the change in electrode polarity is from negative to positive, a cathode root or a group of multiple cathode roots have to form on the plate. This function requires a high restriking voltage to reignite the arc, which is over 150 V when welding aluminium.

For the usual inductive supply, the arc voltage and current wave-forms (*Figure 22.12*) both lag considerably on the open-circuit voltage. As a result, a high restriking voltage is available (*Figure 22.12a*). If the arc fails to reignite because of insufficient restrike voltage, a rectifying arc (*Figure 22.12b*) can occur, with the current flowing only in the negative half-periods. Under l.v. conditions it is possible to secure postive half-period current by use of auxiliary equipment to give, for example, *spark reignition*. The sparks must be properly timed, as otherwise some degree of rectification will occur.

A more precise method of obtaining the electrode-positive half-cycle is the use of the *surge injection* technique. With a surge

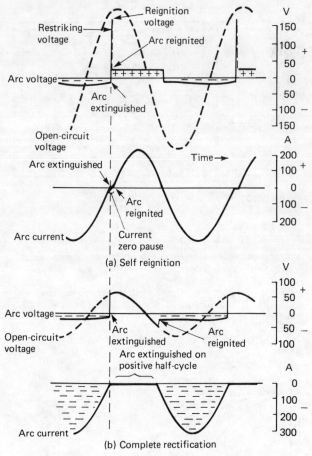

Figure 22.12 Wave-forms of voltage and current for TIG a.c. welding

injector added to a welding transformer the open-circuit voltage can be reduced to 50 V. The basic circuit of the surge injector together with the h.f. arc starter is shown in relationship to the welding circuit in *Figure 22.13*.

The operation of the circuit for starting is as follows. When full open-circuit voltage is applied to the system the relay contact is opened, and the trip unit operates the switch to discharge the surge capacitor into the primary of the step-up transformer. The voltage induced in the secondary builds up until the breakdown voltage of the spark-gap is reached; the capcitor then discharges through the spark-gap into the torch. When the arc is established the voltage applied to the relay falls to arc voltage and the relay contact closes, the surge capacitor being then discharged directly into the arc. The instant of discharge is governed by the trip unit and is so timed as to occur at arc extinction when the polarity is changing to the electrode-positive half-cycle. The surge capacitor, which is charged to a voltage of sufficient amplitude, is then used to provide an artificial restrike voltage.

D.C. component and suppressor capacitor As the tungsten electrode emits electrons when acting as a cathode, there is a difference in the arc voltage in successive half-periods, a consequent difference in the currents, and the appearance of a d.c. component current which may saturate iron cores. To remove this d.c. component it is common practice to insert a capacitor in series with the arc. The value of this capacitor is required to be large, as its capacitive reactance cancels part of the essential inductive reactance. The capacitor comprises an array of reversible electrolytic units totalling, say, 100 000 μF for a 300 A arc. In balancing the current the capacitor takes up a charge and acquires a potential difference (p.d.) of 8–10 V.

It should be noted that, when the arc is starting from cold, until the tungsten is hot only the electrode-positive half-periods are operative. This will tend to charge the capacitor up in a direction so as to oppose the current flow, and hence it may be necessary to short-circuit the capacitor during starting.

22.1.5.7 Developments

Spot welding by the TIG process This is useful where access is available from one side of a lap joint only, or for making spot welds in materials of dissimilar thicknesses, with the thinner material at the top. The equipment is basically the same as for the conventional d.c. electrode-negative system with the addition of a timer and modifications to the welding torch so that pressure can be applied to hold the two sheets to be joined in intimate contact. An arc of predetermined duration is struck between the electrode and the top sheet; this forms a weld pool which penetrates into the bottom sheet, the arc is extinguished and a weld is formed. A crater-filling device is often used to minimise the risk of cracking and also to reduce gradually the size of the weld pool, which helps to minimise the cratering effect. The addition of filler wire, should it be necessary to control weld-pool metallurgy or eliminate the crater effect, is possible though difficult.

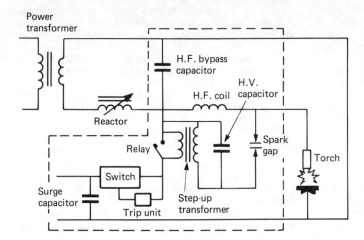

Figure 22.13 Surge injector unit and welding circuit

Pulsed TIG welding The use of a continuous current is, for certain specialised applications, a limitation. A recent development employing a pulsed welding current has increased still further the potential of the TIG system. In pulsed TIG welding a low current is used to maintain the arc and a pulse of current, generally of square-wave form, is superimposed. As the weld progresses, a series of overlapping spot welds is formed.

Programmed machine applications The TIG process is frequently used in fully automatic applications where the complete cycle of gas flow, arc striking, current variation and torch motion is preprogrammed. Various analogue and digital control strategies provide systems that allow the user to program the machine in as flexible a manner as possible. Such systems are employed for small-diameter tube butt welding using orbital welding heads, for tube-to-tube plate welding and for the welding of other small precision components.

22.1.6 Plasma-arc welding

Plasma or constricted-arc welding is essentially a development of TIG welding. There are two basic systems: the plasma-jet or non-transferred plasma, and the plasma-arc or transferred plasma. The essentials of each are shown in *Figure 22.14*. In the plasma-jet

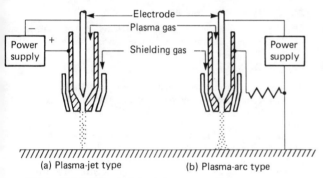

Figure 22.14 The principle of plasma-arc welding

mode the nozzle piece serves as the anode and the plasma jet is forced through the orifice by the pressure of the plasma gas; the effect is similar to a flame. The plasma-arc mode, which is the system most commonly used for welding, involves an arc that is struck between the tungsten electrode and the workpiece and which is constricted by passing through a relatively small orifice. It is usual for the plasma-arc circuit to include provision for a 'pilot' arc, which is a low-current arc struck between the electrode and the copper nozzle, to ionise the plasma gas and allow the main arc to be struck.

When welding is being done with the plasma arc, d.c. electrode-negative is usually employed and a 'keyholing' technique is used. This in effect means that the arc makes a hole through the metal to be joined, and this hole is carried along the joint, surface tension causing the molten metal to flow in behind the hole, forming the weld bead. The use of inert shielding gas, supplied from a nozzle concentric with the plasma orifice, protects the weld bead from atmospheric contamination.

To date, very little work has been carried out on plasma-arc welding at currents above about 400 A, mainly because of the problems associated with the burning out of the copper constricting nozzles. At this current level the process has been successfully used for making single-run square-edge close butt welds in material 12 mm thick.

22.1.6.1 Micro-plasma-arc welding

Micro-plasma-arc welding is a low-current version of the foregoing in which welding is carried out by direct heating rather than by 'keyholing'. The main advantage that this process offers over low-current TIG welding is that the arc column is much 'stiffer' and the process is more tolerant to slight variations in joint fit-up and arc length. Stable arcs at currents as low as 0.5 A that are capable of welding material 0.025 mm thick are claimed by manufacturers of the equipment.

22.1.7 Gas-shielded metal-arc welding

In gas-shielded metal-arc welding a wire is fed at constant speed into an arc formed between the tip of the wire and a plate. A shielding gas is supplied to protect the molten metal from the atmosphere. The current for a given wire diameter is directly proportional to the wire-feed speed. The arc voltage is nominally between 20 and 30 V, being the sum of the anode and cathode drops, and the drop across the arc column.

The power source is conventionally a unit with a flat voltage–current characteristics. This combined with a constant wire-feed speed gives rise to the self-adjusting arc system. If the arc length changes for any reason, the arc voltage will be altered. As the power source has a flat characteristic, the current alters in sympathy with the change in voltage. Any change in current will, as the wire-feed speed is constant, cause an imbalance in the burn-off characteristic which will lead to a readjustment to the correct working voltage.

22.1.7.1 Metal transfer

Two distinct modes of metal transfer across the arc are the low-current short-circuiting mode and the high-current spray or free-flight mode.

Short-circuiting mode In this mode (also termed 'dip transfer welding') the voltage is set low (about 20 V for steel–CO_2) and the resulting current levels are not sufficient to project molten droplets across the arc. Instead, the droplet on the end of the wire grows in size until it bridges the arc gap and an electrical short circuit occurs. The current delivered by the constant voltage rapidly rises; the droplet is pinched off into the weld pool, allowing the arc to re-establish. An adjustable iron-cored inductor in series with the arc is included to limit the rate of rise of current during the short circuit.

The low current levels possible and the small molten weld pools present with short-circuiting transfer make this an attractive technique for the welding of thin materials. Other applications include positional welding and the making of controlled penetration root runs. The short-circuiting technique is especially useful on mild, low-alloy and (to a limited extent) stainless steels; in addition it can be usefully employed on nickel and its alloys, aluminium and its alloys and some copper alloys.

Spray or free-flight mode By increasing the arc voltage of a short-circuiting arc, a type of metal transfer can be obtained whereby a large droplet of molten metal grows on the tip of the electrode until its own surface tension is unable to support it further and it is transferred across the arc by gravity. The frequency at which droplets are transferred can be as low as one per second. By increasing the current the droplet frequency increases, the size of the droplet decreases and metal transfer ceases to be dependent on gravity. The relationship between droplet frequency and current for various wire–gas combinations is shown in *Figure 22.15*. True spray transfer, i.e. metal transfer by arc force rather than gravity, is generally accepted to be established when the diameter of the transferred droplet is smaller than, or equal to, the diameter of the electrode wire.

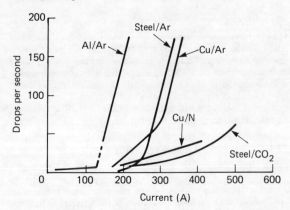

Figure 22.15 Frequency of drop detachment with 1.6 mm wire for various metal–gas systems

This type of metal transfer with its inherent large weld pool is generally only suitable for welding in the flat or horizontal/vertical positions. Exceptions are when the material being welded has a high thermal conductivity, in which case spray-transfer welding can be carried out in all positions.

22.1.7.2 Gases

Commercially available shielding gases for gas-shielded metal-arc welding and the materials on which they are commonly used are listed below:

Non-ferrous metals: Ar, or He, or Ar–He mixture; or N for copper.
Corrosion- and heat-resisting steels: Ar–O_2 (spray transfer only), or Ar–CO_2 mixture.
Mild and low-alloy steels: Ar–CO_2, or Ar–CO_2–O_2, or CO_2.

22.1.7.3 Welding wires

The requirements are similar to those for TIG welding, and one Standard (BS 2901) deals with both processes. The wire size depends on the current, mode of metal transfer, welding position etc. The current density must be high, and the wire diameters common in the UK are 0.6, 0.8, 0.9, 1.0, 1.2, 1.6 and 2.4 mm. A 3.2 mm wire is also available for certain applications such as the high-current welding of aluminium.

When welding with active gases such as CO_2 (which under the influence of the arc dissociates into carbon monoxide CO and atomic oxygen O), it is necessary for the wire to contain elements that combine with oxygen to prevent the formation of porosity. Silicon and manganese are most commonly used, but some wires contain small additions of aluminium, titanium or zirconium.

22.1.7.4 Pulse MIG welding

With the normal metal inert-gas (MIG) process, operating at a high current density of the order of 75 A/mm^2, the metal is transferred across the arc in the form of small droplets. The droplet size is a function of the current, and as the current is decreased the droplets become more coarse. As the force applied to these doplets is a function of the square of the welding current, the direction control of these droplets is dminished; and, if a high current is used for the welding of steel, the weld pool becomes large and fluid. Under these conditions it is not possible to maintain control of the weld pool when welding in any but the flat position. For other welding positions, use may be made of either the dip-transfer or the pulse-transfer technique.

In the pulse-transfer process, a continuous arc current is maintained as a 'background' and short high-current pulses are superimposed. The pulse duration is required to be of the order of 10 ms, with a repetition frequency of 50 or 100 pulses/s. With a 50 Hz supply the former can be produced by a half-wave rectifier and the latter by a full-wave or bridge rectifier. The background supply, from a constant-current 3-phase rectifier or a constant-voltage supply with a series inductor, maintains the arc between pulses and provides an arc voltage that inhibits pulse operation to 7–8 ms in each positive half-period. The conditions are such that, with an appropriate pulse current, one drop per pulse is detached from the electrode wire.

22.1.7.5 Welding with flux-cored wires

The flux-cored wire variant of the gas-shielded metal-arc process has advantages for products that require heavy structural welds that are not conveniently manipulated for fully automatic set-ups, e.g. crane booms and earth-moving machines. The heart of the process is a tubular wire enclosing a flux core. The flux fulfils many of the functions of an MMA coating or submerged-arc flux, and allows much larger and more deeply penetrating single-run welds in a semi-automatic operating mode. The process is fairly tolerant of variations in fit-up, and fills a gap between solid-wire gas metal-arc applications and the heavier automatic submerged-arc process.

22.1.8 Metal cutting and gouging

Metal can be removed from workpieces by use of coated-electrode, plasma-arc and carbon-arc processes.

22.1.8.1 Coated-electrode process

Flux-coated electrodes are available in two types. The first has a specially formulated coating that gives a deeply penetrating and forceful arc, tending to create a weld pool and to blow out the molten metal; a hole or recess is left, which can be carried along the material. When thick material is being cut, a sawing action is employed. The second type of electrode is a hollow electrode through which a high-pressure stream of oxygen is passed. The cutting action is akin to that with oxy-acetylene. Electrode techniques are, in the main, limited to ferrous materials.

22.1.8.2 Plasma-arc process

The plasma-arc process can give neat and accurate cutting. The principles are basically the same as in plasma-arc welding, but the force of the plasma column is increased by higher gas flow through the nozzle and a higher power output. Virtually any conducting material can be cut, at speeds faster than those of other methods: e.g. 8 m/min for aluminium 6 mm thick.

22.1.8.3 Carbon-arc process

In this process an arc between a carbon electrode and the workpiece is used to create a molten pool of metal which is blown away by a jet of compressed air. The method is useful for all grades of ferritic and austenitic steels.

22.2 Resistance welding

Resistance welding is the science of welding two or more metal parts together in a localised area by the application of heat and pressure, the combination ensuring practically molecular continuity at the points of contact. The heat is produced by the resistance set up to the passage of high values of current—predetermined as to density and time interval—through the metal parts held under a pre-set pressure. Copper or copper-alloy electrodes are used to apply pressure and convey the electrical current through the workpieces, these electrodes consisting of dies, wheels or clamps. In special cases copper–tungsten sintered materials are used for electrode facings as well as molybdenum and other special alloys. No extraneous materials such as rods, fluxes, inert gases, oxygen or acetylene are required.

The weld temperature attained depends on the current, the time of application and the resistance. The latter is a function of the electrode material, size and shape, of the pressure applied, and of the resistivity and surface condition of the material being welded. The various factors lead to local material plasticity and intermolecular amalgamation under pressure. The process is analogous to the blacksmith's weld or forge.

The mechanical system used to provide the forging action in resistance welding may be: (a) air or hydraulic cylinders; (b) motor and cam units where the output of a high-speed motor is converted into high torque at low speed by means of a reduction gearbox, and force is transmitted to the moving ram of the welding machine by a cam driven at this low speed (in general practice a clutch is used to engage and disengage the drive to the cam on the welding-ram stroke); (c) foot or hand levers operated through linkage systems; or (d) other means such as electromagnetic forge systems.

22.2.1 Scope

It is possible to weld practically an unlimited number of metals and combinations of metals by one or more of the resistance welding processes. Although certain difficulties are encountered in some combinations because of great differences in metallurgical resistance to intermolecular adhesion, welding temperatures, plastic temperature ranges etc., the field open in practical metal combinations for resistance welding is wide:

(a) Aluminium and its alloys (non-heat-treatable, strain-hardening materials and heat-treatable alloys involving the use of stored energy, 3-phase frequency conversion and secondary rectification machines).
(b) Copper and copper-based alloys.
(c) Low- and high-carbon steels, and low-alloy steels.
(d) Coated steels. Zinc-coated mild steel (where the coating is applied by hot-dip methods or by electroplating), tin plate, terne-plate, tin–terne-plate and cadmium plate can be satisfactorily welded although the surface condition is upset to a certain extent. The contact faces of the welding electrodes also become fouled with the coating material after a number of spots and must be frequently cleaned. Zinc–iron alloy coatings allow more spot welds to be made before it is necessary to remove build-up from the electrode contact faces. Anti-corrosive protective coatings on mild steel must be electrically conductive.
(e) Stainless and high-alloy steels.
(f) Nickel and nickel alloys.
(g) Lead alloys.
(h) Magnesium alloys.
(j) Dissimilar and refractory metals.
(k) Zinc-plated mild-steel sheet coated on one side with a plastic film. Special machines are required for projection welding brackets to the non-plastic side.

The advantages inherent in resistance welding include the following. Less essential skill is required in metal fabrication. Distortion is reduced because of localisation of heating. High production rates and greater uniformity are achievable. Semi-automatic or fully automatic techniques are available. Standard equipment does not require skilled operatives, although experience is necessary in the initial setting up; advanced special-purpose plant does require trained skills.

Resistance welding of tanks and leakproof vessels of light sheet metal can be carried out very rapidly and the welds can be relied upon for gas- and water-tightness. Welds can be made to be almost invisible, products strengthened and reduced in weight, and cheaper materials used, as a result of simpler fabrication. Economies have been made in the substitution of fabrication for castings where machining operations were formerly required. The paramount gain is, however, in time.

22.2.2 Welding equipment

Figure 22.16 shows the essentials of a spot-welder circuit. The three basic units are for heating (transformer), for timing and for applying the necessary electrode force. A normal mains supply is

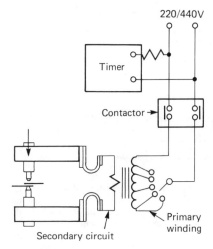

Figure 22.16 Schematic circuit of a spot-welding machine

employed, the transformer having an l.v. tapped secondary winding and electromagnetic or electronic contactor control. The current is conveyed to the work through two copper-alloy electrodes, the lower usually fixed and the upper movable vertically.

22.2.3 Spot welding

Most modern spot-welding machines are air-operated, although machines with mechanical leverage and automatic timing are satisfactory for light-gauge materials when high production rates are not called for. Air-operated machines can produce spot welds at up to 200 per min with consistency and with minimal manual effort on the part of the operator.

The cycle of operations involves four definite stages of time:

(i) *Squeeze time:* the time between the first application of the electrode force and the first application of the welding current.
(ii) *Weld time:* the time during which current flows.
(iii) *Hold time:* the time during which the force remains applied after the current has been switched off.
(iv) *Off time:* the period during which the electrodes are not in contact with the work.

Many machines are arranged for automatic repeat, or automatic work (as for stitch welding); the off time is therefore the time between release and reinitiation of the weld cycle. The four timing periods are programmable within relatively wide limits when modern control methods are used. In addition to the usual single impulse of current, pulsation welding can be employed, a procedure that may be valuable for metallurgical reasons.

22.2.3.1 Timing

In the very simple spot-welding machine where the pressure is obtained by a mechanical leverage system, and where the current is switched on by the contacts connected to the mechanical operating part of the machine, the time of welding current flow is left to the judgment of the operator. *Figure 22.17* illustrates this simple spot welder. Automatic timing, however, is a great

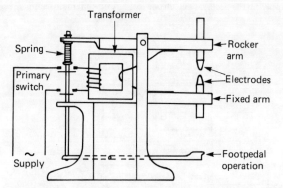

Figure 22.17 Pedal-operated spot welder

advantage, especially with the larger types of machines which are capable of making a spot weld within a fraction of a second; electrically operated contactors have been incorporated in the automatic timing system in this type of machine. In machines using automatic timing equipment, the main current contacts are replaced by an initiating switch which closes the electrically operated contactor coil. With this system it is only necessary to maintain pressure on the foot pedal until the timer, which has been previously set for the prescribed welding period, automatically opens the contactor coil, thus switching off the supply to the primary winding of the welder transformer. This is a great improvement over the operator-controlled timing system, but it is still possible for the operator to release the foot pedal before the pre-set timing period has been completed, and thus inconsistent weld times can still be obtained. Some foot-operated spot-welding machines also incorporate an air-cylinder which comes into use as soon as the electrodes are in contact. This enables the timing unit to be interlocked with the valve controlling the pressure to the air-pressure cylinder, which makes it impossible for the operator to control the length of the welding-current flow time.

Air operation is fast, consistent and extremely versatile in providing variation in stroke, pressure, and 'weld' and 'hold' time. Single-acting cylinders where air pressure is used only for the down stroke and spring pressure for the return stroke, or double-acting air cylinders with power stroke in each direction, are selected for particular applications. In either case, the air-cylinder operation is controlled, usually electrically by a valve, or it is occasionally solenoid operated.

22.2.3.2 Time and sequence control

Weld timers may have a *non-synchronous* or a *synchronous* form. A non-synchronous timer starts and stops the current flow by contactor at instants unrelated to the a.c. wave-form; as a result the current may vary, and the time of repetition of a pre-set sequence be subject to an inaccuracy of plus or minus one half-period of the a.c. supply. A synchronous timer is one in which the operation is directly related to the a.c. wave.

Non-synchronous timers The following are fitted to spot-welding machines, the choice depending on the machine complexity:

(1) Electronic weld or sequence timer comprising four stages: 'squeeze', 'weld', 'hold' and 'off'.
(2) Resistance–capacitance timers as single units or as sequence controls. In this category there are two systems utilised: those which measure the actual timing period necessary to make a particular weld and the alternative system which measures the current–time product or energy put into a weld.
(3) Electromagnetic timers which make use of the interval required to discharge energy stored in the field of a specially designed magnet. A d.c. charging source supplies a current for the system, and the energy stored in the field of the magnet coil is discharged through a parallel resistance when the charging circuit is interrupted.
(4) Electrostatic timers which are based upon the constant time interval required to either charge or discharge a capacitor, assuming a fixed charging current and a specified value of resistance in the charge or discharge path.

Synchronous timers A synchronous timer provides an accurate timing period in that it closes and opens the primary of the welding transformer at the same point of the supply voltage in the making of each weld. It reduces or eliminates electrical transients. Where quality is important, and particularly for such materials as aluminium and its alloys, synchronous timers are usually regarded as essential. Modern 'digital solid-state' controls provide precise and repeatable control of the welding operation.

22.2.3.3 Electrode size

The tip diameter (contact face) controls the size of the spot. If the diameter is too small, the resultant spot weld will be too weak, although it is quite possible that it may be a sound weld metallurgically. Furthermore, a small variation in tip diameter under these conditions may result in a large variation of spot strength.

Conversely, if the tip diameter is too large, the extreme centre of the weld nugget may become too hot; this will result in poor appearance and the weld may be understrength owing to cavitation and blow holes resulting from insufficient pressure. The tip diameter bears a definite ratio to sheet thickness, and this ratio is greater on thin sheets than on thick. The electrode dimensions should comply with the appropriate British Standard. In the welding of sheets of single thickness up to $t=3$ mm, the initial electrode tip diameter D (in mm) should approximate to

$$D = 5\sqrt{t}$$

A greater tip diameter is permissible only if tests prove that the weld strength is not decreased below the desired requirements.

22.2.3.4 Heat balance

Heat balance may be defined as a condition in which the fusion zones of both pieces of material to be welded undergo approximately the same degree of heating and pressure application. This heat balance may be affected by:

(a) Electrode contour.
(b) The shape of the parts to be welded.
(c) The thermal and electrical conductivities of the electrodes.
(d) The relative thermal and electrical conductivities of the materials to be welded.

For example, if it were required to weld an alloy having a high copper content (and thus possessing high thermal and electrical conductivity) to a material of low thermal and electrical conductivity such as 18/8 stainless steel, it would be advisable to use an electrode of much smaller contacting area against the high-conductivity alloy and an electrode of much greater contacting area against the low-conductivity alloy. Occasional use is made of electrodes which are faced with materials such as molybdenum in order to shift the weld-nugget position to a greater depth within the section of the sheet against which the molybdenum-tipped electrode is in contact.

22.2.3.5 Surface indentation marking

It is often desirable to produce spot welds without leaving pronounced electrode indentation marks. Indentation is aggravated by excessive heat and force, and by steeply faced electrodes; it can therefore be mitigated by proper machine control, by careful selection and maintenance of welding electrodes, and by adoption of 'series' or 'indirect' methods of spot welding. Series welding is a method wherein a part of the secondary current bypasses any weld nugget being formed. Usually two welds are made per transformer secondary.

22.2.3.6 Twin-spot welding

When component parts are to be welded to large sheets, considerable variable losses are set up owing to the sheet moving in the inductive area between the machine arms. It is often possible to utilise a twin-spot method of operation where both electrodes are on one side of the material, the current being taken through a copper backing bar or plate below the bottom sheet (*Figure 22.18a*). In this manner the electrodes and secondary connections are located only a few inches apart, thus reducing inductive losses to a minimum and permitting the components to be welded without the necessity of compensating for variable inductive losses.

For heavier materials the 'push–pull' system, with two pairs of electrodes, is used occasionally to permit heating to take place independently in both sheets. Twin-spot series welding can be used for welding mild-steel sheets up to $18+18$ s.w.g. ($1.3+1.3$ mm), but 'push–pull' welding (*Figure 22,18b*) can handle up to $16+16$ s.w.g. ($1.6+1.6$ mm) with greater consistency and spot quality.

22.2.3.7 Portable spot welders

The machines so far described are of the stationary type, but portable gun welders are extensively used where the work to be welded is bulky and where it would not be feasible to carry out the necessary welding on fixed or pedestal-type machines.

Portable gun units are made in a variety of sizes, styles and types. The use of portable welding guns is not confined to very-high-production methods. They have a definite place in moderate- and low-production set-ups, and these guns can be designed and built to accommodate a wide variety of similar or dissimilar work.

Portable welding guns fall into three general classes: pinch guns referred to as scissors and C units, expansion guns, or lever-type guns. Briefly, the portable pinch-type gun can be considered as the most popular type of portable tool and can be used either with or without a fixture. Both the expansion gun and the bar-lever gun, however, require a suitable fixture for their use. Pinch welding guns can be mechanically, pneumatically or hydraulically operated.

One type of portable equipment consists of welding transformer and control gear suspended above the work station with the output from the transformer connected to the portable gun by means of water-cooled twin cables, or a single cable of the 'kickless' type. In some cases the control panel is mounted on an adjacent wall or stanchion and is connected to the movable welding transformer by means of appropriate cable. To keep the guns manoueuvrable the transformer secondary connections are of comparatively small cross-section, and reliance is placed on water cooling. Such machines are of comparatively low electrical efficiency and may have to be rated at several times the power of a comparable fixed machine.

As an alternative to the remote portable welding transformer system, welding guns with built-in transformers can considerably reduce the mains current demand and have a much higher power factor. For spot-welding light work up to 16 s.w.g., simple manual guns with built-in transformers can be used. For this work, water cooling of the electrodes can be omitted, so that the movable part of the equipment is kept light. For heavy-duty work, where continuous operation is likely to be required, and for spot-welding heavier gauges, it is essential to have the electrodes power operated and correctly water cooled.

22.2.3.8 Multi-point welding

The name 'multi-point' or 'multi-spot' is derived from the practice of making a number of spot welds simultaneously. Machines that can do this have many great advantages when large-quantity production runs are undertaken. This type of spot-welding plant is usually a single-purpose, specially designed and constructed machine, and may have from two to as many as 100 or more spot-welding electrodes arranged in the proper positions to make the required number of welds. The pressure

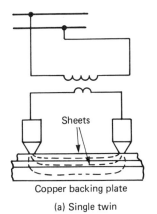
(a) Single twin

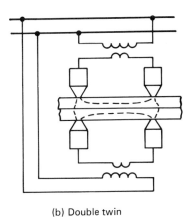
(b) Double twin

Figure 22.18 Twin-spot welding: (a) single twin; (b) push–pull

cylinders, which can be air or hydraulically operated, can be arranged in a single line, in a circle, or in a cluster. Some cylinders may be horizontal, others vertical, while yet others may be mounted at an angle to the vertical. The majority of modern multi-welding machines incorporate a built-in oil-draulic pump unit which supplies high-pressure oil to the cylinder clusters. A number of package-type welding transformers are fitted within the machine frame, in positions adjacent to the points at which welding will be carried out. Each transformer incorporates one, two or more secondaries, permitting two, four or more series welds to be made per transformer. If the machine is designed especially to produce a very large number of welds, then the circuitry is designed to permit all the welding guns to be energised simultaneously so that the work is clamped in many positions, and then groups of transformers are fired in automatic sequence, controlled by separate ignitron contactor panels. Such machines normally operate entirely automatically after the component parts to be welded have been placed in the welding fixture. Exact location of the parts is possible and, in consequence, each assembly is accurately welded. This is a most important factor in the assembly of complex fabricated articles such as motor-car bodies.

As a general rule, the series welding system is adopted for sheet metal up to and including 18+18 s.w.g. In some cases, however, multi-spot welding plant has been designed for welding complex components considerably in excess of 18 gauge and the 'direct' system of welding has been adopted.

In certain cases, large hydraulic presses of the 'up-stroke' type have been converted into multi-spot welding machines by the addition of banks of transformers and air or hydraulic cylinder assemblies, the 'up-stroke' platen being modified to accommodate the copper-alloy component-locating die.

Occasionally these converted 'up-stroke' presses have been used in conjunction with completely interchangeable tooling, which enables the manufacturer to change the machine from, for example, a batch production run of motor-car bonnet lids to offside front doors, with only a tooling change-over time loss of approximately three hours.

Where extremely high production rates are necessary these automatic multi-spot welding machines now form part of an entirely automatic transfer system, so that all component parts are located on a special transfer jig which then moves automatically through a number of special multi-head welders, each machine carrying out its own sequence of welds.

22.2.3.9 Spot welding of aluminium and its alloys

Aluminium and aluminium alloys are extensively welded by the spot process and three types of machines are in use for this work:

(1) Single-phase a.c. machines in ratings up to 600 kVA.
(2) Stored-energy machines, inductive or capacitive.
(3) Three-phase frequency-conversion balanced-load equipment; or machines incorporating a secondary rectifier to provide a d.c. welding current and used, for example, for the welding of heat-treatable light alloys such as duralumin or alclad in aircraft manufacture.

A.C. welding Higher-powered machines are usually controlled electronically, the primary current being initiated and interrupted through an ignitron contactor circuit. Such control panels are usually fully synchronous and incorporate a phase-shifting system for controlling the welding-current amplitude level. For the welding of the commercial light alloys, non-synchronous timing is often used in conjunction with an ignitron contactor for primary current control.

On modern machines, the welding pressure is obtained by means of a pneumatically operated cylinder where the ingoing air pressure is under the control of a reducing valve and gauge assembly, thus making the adjustment to the input pressure extremely easy.

Table 22.2 gives approximate settings for the welding of aluminium alloys on a.c. welding machines. The large currents impose a heavy electrical load on the supply; the load characteristic is single-phase, intermittent and of low power factor, and may adversely affect local voltage regulation.

Energy-storage system In the inductive storage method a transformer core is magnetised by a direct current in the primary, and when it attains a pre-set value it is suddenly interrupted. As a result an impulse with exponential decay occurs in the closed secondary circuit, the energy in the magnetic field being transferred by the secondary current to the weld. This type of machine also incorporates a variable pressure cycle, which is automatically controlled and which permits high initial welding pressure to be applied to the electrodes, followed by a pressure reduction to a much lower value; during this period the weld pulse takes place, and this again is followed automatically by a rapid increase in welding pressure for forging purposes. Machines of this type require a 3-phase rectified supply of about 50 kW, which is balanced and of magnitude considerably less than that required for a.c. welding.

The capacitive storage equipment uses capacitors charged to 1–3 kW d.c. from 3-phase rectifiers. The stored energy is discharged into the primary of a transformer, and again an impulse welding current is induced into the secondary.

Frequency-conversion system This system uses ignitron or thyristor frequency-reduction equipment, so that the welding power factor is raised and the load reduced. Again the demand on the supply is substantially lowered compared with a.c. welding.

Secondary-rectification system A 3-phase transformer has a 3-phase d.c. rectifier in the secondary side. The rectifier bank is quite small and robust in spite of the heavy current demand. Such equipments are suitable for seam, spot and multi-spot welding applications.

22.2.3.10 Stitch welding

Stitch welding consists of a series of overlapped spot welds which appear similar to the line of welds obtained using resistance seam-welding techniques. The main difference is that in the latter case the current is conveyed through the workpieces, using electrode wheels or rolls with the pressure applied continuously

Table 22.2 Setting for a.c. welding of aluminium-alloy sheet

Gauge (s.w.g.)	26	24	22	20	18	16	14	12	10
Time (in 50 Hz periods)	4	6	6	8	8	10	12	12	15
Electrode tip force (kN)	0.9	1.3	1.3	1.8	2.0	2.2	2.7	3.5	3.5
(lbf)	200	300	300	400	450	500	600	800	800
Welding current (kA)	14	16	17	18	21	23	28	32	35

and the current switched on and off at predetermined intervals. In the case of stitch welding, an automatic spot-welding machine arranged for auto-continuous operation at high speeds is utilised. The operator moves the component parts between the electrodes which are reciprocating automatically, and his job is to feed the parts through the electrodes at the correct speed so that spot welds are made consecutively and overlapping. Stitch welding does not supersede seam welding or arc and gas welding, but there are many applications where it has undoubted advantages. Compared with gas or arc welding, stitch welding is quicker and very much cheaper, and the resultant weld does not in the majority of cases need a grinding operation to prepare it for subsequent vitreous enamelling or painting.

An advantage of stitch welding is that it can be applied to workpieces of awkward shape and carried out up to blind corners inaccessible to seam-welding plant. Water- and gas-tight joints can be produced. A typical application is the manufacture of cooker ovens and refrigerator interior cabinets.

If the component parts to be stitch-welded are overlapped by about 10 mm, the method is called 'lap-stitch' welding; if the overlap is about the same as the thickness of the sheet, the method is 'mash-stitch' welding, and the effect is to squeeze the two sheets to a flush joint. Steel sheets can be joined edgewise by running a wire of diameter a little greater than the gap width between the edges, and mash-welding it flat. In this case the lower electrode is a flat table or support.

Resistance welds can be examined using destructive and non-destructive procedures, but (in contrast with, for example, electric arc welding) there are difficulties, particularly with the latter. A sampling procedure, whereby a number of components are tested to destruction, does not provide complete assurance that all welds meet the standard required, for welding conditions can vary from one to another. Fit-up, for instance, can vary through tool wear. Increasingly, therefore, 'in-process' monitoring instruments are being incorporated with the welding-plant controls.

22.2.4 Projection welding

Projection welding is a development of spot welding, with this difference: in spot welding the size and position of the weld are determined by the size of the electrode contacting faces and where they are applied to the workpieces, whereas in projection welding the size and position of the weld or welds are determined in the design of the component to be welded owing to the fact that projections or embossments are raised on the workpieces to form the welded zone. Some of the advantages of projection welding are:

(a) Output is increased for each stroke of the machine, since a number of welds can be made simultaneously.
(b) Electrode life is extended because electrodes with large contacting surfaces may be used, pressure and current being localised by the embossments. Furthermore, it is possible to use special faced electrodes comprising sintered copper–tungsten inserts.
(c) In some cases, components may be projection welded which could not be welded by other resistance-welding techniques.

The principal uses of projection welding are those in which stamped, formed or punched parts are assembled, the projections being formed during the pressing operation. As a general rule, the projections are circular in shape, but they may be of almost any regular design. The circular shape, however, has a number of advantages both in welding and in maintaining the punches and dies used.

When the welding current is forced to pass from one workpiece to the other, a highly localised heat is created at each projection. This, in turn, causes the metal at the projection to rise quickly to a plastic temperature. Under the action of the electrode force the

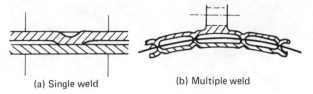

(a) Single weld (b) Multiple weld

Figure 22.19 Projection welds

component parts collapse at the projection positions so that both sheets come into intimate contact.

Figure 22.19(a) shows the conditions that appertain in the making of a single projection weld. Here a projection has been raised on one of the two parts to be welded together and large electrodes, top and bottom, apply the pressure and current, the heat being located by the size and shape of the projection. The number of projection welds that can be made simultaneously is limited only by the gauge of the material and the size of the projection welder available.

Projection welding is an extremely reliable process and involves low maintenance costs for electrodes. However satisfactory a modern spot-welding machine may be, care must always be taken to maintain the tips at a reasonable diameter; otherwise inferior welding will result. Having overcome in projection welding the difficulty of the varying sizes of electrode tips in use, close control has been achieved of the essentials: (1) electrical output; (2) time of current flow; (3) mechanical pressure; (4) area to be welded. As items (1)–(3) are separately adjustable on modern projection-welding machines, and as the area to be welded is predetermined by the projection size, weld consistency can be guaranteed with great accuracy, providing that the material specification remains reasonably constant.

Figure 22.19(b) shows a case in which three thicknesses of metal are to be welded. The projections can be raised on the two outer members as indicated, or they can be formed on both sides of the middle member. Projection welding is also applicable to the use of screw machine parts, the projection being turned on the screw by means of a forming tool. The welding of studs, bolts etc. on to steel plates is a very common application; the bolt ends are turned to produce a conical (internal or external) form.

22.2.4.1 Projection-welding machines

Projection-welding machines operate on principles similar to those of spot-welding plant but are designed for much higher electrode forces because it is usually required to weld several projections simultaneously. The machines, which are rigidly constructed, are pneumatically (or in some cases hydraulically) operated. Very large projection-welding machines are available which can apply a total electrode force of 130 kN and deliver a welding current of 200 kA. As with spot welders, the smaller projection-welding machines can be built with indexing tables for high-speed production.

22.2.5 Seam welding

Seam welding consists fundamentally of making a number of spot welds by means of rotating copper-alloy wheel electrodes without causing the electrodes to open between spots. The electrode wheel applies a constant force to the workpieces and rotates under controlled speeds. Instead of individual welding tips opening and closing at high speeds, as on spot/stitch welding machines, the primary current in the seam-welder transformer is interrupted by an electronic control panel on modern machines in order to produce consistent intermittent secondary current pulses through the workpieces. If the 'off' period of current flow is considerably longer than the 'on' period, then spot welds can be

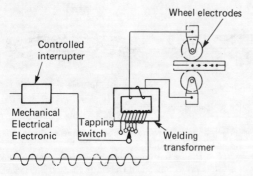

Figure 22.20 The principle of seam welding

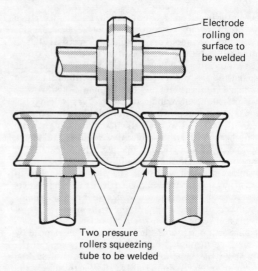

Figure 22.21 Seam welding of small-diameter tubes

produced with unwelded spaces between the spots. On the other hand, if the 'off' period is kept very short, then welds will be made overlapping one another, thus producing a continuous weld line which under the correct conditions of pressure, current and time will result in a pressure-tight seam weld.

The general principles of seam welding are illustrated in *Figure 22.20*. Seam welds cannot be made satisfactorily when uninterrupted welding current is used except where very thin gauges and high speeds are adopted. This is due to the fact that with the uninterrupted current flow system great heat is developed in the component parts to be welded, and this heat travels ahead of the wheels and causes the weld to be unstable. Interruption allows heat to be conducted away by the electrodes.

Resistance seam welding is used mainly for joining sheets of thickness up to 3 mm. Owing to the current shunt loss conditions appertaining when overlapped welds are made on a seam-welding machine, it follows that slightly more current is required under these conditions than would be the case if spaced spot welds were being produced. The selection of the correct type of current interrupter is very important, and on modern seam-welding plant the use of fully synchronous electronic interrupters is now generally accepted to be the most satisfactory method for controlling exactly the period of current flow and 'off' periods between welds. Such control panels generally incorporate a welding-current heat-control system functioning on the phase-shift principle which permits, for all practical purposes, a stepless method of adjustment of welding-current amplitude.

The type of joint normally used in seam welding is the lap joint, and the overlap of the sheets should be at least equal to twice the width of the electrode wheel tread so that the expulsion of metal from the welding joint is avoided. A further method sometimes adopted is 'mash' welding, which is used when flush joints are required. Flat-track electrode wheels are used, and the technique is satisfactory provided that the small overlap is consistently maintained.

In foil butt-seam welding the two sheets to be joined are butted together at their edges, and a narrow strip of foil is run over the butt and mashed flush. With flat steel sheets, the material is held down on a magnetic table. Another system of butt-seam welding has been used for the manufacture of tubes, formed from flat strip by a series of rollers. Welding current is applied at the final closure position, and the welds are subsequently forged by being passed through follower rollers. The welded tube then passes through sizing and straightening rollers and finally through a die. The essentials are shown in *Figure 22.21*.

22.2.5.1 Welding variables

Welding pressure In common with most modern spot-welding plant, seam-welding machines derive their electrode force mainly from pneumatic pressure. This applied force is controllable; its magnitude is related to the thickness of material being welded and, as a general rule, the pressure will be similar or slightly higher than in the case of spot welding.

Welding current The actual current setting required for a given application will depend to a great extent upon the speed of welding selected. It will also bear some relation to the number of pulses of welding current per second. If the component parts are externally spray water-cooled to keep down the spread of heat into the adjacent metal, this will also have a bearing on the magnitude of welding current required. As mentioned previously, shunt loss conditions have also to be compensated for; the magnitude of the welding current in a given case must therefore be determined by trial and error.

Welding speed This may lie between 15 cm/min and 10 m/min. For pressure-tight seams, for which the spot weld must overlap, about 2–3 spots/cm are necessary for 3 mm sheet, and 6 spots/cm for sheet of 22–24 s.w.g.

Roller track width The track width of the electrode wheel very largely determines the resultant width of the weld on the assumption that the conditions of pressure and current have been set accurately. For all general purposes, the track width of the electrode wheel can be the same as the diameter of a spot-welding electrode used for given sheet thicknesses, i.e. approximately equal to the square root of the sheet thickness.

22.2.5.2 Electrode-wheel drive system

Seam-welder electrode wheels are driven by a motor and gear-reducing unit which imparts rotary motion to the electrode wheel in one of three forms: knurl drive, direct gear-shaft drive and friction drive.

The knurl drive system is frequently used when coated metals have to be welded because the knurling on the wheel periphery breaks up the coating on the component surface and minimises 'pick-up' on the welding wheel. This drive system has a great advantage over the gear drive if the appearance of the weld is not of vital importance. The knurl drive is the most positive of the three methods because slipping between the electrode wheels and the parts being welded is avoided. Furthermore, the peripheral speed of the electrode wheel is maintained consistently, irrespective of the diameter of the wheel. Where it is possible to use a knurl drive on both top and bottom electrode wheels, the same peripheral speed on each roller is assured. The knurled wheel automatically maintains the correct contour on the track of the roller and this reduces electrode-wheel maintenance.

Occasionally, a special gear-drive seam welded is equipped with a differential-drive mechanism so that the upper and lower wheels may be individually driven at slightly different speeds to compensate for uneven wheel wear.

22.2.5.3 Mandrel welder

A further type of seam-welding machine is referred to as the mandrel welder; this is used particularly where the welding of small cylinders is involved and where accesss for the welding wheel inside the bore of the cylinder is not permissible. In this case, the component part is loaded on to a mandrel, which can be located on a moving assembly and is driven beneath the idling top electrode wheel. Alternatively, the mandrel can be made stationary and the top electrode wheel assembly can be designed to travel along the length of the stationary component. Many advanced types of modern seam-welding machines are fitted with a means to control the drive to the top or bottom electrode wheel so that the wheel can be stopped and started under predetermined, pre-set conditions. One system uses an interrupted-drive motor which can be stopped and started electronically. Any pre-set pattern of on/off movements can be very accurately controlled and, furthermore, the interruption can be switched out of circuit so that the motor operates as a standard unit providing continuous rotation for the electrode wheels for certain classes of work. For the welding of heat-treatable light alloys and certain nickel-base alloys, the interrupted-drive system is essential. Light alloys, for example, are welded under conditions of variable pressure, and the weld nugget must be forged at the end of the welding-current pulse. It would not be possible to impose a variable pressure cycle system on electrode wheels which were continuously rotating. Such control techniques are now incorporated in straight a.c. seam-welding machines as well as in the 3-phase frequency-conversion equipments used so successfully in the aircraft industry.

22.2.6 Butt and flash welding

Plain butt welding is the simplest form of resistance welding. The opposing faces must be flat and parallel when clamped so that uniform heating at the butt is obtained. As the material heats under the action of current flow and surface pressure, slight collapse and localised swelling take place. Cavities may exist, and the method has never been entirely successful for joining thin sheets. It has to a great extent been superseded by the flash-butt method. *Flash-butt welds* are made on a machine incorporating one stationary platen and one moving platen on which are mounted the flash-welding dies or clamps which securely hold the parts to be welded while serving to conduct the welding current through the components. In general appearance the flash-butt welder is a 'table-top' machine having a movable platen; the dies and clamps slide in very rigid and accurate guides while the transformer is usually mounted in the base of the machine below the flash-welding area. In some large machines, arrangements are made in the design for both platens to be movable, one platen movement controlling the flashing distance and the other being accelerated for the upset force.

The component parts to be flash welded are placed in the dies and clamps, the latter exerting high pressure to avoid slipping and to ensure transfer of maximum welding current through the parts. The components are spaced at a predetermined distance between the welding faces. After being clamped securely in the dies, the two parts are then brought together, the main welding transformer having been energised prior to the components coming into contact. As soon as the parts touch, 'flashing' is established and an intense heat is generated between the two parts.

As flashing continues, the ends of the two pieces reach a very high temperature, until finally they are at a fusion condition and immediately the adjoining metal is at the appropriate plastic temperature. While metal is being expelled by flashing—referred to as 'burn-off'—the clamps are moved towards each other at a carefully controlled and accelerated rate. As the material temperature reaches the molten condition, the pressure on the moving clamps is quickly and greatly increased; this is referred to as 'upset pressure'. This 'upset' action causes the slag and excess

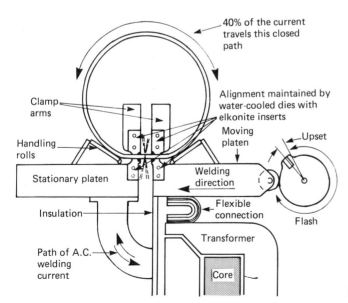

Figure 22.22 Flash-butt welding of steel barrels

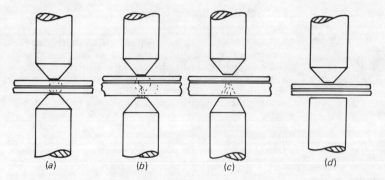

Figure 22.23 Spot-welding electrodes

molten metal to be expelled, while the increased pressure forces the parts together in perfect fusion. After the welding has occurred, the clamps are removed and the parts are taken out of the machine, cleaned and finally machined, leaving no visible evidence of a welding joint.

Figure 22.22 shows a machine for flash-butt welding steel barrels. The cam produces an even movement for the burning-off process and the necessary sudden rise in pressure for the upset.

22.2.7 Resistance-welding electrodes

Resistance-welding electrodes play an important part in the functioning of welding machines, and a large variety of copper alloys have been produced, all with the aim of providing high electrical and thermal conductivity, together with mechanical strength, to withstand pressure and a certain amount of hammering, which is bound to take place in a mechanically operated spot-welding machine.

In the welding of aluminium and light alloys, because of the high conductivity of the material to be welded, it is usual to employ hard-drawn electrolytic copper as this gives the maximum electrical conductivity for the electrodes. In welding ferrous metals, however, it is found that the conductivity properties can be sacrificed to some extent to obtain better mechanical strength.

Spot-welding electrodes are normally manufactured from round copper bar, the shape and taper conforming to BS 4215. Such electrode tips are machined for water cooling, and the water channel should be brought as near as practicable to the face of the electrode. A flow of at least 7 l/min is recommended.

For spot welding two sheets of the same thickness, electrodes with tips of equal diameter should be used (*Figure 22.23a*). For welding sheets of different thicknesses it may be necessary to use electrode tips of different diameters to ensure that the maximum current density is produced at the joint and not midway between the electrodes (*Figure 22.23b* and *c*).

Figure 22.23(b) shows how the maximum current density would be produced in the thicker sheet if tips of equal diameter were used. Usually the tip diameter is made approximately equal to the square root of the sheet thickness. For work where a high finish is required on one sheet, such as in the welding of steel cabinets, it is common practice to use one large-diameter electrode on the exposed surface, thus preventing any indentations from being visible (see *Figure 22.23d*).

Beryllium–copper alloy electrode materials are available for use on nickel–chrome alloys, such as are adopted in the jet-engine industry, and similar materials are used for seam-welding wheel manufacture.

For the spot welding of copper and copper alloys, use is often made of molybdenum-tipped electrodes which enable comparatively low-powered spot-welding machines to weld the high-conductivity copper alloys.

For many projection-welding applications, copper-alloy electrodes are inserted with various grades of sintered copper–tungsten pads which will withstand the high electrode forces imposed upon them, thus reducing the frequency with which the dies have to be remachined. Sintered materials in the form of inserts have to be very efficiently water cooled, and great care has to be taken with this when designing tooling for projection welding.

23 Electrochemical Technology

D Inman,
Imperial College
M Barak,
MSc, DPhil, FRSC, CChem, CEng, FIEE, FRSA
Consulting chemist and engineer

Contents

23.1 Introduction 23/3

23.2 Cells and batteries 23/3
    23.2.1 Redox process 23/3

23.3 Primary cells 23/3
    23.3.1 Leclanché cell 23/4
    23.3.2 Standard cells 23/5
    23.3.3 Alkaline cells 23/5
    23.3.4 Water activated cells 23/6
    23.3.5 Acid cells 23/6

23.4 Secondary cells and batteries 23/7
    23.4.1 Lead/acid cells 23/7
    23.4.2 Nickel/cadmium and nickel/iron alkaline cells 23/8
    23.4.3 Silver/zinc alkali cells 23/10
    23.4.4 Secondary battery technology 23/10
    23.4.5 Lithium cells 23/11
    23.4.6 Sodium/sulphur cells 23/11

23.5 Anodising 23/11
    23.5.1 Process 23/11
    23.5.2 Vats 23/12
    23.5.3 Workpieces 23/12

23.6 Electrodeposition 23/12
    23.6.1 Electroplating 23/12

23.7 Hydrogen and oxygen electrolysis 23/13
    23.7.1 Process 23/13
    23.7.2 Electrolysers 23/14
    23.7.3 Gas purity 23/14
    23.7.4 Plant arrangement 23/14

23.1 Introduction

Electrochemical science is an interdisciplinary subject, (a) unique in its concepts (the chemical interactions between matter and electrons, particularly across interfaces) and (b) general in its applications. Electrochemical reactors use electrical energy to isolate elements from their compounds, as in the electrowinning of metals copper, silver, aluminium or chlorine, or the production of hydrogen and/or oxygen by the electrolysis of water. On the other hand, electrochemical power sources, batteries and fuel cells store chemical energy, which can be instantly converted into d.c. electrical energy by the turn of a switch. Corrosion, an electrochemical process, wastes both energy and materials. Some of the positive applications are discussed below.

Because of the need to conserve conventional sources of energy—coal, oil, gas, nuclear—much attention is being given nowadays to the development of what are generally referred to as 'renewable' sources of energy—solar, wind, wave and tidal, geothermal—but while these may supply relatively low-level local needs, they are unlikely to replace conventional forms in high-level supplies of electrical power. It is worth noting, however, that electrical energy is best stored electrochemically and used directly in electrochemical reactors. Thus, all the branches of electrochemical technology discussed below are likely to be developed much further.

23.2 Cells and batteries

The term 'battery' means an assembly of voltaic primary or secondary cells. Batteries of secondary cells are also known as storage batteries or accumulators. In both types the individual cells consist of a positive and a negative electrode, immersed in an ionically conducting fluid, called the electrolyte, and generally separated by a porous non-conducting diaphragm, called the separator. The electrodes, which must be electrically conducting, may consist of a single rod or plate, or a number of these welded or bolted together in parallel. In some cells (for example, the conventional primary 'dry' cell) the outer metal container may constitute one of the electrodes. The two electrodes have different electrical potentials when immersed in the common electrolyte and the difference between these potentials represents the e.m.f., or open-circuit voltage, of the cell.

In both primary and secondary cells the electrical energy released during discharge is derived from the chemical energy liberated as a result of the chemical reactions taking place in the cell. These reactions involve charged particles in the electrolyte, known as 'ions'. Ions, if positively charged, have a deficiency of electrons, and if negatively charged, carry an excess of electrons. As indicated below, certain ions tend to react with the electrode in their vicinity, causing a transfer of electrons from ions to electrode, or vice versa. If this reaction is allowed to proceed (for example, by closing the external circuit to which the battery is connected), the transfer of electrons from one electrode to the other gives rise to an electric current flowing in the external circuit conventionally from the positive to the negative electrode: the flow of electrons is in the opposite direction. Thus, an *anodic* reaction involving the release of electrons occurs at the *negative* electrode, and a *cathodic* reaction involving the capture of electrons at the *positive* terminal of the battery. In electrochemical reactors the reverse is true, i.e. *anodic* reaction at the *positive* and *cathodic* at the *negative* electrode.

23.2.1 Redox process

The chemical reactions at the electrodes are either 'reduction' or 'oxidation', i.e. 'redox' processes. The basic feature of such reactions is the gain or loss, respectively, of one or more electrons, e.g.

reduction/cathodic reaction $\quad 2H^+ + 2e^- \rightarrow H_2$
$\quad\quad\quad\quad\quad\quad\quad\quad\quad\quad\quad Cu^{2+} + 2e^- \rightarrow Cu$
oxidation/anodic reaction $\quad Zn \rightarrow Zn^{2+} + 2e^-$
$\quad\quad\quad\quad\quad\quad\quad\quad\quad\quad\quad Pb \rightarrow Pb^{2+} + 2e^-$

Here e represents the electron. H, Cu, Zn and Pb represent atoms, or (when charged) ions of hydrogen, copper, zinc and lead. The sign + indicates a deficiency of one electron; 2+ indicates a deficiency of two electrons.

The gaseous hydrogen formed by the first reduction process, if allowed to accumulate, would rapidly polarise the electrode, and the electrochemical reaction would virtually cease. To overcome this, the positive electrodes of many kinds of batteries are selected from substances which readily undergo a depolarising reaction with hydrogen. Typical examples are manganese dioxide, MnO_2, used in the primary dry battery, and lead dioxide, PbO_2, used in the lead/acid storage battery. Negative electrodes must be readily oxidisable, and for these active metals such as zinc, lead, cadmium, iron, magnesium, lithium and sodium may be used.

Electrode potentials and redox reactions are not confined to metals, but include such elements as hydrogen, oxygen, chlorine or fluorine, either in the gaseous form or in combination with some metal in the form of an inorganic salt. Electrode potentials are generally expressed with respect to hydrogen, which for standard conditions is assigned a potential of zero. It is an advantage to have at the two electrodes redox reactions widely spaced on the potential scale to give the highest cell e.m.f. But the choice is restricted by other factors—in particular, the type of electrolyte. Formerly electrolytes were aqueous solutions of salts, acids or bases. Organic electrolytes (e.g. propylene carbonate containing ionic conductors) and molten salts (e.g. LiCl–KCl eutectic, which melts at 352°C) are now being used, so that more active 'anodes' such as lithium or its alloys or sodium, which react with aqueous solutions, can be incorporated.

23.3 Primary cells

Primary cells differ from secondary cells in that the electrochemical reactions are not reversible, or, if so, only to a very limited extent. This may be primarily due to physical factors such as a loss of electrical contact by the chemical products of the discharge. In the case of the primary conventional 'dry' cell, for example, failure often occurs through excessive corrosion of the zinc can which forms one of the electrodes. In secondary cells or accumulators the reactions are readily reversible. The original chemical compounds can be re-formed by passing a direct current through the cell in the reverse direction, and accumulators can generally be submitted to many cycles of discharge and charge in this way.

The simplest primary cell, and one of the earliest (1836), is the *Daniell* cell, named after its inventor. Since many of the principles involved are common to other systems, this cell will be described in some detail. The electrodes are copper and zinc, and the electrolyte is a solution containing sulphate ions. The copper electrode is immersed in a solution of copper sulphate held in an inner porous pot, and the zinc electrode is held in a solution of zinc sulphate or dilute sulphuric acid in the outer glass vessel. The cell is shown diagrammatically in *Figure 23.1*.

The chemical reactions on discharge can be represented as follows:

positive electrode: $\quad Cu^{2+} + 2e^- \rightarrow Cu \quad$ potential $E_+ = 0.34$ V
negative electrode: $\quad Zn \rightarrow Zn^{2+} + 2e^- \quad$ potential $E_- = -0.76$ V
open-circuit e.m.f. $\quad E = E_+ - E_- = 0.34 - (-0.76) = 1.10$ V

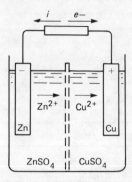

Figure 23.1 Daniell cell

Early forms of this cell were made with both electrodes immersed in dilute sulphuric acid, and had somewhat different reactions: the main reaction at the negative electrode was (as above) the formation of positively charged Zn ions, but in the presence of the acid (H_2SO_4) hydrogen was also produced, by a corrosion reaction. At the positive electrode the reaction was $2H^+ + 2e^- \rightarrow H_2$; and as the discharge reaction proceeded, hydrogen gas accumulated on the electrode surface, polarising the cell and depressing the terminal voltage. Other factors in the voltage 'fall-off' are (a) overvoltages arising from the kinetic limitations of the electrode reactions themselves, (b) the slow diffusive mass transfer of the ions up to the electrode surfaces; and (c) the ohmic resistance of the electrolyte. Substituting sulphuric acid by copper sulphate ensured that copper and not hydrogen ions were discharged at the positive electrode, eliminating the counter-e.m.f. By placing the zinc electrode in its own zinc sulphate electrolyte, the evolution of hydrogen was eliminated; but this made it necessary to enclose the copper electrode and its electrolyte in a porous pot, increasing the ohmic resistance of the cell.

23.3.1 Leclanché cell

In its 'dry' form, the Leclanché cell is the most common. A zinc plate or container forms the negative, a carbon plate or rod the positive electrode. The electrolyte is aqueous ammonium chloride. Through ionisation of the water, hydrogen and hydroxyl ions are produced ($H_2O \rightarrow H^+ + OH^-$). During the discharge the zinc electrode is oxidised in the reaction

$$Zn + 2(OH)^- \rightarrow Zn(OH)_2 + 2e^-$$

and hydrogen migrates to the carbon. Polarisation is avoided by surrounding the carbon by manganese dioxide, which participates in the redox reaction

$$2MnO_2 + 2H^+ + 2e^- \rightarrow Mn_2O_3 + H_2O$$

and the complete cell reaction, in which the electrolyte is also involved is

$$2MnO_2 + 2NH_4Cl + Zn \rightarrow 2MnOOH + Zn(NH_3)_2Cl_2$$

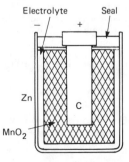

Figure 23.2 'Dry' Leclanché cell

corresponding to an open-circuit e.m.f. of about 1.5 V. The presence of zinc and ammonium chlorides in the electrolyte keeps the acidity at the right level and helps to reduce polarisation of the zinc electrode by the flocculent jelly-like zinc hydroxide that would otherwise coat the surface. In practice, the cells are not strictly 'dry', as the electrolyte is a thick paste. The cell (*Figure 23.2*) comprises (a) a central rod forming the positive terminal, (b) a depolariser of manganese dioxide (mixed with graphite or a highly active form of carbon black known as acetylene black to improve its conductivity) that is liberally moistened by the electrolyte, (c) a thin layer of jellified electrolyte, (d) a hard-drawn zinc outer can forming the negative terminal and (e) a seal, such as a card disc covered by a layer of bitumastic compound. Recent improvements, particularly in high rate performance, have been achieved by the replacement of the bulk of the ammonium chloride in the electrolyte with zinc chloride.

23.3.1.1 Performance

In many applications the use is intermittent. Polarisation develops as the current flows; during rest periods the cell recovers. *Figure 23.3* shows typical curves for a U2 cell discharging continuously through a lamp, the resistance of which varied from 1.32 to 1.03 Ω during the discharge. The cell

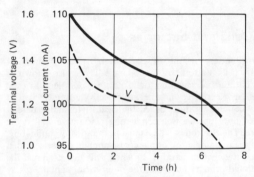

Figure 23.3 Discharge of U2 cell

dimensions are 32 mm diameter, 57 mm height and 0.1 kg mass. Cell behaviour in store and in transit ('shelf life') is lengthened by adding a soluble mercury salt to the electrolyte, which lightly amalgamates the active surface of the zinc container. Shelf life is reduced with high, and extended with low, ambient temperatures.

23.3.1.2 Flat and layer cells

Flat and layer cells have been developed for transistorised electronic equipments. In a duplex type the positive and the negative electrodes are placed on opposite sides of an electrically conducting diaphragm impermeable to the electrolyte. In the layer-built battery the zinc plate serves as the conducting diaphragm as well as the negative plate. On the opposite side this is first coated with a thin adherent layer of carbon mixed with a plastic resin, which functions much as the carbon rod in the cylindrical cell. Against this carbon layer is placed a moulded cake of the usual black depolariser mix and next to it a layer of absorbent material, such as filter paper, impregnated with the electrolyte. Another duplex electrode is placed on top of the first, and in this way a multicell battery of any desired voltage can be built up.

Sealing presents a particular difficulty with layer-built batteries. Obviously, there must be no electrical contact between the electrolytes of neighbouring cells, as this would permit leakage current. Also, some provision must be made for the release of adventitious gas from each cell. One means of sealing is to separate the adjoining duplex electrodes with annular spacing pieces and to cover the edges of the whole battery in wax or a suitable cold-setting plastic resin. Another way is to encapsulate the unit in a tightly stretched plastic stocking which is shrunk into position by gentle heat after application. Yet another method is to enclose the edges of each duplex diaphragm in an annular envelope of rubber or a plastic material, shrunk into position by heat. To reduce the risk of leakage, such envelopes may be cemented to the diaphragm and the assembled unit sealed together or firmly wrapped with plastic tape. Thin strips of rubber or suitable plastic materials in this form are sufficiently permeable to hydrogen to allow the release of gas, and the construction allows for sufficient expansion to release any internal pressures that may develop, without permitting significant leakage of electrolyte.

23.3.2 Standard cells

The Weston standard cell (*Figure 23.4*) has a cadmium/cadmium sulphate negative electrode and an electrolyte of cadmium sulphate. To give added stability, the cadmium is amalgamated

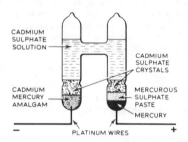

Figure 23.4 Weston standard cell

with mercury and the cell assembled in an H-shaped glass vessel with platinum leads to the terminals. The cadmium and mercury sulphates are usually prepared as thick pastes by digesting fine crystals of the salts with cadmium sulphate solution. Small quantities of sulphuric acid are added to reduce any tendency of the salts to hydrolyse.

When a saturated solution is used for the electrolyte, the cell is termed 'normal' and has an e.m.f. of 1.0183 V at 20°C, and a temperature coefficient over the range 0–40°C given for cell temperature θ by $[-4.06(\theta-20)-0.95(\theta-20)^2]\,10^{-6}$.

23.3.3 Alkaline cells

Three alkaline cells use zinc for the negative electrode.

23.3.3.1 Copper oxide/zinc

The positive electrode or depolariser is CuO, the electrolyte a solution of caustic soda of relative density about 1.2. The cell was formerly common in railway signalling, being cheap to make and able to deliver intermittent currents over considerable periods. The open-circuit e.m.f. is about 1.0 V, and on load 0.5–0.7 V. The reaction at the positive electrode involves the reduction of the copper oxide, while at the negative the zinc is oxidised:

positive: $\quad CuO \rightarrow Cu(OH)_2 \rightarrow Cu^{2+} + 2(OH)^-$

$\quad\quad\quad\quad\quad\quad\quad\quad\quad$ and $\quad Cu^{2+} + 2e^- \rightarrow Cu$

negative: $\quad Zn + 2(OH)^- \rightarrow Zn(OH)_2 + 2e^-$

23.3.3.2 Silver oxide/zinc

The silver oxide/zinc cell has military and other special fields of application. With appropriate modifications this couple will also function as a reversible cell, at any rate for a limited number of cycles. Batteries of cells have outputs four to five times those of any other system for the same weight and volume, and they are particularly in demand where very high currents and a low weight and volume are required, as in torpedo batteries, guided missiles and space satellites. In the primary form the batteries may be stored dry for long periods and brought into action by priming with electrolyte either by some pressure device or simply under the influence of gravity. To reduce polarisation, the electrodes are generally made porous. In one method of manufacture the positive electrode is made by sintering silver powder, the porous compact then being anodised to convert it to silver peroxide (Ag_2O_2). The zinc electrode is made from a paste of zinc oxide and caustic potash solution which is pressed into a screen of a suitable metal, such as silver or silver-plated copper. This is then reduced electrolytically to spongy zinc in dilute alkali solution. The electrolyte is caustic potash solution of about 30% concentration. A thin separator made of cellophane or paper may be used to prevent the plates from coming into contact and causing a short-circuit.

The discharge reactions at the electrodes are analogous to those already described for the copper oxide/zinc cell, but the silver peroxide passes through an intermediate stage to silver oxide (Ag_2O) before being reduced completely to metallic silver. The two separate reduction stages each have their own characteristic potential, so the discharge voltage curve of the cell with freshly prepared electrodes has two plateaus, a short one at about 1.80 V and a much longer one at about 1.50 V. The second portion of the discharge at about 1.50 V per cell or less, depending on the rate, remains remarkably steady until the active materials are exhausted, when the cell voltage falls sharply.

23.3.3.3 Mercury oxide/zinc

The electrolyte is caustic potash, but the elements are merely moistened. Known as the Ruben–Mallory cell, the mercury oxide/zinc cell has very low standing loss and gives steady voltages for long periods on low discharge rates. For such applications as hearing aids it is made in the form of a small metal button; as batteries of 6–8 V the applications are to portable transistorised radio receivers and 'walkie-talkie' sets. Mercuric oxide (HgO), mixed with 5–10% graphite to increase conductivity, acts as the positive depolariser. The mixture is compressed into a pellet, and in the button form this is placed in contact with the steel cup that makes one half of the button. A pellet of compressed zinc powder with a small amount of mercury is in contact with the other metal half of the button, which may be of copper. The separator may be a layer of absorbent paper saturated with the caustic potash solution. The outer surfaces of the two metals are insulated from each other by a plastic grommet, and form the terminals. Such sealed buttons have a capacity of a few milliampere-hours. In other forms the zinc electrode may be a coiled strip or perforated sheet, and the electrolyte is a jelly with gelling agents such as carboxymethylcellulose. The reactions are similar to those of other alkaline couples:

positive: $\quad HgO + H_2O + 2e^- \rightarrow Hg + 2(OH)^-$
negative: $\quad Zn + 2(OH)^- \rightarrow Zn(OH)_2 + 2e^-$

The open-circuit cell e.m.f. is 1.35 V, and the terminal voltage on load is 1.25–1.0 V. Prolonged low-rate discharges are obtainable. Standing losses are very low.

23.3.3.4 Air depolarised

Analogous to the carbon/zinc Leclanché cell, the air depolarised cell uses atmospheric oxygen as depolariser for the porous carbon electrode, which is made by roasting finely divided carbon and charcoal with a binder and then wet-proofing (e.g. by paraffin wax) to render the interior hydrophobic. The carbon electrode is only partially immersed in the electrolyte, and oxygen is readily adsorbed from the air. The negative electrode, of large plates of zinc lightly amalgamated with mercury, is completely immersed. A cell generally consists of two zinc plates flanking one carbon electrode, with capacities up to several hundred ampere-hours.

One range of cells has as electrolyte a solution of caustic soda. The essential reaction at the positive (oxygen) electrode is the reduction of oxygen with the production of $(OH)^-$ ions, and at the negative (zinc) electrode these contribute to the usual oxidation reaction, as in the copper oxide/zinc cell. The open-circuit e.m.f. is about 1.45 V; the terminal voltage on load for small currents is about 1.3 V, reducing to about 1.1 V at the end of the discharge.

The Le Carbone AD cell has as electrolyte a solution of ammonium chloride, which reduces polarisation at the zinc electrode by the action of zinc hydroxide. Large batteries of this type have been used in European railway signalling.

23.3.3.5 Alkaline manganese batteries

Another important recent development, providing up to 50% more power than Leclanché cells of the same volume uses dilute KOH electrolyte in a manganese dioxide–zinc cell with a reverse type of electrode assembly. The zinc anode paste is held in a porous tubular separator surrounding the centrally placed current-collecting nail anode. The cathode mixture of MnO_2 and graphite is highly compressed in the form of annular tablets and these are packed concentrically around the zinc anode. The whole assembly is contained in a thin-walled steel can, with the usual plastic disc-bitumen closure. The extra capacity over standard cells is due to the greater amount of MnO_2—40 to 70%—the absence of solid NH_4Cl and the higher conductivity of the KOH electrolyte.

23.3.4 Water activated cells

Water activated cells are stored dry for long periods, then activated by filling with (or immersion in) salt or fresh water; in the latter case, sodium or potassium chloride is included in the cell to increase its conductivity.

23.3.4.1 Silver chloride/magnesium

Silver chloride/magnesium cells are costly, but can deliver very large currents for short periods, typically for electric torpedoes. When flooded with sea-water, they become immediately active. The positive electrode (silver chloride) is a thin plate obtained by rolling a slab cast from the molten salt or by chloridising thin sheets of silver. The negative electrode is formed from magnesium strip or sheet. The electrodes are separated by rubber bands, absorbent paper, ebonite forks, or glass beads cemented to the electrode surface. Cylindrical cells with coiled plates have been used in batteries for meterological pilot balloons. During discharge, silver chloride is reduced to metallic silver and the chloride ions Cl^- migrate through the aqueous electrolyte to the negative (magnesium) electrode, which is oxidised to magnesium chloride ($MgCl_2 \cdot 6H_2O$), with the transfer of electric charge. The formation of silver raises the conductivity during discharge: this counteracts polarisation and tends to stabilise the terminal voltage. The e.m.f. is about 1.7 V, and the terminal voltage during discharge falls from 1.5 V to 1.0 V.

23.3.4.2 Copper chloride/magnesium

The chemical reactions are similar to those in the foregoing; the cell is cheaper, but the voltage is lower. Batteries of this type have been used in radar-sonde and other meteorological equipments. For higher voltage at low discharge currents a bipolar construction with duplex electrodes has been used. Positive plates are made by pressing the pelleted salt into copper screens or by dipping into molten chloride. Duplex electrodes are made by pressing or welding positive and negative electrodes to opposite sides of a thin copper foil.

23.3.4.3 Lead oxide/magnesium

The lead oxide/magnesium cell is simple, and comprises a fully formed lead dioxide positive plate (as in a lead/acid secondary cell), flanked by a U-shaped magnesium negative plate, with absorbent paper separators. The paper is impregnated with potassium chloride solution, then dried before assembly. The base of the element is left open and the cell is activated by dipping into fresh water for about 30 s. The electrode reactions are the reduction of lead dioxide to lead oxide at the positive, and oxidation of magnesium to magnesium hydroxide at the negative, electrode. A two-cell battery for a pilot balloon lamp gives typically 0.3 A at 3.0 V for 30 min at temperatures down to 0°C. As the discharge reactions are exothermic, batteries can still operate in ambient temperatures of $-40°C$.

23.3.5 Acid cells

Primary cells with acid electrolytes have been developed for special military and meteorological requirements.

23.3.5.1 Lead oxide/zinc (or cadmium)/sulphuric acid

These cells can be stored 'dry' for long periods and activated as required, an advantage in radio-sonde equipment, meteorological balloons, telemetering in experimental guided missiles and similar special applications. Batteries are referred to as 'short-duration reserve'. Lead dioxide electrodes (like those in lead/acid secondary cells) are welded in parallel according to the capacity required, interleaved with sheets of zinc or cadmium, and separated by thick strips of absorbent paper. The sulphuric acid electrolyte, of relative density 1.270 at 15.5°C, can be introduced some hours before use. The chemical reactions involve the reduction of lead dioxide to lead sulphate and oxidation of the negative electrode to zinc (or cadmium) sulphate. The cell e.m.f. is 2.5 V (with zinc) or 2.2 V (with cadmium). Corrosion of the zinc is reduced by amalgamation obtained by adding about 1% of mercuric sulphate to the electrolyte. The zinc couple gives higher voltage and discharge rate but at temperatures below 10°C, the zinc is polarised by the build-up of a reaction product; and at higher temperatures local action is acute, standing loss is high and accelerated activity on discharge can lead to gas polarisation. The cadmium negative is less temperature-sensitive, does not require amalgamation and gives satisfactory performance over the range 0–60°C. Typical voltages for a 3-cell battery at 3 A discharge for 3 min are 6.8–6.5 V (zinc) and 6.3–6.0 V (cadmium).

23.3.5.2 Lead dioxide/lead/perchloric acid

Excellent discharge reactions are obtainable with perchloric (or fluoboric or fluosilicic) acid producing soluble lead compounds. Polarisation by the build-up of reaction products is considerably reduced, so that it is not necessary to increase the surface of the electrodes by making them porous, as in a *secondary* cell. Nearly all of the active electrode material is usable, giving a high output/mass ratio. But for the same reason the cells must be used soon after priming, to avoid high standing loss. The cells are useful where they can be stored dry and primed immediately before use, and where low-mass batteries are required for large currents for relatively short periods. The positive plate is made by electrodepositing a thin non-porous film of lead dioxide on a sheet of nickel, iron or copper from a bath of lead perchlorate, lead nitrate or sodium plumbate. The negative plate may be made by electrodepositing thin layers of lead on a similar metal conductor, or directly from lead sheet. The plates are connected in parallel packs, the interleaved negatives being separated from the positives by spacers allowing rapid ingress of the electrolyte. The latter is held in compartments above the cell and allowed to enter the cell when the cell is primed. During discharge, lead dioxide is reduced to lead monoxide which is at once converted to lead perchlorate; the lead negative is oxidised to the same final product. Batteries have been constructed on the bipolar principle, the lead and lead dioxide being deposited on opposite sides of a nickel sheet. The e.m.f.s with electrolytes of 40–60% concentration range from 2.1 V (perchloric acid) to 1.9 V (fluoboric and fluosilicio acids). Perchloric acid in high concentration can become explosive in the presence of organic materials such as paper, sawdust, etc. Cells with the alternative electrolytes present no such risk, but have lower outputs.

23.4 Secondary cells and batteries

A secondary (or storage or accumulator) cell consists essentially of two electrodes held apart by separators and immersed in an electrolyte, the assembly being fitted into a suitable container. In the lead/acid cell the positive electrode is lead dioxide and the negative is pure lead in spongy form. In the alkaline cell the positive is nickel hydroxide and the negative is iron or cadmium. The electrolyte for the former cell is dilute sulphuric acid and for the latter, dilute potassium hydroxide.

23.4.1 Lead/acid cells

The chemical reactions follow the redox pattern (see Section 23.2.1). The current in the external circuit flows conventionally from the positive to the negative electrode. The two reactions can be written as follows:

positive: $PbO_2 + 4H^+ + SO_4^{2-} + 2e^- \rightarrow PbSO_4 + 2H_2O$
negative: $Pb + SO_4^{2-} \rightarrow PbSO_4 + 2e^-$

The reversible potentials are $E_+ = +1.685$ V and $E_- = -0.356$ V, giving an open-circuit e.m.f. $E = 2.041$ V.

23.4.1.1 Charge and discharge reactions

A secondary cell must be connected to a d.c. supply for charging, positive to positive and negative to negative. During *charging* the reactions are

positive electrode: $PbSO_4 + 2OH^- \rightarrow PbO_2 + H_2SO_4 + 2e^-$
negative electrode: $PbSO_4 + 2H^+ + 2e^- \rightarrow Pb + H_2SO_4$

The fully charged cell has

positive	electrolyte	negative
PbO_2 (lead dioxide)	dilute H_2SO_4	Pb (spongy lead)

and the overall cell reactions are

$$PbO_2 + 2H_2SO_4 + Pb \underset{\leftarrow \text{charge}}{\overset{\text{discharge} \rightarrow}{}} 2PbSO_4 + 2H_2O$$

During *discharge* the active parts of both electrodes are converted to lead sulphate, and the concentration of the electrolyte is reduced by both the removal of sulphate ions and the formation of water. During *charge* the lead sulphate at the negative plate is reduced to spongy lead, and at the positive electrode to lead dioxide with the release of sulphate ions at both plates and an increase in the concentration of the electrolyte. Measurement of the concentration indicates the state of the electrodes: the relative density of the electrolyte is read by a hydrometer.

Typical charge/discharge cell voltages are shown in *Figure 23.5*. Lead sulphate, the product of the discharge reaction, is practically insoluble in the electrolyte, a factor that endows the cell with its high degree of reversibility. During cycling the lead sulphate remains where it is formed, and the structure of the active materials is relatively undisturbed.

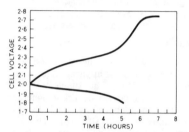

Figure 23.5 Typical charge and discharge voltages of lead/acid cell

23.4.1.2 Materials

Lead dioxide and spongy lead are the active materials, but it is necessary to provide metallic frames or supports for them. The success of the modern secondary cell depends on plate design and manufacture, and on the separators that prevent internal short-circuits and help to retain the active material in position.

Plates. Planté (formed) plate The positive plate has its effective surface area increased tenfold by forming close-pitched fins on the surface of a pure lead plate. The negative plate was commonly of a 'box' form but, see below.

Faure (pasted) plate The active material applied to open-mesh grids cast in antimonial lead is a paste made by mixing lead oxide with water and sulphuric acid. The plate is seasoned, dried and then electrochemically converted to lead dioxide or spongy lead by charging in dilute sulphuric acid. The grid acts both as a support for the active material and as the conductor of the current to and from the active material. The merit of the pasted plate is that the grid can be cast in precisely defined thin sections as low as 1.5 mm or less, and the ratio between active material and carrier grid is high. In addition to the standard gravity die-casting process, in modern developments, thin grids are made by stamping or punching suitable patterns from thin pure lead or lead-calcium alloy sheet. The red lead and litharge originally used for the positive and negative plates have been superseded by a grey oxide containing one-third fine metallic lead particles and two-thirds lead monoxide.

Reference has already been made to the box type of pasted negative plate used with Planté positive plates, in which the paste is held in the grid by thin sheets of perforated lead covering both surfaces. The box negative plate is however costly, both on materials and manufacture and has now been largely superseded by the common pasted negative plate.

Separators In some cells the separators serve only to keep the plates equidistant; in others they act as diaphragms to prevent internal short-circuit or to retain the active material.

Earlier wood separators have been almost entirely superseded by artificially made forms. One of the most successful is a microporous PVC. It has a high degree of diffusability, a low electrical resistance in acid and great durability under normal battery conditions. It is used in all types of lead/acid batteries. Another type of separator is made by sintering fine particles of PVC, and yet another by impregnating an absorbent paper with an acid-resistant resin such as phenol formaldehyde which both stiffens the paper and protects it against attack by the acid.

The profile of the separator is of interest. One product of the chemical reaction at the positive plate is water, the other being lead sulphate, which removes SO_4 ions from the electrolyte. The acid is doubly diluted at the positive plate, whereas at the negative only lead sulphate is formed. It is, consequently, necessary to provide a greater reservoir of acid adjacent to the positive plate, and in cells where the amount of acid is minimised it is usual to make separators with ribs which rest against the plate surface.

Electrolyte Different concentrations of the sulphuric acid electrolyte may be used, depending on the application for which the particular battery has been designed. The voltage of the cell depends on the concentration of the electrolyte, being higher for higher concentrations. Also, the minimum electrical resistivity at 20°C occurs at a concentration of about 31%, equivalent to a density of 1.225; and the minimum freezing point at a relative density of 1.300. Taking these and other factors into account, stationary cells are generally filled with acid of relative density 1.200–1.215; automobile and traction batteries, 1.275–1.285, or in tropical climates 1.240–1.260.

One important operational aspect of the electrolyte concerns *impurities*. Impurities such as chloride and acetic or nitric acid attack the positive grids and are generally kept to a minimum in both the filling-in acid and the 'topping-up' water which is added to replace water lost by evaporation and by electrolysis during charging. Metals such as iron and manganese cause self-discharge of both positive and negative plates, while nickel and copper are 'plated out' on the negative and also cause self-discharge. Limits for various impurities are given in British Standards. For topping up, it is generally advisable to use only distilled water.

23.4.1.3 Construction

Battery design and construction is considerably influenced by the application.

Road vehicle starting, lighting and ignition Faure plates have a special advantage where high currents are required. The larger the plate surface area, the lower the polarisation and the higher the cell voltage. Batteries normally comprise 3-cell (6 V), 6-cell (12 V) and 12-cell (24 V) assemblies, with plates connected in parallel groups in moulded containers.

In the past, batteries were supplied which, after filling with acid, required an extended first charge to reduce the active spongy lead negative plate. More recently, a dry-charged automobile battery has been developed to give about 75% of its nominal capacity shortly after filling, even after lengthy storage. The basic requirement in manufacture is to ensure a high percentage of lead dioxide in the positive plates and a minimum (e.g. 10%) of lead monoxide in the negative plates. This is achieved by the inclusion in the plates of an anti-oxidant, or by drying the plates after formation in an oven in which they cannot come into contact with atmospheric oxygen.

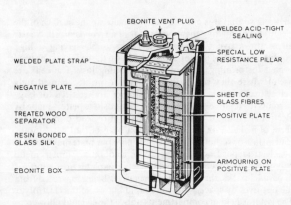

Figure 23.6 Sectional view of traction cell

Traction Batteries for electric vehicles may have a flat-plate or an iron-clad tubular form. The application involves deeper cycling and causes shedding of the active material. In the Faure plate design, 'retainers' (generally of thickly matted glass-wool fibres) are placed in close contact with the surfaces of the positive plates. Batteries of this type (*Figure 23.6*) have lives of 6 years or more. With tubular plates (*Figure 23.7*) tubes of high porosity are fitted: one such plate uses non-woven fibres of a plastics inert to sulphuric acid, such as Terylene. Another type uses an inner woven stocking of glass fibre, strengthened by individual thin-walled perforated PVC tubes. By raising the permeability to acid, these designs have enabled the output/mass and output/volume ratios to be increased by 30% without loss of durability and cycling life.

23.4.2 Nickel/cadmium and nickel/iron alkaline cells

Both of these commercially available alkaline cells have the same electrolyte, dilute potassium hydroxide, and the same positive active material, nickel hydroxide. The nickel/cadmium cell has a negative plate of cadmium with a small proportion of iron. Both have an e.m.f. of about 1.2 V. Cadmium gives the nickel/cadmium cell a lower charging voltage and reduced ohmic resistance, and its characteristics resemble those of the lead/acid cell. Use of the nickel/iron cell is mainly in traction, where the

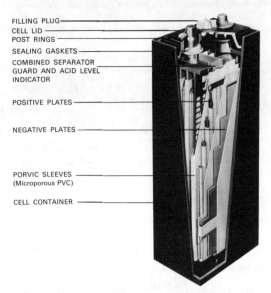

Figure 23.7 Ironclad cell with tubular positive plates

higher charging voltage and internal resistance are less important.

23.4.2.1 Charge and discharge reactions

The chemical reactions are complicated. The following gives a general guide:

positive: $2Ni(OH)_3 - H_2O \rightarrow 2NiOOH + 2H^+ + 2e^- \rightarrow 2Ni(OH)_2$

negative: $Cd + 2OH^- \rightarrow Cd(OH)_2 + 2e^-$

The reversible potentials of these reactions are respectively $E_+ = +0.49$ V and $E_- = -0.81$ V, giving an open-circuit e.m.f. per cell of $E = 1.30$ V. The overall reaction is

$$2NiOOH + 2H_2O + Cd \underset{\leftarrow \text{charge}}{\overset{\text{discharge} \rightarrow}{}} 2Ni(OH)_2 + Cd(OH)_2$$

These reactions apply to the nickel/cadmium cell. In the nickel/iron cell the cadmium (Cd) is replaced by iron (Fe).

The electrolyte is a solution of pure potassium hydroxide (KOH) of relative density about 1.200, a small amount of lithium hydroxide being sometimes added. The electrolyte takes no apparent part in the reactions and its density remains substantially constant. Cells can stand indefinitely in any state of charge, provided that the plates are kept immersed. As in the lead/acid cell, water is lost during gassing on charge and is made up with distilled water. Electrolyte is added only to make good accidental spillage.

Typical charge and discharge characteristics are shown in *Figure 23.8*. The time for a full charge is 7 h, the voltage per cell rising from about 1.4 V at the start to about 1.8 V (nickel/iron) or 1.7 V (nickel/cadmium) after about $5\frac{1}{2}$ h, and then remaining constant for the remainder of the charge. The input (in ampere-

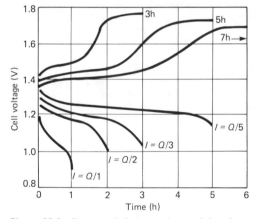

Figure 23.8 Charge and discharge characteristics of low-resistance nickel/cadmium cell

hours) should be 1.4–1.5 times as great as the previous discharge. Filler caps or vents should be kept closed except when topping up or taking gravity readings.

As alkaline cells are not damaged by overcharging, a full normal charge can be given irrespective of the state of the cells.

23.4.2.2 Construction

Nickel hydroxide forms the active material of positive plates, the form being 'tubular' for the nickel/iron and 'pocket' fir the nickel/cadmium cell. Negative plates are of the pocket form for both: the active material is first pelleted, and the pellets then firmly enclosed in pockets of this nickel-plated steel, perforated with many minute holes. In tubular plates helically wound tubes of similar perforated material are constructed, and active material interspersed by layers of thin nickel flake is tamped into them. Groups of plates of the same polarity are bolted or welded to steel terminal pillars. Plates are kept apart by ebonite rod separators. The cell assembly is fitted into a welded sheet-steel container, the terminals being brought out through the lid in suitably insulated glands.

The construction (*Figure 23.9*) produces sturdy, robust cells, unaffected by vibration and shock. Batteries are built up by assembling cells in hardwood crates.

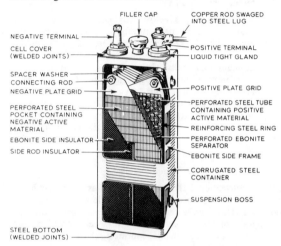

Figure 23.9 Nickel/iron cell

The need to increase the electrical conductivity of the nickel hydroxide positive active material of the alkaline cell was recognised at an early stage of its development. In the tubular plate this was done by introducing small pockets of extremely thin metallic nickel flake at regular intervals in the positive active material as it was tamped into the tube. In the pocket type of plate graphite powder is mixed with nickel hydroxide before it is pressed into a pellet. The same object was achieved in a *sintered* plate developed in the late 1930s. The plate grid or support was prepared in the form of a highly porous sintered nickel plaque and the active materials (nickel hydroxide for the positive plate and cadmium hydroxide for the negative) were deposited in the fine pores from suitable solutions of their salts. In this way the active materials were distributed evenly over a very large conducting surface and relatively high coefficients of use were obtained. By this method also very thin plates could be made in closely controlled thicknesses of 0.75 mm or less, and when these were interleaved with thin separators of woven or felted cloth, close-packed assemblies were produced capable of delivering very high currents, particularly at low temperatures.

Sintered plate batteries have been extensively used in aircraft and in other applications calling for high discharge currents.

It was later discovered that if the amount of electrolyte were reduced to such a point that there appeared to be no free potassium hydroxide, the cell could be submitted to overcharge in a fully sealed condition without any risk of the container bursting. The chemistry of this reaction can be explained simply by saying that any oxygen produced at the positive electrode during charge is immediately absorbed by the cadmium in the negative electrode and this is converted to cadmium oxide. The negative charge is therefore used to reduce this cadmium oxide and no gaseous hydrogen is evolved.

Sealed cells require no topping up, evolve no vapour or spray, and can be installed in equipments without hazard. But occasionally they partially dry out, so it is the custom to fit release valves. Batteries of sealed cells have found application in aircraft.

Development has resulted in small cells of capacities less than 1 Ah. They are made in the form of a large button or in cylindrical shape. The former generally have compressed pellets of the active materials; the latter have thin sintered electrodes. Both have restricted amounts of electrolyte and function as fully sealed cells. They are applied to transistor radio receivers and other electronic devices.

23.4.3 Silver/zinc alkali cells

Silver/zinc alkali cells have recently come to the fore in applications where very high outputs per unit of weight and volume are required. In these respects they give outputs from four to five times those of the equivalent lead/acid and nickel/cadmium alkali cells. Thus, a 100 Ah cell may be only 10 cm × 5 cm in plan and 20 cm in height, and have a mass of 1.5 kg. The silver is costly, and in view of the relatively low reversibility of multicell batteries, silver/zinc batteries have found few commercial applications. They have been used in aircraft, in guided missiles and to supply power to the communication systems in satellites.

23.4.3.1 Charge and discharge reactions

As in the conventional alkaline system, potassium hydroxide of about 1.200 relative density is used as the electrolyte. The silver oxide used as the depolariser in the positive plate passes through two stages of oxidation—AgO and Ag_2O—each of which has a characteristic electrode potential. During discharge the following reactions take place:

positive: (a) $2AgO + H_2O + 2e^- \rightarrow Ag_2O + 2OH^-$
$$E_+ = +0.57 \text{ V}$$
(b) $Ag_2O + H_2O + 2e^- \rightarrow 2Ag + 2OH^-$
$$E_+ = +0.34 \text{ V}$$
negative: $Zn + 2OH^- \rightarrow Zn(OH)_2 + 2e^-$
$$E_- = -1.24 \text{ V}$$

The overall cell reactions are therefore

$$\left. \begin{array}{l} 2AgO + H_2O + Zn \\ Ag_2O + H_2O + Zn \end{array} \right\} \begin{array}{l} \text{discharge} \rightarrow \\ \leftarrow \text{charge} \end{array} \left\{ \begin{array}{l} Ag_2O + Zn(OH)_2 \\ 2Ag + Zn(OH)_2 \end{array} \right.$$

For the former reaction the open-circuit cell voltage is 1.81 V and for the latter it is 1.58 V.

The discharge voltage curves of silver/zinc cells therefore generally show two plateaux, a relatively short one at about 1.8 V and a relatively long one at about 1.5 V. With some pretreatment, as, for example, a very short preliminary high-rate discharge, it is possible almost to eliminate the first plateau, and the discharge voltage then remains fairly steady at about 1.5 V per cell. A curve of this type is shown in *Figure 23.10*.

The nominal cell voltage is 1.5 V, and this is held (at normal discharge rates) substantially constant over most of the discharge period, as shown. High output currents and charge rates can be employed with little sacrifice in effective capacity. Normal ampere-hour efficiency is 90–95% and watt-hour efficiency 80–85%. Practically no gassing takes place. The electrolyte is to a major degree absorbed in the active material and almost completely immobilised.

Owing to the slight solubility in potassium hydroxide of the higher silver oxide, overcharging of reversible silver/zinc batteries should be avoided. In general, charging should be stopped when the voltage reaches 2.1 V per cell and, for this reason, recharging is best carried out with constant-voltage control. If this is not done, short-circuits may develop through the separators. To reduce this risk, manufacturers generally enclose the zinc negative plates in several layers of the separator material, which is usually of a cellulosic base. Single cells of this type have a fair degree of reversibility, but in a multicell battery the irregularity between cells, which is difficult to eliminate, has an adverse effect on reliability. And the greater the number of cells in the battery, the shorter the cycling life becomes.

23.4.4 Secondary battery technology

23.4.4.1 Voltage

The open-circuit voltage of a fully charged secondary cell is the same, however large or small, but varies according to the type, i.e. lead/acid or alkaline. That of a fully charged lead/acid cell is approximately 2 V; that of an alkaline cell is about 1.2 V.

Because of the difference in cell terminal voltages—2.0 V for lead/acid and 1.2 V for Ni/Cd–5 Ni/Cd cells are required to replace batteries with 3 lead/acid cells.

23.4.4.2 Capacity

The capacity of the lead acid battery increases with decreasing rates of discharge (currents) and increasing durations (hours). It is therefore usual to state the rate of discharge for any declared capacity, e.g. at the 10-h rate, the 5-h rate, the 1-h rate and so on. *Figure 23.11* shows families of discharge- and charge-voltage curves for a typical lead acid cell on discharge at constant current in each case. The salient properties are listed in the table.

discharge time (h)	10	7.5	5.0	3.0	2.0	1.0	0.5
current rate (A)	11	14	19	29	40	65	118
mean voltage (V)	2.0	2.0	1.98	1.94	1.90	1.84	1.70
energy (Ah)	110	105	96	88	79	65	59
power (Wh)	220	210	190	170	150	120	100

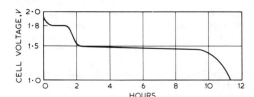

Figure 23.10 Discharge voltage of silver/zinc alkali cell

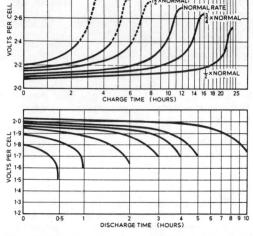

Figure 23.11 Representative curves of charge and discharge

Points to note are: (a) the O/C voltage is about 2.0 V, (b) the discharge voltage curves all start with a relatively flat plateau which ends with a sharp decline, usually referred to as the 'knee' of the curve, beyond which little further capacity is available, (c) the slope of the discharge curves becomes steeper as the rate (current) increases and the knee becomes less prominent, (d) the 'energy' capacity is represented by the product of the current and the duration—Ah—and the 'power' capacity by the product of energy and the mean voltage—Wh.

The theoretical amounts of the active materials consumed per Ah of discharge energy are: PbO_2, 4.45 g, Pb, 3.86 g and H_2SO_4, 3.68 g. The actual amounts are considerably greater as the coefficients of use fall very much below 100%, as indicated below:

	5-h rate	1-h rate	5-min rate
positive, PbO_2	52%	33%	16%
negative, Pb	66%	40%	19%

The capacity is related to the amounts of the active materials and to their porosity and intrinsic surface area. The main product of the discharge in both positive and negative active materials is lead sulphate. This has a very high electrical resistivity. It polarises the active material and, by clogging the pores restricts diffusion of the electrolyte into the reaction zones. Planté positive grids are cast with a developed area at least 10 times the superficial area and the anodically prepared active material—PbO_2 is spread in a thin layer over a large area. In Faure pasted plates on the other hand, the porosity of the active material must be adequate to permit rapid diffusion of the sulphuric acid electrolyte, particularly during discharges at high rates.

Thus, it is found that (a) for stationary batteries of the lead/acid type, on the score of durability and efficiency it is best to use a Planté positive with a large surface area lightly coated with active material; (b) for automobile or diesel engine starter batteries it is best to use a large number of comparatively thin pasted plates; (c) for traction battery service, where lower rates are usual, thicker plates with denser material can be used.

23.4.4.3 Charging

All secondary batteries require a supply of direct current for recharging. The method of charging is important in its effect on battery performance and service life. The three principal methods are as follows.

System control The batteries work in parallel with a generator across the load. Typical applications include motor vehicles, in which the battery forms an essential standstill reserve for ignition, lighting, signalling, etc. Most small motor cars employ a d.c. generator with control gear for cutting out the generator when its speed is too low, and for voltage regulation for operation over the wide range of active speed. Larger cars and road vehicles employ a.c. generators (alternators) with rectifiers and appropriate automatic control equipment.

Manual or semi-automatic control Here the battery is recharged from a separate source, such as a transformer/rectifier. Typical applications are to electric vehicles, in some cases with the transformer/rectifier carried on the vehicle.

Since the performance and life of motive power batteries largely depend on the efficiency of the charging system, these will be described in more detail. Three forms of charger are in general use, the single-step taper charger, the two-step taper charger and the pulse-control charger, of which the Chloride Spegel is a typical example. The voltage-current-time relationships for these three types are shown in *Figure 23.12*. With the single step charger, the voltage is held constant just below the battery gassing point, 2.4 V per cell. When the battery voltage reaches this point, the current is allowed to decay until the battery becomes fully charged. With the two step process, the first stage

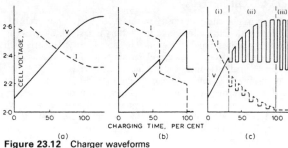

Figure 23.12 Charger waveforms

follows the same line as with the single step; the current is then reduced to a pre-determined value and charging continued for a pre-determined period. The pulse type of charger is devised more to keep the battery fully charged. When the main charge is stopped, the battery receives short pulses of charge, controlled by the decay in the battery voltage during the intervening open-circuit period.

Float-and-trickle control Large stationary batteries in generating stations and (for emergency lighting) in hospitals are connected permanently to a charger, and may also be connected across the load. In normal operation the battery receives a trickle charge to keep it fully charged and ready for intermittent loading. Such a battery requires ventilation and ready access for servicing.

23.4.4.4 Maintenance-free, gas recombination technology

One of the most important developments in storage batteries during the past 20 years or so has been the production of maintenance-free (MF) systems. The primary object was to reduce or eliminate altogether the need to replace water lost through three main factors, (a) evaporation, (b) local electrochemical action at lead negative plates due to the deposition of metals, such as antimony having a lower hydrogen overvoltage than lead and (c) electrolysis of the aqueous electrolyte towards the end of each recharge. With well stoppered cells, (a) presented few problems. So far as (b) was concerned, from the earliest days of manufacture, antimony has been the favourite hardener for the lead grids used to support the active materials and to collect the current. But, because of their different hydrogen over-voltages, when lead and antimony come in contact in the electrolyte, hydrogen is evolved, water in the electrolyte is decomposed, some of the lead active material is oxidized causing a corresponding loss of capacity—the so-called 'standing loss'. Many metals have been tested in lead alloys for this application and the most successful has been lead-calcium alloy, containing 0.08% of calcium and small amounts of tin. With grid cast in this metal, standing losses are negligible. Losses by the third factor (c) proved more difficult to contain, but following the route taken for alkaline cells, the principle of gas re-combination has now been fully established for lead-acid batteries. The main requirements are (a) no free electrolyte—to permit rapid diffusion of oxygen to the negative plate surfaces, the elements should be 'starved' of electrolyte; (b) the bulk of the electrolyte should be retained by the porous separators and (c) there should be excess negative capacity. Oxygen is evolved from the positive group before the negative is fully charged. It diffuses to the negative plates where it is absorbed causing discharge of the active material. The negative plates are, therefore, never fully charged and hydrogen is not evolved. This principle is now being applied to cells of various types from small button cells to batteries for SLI, traction and stationary service. Cells can be fully sealed, though it is usual to incorporate a pressure-controlled safety valve.

23.4.5 Lithium cells

From a laboratory curiosity in the 1960s, the lithium cell has become an active addition to the battery market. Lithium as an electrode gives a high energy/mass ratio and an inherently high power density. The standard electrode potential is over -3.0 V and when allied with, say, a fluorine electrode ($E_0 = +2.87$ V), the resulting cell should theoretically have an O/C voltage of over 6.0 V. Nevertheless, it cannot be used with aqueous electrolytes: non-aqueous electrolytes, both organic at ambient temperatures and anhydrous molten salts at high temperatures, have to be substituted. The following couples have been employed with organic electrolytes: Li/MnO_2, Li/SO_2, $Li/(CF)_n$, Li/MoS_2, Li/Ag_2CrO_4, Li/CuO, Li/FeS_2 and Li/TiS_2. Unfortunately the electrical resistance of these organic electrolytes, even when improved by inorganic salt additives, is relatively high and cells of this type are only suitable for low rate discharges.

With molten LiCl–KCl, the alloys Li–Al and Li–Si have been used as negative electrodes, and FeS and FeS_2 as positive electrodes. Operational temperatures may be as high as 400°C.

The Li/SO_2 and $Li/SOCl_2$ couples have inorganic electrolyte and can be operated at high discharge rates at ambient temperatures. They have, in fact been tested for possible torpedo propulsion.

23.4.6 Sodium/sulphur cells

Another high-temperature contender (250/300°C) is the sodium/sulphur cell, which employs liquid sodium and sulphur as the negative and positive electrodes, respectively, with the solid electrolyte β-alumina which is a sodium ion (Na^+) conductor. In the UK a company (Chloride Silent Power) set up by the Electricity Council and the Chloride Group is exploiting the sodium/sulphur battery for large-scale application.

23.5 Anodising

Anodising is the production of a film of oxide or hydrated oxide on the surface of aluminium or of its alloys, to prevent corrosion of aircraft metal and other surfaces likely to be exposed to the effects of the sea. The anodic film is usually of a light grey colour. When newly formed, it can absorb dye, a property of use for decorative purposes.

23.5.1 Process

The workpiece is made the anode in a bath of chromic or sulphuric acid, through which a direct current is passed at a voltage of about 50 V for the former acid bath and 10 V for the latter. In large installations a current of 1–2 kA may be required from a rectified a.c. supply. The action produces on the aluminium workpiece a semi-insulating film, and the vat voltage is increased as the action proceeds. Non-aluminium parts must be screened by plastics materials. The current density is 0.5–0.7 A/cm^2 of active anode surface. In a typical process about 10 V is initially applied, the current being 1.5 kA; after 30 min the voltage is steadily raised in 5 min to 50 V, and maintained for a further 5 min. The work is then removed from the vat and washed.

The liquid carries the anodising process into every bare recess so long as there is no exclusion by gas. A tube 2 m long and of 5 mm internal diameter can be satisfactorily anodised over its entire surface. The effectiveness of the surface can be tested by applying 50 V between the workpiece and a 25 mm diameter steel ball. Anodising is best carried out at a standard temperature (e.g. 40°C).

23.5.2 Vats

In small baths the cathodes are carbon plates or rods hung from the cathode connection. Current is conveyed to the cathodes and the anode workpiece by round copper rods resting on porcelain insulators attached to the rim of the wooden or metal vat. A typical vat is 6 m × 2 m with a depth of about 2 m, sunk into a concrete floor for convenient access. The cost of filling a bath with diluted acid is considerable, making necessary scrupulous attention to cleanliness and the avoidance of contamination. Precipitates must be allowed to settle or be filtered off. Gentle agitation of the liquid by means of compressed air from a small motor-driven pump is applied to avoid slight pitting of the workpiece at the liquid surface level: this keeps foreign bodies in suspension, so that filtering may be necessary at the end of the day's work.

23.5.3 Workpieces

The pieces to be anodised are wired for connection to the anode rod by aluminium wire if they are small. Larger pieces may be connected by dural rods and clamps. The cross-section of the wires and rods must be adequate for the appropriate current, and that of the connection point to the workpiece must also be sufficient. Multiple connections can be used for this purpose. The action of the bath produces an effervescence of oxygen from the aluminium and dural, and a gas accumulation may exclude workpiece recesses from the anodising process. This can be overcome by tilting the workpiece, or may necessitate two runs with the workpiece in different orientation.

23.5.3.1 Cleaning

Before anodising, workpieces must be cleaned by a process similar to those for the preparation of work for electroplating. Items that have not been heat-treated, welded or riveted are more easily cleaned than fabricated parts or castings. If the workpiece has been previously treated in a salt bath, all trace of the salt must be removed by flushing in running water. Built-up fittings require the joining surfaces to be scratch-brushed or buffed before refitting, and immersed in boiling water immediately prior to anodising. Welded aluminium parts must be kept in contact with boiling water for 30 min to remove flux residues, which lead to pitting. In large anodising shops greasy articles are held in the smoke arising from electrically heated trichloroethylene, which cleans them thoroughly in about 2 min. Before anodising they are dipped in a hot swill to remove traces of the smoke, which would contaminate the anodising bath.

23.6 Electrodeposition

Electrodeposition is carried out in electrochemical reactors that use electrical energy to extract metals from their compounds. The processes include electrowinning, electrorefining, electroplating and electroforming.

Electrowinning A mineral, or a compound prepared therefrom, is decomposed to a metal and (usually) a gas. The prime process in this class is the refining of aluminium: the Hall–Heroult process involves the electrolytic decomposition of alumina (prepared from bauxite) dissolved in molten cryolite (Na_3AlF_6) at about 1000°C. Aluminium is produced as a liquid at the cathode. The primary anode product, oxygen, reacts with the carbon anode to give mainly carbon dioxide, with a small proportion of carbon monoxide. The process is highly energy intensive.

Electrorefining An impure anode (perhaps a cathode from an electrowinning stage) is converted to a pure cathode. Copper is refined in this way. The process is much less energy intensive than electrowinning, mainly because of the lower voltages employed.

Electroplating This is the electrodeposition of a thin protective metal coating on another metal. Protection is obtained with the minimum amount of material and with a low expenditure of energy. The process has imperfections: for example, chromium plating has high porosity and poor adherence, particularly in extreme conditions.

Electroforming The aim in this extension of electroplating is to build up the plate so that the cheap substrate can be dissolved to leave an article of the plate metal which has mechanical integrity. Nickel electroforms are made in this way, but the process is particularly advantageous for refractory metals (e.g. tantalum, niobium). Thus, an article that would be costly and wasteful when made from bulk metal can be prepared directly.

23.6.1 Electroplating

In modern mass production processes automatic plating plants are regularly used for the deposition of nickel, chromium, brass, zinc, silver, copper and cadmium, and, in addition, for depositing composite coats of two metals, such as in nickel and silver plating, brassing and bronzing. In using plants of this kind, the product is extremely consistent, as the plater is relieved from manual duties and left free to devote his attention to the control of essential operations. The plant is usually arranged so that the various vats are placed in line one behind the other in proper sequence. Two conveyors travel slowly over the vats, the articles to be plated being loaded on insulated suspenders which are arranged in two or three rows on cross rods attached to the conveyor chains. The suspended articles pass in succession through cleaning and swilling vats before passing to the vats for deposition, and finally to the swilling and drying-out apparatus after plating is completed.

As each loaded cross rod arrives at a vat or tank it is lifted by an auxiliary fast-moving transfer chain, and lowered into the next vat along the line. After passing through drying apparatus, the suspenders are unloaded from the cross rods, the latter being automatically returned to the loading position for the work to continue. The cycle of operations in the case of automatic nickel-plating plants is usually as follows: hot alkaline electrolytic cleaner used cathodically at about 300 A/m^2; cold water swill; anodic etch in sulphuric acid in the case of iron and steel articles; agitated cold water swill; nickel plating; cold water swill; hot water swill; and drying.

23.6.1.1 Barrel plating

The barrel plating method of electroplating is automatic in so far as the actual deposition of metal is concerned, and the method is extremely efficient and economical within the limits of the capacity of the barrel and the size of the articles. Any difficulties experienced are usually traceable to overloading the barrel, which causes the articles to be carried round in a mass as the barrel revolves, or to attempts to plate articles having awkward shapes which shield one another in the mass and prevent regular deposition.

Plating barrels are available for use with all types of solutions, and excellent results can be obtained with such articles as screws, cycle and car fittings, and hooks and eyes. Some of the early machines were extremely clumsy, and offered great resistance to the flow of current, since when many improvements have been made, particularly in the use of rubber lined tanks, stoneware vats and containers of other non-metallic material.

An important advantage of barrel plating is that, as the articles are receiving the deposit of metal, they rub together and become burnished, while the deposit being consolidated by the rubbing is close-grained and durable. As in so many plating processes, the speed at which a plating barrel should revolve is critical and is governed by the class of article being plated and the diameter of the barrel.

Speeds vary between 30 and 45 rev/min, although in the case of large plating barrels the speeds may be as low as 10–15 rev/min. It sometimes happens that articles all of one shape cannot conveniently be plated in a barrel by themselves, and in such cases scrap metal pieces or steel balls of suitable size can also be put into the barrel.

Although barrel plating can be carried out with some solutions at 6 V, it is usually necessary to use voltages of 10–18 V. In the case of nickel a current of about 45–60 A at 10 V can be passed. A higher current is needed for 'brassing'.

Barrel plating was developed specially for handling small articles in quantities, and in many cases this process has a number of advantages over the original suspended anode method. In one of the best-known makes the anode insulated by a vulcanite cover is situated at the base of the barrel. The barrel itself is made of welded steel internally lined with vulcanite, and is mounted on a swivel arrangement, which permits immediate removal of the work by dumping it into a suitable container for transfer to the drier. Electrical contact is made through an insulated rod passing through the centre of the shaft and directly connected to the anode, the current returning to the plating rectifier by way of the barrel and framework of the machines. An important feature of this type of barrel is that it can be relined with vulcanite at very small cost, and it is also easily and quickly rinsed.

23.6.1.2 Polishing

The speed of a polishing mop or bob is a factor affecting efficiency and economy, and speeds between 30 and 45 m/s are the rule according to the article to be polished.

Variable-speed machines are necessary, so that when a mop or bob becomes worn, say from 30 to 20 cm in diameter, the peripheral speed may be increased. Too low a speed causes a mop to drag and the articles to become heated, and a burnishing rather than a polishing effect is produced. 'Polishing bobs' is the term used for all solid leather wheels, compress wheels, made up of sections of leather, canvas or felt, and solid felt. These are dressed with emery, and the felt bob is the type in most general use. Polishing mops consist of discs of cotton cloth, varying in size from 5 to 40 cm in diameter and held together at the centre by means of washers of leather or fibre.

23.7 Hydrogen and oxygen electrolysis

There are several processes by which oxygen and hydrogen are produced as a main product or by-product. The industrial uses are considerable.

Oxygen is used to a very large extent in metal working and metallurgical industries. In conjunction with acetylene or hydrogen, it is used for welding and cutting; with butane and propane, for tempering steel. Very large (tonnage) quantities are now used in the production of steel, in the gasification of coal and to a lesser extent in the oxidation of olefines. The use for medical purposes and in super atmospheric aviation is relatively small in quantity but of high importance.

In the chemical industry oxygen or oxygen-enriched air is sometimes used in place of air in the oxidation of ammonia to produce nitric acid in highly concentrated form. The electrolysis of water can therefore conveniently be used for the production of hydrogen for the synthesis of ammonia and of oxygen for the subsequent preparation of nitric acid. The nitric acid and ammonia together produce ammonium nitrate, an important fertiliser.

Hydrogen is used in a number of industries. In the food industry the production of margarine and cooking fats from liquid oils such as groundnut, cottonseed, whale, etc., is based upon the partial catalytical hydrogenation of these oils, which converts them into solids at normal temperatures. Sorbitol, a sugar used by diabetics and in the production of synthetic resins, is produced from glucose by hydrogenation. Ammonia, a basic chemical for the fertiliser industry, is produced in very large quantities annually by the catalytic combination under high pressure of hydrogen and nitrogen.

Large quantities are used in the fuel industries, mainly in the processing of mineral oils but also in the production of synthetic liquid fuels from coal.

Many other uses exist in the chemical industry, such as in the production of synthetic solvents and the hydrofining of benzole. In the metallurgical industry hydrogen alone or mixed with nitrogen is used to provide inert atmospheres for the annealing of alloy steels, and in the lamp industry it is used for the reduction of tungsten and molybdenum ores to the metals and in the subsequent manufacturing processes for the production of lamps. Large synchronous machines are sometimes cooled by the circulation of hydrogen in a closed system. Meteorological balloons are filled with the gas.

23.7.1 Process

Water dissociation to give small (equal) concentrations of the hydronium ion (H_3O^+) and the hydroxide ion (OH^-) is in accordance with the equation

$$2H_2O \rightarrow H_3O^+ + (OH)^-$$

Thus, it is possible in principle to decompose water electrolytically in an electrochemical reactor, to give hydrogen at the cathode and oxygen at the anode according to the reactions

cathode: $2H_3O^+ + 2e^- \rightarrow H_2 + H_2O$
anode: $4(OH)^- \rightarrow 2H_2O + O_2 + 4e^-$

Practical reactions are slightly more complex, as the electrolyte added to the water to improve its conductivity plays a role in the reaction.

In simple water electrolysis the hydrogen produced per coulomb is 0.010 45 mg/C. Thus, 1 kA produces 0.42 m^3/h (at 0°C, 760 mm Hg, dry). The corresponding quantity of oxygen is 0.21 m^3/h. Stray current prevents full attainment of these outputs, but the current efficiency can be as high as 98–99%. The specific energy consumption depends on the voltage. The decomposition voltage of water is 1.23 V, but that to operate a hydrogen/oxygen cell exceeds this value, because of the ohmic resistance of the electrolyte and the overpotentials for the evolution of the gases at the electrodes. Solutions of caustic soda or potash are almost invariably used to lower the resistivity, using mild steel cathodes and nickel-plated anodes, which are immune to attack by the electrolyte or the gases. Soda and potash solutions are always prepared from the purest material: the caustic must have limited amounts of chlorides, sulphates and carbonates, which raise the resistance, while the former two may cause corrosion of the electrodes. The water feed must be pure (i.e. a very high resistivity). It is produced in a water still or by treating the raw water in an ion exchange plant, followed sometimes by active carbon treatment to remove oil traces, as when the source is a steam condensate.

23.7.2 Electrolysers

Water electrolysers are distinguished by electrode arrangement and gas pressure.

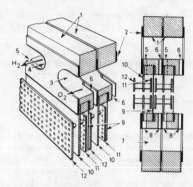

Figure 23.13 Section through typical 'filter-press' electrolyser. 1. Cell frame; 2. joint; 3. oxygen gas channel; 4. hydrogen gas channel; 5. hydrogen gas port; 6. oxygen gas port; 7. electrolyte channel; 8. electrolyte ports; 9. diaphragm; 10. main electrode; 11. auxiliary cathode; 12. auxiliary anode

23.7.2.1 Tank type

The tank or unit-cell electrolyser has unipolar electrodes in one tank and connected in parallel. Normally several tanks are connected in series, facilitating extension and enabling a faulty tank to be by-passed.

23.7.2.2 Filter-press type

The filter-press type has bipolar electrodes arranged to act as a positive on one side and a negative on the other, with terminal connections at the end of the battery. *Figure 23.13* shows a section, the construction having some similarity to a filter press. The cell frames (1) are made of an electrically insulating caustic-resistant material and into them are fitted the diaphragms (9) and main electrodes (10). In order to maintain the optimum distance between the active faces of the electrodes, the main electrodes each carry a perforated auxiliary cathode (11) and a similar auxiliary anode (12). The various ports are sealed off from one another, and the whole assembly of a number of cells is made leak-proof by means of the joints (2). The action is as follows.

Hydrogen evolved at the cathode (11) rises through the gas port (5), carrying with it some of the electrolyte. The mixture passes along the channel (4) common to the pack of cells, thence to an arrangement of gas washing and cooling drums mounted on the electrolyser, from which the gas passes to process. The electrolyte, after cooling, returns by a pipe system through a filter to the electrolyte channel (7). The ports (8) then ensure an electrolyte feed to each cell to replace that taken out with the gas. A similar circulation occurs with the oxygen from the anodes (12) through the ports (6) and channel (3) returning through (7) and (8). The current connections to the electrolyser are to the electrodes at the ends of the pack, which are extended beyond the bottoms of their respective cells for cable or busbar.

It will be seen that, for a given output, a filter press electrolyser is constructed as one unit consisting of a number of cells of an appropriate size, clamped together as a pack, variations in the size and number of electrodes being made to suit the required electrical conditions. A tank type electrolyser, on the other hand, will consist of a number of unit cells connected in series externally, but here again variations in size and number can be made to suit the electrical conditions.

The filter-press electrolyser can produce in a single unit up to 500 m^3/h of hydrogen; the connections are simple; gases can be delivered at normal gasholder pressures without the need for separate boosters; the electrolyte is in an enclosed system and the constant circulation through a filter ensures thorough mixing and freedom from contamination; and the quantity of electrolyte is small compared with the output.

23.7.2.3 Pressure type

The pressure type electrolyser is a development of the filter-press electrolyser in which electrolysis is carried out under pressure. Lower cell voltages are obtained because of decreased blanketing effect of the smaller gas bubbles. The optimum pressure is about 30 atm, giving a 14% power advantage over that achieved at normal atmospheric pressure. Further, the cost of post-compression of the gases is less. The basic design is similar to that of the filter-press type, but adapted to work under pressure, having cells of circular section. Gas purity is a little less than for the unpressurised equipment, but is adequate for most industrial purposes.

23.7.3 Gas purity

The makers of atmospheric pressure filter-press electrolysers guarantee a purity of 99.7% for oxygen and 99.9% for hydrogen. A check is normally kept on the gas purities by drawing off samples into measuring burettes. In the case of oxygen the sample is then passed over copper, which combines with the oxygen and leaves the hydrogen impurity as residue; while in the case of hydrogen, a platinum spiral electrically heated to redness in the gas causes the oxygen impurity to be removed by a combination with twice its volume of hydrogen to form water, leaving the main bulk of hydrogen as residue.

Automatic gas analysers can also be fitted which give a continuous check on gas purities.

23.7.4 Plant arrangement

The electrolytic plant will usually comprise a source of direct current, the electrolyser, a tank for the storage and make-up of electrolyte and a transfer pump, a source of purified water, gasholders and compressors for hydrogen and oxygen, and pressure storage tanks. In the case of pressure electrolysers buffer tanks are provided in place of low-pressure gasholders. The direct connection of electrolysers working at low pressure to compressors is not recommended.

Depending on the required specification of the gases, dryers are installed before the gases are delivered to process. The capacities of gasholders, compressors and high-pressure storage vessels depend upon the user process and its demand cycle. The compressors are normally started and stopped automatically by pressure switches on the high-pressure storage vessels and capacity switches on the gasholders. Complete automatic control of electrolytic plants including variation of output to suit demand is possible but so far has not been attempted.

Because of the explosion hazard, electrolysers are housed in rooms isolated from the electrical conversion equipment by a gasproof wall, intercommunication between the two being by outside doors. Lighting of the electrolyser room can be either by pressurised fittings or by lamps placed outside the windows.

The electrical energy consumption for the production of given quantities of hydrogen and oxygen depends on the type of electrolyser and on the output per unit cell, because the cell voltage increases with the current density at the electrodes. For atmospheric types the operating voltage per cell is 1.9–2.1, corresponding to specific d.c. energy consumptions of 4.5–5.0 kWh/m^3 of hydrogen. The selection of an electrolyser for a given duty is based on the capital cost of the installation and the cost of energy. Thus, where the energy cost is low, it is economic to work at a high current density (and therefore a high specific energy consumption).

24 Microcontrollers and their Applications

J McTaggart
GEC Industrial Controls Limited

D I Alabaster BTech
GEC Electrical Projects Limited

J W Beeston BSc (Eng), PhD
GEC Electrical Projects Limited

S W Bullett BSc
GEC Electrical Projects Limited

A M Plaskett BSc(Eng), CEng, MIEE
GEC Electrical Projects Limited

B F Pope, MSc, AMBIM
GEC Industrial Controls Limited

G S Powell, BA, MBCS
GEC Electrical Projects Limited

Contents

24.1 Introduction 24/3

24.2 General requirements 24/3
    24.2.1 Basic I/O 24/3
    24.2.2 Fast I/O 24/4

24.3 Dedicated controller 24/4

24.4 Operator interface 24/4

24.5 Communications with other controllers 24/4

24.6 User-designer interface 24/4

24.7 Maintenance facilities 24/5

24.8 Applications 24/5
    24.8.1 Rod mill control 24/5
    24.8.2 Mine winder event recorder 24/6
    24.8.3 Distributed control of water supplies 24/8

24.9 Serial communication between microcontrollers 24/11
    24.9.1 Data integrity and error recovery 24/11
    24.9.2 Link statistics and diagnostics 24/11
    24.9.3 Signalling paths 24/11
    24.9.4 Telemetry 24/11

24.1 Introduction

Before present-day industrial control systems are considered it is worth while reflecting briefly on the changes that have been made in the realm of industrial control systems over the past 30 years. The late 1940s and early 1950s marked the era of the conventional types of relays and contactors for any type of sequencing control, and these were supplemented by rotary and magnetic amplifiers for any alternative types of control. At this stage on-line process control was still in its infancy, despite the fact that valves were starting to appear in operational amplifiers.

In the 1950s pressure was exerted on industrialists to increase the efficiency, accuracy and overall performance of plant machinery, and the mid-1950s saw the emergence of the transistor. It was not long before transistors as discrete components were used in operational amplifiers, although such operational amplifier schemes were used in conjunction with the relay and contactor systems for industrial control.

The name 'computer' was soon added to the industrialists' vocabulary and the drive for automation was increased. On-line process control suddenly became widespread in the form of computers constructed of discrete circuits, which were used to supplement the relays of operational amplifiers, and the search increased for more powerful and yet smaller components. The early 1970s saw the large-scale integration of components used in computers, the minicomputer become a reality and hardware logic appear in control schemes, although they still included relays, contactors and operational amplifiers.

Today not only do we have all these technologies, but also additional ones have emerged over the past three or four years. The programmable logic controller (PLC) is now a well-known product to the control engineer and many more programmable-based products are appearing. Specialized microprocessor systems are becoming more commonly available, and so today the control engineer has a multitude of technologies to cope with, as well as specialist programmers and engineers with whom to communicate.

To summarise the situation, a typical control system today could comprise a process control computer is interfaced with an input/output system built up of small-scale integrated transistor logic, which in turn interfaced with a variety of equipments such as relay sequencing, analogue controllers, hard-wired logic, PLCs and, of course, those control functions achieved with a computer or minicomputer.

It is evident that if one technology could be used to fulfil the requirements of a control system, then the task of the control engineer would be greatly simplified, not to mention the improvement for the commissioning and maintenance engineer.

With the introduction of the microprocessor such an objective of using one technology to produce a whole range of functional industrial controllers became a possibility and is now a reality.

The impact of microprocessor-based controllers (microcontrollers) on the whole range of industrial control equipment is illustrated in *Figure 24.1*. This shows that microcontrollers with suitable input and output equipment will be used in place of relays and analogue and hard-wired digital controllers except for very small or very-fast-response systems. They will also perform some of the more mundane tasks which have previously been carried out by process control computers.

A well-designed microcontroller can have advantages over previous forms of control equipment, the most important being as follows.

Cost: The cost of electronic components is low and reducing in comparison with other forms of hardware.

Flexibility: The function of the equipment is determined by its program, so that extensive modifications or improvements to the system can be made without hardware changes or long production delays.

Reliability: Static equipment with reduced wiring and reduced component count all contribute to improved reliability.

One technology for all types of control: Complete control of a plant function within one controller which would previously have been split up because of different hardware types.

Cheap intelligence: A given amount of plant input information can be used for more sophisticated control, plant fault detection, etc.

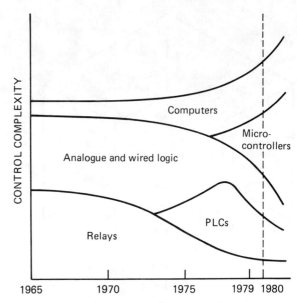

Figure 24.1 Trends in control hardware

However, before these advantages can be realised, there are problems to be overcome, which may be summarised as the need for suitable interfaces between the microprocessor, the plant and the people who work with the equipment.

In addition, as shown in *Figure 24.1*, there is a wide range of system complexity which must be covered economically at all levels.

24.2 General requirements

The necessary range of interfaces is shown in *Figure 24.2*, although not all the interfaces will be required in every system.

To provide illustrative technical detail, features of a typical industrially available microprocessor-based control system (microcontroller) will be described.

The range of plant input/output (I/O) in terms of noise rejection, speed of response and amount of information, is covered by using two I/O highways known as basic I/O and fast I/O channels.

24.2.1 Basic I/O

The input units, in the form of basic I/O cards are designed to accept signals at 110 V a.c., 220 V a.c. or 9–66 V d.c. together with noise spikes up to 2 kV. The output units work at the same signal levels with continuous current ratings of up to 2 A with short-term peaks above this level.

Plant wiring is terminated at terminals which form part of the I/O card.

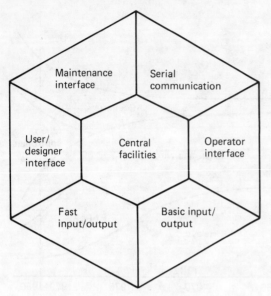

Figure 24.2 Interfaces

The basic I/O highway itself is a pair of ribbon cables which loop between the I/O panels and then to the controller.

All the units include opto-isolation and lamp indication of the state of the signals.

Each card has eight channels, and a total of 512 input or output logic states can be interfaced to the controller via this highway.

Typical uses for basic I/O would be inputs from limit switches and outputs to solenoids and contactors.

24.2.2 Fast I/O

The fast I/O highway is designed for high-speed lower-noise signals and larger total amounts of information than the basic I/O highway. It consists of a printed circuit backplane to which a variety of printed circuit modules can be connected. Ribbon cables from the module rack connect to terminal panels for plant connections.

The range of modules includes analogue input and output (16 channels per card), dedicated controllers, pulse interfaces and low-level digital inputs. Analogue output units work over the range ± 10 V, while input units are available for full-scale ranges of ± 0.5 V, ± 5 V or ± 10 V.

A range of signal conditioning units is available to convert signals from thermocouples, resistance bulb thermometers, strain gauges, CTs, VTs, etc., to the standard level of the analogue input modules. By this means the raw signals may be connected directly to the controller without any interposing equipment.

24.3 Dedicated controller

The dedicated controller is a small but complete microcontroller in itself, including a microprocessor and store, analogue input and output circuitry and a small basic I/O highway. It can in fact be used with a power supply as a small stand-alone system. When plugged into the fast I/O highway, it can carry out dedicated control tasks under the supervision of the main controller. One of its main applications is in position control using coarse/fine resolvers as the transducer. The unit provides resolver excitation and includes standard software for the conversion of the feedback signals into a scaled and datum-corrected position measurement.

This can then be programmed by the user to control analogue or digital outputs to the plant.

24.4 Operator interface

Operator switches, push-buttons and indicators can of course be connected to basic I/O blocks; but where these operator devices are close to the controller, a range of devices is available which connect directly on to the basic I/O ribbon cable. The range includes thumb-wheel switches, decade displays, a contact input unit, a lamp driver and a user-defined 8×8 matrix operator keypad giving a total of up to 64 inputs and 7 outputs, the outputs being in the form of LEDs. If the desk is remote from the main controller, a remote I/O unit can be used which provides a conversion between its basic I/O highway and a serial communications link to the main controller. The combination of ribbon cabling within the desk and serial link to the controller saves enormously on wiring.

One aspect of the use of cheap intelligence mentioned above is the improved presentation of information to operators as video displays and printer output.

Certain controllers include provision for video display by means of a dedicated processor module which plugs into the same highway as the control processor and its store. The video processor produces video output signals to drive black and white or colour industrial monitors. The definition of the formats, which can include tabular displays, bar charts and mimic diagrams, is held in a reserved area of read/write memory while the video processor fetches the data values from the store which it shares with the control processor. Printout of the displayed information is available, and in some systems one of the serial ports on the main processor can be used to drive a serial printer to produce alarm records and production logs.

24.5 Communications with other controllers

Usually controllers include provision for serial links as standard. By this means large amounts of information can be passed as a string of current pulses over two twisted pairs. Standard programs in the controller automatically code the information into serial form so that the link is transparent to the user. A 16 bit check code is included to ensure very high reliability of the communication link.

One important use of the serial link is to provide a connection to a remote I/O unit, one example of which might be an operator's desk, as described above.

The serial links can also be used to transmit data to supervisory computer systems, provided that the computer is programmed to accept the same protocol.

24.6 User-designer interface

As explained above, programmable microcontrollers will be used in place of relays, analogue controllers and other conventional control equipment, often for small systems. It is therefore essential that the control engineer should be able to set the controller up to perform the required function without the need to involve a specialist software expert. Also, the method of programming must allow logic sequencing, closed loop control, data handling and display to be freely combined if the functional advantages of 'one technology' are not to be lost.

A typical control language is based on the familiar ladder diagram representation (*Figure 24.3*) of relay circuits which has

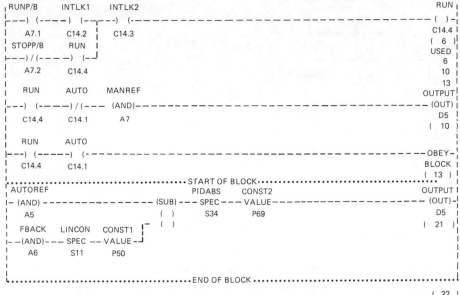

Figure 24.3 Ladder diagram example

been extended to include sequential control, arithmetic, closed loop control, data handling and printout. A portable intelligent programming unit with a video display of the ladder diagram is used to enter the program and features a very high degree of protection against programming errors. The unit can be connected to a domestic cassette recorder to record or reload programs from magnetic tape, or to a printer to produce hard copy of the system program.

When the system is running, the programming unit can be used in monitor mode to display the ladder diagrams complete with a dynamic indication of the state of relay contacts and data values to assist in system commissioning.

24.7 Maintenance facilities

In addition to the monitoring and indication described above, which would assist an engineer to diagnose faults, controllers include comprehensive self-test routines to check the operation of power supply, processor, store and input/output units. These tests are carried out rapidly each time the system is switched on and at a slower rate during normal operation. In the event of failure of units other than the power supply or processor, a simple indication of the faulty unit is given via the programming unit. In addition, the designer can include application dependent checks in the control program so that the controller can also assist in fault finding on the plant as well as within itself.

To summarise, *Figure 24.4* shows a microcontroller configuration including all the above features.

24.8 Applications

Some applications of microprocessor-based control systems follow which illustrate the versatility of this unified approach to industrial control.

24.8.1 Rod mill control

A simplified line diagram of the rod type of rolling mill is shown in *Figure 24.5*. It comprises typically 20 rolling mill stands which reduce the product from a billet approximately 150 mm square to a rod which may be 5 mm in diameter and travelling at 70 m/s. To ensure satisfactory operation, the individual stands need to be controlled so that there is no interstand tension in the rod.

At the early stages of the mill, where the product has a large cross-section so that it will not bend easily, tension-free rolling is achieved by monitoring mill motor drive currents before and after coupling to the next stand and adjusting stand speed to reduce the difference. At the other end of the mill various forms of controlled loop in the rod are used to prevent tension. Often the amount of material stored in the loops is very small, so that fast response control is needed.

Before the use of microcontrollers the mill reference system was a hybrid analogue/digital system: separate analogue controllers were used for tension and loop control, and additional digital units were used for accurate speed indication. A large number of separate operator devices were needed which required an expensive desk and a standing operator. An integrated system including all the above features and some improvements is provided by the scheme shown in *Figure 24.6*.

The mill master system mounted in the operator's control station provides a convenient interface for the operator, allowing him to sit at a much smaller desk. Video displays are used in place of many dedicated instruments, while much of his input is entered by means of function keys and number pad. This system is linked by a serial data link to a number of other units which control the various sections of the mill. These units monitor drive current in analogue form and mill speeds using pulse generators coupled to the mill motors. Analogue-speed references to standard packaged drives are calculated on the basis of initial references from the master system modified by the following control functions.

(1) Cascade control: The operator or the tension and loop controls can modify the speed difference between two stands while preserving all other ratios constant. A new set of matched references are output every 100 ms.

(2) Tension control: The microcontroller control language permits parts of the control programs to be skipped when they are not relevant. This feature is used so that the tension control calculation is only carried out for part of the mill where the nose is passing, thus reducing the load for this function to a low level.

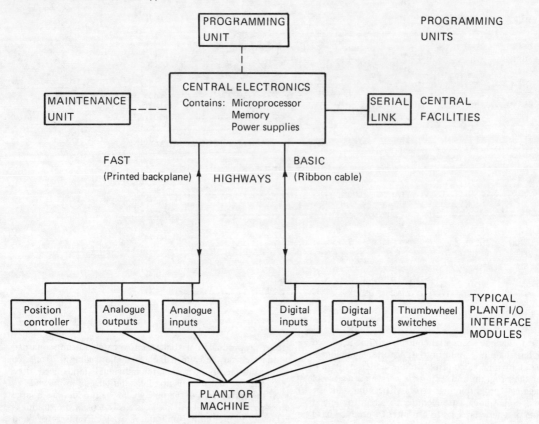

Figure 24.4 Block diagram of typical microprocessor-based control system

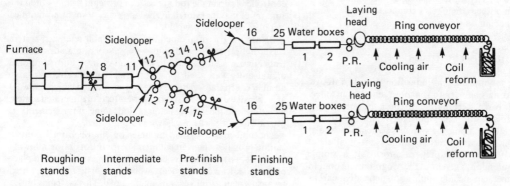

Figure 24.5 Simplified arrangement of typical two-strand rod mill

(3) Digital speed trim: The pulse generators give the system an accurate measure of speed which is used to correct for slow temperature drifts or calibration errors in the analogue drive controls.

(4) Loop control: To achieve the necessary response, dedicated controllers are used to control the stands next to storage loops.

The system can also calculate r.m.s. load on each stand from the current feedback signals to check for thermal overload of the motors. This form of protection can be tailored to meet the machine overload rating far more exactly than thermal overload relays.

24.8.2 Mine winder event recorder

Mine accidents have highlighted the need for a recorder to monitor conditions on winding equipment which could be examined in the event of an accident, similar to the black box recorder on aircraft.

The older recorders used ink and paper to record and ran continuously. They used a very slow paper speed (time base) and recorded very few signals, which made the results difficult to interpret and of limited value. The recorder also needed very regular maintenance.

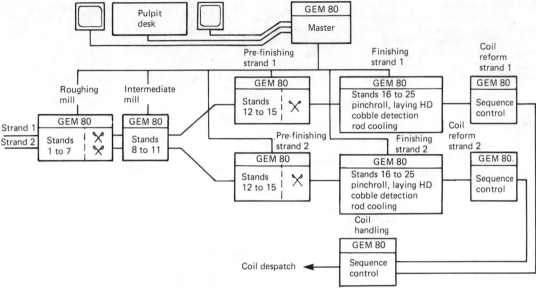

Figure 24.6 Block diagram of systems for typical two-strand rod mill

The ideal recorder must enable the cause of the accident, or dangerous incident, to be determined and must require a minimum of maintenance. It will remember data in a re-useable store and only produce permanent records when they are needed. The aircraft industry uses a 'black box' flight recorder, but analysis of its memory, after an incident, requires work in a specialist laboratory. The mining industry wants the data to be available immediately, so that the record is acceptable to all concerned.

The use of tape recorders is one solution, but they suffer from wear problems. A re-useable memory which will not wear out is required such as a battery-supported semiconductor memory, to print out the data when required. A printer using electrosensitive paper has the best reliability.

To investigate a dangerous incident properly requires a lot of data and, with so much data available, it is evident that the recorder can do more than simply enable dangerous incidents to be investigated. This leads to other information being extracted for the benefit of the user. The mine hoist recorder can be used for: (a) investigating dangerous incidents; (b) giving data on past performance and signalling, when requested; (c) assisting in fault diagnosis, particularly of intermittent faults; (d) monitoring the operation of speed–distance protection contrivances; (e) providing speed–distance protection for the hoist; and (f) giving data on hoist utilisation.

24.8.2.1 The data and method of storing

The recorder records analogue data such as speed and current, counts pulses for depth measurement and for bell ringing codes, and recognises events such as circuit-breakers opening and brakes being applied. These are recorded against time. The information is put into static semiconductor, random access, memories (RAM) in digitally coded form. The system scans the memory sequentially, in what amounts to a continuous loop. The oldest data are overwritten by new, but past history is always available. The size of the memory store and the frequency of the data coming in determine the extent of the past history.

24.8.2.2 The event store

A portion of the store is devoted to 'events', which are inputs going on or off.

Normally the events will include both the shaft signal lamps and the count of the bell rings with which the lamps should tally. In addition, there will be signals from the hoist control (relay) logic and the state of the safety circuits and the various circuit-breakers. There will also be a check on the duty cycle of the hoist. This will be achieved by monitoring the speed and noting the four cardinal points of the duty cycle as shown in *Figure 24.7*.

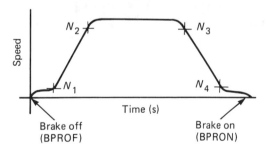

Figure 24.7 Duty cycle check: N_1, low speed, speed rising; N_2, high speed, speed rising; N_3, high speed, speed falling; N_4, low speed, speed falling. The start is indicated by BPROF (brake proving relay off) and the stop by BPRON (brake proving relay on)

If the wind (journey) takes longer than a predetermined time, then a slow wind should be recorded. At speed N_1, the opportunity is taken to check things which could affect safety, such as headgear (end of travel) limit switches being short-circuited (checked from a back pole of the switch). The first time that the brakes are applied, after N_3, is also recorded together with the depth at which that occurred.

In order to compress the event–data–time record, the event inputs, although scanned regularly, are not recorded until a change occurs. When a change occurs, the time and the details of the event are stored. The time is stored as minutes and seconds but the change of each hour is also stored. When printing, time always starts a new line. The events themselves are stored as a number which can be decoded later into the appropriate mnemonic. The mnemonics can be chosen to suit the individual user and can be changed at will.

24.8.2.3 Analogue data

Typically, the 'analogue' signals will be: (*a*) depth of the conveyance (m or ft); (*b*) speed of the conveyance (m/s or ft/s); (*c*) acceleration or deceleration (calculated from speed and time); (*d*) armature current or rotor current (kA); (*e*) field current or stator current (A); (*f*) brake effort (either from brake back-pressure or from a torque transducer). Motor torque can be determined from (*d*) and (*e*).

The analogue data are put into store to both a 0.1 s time base (fast analogue) and a 1 s time base (slow analogue). The part of the program which does this is tied to the time change of 0.1 s.

24.8.2.4 Speed–distance safety shadow

Speed–distance protection curves are held in the memory, and the speed of the hoist is checked against the position of the conveyance in the shaft. This is done for both coal and man winding, throughout the shaft, for both conveyances. Cross-checks are also made between the depth counters for each conveyance. This facility could be used as the electronic winder overspeed protection.

24.8.2.5 Production data (hoise utilization)

For each hour, the store remembers: (*a*) the minutes for which the hoist is stopped; (*b*) the number of times it ran at full speed for more than a predetermined time (fast winds); and (*c*) the number of times it ran continuously for more than a predetermined time, without qualifying as a fast run (slow winds). When the appropriate push-buttons call for a reprint of the data, they are printed for each complete hour during the previous 24 h. In addition, a 24 h summary of the data is printed daily (automatically) at a predetermined time.

24.8.2.6 Program arrangement

An outline of a typical microcontroller program follows. First, all the inputs are read into a store. Then any of the following, which are required, are performed but not necessarily on the same scan. Some must be dealt with more often than others.

Set initial conditions (when starting initially).
Set test conditions (authorised person's tests).
Scale speed and calculate acceleration.
Store 0.1 s time base analogue data.
Store 1 s time base analogue data.
Count bell rings.
Determine N_1, N_2, N_3, N_4.
Store events.
Set time (if required, e.g., after switch-on or change of hour).
Autoprint logic.
Print request logic.
Set store pointers for printing (when required).
Print headings (when required).
Print line of 0.1 s time base analogue data (if required).
Print line 1 s time base analogue data (if required).
Print an event (if required).
Speed–distance shadow.
Hoist utilisation data.

Then the outputs are sent out and the sequence starts again.
Figure 24.8 shows the equipment configuration.

24.8.3 Distributed control of water supplies

24.8.3.1 Introduction

This section describes the application of some of a range of microprocessor-based controllers to the first stages of a potable water storage and distribution scheme.

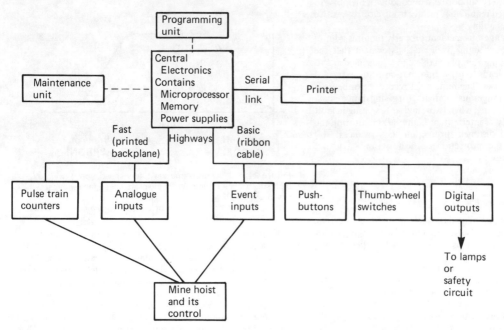

Figure 24.8 Equipment configuration

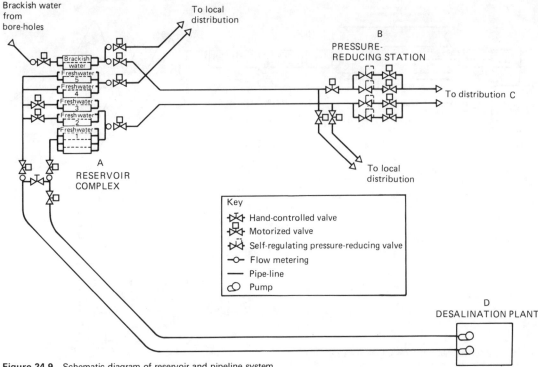

Figure 24.9 Schematic diagram of reservoir and pipeline system

Sea-water is taken and desalinated before being pumped to fresh-water reservoirs forming part of a reservoir complex on high ground. Brackish water, used for blending with the distilled water to improve the taste, is abstracted from bore-holes elsewhere and stored separately in the same complex. Water from these reservoirs is distributed partly via a pressure-reducing station and partly to reservoirs at other locations for further distribution to surrounding villages. *Figure 24.9* shows a schematic diagram of the pipeline and reservoir system.

Controllers interfacing directly to conventional instrumentation have been used at the three stations. These stations are geographically remote from one another, with distances varying from 8 to 17 km, necessitating a telemetry network. The inherent serial communications facility provided by microcontrollers has been used for telemetry by connecting controllers via modems at each end of a data link. The local terrain allows use of a cable link between A and B but a UHF radio link is required between B and D.

The pressure-reducing station at B and the unmanned reservoir station at A are controlled from a master control centre installed at B and a slave control centre installed at D. *Figure 24.10* shows the control arrangement of the entire system. These control centres provide facilities for remote monitoring and control of: (a) reservoir levels; (b) pipe-line pressures and flows; and (c) control valve positions and pump status as well as alarm annunciation.

A printer connected at B is used to give additional logging and event-recording facilities. The most significant feature of this application is the use of the microcontroller as a remote control telemetry system.

24.8.3.2 System description

Each of the three locations has a control panel equipped with a microcontroller system. Apart from the processor, these include, where necessary, analogue input and output (I/O) modules, isolated d.c. I/O blocks for interfacing of digital plant I/O contact input units for push-buttons and switches, mounted on the front of the panel, and lamp driver units to power alarm and status lamps. Moving coil indicators are used to display analogue quantities. A few relays and timers are also included.

Reservoir complex A *Figure 24.9* shows that a number of motorised valves are incorporated in the inlets to and outlets from the reservoirs. Their actuators provide an analogue signal, 4–20 mA, proportional to valve position and also digital signals for 'open', 'closed' 'fault' and 'local control only'.

As may be seen from *Figure 24.9*, the reservoirs are each subdivided into compartments. A pressure bulb and transducer combination in each of these gives an analogue signal proportional to compartment level, and 'pear-drop'-type float switches provide separate high- and low-level alarms.

Control commands for the valves from the microcontroller are produced via standard output blocks and operate on a 24 V d.c. system.

As the control station at A is intended for unmanned operation, very few hand controls are needed on the control panel there. All commands are output via the microcontroller at A except for one back-up loop, to be described below; all control commands for the equipment at A are received via the telemetry link from B. All data collected at A are sent to B via the telemetry link.

It was desired that some degree of control and indication should be maintained in the unlikely event of one of the microcontrollers or the serial links being off-line for any reason. Accordingly, if a high-level float switch operates in any compartment of a reservoir, the associated inlet valve is driven fully closed, regardless of the commands sent by the controller. Similarly, if a low-level float switch operates, the associated inlet valve is driven fully open.

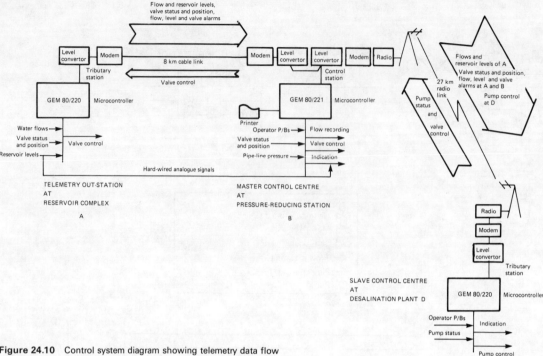

Figure 24.10 Control system diagram showing telemetry data flow

Each of the four outlet pipe-lines is equipped with a flow switch for detection of extra high flow, such as would occur if a pipe-line burst. If this flow persists, then the valve feeding that pipe-line will be driven closed and then locked out, it being necessary to operate a push-button on the control panel to rest. This function is backed up by the controller for extra security.

In addition to the cable link for the telemetry, a further four circuits have been laid between A and B for transmitting reservoir levels, manually selected at A, independently of the microcontrollers. The selected signals are also displayed locally on the control panel.

A further indicator is provided to allow the position of any selected valve to be displayed at A.

Pressure-reducing station B A pair of pipe-lines run from A to the pressure-reducing station at B where each pipe-line divides to pass through two parallel self-regulating pressure-reducing valves each in series with a motorised control valve. Other motorised valves control branches off the main pipe-lines.

The pressure in the pipe-line both upstream and downstream of the pressure-reducing station is continuously measured by a pressure transducer for display on the front of the control panel at B.

All of the control valves at B and at A have their own group of controls on the front of the control panel at B. Each group consists of a valve position indicator, status lamps, control mode selector switch and control push-buttons. Thus, full manual control can be carried out.

All water flows and reservoir levels from A are displayed on the front of the panel, the latter by vane switch relay-type instruments. These are used to initiate stop and start commands for the pumps feeding the reservoirs. When a pump has been registered as stopped via the telemetry link, the corresponding reservoir inlet valve is driven closed.

The indicators used to display the pipe-line pressure downstream of the pressure-reducing station are also vane switch relays. Their high and low contacts provide alarms, and will also close and open, respectively, the motorised valves situated in the pressure-reducing station.

Vane switch relays are used for auto control in preference to incorporating the functions in the microcontroller user program in order, once again, to allow some control to be maintained if the controllers or the serial links should be off-line for any reason.

All alarm lamps and some of the valve status lamps are required to operate in the manner of an alarm annunciator, consequently, the necessary audible alarm device and alarm accept push-button are incorporated, and the microcontroller user program augmented to create the required sequence. All lamps on the control panel can be tested by depressing a lamp test push-button.

Data recording facilities at B include seven chart recorders, for the water flow rate signals from A, and a printer unit for logging and event recording. Reservoir levels, line pressures and integrated flows are logged on demand or at three programmable times per day. Alarms and valve status changes are logged as they occur.

Desalination plant control centre D The control panel at D is very similar to that at B, and displays valve status and position data, water flow rates and pressures, reservoir levels and alarms.

In order to control a valve at A or B from D, the appropriate control mode selector switch at B must be set to 'D control'. This condition is indicated by a lamp at D and the push-buttons can then be used to control the valve.

The alarm annunciator facility at B is duplicated at D: (*a*) alarms will appear at both locations and can be accepted at either; (*b*) all lamps can be tested by depressing a lamp test push-button.

Telemetry One of the features of some controllers is the provision of at least one serial port for high integrity, two-way communication with other controllers. By including modems in the signalling paths and by programming the controllers using the ladder diagram control language, this feature has been used to build-up a telemetry network between the three locations. The controller at B is connected using a multidrop arrangement to the controllers at D and A (see *Figure 24.10*).

24.9 Serial communication between microcontrollers

Special signalling protocols suitable for the specific requirements of a microprocessor-based control system are desirable, to lower processor overheads while at the same time retaining high data integrity and providing informative link statistics and diagnostics.

All communication is two-way between a control station and one or more tributary stations. The control station initiates all transmissions, and tributary stations can only respond to interrogations from the control station. The system can operate either on a point-to-point basis or the link from the control station can be connected to up to seven tributary stations in a multidrop arrangement as shown in *Figure 24.11*.

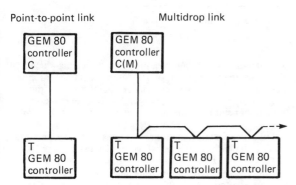

Figure 24.11 Methods of connecting GEM 80 controllers for serial communication: C, control station; C(M), multidrop control station; T, tributary station

One such protocol is known as ESP, which is used with the GEM 80 series of microcontrollers and provides the following facilities.

24.9.1 Data integrity and error recovery

ESP includes a 16-bit cyclic redundancy check code (CRC) at the end of every message. The CRC generator polynomial used is $x^{16}+x^{15}+x^{2}+1$ which detects the following errors: (*a*) any odd number of single errors; (*b*) any error burst covering up to 16-bits; (*c*) all but one in 32 768 of 17-bit burst; (*d*) all but one in 65 536 of longer bursts; and (*e*) any combination of two bursts of 1 or 2 bits. Recovery from transient errors is automatic using retries and timeouts.

24.9.2 Link statistics and diagnostics

ESP control stations maintain a record of the following link performance statistics in the form of monitorable decimal numbers: (*a*) requests; (*b*) attempts and retries; (*c*) framing and overrun errors; (*d*) timeouts; (*e*) corrupt replies; (*f*) negative acknowledgements; (*g*) termination failures; and (*h*) data format faults.

After three consecutive unsuccessful retries by a control station or a period of link inactivity at a tributary station, a link failure is recorded. The system designer can use this information within the control program to take suitable protective action. Separate transmit and receive activity indicators are provided at each serial port which, together with the link statistics, provide the user with powerful fault diagnostic and preventive maintenance aids. This ensures that the user/maintainer can keep that link in peak operating condition.

24.9.3 Signalling paths

The microcontroller serial ports are connected using a 20 mA current loop technique with optical isolation which eliminates any ground loop problems. Where requirements exist for the signalling path to include public telephone lines or radio path transmissions, it is merely necessary to include the appropriate modems. Where modems are not used, the current loop technique allows transmission over relatively large distances (several kilometres). *Figure 24.12* shows maximum permissible signalling rates against link length for various numbers of tributaries in a directly connected multidrop arrangement. The links are wired full duplex (separate transmit and receive loops)

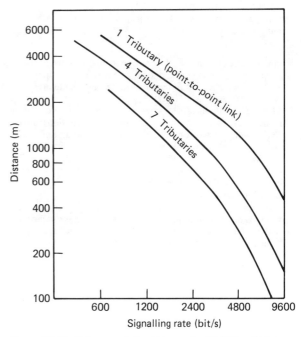

Figure 24.12 Safe transmission distances for GEM 80 serial links

but only operate in half duplex (one direction at a time). Connection is by a twin-twisted pair lightweight cable with overall screen, connected to a termination panel which, in turn, is connected by ribbon cable to its controller.

In a multidrop arrangement a shorting relay is provided on the termination panel to allow power-off or disconnection of any multidropped tributary controller without interruption of the current loops to other controllers on the link.

24.9.4 Telemetry

The inclusion of the appropriate modems in the signalling path of a serial link enables land-line or radio path transmissions to be included in the microcontroller control scheme. In the water

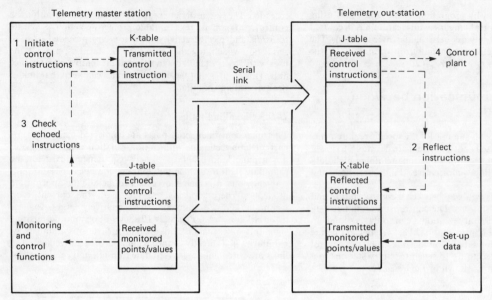

Figure 24.13 Diagram demonstrating sequence of events (shown numbered) for successful transmission of a control instruction

supply control scheme the recommended modems use a frequency shift keying technique capable of data rates up to 1200 baud. This was the limiting factor in choosing the baud rate of the link at 1200 baud.

From calculations of error rates on the radio equipment due to the signalling path, and assuming worst case corruption from the CRC's ability to detect errors, one would expect an undetected error to occur on the link approximately once every 10 years. However, when the possibility of equipment failures or the effect of atmospheric disturbances are taken into account, it is clear that some additional security has to be provided for messages controlling the operation of plant.

In a typical remote control installation, the telemetry outstation is required to retransmit a received message for checking by the originating station before actioning the message.

On the water distribution systems two distinct forms of data are transferred using the serial links: (1) instructions to control the operation of remote plant; and (2) a large quantity of monitored plant parameters. To allow simple programming in the microcontroller control language, these two types of data are allocated dedicated portions of the serial input and output data tables. This allows independent handling by the ladder diagram control program and eliminates problems in deciding which type of message to send as both sorts of data were catered for in one message. A simplified system with control instructions from the master station to the out-station and monitored values/points from the out-station to the master station, is illustrated in *Figure 24.13*.

Because of the large quantities of monitored values/points (35 words of data from A to B and 48 words of data from B to D), not all the data can be sent in one message transmission. A simple system of sequentially multiplexing different blocks of data into the serial output data tables on consecutive message transmissions was devised. Ideally these blocks of data would have been as large as the maximum possible message length of 32 words allowed, thus minimising the time to transmit all the monitored values/points. However, for a given error rate, the information rate drops with increased message length as both the likelihood of a message being hit and the time for retransmission increase. This means that for any level of noise there is an optimum message length. In practice, because error rates in the field were unknown, a message length was chosen to give acceptable response while still coping with a high corruption rate.

25 Analogue, Hybrid and Digital Computation

D J Holding
Department of Electrical and Electronic Engineering and Applied Physics, Aston University
(Sections 25.1–25.4.6)

P J King
Control Systems Engineering, Tosco Corporation
(Sections 25.5–25.17)

P H Hammond, FIEE, MInstMC
Computer Aided Design Centre, Cambridge
(Sections 25.18–25.35)

R K Wood, PEng
Department of Chemical Engineering, University of Alberta
(Sections 25.36–25.41)

Contents

DIGITAL AND PROGRAMMABLE ELECTRONIC SYSTEMS

25.1 Digital systems 25/3
    25.1.1 Digital signals 25/3
    25.1.2 Digital representation of information 25/3
    25.1.3 Textual information 25/3
    25.1.4 Logical information 25/4
    25.1.5 Numerical information 25/4
    25.1.6 Binary number system 25/4

25.2 Processing binary information 25/4
    25.2.1 Logical operations 25/4
    25.2.2 Boolean algebra 25/5
    25.2.3 Logic gates 25/5
    25.2.4 Combinational logic 25/5
    25.2.5 Arithmetic operations 25/6
    25.2.6 Digital multiplexers 25/7
    25.2.7 Encoders and decoders 25/8
    25.2.8 Programmable logic arrays 25/8

25.3 Sequential logic 25/8
    25.3.1 Asynchronous sequential logic 25/9
    25.3.2 Timing problems 25/9
    25.3.3 Synchronous sequential logic 25/10
    25.3.4 Clocked flip-flops 25/10
    25.3.5 Memory registers 25/11
    25.3.6 Shift registers 25/11
    25.3.7 Counters 25/12
    25.3.8 Tri-state logic 25/13
    25.3.9 Memory devices 25/14
    25.3.10 Combinational logic design using memory 25/14
    25.3.11 Sequential logic design using memory 25/16
    25.3.12 Structured design of programmable logic systems 25/17

25.4 Programmable systems 25/17
    25.4.1 The logic design of a digital computer system 25/17
    25.4.2 Architecture 25/18
    25.4.3 Control and timing unit 25/18
    25.4.4 Arithmetic logic unit 25/19
    25.4.5 Memory unit 25/20
    25.4.6 Microprocessors 25/20

DIGITAL COMPUTERS AND COMPUTER PROGRAMMING

25.5 Digital computer structure 25/21

25.6 Interrupts 25/23

25.7 Input/output 25/23

25.8 Instruction execution 25/23

25.9 Computer instruction types 25/24

25.10 Computer instruction format 25/24

25.11 Computer program execution 25/25

25.12 Computer programming 25/25

25.13 Software design and specification 25/26
    25.13.1 Top down design 25/26
    25.13.2 Structured programming 25/26
    25.13.3 Mascot 25/27
    25.13.4 Modular program structure 25/27
    25.13.5 General design procedures 25/27

25.14 Machine language 25/27
    25.14.1 Assembly language 25/27
    25.14.2 High-level languages 25/28

25.15 Program development 25/29
    25.15.1 Program testing 25/29

25.16 High-level language types 25/30
    25.16.1 FORTRAN 25/30
    25.16.2 BASIC 25/31
    25.16.3 Pascal 25/31
    25.16.4 Other high-level languages 25/32
    25.16.5 Merits of high-level and low-level languages 25/32

25.17 Real-time operation 25/32

ANALOGUE COMPUTATION

25.18 Operational amplifier 25/34
    25.18.1 Voltage gain 25/34
    25.18.2 Input resistance 25/34
    25.18.3 Transient response 25/34
    25.18.4 Stability 25/34
    25.18.5 Output impedance 25/34

25.19 Operational amplifier integrated circuits 25/34

25.20 Analogue functional elements 25/34
    25.20.1 Inverting amplifier 25/34
    25.20.2 Summer 25/35
    25.20.3 Integrator 25/35
    25.20.4 Scaling 25/35
    25.20.5 Differentiator 25/36

25.21 Complex functions 25/36

25.22 Voltage follower 25/36

25.23 Function generators 25/37

25.24 Comparator 25/37

25.25 Diode function generator 25/37

25.26 Multifunction converter 25/38

25.27 Analogue computer 25/39

25.28 Applications 25/40

HYBRID COMPUTATION

25.29 Signal switches 25/40

25.30 Sample and hold 25/41

25.31 Digital-to-analogue converter (DAC) 25/41

25.32 Analogue-to-digital converter (ADC) 25/42
    25.32.1 Parallel or flash conversion 25/42
    25.32.2 Successive approximation conversion 25/43
    25.32.3 Dual slope integrating conversion 25/44

25.33 Analogue multiplexers 25/44

25.34 Hybrid computer 25/45

25.35 Hybrid computer applications 25/45

DIGITAL SIMULATION

25.36 Introduction 25/46

25.37 System models 25/47

25.38 Integration schemes 25/48

25.39 Organization of problem input 25/48

25.40 Illustrative examples 25/48
    25.40.1 Example 1 25/48
    25.40.2 Example 2 25/49
    25.40.3 Example 3 25/50

25.41 Conclusions 25/51

DIGITAL AND PROGRAMMABLE ELECTRONIC SYSTEMS

25.1 Digital systems

Digital systems are used to process discrete elements of information. A digital electronic system processes discrete electrical signals. They can be designed to perform simple logic and arithmetic operations and can therefore be used to construct the basic operational parts of a calculator. A digital electronic system can also be used to hold or store discrete elements of information and this gives the system a memory capability. The ability to store information or data and to process the data by logical or arithmetic operations is central to the design of nearly all digital information processing systems including digital computers.

The function of a digital system is determined by the sequence of operations which are performed on the information or data being processed. A digital system can be classified by the way in which its sequence of operations is implemented.

A digital system is considered to be hard wired if the sequence of operations is governed by the physical interconnection of the digital processing elements. For example, in hard-wired logic systems the physical interconnections of the elements govern the routes by which data flow between the processing elements and thus, the sequence of processing operations performed on the data. This type of system is inflexible because the design is specific to a particular processing function: if the processing function is changed, then the processing elements and their interconnections have to be altered.

A digital system is considered to be programmable if a prescriptive program of instructions can be used to control the data-processing function of the system. This type of system usually incorporates a general-purpose processing element which is programmed to implement a specific function in a predetermined way. The coded instructions are normally stored in the memory part of the system and the program forms an integral part of the system.

The ability to define the functions of the digital system by programming introduces considerable flexibility into the system because the programming operation can take place after the general-purpose digital elements have been designed. It also means that identical hardware designs can be used in a number of different applications, the system being tailored to the individual tasks by the applications program.

The digital computer is a very important class of stored program system. The computer is distinguished by the fact that its processing function depends on both the prescriptive sequence of coded instructions and the value of the data being processed. In effect, the program prescribes a number of possible sequences of operations and the conditions under which they may be carried out. The computer, under program control, assesses the data and determines which specific sequence of instruction is to be executed. It is the ability of the computer to take into account the nature of the data being processed, when taking decisions about the type of processing to be performed, which makes the computer such a significant and powerful information-processing device.

All three forms of digital electronic system find widespread application. Hard-wired logic is used extensively to perform combinational and sequential logic operations and to provide the control and interface logic for more complex digital components such as microprocessors and other very large-scale integration (VLSI) devices. Since hard-wired logic can be optimised for a particular application it is also used in high-speed applications such as digital signal processing. Hard-wired logic circuits are often designed as 'random logic' systems and are implemented using discrete logic devices and memory elements which are available as small-scale integration (SSI) circuits.

Modern design techniques make systematic use of a wide range of digital components, including general-purpose devices such as digital multiplexers, read only memories and programmable logic arrays which are available in medium-scale integration (MSI) form. Emphasis is often placed on the overall design of these systems and this has led to the development of 'top-down' or structured methods of logic design.

Where flexibility is required, it is common to use programmable systems particularly in more complex applications. The simpler fixed-function systems are often used in repetitive tasks such as input scanning and data acquisition. They are also used in mass-produced products and as components of larger systems such as telephony equipment. However, many of these applications now involve fully programmable digital computers and make good use of the additional flexibility which these systems provide. The trend towards computer-based solutions is reinforced by the continually increasing computational power of the microprocessor. Such systems are providing economic solutions to design problems in an increasingly wide range of application.

25.1.1 Digital signals

Digital electronic systems operate on physical quantities such as the level of a voltage or current. These quantities or signals are used to represent discrete elements of information. The signals within a digital electronic system are normally allowed to take only two discrete values because two-valued or binary signals can be represented unambiguously and reliably by electrical circuits or magnetic storage devices which have two clearly defined states. These states are often referred to as the binary coefficients 0 and 1, or the logical values 'false' and 'true'. It follows that all discrete elements of information in the system must be represented by groups of binary digits.

The number of binary digits needed to encode the various discrete elements of information in a system has a significant effect on the design of a digital system. The system must be able to move, process and store discrete elements of information using either simultaneous (parallel) operations on a group of binary digits or sequential (serial) operations on a stream of binary digits. Since one binary digit or bit of information can be used to encode only two data objects, it is common to find digital systems operating on groups of 8, 16 or 32 bits of information at once. The range of different objects which can be encoded by a group of n bits is 2^n. Systems which operate on larger groups of bits can therefore process a wider range of encoded information.

25.1.2 Digital representation of information

Digital systems may be used to process textual, numerical or logical information. This information usually takes the form of a sequence or string of alphabetic characters, punctuation symbols, decimal digits and the symbols representing arithmetic operators, logical values and logical operators. It also includes the spaces which mark the boundaries between various words or quantities. This sort of information is represented in a digital system by codes of binary digits. Since the information must be input, processed and output in coded form, the encoding must be reversible and assign a unique representation to each element of information. Each item of information or data whose value can change during processing is known as a variable and can be classified by the type of information which it represents.

25.1.3 Textual information

Character codes are used to represent textual information. These codes are usually based on a group of seven binary bits which have a range of $2^7 = 128$. This is sufficient to encode the fifty-two upper- and lower-case characters of the alphabet, the characters

0–9, and certain punctuation and other symbols as used in text processing. Seven-bit character codes are widely used to represent textual variables and constants and are standardised to allow communication between different systems. The American Standard Code for Information Interchange, known as the ASCII code, is the character code generally used in digital electronic systems.

The seven-bit code is usually augmented by an eighth bit, the parity bit, which is chosen in such a way that the total number of 1 bits in each character code is either even (for even parity coding) or odd (for odd parity coding). The inclusion of a parity bit allows primitive error detection to take place, for if one bit, or an odd number of bits, within a character is changed during data transmission, then the received character will not have the correct parity. The memories of most general-purpose digital systems are configured to hold groups of eight bits of information (1-byte) and can store character codes. Similarly, most computers and microprocessors process 8-bit data and can be used for processing text information.

25.1.4 Logical information

Logical or Boolean variables have two states, FALSE and TRUE and can be represented by a single binary digit. The usual coding assigns the binary value 0 to FALSE and 1 to TRUE. In signal processing systems it is usual to process each logical variable separately, often using hard-wired logic. Individual logical variables can be processed in programmable systems. However, this does not make good use of the processing capability of programmable systems which can process groups of binary signals. In programmable systems such as microprocessors it is common to find that if a group of logical variables has to be processed by identical operations then the operations are performed simultaneously on each member of the group.

25.1.5 Numerical information

Digital systems can manipulate numbers as either text characters or numerical values. In many engineering, scientific and commercial applications, numerical values are processed using arithmetic and logic operations. Within a digital system discrete numerical values are represented using binary number codes which are valid under arithmetical or logical operations. This allows the same system of coding to be used for both the operand and resultand data in any specific processing operation. The coding of numerical data often involves many binary digits because the range of discrete values of each binary number must be sufficient to accommodate the corresponding numerical quantity.

25.1.6 Binary number system

The binary number system is fundamental to the representation of numbers as binary codes. The binary digit can take two values (0 or 1) and is said to be of base 2. Binary numbers consist of the coefficients of successive powers of the binary base and can be used to form both whole numbers and fractions. Thus, the binary number 1110.0101 represents the series:

$$1(2^3) + 1(2^2) + 1(2^1) + 0(2^0)$$
$$+ 0(2^{-1}) + 1(2^{-2}) + 0(2^{-3}) + 1(2^{-4})$$

which has an equivalent decimal value 14.3125. It is usual to include the base as a suffix to a number to distinguish between various number representations.

The range of a multi-digit binary number is defined by the number of digits and this limits the number of discrete values which can be represented by a particular number. For example, the 8-bit number $1110.0101_{(2)}$ is one of only 256 discrete values which can be represented in this specific format if the binary point is considered to have a fixed position.

The conversion from decimal to binary can be performed by the successive division of the whole number and fractional parts of the decimal number by the binary base to yield the coefficients of the corresponding binary number:

$$6.75_{(10)} = \qquad 6.0 + 0.75$$
$$= 3(2^1) + 0(2^0) + 1(2^{-1}) + 0.5(2^{-1})$$
$$= 1(2^2) + 1(2^1) + 0(2^0) + 1(2^{-1}) + 1(2^{-2})$$
$$= \qquad 110.11_{(2)}$$

The binary number system can be used to represent positive values directly. It can also be enhanced to represent negative number by augmenting each binary number by a sign-bit. However, such a number system has two representations of zero $(+0, -0)$ and cannot be used directly in many logical and arithmetical operations.

In most digital systems, positive numbers are represented as in the binary number system and negative numbers are represented using number systems based on complements. For example, in the widely used two's complement ($\bar{2}$) number system, the complement of a binary number of n bits with an integer part of m bits ($m < n$), is calculated by subtracting the binary number from 2^m (or more simply by complementing each of the n bits of the binary number and then adding 1 to the least significant bit):

$$14.3125_{10} = 1110.0101_2 = 1110.0101_{\bar{2}}$$
$$-14.3125_{10} = (10000.)_2 - (1110.0101)_2 = 0001.1011_{\bar{2}}$$

The two's complement number system has a unique representation of zero (0), and is consistent under the arithmetical operations of addition and subtraction: these operations also form the basis for more complex operations such as multiplication and division. This number system is adopted for arithmetic operations and logical manipulations in many digital electronic systems and microprocessors.

25.2 Processing binary information

Each element of information within a digital system is represented as a binary code and is stored, transmitted and processed as a set of binary signals. Within the digital electronic system the binary signals are processed by digital logic circuits which route the binary signals through appropriate combinations of logic gate. Each logic gate implements a primitive binary logic operation and the function of the complete digital logic circuit may be described mathematically in binary logic, which is better known as Boolean algebra. Logic or Boolean algebra is used extensively in the specification and design of digital logic systems, because of the one-to-one correspondence between the logical operators and digital logic gates.

25.2.1 Logical operations

Binary signals are processed by binary logical operations in which each signal carries one bit of information. All binary logic functions can be transformed into operations involving three primitive operators, 'NOT', 'AND' and 'OR'. The most elementary operator is the unary or mondic operator 'NOT', which has only one input signal or operand:

if z is NOT x then $z = 1$ for $x = 0$ and $z = 0$ for $x = 1$

The NOT operator is often called an invertor and its output or resultand is denoted by an inversion symbol, usually a superscript bar or prime:

$\bar{x} =$ NOT x $\qquad$ or $\qquad$ $x' =$ NOT x

The other two pimitive operators, AND and OR, have two inputs and are said to be dyadic. Boolean expressions involving these operators are written with the operator infixed between the inputs:

$$x = x \text{ AND } y \qquad x = x \text{ OR } y$$

The operators AND and OR are often shown symbolically as "." and "+" and are defined in the following truth tables:

AND			OR		
INPUTS		OUTPUT	INPUTS		OUTPUT
x	y	$x \cdot y$	x	y	$x + y$
0	0	0	0	0	0
0	1	0	0	1	1
1	0	0	1	0	1
1	1	1	1	1	1

More complex binary logic functions, such as NAND, NOR and Exclusive OR (XOR) can be expressed in terms of the primitive operators NOT, AND and OR:

$$x \text{ NAND } y = (x \cdot y)'; \qquad x \text{ NOR } y = (x + y)';$$

$$x \text{ XOR } y = (x + y) \cdot (x \cdot y)'$$

25.2.2 Boolean algebra

Boolean algebra provides a set of general rules for logic operations on binary variables and is widely used in the systematic description and manipulation of logic circuits. The rules consist of a set of assertions and theorems deduced from these assumptions. Four rules deal with only one logical variable x and the digits 0 and 1, and each rule states that if the left part of the equation is true then the right part must also be true::

$$x + 0 = x; \qquad x \cdot 1 = x; \qquad x + 1 = 1; \qquad x \cdot 0 = 0$$

The other laws deal with more than one binary variable and each may be verified by substitution of the binary values 0 and 1:

Idempotent Law:

$$x + x = x \qquad x \cdot x = x$$

Complementary Law:

$$x + x' = 1 \qquad x \cdot x' = 0$$

Involution Law:

$$(x')' = x$$

Commutative Law:

$$x + y = y + x \qquad x \cdot y = y \cdot x$$

Associative Law:

$$x + (y + z) = (x + y) + z \qquad x \cdot (y \cdot z) = (x \cdot y) \cdot z$$

Distributive Law:

$$x \cdot (y + z) = x \cdot y + x \cdot z \qquad (x + y)(x + z) = x + y \cdot z$$

Absorption Law:

$$x + xy = x \qquad x \cdot (x + y) = x$$

De-Morgan's Theorem:

$$(x + y)' = x' \cdot y' \qquad (x \cdot y)' = x' + y'$$

These laws are commonly used in the simplification or manipulation of logic functions. For example, De Morgan's Theorem defines the transformation between AND and NAND operations, or between OR and NOR operations and is widely used when implementing logic circuits.

25.2.3 Logic gates

Boolean algebra provides an efficient notation for the specification and design of binary logic systems. The designs can be implemented as digital electronic circuits. Each part of the circuit which performs a logical operation such as NOT, AND or OR is known as logic circuit or gate. The logical functional of such circuits can be shown in a schematic or logic circuit diagram in which each gate forms a logic circuit element and is indicated by the corresponding symbol:

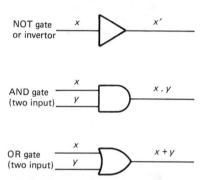

Figure 25.1 Fundamental logic gates

More complex logic functions such as the widely used NAND and NOR or the exclusive-OR, XOR, can be built using combinations of these gates, or may be implemented directly using appropriate circuit elements:

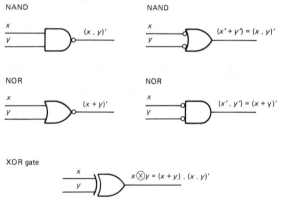

Figure 25.2 Logic gates

25.2.4 Combinational logic

Many digital signal processing systems are required to produce output signals which are simple logical combinations or Boolean functions of the input signals. This type of logical function can be implemented using networks of logic gates in which the signal paths do not form feedback loops. The absence of feedback ensures that the output of the circuit is determined by only the logic function of the circuit and the present set of inputs. Such a circuit is known as a combinational logic circuit.

Combinational logic is used extensively in a wide range of applications. It provides many of the logical and arithmetic functions used in signal processing and computing systems. In programmable systems it is used to decode memory addresses and to provide the signal routing and bus structures necessary to transfer data from one part of the system to another. It is also used to provide much of the control and interface logic within these systems.

The function of a combinational logic circuit can be specified as a Boolean equation in which the output variable is expressed in terms of the input variables. This equation defines the value of the output for each combination of the input variables. The complete set of inputs and their corresponding outputs are often listed in a truth table which can be used as an alternative form of specification for a combinational circuit. For example, the Exclusive OR function of two variables x_1, x_2 generates an output y which can be defined by the truth table in *Table 25.1*.

Table 25.1 XOR truth table

x_1	x_2	y
0	0	0
0	1	1
1	0	1
1	1	0

The reverse process of translating a truth table into a Boolean function generates a set of equivalent expressions, rather than a unique solution. The translation is usually achieved by forming a logical expression of all the input conditions for which the output has the same value, either 1 or 0. These logical expressions, known as the 1st and 2nd canonical forms, can be implemented as two-level logic structures of AND/OR and OR/AND gates respectively. This structure can be seen in the translation of the truth table of the XOR function of *Table 25.1* which yields:

$y = \bar{x}_1 . x_2 + x_1 . \bar{x}_2$ 1st Canonical form

$y = \bar{x}_1 . \bar{x}_2 + x_1 . x_2 = (\bar{x}_1 + \bar{x}_2)(x_1 + x_2)$ 2nd Canonical form

Combinational logic circuits are designed either by determining the minimum set of logic gates necessary to implement a particular function, or by selecting an appropriate range of integrated circuits containing logic gates and more complex function and using them to implement a solution. In general, the various logic reduction techniques provide a good basis for design and for producing designs with reduced numbers of gates or components.

Logic reduction techniques generally examine the complete specification of a system (in Boolean function or truth table form) and reduce the logic by removing redundancy. Boolean functions can be reduced directly using the rules of Boolean logic. Truth tables are less amenable to direct reduction and are often translated into a graphical representation or Karnaugh map in which each cell represents the output or value of a function for a particular combination of inputs. The cells are ordered in Gray code so that inputs of adjacent cells differ in value by only one binary digit. If the value of the function in adjacent cells is the same, then the cells can be combined to effect a reduction in the combinational logic. This technique is shown in *Table 25.2* in which the eight output states in the truth table are translated into a Karnaugh map and are reduced by grouping adjacent cells to give the reduced canonical expressions.

In more complex systems it is possible to reduce the size of the Karnaugh map by entering variables expressions, rather than values, as elements of the Karnaugh maps. Variable entry map reduction techniques have advantages when dealing with complex systems.

Table 25.2 Truth table

x_3	x_2	x_1	y
0	0	0	0
0	0	1	0
0	1	0	0
0	1	1	1
1	0	0	0
1	0	1	1
1	1	0	1
1	1	1	1

Reduced 1st Canonical Form

$y = x_2 x_1 + x_3 x_1 + x_3 x_2$

Reduced 2nd Canonical Form

$y = (x_2 + x_1) . (x_3 + x_2) . (x_3 + x_1)$.

Karnaugh map

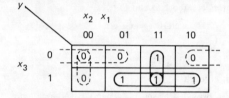

25.2.5 Arithmetic operations

Combinations of logic gates can be used to implement arithmetic operations on binary numbers. The operation of addition is perhaps the most basic operation: it also provides the foundation for more complex operations such as multiplication and division. Arithmetic operations of this type are often used in digital signal processing systems. They are also fundamental to the operation of the arithmetic logic unit (ALU) within the central processing unit (CPU) of computers and microprocessors.

The basic operation of binary addition involves the addition of two one-digit numbers x and y which have a one-digit sum, S, if both x and y are not both equal to one, and a two-digit binary sum, 10, consisting of the one-digit sum S (value 0) and a carry C (value 1) when $x=1$ and $y=1$. The carry C is held over to the next digit position for use in multi-digit operations. This primitive addition operation is defined in the following truth table and can be implemented as shown in *Figure 25.3* by an arithmetic logic circuit known as a binary half-adder:

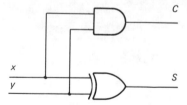

Figure 25.3 Binary half adder

Binary numbers of more than one digit can be added together by a series of operations which form the binary sum of each corresponding pair of digits in turn. Thus, the nth digits of two numbers x and y, x_n and y_n, can be added together with the intermediate carry C_{n-1} from the previous digit position to yield

the S_n and the carry C_n. This operation is defined by
$$S_n = C_{n-1} \otimes (x_n \otimes y_n)$$
$$C_n = C_{n-1}(x_n \otimes y_n) + x_n \cdot y_n$$
and may be verified by substitution. The logic function is known as a binary full adder and can be implemented by a logic circuit formed from two half-adders and an OR gate as shown in *Figure 25.4*.

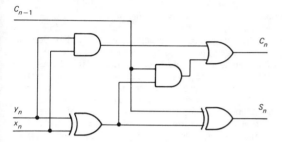

Figure 25.4 Binary full adder

A similar circuit can be used to implement the operation of subtraction in which an intermediate carry or 'borrow' is used. However, many digital systems use two's complement arithmetic and implement subtraction by the addition of the complement of a number.

Processing binary numbers one bit at a time involves a long sequence of operations and is necessarily slow. Speed can be achieved by using parallel binary addition in which two n-bit numbers are added using n binary full-adders connected in cascade with the carry output from each adder forming the carry input to the operation on the next digit. The speed of operation depends primarily on the delay as the carry is propagated through the parallel adder: this delay is reduced in the more sophisticated designs by the use of carry look-ahead mechanisms.

The phenomenon of propagation delays is a characteristic of all digital circuits. In combinational logic circuits timing problems occur when there are different numbers of gates in the various signal paths leading to different propagation delays. This may result in transient changes (glitches) in the specified signal level. Similar problems also arise in high-speed logic circuits in which the lengths of the physical signal paths are unequal. Single transient changes in specified signal levels are known as static hazards, while additional changes are known as dynamic hazards: both can be removed by incorporating additional gates to maintain the continuity of the specified logic function during signal transitions.

25.2.6 Digital multiplexers

The primary function of a combinational logic circuit is to produce an output which is a combinational function of the input variables. If a system has n inputs, there will be 2^n combinations of inputs and each combination will assign a specific value to the output. This type of operation can be performed by a multiplexer, which is a multi-input, single-output combinational logic circuit in which the logic signal or value from a selected input is directed to the output. The signal routing action of the multiplexer is controlled by external logic signals applied to the 'selection control' inputs, which can be considered as the input variables for the device. A set of n selection control inputs can be used to select or address one of a set of 2^n inputs, although in practice, the number of inputs is limited, typically to 4, 8, or 16. Systems with more inputs can be configured using multilevel structures (or trees) of multiplexers. The multiplexer therefore provides a general signal routing capability and its function is controlled externally.

Multiplexers are widely used as general-purpose combinational logic devices. A 2^n to 1 multiplexer can be used to implement any logic function of n input variables using direct mapping, in which the input variables drive the selection control inputs of the multiplexer. A more economical and elegant solution is achieved if a logic function of n variables is implemented using a 2^{n-1} to 1 multiplexer where the $(n-1)$ most significant inputs drive the selector control lines and the nth input is asserted on the selected input. This technique is shown in *Table 25.3*, where two 4 to 1 multiplexers are used to implement a full adder to generate the sum S and carry-out C_0 as a function of the two addends x_1 and x_2 and the carry-in C_I (which is used as the least significant variable). It can be seen that the inputs to channels 0, 1, 2 and 3 of multiplexers S and C are $C_I, \bar{C}_I, \bar{C}_I, C_I$ and O, C_I, C_I, 1 respectively. This function can be implemented by the circuit shown in *Figure 25.5*.

Table 25.3

MULTIPLEXER S				MULTIPLEXER C_0			
Channel	Input variables		Sum S	Channel	Input variables		Carry out C_0
	Selector control	C_1			Selector control	C_1	
	x_1	x_2			x_1	x_2	
0	0 0	0 0	0 1	0	0 0	0 0	0 0
1	0 0	1 1	0 1	1	0 0	1 1	0 1
2	1 1	0 0	0 1	2	1 1	0 0	0 1
3	1 1	1 1	0 1	3	1 1	1 1	0 1

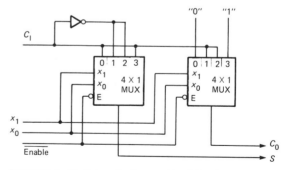

Figure 25.5 Multiplexer implementation of full adder

A multiplexer's signal routing capability is central to its other main role of providing a control element in the data bus structure of computers and other digital systems. In its simplest form a multiplexer can be used to select one of a number of sources and route its data on to a communications bus. In many systems the data bus is time division multiplexed and multiplexers are used to connect signal sources on to the common data bus at appropriate points in time. Multiplexers are commonly used in computer and microprocessor-based systems to connect and route information on common data highways or bus systems and to time division multiplex different classes of information over these bus systems. In many of the above applications demultiplexing devices are also used to separate multiplexed signals and route information to appropriate destinations. A variation of this technique is also involved when a multiplexer is used to convert parallel data to serial data by sequentially switching the parallel inputs on to a single serial output line.

25.2.7 Encoders and decoders

Combinational logic circuits are often required to produce more than one output from a given set of input variables. Since each output can be considered to be a separate function of the inputs, the system can be implemented using a set of multi-input single output combinational logic circuits. These parallel systems would contain much redundant logic and this can be reduced by combining the parallel networks and sharing common sub-functions or elements to produce a multiple-input/multiple-output system.

Multiple-input/multiple-output combinational logic circuits can be classified by the relative number of inputs and outputs. If the number of outputs is greater than the number of inputs the circuit is said to decode the inputs. If the number of inputs is greater than the number of outputs the circuit is known as an encoder. A full decoder is one which generates all 2^n combinations of n inputs. Similarly, a full encoder will generate n outputs from 2^n inputs. Devices of this type are available as MSI circuits and are used within VLSI circuits.

A full decoder can be used as a general-purpose combinational logic element because it generates a complete set of logical combinations, or products, of its inputs. These products can be used to implement any Boolean function expressed in the sum-of-products or first-order canonical form. The resulting circuit consists of a two-level combinational structure in which a decoder generates products (AND functions) and an OR gate is used to sum the relevant products to form the required function. This technique can be extended to multi-input/multi-output combinational logic functions simply by using an OR gate to sum each output. This technique is particularly useful when a large number of simple combinational expressions are to be formed from a set of inputs.

Decoders are used extensively in programmable electronic systems and computers to provide mechanisms for addressing specific devices or elements. For example, the central processing unit (CPU) of a computer or microprocessor will communicate with its immediate access memory using the CPU address bus and data bus. In a typical system many memory devices will be connected to these bus structures and address decoders are used to select a particular device. If the device contains a number of addressable elements, as in programmable interface devices or memory devices, then additional decoders will be used within the device to select a particular register or element.

25.2.8 Programmable logic arrays

In general it is not necessary to generate all combinations of a set of input variables in order to implement the sum-of-products form of a Boolean function, or set of functions. Each function requires sufficient AND gates to form only its constituent products and an OR gate to form the sum of products. Thus, a general-purpose, multi-input/multi-output combination function can be formed from a minimum set of AND gates which provide the necessary products and a set of OR gates which form the required sums.

A programmable logic array (PLA) is a general-purpose combinational logic element in which the inputs are connected via a matrix of links to an array of AND gates and the outputs of the AND gates are connected via a second matrix of links to an array of OR gates which sum the products to provide the outputs of the circuit. The PLA is characterised or programmed by removing unwanted links (or inserting required links) so that the required input variables are linked to the inputs of the appropriate AND gates and the required product terms are linked to the inputs of the appropriate OR gates. The number of gates in the AND and OR arrays is limited and it is usual to minimise the number of product terms before the links are programmed.

Mask-programmed PLAs are programmed or characterised during the manufacturing process and are widely used as customised integrated circuits in products which are to be produced on a large scale. Field programmable PLAs have fusible links which are programmed or 'blown' after manufacture to provide the required gate structure and they are used extensively in prototype designs and in products intended for small- and medium-scale production. The process of determining the pattern of links required to implement a function can be automated and computer programs are available which allow the problem specification to be input in either functional or truth-table form.

25.3 Sequential logic

The introduction of feedback into a digital circuit produces a radical change in the mode of operation of the circuit because the output of the circuit will depend not only on the present external input, but also on the sequence of previous inputs. Logic circuits with feedback are therefore characterised by their sequential behaviour, and they are known as sequential logic circuits. The most important application of sequential circuits is in the construction of the digital memory elements which are widely used in digital computers and many signal processing systems. Sequential circuits are also used to implement specific sequential logic functions for signal processing applications. A wide range of general-purpose sequential logic circuits is available in integrated circuit form.

Sequential logic circuits are formed by introducing feedback into combinational logic circuits. The output of a combinational circuit is defined by the present external inputs and the logic function of the circuit. In a sequential circuit, some of the output states are fed back and used as additional internal inputs to the circuit. This allows information about the current state of the system to be introduced into the inputs which will determine the next state. This mechanism for propagating information between states provides a method for implementing a memory. Equally important, the transition between the current and next state introduces the concept of a sequence of operations. Sequential logic circuits, including memories, are characterised by their current and next state behaviour which is usually defined as a table, sequence, or flow of transitions.

The dynamic behaviour of a sequential circuit depends on the order in which inputs change and on the speed at which the effect of the change of input is propagated through the circuit to external outputs or to new internal inputs. A stable state will be achieved only if the set of feedback logic values or states is the same as the previous set of feedback states. Since the change, or sequence of changes, of internal states and outputs can be affected at any time following a change of input, this type of circuit is known as an asynchronous logic circuit. Asynchronous logic is fast because its speed of operation depends only on gate propagation delays. However, asynchronous logic must be designed carefully in order to eliminate the logic hazards that will occur from unequal path propagation delays and to ensure that the feedback among the logic gates does not introduce stability problems. Asynchronous sequential logic is used in signal processing applications. It is commonly used to handle asynchronous inputs or to link two separate synchronous circuits. It is also used to implement asynchronous subsystems within a system of synchronised or communicating sequential processes.

The timing problems which arise in asynchronous sequential circuits can be reduced by introducing a synchronising signal which allows the logic to switch once only in each clock period. The whole sequential circuit can then be synchronised by distributing a periodic sequence of synchronising clock pulses throughout the system. At appropriate points in the logic circuit, the logic signals and the clock pulses are input to AND gates such that the logic signal is output only when the clock pulse is present. This reduces the signal propagation timing problem to a series of independent discrete events, each of which can be considered separately. Such circuits are known as clocked sequential circuits and although they are not as fast as asynchronous logic circuits, they are easier to design and implement. Synchronous logic circuits are used very widely in digital and programmable systems.

Most modern digital programmable electronic systems consist of a mixture of combinational logic, asynchronous sequential logic and synchronous sequential logic. Both forms of sequential logic are described in the following sections.

25.3.1 Asynchronous sequential logic

Asynchronous sequential logic circuits can be used to implement primitive two-state logic elements or flip-flops which form the basic building blocks for practical memory devices and more advanced sequential circuits.

Figure 25.6 shows a set–reset flip-flop in which the output state Q is fed-back to provide an internal input. This circuit can be

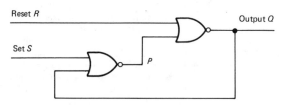

Figure 25.6 Basic RS flip-flop

driven into a stable 'set' state $Q=1$, $(P=0)$, or a 'reset' state $Q=0$, $(P=1)$ by asserting the values $(S=1, R=0)$ and $(R=1, S=0)$ respectively on the set, S, and reset, R, inputs. When the inputs are removed $(S=0, R=0)$ the circuit remains stable in either the 'set' or 'reset' state and simply continues to output the previous state of the flip-flop. Thus, when the inputs are removed, the device acts as a 1-bit memory element or cell and stores the previous state of the flip-flop.

At any point in time the state of an asynchronous sequential circuit is defined by the function of the circuit and the value of the logic signals. When a change of input occurs, the input to the circuit will consist of the current state feedback and the new external input. This complete set of inputs will determine the next state of the system at a time determined by the propagation speed of the signals. There are a number of ways of synthesising or analysing such state machines.

The traditional approach is a direct one which involves an examination of the sequence or flow of state transitions which occur following a change of input. Each state of the system is assigned a binary code (which normally corresponds to the set of binary states of the memory elements). The current and next states of the system are then specified or analysed for various changes in the value of the input variables. For example, if the RS flip-flop of *Figure 25.6* has a present state $Q=0$ and a total state $S=0, R=0, P=1, Q=0$ then, when a new input condition $S=1, R=0$ is asserted, the value of P will change to $P=0$. The new total state of the system $S=1, R=0, P=0, Q=0$ is an unstable and temporary state and will result in the generation of a new output $Q=1$ which will be fed-back as an internal input. However, there will be no further transitions because the signals $S=1, R=0, P=0$ are consistent with the output $Q=1$, and the new state of the system is said to be stable. The complete set of such state transitions can be summarised in a state transition table, *Table 25.4*, which shows the computed next state for all permutations of present state and input. Stable states are shown encircled and unstable states result in transitions to the corresponding next state, as in the transition described above which is shown in the table by an arrow.

Table 25.4 Transition table for a basic RS flip-flop

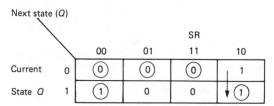

The scope of the system specification must cover all the states of the system, including any unused stable states, because appropriate recovery procedures must be used to guard against the system erroneously entering such states. For example, an RS flip-flop is generally required to have complementary outputs, such that P is the complement of Q. This is true for all the stable states (set, reset, and both undriven states) except the fifth stable state $S=1, R=1, P=0, Q=0$. Additional logic is therefore used to prevent the logic value 1 being asserted on both inputs, S and R, at any instant of time.

25.3.2 Timing problems

The design procedure for asynchronous circuits must include a detailed analysis of the static and dynamic behaviour of the system. Considerable care is needed to ensure that signals are free from noise or momentary changes because their presence in the feedback path could cause erroneous state transitions. Some problems, such as static and dynamic hazards, are common to combinatorial circuits and can be removed by the addition of fault-covering logic. Other problems arise from the sequential nature of the system and are less easily removed. Race conditions may occur when two or more feedback variables change at once in response to the change of an input variable. The order in which the two variables actually change is unpredictable and the resulting state transitions may lead either directly, or indirectly, to the intended stable state (a non-critical race) or to an unintended stable state (a critical race). Another timing problem may occur in asynchronous sequential circuits if the time delays along two or more paths are unequal and are such that one of the feedback signals is returned and used as an input sufficiently quickly to alter the intended order of transition. This type of problem is known as an 'essential hazard' and is corrected by carefully adding an appropriate amount of delay to the effected signal path.

The problems outlined above can be overcome and fast and

reliable asynchronous circuits can be designed and implemented. However, some aspects of the design process require skill, particularly when considering the dynamic behaviour of the system and the associated timing. Timing problems of this type do not occur so readily in synchronous sequential logic and synchronous logic often provides a more appropriate solution to a wide class of application including most programmable systems.

25.3.3 Synchronous sequential logic

The design of sequential logic systems can be simplified if a synchronising signal is used to control the instant at which transitions occur in the state of the system. Synchronisation forms a constraint on asynchronous behaviour and removes many of the timing problems, such as essential hazards, which affect asynchronous systems. This simplifies the task of designing, commissioning and testing this type of circuit. These advantages often outweigh the small penalty which is incurred by the slower speed of operation of the sequential circuit and the added complexity required to generate and distribute a reliable synchronisation signal. Synchronous sequential circuits are widely used in the design and construction of the general-purpose sequential logic elements such as memory registers, shift registers and counters which form the building blocks for many digital signal processing systems. It is also used extensively in the design of more complex systems such as semi-conductor memories and microprocessors.

Synchronous sequential circuits are formed by controlling the instant at which memory elements assume new states. A synchronising signal, which usually consists of a periodic pulse train (or clock sequence), is used to enable the inputs to the memory elements. If the clock period is chosen carefully, the 'current state' output of the memory elements will have propagated throughout the circuit and all binary signals will have reached a steady state before the occurrence of the next clock pulse. When the inputs to the memory elements are enabled by the clock pulse, the appropriate state transitions will take place and signals representing the new or 'next state' of the systems will be propagated in the circuit. The clock therefore constrains each memory element to make a transition only once in each clock cycle. All the memory elements in a circuit are synchronised with each other by the common (or global) clock signal. This allows the dynamic behaviour of the system to be decomposed into a sequence of clock cycles (or discrete steps), each of which can be considered separately. The design and analysis of synchronous circuits is therefore simplified and evolves around consideration of the 'current state' and 'next state' behaviour of the system.

25.3.4 Clocked flip-flops

The basic circuit element in a synchronous sequential circuit is the clocked flip-flop which is used to provide the memory elements which are essential to the sequential operation of the circuit. Clocked flip-flops can be formed from the RS asynchronous flip-flop by introducing additional logic and a clock signal to inhibit the inputs to the flip-flop except when the clock pulse is present. This can be seen clearly in the circuit for a clocked RS flip-flop shown in *Figure 25.7*.

The operation of this circuit can be defined using a characteristic table, *Table 25.5(a)*, which specifies the next state of the circuit for each combination of current states and inputs. This information can also be presented as an excitation table, *Table 25.5(b)*, which specifies the excitation or inputs necessary to bring about a change in state.

The transitions defined in the excitation table may also be entered on a Karnaugh map in which each cell identifies the next state of the system and stable states are shown encircled. (This is

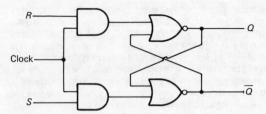

Figure 25.7 Clocked RS flip-flop

Table 25.5 Flip-flop
(a) Characteristic table (b) Excitation table

| Present state | | | Next state | Transition | | Excitation | |
Q	S	R	Q	Present Q	Next Q	S	R
0	0	0	0	0	0	0	0
0	0	1	0	0	0	0	1
0	1	0	1	0	0	1	1
0	1	1	0	0	1	1	0
1	0	0	1	1	0	0	1
1	0	1	0	1	0	1	1
1	1	0	1	1	1	0	0
1	1	1	0	1	1	1	0

analogous to the transition tables used in the analysis of asynchronous circuits, as in *Table 25.4*.)

All sequential circuits must perform a specified function and have satisfactory dynamic behaviour. Close examination of the RS flip-flop described above shows that it fails to provide the specified complementary outputs when both $S=1$ and $R=1$. The circuit also suffers from indeterminate behaviour when the system is in the state $S=1$, $R=1$, $Q=0$ and both inputs are removed $S=0$, $R=0$ for the resultant state will depend on the actual order in which the inputs were removed.

In practice, additional logic is normally added to the primitive clocked RS flip-flop to force the use of complementary inputs as shown in *Figure 25.8(a)*. This circuit, known as the D-type flip-flop or D-type latch can be considered to be a 1-bit memory element with a single 'data' input D. The D-type flip-flop or D-type latch forms one of the basic building blocks for synchronous sequential systems, including semi-conductor memories, and is usually represented schematically as in *Figure 25.8(b)*.

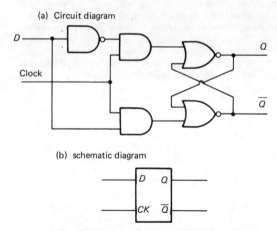

Figure 25.8 D-type flip-flop or latch. (a) Circiut diagram. (b) Schematic diagram

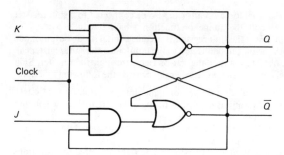

Figure 25.9 Primitive JK flip-flop

The primitive clocked JK flip-flop shown in *Figure 25.9* uses a different technique to overcome the input restrictions which apply to the RS flip-flop. Instead of restricting the inputs, as in the D-type latch, it uses internal feedback to force the flip-flop to oscillate (alternatively output a 0 or 1) while the clock is enabled and the value 1 is asserted on both the J and K inputs. Such a system is deterministic only if the number of state transitions are known during any clock pulse and this prevents the use of the JK flip-flop in its primitive form.

It follows that the synchronising effect of the clock signal is central to the operation of synchronous sequential logic. Particular attention must be given to the distribution of clock signals and to the precise effect of the clock signal on each memory element. For example, the inputs to the D-type flip-flop or latch are enabled while the clock pulse is applied. The clock pulse width should therefore be less than the propagation delay of the device if the inputs are to be disabled before new output signals are generated. Similar restrictions also apply to a primitive clocked JK flip-flop. In practice these problems are removed by using additional logic to provide a more precise method of triggering the flip-flop and controlling the input capture period. In practice the JK flip-flop is improved by connecting two flip-flops in series and applying overall state feedback to produce a 'master-slave' configuration as shown in *Figure 25.10*. The excitation table for this circuit is also presented. The triggering is such that all state transitions (but not input capture) are arranged to coincide with an edge of the clock pulse such as the falling edge. The JK master-slave flip-flop is used in many synchronous logic systems.

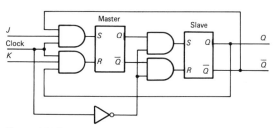

Figure 25.10 JK master-slave flip-flop

Excitation table

Present state	Next state	J	K
0	0	0	X
0	1	1	X
1	0	X	1
1	1	X	0

(Where X can take either the value 0 or 1, and is known as a don't care state.)

An alternative and often preferred approach uses edge-triggered devices in which the inputs are enabled only during a particular transition of the clock signal: only then are the new outputs generated. Edge-triggered devices such as edge-triggered D-type and JK flip-flops give precise and reliable operation provided the input is held steady for a minimum period during and following the enabling clock edge. The widely used D-type edge-triggered flip-flop, including its overriding preset and clear inputs (which force Q to 1 and 0 respectively) is shown in *Figure 25.11*. The chevron notation denotes the edge-trigger clock input.

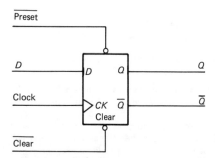

Figure 25.11 Edge-triggered D-type flip-flop

25.3.5 Memory registers

Memory elements are essential to sequential systems and are used extensively in digital signal processing systems. They also form a major part of any programmable electronic system. These systems generally process groups of binary signals in parallel and their memory therefore consists of groups of single-bit memory elements which are synchronised and operate in parallel. These groups (or arrays) of memory cells are known as memory registers and are classified by the number of bits stored in parallel, as in the widely-used 8-bit memory register.

Semiconductor read–write memory is the most widely used form of memory. Each memory register can be considered to consist of a parallel configuration of D-type flip-flops with common preset, clear and clock inputs. Data asserted on the input lines are input or 'written' into the register when the inputs are enabled by the clock or 'write' signal. Following the clock, the register will maintain the new state or 'contents' until data are next written into the register. The output of each cell in the memory register is generally connected to a tri-state buffer so that more than one register can be connected in parallel to a common set of output circuits. When the tri-state buffer is enabled by a 'read' signal, the state or contents of the register are output and asserted on the external circuit.

Figure 25.12 shows, in symbolic form, an 8-bit memory register with clock (write) and output enable (read) control lines. For convenience, identical parallel signal paths such as the data inputs $D_0 \ldots D_7$ are shown as a single line and marked $/n$, where n is the number of times the path is replicated. Common signals, such as clock, are shown once only. Identical parallel hardware is also shown once only and the number of replications can be inferred from the inputs D_{0-7} or the outputs Q_{0-7}. This type of notation, or shorthand, plays an important role in reducing the complexity of symbolic circuit diagrams without removing the function of the circuit; it is often used to describe the complex circuits in microprocessor-based systems.

25.3.6 Shift registers

A shift register is a memory register in which the memory cells are connected in cascade by data paths which can be used to move

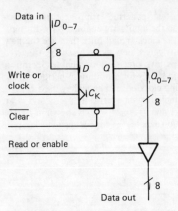

Figure 25.12 Symbolic diagram of 8-bit memory register

data from one cell to another in such a way that the data in the register are shifted synchronously either to the left or to the right. Repeated shift operations can be used to load serial data into a shift register, or to output the contents of the register in a serial manner. Some shift registers have additional data paths which allow data to be loaded or output as parallel data.

A wide variety of shift register is available. Some provide limited functions such as bi-directional shifts. A shift register of this type is shown schematically in *Figure 25.13* in which control inputs are used to configure the register as either a left or right shift register with serial input and either serial or parallel output. Multi-function shift registers which provide a wider range of function are also available. These devices also have external control inputs which can be used to select a particular mode of operation.

External logic circuits can also be connected to a shift register to extend the range of function which it can provide. For example, if the shift register is used in a serial input/output mode with the output connected to the input, then the shift register can be used to rotate data, left or right, through the register. Shift registers can therefore generally provide the following functions:

Bit manipulation (by left or right shift (or rotate) operations).
Serial data input or output (using either right-shift or left-shift operations).
Parallel to parallel operations (with parallel data load and parallel output).
Serial to parallel conversion (using right- or left-shift operations to input serial data followed by a parallel output operation).
Parallel to serial conversion (using a parallel data load followed by right- or left-shift operations to generate serial output).

Shift registeres are widely used in digital signal processing systems. They are also used extensively in programming electronic systems including microprocessors and computers. For example, shift register concepts are normally used within the central processing unit of a computer or microcomputer. In many cases the accumulators and other CPU registers can be used directly as shift registers and operate under software control. As serial-to-parallel and parallel-to-serial convertors, shift registers are incorporated within programmable serial interface devices such as the widely used USART, PIA or SIO components.

25.3.7 Counters

A counter is a sequential circuit which is designed to go through a specific sequence of states and which performs a state transition on each occurrence of an input. A counter can be used directly to count the number of occurrences of an input by converting the

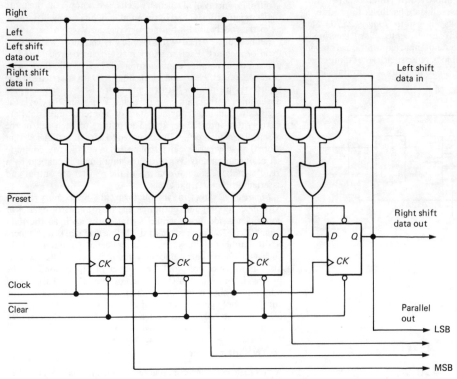

Figure 25.13 Bi-directional shift register

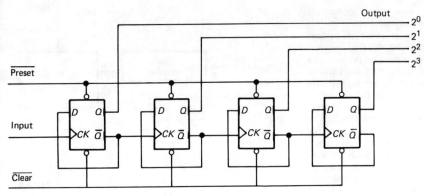

Figure 25.14 4-bit binary asynchronous (ripple) counter

series of inputs into a prescribed sequence of states such that at any point the current state or count may be output or read. The prescribed sequential behaviour of the counter may also be utilised to form a code sequence generator in which code from a prescribed sequence is output on each occurrence of an input.

At any instance of time the output and the next state of the counter will depend on only the current state and the occurrence of an input. If the inputs which activate the counter or the code sequence generator are part of a discrete event system in which significance is attributed to the occurrence of an event rather than the time at which it occurred, then the counter will be an event counter and the sequence generator is said to be event-driven.

However, if the input is generated by a clock, or a similar time-synchronised event, then the counter can be used as a timer and the code sequence generator can be used to generate timing signals or sequences. Event-counters, timers and sequence generators are widely used in signal processing systems and within microprocessors and microprocessor-based systems.

Counters can be divided into two subgroups: asynchronous or ripple counters in which the memory elements do not have a common clock and the signal propagates from memory element to memory element, and synchronous counters in which the memory elements are connected to a common clock and all state transitions are synchronised by the clock signal. A four-bit asynchronous binary counter is shown in *Figure 25.14* in which the frequency of the signal applied to the clock input of the D-type flip-flops is divided by two at each stage of the counter, thus giving a binary count. A synchronous version of a four-bit binary counter is shown in *Figure 25.15*. In this circuit the first or least significant JK flip-flop toggles on each input pulse and the flip-flops representing the higher order binary coefficients toggle only if all lower order binary coefficients have the value 1.

Both the asynchronous and synchronous binary counters described above use direct binary coding of the count or state. This can lead to timing problems in external decoders or other circuits because, in binary-coded systems more than one state transition can occur following a single input. The problem can be avoided by using Gray code counters in which the sequential state or count changes by only one bit in each state transition.

Code sequence generators are generally formed by taking the output of the most significant memory element of a counter and passing the signal via a feedback path to the input of the counter to produce a 'ring' counter. If the complementary state is used as the feedback signal, a twisted ring counter is produced. Both types of ring counter are used to generate the timing sequences for the timing control of microprocessors and of the programmable electronic systems.

25.3.8 Tri-state logic

A tri-state logic gate is a logic gate with an additional input, the mode control input and an output which can assume one of three states. When the device is in the enabled mode, the tri-state logic gate behaves as a conventional logic gate and can output either the value 0 or 1. However, when the gate is not enabled, the output enters the third state and presents a high impedance to any external circuit. In this state the gate cannot drive or load any devices connected to it and is effectively disconnected from any

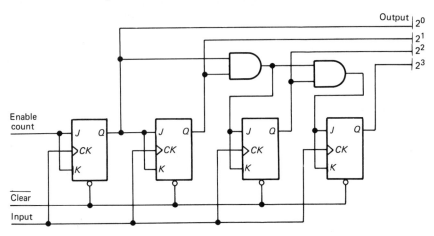

Figure 25.15 4-bit synchronous binary counter

output circuits. Logic circuits with tri-state output drivers are widely used in bus-based logic design.

Most modern digital systems process parallel data and parallel data paths are required to move or transfer this data between various logic gates or devices. The interconnection problem can be reduced by using a common data highway or data bus to connect the units. The logic systems can be connected to a common data bus using tri-state drivers, provided the bus is operated under a protocol such that not more than one driver is enabled at any instant of time. A bus system of this type can be extended by connecting additional systems to the bus without the need to modify existing circuits (subject to the bus driving capability of the tri-state logic). It also means that the direction of data flow on a bus depends on the relative position of the enabled driver and receiving devices. Many logic devices, including some microprocessors, are produced with tri-state output drivers. The symbolic diagram and truth table for a tri-state logic driver are shown in *Figure 25.16*.

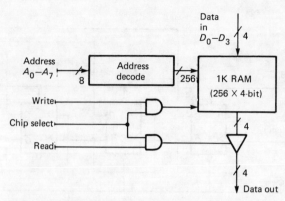

Figure 25.17 Schematic diagram of a linear address 256×4-bit RAM

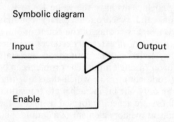

Figure 25.16 Tri-state logic driver

25.3.9 Memory devices

Large general-purpose memories are formed from sets of memory registers which are connected in parallel. The registers share common data input lines and are connected via tri-state drivers to common output lines. Addressing mechanisms are used to select a particular register within a memory unit and each register is allocated a unique binary code or address such that the complete set of addresses forms a continuous sequence. When an address is asserted on the address inputs of a memory device, the address is decoded and another appropriate register select line is enabled. This establishes signal paths so that write (clock) and read (output enable) control signals can reach the selected register. A typical arrangement for a 1 K-bit read–write memory configured from 256 4-bit registers is shown schematically in *Figure 25.17*.

Large memories generally use multi-level address decoding in order to reduce the amount of logic required to select a particular register. In effect the memory unit is reconfigured as a two-dimensional array (or matrix) of registers. The address inputs are split into two orthogonal groups to provide the row address and column addresses of a particular register. Thus a 16 K-bit memory could be configured as 128 rows of 128 bits (or 16 registers of 8-bits). This is shown schematically in *Figure 25.18*.

In the memory register address and data access systems outlined above, the time taken to access an individual register is basically independent of the address of the register. This type of memory is therefore known as a random access memory. The memory registers described above were capable of both read and write operations. Random access read–write memory is commonly called RAM. Typical timing diagrams for write and read operations on RAM memory are shown in *Figure 25.19*. During the write cycle the signals on the data bus are allowed to stabilise before the data are strobed into the selected register by the removal of the write control signal. After the write strobe, data must be held steady for a specified period to ensure that the data are captured without error. Similarly, during a memory read operation the contents of the selected register are output on to the data bus. This output should be held steady for a specified period after the read signal is removed.

Random access read-only memories, or ROM memories, consist of arrays of memory elements in which the data has been pre-loaded or programmed either by electrically programming the ROM (EPROM) or by blowing fusible links (fusible link ROM). A ROM memory cannot perform a 'write' operation. However, the contents of a ROM can be accessed as in the memory read operations described above.

Large arrays of RAM and ROM memory are found in nearly all programmable systems including computers. They are used typically to hold the program code which defines the data processing operations and to hold the data being processed. A RAM memory register or a ROM memory element is also an important general-purpose programmable logic element and can be used individually or in arrays to perform a wide variety of logic functions.

25.3.10 Combinational logic design using memory

A memory can be used as a general-purpose, multi-input/multi-output combinational logic element in which the input variables form the register addresses and the contents of the registers provide the value of the functions. In effect, the memory is used to store the complete truth table of a multi-input/multi-output combinational function. This technique is extremely flexible because the circuit is independent of the function being implemented. A general-purpose circuit can therefore be designed which can be programmed at a later stage to provide a particular application function. For example, a memory device containing 2^n 8-bit registers may be used to implement up to eight separate

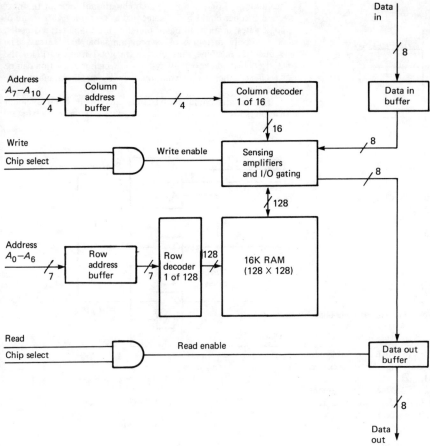

Figure 25.18 Two-level addressed 16 K-bit memory

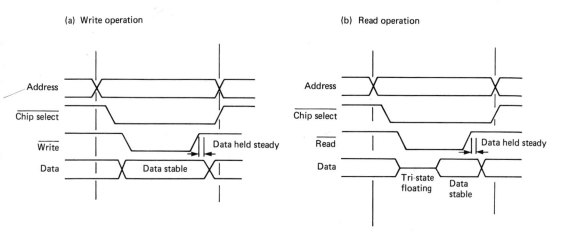

Figure 25.19 Idealised timing diagram for read memory operations. (a) Write operation. (b) Read operation

(a)

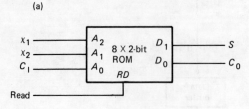

Read

Address			Contents	
x_1	x_2	C_I	S	C_o
0	0	0	0	0
0	0	1	1	0
0	1	0	1	0
0	1	1	0	1
1	0	0	1	0
1	0	1	0	1
1	1	0	0	1
1	1	1	1	1

(b)

Figure 25.20 ROM implementation of full adder. (a) Symbolic circuit diagram. (b) Truth table

combinational functions of n variables. It is usual to implement such designs using non-volatile read-only memory devices such as EPROM and fusible link PROM.

This technique is illustrated in *Figure 25.20* which implements the full adder truth table shown in *Table 25.3* using a memory device consisting of eight two-bit memory registers.

25.3.11 Sequential logic design using memory

A sequential digital system can be considered to be a feedback controlled digital system with a memory or delay capability in which the next-state of the machine is a combinational function of the current states and the external inputs. A generalised system of this type is known as a sequential machine or a finite state machine and is illustrated in *Figure 25.21*.

The model can be used to partition the design activity into a set of processes such as the selection for the memory elements, the design of the combinational logic for the next-state decoder and the design of the output conditioning logic. Various methods can be used to elaborate the design and to select the implementation technology. For example, the combinational logic of the next state encoder could be implemented as a two-level structure of logic gates, or a circuit based on either a multiplexer, a decoder, a read-only memory, or a programmable logic array. The sequential machine thus provides a good framework for the design of a wide variety of systems in which the function can be specified either by conventional or by programmable logic elements.

This design technique can be illustrated by the serial full adder circuit shown in *Figure 25.22*. In this circuit the next-state

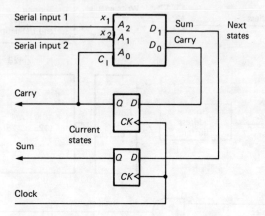

(a)

Address			Contents	
A_2	A_1	A_0	D_0	D_1
0	0	0	0	0
0	0	1	0	1
0	1	0	0	1
0	1	1	1	0
1	0	0	0	1
1	0	1	1	0
1	1	0	1	0
1	1	1	1	1

(b)

Figure 25.22 Serial binary full adder. (a) Schematic diagram. (b) ROM truth table

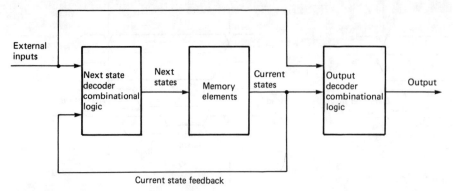

Figure 25.21 Sequential machine

decoder consists of a binary full adder and is implemented by an 8 × 2-bit ROM which has been programmed with the truth table of the full adder, as in *Table 25.5*. (This combinational function could also be performed by the network of logic gates shown in *Figure 25.4* or the multiplexer circuit shown in *Figure 25.2*). Two D-type flip-flops provide the current state memory elements. Serial binary addition is performed by clocking the D-type flip-flops whenever new serial data are present at the inputs to the circuit.

25.3.12 Structured design of programmable logic systems

Structured methods have now been evolved for the design of sequential circuits including asynchronous circuits. The design methods involve the so-called 'top down' approach and start from a specification which includes a statement of the problem. This can be elaborated as an architectural specification which identifies the major components of the system and a specification at an abstract level of an algorithm or procedure for the functional control of the system architecture. These specifications are often made using a formal notation or language, as in the algorithmic state machine method. In particular, the control algorithm may be expressed as a sequence or flow from one assigned state of the algorithm to another. Each stage of the algorithm will include a specification of the actions required to move from the current state to the next state in the algorithm and will include details of the use of inputs and the generation of outputs. During the design process, hardware is selected to elaborate the architectural model and the control algorithm is translated into the sequence of commands or transitions required to make the hardware perform the specified task.

Structured design methods are essentially a systems-level approach to the design of sequential machines. The design is centred on the algorithmic states in the algorithm-part of the specification which 'controls' the system architecture and on the conditions which must be satisfied for a transition from the current state to the next state. The logic conditions for the transitions are then implemented in the 'system controller' part of the design. In many ways this is analogous to the use of a program to control a fully programmable system such as a microprocessor or a computer.

25.4 Programmable systems

The primary function of a digital system is to process signals or data. To carry out this function the system must be able to input signals or data from external circuits or systems; move or transfer the data within the system, store the data before, during and after processing; process the data by logic, arithmetic or bit manipulation operations and output the processed data to external circuits or systems. All these operations can be performed by an appropriate configuration of combinational and sequential logic circuits including memory elements. Logic subsystems can therefore be designed to implement each of the operational functions outlined above and these subsystems can be used as building blocks in advanced digital systems.

The logic subsystems described above can be used in the structured design of signal or data processing systems. Each application will involve the derivation of a specific control structure which is embedded in the logic design of the system. Programmable logic systems of this type can only be designed on a bespoke basis for known applications and this can lead to high design costs and may result in inflexible designs which are difficult to modify. These systems can only be used economically in a limited range of application.

A programmable system overcomes these difficulties by using a general-purpose logic system which can be configured to perform a wide range of operations such as those described above. The hardware is controlled by a program or sequence of instruction codes which define the operations necessary to implement a particular processing function. The instruction codes are normally stored in the memory part of the system and the function of a programmable system can be changed simply by altering the stored program. This type of system can be considered to be composed of two parts—the hardware which is basically independent of the application and the software which defines the application function. It is therefore possible to design a general-purpose programmable system which can be used in a wide range of applications. The hardware part of such a system is invariant and can be produced economically as a standard design.

There are basically two types of programmable system. The fixed function programmable machine is a limited form of programmable system which is constrained to perform a prescribed and fixed sequence of instructions. This type of system does not have the capability under software control to select between two alternative sequences of instruction. The application function of such a system can be altered only by reprogramming the system. A fixed function programmable machine is therefore forced to execute a fixed sequence of instructions in all circumstances and is properly regarded as a programmed machine rather than a computer. These systems can be used in any applications in which the processing function does not depend on the nature or value of the data being processed.

The digital computer is the most powerful and flexible form of programmable system. Its function is again defined by a program, but in the case of the computer the program may specify alternative sequences of instructions and the conditions under which they can be executed. When the program is run or executed, the conditional expressions are evaluated and return a logical result which is used to identify which alternative sequence of instructions is to be executed next. Special hardware is required to carry out this type of operation and a computer is equipped with a logic circuit, known as a status register, which is used to store information about the status of the processor when it has completed the execution of an instruction. This condition or status information can then be used to determine the identity of the next instruction and, when necessary, to effect a change in the value of the program counter which points to the next instruction. It is this mechanism which allows a program to take into account the nature of the information being processed.

For example the conditional expressions in a program may be a function of the data variables. When the program is executed, the actual values of the data variables are used in the evaluation of the conditional expression and the data therefore determines the future sequence of operations. In effect, the computer can be programmed to take a decision about the future courses of action it may take. The result of the decision and the subsequent sequence of actions will depend on the actual value of the data being processed. It is this facility which characterises a proper computing system and which makes the computer such a powerful data processing system. Such systems are providing economic solutions to digital signal and data processing problems in an increasingly wide range of applications.

25.4.1 The logic design of a digital computer system

A digital computer system can be considered to be composed of two logic structures. The first is associated with the flow, storage and processing of data and consists of the data input and output subsystems, the data highways used to move or transfer data within the system, the memory which is used to store the data and the arithmetic logic unit which is used to process the data. The second structure is used to control the data processing operation.

It is primarily responsible for looking after the program part of the system and for ensuring that the instructions are executed in the proper sequence. It is also responsible for the execution of each instruction and controls the data flow and data processing elements so that they perform the required processing operation.

The control structure consists of the memory which is used to store the program code and mechanisms for identifying the location of the next instruction, for fetching and decoding the instruction and for identifying the location of any inputs or stored data which form the operands in a processing operation. It controls all aspects of the execution of the instruction including fetching operand data and, when necessary, taking into account the status of the previous programming operation. It also identifies the destination location of any resultand data generated by the processing operation and stores the resultand or generates an output.

The data processing and program control structures are heavily interconnected and often share common hardware. In particular, the program code and the data being processed are usually stored in an identical binary format. It is usual for code and data to be stored in separate areas or segments within a common memory unit.

25.4.2 Architecture

Computer and microprocessor systems are general programmable systems which perform sequential processing operations. The computer may therefore be represented as a type of sequential state machine which can be programmed to implement either a bounded or an unbounded sequence of state transition. These systems are constructed using a number of subunits as shown in *Figure 25.23*.

The Central Processing Unit (CPU) is the heart of the computer and it contains the control and timing unit which controls the programmed operation of the system and the arithmetic logic unit (ALU) which processes the data. It also contains a number of very important registers such as the Program Counter which points to the next instruction and the Status Register or Flag Register which stores the status information about the result of the previous instruction executed by the CPU.

The Memory Unit provides storage for program code and data. The code and data are always considered to be separate entities although they may share physical memory. However, some processor designs enforce this conceptual separation by providing separate memories for code and data.

The input/output subsystem of a computer provides the interface to external circuits or systems. It also communicates with the operator via a Man–Machine Interface (MMI). The input/output system is used to input program code and data and to output computed results. The processing power of the computer can only be used if the input/output subsystems allow efficient communication between the user or application and the processing system.

There are two commonly used methods for connecting input and output systems to a processor. The most elegant technique treats all input ports and output ports as if they were memory registers in the memory unit. The input and output ports are connected to the address, data and control bus structures as if they were memory elements and are designed to operate to the same electrical and functional specification as a memory register. Data can then be output using a memory reference 'write' at the output address, or input using a memory reference 'read' instruction at the input address. This method, which is known as memory-mapped input/output, is used in a wide range of processor. It gives fast input and output and is compatible with other software data transfer instructions.

An alternative approach connects all inputs and outputs to a separate input/output bus structure which normally consists of a limited number of address lines and the usual control signals. In bus-orientated systems a subset of the memory address lines is used and an additional memory or input/output discriminator signal ($\overline{M}/IO$) is used to generate unambiguous addresses. Input/output mapped input/output is not compatible with memory reference operations and special instructions such as IN and OUT are often used to distinguish this mode of operation.

Many computer systems are configured around a general-purpose data highway. This consists of a common bus structure of CPU address, data and control lines and is used to communicate with all devices external to the CPU including memory, input/output systems and backing stores. The bus-orientated architecture illustrated in *Figure 25.24* provides ease of access to the control, address and data highways which are used to interface any logical system to the CPU. Bus-oriented architectures are therefore used in many designs to provide a flexible and easily expanded computer system.

25.4.3 Control and timing unit

The basic operation of a computer or microprocessor is governed by the control and timing unit which coordinates, synchronises and controls the movement and processing of all information within the system. The unit is driven by an external clock and uses this as a reference to generate the timing and control signals necessary to operate the various logic systems. There are a number of variations on this theme and some designs separate the control and timing unit and use an external timing subsystem to generate the clock signals and multi-phase timing sequences which are required by the control unit and to other logic systems.

The control and timing unit is responsible for controlling the main operational cycle of the processor which is known as the instruction cycle. The instruction cycle can be split into two distinct phases, the instruction fetch and the execution of the instruction. During the instruction fetch the address of the next instruction is obtained from the program counter mechanism

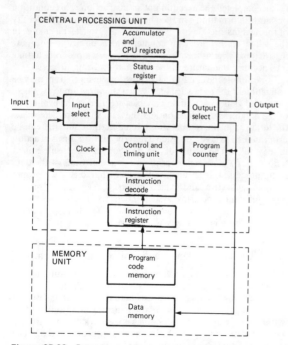

Figure 25.23 Processor architecture

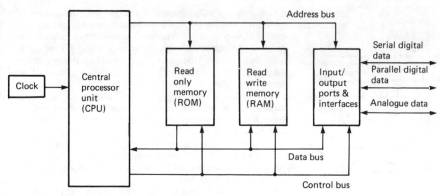

Figure 25.24 Bus-orientated systems architecture

and transferred to the memory address register (MAR). A memory reference operation is then performed on the code part or segment of memory to read the op-code which is the first part of an instruction. The op-code data are transferred via the memory buffer register (MBR) to the Instruction Register where it is decoded and then input to the control and timing unit. The program counter is then updated to point to the next part of the instruction or to the next instruction.

The op-code identifies any further memory reference operations which are required to complete the instruction fetch. The control unit uses the updated program counter to make reference to successive addresses in the code part of memory to fetch any further parts of the instruction, such as the addresses of the operands and resultand and any immediate data values. This information is transferred to various temporary registers in the central processing unit for use during the execute cycle. At the end of the instruction fetch, the central processing unit will contain all the information it requires to control the execution of the instruction and the program counter will be pointing to the next instruction to be fetched (assuming that the execution cycle does not compute a new program counter address). The various logic units used during the instruction fetch cycle are shown in *Figure 25.25* in which the memory and I/O discriminator $\bar{M}/IO$ is used to distinguish between memory reference operations and any operations involving peripheral systems which may use the same address and data bus.

The op-code also defines the sequence of operations necessary to execute the instruction. During the execution part of the instruction cycle the control and timing unit will synchronise the transfer of data within the system and control the operation of the arithmetic logic units. The control unit will access operand data by transferring the operand addresses from the temporary registers to the memory address register to perform memory reference operations. In practice, most modern processors will have a complex data reference pointer which will compute the address of the data object using not only the temporary register but also base or segment registers, offset registers, and index registers according to the addressing mode specified in the instruction. If the computer has a memory-to-memory architecture, then operand data can be transferred direct from immediate access memory to the arithmetic logic unit and resultands can be returned direct to storage in immediate access memory. However, if the computer has a register-to-register architecture, then the operand data are normally transferred to a CPU register before being processed by the arithmetic logic unit and resultand data are held in the accumulator or transferred to another CPU register.

25.4.4 Arithmetic logic unit

The actual data processing operations are performed by the arithmetic control unit (ALU) which operates under the control and timing unit. The ALU is a general-purpose logic system which can normally perform logical, arithmetic and bit manipulation operations. The ALU can perform both monadic (single operand) and dyadic (two operand) operations and, therefore, has two input data paths. The resultand is usually held in a special register, known as the accumulator, which is normally a multi-function register which can participate fully in the processing operations. In some systems the accumulator is used to store one of the operands before a processing operation and is subsequently used to store the resultand. This technique increases the operational speed of the processor; it also removes the need to have two operand registers. However, the need to minimise the number of CPU registers is no longer a major design objective and many modern microprocessors have a number of CPU registers of advanced design which can also act as accumulators.

The arithmetic logic unit also contains the status register which is also known as a flag register or condition code register. This register consists of a number of flip-flops (flags) whose state reflects the result or state of the processing element at the end of the previous processing operation. This is illustrated in *Figure 25.26* which shows in schematic form the structure of a typical

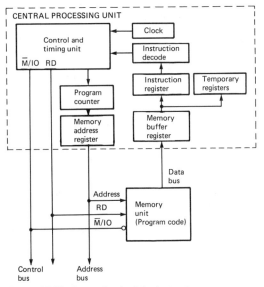

Figure 25.25 Instruction fetch logic structure

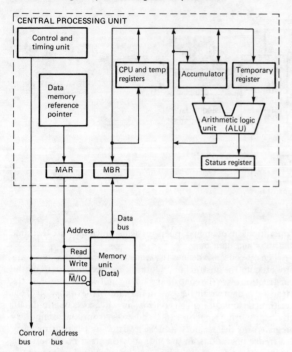

Figure 25.26 Instruction execution logic structure

ALU and the other logical systems associated with the execution part of the instruction cycle.

25.4.5 Memory unit

Computers commonly use two types of memory to hold program code and data. A fast, immediate access memory is used to store program code and the data associated with the program so that it may be readily accessed during execution of the program. This memory normally uses semiconductor random access memory devices such as static or dynamic RAM or ROM memory. In specialist applications, magnetic core memory or bubble memories may also be used. The immediate access memory is normally limited in size by the computer architecture. For example, many simple microprocessors have a 16-bit memory address and this limits the size of the immediate access memory to 64 K registers or elements. More advanced microprocessors commonly use 20-bit or 24-bit memory address systems and may be equipped with a large immediate access memory up to a limit of 1 M or 20 M registers or elements.

Immediate access memory is expensive to implement and is unsuited to the long-term storage of large program or data files. Most computers are therefore equipped with auxiliary or backing stores which are normally sequential access magnetic memory storage systems such as floppy discs, hard discs, Winchester technology discs and magnetic tape or tape cassettes. These systems provide economic storage for the large volumes of data which are commonly used in text processing or data base processing applications.

The immediate access memory is considered to be the primary memory unit of a computer. It consists of a large number of memory registers or elements connected to a common address, data and control bus structures. The memory is normally built using MSI or VLSI memory devices. Many memories require more registers or cells than can be contained within a single memory device and these memories have to be implemented as multi-device memories.

Large multi-device memories can be built in one of two configurations. The particular configuration used depends on the specific design of the memory cell matrix within the memory device. If the memory cell is configured as an array of registers or elements (say 8 K × 8-bit), then a number of such devices could be connected to a common set of address, data and control lines provided each device occupies a different area of the memory address space. This is achieved by decoding the high order address lines to provide chip select inputs for each device such that each device occupies a separate and distinct range of addresses. Thus eight 8 K × 8-bit memory devices could be configured to occupy successive memory addresses to produce a contiguous 64 K byte memory. Within each device the address of the registers is determined by the low order address lines which are common to each device.

An alternative approach can be used if the memory cell matrix is arranged as an array of single bit cells (say 64 K × 1-bit). Devices of this type can be used to construct a large memory by adding similar devices in parallel to make up the desired length of memory word. The devices will all be connected to common address and control lines and each will be connected to an individual data line. All the parallel devices will be selected and accessed simultaneously. Thus eight 64 K × 1-bit memory devices can be used to configure a 64 K byte memory.

In practice, the memory addressing and data bus structures are often integrated by multiplexing both types of signal on to a common bus. This reduces considerably the number of signal paths and the number of pins on the integrated circuits. The control and timing unit is used to synchronise data transfers over the multiplexed bus and generates the synchronisation or strobe signals, such as 'address latch enable' or ALE, which are used to demultiplex the various forms of data on the bus.

The data bus usually has a smaller number of lines than the address bus, and the data bus is often multiplexed with the low order address lines. In more advanced processors with a large number of address and data lines it is common to find that other signals such as status information are multiplexed with the data on to the address bus in order to reduce the number of pins on the CPU. In this case a separate bus controller is often used to demultiplex the complex set of signals.

25.4.6 Microprocessors

Advances in microelectronics and computing science have provided the technologies necessary to construct the complete central processing unit of a computer on a single integrated circuit which is called a microprocessor. These technologies were also used to build large integrated circuit memories. These devices, which were developed in 1971, realised a step change in the cost, performance, power consumption and reliability of a minimum computer system. Further advances in VLSI design led to the development of integrated circuits containing both the CPU and the memory unit; the so-called single chip computer. In effect, these advances in microelectronics had resulted in the miniaturisation of the computer.

The microprocessor can also be viewed as an advanced programmable logic device. Special microprocessor and other advanced programmable systems have been developed to carry out specific computational functions. These processors are often designed to work in conjunction with a general CPU and are known as co-processors. A number of devices such as numeric, text and communication (or local area network) co-processors are available and can be used in the design of powerful processor architectures. These devices are often used in conjunction with programmable interface controllers such as graphics controllers, video interface controllers, and Winchester disc controllers. Such systems make significant demands on immediate access memory and this is often implemented using VLSI dynamic RAM

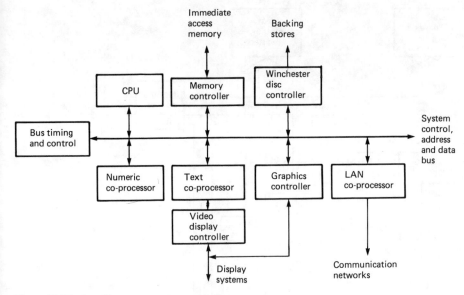

Figure 25.27 A multi-processor system architecture

components which are governed by a programmable memory controller. A typical system architecture of this type is shown in *Figure 25.27* which illustrates the use of programmable systems including microprocessors in the design of an advanced information processing system.

DIGITAL COMPUTERS AND COMPUTER PROGRAMMING

25.5 Digital computer structure

The computer uses a combination of electronic elements to process data in digital (binary number) form. The processing operations performed are specified by a computer program, which consists of a set of logical instructions stored in the computer memory. The instructions are executed in sequence, but the order of processing may be modified according to the result of the instruction. Thus, the computer operation is program controlled. The arrangement of the main functional elements of a typical computer are shown in *Figure 25.28*. They are as follows:

(1) *Memory* The memory is used to store information in the form of binary numbers. These numbers are coded to represent data values (e.g., numbers, logical states, character text) and program instructions. The memory is arranged into words which consist of several binary digits (or bits). The word length or number of bits/word depends on the computer design. Each word can be individually addressed and operated on by the computer.

(2) *Arithmetic unit* This unit can be instructed to perform a variety of digital and logical operations such as addition, subtraction, logical OR, etc. The operand(s) consist of one or two data values extracted from memory.

(3) *Control unit* This generates the necessary switching signals to control the execution of the computer instruction. The operation of the unit consists of several sequential steps (machine cycles); one to fetch the instruction and interpret its function, another to fetch the data, another to operate the arithmetic unit, and finally, another to store the result. The number of steps required depends on the nature of the instruction and the computer architecture.

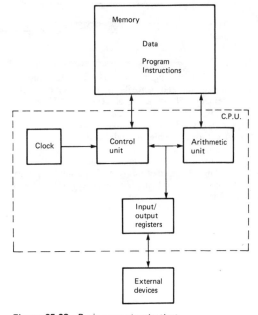

Figure 25.28 Basic computer structure

(4) *Clock* The timing of the computer operations is controlled by pulses generated by the clock.

(5) *Input/output* This consists of the circuitry needed to transfer data to or from the computer to external devices. Data may be transferred either under program control, i.e. by an instruction, or by direct memory access between the memory and the external device.

The functional computing elements of the computer are called the central processor unit (CPU). It includes the arithmetic unit, control unit, clock and buffer registers for data, address and input/output information. The execution of an instruction depends on the correct flow of information in terms of program instructions and data to the various CPU elements which operate

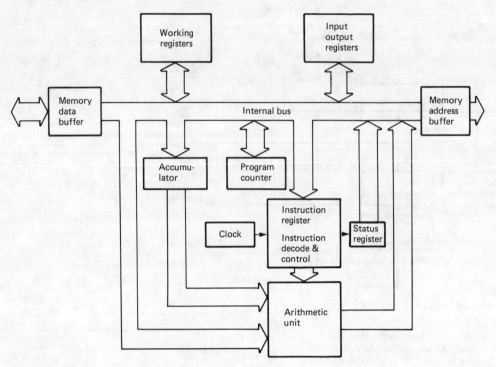

Figure 25.29

upon it or store it. A digital highway or bus is used to route the data, address, instruction and control signal information to the various CPU elements. The structure of a typical CPU and the bus arrangement is shown in *Figure 25.29*.

The computer bus is composed of a large number of parallel lines and the structure of the bus depends on the architecture of the CPU. For example, in some computers the CPU bus lines are dedicated to particular functions, such as an address bus, a data bus and a control signal bus. In other computers a single bus may be time shared between several functions and carry either data or address information at particular times. Some computer architectures may employ more than one bus and information transfer along a bus may be either uni- or bi-directional. The control unit ensures that data and instructions are correctly routed over the bus by setting control signals in the appropriate, precisely timed sequence.

The CPU usually contains a number of registers which hold one word of information. The number of registers available within the CPU is an important feature of computer architecture. Access to registers connected to the digital bus within the CPU is faster than access to the main computer memory, thus registers lead to faster instruction execution and operation.

The sequencing of instructions from memory into the control unit instruction register is supervised by the program counter register. This holds the address of the memory location containing the next instruction to be executed and is incremented to the next instruction address during the instruction fetch cycle. In this way the CPU is forced to execute instructions in strict sequence. However, some instructions are provided which modify the contents of the program counter, either by simply adding a number to branch to a different instruction or by adding a number if a particular condition is satisfied, e.g. if the result of the previous instruction was negative. This feature allows the program sequence to be modified if a specified condition is detected in the data being processed.

The instruction register is an internal register in the control unit which is used to store the current instruction. The control unit decodes the instruction and generates an appropriate and precisely timed sequence of bus signals to execute it. The time reference is provided by the computer clock. The CPU contains a number of additional internal registers which are used to store address or data while they are being written to or read from memory. These include the memory address and memory data buffer registers which store the location and contents of the memory word currently being assessed.

In most computers a number of registers are provided which can be assessed directly from a computer instruction. Some of these registers do not have a defined purpose and are available to the user to provide temporary 'scratch-pad' storage. A number of these registers have specific purposes which are related by the computer hardware to the execution of particular types of instruction. The most important of these registers are the 'accumulator' which is used to store the result of a numerical or logical operation performed by the arithmetic unit and the status or flag register which stores the state of the arithmetic units. The contents of the accumulator may also be used as an operand store for the next instruction in most computer architectures and can be used to store 'working' results of calculations. Since it is overwritten at the end of each instruction it is necessary to store some results in memory if they are required at a later point in the calculation. The index register is used to provide an offset from a base address and facilitates access to memory when the address is indexed. The Memory Data Buffer is a temporary store for the data value currently being read from or written to memory.

Many computers have facilities for using an area of memory as a stack. This is a block of RAM memory which is used on a last in–first out (LIFO) basis for storing data, register values and the contents of the program counter. This facility is particularly useful for storing addresses and register contents on interrupt or before entering a subroutine.

A stack is organised by a stack pointer, which is a CPU register holding the address of the last item placed on to the stack, also

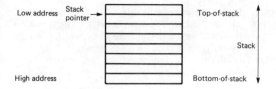

Figure 25.30 Stack layout in RAM memory

called the top-of-stack. Instructions are provided which automatically increment and decrement the stack pointer when data are stored (PUSH) or retrieved (POP) from the stack. A stack grows downwards in the store as shown in *Figure 25.30*.

The stack structure permits a main program to call a subroutine or procedure, which can in turn call other subroutines. The subroutine calls are nested by simply putting return addresses and register contents on to the top of the stack during successive calls. On return from each subroutine the return address, etc. are removed from the top-of-stack in reverse order. The stack can be employed in a similar fashion to store register and address information following an interrupt.

25.6 Interrupts

In many applications the computer must respond rapidly when an external event occurs. This is usually achieved by an interrupt facility. The CPU is provided with an interrupt control line in the digital bus which is connected to the instrument or device which detects the occurrence of the event. When the event occurs, e.g. a key is depressed, the interrupt control line is driven to a specified logical state and the CPU is interrupted. The actions which follow an interrupt vary from computer to computer, but in general terms the following sequence occurs:

(1) At the end of the current instruction, the contents of the program counter and the status register are stored and the interrupt line is disabled.
(2) The program counter is loaded with the address of an interrupt service routine.
(3) The CPU registers needed by the service routine are stored at a predefined location, e.g. in the stack. This may be an automatic hardware facility or may be performed by the interrupt handler software.
(4) The interrupt program is entered.
(5) When the interrupt program is complete, control is returned to the interrupted program by restoring the registers and the contents of the status register and the program counter.

The computer can usually enable and disable interrupts under program control. Most computers have facilities for accepting interrupts from a number of sources via one of several interrupt lines. The CPU is often vectored to one of a number of interrupt entry addresses appropriate to the particular interrupt line which is triggered.

25.7 Input/output

The input or output of data to logic which connects devices external to the computer to the CPU are called input/output channels. There are several ways of implementing input/output: under program control, under interrupt or by direct memory access.

(1) *Program controlled input/putput* The transfer of data between the CPU and the external device is controlled by program instructions which are executed when input or output is required. The data are communicated via an input/output port which consists of a buffer connected to the CPU data bus and logic to identify the device address which is connected to the CPU address lines. In addition, the device logic generates control signals which indicate if the device is operative or if it is busy. This information is held in a status register which can also be assessed by the CPU. Communication with the external device then proceeds under program control in which the CPU executes an instruction using the address of the appropriate device to interrogate the status register for the device. When the status register indicates that valid communication is possible, the program instruction to communicate data to or from the device is executed.

(2) *Interrupt driven input/output* The operation of the CPU on interrupt has been described, in that when an external event driven logic signal is invoked the CPU is forced to execute program at a specified address. Interrupt driven input/output is implemented by connecting the control logic of the external device to an interrupt line so that the device can demand the CPU's attention. It is conventional practice to generate an interrupt when information is input, or when an operation is completed. This allows the CPU to respond immediately and execute a service program to maximise the utilisation of the external device.

(3) *Direct memory access* The use of direct memory access allows the external device to transmit data directly into the computer memory without involving the CPU. The CPU is provided with control facilities which allow an external device (the DMA controller) to gain control of the CPU data bus. The DMA controller then communicates data directly over the bus with a memory location. It is common to communicate blocks of data to or from the device to memory, for example in transferring data from a disc to the CPU. The DMA controller must provide a memory address, the data and bus control signals to effect a data transfer. The controller will also contain a counter to count the number of transfers made within the data block and to increment the memory address. The DMA process is also referred to as cycle-stealing, since it proceeds simultaneously with program execution, the only effect being that the instruction execution time is increased by the number of memory cycles used when a transfer is in progress.

The relative merits of DMA over other means of input/output is that it is fast, uses the minimum amount of computer time per data word transferred and operates autonomously. The loss of instruction execution time is not usually significant unless a very large number of devices are under DMA. The major disadvantage of DMA is that the computer program is not explicitly aware of changes in data or the completion of a DMA transfer and it is usually necessary to make the DMA controller invoke an interrupt to inform the CPU that a data block transfer is complete. The use of program controlled input/output is inefficient since considerable computer time is lost in scanning status registers and waiting for the device to complete the previous data transfer. Interrupt controlled input/output operates asynchronously since it is effectively controlled by the device and thus it is not possible to predict the frequency of input/output scanning. The major disadvantage of interrupt controlled input/output is the high program overhead which results from the execution of the interrupt subroutine each time a data transfer is required. In most computers a combination of program controlled, interrupt driven and DMA based input/output is employed to optimise the operation of the peripheral devices connected to a particular computer.

25.8 Instruction execution

To execute a computer instruction, several sequential steps are required: one to fetch the instruction, another to fetch the data and another to operate the arithmetic unit. These operations are referred to as fetch and execute cycles respectively. These cycles are themselves composed of a number of more primitive operations executed in sequence under the control of the control unit. Typically these most primitive operations consist of the control unit generating switching signals to control data movements and arithmetic unit operations. The duration of each stage in the cycle sequence depends on the type of operation which is being performed. The operation of the control unit may either be asynchronous in which the completion of one operation initiates the next, or synchronous in which the timing is related to a fixed interval.

The functions provided by the arithmetic unit determine the instruction execution time. It is usual to provide a limited number of basic operations which the arithmetic unit can perform and build up more complex operations by calling them in the correct sequence. In principle, all arithmetic operations can be performed if addition and complement operations are implemented. However, a wider range of functions is usually provided to improve instruction execution time.

The control unit controls the sequence of operations required to implement a particular instruction. Each step in the sequence is called a micro-instruction. The micro-instruction is a very primitive operation which directly operates the CPU hardware but does not perform a complete arithmetic or logic instruction. The instructions do not reference any external memory. The fetch sequence and the execute sequences for the various computer instructions are referred to as a micro-program. The micro-program is stored within the control unit in a read-only memory and is usually fixed for a particular type of computer. The micro-program should not be confused with the computer program stored in the main memory.

Modern computers such as the DEC VAX range do allow facilities for adding new instructions by altering or adding to the microprogram. This facility is generally used by the manufacturer to improve the system capability.

25.9 Computer instruction types

The total range of instructions implemented by a particular computer is termed the instruction set or order code. In most computers the instructions can be divided into several classes:

(1) *Load/store* These instructions transfer the contents of memory locations to (load) or from (store) specified registers (e.g., the accumulator).
(2) *Arithmetic* These operations are performed on two operands and include add, subtract, multiply and divide. Negative numbers are usually held in two's complement form.
(3) *Logical* These operations implement boolean operations such as AND, OR, EXCLUSIVE OR (XOR) on two operands.
(4) *Shift/rotate* These involve moving all the bits in a computer word either to the right or the left. There are several possible forms of shift, such as cyclic, logical and arithmetic shifts.
(5) *Input/output instructions* These instructions transfer a binary number to an address in the input/output range. In addition, functions are also provided to test the status of individual input/output units.
(6) *Skip/branch* These instructions consist of a test on the operand (e.g. accumulator contents) against some condition, such as zero, negative, etc. If the condition is met (true) then the program counter address is set to the value specified in the instruction. Skip instructions usually jump over single instruction. Branch instructions usually involve jumping to a program in another part of memory.

While these general types of instruction are supplied with almost all computers, the instructions available with a particular computer depend to a considerable extent, on the design objectives of the particular manufacturer. There is no standardisation of computer instructions at present. The range and capability of the order code functions provided may have a considerable influence on the choice of a computer for a particular application task.

25.10 Computer instruction format

Computer instructions are stored as binary numbers in the memory. The instruction contains the following types of information:

(1) *Operation code* (*op-code*). This part of the instruction defines the action to be taken by the computer, such as add, jump, etc.
(2) *Operand(s)* This specifies the addresses of the data on which the instruction operation is to be performed. The address information is interpreted in a number of ways depending on the addressing mode being employed.

In principle a general computer instruction, e.g. to add two numbers together and store in memory, must specify the following items of information:

(a) The operation code, e.g. add.
(b) The addresses of the two operands in memory.
(c) The address in memory for the result.

This gives a three-address instruction as shown in *Figure 25.31*.

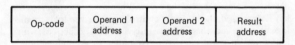

Figure 25.31 Three-address computer instruction

For a reasonable range of memory address (e.g. 16-bits), this implies a long computer word length or the use of several words for each instruction which is inefficient. Thus, the CPU architecture which employs working registers, as shown in *Figure 25.29*, has evolved to allow the size of individual instructions to be reduced. The accumulator or a register is used to store one of the operands and also the result of the instruction, thus reducing the number of addresses required in the instruction to one. Most modern computers employ a one-address instruction as shown in *Figure 25.32*.

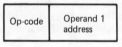

Before execution :   Accumulator = Operand 2
After execution: Accumulator = Result

Figure 25.32 One-address computer instruction

The computer word length is restricted to a limited number of bits, e.g. 8, 16, 32. This, in turn, restricts the address range and number of op-codes which can be included within a single computer word. Thus, it is common for a computer instruction to consist of two or three computer words depending on the word length and computer architecture.

A number of different methods of addressing operands have been developed. These address modes are used to extend the address range of the memory which can be accessed from an instruction. In addition, facilities are provided to efficiently assess areas of memory and thus improve program execution speeds. The address modes commonly encountered for accessing operands are as follows:

(1) *Direct addressing* The number in the address field of the instruction is interpreted as the absolute address in memory of the operand. In limited word length computers this means that only a small area of memory can be assessed directly.

(2) *Implicit or register addressing* In this mode of addressing the instruction specifies the operands as being held in one of a limited number of working registers. An example of such an instruction is, add working register *R* to the accumulator. This mode of addressing has two advantages: it is fast because the operands are already held in the CPU and the restricted address field permits the use of single word instructions.

(3) *Page relative addressing* Paged addressing is used to extend the memory address range which can be assessed using a limited number of instruction address bits. The memory is divided into 'pages' of the same size as the instruction address range (e.g. 256 for an 8-bit address field). The page currently in use is indicated by the contents of a page address register which is set up by a computer instruction. In most paged systems it is possible to access data on one page while executing instructions on another page. Direct addressing of page zero is used to provide an area for global data values.

(4) *Relative addressing* The instruction address field contents are interpreted as a positive or negative binary number which is added to the current contents of the program counter to determine the address required. The philosophy behind this mode of addressing is that most data references, etc. will be fairly close to the instruction being executed. This overcomes the difficulty with paged addressing of accessing data nearby which is located on the next page.

(5) *Indirect addressing* This uses a CPU register to hold the address of the operand. This allows the operand to be located anywhere in memory and be accessed by a one-word instruction. It is commonly used for incrementing an address during successive passes through a loop of instructions. The register must first be loaded with the memory address.

(6) *Indexed addressing* This adds the contents of the address field in the instruction to the contents of an 'index' register in the CPU to define the memory address. This permits the address of the operand to be modified during successive passes through a loop of instructions.

(7) *Immediate mode addressing* In this mode, the address field is a literal. Thus, it is not interpreted as an operand address but is the data value itself. This allows small integer values to be loaded into a register.

(8) *Extended addressing* This is similar to direct addressing, except that the instruction occupies sufficient words to contain the full address.

The preceding addressing modes were concerned with accessing operands. Jump and branch instructions use the operand as an address to change the value of the program counter. These jump instructions employ a variety of addressing modes:

(a) *Extended addressing* This is similar to (8) above. The jump instruction includes sufficient words to fully define the new address.

(b) *Present page addressing* The contents of the address field are interpreted as an address value relative to the start of the current page.

(c) *Relative addressing* The number in the address field is interpreted as a signed number which is added to the current value of the program counter to define the new address.

(d) *Indirect addressing* The contents of a working register are loaded into the program counter.

The address modes discussed are all methods of accessing large memories using a fixed word length machine. A specific computer will employ several of these addressing modes and the use of a wide range of addressing modes provides flexibility to produce efficient programs.

25.11 Computer program execution

The computer instruction has been described as a binary number coded to define the instruction operation, the addressing mode and the operand. The instruction set of a specific computer defines the individual operations that the computer can perform. Each instruction represents one primitive operation from which more complex operations can be constructed by writing a sequence of instructions (a program). The execution of a sequence of instructions is started by setting the program counter to the address of the first instruction. This is usually done by execution of a bootstrap or initialisation program which is initiated by manual action. Instructions are then executed in sequence; that is, after the first instruction is executed, the second follows and this pattern is repeated continually. This sequential behaviour allows for an enormous flexibility in the computation, but at the expense of execution speed. Each instruction takes at most a few microseconds to execute so that in many applications the speed of execution is not critical. It is also important to note that a computer with a complex order code will produce a program of smaller size and hence faster execution time.

At certain points in the program the sequential execution of instructions can be broken, if required, by inserting a branch instruction. These branch instructions modify the contents of the program counter and hence cause a change in the program flow. Branches may be either unconditional, in which case there is only one possible path following the branch, or they may be conditional upon the results of the preceding operation, e.g. if the result is negative, in which case there are two or more alternative paths in the program after the branch point.

A large number of computer programs consist of some frequently used sequences of instructions which are required at several different places in the program. Therefore, it is more convenient to write each sequence of instructions once as a subroutine which can be entered from any point in the program. Subroutines are key elements in a program and the careful use of subroutines can result in a better structured program. Subroutines operate on data passed as 'arguments', or parameters and produce results which may either be data values returned to the main program, or an event such as sending a character to an input/output port. It is also necessary to provide a return address or link to the subroutine so that, on completion, it returns control to the correct point in the program. Parameters can be passed to a subroutine using the stack which provides facilities for entering (PUSH) and extracting (POP) data values into the stack. Within the subroutine the parameters can be accessed by using a copy of the stack pointer. The use of a call instruction to branch to the subroutine automatically enters the return address in the top of the stack for subsequent use on returning from the subroutine.

25.12 Computer programming

A program is a list of computer instructions held in memory and executed in sequence. This list defines the operations which the computer has to perform to execute a particular task. The instructions are held in memory as binary numbers in a coded form which are directly understood and interpreted by the processor. These binary instructions represent the language which the machine uses in all its calculation and processing and correspond to the op-codes in the instruction set. They are the most primitive language which can be used and are very closely related to the architecture and design of the particular computer. A program written using the basic binary instructions is called a machine language program. The use of machine language is tedious and time-consuming for all but the most trivial programming task. To reduce the labour involved in developing programs a number of programming languages have been developed in which the program is generated in an abbreviated symbolic form by the programmer and then converted or translated into machine language using the computer. Programming languages may be either high-level and oriented to the solution of a particular class of problems, or low-level and oriented towards the architecture of a particular machine.

The word 'language' can be defined as 'an artificial system of signs and symbols with rules for forming intelligible communication'. Thus, a computer programming language is a means by which a human being can communicate his requirements to a computer. It is important that a computer programming language obeys strictly defined rules, since there is no place for ambiguity or interpretation when communicating with a machine. In particular, it must use a standard set of symbols, i.e. vocabulary, and there must be a systematic method of using them. This is the syntax of the language, that is, the rules which define which words to use and how to use them. A number of types of computer language are available.

25.13 Software design and specification

The application of computers to a wide range of tasks in the business, production and defence fields has shown that the development of the computer programs or software is not a trivial task. The more complex the task, the more difficult the software is to specify and the more difficult the management of the program writing process. Major cost and time overruns occurred in software development for early computer applications and it is now accepted that proper system design and documentation procedure must be applied to software development. A number of development and specification tools have been developed to support this activity.

25.13.1 Top down design

This is a design procedure which is essentially analytical in that it starts from a stated objective, a specification of system functions and subdivides them into subsidiary functions in increasing levels of detail. This involves a number of stages:

(1) *Goal specification* This is a statement of the purpose of the program and the limitations of its function are defined.
(2) *Specification* This is the second stage of top down design. The goal specification is broken down into specific task modules. The number and types of input/output are defined and the representation of data is specified. The strategy to be adopted in processing the inputs, outputs and data is also defined.
(3) *Pseudocode* The program is written in plain English constructs or a formal notation which describes clearly the program logic. The pseudocode may not prepresent a specific computer language but will contain conditional statements, etc.
(4) *Coding* The pseudocode modules are coded into the computer language chosen.

25.13.2 Structured programming

This is a technique of programming which is aimed at producing a readable, reliable and understandable program. Three basic mechanisms are allowed in structured programming:

IF—THEN—ELSE

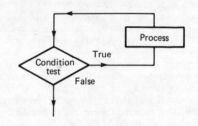

This is the standard conditional structure represented by:

IF Condition = True THEN A
    ELSE B

DO—WHILE

This is the loop structure specified by:

DO process WHILE condition = True
(WHILE condition = True DO process)

DO—UNTIL
(Repeat—Until)

This is the alternative type of loop:

DO process UNTIL condition = True
(REPEAT process UNTIL condition = True)

The structured programming mechanisms are also a compound of the three basic types of building blocks: a process box, a loop mechanism and a binary decision mechanism. Complex tasks are readily decomposed into other simpler tasks, all with a similar level of description. Furthermore, each task has a single input and output and can be readily documented, tested and understood.

An essential element in the structured representation is the elimination of confusing GO TO changes in control. In the structured software, a process is invoked and shown alongside the condition which controls its execution, hence improving documentation. Some languages such as Pascal, Coral 66 and Algol provide constructs and facilities which directly correspond with the structures of structured programming, thereby simplifying the coding and documentation of a design based on this technique.

25.13.3 Mascot

This is a software specification, development and testing procedure developed at the Royal Signals and Radar Establishment. It is based on a diagrammatic description of the interaction of program modules in realtime-systems and consists of three elements:

(1) A root or process that carries out a defined function.
(2) A channel which represents a line of communication between roots (processes).
(3) A pool which represents a data store element.

The use of these symbols in a Mascot Diagram shows the data flow between independent tasks whose function can then be specified using structured programming techniques.

25.13.4 Modular program structure

The structured programming approach leads naturally to a modular approach in program coding. Each program element is constructed as a module which is composed of a separate code and a data segment. The indiscriminate mixing of code and data is bad practice. It is also necessary to divide data segments and define the scope of data objects. Data which are only required by a single program module are local and should be stored in a local data segment. Global data should be held in a separate global data segment as shown in *Figure 25.33*.

This type of structure is implemented directly in block-structured languages such as Pascal, Coral 66 and Algol since each local block and its related data segment are held as a module. These languages also permit the data case to be defined in terms of integer, real, logical, etc. so that additional validity checks on expressions can be implemented by the compiler.

25.13.5 General design procedures

A number of procedures should be employed when writing and designing programs which employ one or more of the techniques described:

(1) The processing should be clearly and unambiguously specified in a number of stages of increasing detail (top down design).
(2) Each stage in the specification should be flowcharted using the rules of structured programming.
(3) When the program is written it should consist of modules which correspond with those of (1) above.
(4) The program modules should be coded independently and interactions with other modules should be defined. All variables should be meaningfully labelled and program source code should be commented.
(5) The program must be documented both for the overall functional level and for each program module.

25.14 Machine language

Machine language or machine code is the most primitive form of computer language. Although the instructions are held in memory in binary number form, the machine language may be specified in either binary, octal or hexadecimal form depending on the particular computer. In most modern computers a PROM resident monitor program is supplied with the computer hardware, to load machine language and provide facilities to enter programs at specific addresses. The loader program will convert

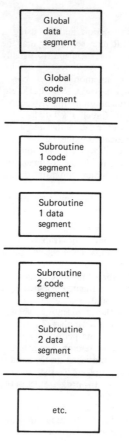

Figure 25.33 Segmented program structure

from the external machine language representation, e.g. hexadecimal, to the internal binary number form.

Machine language suffers from a number of disadvantages:

(1) It is hard to understand and no facilities for explanatory comment are provided.
(2) Instruction codes, operands and data values all have a similar numeric representation.
(3) Addresses are specified in absolute terms. When an error is found which involves adding or deleting program instructions, a large number of address values are likely to need changing.

25.14.1 Assembly language

The difficulty of using machine language led to the introduction of assembly languages in which the computer allocates addresses when the program text (source programmer code) is translated and assembled. Assembly language usually corresponds with instructions on a one-to-one basis. The notation employed in assembly language is symbolic. Instruction op-codes are referred to by mnemonics and addresses are referred to by symbolic labels.

The most commonly encountered format for a line of assembly language is:

LABEL: OP-CODE; OPERAND(S) (or REGISTERS);
       COMMENT

This consists of a number of fields, though the format will vary for different computers, as follows:

(1) LABEL This identifies the instruction address using a symbolic name, e.g. LOOP, TIMER, etc. so that it may be referenced from other instructions. Its inclusion is usually optional.
(2) OP-CODE This consists of a mnemonic which corresponds to the instruction operation, e.g. ADD, JMP, LDX, etc.
(3) OPERAND(S) Depending on the nature of the instruction, the operand may be the symbolic name (label) of a data location or another instruction. It may also refer to one of the computers working registers usually using a predefined name, e.g. B, C, etc.
(4) COMMENT This is provided to allow the programmer to describe the function of the instruction and hence document the program. It is ignored by the assembler program during translation.
(5) *Delimiters* Certain symbols, such as : ; and space are used to define the end of particular fields.

Programming in assembly language is very similar to programming in machine code. The mnemonic codes translate on a one-to-one basis with machine instructions. Thus, the computer programmer has complete control over the program code which is generated. The use of symbolic labels and mnemonics are provided to assist the programmer in producing error-free code and in eliminating the use of absolute addresses.

Advanced assemblers may provide additional features, such as the use of macro instructions. A macro is a number of program instruction lines which are given a name and perform a specific operation, e.g. binary to b.c.d. number conversion. The macro is defined at the beginning of the program and each time the name is referred to in the program text, the appropriate instruction lines are copied into the program. This relieves the programmer of the tedium of writing repetitious groups of instructions.

The task of the assembler program which translates source program in assembly language into object program in machine language is quite straightforward. It frequently requires several scans (passes) of the source code to complete the process, as follows:

(a) *Pass 1* Read the program and expand any macro instructions to lines of assembly code.
(b) *Pass 2* Read the program again and record all symbolic names and note the location address.
(c) *Pass 3* Read the program again and convert op-codes to binary form and replace each operand symbolic name by the appropriate address.

At each stage, error checks are applied to detect that each label is unique and that op-code memories are correct.

An example listing of a portion of an assembler output is shown in *Figure 25.34*. It shows the hexadecimal instruction address and machine code together with the assembly language source program.

25.14.2 High-level languages

In a high-level language it is possible to write programs using statements which are close to those used to express the problem in everyday circumstances. A typical example is the use of a formula to evaluate an algebraic expression. A comparison of high- and low-level language is shown in *Figure 23.35*. A high-

High-level

$P = Q + R$

Low-level (assembly language)

```
Start:  LDA  Q
        MOV  B, A
        LDA  R
        ADD  B
        STA  P
```

Figure 25.35 Low- and high-level languages

OBJECT CODE			SOURCE CODE			
Address (hexa-decimal)	Machine instruction (hexa-decimal)		Label field	Instruction field	Operand address field	Comment
	Byte 1	2 3				
			KBIN:			; KBIN ; Gets one character from keyboard ; Interrogate keyboard status,
0873	DB	ED		IN	OEDH	; Has character been typed?
0875	E6	02		ANI	02H	; Test bit 2 of status port
0877	CA	73 08		JZ	KBIN	; If no character, try again
087A	DB	EC		IN	OECH	; Else yes, input character
087C	C9			RET		;
			PROUT:			; PROUT ; Sends one character to the teletype printer ;
087D	DB	ED		IN	OEDH	; Test printer status, is it busy?
087F	E6	01		ANI	OIH	; Test bit one of status
0881	CA	7D 08		JZ	PROUT	; If printer busy, then try again
0884	78			MOV	A,B	; Else, get character from store
0885	D3	EC		OUT	OECH	; Send character
0887	C9			RET		

Figure 25.34 Assembler output listing example (for Intel 8080)

level language is problem-oriented rather than being machine-oriented and provides a degree of machine independence and portability for programs. This is of great importance because programs may run on any computer with a suitable language compiler, though in practice minor changes may be needed to suit a particular language implementation.

The use of a high-level language allows the programmer to specify the functions to be performed in an abstract form and thus reduce the time to write a program for a given task. It is clear from *Figure 25.35* that the programmer no longer has to be concerned with machine operations or storage allocation. This is a great advantage to the unskilled user but can be a restriction on the expert, particularly if an efficient program is required. The requirement for efficient programs is most necessary in applications where the computer has to perform a task in real-time.

A high-level language program is translated into machine code by a compiler. Each statement in the source program is translated into several lines of object code. The high-level language is general purpose and designed for a range of tasks. Thus, the object code generated will usually be less efficient in terms of memory space required and execution time than the equivalent assembly language program written by a skilled programmer. During compilation, comprehensive error checking of the source code is implemented to ensure that the program conforms to the language syntax. In addition, error checking is also included within the object code to detect run-time errors, such as, array bounds exceeding predefined limits. It is possible to write compilers for one computer which run on another. These are called cross compilers.

An alternative method of implementing a high-level language is to employ an interpreter. In this system the source code text (or some representation of it) is stored in memory. Each line of the source code text is translated and executed in turn by the interpreter program. No complete translation of the program is ever generated; only the current line. This simplifies correction and development of programs since new statements can be entered at any time and execution can proceed without having to compile the program again. However, the repetitive line-by-line translation slows down execution time and tends to make interpretive languages less efficient than compiled ones.

There is a growing tendency to implement languages using a 'semi-compiler'. The source code is compiled to a machine independent intermediate code that is then interpreted. This is an elegant way of allowing the rapid implementation of compilers for particular processors, since all that is needed is to write an interpreter in assembly language for the inetermediate code. An example is the generation of P-code by Pascal compilers which is then interpreted by a P-code interpreter.

25.15 Program development

The conversion of a source program into machine code running in the computer follows an established sequence of operations which are shown in *Table 25.6*. The procedure is similar for both compiled high-level language and assembly language programs, though not all the operations need be involved in some cases. The computer-user develops a number of files during the development process which are usually held on a mass store, e.g. disc, floppy disc, etc., though in the past, punched cord or paper tape was also used to generate the files. A number of development and utility programs then operate on these files until a running program is produced. The following stages are shown in *Table 25.6*.

(1) *Source file generation* The user prepares a source file in the chosen language using a text editor. The editor's function is to accept the sequence of characters entered, prepare a suitably

Table 25.6 Stages in program development

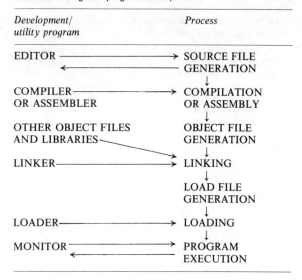

formatted source file and provide facilities of correcting and adding text under the user's direction.

(2) *Compilation or assembly* The compiler or assembler program translates the source file into an object file which contains the corresponding series of machine constructions and data words required to implement the source program. The compiler or assembly program may require several passes to convert the source program into the object code. Any errors detected will be listed as error messages and the user will usually have to edit the source file and report the translation before being able to proceed.

(3) *Linking* The user program may refer to a library of subroutines or several program modules held in a number of object files. The linker combines all these modules in a single executable program and stores them in the load file. The object code may be relocatable in memory, that is, able to run in any area of memory and still be run correctly and correctly address data, in which case, during linking the global variable references are evaluated and instructions are corrected to account for absolute addresses.

(4) *Loader* Finally, the linked object modules are loaded into memory at a defined location by the loader ready for execution. All relocatable instructions in the linked program are modified by the program base address (taken as 0 in a relocatable program) to produce an absolute binary program. The loader may also convert from load file format, e.g., HEX, to binary during the loading process.

(5) *Execution* The program is now resident in memory in the target machine and can be entered via a monitor program which provides facilities for program entry, terminal handling, interrupt handling and limited debugging.

All the above operations may be contiguous but not necessarily. Some of the operations may be carried out on a host computer (or mini- or a mainframe) and the load file only transformed to the final target computer. This procedure is known as cross-compiling.

25.15.1 Program testing

The above procedure does not eliminate all program errors. It does ensure that valid source code is presented to the machine, but any logical errors in the program can only be detected by testing the program in execution. This is usually achieved by processing test data and checking that the program produces the correct result.

The monitor program usually provides a number of features to assist in detecting program errors:

(a) *Memory and Register Examination* Facilities to examine and change the contents of memory and registers at any point in the program assist in testing and identifying errors.
(b) *Breakpoints* This facility allows the programmer to specify any location in the development program as a breakpoint. At this point, control returns to the monitor and the contents of all registers, etc. is preserved. The programmer can now inspect and change memory or register contents before restarting the program under development.

These development facilities are of most use in testing assembly language programs where the programmer is aware of address allocations, etc. For high-level language programs, debugging should take place at the level of the programming language used, e.g. on a statement oriented basis. This is achieved by either requesting a debug mode during compilation which prints out the result of particular tagged statements when they are executed; the other facility is 'program trace', where the execution and result of each statement executed is reported. These facilities slow down program execution and make debugging difficult, particularly for real-time applications.

25.16 High-level language types

There are a number of mature high-level languages available on a number of computers. The selection of a particular language must be made on the basis of the facilities provided and the task to be performed. The syntax of the language to a large extent, determines its ease of use, its ability to implement the logical and arithmetic structures required and the degree to which it allows structured programming and its capability for self-documentation.

The main language features which should be considered in language selection are:

(a) The ability to evaluate complex expressions on numerical, Boolean and other data types (e.g. character string, bit patterns).
(2) The implementation of conditional expressions, in general, *if --- then --- else ---*, forms are preferable to jumps.
(3) The language constructs available to carry out repetitive operations.
(4) The ability to structure programs using subroutines, procedures and block (*begin --- end*) constructs in a nested fashion.
(5) The data structures and data types available. Languages which provide facilities for declaring all data types and provide a wide range of data structures under the control of the programmer are superior both in ease of programming and in providing clear documentation of program.
(6) The compile and run-time error checking provided.
(7) The ability to include machine or assembly code within the high-level language statements and to access computer locations by address.

The more common languages currently available will now be described.

25.16.1 FORTRAN

FORTRAN was invented in 1957 as a language to support scientific computation, hence FORmula TRANslator. Two basic types of variables, namely *real* variables and *integer* variables, are available. Integer variables may assume any integer value between prescribed limits determined by the word length of the computer. For a machine with n-bit words the range is from -2^n to $+(2^n-1)$. Even for a short word length this is generally adequate for operations involving integers. However, the range is inadequate for many calculations and no direct provision is made for handling fractions. To overcome this difficulty, larger machines intended mainly for numerical calculation have special hardware to allow the manipulation of numbers in a *floating point* from $a(2^b)$ where a is always in the range $-\frac{1}{2}$ to $+\frac{1}{2}$ and b is a positive or negative number.

The programmer can choose names for variables subject to the constraint that a variable with a name commencing with one of the letters I, J, K, L, M or N is treated as an integer, while a variable with a name commencing with A to H or O to Z is treated as a *real* variable; that is, one represented in floating-point form. The standard method of input of programs employs punched cards or punched paper tape and there are certain limitations imposed by the keyboards of the machines used for the preparation of input data using the media. The arithmetic operations are represented as

+ addition   * multiplication
− subtraction   / division   ** exponentiation

so that the expression

$$Y = A * B / (C * D) + X ** (P + 1)$$

represents

$$y = (ab/cd) + x^{p+1}$$

If numerical values have been provided for the variables on the right-hand side of the expression, the r.h. side will be evaluated and assigned to the left-hand side.

There are facilities for inputting quantities from paper tape or card reader using the READ statement and for printing results using the PRINT statement.

Conditional statements will allow alternative actions within a program. An example of a conditional statement with alternative actions is

```
       IF (X − Y) 4, 5, 6
  4    Z = Y
       GO TO 7
  5    Z = O
       GO TO 7
  6    Z = X
  7    END
```

The first statement forms the value of $(X − Y)$: if it is negative the statement labelled 4 is obeyed; if zero the statement 5; and if positive the statement 6 is obeyed. The effect is

if $(X − Y) > O$ then $Z = X$;
if $(X − Y) = 0$ then $Z = O$;
if $(X − Y) < O$ then $Z = Y$

An example of a repetitive sequence in a program forming the sum of a set of 100 numbers $A(1), A(2) \ldots A(100)$, is

```
       S = 0
       DO 5   I = 1, 1, 100
  5    S = S + A(I)
```

The first statement sets the value of the variable S to zero. The 'DO' statement causes the execution of all the following statements down to and including that labelled S (only one statement in this case) repeated for values of I starting with 1 and increasing by steps of 1 until 100. The program has the effect of forming the sum $S = a_i$ for all integral values of i between $i = 1$ and $i = 100$.

25.16.1.1 Print-out

FORTRAN provides facilities for printing results with a convenient layout and in a desired format. For example the statements

```
    PRINT 10, L
10  FORMAT (7HLENGTH=, 14)
    PRINT 12, A
12  FORMAT (5HAREA=, F3.2)
```

would result in a print-out

```
    LENGTH=25
    AREA =625.00
```

In terms of modern software concepts, FORTRAN is dated and its capabilities are limited. It provides powerful expression evaluation facilities, but its ability to implement conditional (IF) and repetitive (DO) functions is restricted. The use of subroutines provides a powerful aid to structuring a program, but this is limited to one level (e.g. main program and subroutines). FORTRAN provides only limited characterhandling ability and data types Real, Integer, Logical, and Complex. However, due to its widespread (amost universal) use in numerical computation, FORTRAN remains a major computer language for scientific and engineering purposes.

25.16.2 BASIC

BASIC is a mathematical problem-oriented language which is usually executed interpretively. This eliminates the time-consuming compilation and linking procedures required with most other high-level languages. The execution time of the language is correspondingly slower.

The language was originally designed as a teaching aid and is intended to be easy to learn. In its original form, the language contained only seven basic instructions or commands, but the language has been further developed to widen its scope and application area. Thus, there are now a large and complex set of commands and in many respects a resemblance to FORTRAN in power.

A BASIC statement comprises the primary elements, identifiers and operators. Identifiers consist of a string of characters which represent a value. Identifiers may either be constants which are not changed during the execution of the program, or variables that do change. Real and Integer variables are merged into a single numeric type and arrays may be defined as variables and the whole matrix can be manipulated. Operators consist of the five mathematical functions +, −, *, / and exponentiation, mathematical functions, and relational operators which include = (equal), > (greater than), < (less than), < > (not equal to), etc.

BASIC expressions are composed of a sequence of valid identifiers and operators arranged so that the part of the statement on the r.h.s. of the equals sign is assigned to the value on the l.h.s. The elementary BASIC statements are defined in *Table 25.7*.

Table 25.7 Elementary BASIC statements

Statement	Function
INPUT	Input data from terminal
DIM	Dimension array variable
PRINT	Print variable or text on terminal
LET	Computes and assigns values
GO TO	Transfers control
IF/THEN	Conditional transfer
FOR	Sets up and operates loop
NEXT	Completes loop
REM	Comment (remark)
END	Final program statement

Each BASIC statement has a line number to facilitate editing the program. A line can be added or altered simply by retyping; the lines are re-arranged in numerical order by the interpreter. An example BASIC program is given in *Table 25.8*.

Table 25.8 Simple BASIC program

```
10  INPUT A, C
20  LET B=C−10
30  IF A=0 THEN 70
40  LET D=B/A
50  PRINT D
60  GO TO 80
70  PRINT "DIVISION BY ZERO"
80  END
```

As a computer language, BASIC is easy to learn and convenient for terminal users who have the language interactively for numerical computation. However, it offers no facilities for structuring programming and a restricted set of data types. There are many versions of BASIC now available, particularly for use with microprocessors. These range from 'Tiny Basics', requiring approximately 2 K bytes to larger 12 K byte interpreters. A number of BASIC interpreters have been developed with real-time features for use in control applications. An ANSI standard REAL TIME BASIC has been defined which provides facilities for concurrent operation and communication with external hardware.

25.16.3 Pascal

Pascal is a modern language which was designed by N. Wirth to support the concepts of structured programming. It is derived from Algol and retains the excellent structural features of that language and also provides comprehensive facilities for structuring data. It can be compiled or 'translated' into the primitive Pascal 'P-code' which is then interpreted. A PASCAL program consists of three elements: declarations, block structure and procedural code.

The declarations require the programmer to specify the name and types of all variables that will be used and the names of all labels referenced in transfer of control (GO TO) statements. The extensive use of data declarations may seem tedious, but they force the programmer to be disciplined and explicitly state the use of each variable. Thus, documentation is improved and more vigorous error checking procedures can be applied. Standard data types are Real, Integer, Boolean and Character. In addition, the user may declare new data types of his own. For example, type dayofweek=(mon, tues, wed, thurs, fri, sat, sun). Variables of these types may now be declared, e.g., var today: dayofweek, and today is then restricted to taking only the values specified in dayofweek in an expression, e.g. today=mon.

Pascal statements can consist of simple assignment statements, procedure statements where a sequence of statements is invoked by a single statement and compound statements which consist of a number of statements prefixed by 'begin', and terminated by 'end' which are treated as a single statement entity or block. The scope or validity of variables is restricted to the program block in which they are defined or to any subsidiary blocks.

Pascal also provides powerful features for the execution of repetitions, e.g. while [condition] do [statement] and, for [control var]:= [first] to [last] do [statement].

The selection of alternative actions uses the construct:

If [condition] then S_1
   else S_2

25.16.4 Other high-level languages

There are many high-level languages in use or under development which do not merit a full description here. The more significant ones are as follows.

25.16.4.1 Algol

This language was designed for scientific purposes. It is a block structured algorithmic language well-suited to structured programming and many of the concepts employed in Algol are used in Pascal. It failed, however, to replace FORTRAN in scientific computation, largely due to commercial rather than technical reasons.

25.16.4.2 COBOL (COmmon Business Oriented Language)

A high-level language designed for programming commercial and business applications. It allows statements to be expressed as a subset of English language, allowing programs to be read by persons unfamiliar with the language rules. It makes provision for handling the data structures, such as files and records, normally required in commercial work.

25.16.4.3 PL/1

This language was designed to combine the features of scientific and business languages in a single language of greater generality. Versions of the language are available on a number of computers and adaptations of it, e.g. PL/M, PL/Z, are used by some makes of microprocessor.

25.16.4.4 Coral 66

This language was introduced in the 1960s by the Royal Radar Establishment and is a British military standard for programming real-time control applications. It has the algorithmic features of Algol (block-structure) and also has facilities to allow the programmer to access machine specific facilities either directly or via procedures, as is possible in Assembly Code (e.g. memory address locations). The language does not include standards for input/output or real-time multi-tasking since these are considered to be machine specific. Coral compilers are designed to produce efficient code with an overhead of only approximately 50% over Assembly Code.

25.16.4.5 RTL2

This language was designed by ICL for real-time control and has the algorithmic structure of Algol. It differs from Coral in that multi-tasking and real-time operations are included implicitly within the language structure.

25.16.4.6 ADA

This language has recently been specified by the American Defence Department and its definition has Pascal-like features. It is a language which is intended to be used on 'embedded' computer systems, i.e. those in a real-time environment. It has all the features necessary to support real-time programming and a structure which allows it to be employed on large projects. It differs from other languages in that the total language environment, including the executive software, is specified within the language definition.

25.16.5 Merits of high-level and low-level languages

The choice of a language for a particular task depends on a number of factors—the skill of the programmer, the task requirements and the availability of particular languages on the computer to be used. If a suitable language is available, then it usually takes far less time to write a program in a high-level language than it does in assembly level. However, the compilation process determines such factors as program execution time and storage allocation and frequently means that a less efficient program results. High-level language programs are also self-documenting to some extent. This is less true of low-level languages.

To summarise, assembly language should be used when:

(1) The program function demands that specific computer features or instructions are employed.
(2) Storage allocation or execution speed is critical to the application's success.

High-level languages should be used when:

(a) A language with suitable features is available.
(b) Detailed use of machine features is not required.
(c) Execution speed is less important.
(d) Assembly code inserts can be included to improve execution speed and allow access to machine features.
(e) The application is of moderate complexity.

It should be noted that modern high-level languages are more efficient and there is a trend to designing computer instruction sets to support particular languages. Thus, the necessity to code in assembler is declining.

25.17 Real-time operation

In many applications, particularly those concerned with control of a process or instrument, it is often necessary to arrange that one computer provides resources to control several independent devices. Frequently, the timing of control operations is also crucial for successful control. A typical example could be the updating of a machine control equation every second and the printing of a less important log every minute. If the log takes ten seconds to complete, then it is clear that a single thread of program execution (e.g. log task, control task, log task, etc.) will not allow the critical control task to be executed successfully at all times.

The alternative is to design the computer software to support concurrent operation. That is, to use a separate program to perform each task function and then organise the allocation of processor time to the various device programs according to their timing, computational requirements and the priority of the task. This can only be achieved if the task execution can be related to an external event such as either an interrupt from the device or a clock interrupt. This type of operation is called real-time computing. A feature of real-time operation is that all the tasks must be completed in real-time; this leads to special problems in hardware and software design.

There are a number of methods of achieving concurrent operation, all of which require an interrupt handler or executive program to determine the allocation of computer time:

(1) *Polling* This is not strictly real-time operation and can only be used successfully if the program execution time for servicing each device is much less than the time between device demands. In the polling method, events merely set flags; the processor examines these flags by scanning them in sequence (e.g. polling) to see which ones are set and require servicing.
(2) *Interrupt driven* Each device or process can be connected to an independent interrupt line which, when asserted, causes a

break in the execution of the current program. The processor is forced to execute program at a predefined location to identify the source of the interrupt and determine which program to execute next. It may be necessary to establish the priority of particular tasks following an interrupt to ensure that the most important task is executed first. This will employ polling and the extensive use of a priority system which will increase the program executive overhead. An outline flowchart of an interrupt executive is shown in *Figure 25.36*.

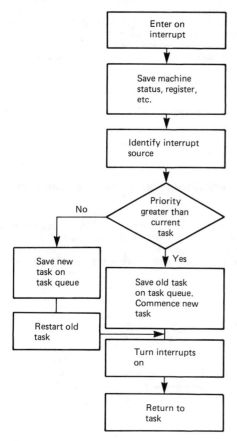

Figure 25.36 A simple interrupt handler

There are several variations in executive structure. For example, each device may be polled, each device may have an interrupt line or each device may be polled at each clock interrupt. Each method has particular advantages in particular applications. In designing a real-time system, it is necessary to define the function of each task and to specify the interactions between tasks. Since all the tasks are performed by one set of hardware, care is required to ensure that the hardware will satisfactorily time-share itself between the tasks, that shared resources are allocated properly, and that the tasks will be executed within the real-time constraints imposed by external events.

ANALOGUE COMPUTATION

Analogue and digital computational methods are now widely used throughout engineering practice. Combined analogue and digital, or hybrid, techniques are also very commonly encountered. Whereas a decade ago analogue and digital computation was not frequently met with in stand-alone functional assemblies (computers) the situation today is changing. The development of large-scale integrated (LSI) circuit technology is creating both analogue and digital functional devices or 'microchips' which are computationally powerful and relatively cheap. This is enabling computing power to be tailored more specifically to engineering requirements and, where this is appropriate, to be distributed physically in engineering systems. Today, though the stand-alone digital computer is available in a very wide range of powers, the analogue computer has almost disappeared, as has the hybrid computer, except for very specific application areas. However, analogue and hybrid computational elements are extensively used within engineering systems as well as pure digital elements, though the relative usage of the three techniques is very much weighted towards digital and this is a process which is likely to accelerate as the speed and economy of digital computation increases.

In a general sense, analogue computation implies the solution of arithmetic, algebraic or differential equations by the representation of mathematical entities by physical quantities. A simple example is the slide rule where, for instance, quantities $\log A$ and $\log B$ are each represented by the lengths between marks on a rule and a summation is performed using the equation

$\log AB = \log A + \log B$

By making the lengths on the rule proportional to the logarithms of the variables, multiplication of quantities A and B can be performed by addition of lengths $\log A$ and $\log B$.

Any physical system governed by a particular set of equations and capable of being disturbed in a controlled manner can be used to solve those equations and, in so doing, can be called an analogue computer.

In engineering practice, analogue computation is today performed almost universally by electronic circuits and the values of variables in equations are almost invariably represented by voltages. Other systems of representation, e.g. by currents and by mechanical quantities are equally feasible and have been employed but the technology which has developed most successfully uses voltage as the computing variable. Values of parameters or coefficients in equations are represented by resistance ratios or by resistance–capacitance products. Inductance could be used equally as a circuit element in analogue computation but the difficulty of manufacturing or simulating pure inductances of precise value has strongly discouraged its use.

An analogue computation is limited in accuracy by the resolution with which the chosen analogue quantity (e.g. voltage) can be measured. Analogue quantities are, in principle, continuous and the limits to resolution are physical effects such as the wire-to-wire spacing of potentiometers or, ultimately, random circuit noise.

Analogue computation is used very commonly for the solution of differential equations. Consider, for example, the third order equation

$$ax + \frac{b\,\mathrm{d}x}{\mathrm{d}t} + \frac{c\,\mathrm{d}^2 x}{\mathrm{d}t^2} + \frac{\mathrm{d}^3 x}{\mathrm{d}t^3} = f(t) \qquad (25.1)$$

where a, b and c are coefficients or parameters, t is the time variable and $f(t)$ is some function of time. A representation of this equation can be set up by commencing with the highest order term and successively integrating it to form the lower order derivatives and, finally, x itself. Examination of the equation will show that the highest derivative, which is to be fed to the beginning of the chain of integrations may be obtained by adding together the variable x and its lower derivatives in proportions determined by the coefficient values.

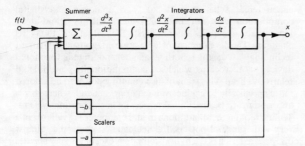

Figure 25.37

Figure 25.37 shows the resulting representation of the third-order differential equation. In an analogue computer the integrators and summer are realised by the use of electronic circuits which form the building blocks of analogue computation. Solutions of the differential equation so represented are obtained by applying a drive voltage having the same form as $f(t)$ and observing or recording the resulting changes in voltages representing x or its derivatives using a cathode ray oscilloscope or other electrical recording instrument.

25.18 Operational amplifier

The basic element of analogue computation is the operational amplifier. An idealised representation of an operational amplifier is illustrated in *Figure 25.38*. The equation governing the operation of the circuit is

$$v_0 = -A(e_1 - e_2)$$

i.e. the output voltage is proportional to the difference between the input voltages.

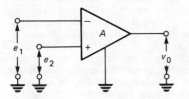

Figure 25.38

Where v_0 is the output voltage (typically in the range $-10 < 0 < 10$ V), e_1 and e_2 are input voltages in the same range as v_0, A is a voltage gain assumed very large, approaching infinity.

A practical operational amplifier fails to meet the ideal characteristics in several ways as follows.

25.18.1 Voltage gain

At zero frequency this will range between 10^4 and 10^6. As the frequency is increased, the gain falls off and phase shift between input and output signals increases.

25.18.2 Input resistance

This depends on the properties of the input transistors. For high values, field effect transistors are employed in the input stages.

25.18.3 Transient response

For large step function changes at the input, i.e. changes comparable with the nominal swing of say ± 10 V, the amplifier will, in general, respond more sluggishly than for small input changes. This is due to overloading of stages in the amplifier. To indicate the performance under these conditions the slewing rate, that is, the maximum rate of change of output voltage, is defined.

25.18.4 Stability

When feedback is introduced between output and input of an operational amplifier, instability may ensue. A correcting network must be incorporated, either within the amplifier or by externally connected components to ensure that the closed loop gain falls to zero at a frequency lower than that at which the closed loop phase shift becomes 360 degrees. Operational amplifiers should be unconditionally stable to avoid unexpected instabilities which may otherwise result in particular computational configurations.

25.18.5 Output impedance

Output impedance of an operational amplifier is low, of the order of tens of ohms or less, since the output is normally derived from an emitter follower stage.

25.19 Operational amplifier integrated circuits

Most modern operational amplifiers are available in the form of LSI circuits on single silicon chips. Dual and quad configurations are also available with two or four amplifiers to each chip. Some specialised amplifiers for high performance applications are made in hybrid circuit form. The circuit layout of a low cost, general purpose IC amplifier is shown in *Figure 29.39*.

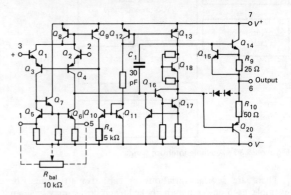

Figure 25.39

Operational amplifiers are available in a very wide range of specifications in terms of bandwidth, slew rates, input current and voltage offsets, etc. Typical specifications for integrated circuit types are indicated in *Table 25.9*.

25.20 Analogue functional elements

By the connection of external components to the basic operational amplifier of *Figure 25.38* various analogue functional elements can be derived.

25.20.1 Inverting amplifier

The inverting amplifier can be implemented by the connection of

Table 25.9 Typical characteristics of integrated circuit operational amplifiers

Specification	Input bias current (ηA)	Input offset current (ηA)	Input offset voltage (μV)	Voltage gain	Common mode rejection factor (dB)	Input resistance (MΩ)	Gain-bandwidth product (MHz)	Slew rate (V/μs)	Output voltage swing (V)
General purpose low cost	500	40	4	50 000	90	0.3	1.0	1.0	±10
Low drift, FET input stage	15×10^{-3}	2×10^{-3}	0.25	100 000	90	10^5	1.0	0.5	±10
Fast slewing	250	50	10	30 000	90	100	20	80	±10
Wide bandwidth	25	25	5	200 000	100	300	100	20	±10

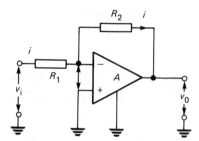

Figure 25.40

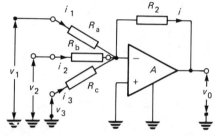

Figure 25.41

two resistors in the current feedback arrangement shown in *Figure 25.40*. Given zero bias current at the amplifier input point the following equations apply

$$\frac{v_1 - e}{R_1} = \frac{e - v_0}{R_2} = i$$

and

$$v_0 = -Ae$$

so that

$$\frac{v_1 + v_0/A}{R_1} = -\frac{v_0/A + v_0}{R_2}$$

When A is very large this approximates to

$$v_0 = -\frac{R_2}{R_1} v_1$$

So that the overall gain v_0/v_1 depends only upon the resistors R_1 and R_2. Note that v_0/v_1 is negative, i.e. voltage inversion takes place. Under these conditions of very high gain, A the summing junction, remains virtually at earth potential since only an infinitesimal change in its potential is sufficient to set up the voltage v_0. The summing junction is commonly referred to as a virtual earth.

25.20.2 Summer

Voltages can be summed by adding them to the summing junction as shown in *Figure 25.41*. Here, for A large and amplifier input current zero, as before, we have

$$i = i_2 + i_2 + i_3$$

and

$$v_0 = -\frac{R_2}{Ra} v_1 - \frac{R_2}{Rb} v_2 - \frac{R_2}{Rc} v_3$$

25.20.3 Integrator

The operation of integrating a voltage input can be achieved by the operational amplifier with capacitive feedback as shown in *Figure 25.42*. Here, with the same constraints as before on amplifier gain A and on amplifier input current

$$i = \frac{v_i}{R} = -C \frac{dv_0}{dt}$$

Integrating both sides yields

$$v_0 = -\int \frac{v_i}{RC} dt$$

where RC is the time constant of the integrator in seconds.

25.20.4 Scaling

Scaling may be achieved by the use of a precision potentiometer as in *Figure 25.43* or *Figure 25.44* where α is the fraction of the input voltage v_i which is applied to the input resistor of the operational amplifier and the potentiometer resistance is small compared with the input resistance. This arrangement can also be used when the latter constraint is not met, by connecting a reference voltage to the input as v_i and adjusting the potentiometer to make v_0 bear the desired ratio to v_i.

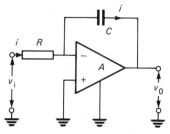

Figure 25.42

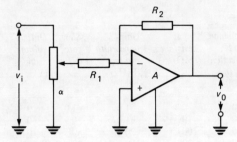

Figure 25.43

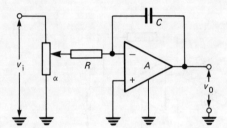

Figure 25.44

25.20.5 Differentiator

In principle, an operational amplifier may be connected as a differentiator as in *Figure 25.45*. The governing equation is

$$v_o = RC \frac{dv_i}{dt}$$

However, this circuit is not normally used because of its propensity to overload on noisy and rapidly changing input signals and thus, to introduce serious errors into computations.

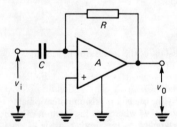

Figure 25.45

25.21 Complex functions

By appropriate resistance and capacitance networks in the input and feedback paths of the operational amplifier, a wide variety of circuit functions may be derived. For example the equation of a so-called exponential or first-order lag is given by the equation

$$T \frac{dv_o}{dt} + v_o = kv_i$$

where k is a constant and T is a time constant.
This equation can be represented by the arrangement shown in *Figure 25.46* where

$$k = \frac{R_2}{R_1} \quad \text{and} \quad t = CR_2 \text{ s}$$

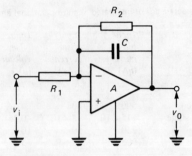

Figure 25.46

The so-called second-order or quadratic lag is given by the equation

$$\frac{1}{\omega_0^2} \frac{d^2 v_o}{dt^2} + \frac{1}{Q\omega_0} \frac{dv_o}{dt} + v_o = kv_i$$

where ω^2 is the square of the natural angular frequency, Q is the magnification factor and k is a constant.
This equation can be solved by the arrangement shown in *Figure 24.47* where

$$2C_2 R_1 = Q\omega_0 \quad \text{and} \quad \frac{1}{\sqrt{C_1 C_2 R_1^2}} = \omega_0$$

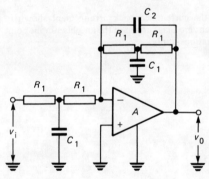

Figure 25.47

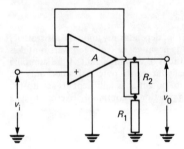

Figure 25.48

25.22 Voltage follower

An operational amplifier with voltage feedback is shown in *Figure 25.48*. The equation governing the behaviour of this arrangement is

$$A \left[v_i - v_o \frac{R_1}{R_1 + R_2} \right] = v_o$$

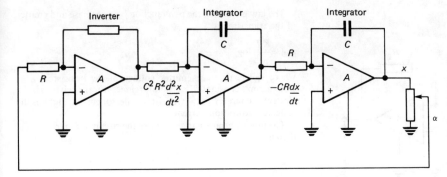

Figure 25.49

for very large gain, A this reduces to

$$v_o = v_i \frac{R_1 + R_2}{R_1}$$

The input impedance of this amplifier is very high (e.g. see *Table 25.9*) and the non-inverting gain is determined entirely by R_1 and R_2. The output impedance is very low.

For the limiting case where R_1 is infinite and R_2 is zero (i.e. output connected directly to inverting input) the closed loop gain is unity so that the output voltage follows the input voltage.

25.23 Function generators

The function $f(t)$ in the Equation (25.1) may have any rational form. Certain well-defined forms of function are steps, ramps, square waves, triangular waves and sinewaves. Function generators which generate such a range of functions are commercially available. In some cases they can conveniently be set up using analogue computational elements. For example a sinewave can be generated by solving the equation having the solution $x = \cos \omega_0 t$ which is

$$\frac{1}{\omega_0^2} \frac{d^2 x}{dt^2} + x = 0$$

This equation is solved by the analogue arrangement shown in *Figure 25.49* where $\omega_0^2 = \alpha/C^2 R^2$. The voltage output at x is a sinewave of frequency depending upon the setting of the coefficient potentiometer. When this is set at zero attenuation ($\alpha = 1$) the angular frequency of the sinewave will be $\omega_0 = 1/CR$ rad/s. Tapping down the potentiometer to reduce α will increase the angular frequency proportionally to $\sqrt{\alpha}$.

25.24 Comparator

It is often necessary to compare two time varying voltages and to derive a signal at the instant when one become algebraically greater than the other. This function is achieved by the use of an operational amplifier operated as a differential amplifier as shown in *Figure 25.50*. When the gain A of the amplifier is assumed infinite the equation governing the operation of this circuit is:

$$v_o = |v_{o\max}| \operatorname{sgn}(v_2 - v_1)$$

where $|v_{o\max}|$ is the magnitude of the saturating value of output voltage and $\operatorname{sgn}(v_2 - v_1)$ is the sign of the algebraic difference between voltages v_2 and v_1.

An essential property of a comparator is speed of response since v_2 and v_1 might be varying rapidly and their point of

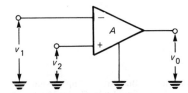

Figure 25.50

equality must be detected without delay. Response times in the order of 10^n s are possible. The output voltage of some comparators is designed to vary between 0 and 5 V, representing logical 0 and logical 1, rather than between negative and positive voltages as indicated by the above equation. The law governing the comparator would then be

$v_0 = 0$ for $\operatorname{sgn}(v_2 - v_1)$ Negative

$v_0 = 5$ V   for $\operatorname{sgn}(v_2 - v_1)$ Positive

Other sign conventions are clearly possible.

25.25 Diode function generator

When it is required to generate arbitrary functions of an analogue variable a piecewise linear approximation to the required curve can be derived by the use of diodes and resistors. In *Figure 25.51* the principle of this method is illustrated. A biased diode D_1 supplies current to the summing point of an operational amplifier from the input voltage v_1. The current via the diode, and hence the operational amplifier output voltage v_0, is zero up to the bias level and then rises linearly with a slope R_2/R_1.

To add further linear segments to the approximated non-linear function, further biassed diodes are connected to the summing point and contribute current as indicated in *Figure 25.51*. In this way a further slope change is introduced at bias level v_b. By connecting a series of biassed diodes to the summing point a

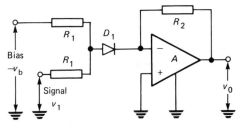

Figure 25.51

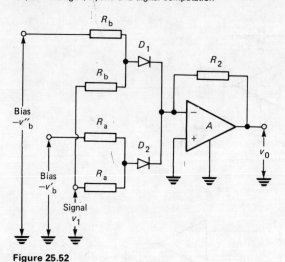

Figure 25.52

series of linear segments can be built up to approximate the desired function as shown for two diodes in *Figure 25.52*.

25.26 Multifunction converter

Several manufacturers offer hybrid devices, which provide a range of non-linear analogue functions. A typical converter might provide

Multiplication
Division
Square
Square root
Exponentiation
Roots
Sine
Cosine
Tan$^{-1}y/x$
$\sqrt{x^2+y^2}$

The basis of one such converter is the high precision logarithmic amplifier. This relies on the fundamental characteristic of a p–n junction that the voltage across the junction is proportional to the logarithm of the current through it.

A logarithmic amplifier can be constructed using an operational amplifier and the p–n junction of a transistor T_1 with shorted collector and base as shown in *Figure 25.53*. The shorting of collector and base is achieved by the use of the virtual earth summing junction of the operational amplifier.

The law governing the p–n junction between base and emitter of the transistor is

$$e_{EB} = \frac{KT}{q} \log_e \frac{I_c}{I_s}$$

where e_{EB} is the base-emitter voltage of the transistor, K is Boltzmann's constant, T is absolute temperature in kelvin, q is the electronic charge, I_c is the transistor collector current and I_s is the collector saturation current.

The input/output characteristic of the circuit in *Figure 25.53* is therefore

$$I_c = \frac{v_1}{R_1}$$

so that

$$v_0 = e_{EB} = \frac{KT}{q} \log_e \frac{v_1}{R_1 I_s}$$

To obtain anti-logarithmic characteristics the p–n junction must be connected on the input side of the operational amplifier as in *Figure 25.54* where the collector to base junction of transistor T_2 is used. Here

$$I_c = v_0/R_2$$

so that

$$v_i = e_{EB} = \frac{KT}{q} \log_e \frac{v_0}{R_2 I_s}$$

or alternatively

$$v_0 = R_2 I_s \text{ anti} \log_e \frac{q}{KT} v_i$$

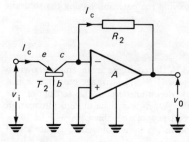

Figure 25.54

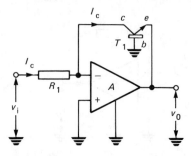

Figure 25.53

Both logarithmic and anti-logarithmic relationships are highly temperature sensitive and special measures must be taken to compensate for this. The use of identical transistor structures arranged in close physical proximity is an effective compensation technique in the generation of functions such as multiplication where both log and anti-log functions are used simultaneously.

A multi-function convertor can be made using logarithmic and anti-logarithmic amplifiers as shown in *Figure 25.55*. In this circuit various functions can be generated by appropriate setting and allocation of the voltages X, Y and Z and of the gain m as indicated in *Table 25.10*. Note that the variables X, Y and Z must always be positive in this arrangement which limits the range of computed functions. Other types of multifunction converter are available which accept inputs and outputs of either sign.

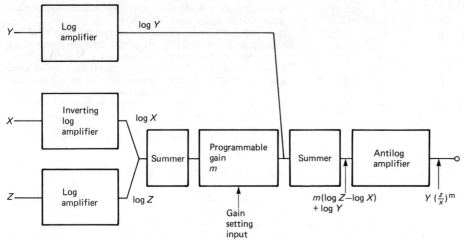

Figure 25.55

Table 25.10 Settings for multi-function converter

Function	X	Y	Z	m	Result
Multiply	Constant level K	Variables to be multiplied		Unity	YZ/K
Divide	Denominator	Constant level	Numerator	Unity	$K \cdot Z/X$
Root	Constant level K_1	Constant level K_2	Operand	Fractional	$(K_2/K_1)m\sqrt{z}$
Power	Constant level K_1	Constant level K_2	Operand	>1	$(K_2/K_1)Z^m$

25.27 Analogue computer

Consider the third-order equation already discussed and illustrated in *Figure 25.37*. Rearranging the differential equation we have

$$\frac{d^3x}{dt^3} = f(t) - c\frac{d^2x}{dt^2} - b\frac{dx}{dt} - ax$$

Assuming a, b and c are fractional, the analogue computer is set up as in *Figure 25.56*. Each operational amplifier introduces sign reversal and unity gain inverters must be used where shown to restore the correct signs to voltage variables. Coefficients are set by potentiometers shown as scalers.

Initial conditions of the variables x, dx/dt and d^2x/dt are set by establishing an initial voltage across the appropriate integrator capacitor as shown in *Figure 25.57*. When the switch is closed the tapping on the potentiometer is adjusted to give the required initial voltage across the capacitor. Each integrator used in the computation is provided with a similar circuit. At the instant of starting the computation all initial condition switches are opened simultaneously. At the same instant the input function $f(t)$ is applied. Records are taken of the variations in variables of interest which follow the initiation of computation.

In the design of a general-purpose analogue computer it is convenient to choose a speed of operation which gives voltage variations which can be recorded using electromechanical

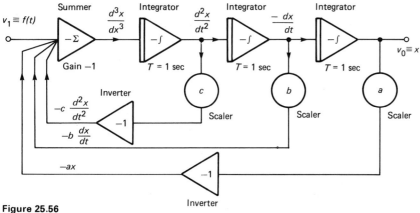

Figure 25.56

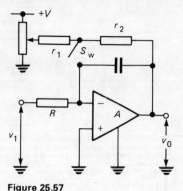

Figure 25.57

recorders so that hard copy can be easily obtained. Under these circumstances resistance values of 0.1–10 M are typical and capacitances may be fixed at $1\,\mu F$. It is common practice to provide fixed inverter amplifiers with selectable gains of 0.1, 1 and 10 and integrators with time constants of 0.1, 1 and 10 s or other convenient values. Precision potentiometers are then used as scalers to generate the correct coefficient values.

When an analogue computer is used to simulate the dynamic behaviour of an engineering system, it is first necessary to formulate the differential equations governing the real system in a way which will allow their representation in the form of *Figure 25.56* where the scaler constants a, b and c, etc. are fractional but where summers and amplifiers with gains up to 10 may also be used in the system. Care must be taken to ensure that transient voltage swings in response to simulated normal disturbances and inputs do not overload the analogue circuits, thereby introducing deviations between the simulated and actual responses. This process is known as scaling of the analogue computer. Another important consideration is the relative speed of response of the simulation and of the real system. This is governed by the value of the time-dependent parameters in the analogue computer, that is, the time constant RC of the integrators. If the analogue is initially set up to run at real time speed, i.e. a response time equal to that of the actual system, then altering each integration time constant by the same factor will change the response time of the simulation by that factor.

Validation of the behaviour of an analogue computer may be done in a variety of ways, depending on the application. Checking the overall response of the analogue computer against a known solution of the equations of motion being simulated is sometimes possible. In other cases it will be necessary to check out the analogue computer in isolated sections of the total configuration where known solution of the equations governing the isolated sections can be defined.

A general purpose analogue computer is provided with a patchboard. Input and outputs of integrators, summers, inverters, function generators, comparators, multi-function converters, scaling potentiometers, etc. are all brought to individual contact points on the patchboard. A required configuration of elements is then assembled by making connections between the appropriate contact points, thus forming a hard wired 'programme' for the solution of the particular problem.

25.28 Applications

Analogue computers are now used largely for teaching purposes though examples, with a large number of operational amplifiers, are sill used in the fields of power systems analysis and aerospace applications. Very often such applications use hybrid computers

(q.v.) which are combinations of analogue and digital machines and are described in a later section.

Analogue computing elements are very commonly employed in systems of real-time computing. Examples are in aerospace where, for instance, a continuous measure of distance flown might be derived by integrating the signal from an airspeed instrument and feeding in a correction for windspeed. Any system requiring computations to be carried out on analogue signals from instruments may employ analogue computing techniques. It must be emphasised, however, that digital computational techniques are replacing analogue techniques in many application areas and that this will be a continuing trend.

HYBRID COMPUTATION

Hybrid computation is the combined use of digital and analogue computing techniques. Digital techniques provide high accuracy solutions of numerical expressions, statistical relationships, matrix equations, etc., whilst analogue techniques provide much lower accuracy but higher speed solutions of differential equations. By combining the techniques, an optimum speed versus accuracy result may be obtained for the solution of a range of engineering problems. A stand-alone computer which combines these techniques and in which separate analogue and digital computers can be identified is called a hybrid computer.

Hybrid computation can also be identified in a more general sense. Dynamic engineering systems, e.g. vehicles, machine tools, manufacturing machinery of all types and chemical and food processing plant have variables such as linear and rotational motion, speed, acceleration, pressure, flow, temperature and a host of others which may be transduced by sensors into analogue form. The actuators for these engineering systems are also usually analogue in principle and include directional control mechanisms, motors, pumps, rams, fluid flow valves, heat sources, etc. In the digital control and instrumentation of such engineering systems the essential functions are the conversion of analogue sensor signals to digital form for processing by digital computing elements of which the output must be converted back to analogue form to drive the actuators. Computation will, in general, occur in both the analogue and digital modalities, i.e. both before and after the conversion processes.

Thus hybrid computation is of very general application in engineering systems, though hybrid computers, in the formal sense of combined analogue and digital calculating, are becoming comparatively rare.

Hybrid computation requires all the elements used in analogue computation together with a range of signal switching, selection and converting elements, and, of course, digital computation equipment and software.

25.29 Signal switches

These are used for the high-speed switching of analogue voltages and a schematic diagram is shown in *Figure 25.58*. An ideal signal switch should provide a zero resistance path when ON and an

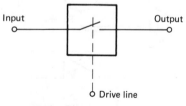

Figure 25.58

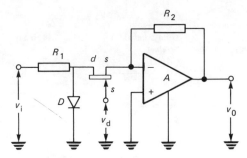

Figure 25.59

infinite resistance path when OFF; switching time should be negligible. The first two requirements are met by a reed relay switch but the switching time of several milliseconds is quite inadequate for most computing requirements and contact bounce is an added defect. Solid state technology now allows very high-speed switches to be manufactured which give good approximations to the ideal characteristics. These are driven on and off by logic voltage levels applied to the drive line. They may use bipolar or field effect transistor (FET) technology. A typical FET current switching arrangement is illustrated in *Figure 25.59*; the diode D holds the source of the switch at about 0.6 V in the OFF state to limit the voltage which may be applied between drain and source. For an N-channel FET as shown in *Figure 25.59*, the drive voltage V_d will be at 0 V for switch ON and a negative voltage for switch OFF.

A wide range of integrated circuit devices is available packaged in a range of options, e.g. SPST, SPDT, DPST, DPDT and dual versions of these configurations.

25.30 Sample and hold

A common requirement is to sample a fluctuating voltage level at a defined instant of time and to store this value for a finite period. The operation of sampling may be purely repetitive, defined by an oscillator or timer, or may be initiated on demand from a digital computer or other source. In either case the basis of the circuit is a signal switch driven by the sample initiating mechanism; a simplified arrangement is shown in *Figure 25.60*. If the switch shown in this circuit is turned on briefly at fixed intervals of time ΔT, the relations between input and output voltage are as illustrated in *Figure 25.61*. Leakage of charge from the capacitor due to switch leakage and voltage follower buffer stage input impedance, gives rise to a droop on the steps of output voltage which must be kept within a specified level by an appropriate choice of switch properties, capacitance value and sampling interval.

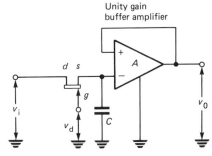

Figure 25.60

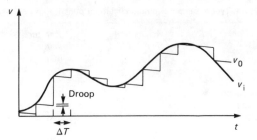

Figure 25.61

In defining the performance of a sample and hold circuit, two time intervals are important, viz

(1) *Acquisition time* The time elapsing from the initiation of switch closure to the point where the output settles to within some specified margin of the input.
(2) *Aperture time* The time taken from the initiation of switch opening to the point where the switch actually opens.

A more complex and commonly encountered sample and hold arrangement is shown in *Figure 25.62*. In this circuit three switches are used and the buffer amplifier is an operational amplifier connected as a non-inverting unity gain device with very high input impedance as in *Figure 25.60*. In operation,

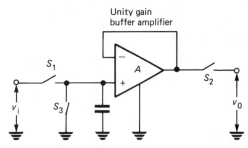

Figure 25.62

switch 1 and switch 2 are initially open and switch 3 is closed. Capacitor C_1 is discharged to zero volts through the switch ON resistance of S3. When S1 closes, S3 simultaneously opens and C charges to the analogue signal voltage v_1. S1 is then opened and C holds the voltage representing the sampled value of v_1. This voltage is established at the output of the unity gain amplifier and when S2 closes the sampled voltage is available for use by subsequent circuits. S3 then closes, S2 opens and the circuit is ready to accept the next sample.

Sample and hold amplifiers are available in both monolithic and hybrid circuit forms. Typical performance figures are shown in *Table 25.11*.

25.31 Digital-to-analogue converter (DAC)

The principle is illustrated in *Figure 25.63*. Switches S1 to S4 represent a binary word, i.e. a closed switch represents 1 and an open switch 0. S1 is driven by the most significant bit (MSB) and S4 by the least significant bit (LSB). For various combinations of switch states *Figure 25.64* shows the relationship between switch state and voltage output from the operational amplifier. It can be seen that the output voltage v_0 is proportional to the number

Table 25.11 Typical sample and hold circuit performance figures

Specification	Acquisition time (μs)	Aperture delay (ns)	Droop ($\mu V/msec$)	Gain-bandwidth product (MHz)	Accuracy
Hybrid					
General purpose	6	100	100	1	0.01%
High speed	1	20	10^4	5	0.01%
Ultra HS	0.025	6	5×10^4	50	0.1%
Monolithic					
General purpose	10	200	100	0.15	0.01%

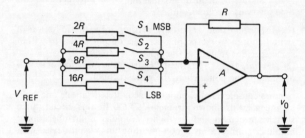

Figure 25.63

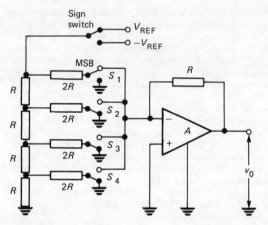

Figure 25.65

Switch state				Decimal value	v_0
S_1	S_2	S_3	S_4		
1	1	1	1	15	$^{15}/_{16} \times V_{REF}$
1	0	0	0	8	$^{1}/_{2} \times V_{REF}$
0	1	0	0	4	$^{1}/_{4} \times V_{REF}$
0	0	1	0	2	$^{1}/_{8} \times V_{REF}$
0	0	0	1	1	$^{1}/_{16} \times V_{REF}$

Figure 25.64

represented digitally by the switch position and that the resolution, set by the least significant bit, is one-sixteenth of the maximum output voltage.

In a practical converter it is necessary to avoid the very wide range of resistance values which would be necessary to carry out a conversion of, say, 12-bits in the type of circuit shown in *Figure 25.63*, where the range of the input resistance would be 2^{n-1} or 2048 to 1. This is done as shown in *Figure 25.65*, which illustrates the so-called current switched mode of conversion in which currents are switched into the virtual earth point of an operational amplifier.

The resistance ladder network gives the same effect as the resistance arrangement in *Figure 25.63* but uses only two resistance values for all digital word lengths. *Figure 26.65* also shows one method of dealing with negative numbers. In this method, the so-called sign magnitude code, a 1 is added to the left of the most significant bit to indicate negative number. This sign bit can be used to operate a switch reversing the sign of the reference voltage V in *Figure 28.65*.

Digital-to-analogue converters are available in both monolithic and hybrid circuit form. In the former configuration 12-bit operation is currently the highest accuracy available; in hybrid form up to 16-bit accuracy devices are available. Some typical specifications are indicated in *Table 25.12*.

Table 25.12 Typical DAC performance figures

Specification	Resolution (word length) bits	Settling time (ns)
Hybrid		
General purpose	12	3000
Fast	10	25
Monolithic		
General purpose	12	200
Fast	8	80

25.32 Analogue-to-digital converter (ADC)

25.32.1 Parallel or flash conversion

The principle of this technique is shown in *Figure 25.66*. An *n*-bit converter requires 2^{n-1} comparators, so for circuit economy the method is normally only used where *n* is small, i.e. accuracy is low. The resistor ladder network generates 2^{n-1} voltage reference levels corresponding to the least significant bit (LSB) at one end of the ladder to $(2^{n-1}) \times$ LSB at the other end. The input voltage v_i is simultaneously compared with all the reference levels and, as v_i passes through a given reference level the output of that particular comparator will make a transition from 0 to 1 or 1 to 0, depending on the direction of change of the input voltage. The comparator outputs are fed to encoding logic which converts the signals into any required digital code convention. Parallel

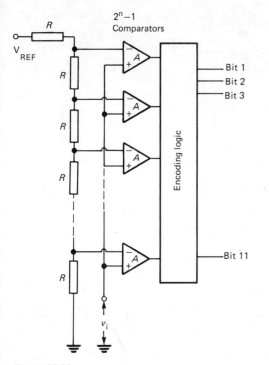

Figure 25.66

25.32.2 Successive approximation conversion

A circuit arrangement for positive input voltages is shown in *Figure 25.67*. At the start of conversion the contents of the output register are set to zero and the shift register then starts with a 1 in its most significant bit (MSB) and zeros in other positions. Since v_0 is initially zero this causes a 1 to appear in the MSB of the output register and the D/A converter outputs an analogue signal of one-half the full-scale output. This analogue level v_0 is compared with the input signal v_i and the comparator will produce output 1 if the input signal exceeds half full-scale DAC output and zero if v_i is less than half full scale. In the former case, gate 1 will turn ON and the output register logic will retain a 1 in the MSB of the output register, in the latter case gate 1 will turn OFF and the output register MSB will be held at zero. The next clock pulse then arrives and the next most significant bit is set in the output register. If the MSB was retained at 1 in the previous state then the converted value of 110 000 (i.e. $\frac{3}{4}$ of the full scale) in the output register is compared with the input voltage v_i and a 1 or 0 appears at the comparator output depending on whether v_i is more than or less than v_0. Gate 2 will detect which state exists and output register logic will set the second most significant bit accordingly before the next lowest position in the output register is tested. In this way all the n-bits in the output register are tested and conversion is complete when the LSB is tested. The contents of the output register are then read into a buffer register for input to the digital machine, the output register is cleared and another conversion operation commences.

converters are very fast since the transmission delays are very small and are in parallel. They may be used for the conversion of video signals in bandwidths of tens of MHz but high accuracy requires a very large number of comparators.

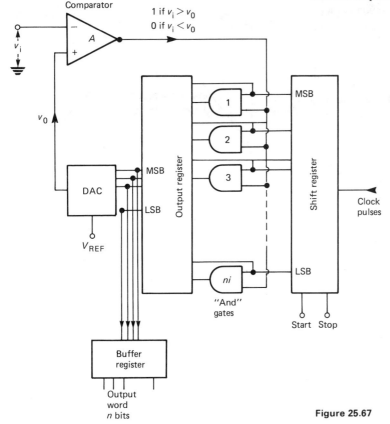

Figure 25.67

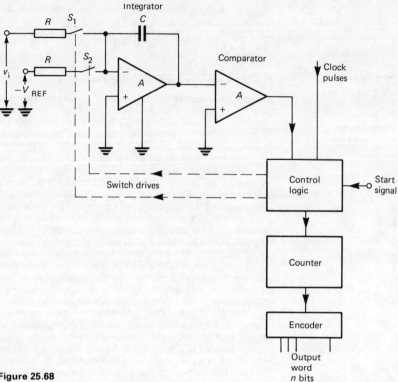

Figure 25.68

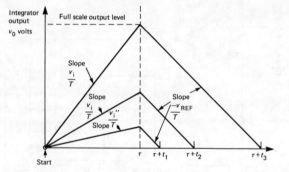

Figure 25.69

The design constraints on an ADC using the principle are

(1) The change in input voltage during one conversion cycle should correspond to less than one significant bit.
(2) For an n-bit output word the conversion time is nt where t is the clock pulse train period.
(3) Conversion speed is limited by the settling time of the bit-setting logic gates.
(4) The accuracy of the conversion is limited by the accuracy of the DAC incorporated in the system.

25.32.3 Dual slope integrating conversion

The arrangement of this converter type is given in *Figure 25.68*. The input voltage v_i and a reference voltage V_{ref} are supplied to an integrator of time constant t via analogue switches S1 and S2. At $t = 0$ a conversion commences by the closure of S1; the integrator output then increases linearly at a rate v_i/t V/s (see *Figure 25.69*). At a predetermined time the switch S1 is opened and S2 closed. Now reference voltage $-V_{ref}$, which is of reverse sign to v_i is applied to the integrator whose voltage decreases at $-V_{ref}/t$ V/s.

The integrator output reaches zero at time $\tau + t_2$ which is detected by the comparator. If this procedure is repeated for a different value of input voltage v_i' or v_i'', the times to zero intercept will be $\tau + t_3$ and $\tau + t_1$ respectively. In fact, $t_2 = (v_i/V_{ref})\tau$ where τ and V_{ref} are constants.

The controller counts clock pulses into a binary counter for the duration t_1, t_2 or t_3 depending upon the input voltage level so that the contents of the counter represent a scaled and digitised version of the input voltage sample v_i.

Suppose that a 12-bit conversion is to be made with a full-scale value of v_i' of 10 V. Then 2^{12} (4096) pulses must be counted in the time interval t_3. For a clock pulse rate of 1 MHz, t_3 must be 4.096 ms and this must equal $v_i'\tau/V_{ref}$. For $V_{ref} = 10$ V, $\tau = 4.096$ ms and the total conversion time is therefore $\tau + t_3$, that is 8.192 ms.

Clearly this method is comparatively slow and is restricted to analogue signal bandwidths of up to about 10 Hz. However, due to the presence of an integrator, the technique is relatively insensitive to noise in the input signal and completely insensitive to interfering frequencies having period of $1/T$ and to their harmonics. ADCs of this type are used where high resolution at low cost and low speed are required. They are available in hybrid circuit form. *Table 25.13* gives typical performance figures for ADCs.

25.33 Analogue multiplexers

This function is required when a number of analogue voltage signals are to be converted by a single ADC. This is a common requirement since the alternative of a separate ADC committed to each analogue signal to be converted is likely to be prohibi-

Table 25.13 Typical ADC performance figures

Specification	Resolution (word length) bits	Conversion time
Hybrid		
Successive approximation	12	2.5 μs
Integrating	12	6 msec
Flash	3	20 nsec
Monolithic		
Low cost	12	24 msec

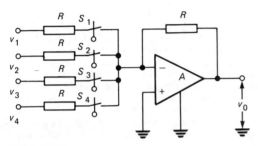

Figure 25.70

tively costly. A typical 4-channel multiplexer is illustrated in *Figure 25.70*.

When one of the switches, i.e. S1 is ON then the voltage v_i appears at the output as v_o since the circuit behaves as a unity gain operational amplifier. By driving the switches sequentially from a clock or on demand from a digital source, any of the for analogue signals can be selected as v_o and routed to an ADC for conversion to digital form.

25.34 Hybrid computer

The block diagram of a typical computer arrangement is shown in *Figure 25.71*. The digital and analogue computers are shown as stand-alone units with data flowing from one to the other via the bold lines and control and timing signals passing via the fine lines. The voltage outputs of the analogue computer are sent to an addressable multiplexer which selects a particular output on command and routes it to a sample and hold circuit where it is held as a voltage level and read by an analogue-to-digital converter ADC. The voltage input to the ADC is converted into a digital word and held in a buffer register at the input of the digital computer to await reading into the computer.

Digital words which are output from the digital machine are held in an output buffer register and then pass to a digital distributor which is addressed by the digital machine and routes the digital word to the correct digital–analogue converter for inputting into the analogue computer, either as a computing variable or as the setting of a coefficient in the analogue computation.

The program in the digital machine governs the computation within the digital machine and also determines data transfers between analogue and digital machines. In this sense the digital computer uses the analogue machine as another peripheral and is in complete control of analogue computing activity. Conventional peripherals may be attached to the digital machine, e.g. disc and tape memory, keyboard input and printer and/or VDU output. The analogue machine is provided with recorders for obtaining hard copy of selected analogue variables and these may themselves be controlled by the digital machine.

25.35 Hybrid computer applications

Hybrid computers can be used in a variety of ways for the solution of engineering problems. Examples are

Repetitive solution of differential equations, e.g. for optimisation of dynamic systems.
Solution of partial differential equations.

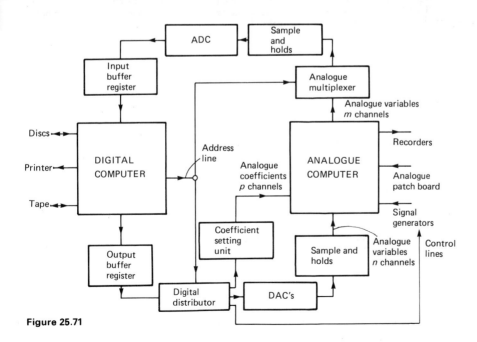

Figure 25.71

Study of behaviour of systems with random disturbances.
Solution of multi-degree of freedom problems.

In all these problems the key decision to be taken is the nature of the division of computational load between the analogue and digital systems. For example, in optimisation studies a given set of differential and algebraic equations will have to be solved for a range of parameter changes. After each parameter change the criterion function will be computed and the direction of further parameter changes will depend on the nature of the changes in the criterion function. In such a problem the differential and algebraic equations may be set up on the analogue machine and the digital machine will compute the criterion function and determine the values of the parameters for the next analogue computation. In this application the analogue machine is used as a differential equation solver and solutions are initiated, assessed and their parameters appropriately modified by the digital machine.

Partial differential equations have more than one independent variable and analogue computers can deal with one only, i.e. time. Consider a simple partial differential equation

$$\partial^2 \phi / \partial x^2 = K(\partial \phi / \partial t)$$

This has two independent variables, a space variable x and the time variable t.

The solution of this equation can be achieved computationally in a number of ways depending upon which of the two variables is treated in continuous form and which in a discrete or finite difference form. By regarding x as discrete, the equation may be solved by holding x at a discrete value and deriving a continuous solution in t, then altering x to its next discrete value and again deriving the continuous solution in t. Four methods of solution are possible, viz

(1) Continuous space, continuous time.
(2) Discrete space, continuous time.
(3) Continuous space, discrete time.
(4) Discrete space, discrete time.

Method 1 is not suitable for analogue, digital or hybrid computation. Methods 2 and 3 lead to hybrid techniques when the discrete variable is handled by the digital machine and the continuous variable by the analogue machine. Method 4 requires a digital computer on its own since both variables are discrete.

In studying the behaviour of systems governed by differential equations when subjected to random disturbances, the hybrid computer can save considerably in time compared with pure analogue or pure digital. One such problem arises in the development of homing missile systems when it is necessary to evaluate the statistics of the miss distance (i.e. the nearest approach of missile and target). Such computations must be carried out for a range of system design parameters and for a range of random variations in such variables as target manoeuvre and homing, radar angular and amplitude noise. In the hybrid solution of such a problem the analogue computer can solve the dynamic and kinematic equations of the missile-target system and the digital computer can vary the statistical parameters of target and homing system and can compute the required statistic of the miss distance.

In the solution of flight equations, there are six degrees of freedom, i.e. three degrees of rotational freedom, roll, pitch and yaw and three degrees of translational freedom represented by trajectory motions in three space coordinates. A major applications area for hybrid computers has been the solution of these equations. The analogue system is used for the rotational equations solution, i.e. the motion of the vehicle about its own axes and the digital system is used for the translational equation solution, i.e. the trajectory of the vehicle in space. This division of computation is well matched to the problem, which demands higher speed and lower accuracy in the vehicle motion solutions and lower speed but higher accuracy in computing trajectories.

Hybrid computers are used to a decreasing extent in industry though they are still employed in aerospace system and electrical power system design to solve dynamic problems. As the speed of digital computation increases and as new digital computing techniques appear, the need for hybrid computers will inevitably decline.

DIGITAL SIMULATION

25.36 Introduction

Simulation studies, previously only performed using an analogue computer, are now conducted by hybrid simulation and digital simulation. Digital simulation, as the name suggests, involves only the use of a digital computer, while hybrid simulation is performed with a hybrid computer which is a combination of an analogue and a digital computer. A hybrid computer requires an interface to implement the analogue-to-digital and digital-to-analogue conversion, to provide digital control of analogue computer operation. Such a combination combines the speed of the analogue computer with the logic, storage and the input/output facilities of the digital machine. Although such a link-up should be ideal, the hybrid computer suffers from the inherent problem of low analogue computer precision and errors arising from data sampling through the interface. Consequently, in order to achieve satisfactory accuracy, it is necessary to use high precision components and clever programming. In the case of digital simulation, the accuracy of the computation can be improved simply at the expense of increased computing time. Hybrid computers suffer from the disadvantage that hardware cost is significantly influenced by the number and complement of analogue components considered to be necessary. This judgement decision, based on an anticipated maximum problem size, will invariably lead to a hardware cost far in excess of that required to handle many simulations.

Hybrid simulation also suffers from the fact that a substantial amount of time is required for problem preparation, scaling and checkout so the cost of performing a simulation may be very expensive. Consequently, relatively expensive hybrid computers will generally only be cost effective when speed of solution is important or perhaps where the simulation needs to be of a highly interactive nature.

The increased availability of digital simulation languages, which do not require that the user be familiar with a programming language, has lead to rapid growth in the use of digital simulation. Use of digital simulation has also increased because of the availability of such software on low-cost minicomputers. Continuous system digital simulation languages are described as being of either the model/block diagram type or the equation-oriented type. Most of the common continuous system simulation languages (CSSL) are of the equation type which employ a problem definition in state equation form. This means that the system is modelled by a collection of first-order ordinary differential equations and linking algebraic equations.

Model-based simulation languages differ in that a wide variety of different types of mathematical models are provided so the user describes his problem by appropriate combinations of the available models. In the case of some languages, the models are simply a defined grouping of differential and algebraic equations, as in an equation-based language, but the equations remain transparent to the user. Regardless of the type of CSSL language, the equations are automatically translated into a FORTRAN program, which is compiled, loaded and executed to produce time histories of the problem variables of interest. The calculated variables of interest can be displayed and/or plotted and/or printed through the use of simple output commands. Use of a host language, such as FORTRAN, not only allows use of existing compilers, but also offers knowledgeable users, full use of FORTRAN functions and procedures in the formulation of the simulation problem. Utilising the conditional logic of the host language allows multiple simulation runs to be easily performed. This facility allows for a series of simulation runs to be performed at selected parameter values or the control of multiple runs to perform even simple parameter optimisation. Changes in parameter values (and initial condition values) are effected without retranslation or recompilation. Typical of the equation-based languages and the organisations supporting these simulation languages are:

(1) CSMP III (IBM Corporation).
(2) ASCL (Mitchell and Gauthier Associates).
(3) CSSL III (Control Data Corporation and Sperry Univac).
(4) DARE (University of Arizona, Digital Equipment Corporation).

25.37 System models

Regardless of the simulation language to be utilised, a necessary pre-requisite is a description of the system of interest by a mathematical model. Some physical systems can be described in terms of models that are of the state transition type. If such a model exists, then given a value of the system variable of interest, e.g. voltage, charge position, displacement, etc. at time t, then the value of the variable (state) at some future time, $t + \Delta t$ can be predicted. The prediction of the variable of interest (state variable), $x(t)$ at time $t + \Delta t$, given a state transition model, S, can be expressed by the state equation.

$$x(t + \Delta t) = S[x(t), t, \Delta t] \tag{25.2}$$

Equation (25.2) shows that the future state is a function of the current state, $x(t)$, at the current time, t and the time increment Δt. Thus, once the model is known, from either empirical or theoretical considerations, Equation (25.2), given an initial condition (value), allows for the recursive computation of $x(t)$ for any number of future steps in time. For an initial value of the state variable, $\bar{x} = x(t_1)$ at time t_1, then

$$x(t_1 + \Delta t) = S[\bar{x}, t_1, \Delta t]$$

then letting $t_2 = t_1 + \Delta t$, Equation (25.2) for the next time step Δt, becomes

$$x(t_2 + \Delta t) = S[x(t_2), t_2, \Delta t]$$

Obviously, this operation is continued until the calculation of the state variable has been performed for the total time period of interest.

Systems of interest will clearly not be characterised only by a single state variable but by several state variables. *Figure 25.72* is a schematic representation of a multi-variable system that has r inputs, n states and m outputs.

In general the simulation will involve calculation of all of the state variables, even though the response of only a selected number of output variables is of interest. For many systems, the output variables may well exhibit a simple one-to-one correspondence to the state variables. As shown by the representation in *Figure 25.72*, the values of the state variables depend on the inputs to the system. For a single interval, between the k and $k+1$ time instants, the state equations for the n state variable system for a change in the jth input ($j \leq r$) $u_j(t)$ would be written as

$$\begin{aligned} x_1(t_k + \Delta t) &= S_1[x_1(t_k), u_j(t_k), t_k, \Delta t] \\ x_2(t_k + \Delta t) &= S_2[x_2(t_k), u_j(t_k), t_k, \Delta t] \\ &\vdots \\ x_n(t_k + \Delta t) &= S_n[x_n(t_k), u_j(t_k), t_k, \Delta t] \end{aligned} \tag{25.3}$$

The system of equations, Equation (25.3), a collection of difference equations, would be used to predict the state variables $x_1, x_2, \ldots, x_n$ at time intervals of Δt from the initial time, t_0 until the total time duration of interest, $T = t_0 + K \Delta t$. For engineering systems, the dependant variable will generally be a continuous variable. In this case the system description will be in terms of a differential equation of the form

$$dx/dt = g(x, t) \tag{25.4}$$

Recalling basic calculus for a small time increment, the left-hand side of Equation (25.4) can be expressed as

$$\lim_{\Delta t \to 0} \frac{x(t + \Delta t) - x(t)}{\Delta t}$$

so for a small time increment, Δt, Equation (25.4) can be written as

$$x(t + \Delta t) = x(t) + [g(x, t)] \Delta t$$

or

$$x(t + \Delta t) = G[x(t), t, \Delta t] \tag{25.5}$$

Equation (25.5) is the form of Equation (25.2), so for a small time increment, a first-order ordinary differential equation can be approximated by a state transition representation.

It thus follows from the preceding discussion, that in digital continuous system simulation, the principle numerical task is the approximate integration of Equation (25.4). For a small time increment, DT, the integration step size, the computation involves the evaluation of the difference equation

$$x(t + DT) = x(t) + [g(x, t)]DT \tag{25.6a}$$

which can be written explicitly as

$$x(t_{k+1}) = x(t_k) + \int_{t_k}^{t_{k+1}} g[x(t_k), t_k] DT \tag{25.6b}$$

where $DT = t_{k+1} - t_k$. The calculation starts with a known value of the initial state $x(0)$ at time t_0 and proceeds successively to evaluate $x(t_1)$, $x(t_2)$, etc. The computation involves successive computation of $x(t_{k+1})$ by alternating calculation of the derivative $g[x(t_k), t_k]$ followed by integration to compute $x(t_{k+1})$ at time $t_{k+1} = t_k + DT$.

Obviously, most physical systems will be described by second- or higher order ordinary differential equations so the higher order equation must be re-expressed in terms of a group of first-order ordinary differential equations by introducing state

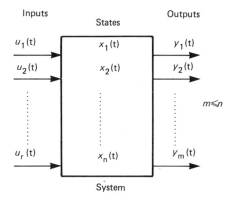

Figure 25.72 Schematic representation of a multi-variable system

variables. For an nth order equation,

$$\frac{d^n z}{dt^n} = f\left[z, \frac{dz}{dt}, \frac{d^2 z}{dt^2} \cdots \frac{d^{n-1} z}{dt^{n-1}}; t\right] \quad (25.7)$$

the approach involves the introduction of new variables as state variables to yield the following first-order differential equations

$$\frac{dx_1}{dt} = x_2$$

$$\frac{dx_2}{dt} = x_3$$

$$\frac{dx_3}{dt} = x_4 \quad (25.8)$$

$$\vdots$$

$$\frac{dx_{n-1}}{dt} = x_n$$

$$\frac{dx_n}{dt} = f(x_1, x_2, x_3, \ldots, x_n; t)$$

It should be noted that this equation can be expressed in shorthand notation as a vector-matrix differential equation. In an analogous manner, Equation (25.3) can be expressed as a vector difference equation. There is no unique approach to the selection of state variables for system representation, but for many systems the choice of state variables will be obvious. In electric circuit problems, capacitor voltages and inductor currents would be logical choices, as would position, velocity and acceleration for mechanical systems.

25.38 Integration schemes

The simple integration step, embodied by the first-order Euler form in Equation (25.6a) only provides a satisfactory approximation of the solution of the differential equation, within specified error limits, for a very small integration step size, DT. Since the small integration interval leads to substantial computing effort and to round-off error accumulation, all digital simulation languages use improved integration schemes. Despite the wide variety of different integration schemes that are available in the many different simulation languages, the calculational approach can be categorised into two groups. The types of algorithm are:

(1) *Multi-step formulas* In such algorithms, the value of $x(t+DT)$ is not calculated by the simple linear extrapolation of Equation (25.6a). Rather than use only $x(t)$ and one derivative value, the algorithms use a polynomial approximation based on past values of $x(t)$ and $g[x(t), t]$, that is at times $t-DT$, $t-2DT$, etc.

(2) *Runge–Kutta formulas* In Runge–Kutta type algorithms, the derivative value used for the calculation of $x(t+DT)$ is not the point value at time t. Instead, two or more approximate derivative values in the interval t, $t+DT$ are calculated and then a weighted average of these derivative values is used instead of a single value of the derivative to compute $x(t+DT)$.

25.39 Organisation of problem input

Most simulation language input is structured into three separate sections, although in some programs the statements can be used with limited sectioning of the program. A typical structure and the type of statements, functions or parts of the simulation program that appear are as follows:

(1) *Initialisation*
   Problem documentation (e.g. name, date, etc.).
   Initial conditions for state variables.
   Parameter values (problem variables that may not be constant, problem time, integration order, integration step size, etc.).
   Problem constants.
(2) *Dynamic*
   Derivative statements.
   Integration statements (including any control parameters not given in the initialisation section).
(3) *Terminal*
   Conditional statements (e.g. total time, variable(s), value(s), etc.).
   Multiple run parameters.
   Output (print/plot/display) option(s).
   Output format (e.g. designation of independent variable; increment for independent variable; dependent variable(s) to be output; maximum and minimum values of variable(s); or automatic scaling; total number of points for the independent variable or total length of time).

It should be understood that the specific form of the statements within each section are not exactly the same for all digital simulation languages. However, from the CSMP simulation programs presented in the next section, with the aid of the appropriate language manual, there should be no difficulty in formulating a simulation program using any CSSL-type digital simulation program.

25.40 Illustrative examples

Simulation programs are presented, using the CSMP language, that would be suitable for investigating system dynamic behaviour. The three different system models, although relatively simple in nature, are typical of those used for system representation.

25.40.1 Example 1

Frequently, it will be found that system dynamic behaviour can be described by a differential equation of the form

$$y^{(n)} + a_1 y^{(n-1)} + a_2 y^{(n-2)} + a_{n-1} y^{(1)} + a_n y$$
$$= b_0 r^{(m)} + b_1 r^{(m-1)} + b_{m-1} r^{(1)} + b_m r \quad (25.9)$$

where

$$y^{(n)} = \frac{d^n y}{dt^n}, \quad r^{(m)} = \frac{d^m r}{dt^m}.$$

Use of CSMP for studying the dynamic behaviour of a system described by a high-order differential equation will be illustrated with a simulation program for the following differential equation.

$$y^{(3)} + 2.5 y^{(2)} + 3.4 y^{(1)} + 0.8 y = 7.3 r \quad (25.10)$$

with initial conditions

$$y^{(2)}(0) = 0; \quad y^{(1)}(0) = -4.2; \quad y^{(0)} = 2.5$$

Development of the simulation program follows logically by rewriting Equation (25.10) as

$$\frac{d^3 y}{dt^3} = -2.5 \frac{d^2 y}{dt^2} - 3.4 \frac{dy}{dt} - 0.8 y + 7.3 r \quad (25.11)$$

$$\left.\frac{d^2 y}{dt^2}\right|_{t=0} = 0; \quad \left.\frac{dy}{dt}\right|_{t=0} = -4.2; \quad y|_{t=0} = 2.5$$

A block diagram showing the successive integrations to solve for the dependent variable, y is given in *Figure 25.73*. As can be seen from the labelling on the diagram, the output of the

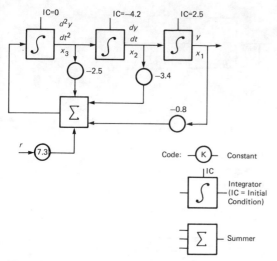

Figure 25.73 CSMP block diagram for a third-order differential equation

integration blocks are successive derivative values and the dependent variable. In fact, the output of each integration block is a state variable. This becomes obvious by introducing new variables, x_1, x_2, x_3 defined as:

$$x_1 = y$$

$$\frac{dx_1}{dt} = x_2$$

$$\frac{dx_2}{dt} = x_3$$

which allows Equation (25.11) to be expressed as

$$\frac{dx_1}{dt} = x_2 \quad (25.12)$$

$$\frac{dx_2}{dt} = x_3$$

$$\frac{dx_3}{dt} = -2.5x_3 - 3.4x_2 - 0.8x_1 + 7.3r$$

with initial conditions

$$x_3(0) = 0, \quad x_2(0) = -4.2, \quad x_1(0) = 2.5$$

A program for solving Equation (25.12) is given as *Figure 25.74*. Examination of the program shows that the value of the forcing function, r, is not constant but varies with time. The variation is provided using the quadratic interpolation function, NLFGEN. Total simulation time is set for 6 min with the interval for tabular output specified as 0.2 min. The time unit is determined by the problem parameters. It is to be noted that the program does not include any specification for the method of integration. The CSMP language does not require that a method of integration be given but a particular method may be specified (cf. Example 2, *Figure 25.77*). If a method is not given, then by default the variable step size 4th-order Runge–Kutta method is used for calculation. The initial step size, by default, is taken as 1/16 of the PRDEL (or OUTDEL) value. Minimum step size can be limited by giving a value for DELMIN as part of the TIMER statement. If a DELMIN value is not given, then by default, the minimum step size is FINTIM $\times 10^{-7}$.

```
LABEL THIRD ORDER DIFFERENTIAL EQUATION
INITIAL
CONSTANT A1=-2.5,A2=-3.4,A3=-0.8,B0=7.3, ...
         X1INIT=2.5,X2INIT=-4.2,X3INIT=0.0
FUNCTION FCHG=(0.5,4.8),(1.0,6.3),(1.5,2.8),(2.0,3.9), ...
              (2.5,4.8),(3.0,3.2),(3.5,2.1),(4.0,5.6), ...
              (4.5,6.8),(5.0,3.7),(5.5,4.6),(6.0,3.4)
DYNAMIC
         R=NLFGEN(FCHG,TIME)
         X1=INTGRL(X1INIT,X2)
         X2=INTGRL(X2INIT,X3)
         X3=INTGRL(X3INIT,DHX3)
         DHX3=A1*X3+A2*X2+A3*X1+B0*R
TERMINAL
TIMER FINTIM=6.0,PRDEL=0.2
PRINT R,X1,X2,X3
END
STOP
ENDJOB
```

Figure 25.74 Simulation program for studying the dynamic behaviour of system described by a third-order differential equation

25.40.2 Example 2

Analysis of the dynamic behaviour of a system described by a transfer function is easily accomplished by digital simulation. Suppose that the transfer function to be simulated is

$$G(s) = \frac{Y(s)}{U(s)} = \frac{3 + 9s^2 + 23s + 15}{4 + 13s^3 + 50s^2 + 56s + 37} \quad (25.13)$$

The simulation block diagram is developed by a two-stage process. First, consider the transfer function

$$\frac{Y(s)^*}{U(s)} = \frac{1}{s^4 + 13s^3 + 50s^2 + 56s + 37} \quad (25.14a)$$

which can be written as

$$s^4 Y^*(s) = -13s^3 Y^*(s) - 50s^2 Y^*(s)$$
$$\qquad -56s Y^*(s) - 37 Y^*(s) + U(s) \quad (25.14b)$$

The block diagram, to obtain the dependant variable $Y^*(s)$ is shown as *Figure 25.75*.

Since the desired dependent variable is in fact $Y(s)$, which is related to $Y^*(s)$ by

$$Y(s) = (s^3 + 9s^2 + 23s + 15)Y^*(s) \quad (25.15)$$

it follows that the complete block diagram for studying system behaviour described by the transfer function $G(s)$ is that given as *Figure 25.76*.

Although the simulation diagrams have used the s domain representation, it should be realised that the calculation obviously is performed in the time domain. This equivalence between the s domain and time domain diagrams is clearly evident by realising that the operator $1/s$ in the Laplace domain is equivalent to integration in the time domain. Consequently a $1/s$ block in an s domain simulation diagram is equivalent to the integration block in a time domain simulation diagram.

In a manner analogous to that used in the previous example, a state variable can be associated with the output of each integrator. If the output of the last integrator is denoted as variable x_1 then the equivalent time domain representation is

$$\frac{dx_1}{dt} = x_2$$

$$\frac{dx_2}{dt} = x_3 \quad (25.16)$$

$$\frac{dx_3}{dt} = x_4$$

$$\frac{dx_4}{dt} = 13x_4 - 50x_3 - 56x_2 - 37x_1 + u$$

$$y = x_4 + 9x_3 + 23x_2 + 15x_1$$

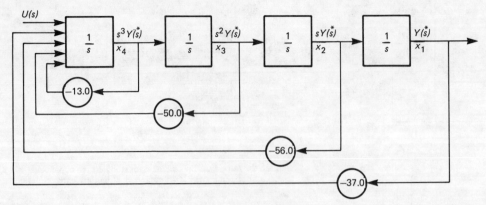

Figure 25.75 Block diagram for denominator of transfer function

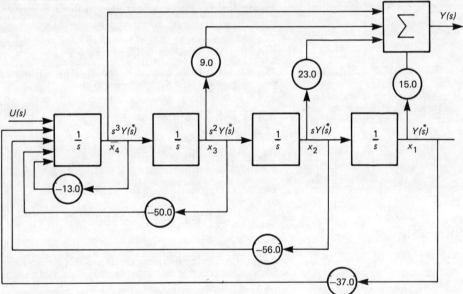

Figure 25.76 Block diagram of transfer function

For the case where the change in the input to the transfer function, u is a pulse at time zero for a duration of 4.5 time units, then a suitable CSMP simulation program is that given in *Figure 25.77*. Examination of the terminal section of the program reveals that tabular output of all the state variables has been required as well as a printer plot of the system output variable, y, versus time.

Low-order transfer functions, particularly for control system description are frequently used. Consequently CSMP provides system defined functions (as do other simulation languages) which require only that the user insert the numerical parameters of the transfer function. For the transfer function

$$G(s) = \frac{Y(s)}{X(s)} = \frac{5s+1}{12s+1}$$

the program statement would be simply

Y = LEDLAG (5.0,12.0,X)

Furthermore, although CSMP is a continuous-system simulation language, digital (discrete/sampled data) control systems can easily be simulated by the use of such functions as zero order hold (ZHOLD), etc.

```
LABEL TRANSFER FUNCTION
INITIAL
CONSTANT A1=-13.0,A2=-50.0,A3=-56.0,A4=-37.0,B1=-9.0, ...
         B2=23.0,B3=15.0,K=3.5
DYNAMIC
    X1=INTGRL(0.0,X2)
    X2=INTGRL(0.0,X3)
    X3=INTGRL(0.0,X4)
    X4=INTGRL(0.0,DX4)
    TR=IMPULS(0.0,20.0)
    U=K*PULSE(4.5,TR)
    DX4=A1*X4+A2*X3+A3*X2+A4*X1+U
    Y=X4+B1*X3+B2*X2+B3*X1
TERMINAL
TIMER FINTIM=6.0,PRDEL=.1,OUTDEL=.15,DELT=0.005
METHOD RKSFX
PRINT X1,X2,X3,X4
PRTPLT Y
END
STOP
ENDJOB
```

Figure 25.77 Transfer function simulation program

25.40.3 Example 3

The final example to be considered is the spring restrained voltmeter system shown schematically in *Figure 25.78*.

The system equations can be written as

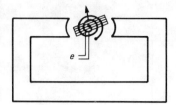

Figure 25.78 Schematic representation of voltmeter

$$e = iR + e_b + L\frac{di}{dt} \quad (25.17)$$

$$e_b = K_v \frac{d\theta}{dt} \quad (25.18)$$

$$K_t i = J\frac{d^2\theta}{dt^2} + K_s \quad (25.19)$$

where e = applied voltage (to be measured by the meter), R = armature resistance, L = armature inductance, i = armature current, e_b = back-emf, J = polar moment of inertia of moving parts, K_s = spring constant, K_t = torque constant, K_v = back-emf constant and θ = angular deflection (needle).

As stated previously in discussing the state variable concept, there is no unique approach to assigning state variables but for this simple system the choice is not difficult. The output of interest is θ, needle angular deflection, so one logical choice of state variable is $\theta(t) = x_1(t)$. Since the equations must be transformed into a system of first-order equations, a further logical choice is $d\theta/dt = x_2(t)$. Then from Equation (25.17), the third state variable is selected as current, that is $i(t) = x_3(t)$.

Substituting for e_b in Equation (25.17) using Equation (25.18) followed by rearrangement of the equations gives the system of state equations as

$$\frac{dx_1}{dt} = x_2$$

$$\frac{dx_2}{dt} = \frac{K_t}{J}x_3 - \frac{K_s}{J}x_1 \quad (25.20)$$

$$\frac{dx_3}{dt} = \frac{e}{L} - \frac{Rx_3}{L} - \frac{K_v}{L}x_2$$

```
LABEL    VOLT METER DYNAMICS
INITIAL
PARAMETER KS=0.15,KT=0.1,KV=0.1,R=8.0,L=0.5,J=1.25E-04, ...
         E=20.0
DYNAMIC
    X1=INTGRL(0.0,X2)
    DX2=(KT/J)*X3-(KS/J)*X1
    X2=INTGRL(0.0,DX2)
    DX3=(E-(R*X3)-(KV*X2))/L
    X3=INTGRL(0.0,DX3)
TERMINAL
TITLE          DYNAMICS OF METER
TIMER          FINTIM=5.0,OUTDEL=0.2
PRTPLT         X1(X2,X3)
END
PARAM          E=40.0
END
PARAM          E=60.0
END
STOP
ENDJOB
```

Figure 25.79 Meter dynamics program

A program that could be used to study the dynamic response of the meter to changes in the applied voltage is given as *Figure 25.79*.

Examination of the TERMINAL section of the program illustrates the feature which allows multiple runs to be performed. The first run uses an applied voltage of 20 V followed by successive runs at voltages of 40 and 60. Note also that the additional feature of PRTPLT which allows tabular output of two variables is employed. Use of TITLE provides labelling for each page of output.

25.41 Conclusions

It must be emphasised that only some of the most elementary features of the CSMP language have been demonstrated. However, it should be possible from these examples to appreciate the scope for the use of digital simulation for the analysis of system transient behaviour. Developments now taking place will make digital simulation even a more powerful aid to the engineer than at present. Most existing simulation languages, such as CSMP, have been developed for use on large computer systems. Now with the increased computing power of microcomputers continuous system simulation languages are becoming available on small desk top computers. Future developments will also see the introduction of simulation languages that allow for inclusion of hardware/instrumentation as part of the simulation exercise by means of analogue-to-digital and digital-to-analogue converters.

26 Electromagnetic Machines

M G Say PhD, MSc, CEng, FRSE, FIERE, ACGI, DIC
Heriot-Watt University

Contents

26.1 Energy conversion 26/3

26.2 Electromagnetic devices 26/3
    26.2.1 Electromagnets 26/3
    26.2.2 Tractive electromagnets 26/4
    26.2.3 Actuators 26/4
    26.2.4 Lifting magnets 26/5
    26.2.5 Crack detectors 26/6
    26.2.6 Separators 26/7
    26.2.7 Clutches 26/8
    26.2.8 Couplings 26/8
    26.2.9 Brakes 26/9
    26.2.10 Magnetic chucks 26/10
    26.2.11 Vibrators 26/10
    26.2.12 Relays and contactors 26/11
    26.2.13 Miniature circuit-breakers 26/12
    26.2.14 Particle accelerators 26/13

26.3 Industrial rotary and linear motors 26/15
    26.3.1 Prototype machines 26/15
    26.3.2 D.C. motors 26/16
    26.3.3 Three-phase induction motors 26/24
    26.3.4 Three-phase commutator motors 26/28
    26.3.5 Synchronous motors 26/30
    26.3.6 Reluctance motors 26/31
    26.3.7 Single-phase motors 26/31
    26.3.8 Motor ratings and dimensions 26/34
    26.3.9 Testing 26/34
    26.3.10 Linear motors 26/39

26.1 Energy conversion

Electromagnetic machines convert electrical into mechanical energy in devices with a limited stroke (actuator, brake, relay etc.) or continuous angular rotation (motor), or linear motion (linear motor).

Mechanical energy involves a force f_m acting over a distance x or a torque M_m acting over an angular displacement θ. Electrical energy involves the displacement of a charge q (a current i for a time t) through a potential difference (p.d.) v. The energies W and corresponding powers $P = dW/dt$ are

Mechanical:
$W_m = f_m x \quad P_m = f_m(dx/dt) = f_m u$ (linear)
$W_m = M_m \theta \quad P_m = M_m(d\theta/dt) = M_m \omega_r$ (rotary)

Electrical:
$W_e = vq \quad P_e = v(dq/dt) = vi$

where $u = dx/dt$ is the translational speed and $\omega_r = d\theta/dt$ is the rotational speed. In an electromagnetic machine the basic physical conversion mechanism between the two forms of energy is the magnetic field, a characteristic property of electric current. The elements of electromagnetic/mechanical conversion are set out in Sections 2.4.1–2.4.3.

26.2 Electromagnetic devices

26.2.1 Electromagnets

Electromagnets for stroke-limited devices (e.g. actuators) are such that estimation of the flux distribution in the airgap (the working region) is difficult. The total magnetomotive force (m.m.f.) produced by a current i in an N-turn coil is $F = Ni$.

26.2.1.1 Coil windings

Most coils for magnetic-circuit excitation are wound by one of the following (usually automated) methods: (1) on a former with end-cheeks; (2) on a bobbin that forms an integral part of the coil and comprises a moulded or fabricated construction of a suitable insulant; (3) by a winding machine that feeds insulated wire into a self-supporting form, with an epoxy-resin binder.

The coil design is based on the provision of the required m.m.f. for a specified voltage (or current), with an acceptable coil temperature rise on a specified duty cycle.

D.C. excitation For direct current (d.c.), the current i at a coil terminal voltage V is determined by the coil resistance $R = V/i = \rho L_{mt} N/a$, where the N turns have a mean turn length L_{mt} and the conductor, of resistivity ρ, has a cross-sectional area a. Then

$a = \rho L_{mt}(Ni)/V = \rho L_{mt} F/V$

for a total m.m.f. F. The current cannot be determined until the cooling conditions are established. Let the conductor current density be J, so that $i = Ja$; then $N = F/Ja$. The total conducting cross-section of the coil is Na and the gross cross-sectional area of the wound coil is Na/k, where k is the *space factor*.

The power taken by the coil is $P = Vi$, and the consequent temperature rise on continuous operation is $\theta_m = P/cS$. Here c is a cooling coefficient representing the power dissipation per unit of the coil surface area S per °C rise of surface temperature above ambient. The value of θ_m for continuously rated coils is usually specified. On intermittent or short-time rating the rise is a function of the thermal capacity of the coil.

A.C. excitation For alternating current (a.c.), the current i at voltage V is determined by the coil impedance $Z = R + j\omega L$ at angular frequency ω. An a.c. coil therefore tends to have fewer turns that one for d.c. Further, the coil inductance L varies widely, depending on the saturation of the ferromagnetic parts and in particular on the length of the airgap. A wide gap increases the magnetic reluctance and reduces L, but as the gap length reduces (e.g. by movement of the working parts) the inductance rises. If $\omega L \gg R$, as is usual, the root-mean-square (r.m.s.) value of the m.m.f. approximates to $F = VN/\omega L$, with L estimated for the range of airgap lengths.

In practice, performance is based on data obtained on test. A particular feature is the double-frequency fluctuation of the mechanical force, which produces a characteristic 'chatter' in the closed position of the device; this may have to be mitigated by means of a shading ring.

26.2.1.2 Coil design

Space factor A simple coil wound from a circular-section wire of diameter d, and insulated to a diameter d_i, will pack down in a manner that is affected by the method of winding, one layer partly occupying the troughs in the layer beneath it; the space factor may then approximate to $k = 0.85(d/d_i)^2$. Conductors of small diameter bed less effectively, and the space factor is reduced.

Cooling coefficient A typical value of the cooling coefficient c is 0.075 W/m² per °C above ambient. However, cooling conditions vary widely with the efficacy of ventilation.

26.2.1.3 Operating conditions

Whether d.c. or a.c. excited, the current in an operating coil is affected by that movement of the working parts that closes or opens the airgap. Let a quiescent spring-loaded relay (*Figure 26.1a*) in the open position be connected to a direct source-voltage V. The coil current begins to rise exponentially, but the armature does not move until the magnetic force exceeds the spring restraint. Thereafter, the shortening gap increases the coil inductance, setting up a counter electromotive force (e.m.f.) and checking the current rise and the attracting force. Finally, the

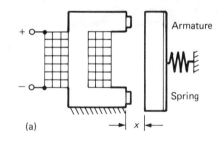

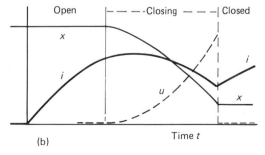

Figure 26.1 Operation of a d.c. relay

armature reaches the closed position at the end-stop, dissipating kinetic energy in noise, bounce and mechanical deformation. The sequence of events, in terms of the gap length x, armature speed u, coil current i and time t, is shown in *Figure 26.1(b)*.

If the coil is energised from an a.c. source there are two further effects: the closing time depends on the instant in the cycle at which the voltage is applied and (more importantly) the operating force fluctuates. Ferromagnetic parts must be laminated to prevent excessive core loss and the counter-effects of eddy currents. Suitable sheet steel for the purpose has a core loss of less than 5 W/kg.

The force fluctuation can be reduced (but not eliminated) by a *shading* ring (*Figure 26.2*) embedded in one of the pole faces flanking the gap. Currents induced in the ring delay part of the pole flux. Thus the combination of shaded and unshaded flux gives a resultant that still fluctuates but does not at any instant fall to zero.

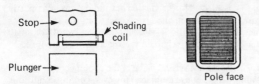

Figure 26.2 Shading coil or ring

26.2.2 Tractive electromagnets

Two forms of tractive electromagnet are shown in *Figure 26.3*. Type (a) usually has cylindrical poles, sometimes with shouldered ends to retain the coils, a rectangular yoke to which the poles are screwed or bolted, and a rectangular armature. Two exciting coils are used; they are connected to give opposite polarities at the respective pole ends. Type (b) has a single coil mounted on a cylindrical core to which the rectangular pole-pieces are attached. In both cases the total airgap length is the sum of the gaps at the respective poles; in some designs, however, the armature is hinged to the pole-piece at one end. In this case the free end forms the major gap.

In (a), let each polar surface have an area of 250 mm² and be required to exert a total force of 1.0 N on the armature when both gaps are 3.0 mm long. Then with *d.c. excitation*, $f = 400\,000 B^2 A$ giving $B = 70$ mT, for which $H = 57\,000$ A-t/m. For a total gap of 6 mm the gap m.m.f. required is 340 A-t. Adding 10% for the iron circuit and 25% for leakage, the total excitation required is about 450 A-t, from which the coil design follows.

With *a.c. excitation* it is necessary to estimate the inductance in the open and closed positions, and to adjust the number of turns for a given operating voltage so that adequate force is available. The change of magnetic flux between the two extreme armature positions is very much less than with d.c. operation, so that for the same (average) force in the open position, that in the closed position is only a little greater.

26.2.3 Actuators

26.2.3.1 D.C. actuators

Three typical arrangements for d.c. actuators are shown in *Figure 26.4*. Form (a) is convenient for small devices as the frame can be bent from strip; it is common for overcurrent and undervoltage relays. Form (b) may have a cast frame, and provides parallel flux paths through the iron. In (c) the cylindrical iron circuit presents a low reluctance, the circuit being completed by a lid attached by studs or screwed into the cylindrical body.

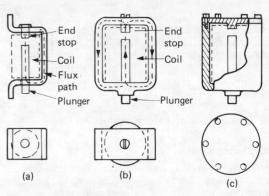

Figure 26.4 D.C. actuators

The iron *end-stop* should project well into the coil to improve flux concentration. It may be integral with the frame or screwed into it (in which case it can be used to locate and secure the operating coil). The plunger passes through the frame at the *throat*, the reluctance of which can be reduced by minimising the annular gap and extending the effective axial length, as shown at (b) and (c).

A typical ironclad actuator in part section is shown in *Figure 26.5*. With the dimensions $a = 220$ mm, $d = 65$ mm, $x = 63$ mm and $y = 150$ mm, the coil may develop about 15 kA-t to give a pull of 400 N across a 25 mm gap in the open position. The brass pin forms a stop, and cushions the plunger by expelling air through the vent.

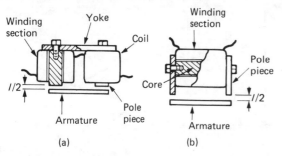

Figure 26.3 Tractive electromagnets

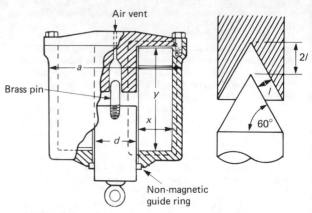

Figure 26.5 Ironclad 'pot' actuator

With a flat-ended plunger the stroke is equal in length to the magnetic airgap. Maximum work (force × displacement) occurs with a short stroke. By using a coned plunger (see *Figure 26.5*), maximum work is obtained with a longer stroke. If the cone angle is 60°, the comparable stroke is twice that for a flat-ended plunger for about the same magnetic pull. It is possible to obtain a wide variety of characteristics by modifying the shapes of the stop and plunger ends.

26.2.3.2 A.C. actuators

A common arrangement for a single-phase device is that shown in *Figure 26.6*. The E-type laminations are clamped. In the plunger, rivets should lie in a line in the flux direction to minimise eddy currents. To keep down the 'holding' current the plunger and stop ends should be flat.

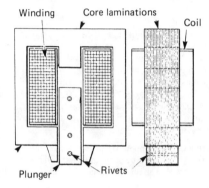

Figure 26.6 Single-phase actuator

Because of the many variables concerned, the design is complicated. An empirical rule is to allow 1.5 mm² of plunger cross-section for every 1 N of force; this corresponds to a peak flux density of 0.8 T in the laminations. The size of the coil (and therefore the main dimensions) may be taken as having a length 2.5–3 times the stroke and a depth equal to the stroke. The number of turns N is estimated from

$$N = V/4.44 f B_m A$$

where $B_m A$ is the peak flux and f is the frequency. Final adjustment of N is made on test; it is reduced if the force is too low.

26.2.3.3 Polyphase actuators

A 3-phase actuator has three limbs. Because of the phasing, the net force on the laminated bar armature assembly is never zero, and shading is not necessary.

The typical unit (*Figure 26.7*) is for operating a brake. It has three limb-coils E connected in star. The armature A is shown in the lifted (energised) condition. The plunger rod, fitted with a piston in the dashpot D, cushions the end of the stroke. A valve in the piston allows unretarded drop-out for quick brake application.

26.2.4 Lifting magnets

Lifting magnets are of use in the handling of iron and steel, as they dispense with hooks and slings. The maximum load of a magnet varies with the material to be lifted. A magnet capable of lifting 1 t of scrap may raise a 20 t load in the form of a thick solid piece with a flat upper surface. As the excitation is limited by

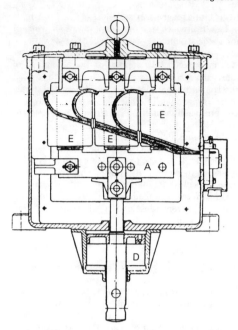

Figure 26.7 Three-phase actuator

temperature rise of the coil, the lifting capacity is also dependent on the duty cycle. For the comparatively arduous conditions normally ruling in industrial use, a robust and weatherproof construction is essential.

26.2.4.1 Circular magnets

The essential features of a circular magnet are shown in *Figure 26.8*. As the magnetic properties of the material lifted and the airgaps between the magnet poles and the material are both arbitrary and subject to wide variation, the design of a lifting magnet is generally based on thermal considerations. A given carcass and winding are assigned an empirically derived power rating such that the temperature rise of the coil is not excessive. The designer's aim is then to secure the maximum effective ampere-turn excitation and working flux density by adjustments of iron and conductor materials and heat dissipation. Allowances in design must be made for the development of adequate pull under conditions of low line voltage (e.g. 80% or less of nominal), and high conductor resistivity when hot.

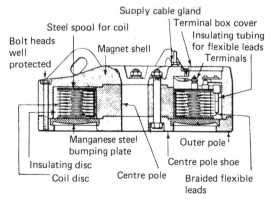

Figure 26.8 Circular lifting magnet

The majority of lifting magnets, except those of small size, have a winding of flat strip, which is more adaptable than wires of circular section to the attainment of a good space factor with the large conductor areas generally necessary. Aluminium is sometimes employed in preference to copper for the advantage of weight economy: the weight of a magnet is important as it represents a useless additional load to be moved every time its crane is operated. The winding in *Figure 26.8* is shown diagrammatically: it comprises a number of flat spirals with heat-resistant insulation.

The general dimensions are such that the diameter of the inner pole-face is about one-third of the overall diameter d. The load lifted is proportional to d^2 and the power rating (in kW) is of the order of $4d^2$ (with d in metres). As regards the load, only in exceptional cases does close and uniform contact occur between magnet and load surfaces. The tendency for flux concentration over small contact regions develops local saturation and increases the effective gap.

26.2.4.2 Rectangular magnets

Materials of regular shape, such as sheets, bars, pipes etc., are well suited to lifting by rectangular magnets. The general construction is similar to that of the circular type, the body being formed of a box-shaped steel casting with a central projection to give the inner polar surface.

The approximate lifting capacities of circular and rectangular magnets are given in *Table 26.1*.

26.2.4.3 Control

Simple on/off switching is not practicable because of the high level of stored magnetic energy. The general control features needed are: (1) discharge resistors connected across the winding just prior to disconnection to reduce contact arcing and limit inductive e.m.f.; (2) auxiliary resistors introduced into the coil circuit after a predetermined time to limit coil temperature rise; (3) reversal of polarity at a low current level to overcome remanence and so release small pieces such as turnings or scrap.

26.2.5 Crack detectors

Electromagnetic crack detection depends on the fact that, in magnetic material, the magnetic susceptibility of a fault is markedly inferior to that of the surrounding material. The success of the whole technique of magnetic crack detection depends largely on the care taken to ensure correct strength and direction of magnetisation. The following methods are used:

(1) *Needle method*. The surface to be tested is explored with a small magnetic needle. This needle carries a pointer which moves over a scale, with a right and left motion as the needle turns on its pivot to align with the field distortion passing beneath it in the direction of the arrow. Thus the fault is detected. The sensitivity is increased by using a mirror and light beam.
(2) *Powder method*. The part to be tested, previously cleaned, is laid across the arms of the machine, and the circuit-closing push-button switch depressed and released quickly. The article is then removed and sprinkled with special powder, the excess of which is blown away or shaken off; it will then be found that the defects are clearly indicated by the magnetic patterns.
(3) *Fluid method*. This resembles the powder method but employs a fluid (e.g. paraffin) containing finely divided magnetic material in suspension.

Each of these techniques can be applied to the detection of cracks or other flaws in parts which have been magnetised. There are two methods of attaining this magnetisation in normal commercial use.

In the first method the part to be tested is placed between the poles of an electromagnet, in which case the direction of the field

Table 26.1 Approximate lifting capacities

CIRCULAR MAGNETS: *Load lifted* (t)

Material	Magnet diameter (m)				
	1.6	1.4	1.2	1.0	0.6
Skull-cracker ball	18	15	10	7	3
Slabs	27	23	16	9	3
Pig-iron	1.3	1.0	0.6	0.3	0.1
Broken scrap	0.8	0.5	0.4	0.3	0.1
Cast-iron borings	1.0	0.7	0.5	0.3	0.1
Steel turnings	0.5	0.3	0.2	0.1	0.05

RECTANGULAR MAGNETS: *Plate area lifted* (m^2)

Plate stack			Magnet dimensions (m)			
Plate thickness (mm)	Longest plate (m)	Maximum no. of plates in stack	0.6 × 0.4	1.0 × 0.4	1.4 × 0.4	2.0 × 0.4
0.4	1.5	80	0.9	2.3	3.5	4.6
1	2.8	20	1.8	4.3	6.5	8.7
3	4.2	10	2.4	5.5	8.3	11
6	6.7	5	2.8	6.5	9.7	13
12	9.5	3	3.2	7.5	11.3	15
25	13.5	2	3.2	7.5	11.3	15

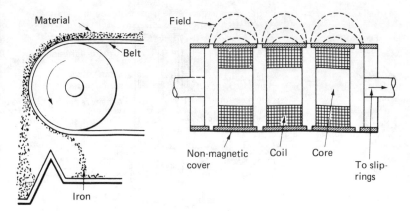

Figure 26.9 Magnetic pulley separator

is from pole to pole. The second method utilises the fact that a concentric magnetic field forms round an electric current. A heavy low-voltage current is passed through the part itself, or through a current-bar adjacent to it or threaded through it.

As only those cracks or flaws will be shown up which cut across the magnetic field, it will readily be understood that the first method is most suited to the detection of transverse cracks, the second to the detection of longitudinal ones. However, apparatus designed for testing by means of the second method may be adapted to the detection of transverse cracks by encircling the part with several turns of cable through which the heavy current is passed.

26.2.6 Separators

The bulk handling of material, particularly where the process involves crushing or grinding, may require the use of a magnetic separator for removing unwanted or tramp iron and steel, or for quickly separating ferrous from non-ferrous scrap metals. Successful operation depends on uniformity of the feed thickness, and often an installation must include a suitable conveyor/feeder.

26.2.6.1 Types

Magnetic pulley This form of separator (*Figure 26.9*) comprises a number of circular cores and poles, the magnetic axis being that of the shaft. Coils encircle the cores, with d.c. (or rectified a.c.) excitation, and set up a magnetic field pattern. Iron attracted to the pulley surface is removed by aid of the conveyor belt, the material being drawn away from the magnetic field region. When the belt speed or width, or the thickness of the feed, is unsuitable for a single pulley, two may be used, one at each end of a short auxiliary belt that receives its feed from the main conveyor.

Drum This has an advantage over the pulley type in respect of its more effective separation. A drum can operate in conjunction with a belt conveyor if placed below the head pulley and a suitable guide. Feed is readily arranged down a chute or directly on to the feeder tray, if one is provided. A common type of feeder has the tray oscillated by an eccentric motion, or vibrated in a straight-line motion, at about 15 Hz.

Suspension A structure resembling a lifting magnet is suspended over a conveyor belt. It operates successfully on feeds containing awkward shapes of tramp iron at a belt speed up to 2.5 m/s. The magnet will not automatically discharge its load, but the large gap can contain a considerable load. The power rating is large.

Disc Most machines utilise rotating discs with a sharp or serrated periphery, set above the conveyor belt and over the magnet. Separation of iron depends on the change of polarity of a given region of the disc as it rotates, so that ferrous particles can be released.

Induction roll A powerful magnet (*Figure 26.10*) is provided with a return path plate. Rollers set between them are magnetised by induction. Material is fed into the top. Non-magnetic pieces fall through under gravity, while ferrous material adheres to the roller and is carried round and detached. Up to eight rollers in tandem may be used.

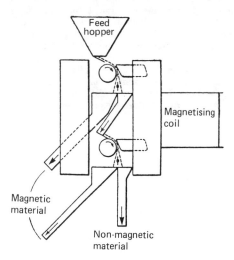

Figure 26.10 Induction roll separator

Wetherill The Wetherill separator has a single magnet unit mounted either side of a conveyor belt on which the material to be treated is passed beneath the upper magnet pole (*Figure 26.11*). Another belt is arranged over the upper pole of each magnet to take off the extracted ferrous material. The success of the separator depends on the shape of the magnet poles: the lower is flat and the upper is arranged with a ridge to concentrate the field. As the material passes under the magnets, each ferrous particle jumps towards the upper pole and is intercepted by the take-off belt, which in turn carries it to the side where it is discharged in a

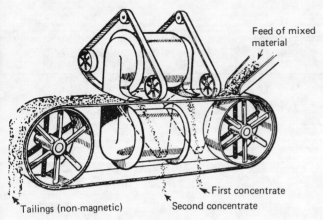

Figure 26.11 Wetherill separator

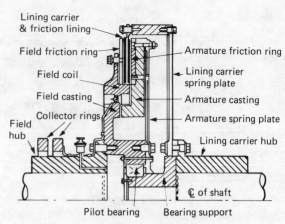

Figure 26.12 Electromagnetic clutch with double friction lining

continuous operation. In practice, several magnets are used; the number of products that can be separated in a single operation is determined by the number of take-off belts, of which there are two per magnet unit.

26.2.6.2 Ore separation

For dealing with material in large lumps the magnetic field must have a deep penetration. This involves widening out the poles. The flux density is inevitably weakened. Thus, while feed depths of 250 mm are usual with a drum of 1 m diameter for the removal of tramp iron, the depth must be cut to, say, 75 mm when feebly magnetic material is operated on.

An important branch of separation deals with the subdivision and concentration of ores, the constituents of which have permeabilities very much lower than that of iron. Data on this point are given in *Table 26.2*. The process may be performed in several ways. A single product can be removed from the bulk; several constituents may be removed, each separately, in a single operation; or the separation may be carried out by a wet process.

The general design for feebly magnetic materials differs from that for the removal of tramp iron, essentially in the necessary flux density. A material with a permeability of 1% of that of iron may require a gap density exceeding 1.6 T, and the field must be divergent. For this purpose the lower pole over which the material passes is made flat; the upper pole, whether fixed or moving, is provided with a concentrating V-edge so that particles travel to it out of the general bulk of the material treated.

26.2.7 Clutches

The conventional clutch consists essentially of two members: the field member, which carries the exciting winding, and the armature member, consisting virtually of a steel ring which becomes attracted to the field member when the winding is energised. The engaging surfaces of these members have a friction lining for taking up the load when the clutch engages, and means are provided for spring disengagement of the armature when the winding is de-energised. As the clutch rotates in operation, it is necessary to employ slip-rings and brushes for the current supply.

A special type of clutch with a double friction lining is shown in section in *Figure 26.12*. In this case the field and armature members rotate together on the same shaft. The other shaft carries on a spring plate the lining carrier member. The two friction surfaces on this member engage between the armature and field members when the field coil is energised. General particulars for representative sizes of this type of clutch are given in *Table 26.3*.

26.2.8 Couplings

Eddy-current couplings resemble induction motors in that they develop torque by 'slip', and the throughput efficiency falls with decrease of speed. In selecting a coupling the critical factors are the speed range and the load torque variation therein.

The essential features are shown in *Figure 26.13*. The outer member (the loss drum) is mounted on the shaft extension of the drive motor, and the inner member (the pole system) on the driven shaft. Operation depends on the induction of current in the loss drum by e.m.f.s resulting from the speed difference between the driving and driven shafts. The two types illustrated are:

(a) *Interdigitate.* This is common for drives transferring up to about 100 kW. The loss drum is of plain ferromagnetic material of low resistivity, normally with forced cooling. The 'claw'-shaped pole structure gives a multipolar field by means of a single exciting coil. There is substantial interpolar leakage flux.

(b) *Inductor.* The toothed rotor produces an alternating flux

Table 26.2 Relative attraction force (iron = 100) of various materials

Apatite	0.2	Dolomite	0.2	Lithium	0.5	Quartz	0.4
Argentite	0.3	Fluorite	0.1	Magnesium	0.8	Rutile	0.4
Biotite	3.2	Franklinite	35.4	Magnetite	40.2	Siderite	1.8
Bornite	0.2	Garnet	0.4	Manganese	8.9	Strontium	3.4
Cerium	15.4	Haematite	1.3	Molybdenite	0.3	Titanium	1.2
Chromium	3.1	Ilmenite	24.7	Palladium	5.2	Tungsten	0.3
Corundum	0.8	Limonite	0.8	Pyrrhoite	6.7	Zircon	1.0

Table 26.3 Clutches with double friction linings

Overall diameter (m)	Max. power per 100 rev/min (kW)	Max. torque (kN-m)	Max. speed (rev/min)	Kinetic energy at 100 rev/min (kJ)	Mass (kg)	Current at 240 V (A)
0.6	33	3	1200	0.5	300	0.9
0.8	95	9	900	1.8	520	1.4
1.0	200	20	700	5	960	2.1
1.2	360	36	600	11	1300	2.6
1.5	670	67	480	31	2200	3.2
1.8	1250	125	400	67	3400	4.2
2.1	1600	160	250	125	4700	4.7

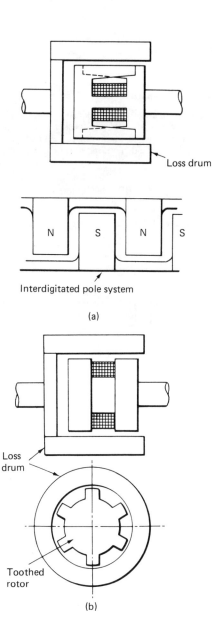

Figure 26.13 Eddy-current couplings

density pattern in the loss drum by the modulation of the airgap permeance. An annular exciting coil, fixed or rotary, causes the two airgaps to have opposite polarity, the flux between them completing its path through the loss drum.

26.2.9 Brakes

The three basic forms of brake are (i) solenoid-operated, (ii) tractive and (iii) a thruster (electrohydraulic). In each case a brake-band or (more commonly) a brake-shoe is pressed against the brake-drum, either by weights or by springs operating through a lever. The use of springs is preferable, especially with large brakes, as the cushioning of the shock due to a falling weight introduces additional problems of design as well as limiting the positions in which the brake may be mounted. The brake is released by the operating force acting against the force due to the resetting spring. The brake is held in the off position for as long as the controlling circuit is energised.

The pressure used on the friction surfaces and the coefficient of friction are of the same order as for clutches. The pressure employed should be such as to give a reasonable rate of wear, and the figure chosen will determine the width of shoe required for a given operating force and wheel diameter. In general practice there are two brake-shoes, each embracing about one-quarter of the wheel circumference.

26.2.9.1 Solenoid brake

The brake is held 'off' by a solenoid/plunger device acting through leverage against spring loading, the latter being adjustable to suit the brake-torque requirements. If the brake is energised only for short periods, with intervening periods of rest (with the brake on), it is usually possible to fit a coil giving more ampere-turns than are obtainable with a continuous rating and thus to use a greater resetting spring pressure, giving increased braking torque.

26.2.9.2 Tractive brake

The example in *Figure 26.14* embodies a tractive electromagnet operating on inner and outer disc armatures A when the magnetising coil B is energised. The mechanical features are the adjusting wedge C, the brake-shoes D, the adjusting nuts E for the outer shoe-lever F, the torque spring G and its adjuster H, the tie-rod J, the terminals K, and the shoe-clamping screws L. Armatures AA rest in slots in the base and tend to remain against the slot abutments as a result of spring pressure and magnetic force. The powerful mainspring forces the armatures AA apart, causing the inner to apply pressure to the right-hand shoe and the

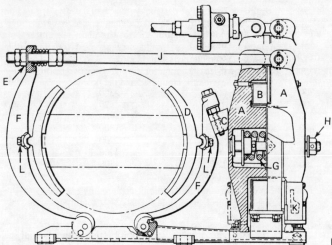

Figure 26.14 Tractive electromagnetic brake

outer to the left-hand shoe through the tie-rod J. When coil B is energised, the armatures AA mutually attract, so releasing the brake.

26.2.9.3 Thruster brake

The thruster brake employs a hydraulic thruster cylinder, with a piston acting under the fluid pressure produced by a small motor-driven pump unit. The power consumption is relatively low, but there is a short time-lag in brake response.

26.2.10 Magnetic chucks

In cases where awkwardly shaped ferrous-metal parts have to be machined in any quantity, the electromagnetic chuck forms a valuable auxiliary to various kinds of machine tools. The chuck contains a number of distributed windings which when energised from a d.c. source produce a concentrated and uniform field at the surface of the chuck, which is ground flat so as to form a suitable base-plate for accurate machining operations. The magnetic pull on ferrous materials in contact with the chuck surface is sufficient to prevent movement under all normal machining stresses.

When the current is switched off, the residual magnetism is in some cases sufficient to prevent easy removal of the part. The usual form of control switch accordingly has a demagnetising position.

The principle can be applied to rotating chucks, in which case slip-rings are necessary to convey exciting current to the windings.

In some cases *permanent-magnet* chucks can be employed. Hold and release of the workpiece are effected by an operating lever which, in the off position, closes the flux paths of the magnets through high-permeability bridges and reduces the flux through the work. With either electro- or permanent-magnet forms, the workpiece may have to be demagnetised after machining.

26.2.11 Vibrators

A vibrator generator develops a vibro-motive force of adjustable magnitude and frequency for the noise, fatigue and vibration testing of small structures and for the assessment of mechanical resonance.

26.2.11.1 Electrodynamic vibrator

Figure 26.15(a) shows the essential features of an electrodynamic vibrator, which are those of a powerful loudspeaker mechanism in which a circular coil, carrying an alternating current and lying in a constant radial magnetic field, develops vibratory force and

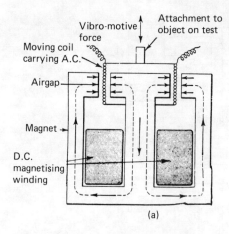

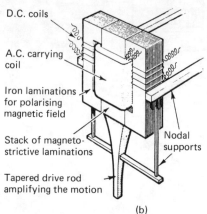

Figure 26.15 Vibrators

displacement of corresponding frequency. A construction of the form shown can be adapted to develop torsional vibration by pivoting the armature centrally.

26.2.11.2 Magnetostrictive vibrator

The magnetostriction effect can be employed by placing the a.c. exciting coil around a stack of magnetostrictive material (*Figure 26.15b*). Mechanical amplification of the very small displacement is provided by a truncated drive rod. Vibrators of this kind are generally fixed-frequency devices, but they are suitable for relatively high frequencies only.

Single-frequency low-power vibrators can be constructed with piezo-electric drive. As large crystals are not readily available, these vibrators are usable only in the ultrasonic frequency range.

26.2.12 Relays and contactors

Relays and contactors, a.c. or d.c. excited, are widely employed for low- and high-power switching. The basic features are shown in *Figure 26.16*.

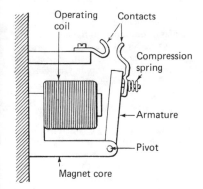

Figure 26.16 Elements of a contactor

26.2.12.1 Contactors

The term 'contactor' applies to power-control devices. For d.c. operation the contactor is made single- or double-pole as required. When the coil is energised, a magnetic field is established across the airgap and the armature is attracted to the pole to close the contacts. The moving contact has a flexible conductor attached to it in order to avoid passing current through the hinge. The destructive effects of d.c. arcs are such as to make necessary an arc shield and magnetic blow-out arrangement. The blowout winding carries the main current and its connection is so arranged that the arc is expelled from the contact region when the contacts separate.

For a.c. service the contactor normally has two or three poles. The magnetic circuit is laminated and the pole-face has a shading coil to reduce 'chatter'. Blow-out coils may not be provided because the principle operates less effectively on a.c.; reliance may be placed on extinction at a current zero. A typical a.c. contactor is illustrated in *Figure 26.17*.

Ratings These have been standardised. The severity of operating conditions varies considerably according to the class of service. Although the cleaning action on the contacts due to frequent operation is desirable in removing cumulative high-resistance films which tend to increase heating, this class of service causes greater contact wear and erosion for a given loading than would occur with less frequent operation. Conversely, very infrequent operation which involves the contacts carrying current for long periods is not onerous from the viewpoint of wear and erosion but is conducive to the formation of high-resistance surface films unless a suitably low temperature is maintained so as to limit the formation of the films. The permissible temperature rise for different types of contact is given in *Table 26.4*. Operation must be satisfactory with the shunt windings at final rated temperature and with reduced operating voltage (80% of normal for d.c., 85% for a.c.).

Table 26.4 Temperature limits for contacts

Type of contact	Temperature rise (°C)
Solid copper in air: standard rating	65
uninterrupted rating	45
Solid copper in oil	45
Laminated copper in air or in oil	40
Solid silver or silver-faced in air	80
Carbon	100

26.2.12.2 Relays

The electromagnetic relay operates one or more sets of contacts by the attraction of a movable armature towards a magnetised core. The representative types in *Figure 26.18* are: (*a*) the

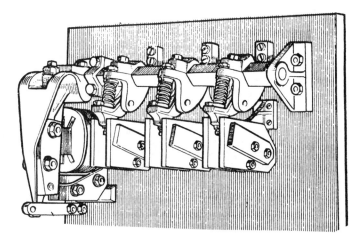

Figure 26.17 Triple-pole a.c. contactor:

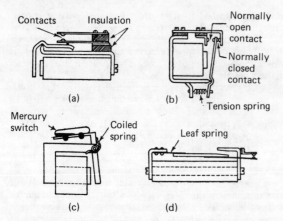

Figure 26.18 Electromagnetic relays

'telephone' type with pivoted armature; (*b*) the 'commercial' version of (*a*); (*c*) a mercury switch with hinged armature; (*d*) a spring-suspended armature. An important feature is the operating time.

High-speed operation may be obtained by one or more of the following methods: (i) lamination of the magnetic circuit to minimise eddy-current delay; (ii) reduction of the mass of moving parts; (iii) use of a large coil power; (iv) reduction of coil inductance.

Low-speed operation, sometimes needed to introduce a time-lag, is obtained by: (i) use of a lag (or *slugging*) coil comprising an additional and separate short-circuited loop or winding; (ii) use of a series inductor or shunt capacitor; (iii) addition of an external time-delay relay.

Design features Contact sets may be normally open or normally closed, and both types may be fitted on the same relay mechanism. The arrangement is determined by the operating sequence required: i.e. make, break, change-over, make-before-break, break-before-make. The contact size and material must be chosen in accordance with the rating and electrical characteristics of the circuits controlled.

Ideally, the contacts should operate cleanly and with no bounce. They should be of adequate size and of the most suitable material. In extremely low-voltage circuits the contact resistance is usually an important consideration and special precautions may also have to be taken to ensure reliable operation under conditions of vibration or shock.

Similarly, in cases of high-current switching it may be necessary to ensure wide separation of the contacts or even to arrange for several gaps to operate in series. In some cases it may be necessary to use arc-suppressing circuits.

The number and type of the contacts and springs determines the switching operation to be performed by the relay; this factor also determines the work to be done by the magnetic circuit. It follows, therefore, that the choice of a suitable coil and iron circuit design is determined by the contact arrangement of any particular relay. Various configurations of magnetic circuits and materials are used in the relays under review, depending upon their particular application. For example, in the high-sensitivity relays, where the airgap has to be kept to a minimum, it is necessary to use materials having a very low residual magnetism and high permeability.

The power required to operate the relay is determined by the spring-set arrangement and the magnetic circuit. The method of construction is important, since it largely determines the safe operating temperature of the winding and this, in turn, governs the coil power and the maximum pull available at the armature.

By increasing the area of the flux path while maintaining the ampere-turns and coil power constant, the total airgap flux, and therefore the armature pull, can be increased and the increased coil area will permit cooler operation of the coil. This may, however, lead to a relay that is physically larger than can be tolerated. In practice, therefore, it is more reasonable to build a relay of a given size and to use other means to amplify the controlling power.

The continuous power input to a given relay coil is limited only by the maximum temperature that the coil insulation can withstand without breakdown. This temperature is governed by the environmental conditions as well as by the coil construction and the quality of the insulating material.

Many of the functions performed by the electromagnetic relay have been taken over by solid-state switching.

26.2.13 Miniature circuit-breakers

The miniature circuit-breaker (m.c.b.) is, for the control of small motors and domestic subcircuits, considered primarily as an alternative to the fused switch. The appropriate British Standard is BS 3871, which lays down specific technical requirements. The usual form of the m.c.b. embodies total enclosure in a moulded insulating material. As the operating mechanism must be fitted with an automatic release independent of the closing mechanism, the m.c.b. is such that the user cannot alter the overcurrent setting nor close the breaker under fault conditions. At the same time the m.c.b. must tolerate harmless transient overloads while clearing short circuits. For most practical conditions, a change-over from time-delay switching to 'instantaneous' tripping at currents exceeding 6–10 times full-load rating is suitable.

26.2.13.1 Tripping mechanisms

Methods of achieving the required operating characteristics can be classified as (i) thermo-magnetic, (ii) assisted thermal and (iii) magnetohydraulic. In the *thermo-magnetic* method the time-delay is provided by a bimetal element, and the fast trip by a separate magnetically operated mechanism based on a trip coil. In the *assisted thermal* method the bimetal is itself subjected to magnetic force. The *magnetohydraulic* mechanism incorporates a sealed dashpot with a fluid and a spring restraint, the dashpot plunger being of iron and subject to the magnetic pull of the trip coil. The essential features are illustrated in *Figure 26.19*.

Thermo-magnetic The bimetal element shown in *Figure 26.19(a)* may carry the line current or, for low current ratings, be independently heated. Its flexure operates the trip latch through a crank. On overcurrent the magnetic force acts directly on the latch bar, with or without the aid of the bimetal deflection.

Assisted thermal The time-delay characteristic is provided by a bimetal element, and instantaneous tripping by magnetic deflection of the bimetal. The operation is shown in *Figure 26.19(b)*. A bar of magnetic material is placed close to the bimetal element, and the magnetic field set up by the current develops a pull on the bimetal such as to increase its deflection and release the trip latch. The magnetic effect is proportional to the square of the current and so becomes significant on overcurrent. However, as the position of the bimetal element on the occurrence of a short circuit is arbitrary, there is no well-defined change-over point at which instantaneous tripping occurs.

The method is cheap and simple, but is difficult to design for low-current (e.g. 5 A) breakers because the operation tends to be sluggish, particularly at fault-current levels that are less than 500 A.

Magnetohydraulic This method, shown in *Figure 26.19(c)*, combines in one composite magnetic system a spring-loaded

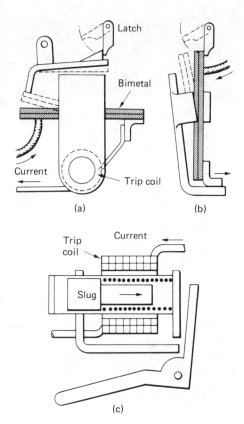

Figure 26.19 M.C.B. trip mechanisms

dashpot with magnetic slug in a silicone fluid, and a normal magnetic trip. When the line current flows, the magnetic field produced by the trip coil moves the slug against the spring towards the fixed pole-piece, so reducing the reluctance of the flux path and increasing the magnetic pull on the trip lever. If it reaches the end of the dashpot, the pull is sufficient to operate this lever and trip the circuit-breaker. On sudden overcurrent exceeding 6–10 times full-load value, there is sufficient pull at the fixed pole-piece to attract the armature of the trip lever regardless of the position of the slug in the dashpot. The characteristic is more definite and satisfactory for low-current ratings than that of the assisted thermal mechanism.

26.2.13.2 Operating features

Thermal operation by bimetal elements implies that the effective current rating is a function of the ambient temperature. It is the practice, if complete ambient compensation is not fitted, to rate m.c.b.s in such a way as to allow for the type of enclosure. With magnetohydraulic devices the tripping is independent of the ambient temperature over a specified range, the small variations due to change of viscosity of the damping fluid being minimised by use of a fluid with a nearly flat viscosity–temperature characteristic.

The combination of thermal and magnetic functions is not easily controlled for low current ratings, and for m.c.b.s with such ratings the tolerances on operation must be wider than they are for larger currents.

Normally, m.c.b.s are suitable only for a.c. circuits. As with all a.c. switchgear, the problems of breaking efficacy are associated not only with the actual short-circuit current but also with its asymmetry and power factor.

As m.c.b.s can be linked to give two- and three-pole versions, so arranged that a fault on one pole will produce complete circuit isolation, the risk of single-phasing in motor control is effectively eliminated. In other directions, however, m.c.b.s cannot necessarily replace fuses: they do not possess the high short-circuit breaking capacity of the modern h.r.c. fuse, nor do they have its inherent fault-energy limitation. If, therefore, conditions are such that back-up protection has to be provided for m.c.b.s, the 'take-over' zone should be of the order of 1.0–1.3 kA.

26.2.14 Particle accelerators

Modern accelerators produce high-energy beams of electrons, ions, X-rays, neutrons or mesons for nuclear research, X-ray therapy, electron irradiation and industrial radiography. If a particle of charge e is accelerated between electrodes of p.d. V it acquires a kinetic energy eV electron-volts (1 MeV $= 1.6 \times 10^{-13}$ J). Accelerators are classified as *direct*, in which the full accelerating voltage is applied between the two electrodes; *indirect*, in which the particles travel in circular orbits and cyclically traverse a region of electric or magnetic field, gaining energy in each revolution; and *linear*, in which the particles travel along a straight path, arriving in correct phase at gaps in the structure having high-frequency excitation, or move in step with a travelling electromagnetic wave.

26.2.14.1 Direct accelerators

The Cockcroft–Walton multiplier circuit has two banks of series capacitors, alternately connected by rectifiers acting as change-over switches according to the output polarity of the energising transformer. The upper limit of energy, about 2 MeV, is set by insulation. A typical target current is 100 μA.

The Van de Graaff electrostatic generator is capable of generating a direct potential of up to about 8 MV of either polarity. It has an endless insulating belt on to which charge is sprayed from 'spray-set' needle-points at about 50 kV. The charge is carried upwards to the interior of the high-voltage (h.v.) electrode, a metal sphere, to which it is transferred by means of a second spray set. H.V. insulation difficulties are overcome by operating the equipment in a tank filled with a high-pressure gas, e.g. nitrogen–freon mixture at 1500 kN/m². In two-stage Van de Graaff generators for higher energies, negative hydrogen ions are accelerated from earth potential to 6 MeV; they are then fired into a thin beryllium foil 'stripper' which removes the electrons from the outer shells of the atom and leaves the remanent ions moving on with little change of energy but with a positive charge. The second stage brings these ions back to earth potential and the total energy gain is 12 MeV. To bring ions on to a small target the accelerating and deflecting fields must be accurately controlled, and scattering limited by evacuating the accelerator tubes to very low pressure. The energies are sufficient for the study of nuclear reactions with the heaviest elements.

26.2.14.2 Indirect (orbital) accelerators

Indirect (orbital) accelerators may have orbits of approximately constant radius with a changing magnetic field (betatrons and synchrotrons) or orbits consisting of a series of arcs of circles of discrete and increasing radii in a constant magnetic field (cyclotrons and microtrons).

Betatron The betatron is unique in that the magnetic field not only directs particles into circular orbits but also accelerates them. The magnet has an alternating field of which only one quarter-period is used. Electrons are accelerated in an evacuated toroidal chamber between the poles of the magnet. They are injected at an energy corresponding to a low magnetic field,

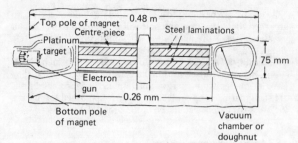

Figure 26.20 Cross-section of a 20 MeV betatron

which bends them in circular orbits round the toroid. A cross-section of the poles and vacuum chamber is shown in *Figure 26.20*. As the magnetic flux through an electron orbit increases during the cycle of alternation, the electron experiences a tangential force, and its gain in energy per revolution is the voltage that would be induced in a loop of wire in the orbit. As the electron gains energy, the magnetic guide field intensity at the orbit increases at a suitable rate. To keep the electron on a constant radius from injection to peak energy requires the rate of change of intensity at the orbit to be half that of the mean flux per unit area within the orbit. At peak energy (or earlier) the electrons are caused to move away from their equilibrium orbit and to strike a target inside the vacuum chamber, producing X-rays or corresponding energy. The output consists of short pulses of radiation whose repetition rate is the frequency of the magnet excitation. Energy limitations are set by the size and cost of the magnet and the radiation loss when a high-energy electron has circular motion.

Synchrotron The synchrotron uses an annular magnetic guide field which increases as the particles gain energy, as in the betatron, so that they maintain a constant orbit radius. Electrons are initially accelerated by the action of central 'betatron bars' which saturate when the main magnetic field corresponds to an energy of 2-3 MeV when electrons travel at a velocity only 1-2% less than the velocity of light. Further gain of energy is produced by radio-frequency (r.f.) power at the frequency of orbital rotation (or a multiple of it) that is fed to resonators inside the vacuum chamber. The particles become bunched in their orbits so that they pass across the accelerating gap in the resonator at the correct phase of the r.f. field. The limitation on electron acceleration is now mainly set by radiation losses due to circular motion.

Protons are injected at about 500 keV, which produces a velocity of only 3% of that of light. Further acceleration changes the frequency of orbital rotation. For a proton synchrotron the magnetic guide field strength and the r.f. power frequency have to be varied accurately over large ranges.

Cyclotron This early form of accelerator consists of a vacuum chamber between the poles of a fixed-field magnet containing two hollow D-shaped electrodes which load the end of a quarter-wave resonant line so that a voltage of frequency 10-20 MHz appears across the accelerating gap between the 'D's. Positive ions or protons are introduced at the centre axis of the magnet and are accelerated twice per rotation as they spiral out from the centre. The relation between particle mass m, charge e, magnetic flux density B and frequency f is $f = Be/2\pi m$. Energy limitation is set by the relativistic increase of mass, which limits the speed of high-energy particles so that their phase retards with respect to the r.f. field.

Synchrocyclotron In this device the energy limitation of the cyclotron can be removed by modulating the oscillator frequency to a lower value as a bunch of particles gains energy.

Microtron In the microtron, or electron cyclotron, electrons are accelerated in a vacuum chamber between the poles of a fixed-field magnet. The orbits consist of a series of discrete circular arcs which have a common tangent at a resonant cavity in which the electrons gain their successive increases of energy from an r.f. electric field. The highest energy achieved with such a machine is 6 MeV, and mean currents are less than 1 μA.

26.2.14.3 Linear accelerators

Indirect accelerators of protons have so far used a resonant cavity in which drift-tube electrodes are introduced that distort the fields and enable particles to be shielded from field reversals. Particles are accelerated between gaps and move between centres of successive gaps in one complete period of oscillation (*Figure 26.21*). Oscillators operating at about 200 MHz and a pulse power of 1-2 MW are used to excite the cavity for some hundreds of microseconds. Injection is by a Cockcroft-Walton or Van de Graaff device.

Figure 26.21 Field resonant cavity for a proton accelerator

An important device for electron acceleration is the travelling-wave accelerator, using megawatt pulses of r.f. power at 3000 MHz. The power is propagated along a cylindrical waveguide loaded with a series of irises. A travelling wave is set up with an axial electric-field component, and correct dimensioning of the iris hole radius a and the waveguide radius b (*Figure 26.22*) enables the propagation velocity and the field-intensity/power-flow relation to be varied. An electron injected along the axis with

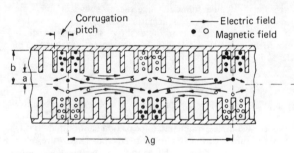

Figure 26.22 Fields in a corrugated waveguide

an energy of the order of 45 keV is accelerated by the axial field, and as its velocity changes it remains in correct phase with the travelling field, the propagation velocity of which is varied to match. A fixed axial field is required to provide for electron focusing.

High-energy machines with low beam-currents have been used in the USA, low-energy machines with high beam-currents in the UK. The 25 MeV Harwell accelerator has a length of 6 m divided into six sections, each fed by a 6 MW klystron amplifier to give a peak beam-current of 1 A and a mean output power of 30 kW.

26.2.14.4 Large machines

The Harwell proton synchrotron gives the particles and energy of 7 GeV in an orbit of radius 19 m within a 7000 t magnet. The magnet takes 10 kA to raise the orbit flux density to about 1.4 T in 0.75 s, to hold this value for 0.25 s and to reduce it to zero in 0.75 s, with a repetition frequency of about two cycles per hour. The inductance of the magnet is about 1.1 H, and to produce a rate of change of current of $10/0.75 = 13.3$ kA/s the magnet supply voltage must be about 14 kV. The peak stored energy is 40 MJ. The supply is from a pair of 3750 kW/75 MW motor/generators through rectifiers.

The new accelerator for CERN near Geneva is to use the existing 25 GeV machine to inject particles into a 300 GeV proton synchrotron with an orbit of diameter about 2.2 km. The magnet will employ superconducting exciting windings giving a flux density of 4–6 T. The design is such that it can be built initially with only alternate magnet sections, and upgraded in energy later without basic alteration of the main structure, possibly to 800 GeV.

26.3 Industrial rotary and linear motors

The elements discussed in Section 2.4.3.6 indicate that there are two methods of developing a mechanical force in an electromagnetic machine:

(1) *Interaction.* The force f_e on a conductor carrying a current i and lying in a magnetic field of density B is $f_e = Bi$ per unit length, provided that the directions of B and i are at right angles; the direction of f_e is then at right angles to both B and i. This is the most common arrangement.

(2) *Alignment.* Use is made of the force of alignment between two ferromagnetic parts, either or both of which may be magnetically excited. The principle is less often applied, but appears in certain cases, e.g. in salient-pole synchronous machines and in reluctance motors.

26.3.1 Prototype machines

Three basic geometries (*Figure 26.23*) satisfy the requirement for the relative orientations of B, i and f_e. For a rotary case the *cylindrical* form is the most common, while the *disc* with its short axial length suits particular applications. The *flat* form is employed for linear motion.

Such machines are almost exclusively *heteropolar* (i.e. have alternate N and S poles). To maintain unidirectional interaction force, the direction of the current in a given rotor conductor must reverse as it passes from an N pole to an S pole region.

26.3.1.1 Heteropolar cylindrical machine

A heteropolar cylindrical machine for a 2-pole magnetic circuit is shown in *Figure 26.24*. The active region is the airgap between stator and rotor. In *Figure 26.24(a)* the stator conductors are arranged (normally in slots) on the surface and are connected so

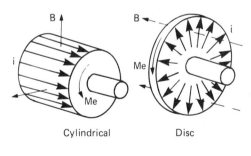

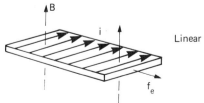

Figure 26.23 Basic geometries for electromagnetic machines

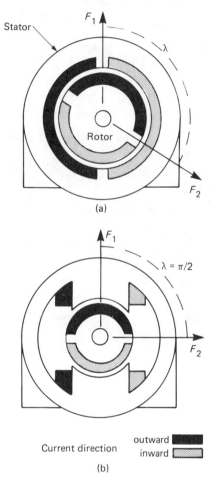

Figure 26.24 Prototype 2-pole cylindrical machines

as to develop the current-sheet pattern indicated, giving rise to a distributed m.m.f. of peak value F_1 on the axis of the winding. A corresponding rotor current-sheet pattern sets up an m.m.f. F_2. If the m.m.f. distributions are assumed to be sinusoidal, the torque on the rotor can be shown to be

$$M_e = kF_1 F_2 \sin \lambda$$

where λ is the *torque angle* between the stator and rotor winding axes, and k is a function of the airgap dimensions. For $\lambda = 0$ the m.m.f.s F_1 and F_2 are in alignment and there is no torque. Displacement of the rotor increases the torque, which reaches a maximum for $\lambda = \pi/2$ rad. For further displacement the torque falls, to become zero again for $\lambda = \pi$ rad.

The machine in *Figure 26.24(b)* has a fixed optimum torque angle $\lambda = \pi/2$ rad. Here the rotor must have a closed winding and be provided with a commutator or alternative switching device so that each conductor, as it passes from an N to an S polar region, has its current automatically reversed. Then the direction of F_2 is fixed for all operating conditions. As the direction of F_1 is also fixed, it is usually developed by salient poles.

26.3.1.2 Types of machine

The three most common machines—synchronous, induction (asynchronous) and commutator—are all heteropolar and have at least one member cylindrical. They are distinguished by the nature of the supply (a.c. or d.c.) and that of the airgap flux (travelling-wave or fixed-axis).

Travelling-wave gap flux The stator current-sheet pattern in *Figure 26.24(a)* is set up by a 3-phase winding ABC as in *Figure 26.25* for a 2-pole machine. With the phase windings excited by balanced symmetrical 3-phase currents of frequency f_1, the

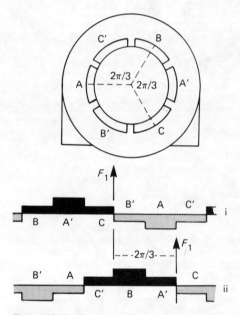

Figure 26.25 Production of a travelling-wave field

sequential cyclic reversal of currents in the displaced windings shifts the current-sheet pattern, as shown for peak current (i) in phase A and (ii) in phase B (one-third of a period later). Thus the stator m.m.f. F_1 produces a travelling wave of m.m.f. and airgap flux (often called a 'rotating field') moving at synchronous speed $n_s = f_1$ (r/s) or angular speed $\omega_1 = 2\pi f_1$ for a 3-phase supply of frequency f_1 to a 2-pole machine; in general $n_s = f_1/p$ and $\omega = 2\pi f_1/p$ for a machine with p pole-pairs.

Let the rotor, rotating at angular speed ω_r, have a 3-phase winding carrying currents of frequency f_2; then it has an m.m.f. F_2 rotating at angular speed $\omega_2 = 2\pi f_2$ with respect to the *rotor* body, and therefore at $\omega_r \pm \omega_2$ with respect to the *stator*. For a steady unidirectional torque to be developed, F_1 and F_2 must rotate in synchronism to preserve unchanging the torque angle λ. Thus $\omega_r \pm \omega_2 = \omega_1$ is the essential running condition.

Synchronous machine The rotor is d.c. excited, so that F_2 is 'fixed' to the rotor body; then $\omega_2 = 0$ and $\omega_r = \omega_1$. The rotor must therefore rotate synchronously with the stator travelling-wave field. The torque angle accommodates to the torque demand up to a maximum for the torque angle $\lambda = \pi/2$ rad. The machine can operate in both generator and motor modes by simple reversal of the torque angle.

Induction machine The rotor winding, isolated and closed, derives its current inductively from the stator. If the rotor spins at synchronous speed, its conductors move with the stator field and no rotor current can be induced. However, if ω_r is less than ω_1 by a fractional 'slip' $s = (\omega_1 - \omega_2)/\omega_1$, the rotor conductors lie in a field changing at slip frequency $s\omega_1$, and currents of this frequency are induced to provide an m.m.f. F_2 travelling around the rotor at this frequency. This gives $\omega_r + \omega_2 = \omega_1$, the required condition. Torque is developed for any slip s other than zero (synchronous speed), with F_1 and F_2 mutually displaced by the torque angle. By driving the machine above synchronous speed the slip and torque are reversed, and the machine generates.

Fixed-axis gap flux In the usual constructional form (*Figure 26.24b*), the gap flux is produced by the stator m.m.f. F_1, generally with the poles salient. The rotor m.m.f. F_2 has the optimum torque angle $\lambda = \pi/2$ rad. As the rotor spins, the current of an individual conductor is reversed as it passes from the outward- to the inward-directed region of the current sheet in the process of *commutation*. In consequence the machine can develop torque at standstill and at any practicable speed.

D.C. commutator machine Both stator and rotor windings are d.c. excited. The torque is smooth and continuous, with simple control of speed and both motor and generator operation. In small d.c. motors the stator may be magnetised by permanent magnets, dispensing with the exciting winding.

Single-phase commutator machine As simultaneous reversal of F_1 and F_2 does not affect the direction of the torque, the d.c. motor can be operated on a 1-phase supply with the stator and rotor windings connected in series. However, the torque has a double-frequency pulsation about a unidirectional mean.

Other forms There are many variants, especially in small and miniature machines. Single-phase induction motors require special starting techniques ('split-phase', 'shaded-pole'). Some operate on alignment torque ('reluctance', 'hysteresis', 'brushless', 'stepper'). A few large homopolar d.c. machines have been devised.

Disc motors The geometry shown in *Figure 26.23* has been applied with permanent-magnet multipolar field systems to machines that must have a very short axial length, e.g. for driving cooling fans for motor vehicles.

Linear motors These are most usually based on the 3-phase induction-motor principle.

26.3.2 D.C. motors

In spite of the fact that a standard d.c. motor costs 1.5–2 times as much as a cage induction motor, and that alternating current is universal for general power distribution, the scope for d.c. motors is still large, particularly for drives requiring speed control or some other special feature. D.C. motors are built in all sizes from fractional-kilowatt up to about 4 MW, the upper limit being imposed by commutation problems.

In addition to the standard types of motor (shunt, series or compound) which are normally fed from a constant-voltage d.c.

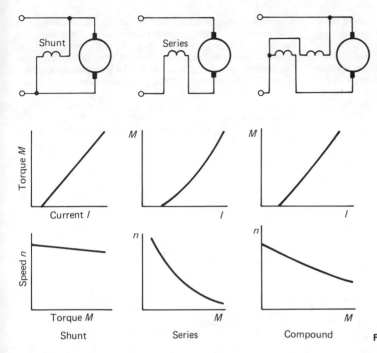

Figure 26.26 D.C. motors: basic characteristics

supply, many modern d.c. motors incorporate thyristors enabling them to operate from a standard a.c. supply. The thyristors rectify the alternating voltage and, by gate control, enable the resulting direct voltage to be varied, thereby giving a wide range of speed control.

26.3.2.1 Characteristics

The connections and basic torque–current and speed–torque characteristics of standard shunt, series and compound motors are given in *Figure 26.26*. Motors with separately excited fields are often used for control purposes.

When connected to a d.c. supply of voltage V, a motor takes a current $I = P/V\eta$ when developing a useful output power P. The efficiency η varies with the rating: typical rated currents are given in *Table 26.6* for a range of operating voltages. The rotational e.m.f. in an armature of resistance r_a over the brushes and carrying a current I_a is $E = V - I_a r_a$. The power-conversion relation is $EI_a = M\omega_r$, where M is the torque and $\omega_r = 2\pi n$ is the angular speed. Then

$$E = V - I_a r_a = 2(p/a)nN\Phi = Kn\Phi \qquad M = K\Phi I_a/2\pi$$

where Φ is the flux. Here $K = 2(p/a)N$ involves the number of pole-pairs (p), the total number of turns (N) on the armature and the number of pairs of parallel paths (a) of the winding. For simple lap and wave windings, $a = 1$ and $a = p$ respectively. In a shunt-connected machine the total input current I is the sum of the armature and field currents, $I = I_a + I_f$. In a series-connected machine the same current I flows in both field and armature windings.

Shunt excitation The field winding has the constant terminal voltage applied to it so that the flux will be approximately constant and the torque will be proportional to the armature current. Speed is proportional to E and is approximately constant since $I_a r_a$ is normally not more than 3–5% of V at full load. In practice, the flux will, owing to armature reaction, be distorted when the machine is loaded, the flux density under the leading pole tips being increased and that under the trailing pole tips decreased. Owing to saturation of the iron in the teeth under the leading pole tips, the increase in density there is less than the decrease in density under the trailing tips, so that there is a net reduction of flux at full load of 2–3%. The drop in speed from no load to full load is therefore less than would be expected from the speed equation—in some cases this action even gives a rising speed characteristic, a disadvantage which can be corrected by the use of a small series field winding. Increase in temperature from cold to hot raises the resistance of the field winding and reduces the current in it, thereby reducing the flux and increasing the speed for a given load.

Motors designed to give a wide range of speed control by variation of the field or to be used in situations where sudden and heavy load fluctuations occur are often fitted with a compensating winding in the pole face to neutralise the effect of armature reaction and prevent flux distortion; such windings are connected in series with the armature so that neutralisation is correct at all loads.

The shunt motor can be used for the drive of machine tools, pumps and compressors, printing machinery and all forms of industrial drive requiring a speed which is approximately constant and independent of the load.

Separate excitation This is applied widely to control motors, particularly where speed variation is required over a considerable range. For a given field voltage, the characteristics resemble those of a shunt motor. Separate excitation (but without the facility of field control) applies to small motors with permanent-magnet field systems.

Series excitation The field m.m.f. is produced by the motor current so that at low currents where the iron is unsaturated Φ is approximately proportional to I, but at high currents (1.5–2 times full-load current) Φ tends to become constant as the iron saturates. The starting torque, when the current is above the full-load value, is thus greater than for a shunt motor with corresponding full-load current and flux. The speed at heavy currents drops to a low value of account of the increase of flux.

The high starting torque and falling speed–torque characteristic make the series motor suitable for driving hoists and cranes, for traction and rope haulage and for driving fans, centrifugal pumps or other apparatus where there is no danger of the motor being run light.

Compound excitation Where a drop in speed between no load and full load greater than that obtainable with a plain shunt motor is required, a series winding may be added to assist the shunt winding, giving a speed–torque characteristic having any desired amount of droop. Instability of a shunt motor due to the weakening of the field by armature reaction can also be cured by the addition of a series winding known as a *series stability* winding. The chief application of the compound motor arises when the motor is used in conjunction with a flywheel—a fairly steep droop to the speed–torque characteristic is then necessary in order to enable the flywheel to give up its stored energy when a sudden load comes on. Compound motors are also used for driving pumps, compressors and other heavy-duty machinery.

If the series winding is arranged to oppose the shunt winding, a motor with a flat or even a rising speed–torque characteristic can be designed, such a motor being known as a differentially compound motor. Such motors are, however, very rarely used.

26.3.2.2 Construction

Motor design aims at economy of materials and the reduction of loss. Further, as most industrial d.c. machines are fed from an a.c. supply through thyristors, an all-laminated magnetic circuit reduces the effect of supply harmonics. There is increasing use of square frames, either of rolled steel or of laminations.

Poles Constructed separately, the pole body and shoe are assembled from laminations, or a solid body is provided with a laminated shoe. The poles are bolted to the yoke and retain the field windings. Commutating poles may be solid or, more usually, laminated. Motors of rating below about 10 kW may have half as many commutating as main poles.

Field windings Main *shunt-field* windings are of circular- or rectangular-section wire, insulated and wound on a former. The whole is then taped, impregnated, slipped on to the pole and held by the pole-shoe. Large machines may have the turns wound on a bobbin of pressboard or of steel lined with micanite. *Series windings* to carry currents exceeding 50 A are more generally of copper strip wound on edge, and a similar construction is used for *compole* windings.

Armature core This is built from core-steel laminations (0.35–0.6 mm), coated on one side with an insulating varnish and bolted or clamped between thick end-plates. For diameters up to about 1 m the stampings may be made in disc form; above this size they are in sectors keyed to the shaft hub. If the core length exceeds about 20 cm, radial ducts are provided, each 5–6 cm. Axial ducts are employed with small machines. Machines of ratings up to 50 kW may have slots skewed to reduce noise.

Commutator and brushgear The commutator is conventionally made by assembling hard-drawn copper sectors interleaved with 0.7–2 mm sheet mica, these separators being 'undercut' by about 1 mm. The brushes, of a suitable carbon/graphite content, are mounted in boxes with spring loading to hold them against the commutator surface with a medium to strong pressure depending on the application. The circumferential brush width is typically 2–3 sector widths (10–20 mm) and about 30 mm axially. One brush-arm per pole is employed except for certain 4-pole wave-wound machines which have two brush-arms in adjacent positions to facilitate maintenance.

Armature winding Almost all motors other than very large machines use a simple 2-circuit wave winding. Conductor wires of section 1 mm² or less are circular and enamel-insulated, the conductors of a coil being taped half-lap before being placed in the slots. Larger machines have former-wound coils of rectangular-section conductors, insulated by half-lapped tape. The coils are assembled from the conductors and taped before insertion. The slots are lined with pressboard, and the two layers separated by a pressboard spacer. Various recent developments in epoxy resins have made possible the use of better insulants at higher temperatures. Typical slot sections are shown in *Figure 26.27*. The coils are contained in the slots by wedges or by steel or glass-cord binding.

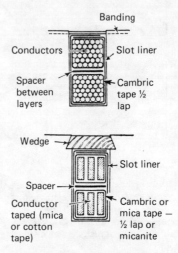

Figure 26.27 Typical slot sections

In small wire-wound armatures the coil ends are soldered direct into grooves in the commutator sectors. For strip windings, the sectors carry 'risers' for connection to the ends of the coils.

Bearings End-shield bearings are usual for ratings up to 250 kW, above which pedestal bearings are employed. Journal bearings are fitted where ball or roller bearings are unsuitable.

Enclosure and ventilation Recent Standards define the conditions to be met by machines for a variety of ambient conditions (e.g. dripproof, splashproof, hoseproof, weatherproof, flameproof).

Cooling air is drawn into the machine directly or through cowls, pipes or screens except in totally enclosed machines, for which there is no communication between the outside air and the interior of the motor. As cooling must then be solely by dissipation from the outside of the carcass, the rating is limited to about 75 kW. Cooling can be improved by shaft-mounted fans, by inlet and outlet pipes or by closed-circuit ventilation. In the latter case the fan-assisted air circulating through the machine is cooled by passing it through an air/air (c.a.c.a.) heat-exchanger or an air/water (c.a.c.w.) exchanger mounted on the motor frame.

26.3.2.3 Commutation

Modern motors can be made to commutate sparklessly up to 1.5–2 p.u. load. To secure this behaviour, commutating poles are fitted to all motors in the integral-kilowatt range.

Compoles Before the commutator sectors connected to a particular armature coil reach a brush, the coil will be carrying current in a certain direction; while the sectors connected to the coil are passing under the brush the coil will be short-circuited by the brushes, and after leaving the brushes the coil will be carrying current in the opposite direction. The current must thus be reversed during the time for which the coil is short-circuited (the time of commutation). At normal commutator peripheral speeds of 10–30 m/s this time will usually lie between 2.5 and 0.2 ms. Owing to the inductance of the coil, the current cannot reverse in this time without some external assistance; it is necessary to induce in the short-circuited coil an e.m.f. to assist the change of current, i.e. an e.m.f. in a direction opposite to that of its e.m.f. when, after leaving the commutating zone, it enters to next pole region. Compoles are therefore fitted to influence the coil-sides undergoing commutation; the compoles have an excitation proportional to the armature current and the polarity of the successive main pole. The arrangement (i) neutralises the main armature m.m.f. in the commutating zone and (ii) produces the necessary commutation flux density there. The compole flux required is proportional to the armature current, and the compole windings are therefore connected in series with the armature.

Commutation in a machine not fitted with compoles can be effected by brush-shifting backward (against the direction of rotation) so that the commutating flux is provided by the succeeding pole. This cannot be done if the motor is required to run in both directions.

Sparking Sparking causes burning and pitting of the commutator surface, so intensifying the trouble. The origins of sparking, and the remedies, are as follows.

Mechanical defects The chief causes are: the sticking of brushes in holders; 'high' or 'low' commutator bars; flats, irregularities or dirt on the commutator surface; badly bedded brushes. The commutator can be cleaned while running with a commutator stone, but irregularities make it necessary to grind the commutator. Small sparks between sectors, starting a few centimetres away from the brush, are probably due to partial short circuits caused by dirt on the mica surfaces. Correct bedding is essential to ensure that the brushes carry current over the whole of their contact surface. It can be carried out by adjusting the brush springs to give a fairly heavy tension and passing, first coarse and finally fine, glass-paper between the brush and the commutator; care must be taken to remove all trace of dust after the operation.

Incorrect brush position The correct position of the brush rocker is usually marked by the manufacturer, but it may tend to move in service. If the marking is obliterated, the correct position can be found by determining the neutral position. If the machine has no compoles the brushes will have to be moved backward from the neutral position by 2 or 3 sectors, the best position being found by trial. With a compole motor the brushes should be almost exactly on the neutral position, although they may be moved forward by a small distance so that the compole flux adds to that of the preceding main pole and prevents any tendency of the speed to rise as the load comes on, the compole winding acting as a series stability winding. Incorrect spacing between adjacent brush arms may occur; this should be checked very carefully by a steel tape, not by counting the sectors.

Winding defects Open- and sort-circuited coils in the armature will cause severe sparking. An open-circuited coil will cause sparks to go all round the commutator with severe burning at the bars connected to the open-circuited coil. A short-circuited coil will cause overheating of the faulty coil and segments and is often due to molten solder falling between the commutator risers. In either case the presence of a fault can be verified by carrying out a drop test.

Incorrect compole excitation The compole strength can be checked by the brush-drop or black-band tests. Incorrect strength can be remedied by inserting or removing thin steel shims between the back of the compole and the yoke: removing a shim increases the airgap and weakens the compole. Weakening can also be obtained by shunting a resistor diverter across the compole winding, but this is not fully effective in transient conditions.

Thyristor-assisted commutation The speed and output limitations imposed by commutation have encouraged attempts to use thyristors to perform the switching function, leaving the brush to act as a simple current collector. Success would enable the voltage per sector (30–40 V peak, 15–20 V mean) to be raised, fewer sectors could be employed and high-power high-voltage d.c. machines achieved. Promising results have been obtained with the arrangement of *Figure 26.28*. The armature coils are connected to two separate commutators with alternately 'live' and 'dead' sectors, the latter shown shaded. In the diagram, brush

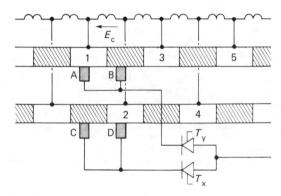

Figure 26.28 Thyristor-assisted commutation

D has just come fully into contact with active sector 2, and thyristor T_x has been turned on; the e.m.f. E_c in the coil that has just been commutated must, acting through brush A, be enough to turn off thyristor T_y. Separation of the brushes into two parts, AB and CD, is necessary to ensure that the part-brush is fully on to an active sector before it begins to carry current; otherwise contact damage occurs. Although the switching procedure is satisfactory, current collection difficulties, causing commutator damage and arising from the transient current changes, have not yet been overcome.

26.3.2.4 Starting

If a d.c. motor is to be started from a constant-voltage supply, its normally low resistance must be augmented to limit the current to a safe value, e.g. 1.5–2 times full-load value. The starting rheostat is cut out as the motor speed rises and the counter-e.m.f. imposes a limit on the current.

Shunt motor For maximum starting torque, the field must be fully established at starting; the starting rheostat is therefore connected only in the armature circuit. The total starting resistance R is determined by the maximum starting current $I = V/(R + r_a)$, where the armature resistance r_a has the typical values:

Motor rating (kW)	1	2	5	10	50	100
Armature resistance (Ω)						
at 110 V	1.4	0.8	0.2	0.1	0.08	0.006
at 230 V	6.0	3.0	0.8	0.5	0.10	0.025
at 460 V	22	13	3.0	2.0	0.50	0.010

Traditional face-plate starters are obsolescent. Usually a fully automatic push-button system is employed.

Series motor The starting rheostat is in series with the motor, the resistance of which is about twice the value given for r_a in the table above. Industrial series motors are often used in cranes and hoists, and the starting resistance is commonly utilised also for speed control.

26.3.2.5 Speed control: standard motors

A prime reason for the continued use of d.c. motors is the possibility of simple and economic speed control over a wide range. Reference to the expression in Section 26.3.2.1 shows that speed can be controlled by varying the applied voltage, the flux or the armature resistance.

Shunt motor In a shunt machine the variation of the supply voltage does not greatly affect the speed because the result is also a comparable change in flux.

Field control The field current (and therefore the flux) is varied by adding resistance into the field circuit (*Figure 26.29*). For a given setting of the field regulator the speed is approximately independent of the load, giving a series of flat speed–torque curves. The upper limit of speed for a standard motor is about 30–50% above normal, fixed partly by mechanical considerations and partly by weak-field flux distortion. However, a 3:1 range can be obtained by suitable design, although very low speeds cannot be obtained in this way. For a given armature current, a flux reduction raises the speed but reduces the torque to yield a *constant-power* characteristic, *P* in *Figure 26.29*. The loss in the field-regulator resistance is small, so that the efficiency of the machine is not affected.

Armature-circuit resistance control The speed for a given value of resistance added into the armature circuit falls with the load, giving a group of speed–torque characteristics (*Figure 26.29*). The flux remains constant so that for a given current the torque will not vary with speed (*constant-torque* characteristic); power output therefore falls proportionately with speed. Owing to the losses in the added resistance the efficiency is low and approximately proportional to the speed: e.g. with a 60% speed drop the efficiency will be a little less than 40%. The resistance required is $R = x(V - I_a r_a)/I_a$ where x is the desired fractional speed reduction and I_a is the armature current at the reduced speed, the latter depending on the type of load.

Diverter control With series armature-circuit resistance control a large resistance is required to obtain low speed on small load, and the machine is unstable in that there is a large change of speed with load. This can be overcome by adding a variable resistor in parallel with the armature circuit. The efficiency is low, and the method is justified only as a temporary measure or with very small motors.

Ward–Leonard control The main d.c. motor M is separately excited with a constant field current, the armature being supplied with a controlled variable voltage obtained from a d.c. generator G driven by a constant-speed motor (*Figure 26.30*). Control of the generator field varies the main motor armature voltage and consequently the armature speed: a range of 25:1 is obtainable, and reversal is possible if the generator field excitation can be reversed. Each setting of the generator field provides for a 'shunt' operating characteristic in the main motor, with a torque proportional to the armature current. The method is economical in energy and is applicable to mine winding gear and machine-tool drives, but it is high in capital cost; consequently for smaller ratings the motor–generator is replaced by a thyristor bank.

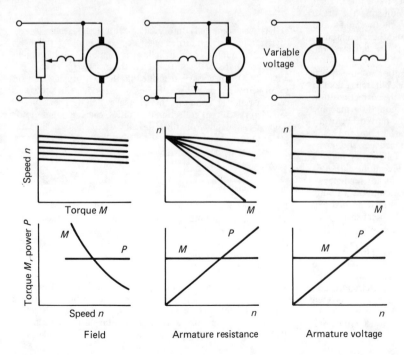

Figure 26.29 Shunt motor: speed control

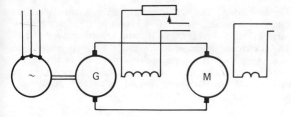

Figure 26.30 Ward–Leonard control

Series motor The three basic methods of speed control are shown in *Figure 26.31*.

Field control This is obtained by a diverter rheostat in parallel with the field circuit. Only on small machines is a continuous speed variation obtainable, and the diverter must generally be varied in one or two discrete steps. An alternative, commonly used with traction motors, is to tap each field winding so that part of the winding is cut out to reduce the field m.m.f. and raise the speed. Both methods can give only a speed rise. In some cases it is possible to arrange the field windings in two groups, which can then be connected in series or parallel, the latter giving 20–30% higher speed for a given current than the former.

Resistance control A variable resistor in series with the motor reduces its terminal voltage and lowers the speed. Although the method involves I^2R loss, it is commonly employed for cranes, hoists and similar plant, the resistance steps being used also for starting.

Series/parallel control If two series motors are connected mechanically to ensure the same speed for each (as is usual in d.c. traction systems) series/parallel voltage control can be obtained by connecting the motors electrically in parallel or in series, the former giving full voltage and the latter one-half voltage to each motor. Intermediate speeds can be obtained by resistance or field control. The full-parallel speed is, for a given motor current, approximately twice that in full-series.

A scheme known as *parallel/series* control is sometimes applied to battery vehicles, the battery being arranged in halves that can be paralleled for starting and low speed, and in series for full speed.

26.3.2.6 Speed control: thyristor-fed motors

A separately excited motor may be fed from a constant-voltage a.c. supply through thyristors, which rectify the current and also (by delaying the commutation angle α by controlled gate signals) furnish a variable-voltage supply to the armature. The field current is obtained from the a.c. supply through semiconductor diodes or thyristors. The main thyristor equipment thus replaces the motor–generator set of the Ward–Leonard speed control in ratings up to about 500 kW and, being static, is more economical and commercially viable even down to fractional-kilowatt sizes.

Connections The choice of thyristor circuit is a compromise between cost (i.e. the fewest thyristors, which with their firing circuits are more expensive than diodes) and operational difficulties arising from harmonic production or poor commutation, both accentuated by the use of a small number of thyristors. For economy, half-controlled circuits in which half the units are thyristors and half are diodes are common, but such circuits cannot regenerate.

The thyristors may be fed direct from an a.c. supply of r.m.s. voltage V_a as in *Figure 26.32*. The mean direct voltage available with zero commutation delay ($\alpha=0$) is $V_{a0}=0.9V_a$ for the 1-phase and $V_{d0}=0.43V_a$ for the 3-phase arrangement. With commutation delayed by angle α the mean direct voltages are

Fully controlled: $\quad V_d = V_{d0}\cos\alpha$

Half-controlled: $\quad V_d = V_{d0}\tfrac{1}{2}(1+\cos\alpha)$

The 1-phase half-controlled arrangement of *Figure 26.32(a)* is very widely used for ratings up to about 5 kW. The current tends to be discontinuous, causing bad commutation, and sufficient inductance should be included in the circuit to avoid this except at large delay angles; a separate inductor may be used or, in the smaller ratings, the motor winding may be designed to have a sufficiently high inductance.

The 3-phase half-controlled circuit (b) has been used for outputs up to about 200 kW but recent practice tends towards the use of the fully controlled circuit from about 25 kW up to 1000 kW. The fully controlled circuit (c) gives rise to a 300 Hz ripple on the direct voltage but the half-controlled circuit gives 150 Hz over most of its range, rising to 300 Hz when α approaches zero.

The transformer-fed networks in *Figure 26.32* give the designer a free hand in choosing the operating voltage of the d.c. motor. The 1-phase centre-tap connection (d) has the advantage over the 1-phase bridge of making regeneration possible.

By connecting in series two 6-pulse units, the supplies to which are obtained respectively from delta- and star-connected windings of a three-winding transformer (*Figure 26.32e*), a 30° phase-shift is obtained giving a 12-pulse operation and a 600 Hz ripple of small amplitude. Such an arrangement is generally desirable for outputs above 1 MW and up to 5 MW (the upper limit for a d.c. motor).

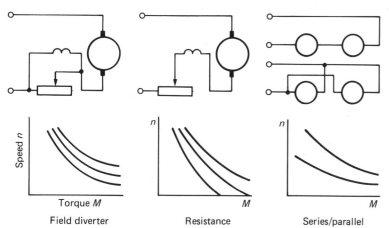

Figure 26.31 Series motor: speed control

Thyristor firing The signals applied to the thyristor gates are usually high-frequency pulses (2–8 kHz). For large units the pulse current may be 1–2 A in amplitude with a rise-time less than 1 μs. The electronic equipment to generate the signals comprises a power supply, timer and phase-shift units, pulse generator and amplifier, and output unit. The control of the gate pulses may be manual, or by signals from a closed-loop control system.

26.3.2.7 Braking

Rheostatic, regenerative and plug braking are applicable, but they cannot hold a motor at rest: for that a mechanical brake is necessary.

Shunt motor *Rheostatic braking* The field connection to the supply is maintained but the armature is disconnected and then reconnected on to a resistor (*Figure 26.33*). The machine generates, dissipating power in the resistor. The braking effect is

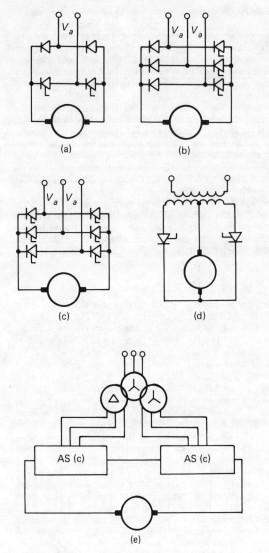

Figure 26.32 Thyristor control

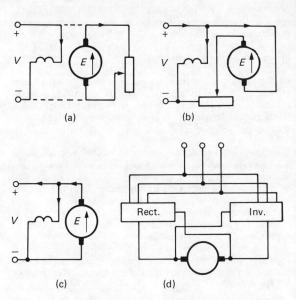

Figure 26.33 Electric braking of a shunt motor: (a) rheostatic; (b) plugging; (c) regeneration; (d) thyristor

The field supply for the motor is generally obtained from the a.c. supply through a 1-phase bridge or centre-tap connection using diodes. If field control is desired in addition to armature voltage control it can be achieved by a conventional field resistor or by using thyristors instead of diodes.

Harmonics In addition to the harmonics on the d.c. side which may interfere with commutation, harmonics also appear on the a.c. side and can cause interference with communication and control circuits and difficulties with other plant connected to the system, particularly overloading of shunt capacitors and possible resonances at 11th and 13th harmonics. For these reasons the rating of plant connected to a single point on the supply is limited by the Supply Authority, typically from 250 kW at 0.4 kV to 3 MW at 33 kV for 6-pulse units, and from 750 kW to 7 MW for 12-pulse units.

Motor construction As a result of harmonics on the d.c. side, a motor may have to be derated by 15–20%. To minimise commutation troubles, compoles and main poles and yokes may all be laminated.

controlled by varying the field current. For a total armature-circuit resistance R, the armature current is $I = E/R = k_1 n\phi/R$ and the braking torque is $M = k_2 EI/n = k_3 n\phi^2/R$. If the excitation is constant, then the braking torque is directly proportional to the speed n and decreases as the motor speed falls.

Plugging The armature connections are reversed, and the motor torque tends to retard the machine and then run it up in the opposite direction. The applied voltage and the armature e.m.f. are additive, so that a resistance of about twice starting value must be included to limit the current. For a total armature-circuit resistance R, the armature current is $I = (V+E)/R = (V+k_1 n\phi)/R$ and the braking torque is $M = (k_4\phi + k_3 n\phi^2)/R$. With constant excitation the braking torque is $k_5 + k_6 n$. Braking by plugging gives a greater torque and a more rapid stop, but current is drawn from the supply during the braking period, and this energy together with the stored kinetic energy has to be dissipated in resistance. A relay must be provided to open-circuit the motor at rest in order to prevent it from running up in reverse.

Regenerative braking If a load (such as a descending hoist) overruns the motor at a speed higher than normal, the counter-e.m.f. E exceeds the terminal voltage V and the machine generates. This is a very convenient method of 'holding' a load, but not at low speeds unless excessive field current is supplied.

With *thyristor-controlled* motors regeneration is possible if fully controlled connections are used. It is necessary to reverse the polarity of the motor terminals relative to those of the thyristor unit. This can be done by reversing either the armature or the field terminals at the moment of entering regeneration, the reversal being effected by conventional reversing contactors. Armature reversal requires about 150 ms but field reversal may require up to 2000 ms. Where a very rapid reversal is required it is necessary to employ separate thyristor units for motoring and regenerating, these being connected in 'anti-parallel'. Only switching of the gate pulses is then required, a process that can be achieved in about 10 ms.

Series motor The electric braking methods are shown in *Figure 26.34*.

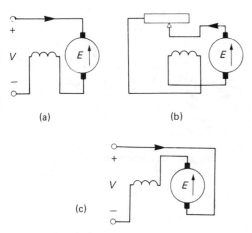

Figure 26.34 Electric braking of a series motor: (a) motoring, $E<V$; (b) rheostatic; (c) plugging, $E>V$

Rheostatic braking The motor acts as a series generator loaded on resistance. It is necessary to reverse the field connections at the instant of changing from motoring to braking. Further, the load resistance must be below a critical value if the machine is to self-excite. In practice the starting resistance, the value of which is well below the critical, is used for braking. The braking torque is approximately $M = knI^2$.

Plugging The conditions are generally similar to those in the shunt machine.

Regenerative braking This is not practicable with series motors. In traction, regeneration is sometimes effected by separately exciting the motors.

26.3.2.8 Design data

Some general and typical data are given here. Small motors up to 5 kW at 1000 rev/min normally have two poles, up to 50 kW four poles and up to 200 kW six poles. Larger motors have more, the number being such that the 'speed frequency' is 20–30 Hz and the pole-pitch between 45 and 55 cm.

The rated output power P is related to the main dimensions (diameter D, length l), the speed n and the specific magnetic and electric loadings B and A by

$$D^2 ln = P/\pi^2 \eta BA$$

where B is the mean gap flux density of value up to about 0.55 T and A varies between 5 and 35 kilo-ampere-conductors per metre of armature periphery.

The maximum safe peripheral speed is 30–40 m/s for the armature and 25–30 m/s for the commutator. The number of slots is about 6 per pole for fractional-kilowatt motors and 10–14 per pole for machines of 100 kW. The total current per slot is about 700 A for the latter rating. The number of slots must suit the winding required. A simple wave winding is used for nearly all industrial motors except in ratings of several hundred kilowatts for which lap (parallel-circuit) windings with equalising connectors may be used. The number of conductors is $Z = E/(p/a)n\phi$; the number of coils and commutator sectors is $2pV/e$, where the mean voltage e between sectors is usually between 12 and 15 V.

The field m.m.f. is 1.15–1.25 times the full-load armature m.m.f., and in designing the magnetic circuit the pole pleakage (about 15%) must be taken into account. The length of the airgap is governed by the requirement that the field distortion shall not be excessive.

26.3.2.9 Disc motors

The disc-armature machine has been developed for special applications such as small pumps, freezer-compressor drives, domestic sewing machines, computer spool drives and light battery vehicles. The basic geometry can be realised in the diagrams shown in *Figure 26.35*. A ring of permanent magnets of

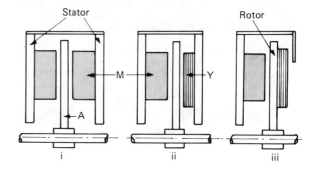

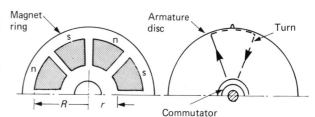

Figure 26.35 Elements of a disc motor

alternate polarity provides the axial field. In (i), two magnet rings M flank the armature A. In (ii), one of the magnet rings is replaced by a flux-conveying steel 'yoke' Y. In (iii), the yoke is carried by the armature, increasing its inertia but reducing the effective airgap length. The wave or lap winding (of which a single turn is shown) has an angular span equal to the angular pole-pitch. A multipolar structure, in which a front-face conductor is paired with another on the back of the armature disc, is employed to reduce the length of the 'overhang'. In small low-inertia motors the winding is punched from a copper or aluminium sheet and then placed on either side of an epoxy-resin-impregnated glass-fabric disc. The resin is then given a completion cure. Larger motors have wire-wound armatures with face or barrel commutators.

The magnets are usually of sector rather than circular shape to increase the working flux density towards the outer radius, where it is the most effective. Consider a machine in which the magnets, set between outer and inner radii R and r, produce in this annulus a uniform mean flux density B. Then for an angular speed n and a current loading A (in A/rad) the e.m.f. in a conductor and the torque developed are respectively

$$E_r = \pi(R^2 - r^2)Bn \quad \text{and} \quad M_e = \tfrac{1}{2}(R^2 - r^2)BA$$

showing the influence of the outer radius in increasing the output. However, R is limited by rotational stress, flexure of the disc and any operational constraints on the armature inertia.

Homopolar motors A homopolar d.c. motor is an embodiment of the Faraday disc. The voltage is inevitably low (less than 200 V) and the current high. Such machines as generators have been built to supply electromagnetic pumps and electrochemical processes requiring very large currents. Homopolar motors have also been built, notably a 2.5 MW 200 rev/min machine for a power-station pump where the motor has a superconducting field system developing a working flux density of nearly 4 T by means of an m.m.f. of 3×10^6 ampere-turns.

26.3.3 Three-phase induction motors

The great majority of industrial, commercial and agricultural electric motors above the fractional-kilowatt size are 3-phase induction machines, on account of their simple, cheap and robust construction and the almost universal availability of 3-phase supplies.

The induction motor has a 'shunt' speed–torque characteristic, the operating speed n falling slightly below the synchronous speed $n_s = f/p$, where f is the supply frequency and p is the number of pole-pairs for which the 3-phase stator winding is arranged. The drop in speed below n_s is the *slip*, given by $s = (n_s - n)/n_s$. The slip increases from nearly zero on no load to 0.03–0.05 on full load. Most industrial motors are 4-pole machines with a synchronous speed of 1500 rev/min, but 2-pole machines ($n_s = 3000$ rev/min) are sometimes of use, and lower speeds obtained with six, eight or more poles may be used in large ratings.

The *cage* motor has a rotor winding internally short-circuited on itself with no external access. The *slip-ring* motor has the polyphase rotor winding brought to three slip-rings so that connection can be made to it for starting or speed control; for normal operation, however, the rotor winding is short-circuited. In each type the stator carries a conventional 3-phase winding, fed from the main supply and generally delta-connected. Almost all induction motors are of the cage type, with slip-ring machines used normally only in ratings above about 100 kW.

26.3.3.1 Operating principle

Currents in the stator windings set up an airgap travelling-wave magnetic field of almost constant magnitude and moving at synchronous speed. The field cuts the rotor conductors at slip speed, inducing a corresponding e.m.f. and causing currents to flow in the short-circuited windings. The interaction of these currents with the travelling-wave field produces torque to turn the rotor in the direction of the field. The magnitude of the rotor currents depends on the slip and on the impedance (comprising resistance, and inductive reactance proportional to slip) of the rotor windings. With the rotor running at synchronous speed, the rotor slip is zero, the rotor inductive reactance vanishes and, as the gap flux does not cut the rotor conductors, the induced e.m.f. is zero; as a consequence there is no rotor current and no torque is developed. Since there must always be a small torque to overcome mechanical loss, the motor cannot quite achieve synchronous speed. As the motor is loaded, it slows so that the slip becomes a small finite quantity, rotor e.m.f. is developed and rotor current flows; the rotor circuits are mainly resistive but a small inductive reactance is introduced. The various interactions yield the torque–speed curve A in *Figure 26.36*. It has been taken into the region of reverse rotation (slip greater than unity), for

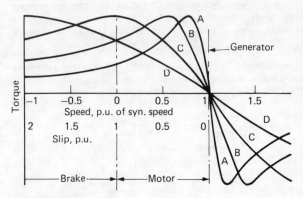

Figure 26.36 Three-phase induction motor: torque–speed characteristics

which the machine acts as a brake. The normal working range of the machine as a motor is the region for small positive slips: here the torque–speed relation is almost linear, corresponding to that of the d.c. shunt motor.

Reversal of the direction of rotation of the motor is obtained by interchanging two of the stator terminal connections, thus reversing the direction of the travelling-wave field.

26.3.3.2 Equivalent circuit

The performance is most readily predicted with the aid of an equivalent circuit (*Figure 26.37*) assembled from the various resistances and inductances (i.e. magnetising and leakage) of the machine, taken as independent of current, frequency and saturation conditions. The essential parameters are as follows, it being assumed for convenience that the rotor and stator windings are identical; all electric-circuit quantities are *per phase*:

V_1 stator applied voltage
E_1, E_2 stator e.m.f., rotor e.m.f. at standstill
r_1, x_1 stator resistance and leakage reactance
r_2, x_2 rotor resistance and leakage reactance at supply frequency (corresponding to standstill, $s = 1$)
r_m, x_m resistance representing core loss, magnetising reactance
I_1, I_2 stator current, rotor current
I_0 no-load current given by $\sqrt{(I_m^2 + I_c^2)}$, where $I_c = E_1/r_m$ and $I_m = E_1/x_m$

The basic equivalent circuit is shown in *Figure 26.37(a)*: it is similar to that of a transformer on short circuit except that the transformation ratio varies with slip (and therefore with the speed). With unity turns-ratio and a division of the rotor parameters by the slip s, equivalent crcuit (*b*) is obtained. For easier calculation the approximate circuit (*c*) can be used, with an error not exceeding 2% or 3% provided that the operating conditions are not abnormal. The rotor resistance r_2/s has been

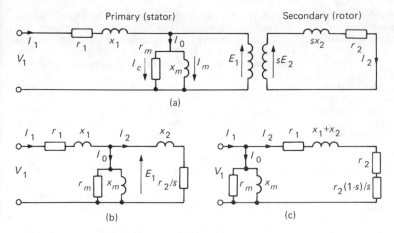

Figure 26.37 Three-phase induction motor: equivalent circuit per phase

split into r_2 for the rotor I^2R loss, and $r_2(1-s)/s$ in which the I^2R value represents the power conversion to the mechanical form.

Equations The following relations can be developed from the approximate equivalent circuit:

Rotor current: $I_2 = V_1/[r_1 + r_2/s)^2 + (x_1 + x_2)^2]^{1/2}$
Stator current: $I_1 = I_2 + I_0$
Power division: rotor input: rotor I^2R: gross output $= 1:s:(1-s)$
Gross torque: $M = V_1 I_2 (r_2/s)/2\pi n_s$

The peak torque, which is independent of the actual value of rotor resistance, is approximately

$M_m = V_1^2/(x_1 + x_2)4\pi n_s$ at $s = r_2/(x_1 + x_2)$

and its value is normally 2–2.5 times the full-load torque.

The losses in the machine comprise the core loss, stator and rotor I^2R loss, and mechanical loss in windage and friction. The no-load input current is 0.25–0.3 of the full-load current, at a lagging power factor in the range 0.15–0.2. At standstill on normal voltage the stator current is 4–6 times full-load current. Typical full-load values for conventional induction motors are:

10 kW motor:
          p.f., 0.87 (4-pole), 0.75 (12-pole); efficiency, 0.85
1000 kW motor:
          0.94   0.91   0.95

26.3.3.3 Construction

As the working flux is alternating, the stator and rotor cores must be laminated, using plates 1.0–1.5 mm thick.

Stator The core comprises annular stampings for small and segmental plates for large machines; it is mounted in a welded steel frame that does not form part of the magnetic circuit.

The voltage for which the stator is wound is normally between 380 and 440 V for motors up to 250 kW. Larger machines are wound for higher voltages, the minimum economic sizes being about 250 kW for 3.3 kV, 400 kW for 6.6 kV and 750 kW for 11 kV.

Rotor Slip-ring rotors may be wound to develop an e.m.f. when stationary of about 100 V for small and up to 1 kV for large machines, with insulation to correspond. The winding of a cage rotor comprises copper or aluminium bars located in slots (usually without insulation) and welded or brazed at each end to a continuous end-ring. The joints between bars and end-ring may prove to be points of weakness unless carefully made. Small cage motors generally have aluminium bars and rings cast into the rotor in one piece, with the end-rings shaped to form simple fan blades. To minimise the magnetising current the airgap is made as small as is mechanically practicable, e.g. from 0.25 mm for small motors up to about 3 mm for large motors.

Enclosure This follows standard practice. The *flameproof* construction is available for motors used in hazardous atmospheres classified as 'Divison 1 Areas' (mines, petroleum plants, etc.): it is a total enclosure with all joints flanged so that any flame generated by an internal explosion will be cooled by its passage through the joint and will not ignite external explosive gases. Cage motors, which have no slip-rings, are well suited to such situations, and they may be used without flame-proofing in 'Division 2 Areas' where flammable gas is not present unless there is some breakdown in the plant.

26.3.3.4 Starting

The factors of importance are (i) the starting torque and (ii) the starting current drawn from the supply. If the motor is to be started on full load, the starting torque must be 50–100% above full-load torque to overcome static friction and to ensure that the motor runs up in a reasonably short time to avoid overheating. Lower values are acceptable if the motor is always to be started on no load, and may be desirable in order to avoid too abrupt a start. The starting current should be as low as practicable to avoid overheating. Occasionally, Supply Authorities limit the starting current that may be drawn from the supply in order to avoid excessive voltage drops interfering with other consumers; it must also be remembered that, since torque is proportional to (voltage)^2, any voltage drop will significantly reduce the available starting torque. The torque and current may be expressed in terms of full-load values, but a more significant comparison between motors of different efficiencies and power factors is had by expressing the starting kVA in terms of the full-load output in kilowatts.

The curves A, B, C, D in *Figure 26.36*, drawn for different resistances, show that the rotor resistance has a major effect on the starting torque.

Slip-ring motor By adding external resistance to the rotor circuit any desired starting torque, up to the maximum-torque value, can be achieved; and by gradually cutting out the resistance a high torque can be maintained throughout the starting period. The added resistance also reduces the starting current so that up to 2–2.5 times full-load torque can be obtained with 1–1.5 times full-load current.

Cage motor For high efficiency, the rotor resistance must necessarily be low, and the starting torque on direct starting (when the motor is switched on to full voltage) is likely to be 0.75–1 times full-load torque with a stator input of 4–6 times full-load current. The table below gives typical values of the ratio (starting kVA)/(full-load kW) for a range of conventional motors with direct starting are:

Range, kW:	1–6	6–40	40–250	250–500	500–1500	1500–4000
kVA/kW:	10	9	8	7.7	7.4	7.2

These values are acceptable to Supply Authorities in most cases.

From the relations already given for the division of the power input to the rotor, the starting torque M_s for a starting current I_s is, in terms of the full-load values M_1 and I_1 and the full-load slip s_1, given by

$$M_s/M_1 = (I_s/I_1)^2 s_1$$

If direct starting is not admissible by reason of the initial current and/or the impulsive torque, then the voltage must be reduced for starting, bearing in mind that torque is proportional to (voltage)^2.

Series resistance A resistor in each line to the stator terminals can reduce the current to any fraction x of the direct-starting value, but the torque will be the fraction x^2. Although cheap and simple, this method is acceptable only for motors that start on no load.

Auto-transformer Usually no more than three tappings (50%, 70% and 80%) are provided. With a tapping giving the fraction x of normal voltage, both current and starting torque are reduced to x times the direct-start values. The cost of the transformer and contactors is high, especially if special connections are used to avoid disconnection of the supply when tap-changing.

Star–delta switch This starts the motor in star connection, and changes it to the normal delta connections as the speed approaches normal. Both current and torque at starting are one-third of their respective direct-start values. All six stator phase-ends must be available. The momentary disconnection can cause significant transient effects during the change-over.

Current displacement This relies on the change of leakage reactance with rotor frequency to give enhanced starting and run-up torque, but with some sacrifice in pull-out torque. The inner cage in (*a*) and the inner region of the deep bar in (*b*) of *Figure 26.38* have a high leakage inductance, and at low speed (i.e. at higher rotor frequency) current is forced mainly into the outer cage in (*a*) and the upper region of the deep bar in (*b*). Thus most of the rotor current flows in the higher-resistance outer cage or in the constricted region of the deep bar.

Solid-state For loads with a rising torque/speed load demand, a 'soft' start can be obtained by including anti-parallel thyristors in the stator circuits, using phase control. At starting, stator voltage control can be obtained by appropriate triggering, or stator impedance control by use of a gapped-core inductor.

26.3.3.5 Speed control

The speed of an induction motor is $n=(1-s)f/p$. The frequency f is normally fixed, the machine is built with p pole-pairs and, as the operating slip s lies generally between the limits 0.03 and 0.05, the motor is a substantially constant-speed machine with a working range as shown on curve A of *Figure 26.36*. Speed variation is often needed and, with some additional cost and complication, can be achieved by varying the slip, the number of poles or the frequency. For small motors a limited control can also be obtained by varying the applied voltage.

Slip control This, applicable only to slip-ring motors, requires connection into the rotor circuit of a device producing an adjustable volt drop or counter-e.m.f. The rotor induced e.m.f. sE_2 must overcome this to enable torque-producing current to flow, so that the slip must change in accordance with the magnitude of the injected volt drop or e.m.f.

Resistance control A variable volt drop is set up in the rotor circuit by variable resistors in each phase, giving a series of torque–speed characteristics (A, B, C, D in *Figure 26.36*). Only the low-slip parts can give normal operation, as the lower speeds may be unstable. The starting resistors, if continuously rated, can be used for speed control. The method is cheap and simple, but results in high I^2R ('slip energy') especially for low speeds. The efficiency is a little less than $1-s$, the speed varies widely with load, and low speeds are not obtainable at low loads. This form of control is used only where small or infrequent speed reductions are called for.

Slip-energy recovery Here the slip energy is not dissipated in resistance but is returned to the supply (constant-torque drive, as with resistance control) or added to the shaft output of the main motor (constant-power drive). Commutator machines have in the past been employed to deal with the slip energy. A modern method (*Figure 26.39*) is to rectify the slip-frequency currents in a diode bridge network; the unidirectional output current is smoothed and passed on to a 3-phase line-commutated inverter at a rate depending on the supply voltage, the rectified direct voltage and the thyristor firing angle. The inverted current has a fixed wave-form and a constant conduction angle of $2\pi/3$ rad. The onset of conduction with respect to the phase-voltage zero is controlled by the firing angle. As power flow through the rectifier is unidirectional, only subsynchronous speeds are feasible.

Pole changing Switching the stator winding to give two (sometimes three) different numbers of pole-pairs gives two (or three)

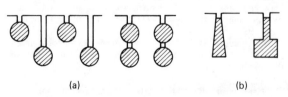

Figure 26.38 High-torque rotor cages: (a) double-cage; (b) deep-bar

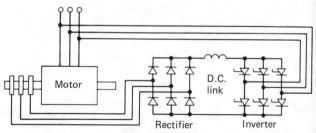

Figure 26.39 Thyristor slip-energy recovery

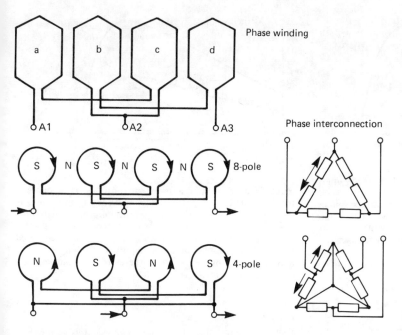

Figure 26.40 Pole change in the ratio 2/1

alternative running speeds. Cage rotors are normally employed, as slip-ring windings must be pole-changed to correspond always to the stator. Pole-changing motors with a 2:1 ratio have been used for many years, a typical arrangement being that in *Figure 26.40*. A more recent innovation is the pole-amplitude-modulated method.

Pole-amplitude modulation The m.m.f. (and flux) distribution around the airgap produced by one phase of a conventional machine can be expressed as $F(\theta) = A \sin p\theta$ where θ represents an instantaneous angular position. If $F(\theta)$ is modulated by making $A = C \sin k\theta$, the m.m.f. becomes

$$F(\theta) = \tfrac{1}{2} C [\cos(p-k) - \cos(p+k)]$$

which is an m.m.f. comprising two superimposed waves of pole-number $p-k$ and $p+k$. One wave can be eliminated by adjusting the chording and relative position of the phase windings so that the machine has p pole-pairs (unmodulated) and either $p-k$ or $p+k$ pole-pairs (modulated). The modulation is effected by reversing half of each phase winding and, in some cases, isolating certain coils. Two basic forms of connection are given in *Figure 26.41*. The coils that are isolated for the unmodulated connection are in the sections A'A, B'B and C'C. For designs that do not need coil isolations the phase terminals during modulation are A'B'C'.

In general, when the distribution of coil groups per pole and per phase is not uniform, some coils are isolated for modulation; with a more uniform distribution, however, a simple reversal of the second half of each phase winding is sufficient. The latter is more suitable for power outputs that are required to be similar at the two speeds.

This simple theory led to the design of successful industrial motors with close speed ratios from, for example, 8/10 or 12/14 poles. The recent development of a more general theory has made possible wide ratios from, for example, 10/2 and 16/4 poles, and also three-speed motors. There is virtually no limit to the size and speed ratio available, even in fractional-kilowatt ratings. The pole-amplitude-modulated (p.a.m.) motor is thus superseding the two-winding change-speed motors commonly used in the past, as the starting torques, power factors and efficiencies of p.a.m. machines are comparable with those of single-speed standard

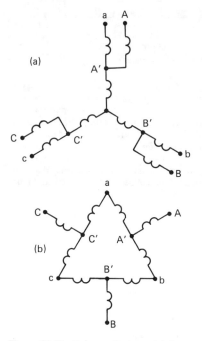

Figure 26.41 Pole-amplitude modulation. Unmodulated/modulated connections: (a) parallel-Y/series-Y; (b) parallel-Y/series-delta. Poles $2p$: supply to ABC with abc joined. Poles $2(p\pm1)$: supply to abc with ABC isolated

machines and have optimum performance on all ranges. The rating for a given frame-size is about 90% of that of a normal single-speed motor.

Frequency variation Development of the static thyristor inverter has made possible the provision from a 3-phase supply of an a.c. source of controllable voltage and of frequency infinitely

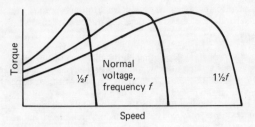

Figure 26.42 Torque–speed relations for variable frequency

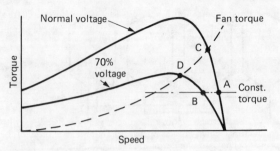

Figure 26.44 Torque–speed relations for variable voltage

variable from zero up to three or more times the supply frequency. Wide speed variation with such a source is possible with a cage motor. To maintain a constant motor flux, the voltage applied to the motor must be proportional to the frequency. The speed–torque characteristics (*Figure 26.42*) show that the peak torque is approximately the same at all operating frequencies.

D.C.-link converter A rectifier converts the a.c. of mains frequency to d.c., and a thyristor inverter converts this to a.c. of the desired frequency in the d.c.-link converter (*Figure 26.43*). The

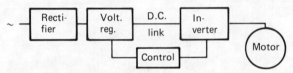

Figure 26.43 Basic d.c.-link converter system

direct voltage is varied by a thyristor chopper instead of by a controlled thyristor equipment in order to mitigate harmonics in the supply. The inverter produces a wave-form with about 20% of 5th and 14% of 7th harmonic, and these may cause small additional losses in the motor; however, the impairment may be reduced by series inductors in the motor circuit. It must be remembered that although the motor has the usual overload capacity, the converting equipment has no such reserve and must be rated for the peak power.

An upper frequency of about 150 Hz is usual, permitting the speed range of a 2-pole machine to be from zero to 9000 rev/min. For small machines and with thyristors having very fast switching characteristics, higher frequencies can be generated and speeds up to 100 000 rev/min achieved.

Cyclo-converter In this alternative equipment, sections of the normal mains-frequency wave are selected and used to build up an outgoing wave of lower frequency, usually not higher than about two-thirds of the mains frequency, in a single static unit. The equipment has an efficiency higher than that of the d.c.-link converter, but a considerably more complicated control network: basically 18 thyristors are required for a 3-phase motor. The cyclo-converter employs natural commutation and may be more reliable than the forced commutation in the d.c.-link converter under abrupt changes of load; it can also be made reversible for regenerative braking without the additional complication that a d.c.-link converter would involve for this duty.

Voltage variation As the peak torque of an induction motor is proportional to the square of the voltage, a limited speed control can be effected by voltage variation. Typical speed–torque curves for a small motor operating at 100% and 70% of rated voltage are given in *Figure 26.44*. For a constant-torque load the speed is reduced from A to B; for a torque proportional to the square of the speed, as for a fan, the greater reduction from C to D is obtained. Greater variation occurs if the motor has a high rotor resistance/reactance ratio, so that the method is ineffective for motors of rating more than a few kilowatts; it is, in fact, used chiefly with small mass-produced motors. The voltage variation may be achieved by means of series rheostats (cheap but wasteful) or by thyristors in the supply circuit (expensive, efficient and harmonic-producing).

26.3.3.6 Braking

Braking may be required to bring a motor and its load to rest rapidly in an emergency or as part of a production process, or to hold the motor at a set speed against gravity, as in a descending hoist. Mechanical brakes must be used if a motor is to be held at rest, but motional braking can be electrical and may not need much auxiliary control equipment.

Plugging If a motor, operating with a small slip s, has a pair of its supply leads interchanged, the direction of its travelling-wave airgap field reverses and the slip becomes $2-s$. A braking torque is developed, retarding the motor towards standstill, $s=1$. The braking torque normally varies from about one-half the value of the starting torque initially, up to the starting torque when the machine comes to rest, and the braking current throughout is approximately equal to the starting current. Unless the motor is disconnected when it stops, it will start up again in the opposite direction.

D.C. injection The 3-phase supply to the stator is disconnected and immediately replaced by a d.c. supply giving a stationary field in the airgap. The rotor conductors, moving in this field, develop e.m.f.s and currents that exert a braking torque, initially about equal to the starting torque but falling with the speed and vanishing as the motor is brought to rest by friction. The direct current is fed into the stator by two terminals, leaving the third isolated if the stator winding is star-connected.

Regeneration When a load overdrives the motor to a speed exceeding synchronous (i.e. with negative slip) as with a descending hoist load, the machine acts as an induction generator and sets up a braking torque (see *Figure 26.36*). Such braking cannot bring the motor to rest but it can limit the speed to a value a little above synchronous, the power from the load being partly returned to the supply. With two-speed pole-change motors a high braking torque can be obtained if, when running at the higher speed, the motor is switched to the larger pole-number.

26.3.4 Three-phase commutator motors

The recovery of slip energy can be achieved in a single machine by incorporating in the rotor a commutator winding. The *Schrage* (rotor-fed) and *doubly fed* (stator-fed) motors have some commercial importance. Both have a 'shunt' speed–torque characteristic and can operate both above and below synchronous speed. A 3-phase commutator motor with a 'series' characteristic is also available.

26.3.4.1 Schrage motor

The Schrage motor has its primary winding on the rotor, connected to the supply through slip-rings and brushes. The rotor also carries a low-voltage commutator winding with conductors located in the same slots as, and above, those of the primary. The secondary winding is on the stator. The primary and secondary windings are similar to those of a conventional induction motor. The brushgear comprises two movable rockers, each fitted with three brush spindles per pair of poles. The two rockers are geared together, each being fitted with a toothed segment. The two segments mesh with pinions fitted to a short shaft to which either a handwheel or a small pilot motor is connected. The gearing is so arranged that the movement of the handwheel or pilot motor causes the brush rockers to move in opposite directions.

The brushes attached to each rocker move over separate portions of the commutator surface to enable these brushes to be placed 'in line' or to be moved in either direction, so that more or fewer commutator sectors are included between a brush on one rocker and the corresponding brush on the other.

The bus-rings of each rocker are connected to the secondary (stator) winding as in *Figure 26.45*, which represents a 2-pole machine.

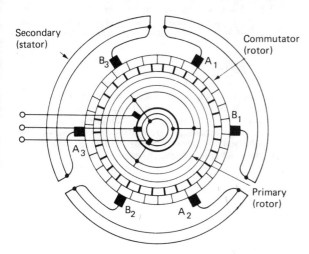

Figure 26.45 Schrage motor: stator and rotor circuits

As the primary winding is on the rotor, e.m.f.s of slip frequency are induced in the secondary (stator) winding. The e.m.f.s at the brushes are also of this frequency. The e.m.f. induced in each coil of the commutator winding is constant at all speeds, and therefore the e.m.f.s injected, via the brushes, into the secondary winding are proportional to the number of commutator sectors included between corresponding brushes on the two rockers, i.e. between the brushes connected to a particular phase of the secondary. Thus, when these brushes are in line, the secondary winding has no e.m.f. injected into it and the motor will run with its natural slip. When the brushes are moved so that the injected e.m.f. opposes the current, the speed is reduced (slip positive); for the opposite movement the speed is increased (slip negative).

Performance In a machine with a synchronous speed n_s and a brush-separation electrical angle θ, the no-load speed is

$$n = n_s(1 - k \sin \tfrac{1}{2}\theta)$$

where k is a constant depending on the numbers of turns in the secondary and commutator windings. A typical relation between n and θ is given in *Figure 26.46(a)* for a motor with a 4:1 speed range—about the practical limit. For a given brush position the speed is nearly constant up to 1.5–2 times full-load torque, as shown in *Figure 26.46(b)*. The speed drop with increasing load is greater than for a plain induction motor because of the brush resistance and the impedance of the tertiary commutator winding. At synchronous speed, when $\theta = 0$ and the secondary is

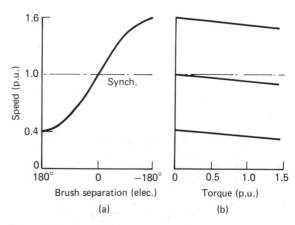

Figure 26.46 Schrage motor: characteristics

short-circuited through the brushes, the overall efficiency is similar to that of an induction motor. At other speeds the efficiency is perhaps 5% lower on account of the tertiary I^2R loss and the stator core loss. The power factor approaches unity at speeds above synchronous as the negative slip results in a capacitive effect. At subsynchronous speeds the power factor falls; however, it can be raised in non-reversing motors by arranging that the brush movement is asymmetric with respect to the 'in-line' position, the axis bisecting a corresponding pair of brushes being progressively displaced in a direction opposite to that of the rotor rotation.

Starting can be effected by direct switching with the brushes set in the lowest-speed position, the starting torque being about 1.5 times full-load value with 1.5–2 times full-load current. Commutation limits the output to about 20 kW/pole and the speed range to 4:1. The maximum output for a motor is thus about 200 kW. With limited speed range, rather higher ratings, e.g. 350 kW and 1.5:1, can be achieved. As the primary winding is supplied through slip-rings the supply voltage is restricted to about 600 V.

26.3.4.2 Doubly fed motor

The stator resembles that of a conventional induction motor; it is fed from the supply at any desired voltage up to 11 kV. The rotor carries a commutator winding and has six (occasionally three) brushes per pole-pair (see *Figure 26.47*). The rotor voltage is necessarily low (200–300 V) and the brushes are connected to the supply through a variable-ratio transformer. The gap flux travels at synchronous speed, and as a result of the commutating function the e.m.f.s at the brushes are of supply frequency at any speed. A variable e.m.f. can thus be obtained from the transformer

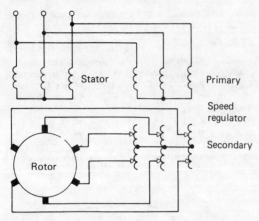

Figure 26.47 Stator-fed 3-phase commutator motor

and injected into the rotor circuits to give speed control from standstill to 1.5–2 times synchronous speed. In practice the variable-voltage transformer is an induction regulator to give smooth speed control. It must be a double regulator to avoid changing the phase angle of the injected voltage.

Performance At the zero-voltage position of the regulator the brushes are short-circuited through the regulator winding and the machine operates as an induction motor, though with a higher rotor effective impedance. Moving the regulator in one or other direction introduces an e.m.f. into the rotor circuit, to give speed–torque relations similar to those of the Schrage motor. The overall power factor tends to be low owing to the magnetisation of the regulator, but it can be improved by special means.

With the regulator in the lowest-speed position the motor can be direct-started to give about 1.5 times full-load torque with 1.5–2 times full-load current. The regulator must carry the slip power, so that speed variation down to zero requires it to be of a physical size comparable to that of the motor. Machines of some thousands of kilowatts can be economically built for speed ranges of ±15% or 20% of synchronous speed.

Unlike the Schrage motor the stator-fed commutator motor is not self-contained, but it can be made in larger ratings and for higher voltages. Again, the simpler brush arrangements makes the machine economic in ratings down to 2 or 3 kW.

26.3.4.3 Three-phase series motor

It is possible to connect the rotor brushes in series with the stator winding to give a machine with a 'series' speed–torque characteristic and with speed variation by moving the brush position. To limit the rotor voltage, however, a transformer is necessary between stator and rotor. If the transformer is that of the induction-regulator type the brushes can be fixed and the speed adjusted by means of the regulator. The series commutator motor is uncommon, but if a steeper characteristic than that furnished by the stator-fed machine is desirable or acceptable, as for fan drives, there is some economy because of the smaller losses and the simpler regulator.

26.3.5 Synchronous motors

Any synchronous generator will operate as a motor and run at precisely synchronous speed up to its pull-out torque of 2–2.5 times full-load torque. Other significant features of the motor are the controllability of its power factor up to unity or leading values, the necessity of a d.c. excitation circuit and the fact that the motor is not inherently self-starting. In ratings above 300–500 kW, however, the synchronous motor, although more expensive than the induction motor, has a higher efficiency and lower running cost and therefore often gives a more economic drive. Except for 3000 rev/min motors the salient-pole construction is generally adopted.

26.3.5.1 Starting

If the motor is always to be started on no load it can be run up to speed by a small pony motor, usually an induction motor, and then allowed to pull into synchronism when the excitation is switched on. If the motor has solid poles and pole-shoes it may be possible to start it by induction-motor action resulting from eddy currents induced in them when the supply is switched on.

Induction start Most synchronous motors are started by use of a cage winding embedded in the pole-faces to give an induction-motor torque when the stator is energised, by direct switching or through an auto-transformer. When the speed approaches synchronous the d.c. excitation is applied and the motor synchronises. During starting, high voltages may be induced in the field winding, and it is usual to short-circuit this winding during the start through a resistor which is disconnected after the machine has pulled into step. The current in the field winding adds significantly to the starting torque.

To ensure that the motor closely approaches synchronous speed at the end of an induction start, the resistance of the cage winding should be low; however, for good torque production at low speeds the cage resistance should be high, so some compromise is required.

With direct switching the starting torque is about one-half of full-load torque with 2–3 times full-load current. For machines rated above 200 kW an auto-transformer is needed for starting.

26.3.5.2 Excitation

The conventional method of excitation is by a shunt-connected d.c. exciter mounted on the motor shaft, control being effected by variation of the exciter field current; the exciter should, however, be disconnected from the motor field winding during starting on account of the alternating currents that would otherwise be induced in it and which could destroy residual magnetism and prevent the build-up of the excitation.

An a.c. mains-fed rectifier with its d.c. output fed to the rotor through the slip-rings could replace the d.c. exciter, but modern practice favours *brushless excitation*, in which an a.c. exciter feeds the rotor field winding through rectifiers, the whole arrangement being incorporated in the rotor. The a.c. exciter field is energised by means of a small permanent-magnet generator to ensure build-up of the excitation under all conditions.

Excitation control can be made automatic, but it is more usually pre-set to give unity or leading power factor at full load. The increased reactive leading power at lighter loads helps to raise the overall system power factor and to improve the transient stability of the motor when it is subject to disturbances.

26.3.5.3 Synchronous-induction motor

Where high starting torques (e.g. 2–2.5 times full-load) are required, the synchronous-induction motor is suitable. It resembles a slip-ring induction motor and is started on resistance; when it is up to nearly synchronous speed, d.c. excitation is switched on to the rotor through the slip-rings and the machine synchronises.

The airgap of the synchronous-induction motor is longer than that of the normal induction motor in order to achieve synchronous stability, and the rotor winding resistance is lower in order to ensure pulling into step. An exciter must also, of course, be provided. Another difficulty is the adaptation of the rotor for the dual purpose of starting and excitation; the direct current is normally fed into two of the slip-rings, the third being isolated so that the winding is not all usefully employed during running. Moreover, the relatively few turns and the large current for which the winding is usually designed necessitate an abnormal design for the exciter, i.e. a low-voltage, high-current machine. The starting performance is similar to that of a slip-ring induction motor, but the running performance is better in that the efficiency is 1–2% higher and the power factor may be made unity or leading. The pull-out torque in the synchronous mode is about 1.5 times full-load torque, but if the machine pulls out it can continue to run as an induction motor with a peak torque up to 2.5 times full-load value.

Where the compromise characteristics of the synchronous-induction motor are inadequate, large salient-pole synchronous motors may be built with a slip-ring pole-face winding, so that the starting and synchronous functions may each be optimised.

26.3.6 Reluctance motors

The reluctance motor is a cheap and reliable synchronous motor that requires no d.c. excitation. Commercial motors are available in ratings of 20 kW or more. The machine has a 3-phase stator winding similar to that of an induction motor, and a rotor without windings. For a given frame the output is about 60% of that of an induction machine, and the motor has a slightly lower efficiency. It has, however, advantages for drives such as the accurate positioning of nuclear-reactor rods, the operation of rotating stores in computers, and in synchronised multi-motor drives.

The essential feature of the rotor is a strong 'saliency' effect obtained in ways such as those illustrated in *Figure 26.48*. The obvious saliency in (a) is, in modern machines, replaced by designs based on studies of flux patterns to give the greatest difference in the reluctance offered respectively in the direct and quadrature axes—a condition of good saliency effect. The rotor iron may be solid but is more usually laminated.

Reluctance torque is maintained only at synchronous speed, so that some form of cage winding must be incorporated for starting. Direct switching is employed, giving starting currents up to 4–6 times full-load current. The effective rotor resistance has an important influence on the starting and pull-in torques.

The requirements for a satisfactory motor are good synchronous performance (efficiency, power factor and pull-out torque), good pull-in torque (especially for high-inertia loads) and stability. Some of these conflict: increasing the ratio of d- and q-axis reluctances gives higher output and pull-out torque, but lower pull-in torque and impaired stability. The design is influenced particularly by the load inertia. For a motor of 5 kW rating typical data are: reluctance ratio, 3–6; efficiency, 70–80%; power factor, 0.6–0.75 lagging; pull-out torque, 2–2.5 p.u.; pull-in torque, 0.9–1.2 p.u.

Motors with change-speed windings are possible, and motors can be built for variable frequency (20–200 Hz) but these are liable to instability at the lower end of the range.

26.3.7 Single-phase motors

Single-phase motors are rarely rated above 5 kW. Fractional-kilowatt motors, most of which are 1-phase, account for 80–90% of the total number of motors manufactured and for 20–30% of the total commercial value. A typical modern home may have 10 or more 1-phase motors in its domestic electrical equipment.

26.3.7.1 Series motor

As the direction of rotation and of torque in a d.c. series motor are independent of the polarity of the supply, such a motor can operate on a.c. provided that all ferromagnetic parts of the magnetic circuit are laminated to minimise core loss.

Universal motor In the fractional-kilowatt sizes the series motor has the advantage, since it is non-synchronous, of being able to run at speeds up to 10 000 rev/min. It is very well adapted to driving suction cleaners, drills, sewing machines and similar small-power rotary devices. Its facility of operating on d.c. and a.c. is not now important, but is the origin of the term 'universal'. The machine has a 'series' speed–torque characteristic, the no-load speed being limited by mechanical losses. The power factor is between 0.7 and 0.9 (mainly the result of armature inductance), but this is of no significance in small ratings. Typical characteristics for a motor for d.c. and 50 Hz supplies of the same nominal voltage are shown in *Figure 26.49*.

In all a.c. commutator motors the commutation conditions are more onerous than on d.c. because the coils undergoing commutation link the main alternating flux and have e.m.f.s induced of supply frequency. The e.m.f.s are offered a short-circuited path through the brushes and contribute to sparking at the commutator. As the e.m.f.s are proportional to the main flux, the frequency and the number of turns per armature coil, these must be limited; a further limit on the current in a short-circuited coil is provided by high-resistance carbon brushes.

Compensated motor Series a.c. commutator motors up to 700–800 kW rating are used in several European railway traction systems. For satisfactory commutation the frequency must be low, usually $16\tfrac{2}{3}$ Hz, and the voltage must also be low (400–500 V), this being provided by a transformer mounted on the locomotive. The inductance of the armature winding is necessarily rather high, so that a compensating winding must be

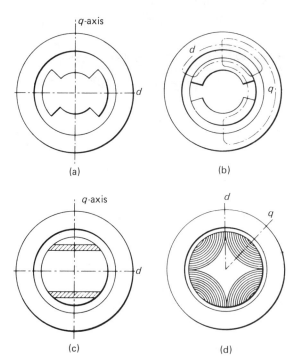

Figure 26.48 Reluctance motors: (a) salient 2-pole; (b) segmental 2-pole; (c) flux-barrier 2-pole; (d) axially laminated 4-pole

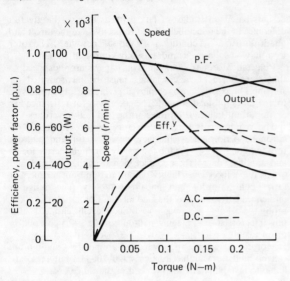

Figure 26.49 Characteristics of a 75 W universal motor

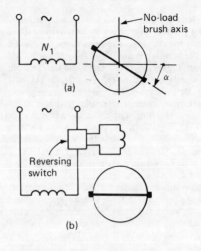

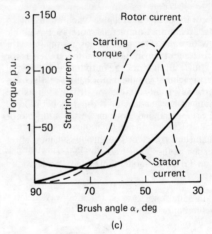

Figure 26.50 Repulsion motor: alternative forms and starting characteristics

fitted to neutralise armature reaction in order to ensure a reasonable power factor.

Motors of this type have been built, of limited output, for operation on modern 50 Hz traction systems but have now been superseded by rectifier- or thyristor-fed d.c. motors.

26.3.7.2 Repulsion motor

The repulsion motor is a form of series motor, with the rotor energised inductively instead of conductively. The commutator rotor winding is designed for a low working voltage. The brushes are joined by a short circuit and the brush axis is displaced from the axis of the 1-phase stator winding (*Figure 26.50*). With non-reversing motors (*Figure 26.50a*) a single stator winding suffices; however, for reversing motors the stator has an additional winding, connected in one or other sense in series with the first winding to secure the required angle between the rotor and effective stator axes for the two directions of rotation, as in *Figure 26.50(b)*.

A stator winding of N_1 turns as in (*a*) can be resolved into two component windings respectively coaxial with and in quadrature with the axis of the rotor winding, and having respectively turns $N_1 \sin \alpha$ and $N_1 \cos \alpha$. Windings (*b*) give the two axis windings directly, although here the turns can be designed for optimum effect. The coaxial winding induces e.m.f.s and currents in the rotor, and these currents lying in the field of the other stator winding develop torque; since both stator and rotor currents are related, the motor has a 'series' characteristic.

When the motor is running, the direct and quadrature axis fluxes have a phase displacement approaching 90°, so producing a travelling-wave field of elliptical form which becomes nearly a uniform synchronously rotating field at speeds near the synchronous. Near synchronous speed, therefore, the rotor core losses are small and the commutation condictions are good.

Small motors can readily be direct-switched for starting, with 2.5–3 times full-load current and 3–4 times full-load torque. The normal full-load operating speed is chosen near, or slightly below, synchronous speed in order to avoid excessive sparking at light load. Repulsion motors are used where a high starting torque is required and where a 3-phase supply is not available. For small lifts, hoists and compressors their rating rarely exceeds about 5 kW.

26.3.7.3 Induction motors

The 1-phase induction motor is occasionally built for outputs up to 5 kW, but is normally made in ratings between 0.1 and 0.5 kW for domestic refrigerators, fans and small machine tools where a substantially constant speed is called for. The behaviour of the motor may be studied by the rotating-field or the cross-field theory. The former is simpler and gives a clearer physical concept.

Rotating-field theory The pulsating m.m.f. of the stator winding is resolved into two 'rotating' m.m.f.s of constant and equal magnitude revolving in opposite directions. These m.m.f.s are assumed to set up corresponding gap fluxes which, with the rotor at rest, are of equal magnitude and each equal to one-half the peak pulsating flux. When the machine is running, the forward field component f, i.e. that moving in the same direction as the rotor, behaves as does the field of a polyphase machine and gives the component torque–speed curve marked 'forward' in *Figure 26.51*; the backward component b gives the other torque component, and the net torque is the algebraic sum. At zero speed the component torques cancel so that the motor has no inherent starting torque, but if it is given a start in either direction a small torque in the same direction results and the machine runs up to near synchronous speed provided that the load torque can be overcome.

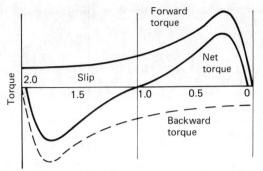

Figure 26.51 Torque components in a simple 1-phase induction motor

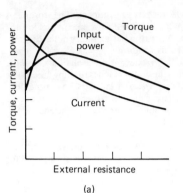

(a)

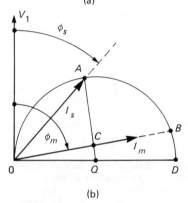

(b)

Figure 26.53 Single-phase induction motor: split-phase resistance-start

The component torques in *Figure 26.51* are, in fact, modified by the rotor current. Compared with the 3-phase induction motor, the 1-phase version has a torque falling to zero at a speed slightly below synchronous, and the slip tends to be greater. There is also a core loss in the rotor produced by the backward field, reducing the efficiency. Moreover, there is a double-frequency torque pulsation generated by the backward field that can give rise to noise. The efficiency lies between about 40% for a 60 W motor and about 70% for a 750 W motor, the corresponding power factors being 0.45 and 0.65, approximately.

The equivalent circuit of *Figure 26.52* is based on the rotating-field theory, using parameters generally similar to those for the 3-phase machine. The e.m.f.s E_f and E_b are generated respectively by the forward and backward field components and are proportional thereto. The respective component torques are proportional to $I_{2f}^2 r_2/2s$ and $I_{2b}^2 r_2/[2(2-s)]$, the net torque being their difference.

Starting To start a 1-phase induction motor, means are provided to develop initially some form of travelling-wave field. The arrangements commonly adopted give rise to the terms 'shaded-pole' and 'split-phase'.

Shaded-pole motor The stator has salient poles, with about one-third of each pole-shoe embraced by a shading coil. That flux which passes through the shading coil is delayed with respect to the flux in the main part of the pole, so that a crude shifting flux results. The starting torque is limited, the efficiency is low (as there is a loss in the shading coil), the power factor is 0.5–0.6 and the pull-out torque is only 1–1.5 times full-load torque. Applications include small fans of output not greatly exceeding 100 W.

Resistance split-phase motor The additional flux is provided by an auxiliary starting winding arranged spatially at 90° (electrical) to the main (running) winding. If the respective winding currents are I_m and I_s with a relative phase angle α, the torque is approximately proportional to $I_m I_s \sin \alpha$. At starting, the main-winding current lags the applied voltage by 70–80°. The starting winding, connected in parallel with the main winding, is designed with a high resistance or has a resistor in series so that I_s lags by 30–40°. The effect of this resistance on the starting characteristic is shown in *Figure 26.53(a)*. With given numbers of turns per winding and a given main-winding resistance, then for a specified supply voltage and frequency there is a particular value of starting-winding resistance for maximum starting torque. The relation can be obtained from the phasor diagram, *Figure 26.53(b)*, in which V_1 is the supply voltage and I_m at phase angle ϕ_m is the main-winding current. The locus of the starting-current phase I_s with change in resistance is the semicircle of diameter

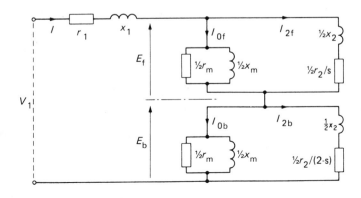

Figure 26.52 Simple 1-phase induction motor: equivalent circuit

OD (which corresponds to zero resistance). The torque is proportional to $I_m I_s \sin(\phi_m - \phi_s)$ and is a maximum for the greatest length of the line AC. From the geometry of the diagram it can be shown that for this condition $\phi_s = \frac{1}{2}\phi_m$.

Direct switching is usual. To reduce loss, the auxiliary winding is open-circuited as soon as the motor reaches running speed. The starting torque for small motors up to 250 W is 1.5–2 times full-load torque, and that for larger motors rather less, in each case with 4–6 times full-load current. The operating efficiency is 55–65% and the power factor 0.6–0.7.

Capacitor split-phase motor A greater phase difference $(\phi_m - \phi_s)$ can be obtained if a series capacitor is substituted for the series resistor of the auxiliary winding. Maximum torque occurs for a capacitance such that the auxiliary current leads the main current by $(\frac{1}{2}\pi + \alpha)/2$. The capacitor size is from 20–30 μF for a 100 W motor to 60–100 μF for a 750 W motor. For economic reasons the capacitor is as small as is consistent with producing adequate starting torque, and some manufacturers quote alternative sizes for various levels of starting torque.

If the capacitor is left in circuit continuously (*capacitor-run*) the power factor is improved and the motor runs with less noise. Ideally, however, the value of capacitance for running should be about one-third of that for the best starting. If a single capacitor is used for both starting and running, the starting torque is 0.5–1 times full-load value and the power factor in running is near unity.

Repulsion-induction motor Machines have been designed to combine the high starting torque capability of the repulsion motor with the constant-speed running characteristic of the induction motor.

Repulsion-start motor This motor has a stator winding like that of a repulsion motor and a lap commutator winding, with the addition of a device to short-circuit the commutator sectors together by centrifugal action when the speed reaches about 75% of normal. The device may also release the brushes immediately thereafter. Thus the commutator rotor winding becomes, in effect, a short-circuited 'induction'-type winding for running. Small motors direct-switched give 3–4 times full-load torque with about 3 times full-load current. A lower starting current is obtained by connecting a graded resistor in series with the stator winding.

Repulsion-induction motor The machine has a repulsion-type stator winding but the change from the repulsion-mode to the induction-mode operation is gradual as the machine runs up to speed. The rotor has two windings in slots resembling those of a double-cage induction motor. The outer slots carry a commutator winding with brushgear, the inner slots contain a low-resistance cage with cast aluminium bars and end-rings, and its deep setting endows it with a high inductance. During acceleration the reactance of the cage falls and its torque increases, tending to counterbalance the falling torque of the commutator winding. At speeds above synchronous the cage torque reverses, giving a braking action which holds the no-load speed to a value only slightly above synchronous speed. The commutation is better than that of a plain repulsion motor, and the motor is characterised by a good full-load power factor (e.g. 0.85–0.9 lagging). With direct switching the starting torque is 2.5–3 times and the current 3–3.5 times full-load value.

26.3.8 Motor ratings and dimensions

Motors of small and medium rating are built to the standards of IEC 72, which list a coherent range of main structural dimensions with centre heights between 56 and 1000 mm. BS

Table 26.5 Rotating machines: recommended ratings and shaft heights

Rating (kW)
0.06, 0.09, 0.12, 0.18, 0.25, 0.37, 0.55, 0.75, 1.1, 1.5, 2.2, 3.7, 5.5, 7.5, 11, 15, 18.5, 22, 30, 37, 45, 55, 75, 90, 110, 132, 150, 160, 185, 200, 220, 250, 280, 300, 315, 335, 355, 375, 400, 425, 450, 475, 500, 530, 560, 600, 630, 670, 710, 750, 800, 850, 900, 950, 1000

Shaft-centre height (mm)
56, 63, 71, 80, 90, 100, 112, 132, 160, 180, 200, 225, 250, 280, 315, 355, 400, 450, 500, 630, 710, 800, 900, 1000

3939 gives the standard ratings for the UK. *Table 26.5* lists data for rotating machines up to 1 MW. Approximate values of rated current for 3-phase, 1-phase and d.c. machines are set out in *Table 26.6* for machines up to 150 kW and for a range of supply voltages. Voltages for d.c. machines correspond to nominal values obtained from rectified a.c. supplies.

26.3.9 Testing

Tests on machines after manufacture or after erection on site are made in accordance with standard Specifications. They cover (i) insulation resistances, (ii) winding resistances, (iii) temperature rise and (iv) losses. Further tests on particular types of machine (e.g. commutation, starting) are required to meet customers' requirements or to obtain design data. Where a batch of similar machines is concerned, a 'type test' on one for detailed performance is usually acceptable.

26.3.9.1 Insulation

'Megger' testing of the insulation resistance between windings, and from windings to frame, must be performed before any live connections are made. The insulation resistance (in MΩ) should not be less than 1, or less than $V/(1+S)$ for a machine of rating S (in kVA), where V is the rated voltage. It may be necessary, if the insulation resistance is low, to 'dry out' the machine. The winding continuity having been checked, an insulation test should immediately precede a h.v. test and also be made prior to energising the machine for the first time.

26.3.9.2 Resistance

Measured resistances of the windings check the design figures and are required for calculating losses: winding temperatures must be noted. An ammeter/voltmeter method is usual; alternatively a bridge method may be used if there are no brush contacts in the circuit. For a commutator winding, the volt drop is taken between the commutator sectors under the brushes, the brush drop being taken separately.

26.3.9.3 Temperature rise

The permissible temperature rise of a machine on rated load depends on the insulation class. Temperatures are measured at a number of points (particularly at or near likely 'hot spots') and, where possible, during a heat run with the machine operating at rated load, until a steady temperature has been reached. The rated load may be a *maximum continuous rating*, or a *short-time rating*, or some special rating based on a duty cycle. The duration of the heat run varies from 2 h for small machines to 8 h or more for large ones. Temperature readings are taken (where feasible) every 15 or 30 min. The following three methods of temperature measurement are in use.

Table 26.6 Approximate rated motor currents (A)

Rating (kW)	Induction motors						D.C. motors			
	3-phase			1-phase						
	350 V	415 V	500 V	240 V	415 V		170 V	290 V	440 V	570 V
0.06	—	—	—	0.7	0.4		0.6	0.4	0.3	0.2
0.09	—	—	—	0.9	0.5		0.9	0.5	0.4	0.3
0.12	—	—	—	1.2	0.7		1.1	0.6	0.5	0.4
0.18	—	—	—	1.5	0.9		1.6	0.9	0.6	0.5
0.25	—	—	—	1.9	1.1		2.0	1.2	0.8	0.6
0.37	1.3	1.1	0.9	2.8	1.6		2.8	1.7	1.1	0.9
0.55	1.8	1.6	1.3	4.0	2.3		4.1	2.4	1.6	1.2
0.75	2.4	2.0	1.7	5.0	2.9		5.3	3.0	2.0	1.6
1.1	3.4	2.9	2.4	7.2	4.1		7.5	4.4	3.0	2.3
1.5	4.5	3.7	3.1	9.3	5.4		10	6.0	4.0	3.1
2.2	6.3	5.3	4.4	13	7.8		15	8.7	5.8	4.4
3.7	10	8.4	7.0	22	13		24	14	9.6	7.4
5.5	13	12	10	32	18		36	21	14	11
7.5	18	16	13	43	25		49	29	19	15
11	28	23	19	62	36		—	42	28	21
15	35	30	25	82	48		—	57	38	29
18.5	45	38	31	100	58		—	70	47	36
22	53	44	37	118	68		—	83	56	43
30	68	56	48	—	—		—	114	76	58
37	85	72	60	—	—		—	138	92	71
45	100	85	71	—	—		—	169	112	86
55	125	105	88	—	—		—	206	137	106
75	165	137	113	—	—		—	280	187	144
90	185	150	125	—	—		—	337	225	173
110	210	175	145	—	—		—	408	270	210
130	245	215	180	—	—		—	480	320	247
150	320	250	210	—	—		—	555	370	285

Thermometer Mercury or alcohol thermometers may be used, the latter being preferable especially on large machines, as eddy currents in the mercury caused by stray fluxes may cause high readings; also mercury from a broken thermometer can cause damage to certain alloys. Good contact must be made between the thermometer bulb and the surface concerned, and the bulb should be well covered by a non-heat-conducting material such as felt or putty.

The thermometers should be located on the stator core and windings and read at intervals throughout the run, and then affixed to the rotor core and windings and to the commutator, if any, as quickly as possible after shut-down. They should be placed where temperatures are likely to be highest; however, only the surface temperature is measured, and not the true hot-spot temperature.

Resistance The resistance–temperature coefficient of the winding material can give an average winding temperature. For copper, the temperature θ_2 corresponding to a measured resistance R_2 is related to the resistance R_1 at θ_1 by

$$\theta_2 = (R_2/R_1)(\theta_1 - 235) - 235$$

Embedded detector Thermocouples or resistance thermometers can be embedded in the core and windings during manufacture, a site between the coil-sides of a double-layer winding being common. At least six detectors, suitably distributed, should be installed. When well sited, detectors can give a closer estimate of hot-spot temperatures than other methods.

Temperature limits The three techniques above do not measure the same quantities, nor do they measure actual hot-spot temperatures. The values must therefore be in most cases significantly lower than the limits appropriate to the insulation class. Typical values for various locations as laid down in BS 2613 are given in *Table 26.7*.

26.3.9.4 Losses and efficiency

Efficiency may be determined by direct output/input ratio, by the total loss and either input or output, or by loss summation. As the efficiency of a large machine may have to be guaranteed within 0.01%, accurate determination is essential.

Output/input Electric power is readily measured, but mechanical power measurement requires some form of dynamometer, often of limited accuracy. Test rigs with instrumentation are used for small motors, but for large machines an adequate estimation of efficiency by this method may be impossible.

Table 26.7 Temperature limits (°C)

Part of machine	Method	Insulation class				
		A	E	B	F	H
A.C. windings:						
above 5000 kVA or core length over 1 m	T, D	60	70	80	100	125
below 5000 kVA	T	60	75	80	100	125
Commutator windings	R	50	65	70	85	105
	T	60	75	80	100	125
Field windings:						
low-resistance	R, T	60	75	80	100	125
other windings	R	50	65	70	85	105
	T	60	75	80	100	125
single-layer windings with bare or varnished metal	R, T	65	80	90	110	135
Permanently short-circuited windings	T	60	75	80	100	125
Iron core in contact with insulated windings	T	60	75	80	100	125
Commutators and slip-rings	T	60	70	80	90	100

Temperature measurement: T, thermometer; R, resistance; D, embedded detector

Back-to-back If two similar machines are available, one as a motor can drive the other as a generator. The net input is then the total loss, which can be accurately measured. This method can be applied for heat runs with comparative economy.

Loss summation A separate determination of each separable loss is made; the items are then summed to give the total loss, and thence the efficiency. The losses to be determined are:

Core   Stator I^2R Rotor I^2R Load (stray)

Brush-contact   Excitation   Friction and windage

Apart from the stray loss, each of these can be determined without loading the machine (although it must be run at normal speed). For uniformity the current losses are found from the measured resistances referred to a standard temperature of 75°C.

Stray losses These are, in practice, largely proportional to (current)^2 and, although insignificant in small machines up to a few kilowatts, they become very important in large machines. It is possible to estimate the stray loss from certain tests, as described later for particular machines; in other cases, however, they must be estimated from experience.

26.3.9.5 H.V. tests

The final test carried out before shipping a machine is the h.v. test in which a specified voltage at a frequency between 25 and 100 Hz is applied for 1 min between windings and earth and between windings. The specified voltage is usually (twice rated voltage + 1000) volts, although certain exceptions to this are given in BS 2613.

The purpose of the test is to ensure that the insulation has a sufficient factor of safety to guard against fortuitous voltage transients which may occur in practice. The test should, however, only be carried out once as repeated applications may damage the insulation. It may be desirable in some cases to repeat the test after the machine has been assembled on site, in which case a voltage of not more than 80% of the original test voltage should be applied.

26.3.9.6 D.C. motors

The following tests are related specifically to d.c. machines.

Armature volt-drop test This is carried out on the armature winding before assembly or if it has developed a fault. Current is fed into the armature by clamped connectors and the voltage between sectors is measured around the commutator. A placing of the connectors a *pole-pitch* apart is suitable for most industrial 2- or 4-pole wave-connected armatures: the direction of the volt drop changes at each pole-pitch. A *diametral* connection is suitable for small lap-connected rotors and for 6-pole wave-connected rotors: the direction of the drop reverses at the lead-in points. With the *bar-to-bar* connection the current is led into adjacent sectors and repeated all round the commutator, and all measured drops should be the same. An alternative test for wave windings is to lead the current into the commutator at sectors separated by the *commutator pitch* of the winding. This is repeated all round the commutator and all readings of volt drop should be the same: this checks each individual coil.

Neutral setting Adjustment of the brush rocker so that the brushes are in the correct neutral position can most conveniently be done by applying about half normal voltage to the field winding with a low-reading voltmeter connected between the positive and negative brushes. The position of the rocker should be adjusted until there is no kick on the voltmeter when making or breaking the field circuit. It is desirable to remove all brushes except the two being used, and these should be bevelled so that they do not cover more than one segment.

If the exact position of a coil on the armature can be observed, the armature can be moved until this coil is symmetrically placed with its centre line opposite the centre line of a pole; the brush rocker should then be moved until a set of brushes stands on the commutator segment connected to this coil and it will then be in the neutral position. Occasionally one of the armature coils is specially marked by the manufacturers to facilitate this method of setting the neutral.

No-load test The power input to the motor running on no load with normal field current and at normal speed gives the sum of the core, friction and windage losses, together with a small armature I^2R loss which can be calculated and deducted. A *shunt* motor is run at normal field current, the speed being adjusted to normal full-load value by varying the applied armature voltage. The core and mechanical loss so determined is the *fixed* or *constant* loss, so called because it changes only slightly with load.

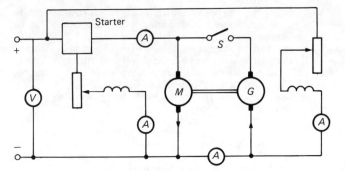

Figure 26.54 D.C. shunt motors: back-to-back test

A *series* motor is run with separate excitation to the field, and an armature voltage that will give (for a specified field current) the speed corresponding to load conditions at rated voltage. The voltage to be applied is equal to the counter-e.m.f. under load conditions, i.e. rated voltage less the volt drop in field and armature resistances. The mechanical loss alone can be estimated by running the motor on no load in normal connection but with a low applied voltage that is adjusted to give the speed required. Then the power input corrected for I^2R loss is substantially the mechanical loss alone.

Loss summation The core, friction and windage losses from the no-load test are added to calculated armature and field I^2R losses (corrected to 75 °C) and to the brush loss taken generally as that due to a total brush volt drop of 2 V. The stray loss is allowed for by a deduction of 1% from the efficiency calculated from the losses mentioned above. If the machine has a compensating winding to counteract the distorting effect of armature reaction, the deduction is 0.5%.

Back-to-back *Shunt motors* Figure 26.54 shows the connections for a back-to-back test for similar machines. The motor M drives the generator G, the excitation of which is adjusted until there is no voltage across switch S; this is then closed. Raising the excitation of G causes it to generate and supply most of the power required for driving M. The two machines operate with both field and armature currents slightly differing in magnitude, but except for small machines the differences are minor.

Series motors Back-to-back tests are more common for series motors, which are not easily loaded with safety. Many series motors are, however, operated in pairs as in traction. The four most common methods are set out in *Figure 26.55*. Method (*a*) is not strictly a back-to-back test as the generator output is dissipated in resistance. Putting the generator field in series with the motor ensures that the generator will always excite. In method (*b*) a variable-voltage booster B is included in the circuit of G to raise its voltage and enable power to be returned to the supply. In both (*a*) and (*b*) it is desirable to boost the supply so that the voltage applied to M can be held at a correct value. Method (*c*) requires an auxiliary drive motor to supply the core and mechanical losses, and a low-voltage booster to supply the I^2R loss. Method (*d*) requires a separate source of excitation, although for small machines the field winding can be connected in series with G and controlled by a diverter.

Commutation The compole setting can be checked by measuring the *volt drop* between the brush and the commutator at three points along the brush width with the machine running on load (*Figure 26.56a*). The drop should be approximately the same at all points, as in *Figure 26.56(b)*; if it is greater at the trailing edge, commutation is being delayed and the compoles are too weak. If it is higher at the leading edge the opposite is the case.

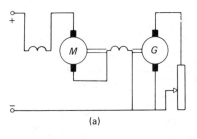

(a)

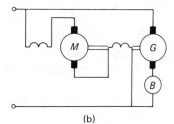

(b)

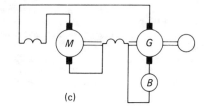

(c)

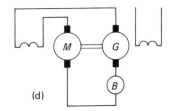

(d)

Figure 26.55 D.C. series motors; back-to-back tests

The *black-band* test is an alternative method. The compole current is varied independently of the armature current, and at each constant armature current the compole excitation is raised and lowered until sparking occurs. In the zone between these limits (*Figure 26.56c*), commutation is 'black', i.e. spark-free. The black band should be symmetrical about the axis of armature current.

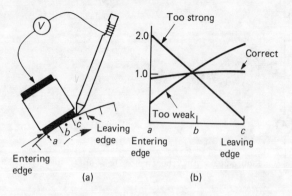

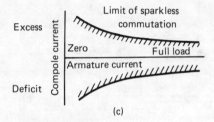

Figure 26.56 Commutation tests

26.3.9.7 Induction motors

The following are related specifically to 3-phase induction motors. Most of the necessary data are obtained from the no-load and short-circuit tests.

No-load The motor is run on no load at rated voltage and frequency, and the current and power input are measured. The input power supplies core and mechanical losses plus the stator I^2R loss; the rotor I^2R loss on no load can be neglected. If required, the motor may be operated at voltages above and below normal (*Figure 26.57*) and the power input and current plotted. Extrapolation of the power curve to zero voltage gives the friction and windage loss.

Short-circuit (*locked-rotor*) The short-circuit current and power are measured by holding the rotor satationary and applying a low voltage to the stator, it being usual to adjust the voltage to obtain full-load current. In the case of a cage motor the starting

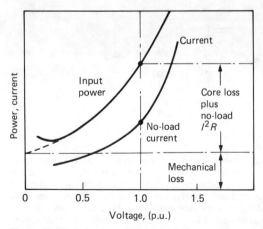

Figure 26.57 Induction motor: no-load characteristics

torque may also be measured. If the motor is connected for star–delta starting, the actual starting current and torque may be measured direct. The short-circuit current at full voltage is calculated by assuming the current to be proportional to voltage, although on some machines the short-circuit current is actually higher because the leakage reactance is reduced by saturation. This is more noticeable on 2- and 4-pole motors designed for a high flux density. The starting torque is approximately proportional to (phase voltage)^2.

The short-circuit current is 4–8 times full-load current, depending on the speed and type of motor. A multipolar motor works on a lower flux density to allow a reasonably good power factor. To this end some overload capacity may be sacrificed and the short-circuit current is therefore low.

Parameters The parameters of the equivalent circuit (*Figure 26.37*) can be evaluated from test results. The no-load test gives r_m and x_m if the stator leakage impedance drop is neglected (which is usually justifiable). Neglecting r_m and x_m in the short-circuit test gives the total motor effective resistance and leakage reactance, r_1+r_2 and x_1+x_2. The stator resistance r_1 can be measured directly and r_2 found; however, it is not possible to separate the two leakage reactances and it is usual to assume that they are equal.

Stray loss Stray loss is included in the short-circuit power. If, as in a slip-ring machine, both r_1 and r_2 (actual and referred) are measurable, the 'true' I^2R loss can be calculated and the stray loss found by subtraction. Where efficiency is calculated by loss summation, a deduction from the calculated efficiency is made (e.g. 0.0625 p.u. at rated load).

Back-to-back For large machines this test is essential. There must be provision for accommodating the speed difference resulting from the positive and negative slips of the motor and generator machines. Methods available are: (i) a close-ratio gearbox where at least one of the machines has a slip-ring rotor for slip control; (ii) a fluid coupling where both are cage machines; (iii) an adaptation of the automobile differential.

In (iii), the test machines are coupled to the 'road-wheel' shafts. The torque shaft, driven slowly by a geared auxiliary motor, depresses the speed of the motoring machine and raises that of the generator. The main supply of rated voltage and frequency to the two stators provides the mechanical, core and stator I^2R loss and magnetising currents. The torque shaft controls the drive-power exchange between the two test machines.

26.3.9.8 Synchronous motors

Relevant procedures are given below. Again use is made of the open- and short-circuit tests, for which the machine is driven at rated speed by an auxiliary motor, preferably calibrated.

Open-circuit The stator is on open circuit and the field current is varied. The stator e.m.f. gives the *magnetisation characteristic* or open-circuit characteristic (o.c.c.) (*Figure 26.58*). The power input to the test machine comprises core and mechanical losses, which can be separated.

Short-circuit The stator windings are short-circuited. The field current is increased to give stator current up to slightly above full-load value. The power input to the test machine comprises the stator I^2R, mechanical and stray losses, the core loss usually being negligible. An approximation to the stray loss is obtained by subtracting from the input power to the calculated I^2R loss and the mechanical loss (from the o.c.c. test).

Excitation The excitation for given load conditions is derived as for a generator (Chapter 11), with reversed current and load

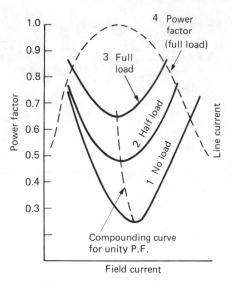

Figure 26.59 Synchronous motor: V-curves

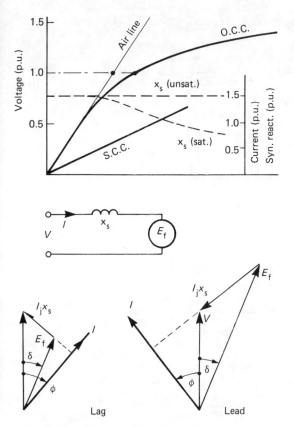

Figure 26.58 Synchronous motor: Open- and short-circuit characteristics, equivalent circuit and phasor diagrams

angle. The appropriate phasor diagrams are given in *Figure 26.58*. Typical V-curves, relating the stator and field currents for specified output powers, are plotted in *Figure 26.59*.

Loss summation The friction and windage and the core losses can be obtained from the open-circuit tests, and the stray loss from the short-circuit test, as already described. For any given load the field current can be determined and the field power obtained therefrom. To the summation must be added the loss in the exciter or excitation circuit, and the efficiency can then be calculated.

Back-to-back If two similar machines are mechanically coupled so that their e.m.f.s are in phase, and they are driven mechanically at normal speed then, if they are connected in parallel electrically, a reactive current can be made to circulate between them by suitable adjustment of the field currents or by a booster transformer fed from an external source. The total electrical and mechanical input represents the total loss, but as the machines are not operating at their rated power factors the results are not valid for calculating full-load efficiencies. The test is, however, convenient for carrying out heat runs.

With the machines coupled so that their e.m.f.s are phase-displaced by an angle equal to twice the full-load load-angle, then full-load active and reactive powers can be circulated under conditions closely simulating normal load operation.

Zero-power-factor A pseudo full-load test can be carried out by loading the machine on inductors so that it operates at normal voltage and current but at a power factor near zero. The airgap flux will be 4–5% above its normal full-load value and the field current will be up by 20–30%. If the machine is used for a heat run, temperatures will be higher than those that would obtain at rated load; however, appropriate adjustments can be made to simulate rated conditions more closely. Results from the zero-power-factor test can be used as described in BS 4296 to predetermine the field current on load.

26.3.10 Linear motors

Linear machines have translational instead of rotary motion. They can be applied to the drive of a conveyor belt, of a traversing crane on a limited track, of liquid metal in the hext-exchanger of a nuclear reactor plant, or of trains on a high-speed railway system. Short-stroke linear machines are suitable for powerful thrusting action.

26.3.10.1 Forms

A linear machine can be regarded initially as resembling a normal rotating machine that has been cut and opened out flat. Of the two elements derived respectively from the stator and rotor, either may move. The member connected to the supply is called the *primary*, the other the *secondary*. In use, either member is fixed as the *stator*, while the other becomes the movable *runner*.

Two forms of linear machine (*Figure 26.60*) are the *flat* machine (geometrically an 'unrolled' rotary structure) and the *tubular* machine, which is equivalent to a flat machine 'rerolled' around the longitudinal axis. For electromagnetic force to be developed, it is necessary to ensure that interaction between the working flux and the working currents should be achieved: the directions of the two components in and around the airgap must be at right angles. In the flat machine (*Figure 26.60a*) the secondary interaction currents are arranged to flow across the element from front to back, with suitable end-connectors to complete the secondary circuits. In the tubular machine the flux enters the pencil-shaped secondary in a radial direction, so that the secondary interaction currents must flow circumferentially.

In both of the shapes in *Figure 26.60* there are certain practical difficulties: (i) how to deal with a linear movement which, if continued, must eventually cause a runner of finite length to part company with its stator; (ii) how to support a very strong magnetic pull tending to cause runner and stator to adhere;

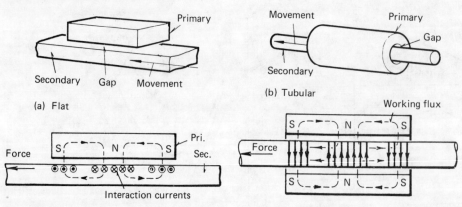

Figure 26.60 Linear motors: (a) flat; (b) tubular

(iii) how to supply electrical energy to a linear-moving runner. In the flat machine, and in the tubular machine if it is not precisely centred, the otherwise unbalanced attraction force must be sustained by some suitable mechanical arrangement. The linear machine is inherently a device lacking some of the symmetry and balance of the normal rotary form.

26.3.10.2 D.C. and a.c. machines

Any arrangement possible in a rotary machine can be realised in the flat form. The outlines in *Figure 26.60* show that the most convenient type is likely to be that corresponding to the cage induction motor, for then the stator can often be made the primary because of the convenience of its supply, while the moving secondary member will correspond to the cage rotor and will require a very simple winding with no supply connections. However, this arrangement is by no means the rule, and much depends on the particular conditions of a given application.

The d.c. and induction forms are the most common. Because of the essential secondary current supply, the d.c. linear motor is usually found as an electromagnetic pump for liquid metal. There is more freedom of shape in the induction machine, as indicated in *Figure 26.61*. The short-primary arrangement (*a*) suits cases in which the total distance to be travelled is great, for it would be uneconomic to wind a long 3-phase primary, with only that part in the neighbourhood of the secondary being effective at any one time. The short-secondary form (*b*) is useful if the total excursion is limited and the moving secondary must be comparatively light. In both (*a*) and (*b*) the secondary is shown as a flattened 'cage'. A conducting sheet or plate, (*c*), is often as effective, and it obviates magnetic attraction. The now indefinite 'gap', which increases the non-useful leakage flux, can be shortened by the use of a double primary, (*d*).

Long-established methods for the design of orthodox rotary machines are not applicable to the radically changed geometry of linear flat or tubular machines. Lacking cylindrical symmetry, the magnetic flux is heavily distorted near the ends of the short element, and as the primary and secondary move relatively to one another, 'dead' regions of the secondary abruptly enter a magnetised gap, and 'live' regions abruptly leave it at the other side. As a result there are important transient effects, the elimination or mitigation of which will impose quite unusual restrictions on the design, affecting not only the length and number of poles of the primary, but also its optimum working frequency.

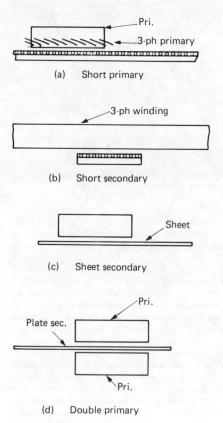

Figure 26.61 Polyphase linear induction motors

26.3.10.3 Duty

There are, in general, three operating duties, which influence the design and construction.

Power (*drive*)   This is concerned with the transport of loads in conveyors, haulers, electromagnetic pumps, travelling cranes and railway traction with acceptable overall *power* efficiency.

Energy (*accelerator*)   Here the duty is to accelerate a mass from rest within a specified time and distance, as in rope-break and car-crash test rigs and the launching of aircraft. The criterion is the *energy* efficiency, i.e. the ratio of the energy imparted to the load and the total primary electrical energy input, a figure that cannot exceed 0.5.

Force (actuator) This develops a thrust at rest, or over a short stroke, as in the operation of stop-valves, impact metal-forming, door closers and small thrustors. The criterion is the *force* per unit electrical power.

26.3.10.4 General principles

The following analysis, applying to power (drive) types of linear motor, outlines in simplified terms the basic principles.

Speed The speed of a d.c. linear machine is associated with the secondary applied voltage and the flux per pole in a way comparable to the relationship between these variables in a rotary machine.

The speed of an induction-type linear machine is associated with the synchronous speed u_1, which is given by $u_1 = f\lambda$ for a supply frequency f, where λ is the wavelength (i.e. the length of a double pole-pitch). The actual speed u differs from u_1 because of the slip. If $\lambda = 1$ m and the supply frequency is 50 Hz, then $u_1 = 50$ m/s $= 180$ km/h. For lower translational speeds on a 50 Hz supply it is necessary to shorten the wavelength. However, if λ is less than about 0.2 m, corresponding to $u_1 = 10$ m/s $= 36$ km/h, the performance is impaired because the pole-pitch is short compared with the gap length. The effect is most significant in machines with an 'open' magnetic circuit (*Figure 26.61c*). Much better low-speed performance can be achieved if a low-frequency supply is available. In some cases in which starting from rest at high translational force is sought, a primary with a graded pole-pitch may be of advantage; alternatively the frequency can be raised during the starting period if the method can be economically justified.

Power If in a secondary member the surface current density is J and the gap flux density is B, the force developed per unit area of gap surface is BJ, and the motional e.m.f. developed at a translational speed u is Bu per unit width. The working voltage to be supplied to maintain motion is $v = Bu + J\rho$ per unit width, where ρ is the resistivity of the secondary surface current conducting path. Where the secondary is a conducting sheet (solid or liquid) the secondary current paths are ill-defined, so that estimation of performance is not straightforward.

In an induction linear machine with a 2-phase or 3-phase winding fed at frequency f and providing a travelling wave of gap flux at synchronous speed $u_1 = f\lambda$, the runner moves in the same direction at a lower speed u, i.e. with a slip given by $s = (u_1 - u)/u_1$. At a point in the gap where the instantaneous gap flux density is B_x, the e.m.f. induced in the secondary is $e_x = B_x s u_1$ per unit width, producing a secondary current of linear density $J_x = e_x/\rho$ (ignoring inductive effects in the current path). The interaction force per unit width is therefore $B_x J_x = B_x^2 s u_1/\rho$. Summed over a wavelength (one double pole-pitch of length λ) the force is $\frac{1}{2}\lambda s u_1 B_m^2/\rho$, where B_m is the peak density of the travelling wave of magnetic field. As the runner moves at speed $u_1(1-s)$, the mechanical power produced per wavelength and per unit width is

$$P_m = \tfrac{1}{2}s(1-s)u_1^2 B_m^2/\rho$$

This is a maximum for $s = 0.5$. The simple analysis applies only to wavelengths remote from the ends of the shorter member of the machine. In general, the precise current circuit of the secondary is somewhat indefinite, there are effects of leakage inductance, and near the ends of the primary block the 'dead' regions of the secondary abruptly enter the magnetised gap while 'live' regions as abruptly leave it. As a result there are transient 'entry and exit' effects, the mitigation of which imposes restrictions on the design that are absent in rotary types.

26.3.10.5 Applications

Some typical applications are described.

Electromagnetic pump The electromagnetic pump utilises the good electrical conductivity of the liquid metal being pumped to establish electromagnetic forces directly within the liquid itself. The liquid is contained in a simple pipe, which can be welded to the rest of the circuit, so that valves, seals and glands are avoided; this is desirable since the low-melting-point metals are highly reactive chemically. Because of the absence of glands and moving parts, the pump reliability should be high and maintenance costs low. In addition, the pump itself is often smaller and the amount of liquid contained sometimes less, which is important when expensive liquids are being handled. Moreover, in favourable cases, particularly with high-conductivity liquid metals, the pump is likely to be cheaper.

The most important applications for circulating liquid metal are in nuclear energy projects where the metal is used as a coolant: sodium, sodium–potassium and lithium are the metals chiefly concerned. Most other industrial applications are restricted to low-melting-point metals, as in the die-casting industry or in chlorine plants for pumping mercury, but a form of liquid-metal pump has been found useful for stirring molten steel in arc-furnaces.

The liquids mentioned divide into two distinct classes: (1) sodium, sodium–potassium and lithium; (2) bismuth, mercury, lead and lead–bismuth, which have much poorer pumping properties. For example, the resistivity of bismuth is nearly eight times that of sodium, its viscosity is five times higher and its density is 11 times higher. A high resistivity lowers the efficiency, while high viscosity and density make the pump appreciably larger, for, to reduce hydraulic losses, the liquid velocity must be low, and this increases the size of the pump duct.

The wide difference in liquid pumping properties influences the type of pump and the efficiency obtainable. Liquids like bismuth usually necessitate conduction pumps, in which current is supplied directly to the liquid through the tube walls. In contrast, sodium can be pumped by conduction or induction.

D.C. conduction pump The general arrangement is shown in *Figure 26.62*. The magnetic field can be produced by a permanent-magnet system or by an electromagnet. A series-excited electromagnet is preferred for large pumps as it is usually smaller and

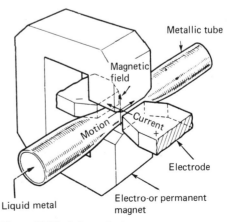

Figure 26.62 D.C. conduction pump

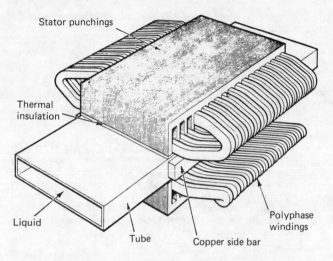

Figure 26.63 A.C. induction pump

cheaper and involves only the provision of a few turns around the pole, as close as possible to the gap, to carry the main operating current. The supply requirements vary from 5 kA at 0.5 V for a pump of 0.05–0.1 m^3/min capacity at a pressure differential of 350 kN/m^2, up to 250 kA at 3–4 V for a nuclear reactor pump delivering 25 m^3/min at 500 kN/m^2. These inconvenient requirements form the major disadvantage of the d.c. pump; the compensating advantages are the minimal insulation levels (desirable if the liquid is hot or radioactive), the accommodating performance which can deal with a wide range of flow-rates and pressures with good efficiency, and the adaptability to a variety of metals, including bismuth.

In d.c. pump design it is important to assess the field distortion due to the current in the liquid metal and to apply compensation if necessary, by returning the current to the electrode between the pole and the pump duct, usually as a pole face winding. It is also important for good efficiency to restrict the useless components of electrode current, which flows in the tube wall and in the liquid outside the pole region. The tube-wall current can usually be limited only by choosing a thin-walled tube of high intrinsic resistivity. The useless current can be substantially reduced by fringing the field to match the natural fringing of the current density between the electrodes. It is also important in design to limit magnetic leakage by appropriately positioning the magnetising turns and by shaping the iron.

The operation can be approximately described in terms of the flow of a quantity q of liquid at velocity u in a tube of width b in the current direction and effective length c in the direction of the liquid movement, the liquid having a resistivity ρ:

Electrodes: voltage $V_0 = (Bu + J\rho)b$;
 current $I_0 = I + I_t + I_s$

Pump: gross pressure $p = (B_m I)c$; gross power $P_0 = pq$

Here B is the flux density, of mean value B_m, and J is the current density. The electrode current I_0 includes the useful current I, and non-useful shunt currents I_t in the tube walls and I_s in the liquid outside the magnetised region. The gross pressure p includes the hydraulic pressure, drop in skin friction, etc.

A.C. induction pump The two basic forms are the flat and the annular. The flat type (*Figure 26.63*) has a straight duct 10–25 mm wide and up to 1 m across. Copper bars are attached to each side of the channel to form the equivalent of the 'end-rings' of a normal cage winding. A flat polyphase winding is placed on each side of the duct. A heat-insulating blanket is usually necessary between duct and winding, which is force-cooled. The annular form is similar except that the cross-section of the duct is an annulus and embraces a radially laminated magnetic core. The polyphase winding comprises a number of pancake coils surrounding the duct and set in comb-like stacks of radial punchings. If the duct is made re-entrant, with inlet and outlet at the same end, it is possible to remove the primary windings for repair and maintenance, should this become necessary, without disturbing the pipework.

Cranes Two linear induction motors can be applied to the traversing of an overhead crane, replacing the conventional rotary cage motor, gearing, drive shaft and control equipment. Maintenance is simplified and, as the linear motors are unaffected by atmospheric conditions, the likelihood of breakdown is reduced. The arrangement for one of the motors is shown schematically in *Figure 26.64*. Each motor comprises a horizontal pack of laminations carrying a 3-phase winding. Motors are located at each end of the crane gantry and directly below the tracks from which the crane is suspended. The track functions as the secondary, the arrangement being basically that of *Figure 26.61(c)*.

Traction Several schemes have been proposed for railway traction. One uses a short-primary form of motor on the locomotive and a fixed 'plate' secondary secured to the track. The moving primary demands either that power be supplied to the moving train or that a prime mover and 3-phase generator be

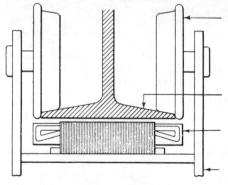

Figure 26.64 Linear traversing crane motor

carried on board. Some recent developments exploit magnetic levitation and the elimination of running wheels.

Stirring The stirring of molten metal in furnaces can be done by external application of short linear induction motors. The process has considerable metallurgical advantage in improving the casting properties of aluminium and copper, and in accelerating the deoxidation processes in steel. The primary of the motor does not come into contact with the melt, is readily controlled and can be arranged for vertical, rotary or horizontal stirring.

Short-stroke devices The linear induction motor offers a compact and readily controllable means for the automation of punching, stamping and impact extrusion. With the wavelength graded on the fixed primaries to raise the 'synchronous' speed as the short secondary runner approaches the workpiece, end speeds of 30 m/s and considerable kinetic energy can be attained. Other industrial applications include the following:

Sliding doors. Here an advantage is that, should the supply fail, the motor does not impede movement of the door by hand.

Goods lifts. The arrangement is that of *Figure 26.64* turned vertically, the linear motor stator providing some of the counterweight.

Tensioning machines. The linear motor gives readily controllable tension for testing ropes, aluminium strip etc.

27 Illumination

R I Bell
Thorn Lighting Ltd

Contents

27.1 Light and vision 27/3

27.2 Quantities and units 27/3

27.3 Photometric concepts 27/4

27.4 Lighting design technology 27/6

27.5 Lamps 27/8
    27.5.1 Incandescent filament lamps 27/8
    27.5.2 Discharge lamps 27/10
    27.5.3 Lamp codes 27/10

27.6 Choice of lamps 27/18
    27.6.1 Colour qualities 27/18
    27.6.2 Operating and replacement costs 27/18
    27.6.3 Developments 27/19

27.7 Lighting design 27/19
    27.7.1 Objectives and criteria 27/19
    27.7.2 Luminaires 27/22
    27.7.3 Types of luminaires (fittings) 27/22
    27.7.4 Installation and maintenance 27/23
    27.7.5 Design 27/23
    27.7.6 Glare 27/24
    27.7.7 Local lighting 27/25
    27.7.8 Practical illumination measurements 27/25

27.8 Applications 27/25
    27.8.1 Factory lighting 27/25
    27.8.2 Domestic lighting 27/30
    27.8.3 Shop lighting 27/31
    27.8.4 Office lighting 27/32
    27.8.5 Floodlighting 27/32
    27.8.6 Street lighting 27/32
    27.8.7 Theatre lighting 27/34
    27.8.8 Television studio lighting 27/37

27.1 Light and vision

Light is electromagnetic radiation of a wavelength capable of causing a visual sensation in the eye of an observer. It is measured in terms of its ability to produce such a sensation.

The *spectral range* of visible radiation is not well defined, and can vary with the observer and with conditions. The lower limit is generally taken to be 380–400 nm (deep-blue radiation) and the upper limit 760–780 nm (deep-red radiation).

The human eye is not equally sensitive to all wavelengths, as shown in *Figure 27.1*: it has a peak sensitivity at 555 nm for normal daytime (*photopic*) vision. The eye contains two distinct types of receptors—'rods' and 'cones'. The cones are related to colour vision, while the rods are more sensitive but permit only black and white vision. In photopic vision the cones are operative; hence, vision is in colour.

At low levels of illumination (luminance below 10 cd/m²) the more sensitive rods begin to take over and the resultant image appears less brightly coloured. Further, the peak sensitivity shifts towards the blue/green region of the spectrum. This condition is known as *mesopic* vision.

At still lower levels (below 10^{-2} cd/m²) vision is almost entirely by rod receptors and the eye is said to be dark-adapted. In this state, known as *scotopic* vision, the sensation is entirely in black and white and the peak sensitivity has moved to 505 nm.

Commission Internationale de l'Eclairage, or CIE, has defined an agreed response curve for the photopically adapted eye, known as the *spectral luminous efficacy* or $V(\lambda)$ function. Luminous flux, which is the rate of flow of light, is radiant power weighted according to its ability to produce a visual sensation by the $V(\lambda)$ function.

The luminous flux emitted by a source of light will vary with direction of emission. The rate of change of luminous flux with solid angle is termed the *luminous intensity*.

Illumination is the process whereby luminous flux is incident upon a solid surface and the corresponding quantity (flux density per unit area) is the *illuminance*.

Light striking a surface can be reflected, transmitted or absorbed according to the nature of the surface, and the fractions of the incident luminous flux thus affected are termed the reflectance, transmittance or absorbtance, respectively.

The brightness of a luminous surface is a subjective sensation having a physical measure termed the *luminance*.

27.2 Quantities and units

Each quantity has a quantity symbol (e.g. I for luminous intensity) and a unit symbol (e.g. cd for candela) to indicate its unit of measurement.

Luminous flux: Φ or F; lumen (lm)   The rate of flow of luminous energy. Quantity derived from radiant flux by evaluating it according to its ability to produce visual sensation. Unless otherwise stated, 'luminous flux' relates to photopic vision as defined by the $V(\lambda)$ function of spectral luminous efficacy.

If K_m is the maximum spectral luminous efficacy (about 680 lm/W at a wavelength of 555 nm), then the luminous flux Φ (in lm) is related to the spectral power distribution $P(\lambda)$ at wavelength λ by

$$\Phi = K_m \int P(\lambda) \cdot V(\lambda) \cdot d\lambda$$

Luminous efficacy (of a source): η; lumen per watt (lm/W)   The quotient of the luminous flux emitted by a source to the input power. It should be noted that for discharge lamps the luminous efficacy may be quoted either for the lamp itself or for the lamp with appropriate control gear. The latter figure will be lower.

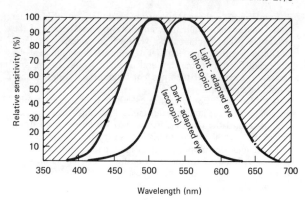

Figure 27.1   Relative spectral sensitivity of human eye

Luminous intensity: I; candela (cd)   The quotient of the luminous flux $\delta\Phi$ leaving the source, propagated in an element of solid angle containing the given direction, by the element of solid angle $\delta\omega$ (see *Figure 27.2*). $I = \delta\Phi/\delta\omega$, and in the limit as $\delta\omega \to 0$,

$$I = d\Phi/d\omega$$

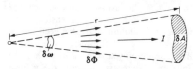

Figure 27.2   Luminous intensity and illumination

Illuminance: E; lux (lx) *or* lumen/metre² (lm/m²)   The incident luminous flux density at a point on a surface. The quotient of the luminous flux incident on an element of surface, by the area of that element. Referring to *Figure 27.2*,

$$E = \delta\Phi/\delta A$$

Note: The term *illuminance* is used for the quantity, while the term *illumination* describes the physical process.

Luminance; L; candela/metre² (cd/m²)   The luminous intensity in a given direction of a surface element, per unit projected area of that element.

For a small element of surface, of area δA (*Figure 27.3*), having an intensity I in a direction at an angle θ to the surface normal n, the luminance in this direction is given by

$$L = I/[\delta A \cos \theta]$$

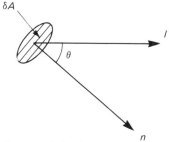

Figure 27.3   Luminance

27/4 Illumination

Luminance is a physical measure of brightness, but it should be noted that, unlike luminance, an observer's assessment of the brightness of an object is subjective. It will depend upon the level of adaptation and other factors. For example, the luminance of a car headlight during the day and at night would be approximately the same, but the apparent brightness during the day would be significantly less.

It is often more convenient in calculations to express luminance in terms of the luminous flux emitted per unit area of surface. This is termed the *luminous exitance M* (in lm/m², sometimes called the apostilb—asb—to distinguish it from the lux).

For a uniform diffuser (such as matt paint, blotting paper, etc.) of reflectance R, the relationships between luminance, luminous exitance and illuminance are

$M = RE$ (in lm/m²)

$L = RE/\pi = M/\pi$ (in cd/m²)

27.3 Photometric concepts

Inverse square law The illuminance E at a point on a surface produced by light from a *point* source varies inversely with the square of the distance r from the source, and is proportional to the luminous intensity I towards that point. Referring to *Figure 27.2*, the illuminance is given by $E = I/r^2$.

Point source Any source can be considered as a point source if its dimensions are small compared with the distance between source and receptor. In practice, the inverse square law can be applied with acceptable accuracy if the largest dimension of the source is not greater than one-fifth of the distance between source and receptor. For longer sources, as with fluorescent luminaires, alternative methods must be employed. One approach is to subdivide the source into small elements, to apply the inverse square law to each, and to sum the contributions. Another method is to use the aspect factor method described in Technical Report No. 11 of the Lighting Division of the Chartered Institute of Building Services.

Cosine law The illumination on a surface is proportional to the cosine of the angle θ between the directions of the incident light and the normal to the surface. This is due to the reduction of projected area as the angle of incidence increases from zero (normal incidence) to 90°. For a point source at distance r, the illumination for angle θ is

$E = E_0 \cos\theta = I \cos\theta / r^2$

where E_0 is the illumination for normal incidence, $\theta = 0$. With the working surface horizontal and the source mounted a distance h above the surface, the illumination on the working surface is

$E = I \cos^3\theta / h^2$

involving θ as the only variable.

Reflection Light falling on a surface may undergo *direct* or *diffuse* reflection. Direct reflection is specular, as by a mirror. Diffuse reflection may be *uniform* or *preferential*: in the former the luminance is the same in all available directions; in the latter there are maxima in certain directions (see *Figure 27.4*). Direct and diffuse reflection may occur together as *mixed* or *spread* reflection.

Examples of reflecting surfaces are: direct—mirror glass, chromium plate; uniform diffuse—blotting paper; preferential diffuse—anodised aluminium, metallic paint.

Reflectance: R The ratio between the reflected luminous flux and the incident luminous flux.

Figure 27.4 Reflection

Figure 27.5 Transmission with partial reflection

Transmission Light falling on a translucent surface undergoes partial transmission (*Figure 27.5*). The transmission may be *direct*, as through clear plate glass; *diffuse*, as through flashed opal glass; or *preferential*, as through frosted glass.

Transmittance: T The ratio between the transmitted luminous flux and the incident luminous flux.

Density (optical): D The common logarithm of the reciprocal of the transmittance:

$D = -\log_{10} T$

Absorption That proportion of light flux falling on a surface which is neither reflected nor transmitted is *absorbed* and, normally, converted into heat.

Absorbtance: A The ratio between the absorbed luminous flux and the incident luminous flux.

Note: The factors R, T and A vary with angle of incidence, nature of the surface, etc., but for any condition $A + T + R = 1$.

Refraction While a light ray is travelling through air, its path is a straight line. When the ray passes from air to glass (or any transparent material, e.g. clear plastics, diamonds, etc.), the ray is, in general, bent at the surface of separation. The path of the ray after bending or refraction from air to glass is always more nearly perpendicular to the bounding surface than is that of the incident ray. The degree to which the ray is bent is dependent upon the type of glass or transparent material, the angle of incidence of the ray and also the colour of the light.

Should the ray, while in the glass, strike another bounding surface, it may again be refracted. In this case the refracted ray may be more nearly parallel to the bounding surface than is the incident ray. If the light ray strikes the bounding surface at any angle above a certain limit (the *critical* angle), it will not be refracted but will be totally reflected.

Both refraction and total internal reflection are used in the design of lighting units, the prismatic types of reflector being typical examples. *Figure 27.6* shows examples of both refraction and total internal reflection as applied to the design of prismatic reflectors.

Polar intensity distributions Utilising the inverse square and cosine laws, it is possible to calculate the direct illuminance at a point from a single luminaire, or an installation, using the 'point-

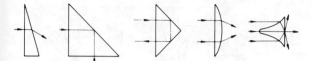

Figure 27.6 Types of refracting and reflecting prismatic glass

by-point' method. The effect of inter-reflected light is not included, as the calculations are too complex to warrant it.

Figure 27.7 shows the intensity distributions for two different interior luminaires:

(a) is for a luminaire with a luminous intensity distribution symmetrical about a vertical axis, and (b) is for a disymmetric luminaire with a luminous intensity distribution symmetrical about two orthogonal vertical planes. These are, respectively, typical of discharge and fluorescent luminaires.

For symmetric luminaires, only one average intensity distri-

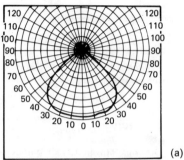

(a)

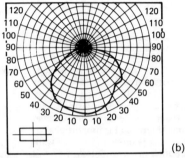

(b)

Luminous intensity (cd/1000 lm)	
Angle (deg)	Mean vertical intensity (cd)
0	234
5	234
10	235
15	236
20	234
25	232
30	230
35	222
40	205
45	180
50	137
55	95
60	66
65	46
70	30
75	19
80	13
85	11
90	9
95	9
100	9
105	10
110	11
115	13
120	15
125	18
130	22
135	28
140	32
145	35
150	34
155	30
160	26
165	20
170	14
175	7
180	4

Luminous intensity (cd/1000 lm)		
Angle (deg)	Transverse plane (T)	Axial plane (A)
0	232	232
5	232	230
10	231	228
15	231	224
20	228	217
25	224	208
30	220	199
35	211	187
40	192	174
45	168	199
50	141	187
55	113	174
60	86	159
65	58	142
70	33	124
75	19	106
80	12	85
85	7	67
90	1	46
95	2	28
100	5	11
105	7	1
110	8	2
115	9	2
120	11	2
125	12	3
130	13	4
135	14	5
140	14	6
145	15	7
150	16	8
155	16	9
160	16	10
165	15	10
170	13	11
175	12	11
180	12	12

Figure 27.7 Luminous intensity distributions for (a) symmetric luminaire and (b) di-symmetric luminaire

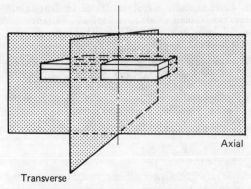

Figure 27.8 The transverse and axial planes in which the transverse and axial polar curves are measured

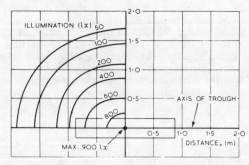

Figure 27.9 Isolux diagram for a trough reflector fitting, 1.8 m long

bution is normally given, and this can be presented graphically on polar co-ordinates or in tabular form (which is easier to use). For disymmetric luminaires two or more distributions are given. The principal ones are the *axial* and *transverse* distributions, which lie in vertical planes down the axis of the luminaire, and at right angles thereto, respectively (*Figure 27.8*).

Many luminaires can accommodate various lamp types without affecting the shape of the intensity distribution. For this reason it is a common practice to quote intensities in candelas per 1000 lamp lumens (cd/klm) rather than in candelas. This permits easy scaling of the data according to the luminous output of the lamps.

For streetlighting and floodlighting luminaires, the main distributions are usually insufficient, and contours of equal intensity (isocandela) are normally published on a convenient Cartesian grid system.

It is important to establish the conditions for which the calculations are to apply, as this will determine whether the lumen output of the lamps is to be 'initial' or 'lighting design', and will permit a realistic maintenance factor (e.g. 0.8) to be employed.

The light output of the luminaire can be obtained from the mean polar curve by the use of Russell Angles (*Table 27.1*) or zone factors (*Table 27.2*). In the Russell angle method, the mean of the luminous intensities at the angles stated is multiplied by 4π to obtain the lumen output. With zone factors, the luminous intensity at each mid-zone angle in zones 10° wide is multiplied by the appropriate factor and the results are added.

Isolux diagrams A convenient way of plotting the illuminance produced by single luminaires or complete installations is by contours of equal illuminance, or isolux contours. Isolux diagrams are frequently used to depict the performance of asymmetric luminaires such as 'wall-washers', and are now often used when the calculations are done by computer. *Figure 27.9* shows a typical isolux diagram for a reflector fitting at a particular mounting height.

27.4 Lighting design technology

Light output ratio: LOR The ratio between the light output of the luminaire measured under specified practical conditions and the sum of the light outputs of individual lamps operating outside the luminaire under reference conditions.

Photometric centre The point in a luminaire or lamp from which the inverse square law operates most closely in the direction of maximum intensity.

Upward [downward] flux fraction UFF [DFF] The fraction of the total luminous flux of a luminaire emitted above [below] the horizontal plane containing the photometric centre of the luminaire. Also known as upper [lower] flux fraction.

Upward [downward] light output ratio ULOR [DLOR] The product of the light output ratio of a luminaire and the upward [downward] flux fraction.

Point luminaire A luminaire, the size of which can be ignored for the purposes of lighting calculations.

Linear luminaire A luminaire, the width of which can be, but the length cannot be, ignored for the purposes of lighting calculations.

Symmetric luminaire A luminaire with a light distribution nominally rotationally symmetrical about the vertical axis passing through the photometric centre.

Table 27.1 Russell angles

Angles from the downward vertical (deg), on *each* side:										
18.2	31.8	41.4	49.5	56.6	63.3	69.5	75.5	81.4	87.1

Table 27.2 Zone factors

Mid-zone angle (deg):	5 / 175	15 / 165	25 / 155	35 / 145	45 / 135	55 / 125	65 / 175	75 / 105	85 / 95
Zone factor:	0.095	0.283	0.463	0.628	0.774	0.897	0.992	1.058	1.091

Disymmetric luminaire A luminaire with a light distribution nominally symmetrical only about two mutually perpendicular planes passing through the photometric centre. Where such a luminaire is linear, the vertical plane of symmetry normal to the long axis is designated the *transverse* plane, and the vertical plane passing through the long axis is designated the *axial* plane (see *Figure 27.8*). The vertical distributions taken in these planes are the transverse and axial distributions, respectively.

Working plane The horizontal, vertical or inclined plane in which the visual task lies.

Reference surface The surface of interest over which the illuminance is to be calculated. A reference surface need not contain the visual task.

Horizontal reference plane A horizontal reference surface. This is usually assumed to be 0.85 m above the floor and to correspond with the horizontal working plane. The horizontal reference plane is also the mouth of the floor cavity (see below).

Plane of luminaires The horizontal plane which passes through the photometric centres of the luminaires in an installation. This is also the mouth of the ceiling cavity.

Floor cavity The cavity below the horizontal reference plane in a room (see *Figure 27.10*). The horizontal reference plane and the floor cavity may be designated by the reference letter F.

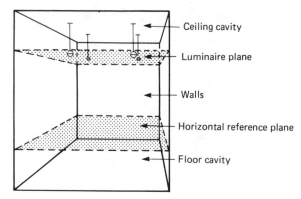

Figure 27.10 Ceiling cavity, walls and floor cavity

Walls The vertical surfaces of a room between the plane of the luminaires and the horizontal reference plane (see *Figure 27.10*). The walls may be designated by the reference letter W.

Ceiling cavity The cavity above the plane of the luminaires in a room (see *Figure 27.10*). The luminaire plane and the ceiling cavity may be designated by the reference letter C.

Distribution factor: DF(S) The distribution factor for a surface S is the ratio between the direct flux received by the surface S and the total lamp flux of the installation. DF(F), DF(W) and DF(C) are the distribution factors for the floor cavity, walls and ceiling cavity, respectively, treated as notional surfaces.

Utilisation factor: UF(S) The utilisation factor for a surface S is the ratio between the total flux received by the surface S (directly and by inter-reflection) and the total lamp flux of the installation. UF(F), UF(W) and UF(C) are the utilisation factors for the floor cavity, walls and ceiling cavity, respectively, treated as notional surfaces.

Direct ratio: DR The proportion of the total downward flux from a conventional installation of luminaires that is directly incident on the horizontal reference plane. The direct ratio is equal to DF(F) divided by the DLOR of the luminaires.

Distributance: D(S) The distributance for a surface S is the ratio between the direct flux received by the surface S and the total emitted flux of the installation. D(F), D(W) and D(C) are the distributances for the floor cavity, walls and ceiling cavity, respectively, treated as notional surfaces.

Utilance: U(S) The utilance for a surface S is the ratio between the total flux received by the surface S (directly and by inter-reflection) and the total emitted flux of the installation. U(F), U(W) and U(C) are the utilances for the floor cavity, walls and ceiling cavity, respectively, treated as notional surfaces.

Zone factor The solid angle subtended at the photometric centre of a lamp or luminaire by the boundary of a zone. The zonal flux is obtained by multiplying the intensity of the lamp or luminaire, averaged over the zone, by the zone factor.

Room index: RI Twice the plan area of a room divided by the wall area (as defined above). The room is taken to have parallel floor and ceiling, and walls at right angles to these surfaces. From any point in the room all of the surfaces in the room should be visible.

Spacing/height ratio: SHR The ratio between the spacing in a stated direction between photometric centres of adjacent luminaires and their height above the horizontal reference plane. It is assumed that the luminaires are in a regular square array unless stated otherwise.

Maximum spacing/height ratio: SHR MAX The SHR for a square array of luminaires that gives a ratio between minimum and maximum direct illuminance of 0.7 over the central region between the four innermost luminaires.

Maximum transverse spacing/height ratio: SHR MAX TR The SHR in the transverse plane for continuous lines of luminaires that gives a ratio between minimum and maximum direct illuminance of 0.7 over the central region between the two inner rows.

Nominal spacing/height ratio: SHR NOM The highest value of SHR in the series 0.5, 0.75, 1.0, etc., that is not greater than SHR MAX. Utilisation factor tables are normally calculated at a spacing: height ratio of SHR NOM.

Maintenance Factor: MF The ratio between the illuminance provided by an installation in the average condition of dirtiness expected in service and the illuminance from the same installation when clean. It is always less than unity.

Uniformity The ratio between the minimum and average illuminance over a given area. For interior lighting it should be not less than 0.8 over the task area. This requirement can be satisfied by ensuring that the spacing/height ratio of an installation does not exceed SHR MAX.

Daylight factor: DF The ratio between the illumination measured on a horizontal plane at a given point inside a building and that due to an unobstructed hemisphere of sky. Light reflected from interior and exterior surfaces is included in the illumination at the point, but direct sunlight is excluded. A directly overcast sky is assumed whose range of luminance from horizon to zenith is about 1:3.

Sky factor: SF The ratio between the illumination on a horizontal plane at a given point inside a building and that due to an unobstructed hemisphere of sky of uniform luminance. This factor may be calculated geometrically.

27.5 Lamps

Light can be produced from electrical energy in a number of ways, of which the following are the most important.

(1) Thermoluminescence, or the production of light from heat. This is the way light is produced from a filament lamp, in which the filament is incandescent.
(2) Electric discharge, or the production of light from the passage of electricity through a gas or vapour. The atoms of the gas are excited by the passage of an electric current to produce light and/or ultraviolet energy.
(3) Fluorescence, a two-step production of light which starts with ultraviolet radiation emitted from a discharge; the energy is then converted to visible light by a phosphor coating within the lamp.

27.5.1 Incandescent filament lamps

Thermoluminescence is the emission of light by means of a heated filament. Although the carbon filament lamp is in this category, the term is normally synonymous with the tungsten filament lamp in its various forms. The most general form is the GLS (general lighting service) lamp (*Figure 27.11*). Light produced

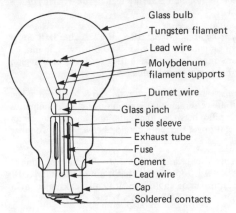

Figure 27.11 Construction of a GLS lamp

from a hot wire increases as the temperature of the wire is raised. It also changes from a predominantly red colour at low temperature to a white which approaches daylight as the temperature is increased.

Of the electrical energy supplied to an incandescent lamp filament, by far the greatest proportion is dissipated as heat, and only a small quantity as visible light (about 95% heat and 5% light; see *Figure 27.12*). Because the quantity of visible light emitted depends upon the filament temperature, the higher the filament temperature the greater will be the visible light output in lumens per watt of electric energy input. Thus, for an incandescent filament, a material is needed that not only has a high melting point, but is also strong and ductile so that it can be

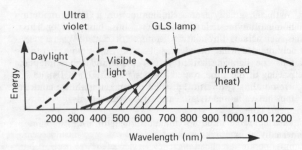

Figure 27.12 Spectral power distribution of daylight and a GLS lamp

formed into wire. At present, tungsten metal is the material nearest to this ideal.

The colour temperature of a normal GLS filament lamp is around 2500°C or 2770 K (0°C = 273 K). At the extremely high temperature of the filament, tungsten tends to evaporate. This leads to the familiar blackening of an incandescent lamp bulb. The evaporation of the tungsten filament can be reduced by filling the lamp bulb with a suitable gas that does not chemically attack the filament. Suitable gases are hydrogen, nitrogen, and the inert gases argon, neon, helium, krypton and xenon. However, gases also cool the filament by conducting heat away from it, and they decrease lamp efficiency. The gas used must therefore be carefully chosen. It should adequately suppress tungsten evaporation without overcooling the filament. In addition, it should not readily pass an electric current, for otherwise arcing may occur which would destroy the lamp.

Argon and nitrogen are the gases most commonly used. Nitrogen will minimise the risk of arcing, but will absorb more heat than argon. Argon is used by itself in general service lamps. A mixture of the two gases is used in incandescent lamps where the tendency for arcing is more likely, such as in projector lamps. In this case the amount of nitrogen present is kept very small—as little as 5%—in order to obtain optimum lamp efficiency.

Not all incandescent lamps benefit from gas filling. Mains voltage 15 W and 25 W lamps are mainly of the vacuum type, whereas lamps of 40 W and above are normally gas filled.

In general service lamps at least one lead is fused to prevent the bulb shattering should an arc occur. Modern fuses are encapsulated in a glass sleeve filled with small glass balls.

27.5.1.1 Coiled and coiled-coil filaments

If a filament is in the form of an isolated straight wire, gas can circulate freely round it. Filament temperature is thus decreased by convection currents, and has to be raised by increasing the electrical power input.

Coiling the wire reduces the cooling effect, the outer surface of the helix alone being cooled by the gas. Further coiling (coiled-coil filament) again reduces the effect of the gas cooling and results in further increase in lamp efficiency of up to 15%.

27.5.1.2 Bulbs

Clear-glass lamp bulbs have smooth surfaces and absorb the smallest possible amount of the light passing through them. The high temperature of the filament results in a high brightness which the bulb does not modify.

Early attempts to reduce glare from an unobscured filament used bulbs externally frosted. These were difficult to keep clean. The drawback is obviated in the modern *pearl* bulb by etching the inside surface instead. The light source appears to be increased in size and to have a larger surface area. The loss of light is negligible.

With the greatly increased illumination levels of modern lighting techniques, a further degree of diffusion is called for. This has been achieved by coating the inside of the bulb with a very finely divided white powder, such as silica or titania. In such lamps the lighted filament is not apparent. The luminous efficiency of the silica coated lamp is about 90% of that of a corresponding clear lamp of equal power rating. Silica coated lamps have a more attractive unlit appearance than either clear or pearl lamps.

In a coloured incandescent lamp the bulb is coated either internally or externally with a filter. All coloured incandescent lamps operate at reduced efficiency. In view of the low proportion of blue light in the spectrum, the efficiency of lamps of this colour is particularly low, as more than 90% of the light is filtered out. It is not possible to obtain a bright saturated blue colour.

27.5.1.3 Caps

The BC or bayonet lamp-cap is the standard in the UK for GLS lamps up to 150 W; it is also widely used in France. It makes lamp replacement easy, but it is not suitable for large currents that would weaken the lamp-holder contact plunger springs; screw caps are therefore used for 200 W rating and upwards. The ES, or Edison screw, cap is used for 200 W lamps and is also the standard cap for smaller lamps in many parts of the world. For 300 W and larger ratings, the GES or Goliath Edison screw, cap is used to ensure adequate mechanical strength. The caps must be so constructed and fitted to the lamp that they will withstand the torsion test specified in BS 161—i.e. 25 lb-in (2.8 Nm) for BC and ES caps, and 45 lb-in (5 Nm) for GES caps, both before and at end of life.

27.5.1.4 Decorative and special-purpose lamps

The incandescent filament lamp in its simplest form is purely a functional light source, but the fact that an integral part of the lamp is a glass bulb enables the manufacturer to adapt this envelope to give some aesthetic appeal. The commonest form is the candle lamp, with glass clear, white or frosted. Other lamps have been marketed which combine the role of light source and decorative fitting by virtue of their bulb shape. They are usually of larger dimensions than conventional lamps. Apart from their attractive shapes, they are made with silica coatings, coloured lacquer coatings and crown silvered tops, and are therefore rather more efficient as light-producing units than a combination of lamp and separate diffuser.

To cater for locations where vibration and shock are unavoidable, special *rough service* lamps are produced which combine filament wire modifications combined with the inclusion of an increased number of intermediate filament supports.

To provide directional beam control a further range of special-purpose lamps is made with blown or pressed paraboloidal bulb shapes coated with an aluminium reflector film. The filament is accurately placed at the focus of the reflector to provide the directional beam. More accurate beam control is provided by the pressed glass versions (PAR lamps).

In some lamps dichroic reflectors are employed. These reflect visible light but transmit infrared radiation. This permits the lamp to have a cooler beam, since the heat radiation is not focused. However, these lamps can be used only in fittings able to dissipate the extra heat transmitted by the reflector. Dichroic coatings are widely used in film projector lamps with integral reflectors, to prevent excessive temperature at the film gate.

27.5.1.5 Efficacy, life and light output

The efficacy of an incandescent lamp is related to the quantity of visible light emitted per unit of electrical power input. Thus, a 100 W incandescent lamp having a total light output of 1200 lm has an efficacy of 1200/100 = 12 lm/W.

A higher filament temperature increases lamp efficacy, but the temperature of a tungsten filament cannot be increased indefinitely, as it will melt catastrophically if the lamp efficacy approaches 40 lm/W.

At high filament temperatures tungsten evaporation—even though it is reduced by gas filling—is more rapid and leads to a shorter lamp life. Thus, the more efficient an incandescent lamp is the shorter is its life.

Variations in supply voltage vary filament temperature, which, in turn, increases or decreases lamp life. *Figure 27.13* shows how the lamp efficiency, life, light output and input power vary with

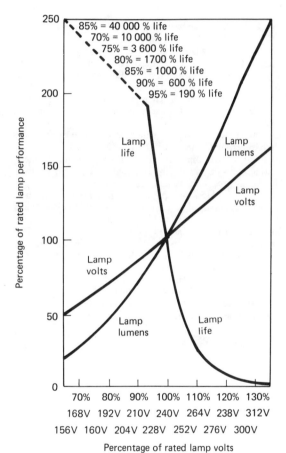

Figure 27.13 Effect of voltage variation on incandescent filament lamps

supply voltage. For example, if a lamp is under-run by 5% below its rated voltage, its life will be nearly doubled (190% of rated 1000 hour life) but the lamp power would be reduced to around 92% of the rating and the light output to less than 85% of the normal lumen output.

27.5.1.6 Tungsten–halogen lamps

If the envelope of a tungsten lamp is made of quartz instead of glass, it can be much smaller, because quartz can operate safely at a higher temperature. As with a glass lamp, tungsten evaporated from the filament will deposit on the quartz envelope, causing it to blacken. However, if a small quantity of iodine is introduced

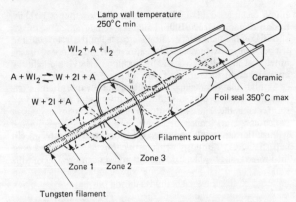

Figure 27.14 Simplified mechanism of the iodine regenerative cycle. W=tungsten; I=iodine; A=inert gas

into the lamp, and if the temperature of the quartz envelope is above 250°C, the iodine combines with the tungsten on the inner face of the quartz to form tungsten iodide, a vapour.

When the tungsten iodide approaches the much hotter filament, it decomposes; the tungsten is deposited on the filament and the iodine is released, to perform its cleaning cycle again. (See *Figure 27.14*.)

Unfortunately, the tungsten is not necessarily redeposited on those parts of the filament from which it originally evaporated. Even so, substantial improvements in life and/or higher filament operating temperatures can be achieved, giving higher lumen outputs compared with the equivalent GLS lamp.

The increase in life is mainly due to the increased gas pressure, which can be employed in a tungsten–halogen lamp to reduce filament evaporation. This, in turn, is only possible because a small outer envelope can be used without risk of bulb blackening.

Tungsten–halogen lamps greatly improve the field of the filament lamp, especially in floodlighting, and photographic and automobile lighting. They provide up to 50% greater lumen output and up to 2000 h of life, depending upon type and application. They have limitations, in that the bulb wall temperature must exceed 250°C but not exceed 350°C, but these requirements are taken care of in fitting design.

27.5.2 Discharge lamps

27.5.2.1 Principle

A discharge lamp consists essentially of a tube of glass, quartz or other suitable material, containing a gas and, in most cases, a metal vapour. The passage of an electric current through this gas/vapour produces light or ultraviolet radiation. Most practical discharge lamps (excluding those used for coloured signs) rely upon discharges in metallic vapours of either sodium or mercury, with an inert gas filling. The nature of the filling, the pressure developed (which ranges from 10^{-5} atm to over 40 atm in different types of lamp) and the current density determine the characteristic radiation produced by the arc. In most lamps the arc tube is enclosed within an outer glass or quartz jacket. This affords protection, can be used for phosphor or diffusing coatings, controls UV emission and, by suitable gas filling, can control the thermal characteristics of the lamp.

All discharge lamps include some mechanism for the production of electrons from the electrodes within the lamp. The commonly used methods are thermionic emission and field emission, and in both cases emissive material such as barium oxide is often contained within the electrode to lower its work function and, hence, reduce energy loss.

When the lamp is put into circuit and an electric field is applied, the electrons begin to accelerate towards the positive electrode, and may collide with gas or metal atoms. These collisions may be elastic, in which case the atom and electron only change their velocities, or inelastic, in which case the atom changes its state. In the latter case, if the kinetic energy of the electron is sufficient, the atom may become excited or ionised. Ionisation produces a second electron and a positive ion, which contribute to the lamp current and which may cause further collisions. Left unchecked, this process would avalanche, destroying the lamp. To prevent this catastrophe some form of electrical control device (such as an inductor) is used to limit the current. This condition is described as an arc discharge. Excitation occurs when the electrons within the atom are raised to an energy state higher than normal (but not high enough to cause ionisation). This is not a stable condition, and the electrons fall back to their previous energy level, with a corresponding emission of electromagnetic radiation (which may be visible, ultraviolet or infrared).

In some lamp types, an inert gas is used to maintain the ionisation process, while it is the metal vapour which becomes excited. The vapour pressure in the lamp affects the starting and running characteristics, and the spectral composition of the emitted radiation.

In most lamps (except the fluorescent lamp) there is a run-up period, during which the metal is vaporised and the pressure increases to its operating condition. In some lamp types this may take 10–15 min. If, once the lamp is run-up, the supply is interrupted, then it will extinguish; and unless special circuits and suitable lamp construction are used, the pressure will be too high for the arc to restrike until the lamp has cooled.

Broadly, practical discharge lamps for lighting are either mercury vapour or sodium vapour lamps, at either high or low pressure.

27.5.3 Lamp codes

When mercury and sodium lamps first became available in the UK, the lamp industry devised a simple code to distinguish between the various types, not only for reference purposes within the industry, but also to assist users in their applications. The code indicates the main characteristics of each type.

27.5.3.1 Mercury lamps

A mercury vapour lamp consists essentially of an arc tube, containing a small quantity of mercury and filled with an inert gas (or gases) to a pressure of from 10^{-5} atm to over 40 atm. The nature of the filling, the pressure developed and the current density determine the characteristic radiation produced by the arc.

In the fluorescent tube a low mercury vapour pressure produces ultraviolet radiation, which is converted to useful visible radiation by the phosphor coating.

In high-pressure mercury vapour lamps the arc tube is normally small, and enclosed within an outer protective envelope. The high-pressure mercury lamp produces a substantial proportion of its radiation within the visible spectrum. The spectral composition of the light may be acceptable, or can be modified by:

(1) Using a tungsten filament within the lamp to act as a ballast and emit useful light.
(2) Applying a phosphor coating to the inside of the outer envelope.
(3) Introducing metal halides into the arc tube in addition to the normal mercury.

27.5.3.2 Classification

Mercury lamps The first letter of the code is M (for mercury lamp). The second specifies the pressure, the material of the arc tube and the loading (in W/cm of arc length). The third and fourth letters indicate special features, and the fifth specifies the correct operating position.

Second letter:

A   medium pressure (about 1 atm), glass, above 10 W/cm (obsolete).
B   high pressure (about 2–10 atm), quartz, 10–100 W/cm.
C   low pressure (about 1/100 000 atm), glass, below 10 W/cm.
D   extra-high pressure (50–200 atm), forced liquid cooling (obsolete).
E   extra-high pressure (20–40 atm), quartz, above 100 W/cm.

Third and fourth letters:

F   fluorescent coating of the outer envelope.
I   halide additive to the arc.
L   linear lamp construction (usually double-ended).
R   internal reflector.
T   tungsten filament/ballast within lamp.
W   Wood's glass outer envelope.

Last letter:

/U universal.   /V vertical,   /D vertical,
                              cap up.   cap down.
/H horizontal.   /BU base up*.   /BD base down*.

* Cap not more than 15° below/above horizontal (*Figure 27.15*).

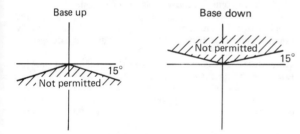

Figure 27.15 Base up and base down burning positions

Sodium lamps The first letter is S (for sodium). Thereafter:
Second and third letters:

LI linear low-pressure lamp with cap at each end.
OX single-ended low-pressure lamp with U-shaped arc tube.
ON high-pressure lamp with sintered aluminium-oxide arc tube.

Last letter:

/T   high pressure, clear glass tubular outer bulb, single-ended.
/TD high pressure, clear quartz tubular outer bulb, double-ended.

27.5.3.3 Low-pressure mercury fluorescent lamps (MCF)

Construction A typical mercury fluorescent tube consists of a glass tube, 38 mm in diameter and up to 2400 mm long, filled with argon or krypton gas at a pressure of approximately 0.005 atm and containing a drop of liquid mercury. A diagram of the tube is shown in *Figure 27.16*. The interior surface of the tube is coated with a fluorescent powder, the phosphor, which converts the ultraviolet light produced by the discharge into visible light. At each end of the tube are electrodes which serve the dual purpose of cathode and anode, for generally these tubes are used on a.c. circuits.

The cathodes of a hot cathode fluorescent lamp consist of coiled-coil, triple-coiled or braided tungsten filaments, coated with a barium oxide thermionic emitter and held by nickel support wires. Cathode shields in the form of metal strips bent into an oval shape surround the cathodes in certain sizes of tube and are supported on a separate wire lead. These shields trap material given off by the cathodes during life and prevent black marks forming at the ends of the tube. The shields also reduce flicker which is sometimes noticeable at the ends of the tubes, and protect the more delicate cathodes by acting as anodes on alternative half periods. The bases of the electrode support wires are gripped in a glass pinch through which the lead wires pass, forming a vacuum-tight glass-to-metal seal.

The lead-in wires are welded to the pins of the bi-pin cap, which is itself sealed to the ends of the glass tube. Some tubes are still available with a B.C. cap, but the bi-pin cap is now British standard.

Principle An external control circuit is required, which causes a preheating current to flow through the electrodes. This causes electrons to be emitted by the emitter coating. Once these have been produced, the control circuit must apply an electric field across the length of the lamp to accelerate the electrons and strike the arc. Once struck, the arc must be stabilised by the control circuit.

Colliding electrons excite mercury atoms, and produce ultraviolet and visible radiation (about 60% of the energy consumed is converted to UV). This radiation, when absorbed by the phosphor on the inside of the glass wall, is converted to visible light.

The colour and spectral composition of radiated light will depend upon the phosphors used. Lamps can be made with a 'white' appearance but with widely different efficacies or colour rendering properties.

Basic starter-switch circuit The basic starter-switch circuit is shown in *Figure 27.17(a)*. The principle of operation is as follows:

(a) When the mains voltage is applied, a glow discharge is created across the bi-metal contacts inside the glow starter (enclosed in a small plastics canister). The contacts warm up and close, completing the starting circuit and allowing a current to flow from the 'L' terminal, through the current limiting inductor through the two tube cathode filaments and back to the 'N' mains terminal.

(b) Within a second or two, the cathode filaments are warm enough to emit electrons; a glow is seen from each end of the tube. At this stage, the starter-switch bi-metal contacts open (because the glow discharge, which caused them to heat and close, ceases when they touch, and they cool and open), and interrupt the preheat current flow. If an inductor (*choke*) ballast (coils of copper wire around a laminated iron core) is

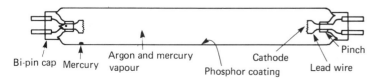

Figure 27.16 Low-pressure mercury vapour fluorescent tube

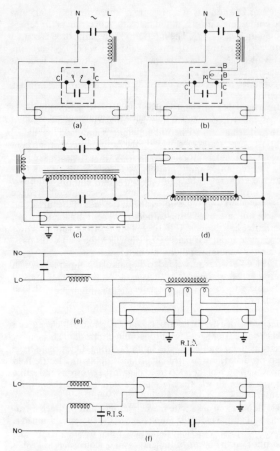

Figure 27.17 Starting methods. (a) Glow starter; (b) Electronic starter; (c) Instant start; (d) Starterless; (e) Semi-resonant circuit

used the magnetic energy stored in the core collapses to produce a high-voltage pulse (600–1000 V) across the fluorescent tube sufficient to strike the arc and set up the electric discharge through the tube.

(c) Once the tube arc has been struck, the current through the tube gradually builds up. This means that the current through the inductor also increases. As this happens, the voltage across the inductor increases and the tube voltage falls. The inductor is so designed that when the tube and inductor current rise to a value determined by the inductor design setting, the circuit stabilises; e.g. for a 1.25 m (4 ft) tube on a 240 V supply, the inductor stabilises the tube current at around 0.4 A, which coincides with about 100 V across the tube. This gives the rated tube power (0.4 A × 100 V = 40 W) at power factor approximately unity for the lamp itself.

Electronic start circuit A recent improvement of the basic starter-switch circuit (*Figure 27.17b*) is the electronic start circuit. It is identical with the starter-switch circuits in all respects but one; the glow starter is replaced by an all electronic starter. In some cases it may be fitted into a conventional glow start cannister, as a direct replacement, and in other more sophisticated luminaires it is a small encapsulated box.

The main advantages of this circuit are that it affords reliable starting, does not shorten tube life (a problem with glow switches at the end of their life) and will last the life of the luminaire without replacement. Some circuits also inhibit the start of faulty lamps after a number of attempts.

Starterless/quickstart circuit See *Figure 27.17(c)*. When the mains supply is switched to this circuit, a very small current flows through the local path of inductor and transformer to neutral. This gives rise to larger currents circulating round the closed loops formed by the transformer secondaries and the tube electrodes, and the electrodes are thereby preheated. As the current drawn is small, it follows that approximately mains voltage appears across the transformer and, hence, across the tube. There is no amplification of voltage or any form of voltage pulse in this circuit.

The presence of an earthed metal surface, such as a reflector or cover plate, close to the tube is essential to ensure satisfactory starting in a switchless start circuit (and in other forms of starterless circuit). A voltage gradient exists across the tube and also between one electrode and the earthed surface. The cloud of electrons emitted from the electrodes by the preheating current is spread along the tube under the influence of the voltage gradient and the tube arc will be struck in this way.

The voltage across the tube and the transformer then falls to the operating value, and the cathode heating is also reduced. Starting time is of the order of 1–2 s.

The capacitor plays no part in the starting or operation of the tube: it is purely for the power factor correction.

Semiresonant start (SRS) This circuit (*Figure 27.17e*) may be regarded as a hybrid of the switch start and switchless start circuits.

In the prestart condition, the circuit is a straight series arrangement comprising two coils, two electrodes and a capacitor, through which a preheating current flows directly the supply is switched on. The coils and the capacitor form a partially tuned circuit in semiresonance, providing an increased voltage across the tube (of the order of 270 V in the case of the 65 W arrangement), sufficient to ionise the vapour filling and start the tube. Starting time is of the order of 2–3 s.

The capacitor fills an essential role in the starting sequence and in operating the tube. Although the circuit in the prestart condition is a series arrangement with one current, when in the running condition it becomes a parallel circuit with a capacitive current improving the overall power factor.

Fluorescent tube replacement It is not easy to determine the end of the useful life of a fluorescent tube. Although failure to start will eventually occur due to exhaustion of the oxide coating on the

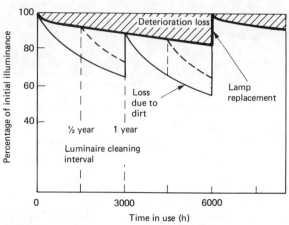

Figure 27.18 Effects of deterioration, cleaning and lamp replacement on illuminance of a fluorescent-tube installation

Table 27.3 Fluorescent tube colours

Tube colour	Light output relative to white (%)	Luminous efficacy† (lm/W)	Colour rendering	Colour appearance	Application and remarks
'White'*	100	45–65	Fair	Intermediate	The best general-purpose tube available, since it combines maximum lumen output with a white appearance. Warmer than daylight, but cooler than incandescence.
'Warm white'*	98	45–65	Fair	Warm	For general lighting where a warmer appearance than white is required. Incandescence effect, but without good red.
'Plus white' 'Colour 84'	95	45–65	Good	Intermediate	For general lighting where reasonably good colour rendering is required covering the complete visual spectrum, particularly for illuminance standards around 500 lx.
'Daylight'* 'Cool white'	94	45–65	Fair	Cool	For general lighting where a cooler appearance than white is required. Daylight effect, but lacking in red.
'Natural'*	70	30–50	Good	Intermediate	For office and shop lighting to give a cool effect. Close to natural daylight but with a flattering red content.
'De luxe warm white' 'Softone 32'	66	20–45	Good	Warm	For office, shop and domestic lighting where a warm effect, similar to incandescent light, is desired.
'Kolor-Rite' 'Trucolor 37'	65	20–45	Very good	Intermediate	Office, studios, showrooms, colleges, hospitals, displays; gives effect of sunny day.
'Northlight' 'Colour matching'	59	20–40	Good	Cool	For displays in lighting where a cool north skylight (winter light) effect is required, with normal red rendering. For colour matching.
'Artificial daylight'	41	20–40	Very good	Cool	Special tube with added ultraviolet to give a very close match to natural daylight. For colour matching cubicles.
'De luxe natural'	49	15–25	Special	Intermediate	For food and supermarket displays with meat or highly coloured merchandise. Combination of good blue and red rendering.

* BS colours.
† Based on total circuit watts.

electrode filaments, it is normally possible to justify replacement of the tube before this stage. As most installations of fluorescent fittings are designed to give a planned illuminance, random replacement of tubes at end of life will result in a non-uniform illuminance, which is uneconomic when related to energy costs and labour costs for replacement.

Figure 27.18 indicates typically the inherent deterioration of the illuminance from the luminaires of a fluorescent tube installation, and the gains that result from regular cleaning and lamp replacement. It is assumed that the use of the installation is 3000 h/year.

Tube colours Table 27.3 shows how the tube colours are graded in terms of lumen output efficacy and colour rendering quality. In choosing a tube colour a choice must be made between light output and colour, since tubes with high lumen output have only modest amounts of blue and red energy, whereas good colour rendering lamps have reduced yellow/green content and the tube lumen output is subsequently reduced.

A recent development in fluorescent tube phosphors is the multi-phosphor. Based on the principle that a mixture of red, green and blue light produces a white light, efficient red, green and blue phosphors are mixed in appropriate proportions to produce a white light when irradiated by ultraviolet light in a fluorescent tube. This produces a high-efficiency tube with good colour rendering for general applications. As the phosphors are costly, the tubes using them are more expensive and are frequently made in a diameter of 25 mm instead of 38 mm.

27.5.3.4 Medium-pressure mercury lamps

Medium-pressure mercury lamps are now virtually obsolete,

because high-pressure mercury lamps are both compatible and superior. The lamp consists of a hard, aluminosilicate glass arc tube inside an outer bulb of soda-lime glass (the gap between the two is filled with nitrogen to a low pressure, in order to obtain the correct amount of heat loss).

27.5.3.5 High-pressure mercury lamps

Construction The lamp consists of a quartz glass (pure fused silica) arc tube, enclosed within a borosilicate outer bulb. The bulb can be tubular, but is normally ellipsoidal to ensure an even outer-bulb temperature. (Ellipsoidal outer bulbs operate at about 550 K, compared with 800 K for the hottest parts of tubular bulbs.)

The arc tube contains a controlled quantity of mercury sufficient to produce the desired pressure at operating temperature and about 0.13 atm of argon, for starting purposes.

Electrodes, in the form of helices of tungsten wire, about a tungsten or molybdenum shank, are fitted at opposite ends of the arc tube. Emissive material is coated onto the electrodes (or may be held inside in pellet form). To assist in starting the lamp, an auxiliary electrode is mounted in close proximity to one of the main electrodes and is connected to the other electrode via a high resistance (10–30 kΩ). The electrodes are sealed into the quartz glass arc tube by means of molybdenum foil (as with tungsten halogen lamps).

The outer bulb (*Figure 27.19*) is filled to a pressure of 0.25–0.65 atm with either nitrogen or a nitrogen–argon mixture. Normally GES or ES lampholders are used.

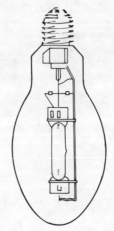

Figure 27.20 MBF lamp

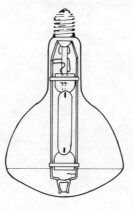

Figure 27.21 MBFR lamp

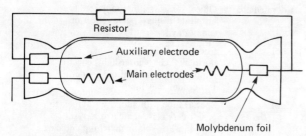

Figure 27.19 Arc tube construction of an MB lamp

High-pressure mercury fluorescent lamp (MBF) The quartz glass arc tube transmits ultraviolet light and enables a phosphor coating to be used on the inside of the outer bulb. The phosphor improves the colour rendering properties and luminous efficacy. Phosphors have upper limits of operating temperature and the use of ellipsoidal bulbs ensures the minimum size of bulb. Improvements in phosphors to give operation at higher temperatures have resulted in smaller bulb sizes.

MBF lamps (see *Figure 27.20*) have efficacies of about 58 lm/W with acceptable colour rendering for factories, storage areas, offices, etc.

High-pressure mercury fluorescent reflector lamp (MBFR) The MBFR lamp (*Figure 27.21*) is identical with the MBF lamp, except that the outer bulb is shaped to form a reflector. Titanium dioxide is deposited inside the conical surface of the outer bulb: this reflects about 95% of all the light in a diffuse manner. The phosphor is applied over this coating. The front face is left uncoated (although it may be etched or have a diffusing coating of silicon dioxide) and about 90% of the light is emitted through this opening. The lamp is less efficient but has a directional light output, suitable for installations where the luminaires have to be mounted high up (as in storage areas, hangars, etc.), and where dirty conditions obtain.

High-pressure mercury-blended lamp (MBT, MBTF) By including a filament within the outer bulb the need for special control gear can be eliminated, as the filament can act as a ballast. The efficacy of the lamp, 12–25 lm/W, is poor compared with an MBF lamp and inductor, but the lamp has two important advantages: (1) it can be fitted into any standard lampholder as a direct replacement for a tungsten lamp of the same rating, and will produce more ultraviolet and visible light; (2) unlike other mercury lamps, it will emit some light from the filament immediately upon switching on.

High-pressure mercury–metal halide lamps The efficacy of high-pressure mercury lamps is lowered because regions of the lamp are at potentials too low for the excitation of mercury. By adding suitable metals with lower excitation potentials to the arc tube, it is possible to increase the light output and improve colour rendering. The only suitable metals are highly reactive. These would damage the quartz glass arc tube and seals. By adding the metals (sodium, thallium, gallium, scandium and others) in the form of halides (usually iodides), these problems can be eliminated. The halides dissociate in the arc itself, but recombine at the arc tube wall.

The metal halide lamps are better than their MBF counterparts, in both colour rendering and efficacy (85 lm/W). They are available with fluorescent coatings (MBIF), in linear versions

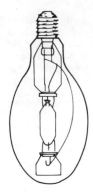

Figure 27.22 MBI/MBIF lamp

Figure 27.23 MBIL lamp

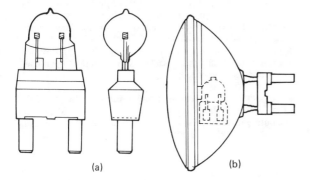

Figure 27.24 CSI/CID lamp. (a) Lamp; (b) Lamp and PAR64 reflector

(MBIL) or compact source versions (CSI and CID), with or without a PAR 64 outer reflector bulb (*Figures 27.22–27.24*).

Operating pressures can be as high as 20 atm, at which the spectrum becomes discontinuous providing an excellent white light for studios, colour TV work, etc.

27.5.3.6 Low-pressure sodium lamp (SOX, SLI)

The low-pressure sodium lamp is characterised by its mono-chromatic yellow light, which consists of two radiation lines (resonant doublet) at 589.0 nm and 589.6 nm. The lines are close to the maximum spectral sensitivity of the eye (555 nm) and the lamp is therefore efficient. Efficacies of over 150 lm/W are typical.

The lamp has poor colour rendering properties. Because the light is monochromatic, it is restricted to applications where the colour of the source and colour discrimination can be sacrificed for high efficacy. For example, the lamp is used for floodlighting and street lighting.

Construction and operation The lamp consists of a long arc tube made of a glass construction, known as ply tubing, filled with low-pressure gas (usually neon + 1% argon) to about 1/150–1/100 atm. The lamp also contains a small quantity of metallic sodium (solid at room temperature), which provides a pressure of about 1/150 000 atm in the lamp when operating. Precautions have to be taken to prevent the sodium from attacking the lamp seals. At each end of the arc tube is a tungsten electrode of coiled-coil, triple-coil or braided construction (similar to those in fluorescent MCF lamps) with an emissive coating (barium oxide or similar material).

The temperature of the arc tube must be maintained at about 270 °C in order to successfully vaporise the correct amount of sodium to give a vapour pressure of about 1/150 000 atm. It is therefore essential to avoid excessive heat loss if this temperature is to be maintained. In early lamp designs a Dewar flask was placed around the arc tube.

Modern lamps have an outer glass envelope enclosing the arc tube, and the space between the two is fully evacuated. In addition, the outer envelope is coated with an infrared reflecting film (bismuth oxide, tin oxide or gold) which transmits light but reflects heat back onto the lamp.

The requirements for efficient operation of a sodium lamp are (1) the arc voltage gradient must be low (long arc tubes for a given power) and (2) the current density must be low (arc tubes must have a large diameter). Sodium vapour readily absorbs light at the resonant doublet wavelengths, and therefore light generated in the centre of the arc is reabsorbed by the vapour; thus, only the outer surface of the arc emits light. This conflicts with the need for a large-diameter arc tube, and demands a compromise.

Sodium, especially hot sodium vapour, is highly reactive and attacks any glass with more than a small proportion of silica in it (i.e. almost all normal glasses and quartz glass). Special glasses have been developed with low silica content which resist the attack of sodium vapour, but they are expensive, are difficult to work and are attacked by moisture. Hence, ordinary soda-lime glass tubing has a coating of this resistant glass flashed onto its inside surface, and the resultant cheap, easily worked material is known as ply tubing.

Unless checked, sodium vapour readily migrates along the lamp to the cooler parts. To prevent this, small protrusions are moulded into the arc tube. They project out from the arc tube and are therefore slightly cooler than the surrounding wall. They act as reservoirs for the sodium metal and help to maintain the correct vapour pressure at all points along the lamp.

Two basic types of lamp construction in common use are:

(1) U tube or SOX lamp (*Figure 27.25*). A long arc tube is folded into a tight U shape. Mutual heating is provided by the two arms of the arc tube, but also each arm absorbs any light from the other arm which may strike it. The two effects are almost self-cancelling, but do produce a slight improvement in efficacy. The resultant lamp is fairly compact, with the

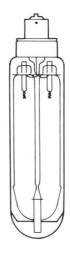

Figure 27.25 SOX lamp

Figure 27.26 SLI lamp

advantage that all of the lamp connections are at one end. A single-bayonet lamp-cap can therefore be used. The lamp efficacy is around 100 lm/W.

(2) Linear or SLI lamp (*Figure 27.26*). A long straight arc tube is used, indented along its length or given a cruciform cross-section. This increases the surface area of the lamp without increasing the cross-sectional area of the arc tube, and has the effect of reducing the lamp length required for a given power. The cruciform section also acts as a reservoir for sodium metal. The lamp efficacy is about 140 lm/W.

When a sodium lamp is first switched on, the sodium is all present as solid metal. An arc discharge cannot therefore occur unless the sodium is first vaporised. For this reason a mixture of neon and argon is used in the lamp. The initial discharge occurs through this gas; the sodium metal is vaporised by the heat from this discharge, and slowly takes over.

The mixture of neon and argon (1%), known as a Penning mixture, reduces the starting voltage required. The energy required to *excite* neon is slightly higher than that to *ionise* argon. Electrons passing through the gas collide with the main gas (neon) and excite the atoms, which may collide with argon atoms giving up their energy by ionising the argon and producing an extra electron. This mixture reduces the starting voltage by 30–50%.

All sodium (low-pressure) lamps when first switched on produce a distinctive red neon glow. Should sodium migrate from any part of the lamp, that part will also exhibit the red neon glow.

The effect of the Penning mixture is to make the starting voltage almost independent of ambient temperature. A low-pressure sodium lamp can therefore be restruck when hot within about 1 min.

Although electrodes are fitted at each end of the arc tube, it is not normal to provide them with a heating current, and the two ends of an electrode are connected together. In the SOX lamp this connection is made within the lamp and a single-bayonet lamp-cap is used. In the SLI lamp two bi-pin lamp-caps are used.

27.5.3.7 High-pressure sodium lamp (SON, SON/R, SON/T, SON/TD)

If the sodium vapour pressure in a low-pressure discharge tube is raised by a factor of 10^5, the characteristic radiation is absorbed and broadened, greatly improving the colour rendering. However, hot sodium vapour is highly reactive, destroying or discolouring conventional arc tube materials; further, to achieve a high vapour pressure the coolest region of the arc tube must have a temperature of 750°C. These phenomena, known in the 1950s, were not exploited until the development of a translucent ceramic material—isostatically pressed doped alumina—capable of operating at temperatures up to 1500°C and of withstanding hot sodium vapour. The difficult process of sealing electrodes into the ends of the arc tube has also been solved: typical methods are (a) brazing a niobium cap to the alumina tube, and (b) using hermetically sealed sintered ceramic plugs holding the electrodes.

Most high-pressure sodium lamps have an arc tube containing metallic sodium doped with mercury and argon or xenon. Radiation of light is predominantly from the sodium. The mercury vapour adjusts the electrical characteristics, and acts to reduce thermal conductivity and power loss from the arc. Argon or xenon aids starting. The arc tube is sealed into an evacuated outer jacket to minimise power loss and to inhibit oxidation of the end-caps, lead wires and sealing medium. Typical arc tube operating temperatures are 700–1500°C, and pressures are 0.13–0.26 atm; efficacies of 100–200 lm/W are achieved.

Starting is effected by high-voltage pulses (2–4.5 kV) to ionise the xenon or argon gas. The ionisation heats the lamp and the sodium vapour discharge takes over. The mercury vapour does not ionise, as its ionisation potential is higher than that of sodium, but its effect is to increase the lamp impedance, raising the arc voltage from about 55 V to 150–200 V. Electrodes are not heated during lamp operation. A hot lamp, after extinction, will not restart until cooled, unless 'hot-restart' ignitors (giving, e.g. 9 kV pulses) are provided.

SON lamp The conventional high-pressure sodium or SON lamp (*Figure 27.27a*) is similar in construction to an MBF lamp. The outer bulb is normally elliptical and is coated with diffusing material.

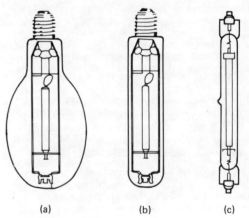

Figure 27.27 SON lamps. (a) SON lamp; (b) SON/T lamp; (c) SON/TD lamp

SON/T lamp This (*Figure 27.27b*) is almost identical with the basic SON lamp, except that the outer glass envelope is clear and tubular.

SON/TD lamp The SON/TD lamp (*Figure 27.27c*) is a linear lamp of double-ended construction (similar to an MBIL lamp or linear tungsten–halogen lamp) with a quartz-glass outer envelope.

In the SON/T and SON/TD lamps the arc tube is in a clear outer bulb, unobstructed by any diffusing coatings. For this reason such lamps are used where precise optical control is required. In contrast, the basic SON lamp is used where a direct view of the arc tube would be unacceptably bright.

SON/R lamp The SON arc tube can be mounted in a reflector bulb (similar to an MBFR) lamp, to produce a lamp with a defined beam, like a spotlight.

General It is interesting to note that, although the low-pressure sodium lamp is more efficient than the high-pressure version, the large dimensions of the former make accurate optical control difficult. Furthermore, the luminaires are large, cumbersome and expensive. High-pressure sodium lamp luminaires are therefore the better choice for many applications, and only in street lighting does low-pressure sodium find a major application. High-pressure sodium lamps are employed for street lighting, flood-lighting and industrial and commercial interiors.

27.5.3.8 Cold-cathode lamps

'Cold-cathode' is the name given to a form of long-life fluorescent lamp, developed originally from neons but sharing the technical advances made in the hot-cathode lamp industry in the post-war years. The production of light is fundamentally the same, by excitation of fluorescent powders by ultraviolet energy from mercury vapour. The term 'cold cathode', as opposed to 'hot cathode', refers to the electrodes, which run at about 200 °C and are in the form of deep-drawn, or otherwise formed, shells of high-purity steel housed at the end of the tube.

The cold-cathode tube is generally made in 22 mm or 25 mm diameter glass, at either 120 mA or 150 mA running current, in a variety of lengths and shapes. Electrodes can be placed at right angles or turned back, or the tubes can be curved and bent to shapes. The maximum straight length generally available is 3 m.

Certain primary colours can be obtained, some vivid hues being produced from the use of neon as a filling gas without any mercury content. General purpose lighting colours are available to follow the normal range. *Table 27.10* shows average lumen outputs for a 3 m tube through the first 15 000 hours of life.

It will be noted that the efficacy of the 25 mm white lamp, although less than standard fluorescent, is 45 lm/W, including gear losses, and the efficacy of 22 mm lamps is 42 lm/W. Reflector lamps, internally coated, with a 180° window are available in straight or curved 25 mm tubes.

The life of cold-cathode tubes is 15 000 h; this is generally held to be the economic life, and at this point the lamps should be replaced. However, there are many instalations which have lasted for more than 20 000 h. When a cold-cathode tube fails, there is no unsightly intermediate stage of partial starting and flickering.

Cold-cathode lamps will start instantly without delay or flicker, as there is no electrode pre-heat time, under a wide range of voltage fluctuation conditions: although reduction in voltage results in reduction in light output, there is not such a sensitive response as in tungsten lamps. Owing to the full part of each half-cycle of current being utilised with the tube in a 'triggered' condition, there is very little stroboscopic effect.

There are a number of types of control gear; desirable standards of design and manufacture are outlined in BS 559. All types are essentially of leakage reactance form. The open-circuit voltage is $1\frac{1}{2}$ or 2 times the operating voltage, and the transformer is capable of operating under secondary short-circuit conditions for several hours.

The most common form is the double-wound transformer (*Figure 27.28a*). The auto-transformer connection (*Figure 27.28b*) is designed to give a slim section, and to provide each tube with

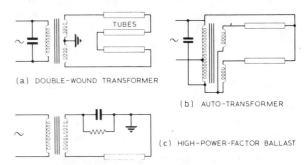

Figure 27.28 Control gear for cold-cathode tubes

an independent secondary winding, so that in the event of tube failure or disconnection, the operation of the remainder is unaffected. Up to four tubes can be operated from one auto-transformer. Both these types require a power factor correction capacitor to raise the p.f. to 0.85. *Figure 27.28(c)* shows a high-power-factor ballast for a single tube, with a series capacitor. This circuit tends to give a peaky waveform.

Simple methods of dimming are possible by progressively reducing the applied voltage to the transformers. Reduction to some 10–15% of the maximum brightness is possible before a slight flicker occurs immediately before extinction. For total control of dimming in the lower register, it is necessary to use a triggered circuit.

Through the rugged construction of the electrode system, cold-cathode tubes withstand continuous vibration. This form of lighting lends itself to decorative treatments to follow architectural features and for siting in coves and cornices for indirect lighting. The long-life lamp renders cold-cathode ideal for lighting where maintenance access is difficult and costly.

27.5.3.9 Electroluminescent devices

Electroluminescence is the emission of light from a semiconductor under the action of an electric field. The process involves heat only as a by-product of the mechanism, which is essentially a 'cold' one. The phenomenon is of commercial interest because of its relatively highly efficient production of visible light.

Table 27.4 Performance of cold-cathode lamps

Power (W)	Current (mA)	Size (mm)	Average output in lumens					
			3500 K	WW	4500 K	Nat	6500 K	DWW
87	120	22	3600	3500	3400	3150	2950	2700
94	150	25	4200	4100	3950	3650	3450	3150

WW = warm white; DWW = de luxe warm white; K = temperature, kelvin.

Characteristics Luminescence decays with time exponentially, so that it is convenient to quote the half-life of an electroluminescent source. The half-life varies from hundreds to millions of hours, depending on the type and purity of the semiconductor materials used. The sources emit light in comparatively narrow spectral bands, producing colour without the use of filters: and sources can be of almost any size, down to areas of less than 0.05 mm². The surface brightness is, however, comparatively low.

In an electroluminescent material two processes must take place: first, an adequate supply of electrons must be available in the conduction band, made possible by using an electric field to raise the energy level of electrons in the valence band; second, the electrons must give up their energy in the form of photons and so return to the valence band. The recombination process is dependent on the 'forbidden gap', the wavelength λ of the emitted radiation being defined by the energy jump E through the relation $\lambda = hc/E$, where h is Planck's constant and c is the free-space velocity of electromagnetic waves. Consequently, materials with a band-gap E between 1.65 and 3.2 eV are capable of producing visible light by electroluminescence, provided that the electron return from the conduction to the valence band is not made in two or more stages by reason of the presence of lattice impurities; but if this is the case, luminescence can be obtained with materials having an intermediate gap energy.

Materials The requirements for an electroluminescent material emitting in the visible spectrum are, generally, that it should have a bandgap of at least 2 eV, be susceptible to both p- and n-type 'doping', and should have either a direct band-gap or an activator system permitting 'steps'. Typical semiconductor materials are ZnS, GaP, GaAs and SiC.

The preparation of luminescent panels based on ZnS involves baking pure zinc sulphide with various dopes such as Mn, Cu and Cl. The sintered mass is ground to a particle size of the order of 10 μm, washed and dried, and mixed with a suitable resin. The suspension is coated on to a glass substrate supporting a very thin layer of Au or SnO, to form one electrode. The other electrode, an evaporated layer of Al on the rear surface of the cell, also acts as a reflector. The small cells are produced from GaP crystals, with areas selected for fabricating electroluminescent diodes. They are cut into 0.5 mm squares, and a p–n junction is formed by alloying a Sn sphere into the GaP, which is doped with Zn to produce the junction. The diode structure is completed by alloying an Au–Zn wire into the GaP to make an ohmic contact, and then connecting a wire to the Sn sphere.

Luminescent panels are, in circuit terms, equivalent to lossy capacitors, and are essentially a.c. devices requiring a minimum drive to about 150 V. A luminance level of 10–20 lm/m² and a half-life of 5000 h are obtainable.

27.6 Choice of lamps

It is advisable to purchase lamps of good manufacture which comply with the appropriate British Standard. When choosing any type of lamp, an inferior one may initially cost less, but the power consumption for a given output may be such that the running and, therefore, the overall cost will be appreciably higher. In addition, the average life may be less, involving extra labour costs for renewals.

Apart from quality the following factors affect lamp choice: (a) economics; (b) load and supply, (c) colour; (d) suitability; (e) life.

The loading of any lighting scheme is dependent primarily upon the efficiency of the lamps used. In cases where the available load is limited owing to the existing plant or cables, high-efficiency discharge lamps should be considered. When calculating loadings for discharge lamps, the losses in the control gear should be allowed for, as also must the reactive current in the cables brought about by low power factors. Although the starting current of high-pressure mercury lamps is about twice the running current, little or no provision need be made for it, as it only lasts for a minute or so. Most discharge sources are made for a.c. working only and the gear is designed for 50 Hz systems. If the supply is not a.c., the adoption of discharge lamps involves the use of special gear, adding to the cost.

Colour composition of the light often has an important bearing on lamp choice. For example, processes involving critical colour discrimination may exclude the use of discharge lamps other than fluorescent and metal halide lamps.

The suitability of a particular light source for any specific task can only be determined by a close study of working conditions, etc. Where cool operation is required, with freedom from harsh shadow and glare, the fluorescent tube is a natural choice. There are, in fact, few applications for which this type of lamp is definitely unsuitable. Lumen output, brightness, size, shape and temperature are factors which must be considered.

27.6.1 Colour qualities

From the colour point of view, light sources can be classified as (a) incandescent lamps and (b) discharge lamps.

In the first group are filament and arc lamps, candles, gas mantles, etc. The spectrum is continuous (i.e. there are no gaps in the range of wavelengths emitted). The distribution of energy in most cases follows very closely the radiation laws of a full radiator. At the temperatures at which filament lamps can be practically worked the peak of the emission curve is in the infrared and the curve slopes down through the visible spectrum from the red end to the violet end, a small amount of ultraviolet light being emitted. Filament lamps have, therefore, a preponderance of red and yellow light, making them unsuitable for accurate colour matching.

In the second group are the gas and vapour discharge lamps. The radiation from a discharge lamp is emitted at certain wavelengths only. The amount of energy radiated in these 'lines' and their position depends upon the gas or vapour used and its operating pressure. In general, as the pressure is increased, the lines broaden out until an almost continuous spectrum is superimposed on the line spectrum.

In the case of the high-pressure mercury lamp, light is emitted in lines at 405, 436, 492, 546 and 577/579 nm. The resultant light may appear white, green, blue or bluer green, depending upon the observer's eye, the distance from the source and the presence of other lamps. Colour rendering of blues, greens and yellows is fairly good under mercury lighting, but the reds seem brownish, owing to the lack of red in the lamp's spectrum.

Colour correction of mercury vapour lamps can be effected by the introduction of metallic halides into the mercury, by blending with tungsten lamps or by the use of fluorescence.

With the low-pressure lamp the colour of the light is controlled almost completely by the fluorescent powders.

Table 27.5 gives the efficacy range and colour properties of discharge and fluorescent lamps. For comparison, the GLS lamp has an efficacy range of 8–18, the tungsten–halogen lamp of 17–25 lm/W; both have very good colour rendering and a warm appearance.

27.6.2 Operating and replacement costs

Operating costs depend upon four factors: (1) power rating, and, in some cases, power factor of lamp and gear; (2) cost of electrical energy; (3) hours of use; and (4) lamp replacement costs (including labour).

Obviously, the more efficient the lamp the lower is the cost of electrical energy to provide a given illumination value. This

Table 27.5 Efficacies and colour properties

Lamp type	Efficacy* (lm/W)	Colour rendering	Colour appearance
Discharge lamps			
SOX	70–135	Nil	Yellow
SLI	65–110	Nil	Yellow
SON	60–110	Fair/poor	Warm
SON/T	55–110	Fair/poor	Warm
SON/TD	55–110	Fair/poor	Warm
SON/R	75–80	Fair/poor	Warm
MBF	35–50	Fair	Intermediate
MBFR	40–45	Fair	Intermediate
MBTF	12–20	Fair	Intermediate
MBI	45–70	Good	Intermediate
MBIF	45–70	Good	Intermediate
MBIL	40–70	Good	Intermediate
Tubular mercury fluorescent lamps†			
NORTHLIGHT, COLOUR-MATCHING	20–40	Very good	Cool
ARTIFICIAL DAYLIGHT	20–40	Very good	Cool
DAYLIGHT, Cool White	45–65	Fair	Intermediate/cool
NATURAL	30–50	Good	Intermediate
Kolor-rite, Trucolor 37	20–45	Very good	Intermediate
Colour 84, Plus-white	45–65	Good	Intermediate
De luxe Natural	15–35	Very good	Intermediate
WHITE	45–65	Fair	Intermediate
WARM WHITE	45–65	Fair	Warm
De luxe Warm White Softane 32	20–45	Good	Warm

* Based on total circuit power.
† BS colours in CAPITALS.

saving in actual running costs should, however, be set against the increased cost of the high efficiency (discharge) lamps and their associated control gear.

Hours of burning influence replacement costs and, consequently, overall running costs. In the case of discharge lamps, not only burning hours but also switching operations determine how long a lamp remains in service.

Lamp replacement costs are too often considered to be the net cost of the replacement lamp. This is not so, labour costs incurred in the actual replacement amounting to a surprisingly high figure. In many installations it has been found economical to adopt a system of 'group' replacement. Two methods are used. The first leaves all premature burnouts in until some 20% of the lamps have failed. With this system a satisfactory uniform illumination over the working plane obviously cannot be guaranteed. The second system caters for the replacement of premature failures individually and then group replacement when 10–20% have failed. Before group replacement is instituted it is necessary to look into the economics of the various systems in detail in order to determine the most suitable arrangement. See Technical Report No. 9, *Depreciation and Maintenance of Interior Lighting*, of the Lighting Division of the CIBS.

27.6.3 Developments

The high-pressure sodium lamp, with a guaranteed life of 8000 h and better colour rendering, is becoming competitive. Metal halide lamps are being improved, and made for smaller power ratings.

Multiphosphor fluorescent tubes of diameter 25 mm (instead of the conventional 35 mm) give excellent colour rendering and high efficacy with rare earth phosphors. The substitution of krypton gas for argon has lowered the power demand without loss of performance; the power can also be reduced by 10% with conventional phosphors, but without improvement in colour rendering.

27.7 Lighting design

27.7.1 Objectives and criteria

To design a lighting scheme the basic objectives must be first established. What sort of tasks will be carried out in the area to be lit? What mood needs to be created? What type of lighting will create a comfortable, pleasant environment? The objectives having been established, they have to be expressed as a series of lighting criteria. For example, what level of illuminance is required? How much glare is acceptable?

The designer then plans a scheme that will best meet the criteria by selecting the appropriate luminaires and considering the practical problems.

27.7.1.1 IES/CIBS code

The Lighting Division of the Chartered Institute of Building Services (formerly the Illuminating Engineering Society) publishes a code of recommendations for the lighting of buildings. It puts forward ideas and methods representing good modern practice and is concerned with both quantity and quality of light.

The Interior Lighting Code deals with: how building design affects lighting; lighting criteria; lighting and energy consumption; design methods; lighting equipment; and methods of maintenance. In addition, there is a schedule giving specific recommendations for a wide variety of areas such as assembly areas, factories, foundries, schools, hospitals, shops and offices.

Table 27.6 Illuminance and glare recommendations for interiors, based upon 1977 IES/CIBS lighting code

Class of visual task	Typical examples	Standard service illuminance (lx)	Reflectances or contrasts unusually low?	Will errors have serious consequences?	Is the area windowless?	Final service illuminance (lx)
Storage areas and plant rooms with no continuous work: *Movement and casual seeing*	Boiler houses, pump houses (25) Stairs, gangways (steelworks) Loading bays (large materials) (25) Mill motor rooms, blower houses (28)	150				150
Rough work: *Rough machining and assembly: Rough tasks with large detail*	Foundries (charging floors) (28) Heavy machinery assembly (25) Bakeries (general lighting) (22) Fire stations (appliance rooms) (22) Paper mills (pulping, general) (25)	300	NO — 300 YES	NO — 300 YES		300
Routine work: *Ordinary tasks with average detail*	Paint shops (rubbing down, spraying) (22) Medium machine and bench work (22) Marking off (structural steel) (28) Printing works (pre-make-ready) (22) Ordinary offices (19)	500	NO — 500 YES	NO — 500 YES	NO YES	500
Demanding work: *Fairly difficult small tasks with small detail*	Woodworking (fine bench and machine work) (22) Pressing, glazing (leather) (22) Final inspection (motor vehicles) (19) Final buffing, polishing (plating) (22) Deep-plan and mechanised offices (19)	750	NO — 750 YES	NO — 750 YES		750
Fine work: *Colour discrimination: Very difficult visual tasks with very small detail*	Inspection (radio, telecomn) (29) Paint works—colour matching (19) Fine bench and machine work, fitting (22) Hosiery, knitwear (linking, running on) (19) Gauge and tool rooms—general (19)	1000	NO — 1000 YES	NO — 1000 YES		1000
Very fine work: *Extremely difficult Visual tasks, with minute detail*	Fine, intricate inspection, gauging (16) Gem cutting, polishing, setting (19) Upholstery (cloth inspection) (16) Hand tailoring (19) Fine die-sinking (19)	1500	NO — 1500 YES	NO — 1500 YES		1500
Minute work: *Exceptionally difficult visual tasks with minute detail*	Inspection (small instruments) (19) Jewellery, watchmaking (finest work) (19) Final inspection, perching (textiles) (19)	3000				3000*

*Localised lighting, if necessary supplemented with optical aids, magnifiers, profile projectors, etc.
Note: The numbers in brackets are the recommended limiting glare indices.

For each entry a standard service illuminance and position of measurement is quoted. The amount of discomfort glare that can be tolerated is given by a limiting glare index. A list of suitable lamp types and their colour appearance is given together with notes on special problems that may be encountered.

The recommended standard service illuminance—in other words, the recommended illuminance that should be provided for a particular application—is a useful guide for the designer. But it is only a recommendation: in some circumstances it should be increased. For example, if errors could have serious consequences in terms of cost or danger, or if unusually low reflectances or contrast are present in the particular task, or if tasks are carried out in windowless interiors where the recommended standard service illuminance is less than 50 lx, the illuminance level should be increased. On the other hand, there are circumstances (e.g. when the duration of the task is unusually short) when the designer would use his judgement to reduce the standard service illuminance. Incorporated into the Code is a flow chart (*Table 27.6*) which helps the designer to take the basic standard service illuminance and make such desirable adjustments.

27.7.1.2 Uniformity

In addition to providing the correct illuminance, the uniformity of the lighting level is also important. The uniformity is expressed as the ratio between the minimum and the average illuminance over the working area. It should not be less than 0.8 in areas where the tasks are performed.

27.7.1.3 Reflectance

Reflectances should also be considered. The effective reflectances of walls in a room should be between 0.3 and 0.8; the ceiling should have a reflectance of 0.6 or greater; and the floor should have a reflectance of between 0.2 and 0.4.

27.7.1.4 Illuminance ratio

The ratio between the illuminance on the wall and that on the task, and between the ceiling and that on the task, is also important. The wall illuminance should be between 0.5 and 0.8, and the ceiling illuminance should be between 0.3 and 0.9, of the task illuminance (see *Figure 27.29*).

For normal working environments, if the reflectances and the ratios between wall illuminance and the task, and between ceiling illuminance and the task, are outside the recommended levels, they will be unacceptable. However, there are exceptions if a particular mood or atmosphere is being created.

27.7.1.5 Modelling and shadow

The direction of light and the size of the luminaires in an interior affect highlights and shadows, influencing the appreciation of shape and texture. The term 'modelling' is used to describe how light reveals solid forms. It may be harsh (the contrast may be excessive and produce deep shadows) or flat (the light provides low contrast with little shadowing). Either extreme occurring in the lighting of a general working area makes vision difficult or unpleasant. Therefore, the aim of good lighting design must be to produce a suitable compromise between the two, although it may not be the same for different applications. Texture and surface details are best revealed by light with fairly strong directional characteristics. Reflections of lamps on polished surfaces may veil what one needs to see, reducing contrast and handicapping one's vision.

27.7.1.6 Glare

A bright object on a dull background is liable to cause glare. *Disability glare* makes it difficult to see contrast or detail. An example is the headlight of an approaching car at night: it is difficult to see past the car to the road beyond. *Discomfort glare* does not perceptibly reduce the ability to see, but it contributes to a feeling of fatigue towards the end of the working day, or when doing difficult visual tasks.

In a conventional lighting scheme disability glare is unlikely. For interiors lit with a regular array of overhead luminaires, the level of discomfort glare can be expressed as a glare index. Glare indices are measured in steps of three units and typically range from about 13 to 28. The Interior Lighting Code shows the limiting glare index acceptable for a particular working environment: for example, a general office would have a limiting glare index of 19.

27.7.1.7 Colour rendering

Generally the apparent colour of objects seen under an artificial lighting installation is of some importance, sometimes it is vital, as for instance in the buying and selling of foodstuffs, the preparation of paints and dyes and the matching of silks, cottons, etc., to fabrics and to one another.

A red article looks red because it reflects red light more strongly than other colours of light, but it can only do so if there is some red light present for it to reflect. Similarly, a green article can only look green if there is some green light present, and so on. It follows that if all colours are to be seen well, the light must contain a mixture of all possible colours of light of roughly equal strength. Such a light will look white, or nearly so. Unfortunately,

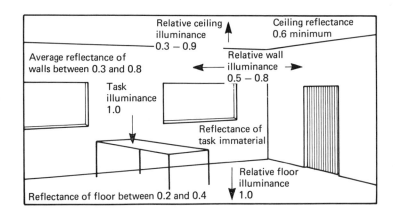

Figure 27.29 Recommended ranges of reflectance and relative illuminance for room surfaces

it is possible for a light to look white even though lacking some of the possible colours (e.g. the light of an ordinary mercury lamp can in certain circumstances look fairly white, though it lacks more colours than it contains, and the poor colour rendering of this type of lamp is well known).

Filament lamps of all types emit all possible colours of light, but in far greater strength at the red end of the spectrum than at the blue end. Thus, red and yellow objects appear in stronger colour than in natural daylight, whereas blues are weak and somewhat muddy in appearance. This, however, is a distortion familiar to most people, who have learned to make allowance for it in choosing decoration materials, etc. The light of a 'deluxe warm white' fluorescent lamp is very similar to that of a filament lamp.

For accurate colour judgement a light similar to north-sky daylight is necessary. An 'artificial daylight' fluorescent lamp has been developed which meets the requirements of BS 950. The 'colour-matching' or 'north-light' type of fluorescent lamp gives a close approximation to north-sky light and is found entirely satisfactory in many industries critically concerned with colour, but this very white light is perhaps too 'cool' in appearance to be used alone in interiors of a more domestic nature, where a 'warm' light is traditional. The other colours of fluorescent lamps are mainly of higher efficiency than the above-named two types, and the choice will lie between them according to the relative importance of their efficiency and their particular colour rendering properties.

27.7.2 Luminaires

Most light sources emit light in nearly all directions; generally, this is both wasteful and visually uncomfortable, and the important functions of most luminaires are: (a) to redirect the light output distribution of a bare electric lamp to give more efficient control of the lamp lumen output in a preferred direction with minimum loss; (b) to reduce glare from the light source; (c) to be acceptable in appearance, and, in some cases, to contribute to the decor of the surroundings.

Other practical aspects which should be considered by the fittings designer should at least be:

(a) *Lamp protection:* The lamp must be mechanically supported and protected, although the extent of this protection will depend on the application for which the luminaire is intended.
(b) *Electrical safety:* The lampholder, associated wiring and control gear (required for use with fluorescent tubes or discharge lamps) must be supported and protected.
(c) *Heat dissipation:* The heat generated by lamps and control gear must be conducted away from the wiring, control gear or other part of the luminaire which can deteriorate owing to overheating.
(d) *Finishing:* Any exposed metal work must be painted or plated to protect it against corrosion. The degree of protection necessary will depend on the application—i.e. special 'proof' fittings for use in corrosive atmospheres are made from glass-reinforced plastics material (GRP).

In addition to the functional aspects of mechanical, electrical, thermal and chemical protection, the designer must use practical techniques suited to low-cost production in either small- or large-scale batch quantities. An allowance must also be made for decorative features. 'Built-in' design extras may be aimed at easing the installation or maintenance of the luminaires, such as through-wiring, end entries or quick-release clips for reflectors and other attachments. Fused terminal blocks may be fitted to give individual protection of each fitting.

27.7.2.1 Light distribution control

A material used to control light either reflects, refracts or absorbs the light that falls on it. The less light absorbed, the more efficient will be the luminaire. Reflecting materials are used to produce large changes in the direction of light, refracting materials to produce smaller changes in direction, and translucent materials to reduce glare.

27.7.3 Types of luminaires (fittings)

Owing to absorption and inter-reflection between adjacent surfaces, no luminaire gives all the light output of the lamp which it surrounds.

The classification of luminaires can be carried out in several ways—according to the light source they use, their field of application or their appearance. More fundamentally, they are classified according to their light distribution, in terms of the proportion of the total light output of the luminaire in the upper and lower hemisphere (up or down) as follows:

Classification	Percentage	
	Upper	Lower
Direct	0–10	90–100
Semidirect	10–40	60–90
General diffusing	40–60	40–60
Semi-indirect	60–90	10–40
Indirect	90–100	0–10

A further classification is by the BZ (British Zonal) system, based on the direct ratio of the luminaire, i.e. the proportion of downward light directly incident upon the working plane. For a particular luminaire this increases with the size of the room in which it is used, and by plotting the direct ratio of luminaires against room size and mounting height a series of characteristic curves results. These have been classified from 1 to 10 (*Table 27.7*).

Designation by type takes the following form.

Industrial These fittings are designed for optical efficiency and durability and cover a wide range of applications.

High bay These are for use when mounting heights of 8–10 m (25–30 ft) or above are encountered. They have a narrow distribution to concentrate light on the working plane.

Batten or spine This term refers to a fluorescent luminaire with no attachments (other than the control gear) in a channel and a bare tube.

Diffuser fittings, controller fittings This term refers to fluorescent fittings with either opal diffuser or prismatic controller attachments.

Troffer Fluorescent or discharge luminaires designed to recess into suspended ceilings and fit their modular size.

Bulkhead These fittings contain tungsten or mercury lamps, or miniature fluorescent tubes. The front is sometimes prismatic (plastics or glass) and the fitting may be mounted vertically or horizontally. They are used in corridors, and over doorways and stairways, etc., and can be used outdoors.

Wellglass This is a lamp surrounded by an enclosure of transparent or translucent glass. It is usually weatherproof and often used outdoors fitted to a bracket.

Table 27.7 Direct ratio limits for the BZ classifications

Classification	Room index								
	0.75	1.00	1.25	1.50	2.00	2.50	3.00	4.00	5.00
BZ1	0.577	0.666	0.724	0.764	0.818	0.852	0.875	0.905	0.924
BZ2	0.517	0.611	0.674	0.719	0.780	0.812	0.847	0.883	0.905
BZ3	0.462	0.558	0.624	0.673	0.739	0.784	0.816	0.858	0.884
BZ4	0.422	0.507	0.576	0.626	0.697	0.746	0.781	0.828	0.858
BZ5	0.367	0.435	0.504	0.554	0.626	0.677	0.716	0.769	0.804
BZ6	0.297	0.358	0.424	0.473	0.545	0.599	0.641	0.699	0.740
BZ7	0.253	0.310	0.374	0.422	0.494	0.550	0.593	0.656	0.700
BZ8	0.204	0.260	0.324	0.371	0.446	0.504	0.550	0.617	0.665
BZ9	0.154	0.208	0.274	0.322	0.400	0.461	0.509	0.581	0.633
BZ10									

Flameproof A flameproof fitting is designed to operate safely in situations where explosive vapours or gases may be present (e.g. coal mines, oil refineries, etc.). The fittings are classified according to the location for which they are designed.

Zone 2 A fitting designed to meet requirements which allow it to be used in certain hazardous areas. It can carry fluorescent, tungsten or mercury discharge lamps.

Decorative Decorative fittings are designed primarily to have a pleasing appearance, efficiency being of secondary importance.

27.7.4 Installation and maintenance

Lighting equipment should be capable of easy installation. This is particularly true of fluorescent fittings, which can be somewhat bulky.

Fixing centres should comply with the appropriate British Standard throughout the range, and the facilities of end entry and through-wiring, where possible, provided. Mains terminal blocks should be readily accessible and of size sufficient to allow looping of cables where possible. Fusing should be readily accessible. Cool wiring boxes on discharge fittings should be provided, as well as adequate earthing of the whole of the components of the fitting.

All attachments, such as diffusers, should be easily removable with safety chains or cords provided as necessary.

Lamp replacement and cleaning should not entail major dismantling of the fitting.

27.7.5 Design

The majority of industrial and commercial lighting problems are solved by the use of general lighting. One method of calculating the illumination derived from a lighting system is known as the 'Lumen' method. Details of this are given below, procedure being grouped into logical steps.

Step (1) Illuminance Illumination values for various tasks are set out in the IES Code V 'Recommendations for good interior lighting'.

Step (2) Luminaire type Consideration must be paid to all relevant factors, i.e. horizontal and vertical illumination requirements, glare prevention, efficiency, appearance, maintenance, economy, etc. For example, ordinary assembly work requires a certain amount of shadow to enable shape to be distinguished easily; therefore indirect lighting would be inappropriate, apart altogether from its relatively high running cost. The obvious choice for this class of work is the dispersive type of reflector.

Step (3) Mounting height The mounting height of fittings is usually dictated by the building construction, but, in general, it is advisable to make the height as great as possible compatible with good maintenance and installation facilities.

Step (4) Room reflectance It is now necessary to calculate, measure (or guess) the effective reflectances of the three main room surfaces: (1) *the ceiling cavity* (area above the luminaires); (2) *the walls* (from the height of the working plane to the height of the luminaires); (3) *the floor cavity* (area below the working plane).

Step (5) Room index From the room dimensions is calculated the 'room index', i.e. the ratio between twice the floor area and the area of the walls measured as the area between the working plane and the luminaires, i.e.

$$\mathrm{RI} = (L \times W)/[(L+W) \times H]$$

where L and W are the room length and width, and H is the mounting height. Results may be rounded to the nearest value in the series 0.75, 1.25, 1.5, 2.0, 2.5, 3.0, 4.0 and 5.0. A room in which L and W are many times greater than H will have a high room index, while if L and W are less than H, it will have a low room index. It is important to realise that the room index is a measure of the relative, not the absolute, dimensions of the room.

Step (6) Utilisation factor and number of luminaires The room index affects the utilisation factor. The larger the room index, the higher the percentage of light reaching the working place. The utilisation factor will also be affected by the reflectances. The higher the room reflectances, the more light will be inter-reflected around the room. Utilisation factors are provided for different combinations of room reflectance and room index. In more recent publications utilisation factors (UF) may have the letters F, W or C in parentheses to indicate whether they refer to the floor, walls or ceiling, respectively, although normally only UF(F) values are published.

The Lighting Division of CIBS has encouraged manufacturers to produce UF tables in a standard format to simplify performance comparisons and to prevent ambiguity. An example is shown in *Table 27.8*.

Allowance must be made for depreciation in light due to dust and dirt on fittings and surroundings: this is in the form of a maintenance factor. For normal interiors the factor is typically 0.8; for dirty locations it may be as low as 0.4.

For installations in high-bay foundries it may be necessary to make additional allowance for light absorption due to dirt and smoke. The average illuminance $E(\mathrm{F})$ over the working plane is found from

$$E(\mathrm{F}) = [\mathrm{UF}(\mathrm{F}) \times n \times N \times F \times \mathrm{MF} \times \mathrm{CF}]/A$$

Table 27.8 Example of a standard utilisation factor (UF(F)) table

Utilisation factors UF(F) $SHR_{nom} = 1.5$

Room reflectance			Room index								
C	W	F	0.75	1.00	1.25	1.50	2.00	2.50	3.00	4.00	5.00
0.70	0.50	0.20	0.43	0.49	0.55	0.60	0.66	0.71	0.75	0.80	0.83
	0.30		0.35	0.41	0.47	0.52	0.59	0.65	0.69	0.75	0.78
	0.10		0.29	0.35	0.41	0.46	0.53	0.59	0.63	0.70	0.74
0.50	0.50	0.20	0.38	0.44	0.49	0.53	0.59	0.63	0.66	0.70	0.73
	0.30		0.31	0.37	0.42	0.46	0.53	0.58	0.61	0.66	0.70
	0.10		0.27	0.32	0.37	0.41	0.48	0.53	0.57	0.62	0.66
0.30	0.50	0.20	0.30	0.37	0.41	0.45	0.52	0.57	0.60	0.65	0.69
	0.30		0.28	0.33	0.38	0.41	0.47	0.51	0.54	0.59	0.62
	0.10		0.24	0.29	0.34	0.37	0.43	0.48	0.51	0.56	0.59
0.00	0.00	0.00	0.19	0.23	0.27	0.30	0.35	0.39	0.42	0.46	0.48

Rating: 65 W, 1500 mm.
Mounted: on ceiling.
Multiply UF values by service correction factors.

where

UF(F) = the utilisation factor for the reference surface S;
N = the total number of luminaires in the installation;
F = the light flux of each lamp (bare);
MF = the maintenance factor of the installation;
CF = the product of any additional correction factors necessary;
n = the number of lamps per luminaire;
A = the area of the working plane.

Alternatively, the number of luminaires required is

$N = [E(F) \times A] / [UF(F) \times n \times F \times MF \times CF]$

Table 27.9 gives utilisation factors (UF(F)) as a guide. More accurate values should be used when possible.

Step (7) Spacing of luminaires and layout For a regular array of luminaires, SHR_{max} is the maximum spacing/height ratio that will provide acceptable uniformity. When disymmetric luminaires are used in long continuous runs, the maximum transverse spacing can be increased to $SHR_{max,tr}$. In addition to these limits, the sum of the transverse and axial SHR values should not exceed twice SHR_{max}. If SHR_{max} is not given, it can not be assumed to be greater than SHR_{nom} (the value for which the UF table is calculated).

SHR_{max} and $SHR_{max,tr}$ provide information only about the maximum spacing/height ratio that will result in acceptable uniformity on the horizontal working plane. In practical installations, obstructions or other factors frequently make closer spacing essential.

The spacing of units is limited by the mounting height, building structure and plant layout. In all cases it is recommended that spacing/height ratios (issued by the fittings manufacturers) be not exceeded if even illumination is desired. Where freedom from shadow is required, closer spacings should be used. In calculating spacings it should be remembered that the mounting height should be taken from the working plane, be it the floor, a desk at 0.75 m or a work bench at 1.0 m.

Ceiling divisions, columns, shafting, ventilation trunking and other obstructions restrict possible layouts of outlets. It is thus desirable to draw a scale plan of the area showing all obstructions and plan the layout on this. In multistorey buildings the units should form, if possible, a symmetrical layout in the ceiling panels formed by the joints.

Especially where fluorescent lighting is concerned, the use of continuous trunking from which the fittings can be suspended and which carries the wiring should usually be considered, as in many cases it can effect considerable economies in installation cost.

Where work benches or machines are located along the outer walls, the distance between the outside rows of fittings and walls should not exceed one-third of the nominal spacing distance. If the wall space is a non-working area, this distance can be increased to one-half the nominal spacing. This should not be exceeded, as it is desirable to have a certain amount of light thrown upon the walls in order to maintain a reasonable brightness throughout the room.

Plant layout may determine the location of outlets, as in the case of the textile industry. What are known as localised general systems are installed in such cases, the outlets being localised in relation to the plant.

An installation will not be satisfactory if the maximum spacing/height ratio is exceeded. Conformity, however, will still be unsatisfactory if there are obstructions (such as large machines); closer spacing is then essential.

27.7.6 Glare

An installation is not completely planned until it has been checked to ensure that the discomfort-glare level is acceptable. Glare calculation is based on CIBS/IES Technical Report No. 10.

The room dimensions, the mounting height of the fittings and the reflection factors of the ceiling and walls are already known. In addition, one has to find (normally from the catalogue) the following data for the luminaire proposed: (a) the BZ (British Zonal) classification; (b) the upward and downward light output ratios; (c) the luminous area (in plan); (d) the total light output of the lamp(s) contained; and (e) the mounting height above eye-level.

By reference to the appropriate table in the Report it is then a fairly simple matter to work out the glare index of the proposed installation and compare it with the limiting glare index for the occupation concerned. If the glare index of the installation is lower than this limit, the lighting will be reasonably comfortable visually; if it is higher, there is the probability that some of the occupants will have reasonable cause for complaint, and the installation should be redesigned (perhaps with a different kind of fitting) to bring the glare index below the limiting value.

It should be noted that although some types of fittings are inherently less glaring than others—in general, the lower the BZ number the better in this respect—the geometry of the installation also has an important effect on its glare index. *Figure 27.30* shows the glare indices computed for three typical classes of fluorescent fittings in an office 5.5 m wide, with light-coloured decoration and with fittings mounted 3 m above the floor. The limiting glare index for a general office (19) is marked. It will be seen that while louvred modular ceiling panels will be satisfactory at any room length, enclosed plastics diffusers viewed sideways-on become unreasonably glaring at a room length greater than about 10 m and batten (bare-lamp) fittings are unacceptable under almost any condition. (NOTE: these curves apply only to the conditions stated. Different room widths, different reflection factors and different mounting heights would all influence the shape and position of the curves.)

The British Lighting Council handbook *Interior Lighting Design* presents a simplified method of calculating glare index, sufficiently accurate for day-to-day purposes not involving compliance with a tight specification. The CIBS is encouraging manufacturers to produce initial glare index tables for these luminaires, as a further simplification, and these should be used when they are available.

27.7.7 Local lighting

Where a system of general lighting proves inadequate for specific tasks, it is necessary to provide a localised light in order to obtain the required type and degree of illumination.

Local lighting should never be used alone, but always as a supplement to general lighting. The methods of application can be classified as follows: local lighting from *fixed points*; from *flexible bracket arms*; by means of *portable lamps*; and by means of *fluorescent lamps*.

The use of fixed points is usually confined to bench or desk where working positions are set. Mains voltage can be used with safety in most cases, as units are mounted out of normal reach. Most manufacturers market special units for this class of work. In some cases projector-type units are to be preferred, as they can be mounted well clear of the working area. A typical example is the lighting of the delivery ends of printing presses.

Flexible-arm brackets find their greatest use with machine tools, although for vice work on intricate parts they are very useful. Units are marketed with one, two or three arms, depending on the use for which they are required. As the wiring is easily accessible to the worker, it is recommended that low voltages be used with this type of fitting. These low voltages (12, 25 and 50 V are standard) can be obtained by means of small transformers in the case of a.c. supplies. Primary windings are arranged for either the 200/260 V or 350/440 V ranges, so that wiring can be arranged from either single-phase or 3-phase supplies. Apart from the safety of low voltages they have the advantage of small fitting size and, also, more robust lamps.

In many works where the assembly of large or intricate parts is undertaken, e.g. aircraft, switchgear assembly, etc., overhead lighting cannot penetrate to all working points. Portable local lights must therefore be used, often taking the form of hand lamps. Here, again, low voltages are recommended. Hand lamps should be fitted with antiglare shields or reflectors. Banks of projector units mounted on portable frameworks are often preferable to hand lamps.

27.7.8 Practical illumination measurements

For all practical illumination measurements the photoelectric light meter has replaced the older visual methods. In its essentials the photoelectric cell consists of a copper, steel or iron plate upon which a selenium film has been formed. Contacts are made on this plate and on the thin transparent special metal coating which covers the oxide or selenium film. When light strikes the boundary surface between the copper oxide or selenium and the transparent coating, an electron flow is inaugurated which passes through the external circuit and deflects a moving-coil microammeter. Current, and, hence, deflection, is almost strictly proportional to the illumination, provided that the instrument resistance is fairly low (not exceeding $200\,\Omega$).

This type of cell has a spectral response different from the human eye and it is sensitive to ultraviolet light. Light meters are usually calibrated with tungsten filament lamps at a colour temperature of 2700 K.

To make allowance for the error in response with other light sources, an appropriate correction factor should be applied. Alternatively, the photocell may be approximately corrected to the relative response of the average human eye by means of a green filter.

An appreciable error in the measurement of general lighting can occur due to the fact that the cell is protected by a glass window or by lacquer. This reflects a proportion of the light which would otherwise fall on the sensitive surface of the cell. Readings, therefore, tend to be low, especially when the light reaches the cell at high angles of incidence. Modern light meters are available with cells which are both colour-corrected and corrected for obliquity (cosine) error, and with ranges of $1-10^5$ lx (lm/m^2).

When taking readings with photocells, it should be remembered that the instrument indicates the illumination on the plane in which the cell's surface lies.

General lighting is usually measured on a horizontal plane with the light unobstructed, so that care should be taken not to let the cell be screened by any part of the body. When testing the light available on a machine, however, the operative should adopt his normal working position. If an average value is required, a number of readings should be taken so that the effects of new or old lamps, which may be in use, are cancelled out.

While light meters are usually housed in robust cases, it should be remembered that the microammeter is relatively sensitive and also that only a small scratch on the cell's surface is sufficient to render it useless. These cells are also damaged by moisture and by temperatures over 70°C.

Apart from checking calculations, the light meter finds great use in determining when the illumination has fallen to such a level as to require reflector cleaning and relamping. By the regular use of a light meter a maintenance engineer can draw up an economic cleaning schedule.

27.8 Applications

27.8.1 Factory lighting

The first statutory provision for industrial lighting were made with the publication of the 1937 Factories Act. The regulations laid down in this Act cover all factories, irrespective of size, working hours, etc. Section 5 of the Factories Act 1961 states:

'(1) Effective provision shall be made for securing and maintaining sufficient and suitable lighting, whether natural or artificial, in every part of a factory in which persons are working or passing.

(2) The Secretary of State may, by regulation, prescribe a standard of sufficient and suitable lighting for factories or for any class or description of factory or parts thereof or for any process.

(3) Nothing in the foregoing provisions in this section or in any regulation made thereunder shall be construed as enabling

Table 27.9 Utilisation factors

Description of fitting, and typical DLOR (%)		Basic DLOR (%)	Ceiling	0.70			0.50			0.30		
			Walls	0.50	0.30	0.10	0.50	0.30	0.10	0.50	0.30	0.10
			Room index				Reflectance					
(M)	Aluminium industrial reflector (72–76)	70	1.0	0.52	0.49	0.45	0.52	0.48	0.45	0.52	0.48	0.45
(T)	High-bay reflector, aluminium (72) or enamel (66)		1.25	0.56	0.53	0.50	0.56	0.53	0.49	0.56	0.52	0.49
			1.5	0.60	0.57	0.54	0.59	0.57	0.53	0.59	0.55	0.53
			2.0	0.65	0.62	0.59	0.63	0.60	0.58	0.73	0.59	0.57
			2.5	0.67	0.64	0.62	0.65	0.62	0.61	0.65	0.62	0.60
			3.0	0.69	0.66	0.64	0.67	0.64	0.63	0.67	0.64	0.62
			4.0	0.71	0.68	0.67	0.69	0.67	0.65	0.69	0.66	0.64
			5.0	0.72	0.70	0.69	0.71	0.69	0.67	0.71	0.67	0.66
(F)	Recessed louvred trough with optically designed reflecting surfaces (50)	50	1.0	0.37	0.35	0.32	0.37	0.34	0.32	0.37	0.34	0.32
			1.25	0.40	0.38	0.35	0.40	0.37	0.35	0.40	0.37	0.35
			1.5	0.43	0.41	0.38	0.42	0.40	0.38	0.42	0.39	0.38
			2.0	0.46	0.44	0.42	0.45	0.43	0.41	0.44	0.42	0.41
			2.5	0.48	0.46	0.44	0.47	0.45	0.43	0.46	0.44	0.43
			3.0	0.49	0.47	0.46	0.48	0.46	0.45	0.47	0.45	0.44
			4.0	0.50	0.49	0.48	0.49	0.48	0.47	0.48	0.47	0.46
			5.0	0.51	0.50	0.49	0.50	0.49	0.48	0.49	0.48	0.47
(F)	Open-end enamel trough (75–85)	75	1.0	0.49	0.45	0.40	0.49	0.44	0.40	0.48	0.43	0.40
(F)	Closed-end enamel trough (65–83)		1.25	0.55	0.49	0.46	0.53	0.49	0.45	0.52	0.48	0.45
(T)	Standard dispersive industrial reflector (77)		1.5	0.58	0.54	0.49	0.57	0.53	0.49	0.55	0.52	0.49
(T)	Enamel deep-bowl reflector (72)		2.0	0.64	0.59	0.55	0.61	0.58	0.55	0.60	0.56	0.54
			2.5	0.68	0.63	0.60	0.65	0.62	0.59	0.64	0.61	0.58
			3.0	0.70	0.65	0.62	0.67	0.64	0.61	0.65	0.63	0.61
			4.0	0.73	0.70	0.67	0.70	0.67	0.65	0.67	0.66	0.64
			5.0	0.75	0.72	0.69	0.73	0.70	0.67	0.70	0.68	0.67
(F)	Shallow fitting with diffusing sides, optically designed downward reflecting surfaces (55)	50	1.0	0.33	0.30	0.27	0.32	0.29	0.27	0.32	0.29	0.27
			1.25	0.36	0.33	0.30	0.35	0.32	0.30	0.35	0.32	0.30
			1.5	0.39	0.36	0.37	0.38	0.35	0.33	0.38	0.34	0.32
			2.0	0.43	0.39	0.37	0.41	0.39	0.36	0.40	0.37	0.36
			2.5	0.45	0.41	0.39	0.43	0.41	0.39	0.42	0.40	0.38
(T)	Industrial reflector with diffusing globe (50)		3.0	0.46	0.43	0.41	0.44	0.42	0.41	0.43	0.42	0.40
			4.0	0.48	0.45	0.44	0.46	0.45	0.43	0.45	0.44	0.42
			5.0	0.50	0.47	0.46	0.48	0.47	0.45	0.47	0.45	0.44
(F)	Recessed (modular) diffuser (43–54)	50	1.0	0.32	0.29	0.26	0.31	0.28	0.26	0.30	0.28	0.26
			1.25	0.35	0.32	0.29	0.34	0.31	0.29	0.32	0.30	0.28
(F)	Shallow ceiling-mounted diffusing panel (40–55)		1.5	0.37	0.34	0.31	0.36	0.33	0.31	0.34	0.32	0.30
			2.0	0.41	0.37	0.35	0.39	0.37	0.34	0.38	0.36	0.34
			2.5	0.43	0.40	0.38	0.42	0.39	0.37	0.40	0.38	0.37
			3.0	0.45	0.42	0.40	0.44	0.41	0.40	0.42	0.40	0.39
			4.0	0.47	0.44	0.43	0.46	0.44	0.42	0.44	0.42	0.41
			5.0	0.49	0.46	0.45	0.47	0.46	0.44	0.46	0.44	0.43
(F)	Suspended opaque-sided fitting, upward and downward light, diffuser or louvre beneath (40–50)	45	1.0	0.41	0.36	0.32	0.37	0.33	0.30	0.34	0.30	0.27
			1.25	0.45	0.41	0.36	0.41	0.37	0.34	0.37	0.33	0.30
			1.5	0.49	0.45	0.40	0.44	0.40	0.37	0.39	0.35	0.33
			2.0	0.55	0.50	0.46	0.48	0.45	0.42	0.42	0.39	0.37
			2.5	0.58	0.53	0.50	0.51	0.48	0.45	0.45	0.42	0.40
			3.0	0.60	0.55	0.53	0.53	0.50	0.48	0.47	0.44	0.42
			4.0	0.63	0.59	0.57	0.55	0.53	0.51	0.48	0.46	0.44
			5.0	0.65	0.62	0.60	0.57	0.55	0.53	0.50	0.48	0.46
(M)	400 W reflectorised mercury lamp	100	1.0	0.66	0.59	0.51	0.64	0.58	0.53	0.64	0.57	0.53
			1.25	0.71	0.65	0.58	0.69	0.64	0.59	0.68	0.62	0.58
			1.5	0.76	0.70	0.64	0.74	0.69	0.64	0.73	0.67	0.63
			2.0	0.84	0.77	0.72	0.80	0.76	0.71	0.79	0.73	0.70
			2.5	0.88	0.82	0.77	0.84	0.80	0.76	0.82	0.78	0.75
			3.0	0.91	0.86	0.82	0.88	0.84	0.81	0.86	0.82	0.79
			4.0	0.95	0.90	0.87	0.92	0.89	0.85	0.90	0.86	0.83
			5.0	0.99	0.94	0.91	0.96	0.93	0.89	0.93	0.89	0.87

Table 27.9 (continued)

Description of fitting, and typical DLOR (%)		Basic DLOR (%)	Ceiling	0.70			0.50			0.30		
			Walls	0.50	0.30	0.10	0.50	0.30	0.10	0.50	0.30	0.10
			Room index				Reflectance					
(T)	Near-spherical diffuser, open beneath (50)	50	1.0	0.43	0.36	0.32	0.38	0.34	0.29	0.31	0.29	0.26
			1.25	0.48	0.41	0.37	0.42	0.38	0.33	0.34	0.32	0.29
			1.5	0.52	0.46	0.41	0.46	0.41	0.37	0.37	0.35	0.32
			2.0	0.58	0.52	0.47	0.50	0.46	0.43	0.42	0.39	0.36
			2.5	0.62	0.56	0.52	0.54	0.50	0.47	0.45	0.42	0.40
			3.0	0.65	0.60	0.56	0.57	0.53	0.50	0.48	0.45	0.43
			4.0	0.68	0.64	0.61	0.60	0.56	0.54	0.51	0.48	0.46
			5.0	0.71	0.68	0.65	0.62	0.59	0.57	0.53	0.50	0.48
(F)	Suspended louvres metal trough, upward and downward light, optically designed reflecting surfaces (47–54)	50	1.0	0.46	0.42	0.40	0.44	0.41	0.39	0.42	0.39	0.37
			1.25	0.49	0.46	0.43	0.47	0.44	0.42	0.45	0.42	0.40
			1.5	0.52	0.49	0.46	0.49	0.47	0.44	0.47	0.44	0.42
			2.0	0.56	0.53	0.51	0.52	0.50	0.48	0.49	0.47	0.45
			2.5	0.58	0.55	0.53	0.54	0.52	0.50	0.51	0.49	0.47
			3.0	0.59	0.57	0.55	0.55	0.53	0.52	0.52	0.50	0.49
			4.0	0.61	0.59	0.57	0.57	0.55	0.54	0.53	0.51	0.50
			5.0	0.63	0.60	0.59	0.58	0.57	0.55	0.54	0.52	0.51
(M)	250 W reflectorised mercury lamp (100)	100	1.0	0.71	0.65	0.60	0.70	0.64	0.60	0.69	0.63	0.60
			1.25	0.77	0.71	0.64	0.76	0.70	0.66	0.76	0.69	0.66
			1.5	0.82	0.77	0.72	0.81	0.76	0.71	0.80	0.74	0.71
			2.0	0.90	0.84	0.80	0.87	0.83	0.79	0.86	0.81	0.78
			2.5	0.93	0.88	0.84	0.90	0.87	0.83	0.90	0.85	0.82
			3.0	0.96	0.91	0.88	0.93	0.90	0.87	0.92	0.88	0.85
			4.0	0.98	0.95	0.92	0.96	0.94	0.91	0.94	0.91	0.89
			5.0	1.00	0.98	0.96	0.99	0.97	0.94	0.96	0.94	0.92
(F)	Enamel slotted trough, louvred (45–55)	50	1.0	0.35	0.32	0.30	0.35	0.32	0.30	0.34	0.31	0.30
(F)	Shallow ceiling-mounted louvre panel (40–50)		1.25	0.38	0.35	0.32	0.38	0.35	0.33	0.38	0.34	0.33
			1.5	0.41	0.38	0.36	0.40	0.38	0.35	0.40	0.37	0.35
			2.0	0.45	0.42	0.40	0.43	0.41	0.39	0.43	0.40	0.39
			2.5	0.47	0.44	0.42	0.45	0.43	0.41	0.45	0.42	0.41
(F)	Louvred recessed (module) fitting (40–50)		3.0	0.48	0.45	0.44	0.46	0.45	0.43	0.46	0.44	0.42
			4.0	0.49	0.47	0.46	0.48	0.47	0.45	0.47	0.45	0.44
			5.0	0.50	0.49	0.48	0.49	0.48	0.47	0.48	0.47	0.46
(F)	Suspended pan-shaped fitting, partly open top, louvred beneath (50)	50	1.0	0.43	0.38	0.35	0.40	0.36	0.34	0.38	0.34	0.32
			1.25	0.47	0.43	0.39	0.44	0.40	0.38	0.42	0.37	0.35
			1.5	0.51	0.47	0.43	0.47	0.44	0.41	0.44	0.40	0.38
			2.0	0.56	0.52	0.49	0.51	0.48	0.46	0.47	0.44	0.42
			2.5	0.58	0.55	0.52	0.53	0.51	0.49	0.49	0.46	0.44
			3.0	0.60	0.57	0.55	0.55	0.53	0.51	0.51	0.48	0.46
			4.0	0.63	0.60	0.58	0.57	0.55	0.54	0.52	0.50	0.48
			5.0	0.65	0.62	0.60	0.59	0.57	0.56	0.53	0.52	0.50
(F)	Plastics trough, louvred (45–55)	50	1.0	0.39	0.34	0.30	0.36	0.32	0.29	0.34	0.31	0.28
			1.25	0.43	0.38	0.34	0.39	0.36	0.33	0.37	0.34	0.31
			1.5	0.46	0.41	0.37	0.42	0.39	0.36	0.39	0.36	0.33
			2.0	0.50	0.46	0.43	0.46	0.43	0.40	0.53	0.39	0.37
			2.5	0.53	0.49	0.46	0.49	0.46	0.43	0.45	0.42	0.40
			3.0	0.55	0.51	0.49	0.51	0.48	0.46	0.47	0.45	0.43
			4.0	0.58	0.54	0.52	0.53	0.51	0.49	0.48	0.47	0.45
			5.0	0.60	0.57	0.55	0.55	0.53	0.51	0.50	0.48	0.47
(F)	Plastics trough unlouvred (60–70)	70	1.0	0.48	0.43	0.38	0.46	0.42	0.38	0.45	0.42	0.38
			1.25	0.52	0.47	0.43	0.50	0.46	0.42	0.49	0.45	0.42
			1.5	0.56	0.51	0.47	0.54	0.50	0.46	0.52	0.48	0.45
			2.0	0.62	0.56	0.53	0.58	0.55	0.51	0.56	0.52	0.50
			2.5	0.65	0.60	0.57	0.61	0.58	0.55	0.59	0.56	0.53
			3.0	0.67	0.63	0.60	0.64	0.61	0.58	0.62	0.59	0.56
			4.0	0.70	0.66	0.64	0.67	0.64	0.61	0.64	0.62	0.59
			5.0	0.73	0.69	0.67	0.69	0.67	0.64	0.66	0.64	0.62

Table 27.9 (continued)

Description of fitting, and typical DLOR (%)	Basic DLOR (%)	Ceiling	0.70			0.50			0.30		
		Walls	0.50	0.30	0.10	0.50	0.30	0.10	0.50	0.30	0.10
		Room index				Reflectance					
(F) Bare lamp on ceiling (F) Batten fitting (60–70)	65	1.0	0.44	0.37	0.33	0.40	0.35	0.31	0.35	0.32	0.29
		1.25	0.49	0.42	0.38	0.45	0.40	0.36	0.39	0.36	0.33
		1.5	0.54	0.47	0.42	0.50	0.44	0.40	0.43	0.40	0.37
		2.0	0.60	0.52	0.49	0.54	0.49	0.45	0.48	0.44	0.41
		2.5	0.64	0.57	0.53	0.57	0.53	0.49	0.52	0.48	0.45
		3.0	0.67	0.61	0.57	0.60	0.57	0.53	0.56	0.52	0.49
		4.0	0.71	0.66	0.62	0.64	0.61	0.57	0.59	0.55	0.52
		5.0	0.74	0.70	0.66	0.68	0.64	0.61	0.62	0.58	0.54
(F) Enclosed plastics diffuser (45–55)	50	1.0	0.40	0.35	0.31	0.37	0.33	0.30	0.33	0.30	0.28
		1.25	0.44	0.39	0.35	0.40	0.36	0.33	0.36	0.33	0.31
		1.5	0.47	0.42	0.38	0.43	0.39	0.36	0.38	0.35	0.33
		2.0	0.52	0.47	0.44	0.47	0.44	0.41	0.41	0.39	0.37
		2.5	0.55	0.51	0.48	0.50	0.47	0.44	0.44	0.42	0.40
		3.0	0.58	0.54	0.51	0.52	0.49	0.47	0.47	0.45	0.43
		4.0	0.61	0.57	0.54	0.55	0.52	0.50	0.49	0.47	0.45
		5.0	0.63	0.59	0.57	0.57	0.55	0.53	0.51	0.49	0.47
(T) Opal sphere (45) and other enclosed diffusing fittings of near-spherical shape	45	1.0	0.36	0.29	0.25	0.31	0.26	0.22	0.26	0.13	0.19
		1.25	0.41	0.34	0.29	0.35	0.30	0.26	0.29	0.26	0.22
		1.5	0.45	0.39	0.33	0.39	0.34	0.30	0.31	0.28	0.25
		2.0	0.50	0.45	0.40	0.43	0.38	0.34	0.34	0.32	0.29
		2.5	0.54	0.49	0.44	0.46	0.42	0.38	0.37	0.35	0.32
		3.0	0.57	0.42	0.48	0.49	0.45	0.42	0.40	0.38	0.34
		4.0	0.60	0.56	0.52	0.52	0.48	0.46	0.43	0.41	0.37
		5.0	0.63	0.60	0.56	0.54	0.51	0.49	0.45	0.43	0.40
(T) Diffuser with open top louvred beneath (30)	30	1.0	0.40	0.34	0.31	0.34	0.30	0.27	0.27	0.25	0.23
		1.25	0.45	0.39	0.36	0.38	0.33	0.31	0.30	0.28	0.26
		1.5	0.49	0.44	0.40	0.41	0.36	0.34	0.32	0.30	0.28
		2.0	0.54	0.50	0.46	0.45	0.41	0.39	0.34	0.33	0.31
		2.5	0.57	0.53	0.50	0.47	0.44	0.42	0.36	0.35	0.33
		3.0	0.60	0.56	0.53	0.49	0.46	0.45	0.38	0.37	0.35
		4.0	0.63	0.59	0.57	0.51	0.49	0.48	0.40	0.39	0.37
		5.0	0.65	0.62	0.60	0.53	0.51	0.50	0.41	0.40	0.38
(T or F) Totally indirect-fitting; based on upward light output ratio 75% (upper and lower walls the same colour)		1.0	0.16	0.15	0.12	0.15	0.12	0.10			
		1.25	0.20	0.19	0.16	0.18	0.15	0.13			
		1.5	0.24	0.23	0.20	0.20	0.18	0.16			
		2.0	0.28	0.27	0.23	0.22	0.20	0.18			
		2.5	0.32	0.31	0.26	0.24	0.22	0.20			
		3.0	0.36	0.35	0.29	0.25	0.23	0.21			
		4.0	0.40	0.38	0.31	0.26	0.24	0.22			
		5.0	0.43	0.40	0.33	0.27	0.25	0.23			
(T or F) As above, but with upper walls the same colour as the ceiling		1.0	0.21	0.17	0.14	0.13	0.11	0.09			
		1.25	0.25	0.21	0.18	0.15	0.13	0.11			
		1.5	0.29	0.25	0.22	0.17	0.15	0.13			
		2.0	0.33	0.30	0.27	0.20	0.18	0.16			
		2.5	0.37	0.34	0.32	0.23	0.21	0.19			
		3.0	0.40	0.38	0.36	0.26	0.24	0.22			
		4.0	0.43	0.42	0.40	0.28	0.27	0.25			
		5.0	0.45	0.44	0.42	0.30	0.29	0.27			
(T or F) Indirect cornices, recessed coves and coffers giving all their light above the horizontal; based on an upward light output ratio of 40%, but details of construction may vary this figure considerably		1.0	0.11	0.09	0.08	0.08	0.07				
		1.25	0.13	0.11	0.09	0.09	0.08				
		1.5	0.14	0.12	0.10	0.10	0.09				
		2.0	0.16	0.14	0.12	0.11	0.10				
		2.5	0.17	0.15	0.14	0.12	0.11				
		3.0	0.18	0.16	0.15	0.12	0.11				
		4.0	0.19	0.18	0.16	0.13	0.12				
		5.0	0.20	0.19	0.17	0.14	0.13				

Table 27.9 (continued)

Description of fitting, and typical DLOR (%)	Basic DLOR (%)	Ceiling	0.70			0.50			0.30		
		Walls	0.50	0.30	0.10	0.50	0.30	0.10	0.50	0.30	0.10
		Room index				Reflectance					
(F) Complete luminous ceiling composed of translucent corrugated strip or individual pan shaped elements; based on ceiling cavity surfaces being white, and cavity width being three times cavity depth		Z.0	0.34	0.31	0.27						
		1.25	0.37	0.34	0.31						
		1.5	0.40	0.36	0.34						
		2.0	0.45	0.42	0.39						
		2.5	0.47	0.44	0.42						
		3.0	0.49	0.46	0.44						
		4.0	0.52	0.49	0.47						
		5.0	0.54	0.51	0.49						
(F) Complete louvred ceiling composed of half-inch translucent plastics cells; based on ceiling cavity surfaces being white, and cavity width being three times cavity depth		1.0	0.37	0.34	0.30						
		1.25	0.39	0.36	0.33						
		1.5	0.41	0.38	0.36						
		2.0	0.44	0.42	0.39						
		2.5	0.46	0.44	0.41						
		3.0	0.48	0.46	0.43						
		4.0	0.50	0.48	0.45						
		5.0	0.51	0.49	0.47						

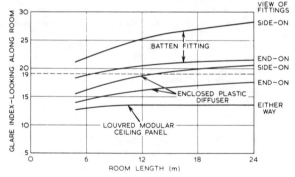

Figure 27.30 Glare indices of three typical fittings in an office 5.5 m wide

directions to be prescribed or otherwise given as to whether any artificial lighting is to be produced by any particular illuminant.'

Subsection 4 deals with natural illumination.

It will be seen that these regulations are of a very general nature, the onus of deciding what is sufficient and suitable lighting being upon the Factory Inspector.

The Fourth and Fifth Reports of the Departmental Committee on Lighting in Factories form the basis of the Factories (Standards of Lighting) Regulations, Statutory Rules and Orders No. 94, February 1941. While this order only applies to factories where persons are regularly employed for more than 48 working hours per week, or in shifts, it can be safely assumed that, for the most part, the standards laid down in the regulations will be looked upon as applicable to all classes of factories.

The regulations, which are due for revision doubtless in an upward direction, are as follows:

REGULATION 2a. 'The general illumination over those interior parts of the factory where persons are regularly employed shall be not less than 6 foot-candles measured in the horizontal plane at a level of 3 feet above the floor [equivalent to 66 lm/m^2 at a level of 0.9 m]:

'The standards specified in this regulation shall be without prejudice to the provision of any additional illumination required to render the lighting sufficient and suitable for the nature of the work.'

While no definite indication is given as to what is 'sufficient', it has become the practice of lighting advisors to regard the Interior Lighting Code of the CIBS as a basis which satisfies present requirements.

REGULATION 2b. 'The illumination over all other interior parts of the factory which persons employed pass shall when and where a person is passing be not less than 0.5 ft-cd measured at floor level.'

This calls for illumination levels which necessitate the use of fairly closely spaced units for passages, stairways, etc., a long-needed regulation.

Subsection (a) of Regulation 3 precludes the use of angle type units below a height of 16 ft (5 m) unless the light is screened from normal view or suitable provision has been made to reduce the surface brightness by means of diffusing glasses.

All general lighting units marketed by reputable firms comply with Regulation 3a, with the exception of certain wide distribution units designed for the illumination of storage areas, etc.

Glare from inadequately screened local lights has a very severe effect on working efficiency, as the source is usually very close to the normal line of sight.

The prevention of dense shadows is of great importance, as a high proportion of industrial accidents is caused by them. Special attention should be given to all obstructions such as shafting, heater pipes, ventilation trunks, stanchions and units located so that shadows are not cast by them on to floor or working points.

27.8.1.1 Lighting requirements for various applications

In the Interior Lighting Code recommendations are given for illumination and limiting glare indices for a very wide variety of interior lighting applications. The illumination values recommended may be regarded as the minimum to satisfy the requirements of good lighting, and aim to provide a standard of visual performance (at least 90% of the maximum possible for the job) considered reasonably adequate, bearing in mind the economics of lighting today. Special circumstances may require a 50% higher illumination than that listed, e.g. if it is a windowless interior, if the materials handled are valuable or dangerous, if the people concerned are mainly over 40 years old, if protective goggles have to be worn or if specially high-speed accurate work has to be done. Conversely, the recommended value is reduced by one-third if critical seeing is required only occasionally and for brief periods.

27.8.1.2 Special requirements

Explosive or inflammable atmospheres Many industrial processes involve the use of explosive or highly inflammable materials or vapours; typical of these are low-flashpoint paint spraying, petrol or oil storage, acetone production, 'dope' shops, etc. Many normally non-explosive materials become highly explosive when in the form of fine dust.

In such locations it is imperative that special precautions be taken to lessen fire or explosion risks. Three methods of lighting are in general use:

(1) Special flameproof units. These units should comply with BS 5501 Part 5, and hold certificates issued by the Mines Department under one of the following groups: Group I— Methane; Group IIA—Petroleum vapour, acetone vapour; Group IIB—Town gas or coke-oven gas containing not more than 60% hydrogen.

Flameproof units can be installed inside the danger area provided that the wiring is carried out in solid drawn conduit or similar, and all joints, switches, etc., are of the flameproof type. NOTE: Flameproof units are not vapour-proof; a small gap is allowed between two broad machined surfaces. If an explosion occurs within the unit, the gas is cooled in its passage to the outer air by the mass of metal and so does not ignite any vapour present in the atmosphere.

(2) Units can be located outside the danger area and project their light through sealed windows into the shop. This method is most suitable for the highly inflammable atmospheres, as no wiring, etc., is brought into the danger area.

(3) If the danger area is small, standard industrial units can be used provided that they are at least 6 m (20 ft) away from the source of gas or vapour. This method is suitable for isolated spray booths.

Food factories Special care must be taken to prevent falling glass contaminating the food. Enclosed or indirect units are well suited for this type of factory. Good general lighting, with an absence of dense shadows, is required so that any perishable particles can be detected and swept away.

Corrosive atmospheres In such locations as plating or accumulator charging shops and many of the chemical industries, precautions must be taken to avoid the corrosion of reflecting surfaces and electrical contacts. Special 'wellglass' units are marketed for this purpose, all metal contacts being protected by glass or china housings. Where enamelled reflectors are used, special care should be taken not to chip them and thus permit the ingress of corrosive fumes.

27.8.2 Domestic lighting

In the average home there are few critical viewing tasks, so that a high standard of general lighting is not required. General lighting should be supplemented by local lighting using floor or table standards in lounges and sitting-rooms or fixed points in kitchens and bathrooms. By this means satisfactory illumination is obtained where it is needed and an interesting brightness pattern is introduced.

The choice of suitable lighting equipment is extremely difficult, owing to the personal likes and dislikes of the individual. All the points looked for in commercial or industrial lighting units should be borne in mind: glare prevention by screening or diffusion; good light control and efficiency; low brightness, especially at normal angles of view; easy to clean and to replace lamps; good wiring; sound mechanical construction.

Many domestic lighting fittings, both cheap and fashionable ones, are designed without much regard to some of these factors.

Entrance or porch lighting The size of lamp required for porch lighting is small (40–60 W), so that it can be burnt for long periods for little cost. The unit used should have a fairly low surface brightness (about 1 cd/cm^2). The lamp should be placed so that it illuminates the caller.

Hall A warm colour of light will convey an atmosphere of welcome. Coves and niches can be used to advantage as illuminated features. Adequate illumination must be provided for the stairs, without causing glare to persons descending them. Dense shadows may be confused with the stair treads or risers, and should be avoided.

Drawing-room, lounge or living-room A medium general-illumination level should be supplemented, at the working points, by portable local lights, but not cause glare or distract other people in the room. A wide, deep shade on a tall stem will often achieve this. BS 4533 and BS 5225 Part 1 specify a performance considered suitable for reading lamps.

The general appearance of a room may be improved if a proportion of light is direct on to the ceiling. Fluorescent lamps, concealed in pelmets, can introduce light into a room unobtrusively and enhance the appearance of curtains. They should give plenty of light locally.

Illuminated features such as china cabinets, flower vases, etc., add interest at very little cost.

Dining-room The dining-table is the main feature and should have the highest illumination. A semi-direct light fitting, whose height can be adjusted, placed centrally over a table is suitable treatment. Here, again, illuminated features can be used to advantage to provide pleasant brightness patterns.

Kitchen Good general and local lighting is required for this room. The location of the general lighting unit should be such as to avoid dense shadows on the cooker, work tables, etc., by the housewife. Fluorescent lamps are recommended for the kitchen; they should normally be placed parallel to, and above, the front edge of the sink and adjacent work surfaces. If tungsten lamps are used, they should be totally enclosed in diffusing glass or plastics. The electrical work should be faultless because of the presence of water and earthed metal.

Bathroom Requirements are much the same as for the kitchen. For shaving mirrors the light should be projected on to the face, not the mirror. Low-brightness panels at the sides are suitable for this, but tubular lamps will serve.

The requirements of the IEE Regulations in respect of bathrooms should be complied with.

Bedroom Two points require special attention—the bedhead and the dressing mirror. Bedhead lighting can take many forms, but the main requirement is the provision of adequate illumination for reading purposes when the occupant is in a comfor-

Table 27.10 Recommended lamp ratings for lighting medium-sized home

Location	W	Location	W
Hall	60	Bedrooms	
Landing	60	(general)	100–150
Stairs	60	Bedhead or	
Passages	60	bedside	60–100
Living-rooms		Mirror	2 × 60
(general)	150–200	Nursery	
Easy chairs		(general)	150
(standard lamp)	150	(table)	100
Writing table	100	Bathroom	100
Sewing table	150	(mirror)	60
Dining-room		Garage (general)	150
(general)	150	Workbench	100
Kitchen (total)	200	Loft, cellar	40–60

NOTE The above powers refer to tungsten filament lamps. If fluorescent lamps are used, from one-third to one-half the above power ratings will be required.

table position. Low-brightness units should be used throughout this room in order to provide a restful atmosphere.

Wardrobes, cupboards, largers, etc. The provision of low-power lamps, controlled by door switches, adds greatly to the amenities of a home. The cost of operation is negligible.

Exterior lighting For most homes the provision of a waterproof bulkhead unit over the rear entrance and a fluorescent lamp in the workshop or garage is well worth the small outlay and running cost they require. A garden can be made very attractive at night by a few weatherproof fittings, strategically placed to illuminate paths and trees.

27.8.3 Shop lighting

27.8.3.1 Shop windows

A shop window has two primary functions—to attract attention and to display to advantage the goods which are to be sold. To do this acceptably, artificial lighting is essential. High brightness, drama and the absence of glare are the three main requirements of good window lighting. To achieve a good brightness level the vertical illumination has to be high and backgrounds should normally be of light-coloured materials, preferably textured or draped.

Tungsten filament lamps are often still the main sources of directed window lighting, especially where a warm colour of light is preferred (as for many foodstuffs) or where a degree of glint is required from the display. Normally they will be housed in a line of concealed reflectors at the top front of the window, but in deep windows extra lighting may be required further back. These reflectors are of two main types, intensive and extensive; the latter are intended for deep windows or where the dressing is high.

For the main shopping streets of medium towns a filament lamp installation, averaging, for each metre of window frontage, 150 W of spotlighting plus 300 W of general lighting, is usually adequate, rather less being required in less important localities, and much more in the shopping centres of cities. Fluorescent tubes averaging some 150 W per metre of frontage will produce about the same effect as 350 W/m with filament lamps.

Fluorescent lamps are found in a great number of shop windows, mainly on account of their high efficiency and coolness in operation. Shadows tend to be very soft, and, when fluorescent lamps are used alone, the display tends to look flat. Almost invariably, therefore, the addition of tungsten filament spot lamps adds interest to the display by creating shadows which show form and texture, and by producing highlights from articles with sheen or gloss.

The treatment of windows displaying highly specular articles such as rings, cut glass, silver plate, etc., should be such that as much sparkle as possible is obtained. For this reason the use of large light sources, e.g. fluorescent lamps, is deprecated, clear filament lamps being the most suitable.

A simple pelmet adequately conceals lamps and reflectors in a single-aspect window, though a neater and more modern appearance may be obtained by a raked-back 'picture frame' just inside the window, lamps being hidden behind the frame. In re-entrant or island windows, the lamps must be hidden from two or more aspects, commonly by using louvred fittings or by recessing them into the ceiling, the latter device being specially applicable to open-back windows.

An illuminated or self-luminous background is very attractive and can easily be obtained by fluorescent tubes concealed at top and bottom of the window near the back. In many cases, strongly coloured fluorescent tubes in these positions are still more effective, provided that the display itself is lit with white light.

Spotlighting equipment must be adjustable both as regards position and direction. Low-voltage lamps require individual transformers. Six socket-outlets per window will generally cater for this as well as for background lighting, etc. To give flexibility a busbar system has been developed into which spotlamps and display fittings can be inserted.

27.8.3.2 Interiors

Shop interior lighting has two primary functions: the first to produce an atmosphere conducive to buying; the second to display the goods offered for sale, both at the point of selling, i.e. on the counter, and in display cases, etc. The first function requires the provision of a lighting system free from glare and harsh shadows and as unobtrusive as possible to avoid distracting the customer's eye from the displays. In fact, the design of a suitable system resolves itself into producing an attractive brightness pattern, and it is the brightness of the special displays which dictates the brightness level of the other parts of the interior.

No relighted interior nowadays is likely to have a lighting load lower than 20 W/m^2 and of this, approximately half is likely to be for highlighting displays, mainly by means of reflector spot lamps or other concentrating lighting equipment. Illumination levels of 500–1000 lx are provided. The remainder of the power, used to light the interior as a whole, is generally accounted for in fittings placed over or in near relation to the counters so that most light falls where it is chiefly required, the circulation areas being lit adequately but at a lower level.

Much use is being made of false ceilings or sections of false ceiling, which may conceal service pipes in addition to being useful platforms in which or on which lighting equipment is mounted. Such lighting equipment should generally be located over the customers' edge of a glass-topped showcase, in order to prevent glare from brilliant reflections in the glass.

Completely luminous ceilings may be used for open display areas. They are not so suitable for lower levels of illumination, on account of the rather monotonous effect then obtained. For self-service shops and supermarkets lines of bare fluorescent lamps are often used for economy and ease of maintenance. Fashion shops make much use of ordinary domestic types of fittings, to obtain the atmosphere they require.

The colour of light for an interior may be vital, or it may be immaterial, depending on the merchandise concerned. Any shop or section of a shop where accurate colour judgement or the ability to match colours is necessary, should use either Colour-matching, Northlight or Artificial Daylight fluorescent tubes, which virtually bring outdoor daylight indoors.

The following notes may help in choosing lamp types for other special purposes.

Food shops (especially meat shops) require a light which shows colours freshly and reds strongly. Deluxe Natural (very strong reds) and Colour 32 have these characteristics; so also to a lesser extent do filament lamps (weak on blues) and Deluxe Warm White (weak on deep reds).

Shops selling fashion wear, fabrics, furnishings and many other classes of merchandise may require light with a colour effect akin to sunlight (Colour 34 and Kolor-rite), or may wish to use the Natural with rather higher efficacy but less attractive colour.

For ironmongery, etc., where the apparent colour of the merchandise is relatively unimportant, the highest lighting efficiency may be obtained by using White or Warm White tubes.

In all cases the addition of some lighting from filament lamps may be desirable to introduce 'flash' (e.g. to make cut meat look juicy), or for its highlighting potentialities; and in some cases to add to the red content of the lighting.

27.8.4 Office lighting

The Offices, Shops and Railway Premises Act 1963 states that 'there must be provision for suitable and sufficient lighting, either natural or artificial, in every part of the premises in which persons work or pass. Windows and skylights used for lighting must be kept clean and free from obstruction so far as reasonably practicable but they can be whitewashed or shaded to mitigate heat or glare. Artificial lighting apparatus must be properly maintained.'

Open general offices are nearly always lighted nowadays by a symmetrical scheme of pendant or recessed fluorescent fittings, or by a luminous or louvred ceiling, giving general lighting which allows the desk layout to be altered as and when desired. The installation scheme, however, should be as flexible as it can reasonably be made, in view of the possibility of partitions being erected subsequently or moved to new positions.

Discomfort glare is more prevalent in offices than in any other interior, owing largely to the widespread use of diffusing fittings whose glaring characteristics have not hitherto been fully realised, especially when they are viewed sideways-on. Future installations designed to comply with the antiglare recommendations of the IES Code should succeed in avoiding this trouble.

In small offices housing up to about five people it is common to find some or all of them seated with their backs to the walls so that most of the light comes to them from the front; typewriters are badly illuminated and there is likelihood of shine from shorthand notebooks, etc. In such places it is advisable to place fittings much closer to the walls than usual. In private offices of managerial type, which are often lighted to a fairly low general level with supplementary lighting for the desk area, fittings of a semidomestic nature are popular.

New drawing offices are frequently lighted to a considerably higher standard than the 500 lx (45 lm/ft^2) recommended in the Code. If parallel, it is usual for fittings to be mounted over the boards, but towards the lower edge if the boards are near-horizontal, and towards the top edge if boards are near-vertical. The fittings used for horizontal boards may with advantage use specular reflecting surfaces, which avoid the draughtsman seeing a bright far side to the reflector, and for vertical boards a diffusing type of fitting is often preferred. If so, the location of the fittings with respect to the boards is not critical.

If fittings are at right angles to boards, it is common to run the left-hand outer row very close to the wall, the other rows being over gangways rather than over the boards. Cross-louvres beneath the lamps make seeing conditions much more comfortable.

27.8.5 Floodlighting

Special weatherproof projector units are used for the floodlighting of building façades. These units have a range of light distribution with beam angles from 10° to 70°. Long-range narrow-beam floodlights often use silvered glass reflectors; but high-purity anodised aluminium reflectors, with specular or diffusing finishes, are used for most applications. Spun reflectors are used for symmetrical beams; trough forms, rolled or pressed, for asymmetrical beams.

Floodlights designed for tungsten halogen lamps are being increasingly used. The tubular form of this lamp produces a fan-shaped beam with good light control in the vertical plane. Colour effects are produced by the use of colour filters with filament lamps or, more efficiently, by the use of coloured discharge or fluorescent lamps.

A good rule-of-thumb to determine how much illumination a floodlighting system should provide is (a) assume that perfectly white building in an isolated position will require 10 lm/m^2, (b) multiply this figure by the reciprocal of the reflection factor of the surface concerned, (c) multiply the result by 2 if the building is not isolated, and there are a few shop and street lights to distract from the floodlighting, or by 3 or 4 in a well-lighted central area. Thus a building surface having a 25% reflection factor in medium-lighted surroundings would require about 10 (100/25)2 = 80 lx for the lighting to be effective.

Floodlighting calculations can be carried out on the 'lumen' basis if the following modifications are made. The coefficient of utilisation is replaced by a 'beam factor' which is the ratio of beam lumens to lamp lumens and varies from 0.25 to 0.55, depending upon the efficiency of the lantern used. A further factor is introduced: the 'waste light factor'. This depends upon the building being illuminated, but general figures are 1.2 for square buildings and 1.5 for irregular or isolated objects. The depreciation factor is usually about 1.5 but may be as high as 2 if the atmosphere is dirty and maintenance difficult.

The formula is therefore:

lumens required
$$= \frac{(\text{lm/m}^2) \times \text{area per fitting (m}^2) \times \text{waste light factor}}{\text{beam factor} \times \text{maintenance factor}}$$

The floodlighting of areas or hoardings can be carried out by means of angle units or projectors. Typical requirements are as follows:

	lm/m^2		lm/m^2
Bathing pools	100–150	Gardens	5–30
Building construction	40–60	Loading docks and quarries	20–40
Building excavation	10–20	Railway yards	5–20
Car parks	5–10		

27.8.6 Street lighting

The technique of street lighting is that of seeing by silhouette or reversed contrast. The aim is to light the road and background against which objects are seen. The actual objects on the road have little light incident on the surfaces normally seen by a road user and therefore appear dark. This technique gives better visibility to distant objects with the very small lamp power economically practicable.

The overall even brightness of the road necessary for good

seeing conditions depends largely on the correct siting of the lanterns and on their light distribution, and on the physical properties of the road surface.

The light which is of most value to the road user to enable him to see at a distance of 30 m or more, and thus to drive in safety, comes from street lanterns even further away and is reflected from the road at glancing angles of incidence. For this reason in the most frequently used lantern, known as semi-cut-off (SCO), the greatest light output occurs at about 75° from the downward vertical. Each lantern gives a characteristic T-shaped reflection of light on the road and the lanterns must be spaced so that these reflections are joined together without dark spaces between.

Owing to the considerable glare caused by the emission of light from the lanterns at angles near the horizontal, an alternative lantern is sometimes used, known as the cut-off (CO) type. The maximum intensity of light is emitted at about 65° and practically no light is emitted at higher angles. This system requires closer spacing of the lanterns to avoid dark patches. BS 1788: 1964 covers the construction and light distribution of street-lighting lanterns. The cut-off of light from the lanterns is such that very little direct light can be seen more than about 75 m away.

The advantages of the two systems are:

Cut-off Less glare. Better performance on matt surfaces.
Semi-cut-off Longer spacing. Greater flexibility in siting. Better performance with smoother surfaces. Better appearance of buildings. Better suited to staggered arrangements.

In built-up areas, where there is a more or less continuous background of tall buildings, other forms of light distribution are permitted. Adjacent railways, docks or airfields may impose cut-off lighting and screened lanterns; the appropriate authorities should be consulted (see CP 1004).

27.8.6.1 British standard code of practice CP 1004

CP 1004 is based on the Report of the Ministry of Transport Departmental Committee on Street Lighting, 1937, which divided roads into two categories for street lighting.

Group A applies to all through-traffic routes. The lighting should be such that vehicles can proceed safely without the use of headlights.

Group B applies to all other roads which it is desirable to light. The standard of lighting should enable pedestrians to see their way safely.

CP 1004: Part 1 deals with general principles; Part 2 with lighting for traffic routes. Group B is at present covered by CP 1004: Part 2A: 1956, which is under revision.

Classification of installations

Group	Where used
A1	Principal urban traffic routes
A2	Generality of urban traffic routes
A3	Minor urban traffic routes and main rural roads
B	Roads carrying local traffic only (e.g. residential and unclassified roads)
C	Roundabouts and complex junctions on one level
D	Bridges and flyovers
E	Tunnels and underpasses
F	Roads with special requirements (Motorways and roads near airfields, railways and docks)
G	Town and city centres

27.8.6.2 Arrangements of lanterns

CP 1004: Part 2: 1963 establishes geometrical relationships between the effective road width (W) and the spacing (S) and the mounting height (H) of lanterns. Within limits S varies inversely as W, for most systems.

There are four general methods of siting street lanterns.

'Staggered' arrangement In this method the placing of lanterns on alternate columns on the opposite side of the carriageway is most common. It is an economic arrangement for lighting a straight road up to a width of about 15 m between kerbs.

'Opposite' mounting This arrangement requires more lanterns, since the spacing cannot be materially increased compared with the staggered arrangement. It gives a more uniform illumination of the road surface, particularly in wet weather, and is thus superior to any of the other methods considered here. It is used on roads which are too wide for the 'staggered' arrangement.

'Central' mounting This method is often adopted to supplement staggered or opposite systems on very wide roads. On dual carriageways with a central reserve, lanterns may be mounted in pairs on a single column.

'Single side' mounting This should not be used on straight roads, but the arrangement is recommended practice for bends with a radius less than $80H$ (e.g. 800 m for $H = 10$ m). The lanterns should be mounted on the outside of the bend.

For example, for a Group A2 installation at a mounting height of 10 m, the minimum lumens in the lower hemisphere from a clean lantern with a lamp having an output equal to the average throughout life should be 8580 lm (CO) or 11 000 lm (SCO). Similarly, the maximum spacing should be 39 m and 47 m and the maximum effective width 9 m and 11.5 m for the cut-off and semi-cut-off installation, respectively.

There is a maximum spacing no matter how narrow the road and a maximum effective width beyond which a satisfactory result cannot be obtained by further reduction of the spacing.

For Group B the mounting height should be between 4 and 4.5 m. The spacing should average 30–37 m and the maximum spacing for occasional spans 45 m. (Cut-off fittings 30 m average.) The total luminous flux provided should lie between 20 and 80 lm per linear metre. This includes some light in the upper hemisphere.

The guiding principle in siting lanterns at junctions, crossings and roundabouts is that there should be a source in such a position as to produce a bright background to any object which the driver may have in his normal field of view (see CP 1004: Part 4: 1967). Thus at a simple crossroads there should always be a lantern on the left-hand side of the opposite road not more than $\frac{1}{3}$ spacing on the major road beyond the crossroads.

It is essential that a confusing array of lights, as seen by the driver, be avoided, and that any corners and kerbs be shown up distinctly.

As 10 m is about the maximum width which can be covered by the bright area from lamps mounted vertically above the kerb line, brackets should be used for wider roads to reduce the distance between the rows of lanterns. The overhang of side-mounted lanterns should be limited to $H/4$.

To reduce the danger of collisions the clearance between column and carriageway should be 2 m or more wherever possible.

27.8.6.3 Special areas

In the case of city centres or places of special interest very high standards of lighting may properly be demanded. In these cases the purpose of the lighting is to enable ordinary daytime seeing conditions to be continued in the dark and the distant view of fast-moving traffic is not so likely to be involved. For such areas special consideration must be given to spacing, mounting height, light output, colour and the design of the equipment, according to the merits of each particular case.

Many towns, especially those catering for holidaymakers, have realised the value of lighting as an additional local attraction, not only for their streets, but also for promenades, gardens, squares, swimming pools, bandstands and other entertainment areas. As the requirement is mainly for area lighting, where illumination is more important than background brightness, close spacing is desirable, mounting height need not be restricted and control of light is of less concern than the appearance of the units both by day and night.

27.8.6.4 Light sources

The street lighting engineer has a wide range of electric lamps from which to choose, and his selection should be based on the suitability and relative economy of a particular type for the purpose.

Tungsten filament lamps are still used extensively and have the advantages of initial cheapness, simplicity and good colour rendering. For a given light output, however, their current consumption is comparatively high. This, and their relatively short rated average life of 1000 h often makes the total annual cost of lighting higher than with other electric sources.

Mercury discharge lamps have proved very satisfactory on account of their long life and superior efficacy. They are particularly effective for the lighting of trunk roads, where good long-distance visibility is more important than natural colour rendering, and their greenish-blue light is often quite acceptable in built-up areas. High-pressure mercury fluorescent lamps provide light of a better colour quality at the expense of some loss of light control.

Low-pressure sodium discharge lamps have also become widely used owing to their exceptional efficacy and long life. The light given by these lamps is an almost pure yellow, which causes colour distortion. This is no bar to their use where good visibility is the major criterion, and their comparatively low brightness greatly assists in reducing glare. Lanterns designed for linear sodium lamps at relatively high mounting heights are particularly suitable for the lighting of through traffic routes. The high-pressure sodium lamp has very good colour and efficacy and should become more widely used.

In 1946 fluorescent lamps were used for the first time for street lighting. The high efficacy, low brightness (which reduces glare), long life and excellent colour rendering make them particularly suitable for shopping centres.

27.8.6.5 Control

Owing to the spread-out arrangement of street-lighting units, the switching of lamps presents serious problems. The standard method has been to utilise time clocks, sometimes with a spring reserve to avoid manual resetting in the event of power failure, and with solar dials which follow the diurnal changes in lighting-up time in various parts of the UK.

This is being superseded by either photoelectric control or a centralised mains injection system.

Street lighting is normally in operation from about 30 min after sunset until 30 min before sunrise. To compensate for the warming-up required by sodium and high-pressure mercury lamps, and to provide adequate street lighting in dull weather, the switch-on and switch-off times are often adjusted to 15 min after sunset and before sunrise, or a 'London Conurbation' solar dial may be used. Some authorities extinguish the lighting at midnight or 0100, or operate it at reduced level until 0400. Lighting throughout the hours of darkness is becoming more general, at least on traffic routes. The annual burning hours thus become approximately 4100 h, which corresponds to the rated average life of sodium lamps, and 80% of the rated average life of high-pressure mercury and fluorescent lamps.

With photoelectric control the lanterns are controlled individually or in groups by a photocell mounted into the lantern or at the top of the column. The cost and reliability of the photocell unit is comparable with that of a time switch. Thus, the lights are switched on and off (at about 50–100 lm/m^2, respectively) according to the actual natural lighting conditions which prevail.

The 'ripple' system of control is a centralised one in that the lights can be operated from a central point at any given time. The principle behind this system is simple, consisting of a superimposed high-frequency ripple which is injected into the mains, this actuating specially tuned relays which are mounted either on the lamp standards or adjacent. Another well-known method of control is by the d.c. bias system, which depends upon the injection of a positive or negative d.c. impulse, of varying duration, into the lighting mains.

27.8.6.6 Maintenance

It is essential to the proper functioning of a street lighting installation that it should be properly maintained. A good guide to maintenance is visual inspection.

Owing to the high labour cost of replacing individual lamps as and when they fail, it is common practice among street lighting authorities (and also in a number of industrial and commercial concerns) to 'group replace' lamps after a predetermined time, corresponding usually to about 80% of nominal lamp life. By this means, the number of casual failures is reduced very greatly, and the saving in labour charges outweighs the loss occasioned by discarding lamps which could have remained alight for some while longer.

27.8.7 Theatre lighting

27.8.7.1 Regulations

Theatres differ vastly in size and shape and in economic considerations; however, they will all have a stage of some sort as well as an auditorium. The combination of actors and scenery on the stage with an audience of hundreds is regarded as a potential source of danger and the aim of the special regulations for this class of buildings is to prevent this state arising. Some County Councils have their own regulations, but where these do not exist a copy of the GLC Regulations and Rules for structure and Equipment in Places of Public Entertainment should be obtained. The GLC Regulations have to be read in conjunction with those of the IEE.

27.8.7.2 Auditorium

This is the audience area for seeing and hearing the show, and is also known to the theatre as 'front of house' or 'F.O.H.', as distinct from the 'rear of house' or 'backstage' inhabited by the actors and all concerned with putting on the show. In a full-scale enterprise these two areas are isolated and have separate electrical intakes. The boundary between auditorium and stage is normally the proscenium opening. There is a tendency to ignore the proscenium nowadays.

It is essential to provide some maintained lighting duplicated by lighting from a secondary source even in the auditorium. In this connection the regulations are more obliging in respect of the theatre than of the cinema. The audience of the latter usually has to find its way in and out during the continuous performance, whereas a theatre show has a well-defined beginning, end and intervals in which the auditorium decorative lighting is used.

A dimmer to fade the auditorium 'house lights' is desirable and it is possible to dim fluorescent lighting sufficiently well for this purpose. However, it is from tungsten lamps that the type of lighting we require is likely to be obtained.

27.8.7.3 Backstage

Dressing rooms should be provided with *filament* lamps either side of the mirrors so that a strong light is obtained for make-up purposes. Socket outlets for portable appliances are necessary.

The areas immediately adjacent to or forming part of the 'off stage' (as distinct from the 'on stage' where the actors play) will require subdued lights for dark scenes and brighter sources for working. Working lights on the stage itself for setting up or changing scenes are known as 'pilots', and circuits for these are included in the stage lighting proper.

All these circuits local to the stage are controlled from the stage manager's corner, the sole exception being the actual stage lighting. No matter how small the hall or stage, one has to decide which side the stage manager will work from. He then has a small desk bracketed from the proscenium wall and this becomes the site for the local switches, signals to orchestra, projection rooms, etc. Light signals 'warning' and 'go' backed up by buzzers all working on 24 V a.c. are usual. It is customary to provide a pair of signal lamps in series. Thus, the stage manager receives direct proof from the lamps in his corner that the outstation is alight.

27.8.7.4 Stage lighting

There are three requirements: (1) the aim of stage lighting is to conceal more than it reveals; (2) glare is of no account as far as the actors are concerned, yet the audience should be unaware of the lighting sources; (3) productions differ so widely and are often put on in such haste that there must be exgensive use of socket outlets in order to reduce the amount of temporary wiring.

The last requirement is overcome to a large extent by making sure that each circuit terminates in a socket outlet and that there are more circuits than the actual equipment provided requires. As far as possible the socket outlets should be the same regulation 15 or 5 A size throughout. The 13 A fused to BS 1363 is not suitable, as local fuses are not only inconvenient, but also may be inaccessible during a production. Switched socket outlets are also unsatisfactory.

As glare is of no account except in opera, where the singers must see the conductor, the stage lighting can be positioned solely to give the best effect from the point of view of the audience.

27.8.7.5 Front-of house lighting

It is logical that the most important stage lighting must come from the same direction as the audience, being positioned in the F.O.H. As the light must be accurately directed on to the stage, the most important piece of equipment is a spotlight with a sharp cut-off.

In the mirror spotlight or profile spot (*Figure 27.31a*), a large solid angle of light is collected by a mirror and directed through a gate, which can be of any profile. The gate is focused by an objective lens. Mirror spots are very efficient: the 250/500 W model is equivalent to 500/1000 W in a standard spotlight. By choosing a suitable focus lens, a wide aperture can be used in the gate and therefore as little light wasted as possible. An iris diaphragm can be fitted in the gate, but, in general, the advantage of this type of optical system is that the beam need not be circular and, in consequence, a rectangular mask with four independently adjustable sides is more popular. This is especially so for the auditorium position, as the beam can be accurately cut off at the edges of stage and proscenium. In the 1000 W range it is possible to have a profiling gate that produces a sharply defined or a soft-edged cut-off at will. Light beams can be 'married' so as to avoid an ugly mosaic of patterns on the stage floor. Every stage lantern must have provision for colour filters.

In the case of the front-of-house, it is sometimes desirable to change the colours of the spots, although these are usually inaccessible once the audience is in position. An electric clock driven colour wheel with a preset position to each of its five colours can be fitted to the Mirror spots and fed from a selector switch. In the professional theatre a more expensive motor-driven solenoid-selected semaphore arrangement is used so that colours can follow in any order.

The spotlights so far described are directed at appropriate parts of the stage, then locked. Usually the beams are crossed to the opposite side of the stages as at A in *Figure 27.34*. Sometimes there is also used, from the auditorium, a spotlight with operator which is known as a 'following spot'. Although this device is not used in straight plays, it is essential for revue and variety acts and at least two socket outlets should be allowed for such lanterns. A special long-throw 600 W mercury iodide lamp in a profile spotlight will suffice in all but the very largest theatres.

27.8.7.6 Lighting on-stage

Two forms of lighting are used on the stage—spot lighting and floodlighting. Although the Mirror spot described earlier is sometimes used on the stage, the standard stage spotlight, at any rate hitherto, has been the Focus lantern. This consists of a lens with a lamp behind it: as the lamp is moved backwards and forwards by means of a knob underneath the lamp house, so the lamp filament is focused and the beam is enlarged or reduced. The beam actually consists of an enlarged image of the lamp filament and, in consequence, may tend to contain some striation. This can be removed by means of a light frost filter. A common spotlight has a 115 mm diameter lens and should be able to contain a 250/500 W lamp. The quality and intensity of the light has been improved by fitting a Fresnel step lens. This removes the striation and because it is short-focus increases the intensity.

By increasing the diameter of the lens, a larger solid angle can be collected and, in consequence, a spotlight known as a Fresnel spot has been produced (*Figure 27.31b*). As with the Mirror spot, this type of lamp, although of 250/500 W, may be regarded as a 500/1000 W lamp. The quality of light is very even and fades away at the edges and is, in consequence, known as a soft edge spot. Focusing is either by a knob underneath or by a lead screw, and the beam ranges from 45° down to 15°. This lantern is thus used as an adjustable floodlight on occasion or may be focused down to give an intense beam to represent sunlight coming through a window, for example. Other types of special lanterns known as Pageant lanterns, acting area floods, etc., have been used on the stage, but these have been replaced nowadays by the three types of spotlight just described. Fresnel spotlights, because they have a slight diffusion in their lenses, may scatter light well outside their main beam. This can be a nuisance, and where this is the case 'barn doors' are fitted. These consist of four independently operated flaps mounted on a turntable which fits in the front runners of the lantern and thus the flaps may be adjusted to intercept stray light in the directions where this is serious.

The floodlights used on the stage consist of an anodised aluminium reflector mounted in a housing with runners on the front to carry colour frames. There are two sizes of floodlights—the 60/200 W flood (*Figure 27.32*) and the 500 W version. The reflector of the flood is often made up into magazine equipment known as a batten (see *Figure 27.33*).

27.8.7.7 Lamps

The lamps used in floods are general-service lamps with clear bulbs in order to allow the reflectors to be effective. Projector lamps with prefocus caps are always used in spotlights to ensure

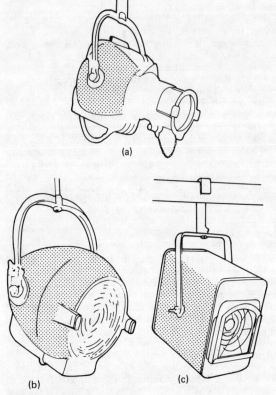

Figure 27.31 Theatre lanterns. (a) Mirror (profile) spot; (b) Fresnel spot; (c) Focus lantern

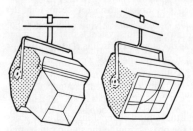

Figure 27.32 200 W flood with and without masking hood

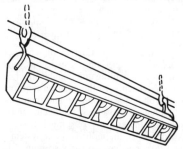

Figure 27.33 Magazine batten

correct focusing. The sizes up to 500 W employ the medium prefocus lampholder, while the 1000 W employ the large prefocus. The common lamp for spotlight work is the class T, which was specially designed for theatre work. This is a lamp with a flat grid filament mounted in a round bulb. This arrangement ensures a filament which is effective in the various optical systems and which can be tilted within the full range of angles required for stage work.

27.8.7.8 Lighting layouts

The chief lighting positions on stage (*Figures 27.34*, *27.35*) are: (a) over and just behind the proscenium at C, known as No. 1 batten;

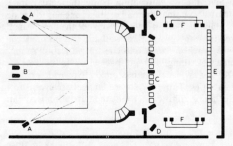

Figure 27.34 Plan of basic stage lighting layout

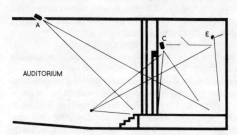

Figure 27.35 Section of basic stage lighting layout

(b) either side of the proscenium high up, known as Perch position at D; (c) a rear position about $2\frac{1}{2}$ m down-stage from the back wall—known as the back batten at E. Depending on the depth of the stage, there may be further intermediate battens, one or two, although it is unlikely to have more than four battens all told and three or even two are quite usual. The number of battens is bound up with a form of masking over the stage, whether there is a ceiling as in the section shown in *Figure 27.35* or the more normal borders.

No. 1 batten will always consist of a mixture of spots and floods. Thus, as in the plan *Figure 27.34* at C, there are five spotlights all on separate circuits, separated by two parallel groups of floods. The individual floods have the advantage that they can be angled in various directions, and thereby distribute the light properly. On the other hand, on the larger stages the magazine batten allows two or three colour circuits to be wired across and colour effects thereby to be obtained on curtains, etc.

The back batten will generally be a magazine batten wired to two or three circuits, as its object is to provide even lighting on the backcloth or on the cyclorama.

At each end of the back batten a mirror spot, fed by separate circuits, can be attached and these are used for lighting the acting area immediately in front of the cyclorama. This is side lighting, but it is the only angle that is effective in this area. Other side lighting comes from high up at the sides of the proscenium in the perch position (*Figure 27.34* at D).

27.8.7.9 *Side lighting*

Side lighting is very important on the stage and should in the main come from high up and be directed to the opposite side of the stage. This is very conveniently done at the perch position, but elsewhere it may be necessary to use a vertical pipe known as a boomerang to which the lanterns are clamped. It would not be usual to fit such lanterns permanently, and it could be wired for a particular production requiring them. It is essential to provide plenty of auxiliary socket outlets on either side of the stage to which portable equipment can be connected (*Figure 27.34* at F). Other portable equipment will be carried on telescopic stands and will consist of floodlights for use behind backings to doorways and windows: also there will be Fresnel spots to provide some light coming through these windows, or to provide highlighting on dancers on the stage.

Intermediate battens, where necessary, consist of a mixture of floodlighting and spotlighting with rather fewer spotlights than No. 1 batten, and rather fewer floods than the back batten.

27.8.7.10 *Lighting control*

Almost all new installations have remote control of lighting from a compartment in the auditorium from which the operator can see the stage. Power electronic devices now make preprogrammed control boards available, to give multiple changes in sequence on command. Control of the light output from one or any combination of lamps by 'dimming' gives fast or slow increase or decrease, and balance of illumination for emphasis, mood and appearance ('stage picture').

27.8.8 Television studio lighting

This must not be confused with film studio lighting, for, although television studios are often converted film studios, the requirements are very different. First, there is far more attempt to make full economic use of the space. Thus, scenery is often rigged overnight, rehearsal and production takes place the following day and then the studio is rerigged the next night for something else. Large productions may be allowed two days for rehearsal or, very exceptionally, a further day. The speed of production is reflected in the rapid means for rigging lighting and feeding the electrical connections. Second, such studios are always provided with a central control board with dimmers, so that the lighting may be composed and adjusted rapidly and effectively. Various versions of stage switchboards specially designed for television are in common use to provide the latter facility.

27.8.8.1 *Lanterns*

The main lanterns in use for lighting are 500 W 150 mm Fresnel spots, 1 or 2 kW 250 mm Fresnel spots, and, for special purposes, 5 kW or even 10 kW Fresnels, although these are rare. Other special-purpose lanterns are Profile spots, particularly the 1 kW following type commonly used in the theatre. In addition, there are arc lanterns, special effects projectors, scene projection machines, etc. However, the backbone of lighting is undoubtedly to be found in the lantern range up to 2 kW; in consequence, the majority of studio circuits are 2 kW. As well as spotlighting there will be an equal lamp power devoted to pure floodlighting, the principal lantern used for this purpose being known as the 'scoop': this takes a 1 kW silica-sprayed lamp to ensure even distribution. Such floodlighting is often known as 'base' or 'fill' light. The floodlighting instruments are often used in pairs or fours at a time. Other forms of multi-unit baselight equipment are used, notably the 10 light with ten 200 W reflector lamps. More recently, designs using tungsten–halogen lamps have been favoured. One solution to the problem is a composite unit with softlight optics at one end, reversible to give Fresnel spot lighting.

27.8.8.2 *Lighting layouts*

As every production may be different from its predecessor and successor, every lighting layout will also be different. There can be no fixed units as in the theatre. This means that every facility for quick rigging of lanterns must be provided. Leaving aside the very small interview studios, where fluorescent lighting with a few spotlights may suffice, there are two main systems of television studio lighting rigging.

27.8.8.3 *Motorised hoist rigging*

The first system was developed by the BBC and employs short barrels 1.2–2.5 m long, each containing two or four socket outlets on separate circuits. Each barrel hangs on two lines from a motorised hoist and arrangement is made to take up the cable slack as the barrel is raised or lowered. The motors with their contactors are busbar-fed in the studio roof and the contactor controls are taken back to a centralised console desk where several hoists may be operated simultaneously. Hoists will only occasionally be used during the time a production is on the air; they are concerned with providing ready access for rigging lanterns and, in consequence, a reasonable level of noise in operation is not considered serious.

The procedure is for the lighting supervisor to make a layout to suit the particular production; the bars required are lowered and the lanterns plugged in and rigged. A large number of bars are used and have roughly $1\frac{1}{2}$ m centres all over the studio, and thus in an average studio there may be 300–500 socket outlets on the bars. Each outlet forms a circuit which is taken back to a patching field. Here it is possible to select the circuits in use and plug them appropriately into the control channels. Thereafter, the lighting will usually be controlled at the lighting console, the results being viewed with the aid of a monitor tube. Various systems of patching have been tried but the most satisfactory seems to be the jack and cord. Thus, one or more circuits can be plugged into a control channel. The drawback of this system is that the number of cords is very great, as the number of circuits is usually four or five times the number of control channels.

If, however, a female jack or a series jack is associated with each control channel only, the panel itself will contain sockets for every circuit and the number of cords is greatly reduced. A further advantage of this latter system is that the circuits can be fed each from its own fuse, and the control channel inserted as a series loop. This means that the phase layout in the studio can be well defined and fixed and the dimmer or other channel controls can be plugged up exactly as required without having any effect on the studio phase areas.

27.8.8.4 *Telescope rigging from grid*

The second main system of rigging, pioneered by Granada Television relies on an all-over theatre-type grid carried above the studio floor. The riggers and electricians can walk all over the studio above the level of the lighting. Carried on channels running the length of this grid are telescopic suspensions, each with its own hand winch, to which lighting units can be attached. The procedure is to run the telescope to the position required, lower it, attach the lamp, drop in a lead, plug up and then raise appropriately. The lead is plugged into a socket outlet up on the grid. These socket outlets on the grid are very much fewer in number than those provided when the motorised bar system is used, because the system does not have to carry so many idle outlets. Because the outlets are accessible at all times, they constitute the patching field itself.

27.8.8.5 Colour television studios

Although most 'monochrome' studios are conceived in terms of 2 kW circuits with 5 kW representing a very small percentage for special work, colour television demands higher powers. Where the hanging bar system is used, the control channels should represent roughly one-third of the total circuits; on the other hand, on the grid system every socket outlet on the grid is represented on the control board as a control channel.

27.8.8.6 Dimmers

The question arises as to what proportion of channels should have dimmers, and experience shows that it is desirable to have as many dimmers as possible; 100% is ideal. All television lighting control is carried out by means of remote-control dimmers so that the control panel can be positioned to allow the operator to take part in the production.

27.8.8.7 Control positions

There remains the question as to where the lighting control should be positioned. Some have been placed in the producer's control room so that the lighting supervisor can see the output of the producer's monitors. He can then connect to his own monitor any picture needing attention. However, the tendency is to separate the lighting supervisor and put him and his control alongside the C.C.U. operator or operators. Much careful planning has gone into the B.B.C. Television Centre to ensure that although the lighting supervisor is not in the producer's control room, he is closely in touch, both visually and physically. The arrangements provide all the advantage of close physical contact without the drawback of the inevitable confusion and noise which results when technical services are mixed up with production control.

28 Environmental Control

D Lush, BSc(Eng), CEng, MIEE, MCIBS
Ove Arup Partnership

Contents

28.1 Introduction 28/3

28.2 Environmental comfort 28/3
    28.2.1 Personal comfort 28/3
    28.2.2 Temperature and humidity 28/3
    28.2.3 Limits 28/4
    28.2.4 Visual and acoustic parameters 28/4
    28.2.5 Machines and processes 28/5
    28.2.6 Safety requirements 28/6

28.3 Energy requirements 28/6
    28.3.1 Steady state loads 28/6
    28.3.2 Dynamic or cyclic loads 28/6
    28.3.3 Intermittent heating and cooling 28/7
    28.3.4 Plant capacity 28/8
    28.3.5 Computer aided design 28/9
    28.3.6 Energy consumption 28/9

28.4 Heating and warm air systems 28/9
    28.4.1 Radiators 28/9
    28.4.2 Convectors 28/9
    28.4.3 Warm air systems 28/9
    28.4.4 Storage heating 28/10
    28.4.5 Air conditioning 28/10
    28.4.6 Cooling plant 28/10

28.5 Control 28/13
    28.5.1 Controllers 28/13
    28.5.2 Time controls 28/16
    28.5.3 Building automation 28/17

28.6 Energy conservation 28/18
    28.6.1 Systems 28/18

28.7 Interfaces 28/22
    28.7.1 Electrical loads 28/22

28.1 Introduction

The environment in which we live and work has two basic elements: the external, over which we have as yet no control; and the internal, which can be maintained at a specified condition according to our needs.

The three major subjects for which internal environmental control is considered are people, machines and processes; and while the conditions for each may differ, there are frequently common requirements. In the case of people the requirements are often referred to as 'comfort' or 'environmental comfort' conditions.

Environmental control is sometimes considered only in terms of the ability of a thermostatic control system to maintain specified conditions, but this is a narrow concept. It is necessary to consider the internal environment in terms of the overall building design, as the selection and construction of the building fabric may have a marked effect on the environmental performance and energy consumption. Equally, the selection of suitable plant and equipment may also affect the same parameters.

In the past the achievement of a specified environment tended to ignore the efficient use of the energy utilised for the purpose. Since the fuel crisis of 1973/74 this attitude has radically altered and energy conservation is considered as a major design parameter in all building services systems. In the UK approximately one-half of the overall annual energy is used in building services, overlapping with the domestic sector, which in its own right also consumes about one-half of the total. As both sectors are directly related to the internal environment, there is considerable potential for energy conservation, by suitable integrated design and selection of equipment.

The use of electrical energy is important in all forms of environmental control whether it be for the supply of thermal power, circulation of air and water, or control. Apart from thermal power, where the choice of fuel is often governed either by its availability or the apparent economics during the design period, electricity will virtually always be involved with the other elements. The use of particular energy sources such as oil, gas, coal, electricity, etc., may be governed by the specific application. The comparison of fuels in terms of economics and costs to the client, as distinct from the primary energy consumption for each fuel, is important and needs to be considered in selecting building services systems, but is an indirect element of environmental control.

28.2 Environmental comfort

The indoor environment should be safe, appropriate for its purpose and pleasant to inhabit. The parameters to be considered include the thermal, acoustic and visual conditions.

28.2.1 Personal comfort

In human terms an individual senses skin temperature, not room temperature, although the latter affects the former. The body loses heat by evaporation of moisture from the skin, convection to the surrounding air and radiation to, or conduction with, cold surfaces. These mechanisms, together with the degree of activity and the type of clothing worn, tend to maintain the skin temperature constant (except for exposed extremities) over a wide range of environmental conditions. However, real comfort occurs in a much narrower range of climate, and individual requirements differ considerably both intrinsically and, again, according to the activity and clothing. The narrow zone of real comfort conditions is often classified as neutral or comfortable. This, is shown in *Figure 28.1* in terms of temperature[1], and

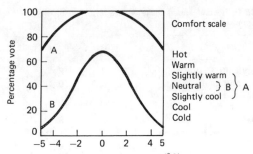

Figure 28.1 Comfort vote for personnel at around the neutral temperature for varying criteria. Curve A is for people giving any of the three central descriptions. Curve B is for the central description alone. © CIBS

illustrates the degree of satisfaction for a group of people in a particular space, about the optimum neutral temperature for the group. The specified space temperature is therefore always a compromise and is only one of the criteria affecting comfort.

Other important parameters that affect comfort are the radiant temperatures of the surfaces surrounding the space, and the relative humidity and air speed in the space, all of which can have a marked effect on the space temperature necessary to provide optimum satisfaction to the occupants. One example of these effects (*Figure 28.2*)[1] illustrates the elevation in space temperature required to compensate for increasing air movement.

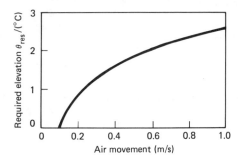

Figure 28.2 Corrections to dry resultant temperature for air movement. © CIBS

28.2.2 Temperature and humidity

The term 'space temperature' has been used so far to avoid confusion. Most people assume that space temperature specified in terms of the dry bulb air temperature defines levels of warmth. The previous comments indicate that this may not be valid, although temperature detectors in common use are mainly calibrated for, and measure, dry bulb air temperature. Alternative temperature indices may provide better definitions of comfort conditions or are used for design calculations: these include equivalent, effective, globe, dry resultant and environmental temperatures. Environmental temperature is used for calculation, and resultant temperature θ_{res} is considered to be a measure of comfort dependent on internal dry bulb temperature θ_{ai}, mean radiant temperature θ_r and speed of air movement u as defined in

$$\theta_{res}=[\theta_r+\theta_{ai}\sqrt{(10u)}]/[1+\sqrt{(10u)}]$$

Table 28.1 Recommended design values for dry resultant temperature

Type of building	θ_{res} (°C)
Assembly halls, lecture halls	18
Canteens and dining rooms	20
Churches and chapels:	
up to 7000 m³	18
>7000 m³	18
Dining and banqueting halls	21
Factories:	
sedentary work	19
light work	16
heavy work	13
Flats, residences and hostels:	
living rooms	21
bedrooms	18
bathrooms	22
Hospitals:	
corridors	16
offices	20
operating theatre suite	18–21
wards and patient areas	18
Hotels:	
bedrooms (standard)	22
bedrooms (luxury)	24
Laboratories	20
Offices	20
Restaurants and tea-shops	18
Schools and colleges	18
Shops and showrooms	18
Swimming baths:	
changing rooms	22
bath hall	26
Warehouses:	
working and packing spaces	16
storage space	13

© CIBS

Extracted from CIBS Guide A1. Part of Table A1.3.

Given that the recommended values of θ_{res} are those in *Table 28.1*, it is possible to adjust room temperature detectors or thermostats to a level suitable for comfort, for any mean radiant temperature and air velocity.

28.2.3 Limits

Limits need to be applied to any specified comfort conditions, particularly in the case of temperature and humidity. Generally, in terms of human comfort, limits of ± 2°C about a specified temperature and a relative humidity (r.h.) of $\pm 10\%$, about a mean of 50%, will be acceptable. Limits more critical than is necessary will create additional and unnecessary costs.

There may also be statutory limits which have to be applied in terms of energy conservation. In the UK since 1974, outside the domestic sector, space temperatures[2] now have an upper limit of 19°C. This is only a partial limit, because the reference covers only the heating. To complete the limit the regulation would have to specify an upper limit of 20°C for heating cycles and a lower limit of, say, 25°C for cooling cycles. Between these two limits there would be neither heating nor cooling input.

The level of humidity in a space can have a considerable impact on comfort, but in a heated building there is a limited range of control over its value. Artifically increasing the air change rate by opening doors and windows is unlikely to be acceptable in winter and is a palliative in summer. Fortunately the band of comfort conditions in humidity terms is fairly broad for most people, and a range of 40–60% r.h. is usually acceptable and often occurs in practice internally. Below approximately 35% r.h., static electricity effects occur and noses and throats may be affected; and above, say, 65% the effect of stickiness may be felt. Air conditioned systems are normally designed to avoid these extremes.

28.2.4 Visual and acoustic parameters

The acoustic and visual impact on comfort conditions is extremely important. The correct level of illuminance for a particular task is important in its own right, but the overall aesthetics are a combination of the lighting level, the lighting source, the architectural finishes and their reflecting properties, and furniture and equipment. These aesthetics contribute to the comfort of the occupant.

Sound, in terms of personal comfort, is also related to the particular task. People's reaction to sound varies according to age and situation. Acoustics is both complex and subjective, and only a broad outline is included for the purpose of defining general criteria for comfort. Because the response of the ear is non-linear and less sensitive at low frequencies, it perceives equal loudness for various combinations of frequency and sound pressure levels, units of loudness being defined in phons. Sound pressure is a fluctuating air pressure sensed by the ear. The fluctuations are minute in relation to atmospheric pressure: sound pressure levels are specified in decibels. The levels are created by the sound power (the power transmitted by the sound waves) which is normally considered only in reference to a sound source. Sound power levels (L) are also referred to in dB.

$$L = 10 \lg(W/W_0) \text{ dB}$$

where W = source power (W) and
W_0 = reference level (normally 1 pW).

Sound pressure is proportional to the square root of sound power.

A series of equal loudness curves (*Figure 28.3*)[1] is split into three sectors defined by A, B and C, which correspond to the sensitivity of the ear under varying conditions and can be

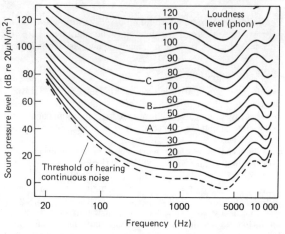

Figure 28.3 Equal loudness level contours. © CIBS

measured by instruments with weighting networks corresponding to these bands. The subjective reactions for comfort in buildings are normally related to the A scale and the noise levels are quoted in dBA. It is common to specify acceptable background noise levels for annoyance and speech intelligibility by means of NR (noise rating) or NC (noise criteria) curves, the former being most commonly used in Europe. Both sets of curves attempt to express equal human tolerances to noise across the audible frequency spectrum and are based on subjective experimental data. Normally the curves specify noise levels between 4 and 8 units below the measured dBA values, although the relationship is not constant. *Figure 28.4* shows the NR curves and *Table 28.2* lists the recommended noise ratings for various situations.

In conditions of adverse background noise the acceptability of differing noise sources may not depend on their absolute level and frequency but on their relationship with one another.

Outside the areas of normal sedentary or light industrial environments noise can rise to levels which may be injurious to health. At 90 dBA or higher, exposure to the noise in confined spaces can be tolerated only for specific periods, e.g. 8 h at 90 dBA, 2 h at 96 dBA and 0.8 h at 100 dBA.

Table 28.2 Recommended noise ratings

Situation	NR value
Concert halls, opera halls, studios for sound reproduction, live theatres (>500 seats)	20
Bedrooms in private homes, live theatres (<500 seats), cathedrals and large churches, television studios, large conference and lecture rooms (>50 people)	25
Living rooms in private homes, board rooms, top management offices, conference and lecture rooms (20–50 people), multipurpose halls, churches (medium and small), libraries, bedrooms in hotels, etc., banqueting rooms, operating theatres, cinemas, hospital private rooms, large courtrooms	30
Public rooms in hotels, etc., ballrooms, hospital open wards, middle management and small offices, small conference and lecture rooms (<20 people), school classrooms, small courtrooms, museums, libraries, banking halls, small restaurants, cocktail bars, quality shops	35
Toilets and washrooms, drawing offices, reception areas (offices), halls, corridors, lobbies in hotels, etc., laboratories, recreation rooms, post offices, large restaurants, bars and night clubs, department stores, shops, gymnasia	40
Kitchens in hotels, hospitals, etc., laundry rooms, computer rooms, accounting machine rooms, cafeteria, canteens, supermarkets, swimming pools, covered garages in hotels, offices, etc., bowling alleys, landscaped offices	45

NR50 and above
NR50 will generally be regarded as very noisy by sedentary workers but most of the classifications listed under NR45 could just accept NR50. Higher noise levels than NR50 will be justified in certain manufacturing areas; such cases must be judged on their own merits.

Notes
1. The ratings listed above will give general guidance for total services noise but limited adjustment of certain of these criteria may be appropriate in some applications.
2. The intrusion of high external noise levels may, if continuous during occupation, permit relaxation of the standards but services noise should be not less than 5 dB below the minimum intruding noise in any octave band to avoid adding a significant new noise source to the area.
3. Where more than one noise source is present, it is the aggregate noise which should meet the criterion.
4. NR ≈ dBA value − 6.

© CIBS

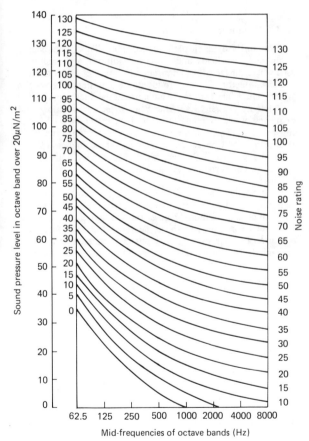

Figure 28.4 NR curves. Each curve is classified by a number corresponding to the speech interference level which was originally defined as the average of the sound pressure levels measured in the octave bands 600 to 1200, 1200 to 2400 and 2400 to 4800 Hz. The maximum permissible loudness level is taken to be 22 units more. Thus NR 30 has a speech interference level of 30 dB and a loudness level of 52 phon; this means that for effective speech communication the loudness level in a space designed to have a background level complying with NR 30 must not exceed 52 phon

Additional factors to be considered for total comfort, particularly over long periods, are the level of CO_2, odours and odour dilution.

28.2.5 Machines and processes

Apart from comfort criteria for personnel, there are two other areas where environmental conditions may be important: machine rooms and process plants. The former covers such spaces as computer rooms and medical machine areas; and the latter, areas such as electronic manufacturing or food processing factories. In many ways the criteria are similar to those for personal comfort, but there may be requirements for closer limits and, in particular, air filtration becomes a significant factor, the number and size of particles being very closely defined according to the process. In certain critical processes the air movement patterns are also specified, laminar flow probably being the most difficult to achieve. While temperatures are often specified with limits of not greater than ±0.25 or ±0.5°C, with humidities to ±2% r.h. or better, two points should be made. Limits specified may be unnecessarily stringent and need to be questioned. As an example, computer rooms in the past needed close limits if the machines were to operate correctly; but nowadays this is not normally necessary. The second point concerns the achievement of conditions throughout the treated space. Strictly, the specified conditions can normally be achieved only at the point of detection and control, the variation throughout the remainder of the space being largely dependent on good plant design and distribution of the heating and cooling media.

28.2.6 Safety requirements

In environmental control the specified conditions and their limits have to be achieved, but generally only during normal periods of occupation. It is therefore necessary to consider the environmental conditions necessary under abnormal circumstances. Examples include (1) emergency or maintained lighting levels when the normal system breaks down or during unoccupied hours, (2) the low limit temperatures to be maintained to prevent freezing or damage to equipment and furniture, and (3) the maintenance of humidity below a specified dewpoint condition to prevent condensation on cold glazing.

28.3 Energy requirements

To achieve comfort conditions energy is required. The selection of electricity, gas, oil, solid fuel or alternative energy sources is a function of the required conditions, the plant selected, the economics, availability of supply, client preference and, to some extent, crystal-ball gazing.

The calculation of the energy necessary to achieve a particular set of environmental conditions is based on a number of concepts ranging from the simple to the complex and covering both the actual loads for heating and cooling the spaces and the plant sizing to deliver these loads. More sophisticated techniques have been developed to improve the accuracy of calculations and predictions in terms of heating or cooling loads and energy consumption. It is necessary to remember that the more precise figures derived from these techniques are valid only so long as the buildings to which they are applied are built with the same precision and with materials having the indices used in the calculations. Care must therefore be exercised to ensure that the calculation procedures do not aim for an order of certainty that cannot be achieved in normal construction.

28.3.1 Steady state loads

In its simplest form the steady state head load for a space within a building may be defined as

$$Q = [\Sigma UA + \tfrac{1}{3}NV]\Delta\theta$$

where the symbols are Q energy required (W); U thermal transmittance of any element surrounding the space (W/m^{2}K); A area of the element (m^2); $\Delta\theta$ temperature differential across the element (°C); V volume of the space (m^3); N number of external air changes per hour (h^{-1}).

The thermal transmittance of a wall, roof, floor, etc., is based on an electric circuit analogue. Normally each wall, etc., is a laminar structure of parallel layers of different materials and air spaces, each having a thermal resistance depending on its composition, thickness and surface properties. Given the thermal resistances (in m^{2}K/W) as R_{si} and R_{so} for the inner and outer surfaces, R_1, R_2,..., for the component layers, and R_a for the air spaces, the thermal transmittance (in W/m^{2}K) is

$$U = 1/[R_{si} + R_1 + R_2 \cdots + R_a + R_{so}]$$

Tabulated U values for a wide range of construction elements[3] are available for different exposures: some are given in *Table 28.3*. There are certain factors for which allowances may have to be made to the tabulated figures. The values quoted are for homogenous areas of construction, whereas in practice edge details and corners affect the figures. In addition, cold bridging can affect the values, i.e. the effect of wall ties between the inner and outer skin of the building or the framing round windows. Some examples of the ventilation rates used in normal calculations are detailed in *Table 28.4*.

The use of this method enables the heating load (or sensible cooling load) to be calculated for steady state conditions based on a minimum (or maximum) outside design condition and the additional assumption that the heating (or cooling) system will operate continuously. In practice systems generally operate intermittently and the building behaves dynamically. It is to cater for these factors that the more sophisticated calculations are introduced.

28.3.2 Dynamic or cyclic loads

To examine the dynamic performance of buildings, i.e. the energy requirements or loads under cyclic conditions, a procedure is used where the factors of admittance (Y), surface factor (F) and decrement factor (f) are introduced, the admittance having the greatest effect. The factors are functions of the thickness, thermal conductivity, density, specific heat capacity, position and frequency of energy inputs of each of the materials used in the construction. These have analogues with reactive loads in electric circuits. Consequently, there are phase changes (ϕ_Y, ϕ_F and ϕ_f) associated with them which, because the fundamental frequency is one cycle per day, are expressed as time lags/leads to the nearest hour.

The use of these factors leads to some complex equations which define the cyclic heat requirements for the building. The admittance can be thought of as the thermal elasticity of the structure, i.e. its ability to absorb heat; the decrement factor is a measure of how a cyclic heat input is attenuated as it passes through the structure; and the surface factor is a measure of how much of the cyclic input at a surface is readmitted to the space.

On thin structures the admittance equals the static U value; on multilayer constructions the admittance is largely determined by the internal layer. Thus, insulation on the inside of a concrete slab gives an admittance close to that of the insulation alone, whereas if the insulation is within or on the outside of the slab, the admittance value is virtually that of the slab alone. Decrement factors range from unity for thin structures of low thermal capacity, decreasing with increasing thickness or thermal capacity. Surface factors decrease with increasing thermal capacity and are virtually constant with thickness. Sample values of these three factors are shown in *Table 28.3*[3].

Other factors affecting the load requirements include environmental temperature, solar gains, internal gains and the latent load for air conditioning plants, i.e. the amount of moisture that has to be removed (or added) to the treated air. Environmental temperature (θ_{ei}) has already been mentioned and it is a concept used in carrying out load calculations, as it defines the heat exchange between a surface and an enclosed space. Its precise value depends on room configuration and the convective and radiant heat transfer coefficients of the surfaces. For the UK and hot climates it may be shown that

$$\theta_{ei} = \tfrac{1}{3}\theta_{ai} + \tfrac{2}{3}\theta_m$$

Table 28.3 Thermal transmittance, admittance, decrement and surface factor for various constructions

Construction (outside to inside)	U value (W/m² K)	Admittance Y value (W/m² K)	ϕ_Y (h)	Decrement f	ϕ_f (h)	Surface factor F	ϕ_F (h)
Brickwork:							
220 mm brickwork, unplastered	2.3	4.6	1	0.54	6	0.52	2
220 mm brickwork, 13 mm dense plaster	2.1	4.4	1	0.49	7	0.53	1
105 mm brickwork, 25 mm air-gap, 105 mm brickwork, 13 mm dense plaster	1.5	4.4	2	0.44	8	0.58	2
105 mm brickwork, 50 mm ureaformaldehyde foam, 105 mm brickwork, 13 mm lightweight plaster	0.55	3.6	2	0.28	9	0.61	1
Concrete brickwork:							
220 mm heavyweight concrete block, 25 mm air-gap (50% plaster), 10 mm plasterboard (on dabs)	2.2	3.2	1	0.38	7	0.62	1
200 mm lightweight concrete block, 25 mm air-gap (50% plaster) 10 mm plasterboard (on dabs)	0.7	2.4	2	0.48	7	0.78	1
Roofs — pitched:							
5 mm asbestos cement sheet	4.9	4.9	0	1.0	0	0.26	0
5 mm asbestos cement sheet, loft space, 10 mm plasterboard	2.4	2.4	0	1.0	0	0.65	0
10 mm tile, loft space, 25 mm glass fibre quilt, 10 mm plasterboard	0.9	1.0	2	1.0	1	0.86	0
Roofs — flat:							
19 mm asphalt, 75 mm screed, 150 mm cast concrete (dense) 13 mm dense plaster	1.8	4.5	1	0.33	8	0.38	2
19 mm asphalt, 13 mm fibreboard 25 mm air-gap, 25 mm glass fibre quilt, 10 mm plasterboard	0.8	0.9	2	0.99	1	0.92	0

Extracted from CIBS Guide Section A3. Sample values from schedules.

© CIBS

where θ_{ai} is internal air temperature and θ_m is mean surface temperature.

Solar gains affect load calculations in two ways. First, there is the effect of solar radiation on the heat transfer characteristics of the building fabric, which is covered by the use of a parameter known as sol-air temperature (θ_{eo}). The definition of θ_{eo} is that temperature which, in the absence of solar radiation, would give the same rate of heat transfer through the wall or roof as exists with the actual outdoor temperature and the incident solar radiation; i.e. it is an artificial outside temperature to take into account the effects of solar radiation. The second factor covers the direct solar radiation gains through windows, some of which have an immediate effect and some of which is absorbed into the internal structure and readmitted subsequently. Both types of gain affect energy consumption, but in terms of load they are ignored for heating calculations and included for air conditioning load purposes.

Internal gains from lights, machines and occupants may be substantial. Again, in heated buildings the gains affect the energy consumed by the environmental plant but are not normally taken into account in calculating the design load (the energy input required for the coldest day). For air conditioning loads the inclusion of these gains is most important.

Latent gains are ignored in heated buildings but for air conditioning a considerable proportion of the maximum cooling load may consist of latent cooling, i.e. the removal of excess moisture from the air because of the use of fresh air for ventilation and internal gains from occupants and, possibly, processes.

28.3.3 Intermittent heating and cooling

The calculation of loads for steady state and cyclic situations leads naturally to consideration of the effects of running the plant intermittently to satisfy only the specified environmental conditions during periods of occupation. Inherent in this are the following:

(1) The preheat period necessary to bring the building up to temperature under varying climatic conditions — in particular, on the coldest day for which the load is calculated.
(2) The thermal response of the building during preheat, which will depend on its construction, insulation and ventilation.
(3) The thermal response of the plant when first switched on.
(4) The ratios between preheat, normal heating and plant-off periods.
(5) Relative running and capital costs.

While the actual calculations can be complex and related to the dynamic states already mentioned, there are some basic points which illustrate the situation. In a steady state or dynamic analysis with continuous plant operation, loads for particular spaces may be calculated and the output equipment sized on this basis. When intermittent plant operation is introduced, the

Table 28.4 Air infiltration rates for heated buildings

Building	Air infiltration rate/h
Assembly hall, lecture halls	$\frac{1}{2}$
Canteens and dining rooms	1
Churches and chapels	$\frac{1}{2}$
Dining and banqueting halls	$\frac{1}{2}$
Flats, residences, and hostels:	
living rooms	1
bedrooms	$\frac{1}{2}$
bathrooms	2
Hospitals:	
corridors	1
offices	1
operating theatre suite	$\frac{1}{2}$
wards and patient areas	2
Hotels:	
bedrooms	1
Laboratories	1
Offices	1
Restaurants and tea-shops	1
Schools and colleges:	
classrooms	2
lecture rooms	1
Shops and showrooms:	
small	1
large	$\frac{1}{2}$
department store	$\frac{1}{4}$
fitting rooms	$1\frac{1}{2}$
Swimming baths:	
changing rooms	$\frac{1}{2}$
bath hall	$\frac{1}{2}$
Warehouses:	
working and packing spaces	$\frac{1}{2}$
storage space	$\frac{1}{4}$

© CIBS

Extracted from CIBS Guide A9 (1975). Extracts from Table A9.1

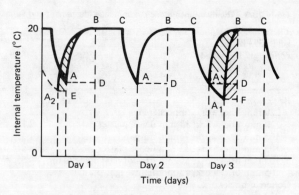

Figure 28.5 Intermittent heating temperature/time curves

situation illustrated in *Figure 28.5* is typical. Here BC represents the period of occupation for which an internal dry bulb temperature of 20°C is required. CA represents the normal temperature decay in the space for a particular set of external weather conditions, and the area ABDA is the energy required to restore the temperature to 20°C at the start of the occupation period.

Because the normal load calculations are based on a constant temperature internally, it is obvious that if the temperature falls as shown, the energy input has to be increased above the steady state requirement, in order to raise the temperature to the required level. Calculating the amount of additional energy becomes part of the load calculations and may be based on variations of the cyclic state techniques already mentioned. The thermal response of the building and the time at which the plant is to be switched on are major factors in the calculations: *Figure 28.5* illustrates some of the variants which have to be taken into account. On the coldest day for which the system is designed the energy input is calculated in terms of the power to satisfy a switch-on time designated by point A. The actual position of A is temperature and time dependent, as indicated by points A_1 and A_2.

Point A_1 represents a situation where the temperature decay has been allowed for longer than that required for A. It is clear that the power input capacity to achieve point B from A_1 must be greater than that for point A, because the differential temperature is greater and the time allowed is reduced. But the actual energy for the purpose is the comparison of areas ABDA and A_1BFA_1. Therefore, apart from the calculations for the load requirements, the economics of plant costs against cost in use (energy costs in this case) have to be evaluated.

The second point A_2 represents the situation after a non-standard shut-down, (e.g. weekends) when the normal load is based on heating from A. Again it is necessary to calculate the load for condition A_2 or to arrange for alternative operation of the plant so that, for example, the temperature is not permitted to drop below A.

28.3.4 Plant capacity

While it is possible to calculate all the heating and cooling loads for achieving environmental comfort conditions in various spaces, including the sizing of individual output terminals, the selection of main plant to supply the necessary energy is another matter. The actual terminal size for a space has to take into account all the elements already considered, which clearly illustrates that capacity is rarely based on the steady state load for continuous plant operation. Economic factors can sometimes inhibit optimum selection. The choice of suitable terminal sizes is reflected in the main plant selection, which has to cover the following points:

(1) Heating plant capacity for intermittently heated buildings is normally based on simultaneous peak loads for all the spaces in the building, with such exceptions as the domestic sector, whereas cooling plant assumes diversity between the peak loads in various conditioned spaces (some diversity is permitted for continuously heated buildings).
(2) Heating or cooling plant capacity should be sufficient to cater for process requirements in addition to the environmental load; e.g. the domestic hot water load may be purely for washing but may also include kitchen or restaurant requirements.
(3) Sizing of source units should be such as to permit efficient operation under part-load conditions.

In respect of (3), particularly for multiple boiler or chiller installations, the environmental load in most cases only reaches the design peak for a small percentage of the total operating hours per annum, and the efficiency of boilers and chillers normally falls as their output decreases from the specified design level. It is therefore important to choose the units so that at, say, 25% of design load the operating source units are matched to the requirement to maintain a high efficiency.

28.3.5 Computer aided design

The use of the computer for environmental comfort design is generally restricted to calculation of loads and annual energy consumptions, and to check whether the summertime temperature in the building rises to a point where air conditioning is essential.

Computer programs enable far more sophisticated techniques to be employed without laborious arithmetic. However, they should be used only as a design tool for the project, which still requires practical knowledge.

The programs for UK use are generally based on detailed extensions[4] of the techniques already outlined. Among the criteria which can be incorporated are the effects of shading provided by building configurations and overhangs, which affect both the load and energy consumption.

Apart from providing the calculated loads, energy consumption programs can provide the annual figures based on hourly weather data over a full year for any location and type of plant, so that a comparison between plants is rapidly available.

28.3.6 Energy consumption

There are three specific points to be identified in considering energy consumption: (1) degree-day figures, (2) energy budgets and (3) energy targets.

The degree-day is a concept that permits energy consumption in a building to be monitored from year to year against a monthly datum. It is normally used only for the comparison of consumption in heated buildings, although a modified form is being considered for use with air conditioning. Degree-days measure the interval for which the outside temperature drops below a specific value (normally 15.5 °C) and the amount by which it does so. The monthly figures are published for various areas. The base of 15.5 °C is used in the UK, but other figures are used elsewhere to broadly represent that outside condition for which no system heating is required to maintain a suitable internal temperature: internal gains, etc., are always assumed to provide a rise of several degrees.

The energy budget for a building is the estimated annual energy consumption of a building. It is necessary for the cost-in-use evaluation.

Energy targets define the design aim for energy use in buildings, related to type, usage and location. The figures may be quoted in terms of power or energy per unit area (e.g. kW/m^2 or kWh/m^2). Buildings are now being designed to meet such energy targets and there is a trend towards making such figures mandatory in certain parts of the world. *Table 28.5* lists the targets for a variety of buildings in various locations.

28.4 Heating and warm air systems

The majority of heating systems in the UK use water as the means of distributing thermal power (with steam as one variant of water). Electricity as a direct source of thermal power is an alternative which has advantages in certain situations but is frequently dismissed on the grounds of comparative energy cost. Water may be utilised for pure heating systems via emitters which produce radiative and convective heating, or for air heating systems where fan assisted devices produce warm air via a heat exchanger. Boilers, by both type and use, are generally known in the engineering professions and are not covered here[5].

28.4.1 Radiators

Radiators are the most common form of heating system. There is a roughly equal split between the radiant and convective heat output. On water systems cast-iron for radiators has been largely

Table 28.5 Energy targets for heated buildings

Building	Consumption per annum	
	(GJ/m^2)	(kWh/m^2)
Offices	0.6–1.2	170–330
Factories	0.7–1.5	190–420
Warehouses	0.6–1.2	170–330
Schools	0.6–1.2	170–330
Shops	0.6–1.2	170–330
Hotels	1.2–2.2	330–610

Notes:
1. The figures are for heated and naturally ventilated buildings.
2. The range of consumption is a function of the thermal insulation, air sealing and efficiency of the heating system.
3. Figures are for conventional hours of occupancy and lighting levels. The factory figures include a 20% allowance for process gain.
4. The figures are for southern England.
5. Factors for modifying the figures in particular circumstances are available (CIBS *Energy Code*, Part 4).

© CIBS

replaced by mild steel. The radiant emission Q is a function of the difference $\Delta\theta$ between the temperature of the ambient air and the mean of the internal liquid, according to the expression

$$Q = k(\Delta\theta)^n$$

where k is a constant depending on dimensions and $n = 1.3$ for radiators. Electric radiators and tubular heaters are rarely used outside domestic or small commercial premises and even then only because no other form of heating is available.

28.4.2 Convectors

Convectors may be natural or fan convectors: both are common. The natural versions may be in upright cases with top and bottom grilles for air circulation, or used as skirting heating. Their emission may be designated in a form similar to that for radiators, but $n = 1.35$ for upright types and 1.27 for skirting versions. Water flow rate can have a considerable effect on emission. Fan convectors provide a form of air heating. They are normally controlled by switching on and off by means of a thermostat. In the emission formula, n is unity.

Convectors using electrical power for the thermal output are unusual compared with the water powered versions.

28.4.3 Warm air systems

Warm air systems vary from domestic units to industrial systems. Some sophisticated versions which duct warm air through a large building complex may be treated as air conditioning systems without cooling and humidification elements, but such systems are rare.

A substantial proportion of heaters, with nil or minimal ductwork, are freestanding and distribute air locally on a recirculation basis without the introduction of fresh air. If the air volume is freely distributed, the complete unit is usually started and stopped by means of a space thermostat, or one mounted in the recirculation air inlet to the unit. Where the air flow is restricted, by manual or thermostatically controlled dampers to control the temperature by varying the air flow, it is necessary to ensure that the unit and fan are switched off when the air volume is reduced to predetermined level.

The primary source of thermal energy for these units may be oil, gas or electricity. The latter is used in domestic units and in some commercial applications.

28.4.4 Storage heating

Underfloor heating can be operated with hot water or electrical energy as the thermal medium. The floor construction, covering, thermal time constant, temperature control and the idiosyncracies of the user make system calculation difficult, and underfloor heating is consequently uncommon. Its future may depend on using low-grade heat spread over large emitting surfaces at temperatures of 23–30 °C, possibly with solar panels as the heat source.

Storage heaters, electrically fed, are employed in domestic and commercial premises, which avoids the need for central heating plant. For adequate thermal capacity the units are unavoidably heavy (60–300 kg), with ratings of 0.5–6 kW. The refractory heat-storage blocks have heating element temperatures up to 900 °C. The heaters are normally run on off-peak supply, emission from the units being regulated to match the periods during which the heat output is required. 'Natural' storage units radiate during the off-peak charging period. They are generally less satisfactory than 'fan assisted' units, which are insulated to restrict output during charging, the output being controlled by timers and thermostats to start and stop the fan, which controls most of the heat output.

The charge during the permitted period should be regulated in accordance with internal and external temperatures. Energy regulators are available for this purpose (see Section 28.5.2).

Other systems are based on large and well-insulated water storage tanks, electrically heated off-peak by boilers or immersion heaters. In some cases the storage temperature may exceed 100 °C, which necessitates a pressurised vessel. Water is circulated (or, for the high-temperature case, injected) into conventional heating or air conditioning systems during the periods required.

28.4.5 Air conditioning

Air conditioning is the filtering, washing, heating and cooling of air to achieve specified temperature and humidity levels. In temperate climates building design and services can usually achieve reasonable comfort conditions without air conditioning, but a system may be found necessary (a) if the extreme conditions are not tolerable, (b) if the building design requires it, (c) if urban noise and dirt have to be reduced, and (d) if internal heat gains (e.g. computer rooms) have to be accommodated.

In an air conditioning system air is moved by fan power through the relevant space, which results in a physical sensation of air movement quite unlike that in normal heated spaces and with simple air heating. Clients and occupants have to be forewarned of this.

28.4.5.1 Systems

Most systems have a section of plant which adjusts the humidity, to ensure that air passed into the conditioned spaces is suitable for the specified humidity (a 'dewpoint' condition of about 10 °C). The dewpoint plant is described in Section 28.5.1.3). After dewpoint treatment, air is ducted at high or low velocity, for which duct size, fan and acoustic treatment differ. Common systems available are given in the following paragraphs.

Constant volume Normally this uses branched ducting, each branch with a reheater controlled by a space temperature detector.

Dual duct Air from the main plant is split into two duct systems: one carries cold and the other carries preheated air. Both ducts traverse the building. Each space has a mixing unit, connected to both ducts and adjusting the hot/cold air ratio in accordance with a room temperature detector.

Variable volume Cold air is distributed to individual spaces through terminals, which throttle the rate of air supply and are controlled by room temperature detectors. Throttling raises the static pressure in the ductwork, the effect being used through a control system to vary the air flow through the supply and extraction fans. Methods of control may be mechanical, or by the several speed control methods applicable to electric motors. As some air is always necessary in a space for ventilation, the terminals do not close in normal operation, and continue to feed in some cooling air. The system is therefore best applied to buildings that require cooling throughout the year, or to systems that incorporate small reheaters (although in the latter case the variable-volume system may be inappropriate). In general, the control of air volume and consequential reduction of air treatment provides a low energy system.

Induction Air is supplied at high velocity to terminal induction units, mainly perimeter mounted, which are fitted with heating and cooling coils. Air is forced out of nozzles in the unit at a velocity high enough to induce entrainment of recirculation air from the space in the discharge jets. Thus, the space air is circulated through the unit, which provides the necessary quantity of conditioned fresh air from the main air plant. The heating and cooling coils are fitted with control valves and sequenced according to the requirements of the space temperature detector, to maintain the correct conditions. An extract system removes the equivalent amount of air to that supplied by the main plant.

Fan coil This has some similarity to the induction system but the air is supplied at relatively low velocity to the units, each of which has its own fan. The coil configuration and control is similar to that for the induction unit, but electrically there is a dual requirement for a distribution system to serve the fractional kilowatt 1-phase fans and to switch them off during plant-off periods.

Reversible heat pump cycle This system is often used when the main plant air is distributed through a ductwork independent of the units, and sometimes with a non-air-conditioned fresh air supply. Each unit contains a reversible cycle compressor so that it can produce either cooling, or heating in a heat pump mode. The energy transfer medium to and from the units is by a circulating water system with the water temperature controlled at approximately 24 °C. Each unit then extracts heat from, or supplies heat to, the circulating water, depending on whether the unit is on the heat pump or chilling cycle mode. The circulating temperature is maintained by sequencing a cooling tower heat exchanger to lower the temperature, or a non-storage calorifier to raise it. The latter may frequently be electrically fed.

Units are generally 'packaged' to include controls. The most common have a 1 kW compressor with a 1-phase motor and a fan of less than 100 W; but 3 kW (1- or 3-phase) units with fans of power more than 100 W are available.

28.4.6 Cooling plant

In essence, cooling plant for comfort conditioning is indicated schematically in *Figure 28.6*, which shows a single machine and tower, but multiple systems are more common. The temperature of the chilled water is controlled by T_{P1} operating the evaporator system and the heat extracted from the primary water appears in the condenser. The cooling tower dissipates this heat and returns the water to the condenser at a fixed temperature dictated by T_C, which by varying the position of valve V_1 controls the amount of tower cooling.

Cooling towers are basically forced or induced draught types, the terms describing the method by which air is drawn past

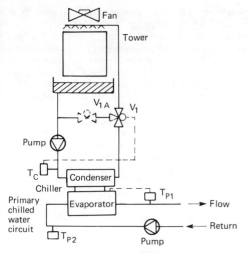

Figure 28.6 Schematic diagram of single chiller and cooling tower. V_{1A} is an alternative valve position for V_1, T_P is an alternative to T_{P1} for specific cases only

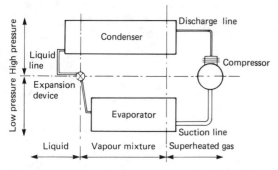

Figure 28.7 Schematic of vapour compression refrigeration cycle.
© ASC Ltd

the sprayed water in the tower for cooling. Air-cooled condensers are also used where, in simple terms, the action of the tower is replaced by air blast cooling. Chillers can be of reciprocating, centrifugal, absorption and screw forms, all of which operate on a refrigeration cycle (*Figure 28.7*). The system circulates a refrigerant which has liquid and gas phases and a boiling point at atmospheric pressure well below that of water. Common refrigerants are the Freon group, chosen according to application and type of chiller. Heat is absorbed by the gaseous and liquid mixture in the evaporator which then becomes a superheated gas: the absorbed heat is extracted from the primary chilled water circuit (*Figure 28.6*), and this is the prime function of the whole system. The gas is then passed through the compressor, which raises the gas pressure and also its boiling (or condensing) point. The condenser then extracts heat from the gas at this higher temperature and condenses it to a liquid again at a relatively high temperature and pressure: this process extracts heat from the refrigerant using the higher-temperature water available in the tower circuit (*Figure 28.6*). The liquid then passes through the expansion valve which reduces both the temperature and pressure of the liquid as it expands to the gaseous and liquid mixture state for the cycle to repeat. The use of the expansion device does not alter the total heat content of the fluid.

Chillers[6] can be made by manufacturers operating internationally, motors may be rated (as in the USA) on a basis of 60 Hz, and may not be suitable for working on a frequency of 50 Hz. Again, 60 Hz control circuitry for one voltage may require modification for 50 Hz and a different voltage. Both input and output are expressed in kilowatts. This may cause confusion, because the output is 3–4 times the input, indicating the 'coefficient of performance' of the system.

28.4.6.1 Chillers

Reciprocating machines or compressors operate on the compression of the refrigerant in a system analogous to that of an internal combustion engine, except that the 'fuel' is contained in a closed loop and an auxiliary motor drives the pistons. Compressors have various numbers of cylinders (up to 16) and methods of control. Machines are available for inputs up to 150–220 kW, corresponding to outputs up to 500–700 kW.

Centrifugal chillers Centrifugal machines use a rotating impeller which performs the compression operation on the gas by centrifugal force. *Figure 28.8* shows how the impeller is included in the system. One- or two-stage compression is normal. The

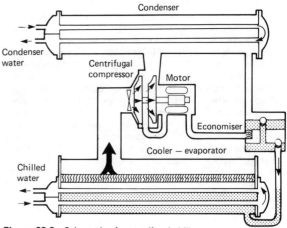

Figure 28.8 Schematic of a centrifugal chiller

control of output is normally by means of inlet guide vanes (not shown in the figure), which are actuated according to the load demand and alter the angle of entry of the gas into, and the performance of, the impeller.

Standard machines are available for outputs of 500–700 kW or greater, but not below 500 kW. Electric motor drives are most common and they may be hermetically sealed into the machine. The electrical rating is from 1/3 to 1/4 of the output rating and for machines above 500 kW input motors, operated economically at 3.3 kV or higher.

Absorption machines Absorption machines also rely on a refrigeration cycle. Analogy with other systems is best considered with the low-pressure section created by a permanently evacuated or high-vacuum system, and compression achieved by heating. The refrigerant fluid is a mixture of water and a fluid (normally lithium bromide) with an ability to absorb water. However, it is the water that acts as the refrigerant, as it will boil at low temperatures in a partial vacuum[6].

Water evaporates more quickly if the surface of a given volume is extended. Rather than using a vessel with a large surface area, this is best achieved by spraying. *Figure 28.9* shows how the heat from the cooling load can be picked up by the action of water boiled in vacuum, the chilled water coils being immersed in the sprayed water. Because of its affinity to lithium bromide the water vapour is carried away from the evaporator to the absorber,

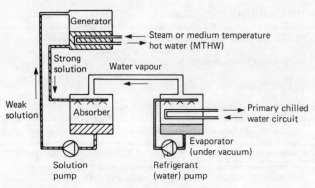

Figure 28.9 Schematic of basic absorption cycle

where it is mixed to provide a lithium bromide solution. If these were the only cycle components, once the lithium bromide had become diluted its capacity for absorbing water vapour would be reduced and an equilibrium position reached whereby no further evaporation of the water and no further useful water cooling could take place.

The weak solution from the absorber can be pumped to a generator where, with the addition of heat, the water vapour can be boiled out of the lithium bromide to produce a strong solution which is returned to the absorber. Here it is sprayed to increase the surface area and, as in the evaporator, increase the capacity to absorb the water vapour from the evaporator. This secondary cycle maintains the absorbent at an operating level, but a water supply is required to replace the water vapour evaporated from the evaporator. If the rejected water vapour from the generator were passed to a fourth vessel, a condenser as shown in *Figure 28.10*, the vapour at a high temperature could be condensed and returned to the evaporator to complete the cycle. In addition to eliminating the need for make-up water, the fourth vessel provides a vacuum-tight system.

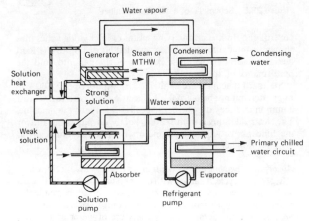

Figure 28.10 Schematic of full absorption cycle

When the lithium bromide solution absorbs water, heat is generated: it consists of the heat of condensation of the absorbed water plus the reaction heat between the lithium bromide and water vapour. To increase the capacity of the lithium bromide to accept the water vapour, it is kept cool by passing the condenser water first through the absorber and then onto the condenser.

Because the generator is hot and the absorber cool, the cycle efficiency can be increased by a heat exchanger which heats the weak solution pumped from the absorber to the generator and cools the strong solution returning.

In the diagrams used to describe the absorption cycle the flow of water vapour is restricted between evaporator and absorber, and between generator and condenser, by the size of the connecting pipes. In practice this is overcome by housing the evaporator and absorber in one shell and generator and condenser in a second. Alternatively, all can be housed in one common shell with a division plate.

The machines are made in two basic size ranges. At the lower end the most common unit is the gas fired domestic refrigerator; and direct gas fired units are also made for commercial use in outputs from 10 to 100 kW. The upper range is for outputs of 350 kW and higher, although 1500 kW is generally the minimum.

Energy for heating cycles associated with absorption machines is normally back-pressure steam from a primary process at 200 kPa (2 bar) or from medium-temperature water at 120°C. The steam system makes efficient use of energy which might otherwise be wasted.

Screw machines These belong to a range of positive displacement compressors, claimed to have advantages over conventional compressors in terms of reduced operating noise, lower operating speed and increased thermal efficiency. The machine essentially consists of two mating helically grooved rotors, a male (lobes) and a female (gullies), in a stationary housing with suitable inlet and outlet gas ports (*Figure 28.11*).

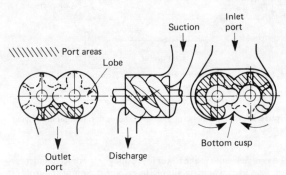

Figure 28.11 Screw chiller operation. © ASHRAE

The flow of gas in the rotors is both radial and axial. Compression is obtained by direct volume reduction with pure rotary motion. For clarity, the description of the four basic compression phases is here limited to one male rotor lobe and one female rotor interlobe space.

(1) *Suction:* As a male lobe begins to unmesh from a female interlobe space, a void is created and gas is drawn in through the inlet port. As the rotors continue to turn, the interlobe space increases and gas flows continuously into the compressor. Just prior to the point at which the interlobe space leaves the inlet port, the entire interlobe space is filled with gas.
(2) *Transfer:* As rotation continues, the trapped gas pocket is moved circumferentially around the compressor housing at constant suction pressure.
(3) *Compression:* Further rotation starts meshing of another interlobe space at the suction end and progressively compresses the gas in the direction of the discharge port. Thus, the volume of the trapped gas within the interlobe space is decreased and the gas pressure consequently increased.

(4) *Discharge*: At a point determined by the built-in volume ratio, the discharge port is uncovered and the compressed gas discharged by further meshing of the lobe and interlobe space. During the remeshing period of compression and discharge, a fresh charge is drawn through the inlet on the opposite side of the meshing point.

Machines are available for outputs ranging from 250 to 2000 kW with both open and hermetic motor drives. The electrical input is from 1/3 to 1/4 of the output.

28.5 Control

The importance of controls to achieve comfort conditions has always varied according to the sophistication of the plant to which they are applied, the conditions specified and the economics of providing them. With cheap energy, simple control systems were the norm and comfort conditions were often of low priority and specified only as minimal levels. Energy is expensive and comfort conditions are considered a more critical factor of human tolerance. Control systems are therefore commonly applied to all types of environmental system, performing the dual role of maintaining comfort conditions and conserving energy. In this combined role it is significant that for heated buildings in the UK a change of 1 °C in normal space temperature will affect the energy consumption by as much as 10%.

The most common parameters considered are temperature, humidity and time. Generically, controls are either electric/electronic or pneumatic. The latter system uses clean dry compressed air as the motive power and is rarely independent of electric/electronic elements, whereas the former system uses electric motive power alone. Pneumatic systems tend to be used where there are many terminal controls, normally on air conditioned systems. Some electric/electronic systems are now competitive with pneumatic terminal controls and the systems described below are based on non-pneumatic systems. Where run and standby air compressors are supplied for pneumatic control systems, the electrical supply must be capable of starting and running both compressors together.

28.5.1 Controllers

Intrinsically, controls operate in a two-position or a modulating mode. The former is recognised by on/off thermostats operating a device such as a fan or two-position control valve; the latter is a combination of detector and controller which can vary the position of the control valve over its full range of travel. The type of controller is selected according to the application.

28.5.1.1 Boilers and chillers

The controls for heating and cooling sources are normally supplied as part of the equipment. When multiple sources are used and sequential control is required to achieve a constant heating or cooling flow temperature, the electrical interlocking requirements are significant. *Figure 28.12* shows an arrangement for boilers in parallel and the interlocking diagram necessary to achieve a constant flow temperature for sequential operation which matches the load requirements. The system permits any boiler to lead the sequence, and various standard safety features are included, but relays, timers, etc., may be replaced by programmable controller software.

28.5.1.2 Heating systems

The control of heating covers the majority of systems in the UK. There are two elements to be considered: central plant control and terminal control. Apart from controlling the flow temperature from the boiler(s), central plant control for radiator and convector systems (the greatest number of heating systems) is carried out mainly by weather compensators. *Figure 28.13* illustrates such a system where the temperature to the load is controlled in accordance with external temperature. In cold weather water is supplied to the load at the boiler flow temperature, i.e. with valve V closed to the bypass and the setting of T_i corresponding to the boiler control temperature. As the outside temperature rises, a signal from T_e, passed via the controller, resets T_i downwards. T_i controls the position of the motorised valve V to mix water from boiler with water from the bypass (which is the return from the load and at a lower temperature than the boiler flow), to correct the temperature. As the temperature T_i decreases, the output from the terminals is reduced and may be matched to the load by selection of the temperature characteristic. *Figure 28.14* curve 1, illustrates a typical characteristic.

The actual characteristic varies according to the system design parameters and to heat losses from the building, and may be varied in a number of ways. The characteristic may be generated and adjusted from point A (curve 2) or point B (curve 3) or points A and B (curve 4). A–C represents the limit condition which is governed by the boiler temperature, and may also be used for warm-up situations during which the compensator system is overridden.

A single compensator is incapable of dealing with varying internal loads or solar gain, and additional controls are employed for this purpose. These may be zonal controls, where an additional thermostat and control valve acts as a local trimming device to detect local gains, or terminal controls on each emitter. The latter permits individual temperature control of each space and might appear to make the compensator redundant. However, the compensator still provides two advantages: heat loss from the distribution mains is reduced as the outside temperature rises, and the reduced circulating temperature prevents individuals from calling for excessive temperatures in mild weather.

The most common terminal control for these systems are thermostatic radiator valves (TRVs), which are self-acting devices requiring no external supply. TRVs have one characteristic which is common to many control systems: they are proportional controllers. This means that they provide their set temperature only at one position of the valve, normally the mid-position as indicated in *Figure 28.15*. A specific change in temperature from

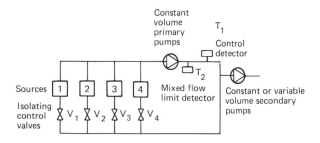

Figure 28.12(a) Flow-controlled multiple modulating boilers in parallel. Sequential control from detector in mixed flow-constant volume through boilers. The system controls any three from four sources ((b) is on next page)

28/14 Environmental control

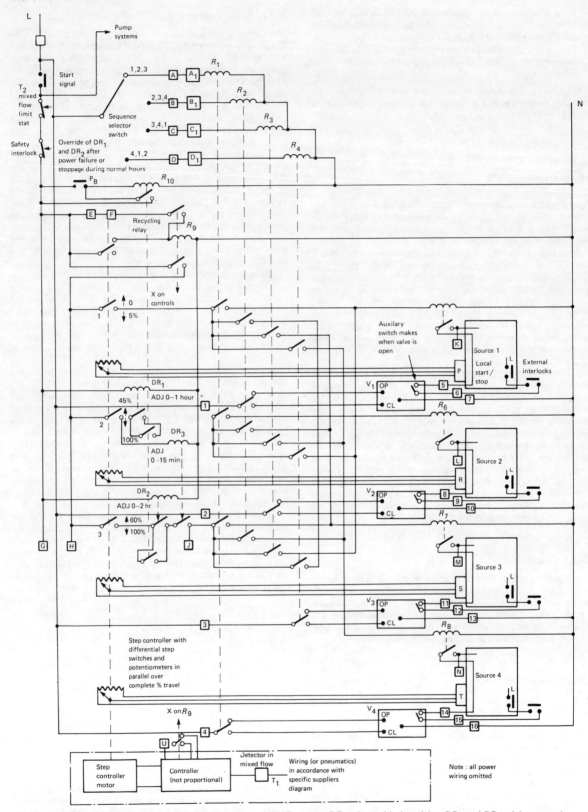

Figure 28.12(b) Interlocking associated with the system of (a). R = relays; DR = relays with time delay; DR_1 and DR_2 = delay start of second and third sources after initial start signal in normal operation steps 2 and 3 make together and DR_3 allows time for the second source to run up before the third source starts. (In the event of failure of a selected source, automatic changeover to the fourth machine can be made using terminals $A-A_1$, $B-B_1$, $C-C_1$, $D-D_1$, E-F, G, H, J, K, L, M and N.)

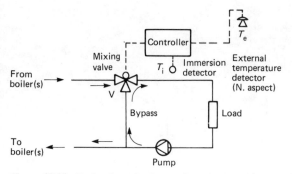

Figure 28.13 Basic schematic of a weather compensated system

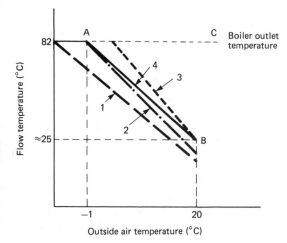

Figure 28.14 Typical compensator characteristics

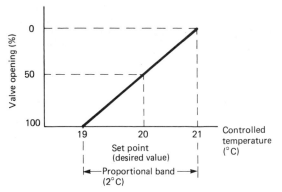

Figure 28.15 Proportional band effect on control setting. The temperature is shown for a set point of 20°C with a proportional band of 2°C

this setting is required for the valve to take up any new position, which means that the valve moves from fully open to fully closed over a range of space temperature and that a particular valve position is related to a specific temperature. The sustained deviation from the set point which this describes is known as 'offset'. The range of temperature over which the valve performs its full travel is the 'proportional band' and TRVs have a fixed band which varies according to source and type. Typically the band is 2–3 °C: thus, if the setting is 20 °C, the space temperature rises to 21 °C with the valve fully open and drops to 19 °C with it fully closed (a 2 °C band). This characteristic is useful in energy conservation terms, as it decreases the internal temperature at times of greatest load, thus reducing energy consumption, always assuming that the 19 °C used in the example is acceptable to the occupants.

Controls for domestic heating systems[7] would not normally include a compensator, and TRVs or room thermostats controlling small motorised valves are more common. The use of a single room thermostat starting and stopping the complete system is common, but should be upgraded for both energy conservation and comfort. Warm air systems may be controlled by thermostatically controlled motorised dampers interlocked with the plant heat exchanger and fan.

28.5.1.3 Air conditioning systems

Air conditioning system controls are similar to those for heating in respect of the final emitters, i.e. a temperature detector-cum-controller, often proportional in character, which modulates either a control valve or a damper (or both in sequence) to maintain the required temperature. If the terminals depend on the sequential operation of heating and cooling, it is again possible to use the width of the proportional band to provide both comfort and energy conservation. The control may be adjusted to provide heating from 19 to 21 °C and cooling from 22 to 24 °C, with a dead zone between 21 and 22 °C with neither heating nor cooling action.

The control of conditions in the main air handling plant is a vital element for the overall space environment and is probably the most complex requirement in the heating, ventilating and air conditioning (HVAC) field. The aim is to achieve a stable dewpoint (fully saturated air) condition, which defines the amount of moisture in the conditioned air. The relation between dry bulb temperature, wet bulb temperature, humidity and total heat (enthalpy) is fully defined in psychrometric charts. Attainment of a particular dewpoint condition is sufficient to provide a specified set of relative humidity conditions in a space, taking into account the moisture pick-up (latent gain) from the occupants. *Figure 28.16* illustrates a typical dewpoint plant which contains basic controls and the necessary override and safety features. The details are related to the systems described in Section 28.4.5.

The dewpoint is controlled by T_1, which sequentially modulates a preheater battery control valve V_1, dampers D_{1a}, D_{1b} and D_{1c}, which operate in parallel, and a cooler battery valve V_2 to maintain a constant saturated temperature condition. The reheater, which is part of many systems, is controlled by the extract temperature detector T_2, modulating the control valve V_3 to maintain a constant space temperature. The low-limit detector T_3 is sometimes employed in the discharge duct to override T_2 and maintain the discharge temperature above a predetermined limit.

The modulation of the dampers is the 'free cooling mode', using air for cooling prior to the use of mechanical cooling. Conditions can occur where the dampers need to be overridden. When the external temperature rises above the return air temperature (or, more precisely, when the enthalpy, or total heat, of the outside air exceeds that of the return air), it is more economic to cool return air than outside air. A detection device is therefore required to measure the total heat. The enthalpy of the outside air can be measured directly (T_{4a}) as being above the design room value, or by dual detectors (T_{4a}/T_{4b}) which compare the room and outside air total heat conditions. In either case, when the room total heat is exceeded by the outside air, a signal from the device drives the dampers to the minimum fresh air position, determined by the

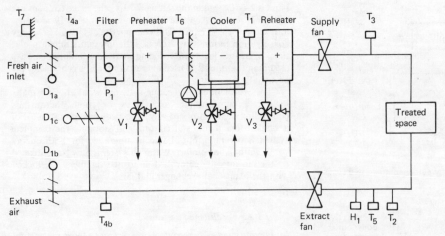

Figure 29.16 Dewpoint schematic. T_1 = dewpoint detector; T_2 = return air temperature detector; T_3 = low-limit detector; T_4 = enthalpy detectors/comparators; T_5 = boost limit thermostat; T_6 = frost protection thermostat; T_7 = free cooling enthalpy detector; H_1 = return air humidity detector; P_1 = differential pressure switch; V_1 = preheater valve; V_2 = cooler valve; V_3 = reheater valve

amount of air necessary to satisfy the fresh air ventilation requirements, which may be a statutory design parameter.

At night or during other shut-down periods the dampers are normally driven to the zero fresh air position. They remain in this position after plant start-up until the space temperature reaches a predetermined level as detected by T_5. During this boost period all the air is recirculated, valves V_1 and V_3 are fully open and the spray coil is de-energised. Thermostat T_6 protects the plant in cold weather conditions. If for any reason the temperature of the preheater drops to approximately 2°C, T_6 operates to shut down the plant and give an alarm.

It is also possible virtually to satisfy the dewpoint condition without mechanical cooling, whenever the enthalpy of the external air is below that approximating to the dewpoint condition. This is accomplished by detector T_7, which holds off the chiller and cooling-tower plants whenever the enthalpy is below the predetermined level.

In rare cases a humidity detector (H_1) mounted in the extract duct from the conditioned space is provided to monitor excessive or reduced latent gains in the space. It then resets the dewpoint detector (T_1) to compensate.

A differential pressure detector (P_1) fitted across the plant filter is a standard control item to provide a warning of high pressure across the filter for both maintenance and energy conservation purposes.

Some interlocking features are desirable on all dewpoint plants. One is associated with fire defence. It is becoming standard practice, sometimes mandatory, to ensure in the event of smoke or fire that the plant shuts down and that firemen can start the extract fan independently of the supply fan, with the dampers run out of sequence. A frequent associated requirement is for recirculation dampers to be fully closed to facilitate smoke exhaust. The basic dewpoint control, and the various additional functions described, illustrate the possible complexity of the interlocking diagrams, which parallel those of the multiple boiler systems described previously. The sustained offset with proportional controllers makes them unsuitable for this application because the dewpoint temperature control is often critical for the comfort conditions required. Thus, floating or two-term controllers are used which do not suffer from offset.

In energy conservation terms, systems which treat *all* the air and bring it to a dewpoint condition are inefficient. Alternatives have to be considered, such as relaxing humidity requirements. In such a case the cooling and heating coils may be controlled in sequence from the detector in the return air duct, and humidity control is only employed at the upper and lower limits of the specified conditions.

28.5.2 Time controls

The use of time switches is generally accepted. On large plants optimum start controls are becoming commonplace. The principle of operation of optimisers is to compare the external conditions with a representative internal condition so that the plant switch-on time is varied to achieve the required internal comfort conditions only at the time of occupation. In contrast, time switches are set to ensure that the conditions can be achieved on the coldest day, which means that in milder weather the building reaches comfort conditions earlier than necessary. Optimisers are therefore used for energy conservation, and can save 7–10% of the energy used with time switch control. The operation is shown graphically in *Figure 28.17*. As in *Figure 28.5*,

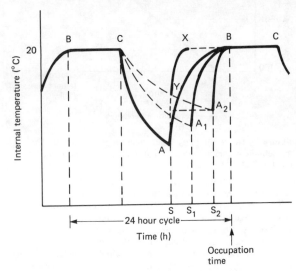

Figure 28.17 Optimised start of heating plant

BC represents the period of occupation and CA the temperature drop for the design coldest day. As the external conditions improve, the decay curve moves to CA_1, CA_2, etc., and the switch-on time is calculated by the optimiser and is delayed, moving from S to S_2, etc. A conventional time switch operating at S when the decay curve is CA_2 would waste energy equivalent to area XYA_2BX. Optimised stop facilities are also available.

28.5.3 Building automation

Building automation, important in building and building services design, is closely related to both comfort and energy conservation. While its original function in buildings was wholly related to monitoring and data collection for HVAC systems, the scope has broadened to include energy conservation, peak-load lopping, lighting control, fire and security alarms, programmed maintenance, etc. Computer-based systems allow all the facilities to be provided, utilising the computer software and data storage capability. Micro-processors bring these automation systems within the reach of a much wider range of buildings.

The systems are often referred to as 'centralised control' but until 1982/83 they virtually never provided the controller functions for HVAC control loops. The latest generation of microprocessor based systems now offers this feature in the form of direct digital control (DDC), which will replace the conventional standalone HVAC controllers. On any but the smallest BAS (which can be a single programmable controller with a number of analogue and digital inputs and outputs), the logic configuration will generally be similar to that shown in *Figure 28.18*. The data gathering panels (DEP) in most modern systems incorporate various levels of software semi-independent of the control processor, including DDC when required.

The main criteria and possible uses are listed briefly below.

Data transmission This may be either multiplexed multicore distribution or single- or 2-wire trunks for pulse-coded messages, the latter for the more sophisticated systems. Where buildings are coupled to one system, British Telecom lines may be used for transmission.

Scanning Typically scan times are between 2 and 30 s, although the point-to-point scan may be much faster. Multiple scan times can be employed, one for analogue signals and others for high- and low-priority alarms.

Hardware and peripherals These include the following:

(1) Data inputs from two-position and analogue devices, and data outputs for control switching and set-point adjustment.
(2) Outstations, which may be relatively simple data processors or semi-intelligent systems.
(3) Intercom, which is normally a feature additional to the transmission system.
(4) Central processor, which contains the memory for automation and alarm functions, often with a back-up power supply.
(5) Operator's keyboard and display unit.
(6) Printer(s) for common or separate logs and alarms.
(7) Visual display units, which may be slides or electronic screens in monochrome or colour.
(8) Permanent displays such as annunciator panels or mimic diagrams.

Software This comprises:

(1) Alarm priorities.
(2) Alarm inhibiting.
(3) Analogue alarms.
(4) Integration, e.g. energy consumption.

(5) Totalisation, e.g. summation of motor run times, etc.
(6) Time switch, including optimised start.
(7) Event initiated sequences, e.g. an alarm which initiates a specific sequence of operations.
(8) Load shedding and lighting control.
(9) Restart after power failure: prevents electrical overload on restart.
(10) Process control, e.g. the use of the centralised system as the controller for individual loops (DDC).
(11) Optimum damper control (see Section 28.5.1.3).
(12) Security, e.g. patrol tours and card entry.
(13) Interlocking, e.g. the use of software in place of conventional relays, timers, etc.
(14) Fire, i.e. alarms and specific event initiated sequences.
(15) Programmed maintenance, i.e. the use of stored data to produce a work schedule for maintenance and service.

An example in block form of a building automation system is shown in *Figure 28.18*.

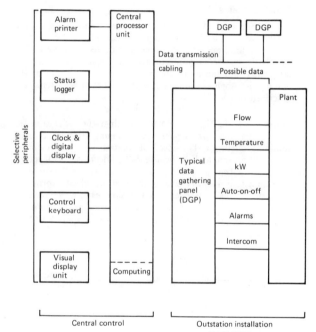

Figure 28.18 Block diagram of a typical building automation system. DGP may be purely for data gathering or may include its own software capability for control functions. Central processor unit may range from a main frame machine to a microprocessor unit

Economics The cost evaluation for a proposed scheme should include consideration of the following:

Savings	Expenditure
Reduction of energy	Capital cost of system
Reduction in maintenance staff	Interest on capital cost
Increased plant life with programmed maintenance	Additional specialist staff
Use of software for interlocking in place of relays, timers, etc.	Maintenance of the automation system
Avoidance of duplicate systems (e.g. fire and security central consoles)	Preparation of detailed programmed maintenance format
	Collection of data and producing particular software

28.6 Energy conservation

The subject of energy conservation and the efficient use of energy is of interest in all branches of engineering design. The provision of satisfactory environmental conditions consumes such a large fraction of the national energy budget that energy conservation cannot be divorced from comfort. The means of reducing energy consumption range from using less energy by more effective building design and more efficient HVAC systems, to lowering comfort standards to reclaiming energy that would otherwise go to waste. Lowering comfort standards is a difficult objective to promote. Heat reclaim methods are numerous and in many cases are applicable also to industrial processes, and they often improve the efficiency of existing systems.

There is also the problem of effective maintenance and servicing of existing installations, which can contribute to both energy conservation and satisfactory environmental comfort standards.

28.6.1 Systems

The importance of heat reclaim systems has always been recognised but in practice their application has been very restricted until recently. Because a heat recovery system is frequently part of a sophisticated installation, it should be integrated into the more comprehensive requirements of the design philosophy. Some of the more common systems are described below.

Thermal wheel These are rotating air-to-air heat transfer devices between two separate air streams in parallel and adjoining ductwork. The speed of rotation will not normally exceed 20 rev/min and the heat recovered decreases with speed. The control of energy transfer is effected by varying the speed or the exhaust air quantity passing through the wheel. Normally, only sensible heat is recovered, but versions exist which reclaim latent heat as well. A standard arrangement is shown in *Figure 28.19*. The temperature of the air supplied to the space (or other

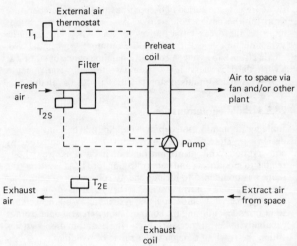

Figure 28.20 Schematic of heat reclaim run-round coil. T_1: T_{2S} and T_{2E} are detectors for different methods of control

Alternatively, a more sophisticated arrangement would be to use a differential thermostat T_{2E}, T_{2S} which runs the pump only when the temperature of the exhaust air is higher than that of the supply air. The capital cost of the plant and the additional pump and fan horsepower must be equated to the energy saved before such a system is adopted.

Cross-flow stationary recuperator (air-to-air heat exchanger) This device is an alternative to the thermal wheel but will provide only sensible heat transfer. The general arrangement is shown in *Figure 28.21*. The control system is similar to the bypass damper arrangement described for the thermal wheel.

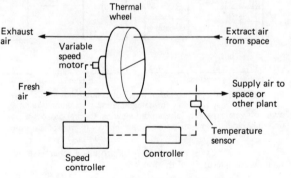

Figure 28.19 Variable speed control of thermal wheel

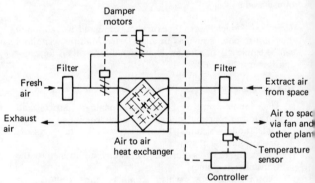

Figure 28.21 Schematic of heat reclaim by air-to-air heat exchanger

elements of the plant) is controlled by varying the speed of the drive motor. Control can also be achieved by bypass dampers to reduce the air passing through the exhaust air section. It is necessary to check that the energy saved by heat transfer is not exceeded by the additional fan power required.

Liquid coupled indirect heat exchanger (run-round coil) This system is simple. The general arrangements are shown in *Figure 28.20*. The pump may be controlled by an externally mounted thermostat T_1, which runs the pump whenever the external temperature is below the design exhaust air temperature.

Heat pump The use of heat pumps in HVAC systems is increasing. The most common form incorporates a refrigerant compressor unit with evaporator and condenser where the functions of the latter two elements may be reversed to give heating or cooling cycles as required. This arrangement is shown in *Figure 28.22*, where the heating cycle mode illustrates the generic definition of a heat pump system. This type of unit is described in Section 28.4.5.

Heat pumps are frequently used purely for heating, but whether in this mode or as reversible units, they are available

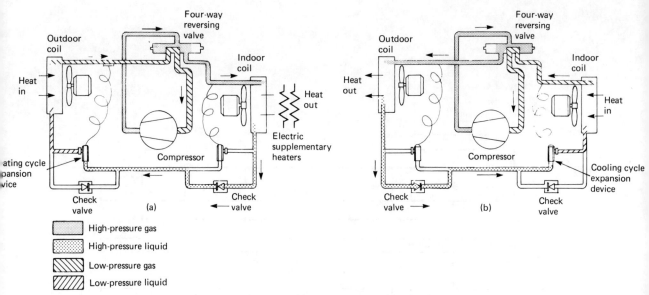

Figure 28.22 Reversible cycle units. (a) Heating cycle (heat pump); (b) Cooling cycle (chiller)

with air-to-air, water-to-water and air-to-water heat transfer. The selection of refrigerant is important to ensure that maximum efficiency is achieved for the specified range of inlet and outlet temperatures.

In some heating applications, particularly where heat is being extracted from outside air, the external coil (acting as the evaporator) will tend to ice up and a defrost control system must be used. This requires an arrangement for allowing hot gas to be passed through the coil for a short period, or the use of separate electric heaters.

Heat pipe This simple device is becoming an accepted heat reclaim component; it contains no moving parts and can be fitted in a manner analogous to a thermal wheel. Essentially, it consists of a sealed and evacuated tube containing a refrigerant, e.g. Freon, and lined with a wick. The action is shown in *Figure 28.23*: (a) shows that the application of heat vaporises the liquid refrigerant, which is then cooled at the top of the tube (giving up its latent heat), absorbed in the wick and returned to the bottom of the tube. This system is utilised in heat reclaim as in (b), where banks of tubes replace the thermal wheel in *Figure 28.19*.

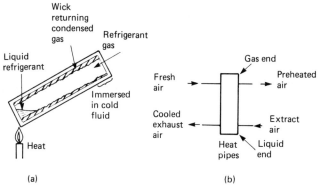

Figure 28.23 Heat pipes for heat reclaim

Double-bundle condenser One common method of heat reclaim normally considered in all air conditioning systems is the double-bundle condenser. *Figure 28.6* shows the conventional condenser, but it is obviously advantageous to use the rejected heat rather than discard it in the cooling tower. The arrangement is shown schematically in *Figure 28.24*, where one tube bundle is

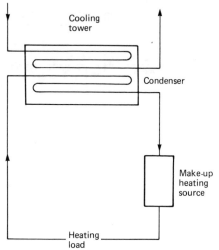

Figure 28.24 Basic schematic of a double bundle condenser for heat reclaim

used for the tower heat rejection circuit and the other for heat reclaim to the heating systems in the building. The system is controlled in sequence so that the tower circuit is brought into use only when there is no requirement for reclaimed heat. Temperatures of 45°C can be attained from the reclaim circuit without difficulty.

28/20 Environmental control

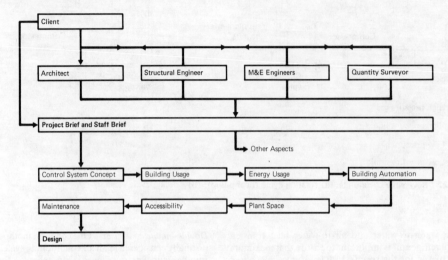

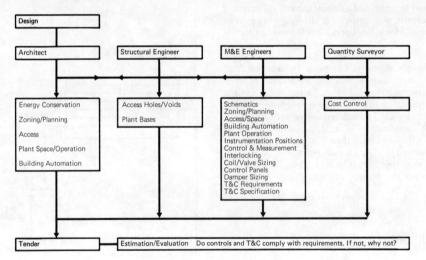

Figure 28.25 Co-ordination for environmental comfort

Energy conservation 28/21

BASIC CONTROL PANEL WIRING DIAGRAM PROGRAMME

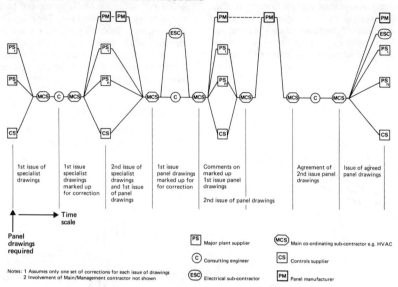

BASIC SITE CO-ORDINATION FOR CONTROLS AND T&C

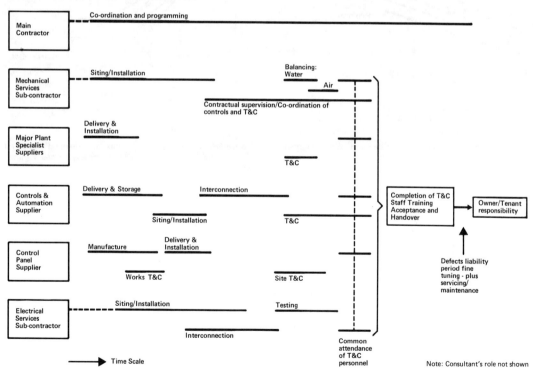

28.7 Interfaces

Plant design for environmental comfort installations often concerns architects and structural engineers initially, electrical and mechanical engineers subsequently, and the occupants of the building eventually. The several interfaces must be recognised and identified if the best efforts of the design team, the client and the occupants are to be realised. The first step towards this is to ensure that all the design disciplines are involved from the concept stage.

For environmental conditioning plant there are many suppliers and contractors, who provide drawings with interfaces between mechanical and electrical elements of the system. The degree of co-ordination and complexity in this context is indicated in *Figure 28.25*, giving the progress from design to commissioning and handover. The diagram specifically relates to controls, but affects all participants in a project. There are four elements of the design and construction process – Conception and Brief, Scheme and Detailed Design, Basic Control Panel Wiring Diagram, and Basic Site Co-ordination for Controls and T&C (Testing and Commissioning).

28.7.1 Electrical loads

For a heavily serviced building the mechanical services engineer has to supply loading data. Figures for the lighting load and for

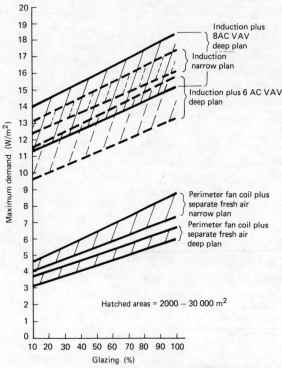

Figure 28.26 Concept design maximum demand figures. Figures include fan and pump power but exclude mechanical cooling load VAV = variable air volume system; AC = air changes per hour

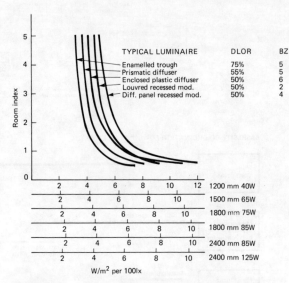

Figure 28.28 Concept design maximum demand figures for lighting schemes with warm white fluorescent tubes. Figures include control gear

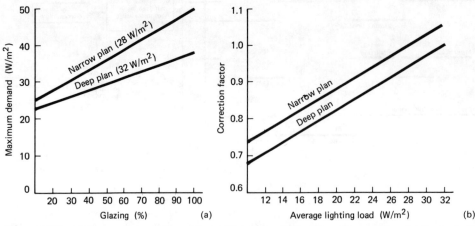

Figure 28.27 (a) Concept design maximum demand figures for refrigeration machinery at given lighting loads; (b) Correction factors for varying lighting loads

the air conditioning (which may be related to that for lighting) have also to be prepared so that electrical cables, substation, transformer(s), switchgear and cabling distribution can be finalised.

Three examples of typical data are given. *Figure 28.26* shows the overall maximum demand for air conditioning for deep- and narrow-plan offices, over a range of floor and window areas: the figures include pumps and fans but not the mechanical cooling load. The refrigeration machinery maximum demand in *Figure 28.27(a)* is assumed proportional to the lighting load, taken as 28 W/m^2 for narrow- and 32 W/m^2 for deep-plan offices; (*b*) gives multiplying factors to be applied to (*a*) for other lighting power levels. *Figure 28.28* gives a range of global figures for particular fluorescent tube lamps in various luminaires, for various length of tube, the figures including control gear.

The programming of projects for site installation does not always conform to the design programme sequences. Control and motor starter panel specifications often need completion to meet delivery dates, with final details of starter ratings and interlocks still unresolved. Mechanical services engineers should be pressed to provide the information on time. The global load figures in *Figures 28.26–28.28* may help in preliminary estimates if no other data are available. Lack of interlock details may create space, delivery and cost problems, but these may be avoided by use of programmable controllers where the software provides the interlocking functions. But this option should be used preferably in its own right, and not just to circumvent proper co-ordination between the design professions.

References

1. CIBS *Guide Book A*, Section A1.
2. *Control of Fuel Electricity* — Statutory Instrument 1974, No. 2160.
3. CIBS *Guide Book A*, Section A3.
4. CIBS *Guide Book A*, Sections A5 and A9.
5. *Fuel Economy Handbook*, National Industrial Fuel Efficiency Service. Graham and Trotman, 1980.
6. *Applied Air Conditioning and Refrigeration*, Second edition, C. T. Gosling. Applied Science.
7. 'Controls for domestic central heating systems'. *British Gas*, October 1978.
8. CIBS *Guide Books B and C*, Sections B3 and C1.

29 Transportation—Roads

D J Barrow
Lucas Electrical Ltd

A Cox
Lucas Electrical Ltd

J A Bailey
GEC Traffic Automation Ltd

B J Prigmore, MA, MSc, DIC, CEng, FIEE
Consultant Electrical Engineer

Contents

29.1 Electrical equipment of road transport vehicles 29/3
    29.1.1 Batteries 29/3
    29.1.2 Charging systems 29/3
    29.1.3 Alternating current generator 29/4
    29.1.4 Ignition systems 29/7
    29.1.5 Starter motor 29/9
    29.1.6 Electronic fuel injection systems 29/10

29.2 Light rail transit 29/12
    29.2.1 Examples 29/12
    29.2.2 Rolling stock 29/12

29.3 Trolleybuses 29/13
    29.3.1 Vehicles 29/13
    29.3.2 Overhead gear 29/15

29.4 Battery vehicles 29/15
    29.4.1 Batteries 29/15
    29.4.2 Motors 29/15
    29.4.3 Control 29/15
    29.4.4 Vehicle and operational details 29/17
    29.4.5 Applications 29/18
    29.4.6 Range and power assessment 29/19

29.5 Road traffic signalling 29/19
    29.5.1 Development 29/19
    29.5.2 Legal requirements 29/20
    29.5.3 Traffic signals 29/20
    29.5.4 Control equipment 29/20
    29.5.5 Central computer control 29/21
    29.5.6 Motorway signals 29/21

29.1 Electrical equipment of road transport vehicles

Within the framework of the modern automobile there is an elaborate mechanism to enable energy to be changed readily from one form to another. Although the electrical equipment is only a small proportion of the whole, its satisfactory operation is essential for the normal running of the automobile.

The chemical energy in the fuel is changed within the engine to provide primarily the propulsion of the automobile. A small portion of the transformed energy is taken to operate the electrical equipment. In its turn the electrical apparatus may have to convert this energy into heat (demister), light (legal requirements) and sound (horn), or back to mechanical energy (starter motor).

In addition, the electrical equipment may be called upon to supply energy when the automobile engine is at rest; so an energy storage system is necessary. To store this energy an accumulator or storage battery is used. Electrical energy is converted into chemical energy and is stored in this form in the battery. *Figure 29.1* shows the various energy conversions concerned.

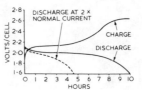

Figure 29.2 Charge and discharge curves of a lead–acid cell

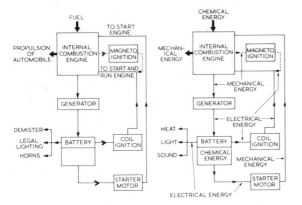

Figure 29.1 Energy transformations in automobile equipment

29.1.1 Batteries

The type of battery used in automobile work has been designed to withstand severe vibration and give maximum capacity for minimum weight. A large discharge current for a short duration is required by the starter motor and the battery must be able to meet this requirement.

The two main types of batteries used in automobile work are the lead–acid and the nickel–alkaline.

The former has been more extensively used for private cars. The type developed has a large number of thin plates of suitable mechanical strength giving a large surface area for a given weight and volume. This construction enables the heavy current demands of the starter motor to be met. The magnitude of the discharge current is a function of the plate area exposed to the electrolyte. The specific gravity of the sulphuric acid used in these batteries is higher than that for stationary batteries: it lies between 1.150 and 1.300. Between these density limits the electrolyte has minimum resistivity—a desirable condition in view of the occasional heavy discharge rates. Too high a density may cause damage to the plates and separators. If the discharged battery is in a healthy condition, it is permissible, in urgent cases, to recharge in less than an hour to about 75% of its fully charged state by giving a boosting charge or a fast rate of charge, provided the battery temperature is not allowed to exceed 80 °C. Typical voltage/time relations for a lead–acid cell are shown in *Figure 29.2*.

The nickel–alkaline battery is of much more robust construction than the lead–acid battery. The plates do not buckle when short-circuited. The electrolyte is a solution of caustic potash, and unlike the lead–acid battery, no indication of the state of charge is given by the specific gravity. The normal working value is 1.200. After about 2 years' service the gravity will fall to 1.160, owing to absorption of impurities from the atmosphere. The action of the battery will become sluggish, and complete renewal of the electrolyte is required.

29.1.2 Charging systems

A generator driven by the automobile engine supplies the energy to charge the battery. If a commutator-type generator is used, an automatic switching device (the cut-out relay) is embodied in the circuit to connect the generator and battery when the generator voltage is of the correct value for charging purposes, and to disconnect the generator when the speed of the automobile engine drops below a certain figure or when the engine is at rest. No cut-out is required in an alternator charging system, since the semiconductor diodes built into the alternator for converting a.c. to d.c. output prevent current reversal.

The automobile driving the generator has a wide speed range, so that the electromotive force produced by the generator, which is proportional to the product of the flux and the speed, would vary considerably unless otherwise prevented. If the required charging current was obtained at low speeds, then the battery would receive a charging current in excess of requirements at the normal driving speeds on the open road.

To keep the generator voltage within the limits set by the battery, a reduction in field current is required for increasing speed. The following systems of control have been adopted for d.c. generators: (1) compensated voltage control; (2) current voltage control.

29.1.2.1 Compensated voltage control

The compensated voltage control system has a single vibrating-contact regulator which carries both a shunt and a series winding. The shunt winding is connected across the generator output and controls a pair of contacts which, in the normally closed position, short out a high-value resistor in the generator field circuit. Ignoring for the moment the compensating action of the regulator series winding, the shunt winding enables a constant voltage to be obtained from the generator. At low speeds the contacts are held together by a spring and generated voltage is applied to the field windings. As speed and voltage rise, the pull of the shunt wound electromagnet overcomes that of the spring and the resistor is inserted into the field circuit, causing the voltage to drop (*Figure 29.3a*), beginning the next contact opening and closing cycle. This vibratory movement of the contacts takes place some 50 times per second and results in a practically constant voltage, the contacts remaining closed for a shorter period of time as speed rises. For applications where the insertion of resistance does not cover requirements over the whole range of speed, a second pair of contacts, which short-circuit the field winding, are arranged to close at the higher speeds, which enables

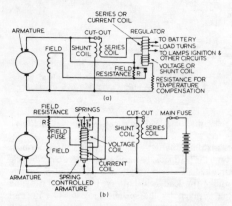

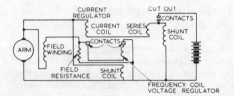

Figure 29.5 Current voltage regulator

Figure 29.3 (a) Voltage regulator with single contact: minimum field current with R in series with field winding; maximum field current with no resistance in field circuit. (b) Voltage regulator with double contact: minimum field current when field winding short-circuited; intermediate field current when R in series with field winding; maximum field current when no resistance in field circuit

constant voltage to be maintained in two stages. *Figure 29.3(b)* shows a barrel-type double-contact voltage regulator.

Constant voltage regulation has to be modified in practice to ensure that current demand by battery charging and other loads is not so high as to overload the generator. In the compensated voltage control system, current control is effected by adding a series winding to the regulator to assist the shunt winding. The voltage characteristic of such a regulator will be a falling one—the voltage will drop with increase of load and thus prevent overloading of the generator. Variation of voltage setting with rising temperature of the voltage coil is prevented by having an armature spring of bimetallic strip type. This temperature compensating device is arranged to give a somewhat higher voltage when cold, to assist in replacing the energy taken from the battery by the starter motor.

The tapering charge characteristic obtained (*Figure 29.4*) means that the compensated voltage control system is unable to utilise those occasions when spare generator capacity is available for increasing the charging rate. This drawback is overcome by current voltage control.

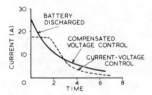

Figure 29.4 Comparison of charging characteristics

29.1.2.2 Current voltage control

A regulator of the current voltage type (*Figure 29.5*) allows the generator to give its safe maximum output into a discharged battery and to continue to do so until the battery approaches the fully charged condition, when the charging current is reduced to a trickle charge. A constant charging current is maintained until a certain battery voltage is reached; then the voltage regulator operates to give constant voltage control. The charging current will decrease until finally a trickle charge is passing through the battery. This type of regulator is used when the vehicle has a heavy electrical load and a more definite regulation is required than that given by the compensated voltage regulator.

In this type of control two separate regulators are used. These are mechanically separate but electrically interlocked. One regulator is shunt wound and so responsive to voltage conditions; the other is series wound and responsive only to current. Their contacts are connected in series with the generator field circuit.

When current from the generator flows to the battery, it passes through the current regulator winding, and on reaching the maximum rated value, it operates the armature of the current regulator, thus inserting resistance into the field circuit of the generator and decreasing the output current. The vibrations of the armature of the current regulator will prevent the rated output current being exceeded. As the battery voltage rises, the voltage regulator takes over, causing the second pair of contacts to operate, thus controlling the output according to the load and the state of the battery.

If, however, the battery is fully charged with little or no load switched on, then as the speed of the dynamo rises, the terminal voltage will increase to the operating value for the voltage regulator, the resistance will be inserted into the generator field circuit by the operation of the voltage regulator armature and the output current will be a function of the potential difference between generator and battery.

29.1.2.3 Cut-out or reverse-current relay

The cut-out relay is a simple automatic relay and is used to connect the battery to the generator when the latter's voltage is just in excess of the battery's. It also disconnects the battery from the generator when a discharge current flows from the battery and the generator tends to run as a 'motor'.

Two coils are provided on the relay: the first, a voltage or shunt coil connected across the generator terminals; and the second, a series coil carrying the generator output current. The fine-gauge wire coil has a large number of turns, but the current is small. This produces the required magnetic effect. When the current flowing in the voltage coil reaches a predetermined value proportional to the generator voltage, the armature is pulled down and the contacts close, allowing current to flow to the battery. This current flows through the series coil and the combined magnetic effect produced by the two coils pulls the contacts close together. The current coil assists the shunt coil as long as a charging current flows. When the generator speed and voltage fall, the current in the series coil decreases, then reverses, so that the resulting magnetic flux produced by the two coils together decreases. With a further fall in generator voltage the magnetic pull exerted on the armature will be overcome by the pull of the spring and the contacts will open, allowing no further discharge current to flow to the generator.

29.1.3 Alternating current generator

The continuing increase in the use of electrically operated components on automobiles calls for higher generator outputs. Higher outputs are also required for vehicles which have long periods of idling or frequent stopping and starting.

The increased currents to meet these needs give rise to

commutation difficulties if the size, weight and speed of the dynamo are to be consistent with a reasonable service life. An alternative to increasing the size of the d.c. generator is to produce alternating current in the generating machine and convert to direct current by means of rectifiers. The a.c. machine can have a higher maximum operating speed, so a higher drive ratio can be chosen which will allow a greater output at lower road speeds.

The alternator has a stationary three-phase output winding wound in slots in the stator so that the generated current can be taken directly from the terminals of the stator winding. There is, therefore, no problem of collecting the heavy current from a commutator. The rotating portion of the alternator carries the field winding, which is connected to two slip-rings with associated brush-gear mounted on the end bracket. The field current is small and can be introduced into the windings without arcing at the brushes. The alternator is not affected by the polarity of the magnetic field or the direction of rotation of the rotor, as in the case of the d.c. machine.

A field pole system of the imbricated type is preferred to salient poles by most alternator manufacturers, as it enables a relatively large number of poles (typically 8–12) to be energised by a single field winding.

The alternator cutting-in speed is about half that of the d.c. generator; consequently, it provides a useful charge at engine idling speed.

Rectifier Rectification of the alternating current output is by means of six silicon diodes in a full-wave three-phase bridge connection. The diodes are usually built into the bracket at the slip-ring end of the machine.

Control Since the alternator is designed to be inherently self-regulating as to maximum current output and since the rectifiers eliminate the need for a cut-out relay, the only form of output control necessary is a voltage regulator in the field circuit.

The conventional type of electromagnetic vibrating-contact voltage regulator, with either single or double contacts, has been used. Also employed to enable higher field current to be used and to give longer contact life has been a vibrating-contact regulator with the contacts connected in the base circuit of a transistor, the transistor being protected by a field discharge diode against inductive voltage surges.

To eliminate all maintenance and wear problems, it is now general practice to use solid state components in all parts of the voltage control unit.

29.1.3.1 Separately excited alternator systems

Fitted to a number of passenger cars and high commercial vehicles are Lucas 10AC and 11AC alternators, which have field systems energised from the battery as opposed to directly from the machines. The fully transistorised control employed for these alternators is shown in *Figure 29.6*. Used in place of the voltage coil and tension spring of a vibrating-contact regulator are two silicon transistors for field switching and a Zener diode (voltage control diode) for voltage reference. The Zener diode is a device that opposes the passage of current until a certain voltage is reached, known as the breakdown voltage. When the ignition is switched on, the base current required to render the power transistor T_2 conducting is provided through resistor R_1; as a consequence, current flows in the collector–emitter portion of T_2, which acts as a closed switch in the field circuit and applies battery voltage to the field winding. Rising voltage generated across the stator output winding is applied to the potential divider (R_3, R_2 and R_4) and, according to the position of the tapping point on R_2, a proportion of this potential is applied to the Zener diode ZD. When the breakdown point of the Zener

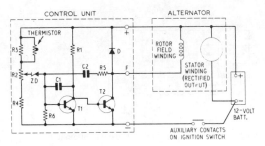

Figure 29.6 Fully transistorised control for a Lucas 10AC or 11AC alternator (positive earth circuit)

diode is reached, the diode conducts and current flows in the base circuit of a driver transistor T_1. The base current of T_2 is reduced and so is the alternator field excitation. To limit power dissipation, it is desirable to switch the voltage across the field winding rapidly on and off instead of using the transistors to provide continuous regulation; this oscillation is achieved by the positive feedback circuit comprising resistor R_5 and capacitor C_2. Transistor T_2 is protected from very high inductive voltage surges by the surge quench diode D connected across the field winding, which also serves to provide a measure of field current smoothing. Radio interference is eliminated by negative feedback provided by capacitor C_1. Resistor R_6 provides a path for any small leakage currents through the Zener diode that would otherwise flow through the T_1 base circuit at high temperatures and adversely affect regulator action. Automatic compensation for changes in ambient temperature is provided by the thermistor connected in parallel with resistor R_3.

Figure 29.7 shows a typical Lucas 10AC or 11AC alternator charging system diagram, including a field switching relay and warning light control unit of 'hot wire' type. The hot-wire resistor

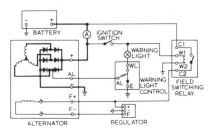

Figure 29.7 A Lucas 10AC or 11AC alternator charging system

is connected to the centre point of one of the pairs of diodes in the alternator; when hot, it lengthens and allows the contacts to open under spring tension, extinguishing the warning lamp.

29.1.3.2 Self-excited alternator systems

The majority of present-day alternators employ what is known as the 9-diode system, and are self-excited at normal operating speeds. The 9-diode arrangement has become popular owing to the simplicity of the warning lamp circuit, which, if combined with an in-built electronic regulator, provides a machine with the minimum of external wiring (*Figure 29.8*). The three field excitation diodes for supplying the field with rectified a.c. are included in the rectifier pack. Operating voltage is built up at starting by energisation of the field winding from the battery through an ignition (or start) switch, a non-charge warning lamp and the voltage regulator.

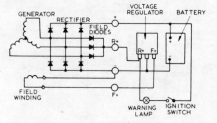

Figure 29.8 Charging system with self-excited alternator

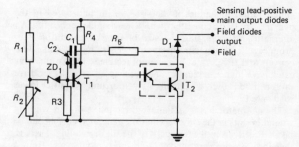

Figure 29.10 Regulator used to sense alternator output terminal voltage

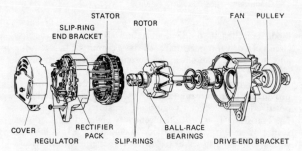

Figure 29.9 Exploded view of typical 9-diode alternator with built-in regulator

Lucas manufacture a series of alternators of this type—all with built-in electronic control units—having outputs ranging from 28 A to 75 A. An exploded view of a typical alternator is shown in *Figure 29.9*.

Control In the simple self-excited system of *Figure 29.8* the regulator controls the voltage at the field diodes' output terminal and therefore controls battery voltage only indirectly. For systems employing high-output alternators or having significant cable resistance between battery and alternator—for example, because the battery is remotely sited—indirect control of battery voltage becomes unacceptable. For such applications, progressive levels of control can be provided as follows: (1) the regulator can be designed to sense and control the voltage at the alternator output terminal; (2) the sensing lead of the regulator can be brought out of the machine to sense and control battery voltage directly. This second method is the most appropriate for remote-battery/high-power systems.

Figure 29.10 shows a Lucas regulator designed to sense alternator output terminal voltage. Field drive transistor T_2 is an integrated Darlington-pair device giving increased gain. This enables the circuit resistor values to be substantially increased, thereby reducing the permanent battery drain (taken through the sensing lead) to an acceptable level. Base drive resistor R_4 is fed from the sensing lead so that, in the event of the latter being disconnected or broken, T_2 is turned off, shutting down the alternator. If R_4 were fed from the field diodes, an open-circuit sensing lead would result in continuous maximum output from the alternator and destruction of the battery.

Figure 29.11 is an example of a regulator designed for remote voltage sensing. It includes the safety feature which affords protection against an open-circuit sensing lead (described above) and, in addition, protection against damage from disconnection of the alternator output lead. In the latter event, since the battery is no longer being charged, the battery/sensing lead voltage falls and the regulator therefore increases the alternator field excitation. This increases the alternator output voltage, which further increases field excitation. Without protection this cumulative effect can result in the destruction of the field drive transistor T_2. Protection is provided by way of resistor R_6 and diode D_2. R_6 is chosen so that under normal regulating conditions insufficient current flows through the potentiometer chain to reverse-bias diode D_2, which is therefore conducting: hence, the regulator controls the sensing lead voltage. The rise in alternator output terminal voltage, which occurs with a break in the output lead, increases the current through R_6, causing D_2 to become reverse-biased. Regulation of the output terminal voltage now takes place at a safe level dictated by R_6, R_1, R_2, R_3 and ZD_1, and T_1 base–emitter voltage. This secondary regulation level is typically 4 V above the normal regulating voltage.

This regulator can be designed to work with a remote network providing battery temperature compensation. The network, which includes a thermistor, is connected in series with the voltage sensing lead and is mounted to sense battery temperature directly.

Lucas alternators incorporating 9-diode systems are also

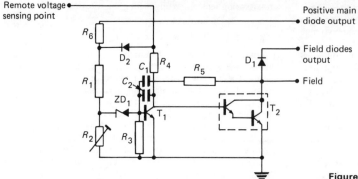

Figure 29.11 Regulator for remote voltage sensing

protected against surge voltages arising from disconnection of the alternator at high output currents, and from switching transients resulting from other equipment (e.g. the ignition system). The protection takes the form of a high-power Zener diode connected between the field diodes' output terminal and earth. Surge voltages are thereby limited to a safe level of typically 35 V. If the Zener diode is overloaded, its normal failure mode is short-circuit. This cuts off the field excitation and switches on the warning lamp. No other damage occurs and the normal charging function of the alternator is restored on the Zener diode's being renewed. A detailed discussion of transient causes and protection is given in SAE paper number 730043, 'Transient overvoltages in alternator systems'.

29.1.4 Ignition systems

The ignition of the compressed charge in a cylinder of an internal combustion engine with petrol as the fuel is brought about by an electric discharge. This discharge takes place across the points of a sparking plug. The voltage required to produce this spark will vary between 7000 and 25 000 V. The equipment may be self-generating, as in the magneto (seldom used on automobiles), or battery powered, as with coil ignition. The high-voltage impulses must be delivered to each cylinder as required and at a precisely controlled time in accordance with engine speed and load. There are several systems how in use to satisfy these requirements.

Contact breaker ignition

The major components of a modern coil ignition system are shown in *Figure 29.12*.

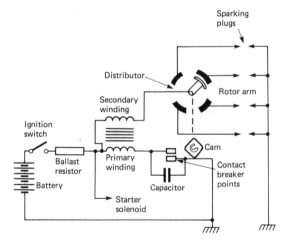

Figure 29.12 Elements of a coil ignition system

Storage battery The source of low-tension current.

Primary winding A coil of a few hundred turns of relatively heavy wire connected to the battery through a contact breaker.

Secondary winding A coil of many thousand turns of fine wire in which the high-tension voltage is induced. One end is usually connected to the more positive end of the primary winding, the other to the sparking plugs through the distributor.

Contact breaker A switch for opening and closing the primary circuit at the required instants. It is operated by a cam driven at half engine speed.

Capacitor Connected across the contact breaker points to suppress arcing.

Rotor arm Together with the distributor head, it forms a rotary switch to distribute the high voltage to each of the cylinders in correct order of firing. The rotor arm may incorporate a resistor to aid radio interference suppression.

Ballast resistor A device to aid starting. It is a resistor of $1-2\,\Omega$ connected in series with the primary winding and arranged to be short-circuited when the starter motor is operated.

Sparking plug The high voltage is led to the insulated central electrode of the plug (one per cylinder) and the spark passes across the gap at the required instant. The gap must be maintained within certain limits, usually 0.3–1.0 mm. In selecting a sparking plug for a given engine, attention must be given to the engine manufacturer's recommendations. The reach of the plug must be such that the correct amount of projection into the combustion chamber is achieved. The correct heat grade is also important: otherwise pre-ignition (overheating) or misfiring (too cold) may occur.

Distributor 'Distributor' is a generic term used to describe the assembly of contact breaker, capacitor, rotor arm and distributor head. It contains a shaft driven at half engine speed on which is mounted a centrifugally operated mechanism coupled to the cam. The mechanism is restrained by two springs and is so designed as to maintain the relationships between spark timing and speed. The cam has as many lobes as there are cylinders.

Usually incorporated is a vacuum control which automatically advances the ignition timing as the engine load falls. Control is achieved by a spring loaded diaphragm which moves in response to changes in depression in the carburettor venturi. As the depression inceases, the diaphragm moves the contact breaker mounting plate contrary to cam rotation and so advances the ignition. Occasionally, more complex vacuum control devices are employed having facilities to retard the ignition timing under certain engine operating modes (usually closed throttle deceleration) to reduce hydrocarbon exhaust emissions.

29.1.4.1 Electronic (contactless) ignition

In contact breaker ignition systems, it is necessary for consistent performance to adjust and—if necessary—to replace the contact points at regular intervals. Failure to do so may result in misfiring and/or excessive levels of exhaust emissions, which in most countries are now controlled by legislation. As a result, electronic (contactless) ignition is used on many vehicles—especially in countries with severe exhaust emission regulations. A general circuit is shown in *Figure 29.13*. The current interrupting function of the contact breaker is now performed by a power transistor, whereas the timing requirement is handled by a sensor in conjunction with a control module.

Many forms of timing transducers are in use. That in *Figure 29.13* is of variable reluctance type. A permanent magnet drives a flux through a magnetic circuit which includes a pickup coil and the distributor shaft. As the shaft revolves at half engine speed, the flux is modulated by a reluctor (replacing the cam of conventional ignition), which is shaped like a cog with as many teeth as there are cylinders. The varying flux generates in the pickup coil a voltage waveform which, after modification by the control module, switches off the power transistor (and therefore the coil primary current), to produce a spark. Timing and

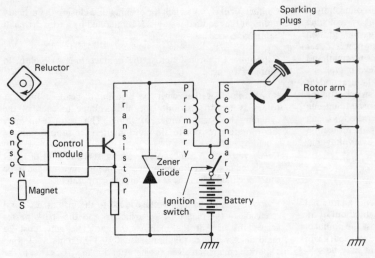

Figure 29.13 Electronic (contactless) ignition system

distribution of the spark are identical with a conventional contact breaker system.

Other forms of timing transducers are:

(1) Photoelectric sensor which is switched by an infrared light beam interrupted by a shutter.
(2) Hall effect device in which a magnetic field is interrupted by a steel shutter. The varying flux in the Hall device causes a proportional voltage change across it.
(3) Radiofrequency differential transformers and eddy current transducers in which the output voltage is varied by rotating magnetic devices.

In its simplest form the electronic control module may be an amplification circuit to drive the switching transistor, and the duty ratio of the sensor signal—equivalent to dwell angle in conventional systems—remains unchanged. More complex systems are in use in which the control module calculates the point of current switch-on in the primary winding according to engine speed. This allows use of special ignition coils having a very short time constant. The module also has a current limiting facility of usually 5–7 A and a device which switches off the current when the engine is stalled. The advantage of such systems is that the ignition spark energy is constant over wide speed and battery voltage ranges. Hence, they are usually referred to as 'constant energy systems'.

29.1.4.2 Electronically timed ignition systems

Contactless ignition systems overcome the problems of contact point degradation but the limitations of spark timing by centrifugal and vacuum operated mechanisms remain. As legislation on exhaust emission levels and energy conservation increases in severity, there is an urgent need to improve the efficiency of internal combustion engines, and a greater accuracy and flexibility in spark timing is required. This is possible with computer control of spark timing. A typical arrangement is shown in *Figure 29.14*.

The timing information for the engine is contained in a semiconductor memory unit. Typically the timing values for over 500 points of the engine speed/load characteristic may be stored permanently in the memory. Sensors placed at the engine crankshaft and in the inlet manifold generate signals in accordance with engine speed, crankshaft position and engine load. These signals are processed by the central processor unit

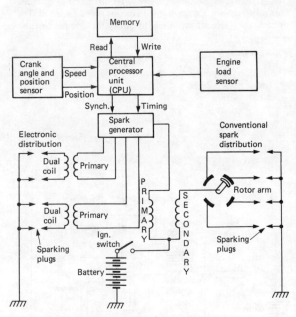

Figure 29.14 Digitally timed ignition (Mechanical and electronic distribution systems are shown)

(microprocessor) into a binary form suitable for addressing the memory. The nearest timing value in the memory for the particular speed and load of the engine is read into the spark generator unit, where, by means of a count-down procedure, the current in an ignition coil primary winding is interrupted at precisely the instant to produce a spark in the correct cylinder.

The spark generator unit may include a constant energy circuit as previously described, although other more complex energy control circuits may be employed and can be programmed by instructions also held in the memory.

Distribution of the high voltage to the sparking plugs may be achieved by a distributor head and rotor arm, or obtained electronically by sequential switching of the primary windings of dual output ignition coils (two for a 4-cylinder engine).

The crankshaft sensors are usually of the eddy current type but occasionally variable reluctance devices are used. Because of their exposed position, a robust construction is essential. A segmented disc fastened either to the rear face of the flywheel or at the crankshaft pulley is used to generate the signals. Where electronic distribution is used, two crankshaft sensors are necessary, the second being required to generate a suitable synchronising signal.

Load signals may be derived from a simple diaphragm operated potentiometer or by a more sophisticated silicon strain-gauge device. Usually the inlet manifold depression is measured rather than that in the carburettor venturi.

More ambitious systems may include programmed timing offsets for ambient temperature, engine coolant temperature, atmospheric pressure and exhaust gas circulation.

29.1.5 Starter motor

A starter motor is required to run the internal combustion engine up to a speed sufficient to produce satisfactory carburation.

The starter motor is mounted on the engine casing and a pinion on the end of the starter motor shaft engages with the flywheel teeth. The gear ratio between pinion and flywheel is about 10:1. A machine capable of developing its maximum torque at zero speed is required. The series wound motor has speed and torque characteristics ideal for this purpose.

The engagement of the pinion with the flywheel is effected in different ways. Perhaps the two most commonly used are the inertia engaged pinion and the pre-engaged pinion methods.

In inertia engagement the drive pinion is mounted freely on a helically threaded sleeve on the armature motor shaft. When the starter switch is operated, the armature shaft revolves, causing the pinion, owing to its inertia, to revolve more slowly than the shaft. Consequently, the pinion is propelled along the shaft by the thread into mesh with the flywheel ring gear. Torque is then transmitted from the shaft to the sleeve and pinion through a heavy torsion spring, which takes up the initial shock of engagement. As soon as the engine fires, the load on the pinion teeth is reversed and the pinion tends to be thrown out of engagement. Inertia drives are usually inboard, i.e. the pinion moves inward towards the starter motor to engage with the ring gear; an inboard is lighter and cheaper than an outboard starter.

To obtain maximum lock torque (i.e. turning effort at zero speed), the flux and armature current must be at a maximum, so resistance in the starter circuit (windings, cables, switch and all connections) must be a minimum; any additional resistance will reduce the starting torque. Generally, the inertia engaged starter motor is energised via a solenoid switch, permitting the use of a shorter starter cable and assuring firm closing of the main starter-switch contacts, with consequent reduction in voltage drop. The use of graphite brushes with a high metallic content also assists in minimising loss of voltage.

While inertia drive has been the most popular method of pinion engagement for British petrol-engined vehicles, the use of outboard pre-engaged drive is increasing. The pre-engaged starter is essential on all vehicles exported to cold climates and for compression ignition engines which need a prolonged starting period.

The simplest pre-engaged type of drive is the overrunning clutch type. In this drive, the pinion is pushed into mesh by a forked lever when the starter switch is operated, the lever often being operated by the plunger of a solenoid switch mounted on the motor casing. Motor current is automatically switched on after a set distance of lever movement. The pinion is retained in mesh until the starter switch is released, when a spring returns it. To overcome edge-to-edge tooth contact and ensure meshing,

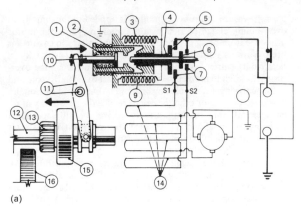

(a)

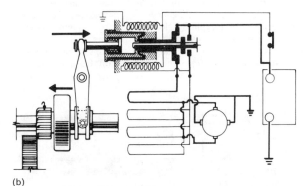

(b)

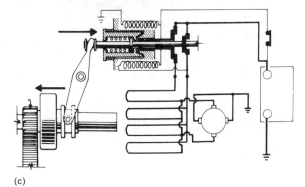

(c)

Figure 29.15 Operation of two-stage switching solenoid. (a) The solenoid is energised in the conventional manner to move the pinion towards the gear ring on the vehicle flywheel:

1. Engagement spring
2. Return spring
3. Solenoid hold-on winding
4. Switch operating spindles (concentric)
5. First set of contacts
6. Second set of contacts
7. Fixed contacts
8. Battery
9. Solenoid operating winding
10. Plunger
11. Operating level and pivot
12. Armature shaft
13. Pinion
14. Field system: four field coils in parallel
15. Roller clutch
16. Gear ring

(b) If tooth-to-tooth abutment occurs, the first set of solenoid contacts close and energise one field coil only, thus giving low power indexing to move the pinion teeth into a meshing position. (c) On full drive engagement, the second set of solenoid contacts close giving full cranking power. If the pinion teeth, on moving forward, can mesh immediately with the gear ring, full drive engagement takes place with the simultaneous closing of both contacts in the final stage.

spring pressure or a rotating motion is applied to the pinion. An overrunning clutch carried by the pinion prevents the motor armature from being driven by the flywheel after the engine has fired. Various refinements may be incorporated, especially in heavy-duty starters. Among these are: a slip device in the overrunning clutch to protect the motor against overload; a solenoid switch carrying a series closing coil and a shunt hold-on coil; an armature braking or other device to reduce the possibility of re-engagement while the armature and drive are still rotating; a two-stage solenoid switch to ensure full engagement of the starter pinion into the flywheel teeth before maximum torque is developed (*Figure 29.15*).

Two other pre-engaged types of starter are used for heavy compression ignition engines—the coaxial and axial types.

The compact size of the coaxial starter is achieved by mounting a two-stage operating solenoid and switching mechanism inside the yoke, coaxial with the armature shaft. When the starter solenoid is energised, the plunger is attracted into the solenoid, which causes the pinion sleeve and integral pinion to move axially along the armature shaft. At the same time the first-stage contacts close, to energise the starter windings through a built-in resistor (*Figure 29.16*). The armature rotates under reduced

29.1.6 Electronic fuel injection systems

To comply with exhaust emission regulations and optimise fuel consumption, the modern petrol engine requires a fuel system of extreme accuracy, reliability and flexibility. To meet this need various electronic fuel injection systems have been developed, the one shown in *Figure 29.18* being typical. The system consists

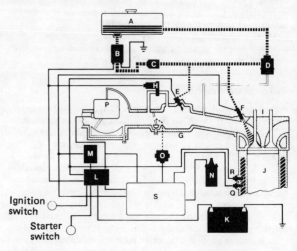

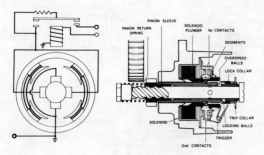

Figure 29.16 Internal wiring and construction of two-stage switching mechanism of coaxial starter

Figure 29.18 Lucas air-flow meter electronic fuel injection system:

A Fuel tank
B Fuel pump
C Fuel filter
D Fuel pressure regulator
E Cold-start fuel injector
F Fuel injector
G Intake manifold
H Extra-air valve
I Cylinder head
J Piston
K Battery
L Relay
M Power resistor
N Ignition coil
O Throttle switch
P Air-flor sensor
Q Coolant temperature sensor
R Thermo-time switch
S Electronic control unit

power and the pinion is driven into engagement by means of the armature shaft helix. When the pinion is almost fully engaged, the second-stage contacts close, to cut out the resistor, which enables full power to be developed.

The axial starter employs a sliding armature, which is moved axially against spring pressure to bring the pinion into mesh. The starter also has a two-stage switching arrangement (*Figure 29.17*) to ensure that pinion/ring gear engagement occurs before maximum torque is developed.

principally of an air-flow meter, engine speed sensor, throttle switch, air and coolant temperature sensors, control unit and fuel injectors. Fuel is delivered to the injectors at a constant pressure, so that the amount of fuel to be injected is determined solely by the time for which the injectors are held open. Both the time of opening of the injectors and the period for which they are held open are determined by the electronic control unit from the various sensor signals it receives.

The basic fuel requirement of the engine is determined from engine load data supplied by the air-flow meter, which is situated between the air filter and the induction manifold. Air flow through the meter deflects a movable flap, which takes takes up a defined angular position depending upon the force exerted on it by the incoming air. A potentiometer operated by the flap converts the angular position to a corresponding voltage. In the control unit this voltage is divided by engine speed to give the air intake per stroke, from which the basic fuel requirement is then derived.

Since the air-flow meter is used to determine the total mass of air drawn into the engine, adjustments to the basic fuel requirement have to be made for variations in air density (which is temperature dependent). An air temperature sensor is incorporated within the air-flow meter and is connected to the control unit for this purpose.

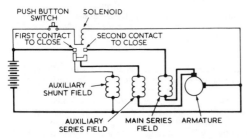

Figure 29.17 Sliding armature or axial-type starter

Adjustments to the basic fuel quantity are also necessary during engine cranking, cold starting and warm-up, idling, full-load operation and acceleration.

Cranking During cranking at any engine temperature, a signal from the starter switching circuit is provided to the control unit in order to increase the 'open' time of all the injectors above that required to supply the basic fuel quantity.

Cold starting and warm-up During cold starting a greatly enriched mixture is required to offset the effect of fuel condensing on the walls of the inlet port and cylinders. This extra fuel is provided by a cold-start injector which delivers a finely atomised spray into the inlet manifold. The cold-start injector operates only when the starter is energised and the thermotime (bimetallic) switch—which is sensitive to coolant temperature—completes the electrical circuit. Additional air required during cold starting and warm-up is controlled by an extra-air valve which bypasses the throttle butterfly. The valve aperture is adjusted by the action of a bimetal strip, which is responsive to the combined temperatures of the engine and an internal heater. The valve becomes fully closed when normal running temperatures are reached. During engine warm-up the control unit steadily decreases fuel enrichment in accordance with signals received from the coolant temperature sensor.

Idling, full-load and accelerating modes In *Figure 29.18* a throttle position switch with two sets of contacts is used to signal engine idling or full-load operating conditions to the control unit and thereby obtain the necessary fuel enrichment. Some applications use a potentiometer instead of a throttle switch. Opening of the throttle is then detected by the control unit as an increasing voltage from the potentiometer, causing acceleration enrichment circuits to be triggered.

29.1.6.1 Closed-loop electronic fuel injection systems

To obtain the very low exhaust emissions required by stringent legislation, a closed-loop electronic fuel injection system may be employed in conjunction with a three-way exhaust catalyst (*Figure 29.19*). The catalyst works at optimum efficiency in converting carbon monoxide and hydrocarbon emissions into carbon dioxide and water, and nitrogen oxides into oxygen and nitrogen when the exhaust gases are from an engine operating near to the stoichiometric air/fuel ratio. The operating condition will be indicated by the amount of oxygen present in the exhaust gases, and is monitored in the closed-loop system by an oxygen (lambda) sensor mounted in the exhaust manifold. The sensor provides a feedback signal to the control unit which continuously adjusts the fuelling level to maintain engine operation at the stoichiometric air/fuel ratio.

29.1.6.2 Digital electronic fuel injection system

Until recently all electronic fuel injection systems have employed analogue computing techniques. This has meant that for reasons of unit size and cost the complexity of the fuelling schedule stored in the analogue control unit has had to be limited and significant compromises made (*Figure 29.20*). In contrast, the latest Lucas

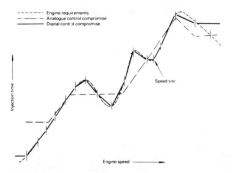

Figure 29.20 Comparison of the fuelling characteristics of analogue and digital systems for a given engine load

electronic fuel injection system employs a digital control unit (incorporating large-scale integrated circuits) in which it has been possible to match very accurately the fuel requirements of an engine under all operating conditions: moreover, this has been achieved in a control unit which has a size readily accommodated on the vehicle.

A key part of the control unit information processing capability is a digital read-only memory (1024 bits) which contains the fuel schedule. The latter is stored as a function of 16 discrete values of engine speed and 8 of load. The fuel requirement at each of these memory sites is identified by an 8-bit number. The fuelling characteristic is smoothed between the points of inflexion (i.e. the discrete value stored at each memory site) by a 32-point interpolation procedure which operates on the load and speed signals to effectively increase the memory size by a factor of 16×16.

The basic steps in the processing of information within the control unit are shown in *Figure 29.21*. The engine speed signal

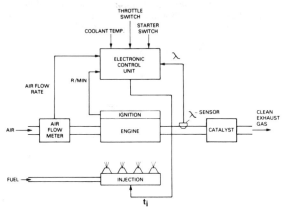

Figure 29.19 Closed-loop electronic fuel injection system

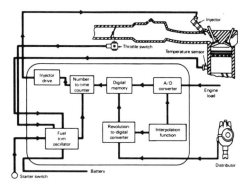

Figure 29.21 Signal processing within the digital electronic fuel injection control unit

(obtained from the ignition system) and the load signal are each converted into a digital word which is modulated by the interpolation function. The modified numbers representing speed and load are then used to select the site in the memory that stores the fuel requirements for these operating conditions. The memory output number (fuel quantity) is fed into a number-to-time counter, where it is stepped to zero by the fuel trim oscillator. The time taken to count down to zero will therefore be proportional to the memory output and will determine the 'open' period of an injector. Signals of air and coolant temperatures and acceleration adjust the fuel quantity read-out of the memory by modifying the frequency of the fuel trim oscillator. The solenoid operated injectors are energised (usually in groups) through power circuits for the countdown period of the number-to-time counter, when fuel is delivered to the engine.

RELEVANT BRITISH STANDARDS

BS2 (SBN 580 00500 3). *Tramway and Dock Rails and Fishplates*
BS 173 (SBN 580 00592 5). *Rotating Electrical Machines for Use on Road and Rail Vehicles*
BS 1727 (SBN 580 06861 7). *Motors for Battery Operated Vehicles* (N.B. not rail vehicles)
BS 2550 (SBN 580 06249 X). *Lead–Acid Traction Batteries for Battery Electric Vehicles and Tracks*
BS 2618 (ISBN 0 580 08473 6). *Electric Traction Equipment*
BS 4846 (ISBN 0 580 07387 4). *Resistors for Traction Purposes* (N.B. not road vehicles)

29.2 Light rail transit

As light rail transit (LRT) is a successor of the tramway, it is conveniently grouped with road transport.

Tramways in the UK and the USA declined between the 1930s and the 1950s in face of competition of buses and private cars. But the establishment of regional transport authorities, confronted with traffic congestion in spite of ambitious building of roads and car parks, turned attention to LRT; this system has more capacity and reliability than a bus service, because it usually has a reserved route or a private right of way, and it needs only modest civil engineering works. The LRT system spans the range between (but excludes) the bus and the full-scale surface or underground railway system.

29.2.1 Examples

The Tyne and Wear Metro (UK) is a form of LRT, but with major civil engineering works and full railway signalling, operated with light twin-articulated vehicles. Blackpool (UK) has a 16-km tramway on a mainly reserved track route, using single- and two-car vehicles. The system in Edmonton (Canada) was opened in 1978 with 7 km of route and 14 two-section articulated vehicles: its success has led to extension. But European cities best exemplify LRT techniques. Amsterdam (Netherlands) has an almost complete 'street Metro' based on the former tramway. Cologne (West Germany) operates 140 km of LRT route, two-thirds on reservation and 5% underground (mainly in the city centre); it is served by 300 large three-section articulated cars.

29.2.2 Rolling stock

Articulated cars, with almost full-width passenger accommodation across the articulations, one-man operated, with slot machines at the steps for ticket purchase (and widespread use of season tickets) are usual. Multiple-unit operation at peak times, and a 2-min service, permit passenger capacities of over 20 000/h per track. A two-section 1-m gauge car may be 20.5 m long, 2.2 m wide; may have seats for 50 and standing room for 107; and may weigh 37.5 t. Larger three-section 1.435-m (standard) gauge cars may be 28 m × 2.35 m and accommodate 270.

29.2.2.1 Mechanical details

Modern bogies usually have fabricated welded steel frames, rubber chevron primary springing, and wheels with rubber sandwiched between wheel centres and tyres for smooth and quiet operation. Disc brakes are usual for final stopping and holding. Service braking is rheostatic: emergency braking is given by adding the action of powerful track magnets. Retardation can attain 2.5 m/s^2 $(0.26g)$.

29.2.2.2 Control equipments

Almost all control equipments incorporate current control for acceleration and braking. The earliest (used today in Belgium and the Netherlands) is based on an American equipment of the 1930s. It gives 99-step controlled current rheostatic control using a combined resistor and accelerator unit in which a servo-motor rotates an arm to cut out rheostats; the same unit gives controlled current rheostatic braking. Another system (Berne, 1976) has series/parallel electropneumatic contactor control with 30 steps plus three weak-field stages. Electronic devices between the controller and the contactors give choice of acceleration, maximum speed and retardation. Many European systems, also Edmonton, have camshaft contactor equipments giving multinotch series/parallel control with bridge transition of series motors: there are usually weak-field stages to give 65 km/h or more. The controllers may be hand operated, joystick operated with a servo-motor as a follower, or may be operated by a small current-setting lever via an electronic comparator controlling the servo-motor which drives the camshaft. The comparators are incorporated into control units which give wheel-slip protection during acceleration and braking.

'Chopper' controls with thyristors, and electronic logic circuits on the control side, are of increasing application. All-electric two-section cars in Zurich (1976) have one 138 kW controlled field separately excited motor on each outer bogie. Control is by a d.c./d.c. chopper for each motor. Nine discrete speeds from 6 to 60 km/h, with nine associated accelerations, are available. Chopper controlled rheostatic braking gives eight discrete speeds, matching downhill speed limits, followed by eight controlled retardations down to standstill. Wheel-slip correction is included. Power/brake changeover is done by electromagnetic contactors.

Experiments with cage-motor induction motor drives were made in Mülheim (West Germany) in 1978. Here a d.c. chopper followed by an invertor gives variable frequency and voltage 3-phase supply to each induction motor, which is operated under controlled slip.

29.2.2.3 Edmonton LRT vehicle

The Edmonton LRT vehicle is a bidirectional two-section standard (1.435 m) gauge (600 V vehicle (*Figure 29.22*) with transverse seats for 64 and a crush capacity of 260. Up to five vehicles may run in multiple, with a passenger capacity of 1300. Each motor bogie has one 600 V 150 kW (1 h) 1200 rev/min series compensated motor with fully laminated construction and Class F insulation, mounted longitudinally and driving both axles via right angle drives. Control is by a 20-step motor driven cam-contactor series/parallel controller giving bridge transition, plus two weak-field steps to 60% and 40% excitation. Acceleration,

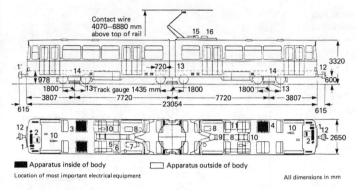

Figure 29.22 The Edmonton LRT vehicle. 1. Master controller; 2. driver's control panel and radio equipment; 3. public address system; 4. SIMATIC control device; 5. battery chargers; 6. battery; 7. motor–generator set; 8. underfloor starting and braking resistors; 9. blower for starting and braking resistors; 10. sander solenoid; 11. controller, reverser, motor cutout switch; 12. coupling; 13. track brake; 14. traction motor; 15. pantograph with servo-motor; 16. overload breaker

fully loaded (45.6 t), is 1.1 m/s². Service speeds are 55–60 km/h: the maximum is about 75 km/h.

Rheostatic braking in 17 steps gives a service retardation of 1.2 m/s². Final stopping is by mechanical spring brake on each motor bogie. Six track magnets, excited from the 24 V auxiliary supply, give emergency braking.

The cam-controller servo-motor is electronically controlled to limit jerk and to give wheel-slip and spin protection, as well as current control for acceleration and braking. Dead-man control is fitted.

29.3 Trolleybuses

Introduced during 1901–1911, early trolleybuses served either as tramway feeders or as alternatives thereto on light-load routes. By the 1930s trolleybuses had replaced some tramways to take advantage of the remaining life of the tramways' power distribution system. Through the 1950s trolleybuses were withdrawn, as public patronage declined with the rising number of private cars, cheap oil, the need to recondition or renew the power distribution system, the higher cost of trolleybuses compared with internal combustion-engined buses, and the wish to rationalise workshop facilities by using a single type of vehicle.

These reasons applied to the UK and the USA; in Europe, where public transport has greater priority and subsidy, trolleybus systems continue to flourish.

Fuel cost has caused a reconsideration of the use of trolleybuses in the UK. Combined battery/trolley vehicles, with contact wires only in the central areas, have been suggested.

29.3.1 Vehicles

The modern European trolleybus is a high-capacity four-wheel single-deck vehicle; alternatively, an articulated vehicle with a two-wheel semi-trailer attached to the main tractor vehicle may be used, the two body sections being joined by a full-width 'concertina' bellows over the articulation. Some have a steered rear axle on the semi-trailer. On other systems the four-wheeler hauls a four-wheel trailer in peak hours, giving high staff productivity. European road regulations do not inhibit public transport.

The four-wheel vehicle (*Figure 29.23*), 11.5 m × 2.5 m overall with 30 seats and room for 60 standing, has a tare weight of 11.8 t, of which the electrical equipment weighs 3.4 t. The fully loaded weight is about 17 t. A back axle ratio of about 10/1, a wheel diameter of 1 m and a 150 kW motor capable of about 3000

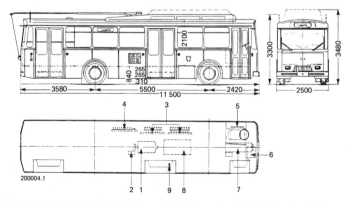

Figure 29.23 Two-axle trolleybus. 1. Traction motor; 2. inductive shunt; 3. contactor bank; 4. 1000 V equipment unit; 5. 24 V control panel; 6. electronic control equipment; 7. motor–compressor–generator set; 8. battery; 9. socket for charging battery. All dimensions are given in millimetres

rev/min, give a maximum speed of 60–65 km/h and an initial acceleration of up to 1.5 m/s². Operation at 600 V is usual, but 1000 V is used on a few long lightly loaded routes or where light rail transit systems share the supply.

29.3.1.1 Mechanical details

The chassis follows bus practice. Air suspension may be used. Owing to the high accelerating torque available from the motor, a heavy-duty rear axle and differential, often based on lorry components, are required. An under-floor motor drives the rear axle via a flexible coupling and a short cardan shaft. Conventional drum brakes, operated by compressed air, give friction braking: service braking is rheostatic, with friction braking added. A hand operated parking brake is available. The all-metal body is usually of all-welded aluminium alloy, with some glass fibre reinforced polyester mouldings. Air operated folding doors give rapid entry and exit, the former often at the front so that the single operator can check or issue tickets.

29.3.1.2 Electrical details

Motors up to 165 kW (1 h) weighing upwards of 1 t for higher-powered and articulated vehicles, down to 100 kW for single vehicles intended for solo operation, are common. They are 4-pole self-ventilated series motors with compoles. The armatures are usually insulated to Class H standards, the fields to Class F. The speed at the 1 h rating is about 1200 rev/min: the maximum allowable speed is maybe 3000–3500 rev/min.

29.3.1.3 Contactor control

Compound motors with a light shunt field to give a range of economic speeds and some regeneration have been tried, but overvoltage problems were severe. Series motors with field weakening, and rheostatic braking to minimise wear on the mechanical brakes, have become usual.

Electromagnetic contactors (*Figure 29.24*) give a dozen or more steps of straight rheostatic control, plus two or three steps by field diversion down to about 50% excitation. Braking is rheostatic, operated by the first depression of the brake pedal, followed by air operated friction brakes on further depression.

For prompt rheostatic braking the motor field is pre-excited from the auxiliary battery. Electronic devices and comparators between the accelerator and brake pedals and the contactors give controlled current and jerk limit for acceleration, and controlled current braking. For off-wire emergency operation or depot manoeuvring at low speeds the auxiliary battery, usually in three or four parallel sections at 24 V, can be connected in series to give 72 or 96 V to supply the traction circuit.

29.3.1.4 Electronic control

Chopper control of trolleybuses is becoming customary. On frequent-stop services it gives 5–15% saving of energy (owing to the avoidance of rheostatic loss), compared with resistive control. In addition, because sustained high currents are not required from the contact wires despite the high motor-starting currents given by the chopper equipment, substation capacities can be reduced. In effect, more or larger vehicles with chopper (rather than rheostatic) control may be put in service before supply reinforcement becomes essential.

Figure 29.25 shows a particularly useful circuit incorporating field current reduction as the motor voltage approaches the supply voltage under the control of the chopper unit. The input bridge ensures that the traction circuit always has the same polarity of supply whatever the overhead wire polarity—a useful feature on rural routes with bot one pair of wires.

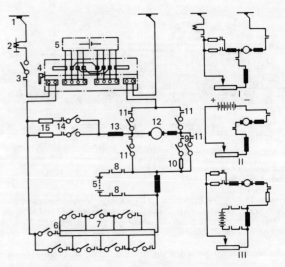

Figure 29.24 Main current circuit. 1. Current collector; 2. overcurrent relay; 3. line contactor; 4. contact wire/battery changeover switch; 5. battery; 6. contactors; 7. starting and braking resistors; 8. motor pre-excitation contactor; 9. braking contactor; 10. additional braking resistor; 11. reversing contactors; 12. traction motor; 13. inductive shunt; 14. shunt contactors; 15. shunt resistor; I. current circuit for motoring with supply from contact wires; II. current circuit for monitoring with supply from battery; III. current circuit for braking

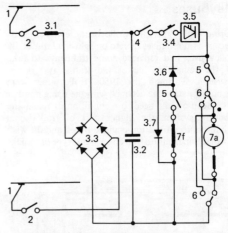

Figure 29.25 Basic circuit 'motoring'. 1. Trolley; 2. main contactor; 3. chopper, comprising 3.1 input choke, 3.2 input capacitor, 3.3 input rectifier bridge, 3.4 protection switch, 3.5 chopper, 3.6 armature freewheel diode, 3.7 field freewheel diode; 4. line contactor; 5. isolating contactor; 6. reversing switch; 7. traction motor; 7a. motor armature; 7f. motor field

The novel feature of the circuit is the inclusion of the motor field winding in the main freewheel diode circuit and the provision of a second freewheel diode for the field winding itself. Inductance is added to the armature circuit. During chopper 'off' periods the full armature current flows through the freewheel path; hence, also through the field winding. During chopper 'on' periods the freewheel circuit carries no current, but the field freewheel diode enables the field current, trapped by the field inductance, to continue flowing. This arrangement enables the mean field current to be nearly equal to the mean armature current, both being approximately constant, owing to the high chopper frequency (400 Hz). However, as full voltage is approached and the chopper 'off' period becomes very short, the freewheel current is insufficient and the field current decreases. Thus arises progressive and smooth field weakening, under electronic control, to the weakest allowable field. The motor voltage has by then increased to about 0.95 of the supply voltage.

This circuit, which necessitates continuous operation of the chopper, is suitable for motors with a low full-voltage full-field speed. A lighter motor, operated on part-voltage full-field, again with continuous operation of the chopper, is sometimes a preferred alternative.

Regenerative braking is not normally installed owing to the risks of overvoltage and a non-receptive line. But instant braking is necessary for safety on road vehicles. Rheostatic braking is obtained from the above circuit by reversing the armature, paralleling the field plus chopper with part of the braking resistor and using the chopper to control the field current. Since the field winding nearly short-circuits part of the braking resistor, this action also decreases the effective braking resistance as the field is strengthened. By this means, constant torque rheostatic braking from 60 to 15 km/h may be obtained. Electronic equipment and comparators between the accelerator and brake pedals and the chopper give controlled current acceleration and braking, and jerk limit at the start of acceleration.

29.3.2 Overhead gear

Since there are no rails to provide an earth return, two trolley wires and two trolley collectors have to be provided. Carbon skid shoes are used, and the design must permit the vehicle to run over a wide range on either side of the trolley wire position to give manoeuvrability. Shoes are preferred to wheels, on account of noise reduction, nearly sparkless collection, fewer dewirements, saving in weight and reduction of radio interference. Effective lubrication of both shoe and trolley wire is necessary.

British trolleybus supplies must be on the d.c. system at voltages not exceeding 600 V. The positive trolley wire must be subdivided into sections of not more than half a mile in length. Under normal conditions the sections are connected together at switch boxes placed by the route. Both positive and negative wires are supported by insulators from the span wire, which is itself broken at an insulator between positive and negative wires. This gives single insulation between the negative wire and earth, double between positive and earth, and triple between positive and negative.

29.3.2.1 Leakage protection

The trolleybus distribution system is free from earth-return problems such as electrolysis, bonding and statutory limits of voltage drop. At the same time the conductivity is inferior to that of the tramway—it may be little more than half. The frame of the trolleybus is insulated from both poles of the supply. Regulations require that careful examination be made every day to ensure that there is no appreciable leakage between any point in the supply circuit and the vehicle frame.

29.3.2.3 Radio interference

The major source of interference with radio receivers by trolleybuses has been found to be the contactor control circuits. Protection is afforded by fitting small inductors in the control circuits as close as possible to the master controller contacts. Slight interference is caused by motor commutation and current collection: the sliding shoe collector is better in this respect than the trolley wheel. Protection can be provided against other sources of interference by fitting two capacitors in series across the supply feeds in the roof of the vehicle with a third connecting their mid-point to earth.

29.4 Battery vehicles

Over many years battery electric vehicles have increased in popularity in the UK. Their characteristics are eminently suitable for services involving frequent stopping and starting and a high rate of acceleration. Briefly, the main sphere of use is for local delivery work such as the door-to-door delivery of milk, bread, laundry, coal, mineral waters and other goods, and specialised work such as refuse collection, streel lighting maintenance and interworks transport. In such services a vehicle may be required to make 200–300 delivery stops, while the total distance is usually between 30 and 50 km daily, and rarely reaches the maximum range of about 70 km. Under these conditions the delivery speed and the road speed are almost unrelated, the chief considerations being acceleration, ease of exit and entry to the vehicle, simplicity of control and reduction of personal fatigue. Efficient service is generally obtained with maximum road speeds of the order of 30–35 km/h.

29.4.1 Batteries

The lead–acid battery is customary. Choice of battery depends on vehicle duty. Thin-plate automotive starter batteries have high energy capacities but short life on deep-cycling duty. Traction batteries have thicker and firmly separated plates: their weight is kept down by using lightweight plastics cases and short intercell connections. 'Light traction' batteries are supposed to withstand 500 charge/discharge cycles, whereas true traction batteries are guaranteed for 1000–1500 cycles, and may last far more. These batteries are most economic (in £/kWh throughput) if they are used almost daily and discharged to between 70 and 90% of their capacity.

29.4.2 Motors

The basic design of d.c. traction motors has not been essentially changed, but the use of improved wires and strips for windings, and class H insulation, have greatly improved thermal transfer and reduced weight.

29.4.3 Control

Until the introduction of semiconductors with high current capacity the accepted methods of speed and torque control were (a) series resistance, (b) series/parallel battery switching, (c) series/parallel motor switching and (d) field control. Almost all controllers provided a stepped variation or, at best, a smoothly variable control over limited ranges of torque and speed. The carbon pile variable resistor was also extensively applied, with compression by hydraulic master and slave cylinders.

29.4.3.1 Series resistance

On small vehicles such as milk prams a single step of resistance usually suffices before the motor is switched on to the full battery, usually 24 V. On larger vehicles and works trucks several steps of series resistance may be used and these are cut out in sequence by a drum or cam controller. The one economical running connection is usually sufficient for low-speed vehicles of the pedestrian-controlled type. In some cases a second economical connection is obtained by field diversion.

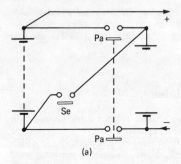
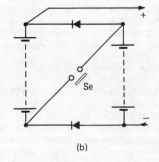

Figure 29.26 Parallel/series battery switching. (a) Contactor; (b) contactor/diodes

29.4.3.2 Parallel/series battery switching

By dividing the battery into halves, and putting them first in parallel to give half-voltage, then in series to give full voltage, and by using two or three resistance steps for each setting, the loss of energy in the resistors may be halved, compared with the previous system. In addition, an economic half-speed running circuit is obtained. The scheme is also less severe on batteries, since the first few current peaks are shared.

A conventional battery switching circuit employs a pair of contactors interlocked (usually mechanically) so that the double-pole contactor (*Figure 29.26a*) is closed for the parallel and the single pole for the series connection: they cannot be closed at the same time. In a circuit which involves no current breaking (*Figure 29.26b*) diodes make the parallel connections, and the contactor for the series connection reverse-biases the diodes so they do not conduct.

A patented diode/contactor arrangement elaborated from the above gives a multivoltage circuit in which, by contrast with the more obvious battery tapping systems, all batteries are used (not necessarily equally) all the time.

29.4.3.3 Field control

Field control is fittingly used in addition to the two previous schemes. By use of a four-pole motor with its field windings in separate pairs, full excitation is obtained with the field windings in series, and half-excitation for increased speeds with the field winding pairs in parallel. Field diversion, for even weaker field and higher speed, may be given by shunting an appropriately low resistance across the paralleled field windings.

29.4.3.4 Solid state switching

With suitable commutation control equipment, thyristors can provide a substantially variable, loss-free motor power control, provided that the pulse rate is not too low. Several forms of pulsed thyristor control are available, differing only in circuit details. The variations are largely associated with the method of providing satisfactory turn-off characteristics for the main circuit thyristors, and correlation of mark/space, repetition frequency and current switching levels to obtain the required motor output characteristics.

Thyristor controllers consist of the following basic elements: (a) main power thyristors, (b) turn-off capacitor, (c) commutator thyristor, (d) function generator, (e) current limiting and fail-safe circuits, (f) manual control device.

29.4.3.5 Chopper control

In the simplified equivalent circuit in *Figure 29.27*, the switch Sw represents the main thyristor in an actual circuit arrangement. In (*a*) for normal motor operation, Sw is closed for a time t_1 and

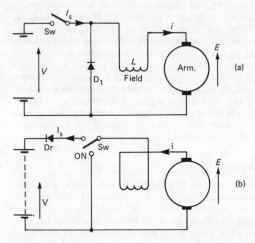

Figure 29.27 Essentials of 'chopper' circuit. (a) For motoring; (b) for regeneration

opened for t_2 (each time being of the order of a few milliseconds), the sequence being repeated with a switching time $T = t_1 + t_2$. During t_1 the current rises at a rate proportional to the difference between the battery voltage V and the motor armature e.m.f. E, such that $L(di/dt) = V - E$, where L is the circuit inductance and resistance is neglected. With Sw now opened for time t_2, the motor current continues through the freewheel diode D_1, driven by the armature e.m.f. E and falling at a rate such that $L(di/dt) = E$. Under steady load conditions the rise Δi during the 'on' time is equal to the fall Δi during the 'off'. As the times are short, then approximately

$$\Delta i = t_1(V-E)/L = t_2 \cdot (E/L)$$

whence $E/V = t_1/(t_1 + t_2) = t_1/T$. The e.m.f. ratio is equal to the ratio between 'on' time and total switching period. For an average motor current i the mean battery supply current is $I_s = i(t_1/T)$ and the mean motor current is $I_m = I_s(T/t_1)$. With $E = V(t_1/T)$ it follows that

$$EI_m = V(t_1/T) \cdot I_s(T/t_1) = VI_s$$

whence the chopper acts like a 'd.c. transformer' in terms of the input and output voltage and current ratios, under steady state conditions and with resistance and other losses neglected.

In operation, t_1/T is increased from zero towards unity to start the motor (with control by the current or voltage limiter). With $t_1/T = 0.95$ or thereabouts, t_2 may be too short to allow the thyristor current to quench. In many equipments a contactor is closed to bypass the thyristor and connect the battery direct to the motor, thus eliminating thyristor forward loss.

If sustained low-voltage high-current operation is required (and it is in such a case that the efficiency of the system is much higher than with resistance control), the mean square motor current is higher than the square of the mean current, which raises the motor I^2R loss. Enhanced cooling or reduced rating are then necessary in chopper control.

Regeneration The basic circuit is rearranged as in *Figure 29.27(b)*. Diode D_r prevents feedback of the supply to the motor. When Sw is 'on', the machine generates, converting kinetic energy into magnetic energy in the inductance; in the 'off' time the machine is connected to the supply, and E together with $L(di/dt)$ drives current through D_r into the battery. Then $V/E = T/t_1$ and $I_s = I_m(t_1/T)$.

Control The function generator in the chopper system is combined with comparators to give current and voltage control, switching off when current or voltage attains a demand value. Both values are increased as the 'accelerator' pedal is depressed. Regeneration, if fitted, is usually current controlled by the initial movement of the brake pedal, which thereafter applies the friction brakes.

The flexibility of electronic control systems allows the use of a controlled field shunt motor, with electronically controlled field current. Full-field acceleration to the running voltage, with power thyristor control (as above) for the armature current, is followed by controlled field weakening to give, e.g., constant current motoring to a preset speed or the weakest allowable field. Regeneration by field strengthening is then easy to arrange. So far, schemes of this nature have been experimental.

Example

Figure 29.28 shows in more detail the scheme for chopper control of a battery electric car.

Motoring The battery motor circuit is completed by firing thyristor Th_1. The current rises, and at a predetermined value is cut off by the switching action of the current monitor C_1. The inductive energy maintains the motor current through the freewheel diode D_1. The current decays, and at a predetermined minimum C_1 switches on Th_1. The motor current fluctuates between the two limits, but battery current flows only while Th_1 conducts. The rate of current fluctuation depends on the armature e.m.f. To switch off Th_1 it is necessary to reduce its current momentarily to a very small value by means of Th_2, C, L, R and D_2. Prior to any current demand, Th_2 is switched on and charges capacitor C so that the potential of the left-hand side is raised to that of the positive battery terminal $B+$ and the right-hand side to that of the negative $B-$: then Th_2 ceases to conduct owing to lack of 'holding' current. When Th_1 is fired, the right-hand side of C is immediately raised to $B+$ and the left-hand side to $2B+$. Current starts to flow from C through inductor L and diode D_2, reducing the voltage across C and increasing the inductive energy in L. When the left-hand side of C has fallen to $B+$, the current in L continues, which causes a continued fall in the voltage of C. When the current in L has ceased, the left-hand side of C will be at a potential $B-$ and diode D_2 becomes reverse-biased. The potentials on C are now $B-$ (left-hand) and $B+$ (right-hand), and remain so until Th_2 again fires. A bleed path through R ensures that leakage currents through D_2 and Th_2 do not raise the left-hand voltage on C when Th_1 conducts for long periods. When Th_2 is fired again, the left-hand side of C is taken up to $B+$ and the right-hand side to $2B+$. The cathode of Th_1 is then positive to its anode and it therefore ceases to conduct. The motor current, which is thus transferred from Th_1 to Th_2, flows through C and lowers the right-hand potential to $B-$. Then, as the motor current diverts through D_1, conduction through Th_2 ceases for lack of holding current. The inductive energy of the motor maintains current through D_1. The circuit is now ready for Th_1 to be fired again.

Regeneration The circuit is set up by changing the armature polarity with respect to the field and by adding diode D_3, accomplished by a simple changeover switch. When Th_1 conducts, the motor is effectively short-circuited and the current builds up rapidly. At maximum current level Th_1 is switched off and the motor current, maintained by the motor inductance, diverts through D_1, supplying a charging current to the battery. The cycle repeats to give a roughly constant average motor current and a constant resulting braking torque.

Conditions are complicated by the transition 'spikes' of voltage. They have a short duration but a large magnitude. The spikes have some effect on the battery, and increase motor core and I^2R losses. They may also impair commutation, particularly at low pulse rate frequency.

The cost of pulsed thyristor equipment is a disadvantage. The reduction of battery energy depends on the type of vehicle and its duty. The overall saving may be about 10% for typical delivery rounds, and rather more for industrial vehicles such as fork-lift trucks.

29.4.4 Vehicle and operational details

29.4.4.1 Charging

Taper chargers are now used extensively for the charging of traction batteries, and are normally provided with germanium or silicon rectifiers. Voltage sensitive thermal relays in conjunction with synchronous timing motors and contactors are used for charge termination. In one form of charge control device a temperature sensitive non-linear resistor, preheated electrically, is cooled by hydrogen from pilot cells when gassing begins. The change in resistance through the cooling of the device causes transductors to regulate the output of the charger.

Improved conductor insulation and high-grade magnetic core material, together with forced cooling, have resulted in weight saving in charger equipments carried on vehicles. A typical

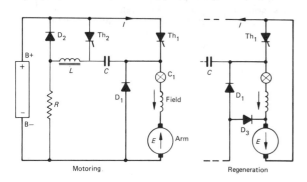

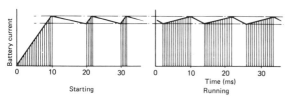

Figure 29.28 Thyristor control of an electric car

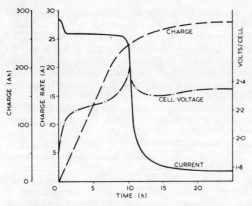

Figure 29.29 Charging characteristics

charger for connection to a 13 A power outlet, and making full use of this current rating up to the start of the gassing period, has the performance shown in *Figure 29.29* when charging a 48 V, 260 A h lead–acid battery.

29.4.4.2 Torque transmission

Most road and industrial battery vehicles are fitted with axles of conventional differential gear design, still the most convenient form of torque transmission to a pair of driving wheels. It has been shown by test that there is scope for improvement in performance by modifying the lubricant viscosity: the axles are normally fitted with extra gear reduction nose-pieces or external chain-and-sprocket reductions, and a fully run-in differential may absorb up to 1.5 kW at maximum road speed. Oils of lower viscosity have been used without excessive gear wear, although the noise is greater.

Hydraulic transmission This has found only limited application. Even simple forms cannot compete as alternatives to thyristor motor control gear for driver-type vehicles, but may be suitable for some industrial applications where the infinitely variable characteristic makes elaborate motor control gear unnecessary.

Motorised wheels Motors built into the wheels are extensively used in fork-lift trucks, sometimes with low-speed designs of reduced length and increased diameter. The powered wheels can be used for steering, thus improving manoeuvrability.

29.4.4.3 Regeneration

In general, the small amount of energy recoverable does not warrant the cost and complexity of the control gear necessary for regeneration, although it reduces brake maintenance (a major operating requirement). Certain industrial vehicles such as fork-lift trucks might benefit from regeneration if enough load lowering duty were involved; this might be possible by use of recirculating ball screws in place of the normal hydraulic lifting rams, enabling the motor to generate on lowering.

29.4.4.4 Tyres

Battery vehicles are affected to a greater extent than other vehicles by the rolling resistance of tyres. Radial ply tyres provide a more supple casing and as a result the loss due to flexing of the walls is less. A saving of 13% in energy consumption (or an increase of 15% in range) can be obtained. An advantage is that radial ply tyres are less sensitive to departures from optimum inflation pressure.

29.4.4.5 Bodywork

Most recent rider–driver battery vehicles use glass-fibre-resin mouldings, which permit improved appearance with much lower tooling cost. Metal inserts are readily moulded into laminates to provide anchorage points or attachments for components. Translucent panels can be moulded into otherwise opaque surfaces.

29.4.5 Applications

29.4.5.1 Pedestrian controlled vehicles

Bakery and dairy delivery vans, 'led' by the driver, have capacities between 0.5 and 1 t and a range up to about 20 km. The steering handle carries running and brake controls and there is a separate parking brake. A typical specification includes a 2 kW motor, 24 V, 100 A h battery, and chain or reduction gear differential drive.

29.4.5.2 Rider–driver vehicles

Invalid carriages with 36 V, 75 A h batteries have speeds up to 20 km/h and can negotiate any normal road. Municipal vehicles (refuse collection, tower wagons and tractors) are usually 4–6 t vehicles with batteries of up to 144 V, 600 A h. The battery weight is useful in adding stability to the tower wagon. The most widely used industrial trucks are platform, elevating and fork-lift trucks, road and rail tractors, mobile cranes, and trucks for carrying hot forgings and molten metal buckets. The capacities range up to 1–2 t. Some lighter types are three-wheelers having the motor and front wheel mounted on a turntable, which makes them exceptionally manoeuvrable.

On elevating, tiering, crane and other types of trucks with power operation, it is usual to employ a separate motor to perform these tasks, power being drawn from the main battery. The location of the battery on the vehicle varies according to the design. On some they are mounted beneath the platform, but in many cases they are mounted on a separate platform or under the driving seat. It is common for these vehicles to work long hours and in such cases two or more sets of batteries are employed, a discharged battery being replaced when necessary. The voltage varies between 24 and 80 V.

When trucks are fitted with a platform for the driver to stand on, steering is usually by means of a tiller, while the controller is hand operated and interlocked with the brake pedal to give a 'dead man' effect. With this design the brakes are held off by brake pedal. When pressure on the pedal is released, the brakes are automatically applied and the electric circuit is broken.

29.4.5.3 Locomotives

Battery shunting locomotives are used by many railways, factories and mines. A typical small locomotive develops a drawbar pull of 1.6 kN (360 lbf) at 6.5 km/h from a 4 kW, 48 V totally enclosed series motor with split field windings for control and worm drive on each of the two axles. Such a locomotive is designed for 0.46, 0.51 or 0.61 m gauge. Larger locomotives may weigh up to 8 t, operate at 120 V, and have two 12 kW series motors with series/parallel control, giving a 1 h tractive effort of 11.5 kN at 8.5 km/h.

29.4.5.4 Electric cars

Several attempts have been made to produce a marketable electric car, but the problem is the inevitable comparison with an equivalent petrol driven car having a speed of well over 100 km/h, a full-tank range of a few hundred kilometres and refuelling in a few minutes. With increasing shortage of liquid fuel, battery electric cars will become attractive for personal transport in and around towns, or, in the absence of rural public transport, for routine trips to the nearest railhead. Electric cars are quiet, non-polluting, easy to start and control, and cheap to run (largely because routine maintenance is negligible). Speeds of 50–60 km/h and ranges of 40–60 km (with 80–100 km possible if battery capacity is maximised) are adequate for such duties.

The growing interest in electric cars led the Electricity Council (UK) to produce a specification for a two-seater and a 40 km range, an acceleration of 1.35 m/s² and a smoothly variable control. Conversion of a small car body was undertaken to meet this specification: its gross weight was 1.3 t, of which the battery accounted for one-quarter. In 1975 the Council acquired 68 commercial electric cars, and used them for evaluation and improvement. Each car weighed 1.13 t (of which 27% was battery weight); had a 48 V, 6 kW series motor; was provided with full- and half-field tappings; and carried a battery with parallel/series switching to provide voltages of 12, 24 and 48 V. The acceleration was 1.3 m/s², the maximum speed 64 km/h and the range 40–90 km.

Conversions of Reliant (three-wheel) and Mini cars have been reported. These have weights in the range 0.7–0.9 t, 4–5 kW series motors, speeds of 45–60 km/h, battery weight fractions of 0.2–0.4 and ranges of 30–60 km. Their overall primary (a.c.) energy consumption is about 350 W h/(t km). Such vehicles give adequate personal mobility without the use of fossil fuel, and may be the first to appear in quantity.

Hybrid petrol or diesel electric vehicles with onboard internal combustion engines in place of battery chargers are advocated by some, as allowing electric operation in town, and sustained operation with the engine running continuously at its optimum conditions (i.e. minimum pollution) for longer runs. Hybrids may find application until a substantial increase in energy/weight ratio is achieved by storage batteries.

29.4.5.5 Electric buses

Two experimental vehicles of the late 1970s demonstrated the feasibility of battery buses. One was a 50-passenger single-deck vehicle, 16 t in weight (including a 4.8 t battery), with chopper control (with regeneration) of a 72 kW, 330 V series motor, 64 km range, 64 km/h speed and 1.0 m/s² acceleration. The other was a 'minibus' for 34 passengers in a short single-deck body, with a 100 kW motor, 150 km cruising range, 80 km/h maximum speed, average acceleration 0.9 m/s² and weight 9.7 t.

29.4.6 Range and power assessment

The range is limited by the energy/mass storage capability. A 1000 kg lead–acid battery operating at a load of 10–15 kW/t stores about 20 kW h, and as the transmission efficiency from motor terminals to wheel treads is about 70%, less than two-thirds of the capability is available at the wheels.

A vehicle of weight G with a (battery/total) weight ratio f, driven against a tractive resistance R, has a range S that is roughly estimated from

$$S = 50(fG/R)$$

Let a vehicle have the following particulars:

G	total weight (kg)	r_r	rolling resistance (N)
u	speed (km/h)	k_r	rolling resistance coefficient (N/kg)
R	tractive resistance (N)	r_a	air resistance (N)
A	frontal area (m²)	k_a	drag coefficient

The tractive resistance R comprises the rolling resistance r_r and the air resistance r_a. The former is $r_r = k_r G$, where k_r has a value in the range 0.1–0.2. The air resistance, proportional to the square of the speed, is

$$r_a = k_a A u^2 / 21$$

The drag coefficient k_a is 1.0 for bluff-fronted vehicles (e.g. vans), about 0.5 for cars and as low as 0.2 for fully streamlined bodies. The total tractive resistance is then $R = (r_r + r_a)$, and can be used to estimate the range S (in km).

Let $G = 8000$ kg, $f = 0.25$, $u = 40$ km/h, $A = 5.0$ m², $k_r = 0.15$ and $k_a = 0.78$. Then $r_r = 1200$ N, $r_a = 300$ N and $R = 1500$ N. Then the range $S = 67$ km.

The power at the wheels (in kW) is $P = Ru/3600$. For the parameters above, $P = 17$ kW. The input to the motor is about $17/0.7 = 24$ kW. Motor inputs of 8 kW/t may be required for a battery vehicle to emulate an internal combustion engined vehicle.

If a vehicle is braked to rest from an appreciable speed, the range S is reduced by reason of the loss of kinetic energy in friction. Some allowance may be made by reducing S by about 0.3 km for each brake stop from 50 km/h, and less for lower speeds in proportion to the square of the speed.

29.5 Road traffic signalling

Road traffic signalling installations cover a range extending from simple fixed-time pedestrian control equipments to vehicle actuated systems which may be co-ordinated by cable linking, electronic clocks deriving their timings from the public mains supply or a centralised computer system.

29.5.1 Development

The earliest signals known in the UK used semaphore arms and gas lighting. Semaphores or mechanical disc signals are still used at roadworks. In the early 1900s colour-light signals were brought into use on tramways and were introduced for the control of road traffic in the UK during the mid-1920s. Road traffic signals were originally operated manually and it was a natural step to mechanically operated signals. These signals normally incorporated a cam mechanism which determined the stop and go periods for the roads under control, the cam taking a fixed time to rotate. Though an economic solution to the problem of accident prevention, they were inefficient for traffic control, as their indication had no relation to the actual traffic conditions. Attempts were made to improve the timings between rush-hour, off-peak and late-night conditions by the use of programmed controllers. These changed their timing in accordance with a programme chart. By means of a synchronised motor the programme was varied throughout the 24 h of a day. This controller still had no regard for actual traffic conditions.

Early attempts at vehicle actuated control of road signals took place in the USA, latterly by electrical contacts or treadles set in the road surface some distance from the signals. The first vehicle actuated signals were introduced into the UK in 1932 with detectors using pneumatic tubes.

Many roads in the centres of cities, such as Oxford Street, were signalled by means of fixed-time signals co-ordinated with the flow of traffic. On early systems this co-ordination was achieved by a linking cable to ensure that the signals operated to a common and synchronised cycle time.

In the mid-1930s traffic controllers adopted a technology based upon relay logic, and the timing of traffic control functions was implemented by means of neon timers (capacitors with resistor charging networks). Later, valve timing circuits using conventional time-base techniques were used.

During the 1960s solid stage logic was introduced into traffic signal controllers, first with discrete transistor ligic with analogue techniques for timing. At a later stage t.t.l. integrated circuits were introduced and many controllers currently in production still use this technology. The new ranges of traffic signal controllers now in production employ microprocessor technology to implement the functional requirements and traffic strategies.

The switching of traffic signal lamps on early controllers was by means of heavy-duty cam contacts. When relay technology was employed, the lamp switching relays were of a type similar to those used for the logic but the contacts were fabricated from a heavy-duty silver–tungsten alloy. More recently with controllers employing solid state logic, plug-in relays have been used with the contacts fabricated from either silver or silver cadmium oxide. In new designs solid state switches will be used for controlling the signal lamps.

29.5.2 Legal requirements

All traffic signal equipment has to conform to statutory requirements and Department of Transport Specifications before it may be installed in the UK. These requirements are defined in a Statutory Instrument, British Standards Specifications and other specifications issued by the Department of Transport.

The major criteria for traffic signals relate to colour, intensity and mounting height of the light signals. The criteria for control equipment are associated with its performance in the presence of hostile environments such as extremes of temperature, vibration and electrical noise.

Prior to the equipment being installed on the public highways, rigorous type approval tests are made by the Department of Transport to ensure that the equipment meets the functional and environmental requirements detailed in the specifications.

29.5.3 Traffic signals

Until the late 1960s the optical systems employed in traffic signals were generally a lens/reflector combination, utilising tungsten filament lamps. The light output from these optical systems, utilising a glass lens with relatively low transmittence, was of the order of 50 cd on axis for the red signals. The current practice is to employ tungsten–halogen lamps with the lens fabricated in either acrylic or polycarbonate, giving light outputs in excess of 400 cd on axis for the red signals, with arrangements made to dim the signal by means of photoelectric cells during the hours of darkness.

One problem with all traffic signals employing reflector systems is associated with 'phantom signals', an apparent illumination of the signal caused by ambient light reflected by the optical system. To overcome this problem, signals have been produced using a system whereby a bundle of fibre-optic light guides is illuminated by a tungsten–halogen lamp: the individual light guides are then distributed across the face of the signal, giving a high-intensity but low-area light source which, when viewed by the road user, presents an even illumination. The small optical area of the light guide gives a significant reduction of phantom signal.

29.5.3.1 Detectors

Early detectors utilised treadles, subsequently pneumatic tubes in frames set in the road surface. The increase of traffic density and vehicle weight resulted in pneumatic tubes having an extremely short life. The inductive loop detector was therefore introduced.

Inductive loop detectors normally consist of three components: (1) a loop laid in the road surface at a depth of 50 mm, (2) a loop feeder and (3) a detector unit. The loop of wire, normally two turns, is laid in the road surface and fed with an alternating current in the frequency range 60–80 kHz. When a vehicle passes over the loop, the inductance is modified and thus the phase of the current is changed. The detector incorporates an electronic circuit responding to the change of phase to operate an output device. The circuit must accommodate small changes in the loop inductance due to environmental changes without giving an output signal.

In addition to normal vehicle detection at intersections, detectors can implement the following traffic functions: (a) speed measuring, (b) vehicle counting, (c) traffic congestion and (d) queue conditions.

Other forms of vehicle detector are also available, two main alternatives in the UK being (1) microwave detectors and (2) axle detectors.

Microwave detectors utilise the Doppler effect as the means of detection: they give an output for an approaching vehicle but ignore vehicles which are moving away. This type of detection is now being brought into service on both pedestrian and intersection traffic control systems.

Axle detectors normally incorporate a special coaxial cable which utilises the piezoelectric effect, whereby pressure by the vehicle wheel produces an electrical signal which can be amplified and made to operate an output relay. Axle detectors in conjunction with inductive loop detectors enable vehicles to be classified by measurement of the number of axles and length of the vehicle.

29.5.4 Control equipment

29.5.4.1 Pedestrian controllers

Pedestrian controllers are normally small pole-mounted equipments using integrated circuits or microprocessor technology and incorporate some measure of solid state lamp switching, based on the use of triacs to flash or to operate the signal lamps. The controllers have a means of adjusting the timings for the signal periods, which may be by means of pin boards or the alteration of data carried in a microprocessor store. The controllers can also be linked either to an adjacent intersection controller or to a centralised computer system.

29.5.4.2 Vehicle actuated intersection controllers

The technology is similar to that used on pedestrian controllers, and new designs have generally used microprocessor techniques.

Intersection controllers incorporate the whole range of functions required by traffic engineers, such as: (a) vehicle actuation, (b) conditional changes, (c) speed measuring, (d) linking by cable or by synchronisation from a mains supply, (e) centralised computer control utilising voice-frequency modems and (f) manual and police facilities.

The normal traffic sequence in the UK for any particular approach or phase is red, red–yellow, green, yellow and then back to red. The red and green periods are capable of independent and conditional timing, depending upon the requirements of the traffic engineer and traffic flow. It is normal for the red–yellow signal to be set at 2 s, for the green signal to have a minimum of 7 s and for the yellow signal to be set at 3 s. In addition, there is usually an all-red period before the signals on another approach change to red–yellow and then green, to allow turning traffic to clear; this all-red period may be extended by queue detectors. In addition, the green signal, when it has reached the end of its minimum period, will be extended if approaching vehicles are still being detected. This green signal will continue until a preset maximum duration is reached and a forced change to the next demanded approach will then be made, but the control will automatically return to give right of way to the approach that has been terminated should demands still exist.

29.5.4.3 Interlocking of signals

To ensure that traffic conflicts cannot occur at intersections (i.e. green signals showing to conflicting approaches), the supply to the signals is interlocked to ensure that a red signal must be shown on a conflicting approach before the opposing green is displayed. On newer equipment additional conflict monitoring is provided as well as relay interlocking. On equipment employing solid state switching, dual conflict monitors are used to ensure safety.

29.5.4.4 Linking of signals

Traffic signals are linked within a particular road or area to ensure signal co-ordination and to maximise traffic flow in the main route. Linking systems are not able to influence the safety timings (red–yellow, minimum green, yellow and all-red conditions) but can call and extend the green signals on any approach, depending upon the pregrouping of the controller phases. Where a relatively simple link between two controllers is less than, say, 100 m, it is usual to employ cable linking to synchronise the two equipments. Over greater distances it is usual to employ mains synchronised clocks in each controller which define time and are therefore able to offset the start time of the main-route green signals by the amount required to ensure a co-ordinated traffic flow over the link. The mains derived time also enables the controller to be fitted with different plans brought in at different times of the day, thereby permitting preferred directions of peak traffic flow. Typically, eight plans will be provided. The controlled equipment is also capable of permitting short periods of free choice within different parts of its cycle, such that with no traffic on the main route the equipment will service demands on a different approach earlier than the programme would normally require.

29.5.5 Central computer control

In the majority of large cities consideration has been given to the improvement of traffic flow by the use of centrally controlled signals. Not only is it possible to predict a reduction in delay time as vehicles pass through a city, but also it is possible to calculate effective energy savings and, hence, lower pollution levels.

The pattern of traffic varies considerably with the time of day. When considered for a complete city, it can be seen that a more structured situation exists: e.g. the pattern of movement of workers from residential areas into both the industrial and city centres is an obvious one early in the morning and in the evening. The random movement of shoppers at mid-day and the need to move emergency vehicles rapidly from one place to another are situations which, if not properly controlled, add significantly to the overall city management. A central computer resembles a 'juke-box' in that 'records' are called upon by time of day and allowed to play until the next required change. The 'record' instructs the intersection controllers, on a second-by-second basis, what indications to give to the traffic. The street controller replies, indicating its actions, so that the computer can check the functioning of the controller and, when necessary, log faults for reporting to the servicing organisation. With the regular pattern of the time-of-day selection of records, manual intervention may become necessary, e.g. to clear a path for fire-fighting vehicles. This occurs in the form of a call for 'waves' of green signals that run away from the fire station, allowing fire appliances to ride on the crest of the green ripple.

High traffic density can also be dealt with, due, e.g., to congestion in the centre of holiday resorts, or the opening of a bridge, or the closing of a road at a railway crossing. A computer can now be used for more significant situations, such as generating restricting rings to prevent congestion in city centres or the provision of special routes for abnormal loads, at the same time considering the overflow of car parks, the monitoring of ramps or (taking a somewhat more sophisticated approach to the time-of-day situation) the selection of plans dependent upon the measurement of traffic densities in various parts of the city.

Most centralised computing systems also have an integrated closed-circuit television facility whereby critical points within the area may be monitored, a system operator being able to take remedial action should he observe congestion occurring.

29.5.6 Motorway signals

Within the UK it is not normal to put standard traffic signals on motorways, the special-purpose roads designated for motorised traffic only. These roads are normally signalled by matrix signals displaying speed restrictions or stop commands, but which may also show symbols indicating the form of hazard.

The signal consists of a matrix sign built up from 50 mm square cells, using a nominal 13×11 matrix. The optical systems are either lamp reflector or fibre optic and each individual cell has a light output of the order of 6 cd.

The matrix sign is backed by yellow lanterns which flash vertically and bring the warning indication on the signal to the attention of the road user. The stop command is implemented by means of red lanterns which flash horizontally.

Motorway signals are co-ordinated from a central computer system with the operation of the signals under the control of the area Police Authority. Signals are set either individually or by specifying the area to be controlled, and the computer system automatically puts out a time and space sequence for the signals to ensure that traffic is not suddenly confronted by a stop signal. The sequences involve the stepped reduction of speed limits by counting down the speed displayed by a signal, e.g. from, say, 60 30, 20 to 'stop', over a period of some 30 s and by spacing the speed restrictions so that 'stop' is preceded by reduced limits on preceding signals.

Motorway signalling systems also have the ability to monitor equipment faults and indicate failures to operators. Automatic operation dependent upon traffic flow or environmental conditions is also possible.

30 Transportation — Railways

W H Whitehouse
British Railways Board

D S Armstrong MIEE, MIMechE
British Railways Research Division

Contents

30.1 Railway electrification 30/3
    30.1.1 Alternating current systems 30/3
    30.1.2 Direct current systems 30/4
    30.1.3 Locomotives 30/4
    30.1.4 Multiple-unit trains 30/5
    30.1.5 Electric braking 30/7
    30.1.6 Track equipment 30/8
    30.1.7 Underground railways 30/8
    30.1.8 Developments 30/9

30.2 Diesel-electric traction 30/9
    30.2.1 Locomotives 30/9
    30.2.2 Design 30/9
    30.2.3 Electrical equipment 30/10

30.3 Railway signalling and control 30/10
    30.3.1 D.C. track circuit 30/11
    30.3.2 A.C. track circuit 30/11
    30.3.3 Jointless track circuit 30/12
    30.3.4 Other forms of track circuit 30/12
    30.3.5 Other means for vehicle detection 30/13
    30.3.6 Colour light signalling 30/13
    30.3.7 Terminology 30/13
    30.3.8 Two-aspect signalling 30/13
    30.3.9 Three-aspect signalling 30/13
    30.3.10 Four-aspect signalling 30/13
    30.3.11 Signalling for junctions 30/14
    30.3.12 Colour-light signals 30/14
    30.3.13 Signal lighting circuits 30/14
    30.3.14 Operating features 30/15
    30.3.15 Signal-aspect control circuits 30/15
    30.3.16 Point operation 30/15
    30.3.17 Modern signal box 30/16
    30.3.18 Interlocking equipment 30/16
    30.3.19 Remote control 30/16
    30.3.20 Train description 30/17
    30.3.21 Automatic warning system 30/17
    30.3.22 Automatic train control systems 30/18
    30.3.23 Signal power supply 30/19
    30.3.24 Immunising of signalling against traction currents 30/19
    30.3.25 Level crossings 30/19

30.1 Railway electrification

Electric traction systems, considered from the point of view of the form in which electrical energy is delivered to the train, fall into two broad divisions: (1) alternating current systems, (2) direct current systems.

For city and suburban services, direct current is most common; for main line services there are many examples of both a.c. and d.c. systems. The use of alternating current has developed mainly in Europe, although there have been installations in the USA. In other parts of the world, notably in Australia, India and South Africa, d.c. systems predominate. The latest trend, however, as shown by French and British Railways electrification plans, is to use the high-voltage a.c. 50 Hz system for any new railway electrifications for both suburban and main line services.

30.1.1 Alternating current systems

A.C. systems can be divided as follows: (a) 3-phase; (b) single-phase, low-frequency ($16\frac{2}{3}$ and 25 Hz); (c) single-phase, standard frequency (50 and 60 Hz).

30.1.1.1 Three-phase

The only 3-phase electrification of importance was in Italy, where the traction supply was obtained from two overhead conductors, the rail providing the third pole. This system has been converted to d.c.

A number of specialised novel vehicles, such as those using magnetic levitation and linear motors for propulsion, use 3-phase power for traction. Power control can be provided from either a trackside or on-board inverter or by voltage control.

30.1.1.2 Single-phase low-frequency

This system is widely used in Europe at $16\frac{2}{3}$ Hz, and to a limited extent in the USA at 25 Hz. Most installations employ 1-phase series motors. Supply is taken from an overhead conductor at voltages up to 16 kV, with the rails as return.

The supply is taken from the overhead line through a circuit-breaker to the primary of a step-down transformer, the secondary of which is tapped for low-voltage feed to the traction motors. *Figure 30.1* is a typical main circuit diagram for a motor-coach with a pair of 1-phase commutator motors.

30.1.1.3 Single-phase standard frequency

The advantage here is the use of main 50 Hz grid systems instead of special low-frequency traction generating stations. Pioneering work was done by the French Railways in 1951 and subsequently. A 50 Hz commutator motor requires a large number of poles and brushes to limit the internal losses. When taking traction power from industrial frequency supplies, d.c. series motors are preferred, fed through a rectifier, using a variable voltage from a tap changer (for rectifiers), or from a fixed-voltage winding on the main transformer (in thyristor control). British rail adopted a 25 kV 50 Hz system (*Figure 30.2*) in 1956. Similar systems are

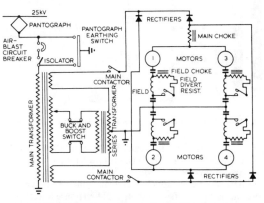

Figure 30.2 Main-circuit diagram for 25 kV locomotive

now operating in many countries, with about 20 000 route km. Some lines, particularly those carrying heavy mineral traffic, use up to 50 kV to reduce the current and permit an increased spacing between feeder stations.

Figure 30.3 shows the basic element of the armature power control circuit of an a.c. thyristor locomotive. The output winding of the transformer is divided into a number of elements (typically two, but higher numbers are also used), which feed asymmetrically controlled thyristor/diode bridges. Each bridge is sequentially advanced to full conduction of the thyristors as increased power is required. In some variants of power circuit (*Figure 30.4*) only one bridge has thyristor control, the others being normal diode bridges. The diode bridges can be switched directly into circuit, the control system using the thyristor bridge to increase power until it is fully conducting. A diode bridge is then added in series, the thyristor bridge being returned to zero

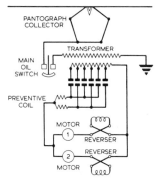

Figure 30.1 Main-circuit diagram for two-motor multiple-unit equipment with single-phase series motors

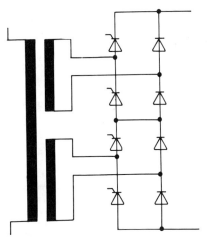

Figure 30.3 Two sequential half-controlled bridges

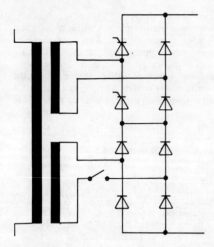

Figure 30.4 One controlled bridge with one switched bridge

conduction so that a steady further increase in voltage can be provided. The fields of the traction motors are normally supplied from a separate thyristor controlled bridge, with arrangements to give the desired series-motor characteristic. Independent control of motor armature and field permits rapid response to transient conditions such as wheel-spin and motor flashover.

The simplest form of regulation uses natural commutation at the end of each half-period to extinguish the thyristor current. The harmonics can be reduced (and power factor increased) by using several semi-controlled bridges in sequence. The a.c. current lags further on the voltage than with a rectifier locomotive, and the power factor is normally lower during thyristor control. Power factor can be improved if the current is forcibly reduced to zero before the end of the natural half-period (a system termed 'sector control'). Thyristor control causes greater distortion of the overhead line voltage than rectifier locomotives, and the disturbance of the supply network may exceed permitted values unless the disturbances are predicted and appropriate measures taken.

The adverse effects of thyristor harmonics can be reduced by using several sequentially controlled bridges or by installing filters on the locomotive.

30.1.2 Direct current systems

Systems operating at 600–1000 V are extensively used for urban and suburban electrification, usually with one live insulated rail and running rail return. Some four-rail systems exist with positive and negative insulated electrical supply rails. Systems working at more than about 1000 V use overhead catenary conductors: typical voltages are 1500 and 3000 V. For heavily trafficked lines a voltage of 6000 V is being considered. The d.c. series motor is universally employed, usually in two- or four-motor sets to meet line-voltage requirements or for series/parallel speed control.

30.1.3 Locomotives

Individual axle drive This is the most common form of arrangement and takes various forms.

Axle-hung geared motor The motor is suspended partly by bearings around the driving axle and partly by a bracket resting on the bogie frame. The motor output is taken to the driving axle via a single reduction gear. This is a simple form of construction but the non-springborne part of the motor mass results in increased dynamic loads on the track.

Axle-hung direct drive The axle can form a driven part of the motor with the electrical armature placed inside (held through the axle bearings) or outside the axle. In one version, a 3-phase induction motor is formed from a wound stator held within a large-diameter tubular axle, the cage winding of the motor being fixed to the inner surface of the axle tube.

Frame-mounted geared motor For express locomotives and for other cases in which riding qualities are important, drives have been devised to reduce the non-spring-borne mass, to raise the mass centre and to allow for relative movement between frame-borne and axle-borne parts.

Spring drive This consists of a hollow shaft surrounding the driving axle and having sufficient clearance therefrom to permit the necessary relative movement between the spring-borne quill and the axle. The quill carries the gearwheel (or a gearwheel at each end) engaging with a pinion on the motor shaft. A twin- or double-armature motor is frequently employed, and to secure flexibility the pinions may be spring cushioned. The chief feature of the drive is the method of connecting the quill to the driving axle, which is accomplished by an arrangement of circumferential springs acting between a spider on the quill and a special spoke arrangement on the driving wheel.

Link drive To avoid the use of springs, forms of quill drive using links (such as the Buchli drive) have been designed. The locomotive frames are within the wheel space and the motor shaft extends over the frame to a pinion engaging with a gearwheel carried in a frame-mounted bearing. The gearwheel is arranged nearly concentric with the driving wheel and outside it; the connection between the two is made by an ingenious gear-link arrangement working on to pins fixed to the driving wheel.

Flexible-disc drive Here the motor is mounted on the bogie frame. The armature shaft is hollow and the drive is taken through a flexible disc coupling at one end by means of a shaft passing through the armature and connecting through another flexible disc to the pinion which is carried in bearings in the gearcase.

Alsthom drive The drive to the wheels is by means of a hollow quill shaft and flexible links, using a 'dancing member' and rubber-bushed bearings in the links. The quill is carried in the motor frame by large diameter taper roller bearings.

Gear-coupled bogie drive (*monomotor*) A motor mounted on the bogie frame can be arranged to drive, via a gear train, all the axles on a bogie. This system is widely used by French Railways. Some units include a gear change (selected when stationary) to provide freight or passenger traction characteristics. Since the motor is mounted above the bogie frame, large power units can be provided and double-armature motors can be used.

Body mounted motors The complete motor may be carried in the body of the locomotive with a suitable cardan shaft drive between motor and axle. The shaft has to accommodate the relative movements between body and axle, these being significantly greater than those between the bogie frame and axle.

Locomotive equipment A locomotive must carry motors, tractive effort and speed control devices, current collection equipment, switches for isolation and overload protection, heating and lighting systems for locomotive and train (if required), power sources for mechanical braking systems, batteries, instrumentation and various signalling and safety devices. It is sometimes capable of independent operation remote from the electrical supply, using a small prime-mover unit.

30.1.3.1 D.C. locomotives

On d.c. locomotives there is almost invariably an even number of traction motors. With two motors (there are usually at least four) the simple well-known series/parallel control can be used.

In this, the motors are first connected in series with starting rheostats across the contact line and rails; the rheostats are then cut out in steps, keeping roughly constant current, until the motors are running in full series; next the motors are rearranged in parallel, again with rheostats; the rheostats are cut out in steps, leaving the motors in full parallel. Some stages of field weakening are generally included. The power input remains roughly constant during the series notching, then jumps to twice this value during the parallel notching. With a four-motor equipment, a series/series–parallel/parallel connection can be used, giving three economical speeds (i.e. running without resistance) unless the line voltage is too high to be applied direct to a motor.

The main circuit contactors performing the functions of motor grouping, field weakening, and cutting out rheostats are operated either electromagnetically (e.g. by solenoid) or electro-pneumatically (electrically controlled compressed air cylinders); further, they may be grouped together and interlocked by a camshaft, or separated and individually controlled.

When a locomotive is fed from a third rail system, the inevitable gaps in the conductor rail at points and crossings (up to 150 m) cause momentary interruptions to the supply, resulting in snatching and surging of the vehicles, particularly with goods trains. This difficulty has been overcome by the Southern Region of British Rail, which has put into service 1100 kW, 600 V d.c. locomotives using a booster motor-generator in the traction motor circuit; a flywheel on the motor-generator set provides sufficient stored energy to enable the locomotive to run over a gap without perceptible reduction in torque.

Proper distribution of the weight of the apparatus in the locomotive body is an important design feature and strongly affects running qualities.

Modern d.c. locomotives usually employ power semiconductor regulators to control the traction motors. These give faster response to load variations and avoid the loss of energy in rheostats.

30.1.3.2 A.C. locomotives

Low-frequency Systems operating at $16\frac{2}{3}$ or 25 Hz normally use 1-phase commutator motors. These require low armature voltages to achieve good efficiency and high power factor. The motors are generally connected in parallel and, except for smaller locomotives, primary tap changing on the main transformer is used to provide a variable voltage supply to the motors. With this type of control, each accelerating notch is an economical running point. This is in contrast to the d.c. locomotive, which cannot run continuously on the resistance starting notches. To avoid jolts and momentary short-circuits, preventive coils connected across successive tappings are used. Various elaborations of the simple transformer tapping system have been devised to reduce the number of tappings without reducing the number of available motor voltages. Thus, by use of a small auxiliary transformer a tapping voltage can be increased or reduced by a required amount, so that six tappings will give 18 running voltages.

Industrial frequency D.C. series motors are fed through a rectifier (or controlled rectifier) which is, in turn, supplied by variable voltage from a tap changer (or fixed voltage) on the main transformer. Semiconductor rectifiers are normally used and give reliable service when suitable protection is provided against the voltage transients which occur due to switching and other causes. Primary protection is provided by an air-blast circuit-breaker mounted on the roof, and in the case of solid state rectifiers, it is usual to protect each string of rectifiers by a fuse. A smoothing inductor is incorporated in the rectifier circuit in order to limit the ripple in the d.c. supply to the motors.

Figure 30.5 shows a typical locomotive layout.

30.1.4 Multiple-unit trains

Multiple-unit trains, used for suburban and city passenger traffic involving frequent services with many stops, are usually operated by d.c. supply. The chief elements of a motor-coach equipment are: two or four motors with nose-suspension drive; starting rheostats; contactor switches; and master controllers to work the train from any required cab. The contactor switches may be operated in the various ways already mentioned in connection with d.c. locomotives. In the present case, however, the control equipment has to be packed into a restricted space, either at one end of the coach behind the driver's cabin or beneath the coach frames. In the latter position it is more exposed to the weather but leaves more passenger accommodation. The tendency is to use under-coach mounted equipment boxes which can readily be removed and replaced, as complete units if necessary.

The essential features of multiple-unit control—that is, the control of the several units in a train from any driving cabin—are shown much simplified in *Figure 30.6*. All motors are notched up by any one master controller by a set of common circuits coupled between coaches. The electromagnetic operation of the contactors is shown. It will be observed that the traction currents are drawn by the motors from local collector shoes, whereas the control current is common to all the coaches. Various safety features are fitted to master controllers, including the well-known 'dead man's handle', which cuts off the power and applies the brakes after a few seconds if the driver releases a button mounted on the controller handle. Automatic acceleration is invariably used on modern motor coach trains; in this arrangement a lock-out relay is used which cuts out a step of starting resistance as soon as the current has dropped to a predetermined value.

30.1.4.1 Control equipment

Electropneumatic camshaft control The most noteworthy feature of this control unit is that the camshaft which controls the operating sequence of the accelerating contactors makes a complete revolution in one direction for the series notches, and a revolution in the reverse direction for the parallel notches, the transfer from series to parallel motor connection being made on a separate pneumatic switch unit. The initial and final positions of the camshaft are identical, so that in the case of a power interruption the equipment is ready for an immediate restart.

It will be seen that this arrangement, which is made possible by a special system of motor and resistance circuits (*Figure 30.7*) introduces each contactor in circuit twice during the accelerating period, once during series and once during parallel notching. Thus, when compared with control units employing separate contact systems for the series and parallel conditions, the number of accelerating notches available in a contactor group of given size and weight is almost doubled, while the simplicity of subsidiary control circuits and of mechanical construction, which are features of the pneumatic camshaft principle, are retained, with low power consumption for control apparatus.

Some equipments have ten series and ten parallel steps (in contrast to the five and four common with simple resistance control), making possible higher average rates of acceleration without wheel slip.

The controller comprises a group of cam-operated contactors. The camshaft is rotated through rack and pinion gear by an air motor (*Figure 30.8*). Energising magnet valve U admits air to the oil reservoir R and exhausts the air cylinder C by means of which

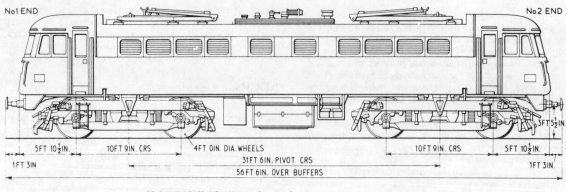

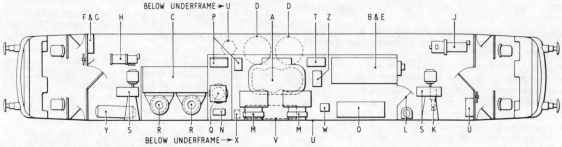

Figure 30.5 British rail 2500 kW, 25 kV a.c. electric locomotive. A. Main transformer; B. Tap changer; C. Rectifier; D. Smoothing inductors; E. Control equipment; F. L.V. fuses; G. Fault indicator; H. Main compressor; J. Exhausters; K. Auxiliary compressor; L. H.V.a compartment door; M. Transformer coolers; N. Hole storage capacitor; O. Motor contactors; P. Braking excitors; Q. Braking resistors; R. Rectifier coolers; S. Motor coolers; T. Auxiliary transformer; U. Battery charger; V. Battery; W. Train heating panel; X. Field divert resistor; Y. Main reservoir; Z. Tap-change inductor

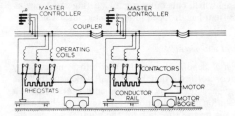

Figure 30.6 Basic multiple-unit control

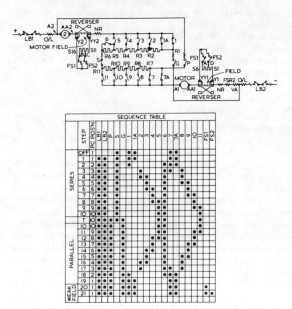

Figure 30.7 Main circuit diagram of 600 V, 2-motor control equipment with 2 field shunt positions

the piston K is moved towards the left at a speed determined by the adjustable orifice N through which the oil must flow: this movement operates the contactors in the sequence required for acceleration in series, and the limit of travel represents the full series position. At this point the motor connections are changed from series to parallel on a separate pneumatic switch, and at the same time the magnet valve U is de-energised so that air is admitted to cylinder C and exhausted from the oil reservoir R, causing the piston to return to its initial position, operating the contactors in the sequence required for parallel acceleration.

Notches are definitely located by means of a star wheel, mounted on the camshaft, with which a pawl engages: a solenoid is provided which, when energised, locks the pawl on any notch so that the camshaft cannot rotate.

Automatic acceleration is carried out by means of a relay operating on the current limit principle, whose contacts control the circuit of a solenoid capable of locking the star wheel pawl on any notch, although air pressure is applied continuously to one cylinder of the air motor. This locking solenoid can also be operated from an alternative circuit controlled by two small contactors, so that acceleration can be carried out a notch at a time, at a rate slower than normal automatic: for this purpose, an inching position for the master controller handle is provided.

Safety devices are provided in the form of voltage relays which ensure that the line breakers can only be closed when the main

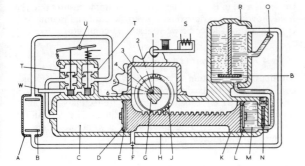

Figure 30.8 Sectional diagram of air motor for contactor group.
A. Expansion chamber; B. Baffle; C. Air cylinder; D. Expander; E. Piston packing; F. Check valve; G. Cam-shaft; H. Air engine cylinder; J. Pinion; K. Air engine piston; L. Piston packing; M. Hydraulic cylinder; N. Timing adjustment; O. Liquid filler plug; R. Liquid reservoir; S. Stop coil; T. Exhaust; U. Magnet valve; W. To air reservoir

voltage is present: also that they are tripped in the event of a power failure.

Electrical interlocks ensure that the motors can only be started from rest on full field, that the reverser can only be thrown when the line breakers are open, and that the series/parallel switch is always in correct relation to the rotation of the camshaft.

The contactors are arranged in two groups on opposite sides of the camshaft: each contactor is an individual unit comprising fixed contact, moving contact with roller and supporting block of moulded insulating compound. The contactors are bolted side by side to steel bars of heavy rectangular section, these bars forming the connecting framework between the two end castings, which carry the camshaft bearings, air engine, locking coil, etc.

The driver's controller will have a small number of positions, and transitions between them will be regulated by the current limit relay. Typical positions are Off, Shunt (motors in series with all resistors in circuit), Series (motors in series with resistors out of circuit), Parallel (motors in parallel with resistors out of circuit), plus a number of Weak Field positions. In urban units the control sequence lasts only for a short time and may be completed before the train has left the station.

Thyristor regulator (chopper) control Regulation of the average voltage applied to the traction motor is available if a thyristor switch is used to connect the supply cyclically to the motor. A thyristor carrying direct current will only switch off if the current is reduced to zero and the thyristor has a reverse voltage applied for a short time. Many circuits are in use to achieve this, the basic concept being that a charged capacitor is used as a temporary source of current, reverse biasing the thyristor to turn it off. The capacitor is switched into circuit using an auxiliary thyristor. A typical circuit is shown in *Figure 30.9*.

When the supply is connected, thyristor T2 is turned on to charge capacitor C via the motor path. When C is charged, the current in T2 falls and it extinguishes. Turning on thyristor T1 applies the full supply voltage to the motor and also allows the charge on C to oscillate (via L and D1) for a half-period, thereby reversing the polarity of the potential across C. When a sufficient current is flowing through the motor, thyristor T2 is turned on. Capacitor C acts as a current source for the motor and applies a reverse voltage to T1. If C has a sufficient charge, T1 will be reverse biased for long enough to be extinguished, thereby blocking the flow of forward current: C will then be charged as in the initial operation. The motor current will decay via the freewheel diode D2. When the motor current has fallen to a selected value, T1 is again fired and the cycle repeated.

By control of the firing pulses to T1 and T2 the mean voltage

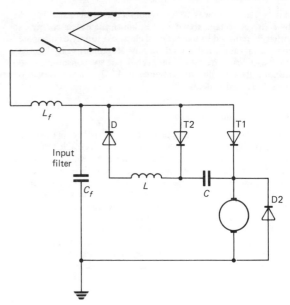

Figure 30.9 Basic circuit of thyristor chopper control

applied to the motor can be regulated as the train speed changes, while maintaining motor current and tractive effort. It is customary to fire T1 at a fixed frequency, so that the input filter $L_f C_f$ can be designed to limit the ripple currents drawn from the traction supply. A fixed frequency also avoids the generation of current at track circuit frequency.

This form of control avoids the energy losses associated with rheostatic control and is particularly valuable where frequent starting is required. It is therefore used widely for urban rapid-transit systems. By reconnecting the elements of the circuit it can use the motor as a generator and convert the mechanical energy of the train during braking into electrical energy and return it to the traction supply. If the supply is not receptive, braking rheostats on the vehicle can be used.

Recent equipments have used gate turn-off thyristors which avoid the need to provide a separate thyristor-switched turn-off circuit.

30.1.4.2 A.C. systems

Multiple-unit trains are used for suburban duties with a similar system of linking the controls of the various units to the single driver's controller. Each power unit has a transformer/regulator/motor drive and both tap change and thyristor regulation are used.

30.1.5 Electric braking

30.1.5.1 Regenerative braking

The braking of heavy trains on long down gradients is materially facilitated by the use of regenerative braking, whereby some of the mechanical energy released by the train in its downward progress can be reconverted into electrical energy and returned to the supply system. For this purpose it must be possible to reconnect the traction motors as generators and control the speed by the braking torque developed. Regeneration is possible with any form of electric traction at the expense of additional weight, cost and complication, except in the case of 3-phase traction with induction motors, for which regeneration is inherent and automatic. Regeneration has been most successfully employed on locomotives which have to negotiate heavy grades. Regenerative braking is also feasible in heavily loaded urban railways. If the railway system is not receptive (detected typically by a rise of voltage above a given limit), the excess energy can either be returned to the national electrical supply or dissipated in rheostats at the railway substations. Regenerative braking can give significant reduction in overall energy consumption on urban railways, and typical figures of 15–20% are achieved on densely loaded systems.

30.1.5.2 Rheostatic braking

An alternative form of electric braking, referred to as rheostatic or electrodynamic braking, is available for electric locomotives and is particularly useful for a.c. locomotives, which cannot be designed for regenerative braking as readily as those operated on d.c. systems. With this type of braking, the energy generated by the motors is dissipated in resistors. Control is usually exercised by separately exciting the motor fields, but some systems use self-excitation of the motors with main resistance switching. Owing to the large amount of energy which may have to be dissipated for considerable periods if descending a steep incline, the braking resistor is usually forced-air cooled, and special types of strip resistor units have been developed for this purpose.

30.1.6 Track equipment

The live conductor for carrying power to the trains is both important and costly.

30.1.6.1 Conductor rail and collector shoe

The d.c. positive conductor rail is a flat-bottomed steel rail of 50–75 kg/m mounted on porcelain insulators at the side of, and about 40 cm spaced from, one of the running rails. The negative return is the track itself unless (as on London Transport lines) circumstances require an insulated rail placed between the two track rails. The collector shoes are of cast-iron, and adequate contact between rail and shoe is effected by the mass of the shoe (10–25 kg). For high-speed running a spring-loaded shoe may be employed in which the pressure is to about 300 N.

The use of an upwards-facing contact surface on the rail permits a simple gravity collector shoe to be used, but the surface is then vulnerable to contamination and ice formation. Alternative systems use a side-contact or inverted rail.

30.1.6.2 Overhead conductor

For 1.5 kV and above, a contact wire, usually of hard-drawn or cadmium copper, is supported above the track by a catenary and dropper wires in such a way that the contact wire is generally level. To maintain the wire tension in varying weather conditions, the catenary and contact wires are anchored via insulators to a structure at the mid-point of the wire length and are then stretched by weights at each outer end. The longitudinal force in catenary and contact wire is about 10 kN. The lateral position of the contact wire alternates from side to side of the track centre line, traversing the active width of the pantograph collector strip during running to equalise the wear of the contact materials.

30.1.6.3 Current collectors (overhead)

A mechanism is required to maintain a current collector in contact with the overhead system. This, normally a pantograph, can be of several forms: a typical version comprises a folded arm which is raised by an air cylinder, carrying a contact head on separate springs. The collector material which slides along the overhead conductor can be of various materials, but metalised carbon is used in Britain. The pantograph frame ensures that the head moves on an approximately vertical path and maintains a constant contact force between head and conductor.

30.1.6.4 Feeder stations (d.c.)

Multipulse rectification is used with star/delta transformer connections and the basic ripple frequency on the d.c. output is 6 or 12 times the a.c. frequency. Mercury arc rectifiers are still in service, but new stations use silicon rectifiers. The d.c. rail (or overhead conductor) is connected as a continuous circuit, energised at each feeder station.

30.1.6.5 Feeder stations (a.c.)

The $16\frac{2}{3}$ and 25 Hz systems use frequency converter stations at the points of connection to the national electrical network. Static frequency converters using thyristors are now in service at ratings of several MVA, displacing the earlier synchronous converters.

Each feeder station at 25 kV supplies about 50 km of route and, to distribute the single-phase load on the main electrical network, different phases are used at succeeding feeder stations. The mid-point between feeders consists of an insulating neutral section through which the trains coast at zero power.

A wide spacing between feeder stations is a particular advantage in countries which do not have an extensive industrial frequency network. In extreme cases, the electrical supply for the railway has to be taken along the railway route as a 3-phase high-voltage system, sometimes using structures common with those which support the overhead contact wire. One alternative feeding method is to use a 25 kV supply from overhead wire to rail, feeding this through auto-transformers fed from a 50 kV longitudinal feeder.

The traction current passing through the locomotive transformer returns to the feeder station via the rails. If no special arrangements are made, the current will enter the earth as well as the rails. This current distribution can induce voltages in parallel electrical conductors such as communication cables. These voltages can reach dangerous values and it is customary to provide a return conductor in parallel with the rails to reduce induction. A further improvement is obtained if current transformers are connected in the energised and in the return conductors, forcing the current into the latter and removing the current from the track rails and earth.

30.1.7 Underground railways

In city centres where a high-capacity transport system is required but where surface routes are not available, railways can be built underground to provide a rapid passenger transport. The cost of tunnelling is very high and this method is only economic for large concentrated passenger flows. The electrical supply is normally by third-rail system at about 750 V d.c.; but where tunnelling costs permit, overhead electrification at higher voltages is provided. Underground railways use the same range of electrical propulsion and control techniques as on surface railways. The acceleration of underground trains is rapid and the separation between stations is small, perhaps 800 m. Under these conditions, if rheostatic camshaft control is used, the starting sequence may be completed before the train leaves the station. Subsequent running is on the natural characteristic of the motors, with weak-field operation to extend the speed range.

In France some underground lines use rubber tyres for propulsion and guidance, running on concrete tracks. This provides high acceleration and braking, contributing to passenger capacity. The rubber tyres have a higher energy loss than steel wheels, and additional ventilation may be necessary if a line is converted to rubber-tyred operation.

30.1.8 Developments

30.1.8.1 Novel systems

Although steel wheels on steel rails is the usual form of railway, other systems, mainly using magnetic levitation, are under examination. In Germany a vehicle using controlled d.c. electromagnets for suspension and guidance, and a 3-phase winding distributed along the track for propulsion, is being tract tested at up to 400 km/h. Japanese Railways are testing a levitation vehicle which uses the repulsion force between a superconducting d.c. magnet on the vehicle and conducting coils fixed to the tract. Two 6 tonne maglev vehicles are used in an airport to railway link at Birmingham International Airport and other maglev vehicles are being evaluated.

30.1.8.2 Three-phase induction motors

The development of high-power inverters using thyristors has permitted the use of induction motors for railway traction. The machines are simpler than d.c. motors, and are of smaller size and weight for a given output. The largest example is the Class 120 locomotives of German Railways. The 15 kV transformer has four traction windings at 1500 V, each of which supplies a controlled rectifier which produces a fixed direct voltage for input to an inverter, which delivers a variable frequency of up to 200 Hz to the four traction motors. The inverters and rectifiers are reversible and allow regenerative or rheostatic braking. With a rating of 5.6 MW the locomotives weigh 80 t.

30.1.8.3 Three-phase synchronous motors

French Railways, after tests with a 4 MW locomotive, have ordered a series of locomotives with synchronous mono-motors. Although the motor has a wound rotor, the inverter can be simplified to give an overall financial advantage. A typical 2.5 MW synchronous mono-motor weighs about 7 tonnes and has a top speed of 1950 rev/min at 130 Hz.

30.1.8.4 High-speed train sets

For high-speed passenger services it is sometimes desirable to use a fixed train formation with a distribution of the traction equipment to more than one vehicle. This applies where a high power output is required but a low axle load is necessary to limit track stress. Examples of this concept are seen in both diesel-electric and all-electric traction. The diesel-electric High Speed Train of British Rail uses a power car at each end of a set of seven or eight passenger vehicles. The power car contains a diesel/alternator set, rectification equipment and d.c. traction motors, within a total weight of 68 t. The 'Train Grande Vitesse' of the French Railways operates (under electric traction conditions) from either a 25 kV, 50 Hz system or a 1500 V d.c. system. This train is provided with a high-voltage busbar along the tops of the coaches, so that the power equipments in both traction vehicles can be supplied from the pantograph on one vehicle.

30.1.8.5 Switched reluctance motors

A reluctance motor with different pole numbers on the rotor (unwound) and the stator (wound) can provide traction if the stator coils are sequentially energised. A Blackpool tram fitted with four of these motors (each rated at 22 kW) is undergoing trials.

30.2 Diesel-electric traction

Petrol and diesel engines have been used for traction purposes, both road and rail, for a number of years, but the extensive application of diesel-electric locomotives to railway service has only occurred since about 1935.

30.2.1 Locomotives

Diesel electric locomotives may be divided into two main groups—main line and shunting; railcars form a third group which is quite small, and in which, on the lighter cars which predominate, mechanical or hydraulic drive is more often employed.

Main line locomotives may again be subdivided, the smaller units ranging from 450 to 750 kW, used for mixed traffic duty; and the larger units ranging from 1000 to 4000 kW for main line service. The larger locomotives are usually arranged for multiple-unit operation and several units can be used together to haul a large train. The units are sometimes distributed throughout the train to reduce drawbar stresses and braking transients, and radio control is sometimes used to give balanced power distribution.

Diesel-electric shunting locomotives have been used for a number of years on British Rail and are mostly in the 250–350 kW range and weigh about 50 tonne.

30.2.2 Design

The diesel engine is primarily a constant-speed, constant-power unit, and it is necessary to convert this to a variable speed, constant power at the wheels of the locomotive. The characteristic of the series d.c. motor is ideal for traction, since it provides a falling torque with increasing speed; by varying the voltage applied to the motor the torque–speed product can be made to match the constant power of the engine. The generator, which becomes in effect a flexible power link between the engine and the motors, provides the necessary current at varying voltage.

The essence of the problem of diesel-electric equipment design is to match the generator output curve to the engine power curve. The typical generator characteristic at constant (maximum) engine speed in *Figure 30.10*(a) is shown by the curve ABCDE.

The voltage–current curve for constant input power at this speed is BGD. Between B and D the two do not match and engine speed falls to give the relation BFD, with loss of power. In a shunting locomotive the speed drop produces output–input balance. Alternatively, the V/I characteristic of the generator

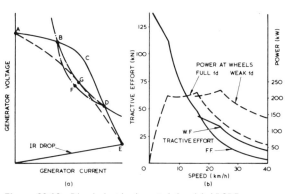

Figure 30.10 Diesel-electric characteristics. (a) ABCDE: generator voltage/current characteristic at constant maximum speed; BGD: Voltage/current characteristic for constant generator input power at constant maximum speed; BFD: characteristic at reduced speed. (b) Performance of 300 kW 50 ton shunting locomotive; Traction, 280 kW; Auxiliaries, 20 kW; Two motors with double reduction gearing

may be made to follow the curve AGE, but full engine power can then only be developed at G, with loss of tractive effort for other operating conditions (*Figure 30.10(b)*).

For main line locomotives it is desirable that full engine power be maintained over a wide range of train speeds. The diesel engine has an optimum operating zone in which the fuel efficiency is highest. For any desired output power there is therefore an optimum diesel engine rotational speed. The load regulation system normally responds to the setting by the driver of a desired engine speed. The excitation of the generator is then adjusted so that the optimum output power is taken from the diesel engine. If the train load reduces, the generator will run at a high-voltage/low-current condition delivering part power.

30.2.3 Electrical equipment

Electrical equipment consists essentially of a main traction generator, one or two auxiliary machines, a number of traction motors and the control equipment necessary for both main and auxiliary circuits. On large main line locomotives the auxiliary supply has to provide for compressors, traction motor blowers, lighting, heating, control and similar requirements. A battery is usually included to provide stand-by power and starting.

30.2.3.1 Generator

The generator armature is usually coupled direct to the engine crankshaft with a single bearing in the generator end housing. The auxiliary generator may be overhung from the main generator or in some cases driven together with an exciter by belts. Since the generator is a variable-voltage machine, the maximum voltage is arranged so as to obtain the most economical use of materials and the lowest I^2R losses. The maximum voltage is usually limited to 1000 and normal operation is around 600 V.

30.2.3.2 Motors

Motors are of series type generally similar to those used on d.c. electric trains and locomotives. They are usually built for operation at the full generator voltage but in some cases may be connected permanently two in series. The size of motor is determined largely by the maximum tractive effort required and bears little relation to the power capacity of the diesel engine.

30.2.3.3 Control

The control equipment on a diesel-electric locomotive is concerned mainly with the following functions.

Engine speed control The varying requirements of railway service make it necessary for the engine to have some speed control so that it can operate when required at reduced speed and power. In its simplest form a hand-operated lever, which alters the governor setting, provides a range of speeds between idling and full speed. No more is required on locomotives of the shunting type.

Generator field control On large locomotives a speed-sensitive device is used to adjust the generator field to match the engine power.

Motor control Motors are normally connected in parallel and this provides the simplest arrangement. In certain cases, however, series/parallel control may be necessary to give the required locomotive characteristics and provision must be made for changing over the motor connections. A reverser is required for changing direction of running, and diverter contactors may be provided if the fields have to be shunted to give top speed running.

Driver's controls On a typical shunting locomotive three control levers are provided on each side of the cab so that the driver has alternative driving positions. A time delay in the operation of the dead-man pedal is introduced so that the driver can change over from one operating position to the other without shutting off power. The engine, or master, control is used to start and stop the engine, and provides for engine testing with the locomotive at rest. A reversing lever with an off position determines the direction of movement of the locomotive. Finally, a power controller gives the driver complete control over all locomotive movement. Initial movement of this control lever prepares the power circuits and brings on generator excitation with the engine running at minimum operating speed. Further movement strengthens the generator field and finally brings the engine up to full speed.

The arrangement for main line locomotives is similar, except that a single driving position at each end of the locomotives is provided, with facilities for multiple-unit operation when double-heading is required.

30.2.3.4 Starting

The usual method is to start the engine from the battery by motoring the main generator, using a special starting winding. Compressed air starting is occasionally employed, but it involves the use of a compressor for charging the air cylinders. Bendix starters, similar to the type used on motor-car engines and operated from a 24 V battery, are used on some American locomotives. In the case of most shunting locomotives it is usual to make provision for tow starting in an emergency if the battery is too far discharged to motor the generator for normal starting. This involves connecting one or more of the motors so that they act as generators to motor the main generator and thus start the engine. Although emergency starting is seldom required, it is a useful provision, since another locomotive is usually available and starting can be effected with very little delay.

30.2.3.5 Mechanical design

The design of diesel-electric locomotives follows very much the same lines as that of normal electric locomotives; main line locomotives, particularly, use axle-hung motors and bogies which are generally similar.

Shunting locomotives up to about 50 t, which seldom require a maximum speed in excess of 40 km/h, have two or three pairs of coupled wheels driven by one, or two, geared motors. Where the top speed is limited, double-reduction gearing enables smaller high-speed motors to be fitted. Larger shunting locomotives, as used in America, and practically all main line locomotives in the 1500/4000 kW class, are of the bogie type using either two- or three-axle bogies.

30.3 Railway signalling and control

The purpose of a railway signalling system is to control the passage of trains such that they may run at any desired speed in safety. Thus, any system adopted must be able to detect the presence of all trains and vehicles and ensure adequate separation.

The basis of modern signalling is the electrical *track circuit* to detect the presence not only of trains but also of single vehicles. A track circuit equipment must detect the presence of a vehicle presenting a maximum resistance of 0.5 Ω between rails, a figure taking account of contact effects due to rusty rail surfaces and other features. There are several forms of track circuit, and their principles are considered first in relation to the simplest d.c. track circuit.

The fundamental philosophy of all signalling systems is that they should *fail safe*, i.e. fail to a more restrictive condition. Solid state devices are used extensively for supervisory and information purposes, and are being introduced for safety functions such as track circuits and the transmission of vital controls.

30.3.1 D.C. track circuit

The track circuit (*Figure 30.11*) is formed by a section of the track normally isolated electrically from adjacent sections by insulated rail joints. Ordinary joints within the section are bridged by

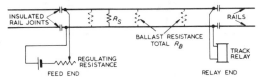

Figure 30.11 D.C. track circuit

metallic bonds. Feed-end and relay-end connections are made to the rails close to the insulated joints. The rail resistance is small enough for the rails to be considered as equipotential busbars between which is a distributed ballast resistance R_B. The ballast resistance is the critical variable in the track circuit performance: it varies from day to day, being low in wet conditions, high in dry and totally frozen conditions. It varies also with the type of track. Modern track is laid on reinforced-concrete sleepers, and it is necessary to ensure that the rail is isolated electrically from the sleeper. The fixings for the rail are metallic and thus the rail has to rest on an insulating pad and the clips holding down the rail are separated from the rail by insulating material. Insulation breakdown can cause a failure, the source of which is difficult to locate. The type of ballast may be significant. Stone ballast (preferably granite) has good resistivity; but in some areas ash ballast is found, and this can make reliable track circuit conditions difficult to achieve. Low-resistance ballast conditions occur where the track is adjacent to the sea or is carried through undersea tunnels. A good track under the worst conditions may have a ballast resistance of 14–20 Ω/1000 m; but on poor track it may be as low as $\frac{1}{2}$ Ω/1000 m.

Track circuit adjustment must take account of variation in the battery voltage. The cell in *Figure 30.11* can be a trickle-charged cell which, if of the lead–acid type, has a voltage range between 1.8 and 2.5 V. If a guaranteed mains supply is available, the battery can be replaced by a transformer and rectifier.

30.3.1.1 Relay

Standard d.c. track relays have a resistance of 4 Ω (though there are many of $2\frac{1}{4}$ or 9 Ω in service). The significant characteristics are: (a) *drop-away voltage*, at which the energised contacts 'break', when, after the relay has been energised, the voltage is reduced; and (b) *pickup voltage*, at which the energised contacts 'make' when, after the relay has been de-energised, the voltage is increased. The drop-away voltage is a minimum of 68% of the pickup voltage. Associated with these voltages is the train-shunt resistance R_S: (1) the *drop shunt* is the maximum resistance that produces track-relay drop-away conditions; and (2) the *pickup shunt* is the minimum resistance such that the relay just picks up.

In practice the pick-up shunt is greater than the drop shunt and as long as this is a practical resistance the track circuit will work. If not, the track circuit will fail to re-energise.

In setting up the track circuit, the regulating resistance is adjusted for maximum battery voltage and maximum ballast resistance such that with a drop shunt of, e.g., 0.5 Ω the track relay is de-energised. Then it is necessary to check that the track relay will re-energise with minimum battery voltage and minimum ballast resistance. If the pickup shunt is relatively high, say 10 Ω, then the conditions only require to become a little more critical for the track circuit to fail to re-energise. In this case, assuming that no fault exists, the selected length of track circuit is too great. On the other hand, if the pickup shunt is relatively low under these conditions, increasing the regulating resistance will economise in power, important where primary cells are used, as it gives them longer operating life.

It is normally possible to work track circuits successfully with ballast resistances down to 2 Ω. The length of track to which this corresponds depends on the ballast conductivity: on good ballast a length of 1–2 km is typical.

30.3.1.2 Fail-safe conditions

To prove the track clear, the track relay must be energised. It will de-energise if (a) the power supply fails; (b) a disconnection occurs in the feed-end or relay-end connections, or in the rails (e.g. a rail removed); (c) a short-circuit occurs on the track. As the de-energised state is taken to mean that a train is present, the fail-safe principle is achieved.

A wrong side-fail condition could occur if the voltage were excessive. The possibility that this condition could occur owing to a voltage from an adjacent track is eliminated where possible by reversing the supply polarity at each pair of insulated rail joints: then, should the joint insulation fail, both track circuits will fail safe. The staggered polarity principle is applied to point work, where the requirements can be complicated.

30.3.1.3 Operation

The d.c. track circuit system is reliable and economic. In certain circumstances relay timings of successive track circuits may become critical. When the supply voltage is high, the track relay may pick up faster than it drops away (which may take as much as 2 s), and in consequence a short train or a single locomotive may be 'lost' for a brief time.

The d.c. track circuit will work satisfactorily in a.c. traction areas. It is necessary to ensure that the track relay is a.c. immune, to guard against false energisation from traction currents. For the same reason it is not possible to use d.c. track circuits in d.c. electrified areas and, hence, a.c. track circuits are used.

30.3.2 A.C. track circuit

The track circuit is fed by a transformer which isolates the power supply from the track. The track relay, a double-element device, has a 'local' winding connection to the same supply as the feed-end equipment (normally at 110 V), and a 'control' winding fed through the track circuit. This makes possible reliable operation with less track power.

30.3.2.1 Track relay

The two-position a.c. relay is an induction device, in principle the same as an induction-disc wattmeter. An aluminium vane lies in the magnetic fields produced by two electromagnets (*Figure 30.12*); one, the 'local' (Q) is energised at voltage V_q from the main transformer, and the other, the 'control' (R) at voltage V_r from the track circuit. The torque is proportional to $V_q V_r \sin \theta$, where θ is

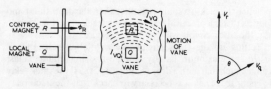

Figure 30.12 Principle of a.c. track relay

the angle between the voltage phasors. If V_q is large, a relatively small V_r can ensure operation of the relay, the angle θ being adjusted by variation of the current-limiting devices.

A three-position relay is needed in point detection circuits, where a point may be in normal, reverse or neither position. One set of relay contacts is closed when V_r leads V_q, and another when it lags. With zero control voltage the relay remains in the central de-energised position.

30.3.2.2 Capacitor feed

In capacitor-fed track circuit a 1:1 isolating transformer feeds the track through a variable capacitor (*Figure 30.13*). The advantage is that there is little loss in the feed and no path for d.c. traction currents.

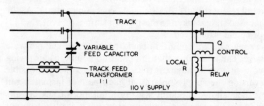

Figure 30.13 Capacitor-fed track circuit

30.3.2.3 Impedance bonds

On electrified lines where both rails are used for traction return currents, impedance bonds are used to provide a low-resistance path for the traction current (d.c.) but a high impedance to track-circuit currents (a.c.). A simple bond is shown in *Figure 30.14(a)*: it consists of a few turns, centre-tapped, on a laminated ferromagnetic core. Balanced traction currents give zero effective m.m.f.; but unbalance up to 20% is mitigated by the provision of an air-gap in the core. The d.c. resistance of an impedance bond may be about 0.4 mΩ, and its impedance of the order of 0.5 Ω at 50 Hz.

30.3.3 Jointless track circuit

In modern rail track designed with continuous welded rail, there are no joints: concrete sleepers and careful installation allow

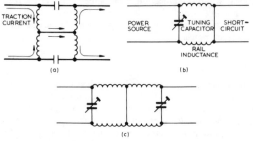

Figure 30.14 Impedance bond and tuned sectionalisation

thermal expansion and contraction stresses to be accommodated. For track circuit sectionalisation without insulated rail joints the simplest method is a short-circuit, but this is impracticable and it is necessary to increase the effective impedance. The short-circuit and rails have inductance that can produce a resonant circuit with a parallel-connected capacitor, as in *Figure 30.14(b)*. With this arrangement no power from the source on the left can reach the receiver on the right. If two adjacent track circuits are required, it is possible to tune from both sides as in (c). The obvious disadvantage is that vehicles near to the short-circuit may not be detected.

One form (*Figure 30.15*) uses a series circuit in resonance at a given frequency. Adjacent track circuits operate at different

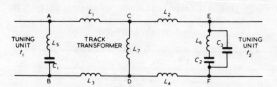

Figure 30.15 Jointless track circuits for adjacent sections

frequencies, and their limits are achieved as follows. At a frequency f_1, components $C_2 L_6$ are in series resonance, short-circuiting C_3 and presenting (ideally) zero impedance across EF, while C_1 is such that the whole circuit is in parallel resonance and of high impedance: thus, the end of track circuit f_1 is defined. At a different frequency f_2 (with $C_2 L_6$ now off-resonance and of high impedance) $C_1 L_5$ are series resonant short-circuiting AB, and C_3 is such that the whole circuit is antiresonant, so defining the limit of track circuit f_2. The rails contribute inductances $L_1 \ldots L_4$ to the network. Transmitter and receiver units are connected to the network through the track transformers.

Operating frequencies are a compromise between capacitor size and maximum track circuit length. With one transmitter and one receiver at opposite ends, track circuits between 50 m and 1000 m in length can be operated at frequencies in the range 1.5–3 kHz. If a track transformer coupled with a transmitter is placed at the centre and a receiver at each end, the length of the track circuit can be doubled.

Another type of jointless track circuit uses frequencies in the range 370–378 Hz: as these are not harmonically related to 50 Hz, they can be applied to a.c. industrial-frequency electrified lines. In this case track circuit energy is detected by a multi-turn coil located between the rails and inductively coupled thereto.

30.3.4 Other forms of track circuit

30.3.4.1 Coded circuits

Coded pulses (on–off or frequency-shift) may be superimposed on track circuits designed for the purpose, and detected by the receivers or by pickup coils at the front of a train. The system can be used for train control.

30.3.4.2 Impulse circuits

Where wheel contact is bad, as on rusted rails or on gradients and curves where locomotives use sanding to improve adhesion, non-detection can be overcome by using high-voltage pulses detectable by the receiver.

30.3.4.3 Audio frequency circuits

In certain areas a track circuit has to work on lines adjacent to both d.c. and 50 Hz a.c. electrified lines. Circuits have been devised using transistor oscillators and receivers tuned to a frequency around 400 Hz (but avoiding mulitples of 50 Hz) by sharply selective electromechanical reed filters.

30.3.4.4 Overlay circuits

In all cases the relay is de-energised in the presence of a train. Sometimes interlocking controls require a train to be detected at a given spot to operate some function. This may be done by providing an 'overlay' circuit which energises when a train is present, additional to the normal track circuit. Frequencies in the range 20–40 kHz can be used, but the attenuation is such that detection is possible only for distances up to about 50 m.

30.3.5 Other means for vehicle detection

Various methods are available to detect the presence of a vehicle at a specific point. Mechanical treadles which are depressed by the wheel rim and close a circuit are common. The most widely used non-mechanical method is the axle counter. The passing of an axle is detected by the change in magnetic flux in the (metallic) wheels, which can be picked up as an impulse and electronically counted. Comparison of the counts at opposite ends of a section gives information as to whether the section is clear or not. The method saves track-circuiting a long stretch of single-track line with sparse traffic.

30.3.6 Colour light signalling

Electric lamps are reliable, and the advantages of colour light signals over the old semaphore system are so great that the system is becoming universal. There is little limitation on the distance from which a colour light signal can be controlled, and it is possible to provide a simple, logical, readily understandable and visible system. In Britain the four basic indications are:

Red: danger.
Yellow: caution (be prepared to stop at the next signal).
Double yellow in vertical axis: preliminary caution (prepare to pass next signal at restricted speed and to find it showing one yellow light).
Green: clear (next signal exhibiting a proceed indication).

A yellow aspect is always required on BR, but is used on the London Transport system only in open sections as repeater signals in fog, and in tunnels.

30.3.7 Terminology

Braking distance The service braking distance is the minimum spacing between the first warning indication that a driver sees and the red aspect to which it refers, such that trains running at their maximum permissible speeds can stop by normal service (not emergency) braking before reaching a signal at danger. The distances adopted by BR for level track are: 2240 yd for 100 mile/h, 1680 for 90, and 1260 yd for 75 mile/h (2000 m for 160 km/h, 1540 for 145, and 1150 m for 120 km/h).

The improved braking performance of the High Speed Train sets, capable of a maximum speed of 125 mile/h (200 km/h), enables them to stop within the same braking distance provided for the locomotive-hauled trains at 100 mile/h.

Sighting distance A driver on approaching a restricted signal takes some action to reduce his train speed at a sighting distance from the restricted signal.

Overlap It has always been British practice to provide 'overlap' against the possible over-run of trains past signals. For level track, overlaps of 275 m (300 yd) are used for two- and three-

aspect systems, and 180 m (200 yd) for four-aspect systems. Recent evaluation of operating performance has resulted in a rationalisation of overlap distances, and for future projects a standard 180 m (200 yd) overlap distance will be provided for all multiple-aspect signals.

Headway This is the separation (in time or distance) between successive trains travelling at the same speed such that the following train can maintain speed.

30.3.8 Two-aspect signalling

The system (*Figure 30.16*) is similar to that given by the lights of a semaphore system at night. Signal B clears from red to green and

Figure 30.16 Two-aspect signalling

its distant or repeater A clears from yellow to green when the rear of the preceding train clears the overlap of signal D, i.e. point X. The two-aspect system is used on lines with low traffic density. The separation of stop signals B and D is governed by the traffic capacity required.

30.3.9 Three-aspect signalling

With greater traffic density the positions of the home signal B and the next distant signal C become close, and the two are combined to give one signal with three indications (stop, caution, proceed: *Figure 30.17*). Each signal is at least full service braking distance

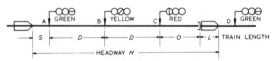

Figure 30.17 Three-aspect signalling

from the next. Signal B clears from red to yellow when the rear of the leading train passes the overlap of signal C. Signal A changes from yellow to green at the same time. The headway can now be defined: if the following train is to maintain its headway it must not be closer to A than its sighting distance. For a train of length L running at speed V, the headway distance H and time T are

$$H = S + 2D + O + L \quad \text{and} \quad T = H/V$$

where S is the sighting distance, D the braking distance and O the overlap. A 3-aspect signalling system is normal on lines with relatively dense traffic with an occasional requirement to run a few trains at 5 min intervals.

30.3.10 Four-aspect signalling

Where there is inadequate braking distance between a yellow and a red aspect, a fourth aspect (double yellow, displayed vertically) is introduced. It is used where a three-aspect system does not give adequate headway in areas of dense traffic; and where it is desirable to relate signal positions to point connections to aid traffic movement with such a requirement dictating signal positions separated less than the braking distance. The principle

Figure 30.18 Four-aspect signalling

is shown in *Figure 30.18*. To get maximum capacity the signal spacing should be regular and one-half of the service braking distance D. Then the headway is

$$H = S + 1\tfrac{1}{2}D + O + L \quad \text{or} \quad T = H/V$$

In practice a four-aspect system gives 25–30% greater capacity than a three-aspect system with similar characteristics. It can give even greater capacity: as $D \propto V^2$ approximately, a train travelling at 0.7 of the maximum line speed requires only one-half of the system braking distance, and provided that the signals are evenly spaced a driver may ignore the double yellow, giving $H = (S + D + O + L)$ and consequently a greater capacity.

If one lamp of a double yellow fails, it leaves the greater restriction of a single yellow, and fails safe.

30.3.11 Signalling for junctions

When there are diverging junctions, it is British practice to tell the driver which route he will take. On lines where speeds exceed 40 miles/h (64 km/h) the routing is given by a junction indicator mounted above the main aspects (*Figure 30.19*). A position 1

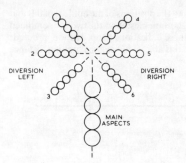

Figure 30.19 Junction indicator

indication tells the driver that he is taking the first road to the left of the main through-route; position 2 is for the second and position 3 for the third road to the left. Positions 4, 5 and 6 serve similarly for roads to the right. The indications are always used with the main signal. The indication is given by a row of five white lights and these are detected before a proceed main aspect is indicated. The detecting relay is set to close limits so that it will be energised for three lamps but will not energise for two lamps. The main (highest speed) route has no indication.

Most diverging junctions have speed restrictions lower than for the main route, and the usual practice is to prevent the main signal and junction indicator from clearing until the train speed has been reduced, by requiring the train to occupy a track circuit approaching the junction signal for a specified time. The timing is calculated to permit the signal to change from red to a proceed aspect at the sighting distance of the junction indicator.

At the entry and exit of large stations and other locations where the line speed is 40 mile/h or less, letter or number indicators can be used to identify the route, with the characters shown by selected lamps in a square array.

At junctions where the approach release from red would be unduly restrictive, preventing the driver from maintaining the full authorised speed over the diverging route, and the permissible speed through the turnout is not less than

(1) 50 mile/h irrespective of the straight route speed.
(2) 40 mile/h where the straight route speed is less than 100 mile/h.
(3) 30 mile/h where the straight route speed is less than 80 mile/h.

the junction signal may display an unrestricted yellow aspect and the signal in rear a flashing yellow aspect. With four-aspect signalling the second signal in rear of the junction signal would display a flashing double yellow aspect for this sequence. The aspect at the junction signal may be released from yellow to a less restrictive aspect in accordance with conditions ahead after the train has passed over the automatic warning system (AWS) inductor.

30.3.12 Colour light signals

The normal pattern is the multi-unit type with a unit for each aspect. Each aspect unit has a double-lens optical system giving an 8° beam angle. The outer lens, about 20 cm diameter, is of clear glass with a downward-deflecting sector for close-up visibility; the 14 cm diameter inner lens is coloured. The units are independent and set about 30 cm apart to avoid false indications, hooded to screen off sunlight, and backed by matt black plates to give a dark background.

30.3.13 Signal lighting circuits

Relays are associated with each signal. For a four-aspect signal there are three controlling relays:

HR: signal control relay, to prove that the conditions are satisfied for the signal to clear from red to any other aspect.
HHR: double-yellow control relay, to prove that it is satisfactory for the signal to clear to double-yellow.
DR: green control relay, to prove that it is satisfactory for the signal to clear to green.

Figure 30.20 gives details of a signal control circuit. Double-filament tripole lamps are used. The main filament (with a guaranteed normal life of 1000 h) is detected by the G(M)ER relays in each case. When a main filament fails, the auxiliary filament is then illuminated; its normal guaranteed life is 250 h but the practice is for maintenance staff to replace lamps within a short while of a main filament failure. It will be seen that an

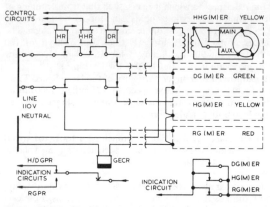

Figure 30.20 Circuits for four-aspect signal

indication circuit is drawn for the red and yellow and green filaments. It is possible to include the double yellow, but in this case it is necessary to prove that filament when the aspect is alight, i.e. when the HHR relay is energised. Thus, any energised contact of the HHG(M)ER relay must be bridged out by a de-energised contact of the HHR relay. This can be expensive in line wires if the controlling relays are an appreciable distance from the signal.

GECR is the signal lamp proving relay which is energised to prove that the signal lamp is alight. The design accommodates a $\pm 10\%$ supply voltage tolerance and the relay must not be operated by the magnetising current of the signal transformers.

30.3.14 Operating features

30.3.14.1 Lamps

Main signal aspect lamps are rated at 12 V with 24/24 W double filaments normally operated at a little less than the rated voltage, which prolongs the filament life without undue sacrifice of light flux. The main filament is placed at the centre of the lantern optics and failure automatically energises the second filament and indicates the changeover.

30.3.14.2 Supply

The guaranteed 110 V a.c. supply is general for comprehensive signalling schemes having a standby power supply and independent of earth. In areas where such provision would be uneconomic, the signals are fed by trickle-charged batteries which take over if the mains supply fails.

The maximum distance of a signal from its controlling relays is dependent on the operating voltage, filament characteristic and cable impedance: it may be 1000–1500 m except in electric traction areas.

The indicator circuit shown in *Figure 30.20* is not usually provided for automatic signals that work with the passage of trains, but is required where a signal is controlled from a signal box and the signalman must make a positive action before the signal will clear.

30.3.15 Signal-aspect control circuits

Typical circuits for the control of a section of four-aspect signalling are shown in *Figure 30.21*. The fundamental control relay is HR: before it can be energised and the signal controls clear from a red aspect, the track ahead must be proved clear. Thus, before, say, signal 1 can clear from red to yellow, it is necessary to prove that no train is between signals 1 and 2 (i.e. track circuits A and B are proved clear) and further that no train is within the overlap of signal 2 (i.e. track circuit C is proved clear).

Apart from auxiliary filament back-up in case of the failure of a main lamp filament, if both filaments fail, then the signal(s) preceding the faulted signal are held at danger.

For signal 1 to clear to double yellow, it is necessary for both relay 1HHR and relay 1HR to be energised; i.e. there must be no train between signals 1 and 2, and 2 must be showing yellow. Thus, 1HHR is fed by an energised contact of 2HR, and the signal lighting circuit ensures that signal 1 can clear by incorporating a contact of 1HR.

Similarly, for signal 1 to clear to green, signal 2 must show double yellow and signal 3 yellow. Thus, 1DR is energised by energised contacts of 2HR (proving that signal 2 can clear from red) and 2HHR (proving that signal 2 can clear to double yellow). It will be seen that relay 2HHR proved relay 3HR to be energised. All these relay controls mean that, if signal 1 clears to green, then the track is clear through to track circuit G and all signals are alight.

In controlled areas involving points, the circuits become much more complicated, and the philosophy of interlocking is involved.

30.3.16 Point operation

In modern power signalling, points are moved by equipment on the ground controlled electrically. There are three basic forms of operation: electric, electropneumatic and hydraulic.

30.3.16.1 Electric

The drive machine is a d.c. split-field motor to reduce the possibility of false operation. The control circuits are preferably four-wire. Control is through a contactor designed to pass a control current up to 10 A at 120 V. There are separate contactor relays for the two operations: when the signalling controls require a point to be moved, then the appropriate contactor is energised first, proving that its complementary contactor is de-energised. The power is then applied to the motor. For passenger train movements it is necessary to ensure both that the point machine is detected in its correct position and that it is mechanically locked in this position. The first operation of the point machine is to unbolt this lock, then to move the point switches in the required direction and rebolt the lock. As this is completed, the point machine windings are cut off by special contacts within the machine and a special electrical snubbing circuit is brought into use by another set of contacts to prevent the motor from overrunning. Also operated by the point machine contacts is a detection circuit which uses a three-position a.c. relay or two neutral polar relays: this circuit detects the point machine normal or reverse setting, including the detection of the mechanical bolt in its locked position. When the detection is in correspondence with the point machine calling, the point contactor is de-energised.

Normal operation takes about 3 s. A clutch is provided to slip if, owing to poor adjustment or an obstruction, the point machine is unable to make its full travel; and to protect the motor, a relay can be used to disconnect the machine on overload sustained for about 10 s. Alternatively, a timing relay operated by the contactor can de-energise the contactor coil after 7 s. Both systems are reset automatically.

30.3.16.2 Electropneumatic

Operation is by compressed air at 300–400 kN/m^2 (40–60 lbf/in^2), but control and detection are electrical and are similar to those for an electrical point machine. The controlling circuit energises a valve, one for normal and one for reverse, to admit compressed

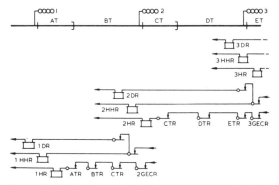

Figure 30.21 Control circuits for four-aspect automatic signals

air to move a piston in a cylinder. This movement is linked to the point mechanism. A mechanical lock is provided: it must be released before the point moves and replaced at the completion of the stroke. This form of point operation requires a supply of compressed air, but it is quick in operation and is applied in marshalling yards, where point operation between wagon cuts must be rapid.

30.3.16.3 Hydraulic clamp lock

The purpose of point (switch) mechanisms is to ensure that the switch mates with the stock rail without gap. A positive and definite method is to clamp the two together by an electrically controlled hydraulically operated unit. Each unit has its own electrohydraulic power pack, and control from the hydraulic unit is by means of valves which actuate two pistons, one for movement to the normal and the second for movement to the reverse position. When the movement is completed, the control circuit is de-energised. The locking of the points is achieved by the clamping mechanism, which will only make in its correct position if the switch is properly against its stock rail. Thus, in operation the clamp is unlocked, and the switches are moved and then clamped. The units include detection contacts which operate a detection circuit similar to that for a point machine.

30.3.17 Modern signal box

With track circuits for train detection it is possible to provide signal boxes which cover large areas of railway. The largest boxes control over 100 route miles of signalling. The advantage is that one man can see at a glance the position of all trains under his control, together with those that shortly will enter his control area. As a result, delays due to late running of trains and indefinite information can be minimised. These signal boxes (which are really control centres) are normally situated at the large railway centres and have been made possible by development of remote control systems which enable outlying equipment centres to be connected to the central signal box at relatively low cost.

At the control centre the track layout is presented on a panel showing the position of point switches, track circuits, controlled signals and other signalling features. Train description equipment is incorporated to identify the many trains in the area. The signalman can control all signals and points. Normally signals are set up by operating a 'route', which normally extends from signal to signal and is usually operated by a push button or switch at the beginning of the route and a push button at the end. This is referred to as the Entrance–Exit (or NX) system.

30.3.18 Interlocking equipment

It is necessary to ensure, before a signal clears, that the route has its points in the correct position and that there is no possibility of a train being signalled in an opposing direction. Equally, once a train has been signalled, there must be no possibility of a point in that signalled route being moved or another movement being signalled against the first-signalled move. In electric signalling the necessary interlocking is obtained by electromechanical relays made to stringent specifications. Some relays control signals and point machines; others detect signals, points and track circuits. These and other relays are electrically so interlocked as to make an unsafe movement impossible.

A signalled route generally extends from one signal to the next. In it there will probably be a number of track circuits, and a point in one of the middle track circuits. The control of that point is fundamental to the safe passage of trains: (1) the point will be locked in position by occupation of the track circuit that includes it; (2) when the signal has been operated and cleared, it must be locked; (3) if after (2) the signal has to be restored to danger (e.g. in emergency), then the point must be kept locked for a time sufficient for an approaching train to stop: the time normally allowed is 2 min minimum and 4 min maximum, depending on the spacing of the signals to the rear; (4) when a train has passed a signal which had cleared, on occupation of the first track circuit and subsequently others, the signal would revert to danger and it would be necessary to ensure that the points are still locked for these movements. But there might be movements in opposite or other directions through these track circuits, in which case point locking may not be necessary. Thus, locking by these track circuits is conditional: it is called route locking and is often designed to incorporate the other features listed above; and it is essentially a part of any electrical interlocking system. Another important feature is that of sectional release, whereby, as a train clears each track circuit and the sections ahead are still locked, the locking of points in those track circuits that have been occupied and cleared can be released.

It is normal to find the interlocking equipment for a number of points and signals grouped in one relay room (called an interlocking) so that they cover a convenient geographical area. The interlocking may be some distance from the control centre. The circuits linking the interlocking and the functions must have high standards of safety. The circuits connecting the interlocking with the control centre need not be of such quality, as they convey controls or commands from the control centre and return indications; contact between two circuits would give a wrong command (which could be embarrassing operationally), but the wrong instruction would be set up only if the safety equipment allowed this to happen. Thus, with large signal boxes it is normal to find a number of interlockings along the line, each connected to the control centre by non-safety remote control equipment to reduce the cost of connections.

All the equipment associated with the interlocking is manufactured to certain standards. For example, the standards adopted in the manufacture of signalling relays are much greater than those adopted for telephone-type relays. Cable and other terminations are specified and so are types of wire and cable. The philosophy of fail-safe is applied so that the possibility of wrong side-failures due to contact between two adjacent circuits is reduced. The additional cost involved in eliminating this possibility altogether is so great that it is not justified.

30.3.19 Remote control

The three systems of remote control currently in use are (1) direct wire; (2) solid state time-division multiplex; and (3) solid state frequency-division multiplex. Multiplex systems share a common circuit, and either time elements are allotted to various channels or the available frequency spectrum of the line circuit is divided into discrete bands with one band allotted to each channel.

30.3.19.1 Direct wire systems

Direct wire systems comprise separate line circuits in a multicore cable with telephone-type relays energised by d.c. A separate circuit is used for each control and indication. The cable is either a paper cored telephone cable or a special microcore cable. As many circuits as possible are contained in one cable, giving minimum first cost and ease of laying.

30.3.19.2 Time-division multiplex

Mass production has led to the development of solid state t.d.m. equipment with the advantage of high coding rates. These systems use audio frequency energy for communication over line circuits. This system results in a fairly quick response. The equipment scans each piece of information in turn, and scanning times of 1 s are achieved. The information to be sent is divided into groups and possibly further subdivided. As each piece of information can only be in one of two states, yes or no, parity checks are sent at the end of each group. Say a group consists of 12 bits: 11 pieces of information could be sent with parity as bit 12. Normally parity adds all the yes bits and ensures that there is an odd number sent in the group, so that if the number of yes bits from 1 to 11 was even, parity would be a yes bit; but if there was an odd number of yes bits, parity would be a no bit. It is thus possible to check small groups of the information sent. If the system shows a failure indicated by parity, or any other such monitoring equipment on a number of successive occasions, then a fault is alarmed. The alarming of a single fault is undesirable as it may have been caused by line interference.

Common systems can be built between the control centre and a number of satellite interlockings.

30.3.19.3 Frequency-division multiplex

The f.d.m. system is made possible by the use of highly selective bandpass filters, which are required to give unusable output signals under fault conditions. Systems acceptable for safety circuits employ electromechanical tuning units that are very highly selective. The frequencies are in the range 400–800 Hz with narrow band separation, as the filters give negligible output if the channel frequency shifts by 0.1 Hz. The operating frequencies must be clear of 50 Hz power harmonics.

30.3.19.4 Choice of system

Cost is the ultimate consideration, in the context of which there are four factors to be considered.

(1) *Reliability* The simpler the system the more reliable it is likely to be: the order is direct wire, f.d.m. and t.d.m.
(2) *Response* Direct wire systems are slightly quicker, but a response time of 1 s is acceptable.
(3) *Flexibility* A direct wire system is flexible and can be worked at any desired power level. An f.d.m. system can be used in a similar manner, but t.d.m. systems require complex equipment and it is desirable to limit the number of stations.
(4) *Security* This is the function of the interlocking equipment. On these grounds there is no difference between the systems. However f.d.m. has been used for security information, which may in some cases be significant.

Direct wire systems are cheap for interlocking near the control centre. At greater distances a saving in cabling cost offsets the cost of multiplex. Transfer of small amounts of information between many points (e.g. lineside to interlocking) is cheaper by f.d.m., while t.d.m. is better suited to the transfer of large quantities of data between a few points (e.g. an interlocking and a control centre).

30.3.20 Train description

A large signal box can have many trains under its control, and identification is essential. The occupation of each track circuit is indicated on the control panel, but this does not identify the train. Each train is allocated a four-unit alphanumeric display (e.g. 1A64; 1 for express passenger, A for destination area, and 64 for a particular train or service number). Such descriptions can be displayed on the control panel at each signal, being progressed automatically with the movement of the actual train. Modern train describers use data processors for this purpose. The equipment takes care of routing at divergences, and it is possible to cancel or interpose a description at any display, a feature essential at places such as termini and freight yards where traffic originates.

30.3.21 Automatic warning system

The standard British Rail system is intermittent, in that the driver receives an audible indication at a point on the approach side of a fitted signal. Signals are fitted only if they convey to the driver information of the aspect displayed at the next signal ahead. In semaphore signal territory only those signals which carry a distant arm are fitted. In colour light signal territory any signal capable of displaying a yellow or double-yellow warning aspect is fitted, also every red/green two-aspect signal interposed between or immediately following a sequence of two or more multiple-aspect signals.

Two different audible sounds can be given to the driver. If the signal ahead is in the full clear or green position, an electric bell rings for approximately 2 s; but if the signal displays any other aspect, an air operated (pressure or vacuum) horn sounds continuously until the driver makes an acknowledgement. If the driver fails to acknowledge, a full brake application is automatically initiated. When the driver makes an acknowledgement, a visual indicator operates to remind him of the fact that he has received and acknowledged this warning, and the indication remains displayed until it is cancelled on approaching the next fitted signal.

30.3.21.1 Track equipment

The link between the track and locomotive or vehicle is magnetic, the track being fitted with two inductors, the first one to be passed over being a permanent magnet and the second an electromagnet. The magnet centres of these two inductors are 76 cm apart. The magnets have a vertical axis, the permanent magnet having its S-pole uppermost and the electromagnet, when energised, having its N-pole uppermost. The electromagnet is only energised when the signal ahead is clear (green). The inductors are usually fitted about 186 m on the approach side of the signal with their upper surface at rail level. They are sufficiently powerful to operate the equipment on the locomotive or vehicle through a 16 cm air-gap at speeds up to 150 miles/h. The permanent inductor has a magnet of Alcomax II, fitted with mild steel spreader plates, top and bottom, to give a projected flux curve with a peak of around mT and of sufficient intensity to operate the vehicle equipment over a horizontal distance of 25 cm measured at a height of 16 cm above the top of the inductor casing. The electro-inductor coil has two windings, and the core is similarly fitted with spreader plates to give a flux curve like that of the permanent inductor. The two coils can be connected in series or parallel for 24 or 12 V d.c., the nominal coil consumption being 9 W.

The electro-inductor is energised through a relay which, in the case of a wire-worked semaphore signal, is controlled (a) by an electrical contact made only when the distant signal lever in the signal box is pulled, and (b) also through a contact on the signal arm, made when the arm is pulled off to an angle of at least 25°. In the case of a colour light signal, the relay controlling the signal to green also energises the electro-inductor with the addition of a contact on a relay made only when the green lamp is actually alright.

30.3.21.2 Rolling-stock equipment

Rolling-stock equipment has various forms for diesel and electric locomotives hauling vacuum-braked trains, diesel multiple-unit trains with vacuum brakes, and diesel or electric multiple-unit trains with air brakes.

On the underside of the locomotive or vehicle is a receiver responding to the magnetic flux from the track inductors. It consists of a permanent-magnet armature, centrally pivoted and lying between soft-iron poles. The armature carries a flexible beryllium–copper contact strip to act as a single-pole changeover switch when the armature moves under the influence of the S- and N-poles of the inductors. Its own magnetism holds it in one or other of the two positions against the vibration of the vehicle. A coil wound round one of the poles is energised by the driver when acknowledging a warning indication, the magnetic field causing the armature to return to the normal position from which it had been moved by the flux from the permanent inductor. If the signal had been clear, the receiver armature after being thrown by the S-pole of the permanent inductor would be automatically restored by the N-pole of the now energised electro-inductor, no acknowledgement by the driver being required.

Power for the automatic warning system (a.w.s.) apparatus is provided from the vehicle battery through a static voltage converter, which serves to isolate the a.w.s. equipment electrically from the driving control and lighting circuits and to provide the correct voltages (12 V for the main circuits and 40 V for the acknowledging or resetting circuits). Normal power consumption is about 4 W with a peak of 12 W for a few seconds on receiving a signal from the track.

The caution indication is a vacuum-operated horn controlled by an electropneumatic (e.p.) valve normally energised. A second e.p. valve, similarly energised and mounted in the same unit, is used to control the a.w.s. brake valve. On passing over a permanent-way inductor with the electro-inductor de-energised, the breaking of the normal receiver contact de-energises the e.p. relay, which, in turn, energises both e.p. valves. The horn immediately sounds and the brake valve admits air to the train pipe, causing a progressive lowering of its vacuum. Within about 3 s the vacuum will have lowered sufficiently for a brake application to be initiated; this progresses to full braking. At the same time a relay passes current to the visual indicator to cancel an indication, if displayed, and change it to a completely black face.

The driver can acknowledge and reset the equipment by pressing a reset plunger.

30.3.21.3 Mechanical safeguards

On double-ended locomotives and multiple-unit trains, means have to be provided to isolate the a.w.s. equipment in all driving cabs except that from which the train is being driven. For vacuum braked stock, a combined electrical and vacuum switch is provided with a loose handle which is carried by the driver. When the driver enters a cab, he inserts the handle into the switch and turns it to close the necessary electrical contacts, which puts supply to the a.w.s. apparatus and opens the a.w.s. brake valve to the train pipe. The handle cannot be removed without first switching off to isolate. To ensure that the driver inserts the loose handle and switches in the a.w.s., additional contacts are provided through which the supply to the main driving controls is taken. The train cannot be moved unless these contacts are made. This switch also carries a second fixed handle, which is sealed in the running position. In case of a failure of the a.w.s. apparatus which would apply a permanent brake, this seal can be broken and the handle moved to isolate the a.w.s. equipment and yet allow the train to be moved.

On air braked trains the same principles apply, but the independent a.w.s. loose handle can be dispensed with and the switching effected by operation of the driver's brake control key.

30.3.22 Automatic train control systems

Problems of high-speed trains include extended braking distances, the presence on the line of trains with a range of lower speeds, and the need of the driver for vital information.

The concept is that a cab signal should not necessarily repeat a ground signal but should give a driver an indication of the permissible speed. Into such a system it is possible to build-in lineside speed restrictions and stop signals. On approaching a point where a driver has to reduce speed, the more sophisticated systems can tell him the maximum permissible speeds at each point as he approaches the restriction or signal. An override device, as in the automatic warning system, can be incorporated to make emergency brake application should a driver exceed the speed permitted and ignore a warning signal.

There are a number of ways in which this type of system can be realised. Some are intermittent, others are continuous or combine both features. Cables can be laid in agreed ways between the rails, and carrier frequencies picked up by sensors on the locomotive. This system can pass a considerable quantity of information between control and locomotive, or vice versa. Another system employs lineside transmitters at specific points. Other systems use sensors to pick up information from coded track circuits or frequency modulated jointless track circuits.

A principal feature of the BR Advanced Passenger Train (APT) is the ability to negotiate curves at substantially higher speeds than are permissible on conventional trains (as indicated by the lineside speed restriction signs which form part of the driver's knowledge of the route). To avoid proliferate lineside indications, the permissible APT speeds throughout the route are given to the driver in the form of a continuous cab display. Advice of a change to a lower speed is given at braking distance plus reaction time from the restriction by enforcement of a cancellable warning, which, if not acknowledged by the driver, automatically applies the brakes. Advice is also given of the commencement of the restriction and of a change to a higher speed. The route information is picked up by the train from passive transponders situated on the track at intervals of 1 km. If the system fails, the driver receives an audible cancellable warning, the cab display is blanked out and the driver reduces speed to that applicable to conventional trains on the route. Thus, while the system is operative, it acts as the driver's authority to drive at APT speeds, whereas when the system is inoperative, the driver must revert to conventional line speeds.

30.3.22.1 Victoria Line (London Transport)

The Victoria Line has coded track circuits and intermittent spot information, a system suitable where all trains are similar. Train movement is controlled automatically. Each train carries an operator, but his function is to control doors and to give the control system the start signal. The normal minimum operating headway is 82 s, based on 30 s station stops. The peak permitted speed is 58 mile/h, but the automatic equipment is set to 48 mile/h. Signalling is closely related to the driving equipment and is dependent on information from lineside units through the track. The two kinds of information passed are coded track circuits and a train command system.

30.3.22.2 Coded track circuits

Coded track circuits are pulsed at 180, 270 and 420 pulses/min. One must be detected for a train to move automatically. Detection of the 420 code allows a train to move at maximum speed; the 270 code restricts the speed to 22 mile/h, motors on, and 180 to the same speed, motors off. Failure to detect code results in emergency brake application.

30.3.22.3 Train command system

It is necessary to have a second system whereby trains are controlled for such features as coasting after acceleration from a station stop, for trains to stop at a danger signal or at a station, and for trains to slow for speed restrictions. This is achieved by a system of command frequency spots.

Coasting and signal stops The outputs of the command frequency spots are fed to the rail at one point and detected in the vicinity by the train equipments. A frequency of 20 kHz is the command to stop for a signal at danger; 15 kHz instructs the motor power to be cut and the train to coast.

Station stops Commands are sited for ideal braking. The commanded speed is 5 mile/h lower for each successive command frequency spot. A train passes the 35 mile/h spot just before entering the platform. The frequencies used are on the basis of 0.1 kHz for each 1 mile/h.

Train-reaction braking The degree of brake application depends on the train speed at the instant that it passes over the command spot compared with the command speed there. If the speed is much lower, no brake application is made.

Emergency operation Should no code be received by the train, the operator can drive it at a speed up to 10 mile/h. If a code is received the operator can drive the train up to 22 mile/h. At each speed an audible signal is given to the driver as he approaches the particular maximum, and an automatic brake application is made if the maximum is exceeded.

Stopping accuracy The automatic braking gives an accuracy within ± 1.5 m.

30.3.23 Signal power supply

As industrial power supply reliability cannot be guaranteed, all signal systems have standby arrangements. For isolated signals and light loads, the standby is in the form of trickle-charged batteries. Where a.c. is essential, inverters have been used. The normal provision for major power signalling systems is a standby diesel-engine/generator set. It is accepted that the set may take up to 7 s to be ready for load. Transfer back to normal supply is done manually. An acceptable main power variation is $\pm 10\%$ in voltage and ± 2 Hz in frequency. The standby equipment is designed to detect wider variations and take over the load in such a case.

A normal power distribution system is based on 650 V, a.c. In densely signalled areas ring mains are used, and the rest on spur mains. Power supplies are sited at suitable points according to loads but on average power supplies are required, with associated standby equipment, every 16 km.

The basic signalling equipment operates at voltages of 110 V a.c. for signals and track circuit equipment, 50 V, 24 V and 12 V d.c. for relays, and 120 V d.c. for point machines. Transformers are provided at suitable intervals to give 110 V from the 650 V main, and the other voltages used are obtained from the 110 V a.c.. Batteries are normally used only on telecommunications equipment and for point machines, the latter because of the costs involved. In certain track layouts it is possible to operate 12 point machines at once and this would require a large transformer rectifier. As it is only an intermittent load a small transformer rectifier and heavy-duty battery are used to overcome this heavy peaking.

All signalling circuits other than track circuits are insulated from earth.

30.3.24 Immunising of signalling against traction currents

A.C. traction systems are unbalanced, and as much as 50% of the return current may reach the substations through the earth, and signalling circuits are subject to induction. So that induced voltages will not exceed the CCITT limits of 60 V normal and 430 V under fault conditions, circuits are restricted to a length of about 2 km, and in the presence of a.c. traction use a.c. immune d.c. relays.

Colour light signals can be worked from relays at distances up to about 1400 m, dependent on the cable size, but this must be restricted to 300 m in electrified areas. Point machines are d.c. operated with a.c. immune motors and contactors, the latter being located near the machines. For track circuits, a.c. feed is used on d.c. electrified lines and d.c. on a.c. lines. A.C. track circuits operating $83\frac{1}{3}$ Hz or at frequencies not a multiple of 50 are acceptable for both forms of traction electrification.

30.3.25 Level crossings

The railway has an obligation to protect road users at level crossings on public highways. Traditionally this was done with gates, but currently there are three main ways of achieving modernised protection.

(1) *Manually controlled barrier crossings* with barriers which close off the whole of the road approaches, the operation of the barriers normally being associated with amber/twin-red aspect road traffic signals and audible warning devices. Many crossings of this type are remotely operated and supervised with the aid of closed-circuit television.

(2) *Automatic half-barriers.* The crossings work automatically with the passage of trains. (Half-barriers are used so that road vehicles cannot be trapped on the crossing.) The controls worked out for the fastest train are: 27 s before the arrival of a train at the crossing the amber aspect of the road traffic signal is shown and an audible warning device is switched on; 3 s later the signal changes to two red lights flashing alternately; 4–8 s from the start of red signals the barriers fall to the horizontal, taking 6–8 s to do so. When horizontal, the audible warning device is switched off. The barriers will be down for not less than 8 s before the train reaches the crossing. The barriers will rise after the train clears the crossing, unless another train is within 37 s of the crossing.

(3) *Automatic open crossings.* As above, except that, being installed on lesser trafficked roads, the system is without barriers. The audible warning device is operative from the onset of the amber aspect until the train clears the crossing.

31 Transportation: Ships

A C Bailey, BSc, FIEE
Senior Principal Surveyor for Electrical and Control Engineering

Contents

31.1 Introduction 31/3

31.2 Regulations 31/3

31.3 Conditions of service 31/3

31.4 D.C. installations 31/3

31.5 A.C. installations 31/3

31.6 Earthing 31/4

31.7 Machines and transformers 31/4
    31.7.1 A.C. generators 31/4
    31.7.2 Voltage build-up 31/5
    31.7.3 Reverse-power protection 31/5
    31.7.4 Single-phasing 31/5

31.8 Switchgear 31/5
    31.8.1 D.C. switchgear 31/5
    31.8.2 A.C. switchgear 31/5

31.9 Cables 31/6

31.10 Emergency power 31/6

31.11 Steering gear 31/6

31.12 Refrigerated cargo spaces 31/7

31.13 Lighting 31/7
    31.13.1 General 31/7
    31.13.2 Navigation lights 31/7

31.14 Heating 31/7

31.15 Watertight doors 31/8

31.16 Ventilating fans 31/8

31.17 Radio interference and electromagnetic compatibility 31/8

31.18 Deck auxiliaries 31/8
    31.18.1 Variable speed 31/8
    31.18.2 Deck auxiliary services 31/8

31.19 Remote and automatic control systems 31/9
    31.19.1 Operational modes of machinery spaces 31/9
    31.19.2 Alarms and safeguards 31/9
    31.19.3 Reliability 31/10

31.20 Tankers 31/11

31.21 Steam plant 31/11

31.22 Generators 31/11

31.23 Diesel engines 31/11

31.24 Electric propulsion 31/11
    31.24.1 Propeller characteristics 31/11
    31.24.2 Turbo-electric systems 31/12
    31.24.3 Diesel-electric systems 31/12
    31.24.4 SCR power system 31/12
    31.24.5 Electromagnetic slip couplings 31/13
    31.24.6 Electromagnetic gearing 31/13

31.1 Introduction

Prior to 1950 electrical installations in ships (other than tankers) were predominantly d.c. This predominance has been reversed because of the higher power requirements of modern ships, for which a.c. systems have lower capital and maintenance costs. The main problems of this change-over have concerned the requirement, based on tradition, for variable-speed deck auxiliaries (winches, capstans and windlasses), pumps and fans. For the latter, makers have accepted single-speed or change-pole 2- or 3-speed induction motor drives, with throttle control additionally in appropriate cases. For deck auxiliaries it was similarly found that for certain trades the change-pole induction machine was adequate, and that Ward Leonard or other sophisticated systems could meet more demanding duty.

The development of North Sea exploration since 1970 led to the introduction of Mobile Offshore Installations generating large electrical power: their systems are invariably a.c., with d.c. conversion for operation of drilling machinery.

The concept of centralised control stations has been fully developed, with control gear, alarms and instrumentation grouped in enclosed, air conditioned and soundproofed rooms. With advances in the automatic control of steam raising plant, including the re-emergence of coal burning, engine rooms unmanned at night and with reduced manning by day have become the norm.

31.2 Regulations

With few exceptions, every seagoing ship must comply with national, international and Classification Society Rules and Regulations. The leading Classification Societies (to which the administration of international requirements is sometimes delegated by the government concerned) include Lloyds Register of Shipping, Bureau Veritas, American Bureau of Shipping, Germanischer Lloyd, Nippon Kaiji Kyokai, Norske Veritas and Registro Italiano Navale. Others are emerging and Unified Requirements are promulgated by the International Association of Classification Societies (IACS).

Ships are to comply with the 1974 Convention for the Safety of Life at Sea (SOLAS) and the 1973 Marine Pollution Convention (MARPOL) together with their respective 1978 Protocols. Passenger ships (carrying more than 12 passengers on international voyages) must have a valid Passenger Safety Certificate and cargo ships a valid Safety Construction Certificate. Validity for Radiotelegraphy or Radiotelephony is also required, as is compliance with specific codes such as those for Gas Carriers and Bulk Chemical Carriers. Conventions and Codes are reviewed by the International Maritime Organisation (IMO), formerly known as IMCO. Revisions have to be accepted by a stipulated number of leading maritime nations.

Electrical construction and performance standards are set by Classification Societies, SOLAS, Government Regulations and (for the UK) by BSI and IEE Regulations for the Electrical and Electronic Equipment of Ships IEE Recommendations for the Electrical and Electronic Equipment of Mobile and Fixed Offshore Installations were published in 1983. The IMO Code For The Construction and Equipment Of Mobile Offshore Drilling Units (MODU Code) was introduced in 1979. The IEE publications include electromagnetic compatibility (EMC). BSI specifications deal with motors, generators, transformers and cables. IEC Publication 92 has the forge of internationally agreed marine standards.

All apparatus should be suitably enclosed, as indicated by IEC 92–201 (e.g. on open decks, IP 56; in galleys, IP 44).

31.3 Conditions of service

While standard industrial designs can in some cases meet the arduous conditions pertaining to ships, each should be specially considered. Ambient temperatures must be taken into account when considering permissible temperature rises and performance; variation in system voltage and frequency is liable to be greater than in shore installations; generators, for example, must deliver rated voltage under tropical conditions, and semiconductor diodes must function over a wide range of temperature; saline atmospheres require special consideration and rust prevention is essential. Especially vulnerable are fine-wire steel springs.

IEC Publication 92–101 specifies for ships on unrestricted service an ambient air temperature of 45 °C for all equipment other than rotating machines in machinery spaces, in galleys and on weather decks. For rotating machines in machinery spaces it is 50 °C. In all other spaces and for vessels on restricted service (i.e. coasters, tugs and harbour craft operating solely in temperate climate) 40 °C is recognised. For electronic devices, semiconductor diodes, etc., much higher ambient temperature conditions may have to be withstood.

Other onerous conditions are vibration, and inclination up to 15° transversely and with rolling up to $22\frac{1}{2}°$. Voltage variation may be $+6$ to -10%, with simultaneous frequency variation of $\pm 2.5\%$. With a.c. generation a momentary voltage dip of up to 15% at the generator is permissible when large motors or groups of motors are switched direct-on-line.

From the point of view of electrical equipment, very severe conditions prevail while a ship is under construction. Welding and painting will be in progress in the vicinity, accompanied by dirt and exposure to the weather.

Skilled maintenance and repairs can be carried out only in ports where suitable facilities exist. This applies particularly to machine windings. Installations must therefore be of a high standard of reliability and suitable for operation over prolonged periods with a minimum of attention. Unlike industrial conditions, in which there are shut-down or reduced load periods, apparatus on essential ship's services may operate continuously for several days.

31.4 D.C. installations

Standard practice is to use parallel-operated level-compounded generators with equaliser connections and reverse-current protection. For small installations 110 V may be used, but 220 V is the general norm. Installations of up to 3000 kW with individual machine ratings up to 1000 kW were formerly common, but the present preference for a.c. means that d.c. systems do not now exceed 1000 kW total with smaller generating units.

31.5 A.C. installations

Tankers and passenger ships of recent construction are almost all equipped with a.c. systems—about 40% with a frequency of 50 Hz; the remainder of 60 Hz. As most marine generators and motors are special to this service, the choice of frequency does not have to be related to particular national supply systems. The frequency of 60 Hz gives the advantage of higher operating speed and lower weight. Motors built for 440 V 60 Hz can operate from 380 V 50 Hz shore supplies and (if some additional heat can be tolerated) on 415 V 50 Hz. However, contactors and voltage operated relays may not always be amenable to such conditions; and 50 Hz motors may not operate satisfactorily on 60 Hz, particularly with centrifugal fan and pump loads.

In some tankers and passenger ships voltages of 3.3 and 6.6 kV have been adopted for primary generation and for some of the larger motors.

31.6 Earthing

Regulations permit isolated or earthed neutral systems, except for tankers, in which earthed systems are forbidden. The choice (where there is one) is almost always for isolation, the risk of overvoltage being accepted to avoid the loss of a vital service, e.g. steering, should one earth fault occur. A single earth fault on an insulated system can be detected, and does not result in an outage unless a second fault occurs; and in any case overvoltages are rare compared with the incidence of earth faults. However, care must be taken when designing the installation to protect electronic circuits, particularly those containing semiconductor diodes, from overvoltage 'spikes', direct or induced.

31.7 Machines and transformers

The construction and performance of rotating machinery are dealt with in BS 2949, and of power transformers in BS 3399,; they are also covered by IEC Publication 92–303. Auto-transformers are permitted only for motor starters, and all power transformers are required to be double-wound.

Because relatively large motors and groups of motors are started direct-on-line, the consequent voltage dip is important because of its effect on the system as a whole and, in particular, on lights, contactors and voltage operated relays. Stipulations are:

(1) A limit of 15% voltage dip at the generator terminals and a recovery to within 3% of rated voltage within 1.5 s, when the generator is subjected to a suddenly applied symmetrical load of 60% rated current at a power factor between zero and 0.4 lagging; the recovery may be increased to 4% within 5 s for emergency generators.

(2) A limit of 20% excess voltage following initial recovery.

(3) Under steady state conditions the system, which may include an automatic voltage regulator, has to be maintained within $2\frac{1}{2}\%$ rated busbar voltage ($3\frac{1}{2}\%$ for emergency sets). The conditions are shown in *Figure 31.1*.

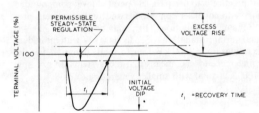

Figure 31.1 Typical voltage response

These requirements are for normal installations. Special conditions apply if the impact load exceeds 60%, or if the deck machinery consists of groups of multi-speed cargo winches liable to be switched simultaneously, or in any other special conditions.

It is common practice in modern ships to use self-excited compounded a.c. generators, or brushless machines with a.c. exciters and shaft mounted rectifiers (*Figure 31.2*). In order not to nullify short-circuit protective gear, such generators must maintain adequate voltage under s.c. fault conditions. BS 2949 specifies a current of at least three times rated value for 2 s unless provision is made for a shorter duration without impairing safety.

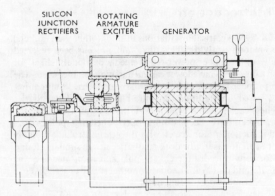

Figure 31.2 Cross-section of brushless a.c. generator

31.7.1 A.C. generators

Self-excited self-regulating a.c. generators are now available in all sizes for marine service, with excitation obtained from a 3-phase exciter through semiconductor rectifiers. The principles are the same whether a slip-ring or a brushless form is applied. Voltage and current transformers connected to the generator output feed field excitation through a 3-phase rectifier to give a compounding effect. A typical arrangement (*Figure 31.3*) combines compounding with closed-loop control: the effect on the excitation due to

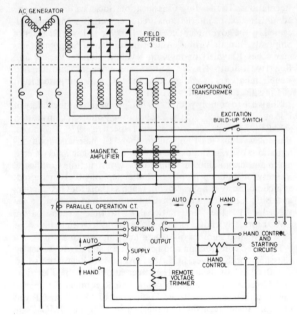

Figure 31.3 Schematic diagram of static-excited a.c. generator

the compounding is supplemented by a fine control, the response of which is determined by the divergence of the generator voltage from a pre-set reference. A transient dip of generator voltage during the first period following a disturbance is inevitable, but no further change due to armature reaction occurs, because the excitation is rapidly corrected by field forcing through the action of the series windings of transformer 2.

Under short-circuit conditions saturation of the static excitation components limits the excitation to about 1.5 times full-load value, and the sustained s.c. current is about 3 times rated value. Variations in field resistance are swamped by the series impedance of the magnetic amplifier. The circuit self-compensates for speed changes, as the amplifier current rises as the frequency falls. The automatic control circuit incorporates a Zener diode which controls a silicon transistor driving a thyristor. The 3-phase rectifier is connected to the generator main field through slip-rings.

An example of a brushless system is given in *Figure 31.4*. Excitation is provided partly by saturable current transformers

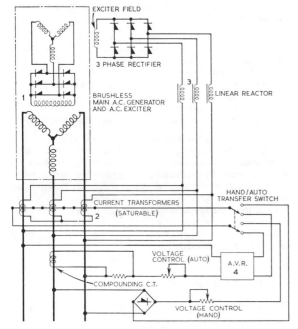

Figure 31.4 Schematic diagram of brushless a.c. generator

and partly by a 3-phase linear inductor, all of small rating and easily accommodated on the switchboard. There is sufficient magnetic remanence in the a.c. exciter to ensure starting. By means of the a.v.r. a steady voltage regulation of $\pm 1\%$ is obtained with high-speed response, typically within 0.5 s.

31.7.2 Voltage build-up

In *Figure 31.3* the no-load terminal voltage is applied to magnetic amplifier 4; its current lags this voltage by nearly a quarter-period, passes through the compounding current transformer 2 and the field rectifier, and supplies the no-load excitation. When the generator is on load, the load current passes through the series coils of current transformer 2; the secondary current is then the phasor sum of the inductor current and a current proportional to the load current. By correct proportioning the static excitation is appropriate for all normal loads, even at low lagging power factors.

With brushless generators two rectifiers in parallel are provided in each phase of the rotating element to provide back-up should one fail. The commonest diode failure is a short-circuit, so that each diode must be fused. A failed diode produces an unbalance and a ripple in the exciter field current. This can be used to detect failure.

31.7.3 Reverse-power protection

Loss of power from the prime mover may be accidental or intentional. If the generator were left connected in parallel with other generators, it would act as a motor and continue rotating, with possible damage to the prime mover. With d.c. generators this is taken care of with a reverse-current tripping relay, but with a.c. a reverse-power relay is necessary. To prevent inadvertent operation of reverse-power relays due to power surges, particularly when synchronising, a time delay feature is incorporated. With diesel engines a fairly coarse setting of the order of 10–15% of full power is suitable, but steam turbines absorb very little power when motored and a fine setting of $2\frac{1}{2}$–3% is necessary.

The alternative to reverse-power relay protection is to provide electrical interlocks or contacts which will respond to predetermined occurrences, such as failure of lubrication, closing of fuel or steam admission valve, operation of overspeed governor or excessive back pressure.

31.7.4 Single-phasing

A common cause of motor burn-out is single-phasing, which can arise from broken or faulty connections or, more commonly, the blowing of one of the 3-phase fuses. For this reason it is strongly recommended that matching filled cartridge fuses be used in preference to rewireable fuses.

Undervoltage releases do not provide protection. Should the open circuit occur on the motor side of the circuit-breaker, the coil will continue to be fed from the supply, and if the open circuit is on the supply side, the coil will be fed by voltage induced in the motor.

A 3-phase motor will not restart with an open-phase connection: out of sight and sound it may remain stalled, and if the overload protection is set too high or with too long a time delay, the motor may burn out. Pilot lights across one phase will give a false indication if the fault is on another phase. The remedy is to set overload protective devices closely, with suitable time lags, or to adopt some form of single-phase protection.

31.8 Switchgear

31.8.1 D.C. switchgear

Open switchboards have all essential switchgear exposed on the front. Some owners prefer dead-front construction. For the open type all parts, front and back, are readily accessible for maintenance, an important consideration when it is remembered that these operations have to be performed on a live board.

Generator switchgear for parallel operation comprises a double-pole circuit-breaker and an interlocked switch for the equaliser connection. Alternatively, the equaliser switch may be combined in a triple-pole circuit-breaker.

Regulations require preferential tripping in large installations so that non-essential loads can be automatically switched off when the generators become overloaded. This can be done in either one stage in a simple installation or in two or three stages in larger systems. A typical arrangement is shown in *Figure 31.5*.

31.8.2 A.C. switchgear

Because of the greater risk of shock in a.c. systems, the open construction is not permitted and switchboards must be 'dead-front'. In the usual construction the switchboard is divided into

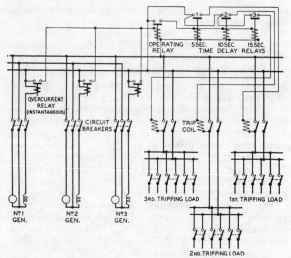

Figure 31.5 Typical diagram of preferential tripping circuits

cubicles. Circuit-breakers can be withdrawn and isolated from the busbars for maintenance and adjustment. Access doors are interlocked to prevent access to live parts. A similar construction is used for control gear.

Preferential tripping as described for d.c. systems is also required. Instrumentation must include a synchroscope; see *Figure 31.6*. Automatic synchronising is now becoming necessary

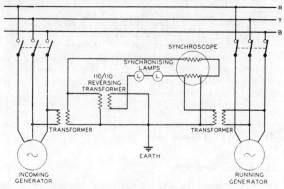

Figure 31.6 Schematic diagram of synchroscope and synchronising lamps for 'lamps bright' system

in installations comprising prime movers which start automatically, and SOLAS 1974 requires automatic start-up and circuit-breaker closure within 45 s to ensure continuity of supply with unmanned installations.

Air-breakers and miniature circuit-breakers are common at 440 V; high fault levels have led to the introduction of current-limiting breakers. At 3.3 kV and above, vacuum contactors are often used; active consideration is given to SF_6.

31.9 Cables

EPRCSP, PVC and XLPE, all with metallic sheath and PVC overall, are most common, being flame retardant as required by SOLAS. MICC and silicone rubber withstand high temperatures and have appropriate applications.

With single-core a.c. cables, separation from steel bulkheads and non-magnetic inserts are required to limit induced currents. Cables to emergency services should not pass through areas of high fire risk such as galleys and machinery spaces; those penetrating fire- and watertight bulkheads should not impair integrity; those for intrinsically safe circuits would be separated from power cables, and identified.

Future regulations may require low flame spread, and low smoke, acid and toxic emission. Development is well advanced; so, too, is the development of cables which retain insulating properties after severe fire damage.

31.10 Emergency power

SOLAS 1974 has extended for all ships the services to be supplied; for cargo ships the period for which this power must be available is stated.

Passenger ships are to have emergency power from generator or battery adequate for 36 h duration. With a generator there must be a transitional battery to supply specified services for 30 min to come into operation automatically.

Cargo ships have similar requirements for 18 h duration but do not need a transitional battery if the emergency source can start and be connected automatically.

Mobile Offshore Installations are covered in the MODU Code.

Both Ni–Ca and lead–acid types of battery are used for emergency or stand-by services to vital circuits such as computer, communication, fire detection, etc., without excessive volt drop, a term interpreted by the British authority as a drop, at the end of the specified emergency period, not exceeding $12\frac{1}{2}\%$ of nominal system voltage. The voltage should be within the limits $+10$ to $-12\frac{1}{2}\%$ from the fully charged condition to that at the completion of the prescribed duty.

Battery installation, ventilation and maintenance correspond to the best practice ashore.

31.11 Steering gear

Electric steering can be either all-electric or electrohydraulic. Its function is to control the position of the rudder through an angle of 35° each side of the central position. Regulations require that the time taken to put the rudder from 35° on one side to 30° on the other is not to exceed 28 s at maximum service speed. The exact position of a rudder, if power-operated, has to be indicated at the main steering station. A usual arrangement is that one quarter-turn of the steering wheel corresponds to one degree of rudder movement. Automatic steering on a set course controlled by a gyro-compass can be superimposed on a power-operated system.

One such system is shown in *Figure 31.7*. Mounted in the

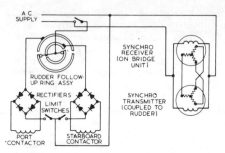

Figure 31.7 Schematic diagram of rudder follow-up system and the synchro transmitter and receiver (Sperry System)

bridge control unit is a follow-up ring assembly consisting of two pairs of silver rings mounted on insulating formers. Carbon brushes on the outer rings are connected to rectifiers which, in turn, are connected to contactor coils. Follow-up rollers make contact with the inner surfaces of the rings, one half-ring of each inner ring being connected to the complete outer ring. The carbon brushes are fixed, but the rings and the bracket carrying the inner contact rollers are free to rotate. The inner roller assembly is rotated by a synchro receiver; the contact rings by the pilot wheel when under hand control or by the gyro-compass transmitter when under automatic control. In this system the transmitter is geared to the rudder so that it adopts a corresponding angular position. When the rudder has reached the angle set by the pilot wheel (or by the compass), the rollers will have caught up with the gap in the ring and the contactors will open.

Electrohydraulic systems usually depend on a continuously running motor coupled to a variable-delivery pump supplying oil pressure to one or other of a pair of hydraulic cylinders. Operation from the bridge is by telemotor control, in which the rudder motion continues until its position coincides with that of the bridge setting.

All-electric systems depend either on a direct-coupled reversing motor or on Ward Leonard control. In the latter the generator voltage may be controlled by a voltage divider or by field windings of opposite polarity. In push-button systems the rudder movement continues so long as the push-button is held closed.

Steering is a vital service. SOLAS now requires duplication of power supplies and control circuits, the latter to be supplied specifically from the associated power circuits. Alarms and transfer switching are to be sited on the navigating bridge. Large ships require steering power to be available, automatically within 45 s, from the emergency source on loss of main power.

31.12 Refrigerated cargo spaces

The system now commonly used consists of batteries of brine cooled pipes over which air is circulated by fans and distributed uniformly to all parts of the hold. The fans vary in size and number according to the holds, and a ship may require up to about 40 fans of 1–8 kW rating. Compressor motors up to about 200 kW at constant speed are commonly required. The fans are located in the holds and their lubrication presents a problem, as the ambient air temperature can vary between tropical conditions at the loading port and approximately $-20\,°C$ when fully refrigerated.

Temperature control of the holds within narrow limits is essential for some cargoes (e.g. bananas), so that provision must be made for accurate and sensitive sensing and control at numerous points. Electrical thermometers have been specially developed to read to 0.1 °C over a range of approximately $-20\,°C$ to $+15\,°C$. A tolerance of $\pm 0.1\,°C$ at freezing point of water is attainable under working conditions.

Container ships with typical installed power of 6 MW 450 V 60 Hz and considerable instrumentation are now common. Up to 3000 containers of rating 3.5 kW may be supplied through flexible connections.

31.13 Lighting

31.13.1 General

For general lighting both tungsten and fluorescent lamps are used. For deck lighting halogen floodlights or high-pressure mercury lamps may be installed. Where colour rendering is not important, sodium lamps, which have high efficiency and long life, are suitable for high-level general lighting. The lighting load for large public rooms can be appreciable, and fluorescent lighting can reduce power and heat, which, in turn, assists air conditioning.

Under the Merchant Shipping Act the Department of Trade requires artificial lighting in crew accommodation to be well diffused, avoiding glare and deep shadows. Bunk lights must be provided, and 25 W or 40 W tungsten or 15 W fluorescent lamps are considered satisfactory. In cold stores separate light fittings of robust construction, with a switch outside the compartment and a pilot light, are required.

For general lighting the level of lighting measured at a height of 0.85 m above floor level and midway between adjacent lamps, and between any lamp and a boundary of the space, is prescribed in precise terms. Good lighting in galleys is important, and brings faster working, fewer mistakes and accidents, and better hygiene.

Statutory regulations (when applicable) require emergency lighting supplied from the emergency source of power to be embodied in the lighting system. All boat stations, lifeboat launching gear and all public and crew areas, alleyways, service spaces, stairways and exits must be provided with emergency lighting.

Voltage variation affects the light flux output and life of both filament and fluorescent lamps.

31.13.2 Navigation lights

The requirements are prescribed in International Regulations for Preventing Collisions at Sea; standards of visibility and lamp types are specified in Department of Trade Regulations. Classification Rules require that navigation lights shall be connected to a distribution board reserved solely for this service, and connected directly or through transformers to the main or emergency switchboard. Each light must have aural and/or visual indication of failure. If only an aural device is fitted, it must be battery operated. If a visual signal is connected in series with the navigation light, there must be means to prevent extinction of the navigation light through failure of the signal lamp. Typical arrangements are shown in *Figure 31.8*. The volt drop across series connected indicators must not exceed 3% of the system voltage. The use of double-filament navigation lights is not permitted.

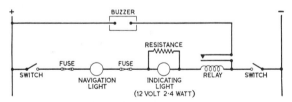

Figure 31.8 Schematic diagram of typical navigation light indicator

31.14 Heating

SOLAS prescribes that electric heaters, where used, must be fixed in position and must not have an element so exposed that clothing, curtains, etc., can be scorched or set on fire.

Except in ships employed solely in the tropics, crew accommodation has to be provided with a heating system of prescribed minimum performance. With certain exceptions a temperature of 19 °C must be maintained in a ship regularly employed otherwise than as a 'home-trade ship', and 15 °C in the case of any other ship, when the outside air temperature is $-1\,°C$.

31.15 Watertight doors

Watertight doors, under certain conditions prescribed by regulations, have to be power operated and to conform to specific requirements, including testing at maker's works in the presence of a surveyor.

They may be either electrically or electrohydraulically operated and controlled, either in groups or individually from a central control position, usually the navigating bridge. In addition, hand operated gear must be provided which can be worked at each door or from a position above the bulkhead deck. At each control position an indicator must show whether the door is open or closed. An audible warning device situated near the door and operated by a time switch giving about 10 s notice functions before it starts to close.

The central control station has overriding features which decide whether the doors are to close; they can be opened locally but re-close automatically. They can also be closed locally independently of the central control.

An essential feature of the powering system is that it should exert an initial high effort to withdraw the door from the grip of door wedges which come into effect in the fully closed position. Thereafter this force diminishes and the rate of opening increases.

31.16 Ventilating fans

When trunking carries ventilating air to accommodation and cabins, any noise arising from fans or fan motors must be avoided. For this reason sleeve bearings are generally preferred. A variable speed or a choice of speed is usually necessary for control purposes.

Regulations require means to be provided for stopping fans from remote positions in the event of fire.

31.17 Radio interference and electromagnetic compatibility

Radio communication is vital and must as far as practicable be interference-free. 'Noise' may originate in the ship's electrical installation, in the rigging, from other radiofrequency apparatus (e.g. public address systems) and from electromedical appliances. The radio installation can also cause interference with other electronic circuits; hence the concept of electromagnetic compatibility.

Suppression of unwanted noise is an economic problem to be shared by the supplier of the radio equipment, the shipbuilder, the electrical contractor and the manufacturers. Usually interference cannot be entirely eliminated, so permissible levels are prescribed. Interference can be alleviated by careful planning of radio installations and aerial systems, by applying certain techniques to the rigging, and by fitting suppression devices to electrical equipment and wiring. Interference arising from the electrical installation can be of two kinds—i.e. radiated and picked up by the aerial or conducted by cables entering the radio cabin. BS 1597 deals with remedies generally considered necessary for the electrical installation and prescribes the permissible level of interfering emissions in the various wavebands. Precautions to be taken in the construction of the rigging and in the installation of cables are included.

This subject is detailed in an Appendix (1982) to the IEE Regulations for the Electrical and Electronic Equipment of Ships.

31.18 Deck auxiliaries

Deck auxiliaries include cargo winches, cranes, capstans, warping winches, windlasses and hatch cover winches. For some classes of ship a variable speed for deck auxiliaries is still preferred, but, in general, 2- or 3-speed change-pole induction motors are suitable. For cranes a high-speed facility is needed for rapid return of the empty hook.

31.18.1 Variable speed

Where variable speed is essential in a.c. systems, slip-ring motors or Ward Leonard control may be provided, the latter with a motor generator for each winch or for a group. With d.c. systems a variety of methods has been used: e.g. series resistance, Ward Leonard, boost and buck, and lowering by dynamic braking or regeneration. Dynamic braking is generally achieved by means of diverters in shunt with the armature giving a potentiometer connection. In nearly all cases load discriminators may be necessary to control speed in relation to load, i.e. to permit of higher speeds with empty or lightly loaded hooks.

31.18.2 Deck auxiliary services

Some of the additional factors concerned with particular applications are given below.

31.18.2.1 Cargo winches

Emergency centrifugal brakes are fitted, where necessary, to prevent excessive speeds when heavy loads are being lowered. Provision must also be made to prevent the load from running back in the event of a power failure. Light-hook speeds are $3-4\frac{1}{2}$ times normal full-load speed.

Serious generator loading problems may be introduced when several winch motors with direct-on-line starting are in use. Large currents at low power factor cause generator and cable heating. *Table 31.1* gives empirical data for the effective current (in per-unit of the full-load current per winch) of a group of n winches ($n > 2$). Load assessment curves are given in the IEE Regulations for the Electrical and Electronic Equipment of Ships.

31.18.2.2 Warping

Warping is frequently done with an additional barrel on cargo winches or windlasses. A flange prevents the hawser from running over the rim, and at the inner end of the frame a projection prevents it from becoming jammed between frame and warp end. If a foot-brake is fitted, its effect should be limited to the full-load torque of the winch, or torque-limiting relays should be fitted.

Table 31.1 Effective group current of winches. Current in p.u. of the full-load current of n similar winch motors in a group, for $n > 2$

Part of system	D.C. motors (series resistance)	A.C. cage motors	A.C. slip-ring motors	A.C. Ward Leonard
Cables and switchgear	$0.33n$	$3.3 + 0.3n$	$1.6 + 0.2n$	$1.2 + 0.15n$
Generators	$2.2 + 0.2n$	$5.0 + 0.4n$	$2.0 + 0.25n$	$2.2 + 0.2n$

31.18.2.3 Capstans

Capstan barrels are normally mounted on a vertical shaft and the motor mounted below deck to leave the deck free.

31.18.2.4 Windlasses

Anchor windlasses are vital to the safety of the ship and have no stand-by. They are subject to Classification and governmental requirements. The cable lifter is shaped to fit the links of the cable, and will normally accommodate four or five links around its circumference, although only two links are actually engaged at one time. The lifter can be declutched so that it runs freely for lowering, the speed being controlled electrically or by band brake. A slipping clutch is fitted to prevent excessive stress which could otherwise occur when heaving or when entering the anchor into the hawsepipe. A crawl speed is necessary to enable the anchor to be housed safely and to allow the motor to stall when it is fully home. It is also necessary while the anchor is still holding.

The overall efficiency of a windlass is about 60% and as much as 30% can be lost in friction unless, in accordance with modern practice, rollers are fitted. Windlasses can perform warping duties and extra control refinements are sometimes incorporated for these. They provide for a pull of about one-third of that of the cable lifters but at a greater speed. This speed is also required for recovering lines cast off from the quayside.

31.18.2.5 Mooring winches

Mooring winches are similar to warping winches except that one end of the line is fixed to the barrel.

The St Lawrence Seaway regulations require short-period stalling while docks are navigated, also constant tensioning against rising and falling tides or lock waters and during rapid loading or unloading.

31.19 Remote and automatic control systems

Since 1960 automation of ship's machinery spaces has increased considerably to the point where it is now generally recognised as an indispensable part of modern propulsion plant. In 1981 about 55% of the ships classed with Lloyd's Register of Shipping were designed to be operated with an unattended machinery space (UMS), and approaching 80% of the ships, including those with a UMS notation, had machinery spaces operated from a centralised control station.

Automation has also become indispensable in such areas as cargo handling, disposal by incineration of obnoxious waste, and pollution prevention. With high discharge rates, critical incinerator temperatures and specified limits of contamination, control systems are required to anticipate demand and give rapid response if catastrophies are to be avoided on board or at the terminal.

The proper functioning of control systems demands careful planning, since their viability may be governed by the nature of the ship's trading, the type of machinery to be controlled and the shipowner's manning and maintenance policies. A full specification is required of the extent of control facilities to be provided to ensure compatibility of controls with machinery, the marine environment and the operating personnel.

31.19.1 Operational modes of machinery spaces

Modern applications of control engineering systems permit two basic operational modes of the machinery space: (1) continuously attended, but with a high degree of remote and automatic controls to allow operation from a centralised control station; (2) periodically unattended machinery spaces so that engineers need not be tied to traditional watch-keeping routines and, for example, may leave the machinery space unattended during the night.

On vessels where centralised control is adopted, it is usual to incorporate all controls, alarms and instrumentation in an enclosed control room which is sound- and vibration-proofed and air conditioned. With the withdrawal of the engineer from the machinery space, it is essential that the controls and instrumentation provided be such that supervision of the machinery plant from the control room is as effective as it would be under direct supervision. The arrangements must give provision for corrective actions to be taken at the control station in the event of faults such as stopping of machinery, starting of stand-by machinery, adjustment of operating parameters, etc. These actions may be effected by either remote or automatic control.

With the advent of microprocessor based control systems the concept of centralised control has been extended to incorporate techniques known as 'totally distributed control'. This enables all machinery within specified areas to be controlled and monitored by one integrated system. To implement this, individual microprocessor based interface units are placed at various locations throughout the machinery space and are interconnected to the central control station by a data link.

Figure 31.9 shows diagrammatically a totally distributed control system. The outstations, interface units, have two functions: (1) to receive data from the various sensors and (2) to output information to the control actuators. The system is arranged so that information is transmitted to the central station only as and when requested, or when a fault condition develops. If the data link between the central station and the outstation is broken, the outstation is capable of continuing operation.

In order to transmit information from the central station to the outstations, and vice versa, a multiplex system has to be used. This arrangement eliminates the need for interconnecting individual signals from each outstation to the central station. Multiplexing is a technique whereby each individual signal is given a specific address which can be transferred from one station to another along the cable in a very short time. The address is recognised only by the unit it is intended for, i.e. it will search many 'go–no go' gates until it finds the one 'go'. The information will then be used by the unit for action. *Figure 31.10* shows diagrammatically how multiplexing functions.

At the control station the complete operation of the system is organised; it consists of the central computers along with visual display units, keyboards, printers, analogue recorders, etc. In early distributed-control systems conventional analogue controllers were used at the outstations with the facility for the set points to be changed from the central station. However, as microprocessor systems have become more reliable, analogue controllers have been replaced by software generated digital control algorithms. Thus, the desired requirement for each control loop is retained in software rather than hardware. These systems are called direct digital control (DDC) systems. All the facilities that were available with the analogue controllers are built into these DDC systems.

31.19.2 Alarms and safeguards

For periodical unattended operation of a machinery space, all controls and safeguards necessary for centralised control are required, but safety actions must be automatic. In addition, it is necessary to extend the machinery space alarm system to the bridge and accommodation areas so that engineering personnel are made aware when a fault occurs. Control of propulsion must also be extended to the bridge to enable the navigating officers to carry out manoeuvres if the need arises. It is important, when an engineer responds to an alarm and enters the machinery space

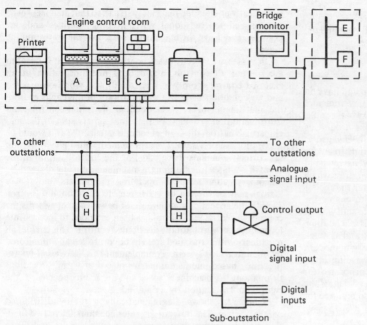

Figure 31.9 Totally distributed control system. A. Operator station alarm; B. Operator station control; C. Multiplexer and control unit; D. Trend recorders; E. Magnetic storage unit and general-purpose computer; F. Extension alarm system; G. Process interface analogue/control and alarms; H. Process interface digital/alarms; I. Multiplexer unit

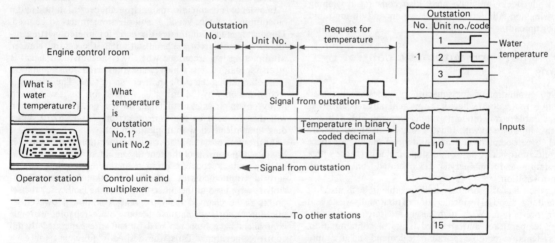

Figure 31.10 Totally distributed control system: form of signal on the data link

alone, that other personnel are aware of his well-being. It is usual to configure the alarm system so that the navigating officer is also made aware of a machinery fault, when it is being attended to and when it is corrected. *Figure 31.11* shows the functioning of an alarm system to meet these requirements.

31.19.3 Reliability

No matter how comprehensive the control and protection systems provided, they are of little value if the equipment used is not reliable. Experience has shown that many items of control equipment do not operate satisfactorily at sea. Control equipment, designed for use ashore, is all too often unsuitable or unreliable when used on board ship.

In order to improve reliability, the majority of Classification Societies have adopted and developed criteria to which equipment should be tested. This is known as Type Approval, a procedure whereby a prototype or a production unit is tested under conditions of environmental and mechanical stress that simulate severe shipboard operating conditions.

Basic minimum requirements specified by Lloyd's Register of Shipping include:

Visual inspection: to ensure that workmanship is good and materials used are adequate for the duty of the equipment.

Performance tests: to ensure that the manufacturer's specified limits of accuracy, repeatability, etc., are fulfilled.

Fluctuations in power supply: voltage variations, steady, $\pm 10\%$ with simultaneous frequency variation of $\pm 5\%$. Tran-

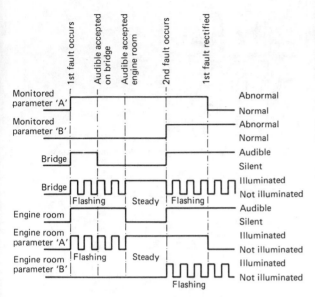

Figure 31.11 Alarm system functional diagram

sient, ±20% voltage with ±10% frequency. For hydraulic and pneumatic systems supply pressure variations of ±20%.

Vibration tests: testing in frequency range 1–13.2 Hz at ±1.0 mm displacement and from 13.2 to 100 Hz at ±0.7 g. Endurance tests carried out at each major resonant frequency for 2 h.

Humidity: 90–100% for 12 h at 55±2 °C, then reduced to 20±5 °C over a period of 1–3 h and remaining at the lower temperature for not less than 6 h. This test repeated over two full cycles.

Dry heat: at manufacturer's stated maximum operating temperature if greater than 55 °C for 16 h.

Inclination test: at least 22½° each side of the vertical in one plane, repeated in a second plane at right angles to the first.

Additionally for electrical equipment, high-voltage and insulation resistance tests are required.

General: Further requirements may include low-temperature and salt mist tests. As applicable, intrinsic safety certification may be called for.

31.20 Tankers

It is of vital importance to take account of bending stresses in the hull resulting from unequal buoyancy. These may arise from ballasting or from different grades of oil, or may occur during loading or discharging. A large number of valves control the filling and unloading of cargo tanks, and these must be operated in logical sequence. Trim and list must also be taken into acccount. There is obviously a fertile field for computer controlled centralised operation. Because of explosion risks, hydraulic power is favoured for valve operation. For instrumentation and control purposes, pneumatic and intrinsically safe electronic circuits are suitable, subject to Classification approval.

31.21 Steam plant

Correct relationship between fuel supply, feed-water supply, temperatures, forced draught, engine load and sea-water temperatures depends on a large number of interdependent controls.

Optimum efficiency is rarely, if ever, achieved without some form of automatic control. The essential factors are: (a) maintenance of steam pressure under varying loads by fuel control; (b) optimum control of combustion air flow and fuel/air ratio.

31.22 Generators

Two or more generators are always provided, but for economy it is undesirable to run more sets than necessary for the prevailing load. Preferential tripping of non-essential loads has already been dealt with. This is only a temporary expedient, and, to prevent overloading, systems are available for automatic starting, stopping and synchronising of sets according to demand. In confined waters it is necessary for safety reasons to have a margin in reserve, and it is therefore usual practice to have two sets on the board. An overriding provision should be included in automatic schemes.

Advances in technology have led in recent years to the introduction of micro-based systems for efficient power management. Escalating fuel costs and the need for reliable unmanned operation have stimulated this development.

31.23 Diesel engines

When bridge control of main propulsion engines is installed, it is necessary to make provision for a repeat start if the first sequence is not completed. After a set number of false starts (usually three) an alarm operates. If acceleration to the firing speed is not reached in about 3 s, a further period of about 4 s is allowed and the complete cycle is then repeated.

For auxiliary generator plant, standby sets can be pre-warmed by continuous circulation of hot water from the main engine cooling system. It is usual to bring in another set before full load is reached. However, a short time delay is advisable to prevent starting on a spurious load demand.

31.24 Electric propulsion

An efficient propeller must run at a relatively low speed, say 80–400 rev/min. If a steam turbine is the prime mover, an efficient turbine speed will be of the order 3000–4000 rev/min, and some form of speed reduction is essential, either by mechanical gearing or by a combination of electric generator and motor. With diesel engines large powers at suitable speeds are available, but for relatively small powers a combination of high-speed engines with electric generators and motors offers advantages. Alternatively, high-speed diesels coupled to mechanical gearing through electric slip couplings can be used.

Admiralty coefficient Using the formula $D^{2/3}U^3/P$, this compares the efficiency of different ships in the same general category on a basis of the displacement D, speed U and power P at the propeller-shaft coupling. A high value of the coefficient indicates a low power and economic propulsion. Ships in different categories cannot be compared because the Admiralty coefficient is affected by the hull proportions—in particular, speed/length ratio $= U/\sqrt{L}$, where L is the length of the hull at the water-line.

Fuel coefficient Another basis of comparison is given by $D^{2/3}U^3/W$, where W is the weight of fuel burnt per day.

31.24.1 Propeller characteristics

Propeller characteristics are important, particularly for turbo-electric drive. A typical full-speed reversal characteristic is given in *Figure 31.12*. The following conditions emerge:

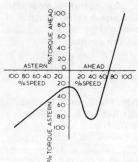

Figure 31.12 Typical speed reversal characteristic of propeller

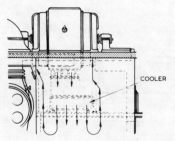

Figure 31.13 Typical cooling system for a turbo-generator

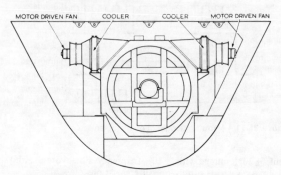

Figure 31.14 Typical cooling system for a propeller motor

(1) On reversal from full power ahead, the propeller speed drops to about 70% when power is cut off and way on the ship is still maintained.

(2) With the ship moving ahead, the power required to reverse the propeller reaches a peak at about 40% of the initial propeller speed. Thereafter less reversing power is called for.

The values quoted, which are typical only, are of great significance in view of the relatively poor torque of synchronous motors when reversing. In the d.c. systems common with diesel-electric propulsion, the d.c. propulsion motor generates, and drives the main generator as a motor. Precautions are necessary to avoid overspeeding of the engines.

Tidal currents may rotate a propeller at crawling speed when the ship is not under power. The motor lubricating system must be adequate for this condition as well as for normal running.

31.24.2 Turbo-electric systems

Turbo-electric propulsion is found in high-powered ships, equipped with two turbo-generators and a synchronous motor. Speed (and therefore frequency) adjustment is obtained by varying the speed of the turbines down to about 25% of normal, at which starting, reversing and manoeuvring are carried out, the motor field being unexcited. The motor operates under these conditions on a cage winding as an induction motor. Considerable slip-ring voltages are induced, and the generator excitation has to be doubled.

Controls are interlocked so that excitation circuits cannot be opened until the turbine controls have been brought to the low-speed setting. Reversing contactors can be opened or closed only when the field circuit is open.

Synchronous generator/motor transmission is analogous to mechanical gearing in that there is a fixed ratio between turbine and motor speeds. The system can operate at unity power factor. It is necessary to co-ordinate the design of a.c. generators and motor to suit the duty with special regard to stability when driving through heavy seas, and the high torque necessary for reversing. Machines are 'stiffer' than those required for land applications, i.e. the ratio between excitation and armature reaction is higher. Special attention must be given to design of stator windings to withstand twice full-load current at about one-quarter normal frequency.

Generators have closed-circuit cooling through a heat exchanger using sea-water (*Figure 31.13*). The propeller motor is generally fitted with a cooling air system using separate fans. If the motor is accommodated in its own compartment, air may be withdrawn through a cooler and re-enter the machine through the endshields (*Figure 31.14*).

Generator and motor losses reduce the efficiency below that of direct or geared drives, so that the turbo-electric system is considered only where there are compensating practical advantages. During the past two decades there has been a steady decline in the gross tonnage of turbo-electric ships.

31.24.3 Diesel-electric systems

Diesel-electric transmission uses constant-speed engines in groups driving d.c. generators connected in series cascade, with d.c. transmission and propulsion motors. Two systems are in use: (1) Ward Leonard control, and (2) constant current control. In (1) the propulsion motor speed is varied by control of its armature voltage; in (2), by its field excitation, the constant current level also being adjustable in accordance with load requirements. The system in either form lends itself to control from the bridge, and is suitable for tugs, single- and double-ended ferries, fishing vessels, ice-breakers, cable-layers, dredgers.

The controlled constant current system is shown in *Figure 31.15*. Use is made of the Kramer 3-winding generator used as an exciter to maintain an approximately constant current in the propulsion circuit. Three winding main generators would be costly, so a 3-winding motor-driven exciter is preferred. The windings comprise a differential series field connected in the propulsion circuit, a self-excited field and a separately excited field. Excitation of the main generators controlled by the exciter automatically varies, so that, for any setting of the controller (which can be on the navigating bridge), the current in the main circuit remains constant. The number of main generators in circuit with either system can be controlled by set-up switches in the main circuit. To disconnect a generator, its excitation is interrupted, and its armature short-circuited and then isolated. The sequence of switching is safeguarded by interlocks. To insert a generator the reverse procedure applies.

31.24.4 SCR power system

The use of silicon controlled rectifiers (thyristors) allows a flexible system without need for d.c. generation. A constant voltage and

31.24.5 Electromagnetic slip couplings

In diesel-engined ships in which the propeller is driven through mechanical reduction gear, up to four engines per shaft can be coupled to the gear through slip couplings. It is necessary to protect the gears from torsional vibration transmitted by the engines, and the slip coupling serves both this purpose and that of a disconnecting clutch, enabling the numbers of engines in service to be altered without stopping the engines already operating. A typical 4-engine arrangement is shown in *Figure 31.17*.

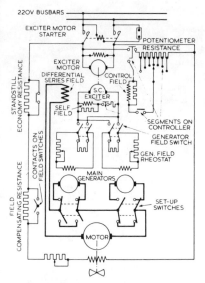

Figure 31.15 Diesel-electric d.c. system with modified Ward Leonard control

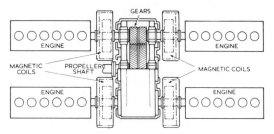

Figure 31.17 Typical four-engine coupling arrangement

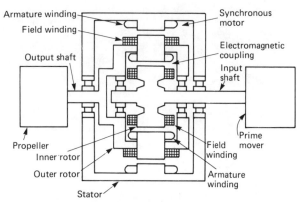

Figure 31.18 Versatile electromagnetic gear

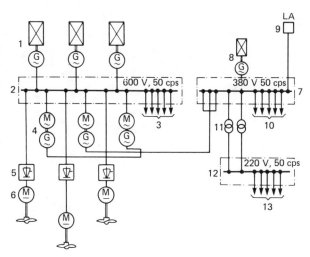

Figure 31.16 Electrical system with thyristor controlled propulsion. 1. Main diesel-generator; 2. Main switchboard, 600 V; 3. Compressors, fans, towing winch, fire extinguishing pump; 4. Converters, 600/380 V; 5. Thyristor rectifiers; 6. Propeller motors; 7. Auxiliary switchboard, 380 V; 8. Auxiliary diesel set; 9. Shore connection; 10. Ship's mains, 380 V; 11. Transformers, 380/220 V; 12. Auxiliary switchboard, 220 V; 13. Ship's mains, 220 V and feeder to emergency switchboard, 220 V

frequency supply is available for electrical auxiliaries such as lighting, pumps and compressors. This a.c. supply is fed to thyristors to give a variable, controllable d.c. voltage for speed control of propulsion motors, bow thrusters, towing winches, and the like.

A common arrangement is shown in *Figure 31.16* and would be found on ice-breakers, offshore supply ships, tugs and other vessels.

Manoeuvring can be carried out by having some engines running ahead and the others astern and selecting the direction required by switching.

A coupling can exert a starting torque from rest, with the engine at full speed, equal to full-load torque. The efficiency is high, the only losses being windage, excitation and the I^2R loss due to slip. Slip varies with speed and rating and is generally 1–2%.

31.24.6 Electromagnetic gearing

Elements of a speed reduction and reverse gear which can also act as an auxiliary a.c. generator, clutch and flexible coupling are shown in *Figure 31.18*. An inner rotor carries the input shaft, field windings and slip-rings. The outer rotor carries the electromagnetic coupling armature winding on the inner side which is connected, through a switching device, to the synchronous motor field windings on the outside, and two sets of slip-rings.

32 Mining

R Hartill, BSc(Hons), CEng, FIEE, FIMEMME, FRSA
Trolex Products Ltd

Contents

32.1 General 32/3
    32.1.1 Load growth 32/3
    32.1.2 Regulations 32/3

32.2 Power supplies 32/3
    32.2.1 Distribution 32/4

32.3 Winders 32/5

32.4 Underground transport 32/7
    32.4.1 Conveyors 32/7
    32.4.2 Rope haulage 32/7
    32.4.3 Locomotives 32/8

32.5 Coalface layout 32/9

32.6 Power loaders 32/9

32.7 Heading machines 32/11

32.8 Flameproof and intrinsically safe equipment 32/11
    32.8.1 Flameproof transformers 32/12
    32.8.2 Flameproof switchgear 32/13

32.9 Gate-end boxes 32/14
    32.9.1 Single point SEL 32/15
    32.9.2 Multi-point SEL 32/16

32.10 Flameproof motors 32/16

32.11 Cables, couplers, plugs and sockets 32/17

32.12 Drilling machines 32/19

32.13 Underground lighting 32/19

32.14 Monitoring and control 32/20
    32.14.1 Computer system 32/20
    32.14.2 Underground transducers 32/21

32.1 General

In order to deep-mine coal the sinking of two vertical shafts or inclined roadways must be established to access the coal seams. Underground roadways are then established to the winning area of the coalface. There are usually two roadways to establish a ventilation system. One shaft and roadway is used to transport the coal, the other usually to transport men and materials to the coalface. One shaft and roadway is referred to as the intake and the other as the return airway, signifying the direction of the ventilating air flow with all precautions taken to separate the air flows to maintain adequate ventilation.

Whenever coal is mined, methane gas is liberated, and the electricity regulations require that the electrical power must be removed from that part of the mine if the methane gas content exceeds $1\frac{1}{4}\%$ by volume. The regulations allow an exception to this rule to permit communications and certain safety monitoring equipment to be maintained even in the heavy concentrations of methane: in this case the equipment must be intrinsically safe—that is, the equipment must be tested and it must be certified that, in the event of open sparking in either normal or faulty condition, insufficient energy would be released to ignite the most easily ignitable methane concentration.

The regulations regarding the use of Electricity in Mines require approved equipment to be used in any part of the mine where gas, though not normally present, is likely to occur in sufficient quantities to be a danger. As it is difficult for the mine manager to state categorically that gas will not occur, it is usual to employ approved equipment in *all* underground situations. In the UK this means general use of equipment certified Flameproof (FLP) or Intrinsically Safe (IS) for Group 1 (methane) gases.

It will be appreciated that the ventilation for the dilution of liberated methane is the first safety measure in the use of electricity in mining, with the added precaution of approved apparatus should concentrations of methane occur.

Ventilation of the mine is normally achieved by reducing the pressure at the surface end of the return shaft (upcast) by means of an axial- or radial-flow fan, one fan working and one fan standby. A typical installation would operate at 10–15 in water gauge (2.5 kPa) and 125–250 m^3/s air flow. Since the fan is running continuously, efficiency is of prime importance: consequently detailed attention is paid to this factor. The prime mover used is a cage or slip-ring induction motor, or a synchronous machine with speed change effected by gear or V-belt drive. More recently variable-speed machines have been used such as a.c. commutator motor, Kramer, cascade arrangements and the pole-amplitude modulated (PAM) motor.

Where underground workings are extensive it may be necessary to provide booster fans in the underground system, usually powered by cage motors up to 400 kW.

In order to ventilate single headways, i.e. roadways being driven, auxiliary ventilation fans are used providing air at the end of the tunnel by ducting. These are smaller machines of 10–35 kW, and where they are exhausting, a situation could exist where gas flows over the fan blades; in this condition it is important to ensure that sparking cannot occur at that point. For this purpose vibration monitoring is extensively used. Two frequency ranges are monitored (500 Hz and 5000 Hz) to detect bearing failure and mechanical out of balance. The equipment will first give a warning and later 'trip the power' if the force on the bearing or out-of-balance exceeds the preset value.

For some years the analogue computer has been used for ventilation calculations for a mine, but is now being superseded by digital computer techniques.

32.1.1 Load growth

All activities at the mine make extensive use of electricity. Steam winders are being eliminated and compressed air as a power

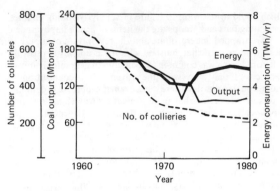

Figure 32.1 Annual coal output and electrical energy consumption (UK)

medium is almost non-existent. Coal is now almost exclusively won by electrically powered machines, and roadways are driven by large roadheading or tunnelling machines with extensive use of electronics to provide protection, control and monitoring. The trend of electricity is shown over the recent past in *Figure 33.1*.

32.1.2 Regulations

Rules were introduced for Electricity in Mines in 1905. These were replaced by regulations in 1913. These regulations have been modified to take account of changes in technology but have well stood the test of time. Following the establishment of the Health and Safety at Work Act, the electricity regulations are being examined and updated. An important step was the Coal Mines (Mechanics and Electricians) General Regulations 1954, which required the certification of engineers and craftsmen at a mine. These regulations also laid down requirements for the qualifications and responsibilities with regard to examination, testing and repair of apparatus. Qualifications required for the Electrical Engineers' Certificate and Electricians' Certificate are laid down in the Coal Mines (Training) Regulations 1967.

Other mining regulations include sections which affect the electrical engineer:

All coal-mining activities are carried out in accordance with the Law relating to Safety and Health in Mines and Quarries Parts 1 and 2. Part 1 consists of the Mines and Quarries Act 1954, Mines and Quarries (Tips) Act 1969 and Mines Management Act 1971. Part 2 contains all the relevant Regulations and Orders under the Mines and Quarries Act, Approved Specifications and Procedures, and Mining Qualifications Board Rules.

The principal Regulations and Orders associated with the installation, testing, maintenance and operation of electrical equipment underground in coal mines are as follows.

Coal and Other Mines (#) Regulations
(Electricity) 1956 (Safety Lamps and Lighting) 1956
(Fire and Rescue) 1956 (General Duties and Conduct) 1956
(Locomotives) 1956 (Shafts, Outlets and Roads) 1960
(Ventilation) 1956 (Mechanics and Electricians) 1965

Coal Mines (#) Regulations
(Explosives) 1961 (Training) 1967

Mining Qualifications Board
Mechanics and Electricians Certification Rules

32.2 Power supplies

The supply of electricity to collieries is given priority by the Electricity Boards because of the high degree of risk to human life which could arise owing to failure of supply. Winding engines, ventilating fans and pumps are the items of prime importance if men are trapped underground, and the restoration of supply to these items, in particular, must be achieved as quickly as possible.

Before 1939 it was common for collieries to generate their own electricity supply, but with the advent of mechanisation of underground operation there was a dramatic increase in demand. It was at this period that mine generation was eliminated and supplies taken from the Electricity Boards. Small generators are, however, being introduced, powered by gas engines or turbines burning methane gas extracted from the mine.

The supply usually takes the form of an Electricity Board primary substation located adjacent to the colliery premises into which a duplicate supply is taken at 33, 66 or 132 kV. Alternatively, the colliery may be fed with at least duplicate supplies from the Electricity Boards 11 kV network.

To comply with the requirements of NCB Mining Department Instruction PI/1957/31, Electricity Boards are required to provide a minimum of two supplies to a colliery, each supply being taken from a separate point in the Electricity Board's network and routed so as to prevent a complete failure due to a common hazard. Each supply must be capable of handling the full output of the colliery (in the case of a large modern mine, 15–20 MVA), the system being capable of handling the fluctuating load of electric winders, etc., which could be 3–4 MW.

32.2.1 Distribution

32.2.1.1 Surface

The supply to the colliery substation from the Electricity Board's primary substation (which is usually adjacent to the colliery) is in most cases 11 kV and consists of a minimum of two feeders. Larger collieries may have three or four feeders; a typical distribution layout for a large colliery would be as shown in *Figure 32.2*.

All incoming cables, circuit-breakers, bus-section switches and metering equipment located within the colliery substation remain the property of the Electricity Board, whereas all outgoing

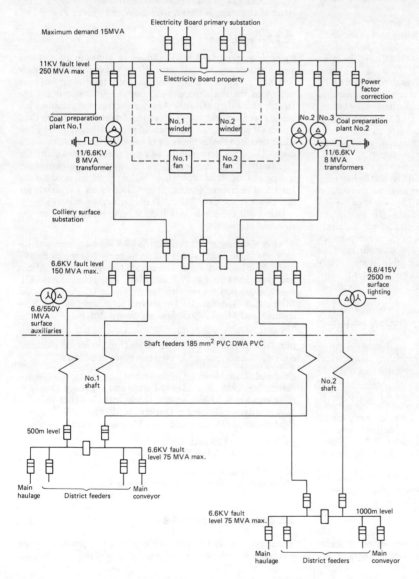

32.2 Typical colliery h.v. distribution

circuit-breakers would be the property of the National Coal Board.

Large drives on the surface such as winding engines and ventilating fans are generally supplied direct from the 11 kV switchboard. To ensure continuity, such supplies are usually duplicated in the form of a ring main or duplicate feeders.

The coal preparation plant, the largest energy consumer at the surface of the mine (usually 1.5–2.0 MVA), is also supplied direct from the 11 kV switchboard by duplicate feeders.

The greatest proportion of load at a colliery is from the induction motors (mainly cage), which leads in some cases to a low power factor on the supply. In the majority of cases this is corrected by manually switched static capacitors on the 11 kV switchboard.

It was the practice to use 3.3 kV as a standard distribution voltage for underground activities. With increase of electrical powered units in operation and increased rating of coalface machines, there may be several 500 kW units being switched direct-on-line at a distance of several miles from the pit bottom; a common distance would be 8 km. The 3.3 kV distribution voltage becomes inadequate and there has been a change to the extensive use of 6.6 kV. While 11 kV is not widely used, at present for underground distribution, flameproof 11 kV switchgear and transformers have been developed and several systems are in operation.

It is envisaged that 11 kV will become the standard underground distribution voltage for systems operating 10 km or more from the source of supply, as in undersea workings.

To provide a 6.6 kV supply for underground distribution, 11/6.6 kV surface installed transformers are used with ratings ranging from 6 to 8 MVA. Further transformation is provided from the 6.6 kV switchboard to provide lower voltages for other surface auxiliaries such as workshops, stores, stockyards, lamp rooms, offices, pithead baths, lighting, etc.

32.2.1.2 Underground

A minimum of two h.v. supplies are provided to underground workings to increase security, being installed in each of two shafts or drifts. The shaft cables are usually 185 mm² 3-core PVC DWA PVC and are secured to the shaft wall by large wooden cable cleats spaced at approximately 25 m intervals.

At the shaft bottom, a main h.v. substation is provided from which all supplies radiate to the various districts of the mine. The supply is taken to the coalface, where it is transformed to the coalface utilisation voltage of 1.1 kV, 3-phase 50 Hz.

Substations are provided at various points along the cable route to provide an h.v. supply for the main coal conveyors/haulages, etc., or an m.v. 1.1 kV supply for smaller drives such as secondary conveyors, haulages, pumps, auxiliary ventilation fans, lighting, etc. It was the practice to use the utilisation voltage of 550 V for underground activities, and this still remains in some parts of the mine; this, however, proved to be inadequate for the large modern machines, e.g. coal winning machines (shearers), roadheading machines and armoured face conveyors.

The 3-phase system at the mine is earthed to its own earth plates at the surface, and is normally maintained at 2 Ω. Earthing resistors are normally included. The practice was generally to limit the earth faults to the full-load current of the transformer, but modern practice is to limit the 6.6 kV systems to 100 A and the older 3.3 kV systems to 150 A. Standard protection of overload, short-circuit and earth leakage is provided on the outgoing feeders.

Improved short-circuit protection is provided near the coalface, referred to as phase sensitive short-circuit protection. This phase sensitive protection was developed in order to permit the through current necessary to start the high-rated (500 kW) machines—which may be five or six times full-load current at a power factor of 0.2—and yet trip the supply on short-circuit of around twice full-load current which is mainly at a power factor of 0.8.

The earth leakage protection near the face is restricted to a prospective earth fault current on the 1100 V and 550 V systems to 750 mA. Two forms are used, one a restricted core balance, the second being multiple earthing. Extensive use of electronics is made to provide the necessary detection.

32.3 Winders

Early colliery winding engines were powered by steam. Owing to the superior efficiency of electric motors and the greater ease in the provision of automatic control, nearly all winding engines in British mines are now driven by electric motors. These can vary from the very small a.c. winders in the range 100–200 kW, to the large d.c. thyristor automatically controlled winders of 4000 kW. Winding engines exist in several different forms (*Figure 32.3*), and whereas the majority of the older designs were of the ground mounted type (with unsightly headgear), new mines generally adopt the tower winder, with its cleaner lines.

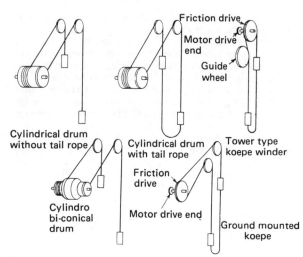

Figure 32.3 Types of winder

Figure 32.4 shows the comparative power–time diagrams for the four different types of winding engines with the same output, net load, depth and decking time (net load, 12 t, depth of shaft, 1000 m; output, 450 t/h). There is little difference between the power requirements for a particular duty for ground and tower mounted Koepe winders, but it is obvious that a comparable Koepe winder does not need such a large motor and the energy consumption is less.

Koepe-type winders are specially suitable where extremely heavy loads are to be handled, owing to the fact that a multi-rope arrangement can be adopted instead of one large single rope. Two or four ropes are generally used, with special devices added to ensure that the ropes share the load equally. With Koepe winders the drive is transmitted from the winder motor by a 'friction' drive; the winding ropes may therefore have a tendency to 'creep' or 'slip'. A 'rope creep compensating' device is provided, which automatically corrects this situation at the end of a winding cycle if it occurs, bringing the depth indicator and other safety devices into line.

Automatic winding techniques have been developed for modern winding engines, and many winders are arranged to wind

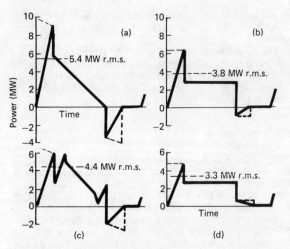

Figure 32.4 Winder power–time diagrams: (a) parallel drum without balance rope; (b) parallel drum with balance rope; (c) bi-cylindro-conical drum; (d) Koepe wheel

coal in the automatic mode. Usually such winders employ skips attached to the winding rope/ropes, instead of cages containing tubs or mine cars. Such skips would hold some 10/12 t and would be loaded automatically from a weigh pocket located at the side of the shaft in the shaft bottom, coal being transferred from the workings to the weigh pocket by conveyor or mine car.

As the skip arrives at the shaft bottom, it is first automatically sensed for being in the correct position and stationary. The weigh pocket door is then operated, and 10/12 t of coal is deposited in the skip in a few seconds. With the skip full and all loading/unloading doors closed, a signal is automatically given to the winding engine to start the wind. The winder is automatically started and accelerated at a predetermined rate to maximum speed. Deceleration commences at a set point in the wind, causing the winder to retard at a predetermined rate to standstill.

With the winder proved stationary and the skip in line, the surface skip door is opened, discharging the coal onto the run-of-mine conveyor at the same time as the shaft-bottom skip is being filled, the whole winding cycle being repeated automatically.

The run-of-mine conveyor transfers the newly won coal to the coal preparation plant, where it is washed, cleaned and loaded into wagons/lorries, etc., for transfer to the customer.

The most common electric winder in the UK mining industry is the a.c. winder employing the slip-ring induction motor with either liquid controller or contactor operated metallic resistors, with a measure of speed and torque control to limit the acceleration or deceleration. Dynamic braking is used on all but the smallest winders; this is compensated to avoid saturation of the machine and ensure control. A typical layout is shown in *Figure 32.5*.

D.C. Ward Leonard winders have been used since the turn of the century. The basic layout is shown in *Figure 32.6*. Closed-loop control was introduced to make the machine start from a signal and automatically come to rest at the surface, i.e. acceleration, deceleration, torque, current control, etc.

In the late 1950s the mercury arc converter replaced the Ward Leonard generator, but was replaced in the early 1970s by thyristor control. The modern machine is now fully automatic, all control and protection being solid state with thyristors in an anti-parallel connection to give complete automatic winding.

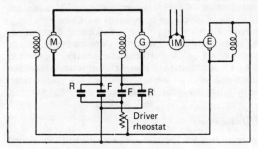

Figure 32.6 Ward Leonard winder control

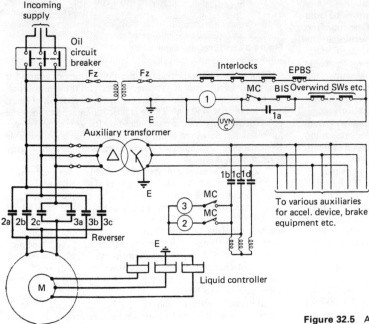

Figure 32.5 A.C. winder control

32.4 Underground transport

Coal mining has two main transport problems. One is to convey the mineral from the coalface to the pit bottom for winding to the surface. Conveyors are the main means for transporting the mineral. The second problem is the transport of men and materials to and from coalface to pit bottom, this is in the main either by rope haulage or by locomotives.

32.4.1 Conveyors

To meet the higher levels of coalface performance, recent advances in the technology of conventional belt conveyor design and belting have resulted in average conveyor capacities in excess of 2000 t/h. Current underground coal transport systems utilising high-capacity belt conveyor, multi-motor drives and booster drives, together with manless transfer points, remote conveyor control techniques and sufficient automatically controlled bunkerage facilities, provide the most efficient system for the tonnage rates now being produced.

A Code of Practice (NCB Mining Department Instruction PI 1979/5) has been introduced, requiring full compliance by 1982, to standardise on the protective devices for the safe operation of all underground roadway belt conveyor drive units. For the additional requirements appropriate to manless operation a memorandum of Guidance Minimum Requirements for Manless Transfer Points on Conveyors has been issued. The provisions of the codes are in amplification of the Health and Safety at Work etc., Act 1974, the Mines and Quarries Act 1954, statutory regulations and mandatory NCB instructions.

The basic type of conveyor drive used underground is the solid mechanically coupled arrangement employing a single flameproof NCB specification 542 or 291 motor up to 70 kW rating supplied at either 550 V or 1100 V from a standard flameproof gate-end box starter fitted with a vacuum contactor. The conveyors are started direct-on-line in sequence, and interlocked with the conveyor pre-start alarm system, signals and drive head protection transducers. Local control of starting may be used.

Main trunk roadway conveyors require higher belt speeds, coal carrying capacities and power. To provide a soft and smooth acceleration to the belt, single- or multi-motor drives with traction-type couplings are used. A limited acceleration period is provided by the fluid coupling. Specially designed fluid couplings with coolers are fitted to keep the starting torque under the fully loaded conveyor conditions to a low level. NCB Specification 625 flameproof motors up to 112 kW rating and flameproof gate-end box starters at 1100 V can be used to supply this type of drive, but a more common arrangement is either 3.3 kV or 6.6 kV flameproof direct-switching vacuum starters supplying higher-rated flameproof motors.

The conveyors are started remotely via a telemetry data transmission link from the surface control room. The high-voltage flameproof starters can be fitted with additional auxiliary equipment to provide, for example:

(1) Manually controlled electrical loop-winch take-up facilities, used for the higher belt tensions and for accommodation of greater amounts of belt associated with longer conveyors.
(2) Integral contactors included in the control gear at 550 V for disc or drum brakes fully interlocked with main drive motors.
(3) Forward reverse contactors at 110 V or 550 V for electrically inserted scoop controlled fluid couplings, the arrangement allowing the main motors to be started first, and for separate insertion of scoops to provide drive to the belt.
(4) Electrical controls for acceleration-torque-limit control (ATLC) fluid couplings, which incorporate a separate hydraulic power pack unit for insertion of scoops at a pre-set rate.

Steel cord belting is now being widely used for drift belt installations and where drive arrangements are similar to those described for main trunk roadway conveyors, utilising ATLC fluid couplings. Microprocessor techniques have been successfully used to control and limit the torque to 150% full-load torque during start-up and acceleration.

At Prince of Wales colliery in the North Yorkshire Area of the NCB the second longest cable belt conveyor in the UK is installed and fully operational. Powered by a single 2240 kW drive unit, the 1706 m conveyor is the sole underground-to-surface conveyor at the mine. Coal is transported to the surface at a rate of 1000 t/h. The vertical lift is 334 m. The conveyor is designed to operate 24 hours a day, with an annual operating time of 4500 h. It is capable of an annual tonnage of 2 million t and the system has the capacity to be extended to a 3000 m length.

The Selby Project (the development of an integrated mine with a capacity of 10 million t/year from five mines) utilises two drift conveyors each capable of dealing with the whole mine's output. The conveyor in the South Tunnel will be 14 800 m in length with a total lift of 1000 m. The belting will be steelchord type SC 7100, 1.3 m wide, with a rated breaking strength of 9054 kN (923 t). The conveyor drive will be a single 2.67 m diameter drum powered by two direct-coupled thyristor controlled NCB type 'E' rationalised winder motors providing an available input to the belt of 10 MW. The speed will be variable up to a maximum of 8.4 m/s (1650 ft/min) and the conveyor will be capable of delivering a maximum 3276 t/h at the surface. The conveyor in the North Tunnel will be a cable belt conveyor and have duties similar to those of the South drift conveyor. The conveyor will be 14 923 m long, with a maximum lift of 1000 m. The drive will consist of twin 6.7 m diameter friction wheels driven by two thyristor controlled type 'D' rationalised winder motors via differential and single reduction gears. Available power input to the drive will be 8.3 MW.

With the introduction of multiplex data transmission equipment and computer controlled coal clearance systems, a much wider range of FLP and IS transducers is now used to protect the conveyor. The range of transducers includes protection against belt slip, motor overheat, belt misalignment, bearing overheat, blocked chute, torn belt, brake overheat and limit switches for brake and scoop 'on' and 'off' proving.

32.4.1.1 Bunkers

Underground horizontal storage bunkers of the moving bed and moving car type with variable speed outfeed metering conveyors are used extensively for capacities up to 500 t. The NCB bunker automation system is now fitted to bunkers of 100 t and over to provide local/remote bunker infeed/outfeed control, using a capacitance probe system, bunker contents and bunker position facilities.

Staple shaft bunkers mounted vertically in the seam incorporating variable speed thyristor controlled outfeed vibrofeeders are used for larger coal storage facilities up to 1000 t.

32.4.2 Rope haulage

Underground steel wire rope haulage systems employ vehicles running on (conventional) two rails, or alternatively on single or double captive rails and a limited use of overhead monorail, for the transport of men and/or materials, with operating speeds between 1.61 km/h and 32 km/h.

The prime mover of the haulage engine is usually a cage or a slip-ring induction motor with a range of 7.5–375 kW at voltages of 550, 1100 and 3300 V 3-phase 50 Hz. All the electrical equipment is certified flameproof to Group 1 requirements.

Motors up to around 75 kW are generally started direct-on-line with a 'soft start' feature provided by a fluid coupling of traction or scoop type, although some 10–50 kW designs use a manually operated friction clutch. Also used are electrical devices which control acceleration by automatically increasing frequency and voltage from zero.

The larger machines use slip-ring motors having FLP rotor resistors and drum-type controllers to provide a variable speed drive. Another speed control for motors of 120 kW and above is the cycloconverter, which uses a thyristor converter controlled by a signal to give a varying frequency/voltage output to supply a purpose-designed cage motor.

Generally haulage systems are operated manually from a position local to the haulage engine, with a guard travelling with the vehicle(s) (either riding or walking) to stop and start the system via hardwire transmitted signals presented in audible and visual form to the operator. For higher speed haulages (above 8 km/h) the NCB type 986 Radio System, which operates on the 'leaky feeder' principle for transmission, is used for signalling from the travelling guard to the haulage engine operator.

For small ratings only, transporting haulages are operated by a man walking with the vehicles, from frequently spaced key operated hardwired connected switches, controlling forward or reverse and brake operating contactors. Because of the unattended (i.e. remote operated) haulage, the system must comply with requirements additional to those of conventionally operated haulages to obtain exemption from mining legislation, which normally requires an operator to be in attendance at the haulage engine.

The NCB type 986 Radio Communication System has been extended to provide control, by a travelling guard, of the speed and direction of the vehicles, in conjunction with the cycloconverter drive. The control transmission uses a coded address binary-function digital signal to switch specific function relays of the haulage drive control system.

32.4.3 Locomotives

In most modern mines the coal is transported from the coalface to the shaft bottom by belt conveyor. There are, however, some mines where the coal is conveyed from the coalface to an inbye loading point, where it is loaded into 4–5 t mine cars.

From the inbye loading point a train of mine cars will be hauled to the shaft bottom by either diesel, battery or trolley locomotives. Currently there are about 650, 75 kW, diesel locomotives operating in British mines and 300 battery locomotives ranging in size from 6 t 22.5 kW up to 40 t 70 kW.

The largest battery locomotives are powered by a 100 cell 550 Ah lead–acid battery with a nominal voltage of 200 V. The battery is contained in a large robust ventilated steel container located in the centre of the double-ended locomotive (*Figure 32.7*), the battery weight being about 4 t.

When a battery change is required, the locomotive enters the battery charging station and positions itself between a pair of charging racks. The batteries are mechanically connected by the use of specially designed links and a racking device is set in motion. The discharged battery is racked onto the empty rack and the charged battery onto the locomotive, the whole process being completed in a few minutes.

Because the large lead–acid batteries on underground locomotives give off large quantities of hydrogen during the charging process, special requirements are laid down (Locomotive Regulations) governing the design of battery charging stations. One of the principal requirements is that charging apparatus must be located on the intake side of the battery charging racks and that ventilation air, having passed over the batteries, is directed into a return airway and does not subsequently ventilate a working face.

Storage battery locomotives for general use underground in coal mines are required by the terms of the Locomotive Regulations to be of a type 'approved by the Minister'. 'Health and Safety Executive—Testing Memorandum TM12' details the test and approval requirements. All electrical equipment used on storage battery locomotives, with the exception of the battery, is required to be certified flameproof.

Most locomotives have two driving motors, one on each of two sets of driving wheels. Series motors are employed with the armatures directly coupled to the driving wheels. Each motor is equipped with a bank of grid resistors controlled through a flameproof contactor and a speed controller. Each driver's cab is equipped with a speed controller, the two electrically interlocked to ensure that only one controller is in operation at any one time.

New locomotives incorporate thyristor chopper control, and methods are being devised to convert some of the existing locomotives to this form of control.

Most battery locomotives in mines are fitted with a battery leakage monitoring device, which consists of an electronic detection unit connected to the battery and an audiovisual alarm unit mounted in the driver's cab. The detector unit has a selector switch which allows the sensitivity at which the unit operates to be varied, a feature necessary to take account of the varying conditions under which the locomotives operate.

Four settings are available, giving battery leakage resistance values of 0.8, 1.3 and 2.6 kΩ. Operation of the alarm indicates that a fault has developed on one pole of the battery and that remedial action should be taken before a second fault develops on the other pole and sets up dangerous circulating currents.

Battery fires underground in coal mines are particularly dangerous and extremely difficult to extinguish once initiated.

Within the UK, trolley-wire locomotive installations have been tried in the past 30 years, but the system has not found universal acclaim, owing principally to the fact that the operating area of the locomotive is restricted to that covered by the trolley wire.

No statutory regulations exist covering the use of trolley locomotives underground in coal mines; consequently, when such installations are considered, special regulations are drawn up to suit each installation.

As coalfaces recede further from the shaft bottom and the need for increased efficiency demands quicker transport of men, materials and minerals to and from the coalface, the advantages of trolley locomotives in certain circumstances have caused the National Coal Board to look at further trolley installations. One such installation is currently operating on a single overhead 500 V d.c. conductor rail-return system, at a distance of 7 km from the shaft bottom, subsequently extending to 10 km.

Work is currently being undertaken on the design and development of a trolley/battery locomotive which will have complete shaft-bottom to coalface capabilities, a distinct advantage over trolley systems. It is envisaged that the new trolley/battery locomotives will be rated at 20 t/120 kW, consisting of four 30 kW motors, and incorporating a 250 V 500 Ah lead–acid battery capable of giving a 75 kW output over a 5 h period. The

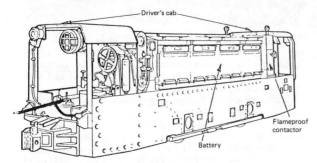

Figure 32.7 Battery locomotive

locomotives will operate on a twin 250 V overhead conductor system, and battery charging will take place while the locomotive is drawing power from the trolley wire. On reaching the end of the trolley wire the locomotive will change to battery power and will carry out excursions away from the trolley area.

The speed at which trolley locomotives operate is 15–20 km/h, compared with the 10–12 km/h of conventional diesel or battery locomotives. With improvements in roadway conditions and better standards of track, speeds of 30 km/h can be expected.

32.5 Coalface layout

When a new coalface is started, two roadways are driven from the main intake and return roadways, to form the intake (main gate) and return (return gate) roadways for the coalface. In each roadway a 6.6 kV DWA PVC 3-core aluminium 185 mm^2 roadway cable is installed. This supply is obtained from a local substation which, in turn, obtains its supply from the pit-bottom substation via the parallel district feeders. The main and return gate roadway cables are supported on special hangers attached to the roadway arches at approximately 2 m spacing. These are adequate to support the cable, but in the event of a roof fall, the additional weight causes the cable supports to give way and allows the cable to fall to the floor.

Each roadway cable terminates into a 6.6 kV flameproof 400 A 150 MVA circuit-breaker incorporating overcurrent, earth leakage and short-circuit protection. This breaker is a semi permanent unit, being moved up periodically by the insertion of 100 m of roadway cable as the coalface advances (*Figure 32.8*).

Advancement of the h.v. circuit-breaker and armoured roadway cable is always carried out with all power isolated, as opposed to all other equipment (flexible wire armoured cables, transformers, contactors, etc.), which is advanced automatically by hydraulic power as the coalface advances.

From the h.v. circuit-breaker a 6.6 kV 50 mm^2 flexible pliable wire armoured (PWA) cable takes the h.v. supply to a flameproof transformer. The PWA cable is supported in loops from a monorail attached to the crown of the roadway arches. Special cable supports with rollers permit automatic advancement.

Flameproof air cooled transformers are used underground. Common ratings in use are 500, 750 and 1000 kVA; they weigh about 5 t.

To permit automatic advancement, the transformer, hydrostatic power pack, flameproof contactors (gate-end boxes) and face signal/communication unit, along with spares container, oil drums, stretchers, first aid and firefighting equipment, etc., are mounted on a robust rail mounted pantechnicon which straddles the main roadway conveyor. The pantechnicon is securely attached to the stage loader conveyor, which, in turn, is attached to the coalface armoured flexible conveyor (AFC).

As the coal is cut by the power loaders, the AFC is pushed forward by hydraulic rams attached to the hydraulic roof supports. This action causes the stage loader (AFC) to advance forward, which, in turn, automatically advances the pantechnicon. When the h.v. PWA cable has been fully extended, power is isolated, allowing the h.v. circuit-breaker and PWA cable to be moved forward, and a new length of 185 mm^2 aluminium PVC DWA PVC roadway cable to be installed.

Armoured roadway cables are installed in 100 or 200 m lengths and are sent into the mine already fitted with 300 A flameproof cable couplers, which are connected on site with copper connecting pins, rubber gasket, bolts and nuts.

PWA 120 mm^2 4-core cable is used to take the m.v. 1100 V supply from the transformer to the bank of flameproof contacters (known in the industry as gate end boxes). Each gate-end box (Section 32.9) is equipped with a 200 A flameproof restrained plug and socket which permits supply to the machine via a 50 mm^2 5-core trailing cable. These cables pass the side of the stage loader to the machines on the coalface. The static parts of power-loader trailing cables are located in specially designed cable troughs attached to the AFC, whereas the part of the trailing cable which flexes backwards and forwards as the power loader moves along the coalface is contained in a robust flexible steel or plastic cable handler. Such handlers also contain a water hose which supplies the machine with water for dust suppression and motor cooling.

A typical coalface could be established as in *Figure 32.8*.

The right-hand (RH) single-ended ranging drum shearer (SERDS) cuts the right-hand end of the face to a distance of about 25–30 m, while the main machine, the double-ended ranging drum shearer (DERDS) cuts the rest of the face.

Attached to the bank of gate-end boxes in the main gate is a flameproof and intrinsically safe face signal and communication unit. Connected to this unit and spaced approximately every 7 m along the stage loader and coalface conveyors are face signal and communication units. Each unit is equipped with a signal push-button and lockout stop key, and every third or fourth unit incorporates a loudspeaker and microphone.

A control point attendant at the face signal and communication unit position controls the stage loader and face conveyor in response to signals transmitted by any of the signal push-buttons. Upon operation of the start button for the stage loader (which, in turn, automatically starts the face conveyor) a seven-second pre-start warning two-tone 'bleep' is transmitted along the whole length of the stage loader and face conveyor, warning faceworkers that the conveyors are about to start.

Operation of a lockout push-button causes the respective conveyor to stop. Should this conveyor be the stage loader, the face conveyor will also stop, as they are connected in sequence.

Should the lockout push-button be latched in the lockout position, the conveyor cannot be started until that push-button has been reset. Each lockout push-button has a specific number which is automatically displayed by a digital readout on the face signal and communication unit whenever a particular lockout is operated. By the use of such communication facilities, the cause, necessary remedial action and subsequent duration of a stoppage in production can be quickly ascertained.

The face signal and communication unit is connected by cable to the main colliery control room on the surface, which permits instant and direct communication between the surface control room and any point along the working face or vice versa.

Each power loader is controlled by an individual operator, sometimes by using radio control. Before the machine can be started, water must be turned on to the pre-start warning water jets positioned on either side of the cutting drum. This condition must persist for approximately 7 s before power can be switched on to the machine, thereby warning by wetting anyone inadvertently in a dangerous position that the machine is about to start.

32.6 Power loaders

The modern coal-getting machine is termed a 'power loader' because it not only cuts coal, but also loads it onto the armoured flexible conveyor (AFC), which is in effect a steel scraper conveyor running the full length of the working face. *Figure 32.9* illustrates a typical modern power loader which has a rotating cutting disc at each end of the machine, mounted on a ranging arm to cater for thicker seams of coal which could be 2–3 m or more. This type of machine is known as a double-ended ranging drum shearer (DERDS). Certain methods of mining call for the use of a machine with only one cutting disc. Such a machine,

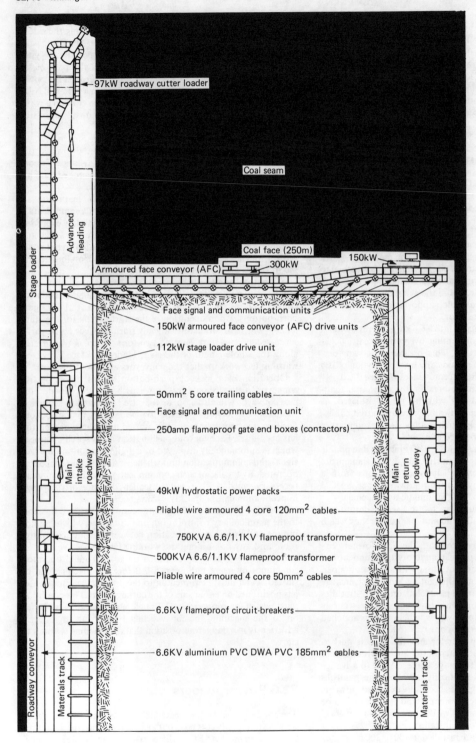

Figure 32.8 Layout of typical 250 m coalface

similar to that illustrated, is termed a single-ended ranging drum shearer (SERDS).

Most power loaders are driven by a single 150 kW motor, operating at 1100 V. Some larger machines, however, have been developed using a single 300 kW or 2×300 kW motor arrangement. The supply is obtained from a gate-end box in the roadway via a flexible trailing cable, which on the coalface is enclosed in a robust flexible cable handling device for protection.

Each complete machine is built up of a number of sections, consisting basically of electric motor, haulage unit and gearhead.

Incorporated in the motor are a reversing isolator, control facilities and fault diagnostic equipment, which are connected to

Figure 32.9 Power loader

the flexible cable by a flameproof plug and socket. A drive shaft protrudes from each end of the motor to transmit power to the adjacent units.

The motor at one end is attached to the haulage unit, which provides the hydraulic power, to haul the machine along the coalface via a driven pinion on the machine, and a static rack attached to the AFC. Speed control is automatic, following the limits determined by the operator who travels with the machine.

Attached to the opposite end of the motor (single-ended machine) is a gearbox, which drives the rotating coal cutting disc. In a double-ended machine a similar gearbox is attached to the haulage unit at the opposite end of the machine.

Radio control techniques have been developed for power loaders, which allow the operator to control the machine from a comparatively safe area, some 15–20 m from the machine.

32.7 Heading machines

One of the principal requirements of modern mining is the ability to drive roadways quickly and safely. This is achieved in the main by the use of heavy-duty roadway cutter loaders (*Figure 32.10*). These machines are equipped with one or two rotary cutting heads, which cut out the stone and shape the profile of the roadway to 4 m × 3.5 m.

Debris from the cutting heads falls onto a rotary scraper conveyor at the front of the machine, which transfers the cut material to a bridge conveyor at the rear of the machine, and then on to the roadway conveyors (*Figure 32.8*).

Roadway machines are fed by flexible trailing cables at 1100 V and typically have a 50 kW motor driving a hydraulic power pack and a 50 kW motor driving each cutting head. The machine traverses on hydraulically powered caterpillar tracks, the whole machine being controlled by an operator sitting in the middle of the machine.

32.8 Flameproof and intrinsically safe equipment

During the process of extracting coal from the seam, methane gas is usually given off. It combines with the normal mine air flow and is eventually disposed of at the surface of the mine through the mine ventilation system. By this means the methane content in the mine air is kept to low and safe proportions.

Figure 32.10 Boom miner with bridge conveyor

Owing to possible malfunctioning of ventilation apparatus, power supply failure or heavy emissions of methane, the methane/air ratio can increase. Between approximately 5 and 15% methane to air, the mixture becomes explosive, the most explosive mixture being 8.3% methane to air. Electrical equipment in which sparking during normal operation may occur is capable of igniting an explosive methane/air mixture and must therefore be given special consideration. For equipment operating at low voltage and current levels, the circuits can be designed such that the energy released at the spark is insufficient to cause ignition. This is generally achieved by the use of non-inductive resistors, non-linear resistors, capacitors, shunt diodes, Zener diodes, fullwave rectifiers, etc., to give 'intrinsically safe' (IS) apparatus.

Apparatus classed as 'intrinsically safe' includes telephones, signals, communications, testing instruments, methanometers, and remote control and monitoring. Since any open sparking produced within such equipment is incapable of igniting an explosive methane/air mixture, no other form of protection is required other than to house the components in a robust enclosure.

In the UK intrinsically safe apparatus is designed to conform to the requirements of BS 1259 (Intrinsically Safe Electrical Apparatus and Circuits) and must be certified by the Health and Safety Executive.

Intrinsically safe equipment is certified as Class 1 for use in mines, and the classification reference and intrinsically safe certificate number along with other specified information must be clearly marked on each item of equipment.

One of the principal requirements of the intrinsically safe certificate is that the equipment must be supplied from an approved source of supply, which can be either a.c. or d.c. The current British Standard covering such supplies is BS 6182 (Intrinsically Safe Power Supplies) and caters for a.c. and d.c. supplies and rechargeable battery units.

There are three categories of d.c. supplies, i.e. 7.5 V, 12 V and 18 V, and two of a.c. supply, i.e. 12 V and 15 V. Rechargeable batteries of 8 V and 14 V are specified, each capable of being charged from the respective source of supply, i.e. the 8 V battery from the 12 V d.c. supply and the 14 V battery from the 18 V d.c. or 15 V a.c. supply.

Power equipment operating on higher current levels which produce sparks during normal operation, and which cannot be designed to be intrinsically safe, must be enclosed in a robust enclosure. If an explosive methane/air mixture exists in a mine roadway and enters the apparatus containing spark producing components, the flame resulting from the ignition will not be transmitted to the ambient atmosphere and so ignite the general body of the mine air. The enclosure in which spark producing components are housed is termed a 'flameproof enclosure'.

Equipment coming within this category covers such items as motors, contactors, switchgear, transformers, lighting fittings, plugs and sockets, cable couplers, etc., and requires to be certified flameproof by the Health and Safety Executive as safe for Group 1 gases (Methane). To meet the H & SE requirements the equipment, for the UK, has since 1971 been required to be designed and tested in accordance with the Group 1 requirements of BS 4683 Parts 1 and 2.

A 'Flameproof Certificate' is issued, the number of which must be permanently displayed on each item of equipment along with other relevant details required by the Standard, i.e. manufacturer's name, type reference, number of BS, etc.

The harmonisation of European Standards resulted in the issue in 1977 by CENELEC (European Committee for Electrotechnical Standardisation) of the EN50 series of standards, i.e. EN50-014 to EN50-020 Electrical Apparatus for Potentially Explosive Atmospheres. Equipment designed and certified to this standard is accepted by the European Community without further testing, etc.

The European Standards of 1977 were issued as British Standards, namely:

Electrical apparatus for potentially explosive atmospheres

CENELEC Standard EN50-	British Standard BS 5501	Subject
014	Part 1	General requirements
015	Part 2	Oil immersion 'o'
016	Part 3	Pressurised apparatus 'p'
017	Part 4	Power filling 'q'
018	Part 5	Flameproof enclosure 'd'
019	Part 6	Increased safety 'e'
020	Part 7	Intrinsic safety 'i'

Gaps associated with any removable covers, doors, motor shafts, etc., must conform to minimum dimensions laid down by the standard. 'Flameproof' apparatus must be designed, installed and maintained at all times within those limits. As a typical example of the tolerances permitted, gaps with a length of 12.5 mm must not exceed 0.4 mm and for 25 mm flanges the maximum is 0.5 mm. During normal maintenance procedures in the mine these gaps are periodically checked by colliery craftsmen using feeler gauges.

32.8.1 Flameproof transformers

Providing sufficient electrical energy at the modern coalface with a demand of some 1–1.5 MVA at a distance of 6–8 km from the shaft bottom and at a depth of 900–1000 m requires an efficient h.v. electrical distribution system.

From the voltage regulation point of view it is essential that the h.v. supply be taken up to the working face, and since this voltage is not suitable for general utilisation on the working face, transformation facilities are needed to provide the coalface utilisation voltage of 1100 V.

Before the advent of flameproof dry-type transformers, standard industrial oil filled transformers were utilised, often with flameproof oil circuit-breakers attached to each end. Mining legislation at that time, however, decreed that such transformers could not be used nearer than 300 yards from the coalface and in no circumstances could they be used in a return roadway. As coalface loading increased in the 1950s and early 1960s as a result of the mechanisation of coal-getting, it became of paramount importance to move the transformer right up to the working face and strengthen the h.v. distribution system. This led to the development of the flameproof air-cooled transformer, which has a flameproof h.v. circuit-breaker on the h.v. side and a flameproof m.v. chamber mounted on the opposite end to house the overcurrent, sensitive earth leakage and short-circuit protection equipment, which on operation causes the h.v. circuit-breaker to trip. Typical transformer ratings would be 500, 750 and 1000 kVA. Comparing these with the old oil filled types (which were usually of the order of 250 kVA) indicates the change which has taken place in the mining industry since nationalisation in 1947.

Figure 32.11 shows a typical 750 kVA flameproof transformer viewed from the h.v. end, the circuit-breaker being a 6.6 kV 400 A 150 MVA SF_6 unit complete with incoming cable adaptor suitable to accept the 6.6 kV 300 A 6-bolt cable coupler attached to the end of the incoming 6.6 kV h.v. cable. The m.v. chamber at the opposite end is approximately $\frac{3}{4}$ size of the h.v. circuit-breaker and is similarly equipped with flameproof adaptors to accommodate the outgoing 1100 V cables.

Figure 32.11 Flameproof transformer

The transformer is equipped with lifting lugs for loading and unloading and adjustable wheels for transportation underground. The total weight would be of the order of 5–5½ tonne.

32.8.2 Flameproof switchgear

Prior to the mid-1960s nearly all the h.v. and m.v. circuit-breakers used in mines were of the oil-break type, usually of the order of 150/200 A 3.3 kV 25 MVA; and although certified flameproof, they constituted a hazard owing to the oil-fire risk. In addition, maintenance of the oil was a problem due to the oil transport and cleanliness, and the disposing of waste oil. With the introduction of no-oil switchgear, maintenance was reduced and the oil-fire risk eliminated.

Modern flameproof mining-type circuit-breakers operate on either the air-break, vacuum interrupter or sulphur hexafluoride gas principle and can be arranged such that they can be utilised as single free-standing units, on a complete switchboard, or mounted on the h.v. end of a flameproof transformer.

Owing to the increase in demand for electrical power underground and the uprating of underground distribution systems, switchgear ratings have also increased. A typical modern flameproof circuit-breaker as shown in *Figure 32.12* would be a 6.6 kV 400 A 150 MVA unit. This circuit-breaker is of the vacuum interrupter type and the illustration shown is a classical example of the construction of flameproof switchgear. The flameproof enclosure is divided into separate flameproof compartments electrically linked by the use of flameproof bushed terminals. In the bottom compartment the circuit-breaker is housed on a withdrawable chassis complete with overcurrent, earth leakage and short-circuit protection.

Two separate compartments are provided in the centre of the circuit-breaker which accommodate the isolator(s) and incoming or outgoing cable terminations. Also incorporated in this section are the isolator and circuit-breaker operating handles, which are mechanically interlocked to ensure that the isolator can be

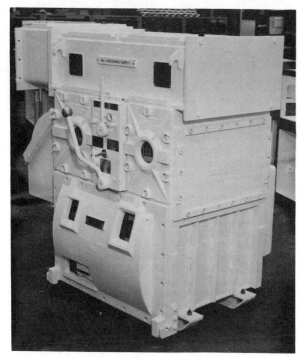

Figure 32.12 Flameproof circuit-breaker (a)

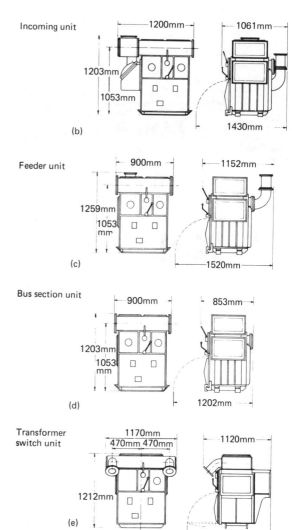

operated only with the circuit-breaker in the 'off' position. Mechanical interlock is also provided between the isolator and circuit-breaker handles and the front cover of the circuit breaker compartment, to prevent access until the interior has been made dead.

With slight modifications to the basic design, flameproof mining-type circuit-breakers can be used in either of four different modes, i.e. incoming unit, feeder unit, bus-section unit and transformer switch unit.

32.9 Gate-end boxes

The control of individual drives in a coal mine, such as conveyors, power loaders, pumps, haulages, etc., is achieved as in any other industry, i.e. by use of contactors.

For mining purposes contactors have to be enclosed in a robust flameproof enclosure or box. In the early days of electricity in mines, such a box would be installed at the end of the roadway leading to the coalface. To use mining parlance, a roadway is a 'gate'; therefore the 'box' installed at the 'gate end' became known as the 'gate-end box'. The term is still used today to refer to a flameproof contactor unit.

Figure 32.13 illustrates a typical flameproof gate-end box

Figure 32.14 Bank of gate-end boxes

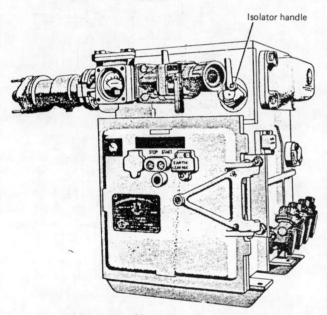

Figure 32.13 Gate-end box

suitable for use on an 1100 V, 3-phase, 50 Hz system, and rated at 250 A. The box can be adapted for use as a single unit or assembled to form a 'bank of panels'. *Figure 32.14* shows a typical bank of panels mounted on a pantechnicon over a roadway conveyor adjacent to a coalface.

A gate-end box consists of upper and lower chambers interconnected by flameproof bushed terminals. The upper compartment contains 300 A throughgoing bus-bars and a 3-phase isolator. The ends of the bus-bar chamber are designed to accept flameproof bus-bar trunking units and links to enable individual panels to be built up into a bank or to accept a bus-bar blank at one end and an incoming cable adaptor at the other to form a single unit. In the lower compartment is housed the contactor of rating 150, 250 or 300 A, and a control unit, all mounted on a removable chassis for ease of maintenance, repair or removal.

A mechanical interlock is provided between the isolator operating handle and the contactor compartment cover to ensure that access to the contactor compartment is prevented until the isolator has been placed in the 'off' position.

Power is transferred to the drive from the gate-end box via a flexible trailing cable which, owing to the hazardous environment of the coalface, is always susceptible to damage. Facilities must therefore be provided to enable the trailing cable to be changed quickly. This facility is provided on the gate-end box and the motor in the form of a 200 A restrained flameproof plug and socket (*Figure 32.15*).

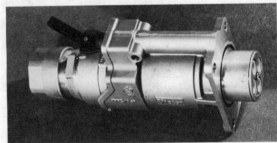

Figure 32.15 Flameproof plug and socket

Taking into consideration the control facilities, sensitive earth leakage protection, static overcurrent protection equipment, etc., the modern gate-end box is a complex piece of electrical equipment which has to work with a high degree of reliability in an atmosphere that can be hot, cold, dry, wet and on occasion subject to considerable vibration. Some of the electrical/electronic components can therefore be subject to abnormal abuse while in service. It is for this reason that the evolution of modern gate-end box has resulted in the majority of the small electrical/electronic components being contained in a 'control unit', which is a plug-in unit on the contactor chassis. This control unit can be quickly changed in the event of trouble, and transported to the surface workshop for overhaul and repair.

Control of the gate-end box can be effected locally or remotely, the selection being by way of a changeover switch on the control unit. In the majority of cases the remote control facility is adopted: it utilises a pilot control core in the 5-core flexible trailing cable and a flameproof starting device at the motor end. The pilot circuit (more correctly termed the intrinsically safe remote control circuit) is designed to NCB Specification P130, based on the following principles:

(1) The circuit is energised from an intrinsically safe constant-voltage transformer within the gate-end box, designed to give a constant 12 or $7\frac{1}{2}$ V secondary output over a wide variation of primary input. One side of the 12 or $7\frac{1}{2}$ V winding is earthed.

(2) A pilot relay is provided in the gate-end box which will operate on half-wave but not on full-wave a.c.

(3) At the far end of the trailing cable a diode is provided along with a start switch across which is connected a 30 Ω resistor.

Figure 32.16 shows a basic arrangement of the contactor coil/pilot circuit within a gate-end box, the associated trailing cable and a face machine with inbuilt start switch. To start the

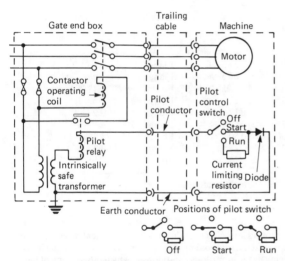

Figure 32.16 Contactor-coil/pilot circuit

machine, the switch is moved to the 'start' position. With only the diode in circuit the pilot relay (PR) energises and the machine starts. The start switch now reverts automatically to the 'run' position inserting the 30 Ω resistor into circuit.

Should a power failure occur with the control switch in the 'run' position, following which the power is restored, the relay PR will not energise at a voltage below 120% of the declared voltage of the incoming supply. Operation of the start button causes PR to energise: it must do so down to 75% of the declared incoming voltage supply. Once energised and with the 30 Ω resistor in circuit, PR must remain energised down to 60% of the declared voltage but under no circumstances continue to operate at 20% or below.

Should a damaged trailing cable result in a pilot core-to-earth fault, full wave a.c. would be applied to PR which (on account of the increased impedance) would de-energise if in operation, or fail to energise upon operation of the start button. This condition, known as pilot core protection (PCP), would be indicated by a lamp at the front of the gate-end box.

Prior to 1959 most transformers used underground to supply coalface equipment had the neutral point of the secondary windings solidly earthed, and earth leakage protection in gate-end boxes operated on the core-balance principle. Trailing cable damage on such systems resulted in very severe incendive arcing which, on the coalface especially, was a very serious hazard. Following an explosion at Walton Colliery near Wakefield in 1959, H.M. Principal Inspector of Mines recommended that a further attempt should be made to devise an electrical protective system capable of eliminating, or at least substantially reducing, the dangers of incendive sparking resulting from damaged trailing cables. This led to the development of the sensitive earth leakage (SEL) circuit, which exists in two forms, single point or multi-point.

32.9.1 Single point SEL (Sensitive Earth Leakage)

The basic principles of single-point earthing systems are similar to those of solidly earthed systems in that a core-balance transformer more sensitive than used on solidly earthed systems is employed. This system is sometimes referred to as sensitive core balance, and the main difference between the two systems is in the method of earthing the neutral point of the transformer secondary winding.

In the single-point system an impedance is inserted between the neutral point and earth of such value as to limit the earth fault current to a maximum of 750 mA (*Figure 32.17*). Although this is the maximum earth fault current permitted, individual earth fault trip circuits are set to trip at between 80 and 100 mA, giving a safety factor of approximately 7 to 1.

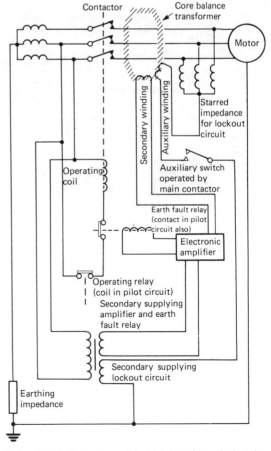

Figure 32.17 Protection unit for high-impedance single-point earthing

The core-balance transformer output under fault conditions is very small and an electronic amplifier is used to control the earth fault relay, which is energised under healthy conditions and de-energises on the occurrence of a fault. This arrangement results in a fail-safe system. The earth leakage relay contacts are inserted in the contactor coil circuit, which opens on the occurrence of a fault.

To ensure that a contactor cannot close onto a system on which an earth fault condition exists, an additional arrangement is provided which is termed the electric lockout circuit. It consists of three impedances, one end of which is connected to the three outgoing phases; the other ends are star-connected and, in turn, are connected to an auxiliary core-balance transformer winding, a pair of contactor auxiliary contacts (closed when the contactor is open) and an intrinsically safe source of supply which has one end of its winding earthed. Should an earth fault develop while the contactor is de-energised, the earth leakage lockout circuit would operate to prevent the contactor coil energising, a condition that would persist as long as the earth fault was in existence.

32.9.2 Multi-point SEL (Sensitive Earth Leakage)

In the multi-point system the transformer secondary is completely insulated from the earth, i.e. it is a free neutral (*Figure 32.18*).

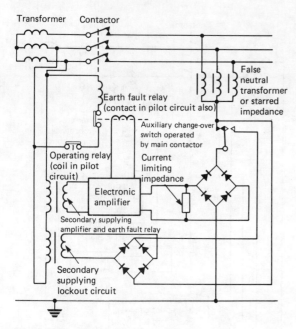

Figure 32.18 Protection unit for multi-point earthing

Each contactor is provided with a false neutral which, similar to the single point system, consists of three impedances connected to the three outgoing phases. The star point is connected via a pair of contactor auxiliary change-over contacts to either the fail-safe earth leakage detection circuit or the earth leakage lockout circuit, depending on whether the contactor is energised or not.

The earth fault current on the multi-point system is limited to 40 mA on a 1100 V system. Since an earth fault on a system supplied from one transformer could cause all gate-end boxes on the system to trip, and in order to keep the maximum earth fault current to 750 mA, the number of gate-end boxes on a system must be limited to $750/40 = 18$.

When a contactor trips on earth leakage on either system, that contactor locks out and displays earth leakage trip conditions, which can only be reset by an authorised craftsman with the appropriate specialised equipment.

32.10 Flameproof motors

Mechanisation of the coal mining industry in the 1950s, followed by further mechanisation and automation in the 1960s and 1970s, completely changed the face of the industry.

Elimination of 'self-generation' at collieries and the introduction of duplicate and more substantial power supplies from the Electricity Boards plus the strengthening of colliery distribution systems made the direct-on-line starting of large cage induction motors possible.

NCB Specifications have been produced covering the majority of FLP motors used in British mines:

National Coal Board Specifications for flameproof motors

NCB Spec.	Motor rating (kW)	Mounting	Voltage (kV)
291	37.5–50	Flange	up to 1.1
420	67.5–90	Flange	up to 1.1
542	2.25–30	Foot/flange	0.11–1.1
625	112	Flange	0.55–1.1
634	112	Flange	2.2–3.3*
635	112	Flange	2.2–3.3†
636	150	Flange	1.1
637	225	Flange	1.1
651	75	Foot	0.55, 1.1, 3.3**

* Single-stack conductors.
†Other than single-stack conductors.
**For booster fans.

Motors associated with coalface equipment need changing more often than those operating in roadways and engine houses, owing to the hazardous conditions in which the equipment has to

Figure 32.19 Flange-mounted flameproof motor

operate on the coalface, physical damage and ingress of water-moisture being the prime cause of failure. To facilitate speed and accuracy in changing and lining up, such motors are designed for flange mounting. *Figure 32.19* shows a typical flange mounted motor complete with flameproof terminal box and flameproof 200 A plug and socket.

32.11 Cables, couplers, plugs and sockets

All cables used in mines for any purpose must conform to the requirements of the relevant NCB Specifications, which are as follows:

National Coal Board Specifications for cables

NCB Spec.	Subject
P115	Shotfiring cables (other than in shafts)
P188	Flexible trailing cables (for coal cutters and similar use)
P295	PVC-insulated wire armoured and sheathed cables
P492	PVC-insulated wire armoured telephone cables
P493	PVC-insulated wire armoured signalling cables
P504	Flexible trailing cables with galvanised steel pliable armouring
P505	Flexible trailing cables (for drills)
P610	Flame retardant properties of flexible trailing cables
P648	Multicore PVC-insulated wire armoured and PVC sheathed 0.6–1.0 kV cables (with special screening for mine winder safety and control circuits)
P653	Flexible multicore screened auxiliary cables with galvanised steel pliable armouring
P656	EPR-insulated wire armoured and PVC sheathed cables

Colliery surface and underground h.v. and m.v. distribution systems utilise, in general, PVC insulated and sheathed mains cables. DWA cables have up to a few years ago been exclusively used for underground systems but recently SWA cables have received favourable consideration, owing to reduced cost and flexibility in handling. Such cables conform to NCB Specification 295 or 656.

Wire armoured roadway cables are usually received at the colliery on 100 m drums with 300 A flameproof couplers already fitted at each end. On completion of assembly the cable coupler is filled with either bituminous compound or a cold-pouring compound, consisting of a bituminous oil and a hardener. A typical 6.6 kV 300 A flameproof cable coupler is shown in *Figure 32.20*, the halves being connected by the use of three connector pins, a rubber sealing gasket and six connecting bolts.

Mobile machines such as power loaders and roadheading machines, which require to move while energised, must be powered by the use of a flexible trailing cable to satisfy the requirements of NCB Specification 188. Several types are available, ranging in sizes from 16 mm² to 95 mm². *Figure 32.21* illustrates a typical trailing cable (type 7). It consists of three power cores, a pilot control core and an earth core. The earth

Figure 32.20 Flameproof cable coupler

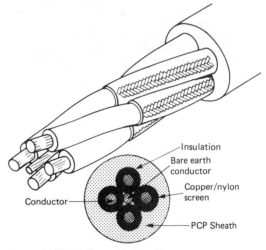

Figure 32.21 Trailing cable (type 7)

core is uninsulated, and is located in the centre of the cable, with the three power cores and the pilot core equally spaced around it. All four cores are insulated with ethylene propylene rubber (EPR); the power cores have an additional copper/nylon screen over the insulation. Overall protection is provided by a tough heavy-duty polychloroprene (PCP) oversheath. A similar cable but of slightly different construction is the type 10 cable shown in

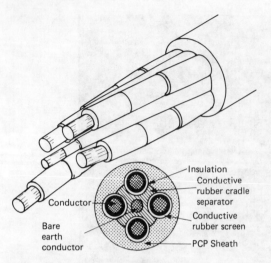

Figure 32.22 Trailing cable (type 10)

Figure 32.22. The three power cores and pilot core are again EPR insulated, but each has a conductive rubber screen, and all four cores are laid up around a conductive rubber cradle separator. The overall protective sheath is PCP.

Since trailing cables associated with mobile machines are susceptible to damage, means must be provided to quickly connect or disconnect the cable from the motor/gate-end box. This facility is provided in the form of a 200 A 1100 V flameproof restrained plug and socket, a typical example of which is shown in *Figure 32.23*. The socket is attached to the motor/contactor and the plug is attached to the cable. A scraper earth facility is provided on the socket which mates with the nose of the plug as it is inserted to maintain the necessary earth connection. Four pins are provided on the plug and socket assembly, three power cores and one pilot control core. The pilot pin is shorter than the power pins, so that on insertion it makes contact after, and on withdrawal it breaks before, the power pins, preventing their making or breaking on load.

Pliable wire armoured cables are used to power permanent and semi-permanent apparatus such as transformers and gate-end box assemblies which move up automatically as the coalface advances. Such cables must conform to NCB Specification 504, and range in size from 10 mm^2 to 150 mm^2 at voltages between 660 V and 6.6 kV. *Figure 32.24* illustrates a typical 6.6 kV pliable wire armoured cable (type 631) which has four cores insulated with EPR. A copper/nylon screen is provided over the insulation on the power cores, and all four cores are laid up around a PCP centre, over which is provided a PCP sheath. Over the inner PCP sheath is a galvanised steel strand armouring and an overall sheath of PCP.

Two, 3-, 4- and 5-core pliable wire armoured cables of similar construction but with much smaller conductor size, e.g. 4 mm^2, are used for control circuits and coalface lighting installation. Such cables are termed types 62, 63, 64, 70 and 71, respectively.

Five-core 6 mm^2 flexible cables are used to power hand-held drilling machines which operate at 125 V 3-phase 50 Hz. Type 43 has three power cores, one pilot core EPR insulated and one earth core conducting rubber covered laid around a conducting rubber cradle separator, screened with conducting rubber and a heavy duty overall sheath of PCP. Type 44 has five EPR insulated cores with the three power cores copper/nylon screened, laid up round a PCP centre with a heavy-duty PCP sheath overall.

Cables for telephone communication are designed to NCB Specification 492, and can be either PVC SWA PVC or PVC

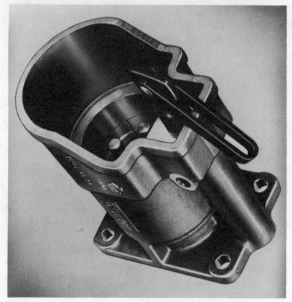

Figure 32.23 Flameproof restrained plug and socket

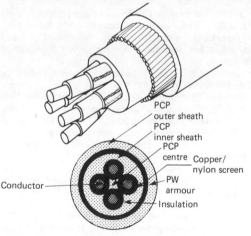

Figure 32.24 Flexible wire-armoured cable

DWA PVC. The cores are laid up in one to 91 pairs of 1.5 mm² conductors and follow a set colour code.

Signal cables conform to NCB Specification 493, and can also be PVC SWA PVC or PVC DWA PVC. These cables have conductor sizes of 1.5 mm², range from 2 to 91 cores and are laid up as single cores to a colour code.

32.12 Drilling machines

Mechanisation and automation have reduced the application of hand-held drilling machines. A typical machine is shown in *Figure 32.25*. Such machines are rated at 1.1 kW 120 V, 3-phase, 50 Hz, and operate from a purpose-designed flameproof drill

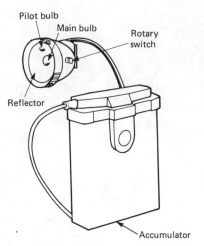

Figure 32.26 Cap lamp

Figure 32.25 Hand-held drilling machine

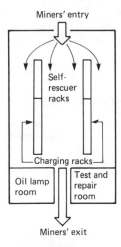

Figure 32.27 Layout of lamproom

panel. Drill panels are approximately the same size and shape as gate-end boxes and are designed so that they can be connected together mechanically. The drill panel contains its own step-down transformer to feed the drilling machine, contactors, protection, etc. The supply to the drill is taken from the drill panel via a 30 A flameproof plug and socket, and a 5-core drill cable, consisting of three power cores, pilot core and earth core.

32.13 Underground lighting

Illumination underground is provided in accordance with the Coal Mines Safety and Health Regulations and to improve environmental conditions. The areas illuminated are pit bottoms, haulage stations, locomotive stations, main trunk conveyors, assembly areas and main roadways where men pass to and from their place of work.

Lighting underground is provided in four ways: (1) by a portable lamp carried by each person underground; (2) by permanent lighting installations supplied by a power transformer at 120 or 240 V 50 Hz; (3) by mains lighting which forms part of a mobile machine; (4) by portable compressed-air turbines.

The cap lamp (*Figure 32.26*) comprises a headpiece provided with a main 4 V, 0.9 or 1.0 A lamp and a 4 V 0.46 A pilot lamp. The headpiece, which can be carried or worn on a special helmet, is connected by flexible cable 2-core 1.0 mm² vulcanised-rubber insulated PCP sheath. The assembly is covered by BS 4945. The battery consists of two lead–acid cells in a polycarbonate moulded case giving 4 V output. A fuse is provided in the battery top to afford protection. The capacity of the batteries is 13 or 16 Ah.

To prevent interference underground, all lamps are locked and sealed before issue at the surface lamproom. *Figure 32.29* shows a typical lamproom layout. The miner enters the lamproom and collects his own personal cap lamp before entering the mine. On his return he places his lamp on the charging rack in the lamproom, when recharging commences automatically. During the period between shifts, the lamps are examined, cleaned and topped up ready for the next period of duty.

Permanent lighting installations below ground are similar to those provided on the colliery surface, viz. filament, fluorescent, discharge, sodium, and mercury lamps, in substantial dustproof fittings outside hazardous areas, and flameproof to Group 1 (Methane) in the hazardous areas. Special attention is paid to the design of electric circuits, proper loading, and fault protection, i.e. earth fault, overload and short-circuit.

Illumination on mobile machines is provided by equipment designed and tested for the automotive industry but fitted into flameproof lights with protection to prevent damage. In mining development work where the heading is progressively moving, the provision of lighting on machines is advantageous.

Figure 32.28

Coalface lighting has had a long history. Numerous approaches and types have been used but have failed for various reasons. The conditions imposed by the Coal Mines Act 1911 limited such installations to 'naked-light mines' only; it was not until 1934 that mains lighting on the face in safety lamp mines was permitted. In 1973 mains fed intrinsically safe lighting was developed using 12 inch 8 W fluorescent tubes on a high-frequency supply. This has resulted in a smaller, lighter, more easily maintained system.

32.14 Monitoring and control

Different types of monitoring and control systems are in use at collieries: either 'surface only' installations, designed as for other industries, or 'surface/underground' systems, which are specifically designed for use underground, requiring Intrinsic Safety.

A typical system for use underground consists of a surface control or central station, a number of two-way transmission links (called 'rings') to convey data over distances up to 8 km, to and from numbers of underground FLP/IS outstations to which are connected transducers for monitoring and control facilities. These systems may perform individual functions, such as coal clearance by remotely controlling conveyors, environmental monitoring, pumping, fan monitoring, etc.; or may be multi-functional, as for a small mine.

A surface control or central station would consist of a mini/micro computer, controlling a number of transmission rings. Transmission rings are usually from 2- to 6-wire time-division multiplex data transmission systems, capable of approximately 500 bit/s, coupled to a maximum of 32 outstations. Some frequency-multiplex systems are used: in this case the signals are passed along armoured low-loss coaxial cable. Audio frequencies, modulated low radio frequencies or modulated VHF frequencies are employed.

The outstations, sited underground near to the plant concerned, are constructed part FLP to give a number of IS supplies from 110/240/550 V, and part non-FLP to house the printed circuits and terminations. The outstations are required to collect information from various transducers and other circuits, pass this information to the transmission rings and receive from them surface information and commands. Action can then be taken by the outstation on these commands, or independently appropriate shut-down and alarm action can be taken through fault-tripping logic, according to the state of the protective transducers and other monitors associated with the plant.

Where possible, all circuits are designed for 'failure to safety'; therefore a.c. signals are preferred, with the final control operator being a relay, supplied through an isolating transformer or 'diode pump' circuit, fed through two high-voltage series capacitors, the relay being energised in the healthy state.

On/off transducer signals are obtained using a half-wave rectifier at the transducer end and an a.c. IS supply to provide open- and short-circuit cable protection.

The surface control (*Figure 32.28*) would be in an air-conditioned room and may house many separate control and monitoring systems, together with a data collection computer system, using larger disc storage. The electrical supply is taken from a source as near to the colliery feed as possible, through CVTs to give as clean and permanent a supply as possible.

32.14.1 Computer system

Hardware Mini- and/or micro-computers; floppy or hard disc loading and data storage; all parity checks utilised; 'watchdog'-type timer system; system scan time with 1 s; 2-colour graphic VDUs displaying mimic representation of system—faults, alarms, etc.— in different areas of the screen; keyboard used to effect system changes and issue commands; printer—300 baud rate to print-out shift analysis of data, breakdowns, etc. Suitable intrinsically safe interface to the underground transmission system. All changeable type ROMS, marked to a standard by the suppliers. Computer interconnections via 20 mA serial loops, terminals, VDUs, etc., connected via either 20 mA loop or VT24.

Software Normally written in assembler of Coral-66 languages. Most systems have executive-type control software to give safer systems and provide rapid response to real-time requirements. Software is written to be as inherently safe as possible and is resident in memory at all times. Some systems perform periodic check sums on ROM and RAM memory, as well as transmitted data. Entry to make system changes or give commands is restricted by level of password. Some diagnostic software is available to check much of the hardware, including on-line checks of transmission data. A software system is supplied to a colliery in a package form, containing an operating system and a set of modules or application programs from which a particular system may be configured. Configuring, carried out in password, is on a questionnaire basis and allows the operator or engineer to add or delete items of plant control facilities—transducers, etc.; change set points, levels and limits of analogue signals; set outstation types; build, using graphics, a representative mimic of the colliery system; select data to be periodically printed; and so on.

32.14.2 Underground transducers

Listed below are some of the many different types, together with a few of the varied monitoring applications of each:

Temperature: bearings, brakes, temperature rise.
Pressure: barometric, fluid, differential (for flow).
Flow: liquids, ventilating air, gas.
Proximity: position and movement (for velocity and acceleration).
Weight, volume, force: belt load, static weighers, chain force, torque, tension.
Vibration: bearings.
Gas and atmosphere analysis: methane, oxygen, carbon monoxide, carbon dioxide, humidity, dust, smoke.
Electrical analysis: power, voltage, current.
Level: coal/stone bunkers, fluid level (e.g. of lubricating oil).

Most transducers, including any electronic circuitry, are designed to be IS 'fail-safe'. With on/off switching, a.c. IS voltages are used, with a half-wave rectifier at the transducer. Analogue signals are 0.4–2.0 V, or 4–20 mA. Underground transducers must be manufactured from approved materials such as brass to eliminate incendive sparking risks.

Analogue temperature Low impedance such as 100 Ω platinum resistance, or thermistors to enable more points to be monitored from one IS supply. Measuring electronics 'fail-safe' throughout, giving variable shutdown/alarm level, or in some cases (such as monitoring of compressors) fixed maximum shutdown levels. Maximum air temperature would be 160 °C and oil 80 °C.

Pressure Measurement from 0 to 0.25 kPa, to measure differential pressures across stoppings to 1500/bar/in^2 for hydraulic pressure. Lower pressures are measured using diaphragm-type transducers; higher pressures are monitored usually by strain-gauge bridges.

# 33

Agriculture and Horticulture

M G Say, PhD, MSc, CEng, ACGI, DIC, FIEE, FRSE
Heriot-Watt University

Contents

33.1 Introduction 33/3

33.2 Supply 33/3

33.3 Installation 33/3
    33.3.1 Earthing 33/3
    33.3.2 Safety 33/3

33.4 Electrical equipment 33/4
    33.4.1 Motors 33/4
    33.4.2 Fans 33/4
    33.4.3 Pumps 33/4
    33.4.4 Feed mills and mixers 33/4
    33.4.5 Elevators, hoists and conveyors 33/5
    33.4.6 Heaters and coolers 33/5
    33.4.7 Auxiliary plant 33/6

33.5 Farming processes 33/6
    33.5.1 Dairying 33/6
    33.5.2 Feed processing 33/6
    33.5.3 Drying 33/6
    33.5.4 Conditioning of livestock housing 33/7
    33.5.5 Irrigation 33/8

33.6 Horticultural processes 33/8
    33.6.1 Greenhouse operation 33/8
    33.6.2 Field operation 33/9

33.7 Information 33/10

33.1 Introduction

The ambient conditions in which electrical energy is applied in farms, market gardens and commercial greenhouses differ radically from those in urban shops, offices and domestic premises. Some farm machines are exposed to the weather; many must operate in the presence of humidity, water, effluent and gaseous contamination; and livestock are highly susceptible to electric shock. In horticulture the cost of fuel, labour, feed and fertilisers, and other main items concerned in crop growth and marketing, has increased greatly of late. The careful use of electricity can be effective and cost-saving as a main energy source and as a control medium. Combination of the power and control facilities enables a measure of automation to be obtained for reducing a number of tedious and lengthy manual tasks.

Apart from scale, the needs of farming and horticulture installations are basically the same in respect of heating, cooling, ventilation, drying, lighting, irrigation, planting, gathering and transport. In farming, the intensive rearing of pigs and poultry, and the milking of dairy cows, have benefited by the application of electrical methods. Semi-automatic planting, propagation, gathering and preparation for the market have contributed to competitive horticulture.

This chapter does not deal with agricultural or horticultural processes in detail, but is concerned with the general electrical applications thereto.

33.2 Supply

In the UK the supply is normally obtained from the 50 Hz a.c. mains at 240 V (1-phase) or 415 V (3-phase), by overhead lines or buried cables. For large installations a higher voltage may be supplied, requiring a local step-down transformer. The supply system impedance at the point of connection may often be such as to limit allowable motor starting currents.

Both overhead line and cable supplies introduce potential hazards. Chance contact with an overhead line may occur with irrigation booms, bale loaders, fork-lift trucks and metal pipes and ladders. Siting metal framed buildings or bale stacks under or too close to an overhead feeder, or shifting tall machinery, may cause serious mishaps. Hazards can be minimised by the adoption of sensible precautions and working procedures. Supply by cable is buried at depths related to voltage and situation with a minimum of 0.45 m; but topsoil can be eroded by heavy rain and worn away by the passage of movable farm machinery. The location of a cable run must be known for due precautions to be taken, for accidents occurring when farm workers strike a cable with a hand- or machine-tool are not uncommon. Damage to a line or cable must be reported at once to the Supply Authority, and repaired only by its engineers.

33.3 Installation

The installation covers lines, cables, conduit, busbar chambers, distribution boards, trunking, control gear, socket outlets and plugs, and all equipment housing the live conductors back to the Supply Authority's metering.

Conduit is liable to suffer from condensation and corrosion, and the 'all-insulated' system is advised. The most suitable wiring (particularly in milking parlours and stables, where the atmosphere is moist and contains ammonia) is of tough rubber insulation with partially embedded braid applied prior to vulcanisation, and with an overall serving of insulating compound. One alternative is polychloroprene sheathed rubber. Another, polythene sheathed rubber insulated cable, does not support combustion, but is liable to damage by rodents unless protected. Cable ends and terminals should be covered with compound to resist corrosion. Where wiring is carried on roof beams, it should be located if possible on the underside, to reduce damage by rodents.

With the all-insulated wiring system, non-metallic junction boxes can be used, sited out of reach of livestock. Socket outlets are normally fixed; they should be of insulated construction with protective flaps. Special sockets and plugs are used for loads in excess of about 4 kW. The round-pin plug/socket has been superseded by the flat-pin type.

Power cables to motors may be carried in galvanised screwed conduit, but are better made by PVC insulated wire armoured and sheathed cable, which is easier to install and is corrosion proof; the combination of wire armouring and PVC sheathing gives ample mechanical and shock protection.

Farm and horticultural buildings are often scattered, lacking covered interconnections. It is then necessary to link buildings electrically by buried cable (an expensive method) or by PVC insulated and sheathed cable suspended by clips from a galvanised steel catenary wire, or a form of cable having a screened catenary wire incorporated in the lay. To obtain adequate height for safety, it may be necessary to suspend the cable between posts fixed to the roofs of the buildings concerned.

Portable and transportable equipment has often to be used in an adverse environment. Low-voltage operation from a step-down transformer, or a double-insulated construction, is desirable, although the flexible cord in the latter is the most vulnerable part. There is an increasing use of transportable generating sets: these can be dangerous and demand effective earthing.

Rewirable or cartridge fuses protect against overcurrent. Main and circuit fuses protect the whole and specific parts, respectively, of the installation. Fused plugs protect the flexible cords to which they are connected. Equipments such as motors for milking machines are usually provided with individual circuits with control switches and fuses. An alternative to the fuse is the miniature circuit breaker, a totally enclosed switch mechanism operable by hand but with automatic opening on overcurrent.

33.3.1 Earthing

Earthing is vital. The Health and Safety at Work Act states that everyone at work bears the responsibility to avoid putting himself and others at risk. Employers not themselves qualified to carry out an installation and earthing system must engage the expertise of those who are so qualified, to install a system to proper standards and in accordance with the relevant Codes of Practice. Supply Authorities can provide this service. Earth continuity conductors in the installation must be connected to a suitable earth. As many water mains are non-metallic, they cannot be relied upon. Earth rods or mats of the 'self-watering' type, or an approved multiple-earthing system, may be employed. Earth leakage circuit breakers can provide protection against the effects (e.g. fire and shock) of earth leakage, especially where there is difficulty in providing an effective earth connection. Where necessary, high-sensitivity current operated earth leakage circuit breakers offer additional protection.

33.3.2 Safety

Overloaded circuits, fuses of incorrect rating, damaged trailing cables, extensions with temporary joints, connection of portable power tools to lighting sockets: these malpractices can be very dangerous—more so in farming and horticulture than in the domestic environment. They contravene a number of Acts and Regulations, including the following:

Statutory Agriculture (Safety, Health and Welfare Provisions) Act, 1956; Agriculture (Stationary Machinery) Regulations, 1959; Factories Act, 1961; Health and Safety at Work etc. Act, 1974.

Non-statutory IEE Regulations for the Electrical Equipment of Buildings.

British Standards: 186 (Plugs, socket outlets and couplers with earthing contacts for 1-phase circuits up to 150 V); 1363 (13 A plugs, switched and unswitched, socket outlets and boxes); 3006 (Milking installations); 4343 (Industrial plugs and socket outlets and couplers); 5502 (Code of Practice for the design of buildings and structures for agriculture, Section 1.5 Services).

Some sensible practices for the safer use of electricity are quoted. In damp conditions (e.g. milk parlours) rubber boots should be worn. When cleansing down with pressure hoses or by steaming, fixed electrical equipment and switches should be protected by plastics sheeting. Switches should be operated with dry hands. Motors and controls should be kept clear of obstruction so that they are accessible in emergency. Installations should be checked for electrical integrity every two or three years. Earth leakage circuit breakers should be readily tested by the means provided. Damage to wiring and plant should be treated as an emergency and repaired forthwith.

33.4 Electrical equipment

Besides motor driven ventilating fans, water pumps, feed mills, cutters, grinders, mixers, conveyors and trucks, electricity is employed for luminaires, heaters, coolers, soil warmers and, in particular, control and automation.

33.4.1 Motors

Either 1-phase or 3-phase cage induction motors are normally employed. They are simple and robust, have a long life and can operate in situations impossible for internal combustion engine drive. It is good practice to choose totally enclosed machines, with external fan ventilation where necessary; however, motors for pumping water from wells may be of the submersible type. Rated speeds are 3–6% lower than the synchronous speed, which is most commonly 1500 rev/min (i.e. a 4-pole motor on a 50 Hz supply). Small loads suitable for 1-phase motors can be driven by the split phase type unless the starting torque has to be higher, in which case capacitance or repulsion machines may be needed. Typical ratings and direct-on-line starting current and torque per-unit values are:

Type	Rating kW	Current p.u.	Torque p.u.
Split phase	0.1–0.5	6–8	1.5
Capacitance start	0.1–5.0	3–5	2–4
Capacitance start/run	0.2–10	3–5	1.5
Repulsion start	0.3–3.5	2–3	5
Three-phase cage	1–350	4–7	2–3

Direct-on-line switching is likely to be restricted to motors of small rating. For 3-phase cage motors a simple starter (e.g. star/delta) will usually suffice. Switches, with standard overcurrent and undervoltage protection, should be sited close to the motor; but if the starter has to be remote, an additional isolator must be fitted at the motor end of the cable run, a requirement particularly important with grain elevators, conveyors and remote-controlled ventilation fans. Resistance speed controllers should never be used, as they constitute a fire risk.

The Agriculture (Stationary Machinery) Regulations require the adequate guarding of motors and associated belt drives.

33.4.2 Fans

Ventilation (with heating) is needed for grain and green crop drying and storage to control temperature and moisture. Typical ratings for a fan air pressure of about 1 kPa are:

Storage mass (t)	200	350	500
Air flow rate (m^3/s)	6	9	12
Motor rating (kW)	12	20	30
Heater rating (kW)	30	45	60

Centrifugal fans are commonly used for grain drying. For low-volume ventilation of moist grain, a centrifugal or an axial flow fan with a motor in the range 0.4–0.75 kW can be applied. Fans mounted in walls or other accessible places must be guarded.

Housed non-ruminants (e.g. pigs and poultry) and ruminants (cows and sheep) require ventilation. In the case of non-ruminants the combination of fan air speed and temperature may be critical.

33.4.3 Pumps

Water is needed for domestic use, for livestock (including drinking and washing), for crop spraying, for irrigation of fields and protected crops, and for the handling and dilution of effluents. There are several Regulations relating to supply and pollution. The best and most reliable source of water is the public main network; but where this is not available within a reasonable and economic distance, alternative sources must be found, such as rivers, bore-holes, wells and springs, with pumping and storage provided.

The water delivery of a constant volume (positive displacement) pump is proportional to speed, while that of centrifugal and axial pumps is proportional to speed squared; the former therefore demands the greater starting torque. A modern bought-in pump set incorporates a matched motor and switching, but ad hoc combinations can be direct coupled or linked by a vee-belt and pulleys. A vee-belt is often used with a reciprocating pump (a low-speed device), and with positive displacement pumps, where it forms a useful clutch in case of pump blockage.

Where the water level of a bore-hole or well is 10 m or more below the ground surface, direct pumping may be difficult. Deep-well reciprocating pumps may then be necessary, in which the pump plunger operates at the lower end of a large-bore delivery pipe carrying a reciprocating actuator rod, driven from the overhead gear mounted at the top of the well. Alternatively, a submersible pump can be used. This consists of a waterproof cage motor of very small diameter, directly coupled to a rotary pump of similar size. The unit, screwed to a delivery pipe, is lowered into the well until it is submerged. The motor is designed with a high current density and is cooled by immersion. A cutout is essential to act in the event of any failure of water flow.

It is not normally permissible to boost the pressure of a mains supply. With alternative supplies it is necessary to provide both pumping and high-level storage. Basically, the method comprises a raised storage tank and a pump with a simple on/off control to keep the tank water level between set limits, by float, pressure or time switching. Where it is not feasible to provide storage at a height suitable for gravity distribution, the pump is coupled to a sealed pressure tank of about 3 m^3 capacity. As water is pumped in, the air in the tank is compressed to about 30 kPa (40 lbf/in^2). A pressure sensor then stops the motor. Draw-off lowers the pressure and the pump is restarted.

33.4.4 Feed mills and mixers

The *hammer* mill consists essentially of a steel cylinder within which steel bar flails are rotated at high speed, reducing input grain or pulse charges to a meal, fine or coarse according to the mesh screen located at the bottom of the mill. In the *plate* mill, grinding takes place between fixed and moving fluted metal plates set vertically, fineness of grinding being varied by adjustment of the spacing between the plates. A *roller* mill comprises a feed hopper and two crusher rollers, one of which is motor driven. Typical motor ratings and outputs, in terms of tonnes per week of feed, are:

Motor rating (kW):	2.5	5	7.5	10	15
Feed (t/week):					
hammer mill (P)*	4	12	17	24	35
hammer mill (A)*	8	18	24	30	50
roller mill	12	30	48	60	—

* P with pneumatic, A with auger conveying.

The mill output may be combined with other ingredients in a mixer. Vertical mixers are used for pig and poultry feeds; but as cattle prefer flaked feed, a horizontal mixer is used for them, as it does less break-up damage to the flakes.

33.4.5 Elevators, hoists and conveyors

The hard manual effort required to lift sacks of corn or bales of hay can be eliminated by motor driven transportable elevators and hoists. A sack or bale *elevator* consists of a steel tray, 4–5 m in length and about 1 m wide, mounted on a transportable undercarriage and pivoted so that one end may be raised by a winch. Endless chains on both sides of the tray are joined at 0.5 m intervals by L section steel cross-bars to form an endless conveyor belt. The chains are driven through a gear-box by a motor. An alternative is a self-contained *hoist*, slung from a suspension hook. Socket outlets must be provided where the elevator or hoist is to be used, and a suitable switch for starting and protection.

Materials handling in horticulture requires attention to site layout and the smooth flow of plants and materials. Several forms of *conveyor* are available. One is the moving belt, capable of operating up a 25° slope. Belts may be of composition or rubber for loads such as soil blocks or seed trays. Heavier loads in packing sheds may be taken by slat or roller conveyors.

33.4.6 Heaters and coolers

33.4.6.1 Space heating

The principal uses of space heating on farms and market gardens are in the storage of potatoes, farrowing pens, incubator and breeder houses, and greenhouses. According to the requirements, the heaters comprise unit or tubular forms, infrared lamps, and floor and soil warming conductors.

Resistance cables with PVC sheathing are manufactured for floor warming, the advantages being that the thermal storage can provide for temporary supply interruption, the system is easily installed, complete protection against shock is obtained and the working surface of the floor is kept free.

Infrared heaters are made either as bright emitters with 250 W lamps or as dull emitters with 300 W sheathed wire elements. The latter are robust and long-lasting. The former, with the lamp in an evacuated glass envelope, require protection against shattering should water drip on to the envelope and (where suspended) a heat resisting cord.

Tubular heaters rated at 200 W/m run should, in general, be waterproof. Unit heaters of 3–4 kW rating are provided with fans to protect potatoes in store from frost.

Heating of the drying and ventilating air is necessary at some stage in most crop conditioning plants. Electric heating is used in installations other than continuous flow systems. The rating is assessed on the basis that 1 kW is required to raise by 1°C the temperature of 0.8 m^3/s of ventilating air. Heater elements, of bare resistance wire on refractory formers or metal sheathed, are assembled in rows and banks within a casing that is normally mounted on the fan intake. A heater is usually connected in sections (e.g. 3, 6 and 12 kW for a 21 kW heater to give seven steps). The heater can be energised only when the fan is running, and is protected by a thermal trip should the fan speed be unduly reduced. Advanced control methods employ automatic sampling of the ambient air temperature and humidity to determine the appropriate combination of heater power and fan speed.

33.4.6.2 Hot water and steam

Dairies need steam for sterilising and hot water for washing. A common steriliser consists of a boiler with immersion-heater elements, feeding steam to a chest in which items of milking equipment are placed, to a churn stool on which churns are inverted, and to a trough to heat water. A typical loading is 5–10 kW. For items that have to be sterilised *in situ* with steam drawn from pipework, the loading may be 15–20 kW. A low-power storage method has been developed, using a thermally insulated cast-iron block with embedded heater elements. Small quantities of steam are flashed by allowing a controlled quantity of water to come into contact with the block. The loading for a unit able to supply steam to a 0.8 m^3 chest twice per day is only 1 kW, the recovery period being 6–7 h.

33.4.6.3 Cooling

Cooling is a compulsory feature of milk production. The former process of cooling milk by copious quantities of water may fail during the summer, making refrigeration necessary. In the *water-plus-refrigerant* plant the water section removes a large fraction of the heat, the refrigerant section completing the cooling down to 5°C. The *refrigerant* cooler employs a surface chilled by brine or other refrigerant; no water is used, so that the plant must have a capacity greater than for the previous method. An alternative is the *immersion* type plant, which uses chilled water in which churns are immersed. The *turbine* cooler comprises an air cooled refrigerator unit mounted above an insulated cabinet containing a galvanised steel tank complete with an evaporator and ice-making coils. A circulating pump delivers the chilled water through a flexible hose from the tank to a turbine type cooler, fitted to the top of the churn in place of the lid. The chilled water rotates stainless steel agitators in the milk held by the churn. The water, after passing through the turbine, runs down the churn in a thin film, thus cooling the milk within; it is then returned by gravity to the recooling tank, where it is chilled for recirculation.

33.4.6.4 Soil warming

The growth of seedlings and plants is dependent on root development, which is, in turn, influenced by soil texture, moisture and temperature. Electric soil warming is applied in greenhouse frames and cloches by burying resistance cable at a suitable depth. The cable is rated at 10–15 W/m of linear run. Assuming a bed 8 m × 2 m to receive 90 W/m^2, the nearest cable is one with a 1500 W rating and a length of 130 m. The cable is laid in loops (*Figure 33.1a*) and connected to a 240 V 50 Hz mains supply. The number of 'passes'—an even number—is 130/8 = 16, with a spacing of 2/16 = 0.125 m = 12.5 cm. The heater conductor of the cable is insulated with polythene, and has a flexible tinned-copper earth screen, sheathed overall with PVC. The loop ends are located by fixtures that hold the cable loops in position and apply tension. Spacers may be necessary at 3 m intervals to avoid hot areas that might occur due to cable displacement. Thermostats are placed in the soil for temperature control: they may be of the rod (bimetal), expansion (capillary tube) or electronic type.

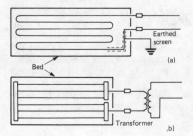

Figure 33.1 Soil warming arrangements: (a) 240 V, (b) 6–12 V

All exposed switchgear must be waterproof, and the whole installation bonded and earthed, including the cable screen.

Early soil warming installations were operated at voltages of 6 or 12 V, a system inherently safe but requiring a step-down transformer. With loads of 12 kW upward the low-voltage cable was laid in grid formation (*Figure 33.1b*), with the wires kept in tension by attaching them to busbars.

33.4.7 Auxiliary plant

Fracture of component parts of farm machinery that are repairable makes *welding* equipment useful. Potato *sorters* are often required in several positions in buildings, and socket outlets should be available for them. Although most tractors have magneto ignition, some are provided with a battery and coil system, so that *battery charging* rectifiers are employed. Further, most farms and horticultural establishments have need of utility vans and lorries with battery drive, and electric fences also rely on batteries. Fluorescent *lighting* equipment is now well established for farm buildings, and in plant growth in greenhouses it may have to be chosen on a basis of its spectral content compared with daylight.

33.5 Farming processes

33.5.1 Dairying

A milking machine consists of a simple rotary vacuum pump, driven by a motor of 0.5–1 kW rating. From the pump, vacuum pipelines run along the length of the cowshed with cocks for the attachment of milking buckets. The vacuum, of the order of 330 mm Hg, is maintained by an automatic control valve near the pump. The buckets are steel containers with airtight lids. A bucket lid is attached by rubber tubes to a cluster of four teat cups, each cup fitted with one vacuum and one milk tube. Vacuum is applied by means of a pulsator on the lid, producing an action on the teat like that of a calf. Milk drawn from the teat is passed into the sealed container. The average time for milking a cow is 4–5 min.

In a *milking parlour* the milking cluster is attached and the milk drawn off into a large glass vessel, where it is weighed and recorded. After milking, the milk is drawn from the vessel by vacuum and passed to a large container sited above the cooler. An advantage is that only a small building is required (the cows entering in batches), and the milk is delivered via the cooler to the churn, and risk of contamination is substantially removed.

33.5.2 Feed processing

Three-quarters of the production costs on livestock are for feed. In place of the purchase of ready-made feed, preparation on the farm has the advantages of control over quality, freshness and mix to exploit variations in commodity prices. The cereal part of the feed has to be disintegrated by milling to make it more

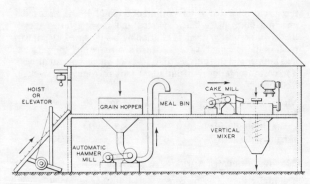

Figure 33.2 Batch food processing plant

digestible. The flaked cereal is then mixed with protein based concentrates to yield a nutrient balanced for optimum performance. Plant for small-scale 'batch' production is shown in *Figure 33.2*. The 'continuous flow proportioning' system is advantageous for throughputs of 800–1500 t/year upward. The prepared feed can be delivered direct to buffer storage bins, and therefrom to the livestock by automatic devices. A schematic diagram is given in *Figure 33.3*.

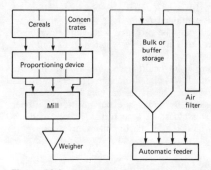

Figure 33.3 Continuous flow milling and mixing plant

Hammer and plate mills are suitable for pig and poultry feeds; plate mills are preferred for cattle feed. In continuous flow production, proportioning of the feed content is usually by variable speed auger conveyors that control the flow rate volumetrically, calibration (in kg/m^3) being necessary for each new batch of ingredients.

33.5.3 Drying

33.5.3.1 Grain

In the UK grain gathered by combine harvester has a high moisture content (e.g. 30%), which must be reduced to 14% in bulk storage and about 16% in sacks. This is achieved by passing the grain through a drier. The typical *in-sack* drier (*Figure 33.4*) dries 40 sacks by means of a single-stage fan and a 48 kW air heater. The sacks are laid on a concrete platform having 0.6 m × 0.4 m openings with large-mesh grids. In a *tray* drier, hot air is blown through the perforations in large metal trays. Combined drier and silo storage is being extensively adopted (*Figure 33.5*). The grain is emptied from the self-tip lorry into the grain pit, hence by elevator through a cleaner into a horizontal elevator and then into a selected silo. Dry heated air is blown upward through the grain in the silo by opening an appropriate hatch.

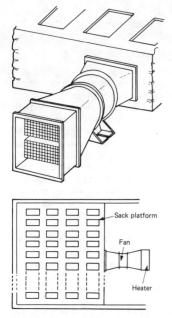

Figure 33.4 In-sack grain drier

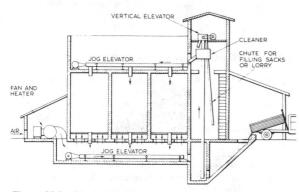

Figure 33.5 Silo drying plant

Grain is discharged by gravity into the lower elevator, raised by a vertical elevator and dispatched through a chute into sacks or lorries.

Large-scale *continuous* driers can be sophisticated and complex. They all depend on the process of blowing or sucking hot air through a 15–30 cm bed of grain. Air speed is a compromise between thermal efficiency and fast drying, and the air temperature is as high as the grain can withstand, bearing in mind the use to which it is to be put. The sections of a continuous drier are:

(1) A compartment to receive the undried grain, kept full.
(2) A drying section supplied with thermostatically controlled hot air.
(3) A cooling section in which ambient air is passed through the dried grain to cool it down to ambient temperature.
(4) Control gear to vary the flow of grain, acting on the output.
(5) A heat source, usually a direct oil fired furnace, the combustion products being blown through the crop.
(6) Fans to move heated air through the crop.
(7) A fan to pass cooling air through the grain before final discharge from the drier.

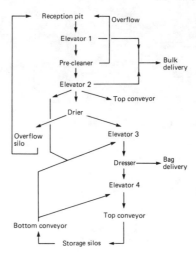

Figure 33.6 Flow diagram of continuous drier with cleaning and conveying stages

Figure 33.6 gives a flow chart representing the possible complexities of the grain movements in a continuous drier. The performance is usually expressed in t/h when extracting 5–6% moisture content from wheat using a hot air temperature of 66°C. The heat requirements render electric heating impractical: for example, a 10 t/h drier would require 1 MW of heater capacity. However, fans and elevators are electrically driven.

33.5.3.2 Hay

After cutting and wilting in the field, hay is baled for transport to a drier. The basic principle of all artificial methods of green crop drying is the removal of moisture by forcing air through the crop. High-temperature drying electrically is not economic. In all-electric driers the air is not heated, and the process is reduced to atmospheric air flow, with mechanical handling plant for transport and conveying.

33.5.4 Conditioning of livestock housing

All intensively housed livestock need effective ventilation to remove moisture, carbon dioxide, excess animal heat, ammonia and noxious organisms. Non-ruminants (e.g. pigs and poultry) respond rapidly in terms of liveweight gain and feed conversion efficiency to the air temperature and speed in their environment. Specific ranges of optimised values of these variables in combination have been evaluated with reference to age, weight, group size and type of flooring. Fan ventilation is therefore coupled to electric heating. On the other hand, ruminants (cattle, sheep) have a wide band of tolerance to air temperature and do not need a temperature controlled environment. Thus, fan ventilation is sufficient.

Some arrangements for fan ventilation are shown in *Figure 33.7*. All have the air entry (or exit) in the roof ridge, but side inlet schemes are also available:

(a) Pressurised loft: useful for intensive stock in fixed positions; not affected by wind.
(b) Recirculation: may be used in pig weaner buildings with dull-emitter heating, and dust filtration.
(c) Ridge extractor: simple, suitable for farrowing houses where sows can tolerate a range of temperature and local heating is provided for piglets.

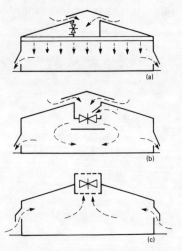

Figure 33.7 Fan systems for livestock housing

Fan ventilation is usually affected by external wind pressure. The design and positioning of the air inlets dominates the air-flow pattern in all cases.

Standard control methods (on/off, manual or timed; thermostat; sequential fan switching; proportional servo) can be applied. For large installations with several houses, central computerised control may be economic, as demands of all the individual houses can be centralised and automatic.

A problem of environment control is provided by *pig rearing*. Piglets at birth have a body temperature of 39°C, a mass of 1 kg and a surface area of low thermal insulation; they are physiologically immature and cannot maintain body temperature unaided. The natural tendency for piglets to lie by the mother sow has led to deaths by crushing. At birth piglets require an air temperature of 27°C, but the sow requires 13–17°C and will reduce feed intake if 17°C is exceeded. The two conflicting demands are met by underfloor heating for the sow, and 'creep heaters' for the piglets some distance from the sow and provided with bright-emitter heat/light, or dull-emitter heaters with a separate 'attraction lamp'.

In the rearing of *poultry* electric *incubators* are in general use, so constructed that sufficient heat storage and insulation are provided as a guard against loss of chicks by temporary mains failure. The loading is quite low, e.g. 3 kW for a 12 000 egg incubator. Heated *brooders* must be provided for chicks not immediately sold as 'day-old'. Brooders are constructed in tiers each holding about 100 chicks, stacked in four or five layers. Young chicks need light for finding food and water for almost all of a 24 h day: the general practice is to cut off light for a 30 min period to accustom chicks to this condition and to reduce 'panic' should there be a supply interruption. After two or three days the chicks learn to find food in the dark, and the lighting period can be shortened. At the same time, the initial brooder temperature of 33°C is also reduced. Subsequently the chicks are transferred to a *hover* house, a hover being a hood raised a few centimetres above ground and electrically heated, under which they find warmth and shelter. After about six weeks the chicks require no further heat and can be removed from the hover house. After transfer to the *laying* houses pullets are sensitive to lighting, which may be increased to about 17 h/day, with an artificial 'dusk' period obtained by switching or dimming fluorescent lamps. This enables eggs to be produced during the winter, when they are most profitable. Battery hens are fed from hoppers by chain conveyors, the feed flow being controlled by sensors and time switches. Water is provided by 'nipple drinkers' or by water troughs. Egg collection and grading systems are also required, and extensive cleaning of the several components of the battery house has to be regularly undertaken.

33.5.5 Irrigation

The area to be irrigated on a given farm depends on its vertical and horizontal distances from the source of water, the required water pressure at the irrigator hydrant connection point, and the characteristic needs of the crops concerned. Application concerns (a) the number of days (e.g. 5) after which an irrigator must be returned to the first site watered to maintain the needs of high-response or valuable crops; (b) the rate at which water can, without loss by run-off, be applied to a site; (c) the specific quantity of water applied (i.e. depth in mm, corresponding to a rainfall record); and (d) the seasonal quantity of water used (in ha-mm). Thus, a reservoir of surface area 0.5 ha and mean depth 3.5 m contains $0.5 \times 3.5 \times 1000 = 1750$ ha-mm, sufficient for a 50 mm depth over a 35 ha area.

Most field irrigators operate mechanically on water pressure, in the production of which electrically driven pumps are usually concerned. When the irrigation site is remote, pumping systems are remote controlled.

Piping is assembled from galvanised steel, copper or plastics pipe lengths, classified according to bore diameter and limit of internal water pressure. Various forms of fixed, oscillating, boom and rotary sprays, and self-erecting spray lines, are available. As irrigation usually proceeds unattended for long periods, protection is required against operational faults, such as failing water pressure or rate of flow.

33.6 Horticultural processes

Electricity is an essential service for the modern commercial nursery, greenhouse, market garden and fruit farm. It provides the means and control of space heating, soil warming, ventilation, lighting, irrigation, materials handling, soil preparation, planting, harvesting, storage and marketing. The basic requirement is to provide an optimum ambient climate for growth, with sophisticated control to ensure that the several factors are maintained and monitored. The cost of equipment is offset by the gain in saleable output. Automatic control makes the 5 day week possible without loss of operating efficiency.

33.6.1 Greenhouse operation

33.6.1.1 Space heating

The temperature in a greenhouse is the basic element in efficient production, representing 20–40% of the direct production cost. Air temperature directly affects plant growth, development, yield and quality. Electric air heating may be (a) by waterproof aluminium cased tubular heaters, normally 50 mm in diameter, 0.6–4.5 m in length, and rating 200 or 250 W/m run; or (b) by heater wires mineral insulated in a copper sheath. The latter are cheaper, but may have a higher installation cost. An alternative is circulation of warmed air through transparent perforated plastics ducts. Typical air temperatures are 15°C during the day and 20°C at night, with a limit of 24°C at which the greenhouse has to be ventilated. The thermal requirements are calculated on a basis of the type of structure, its size and geographical location, and the maximum likely difference between the inside and outside air temperatures. Ventilation is designed to inhibit excessive temperature build-up by 'solar gain' during the summer, to facilitate gaseous interchange within the crop and to purge excessive humidity. The most common ventilation method is by natural convection through vent opening (which can be motorised), but fan ventilation is more sensitive, gives a more even temperature distribution, and is independent of external ambient conditions such as wind velocity and direction.

33.6.1.2 Soil warming

Soil temperature affects germination and development in the root zone of a crop. The ambient air temperature above the soil is not a reliable guide to that in the root zone because of (a) thermal gradient in the soil, (b) evaporative cooling from the soil surface and (c) the effects of irrigation with cold water. Consequently, a thermostatically controlled direct soil warming system, as described in Section 33.4.6.4, is useful. The depth of the heating cable below the soil surface is critical: if too small, there are temperature irregularities across the bed; and if too large, the system is slower to respond and wastes heat. A usual depth/spacing ratio is about 1/1. Advantage may sometimes be taken of off-peak tariffs by including additional thermal storage material in the bed, with an over-riding control to prevent excessive temperature rise.

33.6.1.3 Lighting

Slow growth of crops resulting from the poor level and short duration of natural winter daylight can be overcome with the aid of artificial light sources. Radiation broadly corresponding to daylight is required to promote photosynthesis, and while several electric lamp types have suitable spectra, the illumination levels required are 10 times those for normal visibility. Daylight can be *supplemented* by high-pressure fluorescent or metal halide lamps; it can be entirely *replaced* by low-pressure tubular fluorescent lamps having a suitable phosphor.

33.6.1.4 Irrigation

Modern greenhouse methods require control of the soil moisture. For large-scale operation various forms of automated watering are available by spray lines or 'trickle-and-drip' devices.

33.6.1.5 Cultivation

Electric cultivators have the advantages of light weight, quiet and fumeless working, and easy starting. A typical hand-controlled multipurpose cultivator, driven by a 1 kW motor and fitted with 30 cm slasher blades, can cultivate a 0.6 m strip in one pass; it is reversible, and will spin, weed, hoe and ridge the soil. Compost shredders and mixers, and machines to fill compost into pots, bags and trays, are used to mechanise an otherwise labour intensive operation. Machines should be double-insulated, and when supplied through trailing cables, should employ a 110 V centre-tapped transformer for safety.

33.6.1.6 Sowing and harvesting

Soil-block making greatly reduces the task of planting and pricking out cucumber, lettuce and tomato crops. Soil loaded into a hopper emerges as 40–80 mm blocks at up to 15 000 per hour, with pelleted seed sown into them at the same time. Mechanised planters and harvesters have also been developed for electric drive.

33.6.1.7 Materials handling

Materials handling is an essential function in commercial horticulture. Moving-belt conveyors are usually capable of operating up inclines of 25–30°. Flat belts of rubber or composition are employed for low-profile objects (such as soil blocks or seed trays), and troughed forms for loose compost. Heavier unit loads in cartons or containers in packing sheds require slat or roller conveyors. Monorail transporters in the greenhouse do not obscure light from the growing crops and can permit a useful increase in the space available for plants. While most monorails move their loads by means of a continuous chain, an alternative is provided by small electric tractors towing load-carrying trailers. The shifting of bulk materials between sites can involve walking considerable distances, a labour intensive task that can be relieved by battery electric vehicles, pedestrian- or rider-controlled. Lorry batteries can be used for small trucks, but normally traction batteries are preferred for their longer life of 4–6 years.

33.6.1.8 Storage and marketing

All crops have a natural storage life in uncontrolled ambient atmospheric conditions, and it can be very short. Cool storage slows the inevitable deterioration that follows harvesting. Storage temperature is critical: for a given crop the variation tolerated may be as low as $\pm 1°C$. Further, the humidity must not be so low that the crop wilts or shrivels. The process of storage involves (a) the removal of 'field heat' as rapidly as possible, (b) holding the crop at the prescribed temperature and relative humidity, (c) transport from the store to the buyer in refrigerated transport, (d) further storage if necessary and (e) final display in a refrigerated cabinet until purchased by the consumer. These spoilage-delay processes involve electric or 'ice-bank' cooling equipment. The advent of supermarket chains which, by the nature of their business, require prepacked produce, has made expensive demands on marketing. An automatic continuous flow system is required to process crops from the topping and cleaning input end, through grading, conveying, box making and film wrapping equipment to the dispatching end, with controlled temperature and humidity.

33.6.2 Field operation

33.6.2.1 Irrigation and frost protection

The field-scale watering of market gardens and orchards by spray lines or reciprocating nozzles has the obvious object of making up for soil moisture deficiency. The conditions resemble those described in Section 33.5.5. The more recent practice of applying a continuous spray of water has proved effective in protecting fruit tree blossom from frost damage in the spring and frost-susceptible vegetable crops (e.g. runner beans) in the autumn. The water is electrically pumped; a well-sited air thermostat initiates spraying when the air temperature falls below 0°C. An alternative method is to blanket the crop with a fine mist of water droplets from high-pressure nozzles.

33.6.2.2 Fruit grading and storing

After fruit has been gathered, it has to be picked over and sorted before being sent to market or to store. The fruit is tipped on to a conveyor belt, where damaged items are removed by hand, then passed through a grading machine which sorts the fruit automatically. For marketing the individual items are wrapped, packed into boxes and conveyed to the dispatch department. Storage is illustrated by the treatment of apples: after harvesting they decay by biological action, inhaling oxygen and exhaling carbon dioxide, developing heat in the process. Gas storage equipment retards the decay by placing the fruit in a cold store and allowing the concentration of carbon dioxide to rise no further than a predetermined level. The store is then maintained with a specific proportionality between oxygen and carbon dioxide together with extraction of heat. Typical percentage content is 8 for CO_2 and 12 for oxygen. Cold storage chambers comprise compartments, thermally insulated and gas-tight, into which cases of graded apples are placed. Air chilled by passing over cooling coils is circulated by a fan. Gas indicators are fitted to each compartment. Applies can be stored for longer periods if the oxygen level is lowered, but the cost is greater and the system more complicated; but the method can be economic if employed on a co-operative basis.

33.7 Information

In the UK nationwide technical advisory services for the application and use of electricity in agriculture and horticulture (under the names 'Farmelectric' and 'Growelectric') are available from the Electricity Council's Farmelectric Centre at the National Agricultural Centre, Stoneleigh, Kenilworth, Warwicks. CV8 2LS [Tel.: 0203 58626].

34

HVDC Transmission

A Gavrilović
GEC, Stafford, UK

The author thanks his colleagues for permission to make extensive use of their material: J D Ainsworth, B R Anderson, M H Baker, R Banks, H Gibson, F G Goodrich, C J B Martin, B A Rowe, H L Thanawala, M L Woodhouse

Contents

34.1 Introduction 34/3

34.2 Applications of h.v.d.c. 34/3
    34.2.1 Introduction 34/3
    34.2.2 Types of d.c. interconnection 34/3
    34.2.3 Purposes of transmission interconnections 34/4
    43.2.4 Reasons for choosing h.v.d.c. 34/3
    43.2.5 Application of h.v.d.c. to developing systems 34/4

34.3 Principles of h.v.d.c. converters 34/4
    34.3.1 Converter operation: simplified case of zero commutating inductance 34/4
    34.3.2 Converter operation: practical case of finite commutating inductance 34/5
    34.3.3 Converter operation: converter acting as an inverter 34/6
    34.3.4 12-pulse converters 34/6
    34.3.5 Basic d.c. voltage/d.c. current characteristics 34/6
    34.3.6 Basic principles of control of h.v.d.c. transmission 34/7
    34.3.7 Starting and stopping an h.v.d.c. link 34/7
    34.3.8 Power reversal 34/8
    34.3.9 Isolating a valve group 34/8
    34.3.10 Numerical example 34/8

34.4 Transmission arrangements 34/9
    34.4.1 Bipolar lines 34/9
    34.4.2 Two monopolar lines 34/9
    34.4.3 Cable schemes with sea return 34/9
    34.4.4 Back-to-back arrangement 34/9
    34.4.5 Ground electrodes 34/9
    34.4.6 Sea electrodes 34/10
    34.4.7 Staged construction of h.v.d.c. 34/10
    References 34/10

34.5 Converter station design 34/10
    34.5.1 Valve group arrangements 34/10
    34.5.2 Converter valves 34/11
    34.5.3 Converter transformers 34/11
    34.5.4 AC filters 34/11
    34.5.5 DC smoothing reactor 34/13
    34.5.6 DC isolators 34/13
    34.5.7 Protection 34/13
    34.5.8 Converter station losses 34/13
    34.5.9 Converter station costs 34/13
    34.5.10 Reliability 34/13
    References 34/14

34.6 Insulation co-ordination of h.v.d.c. converter stations 34/14
    34.6.1 Introduction 34/14
    34.6.2 Sources of overvoltages 34/15
    34.6.3 Surge arresters 34/15
    34.6.4 Surge arrester arrangement 34/15
    34.6.5 Safety margins 34/16
    34.6.6 Creepage and clearance 34/14
    34.6.7 Application examples 34/16
    Further reading 34/17

34.7 HVDC Thyristor valves 34/17
    34.7.1 Introduction 34/17
    34.7.2 Thyristor level circuits 34/17
    34.7.3 Voltage rating 34/17
    34.7.4 Current rating 34/18
    34.7.5 Turn-on behaviour 34/20
    34.7.6 Turn-off behaviour 34/21
    34.7.7 Valve arrangements 34/21
    34.7.8 Valve tests 34/23
    References 34/23

34.8 Design of harmonic filters for h.v.d.c. converters 34/23
    34.8.1 Introduction 34/23
    34.8.2 AC harmonic current generation 34/23
    34.8.3 Filtering 34/24
    34.8.4 Harmonic performance evaluation 34/25
    34.8.5 DC filtering 34/26

34.9 Reactive power considerations 34/27
    34.9.1 Introduction 34/27
    34.9.2 Reactive power requirements of h.v.d.c. converters 34/27
    34.9.3 Steady state voltage control and total ratings of reactive equipment 34/27
    34.9.4 Voltage disturbances caused by switching operations and requirements for smooth reactive control 34/27
    34.9.5 Control of temporary overvoltages caused by faults resulting in partial or total loss of d.c. power flow 34/28

34.10 Control of h.v.d.c. 34/28
    34.10.1 Summary of h.v.d.c. controls 34/28
    34.10.2 Pole controls 34/29
    34.10.3 The phase-locked oscillator control system 34/29
    34.10.4 Tapcharger controls 34/30
    34.10.5 Master control 34/30
    34.10.6 Telecommunications 34/30
    34.10.7 Performance examples 34/31
    References 34/32

34.11 AC system damping controls 34/32
    34.11.1 DC link supplies power from dedicated generators or from a very strong system to a small system 34/32
    34.11.2 DC link connecting two systems which are not synchronised but are of similar size 34/33
    34.11.3 DC link connecting two parts of an a.c. system or two separate systems having also a parallel a.c. link 34/33
    References 34/33

34.12 Interaction between a.c. and d.c. systems 34/33
    34.12.1 Study of h.v.d.c. systems 34/33
    34.12.2 Short circuit ratios 34/34
    34.12.3 SCR as a guide to planning 34/35
    34.12.4 System interaction when the a.c. system impedance is high relative to d.c. power infeed (low SCR) 34/35
    34.12.5 'Island' receiving system 34/36
    References 34/36

34.13 Multi-terminal h.v.d.c. systems 34/36
    34.13.1 Series connection 34/36
    34.13.2 Parallel connection 34/37
    34.13.3 DC circuit breakers 34/37
    References 34/38

34.14 Future trends 34/38
    References 34/38

34.15 General references

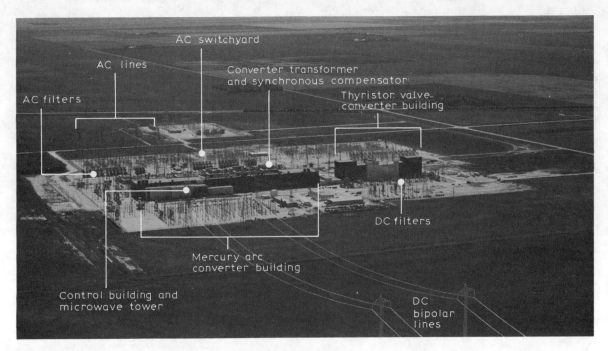

Dorsey Converter Station of the Nelson River h.v.d.c. scheme. The photograph shows valve halls and outdoor equipment of Bipole I rated for 1800 A at ±465 kV and Bipole II rated for 1800 A at ±500 kV. In an emergency the two bipoles can be paralleled on one bipolar line transmitting 3600 A at ±465 kV. Courtesy of Manitoba Hydro.

34.1 Introduction

The first commercial generators were direct current (d.c.) and therefore so were the early distribution systems. As distribution was at relatively low voltages, transmission distances were by necessity very short. The potential benefits of electrical energy were fully recognised and work to improve existing transmission systems was undertaken both in Europe and in America.

In 1883 Nicola Tesla was granted patents for the inventions on which he had worked during the previous ten years relating to polyphase alternating current (a.c.) systems. In May of that year he delivered his classic lecture to the American Institute of Electrical Engineers: 'A New System of Alternating Current Motors and Transformers'. Although, today we cannot visualise life without a.c. electrical systems, they were not immediately or universally accepted. Edison, who was working on a comprehensive d.c. distribution system wrote in 1889 in the Scientific American: 'My personal desire would be to prohibit entirely the use of alternating currents. They are as unnecessary as they are dangerous. I can therefore see no justification for the introduction of a system which has no element of permanency and every element of danger to life and property'.

Elimination of commutators made generators simpler and transformers allowed voltage to be changed easily; the use of higher voltages became practical and transmission over longer distances feasible. Widespread use of a.c. generation and transmission followed the exploitation of the Niagara Falls energy in 1895. Yet engineers continued to seek means of transmitting d.c. at high voltages, because they realised that the cost of overhead lines and cables for high voltage direct current (h.v.d.c.) could be considerably lower than for a.c. at the same power. An a.c. line has three conductors each insulated for the crest value of the alternating phase voltage, but the power transmitted is related to the r.m.s. value; the design of a.c. lines have also to take into account the flow of reactive current. HVDC transmission lines require only two conductors and the normal working voltage equals the rated voltage of the line. However, it was necessary to develop an adequate a.c./d.c./a.c. converter in order to benefit from lower cost of d.c. lines and cables in an a.c. system environment.

The first commercial h.v.d.c. scheme connecting two a.c. systems was a submarine cable link between the Swedish mainland and the island of Gotland. The scheme, rated for 20 MW at 100 kV was commissioned in 1953. Mercury arc valves each rated at 50 kV, 200 A were used as the converting device. Eleven mercury arc valve schemes totalling 6400 MW have since been commissioned. The last scheme to use mercury arc valves was Nelson River Bipole 1 in Canada, rated for 1620/1800 MW at ± 450 kV. It uses the world's largest mercury arc valves, made in the UK, each rated at 155 kV, 1800/2000 A.

In the late 1960s experiments using thyristor valves in mercury arc schemes were carried out in Sweden and in England. In 1970 the Gotland scheme was uprated to 30 MW at 150 kV by the addition of a 50 kV, 200 A thyristor valve bridge. In 1972 the first thyristor scheme, 320 MW back-to-back, was commissioned at Eel River in New Brunswick, Canada. At present (1985) there are some 30 schemes totalling 31 000 MW of installed capacity in service or under construction. The recent increase in the utilisation of h.v.d.c. can be explained partly by its technical attributes and partly by the advantages gained from the interconnection of power systems which it facilitates.

The largest long distance overhead line transmission h.v.d.c. system is the 6300 MW Itaipu scheme in Brazil consisting of two bipoles each rated at ± 600 kV. The largest h.v.d.c. submarine cable scheme is the 2000 MW link between France and England, consisting of two bipoles, each rated for 1000 MW at ± 270 kV. The largest zero distance or 'back-to-back' scheme in Chateauguay in Canada rated for 1000 MW.

34.2 Applications of h.v.d.c.

34.2.1 Introduction

The answer to the question 'Why h.v.d.c.?' was, historically, that high voltage direct current overhead lines and cables are cheaper than those for alternating current for the same power transmission capability, and, provided the transmission distance is more than a critical value (the 'break-even distance'), the savings from using h.v.d.c. lines or cables would more than pay for the a.c.–d.c. converters. The length of submarine cable made a.c. impracticable for some schemes due to the large charging current.

Today, only a few h.v.d.c. links are justified by such simple economics. They are not relevant to back-to-back schemes in which the distance between adjacent a.c. systems is zero, and in most other transmission schemes other attributes of h.v.d.c., its asynchronous nature or its ability to control power, play an equally important part in the choice of transmission.

34.2.2 Types of d.c. interconnection

An h.v.d.c. link is itself asynchronous but it may connect two asynchronous or synchronous a.c. systems and can be further sub-classified by distance according to whether or not there is a h.v.d.c. line or cable between the two converter terminals, as follows:

Asynchronous where h.v.d.c. is the only interconnection between two systems with different frequencies, e.g. 50 and 60 Hz, or two systems at nominally the same frequency but with uncontrolled phase relationships.
Synchronous Where h.v.d.c. link is used within an a.c. system or in parallel with an a.c. interconnection.
Long Distance Point-to-point interconnections by
    (a) Overhead line
    (b) Undersea cable
    (c) Underground cable
    (d) Combination of overhead lines and cables
Zero Distance Back-to-back interconnections.

34.2.3 Purposes of transmission interconnections

A.C. or d.c. interconnections can be classified as follows:

(a) Power transfer exclusively or largely in one direction, normally characteristic of point-to-point applications:
    (i) From remote hydro, thermal or nuclear generation to load areas.
    (ii) From a strong to a weak a.c. system.
(b) Power transfer in either direction, normally characteristic of interconnections between neighbouring a.c. systems, typically of similar strength. Such system interconnections, whether a.c. or d.c., offer one or more of the following benefits:
    (i) Include the links capacity in the spinning reserve for each system, minimising the generating capacity allocated in each system for such duties.
    (ii) Increase the security of supply by offering mutual support.
    (iii) Take advantage of seasonal generation and load pattern differences between the two systems.
    (iv) Take advantage of timing differences between daily load peaks of the two systems.
    (v) Take advantage of different types of generating plant with differing base to peak cost ratios in the two systems.

34.2.4 Reasons for choosing h.v.d.c.

Of the many reasons which may contribute to the choice of the h.v.d.c. as a means of interconnecting two power systems (or elements of power systems), two stand out as being the most important, namely

(a) Frequency or phase angle variation between the two terminals of the interconnection may render an a.c. link impractical. In an extreme case, the a.c. busbars at the terminals of the link may operate at different frequencies. Even if they are synchronised, it does not follow that reliable a.c. transmission can be established, because variations in relative phase angle between the two busbars, caused either by variation in load or by network disturbances, may result in unacceptable power flow severe enough to cause frequent tripping. Thus, it may prove economic to use h.v.d.c. for a zero length (back-to-back) transmission, or in parallel with an existing a.c. transmission path.

(b) The transmission distance may be so long that the cost savings arising from the use of relatively cheaper h.v.d.c. conductor systems is more than sufficient to outweigh the costs of the extra terminal equipment required for h.v.d.c.

Combinations of these two factors constitute more powerful economic pressures than either by itself.

Benefits which h.v.d.c. may provide beyond those provided by an a.c. interconnection can be summarised as follows:

(i) Provide the facility to interconnect two systems which have different operational procedures for frequency or voltage control.
(ii) Provide predetermined and controlled power transfer. Power flow in an a.c. interconnection is controlled by phase relationships which, being relatively uncontrolled, can cause inadvertent overloading or under-utilisation during normal or disturbed operation. In the case of h.v.d.c., two utilities can pre-set the limits of power by which at any time they can assist each other and power will change automatically up to those limits in response to predetermined conditions, such as a frequency change.
(iii) Improve transient stability of the interconnected systems by modulating synchronising or damping power to reduce intermachine swings.
(iv) Avoid excitation of subsynchronous resonance as might occur in the case of series capacitor applications in an equivalent a.c. interconnection.
(v) Distribute the available power more effectively and thus delay the introduction of new power stations and major transmission reinforcements.
(vi) Permit staged development of a country's overall power system in a more controlled and hence less expensive manner by providing the means to utilise generation in geographically separate systems, compared to what could be done by a purely a.c. transmission development.
(vii) H.v.d.c. does not contribute to the a.c. system fault current. The contribution to the system fault current by an a.c. interconnection may necessitate the replacement of the existing switchgear.

Therefore in addition to the use of h.v.d.c. to connect two systems which cannot be synchronised, it should also be considered as one of the possible alternatives whenever enhancements are needed to make an a.c. interconnection attractive: the use of series compensation, variable shunt reactive compensation, phase shift boosters, etc.

34.2.5 Application of h.v.d.c. to developing systems

In developing countries or regions experiencing load growth the integration of h.v.d.c. should be considered at every stage of system development, not merely as an appendage to an otherwise fully designed system.

Two kinds of interconnection are often required: long distance links for bulk power transfer from remote generation, and shorter links (perhaps even of the back-to-back type) to interconnect adjoining relatively large regional systems. For both kinds of interconnection, but particularly in the latter case, to provide a sufficiently secure link using a.c. may require a large installation, typically of multiple e.h.v. lines, which may not be justifiable economically until a much later stage of load growth. The immunity of an h.v.d.c. link from problems arising from variations in the relative phase of the two networks may permit the benefits of interconnection to be realised economically at a much earlier stage of system and load growth than that at which a.c. becomes justifiable. If at some future date the natural system growth justifies an e.h.v. or u.h.v. overlay of a.c. transmission lines, the d.c. interconnection would readily become an integrated part of the combined a.c./d.c. system and, by virtue of its rapid controllability, improve the overall system stability and dynamic performance. Thus the economic and technical advantages of both d.c. and a.c. interconnections can contribute both to the intermediate and to the long term transmission planning.

34.3 Principles of h.v.d.c. converters

34.3.1 Converter operation: simplified case of zero commutating inductance

The standard 'building-block' for h.v.d.c. converters is the 3-phase full wave bridge using six controlled (thyristor) valves, as shown in *Figure 34.1*. This is known as a '6-pulse' converter group or bridge, because there are six valve firing pulses, and six pulses per power frequency cycle in the output, at the d.c. terminals. *Figure 34.2* shows the 'idealised' current and voltage waveforms, neglecting commutation inductance L in *Figure 34.1*, and assuming acceptably smooth direct current output I_d, achieved by the action of the relatively large d.c. smoothing reactor L_d. For this case valve current pulses are 120° long, and their flat-tops have a magnitude equal to I_d. The time at which uncontrolled (diode) valves would commence conduction is used as a reference, and the 'firing delay angle' is defined to be zero at this point on wave. *Figure 34.2* is drawn for the case where the firing time for each valve is delayed by the 'firing delay angle' α relative to diode operation, i.e. $\alpha = 0°$.

All conventional treatments of converter theory make the assumption that the e.m.f. E_l in *Figure 34.1* is sinusoidal; an assumption which is substantially true in practice because of the a.c. harmonic filters which are usually connected at the a.c. terminals of converter stations, preventing the non-sinusoidal

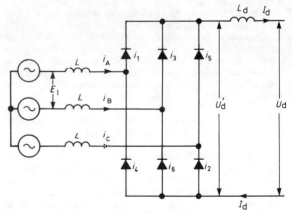

Figure 34.1 Basic 6-pulse converter bridge

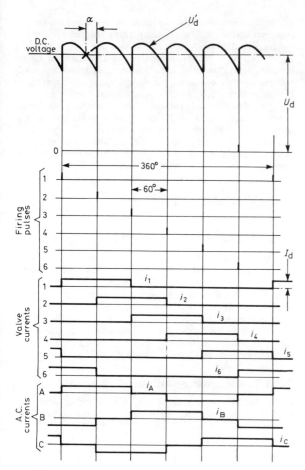

Figure 34.2 Idealised waveform for a 6-pulse converter, neglecting commutation inductance

converter current from appreciably disturbing the shape of the power frequency voltage. The analysis which follows neglects the effects of the current-dependent losses of converters. This is because they are both small and non-linear, making it unreasonably laborious to take them into account unless digital computers are used to carry out the calculations.

Some numerical relationships for this simplified case are

$$U_d = E_l(3\sqrt{2}/\pi)\cos\alpha = 1.35 E_l \cos\alpha \quad (34.1)$$

$$I_l = (\sqrt{2}/\sqrt{3})I_d = 0.816 I_d \quad (34.2)$$

$$I = (\sqrt{6}/\pi)I_d = 0.780 I_d \quad (34.3)$$

where U_d is the d.c. voltage of the 6-pulse bridge, E_l is the commutation e.m.f. (r.m.s. line-line), I_d is the d.c. current, I_l is the r.m.s. a.c. current per phase, and I is the fundamental component of the a.c. current.

Equation (34.1) describes the principal control action of a converter, i.e. by change of firing angle α the d.c. voltage can be changed from maximum positive (rectification) at $\alpha = 0°$, through zero at $\alpha = 90°$ to negative (inversion) for α approaching $180°$. The d.c. voltage, U_d, at $\alpha = 0$ and $I_d = 0$ is termed ideal no-load voltage, U_{dio}; this is a fictitious quantity but it is often used as the basis for further calculations.

$$U_{dio} = 1.35 E_l \quad (34.4)$$

In practice the a.c. connection is via a transformer (not shown in *Figure 34.1*). The transformer rating is defined as $E_l I_t \sqrt{3}$. Although the choice of this definition is arbitrary from the viewpoint of converter operation, it offers the convenience that it would be equally applicable if the transformer were to be utilised for a.c. transmission. With the combined simplifications of zero commutation (leakage) reactance and assuming diode operation ($\alpha = 0$) from equations (34.1) and (34.2) this exhibits its minimum value of 1.047 times d.c. power.

34.3.2 Converter operation: practical case of finite commutating inductance

In a 6-pulse bridge circuit, *Figure 34.1*, the valves 1, 3 and 5 commutate the outgoing direct current I_d between themselves, while the valves 2, 4 and 6 commutate the incoming direct current I_d; the two 3-pulse conversion processes form the 6-pulse bridge conversion. For clarity the *Figure 34.3* is drawn for one half of the 6-pulse bridge, i.e. the commutations between valves 1, 3 and 5 only are shown.

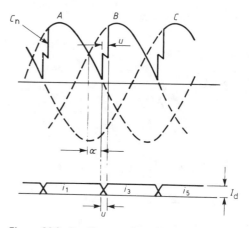

Figure 34.3 Rectifier operation with finite commutating reactance

In practice the converter transformer will have a finite leakage inductance (L in *Figure 34.1*). This causes current waveforms to exhibit more gradual transitions as shown in *Figure 34.3*, i.e. the current requires a finite time to commutate from one valve to the next valve in sequence in that particular row of three valves of the 6-pulse bridge. This is known as commutation overlap time, usually expressed as an angle u in electrical degrees. The value of u increases with increasing d.c. current, reaching typically $25°$ at rated current.

The d.c. voltage U_d is reduced by the value dx due to the commutation notch cn on *Figure 34.3*. (The derivation of the equations used is well documented in text books given as the general references at the end of the chapter.)

Equation (34.1) becomes

$$U_d = 1.35 I_l \cos\alpha - dx \quad (34.5)$$

where

$$dx = \frac{3}{\pi} I_d X_c \quad (34.6)$$

where X_c is the commutating reactance. Usually the commutating (i.e. converter transformer) reactance is expressed in per unit of the converter transformer rating. The equation (34.5) becomes

$$U_d = 1.35E\left(\cos\alpha - 0.5\frac{I_d}{I_{d1}}x_c\right) \quad (34.7)$$

or using equation (34.4)

$$U_d = U_{dio}\left(\cos\alpha - 0.5\frac{I_d}{I_{d1}}x_c\right) \quad (34.8)$$

where I_{d1} is rated direct current.

34.3.3 Converter operation: converter acting as an inverter

This occurs when the firing angle α exceeds 90°. If current flow is to continue, this can only occur as a result of an external power source supporting the direct voltage. An inverter connected to an external circuit composed only of passive components does not conduct, being essentially a provider of back-e.m.f., to be overcome by the d.c. line voltage. The waveforms are generally similar to those above, but d.c. voltage U_d is negative. Thus to reverse power flow in a converter, although d.c. current cannot be reversed, d.c. voltage can be reversed by control action.

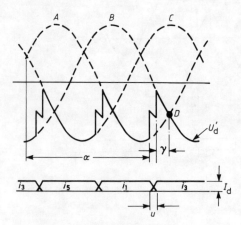

Figure 34.4 Invertor operation

Figure 34.4 shows the inversion process for valves 1, 2 and 3 of *Figure 34.1*. When the valve 3 fires, its current rises to I_d, and valve 1 current falls to zero, in a time u degrees, similarly as for rectifier operation. However, the current is now flowing due to d.c. line voltage and against the (negative) inverter transformer voltage, which acts as 'back-e.m.f.'. The commutation process must be completed before phase A voltage becomes more positive than phase B voltage, point D on *Figure 34.4*. If valve 1 is still conducting at that point it will continue to conduct driven by the sum of d.c. line and the phase A voltages. The inverter is operated so that the commutation process is completed well before point D. The quantity γ is the 'extinction angle' and is the time available for the valve to turn off, i.e. become capable of withstanding the subsequent forward voltage. Valve performance is discussed in Section 34.7.

Small working values of γ (i.e. values of α approaching 180°) lead to low capital cost of valves and transformers, low harmonic generation, low reactive power consumption and low station losses. However, too small a value of γ causes commutation failure. This is usually initiated by disturbances arriving from the a.c. system which distort the waveform at the a.c. terminals of the converter station, resulting in temporarily reduced γ for one or more commutations. A reduction of γ to less than 10° (12°) at 50 Hz (60 Hz) is usually needed before commutation failure becomes likely, but once it has occurred, it may temporarily collapse inverter operation, requiring 100 ms or so before the control system succeeds in restoring normal operation.

A typical running value of γ, which gives reasonable immunity to commutation failure, is 15° to 18° (for a 50 Hz system). The corresponding value of α to produce this is typically about 140°. It is important that inverters are provided with constant-extinction angle (γ) control to prevent commutation failures in normal steady-state operation, as discussed in Section 34.10.

For constant γ operation, equation (34.8) becomes

$$U_d = U_{dio}\left(\cos\gamma - 0.5\frac{I_d}{I_{d1}}x_c\right) \quad (34.9)$$

34.3.4 12-pulse converters

The harmonics produced by a 6-pulse converter are large, requiring expensive filters. They can be reduced by use of a 12-pulse converter as discussed in Section 34.8. The usual arrangement of this for h.v.d.c. uses two 6-pulses bridges connected in series on the d.c. side, with their transformers respectively star-star and star-delta, connected in parallel to the a.c. busbar. As the cancellation of harmonics takes place at the a.c. side of the converter/transformers, the conversion process takes place independently in each 6-pulse bridge.

34.3.5 Basic d.c. voltage/d.c. current characteristics

Figure 34.5 shows these for a converter operating on a zero-impedance a.c. system. Natural boundaries to this occur at zero

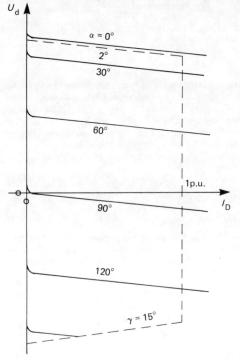

Figure 34.5 Basic firing angle control characteristics (a.c. voltage constant)

d.c. current (because d.c. current cannot reverse) and at $\alpha = 0°$ (firing cannot occur for α negative because this would mean attempting to fire valves when their anode-cathode voltage is negative). Other boundaries are applied by control action:

(a) A minimum limit of $\alpha = 2°$ is applied in practice to ensure reliable firing of each valve.
(b) A minimum γ limit (say at $\gamma_1 = 15°$) prevents commutation failure in normal operation as described above.
(c) A d.c. current limit is applied at the thermal current limit of valves and other components.

Within these boundaries, any desired shape of d.c. voltage/d.c. current characteristic can be obtained by control action, i.e. by change of α, as described later.

34.3.6 Basic principles of control of h.v.d.c. transmission

Figure 34.6 shows a simplified diagram for a two-terminal h.v.d.c. link, with elementary controls.

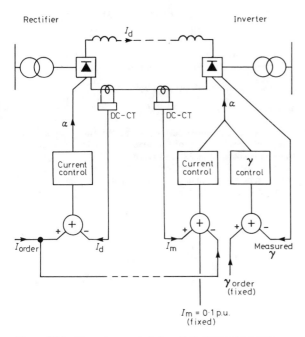

Figure 34.6 Elementary controls for a 2-terminal h.d.v.c. link

At the rectifier a closed-loop current control is provided, which adjusts firing angle α in response to the difference between measured d.c. current I_d, measured by means of a d.c. current transformer, and a current order signal I_o, assumed fixed for the present.

At the inverter, closed-loop γ control is provided, operating similarly but from measured γ, with a fixed reference demanding γ_1 of typically 15° to 18°. A current control loop is also provided, similar to that at the rectifier, supplied with the same current order, but with a 'current margin' signal I_m substracted from it. I_m is typically 0.1 of rated d.c. current, I_{d1}.

Figure 34.7 shows the resulting d.c. voltage/d.c. current characteristics. The rectifier current loop generates the constant-current characteristic BCD. This has a natural transition at B to the $\alpha = 0°$ line AB. The inverter has a constant-gamma characteristic FCE, with a transition at F to a constant-current characteristic FG at I_m below BCD. DC line resistance may be

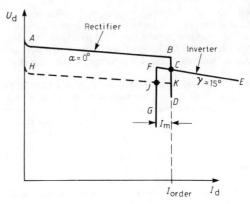

Figure 34.7 DC voltage/d.c. current characteristics for *Figure 34.6*

included with either characteristic for constructing the U_d/I_d diagram. The steady-state working point is at the cross-over, i.e. point C.

Thus in normal operation the rectifier controls current and the inverter controls voltage.

Tapchangers on each converter transformer are often used. These do not have any major control functions; their duty is to optimise working conditions for each converter. The inverter tapchanger is usually arranged to effectively move FCE up or down to obtain rated d.c. voltage; whereas the rectifier tapchanger adjusts AB up or down so that measured α lies within a range of about 5° to 20°.

Changes of a.c. voltage experienced by the converters are effectively corrected in the long term (many seconds) by the tapchangers, but can temporarily shift the characteristics. The only important case is that in which the rectifier a.c. voltage falls significantly (or inverter a.c. voltage rises), so that for example AB moves down to HJK. In the absence of the current loop at the inverter this would cause a complete loss of d.c. current. When this loop exists, the working point moves only to point J, at a d.c. current lower than I_{d1} by the current margin I_m. (The change of working point from C to J, even for a slow a.c. voltage change, would be too abrupt as shown in *Figure 34.7*. More sophisticated characteristics are described in Section 34.10.)

34.3.7 Starting and stopping an h.v.d.c. link

34.3.7.1 Starting

H.v.d.c. convertres can be started and stopped very rapidly if required. However in normal operation this is done relatively slowly to avoid shocks to the a.c. systems.

The normal starting procedure is to first de-block (i.e. initiate firing pulses) at the inverter, with a firing angle of about 160°; as the d.c. voltage is zero, thus causes no current.

The rectifier is then de-blocked, initially at a similar firing angle, which is then slowly reduced over a few hundred milliseconds, raising d.c. voltage until the inverter current rises and the system settles at normal firing angles, with a low current order (0.1 p.u. or less). Current (or power) is then increased slowly over, say, 10 seconds to 10 minutes to the desired final value.

34.3.7.2 Stopping

Whilst in a.c. practice a circuit is invariably taken out of service by opening a switch, on the d.c. side of a converter station the technique for shutting down a 2-terminal scheme is normally to reduce power by control action over a period to suit the needs of the a.c. system, and then to block all valves.

When a converter is shut down, a bypass path is often provided on the d.c. side. For example, normal firing pulses may be blocked and a pair of series connected valves fired to provide a bypass path. This collapses the d.c. voltage and the change in voltage can be detected at the other station and used to initiate its shut down sequence.

If converters at both ends of a link are bypassed whilst high current is flowing in the link, because the resistance of the line circuit is low and the inductance of the d.c. reactors is large, current may take a long time to decay. The d.c. line can be discharged faster by ensuring that one or both of the converter stations remains in inverter mode until current has stopped.

In an emergency, stopping is achieved much more rapidly. A typical method is to separate fault signals according to their origin, into non-urgent stop (from relays detecting persistent commutation failure, asymmetry or mis-fire, undervoltage, abnormal firing angle, etc.) or emergency stop (from relays detecting overcurrent, or flash-overs from differential measurement). Non-urgent stop signals are usually allowed to persist for about 300 ms, and at a rectifier cause forced-retard, i.e. firing angle is forced into inversion at, say, 150°, which will normally stop d.c. current in about 10 ms; at an inverter by-pass operation is caused by blocking normal firing pulses and instead firing a pair of series connected valves in each bridge. The latter does not directly stop d.c. current, but causes zero d.c. voltage, which is detected by the rectifier which then stops current by forced retard after 300 ms.

For an emergency stop signal, full blocking, i.e. suppression of all firing signals is applied within about 2 ms, and the converter group circuit breaker is tripped. Zero a.c. side current usually occurs in less than 10 ms, except for some types of flashover within the station, which the circuit breakers will clear in, say, 3 cycles.

34.3.8 Power reversal

As described earlier, in h.v.d.c. schemes, due to the unidirectional property of the thyristors, the current cannot reverse. Power reversal is achieved by reversing d.c. voltage. This is carried out by changing over the operation characteristics of the converters in the two stations. The effect of this is to change the phase angle of the current at the a.c. busbars by reversing its power component; its reactive component always remains negative.

As for start and stop, power reversal can be done rapidly if desired, within typically 200 ms. However, reversal is normally carried out by reducing power order slowly to zero or a low value, and then reversing the power flow at low current. The power order is then increased to the new desired value.

34.4.9 Isolating a valve group

In some schemes, several valve groups at a converter station are connected in series and it may be desirable to be able to block or deblock, or isolate, one group whilst another remains in service. In such cases it is necessary to provide a bypass path for the current and means to transfer the current between the bypass path and the path through the converters.

The ultimate bypass path is normally a metallic switch in a substantially conventional form, and transfer of current from the converter valves to the bypass switch presents no difficulty as the volt drop in conducting valves will be greater than in the switch. The converter can first either be controlled just into the inverter region, or put into its bypass mode. When the contacts of the switch close, current will transfer and firing pulses can be blocked. The valve group may then be isolated if desired.

To transfer current from the bypass switch to the converter valves a special technique is necessary to extinguish the current in the switch. One alternative is to provide a means to increase the voltage drop in the switch to exceed that in the valve bypass path, to force current to transfer.

34.3.10 Numerical example

Calculations performed for the purpose of equipment steady state ratings are normally carried out by means of computer programs using equations similar to those described previously but taking into account the losses of the converters. For nominal conditions, simplified equations can provide reasonably accurate results. The following takes as an example a 1500 MW Bipolar h.v.d.c. scheme with a rated voltage of ± 500 kV at the rectifier d.c. line terminals. The nominal d.c. resistance of the overhead line is 10.0 ohm. The d.c. voltage (U_{di}) at the inverter d.c. line terminal, for rated d.c. current $I_d = 1500$ A, is

$$U_{di} = 500 - 0.5 \times 1500 \times 10 = 492.5 \text{ kV/pole}$$

and the bipolar powers at the rectifier and inverter d.c. line terminals are

$$P_{dr} = 2 \times U_{dr} \times I_d = 2 \times 500 \times 1500 = 1500 \text{ MW}$$
$$P_{di} = 2 \times U_{di} \times I_d = 2 \times 492.5 \times 1500 = 1477.5 \text{ MW}$$

Each pole consists of two series connected 6 pulse valve groups. Therefore, U_{dr} and U_{di} of each 6 pulse bridge is

$$\tfrac{1}{2} U_{dr} \text{ pole} = 500/2 = 250 \text{ kV d.c.}$$
$$\tfrac{1}{2} U_{di} \text{ pole} = 492.5/2 = 246.52 \text{ kV d.c.}$$

To limit the fault surge current in the thyristor valves to an acceptable level a transformer reactance of 15% is specified. Rated α is specified as 12° and rated γ as 15°.

The valve winding e.m.f. required at the nominal operating point can be calculated using equations (34.7) for the rectifier and equation (34.9) for the converter.

$$E_{lr} = \frac{U_{dr}/2}{1.35 \times \left(\cos \alpha - 0.5 \dfrac{I_d}{I_{dl}} x_{cr} \right)}$$

$$= \frac{250}{1.35 \times (\cos 12° - 0.5 \times 0.15)} = 205 \text{ kV rms}$$

$$E_{li} = \frac{U_{di}/2}{1.35 \times \left(\cos \gamma - 0.5 \dfrac{I_d}{I_{dl}} x_{ci} \right)}$$

$$= \frac{246.25}{1.35 \times (\cos 15 - 0.5 \times 0.15)} = 204.7 \text{ kV rms}$$

The transformer rating can be calculated as

Rectifier: $\sqrt{2} \times E_{lr} \times I_{d1} = 1.41 \times 205 \times 1.5 = 433.6$ MVA

Inverter: $\sqrt{2} \times E_{li} \times I_{d1} = 1.41 \times 204.7 \times 1.5 = 432.9$ MVA

The reactive power absorbed by the six pulse group can be calculated as

$$Q_r = U_{dr} \times I_{dr} \sqrt{\left(\frac{1.35 E_{lr}}{U_{dr}} \right)^2 - 1}$$

$$= 250 \times 1.5 \times \sqrt{\left(\frac{1.35 \times 205}{250} \right)^2 - 1}$$

$$= 178 \text{ MVAr} \quad \text{for the rectifier}$$

and

$$Q_i = U_{di} \times I_{di} \sqrt{\left(\frac{1.35 E_{li}}{U_{di}}\right)^2 - 1}$$

$$= 246.25 \times 1.5 \sqrt{\left(\frac{1.35 \times 204.1}{246.25}\right)^2 - 1}$$

$$= 188 \text{ MVAr} \quad \text{for the inverter}$$

34.4 Transmission arrangements

34.4.1 Bipolar lines

The bipole is the most commonly used arrangement. It consists of one line at positive potential with respect to earth and the other of negative potential, the neutral being solidly grounded in the converter station. Figure 34.8(a) indicates an overhead line, but the same arrangement is used for bipolar cable schemes.

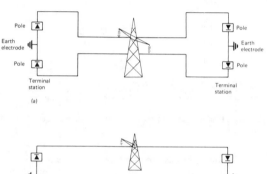

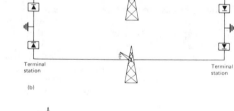

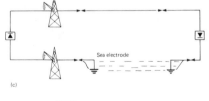

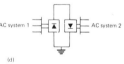

Figure 34.8 Transmission arrangements. (a) Bipolar transmission line; (b) two monopolar transmission lines; (c) cable monopolar transmission with sea return; (d) back-to-back connection

Control is so are arranged that during normal operation the currents in the two poles are balanced so that the current flowing out from the positive pole returns via the negative line, and the current flowing in the earth is negligible. A fault on one pole controls will reduce direct current and voltage of the affected pole to zero in an attempt to clear the fault. In the meantime transmission by the unfaulted pole continues using earth as the temporary return path. Clearly this technique offers the prospect of leads to increased reliability of h.v.d.c. transmission comparable with three phase a.c. transmission, in which phase faults often cause loss of the complete circuit until successful like auto-reclosure.

34.4.2 Two monopolar lines

In circumstances where the probability of line failure arising from environmental conditions is high, two monopolar towers (Figure 34.8(b)) have been used on separate rights-of-way although the system operates as a bipole.

A monopolar line can be operated using earth return with the connection to earth made via a ground electrode. However, ground electrodes have so far only been constructed for use in emergency conditions, being designed to operate only for a matter of hours, or in some cases for a few months.

34.4.3 Cable schemes with sea return

Sea electrodes have been used successfully (Figure 34.8(c)) on several schemes for continuous monopolar operation, low resistivity sea water providing a permanent return path. The resistivity of sea water is in the order of 0.2 ohm-m compared with 10 ohm-m for earth at an ideal land site or 100 ohm-m for fresh water.

34.4.4 Back-to-back arrangement

If there is a need to connect two nearby systems by h.v.d.c., economies can be achieved by combining two converter stations in a back-to-back arrangement (Figure 34.8(d)).

34.4.5 Ground electrodes

The earth provides a readily available medium for the return of direct current. While in certain countries permanent use of the ground in a power circuit is not permitted, its use under emergency conditions such as a line or terminal outage is accepted. A bipolar circuit with ground return thus provides two independent transmission paths.

A typical design of a ground electrode consists of a 3 m annular trench, 250–400 m in diameter, containing coke (Figure 34.9). A steel conductor embedded in the coke is connected to the electrode line by four or more radial insulated conductors. The coke acts as the electrode and the ring diameter is chosen so that the maximum voltage gradient at the ground surface does not exceed $(5 + 0.03\rho)$ volts/m, where ρ is the surface material resistivity in ohm-m. This is a safe value for humans or animals.

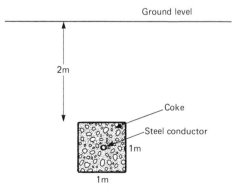

Figure 34.9 Typical ground electrode

A ground electrode requires a damp site of low resistivity both at the surface and in the underlying strata. Very fine soils or sands may be unsuitable because thermal or electrical osmosis could remove the water from the soil-coke surface. A careful survey of neighbouring installations is necessary to identify any electrical conductor which by intercepting the equipotential lines near electrodes would carry a residual d.c. current. This current may cause corrosion where the (anodic) current enters, unless protected by sacrificial anodes or an applied reverse d.c. voltage. Pipelines, railway lines, telephone and power cable sheaths, and power distribution systems using multiple grounding of neutrals may require additional corrosion protection or segmentation into insulated sections.

34.4.6 Sea electrodes

A sea electrode can be relatively simple. One design uses 24 concrete boxes each containing two 1.5 m high silicon cast iron electrodes[1]. The boxes prevent contact with marine life, protect electrodes from silt and damage and restrict the voltage gradient outside to 2.5 V/m. The resistance is 0.02 ohm, well below that of the connecting cables and 35 km overhead line to the converter station (1.1 ohm). An alternative electrode material having good resistance to corrosion is platinized titanium[2].

Electrodes can be built on shore, but assurance of contact with sea water is likely to require a pump installation and consequent maintenance. In both cases annual inspection of the anode electrode for deterioration is necessary. The cathode electrode consists of a simple single rod.

34.4.7 Staged construction of h.v.d.c.

H.v.d.c. systems can be built in stages to suit generation development. For example, the Nelson River Bipole 1 is made up of three 6-pulse groups in series in each pole. Stage 1 was rated at 810 MW at $+320$ kV and -160 kV (*Figure 34.10(a)*) compared to final rating of ± 460 kV (*Figure 34.10(b)*). The low initial voltage implies higher losses until the final voltage was reached. Similarly, Nelson River Bipole II was first used at ± 250 kV and later at ± 500 kV.

If the time between stages of generation development is very long, the cost of operation at low voltage may prove to be too high. In such a case it is possible to start with converters at full voltage but low current, subsequently adding another converter in parallel. The 'parallel' build-up is more expensive than the 'series' build-up from the converter station point of view, but it is more economic from the line loss point of view.

The Pacific Coast Inertie is an interesting example of both series and parallel extensions, illustrating the flexibility of h.v.d.c., *Figure 34.11*. The original scheme was rated for 1420 MW at

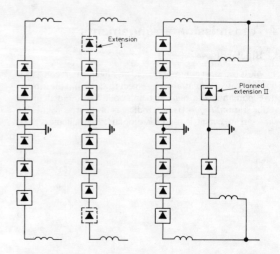

Figure 34.11 Development stages of Pacific Coast Inertie h.v.d.c. scheme

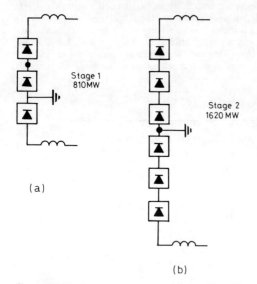

Figure 34.10 Development stages of Nelson River Bipole I scheme

± 400 kV using three mercury arc 6-pulse groups in series per pole. The scheme proved to be capable of higher current rating, giving 1600 MW capability. By restringing parts of the h.v.d.c. line and by adding a 100 kV thyristor 6-pulse bridge to each pole, it was possible to achieve a rating of 2000 MW ± 500 kV. It has now been decided to add in parallel a thyristor converter rated for 500 kV and increase the h.v.d.c. line current to 3100 A, which will give a total external rating of 3100 MW at ± 500 kV, more than twice the original rating of 1420 MW.

References

1 KIMBARK, E. W., *Direct Current Transmission*, Vol 1, Wiley Interscience (1971)
2 THORP and MACGREGOR, Design of the Sea Electrode system, Sardinia–Italian Mainland 200 kV scheme, IEE conference publication 22, London (1966)

34.5 Converter station design

34.5.1 Valve group arrangements

The size and number of converter groups will generally be dictated by the firm power requirements and will in turn dictate the complexity of the d.c. switchgear requirements. Firm power requirements will also influence the selection of which equipment should be switched simultaneously or independently.

Figure 34.12(a) shows the main equipment of a typical converter station arrangement: converter transformers, a.c. filters, valves, d.c. reactors and associated isolators form two separate and independent poles of a bipole. Surge arresters are essential components and are dealt with in Section 34.6. DC filters may be added in overhead transmission line schemes.

In *Figure 34.12(a)*, the loss of any major component, say a

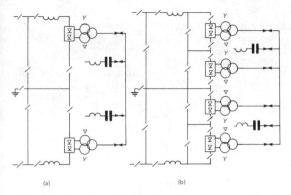

Figure 34.12 Alternative bipolar h.v.d.c. converter connections (a) single 12-pulse groups, (b) series connected 12-pulse groups

Twelve air-cooled air-insulated thyristor valves (three quadrivalves) forming one of four 500 MW 12-pulse poles of the 2000 MW h.v.d.c. submarine link connecting British and French networks (courtesy of Central Electricity Generating Board)

transformer, would lead to the loss of the 12-pulse pole. If power requirements are such that 75% of rated power must be firm then the arrangements in *Figure 34.12(b)* should be considered. Firm power of 75% can be achieved by dividing the pole into two 12-pulse groups. There is no penalty in the thyristor valve arrangement, as the same number of thyristors is required to withstand the desired voltage. However, failure of a d.c. reactor would constitute a loss of 50% power, but no such failure has yet been recorded on any operational h.v.d.c. scheme.

In mercury arc schemes it was economic to use a 6-pulse bridge as the operating unit. In thyristor schemes in most cases, it is more economic to use 12-pulse converter group as the operating unit, and avoid the need for large filter, for 5th and 7th harmonic currents.

34.5.2 Converter valves

The valves are arranged in three-phase bridge circuits as discussed in Section 34.3. The rating of the thyristor valve is flexible and the operating voltage of the valve can be varied by choosing a different number of series connected thyristors. Thyristors connected in parallel have been used, but current ratings of up to 4000 A d.c. bridge current can be accommodated by a single power thyristor of the type now available.

Physically the thyristor valves of a 12 pulse group are usually arranged in stacks with four valves mounted on top of each other to form a quadrivalve, as shown on *Figure 34.13*. This arrangement has the advantage of enabling insulation to ground for the valves operating at the highest voltage potential to be provided by the other valves in the stack (*Figure 34.14*). *Figure 34.13* shows a section through such a valve hall. For electrical power circuit of the equipment shown refer to *Figure 34.20* in Section 34.6. An alternative design from floor supported valves is to hang them from the valve hall ceiling.

34.5.3 Converter transformers

The following transformer arrangements can be used to supply a 12-pulse converter group:

(a) *One* three phase transformer having one line (primary) and two valve (secondary) windings.
(b) *Two* three phase transformers each having one line and one valve winding.
(c) *Three* single phase transformers each with one line and two valve windings.
(d) *Six* single phase transformers each with one line and one valve winding.

Figure 34.15 shows a layout using transformer arrangement (a) in which transformer bushings protrude into the valve hall. This layout has the advantage of eliminating outdoor connections between the transformers and the valve hall which can be a source of radio interference.

Figure 34.14 shows a layout using transformer arrangement (c). It may not be economic to arrange for transformer bushings to protrude into the building as this would necessitate a very long valve hall. Gas insulated busbars could be considered for such a layout.

For very large ratings, arrangement (d) may be used for a 12-pulse group, requiring a much larger area to provide the six a.c. connections between transformers and valve hall.

Because the converter groups are connected in series on the d.c. side, the windings of the outer converter transformers will be biased at a d.c. potential with respect to earth, which is equal to the sum of the voltages of the inner groups. DC potential has a different distribution between oil and paper insulation from a.c. which has to be considered in the design and test. The valve winding line-to-line voltage has a sinusoidal waveshape, but the design must take into account the significant harmonic current content due to the converter action. A major factor in the choice of transformer reactance is the prospective fault current in the thyristor valves during worst case fault conditions. This may dictate the use of a transformer reactance greater than the most economic value for a given transformer design.

34.5.4 AC filters

Filters, to absorb harmonic currents and to provide reactive power, are connected to the same a.c. busbar as the converter

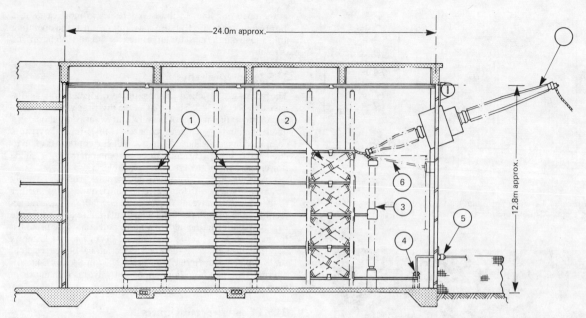

Figure 34.13 Section of valve hall; 1. thyristor quadrivalves; 2. surge arresters-valves; 3. surge arrester-valve group; 4. surge arrester-neutral; 5. through-wall bushing neutral d.c.; 6. earth switch; 7. through-wall bushing-h.v.d.c.

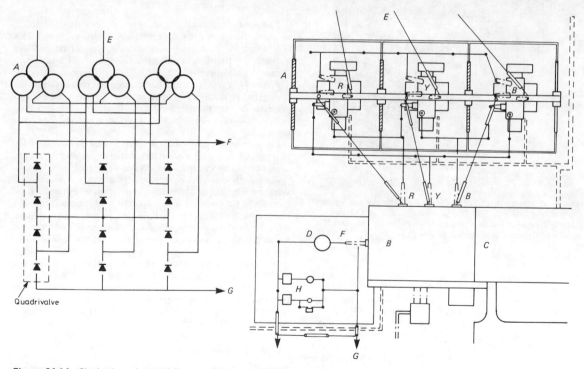

Figure 34.14 Single phase three winding transformer arrangement; A, converter transformers; B, valve hall; C, control building; D, d.c. smoothing reactor; E, a.c. connections; F, h.v.d.c. connection; G, neutral connection; H, d.c. filters

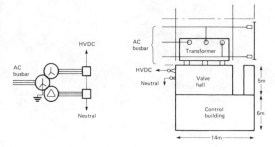

Figure 34.15 Three-phase three-winding transformer arrangement

transformers. It will normally be necessary to split filters into several banks, both for separate maintenance (e.g. capacitor replacement) and to restrict the voltage step at switching. Filter design and reactive power are considered in Sections 34.8 and 34.9.

34.5.5 DC smoothing reactor

The converter valve groups are connected to the d.c. transmission system via a smoothing reactor. In deciding the inductance of this reactor several factors have to be considered. The reactor ensures that the overcurrent transient occurring during an inverter commutation failure or a d.c. line fault is kept within limits acceptable to the valves.

The smoothing reactor exhibits a very low resistance to direct current but provides a high impedance to the characteristic 12-pulse harmonic voltage resulting from converter operation. In the case of transmission schemes employing overhead lines the smoothing reactor acts to filter the harmonics appearing on the d.c. side of the convertors in conjunction with shunt connected capacitors or filters. Unattenuated, these harmonics may cause telephone interference in the area surrounding the d.c. line.

Another important feature of the d.c. smoothing reactor, arising from its high impedance to high frequencies, is that it shields the remaining converter station equipment from being directly exposed to fast voltage transients which can occur on the d.c. line.

The reactor can be placed either in the h.v. or the l.v. connection. By placing the reactor in the l.v. connection savings can be made on reactor insulation costs and this arrangement is feasible for back-to-back schemes. The insulation to ground of the inner valve group would have to be increased however and for this reason, and because protection from fast voltage wavefronts is required for schemes with high voltage lines, it is usual to place the reactors in the h.v. end of transmission schemes.

34.5.6 DC isolators

Most isolators in a two-terminal scheme will be of the conventional slow type. In the few cases where fast operation is required, high-speed isolators which do not have a d.c. current interruption capability will normally be sufficient to provide the switching of lines and valve groups while zero d.c. current is temporary imposed by the converter action.

When series connected 12-pulse groups are used in a pole it is usually necessary to incorporate across each group high speed bypass switches, to assist the blocking and deblocking sequences, and to allow independent operation of the groups.

34.5.7 Protection

The protective functions required of a converter terminal can be divided broadly into three groups.

(a) Conventional protection
 This group covers the standard forms of protection applied to transformers and reactors and would include differential, overcurrent, earth fault, Buchholz etc.
(b) Special power equipment protection
 This group covers special forms of protection which have been developed for converter plant. High speed systems based on fibre optic coupling used with circuit breakers having 2 cycle interruption time, can provide tripping in less than 50 msec. For the capacitor banks used in a.c. filters and static compensators, capacitor unbalance in protection is utilised to detect fuse operations and will have alarm, delayed shutdown, and immediate trip settings.
(c) Protective control equipment
 Special forms of protection for the converter equipment are incorporated as part of the electronic controls for the poles and valve groups. This protection will cover commutation failure, asymmetry, d.c. line or cable fault, d.c. undervoltage, etc. Transient occurrences do not cause shutdown, but if the condition persists for longer than say 300 ms shutdown would be initiated. Asymmetry protection would operate as a result of a converter valve misfire resulting in the generation of disturbed d.c. voltage waveforms. Again, this condition would cause shutdown if it persisted for more than 300 ms. Fault currents on the d.c. line or cable can be limited very quickly (within about 20 ms) by exerting control on the triggering of the valves. For cable schemes this action would be followed by shutdown, but for overhead line schemes, where recovery from the fault may be achieved by temporary reduction of the d.c. current to zero, one or more restarts can be attempted before shutdown is initiated.

These three groups of protection are co-ordinated where appropriate, and are used as back-up to each other to provide a comprehensive protection system. The functions of control and protection are increasingly being coordinated and carried out by microprocessors.

34.5.8 Converter station losses

The high cost of losses can appreciably influence the equipment design (thyristor size, transformer copper size). The total losses of a converter station can be in the region of 0.75% of d.c. power. *Figure 34.16* gives the distribution of the losses between the major items, at full and no load. The transformer, filters and valves account for over 86% of the full-load losses.

Co-ordination of the design of thyristor valves with converter transformers can reduce station losses. Not only will a larger thyristor itself incur a lower current dependent loss, but by its ability to withstand a higher short-circuit current it permits a lower transformer reactance to be used[1]. This in turn favours lower transformer copper loss. Such a choice gives as by-products some overload and a reduction in converter Var consumption capability.

34.5.9 Converter station costs

Typical cost curves for two complete terminal stations are given in *Figure 34.17*. These guidelines will be affected in practice by a.c. voltage, d.c. voltage, location and other matters. Appropriate division of costs between the various components of a station is illustrated in *Figure 34.18*.

34.5.10 Reliability

The term 'reliability' is often used to describe the overall operating performance of a h.v.d.c. scheme which is quantified by its average frequency of failure and average energy availability. The desired performance criteria are usually specified for a

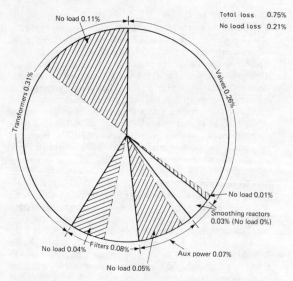

Figure 34.16 Converter station losses, percentage of maximum power transfer

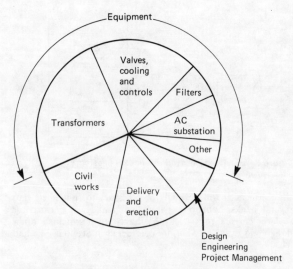

Figure 34.18 Cost division for h.v.d.c. converter stations

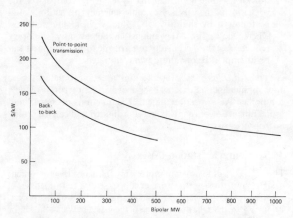

Figure 34.17 Representative d.c. terminal costs 1985 (two terminal, bipolar installed complete)

scheme at the early planning stages to meet the overall requirements of power transmission strategy. The inclusion of financial penalty clauses in scheme contracts has lead to great emphasis being placed on system reliability and availability by system designers.

A quantitative analysis is used to assess the effect of the basic elements in the scheme on the overall performance. Whenever possible equipment with proven reliability is used, but the use of redundant capacity is also extensive. Equipment such as thyristor valves, cooling plant, auxiliary power supplies and control systems usually include redundancy.

The provision of adequate spares to minimise maintenance and repair times contributes to system availability. Spare converter transformers and smoothing reactors may be considered essential due to the long lead time for repair or replacement even though in practice both have proved to be highly reliable items. Indeed there has been no recorded failure of a smoothing reactor on any operational h.v.d.c. scheme.

The energy availability of a bipolar transmission scheme can be maximised if the scheme has the capability to transmit power during forced outages of one pole by operating its remaining pole as a monopole. This can be achieved in the case of converter station pole equipment failures by using its conductors as a metallic return path for the remaining operational pole. The availability can be further increased by utilising an earth return system rated at full current to permit power transfer even during outages of a transmission line conductor.

Emphasis is placed on providing independence between the poles of bipolar schemes such that the number of possible common failure modes is kept to an economic minimum, while still sharing the transmission line and associated d.c. switchgear. Overhead transmission lines usually have both pole conductors on common towers. Experience has shown that common mode failures in this arrangement are unlikely. Even in areas of high lightning activity failures are usually restricted to one pole.

Typical performance targets would be a frequency of failure per pole of a converter station of 1 per year, an availability at full rated power of 98% and scheme total energy availability of 99.25%.

Reference

1 GAVRILOVIČ, A., *HVDC Scheme Aspects Influencing Design of Converter Terminals*, International Symposium HVDC, Rio de Janeiro, March (1983)

34.6 Insulation co-ordination of h.v.d.c. converter stations

34.6.1 Introduction

Insulation co-ordination is the selection of the electric strength of equipment in relation to the voltages to which it may be exposed. Protective devices are chosen to reduce the voltage stresses imposed on the equipment to an economically and operationally acceptable level. The main object of insulation co-ordination for any system, whether a.c. or d.c., is to ensure reliable operation of the scheme at minimum cost.

Generally, an a.c. system is considered to consist of parallel-connected equipments which all have identical insulation levels. A d.c. converter station consists of both series and parallel connected equipment. However, when examining the insulating characteristic of individual a.c. equipment in detail, it is often found that the equipment has been designed and manufactured in discrete units which are connected in series. Obvious examples of this technique are found in shunt capacitor banks, insulators and a.c. circuit breakers which may use several interruptors in series. Less obvious examples include transformers and reactors.

Insulation levels on a.c. systems are relatively higher the lower the a.c. system voltage (facilitating co-ordination between different voltage levels). In a d.c. converter it is generally economical to have relatively lower insulation levels than on the adjacent a.c. system. These lower levels are made possible by close control of the voltages applied during both normal and transient conditions, but this does mean that a.c. system overvoltages can cause significant energy absorption in the converter surge arresters.

34.6.2 Sources of overvoltages

The magnitude and slope of overvoltages arriving at the converter station from the a.c. system will be attenuated by the action of a.c. filters and converter transformer reactance so that overvoltages with fast front times (less than, say, 10 microseconds) do not penetrate the converter from the a.c. system.

Similarly, lightning or other impulsive overvoltages travelling along d.c. overhead line towards the h.v.d.c. converter station will be almost fully reflected at the d.c. smoothing reactor.

Thus, thyristor valves are protected from fast transient overvoltages arising from the a.c. system by the converter transformer, and from the d.c. system by the smoothing reactor. However, in the event of an insulation breakdown within the boundaries established by these protecting inductances, a fast transient overvoltage may occur across the thyristor valves. The most onerous overvoltage occurs if a flashover from an outer (highest d.c. voltage) converter transformer bushing to ground takes place when the converter station has been charged by a switching surge originating in the a.c. system. During such an event the prospective valve overvoltage can be up to twice the switching surge protective level of the valve arrester. It is important to ensure that this surge arrester is able to limit the overvoltage to a level which is safe for the thyristor valve and that its energy absorption capability is adequate for this duty.

34.6.3 Surge arresters

The zinc-oxide non-linear resistor material used in modern surge arresters exhibits a very high impedance at normal applied voltage whilst at a voltage only some 50% higher a very low impedance is provided. The extremely non-linear relationship between voltage and current, shown in *Figure 34.19*, has rendered obsolete the spark gaps which were a feature of previous arresters based on silicon carbide.

Gapless metal oxide arresters are simple in construction, consisting merely of enough resistor blocks connected in series to ensure that the current during normal operating conditions is very small, typically less than one milliampere. The zinc oxide resistor becomes unstable if the continuously applied voltage appreciably exceeds this level, which defines the 'Maximum Continuous Operating Voltage' (MCOV). It may also suffer thermal runaway if the surge energy absorbed is too high (unless the applied voltage is removed after the transient). Its capacity for energy absorption is limited either by the accumulation of the consequences of these two major thermal considerations or in some cases by instantaneous thermal shock.

The protective characteristic of the arrester is the envelope of the discharge voltage (being the maximum voltage developed across the arrester during the passage of a specified current impulse waveform) for various waveshapes. Parallel connected surge arresters can be made to share the total energy to be absorbed in limiting switching surge overvoltages if their VI characteristics are properly matched during manufacture.

34.6.4 Surge arrester arrangement

A typical arrangement of surge arresters is shown in *Figure 34.20*. Arresters are normally connected across each individual thyristor valve, and also across each 6-pulse group. In schemes

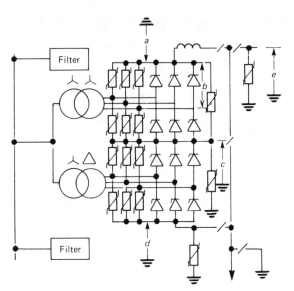

Figure 34.20 Arrangement of surge arresters in one pole of a 500 kV converter

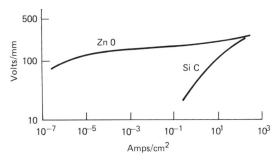

Figure 34.19 Voltage/current characteristic of non-linear resistor material

employing more than one 12-pulse group in series per pole, the point of interconnection between groups is also protected by a surge arrester connected to ground. The d.c. line (and d.c. reactor) is similarly protected by a surge arrester.

The surge arrester protective level applicable to each item of equipment is obtained by examination of the circuit to find the path giving the lowest discharge voltage. Thus, the protective level between the outer converter transformer busbar and ground is determined by the series connection of the valve arrester and the 6-pulse bridge arrester (in the case of two 12-pulse groups in series the protective level of the 12-pulse group arrester must also be added).

Surge arresters are also used on the a.c. system to protect the converter station in the manner shown in *Figure 34.21*. The phase-to-earth insulation is protected by surge arresters which

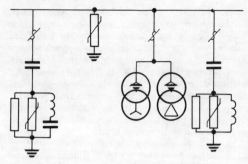

Figure 34.21 Arrangement of surge arresters on the a.c. side of a converter

are often placed close to the transformer terminals. Additional surge arresters may be applied within the a.c. harmonic filters specifically for the protection of filter reactors and resistors.

34.6.5 Safety margins

The source and magnitude of each credible overvoltage can be calculated, knowing the impedance of the circuit elements and the characteristics of arresters. Then the insulation level of the equipment in the converter station can be determined. For most items of conventional equipment the voltage withstand increases as the front time of the applied impulse decreases (*Figure 34.22*).

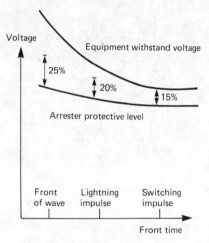

Figure 34.22 Insulation characteristics—conventional equipment

For such conventional plant the safety margin between the protective level of the arrester and the withstand voltage of the equipment is 15% for switching surges. A minimum safety margin of 20% at lightning impulse waveforms and 25% at front of wave (FOW) waveforms is usually applied. The relatively flat protective characteristic of zinc oxide means that the actual margin between the protective level and the equipment withstand at lightning and FOW for most equipment is substantially higher than the minimum recommended. In practice, standard insulation levels (e.g. the IEC 71 series) are used for conventional equipment.

The safety margins applied to conventional equipment have evolved over many years and are intended to take into account both the measuring tolerances and the anticipated deterioration with age of the insulation of the protected equipment and of the surge arrester characteristics. A margin is also necessary to allow for the increase in voltage which may arise as the distance from the surge arrester to the protected equipment increases.

Arresters are placed immediately adjacent to valve terminals, so no allowance is needed for distance effects. The thyristor valves incorporate redundancy which can be restored at regular (e.g. annual) intervals by replacement of any failed thyristors ensuring that the insulating properties of the thyristor valve stay virtually constant throughout its life.

The withstand voltage of the thyristor valve in its off-state is dictated not only by the sum of the withstand voltages of all the thyristors (which is essentially independent of waveform), but also by the interaction between its distributed inrush-limiting reactors and the thyristor grading and damping network. During slow wavefronts only a small proportion of the applied voltage will appear across the reactors. However, as the front time decreases an increasing proportion of the voltage will appear across the reactors, and as a result the valve withstand voltage increases. It is possible to match the valve voltage-time curve to the surge arrester characteristic, achieving the margins listed above for the various impulse waveforms.

34.6.6 Creepage and clearance

The selection of creepages and clearances for the converter station is an important part of insulation co-ordination. The creepage required for a given item of equipment will vary substantially with the environment in which it is required to operate. The creepage length required is proportional to the maximum continuous voltage. In the clean air-conditioned environment provided by a valve hall, a creepage of 14 mm per kV peak will provide satisfactory performance. Under polluted conditions, such as may be present on d.c. overhead lines or on outdoor converter equipment in industrial areas, a creepage distance of 40 mm per kV peak, or sometimes even more may be required to give adequate performance. Clearances within the converter station are determined primarily by lightning impulse and switching impulse withstand requirements.

34.6.7 Application examples

The economic incentive for using a low protective level across the thyristor valves is very strong, since the number of thyristor levels required is directly proportional to this voltage. Therefore, the valve arresters normally exhibit a low protective level, which means that they may be required to absorb large amounts of energy during overvoltages.

With all a.c. harmonic filters connected, recovery from a local three phase short circuit to ground in a weak a.c. system may cause high prospective overvoltages. These overvoltages will have a high content of low order harmonics, and will therefore appear relatively unattenuated on the valve winding side of the converter transformer. The a.c. system phase to ground arresters will limit the peak amplitudes of these recovery overvoltages, but the thyristor valve surge arrester may nevertheless be required to absorb a large amount of energy.

A flashover from the outer converter transformer valve winding busbar to ground, occurring when the d.c. line is charged to overvoltage can also lead to high energy absorption in the thyristor valve surge arrester. During this event the surge arrester absorbs a substantial part of the energy stored in the capacitance of the d.c. line and d.c. filters (where applicable). Although this event is not very likely, the surge arresters across the top three thyristor valves are sometimes specified to be capable of higher energy absorption than the other valve arresters to accommodate it.

Figure 34.23 shows typical lightning and switching impulse

	a	b	c	d	e
BIL	1300	603	650	60	1175
BSL	1172	586	586	35	992

Figure 34.23 Typical lightning and switching impulse levels in a 500 kV d.c. pole

levels in a 500 kV d.c. pole as applicable at various locations, (a) to (e), shown in *Figure 34.20*. It should be noted that the thyristor valves use non-standard insulation levels.

During normal operating conditions the inductor and resistor of an a.c. or d.c. harmonic filter experience only a small fraction of the total line-to-ground voltage. However, a major fraction of any transient overvoltage can appear across the inductor and/or resistor. Filter energisation is an example of a routine event causing an overvoltage across the filter components. If the a.c. system is strong or if several filters are already connected to the busbar, the voltage across the inductor when energised at peak voltage can easily approach the full line-to-ground voltage. By connecting a surge arrester in parallel with the inductor or resistor as shown in *Figure 34.21*, it is possible to utilise components with an insulation level significantly below that applicable to the rest of the a.c. system. However, if it is fitted, such an arrester may be subject to very fast-rising wavefronts, associated with high energy discharge duties. For example, if a flashover occurs from the a.c. busbar to ground, most of the energy stored in the main capacitor will be discharged into the surge arrester. Such an event can lead not only to high energy absorption in the arrester, but also to very high discharge current amplitudes. For example for a filter connected to a 400 kV system the arrester protecting the inductor of the filter may need a co-ordination current of 80 000 A. However, by this means it becomes possible to specify inductors for 650 kV BIL, whereas the a.c. system BIL is 1425 kV.

Further reading

IEC Publication 71-1, 1976 and 71-2, 1976, *Insulation Co-ordination*.
Application Guide for Insulation Co-ordination and arrester protection of HVDC convertor stations, Paper presented by CIGRE WG33-05, *Electra*, No. 96 (1984)

34.7 HVDC thyristor valves

34.7.1 Introduction

Together with the central control system, the valves and their auxiliary cooling and overvoltage protection equipment account for approximately one third of the equipment cost of the converter terminal. In addition, between 30% and 40% of the total station loss is incurred by the valves. At typical capitalised values of US $3000–$8000 per kilowatt, the evaluated cost of losses can approach and sometimes exceed the capital cost of the valves. In such cases, it is economical to invest more in the hardware to reduce the level of losses.

The single most significant variable that influences both the capital cost and the level of losses is the number of series connected thyristors in each valve. The most economic solution is one that uses the minimum number of series levels consistent with reliable long term operation.

A valve is required to act as a switch. It should switch on (turn-on) and switch off (turn-off) efficiently. When off, it should withstand the applied forward and reverse voltages and when on, it should have low resistance.

Unfortunately, thyristors are not perfect switches. At turn-on, they initially have reduced current carrying capacity. At turn-off, the current reverses for a brief period while the thyristor stored charge is extracted and the thyristor's ability to withstand forward voltage is severely limited for some time after negative recovery starts[1,3]. *Table 34.1* summarises how the principal thyristor characteristics are influenced by design parameters. The thyristor designer can trade-off one characteristic against another to achieve the most economic solution for a given application[1]. Close collaboration between the converter valve designer and the thyristor designer is essential in order to achieve the best economic solution.

When properly applied, the thyristor does not suffer from any 'ageing' effects which would cause deterioration of its characteristics with time and it has proved to be a very reliable device provided it is used within its rating. *Figure 34.24* indicates the significant advances which have been achieved in recent years in both voltage and current ratings of thyristors.

34.7.2 Thyristor level circuits

To cater for high voltages for h.v.d.c., each valve is made up of many thyristor levels connected in series.

Figure 34.25 shows the basic electrical circuit of one thyristor level as it may typically be implemented. Many variants are possible but all key features are shown. In some designs one series reactor serves several thyristor levels.

The power thyristor is electrically triggered via its gate and logic unit in response to an optical command received via a fibre optic waveguide from earth potential. Electrical power for energising the electronics is derived from the main damping circuit, local to the thyristor.

The gate and logic unit and the associated power supply are designed to ensure that full and adequate control and protection of the thyristor is afforded, not only under normal steady-state operation, but also under abnormal and disturbed conditions, such as operation with discontinuous current or the loss of a.c. system voltage for up to the maximum a.c. system fault clearance time (typically 200 ms–300 ms).

In series with the thyristor is a saturable reactor to control inrush current and across each thyristor the usual capacitor and capacitor/resistor circuits used for voltage grading and damping purposes.

34.7.3 Voltage rating

The required d.c. voltages can vary from 25 kV for a 50 MW back to back scheme to 600 kV to ground (1200 kV −ve to +ve line) for a 3000 MW, long transmission scheme. Thyristors for h.v.d.c. having a withstand capability in excess of 5 kV are already available and the economic pressure to reduce the number of series levels will continue to force ratings upward.

In the case of long distance schemes, the need to optimise the transmission line costs and losses usually results in a defined optimum transmission voltage for a particular power transfer requirement.

For back-to-back interconnections the designer can choose the operating voltage and current to suit the capability of available thyristors. This usually results in a scheme with low operating voltage and high current. Whatever the application, the general form of the normal valve terminal-to-terminal voltage is similar. The voltage waveshape is complex (*Figure 34.26*), containing fundamental frequency components, fast transient components and a d.c. off-set.

Each valve is protected against overvoltages by a zinc oxide surge arrester, connected directly across its terminals. For maximum economy, the protective level of this arrester should be as low as possible. The achievable protective level depends on the performance requirement of the arrester, which in turn is determined by the system voltage conditions. While the nature of

Table 34.1 Interaction between thyristor parameters

Thyristor parameter	Desired magnitude	Increased by	Reduced by
Current rating	High	increased area increased carrier lifetime packaging (improved cooling) reduced thickness	reduced area reduced carrier lifetime increased thickness
Voltage rating	High	increased resistivity increased thickness shallow impurity gradients uniformity of purity distribution (NTD silicon) edge profiling (reduced peak surface field) edge passivation (stability)	lower resistivity reduced thickness steep impurity gradients less uniform purity distribution steeper edge bevels
Turn-off time	Low	increased carrier lifetime increased thickness emitter geometry	reduced carrier lifetime thinner slices
Stored charge	Low	increased carrier lifetime increased thickness	reduced carrier lifetime thinner slices
Surge current	High	increased area reduced thickness increased carrier lifetime	reduced area increase thickness reduced carrier lifetime
dV/dt withstand	High	shorted emitter pattern density vertical diffusion geometry	trigger sensitivity
Inrush dI/dt capability	High	fast rising gate pulse amplifying gate interdigitated gate	heavy shorted emitter pattern minimal gate area
V_{gt}, I_{gt}	Low	Design for high dV/dt	favourable lateral patterns favourable diffusion profiles
Turn-on delay time	Low	carrier lifetime control	favourable vertical diffusion profile vertical geometry
Forward voltage drop	Low	thicker slices shorter emitter pattern density	thinner slices less dense shorter emitter pattern

transient overvoltages is important because it determines the energy rating of the arrester, it is the ability of the arrester to withstand the peak of the fundamental frequency voltage after a maximum energy surge, which is normally the determining factor. Once the voltage rating of the arrester is chosen, the protective level arises naturally from the zinc oxide voltage-current characteristic and the value of co-ordinating current.

The protective level of the arrester is not a single value but is somewhat dependent on the front time of the incident voltage wave, the value being higher for faster fronted impulses. This must be taken into account in the valve design. In addition, the valve must contain enough thyristors in series to be capable of being tested at values sufficiently above the arrester protective level to give confidence that long service life will be achieved.

In accordance with international standard IEC 700, valves are normally designed for the following test margins with respect to the surge arrester protective levels:

15% for switching impulse
15% for lightning impulse
20% for front-of-wave

Having established the test voltages to be applied, the number of thyristors may be determined. This number is a function of the reverse voltage capability of the thyristors and on the achievable accuracy of the grading circuit.

Thyristors can be damaged if they are exposed to excessive voltage in the forward direction. Further, the forward voltage capability is usually lower than that in the reverse direction. For this reason, protective firing of the thyristors is often employed for overvoltages in the forward direction. *Figure 34.25* shows this feature implemented by a breakover diode. (See Section 34.7.5.) The valve voltage, above which protective self firing may occur, depends on the detailed application but, as a general rule, it will not be below 90% of the surge arrester protective level for switching surge wavefronts (see *Figure 34.27*).

34.7.4 Current rating

For an h.v.d.c. application it is desirable that any credible fault current, arising from insulation failure, can be blocked by the thyristors at the first current zero. It is therefore necessary to ensure that the post fault thyristor junction temperature is below the critical value, above which the thyristor may not be able to block the ensuing recovery voltage.

The problem is a thermal one, bound by two limits: One limit is the temperature above which the thyristors may be unable to block the post fault recovery voltage, the other limit is the ultimate heatsink temperature to which all losses are finally dissipated (e.g. ambient air, river water, etc.).

Figure 34.28 shows a typical breakdown of temperature rise

HVDC thyristor valves **34**/19

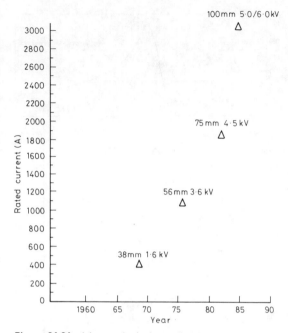

Figure 34.24 Advances in thyristor ratings

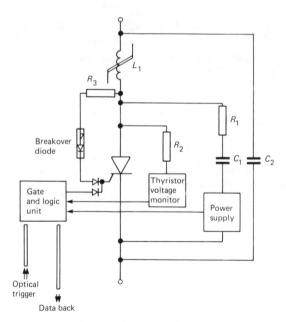

Figure 34.25 Basic circuit of one thyristor level

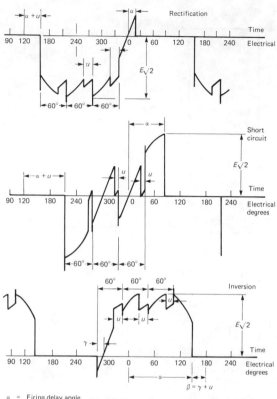

α = Firing delay angle
u = Commutation overlap angle
γ = Safety angle
E = Commutating EMF (RMS line-line)

Figure 34.26 Voltages across a valve operating in a 6-pulse bridge for rectification, zero voltage and inversion

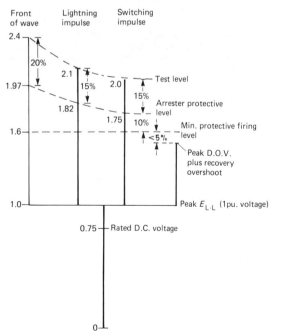

Figure 34.27 Typical insulation coordination for a thyristor valve

between the two limits. In practice, the designer has direct or indirect control over each component of the temperature rise and, for any given application, aims to achieve an optimum economic balance. As can be seen, nearly 60% of the available temperature rise has to be kept in reserve as a margin for transient overloads and fault currents. It is most important to be able to determine the temperature rise resulting from any fault current.

Figure 34.29 indicates how junction temperature follows the current. A short circuit caused by a d.c. side flashover must not be

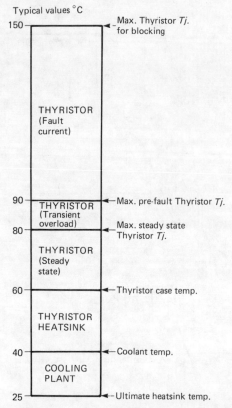

Figure 34.28 Typical breakdown of temperature rise between the coolant and thyristor junction

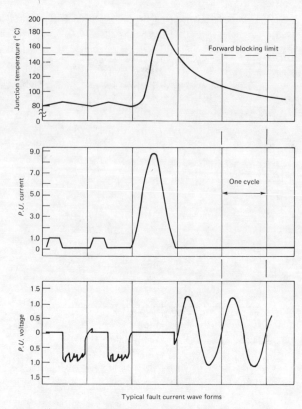

Figure 34.29 Variation of thyristor junction temperature for typical fault current waveforms

allowed to raise junction temperature to a value that would prevent forward blocking after current zero.

There are sophisticated techniques for producing an accurate mathematical model from which the worst case thyristor junction temperature excursion, for any applied current waveform, can be derived[1,2]. Using such models, the sensitivity of junction temperature to changes in fault current produced by changes in transformer leakage reactance can be determined and an optimum value of reactance chosen.

Technically, the method of cooling employed is irrelevant to the performance of the thyristor as it is only a means of achieving a defined steady-state condition from which junction temperature excursions due to transients and faults can commence. Because heat generated at the thyristor junction does not reach the heatsink in times less than one second, the design of heatsink and the method of cooling have no influence on the transient excursions due to faults and disturbances, which rarely last more than a few hundred milliseconds.

Increasingly, the cost of losses is such that it is often justified to select a larger thyristor than could technically satisfy the current requirements of a scheme. Under these conditions, considerations other than overcurrent rating may become limiting (e.g. maximum coolant temperature).

34.7.5 Turn-on behaviour

When thyristors are gated, there is a short, 1 to 2 μs delay before any significant change in impedance takes place. After this initial delay, the impedance of each thyristor collapses rapidly, but the steady-state impedance is not reached until several hundred microseconds later. During this turn-on phase, the thyristors have reduced current carrying capacity and it is necessary to protect them from the prospectively high rates of rise of current arising from the discharge of the circuit stray capacitances. Series connected saturable reactors are used which are active during the initial stages of turn-on but exhibit a low inductance value after conduction is established.

The rate of rise and amplitude of voltage applied to the thyristors, when fast fronted voltage waves appear at the valve terminals are controlled by the reactor and capacitor grading circuit.

Normally, valve turn-on is initiated by the coherent triggering of all thyristors in response to a firing command originating in the central control system. It is clear that if a thyristor is late turning on, it could be destroyed by excessive voltage. Protective self-firing described earlier can be achieved by electronic means or, if very high reliability is required, by use of a break-over diode (BOD) at each thyristor level. The BOD is a voltage sensitive semiconductor switch that operates in response to an over-voltage. It gates the main power thyristor directly and, unlike the electronic alternative, is fully independent of the normal firing system. Because of tolerances, overvoltage firing of the valve is non-coherent and a cascading type of turn-on may take place, and the last level to turn-on may experience increased stress. Being independent, the BOD firing circuit acts as a back up to the normal firing circuits. If normal firing on any one level fails, the BOD responds to the rapid rise in voltage arising from firing of other thyristors in the valve and safely initiates conduction before damage to the thyristor can occur. Efforts to integrate the BOD into the main thyristor, making it self protecting against over-voltages will eventually remove the need for such circuits.

34.7.6 Turn-off behaviour

At turn-off, the switching action of the valve excites an oscillation between the transformer leakage reactance and the circuit stray capacitances. This oscillation cannot be critically damped but the degree of voltage overshoot can be minimised by the correct choice of values for the components of the valve grading network. The thyristors themselves aggravate the recovery overshoot as they do not cease conduction immediately the current reaches zero, but shoftly after. The reverse current flow that is established before turn-off occurs allows energy to be stored in the magnetic fields of the inductive components, and this energy has to be dissipated in the damping circuits.

Immediately after turn-off has occurred, thyristors are unable to support any forward voltage without spontaneously re-conducting. Gradually the thyristors acquire some forward hold-off capability, but it is typically more than one millisecond before their full off-state capability is attained.

The economics of converter design dictate that inverter operation be achieved with the smallest practicable extinction angle margin (gamma) but the value of gamma chosen must not only allow time for the thyristors to recover sufficient forward voltage capability, but must also include additional margin so that temporary minor distortion of the a.c. system voltage waveforms do not result in an unacceptably high incidence of commutation failure. Severe transients, such as those arising from nearby a.c. line faults, will result in loss of margin angle such that commutation failure becomes inevitable. (See Section 34.10.)

Two aspects relating to commutation failure are worthy of mention:

First, a thyristor may be exposed to positive voltage (arising from, for example, disturbed a.c. network voltage) before its recovery process is complete. Forward recovery failure of thyristors may be in itself potentially destructive to the thyristors, particularly when re-application of forward voltage is rapid. The mechanisms leading to failure are complex and are discussed in more detail in References 1 and 3. It is essential for the security and reliability of the system that proper protection is afforded to the thyristors. This usually has to be carried out at each thyristor level.

Second, in marginal cases, some thyristors may re-conduct while others do not. This leaves the valve in a partially blocked state with only a portion of the thyristors supporting the applied voltage. Those thyristors which have successfully blocked are now exposed to a prospective overvoltage, even though the valve terminal voltage may not exceed 1 per unit. The use of individual overvoltage protection, at each thyristor level, and/or whole valve protective firing in response to a data-back signal, ensures that no damaging overvoltage can occur.

34.7.7 Valve arrangements

Oil-cooled, oil-insulated thyristor valves have been used in the past. Air-cooled, air-insulated valves have been used and are still being offered. However, most economic arrangements at present use air-insulation and water cooling.

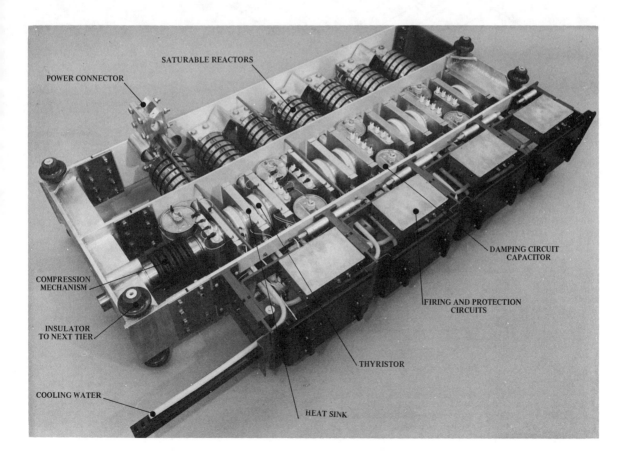

Figure 34.30 Water cooled assembly having seven thyristor levels in series. Courtesy General Electric Co. plc.

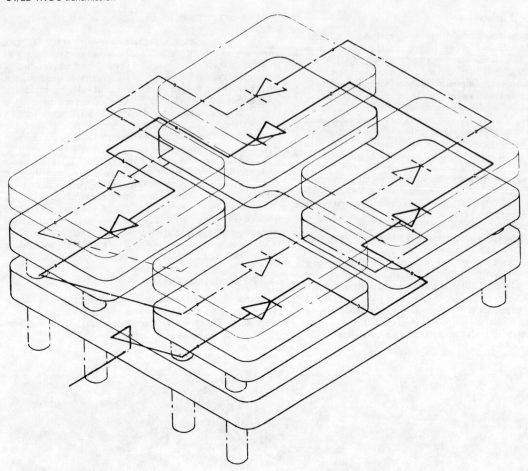

Figure 34.31 Diagrammatic illustration of part of a valve showing four thyristor assemblies connected in series. Each assembly consists of, typically, seven thyristor series levels

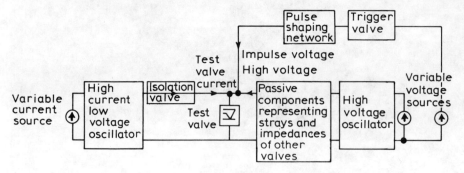

Figure 34.32(a)

The following description of a valve is given as an example of how a 1500 MW ±500 kV d.c. converter is built up. Two 6-pulse bridges connected in series are used to form a 500 kV 12-pulse valve group. Therefore each valve has to be capable of operating in a 250 kV 6-pulse bridge. To achieve this voltage over 100 thyristor levels (*Figure 34.25*) would have to be used, including 2 to 3 levels for redundancy.

Figure 34.30 shows a water-cooled assembly of seven thyristor levels in series, with interleaved heat sinks, held in compression for efficient heat transfer. Replacement of any thyristor is possible without disturbance of the water circuit by jacking against the compression of the spring and lifting out the device. The four assemblies which form one tier are mounted in a structure and supplied with two counter-flow water circuits providing uniform temperature through the valve. Two tiers are diagramatically shown in *Figure 34.31*. Four such tiers are required to form one valve for the example chosen. Four valves are staked on top of each other to form a quadrivalve. Electrical connections form a spiral from ground level (neutral voltage) through four series valves forming a quadrivalve (i.e. four valves associated with one phase of a 12-pulse valve group). Three such quadrivalves form a complete 12-pulse pole contained in one valve hall, rated 500 kV, 1500 amps, 750 MW (see *Figure 34.13*).

34.7.8 Valve tests

It is important that the thyristor valves are fully proven before service. The valve performance for turn-on, turn-off and cascade firing is demonstrated on a complete valve or a portion of the valve with sufficient thyristors in series to achieve realistic conditions. This can be done using a synthetic circuit as shown in the block diagram form in *Figure 34.32(a)* or other suitable circuits always provided that sufficient series connected thyristors are used to obtain representative valve performance.

Figure 34.32(b) shows the voltage response of one level relying on back-up triggering when tested in a 100 level valve and in a proportionality rated 10 level test object. The obvious differences become significantly more pronounced if the reduced scale test object has even fewer series levels.

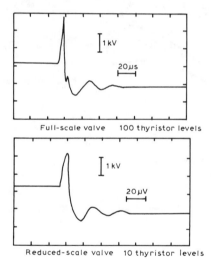

Figure 34.32(b) Voltage response of thyristor level with back-up triggering ($a = 90°$ firing conditions)

There is not yet universal agreement for special valve tests[4,5,6,7]. IRC 700[8] recognises the need for special tests in clause 11 and specifies for test circuits the use of either two 6-pulse bridges connected back-to-back or a suitable synthetic circuit. However, the positive recovery phenomenon discussed above, was not identified and the IEC 700 (1981) recommentation proved to be inadequate[3,9].

References

1 GIBSON, H., BALLAD, J. P. and CHESTER, J. K., *Characterisation, Evaluation and Modelling of Thyristors for HVDC*, IEE Conference Publication No. 255, London, England (1985)

2 CHESTER, J. K., 'A New Technique for Deriving Self-consistent Electrical and Thermal Models of Thyristors During Surge Loops', IEE Conference Proceedings, *Power Electronics, Power Semi-Conductors and their Applications*, London, England (1977)
3 WOODHOUSE, M. L., BALLAD, J. P., HADDOCK, J. L. and ROWE, B. A., 'The Control and Protection of Thyristors in the English Terminal Cross Channel Valves, Particularly During Forward Recovery'. IEE Conference Publication No. 205, *Thyristor and Variable Static Equipment for A.C. and D.C. Transmission*, London, England (1981)
4 EKSTROM, A. and JUHLIN, L. E., *Testing of thyristor valves*, CIGRE 1972 14-03, Paris, France
5 BANKS, R., ROWE, B. A. and NOBLE, R. G., *Testing Thyristor Valves for HVDC Transmission*, CIGRE 1978 14-07, Paris, France
6 DEMAREST, O. M. and STOILS, C. M., *Solid State Valve Test Procedures and Field Correlation*, CIGRE 1978 14-12, Paris, France
7 LIPS, P., THIELE, G., HUYNH, H. and VOHL, P. E., 'Design and Testing of Thyristor Valves for the HVDC Back-to-Back TIE Chateauguay', International Conference on *DC Power Transmission*, June 1984, Montreal, Canada
8 International Electrotechnical Commission, Publication 700, 1981, Geneva, Switzerland
9 KRISHNAYYA, P. C. S., *Important Characteristics of Thyristors, of Valves for HVDC Transmission and Static Var Compensators*, CIGRE 1984 14:10, Paris, France

34.8 Design of harmonic filters for h.v.d.c. converters

34.8.1 Introduction

The a.c./d.c. converter is a source of harmonics, and since excessive levels of harmonic distortion on an a.c. system can lead to a number of undesirable effects (overheating of induction motors, generators and capacitors, telephone interference, etc.) shunt harmonic filters are used on the a.c. terminals of all h.v.d.c. converter stations.

Harmonic distortion on the d.c. line may in some cases cause unacceptable interference in adjacent telecommunication circuits. In order to minimize this interference, harmonic filters are often fitted at the terminations of d.c. overhead lines.

34.8.2 AC harmonic current generation

Due to large inductance of the d.c. smoothing reactor the current conducted by each converter valve consists of a train of nearly flat topped current pulses. Thus, the current at the a.c. terminals of the converters is not sinusoidal.

Figure 34.33 superimposes the a.c. current pulses from two ideal 6-pulse bridge converter circuits connected via star/star and star/delta transformers with zero commutating reactance but with infinite d.c. circuit inductance. A Fourier analysis gives, for a d.c. current of I_d, the following series for the star/star and the star/delta transformer, respectively:

$$i = \frac{2\sqrt{3}}{\pi} I_d [\cos(\omega t) - 1/5 \cos(5\omega t)$$
$$+ 1/7 \cos(7\omega t) - 1/11 \cos(11\omega t) + \ldots]$$

$$i = \frac{2\sqrt{3}}{\pi} I_d [\cos(\omega t) + 1/5 \cos(5\omega t)$$
$$- 1/7 \cos(7\omega t) - 1/11 \cos(11\omega t) + \ldots]$$

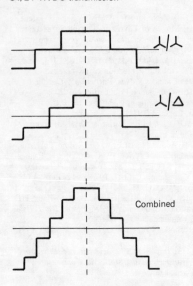

Figure 34.33 Idealised phase current on a.c. side of converter station

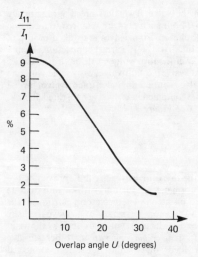

Figure 34.35 11th harmonic as a percentage of fundamental

By addition of these two currents the 5th and 7th harmonics cancel and only harmonics of orders $12k \pm 1$ will enter the a.c. system for a 12-pulse group.

In practice, because the converter transformer reactance is not zero, the current takes a finite time to transfer from one valve to the next. As shown in *Figure 34.34* the resulting current waveform is smoother than that shown in *Figure 34.33* The amplitude of each harmonic component of current depends on the value of overlap angle u which is a function of the commutating reactance, of the firing angle α (or γ) and of the load current. The effect of u

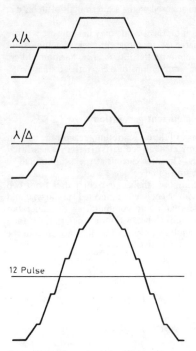

Figure 34.34 Phase current with firing and overlap delays

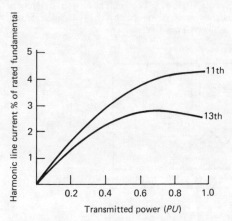

Figure 34.36 Harmonic generation

can be judged from *Figure 34.35*. The 11th harmonic with $u=0$ would be 1/11 (9.1%) of the fundamental, while in practice it is nearer to 4%. *Figure 34.36* shows the variation of characteristic harmonics with d.c. load. Text books should be consulted for a full analysis of harmonic currents.

The theoretical analysis above is valid for balanced a.c. systems and converter operation. In practice, neither the a.c. system nor the converter circuit are perfect, and cancellation of harmonics will be incomplete. The major causes of harmonics other than $12k \pm 1$ are:

1. a.c. system phase unbalance.
2. Unbalance between 6 pulse bridges.
3. Unbalances within 6 pulse gridges.

34.8.3 Filtering

Shunt harmonic filters connected on the converter a.c. busbars provide a low impedance into which most of the harmonic currents are diverted. Shunt filters also generate reactive power at fundamental frequency providing some or all of the reactive power required by the converters.

The most direct method of achieving a low impedance at a given frequency is by means of a tuned filter as shown in *Figure*

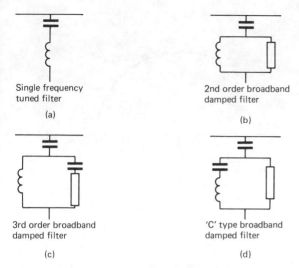

Figure 34.37 Alternative types of harmonic filter

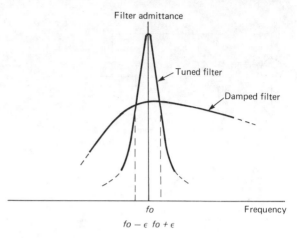

Figure 34.39 Filter performance

34.37(a). The admittance of a tuned filter varies sharply around the resonant frequency as demonstrated in *Figure 34.39*. The sharpness of tuning and the Var rating of the filter must be chosen to achieve the specified performance over the required range of system frequency, temperature and component tolerances.

Each sharply tuned filter is capable of providing significant attenuation at one frequency, but gives virtually no damping at other frequencies. This means that the a.c. busbar voltage during transient phenomena may exhibit ringing at low order non-characteristic frequencies, and that this ringing may persist for a long time. The busbar voltage shown in *Figure 34.40* is typical of the transient occurring during the energisation of a combination of tuned filters and a single high pass damped filter.

It is possible to combine two separate tuned filters in a single filter as shown in *Figure 34.38*. The characteristics can be altered

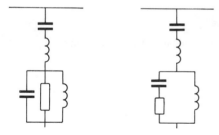

Figure 34.38 Alternative types of double tuned harmonic filters

by increasing the damping thus reducing the sharpness of 'tuning'. Combining the main capacitors of two individual tuned filters to create one double-tuned filter can often give significant cost savings.

The damped broadband filter shown in *Figure 34.37(b)* requires a significantly higher Var rating than a corresponding sharply tuned filter to provide the same harmonic absorption at a given frequency, but *Figure 34.39* shows that damping and filtering is provided over a range of harmonic frequencies.

Damped filters must have a higher Var rating than the corresponding tuned filters to achieve same filtering performance, and their losses are higher. The losses can be reduced significantly by the use of an additional capacitor as

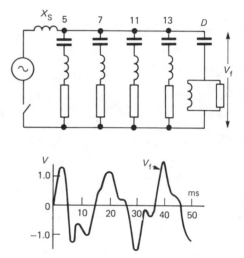

Figure 34.40 Switching transient tuned combination single and high pass damped filter

shown in *Figures 34.37(c)* and *34.37(d)*. In the 3rd order broadband damped filter the fundamental frequency losses in the resistor are reduced by the use of a capacitor C2 in series with the resistor. In the 'C'-type damped filter a capacitor C2 is connected in series with the inductor L and the fundamental frequency losses are minimised by tuning C2 to resonate with the inductor at fundamental frequency. The 'C' type filter attenuates low order non-characteristic harmonics, where the losses in a 2nd order filter would be uneconomically high.

Figure 34.41 shows the busbar voltage following energisation of a combination of a 'C'-type and 2nd order damped filter. The transient can be seen to be virtually damped out within 10–15 ms after energisation.

In general, a combination of tuned and high-pass damped filters will provide the lowest loss solution. However, a combination of damped filters can often provide substantially better service to the a.c. system by preventing resonances.

34.8.4 Harmonic performance evaluation

The harmonic current generated by the converter is injected into the parallel combination of the filter impedance and the a.c.

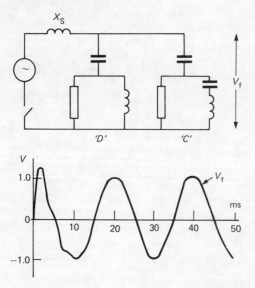

Figure 34.41 Switching transient C-type and second order damped filter

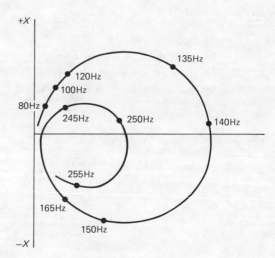

Figure 34.43 Typical supply network impedance diagram

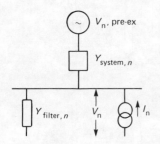

Figure 34.42 Circuit diagram for the calculation of harmonic distortion V_n at harmonic number R_n

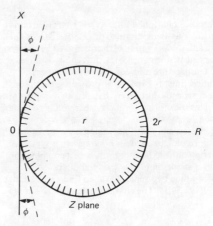

Figure 34.44 AC network impedance

network impedance as shown in *Figure 34.42*. Whilst the impedance of the filters can be determined the harmonic impedance of the a.c. system can vary substantially because of line switching operations and load/generation changes in the network. Variation of system impedance with harmonic number for a typical system condition is shown in the polar R–X diagram of *Figure 34.43*. Information presented in this manner is more informative than that provided in tabular form at integral harmonics because it gives a clear indication of where the resonant frequencies occur, i.e. when the locus crosses the R-axis. When resonance occurs close to a particular harmonic under consideration there will be a rapid change in impedance with frequency, making it difficult to assess performance accurately. Where the supply system has many different operating configurations it can be assumed that the impedance lies within a circle limited by the impedance angles ϕ, or within a segment of a circle of specified radius. The coordinates of such a circle encompass all the anticipated system conditions and enable the determination of the worst case of harmonic distortion to be carried out. A typical example of an harmonic impedance locus specified for filter performance evaluation purposes is given in *Figure 34.44*.

The a.c. network impedance locus is converted into an admittance locus and the worst case of harmonic voltage distortion V_n due to the converter harmonic current is determined by assuming an a.c. system admittance giving the minimum resultant admittance of the parallel combination of the a.c. harmonic filters and a.c. system.

Pre-existing distortion originating in the a.c. network must be added to the voltage distortion caused by converter harmonic currents. Permitted harmonic voltage distortion levels vary from country to country and according to the a.c. network voltage level, but typical values are 1% for odd harmonics, 0.5% for even ones, and 2% for the total root of the sum of the squares.

Filtering performance must normally be achieved for system voltage and frequency variations and ambient temperatures which can persist for long periods. For conditions which can only exist for short times, it is often acceptable to allow the specified harmonic distortion to be exceeded. However, the equipment must be rated for these more arduous conditions.

34.8.5 DC filtering

The converters produce harmonic voltages between the d.c. terminals of the valve groups. In a 12-pulse scheme the lowest order characteristic harmonic is the 12th, but because of inevitable imperfections in the a.c. system and the converter circuit, harmonic voltages of other orders will also be present.

The converter circuit contains a d.c. reactor which exhibits a large impedance at high harmonic orders, minimizing the harmonic current flowing into the d.c. line. Nevertheless, the voltage and current profile along the line should be calculated and the possible induced noise in nearby communication circuits checked. The current and voltage profile is dependent on the earth resistivity and the line characteristics, and it will vary along the line.

If calculation suggests that unacceptable noise is likely to be induced in nearby circuits it may be necessary to provide filters at the converter station terminals. These filters interact closely with the d.c. reactor and the line, but often it is possible to achieve the required performance by means of a simple damped filter.

Figure 34.45 shows a typical arrangement of a d.c. filter on a converter pole.

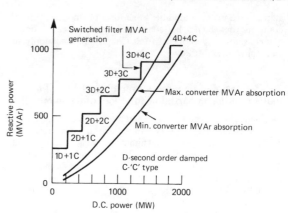

Figure 34.46 Reactive power of h.v.d.c. converters

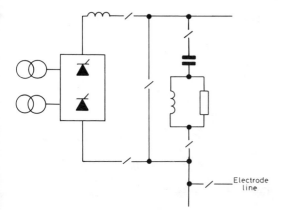

Figure 34.45 Second order damped d.c. filter

34.9 Reactive power considerations

34.9.1 Introduction

Generation and absorption of reactive power constitute major consideration for long h.v.a.c. overhead lines and for much shorter a.c. cables. There is usually a surplus of reactive power at light load and a deficit at heavy load, and it often becomes necessary to provide fixed or variable shunt and/or series compensation which also affects the stable power transmission limit. While h.v.d.c. lines do not consume, generate or transmit reactive power, converters do. At the a.c. terminals of the converter stations, the solution of reactive power requirements of the converters is combined with the reduction of harmonic distortion.

34.9.2 Reactive power requirements of h.v.d.c. converters

The converter absorbs reactive power irrespective of whether it is operating as a rectifier or an inverter as explained in Section 34.3.

At the rectifier end the a.c. system has normally some excess reactive power capability. Therefore, the a.c. filters are usually designed to achieve economically the filtering requirements. This may lead to the fact that the MVar rating of filters is lower than the reactive power consumption of the rectifier, the a.c. system supplying the difference. At the receiving end the situation is normally different. Here, reactive power is required not only for inverter operation but also for loads supplied by the inverter. It is normal to specify that the converter station must as a minimum supply all inverter reactive power requirements. System requirements for reactive power can be supplied by additional shunt capacitors in the converter station or situated elsewhere in the system. In bidirectional schemes each end will in turn act as the receiving or the sending end stations.

Figure 34.46 shows the variation of the reactive power Q with changes in d.c. power for the converter of a 2000 MW scheme. The two continuous curves must allow for all the relevant design and operating tolerances. The nominal reactive power demand of the converters at full load is about 50% of the MW rating for this scheme.

In the example considered in *Figure 34.46*, assuming that the filters are designed to supply all inverter Var, Q_i at rated load, there would be an excess of reactive power supplied to the system at lower powers. This may not be acceptable. In such cases filters are designed so that parts can be switched out at lower loads, reducing Q_f, but always having sufficient filtering capacity. For the example in *Figure 34.19* the filter has been divided into four identical banks for filtering of lower order harmonics (indicated as 1C etc. in *Figure 34.19*) and four banks for filtering higher order harmonics (indicated as 1D etc.).

34.9.3 Steady state voltage control and total ratings of reactive equipment

AC system load flow studies including the effects of d.c. link P and Q should be carried out for all likely network situations in order to determine the limiting levels of Q that the a.c. system would be able to absorb or generate if the network and converter station voltages were to remain within the usual operating limits (e.g. $\pm 5\%$).

The size of harmonic filters necessary from voltage distortion considerations at different power levels should first be determined. Further studies would indicate the required +ve and −ve range of additional reactive compensation taking account of the tolerances of all components. This additional reactive compensation equipment may be switched in steps or may have a proportion under smooth or continuous control, as determined by the following further considerations.

34.9.4 Voltage disturbances caused by switching operations and requirements for smooth reactive control

Step changes of reactive power caused by switching of shunt capacitors and filters have to be limited in order to minimise voltage disturbances to other consumers and to the h.v.d.c. converter itself. The permissible magnitude of voltage step may be appreciable for infrequent occurrences (e.g. 3%) but for frequent events the value may have to be smaller (e.g. 1% or 1.5%). Such step changes may be regularly imposed by filter switchings corresponding to the daily load cycle. If the minimum practical step change is not acceptable some form of smooth acting var compensator should be considered, such as a static var compensator as described in the Chapter 16 on Supply and Control of Reactive Power.

34.9.5 Control of temporary overvoltages caused by faults resulting in partial or total loss of d.c. power flow

DC load rejection resulting, for example, from a d.c. line fault can lead to excessive temporary overvoltage at converter station busbars because the connected filters and capacitors represent a substantial surplus over the reduced var demand of the converter. The worst disturbances for a converter at one end of the d.c. link are usually due to faults in the a.c. system at the other end. This normally results in reduced power flow, the exact nature of it being dependent on whether the converter is operating as a rectifier or an inverter.

Faults in the a.c. system, at the remote rectifier end, cause low d.c. line voltage giving low d.c. power flow. The d.c. link is usually designed to attempt to ride through such temporary system faults. Usually the rectifier-end fault situation is communicated to the receiving end by a telecommunication link, but, in order to discriminate against temporary, rapidly cleared rectifier-end faults, no protective switching action is initiated at the inverter terminal for a few hundred milliseconds. Subsequently the converter is blocked and the filter and other capacitors are switched off to reduce the overvoltage.

The transient load rejection effect on the a.c. system of such inverter blocking should be investigated for the a.c. network conditions of minimum fault level at which full power could be transmitted. As an example, a theoretical temporary overvoltage of about 1.35 p.u. at power frequency, may be expected at the inverter station a.c. busbars for a system short circuit level of 3 times the d.c. power level (short circuit ratio SCR = 3 see Section 34.12), if the a.c. system impedance were a pure reactance. In practice a lower overvoltage will be more usual, due to the resistive effect of network loads. In many urban networks such an overvoltage, even for a fraction of a second, may not be acceptable.

For a rectifier, faults in the remote receiving system have the same effect as a d.c. short circuit. The reactive power demand of the rectifier is then determined by the low voltage current limit (l.v.c.l., see Section 34.10) setting of its current controller. A typical value of l.v.c.l. of 0.3 p.u. would result in reduced var demand compared with the levels provided by the connected filter capacitors, so that there would be some surplus var generation and the rectifier a.c. terminal voltage would rise.

The reduction of such dynamic overvoltages (DOV) due to load rejection cannot generally be achieved by d.c. converter controls themselves. Fast responsive equipment capable of absorbing a large part of the filter Mvar temporarily until the filters can be disconnected is required. Synchronous compensators, even with field forcing, cannot rapidly achieve the high reactive power absorption required. If a.c. system requirements demand faster control of temporary overvoltage, they can be achieved by switching actions, variable static var compensators may have to be used.

Figure 34.47 compares the calculated dynamic overvoltage for a converter station with and without the presence of static compensators following a fault leading to total d.c. load rejection. To achieve the desired control of the temporary overvoltage, the saturated reactor compensator was designed for reactive absorption overcurrents of 3 times its rated current. In *Figure 34.48* the main parameters of a thyristor controlled reactor (TCR) and of a saturated reactor (SR) compensator required to achieve this rating are summarised.

A very large bank of non-linear zinc-oxide type resistors could be considered as an alternative as a voltage limiter of high energy

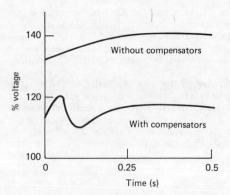

Figure 34.47 Overvoltages due to h.v.d.c. link blocking

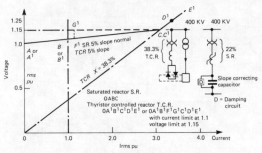

Figure 34.48 V/I characteristics of static variable reactors in overload region

capability suitable for repetitive current surges lasting up to 0.5 second or more. At present this would be expensive.

34.10 Control of h.v.d.c.

34.10.1 Summary of h.v.d.c. controls

Figure 34.49 shows the general arrangement of controls for a typical h.v.d.c. bipole. Progressing backwards from the converter valves these are:

Valve Firing Circuits: These have some protective and monitoring functions, but in normal conditions they act only as an interface between the pole controls and the valves.

Pole Controls: These are the main controls responsible for changing the firing angles of converters in response to various control loops, and are fast (response time typically 5 ms to 50 ms).

Tapchanger Controls: These are relatively slow (about 5 s per step) and act only to optimise working conditions of the converters for minimum reactive power, losses, and harmonics.

Master Control: This is at one station only and controls the whole bipole in response to a power order; it has a slower response than pole controls.

Telecommunication: Transmits current order digitally to the remote station, with a check-back signal on a return channel. It may also carry supervisory and other signals.

AC System Damping Controls: Measurement and feedback of various a.c. system quantities to provide damping to one or both a.c. systems, normally acting via the master control. These are discussed individually in Section 34.11.

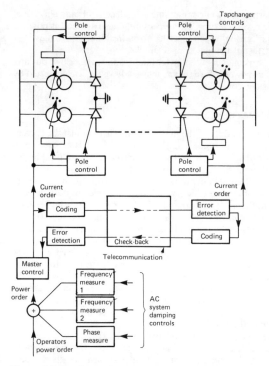

Figure 34.49 Control hierarchy for a bipole h.v.d.c. link

34.10.2 Pole controls

Figure 34.50 shows typical d.c. voltage/current characteristics for the rectifier and inverter respectively as seen from the d.c. line. This diagram effectively summarises the various control loops used in the converters as follows:

Rectifier: AB is from a voltage limit loop. BC is from a minimum alpha limit at about $\alpha = 2°$, CDE is from a constant current loop at a current equal to current order. FG is the low-voltage current clamp (LVCC) characteristic and also acts via the current loop, switched to a fixed current order of 0.3 per unit if d.c. voltage falls below 0.3 per unit.

Inverter: HD is from a constant-extinction angle (γ) loop, typically at $\gamma = 15°$ to $18°$; KD is a 'current-error characteristic' to ensure stable operation near normal voltage[1]. KL is from a constant current loop, at a current lower than current order by the current margin (0.1 per unit). LM is a low-voltage current limit (LVCL) characteristic obtained by compounding a current loop with measured d.c. voltage.

In normal operation the working point is at the cross-over point D of the two characteristics, corresponding to the inverter in constant gamma (γ) control determining d.c. voltage, and the rectifier in constant current control determining d.c. current at the ordered value.

If rectifier a.c. voltage falls relative to inverter a.c. voltage, then BC falls, sweeping the working point along DK and perhaps down KL, in a stable manner, i.e. the inverter takes over current control at a current below order by up to 0.1 p.u. It is not satisfactory to omit slope KD because the transition is then too abrupt. The master control will normally correct the resulting power error by increasing current orders.

34.10.3 The phase-locked oscillator control system

The type of converter control used in the pole controls of all modern h.v.d.c. schemes is the phase-locked oscillator control system[2] giving nominally equidistant pulse firing. The principal reason for this is its freedom from harmonic instability[3] when the converter is connected to a relatively weak a.c. system, as generally applies when the d.c. power forms a substantial part of the a.c. system infeed.

Figure 34.51 shows the principles of the phase-locked oscillator control in simplified form. Its basic components are a

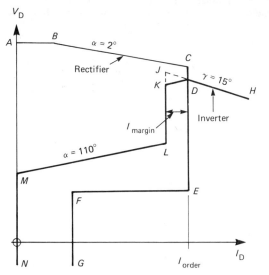

Figure 34.50 Typical V_d/I_d characteristics of rectifier and inverter stations

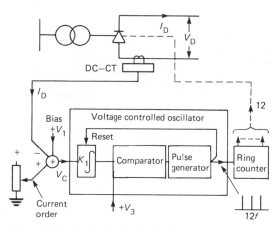

Figure 34.51 Basic phase-locked oscillator control system

voltage-controlled oscillator, and a 12-stage ring counter (for a 12-pulse converter) which feeds the 12 pulses to the respective converter valves.

The oscillator comprises an integrator, comparator, and short pulse generator, and normally runs at 12 times supply frequency, hence valve firing pulses are normally once per cycle per valve, at an accurate $30°$ spacing. The oscillator input has a fixed bias V_1, plus an error signal equal to the difference of the measured quantity for the particular loop (d.c. current as shown) and an order signal. The system settles as for an integral control system, with zero steady-state error, and with firing pulse times (firing angle α) at the correct values to obtain this.

In its recent form, the feedback signal is applied 'raw', i.e. without extra smoothing lags, giving fast response and good stability. This applies not only to approximately smooth signals such as d.c. voltage or current, but also to mark/space type signals such as α or γ. Because this method integrates the input signal, it correctly controls the *mean* value of the controlled quantity. Some methods (including some forms of digital control) do not have this property and tend to respond more to sampled values of the measured quantity near to the normal firing times; this gives excessive response to ripple and sudden disturbances, with poor stability, unless extra smoothing lags are added, which also degrades stability.

All practical control systems must be *multi-loop*, as seen from Figure 34.50, since the operating mode must change according to system conditions at each break point. As an example, operation changes between alpha control ($\alpha = 2°$) to constant-current control at point C of the rectifier characteristic.

Figure 34.52 shows a recent development of multi-loop control based on multiple oscillators[6]. As an example this is

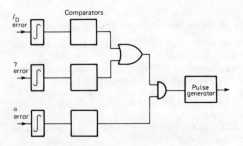

Figure 34.52 Loop control system

shown for three loops, respectively for d.c. current, γ and α. (A scheme may use up to 6 loops.) Each loop has an individual integrator and comparator, and is coupled via OR and AND gates to a common pulse generator; the latter resets all integrators together, and also operates a common ring counter (not shown) as before. This gives extremely fast and precise hand-over between modes, without the extra smoothing lags and amplifier desaturation delays which characterise systems which use hand-over in the pre-oscillator portions of the controls.

The most modern controls include *gamma-balancing* circuits, which equalise gamma values even in conditions of a.c. system unbalance; this gives the control the combined advantages of the equal pulse-spacing method (stable operation on weak a.c. systems) and the individual phase control method (maximum real power and minimum reactive power during unbalanced a.c. system conditions).

Flux-control circuits are included to prevent core saturation instability in converter transformers, and also to reduce the effects of even harmonics from the a.c. system or of a small fundamental frequency component on the d.c. line.

34.10.4 Tapchanger controls

At the inverter station the converter transformer tapchangers are automatically controlled from d.c. voltage to obtain rated d.c. voltage.

At the rectifier the tapchangers are controlled from measured firing angle to hold it within set limits, say 8° to 18°. The lower limit is as low as possible to give minimum reactive power, losses, and harmonics, while preserving a small control range in case of small a.c. voltage dips. The upper limit is also as small as possible without giving tapchanger control instability [i.e. (cos 8° − cos 18°) > tap voltage stepper unit].

The tapchangers are conventional, and rather slow, but any errors in power before they reach steady values are corrected by the master control.

34.10.5 Master control

The principal duty of the master control is to adjust the current orders of the two poles so that measured power is equal to a power order. This is done by direct measurement of total bipole power from the d.c. quantities, subtracting it from the power order signal, and integrating the resulting error signal to generate the 'master current order', which is then applied to both poles in both stations[4]. It is therefore a feedback loop, at a higher level than pole controls, but rather slow (settling time say 100 ms–300 ms, depending on telecommunication rate); the telecommunication gives a direct delay when the remote station happens to be controlling current, and an indirect delay even when the local station is in control because it may then have to wait for the checkback signal (see Section 34.10.6 below). However, the master control contains various other functions including special forms of limit; an example of the latter is in recovering from a zero-voltage fault, when it would otherwise try to generate an infinite current order.

The master control will normally control to constant power in the presence of changes of a.c. voltages and other disturbances, subject to maximum current limits applied locally in pole controls. This applies even if the inverter temporarily takes over current control; the master current order will then be temporarily above the actual current.

Where the rectifier and inverter stations are a long distance apart, a telecommunication system (see below) is required to send current order to the remote station. Although schemes have been operated without telecommunications, optimum performance cannot be obtained without it. Generally the master control is at one end and the remote station acts only as a 'slave' without any master control function. It is not in principle important whether the master control is in the rectifier or the inverter station. Both have been used in practice, though a.c. system damping controls (see Section 34.11) may influence the choice.

The power order signal input will normally be set by an operator, either locally or in a remote control centre. It may be modified by a.c. damping controls.

Back-to-back schemes have both converters in one station, hence do not require a long-distance inter-station telecommunication, but their control and performance is otherwise similar to that for long-distance schemes (except that their response is much faster due to the omission of d.c. line capacitance and telecommunication delays).

34.10.6 Telecommunication

The main duty of this is to send a current order signal to the 'remote' station. Because practical telecommunication media are subject to noise and interference, signals are always sent in digital (binary) form, with error checking code[5]. The coding must generally be much more powerful than normally used for sending computer data, because of the serious effect of an undetected error on power flow.

Detected errors are less of a problem since they are arranged to 'freeze' pole current orders in both stations at the last correct value, using a check-back channel[5]. The latter must itself have error-detection coding; this is usually shared with other signals, e.g. supervisory signals or a.c. network damping signals. Since normal operation is steady state, or with very slow change, the effect of brief errors is then negligible. For permanent telecommunication failure the controls are transferred to a simpler manual mode.

Various media are possible for telecommunication. Direct wires via private land lines or the telephone system have been used, but usually only for emergency operation because of their high cost and poor reliability. Power-line carrier is probably the cheapest method. It would require 2 or 3 repeaters for say 900 km of overhead line. Interference generated by the converters, penetrating via the d.c. reactors, is greater than normally seen on a.c. systems, and requires special measures. Microwave radio has been used on several schemes. It requires rather short repeater intervals of about 40 km since line-of-sight transmission is required, but its bandwidth is very large and can accommodate many telephone and even television channels in addition to the h.v.d.c. control requirements.

Tropospheric scatter radio is another possible method, requiring repeater intervals between 200 km and 500 km depending on transmitter power. It is subject to heavy fading, and requires multiple redundant channels and multiple antennae to give an acceptable basic signal error rate. Optical fibre is being used more and more for medium-distance telecommunications, and has been installed experimentally inside power conductors and overhead earth wires. It is almost immune from interference and has wide bandwidth but would require repeaters about every 30 km at present, though future improvements to this are expected.

The choice of telecommunication will depend on cost and the policy of the utility. Generally a baud rate of 2400 (block period about 50 ms) will be adequate to obtain sufficient bandwidth for damping a.c. system swings up to about 1 Hz.

34.10.7 Performance examples

Figures 34.53–34.56 are oscillograms taken from simulator tests showing the behaviour of a typical h.v.d.c. link for a.c. and d.c. faults. These are all for the relatively onerous case of a weak receiving system (effective short-circuit ratio 2.4, impedance angle 75°), and at rated power.

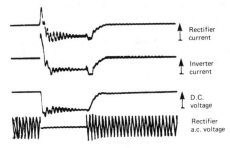

Figure 34.53 3-phase 100% fault at the inverter

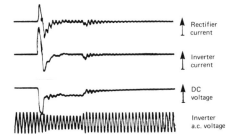

Figure 34.54 1-phase 40% fault at the inverter

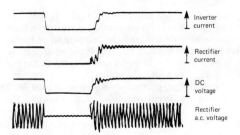

Figure 34.55 3-phase 100% fault at the rectifier

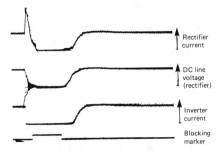

Figure 34.56 DC line fault

Figure 34.53 is for a 3-phase fault close to the inverter a.c. busbar. There is an initial current surge, then the current settles to 0.3 p.u. during the fault, at zero d.c. voltage. After fault removal, most of the recovery occurs in about 100 ms, but the last part of the recovery is relatively slow because of the effect of magnetising inrush current of converter transformers on a.c. voltages.

Figure 34.54 is for a more remote fault in the a.c. system fed by the inverter, which reduces a.c. voltage by about 40%. There is an initial commutation failure, but commutation is then restored at reduced voltage until the fault is removed, with recovery as before.

Figure 34.55 shows a 3-phase fault at the rectifier. DC current falls to zero during the fault, and recovery afterwards is similar to *Figure 34.53*.

Figure 34.56 shows a fault, e.g. due to lightning, about half way along a d.c. overhead line. The fault reduces d.c. voltage to an average of zero, with line oscillations and a rectifier current surge. The rectifier starts to reduce current to 0.3 p.u., but a line fault detector relay operates at 30 ms after the fault, forcing complete

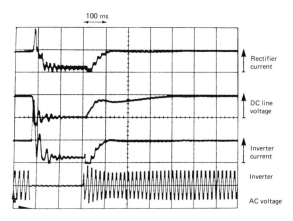

Figure 34.57 3-phase fault at the inverter (ESCR = 1.5)

(temporary) shut-down; by about 40 ms both rectifier and inverter currents are at zero. After waiting about 100 ms for transient line currents to decay, and the fault arc to de-ionise, power is then restored to normal. This simple action appears to be sufficient in most real cases, though if necessary operation can be more complicated, with several re-starts.

To illustrate several of the points discussed above, *Figure 34.57* shows an oscillogram from simulator tests showing the behaviour of an h.v.d.c. link for a 3 phase fault at the inverter, similar to *Figure 34.53* but with a very low value of effective short circuit ratio (see Section 34.12) ESCR = $1.5 \angle 75°$. This gives good performance, only slightly worse than for *Figure 34.53* (SCR = 3), with fast recovery and good stability without using special controls. However, in practice, for an a.c. system so weak, the excessive a.c. voltage changers to which consumers may be exposed following, say, a d.c. line fault, may not be acceptable. In such a case some form of fast reactive control, such as a static var compensator, may have to be used.

References

1. British patent 1 300 226. (Current-error characteristic)
2. AINSWORTH, J. D., 'The phase-locked oscillator—a new control system for controlled static converters', IEEE Trans., vol. PAS-87, no. 3, pp. 859–65 (March 1968)
3. AINSWORTH, J. D., 'Harmonic instability between controlled static converters and a.c. networks', Proc. IEE, vol. 114, no. 7, pp. 949–58 (July 1967)
4. British patents 1 170 249 and 1 258 974. (Master Control and Telecommunication)
5. AINSWORTH, J. D., 'Telecommunication for HVDC', IEE Conference on Thyristor and Variable Static Equipment for a.c. and d.c. transmission, IEE Conference Publication No. 205 (Nov. 1981)
6. AINSWORTH, J. D., 'Developments in the phase-locked oscillator control system for HVDC and other large converters', IEE Conference on a.c. and d.c. Power Transmission, IEE Conference Publication No. 255 (September 1985)

34.11 AC system damping controls

The normal duty of an h.v.d.c. link is to transmit power at a preset level set, usually, by an operator. However, h.v.d.c. is a very powerful control device as it is capable of changing transmitted power in a controlled manner, rapidly and by a large amount and thereby influencing the a.c. system transient performance significantly. Three main system conditions are discussed below.

34.11.1 DC link supplies power from dedicated generators or from a very strong system to a small system

Figure 34.58 shows a block diagram of the Nelson River h.v.d.c. transmission scheme in Manitoba, Canada. The scheme is essentially a 900 km long, 'asynchronous' interconnection from remote hydro generation to an urban industrial load area. The d.c. link controls are arranged to produce a current order in response to four principal signals:

(a) Steady state power order from the system dispatch control.
(b) Transient power order in response to the receiving end load area frequency changes; this provides receiving frequency control by the d.c. link of the form usually carried out by governors with defined droop characteristics.
(c) Transient power order in response to frequency changes at the sending end; this assists the hydro-generator governors in controlling the sending end frequency.

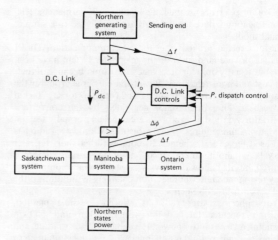

Figure 34.58 Block diagram of the Nelson River transmission scheme

(d) Transient power order in response to the inverter a.c. bus voltage phase angle changes; this signal modulates d.c. power to provide swing damping within the local network and also between the local (Manitoba) power system and the three neighbouring systems of Ontario, Saskatchewan and the USA Northern States Power Pool. The power change required for swing damping is usually quite small, typically less than 10% of rated h.v.d.c. power.

Figure 34.59 illustrates the damping effectiveness of the frequency controls on the speed (frequency) of the receiving end (Manitoba) system equivalent machine and on the sending end generator at the Kettle generating station. *Figure 34.60* shows site measurements of the effect on system frequency oscillations of an inadvertent loss of h.v.d.c. system damping control; before this event the system frequency is well damped, but after the loss of damping, sustained oscillations at about 0.3 Hz are to be observed on the nominal 60 Hz frequency trace. (These appear rather small on a frequency trace, but power oscillations in tie-lines are much larger in proportion.) In this scheme the damping signal (d) is based on measurement of the rate of change of the absolute phase of the inverter a.c. bus voltage using a phase-locked oscillator. Direct derivation of a network frequency signal would require filter circuits which generally reduce the sensitivity and accuracy of the measurement.

Signal (d) is effectively 'a.c. coupled' having no effect in steady state; signals (b) and (c) may be either of this type or may be d.c.

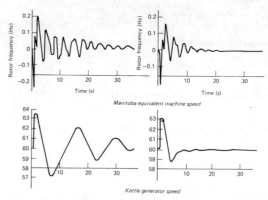

Figure 34.59 Nelson River transmission scheme: rotor frequencies following on disturbance without (on the left) and with d.c. link system damping controls

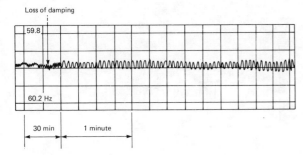

Figure 34.60 Site measurements showing the effect of loss of h.v.d.c. damping controls on a.c. system frequency

coupled to influence steady state operation. Various other signals are also possible for particular systems, for example power in a tie-line.

The use of a damping signal such as (d) above (which is relatively fast) may influence the location of the master control. The average response to modulation at the master control is not much affected by its location (at rectifier or inverter stations) but if it is remote from the source of the fastest damping signal then an extra delay will occur in its loop because of the necessity for sending the damping signal via telecommunication; this is undesirable, hence the master control should be at the end where the highest natural frequency (e.g. 1 Hz) is to be damped. This may require the operator's power order to be sent via a telecommunication channel, but this is rather easy since a relatively slow channel can be used.

34.11.2 DC link connecting two systems which are not synchronised but are of similar size

Care must be taken that the requirements of one system do not adversely affect the other. For example, two systems of similar size may have the same dominant natural frequency, say, in the region of 0.5 Hz to 1 Hz. In this case it will generally be necessary to provide damping signals from both a.c. systems, so that a disturbance in one system is shared between the two; this will slightly disturb the healthy system, but in a heavily damped manner. (Taking a damping signal from one a.c. system only will cause disturbance in the healthy system which may not be damped by the d.c. link.) It should be noted that even if the converter is operated at its maximum rating, a degree of damping can be achieved by power reduction modulation only.

34.11.3 DC link connecting two parts of an a.c. system or two separate systems having also a parallel a.c. link

Figure 34.61 shows a simple diagram illustrating parallel a.c. and d.c. transmission.

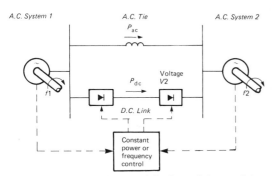

Figure 34.61 Block diagram of simple parallel a.c. and d.c. transmission

The Pacific Intertie on the USA West Coast is an example of a h.v.d.c. link between two a.c. systems which also have long interconnecting a.c. lines. Initially this scheme was operated with constant power control, but in the mid-1970s additional power modulation controls were provided which considerably enhance its use by arranging for the transient stability of the a.c. lines to be improved. In such applications, the amount of power modulation required to achieve beneficial transient stability and swing damping effects is generally small compared with the transmitted power.

There are several examples in Canada, USA and Europe of back-to-back h.v.d.c. links between two large systems where an a.c. interconnection would have been much less advantageous technically and economically. By providing controllable d.c. power flows in the interconnection, even in the event of a major disturbance in either a.c. system, the h.v.d.c. links can help reduce the instabilities, and prevent complete system collapse situations by avoiding the cascading of disturbances which could otherwise occur in uncontrolled a.c. ties.

References

1   *AC Network Stabilisation by DC Links*, CIGRE Paper 32-01, 1970. Paris, France, Influence of d.c. link controls, C. J. B. Martin etc.
2   HAYWOOD, R. W. and RALLS, K. J., *Use of HVDC for Improving AC System Stagility and Speed Control*, Maintoba Power Conference EHV-DC, Winnipeg, Manitoba, June 1971.
3   AINSWORTH, J. D. and MARTIN, G. J. B., *The influence of h.v.d.c. links on a.c. power systems*, GEC Journal of Science and Technology, Vol 44, No 1 (1977)

34.12 Interaction between a.c. and d.c. systems

34.12.1 Study of h.v.d.c. systems

The relationship between an h.v.d.c. link and the a.c. system to which it is connected can be considered in two categories:

(i) The line commutated converter depends for its operation on being supplied with a reasonable sinusoidal voltage. A distorted voltage, or a significant unbalance of the three phase system voltage such as occurs during faults, will detract from the essentially symmetrical rectification and inversion processes.

(ii) The rectifier takes from its a.c. system both power (P) and reactive power (Q). The inverter feeds power to its a.c. system but takes the reactive power from it. If the d.c. link is relatively large compared with the a.c. system to which it is connected, any large changes in P and Q could have significant effects on the system.

AC system disturbance can affect the operation of a small converter but mal-operation of a small converter will have negligible effects on the a.c. system. However, it is not uncommon for an h.v.d.c. link to supply a large proportion of the a.c. system load so that the loss of d.c. power and associated reactive power changes can have a profound effect on the system.

The effects of h.v.d.c. operation and mal-operation can be simulated accurately using loadflow, transient stability and other digital programmes as regards category (ii) above and by the use of an h.v.d.c. simulator as regards category (i) above.

An a.c./d.c. simulator is an important tool in the study of h.v.d.c. In particular its use is important in the development of h.v.d.c. controls and in the study of the inverter recovery after system faults, as illustrated in Section 34.10. It is not necessary to represent the d.c. network in detail for these studies. AC systems can be represented by a Thevenin equivalent of a constant e.m.f. behind an impedance, as shown in *Figure 34.62*. It is important to

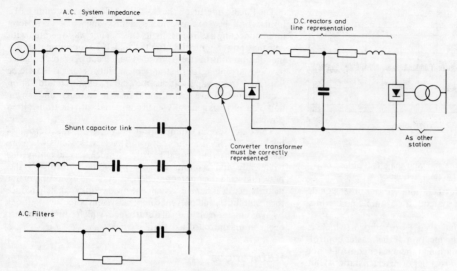

Figure 34.62 System representation for h.v.d.c. simulator

represent adequately both the value of the impedance and its damping. AC filters should be correctly represented as well as any shunt capacitor banks.

Simulator studies are usually concerned with a time scale of up to 400 ms. Sufficiently accurate representation will generally be achieved by representing generators by their subtransient reactances, though more accurate representation is sometimes needed. As far as the converter transformer is concerned, it is important to represent it correctly as the major contribution to the distortion of the a.c. voltage is the transformer inrush current. It is therefore important to represent not only the commutation reactance but also saturated self and mutual reactances.

34.12.2 Short circuit ratios

The higher the system impedance and the lower the system damping for a given h.v.d.c. inverter, the greater the effect of the inverter mal-operation on the a.c. system. It has become customary to refer to relative sizes of the a.c. system and the h.v.d.c. power by the term short circuit ratio, SCR

$$\text{SCR} = \frac{S}{P_d}$$

where S is the minimum a.c. system short circuit MVA at which the maximum d.c. power P_d is transmitted.

S is calculated similarly as in Section 34.12.1 using Thevenin equivalents.

AC harmonic filters must be used in all schemes. At fundamental frequency the filter acts practically as a shunt capacitor. Shunt capacitors increase the impedance at fundamental frequency of essentially inductive systems and to allow for this the term Effective Short Circuit Ratio (ESCR) is used. In admittance form this is defined as for SCR but it is the admittance of the a.c. system plus all filters and capacitor banks additionally connected to the a.c. bars. Note that both SCR and ESCR always have an angle as well as magnitude. Thus for example for a system with SCR $= 3 \angle 80°$ the addition of 0.6 p.u. of capacitors plus filters gives ESCR $= 2.4 \angle 78°$. If the simplification of a system impedance angle of 90° is assumed, then in short-circuit MVA form:

$$\text{ESCR} = \frac{S - (Q_f + Q_s)}{P_d}$$

where Q_f is equal to fundamental frequency MVA of a.c. filters and Q_s is the MVA of any additional shunt capacitors connected to converter station terminals.

Figure 34.63 indicates the definition of SCR and ESCR

SCR = the value of the admittance Y at the fundamental frequency, on the base of the rated d.c. power at the rated a.c. voltage.

ESCR = is defined as SCR except that the admittance includes the admittance of the capacitor, $Y = Y_s + Y_c$

It should be noted that because a.c. systems are largely inductive, it is the change of reactive power which is mainly responsible for the effect of converter behaviour on the a.c. network voltage. Most schemes in the past were designed with transformer reactance of the order of 20% or more to limit the thyristor fault current. The availability of large thyristors and the pressure of cost of losses, are responsible for some schemes being designed using a larger thyristor than necessary in order to reduce the losses. An additional benefit is that it is possible to design the transformer nearer to its economic reactance of typically 15%, as the oversized thyristor has correspondingly larger surge current capability[1].

Lower converter transformer reactance can bring a number of advantages, including:

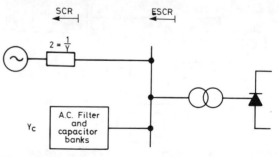

Figure 34.63 Definition of short circuit ratios

(a) Converter reactive power consumption (Q_c) will be reduced.
(b) AC filters and any additional shunt capacitors are normally designed to supply at least all converter reactive power. The total $Q_f + Q_s$ can be reduced, reducing the cost and increasing ESCR.
(c) Temporary overvoltages will be reduced, due to smaller shunt capacitors.
(d) Rating of equipment such as surge arresters may be reduced.

SCR and ESCR represent the a.c. system reasonably accurately for the short timescale considered. It could be argued that overvoltages would be better estimated if the SCR were defined to the base of the rated converter reactive power. There are also some proposals to define SCR to the base of converter transformer rated MVA. It should be emphasised that in the study of a.c./d.c. simulation, not only the a.c. system but also the converters, including their transformers, must be correctly represented. The short circuit ratio concept should be used only to get initial 'feel' and it is best to use SCR to the base of rated power for that purpose. In any case the a.c. system and the d.c. power to be transmitted would be known much before the short circuit level of the converter station equipment is specified.

34.12.3 SCR as a guide to system planning

The following values of SCR can be used to indicate approximately the strength of an a.c. system relative to the d.c. power infeed. (The conditions at the rectifier end are not critical.)

Strong a.c. system SCR > 4

It is unlikely that any special measures would need to be taken to strengthen the system.

Intermediate a.c. system 2 < SCR < 4

Voltage support may have to be provided at the a.c. terminals of the converter station, by for example, static var compensators.

Weak a.c. system SCR < 2

Synchronous compensators may have to be used to strengthen the system in addition to or instead of static var compensators.

34.12.4 System interaction when the a.c. system impedance is high relative to d.c. power infeed (Low SCR)[2]

The important and desired interaction between a.c. and d.c. systems is beneficial. DC can bring power into an a.c. system in a controlled way and can assist the a.c. system by improving its frequency control, stability and damping. The design of the converter system must allow for:

(a) steady state stability
(b) recovery after a.c. or d.c. faults
(c) a.c. and d.c. overvoltages

34.12.4.1 Steady state stability

(a) Power on voltage instability
This is a steady state cumulative type of instability which can occur at low SCR at the inverter end, say less than about 2.5, when constant-power control is used[3]. The simplest cure is to design the master controller to give a not-quite constant power-control when necessary. Other methods are to use control action in the inverter itself or to add a static compensator; neither of these latter two methods may be effective if a.c. system voltage is left slightly low after a system fault. Technically the best solution is either to connect the d.c. converter to a stiffer point in the a.c. system, which may not be

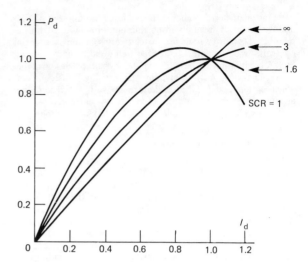

Figure 34.64 Power/d.c. current characteristics of an inverter feeding an a.c. system by finite impedance (commutation reactance 20%, $\gamma = 18°$ fixed)

possible, or to stiffen the system by adding an a.c. line or a synchronous compensator which is expensive[4].

(b) Core saturation instability
This is an instability which involves saturation of converter transformers, and occurs when there is a resonance near to fundamental frequency on the d.c. side between d.c. reactors and d.c. line, with an anti-resonance (or at least a high impedance) near to 2nd harmonic on the a.c. side. The mechanism involves fundamental frequency on the d.c. side, d.c. components in valve windings which cause transformer saturation, and 2nd harmonic on the a.c. side. The control system is also involved. Build up of this type of instability is very slow (several minutes) because of the transformer saturation effect. It can be cured by control system additions which effectively monitor the effect of saturation, and oppose it by feedback in the converter[5].

(c) Subsynchronous instability
This can occur typically at frequencies of the order of 10 Hz to 40 Hz, assuming the worst case of transmission via a long d.c. line or cable. It corresponds to ordinary control loop instability and can be studied by for example the Nyquist Criterion or by eigenvalues and cured by proper choice of control parameters such as loop gain settings at the rectifier and inverter. Synchronous machines with steam turbines exhibit mechanical shaft resonances in this frequency region, hence guaranteed stability is particularly important where these are present in the a.c. networks, to prevent their resonances being excited.

Suitably designed h.v.d.c. controls should not excite subsynchronous instability and in fact can be used to damp such instability[6].

34.12.4.2 Recovery after a.c. and d.c. faults

Satisfactory recovery of the h.v.d.c. link after a major a.c. fault at the inverter can present problems, because the a.c. voltage distortion caused by magnetising inrush current of transformers during recovery may be substantial. The most important criterion is the shock to machines caused by the loss of MW-seconds, which must be minimised to prevent pole-slipping in the a.c. system (transient instability). From the control point of view this requires a control system which re-starts as rapidly as possible on re-establishment of a.c. voltage, which has static and dynamic characteristics chosen to give fast restoration of power, and which gives freedom from commutation failures during recovery. As can be seen from oscillograms in the section on controls, modern controls can cope well even at low values of SCR. The situation can be helped further by suitable design of the converter transformers to give low magnetising inrush current when economic, and of the a.c. filters to give substantial damping at low frequencies of the order of 2nd to 4th harmonic.

Recovery following d.c. faults represent a less severe condition.

34.12.4.3 AC and d.c. overvoltages

Where a converter is connected to a weak a.c. system the effect of sudden blocking of the converter (load rejection) is to cause a substantial a.c. voltage rise. This may be caused for example by an a.c. fault at the remote station, or because of the necessity to cut off direct current to clear a d.c. line fault. Another case is where an a.c. fault near to the converter busbars is removed by a circuit breaker opening; when the a.c. voltage reappears it is likely to be excessive, particularly if for some reason the converter does not commence commutation immediately. The control of these and other overvoltages is discussed in the Section on Reactive Power Considerations.

Overvoltages on the d.c. line are generally lower, at least with a d.c. cable scheme, because of the isolating effect of the d.c. reactors. However, with weak a.c. systems it is advisable to provide each station with voltage limiting loops in its controls; d.c. voltage surges can then be effectively reduced in most conditions, with benefit to cable insulation. A d.c. overhead line is subject to lightning overvoltages, which can be reduced by conventional methods.

34.12.4.4 Conditions at the rectifier end

The above are primarily concerned with the inverter (receiving system) end. The rectifier end source impedance has a much smaller effect on stability and recovery. However, the rectifier end is still liable to moderate a.c. voltage changes caused by disturbances. If the source is an isolated generating station without local loads, the a.c. voltage changes are generally acceptable. If there are local consumers then static compensators may again be a suitable solution to reduce a.c. voltage changes.

34.12.5 'Island' receiving system

It is possible to supply most or all of the system power requirements to an island system by h.v.d.c. The example for such a scheme is the d.c. link supplying power from the Swedish mainland to the island of Gotland[7]. Because the loads are near the inverter station the system impedance will generally have good damping. If there is no generation in the island system, or if the amount of generation is very small, a synchronous compensator must be used to provide sufficient inertia so that temporary faults do not lead to the system breakdown. Any loss of power, arising from, say, a fault on the sending a.c. system, would cause the island system to slow down, but for fault duration less than some defined value, e.g. 200–300 ms, the system should be designed to recover.

References

1 GAVRILOVIC, A., *HVDC Scheme Aspects Influencing the Design of Converter Terminals*, International Symposium on HVDC Technology, Rio de Janeiro (March 1983)
2 AINSWORTH, J. D. and GAVRILOVIC, A., *Interaction between HVDC and a.c. Systems when the d.c. Link is large compared to the a.c. system*, United Nations Seminar on High Voltage Direct Current (HVDC) Techniques, Stockholm (May 1985)
3 AINSWORTH, J. D., GAVRILOVIC, A. and THANAWALA, H. L., *Static and synchronous compensators for h.v.d.c. transmission converters connected to weak a.c. systems*, CIGRE Paper No. 31-01 (August 1980)
4 HAMMAD, A. and KAENFERLE, J., *A New Approach for the Analysis and Solutions of a.c. Voltage Stability—Problems at h.v.d.c. Terminals*, HVDC Symposium, Montreal, Canada (June 1984)
5 AINSWORTH, J. D., *Development in the phase-locked oscillator control systems for HVDC and other large converters*, IEE Conference Publication No. 255, London, England (1985)
6 PIKE, P. J. and LARSEN, E. V., *HVDC system control for damping of subsynchronous oscillations*, IEEE PZS-101, July 1982
7 LISS, G. and SMEDSFELT, S., *HVDC Links for connection to isolated a.c. networks*, United Nations Seminar on High Voltage Direct Current (HVDC) Techniques (May 1985)

34.13 Multi-terminal h.v.d.c. systems

The applicability and flexibility of h.v.d.c. systems can be enhanced in some conditions if several converters are coupled to form a multi-terminal h.v.d.c. system. The earliest application of this philosophy was the paralleling of bipoles I and II of the Nelson River Scheme onto one line in the event of an outage of one of the bipolar transmission lines. The first true multi-terminal h.v.d.c. scheme will be achieved with the construction of a parallel tap (50 MW) on the Sardinia-Italian Mainland h.v.d.c. scheme (200 MW).

Multi-terminal h.v.d.c. systems may be divided into series and parallel types, illustrated in *Figures 34.65* and *34.66* respectively. However, there are many permutations of each type, depending on system requirements.

34.13.1 Series connection

Figure 34.65 shows an example, in which one pole of a 500 kV, 1000 A rectifier supplies two inverter stations in series, respectively of 100 kV, 1000 A, and 400 kV 1000 A. In the case

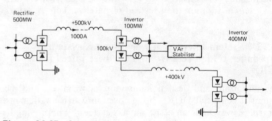

Figure 34.65 3-terminal series scheme

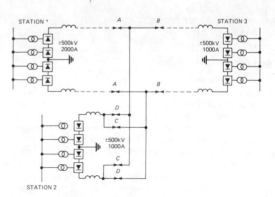

Figure 34.66 3-terminal parallel scheme

shown, one method of control is to operate the rectifier permanently at constant current equal to full rated current, and vary inverter powers individually as required by varying their voltages. This makes the series method expensive in terms of line losses, filter losses, damping losses, and reactive compensation, but does give virtually independent control (including smooth power reversal in any station), and very little interaction between stations during most disturbances. Some economy is obtained by reducing current when total power is less than full load, so that at least one inverter is at its rated voltage, but the sum of the ratings required of each item of plant (including filters, etc.) is always greater than for a parallel scheme.

34.13.2 Parallel connection

The example in *Figure 34.66* shows an arrangement for a 3-terminal parallel scheme intended principally for operation as a 2000 MW rectifier supplying two inverters each of 1000 MW. This is an example of a system of medium control difficulty. Switches A and B are solely to isolate their respective lines in case of permanent d.c. line faults. Switches C and D form in addition a reversing system for station 2. The switches shown may be slow-acting isolators, plain fast switches, or true d.c. circuit-breakers.

Control of parallel rectifiers is easy since by using normal constant current control loops at each rectifier, their currents (and powers) are easily set to any desired values (including zero) and there are no current sharing problems.

The basic problem of control of parallel inverters is that an h.v.d.c. inverter operating in the most efficient mode, of constant extinction angle (γ), has an effective slope resistance which is negative, so that two such inverters in parallel are obviously unstable.

Many ways round this problem have been proposed, such as operating one inverter at constant extinction angle (to control direct voltage), and the other in a constant-current mode; this means that the second inverter requires higher plant rating. However, the system is in a rather delicate state for even minor transients.

Figure 34.67 shows an alternative control characteristic, originally developed to enable parallel operation of the two bipoles of the Nelson River scheme, and which gives stable operation without requiring excessive plant rating. The current order at each station is communicated from a master control. Obviously the actual currents obey Kirchhoff's first law, so current orders should be restricted so that the sum of the rectifier current orders is equal to the sum of the inverter current orders.

Figure 34.67 has been drawn for the normal steady-state case where respective transformer tapchangers have been adjusted to the ideal positions, for the particular a.c. voltages, d.c. currents, etc. The working points of the inverters are right on the constant-

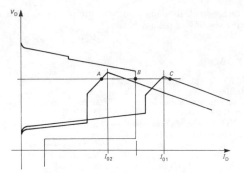

Figure 34.68 Inverter 1 tapchanger slightly low

characteristic (in the ideal positions) at their correct currents, and the rectifiers have the usual small voltage margin in hand (i.e. their firing angles are in the range 2° to 15°).

The example of *Figure 34.68* shows the case in which inverter 1 a.c. voltage becomes relatively slightly low. The working points ABC must still obey Kirchhoff's and Ohm's laws, but are now not quite at their ideal positions, inverter currents being slightly high and low respectively, with inverter 2 operating at slightly high extinction angle. This is a stable operating condition. Powers are corrected in the short term by master control action, and normal (optimum) working conditions are re-established in a longer time by transformer tapchangers.

A commutation failure is a relatively frequent (and inevitable) minor fault, usually caused by a switching transient from the a.c. system at an inverter. Its effect is to cause temporary collapse of inverter operation for say 100 ms to 200 ms, momentarily giving zero d.c. voltage and a.c. current. In a multi-terminal scheme there is a tendency for the whole of the d.c. current to be diverted temporarily into the 'failed' inverter. The rectifier controls will restrain this automatically after a cycle or two; the inverter will not be able to re-commutate (particularly with weak receiving system) until actual inverter current and firing angle fall below rated values.

34.13.3 DC circuit breakers

One of the main advantages of an h.v.d.c. scheme is the controllability conferred by the ability of the converter valves to conduct or block the full load direct current as desired. This capability has meant that the need for true high voltage direct current circuit breakers has been small since by appropriate converter action, ordinary a.c. circuit breakers could be utilised for circuit switching on the d.c. side.

The first application for a d.c. breaker was the metallic return transfer breaker (MRTB). The MRTB is principally introduced to increase the independence between the poles of a bipolar scheme: A station fault requiring the blocking of a pole will divert the direct current from the healthy pole into the ground electrode. If the station fault is permanent, the faulty pole is isolated and bypassed, and transmission continues in the monopolar mode. However, although the metallic return conductor is then in parallel with the ground return, the major part of the current continues to flow into the electrode because of the lower resistance of this path. The MRTB can divert this current into the metallic conductor by developing a significant back e.m.f.

High voltage direct current breakers have now been developed and tested, but have not yet been commercially applied[4]. Such d.c. breakers may give some small reduction in the duration of outages caused by permanent faults in multi-terminal d.c. networks. Since their duty is significantly harder than the transfer duty of the MRTB, they are still costly, and thus justifiable only

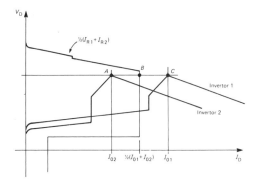

Figure 34.67 Normal steady state operation

when the rather small time improvements are of considerable benefit to the stability of the a.c. system.

References

1 CARCANO, C. et al., *Rebuilding of the h.v.d.c. Sardinia–Corsica–Italy Mainland Link (SACOI): Installation of two new conversion stations and a tapping station in Corsica for Multi-terminal operation*. IEE Conference Publication No. 255 (1985)
2 AINSWORTH, J. D., *Multi-terminal h.v.d.c. Systems*, CIGRE SC.14 Meeting Winnipeg, Canada (June 1977)
3 REEVE, J. et al., *Considerations for Implementing Multi-terminal d.c. Systems, W. F. Long*, IEEE PES Winter Meeting (1985)
4 VITHAYATHIL, J. J., *HVDC Breaker and its Application*, International Symposium in HVDC Technology, Rio de Janeiro, Brazil (March 1983)

34.14 Future trends

HVDC is now a maturing technology which has proved its technical and economic worth in the field. Future trends will be towards reduced complexity and lower cost, the latter achieved through both lower installed equipment cost and reduced running costs.

Very large transmission projects for which the economic transmission voltage lies above 600 kV will probably materialised, prompting the evolution of new insulation techniques for the d.c. side equipment. At lower voltages, developments will be generally confined to the specialist items of equipment such as valves and controls.

Two major developments which have contributed to present-day h.v.d.c. valves are open to further extension.

First, the increased rating per thyristor, which has led to thyristors based on 100 mm diameter silicon, may continue though rather slowly. Although silicon slices up to 150 mm in diameter are being developed, the high quality, high resistivity silicon, which is essential for h.v.d.c. applications, is unlikely to be available in economic quantities even in 125 mm size for several years. Further, the current rating required for most applications can be achieved with 100 mm thyristors without the need for paralleling. It is mainly the quest for higher blocking voltage and the likely increasing economic evaluation of losses which will put pressure on manufacturers to develop larger devices.

Second, the development of light triggering of individual thyristors, initially via auxiliary electronics mounted at valve potential, has already extended to the production of thyristors incorporating light sensitive gates suitable for direct optical triggering. In many applications, this reduces the need for 'local' electronics at each thyristor level, with the prospect of substantial economic advantage. For h.v.d.c., in the first instance, small direct optically triggered 'slave' thyristors have been employed to gate conventional thyristors, and to enable local protective circuits to act directly via the electrical gate. The value of these thyristors to h.v.d.c. is limited by the continuing need for local protective circuits to prevent damage to the thyristors during forward overvoltage and adverse forward recovery conditions. Directly light fired h.v.d.c. thyristors are now being developed, such as illustrated in *Figure 34.69* in which it is planned to achieve full self-protection[1,2]. The thyristor illustrated achieves high di/dt capability for turn-on by using an extended gate structure, is light fired via an amplifying gate arrangement, and is self-protected against overvoltage, and unsafe forward recovery failure by the action of three protective structures seen within the extended arms of the gate. Successful completion of this development should obviate the need for any local electronic circuits but its implementation will have to be justified economically because saving in the cost of supporting electronics will be partly off-set by the greater cost of the more complex thyristor. However, this approach also offers the prospect of an improvement in the reliability of valves since they will then incorporate far fewer components. For any given degree of redundancy, this will permit longer periods of operation between maintenance outages than are practical at present.

The universal adoption in recent schemes of indoor air-insulated valves, rather than outdoor oil or SF_6 insulated, stems largely from the need for regular, perhaps annual, access to thyristor levels which must be replaced or renovated to restore acceptable redundancy. Reducing the component count in a valve through the use of higher voltage self-protecting light-fired thyristors may make the outdoor oil or SF_6 insulated valve attractive once more, at least in regions where extreme climatic conditions are not encountered. However, this approach is unlikely to be adopted until satisfactory operating experience of light-fired self-protecting thyristors has confirmed the high availability and low replacement rates which are expected.

Today's h.v.d.c. inverter is line-commutated, that is it relies on the e.m.f. of the receiving a.c. system to commutate current from one phase to the next against the restraining action of the commutation reactance. Forced commutation inverters, in which the energy for commutation is stored locally, usually in

Figure 34.69 Silicon slice of a direct light-fired, self-protected thyristor. Courtesy of Marconi Electronic Devices Ltd

capacitors, and is released by auxiliary thyristors to force commutation at any desired time, is too expensive in first cost, in losses, and in reduced reliability through complexity, to be a serious competitor in large systems. Such circuits could, in principle, be applied to small schemes feeding isolated a.c. systems, where the advantage of being able to operate without the support of rotating machinery is greatest.

Even in such circumstances, this technique has not yet been employed and it is likely to be supplanted as a contender for such application by the gate turn-off thyristor (GTO). In the GTO, turn-on takes place very much as in a conventional thyristor, but turn-off in face of forward voltage is achieved by applying a substantial reverse gate current of short duration. Thus, turn-off requires local electronic control at each voltage level. A direct light-controlled gate turn-off thyristor is not practicable, though one possible successor to the gate turn-off thyristor, the static induction thyristor[3] does not have optical switching capability. The GTO has higher conduction losses and exhibits a lower rating for a given size of silicon slice, so its use in large schemes is unlikely to be economic in the near future. However, it is becoming available in ratings up to 4.5 kV, capable of turning off 2.5 kA, and having a d.c. rating of about 1 kA. This improves the prospects for small h.v.d.c. systems, capable of starting and operating without the support of independent a.c. generators or VAR control systems. Even for this purpose, the conventional line commutated converter remains a serious contender.

The performance of an h.v.d.c. scheme is critically dependent on the central control system, particularly if the a.c. system at the receiving end is weak. The cost of the controls is a small fraction of the overall cost of a scheme, hence improvement of control performance and operational reliability will continue even where this incurs somewhat higher costs. Some parts of the controls will be implemented on programmable digital systems with internal self checking and, in critical areas, with duplicated circuits and automatic changeover in the event of faults. Because of their relative simplicity and low component count, analogue circuits are likely to be retained in those areas to which they are best suited.

References

1 MELTA, H. and TEMPLE, V. A., *Advanced Light-triggered thyristor*, AC and d.c. Power Transmission Conference, IEE Conference Publication 255, London (1985)
2 TAYLOR, P. D. and FRITH, P. J., *Recent advances in high voltage thyristor design*, AC and d.c. Power transmission Conference, IEE Conference Publication 255, London (1985)
3 NISHIZAURA, J., *New Thyristor Applicable to d.c. Power Transmission*, AC and d.c. Power Transmission Conference, IEE Conference Publication 255, London (1985)

34.15 General references

1 ADAMSON, C., and HINGORANI, N. G. *High Voltage Direct Current Power Transmission*, London, Garraway (1960)
2 CORY, B. J. (ed), *High Voltage Direct Current Convertors of Systems*, Macdonalds (1965)
3 *High Voltage DC Transmission*, IEE Conference Publication no. 22, 1966, London, England
3 KIMBARK, E. W., *Direct Current Transmission*, Vol. 1, Wiley Interscience, (1971)
5 Manitobe Power Conference EHV-DC, Winnipeg, Canada 1971
6 *High Voltage DC and/or AC power Transmission*, IEE Conference Publication 107, London (November 1973)
7 *IEC 146: Semiconductor Converters* (3 parts), International Electrotechnical Commission (1973)
8 URLMANN, E., *Power Transmission by Direct Current*, Springer-Verlag, Berlin (1975)
9 *Incorporating HVDC Power Transmission into Power System Planning*, US Dept. of Energy, Phoenix, Arizona (1980)
10 *Overvoltages and Compensation on Integrated a.c.–d.c. systems*, IEEE Conference Proceedings, Winnipeg, Canada (June 1980)
11 *Thyristor and Variable Static Equipment for a.c. and d.c. Transmission*, IEE Conference Publication No. 205, London, England (1981)
12 *International Symposium on HVDC Technology – Sharing the Brazilian Experience*, Rio de Janeiro (1983)
13 *An annotated Bibliography of High Voltage Direct Current Transmission 1969–1983*, IEEE, Bonnerville Power Administration, DOE/BP-203 (May 1984)
14 ARRILLAGA, J., *High Voltage Direct Current Transmission*, IEE Power Engineering Series 6, Peter Peregrinus, London (1983)
15 International Conference on DC Power Transmission, Montreal, Canada (1984)
16 A.c. and d.c. Power Transmission, IEE Conference Publication No. 255, London (1985)

35 Education and Training

A Sensicle, BSc, CEng, MIEE, MInstMC
Institute of Measurement and Control

Contents

35.1 Introduction 35/3
    35.1.1 Qualifications 35/3
    35.1.2 The three levels of engineer 35/3
    35.1.3 Historical background 35/3
    35.1.4 The present 35/4

35.2 Summary of the three levels 35/4
    35.2.1 Chartered Engineer 35/4
    35.2.2 Technician Engineer 35/4
    35.2.3 Engineering Technician 35/4

35.3 Chartered Engineers 35/4
    35.3.1 Education 35/4
    35.3.2 Training 35/5
    35.3.3 Experience 35/5

35.4 Technician Engineers 35/5

35.5 Engineering Technicians 35/5

35.6 Bridges and ladders 35/6

35.7 Addresses for further information 35/6

35.1 Introduction

35.1.1 Qualifications

This section, although entitled 'Education and Training', provides a background and information on *qualifications* and qualifying routes suitable for those wishing to be electrical engineers. In the United Kingdom (UK) no one word describes the development of electrical engineers to the point where they can be considered to be fully qualified, i.e. the time at which the profession considers a young engineer to have had sufficient education, training and *experience* to practise as an electrical engineer. Engineers and educationists in the rest of Europe generally have a system with less distinction between these three elements of the qualifying route, and there is a tendency for all the elements to be subsumed under a single definition (a single word or statement); in particular, the French use the word *formation* to describe such development. Attempts have been made to introduce an Anglicised version in this country but, even though there have been considerable changes in the approach to engineering education in the UK during the last 50 years, the deeply rooted tradition of identifying education, training and experience separately has remained. Consequently, true to this tradition, much of the information contained in this section relates to these three elements of the qualifying route.

35.1.2 The three levels of engineer

The engineering profession recognises in addition three separate levels of qualification, respectively Chartered Engineer, Technician Engineer and Engineering Technician. We shall describe these separately in relation to electrical engineers, together with the required education, training and experience.

35.1.3 Historical background

The development of electrical engineering education in the UK is complex. Apart from responding to the needs of society, aspects of its structure are deeply rooted in British culture. Consequently it may help to understand the qualifying route if the background to the development of engineering education in Britain is considered.

Unlike engineering education in the rest of Europe, the development of which was rooted in the establishment of prestigious schools and universities concentrating on applied science as long ago as the 18th century, the British system was largely developed from educational establishments called Mechanics Institutes. The universities were places of learning for the arts and natural sciences, where engineering was not studied, as it was considered appropriate only for those of a lower academic ability.

The Mechanics Institutes in Britain had become well established by the middle of the 19th century, there being around 700, of which 200 were in Scotland. Overall some 120 000 students were attending them annually on a part-time basis. These institutes provided the origins of part-time engineering education in Great Britain.

In the university sector, Scotland was slightly in advance of England with regard to engineering education. The first Chair in civil engineering (i.e. *non-military* engineering) was established at Glasgow University in 1840. Shortly afterwards, one was established at University College, London. The first English university to offer a degree which contained engineering was Durham. However, successful students were not formally awarded a degree, though they were entitled to be called civil engineers and unofficially could put 'CE' after their names. Oxford and Cambridge Universities were particularly late in introducing engineering, establishing Chairs in 1875 and 1910 respectively.

35.1.3.1 Professional institutions

The professional engineering institutions, originally established as learned societies where like-minded people could meet, became qualifying bodies through setting standards for engineers in relation to their grades of membership. For many years the only formal evaluation of true professional engineers has been through qualifications awarded by the engineering institutions.

The Institution of of Electrical Engineers (IEE), founded in 1871 and incorporated by Royal Charter in 1921, has been in the forefront of setting standards of education, training and experience for electrical engineers. The IEE first set examinations for those who did not hold exempting academic qualifications in 1914. The examinations were quite demanding, even including a French translation paper.

35.1.3.2 National Certificates and Diplomas

A very important step in the development of engineering education was the introduction of engineering National Certificates and Diplomas, mostly for students undergoing part-time education; they were established and moderated jointly by the Board of Trade and the Institutions of Electrical Engineering and Mechanical Engineering in 1922. These qualifications quickly became established, and until 1957 Higher National Certificates (HNCs) and Diplomas (HNDs) in Electrical Engineering were considered to be the main educational qualification for the professional electrical engineer. After 1957, however, the IEE introduced a restructured examination in three parts (considered equivalent in academic standard to a degree) which became the educational requirement for those seeking corporate membership of the Institution.

Degrees in electrical engineering were quite rare until the 1950s. The UK graduate output just after the Second World War was less than 300. By contrast, during 1982 well over 3000 students were accepted on electrical and electronic engineering degree courses in the UK. The establishment of more universities and polytechnics, and the rapid expansion of technology, has led to this increase in the graduate population of electrical engineers. The increase has not been confined to graduates, however, for many professional engineers have come from the ranks of those with HNC/Ds, this educational route being accepted as normal for professional engineering until the late 1950s.

Normally, those attaining a degree in electrical engineering were considered to be en route to becoming Chartered Engineers. Technician Engineers and Technicians came from the ranks of those who attained HNC/Ds, Ordinary National Certificates/Diplomas (ONC/Ds) and City and Guilds of London Institute qualifications or equivalent.

In the 1950s and early 1960s concern was being expressed about the considerable overlap between the awards in electrical engineering obtainable through the Joint Committee for the Awards of National Certificates and Diplomas administered by the Institutions of Electrical Engineers and Electronic and Radio Engineers, the Department of Education and Science and the Scottish Education Department, and those qualifications issued by the City and Guilds of London Institute and equivalent regional organisations throughout the UK. During the 1960s a Committee was set up under the Chairmanship of Dr Haslegrave to look into the situation regarding Technician Engineer and Technician education, and to make recommendations. As a result, a plan was prepared to rationalise these sectors of education, and eventually the Technician Education Council was established in the 1970s to take over the work of the joint committees and some of the work of the City and Guilds of London Institute.

35.1.4 The present

There have been considerable developments in recent years in qualifications for electrical engineers. It is generally agreed that such qualifications have been rationalised and enhanced in standard and content. Among the factors that have contributed to these changes are the implementation of the IEE's Merriman Report recommendations (1978), the Finniston Inquiry into the Profession of Engineering (1980), the subsequent National Conference on Engineering Education and Training, the establishment of the Engineering Council (1983), the establishment of the Technician Education Council (and its subsequent amalgamation with the Business Education Council to form B/TEC), and the Scottish Vocational Education Council (SCOTVEC).

These developments have resulted in educational qualifications that more closely relate to the needs of the profession and industry and which are identified through three sectors of engineering registration. In electrical engineering, Chartered Engineers are normally educated to at least degree or equivalent standard, Technician Engineers to at least HNC standard, and Technicians to at least ONC or Technicians City and Guilds of London Institute Certificate standard.

In the past, qualification centred on education to the exclusion of training and experience, which are vital components for qualifying as an electrical engineer. Training cannot be defined so exactly as educational qualifications and, traditionally for the engineer, has been integrated with the educational aspects through apprenticeship and on-the-job training. The Industrial Training Act of the 1970s created Industrial Training Boards which evaluated training separately from education and imposed a training levy on employers; the Act increased the awareness of training but possibly contributed to a further tendency to separate it from education. In recent years there has been a greater trend towards comprehensive schemes of education and training at all levels.

The experience required for engineers to be considered qualified has been evaluated by the professional institutions since their inception.

35.1.4.1 Electrical and electronic engineering

Although this text is devoted to electrical engineering, progress in technology has required electrical engineers to have a considerable knowledge of electronic engineering. Consequently some of the references to courses and advisory bodies will touch on the electronic engineering sector.

35.2 Summary of the three levels

A brief summary of the main requirements for the Chartered Electrical Engineer, Electrical Technician Engineer and Electrical Engineering Technician follows.

35.2.1 Chartered Engineer

He or she should have:

An approved degree in electrical engineering or equivalent.
Approved industrial training of at least 2 years.
Professional experience of at least 2 years, tested by interview.

The aggregate of education, training and experience must be at least 7 years and Chartered Electrical Engineers cannot qualify before the age of 25.

35.2.2 Technician Engineer

He or she should have:

An approved electrical engineering HNC/D or City and Guilds of London Institute Full Technological Certificate or equivalent.
Approved industrial training of at least 2 years.
Professional experience of at least 2 years.

Electrical engineers cannot qualify as Technicians until they are at least 23 years of age. Normally education, training and experience cover a period of at least 5 years.

35.2.3 Engineering Technician

He or she should have:

An approved electrical engineering ONC/D or City and Guilds of London Institute Technician Certificate or equivalent.
Approved industrial training of at least 2 years.
At least 2 years experience.

Electrical engineers cannot qualify as Engineering Technicians until they are at least 22 years of age. Normally education, training and experience cover a period of at least 5 years.

35.3 Chartered Engineers

35.3.1 Education

An electrical engineering degree at second-class honours standard accredited by the IEE is the normal main educational requirement.

Entry to degree courses at present requires a broad range of General Certificate of Education (GCE) O-level or equivalent subjects covering both sciences and arts. Mathematics, physics and English are essential. In addition, GCE A-levels or equivalent in mathematics and physics are required. Although it is possible to enter some degree courses with only two A-levels or equivalent, to ensure entry, particularly on a good degree course, it is necessary to have three A-levels with above-average passes. The subject of the third A-level may be a second science, or mathematics, or economics.

35.3.1.1 The degree

There are several ways in which accredited degrees are organised. Full-time degree courses are normally of 3 years duration in England and 4 years in Scotland, and contain no industrial training. Sandwich courses are designed to have a programme that links academic work and periods of industrial training. 'Thin' sandwich courses typically alternate 6 months in industry and college over a 4- or 5-year period, while 'thick' sandwich courses are normally one–three–one or one–four–one arrangements, the academic work being sandwiched between two 1-year periods of industrial training.

Owing to the great variety and the fast rate of change of technology associated with the electrical engineering profession, the IEE is not prescriptive of the content of degree courses. However, for accreditation purposes, electrical engineering courses should contain certain elements (e.g. mathematics), but also an adequate level of specialisation. They should embrace the basic principles of electrical engineering and its application, and an appreciation of the product aspects of industry.

Universities and polytechnics tend to have particular strengths in topics covered and may emphasise these on their courses. With regard to electrical engineering, the topics covered will include the majority of those found in this *Reference Book*. However, additional subjects are studied on all courses to ensure that a good education for the professional engineer is provided. It is quite common, for example, for courses to include management, communication and economics.

Some universities and colleges offer enhanced electrical engineering degree courses, closely integrated with industry, for exceptional engineering students. In general the entry qualifications are higher and excellent passes in three or more A-levels or equivalent are expected. In addition, the universities and colleges have a stringent selection procedure, sometimes in conjunction with industrial sponsors.

The New Opportunity Press in association with the IEE publish an annual guide to IEE Accredited Degree courses.

35.3.2 Training

The training requirement for the Chartered Electrical Engineer is an approved scheme of at least 2 years duration. The IEE has details of approved training schemes. However, many schemes that meet the requirements have not been considered for approval, but would be so considered on application by the individual. Training can be structured in many ways and, where possible, aspiring Chartered Electrical Engineers should seek a 'thin' or 'thick' sandwich course. This is advantageous in many ways. The total course is usually better planned to link the academic work and the industrial training. It enables undergraduates to utilise their time more effectively; the undergraduate with no specific training scheme may not find appropriate experience through the long vacation. Additionally, there are financial advantages, particularly for industrially sponsored students. The IEE can provide details of companies who offer sponsor training. Further information can be obtained from the Engineering Industry Training Board (address given in Section 35.7).

In certain cases, when a graduate follows a non-accredited degree, or an accredited degree not assessed in lieu of training, at least twice as much experience as formal training is required. It is the usual practice of engineering institutions to consider such cases on their merits.

35.3.3 Experience

Experience in a responsible position is required, and guidance on this can be obtained from the IEE. Experience is usually judged through a written submission and an interview.

Chartered Electrical Engineers are normally registered as Chartered Engineers with the Engineering Council (see below).

Although a degree course is the usual way of achieving the educational requirements for the Chartered Electrical Engineer, there are some routes available which are less commonly taken. One of these is through the Engineering Council examinations, another (for those working in the industry and aged 35 or more) is the 'mature candidate' route, which requires a thesis and an oral examination. Information about these routes can be obtained respectively from the Engineering Council and the IEE (addresses given in Section 35.7).

35.4 Technician Engineers

The main educational qualification for a Technician Engineer is through an accredited Business and Technician Education Council (B/TEC) or SCOTVEC course in electrical engineering at a minimum of HNC level. Other courses that normally lead to Technician Engineer qualifications are accredited B/TEC and SCOTVEC National Certificates or Diplomas, and a range of ordinary or unclassified electrical engineering degrees.

HNC/Ds can be entered either through achieving a B/TEC or SCOTVEC National Certificate or Diploma in engineering or, particularly in the case of the HND, through achieving $1\frac{1}{2}$ A-levels or equivalent (i.e. a pass in one A-level and study in another, generally mathematics or physics). Although both B/TEC and SCOTVEC publish syllabuses, there is some variation in the courses owing to college-prepared syllabuses being accredited, in particular by B/TEC. The arrangements for HNC/D courses can vary from college to college. However, an HNC typically takes the form of 2 years of study, and an HND 3 years. They are normally designed to be integrated with industrial training. Information about B/TEC and SCOTVEC courses and the colleges offering them in electrical engineering can be obtained from B/TEC and SCOTVEC (addresses given in Section 35.7).

In addition, the Institution of Electrical and Electronics Incorporated Engineers (IEEIE), which acts as a qualifying body for those wishing to become Technician Engineers, can provide further advice (addresses given in Section 35.7). The IEEIE evaluates industrial training and experience. Normally these elements are each of two years duration, and it is proposed that the combination of education, training and experience for the Technician Engineer cover a period of not less than 5 years. An electrical engineer cannot qualify as a Technician Engineer below the age of 23. In general, aspiring Technician Engineers in their training are required to develop abilities to use electrical engineering equipment, instruments, apparatus and techniques. Training normally includes a period of induction, one of basic training and a period in which the trainee acquires a knowledge of relevant electrical engineering practice and competence in its application.

The IEEIE will assess the experience of a potential Technician Engineer through a written application for the appropriate grade of membership.

35.5 Engineering Technicians

The main *educational* qualification for an Engineering Technician in electrical engineering is through an accredited B/TEC or SCOTVEC course in electrical engineering at a minimum of National Certificate level. Other courses which normally lead to Engineering Technician qualifications are accredited National Diplomas and City and Guilds of London Institute Certificates or equivalent.

National Certificates and Diplomas can be entered through achieving the minimum of three approved Certificates of Secondary Education (CSEs) or four appropriate O-levels respectively or the Scottish equivalents. Among the subjects, it is essential to have mathematics. The arrangements for National Certificate and Diploma courses vary from college to college. However, a National Certificate typically takes three years of part-time study and students are usually 16 years of age on entry to the course. A large proportion of National Certificate courses are undertaken through day release from an appropriate form of employment. National Diplomas take two years from 16+ and usually are full-time courses, but there are arrangements for National Diplomas to be integrated with appropriate employment. In both cases, the employment has a strong *training* orientation.

Certain professional bodies evaluate industrial training and experience for Engineering Technicians, and the Engineering Council can provide information.

The earliest an electrical engineer can qualify as an Engineering Technician is at 21 years of age. The three components over a 5-year period are education, training and experience. Two years approved training is required, and this is counted as two years experience. In addition a further year's experience is necessary.

35.6 Bridges and ladders

The standards and routes described above for the qualification of electrical engineers at the Chartered, Technician Engineer and Engineering Technician levels are the straightforward direct routes. There are bridges and ladders between the various levels of engineering, and between the various courses. In addition, for those who have the training and experience but do not have the appropriate academic qualifications, the appropriate professional engineering institutions offer a mature route for persons aged 35 and over. A useful document for reference is *Policy Statement: Standards and Routes to Registration*, published by the Engineering Council. This describes patterns of study, training and experience, defines the various engineering roles, goes into detail about training and responsible experience and mature routes, and provides information about bridges and ladders. It should, however, be stressed that this document is a general document covering all disciplines in engineering without subject detail. Detailed information about electrical engineering can be obtained from the IEE and the IEEIE.

35.7 Addresses for further information

The Institution of Electrical Engineers (IEE)
Savoy Place
London
WC2R 0BL

The Institution of Electrical and Electronics
Incorporated Engineers (IEEIE)
2 Savoy Hill
London
WC2R 0BS

Business and Technician Education Council (BTEC)
Central House
Upper Woburn Place
London
WC1E 0HH

Scottish Vocational Education Council (SCOTVEC)
38 Queen Street
Glasgow

City and Guilds of London Institute
76 Portland Place
London W1

Engineering Industry Training Board
140 Tottenham Court Road
London W1

The Engineering Council
Canberra House
Maltravers Street
London
WC2R 3ER

36 Standards

J S Cliff, MBE, CEng, FIEE

Contents

36.1 Introduction 36/3

36.2 United Kingdom 36/3
    36.2.1 British Standards 36/3
    36.2.2 Certification and Testing 36/4
    36.2.3 Other standards and certification 36/5
    36.2.4 Microfilm and microfiche indexes 36/5
    36.2.5 Computers 36/5
    36.2.6 Addresses 36/6

36.3 International Standards 36/7
    36.3.1 Addresses 36/8

36.4 USA Standards 36/8
    36.4.1 Addresses 36/8

36.5 World Standards 36/9
    36.5.2 Nordic Regional Standardization 36/9
    36.5.3 Pacific Region 36/10
    36.5.4 Council for Mutual Economic Assistance (CMEA) 36/10

36.6 Further reading 36/11

36.1 Introduction

Before designing any new equipment, particularly for export from the UK, it is necessary to comply with the standards which apply to that country. So far as the UK is concerned most of the equipment will have to comply with the latest British Standards (BS) which is covered in Section 36.2. International Standards are dealt with in Section 36.3, USA standards are covered in Section 36.4, while World Standards are covered in Section 36.5. Europe (EEC & EFTA) is dealt with in Section 36.5.1, the Nordic Region in Section 36.5.2, the Pacific Region in Section 36.5.3, and the CMEA Region in Section 36.5.4. There are also some useful addresses. Further reading is covered in Section 36.6.

Although the prices of standards vary (with a discount for subscribing members), they are fairly expensive for individuals to obtain copies. However, the 1985 BSI Catalogue lists over 135 places in the UK, and approximately 150 places throughout the world, which hold complete sets of BS for reference purposes, and approximately half of the latter places are sales agents for BS. Before purchasing copies of the relevant standards it is worth seeing whether the firm is a subscribing member of the BSI, or seeing a copy of the standard at the nearest library.

36.2 United Kingdom

36.2.1 British Standards

In the UK, the British Standards Institution (BSI) is the main body for the preparation and promulgation of national standards in all fields. It was set up by the professional engineering associates in 1901 as the Engineering Standards Committee, and in 1918 became the British Engineering Standards Association (BESA). A Royal Charter was granted in 1929, followed by a supplementary Charter in 1931, when the present name was adopted.

The BSI is an independent organization and its main function is to draw up voluntary standards by agreement among all the interests concerned and to promote their adoption. The membership of the BSI is divided into several categories such as: individuals, professional firms and partnerships; industrial and commercial firms in both the private and public sectors; associations, institutions and other organizations; local authorities, etc.; their subscriptions being variable with size.

There are now more than 9000 current British Standards (BS) covering almost the whole range of industrial products and processes. The preparation and maintenance of these standards is the responsibility of more than 4000 technical committees of which about 1000 are actively working at any one time.

The BSI provides the secretary, the meeting place, and produces the documents for discussion. The members of the committee usually are nominees of the relevant organizations and are responsible for expressing their organization's views rather than their own, and for keeping their organizations well informed about progress of current projects.

BS may be used to promote standardization at any of the following stages:

1. terminology, symbols,
2. classification,
3. methods of measuring, testing, analysing, sampling, etc.,
4. methods of declaring, specifying, etc.,
5. specifications for materials or products,
6. dimensions, performance, safety, etc.,
7. specifications for processes, practices, etc.,
8. recommendations on product or process applications,
9. codes of practices.

The best known type is the BS specification which lays down a set of requirements to be satisfied by the material, product or process in question and embraces, often by reference, relevant methods by which compliance may be determined.

Like other BS guides or recommendations, BS codes of practice are written in the form of guidance only. Other types of standards are: BS glossaries, BS methods of various kinds, suitable as reference documents to be called up where appropriate in other standards. Drafts for Development (DD) are a comparatively recent addition, and are equivalent to 'provisional' or 'tentative' standards issued in some other countries. They are published where guidance is urgently needed which, though theoretically sound, has not yet been subjected to enough practical application to justify the publication of a BS. They can be converted to a BS of any type, or withdrawn, when sufficient experience has been fed back to the BSI.

The description 'Published Document' (PD) is used for miscellaneous BSI publications containing supplementary information relating to standardization. 'Handbooks' comprise texts taken from a number of separate Standards relating to a particular field. Sectional lists of BS are also available, the one for electrical engineering being number SLZ6.

As the Standards are being continuously up-dated by the relevant technical committees, any list can only be partially up-to-date. Particulars of standards, amendments, and other publications are given in the current 'BSI Catalogue', which is correct to the end of September of the previous year. Any publications issued since that date are given in the current Sales Bulletin (issued every 2 months) and in BSI News (issued monthly). It is important that users of BS should ascertain that they are in possession of the latest amendments or editions.

The BSI library at Milton Keynes has complete sets of BS, international, and overseas national standards (some of which have been translated into English) and related indexes and catalogues for consultation.

A complete set of BS is maintained for reference in the UK at about 280 libraries, technical colleges, etc., and in most countries of the world they are also available. Stocks of BS are held for over the counter sale, and for reference, at Birmingham, Bristol, Dundee, Glasgow, Leeds, Liverpool, Manchester and Norwich. The BSI has recently introduced a scheme for automatically updating an extensive list of classified 'actional lists' from which members can select the 'base units' appropriate to their needs. This involves a Standing Order Scheme (SOS) which is only available to subscribing members.

A withdrawn BS is no longer effective but may be needed for ordering spares, etc. Photocopies of such BS can still be purchased from the BSI. Where a BS has been superseded by a standard with a different BS number this information is shown against the title of the withdrawn BS in the catalogue. Similarly the revision of any BS automatically supersedes all previous editions of that standard, or parts of the standard so revised. Only the current issues are listed and they are considered as authorised BS in the catalogue. Confirmation of a BS, which is a formal procedure following the review of the standard by the technical committee, is indicated on the first line of entry by the inclusion, in brackets, of the year of confirmation. 'Withdrawn' after the title of a standard indicates that the BS is not recommended for the design of new equipment. Current amendments to publications are shown below the titles to which they apply. All amendments to the date of dispatch are included with any publication ordered, but when publications are re-printed all previously issued amendments are incorporated in the text. Amendments may be prefixed by PD, but the latest prefix is AMD.

Also included in the 1985 catalogue, in addition to the list of BS, are

BS.CECC publications
Codes of Practice, covering Building, Electrical Engineering, Civil Engineering, Mechanical Engineering

Automobile services
Marine services
Aerospace services
Aerospace standards withdrawn
Handbooks and Published Documents
Drafts for Development
Education and British Standards Society publications
Public Authority Standards
European Standards (EN)
BSI Certification Trade Marks
Miscellaneous Items
    Numerical list of IEC and 150 standards corresponding to BS. Also includes CISPR and QC standards

BS are publicly available documents voluntarily agreed. However, the publication by the BSI does not ensure their use. A BS becomes binding only if a claim of compliance is made, if it is invoked in a contract or if it is called up in legislation.

Support for the application of BS as agreements produced in the public interest is given by the Restrictive Trade Practices Act 1976. The care exercised in the production of BS is relied upon by users who themselves owe a similar duty to the public. It remains the responsibility of users to ensure that a particular BS is appropriate to their needs. Within their scope national standards provide evidence of an agreed 'state of the art' and may be taken into account by the Courts in determining whether or not someone was negligent.

A BS forms part of the trade description of a product when quoted by a BS number or when compliance with it is claimed. Marking with a BS number constitutes a unilateral claim that the product complies with all the requirements of the BS quoted. The person making the claim is responsible for its accuracy under the Trade Descriptions Act 1968. To support their claims manufacturers may have their products certified as complying with the requirements of the BS.

The existence of a relevant BS facilitates the preparation of contracts, and BS invoked in contracts become legally binding on the contracting parties. Many BS contain options and other matters that need to be clarified by additional contractual provisions, and some BS are deliberately drafted in advisory form, i.e. codes of practice, guides, and recommendations, and are therefore not appropriate for simple reference in contracts.

Reference to BS in regulations may have one of two effects:

1. *Standards made mandatory.* The standard, or part of it, referred to must be followed in order to obey the statutory requirement, i.e. the text of the standard ceases to be voluntary in the context of the legal requirement.
2. *Standards deemed to satisfy.* Here compliance with the standard is indicated as one way of complying with a regulatory requirement. Anyone choosing another method may be required to prove that this solution complies with the regulation.

There are three distinct methods of reference to BS:

1. *Reference to standards by exact identification (strict reference).*
   One or more specific BS are designated in such a way that later revisions of the BS will not be applied unless the regulation is modified. The BS is usually designated by its number and date.
2. *Reference to standards by undated identification (undated reference).*
   One or more specific BS are designated in such a way that later revisions of the BS will be applied without the regulation needing to be changed. The BS is usually designated by its number only.
3. *General reference to standards.*
   Reference is made in a general way to present and future standards, this means that the relevant regulation includes a general clause so that all the present and future standards in a specific field are regarded as meeting the aim of the regulation.

Reference to BS by exact identification (BS number and date) is the method normally followed in the UK, where regulations made under a number of Acts of Parliament invoke some 300 BS.

In July 1982 the Department of Trade on behalf of HM Government issued a document entitled 'Standards, Quality, and International Competition' (Cond 8621) in which the BSI was recognised as the main producer of standards in the UK, and HM Government agreed to maintain the annual grant-in-aid based on the level of contributions by other subscribing bodies, and to support BSI's efforts to achieve international harmonization of standards through ISO, CEE, CCN, CENELEC, etc. in the interests of UK industry and trade, and to encourage fuller participation of public purchasing authorities in the preparation and compliance with BS in their purchasing decisions, quality assurance requirements, and operational procedures. The BSI agreed to review, and where necessary revise, existing BS to ensure that these, and any new BS, will be suitable for reference in Government regulations as unambiguous statements of technical requirements. This may take some time.

36.2.2 Certification and Testing

In addition to producing BS it is also necessary to test and mark various products to ensure that they comply with the relevant BS, and the manufacture is maintained at a reasonable level. The BSI operates three certification marking systems:

1. the kitemark, the BSI's certification trade mark, first used in 1903.
2. the Safety mark introduced in 1974 to provide manufacturers with a means of demonstrating compliance with a BS specifically related to safety.
3. the Registered Firm Symbol operated for manufacturers who produce goods which are not, at present, covered by BS.

In its certification activities BSI does not have, or seek, a monopoly position, but responds to UK needs and aims to provide a service which can be used by industry. Through the Quality Assurance Council the BSI cooperates with other organizations concerned with certification of compliance with standards. The more important of these are

British Electrotechnical Approvals Board (BEAB)
British Approvals Service for Electric Cables (BASEC)
Association of Short-Circuit Testing Authorities (ASTA)
British Approvals Service for Electrical Apparatus in Flammable Atmospheres

The BSI has compiled a register of test houses of assessed capability indicating their fields of testing.

The BSI Certification and Assessment Department is responsible not only for the certification of products but also the assessment of the capabilities of firms in manufacturing and service industries.

The Department also provides services to other certification bodies including the Canadian Standards Association (CSA).

The BSI Test House provides a wide range of testing facilities that are generally available to industry in the UK, and other foreign clients, and embrace electrical, electronic, mechanical, chemical, photometric, motor vehicle component, etc. testing.

The BSI Technical Help to Exporters (THE) service was set up in 1966 to provide technical information and assistance to all sectors of industry engaged in exporting. These services cover the identification, sale and translation of foreign specifications and regulations, and assistance in meeting their requirements, in obtaining test certificates and in arranging for testing and factory investigations in the UK if this is acceptable to the overseas organization concerned.

36.2.3 Other standards and certification

Probably the most important is the standard issued by the Institution of Electrical Engineers (IEE) which is called the 'Regulations for Electrical Installations' (usually called the 'IEE Wiring Regulations') on 30 March 1981. This is the 15th edition and supersedes the 14th edition on the 1 January 1983. The new edition is quite different from its predecessors and now follows the IEC Publication 364 'Electrical Installations of Buildings', which has been adopted by CENELEC for the harmonization of the wiring regulations of the EEC and EFTA countries. In this edition, as far as is practicable, account has been taken of the technical substance of the parts of IEC 364 so far published, and of the corresponding agreements reached in CENELEC. As the IEC and CENELEC work is at a relatively early stage certain parts of this edition use re-arrangements and factual up-dating of the content of the 14th edition as amended to April 1976. The 15th edition will be amended from time to time to take account of further progress in the IEC. The publication of a further edition will be considered when IEC 364, and the corresponding CENELEC work, is completed.

BEAB certifies household appliances and all home laundry equipment, heating and cooking appliances, refrigerators and freezers, and home sound and vision equipment, for compliance with one of two BS:

BS 3456, many sections of which agree with CENELEC HDs or CEE specifications.

BS 415, which technically agrees with IEC 65 or CENELEC HD 19553.

Many large shops in the UK will stock only BEAB-approved appliances.

BASEC provides a certification scheme for manufacturers of electric cables and flexible cords, and that ongoing quality control procedures are sufficient to ensure a consistently high standard. Administration is carried out, under contract, by the BSI.

ASTA operates a certification and product marking scheme for circuit-breakers, fuses, fuse-links, fuse-boards, switches, isolators, starters, transformers, reactors and electrical wiring accessories. It offers a number of classes of certificate, including a certificate of complete compliance with the requirements of a relevant standard.

BASEEFA certifies electrical equipment using any of the recognised forms of explosion protection, and certification of equipment intended for hazardous industrial (non-mining) areas, i.e. for Group II apparatus. Certification is against any recognised standard or where these are not available to standards prepared by BASEEFA.

Most of the above bodies have their own interpretation of their relevant standards.

36.2.4 Microfilm and microfiche indexes

Each new design requires more accurately manufactured components and has to comply with more demanding standards to enable it to be sold to different markets and necessitates more documentation to ensure quality control and standardised methods of production. The increase in the amount of technical information over the past few years has been truly explosive. However, the 'information problem' is not simply the size of the overwhelming amount and choice of data; nor is it only the difficulty of retrieval—a difficulty that is being increasingly overcome by information technology—the major problem for those who endeavour to use this limitless resource is one of *selection*. There are vast stores of information from which any single item can be retrieved in fractions of a second at the touch of a button, but which standard is relevant and which item of information is helpful at a particular time, *that* is the information problem.

There are two ways of solving this problem:

1. by obtaining Product Information and Standards and Regulating Documents from Technical Indexes Ltd,
2. by using INSPEC and the IEE Library, using computer retrieval, see Section 36.2.5.

From the Product Data Book or the Index Book the user is guided by a simple reference directly to the full document in the microfilm file in his office or library. There a standard or regulation that contains all the latest amendments updated every 30 days by Technical Indexes Ltd, can be printed to provide instant paper copies, and you have the benefit of being able to take away a copy while the original document stays in place ready for the next user.

Technical Indexes Ltd collaborate with the BSI. The film cassettes of all the standards fill a small rack about $0.5 \text{ m} \times 1 \text{ m}$, as opposed to the long shelves required by printed standards. It takes a matter of moments to select the right cassette, snap it into the machine, and summon up the relevant standard on the screen. If a paper copy is required the printer will produce one.

The microfilm file is divided into ten sections. Each section containing related documents, and separately priced. The standards themselves are included as complete documents. Any amendments are added to the base documents so that the current standard will always be referred to. Record is kept of the amendments so the actual amendment document can be referred to as well, if required. The 16 mm roll microfilm is loaded into cartridges.

The microfiche file is divided into 35 sections, and consists of $24 \times$ reductions as in frame microfiche with separate prices per section.

There are times when you simply cannot find a particular piece of information. 'Extension 99' is an exclusive enquiry service for your benefit, so simply ring Technical Indexes Ltd and ask for extension 99. Note for UK only: BS micros can only be supplied to members of the BSI, or to Subscribers who undertake to become subscribing members within one month of delivery of the films. The address of Technical Indexes Ltd is Willoghby Road, Bracknell, Berkshire, RG12 4DW.

36.2.5 Computers

INSPEC (Information Services for the Physics and Engineering Communities), is operated by the Institution of Electrical Engineers (IEE), and this service provides access to the published information in Physics, Electrical Engineering, Electronics, Computers and Control Engineering.

INSPEC collects and scans the world's output of scientific and technical literature. All papers relevant to the above fields are selected and processed by a team of subject specialists. For each paper an abstract is stored in a computer file together with indexing and classification codes by which the information may be retrieved. This computer file is currently being added to at the rate of some 100 000 items per year, and forms the source for a wide range of services to meet a variety of needs.

INSPEC Abstracts are available in space-saving microform. Microfilm, in 16 mm open reel or cartridges, 35 mm in open reel only, for years 1898 to the present. In microfiche, in standard 98 frame, $24 \times$ reduction, for years 1974 to the present are available.

Back volumes are available through Dawson's Back Issues, Cannon House, Folkstone, Kent, CT19 5EE.

In addition INSPEC offers a number of printed lists of the Abstract Journals in use, broken down into Physics, Computer and Control, and Electrical and Electronics Abstracts. The eight Key Abstracts journals focus attention on significant technical developments as they are published in the international literature. The full range of Key Abstracts journals covered is: Power Transmission and Distribution, Industrial Power and Control Systems; Communication Technology; Solid State Devices; Electronic Circuits; Systems Theory; Electric Measurement and Instrumentation; Physical Measurement and Instrumentation. In addition INSPEC has recently started 'IT Focus' which covers Information Technology update for Managers, and gives the title of paper, an abstract of the paper, the author, the source of the paper, and entry number in 'IT Focus'. INSPEC also prints on 6 in × 4 in cards, for ease of filing, a Selected Dissemination of Information (SDI) which matches the interest profile of the scientist, engineer or manager against the information added to the INSPEC database each week.

INSPEC is only one of many organisations producing printed cards or articles on electrical engineering subjects, with abstracts or summaries of the various types of information produced throughout the world. Should you have a personal computer and a telephone, or work for a firm which employs such things in, say, the library then you can be in contact with over 1500 online databases, including INSPEC which is typical of the service. You can search over a million references in a few minutes, which are updated monthly so that the most recent information is available and you can find precisely the information you require by direct interrogation of the database using the online computer systems.

If you have not got access to a computer and a telephone they you can use one of the libraries which offer this service, the IEE library being typical. The IEE library, using computer retrieval, now has access to details of over 14 million scientific journal papers, etc. throughout the world, using more than 15 databases worldwide. Charges are based on searched time and computer time used. It is possible to fix a cost limit in advance of each search. Computer searches usually cost between £30–£50, manual searches cost about £10 per hour, the first half hour being free. Photocopies of articles cited in several results can normally be obtained through the IEE library.

To arrange a search write to

Library Information Service
IEE
Savoy Place
London
WC2R 0BL   Tel: 01 240 1871
or
Marketing Department
INSPEC/IEE
Station House
70 Nightingale Road
Hitchin
Herts
SG5 1RJ   Tel: (0462) 53331

36.2.6 Addresses

Head Office—Standards Division
BSI
2 Park Street
London
W1A 2BS   Tel: 01 629 9000
Information Department
Marketing Department
Sales Department
Subscriptions Department   Tel: (0908) 320033
Technical Help to Exporters (THE)
BSI
Linford Wood
Milton Keynes
MK14 6LE
For orders and all enquiries other than subscriptions
                                                  Tel: (0908) 320066

Certification and Assessment Service (C&AS)
Quality Assurance Services
BSI
PO Box 375
Milton Keynes
MK14 6LO   Tel: (0908) 315555

The Inspectorate
Quality Assurance Services
BSI
PO Box 391
Milton Keynes
MK14 6LW   Tel: (0908) 315555

The Test House
BSI
Maylands Avenue
Hemel Hempstead
Herts
HP2 4SQ   Tel: (0422) 3111

BSI
Birmingham Chamber of Commerce and Industry
PO Box 360
75 Harborne Road
Birmingham
B15 3DH   Tel: 021 454 6171

BSI
British Chamber of Commerce, Industry and Shipping
16 Clifton Park
Bristol
BS8 3BY   Tel: (0272) 737373

BSI
Dundee and Tayside Chamber of Commerce and Industry
Panmure Street
Dundee
DD1 1ED   Tel: (0382) 22122

BSI
Glasgow Chamber of Commerce
30 George Square
Glasgow
G2 1EQ   Tel: 041 204 2121

BSI
WIRA House
West Park Ring Road
Leeds
LS16 6QL   Tel: (0532) 780119

BSI
Merseyside Chamber of Commerce and Industry
1 Old Hall Street
Liverpool
L3 9HG   Tel: 051 227 1234

BSI
Manchester Office
3 York Street
Manchester
M2 2AT   Tel: 061 832 3731

The Institution of Electrical Engineers (IEE)
Savoy Place
London
WC2R 0BL

IEE Publishing Department
PO Box 8
Southgate House
Stevenage
Herts
BG1 1HQ

Institution of Electrical Engineers
Station House
70 Nightingale Road
Hitchin
Herts
SG5 1RJ

Institution of Electrical Engineers/Peter Peregrinus Ltd
PO Box 26
Hitchin Herts
SG5 1SA

The British Electrotechnical Approvals Board (BEAB)
9/11 Queens Road
Hersham
Walton-on-Thames
Surrey
KT12 5NA

The British Approvals Service for Electrical Cables (BASEC)
Maylands Avenue
Hemel Hempstead
Herts
HP2 4SQ

The Association of Short Circuit Authorities (Inc) (ASTA)
23/24 Market Place
Rugby
Warwickshire
CV21 3D

The British Approvals Service for Electrical Equipment for Flammable Atmospheres (BASEEFA)
Health and Safety Executive
Harpur Hill
Buxton
Derbyshire

36.3 International Standards

Formed in 1906 the International Electrotechnical Commission (IEC) comprises national electrotechnical committees of more than 40 countries. Approximately nearly 2000 standards or reports compiled by over 80 technical committees, and over 120 sub-committees, that collectively represent some 80% of the world's population that produces and consumes 95% of electrical energy. The national committees include manufacturers, users, trade associations, governments and academic professions.

The IEC issues a yearly catalogue with six supplements per year. Its standards are widely adopted as the basis of national electrotechnical standards so far as local customs and conditions permit. They are also quoted in manufacturer's specifications, and by users when calling for tenders. This widespread adoption facilitates international trade in the electrical and electronic engineering sectors.

The International Conference on Large High-Voltage Electric Systems (CIGRE), which meets in Paris every alternate year, has a number of working groups which produce papers which are discussed and many of these form the basis of work within the IEC. It also produces *Electra*, a monthly magazine. The IEC works in close cooperation with the International Organization for Standardization (ISO) which is mainly concerned with standards in the non-electrical field. By agreement the two bodies do not compete in their activities, but the IEC is responsible for standards for steam and hydraulic turbines which are used almost exclusively for electrical generation.

A recent development in 1982 has been the start of its Quality Assessment System for Electronic Components (IECQ). This is intended initially for mass produced components, such as resistors and capacitors, but many embrace components made for special purposes. Twenty-one member countries representing the major exporting countries of electronic components are members of IECQ.

The BSI is the UK member body of the International Organization for Standardization (ISO) and the European Committee for Standardization (CEN).

The Electrotechnical Council forms the British Electrotechnical Committee, the UK National Committee of the parallel electrotechnical organizations the International Electrotechnical Commission (IEC) and the European Committee for Electrotechnical Standardization (CENELEC). BSI also participates on behalf of the UK in standards work on EURONORMS for the European Coal and Steel Community (ECSC).

Some 600 standards projects related to international work now account for 75% of the total BSI work programme. It is responsible for more than 375 international secretariats but the BSI has to match its work programme to the resources available from the sale of standards, subscription income and government grant. Within these limitations the actual UK participation in international standardization is therefore concentrated on work which will realise a significant return on the effort invested as judged by the potential sales demand for a BS.

Every international committee in which the UK participates has an equivalent committee, usually a BSI committee, which appoints and briefs the UK delegation. Every delegation has a leader who is its principal spokesman at the international meeting, and a rapporteur to assist the leader in checking reports and brief minutes at the meeting, to prepare a brief report for the BSI committee concerned on return. Members invited to participate as delegates in international standards work are chosen for their special knowledge and powers of advocacy, and they are responsible for putting forward the UK viewpoint agreed in the relevant BSI Committee. Five responsible and properly briefed delegates should be the maximum number. If regulatory matters are involved a member of the relevant government department, serving on the BSI committee, should be a member of the delegation.

One of the fundamental goals of international standardization is to bring into use a common set of standards world-wide. So far as electrical products are concerned it has been agreed that the IEC standards are the most suitable. The benefits are two fold:

1. the removal of barriers to international trade,
2. the specification by users ensures that they have a common and valid base for examining and comparing competing products.

Thus many BS are drafted abroad in the IEC.

In 1959 the standardization institutions of western Europe formed the European Committee for Standardization (CEN) which prevented the drifting apart of standards of the EEC and EFTA. Electrotechnical standards were originally the responsibility of CENEL the electrotechnical counterpart of CEN, but with the enlargement of the EEC in 1972 the European Committee for Electrotechnical Standardization (CENELEC) was set up which creates from IEC Published documents Harmonization Documents (HD) and European Standards (EN) generally agreed by both EEC and EFTA countries on the basis of a weighted voting procedure. Once ENs are approved they are printed or endorsed as national standards in the countries concerned.

With the advent of CENELEC there has been a considerable change in the method of drafting British Standards in order to avoid conflict with HDs. When CENELEC begins work on a particular subject a 'Standstill' or 'Status quo' arrangement comes into effect. Changes in national standards relating to that subject cannot then be made until harmonization has been agreed, or permission obtained from CENELEC. It is still possible to have a national standard for a particular type of equipment which is of no interest to other member countries of CENELEC. Individual customers can still obtain equipment to their own specification, but as these will be produced as 'specials' they may suffer from a delivery and/or price penalty.

As the CENELEC documents are based on IEC recommendations the emphasis on these meetings has now increased in importance rather than work in the BSI committees alone.

For many BS there are now corresponding international standards on the same subject in varying degrees of agreement, usually ISO or IEC standards. These are marked in the 1985 catalogue with a symbol, the meaning of which is:

≡ an identical standard in every detail published with dual-numbering
= a technically equivalent standard in which the wording and presentation may differ quite extensively
≠ a related standard covering subject matter similar to that covered by a corresponding international standard or standards. It is emphasised that while the subject matter is similar the standard may deal with it in a different manner.

It may be assumed that for entries in which none of the above references appear, no corresponding international standard exists.

36.3.1 Addresses

International Electrotechnical Commission (IEC)
3 Rue de Varmbe
CH-1211 Geneva 20
Switzerland

International Organization for Standardization (ISO)
Case Postale 56
CH-1211 Geneva 20
Switzerland

International Conference on Large High-Voltage Electric Systems (CIGRE)
112 Boulevard Haussman
F-75008 Paris
France

Electron, as CIGRE above

International Commission for Conformity Certification of Electrical Equipment (CEE)
310 Utrechtsweg
Arnhem
Netherlands

CENELEC General Secretariat
2 rue Briderobe, Bte 5
1000 Brussels
Belgium

Harmonization Documents (HD), European Standards (EN), CECC Specifications, as CENELEC above.

36.4 USA Standards

As the standards in the USA were English, and some countries followed such standards, they have been used by many English speaking engineers throughout the world. At the beginning of 1898, a discussion was organized on the subject of 'Standardization of Generators, Motors, and Transformers' and from this arose the first American Institute of Electrical Engineers (AIEE) product standards committee. The Institute was a prime mover in 1901 for the endorsement of a bill before the US Congress for establishing a national standardizing bureau in Washington DC which ultimately became the National Bureau of Standards. In 1919, as the result of action spearheaded by the AIEE, the American Engineering Standards Committee (AESC) was formed, the latter, in 1926, becoming the American Standards Association (ASA) which later became the American National Standards Institution (ANSI).

ANSI itself does not develop standards but acts as a co-ordinating body for the purpose of encouraging development and adoption of worthwhile standards produced by other organizations and approves these as American National Standards. Thus there are about 35 organizations producing electrotechnical American Standards, produced by such bodies as the Institute of Electrical and Electronic Engineers (IEEE), National Electrical Manufacturers Association (NEMA); Underwriter Laboratories (UL); etc.

Two of the most widely used are:

1. the National Electrical Code, produced by the National Fuse Protection Association (NFPA)
2. the National Electrical Safety Code, produced by a committee under a secretariat of the IEEE.

Thus the standards scene in the USA differs markedly from that in most other industrialized countries.

36.4.1 Addresses

The following are the major organizations concerned with developing and coordinating electrotechnical standards in the USA. They are taken from 'Electrical Standards in World Trade'.

Association of Home Appliance Manufacturers (AHAM)
20 North Wacker Drive
Chicago
IL 60606
USA

Aerospace Industries Association (AIA)
1725 de Sales Street
NW
Washington
DC 20036
USA

American National Standards Institute (ANSI)
1430 Broadway
New York
NY 10017
USA

American Society of Mechanical Engineers (ASME)
345 East 47th Street
New York
NY 10017
USA

American Society for Testing and Materials (ASTM)
1916 Race Street
Philadelphia
PA 19013
USA

American Welding Society (AWS)
2501 NW Seventh Street
Miami
FL 33125
USA

Computer and Business Equipment Manufacturers' Association (CBEMA)
1828 L Street NW
Washington
DC 20036
USA

Electrical Apparatus Service Association (EASA)
1331 Baur Boulevard
St. Louis
MO 63132
USA

Edison Electric Institute (EEI)
1111 19th Street NW
Washington
DC 20036
USA

Electronics Industries Association (EIA)
2001 Eye Street NW
Washington
DC 20036
USA

Electrical Testing Laboratories Inc (ETL)
Industrial Park
Courtland
NY 13045
USA

Institute of Electrical and Electronics Engineers (IEEE)
345 East 47th Street
New York
NY 10017
USA

Illuminating Engineering Society (IES)
345 East 47th Street
New York
NY 10017
USA

Insulated Power Cable Engineers' Association (IPCEA)
PO Box
South Yarmouth
MA 02664
USA

Instrument Society of America (ISA)
400 Stanwix Street
Pittsburgh
PA 15222
USA

National Electrical Manufacturers' Association (NEMA)
2101 L Street NW
Washington
DC 20037
USA

National Fire Protection Association (NFPA)
Batterymarch Park
Quincy
MA 02269
USA

National Standards Association (NSA)
1321 14th Street NW
Washington
DC 20005
USA

Underwriters Laboratories, Inc. (UL)
333T Pfingsten Road
Northbrooke
IL 60022
USA

36.5 World Standards

36.5.1 Europe (EEC and EFTA) Region

The countries of Western Europe have been harmonizing their national standards for a number of years. They formed the European Committee for Standardization (CEN) which prevented the drifting apart of the standards of EEC and EFTA, and CENEL was originally the electrotechnical counterpart of CEN. With the enlargement of the EEC the European Committee for Electrotechnical Standardization (CENELEC) was set up through the joining of CENEL with CENELCOM which had been set up by the original founder members of the EEC. CENELEC creates Harmonization Documents (HDs) and European Standards (ENs) generally agreed by representatives of both EEC and EFTA. CEE originally set up specifications and operated a certification scheme. With the acceptance of the IEC for producing specifications leaving the CEE to administer the certifications scheme through its Certification Body (CB).

CENELEC agreements relating to the common working of cables and cores complying with harmonized standards is the HAR scheme. The CENELEC Electronic Components Committee (CECC) prepares specifications for the quality assessment scheme for EEC and EFTA countries, but this operates in parallel with a world-wide scheme set up by the IEC known as the IECQ for similar components. CENELEC has also produced a number of standards for apparatus for explosive atmospheres, and these are likely to become law in various countries of Western Europe.

36.5.2 Nordic Regional Standardization

This covers the countries of Denmark, Finland, Norway and Sweden, and their export oriented industries already benefit from mutual cooperation in electrical standards through the Nordic Electrotechnical Standards Cooperation Committee (NOREK) which reviews the work of national organizations and formulates a regional policy for presentation to the IEC and CENELEC, and ensures that standards prepared by the IEC can be implemented in the Nordic countries with as few modifications or deviations as possible.

As a general rule all goods made for export to Nordic countries conform to IEC standards, but there is still a significant amount specified in various national standards and tendering for such projects results in extra costs which can often price non-Nordic countries out of a market.

In Denmark the international outlook of the Danish Electrotechnical Committee (DEK) represents Danish standards bodies in the IEC and CENELEC and Electronick Centralen is the supervising inspectorate and they play the same part in the IEC Quality Assessment System for Electronic Components (IECQ). In addition, as a member of the EEC, Denmark complies with Community Directives.

In Finland the electrical standards are prepared by a private organization, the Finnish Electrotechnical Standards Association (SESKO) which has also representatives on IEC, CENELEC, and NOREK committees. All national standards in Finland are ratified by the Finnish Standards Association (SFS) of which SESKO is a member. SFS standards are voluntary but the Electrical Inspectorate (SETI) has decreed some 200 obligatory safety standards mainly for use in homes, offices, shops where the public is admitted. The regulations apply equally to imported equipment.

The Norwegian Electrotechnical Committee (NEK) is a member body of the IEC, CENELEC, and NOREK. NEK is a private organization and its major task is to monitor and take part in IEC and CENELEC committees, submitting proposals to NEK for Norwegian Electrotechnical Standards (NEN) which are mainly voluntary but certain electrical standards, particularly those concerned with safety, are officially enforced under the Norwegian Electricity Act. The majority of NEK Standards are identical to the IEC standards, or translations with very few or no deviations in the technical content.

In addition to representing Sweden on the IEC, CENELEC and NOREK the Svenska Elektriska Kommissionen (SEK) is responsible for all Swedish standards in the electrical fields, and cooperation with the Swedish National standards Organization (SIS). Most of the standards are voluntary, however, electrical domestic equipment is subject to compulsory approval and cannot be sold or installed unless it is approved by the Swedish Institute for Testing and Approval of Electrical Equipment (SEMKO). The main form of Swedish standards in the electrical field are:

endorsement of IEC standards,
reproduction of the English text of the IEC standard,
translation of the IEC standard into Swedish.

In the case of deviations from the IEC standard these are listed in an Annexe forming part of the standard.

Addresses of Standards Organizations and Certification Bodies in the Nordic Countries

General

Nordic Electrotechnical Standards Cooperation Committee (NOREK)

NOREK has no central office as it is a joint body of the national organizations.

Danish Electrotechnical Committee (DEK)
(Dansk Elektroteknisk Komite)
Strandgarde 36
DK-1401 Copenhagen K
Denmark

Damarks Elektriscke Materielkontrol, (DEMKO)
Luska 8
DK-2730 Herleu
Denmark

Finnish Electrotechnical Standards Association (SESKO)
PO Box 134
SF-00211 Helsinki 21
Finland

Finnish Electrical Inspectorate (SETI)
Sarkiniementie 3
SF-00210 Helsinki 21
Finland

Norwegian Electrotechnical Committee (NEK)
(Norsk Elektroteknisk Komite)
Boks 7099 Homansbyen
Oscars Gate 20
Oslo 3
Norway

Norwegian Board for Testing and Approval of Electrical Equipment (NEMKO)
(Norges Elektriske Materiell Kontroll)
Boks 288 Gaustadalleen 30
Blindern
Oslo 3
Norway

Swedish Electrotechnical Commission (SEK)
(Svenska Elektriska Kommissionen)
Box 5177
S-10244 Stockholm
Sweden

Swedish Institute for Testing and Approval of Electrical Equipment (SEMKO)
(Svenska Elektriska Materielkontrollanstaten AB)
Box 30049
S-10425 Stockholm
Sweden

Statens Provingsanstalt
Box 857
S-50115 Boras
Sweden

36.5.3 Pacific Region

There is no regional standards body for Australasia or the Pacific areas, although Australia and New Zealand have begun to introduce harmonized standards. The main difficulty in achieving harmonized standards is the cost of travel between the countries concerned.

The Pacific Area Standards Congress (PASC) is an organization comprising representatives mainly from countries bordering the Pacific ocean. Active members include Australia, Canada, China, Hong Kong, Indonesia, Japan, Republic of Korea, Malaysia, New Zealand, Philippines, South Africa, Thailand, USA.

PASC is not a standards writing body, but endeavours to identify areas where harmonization of existing national standards would be of advantage, and to put forward to the IEC and ISO the concerns of the region and ensure adequate support for the standardization needs of the region.

The most important addresses (apart from the USA, see Section 36.4.1) are:

Standards Association of Australia (SAA)
Standards House
80 Arthur Street
North Sydney
NSW 2060
Australia   Tel: (02) 929 6022

Standards Association of New Zealand (SANZ)
Private Bag
Wellington
New Zealand   Tel: (4) 842108

36.5.4 Council for Mutual Economic Assistance (CMEA)

CMEA covers the following countries: Bulgaria, Cuba, Czechoslovakia, German Democratic Republic (GDR), Hungary, Mongolia, Poland, Romania, Vietnam, Union of Soviet Socialist Republics (USSR), Yugoslavia also has a special agreement on cooperation with CEMA.

CMEA pays great attention to standardization in electrical engineering. It is helped by efforts of the countries to harmonize their national standards and to bring them into line with IEC, ISO documents.

Procedures for the preparation and application of CMEA standards are stipulated in the Statute for CMEA standards, and under the Statute CMEA standards are compulsory for the signatory states. They are applied in contractual and legal relations between CMEA countries and their national economies. The addresses of the members of CMEA are as follows:

Bulgaria
State Committee for Science and Technical Progress
   Standards Office
21 6th September Str.
Sofia 1000, Bulgaria   Tel: 8591

Czechoslovakia
Office for Standards and Measurements
(Urad pr Normalizaci and Mercini)
Vaclavske nam 19
11347 Praha 1   Tel: 262251

German Democratic Republic
Amt fur Standardisierung Messwesen und Warenprufung
   der DDR
Furstenwalder Damm 388
DDR-1162 Berlin, German Democratic Republic

Hungary
Magyar Szabvanyugyi Hivatal
Pf. 24
H-1450 Budapest 9, Hungary   Tel: 183-011

Poland
Polski Komitet Normalizacji, Miar, i Jakosci
Ul. Elektovalna 2
00-139 Warszawa, Poland   Tel: 200241

Romania
Institutl Roman de Standardizare
Bucaresti Sect 1
Str. Roma 32-34
R 71219-RS   Tel: 337660

USSR
USSR State Committee for Standards
Leninky Prospekt 9
117049 Moscow M-49, USSR   Tel: 236 4044

Yugoslavia
Federal Institution for Standardization
Slobodana Penezica
Krcuna 35
11000 Beograd, Yugoslavia   Tel: (011) 644-066

36.6 Further reading

Electrical Standards in World Trade obtainable from the IEC (see Section 36.3.1)
Catalogue of Publications obtainable from IEC (see Section 36.3.1)
ISO Catalogue obtainable from ISO (see Section 36.3.1)
CENELEC Catalogue obtainable from CENELEC (see Section 36.3.1)
BSI Catalogue obtainable from the BSI (see Section 36.2.6)
Regulations for Electrical Installations obtainable from IEE (see Section 36.2.5)

Index

Absorbed dose, unit, 1/3
Absorbtance, 27/4
Absorption, 2/22, 27/4
Absorption factor, 2/24
Acceleration conversion factors, 1/5
Acceleration-error coefficient, 4/9
Acceleration transducers, 6/47
Accelerators, 26/13
Acetate films, 5/36
Acoustic impact on comfort conditions, 28/4
Acrylates, 5/29
Acrylonitrile butadiene styrene, 5/29
Actuators
  a.c., 26/5
  d.c., 26/4
  ironclad pot, 26/4
  polyphase, 26/5
  single-phase, 26/5
  three-phase, 26/5
ADA, 25/32
Adhesives, 5/31
Admittance, 2/3
Admittance operator, 2/4
Agriculture, 33/1
  electrical equipment, 33/4
  installation, 33/3
  power supply, 33/3
  safety measures, 33/3
Air as insulating material, 5/20
Air conditioning, 28/10, 28/21
  control boards, 20/7
  controllers, 28/15
Aircraft cables, 13/12
Alarm systems, 15/29, 17/39, 31/9
Algol, 25/32
Alnico alloys, 5/52
Alternative energy sources, 10/1
Alternating current, 2/3
  maximum, 6/10
  measurement of, 6/8
Alternating voltage
  measurement of, 6/8
  medium and high, 6/10
Alternators, automobile systems, 29/5
  charging system, 29/5
  9-diode, 29/6
  output terminal voltage, 29/6
Aluminium, 5/4
  applications, 5/5
  mechanical properties, 5/5
  protective purposes, 5/6
  resistivity, 5/5
  spot welding, 22/16
Aluminium alloys, 5/5, 5/5
  spot welding, 22/16
Aluminium busbars, 5/5
Aluminium conductor jointing, 13/30
Aluminium conductors, 13/6
Aluminium sheath for cables, 13/15
Ampere (unit), 1/3, 1/24
  definition, 6/4

Ampere law, 1/28
Ampere-turns, 2/4
Amplifiers, inverting, 25/34
  logarithmic, 25/38
  operational, 25/34
Amplitude comparator, 15/22
Amplitude distortion, 6/52
Amount of substance, unit, 1/3
Analogue computation, 25/33
Analogue computers. See Computers
Analogue functional elements, 25/34
Analogue inputs, 17/8
Analogue multipliers, 25/44
Analogue output, 17/8
Analogue-to-digital converters, 25/42
Analogues, 1/13
Analysers, 6/27
Angles (unit), 1/3, 1/4
Angstrom (unit), 1/4
Angular acceleration, unit, 1/3
Angular frequency, 2/4
Angular velocity (unit), 1/3
Angular velocity transducers, 6/49
Anode, 1/25, 2/10
Anodising, 23/11
Arc, 1/25
  interruption, 15/3
Arc chutes, 15/9
Arc control devices, 15/8
Arc extinction, 15/3
Arc mechanism, 15/3
Arc resistance, 15/3, 15/4, 15/27
Arc suppression coil, 15/37
Arcing time, 20/13
Arc (unit), 1/4
Area (units), 1/4
Area conversion factors, 1/5
Arithmetic logic unit (ALU), 25/6, 25/19
Arithmetic operations, 25/6
Arithmetic progression, 1/9
Armature volt-drop test, 26/36
Armature windings, 2/16
Arnold method for measurement of current
  transformers errors, 6/42
Asbestos-cement boards, 5/32
Asbestos-cement compounds, 5/39
Askarels, 5/25
Assembly language, 25/27
Asynchronous sequential logic circuits, 25/9
Atomic mass number, 1/17
Atomic number, 1/17
Atomic structure, 1/17, 1/19
Atoms, 1/17, 1/19
Attentuation coefficient, 3/7
Automatic control, ships, 31/9
Automation, building, 28/17
Automobiles, 29/3
  electrical equipment, 29/3
  ignition systems, 29/7
  standards, 29/12
Availability of plant and equipment, 18/4

Balanced protection, 15/25
Barium titanate, 5/42
Barn (unit), 1/4
Barrel planting, 23/12
Base-load, 18/7
BASIC, 25/31
Batteries, 5/6, 23/3
  alkaline manganese, 23/6
  automobile, 29/3
  charging, 23/11, 29/3, 29/17, 33/6
  cut-out relay, 29/4
  design and construction, 23/8
  electric vehicles, 23/8
  electrolyte, 23/8
  lead-acid, 29/3, 29/15
  nickel-alkaline, 29/3
  secondary, 23/7, 23/10
  trickle-charged floating, 15/29
  see also Cells
Battery vehicles. See Electric vehicles
Becquerel (unit), 1/3
Bessel functions, 1/8
Bimetal elements, 26/12, 26/13
Bimetals, temperature-sensitive, 5/62
Binary full adder, 25/7
Binary information processing, 25/4
Binary number system, 25/4
Binary numeration, 1/12
Binomial coefficients, 1/9
Biological energy conversion, 10/7
Bitumens, 5/26
Block diagrams, 4/5
Boards as insulating materials, 5/32
Bode diagram, 3/24, 4/16, 4/21
Bohr magneton 1/22
Boilers, air-gas flow diagram, 7/11, 7/13
  bi-drum, 7/7
  bi-drum water-tube, 7/9
  burner systems, 7/8
  coal-fired sub-critical, 7/10
  controllers, 28/13
  fluidised bed, 7/7, 7/11
  furnace configurations, 7/8
  heat flux in, 7/6
  industrial, 7/6
  oil-fired shop assembled package, 7/9
  pulverised fuel, 7/14
  recuperative, 7/11
  regenerative, 7/11
  stoker-fired water-tube, 7/8
  triple-pass wet-black shell, 7/7
  see also Steam generating plant
Boolean algebra, 25/5
Boolean variables, 25/4
Boundary conditions, 3/31
Brakes, electromagnetic, 26/9
  solenoid, 26/7
  thruster, 26/10
  tractive, 26/9
Brazing, 5/9
Break frequency, 4/19

Bridge-arm components, 6/37
Bridge Megger tester, 6/11
Bridge networks, a.c., 6/37
   d.c., 6/36
Bridged-T network, 6/41
Bridges, admittance, 6/40
   audiofrequency, 6/38
   audiofrequency tests, 6/45
   capacitance, 6/38
   current comparator, 6/40
   discharge, 6/39
   h.v. capacitance comparator, 6/41
   inductive ratio-arm, 6/40
   low-frequency, 6/38
   mutual inductance, 6/38
   universal, 6/40
Bristol submerged cylinder, 10/5
British Standards. *See* Standards
Brushless excitation sustems, 17/11
Buchholtz relay, 14/19, 15/28
Building automation, 28/17
Bunkers, underground horizontal storage, 32/7
Busbars, 5/5, 20/7
   protective equipment, 15/33
Buses, electric, 29/19

Cable boxes, 14/19
Cable couplers, flameproof, 32/17
Cables, 13/1
   a.c. transmission, 13/21
   aircraft, 13/12
   aluminium sheathed, 13/15
   arc welding, 13/11
   armour, 13/7, 13/15, 32/9, 32/18
   armour losses, 13/29
   categories, 13/3
   combined neutral and earth (CNE) types, 13/16
   conductor constructions, 13/6
   conductor losses, 13/29
   conductor materials, 13/5
   conductor screens, 13/19
   conductors, 13/19
   Consac, 13/16
   construction, belted and screened, 13/14
   continuous ratings, 13/27
   current rating, 13/27, 13/29
   d.c. transmission, 13/25
   dielectric losses, 13/29
   distribution, 13/9, 13/31
   electronic applications, 13/11
   essential components, 13/3, 13/5
   essential requirements, 13/22
   fault location, 13/32
      diagnosis, 13/33
      pinpointing, 13/35
      preconditioning, 13/33
      prelocation, 13/34
   fire hazard, 13/13
   flexible cords, 13/10
   forced cooling, 13/26
   future development, 13/27
   gas pressure, 13/24
   general wiring, 13/7, 13/8, 13/9
   heat sensor, 13/13
   high-pressure oil-filled, 13/24
   instrumentation, 13/10
   insulation, 13/6, 13/19
      alternatives to impregnated paper, 13/22
      electronic cables, 13/11
      paper insulated, 13/14, 13/21
      paper tapes, 13/15
      polymeric, 13/24
      stress distribution, 13/25
      temperature limits, 13/28

Cables, *continued*
   insulation screen, 13/19
   jointing, 13/30
   lead sheathed, 13/15, 13/24
   low-pressure oil filled, 13/23
   mineral insulated metal sheathed, 13/13, 20/9
   mines, 13/12
   Mollerhoj type oil-filled, 13/24
   overload current protection, 13/30
   oversheaths, 13/8
   partial discharges within paper insulation, 13/21
   PE insulated, 13/18
   power distribution. 13/7, 13/17
   pressurised, 13/21
   protective finishes, 13/8
   protective multiple earth (PME) systems, 13/16
   PVC insulated, 13/17
   PVC sheathed, 20/10
   quarries, 13/12
   railways, 13/12
   ratings, 13/26, 13/28
   reactive power, 16/4
   regulations, 13/27
   service, 13/17
   sheath losses, 13/27, 13/29
   sheathing materials, 13/8
   ship's, 13/11, 31/6
   short-circuit ratings, 13/30
   short-time and cyclic ratings, 13/30
   specifications, 32/17
   standards, 13/3, 13/11
   stranded conductors, 13/6
   submarine, 13/25
   supply distribution, 13/7, 13/14
   telephone communications, 32/18
   temperature limits, 13/26, 13/28
   testing, 13/20
   thermocouple, 13/10
   3-conductor split-concentric, 13/17
   trailing, 32/17
   transmission, 13/9, 13/21, 13/31
   treeing, 13/20
   types of, 13/22
   usage groups, 13/3
   voltage drop, 13/30
   voltage range, 13/3
   water cooling, 13/26
   Waveconal, 13/17
   Waveform, 13/17
   wiring system, 13/9
   XLPE insulated, 13/18, 13/22, 13/24
Calibration, 6/3
   cathode-ray oscilloscope, 6/32
   digital voltmeters, 6/26
   galvanometer, 6/53
Calorie (unit), 1/4
Calorific values, 7/3
Candela (luminance), 6/5
Candela (unit), 1/3
Capacitance, 2/4, 2/20, 3/3, 6/5
   calculation of, 2/20
   unit, 1/3, 1/6
Capacitive reactance, 2/21
Capacitor charge and discharge, 2/19
Capacitor commutation, 19/13
Capacitor rating, 20/17
Capacitors, 1/26, 2/4, 2/20, 5/6, 15/14
   connection of, 2/21
   mechanically switched, 16/8
   series, 15/13, 15/14, 16/7, 16/18
   shunt, 15/13
   voltage applied to, 2/21
Capstans, 31/9

Carbon, 5/7
   applications, 5/7
   resistance brazing and welding, 5/9
Carbon brushes, 5/7
Carbon contacts, 5/9
Carbon electrodes, 5/9
Carbon fibres, 5/9
Carbon granules, 5/9
Carnot cycle, 2/10
Cathode, 1/25, 2/10
Cathode-ray oscilloscope, 6/27
   analogue, 6/30, 6/31
   applications, 6/32
   calibration, 6/32
   digital, 6/31
   instrument selection, 6/31
   operational use, 6/31
Cathode-ray tube, 6/28
   brightness control, 6/28
   definition, 6/29
   fluorescent screen, 6/29
   focus control, 6/28
   limiting frequency, 6/29
   maximum recording speed, 6/29
   maximum specific recording speed, 6/29
   performance, 6/28
   specific deflection sensitivity, 6/29
Cause-and-consequence chart (CCC), 17/45
Cavity floor system, 20/11
Cells, 2/11, 23/3
   acid, 23/6
   air depolarised, 23/6
   alkaline, 2/11, 23/5
   copper chloride/magnesium, 23/6
   copper oxide/zinc, 23/5
   flat, 23/4
   hydrogen-oxygen, 2/11
   layer, 23/4
   lead-acid, 2/11, 23/7
   lead dioxide/lead/perchloric acid, 23/6
   lead oxide/magnesium, 23/6
   lead oxide/zinc (or cadmium)/sulphuric acid, 23/6
   Leclanché, 2/11, 23/4
   lithium, 23/11
   mercury oxide/zinc, 23/5
   nickel/cadmium alkaline, 23/8
   nickel/iron alkaline, 23/8
   primary, 2/11, 23/3
   secondary, 2/11, 23/3, 23/7
   silver chloride/magnesium, 23/6
   silver oxide/zinc, 23/5
   silver/zinc alkali, 23/10
   standard, 2/11, 23/5
   water activated, 23/6
   see also Batteries; Fuel cells
Cellulose acetate and triacetate, 5/29, 5/36
Cements, 5/25
CENELEC, 36/9
Central Electricity Authority, 18/14
Central Electricity Board, 18/13
Central processing unit (CPU), 25/6
Ceramics, 5/38, 5/39, 5/52
Characteristic equation, 4/7
Characteristic impedance, 3/7
Charge. *See* Electric charge
Charge carriers, 1/24
Charged particles, 2/22
Charges, at rest, 1/24
   in acceleration, 1/26
   in motion, 1/24
Chemical equivalent, 1/25
Chillers, 28/11
   controllers, 28/13
Chlorosulphonated polyethylene (CSP), 13/7, 13/11, 13/12

Index/3

Chucks, magnetic, 26/10
Circuit breakers, 15/4, 18/15
  air-blast, 15/9
  air-break, 15/9
  arc extinction methods, 15/3
  automatic, 20/4
  auto-reclose, 15/37
  bulk-oil, 15/8
  circuit conditions, 15/4
    asynchronous, 15/5
    capacitive, 15/5
    inductive, 15/4
    resistive, 15/4
  continuously pressurised, 15/10
  cubicle, 15/7
  d.c., 15/9
  fault-voltage operated, 20/22
  flameproof, 32/13
  forms of, 15/8
  hard gas, 15/9
  low-oil-content, 15/8
  miniature, 20/12, 20/13, 26/12
  modular construction, 15/8
  moulded-case, 20/12, 20/13
  oil, 15/8
  plain break, 15/8
  ships, 31/6
  single-pressure, 15/12
  sulphur hexafluoride, 5/22, 15/10
  testing, 15/14, 15/16
  tripping of, 15/28
  truck, 15/7
  vacuum, 5/23, 15/12
  see also Switchgear
Circulating current protection, 15/25
Clock module, 17/8
Closed-loop transfer function, 4/6
Clutches, electromagnetic, 26/8
Coal fuels, 7/4
Coal mining. *See* Mining
Coal stacking and reclaiming equipment, 7/12
Coatings, insulative, 5/45
COBOL, 25/32
Cockerell raft, 10/5
Coercive force, 2/4, 5/46
Coercivity, 2/4
Coil design, 26/3
Coil resistance, 6/54
Coil windings, 26/3
Coils, arc suppression, 15/37
  circuit-breaker trip, 15/28
  crossover or bobbin, 14/7
  disc, 14/7
  helix, 14/7
  induction heating, 21/13
  multilayer helix, 14/7
  multistart interwound helix, 14/7
  transformer, 14/7
Combination, 1/9
Combinational logic, 25/5
Combinational logic circuits, 25/8
Combinational logic design using memory, 25/14
Combined heat and power (CPH) schemes, 8/16, 10/10, 10/12, 10/13
Combustion process, 7/3
Communication(s), radio, 32/8
  serial, 24/11
Communication channels, 17/10
Commutation, 19/3
  capacitor, 19/13
Comparator, 5/7, 25/37
Comparator potentiometer, 6/34
Comparator resistance bridge, 6/37
Compensated voltage control, 29/3

Compensation, application aspects, 16/15
  light flicker, 16/15
  series capacitor, 16/7, 16/18
  shunt, 16/18
  shunt reactor, 16/7
  var equipment developments, 16/18
  variable static shunt, 16/7
Compensation theorem, 3/5
Compensators, harmonic-compensated self-saturated reactor, 16/11
  saturated reactor, 16/12
  static var, 16/10, 16/18
  synchronous, 16/10
  thyristor-controlled reactor, 16/12
  thyristor switched capacitor, 16/15
  variable var, 16/8
Complex functions, 25/36
Complexor, 2/4
Composite materials, sheet insulations, 5/38
Compound, chemical, 1/17
Compressed-air storage plant, 18/8
Computation, 25/1
Computer control, road-traffic signalling, 29/21
Computer elements, 5/11
Computer information systems, 36/5
Computer instruction execution, 25/23
Computer instruction format, 25/24
Computer instruction types, 25/24
Computer program development, 25/29
Computer program execution, 25/25
Computer program testing, 25/29
Computer programming, 25/21, 25/25
  structured, 25/26
Computer programming languages, 25/26
Computer programs, environmental comfort design, 28/9
Computer software design and specification, 25/26
Computers, 24/3, 25/6
  analogue, 25/33, 25/39
  architecture, 25/18
  control and timing unit, 25/18
  digital, 25/3, 25/21
    logic design, 25/17
  hybrid, 24/45
  interrupts, 25/23
  mining, 32/20
Concentration (unit), 1/3
Concrete, 5/39
Condenser, double-bundle, 28/19
Conductance, 2/5
  units, 1/3, 1/6
Conductance instrumentation, 6/11
Conduction, 1/24, 2/8, 2/22
Conductivity, 1/20, 2/5, 2/7
  of aqueous solutions, 2/7
  units, 1/4, 1/6
Conductivity tests, 6/12
Conductors, 1/24, 1/26, 2/19, 5/3
  cables, 13/19
  liquid, 2/7
  overhead lines. *See* Overhead lines
  protective, 20/18
  see also Cables
Connection charges, 18/11
Constant energy systems, 29/8
Constant linkage principle, 2/15
Constitutive equations, 1/28
Contact breaker ignition systems, 29/7
Contactor-coil/pilot circuit, 32/15
Contactor starter, 20/14
Contactors, 26/11
  vacuum, 5/25
Contacts, 5/4, 5/56
  area of application, 5/58
  basic forms of, 5/60

Contacts, *continued*
  heavy duty, 5/62, 5/64
  high voltage-high power, 5/60
  light duty, 5/63
  low voltage, high current, 5/56
  low voltage, low current, 5/56
  make and break, 5/60
  medium duty, 5/62, 5/63
  medium voltage, 5/60
  properties of materials, 5/57, 5/58
Continuity testing, 20/20
Control room, 17/39
  operation and display, 17/30
Control systems, 24/3
  analysis, 4/1
  classification, 4/8
  electric vehicles, 29/15
  environmental comfort, 28/13
  first-order systems, 4/9
  second-order systems, 4/10
  static accuracy, 4/8
  steady state errors due to disturbances, 4/9
  steam-turbines, 17/27
  transient behaviour, 4/9
  water turbines, 17/31
  see also Power system operation and control
Controllers, 28/13
  programmable, 28/22
Convection, 2/8
Convection current, 1/26
Convectors, 28/9
Converters, analogue-to-digital, 25/42
  digital-to-analogue, 25/41
  h.v.d.c., 16/19, 34/1
  multifunction, 25/38
Conversion factors, 1/4
Conveyors, 33/5
  mining, 32/7
Convolution, 4/5
Coolers, 33/5
Cooling, 2/8
Cooling coefficient, 2/8, 26/3
Cooling cycles, 2/9
Cooling time constant, 2/9
Cooling towers, 28/10
Copper, 5/3
  applications, 5/3
  conductivity, 5/3
  mechanical properties, 5/3
  resistivity, 5/3
  welding, 22/4
Copper alloys, 5/3, 5/4
  welding, 22/4
Copper conductors, 13/5
Copper clad aluminium conductors, 13/6
Copper loss. *See* I^2R loss
Copper multilayers, 5/17
Copper-nickel alloys, 5/61
Copper windings, 2/8
Coral 66, 25/32
Cords for lashing bundles of switchboard wiring, 5/38
Core loss, 2/5, 2/13, 6/44
Corner frequency, 4/19
Corona, 5/22
Cosecant, 1/4
Cosine, 1/4
Cosine law, 27/4
Cotangent, 1/4
Coulomb (unit), 1/3
Counters, 25/12
Coupling coefficient, 2/19
Couplings, eddy-current, 26/8
  electromagnetic slip, 31/13
Crack detectors, 26/6
Cranes, linear motors, 26/42

Crop storage and marketing, 33/9
Cross-linked polyethylene (XLPE), 13/7
Crystals, 1/22
CSMP language, 25/48, 25/49, 25/50, 25/51
Cultivation, 33/9
Cunife, 5/54
Current, 1/24, 2/3, 2/5
  unit, 1/3
Current bias, 17/16
Current density, 2/5
  units, 1/4, 1/6
Current generator, 3/3
Current rating, overhead lines, 12/7
Current sensitivity, 6/53
Current-voltage characteristics, 2/6
Current-voltage control, 29/4
Cyclic redundancy check code (CRC), 24/11
Cyclotron, 26/14
Cylinders, dielectric, 5/35

Dairy equipment, 33/5, 33/6
Damping factor, 6/54
Data acquisition system, 6/19, 17/6
Data bases, 17/41
Data integrity and error recovery, 24/11
Data loggers, 6/27
Data recording, 6/51, 24/7
Data storage, 24/7
Data transmission, 17/8, 17/9
D.c. machines, 2/17, 5/11
D.c. precision sources, 6/21
de Broglie, 1/18
Decibel (unit), 1/12
Decibel gain, 1/16
Decimal prefixes, 1/4
Decoders, 25/8
Dedicated controller, 24/4
Deflection amplifiers, 6/30
Degree (unit), 1/4
Density, conversion factors, 1/5
Density (optical), 27/4
Depolariser, 2/11
Derivative control, 4/11
Derivatives, 1/10
Desalination plants, 10/12, 24/10
De Sauty bridge, 6/38
Descartes rule of signs, 4/7
Deuterium, 10/9
Diamagnetic materials, 2/13
Diamagnetism, 1/22, 2/5
Dielectric breakdown, 2/21
Dielectric loss, 2/5, 5/20
Dielectrics
  liquid, 2/21, 5/23
  solid, 2/21, 5/32
  see also Insulating materials
Diesel-electric ship propulsion, 31/12
Diesel engines, 8/22, 10/12, 11/27
  ancillaries, 8/27
  basic classification, 8/25
  combustion chambers, 8/23
  combustion process, 8/23
  cooling systems, 8/26
  exhaust heat dissipation, 8/33
  four-stroke, 8/23
  fuel injection, 8/25
  governors, 8/27
  induction system, 8/26
  inter-cooling, 8/23
  locomotive, 30/9
  lubrication, 8/25, 8/31
  maintenance, 8/31
  marine, 31/11
  mechanical arrangements, 8/25
  monitoring, 8/28

Diesel engines, *continued*
  operating modes, 8/25
  operational aspects, 8/30
  plant layout, 8/32
  pressure charging, 8/23
  primary systems, 8/25
  principal roles, 8/22
  ratings, 8/30
  starting equipment, 8/27
  synchronous speed, 8/25
  theory and general principles, 8/23
  turbochargers, 8/24
  two-stroke, 8/23
  working cycles, 8/23
Differential instruments, 6/21
Differentiator, 25/36
Digital computers. *See* Computers
Digital input modules, 17/7
Digital multiplexers, 25/7
Digital output, 17/8
Digital signals, 25/3
Digital simulation, 25/46
Digital systems, 25/3
  information representation by, 25/3
  programmable, 25/3
Digital-to-analogue converters (DAC), 25/41
Differentiation, 4/5
Dimensional relationships, 6/5
Diode function generator, 25/37
Diodes, 19/11
Direct conversion processes, 10/7
Direct current, 2/3
  measurement of, 6/8
  transmission and interconnection, 18/15
Direct memory access, 25/23
Direct voltage, medium and high, 6/10
  measurement of, 6/8
Disconnectors, 15/13
Discriminating delta connection, 15/26
Discrimination, 20/4
Displacement current, 1/26, 2/5
Distribution systems, 20/3, 32/4
District heating, 10/12
Domains, 1/23
Domestic premises, 20/12
Doubling effect, 2/18
Drilling machines, 32/19
Drives, reversing motor, 19/8
  rolling mills, 19/14
Drying, 33/6
Ductility, 5/45
Dynamic viscosity (unit), 1/3
Dyne (unit), 1/4

Earth current compensation, 15/27
Earth electrode, 20/17
  testing, 20/21
Earth-fault loop impedance, 20/21
Earth faults, 15/27
Earth leakage protection, 15/18, 15/25
Earth resistance measurement, 20/18
Earth resistance tests, 6/12
Earth-rod electrode, 20/18
Earth wire, 15/36
Earthing, 20/17
  designations, 20/19
  high-impedance single point, 32/15
  multi-point, 32/16
  regulations, 33/3
  ship regulations, 31/4
Economic indicators, 18/3
Economics, 18/1
Eddy current, 2/5
Eddy-current couplings, 26/8
Eddy-current losses, 5/43

Edmonton LRT vehicle, 29/12
Education and training, 35/1
  bridges and ladders, 35/6
  Chartered Engineer, 35/4
  Engineering Technician, 35/4, 35/5
  historical background, 35/3
  National Certificates and Diplomas, 35/3
  present situation, 35/4
  professional institutions, 35/3
  qualifications, 35/3
  Technician Engineer, 35/4, 35/5
Electric charge, 2/4
  units, 1/3, 1/6
Electric field, 1/16, 1/24, 1/25, 1/26, 2/5, 2/19, 3/3
  static, 2/22
Electric field intensity, 1/22
Electric field strength, 2/5
  unit, 1/4, 1/6
Electric flux, 2/5
Electric flux density, 2/5, 2/20
  units, 1/6
Electric furnaces. *See* Furnaces
Electric ovens, 21/6, 21/7
Electric potential (unit), 1/3
Electric propulsion of ships, 31/11
Electric space constant, 2/5
Electric strength, 2/5
Electric stress, 2/5
Electric vehicles, 5/6, 29/15
  applications, 29/18
  buses, 29/19
  cars, 29/18
  chopper control, 29/16
  control system, 29/15
  field control, 29/16
  operational details, 29/17
  parallel/series battery switching, 29/16
  range and power assessment, 29/19
  series resistance, 29/15
  solid state switching, 29/16
  see also Batteries
Electrical load. *See* Load
Electrical Standards Laboratories, 6/4
Electrical steels, 5/43
  ancilliary properties, 5/43
  applications, 5/45
  chemistry and production, 5/45
  grain orientation, 5/45
  physical form, 5/45
  test methods, 5/45
Electricity, 1/24
Electricity Act 1947, 18/6
Electricity Act 1957, 18/6
Electricity (Factories Act) Special Regulations 1944, 18/6
Electricity Reorganisation (Scotland) Act 1954, 18/6
Electricity Supply Act 1919, 18/13
Electricity Supply Regulations 1937, 18/6
Electricity supply systems, 18/13
  275 kV network, 18/14
  400 kV network, 18/14
Electrochemical effects, 2/10
Electrochemical equivalents, 2/10
Electrochemical reactions, 23/2
Electrochemical reactors, 23/3
Electrochemical science, 23/3
Electrochemical technology, 23/1
Electrodeposition, 23/12
Electrodes, 1/25, 1/26, 2/10
  installation, 20/18
Electroforming, 23/12
Electroheat, 21/1
  definition, 21/3
Electroluminescent devices, 27/17

Electrolysers, types of, 23/14
Electrolysis, 2/10
  gas purity, 23/14
  hydrogen, 23/13
  oxygen, 23/13
  plant arrangement, 23/14
  process, 23/13
  uses of, 2/10
Electrolytes, 1/25, 2/10
Electrolytic tanks, 1/16
Electromagnetic brakes, 26/9
Electromagnetic clutch, 26/8
Electromagnetic compatibility in ships, 31/8
Electromagnetic crack detection, 26/6
Electromagnetic devices, 26/3
Electromagnetic field, 2/5, 2/22
Electromagnetic force, 2/14
Electromagnetic gearing, 31/13
Electromagnetic induction, 2/5, 2/14
Electromagnetic machines, 2/16, 26/1, 26/3, 26/15
Electromagnetic relays, 26/11
Electromagnetic slip couplings, 31/13
Electromagnetic units, electrostatic, 1/4
Electromagnetic waves, 1/24, 1/28, 2/23
Electromagnets, applications, 26/3
  high-field, 5/10
  tractive, 26/4, 26/9
Electromechanical effects, 2/22
Electromechanical relay, 5/46
Electromotive force (e.m.f.), 2/3, 2/5
  solid-state reference, 6/33
  units, 1/6
Electron atmosphere, 1/24
Electron beam deflection, 6/28
Electron emission, 1/21, 1/25
Electron gas, 1/20
Electron gun, 6/28
Electron spin, 1/22
Electron-volt (unit), 1/4
Electronic analysers, 6/27
Electronic (contactless) ignition systems, 29/7
Electronic devices, 5/11
Electronic oscillators, 6/26
Electronic oscillography, 6/27
Electrons, 1/17, 1/19, 1/20, 2/22
Electroplating, 2/10, 23/12
Electroplating alloys, 5/4
Electrorefining, 2/10, 23/12
Electroslag refining, 21/16
Electrostatic precipitator, 7/14
Electrostatics, 2/19
Electrotechnology, 2/1, 2/3
  circuit phenomena, 2/3
  nomenclature, 2/2
  symbols and units, 2/4
  terminology, 2/3
  thermal effects, 2/6
Electrowinning, 23/12
Element, 1/17
Elevators, 33/5
Emergency supplies, 20/12
  ships, 31/6
Enamels, 5/31
Encapsulation, 5/30
Encoders, 25/8
Energy, 1/13
  units, 1/3, 1/4, 1/6
Energy balance theory, 15/4
Energy bands, 1/22
Energy conservation, 1/13, 28/18
Energy conversion, 1/5, 1/13, 26/3
Energy dissipation, 1/13
Energy meters, 6/26
Energy radiation, 1/13, 1/18

Energy sources, alternative, 10/1
  renewable, 10/3
Energy states, 1/18
Energy transmission, 1/13
Enthalpy-entropy diagram, 7/6
Environmental comfort, 28/3
  computer aided design, 28/9
  control systems, 28/13
  cooling plant, 28/10
  co-ordination for, 28/20
  dynamic or cyclic loads, 28/6
  electrical load calculation for, 28/21
  energy consumption, 28/9
  energy requirements, 28/6
  heating or cooling plant capacity, 28/8
  interfaces, 28/21
  intermittent heating and cooling, 28/7
  limits of, 28/4
  machine rooms, 28/5
  process plants, 28/5
  steady state loads, 28/6
Environmental control, 28/1
  safety requirements, 28/6
Epoxies, 5/30
EPR, 13/12
Equivalent circuits, 2/3, 3/3
Erg (unit), 1/4
Error recovery, 17/41
ESP control stations, 24/11
Ethylene propylene rubber, 13/7, 13/11
Ethylene tetrafluoroethylene, 13/7
Evanohm, 5/61
Event recording, 17/37, 17/39
Excitation potential, 1/20
Exponential functions, 1/6, 1/9

Fabrics, insulating, 5/35
Factorials, 1/9
Factories Act 1961, 27/25
Fans, 33/4, 33/7
  ships, 31/8
Farad (unit), 1/3
Faraday law, 1/28, 2/14
Farming processes, 33/6
Fault conditions, 15/19
  double phase-to-earth, 15/20
  phase-to-phase, 15/19
  single phase-to-earth, 15/19
  3-phase, 15/19
Fault current, 20/4, 20/13
Fault impedance, 15/27
Fault location, protective equipment, 15/35
Feed mills, 33/4
Feed processing, 33/6
Feedback, 4/6
  transfer function, 4/6
Feeders, protective equipment, 15/32
Felici-Campbell bridge, 6/38
F.E.P., 5/36
Ferranti effect, 3/18
Ferrites, 5/47
  magnetically soft, 5/47
    applications, 5/49
    properties of, 5/47
  microwave, 5/49
  permanent magnet, 5/52
  rectangular loop, 5/49
Ferroelectrics, 5/42
Ferromagnetic materials, 2/5, 5/42
Ferromagnetic properties, 2/13
Ferromagnetism, 1/23
Ferroresonance, 3/26
Field-mapping techniques, 1/16
Field strength, 1/28

Fields, 1/16
Films, insulating, 5/35
  see also Thick film; Thin film
Filters, harmonic, 16/17
Final-value theorem, 4/5
Fire-clay refractory ceramics, 5/39
Fire hazards, cables, 13/13
  switchgear, 15/29
Fixed-axis gap flux, 26/16
Flameproof equipment, mining, 32/11
Flameproof motors, 32/16
Flameproof plug and socket, 32/14, 32/18
Flicker, arc furnace induced, 16/16
Flip-flop, 25/9
  clocked, 25/10
Floodlighting system, 27/32
Floor trunking, 20/11
Flow rate, conversion factors, 1/5
Flue-gas wet scrubbing system, 7/14
Fluorescence, 6/29
Fluorinated ethylene propylene, 13/7
Flux, 1/28
  unit, 1/3
Flux change, 2/15
Flux cutting, 2/15
Flux density, 1/28
Fluxes, soldering, 5/62
Fluxmeter, 6/44
Force, 2/5
  (units), 1/3, 1/4, 1/6
Force conversion factors, 1/5
Forced outage, 18/4
Forecasting for operation and maintenance, 18/4
Formed compositions, 5/38
FORTRAN, 25/30, 25/46, 25/47
Forward transfer function, 4/6
Fourier analysers, 6/27
Fourier series, 1/9, 3/13
Fourier's theorem, 2/3
Free space, 1/28, 1/29, 1/30
  propagation, 2/23
Frequency, 2/5
  unit, 1/3
Frequency conversion, 19/9
Frequency decade, 3/24
Frequency effects, 2/7
Frequency indicators, 6/11
Frequency octave, 3/24
Frequency of oscillation, 4/11
Frequency-response methods, 4/14
Frost protection, 33/9
Fruit grading and storing, 33/9
Fuel cells, 2/11, 10/8
Fuel handling and storage, 7/11
Fuel injection, closed-loop electronic, 29/11
  digital electronic, 29/11
  electronic, 29/10
Fuel oil, 7/11, 8/30
Fuels, diesel, 8/25
  fossil, 7/3, 7/13, 18/7
  gaseous, 7/4
  ignition temperatures, 7/3
  liquid fossil, 7/4
  other solid, 7/5
  solid fossil, 7/5
Full-load operating conditions, 29/11
Function generators, 25/37
Furnaces, 21/4, 21/6
  arc, 16/15, 16/16, 21/15
  batch, 21/8, 21/11
  channel induction, 21/14
  construction, 21/8
  conveyor, 21/12
  coreless induction, 21/14
  design patterns, 21/4

Furnaces, *continued*
  direct arc, 21/15
  electron beam, 21/17
  electroslag refining, 21/16
  indirect arc, 21/15
  plasma, 21/19
  salt bath, 21/4
  submerged arc, 21/16
  vacuum arc, 21/16
Fuse action, 15/5
Fuse applications, 15/7
  d.c. circuits, 15/7
  distribution, 15/7
  motors, 15/7
  solid state devices, 15/7
  voltage transformers, 15/7
Fuse characteristics, 15/5
Fuse types, 15/6
Fusegear, 15/5
Fuses, 5/54, 5/62
  composite or duel-element, 5/56
  high-speed, 5/55
  HRC, 20/14, 20/12, 20/13
  'M' effect, 5/56
  major and minor, 15/7
  materials, 5/55
  shaped silver elements, 5/55
  technology, 5/55
Fusible materials, 5/25, 5/26, 5/27
Fusing currents, 2/9

Gain curves, 4/19
Gain margin, 4/16, 4/20
Gall potentiometer, 6/36
Galvanised steel wire (GSW), 13/7
Galvanometer recorders, 6/55
Galvanometers, 6/53
  ballistic, 6/44
  ultraviolet recorder, 6/51
Gas turbines, 18/8, 8/13
  advantages over steam plant, 8/13
  aircraft-type, 8/15
  closed-cycle, 8/13, 8/15
  cogeneration plant, 8/16
  combined-cycle, 8/16
  combustion chamber, 8/14
  compound cycle, 8/15
  disadvantages, 8/13
  efficiency, 8/15
  energy utilisation, 8/16, 8/18
  exhaust gas recovery, 10/12
  exhaust heat recovery, 8/16
  free-piston gas generator, 8/15
  maintenance, 8/31
  open-cycle, 8/13, 8/14
  performance data, 8/17
  power relations, 8/14
  pressure/volume diagram, 8/14
  unit sizes, 8/13
Gases, 1/25
  dielectrics, 2/21
  physical properties of, 1/22
  see also Dielectrics, gaseous
Gate-end boxes, 32/14
Gauss law, 1/28
Gauss-oersted (unit), 1/4
Gauss–Seidel procedure, 3/28
Generating plant, choice of, 18/6
  types of, 18/7
Generator capability chart, 16/4
Generator effects, 2/15
Generators, 5/6, 16/3
  a.c., 5/11, 8/28, 11/1, 29/4, 31/4
    a.c.-exciters with static rectifiers, 11/16
    airgap flux distribution, 11/3

Generators, *continued*
  a.c., *continued*
    ANSI Potier-reactance method, 11/13
    armature leakage reactance, 11/6
    armature reaction, 11/4
    AVR fault protection, 11/21
    brushless excitation, 11/16
    classification, 11/3
    construction, 8/28
    control features, 11/19
    cylindrical-rotor, 11/4, 11/10, 11/11
    d.c. exciters, 11/16
    double-channel AVR, 11/21
    equivalent field m.m.f., 11/5
    excitation control, 11/17
    excitation limits, 11/19
    excitation systems, 11/16, 11/17
    frequency-response tests, 11/8
    insulation, 11/4
    magnetisation (armature-reaction)
      reactances, 11/6
    m.m.f. phasor diagram, 11/12
    negative-sequence reactance, 11/7
    no-load e.m.f., 11/3
    on-load excitation, 11/12
    open-circuit characteristics, 11/9
    operating charts, 11/11
    oscillation frequency, 11/11
    output equation, 11/4
    overall voltage response, 11/21
    overfluxing protection, 11/19
    parallel operation, 8/29, 11/19
    Potier reactance, 11/8
    protection, 8/28
    reactances, 11/6, 11/7
    rotor pole-to-pole leakage, 11/13
    salient-pole, 11/10, 11/12
    salient-pole rotor machine, 11/5
    selection criteria, 8/29
    short-circuit characteristics, 11/9
    short-circuit performance, 8/29
    short-circuit ratio, 11/5
    stator windings, 11/3
    steady load conditions, 11/9
    steady-state operation, 11/9
    sudden 3-phase short circuit, 11/14
    synchronising, 11/10
    synchronous, 11/3
    synchronous reactances, 11/6
    system stability controls, 11/21
    temperature rise, 11/4
    thyristor excitation, 11/17
    time-constants, 11/7
    transformer connections, 11/25
    transient and subtransient reactances, 11/6
    voltage, 8/28
    voltage control, 11/17
    zero-sequence reactance, 11/7
  diesel, 8/33
  economic factors, 8/32
  engine-driven, 11/27
  excitation of large, 17/18
  induction, 11/28
  on-site, 8/33
  parallel operation, 17/26
  powered by diesel engines, 8/22
  protective equipment, 15/30
  salient-pole synchronous, 8/21
  ships, 31/11
  synchronous, 17/10
  thermionic, 2/10, 10/7
  thermoelectric, 10/7
  turbo-. *See* Turbogenerators
  see also Hydro-electric plant
Geometric progression, 1/9
Geothermal energy, 10/4

Germanium diodes, 19/11
Glass, 5/39
  direct resistance heating, 21/4
Glass-bonded mica, 5/32
Glass fabric laminates, 5/34
Glow discharge processes, 21/20
Gouging, 22/12
Grading, 2/22
Grain drying, 33/6
Gray (unit), 1/3
Greenhouse operation, 33/8
Guard ring, 2/20
Gums, 5/26

Hall effect, 2/17
Hall effect instrument, 6/44
Hammer mill, 33/4
Hard-copy devices, 17/39
Hartshorn bridge, 6/38
Harvesting, 33/9
Hay bridge, 6/38
Hay drying, 33/7
Health and Safety at Work Act 1974, 18/6, 33/3
Heat exchangers, 5/6
  air-to-air, 28/18
  liquid coupled indirect, 28/18
Heat pipes, 28/19
Heat pumps, 10/14, 28/18
  1–10 kW (thermal) packaged units, 10/18
  10–100 kW (thermal) packaged units, 10/16
  100 kW–1 MW (thermal) schemes, 10/16
  1–10 MW (thermal) schemes, 10/15
  10–100 W (thermal) modules, 10/18
  100 W–1 kW (thermal) units, 10/18
  air cycle, 10/14
  air/water space heating, 10/17
  applications, 10/19
  dehumidifiers, 10/17, 10/19
  practical cycles, 10/14
  reversible air-to-air space conditioner, 10/16
  size effects, 10/15
  thermodynamics, 10/14
  thermoelectric, 10/15
  vapour compression cycle, 10/14
  water heater, 10/18
Heat reclaim system, 28/19
Heat-resisting alloys, 5/62
Heat sinks, 5/6
Heat transfer materials, 5/4
Heat treatment furnaces, 21/11
Heaters, 33/5
  dielectric, 21/18
  immersion, 21/13
  infrared, 21/6
  resistance, 21/4, 21/7
Heating, 2/8, 28/9
  continuous, 2/8
  controllers, 28/13
  dielectric, 21/17
  direct resistance, 21/3
  indirect resistance, 21/4
  induction, 21/9, 21/10
    billets and slabs, 21/13
    coil design, 21/13
    load coupling, 21/10
    power sources, 21/10
    surface and localised, 21/13
  rapid, 2/8
  ships, 31/7
  space, 33/5, 33/8
  storage heaters, 28/10
  underfloor, 28/10
Heating cycles, 2/9

Heating elements, 5/61
  flexible, 21/7
  for ovens and furnaces, 21/6, 21/9
  metal sheathed, 21/13
  resistance, 21/5, 21/6
  rod or tubular, 21/6
Heating time constant, 2/9
Hectare (unit), 1/4
Helmholtz-Norton equivalent current generator, 3/3
Helmholtz–Morton theorem, 3/6
Helmholz–Thevenin equivalent voltage generator, 3/3
Helmholtz–Thevenin theorem, 3/6
Henry (unit), 1/3
Hertz (unit), 1/3
High-level languages, 25/28, 25/30, 25/32
High-rise buildings, 20/12
High-voltage supplies, installation, 20/3, 34/1
Hoists, 33/5
Horticulture, 33/1
  electrical equipment, 33/4
  installation, 33/3
  power supply, 33/3
  processes, 33/8
  safety measures, 33/3
Hot dry rocks, extracting heat from, 10/4
Humidity control, 28/3
Hybrid computation, 25/40
Hybrid computers. *See* Computers
Hybrid process, 5/16
Hydrocarbons, synthetic, 5/24
Hydrodynamics, of steam generating, 7/5
Hydrogen, as insulating material, 5/23
  electrolysis, 23/13
Hydrogen atom, 1/17, 1/20
Hydrogen/oxygen fuel cell, 10/9
Hydro-Electric Development (Scotland) Act 1943, 18/6
Hydro-electric plant, 18/8, 18/9, 8/18, 17/20
  catchment area, 8/18
  discharge and tail race, 8/21
  economics, 8/21
  electrical plant, 8/21
  generator construction, 11/26
  generator design, 11/26
  generator excitation, 11/26
  mixed thermal and, 18/10
  operational features, 8/21
  pipelines, 8/19
  power station, 8/19
  pumped storage, 8/21, 11/27, 18/3, 18/10
    economics, 8/22
    layout, 8/22
    pump/turbine plant, 8/22
    starting and changeover, 8/22
  reservoir and dam, 8/18
  specific speed, 8/19
  turbines used, 8/19
Hydrogenerators, 11/3
Hydrothermal energy sources, 10/4
Hyperbolic functions, 1/7, 1/9
Hypernik, 2/13
Hysteresis, 2/5, 2/13
Hysteresis effect, 5/42
Hysteresis loss, 2/13, 5/43, 5/49

Idling modes, 29/11
IEE Regulations, 13/27, 18/6, 20/3, 20/18, 20/20, 27/30
Ignition systems, 29/7
  contact breaker, 29/7
  digitally timed, 29/8
  electronic (contactless), 29/7
  electronically timed, 29/8

Ignitron, 19/9
Illuminance, 27/3, 27/4
  and glare recommendations, 27/20
  unit, 1/3
Illuminance ratio, 27/21
Illumination, conversion factors, 1/5
  practical measurements, 27/25
  quantities and units, 27/3
  see also Lamps; Lighting
Immitance, 2/5
Impedance, 2/3, 2/5, 2/7
  intrinsic, 2/23
  measurement, 14/23
Impedance matching, 6/31
Impedance operator, 2/5, 6/37
Impedance time scheme, 15/26
Impulse time scheme, 15/26
Impulse current prelocation method, 13/34
Impulse response, 4/5
Inching control, 20/15
Indicators, direct acting, 6/6
  electrodynamic, 6/9, 6/10
  electrodynamic reactive-power, 6/10
  induced moving-magnet, 6/8
  induction, 6/9, 6/10
  maximum-demand kilovar-hour, 6/17
  maximum-demand kilovolt-ampere-hour, 6/16
  moving coil, 6/8
  moving coil rectifier, 6/9
  moving coil thermocouple, 6/9
  moving iron, 6/8
  multirange, 6/9
  polyphase maximum-demand, 6/16
  split, 6/9
  thermocouple, 6/10
Inductance, 2/5, 2/17, 3/3
  calculation of, 2/18
  concentric cylindrical conductors, 2/18
  long straight isolated conductor, 2/18
  parallel conductors, 2/18
  units, 1/3, 1/6
Induction machine, 2/17
Inductive reactance, 2/18
Inductors, 2/5
  connection of, 2/19
  interphase, 19/5
  voltage applied to, 2/17
  with ferromagnetic circuit, 2/18
Inertia conversion factors, 1/5
Information, logical, 25/4
  numerical, 25/4
  textual, 25/3
Information processing, binary, 25/4
Information representation by digital systems, 25/3
Infrared conveyor ovens, 21/10
Infrared heaters, 21/6
Infrared ovens, 21/8
Initial-value theorem, 4/5
Inorganic materials, methods of manufacture, 5/39
Input/output control, 17/41
INSPEC, 36/5
Inspection, installation, 20/19
Installation, 20/1
  agriculture, 33/3
  electrodes, 20/18
  high-voltage supplies, 20/3
  horticulture, 33/3
  inspection, 20/19
  layout, 20/3
  lighting, 27/23
  low-voltage equipment, 20/6
  regulations, 20/3
  specifications, 20/3

Installation, *continued*
  substations, 20/4
  system supply, 20/3
  testing, 20/20
  transformers, 20/15
Institution of Electrical Engineers (IEE), 35/3
  see also IEE Regulations
Instruction execution logic structure, 25/20
Instruction fetch logic structure, 25/19
Instrumentation, 6/1
  d.c. and industrial-frequency analogue, 6/6
  distortion, 6/21
  electronic, 6/19
  range, 6/21
  selection criteria, 6/20
  sensitivity, 6/20
  switchgear, 8/29
  switchgear testing, 15/15
  terminology, 6/3
  see also Indicators
Insulating materials, 2/20, 5/17
  chemical properties, 5/20
  classification, 5/17
  electrical properties, 5/18
  gaseous, 2/21, 5/20
  mechanical properties, 5/18
  oils, 5/23
  physical properties, 5/18
  properties and testing, 5/17
  radiation effects, 5/41
  representative properties, 5/19
  resistance heaters, 21/4
  temperature index, 5/17
  see also Dielectric
Insulation, 1/30
  cables, 13/6
  elastomeric, 13/6
  impregnated paper, 13/7
  testing, 26/34
  thermoplastic, 13/6
Insulation co-ordination, 15/35
Insulation resistance, 2/5
  testing, 20/21
Insulation tests, 6/11
Insulators, 1/22, 1/24, 1/25, 1/27, 5/17
  overhead lines. *See* Overhead lines
Integral control, 4/11
Integrals, 1/10
Integrated circuits, 25/34
Integrating (energy) metering, 6/12
Integration, 4/5, 25/48
Integrator, 25/35
International System of Units (SI), 1/3, 6/5
Intrinsic impedance, 2/23
Intrinsically safe equipment, mining, 32/11
Invars, 5/62
Inverse square law, 27/4
Inverse transformation, 1/10
Ionisation, 1/18
Ionisation potential, 1/20
Ions, 2/10
I^2R loss, 2/5
Iron-cobalt alloys, 5/51
  24/27% Co, 5/51
  50% Co, 5/51
Iron loss. *See* Core loss
Irradiation effects. *See* Radiation
Irrigation, 33/8, 33/9
Isocyanates, 5/30
Isolux diagrams, 27/6
Isotopes, 1/18

Joule (unit), 1/3
Joule's law, 2/6

K shell, 1/17
Karma, 5/61
Kelvin (temperature), 6/5
Kelvin (unit), 1/3
Kelvin double bridge, 6/36
Kelvin effect, 2/9
Kilogram (mass), 6/5
Kilogram (unit), 1/3
Kirchhoff laws, 3/4, 3/5, 3/6
Kusters method for measurement of transformer errors, 6/42, 6/43

Lacour theorem, 3/6
Lacquers, 5/31
Laminates, 5/32
Lamp bulbs, 27/8
Lamp-caps, 27/9
Lamps, 27/8
  choice of, 27/18
  cold-cathode, 27/17
  colour qualities, 27/18
  colour rendering, 27/21
  construction and operation, 27/15
  decorative, 27/9
  developments, 27/19
  discharge, 27/10
  efficacies and colour properties, 27/19
  floods, 27/35
  fluorescent, 27/10, 27/34
  general lighting service, 27/8
  high-pressure mercury, 27/14
  high-pressure mercury-blended, 27/14
  high-pressure mercury fluorescent, 27/14
  high-pressure mercury fluorescent reflector, 27/14
  high-pressure mercury-metal halide, 27/14
  high-pressure sodium, 27/16
  incandescent filament, 27/8
  linear or SLI, 27/16
  low-pressure sodium, 27/15
  low-pressure sodium discharge, 27/34
  medium-pressure mercury, 27/13
  mercury discharge, 27/34
  mercury vapour, 27/10
  operating and replacement costs, 27/18
  pressed glass, 27/9
  rough service, 27/9
  sodium, 27/11
  special-purpose, 27/9
  tungsten filament, 27/34
  tungsten-halogen, 27/9
  U tube or SOX, 27/15
  ultraviolet, 21/6
  see also Illumination; Lighting
Lancaster flexible bag, 10/5
Lanchester clam, 10/5
Laplace transforms, 1/10, 1/14, 3/21, 4/3, 4/6
Laplacian fields, 1/16
Larsen potentiometer, 6/35
Lasers, applications, 21/20
Lead sheath for cables, 13/15, 13/24
Leclanché cell, 2/11, 23/4
Left-hand rule, 4/15
Legislation, 18/6
Length (units), 1/3, 1/4, 1/6
Length conversion factors, 1/5
Lenz's law, 2/14, 2/15
Lifting magnets, 26/5
Light, 27/3
Light rail transit (LRT). *See* Railways
Lighting, 28/21
  applications, 27/25
  circuits, 20/10
  crops, 33/9
  design codes, 27/19

Lighting, *continued*
  design objectives and criteria, 27/19
  design problems, 27/23
  design technology, 27/6
  dimmers, 27/38
  domestic, 27/30
  equipment, 33/6
  factory, 27/25
  flood, 27/32
  glare calculation, 27/24
  installation, 27/23
  local, 27/25
  maintenance, 27/23
  navigation, 31/7
  office, 27/32
  railway signalling, 30/14
  regulations, 27/29
  requirements for various applications, 27/29
  ships', 31/7
  shop, 27/31
  special requirements, 27/30
  street, 27/32
  television studio, 27/37
  theatre, 27/34
  underground, 32/19
  see also Illumination; Lamps
Lightning, 12/13, 15/35
  protective equipment, 15/36
  surges in transformers, 14/26
Line of flux, 2/5
Line of force, 1/24, 2/5
Linear acceleration (unit), 1/3
Linear constant coefficient ordinary differential equations (LCCDE), 4/3, 4/7
Linear current collection, 5/7
Linear velocity (unit), 1/3
Linkage, 2/5
Linkage change, 2/15
Liquids, 1/17, 1/25
  dielectrics, 2/21
Lithium bromide solution, 28/12
Livestock housing, 33/7
LNG, 7/11
Load-angle limitation, 17/14
Load calculation for environmental comfort, 28/21
Load curves, 18/6
Load forecasts, 18/3
Load shedding, 18/9
  and restoration, 17/36
Loadflow analysis, 3/28
Locking signals, 15/24
Locomotives, a.c., 30/5
  battery-driven, 29/18, 32/8
  d.c., 30/5
  diesel-electric, 30/9
    equipment, 30/10
    mechanical design, 30/10
    starting, 30/10
  drives, 30/4
  mining, 32/8
Logic gates, 25/5, 25/6
Logical operations, 25/4
Logical variables, 25/4
Loss angle, 2/5
Low-level languages, 25/32
Low load adjustment, 6/13
Low-voltage equipment, installation, 20/6
Lubricating oils, 8/31
Lumen (unit), 1/3
Luminaires, classification of, 27/22
  functions of, 27/22
  types of, 27/22
Luminance, 27/3, 27/4
  unit, 1/4
Luminescence, 6/29

Luminous efficacy, 27/3
Luminous exitance, 27/4
Luminous flux, 27/3
  unit, 1/3
Luminous intensity, 1/3, 27/3, 27/5
Lumped resistances, 3/3
Lux (unit), 1/3

'M' effect, 5/56
Machine language, 25/27
Machines, braking and stopping, 20/15
  d.c., 2/17, 5/11
Magnet conductors, stabilisation of, 5/10
Magnet materials, 6/13
Magnetic chucks, 26/10
Magnetic circuit, 2/5, 2/11, 2/12
  composite iron, 2/12
  concentric conductors, 2/12
  long straight isolated wire, 2/12
  toroid, 2/12
Magnetic core, 2/14
Magnetic fields, 1/16, 1/26, 2/5, 2/11, 2/18, 3/3, 5/10, 26/8, 26/41
  static, 2/22
Magnetic field strength, 1/28, 2/5
  units, 1/4, 1/6
Magnetic flux, 1/28, 2/5, 2/11, 2/12
  unit, 1/3
Magnetic flux density, 2/5, 2/14, 6/44
  units, 1/3, 1/6
Magnetic hysteresis, 5/43
Magnetic leakage, 2/5
Magnetic lines of induction, 2/11
Magnetic materials, 2/13, 5/4, 5/42, 26/8
Magnetic measurements, 6/43
  bridges for, 6/45
Magnetic parameters, 6/44
Magnetic potential difference, 2/5
Magnetic saturation, 2/12
Magnetic separators, 26/7
Magnetic space constant, 2/5
Magnetic suspension assembly, 6/13
Magnetisation curves, 2/12
Magnetising force, 2/5
Magnetism, 1/22
  residual, 2/6
Magnetohydraulic tripping mechanism, 26/12
Magnetohydrodynamic generators, 2/17, 5/11, 10/8
Magnetomechanical effects, 2/14
Magnetomotive force (m.m.f.), 1/28, 2/5, 26/3
  units, 1/6
Magneto-plasma-dynamic generator, 10/8
Magnetostriction, 1/23
Magnetostriction oscillator, 6/49
Magnetostriction transducers, 6/48
Magnets, 5/52
  brake, 6/13
  circular, 26/5
  laboratory, 5/11
  lifting, 26/5
  rectangular, 26/6
  superconducting, 5/11
Main equipotential bonding, 20/19
Maintenance, forecasting for, 18/4
  lighting, 27/23
  microprocessor-based, 24/5
  street lighting, 27/34
  transformers, 14/21
Man-machine interface, 17/39
Manganin, 5/61
Mascot Diagram, 25/27
Mass (units), 1/3, 1/6
Mass conversion factors, 1/5
Mass density (unit), 1/3
Material property, 1/28

Materials, 5/1
  conducting, 5/3
  fusible, 5/25, 5/27
  insulating *See* Insulating materials; Dielectrics
  overhead line supports, 12/11
  overhead lines, 12/3
  semi-fluid, 5/25
  see also under specific materials
Materials handling, 33/9
Mathematical models, 4/3
Mathematical symbolism, 1/4, 1/7
Maximum demand instrument, 6/10
Maxwell circulating-current process, 3/4
Maxwell equations, 1/26, 1/28, 2/23
Maxwell–Wien bridge, 6/38
Megger earth tester, 6/12
Melamine-formaldehyde, 5/29
Melting, resistance, 21/13
Memory address register, 25/19
Memory buffer register, 25/19
Memory devices, 25/14
Memory effect alloys, 5/4
Memory management, 17/41
Memory registers, 25/11, 25/12
Memory unit, 25/20
Mesh or resistor-net analogue, 1/16
Mesh-current equations, 3/4
Mesh law, 3/4
Metal cutting, 22/12
  carbon-arc, 22/12
  coated-electrode, 22/12
  plasma-arc, 22/12
Metal melting, 21/13
Metals, 1/25
  electrons in, 1/20
  physical properties of, 1/20
Meter reading, 18/13
Meters, 18/13
  dial testing, 6/15
  energy, 6/26
  induction, 6/12
  polyphase, 6/15
  prepayment, 18/13
  rotating substandard, 6/15
  single-phase, 6/12, 6/14
  single-phase prepayment, 6/13
  single-phase two-rate, 6/13
  single-rate, 6/13
  special, 6/17
  summation metering, 6/17
  temperature error, 6/14
  terminology, 6/3
  testing of single-phase, 6/15
  two-part tariff, 6/14
  voltage and frequency errors, 6/14
4-Methylpentene-1, 5/29
Metre (unit), 1/3, 6/5
Metrology, 6/1
Mica, 5/33
Mica-glass composition, 5/39
Mica-paper materials, 5/32
Micanite, 5/32, 5/35
Microcircuits, thick and thin film, 5/15
Microcontrollers, 24/1
  applications, 24/5
  basic input/output (I/O), 24/3
  communications with other controllers, 24/4
  equipment configuration, 24/8
  fast I/O, 24/4
  general requirements, 24/3
  maintenance facilities, 24/5
  operator interface, 24/4
  serial communication, 24/11
  signalling paths, 24/11
  user-designer interface, 24/4

Microfiche, 36/5
Microfilm, 36/5
Microprocessor-based controllers. *See* Microcontrollers
Microprocessor-based RTI, 17/6
Microprocessors, 24/3, 25/6, 25/20
  architecture, 25/18
  control and timing unit, 25/18
Microwave circuits, 5/17
Microwave power sources, 21/18
Millman theorem, 3/5
Mine winder event recorder, 24/6
Mineral waxes, 5/26
Mining, 32/1
  cables, 13/12
  coalface layout, 32/9
  computers, 32/20
  conveyors, 32/7
  distribution systems, 32/4
  drilling machines, 32/19
  flameproof equipment, 32/11
  gate-end boxes, 32/14
  heading machines, 32/11
  intrinsically safe equipment, 32/11
  load growth, 32/3
  locomotives, 32/8
  monitoring and control, 32/20
  power loaders, 32/9
  power supplies, 32/3
  radio communication, 32/8
  regulations, 32/3
  rope haulage, 32/7
  sensitive earth leakage, 32/15, 32/16
  underground transport, 32/7
  winders, 32/5
Minute (unit), 1/4
Mixers, 33/4
Mole (unit), 1/3
Molecules, 1/17
Mollier diagram, 7/6, 8/3
Molybdenum disilicide, 21/19
Momentum conversion factors, 1/5
Motional effect, 2/15
Motor control, remote, 20/15
Motor control gear, 20/14
Motor protection, 20/15
Motors, 5/6
  cage, 26/26
  capacitor split-phase, 26/34
  commutation, 26/37
  d.c., 26/16
    braking, 26/22
    characteristics, 26/17
    commutation, 26/18
    commutator, 26/16
    compoles, 26/18
    compound excitation, 26/18
    connections, 26/21
    construction, 26/18
    design data, 26/23
    harmonies, 26/22
    separate excitation, 26/17
    series, 26/17, 26/20, 26/23
    shunt, 26/17, 26/19, 26/20, 26/22
    sparking, 26/19
    speed control, 26/21
    starting, 26/19
    testing, 26/36
    thyristor-assisted commutations, 26/19
    traction, 29/15
  disc, 26/23
  flameproof, 32/16
  fuses, 15/7
  induction, 26/32, 26/38, 33/4
  industrial, 26/15

Motors, *continued*
  linear, 26/15, 26/39
    applications, 26/41
    d.c. and a.c., 26/40
    general principles, 26/41
    operating duties, 26/40
  losses and efficiency, 26/35
  protective equipment, 15/33
  ratings and dimensions, 26/34
  reluctance, 26/31
  repulsion, 26/32
  repulsion-induction, 26/34
  resistance split-phase, 26/33
  rotary, 26/15
  Schrage, 26/29
  series, 26/37
  shaded-pole, 26/33
  shunt, 26/37
  single-phase, 26/31
  single-phase commutator, 26/16
  slip-ring, 26/25
  speed control, 19/14, 26/20, 26/21, 26/26
  starter, 29/9
  synchronous, 26/30, 26/38
  synchronous-induction, 26/30
  testing, 26/34
  3-phase commutator, 26/28
  3-phase induction, 26/24, 30/9
Motorway signals, 29/21
Moulded compositions, 5/38
Multifunction converters, 25/38
Multilayered conductors, 5/17
Multimeters, digital, 6/23
Multi-motor starter boards, 20/6
Multiplexers, 25/7
  analogue, 25/44
Mumetal, 2/13
Mutual impedance, 3/4
Mutual inductance, 2/18, 3/3
  calculation of, 6/4

National Certificates and Diplomas, 35/3
Natural resins, 5/26
Navigation lights, 31/7
Needle gaps, 5/22
Negative sequence, 15/28
Neodymium iron boron, 5/53
Neper (unit), 1/12
Network analysers, 6/27
Network analysis, 3/1
  basic, 3/3
  power-system, 3/27
Network configuration, 3/3, 3/4
Network elements, 3/12
Network equations, 3/7
Network laws, 2/4
Network parameters, 3/3, 3/8
Network solution, 3/4, 3/28
Network theorems, 3/5
Network topology, 3/7
Network transients, 3/18
Network voltage considerations, 16/5
Networks, 2/3
  bridge. *See* Bridge networks
  equal-area criterion, 3/33
  fault-level analysis, 3/29
  line transmission, 3/17
  multi-machine system, 3/33
  multiport, 3/7
  non-linearity, 3/25
  non-sinusoidal alternating quantities, 3/13
  Π, 3/7
  power, 3/11, 3/14
  power limits and stability, 3/31
  resonance, 3/13

Networks, *continued*
  resonant, 15/14
  sinusoidal alternating quantities, 3/10
  steady-state a.c., 3/10, 3/31
  steady-state d.c., 3/9
  symmetrical components, 3/16
  system-fault analysis, 3/30
  system functions, 3/21
  T-, 3/6
  3-phase systems, 3/14
  two-port, 3/6, 3/7, 3/8
  with surface mounted devices, 5/16
  with wire-bonded chips, 5/17
Neutral displacement, 15/28
Neutron, 1/17
Neutron chain reaction, 9/3, 9/4
Newton (unit), 1/3
Newton-Raphson procedure, 3/29
Nichols diagram, 3/25, 4/20, 4/21
Nickel, welding, 22/4
Nickel alloys, welding, 22/4
Nickel–iron alloys, 5/50, 5/62
  30% Ni, 5/50
  36% Ni, 5/50
  50% Ni, 5/50
  80% Ni, 5/50
  forms of supply, 5/51
Nitrogen as insulating material, 5/22
Nitrogen oxides, reduction of, 7/15
Nodal-admittance equations, 3/27
Node law, 3/4
Node-voltage equations, 3/5
Noise ratings for environmental comfort, 28/5
Non-electrolytes, 1/25
Non-metals, physical properties of, 1/21
Normal-frequency recovery voltage, 15/4
North of Scotland Hydro-Electric Board, 18/14
Norwegian wave energy device, 10/5
Nuclear fission, 9/3
Nuclear fission-fusion hybrid system, 10/10
Nuclear fuel cladding, 5/64
Nuclear fuel cycle, 9/17
Nuclear fuel element fabrication, 9/18
Nuclear fuel reprocessing plant, 9/18
Nuclear fuels, 5/64, 9/3
Nuclear fusion, 10/9
Nuclear power stations, 18/8
Nuclear pressure vessels, 5/65
Nuclear radiation sensors, 6/51
Nuclear reactor coolants, 5/65
Nuclear reactor materials, 5/63
Nuclear reactor moderators, 5/65
Nuclear reactor shield, 5/65
Nuclear reactors, 9/1
  advanced gas-cooled, 9/4, 9/10
  as energy source, 9/3
  boiling water, 9/4, 9/7
  breeder, 9/4
  CANDU, 9/12, 9/13
  components, 9/4
  fast breeder, 9/15
  graphite moderated, 9/9
  heavy water, 9/12
  high temperature, 9/10, 9/12
  light water, 9/4
  liquid metal fast breeder, 9/15
  MAGNOX, 9/9, 9/10
  pressurised-water, 9/4, 9/5
  shielding, 9/4
  ship propulsion, 9/17
  steam generating heavy water, 9/13, 9/14
  types of, 9/4
Nuclear waste disposal, 9/18
Nylon, 5/37
Nyquist diagram, 3/24, 4/15

Ocean thermal energy conversion (OTEC), 10/5
Offices, Shops and Railway Premises Act 1963, 27/32
Ohm (unit), 1/3
Ohm's law, 3/5
Oil fuels, 7/4
Oils as insulating materials, 5/23
Open-loop transfer function, 4/6
Open-top trunking, 20/11
Opposed voltage schemes, 15/25
Optics, 1/18
Ore separation, 26/8
Organic materials, moulding and forming, 5/29
Oscillating water column, 10/5
Oscillators, 6/26
  magnetostriction, 6/49
Oscillography, electronic, 6/27
Outage, forced, 18/4
  scheduled, 18/4
Overcurrent protection, 15/18, 15/23
Overhead gear, trolleybuses, 29/15
Overhead lines, 12/1
  conductors, 5/5, 12/3
    bundle, 12/7
    creep coefficients, 12/6
    fittings, 12/6
    materials, 12/3, 12/5
    mechanical characteristics, 12/3
    nomenclature, 12/3
  sag and tension, 12/4
  current rating, 12/7
    electrical characteristics, 12/7
    electrical parameters, 12/7
    insulation flashover, 12/13
  insulators, 12/8
    pollution control, 12/10
    selection, 12/10
    types, 12/9
  lightning performance, 12/13
  mechanical loadings, 12/14
  radio interference, 12/7, 12/8
  supports, 12/11
    configurations, 12/11
    foundations, 12/12
    materials, 12/11
    tower geometry, 12/12
  voltage distribution over insulator strings, 12/10
  voltage gradient effects, 12/8
  voltage regulation, 12/7
  see also Transmission lines
Overvoltage protection, 15/35
Oxygen, electrolysis, 23/13

Paints, 5/31
Papers, insulating, 5/35, 13/7, 13/14, 13/15, 13/21
Parallel-generator theorem, 3/5
Parallel T network, 6/41
Paramagnetic materials, 2/13
Paramagnetism, 1/23, 2/13
Particle accelerators, 26/13
Pascal (unit), 1/3
Pascal language, 25/31
Pauli exclusion principle, 1/18
Peak-load, 18/7
Peltier coefficient, 2/9
Peltier device, 10/15
Peltier effect, 2/9
Period, 2/5
Permalloy C, 2/13
Permanent magnet materials, 5/51
  bonded, 5/53
  characteristics of, 5/53

Permanent magnet materials, *continued*
  ferrites, 5/52
  names and applications, 5/54
Permanent magnets, 2/13
  miscellaneous types, 5/54
Permeability, 1/27, 2/5, 2/13, 5/43, 5/46
  units, 1/4, 1/6
Permeance, 2/5
  units, 1/6
Permittivity, 1/26, 1/27, 2/5, 2/20, 5/20, 5/42
  units, 1/4, 1/6
Permutation, 1/9
Personal comfort, 28/3
Phase angle, 2/5
Phase coefficient, 3/7
Phase comparator, 25/22
Phase curves, 4/19
Phase distortion, 6/53
Phase grouping, 20/15
Phase margin, 4/16, 4/20
Phase-phase fault, 25/27
Phase sequence indicator, 6/11
Phasor, 2/5
Phenol-formaldehyde resins, 5/29
Phenol-furfural, 5/29
Phosphorescence, 6/29
Photodiodes, 6/50
Photoelectrochemical conversion, 10/6
Photoemissive sensors, 6/50
Photojunction, 6/50
Photometry, 27/4
Photoresistive sensor, 6/50
Photo sensors, 6/50
Photosynthesis, 10/7
Phototransistors, 6/50
Photovoltaic conversion, 10/6
Photovoltaic sensors, 6/50
Physical constants, 1/23
Physical properties, of gases, 1/22
  of metals, 1/20
  of non-metals, 1/21
Physical quantities, 1/12
Pilot circuits, 15/21
Pilot wires, 15/26
PL/1, 25/32
Planck constant, 1/18
Plane angle (unit), 1/3
Plane wave transmission, 1/30
Planning, 18/1
Plant scheduling, 18/9
Plasma furnaces, 21/19
Plasma torches, 21/19
  direct-coupled, 21/19
  induction-coupled, 21/19
Plastics, 5/38
Plutonium, 9/3, 9/15
p–n junctions, 6/47, 19/11
Point source, 27/4
Polar intensity distributions, 27/4
Polarisation, 2/6, 2/10, 2/20
Polarisation current, 1/26, 2/20
Polarity testing, 20/21
Polishing, 23/13
Pollution, overhead lines, 12/10
  steam generating plants, 7/13
Polyacetal, 5/29
Polyamides, 5/29, 5/37
Polyaralkyl ether/phenols, 5/30
Polycarbonate, 5/29, 5/37
Polychlorinated biphenyls, 5/25
Polychloroprene (PCP), 13/7, 32/17, 32/18
Polyesters, 5/29, 5/30
Polyethylene, 5/27, 5/37, 13/7, 13/18, 13/22
Polyethylene terephthalate, 5/29, 5/37
Polyimide/FEP tapes, 13/7
Polyimides, 5/30, 5/37

Polynomial equation, 4/8
Polyphenylene oxide, 5/29
Polypropylene, 5/29, 5/37, 13/7
Polystyrene, 5/29
Polytetrafluoroethylene (PTFE), 5/27, 5/36, 13/7, 13/12
Polytrifluorochloroethylene (PTFCE), 5/36
Polyurethanes, 5/30
Polyvinylacetate, 5/29
Polyvinyl chlorides, 5/29, 5/37, 13/7, 13/10, 13/13, 13/17, 20/9, 20/10, 32/17, 33/3
Polyvinyl fluoride, 5/37
Polyvinylidene fluoride, 5/37
Porcelain, 5/39
Position-error coefficient, 4/8
Positive sequence, 15/28
Potential, 1/28, 2/6
 units, 1/6
Potential difference, 2/6
Potential gradient, 2/6
Potentiometer transfer functions, 4/6
Potentiometers, a.c., 6/35
 comparator, 6/34
 d.c., 6/33
 pulse width modulation, 6/35
Potentiometric recorders, 6/55
Potted circuits, 5/30, 5/31
Power, 2/6
 units, 1/3, 1/6
Power angle characteristic, 16/7
Power circuits, 20/11
Power conversion factors, 1/5
Power factor, 2/6
 zero, 26/39
Power-factor correction, 20/16, 22/6
Power-factor indicators, 6/10
Power measurement, 6/10
Power ratio, 1/12, 1/16
Power series, 1/9
Power stations, fossil-fuelled, 18/7
 parallel operation, 17/16
Power supply, agriculture, 33/3
 horticulture, 33/3
 mining, 32/3
 railway signalling, 30/19
Power swing, 15/27
Power system operation and control, 17/1
 automatic control, 17/42
 centralised control, 17/3, 17/38
  hard copy, 17/39
  hardware configuration, 17/38
  software configuration, 17/39
 decentralised control, 17/3, 17/10, 17/24, 17/35, 17/37
 electronic turbine controllers, 17/24
 impact of system control, 17/45
 load controller with frequency response, 17/30
 objectives and requirements, 17/3
 reliability, 17/42
 security, 17/45
 slip stabilisation, 17/17
 system description, 17/4
Power transmission, 5/11
Pre-arcing time, 20/13
Precious metals, 5/62
Predominant time constant, 4/11
Preece's law, 2/9
Pressboard as insulating material, 5/32
Presspaper, 5/35
Pressure (unit), 1/3
Pressure conversion factors, 1/5
Pressure-reducing station, 24/10
Pressure transducers, 6/47
Printer, 17/8
Programmable logic arrays, 25/8

Programmable logic controller (PLC), 24/3
Programmable logic systems, 25/17
Progression, 1/9
Propagation coefficient, 3/7
Proportional-plus-derivative control, 4/11
Proportional-plus-integral-plus-derivative control (PID), 4/12
Protection, accelerated distance, 15/28
 balanced, 15/25
 circulating current, 15/25
 distance (impedance), 15/26
 earth leakage, 15/18, 15/25
 efficacy of, 15/29
 motor, 20/15
 negative-phase-sequence, 15/28
 overcurrent, 15/18, 15/23
 overvoltage, 15/35
 rectifiers, 19/11
 substation, 17/37
 thyristors, 19/13
 transformers, 18/15
 voltage, 15/26
Protective equipment, 15/20, 18/15
 application of, 15/29
 busbars, 15/33
 commissioning, 15/34
 conductors, 20/18
 equipment elements, 15/20
 fault location, 15/35
 feeders, 15/32
 generators, 15/30
 lightning, 15/36
 modes of operation, 15/20
 motors, 15/33
 reactors, 15/31
 rectifiers, 15/34
 solid state, 15/23
 testing, 15/34
 transformers, 15/30, 20/16
Proton, 1/17, 2/22
Proton synchrotron, 26/15
Proximity effect, 2/8
Public buildings, 20/12
Public Utilities and Street Works Act 1950, 18/6
Pulsational effect, 2/15
Pulse controllers, 17/37
Pulse echo prelocation method, 13/34
Pulse inputs, 17/8
Pulse width modulation potentiometer, 6/35
Pumped storage, 8/21, 11/27, 18/3, 18/10
 economics, 8/22
 layout, 8/22
 pump/turbine plant, 8/22
 starting and changeover, 8/22
Pumps, 33/4
 a.c. induction, 26/42
 d.c. conduction, 26/41
Pyrometers, 6/47

Quality assurance, data, 6/3
Quantity, 2/6
Quarries, cables, 13/12

Radial-feeder arrangement, 20/3
Radian (unit), 1/3
Radiation, 2/8
 common form of, 5/40
 effects of, 5/41
 effects on insulating materials, 5/41
Radiation activity (unit), 1/3
Radiation sensors, 6/51
Radiators, 28/9
Radiofrequency power sources, 21/17

Radio interference, overhead lines, 12/7, 12/8
 ships, 31/8
 trolleybuses, 29/15
Railways, 30/1
 a.c. systems, 30/3
  single-phase, 30/3
  3-phase, 30/3
 cables, 13/12
 conductor rail, 30/8
 current collectors, 30/8
 d.c. systems, 30/4
 developments, 30/9
 diesel-electric traction, 30/9
 electric braking, 30/7
 electrification, 30/3
 fail-safe conditions, 30/11
 feeder stations, 30/8
 high-speed train sets, 30/9
 light rail transit, 29/12
 linear motors, 26/42
 multiple-unit trains, 30/5
  a.c. systems, 30/7
  control equipment, 30/5
  thyristor regulator (chopper) control, 30/7
 overhead conductor, 30/8
 track circuit, a.c., 30/11
  capacitor feed, 30/12
  d.c., 30/11
  impedance bonds, 30/12
  jointless, 30/12
  miscellaneous forms, 30/12
 track equipment, 30/8
 underground, 30/8
 signalling and control, 30/10
  2-aspect, 30/13
  3-aspect, 30/13
  4-aspect, 30/13
  automatic train control systems, 30/18
  automatic warning system, 30/17
  coded track circuits, 30/18
  colour light, 30/13, 30/14
  immunising against traction currents, 30/19
  interlocking equipment, 30/16
  junctions, 30/14
  level crossings, 30/19
  lighting circuits, 30/14
  point operation, 30/15
  power supply, 30/19
  remote control, 30/16
  signal boxes, 30/16
  terminology, 30/13
  train identification, 30/17
  typical circuits, 30/15
  vehicle detection, 30/13
 see also Locomotives
Ramp voltage generator, 6/30
Random access road-only memories (ROM), 25/14
Random access read-write memory (RAM), 25/14
Rare earth cobalt, 5/53
Rare metals, 5/62
Reactance, 2/6
Reactance transducers, 6/49
Reactive compensation, 16/6
Reactive power, 16/1
 control, 16/3
 general considerations, 16/3
 management, 16/5
  by consumer, 16/6
  by supply authority, 16/5
Reactive power devices, 16/3
Reactors, 2/6, 15/13
 air insulated, 15/13
 mechanically switched, 16/8

Reactors, *continued*
  oil immersed, 15/13
  protective equipment, 15/31
  saturated, 16/11
  self-saturated, 16/12
  series, 15/13
  shunt, 15/13, 15/14
  thyristor controlled, 16/14, 16/16
  Treble Tripler, 16/11, 16/13
  Twin Tripler, 16/11
Real-time operation, 25/32
Reciprocity theorem, 3/5
Recorders, analogue, 6/55
  digital, 6/55
  potentiometric, 6/35
  strip-chart, 6/55
  ultraviolet, 6/51
  X-Y, 6/54
Recovery rate theory, 15/4
Rectification, controlled, 19/8
  full-wave, 19/3, 19/4
  half-wave, 19/3
  inversion, 19/8
  multiphase, 19/4
  process of, 19/3
  single-phase, 19/3
  voltage control, 19/8
Rectifier circuits, 19/3
Rectifier elements, freewheel, 19/3
  ideal, 19/3
  practical, 19/3
Rectifiers, Bridge connection, 19/4
  connections, 19/5
  current/voltage characteristics, 19/3
  current-voltage relations, 19/7, 19/10
  monocrystalline, 19/10
  multi-anode mercury arc, 19/6
  overlap, 19/7
  polycrystalline, 19/10
  protection, 19/11
  protective equipment, 15/34
  selenium, 19/10
  short-circuit conditions, 19/7
  silicon controlled, 19/12
  3-phase 3-pulse, 19/4
  3-phase 6-pulse bridge, 19/5
  voltage-current relations, 19/4
  welding, 22/5
Recuperator, cross-flow stationary, 28/18
Redox process, 23/3
Reflectance, 27/4, 27/21
Reflection, 27/4
Reflection factor, 2/24
Refraction, 27/4
Refrigeration, 28/11, 28/22
  ships, 31/7
Refrigerator, 2/10
Register, 6/13
Relative attraction force, 26/8
Relay steels. *See* Soft iron
Relays, 15/21
  a.c. track, 30/11
  admittance, 15/32
  beam, 15/23
  Buchholz, 15/28
  characteristics of, 15/27
  cut-out or reverse-current, 29/4
  d.c., 26/3
  d.c. track, 30/11
  definite time, 15/23
  directional, 15/22, 15/24, 15/26
  distance, 15/27
  double-input, 15/22
  electromagnetic, 26/11
  electromechanical, 5/46
  graded time lags, 15/24

Relays, *continued*
  induction, 15/22
  multiple-input, 15/22
  overcurrent, 15/23, 15/24
  rotary, 15/22
  single-input, 15/22
  spring-loaded, 26/3
  types of, 15/22
  see also Protective equipment
Reliability, power supply, 18/4
  ships, 31/10
  system control, 17/42
Reluctance, 2/6, 2/12
  units, 1/6
Reluctance motors, 2/16
Reluctance torque, 2/16
Remanence, 2/6
Remanent flux density, 2/6, 2/13
Remote control, ships, 31/9
Remote terminal units (RTUs), 17/6
Reserve capacity, 18/5
Residual compensation, 15/27
Residual-current devices, 20/22
Residual magnetism, 2/6
Resins, 5/26, 5/27, 5/33
Resistance, 1/25, 2/6, 3/3
  a.c., 2/7
  d.c. or ohmic, 2/7
  testing, 26/34
  units, 1/3, 1/6
Resistance alloys, 5/61
Resistance heating alloys, 5/61
Resistance instrumentation, 6/11
Resistance materials, 5/4
Resistance-temperature coefficient, 2/7, 26/35
Resistivity, 2/6, 2/7
  resistance materials, 21/5
  temperature variation, 21/3
  units, 1/4, 1/6
Resistor networks, 5/16
Resistors, 2/6
  linear, 2/7
  non-linear, 2/8
Resonance, synchronous/subharmonic, 16/18
Resonant link, 15/14
Revenue, 18/10
Rheostat, 2/6
Ring-main arrangement, 20/3
Rise time, 4/10
Road traffic signalling, 29/19
  computer control, 29/21
  control equipment, 29/20
  detectors, 29/20
  development, 29/19
  linked, 29/21
  motorways, 29/21
  phantom signals, 29/20
  statutory requirements, 29/20
Road transport, 29/1
Rods, dielectric, 5/35
Rolling mills, microcontrollers, 24/5
Root-locus method, 4/12
Root mean square (r.m.s.), 2/6, 3/10, 6/19
Rotating-field theory, 26/32
Rotor bearings, 6/13
Rotor-current limitations, 17/13
Routh array, 4/7
RTL2, 25/32
Rubber, 13/7
Russell angles, 27/6

Salt baths, 21/4
Salter duck, 10/5
Sample and hold requirement, 25/41
Saturation curve, 15/34

Saw-tooth generator, 6/30
Scale shapes, 6/7
Scaling, 25/35
Scheduled outage, 18/4
Scheduling, 17/41
Schering bridge, 6/38
Schrage motor, 26/29
Secant, 1/4
Second (time), 6/5
Second (unit), 1/3, 1/4
Seebeck coefficient, 2/9, 2/10, 10/7
Seebeck effect, 2/9
Self-impedance, 3/4
Self-inductance, 2/19
Self-induction, 2/17
Semiconductor converter equipment, 19/12
Semiconductor diodes, 6/47, 19/11
Semiconductor sensors, 6/46
Semiconductors, 1/22, 5/12, 29/15
  manufacture, 21/13
Semi-fluid materials, 5/25
Semi-plastic compounds, 5/25
Sensitive earth leakage, 32/15, 32/16
Sensitivity ratios, 6/53
Separation processes, 5/11
Separators, magnetic, 26/7
Sequential logic circuits, 25/8
Sequential logic design using memory, 25/16
Series, 1/9
Series aiding connection, 2/19
Serial communication, 24/11
Series inductor termination, 2/24
Series opposing connection, 2/19
Settling time, 4/11
Shearers, double-ended ranging drum, 32/9
  single-ended ranging drum, 32/10
Sheets as insulating materials, 5/32
  flexible, 5/35
Shift registers, 25/11, 25/12
Ships, 31/1
  a.c. installations, 31/3
  automatic control, 31/9
  cables, 31/6, 13/11
  circuit breakers, 31/6
  conditions of service, 31/3
  d.c. installations, 31/3
  deck auxiliaries, 31/8
  diesel-electric propulsion, 31/12
  earthing, 31/4
  electric propulsion, 31/11
  electromagnetic compatibility, 31/8
  emergency power, 31/6
  fans, 31/8
  generators, 31/4, 31/11
  heating, 31/7
  lighting, 31/7
  navigation lights, 31/7
  nuclear propulsion, 9/17
  operational modes of machinery spaces, 31/9
  propeller characteristics, 31/11
  radio interference, 31/8
  refrigerated cargo spaces, 31/7
  regulations, 31/3
  reliability, 31/10
  remote control, 31/9
  reverse-power protection, 31/5
  rotating machinery, 31/4
  safety actions, 31/9
  single-phasing, 31/5
  steam plant, 31/11
  steering gear, 31/6
  switchgear, 31/5
  tankers, 31/4, 31/11
  transformers, 31/4
  turbo-electric systems, 31/12
  ventilation, 31/8

Ships, *continued*
  voltage build-up, 31/5
  watertight doors, 31/8
Shock-ionisation, 1/25
Short-circuit current and power measurement, 26/38
Short-circuit ratio (SCR), 16/3, 16/4
Shunt capacitor termination, 2/24
Shunt inductor termination, 2/24
SI system of units, 1/3, 6/5
Siemens (unit), 1/3
Signal amplifiers, 6/30
Signalling, railway. *See* Railways
  road traffic. *See* Road traffic signalling
Silicon, 5/12
  crystal structure, 5/12
  device manufacture, 5/14
  electrical conduction, 5/12
  extrinsic conduction, 5/13
  N-type conduction, 5/13
  P-type conduction, 5/13
  single crystal growth, 5/14
Silicon diodes, 19/11, 22/5
Silicon junction, 6/48
Silicon steels, 5/43
Silicone elastomers, 5/37
Silicone fluids, 5/25
Silicone rubber, 13/7
Silicone varnishes, 5/31
Silicones, 5/29
Silver fuse elements, 5/55
Simulation languages, 25/46
Simulation programs, 25/48
Simulation studies, 25/46
Sine, 1/4
Single-phase commutator machine, 2/17
Sinusoidal wave-forms, 3/10
Skin effect, 2/8
Sleevings, tubular, 5/38
Slip-rings, 5/7
Small-scale integration (SSI) circuits, 25/3
Soft irons, 5/45
  ageing, 5/47
  applications, 5/46, 5/47
  composition, 5/45
  grades, 5/46
  heat treatment, 5/47
  magnetic characteristics, 5/46
  metallurgical state, 5/46
  physical forms, 5/46
  specifications, 5/46
  test methods, 5/47
Soil resistivity, 20/17, 20/18
Soil warming, 33/5, 33/9
Solar cells, 10/6
Solar collectors, 10/6
Solar energy, 10/6
Solar ponds, 10/6
Solar power, 18/8
Solar thermal power generation, 10/6
Solders, 5/62
Solenoid, 2/14, 2/18
Solid angle (unit), 1/3
Solid-state protective equipment, 15/23
Solids, 1/17
  dielectrics, 2/21
Sound level measurements, in transformers, 14/26
Sound-phase compensation, 15/27
South of Scotland Electricity Board, 18/14
Sowing, 33/9
Space factor, 26/3
Spark, 1/25
Specific apparent power, 5/43
Specific heat capacity (unit), 1/4
Specific heat dissipation, 2/8

Spectral luminous efficacy, 27/3
Spectrum analysers, 6/27
Speed control, 19/14, 26/20, 26/21, 26/26
Speed-distance protection curves, 24/8
Speed governors, 17/24, 17/25
Speed-measuring equipment, 17/34
Speed of light, 6/5
Springs, copper, 5/4
SQUID (superconducting quantum interference device), 5/11
Stability, 4/7
Stability criteria, 4/15, 4/16
Stability investigations, 17/21
Stability ratio, 15/26
Stacking factor, 5/45
Standards, 36/1
  agriculture, 33/4
  automobile, 29/12
  British, 18/6, 36/3
  cables, 13/3, 13/11
  derived electric and magnetic, 6/5
  electrical, 6/4, 36/1
  fundamental, 6/5
  horticulture, 33/4
  magnetic, 6/4
  modern, 6/4
  physical reference, 6/4
  previous, 6/4
  street lighting, 27/33
  switchgear, 15/3
  transformers, 14/3
Standby plant, 18/9
Standby systems, 20/12
Star-delta transformation, 3/5
Starter motors, 29/9
Starting, cold, 29/11
Static balancer, 14/20
Static excitation systems, 17/11, 17/20
Stator-current limitation, 17/14
Steady-state conditions, 2/3, 3/5
Steam generating plant, 7/1
  chemical energy sources, 7/3
  combustion process, 7/3
  design options and objectives, 7/8
  flow velocities in, 7/7
  fuel handling and storage, 7/11
  heat recovery, 8/16
  hydrodynamics, 7/5
  pollution control, 7/13
  thermodynamics, 17/5
  types of, 7/6
  water-tube, 7/6
  see also Boilers
Steam plant, ships, 31/11
Steam turbines, 8/3
  air-cooled condenser, 8/11
  auxiliaries, 8/11
  back-pressure (total energy) cycle, 8/3
  back-pressure, 10/11
  blade types, 8/5
  boiling-water reactor cycle, 8/3
  combined cycle, 8/4
  condenser arrangements, 8/7, 8/9
  construction, 8/7
  control system, 17/27
  cooling system, 8/10
  cycles and types, 8/3
  dry cooling tower with jet condenser, 8/11
  exhaust arrangements, 8/7
  extraction cycle, 8/3
  fluids alternative to steam, 8/5
  feed water plant, 8/11
  form and arrangement, 8/7
  foundations, 8/9
  governing and protection system, 8/9

Steam turbines, *continued*
  lubrication, 8/9
  non-reheat cycle, 8/3
  operation and control, 8/12
  part load, 8/6
  pass-out condensing, 10/11
  pressurised-water, 8/3
  reheat cycle, 8/3
  sea-water system, 8/10
  size effects, 8/6
  support plant, 8/9
  technology, 8/5
  wet cooling tower system, 8/10
Steatite, 5/39
Steels, corrosion-resisting, 22/4
  electrical. *See* Electrical steels
  heat-resisting, 22/4
  relay. *See* Soft iron
Steering gear, ships, 31/6
Stefan law of heat radiation, 2/8
Step function response, 6/53
Stepped time scheme, 15/26
Steradian (unit), 1/3
Strain gauges, 6/47
Stress (unit), 1/3
Strip-chart recorders, 6/55
Strips, flexible, 5/35
Stroboscope, 6/49
Structure of matter, 1/16
Substations, 18/15, 20/4
  automatic, 17/35
  automation, hardware configuration, 17/35
    software configuration, 17/36
  equipment, 17/6
  installation, 20/4
  packaged, 20/6
  protection, 17/37
Sulphur dioxide removal, 7/14
Sulphur hexafluoride, 15/10, 5/22
Summators, 6/18
Summer, 25/35
Superconducting alloys, 5/4
Superconducting coils, 5/11
Superconductivity, 1/21
Superconductors, 5/9
  materials and their properties, 5/10
  varieties of, 5/9
Superposition, 3/4, 3/5
Supertransductor, 5/11
Supplementary bonding, 20/19
Surface insulation resistance, 5/43
Surge reflection, 2/23
Susceptance, 2/6
Swing curves, 3/33, 3/34
Switches, 15/13
  earthing, 15/13
  intermediate, 20/11
  isolating, 20/14
  limit, 20/15
  master control, 20/11
  signal, 25/40
  time, 28/16
  two-way, 20/10
Switchgear, 5/6, 8/29, 15/1, 20/12
  a.c., 31/5
  control gear, 8/30
  d.c., 31/5
  disconnectors, 15/13
  fault considerations, 8/29
  fire hazard, 15/29
  flameproof, 32/13
  instrumentation, 8/29
  metal clad, 15/8
  metering, 8/29
  mounting, 15/7

Switchgear, *continued*
　outdoor, 15/8
　planning, 8/29
　protection, 8/30
　ships, 31/5
　standards, 15/3
　testing, 15/14
　　direct, 15/14
　　field testing, 15/16
　　miscellaneous tests, 15/18
　　single-phase and unit, 15/15
　　synthetic testing, 15/16
　use of term, 15/7
　see also Circuit breakers
Switching interlock supervision, 17/36
Switching programs, 17/36
Synchronism indicator, 6/11
Synchronous compensators, 11/3, 11/28
Synchronous machines, 2/17, 3/30, 3/33
　control characteristics, 17/14
　control quality, 17/19
　excitation systems, 17/10
　limiting excitation, 17/12
　models of, 17/21
　power transmission between, 17/17
Synchronous sequential logic, 25/10
Synchroscope, 6/11
Synchrotron, 26/14
Synthetic resins, 5/27
Synthetic waxes, 5/26
System design, 18/5
System models, 25/47

Tangent, 1/4
Tanks, electrolytic, 1/16
Tapes, pressure-sensitive adhesive, 5/35
Tariffs, 18/10, 18/11, 18/12
Telecontrol systems, 17/6
Telemetry, 17/6, 24/10, 24/11
Telephone communications, cables, 32/18
Temperature, characteristic, 1/23
　unit, 1/3
Temperature control systems, 19/14, 28/3
Temperature rise, measurement of, 2/9
　testing, 26/34
Temperature-sensitive bimetals, 5/62
Temperature transducers, 6/46
Tesla (unit), 1/3
Testing, cables, 13/20
　diagnostic, 6/32
　earth electrode, 20/21
　high voltage, 26/36
　installation, 20/20
　insulation, 26/34
　insulation resistance, 20/21
　protective equipment, 15/34
　regulations, 20/22
　resistance, 26/34
　switchgear. *See* Switchgear
　temperature rise, 26/34
　transformers. *See* Transformers
Thermal capacity (unit), 1/4
Thermal conductivity (unit), 1/4
Thermal expansion, controlled expansion
　　alloys, 5/61
Thermal quantities, conversion factors, 1/5
Thermal wheel, 28/18
Thermionic generator, 2/10, 10/7
Thermistors, 6/46
Thermocouple cables, 13/10
Thermocouple indicators, 6/10
　moving coil, 6/9
Thermocouple sensors, 6/46
Thermocouples, 5/62, 26/35
Thermodynamics of steam generating, 7/5

Thermoelectric devices, 2/10
Thermoelectric effect, 2/9
Thermoelectric generators, 10/7
Thermoluminescence, 27/8
Thermo-magnetic tripping mechanism 26/12
Thermometers, 26/35
　optical, 6/47
　platinum resistance, 6/46
　radiation, 6/47
Thermo-plastic materials, 5/25
Thermo-plastic synthetic resins, 5/27
Thermo-setting synthetic resins, 5/29
Thermostatic bimetals, 5/64
Thick films, common alloys, 5/15
　design and manufacture, 5/16
　microcircuits, 5/15
　properties of, 5/15
Thin films
　design and manufacture, 5/16
　microcircuits, 5/15
　properties of, 5/16
Thomson coefficient, 2/9
Thomson effect, 2/9
Thorium, 9/15
Thyratron, 19/9
Thyristors
　capacity compensator, 16/15
　characteristics, 19/12
　chopper control, 30/7
　commutation, 26/19
　drives, 16/16
　excitation systems, 11/17
　firing, 26/22
　gate turn-off, 16/19
　modular assemblies, 16/14
　protection, 19/13
　reactor compensators, 16/12
　solid state switching, 29/16
　speed control, 19/14, 26/21
　synchronous compensators, 11/28
　temperature control, 19/14
　3-phase a.c./a.c. controller, 19/13
　3-phase a.c./d.c. converter, 19/14
　transformer tap changing, 16/18
Tidal energy, 10/7, 18/8
Time (units), 1/3, 1/6
Time constant, 1/7, 2/6, 4/11
Time controls, 28/16
Time/current characteristics, 20/15
Timing problems, 25/9
Toroid, 2/12, 2/18
Torque, 2/16, 6/7
　/units, 1/4, 1/6
Torque angle, 26/15
Torque conversion factors, 1/5
Total energy system, 8/33, 10/13
Total operating time, 20/13
Town and Country Planning Acts, 18/6
Traceability, data, 6/3
Tracking, 5/20
Tractive electromagnets, 26/9
Training. *See* Education and training
Tramways, 30/9
Transducer effect, 5/42
Transducers, 6/45
　acceleration, 6/47
　angular velocity, 6/49
　magnetostriction, 6/48
　photo, 6/50
　pressure, 6/47
　radiation, 6/51
　reactance, 6/49
　temperature, 6/46
　underground, 32/21
Transfer function, 4/3, 4/17
　poles of, 4/5

Transform theorems, 4/5
Transformer effect, 2/15
Transformer oil, 5/24
Transformers, 5/6, 14/1, 18/15
　air-insulated air-cooled, 14/17
　auto, 14/20
　auto-starter, 14/20
　breather, 14/19
　Buchholtz relay, 14/19
　cable boxes, 14/19
　capacitor divider voltage, 6/43
　coil types, 14/7
　connections, 14/12
　　generator, 11/25
　　parallel operation, 14/19
　　phase conversion, 14/13
　　Scott connection, 14/13
　cooling, 14/17
　current, 6/41, 15/20, 15/21
　current testing, 15/34
　efficiency, 14/12
　fittings, 14/18
　flameproof, 14/21, 32/12
　ideal, 2/16
　installation, 20/15
　installation precautions, 20/16
　insulating oil, 14/21
　insulation, 14/8
　insulation maintenance, 14/22
　lightning surge, 14/26
　line-drop compensation, 14/17
　magnetic circuit, 14/3
　　characteristics, 14/5
　　core loss in, 14/3
　　core steel, 14/3
　　design, 14/3
　maintenance, 14/21
　measuring (instrument), 6/41
　mining, 14/21
　oil conservator, 14/19
　oil gauge, 14/19
　oil-immersed air-cooled, 14/18
　oil-immersed water-cooled, 14/18
　oil temperature indicator, 14/19
　overload capability, 14/18
　parallel operation, 14/19, 15/24
　power, 14/3
　protective equipment, 15/20, 15/21, 15/30,
　　　18/15, 20/16
　purchasing specifications, 14/27
　quadrature booster, 14/15
　reactive power, 16/4
　regulation, 14/12
　relief or explosion vents, 14/19
　small, 14/21
　sound level measurement, 14/26
　special types, 14/20
　standards, 14/3
　static balancer, 14/20
　summation, 6/17
　surge protection, 14/26
　surge voltage distribution, 14/10
　tap changing, 16/18, 17/37
　tap changing control, 14/16
　tap changing under load, 14/15, 14/22
　tapping switches (off-load), 14/19
　terminals and bushings, 14/19
　testing, 14/22
　　applied high-potential, 14/24
　　core loss and magnetising current, 14/24
　　impedance measurement, 14/23
　　impulse, 14/25
　　induced overvoltage, 14/24
　　insulation resistance, 14/23
　　load loss, 14/23
　　noise, 14/26

Transformers, *continued*
  testing, *continued*
    ratio, polarity and interphase connections, 14/22
    short-circuit, 14/23
    switching surge, 14/26
    temperature, 14/25
    winding resistance, 14/22
  3-winding, 14/4
    external loads, 14/14
    harmonic suppression, 14/14
    impedance characteristics, 14/14
  voltage, 6/42, 15/20, 15/21
    fuses, 15/7
  welding sets, 14/21, 22/5
  windings, 14/6
    cooling, 14/11
    design, 14/9
    impedance voltage, 14/12
    load losses, 14/11
    short-circuit conditions, 14/12
    temperature indicator, 14/19
Transient conditions, 3/5
Transient current, 2/3
Transient recovery voltage, 15/4
Transient stability, 16/6
Transmission (light), 27/4
Transmission lines, 2/23
  reactive power, 16/4
  reflection of surges, 2/23
  using shunt compensator, 16/8
  voltage/load characteristics, 16/5
  with multiple static var compensators, 16/8
Transmission systems, general, 16/18
  long-distance, 16/18
Transmittance, 27/4
Transportation. *See* Automobiles; Electric vehicles; Railways; Road transport; Ships; Trolleybuses
Travelling-wave gap flux, 26/16
Treeing, 13/20
Trigonometric functions, 1/4, 1/7
Trigonometric relations, 1/4, 1/7
Trip coils, 15/28
Tripping and operating circuits, 15/28
Tripping mechanisms, 26/12
Tri-state logic, 25/13
Tritium, 10/9
Trolleybuses, 29/13
  contactor control, 29/14
  electrical details, 29/14
  electronic control, 29/14
  leakage protection, 29/15
  mechanical details, 29/14
  overhead gear, 29/15
  radio interference, 29/15
  vehicles, 29/13
Trunking, 20/11
Tubes, dielectric, 5/35
  flexible, 5/38
Turbines, 8/1
  steam. *See* Steam turbines
  water. *See* Hydro-electric plant; Water turbines
  wind, 10/3
Turbo-electric ship propulsion, 31/12
Turbogenerators, 11/3, 11/21
  cooling, 11/23
  main dimensions, 11/22
  rotor body, 11/22
  rotor winding, 11/22
  stator casing, 11/23
  stator core, 11/23
  stator winding, 11/23
Tyne and Wear Metro, 29/12

Ultraviolet processes, 21/6
Ultraviolet recorder, 6/51
Under-carpet system, 20/12
Underfloor trunking, 20/11
Unit load demand signals, 17/30
Units, auxiliary, 1/4
  base, 1/3
  derived, 1/3
  relation between SI, e.s. and e.m. units, 1/6
  supplementary, 1/3
Unity-feedback systems, 4/8, 4/9, 4/12
Uranium, 9/3, 9/15
Uranium enrichment plant, 9/18
Urea-formaldehyde, 5/29

Vacuum, 1/25
  as interrupting medium, 5/23
Vacuum arc interruption, 15/4
Vacuum circuit-breaker. *See* Circuit-breakers
Vacuum furnaces, 21/4
Vacuum technology, 5/23
Valence electrons, 1/18
Variable reactive power, 15/14
Variable speed systems, 31/8
Varnishes, 5/31
  air-drying, 5/31
  baking, 5/31
  properties of, 5/31
  silicone, 5/31
  solventless, 5/31
Vehicles, battery-driven. *See* Electric vehicles
Velocity conversion factors, 1/5
Velocity-error coefficient, 4/9
Velocity feedback, 4/11
Ventilation, 33/4, 33/7
  ships, 31/8
Ventilation control boards, 20/7
Very large-scale integration (VLSI) devices, 25/3
Vibrators, 26/10
  electrodynamic, 26/10
  magnetostrictive, 26/11
Vicalloy, 5/54
Viscosity conversion factors, 1/5
Vision, 27/3
Visual impact on comfort conditions, 28/4
Volt (unit), 1/3
Voltage, 2/3, 6/5
  a.c./d.c. transfer, 6/5
  applied to resistor, 2/6
Voltage control, 17/37
  compensated, 29/3
  rectification, 19/8
Voltage converter, a.c./d.c., 6/19
Voltage-current characteristics, 2/6, 19/4
Voltage distribution over insulator strings, 12/10
Voltage fluctuations, 16/16
Voltage follower, 25/36
Voltage generator, 3/3
Voltage gradients, 2/6, 12/8
Voltage protection, 15/26
Voltage ratio, 1/16, 20/15
Voltage regulation, 12/7, 17/11, 17/19
  automatic, 17/12, 17/13
  overhead lines, 12/7
Voltage sensitivity, 6/53
Voltage surge arrester, 15/36
Voltage surge attenuation, 15/36
Voltmeters, digital, 6/21
  calibration, 6/26
  charge balancing, 6/22
  common-mode interference signals 6/25
  dual slope, 6/22
  floating-voltage measurement, 6/25

Voltmeters, digital, *continued*
  Tripping and operating circuits, 15/28
  input and dynamic impedance, 6/23
  instrument selection, 6/25
  linear ramp, 6/21
  multislope 6/23
  mixed techniques, 6/23
  noise limitation, 6/25
  series-mode (normal) interference signals, 6/25
  successive-approximation, 6/22
  voltage-frequency, 6/22
  electronic analogue, 6/19
  electrostatic, 6/10
Volume conversion factors, 1/5
Vulcanised fibre, 5/35

Wagner earth, 6/39
Wall diagram, 17/39
Ward-Leonard control, 26/20, 32/6
Warm air systems, 28/9
Warping, 21/8
Water dissociation, 23/13
Water electrolysers, 23/14
Water heating, direct resistance, 21/4
Water-power stations. *See* Hydro-electric plant
Water supplies, distributed control of, 24/8
Water turbines, 8/18
  bulb, 8/20, 8/21
  control system, 17/31
  Deriaz, 8/20
  design, 11/26
  impulse (Pelton), 8/19
  propeller (Kaplan), 8/20
  reaction (Francis), 8/20
Watt (unit), 1/3
Wattmeters, 6/15
  compensated, 6/10
  digital, 6/26
Wave energy, 10/5
Wave forms, 1/10, 2/6
  periodic response, 6/51
Waveguide, corrugated, 26/14
Wave mechanics, 1/18
Waxes, 5/26
Wayne Kerr bridge, 6/40
Weber (unit), 1/3
Weight (unit), 1/3
Welding, 5/9, 22/1
  arc, 22/3
  cables, 13/11
  electrodes, 22/3
  butt, 22/19
  copper and copper alloys, 22/4
  corrosion-resisting steels, 22/4
  electroslag, 22/8
  equipment, 33/6
  flash, 22/19
  gas-shielded metal-arc, 22/11
    flux-cored wires, 22/12
    gases, 22/12
    metal transfer, 22/11
    welding wires, 22/12
  heat-resisting steels, 22/4
  manual metal-arc, 22/3, 22/5
  metal inert-gas, 22/12
  micro-plasma-arc, 22/11
  nickel and nickel alloys, 22/4
  plasma-arc, 22/11
  power supply, 22/5
  projection, 22/17
    electrodes, 22/20
  pulse MIG, 22/12
  rectifiers, 22/5

Welding, *continued*
  resistance, 22/12
    electrodes, 22/20
    equipment, 22/12
    scope of, 22/13
  seam, 22/17
    electrode-wheel drive system, 22/18
    mandrel, 22/19
    process variables, 22/18
  spot, 22/13
    aluminium and its alloys, 22/16
    electrode size, 22/14
    electrodes, 22/20
    heat balance, 22/14
    multi-point, 22/15
    portable, 22/15
    surface indentation marking, 22/15
    TIG process, 22/10
    time and sequence control 22/14
    timing, 22/14
    twin-spot, 22/15
    stitch, 22/16

Welding, *continued*
  submerged-arc, 22/7
  transformers, 14/21, 22/5
  tungsten inert-gas, 22/8
    a.c. process, 22/9
    applications, 22/9
    capacitor, 22/10
    d.c. process, 22/9
    electrical requirements, 22/9
    electrodes, 22/8
    gases, 22/9
    programmed, 22/11
    pulsed, 22/11
    spot, 22/10
Weston Cell, 2/11, 6/33, 23/5
Wetherill separator, 26/7
Wheatstone bridge, 6/36, 6/47
Winches, cargo, 31/8
  mooring, 31/9
Wind energy, 10/3, 18/8
Wind turbines, 10/3
Winders, 32/5

Winders, *continued*
  Koepe-type, 32/5
  types of, 32/5
  Ward Leonard control, 32/6
Windings, copper, 2/8
Windlasses, 31/9
Wire coverings, 5/38
Wiring regulations. *See* Regulations
Wiring systems, 20/8
  plastic conduit, 20/9
  steel conduit, 20/8
  trunking, 20/9
  see also Cables

XLPE, 13/18, 13/22, 13/24
X-Y recorder, 6/54

Zener diodes, 6/33, 6/50
Zone factors, 27/6
Z-transforms, 4/4